1923 CATALOG

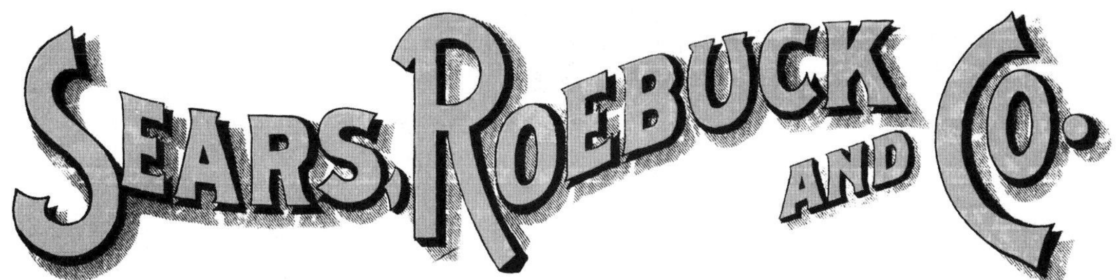

THE THRIFT BOOK OF A NATION

Sears, Roebuck and Co.

DOVER PUBLICATIONS
Garden City, New York

Bibliographical Note

This Dover edition, first published in 2023, is an abridged republication
of the work originally published by Sears, Roebuck and Co., Chicago, in 1923.

International Standard Book Number

ISBN-13: 978-0-486-85116-7
ISBN-10: 0-486-85116-8

Manufactured in the United States of America
85116801 2023
www.doverpublications.com

1923 CATALOG

THE THRIFT BOOK OF A NATION

Truly Charming Dresses VERY Reasonably Priced
31L6280
All Silk Taffeta
$14.95
31L6285
French Ratine
$4.98
31L6290
All Silk Canton Crepe and Silk Faille
$13.95
31L6275
Tissue Gingham
$4.48
What is more satisfactory for a spring and summer frock than all silk taffeta? The model pictured is developed in excellent quality all silk taffeta, and trimmed around the full skirt with bands of organdie embroidery. The collar and cuffs are of embroidery to match. Waist has panel front and back; sides and sleeves are joined to a Jap silk lining.
WOMEN'S REGULAR SIZES—32 to 44 inches bust measure. Give measurements. Shipping weight, 1¾ pounds.
31L6280—Navy blue.
31L6281—Brown.
31L6282—All black.
$14.95
The unique and attractive "Deauville" One-Piece Dress, made of washable, fancy woven French finish ratine. Collar and sleeves are of white ratine, embellished with colored embroidery to match material of dress. Front of skirt is embroidered in white. Dress fastens with white loops and self covered buttons. Tie sash.
WOMEN'S REGULAR SIZES—32 to 44 inches bust measure. Give measurements. Shipping weight, 2 pounds.
31L6285—Tangerine (burnt orange).
31L6286—Cadet blue.
31L6287—Rose.
$4.98
The charming and youthful style of this chic washable frock is sure to please you. Made of superior fancy checked tissue gingham, combined with plain organdie to harmonize. Note the graceful deep organdie collar, front and back, edged with tissue piping; the fancy cuffs and gathered panels of organdie, trimmed with circles of checked tissue. Belt is button trimmed. Dress fastens at shoulders.
WOMEN'S REGULAR SIZES—32 to 44 inches bust measure. Give measurements. Shipping weight, 1¾ pounds.
31L6275—Brown and white check.
31L6276—Blue and white check.
$4.48
Paris furnished the inspiration for this charming and original model. The chic mandarin blouse is made of embossed silk faille, with floral design in self color. The graceful elbow length butterfly sleeves are slashed and faced with contrasting silk crepe. Skirt is of plain all silk Canton crepe, with fashionable side draping as pictured. Girdle is trimmed with large filigree silver color metal buckle ornaments. Fastens at side.
WOMEN'S REGULAR SIZES—32 to 44 inches bust measure. Give measurements. Shipping weight, 1¾ pounds.
31L6290—Black.
31L6291—Brown.
$13.95
SIZES Dresses offered on this page can be furnished in Women's Regular Sizes, 32 to 44 in. bust measure, proportionate waist measure and front length of skirt, 30 to 37 inches, with basted hem. When ordering give bust and waist measures; also front length of skirt.

HERE ARE THE NEW STYLE DRESSES *for* *Spring and Summer*

Fascinating Fashions in delightful variety at prices much lower than you expected to pay

You will find in our Dress Department a style to charm every taste at a price to suit every purse, and whatever the price of the dress you decide to buy, you will receive *exceptional value* for every penny you spend.

There are so many grades and qualities of merchandise that sometimes it is difficult, in a printed description, to show the real value of what we offer; but there is no risk in buying from this catalog. Our reputation deserves your confidence and our guarantee protects you. We do not send out anything that is unworthy of the trust reposed in us by our customers.

We welcome honest competition. But, when you compare our offerings with others, we ask that you base your final judgment on the *quality* and *value* of your purchase. The name of Sears, Roebuck and Co. means *justice, fair dealing and economy* to more than eight million satisfied customers and loyal friends.

Our Guarantee Means Just What It Says—Here It Is:

WE GUARANTEE

that each and every article in this catalog is exactly as described and illustrated.

We guarantee that any article purchased from us will satisfy you perfectly; that it will give the service you have a right to expect; that it represents full value for the price you pay.

If for any reason whatever you are dissatisfied with any article purchased from us, we expect you to return it to us at our expense.

We will then exchange it for exactly what you want, or will return your money, including any transportation charges you have paid.

SEARS, ROEBUCK AND CO.

31L6295
All Silk Georgette Crepe and Spanish Lace
$11.75

A Handsome Dress for Formal Occasions When You Want to Look Your Best

This stunning Paris model, charming and exclusive in style, is a dress that will delight the woman of discriminating taste. It is made of fine **all silk Georgette crepe,** pleasingly combined with rich Spanish lace over a foundation of Jap silk. The blouse has a wide hemstitched panel of Georgette, front and back, loosely draped over a Jap silk bodice. The chic bell sleeves and gathered skirt are of Spanish lace, and over the right side of skirt a long panel of Georgette floats gracefully. The wide crushed girdle is of shimmering cloth of gold, veiled by Spanish lace and adorned with a handsome spray of chenille flowers and tiny Parisian buds of metal cobwebbing. Dress closes invisibly at side. A truly superb frock, combining consummate skill in designing with a very reasonable price!

WOMEN'S REGULAR SIZES—32 to 44 inches bust measure, proportionate waist measure and front length of skirt, 31 to 37 inches, with basted hem. **Give bust and waist measures; also front length of skirt.** Shipping weight, 1½ pounds.

31L6295—Brown.
31L6296—Black. **$11.75**

You will be astonished at this remarkable value! Just think of being able to buy a smart one-piece dress, up to the minute in style, made of durable **all wool French serge** for such a low price as this! The dress is trimmed on front of waist, sleeves and on the triple panels, placed one over the other, at each side of skirt with embroidery banding in artistic contrasting colors. The panels just mentioned hang in uneven lengths, loosely extending to below the hem. Another smart feature is the novelty metal girdle which holds the fullness at waistline. Dress fastens at side. **Positively the best value you ever bought for the money!**

WOMEN'S REGULAR SIZES—32 to 44 inches bust measure. **Give measurements.** Shipping weight, 2 pounds.

31L6300—Navy blue.
31L6301—Brown.
31L6302—Black. **$5.98**

Here is a stunning frock of novel design and beautiful material which you will surely admire. It is made of high grade **all silk taffeta** with an original and charming decorative treatment of the fashionable allover embroidery. This handsome embroidery scheme is carried out in contrasting color on the sleeves and on the wide panel down the entire front of dress. The collar and revers are also embroidered and the vestee of contrasting silk Canton crepe is trimmed with tiny novelty buttons. Slashed cuffs are of plain taffeta. Model has elastic belt concealed by a girdle of taffeta.

WOMEN'S REGULAR SIZES—32 to 44 inches bust measure. **Give measurements.** Shpg. wt., 1¾ lbs.

31L6305—Black.
31L6306—Navy blue.
31L6307—Brown. **$11.75**

Tasteful simplicity is the keynote of this fascinating frock of fine **all silk Canton crepe.** It is a model calculated to appeal to women of refinement; designed with a graceful bertha collar of Brussels net trimmed with fine Valenciennes lace insertion and edging. The blouse shirred on an elastic belt and over this a wide girdle of self material lavishly adorned with multi-colored embroidery in artificial silk and dull gold thread. The plain skirt falls in rippling fullness. **A lovely frock for less than you expected to pay.**

WOMEN'S REGULAR SIZES—32 to 44 inches bust measure. **Give measurements.** Shipping weight, 1¾ pounds.

31L6310—Cinnamon.
31L6311—Navy blue.
31L6312—Black. **$9.98**

SIZES Dresses offered on this page can be furnished in **Women's Regular Sizes,** 32 to 44 inches bust measure, proportionate waist measure, and front length of skirt, 31 to 37 inches, with basted hem. **When ordering give bust and waist measure; also front length of skirt. See Page 557 for measuring instruc io**

DRESSES *of* RARE BEAUTY

Chosen for YOU and offered at Money Saving Prices

Selected from among thousands at our Fifth Avenue Establishment in NEW YORK—the Fashion Center of the World

31L6315
All Silk Canton Crepe
$13.95

31L6320
All Silk Figured Crepe de Chine
$14.50

Designed expressly for women of taste, ho prefer styles that are not common-ace, this superb, richly beaded Paris odel frock elicits admiration from all who hold it. It is made of an exquisite ality **all silk Canton crepe.** The style is the st word and the workmanship beautiful in every tail. The charming mandarin style blouse is elaborately nented with bead embroidery in brilliant contrasting rs. This decorative treatment is carried out on the t and on the girdle, which has a buckle effect of solid ding matching the design on blouse. Note the charming ace of the butterfly wing sleeves. The model has elastic lt at waistline concealed by the girdle. An exclusive yle, appealing to women of taste and one of those onderful values that have made the Sears-Roebuck dress epartment famous.

WOMEN'S REGULAR SIZES—32 to 44 inches bust easure. **Give measurements.** Shipping weight, 2 lbs.

31L6315—Cocoa brown.
31L6316—Navy blue.
31L6317—All black

$13.95

Hats on this page may be found in our Millinery Section, pages 83 to 109, inclusive.

SIZES Dresses offered on this page can be furnished in **Women's Regular Sizes,** 32 to 44 inches bust measure, proportionate waist measure, and front length of skirt, 31 to 37 inches, with basted hem. **When ordering give bust and waist measures; also front length of skirt. See page 557 for measuring instructions.**

A modish style that is grace itself, designed to please the well dressed American woman who wears only the latest and best. This thoroughly smart frock is made of superior quality **all silk crepe de chine** in pleasing Paris novelty figured pattern. It has the popular mandarin blouse with short sleeves and collarless neck edged with contrasting piping. The skirt is made with a graceful double pointed gathered tunic falling in pretty fullness over a silk lining to which the lower part of skirt is attached. Girdle is of self material trimmed with plain crepe de chine to match piping and a chic corsage bouquet of silk fruit and blossoms. A beautiful dress worth every cent of our low price.

WOMEN'S REGULAR SIZES—32 to 44 inches bust measure. **Give measurements.** Shipping weight, 1¾ pounds.

31L6320-Tan ground, Blue and Red figure.
31L6321-Blue ground, Tan and Red figure. **$14.50**

CORRECT TOGS
for Outing and Sports Wear

31L6325
All Wool Homespun
$9.98
31L6330
Hill's Jean Sport Suit
$2.98
31L2960
All Wool Tweed Three-Piece Outfit Coat, Skirt and Knickers 7 to 14 Years
$9.98
Separate Knickers
$2.59
31L6335
Hill's Jean Knicker Outfit
$3.75
31L6340
Khaki Riding Suit
$8.95
Khaki Leggings
98¢

We picture here one of the new Sport Outfits which are all the rage for spring and summer wear. Plain sleeveless jacket and checked plaited skirt are both made of **all wool homespun.** Jacket has tuxedo collar and revers also pocket trimming of checked material. Buttoned belt. Skirt is stylishly knife and box plaited and has separate belt of plain material. A stunning style and a big value!

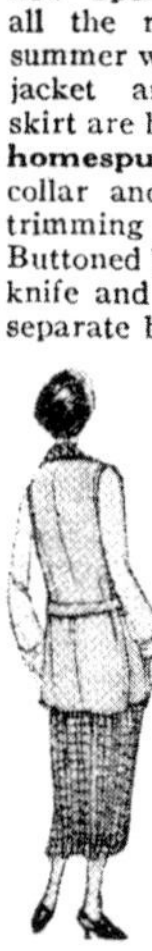

WOMEN'S SIZES—32 to 44 inches bust measure. Misses 'sizes, 14, 16, 18 and 20, years. **Give measurements.** Shipping weight, 2 lbs.

31L6325 Brown and tan check.

31L6326 Copenhagen blue and tan check. **$9.98**

Women's and Misses' Two-Piece Outfit, consisting of blouse and plain tailored skirt. Made of best quality, extra durable, washable **jean.** Blouse has sailor tie, two pockets and buttons at sides. Note the plaited bellows sleeves finished with buttoned tabs. Plain skirt attached to a waistband. Natty outing suit which is a real bargain!

WOMEN'S SIZES—32 to 44 inches bust measure. Misses' sizes, 14, 16, 18, and 20, years. **Give measurements.** Shpg. wt., 1¾ lbs.

31L6330 Khaki jean.

31L6331 White jean.

$2.98

Girls' Stylish **Three-Piece Outfit,** consisting of smart belted jacket, full plaited skirt and a pair of knickers—all made of high grade **all wool tweed.** Jacket is unlined, has plaited back and two bellows pockets with buttoned flap. Skirt is knife and side plaited and attached to a waistband. Knickers may be purchased separately for $2.59. Have cuffs at knees, and belt of self material.

SIZES—7 to 14. **State size.** Shipping weight, outfit, 1½ pounds; knickers, 1 pound.

31L2960—Gray mixture outfit.
31L2961—Tan mixture outfit. **$9.98**
31L2962—Gray Knickers.
31L2963—Tan Knickers. **2.59**

Women's and Misses' Two-Piece Outfit. Jacket and full cut knickers made of superior quality, iron wearing, washable **jean.** A fine suit for sports, camping, or hiking. Jacket has plaited back, two buttoned pockets and belt, as pictured. Well tailored knickers, buttoning at sides; two slash pockets; buttoned cuffs at knees. A dandy suit and a splendid value!

WOMEN'S SIZES—32 to 44 inches bust measure. Misses sizes, 14, 16, 18, and 20. **Give measurements.** Shpg. wt., 1½ lbs.

31L6335 Khaki jean. **$3.75**

Women's and Misses' Swagger London Model Two-Pi[ece] Riding Habit, made of stro[ng] durable **khaki cloth.** Co[n]sists of a finely tailored coat smart semi-fitted belted style and separate breeches—breeches fasten at both sides at hips and are finished at knees with eyelets and laces.

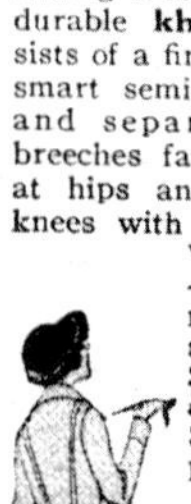

WOMEN'S SIZES—32 to 44 in. bust measure. Misses sizes, 14 to 20 years. **State bust, waist and hip measures.** Shipping weight, 4 pounds.

31L6340
Khaki. **$8.95**

Khaki Cloth Leggings to match Riding Suit, described above. Shipping weight, 1½ pounds.

31L6341
Khaki. **98c**

Outing Suits shown on this page come in Women's sizes, 32 to 44 inches bust measure; Misses' sizes, 14 to 20 years. Knickers, Breeches and Skirts, 22 to 34 inches waist measure. Front length of skirts, 32 to 38 inches. **Be sure to state size.** See page 557 for measuring instructions.

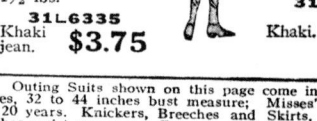

Smart Washable Frocks
Practical and Inexpensive
A charming frock reflecting the spirit of springtime. Made of high grade sheer washable dotted **Swiss** in fast colors. Trimmed with Venise lace on the white organdy collar, the novelty cuffs and also on the two graceful loose hanging panels which adorn the skirt. Note the chic velvet ribbon bows which trim collar and cuffs. Fastens in front. Waistline is shirred on elastic belt, concealed by the tie girdle. A beautiful dress for little money.
WOMEN'S REGULAR SIZES—32 to 44 inches bust measure. **Give measurements.** Shpg. wt., 1¾ lbs.
31L6365—Rose.
31L6366—Copenhagen blue.
31L6367 Orchid. **$4.79**
Ratine
31L6350
Gingham
$3.29
31L6345
Ratine
$4.59
31L6365
Dotted Swiss
$4.79
31L6355
Voile
$2.98
Here is a pleasing novelty in a stylish one-piece slip-over dress for spring and summer. Made of fancy woven washable cotton **ratine** and tastefully adorned with contrasting embroidery on the short sleeves and collar and also down the front. Has girdle of self material and closes invisibly in front. A stylish dress and an exceptional value.
WOMEN'S REGULAR SIZES—32 to 44 inches bust measure. **Give measurements.** Shipping weight, 1¾ pounds.
31L6345—Peach.
31L6346—Heliotrope.
31L6347—Blue.
$4.59
You'll be delighted with the chic style of this neat semi-tailored dress for spring and summer. It is made of high grade cool crisp checked **gingham**, and trimmed on the front and the short sleeves with a band of contrasting linene and real hand embroidery. Back is also elaborated with hand embroidery and two full length loose straps of gingham edged with linene and tacked to skirt at the hem. Fastens in front with large pearl buttons. Buckled belt of patent leatherette included. A splendid dress.
WOMEN'S REGULAR SIZES — 32 to 44 inches bust measure. **Give measurements.** Shpg. wt., 1¾ lbs.
31L6350 — Black and white check.
31L6351 — Red and white check.
31L6352—Brown and white check.
$3.29
A Smart Dress of good quality figured cotton **voile** trimmed down each side of front with wide loose bands of filet lace piped with voile. Collarless neck; short slashed sleeves tying with chic voile bows; fastens invisibly in front. Sash ties at back. An unsurpassed value.
WOMEN'S REGULAR SIZES —32 to 44 inches bust measure. **Give measurements.** Shipping weight, 1¾ pounds.
31L6355—Navy blue.
31L6356—Brown.
31L6357—Black. **$2.98**
Hats on this page may be found in our Millinery Section, pages 83 to 109, inclusive.
See page 557 for measuring instructions.
31L6360
Linene
$2.95
A one-piece dress of fetching style, made of high grade washable fast color cotton **linene** most attractively trimmed with contrasting embroidery in a charming design. The front displays two panels, trimmed with embroidery hanging gracefully over the skirt, held by tie sash of self material. Fastens in front invisibly. Special value.
WOMEN'S REGULAR SIZES—32 to 44 in. bust measure. **Give measurements.** Shipping weight, 1¾ pounds.
31L6360—Rose.
31L6361—Copenhagen blue.
31L6362—Reseda green. **$2.95**
SIZES Dresses offered on this page can be furnished in **Women's Regular Sizes**, 32 to 44 inches bust measure, proportionate waist measure, and front length of skirt, 31 to 37 inches with basted hem. **When ordering give bust and waist measures; also front length of skirt.**

LOVELY FROCKS
Reasonably Priced

Our Guarantee will interest you
Be sure to read it!

It's on page 1

Hats on this page may be found in our Millinery Section, pages 83 to 109 inclusive.

31L6370
Voile
$4.98

Charming taste, dainty material and low price make this dress one of our most alluring models for spring and summer. Made of light cool washable cotton **voile,** elaborated on the tuxedo collar, revers and chic wing panels on the skirt with white honeycomb embroidery, and plaited ruffles of white voile. White voile cuffs, ruffled to match, and edged with voile piping, finish the sleeves. Vestee of white voile. Dress has elastic belt, concealed by a tie sash. Fastens in front invisibly.

WOMEN'S REGULAR SIZES—32 to 44 inches bust measure. **Give measurements.** Shipping weight, 1¾ pounds.

31L6370—Blue with white.
31L6371—Rose with white.
31L6372—Reseda green with white. **$4.98**

31L6375
All Silk Taffeta
$10.95

Handsome One-Piece Dress of **all silk taffeta,** richly embroidered in beautiful Bulgarian colors on front of blouse, cuffs and graceful shirred panels which hang loosely over the skirt, extending below the hem. Elastic waistline. Fastens at shoulders. **A wonderful value.**

WOMEN'S REGULAR SIZES—32 to 44 inches bust measure. **Give measurements.** Shipping weight, 1¼ lbs.

31L6375—Navy blue.
31L6376—Brown.
31L6377—Black. **$10.95**

31L6380
All Wool Jersey
$9.45

A stunning Jacquette Blouse Dress of fine **all wool jersey** cloth. Blouse is elaborated with harmonizing **matelasse** embroidery in allover design. Fastens with large ornamental buttons. Plain collar and belt. Full accordion plaited skirt attached to a body lining. **An exclusive design and a marvelous value.**

WOMEN'S REGULAR SIZES—32 to 44 inches bust measure. **Give measurements.** Shipping weight, 1¼ lbs.

31L6380—Sand and mahogany.
31L6381—Brown and sand.
31L6382—Navy blue and gray. **$9.45**

31L6385
Voile
$4.50

A truly exquisite dress of extra fine cotton **voile** attractively trimmed with pin tucks on front of blouse, also at each side of skirt. A charming style feature is the deep bertha collar of pin tucked net and lace insertion and edging in filet pattern. Large tie sash is of voile. Dress fastens at side. A bargain at this low price.

WOMEN'S REGULAR SIZES—32 to 44 in. bust measure. **Give measurements.** Shpg. wt., 1¼ lbs.

31L6385—Rose.
31L6386—Jade green.
31L6387—Blue. **$4.50**

SIZES Dresses offered on this page can be furnished in Women's Regular Sizes—32 to 44 inches bust measure, proportionate waist measure, and front length of skirt, 31 to 37 inches, with basted hem. **When ordering give bust and waist measures; also front length of skirt.** For measuring instructions see page 557.

Dresses of Fascinating Style
designed for well-dressed American Women
Hats on this page may be found in our Millinery Section, pages 83 to 100, inclusive.
31L6390
Voile
$2.95
31L6395
All Silk Canton Crepe with All Silk Crepelette
$12.95
31L6400
All Silk Charmeuse with Matelasse Waist
$11.95
31L6405
Pongee
$5.98
This is one of the astonishing values that have made thousands of friends for Sears, Roebuck and Co. among thrifty women. It is a chic dress made of good quality figured cotton voile. The trimming consists of gathered ruffles of self material on the neck and short sleeves and the same decorative idea is carried out on front of blouse and sides of skirt. Dress fastens at side and has sash of the material tying in a bow at back. You never received such good quality for so little money.
WOMEN'S REGULAR SIZES—32 to 44 inches bust measure. Give measurements. Shipping weight, 1¾ lbs.
31L6390—Navy blue.
31L6391 Black. $2.95
Two-Tone Dress. One of our most attractive models. Skirt is made of knitted artificial silk, called crepelette; has a pleasing crepe texture and high luster. Upper part of dress is of all silk Canton crepe in a harmonizing shade. Blouse has collarless neck and is trimmed on the short sleeves and on neck and pockets with colorful embroidery banding worked in silk and gold thread on net. Elastic belt at waistline concealed by tie sash of crepelette.
WOMEN'S REGULAR SIZES—32 to 44 inches bust measure. Give measurements. Shipping weight, 2 lbs.
31L6395—Brown and tan.
31L6396—Blue and gray. $12.95
SIZES Dresses offered on this page can be furnished in Women's Regular Sizes, 32 to 44 inches bust measure, proportionate waist measure, and front length of skirt, 31 to 37 inches, with basted hem. When ordering give bust and waist measures; also front length of skirt.
A handsome dress of rare beauty and splendid quality, made of fine lustrous all silk charmeuse. The bodice, both front and back, and the huge flared cuffs are of rich, tinsel thread matelasse embroidery, one of the season's most favored decorative treatments. The skirt and the wide girdle are of plain charmeuse and the girdle has a long fringed streamer held by a metal buckle ornament. Dress fastens at side. Our price for this dress is very low, considering the superb quality and style.
WOMEN'S REGULAR SIZES—32 to 44 inches bust measure. Give measurements. Shipping weight, 1¾ pounds.
31L6400—Black.
31L6401—Brown. $11.95
You will surely admire the style of this practical, good looking dress for spring and summer, and you will be pleased, too, with the serviceable material. It is made of highly mercerized cotton Rajah pongee, cool, light weight and durable. Front of blouse is designed with three panels, adorned with contrasting embroidery and slashed to disclose the foundation beneath of voile to match color of embroidery. The skirt has graceful loose hanging straps tacked at hem and trimmed with loops and buttons like slashed sleeves. Elastic belt concealed by tie sash. Big value!
WOMEN'S REGULAR SIZES—32 to 44 inches bust measure. Shipping wt., 1¾ lbs.
31L6405—Natural tan with brown.
31L6406—Lavender with white.
31L6407—Dark brown with tan. $5.98
SEARS, ROEBUCK AND CO. 101
7

Practical, comfortable, good looking Middy Blouse for women and misses. Made of iron wearing regulation white **jean.** This blouse has a square sailor collar of all wool navy blue flannel trimmed with rows of white braid. Collar is detachable, making it very easy to launder. Buttoned cuffs are also trimmed with white braid. Embroidered emblem on sleeve. Sailor tie drawn through embroidered loop. Breast pocket. Deep yoke front and back.

WOMEN'S AND MISSES' SIZES—34 to 44 inches bust measure. GIRLS' SIZES—6 to 14 years. **State size.** Shipping weight, 12 ounces.

WOMEN'S SIZES.

27L6145—White with blue trim. **$1.98**

GIRLS' SIZES.

27L6146—White with blue trim. **1.89**

Admiral BRAND S.R.AND CO.

Middy Blouses

For Other Descriptions See Opposite Page.

Women's and misses' nicely made, comfortable, becoming, regulation Middy Blouse of durable long wearing white **jean.** Has sailor collar trimmed with rows of white braid and buttoned band cuffs trimmed to match. Breast pocket and sailor tie as pictured.

WOMEN'S AND MISSES' SIZES—34 to 44 inches bust measure. GIRLS' SIZES—6 to 14 years. **State size.** Shipping weight, 12 ounces.

WOMEN'S SIZES.

27L6155—White with blue trim.
27L6156—All white. **98c**

GIRLS' SIZES.

27L6157—White with blue trim.
27L6158—All white. **98c**

Outing and Utility Waists

Smart, Well-made and Moderately Priced

A becoming middy blouse for misses. Made of high grade **jean.** Easy to launder. Blouse has sailor collar and buttoned band cuffs trimmed with naval braid. Left sleeve has embroidered emblem. Tie included. Breast pocket. Wide attached button trimmed belt.
MISSES' SIZES—14 to 20 years; bust measures, 32, 34, 36 and 38 inches. **State size.** Shpg. wt., 12 oz.
27L6165—White with blue trim.
27L6166 All white. **$1.29**

Girls' Middy Blouse of extra strong quality Hill's khaki **jean.** Middy has a sailor collar, buttoned cuffs, and a wide, tailored belt which fastens with buttons as pictured. A sailor tie is included. It is a special value at this price and a very high quality garment.
GIRLS' SIZES—7 to 14 years. **State size.** Shipping weight, 1½ pounds.
31L02811 Khaki. **98c**

The celebrated "Admiral" line of middies has been developed by us under most rigid specifications. You will find them better made, of finer materials; you will find that they fit better and are lower priced than middies you can purchase elsewhere.
The "Admiral" label on your middy assures you of the highest grade of **jean cloth, all wool flannel** or **middy serge** obtainable; the famous non-rip placket cuff; double thread stitching throughout; finest quality embroidered navy emblems, and painstaking care in construction of every garment.

Descriptions of Middies Shown on Opposite Page.

Women's khaki tan middy blouse made of high grade durable khaki **jean.** Blouse has sailor collar, buttoned cuffs and breast pocket. Tie illustrated is included. **A well tailored garment at a bargain price.**
WOMEN'S AND MISSES' SIZES—34 to 44 inches bust measure. **State size.** Shipping weight, 12 ounces.
27L6130 Tan khaki. **$1.19**

Smart, man tailored middy blouse for sport or general wear, made of high grade light weight **all wool flannel.** Middy has sailor collar and buttoned band cuffs trimmed with white naval braid. Sailor tie drawn through an embroidered loop. **Big value.**
WOMEN'S AND MISSES' SIZES—34 to 46 inches bust measure. GIRLS' SIZES—6 to 14 years. **State size.** Shipping weight, 12 ounces.

WOMEN'S SIZES.
27L6135—Scarlet.
27L6136—Navy blue. **$3.98**
27L6137—Green.

GIRLS' SIZES.
27L6138—Scarlet.
27L6139—Navy blue. **$3.85**

Middy blouses are very stylish this year for spring and summer wear. Here is a regulation middy made of extra fine durable white **jean,** with braid trimmed sailor collar and buttoned cuffs, buttoned turned up border at bottom and breast pocket. The sailor tie pictured is included. **A very special value.**
WOMEN'S AND MISSES' SIZES—34 to 44 in. bust measure. **State size.** Shpg. wt., 12 oz.
27L6140—White with blue trim.
27L6141—All white. **$1.39**

Girls' good looking, stylish middy blouse. Extra well made of superior quality durable white **jean.** Blouse has sailor collar and buttoned cuffs, trimmed with white naval braid. Tie pictured is included. Breast pocket. Turned up buttoned band at bottom.
SIZES—6 to 14 years. **State age.** Shipping weight, 12 ounces.
27L6150—White with blue trim.
27L6151—All white. **$1.29**

Girls' and misses' middy blouse of good, strong quality tan khaki **jean.** Has sailor collar, buttoned cuffs and wide belt which fastens with buttons, as pictured. Red sailor tie included. **A special value.**
MISSES' SIZES—16 to 20 years; bust measure, 34, 36 and 38 inches. **State age and bust measure.** Shipping weight, 12 ounces. **$1.48**
27L6160—Tan khaki.
Khaki middy blouse, same as above, for girls. SIZES—6 to 14 years. **State age.** Shipping weight, 12 ounces.
27L6161—Tan khaki. **$1.35**

A well tailored, good looking, outing or camping shirt for women or misses. This mannish, comfortable garment is made of durable quality genuine khaki **jean.** This fabric is far superior to the ordinary khaki cloth. The shirt has attached collar, buttoned cuffs and breast pocket. Tie pictured is included.
MISSES' SIZES—16 to 20 years; bust measures, 34, 36 and 38 inches. WOMEN'S SIZES—34 to 44 inches bust measure. **State size and bust measure.** Shipping weight, 12 ounces.
27L6175—Khaki. Misses' sizes, 16 to 20 years.
27L6176—Khaki. Women's sizes, 34 to 44 inches bust measure. **$1.19**

Knickers are no longer a novelty. They have so many practical uses they are an essential part of every woman's wardrobe. The knickers we show above are very good fitting and whether you order them in iron wearing **khaki jean cloth, all wool tweed** or fine quality **linen,** you will receive a wonderful value. Knickers button at sides, with button cuffs at knee, have metal belt buckle and handy pockets. Waist measures, 25 to 32 inches only. **State waist measure.** Shipping weight, 2 pounds.
31L8002 Khaki Jean. **$1.79**
31L8000—Gray All Wool Tweed. **2.98**
31L8001 White All Linen. **2.98**

A simple, practical, tailored shirt waist for general wear during the spring and summer. Made of reliable quality cotton **pongette,** in dark colors which will not readily show soil. The fabric will give splendid wear and is no trouble to launder. Waist has hemstitched collar and panel effect front through which it fastens visibly between two groups of pin tucks. The long sleeves are finished with tailored cuffs. An excellent value.
WOMEN'S AND MISSES' SIZES—34 to 46 inches bust measure. **State size.** Shipping weight, 10 ounces.
27L6180—Black.
27L6181—Navy blue. **98c**

Correct Coat Styles *for* STOUT FIGURES

Hats shown on this page may be found in our Millinery section, pages 83 to 109 inclusive.

17L4680
All Wool Velour
Unlined
$14.95
Silk Lined
$18.50

Trimline BRAND S.R. and Co

Our "Trimline" garments produce the "slenderizing" effect so much desired by women of full figure.

For other garments for stout figures see following pages:

Dresses....16, 17 and 18
House Dresses.56 and 57
Kimonos....57
Skirts....40 and 41
Suits....34 and 35
Waists....54 and 55
Corsets....117
Underwear...234 and 236
Muslin Wear.154 and 155

17L4685
All Wool Poiret Twill
Peau de Cygne Lining
$29.95

17L4675
All Wool Polo Coat
$12.50
All Wool Serge
$12.50

17L4690
All Wool Bolivia
Silk Lined
$29.50

Stout women will admire the trim beauty and slenderizing lines of this fashionable Dress Coat. It is made of good quality spring weight **all wool velour,** which is both fashionable and serviceable. Coat is cut on stylish loose fitting lines with a tie girdle of self material. The adjustable collar and the back of coat are elaborated with embroidery stitching in a tasteful design, as pictured. The back has box plaits, at each side and the two pockets in front are stitched like collar. Length, 47 inches. A stylish, well tailored coat economically priced.

BUST MEASURES—41 to 53 inches. **State size.** Shipping weight, 4¼ pounds.

Unlined.
17L4680—Brown.
17L4681—Navy blue. **$14.95**

Lined With Fancy Silk.
17L4682—Brown.
17L4683—Navy blue. **$18.50**

Beautiful quality all wool Bolivia is the material used for this up to date and stylish coat, designed expressly for women of stout figure. The coat is an ideal garment for dress wear during the spring and summer, cut with graceful fullness and displaying the new and fashionable Poiret sleeves. The collar, sleeves and sides of coat are adorned with crisscross embroidery stitching. Collar has a stitched border of self material and may be fastened up at neck. The sides are finished with a stitched plait, trimmed with self covered buttons. Two pockets. Length, 46 inches. Lined throughout with all silk satin. A handsome coat combining appropriate style with economy of cost.

BUST MEASURES—41 to 53 inches. **State size.** Shipping weight, 4½ pounds.

17L4690—Tan.
17L4691—Navy blue. **$29.50**

A coat of pleasing style designed to enhance the appearance of women of full figure. Made of beautiful quality **all wool Poiret twill,** a favorite material this season, recommended for practical service and fine appearance. Coat is a graceful, loose fitting model with two box plaits in back. The collar, stitched cuffs and box plait are adorned with embroidery stitching in self color. Two pockets. Tie sash of self material. Length, 48 inches. Richly lined throughout with silk peau de cygne. A wonderful value.

BUST MEASURES—41 to 53 inches. **State size.** Shipping weight, 4½ lbs.

17L4685
Black.

17L4686
Navy blue.

$29.95

Appropriate and becoming style, faultless tailoring and skillful designing make this coat a splendid selection for stout women. It is developed in your choice of good quality **all wool material** or spring weight **all wool serge.** The coat is cut on graceful lines, fitting loosely and held by a buttoned belt of self material. The collar, cuffs, pockets and back of coat are elaborated with rows of embroidery stitching in harmonizing color, as pictured. Length, 46 inches. Unlined. This is a coat of good reliable quality which is a sterling value at the low price quoted.

BUST MEASURES—41 to 53 in. **State size.** Shpg. wt., 4 lbs.

All Wool Polo Coat.
17L4675—Tan.

All Wool Serge.
17L4676
Navy blue.
17L4677—Black.

$12.50

EVERDRY BRAND S.R. AND CO
Rain Coats
Rainwear for Women, Misses and Girls
17L5967 Fancy Mixture $2.95
17L5985 Schappe Silk $8.98
17L5972 Mercerized Cantona $5.49
17L5970 Tweed $4.69
17L5980 Fancy Mixture $3.98
17L5965 Sateen $1.98
17L5975 Tweed $6.49
Girls' Serviceable Well Tailored Everdry Raincoat, made of showerproof rubberized cotton **suiting.** Has adjustable collar which may be turned down if preferred; button trimmed cuffs and flap pockets and all around belt. An excellent storm coat very moderately priced.
AGES—6 to 15 years. (See scale of sizes). **State age.** Shipping weight, 2½ pounds.
17L5967—Oxford gray.
17L5968—Tan. **$2.95**
A Smart Everdry Raincoat of splendid style and appearance, for the well dressed woman. Coat is made of woven **Schappe silk,** a brilliant fabric 50 per cent silk, balance mercerized cotton. Looks just like an all silk coat and has rubberized under surface which makes the coat a perfect protection against the rain. Coat is a full loose model with deep armholes, adjustable tabs on sleeves, all around belt and buttoned pockets. Collar may be fastened up at neck or turned down. Inside seams are sewed, strapped and cemented. Length, 48 inches.
WOMEN'S AND MISSES' SIZES—34 to 46 in. bust measure. **State size.** Shipping wt., 2¼ lbs.
17L5985—Navy blue.
17L5986—Gray.
17L5987—Tan. **$8.98**
A Serviceable Light Weight Everdry Raincoat, made of good quality cotton **suiting** with showerproof rubberized back. This coat has raglan sleeves with adjustable buttoned tabs, all around belt, two large patch pockets with buttoned flap. Lapels may be fastened up at neck when desired. An excellent protective coat at a very low price. Length, 48 inches.
WOMEN'S AND MISSES' SIZES—34 to 46 inches bust measure. **State size.** Shipping weight, 2¼ pounds.
17L5980—Gray.
17L5981—Tan. **$3.98**
GIRLS' SCALE OF SIZES.
Ages, years.......... 4 6 8 10 12 14 15
Length, inches...... 31 33 35 37 39 42 45
Do not fail to state age when ordering.
Schoolgirls' Everdry Raincape with attached Billie Burke hood, made of lustrous rubberized **sateen.** Will keep out the damp as the fabric is showerproof. Hood is shirred on elastic. Cape has arm vents and fastens with buttons. **Built on our own rigid specifications that insure extra fullness and real rain protection.**
AGES—4 to 14 years. (See scale of sizes). **State age.** Shipping weight, 2½ pounds.
17L5965 Navy blue. **$1.98**
Girls' Natty Everdry Storm Coat, of durable quality rubberized two-thirds wool **tweed.** Looks like an all wool material and is impervious to moisture. This is a loose fitting, belted coat with adjustable collar and tabs on sleeves. Two patch pockets with buttoned flap. A well made storm coat of fine appearance, moderately priced.
AGES—6 to 15 years. (See scale of sizes.) **State age.** Shipping weight, 2½ pounds.
17L5970—Gray.
17L5971—Tan. **$4.69**
A Practical Well Tailored Everdry Raincoat, made of moisture defying showerproof rubberized **Cantona cloth,** a mercerized cotton fabric which will give excellent service. Coat has adjustable collar and tabs on cuffs, all around buttoned belt and patch pockets with flaps. A fine protection in stormy weather and a big value. Length, 48 inches.
WOMEN'S AND MISSES' SIZES—34 to 46 inches bust measure. **State size.** Shipping weight, 2½ pounds.
17L5972—Navy blue.
17L5973—Tan. **$5.49**
A Swagger Everdry Raincoat, one of the approved mannish London models, made of good quality showerproof rubberized two-thirds wool **tweed.** Has all around belt, holding fullness at waistline, adjustable collar which buttons up at neck, also adjustable tabs and buttons on sleeves. Two pockets with vents for the hand to go through. A high grade serviceable stylish coat which is a splendid value at the price. Length, 48 inches.
WOMEN'S AND MISSES' SIZES—34 to 46 inches bust measure. **State size.** Shipping weight, 3½ pounds.
17L5975—Gray.
17L5976—Tan. **$6.49**
SEARS, ROEBUCK AND CO. 21L 77

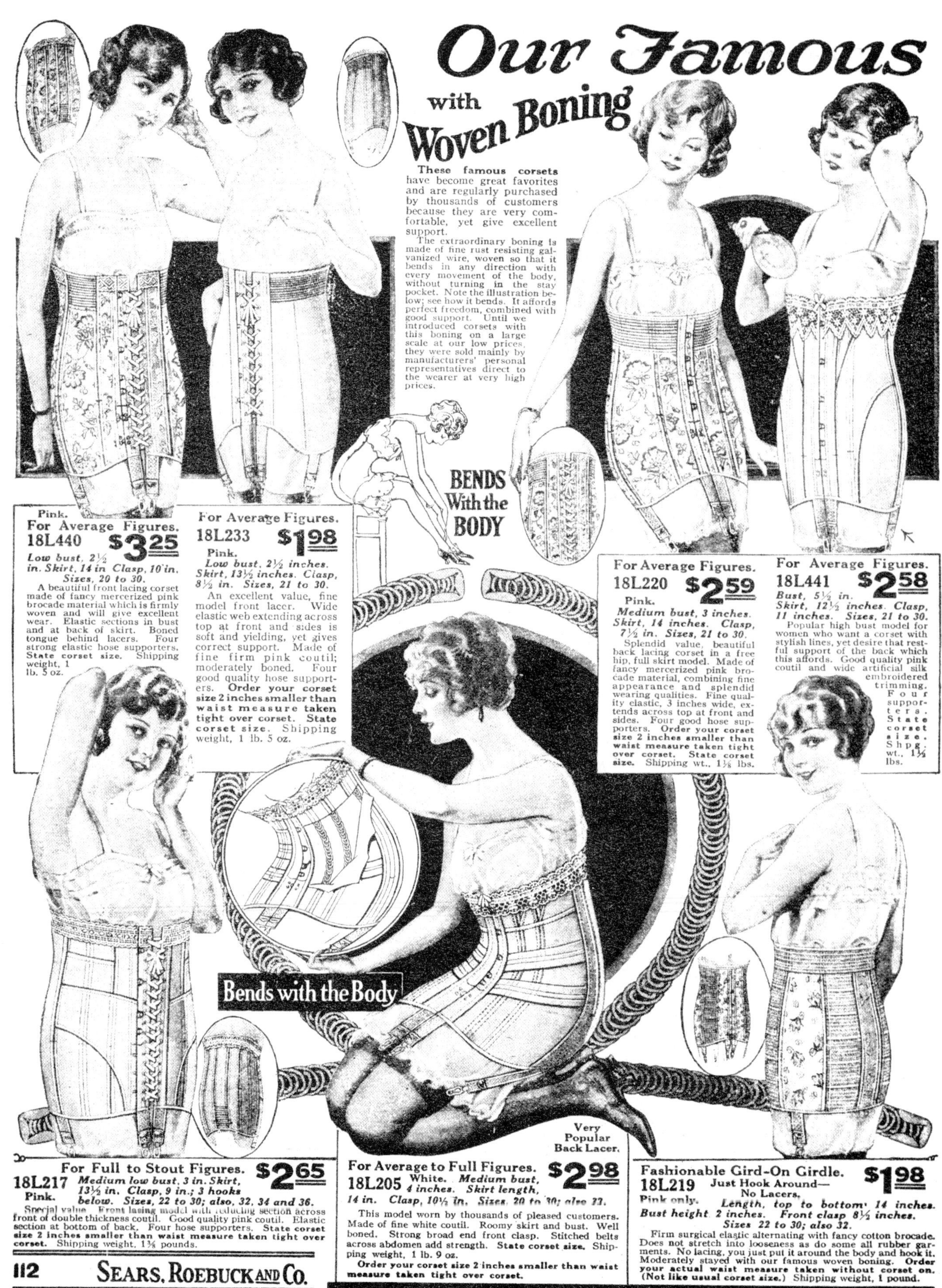
Our Famous
with
Woven Boning
These famous corsets have become great favorites and are regularly purchased by thousands of customers because they are very comfortable, yet give excellent support.
The extraordinary boning is made of fine rust resisting galvanized wire, woven so that it bends in any direction with every movement of the body, without turning in the stay pocket. Note the illustration below; see how it bends. It affords perfect freedom, combined with good support. Until we introduced corsets with this boning on a large scale at our low prices, they were sold mainly by manufacturers' personal representatives direct to the wearer at very high prices.
BENDS With the BODY
Pink.
For Average Figures.
18L440 $3.25
Low bust, 2½ in. Skirt, 14 in Clasp, 10 in. Sizes, 20 to 30.
A beautiful front lacing corset made of fancy mercerized pink brocade material which is firmly woven and will give excellent wear. Elastic sections in bust and at back of skirt. Boned tongue behind lacers. Four strong elastic hose supporters. State corset size. Shipping weight, 1 lb. 5 oz.
For Average Figures.
18L233 $1.98
Pink.
Low bust, 2½ inches. Skirt, 13½ inches. Clasp, 8½ in. Sizes, 21 to 30.
An excellent value, fine model front lacer. Wide elastic web extending across top at front and sides is soft and yielding, yet gives correct support. Made of fine firm pink coutil; moderately boned. Four good quality hose supporters. Order your corset size 2 inches smaller than waist measure taken tight over corset. State corset size. Shipping weight, 1 lb. 5 oz.
For Average Figures.
18L220 $2.59
Pink.
Medium bust, 3 inches. Skirt, 14 inches. Clasp, 7½ in. Sizes, 21 to 30.
Splendid value, beautiful back lacing corset in a free hip, full skirt model. Made of fancy mercerized pink brocade material, combining fine appearance and splendid wearing qualities. Fine quality elastic, 3 inches wide, extends across top at front and sides. Four good hose supporters. Order your corset size 2 inches smaller than waist measure taken tight over corset. State corset size. Shipping wt., 1⅛ lbs.
For Average Figures.
18L441 $2.58
Bust, 5½ in. Skirt, 12½ inches. Clasp, 11 inches. Sizes, 21 to 30.
Popular high bust model for women who want a corset with stylish lines, yet desire that restful support of the back which this affords. Good quality pink coutil and wide artificial silk embroidered trimming. Four supporters. State corset size. Shpg. wt., 1⅛ lbs.
Bends with the Body
Very Popular Back Lacer.
For Full to Stout Figures.
18L217 $2.65
Pink.
Medium low bust, 3 in. Skirt, 13½ in. Clasp, 9 in.; 3 hooks below. Sizes, 22 to 30; also, 32, 34 and 36.
Special value. Front lacing model with reducing section across front of double thickness coutil. Good quality pink coutil. Elastic section at bottom of back. Four hose supporters. State corset size 2 inches smaller than waist measure taken tight over corset. Shipping weight, 1⅜ pounds.
For Average to Full Figures.
18L205 $2.98
White. Medium bust, 4 inches. Skirt length, 14 in. Clasp, 10½ in. Sizes, 20 to 30; also 33.
This model worn by thousands of pleased customers. Made of fine white coutil. Roomy skirt and bust. Well boned. Strong broad end front clasp. Stitched belts across abdomen add strength. State corset size. Shipping weight, 1 lb. 9 oz.
Order your corset size 2 inches smaller than waist measure taken tight over corset.
Fashionable Gird-On Girdle.
18L219 $1.98
Just Hook Around—No Lacers.
Pink only.
Length, top to bottom 14 inches. Bust height 2 inches. Front clasp 8½ inches. Sizes 22 to 30; also 32.
Firm surgical elastic alternating with fancy cotton brocade. Does not stretch into looseness as do some all rubber garments. No lacing, you just put it around the body and hook it. Moderately stayed with our famous woven boning. Order your actual waist measure taken without corset on. (Not like usual corset size.) Shipping weight, 1 pound.

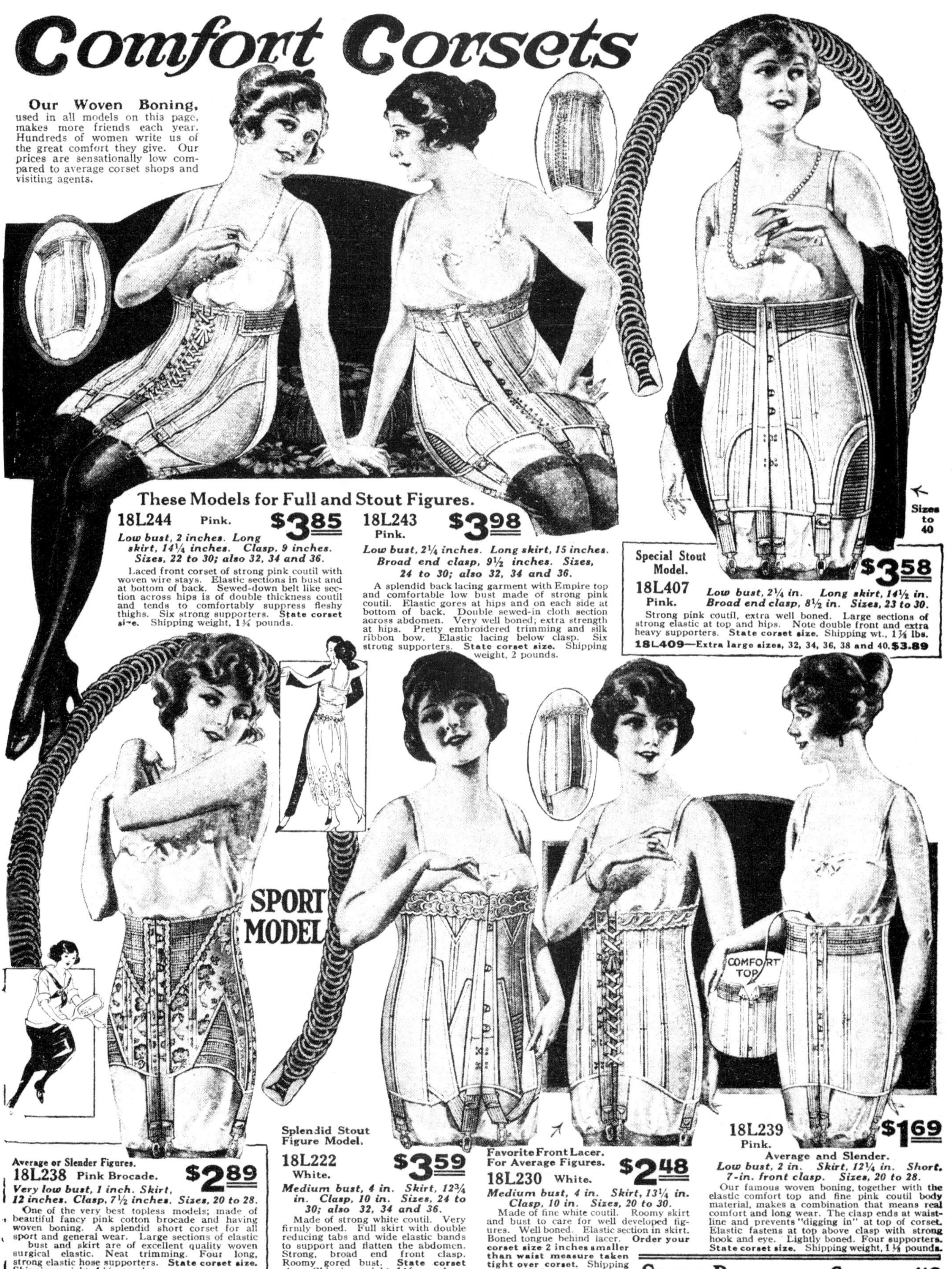
Comfort Corsets
Our Woven Boning, used in all models on this page, makes more friends each year. Hundreds of women write us of the great comfort they give. Our prices are sensationally low compared to average corset shops and visiting agents.
These Models for Full and Stout Figures.
18L244 Pink. $3.85
Low bust, 2 inches. Long skirt, 14¼ inches. Clasp, 9 inches. Sizes, 22 to 30; also 32, 34 and 36.
Laced front corset of strong pink coutil with woven wire stays. Elastic sections in bust and at bottom of back. Sewed-down belt like section across hips is of double thickness coutil and tends to comfortably suppress fleshy thighs. Six strong supporters. State corset size. Shipping weight, 1¾ pounds.
18L243 Pink. $3.98
Low bust, 2¼ inches. Long skirt, 15 inches. Broad end clasp, 9½ inches. Sizes, 24 to 30; also 32, 34 and 36.
A splendid back lacing garment with Empire top and comfortable low bust made of strong pink coutil. Elastic gores at hips and on each side at bottom of back. Double sewed-in cloth section across abdomen. Very well boned; extra strength at hips. Pretty embroidered trimming and silk ribbon bow. Elastic lacing below clasp. Six strong supporters. State corset size. Shipping weight, 2 pounds.
Sizes to 40
Special Stout Model.
18L407 Pink. $3.58
Low bust, 2¼ in. Long skirt, 14½ in. Broad end clasp, 8½ in. Sizes, 23 to 30.
Strong pink coutil, extra well boned. Large sections of strong elastic at top and hips. Note double front and extra heavy supporters. State corset size. Shipping wt., 1⅜ lbs.
18L409—Extra large sizes, 32, 34, 36, 38 and 40. $3.89
SPORT MODEL
Average or Slender Figures.
18L238 Pink Brocade. $2.89
Very low bust, 1 inch. Skirt, 12 inches. Clasp, 7½ inches. Sizes, 20 to 28.
One of the very best topless models; made of beautiful fancy pink cotton brocade and having woven boning. A splendid short corset for all sport and general wear. Large sections of elastic bust and skirt are of excellent quality woven surgical elastic. Neat trimming. Four long, strong elastic hose supporters. State corset size. Shipping weight, 1⅛ pounds.
Splendid Stout Figure Model.
18L222 White. $3.59
Medium bust, 4 in. Skirt, 12¾ in. Clasp, 10 in. Sizes, 24 to 30; also 32, 34 and 36.
Made of strong white coutil. Very firmly boned. Full skirt with double reducing tabs and wide elastic bands to support and flatten the abdomen. Strong, broad end front clasp. Roomy gored bust. State corset size. Shipping weight, 1¾ pounds.
Favorite Front Lacer. For Average Figures.
18L230 White. $2.48
Medium bust, 4 in. Skirt, 13¼ in. Clasp, 10 in. Sizes, 20 to 30.
Made of fine white coutil. Roomy skirt and bust to care for well developed figures. Well boned. Elastic section in skirt. Boned tongue behind lacer. Order your corset size 2 inches smaller than waist measure taken tight over corset. Shipping weight, 1½ pounds.
COMFORT TOP
18L239 Pink. $1.69
Average and Slender.
Low bust, 2 in. Skirt, 12¼ in. Short. 7-in. front clasp. Sizes, 20 to 28.
Our famous woven boning, together with the elastic comfort top and fine pink coutil body material, makes a combination that means real comfort and long wear. The clasp ends at waistline and prevents "digging in" at top of corset. Elastic fastens at top above clasp with strong hook and eye. Lightly boned. Four supporters. State corset size. Shipping weight, 1⅛ pounds.
SEARS, ROEBUCK AND CO. 113

New York's Most Popular Bags

BIG SPECIAL VANITY

$1.98

18L802—Sensational value! Popular new Tray Vanity Box. Articles in compartment under tray concealed when vanity is open. Made of glossy black artificial patent leather with smart double handles. Large good quality mirror. Attractive fancy gold color linings with coin purse and fittings to match. Small comb. Size, 7¼x5⅛x3⅛ inches. Shipping weight, 1 lb. 9 oz.

Billie Burke Beauty Box

$3.19

18L803—Stunning Vanity New York's Latest Creation. Attached coin purse and serviceable fittings. Long beveled mirror. Plenty of room, yet very compact. Can be had in genuine glossy black patent leather or in a very good quality brown calf leather, in a beautiful hand tooled effect, as illustrated. **Be sure to state choice.** Size, 6x3½x2½ inches. Shipping weight, 14 oz.

Distinctive Swagger Vanity Box.

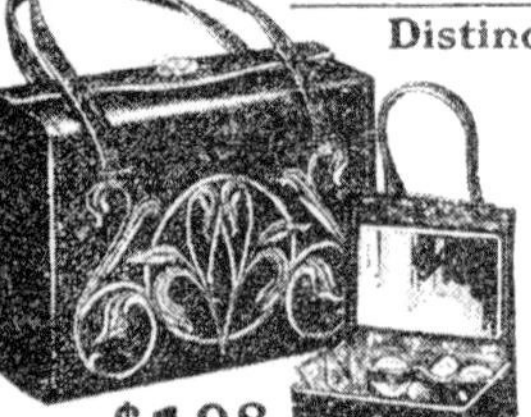

18L841

Made of good quality leather in beautiful hand tooled effect. Durable all leather double strap handles. Serviceable fittings. Finely lined. Good large mirror. Coin purse. **Colors:** Black or brown. **State color.** Size, 6½x4¾x2 inches. Shpg. wt., 1 lb. 6 oz.

$1.98

Our Dollar Special

$1.00

18L833

Challenge value in new shape Vanity Box. Medium quality leather in attractive tooled effect. Mirror, attached coin purse and small comb. Size, 5¾x5¼x2 in. **Colors:** Black or dark brown. **State color.** Shipping weight, 1 pound.

Large Vanity 93c

18L854—Startling bargain offer. Extremely popular double handle Vanity of glossy black artificial patent leather. Large mirror on flap. Fitted with small comb and powder box. Neatly lined. Large roomy size, 7½x2½x5½ inches. Shipping weight, 1¼ pounds.

$3.48

18L843—Very special value bag in new flat shape style. Good quality leather in tooled effect. Special feature large beveled mirror, 7x5 inches on inner flap. Comes in bronze color (the fashionable greenish brown shade). Size, 7⅛x5¾ inches. Shipping weight, 1¼ pounds.

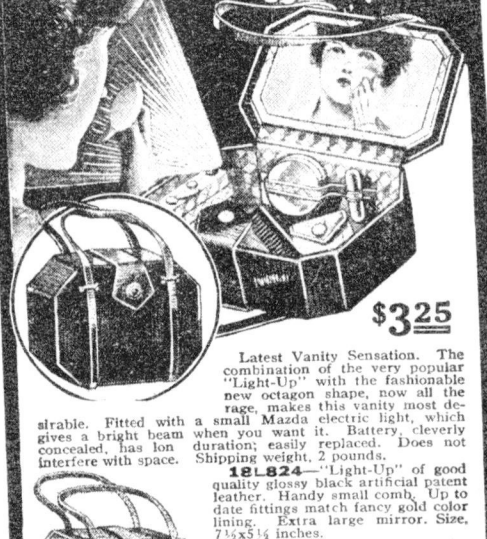

Real Pin Seal

$5.45

18L898—Our greatest offer in a full size bag. Tailor made, with very roomy pockets, one on gold colored metal frame. Pocket for powder puff, etc., on flap opposite mirror. Good quality moire lining. Large beveled mirror. Though dainty in appearance, genuine pin seal gives fine service. A high grade gift. Size, 8x5 inches. Black only. Shipping weight, 1 lb. 1 oz.

New "Light up" Box

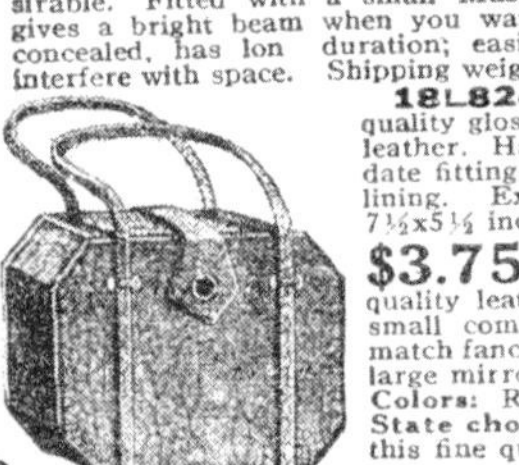

$3.25

Latest Vanity Sensation. The combination of the very popular "Light-Up" with the fashionable new octagon shape, now all the rage, makes this vanity most desirable. Fitted with a small Mazda electric light, which gives a bright beam when you want it. Battery, cleverly concealed, has lon duration; easily replaced. Does not interfere with space. Shipping weight, 2 pounds.

18L824—"Light-Up" of good quality glossy black artificial patent leather. Handy small comb. Up to date fittings match fancy gold color lining. Extra large mirror. Size, 7½x5¼ inches.

$3.75 18L821—"Light-Up" in very good quality leather. Fitted with handy small comb. Up to date fittings match fancy gold color lining. **Extra large mirror.** Size, 7½x5½ inches. **Colors:** Rich dark brown or black. **State choice.** Priced very low for this fine quality.

Drop Mirror Bag With Cord Handle.

$1.98

18L782—Special bargain. Attractive Drop Mirror Bag with new style cord handle and leather lined flap, edged with 14-karat gold clips. Good quality dark brown leather in hand tooled effect. Three deep pockets, one on clasping gold plated metal frame. Handkerchief pocket on inside flap. Large mirror. Size, 7¼x4½ inches. Shippin weight, 12 ounces.

$1.25

18L848

Popular Envelope Shape Bag. Very smart. Made of good quality leather, ornamented with beautiful hand tooled effect design. Three large deep roomy pockets. Long mirror in inner pocket. Attractively lined. Size, 8x5½ inches. **Colors:** Dark brown or black. **State color.** Shipping weight, 1 pound.

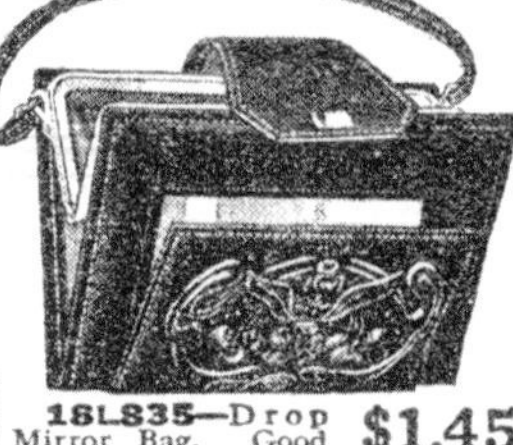

$1.69 18L874 Fine cowhide leather in hand tooled effect. Silverlike edge trimming. Leather lined flap. Deep pockets, one on clasping metal frame. Mirror. Size, 6½x5¼ inches. **Colors:** Rich dark brown or black. **State color.** Shpg. wt., 14 oz.

18L835—Drop Mirror Bag. Good quality leather, hand tooled effect. Large mirror concealed under flap. Three deep pockets, one on metal frame, opens wide. Size, 7x5¾ in. **Colors:** Dark brown or black. **State color.** Shipping weight, 12 oz. $1.45

$1.10

18L834—Bargain value. Popular style B. with special feature of extra large size mirror, 6½x4 inches. Made of medium quality leathe with beautiful design in hand tooled effect. Thr. deep pockets. Neatly lined. Size, 7¼x4¾ inches. **Colors:** Dark brown or black. **State color** Shipping weight, 1 pound.

Shopping Bag That Folds

39c

18L862—Big offer in popular Folding Carryall Bag. Ideal for everyday use. Small change pocket with snap fastener. Made of durable black artificial leather. Size, folded, 10¾x9 inches; unfolded, 16¾x12 inches. Shipping weight, 12 ounces.

59c

18L853—Splendid value bag for such a low price. Good leather in tooled effect. Three deep pockets, one on metal frame. Size, 5½x4⅝ in. Black only. Shipping weight, 8 ounces.

$3.48

New Drop Mirror Style

18L829—Entirely different and a big bargain. Large new drop mirror style swagger bag. Good quality leather in popular fluffed alligator grain. Large roomy pockets. Coin purse. Shirred pocket opposite large mirror inside flap, for powder puff, etc. Neatly lined. Size, 8x5¾ inches. **Colors:** Gray or dark brown. **State color.** Shipping weight, 1¼ pounds.

New Shapes and Colors

Imported Beaded Bag.
18L807 $2.25
Bargain in popular framed Bead Bag. Small size bright beads in attractive floral design. Rich gunmetal finish frame, attractive handle to match. Nicely lined. A wonder at this price. Size, 6⅜x5⅞ inches. Shipping weight, 12 ounces.

$2.89
18L811—Very Smart Up to Date New York Style Bag. Made of a good quality leather, attractively set off with hand tooled effect design. Large beveled mirror concealed under flap. Three deep roomy pockets. Coin purse in inner pocket. Good lining. Excellent value. Size, 6¾x5½ inches. **Colors:** Rich dark brown or black. **State color.** Shpg. wt., 14 oz.

18L837—Exceptional Bargain! $1.98
Fashionable Bag of good quality leather. Large mirror and handkerchief pocket concealed under leather lined flap. Three deep roomy pockets. Serviceable lining. Splendid size and shape, 7x5½ inches. **Colors:** Dark brown or black. **State color.** Shipping weight, 1⅛ lbs.

$1.95
18L840
For street or dress wear. Holds powder puff, etc. Clasping pocket for change, keys, etc. Good leather. Closed, 4¾x3 in. **Colors:** Rich dark brown, navy or black. **State color.** Shpg. wt., 7 oz.

Hand Laced Genuine Calf.

18L823 $3.25
Hand Laced Bag of fine quality **genuine calf leather** in beautiful brown shade. Bargain price! Hand tooled effect design. Three deep pockets, one on clasping metal frame is leather lined. Beveled mirror. Leather lined flap. Size, 6x6¼ inches. Shipping weight, 12 oz.

Real Cowhide.
18L847—This is one of our greatest bargains. $1.98
Stunning tooled effect bag of genuine cowhide with soft suede finish. All leather gussets. Three wide opening pockets, one on clasping metal frame. Long mirror in pocket. Size, 8x5½ inches. Comes in rich shaded brown color. Shipping weight, 12 ounces.

95c
18L830—Velveteen Bag with fancy metal frame at a sensationally low price. Roomy size. Length, 7½ inches. Wide opening frame. Attached mirror. Good plain lining. Strong chain handle. Tassel trimming. **Color,** black. Shipping weight, 12 oz.

BOSTON BAGS

The Better Kind.

18L899—Everybody—men, women, school children—should buy these handy Bags of good quality split cowhide, with sewed half round leather covered steel frame. Strap fastener. A good leather bag, sure to give service. We do not sell the cheaply made, flimsy quality of leather Boston bags. Height, 9 inches; width, 5¾ inches; length, 14 inches. Colors: Black or tan. State color. Shipping weight, 4½ pounds.
$1.59

Special Bargain.

18L858 98c
Again we offer this unusual value in New York's popular flat shape Bag, so much in favor with our customers. Hand tooled effect on improved grade genuine leather in dark brown shade. Three deep pockets, large one on metal frame. Mirror in separate pocket. Size, 7⅜x5¼ inches. Shpg. wt., 12 oz.

Fashionable Girdles and Belts

Smartest Fashions for Spring and Summer.
Belts and Girdles. More popular than ever. Worn over dresses, coats and sweaters by both women and children. Here again we demonstrate our leadership with big money saving values and the extra large assortment of the latest, beautiful styles.

Our Finest Girdle.

18L732—Hammered bronze effect, on beautiful dark blue pearl like slides, artistic narrow metal strips. Goes well with any color. Length, 54 inches. Fits all sizes. Shipping weight, 5 ounces. $1.35

18L729—Smart Girdle. Light weight silverlike links and bright color celluloid slides with beautiful filigree design ornaments. Fits all sizes. **Colors:** Black, red or green. **State color.** Shpg. wt., 4 oz. 69c

18L727—Very Popular Girdle. Charming new style. Novel silverlike ornaments, with beautiful strips to resemble pearl. For dresses, suits, coats, etc. Easily adjusted. Length, about 54 inches. Shipping weight, 5 ounces. $1.19

18L744—Wide Belt with large fancy double disc buckle. Made of good quality patent leather, perforated in neat pattern. Width, 1½ inches. Black only. Sizes, 26 to 40. **State size.** Shipping weight, 5 oz. 50c

18L704—Splendid quality tailored white kid (sheepskin) Belt. Neat pearl buckle. Width, ⅝ inch. Sizes, 26 to 40 inches. Special value. **State size.** Shipping weight, 2 ounces 35c

Our $1.00 Special

18L734—Stunning Girdle. Glossy celluloid rings and slides with beautiful nickel filigree design. Large clasp. 32 or 40-inch length. **Colors:** Black or red. **State length and color.** Shipping weight, 5 oz.

33c

18L736—Special low price on attractive New York Girdle. Celluloid and mounted with silver color ornaments. **Colors:** Red, black or green. **State color.** Shipping weight, 4 oz.

18L705—Genuine patent leather. Width, ⅝ inch. Sizes, 26 to 40. **Colors:** Black or red. **State color and size.** Shpg. wt., 2 oz. 14c

18L716—Dull finish kid (sheepskin). Width, ⅝ inch. **Colors:** Black or dark brown. Sizes, 26 to 40. **State color and size.** Shpg. wt., 2 oz. 15c

19c

18L738—Very special value. New style Girdle. Celluloid with nickel rings. The three large pieces are mounted with silver color ornaments. **Colors:** Red, black or green. Length, 50 inches. **State color.** Shipping weight, 3 ounces.

38c

18L740—Unusual bargain, very smart Girdle. Pretty silverlike mountings on six glossy celluloid pieces. Easily adjusted. Length, 55 inches. **Colors:** Red, black or green. **State color.** Shpg. wt., 4 oz.

18L707—Stylish belt of good patent leather. Width, 1 inch. Neat eyelet metal buckle. Sizes, 26 to 40. **Colors:** Black or red. **State size and color.** Shipping weight, 2 ounces. 18c

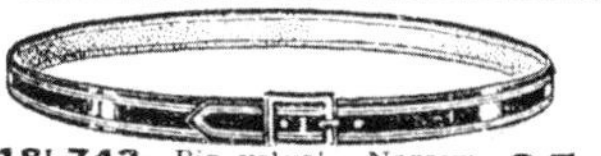

18L742—Big value! Narrow Belt of genuine black patent leather, trimmed with two rows of glossy artificial silk covered cord. Either red or green trimming. Width, ⅝ inch. Sizes, 26 to 42. **State size and color.** Shipping weight, 2 ounces. 25c

Fine Finger Purses

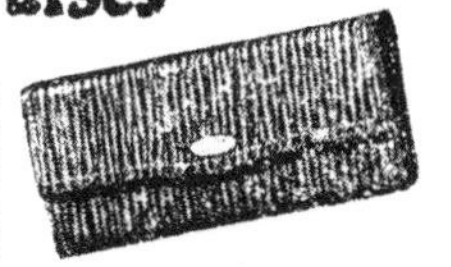

18L788 98c
Popular Finger Purse of good leather. Three pockets, one on clasping metal frame. Leather lined flap. Size, 7x3½ inches. Black or brown. **State color.** Shipping weight, 9 ounces.

18L879—Tool-ed Effect Finger 89c
Purse of good leather with strap at top. Three pockets, one on clasping metal frame. Size, 6x3½ inches. **Colors:** Bronze or black. **State color.** Shipping weight, 9 ounces.

18L820—Handy Coin Purse of fine quality leather. Leather lined. Two pockets. Black only. Size, 4¼x2½ inches. Shipping weight, 2 ounces. 21c

EVER TRY THIS? Ever buy the other fellow's goods and ours for comparison? Then judge the value each gives for the money. You will try a long time before you find us beaten.

For the Children

18L861 38c
Special value in Children's Vanity. Will delight the hearts of all little ones. Made of glossy black artificial leather. Size, 5½x1⅝ in. Shpg. wt., 9¾ oz.

18L784 29c
Low price in Children's Fancy Design Velveteen Bag. Made to look like a beaded bag. Mirror. 4¼x4½ in. Shpg. wt., 5 oz.

Charming Summer Styles
Ages, 6 Months, 1 and 2 Years. Average length, 20 inches.
98c
38L5228—White. Ages, 6 months, 1 and 2 years. State age. Cool dress for hot days, as it is made in low neck, short sleeve style. Made of good quality lawn. Front yoke of lace and embroidery. Silk ribbon rosette. Sleeves lace edged. Bottom of skirt trimmed with embroidery and lace insertions and lace edge. Sweep, 45 inches. Shipping weight, 3 ounces.
HAND EMBROIDERED YOKE
59c
38L5202—White. Ages, 6 months, 1 and 2 years. State age. Little Tots' Walking Length Dress, made of standard quality nainsook. Machine embroidered lawn front yoke. Neck and sleeves lace trimmed. Gathered back. Serviceable and easy to launder. Sweep, about 45 in. Shipping weight, 4 oz.
$1.48
38L5208—White. Ages, 6 months, 1 and 2 years. State age. Beautiful Lace Trimmed Dress of fine quality nainsook. Round hand embroidered front yoke trimmed with lace insertion and silk ribbon rosette. Bottom of skirt smartly trimmed with embroidery and rows of lace insertions and finished with lace edge. Sweep, 48 inches. Shipping weight, 3 oz.
$1.58
38L5229—White. Ages, 1, 2 and 3 years. State age. Beautiful Slipover Dress, made of fine quality lawn. Low neck and short sleeves finished with neat machine embroidered scalloped edge. Neatly trimmed with fine pin tucks and beautiful embroidery work at neck and bottom of skirt which looks like handwork. Silk ribbon sash inserted through eyelets; finished with bow in front. Hemstitched hem. Sweep, 44 inches. Shipping weight, 3 oz.
69c
38L5219—White. Ages, 6 months, 1 and 2 years. State age. Cute slipover style for little tots. Made of cotton poplin. Round neck, kimono sleeves, and bottom of skirt finished with machine embroidered scalloped edge. Artistic colored embroidered design on front of dress. Buttons down back. You will want several of these for the hot weather. Sweep, 42 inches. Shipping weight, 4 oz.
55c
38L5205—White. Ages, 6 months, 1 and 2 years. State age. Unusual value in Little Tots' Walking Length Dress. Made of standard quality nainsook. Round front yoke made of assorted patterns embroidery. Neat embroidery edge attached to bottom of skirt with row of veining. Lace edged neck and sleeves. Sweep, 43 inches. Shipping weight, 3 ounces.
98c
38L5227—White. Ages, 6 months, 1 and 2 years. State age. Cool and becoming Dress for warm days. Made in low neck, short sleeve style of attractive pattern lawn embroidery flouncing. Front yoke smartly trimmed with lace and wide embroidery insertions. Tucked back yoke. Lace edge finishes neck and short sleeves. Ribbon rosette. Sweep, 44 inches. Shipping weight, 3 ounces.
59c
38L5250—White. Ages, 6 months, 1 and 2 years. State age. Baby will be happier if kept cool and comfortable in one of these square neck, short sleeve dresses. Made of standard quality nainsook. Neck and sleeves daintily finished with an embroidery edge. Gathered back and front. Not bulky on the baby and easy to launder. Sweep, 45 inches. Shpg. weight, 4 ounces.
79c
38L5226—White. Ages, 6 months, 1 and 2 years. State age. Little Tots' Walking Length Dress, made of fine quality nainsook. Front yoke attractively trimmed with lace insertions. Bottom of skirt trimmed with lace insertions and lace edged lawn ruffle. Sweep, 43 inches. Shipping weight, 4 ounces.
39c
38L5200—White. Ages, 6 months, 1 and 2 years. State age. A remarkable value in Babies' Plain Bishop Style Dress, made of standard quality nainsook. Lace edged neck and sleeves. Gathered front and back. A full sized garment. Easy to launder. Sweep, about 45 inches. Shpg. wt., 4 oz.
89c
38L9497—Cream-white. Ages, 6 months, 1 and 2 years. State age. A Serviceable Walking Length Gertrude Style Underskirt, made of about one-fourth wool and three-fourths cotton flannel. Neat hemstitched hem. Neck and armholes finished with shell crocheted edges. Buttons at shoulders. Average sweep, 40 inches. Shipping weight, 4 ounces.
29c
38L9491—White. Ages, 6 months, 1 and 2 years. State age. Gertrude Style Walking Length Underskirt of nice quality flannelette. Neck, armholes and bottom finished with shell crocheted edge. Buttons at shoulders. Average sweep, 38 inches. Shipping weight, 4 ounces.
39c
38L5301—White. Ages, 6 months, 1 and 2 years. State age. Easy to dress the little tot with one of these Slipover Style Underskirts. No buttons or buttonholes to try the mother's patience. Garment simply slips over the head. Well made of cambric finished muslin. Trimmed at bottom with pin tucks and embroidery ruffle. Sweep, about 40 inches. Shpg. wt., 3 oz.
49c
38L5224—Pink plaid.
38L5225—Blue plaid.
Ages, 1, 2 and 3 years. State age. One of the biggest bargains of the season in a Little Tots' Colored Dress, made of good quality small plaid gingham. Made on yoke, front and back. Buy several for everyday wear. Shipping weight, 5 ounces.
65c
38L5306—White. Ages, 6 months, 1 and 2 years. State age. Beautiful Underskirt for the walking tot. Well made of good quality nainsook and attractively trimmed at bottom with Valenciennes lace insertions, strip of hemstitched lawn, and lace edged lawn ruffle. Buttons down back. Average sweep, 41 inches. Shipping weight, 3 ounces.
57c
38L5305—White. Ages, 6 months, 1 and 2 years. State age. Another attractive Walking Length Underskirt of good quality nainsook. Effectively trimmed at bottom with two rows of machine hemstitching and dainty embroidery ruffle. Buttons down back. Sweep, 41 inches. Shipping weight, 3 ounces.

Infants' Long Dresses and Sets
GOOD VALUES FOR LITTLE MONEY.
HAND EMBROIDERED YOKE
Infants' Set—26 Inches Long.
38L5094 98c White dress.
Many mothers prefer a Lawn Embroidery Dress and we have scoured the market for this wonderful value. Made of good quality lawn embroidery flouncing. Round machine embroidered front yoke. Lace edged neck and sleeves. Sweep, 44 inches. Shpg. wt., 4 oz.
85c
38L5095—White underskirt.
Lawn Embroidery Flounce Underskirt on waist, to match 38L5094 Dress. Sweep, about 42 in. Shpg. wt., 4 oz.
38L5094 DRESS
38L5095 UNDERSKIRT
38L5074 DRESS
38L5075 UNDERSKIRT
53c
38L5003—White.
Infants' size only.
Practical Long Dress, well made of standard quality nainsook. Round embroidered front yoke. Gathered back. Lace edged neck and sleeves. Length, 26 inches. Sweep, 42 inches. Shipping wt., 4 oz.
69c
38L5006—White.
Infants' size only.
Here is a wonderful opportunity to buy several Infants' Nainsook Dresses, with dainty hand embroidered yoke in assorted designs, at a price which means a great saving. Lace edged neck and sleeves. Gathered back. Length, 26 inches. Sweep, about 44 inches. Shipping weight, 4 ounces.
59c
38L5007—White.
Infants' size only.
This practical Dress is one of our biggest sellers. Made of standard quality nainsook. Embroidered front yoke. Embroidery edge attached to bottom of skirt with row of veining. Lace edged neck and sleeves. Length, 26 inches. Sweep, 42 inches. Shipping weight, 4 ounces.
Infants' Set—26 Inches Long.
38L5074 $1.39 White dress.
Fancy Lace Trimmed Nainsook Dress. Embroidery front yoke with lace insertion. Silk ribbon rosette. Lace and embroidery trimmed skirt, finished with tucked and lace edged lawn ruffle. Sweep, about 45 inches. Shipping weight, 4 ounces.
38L5075 98c White underskirt.
Gertrude Style Nainsook Underskirt, trimmed to match 38L5074 Dress. Sweep, about 43 inches. Shpg. wt., 4 oz.
39c
38L5001—White.
Infants' size only.
Inexpensive Bishop Style Dress made of standard quality nainsook. Gathered at neck both front and back. Neck and sleeves finished with narrow ruffle. Easy to launder. Length, 26 inches. Sweep, about 42 inches. Shipping weight, 4 ounces.
HAND EMBROIDERED YOKE
38L5013 DRESS
38L5015 UNDERSKIRT
Infants' Set—26 Inches Long.
38L5013 $1.48 White dress.
Our Best Long Dress, beautifully made of fine quality nainsook. Daintily hand embroidered front yoke joined to skirt with lace insertion. Silk ribbon rosette. Bottom of skirt elaborately trimmed with embroidery insertion and several rows of Valenciennes lace and lace edge. Lace edged neck and sleeves. Sweep, 44 inches. Shpg. wt., 4 oz.
38L5015 98c White underskirt.
Nainsook Underskirt, lace and embroidery trimmed, to match Dress 38L5013, made on waist. Sweep, about 43 inches. Shipping weight, 4 oz.
38L9426 89c Cream-white.
Infants' size only.
Babies' Gertrude Style Part Wool Long Flannel Underskirt. Made of about one-fourth wool and three-fourths cotton. Hemstitched hem. Neck and armholes finished with shell crocheted edge. Average length, 25 inches; sweep, 46 inches. Shipping wt., 4 oz.
89c
38L5016—White.
Infants' size only.
Long Dress made of fine quality nainsook. Bottom of skirt attractively trimmed with rows of lace insertion and lace edged lawn ruffle. Neat machine embroidered front yoke. Lace edge finishes neck and sleeves. Length, 26 inches. Sweep, 44 in. Shpg. wt., 4 oz.
39c
38L5104—White.
Infants' size only.
Babies' Long Underskirt made of standard quality nainsook. Buttons on one shoulder, making it convenient to dress baby. Bottom of skirt trimmed with clusters of tucks and narrow embroidery edge. Length, 26 inches. Sweep, 42 inches. Shipping weight, 3 ounces.
38L5092 DRESS
38L5093 UNDERSKIRT
Infants' Set—26 Inches Long.
38L5092 98c White dress.
A pretty Dress for the new arrival, inexpensively priced. Made of fine quality nainsook. Round front yoke attractively made of lace and embroidery insertions. Bottom of skirt neatly trimmed with embroidery and lace insertions and lace edge. Lace edged neck and sleeves. Sweep, 43 inches. Shpg. wt., 4 oz.
79c
38L5093—White underskirt.
Nainsook Underskirt, trimmed to match 38L5092 Dress; made on waist. Sweep, 41 inches. Shipping weight, 4 ounces.
For Babies' Shoes See Pages 196 & 197.
45c
38L9692—White.
Infants' size only.
Infants' Long Nightgown. Made of good quality flannelette. Turndown collar, cuffs and bottom trimmed with white shell crocheted edge. Buttons in front. Average length, 26 inches. Sweep, 39 inches. Shipping weight, 5 ounces.
55c
38L9689—White.
Ages, 6 months, 1 and 2 years. State age.
Infants' Nightgown of good quality flannelette. Draw string at the bottom, which insures babies' feet being always covered. Neck and sleeves edged with braid. Well made. Roomy sizes. Shipping weight, 6 ounces.
39c
38L9422—White.
Infants' size only.
Babies' Good Quality Gertrude Style Long Flannelette Underskirt. Buttons at shoulders. Trimmed with fancy machine stitching and shell crocheted edges. Average length, 26 in. Sweep, about 40 in. Shipping weight, 5 ounces.
38L5105—White. 49c
Infants' size only.
Babies' Long Underskirt made of standard quality nainsook. Buttons on one shoulder only, making it convenient to dress baby. Lace edged lawn ruffle joined to bottom of skirt with row of lace insertion. Wonderful value. Length, 26 inches. Sweep, 42 inches. Shipping weight, 3 ounces.
38L9451—White.
29c Pinning Blanket of flannelette on cambric waistband. Neatly hemmed and finished with white shell crocheted edge at bottom. Average length over all, 29 inches. An excellent value. Sweep, 30 inches. Shipping weight, 5 ounces.
SEARS, ROEBUCK AND CO. 137

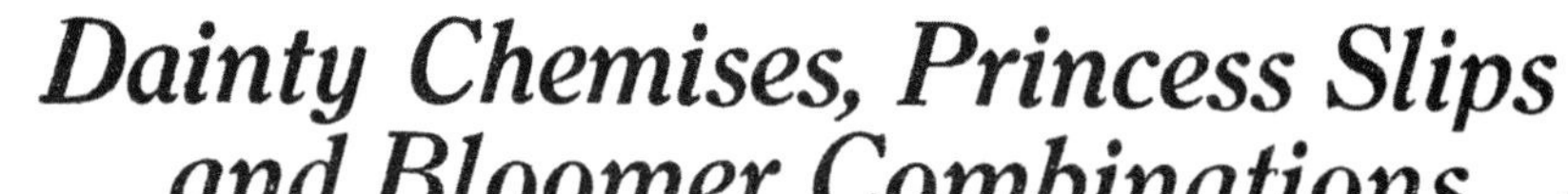

Dainty Chemises, Princess Slips and Bloomer Combinations

Attractive Styles—Priced Low

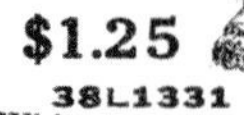

$1.25

38L1331
White.
Sizes to fit 34 to 44 inches bust measure. State size.
A Beautiful Lace Trimmed Chemise. Made in round neck style, of fine quality **nainsook.** Lace insertions and embroidery medallion make a very pretty top. Arm openings, neck and bottom are edged with Valenciennes lace. Silk ribbon draw string. A well made garment that will satisfy. Shipping weight, 4 ounces.

89c

38L1337—White.
Sizes to fit 34 to 44 inches bust measure. State size.
Women's Bloomer Combination. Made of standard quality **nainsook.** Bodice top style. Dainty Lorraine embroidery and hemstitching in front beautify this garment. Neatly shirred in front. Nainsook shoulder straps are hemstitched. Elastic at knees with neat lace edged ruffle. Open crotch. Shpg. wt., 6 oz.

38L1311 White. **$1.89**
Sizes to fit 34 to 44 inches bust measure. State size.
Women's Princess Slip. Made of standard quality **nainsook.** Front of corset cover and deep flounce made of pretty pattern embroidery. Ribbon run beading at waist and ribbon draw at neck. A wonderful garment for this price. Shipping weight, 7 oz.

38L1318 White. **$1.19**
Sizes to fit 34 to 44 inches bust measure. State size.
Women's Princess Slip made of standard quality **nainsook,** with pretty embroidery around bodice and at bottom of skirt. Finished with row of Valenciennes lace with draw string. Silk ribbon shoulder straps. Hemstitched at waistline with draw string running through. A very serviceable garment. Shpg. wt., 6 oz.

38L1329 White. **98c**
Sizes to fit 34 to 44 inches bust measure. State size.
Women's Bloomer Combination. Made of standard quality **nainsook.** Bodice consists of ribbon run embroidery, lace insertion and edge. Neatly shirred. Silk ribbon draw strings. Embroidery shoulder straps. Elastic at knees. Lace edged ruffles. Open crotch. Shipping weight, 6 ounces.

38L1336 White. **98c**
Sizes to fit 34 to 44 inches bust measure. State size.
Women's Bodice Top Step-In Combination. Made of better quality **nainsook,** prettily trimmed with lace insertions and corded organdy medallion. Shoulder straps and bottom edged with dainty lace. Elastic at waistline. Beautifully made and will please the most exacting. Shipping weight, 5 ounces.

38L1335 White. **$1.39**
Sizes to fit 34 to 44 inches bust measure. State size.
Women's Elaborate Bodice Top Chemise. Made of standard quality **nainsook.** Top consists of beautiful pattern shadow lace inserts with ribbon draw. Has two lace insertions down the front. Bottom finished with row of neat lace. Lace shoulder straps. A decidedly stylish and up to the minute garment. Shipping weight, 5 ounces.

49c

38L1301
White. **Sizes to fit 34 to 44 inches bust measure. State size.**
Women's Round Neck Chemise. Made of standard quality **nainsook.** Lace trimmed neck and armholes. Ribbon draw string. Blue shirring in front. A neat and serviceable garment. Shpg. wt., 5 oz.

STOUT SIZES. 59c
38L1313—White.
Sizes to fit 46 to 54 inches bust measure. State size.
Same style as above for stout women. Shipping weight, 10 oz.

38L1309 White. **69c**
Sizes to fit 34 to 44 inches bust measure. State size.
Women's Chemise. Made in bodice top style of standard quality **nainsook.** Top is combination of pretty lace and embroidery insertion. Neatly shirred in front. Shoulder straps are of lace. A dainty, serviceable garment that will please any woman. Shipping weight, 5 ounces.

38L1332 White. **98c**
Sizes to fit 34 to 44 inches bust measure. State size.
One of our Prettiest Chemises. Made of standard quality **nainsook** in round neck style. Bodice trimmed with lace, embroidered medallions and ribbon bows, making a very beautiful combination. Neck, shoulder straps and bottom are finished with pretty lace edge. A chemise that will please you. Shipping weight, 5 ounces.

Hand Embroidered.
38L1333 White. **89c**
Sizes to fit 34 to 44 inches bust measure. State size.
Women's Beautiful Chemise. Hand embroidered in dainty colors. Made in bodice top style of standard quality **nainsook** and trimmed with lace insertion and edging. Shoulder straps are of lace. Shirred in front. Hemstitched at bottom. A surprising value. Shipping weight, 5 ounces.

STOUT SIZES. 98c
38L1334—White.
Sizes to fit 46 to 54 inches bust measure. State size.
Same style as above in extra sizes. Shpg. wt., 10 oz.

Costume Slips and Union Suits

Cool, Dainty, Serviceable

38L1623—White.
Sizes to fit 34 to 44 inches bust measure. State size. **98c**

Women's Built-Up Shoulder Open Front Union Suit. Buttons down front. Made of good quality striped **madras.** Has draw string around neck. Knitted insert in back and sure lap flap seat. A very serviceable and exceedingly neat style. Shipping weight, 6 ounces.

38L1320—White.
Sizes to fit 34 to 44 inches bust measure. State size. **89c**

A neat and serviceable Bodice Top Princess or Costume Slip. Made of **standard quality nainsook.** This garment is gathered at hips and has row of hemstitching around top and on shoulder straps. Ribbon draw. Shadow proof skirt has 20-inch hem all around. Well made and priced exceptionally low. Shipping weight, 7 ounces.

38L1328—Black.
38L1330—White. **$1.15**
Sizes, 34 to 44 inches bust measure. State size.

Same style as above, except without 20-inch hem. Made of good quality sateen. Shipping weight, 12 ounces.

38L1321—Navy.
38L1322—White. **$2.98**
(Shadow proof).
Sizes to fit 34 to 44 inches bust measure. State size.

Women's Dainty Silk Costume Slip. Made of **tub silk** in bodice top style, with hemstitching and draw string. Hips gathered to insure good fit. The skirt of the white slip is lined with nainsook, making it shadow proof. A well made, high grade slip for dress up wear. Shipping weight, 4 oz.

38L1323
Navy. **$1.98**
38L1324—Flesh.
Sizes, 34 to 44 inches bust measure. State size.

Same style as above, made of very fine cotton charmeuse (Cotton satin). Shpg. wt., 8 oz.

Nainsook Princess Slip.
38L1327
White. **98c**
Sizes to fit 34 to 44 inches bust measure. State size.

Women's Bodice Top Costume or Princess Slip. Made of standard quality **nainsook.** Prettily trimmed with fine Lorraine embroidery work, in light colors and hemstitching around bodice and on shoulder straps. Skirt has 20-inch hem all around, which makes the garment shadow proof. Taken in and gathered at hips to give perfect fit. Shipping weight, 7 ounces.

38L1319
White. **$1.59**
Sizes to fit 34 to 44 inches bust measure. State size.

A very pretty Costume or Princess Slip. Made of a better quality **nainsook** in dainty bodice top style. Shoulder straps and top are made of good quality laces and insertions. Two attractive medallions trim the front. Skirt has 20-inch hem, making it shadow proof. Gathered at hips, insuring comfort and fit. Style, beauty, quality and price are all combined here. Shipping weight, 5 ounces.

38L1601—White.
Sizes to fit 34 to 44 inches bust measure. State size. **89c**

Excellent Quality Bodice Top Union Suit. Step-in style. Cool and serviceable. Made of good quality striped madras. Shoulder straps of same material. Hemstitched at top. Knitted ribbed insert in back. Sure lap flap seat. Draw string at bodice. Shipping weight, 5 ounces.

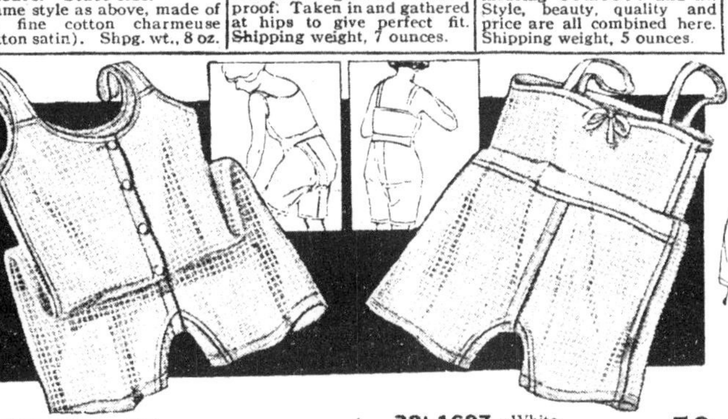

38L1606—White.
Sizes to fit 34 to 44 inches bust measure. State size. **79c**

A remarkable value in a Women's Round Neck Union Suit. Made of crossbar **nainsook.** This material is known for its good wearing qualities. Buttons down the front. Has knitted ribbed insert in back. Sure lap flap seat. Draw string. Well made and priced very low. Shipping weight, 6 ounces.

38L1607—White.
Sizes to fit 34 to 44 inches bust measure. State size. **59c**

Bodice Top Step-In Union Suit. Made of soft pajama check crossbar **nainsook,** which will wear exceedingly well. Shoulder straps of same material. Draw string at top. Knitted insert in back. Sure lap flap seat. Inexpensive yet serviceable. Shipping weight, 5 ounces.

38L1622—White.
Sizes to fit 34 to 44 inches bust measure. State size. **$1.00**

A very attractive and dainty Step-In Union Suit, well finished, with hemstitching at top and side openings. Made from fine mercerized shadow striped **lingerie cloth.** Fancy lingerie tape shoulder **straps.** Closed crotch has insert of **soft knitted** fabric insuring comfort to the **wearer.** Shipping weight, 5 ounces.

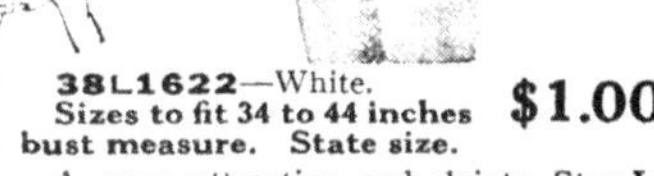

Schoolgirls' Useful Garments

Aprons—Princess Skirts—Sweaters

Full Sizes *Well Made*

38L2058—White.
Ages, 7 to 16 years. State age. **98c**
Schoolgirls' Princess Slip, made of standard quality **nainsook**. Smartly trimmed with colored machine embroidered design, looks like handwork. Wide embroidery flounce (assorted designs) at bottom. Embroidery trimmed neck and armholes. Ribbon drawstring. Buttons down back. Shipping weight, 10 ounces.

ALL WOOL

38L7533
Copenhagen blue, white trim.
38L7534 **$2.98**
Brown, peacock trim.
Sizes, 7 to 14 years. State size.
Schoolgirls' Medium Weight Jersey Tuxedo Sweater Coat. Made of good grade all wool worsted yarns. This attractive garment is well made throughout and is finished neatly with contrasting color on cuffs, lapels and sash belt. Just the right sweater for spring or summer wear. Shpg. wt., 1¼ lbs.

38L2055—Blue and white stripe. **48c**
Ages, 7 to 16 years. State age.
Schoolgirls' Blue and White Striped Gingham Underskirt made on white muslin waist. Buttons down the back. Bottom of skirt finished with a ruffle. A colored skirt saves washing. Shipping weight, 6 ounces.

38L2060 White **69c**
Ages, 7 to 16 yrs. State age.
Exceptional value in a Schoolgirls' Practical Underskirt. Slips over the head. Makes it easy to dress, as there are no buttons or buttonholes. Made of standard quality **nainsook** with colored machine stitched hems and ruffle at bottom of skirt. Shipping weight, 6 ounces.

38L7530 Navy blue. **$1.79**
38L7531—Brown.
38L7532—Honey dew (light apricot).
Sizes, 7 to 14 years State size.
Schoolgirls' Slipover Sweater. Made of fine quality all wool worsted yarns. Knitted in fancy stitch with drop stitch stripe. V shape neck. Two pockets and belt. A very serviceable and inexpensive sweater. Can be worn in place of a waist or blouse. Shipping weight, 1¾ pounds.

38L7512 Maroon. **$2.98**
38L7513—Peacock blue.
Ages, 7 to 14 years. State age.
Schoolgirls' Medium Weight All Wool Sweater Coat. Here is an exceptional value. Made with full belt, attractive Dutch collar and two pockets. We are asking a very low price for this practical well made garment. Shpg. wt., 1¾ lbs.

ALL WOOL

38L2061 White. **79c**
Ages, 7 to 16 years. State age.
Schoolgirls' Standard Quality Nainsook Princess Slip. Trimmed at bottom with neat embroidery ruffle. Ribbon drawn embroidery edge in neck. Buttons down back. Wonderful value. Shipping weight, 6 ounces.

DOUBLE BOTTOM SHADOW PROOF

38L2062—White. **89c**
Ages, 7 to 16 years. State age.
Schoolgirls' Fancy Princess Slip, made of standard quality **nainsook**. Attractively trimmed in front with embroidery medallion, outlined with lace insertion. Ribbon drawn lace edging around neck. Bottom of skirt trimmed with a tucked and lace edged lawn ruffle. Buttons down back. Shipping weight, 7 ounces.

38L4702—Solid blue. **69c**
38L4703—Solid pink.
Ages, 7 to 14 years. State age.
Schoolgirls' Slipover Apron. Made of standard quality **percale**. This apron is gathered at neck and has a large gathered pocket, which gives a very attractive appearance. Neck and short sleeves trimmed with rickrack braid. Sash back. A stylish and useful garment. Shipping wt., 10 oz.

38L4725—Black sateen, red trim. **98c**
Ages, 7 to 14 years. State age.
Schoolgirls' Colonial Style Apron. Trimmed with red rickrack braid. Has large sash which ties in back, and two pockets. An excellent garment for everyday wear, and is both neat and serviceable. Saves washing. Shipping weight, 10 ounces.
38L4726—Dark patterns percale. **75c**
Ages, 7 to 14 years. State age.
Same style as above in dark patterns percale. Trimmed with white rickrack braid. Shipping weight, 8 ounces.

38L4701—Blue and white check gingham. **75c**
38L4713—Dark patterns percale.
Ages, 7 to 14 years. State age.
Schoolgirls' Apron in a good practical coverall style. Has long sleeves and one pocket. Ties in back with sash tie strings in large bow. Collar, cuffs and pockets finished with white binding. Shipping wt., 8 oz.

38L2063—White. **89c**
Sizes, 12, 14, 16 and 18 years only. State age.
There has been a great demand for Schoolgirls' Shadowproof Costume or Princess Slips. Made of standard quality **nainsook**. Simple slips over the head; no buttons buttonholes. Gathered at hips. Makes dressing easy. Colored machine embroidered work on front yoke. Hemstitched bodice. Double panel bottom. Shpg. wt., 9 oz.

Boy's Furnishings

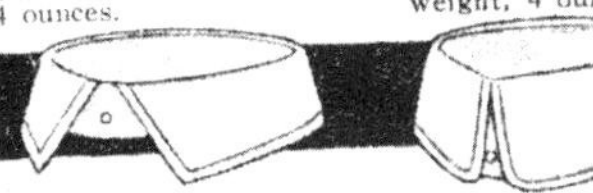

89c For 6 33L8017 Boys' Soft Collars. Good quality self figured madras. Buttons on to inside band. Back, 1½ inches. Points, 2½ inches. Half sizes, 12 to 14, neck measurement. **State size.** Shipping weight, 14 ounces.

59c For 3 33L8018 Boys' Fine Quality Plain Pique Soft Collars. Round corners. Front, 2¼ inches; back, 1⅝ inches. Half sizes, 12 to 14, neck measurement. **State size.** Shipping weight, 4 ounces.

89c For 6 33L8157 Laydown Effect Laundered Collars. Points, 2½ in.; back, 1¾ in. Half sizes, 12 to 14 neck measurement. **State size.** Shipping weight, 11 ounces.

89c For 6 33L8159 Popular Shape Laundered Collars. Front, 2 inches; back, 1⅝ inches. Half sizes, 12 to 14 neck measurement. **State size.** Shipping weight, 11 oz.

Plain Color Pajamas for Boys Are Very Popular.

$1.19 A Pajama for the boy made just like his dad's. Made of excellent quality cotton pajama cloth in plain colors. Trimmed with pearl buttons and artificial silk frog loops as shown in illustration. One pocket. Made over generous dimensions insuring plenty of room.
33L1084—White.
33L1086—Blue.
33L1087—Tan.
Ages, 6 to 16 years. **State age.** Shipping weight, 12 ounces.

Ideal Summer One-Piece Sleeping Suit.

$1.19 This One-Piece Sleeping Suit is one of the best and most comfortable ever manufactured. Made of excellent quality plain color cotton pajama cloth with neat contrasting colored trimming, as illustrated, making a very attractive garment. Button-through flap seat. Fine quality pearl buttons.
33L1080—White.
33L1081—Blue.
33L1083—Tan.
Ages, 6 to 16 years. **State age.** Shipping weight, 11 ounces.

Boys' Flannelette Nightshirt.

89c Boys' Medium Weight Flannelette Nightshirt in neat striped patterns. Flat collar. One pocket. For cool chilly nights.
33L1091 — Striped patterns.
Ages, 6, 8, 10, 12, 14 and 16 years. **State age.** Shpg. wt., 14 oz.

White Muslin Nightshirt.

75c Boys' Good Quality White Muslin Nightshirt. Made collarless style and trimmed with pearl buttons. Well made throughout over large roomy patterns. Our price is exceptionally low for such high quality.
33L1082—White.
Ages, 6, 8, 10, 12, 14 and 16 years. **State age.** Shipping weight, 11 ounces.

Genuine Cowhide.

39c Genuine Cowhide Leather Belt. Fancy grained with creased edges and lined with leather. Nickel plated self-adjusting lever buckle. Width, 1 inch.
33L8885—Black.
Sizes, 24, 26, 28 and 30 in. waist. **State size.** Shipping weight, 3 ounces.

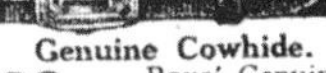

Fancy Initial.

39c Split leather belt, fancy grained, leather lined. Nickel plated roller buckle. Pierced initial. Width, about 1 inch.
33L8880—Black.
33L8881—Brown.
Sizes, 24, 26, 28 and 30 in. **State size and initial.** Shpg. wt., 3 oz.

Priced Very Low.

19c Boys' Fancy Embossed Split Leather Strap Belt. Nickel plated tongue buckle. Width, abt., 1 in.
33L8894—Black.
33L8895—Cordovan.
Sizes, 24, 26, 28 and 30 in. waist. **State size.** Shipping weight, 3 ounces.

Genuine Cowhide.

39c Boys' Genuine Cowhide Bridle Leather Strap Belt. Fancy embossing. Nickel plated self adjusting roller buckle. Width, about ¾ inch.
33L8882—Black.
33L8883—Cordovan.
Sizes, 24, 26, 28 and 30 in. waist. **State size.** Shipping weight, 3 oz.

69c For 12 33L9300 Standard Quality White Cotton Handkerchiefs. Hemstitched border. Put up in a dustproof envelope. Shipping weight, 9 ounces.

31c For 3 33L9302 Soft Finish White Cotton Handkerchiefs with colored hemstitched border. Three assorted borders in a dustproof envelope. Shipping wt., 3 oz.

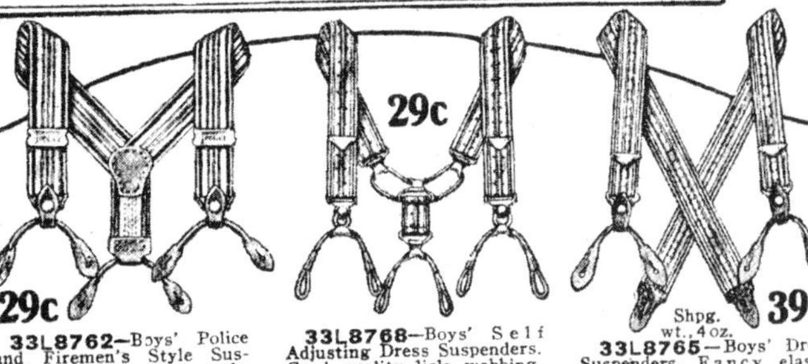

29c

29c 33L8762—Boys' Police and Firemen's Style Suspenders. Good elastic webbing. Strong leather ends. Nickel plated trimmings. Length, 30 in. Shpg. wt., 4 oz.

33L8768—Boys' Self Adjusting Dress Suspenders. Good quality lisle webbing. Nickel plated trimmings. Cord ends. Length, 30 in. Shipping weight, 4 ounces.

39c Shpg. wt., 4 oz. 33L8765—Boys' Dress Suspenders. Fancy elastic lisle webbing. Colored leather ends. Brass plated trimmings. Length, 30 inches.

39c For 3 33L9304 Boys' Fine Quality White Cotton Handkerchiefs with neat design colored initial. Hemstitched border. **State initial.** Shipping weight, 3 oz.

31c For 3 33L9303 Good quality white cotton handkerchiefs with neat printed scenes of outdoor life. Hemstitched border. Shipping weight, 3 ounces.

Windsor Ties.

19c 33L8578 Boys' Good Quality Silk Windsor Tie. A bargain at our price. Always looks neat. Comes in navy blue, brown, red, black or white. **State color.** Shipping weight, 1 oz.

33L8574
Scotch plaid.
Same as above, but in Scotch plaid effects only.

Reversible.

19c 33L8590 Fancy patterns.

33L8592—Plain colors.

Boys' Good Quality Silk and Cotton Mixed Reversible Four-In-Hand Tie. Can be worn on either side, giving double wear. A neat serviceable tie. Colors: Navy blue, red, brown, purple, gray, green or black. **State color.** Shpg. wt., 1 oz.

Fancy Four-In-Hand.

48c 33L8596 Fancy patterns.

Boys' Better Quality Four-In-Hand Tie. Made of heavy quality silk and cotton neckwear material in the very latest patterns. Attractive designs in ground colors: Navy blue, red, brown, purple, gray or green. **State color.** Shipping wt., 1 oz.

Striped Four-In-Hand.

48c 33L8593 Striped patterns.

Boys' Better Quality Four-In-Hand Tie in neat striped effects. Heavy silk and cotton mixed neckwear material. Rich colorings of navy blue, red, brown, purple, gray or green. **State color.** Shipping weight, 1 ounce.

Fancy Four-In-Hand.

29c 33L8573 Fancy patterns.

Boys' Good Quality Silk and Cotton Mixed Four-In-Hand Tie in very neat patterns. A very popular tie. Comes in ground colors: Navy blue, red, brown, purple, gray or green. **State color.** Shipping weight, 1 oz.

Boys' Knit Tie.

39c 33L8597 Fancy stripes.

Fine Quality Knitted Tie of artificial silk and mercerized cotton. Colors: Navy blue, red, brown, purple, green or black with contrasting bias cross stripes. **State color.** Shipping wt., 1 oz.

Four-In-Hand Tie.

25c 33L8571 Plain colors.

Boys' Plain Color Four-In-Hand Tie. Good quality silk and cotton mixed poplin. Comes in navy blue, light blue, red, brown, purple, gray, lavender, green, black or white. **State color.** Shpg. wt., 1 oz.

All Silk.

39c 33L8580 Plain colors.

Better Quality Heavy All Silk Windsor Tie. Plain colors: Navy blue, red, brown, black or white. **State color.** Shpg. wt., 1 oz.

33L8581
Scotch plaid.
Same. Scotch plaid effects only.

These Lowest Prices Are Proof
That We Make Your Dollar Go Further
The Famous Brooklyn Bridge. One of the greatest engineering feats of modern times, connects New York City and Brooklyn.
Just as famous, in their way, are our overalls and raincoats. They're sturdily made for real hard wear.
OUR BEST OVERALLS
Tan Bombazine—Diagonal Print Weave.
Looks Like Gabardine.
40L3356.....$3.89
40L3966—Hat to match..............48c
Handsome Brown TWEED.
40L3358.....$4.98
Dark Gray Pincheck.
40L3359.....$3.45
40L3959—Hat to match..............48c
Coat sizes, 5 to 17 years. State size. Hat sizes, 6⅜ to 7⅛. State size.
Waterproof Rubber Coats.
40L3355—Black Coat. Dull Finish. Lined......$2.98
40L3955—Hat to match............49c
40L3357—Maroon. Better Quality Coat....$3.95
40L3957—Hat to match............55c
Slicker Coat.
40L3351—Black.
40L3353 Olive Drab..$2.49
40L3951—Hat to match black coat........45c
40L3953—Hat to match drab coat.........45c
Coat sizes, 6 to 17 years. State size. Hat sizes, 6½ to 7⅛. State size.
Shipping weight, coat, 3½ pounds; hat, 8 ounces.
Real Work Overalls.
FOR BOYS.
HEAVY WEIGHT White Back Indigo BLUE DENIM.
40L3126 Sizes, 3 to 8 yrs. $0.98
40L3128 Sizes, 9 to 17 yrs. 1.25
Jacket to Match.
40L3129 Sizes, 9 to 17 yrs. $1.25
High grade garments, strongly made of heavy weight white back denim that will give long, satisfactory wear. Triple stitched inseams. All points of strain reinforced. Two regular pockets in back, two full swing pockets in front; all have strongly bar tacked corners. Cut over big, roomy patterns. High grade garments sure to please. State size. Shipping weight, each garment, 1¼ pounds.
OUR BEST ALL-OVER SUIT
These Overalls Will Last Longer.
The Knees Are Doubled.
Made of Double and Twist BLUE DENIM.
40L3131 Sizes, 6 to 14 years. 95c
The double knees on these overalls will add months of hard service to the life of the garment. Medium weight double and twist blue denim of fast color. Sewed throughout with strong, heavy thread and all seams are double stitched. Reinforced at strain points to prevent ripping. Garment is cut full and roomy and can be worn with comfort over other clothes. Big, generous pockets and high bib front. Patent buttons, will not tear out. Adjustable suspenders. A real work garment, strong and durable. State size. Shpg. wt., 1⅛ lbs.
Handy Overall Suits.
Cover Him Up From Head to Toe. Good Quality KHAKI.
40L3114—Sizes, 3 to 10 years..........98c
Genuine Stifel INDIGO BLUE DRILL.
40L3116—Sizes, 3 to 10 years..........98c
These practical suits can be worn alone and they are made big enough to be worn over other garments. Made of strong washable materials. Three pockets and other points of strain strongly bar tacked. All important seams double stitched. Closes in front with patented buttons. State size. Shipping weight, 1 pound.
Our Best All-Over Suit.
Medium Heavy Weight Double and Twist BLUE DENIM.
40L3132—Sizes, 6 to 10 years.........$1.69
40L3134—Sizes, 11 to 14 years........ 1.89
Medium Heavy Weight KHAKI TWILL.
40L3136—Sizes, 6 to 10 years.........$1.69
40L3138—Sizes, 11 to 14 years........ 1.89
Our Highest Grade One Piece Overall Suit. Strongly made for real service. Seams reinforced and pockets and other strain points bar tacked to prevent ripping. Cut full and roomy and can be comfortably worn over other clothes. Materials are of tested quality and guaranteed fast color. State size. Shipping weight, 1½ pounds.
For Men's and Youths' Overalls See Pages 268, 289, 290 and 291.

Playtime Overalls.
Durable and Priced Low.
Good Quality INDIGO BLUE DRILL.

40L3103 — **50c**
Sizes, 3 to 8 years.
Low priced overalls that will give good service. Made of good wearing Indigo Blue Drill of medium weight, with dotted white stripes. Double stitched seams. Corners of pockets and other strain points are securely bar tacked. **State size.** Shipping weight, 14 ounces.

Economical Play Garment.
Fast Color Trimmings.
Genuine Stifel INDIGO BLUE Shadow Stripe DRILL.

40L3608 — **45c**
Sizes, 2 to 7 years.

Good Weight Washable KHAKI DRILL.

40L3610 — **45c**
Sizes, 2 to 7 years.
Slip it on over the dressy clothes when on visits, at picnics, etc., or it can be worn with blouse only. Strongly made of durable materials. **Double stitched** seams. **State size.** Shipping weight, 8 ounces.

BIG PAL SUITS

Easy to Slip Into—Easy to Take Off.
Durable FAST COLOR KHAKI.

40L3611 — **89c**
Sizes, 3 to 8 years.
40L3613 — **98c**
Sizes, 9 to 12 years.

Strong INDIGO BLUE DENIM.
40L3629—Sizes, 3 to 8 years. 89c
40L3630—Sizes, 9 to 12 years. 98c

Nothing pleases the boy more than to dress and act like his dad. He'd be just tickled to do those little chores around the house if he had one of these suits to wear. Made big and roomy so they can be worn with comfort over his other clothes. They make practical play garments, too. Material is firmly woven. Seams are reinforced. Riveted buttons. **State size.** Shipping weight, 1¼ pounds.

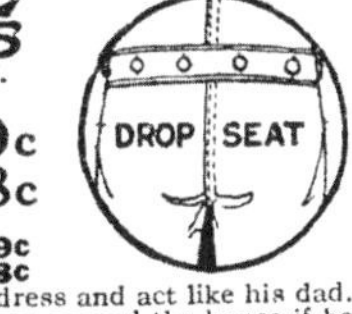

Lowest Prices —and Better Quality too!

For Dad's Helper.
Medium Heavy Weight BLUE DENIM OVERALLS.

40L3110 — **95c**
Sizes, 6 to 17 years.
Strongly made of medium heavy weight double and twist indigo blue denim. **Triple stitched legs;** pockets are reinforced. Plenty of roomy pockets that will be found very handy. Attached suspenders. **State size.** Shipping weight, 1¼ pounds.

Double and Twist BLUE DENIM JACKET. To Match Above Overalls.

40L3112 — **95c**
Sizes, 9 to 17 years.
Very convenient to slip on when working around the yard or the garage. Strongly made, sleeve and shoulder seams **triple stitched;** pockets reinforced to prevent ripping. **State size.** Shipping weight, 1 pound.

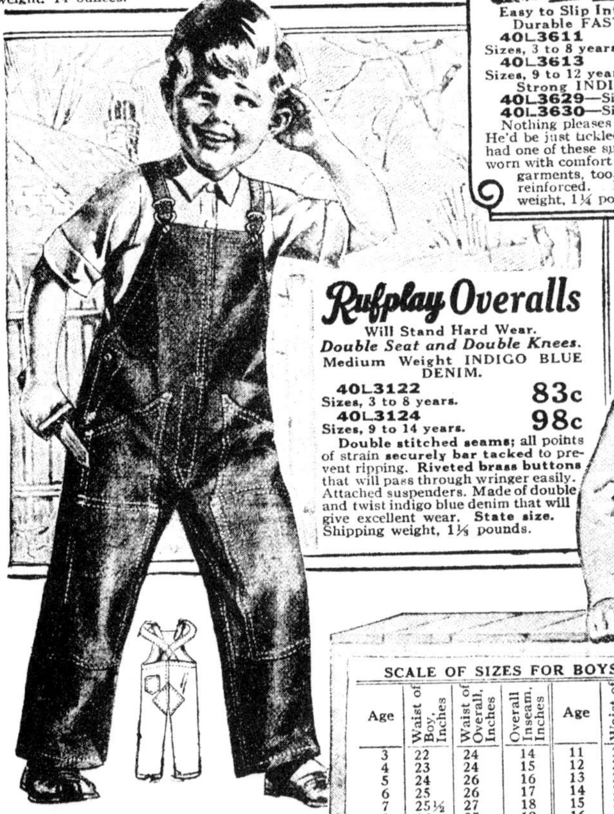

Rufplay Overalls

Will Stand Hard Wear.
Double Seat and Double Knees.
Medium Weight INDIGO BLUE DENIM.

40L3122 — **83c**
Sizes, 3 to 8 years.
40L3124 — **98c**
Sizes, 9 to 14 years.

Double stitched seams; all points of strain **securely bar tacked** to prevent ripping. **Riveted brass buttons** that will pass through wringer easily. Attached suspenders. Made of double and twist indigo blue denim that will give excellent wear. **State size.** Shipping weight, 1⅛ pounds.

Have You Tried Our Famous
Rufplay Rompers
for the Little Fellows?
See Page 168.

Every Garment Made Over Full and Roomy Patterns.

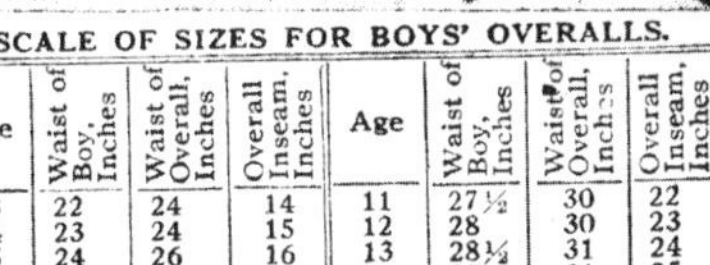

SCALE OF SIZES FOR BOYS' OVERALLS.

Age	Waist of Boy, Inches	Waist of Overall, Inches	Overall Inseam, Inches	Age	Waist of Boy, Inches	Waist of Overall, Inches	Overall Inseam, Inches
3	22	24	14	11	27½	30	22
4	23	24	15	12	28	30	23
5	24	26	16	13	28½	31	24
6	25	26	17	14	29	31	25
7	25½	27	18	15	29½	32	26
8	26	27	19	16	30	32	27½
9	26½	28½	20	17	30½	33	28½
10	27	29	21				

40L3126, 40L3128, 40L3131, 40L3134 and 40L3138 have longer inseams, allowing for turn up bottoms.

Medium Weight Double and Twist BLUE DENIM.
Sure to Satisfy.

40L3101 — **59c**
Sizes, 3 to 8 years.
40L3102 — **79c**
Sizes, 9 to 14 years.

This garment is a big value at our price. Made of double and twist medium weight blue denim that will give good service. Seams are double stitched and pocket corners are bar tacked. Adjustable suspenders. Patent buttons will not tear out. **State size.** Shipping weight, 14 ounces.

New Styles for Boys

All Caps on this page have Canvas Visors.

All Caps on this page have Canvas Visors.

Stitched Throughout.
98c Boys' Cloth Hat. Leather shield protector. Made of a good quality wool mixed tweed. Snap crown. Twill lining and sweatband.
93L4748—Gray mixture.
93L4749—Brown mixture.
Sizes, 6⅝ to 7⅛. State size. Shpg. wt., 1¼ lbs.

A Snappy Style.
59c Boys' Eight-Quarter Golf-Style Cap. Made of wool mixed shepherd check or serge. Taped seams. Leather sweatband.
93L4766—Gray and black.
93L4767—Navy blue serge.
Sizes, 6⅝ to 7⅛. State size. Shipping weight, 14 ounces.

Comfort Hat.
23c Boys Inexpensive Work or Play Hat. Woven from a good quality peanit straw. A very good value at our low price.
93L4750—Natural.
Sizes, 6⅝ to 7⅛. State size. Shpg. wt., 1 lb.

The Latest Style and All Wool.
89c Splendid One-Piece Plaited Golf Style Cap for Boys. Made of a good all wool serge or tweed cloths. Leather shield protector. Good quality twill lining.
93L4782—Navy blue serge.
93L4783—Gray tweed.
Sizes, 6⅝ to 7⅛: State size. Shipping weight, 1 pound.

All Wool.
98c Boys' One-Piece Plaited Golf Style Cap of a good all wool cloth. Leather shield protector. Twill lining.
93L4786—Gray mixture.
93L4787—Brown mixture.
Sizes, 6⅝ to 7⅛. State size. Shipping wt., 1 lb.

It's Rubberized.
69c Boys' Good Quality Rubberized Cotton Poplin Cloth Hat. Taped seams and leather sweatband. Great protection in bad weather.
93L4758—Blue.
93L4759—Olive tan.
Sizes, 6⅝ to 7⅛. State size. Shpg. wt., 14 oz.

All Wool.
69c Boys' One-Piece Golf Style Cap of all wool suitings or serge cloth. Good quality twill lining. Leather shield protector.
93L4780—Gray mixture.
93L4781—Brown mixture.
93L4769—Navy blue serge.
Sizes, 6⅝ to 7⅛. State size. Shipping weight, 1 pound.

Wool Felt.
98c Boys' Mannish, Stylish, Durable Wool Felt Telescope Style Hat. Crown, 3½ inches high. Welt edge brim, 2¼ inches wide.
93L6010—Black.
93L6011—Gray.
93L6012—Navy blue.
93L6013—Brown.
Sizes, 6⅝ to 7. State size. Shipping weight, 1¼ pounds.

The Buddy Junior.
$1.39 Boys' Smart Military Style Hat. A good quality wool felt is used in manufacturing this hat. Crown is about 4¾ in. high. Flat set brim, 2¾ in. wide.
93L6036—Army drab.
Sizes, 6⅝ to 7⅛. State size. Shipping weight, 2¼ pounds.

The Trooper.
$1.15 New style trooper shape. Made of a fine quality wool felt. Crown about 5⅜ in. high. Welt edge brim, 2¼ in. wide.
93L6025—Black.
93L6026—Brown.
93L6027—Green.
93L6028—Navy blue.
Sizes, 6⅝ to 7⅛. State size. Shipping wt., 1⅞ lbs.

The Sailor Soldier.
79c Little Fellows' Extremely Stylish Middy Hat. The style, quality and workmanship are beyond comparison. Made of a fine quality wool mixed tweed in neat designs. Silk faced serge hat lining.
93L4725—Gray mixture.
93L4726—Brown mixture.
Sizes, 6¼ to 6⅞. State size. Shpg. wt., 14 oz.

Cloth Rah-Rah.
69c Little Fellows' Extremely Snappy and Stylish Rah-Rah Hat. Made of a fine quality shepherd check cotton cloth. Good quality cloth lining. A very good value.
93L4745—Gray and black check.
93L4746—Gray and brown check.
Sizes, 6¼ to 6⅞. State size. Shpg. wt., 14 oz.

ALL WOOL

Two Special Values for Little Fellows

Eight-Quarter Top.
49c Little Fellows' Eight-Quarter Golf Style Cap. Made of an assortment of all wool suiting or serge cloths. Taped seams. Leather sweatband.
93L4762 — Assorted mixtures.
93L4768—Navy blue serge.
Sizes, 6¼ to 6⅞. State size. Shpg. wt., 10 oz.

One-Piece Top.
75c Little Fellows' One-Piece Golf Style Cap. Made of an all wool serge or tweed cloths. Good quality twill lining. Leather shield protector.
93L4770—Navy blue.
93L4774—Gray tweed.
Sizes, 6¼ to 6⅞. State size. Shpg. wt., 11 oz.

Jack Tar.
69c An exact duplicate of the U. S. Middy Hat. Little Fellows' Naval Style Hat. Made of a good quality regulation U. S. cotton drill. Stitched brim. Taped seams.
93L4714—White.
Sizes, 6¼ to 6⅞. State size. Shpg. wt., 14 oz.

Unusual Value.
89c Little Fellows' Very Smart and Snappy Rah-Rah Hat. Box plaited crown. Made of good quality wool mixed tweed. Good quality cloth lining.
93L4740—Gray mixture.
93L4741—Brown mixture.
Sizes, 6¼ to 6⅞. State size. Shipping wt., 1 lb.

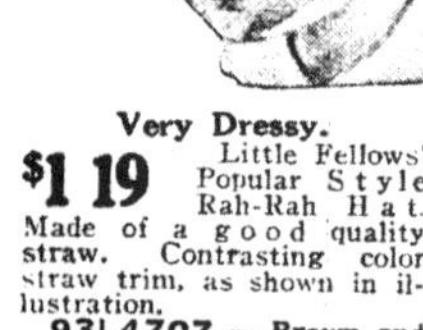

Very Dressy.
$1.19 Little Fellows' Popular Style Rah-Rah Hat. Made of a good quality straw. Contrasting color straw trim, as shown in illustration.
93L4707 — Brown and sand.
93L4708 — Navy blue and white.
Sizes, 6¼ to 6⅞. State size. Shipping wt., 1¼ lbs.

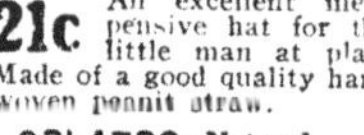

Here's Comfort.
21c An excellent inexpensive hat for the little man at play. Made of a good quality hand woven peanit straw.
93L4730—Natural.
Sizes, 6⅛ to 6⅝. State size. Shipping weight, 1 lb.

This Is Cute.
39c Little Fellows' Rah-Rah Hat. Made of a good quality cotton cloth. Taped seams. Contrasting trim, as illustrated.
93L4737—Solid white.
93L4738—White with navy blue trim.
93L4739—White with brown trim.
Sizes, 6¼ to 6⅞. State size. Shipping weight, 10 ounces.

Very Attractive.
75c A most wonderful offer in a Little Fellows' Rah-Rah Hat. Made of a good quality artificial silk and cotton mixed cloth. Taped seams. An excellent hat for summer wear.
93L4716—Gray and black check.
Sizes, 6¼ to 6⅞. State size. Shipping wt., 10 oz.

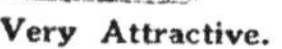

Good Quality.
59c Little Fellows' Good Quality Straw Rah-Rah Hat. Made of an excellent quality straw braid. A fine quality at an exceptionally low price. Very stylish. Will surely please you.
93L4710—Black.
93L4711—White.
93L4712—Brown.
Sizes, 6¼ to 6⅞. State size. Shpg. wt., 1¼ lbs.

Big Value.
79c Little Fellows' Rah-Rah Hat. A good quality straw used throughout in making this hat. Has the new style welt crown, as illustrated.
93L4703—Black.
93L4704—White.
93L4705—Brown.
Sizes, 6¼ to 6⅞. State size. Shpg. wt., 1¼ lbs.

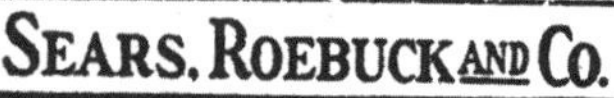

Made *for* Hard Service
Priced to Save You Money

-It sure is easy to order the right size! See Scale Below

Remember—*Every suit we offer is made over full true to size patterns (not skimpy) and no allowances need be made.*

For centuries the elephant has been known as the strongest of beasts. Gentle, willing, a hard worker and long lived, there is no task that this great animal will not undertake.

For strength, long wear and service these suits have a nation wide reputation. Nothing cheap, but the price. And, we guarantee them fully.

Dark Blue Serge.
Dandy Suit for School Wear.
No Better Value Anywhere.
40L3201
Sizes, 8 to 15 years.

$4.75

Dark blue serge in the all round belted model which is always in demand. Material is a firm finished serge weave fabric, about 40 per cent wool. Will stand lots of rough usage. Roomy side pockets with flap and breast pocket. Twill lining in coat. Full lined pants. **State size.** Shpg. wt., 3¼ lbs.

Golf Style Cap to Match.
40L3901
Sizes, 6½ to 7⅛. Shpg. wt., 10 ounces.

69c

RUFFOSUITS
FOR ROUGH WEAR

DOUBLE SEAT and DOUBLE KNEES.

"As Tough as an Elephant's Hide."

40L3215—**Dark Olive Gray.**
40L3217 — Rich Dark Brown.

$6.39

Sizes, 6 to 17 years.
We know how hard boys are on clothes—climbing trees, jumping fences, always dashing around. Here is a suit as "boyproof" as it can possibly be made. The material is a strong, firm cassimere, about 40 per cent wool. **Double seat and knees are strongly sewed on the inside.** Seams are double stitched and **bar tacked** where the strain comes. **Durable** lining in coat and pants. **State size.** Shpg. wt., 3⅝ lbs.

A Dandy Suit for General Wear.
Bluish Gray Cassimere.
40L3202
Sizes, 5 to 15 years.

$3.98

A smart looking well made suit in a pleasing bluish gray mixture. A style that will appeal to those who like a neat, plain suit. The material has a soft finish that will not wear shiny. Seams are double stitched and all belt loops, pockets and other strain points are bar tacked to prevent ripping. A durable suit that will stand up well under the wear that boys give their clothes. About one-third wool. Coat has good quality lining. Full lined pants. **State size.** **Shipping weight, 3¼ pounds.**

A New Model.
Venetian Bound Seams—The Latest Feature in Boys' Suits.
40L3203
Sizes, 5 to 15 years.

$3.35

This is really a wonder value. New spring model with yoke and plaits in front and back. The inside of the coat has seams bound with lustrous Venetian like some late style men's suits. Pants are unlined. Strongly sewed and finished throughout. Dark brown Cassimere, about one-third wool. **State size.** Shpg. wt., 2¼ lbs.

Genuine Crompton.
"All Weather" Corduroy.

40L3273 Dark Drab. **$6.75**
40L3277 Golden Brown. **7.45**
Sizes, 5 to 17 yrs.

This suit, made of the famous Crompton corduroy, known throughout the country, is ideal for school wear. Material is specially treated to shed water. Coat has strong twill lining. Back has yoke and inverted plait, very stylish. Full lined pants. **State size.** Shipping weight, 4 pounds.

40L3275—Dark Drab Corduroy. Style as above.
Sizes, 5 to 17 years.

$4.95

Made of strong corduroy, but not waterproofed. Knickerbocker pants. An excellent value. **State size.** Shipping wt., 3½ lbs.

Look at Our Simple Scale of Measurements.

Simply take the boy's chest measure over his blouse or shirt and order a suit of the corresponding age, as shown in the table below:

Chest, Inches	22	23	24	24½	25	26	26½
Order Size	3	4	5	6	7	8	9

Chest, Inches	27	27½	28	29	30¼	32	33	34
Order Size	10	11	12	13	14	15	16	17

See Simple Measuring Instructions on page 557.

Examples:—
If your boy's chest measures 30 inches, our size 14 suit will fit him. If your boy's chest measures 31 inches, our size 15 suit will fit him. If your boy's chest measures 27¼ inches, our size 11 suit will fit him.

Of Course We Guarantee Correct Fit.

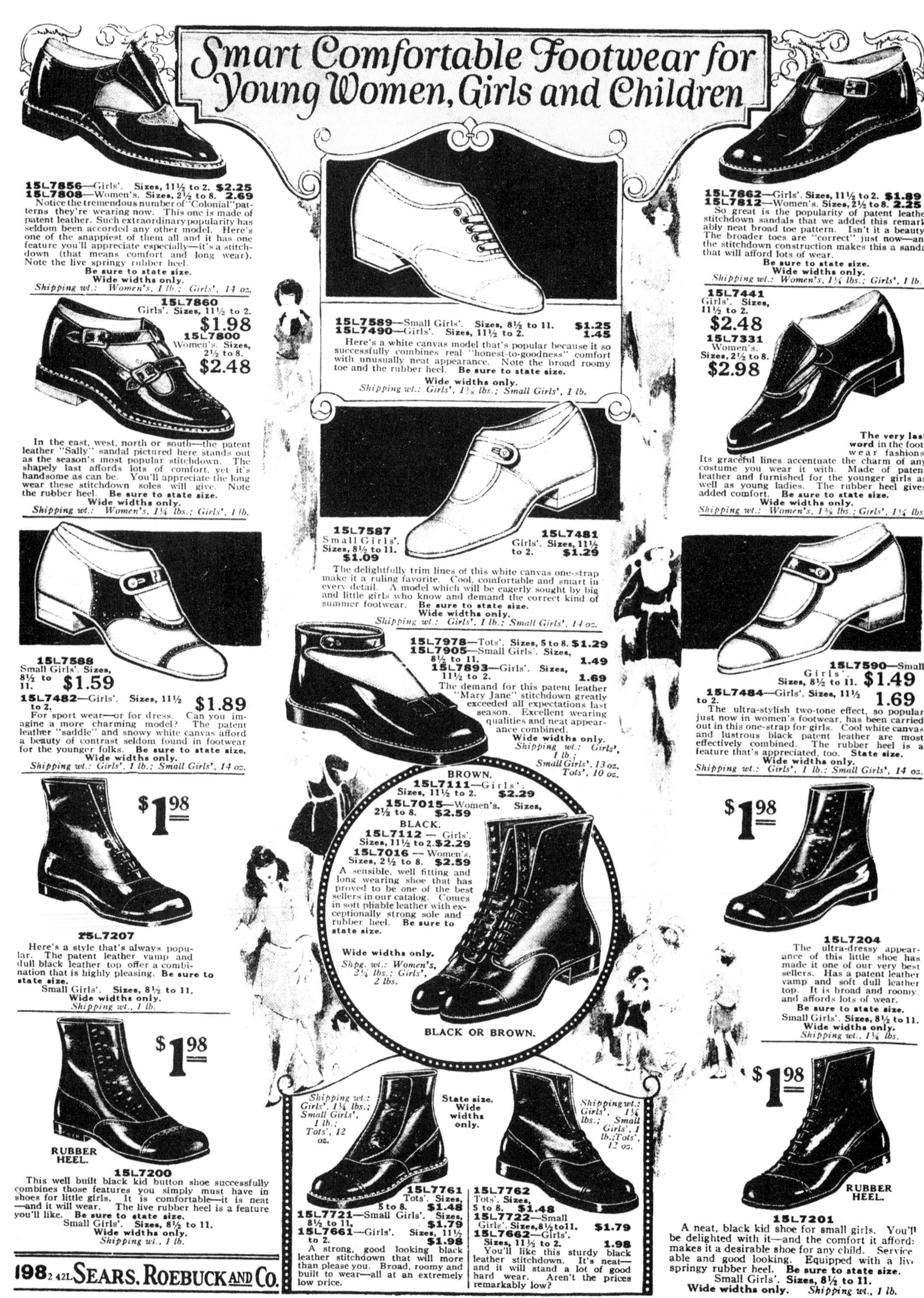
Smart Comfortable Footwear for Young Women, Girls and Children
15L7856—Girls'. Sizes, 11½ to 2. $2.25
15L7808—Women's. Sizes, 2½ to 8. 2.69
Notice the tremendous number of "Colonial" patterns they're wearing now. This one is made of patent leather. Such extraordinary popularity has seldom been accorded any other model. Here's one of the snappiest of them all and it has one feature you'll appreciate especially—it's a stitchdown (that means comfort and long wear). Note the live springy rubber heel.
Be sure to state size.
Wide widths only.
Shipping wt.: Women's, 1 lb.; Girls', 14 oz.
15L7589—Small Girls'. Sizes, 8½ to 11. $1.25
15L7490—Girls'. Sizes, 11½ to 2. 1.45
Here's a white canvas model that's popular because it so successfully combines real "honest-to-goodness" comfort with unusually neat appearance. Note the broad roomy toe and the rubber heel. Be sure to state size.
Wide widths only.
Shipping wt.: Girls', 1⅛ lbs.; Small Girls', 1 lb.
15L7862—Girls'. Sizes, 11½ to 2. $1.89
15L7812—Women's. Sizes, 2½ to 8. 2.25
So great is the popularity of patent leather stitchdown sandals that we added this remarkably neat broad toe pattern. Isn't it a beauty? The broader toes are "correct" just now—and the stitchdown construction makes this a sandal that will afford lots of wear.
Be sure to state size.
Wide widths only.
Shipping wt.: Women's, 1¼ lbs.; Girls', 1 lb.
15L7860
Girls'. Sizes, 11½ to 2.
$1.98
15L7800
Women's. Sizes, 2½ to 8.
$2.48
In the east, west, north or south—the patent leather "Sally" sandal pictured here stands out as the season's most popular stitchdown. The shapely last affords lots of comfort, yet it's handsome as can be. You'll appreciate the long wear these stitchdown soles will give. Note the rubber heel. Be sure to state size.
Wide widths only.
Shipping wt.: Women's, 1¼ lbs.; Girls', 1 lb.
15L7441
Girls'. Sizes, 11½ to 2.
$2.48
15L7331
Women's. Sizes, 2½ to 8.
$2.98
The very last word in the footwear fashions. Its graceful lines accentuate the charm of any costume you wear it with. Made of patent leather and furnished for the younger girls as well as young ladies. The rubber heel gives added comfort. Be sure to state size.
Wide widths only.
Shipping wt.: Women's, 1⅜ lbs.; Girls', 1¼ lbs.
15L7587
Small Girls'. Sizes, 8½ to 11.
$1.09
15L7481
Girls'. Sizes, 11½ to 2. $1.29
The delightfully trim lines of this white canvas one-strap make it a ruling favorite. Cool, comfortable and smart in every detail. A model which will be eagerly sought by big and little girls who know and demand the correct kind of summer footwear. Be sure to state size.
Wide widths only.
Shipping wt.: Girls', 1 lb.; Small Girls', 14 oz.
15L7588
Small Girls'. Sizes, 8½ to 11. $1.59
15L7482—Girls'. Sizes, 11½ to 2. $1.89
For sport wear—or for dress. Can you imagine a more charming model? The patent leather "saddle" and snowy white canvas afford a beauty of contrast seldom found in footwear for the younger folks. Be sure to state size.
Wide widths only.
Shipping wt.: Girls', 1 lb.; Small Girls', 14 oz.
15L7978—Tots'. Sizes, 5 to 8. $1.29
15L7905—Small Girls'. Sizes, 8½ to 11. 1.49
15L7893—Girls'. Sizes, 11½ to 2. 1.69
The demand for this patent leather "Mary Jane" stitchdown greatly exceeded all expectations last season. Excellent wearing qualities and neat appearance combined.
Wide widths only.
Shipping wt.: Girls', 1 lb.; Small Girls', 13 oz. Tots', 10 oz.
15L7590—Small Girls'. Sizes, 8½ to 11. $1.49
15L7484—Girls'. Sizes, 11½ to 2. 1.69
The ultra-stylish two-tone effect, so popular just now in women's footwear, has been carried out in this one-strap for girls. Cool white canvas and lustrous black patent leather are most effectively combined. The rubber heel is a feature that's appreciated, too. State size.
Wide widths only.
Shipping wt.: Girls', 1 lb.; Small Girls', 14 oz.
$1.98
15L7207
Here's a style that's always popular. The patent leather vamp and dull black leather top offer a combination that is highly pleasing. Be sure to state size.
Small Girls'. Sizes, 8½ to 11.
Wide widths only.
Shipping wt., 1 lb.
BROWN.
15L7111—Girls'. Sizes, 11½ to 2. $2.29
15L7015—Women's. Sizes, 2½ to 8. $2.59
BLACK.
15L7112 — Girls'. Sizes, 11½ to 2. $2.29
15L7016 — Women's. Sizes, 2½ to 8. $2.59
A sensible, well fitting and long wearing shoe that has proved to be one of the best sellers in our catalog. Comes in soft pliable leather with exceptionally strong sole and rubber heel. Be sure to state size.
Wide widths only.
Shpg. wt.: Women's, 2¼ lbs.; Girls', 2 lbs.
BLACK OR BROWN.
$1.98
15L7204
The ultra-dressy appearance of this little shoe has made it one of our very best sellers. Has a patent leather vamp and soft dull leather top. It is broad and roomy, and affords lots of wear.
Be sure to state size.
Small Girls'. Sizes, 8½ to 11.
Wide widths only.
Shipping wt., 1¼ lbs.
$1.98
RUBBER HEEL.
15L7200
This well built black kid button shoe successfully combines those features you simply must have in shoes for little girls. It is comfortable—it is neat—and it will wear. The live rubber heel is a feature you'll like. Be sure to state size.
Small Girls'. Sizes, 8½ to 11.
Wide widths only.
Shipping wt., 1 lb.
Shipping wt.: Girls', 1¼ lbs.; Small Girls', 1 lb.; Tots', 12 oz.
State size. Wide widths only.
Shipping wt.: Girls', 1¼ lbs.; Small Girls', 1 lb.; Tots', 12 oz.
15L7761
Tots'. Sizes, 5 to 8. $1.48
15L7721—Small Girls'. Sizes, 8½ to 11. $1.79
15L7661—Girls'. Sizes, 11½ to 2. $1.98
A strong, good looking black leather stitchdown that will more than please you. Broad, roomy and built to wear—all at an extremely low price.
15L7762
Tots'. Sizes, 5 to 8. $1.48
15L7722—Small Girls'. Sizes, 8½ to 11. $1.79
15L7662—Girls'. Sizes, 11½ to 2. 1.98
You'll like this sturdy black leather stitchdown. It's neat—and it will stand a lot of good hard wear. Aren't the prices remarkably low?
$1.98
RUBBER HEEL.
15L7201
A neat, black kid shoe for small girls. You'll be delighted with it—and the comfort it affords makes it a desirable shoe for any child. Serviceable and good looking. Equipped with a live springy rubber heel. Be sure to state size.
Small Girls'. Sizes, 8½ to 11.
Wide widths only. Shipping wt., 1 lb.
198 2 42L SEARS, ROEBUCK AND CO.

Women's Slippers
WINE OR BROWN.
15L800—Wine.
15L801—Seal brown.
98c
An attractive Felt Moccasin, trimmed with ribbon and pompon to match the felt. Has a soft padded chrome leather sole. One of our biggest values.
Be sure to state size.
Sizes, 3 to 8. No half sizes.
Wide widths only. Shipping wt., 10 oz.
15L818 $1.49
Here is a style that can't be beat for comfort. It is made of soft and pliable black leather and has a soft padded leather sole. Can be used either as a boudoir or traveling slipper. Can be folded in a grip very easily.
Be sure to state size.
Sizes, 3 to 8.
No half sizes.
Wide widths only.
Shipping wt., 12 oz.
BLACK LEATHER.
15L955 $1.49
Women's Boudoir Slipper, made of soft black kid finish leather and has a flexible hand turned sole. The upper is reinforced where foot goes into slipper.
Be sure to state size.
Sizes, 2½ to 8.
Wide widths only.
Shipping wt., 1 lb.
BLACK LEATHER.
WINE OR BROWN.
15L850—Wine.
15L851—Brown.
$1.49
An attractive Felt Juliet, neatly trimmed with ribbon. Has serviceable, flexible leather sole, and heel has a rubber top lift.
Be sure to state size.
Sizes, 3 to 8. No half sizes.
Wide widths only.
Shipping wt., 1⅛ lbs.
BLACK.
15L862
59c
Women's Black Felt Everett Style Slipper with hair felt stitchdown sole.
Be sure to state size.
Sizes, 3 to 8. No half sizes.
Wide widths only.
Shipping wt., 12 oz.
BLACK OR GRAY.
15L855—Black.
15L856—Gray.
98c
Serviceable Everett Slipper, made of black or gray felt. Has a desirable leather stitchdown sole which makes slipper flexible and comfortable. State size.
Sizes, 3 to 8. No half sizes. Wide widths only.
Shipping wt., 1 lb.
OLD ROSE OR LIGHT BLUE.
15L802—Old rose.
15L803—Light blue.
98c
Women's dainty Felt Moccasins with fawn color tongues, and ribbon and pompons to match. Have padded chrome leather soles.
Sizes, 3 to 8. No half sizes.
Wide widths only.
Shipping wt., 10 oz.
BLACK LEATHER
15L4550
$1.65
Here is a style that can't be beat for comfort. Made of soft black leather with a soft padded leather sole and is very flexible. It can be used for any purpose and can be carried conveniently for traveling without being injured in any way.
Sizes, 5 to 12, including half sizes. Wide widths only.
Shipping wt., 12 oz.
Men's Leather Slippers—
BLACK OR GRAY.
15L816—Black.
15L817—Gray.
79c
Black or Gray Felt Everett Slipper. Has padded chrome leather sole and heel which makes a regular cushion for the feet affording greater flexibility. It can't be beat for the price. Be sure to state size.
Sizes, 3 to 8. No half sizes.
Wide widths only.
Shipping wt., 12 oz.
BROWN.
15L4512
A lighter weight "Romeo" Brown Kid Slipper. We believe that nothing could be more comfortable than this soft, lightweight slipper, and certainly the price is very reasonable. Be sure to state size.
$2.49
Sizes, 5 to 12. Wide widths only.
Shipping wt., 1¾ lbs.
BROWN OR BLACK.
15L4501-Brown.
15L4502-Black.
$1.98
This neat and comfortable House Slipper comes in brown or black Genuine Kid leather with sewed all leather sole. There is lots of room and comfort in this shoe.
Be sure to state size.
Sizes, 5 to 12.
Wide widths only.
Shipping wt., 1½ lbs.
BROWN.
15L4509
Genuine Glazed Kidskin.
$2.60
This Brown Genuine Kidskin Slipper is one of our best models. The pattern is cut a trifle higher than most slippers and the sole is somewhat heavier. You will note that the vamp is all one piece.
Be sure to state size.
Sizes, 5 to 12. Wide widths only.
Shpg. wt., 1½ lbs.
Men's Felt Slippers
BLACK.
BLACK.
15L4562
98c
This neat House Slipper is made of black felt with collar of the same material. Has two soles, one of hair felt and an outside sole of light split leather. It is of the stitchdown welt construction—a feature which makes it flexible.
Be sure to state size
Sizes, 5 to 12. No half sizes.
W.de widths only. Shipping wt., 1 lb.
15L4553
79c
This low Felt Slipper is made unusually attractive by the "homey" design on its vamp. It comes in black felt with felt sole and has no heel. It's comfortable, durable and very low priced at 79c. A dandy for gift purposes, because of its strikingly attractive appearance and the great comfort that it will afford any man. It is one of our most popular patterns.
Be sure to state size.
Sizes, 5 to 12. No half sizes.
Wide widths only. Shipping wt., 10 oz.
15L4506
$2.79
A Brown Genuine Kid Extra Heavy Soled Slipper. Rubber heel and smooth innersole. A high grade slipper.
Be sure to state size.
Sizes, 5 to 12. Wide widths only.
Shipping wt., 1½ lbs.
BROWN.

SHOES FOR
BROWN.
15L6010—Boys'. Sizes, 1 to 5½. $1.98
15L6210—Small Boys'. Sizes, 9 to 13½. Broad Round Toe. 1.79
A very dressy dark brown stitchdown Oxford for boys. It is made over a new style broad toe last and is cut from the best materials. It insures plenty of wear. In addition to the neatness and style of this splendid oxford, the stitchdown construction and the rubber heel are features which assure comfort.
Be sure to state size.
Wide widths only
Shipping wt. Boys', 1⅝ lbs. Small Boys', 1¼ lbs.
BROWN.
Shape of Small Boys' sizes.
15L5037—Boys'. Sizes, 1 to 5½. $2.65
15L5438—Small Boys'. Sizes, 9 to 13½. 1.98
This splendid shoe has been so popular for boys that we have added the same shoe to the line, but have made it with a broad round toe, for the small boys. It is made of stylish dark brown soft chrome tanned upper leather with sturdy oak tanned leather sole and springy rubber heel. This shoe has comfort, style and long life and is very low priced.
Be sure to state size.
Wide widths only.
Shipping wt.: Boys', 2¼ lbs.; Small Boys', 1⅞ lbs.
Be sure you order right size. Measuring Instructions are on page 557.
BROWN.
15L5028—Boys'. Sizes, 1 to 5½. $2.59
15L5445—Small Boys'. Sizes, 9 to 13½. 1.98
A very sensible Blucher Type Shoe for boys and small boys, full of good strong wear. Soft chrome tanned dark brown leather uppers and good springy rubber heel. The sole is stoutly nailed and sewed to the uppers and won't come loose. Be sure to state size.
Wide widths only.
Shipping wt.: Boys', 2¼ lbs.; Small Boys', 1⅞ lbs.
15L5035 $2.59
This is one of our most popular dull black leather shoes for boys. It is thoroughly sensible in every way with its broad roomy round toe and yet it preserves the neatness and style which most boys want. The comfort is completed with the addition of a live springy rubber heel and the sole is stoutly nailed and sewed to the uppers so it won't come loose.
Be sure to state size.
Sizes, 1 to 5½. Wide widths only.
Shipping wt., 2½ lbs.
Rubber Heels.
BROWN OR BLACK.
15L6950—Brown.
15L6952—Black. $1.79
These Brown or Dull Black Leather Shoes are built especially for little feet. They are strongly made in every way and the uppers are soft and pliable and comfortable. Most important is the fact that there is plenty of room for small toes to spread out and little feet to grow correctly. These shoes have very low rubber heels which make walking easier.
Be sure to state size.
Sizes, 6 to 9. Wide widths only.
Shipping wt., 1½ lbs.

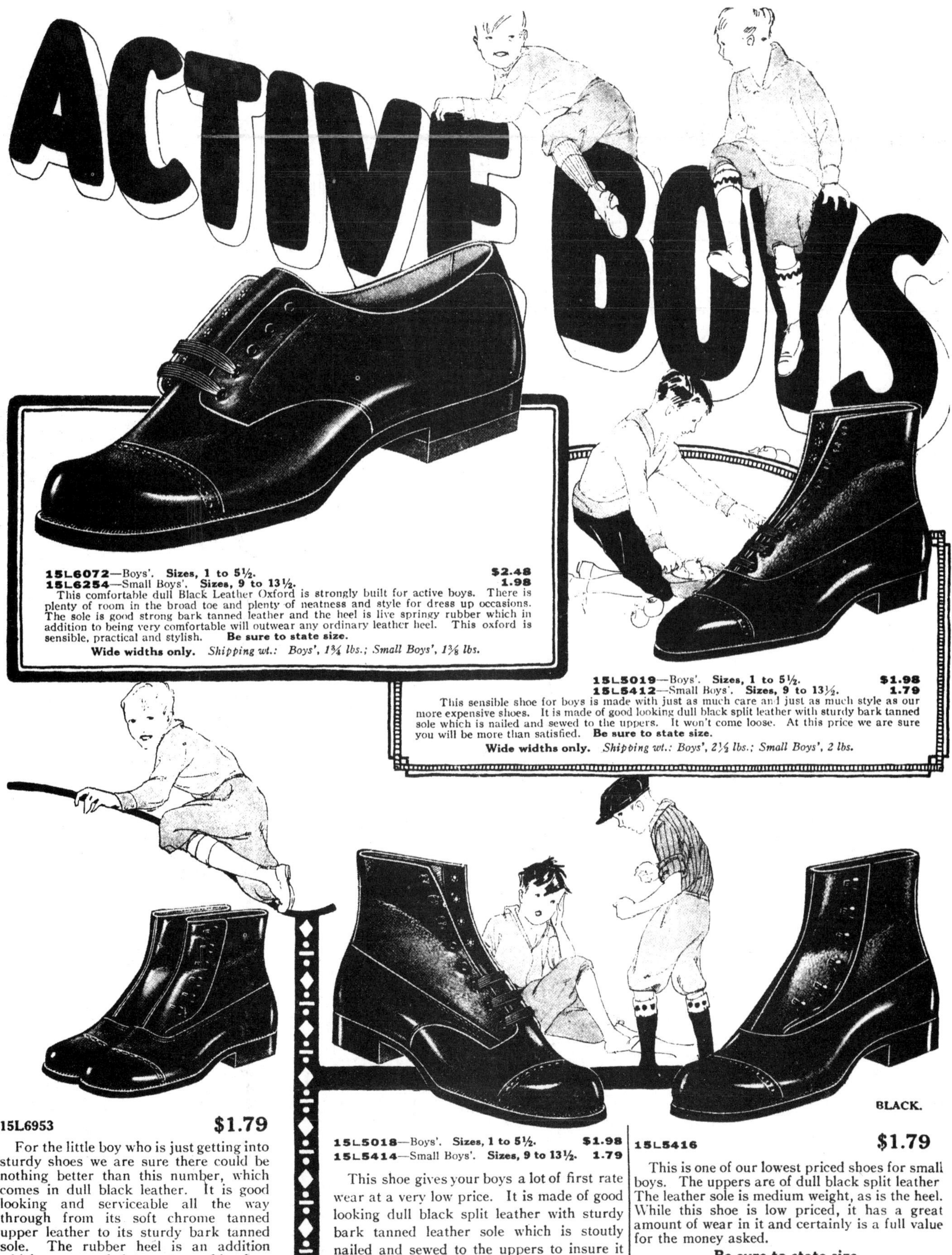
ACTIVE BOYS
15L6072—Boys'. Sizes, 1 to 5½. $2.48
15L6254—Small Boys'. Sizes, 9 to 13½. 1.98
This comfortable dull Black Leather Oxford is strongly built for active boys. There is plenty of room in the broad toe and plenty of neatness and style for dress up occasions. The sole is good strong bark tanned leather and the heel is live springy rubber which in addition to being very comfortable will outwear any ordinary leather heel. This oxford is sensible, practical and stylish. Be sure to state size.
Wide widths only. Shipping wt.: Boys', 1¾ lbs.; Small Boys', 1⅜ lbs.
15L5019—Boys'. Sizes, 1 to 5½. $1.98
15L5412—Small Boys'. Sizes, 9 to 13½. 1.79
This sensible shoe for boys is made with just as much care and just as much style as our more expensive shoes. It is made of good looking dull black split leather with sturdy bark tanned sole which is nailed and sewed to the uppers. It won't come loose. At this price we are sure you will be more than satisfied. Be sure to state size.
Wide widths only. Shipping wt.: Boys', 2½ lbs.; Small Boys', 2 lbs.
15L6953 $1.79
For the little boy who is just getting into sturdy shoes we are sure there could be nothing better than this number, which comes in dull black leather. It is good looking and serviceable all the way through from its soft chrome tanned upper leather to its sturdy bark tanned sole. The rubber heel is an addition which we are glad to put on this shoe, because it means so much for comfort and wear. There are no cramped toes in this little shoe. Be sure to state size.
Sizes, 6 to 9. Wide widths only.
Shipping wt., 1⅜ lbs.
15L5018—Boys'. Sizes, 1 to 5½. $1.98
15L5414—Small Boys'. Sizes, 9 to 13½. 1.79
This shoe gives your boys a lot of first rate wear at a very low price. It is made of good looking dull black split leather with sturdy bark tanned leather sole which is stoutly nailed and sewed to the uppers to insure it from coming loose. There is plenty of comfort and plenty of service in this shoe.
Be sure to state size.
Wide widths only.
Shipping wt.: Boys', 2½ lbs.; Small Boys', 2 lbs.
BLACK.
15L5416 $1.79
This is one of our lowest priced shoes for small boys. The uppers are of dull black split leather The leather sole is medium weight, as is the heel. While this shoe is low priced, it has a great amount of wear in it and certainly is a full value for the money asked.
Be sure to state size.
Sizes, 9 to 13½. Wide widths only.
Shipping wt., 1⅞ lbs.

Do Your Own Shoe Repairing

Have a Repair Outfit right in your home. It's convenient and an inexpensive way of taking care of the family's footwear. When a shoe sole wears through you want it repaired at once, for delay means it will cost you more. Have an outfit of your own and you can do this repairing easily, economically and **without delay.**

The outfit listed here is offered at \$2.39—that's about what it would cost now to have two pairs of shoes equipped with new soles and heels. Why not do your own shoe repairing? It pays, especially when you can buy a good outfit like this at such a low price. It will pay for itself in a few weeks' time. You can do the work easily if you have the proper tools. The outfit illustrated here is complete and reliable and every tool is guaranteed to give complete satisfaction.

76L9950¼—14-Inch Stand. Shipping wt., 18 lbs. Complete outfit................ **\$2.39**

76L9951¼—23-Inch Stand. Shipping wt., 22 lbs. Complete outfit................ **\$2.69**

This Outfit Contains

1 Extra Heavy Guaranteed Iron Stand.
4 Extra Heavy Guaranteed Lasts.
1 Shoemakers' Drop Forged Steel Hammer.
1 Steel Shoe Knife.
1 Patent Peg Awl Handle.
1 Peg Awl.
1 Sewing Awl Handle.
1 Sewing Awl.
1 Wrench for Peg Awl Handle.
1 Ounce Leather Cement.
2 Ounces Rubber Cement.
1 2-Ounce Ball of Thread.
1 Ball of Cobblers' Wax.
1 Pound of Heel Nails.
½ Pound 4-8 Tapping Nails.
½ Pound 5-8 Tapping Nails.
4 Pairs Heel Plates.
6 Shoe and Harness Needles.
1 Copy of Directions for Shoe Repairing.

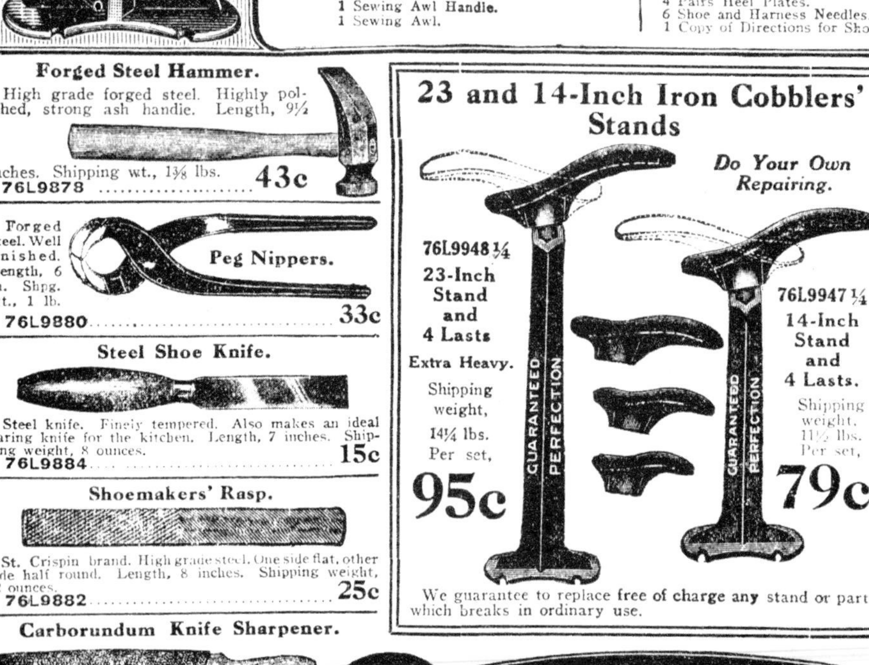

Forged Steel Hammer.

High grade forged steel. Highly polished, strong ash handle. Length, 9½ inches. Shipping wt., 1⅜ lbs.
76L9878 .... **43c**

Peg Nippers.

Forged steel. Well finished. Length, 6 in. Shpg. wt., 1 lb.
76L9880 .... **33c**

Steel Shoe Knife.

Steel knife. Finely tempered. Also makes an ideal paring knife for the kitchen. Length, 7 inches. Shipping weight, 8 ounces.
76L9884 .... **15c**

Shoemakers' Rasp.

St. Crispin brand. High grade steel. One side flat, other side half round. Length, 8 inches. Shipping weight, 12 ounces.
76L9882 .... **25c**

Carborundum Knife Sharpener.

For shoemakers' and kitchen use. One of the simplest and most effective sharpeners on the market. Has two sharpening surfaces, for fine or heavy work. Shipping weight, 8 ounces.
76L9886 .... **35c**

Baker's Celebrated "Hold Fast" Nails.

In ½-Pound Packages.

Baker's Patent Wire Clinch Nails. Sizes, 3-8, 3½-8, 4-8, 4½-8, 5-8, 5½-8, 6-8, 6½-8 and 7-8. **State size.** Shipping wt., 1½ pounds.
76L9861
Per ½-lb. package........ **10c**

23 and 14-Inch Iron Cobblers' Stands

Do Your Own Repairing.

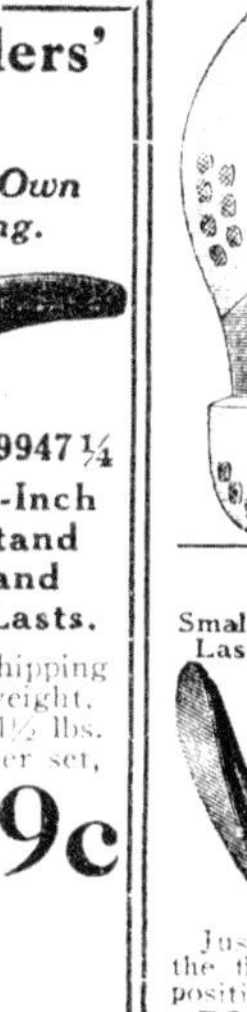

76L9948¼
23-Inch Stand and 4 Lasts
Extra Heavy.
Shipping weight, 14¼ lbs.
Per set,
95c

76L9947¼
14-Inch Stand and 4 Lasts.
Shipping weight, 11½ lbs.
Per set,
79c

We guarantee to replace free of charge any stand or part which breaks in ordinary use.

Sole Savers.

These small plates make an ideal protection for both soles and heels. For men's, women's or children's shoes they are just the thing. Stamped from the best sheet steel and will automatically clinch when driven into the leather. Any number of plates can be attached as close together as you wish, so that the entire wearing surface of heel and sole is protected. Because of their small size "Sole Savers" do not interfere with the flexibility of the leather. Make your shoes wear longer. Buy SOLE SAVERS today. Shpg. wt., per pkg., 3 oz.

Actual Size

76L9937—SOLE SAVERS.
1 package (24 plates).. **10c** | 3 packages (72 plates).. **25c**

Handy Shoe Repair Last.

Something new for the family repairman. It combines three lasts into one. Being constructed in one piece there are no parts to become mislaid, neither is it necessary to fasten it to a table or block of wood. Note the illustration—there's a small last for children's shoes, a large last for men's and women's shoes and a last for repairing the heels.

Just set it on a table, or on a block of wood or on the floor. It is perfectly rigid in any one of three positions. Shipping weight, 8 pounds.
76L9946¼ **55c**

You Don't Have to Be a Cobbler to Repair Shoes!

Pay for this outfit from the saving you will make on the first two or three pairs of shoes you repair. **All tools are of a high grade quality.**

1 Handy Shoe Repair Last (3 lasts combined in one, will fit all sizes in shoes).
1 Cast Iron Hammer.
1 Steel Shoe Knife.
1 Patent Peg Awl Handle.
1 Peg Awl Wrench.
1 Peg Awl Blade.
½ Lb. Iron Heel Nails, size ⅝.
¼ Lb. Clinch Nails, size ⅜.
¼ Lb. Clinch Nails, size ⅘.
1 Copy of Directions for Shoe Repairing.

76L9952¼—Complete outfit.
(Shipping weight, 11 pounds) **97c**

Steel Heel and Toe Plates.

No. 3.

No. 2.

No. 1.

8c Per Box,

Crescent Heel and Toe Plates, cut from sheet steel. Will not break. No. 1 for women's French heels and children's shoes, also used for toe plate. No. 2 for women's Cuban or military heels and children's shoes. No. 3 for men's and boys' shoes. Illustrations show exact size of plates. **State size.** Shipping weight, 10 oz.

76L9931 — Women's No. 1. Per package of 1½ dozen pairs........**8c**
76L9928 — Women's No. 2. Per box of 1½ dozen pairs........**8c**
76L9929—Men's No. 3. Box of 1 doz. pairs.. **9c**

COBBLERS—Save Money Here!

Shinola White Cake.

A canvas cleaner put up in a convenient form. When applied with a damp brush or sponge leaves a clean, white surface. More economical than liquid cleaner because it goes farther. Also good for Suede, Nu-Buck and other white leathers which have a nap. Shipping wt., 5 oz.

76L9772 ....**8c**

SHINOL'A HOME SET

This wonderful polishing set consists of a soft fleecy lambs wool polisher and a genuine bristle dauber. Polisher about 8 inches long. Shipping weight, 9 ounces.

76L9779 ....**29c**

Shinola Paste Polish.
Contents, 1¾ Ounces.

This widely advertised paste polish is very popular. It produces a lustrous polish in either black, light tan, dark brown or ox blood. Opens easily and cleanly by reason of the handy key opener on each box. Regular size. Shipping weight, 5 oz.

76L9725—Black.
76L9726—Light Tan.
76L9727—Dark Brown.
76L9728—Ox Blood.

Per box **8c**

Dri-Foot Oil.

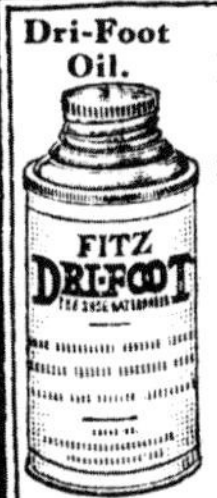

Dri-Foot Oil applied to shoes will make them resist water. It goes into the pores of the leather, making it soft and pliable, thus making it wear longer and turn water as much as any oil can. Good for both black and tan shoes. Shipping weight, 1 lb.

76L9778—Per ½-pint can...**29c**

Jackie White.
For White Canvas.
Contents, 5 Fluid Oz.

Large size bottle of White Cleaner for white canvas and leather shoes which have a nap, such as White Buck, Nu-Buck, Suede, etc. Shipping wt., 1½ lbs.

76L9762
Per bottle,
17c

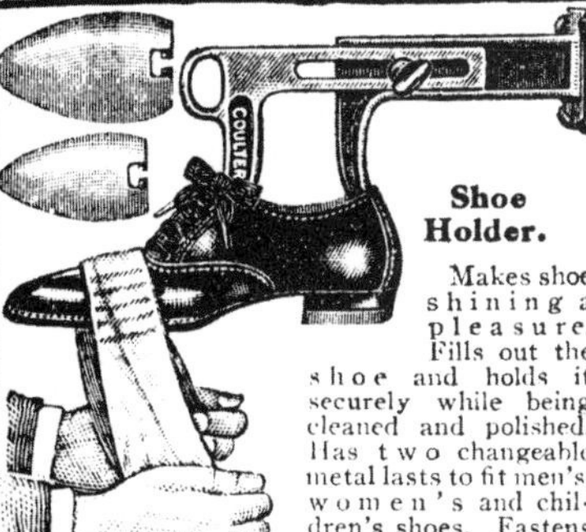

Shoe Holder.

Makes shoe shining a pleasure. Fills out the shoe and holds it securely while being cleaned and polished. Has two changeable metal lasts to fit men's, women's and children's shoes. Fastens to wall with detachable wall bracket and can be taken down instantly when not in use. Shipping weight, 4 pounds.

76L9945 ....**$1.19**

Shoe Dye.
Contents, 4 Fluid Oz.

A permanent dye for all leather goods. Will not rub off or injure the finest leather. Shipping wt., 1 lb.

76L9752
Black.
76L9753
Brown.
76L9754
Ox Blood.
Per bottle,
20c

Black Olo.
Contents, 5 Fluid Ounces.

Black Olo is a black liquid shoe dressing which produces a durable shine without rubbing or polishing. It will not harden or crack the leather. It is also useful for many articles besides shoes, such as rubbers, shopping bags and black kid gloves, etc. Shipping wt., 1¼ pounds.

76L9755
Per bottle....**20c**

Shoe Stretcher.

Wood Shoe Stretcher. Made in four sizes: No. 0, men's large size; No. 1, men's medium size; No. 2, women's size, and No. 3, children's size. **State size.** Has corn and bunion attachments. Shipping weight, 1¾ pounds.

76L9856 ....**75c**

Ice Creepers.

Especially made to use with rubber footwear. Each pair has adjustable straps. Easily attached to boots, rubbers, etc. Shipping weight, per pair, 8 ounces.

76L9940

Per pair............**$0.35**
Dozen pairs.......... **4.00**

A Stretcher That Touches the Spot.

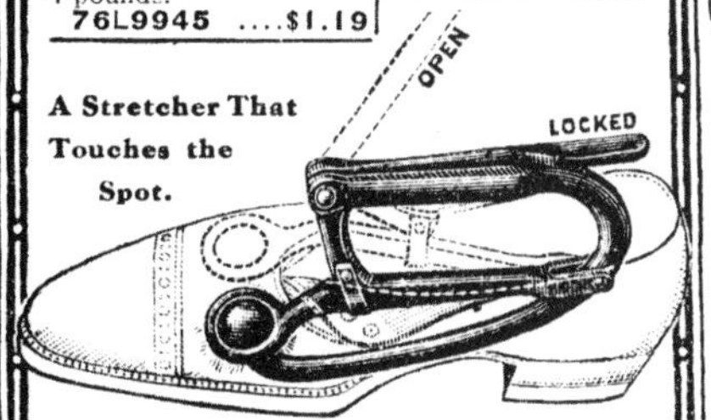

Stretches the leather and relieves the pressure at any part of the shoe without stretching the entire shoe. A great aid to those suffering from corns and bunions. Made of cast iron, japanned finish. Shipping weight, 1 pound 10 ounces.

76L9855 ....**89c**

"FITS 'EM ALL."

The Ace Shoe Tree.

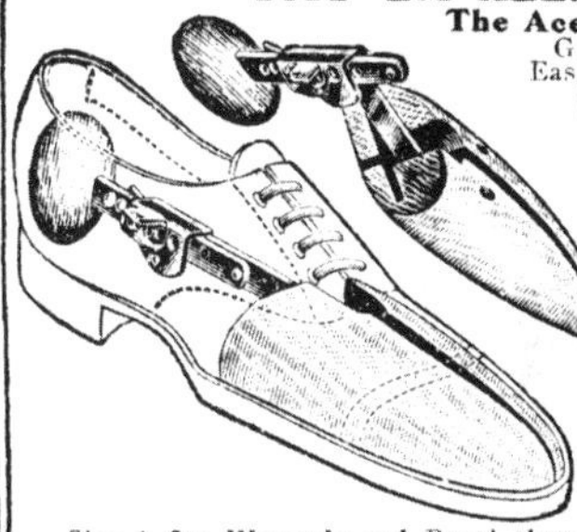

Greatly improved. Easy, quick adjustment. One size in this tree will fit as many as six different size shoes in any width. The fore part of this tree is split in two parts and will conform to the shape of the toe when placed inside. If you want your shoes to wear longer and look better, order a pair of these trees.

Size 1 fits Women's and Boys' shoes, sizes 2 to 5, in any width.
Size 2 fits Women's shoes, sizes 5 to 8, in any width.
Size 3 fits Men's and Boys' shoes, sizes 5 to 8, in any width.
Size 4 fits Men's shoes, sizes 8 to 14, in any width.
Shipping weight, 1½ pounds.

76L9857—Per pair..............................**95c**

Metal Patches.

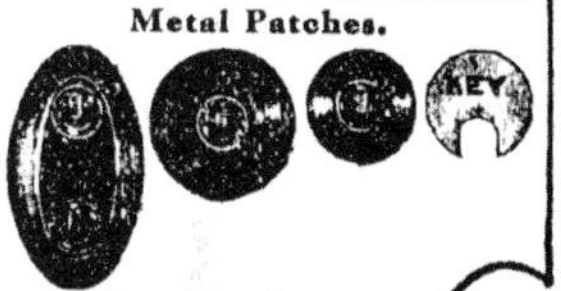

For mending holes in tops of rubber, leather and canvas footwear. Absolutely watertight; instantly applied; will not peel off. Can be used over and over again. Shipping weight, 5 ounces.

76L9944—Per card of 3 patches, with key......**35c**

Metal Heel Stiffeners.

Prevent boots and shoes from running over. Shipping weight, per dozen pairs, 1¼ pounds.

76L9936—Per doz. pairs, **45c**

Round Laces, Best Grade, Highly Mercerized.

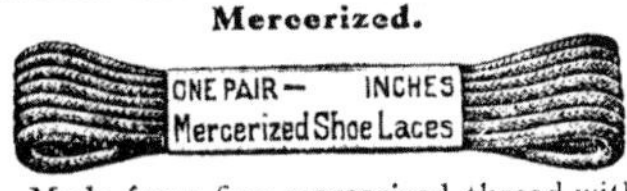

Made from fine mercerized thread with tapered metal tips "that won't come off," and will go through eyelets easily. Shipping weight, per dozen pairs, 6 ounces.

76L9737—White.
76L9738—Black.
76L9739—Dark Brown.
27 inches long. For children's shoes and men's and women's oxfords.
Per pair **5c**
Per dozen pairs..................... **55c**

76L9740—Black.
76L9741—Dark Brown.
40 inches long. For men's regular height shoes.
Per pair **5c**
Per dozen pairs..................... **55c**

76L9742—Black.
76L9744—Dark Brown.
54 inches long. For women's medium high shoes.
Per pair **7c**
Per dozen pairs..................... **75c**

76L9746—Black.
76L9748—Dark Brown.
72 inches long. For women's extra high shoes.
Per pair **$0.09**
Per dozen pairs.................. **1.00**

Fine Quality Leather Shoe Laces.

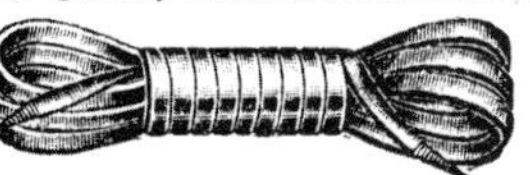

Commonly known as porpoise laces. Fine quality; strongly made with spiral tip. Colors, black or brown. Shipping weight, per dozen pairs, 9 ounces.

76L9806—Brown. 36 inches long.
Per pair **6c**
Per dozen pairs.................. **65c**
76L9807—Black. 36 inches long.
Per pair **6c**
Per dozen pairs.................. **65c**
76L9808—Black. 45 inches long.
Per pair **8c**
Per dozen pairs.................. **90c**

Rawhide Laces.

Full grain, 3/16 inch wide. Extra quality. Used in heavy shoes and Hi-Cut boots where a strong and durable lace is necessary. Shipping weight, per dozen pairs, 1 pound.

76L9809—36 inches long.
Per pair **$0.09**
Per dozen pairs................ **1.00**
76L9810—54 inches long.
Per pair **$0.13**
Per dozen pairs................ **1.50**
76L9811—72 inches long.
Per pair **$0.17**
Per dozen pairs................ **2.00**

Waxed Shoe Laces.

(Commonly Called Waterproof.)

Made from high grade thread, closely woven; about 3/16 inch wide, 1 yard long. Similar to U. S. Army Lace. Color, Black. Shipping weight, per dozen pairs, 8 ounces.

76L9805

Per pair **5c**
Per dozen pairs................................ **35c**

Men's, Women's, Boys', Girls', and Children's Flat Tubular Laces.

These laces are guaranteed 88-thread and full length. Shipping weight, per dozen pairs, 6 ounces.

76L9792—Black.
76L9793—Brown.
27 inches long. For children's shoes and men's and women's oxfords.
Per dozen pairs.................**$0.19**
Per gross laces (72 pairs) **1.10**

76L9794—Black.
76L9795—Brown.
36 inches long. For boys', girls' and men's shoes.
Per dozen pairs **$0.25**
Per gross laces (72 pairs) **1.35**

76L9796—Black.
76L9797—Brown.
45 inches long. For men's and women's shoes.
Per dozen pairs **$0.30**
Per gross laces (72 pairs) **1.65**

76L9798—Black.
76L9799—Brown.
54 inches long. For women's shoes.
Per dozen pairs **$0.35**
Per gross laces (72 pairs) **1.90**

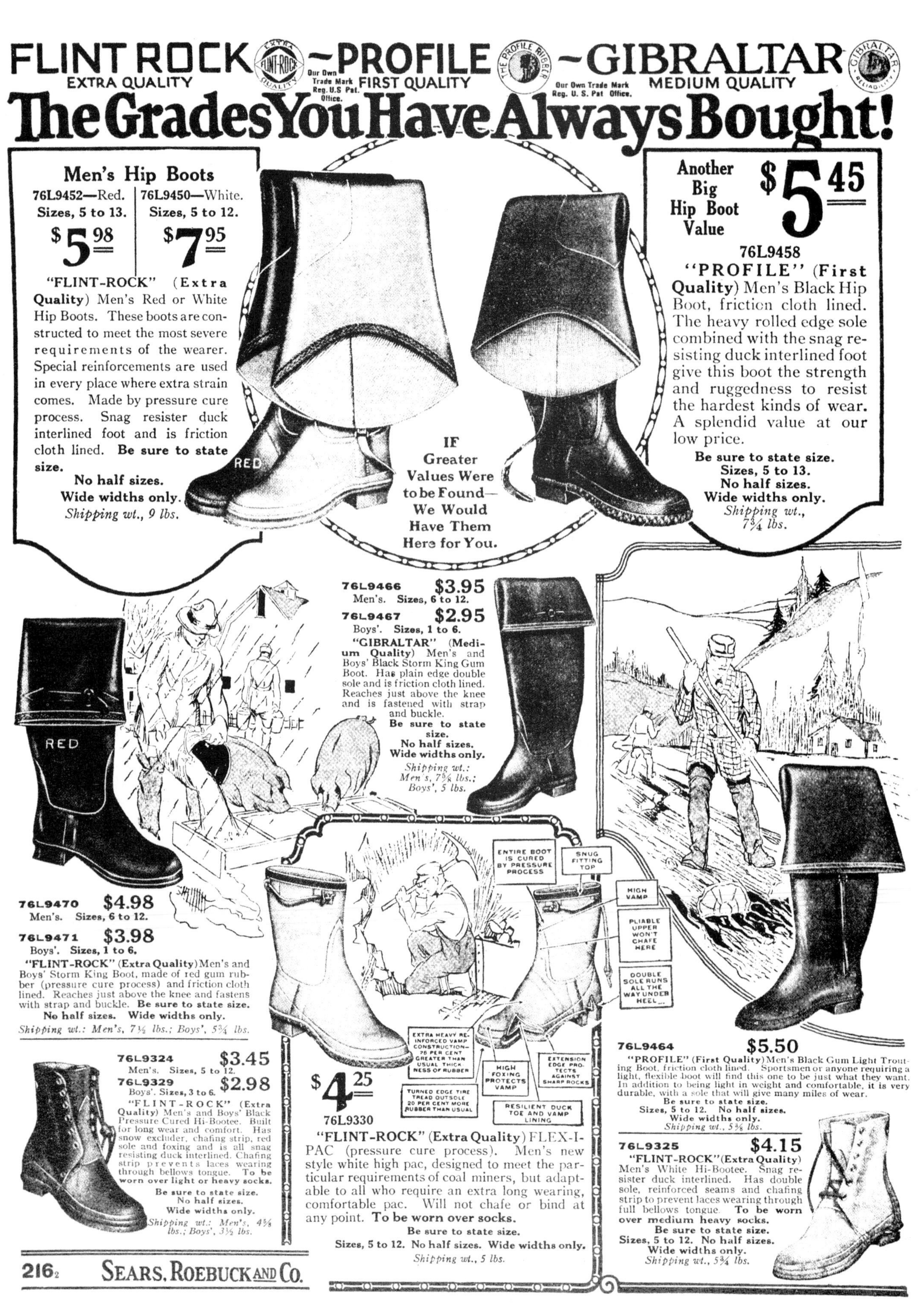

FLINT ROCK
EXTRA QUALITY
~PROFILE
Our Own Trade Mark Reg. U.S. Pat. Office.
FIRST QUALITY
~GIBRALTAR
Our Own Trade Mark Reg. U. S. Pat. Office.
MEDIUM QUALITY
The Grades You Have Always Bought!
Men's Hip Boots
76L9452—Red. Sizes, 5 to 13. $5.98
76L9450—White. Sizes, 5 to 12. $7.95
"FLINT-ROCK" (Extra Quality) Men's Red or White Hip Boots. These boots are constructed to meet the most severe requirements of the wearer. Special reinforcements are used in every place where extra strain comes. Made by pressure cure process. Snag resister duck interlined foot and is friction cloth lined. Be sure to state size.
No half sizes.
Wide widths only.
Shipping wt., 9 lbs.
RED
IF Greater Values Were to be Found—We Would Have Them Here for You.
Another Big Hip Boot Value
$5.45
76L9458
"PROFILE" (First Quality) Men's Black Hip Boot, friction cloth lined. The heavy rolled edge sole combined with the snag resisting duck interlined foot give this boot the strength and ruggedness to resist the hardest kinds of wear. A splendid value at our low price.
Be sure to state size.
Sizes, 5 to 13.
No half sizes.
Wide widths only.
Shipping wt., 7¾ lbs.
76L9466 $3.95
Men's. Sizes, 6 to 12.
76L9467 $2.95
Boys'. Sizes, 1 to 6.
"GIBRALTAR" (Medium Quality) Men's and Boys' Black Storm King Gum Boot. Has plain edge double sole and is friction cloth lined. Reaches just above the knee and is fastened with strap and buckle.
Be sure to state size.
No half sizes.
Wide widths only.
Shipping wt.: Men's, 7⅝ lbs.; Boys', 5 lbs.
RED
76L9470 $4.98
Men's. Sizes, 6 to 12.
76L9471 $3.98
Boys'. Sizes, 1 to 6.
"FLINT-ROCK" (Extra Quality) Men's and Boys' Storm King Boot, made of red gum rubber (pressure cure process) and friction cloth lined. Reaches just above the knee and fastens with strap and buckle. Be sure to state size.
No half sizes. Wide widths only.
Shipping wt.: Men's, 7½ lbs.; Boys', 5¾ lbs.
ENTIRE BOOT IS CURED BY PRESSURE PROCESS
SNUG FITTING TOP
HIGH VAMP
PLIABLE UPPER WON'T CHAFE HERE
DOUBLE SOLE RUNS ALL THE WAY UNDER HEEL
EXTRA HEAVY REINFORCED VAMP CONSTRUCTION—25 PER CENT GREATER THAN USUAL THICKNESS OF RUBBER
HIGH FOXING PROTECTS VAMP
EXTENSION EDGE PROTECTS AGAINST SHARP ROCKS
TURNED EDGE TIRE TREAD OUTSOLE 20 PER CENT MORE RUBBER THAN USUAL
RESILIENT DUCK TOE AND VAMP LINING
76L9324 $3.45
Men's. Sizes, 5 to 12.
76L9329 $2.98
Boys'. Sizes, 3 to 6.
"FLINT-ROCK" (Extra Quality) Men's and Boys' Black Pressure Cured Hi-Bootee. Built for long wear and comfort. Has snow excluder, chafing strip, red sole and foxing and is all snag resisting duck interlined. Chafing strip prevents laces wearing through bellows tongue. To be worn over light or heavy socks.
Be sure to state size.
No half sizes.
Wide widths only.
Shipping wt.: Men's, 4⅝ lbs.; Boys', 3½ lbs.
$4.25
76L9330
"FLINT-ROCK" (Extra Quality) FLEX-I-PAC (pressure cure process). Men's new style white high pac, designed to meet the particular requirements of coal miners, but adaptable to all who require an extra long wearing, comfortable pac. Will not chafe or bind at any point. To be worn over socks.
Be sure to state size.
Sizes, 5 to 12. No half sizes. Wide widths only.
Shipping wt., 5 lbs.
76L9464 $5.50
"PROFILE" (First Quality) Men's Black Gum Light Trouting Boot, friction cloth lined. Sportsmen or anyone requiring a light, flexible boot will find this one to be just what they want. In addition to being light in weight and comfortable, it is very durable, with a sole that will give many miles of wear.
Be sure to state size.
Sizes, 5 to 12. No half sizes.
Wide widths only.
Shipping wt., 5⅜ lbs.
76L9325 $4.15
"FLINT-ROCK" (Extra Quality) Men's White Hi-Bootee. Snag resister duck interlined. Has double sole, reinforced seams and chafing strip to prevent laces wearing through full bellows tongue. To be worn over medium heavy socks.
Be sure to state size.
Sizes, 5 to 12. No half sizes.
Wide widths only.
Shipping wt., 5¾ lbs.
216₂ SEARS, ROEBUCK AND CO.

FLINT ROCK ~ PROFILE ~ GIBRALTAR

FLINT ROCK — EXTRA QUALITY · PROFILE — FIRST QUALITY · GIBRALTAR — MEDIUM QUALITY

Prices Reduced!-Quality Unchanged!

Just Like Dad's

76L9424 Boys'. Sizes, 1 to 6. **$2.63**

76L9425 Small Boys'. Sizes, 11 to 13. **1.98**

The same rugged wearing qualities in this Black Rubber Boot for Boys as found in our Men's Boot, 76L9421, at the right. Clean, fresh new goods in our standard "Gibraltar" (medium quality) brand at a price that defies competition. Compare the price—the saving will surprise you.

Be sure to state size.
No half sizes. Wide widths only.
Shipping wt.: Boys', 3½ lbs.; Small Boys', 2⅝ lbs.

World's Greatest Boot Values!

All Clean, New Stock Fresh From the Mill.
Lowest Price in Over 10 Years.

$2.98 — **76L9421** Short Boot.
$4.48 — **76L9461** Hip Boot.

A most thorough search of the market, aided by our large well known purchasing power, enables us to do almost the impossible. These are our regular "GIBRALTAR" (Medium Quality) Men's Black Rubber Boots with snag resisting duck interlined foot. Heavy soles and friction cloth lined throughout.

Be sure to state size.
Sizes, 5 to 13. No half sizes. Wide widths only.
Shipping wt.: Short Boots, 5¾ lbs.; Hip Boots, 7 lbs.

Sears, Roebuck and Co.
Hart, Mich.

Dear Sirs:
I sent for a pair of boots Wednesday night and asked you to have them here so I could hunt the following Sunday. They arrived Saturday morning and are better boots for $3.98 than a friend of mine has which cost him $7.50 here.
You can look for more orders from here. Thanking you for your very prompt service, I am,
Very respectfully yours,
DON MAXSON.

Long Wear —Good Looks.

76L9402 Men's Red Short Boot. Sizes, 5 to 13. **$3.95**

76L9403 Boys' Red Short Boot. Sizes, 1 to 6. **$3.25**

FLINT ROCK (Extra Quality). There is a reason for the popularity of this Red Boot—it is the decidedly superior service it gives the wearer. Highest grade rubber and other materials are perfectly vulcanized by the famous pressure curing process. The entire boot is friction cloth lined and the foot is duck interlined, making it snag resisting. **Be sure to state size. No half sizes. Wide widths only.**
Shipping wt.: Men's, 7 lbs.; Boys', 5⅛ lbs.

76L9400 **$5.25**
FLINT ROCK (Extra Quality). Miners, dairymen, policemen, construction men and farmers like this white boot for its cleanliness in color and because they know it will "wear like iron." Pressure cured, stout uppers, tough tire tread sole and snag resisting duck interlined foot.

Be sure to state size.
Sizes, 5 to 12.
No half sizes.
Wide widths only.
Shipping wt., 7 lbs.

76L9410
$3.98
This **FLINT ROCK (Extra Quality)** Men's Black Rubber Boot is the result of several years continued effort to discover a way to produce a tough and sturdy sole and upper that would resist to a greater degree the severe wear encountered on concrete floors and in mining and irrigation work, involving contact with alkalies and similar destructive substances. **Be sure to state size.**
Sizes, 5 to 13. No half sizes. Wide widths only.
Shipping wt., 6¾ lbs.

Think of It!

First Quality Rolled Edge Boot

at Only **$3.65**

76L9415 **$3.65**

PROFILE (First Quality). The heavy rolled edge sole on this black rubber boot is just as durable as it looks. The foot is duck interlined, making it snag resisting. Friction cloth lined throughout.

Be sure to state size.
Sizes, 5 to 13.
No half sizes.
Wide widths only.
Shipping wt., 6⅛ lbs.

76L9428 Men's. Sizes, 5 to 11. **$3.29**
76L9430 Small Boys'. Sizes, 11 to 13. **$1.98**
76L9431 Women's. Sizes, 2½ to 8. **$2.39**
76L9432 Girls'. Sizes, 11 to 2. **$1.98**
76L9433 Children's. Sizes, 5 to 10½. **$1.75**

"GIBRALTAR" (Medium Quality). Light Weight, Bright Finish Pebble Leg Black Boot. Sole has plain edge. Men's and small boys', friction cloth lined and does not come in half sizes. Women's, girls' and children's, fleece cloth lined and come in both full and half sizes.

Be sure to state size.
Wide widths only.
Shipping wt.: Men's, 2½ lbs.; Small Boys', 2⅜ lbs.; Women's, 2¼ lbs.; Girls', 1⅞ lbs.; Children's, 1⅝ lbs.

Knit Boot Socks.

76L9890 **45c**

Very Heavy Tufted Knitted Socks with elastic ribbed anklets. About 55 per cent wool, balance cotton. Specially suitable for wearing with rubber boots, lumbermen's overs, high and low pacs, heavy work shoes, etc.

Be sure to state size.
Sizes, 6 to 11.
No half sizes.
Shipping wt., 6 oz.

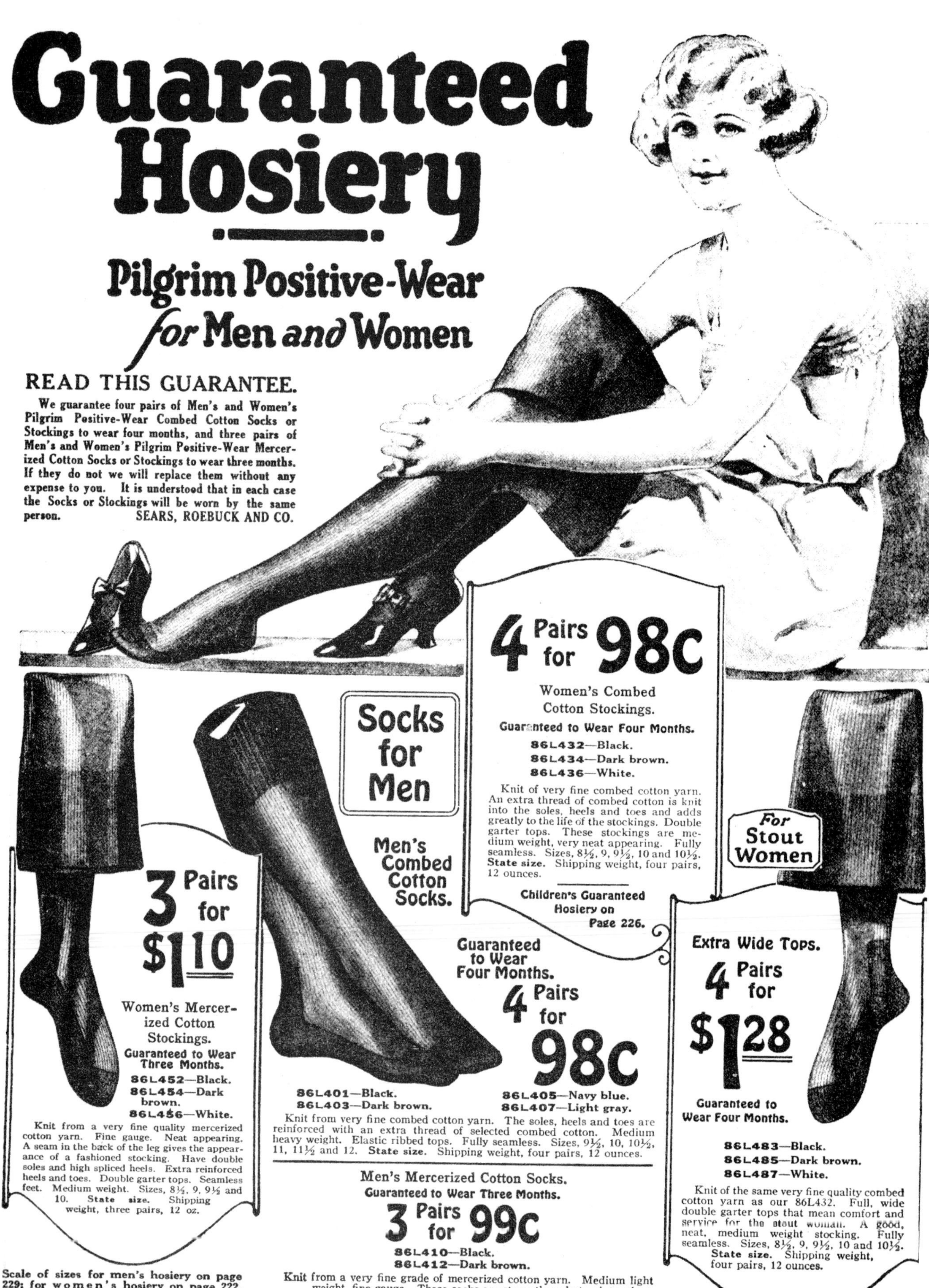
Guaranteed Hosiery
Pilgrim Positive-Wear for Men and Women
READ THIS GUARANTEE.
We guarantee four pairs of Men's and Women's Pilgrim Positive-Wear Combed Cotton Socks or Stockings to wear four months, and three pairs of Men's and Women's Pilgrim Positive-Wear Mercerized Cotton Socks or Stockings to wear three months. If they do not we will replace them without any expense to you. It is understood that in each case the Socks or Stockings will be worn by the same person. SEARS, ROEBUCK AND CO.
4 Pairs for 98c
Women's Combed Cotton Stockings.
Guaranteed to Wear Four Months.
86L432—Black.
86L434—Dark brown.
86L436—White.
Knit of very fine combed cotton yarn. An extra thread of combed cotton is knit into the soles, heels and toes and adds greatly to the life of the stockings. Double garter tops. These stockings are medium weight, very neat appearing. Fully seamless. Sizes, 8½, 9, 9½, 10 and 10½. State size. Shipping weight, four pairs, 12 ounces.
Children's Guaranteed Hosiery on Page 226.
Socks for Men
Men's Combed Cotton Socks.
For Stout Women
3 Pairs for $1.10
Women's Mercerized Cotton Stockings.
Guaranteed to Wear Three Months.
86L452—Black.
86L454—Dark brown.
86L456—White.
Knit from a very fine quality mercerized cotton yarn. Fine gauge. Neat appearing. A seam in the back of the leg gives the appearance of a fashioned stocking. Have double soles and high spliced heels. Extra reinforced heels and toes. Double garter tops. Seamless feet. Medium weight. Sizes, 8½, 9, 9½ and 10. State size. Shipping weight, three pairs, 12 oz.
Guaranteed to Wear Four Months.
4 Pairs for 98c
86L401—Black.
86L403—Dark brown.
86L405—Navy blue.
86L407—Light gray.
Knit from very fine combed cotton yarn. The soles, heels and toes are reinforced with an extra thread of selected combed cotton. Medium heavy weight. Elastic ribbed tops. Fully seamless. Sizes, 9½, 10, 10½, 11, 11½ and 12. State size. Shipping weight, four pairs, 12 ounces.
Extra Wide Tops.
4 Pairs for $1.28
Guaranteed to Wear Four Months.
86L483—Black.
86L485—Dark brown.
86L487—White.
Knit of the same very fine quality combed cotton yarn as our 86L432. Full, wide double garter tops that mean comfort and service for the stout woman. A good, neat, medium weight stocking. Fully seamless. Sizes, 8½, 9, 9½, 10 and 10½. State size. Shipping weight, four pairs, 12 ounces.
Men's Mercerized Cotton Socks.
Guaranteed to Wear Three Months.
3 Pairs for 99c
86L410—Black.
86L412—Dark brown.
Knit from a very fine grade of mercerized cotton yarn. Medium light weight, fine gauge. These socks are strengthened at points where the wear is greatest—heels, toes and soles—with a two-thread mercerized cotton yarn. Elastic ribbed tops. Fully seamless. Sizes, 9½, 10, 10½, 11 and 11½. State size. Shipping weight, three pairs, 10 ounces.
Scale of sizes for men's hosiery on page 229; for women's hosiery on page 222.

INFANTS' HOSIERY

Socks for the Very Wee and their Somewhat Older Brothers and Sisters

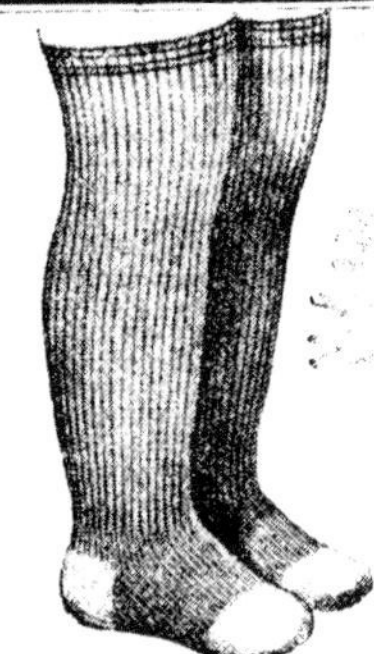

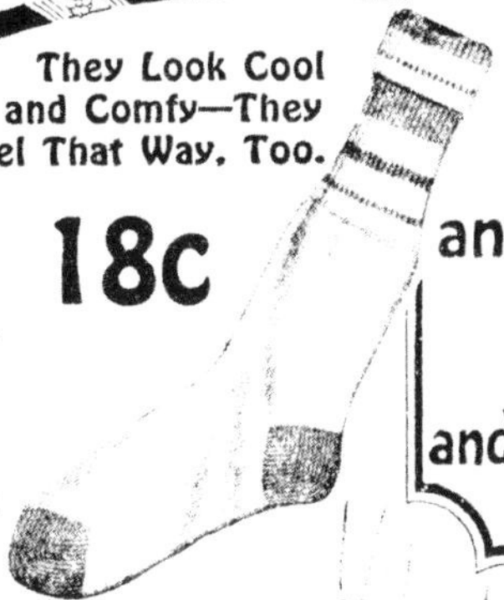

They Look Cool and Comfy—They Feel That Way, Too.

Fine White Mercerized Cotton Socks. Color Striped. Turndown Tops.

18c

86L2800—Blue stripes.
86L2801—Pink stripes.
86L2802—Yellow stripes.

Every little miss has dainty dresses and every "youngest son" has summery wash suits that call for just such pretty little socks. Reinforced heels and toes. Sizes, 4½ to 8½. **State size.** Shpg. wt., 2 oz.

Stockings About One-Third Wool Quite Ably Keep Tiny Feet Warm.

25c

86L2752—Black.
86L2754—Cream-white.

Even in the hot days of summer care must be taken to protect baby from cold. Especially must his feet be kept warm—and the stocking that will do it, and do it well, is this one. It's made of about one-third wool and two-thirds cotton. Elastic ribbed legs. Seamless flat knit feet have reinforced heels and toes. Medium weight. Sizes, 4½, 5, 5½ and 6. **State size.** Shipping weight, 2 ounces.

Wide Ribbed Effect Mercerized Cotton Socks. Turndown Tops.

22c

86L2803—Black.
86L2804—White.
86L2805—Romper blue.
86L2806—Champagne.

As to style, socks for tots pattern after the fancy hose Dad and Mother wear. They're suitable for dress or play. Reinforced heels and toes. Sizes, 4½ to 8½. **State size.** Shipping weight, 2 ounces.

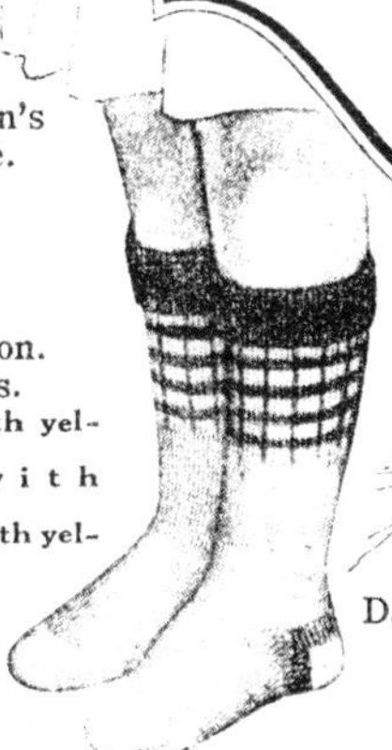

Gayly Striped Tops—the Season's Favored Mode.

29c

Mercerized Cotton. Turndown Tops.

86L2811—Blue with yellow stripes.
86L2812—Tan with brown stripes.
86L2813—Black with yellow stripes.

The striped tops would capture any mother's fancy, they form such an unusually pretty contrast to the body of the sock. Cute as can be. Reinforced heels and toes. Sizes, 4½ to 8½. **State size.** Shipping wt., 2 oz.

Daintily Becoming to Little Tots.

24c

86L2807—Sky blue tops.
86L2808—Pink tops.
86L2809—Dark brown tops.
86L2810—Yellow tops.

These White Mercerized Cotton Socks are among the very latest and prettiest of socks for the kiddies. The ribbed effect turndown tops are very prettily colored. Reinforced heels and toes. Sizes, 4½ to 8½. **State size.** Shipping weight, 2 ounces.

Three-Quarter Length Socks for Children Are Becoming More and More Popular.

You'll Find Several Delightful Styles on Pages 226 and 227.

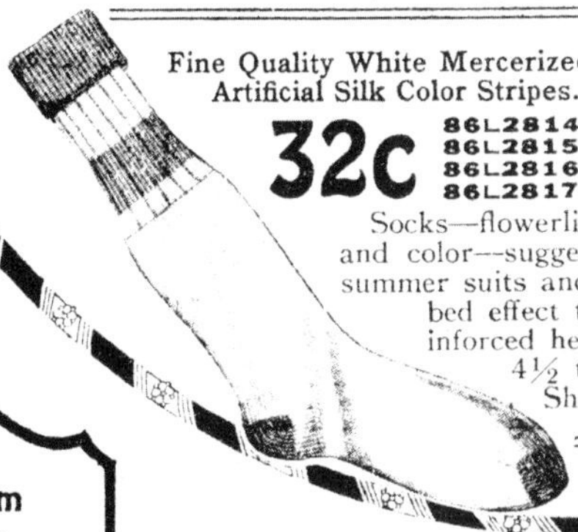

Fine Quality White Mercerized Cotton Socks With Artificial Silk Color Stripes. Turndown Tops.

32c

86L2814—Sky blue stripes.
86L2815—Pink stripes.
86L2816—Brown stripes.
86L2817—Lavender stripes.

Socks—flowerlike in their daintiness and color—suggestive of pretty little summer suits and frocks. Wide ribbed effect turndown tops. Reinforced heels and toes. Sizes, 4½ to 8½. **State size.** Shipping weight, 2 oz.

Though the Weather Be Warm All Wool Stockings for Baby Are Often Advisable.

44c

Silk Tipped Heel and Toe.

86L2774—Black.
86L2777—White.
86L2778—Light sand color.

A chance summer breeze or draught that wouldn't affect an older child at all might be the means of a cold for baby if his feet are not well protected. There's warmth and wear in these stockings. They're knit from very fine quality Australian wool. Medium weight. Fine ribbed elastic knit legs with fully seamless flat knit feet. Heels and toes are tipped with silk. Sizes, 4, 4½, 5, 5½, 6 and 6½. **State size.** Shipping weight, 2 ounces.

ALL WOOL

Combed Cotton Stockings.

3 Pairs for 40c

86L2706—Black.
86L2707—White.
86L2708—Dark brown.
86L2709—Pink.
86L2710—Light blue.

Knit from very good quality combed cotton yarn. Medium weight, with elastic ribbed legs and flat knit seamless feet. Reinforced heels and toes. Sizes, 4½, 5, 5½ and 6. **State size.** Shipping weight, 3 pairs, 6 ounces.

22c

Medium Weight Mercerized Cotton Stockings.

86L2716—Black.
86L2717—White.
86L2721—Light blue.
86L2722—Pink.
86L2723—Dark brown.

Medium weight. Knit of mercerized cotton yarn. Fine elastic ribbed legs with flat knit feet and reinforced heels and toes. Sizes, 4½, 5, 5½ and 6. **State size.** Shipping weight, 2 ounces.

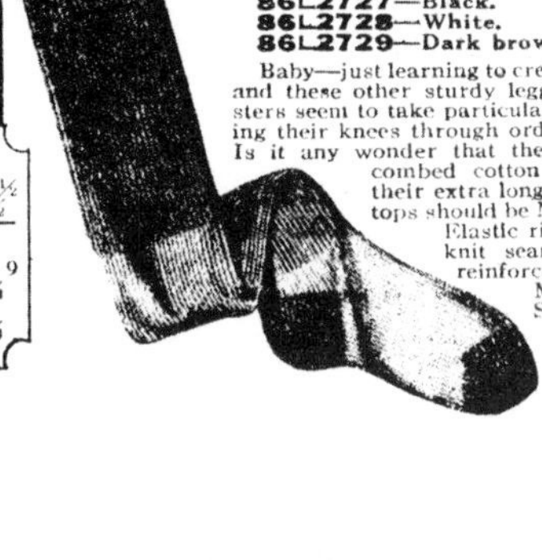

20c

Double Tops. Mother's the One Who Realizes Their Advantage.

86L2727—Black.
86L2728—White.
86L2729—Dark brown.

Baby—just learning to creep on all fours—and these other sturdy legged little youngsters seem to take particular delight in poking their knees through ordinary stockings. Is it any wonder that these good quality combed cotton stockings with their extra long wearing double tops should be Mother's choice? Elastic ribbed legs. Flat knit seamless feet with reinforced heels and toes. Medium weight. Sizes, 4½, 5, 5½, 6 and 6½. **State size.** Shpg. wt., 2 oz.

The Dressiest Little Stocking We Sell. Artificial Silk Plated.

40c

86L2780—Black.
86L2781—White.
86L2782—Dark brown.

Soft, finely knit, rich looking stockings that will suit their tiny wearer to perfection. Knit of a fine quality artificial silk plated yarn. Reinforced mercerized cotton heels and toes. Sizes, 4½ to 6½. **State size.** Shipping weight, 2 oz.

SCALE OF SIZES FOR INFANTS.

Size of Shoe	1	2	3	4–5	6–7	8–8½
Size of Hosiery	4	4½	5	5½	6	6½

SCALE OF SIZES FOR SMALL CHILDREN.

Size of Shoe		4 to 5	6 to 7	8 to 9
Size of Hose		5½	6	6½
Size of Shoe	10 to 11	12 to 13	1 to 2	3
Size of Hose	7	7½	8	8½

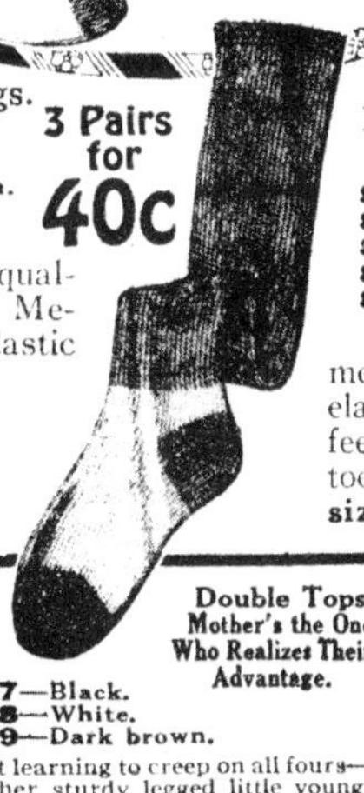

$1.25
CREAM COLOR.
Elastic Ribbed Cotton.

16L5001—Short sleeves.

16L5002 — Long sleeves.

Serviceability and comfort are virtues every man expects his underwear to possess, but when in addition to these he gets quality at a low price, he considers himself fortunate indeed. Knit of extra good grade cotton yarn. Light weight, ankle length. Sizes, 34 to 46 inches chest measure. **State size.** Shpg. wt. 11 oz.

$1.65
WHITE.
Elastic Ribbed, Fine Combed Cotton. Drop Seat.

16L5007—Short sleeves, ankle length.

A Drop Seat as well as any other style can be designed improperly—we've seen a lot of them—we've selected this one because it's a real one, that will afford real comfort. Light weight. Sizes, 34 to 52 inches chest measure. **State size.** Shipping wt., 11 ounces.

For the Men of Average Build.

$1.48
CREAM COLOR.
Long Sleeves, Ankle Length.

16L5011

A suit that will live up to your ideas of what a union suit should be! Knit of a very fine combed cotton yarn. Light weight. Sizes, 34 to 52 inches chest measure. **State size.** Shipping weight, 11 ounces.

$1.48
WHITE.
Short Sleeves, Three-Quarter Length Legs.

16L5260

A favorite of the man desiring a union suit that will not show at the ankle. Legs extend below the knee, just inside cuff of socks. Knit of a fine grade combed cotton yarn. Elastic ribbed. Light weight. Sizes, 34 to 46 inches chest measure. **State size.** Shpg. wt., 11 oz.

$1.48
CREAM COLOR.
Short Sleeves, Ankle Length.

16L5010

Knit of an extra good grade combed cotton yarn. Elastic ribbed. Light weight. Made to fit RIGHT and give you comfort. Sizes, 34 to 52 inches chest measure. **State size.** Shipping weight, 11 ounces.

Short Stout Men.

$1.48
CREAM COLOR.
Short Sleeves, Ankle Length.

16L5037

Knit of a fine combed cotton yarn. Elastic ribbed. Light weight. Made on a special pattern so that it will really fit the short stout man. Sizes, 38 to 52 inches chest measure. **State size.** Shipping weight, 11 ounces.

Tall Slim Men.

$1.48
CREAM COLOR.
Short Sleeves, Ankle Length.

16L5012

Knit of a fine quality combed cotton yarn. Elastic ribbed, light weight. Nowhere could the tall slim man buy a more comfortable, better fitting suit. Sizes, 34 to 46 inches chest measure. **State size.** Shipping weight, 11 ounces.

98c
CREAM COLOR.
Youths' Elastic Ribbed Combed Cotton Union Suit.

16L7525

You young chaps who insist on quality and style in your underwear as well as in your outer clothing will like this suit especially. Made of a fine grade of combed cotton yarn. Light weight. Short sleeves, ankle length. Sizes, 32, 34 and 36 inches chest measure. **State size.** Shipping weight, 7 ounces.

69c
WHITE.
Youths' Nainsook Athletic Style Union Suit.

16L7526

A nainsook athletic suit is always popular with the younger fellows, for a cooler, more comfortable suit for the warm weather cannot be bought. Made of a good light weight nainsook. Sleeveless. Knee length loose knees. Elastic knit ribbed band across back. Sizes, 32, 34 and 36 inches chest measure. **State size.** Shipping weight, 7 ounces.

A Correct Size for comfort and better wear

88c
Each Suit

78c
CREAM COLOR.

Elastic Ribbed Cotton. Ankle Length.

16L5240 — Long sleeves.

16L5241 — Short sleeves.

Once you've worn this suit, you'll be as willing to vouch for it as we are now. There isn't a better suit in the country for the money. Knit of good quality cotton yarn. Light weight. Sizes, 34 to 46 inches chest measure **State size.** Shipping weight, 12 ounces.

$1.98
CREAM COLOR OR WHITE.
Elastic Ribbed Cotton Lisle. Ankle Length. Short Sleeves.

16L5255—Cream color.
16L5256—White.

Lisle yarn is cotton that has been put through a special twisting process, Cotton so treated is better looking, finer, and gives more wear than ordinary cotton. In underwear, it is used only for the finest of suits—and here is one of them. It's light in weight and comes in sizes 34 to 46 inches chest measure. Every man that buys this suit will reorder—we're sure of that. **State size.** Shpg. wt., 12 oz.

$1.65
WHITE.
Flat Knit Cotton. Short Sleeves, Three-Quarter Length Legs.

16L5225

To insure comfort, the legs of this very high grade cotton suit extend below the knee. The socks will extend several inches over the cuff of the suit, thus keeping the leg free from contact with the garter and at the same time giving you the advantage of a garment that does not show at the ankle. Very light weight. Sizes. 34 to 50 inches chest measure. **State size.** Shipping weight, 12 oz.

UNDERWEAR BOUGHT HERE FITS RIGHT

All our underwear, **no matter how low the price**, is made to conform to the same size specifications that we have lived up to all through our years of experience—sizes that we've proved correct.

This is surely something to brag about, for there is a great deal of underwear on the market today in which materials have been skimped in order to make tempting bargain prices. WE can make bargain prices and DO—but the quality and amount of material used is never sacrificed.

Mesh Weave

79c
CREAM COLOR.
Good Quality Mesh Weave Cotton.

16L5234—Short sleeves, ankle length.

Let your first purchase for summer be this mesh weave union suit. Cool underwear makes for comfort when the hot days come. Flat knit of good quality cotton. Light weight. Sizes, 34 to 46 inches chest measure. **State size.** Shpg. wt., 12 oz

$1.08
WHITE.
Fine Quality Mesh Weave Combed Cotton.

16L5198 — Short sleeves, ankle length.

When the mercury climbs to "90° in the shade," this flat knit mesh weave union suit will be a great help in making you forget the heat. Fine combed cotton. Light weight. Sizes, 34 to 46 inches chest measure. **State size.** Shpg wt., 12 oz.

88c
CREAM COLOR.
If It's Value You Seek You'll Find It Here.
Elastic Ribbed Cotton.

16L5253—Short sleeves.
16L5254—Long sleeves.

When you buy this suit, show it to the women folk of your family. They'll pronounce it a bargain sure, for women always recognize value. A Big Buy for Eighty-Eight Cents. Knit of a good grade of cotton yarn. Light weight. Ankle length. Sizes, 34 to 46 inches chest measure. **State size.** Shipping weight, 12 oz.

69c
CREAM COLOR.
Flat Knit Cotton.

16L5235—Short sleeves, ankle length.

No matter how low our price you can always feel certain that you are buying underwear that will compare with higher priced merchandise sold elsewhere. Flat knit of good cotton yarn. Elastic ribbed cuffs and anklets. Light weight. Loose fitting. Sizes, 34 to 46 inches chest measure. **State size.** Shipping weight, 12 ounces.

88c
CREAM COLOR.
Flat Knit Cotton.

16L5231—Short sleeves. ankle length.

This flat knit suit has an enviable reputation for comfort and coolness Men like the cut and fit of it—they like the softness of the material. The sizes are full and roomy. No skimping of materials any place. Good quality cotton. Elastic ribbed cuffs and anklets. Medium weight. Loose fitting. Sizes, 34 to 46 inches chest measure **State size** Shipping weight, 12 ounces.

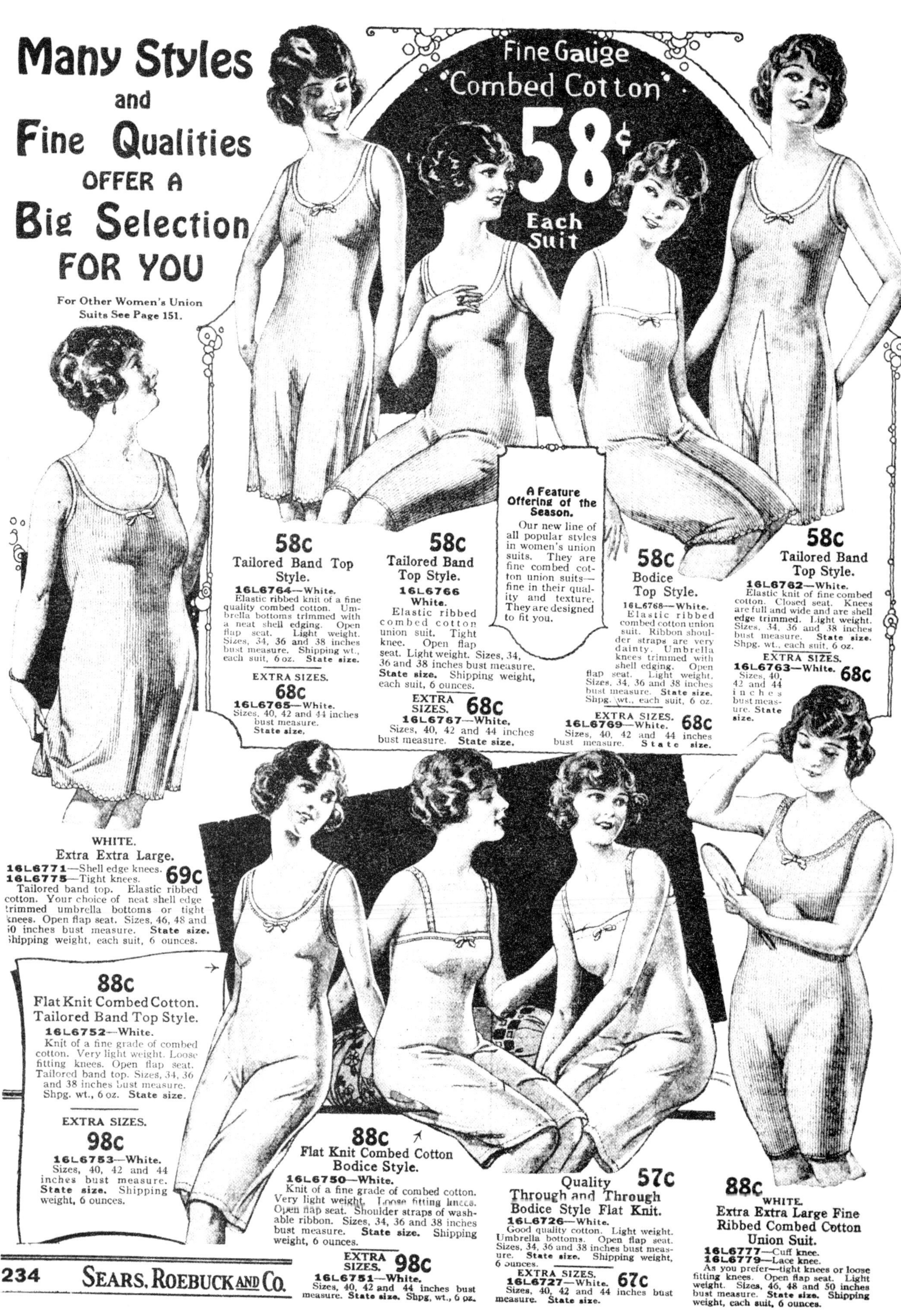
Many Styles
and
Fine Qualities
OFFER A
Big Selection
FOR YOU
For Other Women's Union Suits See Page 151.
Fine Gauge
Combed Cotton
58¢
Each Suit
A Feature Offering of the Season.
Our new line of all popular styles in women's union suits. They are fine combed cotton union suits—fine in their quality and texture. They are designed to fit you.
58c
Tailored Band Top Style.
16L6764—White.
Elastic ribbed knit of a fine quality combed cotton. Umbrella bottoms trimmed with a neat shell edging. Open flap seat. Light weight. Sizes, 34, 36 and 38 inches bust measure. Shipping wt., each suit, 6 oz. State size.
EXTRA SIZES.
68c
16L6765—White.
Sizes, 40, 42 and 44 inches bust measure. State size.
58c
Tailored Band Top Style.
16L6766 White.
Elastic ribbed combed cotton union suit. Tight knee. Open flap seat. Light weight. Sizes, 34, 36 and 38 inches bust measure. State size. Shipping weight, each suit, 6 ounces.
EXTRA SIZES. 68c
16L6767—White.
Sizes, 40, 42 and 44 inches bust measure. State size.
58c
Bodice Top Style.
16L6768—White.
Elastic ribbed combed cotton union suit. Ribbon shoulder straps are very dainty. Umbrella knees trimmed with shell edging. Open flap seat. Light weight. Sizes, 34, 36 and 38 inches bust measure. State size. Shpg. wt., each suit, 6 oz.
EXTRA SIZES. 68c
16L6769—White.
Sizes, 40, 42 and 44 inches bust measure. State size.
58c
Tailored Band Top Style.
16L6762—White.
Elastic knit of fine combed cotton. Closed seat. Knees are full and wide and are shell edge trimmed. Light weight. Sizes, 34, 36 and 38 inches bust measure. State size. Shpg. wt., each suit, 6 oz.
EXTRA SIZES.
16L6763—White. 68c
Sizes, 40, 42 and 44 inches bust measure. State size.
WHITE.
Extra Extra Large.
16L6771—Shell edge knees. 69c
16L6775—Tight knees.
Tailored band top. Elastic ribbed cotton. Your choice of neat shell edge trimmed umbrella bottoms or tight knees. Open flap seat. Sizes, 46, 48 and 50 inches bust measure. State size. Shipping weight, each suit, 6 ounces.
88c
Flat Knit Combed Cotton. Tailored Band Top Style.
16L6752—White.
Knit of a fine grade of combed cotton. Very light weight. Loose fitting knees. Open flap seat. Tailored band top. Sizes, 34, 36 and 38 inches bust measure. Shpg. wt., 6 oz. State size.
EXTRA SIZES.
98c
16L6753—White.
Sizes, 40, 42 and 44 inches bust measure. State size. Shipping weight, 6 ounces.
88c
Flat Knit Combed Cotton Bodice Style.
16L6750—White.
Knit of a fine grade of combed cotton. Very light weight. Loose fitting knees. Open flap seat. Shoulder straps of washable ribbon. Sizes, 34, 36 and 38 inches bust measure. State size. Shipping weight, 6 ounces.
EXTRA SIZES. 98c
16L6751—White.
Sizes, 40, 42 and 44 inches bust measure. State size. Shpg. wt., 6 oz.
Quality 57c
Through and Through Bodice Style Flat Knit.
16L6726—White.
Good quality cotton. Light weight. Umbrella bottoms. Open flap seat. Sizes, 34, 36 and 38 inches bust measure. State size. Shipping weight, 6 ounces.
EXTRA SIZES.
16L6727—White. 67c
Sizes, 40, 42 and 44 inches bust measure. State size.
88c
WHITE.
Extra Extra Large Fine Ribbed Combed Cotton Union Suit.
16L6777—Cuff knee.
16L6779—Lace knee.
As you prefer—tight knees or loose fitting knees. Open flap seat. Light weight. Sizes, 46, 48 and 50 inches bust measure. State size. Shipping weight, each suit, 6 ounces.

Perfect Fitting Union Suits Form a Perfect Foundation for Fit in Stylish Outer Apparel
88c
Pilgrim Princess.
Tailored Band Top Style.
16L6854—White.
Tailored to fit of a very select combed cotton yarn. Open flap seat. The knees are trimmed with a neat shell edging. Elastic ribbed. Sizes, 34, 36 and 38 inches bust measure. State size. Shipping weight, each suit, 6 oz.
EXTRA SIZES.
16L6855—White. 98c
Sizes, 40, 42 and 44 inches. State size.
85c
Pilgrim Princess.
Tailored Band Top Style.
16L6848–White.
Tailored to fit of the very finest combed cotton yarn. Tight knees. Open flap seat. Elastic ribbed. Sizes, 34, 36 and 38 inches bust measure. State size. Shpg. wt., each suit, 6 ounces.
EXTRA SIZES. 95c
16L6849—White.
Sizes, 40, 42 and 44 inches. State size.
Pilgrim Princess
Light Weight Union Suits Are the Best Made.
Because—They're tailored to fit the figure. No binding or bunching.
Because—The best combed cotton yarn is used in the making. Added style and neatness.
Because—They have many special features that guarantee comfort and long wear.
88c
Pilgrim Princess.
Bodice Top Style.
16L6850–White.
Like all Pilgrim Princess Union Suits, this is tailored to fit of the finest combed cotton yarn. Open flap seat. Shell edge trimmed knees. Silk faced ribbon shoulder straps. Elastic ribbed. Sizes, 34, 36 and 38 inches bust measure. State size. Shipping weight, each suit, 6 ounces.
EXTRA SIZES. 98c
16L6851—White.
Sizes, 40, 42 and 44 in. State size.
98c
Pilgrim Princess.
Closed Seat Style.
16L6852—White.
Designed properly and cut with full bottoms that are trimmed with a very neat shell edge. Tailored to fit of the most select combed cotton yarn. Elastic ribbed. Sizes, 34, 36 and 38 inches bust measure. State size. Shipping weight, each suit, 6 oz.
EXTRA SIZES. $1.08
16L6853—White.
Sizes, 40, 42 and 44 inches. State size.
48c
It's a Real Value and a Popular Style as Well!
16L6792—White.
Tailored band top. Knit of a good grade of cotton yarn. Elastic ribbed. Umbrella bottoms are trimmed with a neat shell edge. Open flap seat. Light weight. Sizes, 34, 36 and 38 inches bust measure. State size. Shipping weight, each suit, 6 ounces.
EXTRA SIZES. 56c
16L6793—White.
Sizes, 40, 42 and 44 inches. State size.
44c
Closed Seat.
16L6744—White.
Knit of a good cotton yarn. Elastic ribbed. Low neck, sleeveless. Lace trimmed knees made full and wide. Light weight. Sizes, 34, 36 and 38 inches bust measure. State size. Shipping weight, 6 ounces.
EXTRA SIZES.
52c
16L6745—White.
Sizes, 40, 42 and 44 in. State size.
55c
Something New! A Union Suit That Doesn't Reach the Knees.
16L6758—White.
Many women prefer union suits that do not extend below the knees. Knit in the popular tailored band top style of fine cotton. Close fitting knees. Open flap seat. Light weight. Sizes, 34, 36 and 38 inches bust measure. State size. Shipping weight, 6 ounces.
EXTRA SIZES. 65c
16L6759—White.
Sizes, 40, 42 and 44 inches. State size.
39c
Elastic Ribbed Cotton Union Suit.
16L6739—White.
Low neck, sleeveless. Knee length. Close fitting knees. Open flap seat. Full sizes. Light weight. Sizes, 34, 36 and 38 inches bust measure. State size. Shipping weight, 6 ounces.
EXTRA SIZES. 47c
16L6742—White.
Sizes, 40, 42 and 44 inches. State size.
35c
Elastic Ribbed Cotton Union Suit.
16L158—White.
Low neck, sleevelees. Umbrella bottoms are trimmed with domestic lace. Open flap seat. Full sizes. Light weight. Sizes, 34, 36 and 38 inches bust measure. State size. Shipping wt., 6 oz.
EXTRA SIZES. 43c
16L159—White.
Sizes, 40, 42 and 44 inches. State size.

QUALITY

Sturdy Oak Work Shirts are far superior to the ordinary work shirt generally sold for much higher prices. The quality is there, they are cut full and roomy, the buttons are sewed on to stay, all seams are double stitched to give added reinforcement.

TRADE MARK STURDY OAK

An Ideal Chambray Work Shirt.

89c Made of good quality, medium weight, fine yarn chambray, over large, roomy dimensions. Two large button-through pockets. Interlined collar and faced sleeves. Fine quality buttons to match. Make and finish are the very best.

33L650 Plain blue **33L651** Plain gray. **33L652** Plain striped.
Sizes, 14½ to 17. **State size.** Shipping weight, 14 oz.

Blue Chambray.

89c Our Popular Two-Pocket Blue Chambray Work Shirt. Good quality medium weight chambray, cut over large roomy dimensions with two large button-through pockets. Has interlined collar and faced sleeves and is trimmed with fine quality buttons to match.

33L628—Plain blue.
Sizes, 14½ to 17. **State size.** Shpg. wt., 14 oz.

Medium Weight.

89c A Two-Pocket Work Shirt. Made of good quality medium weight shirting cloth over large roomy dimensions. Two large button-through pockets. Has interlined collar and faced sleeves. Fine quality buttons to match.

33L655—Khaki tan.
33L656—Dark indigo blue.
Sizes, 14½ to 17. **State size.** Shipping wt., 14 oz.

Blue Chambray.

95c Men's Coat Style Shirt with Hi-Band Collar for work and semi-dress. Made of extra quality medium weight fine yarn chambray, cut over our large dimensions. Has interlined collar and faced sleeves. Large pocket. Fine quality buttons.

33L641—Plain blue.
Sizes, 14½ to 17. **State size.** Shipping weight, 14 ounces.

Coat Style.

98c Men's Coat Style Work Shirt. Made of excellent quality medium weight fine yarn chambray. Large roomy dimensions. Made coat style with center plait all the way down and has faced sleeves.

33L674—Plain blue.
33L678—Plain gray.
Sizes, 14½ to 17. **State size.** Shipping weight, 14 ounces.

We believe you will find Our "Sturdy Oak" Line of Work Shirts to be far superior to work shirts that are being sold at much higher prices elsewhere. Built for comfort and long wear.

CUT OVER EXTRA FULL PATTERNS

Slim

Extra Size

For the Tall Man.

95c Made of excellent quality closely woven fine yarn chambray. Made 39 inches long with 36-inch sleeves. Has interlined collar and faced sleeves. Buttons to match.

33L703—Plain blue.
33L704—Plain gray.
Sizes, 14½ to 17. **State size.** Shipping wt., 14 oz.

Extra Size Chambray.

95c Good quality medium weight chambray shirt, in extra large dimensions. Interlined collar and faced sleeves. All principal seams double stitched.

33L684—Plain blue.
33L685—Plain striped.
Sizes, 17½ to 20. **State size.** Shipping weight, 1 lb.

Extra Size Khaki.

$1.05 Khaki Color Twill Work Shirt. Made of an excellent quality medium weight soft finish khaki twill. Cut extra large. Interlined collar and faced sleeves. Button-through pockets.

33L688—Khaki tan.
Sizes, 17½ to 20. **State size.** Shipping wt., 1 lb.

OUR PRICE AND QUALITY COMBINATIONS ARE HARD TO BEAT

MATERIALS AND WORKMANSHIP THE BEST

A Leader.

79c Made of medium weight khaki color shirting in full size dimensions. Principal seams are double stitched. Faced sleeves. Double yoke shoulders and extension neckband. Interlined collar. Buttons to match.

33L666—Khaki tan.
Sizes, 14½ to 17. **State size.** Shipping weight, 14 ounces.

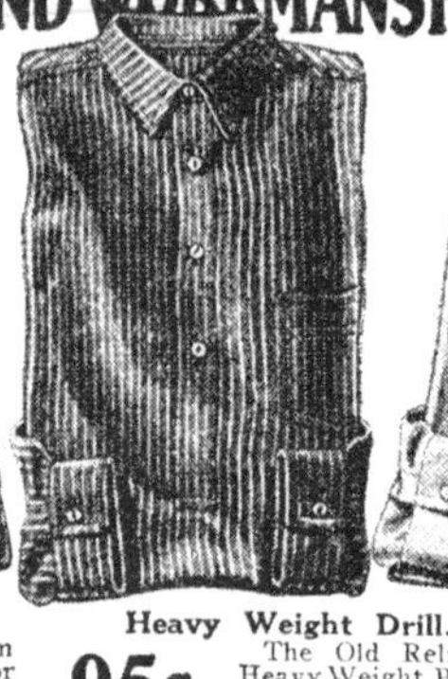

Heavy Weight Drill.

95c The Old Reliable Heavy Weight Black and White Drill Work Shirt. Large roomy dimensions. Principal seams are double stitched. Collar is interlined and sleeves are faced. Large pocket.

33L677—Black with white stripes.
Sizes, 14½ to 17. **State size.** Shipping weight, 14 ounces.

Good Quality Chambray.

69c EACH Good Quality Chambray Work Shirt at an exceptionally low price. Cut over full size dimensions. Principal seams are double stitched and sleeves are faced. Has double yoke shoulders and extension neckband and large pocket.

33L664—Plain blue.
33L665—Plain tan.
33L659—Plain striped.
Sizes, 14½ to 17. **State size.** Shipping weight, 14 ounces.

Two-Pocket Twill.

95c One of our most popular Work Shirts. Made of excellent quality, medium weight, soft finish khaki colored twill with two large button-through pockets and made over large roomy dimensions. Has interlined collar and faced sleeves. Buttons to match. Will wash exceptionally well. A shirt that has been used for work, outing, and semi-dress, and has proved to be one of our most popular sellers. We believe you will be pleased with this shirt at our low price.

33L686—Khaki tan.
Sizes, 14½ to 17. **State size.**
Shipping weight, 14 ounces.

WHAT IS YOUR SIZE?

Fine Yarn Chambray.

79c Medium Weight Fine Yarn Chambray Work Shirt of excellent quality. Cut over large roomy dimensions with large armholes and big cuffs to insure plenty of room. All principal seams are double stitched. Has faced sleeves and large pocket. Interlined collar. Fine quality buttons to match. A quality shirt at a very low price.

33L669—Plain blue.
33L670—Plain gray.
33L671—Plain striped.
Sizes, 14½ to 17. **State size.** Shipping weight, 14 ounces.

HERCULES GUARANTEED WORK SHIRTS

$1.00 EACH

"Hercules" Blue Chambray.

$1.00 Our "Hercules" Work Shirt, made of extra heavy weight blue chambray. This is the finest quality heavy weight chambray on the market today and will give exceptional wear. All our "Hercules" features make this a wonderful shirt and we feel sure that you will be more than satisfied.

33L620—Blue.

Sizes, 14½, 15, 15½, 16, 16½ and 17. **State size.** Shipping weight, 15 ounces.

"Hercules" Polka Dot.

$1.00 Our "Hercules" Polka Dot Work Shirt is made of exceptionally fine indigo blue shirting. These shirts are best by test, and in case you need an extra service shirt you can order this one, knowing it will give more service than any polka dot shirt on the market today at this low price.

33L621 — Blue with White Polka Dots.

Sizes, 14½ to 17. **State size.** Shipping weight, 15 ounces.

"Hercules" Khaki Jean.

$1.00 Our "Hercules" Khaki Jean Work Shirt. Closely woven cloth of firm texture and exceptional wearing qualities. If your desires are for a khaki work shirt, look no further, as this is the shirt to buy. It has wonderful wearing qualities—test them yourself.

33L622—Khaki.

Sizes, 14½ to 17. **State size.** Shipping weight, 15 ounces.

"Hercules" Hickory Stripe.

$1.00 Our "Hercules" Hickory Stripe Work Shirt. A wear resisting woven piece of goods that is a wonder. Comes in neat blue and white stripes. Hickory stripe shirts have been known since work shirts were first worn, but this is the best obtainable at any price. Try it and be convinced.

33L624—Hickory stripe.

Sizes, 14½ to 17. **State size.** Shipping weight, 15 ounces.

"Hercules" Black Sateen.

$1.00 Our "Hercules" Black Sateen Work Shirt. Heavy weight, fine quality lustrous sateen, the best we could buy. It's a beauty, just a little more luster, just a little finer material, and just a little more wear than any other black sateen work shirt sold. Our guarantee protects you.

33L623—Black.

Sizes, 14½ to 17. **State size.** Shipping weight, 15 ounces.

"Hercules" Chambray.

$1.00 Our "Hercules" Work Shirt, made of extra quality closely woven fine yarn chambray, a medium weight cloth that is used in work shirts of the highest grade only and makes a shirt that will please the most critical.

33L625—Blue.
33L626—Gray.
33L627—Plain stripes.

Sizes, 14½, 15, 15½, 16, 16½ and 17. **State size.** Shipping weight, 15 ounces.

"Hercules" Extra Size.

$1.10 Our "Hercules" Extra Size Work Shirt. Extra large dimensions that are guaranteed to fit the bigger men. Made of extra quality closely woven fine yarn chambray, a cloth that will stand lots of wear and tear.

33L629—Blue.
33L630—Gray.

Sizes, 17½, 18, 18½, 19, 19½ and 20. **State size. Shipping weight, 1 pound.**

HERCULES GUARANTEED WORK SHIRTS

This Label Sewed in Every Shirt. Sold Exclusively by Sears, Roebuck and Co.

STATE SIZE

TEN REASONS WHY

1. Made over large, roomy dimensions. Plenty of room for action.
2. Large curved armholes and big cuffs. They cannot bind.
3. Both collars and cuffs are interlined. Adds greatly to the appearance.
4. Non-rip continuous faced sleeves. Guaranteed not to pull out.
5. All principal seams are double stitched. Guaranteed not to rip.
6. Two large button-through pockets. The big kind you'll like.
7. Double yoke shoulders and extension neckband. Built for service.
8. Vegetable ivory buttons that will not chip or crack. Guaranteed to stay on.
9. Only finest quality materials used. Chosen for their long wearing qualities.
10. The very best work shirt we know how to make. Priced to defy competition.

Our Famous Bond Street HATS

A Popular Sailor.

$1.95 Men's Sailor Style Hat, of fine quality sennit braid, hand finished throughout. Fitted with Bon Ton Ivy adjustable cushioned sweatband. Crown, about 3⅜ inches high. Brim, about 2¼ inches wide.

93L4872—Sizes, 6¾ to 7½. **State size.** Shipping weight, 2¼ pounds.

Japanese Panama.

$1.95 Men's Fine Quality Hand Woven Japanese Panama. Telescope style. Stylish and serviceable. The crown is about 3¾ inches high. Brim, about 2½ inches wide.

93L4886

Sizes, 6⅝ to 7½. **State size.** Shipping weight, 2⅜ pounds.

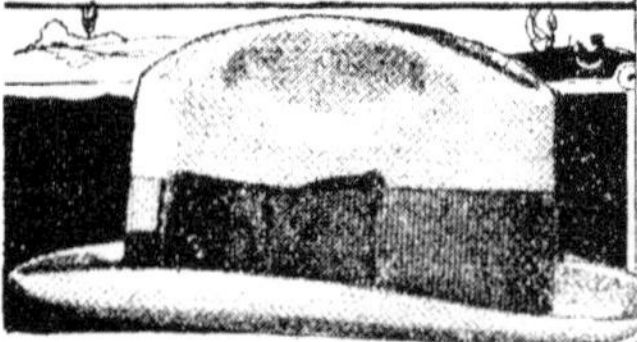

The New Broadway Alpine.

$1.95 Men's Fine Quality Hand Woven Japanese Panama. Fedora or Alpine style. Comfortable and dressy. The crown is about 5½ inches high. Brim, about 2⅜ inches wide.

93L4874

Sizes, 6¾ to 7½. **State size.** Shipping weight, 2 pounds 7 ounces.

King Kumfort.

55c A practical, durable light weight Curaco Panama Work Hat at a very low price. Hand woven Curaco panama straw. Guaranteed to give satisfaction. A hat that is suitable for street wear as well as all kinds of outdoor work. Brim is wide, giving ample protection from the torrid summer sun.

93L4880

Sizes, 6¾ to 7½. **State size.** Shipping weight, 1¼ pounds.

The Swagger.

$4.85 Bond Street De Luxe Welt Edge Style Fine Quality Fur Felt Hat. Newest shades. Crown, about 5½ inches high. Welt edge brim, about 2⅜ inches wide. One of the smartest shapes on the market today. Truly a remarkable value. Being a Bond Street Hat, it is one of the best that money can buy.

93L6331—Gray with black band.
93L6332—Light beaver.
93L6333—Black.

Sizes, 6¾ to 7¾. **State size.** Shipping wt., 2½ lbs.

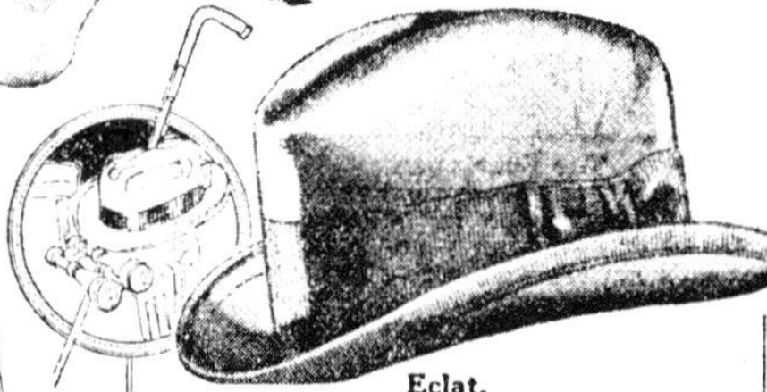

Eclat.

$4.95 Bond Street De Luxe Light Weight Fedora Style. Made of a fine quality light weight fur felt, and is one of the season's leading styles. Fine quality silk lining. Crown, about 5¾ inches high. Brim, about 2⅜ inches wide.

93L6350—Sand tan with brown trim.
93L6351—Seal brown with brown trim.
93L6352—Medium gray with dark trim.
93L6353—Black.

Sizes, 6¾ to 7½. **State size.** Shipping weight, 2½ pounds.

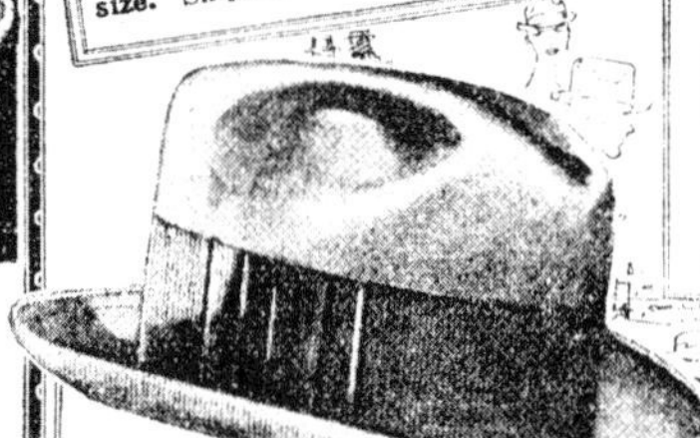

The Campus.

$4.85 Bond Street De Luxe Fedora Style Fur Felt Hat. A very stylish hat in the season's most popular shades. The crown is about 5½ inches high. Bound brim, about 2⅜ inches wide.

93L6310—Beaver.
93L6311—Steel gray with black band.
93L6312—Black.

Sizes, 6¾ to 7¾. **State size.** Shipping weight, 2½ pounds.

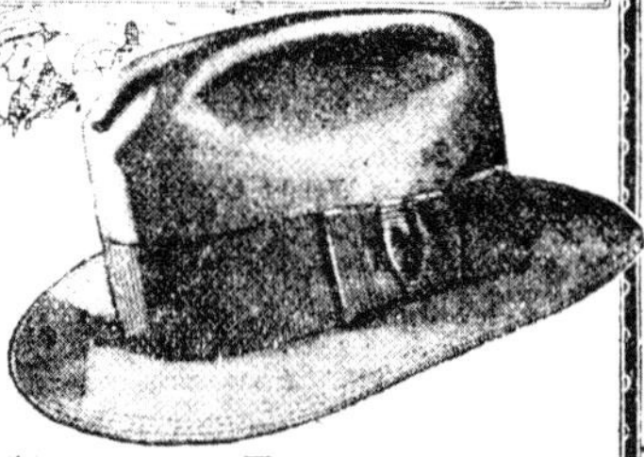

The Beverly.

$4.85 Bond Street De Luxe Trooper Style Fine Quality Fur Felt Hat. An extremely stylish and up to date shape. The crown is about 5¼ inches high. Bound brim, about 2⅛ inches wide.

93L6380—Medium green.
93L6381—Seal brown.
93L6382—Black.

Sizes, 6¾ to 7¾. **State size.** Shipping weight, 2½ pounds.

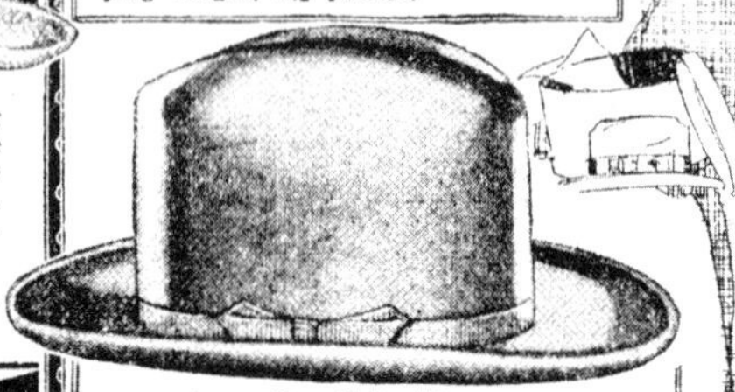

Our Columbia.

$4.95 Bond Street De Luxe Columbia Style Fine Quality Fur Felt Hat. Crown is about 5⅝ inches high. Raw edge brim, about 3 inches wide.

93L6304—Black.
93L6305—Nutria tan.

Sizes, 6¾ to 7¾. **State size.** Shipping weight, 2 pounds 7 ounces.

Our Big Four.

$6.45 Bond Street De Luxe Big Four Style Fine Quality Fur Felt Hat. The crown is about 6¼ inches high. Raw edge brim, about 4 inches wide. For the man desiring a large shaped hat.

93L6323—Black.
93L6324—Nutria tan.

Sizes, 6¾ to 7¾. **State size.** Shipping weight, 2 pounds 9 ounces.

Our Carlsbad.

$6.95 Bond Street De Luxe Carlsbad Style. Extra Fine Quality Fur Felt Hat. The crown is about 7 inches high. Raw edge brim, 4 inches wide. Our Best Quality Carlsbad Style which we feel sure will please you.

93L6360—Black.

93L6361—Nutria tan.

Sizes, 6¾ to 7¾. **State size.** Shipping weight, 2 pounds 9 oz.

Palm Beach Special.

$2.25 Men's Extremely Smar Sailor Style Hat. Made of fine quality imported fancy c natural Swiss braid in new box edge brim, hand finished throughout. Fitted with famous Bon Ton Ivy adjustable cushioned sweatband. Crown, about 3⅜ inches high Brim, about 2¼ inches wide.

93L4875—Bronze tan.

Sizes, 6¾ to 7½. **State size.** Shipping weight, 2¼ pounds.

Miami Beach.

$4.45 Men's Fine Quality Hand Woven South American Panama. Fedora or Alpine style. Fancy navy blue band. Crown, about 5½ inches high. Brim, about 2⅜ inches wide.

93L4887

Sizes, 6¾ to 7½. **State size.** Shipping weight, 2½ pounds.

$5.95 **93L4888**—Same style as above, except in finer quality.

The New Optimo.

$4.35 Fine Quality Hand Woven South American Panama Hat Crown, about 4 inches high Brim, about 2½ inches wide.

93L4890

Sizes, 6¾ to 7½. **State size.** Shipping weight, 2⅜ pounds.

$5.85 **93L4899**—Same style as above, except in finer quality.

Men's Hat and Cap Sizes, 6¾ to 7¾. Boys' Hat and Cap Sizes, 6⅛ to 7⅛.

Hat Sizes	Measures Around Head, In.	Hat Sizes	Measure Around Head, In.
6	19	7	22¼
6⅛	19⅜	7⅛	22½
6¼	19¾	7¼	23
6⅜	20¼	7⅜	23⅜
6½	20¾	7½	23¾
6⅝	21	7⅝	24
6¾	21½	7¾	24½
6⅞	21⅝		

TO MEASURE FOR HAT SIZE.

When you order measure your head above illustrated and send us either the measurement in inches or compare the number of inches your head measures with this scale of hat sizes and send us the hat size you wear.

WE ARE ALWAYS IMPROVING QUALITY
WHAT IS YOUR SIZE?
All Wool Tweed.
$1.69 Men's One-Piece Top Inverted Plait Golf Style Cap of an excellent quality all wool overplaid tweed. Finest quality silk faced cap lining. Leather shield protector. Flexible indestructible canvas visor.
93L4825 Gray mixture.
93L4826 Brown mixture.
Sizes, 6¾ to 7¾. State size. Shipping weight, 1 lb.
A Popular Style.
$1.25 Men's One-Piece Top Plaited Back Golf Style Cap. Made of good quality wool mixed cloth. Good grade twill cap lining. Leather shield protector. An extremely stylish and well tailored cap.
93L4835—Gray mixture.
93L4836—Brown mixture.
Sizes, 6¾ to 7¾. State size. Shipping weight, 1 pound.
A Fine Selection.
89c Men's One-Piece Golf Style Cap. Made of a good quality all wool serge, tweed or shepherd check cloth. Good quality twill lining. Leather shield protector.
93L4812—Navy blue serge.
93L4815—Gray tweed.
93L4810—Black and gray check.
Sizes, 6¾ to 7¾. State size. Shipping weight, 1 pound.
The Latest Style.
$1.89 Men's Snappy One-Piece Top Plaited Back Golf Style Cap. Made of the finest all wool novelty overplaid cap cloths. Good quality silk faced cap lining. Leather shield protector. Flexible indestructible canvas visor. Quality caps have never been priced so low.
93L4837—Gray.
93L4838—Brown.
Sizes, 6¾ to 7¾. State size. Shipping weight, 1 pound.
MEN'S CAPS
Newest Patterns.
$1.15 Men's One-Piece Unlined Golf Style Cap. Made of good quality wool mixed cloth. The very latest colors and patterns have been selected. Leather sweatband. Good quality canvas visor.
93L4822—Gray mixture.
93L4823—Brown mixture.
Sizes, 6¾ to 7¾. State size. Shipping weight, 1 pound.
All Wool.
79c Men's One-Piece Top Golf Style Cap. Made of good quality all wool cloth. Good quality twill lining. Leather shield protector. A most unusual value at our price.
93L4820—Gray mixture.
93L4821—Brown mixture.
Sizes, 6¾ to 7¾. State size. Shipping weight, 1 pound.
All Wool Tweed.
$2.95 Bond Street De Luxe Fedora or Alpine Style Non-Shrinkable Cloth Hat. Stitched throughout. Made of fine quality all wool tweed. Good quality silk faced hat lining. Leather sweatband.
93L4850—Gray mixture.
93L4851—Light brown mixture.
Sizes, 6¾ to 7½. State size. Shipping weight, 1¼ pounds.
Sport Hat.
65c Men's Screen Style Hat. Made of good quality cotton twill cloth. Taped seams. Leather sweatband. A practical hat for all outdoor wear, such as camping, fishing, tennis and golfing.
93L4860—Brown.
93L4861—White.
Sizes, 6¾ to 7½. State size. Shipping weight, 1 pound.
Stitched Throughout.
98c Men's Fedora or Alpine Style Cloth Hat. Stitched throughout. Made of good quality wool mixed tweed. Excellent quality hat lining and sweatband. Leather shield protector.
93L4840—Gray mixture.
93L4844—Brown mixture.
Sizes, 6¾ to 7½. State size. Shipping weight, 1¼ pounds.
A Smart Style.
$1.25 Men's Eight-Quarter Top Taped Seam Golf Style Cap. Overplaid patterns. Made of fine quality wool mixed cloth. Leather sweatband. Canvas visor. A cap that stands out among others for its style, quality and workmanship.
93L4807—Gray mixture.
93L4808—Brown mixture.
Sizes, 6¾ to 7¾. State size. Shipping weight, 1 pound.
All Wool.
$1.39 Men's One-Piece Golf Style Cap. Made of good quality all wool cloth in the newest patterns. Good quality silk faced serge cap lining. Leather shield protector in front. Good quality indestructible visor.
93L4805—Gray mixture.
93L4806—Brown mixture.
Sizes, 6¾ to 7¾. State size. Shipping weight, 1 pound.
YOUTHS' CAPS
A Big Value.
All Wool Golf Style Caps.
89c Youths' One-Piece Top Golf Style Cap. Made of good quality all wool serge and tweed cloths. Durable twill lining. Leather shield protector. Good quality canvas visor.
93L4771—Navy blue.
93L4775—Gray tweed.
Sizes, 6½ to 7⅝. State size. Shipping weight, 15 ounces.
98c Youths' One-Piece Top Plaited Back Golf Style Cap. Made of good quality all wool cloth. Durable twill lining. Leather shield protector. Good quality canvas visor.
93L4795—Gray mixture.
93L4796—Brown mixture.
Sizes, 6½ to 7⅝. State size. Shpg. wt., 15 oz.
Palm Beach Cloth.
$1.00 Men's Eight-Quarter Top Golf Style Cap. Made of a fine quality Palm Beach cloth. Taped seams. Leather sweatband. Canvas visor. An ideal cap for summer wear as it is very light in weight.
93L4813—Tan.
93L4816—Gray.
Sizes, 6¾ to 7¾. State size. Shipping weight, 1 pound.
Crusher Style.
69c Made of good quality rubberized waterproof cotton poplin cloth. Taped seams. Leather sweatband.
93L4865—Olive tan.
93L4866—Navy blue.
Sizes, 6¾ to 7½. State size. Shipping weight, 1 pound.
79c Youths' One-Piece Golf Style Cap. Made of fine quality all wool cloth. Good quality twill lining. Leather shield protector. Good quality canvas visor.
93L4772—Gray mixture.
93L4773—Brown mixture.
Sizes, 6½ to 7⅝. State size. Shipping weight, 15 ounces.
$1.25 Our Best Youths' One-Piece Top Plaited Golf Style Cap. Good quality all wool tweed overplaid cloth. Splendid quality. Cotton twill cap lining. Leather shield protector. Canvas visor.
93L4790—Gray mixture.
93L4791—Brown mixture.
Sizes, 6½ to 7⅝. State size. Shipping wt., 15 oz.
SEARS, ROEBUCK AND CO. 255

COTTON WORK GLOVES

6-Ounce Material.

$1.19 For 12 Pairs. **33L3811** Men's Medium Weight Canton Flannel Gloves. Made of 6-ounce material. Knitted wrists. Nap inside. Shipping weight, twelve pairs, 1⅝ pounds.

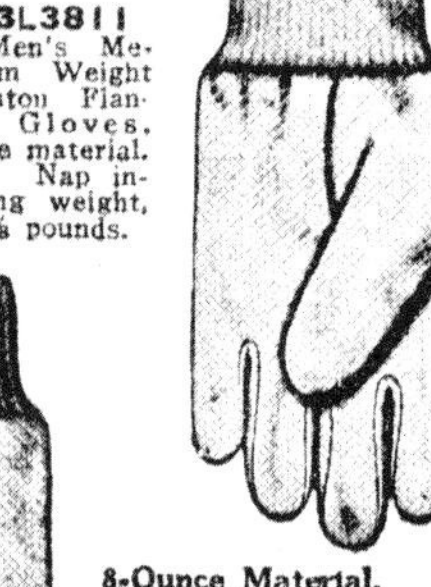

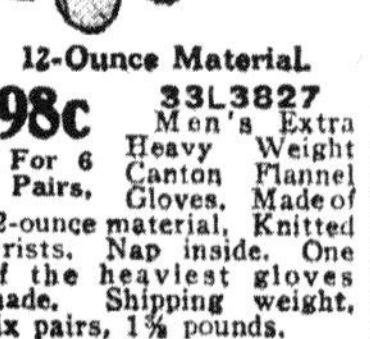

8-Ounce Material.

79c For 6 Pairs. **33L3819** Knitted Wrists. Men's Heavy Weight Canton Flannel Gloves. Made of 8-ounce material. Nap inside. Shpg. wt., six pairs, 1⅛ pounds.

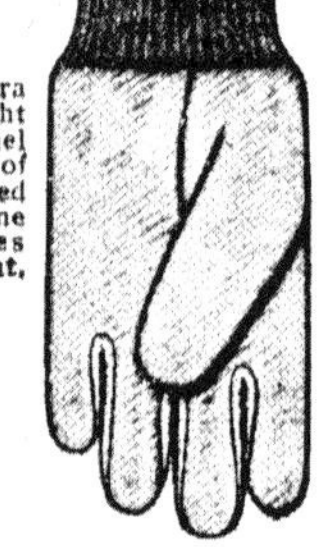

12-Ounce Material.

98c For 6 Pairs. **33L3827** Men's Extra Heavy Weight Canton Flannel Gloves. Made of 12-ounce material. Knitted wrists. Nap inside. One of the heaviest gloves made. Shipping weight, six pairs, 1⅞ pounds.

Napa Sheepskin.

$1.15 For 3 Pairs. **33L3840** Men's Medium Napa Sheepskin Leather Work Gloves. Have good quality knitted wrists. One of the most durable work gloves made. Shipping weight, 1⅛ pounds.

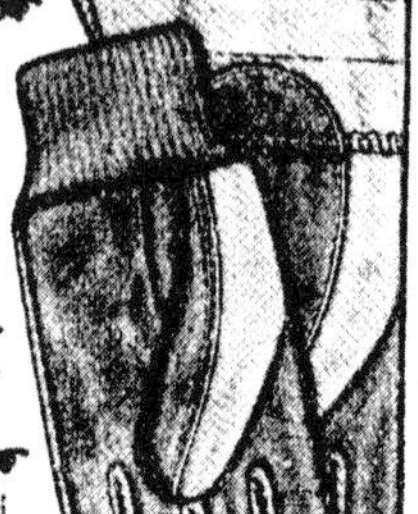

Gauntlets.

89c For 3 Pairs. **33L3832** Men's Heavy Weight 8-Ounce Canton Flannel Gauntlets. Lined canvas cuffs, leather faced palms and thumbs. Reinforced leather pull on cuffs. Shpg. wt., three pairs, 1 pound.

Knitted Wrists.

85c For 3 Pairs. **33L3830** Same as above, but with knitted wrists. Shpg. wt., 1⅛ pounds.

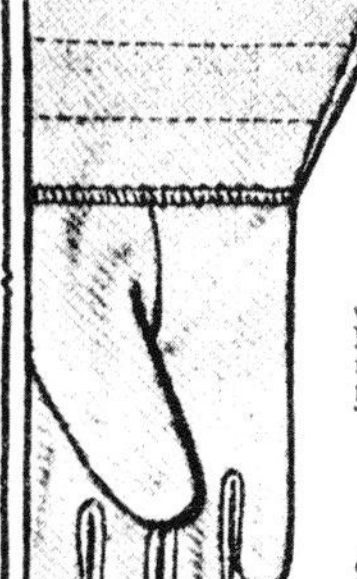

8-Ounce Material.

98c For 6 Pairs. **33L3829** Men's Heavy Weight Canton Flannel Gauntlets, with lined canvas cuffs. Made of 8-ounce material. Nap inside. Shipping weight, 1½ lbs.

83c For 6 Pairs. **93L3878** Boys' or Women's Medium Weight Canton Flannel Gloves, with lined canvas cuffs. 7 oz. material. Nap inside. Shpg. wt., 1 lb.

BETTER QUALITY LOWER PRICES

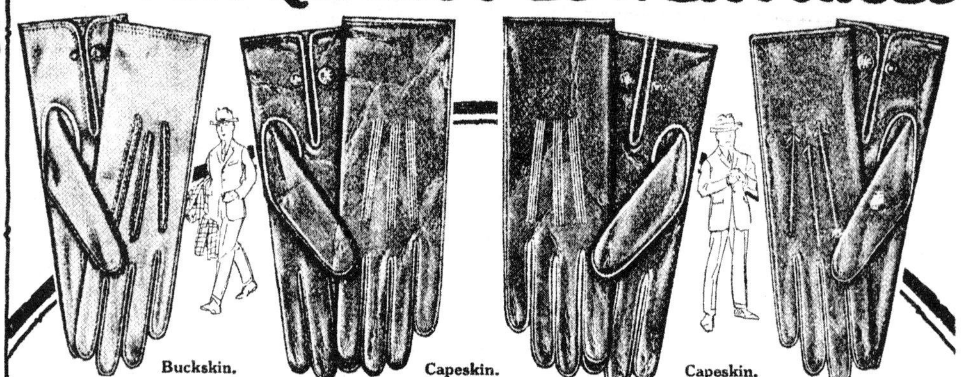

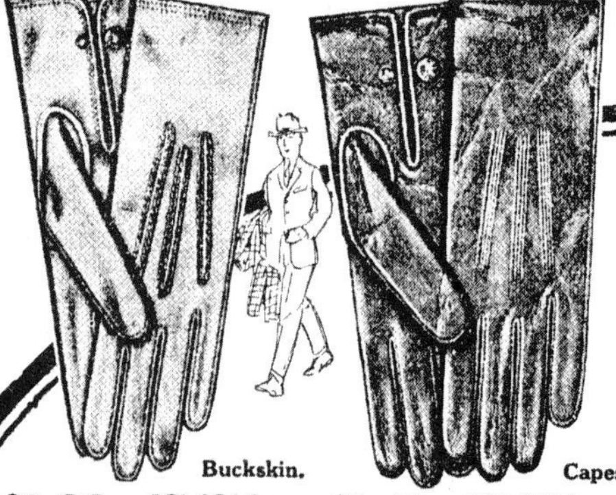

Buckskin. **$2.95** **33L4015** Gray. Men's Unlined Plymouth Buckskin Dress Gloves of fine quality. Embroidered backs. Outseam sewed. Sizes, 7 to 10½. State size. Shpg. wt., 4 oz.

Capeskin. **$1.49** **33L4000** Brown. Men's Excellent Quality Unlined Domestic Capeskin Dress Gloves. Selected skins. Outseam sewed. Stitched backs. Sizes, 7 to 10½. State size. Shpg. wt., 4 oz.

95c **33L4002** Brown. Men's Good Quality Unlined Domestic Capeskin Dress Gloves. Half pique sewed. Stitched backs. Sizes, 7 to 10½. State size. Shpg. wt., 8 oz.

Capeskin. **$1.98** **33L3998**—Gray. **33L3999**—Brown. Men's Fine Imported Capeskin Unlined Dress Gloves. Outseam sewed. Spear point backs. Sizes, 7, 7½, 7¾, 8, 8¼, 8¾, 9, 9½, 10 and 10½. State size. Shpg. wt., 7 oz.

$2.48 "Indestructo" Buckskin. **33L4062** Buckskin. Our Men's "Indestructo" Grained Tanned Buckskin Driving Gloves. Half pique sewed. Stitched backs. Adjustable leather straps. Sizes, 7½ to 10½. State size. Shipping weight, 9 ounces.

Suede. **$1.89** **33L4012** Brown. **33L4013** — Gray. Men's Fine Quality Unlined Suede "Unfinished Kid" Dress Gloves. Sizes, 7, 7½, 7¾, 8, 8¼, 8½, 8¾, 9, 9½, 10 and 10½. State size. Shipping weight, 7 ounces.

YOU CAN AFFORD THESE BETTER QUALITIES

We sell the better grades for less money than others usually ask for the lower grades.

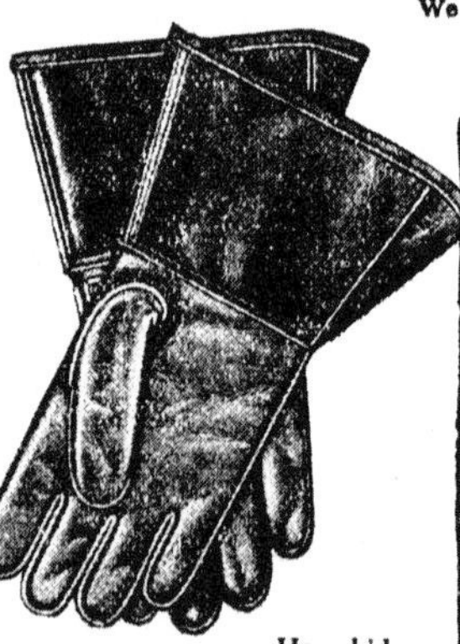

Horsehide. **$1.49** **33L4125**—Black. Men's Good Quality Horsehide Motor Gauntlets. Made of soft, pliable medium weight horsehide leather with large cuffs. Outseam sewed. Adjustable leather straps on backs. Sizes, 8 to 10½. State size. Shipping weight, 1 pound.

Capeskin. **98c** **33L4123**—Black. **33L4124**—Brown. Men's Good Quality Leather Motor Gauntlets. Made of soft, pliable medium weight capeskin leather. Outseam sewed. Adjustable leather straps on backs. Fit and service guaranteed. Sizes, 8, 8½, 9, 9½, 10, 10½. State size. Shipping weight, 1 pound.

Horsehide. **$1.98** **33L4117**—Black. Men's Finest Chrome Tanned Horsehide Motor Gauntlets. Extra large paneled or folding cuff. Can be rolled up and tucked into pocket. Outseam sewed. Adjustable leather straps on backs. Sizes, 8 to 10½. State size. Shipping weight, 1 pound.

HERE ARE GLOVES THAT BOYS LIKE

Capeskin. **33L3921** **49c** Brown. Something every boy desires is this Boys' Capeskin Scout Gauntlet. Black imitation leather gored cuffs. Ages, 6 to 14 years. State age. Shpg. wt., 8 ounces.

Imported Capeskin. **79c** **33L3910** Brown. A good value in Boys' Imported Capeskin Dress Gloves. Stitched backs. Good looking and durable. Ages, 6 to 14 years. State age. Shipping weight, 4 ounces.

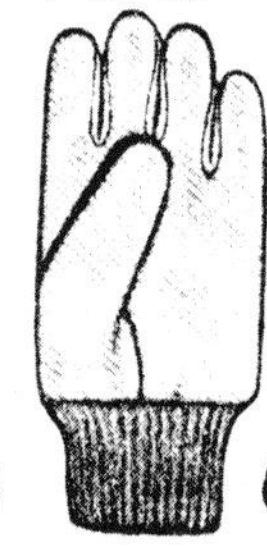

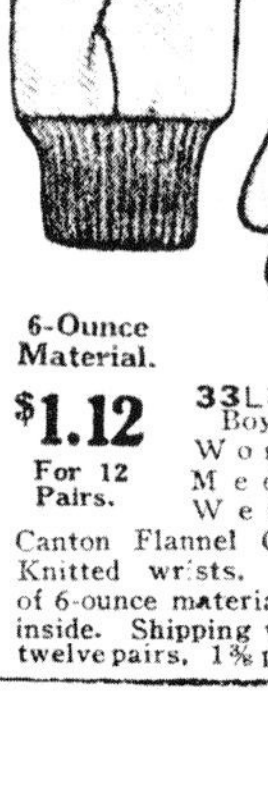

6-Ounce Material. **$1.12** For 12 Pairs. **33L3838** Boys' or Women's Medium Weight Canton Flannel Gloves. Knitted wrists. Made of 6-ounce material. Nap inside. Shipping weight, twelve pairs, 1⅝ pounds.

Split Leather. **49c** **33L3929** Boys' Good Quality Split Leather Gauntlets. Embossed leather cuff with star and fringe, just the thing for the youngster. Ages, 6 to 14 years. State age. Shipping weight, 10 ounces.

STATE SIZE

Horsehide. **79c** **33L3928** Boys' Good Quality Chrome Tanned Horsehide Gauntlets. Handsome embossed leather cuff gored and fringed. Ages, 6 to years. State age. Shipping weight, ounces.

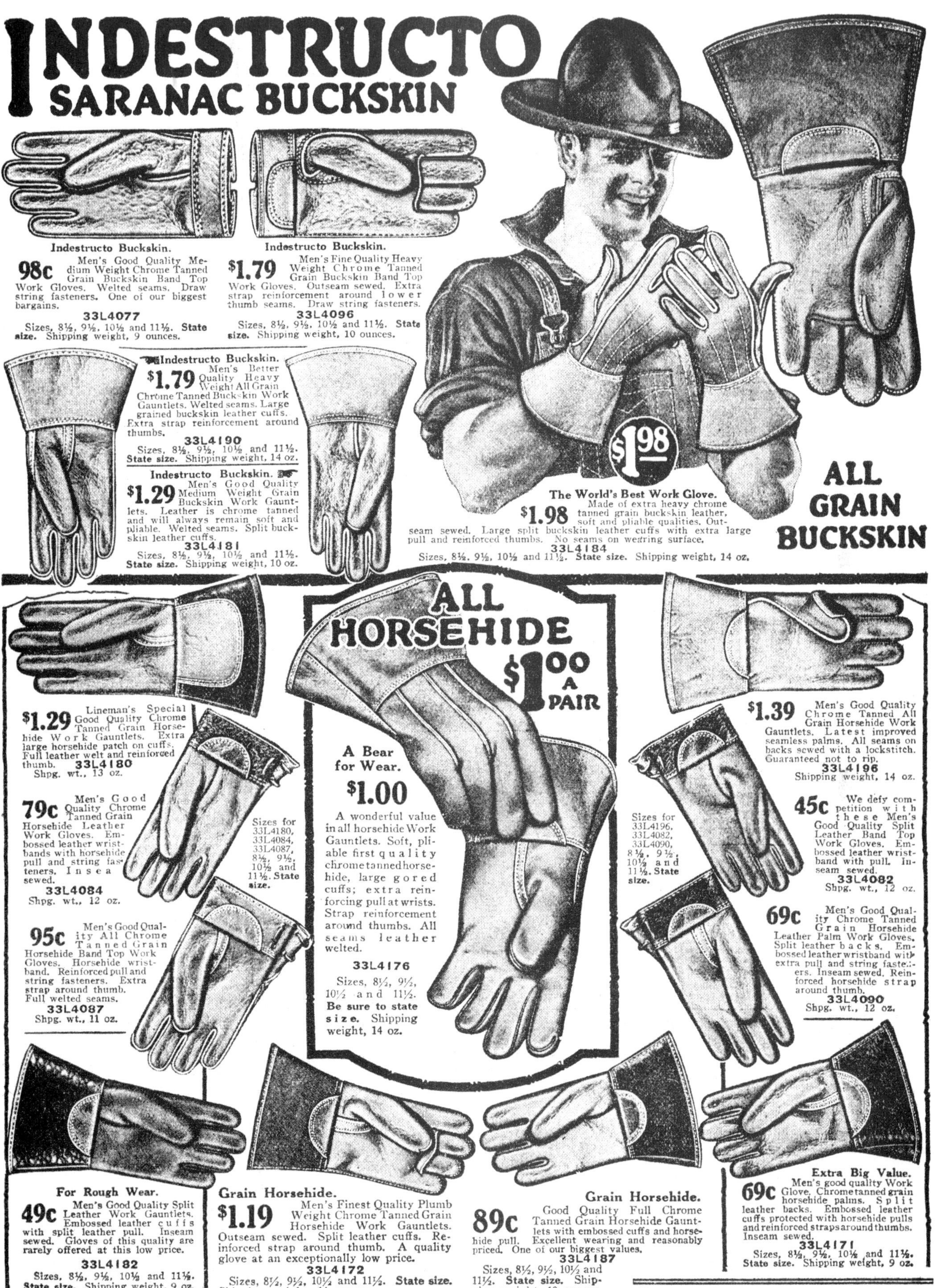
INDESTRUCTO
SARANAC BUCKSKIN
Indestructo Buckskin.
98c Men's Good Quality Medium Weight Chrome Tanned Grain Buckskin Band Top Work Gloves. Welted seams. Draw string fasteners. One of our biggest bargains.
33L4077
Sizes, 8½, 9½, 10½ and 11½. State size. Shipping weight, 9 ounces.
Indestructo Buckskin.
$1.79 Men's Fine Quality Heavy Weight Chrome Tanned Grain Buckskin Band Top Work Gloves. Outseam sewed. Extra strap reinforcement around lower thumb seams. Draw string fasteners.
33L4096
Sizes, 8½, 9½, 10½ and 11½. State size. Shipping weight, 10 ounces.
Indestructo Buckskin.
$1.79 Men's Better Quality Heavy Weight All Grain Chrome Tanned Buckskin Work Gauntlets. Welted seams. Large grained buckskin leather cuffs. Extra strap reinforcement around thumbs.
33L4190
Sizes, 8½, 9½, 10½ and 11½. State size. Shipping weight, 14 oz.
Indestructo Buckskin.
$1.29 Men's Good Quality Medium Weight Grain Buckskin Work Gauntlets. Leather is chrome tanned and will always remain soft and pliable. Welted seams. Split buckskin leather cuffs.
33L4181
Sizes, 8½, 9½, 10½ and 11½. State size. Shipping weight, 10 oz.
$1.98
The World's Best Work Glove.
$1.98 Made of extra heavy chrome tanned grain buckskin leather, soft and pliable qualities. Outseam sewed. Large split buckskin leather cuffs with extra large pull and reinforced thumbs. No seams on wearing surface.
33L4184
Sizes, 8½, 9½, 10½ and 11½. State size. Shipping weight, 14 oz.
ALL GRAIN BUCKSKIN
ALL HORSEHIDE
$1.00 A PAIR
$1.29 Lineman's Special Good Quality Chrome Tanned Grain Horsehide Work Gauntlets. Extra large horsehide patch on cuffs. Full leather welt and reinforced thumb.
33L4180
Shpg. wt., 13 oz.
79c Men's Good Quality Chrome Tanned Grain Horsehide Leather Work Gloves. Embossed leather wristbands with horsehide pull and string fasteners. Inseam sewed.
33L4084
Shpg. wt., 12 oz.
Sizes for 33L4180, 33L4084, 33L4087, 8½, 9½, 10½ and 11½. State size.
95c Men's Good Quality All Chrome Tanned Grain Horsehide Band Top Work Gloves. Horsehide wristband. Reinforced pull and string fasteners. Extra strap around thumb. Full welted seams.
33L4087
Shpg. wt., 11 oz.
A Bear for Wear.
$1.00
A wonderful value in all horsehide Work Gauntlets. Soft, pliable first quality chrome tanned horsehide, large gored cuffs; extra reinforcing pull at wrists. Strap reinforcement around thumbs. All seams leather welted.
33L4176
Sizes, 8½, 9½, 10½ and 11½. Be sure to state size. Shipping weight, 14 oz.
$1.39 Men's Good Quality Chrome Tanned All Grain Horsehide Work Gauntlets. Latest improved seamless palms. All seams on backs sewed with a lockstitch. Guaranteed not to rip.
33L4196
Shipping weight, 14 oz.
Sizes for 33L4196, 33L4082, 33L4090, 8½, 9½, 10½ and 11½. State size.
45c We defy competition with these Men's Good Quality Split Leather Band Top Work Gloves. Embossed leather wristband with pull. Inseam sewed.
33L4082
Shpg. wt., 12 oz.
69c Men's Good Quality Chrome Tanned Grain Horsehide Leather Palm Work Gloves. Split leather backs. Embossed leather wristband with extra pull and string fasteners. Inseam sewed. Reinforced horsehide strap around thumb.
33L4090
Shpg. wt., 12 oz.
For Rough Wear.
49c Men's Good Quality Split Leather Work Gauntlets. Embossed leather cuffs with split leather pull. Inseam sewed. Gloves of this quality are rarely offered at this low price.
33L4182
Sizes, 8½, 9½, 10½ and 11½. State size. Shipping weight, 9 oz.
Grain Horsehide.
$1.19 Men's Finest Quality Plumb Weight Chrome Tanned Grain Horsehide Work Gauntlets. Outseam sewed. Split leather cuffs. Reinforced strap around thumb. A quality glove at an exceptionally low price.
33L4172
Sizes, 8½, 9½, 10½ and 11½. State size. Shipping weight, 14 ounces.
Grain Horsehide.
89c Good Quality Full Chrome Tanned Grain Horsehide Gauntlets with embossed cuffs and horsehide pull. Excellent wearing and reasonably priced. One of our biggest values.
33L4187
Sizes, 8½, 9½, 10½ and 11½. State size. Shipping weight, 10 ounces.
Extra Big Value.
69c Men's good quality Work Glove. Chrome tanned grain horsehide palms. Split leather backs. Embossed leather cuffs protected with horsehide pulls and reinforced straps around thumbs. Inseam sewed.
33L4171
Sizes, 8½, 9½, 10½ and 11½. State size. Shipping weight, 9 oz.

TESTED MERCHANDISE
IS DEPENDABLE MERCHANDISE

By testing, we can guarantee quality and service. Every price we quote therefore means unusual value.

Grain Cowhide Sterling Silver Buckle $1.35

$1.35 A very unusual value. Genuine Cowhide Leather Belt with sterling silver pierced initial buckle of neat design. Belt is fancy grained and leather lined. A high grade belt and buckle for well dressed men at a very low price.
33L8851—Black.
Sizes, 30 to 44 inches waist measure. **State size; also initial wanted.** Shpg. wt., 4 oz.

Our Biggest Value.

48c **33L8664—1⅜ inches. Lisle. 33L8718—1⅛ inches. Lisle.**

Fine Quality Fancy Lisle Crossback Dress Suspenders in two widths. Fancy elastic lisle webbing with stitched colored leather ends to match. All brass trimmed. Very high grade dress suspenders. Length, 38 inches. Shipping weight, 4 ounces.

Genuine Cowhide. Sterling Front Buckle.
95c Genuine Cowhide Belt. Leather lined, with sterling silver front self adjusting Giant Grip buckle. Belt is fancy grained, with creased edges and buckle of a very attractive design. For such high quality our price is exceptionally low.
33L8835—Black.
Sizes, 30 to 44 inches waist measure. **State size.** Shipping weight, 4 ounces.

Genuine Cowhide. Gold Plated Buckle.
95c Genuine Cowhide Belt. Leather lined, with 14-karat gold plated self adjusting Giant Grip buckle. Belt is fancy grained, with creased edges. Very attractive buckle. A very good value at our price.
33L8832—Brown.
Sizes, 30 to 44 inches waist measure. **State size.** Shipping weight, 4 ounces.

Genuine Cowhide. Silver Plated Buckle.
89c Genuine Cowhide Leather Belt, fancy grained, with creased edges. Leather lined. Silver plated self adjusting initial buckle in a very attractive design.
33L8817—Black. 33L8818—Brown.
Sizes, 30 to 44 inches waist measure. **State size; also initial.** Shpg. wt., 4 oz.

Silver Plated Initial Tongue Buckle.
89c Fancy Grained Cowhide Bridle Leather Strap Belt with silver plated initial tongue buckle. One of the best wearing belts on the market and a wonderful value at our price.
33L8820—Black.
Sizes, 30 to 44 inches waist measure. **State size; also initial wanted.** Shipping weight, 4 ounces.

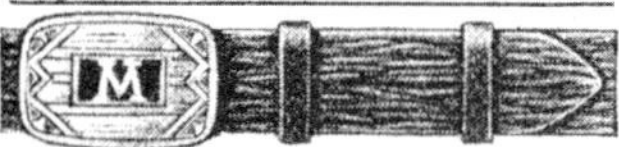

Cowhide Belt. Nickel Silver Buckle.
48c Genuine Fancy Grained Cowhide Bridle Leather Strap Belt. Attractive design nickel silver self adjusting Giant Grip buckle with neat pierced initial. Usually sells for double our price.
33L8864—Black. 33L8865—Brown.
Sizes, 30 to 44 inches waist measure. **State size; also initial.** Shpg. wt., 4 oz.

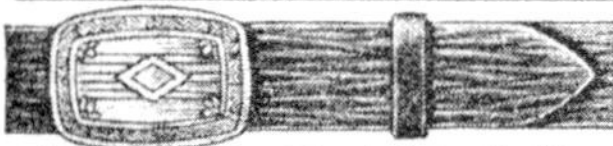

Cowhide Belt. Giant Grip Buckle.
39c Genuine Fancy Grained Cowhide Bridle Leather Strap Belt with nickel plated self adjusting Giant Grip buckle.
33L8842—Black.
Sizes, 30 to 44 inches waist measure. **State size.** Shipping weight, 4 ounces.

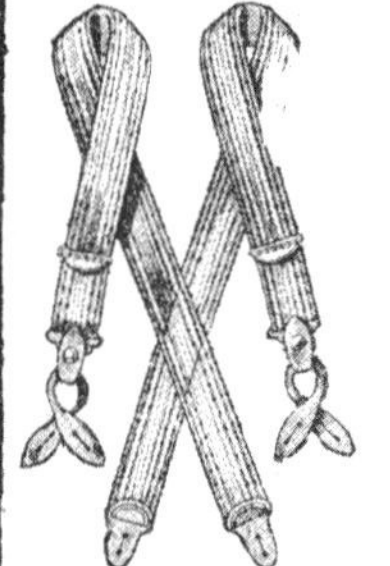

25c **33L8711** Good Quality Fancy Lisle Crossback Suspenders. Non-rusting brass plated buckles and trimmings. Leather ends. Length, 38 inches. Truly a great value at this low price. Shipping weight, 4 oz.

79c **33L8735** Fine Quality Lisle Dress Suspenders. Plain color. Artificial silk elastic lisle webbing with stitched colored leather ends to match. All brass trimmed. Length, 38 inches. Shpg. wt., 4 oz.

69c **33L8700** Fine Quality Lisle Crossback Dress Suspenders. Fancy elastic lisle webbing with non-rusting brass trimmings, stitched colored leather ends. Lengths, 38 or 40 in. **State length.** Shpg. wt. 4 oz.

39c **33L8608** Good Quality Police and Fireman's Style Suspenders. Cushion back elastic webbing with strong cowhide ends. Nickel plated trimmings. Length, 38 in. Shipping weight, 7 ounces.

48c **33L8650** Fine Quality Police and Fireman's Style Suspenders. Heavy cushion back elastic webbing with strong cowhide ends. Nickel plated trimmings. Length, 38 in. Shpg. wt., 7 oz.

59c **33L8625** Our Best Quality Police and Fireman's Style Suspenders. Long wearing cushion back elastic webbing with strong cowhide ends. Brass trimmings. Length, 38 in. Shipping weight, 7 oz.

69c **33L8626** Dress Suspenders of fine quality elastic webbing overshot with artificial silk, having the appearance of all silk suspenders. Stitched colored leather ends. Brass trimmed. Length, 38 in. Shpg. wt., 4 oz.

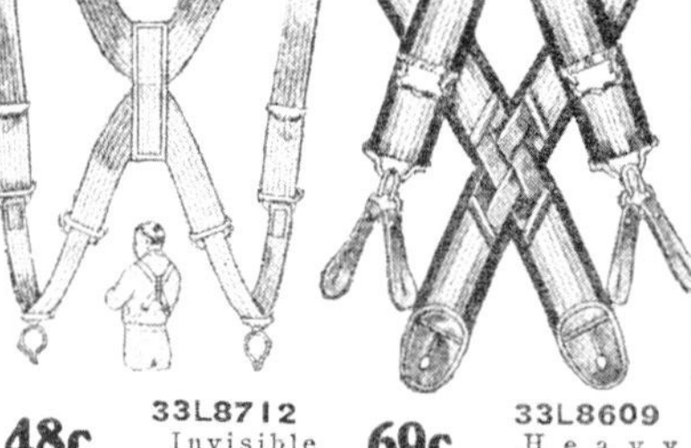

48c **33L8712** Invisible Dress Suspenders. To be worn under the shirt. Light weight elastic webbing with non-rusting brass trimmings. Serviceable as well as comfortable. Length, 38 inches. Shipping weight, 3 oz.

69c **33L8609** Heavy Weight Adjustable Crossback Suspenders for heavy wear. Strong 2-inch cushion back elastic webbing with nickel plated trimmings. Heavy leather ends. Length, 38 inches. Shipping weight, 7 ounces.

Fancy Nickel Plated Tongue Buckle.
19c A good leather belt at a low price. Good quality fancy grained split leather strap belt. Fancy nickel plated tongue buckle.
33L8802—Black.
33L8803—Brown.
Sizes, 30 to 44 inches waist measure. **State size.** Shipping weight, 4 ounces.

Cowhide Bridle Leather Strap Belt.
39c Genuine Cowhide Bridle Leather Strap Belt. Smooth finish with creased edges. Heavy steel tongue buckle. A very neat, long wearing belt.
33L8838—Black.
33L8839—Cordovan color.
Sizes, 30 to 44 inches waist measure. **State size.** Shipping weight, 4 ounces.

Fancy Grained Cowhide Tubular Belt.
79c Genuine Cowhide Tubular Belt, fancy grained, with creased edges. Fancy silver plated tongue buckle. A neat serviceable belt.
33L8847—Black.
Sizes, 30 to 44 inches waist measur **State size.** Shipping weight, 4 ounce

Leather Belt. Giant Grip Buckle.
29c A Leather Belt with a Giant Grip buckle is a wonderful value at this price. Fancy grained split leather with creased edges. Nickel plated, self adjusting buckle.
33L8807—Black.
33L8808—Brown.
Sizes, 30 to 44 inches waist measure. **State size.** Shipping weight, 4 ounces.

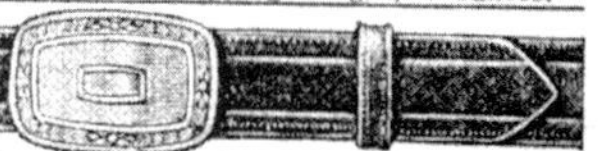

Genuine Cowhide. Giant Grip Buckle.
48c Genuine Cowhide Bridle Leather Strap Belt. Smooth finish with stitched edge effect. Fancy design nickel plated Giant Grip self adjusting buckle that absolutely will not slip.
33L8840—Black.
Sizes, 30 to 44 inches waist measure. **State size.** Shipping weight, 4 ounces.

Genuine Cowhide Narrow Belt.
45c Genuine Cowhide Bridle Leather Strap Belt in the popular narrow width. Smooth finish with stitched edge effect. Fancy nickel plated self adjusting roller buckle. Width, about ¾ inch.
33L8872—Black.
Sizes, 30 to 44 inches waist measure. **State size.** Shipping weight, 3 ounces.

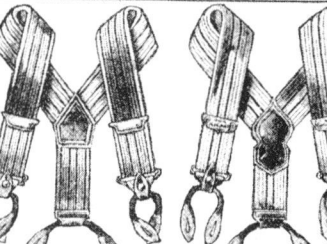

Washable Rubber Belt.
19c Exceptional value in Rubber Belts. Made of good quality interlined rubber with self adjusting nickel plated buckle. Can easily be washed and will give wonderful service.
33L8829—Black.
33L8830—Brown.
Sizes, 30 to 44 inches waist measure. **State size.** Shipping weight, 6 ounces.

33L8671 "Guyot" Style Dress Suspenders. Elastic in black ends only. Nickel plated brass trimmings. Length, 38.

33L8669 Self Adjusting Fancy Lisle Webbing Dress Suspenders. Brass trimmings. Cord ends. Length, 38.
45c **33L8670** Genuine President Suspenders. Nickel plated trimmings. Length, 36 inches.

FINE PIPES and MEN'S PURSES

Men like our pipes. They are real Men's Pipes, made to give honest-to-goodness smokes. They are made by the best known pipe makers, and smokers will at once recognize their favorite brands and trade marks on our pipes. Compare our values! Our great buying power enables us to quote low prices on the popular pipes.

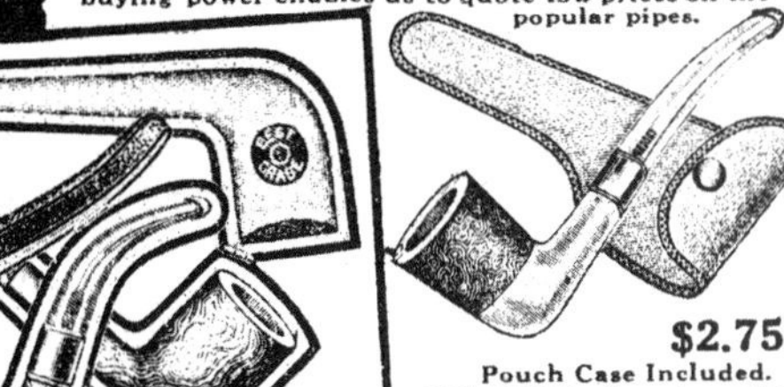

$1.98

18L4001—Value extraordinary! Very high grade pipes, made of genuine French briar, fitted into fine silk plush lined cases. Shapes of several patterns, all with large size highly polished bowls and bits of the popular tasteless Redmanol. A fortunate purchase enables us to offer these fine pipes at less than half their usual selling price. **State choice of straight or bent style.** Shipping weight, 8¼ ounces.

$2.75

Pouch Case Included.

18L4026—Bargain value, popular half bent style. Bowl made of genuine French briar wood. Clear Redmanol base and stem which is absolutely tasteless. Gold plated band. Length, about 5 inches. Soft leather pouch case. Shipping weight, 8¼ ounces.

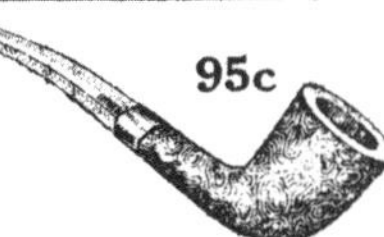

95c

18L4015

Popular thin model with good size bowl of genuine French briar with Bakelite bit and nickel silver band. Length, 5¼ inches. Shipping weight, 7¼ oz.

79c

18L4018

Very popular London shape pipes of selected dark finish genuine briar. Patented "Nuvo" flush fitting bits. Guarantee with each pipe against bowls cracking or burning out or mouthpiece breaking at point entering bowl. Unusual value. Length of straight pipe, 5¾ inches. **State choice of straight or bent shape.** Shipping weight, 7¼ ounces.

Absorbo Lined.

73c

18L4007

Absorbo Lined Pipe with metal cover top. Cherry wood bowl with screw cleaning socket. Horn screw bit. Needs no breaking in. Cool smoker. Length, 5¼ inches. Shipping weight, 8¼ ounces.

The Wellington

18L4035—Extra large size..... **75c**
18L4031—Big size.......... **55c**
18L4028—Medium size....... **33c**

Famous the world over for a cool and comfortable smoke. The well collects the saliva and keeps the tobacco dry to the last puff. Fitted with special mouthpiece. Shipping weight, 8, 7½, 7¼ ounces, respectively.

Famous Perry Pipe

State choice of bent or straight style.

Shipping weight, 8 ounces.

$2.19 EACH **18L4006**

Finest briar used in the "New Improved Perry Pipes." Made with less parts and give even a cooler and cleaner smoke than before. The smoke is purified and cooled as it filters through the discs. Nicotine and tobacco oil fall between discs and are removed when stem is pulled out. We do not handle the highest priced four-disc Perry Pipes which sell at $3.00 and $4.00, preferring to offer our customers the same large up to date shapes in the second quality at a great saving in price. The only difference consists of slight marking in the briar bowls, markings which the average person would seldom notice and which, in our opinion, in no way impair the looks or smoking quality of the pipes. Length, 5¼ inches.

83c **18L4014—Perry Cigarette Holder.** Patented aluminum discs filter the smoke and keep nicotine and oil from the mouth. Spear point holds cigarette securely. When pulled out by stem it is used to remove "butts." Made of tasteless hard rubber. Length, 3⅞ inches. Shipping weight, 1½ ounces.

Cigar and Cigarette Holders

18L4148 45c

Cigar Holder of tasteless Redmanol, nicely finished. Length, 2 inches. Shipping weight, 1¼ ounces.

18L4151 43c

Cigarette Holder of tasteless Redmanol. Nicely finished. Length, 3 inches. Shipping weight, 1¾ ounces.

18L4259 72 for 9c

Our special Pipe Cleaners. Something every pipe smoker should have. Covered wire. The best cleaners we ever saw. Shipping weight, 2 oz.

$1.79

18L966⅓—Exceptional value Bill Fold and Card Case. Made of splendid quality genuine pigskin in a rich looking dark russet brown finish. Does not show soil, folds thin, does not bulk in pocket. Compartments for bills, tickets, stamps, etc. Fine workmanship throughout. Size, closed, 4¾x3¼ inches. Shipping weight, 2 oz. **(Print name if wanted.)**

59c

18L991⅓—Genuine long wearing Horsehide Bill Fold and Card Case. In addition to the large compartment for bills, has three other compartments, also small calendar. Color, rich dark brown. Size, closed, 4⅝x3 inches. Shipping weight, 2 ounces. **(Print name if wanted.)**

59c

18L993⅓—Fine quality genuine Pigskin Six-Hook Key Purse. Folds very thin. Keeps keys from punching holes in your pockets. Easy to select wanted key. Closes with strong snap fastener. Color is a rich dark russet brown. Size, closed, 2½x4 inches. Shpg. wt., 1¼ oz. **(Print name if wanted.)**

YOUR NAME in Gold Letters

EVERY man enjoys the sight of his name on his personal things, especially when set permanently and in such handsome style as on these selected purses. Gold stamping has heretofore been done only at high prices. We are furnishing these seven new style purses, expertly stamped, at prices much less than the usual prices for purses alone. Print name wanted like this:

J.T.WALTER

Do not write it like this:

J.T. Walter

$2.25 OUR BEST

Something Very Fine.

18L986⅓—Mounted edges on fine quality genuine lustrous pinseal make the very smartest of the new folds. Stylish size, 4⅜x3⅛ inches, closed. Pockets for bills, tickets, pass cards, stamps, etc. Gold plated mounts will not come off. Smooth leather facings. High grade article at special price. Color, black. **(Print name if wanted.)** Shipping weight, 3½ ounces.

$1.19

18L989⅓ — Good quality Genuine Leather Bill Fold and Card Case. Especially fitted for holding photographs. Black only. Size, closed, 4¾x3¼ inches. Shipping weight, 2¾ oz. **(Print name if wanted.)**

98c

18L992⅓ — Extra quality, genuine Cordovan Leather Bill Fold and Card Case. Very well made and will give excellent service. Has four compartments in addition to the large one for bills. Rich looking dark brown color. Size, closed, 4⅝x3⅛ inches. Shipping weight, 2¼ ounces. **(Print name if wanted.)**

The Name in Gold.

The real personal touch in fine 22-karat gold leaf. Perfect gifts for "him."

Novel Purses for Coins and Bills

65c

18L948—For a very handy purse which closes flat, we recommend this unusual value. Fine quality dark Leather Coin Purse with special pocket for bills. Large pocket for small change. Purse locks with snap. Size, 3¼x3¼ inches. Shipping weight, 1½ ounces.

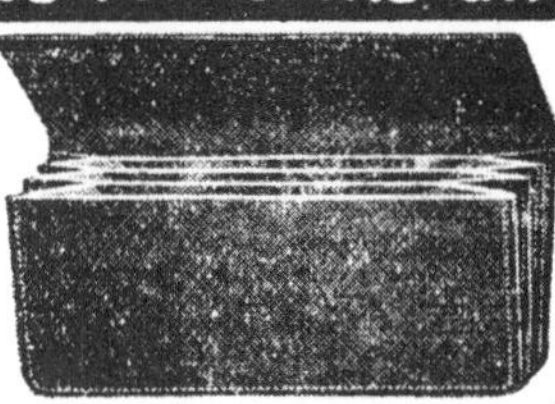

$1.29 **18L953⅓**—Large Document Case. Good black leather. Size, 10⅝x4⅜ in. Four large compartments and three small ones. Name printed in gold letters, 25c extra. Name and address, 50c extra. Shpg. wt., 9¼ oz. **(Print name carefully).**

45c **18L973**—Combination Bill Fold and Coin Purse. Good quality black leather. Three large pockets for coins. Size, closed, 3¼x2¾ inches. Shipping weight, 1¾ ounces.

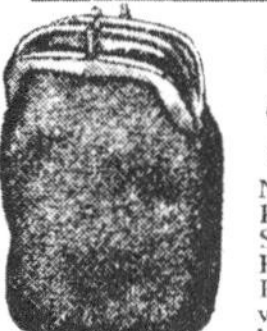

24c

18L949

Nickel Plated Steel Frame Purse with two large pockets. Good quality tan leather. One pocket for change and one for bills. Length, 5 inches. Shipping weight, 2¼ ounces.

39c

18L983—Big Value Bill Fold. Made of medium quality brown leather in alligator grain. Compartments for bills, cards, tickets, etc.; also small calendar. Size, closed, 3x4⅜ inches. Shipping weight, 2½ ounces.

21c

18L984—Combination Bill Fold and Coin Purse. Medium quality alligator grain brown leather. Double snap fasteners. Folds very thin. Size, closed, 2¾x3¾ inches. Shipping weight, 1½ ounces.

39c **18L944** Splendid Value Leather Coin Purse. Two pockets, one for bills and one for small change. Nickel plated frame. Closed, 3 inches long. Dark leather. Shipping weight, 2¼ oz.

33c **18L946**—Bargain value in a good quality Leather Purse with nickel plated metal frame. Well finished. Two pockets. Size, 4x3¼ inches. Dark leathers. Shipping weight, 2 ounces.

"La Belle" Wardrobe and Dress Trunks

Fiber Covered Wardrobe Trunk.

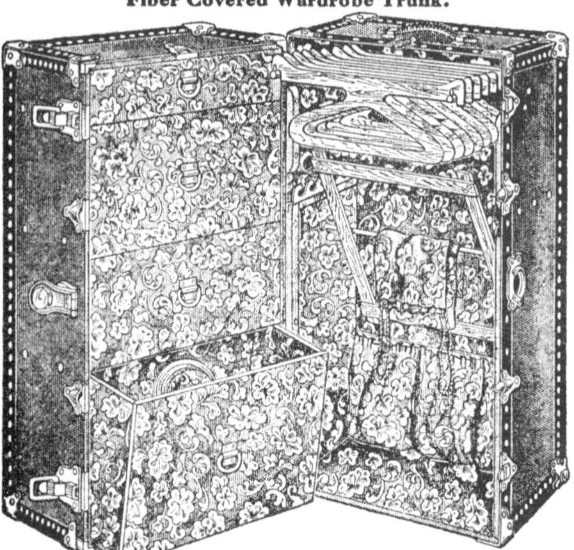

Box of basswood, covered with hard rolled fiber and bound with heavy vulcanized fiber, trimmed with heavy gauge, brassed hardware. A very substantially built, low priced wardrobe, lined with fancy cretonne cloth.

"A place for everything and everything in its place." There are ten hangers for suits, dresses, coats, waists, skirts, etc.; drawers for handkerchiefs, gloves, ties, collars, shirts, lingerie, etc.; space for hats; pockets for shoes and slippers; laundry bag for soiled garments. Small articles of wearing apparel can readily be found without going to the trouble of unpacking. A modern wardrobe closet to accompany you everywhere, and the amount of clothing this trunk holds without crowding is indeed remarkable.

10L9500¼—Size, 40x21½x22 inches. Shipping weight, 75 pounds............**$25.00**

10L9501¼—Same as above only three-quarter size, 40x21½x18 inches. Eight hangers. Shipping wt., 70 lbs..**$23.50**

10L9502¼—Steamer Trunk, made as above except that it has only six hangers; one divided tumbler or drop drawer, two small drawers, and is made in size 40x21½x13¾ inches. Shipping weight, 65 pounds.............**$19.75**

Low Priced Metal Covered Trunk.

10L9550¼—**Very good quality box of well seasoned lumber with one slat all around and three on the top; metal covered, good hardware, strong lock.** Inside is fitted with one dress tray with covered hat box and is neatly lined. Strong enough for all practical purposes, **including those of travel, and will be found especially suitable for home use.**

Size	Shipping weight	Price
26x17½x21 inches.	35 pounds	$6.75
30x19 x22 inches.	45 pounds	7.50
34x20½x23 inches.	55 pounds	8.25

Metal Covered and Metal Bound Dress Trunk.

10L9504¼—**Secured with two strong locks, spring catch in the center and two good leather straps.** All trimmings are of steel, brass finished; edges and top slats in front are protected with heavy steel valance clamps; the wide center band of metal is double studded and the metal bound edges are also nail studded, making a very strong, well wearing trunk that will show its worth in long, hard service. Attractively lined inside and fitted with a deep top tray with full covered lid.

Size	Shipping weight	Price
32x19½x22 inches.	52 lbs.	$13.50
36x21 x23½ inches.	64 lbs.	14.25
40x21½x24 inches.	71 lbs.	15.00

Basswood Dress Trunk With Two Locks, Fiber Covered and Studded.

10L9518¼—**Deserving of special notice is this fine trunk. Away in the lead so far as price goes, and the quality is of high rating.** We feel secure in recommending this number to all who look for something out of the ordinary in value. People who perhaps are planning a trip of much importance will find added pleasure in the occasion by the possession of one of these rare bargains. The selected quality of basswood is covered with heavy hard rolled fiber, studded with brass finished saddle nails. Trimmings are of heavy brassed steel. Two Excelsior style locks and a center draw bolt. Attractively lined inside; deep, full covered and divided top tray; extra dress tray.

Size	Shipping weight	Price
32x21 x23 inches.	60 pounds	$13.75
36x22 x23 inches.	72 pounds	15.25
40x22½x24 inches.	75 pounds	16.75

Fiber Covered and Studded Basswood Steamer Trunk.

Fitted With Excelsior Style Lock and Two Draw Bolts.

10L9519¼—**Of heavy hard rolled fiber over basswood of selected quality and studded with brassed saddle nails;** heavy, brass plated steel trimmings give protection from rough handling. Inside is neatly lined and fitted with deep top tray, fully covered and divided. A trunk of unusually good value and from which years of good service may be expected. Matches 10L9518¼ at left.

Size	Shpg. wt.	Price
32x19x12 inches.	40 lbs.	$13.65
36x21x12 inches.	48 lbs.	14.45
40x22x12 inches.	51 lbs.	15.25

Metal Covered Dress Trunk With Hardwood Slats.

10L9522¼—**Seasoned basswood box, metal covered and heavily slatted with hardwood, reinforced and held securely at all corners with heavy steel bumpers, brass plated.** An unusually strong, durable and attractive, massive appearing piece of luggage. Fitted with one good lock and two lever draw bolts that hold the cover and body rigidly together. Metal center band, double nailed, and metal bound edges; two good leather straps. Neat inside lining with dress tray and covered top tray. All trimmings are of steel, brass plated.

Size	Shipping weight	Price
32x20 x23 inches.	60 lbs.	$15.50
36x21 x23½ inches.	70 lbs.	16.25
40x21½x24 inches.	75 lbs.	17.00

"Douglas" Vulcanized Fiber Wardrobe Trunks.

These trunks are of good appearance and of that ruggedness which resists wear. The interiors are fitted in a pleasing and convenient manner.

10L9508¼—"Regular Douglas." Of three-ply veneer, covered and lined with vulcanized fiber, making five-ply construction. Edges bound with vulcanized fiber and closely nailed with shot head tacks. Hardware and trimmings of good quality steel, brass plated. A strong, well finished trunk that will withstand hard service. Cloth lined; ten garment hangers of five-ply basswood veneer; retainer that holds garments in position; pull out trolley, shoe box; plush lined open top; set of cord hangers; five drawers with new locking device that locks them all at once. Third drawer has removable hat form; second drawer is convertible for large hats. Size, 40x21½x22 inches. Shipping weight, 85 pounds........**$37.50**

10L9523¼—"Extra Douglas." Same as above but larger size, 43 inches high; full studded; rounded edges and corners; extra heavy hardware. A very strongly made and massive looking trunk. Shipping weight, 95 lbs.**$45.50**

10L9524¼—"Special Douglas." 42 inches high, with rounded edges and corners; heavy hardware; iron holder, ironing board; patent lock that fastens the trunk at top and bottom by means of a sliding bar. The lock must be turned upward, after unlocking, before trunk can be opened; after closing, the lock must be turned down again before the trunk can be locked. Shpg. wt., 95 lbs.**$48.50**

Your Choice of Three Popular Designs.

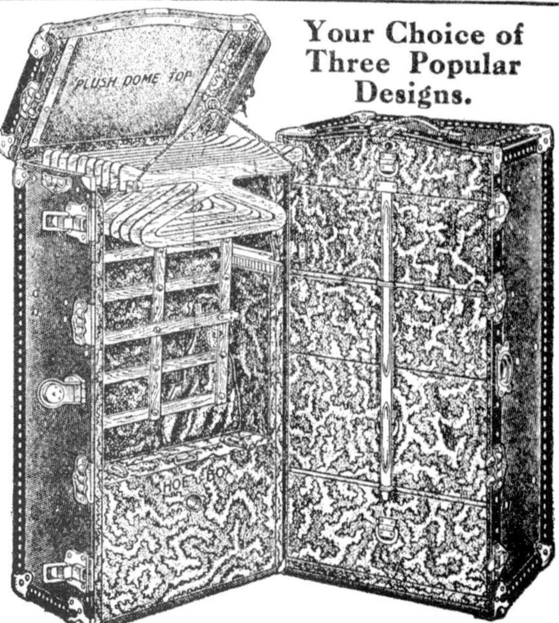

Strongly Constructed—Low in Price

Popular General Purpose Trunk.

Sturdy Metal Covered Dress Trunk of Medium Price.

10L9540¼—Strongly built of well seasoned basswood, metal covered and bound, to withstand the knocks of travel. Corners well protected with heavy iron angles, japanned. Hardwood slatted and strapped, as shown; nicely finished and lined inside; plenty of room for storage; two-compartment tray, one section of which forms a covered hat box.

Size, 32x19½x21½ in.	Shipping weight, 45 pounds........	$ 9.75
Size, 36x21½x23½ in.	Shipping weight, 53 pounds........	10.50
Size, 40x22 x24 in.	Shipping weight, 65 pounds........	11.25

Fiber Covered and Studded Basswood Wardrobe Trunk.

Folks who spend a great deal of time traveling will appreciate the sturdiness and strength of this trunk. The heavy vulcanized fiber binding and metal braces at all edges and corners reinforce and permanently hold the well constructed body solidly together. The heavy brass finished steel lock and draw bolts add security. The feeling comes that so sturdy a trunk must give many years of good service, and this is a fact which fortunate buyers will prove. Rough handling will have little damaging effect on a trunk of this kind.

The interior is lined with figured cretonne and fitted with four drawers (one of which has a removable hat form), ten clothes hangers for garments of various kinds, clothes retainer, three shoe pockets and laundry bag; open top.

10L9515¼—Full size, about 40x21½x22 inches. Shpg. weight, 85 pounds...**$31.50**

10L9516¼—Three-quarter size, about 40x21½x18 in. Eight hangers only. Shipping weight, 75 pounds **$28.50**

10L9517¼—Steamer Trunk. Size, about 40x21½x13¾ inches. Six hangers only. Shpg. wt., 65 lbs....**$25.50**

Some Splendid Values on This Page.

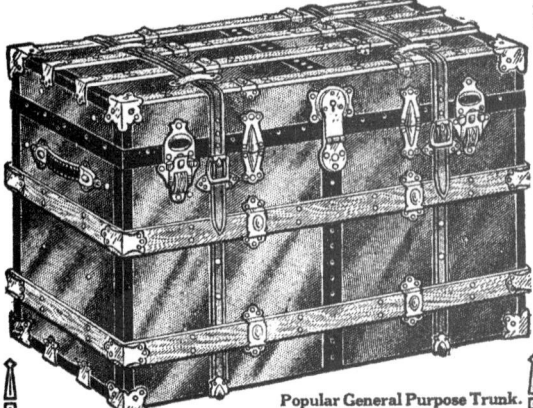

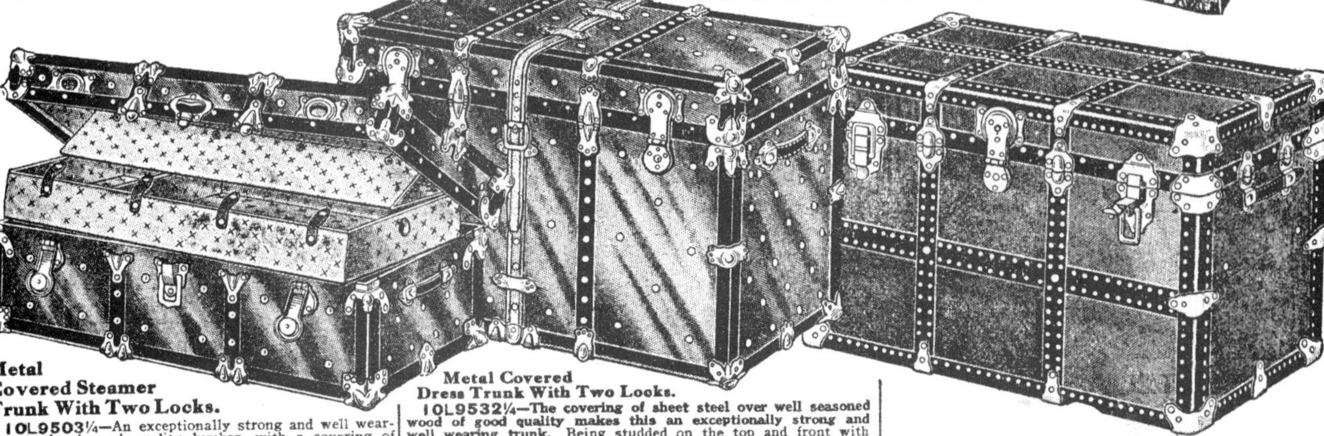

Metal Covered Steamer Trunk With Two Locks.

10L9503¼—An exceptionally strong and well wearing trunk of good quality lumber, with a covering of sheet steel. The top and front are studded with saddle nails and the ends and back with steel tacks, brass plated. It has two heavy fiber center bands. These features combine to make a trunk of sterling wearing quality and high class, massive appearance. The interior is neatly lined and has a deep, divided tray. Matches 10L9532¼ at right.

Size, 32x19x12 in.	Shpg. wt., 40 lbs.........	$ 9.75
Size, 36x21x12¾ in.	Shpg. wt., 48 lbs.........	10.50
Size, 40x22x13½ in.	Shpg. wt., 51 lbs.........	11.25

Metal Covered Dress Trunk With Two Locks.

10L9532¼—The covering of sheet steel over well seasoned wood of good quality makes this an exceptionally strong and well wearing trunk. Being studded on the top and front with saddle nails and on the ends and back with brass finished steel tacks, it has that massive and sturdy appearance so desirable in a much used piece of luggage. The two center bands are of heavy fiber; all trimmings are of brass plated steel. Fitted with two good locks and a heavy leather center strap. Inside is attractively lined and has a deep divided tray, one section of which may be used for a hat box.

Size, 32x20½x22 inches.	Shpg. wt., 45 lbs...........	$10.50
Size, 36x21½x23½ inches.	Shpg. wt., 50 lbs...........	11.25
Size, 40x22½x23½ inches.	Shpg. wt., 55 lbs...........	12.00

Five-Ply Fiber Covered Dress Trunk.

10L9526¼—A high class, strong and serviceable piece of luggage. Constructed of three-ply veneer, covered and interlined with vulcanized fiber, making a five-ply trunk. All edges are rounded; center bands and binding of first quality heavy vulcanized fiber, riveted at all points that must stand the heaviest wear. Strong snap lock, draw bolts and trimmings of steel, brass finished. Inside is lined with figured cloth and fitted with deep top tray containing removable hat form and extra dress tray.

Size, 36x22 x23 in.	Shpg. wt., 72 lbs...........	$25.50
Size, 40x22½x24 in.	Shpg. wt., 78 lbs...........	27.00

"Logan" Extra Well Made Vulcanized Fiber Wardrobe Trunks. Interlined with vulcanized fiber, making five-ply construction.

Great care is taken in the making of these trunks, resulting in a finished article that is hard to equal in appearance and service value.

Three Very Desirable Trunks Here.

10L9512¼

"Regular Logan." Open, plush lined, dome top; ten garment hangers, shoe box, ironing board, laundry bag, five drawers, one of which is fitted with iron holder and another as a hat box; drawer locking device that locks all drawers at once; set of cord hangers for hanging garments in closet; the trunk is fitted with an extra strong lock that fastens it at top and bottom by means of a sliding bar. To operate: Unlock in the usual way, then turn lock upward to the left. To lock: Close trunk and turn lock downward to the right, then snap lock. Trimmings of heavy rolled steel, brass finished. Size, 42x22x21½ in. Shipping weight, 90 pounds.....**$54.95**

10L9525¼—"Extra Logan." Larger and heavier trunk, with steel bolt through center, to which is connected the improved double drawer locking device, operated from center of trunk and making it strong and rigid. Spring clothes retainer with dust cloth. Rounded edges and corners; heavy hardware. Shipping weight, 110 pounds **$65.95**

10L9530¼—"Special Logan." Extra large size, with well rounded edges and corners; heavy brass hardware; very fine quality interior fittings; six drawers, twelve hangers. Spring clothes retainer with dust cloth. A high class trunk in every way, of the same general construction as above, but with these special features added. Shipping weight, 115 pounds **$79.95**

Five-Ply Fiber Covered Steamer Trunk.

10L9527¼—Three-ply veneer construction, covered and lined with vulcanized fiber, making a five-ply trunk. Popular round edge style, with center bands and binding of heavy vulcanized fiber, first quality; strongly riveted by hand where the hardest knocks usually strike. Draw bolts and trimmings of steel, brass plated; good snap lock. Lined with figured cloth; deep divided top tray.

Size, 36x21 x12 in.	Shipping weight, 47 pounds..	$23.50
Size, 40x21½x12 in.	Shipping weight, 53 pounds..	24.95

Combination
Raincoats
and
Topcoats
Tweeds, Cashmeres
Gabardines and
Whipcords
TAKE QUALITY INTO ACCOUNT and you will appreciate our prices more than ever. Careful buying, supported by laboratory tests, make our values better.
Cloth samples sent on request.
Cravenetted Whipcord or Gabardine
Rubberized Cashmere
Rubberized Tweed
Rubberized Tweed
45L7618—Olive Drab Cravenetted Gabardine. About Two-Thirds Wool.
45L7619—Mixed Brown and Tan Cravenetted Whipcord. About One-Third Wool.
EACH $15.85
EXTRA FINE QUALITY SHOWER PROOF GABARDINE OR NEW WHIPCORD TOPCOATS. One of those classy and dressy double breasted models, with raglan shoulders, wide box plait down the center of back and belt all around. Made of light weight closely woven textures, yet they're warm and comfortable when it's too cool to be without a topcoat and not quite cold enough for an overcoat. Both coats are rainproof in all ordinary showers. Other desirable features are the stylish patch pockets and the convertible collar. Shrewd buyers choose these exceptional values. Order now so you'll be prepared for rainy or chilly weather. Length, 44 inches. SIZES—34 to 44 inches chest. State chest measure taken over vest. Shipping weight, 45L7618, 4 lbs.; 45L7619, 4½ lbs.
Shipping weights are based on an average number of shipments. Weight of shipment you receive may, therefore, vary a little from weight specified in description.
45L7614 Brown and Gold Heather.
45L7615 Blue and Gold Heather.
EACH $13.75
YOUNG MEN'S SLIP-ON SPORT MODEL. Cut along snappy lines. Double breasted style, with raglan shoulders, belt all around, large patch pockets and back with inverted half length plait—all features which add class and style. Material is a fine TWEED, about two-thirds wool. Good quality woven plaid lining and rubber interlining. Convertible collar. Your wardrobe isn't complete without this coat. The old saying is, "If you actually need a certain thing—you'll pay as much or even more in other ways if you don't get it." This coat may prevent you from taking cold due to rain soaked clothes. Certainly, it will protect your clothing. Length, 45 inches. SIZES—34 to 44 inches chest. State chest measure taken over vest. Shipping weight, 5½ pounds.
45L7608 Brown Heather.
45L7609 Blue and Brown Heather.
EACH $9.75
BOTH A GOOD LOOKING TOPCOAT AND A DEPENDABLE RAINCOAT. It's good looking because of the richly colored heather effect, the handsome patch pockets and the all around belt. And dependable for rainy weather because it has an interlining of rubber. Outer material is a closely woven TWEED, about two-thirds wool. Body and sleeve linings are of neatly patterned woven plaid cotton material. Collar is of the convertible style; that is, it can be turned up and buttoned for stormy weather. Large ventilation eyelets under the arms allow free air circulation and the vent in the back permits freedom in walking. A good looking topcoat for dress wear and a raincoat for the rainy days. Length, 44 inches. SIZES—34 to 44 inches chest. State chest measure taken over vest. Shpg. wt., 5½ lbs.
45L7610—Dark Gray.
45L7611—Black.
45L7612—Olive Drab.
EACH $9.95
EXTRA LONG CONSERVATIVE MODEL COMBINATION RAINCOAT AND TOPCOAT. Loose fitting single breasted style, to suit the man who prefers a plain back, full length coat. Outer fabric is a fine twilled CASHMERE, about one-half wool. Has woven plaid cotton body lining and rubber interlining. Edges are double stitched and all seams are full strapped and cemented. The collar is convertible and can be turned up in rainy or chilly weather. Sleeves have an adjustable tab. Pockets are of the handy slash type. We urge you conservative men to choose this practical topcoat. At only $9.95 it is unusually low priced for so desirable a coat. Lengh, 48 inches. SIZES—34 to 48 inches chest. State chest measure taken over vest. Shipping weight, 5½ pounds.
264 SEARS, ROEBUCK AND CO.

Combination
Raincoats
and
Topcoats
Cotton and Wool Twills
Tweeds, Velours
and Whipcords
If interested, you'll find Youths' Suits on pages 270 and 271, and Youths' Overalls on page 268. They're specially made to fit boys of high school age.
We'll mail you cloth samples if you write for them.
Rubberized Cotton Tweed
Rubberized Tweed
Cravenetted Whipcord
Rubberized Suede Mole Cloth
45L7600 Brown Heather. 45L7601 Gray-Green Heather. $4.98 EACH
MEN'S AND YOUNG MEN'S LOW PRICED COMBINATION RAINCOATS AND TOPCOATS. Why ruin your clothes when you can buy this coat for only $4.98? Three-button single breasted style with belt all around, handy slash pockets, convertible collar and adjustable tab on sleeves. Between the outer texture of closely woven tweed and the printed plaid cotton lining is an interlining of rubber; all three being closely compressed. Length, 44 inches. SIZES—34 to 48 inches chest measure. State chest measure taken over vest. Shipping weight, 5¼ lbs.
YOUTHS' SIZES.
45L7602 Brown Heather. 45L7605 Gray-Green Heather. $4.48 EACH
Same style and make as above. Av. length, 41 inches. SIZES—32, 34 and 36 inches chest. State chest measure taken over vest. Shpg. wt., 5 lbs.
45L7603 Gray-Green Heather. 45L7604 Dark Gray. $5.95 EACH
TWO-IN-ONE RAINCOAT AND TOPCOAT with stylish patch pockets. It's an ideal coat for rainy weather and a mighty comfortable topcoat for dress wear, too. Made from a strong three-ply fabric, the outside of which is a wool mixed tweed, about one-third wool, and the lining a neatly printed plaid cotton material. Between these is an interlining of rubber which cements the two together and makes the combined fabric both wind and rain proof. This coat is of the three-button single breasted style with belt all around. Has convertible collar and adjustable tab on sleeves. Free air circulation is provided for by ventilation eyelets under each arm. Vent in back. Inside seams are neatly strapped and cemented. You'd better get one of these coats if you want a serviceable low priced coat. Length, 44 inches. SIZES—34 to 44 inches chest. State chest measure taken over vest. Shipping weight, 5¼ lbs.
45L7606—Mole Brown. $8.95
DRESSY TOPCOAT AS WELL AS AN EVER READY RAINCOAT. Designed especially for young fellows who favor smart features, such as the double breasted style with belt all around, yoke back with inverted plait and patrol style pockets. You'll be proud to wear this medium heavy weight topcoat at all times when it's a bit chilly outdoors. Then, too, you needn't dread the rain, because you're always prepared when you wear this combination raincoat and topcoat. Material is a smooth finish strong wearing fabric known as cotton velour, which has the appearance of genuine chamois leather. Lined throughout with good quality dark tan bombazine and has rubber interlining. An unusually fine garment at our low price of $8.95. Length, 44 inches. SIZES—34 to 44 inches. State chest measure taken over vest. Shipping weight, 6 pounds.
Shipping weights are based on an average number of shipments. Weight of shipment you receive may therefore vary a little from weight specified in description.
In all coats on this page, except 45L7616, the material, lining and interlining are rubberized and so closely and tightly compressed as to make one solid fabric. Seams are full strapped and cemented.
45L7616—Olive and Tan Heather. $10.75
POPULAR CRAVENETTED WHIPCORD TOPCOAT. A light weight double service coat for rain or shine. Cravenetted by the special Cravenette process, making it rainproof in all ordinary showers. Made from a closely woven cotton whipcord, a fabric that will give you long wear and always look good. All around belted model, as illustrated above, with raglan shoulders, inverted plait in back, convertible collar and slash pockets. Compare this handsome high quality coat with whipcord topcoats elsewhere and you will appreciate what an exceptional value we here offer you. We're quite sure you wouldn't take twice the amount you paid for this coat, after you have worn it a few times. You'll be just as proud of it as we are. Length, 44 inches. SIZES—34 to 44 inches chest. State chest measure taken over vest. Shipping weight, 4½ pounds.
SEARS, ROEBUCK AND CO. 265

Rubber Surfaced Clothing
Shipping weights are based on an average number of shipments. Weight of shipment you receive may, therefore, vary a little from weight specified in description.
Shpg. wt., 4¼ lbs.
SIZES—34 to 48 inches chest. State chest measure taken over vest.
41L893—Black and Drab Reversible. $5.85
Practical Two-In-One Reversible Waterproof Coat. Worn either side out. Soft and flexible. One side is dull finish black rubber surfaced, other is drab cotton Asia cloth. Double back, double arm shields, deep slash pockets, strongly reinforced buttonholes and firmly cemented seams. Long vent in back. Decidedly practical and popular raincoat and an unusual value. Length, 52 inches.
41L965 Black. 89c
Dull Finish Black Rubber Coated Waterproof Nobby Hat. Soft, pliable brim may be worn turned up or down. Very serviceable and a great favorite. Has tan cotton lining. Band and seams are cemented, strapped and vulcanized. SIZES—6¾ to 7¾. State size. Shipping wt., 8 ounces.
Shpg. wt., 8 oz.
41L960 Black. 69c
Soft and Flexible Sou'wester Waterproof Hat. Standard style. Dull finish black rubber, lined with white sheeting. Stitched down brim. Strapped and cemented seams. SIZES—6¾ to 7¾. State size.
Shipping weight, 4¼ lbs.
State chest measure taken over vest.
41L899 Black. $3.79
Unusually Low Priced Soft and Pliable Dull Finish Black Rubber Surfaced Waterproof Coat. Lined throughout body and sleeves with good quality white sheeting. Has vulcanized one-seam back, two lower flap pockets and ventilation eyelets under arms. Closes with five patent metal clasp fasteners. Good value at a low price. Length, 48 inches. SIZES—34 to 48 inches chest.
41L905 Black. $4.98
Brakemen's, Teamsters' and Taxi Drivers' 36-Inch Length Waterproof Coat. Dull finish black rubber surface; frictioned body and sleeve lining penetrated with rubber. Double cape back across shoulders and double arm shields. Large ventilation eyelets. Two pockets. Corduroy tipped collar. Coat closes with four reinforced patent metal snap fasteners. Length, 36 inches. SIZES—34 to 48 inches chest. State chest measure taken over vest. Shpg. wt., 4½ pounds.
SIZES—34 to 48 inches chest. State chest measure taken over vest.
Shpg. wt., 4 lbs.
Length, 52 inches. SIZES—34 to 48 inches chest. State chest measure taken over vest. Shpg. weight, 6½ lbs.
41L894 Tan Covert. $8.95
Extra Heavy and Extra Long Waterproof Coat. Unusually good quality and decidedly practical. Outside material is a closely woven cotton covert cloth. Lined throughout with heavy weight brown sheeting and has waterproof rubber interlining. Fly front. Full strapped and cemented seams. Has high standing collar, two lower pockets with flaps and ventilation eyelets under arms. Well made to stand extra hard wear. Length, 52 inches. SIZES—34 to 48 inches chest. State chest measure taken over vest. Shpg. wt., 5¾ lbs.
41L935 Black. $7.95
Firemen's Extra Heavy Double Coated Waterproof Coat of heavy cotton jeans cloth, thoroughly coated with dull finish black rubber on both sides. Long fly front. Large corduroy faced collar, adjustable strap and buckle on sleeves, one inside pocket, reinforced armpits, and vent in back. Seams are strongly sewed and vulcanized. Length, 48 inches. SIZES—34 to 48 inches chest. State chest measure taken over vest. Shpg. wt., 6¾ lbs.
41L940 Black. $8.75
Chicago Police Coat. Regulation model, Extra long and positively waterproof. Well made of heavy weight cotton jeans cloth and heavily coated with dull finish black rubber. Tan jeans cloth inside body and sleeves. Made according to specifications of the Chicago Police Department. Has large double storm cape extending all around over shoulders and well down over arms. Large ventilation holes under cape in back. One large inside pocket and one club or billy pocket. Vent in back. Has shield with loops for attaching star. Length, 54 inches. Shipping weight, 5⅝ pounds.
41L915 Black.
SIZES—34 to 48. Length, 52 inches. $5.98
Extra Long Medium Weight Dull Finish Black Rubber Surfaced Waterproof Coat. Soft and pliable. Lined with fancy sheeting. Double yoke back with large ventilation holes. Corduroy tipped collar, two pockets, double arm shields, and vent. Vulcanized seams.
41L891 Drab. $9.95
Heavy Weight Extra Long Double Texture Waterproof Storm Coat. Extra fine quality. Excellently made of finely woven drab color cotton jeans cloth; body and sleeves lined with cotton bombazine to match outside; interlined with rubber. Large double storm cape covers back, shoulders and chest. Triple fly front closes with four patent metal snap fasteners. Corduroy collar has throat tab and sleeves have adjustable tab, all of which give added comfort and protection against rain. Full strapped and cemented seams. Long vent in back. Large ventilation eyelets under arms. Especially suitable for teamsters and others who are regularly exposed to hard and steady rain and windstorms.
266 SEARS, ROEBUCK

ON SCIENTIFIC TEST is assured in buying clothing from us. That makes our low prices doubly attractive.
Oiled Slicker Clothing
41L1015—Black.
41L1016—Yellow.
EACH $2.98
Extra Long Triple Fly Front Waterproof Oiled Slicker Coat. An exceptionally practical low priced general purpose coat for all sorts of stormy and rainy weather. Made double throughout. Has rain excluding wristlets and high standing cloth faced collar with tab to button around throat. Two large patch pockets with flap. Patent buttons. Average length, 54 inches. SIZES—36 to 48 inches chest. State chest measure taken over vest. Shipping weight, 5¼ lbs.
EACH 39c
41L970 Black. 41L971 Yellow.
Sou'wester Waterproof Oiled Slicker Hat. Standard for years. Has chin strap and ear laps. Stitched down brim. Soft cotton flannel lining. SIZES—6¾ to 7½. State size. Shpg. wt., 8 oz.
30L975—Black.
30L976—Yellow.
EACH 33c
Slicker Oil Compound. High quality. Highly recommended for recoating and preserving oiled slicker clothing. ½ pint in each can. Shipping weight, 2½ pounds.
Practical and Comfortable Three-Quarter Length Waterproof Coat.
41L1010—Black.
41L1011—Yellow.
EACH $2.58
Three-Quarter Length Waterproof Oiled Slicker Coat. Especially suitable for brakemen, fishermen, taxi drivers, etc. Made double throughout. Has fly front, and rain excluding wristlets. Two large patch pockets and high standing collar faced with flannel to protect neck. Patent buttons. Average length, 38 inches. SIZES—36 to 48 inches chest. State chest measure taken over vest. Shpg. wt., 3¾ lbs.
Reliable Quality At a Low Price.
Shipping weights are based on an average number of shipments. Weight of shipment you receive may therefore vary a little from weight specified in description.
41L1035 Black.
41L1036 Yellow.
EACH $4.48
Extra Long Triple Fly Front Waterproof Oiled Slicker Coat. Unusually well made, excellent quality and guaranteed to withstand heaviest rains. Large cape extends around front and back. Triple shoulders. Extra large standing collar with corduroy facing and large throat tab which buttons. Waterproof wristlets and triple elbows. Two large patch pockets with flaps. Coat closes with five patent snap fasteners. Average length, 54 inches. SIZES—36 to 48 inches chest. State chest measure taken over vest. Shpg. wt., 5⅝ lbs.
41L1030 Black Jacket.
41L1031 Black Pants.
41L1032 Yellow Jacket.
41L1033 Yellow Pants.
EACH GARMENT $2.69
Waterproof Oiled Slicker Suit. Shoulders, elbows and fly front of jacket are triple thickness. Corduroy faced collar. Large cape around back. Rain excluding wristlets. Pants made apron style with triple seat and triple front. Attached adjustable suspenders. Average length of jacket, 30 inches. SIZES—Jacket, 36 to 48 inches chest; pants, 32 to 44 inches waist. State chest measure of jacket, taken over vest, and waist measure of pants. Shpg. wt. of suit, 6½ lbs.; jacket, 3½ lbs.; pants, 3 lbs.
41L1025 Black.
41L1026 Yellow.
EACH $3.98
Extra Long Triple Fly Front Waterproof Oiled Slicker Pommel Riding or Walking Coat. Extends over entire saddle and is easily adjusted for use by buttoning around legs at bottom. When not used for riding can be worn regular coat style. Double throughout body and sleeves. Has extra large pocket on right side. High standing collar, faced with flannel to protect neck. Rain excluding wristlets. Average length, 58 inches. SIZES—36 to 48 inches chest. State chest measure taken over vest. Shpg. wt., 6¼ lbs.
41L1000—Black Jacket. | 41L1002—Yellow Jacket.
41L1001—Black Pants. | 41L1003—Yellow Pants.
EACH $1.69
Low Priced Pliable Waterproof Oiled Slicker Suit. Large and roomy. Great protection against rain. Jacket has triple fly storm front and double throughout balance of suit. High standing collar. Pants made apron front style with attached adjustable suspenders and patent never-come-off buttons. Average length of jacket, 30 inches. SIZES—Jacket, 36 to 48 inches chest; pants, 32 to 44 inches waist. State chest measure of jacket, taken over vest, and waist measure of pants. Shipping weight of suit, 5¾ pounds; jacket, 2¾ pounds; pants, 3 pounds.
SEARS, ROEBUCK AND CO. 267

45L7036—Dark Blue (Striped).
45L7038—Dark Brown (Striped).
45L7040—Black (Striped). **$17.95**

RICHLY STRIPED ALL WOOL WORSTED MODEL. It has the style men like—really it's exceptionally good looking. Unusually well made from medium weight ALL WOOL WORSTED, having just enough stripes to make it rich looking. You ought to see this texture. You'd say, "That's the strongest ever and should give lasting service." It's one of those hard surfaced, closely woven wool worsteds that does not wear threadbare. The coat is cut along semi-form fitting lines, single breasted, three-button style. Lined throughout body with durable alpaca. Vest is of the five-button style, having the usual pockets and adjustable strap and buckle in back. The trousers are well proportioned and have plain bottoms. If you want dignified style, it's to be had in this model. **Be sure you state measurements in the order.** Shipping weight, 5¾ lbs.

SIZES OF SUITS LISTED ABOVE.

Chest meas.	34	36	37	38	39	40	42	44
Waist meas.	29-31	31-33	32-34	33-35	34-37	35-38	37-40	39-42
Inseam meas.	Ranges from 29 to 36 inches.							

For Extra Large and Stout Suits see page 279.

Shipping weights are based on an average number of shipments. Weight of shipment you receive may therefore vary a little from weight specified in description.

*BUY QUALITY. Our economical buying and selling methods take care of **Price**. You can afford **BETTER** qualities at our prices.*

Cloth Samples of suits listed above sent free, on request.

Whenever You Say We'll Send This

Clothing Sample Folder FREE

Write Today!

PICTURED above is our latest Sample Folder of Men's and Young Men's Suits—a style authority for spring and summer, 1923. From it you can get approved new styles, correct fit, new, dependable ALL WOOL materials, attractive patterns, long wear and comfort. It shows actual cloth samples—sixteen of them—beautiful soft colored, rich fabrics of various popular shades, made in six selected models. A choice of thirty-three suits in all! Every suit is carefully made, strongly sewed and priced lower than equal quality elsewhere. Remember, it's not what you pay that counts—it's what you get for what you pay. Your money back without question if not entirely satisfied. Don't put it off—find out NOW! Write TODAY.

Ask for Clothing Sample Folder 8292GCL

Suits and Trousers for Large Men

SIZES OF SUITS ON THIS PAGE

Chest Measure........	40	42	44	46	48	50
Waist Measure........	39-42	41-44	43-46	42-48	44-50	46-52
Inseam Measure.......	Ranges from 29 to 36 inches.					

For other garments for Large Men see following pages:
Underwear, pages 230 and 233.
Men's Furnishings, pages 250 and 251.

Cloth Samples of Suits Sent on Request.

Back View of Suits on This Page.

Extra Fine Hand Tailored ALL WOOL Serge Suits.

45L7061—Navy Blue. Medium Weight. Fine Quality. **$24.85**
45L7063—Navy Blue. Medium Heavy Weight. Finest Quality. **29.95**
45L7065—Dark Gray. Medium Heavy Weight. Finest Quality. **29.95**

SPRING AND SUMMER SUITS OF SPECIALLY HIGH QUALITY WORSTED SERGE. Well tailored in every respect, in a conservative model that is very becoming to both large and stout men. It is a most comfortable model and we know you will be pleased with the way it fits. These ALL WOOL WORSTED SERGES are fast color and are noted for their lasting service and shape retaining qualities. Coat is made with closed back (no vent) and is full lined. Vest is of the regular five-button style. Trousers have plain bottoms (no cuffs). **State your measurements.** See size scale above. Shipping weight, 6¼ pounds.

Unusually High Quality Hand Tailored All Wool Worsted Suits.
(See illustration above).

45L7067—Dark Brown (Striped). Medium Weight. **$27.75**
45L7069—Black (Striped). French Back. Heavy Weight. **29.95**

YOUR SIZE AND FIT IN GOOD LOOKING AND LONG LASTING ALL WOOL WORSTED SUITS. Two handsome patterns that will render almost unlimited service and will not wear threadbare. 45L7067 is of a striped effect with narrow silk threads of lighter shades and contrasting colors woven into the dark brown background. 45L7069 is a French back worsted in a subdued silk striped pattern. Both are especially well tailored in the dignified model illustrated above. They are cut full size, yet they embody good style and forceful character. Single breasted three-button coat with closed back; lined with durable alpaca. Regular five-button vest with adjustable strap and buckle in back. Plain bottom trousers (no cuffs). **State measurements.** See size scale above at left. Shipping weight, 6¼ lbs.

Exceptionally Well Made Part Wool Worsted or All Wool Serge Trousers.
(See illustration at right).

45L7550—Gray and Black Striped Worsted. About One-Third Wool. **$3.79**
45L7552—Navy Blue All Wool Serge. **5.98**
45L7553—Dark Gray All Wool Serge. **5.98**

EXCELLENT VALUES IN MEDIUM WEIGHT WORSTED OR SERGE TROUSERS for large and stout men who usually have trouble in securing proper fitting clothing. These trousers are correctly proportioned to fit extra large men. Strongly sewed and stitched to give long satisfactory service. They are made only with plain bottoms (no cuffs). All have the usual pockets, suspender buttons and belt loops. You'll find that these trousers will fit you and hang properly. **State your waist and inseam measures.** Shipping weight, 2⅜ pounds.

Shipping weights are based on an average number of shipments. Weight of shipment you receive may, therefore, vary a little from weight specified.

THE QUALITY AND SERVICE YOUR DOLLAR BUYS determines how low or how high a price may be. Big value is assured on all clothing in this catalog by our careful merchandising backed by scientific tests.

SIZES **44 to 50 inches waist and 30 to 36 inches inseam. State waist and inseam measures.**

45L7550, 45L7552 and 45L7553.

See Page 557 for Measuring Instructions.

COATS

Light Weight. Unlined

Serges and Alpacas

45L7340—Navy Blue. All Wool **$5.95**
45L7342—Navy Blue. About One-Half Wool and One-Half Cotton **$4.45**
GOOD QUALITY LIGHT WEIGHT SERGE UNLINED COATS. Three large outside pockets and one inside pocket. Very moderately priced and guaranteed to give satisfactory service. **SIZES —34 to 44 inches chest. State chest measure.** Shipping weight, 1½ pounds.

45L7344—Black Alpaca. Good Quality..........
45L7346—Medium Gray Alpaca. Good Quality.. } **$3.98** EACH
45L7348—Black Alpaca. Fine Quality............
45L7350—Medium Gray Alpaca. Fine Quality... } **$5.48** EACH
FEATHER WEIGHT GRAY OR BLACK ALPACA UNLINED COATS. Two lower and one upper patch pockets and one inside pocket—roomy and well sewed. Decidedly practical and serviceable. **SIZES —34 to 44 inches chest. State chest measure.** Shipping weight, 1½ pounds.

Wear an Unlined Coat—either serge or alpaca—and save the wear and tear of the coat of your regular suit. Extremely comfortable. Just the garment for dentists, doctors, other professional men, clerks and others who work in offices, stores or factories.

Hiking and Riding Suits and Breeches

SIZES of suits and breeches listed at right: Coats, 34 to 44 inches chest; breeches, 30 to 42 inches waist and 24 to 28 inches inseam. **State chest measure of coat; give waist and inseam measures of breeches as shown at bottom of page.**

41L7640—Olive Drab Khaki. Medium Weight............ **$1.98**
41L7641—Olive Drab Moleskin Cloth. Heavy Weight............ **$2.98**
41L7642—Olive Drab Thickset Corduroy. Heavy Weight............ **$3.39**
HIKING AND RIDING BREECHES that will give unusually long service. They are reinforced at seat, as shown in small illustration above at left. Lace at calves. Have two side pockets, two hip pockets with flap and button, watch pocket, and belt loops. Well made and neatly finished. **SIZES —28 to 42 inches waist; comes in 26 inches inseam only. State waist measure.** Shipping weight, 2 pounds.

41L7630—Hiking and Riding Suit........ **$10.98**
41L7631—Breeches only........ **4.48**
MADE FROM OLIVE DRAB THICKSET CORDUROY. Coat is Norfolk style with all around belt. It is unlined and has two roomy patch pockets with flap and button. Breeches are tailored to fit correctly and have double seat and calf reinforced with many rows of stitching. They lace at calves. See small views above at right. Practical for riding or hiking. **State measurements.** See sizes above. Shipping weight: Suit, 5 pounds; breeches, 2⅛ pounds.

41L7632—Hiking and Riding Suit................ **$10.98**
41L7633—Breeches only................ **4.48**
FINE QUALITY OLIVE DRAB GABARDINE SUIT. Well made from a strong medium weight cotton gabardine. Same style as the corduroy suit above. **State measurements.** See sizes above. Shipping weight: Suit, 4½ pounds; breeches, 2 pounds.

41L7634—Hiking and Riding Suit................ **$6.69**
41L7635—Breeches only................ **2.79**
GOOD WEIGHT KHAKI SUIT in an olive drab shade. An ideal suit for hiking, riding, motorcycling and general outdoor wear. It is of the Norfolk style with all around belt and patch pockets. Tailored breeches have double seat and calf. They lace at calves. See small views above. Usual pockets; hip pockets with tab to button. **State measurements.** See sizes above at left. Shipping wt.: Suit, 4 pounds; breeches, 2⅛ lbs.

41L7636—Hiking and Riding Suit................ **$7.98**
41L7637—Breeches only................ **3.79**
STRONG DRAB MOLESKIN CLOTH SUIT. A medium weight material that will wear almost like leather. Same style as the khaki suit above. **State measurements.** See sizes above at left. Shipping weight: Suit, 4¼ pounds; breeches, 2 pounds.

HOW TO SECURE PROPER LENGTH BREECHES FOR SUITS LISTED ABOVE.

If your trousers inseam measures, inches....	28 to 30	31 or 32	33 to 36
Order breeches inseam, inches..........	24	26	28

SIZES All trousers on this page, except 45L7526, are made in sizes 30 to 42 inches waist and 29 to 34 inches inseam. 45L7526 comes in sizes 28 to 42 inches waist and 28 to 34 inches inseam. **Be sure that you state waist and inseam measures in your order.**

Shipping weights are based on an average number of shipments. Weight of shipment you receive may therefore vary a little from weight specified in description.

Young Men's Trousers For Spring and Summer 1923

Cassimeres, Tweeds, Panamas, White Ducks

PROUD, INDEED, are we of these trousers, because they're well made from selected fabrics, and have neat and trim proportions that will appeal to the tastes of discriminating young men. All of these trousers have stylish narrow cuffs, belt loops, and the usual number of pockets. The pockets are well shaped and double stitched; waist lining is strongly sewed; buttons are securely sewed on; in fact, they are well made in every respect. You will find it easy to select from this page just the trousers you want and at a price you want to pay.

45L7527—Dark Blue.
45L7528—Dark Brown.
45L7529 Dark Gray. **$2.98** EACH

ATTRACTIVE SHADES OF NEATLY STRIPED HALF WOOL CASSIMERE. A medium heavy weight soft finish cassimere, about one-half wool. Has subdued thread stripes in harmonizing shades and spaced about ¼ inch apart. Price considered, these smart looking trousers will give you good service and great satisfaction. **State waist and inseam measures.** Shipping weight, 2 pounds.

45L7531—Medium Brown Heather Mixture.
45L7532—Olive Gray Heather Mixture. **$3.79** EACH

CLASSY AND DRESSY ALL WOOL TWEED TROUSERS. Those who prefer snappy looking trousers will be more than pleased with these fine quality medium heavy weight ALL WOOL tweeds. The illustration will give you a good idea as to the pattern of these handsome brown and olive gray heather mixtures. Two of our best values. **State waist and inseam measures.** Shipping wt., 2½ lbs.

VALUE IS MEASURED BY BOTH QUALITY AND PRICE. Some clothing on the market is made to sell at a price regardless of service. Our methods insure ***serviceable*** *clothing at rock bottom prices.*

45L7526 White Duck. **$1.89**

GOOD GRADE WHITE DUCK TROUSERS. Every young man should have at least two pairs of these trim and stylish looking white duck trousers for summer wear. There's nothing better for outing purposes—for picnics, excursions, social functions, tennis, golf, etc. They're easily laundered, and that's why we say that you should have at least two pairs, so that one will always be ready for wear. Nicely made, well fitting, comfortable, and always look cool. **SIZES—28 to 42 inches waist and 28 to 34 inches inseam. State waist and inseam measures.** Shipping weight, 1¾ pounds.

45L7537 — Medium Gray. (Plain.)
45L7538—Sand Color. (Plain.) **$3.48** EACH

STYLISH AND COMFORTABLY COOL SUMMER TROUSERS. Two popular shades of panama cloth, a hard finished fabric made of mohair (luster wool) and cotton—about one-half of each. It is light in weight and very durable, thus being ideal for hot summer wear. Either of the two very attractive shades—in plain colors—go well with a dark coat. They're just the kind of trousers up to date young fellows demand. **State waist and inseam measures.** Shipping weight, 1¼ pounds.

Unusually Fine Quality Trousers—Great Values.

45L7534—Dark Brown (Striped.)
45L7535—Dark Olive (Striped.)
45L7536—Medium Gray (Striped.) **$3.98** EACH

OUR HIGHEST QUALITY CASSIMERE TROUSERS FOR YOUNG MEN. Seldom do you find trousers as serviceable and strikingly looking as these offered at our price. Handsome shades of extra fine ALL WOOL CASSIMERE, richly striped in subdued contrasting colors. They are strongly made—especial attention being given to parts which are subject to greatest strain and wear. We consider these trousers most unusual values. Take advantage of our low price now. **State waist and inseam measures.** Shipping weight, 2¼ pounds.

Work and Outing Pants

How They Are Made. Materials are put through a **strict** laboratory test; all seams are sewed throughout with strong, heavy thread; pants are reinforced at strain points; buttons are put on to stay; lining will not creep above waistband; pockets are large and strong, and every pair is priced to give you the greatest possible value for your money.

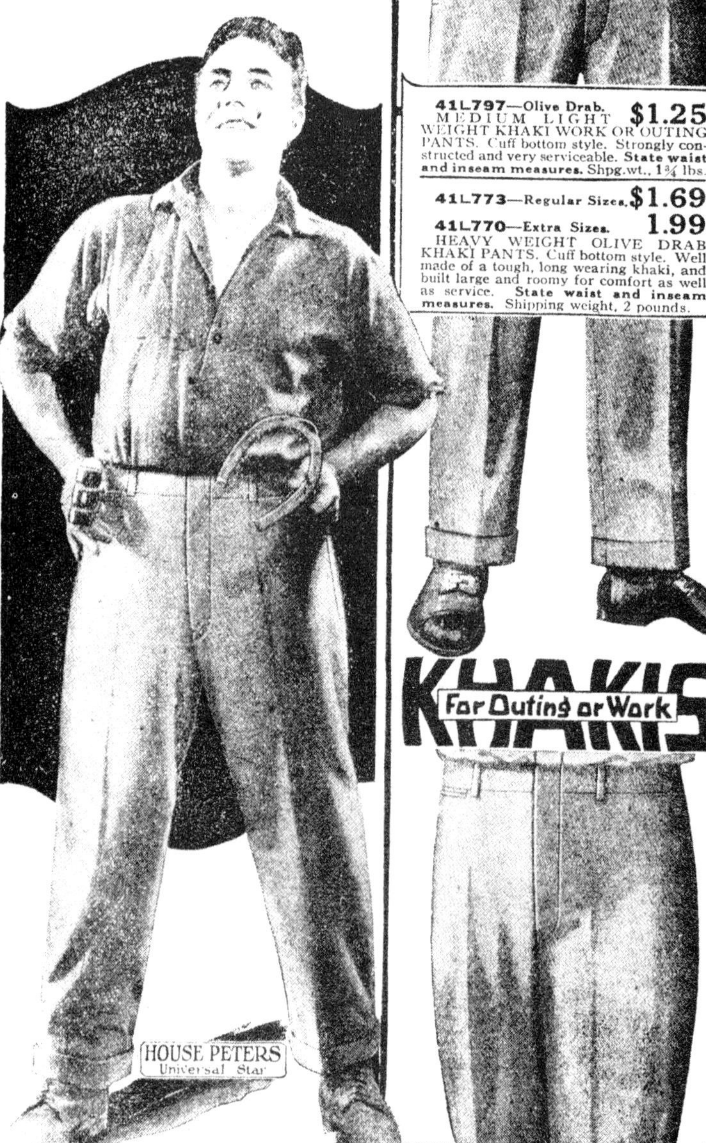

41L797—Olive Drab. MEDIUM LIGHT WEIGHT KHAKI WORK OR OUTING PANTS. Cuff bottom style. Strongly constructed and very serviceable. **State waist and inseam measures.** Shpg. wt., 1¾ lbs. **$1.25**

41L773—Regular Sizes. **$1.69**
41L770—Extra Sizes. **1.99**
HEAVY WEIGHT OLIVE DRAB KHAKI PANTS. Cuff bottom style. Well made of a tough, long wearing khaki, and built large and roomy for comfort as well as service. **State waist and inseam measures.** Shipping weight, 2 pounds.

KHAKIS For Outing or Work

41L786—Olive Drab. **$2.19**
EXTRA FINE HEAVY WEIGHT KHAKI WORK OR OUTING PANTS. Extra well made of exceptionally strong, finely woven khaki twill. One of the strongest cotton cloths in our work pants line. Illustration represents material and shows cuff bottom style in which these pants are made. Strongly sewed and finished throughout. Will give unusually good service. **State waist and inseam measures.** Shipping weight, 2⅛ pounds.

See Page 557 for Measuring Instructions.

41L777—Olive Drab. **$1.98**
GOOD QUALITY MEDIUM WEIGHT MOLESKIN CLOTH WORK PANTS. Noted for their excellent wearing qualities. Seat and crotch are strongly sewed and taped. Decidedly practical, as the olive drab shade does not show dirt readily. Priced extremely low for pants of this quality. Cuff bottoms. **State waist and inseam measures.** Shipping weight, 2¼ pounds.

41L778—Regular Sizes. **$2.89**
41L776—Extra Sizes. **3.19**
HEAVY WEIGHT OLIVE DRAB MOLESKIN CLOTH GUARANTEED WORK PANTS. Our well known Big Chief Brand. These are our heaviest weight moleskin cloth pants, extra well made from one of the strongest and most serviceable materials used for this purpose. Seat and crotch are strongly sewed and taped. Have tough and long wearing pockets made from extra heavy drill. Patent buttons, securely fastened. All seams are sewed with good strong thread. Cuff bottoms. **State waist and inseam measures.** Shipping weight, 2½ pounds.

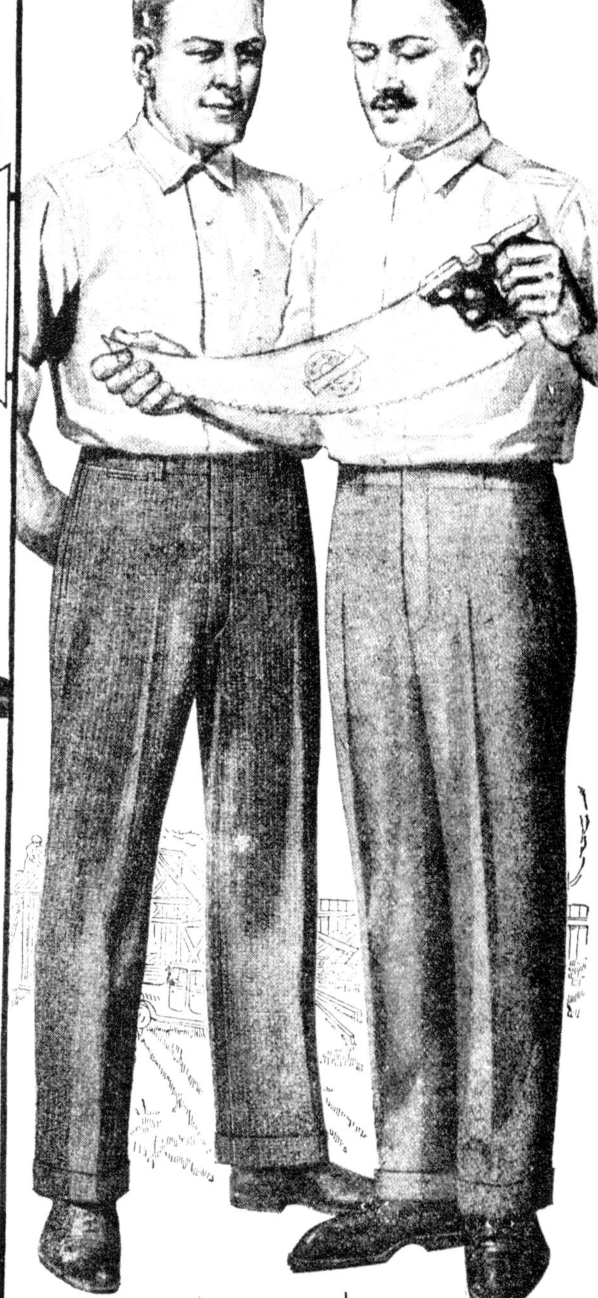

41L802—Dark Gray (Striped).
41L803—Grayish Tan (Striped). **$1.48**
LOW PRICED LIGHT WEIGHT WASHABLE COTTON WORK PANTS. Cool and comfortable for hot weather wear. Strongly made of a good wearing cotton Daytona fabric in either dark gray or grayish tan with narrow stripes, as illustrated above. Wash easily and well. Neatly made with cuff bottoms, two side pockets, two hip pockets, watch pocket, belt loops and sewed on suspender buttons. For general purpose wear these light weight work pants will be most useful. **State waist and inseam measures.** Shipping weight, 1½ lbs.

41L790 Olive Drab. **$1.98**
EXTRA HEAVY STRONG AND SERVICEABLE COTTON WHIPCORD CLOTH WORK OR OUTING PANTS. Especially desirable for drivers, mechanics, shop men, repair men and for general work purposes or rough outing wear. Made with cuff bottoms. **State waist and inseam measures.** Shipping weight, 2¼ pounds.

SIZES The various Work and Outing Pants on this page come in sizes 30 to 42 inches waist and 30 to 36 inches inseam. We also list 41L770 and 41L776 in extra sizes from 44 to 50 inches waist and 30 to 36 inches inseam. (See pages 286 and 287 for other work pants.) **When ordering be sure to state waist and inseam measures wanted.**

KEEP QUALITY IN MIND AS WELL AS PRICE when you are trying to make your money do its full duty. Of course, our prices are low, but the quality our constant testing assures makes our values real.

Shipping weights are based on an average number of shipments. Weight of shipment you receive may therefore vary a little from weight specified in description.

Guaranteed Overalls

Some of the Features Embodied in Our "S. R. Best" Brand Overalls and Jackets

Made from Extra Heavy Weight White Back Indigo Blue Denim, a firmly woven, heavy weight cotton material, and one of the best overall fabrics on the market. Cut over EXTRA LARGE, ROOMY PATTERNS to allow for clothing worn underneath, insuring ease and comfort to the wearer. Reinforced at all strain points. Triple stitched throughout with heavy thread so they will not rip. New flexible buttons that will not pull off.

Solid High Back Style Apron Overalls.

41L705—Regular Sizes...... **$1.69**
41L720—Extra Sizes........ **1.94**

SOLID HIGH BACK, as shown by small illustrations, gives added protection to clothing worn underneath. All pocket corners are strongly bar tacked so they will not rip. Triple stitched seams. A full and roomy garment that will give complete satisfaction. Furnished in regular and extra sizes. **State waist and inseam measures.** Shipping weight, regular sizes, 2 pounds; extra sizes, 2¼ pounds.

Coat Style Jacket.

41L709—Regular Sizes...... **$1.69**
41L723—Extra Sizes........ **1.94**

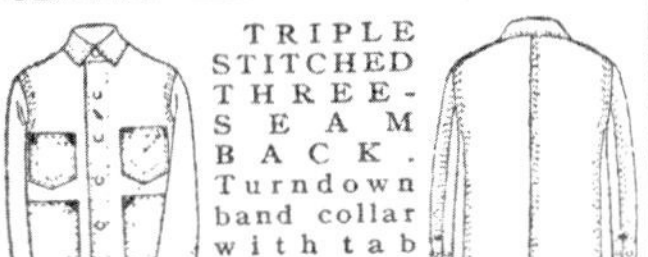

TRIPLE STITCHED THREE-SEAM BACK. Turndown band collar with tab and set-in sleeves with adjustable cuffs which button permit wearer to tighten or loosen the collar and cuffs. A high grade coat style jacket. Very comfortable and serviceable. Four extra large pockets, bar tacked and strongly stitched. All seams are strongly sewed so as to give utmost wear. Furnished in regular and extra sizes. **State chest measure.** Shipping weight, regular sizes, 1⅝ pounds; extra sizes, 2 pounds.

Double Front and Double Seat Band Top Overalls.

41L706—Regular Sizes..... **$1.79**
41L719—Extra Sizes........ **2.09**

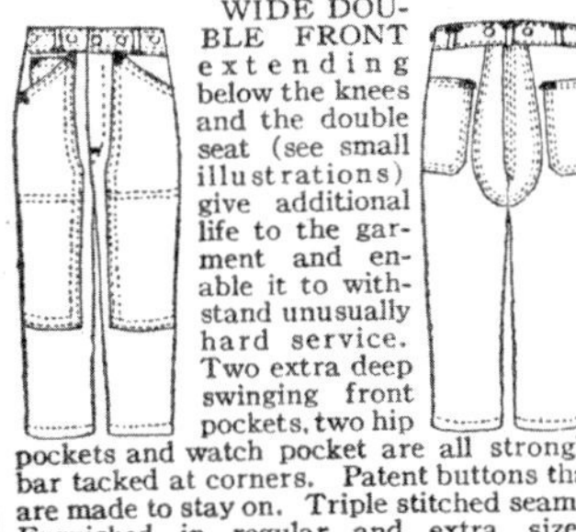

WIDE DOUBLE FRONT extending below the knees and the double seat (see small illustrations) give additional life to the garment and enable it to withstand unusually hard service. Two extra deep swinging front pockets, two hip pockets and watch pocket are all strongly bar tacked at corners. Patent buttons that are made to stay on. Triple stitched seams. Furnished in regular and extra sizes. **State waist and inseam measures.** Shipping weight, regular sizes, 2¼ pounds; extra sizes, 2½ pounds.

Detachable Suspender Style Apron Overalls.

41L707—Regular Sizes..... **$1.69**
41L721—Extra Sizes........ **1.94**

MANY PREFER THE DETACHABLE SUSPENDER STYLE. Made with elastic inserts, allowing full play at the shoulders. An exceptionally well made, comfortable garment. Strongly sewed throughout. Triple stitched seams. Furnished in regular and extra sizes. **State waist and inseam measures.** Shipping weight, regular sizes, 2 pounds; extra sizes, 2¼ pounds.

Double Knee and Front Apron Overalls.

41L710—Regular Sizes..... **$1.98**
41L724—Extra Sizes........ **2.28**

FOR EXTRA HARD WEAR. Have broad double front extending below knees. Wide detachable suspenders with elastic inserts that allow play at the shoulders. Extra strong pockets bar tacked at corners. Two front pockets, deep and roomy. Reinforced at all strain points. Furnished in regular and extra sizes. **State waist and inseam measures.** Shipping weight, regular sizes, 2¼ pounds; extra sizes, 2½ pounds.

California Style Band Top Overalls.

41L708—Regular Sizes..... **$1.69**
41L722—Extra Sizes........ **1.94**

POPULAR STYLE OVERALLS, made with strap and buckle in back, which enables wearer to tighten garment at waist and hip. Full triple stitched throughout. All pocket corners are strongly bar tacked and greatest strength embodied where most needed. You'll derive a great deal of comfort and lasting service from this garment. Furnished in regular and extra sizes. **State waist and inseam measures.** Shipping weight, regular sizes, 1⅞ pounds; extra sizes, 2⅛ pounds.

SIZES Overalls furnished in **Regular Sizes,** from 30 to 44 inches waist and 30 to 36 inches inseam. **Extra Sizes,** 46 to 56 inches waist and 30 to 36 inches inseam. Jackets furnished in **Regular Sizes,** from 34 to 46 inches chest and **Extra Sizes** from 48 to 58 inches chest. **When ordering be sure to state waist and inseam measures of overalls and chest measure of jacket.**

Our Own Brand Registered in U. S. Patent Office.

The Better Kind

Large and Strongly Made

Extra Heavy Weight White Back Indigo Blue Denim

RARELY WILL YOU NEED OUR GUARANTEE, but when you do it's right here and means what it says.

Shipping weights are based on an average number of shipments. Weight of shipment you receive may therefore vary a little from weight specified.

Have Better Clothes—and More of Them!

It's Easy to be Well Dressed when you Make your own Clothes

Every woman can have more and better clothes for less money if she makes them herself. This is the secret of being well dressed. When you see a pretty dress or frock that appeals to you, a few yards of material, a pattern and your sewing machine will enable you to duplicate it for your own wardrobe. You don't have to be a designer or skilled dressmaker! Any woman can easily make the pretty things her heart desires if she has one of the efficient and easy running machines we offer on the following pages. The simple instructions we furnish with each machine will show you how easy it is to make all the dainty and attractive garments you have so often admired. Order your sewing machine today and resolve to be the best dressed woman in town.

Attachments

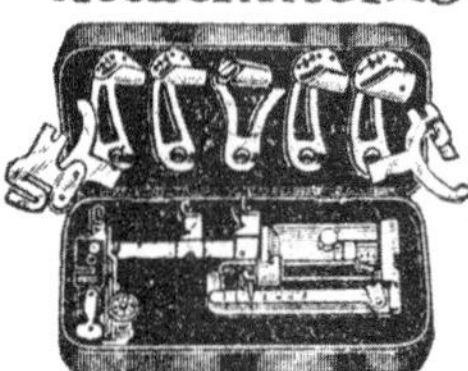

This high grade set of attachments furnished with every Franklin machine. Guaranteed by the Greist Co., the makers, who supply to most manufacturers of high grade sewing machines. Set consists of tucker, ruffler, shirring blade, under braider, short presser foot, binder, bias cutting gauge and set of four hemmers, different widths.

The Genuine Franklin Head

Here is the wonderful Franklin head—the finest and most efficient vibrator type sewing machine we know how to build. The Franklin embodies every desirable and up to date feature, and is as near perfect as human skill and modern machinery can make it. When you buy a Franklin you have the satisfaction of knowing you have the best—a sewing machine that will give many years of faithful service—one that will measure up to the very best sewing machines on the market regardless of make, name or price.

The Franklin is a high arm, double thread, lockstitch type machine. The head is very simple in construction, and has the fewest possible points of friction. This insures a smooth, light running machine. It is a fast worker, and is self threading at every point except the eye of the needle.

Points of Franklin Superiority.

1. *Thumbscrew needle clamp—no screwdriver needed.*
2. *Automatic shuttle ejector—saves time and trouble.*
3. *Presser foot fits solidly against the bar. This eliminates bending and breaking needles.*
4. *Belt on outside—easy to raise head for oiling.*
5. *Automatic bobbin winder.*
6. *Independent positive cam take-up and disc tension—insures perfect stitch on any kind of material.*
7. *Automatic tension release—avoids bending needle or breaking thread when drawing work from under presser foot.*
8. *Extra strong and large feed—direct and positive. Handles materials, from lightest chiffon to heaviest woolens*

The Sit-Right Feature.

Every woman will appreciate the wonderful Sit-Right construction of Franklin drop head models. (See picture above.)

Franklin drop head models are so designed that the needle is directly in front of the operator. No necessity of leaning to one side in order to guide the work when sewing. There is no cramping of arms or body. You can sew for hours on a Sit-Right Franklin without fatigue because you sit in a natural, upright position directly in front of your work. You'll never know real sewing comfort until you have tried the Sit-Right way. (See picture above.)

Accessories

The accessories furnished without extra charge consist of one quilter, five bobbins (and one in the machine), one hemmer screw, large nickel plated screwdriver, small nickel plated shuttle screwdriver, oil can filled with oil, foot hemmer, package of six needles, assorted sizes (and one in the machine).

The Guarantee that Stands the Acid Test

20 years old and nothing worn out but Needles

Certificate of Guaranty. This is to Certify that the Machine ... Sears Roebuck ...

Factoryville Pa 2-24-21
Sears Roebuck & Co
Chicago, Ill.
Gentlemen—
Guess you might as well have this Guaranty as its days of usefulness are over—The machine is still in use
Nothing worn out but needles—
Yours truly,

WHEN you buy a sewing machine from Sears, Roebuck and Co. you are protected by the most liberal guarantee ever made. Every machine we sell is positively guaranteed for twenty years. This means that if any defect in material or workmanship develops within this time we will replace the defective part without charge. We can make this guarantee with perfect assurance because we know our machines are honestly built to give service. None but the finest materials are used, and every part is so carefully fitted and tested that with ordinary care the machines should last much longer than twenty years and they do. The best proof we could offer of the durability and serviceability of our sewing machines are the hundreds of letters we have like the one reproduced at the left, which tell of machines our customers have used over twenty years, and which are still giving satisfactory day to day service.

The illustration at the left is a facsimile of a letter received from a customer in February, 1921, along with the old sewing machine guarantee certificate which had been issued in August, 1900, over twenty years before. The machine was still in use—"nothing worn out but needles."

Look at the Low Price!

Closed View.

On This High Grade Franklin $29.95

Quality First.
Perfect Sewing in Perfect Comfort.

When you sew the Sit-Right Franklin way, you sit in an easy, natural position with your feet straight in front of you and your work squarely before you. The sidewise twist to watch the work produces an unusual muscular effort to which a woman's delicate organism is keenly susceptible. All this sidewise strain compelled by other machines is entirely done away with when you sew the Sit-Right Franklin way. If you wish to accomplish the best results in sewing, and with most comfort, we believe that you will appreciate the Sit-Right way of sewing. If you never have used the Sit-Right Franklin you really do not know the full satisfaction of sewing in perfect comfort.

Where else can you get such a splendid value as this for $29.95? Here is a genuine Sit-Right Franklin, the same make as tens of thousands which we have sold in years gone by and which are today giving universal satisfaction. You cannot buy greater sewing comfort or greater sewing efficiency than we offer in this wonderful Franklin at $29.95. Even though you paid two or three times as much for some other make you wouldn't be getting any greater value.

Open View.

A Beautiful Machine

Our cabinetmakers, skilled in their profession, have done as much to make the Sit-Right Franklin the standard of its type as have our expert mechanics. The two together produce a machine unrivaled in beauty and unsurpassed in mechanical perfection.

This six-drawer model is made of genuine quarter sawed oak, and is finished in a rich golden color. The drawers are enclosed in cases and have turned wooden knobs, finished to correspond with the drawers.

The head supplied is the well known Franklin, illustrated and described on the preceding page.

The stand is the celebrated Franklin wide ball bearing stand which permits of the comfortable Sit-Right sewing position.

26L65—Six-Drawer Sit-Right Franklin Ball Bearing Drop Head Sewing Machine, fitted with automatic lift, quarter sawed oak woodwork. Complete with attachments, accessories and instruction book. Shipping weight, 120 pounds *$29.95*
Shipped from BUFFALO, N. Y., or CHICAGO, ILL., whichever city is nearer you.

New Exclusive Franklin Design

Price Reduced to $32.95

Closed View.
This Design Patented by Sears, Roebuck and Co. Patent Allowed, April 11, 1922.

If you want the very latest development in drop head sewing machine design, this beautiful and up to date Franklin will delight you. The woodwork is an exclusive Franklin feature, this design being patented by Sears, Roebuck and Co. The arrangement is convenient, unique, and is the first real improvement in sewing machine woodwork construction since the change from the old box top to the drop head cabinet.

And, of course, it's a Sit-Right—the scientifically designed sewing machine that enables you to work in a natural, upright position without tiresome strain and unnecessary fatigue. You will find sewing a delight on this beautiful, easy running and comfortable machine.

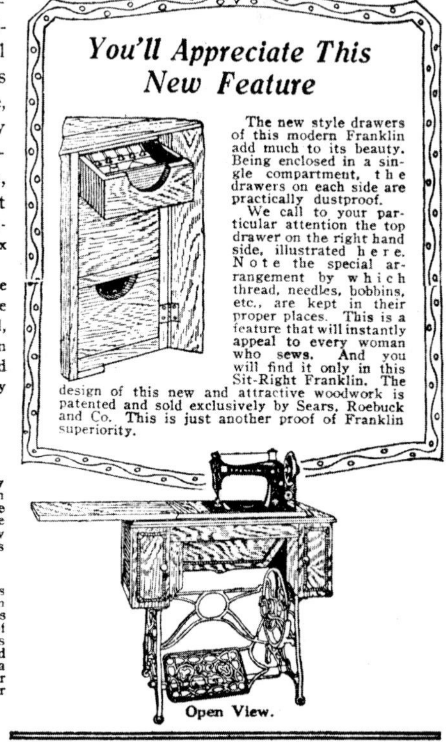

You'll Appreciate This New Feature

The new style drawers of this modern Franklin add much to its beauty. Being enclosed in a single compartment, the drawers on each side are practically dustproof.

We call to your particular attention the top drawer on the right hand side, illustrated here. Note the special arrangement by which thread, needles, bobbins, etc., are kept in their proper places. This is a feature that will instantly appeal to every woman who sews. And you will find it only in this Sit-Right Franklin. The design of this new and attractive woodwork is patented and sold exclusively by Sears, Roebuck and Co. This is just another proof of Franklin superiority.

Open View.

No Finer Machine Made.

We consider this the finest Franklin we have ever made, and we are sure you cannot get a more efficient, durable or more handsomely designed and finished machine, no matter where you go or what you pay. But we don't ask you to take our word for this. Order one of these Franklins today for ninety days' trial. When you get it, put the machine to any test you want to make; compare it with any other sewing machine on the market, regardless of make, name or price. If it does not fully measure up to your expectations, or if you are dissatisfied with it in any way, just send it back at our expense, and we will return every cent you have paid, including transportation charges. You are to be the sole judge of quality and value.

Like all our machines, this Franklin is backed by our iron clad twenty-year guarantee. This is more than a guarantee. It is our pledge of supreme quality, and an expression of our confidence as manufacturers in the durability which we have built into every part of the machine.

It's Easy to Sew on This Sit-Right Franklin.

We send a complete book of instructions with every Franklin machine. With the aid of these instructions even an inexperienced operator can secure good results. The operation of the machine itself, and all the uses of the various attachments, are treated in detail. This will show you how easy it is to be your own dressmaker, and thus have all the pretty clothes your heart desires.

Compare Our Price.

A comparison of our price with those charged by others will show you what a really big value we are offering in this new Franklin. Just think of it! Only $32.95 for this handsome, efficient and up to date machine. Machines of similar quality, even without the many special features you get in the Franklin, are being sold through agents and dealers for from $50.00 to $75.00 and more. Isn't this a saving worth while? Send your order today and see for yourself what a wonderful value you can get here for your money.

26L18—Six-Drawer Sit-Right Franklin Ball Bearing Drop Head Sewing Machine. Automatic lift, quarter sawed oak woodwork, complete with attachments, accessories and instruction book. Shipping weight, 120 pounds *$32.95*
Shipped from BUFFALO, N. Y., or CHICAGO, ILL., whichever city is nearer you.

Our Notions are of Best Quality

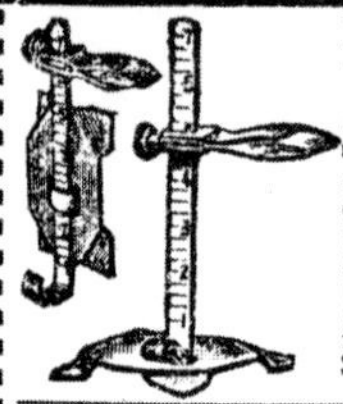

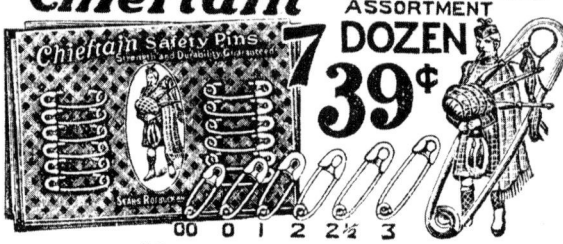

The Handy Machine Drawer Skirt Marker.

Made of nickel plated steel, with folding measuring bar. Marker can be put in any machine drawer or sewing basket. Fitted with screw attachment which holds the chalk and gauges the size. Adjustment from 1 to 7 inches. Shpg wt., 12 oz.

25L5177 .... **25c**

Rustproof Hooks and Eyes.

Made of brass wire. Each card contains 24 each of hooks and eyes and loops. Comes in black or white in sizes 0, 1, 2 or 3. **State size and color desired.** Shipping weight, 2 ounces.

25L5184—Per card **5c**

A. C. E. Snap Fasteners.

Our Own Trade Mark, Reg. U. S. Pat. Off.

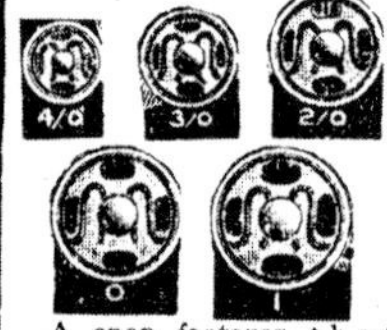

A snap fastener that will give you perfect satisfaction and is dependable in every way. Snaps easily. Made of brass with a rolled back that will not cut the thread. Has been thoroughly tested and will give you all the service you can expect. Colors: Black or white. Sizes, 4/0, 3/0, 2/0, 0, or 1. **State color and size.** Shipping wt., 2 oz.

25L5174—6 dozen for **15c**

Steel Sewing Needles.

English Made Egg Eye Steel Needles. Fifteen to a paper. Sizes: 3 to 7; 4 to 8; 5 to 9; or solid sizes, 1, 2, 3, 4, 5, 6, 7, 8, 9 or 10. **State size.** Shipping wt., 2 ounces.

25L5181

Per paper **4c**

Steel Darning Needles.

Steel Darning Needles packed ten needles to a paper. Sizes, assorted, from 14 to 18. Shipping weight, 2 oz.

25L5198

Per paper **5c**

"Chieftain" SAFETY PINS ASSORTMENT

7 DOZEN 39¢

00 0 1 2 2½ 3

"Chieftain" Safety Pin Assortment.

This is the biggest safety pin value that could be offered. In this assortment you will have the safety pin you want when needed. These are high grade brass safety pins, with guarded coil and are either gold plated or nickel plated. Assortment contains the following, 1 dozen each, sizes 00 and 0, gold plated; ½ dozen each, sizes 00 and 0, and 1 dozen each, sizes 1, 2, 2½ and 3, nickel plated safety pins. Put up in a neat package. Shipping weight, per assortment, 7 dozen, 6 ounces.

25L5165—Per assortment of **7** dozen **39c**

STEWARTS DUPLEX

Stewart's Duplex Safety Pins.

The Duplex Safety Pin is considered the most reliable made. The double sided shield prevents the material from getting caught in the pin. Fastens from either side. Can be locked in the dark as the tongue in the head prevents the pin from going through. Made of brass wire, nickel plated. Put up 1 dozen on a card in the sizes listed below. **State size.** Shipping weight, 2 oz.

25L5163

Size No.	0	1	2
Per card of 1 dozen	**5c**	**5c**	**5c**
Size, No.	2½	3	3½
Per card of 1 dozen	**6c**	**8c**	**10c**

"GRIP-IT."

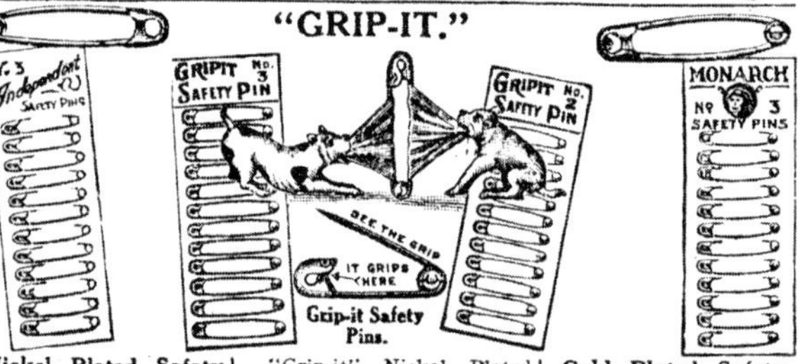

Grip-it Safety Pins.

Nickel Plated Safety Pins.

Nickel Plated Safety Pins, made of brass wire. Reliable in every way. Open or close from either side. **State size.** Shipping weight, per 3 dozen, 4 ounces.

25L5162

Size No. 1. **3** doz. for **8c**
Size No. 2. **3** doz. for **10c**
Size No. 3. **3** doz. for **13c**

"Grip-it" Nickel Plated Brass Safety Pins are a little higher in cost than the ordinary, but we think they are one of the best safety pins made and will prove cheaper in the end. The lock in the shield prevents the pin from pulling out. Shpg. wt., per doz., 2 oz.

25L5167

Size No. 1. Per dozen .. **6c**
Size No. 2. Per dozen .. **7c**
Size No. 3. Per dozen .. **9c**

Gold Plated Safety Pins.

Gold Plated Safety Pins, made of brass wire. **State size.** Shipping weight, 2 ounces.

25L5168

Size	Per Doz.
No 00	**7c**
No. 0	**7c**
No. 1	**8c**
No. 2	**9c**
No. 2½	**10c**
No. 3	**10c**

Spool Holder.

Made of nickel plated steel with a velveteen covered pin cushion. Has room for six spools of thread. Shipping weight, 4 ounces.

25L5247 ..... **22c**

Nickel Plated Crochet Hooks.

6 7 8 9 10 11 12 13 14

Made of steel and nickel plated. Has cap to protect the point. Sizes, 6 to 14. **State size.** Shipping weight, 1 ounce.

25L5169—Each **5c**

Tape Measure.

Good grade non-stretchable cotton, 60 inches long. Inches printed on one side and yards and fractions on the other side. Shipping weight, 2 ounces.

25L5216 .... **8c**

Dressing Table Set.

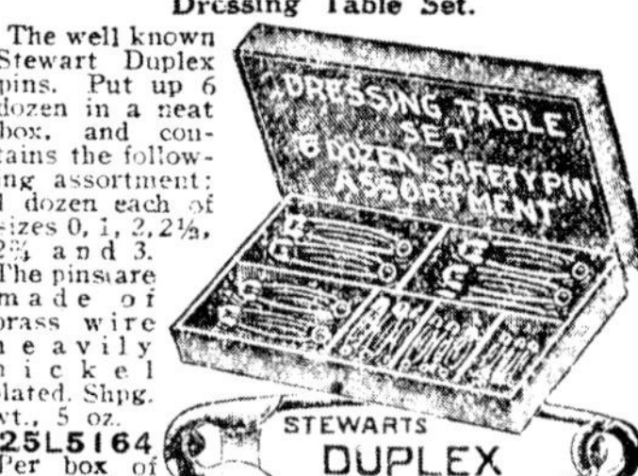

The well known Stewart Duplex pins. Put up 6 dozen in a neat box, and contains the following assortment: 1 dozen each of sizes 0, 1, 2, 2½, 2¾ and 3. The pins are made of brass wire heavily nickel plated. Shpg. wt., 5 oz.

25L5164

Per box of **6** doz. **35c**

Flora McDonald Needle Assortment.

These are the well known Flora McDonald Needles, stuck in flannel cloth. Assortment consists of 1 package each of sizes, 3 to 7, 4 to 8, 5 to 9 sewing needles, 12 needles to a package; 1 package of 8 crewel needles, sizes, 3 to 7; 2 packages darners, 1 fine, assorted 2 to 5, and 1 package long yarn darners assorted 18 to 3, 6 needles to the package. Shipping weight, 2 ounces.

25L5180—Per assortment of **6** pkgs **25c**

WALDES Koh-i-noor

Waldes "Koh-i-noor" Snap Fasteners.

The famous Koh-i-noor Snap Fastener needs no introduction. It is "the only snap fastener." It is a fastener that you do not have to watch. A snap you can depend upon. Koh-i-noor alone contains the patent curved Waldes spring, the ball shape socket and the sunken holes necessary for security. Made of all brass and rustproof. Comes in sizes 4/0, 3/0, 2/0, 0, 1 and 2. In colors black or white. **State color and size.** Shipping wt., 1 oz.

25L5173—Per card of **12** fasteners **8c**

Imported Needle Case.

New and handy metal needle case, divided into five compartments, containing fifty needles of five different sizes, **either** 4 to 8 or 6 to 10. One size to each compartment, and all you have to do is turn the indicator to the size desired and shake the needle out. You always have the right needle in the right place. **State size, either 4 to 8 or 6 to 10.** Shipping weight, 2 oz.

25L5295—Each **19c**

Good Quality Pins.

3 FOR PAPERS 25c

Good Quality Brass Pins. Extra sharp points. Sizes: Small, medium or large. 360 pins to the paper. **State size.** Shipping weight, 3 ounces.

25L5187—3 papers for **25c**

Adamantine Pins.

FULL COUNT ADAMANTINE 200 PINS

Adamantine or Dressmakers' Pins. Put up 200 pins to the paper, in medium size only. Shipping weight, 12 oz.

25L5185

12 papers **23c**

Sharp Point Pins.

PERFECT HEADS 160 PINS 160 NEEDLE POINTS

Good Quality Sharp Point Brass Pins. 160 pins to the paper. Sizes: Small, medium or large. **State size.** Shipping weight, 2 ounces.

25L5183

3 papers for **12c**

"Day by Day." Imported Needle Case.

A very substantial Needle Case containing various kinds of needles for everyday use. Contains the following: 80 egg eye sharps, in sizes 3 to 7, 4 to 8, 5 and 6, and 54 assorted crewel, darning, straw, rug and chenille needles for art work. Shipping wt., 5 oz.

25L5292 .... **49c**

Imported Needle Book.

This Handy Needle Book contains the following first quality needles; 60 assorted sharps and betweens, 12 darning needles and an assortment of 22 straw, embroidery, rug, chenille, carpet and yarn needles and one bodkin. All stuck in flannel. Shipping weight, 2 ounces.

25L5294 .... **59c**

Imported Needle Book.

A needle book that contains the most used needles for everyday purposes. It contains the following assortment: 24 assorted sizes sewing needles, 8 straw, embroidery, rug and chenille needles, 8 long darning needles, 3 carpet needles and one bodkin. All stuck in flannel. Shipping weight, 2 ounces.

25L5291 .... **35c**

Imported Needle Case.

This High Grade Needle Case is filled with first quality English gold eye needles. Contains the following assortment: 200 sharps, assorted sizes, 4 to 10; 66 assorted crewel, darning chenille and rug needles. Shipping wt., 6 oz.

25L5293

Each **79c**

Our Prices are the Lowest

Sewing Companion.

A very convenient and handy article to have in your kitchen or sewing room. Has room for four spools of thread and other sewing articles. Also fitted with mirror and pincushion. Made of gilt steel plate. Shipping weight, 9 ounces.
25L5160 Each........ **21c**

Imported Pin Cubes.

About 64 glass headed pins to each cube. Colors: Black, white or assorted colors. **State choice.** Shipping weight, 2 ounces.
25L5159—Per cube **12c**

Nickel Plated Thimble.

Made of brass. Nickel plated. Sizes: 8, 9, 10 and 11. Size 11 is the largest. **State size.** Shipping weight, 2 ounces.
25L5192.......... **3c**

Nickel Thimbles.

Imported Nickel Thimbles, very strong and durable. Come in sizes 8, 9, 10 and 11. Size 11 is the largest. **State size.** Shipping weight, 2 ounces.
25L5191.......... **6c**

Thimble Set.

This set consists of three nickel plated brass thimbles, three of different styles. One size in a box. Comes in sizes 8, 9, 10 or 11. Size 11 is the largest. **State size.** Shipping weight, 2 ounces.
25L5190—Per set of 3 thimbles....... **15c**

For Silver and Gold Thimbles see page 393.

BIAS SEAM TAPES

Extra Fine Lawn.

An exceptionally fine grade Lawn Bias Seam Tape. The kind that is used on fine lingerie, baby clothes, etc. All seams are carefully [illegible]ned and the edges come together evenly. Put up 6 yards to the piece, in white only. Each piece is put in a glassim envelope, so it reaches you in perfect condition. Shpg. wt., 2 oz.
25L5301—Width, about ¼ inch. Per piece........ **6c**
25L5302—Width, about ⅜ inch. Per piece........ **7c**
25L5303—Width, about ½ inch. Per piece........ **9c**
25L5304—Width, about ⅝ inch. Per piece........ **12c**

Cambric Bias Seam Tape.

Made of good quality cambric. The well known F. A. quality. Put up six yards to the piece. White only. Shipping wt., 2 oz.

	Width, About	Per Piece
25L5308	¼ inch	5c
25L5309	⅜ inch	6c
25L5310	½ inch	7c
25L5311	⅝ inch	8c

Lawn Bias Seam Tape.

Good grade Lawn Bias Seam Tape. The well known F. A. quality, six yards to the piece. Color, white only. Shipping weight, 2 oz.

	Width, About	Per Piece
25L5297	¼ inch	5c
25L5298	⅜ inch	6c
25L5299	½ inch	7c
25L5300	⅝ inch	8c

Colored Percale.

Good grade Percale Bias Seam Tape, put up 6 yards to the piece. Width, about ½ in. Colors: Light blue, pink, red, navy blue, nile green, Alice blue, tan or black. **State color.** Shipping weight, 2 ounces.
25L5312
Per 6-yard piece........ **10c**

Duo-Fold Bias Edging.

Used as a bias tape and trimming combined. Made of two pieces of high grade lawn, one white and one colored, folded so that the colored lawn will show as a trimming. White with the following colored edges: Pink, light blue, navy blue, cardinal, Alice blue, old rose or black. **State color.** Put up 3 yards to the piece. Width, about 7/16 inch. Shpg. wt., 1 oz.
25L5315—Per piece of 3 yards...... **13c**

Striped Percale.

Good grade white percale with the following color stripes: Pink, light blue, navy blue, red, black or lavender. **State color.** Six yards to the piece. Width, about ½ inch. Shipping weight, 2 ounces.
25L5313
Per 6-yard piece...... **9c**

Combination Spool Holder and Pincushion.

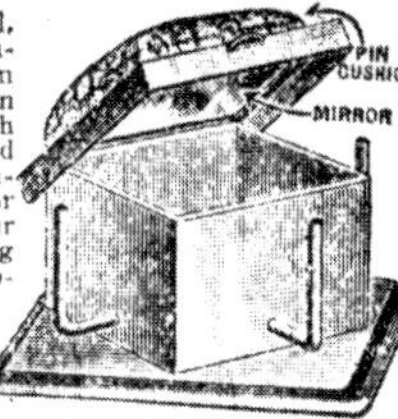

Made of metal, aluminum color finish with pincushion top and mirror in the cover which is hinged and makes a very convenient article for keeping your everyday sewing necessities. Shipping wt., 6 oz.
25L5161
25c

Stocking Darner.

One of the most practical stocking darners made. It has a metal ring that will always hold the stocking in place and thereby relieve the strain on the hand. Made of hardwood in either natural or black enameled finish. **State color.** Shipping weight, 4 ounces.
25L5220 **13c**

Stocking Darner.

Made of natural color polished hardwood. Shipping wt., 5 ounces.
25L5204 **5c**

Mending Tissue.

Tailors' Mending Tissue for mending torn clothing. Size of sheets, 4x36 inches. Shipping wt., 2 oz.
25L5207
Per piece **6c**

"Holdzit" Shirt Waist and Bloomer Bands.

Used in replacing worn out elastic in blouses, petticoats, bloomers, camisole, etc. No sewing—just slip through. The elastic is good grade ⅜-inch white double stretch elastic. Removable for laundering. Comes in sizes, small (22 inches); medium (26 inches); large (30 inches), or extra large (34 inches). **State size.** Shipping weight, each, 2 ounces.
25L5195—Each **8c**

"Holdzit" Elastic Bloomer Knee Bands.

For replacing worn out elastic in bloomer knees. Shipping weight, per pair, 2 ounces.
25L5196—Children's sizes, 9 or 11 inches long. **State size.**
Per pair........ **8c**
25L5197—Women's sizes, 14 or 16-inches long. **State size.**
Per pair........ **9c**

Superfine Tape

A white cotton binding tape of good quality. **State width desired.** Shpg. wt., 2 oz.
25L2918
Width, ¼ inch, 10 yards for........ **10c**
Width, ⅜ inch, 10 yards for........ **14c**
Width, ½ inch, 10 yards for........ **17c**
Width, ⅝ inch, 10 yards for........ **20c**
Width, 1 inch, 10 yards for........ **22c**

Lingerie Braid.

Highly mercerized cotton lingerie braid with self-threading bodkin. For women's undergarments. Width, about ¼ inch. Colors: White, pink or light blue. **State color.** Sold in 6-yard lengths only. Shipping weight, 2 ounces.
25L5200—6 yards for........ **7c**

Artificial Silk Lingerie Braid.

Comes with self-threading bodkin. Width, about ¼ inch. Colors: White, pink or light blue. Sold in 5-yard pieces only. **State color desired.** Shipping weight, 2 ounces.
25L5208—Per 5-yard piece........ **12c**

Artificial Silk Lingerie Braid.

White with woven colored edges. Colors: White with pink edge; white, with light blue edge, or white with cardinal edge. **State color desired.** Width, about 5/16-inch. Bodkin attached. Shipping weight, 2 ounces.
25L5203
Per 3-yard piece........ **8c**

Simplicity Plaiter.

"Simplicity" Is Our Own Trade Mark, Registered in U. S. Patent Office.

Made of heavy tinned plate. Handy and easy to operate. Every woman knows the advantage of having a simple plaiter in the home. Directions are given with each plaiter. You can do box plaiting, ruffling and ruching in a great variety of patterns and can do it very quickly. Size, about 13¾x7½ inches. Shipping weight, 1 pound.
25L5219...... **25c**

Cotton Tape.

Good grade. White only. Comes in 3-yard lengths. **State width desired.** Shipping weight, 2 ounces.
25L5215
Width, about ¼ inch, 6 pieces for........ **10c**
Width, about ⅜ inch, 6 pieces for........ **14c**
Width, about ½ inch, 6 pieces for........ **16c**
Width, about ⅝ inch, 6 pieces for........ **20c**

Lingerie Ribbon.

Made of silk with a cotton warp, ¼-in., for draw strings, and 1-in. for shoulder straps. Colors: White, pink or light blue. **State color and width.**
25L3072
Width, about ¼ in. Per yard... **4c**
Width, about 1 in. Per yard **8c**

Van Heusen Collar Bands for Men's Shirts.

One of the best collar bands made. The material is a single ply white cotton cloth, the same as used in the well-known Van Heusen collars. Non-shrinkable and correctly sized. Ready to sew on. Comes in ½ and full sizes from 13½ to 17½. **State size.** Shipping weight, 2 ounces.
25L5221-2 for........ **25c**

Collarbands for Men's Shirts.

Correctly sized and ready to sew on. Made of good quality white cotton cloth. Come in half and full sizes from 13½ to 17 inches. **State size desired.** Shipping wt., 2 oz.
25L5205-4 for........ **25c**

Fancy Plaited Sateen Skirt Flouncings.

Plaited Skirt Flouncings. Here are two exceptionally pretty designs of plaited flouncings. Made of a heavy sateen underskirt material. You will find that you can make over an old skirt so that it is as good as new simply by sewing this flouncing on to the bottom. It comes in two patterns, as shown in illustration. Colors: Black, navy blue, emerald green or medium brown. **State color and mention number indicating pattern wanted.** Width, about 12 in.
25L2900—Per yard........ **45c**

Skirt Belting.

Boned Curved Skirt Belting. Made of cotton grosgrain with bone supports. Colors: White or black. **State color and width.**
25L2915
Width, about 1½ inches. Yard........ **7c**
Width, about 2 inches. Yard........ **8c**
Width, about 2½ inches. Yard........ **10c**

Curved Belting.

Cotton Grosgrain Curved Inside Skirt Belting. Curved to take the shape of waistline of skirt properly. Colors: White or black. **State color and width.**
25L2913
Width, about 1½ inches. Per yard........ **6c**
Width, about 2 inches. Per yard........ **7c**

"Say it with a Diamond"

Unless otherwise stated all jewelry shown on this page is 10-karat solid gold. Each article is set with genuine regular cut diamond and is accompanied by our certificate of guarantee. Pendants are furnished with 15-inch soldered link chains; can be furnished with 18-inch chains for 40 cents extra. For ring measuring chart, see page 399. Brooches and Bar Pins have patent safety catches. With each scarf pin we furnish patent safety holder without extra charge.

On any article on this page priced under $50.00, shipping weight is 4 ounces; if priced over $50.00, it will be sent by express collect if there is an express office at your station. If not we will send ring by registered mail.

4L3500
Fancy Lacy Perforated Platinum Mounting, five of our first quality diamonds, total weight, 25/100 carat. Certificate "B" sent. Shipped by sealed express.... **$75.00**

4L3501 Fancy Perforated Platinum Mounting, with seven of our first quality diamonds; total weight, 28/100 carat. Certificate "B" sent.... **$81.50**

Fancy Openwork Platinum Engraved Mounting, with one 63/100-carat diamond, and six small diamonds in crown of ring. Certificate "A" sent.
4L3503—(2d)..$275.10
4L3505—(1st). 299.65

Fancy Delicate Openwork Platinum Mounting, heavy weight, with 44/100-carat diamond. Certificate "A" sent.
4L3507—(3d) $168.10
4L3509—(2d) 189.35
4L3511—(1st) 205.00

Fancy Lacy Open Work Platinum Mounting, with 32/100-carat diamond. Certificate "A" sent.
4L3513—(3d) $120.85
4L3515—(2d) 133.95
4L3517—(1st) 143.65

4L3520
14-Karat Green Gold Engraved Mounting, platinum top. Synthetic blue sapphire and two small genuine diamonds. Certificate "B" sent. Shpg. wt., 4 oz. **$30.25**

Solid Platinum Engraved Mounting, with 37/100-carat diamond. Certificate "A" sent.
4L3519—(3d) $115.45
4L3521—(2d) 131.75
4L3523—(1st) 143.85

Openwork Platinum Mounting, with 44/100-carat diamond. Certificate "A" sent.
4L3525—(3d) $167.80
4L3527—(2d) 189.00
4L3529—(1st) 204.70

Fancy Openwork Platinum Engraved Mounting, with 60/100-carat diamond. Certificate "A" sent.
4L3531—(3d) $172.75
4L3533—(2d) 203.15
4L3535—(1st) 221.80

Solid Platinum Engraved Mounting, with 50/100-carat diamond. Certificate "A" sent.
4L3537—(3d) $172.10
4L3539—(2d) 202.35
4L3541—(1st) 221.00

Solid Platinum Engraved Mounting, with 25/100-carat diamond. Certificate "A" sent.
4L3543—(3d) $66.40
4L3545—(2d) 75.00
4L3547—(1st) 81.25

4L3549
14-Karat Solid Gold Initial Ring, yellow and green gold finish, with 5/100-carat diamond. Engraved with one, two or three letters. **State letters.** Price includes engraving. Certificate "B" sent. Shipping wt., 4 oz... **$14.00**

4L3551
18-Karat White Gold Hand Chased Mounting, with 12/100-carat diamond, our second quality. Certificate "A" sent. Shipping weight, 4 ounces.... **$26.50**

4L3553
18-Karat White Gold Fancy Openwork Mounting, with 28/100-carat diamond, our second quality Certificate "A" sent. **$76.00**

4L3557
18-Karat White Gold Mounting, with 18/100-carat diamond, our second quality. Certificate "A" sent. Shipping weight, 4 ounces.... **$43.00**

4L3559
18-Karat White Gold Perforated Mounting, with two diamonds; total weight, 15/100 carat. Certificate "B" sent. Shipping weight, 4 ounces..... **$39.75**

4L3573
14-karat white gold, sautoir style, with genuine black onyx and small diamond. Round cord 22 inches long. Shipping weight, 4 ounces.
$19.00

4L3575
14-karat solid gold, platinum top, with 5/100-carat diamond; whitegoldchain. Shpg. wt., 4 oz... **$21.25**

4L3561
14-karat solid gold, genuine black onyx top, set with diamond. Certificate "B" sent. Shipping wt., 4 oz... **$7.70**

4L3563
14-karat, white gold top, genuine black onyx, set with diamond. Certificate "B" sent... **$8.25**

4L3565
14-karat, green gold genuineblackonyx, set with diamond. Certificate "B" sent. Shpg. wt., 4 oz.... **$10.50**

4L3567
14-karat, with 18-karat white gold top, genuine black onyx, set with diamond. Certificate "B" sent. Shipping weight, 4 oz... **$10.75**

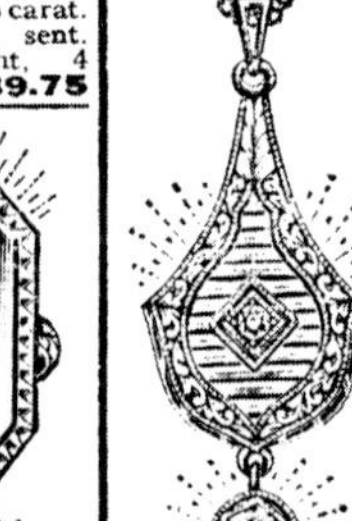

4L3569
18-karat white gold, hand engraved, two small regular cut diamond sets; white gold chain. Shipping wt., 4 oz...... **$21.85**

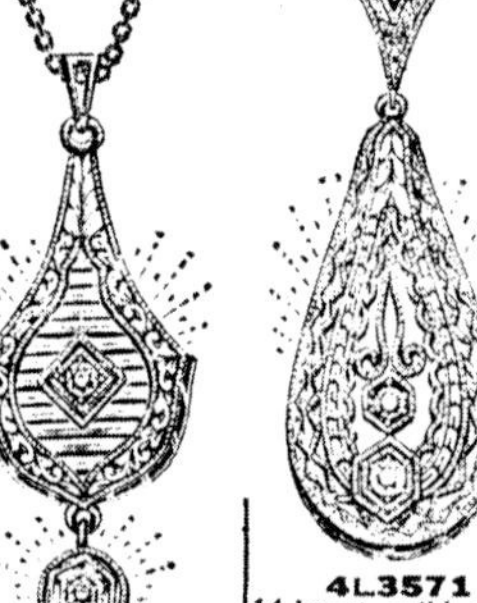

4L3571
14-karat solid white gold, platinum top, silk cord sautoir, 23 inches long. Pendant with two genuine regular cut diamonds and one blue synthetic sapphire. Shpg. wt., 4 oz... **$35.00**

4L3577
14-Karat Green Gold Mounting, with small diamond set in 18-karat white gold. Certificate "B" sent. Shipping weight, 4 ounces.... **$18.95**

4L3579
14-Karat Green Gold Mounting, with engraved platinum top. Synthetic blue sapphire, and two small genuine diamonds. Certificate "B" sent. Shipping weight, 4 ounces..... **$22.00**

4L3581
14-Karat Solid Gold Consistory Emblem Ring, enameled in colors, white gold eagles, with 12/100-carat diamond. Certificate "B" sent. Shipping weight, 4 ounces..... **$40.50**

4L3583
14-karat white gold, platinum top, 5/100-carat diamond set; white gold chain. Shipping weight, 4 oz.... **$15.85**

4L3585
18-karat solid white gold, small diamond; white gold chain. Shpg. wt., 4 oz. **$9.95**

4L3587
14-karat solid gold, white goldtop, 5/100-carat diamond; white gold chain. Shpg. wt., 4 oz. **$15.95**

4L3589
14-Karat Solid Gold Cluster Solitaire, with seven small genuine diamonds in white gold top. Certificate "B" sent. Shipping wt., 4 oz. **$29.75**

4L3591
14-Karat Solid Gold Cluster Solitaire, has seven small fine diamonds, set in all platinum top. Certificate "B" sent. Shipping wt., 4 oz..... **$43.00**

4L3593
18-Karat White Gold Cluster Solitaire, platinum top, with seven fine diamonds. Certificate "B" sent. Shipping weight, 4 ounces..... **$38.45**

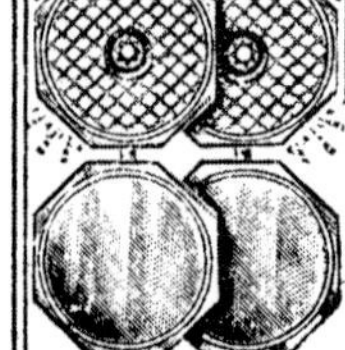

4L3609
14-karat solid gold, platinum top, with 12/100-carat diamond. Shipping weight, 4 ounces.. **$27.50**

4L3615
14-karat solid gold, white gold top, engine turned, with 3/64-carat diamond. Total weight in pair, 5/100-carat. Shipping weight, 4 ounces. **$19.25**

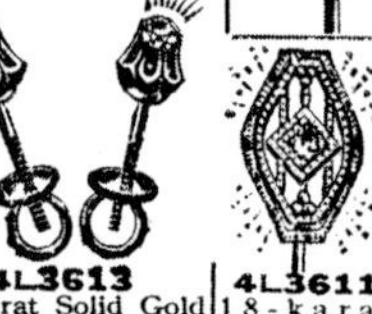

4L3613
14-Karat Solid Gold Earscrews, two 1/16-carat first quality diamonds; total weight, 12/100 carat. Shipping wt., 4 oz. Per pair. **$23.25**

4L3611
18-karat white gold top, with 1/64 or 2/100-carat diamond. Shpg. wt., 4 oz.. **$4.75**

4L3607
Solid gold, green gold rim, with 5/100-carat diamond. Shpg. wt., 4 oz.. **$9.25**

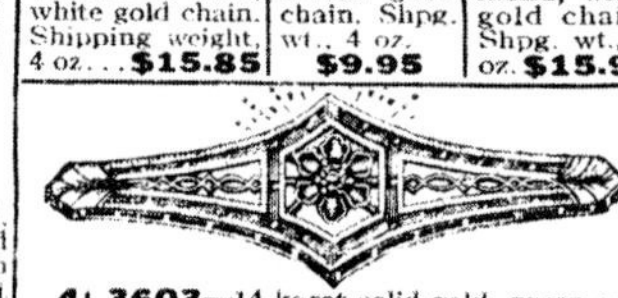

4L3603—14-karat solid gold, green color finish, hand engraved, lacy effect, with 1/32 or 3/100-carat diamond. Shpg. wt., 4 oz... **$8.45**

4L3599
14-karat solid gold, 18-karat white gold trimmed top, hand engraved 3/100-carat diamond. Shipping wt., 4 oz. **$7.95**

4L3601
Solid gold, bright polish, with small brilliant diamond. Shpg. wt., 4 oz..... **$5.00**

4L3597
14-Karat Solid Gold Earscrews, two 3/16-carat first quality diamonds; total weight, 37/100-carat. Shipped by sealed express.. **$88.70**

4L3605—14-karat solid gold, with platinum top, lacy effect, with our first quality 1/16 or 5/100-carat diamond. Shpg. wt., 4 oz......... **$20.00**

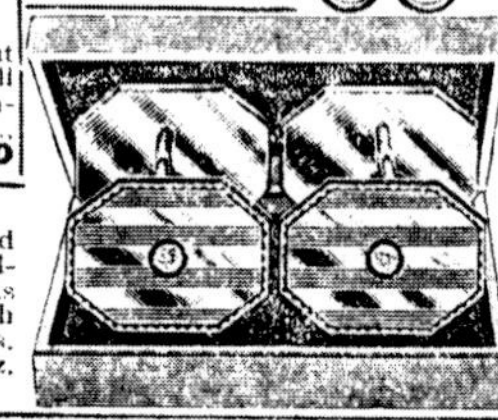

4L3595
14-Karat Solid Gold Bright Polish Cuff Links for soft cuffs, with small diamonds. Shpg. wt., 4 oz.
$17.00

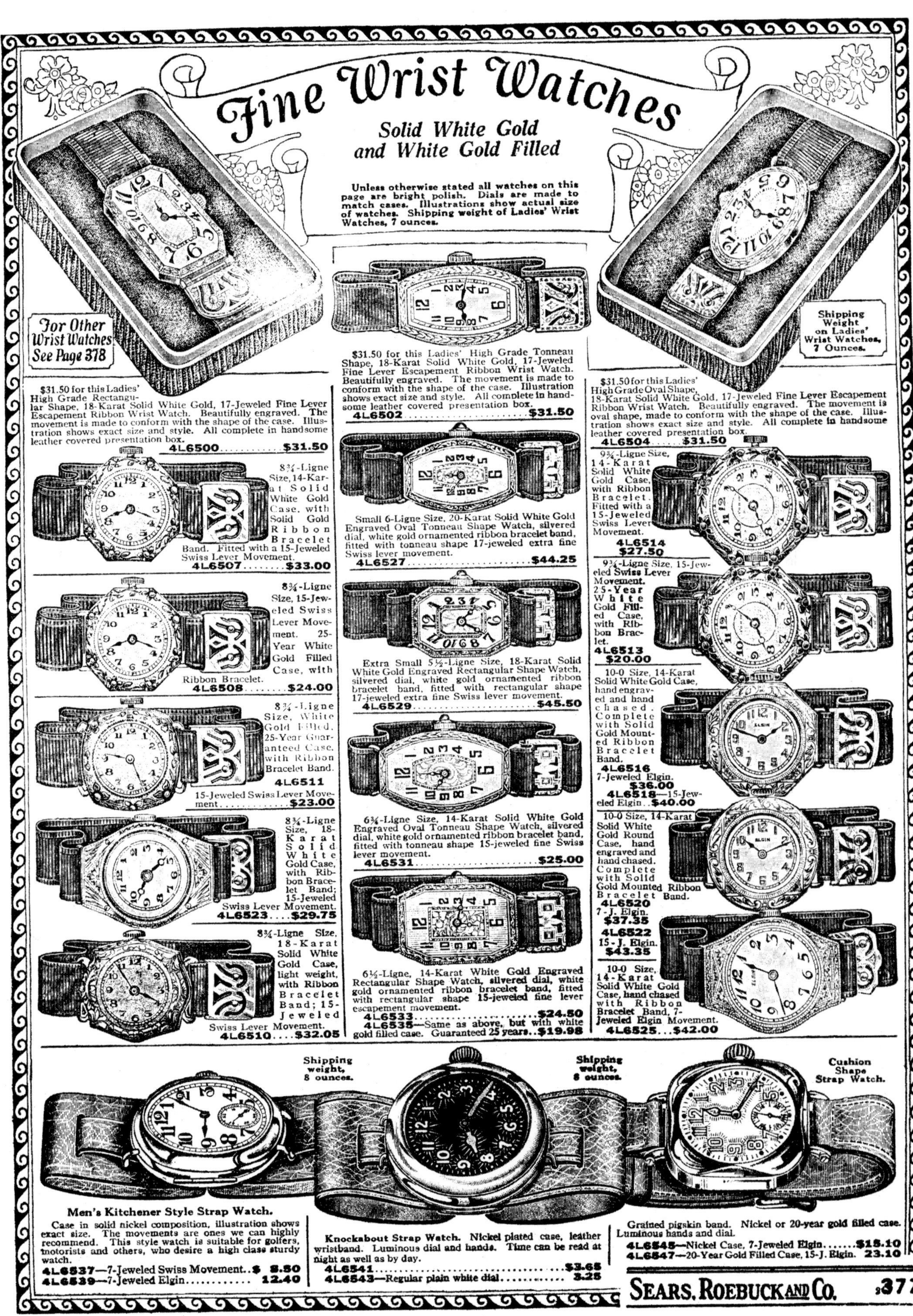

Fine Wrist Watches
Solid White Gold and White Gold Filled
Unless otherwise stated all watches on this page are bright polish. Dials are made to match cases. Illustrations show actual size of watches. Shipping weight of Ladies' Wrist Watches, 7 ounces.
For Other Wrist Watches See Page 378
Shipping Weight on Ladies' Wrist Watches, 7 Ounces.
$31.50 for this Ladies' High Grade Rectangular Shape, 18-Karat Solid White Gold, 17-Jeweled Fine Lever Escapement Ribbon Wrist Watch. Beautifully engraved. The movement is made to conform with the shape of the case. Illustration shows exact size and style. All complete in handsome leather covered presentation box.
4L6500 .... $31.50
$31.50 for this Ladies' High Grade Tonneau Shape, 18-Karat Solid White Gold, 17-Jeweled Fine Lever Escapement Ribbon Wrist Watch. Beautifully engraved. The movement is made to conform with the shape of the case. Illustration shows exact size and style. All complete in handsome leather covered presentation box.
4L6502 .... $31.50
$31.50 for this Ladies' High Grade Oval Shape, 18-Karat Solid White Gold, 17-Jeweled Fine Lever Escapement Ribbon Wrist Watch. Beautifully engraved. The movement is oval shape, made to conform with the shape of the case. Illustration shows exact size and style. All complete in handsome leather covered presentation box.
4L6504 ..... $31.50
8¾-Ligne Size, 14-Karat Solid White Gold Case, with Solid Gold Ribbon Bracelet Band. Fitted with a 15-Jeweled Swiss Lever Movement.
4L6507 .... $33.00
8¾-Ligne Size, 15-Jeweled Swiss Lever Movement. 25-Year White Gold Filled Case, with Ribbon Bracelet.
4L6508 .... $24.00
8¾-Ligne Size, White Gold Filled, 25-Year Guaranteed Case, with Ribbon Bracelet Band.
4L6511
15-Jeweled Swiss Lever Movement $23.00
8¾-Ligne Size, 18-Karat Solid White Gold Case, with Ribbon Bracelet Band; 15-Jeweled Swiss Lever Movement.
4L6523 .... $29.75
8¾-Ligne Size, 18-Karat Solid White Gold Case, light weight, with Ribbon Bracelet Band; 15-Jeweled Swiss Lever Movement.
4L6510 .... $32.05
Small 6-Ligne Size, 20-Karat Solid White Gold Engraved Oval Tonneau Shape Watch, silvered dial, white gold ornamented ribbon bracelet band, fitted with tonneau shape 17-jeweled extra fine Swiss lever movement.
4L6527 .... $44.25
Extra Small 5½-Ligne Size, 18-Karat Solid White Gold Engraved Rectangular Shape Watch, silvered dial, white gold ornamented ribbon bracelet band, fitted with rectangular shape 17-jeweled extra fine Swiss lever movement.
4L6529 .... $45.50
6¾-Ligne Size, 14-Karat Solid White Gold Engraved Oval Tonneau Shape Watch, silvered dial, white gold ornamented ribbon bracelet band, fitted with tonneau shape 15-jeweled fine Swiss lever movement.
4L6531 .... $25.00
6½-Ligne, 14-Karat White Gold Engraved Rectangular Shape Watch, silvered dial, white gold ornamented ribbon bracelet band, fitted with rectangular shape 15-jeweled fine lever escapement movement.
4L6533 .... $24.50
4L6535—Same as above, but with white gold filled case. Guaranteed 25 years.. $19.98
9¾-Ligne Size, 14-Karat Solid White Gold Case, with Ribbon Bracelet. Fitted with a 15-Jeweled Swiss Lever Movement.
4L6514
$27.50
9¾-Ligne Size, 15-Jeweled Swiss Lever Movement. 25-Year White Gold Filled Case, with Ribbon Bracelet.
4L6513
$20.00
10-0 Size, 14-Karat Solid White Gold Case, hand engraved and hand chased. Complete with Solid Gold Mounted Ribbon Bracelet Band.
4L6516
7-Jeweled Elgin.
$36.00
4L6518—15-Jeweled Elgin.. $40.00
10-0 Size, 14-Karat Solid White Gold Round Case, hand engraved and hand chased. Complete with Solid Gold Mounted Ribbon Bracelet Band.
4L6520
7-J. Elgin.
$37.35
4L6522
15-J. Elgin.
$43.35
10-0 Size, 14-Karat Solid White Gold Case, hand chased with Ribbon Bracelet Band, 7-Jeweled Elgin Movement.
4L6525 ... $42.00
Shipping weight, 8 ounces.
Shipping weight, 8 ounces.
Cushion Shape Strap Watch.
Men's Kitchener Style Strap Watch.
Case in solid nickel composition, illustration shows exact size. The movements are ones we can highly recommend. This style watch is suitable for golfers, motorists and others, who desire a high class sturdy watch.
4L6537—7-Jeweled Swiss Movement.. $ 8.50
4L6539—7-Jeweled Elgin 12.40
Knockabout Strap Watch. Nickel plated case, leather wristband. Luminous dial and hands. Time can be read at night as well as by day.
4L6541 .... $3.65
4L6543—Regular plain white dial 3.25
Grained pigskin band. Nickel or 20-year gold filled case. Luminous hands and dial.
4L6545—Nickel Case, 7-Jeweled Elgin $15.10
4L6547—20-Year Gold Filled Case, 15-J. Elgin. 23.10
SEARS, ROEBUCK AND CO.
377

Elgin, Hampden and Buren Watches

12-Size Open Face and Hunting Case Style Gold Filled Watches

Watch cases with engraved designs come in assorted patterns, and the watch we send you may not be exactly like the illustration, but we always endeavor to send a design as near like the illustration as possible. The gold filled cases shown in our catalog are guaranteed. The maker's name and term of guarantee plainly stamped on the inside lid of each case. We sell cases manufactured by recognized reputable manufacturers only. All watches sent in neat presentation case. **When ordering watches state catalog number and movement wanted.**

Shipping weight, any watch on this page, 7 ounces.

Illustrations show actual size. Unless otherwise stated all watches on this page are bright polish.

Gold filled, guaranteed for 25 years, 12-size, new green color gold, plain satin finish, open face, the new cushion shape. Monogrammed with any two or three letters, as illustration shows. **State monogram and movement wanted.** Prices quoted include monogram.

4L4600—7-J. Elgin. **$21.55**
4L4604—15-Jeweled Elgin or Hampden movement........**$25.50**
4L4609—17-Jeweled Elgin or Hampden. Unadjusted..**$28.00**

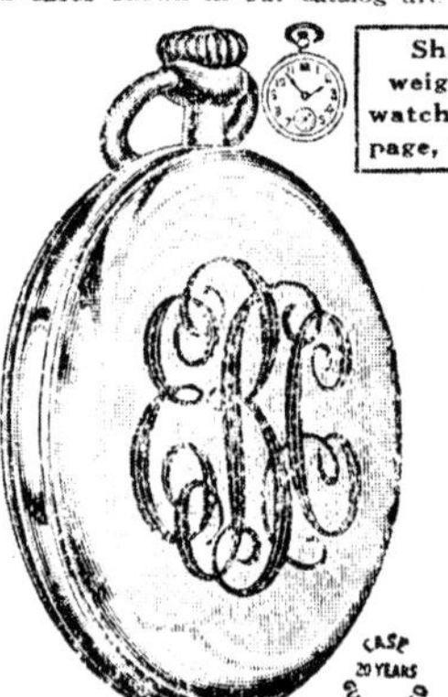

Gold filled, 20-year guaranteed, 12-size, open face, screw back and screw bezel, monogrammed case. Engraved with any two or three letters, as you desire. **State letters and movement wanted.** Prices quoted include monogram.

4L4610—7-J. Elgin. **$14.85**
4L4614—15-Jeweled Elgin or Hampden..........................**$18.80**
4L4618—17-Jeweled Elgin or Hampden. Unadjusted.**$21.25**

Gold filled, 25-year guaranteed, 12-size, open face, screw bezel, solid back, swing ring case. Hand engraved with any two or three-letter monogram, as you desire. **State letters and movement wanted.** Prices quoted include monogram.

4L4622—7-J. Elgin.....**$17.70**
4L4624—15-Jeweled Elgin or Hampden..........................**$21.65**
4L4626—17-Jeweled Elgin or Hampden. Unadjusted.....**$24.10**

14-karat solid gold case, 12-size open face, hinge bezel and back, with inside protecting cap. Any two or three-letter monogram. **State letters and movement wanted.** Prices quoted include monogram.

4L4632—15-Jeweled Elgin or Hampden..........**$38.65**
4L4634—17-Jeweled Elgin or Hampden. Unadjusted.**$41.10**

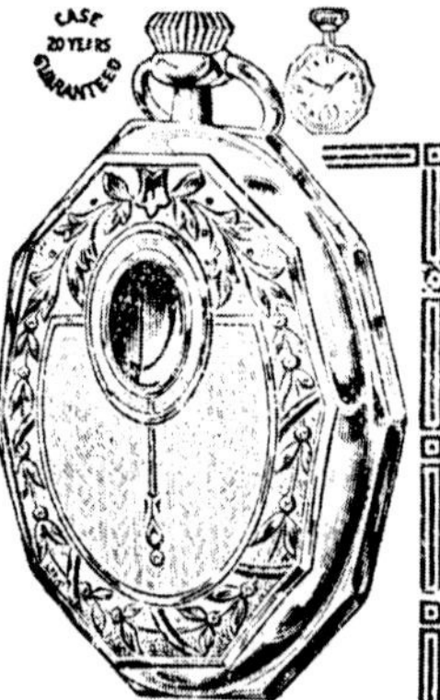

Gold filled, 20-year guaranteed, open face, screw back and screw bezel, new decagon shape 12-size case. Fitted with the movements listed below. **State catalog number and movement wanted.**

4L4640—7-Jeweled Elgin...............**$16.05**
4L4644—15-Jeweled Elgin or Hampden....**$20.05**
4L4645—17-Jeweled Elgin or Hampden. Unadjusted..............**$22.40**

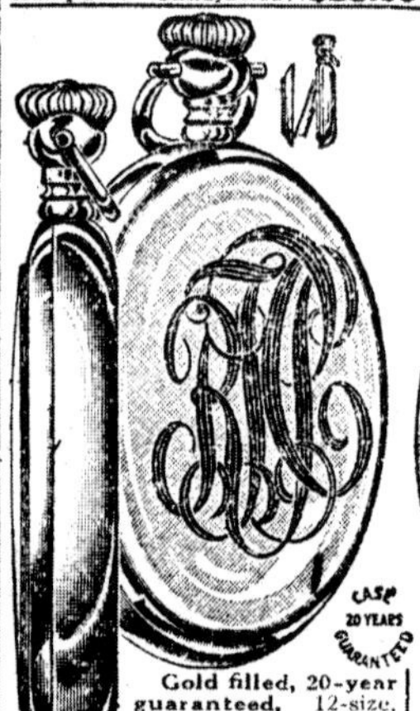

Gold filled, 20-year guaranteed, 12-size, open face, solid back swing ring style case. Engraved two or three-letter monogram. **State letters and movement wanted.** Prices quoted include monogram.

4L4648—7-J. Elgin. **$16.25**
4L4652—15-Jeweled Elgin or Hampden..........**$20.20**
4L4656—17-Jeweled Elgin or Hampden. Unadjusted..............**$22.65**

Gold filled, 25-year guaranteed, 12-size, open face, new green gold color satin finish case. **State catalog number and movement wanted.**

4L4662—7-J. Elgin...**$17.95**
4L4666—15-Jeweled Elgin or Hampden...............**$21.90**
4L4670—17-Jeweled Elgin or Hampden. Unadjusted...**$24.35**

Gold filled, 20-year guaranteed, open face, new decagon shape, plain polished, hand engraved monogram, two or three letters. **State letters and movement wanted.** Price includes monogram. This watch is perfect in size and splendidly constructed in every detail.

4L4674—7-Jeweled Buren Swiss..................**$12.25**
4L4676—15-Jeweled Altrue Buren Swiss.............**$14.15**
4L4678—7-J. Elgin... **15.35**
4L4680—15-Jeweled Elgin or Hampden...............**$19.35**

White gold filled, guaranteed for 25 years, 12-size fancy dial. This case is made to imitate platinum. Any two or three-letter monogram. **State letters and movement wanted.** Prices include monogram. Illustration shows the fancy dial made of metal, gilt background, with raised silverlike figures.

4L4682—15-Jeweled Elgin..............................**$26.55**
4L4684—17-Jeweled Elgin. Unadjusted..................**$28.95**

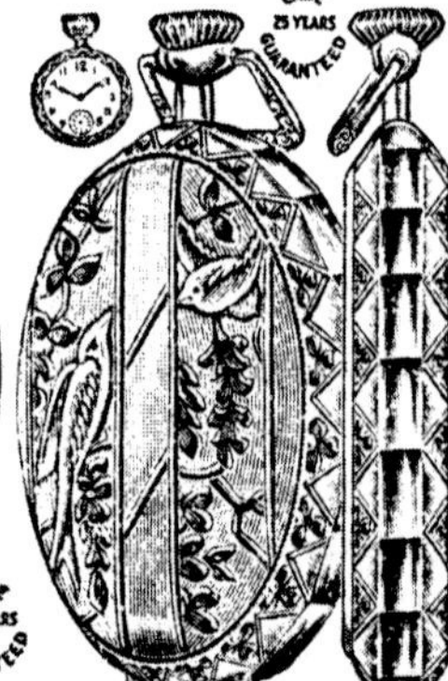

Gold filled, guaranteed for 25 years, 12-size, open face, with the green gold finish. **State movement.**

4L4688—7-Jeweled Elgin..............**$17.65**
4L4692—15-Jeweled Elgin or Hampden..**$21.65**
4L4696—17-Jeweled Elgin or Hampden. Unadjusted..............**$24.10**

Elgin and Hampden Movements—Solid Gold and Gold Filled Cases

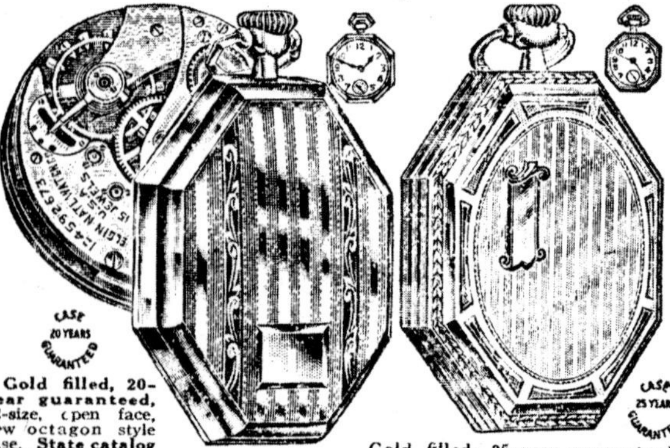

Gold filled, 20-year guaranteed, 12-size, open face, new octagon style case. **State catalog number and movement wanted.**

4L4700—7-J. Elgin.....**$18.40**
4L4703—15-Jeweled Elgin or Hampden.................. **22.40**
4L4705—17-Jeweled Elgin or Hampden. Unadjusted.... **24.85**

Gold filled, 25-year guaranteed, green gold finish, 12-size, open face case. **State catalog number and movement wanted.**

4L4706—7-J. Elgin.....**$20.75**
4L4710—15-J. Elgin or Hampden.......................... **24.70**
4L4711—17-J. Elgin or Hampden. Unadjusted.**$27.15**

Gold filled, 25-year guaranteed, 12-size case with two or three-letter monogram; prices quoted include engraving. New green color gold, satin finish. **Mention letters and movement wanted.**

4L4714—15-Jeweled Elgin or Hampden...........**$25.15**
4L4716—17-Jeweled Elgin or Hampden. Unadjusted.....**$27.60**

14-karat solid gold, 12-size, open face, plain polish with chased border. Any two or three letters, as illustration shows. The price includes monogram. **State letters and movement wanted.**

4L4720—15-Jeweled Elgin or Hampden... **$32.70**
4L4724—17-Jeweled Elgin or Hampden. Unadjusted.............................. **35.10**

Elgin, Hampden and Buren Watches

Open Face and Hunting Case Style Gold Filled Watches.

Prices quoted are for watch case and movement complete.

Watch cases with engraved designs come in assorted patterns, and the watch we send you may not be exactly like the illustration.
When ordering watches state catalog number and movement wanted.

We will always endeavor to send a design as near like the illustration as possible.

Gold filled, 20-year guaranteed case, 12-size, beautifully engraved, open face, screw back and screw bezel. Fitted with the following movements. **State catalog number and movement.**

4L4801—7-Jeweled Elgin............**$15.00**
4L4803—15-Jeweled Elgin or Hampden..........**$19.50**
4L4805—17-J. Elgin or Hampden, Unadjusted.**$21.95**

Gold filled, 10-year guaranteed, green gold finish, 12-size, open face, screw back and bezel case. Fitted with the following movements. **State catalog number and movement wanted.**

4L4811—7-Jeweled Buren..........**$10.15**
4L4813—15-Jeweled Buren..........**$12.05**
4L4815—7-Jeweled Elgin..........**$13.25**

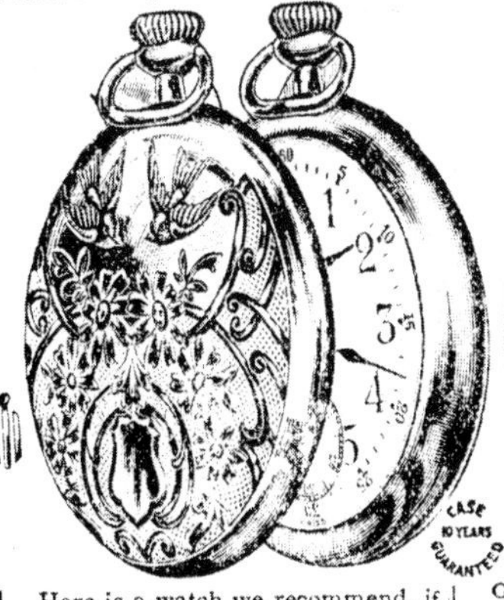

Here is a watch we recommend, if you desire a modest priced attractive proposition. The case is **gold filled,** beautifully engraved, 12-size, **guaranteed for 10 years.** Illustration gives you an idea of the engraving and front view of the watch.

4L4819—Fitted with 15-Jeweled Altrue Buren Movement...**$12.05**

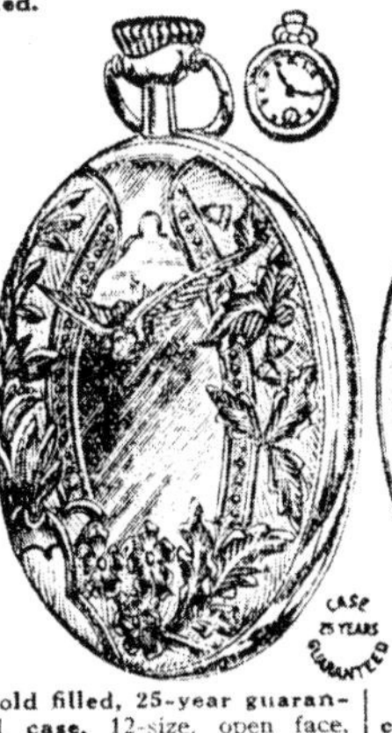

Gold filled, 25-year guaranteed case, 12-size, open face, screw back and screw bezel, beautifully engraved. Fitted with the following movements. **State catalog number and movement wanted.**

4L4823—7-Jeweled Elgin..........**$16.85**
4L4825—15-Jeweled Elgin or Hampden..........**$20.80**
4L4827—17-J. Elgin or Hampden, Unadjusted..**$23.25**

Gold filled, 25-year guaranteed case, hand monogrammed with any two or three letters. Open face, 12-size, screw back and screw bezel. **New green gold finish. Mention letters and movement wanted.** Prices quoted include monogram.

4L4833—7-Jeweled Elgin..........**$18.00**
4L4835—15-Jeweled Elgin or Hampden..........**$21.95**
4L4837—17-Jeweled Elgin or Hampden, Unadjusted..**$24.40**

Shipping weight on all watches on this page, 8 oz.

Gold Filled, 20-Year Guaranteed Case, 12-Size, Hunting Style, Hand Engraved Monogram. **State letters and movement wanted.**

4L4843—7-Jeweled Elgin..........**$19.67**
4L4845—15-Jeweled Elgin or Hampden..........**$23.65**
4L4847—17-J. Elgin or Hampden, Unadjusted.**$26.10**

Gold Filled, 10-Year Guaranteed Case, 12-Size, Hunting Style.

4L4849—7-Jeweled Buren..........**$11.50**
4L4851—15-Jeweled Buren..........**$13.40**
4L4853—7-Jeweled Elgin..........**$13.75**

Shipping weight on all watches on this page, 8 oz.

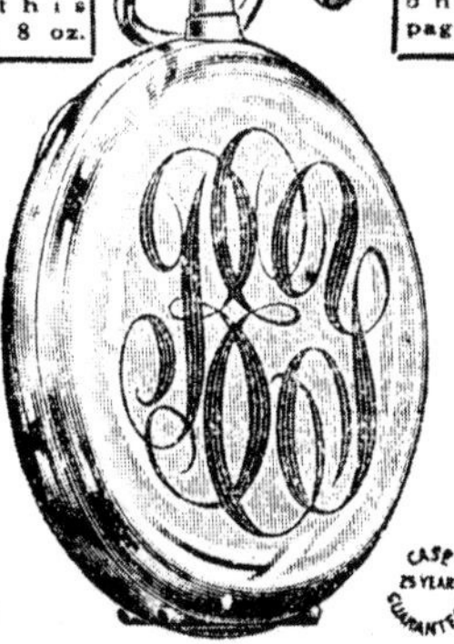

Gold Filled, 25-Year Guaranteed Case, 12-Size, Hunting Style, Hand Engraved Monogram. **State letters and movement wanted.**

4L4855—7-J Elgin..**$18.50**
4L4857—15-Jeweled Elgin or Hampden..........**$23.00**
4L4859—17-Jeweled Elgin or Hampden, Unadjusted...**$25.75**

Gold Filled, 20-Year Guaranteed Case, 12-Size, Hunting Style, engraved and engine turned. **State catalog number and movement wanted.**

4L4861—7-Jeweled Elgin..........**$18.60**
4L4863—15-Jeweled Elgin or Hampden..........**$22.50**
4L4867—17-J. Elgin or Hampden, Unadjusted.**$25.25**

Gold Filled, 10-Year Guaranteed Case, Hunting Style, 12-Size. **State catalog number and movement wanted.**

4L4869—7-Jeweled Swiss..........**$11.50**
4L4871—15-Jeweled Swiss..........**$13.40**
4L4873—7-Jeweled Elgin..........**$13.50**
4L4875—15-Jeweled Elgin or Hampden..........**$17.50**

$23.30

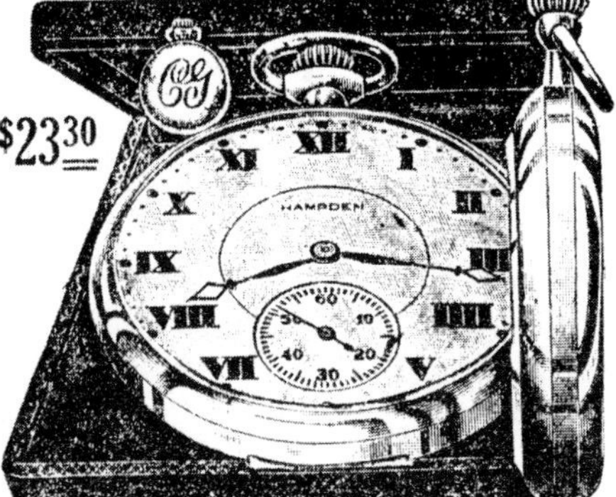

Dueber-Hampden Complete Watches, 12-size, open face cases. Prices include monogram. **State letters wanted.**

4L4877—**Gold filled, 25-year case** with 17-jeweled Hampden Movement..........**$23.30**
4L4879—**14-karat solid gold case** with 17-jeweled Hampden. Luminous hands and dial..........**$38.50**

This 23-Jeweled Special Railway Watch for $26.25.

The movement is guaranteed by the Dueber Hampden Watch Company of Canton, Ohio. 18-size, has 23 jewels, extra fine finished, accurately adjusted to temperature, isochronism and five positions, a watch that you can depend on, a watch of rare accuracy, much used and very popular with railway men. We know of no watch that will give better general satisfaction. Solid nickel composition case, open face, screw back and screw bezel, dust and damp proof, just the kind of case to give a watch good protection.

If you desire, we can supply the same movement fitted in a 20-year gold filled case, plain polished, for $31.85.

4L4881—23-Jeweled Special Railway Movement in nickel composition case..........**$26.25**
4L4883—23-Jeweled Special Railway Movement in 20-year gold filled case..........**31.85**

$26.95

Gold Filled Case, Guaranteed for 20 Years, 12-size, plain polished. Extra thin model. 17-jeweled Elgin, adjusted grade movement. Monogrammed as desired. **State letters.**

4L4885..........**$26.95**

ELGIN, HAMPDEN AND BUREN

Movements—16-Size Gold Filled Cases

Watch cases with engraved designs come in assorted patterns, and the watch case we send you may not be exactly like the illustration. We will always endeavor to send a design as near like the illustration as possible. Illustrations show actual size. When ordering watches be sure to state catalog number and movement.
All watches on this page are bright polish unless otherwise stated.

Gold Filled, 20-Year Guaranteed, 16-Size, Open Face Screw Back and Screw Bezel Case. Hand engraved monogram, two or three letters. **State letters.** Shipping weight, 8 ounces.

4L6103—7-Jeweled Elgin... **$14.35**
4L6105 — 15-Jeweled Elgin or Hampden **$19.20**
4L6107—17-Jeweled Unadjusted Elgin or Hampden **$22.00**
4L6109—17-Jeweled Adjusted Elgin or Hampden... **$27.25**

Gold Filled, 25-Year Guaranteed, new green color, gold satin finish, hand engraved and hand chased 16-size case. Monogrammed as illustration shows with any two or three letters. Prices quoted include monogram.
Be sure to mention letters wanted. Shpg. wt., 8 oz.

4L6113—15-Jeweled Elgin or Hampden........ **$23.00**
4L6117—17-J. Elgin or Hampden Unadjusted........ **25.80**
4L6118—19-Jeweled B. W. Raymond Elgin........ **41.05**

Gold Filled, 20-Year Guaranteed, 16-Size, Open Face, Screw Back and Screw Bezel Style, Monogrammed Case with high grade movements that meet the most exacting railway specifications. **Be sure to mention letters wanted.** Shipping weight, 8 ounces.

4L6119—19-J. B. W. Raymond Elgin.. **$35.00**
4L6121—Same movement, fitted in a 14-karat solid gold case... **$63.00**
4L6123—21-J. Father Time Elgin...... **$40.00**
4L6125—Same movement, fitted in a 14-karat solid gold case... **$68.75**
4L6127—23-J. Veritas Elgin.......... **$49.50**
4L6134—Same as above, but fitted in a 14-k. solid gold case... **$80.00**

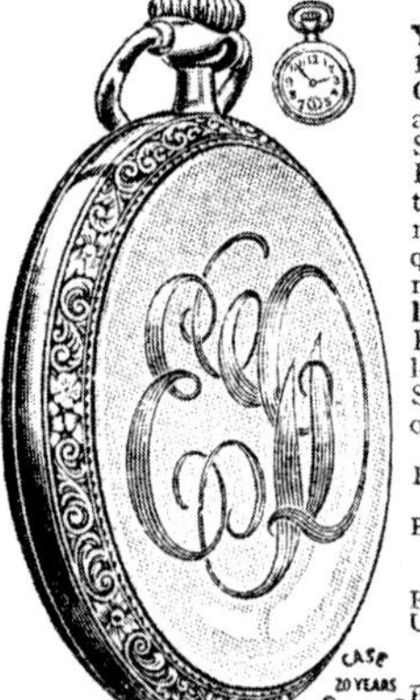

Gold Filled, 20-Year Guaranteed, 16-Size, Open Face Case, Screw Back and Screw Bezel Style, Plain Polish. Engraved with any two or three-letter monogram. Prices quoted include monogram. **State letters wanted.** Fitted with the following movements. Shipping weight, 8 ounces.

4L6135—7-J. Elgin..... **$15.87**
4L6139—15-J. Elgin or Hampden **$20.73**
4L6143—17-J. Elgin or Hampden Unadjusted. **$23.53**
4L6144—17-J. Elgin or Hampden Adjusted... **$28.75**

Gold Filled, 25-Year Guaranteed, 16-Size, Hunting Style, Beautifully Engraved Case. Shipping weight, 8 ounces.

4L6163
15-Jeweled Elgin or Hampden.
$25.88

4L6167
17-Jeweled Unadjusted Elgin or Hampden.
$28.70

Gold Filled, Guaranteed for 25 Years, 16-Size, Open Face, Screw Back and Screw Bezel Case. Fitted with the following high grade movements. All complete. Prices quoted include two or three-letter monogram. **State letters wanted.** Shipping weight, 8 ounces.

4L6150—15-J. Elgin or Hampden. **$20.75**
4L6151—17-J. Elgin or Hampden. Adjusted **$28.75**
4L6153—19-J. B. W. Raymond Elgin... **$38.80**
4L6155—21-J. Father Time Elgin. **$43.60**

Gold Filled, 20-Year Guaranteed, 16-Size, Open Face, Screw Back and Screw Bezel Case. Will meet the most exacting railway specifications. Shipping weight, 8 ounces.

4L6145—19-Jeweled 16-Size Railway Hampden........ **$35.70**
4L6147—21-Jeweled William McKinley Hampden....... **$40.20**
4L6149—23-Jeweled No. 104 Hampden............... **$48.40**

Gold Filled, 10-Year Guaranteed, 16-Size, Hunting Style Beautifully Engraved Case. Shipping weight, 8 ounces.

4L6191—7-Jeweled Buren........ **$11.55**
4L6192—15-Jeweled Buren........ **$13.45**

Gold Filled, 20-Year Guaranteed, 16-Size, Open Face, Screw Bezel and Solid Back Case. Swing ring style. Two or three letters. **State letters wanted.** Shipping weight, 8 ounces.

4L6181—17-J. Unadjusted Elgin or Hampden.......... **$23.43**
4L6183—17-J. Adjusted Elgin or Hampden.......... **28.66**
4L6185—19-J. B. W. Raymond Elgin.......... **37.50**
4L6187—21-J. Father Time Elgin.......... **42.00**

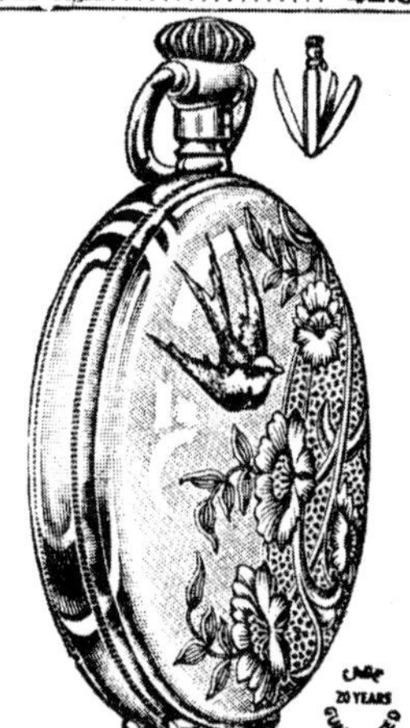

Gold Filled, 20-Year Guaranteed, 16-Size, Hunting Style Case. Shipping weight, 8 ounces.

4L6193—15-Jeweled Buren Movement............ **$18.70**
4L6195—7-Jeweled Elgin or Hampden Movement....... **$18.95**
4L6197—15-Jeweled Elgin or Hampden Movement....... **$23.85**
4L6199—17-Jeweled Elgin or Hampden Unadjusted....... **$26.65**

Gold Filled, 25-Year Guaranteed 16-Size, Hunting Style, Monogrammed Case. Hand engraved monogram of two or three letters. **State letters wanted.** Shipping weight, 8 ounces.

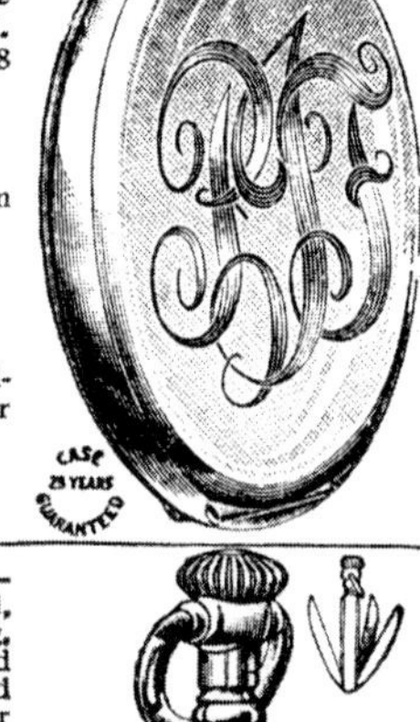

4L6157
15-Jeweled Elgin or Hampden.
$24.55

4L6159
17-Jeweled Unadjusted Elgin or Hampden.
$27.35

Gold Filled, 20-Year Guaranteed, 16-Size, Hunting, Monogrammed Case. Engraved monogram, two or three letters. **State letters wanted.** Shipping weight, 8 ounces.

4L6169
15-Jeweled Buren Movement.
$17.40
4L6171
7-Jeweled Elgin.
$17.68
4L6173
15-Jeweled Elgin or Hampden.
$22.55
4L6175
17-Jeweled Unadjusted Elgin or Hampden.
$25.35

Gold Filled, 25-Year Guaranteed, 16-Size, Open Face, Screw Back and Screw Bezel, Monogrammed case. Hand engraved monograms, two or three letters. **State letters wanted.** Shipping weight, 8 oz.

4L6205
15-Jeweled Elgin or Hampden.
$20.70
4L6207
17-Jeweled Unadjusted Elgin or Hampden.
$23.50
4L6209
17-Jeweled Adjusted Elgin or Hampden.
$28.75

New York Standard Thin Model Watches

We show the choice examples made by the well known New York Standard Watch Company, of Jersey City, New Jersey. For a low priced watch, we know of no timepiece made in the United States that will give better results. This great company guarantee their watches to give the satisfaction you have a right to expect. The cases that we illustrate on this page are all open face. The illustrations at the bottom of the page give an idea of the screw back and screw bezel style, and the open face screw bezel with jointed back.

7 and 15-Jewel.

New 12-Size, 7-Jeweled New York Standard Movement. Bridge model, splendidly damaskeened, pendant winding and setting, exposed winding wheels, lever escapement, Breguet hairspring, true timing screws, 7 genuine jewels.

New 12-Size, 15-Jeweled New York Standard Movement. Bridge model, beautifully damaskeened and nicely finished throughout, patent regulator, true timing screws, pendant winding and setting, exposed winding wheels, lever escapement, Breguet hairspring, 15 genuine jewels.

New 16-Size, 7-Jeweled New York Standard Movement. Bridge model, splendidly damaskeened, pendant winding and setting, exposed winding wheels, lever escapement, Breguet hairspring, true timing screws, 7 genuine jewels. The 16-size is a size larger than the 12-size.

New 16-Size, 15-Jeweled New York Standard Movement. Bridge model, beautifully damaskeened and well finished throughout, patent regulator, true timing screws, pendant winding and setting, exposed winding wheels, lever escapement, Breguet hairspring, 15 genuine jewels, a movement that will satisfy you with its accuracy. The 16-size is a size larger than the 12-size.

Solid Nickel Composition, Plain Polished, Open Face, Thin Model, 12 and 16-Size, 7-Jeweled New York Standard Watches, white porcelain dials. Shipping weight, 8 ounces.
4L6900—12-size, complete. **$6.85**
4L6902—16-size, complete. **6.55**

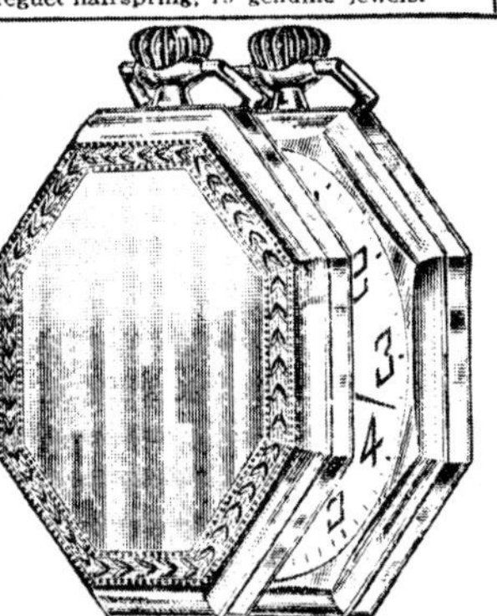

Solid Nickel Composition, or 10-year Gold Filled Fancy Thin Octagon Shape, 12-Size, 7-Jeweled New York Standard Watch. Has silvered metal dial. The nickel case resembles genuine platinum. Illustration shows back and front view of watch. Front and back snap on by friction. Shipping wt., 7 oz.
4L6904—Nickel Composition Case.......... **$8.40**
4L6905—10-Year Guaranteed Case.......... **10.65**

12-size, open face, solid nickel with rolled gold plated rim around edge of watch, screw back and screw bezel style, with a 7-jeweled New York Standard Movement. Has appearance of being a high grade and costly timepiece. The white composition metal contrasting with back and bezel, the rolled gold plated center, creates a beautiful combination of metal colors. Illustration shows back and front view. Shipping weight, 7 ounces.
4L6906—Complete.......... **$8.00**

16-Size, Open Face, Screw Back and Screw Bezel, Solid Nickel, 7-Jeweled New York Standard Watch, with rolled gold plated rim around edge of watch. White enameled dial. This watch has every appearance of being a high grade and costly timepiece, back and bezel, with the rolled gold plated center and crown make a beautiful combination of colors. Illustration shows back and front view of the watch. Shipping weight, 8 ounces.
4L6908—Complete.......... **$7.80**

16-Size, 10-Year Gold Filled Engraved, 7-Jeweled New York Standard Watch. Dial is white enamel. Illustration shows the back and front view of the watch. Shipping weight, 8 ounces.
4L6910—Complete.......... **$8.90**

12-Size, 10-Year Gold Filled, Engraved, 7-Jeweled New York Standard Watch. White enameled dial. Illustration shows the back and front view of the watch. Shipping weight, 7 ounces.
4L6912—Complete.......... **$9.20**

12-Size, 10-Year Gold Filled, Engine Turned Style, 7-Jeweled, New York Standard Watch. White enameled dial. Illustration shows the back and front view of the watch. Shipping weight, 7 oz.
4L6914—Complete.......... **$8.90**

12-Size, 20-Year, Engraved, Gold Filled, 15-Jeweled New York Standard Watch, Silvered metal dial. Illustration shows back and front view of the watch. Shpg. wt., 7 oz.
4L6916—Complete.......... **$13.45**

12-Size, 25-Year, White Gold Filled, 15-Jeweled New York Standard Watch. White gold resembles genuine platinum. Silvered metal dial. Illustration shows back and front view of the watch. Shipping weight, 7 ounces.
4L6918—Complete.......... **$17.40**

12-Size, 25-Year, Gold Filled, Plain Polished, Green Gold Finish, 15-Jeweled New York Standard Watch. Silvered metal dial. Illustration shows back and front view of the watch. Shipping weight, 8 ounces.
4L6920—Complete.......... **$14.75**

16-Size, 20-Year, Gold Filled, Engraved, 15-Jeweled New York Standard Watch. Silvered metal dial. Illustration shows back and front view of the watch. Shipping weight, 8 ounces.
4L6922—Complete.......... **$13.70**

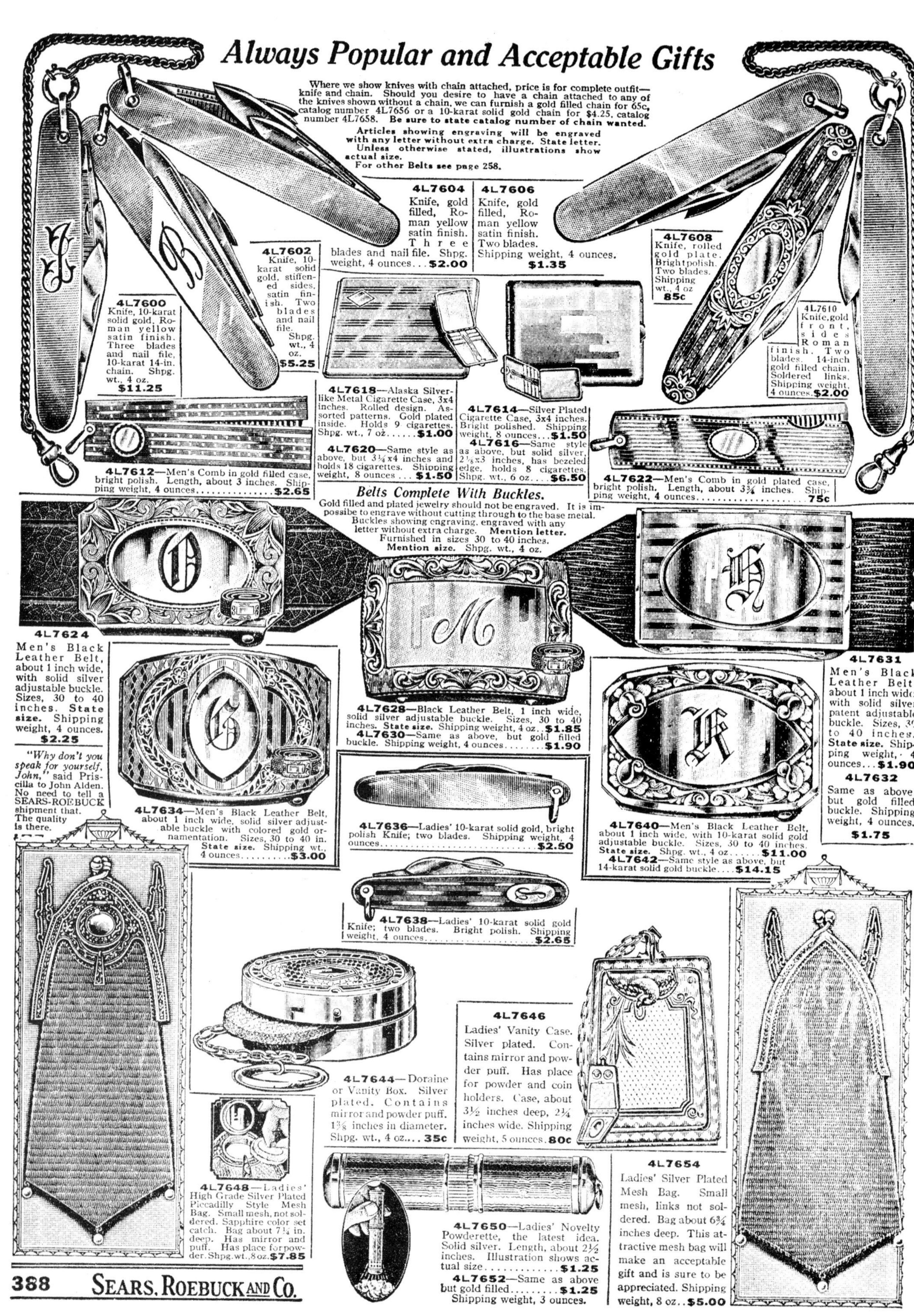
Always Popular and Acceptable Gifts
Where we show knives with chain attached, price is for complete outfit—knife and chain. Should you desire to have a chain attached to any of the knives shown without a chain, we can furnish a gold filled chain for 65c, catalog number 4L7656 or a 10-karat solid gold chain for $4.25, catalog number 4L7658. Be sure to state catalog number of chain wanted.
Articles showing engraving will be engraved with any letter without extra charge. State letter. Unless otherwise stated, illustrations show actual size.
For other Belts see page 258.
4L7604 Knife, gold filled, Roman yellow satin finish. Three blades and nail file. Shpg. weight, 4 ounces... $2.00
4L7606 Knife, gold filled, Roman yellow satin finish. Two blades. Shipping weight, 4 ounces. $1.35
4L7602 Knife, 10-karat solid gold, stiffened sides, satin finish. Two blades and nail file. Shpg. wt., 4 oz. $5.25
4L7600 Knife, 10-karat solid gold, Roman yellow satin finish. Three blades and nail file, 10-karat 14-in. chain. Shpg. wt., 4 oz. $11.25
4L7608 Knife, rolled gold plate. Bright polish. Two blades. Shipping wt., 4 oz 85c
4L7610 Knife, gold front, sides Roman finish. Two blades. 14-inch gold filled chain. Soldered links. Shipping weight, 4 ounces. $2.00
4L7618—Alaska Silver-like Metal Cigarette Case, 3x4 inches. Rolled design. Assorted patterns. Gold plated inside. Holds 9 cigarettes. Shpg. wt., 7 oz...... $1.00
4L7620—Same style as above, but 3¼x4 inches and holds 18 cigarettes. Shipping weight, 8 ounces... $1.50
4L7614—Silver Plated Cigarette Case, 3x4 inches. Bright polished. Shipping weight, 8 ounces... $1.50
4L7616—Same style as above, but solid silver, 2⅝x3 inches, has bezeled edge, holds 8 cigarettes. Shpg. wt., 6 oz... $6.50
4L7612—Men's Comb in gold filled case, bright polish. Length, about 3 inches. Shipping weight, 4 ounces.................. $2.65
4L7622—Men's Comb in gold plated case, bright polish. Length, about 3¾ inches. Shipping weight, 4 ounces.................. 75c
Belts Complete With Buckles.
Gold filled and plated jewelry should not be engraved. It is impossibe to engrave without cutting through to the base metal. Buckles showing engraving, engraved with any letter without extra charge. Mention letter. Furnished in sizes 30 to 40 inches. Mention size. Shpg. wt., 4 oz.
4L7624 Men's Black Leather Belt, about 1 inch wide, with solid silver adjustable buckle. Sizes, 30 to 40 inches. State size. Shipping weight, 4 ounces. $2.25
4L7628—Black Leather Belt, 1 inch wide, solid silver adjustable buckle. Sizes, 30 to 40 inches. State size. Shipping weight, 4 oz.. $1.85
4L7630—Same as above, but gold filled buckle. Shipping weight, 4 ounces........ $1.90
4L7631 Men's Black Leather Belt, about 1 inch wide, with solid silver patent adjustable buckle. Sizes, 30 to 40 inches. State size. Shipping weight, 4 ounces... $1.90
4L7632 Same as above, but gold filled buckle. Shipping weight, 4 ounces. $1.75
"Why don't you speak for yourself, John," said Priscilla to John Alden. No need to tell a SEARS-ROEBUCK shipment that. The quality is there.
4L7634—Men's Black Leather Belt, about 1 inch wide, solid silver adjustable buckle with colored gold ornamentation. Sizes, 30 to 40 in. State size. Shipping wt., 4 ounces.......... $3.00
4L7636—Ladies' 10-karat solid gold, bright polish Knife; two blades. Shipping weight, 4 ounces.............. $2.50
4L7640—Men's Black Leather Belt, about 1 inch wide, with 10-karat solid gold adjustable buckle. Sizes, 30 to 40 inches. State size. Shpg. wt., 4 oz...... $11.00
4L7642—Same style as above, but 14-karat solid gold buckle.... $14.15
4L7638—Ladies' 10-karat solid gold Knife; two blades. Bright polish. Shipping weight, 4 ounces.................. $2.65
4L7646 Ladies' Vanity Case. Silver plated. Contains mirror and powder puff. Has place for powder and coin holders. Case, about 3½ inches deep, 2¼ inches wide. Shipping weight, 5 ounces. 80c
4L7644—Doraine or Vanity Box. Silver plated. Contains mirror and powder puff. 1⅞ inches in diameter. Shpg. wt., 4 oz.... 35c
4L7648—Ladies' High Grade Silver Plated Piccadilly Style Mesh Bag. Small mesh, not soldered. Sapphire color set catch. Bag about 7¼ in. deep. Has mirror and puff. Has place for powder. Shpg. wt., 8 oz. $7.85
4L7650—Ladies' Novelty Powderette, the latest idea. Solid silver. Length, about 2½ inches. Illustration shows actual size.............. $1.25
4L7652—Same as above but gold filled......... $1.25
Shipping weight, 3 ounces.
4L7654 Ladies' Silver Plated Mesh Bag. Small mesh, links not soldered. Bag about 6¾ inches deep. This attractive mesh bag will make an acceptable gift and is sure to be appreciated. Shipping weight, 8 oz.. $5.00
388 SEARS, ROEBUCK AND CO.

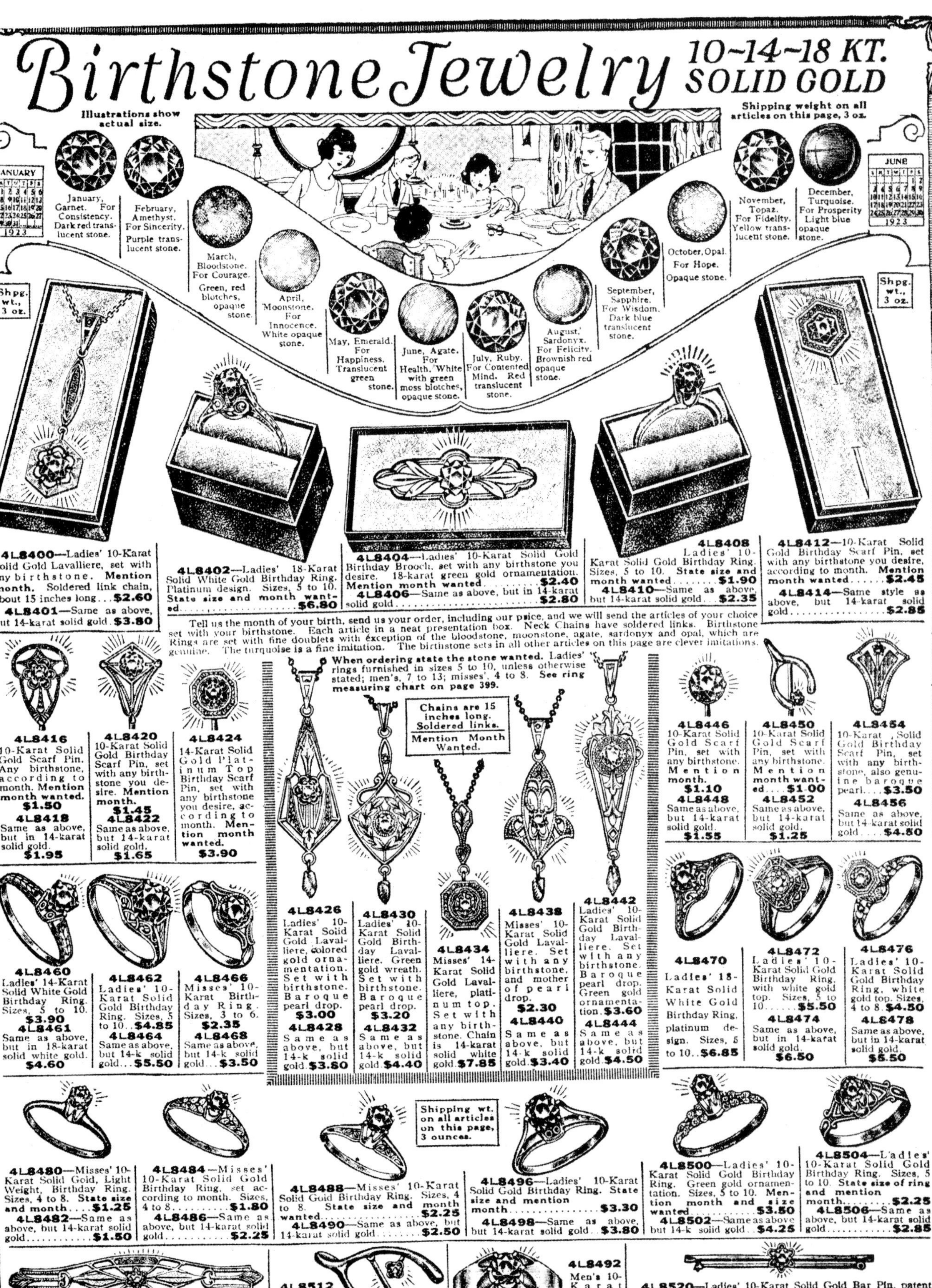

Birthstone Jewelry 10-14-18 KT. SOLID GOLD

Illustrations show actual size.

Shipping weight on all articles on this page, 3 oz.

JANUARY 1923 — JUNE 1923

January, Garnet. For Consistency. Dark red translucent stone.

February, Amethyst. For Sincerity. Purple translucent stone.

March, Bloodstone. For Courage. Green, red blotches, opaque stone.

April, Moonstone. For Innocence. White opaque stone.

May, Emerald. For Happiness. Translucent green stone.

June, Agate. For Health. White with green moss blotches, opaque stone.

July, Ruby. For Contented Mind. Red translucent stone.

August, Sardonyx. For Felicity. Brownish red opaque stone.

September, Sapphire. For Wisdom. Dark blue translucent stone.

October, Opal. For Hope. Opaque stone.

November, Topaz. For Fidelity. Yellow translucent stone.

December, Turquoise. For Prosperity Light blue opaque stone.

Shpg. wt., 3 oz.

Shpg. wt., 3 oz.

4L8400—Ladies' 10-Karat Solid Gold Lavalliere, set with any birthstone. **Mention month.** Soldered link chain, about 15 inches long...**$2.60**
4L8401—Same as above, but 14-karat solid gold. **$3.80**

4L8402—Ladies' 18-Karat Solid White Gold Birthday Ring. Platinum design. Sizes, 5 to 10. **State size and month wanted**...**$6.80**

4L8404—Ladies' 10-Karat Solid Gold Birthday Brooch, set with any birthstone you desire. 18-karat green gold ornamentation. **Mention month wanted**...**$2.40**
4L8406—Same as above, but in 14-karat solid gold...**$2.80**

4L8408 Ladies' 10-Karat Solid Gold Birthday Ring. Sizes, 5 to 10. **State size and month wanted**...**$1.90**
4L8410—Same as above, but 14-karat solid gold...**$2.35**

4L8412—10-Karat Solid Gold Birthday Scarf Pin, set with any birthstone you desire, according to month. **Mention month wanted**...**$2.45**
4L8414—Same style as above, but 14-karat solid gold...**$2.85**

Tell us the month of your birth, send us your order, including our price, and we will send the articles of your choice set with your birthstone. Each article in a neat presentation box. Neck Chains have soldered links. Birthstone Rings are set with fine doublets with exception of the bloodstone, moonstone, agate, sardonyx and opal, which are genuine. The turquoise is a fine imitation. The birthstone sets in all other articles on this page are clever imitations.

When ordering state the stone wanted. Ladies' rings furnished in sizes 5 to 10, unless otherwise stated; men's, 7 to 13; misses', 4 to 8. **See ring measuring chart on page 399.**

Chains are 15 inches long. Soldered links. Mention Month Wanted.

4L8416 10-Karat Solid Gold Scarf Pin. Any birthstone, according to month. **Mention month wanted. $1.50**
4L8418 Same as above, but in 14-karat solid gold. **$1.95**

4L8420 10-Karat Solid Gold Birthday Scarf Pin, set with any birthstone you desire. **Mention month. $1.45**
4L8422 Same as above, but 14-karat solid gold. **$1.65**

4L8424 14-Karat Solid Gold Platinum Top Birthday Scarf Pin, set with any birthstone you desire, according to month. **Mention month wanted. $3.90**

4L8426 Ladies' 10-Karat Solid Gold Lavalliere, colored gold ornamentation. Set with birthstone. Baroque pearl drop. **$3.00**
4L8428 Same as above, but 14-k solid gold. **$3.80**

4L8430 Ladies' 10-Karat Solid Gold Birthday Lavalliere. Green gold wreath. Set with birthstone. Baroque pearl drop. **$3.20**
4L8432 Same as above, but 14-k solid gold. **$4.40**

4L8434 Misses' 14-Karat Solid Gold Lavalliere, platinum top. Set with any birthstone. Chain is 14-karat solid white gold. **$7.85**

4L8438 Misses' 10-Karat Solid Gold Lavalliere. Set with any birthstone, and mother of pearl drop. **$2.30**
4L8440 Same as above, but 14-k solid gold. **$3.40**

4L8442 Ladies' 10-Karat Solid Gold Birthday Lavalliere. Set with any birthstone. Baroque pearl drop. Green gold ornamentation. **$3.60**
4L8444 Same as above, but 14-k solid gold. **$4.50**

4L8446 10-Karat Solid Gold Scarf Pin, set with any birthstone. **Mention month. $1.10**
4L8448 Same as above, but 14-karat solid gold. **$1.55**

4L8450 10-Karat Solid Gold Scarf Pin, set with any birthstone. **Mention month wanted...$1.00**
4L8452 Same as above, but 14-karat solid gold. **$1.25**

4L8454 10-Karat Solid Gold Birthday Scarf Pin, set with any birthstone, also genuine baroque pearl...**$3.50**
4L8456 Same as above, but 14-karat solid gold...**$4.50**

4L8460 Ladies' 14-Karat Solid White Gold Birthday Ring. Sizes, 5 to 10. **$3.90**
4L8461 Same as above, but in 18-karat solid white gold. **$4.60**

4L8462 Ladies' 10-Karat Solid Gold Birthday Ring. Sizes, 5 to 10...**$4.85**
4L8464 Same as above, but 14-k solid gold...**$5.50**

4L8466 Misses' 10-Karat Birthday Ring. Sizes, 3 to 6. **$2.35**
4L8468 Same as above, but 14-k solid gold...**$3.50**

4L8470 Ladies' 18-Karat Solid White Gold Birthday Ring, platinum design. Sizes, 5 to 10...**$6.85**

4L8472 Ladies' 10-Karat Solid Gold Birthday Ring, with white gold top. Sizes, 5 to 10...**$5.50**
4L8474 Same as above, but in 14-karat solid gold. **$6.50**

4L8476 Ladies' 10-Karat Solid Gold Birthday Ring, white gold top. Sizes, 4 to 8. **$4.50**
4L8478 Same as above, but in 14-karat solid gold. **$5.50**

Shipping wt. on all articles on this page, 3 ounces.

4L8480—Misses' 10-Karat Solid Gold, Light Weight, Birthday Ring. Sizes, 4 to 8. **State size and month**...**$1.25**
4L8482—Same as above, but 14-karat solid gold...**$1.50**

4L8484—Misses' 10-Karat Solid Gold Birthday Ring, set according to month. Sizes, 4 to 8...**$1.80**
4L8486—Same as above, but 14-karat solid gold...**$2.25**

4L8488—Misses' 10-Karat Solid Gold Birthday Ring. Sizes, 4 to 8. **State size and month wanted**...**$2.25**
4L8490—Same as above, but 14-karat solid gold...**$2.50**

4L8496—Ladies' 10-Karat Solid Gold Birthday Ring. **State size and mention month**...**$3.30**
4L8498—Same as above, but 14-karat solid gold...**$3.80**

4L8500—Ladies' 10-Karat Solid Gold Birthday Ring. Green gold ornamentation. Sizes, 5 to 10. **Mention month and size wanted**...**$3.50**
4L8502—Same as above but 14-k solid gold. **$4.25**

4L8504—Ladies' 10-Karat Solid Gold Birthday Ring. Sizes, 5 to 10. **State size of ring and mention month**...**$2.25**
4L8506—Same as above, but 14-karat solid gold...**$2.85**

4L8508—Ladies' 10-Karat Solid Gold Bar Pin, set with any birthstone you desire. 18-karat green gold ornamentation. **Mention month wanted**...**$3.10**
4L8510—Same as above, but in 14-karat solid gold...**$3.80**

4L8512 Ladies' 10-Karat Gold Birthday Brooch, set with any birthstone. **State month wanted**...**$1.25**
4L8514—Same as above, but 14-karat solid gold...**$1.75**

4L8492 Men's 10-Karat Solid Gold Birthday Ring. **State size and month**...**$5.90**
4L8494—Same as above, but 14-karat solid gold...**$6.90**

4L8520—Ladies' 10-Karat Solid Gold Bar Pin, patent safety catch, set with any birthstone you desire, according to month. **Mention month wanted**...**$1.95**
4L8522—Same as above, but in 14-karat solid gold...**2.35**

Senorita Pearl Necklaces

4L10406—Fine Quality, C Luster, Artificial Pearl Necklace. Graduated size. 15 in. long. Clasp is sterling silver, imitation diamond set. Shipping weight, 3 ounces.....**$2.50**
4L10408—Same as above, but 18 inches long.......................... **2.75**
4L10410—Same as above, but 24 inches long.......................... **3.50**

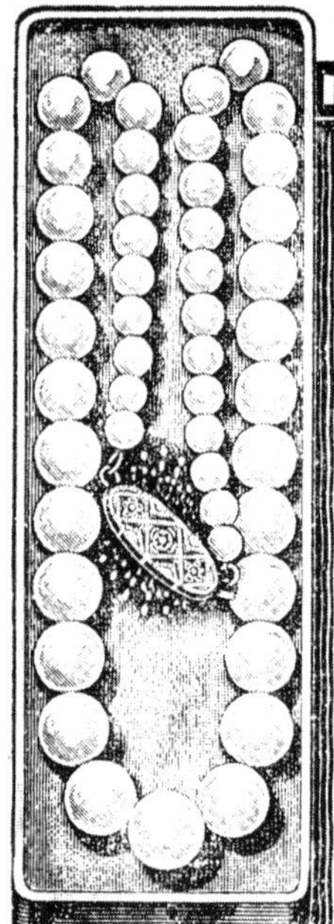

Senorita Artificial Pearls are divided into classes—Superior Quality and Fine Quality. Both classes are guaranteed unbreakable by any ordinary use or wear. **Superior Quality Senorita Pearls** are created by applying a beautiful pearly substance on the outside of the beads, coat on coat, then dried, baked and so treated by secret process as to make them practically impervious to wear and giving them charm and beauty comparable only with true Oriental natural pearls worth thousands of dollars. Only experts by careful tests would discover the difference and then only in structure. The manufacturer guarantees that they will not peel, even if boiled in water, and will give the satisfaction you have the right to expect of the highest grade pearls manufactured and sold under various names at four to six times our price.

Our Fine Quality Senorita Pearls. In this class the pearly substance is applied on the inside of the bead, then filled with wax to insure permanency. This process is not as costly as the process used in our Superior Quality, but produces a gem that will wear indefinitely, of great beauty, and a very near approach to genuine Orientals. Pearls are graded according to luster, the quality of the beads depending on the luster. We quote in each class three lusters—"A" luster, "B" luster and "C" luster. "A" luster being the finest.

Our Guarantee. If the string of pearls that you buy from us does not measure up with your expectations as to appearance and wear, return them and we will exchange them or return your money, together with transportation charges.

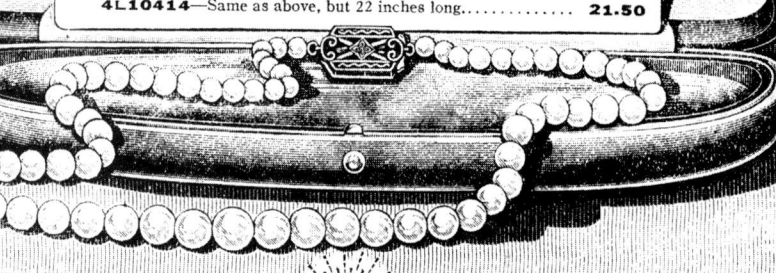

4L10412—Superior Quality, A Luster. Our Highest Grade Solid, Heavy Artificial Pearl Bead Necklace. Luster A, our finest luster. Graduated size. 16 inches long. 14-karat solid white gold clasp. Shipping weight, 3 ounces..**$17.50**
4L10414—Same as above, but 22 inches long............. **21.50**

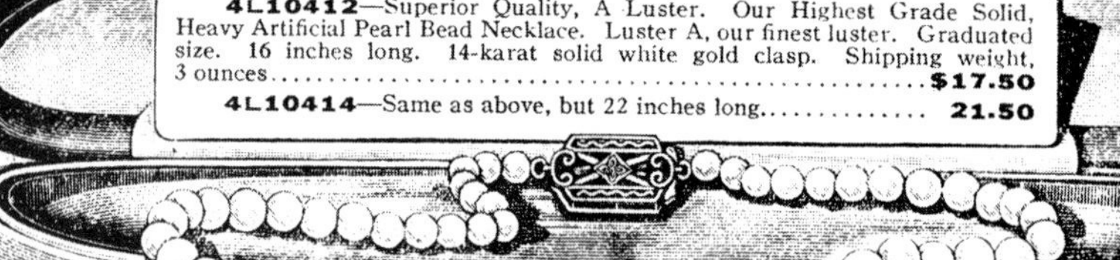

Superior Quality, C Luster. Our Highest Quality Solid, Heavy Artificial Pearl Bead Necklace. Graduated size. 18-karat solid white gold clasp, set with genuine regular cut diamond. Shipping weight, 3 ounces.
4L10400—18 in. long........**$12.00**
4L10402—24 in. long........**$13.50**
4L10404—30 in. long........**$14.95**

Superior Quality, C Luster. Our Highest Quality Solid, Heavy Artificial Pearl Bead Necklace. Graduated size. 10-karat solid gold spring ring clasp. Shipping weight, 3 oz.
4L10416—18 in. long..........**$4.50**
4L10418—24 in. long..........**$5.50**
4L10419—30 in. long..........**$6.50**

4L10424—Fine Quality, A Luster, Artificial Pearl Bead Necklace. Graduated size, 16 inches long. Clasp is 10-karat solid white gold, genuine diamond set.................. **$ 8.50**
4L10426—Same as above, but 18 inches long.................. **9.50**
4L10428—Same as above, but 24 inches long.................. **10.50**
Shipping weight, 3 ounces.

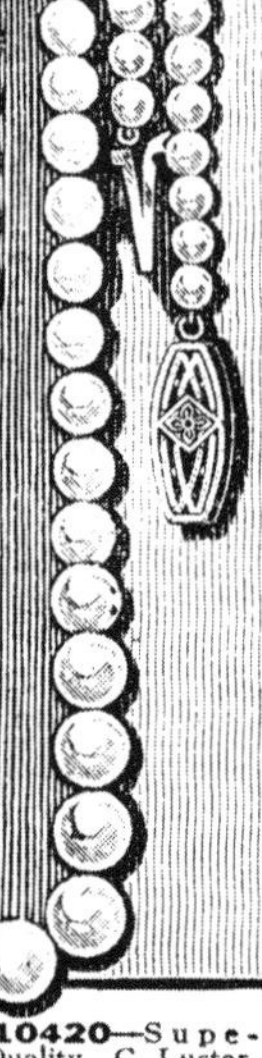

4L10420—Superior Quality, C Luster. Our Highest Grade Solid, Heavy Artificial Pearl Bead Necklace. Luster C, which is a fair luster. Graduated size, 16 inches long. 14-karat solid white gold clasp. Shipping wt., 3 oz...**$11.50**
4L10421—Same as above, but 18 inches long..........**$12.50**
4L10422—Same as above, but 22 inches long..........**$15.00**

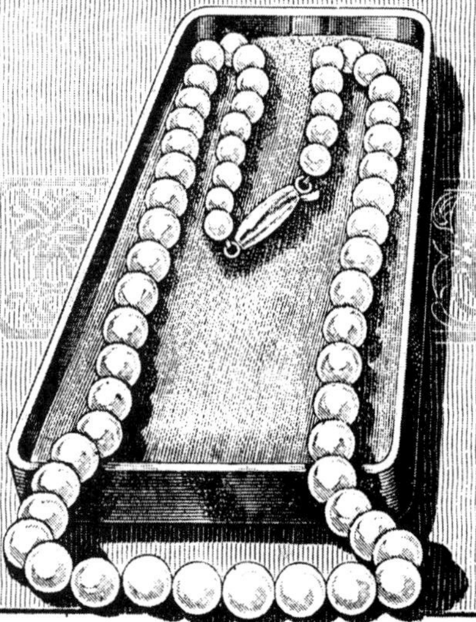

4L10430—Fine Quality, C Luster, Artificial Pearl Bead Necklace. Graduated size. 16 inches long. Clasp is 10-karat solid gold. Shipping wt., 3 oz.......**$2.25**
4L10432—Same as above, but 18 in. long... **2.50**
4L10434—Same as above, but 24 inches long.....**$3.25**

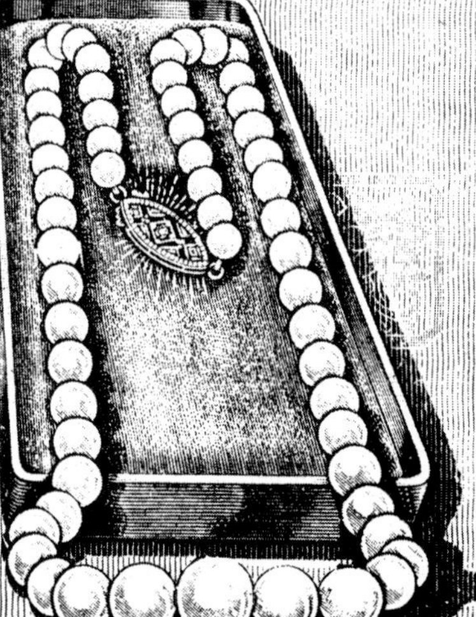

4L10439—Fine Quality, B Luster, Artificial Pearl Bead Necklace. Rosa tint, graduated size, 18 inches long; clasp is 14-karat solid white gold, set with genuine rose diamond. Shipping weight, 3 ounces.............................**$3.75**
4L10441—Same as above, but 24 inches long.............................**$4.50**

60 Inches Long.

The Very Latest in Pearl Bead Necklaces, Called the Opera Length. These are fine artificial pearl bead necklaces, worn in double strand around the neck as illustration shows. Your choice of indestructible or waxed filled. Beads are uniform size and have a beautiful luster. Necklaces are 60 inches long. Small illustration shows actual size of beads. Shipping weight, 3 ounces.
4L10443—Indestructible beads, not a waxed filled bead, but a solid bead with a fine pearly coating **$10.28**
4L10445—Waxed filled beads of beautiful color.....**$2.75**

Genuine Amber, Fancy Bead Necklaces and Earrings

4L10700
Genuine Imported Amber Bead Necklace. Graduated style. Illustration in center shows actual size of beads. Length, about 16 inches. Amber screw clasp. Shipping weight, 4 ounces...... **$3.85**

4L10702
Genuine Imported Amber Bead Necklace. Graduated style. Illustration in center shows actual size of beads. Length, about 16 inches. Amber screw clasp. Shipping weight, 4 ounces...... **$2.50**

4L10704
Genuine Imported Amber Bead Necklace. Graduated style. Illustration in center shows actual size of beads. Length, about 25 inches. Amber screw clasp. Shipping weight, 4 ounces.. **$4.75**

4L10706
Genuine Imported Amber Bead Necklace. Graduated style. Illustration in center shows actual size of beads. Length, about 25 inches. Amber screw clasp. Shipping weight, 4 ounces.. **$6.25**

The enlarged illustrations below show actual size of beads.

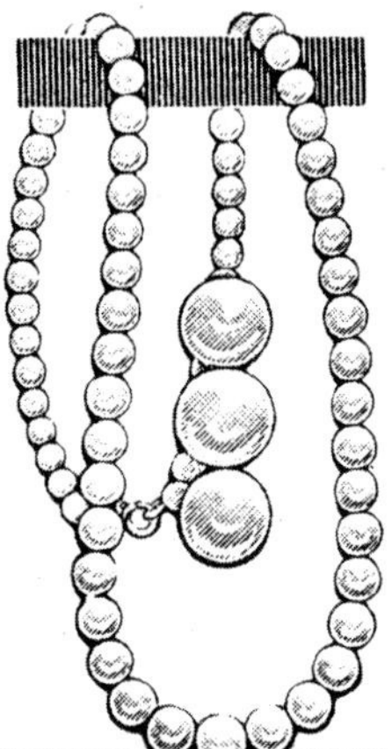

4L10708 — Artificial Pearl Bead Necklace. Graduated style. Length, about 24 in. Gold plated clasp. Illustration in center shows actual size of beads. Shipping weight, 4 oz.. **75c**

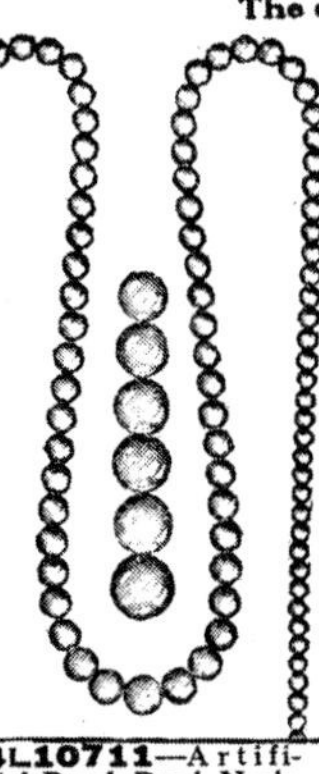

4L10711—Artificial Pearl Bead Necklace. Fine quality small size beads, graduated style, not wax filled, but practically indestructible, coated on the outside with a fine pearly substance. Length, about 16 in. Clasp is 14-karat solid white gold. Illustration of section in center shows actual size of beads. Shpg. wt., 3 oz. **$4.50**

4L10713
Gold Filled Earrings, for unpierced ears. Fine enamel pearl knob and tassel. Shipping weight, 3 ounces.
Pair, **$3.95**

4L10715
Genuine Red Coral Bead Necklace. Illustration shows actual size of beads. Graduated style. Gold plated clasp. Length, about 16 inches. Shpg. wt., 3 ounces. **$1.00**

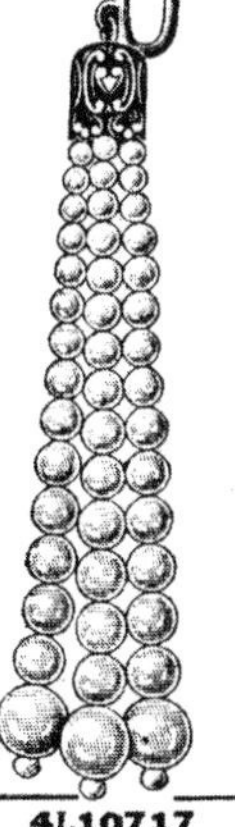

4L10717
Gold Filled Earrings, for unpierced ears. Fine enamel pearl knob and tassel. Shipping wt., 3 ounces.. **$5.00**

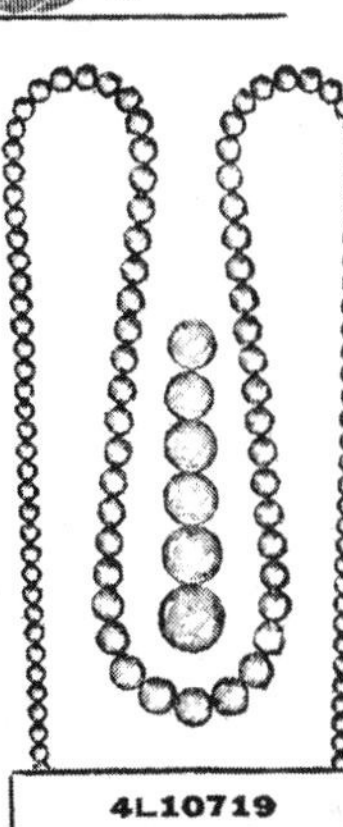

4L10719
Artificial Pearl Bead Necklace. Fine quality; not coated; small size beads; fine Oriental luster; graduated style. Length, about 16 in. Clasp is 10-karat solid white gold. Illustration shows actual size of beads. Shipping wt., 3 ounces...... **$4.50**

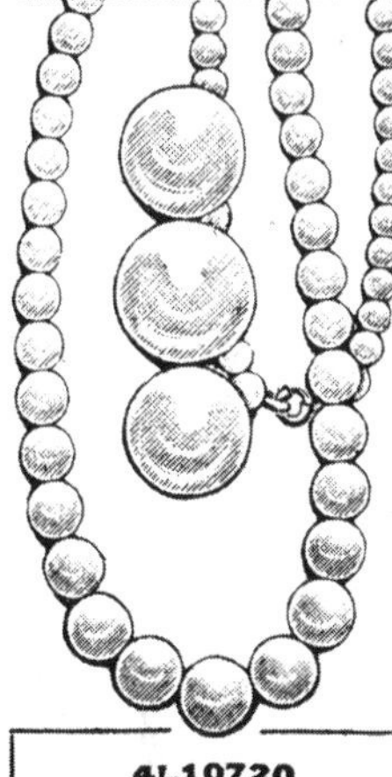

4L10720
Good Quality Artificial Pearl Bead Necklace. Graduated style. Length, about 27 inches. Gold plated clasp. Illustration in center shows actual size of beads. Shipping weight, 4 ounces............ **98c**

Good Quality Glass Bead Necklace. Length, about 24 inches. Graduated style. Gold plated clasp. Shipping weight, 5 ounces.
4L10726 — Amber (yellow) color..... **$1.25**
4L10728 — Amethyst (purple) color..... **$1.25**

Good Grade Colored Glass Bead Necklace. Graduated style. Fancy clasp. Length, about 26 inches. Shpg. wt., 5 oz.
4L10730 — Amethyst (purple) color **87c**
4L10732—Amber (yellow) color...... **87c**

4L10735
Good Quality Imitation Black Jet and Crystal White Glass Bead Necklace. Length, about 37 inches, including tassel. Shipping weight, 5 ounces......... **95c**

4L10737
Good Quality Imitation Black Jet Glass Bead Necklace. Length, about 35 inches, including tassel. Shipping weight, 5 ounces.
$1.10

Good Quality Glass Bead Necklace. Length, about 37 inches, including tassel. Shipping weight, 5 ounces.
4L10739 — Light red color...... **$1.10**

4L10741—Good Quality Glass Bead Necklace. Ruby (red color) and crystal combination. Length, about 27 inches. Graduated style, gold plated clasps. Shipping wt., 5 oz.. **$1.00**

4L10743 — Good Quality Glass Bead Necklace. Bright cherry red color. Length, about 33 inches, not including tassel. Shipping weight, 5 ounces.......... **$1.75**

Large illustration in center shows actual size of beads.

Seamless Wedding Rings

Shipping weight of rings, except 4L12620, 3 ounces.

Prices Include Engraving. Illustrations Show Exact Size.

2¾ dwt. Solid gold. English style.
4L12600 14-karat... **$3.00**
4L12602 18-karat... **$3.75**

4L12616 22-karat, 3 dwt. Solid gold. English style. **$4.95**

4L12610–10-k. solid gold, 3 dwt. **$2.50**
4L12612–14-k. solid gold, 3 dwt. **$3.30**
4L12614–18-k. solid gold, 3 dwt. **$4.15**

4L12618 22-karat, 4 dwt. Solid gold. English style. **$6.50**

4L12626–10-k. solid gold, 2 dwt... **$1.75**
4L12628–14-k. solid gold, 2 dwt... **$2.25**
4L12630–18-k. solid gold, 2 dwt... **$2.75**

4L12619 14-Karat Solid Green Gold English Style Wedding Ring. Hand engraved.... **$3.50**

Bride and Groom Ring Set The Latest Idea

4L12620 — Bride and Groom Set. The bride's ring is made of 18-karat solid white gold, English style, hand engraved, platinum design. The groom's ring is 18-karat solid gold, English style, bright polish yellow gold. Shipping weight, 4 ounces. Complete set.............. **$10.00**

Hand Engraved Solid Green Gold English Style Wedding Ring.
4L12622 14-karat solid gold. **$3.90**
4L12624 18-karat solid gold. **$4.50**

Solid Platinum Wedding Ring, engraved, three, five, or ten small regular cut diamonds according to choice. **Be sure to state correct size wanted in 4L12615 to 4L12619.**
4L12615 –With three diamonds. **$32.25**
4L12617–With five diamonds.... **$38.45**
4L12619–With ten diamonds.... **$51.50**

4 dwt. Solid gold. English style.
4L12638 14-karat... **$4.50**
4L12640 18-karat... **$5.50**

5 dwt. Solid gold. English style.
4L12642 14-karat... **$5.45**
4L12644 18-karat... **$6.75**

Light Weight Solid Gold Wedding Ring. English style.
4L12632–14-karat solid gold..... **$2.55**
4L12634–18-karat solid gold..... **$3.05**
4L12636–22-karat solid gold..... **$3.50**

4L12646 22-karat, 4 dwt. Solid gold. Oval style...... **$6.50**

State Size. **State Lettering.**

We sell only 10, 14, 18 and 22-karat, and guarantee them in every respect. Positively only fine quality pure gold used. Made from one piece. The true wedding ring should be a continuous circle without joint or seam. We cannot fill your order for a ring unless you let us know the size wanted. Sizes, 5 to 13. Misses' rings, sizes 5 to 10. See page 399 to find the exact size you want. Ring 4L12600 shows how the wedding rings are engraved.

Hand engraved. White gold. English style.
4L12648 14-karat........ **$3.85**
4L12650 18-karat........ **$5.00**

4L12652—Solid Platinum English Style Wedding Ring, hand engraved.... **$18.00**

4L12654 — 18-Karat Solid White Gold English Style Wedding Ring, hand engraved. **$6.50**

4L12656 18-Karat Solid White Gold Flat Style Wedding Ring, hand engraved. Heavy weight.......... **$8.00**

4L12658 10-Karat Solid Gold Ring, oval band. Sizes, 5 to 10... **$1.15**

4L12660 Light Weight 10-Karat Solid Gold English Style Wedding Ring. Sizes, 5 to 13. **$1.50**

4L12662 10-Karat Solid Gold Band Ring. Sizes, 5 to 10. **$1.45**

4L12664 10-Karat Solid Gold Ring. Sizes, 5 to 13... **$1.60**

Ladies' and Misses' Solid Gold Rings

4L12674 10-Karat Solid Gold Ring, genuine red coral cameo. Sizes, 4 to 8. **$3.00**

4L12676 10-karat solid gold, colored gold ornamentation, genuine red coral cameo set. Sizes, 5 to 10..... **$2.55**

4L12678—10-Karat Solid Gold Initial Ring, black enameled top. **State initial desired.** Sizes, 5 to 10. **$2.60**
4L12680—Same as above, but 14-karat solid gold.............. **$3.75**

4L12682 10-karat solid gold, set with genuine red coral cameo. Sizes, 5 to 10.... **$4.20**

4L12684 10-karat solid gold, hand chased, Roman finish, set with genuine red coral cameo. Sizes, 5 to 10........ **$5.40**

4L12696 10-Karat Solid Gold Ring, ruby red color set. Sizes, 5 to 10. **$2.45**

4L12698 10-Karat Solid Gold Ring, set with genuine pearl. Sizes, 4 to 8.... **$2.50**

4L12700 14-Karat Solid White Gold Ring, amethyst purple color set. Sizes, 4 to 8... **$2.75**

4L12702 10-Karat Solid Gold Ring, synthetic pink sapphire set. Sizes, 4 to 8. **$3.00**

4L12666 10-Karat Solid Gold Ring, genuine changeable opal. Sizes, 5 to 10...... **$1.10**

4L12668 10-Karat Solid Gold Ring, ruby red color set. Sizes, 5 to 10. **$1.35**

4L12686 10-karat solid gold, genuine pink shell cameo. Sizes, 5 to 10. **$1.65**

4L12688 Misses' 10-Karat Solid Gold Ring, genuine pink shell cameo. Sizes, 5 to 10....... **$1.80**

4L12690 — Ladies' 10-karat solid gold, genuine pink shell cameo set. Sizes, 5 to 10. **$1.70**

4L12692 Ladies' 10-karat solid gold, set with genuine pink shell cameo. Sizes, 5 to 10........ **$3.35**

4L12694 Ladies' 10-karat solid gold, genuine pink shell cameo, colored gold ornamentation. Sizes, 5 to 10...... **$4.80**

4L12704 10-Karat Solid Gold Ring, set with genuine bloodstone, green with red spots. Sizes, 4 to 8. **$3.00**

4L12706 10-Karat Solid Gold Ring, set with genuine reddish brown sardonyx. Sizes, 4 to 8...... **$3.00**

4L12671 Misses' 10 Karat Solid Gold Ring, white gold ornamented, set with blue color sapphire. Sizes, 5 to 10...... **$2.10**

4L12672 10-Karat Solid Gold Ring, set with genuine dark red garnet. Sizes, 5 to 10. **$1.55**

Merchandise Up to Your Expectations and Even Better is assured in buying from these pages. We do not have the opportunity to meet you personally so must depend on our merchandise to speak for itself. If ever we disappoint you, let us make good on our guarantee.

Shipping weight of rings, except 4L12620, 3 ounces.

4L12708 10-Karat Solid Gold Ring, set with turquoise light blue color set. Sizes, 4 to 8. **$3.50**

4L12710 10-Karat Solid Gold Ring, set with genuine cultured pearl. Sizes, 5 to 10. **$3.25**

Where we describe a setting as ruby red color, sapphire blue color, amethyst purple color, etc., we wish it understood that these are the finest artificial stones made to imitate the gem mentioned.

4L12714 10-Karat Solid Gold Ring, set with genuine bloodstone, green with red spots. Sizes, 5 to 10...... **$1.60**

4L12716 10-Karat Solid Gold Ring, ruby red color set. Sizes, 5 to 10. **$1.70**

4L12718 Misses' 10-Karat Solid Gold Ring, genuine pearl set. Sizes, 5 to 10............ **$2.50**

4L12720 10-Karat Solid Gold Ring, set with reddish brown sardonyx carbuncle. Sizes, 4 to 8. **$2.10**

4L12722 10-Karat Solid Gold Ring, ruby doublet red color set. Sizes, 4 to 8. **$2.50**

4L12724 10-Karat Solid Gold Ring, set with red synthetic rubies. Sizes, 5 to 10........ **$1.95**

4L12726 10-Karat Solid Gold Ring, set with synthetic red ruby. Sizes, 5 to 10...... **$1.85**

4L12728 14-Karat Solid Gold Ring, set with golden yellow sapphire. Sizes, 5 to 10..... **$3.25**

4L12730 14-Karat Solid Gold Ring, blue synthetic sapphire set. Sizes, 5 to 10...... **$4.50**

10 and 14 Karat Seal, Stone Set and Initial Rings

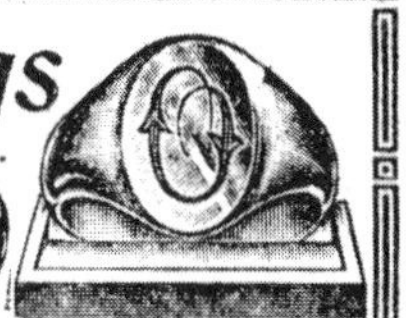

Unless otherwise stated, all rings on this page are bright polish. **Shipping weight of rings, 3 ounces.**

4L12900—10-Karat Solid Gold Initial Ring, black enamel top with gold encrusted initial. **$4.85**

4L12901—Same style as above but in 14-karat solid gold. **$6.75** Sizes, 6 to 13. **State size and initial wanted.**

4L12902 10-Karat Solid Gold Seal Ring, bright polish. Sizes, 4 to 8. **State size. $1.35**

4L12904 10-Karat Solid Gold Seal Ring. Sizes, 4 to 8. **State size. $1.75**

4L12906 10-Karat Solid Gold Seal Ring, bright polish. Sizes, 4 to 8. **State size. $1.85**

4L12908 10-Karat Solid Gold Ring, bright polish, engraved with any letter. Sizes, 5 to 10. **$2.15**

4L12910 10-Karat Solid Gold Seal Ring, bright polish. Sizes, 5 to 10. **$2.35**

4L12912 10-Karat Solid Gold Seal Ring, bright polish. Sizes, 5 to 10. **$2.55**

4L12914 10-Karat Solid Gold Seal Ring, bright polish. Sizes, 5 to 10. **State size $2.65**

4L12916 14-Karat Solid Gold Seal Ring, bright polish. Sizes, 7 to 13. **State size.....$3.30**

4L12918 10-Karat Solid Gold Seal Ring, green gold leaf. Sizes, 5 to 10. **State size. $2.00**

4L12920 14-Karat Solid Gold Seal Ring, bright polish. Sizes, 5 to 13. **State size $3.25**

4L12922 Men's 10-Karat Solid Gold Seal Ring, bright polish. Sizes, 7 to 13. **State size $4.20**

4L12924 Men's 10-Karat Solid Gold Seal Ring, bright polish. Sizes, 7 to 13. **State size $4.15**

4L12926 14-Karat Solid Gold Seal Ring, bright polish. Sizes, 7 to 13. **State size $4.25**

4L12928 10-Karat Solid Gold Seal Ring, bright polish, heavy weight. Sizes, 7 to 13. **State size. $4.60**

4L12930 14-Karat Solid Gold Seal Ring, heavy weight, bright polish. Sizes, 7 to 13. **State size. $5.00**

4L12932 14-Karat Solid Gold Seal Ring, bright polish. Sizes, 7 to 13. **State size $5.25**

4L12934 10-Karat Solid Gold Roman Finish, set with small genuine regular cut diamond; one or two letters engraved. **State letters.** Sizes, 5 to 13. **State size...$8.85**

Seal Rings in 10 and 14-Karat Solid Gold. Hand Engraved Initials and Hand Chased Sides.

4L12936 Seal Ring, 10-karat solid gold, green color or dull finish seal. Sizes, 5 to 10. **State size........$2.55**

4L12938 Same as above, but 14-karat solid gold. **$4.25**

4L12940 Seal Ring, 10-karat solid gold, green color or dull finish seal. Sizes, 5 to 13. **State size.......$3.40**

4L12942 Same as above, but 14-karat solid gold. **$4.00**

4L12944 Seal Ring, 10-karat solid gold, satin finish. Sizes, 7 to 12. **State size........$3.50**

4L12946 Same as above, but 14-karat solid gold. **$4.50**

4L12948 Seal Ring, 10-karat solid gold, green color or dull finish seal. Sizes, 7 to 12. **State size.........$4.25**

4L12950 Same as above, but 14-karat solid gold. **$5.00**

4L12952 Seal Ring, 10-kara solid gold, green color or dull finish seal. Sizes, 5 to 10. **State size..........$4.35**

4L12954 Same as above, but 14-karat solid gold. **$5.75**

4L12956 Seal Ring, 10-karat solid gold, green color or dull finish seal. Sizes, 7 to 12. **State size........$4.90**

4L12958 Same as above, but 14-karat solid gold. **$5.85**

4L12960 Seal Ring, 10-karat solid gold, green color or dull finish seal. Sizes, 7 to 12. **State size........$6.20**

4L12962 Same as above, but 14-karat solid gold. **$7.50**

4L12964 Seal Ring. 10-karat solid gold, green color dull finish seal. Sizes, 7 to 12. **State size. $6.50**

4L12966—Same as above, but 14-karat solid gold.... **$7.75**

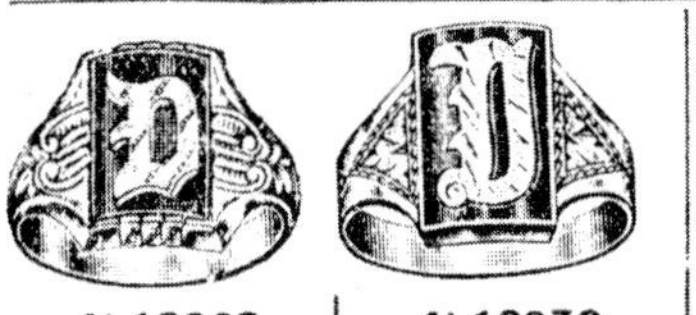

4L12968 Men's 10-Karat Solid Gold Ring, Initial on genuine black onyx. Sizes, 5 to 13. **State size..........$3.50**

4L12970 Men's 10-Karat Solid Gold Ring, Initial on genuine black onyx. Sizes, 5 to 13. **State size.........$6.00**

Men's Fine Quality 10-Karat Solid Gold Initial and Emblem Rings.

We supply the following catalog numbers, 4L12968, 4L12970, 4L12972 and 4L12974, in any initial or any of the following emblems on these rings: Masonic, Odd Fellows, Knights of Pythias, Modern Woodmen, Eagles, Woodmen of the World and Elks. **Be sure to state emblem or initial desired.** The initials on these rings are solid gold set on genuine black onyx top.

ILLUSTRATIONS SHOW EXACT SIZE OF RINGS.

When ordering rings, **be sure to mention size wanted.** See ring measuring chart on page 399. Rings showing engraving will be engraved with any letter without extra charge. **Mention letter.** Where we describe a setting as ruby red color, sapphire blue color, amethyst purple color, etc., we wish it understood that these settings are clever imitations. Where the stone is genuine we so state in the description. Where described as doublets, as the name implies, the setting is made of two pieces, genuine stone fixed on front and a colored glass like material on the back.

4L12972—Men's 10-karat Solid Gold Ring, initial on genuine black onyx. Sizes, 7 to 13. **State size. $8.75**

4L12974—Men's 10-Karat Solid Gold Ring, Initial on genuine black onyx. Initial set with 6 rose diamonds. Sizes, 7 to 13. **State size........$17.00**

4L12976 14-Karat Solid Gold Ring, set with genuine bloodstone; green color with red spots. Sizes, 5 to 10. **State size. $3.50**

4L12978 Men's 10-Karat Solid Gold Ring, set with genuine moss agate with green moss in it. Sizes, 7 to 12. **State size $3.65**

4L12980 10-Karat Solid Gold Ring, set with genuine purple amethyst. Sizes, 5 to 10. **State size. $6.85**

4L12982 Men's 10-Karat Solid Gold Ring, set with genuine dark red garnet carbuncle. Sizes, 7 to 12. **State size. $5.50**

4L12984 Men's 10-Karat Solid Gold Ring, ruby red color set. Sizes, 7 to 13. **State size $5.75**

4L12986 Men's 10-Karat Solid Gold Ring, ruby red color set. Sizes, 7 to 13. **State size $5.25**

4L12988 Men's 14-Karat Solid Gold Ring, set with genuine reddish brown sardonyx. Sizes, 7 to 13. **State size. $6.50**

4L12990 Men's 10-Karat Solid Gold Ring, set with genuine bloodstone; green stone with red spots. Sizes, 7 to 12........**$5.75**

4L12992 Men's 10-Karat Solid Gold Ring, set with genuine sardonyx; reddish brown stone. Sizes, 7 to 12. **State size. $5.50**

4L12994 14-Karat Solid Gold Ring, set with genuine bloodstone. Green stone with red spots. **State size....$7.25**

4L12996 10-Karat Solid Gold Ring, set with genuine sardonyx reddish brown stone. **State size.........$6.25**

4L12998 10-Karat Solid Gold Ring, set with genuine bloodstone. Green stone with red spots........**$5.35**

4L13001 Men's 10-Karat Solid Gold Ring, set with genuine dark red garnet.........**$6.50**

4L13002 Men's 10-Karat Solid Gold Ring, genuine amethyst purple color set.......**$4.50**

4L13004 Men's 10-Karat Solid Gold Ring, set with genuine bloodstone; green stone with red spots........**$7.75**

4L13006 10-Karat Solid Gold Ring, set with genuine purple amethyst. **State size....$8.50**

4L13008 Men's 10-Karat Solid Gold Ring, set with genuine black onyx. **State size...$9.00**

4L13010 Ladies' 14-Karat Solid Gold Ring, colored gold ornamentation, set with genuine pink shell cameo. **$4.95**

4L13012 14-Karat Solid Gold Ring, 18-karat solid white gold border, set with genuine cornelian cameo; brown stone with white head.......**$5.25**

4L13014 14-Karat Solid Gold Ring, 18-karat solid white gold border, set with genuine cornelian cameo; brown stone with white head........**$6.50**

4L13016 Ladies' 14-Karat Solid Gold Ring, set with genuine pink shell cameo.. **$5.75**

4L13018 14-Karat Solid Gold Ring, white gold border, set with genuine cornelian cameo; brown stone with white head.. **$8.00**

4L13020 14-Karat Solid Gold Ring, white gold border, set with genuine red coral cameo. **$8.90**

4L13022 Ladies' 14-Karat Solid Gold Ring, set with genuine pink coral cameo, white gold ornamentation.... **$10.00**

4L13024 Ladies' 14-Karat Solid Gold Ring, set with genuine pink coral cameo, colored gold ornamentation... **$9.50**

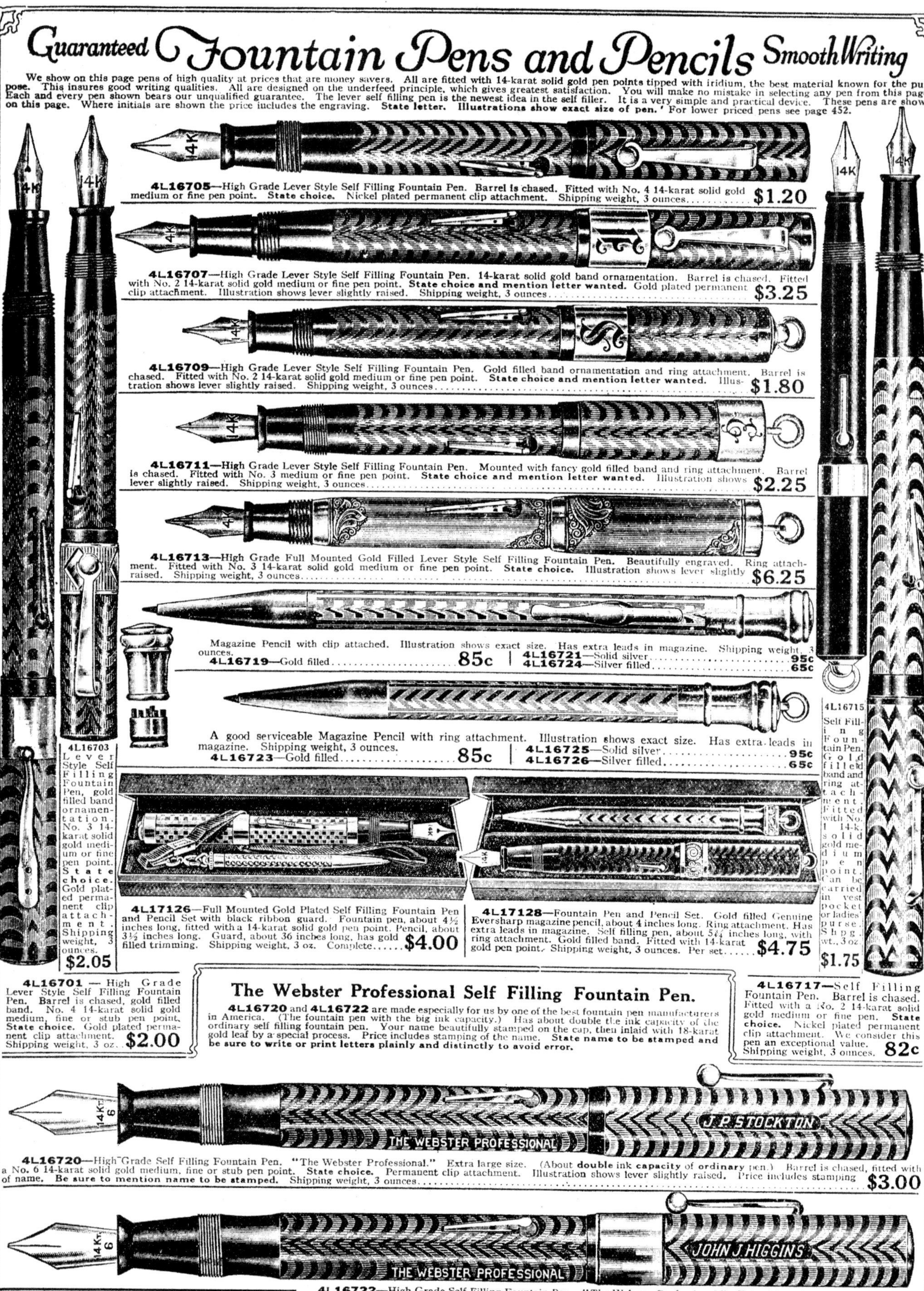

Guaranteed Fountain Pens and Pencils Smooth Writing
We show on this page pens of high quality at prices that are money savers. All are fitted with 14-karat solid gold pen points tipped with iridium, the best material known for the purpose. This insures good writing qualities. All are designed on the underfeed principle, which gives greatest satisfaction. You will make no mistake in selecting any pen from this page. Each and every pen shown bears our unqualified guarantee. The lever self filling pen is the newest idea in the self filler. It is a very simple and practical device. These pens are shown on this page. Where initials are shown the price includes the engraving. State letter. Illustrations show exact size of pen. For lower priced pens see page 452.
14K
4L16705—High Grade Lever Style Self Filling Fountain Pen. Barrel is chased. Fitted with No. 4 14-karat solid gold medium or fine pen point. State choice. Nickel plated permanent clip attachment. Shipping weight, 3 ounces. $1.20
4L16707—High Grade Lever Style Self Filling Fountain Pen. 14-karat solid gold band ornamentation. Barrel is chased. Fitted with No. 2 14-karat solid gold medium or fine pen point. State choice and mention letter wanted. Gold plated permanent clip attachment. Illustration shows lever slightly raised. Shipping weight, 3 ounces. $3.25
4L16709—High Grade Lever Style Self Filling Fountain Pen. Gold filled band ornamentation and ring attachment. Barrel is chased. Fitted with No. 2 14-karat solid gold medium or fine pen point. State choice and mention letter wanted. Illustration shows lever slightly raised. Shipping weight, 3 ounces. $1.80
4L16711—High Grade Lever Style Self Filling Fountain Pen. Mounted with fancy gold filled band and ring attachment. Barrel is chased. Fitted with No. 3 medium or fine pen point. State choice and mention letter wanted. Illustration shows lever slightly raised. Shipping weight, 3 ounces. $2.25
4L16713—High Grade Full Mounted Gold Filled Lever Style Self Filling Fountain Pen. Beautifully engraved. Ring attachment. Fitted with No. 3 14-karat solid gold medium or fine pen point. State choice. Illustration shows lever slightly raised. Shipping weight, 3 ounces. $6.25
Magazine Pencil with clip attached. Illustration shows exact size. Has extra leads in magazine. Shipping weight, 3 ounces.
4L16719—Gold filled. 85c
4L16721—Solid silver. 95c
4L16724—Silver filled. 65c
A good serviceable Magazine Pencil with ring attachment. Illustration shows exact size. Has extra leads in magazine. Shipping weight, 3 ounces.
4L16723—Gold filled. 85c
4L16725—Solid silver. 95c
4L16726—Silver filled. 65c
4L16703 Lever Style Self Filling Fountain Pen, gold filled band ornamentation. No. 3 14-karat solid gold medium or fine pen point. State choice. Gold plated permanent clip attachment. Shipping weight, 3 ounces. $2.05
4L16715 Self Filling Fountain Pen. Gold filled band and ring attachment. Fitted with No. 1 14-k. solid gold medium pen point. Can be carried in vest pocket or ladies' purse. Shpg. wt., 3 oz. $1.75
4L17126—Full Mounted Gold Plated Self Filling Fountain Pen and Pencil Set with black ribbon guard. Fountain pen, about 4½ inches long, fitted with a 14-karat solid gold pen point. Pencil, about 3½ inches long. Guard, about 36 inches long, has gold filled trimming. Shipping weight, 3 oz. Complete. $4.00
4L17128—Fountain Pen and Pencil Set. Gold filled Genuine Eversharp magazine pencil, about 4 inches long. Ring attachment. Has extra leads in magazine. Self filling pen, about 5¼ inches long, with ring attachment. Gold filled band. Fitted with 14-karat gold pen point. Shipping weight, 3 ounces. Per set. $4.75
4L16701 — High Grade Lever Style Self Filling Fountain Pen. Barrel is chased, gold filled band. No. 4 14-karat solid gold medium, fine or stub pen point. State choice. Gold plated permanent clip attachment. Shipping weight, 3 oz. $2.00
The Webster Professional Self Filling Fountain Pen.
4L16720 and 4L16722 are made especially for us by one of the best fountain pen manufacturers in America. (The fountain pen with the big ink capacity.) Has about double the ink capacity of the ordinary self filling fountain pen. Your name beautifully stamped on the cap, then inlaid with 18-karat gold leaf by a special process. Price includes stamping of the name. State name to be stamped and be sure to write or print letters plainly and distinctly to avoid error.
4L16717—Self Filling Fountain Pen. Barrel is chased. Fitted with a No. 2 14-karat solid gold medium or fine pen. State choice. Nickel plated permanent clip attachment. We consider this pen an exceptional value. Shipping weight, 3 ounces. 82c
14Kt 6
J.P. STOCKTON
THE WEBSTER PROFESSIONAL
4L16720—High Grade Self Filling Fountain Pen. "The Webster Professional." Extra large size. (About double ink capacity of ordinary pen.) Barrel is chased, fitted with a No. 6 14-karat solid gold medium, fine or stub pen point. State choice. Permanent clip attachment. Illustration shows lever slightly raised. Price includes stamping of name. Be sure to mention name to be stamped. Shipping weight, 3 ounces. $3.00
JOHN J. HIGGINS
THE WEBSTER PROFESSIONAL
4L16722—High Grade Self Filling Fountain Pen. "The Webster Professional." Extra large size. (About double ink capacity of ordinary pen.) Barrel is chased, gold filled band. Fitted with a No. 6 14-karat solid gold medium, fine or stub pen point. State choice. Permanent clip attachment. Illustration shows lever slightly raised. Price includes stamping of name. Be sure to mention name to be stamped. Shipping weight, 3 ounces. $3.25

Eight-Piece Tea Set. Alaska Silverlike Metal. Six teaspoons, butter knife and sugar shell. Shipping weight, 12 oz.
5L1309—Laval Pattern.
5L1319—Colonial Pattern.
5L1329—Tipped Pattern.
5L1389—Brynathyn Pattern.
Per set **90c**

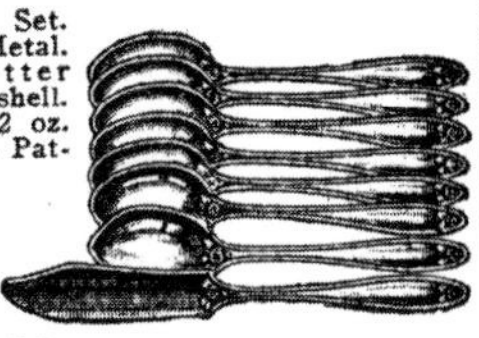

Eight-Piece Set in cloth lined paper box. Shipping weight, 1¼ pounds.
5L1306—Laval Pattern.
5L1316—Colonial Pattern.
5L1326—Tipped Pattern.
5L1336—Brynathyn Pattern.
Per set **$1.15**

Guaranteed to Give Lifetime Service.
Alaska Silverlike Metal Tableware Looks Like Solid Silver.

In color and appearance it is difficult to distinguish from solid silver. Yet Alaska Silverlike Metal Tableware contains no silver. It is not plated. It is made of the same metal throughout and, with ordinary use and care, will last a lifetime.

Like sterling silver, Alaska Silverlike Metal Tableware takes a high polish. It is easy to clean, but, like solid silver or silver plate, it should not be left in vinegar or foods which contain acid or salt.

GUARANTEE—We guarantee Alaska Silverlike Metal Tableware for the term of your natural lifetime. You make no mistake when you buy Alaska Silverlike Metal Tableware.

TIPPED PATTERN
LAVAL PATTERN
COLONIAL PATTERN
BRYNATHYN PATTERN

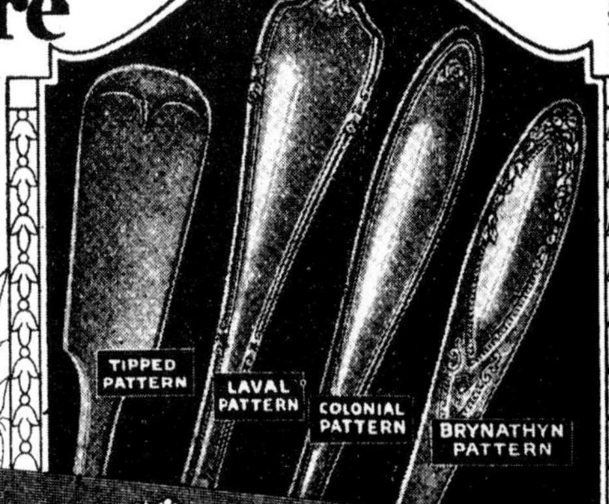

ENGRAVING—On account of the hardness of the metal we do not engrave Alaska Silverlike Metal Tableware.

Article	Shipping Weight		Laval Pattern	Colonial Pattern	Tipped Pattern	Brynathyn Pattern
Teaspoons. Per set of six	8 oz.	$0.58	5L1300	5L1310	5L1320	5L1330
Tablespoons. Per set of six	15 oz.	1.16	5L1301	5L1311	5L1321	5L1331
Medium Forks. Per set of six	1¼ lbs.	1.16	5L1302	5L1312	5L1322	5L1332
Medium Knives. Per set of six	1½ lbs.	1.60	5L1303	5L1313	5L1323	5L1333
Sugar Shell. Each	3 oz.	.17	5L1304	5L1314	5L1324	5L1334
Butter Knife. Each	3 oz.	.18	5L1305	5L1315	5L1325	5L1335

5L1346—Salt Shaker. Shipping weight, 2 ounces. Each **17c**
5L1347—Pepper Shaker. Shipping weight, 2 ounces. Each **17c**

For $5.15 or $6.00 We Furnish 28-Piece Tableware Set in Four Patterns, With Cloth Lined Box or Without Box. 6 Teaspoons, 6 Tablespoons, 6 Knives, 6 Forks, 1 Sugar Shell, 1 Butter Knife and Pair of Salt and Pepper Shakers.

5L1307—28-Piece Set, Laval Pattern, without box **$5.15**
5L1317—28-Piece Set, Colonial Pattern, without box **5.15**
5L1327—28-Piece Set, Tipped Pattern, without box **5.15**
5L1337—28-Piece Set, Brynathyn Pattern, without box **5.15**
Shipping weight, 4 pounds.

5L1308—28-Piece Set, Laval Pattern, with box **$6.00**
5L1318—28-Piece Set, Colonial Pattern, with box **6.00**
5L1328—28-Piece Set, Tipped Pattern, with box **6.00**
5L1338—28-Piece Set, Brynathyn Pattern, with box **6.00**
Shipping weight, 6 pounds.

A Set With Your Initial

This illustration shows the style of lettering on this set. You may have **any initial you desire.**

We ship each 28-Piece Set of Initialed Alaska Metal Tableware in this attractive case, fitted with pull drawer.

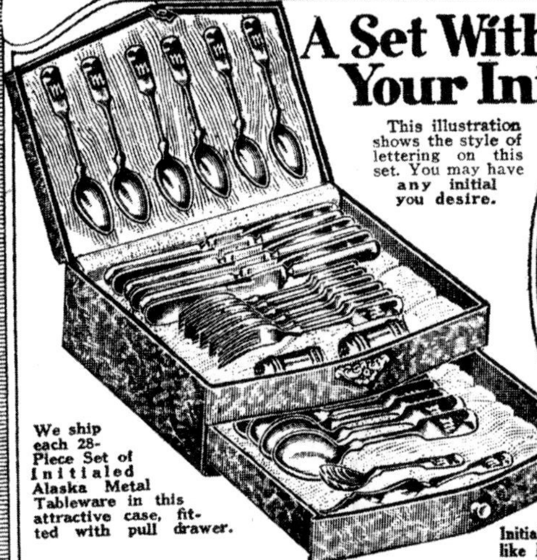

Initialed Alaska Silverlike Metal Tableware.

Each piece comes stamped with the initial you desire, except the knives and salt and pepper shakers. **Write plainly and distinctly the letter wanted,** so that no mistake can be made.

Like sterling silver, Alaska Silverlike Metal Tableware takes a high polish. It is easy to clean, but, like solid silver or silver plate, it should not be left in vinegar or foods which contain acid or salt.

Initialed Alaska Tableware. Twenty-eight pieces. 6 Teaspoons, 6 Tablespoons, 6 Knives, 6 Forks, 1 Sugar Shell, 1 Butter Knife and pair of Salt and Pepper Shakers. Without case. Shipping weight, 3 pounds.
5L1481—Per set of 28 pieces, without case **$5.25**

Initialed Alaska Tableware. Twenty-eight pieces. Complete in cloth lined box, as illustrated. Shipping weight, 6 pounds.
5L1480—Per set, in case **$6.50**

	Order by Number. Any Initial You Desire. State Initial Wanted.		Shpg. Wt.
5L1482	Teaspoons. Set of six	$0.60	8 oz.
5L1483	Tablespoons. Set of six	1.20	11 oz.
5L1484	Medium Forks. Set of six	1.20	11 oz.
5L1344	Plain Knives. Set of six	1.55	1½ lbs.
5L1488	Sugar Shell. Each	.18	4 oz.
5L1489	Butter Knife. Each	.20	4 oz.

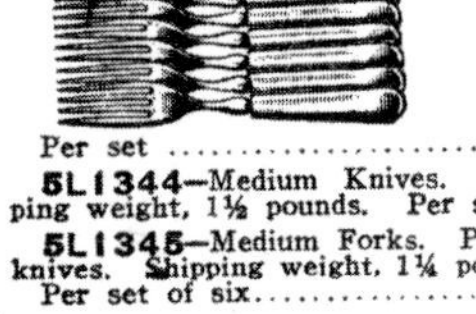

5L1343 — Knife and Fork Set. Six Alaska Silverlike Metal plain handle medium knives, 9⅛ inches long, and six Alaska Silverlike Metal plain handle forks, 7¼ inches long. Shipping weight, 2 pounds 9 ounces.
Per set **$3.10**

5L1344—Medium Knives. Plain handle. Shipping weight, 1½ pounds. Per set of six **$1.55**

5L1345—Medium Forks. Plain handle, to match knives. Shipping weight, 1¼ pounds.
Per set of six **$1.55**

Electroline.

This modern silver cleaner and preserver of silverware comes in paste form. 4½-ounce size. Shpg. wt., 1 lb.
5L574 .... **22c**

Silver Polish.

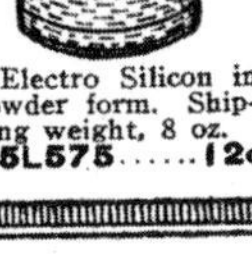

Electro Silicon in powder form. Shipping weight, 8 oz.
5L575 ..... **12c**

Genuine Rogers Nickel Silver Tableware

Genuine Rogers Nickel Silver Tableware is a composition metal, the same metal through and through. It resembles silver, though it has no silver in it, and like solid silver it should not be left standing for any length of time in fatty or acid foods. With proper care it will keep bright looking.

Order by Number	Shpg. Wt.		Plain Pattern	Montrose Pattern
Teaspoons. Set of six	8 oz.	$0.50	5L1902	5L2002
Dessert Spoons. Set of six.	12 oz.	.95	5L1908	5L2008
Tablespoons. Set of six	15 oz.	1.00	5L1914	5L2014
Medium Forks, flat handle. Set of six	15 oz.	1.00	5L1918	5L2018
Sugar Shell. Each	3 oz.	.16	5L1924	5L2024
Butter Knife. Each	3 oz.	.17	5L1926	5L2026
Medium Knives, round handle. Set of six	1½lbs.	1.55	5L1932	5L2032

Montrose Pattern
Plain Pattern

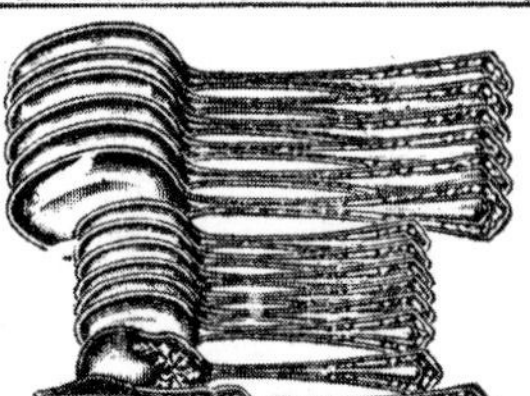
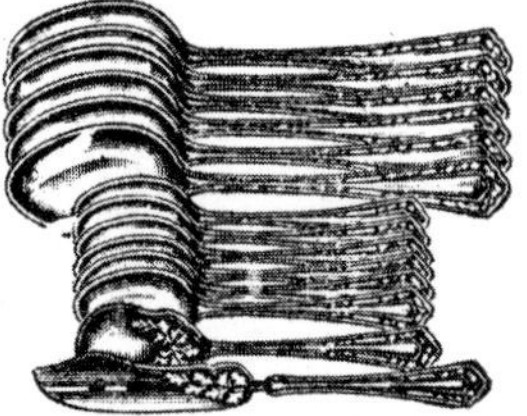

Fourteen-Piece Dinner Set. Rogers Nickel Silver. Six teaspoons, six tablespoons, sugar shell and butter knife. Shipping weight, 2 pounds.
5L1934—Plain Pattern.
5L2034—Montrose Pattern.
Per set **$1.80**

Fourteen-Piece Set in cloth lined paper box. Shipping weight, 2¾ lbs.
5L1935—Plain Pattern.
5L2035—Montrose Pattern.
Per set, in presentation box **$2.10**

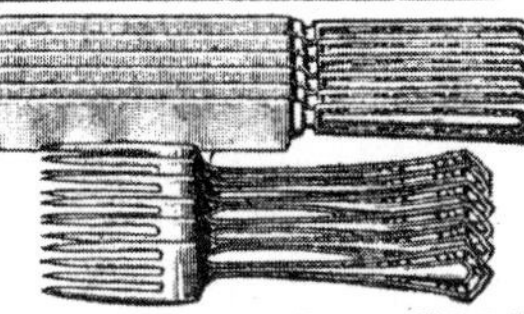

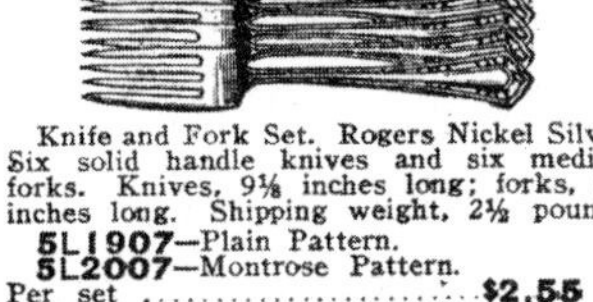

Knife and Fork Set. Rogers Nickel Silver. Six solid handle knives and six medium forks. Knives, 9⅛ inches long; forks, 7½ inches long. Shipping weight, 2½ pounds.
5L1907—Plain Pattern.
5L2007—Montrose Pattern.
Per set **$2.55**

Knife and Fork Set in cloth lined paper box. Shipping wt., 2½ lbs.
5L1905—Plain Pattern.
5L2005—Montrose Pattern.
Per set in box **$2.84**

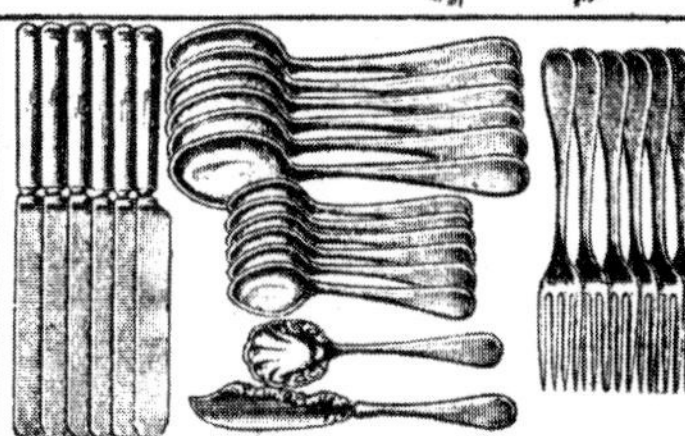

Twenty-Six Piece Genuine Rogers Nickel Silver Dinner Set. Six medium knives, six medium forks, six teaspoons, six tablespoons, sugar shell and butter knife. Shipping weight, 3 pounds.
5L1980—Plain Pattern.
5L2080—Montrose Pattern.
Per set **$4.35**

Twenty-Six Piece Set in cloth lined box. Same case as shown in 5L1308. Shpg. wt., 6 lbs.
5L1982—Plain Pattern.
5L2082—Montrose Pattern.
Per set, in presentation box **$5.20**

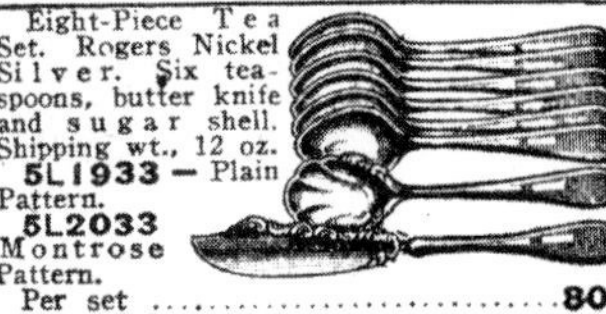

Eight-Piece Tea Set. Rogers Nickel Silver. Six teaspoons, butter knife and sugar shell. Shipping wt., 12 oz.
5L1933 — Plain Pattern.
5L2033 Montrose Pattern.
Per set **80c**

Eight Piece Tea Set in cloth lined paper box. Shipping weight, 1¼ pounds.
5L1930—Plain Pattern.
5L2030—Montrose Pattern.
Per set **$1.05**

Low Priced Silver Plated Tableware

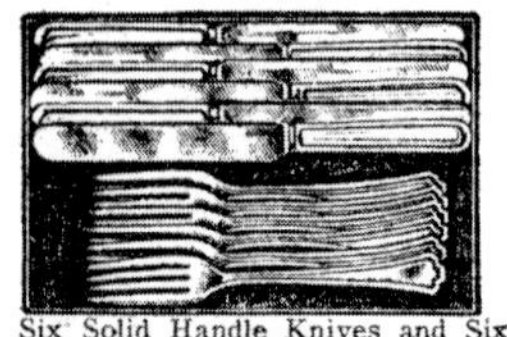

Six Solid Handle Knives and Six Flat Handle Medium Forks. Shipping weight, 2¾ pounds.
5L1735—Chatham Pattern.
5L1835—Manchester Pattern.
Per set **$3.70**

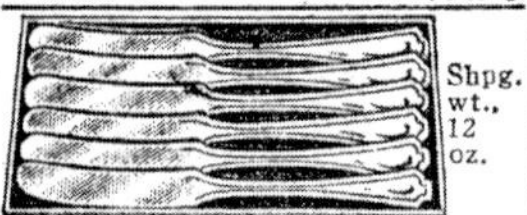

Shpg. wt., 12 oz.

Butter Spreaders.
5L1781—Chatham Pattern.
5L1881—Manchester Pattern.
Per set of six **$1.65**

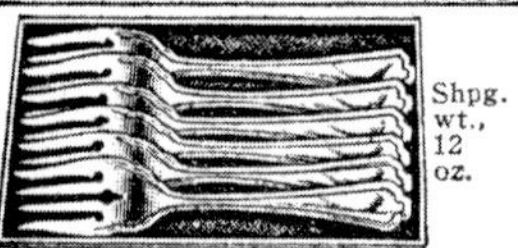

Shpg. wt., 12 oz.

Salad Forks.
5L1737—Chatham Pattern.
5L1837—Manchester Pattern.
Per set of six **$2.40**

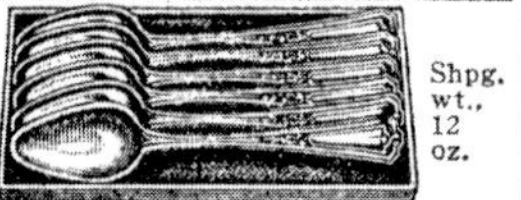

Shpg. wt., 12 oz.

Orange Spoons.
5L1742—Chatham Pattern.
5L1842—Manchester Pattern.
Per set of six **$1.79**

Shpg. wt., 8 oz.

Gravy Ladle.
5L1748—Chatham Pattern.
5L1848—Manchester Pattern.
Each **75c**

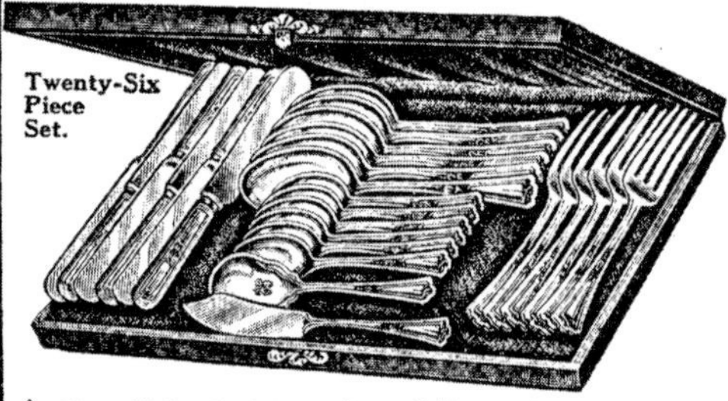

Twenty-Six Piece Set.

Six solid handle knives, six forks, six teaspoons, six tablespoons, sugar shell and butter knife. Shipping wt., 3 lbs.
5L1782—Chatham Pattern.
5L1882—Manchester Pattern.
Per set (without box) **$6.75**

A Twenty-Six Piece Set. Same as 5L1882 but with box, as illustrated, paper covered, cloth lined. Shipping weight, 5 pounds.
5L1780—Chatham Pattern.
5L1880—Manchester Pattern.
Per set........ **$7.65**

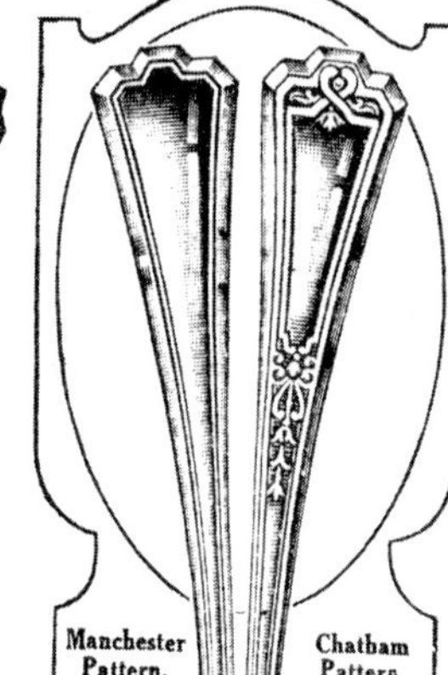
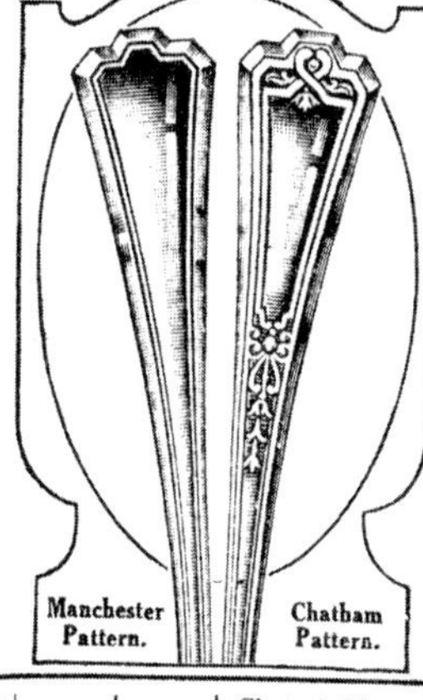
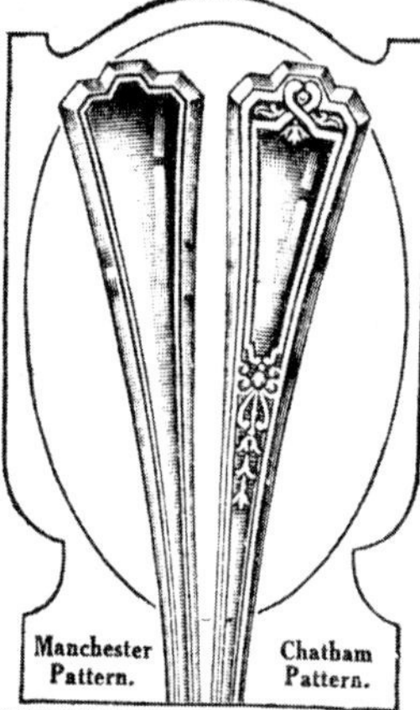

Manchester Pattern. **Chatham Pattern.**

Children's Three-Piece Set.
Consisting of knife, fork and spoon. Shipping wt., 8 ounces.
5L1776—Chatham Pattern.
5L1876—Manchester Pattern.
Per set...... **75c**

We charge 3 cents each for engraving script letters and 5 cents each for engraving Old English letters. **State initial and style of engraving desired.**

A Silver Plated Ware that will give you good service, far beyond what our low prices may lead you to expect. Order by Number	Shipping Weight	Price	Chatham Pattern, Gray Finish	Manchester Pattern, Bright Finish
Teaspoons. Set of six	8 oz.	$0.39	5L1702	5L1802
Tablespoons. Set of six	1 lb.	1.78	5L1714	5L1814
Medium Forks. Set of six	1 lb.	1.78	5L1720	5L1820
Sugar Shell. Each	4 oz.	.29	5L1724	5L1824
Butter Knife. Each	5 oz.	.32	5L1726	5L1826
Soup Spoons. Set of six	1 lb.	1.78	5L1728	5L1828
Solid Handle Knives. Set of six	1¾ lbs.	1.97	5L1732	5L1832
Individual Salad Forks. Set of six	12 oz.	2.40	5L1737	5L1837
Cream Ladle. Each	5 oz.	.60	5L1746	5L1846
Gravy Ladle. Each	8 oz.	.75	5L1748	5L1848
Pickle Fork. Each	5 oz.	.43	5L1752	5L1852
Fruit Knives. Set of six	12 oz.	1.75	5L1760	5L1860
Berry Spoon. Each	8 oz.	.88	5L1762	5L1862
Berry Spoon, gold plated bowl. Each	8 oz.	1.18	5L1763	5L1863
Cold Meat Fork. Each	6 oz.	.65	5L1764	5L1864
Cold Meat Fork, gold plated tines. Each	6 oz.	.90	5L1765	5L1865
Iced Tea Spoons. Set of six	1 lb.	1.47	5L1771	5L1871
Butter Spreaders. Set of six	12 oz.	1.65	5L1781	5L1881

Berry Spoon.
Shipping weight, 8 ounces.
5L1762—Chatham Pattern.
5L1862—Manchester Pattern.
Each **88c**

Pickle Fork.
Shipping weight, 5 ounces.
5L1752—Chatham Pattern.
5L1852—Manchester Pattern.
Each **43c**

Shpg. wt., 1 lb.

Iced Tea Spoons.
5L1771—Chatham Pattern.
5L1871—Manchester Pattern.
Per set of six **$1.47**

Shpg. wt., 12 oz.

Fruit Knives.
5L1760—Chatham Pattern.
5L1860—Manchester Pattern.
Per set of six **$1.75**

Cold Meat Fork.
Shipping weight, 6 ounces.
5L1764—Chatham Pattern.
5L1864—Manchester Pattern.
Each **65c**

Shpg. wt., 5 oz.

Cream Ladle.
5L1746—Chatham Pattern.
5L1846—Manchester Pattern.
Each **60c**

Children's Sets

Children's Three-Piece Set. Silver plated. Dull finish. Consisting of knife, 7½ inches long; fork, 6 inches long, and spoon, 5¼ inches long. Shipping weight, 10 ounces.
5L804—Per set.... **$1.82**

Children's Three-Piece Set. Silver plated. Gray finish. Knife is 7¼ inches, fork is 6 inches and teaspoon is 5⅜ in. long. Shipping weight, 8 oz.
5L808—Per set........ **75c**

Children's Three-Piece Set. Silver plated. Gray finish handle. Knife is 7¾ inches, fork is 6 inches and teaspoon 5 inches long. Shpg. wt., 12 oz.
5L823—Per set........ **89c**

Babies' Sets

Baby Cup. Silver plated satin finish. Engraved "Baby." Height, 2⅛ in. Shipping weight, 14 ounces.
5L815 .. **90c**

Baby Cup. Silver plated. Satin finish. Hand engraved ornament. Gold plated inside. Height, 2⅛ inches. Shipping weight, 12 oz.
5L803 .. **73c**

Baby Spoon. Guaranteed for fifty years. Curved handle. Salem silver plate. Length, 3½ inches. Shipping weight, 3 ounces.
5L800 .... **60c**

Baby Spoon. Silver plated. Gray finish curved handle. Length, 3½ inches. Shipping weight, 3 oz.
5L801 .... **32c**

Babies' Silver

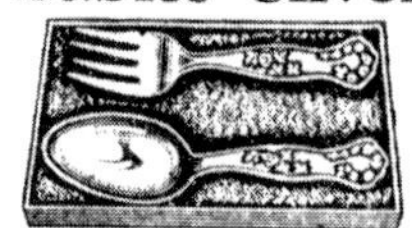

Baby Set.
Guaranteed 30 years. Baby Spoon and Fork. Silver plated. Gray finish. Spoon, 3⅝ inches long. Shipping weight, 6 ounces.
5L811—Per set..... **$1.10**

Baby Cup. Silver plated on nickel composition metal. Bright finish. Gold plated inside. Height, 2⅞ inches. Engraved with one Old English initial. **State initial.** In box. Shipping weight, 14 ounces.
5L809 .... **$1.65**

Baby Set. Silver plated. Cup bright finish, gold plated inside; 2 in. high. Spoon, 3¼ in. long. In box. Shpg. wt., 1¼ lbs.
5L810—Per set.. **$2.50**

Baby Plate. Silver plated. Bright finish center and satin ornamented border. Jack and Jill, Little Bo Peep designs, etc. Diameter, 7½ inches. Shpg. wt., 1 lb.
5L820 .... **$2.94**

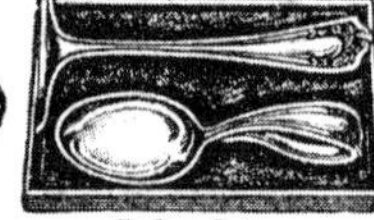

Baby Set. Guaranteed 50 years. Baby Spoon and Food Pusher. Silver plated. Gray finish. Length of spoon, 3⅜ inches. Shipping weight, 6 ounces.
5L817—Per set.... **$1.20**

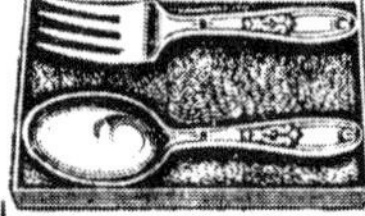

Baby Set. Silver Plated Spoon and Fork. Flat handle. Gray finish. Length of spoon, 4⅛ in. Shpg. wt., 6 oz.
5L806—Per set **56c**

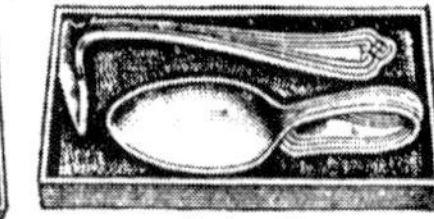

Baby Set. Guaranteed 30 years. Baby Spoon and Food Pusher. Silver plated. Gray finish. Length of spoon, 3⅝ inches. Shipping weight, 6 ounces.
5L825—Per set.. **$1.10**

Silver Plated Communion or Altar Service

Collection Plate. Silver plated. Bright polish. Diameter, 9¼ inches. Cloth lined center. Shpg. wt., 1¼ lbs.
5L7440 .... **$3.35**

Goblet. Silver plated (½ pint).
5L7432
$3.40
Gold plated inside (½ pint). Silver plated on white metal. Shpg. wt., 1 pound.
5L7434
$4.10

Individual Communion Service. Silver plated. Bright polish. Thirty-six glasses in rack. Diameter, 13½ inches. Shipping weight, 12 pounds.
5L7438—Complete.. **$18.95**
Extra Glasses only. Shipping wt., per dozen, 1½ lbs.
5L7439—Per dozen, **$1.48**

Flagon. Silver plated on white metal (3½ pts). Shpg. wt., 7 lbs.
5L7426 .. **$12.20**

Communion Flagon or Filler. Silver plated on white metal (1½ pints). Shipping weight, 2⅝ lbs.
5L7425 ...... **$6.30**

For other ecclesiastical goods see page 445.

Individual Communion Set. Silver plated. Bright polish. Thirty-six glasses in rack. Diameter, 13½ inches. Shipping weight, 12 pounds.
5L7441—Complete. **$16.45**
Extra Glasses only. Shipping weight, per dozen, 1½ pounds.
5L7439—Per dozen.. **$1.48**

Plate. Diameter, 9 inches. Silver plated on white metal. Shipping weight, 1¼ pounds.
5L7436 .... **$3.30**

Baptismal Bowls. Silver plated (1 quart).
5L7428 ...... **$5.80**
Silver plated. Gold plated inside (1 quart). Shipping weight, 2½ pounds.
5L7430 .... **$6.65**

PEARL HANDLE TABLEWARE

We offer here our Pearl Handle Tableware. The pearl used in these articles is of a beautiful luster, well shaped and finished. The blades, tines and bowls of these pieces are silver plated with the exception of the bread knife, which is good quality steel. The pearl handles are made secure by a solid silver ferrule which extends well up on the pearl, making a very neat appearing, well finished article. These pieces all come in a neat, lined paper box. You make no mistake in buying our Pearl Handle Ware if you wish a very attractive, durable gift.

Pickle Fork. Pearl handle. Solid silver ferrule, silver plated tines. Entire length, about 7⅝ in. Shpg. wt., 8 oz. **5L606 $1.04**

Berry Spoon. Pearl handle. Solid silver ferrule, silver plated bowl. Entire length, 8½ inches. Shipping weight, 12 ounces. **5L605 $1.74**

Cold Meat Fork. Pearl handle. Solid silver ferrule, silver plated tines. Entire length, 8⅞ inches. Shipping weight, 12 ounces. **5L613...$1.69**

Tea Ball. Pearl handle. Solid silver ferrule. Entire length, 6¾ inches. Silver plated ball. Shpg. wt., 8 oz. **5L608 $1.14**

Sugar Shell. Pearl handle. Solid silver ferrule, silver plated bowl. Length, 6½ inches. Shipping weight, 9 oz. **5L611 $1.00**

Pie or Cake Server. Pearl handle. Solid silver ferrule, silver plated blade. Entire length, about 9⅛ inches. Shipping weight, 14 ounces. **5L612 .... $1.74**

Knife and Fork Set. Pearl handles. Solid silver ferrules and silver plated blades. Length of knives, 8⅝ inches; forks, 7¼ inches. Shipping weight, 3½ lbs. **5L601**—Set of six knives and six forks **$17.75**

Cream Ladle. Pearl handle. Solid silver ferrule, silver plated bowl. Length, 6¾ inches. Shipping wt., 10 oz. **5L604 $1.05**

Cheese Knife. Pearl handle. Solid silver ferrule, silver plated blade. Entire length, 6⅛ in. Shpg. wt., 8 oz. **5L603 80c**

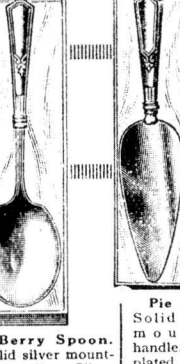

Gravy Ladle. Pearl handle. Solid silver ferrule, silver plated bowl. Length, 8¾ inches. Shipping weight, 12 ounces. **5L609...$1.74**

Bread Knife. Pearl handle. Solid silver ferrule, steel blade. Length of knife, about 11½ inches. Shipping weight, 14 ounces. **5L610...$1.70**

Butter Spreaders. Pearl handles. Solid silver ferrules, silver plated blades. Entire length, 5¾ inches. Shipping weight, 1 pound. **5L600**—Set of six **$6.98**

Fruit Knives. Pearl handles. Solid silver ferrules, silver plated blades. Entire length, 6¼ inches. Shipping weight, 1 pound. **5L602**—Set of six **$6.98**

Butter Knife. Pearl handle. Solid silver ferrule, silver plated blade. Entire length, 7½ inches. Shipping weight, 8 oz. **5L607 ...... $1.04**

SOLID SILVER HANDLE TABLEWARE

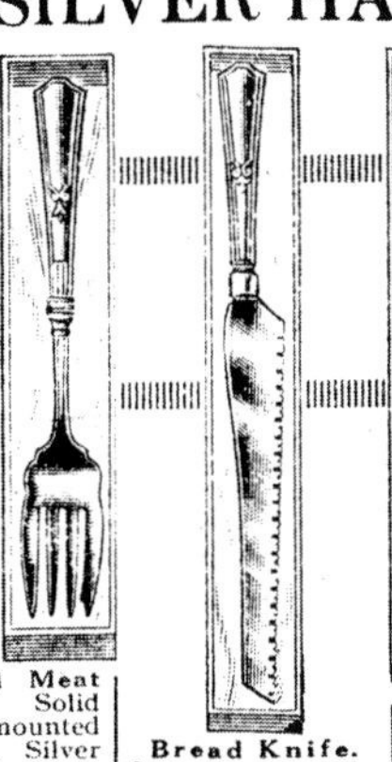

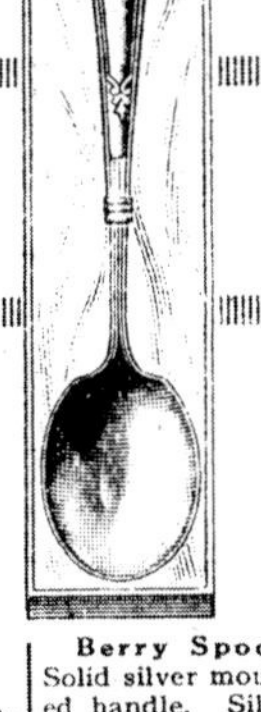

Pickle Fork. Solid silver mounted handle. Silver plated tines. Length, about 7 inches. Shipping weight, 6 oz. **5L701 .... $1.00**

Tea Ball. Solid silver mounted handle. Silver plated ball. Length, 7¼ inches. Shipping weight, 6 ounces. **5L707 .... $1.45**

Cream Ladle. Solid silver mounted handle. Silver plated bowl. Length, 7 inches. Shipping weight, 7 ounces. **5L702 .... $1.49**

Salt and Pepper Set. Solid silver. Height, 3 inches. Shipping weight, 1 pound. **5L703 $3.57**

Salt and Pepper Shaker Set. Solid silver. Height, 1⅝ in. Shpg. wt., 5 oz. **5L713 86c**

Cake Server. Solid silver mounted handle. Silver plated blade. Length, 10 in. Shipping wt., 8 ounces. **5L705 $1.49**

Cold Meat Fork. Solid silver mounted handle. Silver plated tines. Length, about 9 inches. Shpg. wt., 8 oz. **5L710 $1.79**

Bread Knife. Solid silver mounted handle. Steel blade. Lgth., abt. 13 in. Shpg. wt., 10 oz. **5L712 $1.66**

Berry Spoon. Solid silver mounted handle. Silver plated bowl. Length, about 9¼ in. Shpg. wt., 10 oz. **5L711 $2.00**

Pie Knife. Solid silver mounted handle. Silver plated blade. Length, 9¾ in. Shipping weight, 8 oz. **5L709 $1.49**

Gravy Ladle. Solid silver mounted handle. Silver plated bowl. Length, about 8¾ in. Shpg. wt., 8 oz. **5L706 $1.96**

Butter Knife. Solid silver mounted handle. Silver plated blade. Length, about 7¾ inches. Shipping weight, 6 ounces. **5L700 .... $1.32**

Sugar Shell. Solid silver mounted handle. Length, 7 inches. Silver plated bowl. Shipping weight, 6 oz. **5L708 .... $1.35**

Cheese Knife. Solid silver mounted handle. Silver plated blade. Length, 5⅞ inches. Shipping weight, 5 ounces. **5L704 .... 82c**

Individual Salt and Pepper Set. Solid silver. Three salt and three pepper shakers. Height, 1¾ inches. Shipping wt., 1 lb. **5L716 .... $4.77**

SOLID SILVER TABLEWARE

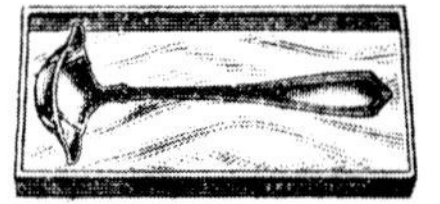

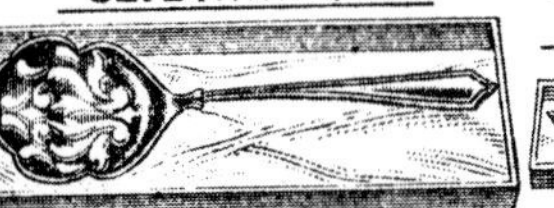

Sugar Shell. Solid silver. Length, 5½ inches. Shipping weight, 5 ounces. **5L723 .... $1.40**

Cream Ladle. Solid silver. Length, 5⅛ inches. Shipping weight, 8 ounces. **5L719 .... $1.40**

Cold Meat Fork. Solid silver. Length, 7⅜ inches. Shipping weight, 12 ounces. **5L715 .... $3.57**

Pickle Fork. Solid silver. Length, 5⅝ inches. Shipping weight, 5 ounces. **5L721 .... $1.38**

Baby Spoon. Sterling silver, curved handle. Length, 2¾ in. Shipping weight, 3 oz. **5L724 ...... $1.25**

Tomato Server. Solid silver. Length, 7¼ inches. Shipping weight, 12 ounces. **5L718 .... $3.46**

Berry Spoon. Solid silver. Length, 7⅜ inches. Shipping weight, 12 ounces. **5L717 .... $4.18**

Butter Knife. Solid silver. Length, 6⅞ inches. Shipping weight, 6 ounces. **5L722 .... $2.38**

Friendship Spoon. Solid silver. Length, 5⅜ inches. Shipping weight, 4 ounces. **5L720 .... $1.19**

Gravy Ladle. Solid silver. Length, 5¾ inches. Shipping weight, 10 ounces. **5L714 .... $3.18**

RELIGIOUS ARTICLES

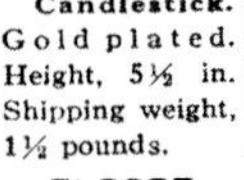

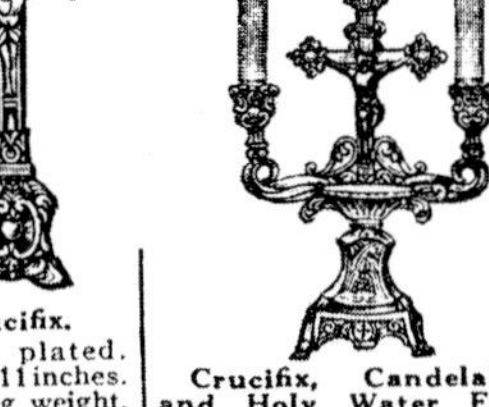

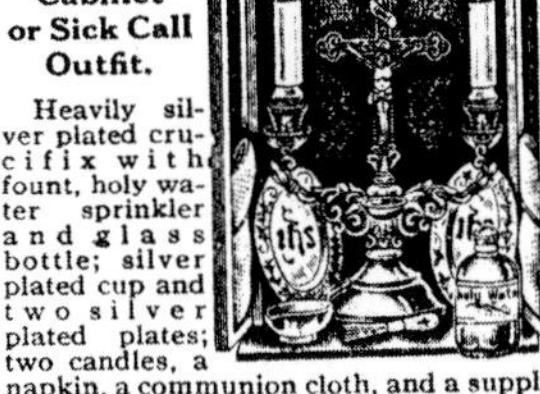

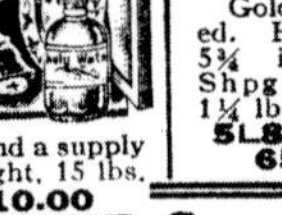

Crucifix. Gold plated. Height, 9½ in. Shipping wt., 2½ lbs. **5L8636 $1.35**

Candlestick. Gold plated. Height, 5½ in. Shipping weight, 1½ pounds. **5L8635 60c**

Crucifix. Gold plated. Height, 11 inches. Shipping weight, 3 pounds. **5L8637 $1.58**

Crucifix, Candelabrum and Holy Water Fount. Gold plated. Height, 11½ inches. Shipping wt., 6 lbs. **5L8632 ...... $3.15**

Viaticum Cabinet or Sick Call Outfit. Heavily silver plated crucifix with fount, holy water sprinkler and glass bottle; silver plated cup and two silver plated plates; two candles, a napkin, a communion cloth, and a supply of fine cotton. Shipping weight, 15 lbs. **5L8638 .... $10.00**

For other Ecclesiastical Goods, see page 445.

Saint Anthony and Child. Gold plated. Height, 5¾ inches. Shpg. wt., 1¼ lbs. **5L8643 65c**

Blessed Virgin. Gold plated. Height, 5¾ inches. Shpg. wt., 1¼ lbs. **5L8644 59c**

Saint Joseph and Child. Gold plated. Height, 5¾ inches. Shpg. wt., 1¼ lbs. **5L8645 65c**

Holy Water Fount. Gold plated. Length, 5⅛ inches. Shpg. wt., 1½ lbs. **5L8641 68c**

Useful Gifts of Unquestionable Value

The four Shaving Stands on this page are silver plated and then lacquered to prevent tarnishing.

Shaving Stand Outfit. Silver plated, gray finish. Extreme height, 28 inches. Beveled mirror, 8¾ inches in diameter. Complete as shown in illustration. Shpg. wt., 11 lbs.
5L8765 .... **$7.95**

Shaving Stand Outfit. Silver plated, with white glass mug, brush with celluloid handle. Beveled glass mirror is 6¾ inches in diameter. Extreme height, 21 inches. Shipping weight, 6 pounds.
5L8767 .... **$3.95**

Ash Tray and Match Holder. Mahogany base. Silver plated trimmings. Glass lining. Height, 6¼ inches, including handle. Diameter of dish, 5 inches. Shipping weight, 3 lbs.
5L8776 .... **$5.50**

Shaving Mug and Brush. Silver plated, bright polished holder, with white glass container. Height, 2½ inches. Brush with wood handle. Shipping weight, 2 pounds.
5L8787 .... **$1.65**

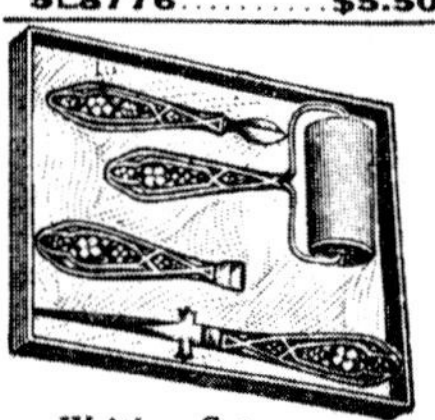

Writing Set. Silver plated. Opener, eraser, seal and blotter. Seal engraved with one Old English letter. **State initial wanted.** Shipping weight, 1½ pounds.
5L8607 .... **$2.25**

Military Set. Silver plated. Military brushes measure 2¾x4½ inches. Comb measures 7½ inches long. In box. Shipping weight, 2 pounds.
5L8772—Per set. **$4.25**

Jewel Case. Gold plated, cloth lined. Height, 4¾ inches; length, 6 inches; width, 4 inches. Shipping weight, 3 pounds.
5L8611 .... **$1.45**

Toilet Set. Three-piece. Silver plated. Mirror, 11 inches long, 4¾-inch beveled glass. Brush, 9¼ inches long. Comb, 7½ inches long. Shipping weight, 3¼ lbs.
5L8608—Per set **$5.95**

Jewel Case. Gold plated, cloth lined. Height, 3¼ inches; length, 4¾ inches; width, 3 inches. Shipping wt., 1½ pounds.
5L8610 ...... **88c**

Hairpin Box. Gold plated top, glass container. Height, 2½ inches; length, 3¼ inches; width, 2¼ inches. Shpg. wt., 1 lb.
5L8614..**59c**

Jewel Case. Gold plated, cloth lined. Height, 3 inches; length, 2¾ inches; width, 1⅞ inches. Shipping weight, 1 lb.
5L8615..**56c**

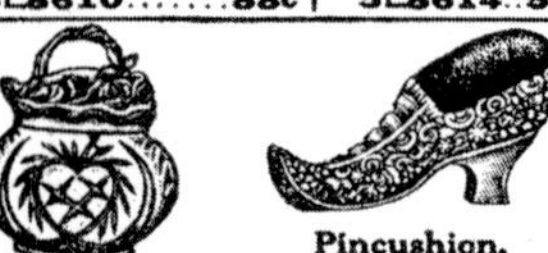

Powder Puff Jar. Gold plated top with glass holder. Height, 5 inches. Shpg. wt., 2¼ lbs.
5L8617..**75c**

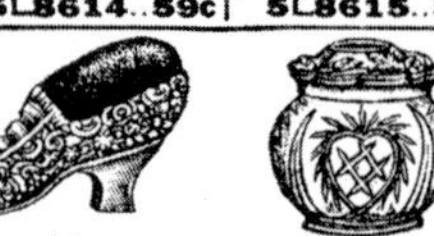

Pincushion. Gold plated, cloth cushion. Length, 4¾ inches; height, 3¼ in. Shipping weight, 1 lb.
5L8613 ..... **55c**

Hair Receiver. Gold plated top, glass container. Height, 4 inches. Shipping weight, 2 pounds.
5L8618..**75c**

Complete Shaving Outfit. Silver plated. Military brush, 3x5 inches. Comb. Pierced design mug with white glass lining. Shaving brush, can of talcum powder, can of shaving soap. In box. Shipping weight, 6½ pounds.
5L8782—Per set **$7.30**

Military and Shaving Set. Silver plated. Consists of comb, military brush, 2¾x4½ inches, shaving brush and mug with glass lining. Shipping weight, 5 lbs.
5L8785—Per set **$4.95**

Military Set. Silver plated. Clothes brush measures 2¼x7 inches. Two military brushes, 3x4½ inches. Comb measures 7½ inches long. In box. Shipping weight, 4 pounds.
5L8769—Per set **$7.85**

Eleven-Piece Set. Silver plated. Mirror with beveled glass. Hair brush, comb, nail file, nail polisher, manicure scissors, cuticle knife, shoe horn, buttonhook, pomade jar and salve jar. In cloth lined box. Shipping weight, 8 pounds.
5L8605—Per set **$9.45**

Eleven-Piece Set. Silver plated. Floral design. Mirror with 4¾-inch beveled glass. Hair brush, comb, nail polisher, two jars, nail file, manicure scissors, shoe horn, buttonhook and cuticle knife. Shpg. wt., 12½ lbs.
5L8604—Per set **$12.75**

Eleven-Piece Set. Silver plated. Mirror, 10¾ in. long, with 4¾-inch beveled glass. Hair brush is 9¼ in. long. Comb, nail polisher, puff jar, pomade jar, buttonhook, cuticle knife, nail file, manicure scissors and shoe horn. In presentation case. Shipping weight, 13 pounds.
5L8602—Per set **$15.50**

Shaving Stand Outfit. Silver plated, with white glass shaving mug. Brush with wood handle. Has 5¾-inch beveled glass mirror. Height, 14 inches. Shipping weight, 5 pounds.
5L8768 .... **$2.85**

Shaving Stand. Silver plated stand. 6¾-inch beveled mirror. Glass mug. Extreme height, 21 inches. Shipping weight, 7 pounds.
5L8766 .... **$5.50**

Shaving Mug and Brush. Silver plated, gray finish holder with white glass mug. Height, 2¾ inches. Good brush with celluloid handle. Shipping weight, 2 pounds.
5L8786 .... **$2.65**

Smokers' Set. Silver plated. Cigar jar is 3¼ inches high. 7½-inch tray. Cigar jar, ash tray and match holder are gold plated inside. Shipping weight, 2¼ pounds.
5L8774
Per set **$3.95**

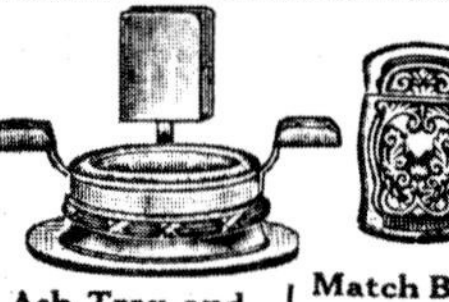

Ash Tray and Match Holder. Silver plated, with 3⅜-inch glass lining. Height, 4¼ inches. Shipping weight, 2 pounds.
5L8781
$1.85

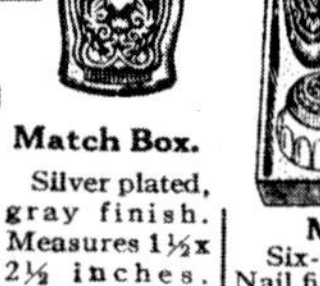

Match Box. Silver plated, gray finish. Measures 1½x2¼ inches. Shipping weight, 5 oz.
5L8779
75c

Manicure Set. Six-piece. Silver plated. Nail file, manicure scissors, buttonhook, cuticle knife, nail polisher and pomade jar. In presentation box. Shipping weight, 2¼ lbs.
5L8606—Set. **$2.95**

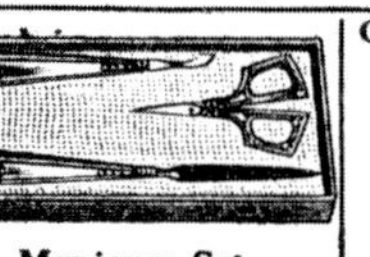

Manicure Set. Three pieces. Solid silver handles. Length of file, 7⅛ inches. Shipping weight, 1 pound.
5L8621
In box **$3.75**

Oval Picture Frames.

Silver plated. Holds picture 2½x4 in. Shpg. wt., 8 oz.
5L8629
50c

Silver plated. Holds picture 5x6¼ in. Shpg. wt., 14 oz.
5L8630
78c

Silver plated. Holds picture 6x7½ inches. Shipping wt., 1¼ lbs.
5L8631
95c

Sewing Set. Solid silver thimble, scissors handles and emery top. Mention size of thimble. Shpg. wt., 6 oz.
5L8624 .... **$1.89**

Book Mark. Silk ribbon with gold filled bangles. Shipping wt., 2 oz.
5L8628..**59c**

Ink Blotter. Solid silver handle. Length over all, 5 inches. Shipping weight, 4 ounces.
5L8626 .... **98c**

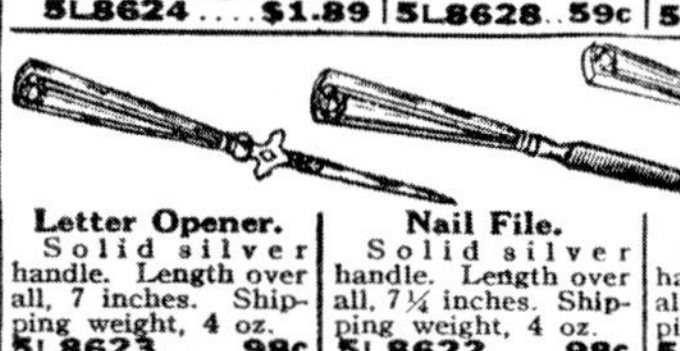

Letter Opener. Solid silver handle. Length over all, 7 inches. Shipping weight, 4 oz.
5L8623 .... **98c**

Nail File. Solid silver handle. Length over all, 7¼ inches. Shipping weight, 4 oz.
5L8622 .... **98c**

Darner. Solid silver handle. Length over all, 6½ inches. Shipping weight, 6 oz.
5L8625 .... **98c**

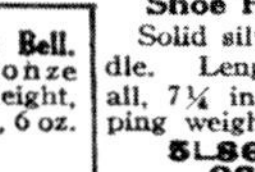

Table Bell. Bronze finish handle. Height, 3 inches. Shpg. wt., 6 oz.
5L8627
$1.10

Shoe Horn. Solid silver handle. Length over all, 7¼ in. Shipping weight, 4 oz.
5L8619
98c

Tooth Brush. Solid silver handle. Length over all, 6¼ in. Shipping weight, 4 oz.
5L8620
98c

"As Automatic as the Sunrise."

The National Call

Automatic 8-Day Alarm Clocks

Our own trade mark registered in U. S. Patent Office.

Remember that you can purchase these famous clocks only from us.

Our New Improved Plain Dial Automatic EIGHT-DAY NATIONAL CALL ALARM CLOCK.

Here's a sturdy, dependable watchman, making the rounds twice each day and sounding the alarm automatically whenever you wish. His voice is clear and vibrant; and if you wish it so, insistent; also the alarm can be short or long as you please; or, he can be silenced by the mere pressure of a button. Eight days with one winding. Graceful hands and large clear figures tell time quickly. Set it once and it will ring and shut off automatically every day at the time set. The dial is 4¼ inches and the height 6 inches. Solid brass and polished nickel parts throughout. Practically rustproof. Seamless. Dustless. A stanch timepiece that will give long and excellent service. Shipping weight, 2⅜ pounds.

5L8512 .... **$2.75**

Our New Improved Automatic LUMINOUS EIGHT-DAY NATIONAL CALL ALARM CLOCK.

Though dark be the night, the luminous Automatic National Call stands with lighted hands pointing clearly to the luminous figures. Set it once and it will ring at the time set automatically and shut off automatically every day without additional attention, except winding once every eight days. The alarm can be short or long, or, by a touch of the finger on the button at the top of the clock, it can be shut off instantly. If you wish, this new improved National Call can be set to silence and it will not alarm at all. Dial, 4¼ inches. Height, 6 inches. Solid brass and polished nickel parts. Seamless. Shipping weight, 2⅜ lbs.

5L8514 .... **$3.75**

OUR NATIONAL ALARM CLOCKS.

These clocks are manufactured expressly for us under our own specifications and represent the highest degree of perfection in American alarm clock making. They run for 30 hours with one winding and will surely alarm at time set. Run for years if properly used. Clocks that we recommend, and will give the satisfaction that you have a right to expect from a high grade article. Every one properly and carefully tested at the factory and again before we ship.

One-Day Nickel Plated Buzzer Alarm Clock. Height, 5¼ inches; 3¾-inch dial. Instead of having the loud ringing bell that startles you, have a clock where the hammer strikes on the brass case, and while it is loud enough to awaken you, it does not startle you. We recommend this clock to nervous people. Shpg. wt., 1½ lbs.

5L8510 .... **$1.45**

The Success and the Pet.

The National Success runs 30 hours. Stands 5¾ in. high and has 4-inch plain white dial with clear numerals. Bell on back of the clock. Shpg. wt., 2¼ lbs.
5L8518 **$1.55**

The National Pet runs 30 hours. Stands 3¾ inches high. Dial is 2¼-in. in diameter. Plain white dial, clear numerals. Has alarm bell on the back. Will awake any ordinary light sleeper. Shipping weight, 1 pound.
5L8519 **$2.15**

The Luminous National Pet runs 30 hours. Luminous figures and hands. The time can be read both night and day. Bell on back will wake any ordinary light sleeper. Stands 3¾ inches high. Dial is 2¼ inches in diameter. Shipping weight, 1 pound.
5L8522 **$2.95**

The Luminous National Success runs 30 hours. Stands 5¾ inches high. Dial is 4 inches in diameter. The time can be read at night as well as in the day time. Alarm bell on back of clock. Shipping weight, 2¼ pounds.
5L8521 **$2.15**

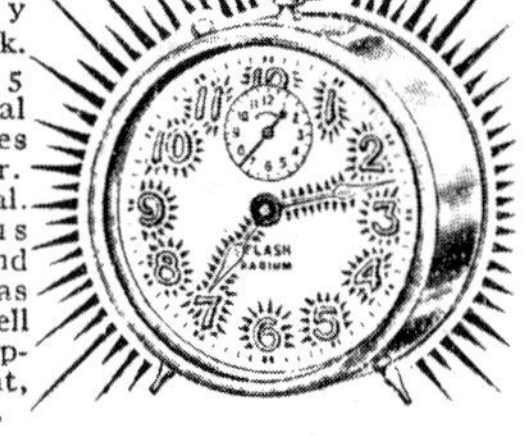

Luminous One-Day Alarm Clock. Height, 5 inches. Dial is 3⅝ inches in diameter. White dial. Luminous numerals and hands. Has 3⅝-inch bell on back. Shipping weight, 2¼ pounds.

5L8507—White Dial **$2.45**

One-Day Reliable Nickel Plated Alarm Clock. Height, 6¼ inches; 3½-inch dial. Alarms for one-half minute. Shipping weight, 2 pounds.

5L8520—With plain dial **$0.89**
5L8511—With luminous dial **1.65**

Small Luminous Black Dial One-Day Intermittent Alarm Clock. Height, 2¾ inches. Has a 2-inch dial; 1¾-inch bell on back. Shipping weight, 1¼ pounds.

5L8508 .... **$3.50**

Small One-Day Nickel Plated Repeating Alarm Clock. Height, 2¾ inches; 2-inch dial. Alarms and repeats at intervals for four minutes. Shipping weight, 1½ pounds.

5L8557 .... **$2.30**

One-Day Nickel Plated Repeating Alarm Clock. Stands 5¾ inches high. Dial is 4⅞ inches in diameter. Rings at intervals for 5 minutes on 4¼-inch bell on back. Made by the celebrated Ansonia Clock Company. Shipping weight, 2 pounds.

5L8530 .... **$2.25**

BEDROOM AND DRESSER CLOCKS.

Mahogany Finish Eight-Day Desk or Dresser Clock. Height, 4⅝ in.; 2⅜-inch dial with beveled glass. Does not strike or alarm. Shipping weight, 2½ pounds.

5L8536 .... **$3.45**

One-Day Desk or Dresser Clock. Height, 2⅞ inches. Dial, 1½ in. Has polished brass hexagon case. Does not alarm. Shpg. wt., 8 oz.

5L8515—With plain dial **$1.45**
5L8516—With luminous dial **$2.15**

Mahogany Finish One-Day Desk or Dresser Clock. Height, 3⅝ inches; width, 6½ inches; 2-inch dial. Does not alarm. Shipping weight, 2 lbs.

5L8517 ...... **$2.75**

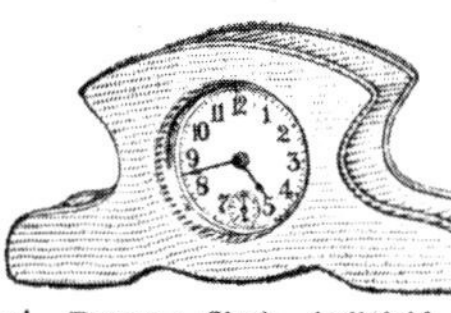

Luminous Dial One-Day Midget Clock (celluloid, ivory color). Height, 3 inches. Fitted with luminous dial and watch movement. Does not alarm. Shipping weight, 8 oz.

5L8509 .... **$1.95**

Same as above, but with plain white dial.

5L8547 .... **$1.25**

Dresser Clock (celluloid, ivory color). Height, 3 in.; width, 6⅛ inches. Does not alarm. Shipping weight, 1¼ pounds.

5L8539 .... **$2.19**

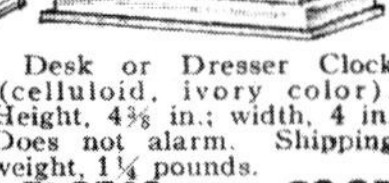

Desk or Dresser Clock (celluloid, ivory color). Height, 4⅜ in.; width, 4 in. Does not alarm. Shipping weight, 1¼ pounds.

5L8549 .... **$2.25**

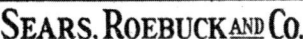

Designs for Every Purpose—Eight-Day Clocks

Eight-Day Black Enameled and Oak Mantel Clocks—beautifully finished in every detail. Similar designs found only **in high grade shops at much higher prices.** We handle only makes of highest quality. You will find a complete description under each illustration. Unless otherwise stated in description these clocks strike the hours and half hours. Full directions for setting up and regulating accompany each clock.

Black Enameled Wood. Bronze finish metal ornaments and columns, side ornaments and feet. Gilt scroll decorations, 4⅝-inch dial and Arabic figures. Runs eight days with one winding. Strikes hours on gong and half hours on a cup bell. Stands 11 inches high and is 18 inches wide. Shipping weight, 18 pounds.
5L9100¼..............**$7.95**

Black Enameled Wood Mantel Clock. Runs eight days with one winding. Bronze finish columns, side ornaments and feet. Column caps and bases gilt. 4½-inch dial. Strikes the hours on gong and half hours on cup bell. Stands 10¾ inches high and is 14 inches wide. Shipping weight, 18 pounds.
5L9102¼.....**$6.60**

Black Enameled Wood Mantel Clock. Runs eight days with one winding. Variegated red sides. Imitation white marble columns. Gilt metal ornamentations. Gong and bell strike. Height, 12¾ inches; width, 18½ in.; 4⅝-inch dial. Shpg. wt., 18 lbs.
5L9104¼ **$9.48**

Black Enameled Mantel Clock. Runs eight days with one winding. Bronze finish, side ornaments and feet. Green columns with gilt caps and base, variegated green top molding. 4½-inch dial. Stands 9¾ inches high and is 14 inches wide. Strikes the hours and half hours. Shipping weight, 16 pounds.
5L9106¼..............**$6.35**

Black Enameled Wood Mantel Clock. Runs eight days with one winding. Ornaments of gilt and bronze variegated red top molding. Green color columns. Stands 10 inches high, and is 14½ inches wide with 4¾-inch dial. Strikes the hours on gong and half hours on cup bell. Shipping weight, 16 pounds.
5L9144¼........**$6.25**

Dark Mahogany Finish Wood Mantel Clock. Runs eight days with one winding. Lever escapement (no pendulum), does not strike. Stands 8¼ inches high, 15¼ inches wide, 2⅞ in. deep, and has a 4½-inch dial. Shipping wt., 6 lbs.
5L9148¼.......**$4.50**

Oak Mantel Clock. Golden finish. Runs eight days with one winding. Very massive. Height, 10¾ inches; width, 19¼ inches at base; 5⅝-inch dial. Gong strike. Shipping weight, 15 pounds.
5L9116¼........**$13.25**

Oak Mantel Clock. Golden finish. Runs eight days with one winding. Height, 13 inches; width, 8¾ inches; 4⅝-inch dial. Bell and gong strike. Shipping weight, 10 pounds.
5L9142¼
Each **$7.45**

Black enameled wood, eight-day Mantel Clock. Bronze finish ornaments, and gilt column caps and base. Variegated green top molding. Strikes the hour on gong and half hour **on cup** bell. Stands 10¾ inches high, 18 inches wide; 4¾-inch dial. Shipping **weight**, 18 pounds.
5L9146¼.....................**$6.75**

Genuine Solid Mahogany Candlesticks.

Mahogany Candlesticks. Height, 10⅛ inches. Shipping weight, 3 pounds.	Mahogany Candlesticks. Height, 8½ inches. Shipping weight, 3 lbs.	Mahogany Candlesticks. Height, 8½ inches. Shipping weight, 2½ lbs.	Mahogany Candlesticks. Height, 8⅛ inches. Shipping weight, 3 pounds.
5L8506	**5L8505**	**5L8504**	**5L8513**
Per pair **$2.48**	Per pair **$3.50**	Per pair......**$1.95**	Per pair . . .**$2.98**

Oak Mantel Clock. Runs eight days with one winding. Height, 10 inches; length of base, 11 inches. Strikes the hour on soft tone gong and half hour on cup bell. Has 4⅝-inch dial. Shipping weight, 13 pounds.
5L9223¼ **$4.75**

Oak Mantel Clock. Runs eight days with one winding. Strikes the hour on a soft tone Cathedral Gong and half hour on cup bell. Has oak columns set in gilt. Metal cap and base. Height, 10 inches, width of base, 16 inches, with 4⅝-inch dial. Shipping weight, 15 pounds.
5L9225¼..............................**$5.75**

Hanging Clock. Runs eight days with one winding. Oak finish; 27 inches high and 14⅝ inches wide; 5¼-inch dial. Fitted with strike attachment. Shipping weight, 16 pounds.
5L9130¼.........**$5.85**

Shelf Clock. Runs eight days with one winding. Oak finish; 22½ inches high and 14½ inches wide; 5¼-inch dial. Shipping wt., 16 lbs.
5L9124¼
Wire bell strike............**$4.20**
5L9126¼—Wire bell strike, with alarm..........**$5.10**
5L9128¼—Gong bell strike, with alarm..........**$5.50**

Church, School, Shop or Factory Clock. Runs eight days with one winding. Oak finish. The kind and make of clock we can recommend. 23¾ inches high. 9¼-inch dial. Shipping wt., 16 lbs.
5L9134¼
Time only...........**$5.60**
5L9136¼—Time with calendar attachment. .**$5.90**
5L9138¼—Time strike on wire gong.........**$6.10**

Oak Front Kitchen Eight-Day Clock, strikes the hours and one-half hours. Fancy embossed case. Stands 22¾ inches high and 16½ inches wide; 5¼-inch dial. Fitted with thermometer for the temperature, and also a barometer that predicts the changes in the weather. Shipping weight, 16 pounds.
5L9122¼ **$5.45**
5L9123¼—Same as above, but with alarm attachment **$5.95**

Shelf Eight-Day Clock, oak finish; 21 inches high and 12 inches wide; 5¼-inch dial. Strikes the hours and half hours. Fitted with alarm attachment. The decorations on this clock are very simple in character. Shipping weight, 14 lbs.
5L9120¼..........**$5.40**

Beautiful Eight-Day Clocks With Dual Chimes

Our new line of genuine mahogany, mahogany veneer and mahogany finish Dual Strike Chime Tone Mantel Clocks. We show nine numbers: 5L9200¼ to 5L9215¼. The dual chime tone is a brand new improvement. The clock chimes by giving two strokes on chime bars with the true chime tone, at the half hour and hour intervals. At 12 o'clock it strikes two notes twelve times.

$11.00 · $9.25 · $8.75 · $9.25

The Sanford.
Mahogany finish. Stands 9¾ inches high and is 21 inches wide. Dial, 4¾ inches in diameter, of porcelain, brass sash fitted with bullseye glass. Runs eight days with one winding. Dual strike chime tone attachment. Strikes the half hour and hour intervals on two-tone chime bars. Shipping weight, boxed, 17 pounds.
5L9204¼ $11.00

The Lakewood.
Mahogany finish. Stands 9¼ inches high and is 17 inches wide. Dial, 4¾ inches in diameter, of porcelain, brass sash with bullseye glass. Runs eight days with one winding. Dual strike chime tone attachment. Strikes the half hour and hour intervals on two-tone chime bars. Shipping weight, boxed, 17 pounds.
5L9210¼ $9.25

The Larchmont.
Mahogany finish. Stands 9¼ inches high and is 18 inches wide. Dial, 4¾ inches in diameter, of porcelain, brass sash fitted with bullseye glass. Runs eight days with one winding. Dual strike chime tone attachment. Strikes the half hour and hour intervals on two-tone chime bars. Shipping weight, boxed, 19 pounds.
5L9212½ $8.75

The Flossmore.
Mahogany finish. Stands 9¾ inches high and is 16½ inches wide. Dial, 4¾ inches in diameter, of porcelain, brass sash fitted with bullseye glass. Runs eight days with one winding. Dual strike chime tone attachment. Strikes the half hour and hour intervals on two-tone chime bars. Shipping weight, boxed, 16 pounds.
5L9214¼ $9.25

Genuine Mahogany Clocks With Dual Chimes

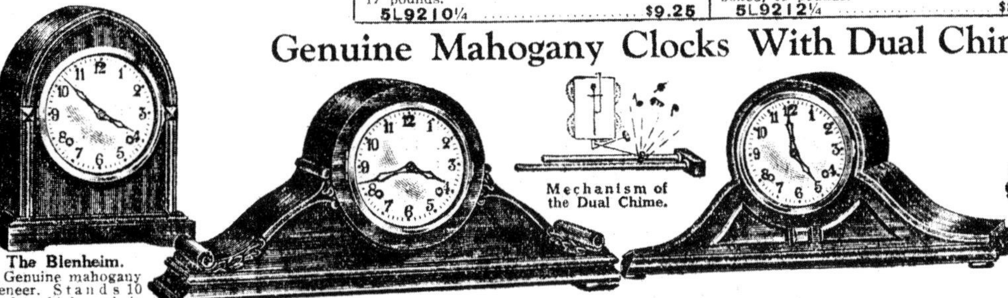

Mechanism of the Dual Chime.

The Blenheim.
Genuine mahogany veneer. Stands 10 inches high and is 8⅝ inches wide. Dial, 4¾ inches in diameter, of porcelain, brass sash fitted with bullseye glass. Runs eight days with one winding. Dual strike chime tone attachment. Strikes the half hour and hour intervals on two-tone chime bars. Shipping weight, boxed, 14 pounds.
5L9200¼ $9.85

The Rosedale.
Genuine mahogany. Stands 9 inches high and is 21 inches wide. Dial, 4¾ inches in diameter, of porcelain, brass sash fitted with bullseye glass. Runs eight days with one winding. Dual strike chime tone attachment. Strikes the half hour and hour intervals on two-tone chime bars. Shipping weight, boxed, 21 pounds.
5L9202¼ $18.23

The Hollywood.
Genuine mahogany. Stands 8¾ inches high and is 18½ inches wide. Dial, 4¾ inches in diameter, of porcelain, brass sash fitted with bullseye glass. Runs eight days with one winding. Dual strike chime tone attachment. Strikes the half hour and hour intervals on two-tone chime bars. Shipping weight, boxed, 18 pounds.
5L9206¼ $16.25

The Marlboro.
Genuine mahogany. Stands 8¾ inches high and is 17¾ inches wide. Dial, 4¾ inches in diameter, of porcelain, brass sash fitted with bullseye glass. Runs eight days with one winding. Dual strike chime tone attachment. Strikes the half hour and hour intervals on two-tone chime bars. Shipping weight, boxed, 17 lbs.
5L9208¼ $13.50

Reliable Eight-Day Clocks

The Ambassador.
Mahogany Finish Eight-Day Dual Chime Clock. Strikes the hours and half hours on two-tone chime bars. Stands 9½ inches high, is 17½ inches wide and has 4¾-inch dial. Shipping wt., 17 lbs.
5L9215¼ $10.00

Mahogany Finish Mantel Eight-Day Clock. Height, 11 inches; width, 17 inches; 6-inch dial with bullseye glass in brass sash. Bell and gong strike. Shipping weight, 14 lbs.
5L9218¼ $7.45

Mahogany Finish Eight-Day Mantel Clock. Height, 9½ inches; width, 19 inches; 5¼-inch dial with bullseye glass. Strikes the hours and half hours on gong. Shipping weight, 15 lbs.
5L9219¼ $7.95

Mahogany Finish Mantel Eight-Day Clock. Has 5⅝-inch dial, is 10¾ inches high and 9¼ inches wide. Strikes the hours on gong, and half hours on cup bell. Shipping weight, 10 lbs.
5L9220¼ $6.35

A Cottage Kitchen Clock. Runs eight days with one winding. Solid oak. Height, 15¼ inches. Width, 10¼ inches. Glass panel door, 4-inch dial. Strikes hours and half hours on a wire gong. Shipping weight, 15 pounds.
5L9132¼ $4.75

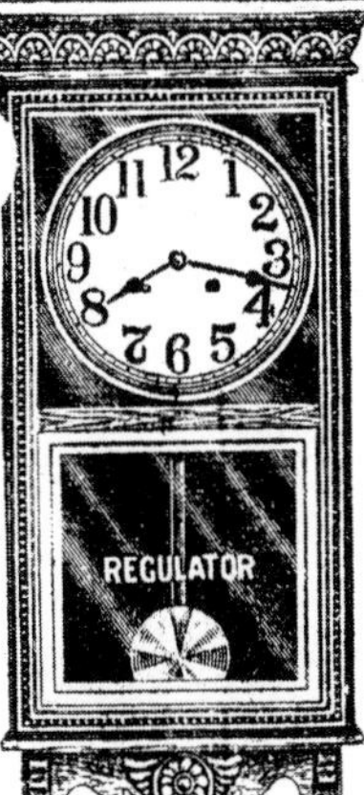

Hanging Wall Regulator Clock, oak, 35 inches high; 12-inch dial. Shipping weight, 32 pounds. Not mailable.
5L9232¼ — Time only $7.50
5L9234¼ — Time and strike $8.50
5L9236¼ — Time, strike and calendar attachment $9.25

Shelf Eight-Day Clock, mahogany or oak finish; 19 inches high and 15¼ inches wide; 5¼-inch dial. Strikes the hours and half hours on soft tone gong. Shipping wt., 15 lbs.
5L9226¼ — Mahogany finish $4.98
5L9228¼ — Oak finish 4.98

Hanging Eight-Day Clock, mahogany finish; 22 in. high, 14 in. wide; 5¼-in. dial. Soft tone wire gong upon which the hours and half hours are sounded. Neat and dignified in design. Shipping weight, 14 pounds.
5L9230¼ $5.65

Hardwood Clock Shelf. Fits shelf clocks only; 16½ inches long and 5 inches wide. Shipping wt., 1⅞ lbs.
5L8595 .... 68c

Hanging Wall Eight-Day Clock, oak finish; 32 in. high, 12⅛ in. wide; 7-in. dial. Strikes hours and half hours on a gong. Shpg. wt., 21 lbs.
5L9224¼ $8.75

A Cottage Kitchen Clock. Runs eight days with one winding. Mahogany finish. 4-inch dial. Height, 15½ inches. Width, 10⅝ inches. Strikes the hours and half hours on a wire gong. Shipping weight, 14 pounds.
5L9140¼ $4.75

Magnifying Glasses.

Magnifiers of high power are necessarily made with lenses of small diameters and short focal lengths. This gives a narrow field of view and ample working distance. The purchaser must choose between a low power with wide field and ample working distance or a high power with narrow field and short working distance. The choice will usually be governed by the character of the work for which the glass will be used, but consideration should also be given to the fact that magnifiers with large lenses and of low power are easier to use and generally prove more satisfactory to those who do not use a glass often enough to become expert.

Single Lens Magnifiers.

Mounted in imitation rubber oval folding pocket cases. Very powerful. Shpg. wt., any size, 2 oz.

	Diameter of Lens	Magnifying Power	
5L9700	1 in.	4 times	$0.98
5L9702	1¼ in.	3½ times	1.12
5L9704	1½ in.	3 times	1.48

Double Lens Magnifiers.

Two lenses of different foci mounted in imitation rubber oval folding pocket cases. The lenses may be used separately or combined, thus affording three degrees of magnifying power. Shipping wt., any size, 3 oz.

	Diameter of Lenses, Inches	Magnifying Power, Times	
5L9706	⅞ and 1	4 to 9	$1.40
5L9708	1 and 1¼	3½ to 8	1.67
5L9710	1¼ and 1½	3 to 6	2.48

Three lenses of different foci mounted in imitation rubber bellows shape folding pocket case. The lenses may be used singly or in various combinations, thus affording seven different magnifications. Diameters of lenses, ½, ⅝ and ¾ in. Magnifying power, 7 to 30 times. Shipping weight, 3 ounces.

5L9712 ....$1.68

Tripod Microscope.

This instrument is made with two extra quality double convex lenses, separated by a diaphragm ring, thus forming a powerful double magnifier. Diameter of lenses, 1¾ inches; diaphragm opening, ¾ inch. Magnifying power, 3 times. Lacquered brass mounting. Made in France. Shpg. wt., 8 oz.

5L9758 ....75c

Folding Pocket Magnifier.

Constructed of brass, ¾-in. lenses, very compact when folded. Generally used for counting the threads in fabrics. Shpg. wt., 4 oz.

5L9762 ....65c

READING GLASSES

Make it easy to read the smallest type and when looking at pictures. They add greatly to the beauty of photographs. Also very convenient for examining any small object for which a magnifying glass is ordinarily used. The smaller sizes have the highest magnifying power, but with the larger sizes a greater area can be seen at one time. AS BURNING GLASSES these instruments are very powerful.

Extra Quality Reading Glasses.

Best grade imported reading glasses, with very powerful lenses of extra quality. Strong polished nickel plated rims. Black wood handles. Very finely made and beautifully finished throughout. Instruments that we can highly recommend.

	Diameter	Shpg. Wt.	
5L9714	2 in.	9 oz.	$1.15
5L9716	2½ in.	10 oz.	1.25
5L9718	3 in.	12 oz.	1.60
5L9720	4 in.	1 lb. 3 oz.	2.90
5L9722	5 in.	1½ lbs.	4.75

Good Reading Glasses at Very Low Prices.

Imported reading glasses. Nickel plated rims. Black wood handles. Remarkably inexpensive, yet very serviceable reading glasses.

	Diameter	Shpg. Wt.	
5L9724	2 in.	12 oz.	$0.50
5L9726	2½ in.	14 oz.	.68
5L9728	3 in.	1⅛ lbs.	.85
5L9730	4 in.	1½ lbs.	1.50
5L9732	5 in.	1 lb. 13 oz.	2.50

POCKET COMPASSES

Explorers' Compass With Luminous Indicators.

The Compass is made with a special hand. The user, after getting his bearings, revolves the bottom of the compass until the hand points exactly in the direction he wishes to go. This special direction hand, the north-indicating needle, and the main points on the dial are **prepared with a radioactive compound, enabling them to be seen at night.** The case is made of brass, silver finish metal dial, with full circle divisions. Jeweled cap to needle and sliding stop. Diam., 2 in. Shpg. wt., 4 oz.

5L9744 ....$1.25

Explorers' Day Compass, 75c.

Exactly like 5L9744, except that indicators and direction points on dial are not luminous, measuring 1⅞ inches. Shipping weight, 4 ounces.

5L9746 ....95c

Fine Jeweled Compass, $3.25.

High Grade Pocket Compass. Lacquered brass case, 2⅝ in. in diameter, with cap cover; heavy beveled glass, automatic stop and best grade **Jewel mounted English bar needle.** Bottom of compass is oxidized in black with white lettering, and the full circle divisions are engraved on a silver finish metal dial **raised to level of needle.** Shipping weight, 6 ounces.

5L9748 ....$3.25

High Grade Compass, $2.85.

Compact Compasses. Made like a watch with **nickel plated dustproof case;** spring hinged cover, opened by pressing on stem; best jewel mounted English bar needle; automatic stop and heavy circle division glass. The full circle divisions are engraved on a silver finish metal dial. Diameter, 2 in. Shpg. wt., 4 oz.

5L9750 ....$2.85

Our 35-Cent Special.

Pocket Compass, watch style, open face, bevel edge glass, brass case. Paper dial. Diameter, 1½ in. Shipping weight, 2 oz.

5L9752 ....35c

Jeweled Compass, $1.20.

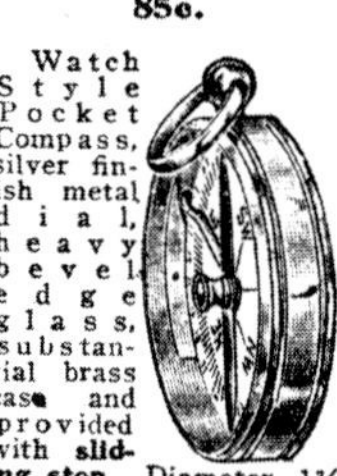

Watch Style Pocket Compass, finished and lacquered brass case with hinged cover, heavy beveled edge glass, **silver finish metal dial** with full circle divisions and **sliding stop.** Very sensitive needle mounted with **jeweled cap.** Diameter, 2 inches. Shpg. wt., 5 oz.

5L9754 ....$1.20

Good Compass for 85c.

Watch Style Pocket Compass, silver finish metal dial, heavy bevel edge glass, substantial brass case and provided with **sliding stop.** Diameter, 1¾ inches. Shpg. wt., 3 oz.

5L9756 ....85c

Wind, Dust and Sun Glasses.

High Grade Imitation All Shell Goggles with 1¾-inch lenses in amber or smoke glass. In individual cardboard box. Shipping weight, 1 lb.

5L9783—Amber glass 50c
5L9784—Smoke glass 50c

Men's and Women's Goggles.

Men's and Women's High Grade Imitation All Shell Goggles with straight temples, 1¾-inch lenses in amber or smoke. In individual cardboard box. Shipping weight, 1 lb.

5L9785—Amber glass 75c
5L9786—Smoke glass 75c

Dust and Foundry Goggles.

Rustproof White Metal Goggles with ventilated leather side shields, flexible cable earbows and the adjustable bridge, 1⅞-in. lenses in amber or clear glass. With metal case. Shpg. wt., 1 lb.

5L9787—Amber glass $1.00
5L9788—Clear glass 1.00

Good Grade Leather Dust Goggles, velvet bound, 1¾-inch lenses in amber or clear. Ventilated round eye cups. Shipping wt., 8 oz.

5L9789—Amber glass 20c
5L9790—Clear glass 18c

Low Priced Leather Dust Goggles with 1¾-inch amber or clear lenses. Ventilated round eye cups. Shipping weight, 8 ounces.

5L9791—Amber glass 18c
5L9792—Clear glass 15c

Dustite Respirator.

Has passed the test and secured the approval of the UNDERWRITERS' LABORATORIES for the protection of workers in dusty conditions. No foundryman, thresherman, tobacco worker, miller or farmer should be without one. Filters and cleans the air you breathe. Light, comfortable and easily cleaned. All parts interchangeable. Made of high grade white rubber, with dry filter of pure cotton, instantly and easily renewable. Complete with one dozen filters and elastic band to hold it on. Shipping weight, 7 ounces.

5L9766 ....$1.50
5L9768—Extra Cotton Filters. Shipping weight, 2 ounces. Per 50 45
Respirator—wet sponge type with relief valve. Shipping weight, 7 oz.
5L9769 ....$2.25

The Gem Respirator or Dust Protector.

Protects nose and mouth from intrusion of dust, which is so injurious to head and lungs. Recommended for anyone whose duties call him into dusty places. Made of leather, with aluminum nose cap. Filtration effected through cloth filter between two wire mesh screens. Interchangeable filters, adjustable to anyone, easy to breathe through (weighs only 1 ounce). Complete with one dozen cloth filters, two extra wire screens and elastic band to hold it on. Shipping weight, 7 ounces.

5L9770 ....$1.10

Gibbs Respirator or Dust Protector, all metal, chamois bound. Shipping weight, 7 ounces.

5L9771 ....$1.00

MICROSCOPES

Microscope.

A Brass Compound Microscope with jointed black enameled metal base. Fitted with rack and pinion focusing adjustment and triple lens objective. Any of the lenses may be removed by unscrewing. It stands 9¼ inches high and when all three lenses are used has a magnifying power of 110 times. This is a well made, practical instrument. Packed in wooden box with tweezers, one mounted and two unmounted slides. Shpg. wt., 3¼ lbs.

5L9772—Complete $17.50

Microscope.

A Brass Compound Microscope with jointed black enameled metal base. Has triple lens with a magnifying power of 90 times. Stands 8 inches high. Fitted with eccentric focusing adjustments. Has swinging mirrors for illuminating objects. Packed in wooden box with one mounted and two unmounted slides. Shpg. wt., 3 lbs.

5L9774 Complete, $12.50

Microscope.

Brass Compound Microscope with jointed black enameled metal base. This instrument is the same as our 5L9774 except that it is not equipped with the eccentric focusing adjustment. It stands 7¾ inches high, comes packed in wooden box with tweezers, one mounted and two unmounted slides. Shpg. wt., 2¾ lbs.

5L9776 ....$9.85

Compound Microscope.

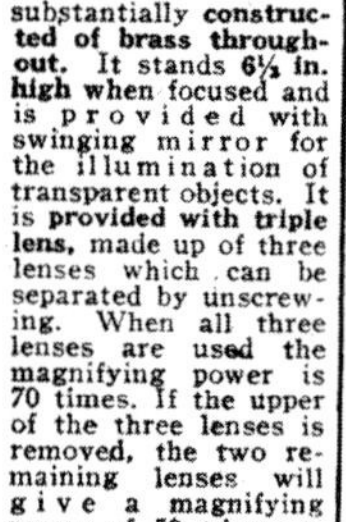

Made in Paris and substantially **constructed of brass throughout.** It stands **6½ in. high** when focused and is provided with swinging mirror for the illumination of transparent objects. It is **provided with triple lens,** made up of three lenses which can be separated by unscrewing. When all three lenses are used the magnifying power is 70 times. If the upper of the three lenses is removed, the two remaining lenses will give a magnifying power of 50 times. When using only the lower lens, the magnifying power is 30 times. Packed in a polished wood case, complete with brass tweezers, two plain glass slips and one prepared object. Shipping weight, 1½ lbs.

5L9778 ....$4.50

Microscope.

A Single Lens Brass Microscope, upright model. Stands 6 in. high. Has magnifying power of twenty-five times and is equipped with swinging mirror. Comes packed in wooden box with tweezers, one mounted and two unmounted slides. Shpg. wt., 1¼ lbs.

5L9780 Complete $3.95

Watchmakers' Eye Glass.

A well constructed glass, aluminum frame with fine, clear lens, ⅞ inch in diameter, 2-in. focus. Shpg. wt., 6 oz.

5L9764 ..45c

Pocket Microscope.

For the Young Natural History Student.

A small brass microscope that can easily be carried in your pocket. Stands 2½ in. high. Has single lens ⅝ in. in diameter. Very convenient for studying plant and insect life. Shpg. wt., ½ lb.

5L9782 ...58c

THERMOMETERS

Outdoor Thermometers.

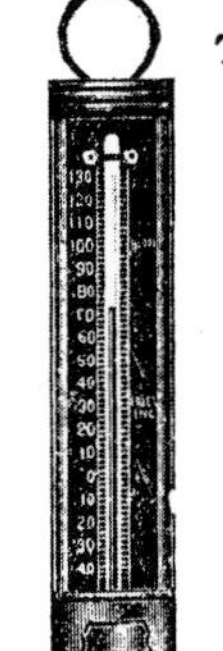

Japanned Metal Case. Mercury. Very accurate. White figures and graduations on black oxidized scale. Thoroughly seasoned tube of large size, guaranteed. Length, 9¾ inches. Shipping wt., 1¾ lbs.
5L9400
$1.50

Japanned Metal Case Thermometer. Spirit. Extra quality. Tube of standard size. White figures and graduations on black scale. Length, 7¼ in. Shpg. wt., 1 lb.
5L9402
65c

Japanned Metal Case Thermometer. Red spirit. Ordinary grade. Black figures on light metal scale. Length, 7¼ in. Shipping wt., 14 oz.
5L9404
25c

Storm Glass Thermometers.

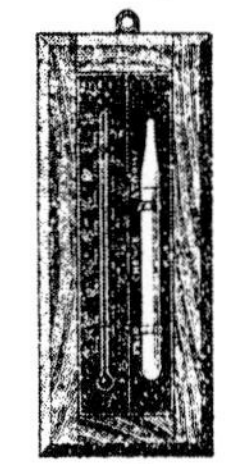

Storm Glass and Thermometer. Sold at a very low price. Has white figures and lines on black scale, set in a wooden frame. Red spirit. Length, 8¼ inches. Shipping weight, 1 pound.
5L9406 ...... **50c**

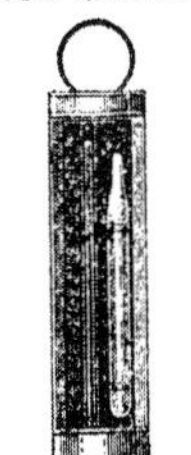

Storm Glass and Thermometer. Copper plated steel case. Black scale with white figures. Good grade spirit thermometer and medium size storm glass. Serviceable for outdoor use. Length, 7¾ inches. Shipping wt., 1½ pounds.
5L9408 ...... **80c**

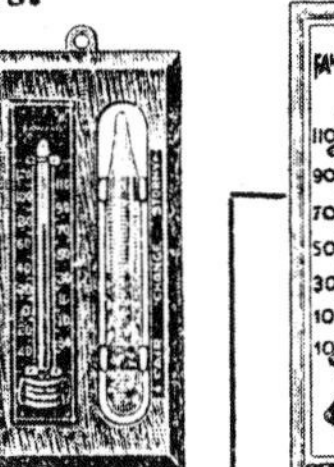

Storm Glass and Thermometer. Oak back, 3⅞x7⅞ inches. Good spirit thermometer with black scale and white figures. Extra large storm glass. Shipping weight, 2 pounds.
5L9410 $1.30

Distance Reading Thermometers.

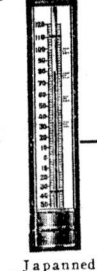

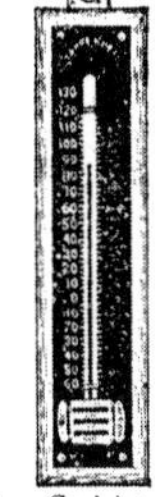

Hardwood, mahogany finish. Accurate. Black scale on white enameled metal. Magnifying spirit tube. Registers from 120 degrees above to 20 degrees below zero. Length over all, 7½ in. Shipping weight, 1 pound.
5L9412
65c

Cabinet Thermometer. Red spirit. Wood back, birch finish. Black scale with white figures and lines. Length, 8 in. Shipping weight, 1 lb.
5L9414
45c

Hardwood, mahogany finish. Etched black scale on silvered background. Spirit tube. Accurate. Registers from 120 degrees above to 50 degrees below zero. Length over all, 9½ inches. Shipping weight, 1 pound.
5L9416
75c

Metal Outside Thermometer. Spirit tube, magnifying. Easily read black figured scale on white enamel. Metal attachment for outdoor use. Registers from 120 above to 50 degrees below zero. Length over all, 8½ in. Shpg. wt., 1 lb.
5L9430
85c

Metal Window Thermometer. Spirit tube. Black figures on finished zinc will resist action of weather. Intended to be mounted on window frame, outside. Read from within the house. Two metal brackets for attaching. Length, 8 in. Shipping weight, 1 pound.
5L9432
95c

Outside Glass Spirit Tube Thermometer, black figures on an opaque background. Resists the action of the weather. Plain reading; high grade. Registers 120 above to 20 degrees below zero. Metal brackets for attaching. Length, 7½ in. Shipping wt., 1¼ lbs.
5L9434
$1.75

Cylinder Shape Glass Window Thermometer. Spirit tube, plain reading. Black figures on opaque glass. Metal case and cap. Metal attachment for outside use. Registers 120 degrees above to 40 below zero. Length over all, 8½ inches. Shipping wt., 1½ lbs.
5L9436
$1.25

Certified Fever Thermometers.

A fever thermometer to have any practical value must be accurate and thoroughly dependable. It must register quickly and exactly. It must be easily read. The mercury should shake down readily. These thermometers are certified by the makers and further guaranteed by us to be of this high quality. There are many cheaper fever thermometers on the market, but we do not handle them.

Shpg. wt., any fever thermometer, 10 oz.

Certified Fever Thermometer. Same as 5L9420, but in twisted aluminum case with chain and clasp. Registers in two minutes.
5L9418 .... **89c**

5L9419—Same as above, one minute registering **$1.20**

Certified Fever Thermometer. Magnifying tube. Length, 4 inches. Hard rubber case. Very accurate. Registers in two minutes.
5L9420 .... **85c**

Certified Fever Thermometer. Same as 5L9420, but registers in one minute.
5L9422 .... **$1.10**

Certified Fever Thermometer. Same as 5L9420, but registers in one-half minute.
5L9424 .... **$1.15**

Certified Rectal Fever Thermometer. Registers in one minute. In black rubber case. Shipping weight, 10 ounces.
5L9426 .... **$1.50**

Certified Veterinary's Thermometer is placed in a hard rubber case for protection. A high grade, accurate instrument. Shipping weight, 10 ounces.
5L9428 .... **$1.50**

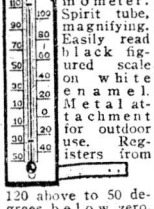

Distance Reading Thermometer. White enameled on steel. For inside or outside use. Spirit tube, black scale on white enamel. Registers from 120 degrees above to 60 degrees below zero. Length over all, 8 inches. Shipping wt., 1 pound.
5L9438 .... **95c**

Extra High Grade White Enameled Thermometer

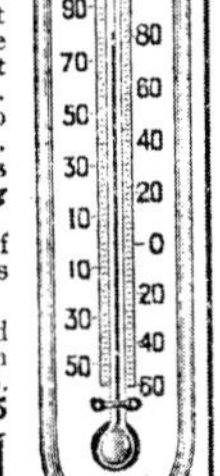

This fine enameled thermometer will stand up indefinitely. The large black figures on a pure white ground stand out plain and clear, and they stay that way because they are protected by an absolutely weatherproof transparent glaze. The surface is just like a piece of porcelain and the figures are under the glaze. It cannot rust. It cannot corrode or fade nor deteriorate in any way. The large column of red spirit is easy to see. It can be read quickly and easily, even in poor light. **This thermometer is accurate. There is a satisfaction in having a thermometer you can believe.**

The back of this instrument is made of metal. It is 10¼ inches long and 2¼ inches wide.

5L9440—Extra High Grade Enameled Thermometer. As illustrated, with scale reading to 60 degrees below zero. Shipping weight, 2¼ pounds **$2.75**

Oven Thermometers.

Hanging Oven Thermometer. Registers as illustrated. White scale on black background. Length over all, 6 in. Directions included. Shipping weight, 8 oz.
5L9442
$1.25

This instrument is placed directly in the oven, thus giving very accurate results. Directions with each thermometer, showing proper temperatures for bread, roasts, pastry, custards, etc. Mercury tube. Length, 5 inches. Shpg. wt., 1 lb. 2 oz.
5L9444 ...... **$1.50**

House Thermometers.

Distance Reading Thermometer at a low price. Well seasoned wood. Spirit tube. Black printed scale. Registers from 120 degrees above to 50 degrees below zero. Length over all, 7¾ in. Shpg. wt., 12 oz.
5L9446 ..... **25c**

Distance Reading Thermometer at a low price. Well seasoned wood. Spirit tube. Black printed scale. Registers from 120 degrees above to 50 degrees below zero. Length over all, 9¼ in. Shpg. wt., 12 oz.
5L9448 ..... **35c**

5L9472

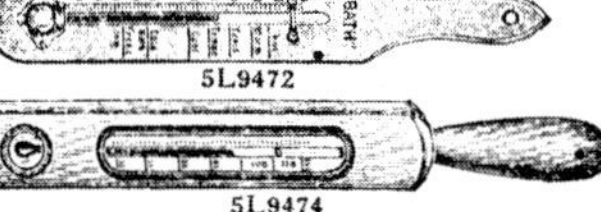

5L9474

Bath Thermometer, plain seasoned wood, black printed scale. Spirit tube, magnifying. Registers from 10 degrees to 120 degrees above zero. Length over all, 10 inches. Shipping weight, 1 pound.
5L9472 .... **27c**

Bath Thermometer, well seasoned wood. Spirit tube, magnifying. Registers from 10 degrees to 120 degrees above zero. Thermometer protected, as illustrated, against breakage. Length over all, 10 inches. Shipping weight, 1 pound.
5L9474 .... **45c**

Dairy Thermometers.

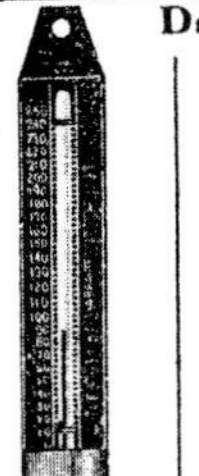

Flange Dairy Thermometer. Spirit tube. Black oxidized brass scale. Stamped white figures and lines. Sliding guard for bulb. Correct and carries maker's guarantee. Length, 8 in. Shpg. wt., 1 lb.
5L9450
79c

Flange Dairy Thermometer. Spirit. Nickel plated brass scale with stamped figures and lines. Sliding guard for bulb. Length, 8 in. Shpg. wt., 1 lb.
5L9452
59c

American Made. Dairy Thermometer. Mercury. **All glass.** Very accurate. Extra large figures; red lettering for scalding, cheese, churning and freezing points. Weighted with shot, stands upright in cream. Length, 8 in. Not to be compared with the cheap imported thermometers. The price is higher than the imported thermometers, but it is well worth the difference. Shpg. wt., 1 lb.
5L9454
65c

High Grade Incubator Thermometers and Hygrometer.

Incubator Thermometer. Same quality as 5L9462, but triangular in shape and without legs. Length, 4 inches. Shipping weight, 9 ounces.
5L9456 .... **49c**

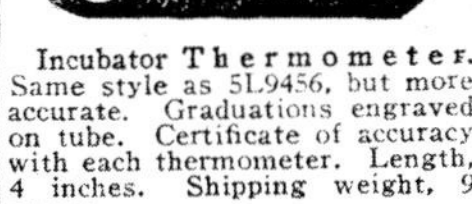

Incubator Thermometer. Same style as 5L9456, but more accurate. Graduations engraved on tube. Certificate of accuracy with each thermometer. Length, 4 inches. Shipping weight, 9 ounces.
5L9458 .... **65c**

Incubator Mercury Thermometer. An all around thermometer. Black scale on silvered background. Length over all, about 5 in. Shpg. wt., 1 lb.
5L9460 .... **65c**

Incubator Thermometer, with folding brass legs. Mercury. Extra large bulb. White graduations on black oxidized plate. Very sensitive. Length, 4 inches. Shipping weight, 9 ounces.
5L9462 .... **55c**

Incubator Hygrometer. A wet bulb thermometer, graduated with humidity percentages. Shows exact humidity by direct reading and is scientifically accurate. This instrument enables the user to maintain the proper degree of moisture in incubator with absolute accuracy. Shipping weight, 1¼ pounds.
5L9466 .... **90c**

Extra Wicks for Incubator Hygrometer. Shpg. wt., 1 oz.
5L9468 .... **8c**

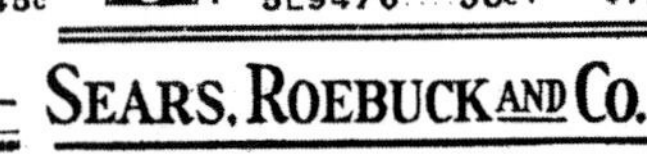

Brooder Thermometer. Mercury. Upright pattern. Large bulb. White figures on black scale. Length, 4 in. Shipping wt., 9 ounces.
5L9464 **35c**

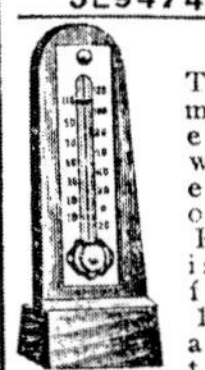

Brooder Thermometer. Spirit tube. Etched white scale on black background. Lgth. over all, 7 in. Shpg. wt., 5 oz.
5L9470
45c

Desk Thermometer, weathered oak. Registers from 120 above to 20 degrees below zero. 5¾ in. high over all. Black lettered scale on silvered background. Spirit tube. Distance reading. Shpg. wt., 1 pound.
5L9476 ... **50c**

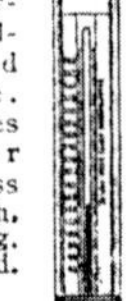

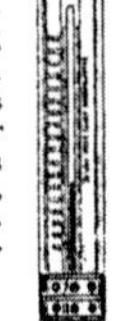

Home Candy Making Thermometer. Mercury. Book of candy recipes with each thermometer. Silver coated copper case. Black figures on silver plated brass scale. Length, 8 in. Shpg. wt., 1 pound.
5L9478
$1.50

Telescopes *and* Binoculars

Combined Telescope and Microscope, $6.00.

A high grade telescope, made by Vion of Paris. Especially suitable for Boy Scout use. Finished throughout in a hard glossy black enamel, a finish which is not only pleasing in appearance but very practical and durable. Vion is noted for the fine optical quality of his telescope lenses and the lenses used in this telescope maintain the usual Vion high standard. Length, extended, about 15⅝ inches; closed, about 5¾ inches. Object glass, 10 lignes or ⅞ inch in diameter; magnifying power, 12 times. Width of field, 40 yards at 1,000 yards distance.

So constructed that first draw may be unscrewed and used as a microscope having a magnifying power of 10 times. It is fitted with holder for glass slips or prepared objects, and we send with each telescope one prepared object and two plain glass slips. Shipping weight, 1¾ pounds.

5L9500—In tan leather pocket case.......... **$6.00**

High Grade Achromatic Telescopes, $4.75 to $8.00.

Real telescopes, made in Paris. Constructed throughout of brass; bodies covered with fine leather, morocco grained; brass draw tubes, highly burnished; trimmings all lacquered. Brass cap for lens and dustproof sliding cover for eyepiece.

5L9502—Achromatic Telescope. Diameter, object glass, 10 lignes or ⅞ inch; length, extended, about 14½ inches; closed, about 5¼ inches; magnifying power, 8 times. Width of field, 30 yards at 1,000 yards distance. Shpg. wt., 1½ lbs..... **$4.75**

5L9504—Achromatic Telescope. Diameter, object glass, 12 lignes or 1₁⁄₁₆ inches; length, extended, about 16⅛ inches; closed, about 6 inches; magnifying power, 11 times. Width of field, 35 yards at 1,000 yards distance. Shpg. wt., 1¾ lbs **$5.50**

5L9506—Achromatic Telescope. Diameter, object glass, 14 lignes or 1¼ inches; length, extended, about 17¼ inches; closed, about 6⅝ inches; magnifying power, 13 times. Width of field, 30 yards at 1,000 yards distance. Shipping weight, 1 pound 14 ounces.......... **$6.75**

5L9508—Achromatic Telescope. Diameter, object glass, 16 lignes or 1⅜ inches; length, extended, about 23¾ inches; closed, about 8¼ inches; magnifying power, 16 times. Width of field, 22 yards at 1,000 yards distance. Shipping weight, 2¼ pounds.......... **$8.00**

Hinged Cap Telescope, $10.00.

Made by Vion, the Paris maker, who is celebrated for the exceptionally high quality of his lenses. Fitted with special hinged cap and sliding cover for the eyepiece. Draw tubes and other metal parts are oxidized by a special process, which produces as fine a finish as is used on any optical instrument. Leather covering, morocco grained. Diameter, object glass, 14 lignes or 1¼ inches; length, extended, about 17 inches; closed, about 6¾ inches; magnifying power, 15 times. Width of field, 30 yards at 1,000 yards distance. Shipping weight, 1 pound 13 ounces.

5L9510.................... **$10.00**

Using Telescopes at Target Practice.

We give below the distances at which a 22-caliber bullet hole can be seen in a black bullseye on a white target with the various telescopes we list:

Telescopes 5L9502, 5L9504, 5L9506 and 5L9510	50 yards
Telescopes 5L9508, 5L9512 and 5L9514	75 yards
Telescope 5L9522	100 yards
Telescope 5L9518	125 yards
Telescope 5L9518 with Celestial Eyepiece 5L9520 (showing target inverted)	200 yards

Sunshade Telescopes, $13.50 and $18.50

Fine achromatic telescopes, made of brass throughout with burnished draw tubes and lacquered trimmings. Bodies covered with fine leather, morocco grained. Provided with sunshade, which can be extended forward to shade the object glass from the direct rays of the sun.

5L9512—Diameter, 19 lignes or 1₁₁⁄₁₆ inches; length, extended, about 29⅛ in.; closed, about 9¾ inches; magnifying power, 18 times. Width of field, 17 yards at 1,000 yards distance. Shipping weight, 4½ pounds.. **$13.50**

5L9514—Diameter, 22 lignes or 1₁⁵⁄₁₆ inches; length, extended, about 36½ in.; closed, about 10¾ inches; magnifying power, 23 times. Width of field, 16 yards at 1000 yards distance. Shipping weight, 5¼ pounds.. **$18.50**

Clamp for Telescopes or Binoculars.

To get the best results, a rigid support for the instrument is essential. This clamp is adjustable for any size telescope we list. Fine on a camera tripod. Shipping weight, 1 pound.

5L9516
$7.50

Extra High Grade Telescope, $34.00

The most powerful terrestrial telescope we handle. A high grade telescope in every respect. **It is the lenses** which make this instrument so much superior to ordinary telescopes, these lenses being especially ground from the finest optical glass, very carefully centered and accurately adjusted, combining to the greatest possible extent the finest definition and highest magnifying power. **For astronomical work** this telescope, with the celestial eyepiece listed at the right, is an ideal instrument. **For observation of the sun** a dark glass is mounted in the slide cover of the eyepiece. Magnifying power, 45 times. Draw tubes, trimmings and all exposed metal parts made with fine gunmetal finish, the very best and most expensive finish known for optical instruments. **This fine steel blue gunmetal finish will never tarnish nor rust,** and the draw tubes always work smoothly and easily. Body of instrument covered with fine leather, morocco grained. **This telescope is made with sunshade,** and instead of the ordinary cap it is provided with hinged metal cover which affords perfect protection to the object glass. **Length, extended, about 41½ inches; closed, about 12½ inches. Weight, 3½ pounds.** Diameter of object glass, 25 lignes or 2⅛ inches. Magnifying power, 45 times. Shipping weight, 6 pounds.

5L9518 .... **$34.00**

Celestial Eyepiece, to fit Telescope 5L9518, for astronomical work, increasing power to 68 times. With all celestial eyepieces the image **is seen inverted,** a matter of no consequence in astronomical observations, but which, of course, renders such eyepieces unsuitable for terrestrial work. Shipping weight, 1 pound 2 ounces.

5L9520.. **$10.00**

Vion's High Power Double Achromatic Telescope, $27.50

Made with double achromatic eyepiece and double achromatic object glass. Special construction of objective and eyepiece makes possible an extraordinarily high magnifying power in a comparatively small and compact instrument. Although object glass measures only 19 lignes (1₁₁⁄₁₆ inches) in diameter, and total length of instrument extended is only about 20¾ inches, the magnifying power is forty times. Width of field, 17 yards at 1,000 yards distance. Draw tubes finished in dead black. Body covered with fine leather, morocco grained. Provided with leather caps for each end and shoulder strap with loop for attaching to the body of the instrument. Length, extended, about 20¾ inches; closed, about 8 inches; object glass, 19 lignes or 1₁₁⁄₁₆ inches, magnifying power, 40 times. Width of field, 17 yards at 1,000 yards distance. Shipping weight, 3 pounds.

5L9522 .... **$27.50**

HIGH GRADE BINOCULARS

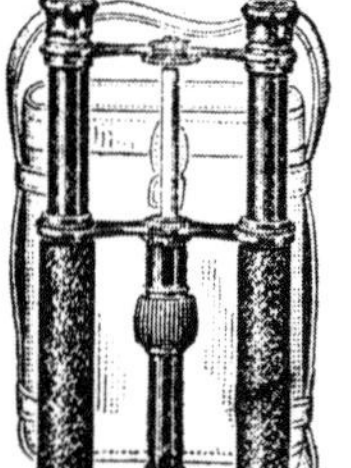

Seven-Power Petit Binocular Telescope, $31.50

Maker—L. Petit, Paris. **Size,** about 6¾ inches long, extended, 4¾ inches closed, 3½ inches wide, ¾ inch thick. Weight, 11 ounces. **Object glasses**—6 lignes in diameter. **Field of view**—70 yards across at 1,000 yds. distance. Convenient and very effective. High class in every detail. Fitted with the highest grade achromatic lenses. Have unusual clearness and perfect definition. Rapid draw style of construction, and adjustable for pupilary distance. Furnished in a strong leather covered case, measuring 1¾x4x5⅛ inches with shoulder straps. All metal parts of brass with glossy black enameled finish. Covering, pebble grained morocco leather. Shipping weight, 2½ lbs.

5L9524 With leather case.... **$31.50**

Oigee Prism Binoculars.

Oigee Binoculars are unsurpassed as regards the well known superior qualities of prism binoculars and are unparalleled in certain mechanical details embodied in their construction, and protected by patent. These glasses, with the central focusing arrangement, allow for exceedingly rapid adjustment, as in the case of moving objects. In addition to this, the right hand eyepiece is made movable, thus allowing an adjustment to equalize the difference in strength of the two eyes. Size, closed, 4 inches high. Field of vision, 115 yards across at 1,000 yards distance. Magnifying power, 8 times. Covering, black. Shipping weight, 4¾ pounds.

5L9526—Complete with leather case..... **$35.00**

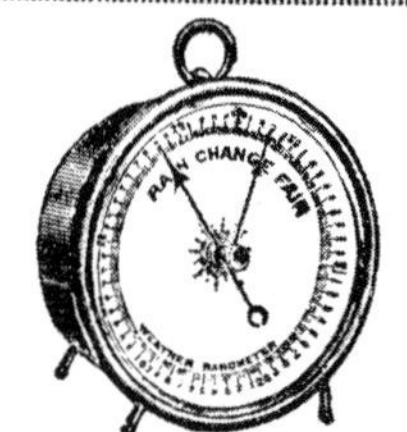

Aneroid Weather Barometer. Foretells the Weather.

A registering barometer in which the indicating hand is the darkened hand which responds to atmospheric pressure or change. The lighter hand is the set hand. To know of a change, turn the brass knob so that the hand covers the black one. A change in atmosphere will be indicated by the movement of the black hand. General directions accompany each instrument. This type of instrument is the most scientific for the purpose intended that we know of. Dial is 5 in. in diameter, brushed brass jacket. Compact and substantially made. Shipping weight, 3¾ pounds.

5L9760 .... **$7.98**

Oigee Prism Binoculars.

These binoculars are of the same make and general construction as our 5L9526 shown on this page, with the exception of the magnifying power, which is 6 times instead of 8. The same general advantages of rapid adjustment, the large range of vision and clearness of detail are all represented in this glass, as is the right hand eyepiece, made movable for equalizing the difference in strength of the two eyes. Field of vision, 100 yards across at 1,000 yards distance. Magnifying power 6 times. Shipping weight, 4¼ pounds.

5L9530
Complete with leather case.......... **$31.50**

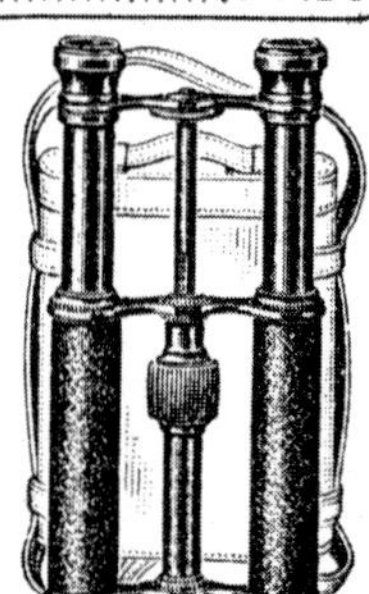

Eight-Power Petit Binocular Telescope, $32.50.

Exactly the same style as 5L9524 Seven-Power Petit Binocular Telescope, but larger, measuring about 7½ inches long extended, 5¼ inches closed, 3¾ inches wide and 1 inch thick; object glasses, 8 lignes or ¾ inch in diameter. This instrument has a more brilliant field than the seven-power glass, owing to the greater diameter of the lenses, and the magnifying power is 25 per cent greater. While larger than the seven-power glass, it is still of very convenient size. Shipping weight, 3 pounds.

5L9532—With strong leather covered case...... **$32.50**

High Power Field and Opera Glasses

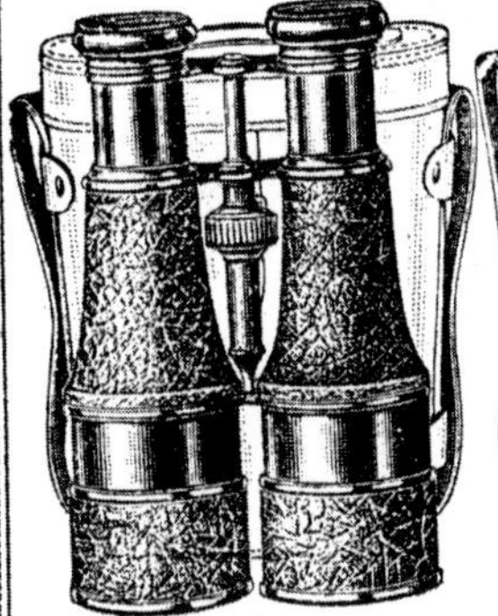

Petit Standard Field Glass.
Magnifying Power—6 times. Width of field, 55 yards at 1,000 yards distance. **Size**—Extended, 7¾ in.; closed, 6 inches. Weight, 2 pounds. **Object Glasses**—26 lignes or 2 3/16 inches in diameter. **Finish**—All metal parts, glossy black enamel; covering, black genuine morocco leather. **Case**—Good quality leather covered, with shoulder strap. In both optical qualities and mechanical construction this instrument is beyond criticism. We recommend it as a high grade, finely made glass that will give satisfaction. Shipping weight, 5 lbs. **$23.00**
5L9600

High Power Field Glass.
Magnifying Power—6½ times. Width of field, 40 yards at 1,000 yards distance. **Size**—Extended, 9 in.; closed, 7 9/16 in. Weight, 1 pound 11 ounces. **Object Glasses**—21 lignes or 1⅞ inches in diameter. **Finish**—All metal parts, glossy black enamel; covering, black genuine morocco leather. **Case**—First quality leather covered, with shoulder strap. This is an ideal instrument for use where the distances are great. Of high grade optical and mechanical construction throughout, an instrument that we can highly recommend. Shipping weight, 4½ pounds. **$20.50**
5L9602

Chevalier Field Glass.
Magnifying Power — 4 times. Width of field, 76 yards at 1,000 yards distance. **Size**—Extended, 6¼ inches; closed, 5¼ inches. Weight, 1¼ pounds. **Object Glasses**—24 lignes or 2⅛ inches in diameter. **Finish**—Draw tubes, glossy black enamel; other metal parts, glossy black enamel and nickel plated; covering, fine black pebble grained leather. **Case**—Covered with artificial leather, with shoulder strap. A well made, serviceable field glass. Shipping weight, 4 pounds.
5L9604 . . . . **$11.50**

Grammont Field Glass.
Magnifying Power—5½ times. **Size**—Extended, 7¾ inches; closed, 6⅝ in. Weight, 1⅜ lbs. **Object Glasses**—26 lignes or 2 3/16 inches in diameter. **Finish**—All metal parts, glossy black enamel; covering, fine quality smooth tan leather. **Case**—Extra quality, covered with smooth tan leather, velveteen lined, with shoulder strap. A large, powerful glass, suitable for general use at the seashore, in the mountains or on the farm. Shipping weight, 4½ pounds.
5L9606 . . . . . **$16.00**

Grammont Field Glass.
Magnifying Power—4 times. Width of field, 76 yards at 1,000 yards distance. **Size**—Extended, 7⅛ inches; closed, 6⅛ inches. Weight, 1 pound 7 ounces. **Object Glasses**—24 lignes or 2⅛ inches in diameter. **Finish**—All metal parts, glossy black enamel; covering, fine black pebble grained leather. **Case**—Leather covered, with shoulder strap. We recommend this instrument to anyone desiring a glass with fine optical qualities at a moderate price. Shpg. wt., 4 lbs. **$12.50**
5L9608

Petit Rapid Draw Field Glass, $18.75.
Magnifying Power — 5 times. Width of field, 50 yards at 1,000 yards distance. **Size** — Extended, 6 9/16 inches; closed, 4 9/16 in. Weight, 1⅛ pounds. **Object Glasses**—17 lignes or 1½ inches in diameter. **Finish** — Draw tubes, dull black enamel; other metal parts, glossy black enamel; covering, black genuine morocco leather. **Case** — First quality leather covered, with shoulder strap.
This field glass is of the "rapid draw" style of construction and may be instantly extended to an exact focus by grasping the disc on the top of the crossbar between the two barrels and pulling them straight out. Shipping weight, 2¼ lbs.
5L9610 . . . . **$18.75**

We state diameters of object glasses in lignes, which is the unit of measurement used in the French optical trade.

10 lignes—About	⅞ in.	17 lignes—About	1½ in.
11 lignes—About	1 in.	19 lignes—About	1 11/16 in.
12 lignes—About	1 1/16 in.	21 lignes—About	1⅞ in.
13 lignes—About	1⅛ in.	22 lignes—About	1 15/16 in.
14 lignes—About	1¼ in.	24 lignes—About	2⅛ in.
15 lignes—About	1 5/16 in.	26 lignes—About	2 3/16 in.
16 lignes—About	1 7/16 in.		

Measurements are made with the object glasses taken out of the instruments, so the figures given in our descriptions represent the full diameters of the glasses.

Chevalier Field Glass, $8.50.
Magnifying Power — 3½ times. Width of field, 70 yards at 1,000 yards distance. **Size**—Extended, 5⅞ in.; closed, 4⅛ in. Wt., 15 ounces. **Object Glasses** — 19 lignes or 1 11/16 inches in diameter. **Finish**—Metal parts, glossy black enamel; covering, black pebble grained leather. **Case**—Imitation leather, with shoulder strap. A well made, compact, serviceable instrument of the tourist type, with short bodies and long draw tubes. Shipping weight, 2¾ pounds.
5L9612 **$8.50**

Petit Extra Power Field Glass, $21.00.
Magnifying Power — 5 times. Width of field, 55 yards at 1,000 yards distance. **Size** — Extended, 5¼ inches; closed, 4⅝ inches. Weight, 1⅜ pounds. **Object Glasses**—21 lignes or 1⅞ inches in diameter. **Finish**—All metal parts, glossy black enamel; covering, black genuine morocco leather. **Case**—First quality leather covered, with shoulder strap.
This is a very compact instrument. Especially suited for tourists' use. Shpg. wt., 2½ lbs. **$21.00**
5L9614

Chevalier Field Glass.

Magnifying Power—3 times. Width of field, 130 yards at 1,000 yards distance. **Size**—Extended, 4⅛ inches; closed, 3¼ in. Wt., 10½ oz. **Object Glasses**—19 lignes or 1 11/16 inches in diameter. **Finish**—Metal parts, glossy black enamel; covering, black pebble grained leather. **Case**—Black cloth in artificial leather, with shoulder strap. A good low priced field glass. Shipping weight, 1⅞ pounds.
5L9616 . . . . **$7.75**

Extra Brilliant Field Glass.

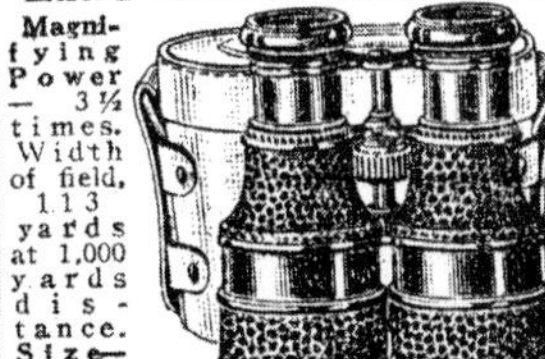

Magnifying Power — 3½ times. Width of field, 113 yards at 1,000 yards distance. **Size**—Extended, 4½ in.; closed, 3⅜ in. Wt., 1½ lbs. **Object Glasses**—24 lignes or 2⅛ in. in diameter. **Finish**—All metal parts, glossy black enamel; covering, black genuine morocco leather. **Case**—First quality leather, with shoulder strap.
Has extreme brilliancy of illumination, very large field and fine definition, qualities possible only in a glass of the most perfect optical construction. Shpg. wt., 3½ lbs.
5L9618 . . . . **$20.00**

Chevalier Field Glass, $8.50.
Magnifying Power — 3½ times. Width of field, 76 yards at 1,000 yards distance. **Size**—Extended, 5⅛ in.; closed, 4⅛ in. Wt., 12½ ounces. **Object Glasses**—19 lignes or 1 11/16 inches in diameter. **Finish**—Draw tubes, dead black enamel; other metal parts, glossy black enamel, with two narrow gold plated bands; black pebble grained leather covering. **Case**—Artificial leather, with shoulder strap.
Extra large eyepieces make this an exceptionally effective glass. Shipping wt., 2 lbs. 9 oz. **$8.50**
5L9620

Grammont Opera Glass, $4.85.
Size—Extended, 3⅛ in.; closed, 2⅜ in. **Object Glasses**—13 lignes or 1⅛ inches in diameter. **Finish**—Draw tubes, gold plated; other metal parts, glossy black enamel; covering, dark green leather with two ornamental beaded gold plated bands.
A well made, serviceable instrument, with achromatic lenses. Shipping wt., 1½ lbs.
5L9636—With black leather case **$4.85**

High Grade Opera Glasses

Petit Folding Opera Glass.
Size—Extended, 3⅝ inches; closed, 2⅜ inches. **Object Glasses**—11 lignes or 1 inch in diameter. **Finish**—Black enameled metal parts and covered with black leather.
A very thin flat model general purpose glass as well as for the opera. Shipping weight, 1 pound.
5L9626 . . . . **$10.75**

Petit Special Opera Glass, $13.50.
Size—Extended, 3 inches; closed, 2⅜ inches. **Object Glasses**—17 lignes or 1½ inches in diameter. **Finish**—Draw tubes and center post, polished aluminum; other metal parts, glossy black enamel; covering, first quality black morocco leather.
Field of view is large and brilliant, magnifying power is high and the extra large fine lenses result in unusually good definition. Shpg. wt., 1½ lbs.
5L9628—With fine leather case **$13.50**

Chevalier Opera Glass, $4.00.
Size—Extended, 3 in.; closed, 2⅜ inches. **Object Glasses** —13 lignes or 1⅛ inches in diameter.
Finish—Draw tubes, gold plated; other metal parts, glossy black enamel; covering, black pebbled leather. Fitted with genuine achromatic lenses. Shipping weight, 1½ lbs.
5L9622
With leather case **$4.00**

Hearing Instruments.
Dr. Fossgate's Vibro-Phones, $2.48, $3.50 and $3.98.

An improved form of Conversation Tube, with an internal diaphragm for dividing and intensifying the sound waves. Very finely made; spiral spring lining, hard rubber earpiece and special metallic mouthpiece finished in black enamel. Length over all, 40 inches. Diameter of mouthpiece, 2½ inches. Weight, 9 ounces. Shipping weight, 1¼ pounds.
5L9734—Vibro-Phone, covered with black mohair. **$3.50**
5L9736—Vibro-Phone, covered with black silk . . **3.98**
5L9738—Vibro-Phone, covered with black cotton fabric **2.48**

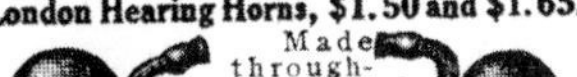

London Hearing Horns, $1.50 and $1.65.

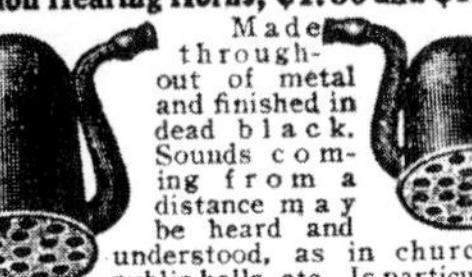

Made throughout of metal and finished in dead black. Sounds coming from a distance may be heard and understood, as in churches, public halls, etc. Is particularly adapted to those who are only moderately deaf, but for those who are very deaf we recommend the conversation tubes.
5L9740—London Hearing Horn, black finish, 2½ inches high. Shipping weight, 6 ounces **$1.50**
5L9742—London Hearing Horn, black finish, 4 inches high. Shipping weight, 8 ounces. Each **$1.65**

Petit Opera Glass, $9.75.
Size—Extended, 3⅜ in.; closed, 2½ inches. **Object Glasses** —15 lignes or 1 5/16 in. in diameter. **Finish**—All metal parts, fine glossy black enamel; covering, first quality black morocco leather. Made with extra fine achromatic lenses and best workmanship throughout, of fine optical qualities. Shpg. wt., 1¾ lbs. **$9.75**
5L9634—With leather case . . . . . .

WITH BOOKS AND AUTHORS

The Six Best Sellers

Here are six of the most popular new fiction titles. Bound in cloth. Average 300 pages. Size, 5½x7¾ inches. Shipping weight, each, 1¼ pounds.

The Country Beyond. By James Oliver Curwood. 3L1061 .... $1.68

Certain People of Importance. By Kathleen Norris. 3L1062 .... $1.58

Flowing Gold. By Rex Beach. 3L1063 .... $1.68

Cappy Ricks Retires. By Peter B. Kyne. 3L1064 .... $1.68

The Prairie Child. By Arthur Stringer. 3L1065 .... $1.68

Wanderer of the Wasteland. By Zane Grey. 3L1106 .... $1.68

Mary J. Holmes' Books.

Bound in cloth. Size, 5⅛x7⅛ inches. Average 250 pages. Shpg. wt., each, 1⅛ pounds.

Aikenside, Bad Hugh, Cousin Maude, Darkness and Daylight, Dora Deane, Edith Lyle's Secret, English Orphans, The, Ethelyn's Mistake, Family Pride, Homestead on the Hillside, Leighton Homestead, Lena Rivers, Maggie Miller, Marian Grey, Meadowbrook Farm, Mildred, Millbank, Miss McDonald, Rosamond, Rose Mather, Tempest and Sunshine, Woman Against Woman

3L86—Each, 48c; any two for .. 94c

Mrs. Southworth's Books.

Bound in cloth. Size, 5⅛x7⅜ inches. Average 250 pages. Shipping wt., each, 1⅛ lbs.

Beautiful Fiend, A; Bridal Eve, The; Bride of Llewellyn, The; Bride's Dowry, The; Bride's Fate, The; Broken Engagement; Capitola's Peril; Changed Brides; Christmas Guest, The; Cruel as the Grave; Curse of Clifton; Deserted Wife; Discarded Daughter; Fair Play; Gypsy's Prophecy; Haunted Homestead; Hidden Hand; How He Won Her; India, or the Pearl of Pearl River; Ishmael; Lost Heiress, The; Lost Heir of Linlithgow; Love's Labor Won; Maiden Widow; Missing Bride, The; Mother-in-Law, The; Mystery of Raven Rocks; Noble Lord, A; Prince of Darkness and Artists' Love; Self Raised, or From the Depths; Three Beauties, The; Three Sisters, The; Two Sisters, The; Tried for Her Life; Unknown, The; Victor's Triumph; Vivian; Widow's Son; Wife's Victory

3L88—Each, 48c; any two for 94c

Rosa N. Carey's Books.

Bound in cloth. Size, 5⅛x7⅛ inches. Average 250 pages. Shipping weight, each, 1⅛ pounds.

Aunt Diana, Averil, Esther, Herriot's Choice, Lover or Friend, Mary St. Johns, Merle's Crusade, Not Like Other Girls, Only the Governess, Our Bessie, Queenie's Whim, Wee Wifie, Wooed and Married

3L1714—Each, 48c; any two .. 94c

Old Favorite Library

A library of popular and standard works of some of the world's best authors. Bound in cloth. Size, 5⅛x7⅛ inches. Average 250 pages. Shipping weight, each, 1⅛ pounds.

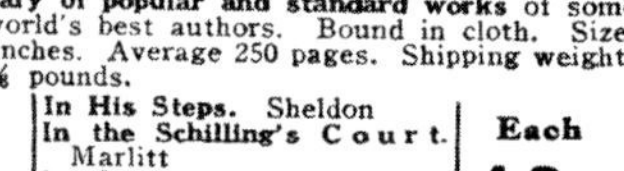

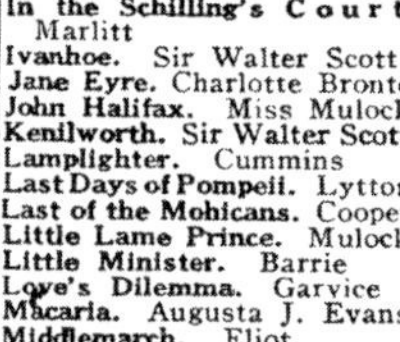

Each 48c

Alice in Wonderland. Carroll
Andersen's Fairy Tales.
Alhambra. Irving
Beulah. Wilson
Black Beauty. Sewell
Black Rock. Ralph Connor
Budge and Toddie. John Habberton
Children of the Abbey. Roche
Child's History of England. Charles Dickens.
Count of Monte Cristo. Alexander Dumas
David Copperfield. Dickens
Deerslayer. J. F. Cooper
East Lynne. Mrs. Wood
Edmund Dantes. Dumas
Emerson's Essays
Felix Holt. Eliot
Fiery Ordeal, The. Clay
First Violin. Fothergill
Green Mountain Boys. D. P. Thompson
Grimms' Fairy Tales.
Hans Brinker. M. Mapes Dodge
Hardy Norseman. Lyall
Helen's Babies. Habberton.
Hypatia. Chas. Kingsley
Inez. Augusta J. Evans
In His Steps. Sheldon
In the Schilling's Court. Marlitt
Ivanhoe. Sir Walter Scott
Jane Eyre. Charlotte Bronte
John Halifax. Miss Mulock
Kenilworth. Sir Walter Scott
Lamplighter. Cummins
Last Days of Pompeii. Lytton
Last of the Mohicans. Cooper
Little Lame Prince. Mulock
Little Minister. Barrie
Love's Dilemma. Garvice
Macaria. Augusta J. Evans
Middlemarch. Eliot
Mysterious Island. Verne
Oliver Twist. Dickens
Olivia. Garvice
Only One Sin. Clay
Pathfinder. J. F. Cooper
Pilgrim's Progress. Bunyan
Pioneers. J. F. Cooper
Prairie. J. F. Cooper
Prince of the House of David. Rev. J. H. Ingraham
Robinson Crusoe. Defoe
Romance of Two Worlds. Marie Corelli
St. Elmo. Augusta J. Evans
Samantha at Saratoga.
Scarlet Letter. Hawthorne
Scottish Chiefs. Jane Porter
Second Wife, The. Marlitt
Silas Marner. Eliot
Sketch Book. Irving
Spy, The. J. F. Cooper

Each 48c

Stepping Heavenward. Prentiss
Sweet Girl Graduate. Mrs. L. T. Meade
Swiss Family Robinson.
Tales of Sherlock Holmes. Doyle
Tales From Shakespeare. Charles and Mary Lamb
Tale of Two Cities. Dickens
Thaddeus of Warsaw. Porter
Thelma. Corelli
Three Guardsmen. Dumas
Tom Brown at Oxford. Hughes
Tom Brown's School Days. Thomas Hughes
Treasure Island. Stevenson
Twenty Thousand Leagues Under the Sea. Jules Verne
Two Orphans. D'Ennery
Uncle Tom's Cabin. Stowe.
Vendetta. Corelli
Weaker Than Woman. Clay
We Two. Edna Lyall
Wide, Wide World. Warner
Woman's Error. Clay
Woman's War. Corelli
Won by Waiting. Lyall
Wormwood. Marie Corelli

3L125—Each ...... 48c; any two for .... 94c; any three for $1.39

The Lambskin Library.

Do you know of a nicer gift to anyone than a book, especially a leather bound volume? Perhaps you have wanted to use this kind of a gift, but the price was a little high. Did you ever want to own a collection of good books in a handsome binding to ornament your library? We offer you here an opportunity to buy the best of both modern fiction and the more serious books, bound in genuine red leather, decorated in gold, for 85 cents—less than you have paid for cloth bound books. Don't fail to take advantage of this chance. Size, 4¾x7¼ inches. Average 350 pages. Shipping weight, 1¼ pounds.

Adventures in Contentment. David Grayson
Alice in Wonderland. Lewis Carroll
Autobiography of Benjamin Franklin.
Black Beauty. Anna Sewell
Bob, Son of Battle. Alfred Ollivant
Casuals of the Sea. William McFee
Cheerful by Request. Edna Ferber
Dracula. Bram Stoker
Further Side of Silence. Sir Hugh Clifford
Gold. Stewart Edward White
Impressions of Theodore Roosevelt. Abbott
Ivanhoe. Sir Walter Scott
Lord Jim. Joseph Conrad
Lorna Doone. Part I. R. D. Blackmore
Lorna Doone. Part II. R. D. Blackmore
Magnificent Ambersons, The. Tarkington
Mother. Kathleen Norris
Octopus, The. Frank Norris
Pit, The. Frank Norris
Pieces of Eight. Richard Le Gallienne
Riverman, The. Stewart Edward White
Romola. George Eliot
Ruggles of Red Gap. Harry Leon Wilson
Stamboul Nights. Harry Griswold Dwight
Story of Gosta Berling. Selma Lagerlof
Story of My Life. Helen Keller
Tale of Two Cities, A. Dickens
Tales From Shakespeare. Lamb
Three Musketeers, The. Part I. Dumas
Three Musketeers, The. Part II. Dumas
Trimmed Lamp, The. O. Henry
Two Years Before the Mast. Dana
Up From Slavery. Booker T. Washington

3L85—Per volume 85c

The Best of the New Fiction.

We have selected from the most recent publications books that we believe will be discussed wherever men and women gather, who are interested in literature or in life and its portrayal. Every volume bound in cloth, averaging 380 pages. Size, 5½x7¼ inches. Shipping wt., ea., 1½ lbs.

3L943—The Pelham Affair. Louis Tracy $1.68
3L944—The Red Marshall. Gordon Casserly 1.68
3L1118—On Tiptoe. Stewart E. White 1.58
3L5551—Carmac's Folly. Gilbert Parker 1.68
3L5554—The Tale of Triona. William J. Locke 1.68
3L5555—Viola Gwyn. George Barr McCutcheon 1.68
3L5574—Fair Harbor. Joseph C. Lincoln 1.68
3L5573—The Shadow of the East. By the author of "The Sheik." 1.68
3L5559—The Breaking Point. Mary Roberts Rinehart 1.68
3L5562—Babbitt. Sinclair Lewis 1.68
3L5563—This Freedom. A. S. M. Hutchinson 1.58
3L5564—The Evil Shepherd. E. Phillips Oppenheim 1.58
3L5565—Peregrine's Progress. Jeffery Farnol 1.58
3L1113—Command. Wm. McFee 1.68
3L1115—Glimpses of the Moon. Edith Wharton 1.68
3L1116—Skippy Bedele. Owen Johnson 1.58

"My Passion for Him Was But a False Glamour."

The Love Story of a Movie Star.

The True Biography of a Famous Screen Actress.

Here is not only a story that makes a searching analysis of the heart of a woman in love, but its scenes are laid so vividly and so intimately in the "behind the scenes" atmosphere of the moving picture studios, that the reader forgets herself at times and believes that she herself is living the romantic exciting and soul testing experiences of the heroine of this wonderful tale.

This book, as no other story ever has done, lays bare all sides of the life in the moving picture colonies, where the stars we know so well in the pictures are "off stage." What are these beings, chosen of the gods, like in their homes, in their associations with each other, in their loves, in their hatreds, in their inspired moments, in their unworthy temperamental outbursts? Aside from the absorbing charm of the story what makes this book notable is that it is actually true. It is an actual human document. It is the authentic autobiography of one of the most famous of the moving picture actresses.

Who is the lovely heroine in real life? Which one of the stars whose face is known almost the world over is Nella? This same lovely heroine perhaps has thrilled you and entertained you many times at your favorite theater. So lifelike is the picture she draws of herself, so vividly is portrayed her striking personality, that every reader is convinced he recognizes her. Will Nella eventually reveal her identity to the world? Who knows? Stranger things have happened.

Over 300 pages. Bound in cloth. Size, 5½x7¾ inches. Shpg. wt., 1⅝ lbs.

3L1047—The Love Story of a Movie Star 89c

Special Bargains

58c EACH

Reprint Fiction Sold Elsewhere for 75c and $1.00.

Here are some of the best stories you ever read at a really low price. The cream of the work of Mary Roberts Rinehart, Eleanor Porter, Florence Barclay and others who have been acknowledged as story tellers of the first class. All bound in cloth with colored jackets. Size, 5½x7¾ inches. Average 250 to 400 pages each. Shipping weight, 1¼ lbs.

Mary Roberts Rinehart.
3L075—Dangerous Days.
3L300—Love Stories.
3L138—The Amazing Interlude.

Eleanor H. Porter.
3L2436—Just David.
3L415—Oh, Money, Money.

3L686—Tangled Threads.
3L687—Across the Years.
3L688—The Tie That Binds.
3L453—Dawn.
3L429—Six-Star Ranch.
3L1291—The Road to Understanding.

Grace Miller White.
3L2233—Tess of the Storm Country.
3L510—The Shadow of the Sheltering Pines.
3L689—Storm Country Polly.

Temple Bailey.
3L494—The Tin Soldier.
3L1262—The Glory of Youth.
3L1264—Contrary Mary.
3L1263—Mistress Anne.

Elmer R. Gregor.
3L677—Son of Rolling Thunder.

Florence L. Barclay.
3L051—The Mistress of Shenstone.
3L1268—Through the Postern Gate.
3L574—The White Ladies of Worcester.
3L805—The Upas Tree.

Edgar Rice Burroughs.
3L2470—Jungle Tales of Tarzan.
3L610—Tarzan the Untamed.

Robert W. Chambers.
3L1048—In Secret.

Each 58c

Are You at Ease Among Strangers?

People who are well poised on all occasions make friends, win social prestige and enviable business success.

People of culture and good breeding can be recognized at once. They are calm, well poised, at ease. They have **dignity and self assurance.** Why? Because they are fortified with the knowledge of what to do, what to say and how to act on every occasion.

But there are other persons who are never at ease before people. They are awkward, self conscious, fearful all the time that they will do something that is not right. These people are afraid to accept invitations because they don't know what to wear. They are afraid to go to receptions, balls, informal gatherings, because they do not know how to act. They have a haunting fear of making social blunders. They don't know how to acknowledge formal invitations, they do not know the social customs that every man and woman should know. What is the result? Such people are ostracized. They offend, by their ignorance and social mistakes, people who would be friends to them and would help them in times of need.

Do You Stand in Your Own Light?

Which kind are you? **Are you the sort that people remark about:** "He is impossible—he needs the rough corners knocked off?" Don't be of this kind when it is so easy to be cultivated, poised, a real gentleman or a real lady. Don't ever be caught wanting on any occasion, for it is now so simple to be well bred and at ease among people. With the aid of **"Etiquette Made Easy,"** the authoritative guide to correct manners, you need never be at a loss to know what to do on all occasions. This book will give you the key to what is proper to do, to write, to wear, to say, no matter where you are or what situation may arise. Everyone who is concerned about making a good impression on people, or values the good regard of friends, certainly should have "Etiquette Made Easy" close at hand.

"Etiquette Made Easy" is a complete treatment of this highly important subject in a handy, readable form. This book gives you everything that the larger encyclopedias on etiquette give you, but in a delightful, concise yet comprehensive one-volume edition. 150 pages. Cloth. Size, 5⅛x7½ inches. Shipping weight, 1 pound.

3L3500—"Etiquette Made Easy".................................. **89c**

Dance Well and Be Popular

Some people seem to be born good dancers; they are light on their feet and naturally fall into all the new steps. Everyone wants to dance with them and their programs are filled before the evening has hardly begun. **Do you envy these people?** You need not do so. For you can be just as popular. Don't be a wall flower, for good dancing is a magic wand which sweeps you to immediate social popularity.

Get this marvelous, up to the minute book, "Dancing Made Easy," which will give you all the latest steps so simply explained that you can learn them easily in one evening. You don't need to take expensive lessons from some instructor. This remarkable book sets forth new steps that are being danced, with illustrations and diagram charts. It will teach you the **latest variations of jazz,** which, when danced properly, are pretty and graceful; the toddle, the camel walk, the canter, the pivot, the one-step, the fox trot, two-step, waltz—in fact, all the up to the minute as well as the older established society dances. **Remember,** too, that here every dance is carefully explained with pictorial diagrams that make the dances so easy to learn from this book. Be the first to introduce in your crowd the new steps. 278 pages. Bound in cloth. Size, 5⅛x7½ inches. Shipping weight, 1 lb.

3L3997—"Dancing Made Easy".............. **89c**

Can You Entertain Successfully?

"Don't we always have a good time at Mary's?" How often have you uttered such words on coming away from a party at the house of a friend where you always have a thoroughly enjoyable evening. Why is it you eagerly accept this friend's invitations and happily look forward to spending an evening at her house?

Perhaps you try to analyze to see why her parties are such a success. Your other friends say "It is a gift; Mary has a knack of entertaining." But it is not a gift—it is just careful planning. She has learned how to entertain, and you can do the same.

It Is Just Knowing How.

Your parties at which there seemed to be something lacking can be made just as entertaining, just as charming as those of your successful friend. Don't have that uncertain feeling after your guests have gone, "I wonder if everybody really had a good time?" **Make sure** your guests have a good time. "Entertaining Made Easy" will tell you how to make your entertainments so attractive that you and your family will be **most popular.** This marvelous book reveals everything you must know to entertain successfully. It will show you all the latest unique "stunts" that will make your parties distinctive, more amusing than ordinary parties. Order this book today and entertain at your next party in the newest up to date way. 164 pages. Bound in cloth. Size, 5⅛x7½ inches. Shipping weight, 1 lb.

3L3900—"Entertaining Made Easy".............................. **89c**

An Inspiring Edition of The Holy Bible at an Unusually Low Price

Profusely Illustrated—Full Divinity Circuit

A revival of Bible reading is sweeping the country. This renewed interest in the Bible, which had its source during the war, has grown stronger and stronger. Today the Bible stands glorified—greater than at any time in its centuries of history. Now you will want the best of Bibles for your home. Have you a suitable Bible for all your family, from the young members to grandfather and grandmother, a Bible which all can use with comfort and joy and which fulfills the utmost service?

Here is a massive, dignified, wonderful Bible, expressive to the very highest point of the spirit of the Holy Book. It is superbly and supremely beautiful, heart reaching and inspiring and it will be the most loved by every member of the family of any edition of the Bible you ever owned. Here is the Bible that you will be glad to see on your parlor or living room table and which you will come to look upon as a valued family possession.

An Ideal Home *and* Old Folks Bible

This is an ideal Bible for the home—a family Bible in all the term implies. This Bible is most suitable for the old folks and for those who like large, clear print for most comfortable reading. It is printed in beautifully clear extra large pica type, which is the largest type ever used in a Bible of this size. The pages are printed in two columns with a short type line which is an additional aid to eyes that are not strong. The paper of a soft, pleasing color tone and dull finish is most restful to the sight. The fine heavy paper makes the pages easy to turn and an older person whose hands are a bit unsteady can pass through this Bible to and fro with the greatest comfort.

This Bible is again particularly fine and beautiful, for throughout its pages is a wealth of illustrations, the work of the greatest artists of all ages, that breathe pictorial life into the sacred writings.

This superb Bible is 7 inches wide by 9½ inches high and 2½ inches thick. It contains 1,935 pages and 664 illustrations. It is handsomely bound in the best quality black artificial leather with full divinity circuit (overlapping covers), semi-flexible style, beautifully morocco grained, so that it is almost impossible to tell it from genuine leather. The words "Holy Bible" are stamped on side and back in real gold leaf. It has burnished red edges, round corners, headband and silk ribbon marker. No money has been spared to make it the best Bible for the home ever published.

This wonderful home Bible contains a complete dictionary of the Bible and of Scriptural proper names. It combines a Family Reading Bible, and an illustrated treasury of Bible art and pictures. It is, in short, an ideal home Bible with 60,000 references for every member of the family.

Sets New Price Standard *in* Bibles

With all these rare features that make this edition such a fine family Bible, our price is only about one-half that you usually pay for a Bible of this quality, size and completeness. When you see this handsome, massive, dignified Bible so full of character on your parlor table you will congratulate yourself on your choice both because it is so appropriate for all the family and also because it was such an excellent purchase from a money standpoint. We heartily recommend and urge you to purchase this Bible for your home.

60L822—The Holy Bible, containing 1,935 pages and 664 illustrations. Bound in fine morocco grained black artificial leather, stamped in real gold leaf. Shipping weight, 10 pounds. Only.................. **$3.98**

Medical Books

Confidential Chats With Girls.

By William Lee Howard, M. D. These chats are exactly what the title indicates—just plain explanations and advice about the body, nerves, brain and sex laws. A good, clean book, which presents knowledge bearing on sex and eugenics in the right way. An essential part of the education of the young girl. Cloth. Size, 5½x7½ in. 128 pages. Shipping weight, 1 pound.

3L3762 .... 89c

Confidential Chats With Boys.

By William Lee Howard, M. D. The plan of this book is to give the boy and youth the information which will enable him to conserve his energies, increase his vigor and learn what his mental endowment and physical powers are. Don't keep your boy in darkness on these most important subjects. A little plain speaking and the right knowledge of intimate facts may preserve both the physical and mental health of your boy. Bound in cloth. Size, 5½x7½ inches. 162 pages. Shipping weight, 1 pound.

3L3761 .... 89c

Plain Facts on Sex Hygiene.

By Wm. Lee Howard, M. D. Deals with the facts of sex hygiene that should be a part of the knowledge of every man and woman. Scientifically correct. Prurient prudery and false social conditions have kept us in ignorance. Here are instructions for avoiding the great peril that is over the land. Dr. Howard fixes your attention upon his revelations; uses plain words, shuns no details, and presents the facts in a way that all can understand. 171 pages. Cloth. Size, 5¼x7½ in. Shipping weight, 1 pound.

3L3754 .... 89c

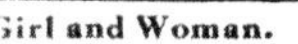

Girl and Woman.

A book for mothers and daughters, **by Caroline Wormiley Latimer, M. D., M. A., with an introduction by Howard A. Kelly, M. D., of Johns Hopkins University.** This book is a notable effort to lead girls safely from girlhood to womanhood, over that most important period in their life where simple precautions may safeguard their future health. 330 pages. Bound in cloth. Size, 5¼x8 in. Shipping weight, 1⅛ pounds.

3L3760 .... $1.68

Be as Young at Sixty as at Twenty.

By Geo. W. Bacon. The haunting fear of all of us is to get old, to lose our freshness, our ambitions, our full powers, our vivacity, our strength. It is difficult today not to show wear and tear in the high speed civilization in which we are living. This book, "Keeping Young and Well," is invaluable to a person who wishes to keep his vigor as long as possible, who wishes to be young at sixty. It will teach him how to absorb the shocks of the daily round of duty and trouble and toil. 177 pages. Bound in cloth. Size, 5½x7½ in. Shipping weight, 14 ounces.

3L3771 .... 48c

The Care and Feeding of Children. New Tenth Edition, Revised and Enlarged. **By L. Emmett Holt, M. D., LL.D.** Children's diseases, ailments, diet, clothing, exercises, bathing, etc. 239 pages. Cloth binding. Size, 4⅞x6⅞ in. Shpg. wt., 1 lb.

3L3537 .... 68c

Mother's Guide to Child Training.

A practical treatise on the rearing and training of children from infancy to maturity. It is a self-instructor which clearly and correctly points the way to successful motherhood and the proper upbringing of children. 442 pages. Bound in cloth. Size 5¼x7¾ inches. Shipping weight, 1½ pounds.

3L3773 .... 98c

Partial List of Contents.

Bronchitis
Catarrh
Asthma
Grip
Pneumonia
Consumption
Bright's Disease
Typhoid Fever
Tonsilitis
Diabetes
Scrofula
Hives
Earache
Rheumatism
Venereal Diseases
Malaria
Bad Complexion
Measles
Diphtheria
Scarlet Fever
Chicken Pox
Mumps
Female Diseases
Whooping Cough
Indigestion
Neuralgia
Heart Disease
Constipation
Convulsions
Boils, etc.

Dr. Evans' How To Keep Well — A HEALTH BOOK FOR THE HOME

The World's Great Medical Book.

A friend in need is a friend indeed. Dr. Evans and his great book represent just such a friend. There is a time—an emergency—in every man's and woman's life when such a book as this is priceless. Such an emergency never gives you warning. Perhaps you or some member of your family will be suddenly stricken—of all times—in the dead of night; it may be hours before you can get a doctor. Perhaps you live miles from a railroad or town and no doctor is available. This great book may mean the difference between life and death. There are certain diseases that must be treated immediately, such as croup and tetanus. What would you do if some member of your family were suddenly taken sick with one of these diseases or one of many others where not a moment was to waste? Would you know what to do in the emergency to relieve the patient or perhaps save his life? Can you administer first aid? Can you stop the flow of blood from a wound till the doctor arrives?

Why will you go on day after day, month after month, trusting to luck that nothing will happen when something may happen at any moment? Are you fair to yourself and to your family? You would never forgive yourself if something happened and you were unable to take the necessary measures to comfort the patient or save his life. Don't go another day without this invaluable book which tells you just what to do in every emergency, in sickness or accident, before the doctor arrives. For the sake of yourself, your family, your loved ones, buy this book.

This book will tell you how to win back your dearest possession—your health—without which nothing counts in this world. Simple in language and rich in information, this book tells you how to escape sickness and how best to recover quickly from the illness that is curable. In it nearly every known disease is frankly and ably discussed.

The book consists of thousands of questions asked Dr. Evans by thousands of people seeking the road to health, together with the answers by the physician. So varied are the topics covered that without doubt the very question that you want to ask about yourself or some one in your family is here answered authoritatively in Dr. Evans' simple, direct manner.

"HOW TO KEEP WELL" is a great big book, printed in large easily read type on book paper, and handsomely bound in cloth. This massive volume measures 10 inches high and 7 inches wide. It contains 1,040 pages, and over 400 practical and helpful illustrations with complete descriptions. A copy of this wonderfully invaluable book should be in your home no later than the time it takes for you to order and for us to get it to you. Shipping weight, 3½ pounds. Former retail price, $3.00.

3L5946—Our new special low price **$1.25**

A Complete and Authoritative Library on Sex Knowledge.

By Dr. William Lee Howard. Sex knowledge, judiciously imparted, is the greatest insurance toward the development of sterling manhood and womanhood. It is the duty of every man and woman to be informed upon the subject of sex and the care of the body. It is a subject vitally important to every individual. Dr. Howard presents facts as they are, and speaks from a knowledge gained through years of study and experience in the great hospitals of the world.

Dr. Howard says, "I want to make clear to the world through these books that right sex living is of as much, if not more, importance than right physical living; that sex laws must be as well known and observed as those of eating, sleeping and the general rules for bodily health."

By special arrangement with the publishers, we are able to offer these three important books, when purchased as a set, for only $2.50. The three volumes of this set are bound in brown cloth with the title stamped on the side and back. Each book measures 5x7⅝ inches. 535 pages. Weight of set, packed, 2½ lbs.

Facts for the Married.

This book is written in the form of consultations, which contain the answers to the questions that husbands and wives want to ask.

Plain Facts on Sex Hygiene.

A plain unvarnished statement of the possible menace to the mental, physical and moral health of those ignorant of sex hygiene.

Sex Problems Solved.

Shows necessity of knowledge of sex activity and its relation to brain, morals and intellect. A book that will remove some of your own perplexities.

3L3765—Complete set of **3** volumes **$2.50**

The Prospective Mother.

By J. Morris Slemons, M. D. A book for women during pregnancy. Many puzzling questions occur to the woman who is about to become a mother. Since these questions are reasonable and natural, they should be answered. There are few such questions that are not fully answered in this helpful book. Endorsed by eminent physicians as a trustworthy guide. Fifteenth printing. 343 pages. Cloth. Size, 5⅛x7¾ inches. Shpg. wt., 1¼ pounds.

3L3769 .... $1.58

Facts for the Married.

By William Lee Howard, M. D. This book is written in the form of consultations. Each consultation takes up with the husband and wife a different phase of health, sex relations and habits, and answers a multitude of questions confronting married people about which there are at present many hazy or wrong ideas. The principles of complete manhood or womanhood are moral, physical and sex health. 161 pages. Cloth. Size, 5⅛x7⅝ inches. Shpg. wt., 1 lb.

3L3758 .... 89c

Sex Problems Solved.

By William Lee Howard, M. D. It makes clear in plain, simple language the problems that are such a cause of worry and distress in later life, based upon the latest scientific knowledge and a professional experience of many years. Its rules, if followed, will aid greatly in preserving the energy and vitality of youth. What part of your hidden forces has been keeping you from health, mental peace and success? Read this book and find out. A book that will help worried men and women. 204 pages. Cloth. Size, 5x7⅝ inches. Shipping weight, 1 pound.

3L3756 .... 89c

Confidential Talks With Husband and Wife.

By Lyman B. Sperry, A. M., M. D. A book of information and advice for the married and marriageable. It states plainly and fully the main physical facts that must needs be known in order to preserve health and promote happiness. 228 pages. Cloth. Size, 5x7½ inches. Shipping weight, 1⅛ pounds.

3L3752 .... $1.18

Mental Healing—A Vital Subject

By Villette H. White. Mental healing is the most vital subject before the people today, because we all realize its common sense possibilities. As a builder of health and of happiness, of contentment, this book is invaluable. This book contains simple and practical methods for acquiring mental self-mastery over mind and body which is so needful to the seeker after health and success. For one who rightly looks upon his health as his most prized possession, this book will be a constant companion. 196 pages. Bound in cloth. Size, 5⅛x7½ inches. Shipping weight, 1 pound.

3L6100 .... 48c

Diet and Health With Key to the Calories. A sane and sensible method of regulating diet, with tables and diets indicated in such a way that you can lose weight or gain it by following these instructions. Cloth. 130 pages. Size, 5¼x7¾ inches. Shipping weight, 1 pound.

3L3804 .... 89c

Beauty Culture.

By William A. Woodbury. Written in simple language, concerning the care of the person. Contains instructions on manicuring, chiropody, hair dressing, facial massage, reduction, removal of facial blemishes, etc. For both men and women. Bound in cloth. 377 pages. Illustrated. Size, 5⅛x7½ inches. Shipping weight, 1 pound.

3L3753 .... $1.35

Walter Camp's New Way to Keep Fit Through Ten Minutes Daily Fun!

Walter Camp, Yale's celebrated football coach and one of the highest authorities in America on health and physical perfection, has been teaching men and women everywhere how to keep fit—"on edge"—full of bounding health and youthful vitality and how to **enjoy** doing it.

Mr. Camp says that civilized, indoor man is a "captive animal," just as much as a tiger in a cage. But the **tiger** instinctively knows how to take the kind of exercise he needs to keep fit. He stretches, turns and twists his **trunk muscles**—the very muscles that tend to become weak and flabby in indoor men and women. And so the tiger, even in close confinement, is never "run-down," never nervous—never suffers from dyspepsia, constipation, insomnia, overweight and other sedentary ills.

The "Daily Dozen," Mr. Camp's now famous little exercises, supply exactly the right movements to put those vitally important trunk muscles in the pink of condition and keep them there.

Not Ordinary Exercises.

These are not the usual tiresome gymnastics or calisthenics. Neither do they **require hours** of valuable time. They are simple stretching, turning and flexing motions scientifically designed to tone up and strengthen every organ of the body and to properly circulate the life giving blood in every part. If you are too stout, these movements will make you slender again. If you are thin and anemic, they will quickly build up your body to normal, ideal weight. Only ten minutes a day devoted to them and you—anyone—can keep yourself healthy, happy and full of "pep."

These easy movements are enjoyable, but the final touch—the addition that has made them irresistible fun—that makes it hard for you **not** to join in—is the jolly, catchy music. With Mr. Camp's special permission, all twelve of the "Daily Dozen" have been set to spirited music—on phonograph records that can be played on any disc machine.

In addition a chart is furnished for each exercise, showing by actual photographs the exact movements to make. A clear, brisk voice on the record gives the "commands," and then, when the lively music strikes up, you've just **got** to follow along. It's fun—a regular frolic—but these simple movements are all you need to keep your whole body in splendid condition. And they take only ten minutes a day!

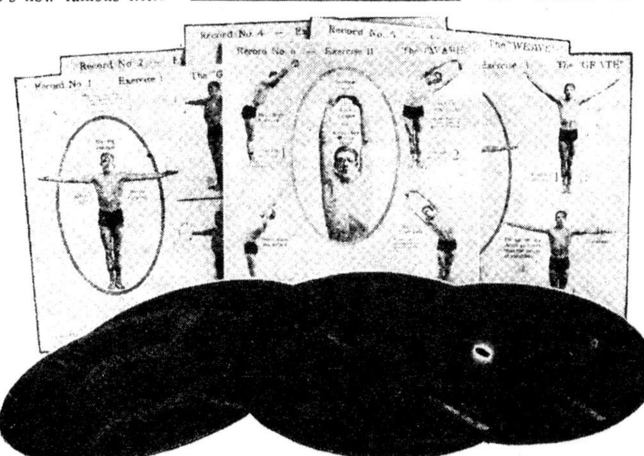

At Much Less Than Original Price.

Walter Camp's improved system of health building now includes the entire "Daily Dozen" exercises, set to specially selected music, on five special 10-inch double disc phonograph records; a beautiful record album; twelve handsome charts printed in two colors, with over 60 actual photographs illustrating each movement, and a little book by Mr. Camp explaining the new principles of his famous system.

By a special arrangement we are able to offer it to you at a BIG REDUCTION. The regular price is $10.50. (Formerly $15.00.) Our price is only $8.75, and you get the same complete course. Send for it today and quickly gain new life and health and "pep."

60L409—Walter Camp's Daily Dozen Health Building Exercises. Shipping weight, 5½ pounds.................... **$8.75**

60L205—Walter Camp's Daily Dozen Health Building Exercises with Special Portable Phonograph which will play 10 and 12-inch records of any make. Shipping wt., 20 lbs..... **$18.25**

Fat? Three internationally famous health experts show you how to easily lose a pound a day

Three internationally famous health experts have collaborated to produce perhaps the most remarkable weight reducing system yet devised. It is really more like play than anything else—purposely so, because these three great authorities know the wonderful health value of the play spirit. But it has marvelous results—10, 20, 50 pounds quickly taken off. At the same time it tones and strengthens the entire system.

You—anyone—can follow this wonderful new system in your own home, without outside advice or help. Thousands of men and women have already been immensely benefited by it. Not only have they lost their unsightly, burdensome fat, but they have also acquired new health, new energy, new vitality. Gone is that continual tired feeling, that sluggishness, that shortness of breath at the least exertion. Gone are constipation, headaches, insomnia, nervousness. And these folks have regained, perhaps to a greater degree than ever before, lithe, slender, well proportioned figures—graceful, supple, full of new life and vigor.

Science tells us fat forms only when the blood is too sluggish to carry it off. From their years of practical experience and wide study, these three famous health building experts—Messrs. Camp (American), Parnet (French), and Mueller (Danish) have been able to devise certain easy, simple, stretching and turning movements which circulate the blood at will in any and every part of the body, carrying off not only the congested fat, but also all accumulated poisons and impurities.

The simple movements are rhythmical and pleasant in themselves, but these wise specialists have made them irresistibly attractive by placing the entire course on phonograph records and accompanying the movements with lively, catchy music.

The new system is not intended to build athletes. It is designed to produce and does produce only perfectly normal, supremely healthy human beings, with all the charm and elegance of appearance and movement which only a slender, graceful, supple figure can give.

Reduces Any Part or Parts.

By this wonderful new system not only can you quickly reduce your entire physique, but directions are also given how to take off flesh just where you wish—on abdomen, bust, hips, thighs, buttocks, arms or legs—without affecting other parts that may now be normal.

You also learn how to tone and strengthen every important internal organ—heart, lungs, liver, and other abdominal organs, the spine and the pelvic region.

Save at Our Low Price.

This marvelous new Musical Weight Reducing System includes five special 10-inch double disc phonograph records, playable on any disc phonograph; a handsome record album, and a large, profusely illustrated booklet with 82 actual photographs, giving the simple instructions accompanying the ten easy lessons and many useful health hints.

The system sells regularly for $7.50. As usual, we are able to offer our customers a nice saving. From us you can get the same complete course for only $6.45.

Why be fat when it costs so little to quickly win a perfect figure—ideal youthful slenderness and grace? Send today.

60L410—Musical Weight Reducing Course. Shipping weight, 5½ pounds.............................. **$6.45**

60L425—Musical Weight Reducing Course with Special Portable Phonograph which will play 10 and 12-inch records of any make. Shipping weight, 20 pounds. **$15.95**

BOOKS FOR THE AUTOMOBILIST

How to Take Care of an Automobile at Small Expense. Repairs and How to Make Them.

By A. Frederick Collins. Here is just the book for the person who owns a car and to whom money is an object. Many times cars are sent to the repair shop to have some small repair made and often the result is a large bill. This book will show you how to save this expense by teaching you how to make your own repairs. It tells you all about the parts of a car, how to take care of your car, what you can fix on your car, how the engine works, how the carburetor works, how the magneto, ignition, starting, lighting and oiling systems work, all about timing, valve setting, valve grinding, etc., and how to run your car at the least cost. 206 pages. Illus. Cloth. Size, 5¼x7⅝ in. Shpg. wt., 1¼ lbs.

3L4846—Only $1.18

Tires and Vulcanizing.

By Henry H. Tufford, Former Chief Vulcanizing Instructor, U. S. A., A. S. M. S., St. Paul. This new book gives complete working methods for all kinds of tire repairs and vulcanizing on both fabric and cord tires; also all kinds of tube repairs. The processes described have all been proved and tested in service. They are selected for their practical value. This book covers rubber, cotton and compounds; tire injuries, abuses and adjustments; repair materials and tools; shop equipment, steam and curing; service and rims; repair data; business methods and accounting; glossary. An invaluable book for both the worker in the garage, as well as the owner. Contains 410 pages and 157 illustrations. Bound in cloth. Size, 4¾x7 in. Shipping weight, 1¼ lbs. Retail price, $2.00

3L4669—Only $1.60

The Model T Ford Car, Including Fordson Farm Tractor and F. A. Lighting and Starting System.

By Victor W. Page, M. E. New revised edition. This is the most complete and practical instruction book ever published on the Ford car and Fordson tractor, explaining the operating principles of all parts of the Ford automobile, with complete instructions for driving, maintenance and repairing. A book that will be invaluable to all Ford owners. All parts of the Ford model T and Fordson tractor are described and illustrated. 410 pages. Illustrated by 153 specially made diagrams and distinctive original photographs of actual parts in correct proportion. Cloth. Size, 5¼x7½ inches. Shipping weight, 1½ lbs.

3L4884—Only $1.49

Automobile Service Station Manuals. New—Complete—Authoritative. By H. P. Manly.

Automobile Ignition.

Covers the operation, upkeep, care and repair of the battery and magneto ignition and gives instructions for adjustment, timing, testing and locating troubles. Includes trouble location tables, and explains the use of testing equipment. Gives complete descriptions with internal wiring of all makes of ignition equipment. Contains 439 pages, 139 illustrations and 98 wiring diagrams. Bound in artificial leather, with gilt stamping. Size, 4⅝x6⅝ inches. Shpg. wt., 1¼ lbs. Retail price, $2.00.

3L4631—Only $1.60

Automobile Starting and Lighting.

A practical explanation of the construction and operation of electrical and lighting equipment, showing internal and external wiring diagrams, with upkeep, care and repair of all parts. Full details of all the various makes commonly used. Cross index. 375 pages, 296 illustrations. Bound in artificial leather, with gilt stamping. Size, 4⅝x6⅝ inches. Shipping weight, 1¼ pounds. Retail price, $2.00.

3L4630—Only $1.60

Automobile Battery—Care and Repair.

A practical manual on the care and repair of the automobile type of modern lead-acid storage battery used with starting, lighting and ignition systems. Fully describes battery construction and action, effect of starting and lighting trouble, charging methods, battery diseases with their remedies, repair methods and shop equipment, service stations, reference tables. Contains 335 pages, 160 illustrations. Bound in artificial leather with gilt stamping. Size, 4⅝x6⅝ inches. Shipping wt., 1¼ lbs. Retail price, $2.00.

3L4629—Only $1.60

Starting and Lighting Troubles—Remedies and Repairs.

This complete manual contains 459 pages, 90 illustrations and 105 internal wiring diagrams. It explains simply and accurately methods for the location, remedy and repair of all forms of electric starting and lighting trouble, including charts for quickly locating troubles and internal wiring diagrams of the complete equipment of all makes and models, both past and present. Covers testing and repairing, wiring dynamos, regulation cut-outs, starters and batteries. Size, 4⅝x6⅝ in. Bound in artificial leather with gilt stamping. Shpg. wt., 1⅛ lbs. Retail price, $3.00.

3L4628—Only $2.25

The New Dyke Course of Automobile and Gasoline Engine Self Instruction. **$10.98**

New 13th Edition. Entirely Rewritten, Rearranged, Illustrated and Enlarged.

The Dyke course not only teaches you the principle and construction of the automobile, but teaches you the gasoline engine as well, and when you master the automobile and gasoline engine with the Dyke course you will understand all automobiles and all gasoline engines, either for automobile, marine or stationary work. If you are really ambitious, there is no quicker, cheaper or better way than this Dyke course. It will surely teach you. The encyclopedia is simple, the diagrams and charts are large and clear and, above all, the demonstrating models will show you the practical workings of the engine. If you are an owner of an automobile, if you are a chauffeur, or desire to become one, if you are desirous of obtaining a position in an automobile factory, garage or repair shop, you will find this course invaluable. With the Dyke course you learn the principle and construction of not only one car, but you learn the principle and construction of all cars.

What Dyke's New Course of Automobile and Gasoline Engine Self Instruction Contains.

Dyke's New and Greater Automobile and Gasoline Engine Encyclopedia.

13th Edition. Just Out. 1238 pages. 4143 illustrations. Mr. Dyke has devoted almost two years' time on rewriting and illustrating this new volume. **Thoroughly practical. Written in simple language so as to be easily understood.** Most of the pages are double column. Bound in cloth. Large size book measures 7x10 inches. It progresses in easy steps from one part to another until finally you are taught the operation of this wonderful power plant as a whole, then how to locate, remedy and repair trouble, etc. Each step is carefully explained by specially prepared charts. A dictionary giving the meaning of the words and terms used will be found in the back. Explains everything you can possibly think of. Tells how to build a repair shop for the home or business; how to make repairs; how to increase the power of your engine; how to overhaul engines; how to drive the different makes of cars; how to figure horse-power; how to start in the automobile business, etc. Contains hundreds of questions and answers on the automobile and its troubles. To this latest edition have been added simplified setting of valves and timing of magnetos, all new and up to date electric starters, generators and lighting systems, operation, care and repair of trucks and tractors, airplanes, airplane engines and motorcycles; also complete instructions for the Ford.

Dyke's Four and Six-Cylinder Demonstrating Models.

Demonstrating models made of stout cardboard with all moving parts made of real metal. These models represent the connecting link between study and practice.

Dyke's Four-Cylinder Engine Demonstrating Model. Size, 9⅛x11 inches. With this model you can see each valve, piston, etc. Makes clear the eight-cylinder principle.

Dyke's Six-Cylinder Engine Demonstrating Model. Size, 11⅜x10⅞ inches. Shows crankshaft, piston and connecting rod side. Also makes clear the twelve-cylinder principle.

Charts. Size, 17½x10 inches. With the demonstrating models we also send charts showing the different parts, such as the clutch, gear box, drive shaft, rear axle, electric starting motor, electric generator, a modern ignition system, inlet and exhaust manifold.

3L4885—Complete Course of Automobile and Gasoline Engine Self Instruction, Consisting of Dyke's Encyclopedia, two Demonstrating Models (one 4-cylinder and one 6-cylinder), and a set of Progressive Charts. Shipping weight, 9 pounds $10.98

3L4868—Dyke's New and Greater Automobile and Gasoline Engine Encyclopedia only. Shipping weight, 4 pounds $5.65

Automobile Painting.

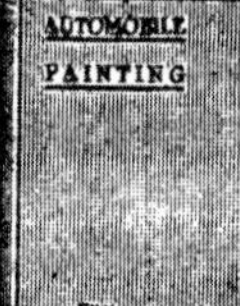

By F. N. Vanderwalker. A practical instruction book, covering every detail of the latest and best methods, specially designed for the average painter. The painting of new cars and the repainting of old ones by several different methods are fully described. Working methods for priming, surfacing, rubbing and varnishing are made perfectly plain as well as factory processes in spraying, dipping, flowing on and baking. The subjects of other chapters are carriage and wagon painting, initials and monograms, color schemes and automobile paint shop plans. This book is invaluable to the painter as well as the man who wishes to paint his own car. 210 pages. 36 illustrations. Bound in cloth. Size, 5¼x7½ inches. Shipping wt., 1¼ pounds. Retail price, $1.50.

3L4627—Only $1.18

Automobile Upkeep and Care.

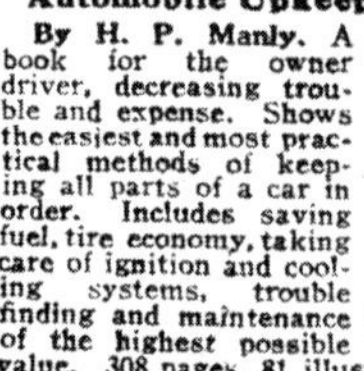

By H. P. Manly. A book for the owner driver, decreasing trouble and expense. Shows the easiest and most practical methods of keeping all parts of a car in order. Includes saving fuel, tire economy, taking care of ignition and cooling systems, trouble finding and maintenance of the highest possible value. 308 pages, 81 illustrations. Bound in cloth. Size, 4⅝x7 inches. Shpg. wt., 1¼ pounds. Retail price, $1.50.

3L4626—Only $1.18

The Motorcycle Handbook.

By H. P. Manly. A complete practical instruction book, covering the construction, operation, care and repair of all modern types of motorcycles, including 2 and 4-cycle types, and multi-cylinder designs. Trouble findings and their remedies, lubrication, ignition, starters, frames, wheels and tires. 320 pages. 183 illustrations. Bound in cloth. Size, 4⅝x7 inches. Shipping weight, 1⅛ lbs. Retail price, $1.50.

3L4625—Only $1.18

Automobile Encyclopedias.

The Modern Gasoline Automobile. Its Design, Construction, Operation and Maintenance.

Revised and enlarged 1921 edition. By Victor W. Page, M. E. The most complete, up to date work on the gasoline automobile ever published. Illustrated by 1,000 specially made detailed illustrations and diagrams and 12 large folding plates. Invaluable to mechanics, repairmen, automobile draftsmen and motorists. Contains 1,032 pages. Bound in cloth. Size, 6x8¾ inches. Shipping weight, 3½ lbs. Retail price, $4.00.

3L4859—Only $2.98

Automobile Repairing Made Easy.

By Victor W. Page, M. E. A comprehensive, practical exposition of every phase of automobile repairing practice. Outlines every practice of motor car restoration; tells how to overhaul all parts of the automobile. Gives plans for work shop construction with suggestions for equipment. Contains 1,060 pages, 1,000 illustrations and 11 folding plates. Size, 5¾x8 inches. Shpg. wt., 3½ pounds. Retail price, $4.00.

3L4675—Only $2.98

Practical Gas and Oil Engine Handbook.

By L. E. Brookes. A practical manual on the care and maintenance of gas and oil engines. This book is invaluable as a guide in the construction, operation and management of stationary, portable and marine oil engines with special reference to the Diesel and other new oil engines. Contains 270 pages, 81 illustrations. Bound in cloth. Size, 4⅛x6¾ inches. Shipping weight, 1⅛ lbs. Retail price, $1.50.

3L4624—Only $1.15

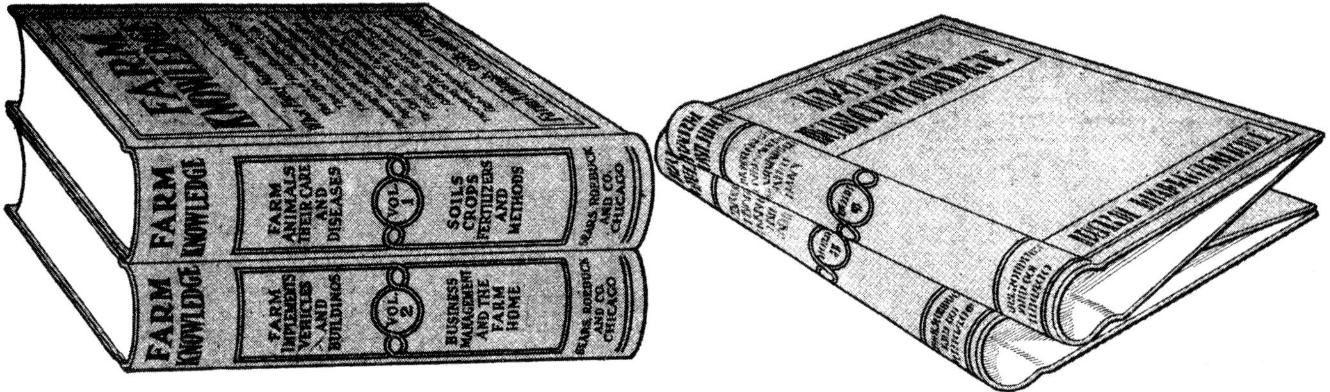

YOUR ONLY LOSS IS TWO EMPTY COVERS AND YOU SAVE $14.05

You Can Now Afford This Great Farm Book

Farm Knowledge Now Bound in Two Handsome Volumes is Offered at 1/4 Former Price

You know what Farm Knowledge is and means. It is the one big outstanding encyclopedia giving reliable and authoritative information on every subject of interest and help to the farmer and agriculturist. Farm Knowledge reveals the great secrets of successful farming, eliminating all guesswork and waste.

No matter what it is on the farm you wish to know about or on which you need help, Farm Knowledge will furnish you with the practical detailed information, written by the best known and most esteemed authorities in the United States—specialists on the particular subjects they write about. No matter which way you turn on your farm, Farm Knowledge is waiting to help you solve your problems, standing at your elbow and guiding you in the best methods to get the most profits and happiness from your farm.

Farm Knowledge will tell you more about the farm than you can get from any other source. It will tell you how to rotate your crops, which crops to raise on every kind of soil. It will tell you all the latest and best methods in growing corn, wheat, oats and all other crops. It will tell you about all breeds of cows, pigs, sheep, horses, chickens and other domestic animals, their characteristics and the advantages of each breed for use on your farm; also it will tell you about their diseases and how to cure them. It will tell you about all fruits and the advantages of the different kinds, the care of the trees, vines and bushes, how to get the maximum yield and the best quality. Farm Knowledge will tell you how to prepare and market your products. It will tell you the machinery to use on your farm and give you invaluable information on thousands of subjects of great interest. You should not fail to have

FARM KNOWLEDGE

This famous work up to this time has been sold at four times the present price. We wished, however, to put this valuable work into the hands of the largest number of farmers possible. To do this meant that it would be necessary to make a worth while cut in price, that is, to bring the price down so far that no one could say he could not afford it. This we have more than done, cutting the price from about $20.00 to less than $5.00.

How was this remarkable reduction made possible? Not by cutting down the contents of Farm Knowledge or leaving out some of its valuable features, but (1) by selling it on a cash basis, thus saving the expense of collecting and handling installment accounts; (2) by using lighter and less expensive paper, which in every way is as practicable and gives as good a reading surface as the costlier paper used in the higher priced edition; (3) by binding Farm Knowledge in two volumes instead of in four volumes, thus saving the cost of material and labor required in binding the two extra books; (4) by using a high grade regular book cloth of splendid wear resisting qualities for the binding instead of a fancy crash buckram, and printing the covers in one color instead of in three colors, as in the higher priced edition. All these have meant an enormous saving, as you can realize.

These two volumes are fine looking books and you will be highly pleased with them. The contents are exactly the same as in the higher priced four-volume set. Page for page, picture for picture, line for line, they are the same. Each volume of the four-volume set contains 550 pages, while each volume of the two-volume set contains 1,100 pages. Both sets contain 2,200 pages and 3,000 illustrations.

Farm Knowledge will be the greatest set of books you ever got hold of on farming subjects. This set will be a revelation to you in the information it will give you. At the price we offer it every farmer and every man interested in farming can afford and should have a set, for many times over he will save the price of the set in money he makes on his farm and the greater crops he obtains because of the use of these books.

Send us $4.95 today, and we will ship you at once this handsome and valuable set. You will never regret your purchase. Farm Knowledge is worth fifteen hired men on the farm in the time and labor it will save you. If, after you have received the set, it does not thoroughly please you, you need not keep the books. Return them to us collect and you are not out a single cent, as we will send back to you all you have paid, including shipping charges. Shipping weight, 10 pounds.

60L969—Farm Knowledge.......................... $4.95

$4.95 PAYS FOR COMPLETE SET

School Bags—Tablets—Students' Note Books

Ruled Pencil Tablet. Tops in assorted designs. Average, 60 sheets (120 pages). A big value for school children. Size, 5⅞x8⅞ inches. Weight of eight, 2½ lbs.
3L12592
8 for 33c

Ink Tablet. Size, 5x8 in. Average 60 sheets (120 pages) ruled white paper.
3L12593
8 for 35c

Ink Tablet. Size, 8x10 in. Average 24 sheets (48 pages).
3L12594
8 for 35c
Weight of eight, either size, 2½ pounds.

Ink Tablet. Size, 5x8 inches. Average 40 sheets (80 pages) ruled plate paper.
3L12602
6 for 26c

Letter Size Ink Tablet. Size, 8x10 inches. Average 16 sheets (32 pages).
3L12603
6 for 26c
Weight of six, either size, 1½ pounds.

Ink and Pencil Tablets

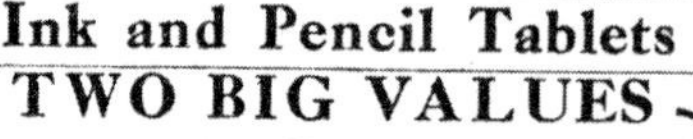

SIX PENCIL TABLETS FOR 25 CENTS.
Pencil Tablet Assortment. Full colored covers. Each tablet averages 50 sheets (100 pages) fair quality ruled pencil paper. Size, 5⅞x8¾ inches. Shpg. wt. of six, 2 lbs.
3L12605—Price, 6 for 25c

SIX INK TABLETS FOR 25 CENTS.
Six Ruled Ink Tablets. Each tablet contains an average of 40 sheets (80 pages) fair quality paper. Assorted colored cover design. Size, 5x8 inches. Shipping weight of six, 1¼ pounds.
3L12604–6 for 25c

Ink Tablet. Size, 5x8 in. Average 70 sheets (140 pages). Good grade ruled smooth plate finish paper.
3L12598
3 for 23c

Letter Size Tablet. Size, 8x10 in. 28 sheets (56 pages).
3L12599
3 for 23c
Weight of three, either size, 1 lb.

Ink Tablet. Size, 5x8 in. Average 50 sheets (100 pages) unruled good quality white cloth finish paper.
3L12608
2 for 17c

Ink Tablet. Size, 8x10 in. Average 20 sheets (40 pages).
3L12609
2 for 17c
Weight of two, either size, 1 lb.

Shannon Linen Finish Ink Tablet. A laid paper commonly known as Irish linen. Size, 5x8 inches. Average 90 sheets (180 pages).
3L12606
Ruled.
2 for 15c
3L12607
Unruled.
2 for 15c
Shipping wt. of two, 1¼ pounds.

A Supply of Writing Essentials.
Pencils—Penholders—Pens—Eraser.

Consists of one cork tip penholder, two good hexagon lead pencils and two high grade round pencils, all with inserted erasers, six pens (assorted styles) and a combination ink and pencil eraser. Shipping wt., 6 ounces.
3L13482
Outfit 29c

School Children's Outfit.

Consists of two pencil tablets, one ink tablet, one composition book, one ruler, a good pencil and a metal pencil clip. Shpg. wt., 1¼ lbs.
3L12620
All for only 25c

SCHOOL BAGS AND STUDENTS' CASE

Very High Grade Corduroy School Bag.

An unusually strong and serviceable bag. Size, 13x10 in. Made of tan corduroy, lined with canvas. Deep gussets. Reinforced seams. Wide flap with leather straps and buckle fasteners. Lunch pocket on outside, 6¼x6¾ inches. Leather strap with buckle and snap fasteners. Shpg. wt., 12 oz.
3L17936 .... $1.38

Waterproof School Bag.

A large bag of water tight black fabricoid lined with canvas duck. Size, 10x13½ inches. Gusseted. All seams and edges reinforced. Lunch pocket on the outside and extra pocket on flap for pencils, etc. Leather strap and buckle fasteners. Shpg. wt., 10 oz.
3L17935 .... 98c

Canvas School Bag.

A school bag made of brown canvas duck. Size, 13½x10 in. Deep gussets. Reinforced seams. Has wide flap with leather straps and buckle fasteners. Lunch pocket on outside, 6½x6 in. Leather shoulder strap with buckle. Shipping weight, 8 ounces.
3L17938 .... 49c

Students' High Grade Carrying Case With Lock.

Attractively made brown fiber carrying case. Very durable. Size, 13½x10x5¼ in. Two metal catches with lock and key. Heavily padded handle with metal fittings. Leather corners, riveted to case with heavy metal studs. Just the thing for an outing. Shipping weight, 3½ pounds.
3L17945 .... $1.48

Crayons.

Box of Sixteen Crayola Wax Crayons, assorted colors. Waterproof. Each crayon 3⅝ inches long. Weight of two boxes, 10 ounces.
3L9657
2 boxes for 35c

Box of Eight Wax Crayons, assorted colors. Waterproof. Each crayon 3½ inches long. Weight of two boxes, 8 oz.
3L9545–2 boxes for 17c

Little Artists' Crayons.

Twenty-four 3½-in. colored crayons. Shpg. wt. of two boxes, 12 ounces.
3L9544
2 boxes for ... 29c

Box of Six Colored Crayon Pencils, assorted colors, 4⅜ inches long, wood covered. Weight of three boxes, 9 oz.
3L9542–3 boxes for 29c

Box of Twelve Colored Crayon Pencils, all different colors. 4⅜ inches long, wood covered. Shipping weight, 5 ounces.
3L9543 .... 19c

Loose Leaf Note Books and Binders

Loose Leaf Binder.

Black board covers in imitation of cloth with red cloth back. Without rings. Can be used either with shoe lace or brass paper fasteners. Durable and inexpensive. An ideal book for school and office use, 8⅛x10¾ in. Shipping weight, 1½ lbs.
3L9688
Complete with 45 sheets of good quality ruled paper 25c

Loose Leaf Binder With Rings.

A popular style with students. Heavy board covers, with cloth back. Size, 8⅛x11 inches. The binding device consists of two interlocking metal rings. Label for name and address on front cover. Shipping wt., complete with 45 sheets good quality paper, 1¾ lbs.
3L9698 .... 32c

Same as above, but made of good grade artificial leather. Complete with filler.
3L9699 .... 59c

Loose Leaf Note Book.

Heavy board covers with cloth back. Metal device of simple construction for locking sheets, one movement opening the rings. Size, 8⅛x11 inches. Shipping weight, 1¾ lbs.
3L9694
Complete with 45 sheets good quality ruled white paper 39c

Same as 3L9694, but bound in flexible black cloth artificial leather.
3L9695 .... 75c

3L9689—Extra Sheets for above, per package of 45 sheets. Wt., 12 oz. 10c

Scholars' Companions.

Made of artificial leather with blotter pad in cover, button clasp fasteners. Contains four round pencils, one hexagonal pencil, all metal tipped with inserted eraser, one sharpener, six pens in tube, one pencil holder with pencils and one penholder. Size, 3¾x8¼ in. Shpg. wt., 9 oz.
3L13574 .... 49c

An assortment consisting of three pencils (good quality), two with erasers in metal tips, one holder, with two pencils, one in each end, a penholder and an eraser. In handy folding cloth box, made to resemble leather, with a button clasp. Size, 2¼x9⅜ inches. Shipping weight, 10 ounces.
3L13571 .... 25c

Pastel Crayons.

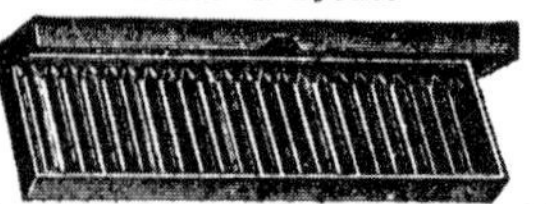

A fine quality set of twenty-four pastel crayons, all different shades, each 3 in. long. Really artistic work can be done with these crayons, which can be blended and shaded. In cardboard box. Size of box, 3⅛x8⅞ inches. Shipping weight, 10 ounces.
3L9585 .... 45c

SEE PAGES 500-501-502
FOR

Drawing Materials	Geographical Globes
Artists' Supplies	Typewriter Supplies
Draughting Materials	Rubber Type Outfits

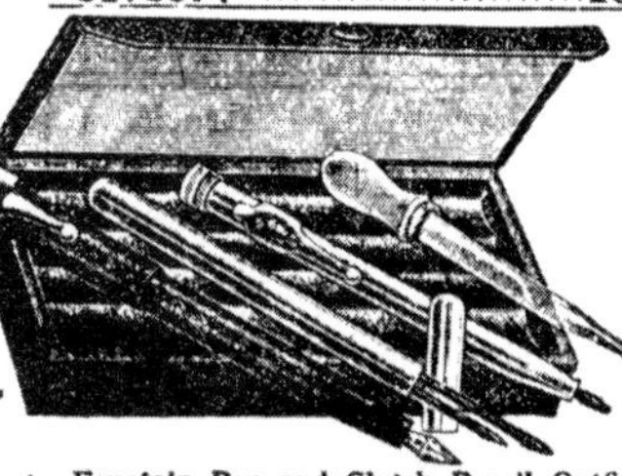

Folding Lunch Box.

Folding Lunch Box. Size, 8x4x4 in. Cloth. Hinged top, leather handle and hinged clasp. Shipping weight, 10 oz.
3L17941 ..... 48c

Compass and Pencil.

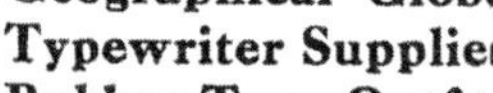

A serviceable metal compass with adjustable pencil. Size, 5 inches. Shipping weight, 4 ounces.
3L14294 .... 19c

Compass and Divider.

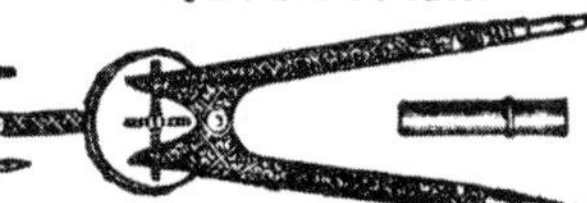

A practical metal compass regulated by spring and screw adjustment with box of extra leads. Shipping weight, 4 ounces.
3L14295 .... 39c

Fountain Pen and Clutch Pencil Outfit. Contains: **A good quality fountain pen** fitted with iridium tipped **14-karat** gold pen. Length of pen, 7 inches. **A metal clutch pencil** with heavy lead. Length, 4½ inches. **A tube** containing 2 extra leads. **Ink dropper.** In box with button clasp. Box, 6x3x⅞ in. Shpg. wt., 6 oz.
3L15360—Complete outfit $1.48

How to Order and Other Information

Order under one name

If possible, have all the members of the family order under one name—**the name of the head of the family.** This name should always be written plainly and always the same way. For example: If the name of the head of the family is J. P. Thompson, sign the two initials and the name every time. Don't sign the order simply J. Thompson. If you have no middle name, please write the first name in full. For example: John Thompson, not merely J. Thompson. When we receive all orders from the same family under one and the same name, the keeping of our records is simpler and prevents mistakes and delays.

Order blanks

Order blanks are enclosed in this catalog. Additional blanks, if wanted, will be sent upon request. If at any time you have no order blanks, write your order on any paper.

Write in any language

We can read it. We receive orders in **all languages**; all are handled with the same promptness.

Necessary information

Give name and number of article in catalog; also size and color where necessary. It is also advisable to check your order carefully to see that the necessary information is correctly stated before enclosing your order in the envelope.

How to send money

We require cash with order. You are perfectly safe in sending cash with order, for our guarantee protects you. If you are not satisfied with the goods you receive we will exchange them for other goods you want, or return your money, together with all transportation charges you paid. You can send the money to us in any of the following ways:

1—Postoffice money order.
2—Express money order.
3—Bank draft.
4—Cash by registered mail.
5—Your personal check.

When goods are to be shipped by parcel post, be sure to include **additional** money to pay for postage.

If you live on a rural route you can give the letter containing your order and money to your carrier and he will buy a money order for you at the postoffice and enclose it in the envelope with your order and mail it to us.

Change of address

If you change your postoffice address, street address, rural route, or box number, please let us know at once. In notifying us be sure to give your old address as well as your new one. This will enable us to send catalogs or letters to the correct address and thus avoid inconvenience to you.

Transportation charges

All transportation charges are to be paid by the customer.

When goods are to be shipped by parcel post be sure to include **additional** money to pay for postage.

When goods are to be shipped by freight or express and there is no freight or express agent at your shipping point, you must send additional money to prepay the freight or express charges. If there is an agent you can pay the charges when shipment reaches you. It is only necessary to prepay freight or express charges when there is no agent at your station. See pages 457 and 458 for express and freight rates.

Freight is the cheapest

Parcel post and express rates are low, but the cheapest way of shipping is by freight. The biggest savings are made by our customers who plan their purchases in advance. Instead of having small orders shipped to them by express or parcel post, they figure out all the supplies they will need for two or three months and order them all at once, shipped by freight. In this way they make a considerable additional saving on the larger order.

If you order goods sent by freight or express be sure to give your shipping point if it is different from your postoffice.

Factory shipments

In order to make our prices as low as we do we find it necessary to ship many of the heavy, bulky articles we sell direct from the various factories where they are made, or from a warehouse, thus saving our customers freight and cartage to our store, double handling and other expenses. The descriptions tell you when goods are shipped from factory or warehouse. By far the greater part of our merchandise is shipped direct from our store.

When you don't tell us how to ship

In this case we will consider that you have left it to our judgment and we will ship your goods the way it will cost you the least.

A Word About Prices

Whenever market values decline, we immediately reduce our prices and return the difference to our customers on all merchandise purchased from this catalog after the decline. In extreme cases we may be compelled to advance our prices when market conditions make such advances necessary, but only extreme conditions could force us to take such action. It may also become necessary, in exceptional cases, to limit the amount of any commodity that is ordered to such quantity as would ordinarily be purchased by the consumer.

Refer to the Index

To find what you want refer to the index, pages 459 to 470. For Parcel Post, Express and Freight Rates see pages 456, 457 and 458.

Come and Visit Us

Whenever you come to our city we invite you to visit us. We'll show you through our great buildings. You will not be asked to purchase anything, but if you wish to make out an order while you are here you may do so, just as you would at home, and the goods will be shipped in the usual way.

HOW to RETURN GOODS
by Parcel Post, Freight or Express

If you ever have occasion to return any merchandise, we can give you **better service** if you follow these instructions:

IF YOU RETURN GOODS BY PARCEL POST, PASTE OR TIE AN ENVELOPE CONTAINING YOUR LETTER AND THE BILLS TO THE OUTSIDE OF THE PACKAGE and put a 2-cent stamp on the envelope in addition to the postage on the package. In this way we will receive your letter and the package at the same time and can attend to your wishes promptly.

The Way to Return Goods to Us by Parcel Post.

Fasten the Envelope on Outside of Package.

Before returning merchandise by freight please write us. We ask this so we may tell you whether we want the goods sent to our store or to the factory. Also do not, under any circumstances, return heavy merchandise by express, unless we request you to do so. If we have asked you to return merchandise by freight or express, mail us a letter enclosing the receipt the agent gives you, as soon as the goods have been shipped.

In any case, whether you return merchandise by freight, parcel post or express, be sure to write your name on the package, so that we may know who sent it. Put the word "From" in front of your name.

We are anxious to improve our merchandise and our service in any possible way, and will appreciate any suggestions from our customers. It will help us very much if you will tell us just WHY goods are unsatisfactory.

Rates for Parcel Post Shipments

Your postmaster will tell you the parcel post zone in which your postoffice is located, measuring from our store.

All merchandise shipped by mail takes parcel post rates. Packages up to **4 ounces** in weight are carried at the rate of 1 cent an ounce, regardless of distance. Packages over 4 ounces are charged for by the pound. The rate per pound varies according to the distance, which is measured by the Government zone system, each zone covering a certain number of miles from point of shipment. Distances and rates are shown in the table below. **Packages carried by parcel post are handled just like any other mail matter.**

They are delivered to your box by your rural mail carrier if you live on a rural route, or delivered to your door if you live in a city where there is carrier service, or delivered to your local postoffice if you live where there is no carrier service.

Loaded or primed cartridges or shells, other explosives, inflammable articles and poisons cannot be shipped by parcel post, nor articles measuring more than 7 feet in length and girth combined.

RATE TABLE FOR PARCEL POST SHIPMENTS

This table shows the charges when shipping by parcel post, according to the weight of the packages and according to distance by zones.

Weight of Package	LOCAL ZONE For Shipments From Our Store to Customers Within Local Zone — Charges Required	ZONES 1 & 2 Not Over 150 Miles From Our Store — Charges Required	ZONE 3 151 to 300 Miles From Our Store — Charges Required	ZONE 4 301 to 600 Miles From Our Store — Charges Required	ZONE 5 601 to 1,000 Miles From Our Store — Charges Required	ZONE 6 1,001 to 1,400 Miles From Our Store — Charges Required
Over 4 oz. up to 1 lb.	5c	5c	$0.06	$0.07	$0.08	$0.09
Over 1 lb. up to 2 lbs.	6c	6c	.08	.11	.14	.17
Over 2 lbs. up to 3 lbs.	6c	7c	.10	.15	.20	.25
Over 3 lbs. up to 4 lbs.	7c	8c	.12	.19	.26	.33
Over 4 lbs. up to 5 lbs.	7c	9c	.14	.23	.32	.41
Over 5 lbs. up to 6 lbs.	8c	10c	.16	.27	.38	.49
Over 6 lbs. up to 7 lbs.	8c	11c	.18	.31	.44	.57
Over 7 lbs. up to 8 lbs.	9c	12c	.20	.35	.50	.65
Over 8 lbs. up to 9 lbs.	9c	13c	.22	.39	.56	.73
Over 9 lbs. up to 10 lbs.	10c	14c	.24	.43	.62	.81
Over 10 lbs. up to 11 lbs.	10c	15c	.26	.47	.68	.89
Over 11 lbs. up to 12 lbs.	11c	16c	.28	.51	.74	.97
Over 12 lbs. up to 13 lbs.	11c	17c	.30	.55	.80	1.05
Over 13 lbs. up to 14 lbs.	12c	18c	.32	.59	.86	1.13
Over 14 lbs. up to 15 lbs.	12c	19c	.34	.63	.92	1.21
Over 15 lbs. up to 16 lbs.	13c	20c	.36	.67	.98	1.29
Over 16 lbs. up to 17 lbs.	13c	21c	.38	.71	1.04	1.37
Over 17 lbs. up to 18 lbs.	14c	22c	.40	.75	1.10	1.45
Over 18 lbs. up to 19 lbs.	14c	23c	.42	.79	1.16	1.53
Over 19 lbs. up to 20 lbs.	15c	24c	.44	.83	1.22	1.61
Over 20 lbs. up to 21 lbs.	15c	25c	.46	.87	1.28	1.69
Over 21 lbs. up to 22 lbs.	16c	26c	.48	.91	1.34	1.77
Over 22 lbs. up to 23 lbs.	16c	27c	.50	.95	1.40	1.85
Over 23 lbs. up to 24 lbs.	17c	28c	.52	.99	1.46	1.93
Over 24 lbs. up to 25 lbs.	17c	29c	.54	1.03	1.52	2.01
Over 25 lbs. up to 26 lbs.	18c	30c	.56	1.07	1.58	2.09
Over 26 lbs. up to 27 lbs.	18c	31c	.58	1.11	1.64	2.17
Over 27 lbs. up to 28 lbs.	19c	32c	.60	1.15	1.70	2.25
Over 28 lbs. up to 29 lbs.	19c	33c	.62	1.19	1.76	2.33
Over 29 lbs. up to 30 lbs.	20c	34c	.64	1.23	1.82	2.41
Over 30 lbs. up to 31 lbs.	20c	35c	.66	1.27	1.88	2.49
Over 31 lbs. up to 32 lbs.	21c	36c	.68	1.31	1.94	2.57
Over 32 lbs. up to 33 lbs.	21c	37c	.70	1.35	2.00	2.65
Over 33 lbs. up to 34 lbs.	22c	38c	.72	1.39	2.06	2.73
Over 34 lbs. up to 35 lbs.	22c	39c	.74	1.43	2.12	2.81
Over 35 lbs. up to 36 lbs.	23c	40c	.76	1.47	2.18	2.89
Over 36 lbs. up to 37 lbs.	23c	41c	.78	1.51	2.24	2.97
Over 37 lbs. up to 38 lbs.	24c	42c	.80	1.55	2.30	3.05
Over 38 lbs. up to 39 lbs.	24c	43c	.82	1.59	2.36	3.13
Over 39 lbs. up to 40 lbs.	25c	44c	.84	1.63	2.42	3.21
Over 40 lbs. up to 41 lbs.	25c	45c	.86	1.67	2.48	3.29
Over 41 lbs. up to 42 lbs.	26c	46c	.88	1.71	2.54	3.37
Over 42 lbs. up to 43 lbs.	26c	47c	.90	1.75	2.60	3.45
Over 43 lbs. up to 44 lbs.	27c	48c	.92	1.79	2.66	3.53
Over 44 lbs. up to 45 lbs.	27c	49c	.94	1.83	2.72	3.61
Over 45 lbs. up to 46 lbs.	28c	50c	.96	1.87	2.78	3.69
Over 46 lbs. up to 47 lbs.	28c	51c	.98	1.91	2.84	3.77
Over 47 lbs. up to 48 lbs.	29c	52c	1.00	1.95	2.90	3.85
Over 48 lbs. up to 49 lbs.	29c	53c	1.02	1.99	2.96	3.93
Over 49 lbs. up to 50 lbs.	30c	54c	1.04	2.03	3.02	4.01
Over 50 lbs. up to 51 lbs.	30c	55c	1.06			
Over 51 lbs. up to 52 lbs.	31c	56c	1.08			
Over 52 lbs. up to 53 lbs.	31c	57c	1.10			
Over 53 lbs. up to 54 lbs.	32c	58c	1.12			
Over 54 lbs. up to 55 lbs.	32c	59c	1.14			
Over 55 lbs. up to 56 lbs.	33c	60c	1.16			
Over 56 lbs. up to 57 lbs.	33c	61c	1.18			
Over 57 lbs. up to 58 lbs.	34c	62c	1.20			
Over 58 lbs. up to 59 lbs.	34c	63c	1.22			
Over 59 lbs. up to 60 lbs.	35c	64c	1.24			
Over 60 lbs. up to 61 lbs.	35c	65c	1.26			
Over 61 lbs. up to 62 lbs.	36c	66c	1.28			
Over 62 lbs. up to 63 lbs.	36c	67c	1.30			
Over 63 lbs. up to 64 lbs.	37c	68c	1.32			
Over 64 lbs. up to 65 lbs.	37c	69c	1.34			
Over 65 lbs. up to 66 lbs.	38c	70c	1.36			
Over 66 lbs. up to 67 lbs.	38c	71c	1.38			
Over 67 lbs. up to 68 lbs.	39c	72c	1.40			
Over 68 lbs. up to 69 lbs.	39c	73c	1.42			
Over 69 lbs. up to 70 lbs.	40c	74c	1.44			

Within Local Zone and Zones 1, 2 and 3, packages up to 70 pounds in weight are carried. The limit of weight for all other zones is 50 pounds. Articles measuring more than 7 feet in length and girth combined, explosives, inflammable articles and poisons cannot be shipped by parcel post.

Books **Parcel post rates apply to books as follows:** All books up to and including 8 ounces in weight will be carried at the rate of 1 cent for 2 ounces to any part of the United States, regardless of distance, and all books over 8 ounces in weight will take the regular parcel post rates according to weight and zone.

How to Return Goods to Us by Parcel Post.

WHEN YOU RETURN GOODS BY PARCEL POST, PUT THE LETTER YOU WRITE AND THE BILLS FOR THE GOODS IN AN ENVELOPE AND PASTE OR TIE THE ENVELOPE SECURELY TO THE OUTSIDE OF THE PACKAGE. In addition to the postage you put on the package, put 2 cents postage on the envelope.

THE WAY TO RETURN GOODS TO US BY PARCEL POST.

About Transportation Charges.

When goods are to be shipped by parcel post, do not send stamps to pay the postage for shipping package, but add the amount for postage to the amount of the merchandise and include in the remittance you send us. This charge for mailing must be paid in advance, as no provision has been made for the collection of mailing charge on delivery.

When goods are to be shipped by freight or express and there is no freight or express agent at your shipping point, you must send money to prepay the transportation charges. If there is an agent you can pay the transportation charges when shipment reaches you. It is only necessary to prepay freight or express charges when there is no agent at your station. Read pages 457 and 458 for complete information about freight and express rates and charges.

Throughout our catalogs you will find the shipping weight is given in the description of merchandise. Occasionally, according to the nature of the merchandise, we are obliged to give the actual weight. In such cases a few ounces extra in weight must be allowed for wrapping and packing, according to the nature of the goods.

Express Rates

Tables showing the Express Rates per 100 pounds on goods shipped from Chicago to a number of cities in each state, these cities being used as a basis for figuring rates for all the towns in the immediate vicinity of each city.

Your express agent will tell you the exact rates from Chicago to your home town and give you full information in reference to their delivery service. If there is no express agent at your nearest railroad station and you desire your order sent by express, it is necessary that you show the nearest town at which there is an express agent.

HOW TO FIGURE EXPRESS CHARGES. First estimate the weight of goods you are ordering; then find the rate per 100 pounds by express to your nearest city in the Table of Rates below; then consult the Scale of Express Charges, following the line for the weight of your goods to the column headed by your express rate per 100 pounds, and the amount shown will be the express charges.

If the exact rate per 100 pounds to your town is not shown in any of the headings of this scale, take the rate shown for the town nearest you and the express charges will be about the same as to your town.

The table of Express Rates also shows the parcel post zone of the various cities named below, enabling you to make an approximate comparison between the express charges and the parcel post charges to the zone in which you live. For Parcel Post Rates see page 456.

EXPRESS RATES PER 100 POUNDS TO CITIES IN EACH STATE.

From Chicago to	Parcel Post Zone	Express, per 100 Pounds
ALABAMA—		
Anniston	4	$3.67
Birmingham	4	3.53
Brewton	5	4.16
Dothan	5	4.22
Florence	4	3.33
Huntsville	4	3.33
Mobile	5	4.30
Montgomery	5	3.81
York	5	3.67
ARKANSAS—		
Ft. Smith	4	4.22
Hamburg	5	3.67
Harrison	4	3.74
Helena	4	3.33
Jonesboro	4	3.11
Little Rock	4	3.88
Texarkana	5	4.57
COLORADO—		
Craig	6	11.78
Denver	5	6.02
Durango	6	8.38
Grand Junction	6	7.82
Las Animas	5	5.75
Leadville	5	7.13
Sterling	5	5.54
Trinidad	5	6.24
ILLINOIS—		
Bloomington	2	1.29
Cairo	4	2.04
Danville	2	1.11
East St. Louis	3	1.91
Effingham	3	1.42
Galena	2	1.42
Joliet	1	.92
Kankakee	1	1.11
La Salle	2	1.29
Moline	2	1.42
Mt. Vernon	3	1.61
Peoria	2	1.42
Quincy	3	1.91
Rockford	2	1.29
Springfield	3	1.61
Waukegan	1	.92
INDIANA—		
Bloomington	3	1.59
Evansville	3	1.94
Ft. Wayne	2	1.59
Gary	1	.97
Indianapolis	2	1.59
Logansport	2	1.45
New Albany	3	1.94
Richmond	3	1.94
South Bend	2	1.25
Terre Haute	3	1.59
IOWA—		
Burlington	3	1.94
Cedar Rapids	3	1.94
Council Bluffs	4	3.33

From Chicago to	Parcel Post Zone	Express, per 100 Pounds
IOWA—Continued.		
Davenport	2	$1.59
Decorah	3	2.08
Des Moines	3	2.70
Dubuque	3	1.59
Ft. Dodge	4	2.91
Mason City	3	2.77
Ottumwa	3	2.36
Rock Rapids	4	3.60
Sioux City	4	3.33
KANSAS—		
Concordia	4	4.08
Dodge City	5	5.13
Great Bend	5	4.57
Leavenworth	4	3.11
Pittsburg	4	3.60
St. Francis	5	5.27
Sharon Springs	5	5.27
Topeka	4	3.33
Wichita	4	4.57
KENTUCKY—		
Ashland	4	2.56
Bowling Green	4	2.14
Frankfort	3	2.14
Hickman	4	2.70
Louisville	3	1.94
Newport	3	1.94
Pikeville	4	3.11
Williamsburg	4	2.97
MICHIGAN—		
Adrian	3	1.59
Alpena	4	3.05
Bad Axe	3	2.42
Cheboygan	4	3.05
Detroit	3	1.94
Escanaba	3	2.63
Houghton	4	3.47
Lansing	3	1.80
Muskegon	2	1.94
Niles	2	1.25
Saginaw	3	2.28
Sault Ste. Marie	4	3.60
Traverse City	3	2.42
MINNESOTA—		
Bemidji	4	4.16
Brainerd	4	3.88
Breckenridge	4	4.22
Chisholm	4	4.02
Crookston	4	4.44
Duluth	4	3.53
International Falls	4	4.78
Mankato	4	2.97
Pipestone	4	3.74
St. Paul	4	2.97
Winona	3	2.56
MISSISSIPPI—		
Biloxi	5	4.30
Clarksdale	4	3.33
Corinth	4	3.33

From Chicago to	Parcel Post Zone	Express, per 100 Pounds
MISSISSIPPI—Cont'd.		
Greenville	4	$3.67
Hattiesburg	5	4.02
Jackson	5	3.88
Meridian	5	3.67
Natchez	5	4.22
West Point	4	3.53
MISSOURI—		
Caruthersville	4	2.70
Hannibal	3	2.14
Jefferson City	3	2.84
Kansas City	4	3.11
Maryville	4	2.97
Milan	3	2.70
Nevada	4	3.60
St. Louis	3	2.14
Salem	4	2.84
Springfield	4	3.33
MONTANA—		
Billings	6	7.69
Glasgow	6	6.79
Glendive	5	6.10
Great Falls	6	8.87
Havre	6	7.90
Plentywood	5	6.24
NEBRASKA—		
Chadron	5	5.41
Grand Island	4	4.22
Laurel	4	4.02
Lincoln	4	3.60
McCook	5	4.71
North Platte	5	4.91
Omaha	4	3.33
Sidney	5	5.41
Valentine	5	4.85
NORTH DAKOTA—		
Bismarck	5	5.27
Bowman	5	5.75
Devils Lake	5	4.91
Dickinson	5	5.61
Fargo	4	4.22
Grafton	5	4.91
Minot	5	5.47
Oakes	4	4.85
Williston	5	5.88
OHIO—		
Cambridge	4	2.28
Chillicothe	3	2.28
Cincinnati	3	1.94
Cleveland	3	2.14
Columbus	3	2.28
Dayton	3	1.94
Ironton	4	2.56
Lima	3	1.80
Mansfield	3	2.14
Marietta	4	2.42
Massillon	4	2.28
Toledo	3	1.80
Youngstown	4	2.42

From Chicago to	Parcel Post Zone	Express, per 100 Pounds
OKLAHOMA—		
Bartlesville	4	$4.08
Enid	5	4.85
Forgan	5	5.61
Muskogee	4	4.36
Oklahoma City	5	4.99
Poteau	4	4.22
Woodward	5	5.41
SOUTH DAKOTA—		
Aberdeen	4	4.57
Deadwood	5	5.47
Dupree	5	5.19
Hot Springs	5	5.75
McIntosh	5	5.19
Mitchell	4	4.08
Pierre	5	4.71
Sioux Falls	4	3.60
Watertown	4	4.02
TENNESSEE—		
Chattanooga	4	3.11
Cookeville	4	2.77
Jackson	4	2.91
Johnson City	4	3.53
Knoxville	4	3.53
Lewisburg	4	2.91
Memphis	4	3.11
Nashville	4	2.63
Union City	4	2.70
UTAH—		
Lucin	6	9.84
Marysvale	6	10.32
Milford	6	10.81
Price	6	9.01
Salt Lake City	6	9.35
WISCONSIN—		
Ashland	4	3.33
Eau Claire	3	2.56
Fond du Lac	2	1.59
Green Bay	3	2.22
Hudson	4	2.84
La Crosse	3	2.08
Madison	2	1.80
Marinette	3	2.63
Milwaukee	2	1.45
Prairie du Chien	3	2.08
Rhinelander	3	2.77
Superior	4	3.53
Wisconsin Rapids	3	2.28
WYOMING—		
Casper	5	6.85
Cheyenne	5	5.96
Cody	6	7.96
Kemmerer	6	8.52
Lander	6	7.62
New Castle	5	6.16
Rawlins	5	7.27
Sheridan	5	6.72

SCALE OF EXPRESS CHARGES BASED ON THE RATE PER 100 POUNDS.

Rate per 100 lbs.	$0.92	$0.97	$1.03	$1.11	$1.25	$1.29	$1.42	$1.45	$1.59	$1.61	$1.80	$1.91	$2.04	$2.08	$2.14	$2.22	$2.28	$2.36	$2.42	$2.56	$2.63
CHARGES ON																					
Package of 5 lbs.	.34	.38	.38	.39	.39	.36	.36	.40	.40	.37	.42	.38	.39	.43	.43	.44	.44	.44	.44	.45	.45
Package of 10 lbs.	.37	.40	.42	.42	.43	.41	.42	.45	.47	.43	.48	.47	.48	.52	.53	.53	.54	.54	.55	.57	.57
Package of 15 lbs.	.39	.44	.44	.45	.48	.46	.47	.52	.53	.51	.57	.54	.57	.60	.60	.62	.64	.66	.66	.68	.69
Package of 20 lbs.	.43	.47	.48	.50	.53	.51	.53	.57	.59	.57	.64	.63	.65	.69	.71	.72	.73	.74	.76	.79	.81
Package of 25 lbs.	.46	.50	.52	.54	.57	.55	.59	.62	.66	.63	.71	.71	.74	.78	.79	.82	.83	.84	.86	.89	.92
Package of 30 lbs.	.50	.53	.55	.57	.60	.61	.64	.68	.72	.70	.78	.79	.83	.86	.88	.89	.93	.95	.97	1.01	1.02
Package of 35 lbs.	.52	.57	.58	.60	.67	.65	.70	.73	.78	.77	.86	.87	.91	.96	.97	1.00	1.02	1.06	1.07	1.12	1.15
Package of 40 lbs.	.55	.59	.62	.66	.71	.71	.75	.79	.84	.83	.93	.96	1.00	1.03	1.07	1.10	1.12	1.15	1.17	1.23	1.26
Package of 45 lbs.	.59	.62	.66	.69	.74	.75	.80	.84	.89	.89	1.00	1.02	1.09	1.12	1.15	1.20	1.22	1.25	1.27	1.35	1.37
Package of 50 lbs.	.62	.66	.69	.72	.79	.80	.87	.89	.97	.96	1.07	1.11	1.17	1.21	1.25	1.27	1.31	1.35	1.39	1.45	1.49
Package of 60 lbs.	.68	.72	.76	.81	.88	.90	.98	1.01	1.10	1.09	1.22	1.27	1.38	1.39	1.42	1.47	1.51	1.55	1.59	1.68	1.71

Rate per 100 lbs.	$2.70	$2.77	$2.84	$2.91	$2.97	$3.05	$3.11	$3.33	$3.47	$3.53	$3.60	$3.67	$3.74	$3.81	$3.88	$4.02	$4.08	$4.16	$4.22	$4.30	$4.36
CHARGES ON																					
Package of 5 lbs.	.45	.47	.47	.47	.47	.48	.48	.50	.50	.50	.52	.52	.52	.52	.53	.53	.53	.54	.54	.54	.54
Package of 10 lbs.	.58	.58	.59	.59	.60	.60	.62	.64	.66	.67	.67	.68	.68	.69	.69	.71	.72	.72	.73	.73	.74
Package of 15 lbs.	.69	.71	.72	.73	.73	.74	.76	.79	.82	.82	.83	.84	.86	.86	.87	.89	.89	.92	.93	.95	.95
Package of 20 lbs.	.82	.83	.84	.86	.87	.88	.89	.95	.97	.98	1.00	1.01	1.02	1.03	1.06	1.08	1.10	1.11	1.12	1.13	1.15
Package of 25 lbs.	.93	.96	.97	.98	1.00	1.02	1.03	1.10	1.12	1.13	1.16	1.17	1.20	1.21	1.23	1.26	1.27	1.30	1.31	1.34	1.35
Package of 30 lbs.	1.06	1.07	1.10	1.11	1.13	1.15	1.17	1.23	1.27	1.30	1.31	1.35	1.36	1.39	1.40	1.44	1.47	1.49	1.51	1.52	1.55
Package of 35 lbs.	1.16	1.20	1.22	1.25	1.26	1.29	1.31	1.39	1.44	1.45	1.49	1.51	1.54	1.55	1.58	1.64	1.65	1.68	1.70	1.73	1.75
Package of 40 lbs.	1.29	1.31	1.35	1.37	1.40	1.42	1.45	1.54	1.59	1.63	1.65	1.68	1.77	1.73	1.76	1.81	1.84	1.86	1.90	1.93	1.95
Package of 45 lbs.	1.40	1.44	1.47	1.50	1.52	1.56	1.59	1.69	1.75	1.78	1.81	1.84	1.86	1.90	1.94	1.99	2.03	2.07	2.09	2.12	2.14
Package of 50 lbs.	1.52	1.55	1.59	1.63	1.66	1.69	1.73	1.83	1.90	1.94	1.97	2.00	2.04	2.08	2.10	2.18	2.22	2.24	2.28	2.32	2.36
Package of 60 lbs.	1.76	1.80	1.84	1.89	1.93	1.97	2.00	2.13	2.22	2.26	2.31	2.34	2.38	2.42	2.47	2.55	2.60	2.63	2.67	2.72	2.76

Rate per 100 lbs.	$4.44	$4.57	$4.71	$4.78	$4.85	$4.91	$4.99	$5.13	$5.19	$5.27	$5.41	$5.47	$5.54	$5.61	$5.75	$5.88	$5.96	$6.02	$6.10	$6.16	$6.24
CHARGES ON																					
Package of 5 lbs.	.55	.55	.57	.57	.57	.57	.58	.58	.58	.59	.59	.59	.60	.60	.60	.62	.62	.62	.64	.64	.64
Package of 10 lbs.	.74	.76	.78	.79	.79	.81	.81	.82	.83	.83	.84	.86	.86	.87	.88	.89	.89	.92	.92	.93	.93
Package of 15 lbs.	.96	.98	1.00	1.01	1.02	1.02	1.03	1.07	1.07	1.08	1.11	1.11	1.12	1.13	1.15	1.17	1.20	1.20	1.21	1.22	1.23
Package of 20 lbs.	1.16	1.20	1.22	1.23	1.25	1.26	1.27	1.30	1.31	1.34	1.36	1.37	1.39	1.40	1.42	1.45	1.47	1.49	1.50	1.51	1.52
Package of 25 lbs.	1.37	1.40	1.44	1.45	1.47	1.49	1.51	1.54	1.55	1.58	1.61	1.63	1.65	1.66	1.69	1.73	1.75	1.76	1.79	1.80	1.81
Package of 30 lbs.	1.56	1.61	1.65	1.68	1.69	1.71	1.73	1.78	1.80	1.81	1.85	1.89	1.90	1.93	1.97	2.00	2.03	2.05	2.07	2.09	2.10
Package of 35 lbs.	1.78	1.83	1.86	1.90	1.93	1.94	1.97	2.03	2.04	2.07	2.12	2.13	2.17	2.19	2.23	2.28	2.32	2.33	2.36	2.38	2.41
Package of 40 lbs.	1.98	2.04	2.09	2.12	2.14	2.18	2.21	2.26	2.28	2.32	2.37	2.39	2.42	2.46	2.51	2.56	2.60	2.62	2.65	2.67	2.70
Package of 45 lbs.	2.19	2.24	2.32	2.34	2.37	2.39	2.44	2.49	2.52	2.56	2.62	2.65	2.68	2.72	2.77	2.84	2.87	2.90	2.94	2.96	3.00
Package of 50 lbs.	2.38	2.46	2.52	2.56	2.60	2.63	2.66	2.73	2.77	2.80	2.87	2.91	2.94	2.97	3.05	3.11	3.15	3.19	3.21	3.25	3.29
Package of 60 lbs.	2.80	2.89	2.96	3.01	3.05	3.09	3.14	3.21	3.25	3.30	3.38	3.43	3.47	3.50	3.59	3.67	3.72	3.75	3.75	3.84	3.88

Rate per 100 lbs.	$6.72	$6.79	$6.85	$7.13	$7.27	$7.35	$7.62	$7.69	$7.82	$7.90	$7.96	$8.38	$8.52	$8.87	$9.01	$9.35	$9.84	$10.32	$10.81
CHARGES ON																			
Package of 5 lbs.	.67	.67	.67	.68	.69	.69	.71	.71	.72	.72	.72	.74	.74	.78	.78	.79	.82	.84	.87
Package of 10 lbs.	.98	.98	1.00	1.02	1.03	1.03	1.07	1.08	1.10	1.10	1.11	1.15	1.16	1.20	1.21	1.25	1.29	1.35	1.39
Package of 15 lbs.	1.30	1.31	1.31	1.36	1.39	1.40	1.44	1.44	1.47	1.49	1.49	1.55	1.56	1.63	1.65	1.69	1.78	1.84	1.92
Package of 20 lbs.	1.63	1.64	1.65	1.70	1.73	1.75	1.80	1.81	1.84	1.85	1.86	1.95	1.98	2.05	2.08	2.14	2.24	2.34	2.44
Package of 25 lbs.	1.94	1.95	1.97	2.04	2.08	2.09	2.17	2.18	2.22	2.23	2.24	2.36	2.38	2.48	2.51	2.60	2.72	2.84	2.96
Package of 30 lbs.	2.26	2.27	2.31	2.38	2.42	2.44	2.52	2.55	2.60	2.61	2.63	2.76	2.80	2.90	2.94	3.05	3.19	3.34	3.48
Package of 35 lbs.	2.58	2.61	2.62	2.72	2.77	2.80	2.90	2.91	2.96	3.00	3.01	3.16	3.20	3.33	3.38	3.49	3.67	3.84	4.01
Package of 40 lbs.	2.90	2.92	2.95	3.06	3.11	3.15	3.25	3.29	3.34	3.36	3.39	3.57	3.62	3.75	3.81	3.94	4.15	4.33	4.54
Package of 45 lbs.	3.21	3.24	3.28	3.39	3.47	3.49	3.62	3.64	3.72	3.74	3.77	3.97	4.02	4.18	4.25	4.40	4.61	4.84	5.05
Package of 50 lbs.	3.53	3.57	3.60	3.74	3.81	3.84	3.98	4.02	4.08	4.12	4.16	4.36	4.44	4.60	4.67	4.85	5.09	5.33	5.57
Package of 60 lbs.	4.17	4.21	4.26	4.42	4.50	4.55	4.71	4.75	4.84	4.88	4.91	5.17	5.25	5.46	5.54	5.75	6.05	6.34	6.63

FREIGHT RATES

Table showing the freight rates per 100 pounds on goods shipped from Chicago to a number of cities in each state, these cities being used as a basis for figuring rates for other towns located in the same part of the state.

Please consult the following Table of Rates, considering it in connection with the classification of merchandise, and you will be able to determine easily, almost exactly, what the freight will amount to on any goods. If your station is not named in this table, the rate to your station will be the same or a few cents per 100 pounds more or less than the rate to the nearest city named, or you may ask your freight agent.

HOW TO SAVE ON FREIGHT CHARGES.

In having goods shipped by freight, it pays you to make up an order of 100 pounds or more, because railroad companies charge as much for a shipment weighing less than 100 pounds as they do for one weighing 100 pounds.

If there is no freight agent at your shipping point you must include sufficient money to prepay freight charges. If there is an agent you pay the freight charges when shipment reaches you. When you leave the method of shipment to us we always choose the cheapest.

CLASSIFICATION OF MERCHANDISE FOR FREIGHT SHIPMENTS.

You will notice in the Freight Table below there are four different rates quoted to each city, **first class, second class, third class and fourth class.** This is because the railroad companies base their rates on the kind of merchandise to be carried. In order, therefore, to give you an idea as to the rates which any kind of merchandise will take to your nearest city, we tell you here the kind of merchandise which takes the different classes of rates.

EXPLANATION.

- **1** means 1st class rates.
- **2** means 2d class rates.
- **3** means 3d class rates.
- **4** means 4th class rates.
- **1½** means 1 and ½ times 1st class rates.
- **D1** means 2 times 1st class rates.
- **3x1** means 3 times 1st class rates.
- **4x1** means 4 times 1st class rates.
- R25 means 15 per cent lower than 2d class rates.
- R26 means 20 per cent lower than 3d class rates.
- E means East of Chicago and North of Ohio River.
- W means West, Northwest and Southwest of Chicago.
- S means South of Ohio River and East of Mississippi River.

	Class
Ammunition (S)	1
Ammunition (W and E)	2
Animal Traps (W and S)	2
Animal Traps (E)	3
Anvils	4
Automobile Repairs	1
Awnings	1
Baby Carriages (E and W)	1½
Baby Carriages (S)	1
Bedsteads, Iron and Wood	2
Bedsteads, Brass	1
Bicycles	1½
Books	1
Boots and Shoes	1
Brooders	1
Cameras (E)	1
Cameras (W and S)	D1
Camera Supplies	1
Carpets	1
Chinaware	2
Cigars	1
Clocks	1
Clothing	1
Commodes	1
Couches, Sanitary (E)	1
Couches (Sanitary) (W and S)	2
Cream Separators	2
Cribs	2
Crockery	3
Curtains	1
Dress Forms (W)	1
Dress Forms (E and S)	1½
Drugs	1
Drums (Musical)	3x1
Dry Goods	1
Electrical Goods	1
Farm Bells	2
Fire Arms	1
Forges (W and S)	2
Forges (E)	1
Glassware (E)	R25
Glassware (W and S)	2
Go-Carts, Collapsible (W)	1
Go-Carts, Collapsible (E and S)	2
Graniteware	1
Graphophones	1
Grindstones with frames (W)	3
Grindstones with frames (E)	R26
Grindstones with frames (S)	4
Hardware (W and S)	2
Hardware (E)	3
Horse Clipping Machines	2
Incubators (W and S)	1
Incubators (E)	2
Iron Kettles (E)	R25
Iron Kettles (W and S)	3
Ladders (W and S)	1
Ladders (E)	2
Lamps	1
Linoleum	2
Matting	1
Milk Cans (E)	1½
Milk Cans (W and S)	1
Morris Chairs	2
Music Cabinets	1
Notions	1
Oilcloth	2
Oils, Lubricating (E)	2
Oils, Lubricating (W and S)	3
Paints (E)	3
Paints (W and S)	4
Pictures	1
Plumbing Material	2
Prepared Roofing (E)	R26
Prepared Roofing (W and S)	4
Pumps (Hand) (S and W)	3
Pumps (Hand) (E)	R25
Pulley Blocks (E and S)	3
Pulley Blocks (W)	2
Reed Chairs (S)	D1
Reed Chairs (E and W)	3x1
Rubber Goods	1
Rugs	1
Sewing Machines (W and E)	1
Sewing Machines (S)	2
Sleds, Children's (E and W)	D1
Sleds, Children's (S)	1½
Sporting Goods	1
Stationery	1
Tents	1
Tinware	1
Tools (W and S)	2
Tools (E)	3
Toys	1
Trunks (W and S)	1
Trunks (E)	1½
Vacuum Cleaners	1
Valises (E)	1½
Valises (W and S)	1
Vises (E)	R26
Vises (W and S)	2
Wagons, Children's	1
Wagon Wheels, Wooden	1
Wall Paper	2
Washing Machines, Hand	2
Wheelbarrows	2
Wringers	2

The following kinds of merchandise are shipped from our various factories and warehouses, and not with other items from Chicago, and are classified for freight shipment for your information only, as they cannot be used in connection with the rates shown below.

	Class
Bathtubs, Cast Iron, Enameled	1
Bathtubs, Sheet Steel	1½
Benches, Piano	1
Bob Sleds	2
Boilers, Heating (E and W)	4
Buffets	1
Buggies (E)	1½
Buggies (W and S)	1
Cabinets, Kitchen (E and S)	1½
Cabinets, Kitchen (W)	1
Chiffoniers	1
Chifforobes (E and W)	1½
Chifforobes (S)	1
China Closets (E and S)	1½
China Closets (W)	D1
Church and School Bells (E)	R25
Church and School Bells (W and S)	3
Concrete Block Machines (E and S)	1
Concrete Block Machines (W)	2
Couches, Upholstered	1
Cupboards, Kitchen	2
Desks (E)	1½
Desks (W and S)	1
Disc Harrows (E and W)	2
Disc Harrows (S)	1
Dressers	1
Dressers, Vanity	1½
Farm Implements (E)	R25
Farm Implements (W and S)	3
Farm Wagons	1
Gasoline Engines (E)	2
Gasoline Engines (W)	1
Gasoline Engines (S)	3
Iron Safes	3
Oil Stoves	1
Pianos and Organs	1
Radiators, Heating	3
Refrigerators	2
Road Carts	D1
Row Boats (E and S)	4x1
Row Boats (W)	3x1
Sideboards	1
Steel Roofing	4
Steel Ceiling and Siding (E)	R26
Steel Ceiling and Siding (W)	4
Steel Ceiling and Siding (S)	3
Stoves and Ranges	3
Tables, Dining	2
Wagon Wheels, Iron (E)	R25
Wagon Wheels, Iron (W and S)	3
Washing Machines, Power	1
Windmills (E)	R25
Windmills (W and S)	3
Wire (Barbed)	4
Wire Fencing (E)	R26
Wire Fencing (W)	3
Wire Fencing (S)	5

HERE ARE THE RATES PER 100 POUNDS TO DIFFERENT TOWNS IN EACH STATE.

From Chicago to	1st Class Freight, per 100 Pounds.	2d Class Freight, per 100 Pounds.	3d Class Freight, per 100 Pounds.	4th Class Freight, per 100 Pounds.
ALABAMA—				
Anniston	$2.32	$1.98	$1.69	$1.37
Birmingham	2.18	1.84	1.55	1.25
Brewton	2.48	2.18	1.86	1.52
Dothan	2.65	2.29	1.96	1.59
Florence	1.92	1.66	1.36	1.07
Huntsville	1.92	1.66	1.36	1.07
Mobile	1.98	1.70	1.51	1.27
Montgomery	2.18	1.92	1.64	1.28
York	2.40	2.04	1.67	1.37
ARKANSAS—				
Ft. Smith	2.04	1.72	1.40	1.20
Hamburg	2.13	1.80	1.47	1.25
Helena	1.59	1.37	1.21	1.02
Jonesboro	1.73	1.45	1.18	1.01
Little Rock	1.91	1.61	1.30	1.12
Texarkana	2.18	1.83	1.49	1.28
COLORADO—				
Craig	5.61	4.79	3.99	3.26
Denver	2.74	2.21	1.67	1.30
Durango	4.71	3.97	3.15	2.36
Grand Junction	3.98	3.35	2.78	2.24
Las Animas	2.74	2.21	1.67	1.30
Leadville	3.98	3.35	2.76	2.10
Sterling	2.64	2.21	1.67	1.28
Trinidad	2.74	2.21	1.67	1.30
ILLINOIS—				
Bloomington	.60	.51	.40	.30
Cairo	.89	.75	.60	.45
Danville	.60	.51	.40	.30
East St. Louis	.79	.67	.53	.40
Effingham	.73	.62	.49	.37
Galena	.67	.57	.45	.34
Joliet	.42	.36	.28	.21
Kankakee	.48	.41	.32	.24
La Salle	.54	.46	.36	.27
Moline	.67	.57	.45	.34
Mt. Vernon	.79	.67	.53	.40
Peoria	.66	.56	.44	.33
Quincy	.79	.67	.53	.40
Rockford	.53	.45	.36	.27
Springfield	.71	.61	.48	.36
Waukegan	.42	.36	.28	.21
INDIANA—				
Bloomington	.75	.64	.51	.38
Evansville	.81	.69	.54	.41
Ft. Wayne	.64	.54	.43	.32
Gary	.41	.35	.27	.21
Indianapolis	.71	.61	.48	.36
Logansport	.59	.50	.39	.30
New Albany	.82	.70	.55	.41
Richmond	.75	.64	.51	.38
South Bend	.53	.45	.36	.27
Terre Haute	.69	.59	.46	.35
IOWA—				
Burlington	.73	.62	.49	.37
Cedar Rapids	.79	.64	.48	.35
Council Bluffs	1.22	.99	.69	.49
Davenport	.67	.57	.45	.34
Decorah	.92	.76	.61	.41
Des Moines	.92	.73	.55	.42
Dubuque	.69	.59	.46	.35
Ft. Dodge	.99	.79	.60	.45
Mason City	.92	.76	.61	.41
Ottumwa	.84	.69	.54	.41
Rock Rapids	1.22	.99	.69	.49
Sioux City	1.22	.99	.69	.49
KANSAS—				
Concordia	1.91	1.59	1.18	.90
Dodge City	2.46	2.12	1.65	1.27
Great Bend	2.21	1.86	1.45	1.18
Leavenworth	1.22	.99	.69	.49
Pittsburgh	1.42	1.28	.92	.69
St. Francis	2.59	2.21	1.67	1.30
Sharon Springs	2.58	2.18	1.65	1.27
Topeka	$1.52	$1.25	$0.90	$0.66
Wichita	1.98	1.69	1.31	1.00
KENTUCKY—				
Ashland	.92	.78	.62	.46
Bowling Green	1.57	1.34	1.12	.90
Frankfort	1.12	.96	.78	.61
Hickman	1.29	1.11	.98	.83
Louisville	.84	.71	.57	.43
Newport	.92	.78	.62	.46
Pikeville	1.68	1.45	1.13	.83
Williamsburg	1.64	1.40	1.17	.97
MICHIGAN—				
Adrian	.75	.64	.50	.38
Alpena	1.06	.90	.72	.53
Bad Axe	1.03	.88	.68	.52
Cheboygan	1.06	.90	.72	.53
Detroit	.79	.67	.53	.40
Escanaba	.92	.76	.61	.43
Houghton	1.16	.96	.76	.49
Lansing	.79	.67	.53	.40
Muskegon	.76	.65	.51	.38
Niles	.54	.46	.36	.27
Saginaw	.85	.73	.57	.43
Sault Ste. Marie	1.15	.99	.73	.58
Traverse City	.97	.82	.66	.48
MINNESOTA—				
Bemidji	1.70	1.43	1.14	.78
Brainerd	1.47	1.23	.98	.69
Breckenridge	1.66	1.39	1.11	.76
Chisholm	1.40	1.18	.95	.63
Crookston	1.80	1.50	1.20	.86
Duluth	.99	.84	.67	.43
International Falls	1.84	1.55	1.24	.85
Mankato	.99	.84	.66	.42
Pipestone	1.27	1.03	.72	.51
St. Paul	.92	.76	.61	.41
Winona	.76	.64	.51	.38
MISSISSIPPI—				
Biloxi	1.98	1.70	1.51	1.27
Clarksdale	1.71	1.47	1.30	1.10
Corinth	1.47	1.26	1.12	.94
Greenville	1.71	1.47	1.30	1.10
Hattiesburg	1.91	1.65	1.45	1.22
Jackson	1.82	1.56	1.38	1.17
Meridian	1.82	1.56	1.38	1.17
Natchez	1.91	1.65	1.45	1.22
Westpoint	1.71	1.47	1.30	1.10
MISSOURI—				
Caruthersville	1.62	1.39	1.16	.92
Hannibal	.81	.69	.54	.41
Jefferson	1.08	.88	.64	.45
Kansas City	1.22	.99	.69	.49
Maryville	1.22	.99	.69	.49
Milan	1.22	.99	.69	.49
Nevada	1.40	1.24	.85	.57
St. Louis	.79	.67	.53	.40
Salem	1.25	1.03	.78	.63
Springfield	1.25	1.10	.77	.57
MONTANA—				
Billings	3.75	3.14	2.54	2.04
Glasgow	3.38	2.88	2.33	1.77
Glendive	3.12	2.66	2.13	1.59
Great Falls	3.90	3.33	2.73	2.15
Havre	3.75	3.21	2.61	2.04
Plentywood	3.26	2.78	2.24	1.68
NEBRASKA—				
Chadron	2.66	2.21	1.69	1.36
Grand Island	1.92	1.59	1.19	.93
Laurel	1.70	1.41	1.03	.78
Lincoln	1.29	1.06	.75	.55
McCook	2.33	1.95	1.47	1.17
North Platte	2.37	1.99	1.51	1.19
Omaha	1.22	.99	.69	.49
Sidney	2.65	2.21	1.67	1.29
Valentine	2.34	1.95	1.47	1.16
NORTH DAKOTA—				
Bismarck	2.44	2.06	1.63	1.19
Bowman	2.89	2.43	1.93	1.37
Devils Lake	$2.38	$2.01	$1.57	$1.14
Dickinson	2.85	2.41	1.95	1.47
Fargo	1.78	1.49	1.18	.82
Grafton	2.12	1.78	1.40	1.02
Minot	2.67	2.27	1.81	1.35
Oakes	1.84	1.52	1.13	.85
Williston	3.03	2.53	2.07	1.54
OHIO—				
Cambridge	.91	.78	.61	.46
Chillicothe	.87	.74	.59	.44
Cincinnati	.81	.69	.54	.41
Cleveland	.85	.73	.57	.43
Columbus	.84	.71	.56	.42
Dayton	.79	.67	.53	.40
Ironton	.92	.78	.62	.46
Lima	.73	.62	.49	.37
Mansfield	.82	.70	.55	.41
Marietta	.95	.81	.64	.48
Massillon	.87	.74	.59	.44
Toledo	.76	.65	.51	.38
Youngstown	.92	.78	.62	.46
OKLAHOMA—				
Bartlesville	1.91	1.67	1.27	1.04
Enid	2.28	1.96	1.57	1.33
Forgan	2.59	2.21	1.83	1.55
Muskogee	2.05	1.83	1.37	1.06
Oklahoma City	2.28	1.96	1.63	1.33
Poteau	2.09	1.76	1.43	1.22
Woodward	2.28	1.96	1.63	1.45
SOUTH DAKOTA—				
Aberdeen	1.74	1.45	1.02	.76
Deadwood	2.95	2.46	2.05	1.67
Dupree	2.83	2.36	1.93	1.49
Hot Springs	2.82	2.39	1.98	1.61
McIntosh	2.52	2.10	1.74	1.24
Mitchell	1.51	1.36	.97	.69
Pierre	1.86	1.58	1.22	.92
Sioux Falls	1.27	1.03	.72	.51
Watertown	1.46	1.25	.97	.69
TENNESSEE—				
Chattanooga	1.88	1.61	1.35	1.07
Cookeville	2.39	2.06	1.78	1.44
Jackson	1.35	1.16	1.03	.87
Johnson City	1.88	1.64	1.27	.93
Knoxville	1.88	1.61	1.35	1.07
Lewisburg	1.89	1.63	1.36	1.15
Memphis	1.47	1.26	1.12	.94
Nashville	1.65	1.41	1.18	.95
Union City	1.16	1.00	.88	.75
UTAH—				
Lucin	4.13	3.57	2.99	2.49
Marysvale	4.68	4.03	3.43	2.86
Milford	5.10	4.43	3.68	3.11
Price	3.98	3.35	2.78	2.24
Salt Lake City	3.98	3.35	2.78	2.24
WISCONSIN—				
Ashland	.99	.84	.67	.43
Eau Claire	.92	.76	.61	.41
Fond du Lac	.62	.52	.43	.31
Green Bay	.66	.55	.45	.31
Hudson	.92	.76	.61	.41
LaCrosse	.76	.64	.51	.38
Madison	.60	.51	.40	.30
Marinette	.66	.55	.45	.35
Milwaukee	.52	.45	.35	.27
Prairie du Chien	.76	.64	.51	.38
Rhinelander	.92	.76	.61	.41
Superior	.99	.84	.67	.43
Wisc. Rapids	.76	.64	.51	.36
WYOMING—				
Casper	3.74	3.10	2.46	1.95
Cheyenne	2.74	2.21	1.67	1.30
Cody	4.21	3.57	2.82	2.26
Kremmerer	3.98	3.35	2.78	2.24
Lander	4.20	3.50	2.78	2.22
New Castle	3.35	2.90	2.34	1.90
Rawlins	3.77	3.17	2.63	2.10
Sheridan	3.75	3.14	2.54	2.04

ITEMIZED INDEX

A

ACCELERATORS ... 755
ACCORDIONS ... 515-517
ADAPTERS
Film Pack Adapters ... 554
ADJUSTERS
Carburetor Adjusters ... 750, 755
Casement Adjusters ... 835
ADZES ... 862
AERATORS ... 832
AERIAL GOODS ... 809
AFGHANS ... 134
AGRICULTURAL IMPLEMENTS AND MACHINERY 818-820, 842-844, 876-890, 983, 904-909
AIR MOISTENERS ... 694
ALBUMS
Phonograph Record Albums ... 548
Photograph Albums ... 556
Post Card Albums ... 556
ALCOHOL
Denatured Alcohol ... 956
ALTAR SERVICES ... 418
ALUM ... 484, 553
ALUMINUMWARE ... 826-827
Aluminum Cleaners ... 826, 829
AMMETERS ... 749
AMMONIAC ... 813, 853
AMMUNITION ... 790-792
ANCHORS
Boat Anchors ... 780
Decoy Anchors ... 793
Embroidered Anchors ... 366
Stanchion Anchors ... 891
ANNOUNCEMENTS
Birth Announcements ... 482
ANTI-OXIDIZERS ... 471
ANTI-RATTLERS
Ball Sockets, Auto ... 760
Shaft Bolt Holders ... 848
ANTISEPTICS ... 482, 484, 487
ANVILS
Blacksmiths' Anvils ... 875
Jewelers' Anvils ... 471
Vise and Anvil Combinations ... 872, 874
APRONS
Blacksmiths' Aprons ... 874
Buggy Storm Aprons ... 851
Clothespin Aprons ... 823
Girls' Aprons ... 143, 162
Massage Aprons ... 803
Mechanics' Aprons ... 855
Sanitary Aprons ... 370
Shampoo Aprons ... 803
Stamped Aprons ... 362
Waterproof Aprons, Men's ... 552, 832
Waterproof Aprons, Women's ... 82
Window Aprons ... 967
Women's Aprons ... 82, 155
AQUARIUMS ... 723, 726
ARCH SUPPORTS ... 472
ARMBANDS ... 243
ARNICA ... 484
ARSENICAL COMPOUND ... 476
ARRESTERS
Lightning Arresters ... 808
ARTGUM ... 500
ARTISTS' MATERIALS ... 454, 500-502
ASAFETIDA ... 484
ASBESTOS
Asbestos Lighting Rings ... 687
Asbestos Mats, Pads ... 338
Asbestos Roof Cement ... 935
Asbestos Roof Paint ... 935
ASPHALT
Asphalt Roofing ... 960-963
Asphalt Roof Paint ... 935
ASPIRIN ... 484
ATHLETIC GOODS ... 766, 769-771
ATLASES ... 436
—Or Write for School Furniture Catalog.
ATOMIZERS ... 481, 502
ATTACHMENTS
—See Name of Article for Which Attachments Are Wanted.
AUGERS
Auger Bits ... 864-865
Boring Machine Augers ... 865
Hollow Augers ... 864
Nut Augers ... 864
Post Hole Augers ... 710, 843
Well Boring Augers ... 710
AUTOHARPS ... 513
AUTOMOBILE ACCESSORIES 738-761, 815, 955-957
—Or Write for Auto Supply Catalog.
Auto Repair Parts for Ford Cars ... 761
Auto Bodies ... 815
Auto Camping Supplies ... 782-785
Auto Oils and Greases ... 956-957
AUTOMOBILES
Children's Automobiles ... 496-497
AWLS
Belt Awls ... 855
Chalk Line Awls ... 864
Harness Awls ... 930
Sewing Awls ... 930
AWNINGS ... 781
AXES ... 793, 856
AXLES
Axle Parts, Auto ... 760-761
Buggy Axles ... 848
Wagon Axles ... 848, 850

B

BABBITT METAL 853
Babbitting Jigs ... 755
BABY GOODS
Babies' Baskets, Bassinets ... 482, 494, 667-668
Babies' Blankets ... 347
Babies' Carriages ... 670-671
Babies' Chairs ... 668-669
Babies' Cradles, Cribs ... 667
Babies' Flannels ... 328, 351
Babies' Jewelry ... 373, 392, 408-409
Babies' Moccasins, Shoes ... 138, 196-197
Babies' Nursery Supplies 134-135, 482-483, 494, 667-669
Babies' Silverware ... 412-419
Babies' Wear ... 134-143, 225, 237, 372
BACKBANDS ... 922
BACKS
Cushion Seat Backs ... 851
Sink Backs ... 702
Stove Fire Backs ... 823
BAGS
Band Instrument Bags ... 523
Banjo Bags ... 514
Beaded Bags ... 127
Bicycle Tool Bags ... 738
Boston Bags ... 127
Brief Bags ... 525
Clothes Bags ... 367
Diaper Bags ... 135
Drum Bags ... 522
Dunnage Bags ... 784
Feed Bags ... 924
Fish Bags ... 779
Game Bags ... 795
Golf Bags ... 765
Guitar Bags ... 514
Gun Shell Bags ... 793, 795
Hand Bags ... 126-127
Hot Water Bags ... 480-482
Ice Bags ... 480
Jelly Strainer Bags ... 831
Laundry Bags ... 363
Mandolin Bags ... 514
Masons' Tool Bags ... 841
Mesh Bags ... 388
Mothproof Bags ... 367
Music Bags ... 524-525
Music Stand Bags ... 515
Reel Bags ... 774
Saddle Bags ... 924
School Book Bags ... 454
Sleeping Bags, Babies' ... 137, 141
Sleeping Bags, Campers' ... 784
Snare Drum Bags ... 522
Striking Bags ... 770
Traveling Bags ... 260-261
Ukulele Bags ... 514
Urinal Bags ... 481
Vanity Bags ... 126-127, 489
Violin Bags ... 508
Water Bags ... 784
BAITS
Animal Baits ... 796
Fish Baits ... 776-777
BALANCES
Sash Balances ... 834
Spring Balances ... 775, 833
BALLOONS ... 496
BALLS
Ball Bearing Balls ... 737, 767
Ball Bearings, Hub, Shaft ... 754, 760
Baseballs ... 762
Basket Balls ... 771
Billiard Balls ... 797
Closet Tank Balls ... 699
Footballs ... 771
Golf Balls ... 765
Hand Balls ... 762
House Play Balls ... 494, 497
Indoor Balls ... 763
Moth Balls ... 478
Ox Horn Balls ... 838
Playground Balls ... 763, 771
Pool Balls ... 797
Rubber Bibb Balls ... 706
Shake Balls ... 797
Tea Balls ... 419, 421
Tennis Balls ... 764
Volley Balls ... 771
BALMS ... 484
BALUSTERS
Baluster Stock ... 967
BAND INSTRUMENTS ... 519-522
Jazz Bands ... 513
BANDAGES
Elastic Bandages ... 474
Gauze Bandages ... 483
Rubber Bandages ... 474
BANDANAS ... 242
BANDEAUX
Bust Bandeaux ... 124-125, 156
Hair Bandeaux ... 130
Hat Bandeaux ... 102
BANDS
Abdominal Bands ... 473
Accouchement Bands ... 135
Babies' Bands ... 157, 237
Backbands ... 922
Bellybands ... 922
Bloomer Knee Bands ... 369
Breast Bands ... 135
Chin Bands ... 480
Elastic Arm Bands ... 243
Elastic Waist Bands ... 369
Feather Bands ... 106-107
Horse Tail Bands ... 928
Nosebands, Bridle ... 919
Oil Tank Bands and Bolsters ... 894
Poultry Leg Bands ... 839
Rubber Bands ... 453
Shirt Collar Bands ... 369
Sweat Bands ... 764
Transmission Brake Bands ... 760
Wrist Watch Bands ... 358, 378
BANJOS and SUPPLIES ... 512, 514
BANKS ... 495
BARBERS' SUPPLIES ... 337, 799-803
BARLEY
Babies' Barley ... 482
BARNS AND BARN EQUIPMENT ... 891, 971
—Or Write for Book of Barns.
Barn Hardware ... 836
BAROMETERS ... 428
BARRELS
Oak Barrels (Kegs) ... 830
Steel Oil Barrels ... 957
BARRETTES ... 131, 408
BARROWS ... 843
BARS
Clothes Hanger Bars ... 367
Crowbars ... 857
Handle Bars ... 738
Iron Bars (Bar Iron) ... 847
Towel Bars ... 934
Trapeze Bars ... 766
Wrecking Bars ... 857
BASEBALL GOODS 762-763, 766
BASEBOARDS ... 967
BASE SHOES ... 967
BASINS
Canvas Basins ... 784
Wash Basins (Lavatories) ... 698-699
Wash Basins (Pans) ... 825-826
BASKET BALL GOODS ... 766, 769, 771
BASKETS
Babies' Baskets ... 482, 668
Basket Ball Goals ... 771
Cake Baskets, Silver ... 420-421
Canvas Baskets ... 780
Feed Baskets ... 839
Fish Baskets ... 780
Flower Baskets, Silver, Glass ... 420
Fruit Baskets, Silver, Glass ... 420-421
Household Baskets ... 822
Wire Nest Baskets ... 839
BASS VIOLS AND SUPPLIES ... 508
BASSINETS ... 494, 667-668
BATH CABINETS ... 480
BATHING GOODS ... 768
BATHROOM OUTFITS ... 696
Bathroom Trimmings ... 934
BATHTUBS
Babies' Bathtubs ... 482, 668
Folding Bathtubs ... 481, 668, 702
Porcelain Bathtubs ... 697
Steel Bathtubs ... 697
BATISTE ... 304-305, 321-324, 326
BATS ... 763
BATTENS
Barn Battens ... 964
BATTERIES
Dry Batteries ... 813
Farm Plant Batteries ... 674
Flash Light Batteries ... 812
Gravity Batteries and Supplies ... 813
Medical Batteries ... 810
Multiple Batteries ... 813
Radio Batteries ... 809
Starting Batteries and Accessories ... 748-749, 756-757
Storage Batteries ... 749
BATTING
Cotton Batting ... 333
Wool Batting ... 333
BAY RUM ... 490
BEADERS
Carpenters' Beaders ... 860
BEADING
Embroidery Beading ... 352-353
Lace Beadings ... 354-355
Wall Beadings—Write for Wall Paper Sample Book.
BEADS
Beads (Necklaces) ... 373, 392-395, 409
Children's Beads ... 494-495, 497
Prayer Beads ... 390, 445
Trimming Beads ... 361
BEAMS
Beams (Lumber) ... 965
Scale Beams ... 833
BEARINGS
Ball Bearings (Balls) ... 737
Ball Thrust Bearings ... 760
Hub Bearings, Auto ... 754, 760
Shaft Bearings, Auto ... 760
BEARS ... 494
BEATERS
Cream Beaters ... 726, 831
Cymbal Beaters ... 523
Drum Beaters ... 523
Egg Beaters ... 726, 831
BEDPANS ... 483, 727
BEDS
Auto Beds ... 784
Babies' Swing Beds ... 494
Brass Beds ... 656
Camp Beds ... 785
Children's Beds ... 667
Cot Beds ... 666-667
Cot Beds, Outdoor ... 785
Davenport Beds ... 607E-607H, 664
Folding Beds ... 653
Steel Beds ... 654-655
Wood Beds ... 644-651
BEDSPREADS 348-350
Bedspread Sets ... 349
Bedspread Sets, Lace ... 582
Bedspread Sets, Stamped ... 362
BEE KEEPERS' SUPPLIES ... 895
BEEF EXTRACTS ... 483
BELLS
Bicycle Bells ... 737
Church Bells ... 889
Cow Bells ... 838
Dinner Bells ... 831
Door Bells ... 834-835
Door Bells, Electric ... 813
Drummers' Bells ... 522
Extension Bells—Write for Electric Goods Catalog.
Factory Bells ... 889
Farm Bells ... 838
Hand Bells ... 831
Orchestra Bells ... 519
Schoolhouse Bells ... 889
Sheep Bells ... 838
Sleigh Bells—Write for Harness Catalog.
Table Bells ... 422
Telephone Bells—Write for Electrical Goods Catalog.
Trip Gong Bells ... 855
Turkey Bells ... 838
BELLYBANDS ... 922
BELTING
Canvas Belting ... 854
Leather Belting ... 854
Link Chain Belting ... 854
Rubber Belting ... 818, 854
Skirt Belting ... 369
Solid Woven Belting ... 854
BELT LACING MACHINES ... 854
BELTS
Abdominal Belts ... 473
Athletic Belts ... 769
Bathers' Belts ... 768
Boys' Belts ... 163
Cartridge Belts ... 795
Children's Belts ... 127
Fan Belts, Auto ... 754
Girls' Belts ... 127
Gun Shell Belts ... 795
Holster Belts ... 795
Hunters' Belts ... 795
Life Preserver Belts ... 780
Linemen's Belts ... 814
Line Shaft Belts ... 818
Machine Belts and Accessories ... 854-855
Men's Belts, Sets ... 258, 388
Money Belts ... 795
Reducing Belts ... 473
Sewing Machine Belts ... 312
Women's Belts ... 127
Women's Girdle Belts ... 127, 366
Women's Sanitary Belts ... 371
BENCHES
Dressing Table Benches 644-645, 649
Piano Benches ... 523
Work Benches ... 863
BENDS
Closet Bends ... 704
Pipe Bends ... 704, 707
BEVELS
Sliding T Bevels ... 863
BEVERAGES ... 478
BIBBS ... 706
Bibb Seat Grinders ... 706
BIBLES ... 431, 444-445
BIBS
Babies' Bibs ... 135, 141
Barbers' Bibs ... 803
BICYCLES AND SUPPLIES ... 732-738
BILLFOLDS ... 259
BILLIARD TABLES
—Write for Prices.
Billiard Supplies ... 797
BINDERS
Abdominal Binders ... 135
Babies' Binders ... 157
Breast Binders ... 135
Loose Leaf Binders ... 454
Music Binders ... 515
BINDER TWINE ... 887
BINDING
Carpet Binding ... 605
Linoleum Binding ... 605
Matting Binding ... 605
Oilcloth Binding ... 605
Seam Binding ... 369
BINOCULARS ... 428
BINS
Feed Bins ... 901
Flour Bins ... 831
Grain Bins — Write for Farm Implement Catalog.
Storage Bins ... 901
BIRDCAGES ... 837
BITS
Auger Bits ... 864-865
Bit and Brace Sets ... 864
Boring Machine Bits ... 865
Brace Bits ... 864-865
Countersink Bits ... 864, 869
Drill Bits ... 864, 869
Drill Bits, Jewelers' ... 471
Expansive Bits ... 865
Gimlet Bits ... 864
Horse Bits ... 931
Mule Bits ... 931
Pony Bits ... 931
Reamer Bits ... 864, 869
Screwdriver Bits ... 864
Ship Auger Bits ... 865
Ship Auger Car Bits ... 865
Stallion Bits ... 931
BLACKING
Shoe Blacking ... 213
Stove Blacking ... 689
BLACKLEG OUTFITS ... 477
BLACKSMITHS' TOOLS AND OUTFITS ... 846-847, 867-875
BLADDERS
Basket Ball Bladders ... 771
Football Bladders ... 771
Striking Bag Bladders ... 770
Volley Ball Bladders ... 771
BLADES
Band Saw Blades ... 852, 866
Bracket Saw Blades ... 859
Buck Saw Blades ... 857
Butchers' Saw Blades ... 806
Compass Saw Blades ... 858
Coping Saw Blades ... 859
Cultivator Blades ... 876
Dehorning Saw Blades ... 838
Drag Saw Blades ... 819
Floor Scraper Blades ... 861
Hack Saw Blades ... 873
Harness Awl Blades ... 930
Hoof Shear Blades ... 873
Jewelers' Saw Blades ... 471
Kitchen Saw Blades ... 806
Mouth Float Blades ... 477
Road Grader Blades ... 888
Safety Razor Blades ... 800-
Tree Pruner Blades ... 840
Turning Saw Blades ... 859
BLANKETS
Babies' Pinning Blankets ... 137
Babies' Receiving Blankets ... 134-135
Bath Robe Blankets ... 347
Bed Blankets ... 346-347
Camp Blankets ... 346, 784
Cow Blankets ... 932-933
Crib Blankets ... 347
Horse Blankets ... 932-933
Saddle Blankets ... 929
Stable Blankets ... 933
Steamer Blankets ... 929
Waterproof Blankets ... 784
BLINDS
Horse Blinds ... 920
Window Blinds ... 583-586
BLOCKS
Aerial Connector Blocks ... 809
Drummers' Blocks ... 522
Pillow Blocks ... 855
Post Blocks ... 855
Pulley Blocks ... 843-845
Saw Setting Blocks ... 859
Snatch Blocks ... 844
Tackle Blocks ... 843, 845
Toy Blocks ... 494
BLOOMERS
Children's Bloomers ... 142
Girls' Bloomers ... 142, 160-161
Gymnasium Bloomers ... 769
Sanitary Bloomers ... 371
Women's Bloomers ... 146-147, 149, 155
BLOTTERS
Photo Blotters ... 555
Roller Blotters ... 422
BLOUSES
Boys' Blouses ... 178-180
Gymnasium Blouses ... 769
Knit Blouses ... 146-147
Middy Blouses ... 24, 50-51
Women's Blouses ... 44-55
BLOW TORCHES ... 705
BLOWERS
Blacksmiths' Blowers ... 874
Insecticide Blowers ... 903
Blowpipes ... 471
BLUE VITRIOL ... 813
BOARDS
Base Boards ... 967
Boards (Lumber) ... 964-965
Clapboards ... 964
Drawing Boards ... 501-502
Game Boards ... 495
Honey Boards ... 895
Ironing Boards ... 823
Ironing Boards, Built-In —Write for Mill Work Catalog.
Ouija Boards ... 495
Paste Boards—Write for Wall Paper Sample Book.
Plaster Boards ... 969
Print Trimming Boards ... 555
Running Boards, Auto ... 754, 759
Sink Drain Boards ... 702
Stove Boards ... 823
Wagon Sand Boards ... 850
Wall Boards ... 969
Wash Boards ... 822
BOATS ... 780
BOBBINETS ... 570
BOBBINS ... 312
BODIES
Automobile Bodies ... 815
Buggy Bodies ... 850
BOILERS
Agricultural Boilers ... 892
Boiler Coverings ... 703
Boiler Fittings ... 694
Coffee Boilers ... 828
Dairy Boilers ... 893
Double Boilers ... 826-828
Ham Boilers ... 829-830
Range Boilers ... 703
Steam Heat Boilers ... 694
Stock Feed Boilers ... 892
Wash Boilers ... 822, 829
Water Heat Boilers ... 694
BOLSTERS
Bed Bolsters ... 657
Couch Bolsters ... 665
Oil Tank Bolsters and Bands ... 894
Wagon Bolsters ... 849-850
BOLSTER STAKES ... 848
BOLTS
Carriage Bolts ... 846
Chain Bolts ... 835
Connecting Rod Bolts ... 755
Door Bolts ... 834-837
Eye Bolts ... 844
Foot Bolts ... 834
Ironwork Bolts ... 846
Lag Bolts ... 847
Machine Bolts ... 847
Mower Guard Bolts ... 876
Plow Bolts ... 846
Rod Track Bolts ... 891
Stall Hook Bolts ... 891
Stove Bolts ... 846
Tire Bolts ... 846
Wagon Box Bolts ... 848
Window Bolts ... 834, 837
BONE
Steamed Bone ... 476
BONNETS
Babies' Bonnets ... 138
Children's Bonnets ... 138
BOOKMARKS ... 422
BOOKCASES ... 625
Book Ends ... 723
Bookstands ... 618-620, 625
BOOKS ... 430-450
Including the following:
Accounting ... 439
Adventure ... 432-433, 436
Agricultural ... 449
Algebra ... 438
Amusements ... 431, 435
Arithmetic ... 438-439
Atlases ... 436
Automobiles ... 448
Baseball Rules ... 762
Basket Ball Rules ... 771
Bibles ... 431, 444-445
Bible Stories ... 445
Biographies ... 430, 436
Birds ... 440
Blacksmithing ... 447
Blotter Books, Photo ... 555
Bookkeeping ... 439
Boxing ... 435
Boys' Books ... 434
Bricklaying ... 447
Building ... 446
Butterfly Guides ... 440
Canning ... 450

Order Blank Enclosed in This Catalog.

BOOKS—Continued.
Candy Making ... 450
Card and Other Games ... 435
Care of Children ... 442
Carpentry ... 446
Cartoons ... 434
Cements ... 447
Checkers ... 435
Chess ... 435
Chord Books ... 524
Children's Books ... 430, 434-435
Civil Service ... 438
Concretes ... 447
Conundrums ... 434
Conversation ... 439
Cooking ... 450
Correspondence ... 435, 438, 450
Cowboy Life ... 436
Crocheting ... 361
Dancing ... 431
Detective Stories ... 433, 436
Dialogs ... 435
Dictionaries ... 440
Drawing ... 438, 447, 501
Dreams ... 435
Dressmaking ... 437, 450
Educational ... 438-439
Electrical ... 446
Encyclopedias ... 441, 445-449
Engineering ... 446
Engines ... 447-448
Entertainment ... 431, 435
Etiquette ... 431, 450
Exposure Tables ... 549
Fairy Tales ... 430
Fancywork ... 361, 365
Farm Knowledge ... 449
Fiction ... 430, 432-433
Flowers ... 440
Foreign Languages ... 438-440
Fortune Telling ... 435
French ... 438, 440
Frontier Stories ... 432, 436
Games ... 435
Geometry ... 438
German ... 440
Grammar ... 439
Hand Ball Rules ... 762
Health ... 442-443
Heating Systems ... 447
History ... 436
Home Study ... 446-447
House Wiring ... 446
How to Reduce (Records) ... 443
Humor ... 434-435
Hygienics ... 442
Hypnotism ... 435
Indian Stories ... 436
Indoor Ball Rules ... 763
Jiu Jitsu ... 435
Jokes ... 434
Keys of Heaven ... 445
Keys to the Bible ... 445
Knitting ... 365
Latin ... 439
Law ... 439
Lettering ... 446
Letter Writing ... 438, 450
Loose Leaf Books and Fillers ... 450, 454
Love Letters ... 435
Love Making ... 435
Machinists' Books ... 447
Made Easy Series ... 431, 435, 438-439, 450
Magic ... 435
Magnetism ... 439
Manuals of Devotion ... 445
Mathematics ... 438-439
Mechanics ... 446-447
Medical ... 442
Memorandum Books ... 450
Memory Helps ... 438-439
Mental Healing ... 442
Mental Training ... 438-439
Millwrights ... 447
Mortars ... 447
Motorcycles ... 448
Motors ... 446
Moving Picture Novels ... 430
Music Dictionaries ... 524
Music Instructors ... 519, 524
Mystery Stories ... 433
Names, Dates and Numbers ... 435
Negative Books ... 555
New Books ... 430
Painting ... 446, 448
Palmistry ... 435
Penmanship ... 438
Phono-Bretto (Words of Records) ... 544
Phonograph Exercises ... 443
Plumbing ... 447
Prayer Books ... 445
Preserving ... 450
Radio ... 447
Reading and Recitations ... 435
Recent Publications ... 430
Recipes ... 450
Religious Books ... 431-432, 444-445
Riddles ... 435
Score Books ... 762, 771
Sewing ... 450
Sex Hygiene ... 442
Shorthand ... 439
Sketch Books ... 502
Sleight of Hand ... 435
Song Books ... 435, 525
Spanish ... 438, 440
Speaker ... 435
Speechmaking ... 435
Spelling ... 439
Steel Square ... 446
Stone Masonry ... 447
Storage Batteries ... 446
Stump Speeches ... 435
Success ... 439
Tatting ... 361
Tennis Rules ... 764
Testaments ... 444
Tire Vulcanizing ... 448
Toasts ... 435
Tractors ... 448
Tree Guides ... 440
Tricks ... 435
Volley Ball Rules ... 771
Welding ... 447
Western Stories ... 432, 436
Wireless Telegraphy ... 447
Wiring ... 446
Working Models ... 447-448
Writing Courses ... 438-439

BOOTEES ... 138

BOOTS
Ankle Boots, Horse ... 924
Interfering Boots ... 928
Rubber Boots ... 216-217
Bordeaux Mixture ... 476

BORDERS
Felt Robe Borders—Write for Harness Catalog.
Oilcloth Wall Borders—Write for Wall Paper Sample Book.
Rug Borders ... 594

BORIC ACID ... 484

BORING MACHINES ... 865
Angular Boring Attachments ... 865
Boring Sets ... 864
Well Boring Outfits ... 710

BOTTLES
Babies' Bottles ... 482
Cream Bottles ... 832
Crown Cap Bottles ... 830
Hot Water Bottles, Metal ... 481
Hot Water Bottles, Rubber ... 480-482
Hot Water Bottles, Stoneware ... 727
Insulated Bottles ... 784
Milk Bottles ... 832
Milk Tester Bottles ... 832
Oil Bottles, Cruets ... 724
Shake Bottles ... 797
Perfume Bottles ... 490
Vacuum Bottles ... 765
Vinegar Bottles, Cruets ... 724
Water Bottles ... 724

BOTTLING SUPPLIES ... 830

BOUILLON ... 483

BOUQUETS ... 130

BOWLS
Baptismal Bowls ... 418
Berry Bowls, Sets ... 724-725
China Bowls ... 714, 716-721
Closet Bowls ... 698-699
Fish Bowls ... 723, 726
Flower Bowls ... 723
Fruit Bowls ... 420-421, 723
Lavatory Wash Bowls ... 698-699
Mixing Bowls ... 726
Nut Bowls, Silver ... 420-421
Oyster Bowls ... 714, 716-721
Pickle Bowls ... 421
Salad Bowls ... 714, 716-721, 723
Soup Bowls ... 714, 716-721
Stock Watering Bowls ... 836, 891
Sugar Bowls, Sets ... 420-421, 714, 716-721, 724-725
Wash Bowls (Pans) ... 825-826
Wash Bowl and Pitcher Sets ... 727, 825

BOWS
Bass Viol Bows ... 508
Cello Bows ... 508
Violin Bows and Repairs ... 508-509

BOXES
Battery Boxes, Auto ... 756
Cash Boxes ... 500, 837
Cedar Boxes ... 613
Deed Boxes ... 500, 837
Doraine Boxes ... 388, 409
Egg Boxes ... 839
Farm Wagon Boxes ... 815
Feed Boxes ... 836
Fireproof Boxes ... 837
Flour Boxes ... 831
Grit Boxes ... 839
Hairpin Boxes ... 422
Hand Cart Boxes ... 893
Honey Boxes ... 895
Ice Boxes ... 638-639
Jewel Boxes ... 422, 491
Leader Boxes ... 774
Lunch Boxes ... 454, 765, 831
Mail Boxes ... 838
Match Boxes ... 422, 784
Miter Boxes ... 859
Outlet Boxes ... 731
Paint Boxes ... 495, 502
Pencil Boxes ... 454
Powder Puff Boxes ... 422, 491
Razor Boxes ... 802
Recipe Boxes ... 450
Shirt Waist Boxes ... 613
Soap Boxes ... 482, 491
Tackle Boxes ... 780
Tool Boxes ... 855
Tool Boxes, Auto ... 756
Vanity Boxes ... 126-127, 388, 409, 488-489

BOXING GOODS ... 766, 770

BRACELETS ... 391-392, 397

BRACES
Ankle Braces ... 472, 474
Auto Body and Fender Braces ... 760
Bit Braces ... 864
Body Braces ... 473
Brace and Bit Sets ... 864
Drill and Brace ... 864
Ratchet Bit Braces ... 864
Screen Door Braces ... 835
Shoulder Braces ... 473
Socket Braces, Auto ... 747
Wagon Box Braces ... 848

BRACKETS
Banjo Brackets ... 514
Barn Door Track Brackets ... 836
Carrier Track Brackets ... 844, 891
Curtain Rod Brackets ... 586-587
Electric Brackets ... 731
Gas Brackets ... 731
Roof Brackets ... 939
Screen Door Brackets ... 835
Shelf Brackets ... 834, 837
Shingling Brackets ... 863
Sink Brackets ... 702
Spotlight Brackets ... 751
Telegraph Brackets ... 814
Telephone Brackets ... 814
Window Screen Brackets ... 835

BRADS ... 835

BRAIDS
Binding Braids ... 366
Featherstitch Braids ... 366
Finishing Braids ... 366
Lingerie Braids ... 369
Middy Braids ... 366
Ric-Rac Braids ... 361
Romper Braids ... 366
Rug Braids ... 363
Soutache Braids ... 366
Tinsel Braids ... 366
Trimming Braids ... 361, 366

BRAKES
Auto Brakes and Accessories ... 753, 760
Bicycle Brakes ... 737
Wagon Brakes ... 815

BRANCHES
Soil Pipe Branches ... 704

BRASSIERES ... 124-125, 156
Corset Brassieres ... 114, 120

BREAD MIXERS, KNEADERS, RAISERS ... 828, 831

BREAST PUMPS ... 480

BREECHES
Riding Breeches ... 282

BREECHING ... 922

BRIDGES
Banjo Bridges ... 514
Billiard Bridges ... 797
Cello Bridges ... 508
Gig Pad Bridges ... 928
Guitar Bridges ... 514
Mandolin Bridges ... 514
Violin Bridges ... 509

BRIDGING ... 964

BRIDLES
Bridle Repairs ... 919-921
Driving Bridles ... 919-920
Pony Bridles ... 919
Riding Bridles ... 919
Stallion Bridles ... 919

BRIEF CASES ... 525

BRONZE
Aluminum Bronze ... 942
Artists' Bronze ... 502
Bronze Liquids ... 942
Bronze Outfits ... 942
Bronze Powders ... 942
Gold Bronze ... 942

BROOCHES ... 373, 376, 389, 391, 396-397, 407, 409
Birthday Brooches ... 389, 407
Cameo Brooches ... 407
Emblem Brooches ... 373, 402

BROODERS ... 900-901

BROOMS
House Brooms ... 824
Stable Brooms ... 929
Street Brooms ... 929
Whisk Brooms ... 479, 803

BRUSHES
Artists' Brushes ... 502
Bath Brushes ... 479
Bath Spray Brushes ... 934
Bee Brushes ... 895
Bronzing Brushes ... 502, 942
Brush and Comb Sets ... 422
Brush and Comb Sets, Babies' ... 482
Brush Attachments, Motor ... 706
Brush, Comb and Mirror Sets ... 422
Clarinet Brushes ... 523
Clothes Brushes ... 479, 491
Complexion Brushes ... 480
Cream Separator Brushes ... 909
Dusting Brushes ... 479, 824
Enameling Brushes ... 942
File Brushes ... 872
Floor Brushes ... 824
Floor Waxing Brushes ... 943
Flue Brushes ... 705
Flute Brushes ... 523
Generator Brushes ... 757
Gun Brushes ... 793
Hair Brushes ... 485, 491
Hair Brushes, Babies' ... 482
Hair Brushes, Barbers' ... 803
Hand Brushes ... 479
Hat Brushes ... 491
Horse Brushes ... 929
Kalsomine Brushes ... 940-941, 952
Kettle Brushes ... 479
Lighter Tip Brushes ... 796
Military Brushes ... 422
Milk Bottle Brushes ... 832
Milk Bottle Brushes, Babies' ... 482
Milk Can Brushes ... 832
Neck Brushes ... 803
Paint Brushes ... 940-941, 951
Paint Brushes, Children's ... 495
Paint Brush Renewers ... 954
Paperhangers' Brushes ... 940
Paste Brushes ... 940
Photographers' Brushes ... 556
Piccolo Brushes ... 523
Plasterers' Brushes ... 941
Rifle Brushes ... 793
Roof Paint Brushes ... 935
Rubber Brushes ... 480
Scrub Brushes ... 824
Shampoo Brushes ... 803
Shaving Brush and Mug Sets ... 422
Shaving Brushes ... 801
Shoe Brushes ... 479
Starter Brushes ... 757 —Or Write for Auto Supply Catalog.
Stencil Brushes ... 840, 952
Street Brushes ... 929
Tooth Brushes ... 422, 487
Varnish Brushes ... 940-941, 944
Watch Brushes ... 471
Water Closet Brushes ... 479, 934
Whitewash Brushes ... 940-941
Window Brushes ... 824
Wire Brushes, Hair ... 485
Wire Brushes, Painters' ... 941

BUCKETS
Chain Pump Buckets ... 709
For Other Buckets—See "Pails."

BUCKLES
Cinch Buckles ... 924
Dress Buckles ... 366
Harness Buckles ... 926
Line Buckles ... 926
Trace Buckles ... 926

BUFFERS
Buffing Compounds ... 853
Buffing Wheels ... 853
Door Buffers ... 835
Finger Nail Buffers ... 491
Horseshoers' Buffers ... 873

BUFFETS ... 628-633

BUGGIES
Babies' Buggies ... 670-671
Doll Buggies ... 497
Runabouts ... 816-817
Top Buggies ... 816-817

BUGGY REPAIRS ... 846-851

BUGLES ... 520

BUILDING MATERIAL AND MILL WORK ... 958-969 —Or Write for Mill Work Catalog.

BUILDINGS
Already Cut Buildings and Modern Homes ... 971-973
Sectional Buildings ... 974

BULBS
Atomizer Bulbs ... 481
Auto Lamp Bulbs ... 751, 756
Bicycle Lamp Bulbs ... 737
Flash Light Bulbs ... 812
House Lamp Bulbs ... 727, 731
Thirty-Two Volt Bulbs ... 674

BUNCHERS ... 887

BUNGALOWS ... 972-974

BUNION RELIEFS ... 472

BURLAP ... 338
Drapery Burlap ... 558

BURNERS
Incense Burners ... 478, 490
Lamp Burners ... 730
Rubbish Burners—Write for Stove Catalog.

BURRS
Feed Grinder Burrs ... 820, 889
Grist Mill Burrs ... 839
Rivets and Burrs ... 846-847

BUSHINGS
Bushing Tools, Auto ... 756
Iron Bushings ... 707

BUST FORMS ... 480

BUTTER MAKING SUPPLIES ... 478, 726, 832-833

BUTTONHOOKS ... 491

BUTTONS
Cloth Covered Buttons ... 372
Cloth Covered Buttons—Made to Order ... 305, 313
Coat Buttons ... 372
Collar Buttons ... 406
Cuff Buttons ... 373, 376, 404-406, 409
Door Buttons ... 836
Emblem Cuff Buttons ... 404
Emblem Lapel Buttons ... 373, 402, 409
Gilt Buttons ... 372
Jet Buttons ... 372
Pants Buttons ... 372
Pearl Buttons ... 372
Push Buttons ... 813, 834
Stock Marking Buttons ... 838
Tufting Buttons ... 851
Vegetable Ivory Buttons ... 372

BUTTS ... 834-835

BUZZERS
Bell Buzzers ... 813
Radio Buzzers—Write for Radio Catalog.

C

CABINETS
Bath Cabinets ... 480
China Cabinets ... 628-632, 638
Kitchen Cabinets ... 640-643
Medicine Cabinets ... 627
Music Cabinets ... 626
Phonograph Record Cabinets ... 626
Piano Roll Cabinets ... 626
Recipe Cabinets ... 450
Sewing Cabinets ... 626
Viaticum Cabinets ... 419

CABLE
Armored Ignition Cable ... 749
Hay Carrier Cable ... 844-845
Lightning Rod Cable ... 888
Manila Cable ... 845
Radio Cable ... 809
Tire Locking Cable ... 743
Wire Cable ... 844-845

CAGES
Bird Cages ... 837
Brooder Cages ... 901

CAKE DECORATIONS ... 492
Cake Topping ... 492

CALICOES ... 316, 318

CALIPERS ... 869

CALKING ... 780

CALKS
Boot Screw Calks ... 857
Horseshoe Calks ... 875

CALLS
Bird Calls ... 793
Dog Calls ... 795
Game Calls ... 793

CALOMEL ... 484

CAMBRICS ... 327, 330

CAMERAS AND SUPPLIES ... 549-556

CAMISOLES ... 146-147, 149
Camisole Straps ... 359

CAMPHOR
Camphor Balls ... 478
Camphor Ointment ... 484
Camphorated Oil ... 484
Gum Camphor ... 484
Spirits of Camphor ... 484

CANDELABRA ... 419

CANDIES ... 478, 492

CANDLES
Cake Candles, Holders ... 492
Disinfecting Candles ... 483
Fancy Candles ... 723
Household Candles ... 784

CANDLESTICKS
Altar Candlesticks ... 419
Crucifix Candlesticks ... 419
Ivory Finish Candlesticks ... 491
Mahogany Candlesticks ... 424, 723
Polychrome Candlesticks ... 723
Silver Candlesticks ... 420-421
Tent Candlesticks ... 784

CANE
Chair Cane ... 837

CANNING OUTFITS ... 726, 830
Canning Supplies ... 478, 827-830
Can Openers ... 831

CANOES ... 780

CANOPIES
Hammock Couch Canopies ... 787

CANS
Cream Setting Cans ... 832
Milk Cans ... 832
Oil Cans ... 825
Oiler Cans ... 855
Tin Packing Cans ... 830

CANULAS ... 477

CANVAS
Army Duck ... 331
Binder Canvases ... 877
Boiler Canvases ... 703
Canvas Duck ... 331
Haystack Canvases ... 781
Machine Canvases ... 781
Sketching Canvas ... 502
Wagon Canvases ... 781
Wagon Cover Canvas ... 851

CAPES
Girls' Capes ... 77-81
Marabou Capes ... 359
Misses' Capes ... 60, 62-63, 73-75
Rain Capes ... 77
Women's Capes ... 60, 62-63

CAPO D'ASTROS ... 514

CAPONIZERS ... 476

CAPPING MACHINES ... 830

CAPS
Auto Caps, Women's, Girls' ... 91, 356
Babies' Caps ... 138
Baseball Caps ... 762
Bathing Caps ... 768
Boudoir Caps ... 359
Boys' Caps ... 176, 255
Children's Caps ... 138
Cushion Bolt Caps ... 797
Ear Caps, Children's ... 138
Eaves Trough End Caps ... 959
Headlight Caps, Men's ... 796
Hunters' Caps ... 794
Ice Caps ... 480
Jar Caps ... 478
Knee Caps ... 474
Men's Caps ... 243, 255
Metal Caps, Bottle ... 830
Milk Bottle Caps ... 832
Mouthpiece Caps ... 523
Pipe Caps ... 707
Pipe Driving Caps ... 710
Pipe Ventilating Caps ... 704
Pole Caps ... 848
Radiator Caps, Auto ... 754
Tire Valve Caps ... 736, 743

CAPSULES ... 484

CARAFES ... 765

CARBIDE ... 796
Carbide Containers ... 796
Carbide Lamps and Accessories ... 796

CARBINES ... 789

CARBOLIC
Crude Carbolic Oil ... 477

CARBON
Carbon Paper ... 500
Carbon Removers ... 752
Carbon Scrapers ... 747

CARBURETORS ... 755

CARDS
Birth Announcement Cards ... 482
Birthday Post Cards ... 478
Correspondence Cards ... 451
Cotton Cards ... 837
Developing Post Cards ... 552
Fortune Telling Cards ... 495
Playing Cards ... 495
Sewing Cards ... 495
Wool Cards ... 837
Also see "Post Cards."

CARPETS
Auto Carpet ... 758
Carpet (by the yard) ... 588-592
Carpet Sweepers ... 811, 824
Carpet Warp• ... 334
Floor Carpet (Rugs) ... 588, 591-593, 596-604
Hall Runners ... 591, 593
How to Order Carpets ... 588
Stair Runners ... 591, 593

CARPET STRIP ... 967

CARRIAGES
Baby Carriages ... 670-671
Doll Carriages ... 497

CARRIERS
Baggage Carriers, Covers, Auto ... 746
Egg Carriers ... 839
Feed Carriers—Write for Farm Implement Catalog.
Fish Carriers ... 775, 780
Game Carriers ... 793, 795
Hay Carriers and Accessories ... 844 —Or Write for Farm Implement Catalog.
Litter Carriers and Attachments ... 891
Parcel Carriers, Bicycle ... 738
Shuttle Carriers ... 312
Trace Carriers ... 923, 927

CARTONS
Egg Cartons ... 839

CARTRIDGES ... 791

CARTS
Babies' Carts ... 673
Barrel Carts ... 893
Boys' Carts ... 494, 496-497
Dog Carts ... 911
Dump Carts ... 893
Goat Carts ... 911
Hand Carts ... 893
Harrow Carts ... 882
Platform Carts ... 893
Pony Carts ... 817
Road Carts ... 817
Tea Carts ... 618, 620, 626

CARVING SETS ... 413, 417, 804

CASCARA ... 484

CASES
Banjo Cases ... 514
Brief Cases ... 525
Camera Cases ... 551
Card Cases ... 259
Cigarette Cases ... 388
Clarinet Cases ... 523
Clarinet Reed Cases ... 523
Cornet Cases ... 523
Drum Cases ... 522
Egg Cases ... 839
Guitar Cases ... 514
Gun Cases ... 793
Honey Shipping Cases ... 895
Jewel Cases ... 422, 491
Knife Cases ... 798
Mandolin Cases ... 514
Manicure Cases, Sets ... 422, 490
Music Cases ... 524-525
Music Stand Cases ... 515
Office Toilet Cases ... 627
Phonograph Record Cases ... 548
Powder Cases (Powderettes) ... 388
Racket Cases ... 764
Razor Cases ... 802
Rifle Cases ... 793
Rosary Cases ... 390
Saxophone Cases ... 523
Saxophone Reed Cases ... 523
School Book Cases ... 454
Snare Drum Cases ... 522
Suit Cases ... 260-261
Toilet Cases, Sets ... 422
Tool Cases ... 855
Trombone Cases ... 523
Trumpet Cases ... 523
Ukulele Cases ... 514
Vacuum Bottle Cases ... 765
Vanity Cases ... 126-127, 388, 409, 489
Violin Cases ... 508

CASHBOXES ... 500, 837

CASINGS
Door Casings ... 967
Inner Casings, Auto ... 739
Pillow Casings ... 329, 331
Tire Casings, Auto ... 740-741
Tire Casings, Bicycle ... 735-736
Tire Casings, Motorcycle ... 738
Tire Casing Repairs ... 736, 742
Window Casings ... 967

CASSEROLES ... 420-421, 725-726

CASTERS
Bed Casters ... 837
Box Casters ... 837
Cruet Casters ... 725
Furniture Casters ... 837
Stove Casters ... 837
Truck Casters ... 837

CASTINGS
Gate Castings ... 898

CASTORIA ... 482

CATCHES
Cupboard Catches ... 834
Door Catches ... 835-836
Elbow Catches ... 834
Screen Door Catches ... 835
Transom Catches ... 835
Window Catches ... 834

CATHARTICS ... 484

CATHETERS ... 480

CEDAR CHESTS ... 613

CEILING
Metal Ceiling ... 958
Steel Ceiling ... 958
Wood Ceiling ... 964

See Page 557 for Measuring Instructions.

CELLOS AND SUPPLIES 508
CELLULOID 745
CEMENT MACHINES 970
Cement Workers' Tools 841-842

CEMENTS
Asbestos Cement—Write for Heating Catalog.
Auto Top Cement .. 745
Bicyle Rim Cement .. 736
Boat Cement 780
Celluloid Cement 745
Chimney Flashing Cement 935
China Cement 478
Iron Cement 752
Jewelers' Cement .. 471
Kettle Cement 478
Leather Cement 211
Linoleum Cement 605
Liquid Glue Cement 478, 954
Pipe Joint Cement .. 707
Radiator Cement, Auto 752
Roof Cement 935
Rubber Cement 211, 736, 742
Stove Repair Cement 689
Tire Cement .. 736, 742

CENTERPIECES .. 363
CESSPOOLS 704

CHAINS
Brass Chains 837
Breast Chains 927
Coil Chains, Wire .. 837
Cow Chains 838
Dog Chains 795
Door Chains 835
Eyeglass Chains 405
Halter Chains 927
Hame Chains 927
Hammock Chains 786-787
Heel Chains 927
Jack Chains 837
Kennel Chains 795
Log Chains 856
Neck Chains 390
Neck Chains With Pendant 373, 376, 389, 391-393, 396-397, 409
Picket Chains 838
Picture Chains 586
Pole Chains 848
Pump Chains 709
Speedometer Chains .. 753
Sprocket Chains, Bicycle 738
Sprocket Chains, Motorcycle 738
Stallion Lead Chains .. 927
Tire Chains and Accessories, Auto 743
Trace Chains 927
Wagon Stay Chains .. 848
Watch Chains, Men's 386-387, 409
Watch Chains, Women's 390, 409

CHAIRS
Arm Chairs 607A-608, 610-611, 614-615, 618-621, 628-631, 636-637, 672
Babies' Chairs 494, 668-669
Bedroom Chairs .. 644-651
Camp Chairs 784
Children's Chair 668
Commode Chairs 668-669
Dining Chairs 628-632, 636-637
Dressing Table Chairs 644-651
Fiber Chairs 618-619, 621
Folding Chairs 637, 784, 787
Folding Chairs, Auto 746
High Chairs 668 669
Invalids' Chairs ...... 483 —Or Write for Invalid Chair Catalog.
Kitchen Chairs 636-637
Lawn Chairs 618-621, 637, 784, 787
Morris Chairs 615
Nursery Chairs 668-669
Office Chairs 672
Parlor Chairs 607A-608, 610-611, 614-615, 618-621
Porch Chairs 618-621, 637, 784, 787
Reed Chairs 620
Rocking Chairs—See "Rockers."
Swivel Chairs 672

CHALK
Billiard Chalk 797
Carpenters' Chalk .. 864

CHALKLINES 864

CHAMBERS
Bedroom Chambers 727, 825

CHAMBRAYS 316-317, 321-322

CHAMOIS
Cleaning Chamois 478, 753
Face Chamois 489

CHANDELIERS
Electric Chandeliers 731
Gas Chandeliers 731

CHAPPED HAND PREPARATIONS 484, 489
CHAPS 924

CHARCOAL
Charcoal and Sulphur 796
Charcoal Tablets 484

CHARMEUSE 296-298, 313, 319
CHARMS 373, 403

CHARTS
Fingerboard Charts 508-509, 514

CHECKS
Dress Checks 298, 300-304, 314, 316-326
Door Checks ... 834-835
Overchecks 923
Side Checks 923
Trade Checks 797

CHEEKS
Bridle Cheeks 920

CHEESECLOTHS . 331

CHEMICALS
Closet Chemicals 478, 483, 934
Household Chemicals 478, 483-484
Photo Chemicals ... 553
Poultry Chemicals .. 476
Stock Chemicals ... 477

CHEMISES 146-147, 150, 154-155
CHENILLE 366
CHESS 495

CHESTS
Cedar Chests 613
Tool Chests 855

CHEWING GUM .. 492
CHIFFONIERS 651-652
CHIFFONS 356
CHIFFORETTES 644-650, 652
CHIFFOROBES 644-645, 648-649, 653

CHIMES
Auto Chimes ... 746, 754
Sleigh Chimes—Write for Harness Catalog.

CHIMNEYS
Lamp Chimneys 729-730
Lantern Chimneys .. 825

CHINAWARE 714-721, 723

CHISELS
Calking Chisels 704
Carpenters' Chisels 861
Cold Chisels 873
Plumbers' Chisels .. 704

CHOCOLATES 478, 492
CHOCOLATE SETS 420

CHOKERS
Fur Chokers 43
Marabou Chokers ... 359

CHOPPERS
Food Choppers and Attachments .. 830-831

CHUCKS
Drill Chucks 868
Lathe Chucks 868

CHURNS 726, 832-833
CIGARS 493

CINCHES
Saddle Cinches 924

CIRCULATORS
Water Circulators, Auto 754

CLAMPS
Cabinet Makers' Clamps 864
Cable Clamps .. 844-845
Carrier Track Clamps 891
Clamp and Drill 869
Cue Tip Clamps 797
Flooring Clamps 863
Garden Hose Clamps 840
Harness Makers' Clamps 930
Hydrant Clamps 706
Quilt Frame Clamps 864
Screw Clamps 864
Splicing Clamps 871
Stanchion Clamps .. 891
Suction Hose Clamps 854
Telescope Clamps .. 428

CLAPBOARDS ... 964
CLARINETS 519

CLASPS
Horse Tail Clasps .. 928
Lingerie Clasps 403, 409
Tie Clasps . 373, 405, 409

CLAY
Complexion Clay ... 489

CLEANERS
Aluminum Cleaners 826, 829
Auto Body Cleaners 753
Clarinet Cleaners .. 523
Closet Bowl Cleaners 478, 934
Drainpipe Cleaners 478, 934
File Cleaners 872
Flue Cleaners 705
Flute Cleaners 523
Grain Cleaners 889
Gun Cleaners 793
Kettle Cleaners 826, 829
Leather Cleaners 213
Paint Cleaners 951
Phonograph Record Cleaners 548
Piccolo Cleaners 523
Pipe Cleaners 259
Scalp Cleaners 486
Shade Cleaners 942
Shoe Cleaners 213
Vacuum Cleaners and Attachments 811, 824
Varnish Cleaners 951
Wallpaper Cleaners 942
Windshield Cleaners . 744

CLEATS
Porcelain Cleats ... 731

CLEAVERS ... 805-806

CLEVISES
Hame Clevises 927
Implement Clevises 848
Pipe Lifting Clevises 710
Trace Clevises 927
Wagon Clevises 848

CLIMBERS
Linemen's Climbers 814

CLIPPERS
Bolt Clippers 846
Dehorning Clippers .. 838
Dog Clippers 929
Fingernail Clippers .. 490
Hair Clippers 802
Horse Clippers 933
Neck Clippers 802

CLIPPING MACHINES AND ACCESSORIES 929, 933

CLIPS
Auto Hood Clips ... 752
Cable Clips 844-845
Film Clips 552
Hair Bow Clips 358
Hame Clips 926-927
Harness Clips .. 926-927
Harrow Tooth Clips 876, 882
Napkin Clips 421
Radio Helix Clips .. 808
Rope Clips 927
Singletree Clips 848
Trace Clips 927
Trousers Clips 736
Wire Clips 871

CLOAKS
Babies' Cloaks 139

CLOCKS
Alarm Clocks .. 423-424
Auto Clocks 746
Dresser Clocks 423
Luminous Clocks 423
Mantel Clocks .. 424-425
Wall Clocks 424-425

CLOSETS
Portable Indoor Closets 483
Water Closet Outfits 698-699

CLOTHESLINES, REELS AND PROPS 823
CLOTHESPINS ... 823

CLOTHING
Athletic Clothing ... 769
Boys' Clothing 165-166, 168-169, 172-180, 269-271
Children's Clothing (Ready to Sew) ... 372
Corduroy Clothing, Men's, Boys' 177, 282, 287, 794
Girls' Clothing 4, 22-27, 50-51, 77-81
Hunters' Clothing ... 794
Little Fellows' Clothing . 165-166, 168-180
Men's Clothing 264-265, 272-288, 794
Misses' Clothing 4, 19-23, 28-29, 33, 42-53, 60-75, 77
Oiled Slicker Clothing 166, 267, 794
Palm Beach Clothing, Men's, Boys' 165, 179, 281
Riding Clothing, Women's, Misses' 4, 39
Women's Clothing 1-21, 28-71, 76-77
Work Clothing, Men's, Boys' 166-167, 268, 286-291

CLOTH SAMPLES

Men's and Young Men's Spring and Summer Suits.

Your choice of 16 samples 8 models 33 suits.

Write for Sample Folder 8292GCL, sent postpaid on request.

See page 278 of this catalog.

CLOTHS
Auto Top Cloths 851
Bandage Cloths 331, 483
Billiard Table Cloths 797
Buggy Cushion Cloths 851
Buggy Top Cloths .. 851
Diaper Cloths .. 328, 332
Dish Cloths 367
Dress Cloths—See "Dress Goods."
Dusting Cloths . 367, 824
Emery Cloth 861
Hair Cutting Cloths 803
Lunch Cloths 342-343
Oil Cloths 334, 561
Oil Cloths, Wall—Write for Wall Paper Sample Book.
Silence Cloths 338
Tablecloths 338-343
Tablecloths (By the Yard) 338-339, 341-343
Tent Sod Cloths 782
Terry Cloths 560
Tracing Cloths 500
Wash Cloths 336
Wash Cloths, Babies' 135
Waterproof Duck Cloths 851
Window Shade Cloths 586

CLUBS
Golf Clubs 765
Indian Clubs 769
Police Clubs 792

CLUSTERS
Electric Plug Clusters 731

CLUTCHES
Clutch Release 760
Speedometer Clutches 753

COAL TAR 935
COASTERS 494

COATS
Automobile Coats, Men's 266
Babies' Coats 139
Barrow Coats 137
Children's Coats 139
Corduroy Coats, Men's, Boys' 794
Cravenette Coats, Men's 264-265
Cravenette Coats, Women's, Misses' 68-69
Firemen's Rubber Coats 266
Fishing Coats .. 267, 794
Girls' Coats 78-81
Hunters' Coats 794
Junior Misses' Coats 72-75
Misses' Coats 60-71
Office Coats 282
Oiled Slicker Coats 166, 267, 794
Police Rubber Coats 266
Pommel Coats 267
Raincoats, Men's, Boys' 166, 264-265
Raincoats, Women's, Girls' 68-69, 77
Riding Coats, Men's . 267
Rubber Coats, Men's, Boys' 166, 266
Sweater Coats, Boys' .. 243
Sweater Coats, Children's 141
Sweater Coats, Men's 243
Sweater Coats, Women's, Girls' 146-147, 162
Waterproof Coats, Men's, Boys' 166, 266-267, 794
Waterproof Coats, Women's, Girls' 68-69, 77
Women's Coats 60-71, 76

COBBLERS' SUPPLIES AND TOOLS 211-213
COCKEYES 926
COCOA BUTTER 484, 802

COCKS
Air Cocks 707
Ball Cocks 699
Basin Cocks 706
Bath Cocks 706
Gauge Cocks 707
Hydrant Cocks 706
Sill Cocks 706
Steam Cocks 706
Stop Cocks 706

COFFEEPOTS 420-421, 826-829
Coffeepots (Percolators) 810, 826-827, 829

COFFEE SETS 420-421

COILS
Coil Parts 757
Ignition Coils 749
Make and Break Coils 813
Radio Tuning Coils 807
Spark Coils 750, 757, 813

COLANDERS .. 826, 828
COLLARBANDS ... 369

COLLARS
Boys' Collars 163
Cable Collars 844
Cat Collars 795
Dog Collars 795
Horse Collars 925
Men's Collars .. 240-241
Mule Collars 925
Shafting Collars 855
Women's Collars, Sets 360

COLOGNES 490

COLORS
Butter Colors 478
Cheese Colors 478
Colors in Oil (Paint) 943, 954
Embroidery Silk Colors 365
Graining Colors 943
Ground Colors 943-944
Oil Colors, Artists' . 502
Paint Color Samples 947, 950
Water Colors, Artists' 502
Water Colors, Photo 556

COLUMNS
Emery Wheel Columns 853
Porch Columns 967

COMBINATIONS
Girls' Combinations 143, 161
Women's Combinations 150

COMBINETS 825

COMBS
Babies' Combs 482
Barbers' Combs 803
Bobbing Combs 803
Cattle Combs 929
Clipping Machine Combs 933
Comb and Brush Sets 422
Comb and Brush Sets, Babies' 482
Comb, Brush and Mirror Sets 422
Curry Combs 929
Dandruff Combs 485
Fancy Hair Combs 130, 397
Girls' Puff Combs ... 131
Graining Combs 943
Hair Combs 485, 491
Hair Tonic Combs ... 486
Pocket Combs .. 388, 485

COMFORTERS 347
Comforter Coverings 318, 333

COMMODES 652, 727, 825
COMMUNION SETS 418

COMPASSES
Children's School Compasses 454
Machinists' Compasses 869
Needle Compasses .. 426

COMPLEXION PREPARATIONS 487-490
CONCRETE MACHINERY .. 970

CONDENSERS
Radio Condensers 808-809

CONDUCTORS
Pump Water Conductors 710
Rain Water Conductors 959

CONDUITS 731

CONFINERS
Bust Confiners 124-125, 156
Hip Confiners 370

CONNECTIONS
Auto Pump Connections 744
Steel Tank Connections 894

CONTRACTORS
Tire Rim Contractors 743

CONTROLS
Filament Controls ... 809

COOKERS
Camp Cookers 785
Fireless Cookers 828
Small Cookers (for Stove) 829
Steam Cookers 827
Stock Food Cookers 892-893
Waterless Cookers 827

COOKING UTENSILS 725-726, 826-831

COOLERS
Iceless Coolers 710
Milk Coolers 832
Water Coolers . 726, 831

COOPS
Poultry Brood Coops 839, 974

COPPER
Battery Copper 813
Copper Sulphate . 476, 813

COPPERAS 477

COPPERS
Soldering Coppers .. 853
Soldering Coppers, Jewelers' 471

CORDS
Curtain Cords and Tassels 575
Electric Cords 731, 810, 813
Lamp Wiring Cords, Auto 751
Sash Cords 834
Telephone Receiver Cords 814
Trimming Cords 366
White Cotton Cords . 479
Window Shade Cords and Tassels 585
Wrapping Cords 823

CORDUROY 305
CORERS 830
CORKS 479
Vacuum Bottle Corks 765
CORKSCREWS 831
CORN REMEDIES 472

CORNERS
Brass Box Corners 837
Corner Irons 846
Stair Corners 586

CORNETS 520
Cornet Supplies 523

CORN GROWERS' IMPLEMENTS 839, 842, 878-886, 888-889

CORSET COVERS
Muslin Corset Covers 149, 155
Silk Corset Covers 146-147

CORSETS
Girls' Corsets 120
Maternity Corsets .. 156
Misses' Corsets 120
Nursing Corsets 156
Women's Corsets 111-123

COSTUMERS 624

COTS
Babies' Cots 667
Folding Cots 664-666, 785
Outdoor Sleeping Cots 785

COTTAGES 972-974
COTTER PINS . 747, 846

COTTON
Absorbent Cotton 483
Cotton Battings 333
Cotton Waste 705
Crochet Cotton 364
Darning Cotton 364
Embroidery Cotton .. 364
Knitting Cotton 364
Marking Cotton 364
Slipper Cotton 364

COUCHES
Fiber Couches 618
Hammock Couches and Fittings 787
Steel Couches 664-666
Upholstered Couches 610-611

COUGH RELIEFS .. 484
COULTERS 877

COUNTERSHAFTS
Belt Countershafts 853
Lathe Countershafts 853
Shearing Machine Countershafts 933

COUNTERSINKS 864, 869
COUNTING MACHINES 869

COUPLINGS
Hose Couplings . 840, 854
Line Shaft Couplings 855
Pipe Couplings 707
Pump Rod Couplings 710
Pump Tubing Couplings 708
Screen Couplings 835

COVERS
Billiard Table Covers 797
Catch Basin Covers 709
Cattle Covers ... 932-933
Chamber Covers 825
Cistern Covers 709
Corset Covers 146-147, 149, 155
Couch Covers 567
Diaper Covers 134
Door Covers, Auto .. 759
Feed Cooker Covers 892
Garment Covers 367
Gun Covers 793
Haycock Covers 781
Horse Covers 932-933
Ironing Board Covers 333
Kettle Covers 827, 831
Luggage Carrier Covers, Auto 746
Machine Covers 781
Mattress Covers 349
Pillow Covers 363
Pillow Covers, Babies' 134
Pulley Covers 854
Racket Covers 764
Range Boiler Covers 703
Rifle Covers 793
Seat Covers, Auto .. 759
Stack Covers 781
Table Covers 334, 342-343, 362-363, 568
Tire Covers, Auto .. 744
Top Covers, Auto 745, 758, 815
Wagon Covers 781
Watch Movement Covers 471

CRACKERS
Nut Crackers 831

CRADLES
Babies' Cradles 667
Grain Cradles 841

CRANKS
Starting Cranks, Auto 761

CRASH .. 327, 335, 361
CRATES 839

CRAYONS
Marking Crayons 862
Pastel Crayons . 454, 502
School Crayons 454, 502
Wax Crayons 454

CREAM SEPARATORS 904-909
Dilution Separators 832
Double Can Separators 832

CREAMER SETS 420-421, 714, 724-725

CREAMS
Almond Creams 489
Cold Creams 489
Dental Creams 487
Deodorant Creams .. 489
Face Creams 489, 802
Freckle Creams 489
Manicure Creams 490
Massage Creams 489, 802
Shaving Creams 802

CREAM TARTAR .. 484
CREELS 780

CREEPERS
Babies' Creepers ... 140
Babies' Creepers (Ready to Sew) .. 372
Ice Creepers 213

CREOSOTE
Creosote Oil 935
Creosote Stain . 946-947

CREPES 295-302, 304-305, 315, 320-324
CRETONNE 315, 319, 333, 558-559, 561

CRIBS
Children's Cribs ... 667
Corn Cribs—Write for Farm Implement Catalog.

CROCKERY 726
CROKINOLE 495
CROQUET SETS ... 787

CROSSBARS
Wagon Crossbars ... 850

CROSSES 373, 390, 393
CROWBARS 857
CROWNPIECES 921
CRUCIBLES 853
CRUCIFIXES . 419, 445

CRUETS
Cruet Sets 725

CRUETS—Continued
Oil Cruets ... 724
Salt and Pepper Cruets ... 419, 421, 723-724, 826
Vinegar Cruets ... 724
CRUPPERS ... 923
CRUSHERS
Clod Crushers ... 882
Corn and Cob Crushers ... 819-820, 889
CRUTCHES ... 483
Crutch Tips ... 483
CRYSTALS
Watch Crystals ... 471
CUES AND SUPPLIES ... 797
CUFFS
Leather Riding Cuffs ... 924
Women's Cuff and Collar Sets ... 360
CULTIVATORS 842, 880-882, 884-886
Cultivator Attachments and Repair Parts ... 876, 884-886
CUPBOARDS
Kitchen Cupboards .. 639
CUPS
Alcohol Cups ... 471
Aluminum Cups ... 784
Babies' Cups ... 418, 482, 494, 723
Collapsible Cups ... 784
Cups and Saucers ... 714, 716-721
Custard Cups 722, 726
Egg Cups ... 714
Enameled Cups ... 828
Grease Cups ... 706
Measuring Cups ... 827, 831
Nut Cups ... 492
Oil Cups ... 706
Oil Cups, Jewelers' 471
Palette Cups ... 502
Shaving Cups ... 802
Soap Cups ... 934
CURLERS
Hair Curlers ... 131, 486, 810
CURRYCOMBS ... 929
CURTAINS
Auto Curtains and Lights ... 745, 758
Baby Carriage Curtains ... 671
Curtain Materials ... 558-563, 568-572
Folding Bed Curtains ... 653
Panel Curtains 574-575
Porch Curtains ... 781
Sash Curtains 574, 582
Window Curtains ... 558, 565, 571-582
CUSHIONS
Air Cushions ... 481
Auto Cushions ... 759
Billiard Table Cushions ... 797
Boat Cushions ... 780
Buggy Cushions ... 851
Heel Cushions ... 472
Implement Seat Cushions ... 876
Piano Bench Cushions ... 523
Pincushions ... 368-369, 422, 491
Sofa Cushions 363, 657
CUSPIDORS ... 825
CUT GLASS ... 722-725
CUTICLE SPECIALTIES ... 490
CUTLERY
Kitchen Cutlery 804-806
Silver Cutlery 411-419
Steel Cutlery ... 798-799, 802-806
CUT-OFFS
Rain Water Cut-Offs ... 959
CUT-OUTS
Muffler Cut-Outs ... 746, 754
CUTTERS
Blacksmiths' Cutters ... 873, 875
Bolt Cutters ... 846
Bone Cutters ... 893
Cigar Cutters ... 798
Clipping Machine Cutters ... 933
Disc Sharpener Cutters ... 876
Emery Wheel Dresser Cutters ... 853
Ensilage Cutters—Write for Farm Implement Catalog.
Expansive Bit Cutters ... 865
Feed Cutters and Attachments—Write for Farm Implement Catalog.
Fiber Needle Cutters ... 535
Gasket Cutters ... 855
Glass Cutters ... 941
Hoof Cutters ... 873
Kraut Cutters ... 831
Lace Leather Cutters ... 855
Meat Cutters ... 830
Molar Cutters, Veterinary ... 477
Parchment Paper Cutters ... 478
Pastry Cutters ... 831
Pipe Cutters ... 705
Plane Cutters ... 860
Plow Fin Cutters ... 877
Poultry Food Cutters ... 893
Root Cutters—Write for Farm Implement Catalog.
Saw Gummer Cutters ... 852
Screw Cutters ... 870
Slaw Cutters ... 831
Stalk Cutters—Write for Farm Implement Catalog.
Wire Cutters ... 871
CYCLOMETERS ... 736
CYLINDERS
Pump Cylinders ... 710
CYMBALS ... 522

D

DAGGERS ... 784, 799
DAIRY BARN EQUIPMENT ... 891
DAIRY SUPPLIES ... 477-478, 832-833, 891, 904-909
Dairymen's Instruments ... 477
DAMASKS ... 338-339, 341-343
DAMPERS
Furnace Damper Attachments ... 694
DARNERS ... 369, 422
DARTS
Air Rifle Darts ... 792
DATERS ... 500
DAVENPORTS ... 607A-607H, 610-612, 664
DAVID BRADLEY IMPLEMENTS AND REPAIRS ... 876-890
DECALCOMANIA .. 951
DECORATIONS
Cake Decorations ... 492
DECOYS ... 793
DEHORNERS ... 838
DEES
Harness Dees ... 927
DENIMS
Art Denims ... 558, 561
Overall Denims ... 317
DENTAL PREPARATIONS ... 487
DEODORANTS ... 489
DEPILATORIES ... 486
DESKS
Bookcase Combination Desks ... 625
Folding Desks ... 626
Office Desks ... 672
Parlor Desks ... 619, 625
School Desks ... 673
Writing Desks ... 619, 625-626
DESK SETS ... 422
DETECTORS
Radio Detectors ... 809
DEVELOPING, PRINTING AND ENLARGING ... 549
Developing Supplies ... 549, 552-553, 555
DIAMONDS ... 374-376, 400
Imitation Diamonds 396
DIAPERS ... 135
Diaper Covers, Pants 134
DIAPHRAGMS
Phonograph Diaphragms ... 548
DICTIONARIES ... 440
Foreign Dictionaries 440
Music Dictionaries .. 524
DIES
Dies and Stocks, Pipe Threading ... 705
Dies and Stocks, Screw Cutting ... 870
Stock Marking Dies 838
DIGESTIVES ... 484
DIGGERS
Post Hole Diggers ... 710, 843
Potato Diggers ... 886
Well Diggers ... 710
DILATORS
Cow Teat Dilators .. 477
DIMITIES ... 324, 326
DIMMERS ... 751
DINNERWARE
China Dinnerware ... 714-721
Silver Dinnerware ... 411-421
DIPPERS ... 826, 828
DIPS
Stock Dips ... 477
DISCS
Harrow Discs ... 876
DISGORGERS ... 775
DISHES
Babies' Dishes ... 482, 494, 723
Baking Dishes ... 420-421, 725-726
Biscuit Dishes ... 420
Bonbon Dishes 421, 723
Bone Dishes ... 716-721
Butter Dishes ... 421, 714, 716-721
Celery Dishes ... 724
Cereal Dishes ... 714, 716-721
Cheese and Cracker Dishes ... 420, 724
China Dishes ... 714-721
Fern Dishes ... 723
Fruit Dishes ... 723
Glass Cooking Dishes ... 420-421, 725-726
Glass Dishes ... 722-726
Jelly Dishes ... 421, 723
Lemon Dishes ... 421
Marmalade Dishes .. 421
Mayonnaise Dishes ... 723-724
Paper Dishes ... 478
Pickle Dishes ... 421, 716-721
Pudding Dishes ... 726
Salad Dishes ... 714, 716-721
Sauce Dishes ... 714, 716-721
Silver Dishes ... 420-421
Soap Dishes ... 727, 825, 828, 934
Vegetable Dishes ... 420, 714, 716-721
DISINFECTANTS ... 476-477, 483
DISTRIBUTORS
Battery Distributors 750 —or Write for Auto Supply Catalog.
DITCHERS ... 888
DIVIDERS
Mechanics' Dividers 869
School Compass Dividers ... 454
DOILIES ... 363
DOLLS ... 494, 498-499
Doll Supplies ... 499
DOMES
Electric Domes ... 731
DOMINOES ... 495
DOORS
Ash Pit Doors ... 823
Door Trim ... 967
Front Doors ... 967
Fuel Doors ... 823
Garage Doors—Write for Millwork Catalog.
Headlight Doors ... 756
Inside Doors ... 967
Screen Doors ... 968
Storm Doors ... 968
Tent Doors ... 782
DOOR HARDWARE ... 834-837
DOUBLETREES ... 815, 849-850
DOUCHES ... 480-481
DOWN ... 657
DRAGS
Farm Drags—Write for Farm Implement Catalog.
DRAG SAWS ... 819
DRAINS ... 704
Drain Boards ... 702
Drain Pipe Cleaners 934
DRAPERY GOODS ... 558-582
DRAWERS
Babies' Drawers ... 134
Children's Drawers 142
Girls' Drawers, Muslin ... 142, 161
Men's Drawers ... 233
Women's Drawers, Knit ... 236
Women's Drawers, Muslin, Silk .. 149, 154
DRAWING INSTRUMENTS ... 454, 500-502
DRAWING KNIVES ... 861
DRESSERS
Bedroom Dressers ... 644-652
Bibb Seat Dressers 706
Emery Wheel Dressers ... 853
DRESSES
Babies' Dresses 136-137
Children's Dresses ... 136, 142, 144-145
Children's Dresses (Ready to Sew) ... 372
Confirmation Dresses ... 25
Girls' Dresses ... 22-27
House Dresses ... 56-59
Junior Misses' Dresses ... 22-23
Maternity Dresses ... 10
Misses Dresses ... 19-23
Mourning Dresses .. 11
Nurses' Dresses ... 57
Women's Dresses ... 1-21, 56-59
DRESS FORMS ... 361
DRESS GOODS
Apron Gingham ... 316
Baby Flannels 328, 351
Batiste ... 304-305, 321-324, 326
Beach Suitings ... 314
Brilliantines ... 303
Broadcloths ... 297, 304
Brocades ... 313
Butcher Linens ... 327
Calicoes ... 316, 318
Cambric ... 327, 330
Cambric Linings ... 313
Canton Flannels ... 332
Canton Silks ... 296-297, 299-301, 305
Canvas (Duck) ... 331
Challis ... 302, 333
Chambrays ... 316-317, 321-322
Charmeuse ... 296-298, 313, 319
Checks (Dress Goods) ... 301-304, 314, 316-326
Checks (Silks) 298, 300
Cheviot Shirtings ... 317
Chiffons ... 356
Chiffon Poplins ... 305
China Silks ... 298
Coating Cloths ... 298, 302-305, 314-315
Corduroy ... 305
Costume Cloths 325-326
Cotton Flannels ... 328, 332
Crash ... 327, 361
Crepe de Chine ... 295-297, 301, 305
Crepe de Meteor ... 298
Crepe Georgette ... 296
Crepes ... 295-302, 304-305, 320-324
Cretonne 315, 319, 333
Diaper Cloth ... 328, 332
Dimities ... 324, 326
Dotted Swiss Muslin ... 323-324
Dress Linens ... 327
Dress Patterns (Paper) ... 292-293
Dress Trimmings ... 356-359, 361, 366
Drill ... 316, 331
Duck ... 331
Eden Cloth ... 328
Eiderdown ... 303
Embroidered Flannels ... 351
Embroidered Swisses ... 351-353
Embroideries ... 351-353, 366
Flannelettes ... 328
Flannels ... 328, 332
Flaxons ... 320
Foulards ... 296
French Serges ... 303
Georgette Crepes ... 296
Ginghams ... 300, 316, 319
Habutai Silks ... 298
Henriettas ... 303
Hickory Shirtings ... 317
India Linons ... 324, 326
Indian Head Muslin .. 325
Infants' Cloth ... 326
Japanese Silk ... 298
Jean ... 315, 325, 331
Jersey Cloths .. 296, 298
Khaki Cloths ... 317
Kimono Cloth ... 300, 315, 320
Laces ... 351, 354-355
Lawns 323-324, 326-327
Linen Finish Suitings ... 314, 319
Linens ... 327
Lingerie Cloth ... 295, 298, 300-301, 305, 313-315, 319-320, 323, 325-326
Linings ... 301, 305, 313, 328, 331
Linons, India .. 324, 326
Longcloth ... 330
Madras .. 299, 317-318, 322, 325
Messalines 295-296, 301
Middy Cloths .. 315, 325
Mohair ... 303
Moonglo Silks ... 298
Mulls ... 326
Muslins ... 323-325, 330
Nainsooks ... 330
Nets (Dress) ... 356
Novelty Suitings ... 302, 304
Nurses' Costume Cloths ... 325-326
Nurses' Ginghams .. 316
Organdies ... 320, 323-324, 326
Ottoman ... 302
Outing Flannels ... 328
Overall Goods .316-317
Pajama Checks ... 325
Panamas ... 304
Peau de Soie ... 298
Percales ... 318-319
Permalawns ... 326
Persian Lawns 324, 326
Pique ... 325
Plaids ... 300, 304, 314, 316-322
Plisses ... 305, 320
Poiret Twill ... 302
Pongees ... 294, 298-301, 305, 315, 327
Poplins ... 304-305, 325
Prints ... 316, 318
Radium Silk ... 294
Ratine ... 295, 297-298, 301-302, 315, 320-321, 324
Rice Cloth ... 315
Romper Cloths 316, 319
Sacking Broadcloth 304
Sateens ... 313, 324-325, 333
Satins ... 295-299
Seersucker ... 316
Serges ... 303, 315
Shadow Checks ... 304
Shaker Flannels ... 328
Shantung Silks ... 298
Shepherd Checks ... 303, 314
Shirtings ... 294, 298-299, 305, 316-317
Sicilians ... 303
Silks ... 294-301, 305
Silk Shirtings ... 294, 298-299
Skirtings ... 296, 299, 301-305, 314-315
Soiesette ... 314
S. R. C. Cloth .300-301
Storm Serges ... 303
Stripes (Dress Goods) 304, 316-320, 322, 324
Stripes (Silks) ... 294-295, 297-299, 301
Suitings ... 302-304, 314-315, 319-321, 325, 327
Suitings, Wash ... 314, 319, 325, 327
Swisses ... 323-324
Taffetas .. 294-298, 301
Tricollette ... 297
Tricotines ... 302
Tub Silks .. 294-299, 305
Tussahs ... 300-301, 305
Tweeds ... 302, 304
Twills ... 302, 316-317
Utility Cloth ... 317
Velours ... 302
Voiles ... 300-301, 319-323, 325-326
Waistings ... 296-299, 326-327, 330, 355-356, 361
DRESS SETS (Jewelry) ... 405-406
DRESSING
Auto Top and Seat Dressing ... 955
Belt Dressing ... 854
Hoof Dressing ... 957
Leather Dressing ... 955, 957
Shoe Dressing ... 213
Stove Dressing ... 689
Strop Dressing ... 802
Waterproof Dressing ... 955
DRESSMAKERS' SUPPLIES 292-293, 305-313, 328, 331, 357-359, 361, 364, 366, 368-369, 371-372, 437, 450
DRILLS
Automatic Drills ... 865, 869
Bench Drills ... 869
Bit Stock Drills ... 864, 869
Breast Drills ... 869
Brick Drills ... 841
Chain Drills ... 869
Clamp and Drill ... 869
Drill Attachments ... 865, 868-869, 871
Drill (Carriage Cloth) ... 851
Drill Sleeves ... 705
Drill, Vise and Anvil ... 872, 874
Grain Drills and Attachments ... 883
Hand Drills ... 865, 869
Jewelers' Drills ... 471
Post Drills ... 868
Power Drills ... 868
Ratchet Drills ... 705, 864
Round Shank Drills 869
Sand Pump Drills ... 710
Stone Drills ... 841
Straight Shank Drills ... 869
Twist Drills ... 869
Wood Brace Drills .. 864
DRUGS AND DRUG SUNDRIES .472-493
DRUMS
Bass Drums ... 522
Chinese Drums ... 522
Drum Heads ... 523
Drummers' Traps ... 522
Minstrel Drums ... 522
Orchestra Drums ... 522
Snare Drums ... 522
DRY CELLS .. 812-813
DRYERS
Artists' Dryers ... 502
Fruit Dryers—Write for Farm Implement Catalog.
DRYING FRAMES ... 134, 823
DUCKS
Canvas (Duck) .331, 851
Decoy Ducks ... 793
Oiled Duck ... 851
DUMBBELLS ... 769
DUSTERS
Dustless Dusters ... 824
Feather Dusters ... 479
Insecticide Dusters 903
Neck Dusters ... 803
Painters' Dusters ... 940-941
Wool Dusters ... 479
DUSTPANS ... 824
DUTCH OVENS ... 829
DYES
Auto Top Dyes ... 955
Curtain Dyes ... 479
Fabric Dyes ... 479
Hair Dyes ... 486
Hat Dyes ... 479
Leather Dyes ... 213
Shoe Dyes ... 213
Soap Dyes ... 479

E

EAGLES
Embroidered Eagles .366
EARRINGS
373, 376, 391, 395, 397, 408-409
EARSCREWS
373, 376, 391, 395, 397, 408-409
EASELS ... 502
EAVES TROUGH AND FITTINGS 959
EDGERS
Sidewalk Edgers ... 841
Turf Edgers ... 841
EDGINGS
Curtain Edgings ... 558, 568
Embroidery Edgings ... 351-353, 366
Lace Edgings ... 351, 354-355
EGGS
Egg Beaters ... 831
Egg Carriers ... 839
Egg Cups ... 714
Egg Preservers ... 478
Eggs (Fish Bait) ... 777
Egg Testers ... 839
Nest Eggs ... 476
EIDERDOWN ... 303
ELASTICS
Bandages, Elastic . 474
Garter Elastic . 370-371
Hat Elastic ... 371
Stockings, Elastic . 474
Truss Elastics ... 475
ELBOWS
Eaves Trough Elbows ... 959
Pipe Elbows ... 704, 707
Radiator Elbows—Write for Heating Catalog.
Spraying Elbows ... 903
Stovepipe Elbows ... 689, 785
ELECTRIC GOODS
Electric Appliances, Household 810-811 —or Write for Electric Goods Catalog.
Electric Appliances, Medical ... 810-811
Electric Cream Separators ... 909
Electric Flashlights 812
Electric Incubators, Brooders ... 900
Electric Ironing Machines ... 711
Electric Light and Power Plants and Accessories ... 674
Electric Lighting Fixtures and Wiring Sundries ... 731
Electric Motors 811, 813
Electric Sewing Machines ... 312
Electric Toys ... 496
Electric Vacuum Cleaners ... 811
Electric Washing Machines ... 712
Electric Water Supply Outfits ... 695
Electric Wire and Wiring Supplies ... 809, 813-814
ELEVATORS
Sacking Elevators ... 888-889
Wagon Box Elevators ... 888
EMASCULATORS .477
EMBLEMS
Embroidered Emblems ... 366
Fraternity Emblems (Jewelry) 373, 376, 391, 398-399, 402-404, 407
Radiator Emblems, Auto ... 752
EMBOSSERS ... 450
EMBROIDERIES ... 351-353, 366
EMERY
Emery Cloth ... 861
Emery Paper ... 933
Emery Wheel Dressers ... 853
Emery Wheel Stands ... 853, 868
EMULSIONS ... 484
ENAMELED WARE ... 825, 827-829
ENAMELS
Aluminum Enamels 942
Auto Body Enamels ... 950, 955
Bathtub Enamels ... 953
Bicycle Enamels ... 736
Carriage Enamels ... 950, 955
Engine Enamels ... 955
Furniture Enamels ... 950, 953
Gold Enamels ... 942
Radiator Enamels ... 942, 953
Radiator Enamels, Auto ... 955
Stovepipe Enamels 953
Wall Enamels ... 950, 953
White Enamels (and Undercoat) ... 953
Wire Screen Enamels ... 953
Woodwork Enamels ... 950, 953
ENCYCLOPEDIAS
Bible Encyclopedias 445
Britannica ... 441
Builders' Encyclopedias ... 446
Farm Knowledge .. 449
Mechanics' Encyclopedias ... 446-448
ENDS
Shaft Ends ... 848
Zinc Matting Ends 605
ENGINES
Gasoline Engines and Outfits 818-819
Engine Accessories, Auto ... 752, 755, 761
Kerosene Engines 820
Motor Boat Engines and Fittings ... 821
Oil Engines ... 820
Portable Engines ... 820
Rowboat Engines ... 821
Toy Engines ... 496
ENLARGING, DEVELOPING AND PRINTING ... 549
ENVELOPES ... 451
Negative Envelopes 555
EPSOM SALTS .477, 484
EQUALIZERS
Plow Equalizers ... 849
Wagon Equalizers .849
ERASERS ... 453, 500
ESCUTCHEONS
Keyhole Escutcheons 837
EVAPORATORS—Write for Farm Implement Catalog.
EVENERS
Buggy Eveners ... 849
Plow Eveners ... 849
Wagon Eveners ... 849-850
EXERCISERS
Athletic Exercisers 766
Babies' Exercisers ... 494, 669
Phonograph Exercises ... 443
Poultry Exercisers .839
EXPANDERS
Roller Tube Expanders ... 705
EXPOSURE GAUGES, TABLES ... 549
EXPRESS RATES ... 457
EXTINGUISHERS
Fire Extinguishers ... 710, 753
EXTRACTORS
Honey Extractors ... 895
EXTRACTS
Beef Extracts ... 483
Flavoring Extracts 478
EYEGLASSES
Jewelers' Eyeglasses ... 426, 471

F

FABRICS
Auto Top Fabrics 851
Buggy Top Fabrics 851
FANS
Fan Belts ... 754
Fan Parts, Auto ... 761
FARM HOUSES AND BUILDINGS ... 971-974

Order Blank Enclosed in This Catalog.

FARM IMPLEMENTS AND MACHINERY 818-820, 842-844, 876-890, 893, 904-909 —Or Write for Farm Implement Catalog.
Farm Implement Repairs 876-877

FARM KNOWLEDGE—(An Encyclopedia for Farmers) 449

FARM LIGHT AND POWER PLANTS 674

FASTENERS
Bed Clothes Fasteners 135
Casement Fasteners 835
Door Fasteners 835-836
Hair Bow Fasteners 131, 358
Hame Fasteners 927
Rug Fasteners 586
Sash Fasteners 834
Snap Fasteners 368

FAUCETS
Faucet Attachments, Hose 711
Plumbing Faucets 701, 703, 706
Steel Barrel Faucets 957

FAVORS 492

FEATHERS
Feather Capes 359
Feather Trimmings 357
Hat Feathers 106
Pillow Feathers 657

FEED CARRIERS Write for Dairy Barn Equipment Catalog.

FEEDERS
Feed Bags 924
Feed Baskets 836
Feed Boxes 836
Hog Feeders 892
Poultry Feeders 839, 901

FEEDS
Stock Feed 477

FEET
Presser Feet 312
Stocking Feet 222-223, 226

FELLOES 850
Felloe Boring Machines 866

FELT
Builders' Deadening Felt 959
Sadiron Felt 823
Table Felt 338
Tarred Felt 959
Wool Felt 770

FENCING
Farm Wire Fencing 896-899
Fence Pickets 964
Fence Posts .... 898 —Or Write for Lumber List.
Fence Making Tools 843, 899
Ornamental Wire Fencing 898
Steel Picket Fencing 898

FENDERS
Auto Fenders 754, 815
Stirrup Fenders 924

FERNERIES ..618-620

FERNS 723

FERRULES
Clean-Out Ferrules 704
Combination Ferrules 704
Neckyoke Ferrules 848
Singletree Ferrules 848

FERTILIZERS 476
Fertilizer Attachments 883
Fertilizer Sowers 883

FIDDLES 506-507

FIELD GLASSES . 429

FIFES 519

FIFTH WHEELS . 848

FIGURES
Aluminum Figures 837
Brass Stencil Figures 840
Steel Figures 859
Stock Marking Figures 838

FILERS
Saw Filers 852, 859

FILES
Auger Bit Files 872
Bill Files 500
Finger Nail Files 422, 490-491
Hand Saw Files 872
Jewelers' Files 471
Letter Files 500
Mill Files 872
Mower Knife Files 844
Needle Files 872
Rat Tail Files 872
Saw Filer Files 852
Sickle Files 844
Taper Files 872
Veterinary Files 477

FILING GUIDES .. 859

FILLERS
Battery Fillers 749
Crevice Fillers 943
Tire Tread Fillers 742
Wood Fillers 942

FILLET 967

FILMS 554
Film Developing 549
Film Packs 554

FILTERS
Ray Filters 549
Water Filters 726

FINGERBOARDS
Guitar Fingerboards 514
Violin Fingerboards 509

FINISHES
Floor Finishes 942-943
Hard Oil Finish 945
Stove Oil Finish 689

FIREARMS 788-789, 792
Firearm Repairs—Write for Sporting Goods Catalog.

FIREBACKS 823

FIRE EXTINGUISHERS . 710, 753

FIRELESS COOKERS 828

FIREPLACES AND FITTINGS—Write for Millwork Catalog.

FISHING TACKLE 772-780

FIXTURES
Bathroom Fixtures 934
Electric Lighting Fixtures and Wiring Sundries 731
Gas Fixtures 727, 731
Grindstone Fixtures 856
Stable Fixtures 836, 891

FLAGONS 418

FLAGS 770

FLAGEOLETS 519

FLANGES
Floor Flanges 707
Roof Flanges 704
Stall Floor Flanges 891

FLANNELETTES .. 328

FLANNELS .. 328, 332
Embroidered Flannels 351

FLASHLIGHTS 812

FLATIRONS 823
Electric Flatirons ... 810
Flatiron Holders 823

FLIES
Fishing Flies, Books 776-777
Tent Flies 782

FLINTS
Lighter Flints 796

FLOATS
Fish Line Floats 775
Horse Mouth Floats 477
Plasterers' Floats 841
Tank Floats 710

FLOORING 964-965

FLOSS 364-365

FLOUNCINGS 351-355
Plaited Skirt Flouncings 369

FLOUR
Paste Flour 954

FLOWERS
Cake Decorations 492
Hat Trimmings 102-109
Table Decorations 723
Wax Flowers 130

FLUTES 519

FLYNETS 932

FLY SWATTERS, POISONS, ETC. 476-477

FOBS 387

FOLDERS
Birth Anouncement Folders 482
Photo Folders 555

FOLIOS
Music Folios 515, 525
Music Folios (Bags) 524

FOODS
Babies' Foods 482
Dog Foods 795
Invalids' Foods 482-483
Plant Foods 476

FOOD CHOPPERS AND ACCESSORIES 830-831

FOOD PUSHERS 412, 414-418

FOOTBALL GOODS 766, 769, 771
Boys' Footballs 771

FOOT REMEDIES . 472

FOOTSTOOLS 607B

FOOTWEAR 766

FORCEPS
Veterinary Forceps .. 477

FORGES 874
Forge Outfits . 867, 874

FORKS
Alfalfa Forks 844
Babies' Forks 412-414, 416
Barley Forks 844
Bicycle Repair Forks 738
Cake Forks 412-414
Coke Forks 843
Cold Meat Forks 412-419
Crupper Forks 923
Fork and Knife Sets 411-419, 804
Fork and Knife Sets, Children's 804
Fork and Spoon Sets, Babies' 412-414, 416, 418
Fork, Knife and Spoon Sets 411-418
Fork, Knife and Spoon Sets, Children's 412-418
Grapple Forks 844
Harpoon Hay Forks 844
Hay Forks 844
Header Forks 844
Manure Forks 844
Oyster Forks ... 415-417
Pickle Forks 412-419
Pot Forks 805
Salad Forks 412-418
Scoop Forks 843
Spading Forks 843
Table Forks, Silver 411-418

FORMALDEHYDE 476, 484
Formaldehyde Candles 483

FORMS
Bust Forms 480
Dress Forms 361

FORTUNE TELLERS 495

FOULARDS 296

FOUNDATIONS
Hair Foundations ... 131
Wax Comb Foundations 895

FOUNTAINS
Poultry Fountains 839
Stock Watering Fountains 892

FOUNTS
Holy Water Founts 419

FRAMES
Clothes Drying Frames 134, 823
Door Frames 967
Jewelers' Saw Frames 471
Photo Frames 422, 491
Printing Frames 555
Saw Frames—Write for Farm Implement Catalog.
Screen Frames 835
Window Frames 966

FREEZERS 831

FREIGHT RATES .. 458

FRETS
Guitar Frets 514

FRINGES
Curtain Fringes 558, 571
Dress Fringes 366
Fancywork Fringes 361, 366
Rug Fringes 605
Window Shade Fringes 583-585

FROGS
Artificial Frogs 776
Braid Loop Frogs 366

FRONTS
Buggy Storm Fronts 851
Water Fronts for Our Stoves—Write for Stove Repair List.

FRYING PANS 826, 829

FUEL
Carriage Heater Fuel 929
Charcoal Iron Fuel .. 823

FUMIGATORS 483-484

FUNGICIDES 476

FUNNELS
Conductor Funnels .. 959
Gasoline Funnels ... 752
Glass Funnels 552

FURNACES
Butchers' Furnaces .. 892
Heating Furnaces 692-693
Soldering Furnaces 704, 853

FURNITURE
Bedroom Furniture 617, 624, 644-667
Camp Furniture 784-785
Cane Furniture 607D, 614-615
Dining Room Furniture 628-638
Fiber Furniture 618-619, 621, 669
Kitchen Furniture 636-643
Lawn Furniture 618-621, 784-787
Library Furniture 608-609, 618-620, 622-623, 625
Nursery Furniture 667-669
Office Furniture 627, 672
Parlor Furniture 607A-612, 614-620, 622-626
Porch Furniture 618-621, 784-787
Reed Furniture 620, 668-669
Rustic Furniture 621
School Furniture and Supplies 637, 672-673

FURRING STRIPS 964

FURS 43

FUSES 814

G

GAGS
Mouth Gags, Stock 477

GALENA 809

GALLOONS .. 352, 355

GAMBRELS
Butchers' Gambrels 806
Skinning Gambrels .. 796

GAMES
Card Games 495
Parlor Games 495

GANGS
Cultivator Gangs ... 884
Worm Gangs 775

GARAGES 974
Garage Hardware ... 836
Garage Machine Tools 868
Garage Supplies 749, 752-753
Garage Tanks and Pumps 957

GARTERS
Boys' Garters .. 370-371
Girls' Garters .. 370-371
Men's Garters 241, 243
Women's Garters .. 370

GASKETS
Cork Gaskets 754
Cylinder Head Gaskets 755
Phonograph Gaskets 548
Pipe Union Gaskets 707
Water Gauge Gaskets 707

GAS LIGHTERS 796

GAS LOGS — Write for Millwork Catalog.

GASOLINE
Gasoline Engines and Outfits 818-819
Gasoline Tanks, Auto 752
Gasoline Storage Tanks, Pumps 957

GAS PLATES 689

GATES
Driveway Gates 898-899
Gate Castings 898
Gate Hangers, Hinges 836
Porch Gates 668
Walk Gates 898-899

GAUGES
Air Gauges, Tire ... 744
Battery Gauges 749
Butt Gauges 863
Gasoline Gauges 755
Jointer Gauges 860
Mainspring Gauges 471
Marking Gauges 863
Mortise Gauges 863
Oil Gauges, Auto ... 755
Photo Exposure Gauges 549
Plane Gauges 860
Pressure Gauges 706
Saw Tooth Gauges ... 859
Scissors Gauges 803
Screw Pitch Gauges 869
Skirt Gauges 368
Surface Gauges 869
Water Gauges 707

GAUNTLETS
Boys' Gauntlets ... 256
Men's Gauntlets 256-257
Riding Gauntlets ... 924
Women's Gauntlets 132-133

GAUZE
Aseptic Gauze 483

GEARS
Ratio Gears 760
Speedometer Gears .. 753
Steering Gear, Auto 761

GENERATORS
Auto Generators—Write for Auto Supply Catalog.
Gasoline Lantern Generators 796
Generator Brushes .. 757
Telephone Generators — Write for Electrical Goods Catalog.

GEORGETTE CREPE 296

GIANT STRIDES . 767

GIMLETS 864

GIMPS 561, 851

GINGHAMS 300, 316, 319

GIRDLES
Belt Girdles ... 127, 366
Corset Girdles 112, 120-123

GIRTHS
Saddle Girths 924

GLAROSCOPES ... 745

GLASS
Art Glass (Vitrophane) 942
Ground Glass 555

GLASSES
Amber Glasses 746
Communion Glasses 418
Eye Glasses, Jewelers' 426, 471
Field Glasses 429
Goblets 722-723
Iced Tea Glasses 722-723
Lemonade Glasses 722-723
Looking Glasses — See "Mirrors."
Magnifying Glasses 426
Measuring Glasses .. 552
Milk Glasses 722
Opera Glasses 429
Reading Glasses 426
Sherbet Glasses 722-723
Storm Glasses 427
Sundae Glasses 722
Watch Glasses 471
Water Glasses, Sets 722-726

GLASSWARE
Lighting Glassware 727-731
Table Glassware 722-726

GLAUBER SALTS 477

GLIDES
Furniture Glides 837

GLOBES
Atlas Globes .... 500 —Or Write for School Furniture Catalog.
Fish Globes ... 723, 726
Gasoline Lantern Globes 796
Oil Lamp Globes ... 729
Oil Lantern Globes .. 825

GLOVES
Baseball Gloves 762
Bee Keepers' Gloves 895
Boxing Gloves 770
Boys' Gloves 256
Canton Flannel Gloves 256
Hunters' Gloves 793
Husking Gloves 839
Men's Gloves ... 256-257
Rubber Gloves 480
Women's Gloves 132-133, 256
Work Gloves ... 256-257

GLUE
Canoe Glue 780
Glue (Mucilage) 478, 556
Ground Glue 954
Liquid Glue ... 478, 954

GLYCERIN 484
Glycerin and Rose Water 489
Glycerin Suppositories 482, 484

GOALS 771

GOBLETS
Communion Goblets 418
Glass Goblets .. 722-723

GO-CARTS
Babies' Go-Carts ... 670
Doll Go-Carts 497

GOGGLES 426, 746

GOLF GOODS 765-766

GONGS
Chinese Gongs 522
Trip Gongs 855

GOPHER KILLERS 796

GRADERS
Grain Graders 889
Road Graders 888
Seed Corn Graders ... 883

GRADUATES 552

GRAIN BINS—Write for Farm Implement Catalog.
Grain Measures 839

GRAINERS 943
Graining Colors 943

GRANITEWARE 825, 827-829

GRAPHITE
Graphite Grease 957

GRASS CATCHERS 841

GRATERS 831

GRATES
Feed Cooker Grates .. 892
Fireplace Grates—Write for Millwork Catalog.

GRAVY BOATS 420, 714, 716-721

GREASE
Axle Grease 957
Cup Grease 956-957
Differential Grease .. 956
Gun Grease 793
Rifle Grease 793
Transmission Grease 956-957

GREASE CUPS 706

GREASE GUNS 753-754

GRIDDLES 826, 829

GRID LEAKS, CONDENSERS ... 808

GRIDS
Camp Grids 785

GRILLS
Electric Grills 810

GRINDERS
Coffee Grinders 831, 839
Corn Grinders and Attachments 819-820, 839, 888-889
Engine and Feed Grinder Outfits 818-820
Feed Grinders and Attachments 819-820, 839, 888-889
Grain Grinders 818-820, 831, 839, 888-889
Knife Grinders 706, 856
Meat Grinders 830
Shearing Knife Grinders 933
Sickle Grinders 844
Tool Grinders 706, 853, 856, 859, 868
Valve Grinders 752, 755

GRINDSTONES AND FIXTURES 706, 856, 859
Grindstones (Wheels) 853, 856

GRIPS
Handle Bar Grips ... 738
Pipe Grips 705
Spoke Nipple Grips .. 737
Wire Grips 871

GROOVERS
Sidewalk Groovers .. 841

GROUND BONE ... 476

GROUND COLORS 943-944

GROUND OUTFITS 809

GROUNDS (Lumber) 964

GUARDS
Babies' Guards 494, 668-669
Cattle Fly Guards ... 932
Grain Saving Guards 876
Harrow End Guards 882
Harvester Guards ... 876
Mower Guards 876, 887
Mud Guards 738
Pants Guards 736
Pea Harvesting Guards 876
Razor Guards 802
Ribbon Guards 387, 390, 409
Shin Guards 763

GUIDES
Fishing Rod Guides 775
Piston Guides 755
Saw Filing Guides .. 859
Saw Guides 859

GUIMPES 371

GUITARS 510
Guitar Supplies 514
Hawaiian Guitars, Sets 510, 514

GUM
Chewing Gum 492

GUMMERS
Saw Gummers 852

GUNS
Grease Guns 753-754
Oil Guns 753-754
Pop Guns 496
Shotguns 788
Spray Guns 903

GUNSTOCKS 793

GUT
Fish Line Gut 774
Stringed Instrument Gut 508-509

GUTTERS 959

H

HABITS
Riding Habits, Skirts, Suits, Etc. 4, 39

HAFTS
Awl Hafts 930

HAIR GOODS
Including Switches, Transformations, Puffs, Toupees and Wigs ... 128-129, 131
Hair Ornaments 130-131, 397, 408
Hair Preparations 486, 490

HAIRPINS 131

HAIRSPRINGS 471

HALTERS 920-921
Halter Accessories 926-928

HAMES 926
Hame Repairs . 926-927

HAMMERS
Ball Pein Hammers 873
Blacksmiths' Hammers 873, 875
Brick Hammers 841
Cobblers' Hammers .. 212
Horseshoers' Hammers 873
Jewelers' Hammers 471
Machinists' Hammers 873
Nail Hammers 863
Orchestra Bell Hammers 519
Plow Hammers 873
Ripping Hammers ... 863
Riveting Hammers 847, 873
Sledge Hammers 873
Stone Hammers 841
Tubephone Hammers 519
Tuning Hammers ... 515
Xylophone Hammers 519

HAMMOCKS 786-787
Hammock Fittings 786-787

HANDBAGS .. 126-127

HANDBALLS 762
Handball Rules 762

HANDCUFFS 792

HANDKERCHIEFS
Bandanas 242
Boys' Handkerchiefs 163
Children's Handkerchiefs 357
Men's Handkerchiefs 242-243
Silk Handkerchiefs 242-243
Women's Handkerchiefs 357

HANDLE BARS ... 738

HANDLES
Adze Handles 862
Auger Handles 864
Awl Handles 930
Ax Handles 856
Bag Handles 261
Box Handles 837
Broom Handles 929
Cant Hook Handles 856
Carpenters' Saw Handles 859
Chest Handles 837
Chisel Handles 861
Compass Saw Handles 858
Crosscut Saw Handles 857
Door Handles 834, 836
Door Handles, Auto 759
Drawer Handles 834, 837
File Handles 872
Forceps Handles ... 477
Hammer Handles 863, 873
Hammer Handles, Jewelers' 471
Hatchet Handles ... 862
Hoe Handles 843
Interchangeable Tool Handles 872
Manure Fork Handles 844
Mattock Handles ... 843
Meat Saw Handles .. 806
Panel Saw Handles . 859
Pick Handles 843
Pitchfork Handles .. 844
Post Maul Handles 843

HANDLES — Continued.
Razor Handles......802
Sadiron Handles....823
Scythe Handles....841
Shovel Handles and Heads......843
Sledge Handles.....873
Soldering Copper Handles.......853
Suitcase Handles....261
Wrench Handles....747

HANGERS
Barn Door Hangers 836
Clothes Hangers......367, 482, 837
Dumbbell Hangers..769
Eaves Trough Hangers.......959
Gate Hangers.......836
Indian Club Hangers.......769
Line Shaft Hangers 855
Pipe Hangers.......707
Stanchion Hangers..891
Seed Corn Hangers..839
Storm Sash Hangers 835
Towel Hangers......934
Track Hangers......891
Window Screen Hangers......835

HARDANGER CLOTH......361

HARDIES......875

HARDWARE..822-875
Barn Door Hardware.....836
Bathroom Hardware.....834, 934
Builders' Hardware.....834-836
Cabinet Hardware...837
Cellar Window Hardware.....834
Dairy Hardware 832-833
French Window Hardware.....835
Garage Door Hardware.....836
Harness Hardware.....926-928, 931
Screen Door Hardware.....835
Vehicle Hardware.....846-851

HARMONICAS....518

HARNESS
Dog Harness...795, 911
Goat Harness.......911
Harness Repairs.....919-923, 926-927, 931
Horse or Mule Harness.......910-915
Plow Harness.......912
Pony Harness...910-911
Web Harness........911

HARROWS.....842, 880-882, 886
Harrow Attachments 881
Harrow Repair Parts.....876, 881-882
Harrow Riding Attachments.......878

HASPS
Hasps, Hooks and Staples......836
Hinge Hasps.......836

HATCHETS......862
Hunters' Hatchets.......793, 856

HATS
Auto Hats, Women's.......91
Boys' Hats...138, 176
Girls' Hats 98-101, 109
Hat and Scarf Sets... 91
Men's Hats.....243, 252-255
Misses' Hats.....97-99, 107-109
Oiled Slicker Hats.....166, 267
Rubber Hats, Men's, Boys'.....166, 266
Straw Hats, Men's, Boys'.....176, 254
Untrimmed Hats.....102-109
Waterproof Hats, Men's, Boys'.....166, 266-267
Women's Hats.....83-99, 102-109
Women's Hat Trimmings.......102-109

HAWKS
Plasterers' Hawks..841

HAY CARRIERS AND ACCESSORIES.......844
—Or Write for Farm Implement Catalog.
Hay Stacking Outfits.......844

HEADLIGHTS
Auto Headlights and Accessories 751, 756
Bicycle Headlights..737
Carbide Cap Lights 796

HEAD SETS (Radio).......808

HEADSTALLS.919-920

HEATERS
Automobile Heaters 929
Camp Heaters......785
Carriage Heaters....929
Gas Radiator Heaters—Write for Stove Catalog.
Heating Stoves......679
Heating Stove Repairs for Our Stoves—Write for Stove Repair List.
Range Boiler Heaters......684-685, 703
Sadiron Heaters.....829
Tank Heaters (Stock)—Write for Farm Implement Catalog.
Water Heaters.......684-685, 703

HEATING PADS...810

HEATING PLANTS.......692-694

HEELS
Heel and Sole Sets.212
Leather Heels.......211
Rubber Heels.......211

HENRIETTAS....303

HERBS.......484

HILLERS
Potato Hillers......886

HINGES
Box Hinges.........837
Brass Hinges.......837
Cupboard Hinges...834
Door Hinges...834, 836
Gate Hinges........836
Hasp Hinges........836
Screen Door Hinges 835
Spring Hinges......834
Strap Hinges.......836
T Hinges...........836
Transom Hinges....835

HITCHES
Rope Hitches.......844

HIVES.......895

HOBBLES
Cow Hobbles.......838
Horse Hobbles.....922

HODS......823

HOES.......842-843

HOISTS.......843, 845

HOLDBACKS.....848

HOLDERS
Baggage Holders, Auto.......746
Babies' Bib Holders.392
Cake Candle Holders.492
Camera Plate Holders.......549
Chalk Holders......797
Cigar Holders......259
Cigarette Holders...259
Clarinet Reed Holders.......523
Cymbal Holders.......522-523
Door Holders.......834
Extension Bit Holders.......864
Fruit Jar Holders..830
Harmonica Holders.518
Hog Holders........838
Horse Tail Holders.928
Ladder Rung Holders.......939
Mop and Brush Holders.......824
Oil Can Holders....855
Pipe Holders.......710
Plate Holders......549
Post Card Holders..555
Sadiron Holders....823
Sash Holders.......834
Saw Tooth Holders 852
Saxophone Reed Holders.......523
Seed Corn Holders.839
Shaft Bolt Holders.848
Shoe Holders.......213
Spool Holders..368-369
Spoon Holders......421
Starting Handle Holders, Auto.....759
Strop Holders......802
Toilet Paper Holders.......934
Tooth Brush Holders.......934
Toothpick Holders 421
Towel Holders......934
Tumbler Holders....934
Umbrella Holders...851

HOLLOW HANDLE TOOL SETS......859

HOLSTERS......795

HONDAS......924

HONES.......798, 801

HONEY MAKING SUPPLIES......895

HOODS
Auto Hoods.......761
Babies' Hoods.....138

HOOKS
Bathroom Hooks....934
Belt Hooks.....854-855
Bird Cage Hooks....837
Bush Hooks.........841
Button Hooks.......491
Cant Hooks.........856
Chain Hooks........856
Clothesline Hooks..823
Coat Hooks.........837
Conductor Pipe Hooks.......959
Crochet Hooks......368
Curtain Rod Hooks 586
Fish Hooks.......774-777
Floor Hooks........844
Gaff Hooks.........774
Gate Hooks.........836
Grass Hooks........841
Hame Hooks.......927
Hammock Hooks.......786-787
Harness Hooks.....926
Harness Room Hooks.......928
Hat Hooks.........837
Hog Hooks.........806
Hooks and Eyes....835
Hooks and Eyes, Dress.......368
Hooks and Staples..836
Hook, Hasp and Staple.......836
Husking Hooks.....839
Ladder Hooks.......939
Manure Hooks.......844
Mud Hooks, Auto..743
Paint Pot Hooks....939
Picture Hooks......586
Potato Hooks.......844
Screw Hooks........837
Singletree Hooks...848
Spoon Hooks....774, 776
Switch Hooks—Write for Electrical Goods Catalog.
Tassel Hooks......586
Trace Hooks.......927
Track Hanging Hooks.......844

HOOPS
Embroidery Hoops.361

HOPPERS
Closet Hoppers.....698-699
Poultry Feeder Hoppers.......901

HOREHOUND....492

HORNS
Alto Horns........521
Auto Horns and Attachments...746, 754
Baritone Horns.....521
Bass Horns.........521
Bicycle Horns......737
Hearing Horns......429
Shoe Horns.........422

HORSES
Stitching Horses....930

HORSESHOES....875
Horseshoeing Outfits and Tools.....873-875
Pitching Horseshoes.769

HOSE
Garden Hose and Fittings.......840
Gas Stove Hose.....689
Oil Hose...........957
Pump Hose, Auto...744
Radiator Hose, Auto.......752, 754
Spray Hose.........903
Steam Hose.........854
Suction Hose.......854

HOSIERY....220-229, 762, 765, 769

HOT WATER HEATING PLANTS....694

HOUNDS
Wagon Hounds.849-850

HOUSES
Already Cut Buildings.......971-973
Garages............974
Hog Houses—Write for Book of Barns.
Poultry Houses....974 —Or Write for Book of Barns.
Ready Made Houses.974
Smokehouses.......893

HOUSINGS......922

HOW TO ORDER..455

HUBS
Bicycle Hubs.......737
Hub Parts, Auto.......754, 761
Wagon Hubs........850

HUMIDIFIERS....694

HUMIDORS......493

HUNTERS' SUPPLIES...784, 788-796, 799

HUSKING GLOVES 839

HYDRANTS......707

HYDROMETERS.. 749

HYGROMETERS..427

I

ICE BOXES...638-639

ICELESS COOLERS......710

ICE CAPS......480

IMITATION DIAMONDS.396, 398

IMPLEMENTS AND FARM MACHINERY 818-821, 842-844, 876-890, 904-909 —Or Write for Farm Implement Catalog.

INCENSE.....478, 490
Incense Burners.......478, 490

INCREASERS
Pipe Increasers...704

INCUBATORS....900

INDIA LINONS.....324, 326

INDIAN CLUBS...769

INDICATORS
Speed Indicators....869
Trombone Position Indicators.......523
Umpires' Indicators.763

INDOOR BALL GOODS.....763, 766

INFLATERS
Bladder Inflaters..771

INJECTORS
Automatic Injectors.706

INKS
Drawing Inks.......500
Fountain Pen Inks 452
Indelible Inks.....500
Stencil Inks.......840
Writing Inks.......452

INKWELLS.......500

INNER CASINGS..739

INNER TUBES
Auto Tire Tubes.....739, 743
Bicycle Tire Tubes..736
Motorcycle Tire Tubes.......738

INSECTICIDES 476-478

INSERTIONS..351-355

INSOLES.......211

INSTRUCTORS
Music Instructors.....519, 524

INSULATORS
House Wiring Insulators.......731
Radio Insulators....809
Telephone Insulators.......814

INTENSIFIERS
Spark Plug Intensifiers.......750

INVALID CHAIRS..483 —Or Write for Invalid Chair Catalog.

IRON
Bar Iron...........847
Perforated Strap Iron.......707
Tire Iron..........847

IRONING MACHINES......711
Ironing Boards.....823
Ironing Boards, Built-In—Write for Mill Work Catalog.
Ironing Board Covers.......333

IRONS
Charcoal Irons......823
Corner Irons.......846
Curling Irons..131, 810
Gas Irons..........823
Neckyoke Irons.....848
Plane Irons........860
Pool Pocket Irons...797
Sadirons...........823
Sadirons, Electric...810
Saw Frame Irons—Write for Farm Implement Catalog.
Singletree Irons....848
Soldering Irons.....853
Soldering Irons, Jewelers'.......471
Tailors' Goose.....823
Tie Irons..........838
Tuyere Irons.......874
Waffle Irons.......829
Wagon Irons........848
Yarning Irons......704

IRREGULAR CURVES.......500

IVORY FINISH TOILET ARTICLES.......491

J

JACKETS
Babies' Jackets....141
Oiled Slicker Jackets 267
Overall Jackets.....167, 268, 289-291

JACKS
Auto Jacks.........742
Jack Screws........845
Ladder Jacks.......939
Pump Jacks.........818
Radio Jacks........808
Wagon Jacks........845

JAMBS
Door Jambs.........967
Window Jambs.......966

JARDINIERES......723

JARS
Battery Jars.......813
Candy Jars.....723-724
Cereal Jars, Sets...726
Fruit Jars.........726
Horseradish Jars.......421, 714
Insulated Jars.....784
Jam Jars......723-724
Marmalade Jars.......421, 723-724
Mustard Jars..421, 714
Powder Puff Jars...422
Preserve Jars......726
Slop Jars..........727
Spice Jar Sets.....726
Vacuum Food Jars..765

JEWS' HARPS.....515

JAZZBO.......513

JERSEYS
Athletic Jerseys....769
Men's Jerseys......243

JEWELRY.....373-409
Babies' Jewelry.....373, 392, 408-409
Birthday Jewelry.....389, 392, 398
Black Jewelry.391, 401
Diamond Jewelry.....374-376, 400
Imitation Diamond Jewelry.....396, 398

JEWELERS' TOOLS AND SUPPLIES.....426, 471, 870, 872

JOINTERS
Plow Jointers.....877-878, 880
Saw Jointers.......859
Sidewalk Jointers...841

JOINT RUNNERS..704

JOINTS
Clarinet Tuning Joints.......523
Insulating Joints....731
Swivel Joints.......753

JOISTS.......965

JUGS
Communion Jugs....418
Food Jugs.....765, 784

JULIETS......192, 199

JUMPERS
Overall Jumpers, Men's, Boys'.....167, 268, 289-291

K

KALSOMINES 950, 952

KEGS.......830

KEROSENE........957
Kerosene Engines...820

KETTLES
Caldron Kettles....892
Cooking Kettles.826-830
Copper Kettles..829-830
Iron Kettles....829, 892
Milk Kettles.......826
Preserving Kettles.......826-830
Steamer Kettles.827, 829
Tea Kettles.....826-829

KEYS
Handcuff Keys......792
Roller Skate Keys..767
Telegraph Keys—Write for Radio Catalog.
Tuning Keys........514
Watch Keys.........471

KILLERS
Gopher Killers......796
Vermin Killers.....476-478, 483-484

KIMONOS.......57

KITCHEN CABINETS.....640-643
Kitchen Cupboards, Safes.......639
Kitchen Utensils.......726, 826-831

KITS
Camp Cooking Kits 785
Tool Kits, Auto....747
Vacuum Lunch Kits 765
Violinists' Emergency Kits.......509

KNAPSACKS......784

KNEADERS.......831

KNEE CAPS......474

KNICKERBOCKERS
Boys' Knickerbockers.......178-180
Girls' Knickers.......4, 24, 51
Golf Knickerbockers 765
Women's Knickers.......39, 51

KNITTED GOODS
Babies' and Children's Knitted Goods.....134, 138, 141, 147, 157
Women's and Girls' Knitted Goods.....146-147, 149, 160, 162

KNIVES
Blacksmiths' Knives.......873
Boning Knives......806
Bread Knives...419, 805
Butchers' Knives.......804, 806
Butter Knives...411-419
Butter Spreaders.......412-419
Cake Knives.....412-414, 417, 419
Carving Knife and Fork Sets.....413, 417, 804
Carving Knives 805-806
Castrating Knives...477
Cheese Knives......419
Cobblers' Knives....212
Corn Knives........844
Cuticle Knives.....491
Dehorning Clipper Knives.......838
Drawing Knives.....861
Fish Knives........775
Food Chopper Knives.......830
Fruit Knives.....413, 415-419
Glaziers' Knives....941
Grapefruit Knives.......804-805
Harness Makers' Knives.......930
Hay Knives.........844
Horseshoers' Knives.873
Hunting Knives.784, 799
Kitchen Knives.....805
Knife and Chain Sets......388, 798
Knife and Fork Sets....411-419, 804
Knife and Fork Sets, Campers'....785
Knife and Fork Sets, Children's...804
Knife, Fork and Spoon Sets..411-418
Knife, Fork and Spoon Sets, Children's......412-418
Linoleum Knives.. 805
Mower Knives, Heads......876, 887
Palette Knives.....502
Painters' Knives...941
Paperhangers' Knives.......941
Paring Knives..804-805
Penknives..........798
Penknives, Gold.388, 409
Pie Knives.....412-414, 417, 419
Pocket Knives. 798-799
Pruning Knives.799, 805
Putty Knives.......941
Skinning Knives.799, 806
Slicing Knives.805-806
Steak Knives.......806
Sticking Knives....806
Table Knives, Silver.......411-418
Table Knives, Steel 804
Tool Knives........799
Tree Pruning Knives 840

KNOBS
Base Knobs.........835
Brass Knobs....834, 837
Curtain Pole Knobs 587
Door Knobs.....834-835
Furniture Knobs.......834, 837
Glass Knobs.......834
Insulating Knobs—Write for Radio Catalog.
Kettle Cover Knobs 831
Percolator Knobs....826
Radio Knobs........808

L

LABELS
Stock Marking Labels.......838

LACES
Curtain Laces..558, 571
Lace Panels........568
Laces and Embroideries.......351-355
Shoe Laces.........213
Window Shade Laces 583

LACING
Belt Lacing Machines.......854
Leather Belt Lacing 854
Steel Belt Lacing...855
Wire Belt Lacing....855

LACQUER
Automobile Lacquer.955
Chinese Gloss Lacquer.......944, 950

LADDERS.......939
Ladder Hooks, Jacks, Platform and Rung Replacers.......939

LADLES
Butter Ladles......832
Cream Ladles.......412, 414-419
Gravy Ladles...412-419
Plumbers' Ladles....704
Soup Ladles........828

LAMBREQUINS...574

LAMPBLACK 952, 954

LAMPS
Alcohol Lamps......471
Arc Lamps..........727
Auto Lamps and Parts.......751, 756
Auto Stop Lamps...751
Bicycle Lamps......737
Boudoir Lamps 491, 723
Camp Lamps....784, 796
Carbide Lamps and Accessories......796
Darkroom Lamps....552
Desk Lamps, Electric.......731
Dome Lamps, Electric.......731
Dome Lamps, Oil.......728-729
Electric Lamps, Farm Lighting....674
Electric Lamps, House.......731
Electric Lamps (Bulbs), House.......727, 731
Electric Lamps, Vehicle.......812
Electric Lamp Outfits, Miniature....812
Flash Lights.......812
Floor Lamps, Electric.......731
Gas Lamps, House.......727, 731
Gasoline Lamps.......729, 796
Oil Lamps......728-730
Piano Lamps, Electric.......731
Tungsten Lamps.......727, 731

LANTERNS
Carbide Lanterns and Accessories.......796
Dark Lanterns......792
Dash Lanterns......825
Electric Lanterns...812
Gasoline Lanterns...796
Oil Lanterns.......825

LARIATS.......924

LASHES
Whip Lashes.......930

LASTS
Shoe Repair Lasts...212

LATCHES
Barn Door Latches..836
Casement Latches...835
Door Latches.......835
Gate Latches.......836
Knob Latches.......835
Night Latches......835
Thumb Latches......835

LATH
Metal Lath—Write for Roofing Catalog.
Wood Lath.......964

LATHES AND ACCESSORIES.......868 —Or Write for Wood and Metal Working Machinery Catalog.
Boys' Lathes.......859
Lathe Sets.........853

LATIGOS.......924

LAVALLIERES.....373, 376, 389, 391, 393, 396-397, 409

LAVATORIES 698-699
Lavatory Fittings...704

LAWNS.....323-324, 326-327

LAXATIVES.......477, 482, 484

LAYETTES...157-159

LEAD
Arsenate of Lead...476
Bar Lead...........793
Pencil Lead........453
Pig Lead...........704
Sheet Lead.........704
White Lead in Oil..954

LEADERS
Gut Leaders........774
Wire Leaders.......774

LEADS
Battery Leads......757
Bull Leads.........838
Cattle Leads.......838
Dog Leads..........795
Stallion Leads.....927

LEADUPS
Breeching-Leadups..921

LEATHER
Artificial Leather.......561, 851
Harness Leather....930
Lace Leather.......855
Pool Table Leathers 797
Pump Leathers.708, 710
Shoe Leather.......211

LEAVES
Album Leaves......556
Loose Leaves.......454
Senna Leaves.......484

LEGGINGS
Boys' Leggings....215
Elastic Leggings....474
Girls' Leggings.......4
Men's Leggings.....215
Women's Leggings.....4

See Page 557 for Measuring Instructions.

LEMON SQUEEZERS ... 831
LEMONADE SETS ... 722
LENSES
Headlight Lenses ... 751, 756
Photo Lenses ... 549
LETTERING
Sign Lettering ... 446
Window Shade Lettering ... 585
LETTER OPENERS ... 422
LETTERS
Aluminum Letters ... 837
Brass Stencil Letters ... 840
Harness Letters ... 922
Rubber Type Letters ... 500
Steel Letters ... 859
Stock Marking Letters ... 838
LEVELERS
Cultivator Levelers ... 884
LEVELS
Level and Square ... 863
Leveling Instruments ... 862
Mechanics' Levels ... 862
LEVERS
Radio Switch Levers—Write for Radio Catalog.
LICE KILLERS ... 476, 839
LICORICE
Licorice Powders ... 484
Licorice Sticks ... 484
LIDS
Kettle Lids ... 827, 831
LIFE PRESERVERS ... 780
LIFTERS
Transom Lifters ... 835
Valve Lifters ... 752, 755
LIFTS
Heel Lifts ... 211
Safety Lifts ... 843, 845
Sash Lifts ... 834
LIGHT AND POWER PLANTS ... 674
LIGHTERS
Asbestos Lighting Rings ... 687
Automatic Gas Lighters ... 796
LIGHTNING RODS AND FITTINGS ... 888
Lightning Arresters ... 808
LIGHTS
Auto Headlights ... 751, 756
Auto Spotlights ... 751
Auto Stop Lights ... 751
Cap Lights ... 796
Carbide Lights ... 796
Curtain Lights, Auto ... 745, 758
Electric Lights (Farm Plants) ... 674
Electric Lights (Fixtures) ... 731
Fishing Lights ... 780, 796
Flash Lights ... 812
Hunters' Lights ... 796
Miners' Lights and Attachments ... 796
Tent Lights ... 784
Torch Lights ... 780, 825
LIME
Lime and Sulphur ... 476
LIME JUICE ... 478
LINCRUSTA—Write for Wall Paper Sample Book.
LINEMEN'S TOOLS AND SUPPLIES ... 814, 871
LINENS
Art Linens ... 327, 335, 337, 341-342, 361
Bed Linens ... 327, 329-331
Dress Linens ... 327
Hotel Table Linens ... 339, 342
Table Linens ... 338-343
LINES
Chalk Lines ... 864
Clothes Lines ... 823
Fishing Rod Lines ... 774
Harness Lines ... 921
Plow Lines ... 921
Set Lines ... 775, 779
Trot Lines ... 775, 779
Web Lines ... 921
Wire Clothes Lines ... 823
LINESHAFTS ... 818, 855
LINIMENTS ... 484
LININGS
Brake Band Linings ... 753, 760
Carpet Linings ... 605
Coat Linings ... 301, 313, 328
Dress Linings ... 301, 305, 313
Hat Linings ... 102
Plush Robe Linings—Write for Harness Catalog.
Transmission Band Linings ... 760
LINKS
Cable Chain Links ... 856
Chain Links, Bicycle ... 738
Chain Links, Motorcycle ... 738
Cuff Links ... 373, 376, 404-406, 409
Emblem Cuff Links ... 404
Skid Chain Links ... 743
Speedometer Links ... 753
LINOLEUMS ... 588, 594-596
LINSEED OIL ... 949
LIPSTICKS ... 489
LISTERINE ... 484, 487
LISTERS ... 886
LITTER CARRIERS ... 891
LITTLE JOE ... 513
LOCKETS ... 403
Lockets With Chains ... 392
Scapular Lockets ... 390
LOCKNUTS ... 707
LOCKS
Box Locks ... 837
Cupboard Locks ... 837
Dog Collar Locks ... 795
Door Locks ... 834-835
Drawer Locks ... 837
Mail Box Locks ... 838
Oar Locks ... 780
Padlocks ... 837-838
Sprocket Locks ... 737
Steering Wheel Locks ... 754
Suit Case Locks ... 837
Tire Cable Locks ... 743
Trunk Locks ... 837
Wardrobe Locks ... 837
Window Locks ... 834-835
LOGS
Gas Logs—Write for Mill Work Catalog.
LOOPS
Curtain Loops ... 575
Frog Loops ... 366
Harness Loops ... 926-928
Portiere Loops ... 575
Track Anchor Loops ... 891
LOOSE LEAF BOOKS AND FILLERS ... 450, 454
LOTIONS ... 489-490
LOUNGES ... 610-611, 618
LOZENGES ... 484, 492
LUBRICANTS
Babbitt Metal ... 853
Phonograph Lubricants ... 548
(Also see Oils and Greases.)
LUBRICATORS
Engine Lubricators ... 706
Spring Lubricators, Auto ... 747
LUMBER ... 964-965
LUNCH SETS ... 341, 363
LYSOL ... 484

M

MADRAS ... 299, 317-318, 322, 325
Curtain Madras ... 559-560
MAGAZINES
Pistol Magazines ... 792
Rifle Magazines ... 789
MAGNAVOX ... 809
MAGNESIA ... 484
MAGNETOS ... 813
Auto Magneto Parts ... 750
—Or Write for Auto Supply Catalog.
MAGNETS ... 814
MAGNIFIERS ... 426
MAILBOXES ... 838
MAINSPRINGS
Clock Mainsprings ... 471
Phonograph Mainsprings ... 548
Watch Mainsprings ... 471
MALINES ... 356
MALLETS
Carpenters' Mallets ... 861
MALTED MILK ... 482
MANDOLINS ... 511
Banjo Mandolins ... 511
Mandolin Supplies ... 514
MANDRELS
Saw Mandrels ... 852
MANGERS ... 891
MANICURE ARTICLES ... 422, 485, 490-491
Manicure Specialties ... 490
MANTELS
(Brick, Tile or Wood)—Write for Mill Work Catalog.
MANTLES
Gas Mantles ... 727
Gasoline Mantles ... 727, 730, 796
Kerosene Mantles ... 729
MANURE SPREADERS ... 890
MAPS ... 436
—Or Write for School Furniture Catalog.
MARABOU ... 357
Marabou Capes ... 359
MARBLES ... 496
MARGUERITES ... 141
MARKERS
Billiard Markers ... 797
Napkin Markers ... 421
Poultry Markers ... 839
Skirt Markers ... 368
Stock Markers ... 838
Tennis Court Markers ... 764
MARQUISETTES
Curtain Marquisettes ... 558, 562-563
MARSHMALLOWS ... 492
MARTINGALES ... 921
MASHERS
Potato Mashers ... 831
MASKS
Baseball Masks ... 763
Complexion Masks ... 480
MASSAGERS
Body Massagers ... 811
MASTIC ... 736, 742
MATCHBOXES ... 422, 784
MATS
Auto Mats ... 744, 758
Bath Mats ... 593
Door Mats ... 605
Drain Board Mats ... 701
Floor Mats ... 592, 596
Fur Mats, Auto—Write for Harness Catalog.
Landing Mats ... 605
Table Mats ... 338, 363
MATTINGS ... 596, 605
MATTOCKS ... 843
MATTRESS COVERS ... 349
MATTRESSES
Bassinette Mattresses ... 668
Bed Mattresses ... 658-661
Camp Mattresses ... 785
Cot Mattresses ... 661, 666
Couch Mattresses ... 664-665
Cradle Mattresses ... 661
Crib Mattresses ... 661, 667
Davenport Mattresses ... 661, 664
Folding Bed Mattresses ... 661
Pad Mattresses ... 661, 664-667
MAULS
Post Mauls ... 843
Woodchoppers' Mauls ... 856
MEASURES
Acid Measures ... 832
Gasoline Measures ... 752
Graduates ... 552
Grain Measures ... 839
Measuring Cups ... 827, 831
Quart Measures ... 831
Tape Measures ... 368, 862
MEDICINES ... 484
Poultry Medicines ... 476
Stock Medicines ... 477
MENDING OUTFITS
Auto Top Mending Outfits ... 745
Mending Tissue ... 369
Tire Mending ... 736, 742
MENTHOL
Menthol Ointment ... 484
MESSALINES ... 295-296, 301
METAL WORKING MACHINERY ... 868
—Or Write for Wood and Metal Working Machinery Catalog.
METERS
Air Pressure Meters ... 744
Battery Testing Meters ... 749
Exposure Meters ... 549
METRONOMES ... 523
MICA
Mica Compounds ... 760
Tire Mica ... 742
MICROMETERS ... 869
MICROSCOPES ... 426
MIDDIES ... 24, 50-51
MILK
Malted Milk ... 482
Milk Fever Outfits ... 477
Milk of Magnesia ... 484
Sugar of Milk ... 482
Milk Shakers ... 826
Milk Testing Outfits ... 832
MILLINERY ... 83-109
MILLS
Cane or Cider Mills—Write for Farm Implement Catalog.
Coffee Mills ... 831, 839
Fanning Mills ... 889
Grinding Mills and Attachments ... 888-889
Hand Grist Mills ... 831, 839
Power Grist Mills ... 839
MILL SUPPLIES ... 854-855
MILL WORK ... 966-967
—Or Write for Mill Work Catalog.
MINCERS ... 830-831
MINNOWS ... 776-777
MINTS ... 484, 492
MIRRORS
Auto View Mirrors ... 746, 759
Bathroom Mirrors ... 627, 934
Folding Mirrors ... 627
Hall Mirrors ... 627
Hand Mirrors ... 485, 491
Headlight Mirrors ... 756
Mirror, Brush and Comb Sets ... 422
Polychrome Mirrors ... 627
Shaving Mirrors ... 422, 485, 802
Stand Mirrors ... 485
Wall Mirrors ... 627
MITERBOXES ... 859
MITTENS
Hunters' Mittens ... 793
MITTS
Baseball Mitts ... 763
Scouring Mitts ... 367
Striking Bag Mitts ... 770
MIXERS
Bread Mixers ... 831
Concrete Mixers ... 970
Feed Mixers ... 970
MOCCASINS
Babies' Moccasins ... 138, 196
Men's Moccasins ... 766
Women's Moccasins ... 199, 766
MODERN HOMES ... 972-973
MODULATORS
Phonograph Modulators ... 548
MOISTENERS
Air Moisteners ... 694
MOLASSES
Stock Feed Molasses ... 477
MOLDINGS
Picture Moldings ... 967
Picture Moldings (Finished)—Write for Wall Paper Sample Book.
Porch Moldings ... 967
Window Moldings ... 967
MOLDS
Concrete Molds ... 970
MOPS
Polishing Mops ... 824
Scrubbing Mops ... 824
MORTISING MACHINES ... 866
MOSQUITO PREPARATIONS ... 775
MOTHBALLS ... 478
MOTOMETERS ... 752, 754
MOTORCYCLE SUPPLIES ... 738
MOTORS
Air Motors ... 706
Auto Motor Accessories ... 752, 754-755, 761
Electric Motors ... 811, 813
Steam Motors ... 706
Water Motors and Attachments ... 706
MOUNTS
Photo Card Mounts ... 555-556
MOUTH ORGANS ... 518
MOUTHPIECES
Clarinet Mouthpieces ... 523
Cornet Mouthpieces ... 523
Fife Mouthpieces ... 519
Saxophone Mouthpieces ... 523
Telephone Mouthpieces ... 814
MOWERS
Hay Mowers ... 887
Mower Repairs ... 876, 887
Lawn Mowers ... 841
MUCILAGE ... 478, 556
MUFFLERS
Auto Mufflers ... 759, 761
Muffler Cut-Outs ... 746, 754
MUGS
Aluminum Mugs ... 784, 802
Coffee Mugs ... 714
Enameled Mugs ... 828
Shaving Mugs ... 802
Shaving Mug and Brush Sets ... 422, 800
Silver Mugs ... 418
MUSIC
Chord Books ... 524
Folios ... 515, 524-525
Instructors ... 519, 524
Organ Rolls ... 515
Phonograph Records ... 544-547
Piano Rolls ... 503-504
Sacred Music ... 503-505, 525, 544-545
Sheet Music ... 504-505
Song Books ... 435, 525
MUSLINS ... 323-325, 330
Curtain Muslin ... 563
MUSLINWEAR ... 124-125, 136-137, 142-143, 148-155, 160-162
MUTES
Cornet Mutes ... 523
Violin Mutes ... 509
MUZZLES
Dog Muzzles ... 725

N

NAILS
Chair Seat Nails ... 837, 851
Horseshoe Nails ... 875
Roofing Nails ... 958, 960
Screen Numeral Nails ... 835
Shoe Nails ... 212
Upholstery Nails ... 837, 851
Wire Nails ... 835, 863
NAIL PULLERS ... 863
NAIL SETS ... 863
NAINSOOKS ... 330
NAPKINS
Napkins for Embroidering ... 341
Paper Napkins ... 450
Sanitary Napkins ... 370-371, 483
Table Napkins ... 339-343
NAPPIES ... 714, 726
NEATSFOOT ... 957
NECKLACES ... 373, 376, 389, 391-397, 409
NECKS
Violin Necks ... 509
NECKTIES
Boys' Neckties ... 163
Men's Neckties ... 240-241, 243
NECKYOKES ... 815, 849-850
NEEDLES
Assorted Needles in Books ... 368
Bead Work Needles ... 361
Billiard Cloth Needles ... 797
Crochet Needles ... 368
Darning Needles ... 368
Embroidery Needles ... 361
Harness Needles ... 930
Knitting Needles ... 365
Phonograph Needles ... 535
Sewing Awl Needles ... 930
Sewing Machine Needles ... 312
Sewing Needles ... 368
NEGATIVES
Negative Printing and Enlarging ... 549
Negative Supplies ... 549, 552-553, 555
NEST EGGS ... 476
NESTS
Poultry Trap Nests ... 901
Wire Nests, Hens' ... 839
NETS
Curtain Nets ... 568-572
Dip Nets ... 777
Dress Nets ... 356
Fish Alive Nets ... 779-780
Fish Nets ... 777-780
Hair Nets ... 129
Horse Fly Nets ... 932
Landing Nets ... 777
Maline Nets ... 356
Minnow Nets ... 777-780
Mosquito Nets, Face ... 775
Tennis Nets ... 764
Trammel Nets ... 779
Volley Ball Nets ... 771
NETTING
Fish Netting ... 778-779
Minnow Netting ... 778
Mosquito Netting ... 331
Poultry Netting ... 899
Rabbit Netting ... 899
Seine Netting ... 778
Wire Screen Netting ... 835
NEWELS ... 967
NICOTINE
Nicotine Solutions ... 476
NIGHTGOWNS
Babies' Nightgowns ... 137, 141
Children's Nightgowns ... 142
Girls' Nightgowns ... 142, 160
Women's Nightgowns ... 152-156
NIGHTSHIRTS
Boys' Nightshirts ... 163
Men's Nightshirts ... 249
NIPPERS
Blacksmiths' Nippers ... 873
Finger Nail Nippers ... 490
Peg Nippers ... 212
Wire Nippers ... 871
NIPPLES
Hose Nipples ... 840
Iron Pipe Nipples ... 707
Nursing Bottle Nipples ... 482
NITRATES ... 476
NOSEBANDS ... 919
NOSING
Stair Nosing ... 586
NOTIONS ... 361, 364-366, 368, 368-369, 371-372
NOZZLES
Garden Hose Nozzles ... 840
Paint Nozzles ... 903
Spray Hose Nozzles ... 903
Whitewash Nozzles ... 903
NUMBERS
House Numbers ... 837
NURSERS ... 482
NUTCRACKERS ... 831
NUTS
Blank Nuts ... 846
Castellated Nuts ... 747
Fingerboard Nuts ... 509
Locknuts ... 707
Threaded Nuts ... 846

O

OAKUM ... 704
OARS AND FITTINGS ... 780
OCARINAS ... 513
OCHRE IN OIL ... 954
ODD JOBS ... 862
OFFICE FURNITURE AND SUPPLIES ... 450-454, 500, 627, 672-674
OFFSETS
Soil Pipe Offsets ... 704
OILCLOTHS
Oilcloth Sets, Table ... 363
Shelf Oilcloth ... 334
Table Oilcloth ... 334
Upholstering Oilcloth ... 334, 561
Wall Oilcloth and Borders—Write for Wall Paper Sample Book.
OIL CUPS ... 706
OILERS
Auto Spring Oilers ... 747, 752
Hog Oilers ... 892
Motor Oilers, Auto ... 755
Oiler Cans ... 855
Pump Oilers ... 855
Watch Oilers ... 471
Wheel Oilers ... 848
OIL GUNS ... 753-754
OILS
Animal Bait Oil ... 796
Artists' Oils ... 502
Auto Engine Oil ... 956-957
Camphorated Oil ... 484
Carbolic Oil ... 477
Castor Oil ... 484
Clock Oil ... 471
Coal Oil ... 957
Cocoanut Oil ... 486
Cod Liver Oil ... 484
Cream Separator Oil ... 957
Creosote Oil ... 935
Cylinder Oil ... 957
Differential Oil ... 956
Dustless Mop Oil ... 942
Engine Oil ... 957
Finishing Oil ... 945
Fish Bait Oil ... 775
Floor Oil ... 942-943
Gasoline Engine Oil ... 957
Gun Oil ... 793
Harness Oil ... 957
Harvester Oil ... 957
Hog Oil ... 892
Hone Oil ... 793
Household Oil ... 942
Incubator Oil ... 957
Kerosene Oil ... 957
Lantern (Signal) Oil ... 792
Linseed Oil ... 949
Mineral Oil ... 484
Mop Oil ... 824, 942
Neatsfoot Oil ... 957
Olive Oil ... 484
Phonograph Oil ... 548
Reel Oil ... 793
Sawmill Oil ... 957
Sewing Machine Oil ... 312
Shoe Waterproofing Oil ... 213
Slicker Oil ... 267
Stove Oil Finish ... 689
Strop Oil ... 793
Tattoo Oil ... 838
Thresher Oil ... 957
Tractor Oil ... 956-957
Transmission Oil ... 956
Watch Oil ... 471
OILSTONES ... 861
OINTMENTS ... 484
OPENERS
Can Openers ... 831
Furrow Openers ... 876
Letter Openers ... 422
OPERA GLASSES ... 429
OPTICAL GOODS ... 426, 428-429
ORGANDIES ... 320, 323-324, 326
ORGANS
Church Organs ... 526
Mouth Organs ... 518
Parlor Organs ... 526
Roller Organs ... 515
ORNAMENTS
Cake Ornaments ... 492
Hair Ornaments ... 130-131
OSTRICH ... 106-107, 357
OUIJA BOARDS ... 495
OUTLET BOXES ... 731
OVENS
Dutch Ovens ... 829
Portable Ovens ... 686
Stovepipe Ovens ... 685
OVERALLS
Boys' Overalls ... 166-167
Men's Overalls ... 289-291
Youths' Overalls ... 268
OVERCHECKS ... 923
OVERCOATS
Boys' Overcoats ... 175
Men's Overcoats ... 264-265
OVERS ... 219
OVERSHOES ... 218-219
OXFORDS
Boys' Oxfords ... 195, 204, 206, 208-209, 215
Canvas Oxfords ... 184-185, 198, 204, 214-215
Children's Oxfords ... 195, 215
Girls' Oxfords ... 183-185, 195, 198, 214-215
Golf Oxfords ... 766
Gymnasium Oxfords ... 766
Men's Oxfords ... 200-201, 204-206, 215
Tennis Oxfords ... 766
Women's Oxfords ... 182-189, 191-193, 195, 214-215

P

PACKERS
Soil Packers ... 882
PACKING
Sheet Packing ... 854
PACKS
Film Packs ... 554
Pack Sacks ... 784
PACS ... 216
PADDING
Table Padding ... 338
PADDLES
Canoe Paddles ... 780
Paint Paddles ... 939
PADLOCKS ... 837
Mail Box Padlocks ... 838
Sprocket Padlocks ... 737
Tire Cable Padlocks ... 743
PADS
Asbestos Table Pads ... 338
Bassinette Pads ... 668
Bolster Pads ... 665
Bunion Pads ... 472
Callous Pads ... 472
Clarinet Key Pads ... 523
Corn Pads ... 472
Cot Pads (Mattresses) ... 661, 666

PADS—Continued.
Couch Pads (Mattresses) 664-665
Crib Pads (Mattresses) 661, 667
Door Pads, Auto 759
Drummers' Practice Pads 522
Elbow Pads 771
Electric Heating Pads 810
Gun Recoil Pads 795
Harness Pads 922, 925
Heel Pads 472
Horse Collar Pads 925
Knee Pads 771
Mattress Pads (Protectors) 134, 349, 480, 482-483
Pedal Pads 759
RecordCleaningPads .548
Sadiron Pads 823
Sanitary Pads 370-371, 483
Scissors Pads 803
Seat Pads, Auto 745
Shoulder Pads, Violin 509
Skate Strap Pads 767
Sliding Pads 763
Stair Pads 605
Window Seat Pads 613
Writing Pads 454

PAILS
Chamber Pails 727, 825
Coal Pails 823
Dinner Pails 826, 831
Folding Pails 780, 784
Milk Pails 832
Minnow Pails 780
Water Pails 822, 825-826, 828
Wringer Pails 824

PAINTERS' TOOLS AND SUPPLIES
935, 939-944, 951-953

PAINTS
Automobile Paint 950, 955
Barn Paint 946-947
Bronze Paint 942
Buggy Paint 950, 955
Children's Paints 495, 502
China Painting Outfits 502
Cold Water Paint 954
Concrete Paint 935
Enamel Paints 942, 950, 953, 955
Flat Finish Paint 950-951
Floor Paint 943, 947
House Paint 947-949
Implement Paint 946
Ironwork Paint 935
Paint Cleaners 951
Paint Colors, Oil 954
Paint Removers 954
Paint Undercoats 935, 943-944
Painting Outfits 940, 955
Porch Floor Paint 943, 947
Porch Furniture Paint 947, 953
Poultry Lice Paint. 476
Roof Paint 935
Wagon Paint 946

PAJAMAS
Boys' Pajamas 163
Children's Pajamas. 142
Men's Pajamas 249

PALETTES 502

PANELS
Lace Panels 568
Radio Panels 807

PANS
Baking Pans 826-829
Bed Pans 483, 727
Bread Pans 726, 826, 831
Bread Raising Pans 828
Cake Pans 826-827, 831
Dish Pans 826-828
Douche Pans 483
Drip Pans 826, 829
Dust Pans 824
Frying Pans 826, 829
Muffin Pans 826, 829, 831
Patty Pans 831
Pie Pans 827-828
Preserving Pans 826-830
Pudding Pans 827-828
Rinsing Pans 826-828
Roasting Pans 829
Sauce Pans 826-828

PANTS
Athletic Pants 769
Basket Ball Pants 769
Boys' Knee Length Pants 178-180
Boys' Long Pants. 269
Cowboys' Riding Pants 924
Diaper Pants 134
Football Pants 769
Golf Pants 765
Gymnasium Pants 769
Hunters' Pants 794
Little Fellows' Pants 178-179
Men's Pants 279, 282-288, 794
Oiled Slicker Pants. 287
Riding Pants 282
Running Pants 769
Work Pants 286-288

PAPER
Blue Print Paper 500
Building Paper 959, 969
Carbon Paper 500
Charcoal Paper .500, 502
Crepe Paper 450
Drawing Paper 500
Emery Paper 933
Loose Leaf Paper 454
Paper Lunch Sets 478
Parchment Paper 478
Photo Blotting Paper 555
Photo Paper 549, 552-553
Sandpaper 861
Shelf Paper 450
Toilet Paper 479
Tracing Paper 500
Typewriter Paper 500
Wall Paper 936-938
Waxed Paper 450
Writing Paper 451

PAPERHANGERS' TOOLS AND SUPPLIES 939-941, 954

PARCEL POST RATES 456

PARCHEESI 495

PARCHMENT 478

PARERS
Fruit Parers 830
Hoof Parers 873

PARIS GREEN 476

PARTITIONS
Stall Partitions 891

PASTEBOARDS AND TABLES—Write for Wall Paper Sample Book.

PASTES
Photo Paste 556
Poultry Lice Paste. 476
Shoe Paste 213
Soldering Paste 853
Tooth Paste 487
Wall Paper Paste 954
Wood Filler Paste. 942

PATCHES
Auto Tire Patches. 742
Auto Top Patches 745
Boot Top Patches 213
Inner Tube Patches 736, 742

PATENT BARLEY 482

PATTERNS
Fancywork Patterns 361
Paper Dress Patterns 292-293
Stamping Pattern Outfits 361

PAULINS 781

PEANUTS
Candied Peanuts 492
Salted Peanuts 492

PEDALS
Bicycle Pedals 737

PEDESTALS
Parlor Pedestals 624
Stove Leg Pedestals 689

PEDOMETERS 784

PEGS
Cello Pegs 508
Iron Tent Pegs 783
Quoit Pegs 769
Violin Pegs 509

PEGWOOD 471

PENCILS
Carpenters' Pencils 862
Charcoal Pencils 502
Crayon Pencils 454
Drawing Pencils 500
Eyebrow Pencils 489
Indelible Pencils 453
Ink Pencils 453
Lead Pencils 453, 500
Lead Pencils, Gold, Silver 391, 409-410
Magazine Pencils 391, 409-410, 453
Styptic Pencils 802
White Pencils .500, 556

PENDANTS
Electric Pendants 731
Gas Pendants 731

PENKNIVES 798
Gold Penknives, Sets 388, 406, 409

PENS
Fountain Pens, High Grade 410
Fountain Pens, Low Priced 452
Pen and Pencil Outfits 410, 454
Steel Pens 450

PEPPERMINT 484

PEPSIN 484

PERCALES 318-319, 334

PERCOLATORS 810, 826-827, 829

PERFUMES 490
Breath Perfumes 492

PERGOLAS—Write for Mill Work Catalog.

PESSARIES 473

PETTICOATS
Girls' Petticoats 136, 143-145, 160-162
Women's Petticoats 146-149, 154

PHONOGRAPHS AND SUPPLIES 535-548
Phonograph Exercises 443
Phonograph Toys 548
Phono-Bretto (Words of Records) 544

PHONO-HARPS 513

PHOSPHATES
Acid Phosphate 476
Phosphate of Soda 484
Phosphate Rock 476

PHOTO SUPPLIES 549-556
Photo Developing, Printing and Enlarging 549

PIANOS 527-534

PICCOLOS 519

PICKETS
Fence Pickets, Wood 964

PICKS
Autoharp Picks 513
Dirt Picks 843
Guitar Picks 514
Mandolin Picks 514
Ukulele Picks 514
Zither Picks 513

PICTURES
Framed Pictures 479

PILLOW BLOCKS 855

PILLOWCASES 319, 344-345
Babies' Pillowcases .134
Embroidered Pillowcases 362
Stamped Pillowcases .362

PILLOWCASING 329, 331

PILLOWS
Babies' Pillows .134, 667
Bed Pillows 657
Boat Pillows 780
Sofa Pillows 363, 657

PILLS 484
Farm Stock Pills 477

PINCERS
Carpenters' Pincers 871
Horseshoers' Pincers 873

PINCUSHIONS 368-369, 422, 491

PINE TAR 957

PINS
Babies' Pins 373, 392, 409
Baby Pins 392, 397, 408-409
Bar Pins 373, 376, 389, 391, 396-397, 408-409
Birthday Scarf Pins .389
Braid Pins 130
Breast Pins 373, 376, 389, 391, 396-397, 407-409
Closet Pins 367
Clothes Pins 823
Collar Pins, Men's 373, 405, 409
Collar Pins, Women's .392, 397, 408-409
Common Pins 368
Cotter Pins 747, 846
Cuff Pins 392, 397, 408-409
Drapery Pins 586
Emblem Pins 373, 402, 407
End Pins 509, 514
Guitar Bridge Pins 514
Hairpins 131
Hair Trimming Pins 130-131, 397
Hat Trimming Pins 106-107
Horse Blanket Pins .928
Husking Pins 839
Knitting Pins 365
Mourning Pins 369
Rolling Pins 831
Safety Pins 368
Scarf Pins 373, 376, 389, 391, 396-397, 406, 409
Tuning Pins 513
Waist Pins 392, 397, 408-409

PIPE
Conductor Pipe and Fittings 959
Furnace Smoke Pipe and Fittings 692
Gas and Water Pipe and Fittings 707
Lead Pipe and Fittings 704
Soil Pipe and Fittings 704
Stovepipe 689, 785

PIPES
Blow Pipes 471
Sprayer Pipes 903
Tobacco Pipes 259
Tuning Pipes 509, 514
Uterine Supporter Pipes 473

PIPETTES 832

PIPING
Chain Trace Piping 922

PISTOLS 792

PISTONS
Engine Pistons, Auto 755
Piston Rings, Auto 752, 755

PITCHERS
Communion Pitchers .418
Cream Pitchers, Sets, China 714, 716-721
Cream Pitchers, Sets, Silver 420-421
Milk Pitchers 714, 716-721
Pitcher and Tumbler Sets 722, 724-725
Pitcher and Wash Bowl Sets 727, 825
Syrup Pitchers 826
Syrup Pitchers, Cut Glass 724
Syrup Pitchers, Silver 420-421
Water Pitchers 826
Water Pitchers, Silver 420-421

PITCHFORKS 844

PITHWOOD 471

PLAIDS
300, 304, 314, 316-322

PLAITERS 369

PLAITINGS 360, 369

PLANES 860

PLANKS
Extension Planks 939
Lumber 964-965

PLANTERS
Corn Planters and Attachments 883
Corn Planters, Hand 842
Corn Planter Repair Parts 876
Cotton Planters 883
Potato Planters 886
Potato Planters, Hand 842

PLANTS
Artificial Plants 723

PLASTER
Plaster Finish 969
Wall Plaster 969

PLASTERBOARD 969

PLASTERS
Adhesive Plasters 478, 483
Belladonna Plasters 484
Billiard Table Plasters 797
Corn Plasters 472
Toupee Plasters 129

PLATEAUS
Mirror Plateaus 723

PLATES
Babies' Plates 418, 482, 494, 723
Baseball Shoe Plates 762
Bread and Butter Plates 714, 716-721
Breakfast Plates 714, 716-721
Cake Plates 724, 726
Ceiling Plates—Write for Heating Catalog.
Collection Plates 418
Communion Plates 418
Dinner Plates 714-721
Dry Plates, Photo 554
Floor Plates—Write for Heating Catalog.
Food Chopper Plates .830
Fruit Plates 724
Gas Hot Plates 689
Guard Plates, Mower and Binder 876
Heel Plates 212
Meat Plates (Platters) 714, 716-721
Mending Plates 846
Oar Plates 780
Pie Plates 714, 716-721, 723, 725
Pie Plates (Pans) 725-726, 827-828
Pie Plates (Pans), Silver 420-421
Push Plates 834
Sandwich Plates .420-421
Screw Plates 870
Squeegee Plates 555
Soup Plates 714, 716-721
Step Plates, Auto 744
Tea Plates 714, 716-721
Tennis Court Plates 764
Toe Plates 212

PLATFORMS
Ladder Platforms 939
Striking Bag Platforms and Attachments 770

PLATTERS 714, 716-721

PLAYER PIANOS
527, 529-530, 532-534

PLAYGROUND EQUIPMENT 767
—Or Write for Sporting Goods Catalog.

PLIERS 871
Cutting Pliers 871
Fencing Pliers 843
Jewelers' Pliers 471
Pipe Pliers 871
Tire Chain Pliers 743

PLOWS
Farm Plows 878-880
Light Plows 842
Plow Attachments 877-878
Plow Repairs 876-880
Plow Riding Attachments 878
Tobacco Plows 842
Tractor Plows 880

PLOWSHARES 842, 877-880

PLUGS
Cast Iron Plugs 707
Electric Plugs 731
Rubber Plugs, Basin .934
Screw Cap Plugs 751
Spark Plugs 750, 757
Tire Repair Plugs 736

PLUMB AND LEVEL 862

PLUMBAGO
Plumbago Grease 957

PLUMBING GOODS 695-707, 934

PLUMES 106-107

PLUNGERS
Water Closet Plungers 934

PLUSH
Drapery Plush 561
Robe Lining Plush—Write for Harness Catalog.

PNEUMATIC WATER SUPPLY SYSTEMS 695

POCKETBOOKS 127, 259

POCKETS
Canvas Wall Pockets 784
Pool Table Pockets .797
Razor Pockets 802
Saddle Pockets 924

POINTERS
Pencil Pointers 450
Spoke Pointers 864

POINTS
Cue Points 797
Drill Points 865, 869
Drive Well Points 710
Switch Lever Points .808
Tooth Saw Points 852

POISONS
Fly Poisons 476
Rat Poisons 478

POKES
Cow Pokes 838

POLES
Buggy Poles 849-850
Clothes Poles 823
Curtain Poles 587
Sprayer Poles 903
Tennis Net Poles 764
Vaulting Poles 769
Wagon Poles 849-850

POLISH
Aluminum Polish 826
Auto Polish 753, 955
Fingernail Polish 490
Floor Polish 942-943
Furniture Polish 942-943
Piano Polish 942
Shoe Polish, Outfits 213
Silver Polish 411
Squeegee Plate Polish 555
Stove Polish 689

POLISHING HEADS 853

POLYCHROME GOODS 627, 723

POMADES 486

PONCHOS 784

PONGEES 294, 298-301, 305, 315, 327

POPGUNS 496

POPLINS 304-305, 325

PORCELAINS
Liquid Porcelains 953
Spark Plug Porcelains 757

PORCH
Porch Awnings 781
Porch Curtains 781
Porch Floor Coverings 596
Porch Furniture 618-621, 784-787
Porch Material (Mill Work) 967

PORK RIND 776-777

PORTABLE HOUSES 974

PORTFOLIOS 525

PORTIERES 564-566

POSTAGE RATES—(PARCEL) 456

POST BLOCKS 855

POST CARDS
Birth Announcement Post Cards 482
Birthday Post Cards .478
Developing Post Cards 552

POSTS
Cedar Fence Posts—Write for Lumber List.
Radio Binding Posts .808
Steel Fence Posts 898

POTHOOKS 939

POTS
Bean Pots 421, 726
Chamber Pots 727, 825
Coffee Pots 826-829
Coffee Pots (Percolators) 810, 826-827, 829
Coffee Pots, Silver 420
Iron Cooking Pots 829
Melting Pots 704
Mustard Pots 421, 714
Tea Pots 723, 726, 826-829
Tea Pots, Silver 420

POULTRY SUPPLIES
476, 839, 893, 898-901
Poultry Houses 974
—Or Write for Book of Barns.

POWDERS
Bronze Powders 502, 942
Cleaning Powders 478
Closet Bowl Powders 478
Developing Powders 553
Face Powders 488
Foot Powders 472
Hog Powders 477
Insect Powders .476, 478
Intensifying Powders 553
Licorice Powders 484
Plant Powders 476
Poultry Lice Powders 476
Reducing Powders 553
Sachet Powders 490
Seidlitz Powders 484
Shaving Powders 802
Talcum Powders 482, 488
Tire Powders 742
Toning Powders 553
Tooth Powders 487

POWER DRIVE 909

PREPARATIONS
Chapped Hand Preparations 484, 489
Complexion Preparations 487-490
Dental Preparations 487
Hair Preparations 486, 490
Manicure Preparations 490
Mosquito Preparations 775
Rust Preparations 689, 793

PRESERVERS
Egg Preservers 478
Fence Post Preservers 935
Leather Preservers 213, 957
Life Preservers 780
Roof Preservers 935, 946
Shingle Preservers 935, 946
Tennis Gut Preservers 764
Wood Preservers 935, 946

PRESSER FEET 312

PRESSES
Cider Presses—Write for Farm Implement Catalog.
Fruit Presses .... 830-831 —Or Write for Farm Implement Catalog.
Hay Presses—Write for Farm Implement Catalog.
Honey Box Presses 895
Initial Letter Presses 450
Jelly Presses 830
Lard Presses 830
Racket Presses 764
Vegetable Presses 830-831
Wine Presses .... 830 —Or Write for Farm Implement Catalog.

PRINTING, DEVELOPING AND ENLARGING 549

PRINTING OUTFITS
Photo Printing Outfits 553, 555
Rubber Type Outfits .500

PRINTS
Dress Prints 316, 318

PROPS
Clothesline Props 823

PROTECTORS
Auto Seat Protectors 759
Bunion Protectors 472
Catchers' Protectors .763
Dust Protectors 426
Lap Protectors 134-135
Line Protectors—Write for Electric Goods Catalog.
Mattress Protectors 134-135, 349, 480, 482-483
Stovepipe Hole Protectors 783, 785
Table Protectors 338

PROTRACTORS 500, 869

PRUNERS 840
Pruning Knives .799, 805

PUFFS
Powder Puffs 482, 488

PULLERS
Nail Pullers 863
Stump Pullers and Supplies—Write for Farm Implement Catalog.

PULLEYS
Awning Pulleys 845
Belt Pulleys 855
Corn Sheller Pulleys 888
Cream Separator Pulleys 909
Drag Saw Pulleys 819
Friction Clutch Pulleys 819, 909
Hay Carrier Pulleys 844
Power Pulleys, Auto 854
Rope Pulleys 844
Sash Pulleys 834
Screw Pulleys 845
Tackle Pulleys .843, 845

PULLEY COVERS 854

PULLS
Auto Wheel Pulls 754
Chest Pulls 766
Door Pulls 834, 836
Drawer Pulls 834
Glass Pulls 834
Window Shade Pulls .585

PULVERIZERS (SOIL) 882

PUMICE STONE 954

PUMPS
Babies' Pumps 196-197
Barrel Pumps 710, 902, 957
Breast Pumps 480
Chain Bucket Pumps 709
Children's Pumps 197-198, 214
Cistern Pumps 708, 710
Closet Cleaning Pumps 934
Electric Pumps 695
Engine Pumping Outfits 818
Force Pumps 708-710
Girls' Pumps 183, 185, 198, 214
Hydraulic Ram Pumps 709
Inflater Pumps 771
Oil Tank Pumps 957
Player Piano Pumps .504
Pump Jacks 818
Pump Standards 708
Pumping Outfits, Water Supply Systems 695
Sand Pumps 710
Sink Pumps and Outfits 702, 708
Spraying Pumps .902-903
Tank Pumps 710
Tire Pumps 736, 744
Well Pumps 710
Windmill Pumps 708-709
Women's Pumps 183-186, 188-189, 198, 214

PUNCHES
Center Punches 869
Drive Punches 847
Hand Punches 872
Hollow Punches 847
Leather Punches 847, 930
Pin Punches 869
Poultry Punches 839
Prick Punches 869
Spring Punches 847
Stock Marking Punches 838

PUNCHING BAGS AND ATTACHMENTS 770

PUPPY CAKES 795

PURSES
Key Purses 259
Knife Purses 798
Men's Purses 259
Razor Purses 802
Women's Purses 127

PUSH BUTTONS, PLATES 813, 834

PUTTEES 215

PUTTY
Glaziers' Putty 954
Stove Putty 689
Wafer Putty 942

See Page 557 for Measuring Instructions.

Q

QUICKSHIFTERS..848
QUILTS 347-350
QUININE 484
QUIRTS 930
QUOITS 769

R

RACKETS 764
Racket Restringing.764
RACKS
Band Instrument Music Racks.....523
Book Racks....618-620
Clothes Drying Racks........823
Coat and Hat Racks 624, 627
Hall Racks....624, 627
Negative Drying Racks.......552
Oil Tank Foot Rests and Racks.......894
Parchment Paper Racks..........478
Towel Racks.......934
RADIATOR COMPOUNDS ...752
RADIATORS
Auto Radiators and Accessories..752, 754
Hot Water Radiators 694
Steam Radiators.....694
Stovepipe Radiators—Write for Stove Catalog.
RADIO APPARATUS 807-809
—Or Write for Radio Catalog.
RAFTERS 965
RAILS
Auto Robe Rails...759
Chair Rails (Finished)—Write for Wall Paper Sample Book.
Harrow Guard Rails............882
Plate Rails........967
Plate Rails (Finished)—Write for Wall Paper Sample Book.
Porch Rails.......967
RAINCAPES 77
RAINCOATS
Boys' Raincoats....166
Girls' Raincoats.... 77
Men's Raincoats 264-265
Misses' Raincoats.......68-69, 77
Women's Raincoats 68-69, 77
RAKES
Garden Rakes......842
Hay Rakes....887, 890
Lawn Rakes........841
RAMS
Hydraulic Rams....709
RANGES
Coal and Wood Ranges......675-677, 680-683
Combination Ranges 678-679
Gas Ranges.....689-691
Oil Ranges.....686-688
RASPS
Horse Rasps.......873
Shoemakers' Rasps.212
Veterinary Rasps...477
Wood Rasps 872
RATCHETS
Drill Ratchets......705
RATES
Express Rates......457
Freight Rates......458
Parcel Post Rates..456
RATTLES
Babies' Rattles.482, 494
RAWHIDE 855
RAW PHOSPHATE ROCK 476
RAZORS 799-800
Razor Outfits.......800
REACHES
Wagon Reaches....850
READY CUT HOUSES 972-973
REAMERS
Bushing Reamers..756
Burring Reamers...705
Iron Reamers..864, 869
Pipe Reamers.......705
Wood Reamers......864
REBABBITTING JIGS 755
RECEIVERS
Ash Receivers......422
Hair Receivers.422, 491
Radio Receivers.807-808
Telephone Receivers.814
RECEPTACLES
Electric Plug Receptacles 731
—Or Write for Electric Goods Catalog.
RECIPES 450
RECORDS
Phono-Bretto (Words of Records).......544
Phonograph Records 544-547
Phonograph Record Exercises 443
RECTIFIERS 809
REDUCERS
Double Chin Reducers 480
Pipe Reducers......707
REDUCING GARMENTS 112-113, 115-119, 122
REED INSTRUMENTS 515-519
REEDS
Clarinet Reeds......523
Saxophone Reeds....523
REELS
Chalk Line Reels..864
Clothes Line Reels..823
Eyeglass Reels......405
Fishing Reels.......773
Hose Reels..........840
Tennis Net Reels...764
REFLECTORS
Auto Lamp Reflectors 756
REFRIGERATORS 638-639
REGISTERS
Tallying Registers.869
Umpires' Registers..763
Ventilating Registers 823
REGULATORS
Auto Lamp Regulators 756
Furnace Regulators.694
Motor Oil Regulators.755
REINS 921
Toy Reins 494
RELINERS
Tire Reliners..739, 743
REMEDIES
Dandruff Remedies.486
Family Remedies....484
Foot Remedies......472
Poultry Remedies...476
Stock Remedies.....477
RENNET 478
REPAIRERS
Rim Repairers......848
Spoke Repairers....848
REPAIRS
—See Name of Article for Which Repair Is Wanted.
REPRODUCERS
Phonograph Reproducers 548
RESPIRATORS 426
RESTS
Chin Rests.........509
Oil Tank Foot Rests and Racks.........894
Soil Pipe Rests....704
Shoulder Rests.....509
Stove Leg Rests....689
RETOUCHING OUTFITS 555
REVOLVERS 792
Revolver Magazines.792
RHEOSTATS 807
RIBBONS 358-359, 369
Locket Ribbons.390, 409
Typewriter Ribbons 500, 674
Watch Ribbons.....358, 378, 387, 390, 409
RICERS 831
RIDING HABIT OUTFITS, ETC. 4, 39
Riding Wear, Men's.282
RIFLES 789
Air Rifles.........792
Rifle Chambers.....793
Rifle Magazines....789
RIMS
Auto Rims and Rim Parts—Write for Auto Supply Catalog.
Bicycle Rims.......737
Buggy Rims.........850
Truck Rims.........850
Wagon Rims.........850
RINGERS
Hog Ringers........838
Telephone Ringers—Write for Electric Goods Catalog.
RINGS
Babies' Rings.......373, 392, 409
Birthday Rings.......389, 392, 398
Bone Rings.........361
Boys' Rings........401
Bull Rings.........838
Cameo Rings......373, 398, 400-401
Curtain Pole Rings.586
Diamond Rings.......374-376, 400
Emblem Rings 376, 391, 398-399, 402
Fruit Jar Rings....478
Girls' Rings.......373, 389, 400
Grinding Rings.....888
Harness Rings.926-928
Hitching Rings.....836
Hog Rings..........838
Initial Rings......373, 376, 400-401
Lighting Rings.....687
Men's Rings.......374, 389, 391, 396, 398, 400-402
Napkin Rings.......421
Neckyoke Rings.....848
Piston Rings..752, 755
Seal Rings.......373, 392, 401
Separator Bowl Rings..............905
Silver Rings..390, 398
Singletree Rings...848
Split Rings........775
Swinging Rings.....766
Teething Rings.482, 494
Wedding Rings......400
Window Shade Rings..............585
Women's Rings.......373-376, 389, 391, 396, 398-401
RINSERS 826-828
RIVETING MACHINES 847
Rivet Sets..........847
RIVETS
Brake Band Rivets.753
Clinch Rivets......847
Copper Rivets and Burrs..............847
Hame Rivets........927
Oval Head Rivets...846
Tubular Rivets.....847
Wagon Box Rivets.846
ROAD MACHINERY 888
ROASTERS
Double Roasters....829
ROBES
Auto or Buggy Robes 929
Baby Carriage Robes 134-135, 347
Bath Robes.........249
Robe Borders—Write for Harness Catalog.
ROCKERS
Bedroom Rockers.......617, 644-651
Children's Rockers 668-669
Fiber Rockers.......618-619, 621, 669
Morris Rockers.....617
Parlor Rockers.607A-608, 610-611, 614-621
Porch Rockers..618-621
Reed Rockers..620, 669
Sewing Rockers.....616
RODS
Casement Rods......835
Cleaning Rods, Gun, Rifle 793
Connecting Rods, Auto 755
Curtain Rods.......587
Drum Rods..........523
Fishing Rods and Fittings......772-777
Ground Rods........814
Guy Rods...........814
Lightning Rods and Fittings..........888
Portiere Rods......587
Pump Rods..........710
Sash Rods..........587
Stair Rods.........586
Wagon Box Rods....848
ROLLER SKATES AND SUPPLIES ..767
ROLLERS
Barn Door Rollers..836
Graining Rollers...943
Land Rollers.......882
Print Rollers......555
Spread Rollers.....927
Wall Paper Rollers—Write for Wall Paper Sample Book.
Window Shade Rollers 586
ROLLING PINS 831
ROLLS
Bolster Rolls.......657
Music Rolls (Cases).525
Organ Rolls........515
Player Piano Rolls 503-504
Razor Rolls........802
Skate Rolls........767
ROMPERS 140, 168-169
Children's Rompers (Ready to Sew)...372
Romper Cloths.316, 319
ROOFING
Asphalt Roofing.......960-963
Composition Roofing 963
Shingle Roll Roofing 961
Slate Surfaced Roofing 960, 962
Steel Roofing.......958
ROOTBEER 478
ROPES
Anti-Rust Ropes....793
Artificial Silk Rope.364
Clothes Line Ropes 823
Halter Ropes...920-921
Hay Carrier Ropes 844-845
Hammock Ropes.....786
Iron Ropes.........845
Jump Ropes.........497
Lariat Ropes.......845
Manila Ropes..845, 854
Sash Ropes.........834
Steel Ropes.......844-845
Striking Bag Ropes.770
Transmission Ropes 845, 854
Wire Ropes....844-845
ROSARIES ...390, 445
ROSETTES
Bridle Rosettes....928
ROSIN
Violin Bow Rosin..509
ROUGE 489
ROWBOATS 780
RUBBER
Tire Patching Rubber 736, 742
RUBBERS
Band Rubbers.......453
Bobbin Winder Rubbers 312
Chair Rubbers......483
Crutch Rubbers.....483
Jar Rubbers........478
Pedal Rubbers..737, 759
Rubber Footwear 216-219
RUCHINGS 360
RUFFLERS 312
RUGS
Auto Rugs..........758
Bath Rugs..........593
Carpet Rugs......588, 592-593, 596-604
Fiber Rugs 592, 596-597
Grass Rugs.........596
Hall Runners.591, 593
Linoleum Rugs 594, 596
Made to Order Rugs.588
Rag Rugs...........593
Remnant Rugs.......588
Rug Borders........594
Stair Runners.591, 593
Stove Rugs.........594
RULES
Architects' Rules...862
Blacksmiths' Rules.869
Caliper Rules......862
Carpenters' Rules...862
Rules for Games—See "Books."
Slide Rules....501, 862
Steel Rules........869
Zigzag Rules.......862
RUNABOUTS..816-817
RUNGS
Rung Holders.......939
RUNNERS
Asbestos Joint Runners 704
Hall and Stair Runners.......591, 593
Table Runners.....342, 362-363, 560, 568
RUST PREPARATIONS......689, 793

S

SACHETS 490
SACQUES
Babies' Sacques.....141
Women's Sacques....57
SADDLES
Bicycle Saddles.....737
Harness Saddles....922
Riding Saddles.916-918
Saddle and Breeching922
SADIRONS 823
Electric Sadirons....810
Sadiron Holders.....823
SAFES
Fireproof Safes.....673
Kitchen Safes......639
SAL AMMONIAC 813, 853
SALT SETS 419, 421, 723-724, 826
SALTS
Medicinal Salts, Household 484
Medicinal Salts, Stock 477
SALVES 484
SANDALS
Babies' Sandals 196-197
Boys' Sandals......195
Children's Sandals 195-197
Girls' Sandals 195, 198
Women's Sandals 195, 198
SANDBOARDS 850
SANDPAPER 861
SASH 966, 968
Sash Hardware.834-835
SASHES
Ribbon Sashes......359
SASSAFRAS 484
SATCHELS
Book Satchels......454
Music Satchels.....524
SATEENS
313, 324-325, 333
SATINS 295-299
SAUCEBOATS 714, 716-721
SAUCEPANS ..826-828
SAUTOIRS 376
SAVINGS BANKS.495
SAWS
Back Saws..........859
Band Saws....852, 866
Bracket Saws.......859
Buck Saws..........857
Butchers' Saws.....806
Circular Saws......852
Compass Saws.......858
Coping Saws........859
Crosscut Saws......857
Dehorning Saws.....838
Double Edge Saws..840
Drag Saw Outfits...819
Felling Saws.......857
Hack Saws..........873
Hand Saws..........858
Ice Saws...........857
Jewelers' Saws.....471
Keyhole Saws.......858
Kitchen Saws.......806
Mill Saws..........852
Miter Saws.........859
Nail Cutting Saws..858
Panel Saws.........858
Power Saws 852, 866, 868
Pruning Saws.......840
Rip Saws......852, 858
Saw Frames—Write for Farm Implement Catalog.
Saw Machines.866, 868
Saw Rigs......819-820
Saw Sets......852, 859
Scroll Saws........859
Tooth Saws.........852
Turning Saws.......859
SAXOPHONES 521
SCALDERS
Hog Scalding Tanks.893
SCALES
Beam Scales........833
Counter Scales.....833
Fishermen's Scales..775
Household Scales...833
Photographers' Scales 552
Pit and Platform Scales 833
Triangular Scales...501
SCALERS
Fish Scalers.......775
SCARFING
Dresser Scarfing.342, 361
SCARFS
Dresser Scarfs.......342, 362-363
Fur Scarfs..........43
Piano Scarfs........568
Scarf and Cap Sets... 91
Stamped Scarfs.....362
Table Scarfs......342, 362-363, 560, 568
Women's Scarfs 146-147, 357
SCENTS
Animal Scents......796
Toilet Perfumes....490
SCHOOL FURNITURE...637, 672-673
SCISSORS
Barbers' Scissors...803
Buttonhole Scissors.803
Cuticle Scissors 490-491
Dressmakers' Scissors..............803
Embroidery Scissors.803
Nail Scissors 490-491, 803
Paperhangers' Scissors—Write for Wall Paper Sample Book.
Pocket Scissors....803
Sewing Scissors....803
Tailors' Scissors...803
Upholsterers' Scissors 803
SCOOPS
Flour Scoops.......831
Furnace Scoops.....843
Grain Scoops..840, 843
Vegetable Scoops...843
SCORE BOOKS 762, 771
SCRAPERS
Bearing Scrapers...747
Cabinet Scrapers...861
Carbon Scrapers....747
Ditching Scrapers...888
Floor Scrapers.....861
Flue Scrapers......705
Foot Scrapers......837
Harrow Scrapers.........880-881
Hog Scrapers.......806
Paint Scrapers.....954
Road Scrapers......888
SCREENING
Wire Cloth Screening 835
SCREENS
Door and Window Screens 968
Folding Screens....626
Screen Door Hardware 835
Screen Frames......835
SCREWDRIVERS
Auto Screwdrivers..747
Automatic Screwdrivers and Attachments 871
Bicycle Screwdrivers 738
Jewelers' Screwdrivers 471
Mechanics' Screwdrivers 871
Ratchet Screwdrivers 871
Screwdriver Bits....864
Sewing Machine Screwdrivers 312
SCREWEYES 837
SCREWHOOKS ...837
SCREWPLATES ...870
SCREWS
Bench Screws.......864
Cap Screws.........847
Drain Pipe Screws..707
Harness Pad Screws 927
Jack Screws........845
Saw Handle Screws.859
Set Screws.........847
Skein Screws.......847
Stop Bead Screws...834
Wood Screws...833, 835
SCRIMS...558, 562-563
SCYTHES 841
SEARCHLIGHTS
Auto Searchlights..751
Bicycle Searchlights 737
Hunters' Searchlights 796
SEATS
Bathtub Seats......934
Boat Seats.........780
Chair Seats........837
Children's Toilet Seats ...494, 668-669
Folding Seats, Auto 746, 851
School Seats.......673
Wagon Seats........815
Water Closet Seats.698
SECTIONAL HOUSES 974
SECTIONS
Binder Sections....876
Honey Sections.....895
Mower Sections.876, 887
Sprouter Sections...901
Stall End Sections..891
SEEDERS
Bag Seeders........842
Broadcast Seeders...883
Drill Seeders..883, 886
Endgate Seeders....883
Grape Seeders......830
SEINES 778
SENNA 484
SEPARATORS
Cream Separators, ...904-909
Cream Separators, Dilution 832
Cream Separators, Double Can.........832
Cream Separators, Electric 909
Cream Separator Repairs 905
Grain Separators and Attachments.889
SEPTIC TANKS ...695
SERGES 303, 315
SERVERS
Cake Servers......412-414, 417, 419
Cheese Servers.....419
Cold Meat Servers 413-419
Jelly Servers...412-414
Pie Servers......412-414, 417, 419
Tomato Servers.....412-413, 417, 419
SERVIETTES 370-371, 483
SETS
Lathe Sets.........853
Nail Sets..........863
Rivet Sets.........847
Saw Sets......852, 859
SETTEES
Fiber Settees 618-619, 621
Folding Settees....787
Lawn Settees 618-621, 787
Parlor Settees 607D, 608
Porch Settees 618-621, 787
Reed Settees.......620
SEWAGE DISPOSAL SYSTEMS... 695
SEWING MACHINES 306-312
Sewing Machine Supplies and Repairs. 312
Sewing Sets........422
SHADES
Electric Lamp Shades 731
Eye Shades.........764
Gas Lamp Shades...729
Oil Lamp Shades...729
Porch Shades.......781
Window Shades 583-586
SHAFTING
Line Shafting Outfits..............818
Steel Shafting.....855
SHAFTS
Axle Shafts, Auto 753, 760
Buggy Shafts..849-850
Crank Shafts, Auto 755
Drive Shafts, Auto 753, 760
Shaft Ends.........848
Speedometer Shafts and Fittings......753
Wagon Shafts......850
SHAKERS
Milk Shakers.......826
Salt and Pepper Shakers 411, 419, 421, 723-724, 826
Sugar Shakers......421
SHAMPOO ACCESSORIES.486-487, 803
SHAMS 362-363
SHARES
Lister Shares..876, 886
Plowshares 842, 877-880
SHARPENERS
Harrow Disc Sharpeners 876
Knife Sharpeners ...212, 798, 803, 805
Pencil Sharpeners..450
Razor Sharpeners 800-801
Scissors Sharpeners 803
SHAVING ACCESSORIES 422, 485, 490, 799-802
SHAVINGS
Steel Shavings 861, 954
SHAWLETTES 146-147
SHAWLS
Babies' Shawls.....147
Women's Shawls....147
SHEARING MACHINES 933
SHEARS
Barbers' Shears....803
Dressmakers' Shears 803
Grass Shears.......841
Hedge Shears.......840
Hoof Shears........873
Horse Shears.......929
Pruning Shears.....840
Sheep Shears.......929
Tailors' Shears.....803
Tinners' Shears....853
Upholsterers' Shears 803
Wall Paper Shears—Write for Wall Paper Sample Book.
SHEATHING..959, 964
SHEATHS
Ax Sheaths.........793
Carbine Sheaths....793
Rifle Sheaths......793
SHEETINGS
Bed Sheetings 327, 329, 331
Waterproof Sheetings 134-135, 334, 480, 482-483
SHEET MUSIC 504-505
SHEET PLASTER. 969
SHEETS
Bed Sheets....344-345
Bed Sheets, Waterproof ..480, 482-483
Campers' Waterproof Sheets.....784
Crib Sheets........134
Crib Sheets, Waterproof 134-135
Stable Sheets......933
SHELLAC 954
Artists' Shellac....502
Shellac Substitute..954
SHELLERS
Corn Shellers and Attachments 839, 888

SHELLS
Gun Shells, Loaded 790
Sugar Shells 411-419, 805
Telephone Receiver Shells 814

SHELVES
Clock Shelves 425
Glass Shelves 934

SHIELDS
Buckle Shields 928
Buggy Storm Shields 851
Bunion Shields 472
Corn Shields 472
Dress Shields 371
Erasing Shields 500
Nursing Shields 156
Sun Shields, Auto 746
Wind Shields, Auto 815

SHINERS 777

SHINGLES
Slate Surfaced Shingles 960
Wood Shingles 964

SHIPLAP 964

SHIRTINGS 294, 298-299, 305, 316-317

SHIRTS
Athletic Shirts 769
Babies' Shirts 237
Boys' Shirts 164-165
Flannel Shirts 244-245
Khaki Shirts, Women's, Girls' 24, 51
Men's Shirts 244-248, 250-251
Night Shirts 163, 249
Work Shirts 165, 250-251

SHIRTWAISTS 49, 51-55

SHOCK ABSORBERS 760

SHOEMAKERS' SUPPLIES 212-213

SHOES
Army Style Shoes 202, 206
Athletic Shoes 766
Auto Brake Shoes 760
Auto Tire Shoes 742
Babies' Shoes 196-197
Baseball Shoes 766
Basket Ball Shoes 766
Bathing Shoes 768
Bowling Shoes 766
Boxing Shoes 766
Boys' Shoes 195, 202-203, 206-210, 214-215
Camp Shoes 766
Canvas Shoes, Men's, Boys' 214-215
Canvas Shoes, Women's, Girls' 184-185, 198, 214-215
Children's Shoes 195-198
Congress Shoes, Men's 200, 206
Corn Planter Shoes 876
Fencing Shoes 766
Girls' Shoes 183, 195, 198
Golf Shoes 766
Gymnasium Shoes 766
Handball Shoes 766
Indoor Ball Shoes 766
Jumping Shoes 766
Men's Shoes 200-207, 210, 214-215
Running Shoes 766
Shoe Polishing Outfit 213
Shoe Size Chart 557
Small Boys' Shoes 195, 202-203, 206-210, 214-215
Tennis Shoes 766
Women's Low Shoes 181-193, 195, 198, 214-215
Women's Shoes 182-183, 186-195, 198, 214-215
Work Shoes, Men's, Boys' 202-203, 210, 214
Work Shoes, Women's 194
Wrestling Shoes 766

SHORTENERS
Fishing Rod Shorteners 775

SHOT
Air Rifle Shot 792
Split Shot 775

SHOTGUNS 788

SHOTS
Outdoor Shots 769

SHOVELS 843
Coal Shovels 843
Cultivator Shovels 876, 884
Drain Shovels 843
Grain Shovels 843
Snow Shovels 823
Stove Shovels 843

SHOWERS
Bath Showers 934
Electric Showers 731

SHRINKERS
Tire Shrinkers 875

SHUTOFFS
Brass Shutoffs 903

SHUTTLES
Sewing Machine Shuttles 312
Tatting Shuttles 361

SICKLES
Hand Sickles 841
Mower Sickles 876

SIDEBOARDS 628-633

SIDING
House or Barn Siding 964
Metal Siding 958

SIFTERS
Flour Sifters 831

SIGHTS
Gun Sights 792
Level Sights 862
Rifle Sights 790

SIGNALS
Auto Stop Signals 751

SILENCERS
Steering Rod Silencers 760

SILICON 809

SILKOLINES 558

SILKS
Crochet Silks 365
Dress Silks 294-301, 305
Embroidery Silks 364-365
Knitting Silks 365
Sewing Silks 364

SILLS
Window Sills 967

SILOS—Write for Lumber List.

SILVERTONES (Phonographs) 535-543

SILVERWARE 411-422
Pearl Handle Silverware 419
Sterling Silverware 419

SINGLETREES 849-850

SINKERS
Line Sinkers 775

SINKS AND FITTINGS 700-701

SIZING
Burlap Sizing 954
Kalsomine Sizing 952
Painters' Sizing 951
Wallpaper Sizing 954

SKATES 767

SKEINS
Wagon Skeins 848

SKILLETS 826, 829

SKIRTINGS 296, 299, 301-305

SKIRTS
Babies' Skirts 136-137
Girls' Skirts 24
Misses' Skirts 42
Riding Skirts 39
Women's Skirts 36-42, 46-49

SLATS
Steel Bed Slats 663

SLEDGES
Blacksmiths' Sledges 873
Stone Sledges 841

SLEEPING OUTFITS
Babies' Sleeping Bags 137, 141
Boys' Sleeping Outfits 163
Campers' Sleeping Bags, Waterproof 784
Children's Sleeping Suits 142, 160
Girls' Sleeping Suits 142, 160

SLEEVES
Ratchet Drill Sleeves 705

SLICERS
Fruit Slicers 831
Vegetable Slicers 831

SLICKERS
Oiled Slickers 166, 267, 794

SLIDES
Harness Slides 927
Playground Slides 767
Shuttle Slides 312

SLINGS
Wagon Slings 844

SLIPPERS
Babies' Slippers 196-197
Bathing Slippers 768
Boudoir Slippers 199
Boys' Slippers 214
Felt Slippers 199
Girls' Slippers 183, 195, 198, 214
Men's Slippers 199
Women's House Slippers 186, 199, 214
Women's Slippers (Low Shoes) 181, 183-193, 195, 198, 214

SLIPS
Babies' Pillow Slips 134
Babies' Slips 136-137
Bed Pillow Slips 319, 344-345, 362
Princess Slips 150-151, 154, 162
Sofa Pillow Slips 363
Stamped Pillow Slips 362

SLITTERS
Cow Teat Slitters 477

SLUGSHOT 476

SMOKEHOUSES 893

SMOKERS' ARTICLES 259, 422, 493

SMOKERS
Animal Smokers 796
Bee Smokers 895

SNAPS
Bull Snaps 838
Cattle Leader Snaps 838
Cuff Snaps 405, 409
Dress, Waist and Collar Snaps 368
Harness Snaps 928
Rope Snaps 928
Set Line Snaps 775, 779
Whip Snaps 930

SNARES
Drum Snares 523

SNATHS 841

SNIPS
Pocket Snips 853
Tinners' Snips 853

SOAPS
Babies' Soaps 482
Dye Soaps 479
Harness Soaps 957
Pumice Soaps 487
Shampoo Soaps 486-487
Shaving Soaps 802
Toilet Soaps 487
Tooth Soaps 487

SOCKETS
Anti-Rattler Sockets 760
Curtain Pole Sockets 587
Electric Lamp Sockets, Miniature—Write for Electric Goods Catalog.
Electric Sockets 731
Headlight Sockets 756
Hex Sockets 747
Radio Tube Sockets 809

SOCKS
Babies' Socks 225
Boys' Socks 229
Children's Socks 225, 227
Guaranteed Socks 224
Lumbermen's Socks 217
Men's Socks 224, 228-229
Polar Socks 211
Rockford Socks 228-229
Silk Socks 229

SODA
Nitrate of Soda 476
Phosphate of Soda 484
Soda Mints 484

SOFAS 612, 618

SOILPIPE AND FITTINGS 704

SOLDER
Household Solder 829
Jewelers' Solder 471
Plumbers' Solder 704
Silver Solder 852
Solderine 478
Soldering Fluid 471
Tinners' Solder 853
Wire Solder 853

SOLDERING OUTFITS 840, 853

SOLES
Leather Soles 211
Slipper Soles 211
Sole and Heel Sets 212
Waterproof Soles 211

SOLVENTS
Lead Solvent 793

SONGS
Song Books 435, 525

SONG-O-PHONES 513

SOUNDERS
Telegraph Sounders—Write for Radio Catalog.

SOU'WESTERS 166, 266-267

SOWERS
Fertilizer Sowers 881
Grass Sowers 883
Lime Sowers 881

SPADES 843

SPARK PLUGS AND ACCESSORIES 750, 757

SPATULAS 805

SPEARS
Fish Spears 774
Frog Spears 774

SPECULUMS 477

SPEED INDICATORS 869

SPEEDOMETERS 754
Speedometer Parts 753

SPITTOONS 825

SPLICERS
Hose Splicers 840

SPLICES
Trace Splices 927

SPOKES
Bicycle Spokes 737
Buggy Spokes 850
Spoke Pointers 864
Spoke Repairers 848
Wagon Spokes 850

SPOKESHAVES 861

SPONGES
Babies' Sponges 478, 482
Bath Sponges 478, 480, 482
Cleaning Sponges 478, 753, 952

SPOONS
Babies' Spoons 412-419
Basting Spoons 805
Berry Spoons 412-419
Dessert Spoons 411-412, 414, 417
Friendship Spoons 419
Iced Tea Spoons 412-418
Jelly Spoons 412-414
Kitchen Spoons 805, 828
Orange Spoons 413, 415-418
Soup Spoons 412-418
Sugar Spoons 411-419, 805
Tablespoons, Aluminum 805
Tablespoons, Silver 411-418
Tablespoons, Tinned 805
Teaspoons, Aluminum 805
Teaspoons, Silver 411-418
Teaspoons, Tinned 805

SPORTING GOODS 762-796

SPOTLIGHTS 751

SPRAYERS
Barrel Pump Sprayers 902
Bucket Sprayers 902-903, 954
Compressed Air Sprayers 902-903
Hand Sprayers 903
Insect Sprayers (Household) 476
Medicine Sprayers 481
Perfume Sprayers 481
Power Sprayers 902
Shower Bath Sprayers 934

SPRAYS
Bath Sprays 934
Spraying Materials 476-477
—Or Write for Prices.

SPREADERS
Butter Spreaders 412-419
Harness Spreaders 928
Manure Spreaders and Attachments 890

SPRING BALANCES 775, 833

SPRINGS
Auto Springs 738, 760
Bed Springs 662-663
Bird Cage Springs 837
Bolster Springs 849
Buggy Springs 848
Coil Springs 846
Door Springs 834-835
Hair Clipper Springs 802
Hair Springs, Watch 471
Hammock Springs 787
Horse Clipper Springs 929
Mainsprings, Clock 471
Mainsprings, Watch 471
Phonograph Springs 548
Seat Springs, Wagon 848
Shock Absorbers 760
Springs and Parts for Ford Cars 760-761
Steering Springs, Auto 760
Top Springs, Auto 759
Tree Pruner Springs 840
Wagon Springs 848

SPRINKLERS 840

SPROUTERS 901

SPURS
Drum Spurs 523
Linemen's Spurs 814
Riding Spurs 931

SQUARES
Square and Level 863
Steel Squares 863
T Squares 501
Try Squares 863

SQUEEZERS
Lemon Squeezers 831

STABILIZERS 760

STACKERS
Cob Stackers 888
Hay Stackers 844, 890

STAGE SUPPORTS 939

STAINS
Shingle Stains 946-947
Varnish Stains 942, 944, 950
Wood Dye Stains 942

STAIR WORK
—Write for Mill Work Catalog.
Stair Carpets 589-591
Stair Fittings 586

STAKES
Bolster Stakes 848
Riveting Stakes 471

STALLS
Steel Stalls and Fittings 891

STAMPED GOODS 362
Stamping Outfits 361

STAMPS
Dating Stamps 500
Rubber Stamp Outfits 500
Tool Stamps 859

STANCHIONS
Cattle Stanchions and Fittings 891

STANDS
Bicycle Stands 738
Book Stands 618-620, 625
Carrier Stands 738
Coat and Hat Stands 624
Cobblers' Stands 212
Drum Stands 522
Emery Wheel Stands 853, 868
Hammock Couch Stands 787
Ink Stands 500
Jardiniere Stands 624
Music Stands 515
Orchestra Bell Stands 519
Parlor Stands 624
Plant Stands 618-620
Sewing Stands 626
Shaving Stands 422
Shoe Polishing Stands 652
Wash Stands 652, 825

STAPLES
Gate or Door Staples 836
Hame Staples 927
Marking Tape Staples 764
Netting Staples 898
Screen Cloth Staples 835
Wire Fence Staples 898

STARS
Embroidered Stars 366
Officers' Stars 792

STATIONERY 450-454, 500

STATUETTES 419

STEAMERS
Steamer Kettles 827, 829

STEAM FITTINGS 706-707

STEAM HEATING PLANTS 694

STEARATE OF ZINC 482

STEEL
Sheet Steel 958
Steel Wool and Shavings 829, 861, 954

STEELS
Butchers' Steels 806
Capping Steels 830
Kitchen Steels 805

STEERING DEVICES 760

STEMS
Handle Bar Stems 738

STENCILS
Brass Stencils 840
Fancywork Stencils 361
Wall Stencils 952

STEPLADDERS 939

STEPPING 964

STICKPINS 373, 376, 389, 391, 396-397, 406, 409

STICKS
Drum Sticks 523
Shaving Sticks 802

STIFFENERS
Heel Stiffeners 213

STILETTOS
Embroidery Stilettos 361

STIRRUPS 924

STITCHING HORSES 930

STOCKINGS
Athletic Stockings 762, 769
Babies' Stockings 225
Baseball Stockings 762
Boys Stockings 226-227
Elastic Stockings 474
Girls' Stockings 226-227
Golf Stockings 765
Guaranteed Stockings 224, 226
Rockford Stockings 223
Silk Stockings 220-221
Stocking Feet 222-223, 226
Women's Stockings 220-224

STOCKS
Gun Stocks 793
Jewelers' Drill Stocks 471
Pipe Threading Stocks and Dies 705
Screw Cutting Stocks and Dies 870
Whip Stocks 930

STOGIES 493

STONEBOAT HEADS 893

STONERS
Cherry Stoners 830

STONES
Corundum Stones (Wheels) 853, 856
Grindstones 706, 853, 856
Jewelers' Jobbing Stones 471
Oil Stones 861
Sickle Grinder Stones 844
Scythe Stones 841

STOOLS
Band Saw Stools 866
Bath Stools 934
Folding Stools 785, 851
Foot Stools 607B
Kitchen Stools 641
Office Stools 672
Stepladder Stools 939
Window Stools 967

STOPLIGHTS 751

STOPPERS
Basin Stoppers 934
Bathtub Stoppers 934
Vacuum Bottle Stoppers 765

STOPS
Bench Stops 863
Door Stops 967
Water Pipe Stops 706
Window Stops 967

STOVEBOARDS 823

STOVEPIPE 689, 785

STOVES
Brooder Stoves 901
Camp Stoves 785
Cooking Stoves 675-691
Electric Stoves 810
Gas Stoves 689-691
Heating Stoves 679
Laundry Stoves 684-685
Oil Stoves 686-688
Stove Repairs for Our Stoves—Write for Stove Repair List.
Tank Heating Stoves 684-685
—Or Write for Farm Implement Catalog.

STOVEWARE 826-829, 831

STRAIGHTEDGES
—Write for Wall Paper Sample Book.

STRAIGHTENERS
Hair Straighteners 486
Heel Straighteners 472

STRAINERS
Eaves Trough Strainers 959
Hose Strainers 854
Jelly Strainers 831
Milk Strainers 832
Sprayer Strainers 903
Vegetable Strainers 827-828
Well Pipe Strainers 710

STRAPS
Ankle Straps 472
Barbers' Chair Straps 802
Bedclothes Straps 135
Carrying Straps 929
Cow Bell Straps 922
Creel Straps 780
Harness Straps 921-923
High Chair Straps 134
Hitching Straps 921
Latigo Straps 924
Linemen's Straps 814
Pipe Straps 707
Singletree Straps 848
Spur Straps 931
Tennis Net Straps 764
Trunk Straps 922
Watch Straps, Men's 382
Wrist Straps 771

STRETCHERS
Curtain Stretchers 823
Fence Stretchers 843, 899
Folding Stretchers (Cots) 78
Fur Stretchers 796
Shoe Stretchers 213
Wire Stretchers 843, 899

STRIDES
Giant Strides 767

STRINGERS
Fish Stringers 775

STRINGS
Cotton String 479
Musical Instrument Strings 508-509, 513-514
Sticky Fly String 476
Wrapping String 823

STRIPS
Carpet Strips 967
Furring Strips 964
Panel Strips—Write for Wall Paper Sample Book.
Wall Board Panel Strips 969
Weather Strips 835

STROPS 800-801

STUDDING 965

STUFFERS
Sausage Stuffers 830

STUMP PULLERS AND ACCESSORIES
—Write for Farm Implement Catalog.

SUGAR SETS 420-421

SUITCASES 260-261

SUITINGS 302-304, 314, 319-321, 325, 327

SUITS
Bathing Suits 768
Boys' Knee Pants Suits 165, 172-175, 177, 179-180
Boys' Long Trouser Suits 270-271
Boys' Sleeping Suits 163
Hunting Suits 794
Khaki Suits, Men's, Boys' 168-171, 179-180, 282, 794
Knicker Suits, Women's, Girls' 4, 23-24, 26-27
Little Fellows' Suits 165, 170-175, 177, 179-180
Men's Suits 272-282, 794
Misses' Suits 4, 28-29, 33
Oiled Slicker Suits 267
Overall Suits, Men's, Boys' 166-167, 268, 289-291
Palm Beach Suits, Men's Boys' 165, 281
Play Suits, Children's 168-169, 179
Riding Suits, Men's 282
Riding Suits, Women's, Misses' 4
Union Suits—See Union Suits.
Wash Suits, Boys' 165, 170-171, 175
Wash Suits, Children's (Ready to Sew) 372
Women's Suits 4, 28-35

SULKIES
Babies' Sulkies 670
Dog Sulkies 911
Goat Sulkies 911
Road Sulkies 817

SULPHUR 476-477
Sulphur Candles 483
Sulphur Lozenges 484
Sulphur Ointments 484
Sulphur and Charcoal 796

SUNSHIELDS
Auto Sunshields 746

SUPERS
Bee Hive Supers 895

SUPPORTS
Abdominal Supports 135, 473
Ankle Supports 472, 474
Arch Supports 472
Athletic Supports 474, 769, 771
Belt and Hose Supports 370
Bust Supports 135
Diaper Supports 135
Elastic Supports 473-474
Hip Supports 370
Hose Supports 135, 143, 370-371
Radius Rod Supports 760
Shelf Supports 834
Stage Supports 939
Uterine Supports and Fittings 473
Wrist Supports 771

SUPPOSITORIES 482, 484

SURCINGLES 924

SUSPENDERS 163, 258
Swimming Suspenders 768

SUSPENSORIES 474

SWAGES
Saw Swages 852

SWATTERS
Fly Swatters 476

SWEATBANDS 764

See Page 557 for Measuring Instructions.

SWEATERS
Babies' Sweaters 141
Boys' Sweaters 243
Children's Sweaters 141
Girls' Sweaters 162
Men's Sweaters 243
Women's Sweaters 146-147

SWEEPERS
Carpet Sweepers 811, 824
Dustless Sweepers 824
Vacuum Sweepers 811, 824

SWEEPS
Cultivator Sweeps 842, 876

SWINGS
Babies' Swings 494
Children's Swings 669
Fiber Hammock Swings 621
Gymnasium Swings 766
Lawn Swings 786
Playground Swings 767
—Or Write for Sporting Goods Catalog.
Porch Swings 621, 786

SWITCHBOARDS 814

SWITCHES
Battery Switches 813-814
Carrier Track Switches 891
Hair Switches 128-129
House Wiring Switches 731
Lighting Switches, Auto 751, 756
Telegraph Switches 814
Telephone Switches 814

SWIVELS
Bait Swivels 775
Gag Swivels 927
Rope Swivels 838
Striking Bag Swivels 770
Swivel Joints 753

SYRINGES
Babies' Syringes 482
Bulb Syringes 480-482
Fountain Syringes 480-481
Hard Rubber Syringes 481
Internal Bath Syringes 481
Veterinary Syringes 477

SYRUPS
Medicinal Syrups 484
Syrup Making Supplies 892
—Or Write for Farm Implement Catalog.

T

T BEVELS 863

T SQUARES 501

TABLECLOTHS 338-343
Tablecloths (by the yard) 338-339, 341-343
Table Oilcloths 334, 363

TABLES
Babies' Dressing Tables 668-669
Bedroom Tables 624
Billiard Tables — Write for Prices.
Camp Tables 784-785
Card Tables 624
Davenport Tables 622-623, 626
Dining Room Tables 628-632, 634-635
Dressing Tables 644-645
Fiber Tables 618-619
Folding Tables 624, 626, 784-785
Kitchen Tables 643
Library Tables 608, 619-620, 622-623, 625
Office Tables 672
Parlor Tables 607D, 618-620, 622-626
Paste Tables—Write for Wall Paper Sample Book.
Pool Tables—Write for Prices.
Porch Tables 618-621
Reed Tables 620
Serving Tables 618, 620, 626, 628-631
Sewing Tables 624
Steel Top Tables 643
Telephone Tables and Stools 624

TABLETS
Butter Color Tablets 478
Cheese Color Tablets 478
Foot Tablets 472
Medicinal Tablets 484
Rennet Tablets 478
Water Color Tablets 502
Worm Tablets 477
Writing Tablets 454

TABLEWARE
China Tableware 714-721, 723
Glass Tableware 722-726
Silver Tableware 411-419
Steel Tableware 804-805

TACKLE
Fishing Tackle 772-780
Tackle Blocks 845

TACKS
Bill Posters' Tacks 837
Brass Head Tacks 837
Carpet Tacks 837
Double Pointed Tacks 835
Gilt Head Tacks 837
Household Tacks 837
Thumb Tacks 500
Upholsterers' Tacks 837, 851

TAFFETAS 294-298, 301

TAILORS' GOOSE 823

TAILPIECES
Stringed Instrument Tailpieces 509, 514

TALCUMS
Babies' Talcums 482
Face Talcums 482, 488

TALKING MACHINES 535-543

TAMBOURINES 522

TAMERS
Hog Tamers 838

TAMPERS 841

TANGLEFOOT 476

TANKS
Expansion Tanks—Write for Heating Catalog.
Gasoline Tanks 957
Gasoline Tanks, Auto 752
Harness Dipping Tanks 957
Hog Scalding Tanks 893
Oil Wagon Tanks 894
Pneumatic Tank Outfits 695
Pressure Tanks—Write for Plumbing Catalog.
Range Boiler Tanks 703
Septic Tanks 695
Steel Tanks 894
Storage Tanks 894
Thresher Tanks 894
Wagon Tanks 894
Water Closet Tanks 698-699
Wood Tanks 894
—Or Write for Farm Implement Catalog.

TANNING COMPOUNDS 796

TAPES
Adhesive Tape 478, 483
Bias Seam Tapes 369
Cotton Tapes 369
Mechanics' Tapes 862
Steel Tapes 862
Tennis Court Tapes 764
Tire Tapes 736, 742
Wall Board Tapes 969
Winding Tapes 763

TAPESTRY 561

TAPIDEROS 924

TAPS
Blacksmiths' Taps 870
Calk Taps 875
Pipe Taps 705
Shoe Taps 211

TAR
Coal Tar 935
Pine Tar 957

TARGETS 792-793

TASSELS
Curtain Tassels 575
Dress Trimming Tassels 366
Window Shade Tassels 585

TEA BALLS 419, 421

TEAKETTLES 826-829

TEAPOTS 420-421, 723, 726, 826-829

TEA SETS
China Tea Sets 723
Glass Tea Sets 722
Linen Tea Sets 341
Silver Tea Sets 420-421

TEAS
Herb Teas 484

TEDDY BEARS 494

TEES
Drainage Tees 707
Laundry Tub Tees 707
Malleable Tees 707
Reducing Tees 707
Soil Pipe Tees 704

TEETER TOTTERS 767

TEETH
Harrow Teeth 842, 876, 882
Hay Rake Teeth 887
Weeder Teeth 886

TELEGRAPH INSTRUMENTS AND SUPPLIES 807-809
—Or Write for Radio Catalog.

TELEPHONES AND SUPPLIES 814
—Or Write for Electrical Goods Catalog.

TELESCOPES 428

TENDERS
Babies' Tenders 494, 668-669

TENNIS GOODS 764, 766
Toy Tennis Sets 495

TENONING MACHINES 866

TENTS 782-783
Auto Tents 782-783
Cot Tents — Write for Prices.
Tent Fittings 782-783
Tenting Supplies 784-785

TERRETS
Harness Terrets 926

TESTAMENTS 444

TESTERS
Battery Testers 749
Egg Testers 839
Milk Testers 832
Scale Testers 833
Spark Plug Testers 750
Tire Testers 744

THERMOMETERS 427
Bath Thermometers 135
Fever Thermometers 427, 483
Hot Water Thermometers—Write for Heating Catalog.
Photographers' Thermometers 552

THERMOSTATS 694

THIMBLES
Banjo Thimbles 514
Gold Thimbles 393
Nickel Thimbles 369
Silver Thimbles 393

THREAD
Basting Thread 364
Carpet Thread 364
Crochet Thread 364-365
Darning Thread 364
Embroidery Thread 364-365
Harness Thread 930
Knitting Thread 364-365
Sewing Thread 364
Silk Thread 364
Waxed Thread 930

THROWS
Women's Throws 357

TICKING 332

TIES
Cattle Ties 838
Middy Ties 360
Neckties 163, 240-241, 243
Rope Ties 920-921

TIGHTS
Athletic Tights 769

TIMBERS 965

TIMERS
Ignition Timers 757

TIN
Block Tin 853
Valley Tin 958

TINNERS' SUPPLIES 853

TIPS
Carbide Light Tips 796
Chair Tips 483
Crutch Tips 483
Cue Tips 797
Fishing Rod Tips 772, 775
Ostrich Tips 106
Violin Bow Tips 509

TIRES
Auto Tires 740-741
Bicycle Tires 735-736
Inner Tires 739
Motorcycle Tires 738
Tire Iron 847
Tire Repairs 736, 739, 742-743
Tire Shrinkers 875

TISSUE
Mending Tissue 369

TOASTERS 831
Electric Toasters 810

TOBACCO 493
Tobacco Dust 476

TOILETS 483, 698-699

TOILET SETS
Babies' Toilet Sets 482
Manicure Sets 422, 490
Men's Toilet Sets 422, 485
Women's Toilet Sets 422

TOILETWARE 727, 825

TOILET WATERS 490

TONGS
Blacksmiths' Tongs 873
Hog Tongs 838
Ice Tongs 857
Sugar Tongs 412, 417

TONGUES
Wagon Tongues 849-850

TONICS
Hair Tonics 486
Household Tonics 484
Stock Tonics 477

TOOLS
Auto Tools 742-744, 747, 752-753, 755-756
Bicycle Tools 736-738
Blacksmiths' Tools 846-847, 867-875
Bricklayers' Tools 841
Butchers' Tools 806
Calking Tools 704
Carpenters' Tools 843, 858-866
Cement Workers' Tools 841-842
Cobblers' Tools 212
Farming Tools 841-844
Fencing Tools 843, 871, 899
Garage Machine Tools 868
Garden Tools 840-843
Gasfitters' Tools 705, 853, 871-872
Glaziers' Tools 941
Graining Tools 943
Gunsmiths' Tools 870, 872
Harness Makers' Tools 930
Horseshoers' Tools 873-875
Jewelers' Tools 471, 870, 872
Linemen's Tools and Supplies 814, 843, 871
Lumbermen's Tools 856-857
Machinists' Tools 868-873
Masons' Tools 841
Painters' Tools 935, 939-944, 951-954
Paperhangers' Tools 939-941
Piano Tuners' Tools 515
Plasterers' Tools 841, 935, 941
Plumbers' Tools 704-705
Pump Repairers' Tools 710
Steamfitters' Tools 705, 868-873
Tinners' Tools 853
Tire Tools, Auto 743
Tool Sets, Hollow Handle 859
Watchmakers' Tools 471

TOOTHPICKS 479

TOOTH PREPARATIONS 487

TOPCOATS 175, 264-265

TOPS
Auto Tops, Repairs 745, 758, 815, 851
Buggy Tops 851
Hame Tops 926
Percolator Tops 826
Play Tops 496
Wagon Tops 851

TORCHES
Blow Torches 705
Fumigating Torches 483
Gasoline Torches 825

TOUPEES 129

TOWELINGS 335
Embroidery Towelings 335

TOWELS 336-337, 343, 362
Babies' Bath Towels 135
Embroidered Towels 362
Paper Towels, Racks 479
Sanitary Towels 370-371, 483
Stamped Towels 362

TOWEL BARS 934

TOWERS
Windmill Towers 894

TOYS 494-497

TRACES
Cable Traces 923
Chain Traces 927
Harness Traces 923
Rope Traces 923
Trace Carriers 923

TRACING PAPER 500

TRACKS
Barn Door Tracks 836
Hay Carrier Tracks 844
Litter Carrier Tracks and Fittings 891

TRACTORS
Tractor Plows 880

TRANSFERS
Decalcomania Transfers 951

TRANSFORMERS
Bell Ringing Transformers 813
Radio Transformers 807, 809

TRANSMISSION
Transmission Grease 956-957
Transmission Oil 956
Transmission Parts 760-761

TRANSMITTERS
—Write for Electric Goods Catalog.

TRAPEZE 766

TRAPS
Animal Traps and Accessories 796
Bee Traps 895
Closet Traps 704
Drain Traps 704
Drum Traps 704
Drummers' Traps 522
Fish Traps 780
Laundry Tub Traps 704
Minnow Traps 780
Mole Traps 796
Mouse Traps 796
Rat Traps 796
Sink Traps 704

TRAYS
Ash Trays, Sets 422
Bread Trays 420-421
Celery Trays 724
Comb and Brush Trays 934
Cooky Trays 420
Crumb Tray and Scraper Sets 420-421, 479
Developing Trays 552
Dresser Trays 491
Fixing Trays 552
Manicure Trays 491
Sandwich Trays 420-421, 724
Serving Trays 421, 479, 724, 826, 829
Serving Trays (Wheeled) 618, 620, 626
Soap Trays 825, 828, 934
Spoon Trays 420

TREADS
Rubber Stair Treads 605

TREES
Shoe Trees 213, 367

TRESTLES — Write for Wall Paper Sample Book.

TRIANGLES
Drawing Triangles 501
Musical Triangles 522
Pool Ball Triangles 797

TRICOTINES 302

TRIMMERS
Paperhangers' Trimmers —Write for Wall Paper Sample Book.
Shears 803

TRIMMINGS
Brass Box Trimmings 837
Buggy Trimmings 851
Curtain Pole Trimmings 586
Dress Trimmings 356-359, 361, 366
Harness Trimmings 928
Hat Trimmings 102-109
Marabou Trimmings 357
Ostrich Trimmings 106-107, 357
Window Shade Trimmings 583-585

TRIP GONGS 855

TRIPODS 549

TROCARS 477

TROMBONES 521

TROUGHS
Eaves Troughs and Fittings 959
Hog Scalding Troughs 893
Oiling Troughs 848
Poultry Troughs 839, 901
Stock Feeding Troughs 892, 894
Watering Troughs 836, 892, 894

TROUSERS
Boys' Long Trousers 269
Boys' Short Trousers 178-180
Men's Trousers 279, 283-285
Work Trousers 286-288

TROWELS
Bricklayers' Trowels 841
Cement Workers' Trowels 841
Garden Trowels 842
Masons' Trowels 841
Painters' Trowels 935
Plasterers' Trowels 841

TRUMPETS 520

TRUCKS
Auto Truck Bodies 815
Binder Trucks 882
Corn Sheller Trucks 888
Cultivator Trucks 881, 885
Engine Trucks 818-819
Farm Trucks 815
Hand Trucks 840
Harrow Trucks 881-882
Tongue Trucks and Attachments 881-882

TRUNKS 262-263
Athletic Trunks 769

TRUSSES
Axle Trusses, Auto 760
Trusses (Rupture) 475

TUBES
Conversation Tubes 429
Inner Tubes, Auto 739, 743
Inner Tubes, Bicycle 736
Inner Tubes, Motorcycle 738
Milking Tubes 477
Sprayer Tubes 903
Uterine Supporter Tubes 473
Vacuum Tubes 807
Water Gauge Tubes 707

TUBE EXPANDERS 705

TUBINGS
Chain Pump Tubing 709
Gas Tubing 689
Oil Tubing 957
Pillow Tubing 329
Pump Tubing, Auto 744
Syringe Tubing 480-481
Wood Pump Tubing 708

TUBS
Babies' Bathtubs 482, 668
Butter Tubs, Silver 421
Folding Bathtubs 481, 668, 702
Foot Tubs 825
Laundry Tubs 703
Porcelain Bathtubs 697
Steel Bathtubs 697
Wash Tubs 822

TUCKERS
Drum Head Tuckers 523
Sewing Machine Tuckers 312

TUGS 923

TUMBLERS
Bathroom Tumblers 934
Iced Tea Tumblers 722-723
Lemonade Tumblers 722-723
Tumbler and Pitcher Sets 722, 724-725
Water Tumblers 722-724, 726

TUNERS 509, 513-515

TURNERS
Cake Turners 826

TURNS
Bell Turns 834
Cupboard Turns 834

TURPENTINE 949, 954
Artists' Turpentine 502
Turpentine Substitute 949, 954

TUYERES 874

TWEEZERS
Hair Tweezers 490, 802
Jewelers' Tweezers 471

TWINE
Binder Twine 887
Fish Net Twine 779
White Cotton Twine 479
Wrapping Twine 823

TWIST (Thread) 364-365

TYPE
Rubber Type 500

TYPEWRITERS 674
Typewriter Supplies 500, 674

U

U's
Cast Brass U's 903

UKULELES 513

UMBRELLAS 110
Wagon Umbrellas and Holders 851

UNDERCOATS
Enamel Undercoats 953
Graining Undercoats 943
Kalsomine Undercoats 952
Paint Undercoats 935, 944
Varnish Undercoats 944

UNDERSKIRTS
Babies' Underskirts 136-137
Girls' Underskirts 136, 143-145, 160-162
Women's Underskirts 146-149, 154

UNDERWAISTS
Babies' Underwaists 143
Children's Underwaists 143, 161, 238, 370-371

UNDERWEAR
Babies' Underwear 136-137, 157, 237
Boys' Underwear 238-239
Girls' Knit Underwear 238-239
Girls' Muslin Underwear 142-143, 160-162
Men's Underwear 230-233
Women's Knit Underwear 234-236
Women's Muslin Underwear 148-151, 154-155
Women's Silk Underwear 146-147

UNIFORMS
Baseball Uniforms 762
Nurses' Uniforms 57

UNIONS 707

UNION SUITS
Boys' Union Suits 238-239
Combinations, Women's, Girls 143, 150
Girls' Union Suits 238-239
Men's Union Suits 230-233
Waist Union Suits 238-239
Women's Knit Union Suits 234-235
Women's Muslin Union Suits 151

UPHOLSTERERS' SUPPLIES 332, 334, 561, 803, 837, 851
Auto Upholstery 759

UPSETTERS
Saw Upsets 852
Tire Upsetters 875

URNS
Marble Urns 723

V

VACCINATING OUTFITS 477

VACUUM BOTTLES 765

VACUUM CLEANERS AND ATTACHMENTS 811, 824

VACUUM TUBES 807

VALANCES 564, 566

VALLEYS 958

VALVES
Air Valves — Write for Heating Catalog.
Angle Valves 706
Check Valves 706
Closet Tank Valves 699
Cut-Out Valves 746
Engine Valves, Auto 755
Gate Valves 706
Globe Valves 706
Pump Foot Valves 710
Radiator Valves 706
Tank Float Valves 710
Tire Valves and Parts, Auto 743
Tire Valves and Parts, Bicycle 736
Whistle Valves 706

VANITIES 126-127, 489

VARIOCOUPLERS 807

VARIOMETERS 807

VARNISH
Artists' Varnish 502
Auto Varnish 955
Carriage Varnish 955
Church Seat Varnish 945
Colored Varnish 944, 950
Fishing Rod Varnish 775
Floor Varnish 942, 944-945
Furniture Varnish 942, 944-945, 950
Interior Varnish 944-945
Linoleum Varnish 945
Oilcloth Varnish 945
Outside Varnish 944-945

Tile Paper Varnish 945
Varnish Cleaners 951
Varnish Removers 954

VASES
Glass Vases 723-724
Silver Vases 421

VEGETALS 490

VEILINGS 356

VEILS
Beekeepers' Veils 895
Women's Veils 356-357

VEINING 352-353

VELOCIPEDES 494, 496

VELOURS 302, 561

VENTILATORS
Building Ventilators 958
Register Ventilators 823
Tent Ventilators 783

VESTEES
Women's Vestees 360

VESTS
Babies' Undervests 237
Girls' Undervests 238
Hunters' Shell Vests 794
Women's Knit Undervests 236
Women's Silk Undervests 149

VETERINARY SUPPLIES 427, 477

VIATICUMS 419

VIBRATORS
Medical Vibrators 811
Song-O-Phone Vibrators 513
Vibrator Parts 757

VIOLINS AND ACCESSORIES 506-509

VIOLONCELLOS AND SUPPLIES 508

VISES
Blacksmiths' Vises 872, 874
Jewelers' Vises 471, 872
Metal Workers' Vises 872
Pipe Vises 705, 872
Saw Vises 859
Vise Boxes and Screws 872
Vise Jaws 705
Woodworkers' Vises 863, 872

VISORS
Windshield Visors 746

VITRIOL 476, 813

VITROPHANE 942

VOILES 300-301, 319-323, 325-326
Curtain Voiles 562-563

VOLLEY BALL GOODS 766, 769, 771

VOLTMETERS 749

VULCANIZERS 742

W

WAFFLE IRONS 829

WAGONS
Boys' Wagons and Attachments 494, 496-497
Farm Wagons (Trucks) 815
Spring Wagons 817
Tea Wagons 618, 620, 626
Wagon Boxes 815
Wagon Repairs 846-851

WAISTINGS 295-299, 324, 326-327, 330, 355-356, 361

WAISTS
Boys' Waists 178-180
Corset Waists 120, 123
Garter Waists 238, 370-371
Waist Fronts 360
Women's Waists 44-55

WALKERS
Babies' Walkers 494, 669

WALLBOARD 969

WALLETS 259
Fishermen's Wallets 777

WALLPAPER 936-938
Wallpaper Cleaners 942

WARDROBES 653
Babies' Wardrobes 669
Wardrobe Trunks 262-263

WARMERS
Auto Warmers 929
Bed Warmers 480-482
Carriage Warmers 929
Foot Warmers 480-482, 829, 929

WARP
Carpet Warp 334
Crochet Warp 334

WASHBOARDS 822

WASHBOILERS 822, 829

WASHERS
Auto Lock Washers 747
Bibb Washers 706
Bolt Washers 846
Clothes Washers (Tin) 822
Felt Washers 754
Leather Axle Washers 848
Lock Washers 846
Rubber Hose Washers 840

WASHING MACHINES 711-713
Window Washing Machines 903

WASHSTANDS 652, 825

WASHTUBS 822

WASTE
Cotton Waste 705

WATCHES
Boys' Watches 382
Luminous Watches 377, 379, 382
Men's Watches 377, 379-385
Railroad Watches 381, 383-384
Stop Watches 382
Watch Repairs 471
Women's Watches 377-379

WATCHMAKERS' SUPPLIES 426, 471

WATER CLOSETS 698-699

WATER COLORS 495, 502, 556

WATER CONDUCTORS 710, 959

WATERERS
Poultry Waterers 839
Stock Waterers 836, 892, 894

WATERGLASS 478

WATERLESS COOKERS 827

WATERPROOFING
Canvas Waterproofing 783, 794
Fish Line Waterproofing 775
Leather Waterproofing 213, 955
Shoe Waterproofing 213
Slicker Waterproofing 267

WATER SETS 724-725

WATER SUPPLY SYSTEMS 695

WAVERS
Hair Wavers 131, 486, 810

WAX
Auto Polishing Wax 753
Floor Wax 943
Fruit Jar Wax 478
Furniture Wax 943
Harness Makers' Wax 930

WEANERS
Calf Weaners 838

WEATHERBOARDS 964

WEATHER INDICATORS 427, 479, 495

WEATHERSTRIPS 835

WEBBING
Elastic Webbing 371, 475

WEDGES 856

WEEDERS 842, 886 —Or Write for Farm Implement Catalog.

WEIGHTS
Scale Test Weights 833

WELDING COMPOUNDS 875

WELL BORING SUPPLIES 710

WHEELBARROWS 843

WHEELS
Bicycle Wheels 737
Buffing Wheels 853
Buggy Wheels 849
Corundum Wheels 853, 856
Fifth Wheels 848
Gauge Wheels, Digger 886
Gauge Wheels, Plow 877-878
Glass Cutter Wheels 941
Light Wagon Wheels 849
Saw Gummer Wheels 853
Steel Wagon Wheels 848, 850
Steering Wheels, Auto 754
Well Wheels 845
Wheel Accessories, Auto 754, 761

WHETSTONES 841

WHIPLASHES 930

WHIPS 930
Dog Whips 795

WHIPSTOCKS 930

WHISTLES
Auto Whistles 746, 754
Dog Whistles 771, 792, 795
Police Whistles 792
Referees' Whistles 771, 792
Song Whistles 522
Steam Whistles 706

WHITE LEAD 954

WHITEWASH
Disinfecting Paint 476, 954
Whitewashing Machines 954

WICKS
Hygrometer Wicks 427
Oil Stove Wicks 688

WIGS 129
Doll Wigs 499

WINDMILLS AND TOWERS 894

WINDOWS 966
Screen Windows 968
Storm Windows—Write for Mill Work Catalog.
Tent Windows 783
Window Hardware 834-835
Window Trim 967

WINDOW SHADES 583-586
Window Shade Cleaners 942

WINDROWERS 887

WIND SHIELDS 815
Wind Shield Accessories 744-746

WINGS
Bathers' Wings 768

WIPES
Gun Wipes 793

WIRE
Aerial Wire 809
Barbed Wire 898
Bell Wire 809
Copper Wire 837
Gold Plated Wire 471
Insulated Wire 731, 809, 814
Iron Wire 837
Lamp Wire, Auto 751
Picture Wire 586
Planter Wire 883
Rubber Covered Wire 731, 814
Screen Wire 835
Steel Wire 837
Telegraph Wire 814
Telephone Wire 814
Tinned Wire 895

WIRE FENCING 896-899

WIRELESS APPARATUS 807-809 —Or Write for Radio Catalog.

WIRING
Auto Wiring Sundries 749-751, 757
Electric Wiring Sundries 731, 809, 813-814

WITCH HAZEL 484

WOOD AND METAL WORKING MACHINERY 866, 868 —Or Write for Wood and Metal Working Machinery Catalog.

WOOL
Steel Wool 829, 861

WORKERS
Butter Workers 832

WRAPPERS
Babies' Wrappers 141
Sheet Music Wrappers 525
Women's Wrappers 56-59

WRAPS
Women's, Misses' Wraps (Coats) 60, 62-63, 65, 68, 75

WREATHS
Bridal Wreaths 130
Confirmation Wreaths 130
Hat Trimming Wreaths 102-109

WRENCHES
Alligator Wrenches 871
Banjo Wrenches 514
Bicycle Wrenches 738
Brake Wrenches, Auto 760
Cap Screw Wrenches, Auto 756
Chain Wrenches 705
Connecting Rod Wrenches 756
Drill Wrenches 870
Monkey Wrenches 871
Nut Wrenches 705, 871
Nut Wrenches, Auto 747
Pipe Wrenches 705
Ratchet Wrenches 871
Rim Brace Wrenches 747
S Wrenches 871
Socket Wrenches 738, 747, 756, 871
Spark Plug Wrenches 756
Tap Wrenches 870
Tappet Wrenches 747
Thread Cleaning Wrenches 871

WRINGERS
Clothes Wringers 822
Mop Wringers 824

WRISTMACHINES 766

WRITING SETS 422, 454

Y

Clean-Out Y 704

YARDS
Babies' Folding Yards 668

YARNS
Crochet Yarns 364-365
Embroidery Yarns 364-365
Wool Yarns 365

YELLOW OCHRE 954

YOKES
Embroidery Yokes 351

Z

ZINCS
Battery Zincs 813
Zinc Matting Ends 605
Zinc Ointments 484
Zinc Stearate 482

ZITHERS 513

Lines of Merchandise for Which We Issue Special Catalogs, Sample Books and Circulars

Automotive Supplies.
Barns and Barn Equipment.
Bicycles.
Buildings — Already Cut and Fitted.
Buildings—Sectional.
Building Material and Mill Work—Mantels, Consoles and Fireplace Furnishings.
Circular Saws.
Cloth Samples of Men's and Young Men's Suits — Write for Cloth Sample Folder 8292GCL.
Concrete Block Machinery and Molds.
Dairy Barn Equipment.
Electric Appliances—Grills, Percolators, Toasters, Fans, Flatirons, Radiators, Etc.
Electric Light Fixtures, Wiring Sundries and Portable Lamps.
Electric Light and Power Plants—Write to our Chicago store for catalog.
Farm Implements.
Fencing—Ornamental.
Fertilizers—Raw Phospate Rock, Etc.
Garages—Wood and Steel, Already Made.
Gas and Electric Light Fixtures.
Groceries and Laundry Supplies—Write to our Chicago store for catalog.
Gun Repairs.
Heating Plants—Furnaces, Hot Water, Steam and Warm Air and Piping.
Hydraulic Rams.
Invalid Chairs.
Lamps—Electric. Table and Floor.
Lighting Fixtures.
Lightning Rods.
Litter Carriers.
Lumber, Lath and Shingles.
Metal Wagon Wheels.
Modern Homes.
Organs—Church and Home.
Paper Veneer.
Pianos and Player Pianos.
Pipeless Furnaces.
Playground Equipment.
Plumbing Instruction Book.
Plumbing Supplies.
Portable Buildings.
Poultry Supplies.
Radio (Wireless) Apparatus.
Roofing, Wall Board, Sheet Plaster and Metal Ceilings.
Samples—Note on Merchandise Pages of This Big General Catalog Whether Samples Are Furnished.
School Furniture and Supplies.
Sectional Buildings.
Sprayers and Spraying Materials.
Stove Repair Price List.
Summer Cottages.
Switchboards—Telephone.
Telephone Parts and Line Materials.
Typewriters.
Vitrophane.
Wall Paper Sample Book and Finished Moldings.
Water Supply Outfits.
Wigs and Toupees for Men and Women.
Winter Footwear.
Wood and Metal Working Machinery.

 Send a Post Card or Letter for the Special Catalog, Sample Book or Circular You Want.

WATCHMAKERS' TOOLS AND MATERIALS.

Should you desire to order balance staffs, hole jewels, cap jewels, mainsprings, hairsprings, etc., send a sample to us and state name of watch. Be sure to enclose in your letter or package your name and address. When ordering hairsprings **be sure to state size and strength.**

WE BUY OLD GOLD. We buy old gold and pay the highest market price, namely: 18-karat gold, 72c; 14-karat gold, 56c; and 10-karat gold, 40c per pennyweight. In all cases we hold old metal until we are advised by customer that estimate of value is satisfactory.

SHIPPING WEIGHTS—We give the weights of the articles on this page. Where no weight is given, the shipping weight is 4 ounces.

FOR THE ACCOMMODATION OF OUR CUSTOMERS WE REPAIR WATCHES THAT HAVE BEEN PURCHASED FROM US.

Should you send your watch to us, we will examine it and write you the cost of repairing. When we receive your favorable reply, with the amount of charges, we will put the watch in perfect condition. It takes ten to fifteen days to repair and regulate a watch properly.

Jewelers' Complete Tool Set, 48 Pieces in All, for Only $15.98

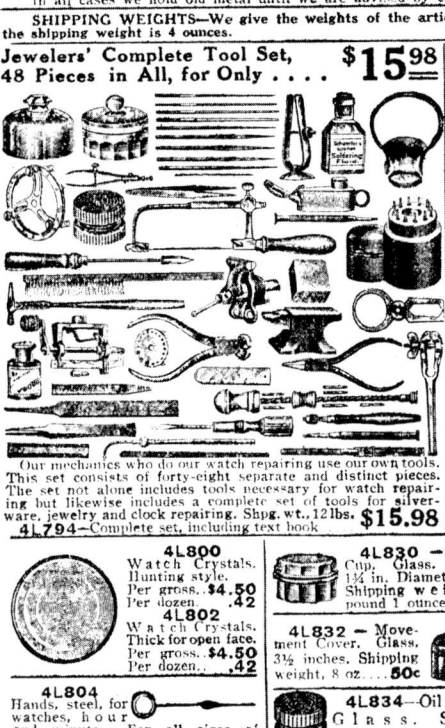

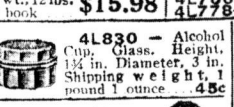

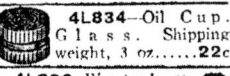

Our mechanics who do our watch repairing use our own tools. This set consists of forty-eight separate and distinct pieces. The set not alone includes tools necessary for watch repairing but likewise includes a complete set of tools for silverware, jewelry and clock repairing. Shpg. wt., 12 lbs. **$15.98**

4L794—Complete set, including text book

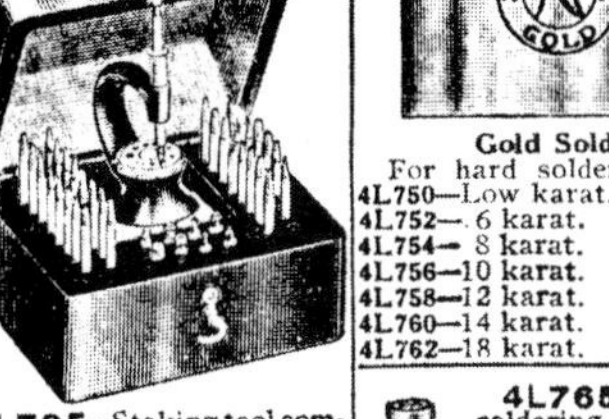

4L795—Staking tool complete in box, 32 punches and 8 stumps. Shipping weight, 4 pounds **$8.46**

4L797—Staking tool complete in box, 24 punches and 4 stumps **$6.58**

Gold Solder.

For hard soldering.

4L750	—Low karat.	Dwt.	$0.23
4L752	—6 karat.	Dwt.	.38
4L754	—8 karat.	Dwt.	.47
4L756	—10 karat.	Dwt.	.66
4L758	—12 karat.	Dwt.	.76
4L760	—14 karat.	Dwt.	.86
4L762	—18 karat.	Dwt.	1.08

4L765—Hard soldering solution. Serves as anti-oxidizer, pickel, hard soldering fluid. Contents, 3 ounces. Shipping weight, 11 ounces **47c**

Watchmakers', Jewelers' and Silversmiths' Outfit for $7.74

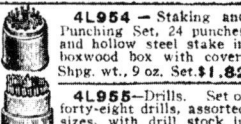

Consisting of twenty-four separate tools and appliances. To this complete set of tools is likewise added one text book. This book gives much information regarding watch repairing, stone setting and other valuable pointers. Weight, complete, 4 pounds.

4L798 Complete . . . **$7.74**

Breguet and Flat Hairsprings.

4L770—Elgin Breguet . . **47c**
4L772—Elgin Flat **22c**
4L774—Waltham Breguet **55c**
4L776—Waltham Flat **31c**
4L786—Hamilton Breguet **47c**
4L778—Hampden Breguet **47c**
4L780—Hampden Flat . . . **30c**
4L782—Illinois Breguet . **59c**
4L784—Illinois Flat **25c**
4L790—New York Standard Breguet **28c**
4L792—New York Standard Flat . **24c**

4L954—Staking and Punching Set, 24 punches and hollow steel stake in boxwood box with cover. Shpg. wt., 9 oz. Set . **$1.82**

4L955—Drills. Set of forty-eight drills, assorted sizes, with drill stock in boxwood box. Shipping weight, 9 ounces. Per set **$1.80**

4L800 Watch Crystals. Hunting style. Per gross . . **$4.50** Per dozen . . . . **.42**

4L802 Watch Crystals. Thick for open face. Per gross . . **$4.50** Per dozen . . . . **.42**

4L804 Hands, steel, for watches, hour and minute. For all sizes of American and imported watches. Not less than lots of one dozen sold. Dozen pairs **40c**

4L806 Hands for clocks, all lengths. Dozen pairs **32c**

4L808 Hands, steel, second, for all sizes American and imported watches. Not less than lots of 1 dozen sold. Dozen **18c**

4L810—Roller Jewels or Ruby Pins for Elgin, Waltham, Hampden or New York Standard. All sizes. Not less than lots of 1 dozen sold. Dozen **50c**

A
B
C

4L812—Mainsprings for watches. All styles and sizes. Not less than ¼ doz. sold. Doz . . **$1.48**

4L814—Mainspring for clocks. 1-day. **Mention width wanted . 14c**

4L816 Mainspring for clocks, 8-day. Mention width wanted. Shipping weight, 13 oz. Each **34c**

4L818—Elgin Balance Staffs. All sizes. Not less than ¼ dozen sold. Dozen **$1.12**

4L820—Waltham Balance Staffs. All sizes. Not less than ¼ dozen sold. Dozen **$1.12**

4L824—Hampden, Springfield, Seth Thomas, Plymouth, New York Standard, Trenton or Rockford Balance Staffs. All sizes. Dozen **$1.12**

4L826—Balance Hole Jewels, cock and foot, for Elgin, Waltham, Hampden or New York Standard, for all sizes. Not less than ¼ dozen sold. Dozen **$1.18**

4L828—Balance Cap or End Stones for Elgin, Waltham, Hampden or New York Standard. All sizes. Not less than ¼ dozen sold. Dozen **76c**

4L830—Alcohol Cup. Glass. Height, 1¾ in. Diameter, 3 in. Shipping weight, 1 pound 1 ounce . . . **45c**

4L832—Movement Cover. Glass, 3½ inches. Shipping weight, 8 oz . . . **50c**

4L834—Oil Cup. Glass. Shipping weight, 3 oz **22c**

4L836—Watch or Clock Oil. **State which wanted.** Shpg. wt., 5 oz. Bottle **20c**

4L838—Watch Oiler. Nickel plated. Shipping wt., 2 oz . . **13c**

4L840—Poising Tool, as illustrated, for poising and truing watch wheels. Shipping weight, 4 ounces . . . **78c**

4L843 Blow Pipe, with ball, 8 or 10 inches. **State length.** Each **35c**

4L844 Blow Pipe, plain, 8 or 10 inches. **State length.** Shpg. wt., 4 oz. Each **18c**

4L846—Watch Brush, 3-row. Shipping weight, 5 ounces **48c**

4L850—4-row. Shipping weight, 5 ounces **58c**

4L856 Pliers, flat. Swiss make, 4-in. Shipping weight, 3 ounces **52c**

4L858—Pliers, round. Swiss make, 4-inch **52c**

4L860—Pliers, end cutting. Swiss make, 4-in. Shipping weight, 4 ounces **90c**

4L862—Pliers, side cutting. Swiss make, 4-inch. Shipping weight, 3 ounces **90c**

4L864—Alcohol Spheric Lamp, glass bulb, nickel plated base. Height, 5½ inches. Shipping weight, 6 ounces **76c**

4L865—Alcohol Lamp, glass bulb and cover, 4 inches high. Shipping weight, 8 oz . . **37c**

4L866—Pendant Sleeve Driver, with nine prongs. Fits all sizes and styles of pendant sleeves. Shipping weight, 3 oz **94c**

4L870 One dozen assorted Clock Drills. Shipping weight, 2 ounces. Dozen . . . . **33c** Not less than 1 dozen sold.

4L872 Drill Stock. Patent geered with adjustable split chuck; top of drill unscrews and has receptacle for holding drills; 10½ inches long. Shipping weight, 1 pound **$1.10**

4L876—Eyeglass. Hard rubber with coil spring; 2 to 5-inch focus. Shpg. wt., 4 oz **98c**

4L878—Eyeglass. Plain hard rubber without spring. 2 to 5-inch focus. Shipping weight, 4 ounces . . . **60c**

4L882—Eyeglass. Double lens. Very powerful, used for very accurate work. Shipping weight, 4 oz . . **98c**

4L884—English Cutting Broaches. Set of 4. Shipping wt., 2 ounces **32c**

4L886—Files, Needle. Length of file complete, 4 inches. Shipping weight, 3 ounces. Set of **6** **95c**

State Length. **4L890**—Flat Files.

3-inch cut 3 . . . . **30c**
4-inch cut 3 . . . . **34c**
5-inch cut 3 . . . . **44c**

4L892 Half Round Files. **State length.**

3-inch cut 3 . . . . **30c**
4-inch cut 3 . . . . **40c**
5-inch cut 3 . . . . **49c**

Shipping weight of files, 5 ounces.

4L894 Files, Screw Head. For filing slots in screw heads. Length, 3½ inches. Shipping weight, 5 ounces **32c**

4L896 Pivot File. Square. 2 inches. Shipping weight, 2 ounces **46c**

4L897 Hammers, Swiss. **State size.**

2 inches . . **30c** | 2½ inches . . **35c**
2¼ inches . . **30c** | 2¾ inches . . **35c**

4L900—Handles for hammers. Maple wood **10c**

4L903 Adjustable roller remover. Nickel plated. Size, 2½-inch. Shipping weight, 3 oz **$1.23**

4L906—Movement Rests. Made of hardwood. Shipping weight, 1 ounce. Set of **6** **38c**

4L912—Tweezers. Fine points, nickel plated **15c**

4L914 Tweezers, hollow handle, Boley make, very light, with fine points, for hairsprings and other fine work **27c**

4L916—Tweezers. Medium points, nickel plated **15c**

4L918—Tweezers with hand remover on opposite end **30c**

4L919 Tweezers, hand remover **54c**

Gold plated wire, round or square; in ½-ounce coils.

4L920—First Quality Round Wire. Sizes, 16 to 21-gauge. Per ounce **85c**

4L924—Second Quality Round Wire. Sizes, 16 to 21-gauge. Per ounce **75c**

4L928—First Quality Square Wire. Sizes, 18 to 22-gauge. Per ounce **85c**

4L930—Second Quality Square Wire. Sizes, 18 to 22-gauge. Per ounce **75c**

4L934—Jobbing Stones, assorted. Containing all colors and sizes in imitation of genuine. Per gross **85c**

4L935—Anvil with hub. Nickel plated. ⅝x1¾-in. Shipping weight, 8 ounces **94c**

4L937—Hand remover with self acting plunger. 4-inch. Shipping weight, 3 ounces **$2.61**

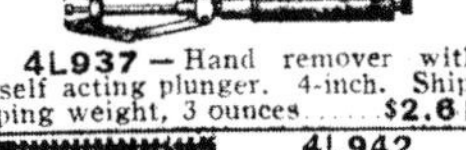

4L942 Gauge for watch mainsprings, with gauge for measuring thickness. Length, 4¼ inches **$1.28**

4L944 Winder, mainspring. Swiss. Length, 3½ inches . . **$1.12**

4L945—Soft Solder. Per bunch **10c**

4L950—Soldering Copper, small, for jewelers. Shipping wt., 6 ounces **28c**

4L951—Soldering Fluid. Shipping wt., 7 oz. Bottle **15c**

4L952 Anti-oxidizer to retain color of metal when hard soldering. Shipping weight, 7 oz . . . **15c**

4L960 Saw Frame. Extra quality. Shipping weight, 7 ounces . . . **84c**

4L961—Saw Blades. (Not less than 1 dozen sold.) Dozen . . **18c**

4L963 Pin Vise, small, adjustable **26c**

4L964 Stake, riveting. Hard steel. Each **33c**

4L965—Pin Vise. Hollow handle. Extra quality **90c**

4L966 Jeweled truing Caliper. Nickel plated jeweled end for balance truing, other end for poising. Shipping weight, 4 oz . . . **$1.68**

4L967—Calipers, with bar, 3¼ inches long. Plain brass. 2 ounces **54c**

4L968 Vise, 1¼-inch steel jaws, clamp vise, handy to adjust to any work bench. Shipping wt., 3 lbs **94c**

For other Jewelers' Vises see page 872.

4L970 Screwdrivers. Set of seven, nickel plated. Shipping wt., 5 oz. Set of **7** **96c**

4L972 Metal Head Screwdriver, in small, medium or large size. **State size wanted.** Each **15c**

4L975—Screwdriver. Adjustable, with four different size blades. Complete **28c**

4L976 Clock Screwdriver, 3-inch blade. Shipping weight, 2 ounces **38c**

4L977 Square Oil Stone Slip; hard; 2¾-inch. Shipping weight, 1 oz **32c**

4L978—Pegwood. Per bundle **10c**

4L980—Pithwood. Per bundle **10c**

4L982 Patent Key. Fits any key wind watch **20c**

4L984—Jewelers' Cement. For cementing china, glass, ivory, beads, pearls and jewelry. Shipping wt., 5 oz. Bottle **24c**

4L986—Granite Hold Fast Cement. Shipping weight, 5 oz. Bottle . . **18c**

Arch Supports and Foot Needs

How to Wear an Arch Support.
The wearing of arch supports at first may cause you some discomfort. Do not attempt to wear them for more than an hour or so each day for the first few days, gradually increasing the period as the soreness leaves the feet, which will occur as the feet gradually conform to the shape of the supports.

Treatment of the Feet—Arch supports serve a special purpose. Many people, especially those walking or standing a greater part of the day, suffer with broken down or weakened arches. A good arch support will hold up this weak or broken arch, help to bring it back to its normal condition and relieve the trouble.
Especially beneficial for persons of heavy weight, for policemen, motormen, workmen and all who are obliged to stand for several hours per day.
Consult your physician before ordering to see that you order the proper support. They are all of first class material, the principal difference being in the shape, curvature, springs or special form for certain foot troubles.

How to Order Foot Supplies.
In ordering arch supports follow these directions carefully unless you are sure of your size and the style you want.
Place your foot on a piece of blank paper and with a pencil held perfectly upright draw the outline of each foot, pressing the weight of your body on the feet. These outlines will enable us to select the correct size. Mail us the pencil drawing with your order and, if possible, state size of shoe worn. **When ordering any foot supply, be sure to state size of shoe and if for man or woman.**

Metal Arch Supports

"Harvard" Arch Support. A reinforced light weight support for heavy people. Plate is nickel silver with grain leather top. (See instructions "How to Order" at top of page.) Shipping weight, 11 ounces.
8L2702 **$1.69**

Reinforced Arch Support. Similar in construction to our "Harvard," but made of nickeled steel. We do not recommend this support, but offer it for those who desire a low priced article. (See instructions "How to Order" at top of page.) Shipping weight, 11 oz.
8L2704 **89c**

"Dixie" Combination Support. For arch and metatarsal. Especially designed to slightly raise the bones and relieve the ligaments causing metatarsalgia or Morton's toe (sharp pains under the ball of the foot). Also has features of arch support for broken down or fallen arches. Genuine nickel silver with grain leather top. (See instructions "How to Order" at top of page.) Shipping weight, 11 ounces.
8L2701 **$2.98**

"Cooper" Arch Support. Made to relieve broken down or fallen arches and for tired, aching feet, weak ankles, pains up and down the leg and callouses brought about by fallen arches. Prevents flat foot by bridging the weight of the body from heel to toe. Has two plates to give spring to the foot. Made of grain leather and nickel silver plates. (See instructions "How to Order" at top of page.) Shipping weight, 11 oz.
8L2708 **$1.98**

"Comfort" Arch Support. A popular arch support with wide and comfortable plate. Light, but strong. Made of nickel silver. Good quality grain leather tops. (See instructions "How to Order" at top of page.) Shipping weight, 11 ounces.
8L2705 **$1.39**

"Sniffen" Arch Support. Similar to our "La Salle," but with sponge rubber cushion in place of felt. Will not last as long as the "La Salle," but offered at a much lower price. (See instructions "How to Order" at top of page.) Shpg. wt., 11 oz.
8L2707 **$1.59**

"La Salle" Arch Support. Provides a felt cushion, leather covered, under the heel of the foot, relieving the pressure and giving relief to a sensitive heel. Plates made of nickel silver, with grain leather top. (See instructions "How to Order" at top of page.) Shipping wt., 11 oz.
8L2700 **$2.39**

Leather Arch Supports

No metal, light weight, easy to adjust and easy to wear. Not recommended for people with serious arch trouble, but offered for those needing temporary relief.

Combination Arch Support.

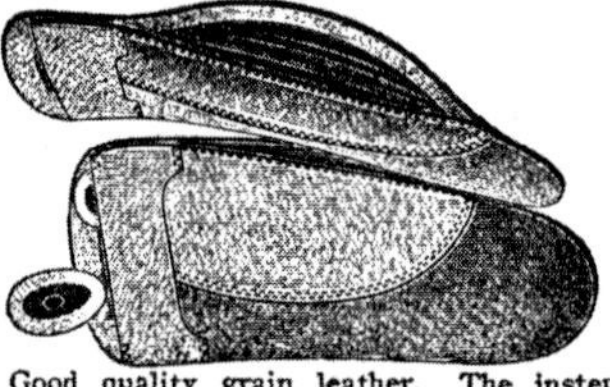

Good quality grain leather. The instep can be raised as desired by inserting small pads. Also has the anterior metatarsal adjustment (for pains at ball of foot). (See instructions "How to Order" at top of page.) Shipping weight, 8 ounces.
8L2729—Per pair **$1.59**

"No-Metal" Arch Support.

Good quality grain leather. Raise the instep to suit yourself. Elevating pads to fit in pocket in support. (See instructions "How to Order" at top of page.)
8L2709 Per pair, **98c**

Ankle Supports

Reinforced Leather.

Good quality soft sheepskin, without stays. Favorite with athletes. Helps to prevent turning of the ankle. Not made in children's size. **Mention size shoe you wear.** Shpg. wt., 5 oz.
8L2786—For boys. Sizes, 1 to 5 shoe. Per pair **98c**
8L2787—For men. Sizes, 6 to 11 shoe. Per pair **98c**
8L2788—For women. Sizes, 2 to 7 shoe. Per pair **98c**

Leather Ankle Supports.

Good grade tanned sheepskin leather, soft and pliable, reinforced with removable stays over both sides of ankle, giving extra strength. Any one stay can be removed to ease pressure at any point. Give size shoe worn. Shpg. wt., 7 oz.
8L2780—For men. Sizes, 6 to 10. Per pair **$1.39**
8L2781—For women. Sizes, 3 to 7. Per pair **1.39**
8L2782—For boys. Sizes, 1 to 5. Per pair **1.39**

REMOVE CORNS WITH DIXIE

Why suffer untold agony with painful corns when Dixie will relieve in a short time? Made only for us under special formula. Has proved to be a very satisfactory remedy for corns. A trial will convince. Dixie liquid for soft corns; Dixie plaster for hard, deep seated corns.

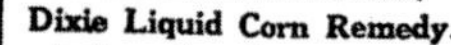

Dixie Liquid Corn Remedy. Apply a few drops to corn each night for four or five nights, then soak foot and corn can be removed, roots and all. Shipping weight, 5 oz.
8L2717—Per bottle **19c**

Dixie Corn Plaster. Apply plaster and wear two days, then soak foot and remove corn. One application is usually enough. Shipping weight, 2 oz.
8L2718—Per package of 6 plasters **19c**

Dixie Foot Powder. U. S. Army Formula. For sore, tired, aching, perspiring or swollen feet. Simply sift a little into the shoes. Shipping weight, 6 ounces.
8L2719—Per 4-ounce carton...... **19c**

Dixie Foot Tablets. Used to relieve tired and painful feet. Dissolve in pail of warm water and bathe feet each night until soreness is gone. Shipping weight, 4 ounces.
8L2732 .... **19c**

Bunion Protectors

Dixie Soft Rubber Protector. To be worn under the stocking. Thick wall fits around the bunion, giving comfort and protection. **State whether for man or woman; for right or left foot.** Shipping weight, 4 ounces.
8L2774—Each **39c**

Dixie Leather Protector. Keeps the shoe from rubbing and allows the bunion to heal. Adjusts itself to foot. **State whether for man or woman and for right or left foot.** Shipping weight, 4 ounces.
8L2768—Each **39c**

DIXIE PADS for Corns, Bunions, Callouses.

A new discovery for removal and relief of corns, bunions and callouses. Simple and efficient. Apply the pad directly over the corn, bunion or callous and Dixie will do the rest. Shipping weight, 2 ounces.

Corn Pads.	Bunion Pads.	Callous Pads.
8L2765	8L2766	8L2767
12 in pkg.	6 in pkg.	6 in pkg.
19c	19c	19c

Coe's Bunion Relief.

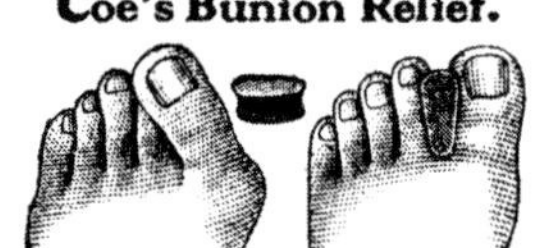

Bunions are caused by wearing too narrow or pointed shoes. This small, soft, spongy rubber fits between the first and second toes, and if shoe is wide enough it will assist in gradually straightening the curved toe to its natural position. Medium and large. **Specify size.** Shipping weight, 2 oz.
8L2770—Each **15c**

Johnson and Johnson Corn and Bunion Shields. White felt with gummed side. Can be cut to fit. Shipping weight, 3 oz.
8L2734 Corn Shields. 12 for **13c**
8L2735 Bunion Shields. 6 for **13c**

Soft Corn Pad. For corns between toes. Soft red sponge rubber to fit between the toes. Hole allows for corn, thus relieving the pressure. Shipping weight, 3 ounces.
8L2773 .... **15c**

Air Cushion Arch Supports

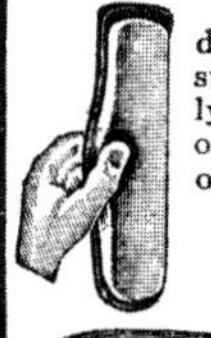

For athletes and all others desiring a resilient, springy support for the arch. Especially helpful for people who are on their feet a great portion of the time.

Nerve Ease Longitudinal.

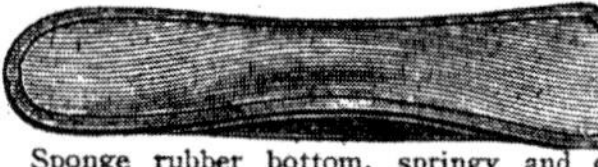

Sponge rubber bottom, springy and durable, covered with a leather top. When walking a suction effect is produced, ventilating the feet. Helps to bring back that quick and snappy step. (See instructions "How to Order" at top of page.) Shipping weight, 8 ounces.
8L2725 .... **$1.19**

Nerve Ease Transversal.

Relieves pain at metatarsal joints (at ball of foot) and helps remove callouses. (See instructions "How to Order" at top of page.)
8L2726 .... **69c**

Heel Reliefs

Keep Your Heels Straight.

Worn inside shoe to prevent the heels from wearing down on the side. **Give size of shoe worn and state whether for man or woman.** Shipping weight, 3 oz.
8L2759—Per pair **15c**

Nerve Ease Heel Cushion.

Made of resilient rubber with leather top. Relieves pressure on heel. Acts as rubber heel inside of shoe. **State size of shoe.** Shipping weight, 7 ounces.
8L2712—Per pair **19c**

Inside Heel Grip.

For inserting inside of low shoes to prevent the shoe from slipping off the heel of the foot. Include a pair with each order for low shoes. Made of leather, in black, brown or white. **Specify color.** Shipping weight, 2 ounces.
8L2750—Pair **13c**

Shoulder Braces and Abdominal Belts

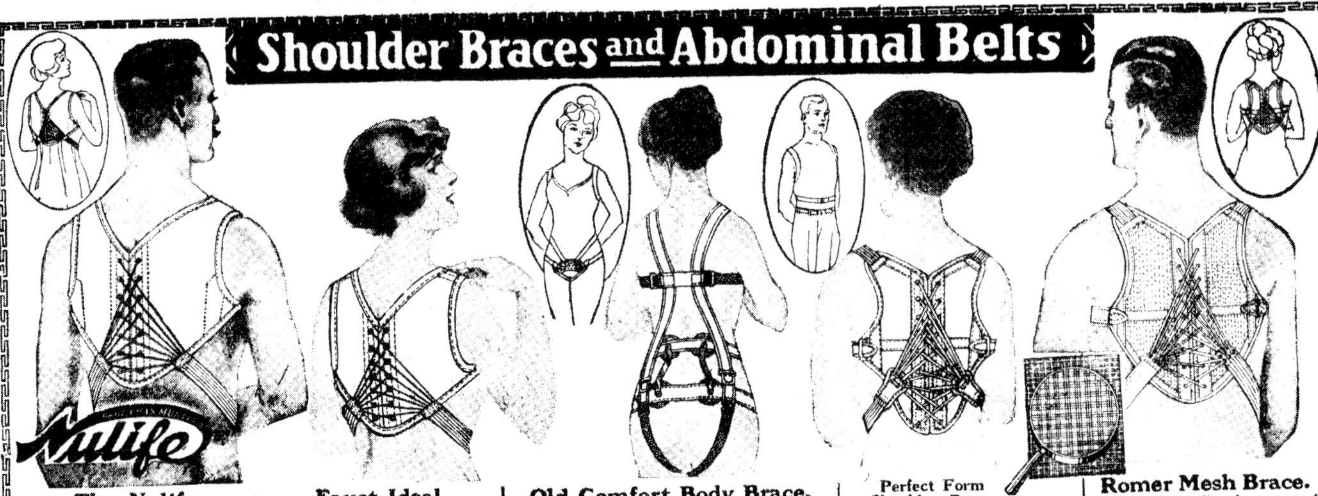

The Nulife Shoulder Brace.

Keeps the shoulders back, the head erect, thereby expanding the lungs and causing deeper and fuller breathing. Sizes, 24 to 44 inches. **State size.** Shipping weight, 8 ounces
8L3725........**$1.69**
NOTE—Have another person take your chest measure all around the body, over the underwear, and up under the arms with your chest fully expanded.

Faust Ideal Shoulder Brace.

For men, women and children. Made of light weight, washable material. Strong, cool and comfortable. An especially good brace for children. Sizes, 24 to 44 inches. **State size.** Shipping weight, 8 ounces.
8L3727........**98c**

Old Comfort Body Brace.

A combination shoulder brace and abdominal supporter. The shoulder straps and waist belt are elastic, but the hip belt and understraps are non-elastic. The nickel plated brass plates are large and comfortable. Sizes, 30 to 52 inches only. Adjustable 6 to 8 inches. Extra set of understraps furnished. Shipping weight, 1¼ pounds.
HOW TO ORDER: State size around body 2 inches below top of hip bones.
8L3700........**$3.79**

Perfect Form Shoulder Brace.

Light, comfortable, strong and washable. Reinforced at all seams. The shoulder straps are adjustable under arms. Sizes, 24 to 44 inches only. **State size.** Adjustable 3 to 5 inches. Shipping weight, 8 oz.
8L3734........**$1.39**

Romer Mesh Brace.

Made of ventilated open mesh cloth; strong, light and comfortable. Adjustable shoulder straps that can be set to the measurements required. Sizes, 24 to 44 inches only. **State size.** Adjustable 3 to 5 inches. Shipping weight, 8 oz.
8L3736........**$1.69**
NOTE—Have another person take your chest measure all around the body, over the underwear, and up under the arms with your chest fully expanded.

Elastic Abdominal Belts

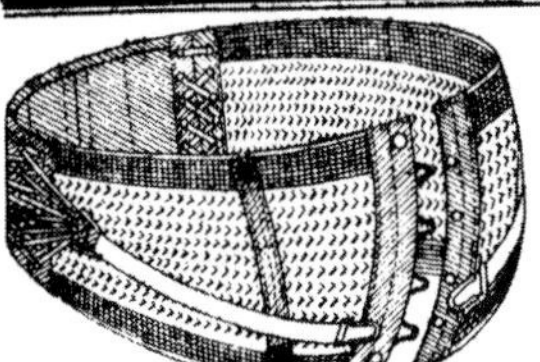

Front Clasp Elastic Supporter.

A comfortable adjustable elastic belt and abdominal supporter for men and women. Made of good quality elastic webbing, carefully woven by machine. Has lace adjustments with lift up strap. Complete with eyelet attachments for understraps. Front reinforced with corset stays. Fastens at front with regular corset clasps. **State largest abdominal measurement.** We allow for stretching. Shipping weight, 13 ounces.

8L3684—Sizes, 30 to 38 inches only. Height at front, about 8 inches....**$3.75**
8L3685—Sizes, 40 to 50 inches only. Height at front, about 9 inches....**$3.98**

Higgins' Elastic Abdominal Supporter.

Popular style elastic supporters. For men and women. Made of good quality elastic webbing. Has additional feature of front corset stays reinforced. **Give largest measurements around abdomen.** We allow for stretching. Shipping weight, 13 ounces.

Higgins' Standard Elastic Supporter.
8L3695—Sizes, 30 to 38 inches only. 8-inch front........**$2.47**
8L3696—Sizes, 38 to 54 inches only. 9-inch front........**$2.89**

Higgins' Best Elastic Supporter.
8L3698—Sizes, 30 to 38 in....**$2.98**
8L3699—Sizes, 38 to 54 in....**3.45**

Hoffman Elastic Abdominal Supporters.

For men and women. Made of elastic webbing woven with cotton. Supports the abdomen. Laces in back. Carried in two grades. **State largest measurement around abdomen.** We allow for stretching. Shipping wt., 14 oz.

Standard Hoffman Elastic Supporter.
8L3653—Sizes, 30 to 38 in....**$1.69**
8L3654—Sizes, 38 to 52 in....**1.89**

Our Best Hoffman Elastic Supporter.
8L3658—Sizes, 30 to 38 in....**$1.98**
8L3659—Sizes, 38 to 52 in....**2.47**

"Faust Understrap" Elastic Supporter.

Our Best Elastic Supporter.

In our opinion the most practical abdominal supporter sold. Made of best quality hand woven elastic webbing, reinforced at top and bottom. Laces in back. Has fancy leather covered corset stays. Bottom of belt and understraps at front padded with chamois. Patented understraps keep the garment in correct position. Carried in silk woven and mercerized. **Give measurement around body at largest part of abdomen.** Shipping weight, 14 ounces.

Silk Hand Woven.
8L3676—Sizes, 30 to 38 in. About 8-inch front........**$8.98**
8L3677—Sizes, 40 to 54 in. About 9-in. front. Each....**$9.98**

Mercerized Hand Woven.
8L3678—Sizes, 30 to 38 in. About 8-inch front. Each....**$7.47**
8L3679—Sizes, 40 to 54 in. About 9-inch front. Each....**$7.98**

Non Elastic Abdominal Belts

Moleskin Cloth Supporters.

Worn by both men and women to support the abdomen. Made of good quality moleskin cloth. Have reinforced corset stays and straps for adjusting on both sides. Has eyelets for attaching understraps. Made in two grades. Sizes, 30 to 54 inches. **Give measurement around body at largest part of abdomen.** Shipping weight, 14 ounces.

"Chicago" Moleskin Cloth Supporter.
8L3668—Made of standard quality moleskin cloth........**$1.98**

Our Best Moleskin Cloth Supporter.
8L3669—Made of extra fine quality moleskin cloth........**$2.98**

Walter's Mesh Cloth Lift Up Supporters.

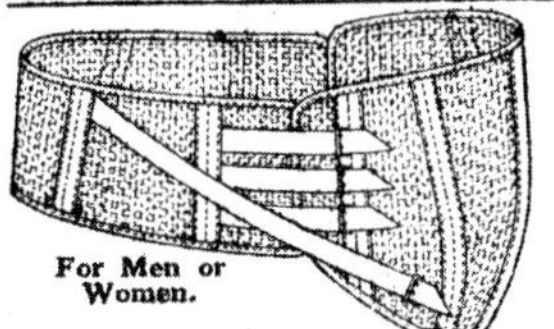
For Men or Women.

For men and women. Woven open mesh cloth. Soft, light, cool and comfortable. Non-elastic. Reinforced with corset stays. Washable. Adjusting straps on each side. Lift up straps give extra support. Carried in two grades. Sizes, 28 to 54 inches only; widths in proportion. **Give measurements around body at largest part of abdomen.** Shpg. wt., 10 oz.

Walter's Standard Mesh Supporter.
8L3656—Made of mesh cloth (double thickness)........**$1.98**

Walter's Best Mesh Supporter.
8L3657—Made of fine quality mesh cloth. Stays are covered with white leather. Each........**$2.69**

"Uplift" Abdominal Belt and Supporter.

An ideal garment for men or women. Fastens in front with special hooks. Laces on both sides. Easily adjusted. Adds grace to the form and supports the abdomen. Light, cool and comfortable. Washable. Can be worn with or without a corset. Sizes, 30 to 52 inches only. **Give measurement around body at largest part of abdomen.** Shipping weight, 10 ounces.
8L3693........**$1.79**

For other Accouchement Bands and Abdominal Supports see page 135.

"Faust Understrap" Non-Elastic Supporter.

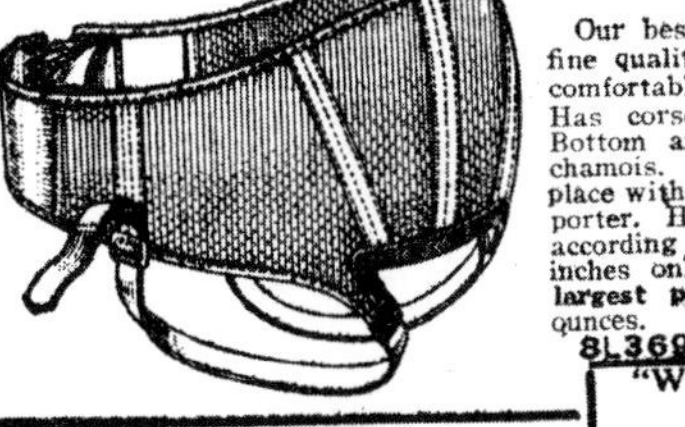

Our best non-elastic supporter. Made of extra fine quality mesh cloth. Light, strong, cool and comfortable. Easily washed. Laces in back. Has corset stays covered with white leather. Bottom and understrap at front padded with chamois. Special understrap feature holds belt in place without discomfort. We recommend this supporter. Height at front varies from 8 to 10 inches according to size of supporter. Sizes, 30 to 54 inches only. **Give measurement around body at largest part of abdomen.** Shipping weight, 10 ounces.
8L3694........**$3.67**

Faust Lift Up Silk Supporter.

For men or women who desire an extra fine quality silk hand woven elastic supporter reinforced at top and bottom. Leather covered, front corset stays. Laces in back. Has patent lift up supporting feature. **State size.** Shipping weight, 12 ounces.

Sizes 30 to 38 inches, about 8-inch front. 8L3667....**$6.98**
Sizes 40 to 54 inches, about 10-inch front. 8L3673....**$7.98**

Uterine Supporters.

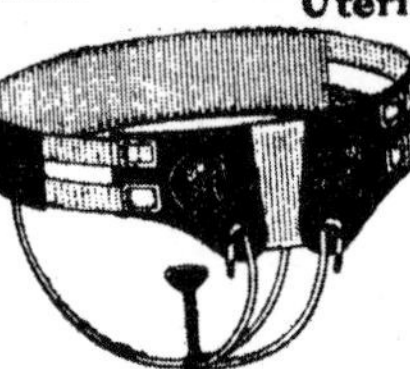

8L3660—Uterine Supporter, complete with rubber tubing and hard rubber cup pessary. Sizes, 28 to 48 inches only. **Give largest abdominal measurement.** Shipping weight, 11 oz....**$1.98**

8L3674—Hard Rubber Pessaries. Shipping wt., 4 oz....**47c**

8L3687—Rubber Tubes. Shipping weight, 4 ounces. Per pair....**21c**

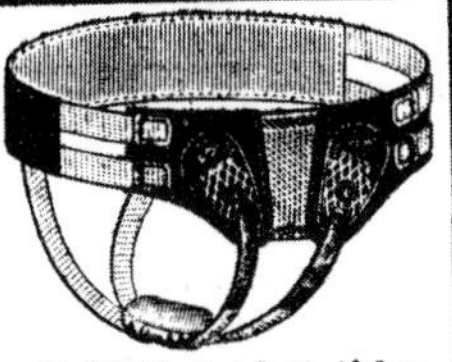

8L3661—London Abdominal Supporter. Complete with small sheepskin pad. Sizes, 30 to 42 inches only. **Give largest abdominal measurement.** Shpg. wt. 11 oz....**$1.89**

"Wrap Around" Abdominal Supporter.

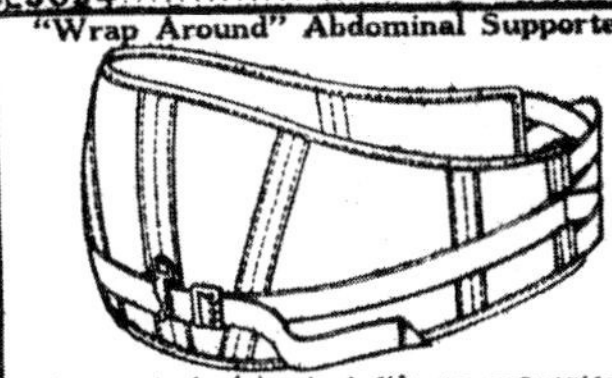

A practical abdominal lift up supporter for everyday use, made of light weight strong cloth. Cool and comfortable. Quickly and easily adjusted. Sizes, 30 to 52 inches. **State size.** Shipping weight, 9 ounces.
8L3666........**$1.47**

BUG KILLERS

Paris Green.

Fine grade Paris Green for killing leaf eating insects. Meets all the requirements of the Insecticide Board at Washington, D. C. Full directions on each package. **Unmailable.**

8L502 — 1 pound ......(Shpg. wt., 1¾ lbs.)......$ 0.36
8L503 — 2 pounds......(Shpg. wt., 2¾ lbs.)...... .69
8L504 — 5 pounds......(Shpg. wt., 5¾ lbs.)...... 1.69
8L505¼— 14 pounds......(Shpg. wt., 15¾ lbs.)...... 4.57
8L506¼—100 pounds......(Shpg. wt., 105 lbs.)...... 29.95

Arsenate of Lead (Powdered).

Of guaranteed quality, for killing leaf eating insects. Full directions on package. Meets all Insecticide Board requirements. **Unmailable.**

Powder Form.

8L574 — 1 pound ......(Shpg. wt., 2 lbs.)......$ 0.36
8L575 — 5 pounds......(Shpg. wt., 7⅛ lbs.)...... 1.67
8L576¼— 10 pounds......(Shpg. wt., 14½ lbs.)...... 2.98
8L577¼— 25 pounds......(Shpg. wt., 31½ lbs.)...... 6.89
8L578¼—100 pounds......(Shpg. wt., 114 lbs.)...... 22.95

Powd. Bordeaux Mixture.

16 Per Cent Strength.

A standard treatment of many fungus diseases of orchard and garden plants.

Special Notice— There are several strengths of Bordeaux Mixture on the market. We sell only the 16 per cent metallic copper strength, the highest obtainable, and popular strength. We do not compete with the lower grades. Directions on each package. **Unmailable.**

	Weight	Shpg. Wt.	
8L520	1 lb.	1½ lbs.	$0.29
8L521	5 lbs.	6½ lbs.	1.29
8L522¼	10 lbs.	14¾ lbs.	2.47
8L523¼	25 lbs.	31 lbs.	4.98

Solution or Powder Lime-Sulphur

We list below a solution of full 31° Baume strength; also Dry Powder which can be diluted with water to make the solution. Two pounds of powder are equivalent to approximately 1 gallon solution. The Lime-Sulphur for spraying is effective against most fungus growths and may be mixed with Arsenate of Lead to make a combined fungicide and insecticide. Powdered Lime-Sulphur may be used to make a dip for controlling sheep scab. Shipping weight of solution, 12 pounds to gallon. **Unmailable.**

Lime-Sulphur Solution.

8L509⅓—47 gallons....$10.98
8L517⅓—25 gallons.... 7.50
8L515¼— 1 gallon..... .79

10-barrel lots, $9.95 per barrel.

Barrels and one-half barrels shipped from warehouse in TOLEDO, OHIO, ROCHESTER, N. Y., and KANSAS CITY, MO.

Lime-Sulphur Dry Powder.

8L526¼—5 pounds. Shipping weight, 7¾ pounds......$1.00
8L527¼—10 pounds. Shipping weight, 12½ pounds.$1.89
8L541⅓—100 pounds. Shipping weight, 102 lbs...$11.35

Blue Vitriol.

Known as Copper Sulphate or Blue Stone. Used in the manufacture of Bordeaux Mixture and other sprays for certain plant diseases; also certain smuts on wheat, rye, etc. **Unmailable.**

8L550 — 5 - pound package. Shipping wt., 6¼ pounds.......59c

8L551 — 10 - pound package. Shipping wt., 11⅝ pounds....$1.15

Powdered Bordeaux Arsenate.

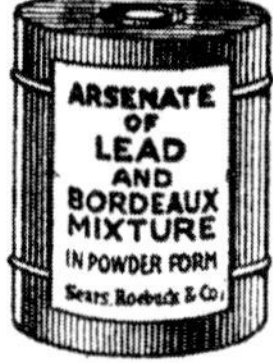

Adhesive poison for killing leaf eating bugs and worms. Will prevent blight on potato vines. Full directions on each package. **Unmailable.**

	Weight	Shpg. Wt.	
8L535	1 lb.	1¾ lbs.	$ 0.35
8L536	5 lbs.	6½ lbs.	1.54
8L537¼	10 lbs.	12½ lbs.	2.79
8L538¼	25 lbs.	28 lbs.	6.37
8L539¼	50 lbs.	56 lbs.	11.97

Tree Tanglefoot.

A sticky material especially adapted for protecting trees against climbing and creeping insect pests, by applying it in bands around the trunks of the trees. Non-injurious. Will remain sticky three months or longer. Keeps perfectly in original packages.

8L542—1-pound tins..$0.39
8L543—5-pound tins.. 1.89

Shipping weights: 1-lb. can, 1¾ lbs.; 5-lb. can, 6 lbs.

Black Leaf "40."

A concentrated 40 per cent Nicotine Solution for destroying aphis, thrips, leaf hoppers and similar sucking insects on rose bushes, shrubs, flowers, vegetables, fruits and plants. Directions on each package. **Unmailable.** Shipping weights, 1 oz., 8 oz.; ½ pound, 1 pound.

8L558—1-ounce bottle, making 6 gallons spray......$0.29
8L556—½-pound tin. 1.19

Formaldehyde.

Extensively used for treating seed potatoes, seed and soil to prevent diseases. Has the approval of the United States Department of Agriculture for use in destroying certain smuts of wheat, oats, barley and rye. U.S.P. strength. **Unmailable.**

8L530—1 - pound (about 15 fluid ounces.) Shipping weight, 3 pounds.......47c
8L531¼—1-gallon jug. Shipping weight, 20 pounds $3.47

Hammond's Slug Shot.

Thoroughly reliable for killing body slugs, raspberry or currant worms. May be used as a dust powder or mixed with water and used as a spray. Does not injure foliage and acts in some measures as a fungicide. Full directions on each package.

8L571—1-pound carton. Shipping weight, 1½ pounds..........19c
8L572—5-pound bag. Shipping weight, 5¾ pounds..........50c
8L573—10-pound bag. Shipping weight, 11¼ pounds..........98c

Tobacco Dust.

Effective against plant lice on currant and berry bushes. Can be used either as a powder or mixed with water as a spray for the control of certain aphids or lice.

	Weight, Lbs.	Shpg. Wt., Lbs.	
8L518	5	5¾	$0.39
8L524¼	25	25⅝	1.39
8L529¼	100	107	4.79

Daisy Fly Killer.

The most popular fly killer on the market, because it does not spill, is not sticky, is not unsightly, is odorless and, above all, very effective. Made of metal, lithographed with daisies in colors, designed to attract the flies, which suck the poison through the small wick in the center of each flower. Will last all season. Shipping weight of three, 10 ounces.

8L2051—3 Fly Killers for..........45c

Fly Preparations

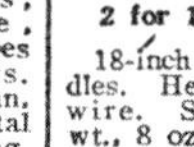

J. O. Bug and Insect Powder. Effective against flies, lice, roaches, mosquitoes and other insects. Harmless to man, fowl or beast. Metal bellows for applying. Shipping weight, 4 oz.

8L2055.......19c

Fly Swatters.

Every home needs fly swatters. Made of fine wire, edges bound, so that they will not mar woodwork or walls.

2 for 15c. 18-inch handles. Heavier wire. Shpg. wt., 8 oz.

8L2053 2 for.....15c

3 for 13c. 14-inch handles. Durable wire. Shpg. wt., 12 oz.

8L2054 3 for....13c

Sticky Fly String. Thirty feet of sticky cotton twine. Pull out a few inches and hang from the ceiling. When full of flies cut off the string and pull more out. Very effective. Shipping weight, 6 ounces.

8L2043 3 for..........10c

Fly-Death.

A Fly Killing Spray for the Home.

One of the quickest, cleanest and most pleasant methods of ridding your home of obnoxious insects—flies, mosquitoes, bedbugs, roaches, moths, etc. Non-poisonous to other than insect life. Pleasant odor. Merely spray around room with sprayer listed below.

8L2060—1 pint Fly-Death, liquid only. Shipping weight, 1⅛ pounds..........49c
8L2061—1 quart Fly-Death, liquid only. Shipping weight, 3 pounds..........79c
8L2062¼—1 gallon Fly-Death, liquid only. Shipping weight, 9 pounds..........$1.98

42L1031—Hand Sprayer, will hold 1 pint of Fly-Death. Shipping weight, 2 pounds...30c

Vegetable Fertilizers

Chemical Fertilizers for Garden, Flowers and Lawn.

Chemically prepared for use where plants do not receive the necessary nourishment from the soil. A soluble, quickly available fertilizer which will maintain healthy, vigorous plants and produce larger crops, more flowers and improve the growing conditions of your lawn. **Shipped from CHICAGO, ILL., or PHILADELPHIA, PENNA.**

8L702—5-lb. bag. Shpg. wt., 5¾ lbs.39c
8L701¼—10-pound bag. Shipping weight, 11¼ pounds..........59c
8L700¼—25-pound bag. Shipping weight, 26 pounds..........98c
8L704¼—100-pound bag. Shipping wt., 102 pounds..........$3.79

Plant Life.

Promotes growth, health and vigor. A life giver to potted plants or flowers in the house or garden. A real food easily absorbed. Dry, rich and lasting. Convenient, economical and easy to use.

8L706—Per 6-ounce can. Shipping wt., 10 ounces..........19c

Raw Phosphate Rock

Phosphated **Unphosphated**

Some very outstanding results have been accomplished in the corn belt states by the application of Ground Raw Rock Phosphate, which proves conclusively that these soils need phosphorus. Maximum crops are not possible without the application of this element, and as Ground Raw Rock Phosphate is the cheapest and most natural form in which phosphorus can be supplied, its value to the farmer is unlimited. Endorsed by Experimental Stations. A trial will convince you of its value. Easily applied.

Reputation Raw Rock Phosphate is guaranteed to contain the equivalent of 14 per cent phosphorus, and to be of 95 per cent fineness through a 100-mesh screen. It is shipped in carload lots only, either in bulk or bags, direct from the rich Tennessee phosphate mines. We can make immediate shipment or on specified future dates. Write for prices and information, addressing the Fertilizer Department, Sears, Roebuck and Co.

Farm Fertilizer

Acid Phosphate and Nitrate of Soda.

Every farmer is acquainted with the value and use of acid phosphate and nitrate of soda in the growing of crops. We are able to give you very prompt shipment and rock bottom prices. For prices and particulars write Fertilizer Department, Sears, Roebuck and Co., Chicago, Ill.

Sheep Manure.

Contains 2 per cent ammonia, 1½ per cent phosphoric acid and 1¼ per cent soluble potash. **Shipped from warehouse in NORTHERN ILLINOIS only.**

8L705⅓—100-pound bag. Shipping wt., 102 pounds..........$ 1.98
8L707⅓—Ton lots.......... 34.75

Steamed Bone.

Contains 3 per cent ammonia and 24 per cent total phosphoric acid. **Shipped from SOUTHERN or NORTHEASTERN ILLINOIS or PHILADELPHIA, PENNA.**

8L711⅓—100-pound bag. Shipping wt., 102 pounds..........$ 2.49
8L710⅓—Ton lots.......... 39.75

Whiticide

Disinfecting White Water Paint.

A non-poisonous disinfecting white water paint for poultry house, stables and cellars to destroy mites, nits and animal parasites. Mixed with water and easily put on with brush or spray. Cover walls and ceilings. Doesn't scale or rub off.

8L647—5-lb. package. Shipping weight, 6¼ lbs.......69c
8L648¼—25-pound package. Shipping weight, 27½ pounds..........$2.69

Poultry Needs

It Pays to Caponize

Relative Sizes. Capons and Non-Caponized.

With Dr. Stanley's Improved Caponizing Instruments.

Caponizing is practiced and endorsed by the leading poultrymen of the country. This simple operation enables one to market birds at a season when most other fowls are off the market; and, due to the scarcity of fowls at these seasons, their added size and quality of the meat, bring fancy prices and are always in great demand.

This set is of simple design, safe, easy to use and very effective and, with the full directions furnished, enables all poultry raisers to easily raise capons. Shpg. weight, 10 oz.

8L1939—Caponizing Set and directions..........$2.98

Sodium Fluoride Lice Powder

Recommended by the U. S. Department of Agriculture as a lice destroyer. Deadly and sure. Becoming more popular each season. One pound, applied dry by "pinch" method, will treat 100 fowls. Directions on package. Shipping weights, 1 pound, 1⅜ lbs; 5 pounds, 5¾ lbs.

87L633—1-pound carton..........$0.29
87L634—5-pound package.......... 1.25

Silver Quill Lice Paste

An efficient remedy for destroying lice and nits (eggs of lice) on chickens. For scaly legs and head lice. Full directions with each tube. Shipping weight, per tube, 7 ounces.

87L628—One tube..........33c
87L629—Three tubes..........89c

White Diarrhoea Remedy

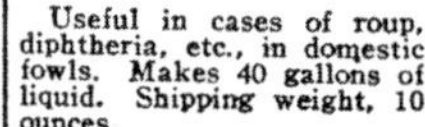

Safe, reliable, carefully made remedy. Directions on package. Shipping weight, 2 oz.

8L915—Per box of 40 tablets to make 10 gallons solution.........33c

Roup Remedy

Useful in cases of roup, diphtheria, etc., in domestic fowls. Makes 40 gallons of liquid. Shipping weight, 10 ounces.

8L914—Per pkg....33c

Water Glass for preserving eggs. See page 478 for description.

Nest Eggs

Every nest should have a nest egg. These eggs give excellent results, are made of a special composition containing lime and are not easily broken. Twelve eggs in a carton. Shipping weight, 2⅛ pounds.

8L901—12 Nest Eggs for 29c

Farm Chemicals

Stock Supplies

Glauber Salts.

Valuable as a good stock regulator and laxative.

8L785⅓—100-pound bag. Shipping weight, 110 pounds $2.89

8L786⅓—350-pound barrel. Shipping weight, 375 pounds $7.89

Shipped from warehouse in NORTHEASTERN ILLINOIS or EASTERN PENNSYLVANIA.

Commercial Sulphur.

Excellent as a stock conditioner or for making spraying mixtures.

8L754⅓—100-pound bag. Shipping weight, 102 pounds $2.69

Shipped from warehouse in NORTHEASTERN ILLINOIS.

Epsom Salts.

A good bowel regulator for stock. Keeps perfectly. A small quantity should be given regularly with the feed.

8L758⅓—100-pound bag. Shipping weight, 102 pounds $3.49

Granular Copperas.

A chemical largely employed around farms.

8L762⅓—100-pound bag. Shipping weight, 110 pounds $2.75

Shipped from warehouse in NORTHEASTERN ILLINOIS, NORTHERN OHIO or EASTERN PENNSYLVANIA.

Reputation Worm Powder

(With Santonin.)

Worms devour your feed, make your hogs subject to disease and rob you of your profits. Our Worm Powder with Santonin is a reliable and dependable preparation. One pound is sufficient for thorough treatment of fifteen hogs, thirty shoats or sixty pigs. Perfectly safe and easily administered. Full directions. Shipping weight, 1¾ pounds.

8L850—Per 1-pound carton $5.98

Dr. Edwards Hard Milker Outfit

This compact set contains teat slitter, teat expander, teat dilator, milk tube and a cake of antiseptic soap. For treating hard milkers. Full directions. Shpg. wt., 6 oz.

8L1879 .... $2.39

Milk Fever Outfit.

For Sterilized Air Treatment.

Milk fever, occurring at calving time and when the milk flows most freely, is an acute disease and very often fatal unless treatment is given at once. When treatment is properly given in time over 90 per cent of the cows recover. Simply inject sterilized air into the udder. Shipping weight, 14 ounces.

8L1895 .... $2.49

Self Retaining Milking Tubes.

Excellent for sore, obstructed teats and hard milkers. Nickel plated. Four assorted lengths to set. Safe and easy to use. Shipping weight, 2 ounces.

8L1975—Set of 4 tubes 32c

Cow Teat Slitter.

Nickel plated with steel blade. For hard milkers and cutting strictures. Length, 7¾ in. Ring handle for holding. Shipping weight, 4 ounces.

8L1985 .... $1.59

Cattle Trocar and Canula.

For draining abscesses and reducing bloat. Black finish wooden handle. Shpg. wt., 9 oz.

8L1973 .... 59c

8L1974—Extra Canula. Brass, nickel plated. Shipping weight, 2 ounces 39c

Self Retaining Teat Dilator.

For expanding the milk canals of hard milkers. Brass, nickel plated. Shpg. wt., 1 oz.

8L1977 .... 19c

Reputation Fly Killer

Mr. Dairyman!

Please Read About This Product. Kills the fly, does not only drive it away. Harmless, stainless and just what you want. Highly endorsed by all users.

BEFORE

AFTER

The Reputation Fly Killer was a new product last season, but its popularity and the very satisfactory results obtained by our customers proves it, in our opinion, the best fly killer on the market. The testimonials from the users of this product prove to us that it more than fulfills every claim made for it and that we now offer you a really reliable and satisfactory fly exterminator.

Reputation Fly Killer is a harmless, colorless, liquid spray with a pleasant odor, which will not blister, stain or discolor animals of any kind. Especially adapted to "Prize" or "Show" stock, as no discoloration results.

Reputation Fly Killer is deadly to the tormenting "horn fly" or "belly fly" or other insects which cause nervousness, thereby reducing the flow of milk. It also keeps flies off much longer than a fly spray. Buy this on our recommendation. We know you will be more than pleased. Shipping weight, per gallon, 8½ pounds.

8L604¼— 1-gallon can $1.39

8L605¼— 5-gallon can 5.98

8L606¼—10-gallon can 10.98

Reputation Fly Spray.

An Excellent Fly Preventive.

Biting, stinging, bothersome flies greatly torment your stock, make them nervous, thereby reducing the flow of milk. Reputation Fly Spray will keep flies off animals from four to five hours. Very effective, easy to apply and pleasing odor. Spray and quiet your cows, thereby increasing the milk flow. No more injuries from kicking.

8L620¼— 1-gallon can $0.98

8L621¼— 5-gallon can 3.98

8L622¼—10-gallon can 6.98

Shipping weight, per gallon, 8½ pounds.

Stock Feeding Molasses

Each barrel, about 54 gallons. Shpg. wt., each barrel, 700 lbs.

Molasses has enabled feeders to market steers and hogs early for top prices. It gives gloss and finish, thereby raising the grade.

An appetizing, energy producing food, and when used with roughages like straw, hay or corn fodder, or in a slop, these feeds are made more palatable and digestible and are turned more easily into weight and energy. Increases the milk flow in cows. Horses do more work. Hogs and steers gain weight more rapidly.

It has been found that when replacing not more than half the corn in a ration, molasses has slightly more feeding value, pound for pound, than this grain. A great aid to the present day feeder in utilizing cheap foods. **Shipped from the following distributing stations:**

8L836⅓—Cedar Rapids, Iowa $10.39

8L842⅓—Philadelphia, Penna., or New York, N. Y. $6.98

8L838⅓—Chicago, Ill. $9.89

8L841⅓—Savannah, Ga., or Norfolk, Va. 8.79

In 5-barrel lots, 25 cents per barrel less. Carloads, 50 cents per barrel less. Carload, 60 barrels.

Santonin Worm Tablets for Hogs

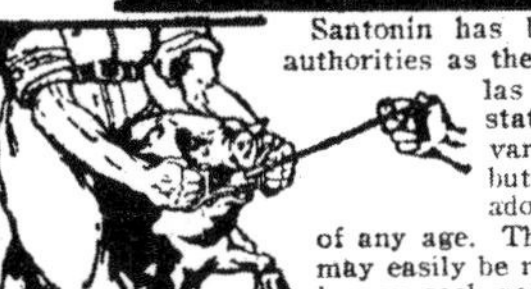

Santonin has been recommended by state and government authorities as the best product to get rid of worms. The formulas listed below have been approved by prominent state experiment stations. In the past we have listed various tablets made for hogs of different ages, but for your convenience and economy we have adopted a new tablet which may be used for hogs of any age. These tablets are creased in the center so dosage may easily be regulated. Full directions as to dosage and feeding on each package. Shipping weight, 10 ounces.

Without Areca Nut or Oil of Wormseed.

Each tablet contains 3 grains santonin, 3½ grains calomel and 7 grains soda bicarbonate.

8L882—25 tablets $4.39

With Areca Nut and Oil of Wormseed.

Each tablet contains 3 grains each of santonin and calomel and 7 grains soda bicarbonate with areca nut and oil of wormseed.

8L883—25 tablets $4.79

Crude Carbolic Oil.

Valuable for prevention of disease in the chicken yard, barns, horse stalls, hog pens, etc. Keep a supply always on hand. Unmailable.

8L630¼—1-gallon can. Shipping weight, 10 pounds 98c

8L631¼—5-gallon can. Shipping weight, 48½ pounds $3.67

Mouth Gag.

A cast iron nicely finished gag for shotes or hogs. Excellent for giving our worm tablets. Two handles, making it easy to use. Size, about 9½x4¼ inches. Shipping weight, 1 pound.

8L1884 .... 47c

Veterinary Instruments

Combination Horse Mouth Float.

Easy to operate. Has straight attachment for lower molars and angle head for uppers. Nickel plated steel shanks. Black enameled wooden handle. Shpg. wt., 2 lbs.

8L1862 .... $2.39

Extra Steel Blades.

For above instruments. Rasp on one side, file on the other side. Length, 3¼ in. Shpg. wt., 3 oz.

8L1863 .... 21c

Nickel Plated Brass Syringes.

Tight fitting plunger; two pipes, one for giving medicine, one for washing cuts and wounds. Brass, heavily nickel plated. Shipping weight, 1 pound.

8L1918—Capacity, 2 ounces $0.98

8L1919—Capacity, 4 ounces 1.35

Double Blade Castrating Knives.

Metal handle, steel blade and hook, spring back.

8L1893 .... $1.49

One-Blade Knife, similar to above, but without hook. Steel blade. Shpg. wt., 2½ oz.

8L1892 .... 89c

Emasculators.

Emasculators, different shaped cutters for different methods. Made of steel, polished and nickel plated. Handles, dull finish.

Safety First Roller Emasculator.	Eclipse Emasculator.	Dr. Edward's Improved Emasculator.
Shipping weight, 1¾ pounds.	Shpg. wt., 1½ lbs.	Shipping weight, 1⅞ pounds.
8L1880 $9.98	8L1890 Straight $6.98	8L1878 $9.98
	8L1891 Curved for standing operation. $6.98	

Molar Cutters and Forceps.

Made of tool steel, polished and nickel plated. Length, without handles, 10½ in. Shpg. wt., 3 lbs.

8L1853	8L1854	8L1850
Closed for trimming.	Open for cutting.	Forceps for extracting.
$6.98	$6.98	$6.98

For handles see 8L1856.

Instrument Handles.

Fit any of the above molar cutters or forceps. Polished, nickel plated. Corrugated hardwood handles, 10½ inches long. Shipping weight, 2½ pounds.

8L1856 .... $2.89

Blackleg

Vaccinate your young cattle before you turn them on pasture in the spring and protect them against Blackleg. Pellet set consists of a nickel plated injector and vial containing ten Blackleg Pellets, sufficient to vaccinate ten cattle. The injector has nickel plated safety cap which fits over the needle to protect it when not in use. Full directions for use.

Blackleg Pellet Outfit.

8L1871—Complete outfit with 10 pellets. Shipping weight, 8 ounces $1.98

8L1872—Vial containing 10 pellets. Shipping weight, 1 ounce 83c

8L1873—Vial containing 50 pellets. Shipping weight, 2 ounces $3.69

Blackleg Filtrate Outfit.

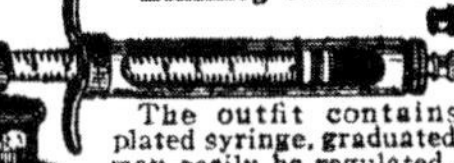

The outfit contains nickel plated syringe, graduated so doses may easily be regulated, and two hypodermic needles. Bottle containing ten doses of Blackleg Filtrate and full directions.

8L1864—Complete outfit, syringe, hypodermic needles and bottle of Filtrate. Shipping weight, 1½ pounds $4.79

8L1865—Bottle containing ten doses of Blackleg Filtrate. Shipping weight, 10 oz $1.98

Reputation Dip

Disinfect your barns and yards by spraying, and your stock by dipping. A reliable coal tar animal dip and disinfectant for extermination of parasites, lice, ticks, fleas and many other insects. One gallon makes 65 gallons of dip. Costs about 2 cents a gallon, ready to use. Directions on package.

8L610¼—1-gallon can. Weight, 10½ pounds $1.25

8L611¼—5-gallon can. Weight, 53½ pounds $4.79

8L612¼—10-gallon can. Weight, 95 pounds $8.98

8L613⅓—50-gallon barrel. Weight, 350 pounds $24.98

Fifty-gallon barrels shipped from factory in NORTHEASTERN ILLINOIS.

Hercules Mouth Speculum

A strong, malleable iron and steel speculum, polished and nickel plated. Safety for the operator, coupled with every advantage offered in a speculum. A single pull on chain closes it entirely. Can be taken apart and carried in grip. Easily cleaned. Shipping weight, 7 pounds.

8L1886 .... $8.98

Solid Brass Barrel Syringe.

Solid brass, full capacity syringe with two extension pipes. Shipping weights, 8 ounces, 1½ pounds; 18 ounces, 6 pounds.

8L1950— 8 ounces $2.67

8L1952—18 ounces 4.47

Forceps.

Nickel Plated Tool Steel.

8L1842—Length, 13¼ inches. Shipping weight, 1½ pounds $3.98

8L1841—Length, about 9½ inches. Shipping weight, 1½ pounds $3.59

Improved Pig Forceps.

Malleable iron, nickel plated, dull finish. Will not injure sow. Blades easily separated for insertion. Shipping weight, 1¾ pounds.

8L1963 .... $2.98

Reliable Pig Forceps.

Malleable iron, nickel plated, dull finish. Use on large or small sows. Shipping weight, 2 pounds.

8L1965 .... $1.25

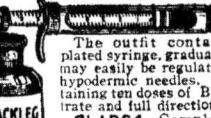

RUBBER GOODS

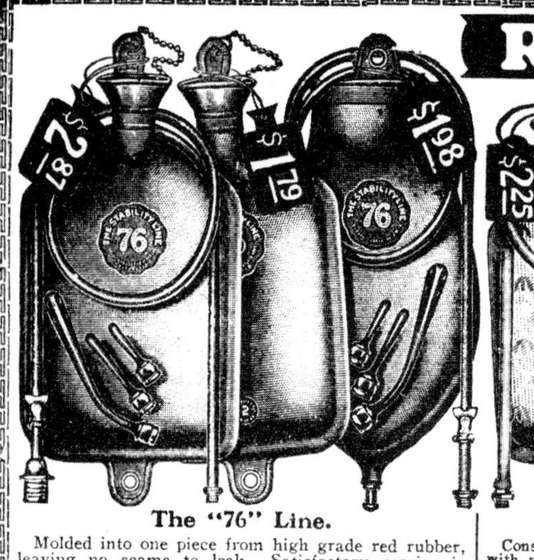

The "76" Line.

Molded into one piece from high grade red rubber, leaving no seams to leak. Satisfactory service is assured. Capacity, each article, full 2 quarts. Equipped as illustrated. Shipping weight, 1⅝ pounds.

Combination Syringe and Water Bottle.	Water Bottle.	Fountain Syringe.
8L2376 ....$2.87	8L2434 $1.79	8L2308 $1.98

The "77" Line.

Constructed in one piece from good quality red rubber, with no seams or joints to leak. Faint leaf design embossed in rubber. Equipped as illustrated. Shipping weight, 1⅝ pounds.

Combination Syringe and Water Bottle.	Water Bottle.	Fountain Syringe.
Capacity, about 2 qts. 8L2356 ... $2.25	Capacity, about 2 quarts. 8L2406 $1.33	Capacity, about 2 quarts. 8L2306 $1.57

Our High Grade Line.

In our opinion the best that money can buy. Handmade articles of fine quality red rubber. Efficient and lasting service are the special features. Soft and pliable. Capacity, water bottle and combination, 2 qts.; fountain syringe, 3 qts. Equipped as illustrated. Shipping wt., 1⅝ lbs.

Fountain Syringe.	Water Bottle.	Combination Syringe and Water Bottle.
8L2303 $2.79	8L2403 $1.98	8L2353 ....$3.25

White Seamless Irrigator.

Metal, welded into one piece, then enameled white. Can be scalded to keep clean and sanitary. Shipping weight, 2¾ pounds.

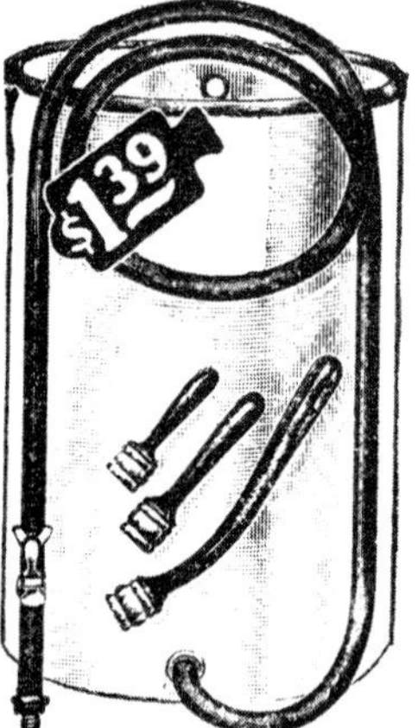

SCREW PIPE SET.

Heavy red rubber tubing with pipes, as illustrated.

8L2335 2-Quart Size **$1.39**

8L2336 3-quart size... **$1.59**

SLIP PIPE SET.

Irrigator as above, but with lighter weight tubing and equipped with slip pipes.

8L2337 2-quart size **$1.19**

Women's Bulb Syringes

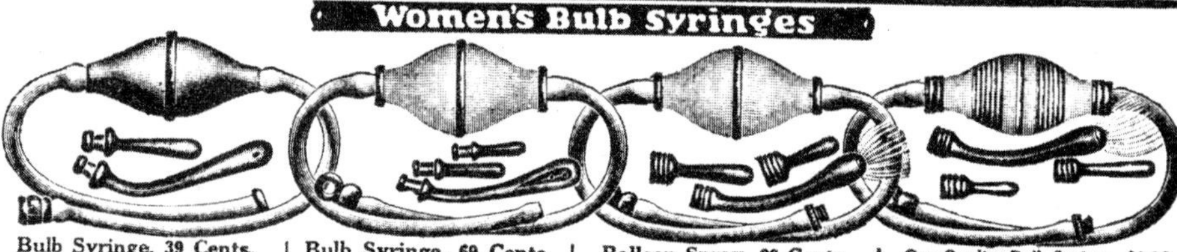

Bulb Syringe, 39 Cents. Medium size bulb of red rubber. Two slip pipes, as illustrated. Shipping weight, 10 ounces.
8L2343 ....**39c**

Bulb Syringe, 69 Cents. Good Grade Chocolate Color Rubber Bulb Syringe. Three polished black hard rubber slip pipes and rapid flow tubing. Shipping wt., 10 ounces.
8L2345 ....**69c**

Balloon Spray, 98 Cents. Red rubber bulb with rapid flow rubber tubing. Three polished hard black rubber screw pipes, including vaginal balloon spray. Shipping weight, 10 ounces.
8L2344 ....**98c**

Our Quality Bulb Syringe, $1.39. High grade red rubber bulb, large size, with rapid flow tubing. Black hard rubber screw pipes, including vaginal balloon spray, as illustrated. Shipping weight, 10 ounces.
8L2348 ....**$1.39**

Breast Pump.

A popular style pump—standard for years. Red rubber bulb and heavy glass bell shaped end. Shipping weight, 12 ounces.
8L2466 ....**33c**

Superservice Breast Pump.

As the name implies, this is a high grade breast pump. Made from excellent material into an article that is practical and durable. Bulb so patterned as to cause an even suction. Shpg. wt., 14 oz.
8L2463 ....**59c**

Combination Sets.

Five feet of good quality extra heavy rubber tubing, three hard rubber screw pipes, one connection cap and one shut off.
8L2468—Red..................**59c**
Similar to above, but lighter weight tubing and slip pipes.
8L2470......................**33c**
Shipping weight, 8 ounces.

Turkish Bath Cabinet.

$7.98

This cabinet is made with full steel support construction, one thickness of black waterproof cloth. Outside is made of neatly printed material. Patented opening top. Alcohol heater or vaporizer included. Size, set up, 26¾x29½x41 inches. Unmailable. Shipping weight, 22 pounds.
8L4005¼.............**$7.98**

Rubber Household Gloves

Carried in all sizes, 7 to 10, including half sizes. All gloves reinforced at wrist. Rubber gloves should be worn loose.

How to Measure—Hold hand out flat with fingers touching, thumb raised; draw tape close, but not tight, as shown in illustration (do not include thumb), and add 1 inch, which will be your correct glove size. For example, if your hand measures 7 inches in this manner your glove size will be 8. **State nearest size.** Shipping weight, 8 ounces.

Daisy Gloves. Red Rubber. 8L2498 ....**39c**

Seamless Gloves. Chocolate Rubber. 8L2497 ....**67c**

Purity Gloves. Tan Rubber. 8L2495 **89c**

Cloth Inserted Ice Cap.

A red rubber cloth inserted ice cap. Has four loops for attaching to body. We recommend this ice cap for service and quality. Shipping weight, 11 oz.
8L2516 ....**$1.39**

Rubber Sponges.

Big Size, **39c**

Impart a gentle friction to the skin so pleasant while bathing. Many other uses. Shipping weight, 2 oz.
8L2526—Size, 5½x4 x2 in...**25c**
8L2527—Size, 7 x4½x2 in...**39c**

Electricians' Gloves.

Extra heavy red rubber, for electricians and workers in factories. Can be used to protect hands when spraying trees. Large sizes only, 9, 10 and 11 inches over knuckles. **State size.** Shipping weight, per pair, 1 pound
8L2501.....................**$1.98**

Waterproof Sheeting or Blanket.

Heavy black rubber sheeting, impervious to moisture. Durable, strong and wear resisting. Used as a protective covering for bedding, etc., in outdoor sleeping porches or camps. Can also be used in the nursery or sick room. Size, 45x71 inches. Shipping weight, 2⅜ pounds.
8L2512.....................**$2.39**

Complexion Brushes.

For aiding removal of blackheads, roughness and dead cuticle. For use wet or dry. Oval shape, fine quality red rubber with heavy rubber teeth. Size, 5¾x3¼ inches.
8L2460 ....**34c**
Similar to above, but not so good quality.
8L2461 Shipping weight, 4 ounces. **19c**

Complexion Mask.

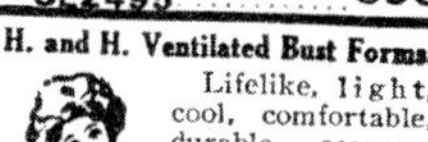

Often called beauty masks. A light weight rubber mask and preferred to the heavier type. Used to smooth wrinkles. Fine quality rubber. Shaped to the face, with cut-outs for eyes, nose and mouth. Strings for tying. Shipping weight, 5 ounces.
8L2492..........**$1.39**

Chin Band.

Used for the reduction of double chin and overfatness under the chin. Easy to use. Made of soft rubber and can be washed. Shipping weight, 3 ounces.
8L2464..........**59c**

H. and H. Ventilated Bust Forms.

Lifelike, light, cool, comfortable, durable, economical and cleanly. Can scarcely be detected from natural bust. Enclosed in cloth cover, lace trimmed; will not get out of place. Expanded by light resilient filling, which can be removed and washed and returned in a few minutes. Pin punctures have no effect. Shipping weight, 8 ounces.
8L2475—Round style. Per pair **$1.98**
8L2476—Oblong style. Per pair **2.47**

Rubber Catheters.

Catheter. Sizes, 12, 14, 16, 18, 20 and 22, French scale. Send size in American or English scale and corresponding size will be shipped. If no size is given medium will be sent. Shipping weight, 3 ounces.
8L2510.....................**19c**

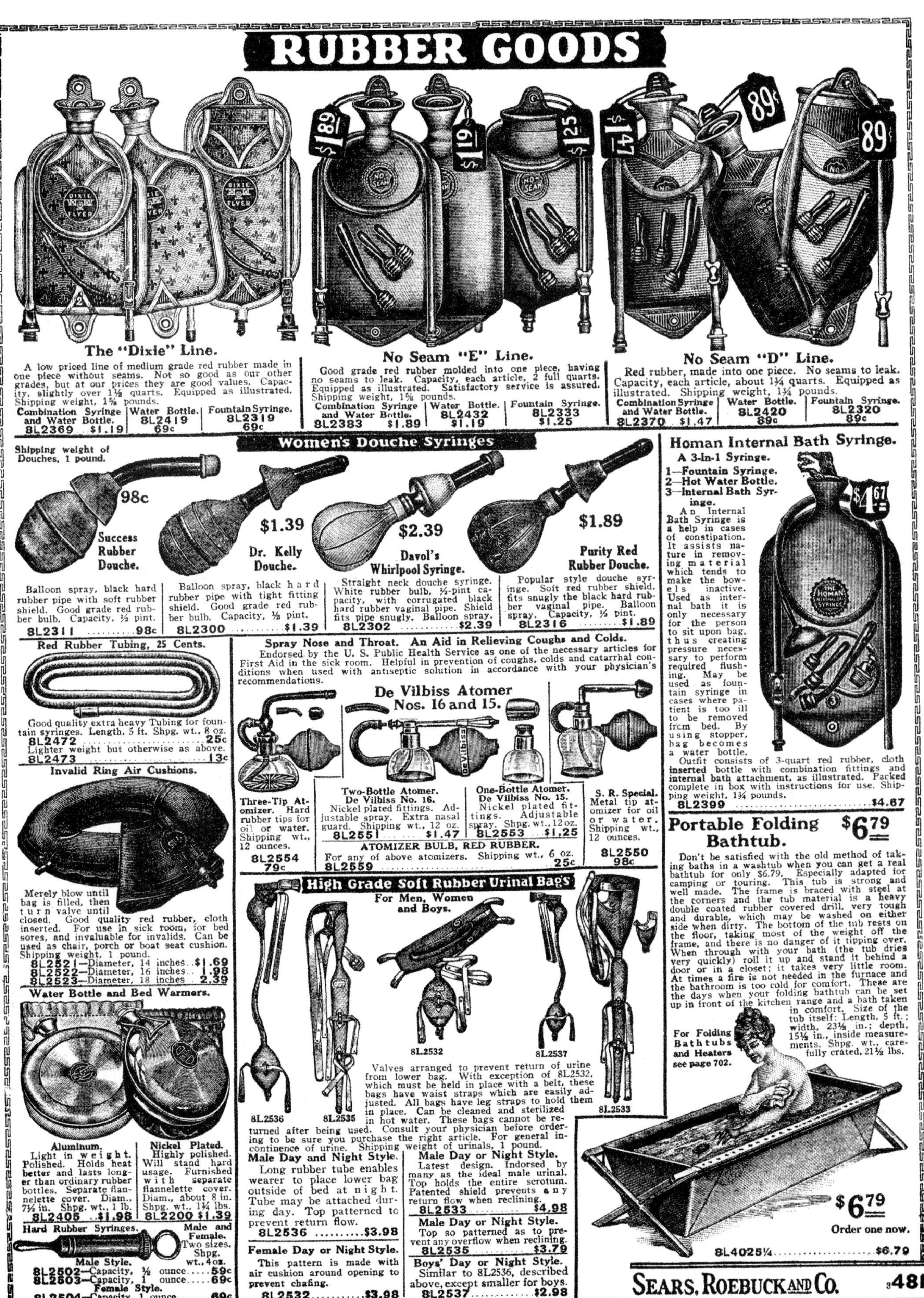

RUBBER GOODS

The "Dixie" Line.

A low priced line of medium grade red rubber made in one piece without seams. Not so good as our other grades, but at our prices they are good values. Capacity, slightly over 1½ quarts. Equipped as illustrated. Shipping weight, 1⅝ pounds.

Combination Syringe and Water Bottle.	Water Bottle.	Fountain Syringe.
8L2369 $1.19	8L2419 69c	8L2319 69c

No Seam "E" Line.

Good grade red rubber molded into one piece, having no seams to leak. Capacity, each article, 2 full quarts. Equipped as illustrated. Satisfactory service is assured. Shipping weight, 1⅝ pounds.

Combination Syringe and Water Bottle.	Water Bottle.	Fountain Syringe.
8L2383 $1.89	8L2432 $1.19	8L2333 $1.25

No Seam "D" Line.

Red rubber, made into one piece. No seams to leak. Capacity, each article, about 1¾ quarts. Equipped as illustrated. Shipping weight, 1¾ pounds.

Combination Syringe and Water Bottle.	Water Bottle.	Fountain Syringe.
8L2370 $1.47	8L2420 89c	8L2320 89c

Women's Douche Syringes

Shipping weight of Douches, 1 pound.

Success Rubber Douche. 98c

Balloon spray, black hard rubber pipe with soft rubber shield. Good grade red rubber bulb. Capacity, ½ pint.
8L2311 .... 98c

Dr. Kelly Douche. $1.39

Balloon spray, black hard rubber pipe with tight fitting shield. Good grade red rubber bulb. Capacity, ⅓ pint.
8L2300 .... $1.39

Davol's Whirlpool Syringe. $2.39

Straight neck douche syringe. White rubber bulb, ½-pint capacity, with corrugated black hard rubber vaginal pipe. Shield fits pipe snugly. Balloon spray.
8L2302 .... $2.39

Purity Red Rubber Douche. $1.89

Popular style douche syringe. Soft red rubber shield, fits snugly the black hard rubber vaginal pipe. Balloon spray. Capacity, ½ pint.
8L2316 .... $1.89

Homan Internal Bath Syringe.

A 3-In-1 Syringe.
1—Fountain Syringe.
2—Hot Water Bottle.
3—Internal Bath Syringe.

An Internal Bath Syringe is a help in cases of constipation. It assists nature in removing material which tends to make the bowels inactive. Used as internal bath it is only necessary for the person to sit upon bag, thus creating pressure necessary to perform required flushing. May be used as fountain syringe in cases where patient is too ill to be removed from bed. By using stopper, bag becomes a water bottle.

Outfit consists of 3-quart red rubber, cloth inserted bottle with combination fittings and internal bath attachment, as illustrated. Packed complete in box with instructions for use. Shipping weight, 1¾ pounds.
8L2399 .... $4.67

Red Rubber Tubing, 25 Cents.

Good quality extra heavy Tubing for fountain syringes. Length, 5 ft. Shpg. wt., 8 oz.
8L2472 .... 25c
Lighter weight but otherwise as above.
8L2473 .... 13c

Invalid Ring Air Cushions.

Merely blow until bag is filled, then turn valve until closed. Good quality red rubber, cloth inserted. For use in sick room, for bed sores, and invaluable for invalids. Can be used as chair, porch or boat seat cushion. Shipping weight, 1 pound.
8L2521—Diameter, 14 inches .. $1.69
8L2522—Diameter, 16 inches .. 1.98
8L2523—Diameter, 18 inches .. 2.39

Water Bottle and Bed Warmers.

Aluminum. Light in weight. Polished. Holds heat better and lasts longer than ordinary rubber bottles. Separate flannelette cover. Diam., 7½ in. Shpg. wt., 1 lb.
8L2405 .. $1.98

Nickel Plated. Highly polished. Will stand hard usage. Furnished with separate flannelette cover. Diam., about 8 in. Shpg. wt., 1¼ lbs.
8L2200 $1.39

Hard Rubber Syringes.

Male and Female. Two sizes. Shpg. wt., 4 oz.

Male Style.
8L2502—Capacity, ½ ounce 59c
8L2503—Capacity, 1 ounce 69c
Female Style.
8L2504—Capacity, 1 ounce 69c

Spray Nose and Throat. An Aid in Relieving Coughs and Colds.

Endorsed by the U. S. Public Health Service as one of the necessary articles for First Aid in the sick room. Helpful in prevention of coughs, colds and catarrhal conditions when used with antiseptic solution in accordance with your physician's recommendations.

De Vilbiss Atomer Nos. 16 and 15.

Three-Tip Atomizer. Hard rubber tips for oil or water. Shipping wt., 12 ounces.
8L2554 79c

Two-Bottle Atomer. De Vilbiss No. 16. Nickel plated fittings. Adjustable spray. Extra nasal guard. Shipping wt., 12 oz.
8L2551 ...... $1.47

One-Bottle Atomer. De Vilbiss No. 15. Nickel plated fittings. Adjustable spray. Shpg. wt., 12 oz.
8L2553 ...... $1.25

S. R. Special. Metal tip atomizer for oil or water. Shipping wt., 12 ounces.
8L2550 98c

ATOMIZER BULB, RED RUBBER.
For any of above atomizers. Shipping wt., 6 oz.
8L2559 .... 25c

High Grade Soft Rubber Urinal Bags

For Men, Women and Boys.

8L2536 8L2535 8L2532 8L2537 8L2533

Valves arranged to prevent return of urine from lower bag. With exception of 8L2532, which must be held in place with a belt, these bags have waist straps which are easily adjusted. All bags have leg straps to hold them in place. Can be cleaned and sterilized in hot water. These bags cannot be returned after being used. Consult your physician before ordering to be sure you purchase the right article. For general incontinence of urine. Shipping weight of urinals, 1 pound.

Male Day and Night Style.
Long rubber tube enables wearer to place lower bag outside of bed at night. Tube may be attached during day. Top patterned to prevent return flow.
8L2536 .... $3.98

Female Day or Night Style.
This pattern is made with air cushion around opening to prevent chafing.
8L2532 .... $3.98

Male Day or Night Style.
Latest design. Indorsed by many as the ideal male urinal. Top holds the entire scrotum. Patented shield prevents any return flow when reclining.
8L2533 .... $4.98

Male Day or Night Style.
Top so patterned as to prevent any overflow when reclining.
8L2535 .... $3.79

Boys' Day or Night Style.
Similar to 8L2536, described above, except smaller for boys.
8L2537 .... $2.98

Portable Folding Bathtub. $6.79

Don't be satisfied with the old method of taking baths in a washtub when you can get a real bathtub for only $6.79. Especially adapted for camping or touring. This tub is strong and well made. The frame is braced with steel at the corners and the tub material is a heavy double coated rubber covered drill, very tough and durable, which may be washed on either side when dirty. The bottom of the tub rests on the floor, taking most of the weight off the frame, and there is no danger of it tipping over. When through with your bath (the tub dries very quickly) roll it up and stand it behind a door or in a closet; it takes very little room. At times a fire is not needed in the furnace and the bathroom is too cold for comfort. These are the days when your folding bathtub can be set up in front of the kitchen range and a bath taken in comfort. Size of the tub itself: Length, 5 ft.; width, 23½ in.; depth, 15⅛ in., inside measurements. Shpg. wt., carefully crated, 21½ lbs.

For Folding Bathtubs and Heaters see page 702.

$6.79 Order one now.

8L4025¼ $6.79

Nursery Supplies

My Baby's Castile Soap.

A fine imported pure olive oil Castile Soap. Made in sunny Spain, carefully prepared from pure olive oil. Cleansing and healing, non-irritating and will not chafe. The best soap for baby's tender skin. Each cake weighs 4 ounces. Shpg. wt., 3 cakes, 1 lb.
8L2155—3 cakes for............**49c**

Baby Bath Sponges. Very soft. Known as silk sponges. Shipping weight, 1 ounce.
8L2116—Large size............**49c**
8L2117—Medium size............**27c**

Baby Bath Tub. White enameled. Edge enameled blue. A pleasing combination. No sharp edges to injure baby. Size, about 20x15¾x5 inches. Shipping weight, 10½ pounds.
8L2168¼....................**$1.98**

For other Nursery Supplies see pages 668 and 669.

Individual White Pyralin Pieces.

Brush.	Soap Box.	Comb.
Length, 4¾ inches. Soft bristles. Shpg. wt., 2 oz.	Will hold cake of baby soap. 2½x3⅜ inches. Shpg. wt., 5 oz.	Length, 4¼ in. Shpg. wt., 2 oz.
8L2236	**8L2240**	**8L2238**
35c	**19c**	**14c**

Baby Basket

Straw chip and willow basket. Wood and plaited willow bottom. Size, about 16x13x4½ inches. Shipping wt., 3 lbs.
8L2131............................**98c**

For larger Baby Baskets see page 668.

Babies' Waterproof Crib Sheet.

Use a White Rubber Coated Sheet to Protect Baby's Mattress. The edges of these sheets are turned and eyelets are fitted so that the sheet can be fastened over the mattress. A very serviceable article. Size, 35x50 inches. Shpg. wt., 1½ lbs.
8L2619......................**$1.39**

Laxatives.

Fletcher's Castoria. Laxative for children. Shpg. wt., each, 14 oz.
8L2250 Each.....**29c**
8L2251 3 bottles. **79c**

Glycerin Suppositories.
8L2252 1 dozen to bottle. Infant size. Shipping wt., 10 oz........**19c**

Red Rubber Syringes.

Eye, Ear and Ulcer Syringe. Often used as a rectal syringe for infants, on account of its soft rubber tip. Capacity, 1 ounce. Shpg. wt., 2 oz.
8L2625..15c

Infants' Syringe with hard rubber rectal pipe. Cap., 1½ oz. Shipping wt., 2 ounces.
8L2112 **15c**

Infants' large size with hard rubber rectal pipe. Cap., 3½ oz. Shpg. wt., 8 oz.
8L2114 **39c**

Birth Announcements. / Folders and Post Cards.

Folders, 6 for 39c. Our new quality ribbon tied announcement folders. Pink or blue with borders of darker shade. Embossed letters. Size, 4⅛x2⅞ inches. Envelope for each. Six folders per package. Shpg. wt., 2 oz.
8L2181—Blue, decorated..............**39c**
8L2182—Pink, decorated..............**39c**

Folders, 10 for 33c. A popular announcement. Neatly designed. Embossed letters. Size, 3x2 inches. Ten folders in package. Envelope for each. Shipping weight, 2 ounces.
8L2178—Pink, decorated..............**33c**
8L2179—Blue, decorated..............**33c**

Post Cards, 10 for 15c. Ten gilt edge high quality cards. Assorted designs. Shipping weight, 2 ounces.
8L2180—Envelope of **10**...........**15c**

All weights given on this page are approximate and may vary a trifle.

Baby's Hand Decorated Sets

Beautiful designs, assorted; on white pyralin pieces. Each piece in separate compartment. Makes a splendid gift for the new baby. Shipping weight, 7 oz.

Four-Piece Set.	Two-Piece Set.	Three-Piece Set.	Five-Piece Set.
Comb, Brush, Rattle and Soap Box.	Comb and brush.	Comb, Brush and Rattle.	Comb, Brush, Rattle, Soap Box and Powder Box.
8L2109...98c	**8L2107..39c**	**8L2108..69c**	**8L2111.$1.25**

Baby Rattles

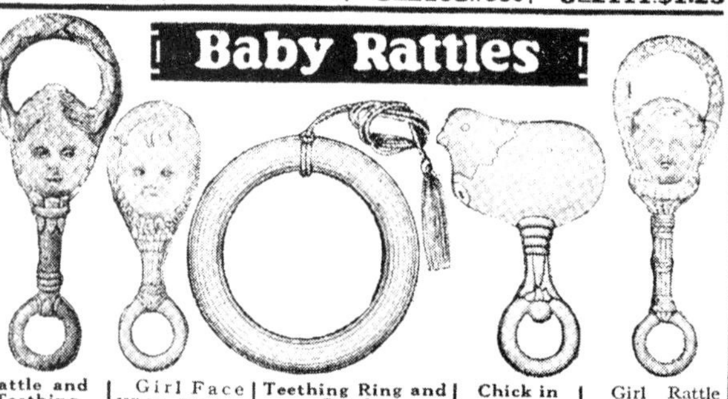

Rattle and Teething Ring.	Girl Face White Rubber Rattle	Teething Ring and Rattle.	Chick in Egg.	Girl Rattle
Can be sterilized in boiling water. Red rubber. Lgth., about 8 inches. Shpg. wt., 3 oz.	with teething ring. Length, 4¾ inches. Shipping weight, 3 ounces.	Every baby needs a teething ring. Can be washed or sterilized. Easily cleaned. Large size, 3¼ inches in diameter. Complete with cord and tassel. Shipping wt., 4 oz.	White rubber rattle and teething ring. Length, 4½ in. Shipping weight, 3 oz.	with loop. White rubber rattle with teething ring. Length, 7 in. Shpg. wt., 3 oz.
8L2187 **39c**	**8L2171** **10c**	**8L2247...19c**	**8L2174** **12c**	**8L2172** **19c**

Foods for Infants

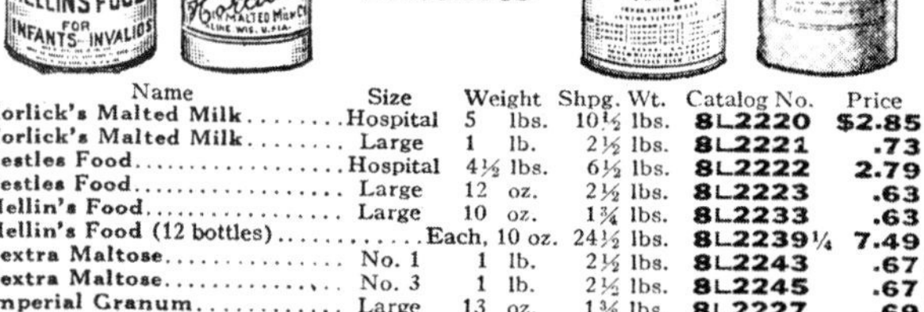

Name	Size	Weight	Shpg. Wt.	Catalog No.	Price
Horlick's Malted Milk........	Hospital	5 lbs.	10½ lbs.	**8L2220**	**$2.85**
Horlick's Malted Milk........	Large	1 lb.	2½ lbs.	**8L2221**	**.73**
Nestles Food................	Hospital	4½ lbs.	6½ lbs.	**8L2222**	**2.79**
Nestles Food................	Large	12 oz.	2½ lbs.	**8L2223**	**.63**
Mellin's Food...............	Large	10 oz.	1¾ lbs.	**8L2233**	**.63**
Mellin's Food (12 bottles)...........	Each, 10 oz.		24½ lbs.	**8L2239¼**	**7.49**
Dextra Maltose..............	No. 1	1 lb.	2½ lbs.	**8L2243**	**.67**
Dextra Maltose..............	No. 3	1 lb.	2½ lbs.	**8L2245**	**.67**
Imperial Granum.............	Large	13 oz.	1⅜ lbs.	**8L2227**	**.69**
Robinson Barley.............	Large	1 lb.	1¾ lbs.	**8L2228**	**.53**
Sugar Milk, U. S. P. Mallinckrodt		1 lb.	1¾ lbs.	**8L2232**	**.47**

Baby's Own Hot Water Bottle.

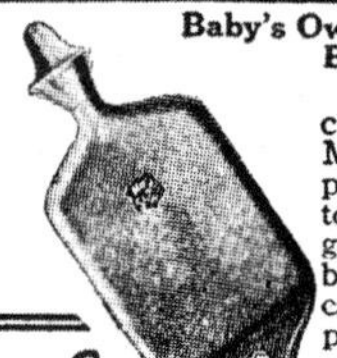

Keeps baby's crib warm. Made in one piece, no seams to leak. Good grade red rubber. Full 1 pint capacity. Shipping wt., 8 oz.
8L2124
69c

Clothes Hangers for Baby's Clothes.

Nicely Painted Hangers with beautiful baby faces decorated on side. Size of hanger, 10¼ in. over all. Shpg. wt., 1 lb.
8L2249—Two per box—pink and blue...........................**67c**

My Baby's Borated Talcum

An exceptionally high grade, pure borated Talcum Powder. Especially prepared for use on baby's tender skin. Snow white in color, made from finest grade imported talcum powder, delightfully perfumed, and with just the right quantity of boric acid, stearate of zinc, calcined magnesia and starch to make it a soft, fluffy hygienic powder for toilet and nursery use. For use after baby's bath, for chafing and toilet. Shipping weight, each, 6 oz.
8L2119
4-ounce can..........**19c**
8L2205—3 for..**55c**

Other Baby Powders.

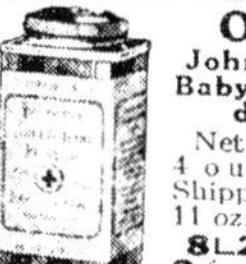

Johnson's Baby Powder.	Mennen's Talcum Powder.	Stork Talcum Powder.
Net weight, 4 ounces. Shipping wt., 11 oz.	Net weight, 5¾ ounces. Shipping wt., 14 oz.	Net wt., 3½ ounces. Shpg. wt., 11 oz.
8L2157 **2** for...**39c**	**8L2120** **2** for...**39c**	**8L2184** **2** for.**39c**

Zinc Stearate.

Stearate of Zinc. A dusting powder or dry antiseptic. Helpful for baby in cases of irritation and used to prevent chafing. 1-ounce carton. Shipping weight, 6 ounces.
8L2110..17c

Babies' Powder Puff. Soft and fluffy. All wool. Diameter, 3 inches. Shpg. wt., 2 oz.
8L2244..25c

Wide Mouth Nursing Bottle

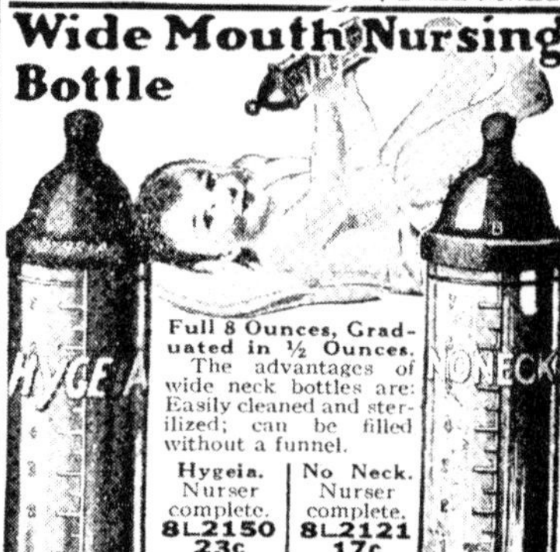

Full 8 Ounces, Graduated in ½ Ounces. The advantages of wide neck bottles are: Easily cleaned and sterilized; can be filled without a funnel.

Hygeia.	No Neck.
Nurser complete.	Nurser complete.
8L2150 **23c**	**8L2121** **17c**
Nipple only.	Nipple only.
8L2153 **2** for.**23c**	**8L2122** **9c**

Shipping weight: Nursers, 1¼ lbs.; Nipples, 2 oz.

Nipples for Other Bottles.

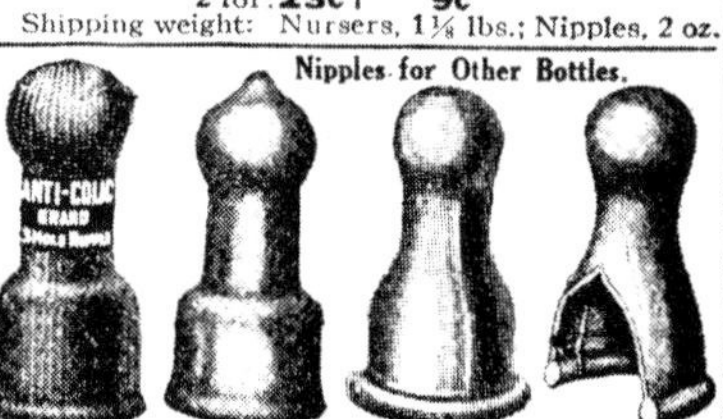

Anti-Colic.	Tiptop.	Transparent Ball Top.	Feed-Rite.
(So called.) Well known Black rubber.	Good quality black rubber. Reinforced ball top to prevent collapsing.	Made of good grade rubber.	The nipple with the cross. Non-collapsible. **Red rubber.**
8L2127 **3** for **13c**	**8L2125** **4** for....**17c**	**8L2126** **3** for **13c**	**8L2128** **5** for...**33c**

Shipping weight, nipples, 1 ounce.

Oval Nursing Bottle. Good grade. Capacity, 8 ounces. Graduated. Shipping weight, 2¾ pounds.
8L2129—3 for....**16c**

Bottle Brush. The tuft on end cleans corners. Lgth., 11 in. Shpg. wt., 3 oz.
8L2123.....**10c**

Babies' Unbreakable Aluminum Set With Alphabet.

Plate, 6¾ inches; saucer, 4 inches; cup, 2¼ inches wide. Shipping weight, 9 ounces.
8L2262..25c

Sickroom Supplies

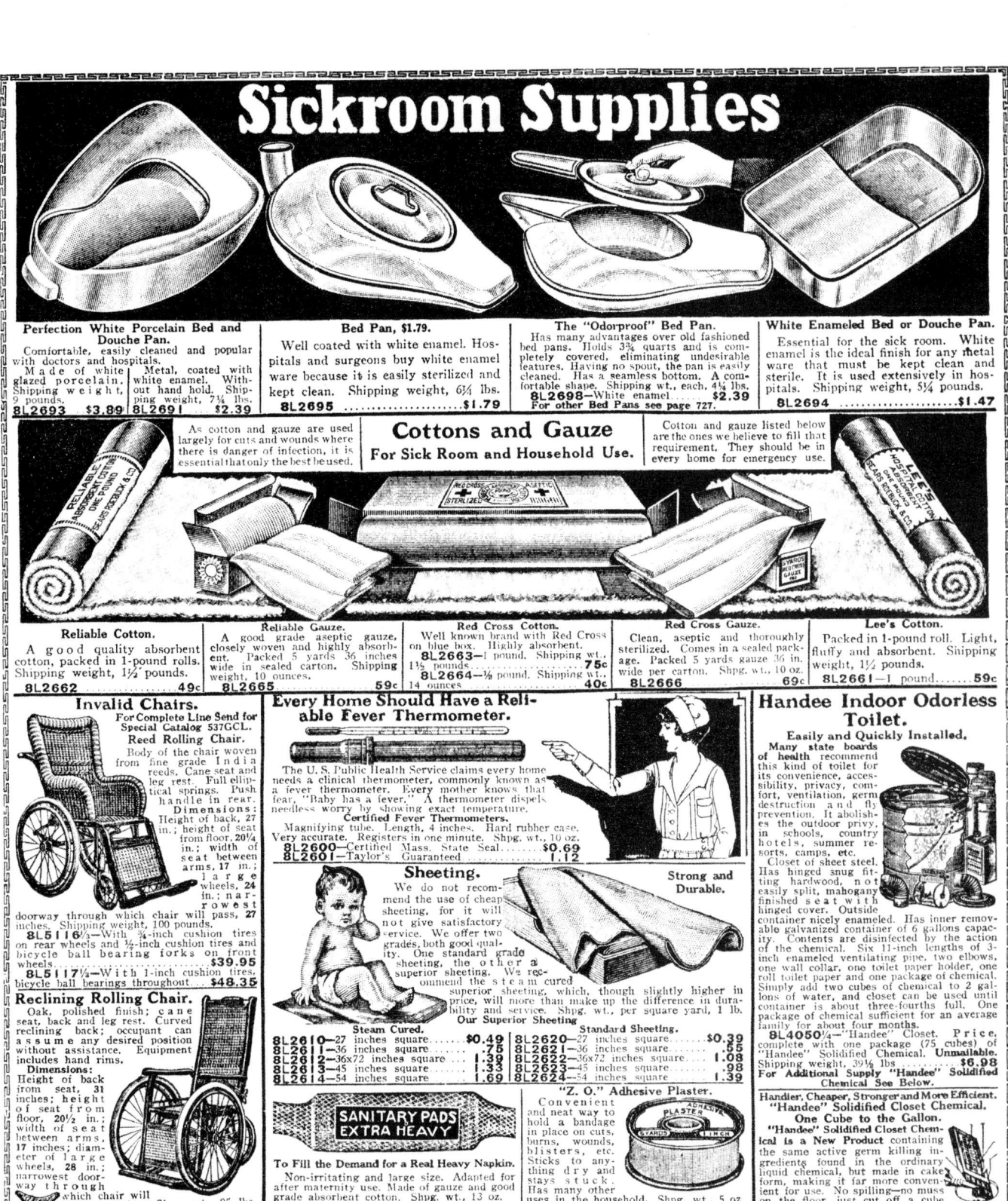

Perfection White Porcelain Bed and Douche Pan.
Comfortable, easily cleaned and popular with doctors and hospitals.
Made of white glazed porcelain. Shipping weight, 9 pounds.
8L2693 $3.89
Metal, coated with white enamel. Without hand hold. Shipping weight, 7¼ lbs.
8L2691 $2.39

Bed Pan, $1.79.
Well coated with white enamel. Hospitals and surgeons buy white enamel ware because it is easily sterilized and kept clean. Shipping weight, 6¼ lbs.
8L2695 .... $1.79

The "Odorproof" Bed Pan.
Has many advantages over old fashioned bed pans. Holds 3¾ quarts and is completely covered, eliminating undesirable features. Having no spout, the pan is easily cleaned. Has a seamless bottom. A comfortable shape. Shipping wt., each, 4¼ lbs.
8L2698—White enamel...... $2.39
For other Bed Pans see page 727.

White Enameled Bed or Douche Pan.
Essential for the sick room. White enamel is the ideal finish for any metal ware that must be kept clean and sterile. It is used extensively in hospitals. Shipping weight, 5¼ pounds.
8L2694 .... $1.47

Cottons and Gauze
For Sick Room and Household Use.

As cotton and gauze are used largely for cuts and wounds where there is danger of infection, it is essential that only the best be used.

Cotton and gauze listed below are the ones we believe to fill that requirement. They should be in every home for emergency use.

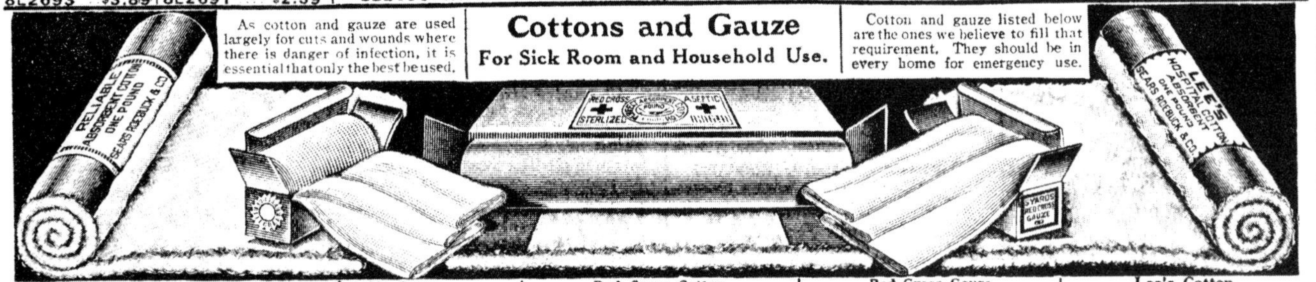

Reliable Cotton.
A good quality absorbent cotton, packed in 1-pound rolls. Shipping weight, 1½ pounds.
8L2662 .... 49c

Reliable Gauze.
A good grade aseptic gauze, closely woven and highly absorbent. Packed 5 yards 36 inches wide in sealed carton. Shipping weight, 10 ounces.
8L2665 .... 59c

Red Cross Cotton.
Well known brand with Red Cross on blue box. Highly absorbent.
8L2663—1 pound. Shipping wt., 1½ pounds 75c
8L2664—½ pound. Shipping wt., 14 ounces 40c

Red Cross Gauze.
Clean, aseptic and thoroughly sterilized. Comes in a sealed package. Packed 5 yards gauze 36 in. wide per carton. Shpg. wt., 10 oz.
8L2666 .... 69c

Lee's Cotton.
Packed in 1-pound roll. Light, fluffy and absorbent. Shipping weight, 1½ pounds.
8L2661—1 pound....... 59c

Invalid Chairs.
For Complete Line Send for Special Catalog 537GCL.

Reed Rolling Chair.
Body of the chair woven from fine grade India reeds. Cane seat and leg rest. Full elliptical springs. Push handle in rear. **Dimensions:** Height of back, 27 in.; height of seat from floor, 20¼ in.; width of seat between arms, 17 in.; large wheels, 24 in.; narrowest doorway through which chair will pass, 27 inches. Shipping weight, 100 pounds.
8L5116⅓—With ⅜-inch cushion tires on rear wheels and ½-inch cushion tires and bicycle ball bearing forks on front wheels $39.95
8L5117⅓—With 1-inch cushion tires, bicycle ball bearings throughout $48.35

Reclining Rolling Chair.
Oak, polished finish; cane seat, back and leg rest. Curved reclining back; occupant can assume any desired position without assistance. Equipment includes hand rims. 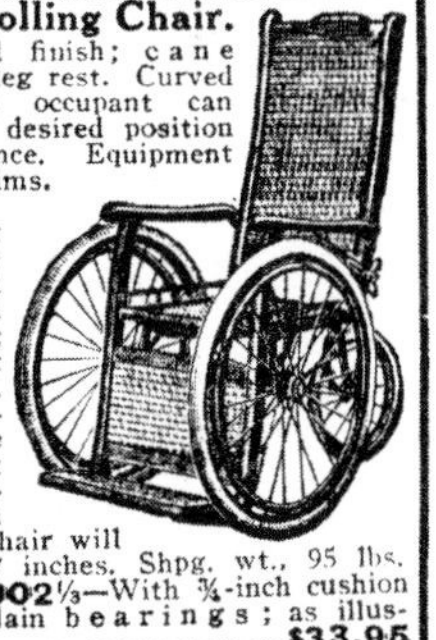
Dimensions: Height of back from seat, 31 inches; height of seat from floor, 20½ in.; width of seat between arms, 17 inches; diameter of large wheels, 28 in.; narrowest doorway through which chair will pass, 27 inches. Shpg. wt., 95 lbs.
8L5002⅓—With ⅜-inch cushion tires; plain bearings; as illustrated $33.95
8L5003⅓—With 1-inch cushion tires, bicycle ball bearings. $40.95

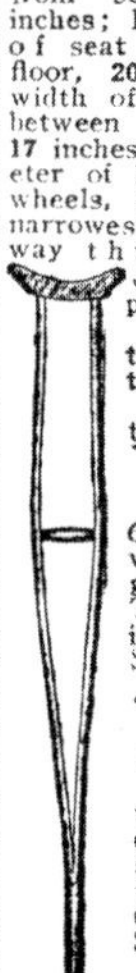

Crutches.
Shipped from stock. Sizes, 36 to 60 inches. Even sizes only. Maple wood with hardwood top and handgrip. Fitted with slip rubber tips. Take measure from armpit to floor in standing position and add 2 inches. Shipping weight, 4 pounds.
8L3853¼—Pair $1.59

Crutch or Chair Rubber Tips.
Bailey's "Won't Slip." 
Construction of bottom tends to minimize the danger of slipping on smooth or polished surfaces. Sizes given are diameter end of crutch. Shipping weight, 3 oz.
8L3866—¾-inch. Pair 19c
8L3867—⅞-inch. Pair 21c
8L3868—1 -inch. Pair 23c

Every Home Should Have a Reliable Fever Thermometer.
The U. S. Public Health Service claims every home needs a clinical thermometer, commonly known as a fever thermometer. Every mother knows that fear, "Baby has a fever." A thermometer dispels needless worry by showing exact temperature.
Certified Fever Thermometers.
Magnifying tube. Length, 4 inches. Hard rubber case. Very accurate. Registers in one minute. Shpg. wt., 10 oz.
8L2600—Certified Mass. State Seal........ $0.69
8L2601—Taylor's Guaranteed 1.12

Sheeting.
Strong and Durable.
We do not recommend the use of cheap sheeting, for it will not give satisfactory service. We offer two grades, both good quality. One standard grade sheeting, the other a superior sheeting. We recommend the steam cured superior sheeting, which, though slightly higher in price, will more than make up the difference in durability and service. Shpg. wt., per square yard, 1 lb.

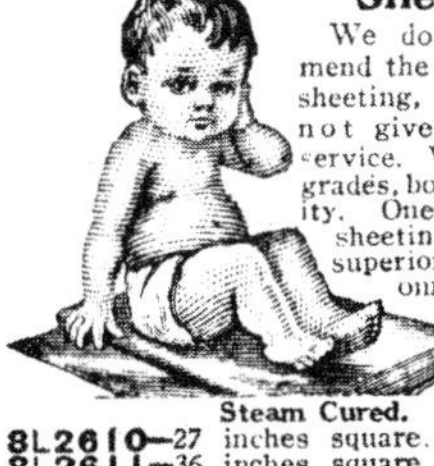

Steam Cured. Our Superior Sheeting		Standard Sheeting.	
8L2610—27 inches square	$0.49	8L2620—27 inches square	$0.39
8L2611—36 inches square	.75	8L2621—36 inches square	.55
8L2612—36x72 inches square	1.39	8L2622—36x72 inches square	1.08
8L2613—45 inches square	1.33	8L2623—45 inches square	.98
8L2614—54 inches square	1.69	8L2624—54 inches square	1.39

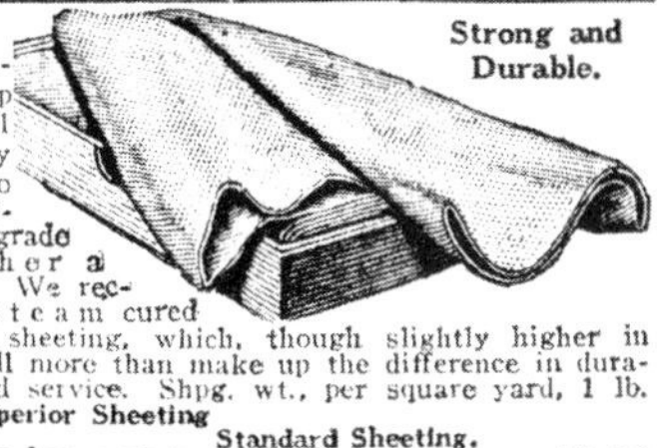

SANITARY PADS EXTRA HEAVY
To Fill the Demand for a Real Heavy Napkin.
Non-irritating and large size. Adapted for after maternity use. Made of gauze and good grade absorbent cotton. Shpg. wt., 13 oz.
8L2644—12 pads for 49c

"Z. O." Adhesive Plaster. 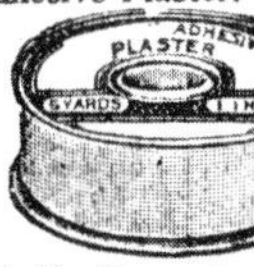
Convenient and neat way to hold a bandage in place on cuts, burns, wounds, blisters, etc. Sticks to anything dry and stays stuck. Has many other uses in the household. Shpg. wt., 5 oz.
8L2675 .... 29c

Gauze Bandage. 
Ten yards of plain gauze bandage for dressing wounds. Put up by Johnston & Johnston, makers of high quality surgical goods. Carefully wrapped at factory. Shpg. wt., 2 oz.
8L2673
Width, 2½ inches............ 13c
8L2671—1½ in., 2 for.... 17c

Bouillon Cubes. 
Each cube makes a cup of appetizing bouillon. Just add boiling water. Shpg. wt., 3 oz.
8L2650—Tin box of 12 cakes 23c

Beef Extract. 
Highest quality. Shipping weight, 4 oz.
8L2652
Per jar 39c

Formaldehyde Torch.
For fumigation purposes about the home. One candle sufficient to thoroughly disinfect 700 cubic feet. Directions on package. Shipping weight, 12 ounces.
8L652 .... 33c

Sulphur Candles.
For destroying most vermin. Candle contains about 1 pound of sulphur, sufficient for 500 cubic feet. Shiping weight, 1⅛ lbs.
8L658........ 25c

Handee Indoor Odorless Toilet.
Easily and Quickly Installed. 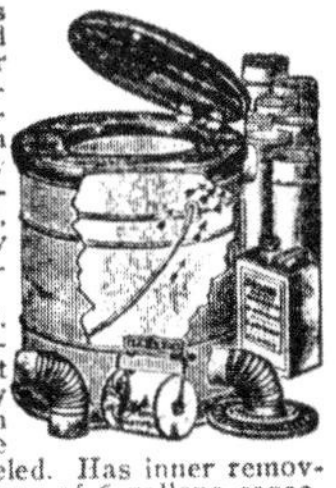
Many state boards of health recommend this kind of toilet for its convenience, accessibility, privacy, comfort, ventilation, germ destruction and fly prevention. It abolishes the outdoor privy, in schools, country hotels, summer resorts, camps, etc.
Closet of sheet steel. Has hinged snug fitting hardwood, not easily split, mahogany finished seat with hinged cover. Outside container nicely enameled. Has inner removable galvanized container of 6 gallons capacity. Contents are disinfected by the action of the chemical. Six 11-inch lengths of 3-inch enameled ventilating pipe, two elbows, one wall collar, one toilet paper holder, one roll toilet paper and one package of chemical. Simply add two cubes of chemical to 2 gallons of water, and closet can be used until container is about three-fourths full. One package of chemical sufficient for an average family for about four months.
8L4050¼—"Handee" Closet. Price, complete with one package (75 cubes) of "Handee" Solidified Chemical. **Unmailable.** Shipping weight, 39½ lbs $6.98
For Additional Supply "Handee" Solidified Chemical See Below.

Handier, Cheaper, Stronger and More Efficient.
"Handee" Solidified Closet Chemical.
One Cube to the Gallon. 
"Handee" Solidified Closet Chemical is a New Product containing the same active germ killing ingredients found in the ordinary liquid chemical, but made in cake form, making it far more convenient for use. No spilling—no muss on the floor, just cut off a cube and everything is ready. Far superior to liquid form.
Packed in cakes, each cake marked into 75 squares for easy cutting. To use, simply cut off one cube for each gallon of water. No need to guess. Always the correct dilution. Not too strong—not too weak—but just right. Prevents waste through leakage, spilling or excessive quantities. Saves the cost of can container, and about 7 pounds freight.
We strongly recommend **"Handee" Solidified Closet Chemical** for use in chemical toilets. Shpg. wt., 3½ lbs. **Mailable.**
8L4053¼—75-cube cake $1.19

"Handee" Liquid Chemical.
Destroys most disease germs and keeps the closet sanitary. **Unmailable.** Shipping weight, 10½ pounds.
8L4054¼—Per gallon........... $1.39
For other Closet Outfits see pages 698 and 699.

Everyone Admires Beautiful Hair

7 Sutherland Sisters'

Hair Preparations.

Seven Sutherland Sisters' hair preparations have been on the market and widely advertised for many years. Directions for applying with each package.

Hair Tonic. Free from injurious substances. Full directions for use with each package. Shipping wt., 2 pounds.

8L3249 $1.00 size. **75c**

Hair and Scalp Cleaner. An easily applied shampoo for keeping the hair and scalp clean and healthy and the hair beautiful. Aids in removing dandruff. Shipping weight, 7 ounces.

8L3235 50c size..... **39c**

Colorator. For changing gray, bleached and faded hair to a natural color. Easy to apply. Shades, as follows: Black, chestnut, dark brown, auburn, medium brown, light brown, ash blonde or gold blonde. **State color wanted.** Shipping weight, 2¼ pounds.

8L3247 $1.00 size.. **79c**

Other Hair Colorators.

Nazimova Colorator.
In black, medium brown or dark brown. State color.
8L3205—$1.00 size **$0.79**

Clay's Hair Shader.
For restoring hair to its natural color.
8L3218—$1.00 size .... **.59**

Lotus Color Restorer.
Light, medium or dark. State color.
8L3225—$1.50 size **1.19**

Walnutta Hair Stain.
8L3217—75-cent size .... **.48**
Shipping weight, Colorators, 1½ pounds.

Powdered Henna Leaves.
Used in many beauty parlors for coloring or shading the hair.
8L3209—¼-lb. package. Shpg. wt., 8 oz **33c**

Hair Curling Fluid.

Assists in making hair curly and wavy. Especially good for keeping hair curly during damp weather. Easily applied. Hair can be let down any time after three hours. Perfumed. About 2½-oz. bottle. Shipping weight, 12 ounces.

8L3221 .... **25c**

Eyebrow and Eyelash Cream.

A popular cream used for beautifying the eyebrows and eyelashes. Helps to keep straggling hairs in place and aids in growth of eyelashes. Apply with finger tips. Shipping weight, 6 ounces.

8L3208 .... **39c**

Eyebrow Tinting Outfit.
For darkening eyebrows or lashes. Complete outfit with brush. Brown or black. **State color.** Shipping weight, 4 ounces.

8L3320........... **47c**

"TAROLA" Shampoo Soap.

A Superior Grade Transparent Tar Soap.

Tarola cleans the hair and scalp thoroughly and aids in preventing the hair from falling out. Tarola makes an ideal shampoo for men, women and children. Each cake foil wrapped. We recommend this soap. Shipping weight, three cakes, 1 pound.

8L4908—3 cakes for **55c**

Cup and Comb for Applying Tonic.

This device has proved very satisfactory for applying hair tonics. The liquid will flow through the teeth of comb directly to the scalp, thus allowing the preparation to reach the roots of the hair. It is easy to use and there is nothing to get out of order. Each appliance in box. Shipping weight, 10 ounces.

8L2474 .... **39c**

Princess Hair Tonic and Colorator.

Beautiful hair adds much to the charm of any woman. The beauty and luster of the hair depend largely upon the condition of the scalp.

Princess Hair Tonic.
A delightful preparation for helping to make the hair grow. Unless the hair follicles are absolutely dead, it is possible to materially stimulate the scalp and help nature to produce a growth of hair. Shipping wt., one bottle, 1 pound; three bottles, 5 pounds.

8L3200—12-ounce bottle **$0.79**
8L3207—3 bottles **2.25**

Princess Hair Colorator.
A reliable product used to restore gray hair to its natural color. Easy to apply and colors evenly. Make various applications to obtain the tint desired. Shipping weight, 1 pound.

8L3206—4-ounce bottle **79c**

Princess Cocoanut Oil Shampoo.

36c

The Proper Way to Shampoo.

We Recommend This Fine Quality Cocoanut Oil Shampoo.

FOR WOMEN—The beauty and luster of your hair depend upon the condition of your scalp. Through its roots the hair receives nourishment to give it life. Princess Shampoo applied to the scalp helps to bring out this natural luster and beauty. It leaves the hair soft and silky and free from impurities.

FOR CHILDREN—Children's hair is naturally beautiful, but it needs regular washing to keep it so. Ordinary soap is not beneficial to the child's tender scalp and tends to leave the hair brittle.

FOR MEN—What is more refreshing to a man, especially on a hot day, than a good shampoo? Two teaspoonfuls of this shampoo will make an abundant lather which rinses out easily, bringing out dust and dirt, together with loose dandruff and excess oil.

Princess Shampoo is clear, pure, nicely perfumed, and free from grease and oil. Cannot injure the scalp. Shipping weight, 1½ lbs.

8L3248—5-ounce bottle **36c**

MULSIFIED COCOANUT OIL SHAMPOO

Makes a rich, creamy lather and cleanses hair thoroughly, removing dust, dirt and dandruff. It brings out the real life and luster, and makes the hair soft, fresh and luxuriant. Shipping weight, 1⅛ pounds.

8L3226 4-oz. bottle. **39c**

Lady Janis' Cocoanut Oil Shampoo.

Your hair can be kept beautiful with just a little care and attention. This nicely perfumed creamy lather shampoo will thoroughly cleanse the scalp of dirt and dust, and leave the hair nice and soft and glossy. Shipping weight, 1⅛ lbs.

8L3243 4-oz. bottle. **25c**

Which type are you?

BLOND — TITIAN — BRUNETTE

JORO — The individual shampoo, Blonde Brunette Titian

Joro Shampoo.

Beautiful hair is your birthright. It is yours to enjoy, yours to take pride in. Joro is a shampoo for three basic types. Joro brings out the true radiance of your hair.

Joro Blonde brings out the natural glint and adds exquisite touch of spun gold as it cleanses.

Joro Brunette beautifies the color and gives gloss to brunette hair.

Joro Titian preserves the true luster of titian and auburn hair, whether natural or not.

Joro is put up in cake form and is used like an ordinary shampoo. It lathers freely and rinses absolutely, leaving the hair clean, silky and lustrous. Shipping weight, 5 ounces.

8L4923—Blonde. 50c size cake... **39c**	8L4922—Brunette. 50c size cake... **39c**	8L4924—Titian. 50c size cake... **39c**

Wildroot Hair Preparations.

Everyone can have clean, fluffy, silky hair and a soft, white, healthy scalp with these Wildroot preparations that contain only the purest ingredients.

Wildroot Hair Tonics.
Even though your hair is dull, lifeless, hard to do up—or even full of dandruff—Wildroot Hair Tonic will bring out the true loveliness and keep it healthy.

8L3239—$1.00 size. Shipping wt., 2 pounds **69c**
8L3201—50c size. Shipping wt., 1⅛ pounds **36c**

Quinine Bouquet Tonic.
A new Wildroot product. Red color and a fine perfume. Shpg. wt., 1 lb.
8L3231—8-ounce bottle **59c**

Wildroot Shampoos.

Wildroot Shampoo Soap.
8L4907—Shipping weight, 6 oz. Cake **19c**

Wildroot Taroleum Shampoo.
A fine quality tar shampoo. Shipping weight, 1 pound.
8L3203—50-cent size.......... **36c**

Wildroot Cocoanut Oil Shampoo.
8L3240—50c size. Shipping weight, 1 pound **36c**

Ker-ene

Deodorized Kerosene. For Hair and Scalp Treatment.

Used for dandruff, falling hair and scalp conditions generally. Kerene is kerosene without the unpleasant odor and grease.

8L3242—4-ounce bottle. Shipping weight, 10 oz. Regular 50-cent size **39c**
8L3244—10-ounce bottle. Shipping weight, 1 lb. Regular $1.00 size **79c**

Other Hair Tonics.

Danderine.
8L3233—$1.00 size **$0.75**

Pinaud's Eau de Quinine.
8L3212—8-ounce bottle **1.25**

K. D. X. Dandruff Remover.
8L3216—$1.00 size .... **.79**

Herpicide.
8L3210—$1.25 size .... **.83**

Ferron's French Tonique.
8L3220—$1.50 size .... **.89**

Berriaults Hair Bitters.
8L3215—$2.00 size **1.69**

Shipping weight, hair tonics, 1½ pounds.

Popular Hair Dressings

Busse's Hair Luster.

Gives that soft, glossy, well groomed appearance. Holds the hair in place. For men, women and boys. Harmless and greaseless. Shipping weight, small size, 10 ounces; large size, 16 ounces.

8L3224—$1.00 size Liquid.. **79c**
8L3223—50c size Liquid... **39c**
8L3219—2-ounce jar Paste. Shpg. wt., 10 oz. **59c**

Hair Pomade.
A paste for straightening kinky or curly hair. Shipping weight, 13 ounces.
8L3241 .... **19c**

Lady Janis Depilatories.

Simple and most effective for safely removing superfluous hair.

Safe and Easily Applied.
A Lady Janis depilatory may be used by any woman with perfect confidence. Easily applied; removes every trace of hair, leaving the skin smooth and white. Unlike shaving, does not encourage the further growth of hair.

Powder Depilatory.
The perfumed powder depilatory enjoys the larger sale and is preferred by many for removing hair from under the arms. With each bottle is a small celluloid cup to use in mixing. Bottle has handy patent airtight cap. Full directions with each package. Shpg. wt., 8 oz.
8L3246 .... **79c**

Liquid Depilatory.
Liquid depilatory is preferred by many on account of no mixing being required. Each bottle fitted with rubber cork. Bottle must be kept airtight when not in use or liquid will lose its strength. Shipping weight, 8 ounces.
8L3245 .... **79c**

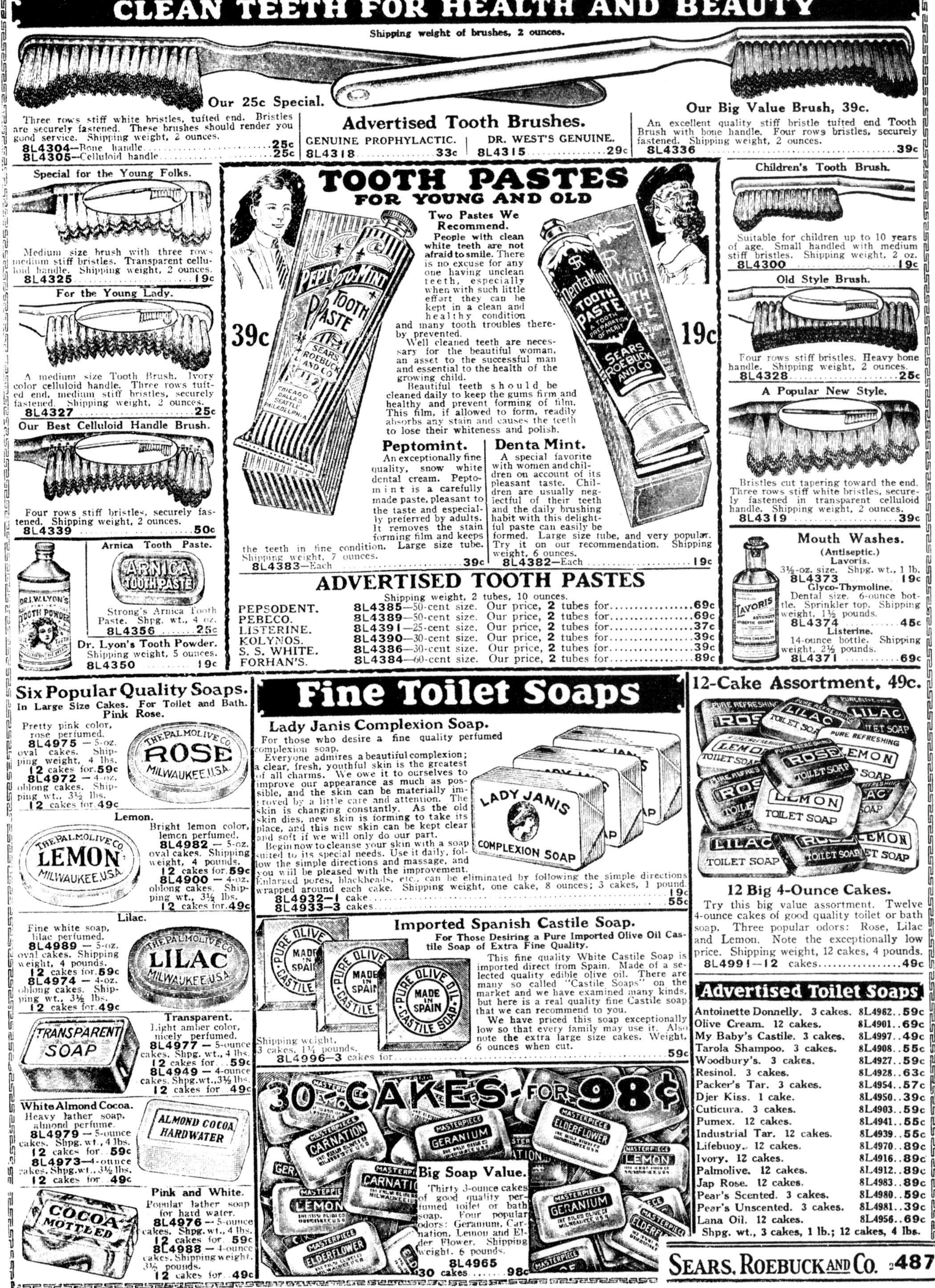

CLEAN TEETH FOR HEALTH AND BEAUTY
Shipping weight of brushes, 2 ounces.
Our 25c Special.
Three rows stiff white bristles, tufted end. Bristles are securely fastened. These brushes should render you good service. Shipping weight, 2 ounces.
8L4304—Bone handle.....25c
8L4305—Celluloid handle.....25c
Advertised Tooth Brushes.
GENUINE PROPHYLACTIC.
8L4318.....33c
DR. WEST'S GENUINE.
8L4315.....29c
Our Big Value Brush, 39c.
An excellent quality stiff bristle tufted end Tooth Brush with bone handle. Four rows bristles, securely fastened. Shipping weight, 2 ounces.
8L4336.....39c
Special for the Young Folks.
Medium size brush with three rows medium stiff bristles. Transparent celluloid handle. Shipping weight, 2 ounces.
8L4325.....19c
For the Young Lady.
A medium size Tooth Brush. Ivory color celluloid handle. Three rows tufted end, medium stiff bristles, securely fastened. Shipping weight, 2 ounces.
8L4327.....25c
Our Best Celluloid Handle Brush.
Four rows stiff bristles, securely fastened. Shipping weight, 2 ounces.
8L4339.....50c
Arnica Tooth Paste.
ARNICA TOOTH PASTE
Strong's Arnica Tooth Paste. Shpg. wt., 4 oz.
8L4356.....25c
DR. I. W. LYON'S TOOTH POWDER
Dr. Lyon's Tooth Powder.
Shipping weight, 5 ounces.
8L4350.....19c
TOOTH PASTES
FOR YOUNG AND OLD
Two Pastes We Recommend.
People with clean white teeth are not afraid to smile. There is no excuse for any one having unclean teeth, especially when with such little effort they can be kept in a clean and healthy condition and many tooth troubles thereby prevented.
Well cleaned teeth are necessary for the beautiful woman, an asset to the successful man and essential to the health of the growing child.
Beautiful teeth should be cleaned daily to keep the gums firm and healthy and prevent forming of film. This film, if allowed to form, readily absorbs any stain and causes the teeth to lose their whiteness and polish.
PEPTO MINT TOOTH PASTE SEARS ROEBUCK AND CO CHICAGO DALLAS SEATTLE PHILADELPHIA
39c
DENTA MINT TOOTH PASTE A TOOTH PREPARATION OF MERIT SEARS ROEBUCK AND CO
19c
Peptomint.
An exceptionally fine quality, snow white dental cream. Peptomint is a carefully made paste, pleasant to the taste and especially preferred by adults. It removes the stain forming film and keeps the teeth in fine condition. Large size tube. Shipping weight, 7 ounces.
8L4383—Each.....39c
Denta Mint.
A special favorite with women and children on account of its pleasant taste. Children are usually neglectful of their teeth and the daily brushing habit with this delightful paste can easily be formed. Large size tube, and very popular. Try it on our recommendation. Shipping weight, 6 ounces.
8L4382—Each.....19c
ADVERTISED TOOTH PASTES
Shipping weight, 2 tubes, 10 ounces.
PEPSODENT. 8L4385—50-cent size. Our price, 2 tubes for.....69c
PEBECO. 8L4389—50-cent size. Our price, 2 tubes for.....69c
LISTERINE. 8L4391—25-cent size. Our price, 2 tubes for.....37c
KOLYNOS. 8L4390—30-cent size. Our price, 2 tubes for.....39c
S. S. WHITE. 8L4386—30-cent size. Our price, 2 tubes for.....39c
FORHAN'S. 8L4384—60-cent size. Our price, 2 tubes for.....89c
Children's Tooth Brush.
Suitable for children up to 10 years of age. Small handled with medium stiff bristles. Shipping weight, 2 oz.
8L4300.....19c
Old Style Brush.
Four rows stiff bristles. Heavy bone handle. Shipping weight, 2 ounces.
8L4328.....25c
A Popular New Style.
Bristles cut tapering toward the end. Three rows stiff white bristles, securely fastened in transparent celluloid handle. Shipping weight, 2 ounces.
8L4319.....39c
Mouth Washes.
(Antiseptic.)
Lavoris.
3½-oz. size. Shpg. wt., 1 lb.
8L4373.....19c
Glyco-Thymoline.
Dental size. 6-ounce bottle. Sprinkler top. Shipping weight, 1½ pounds.
8L4374.....45c
Listerine.
14-ounce bottle. Shipping weight, 2½ pounds.
8L4371.....69c
LAVORIS
Six Popular Quality Soaps.
In Large Size Cakes. For Toilet and Bath.
Pink Rose.
Pretty pink color, rose perfumed.
8L4975 — 5-oz. oval cakes. Shipping weight, 4 lbs.
12 cakes for.59c
8L4972 — 4-oz. oblong cakes. Shipping wt., 3½ lbs.
12 cakes for.49c
THE PALMOLIVE CO. ROSE MILWAUKEE, U.S.A.
Lemon.
Bright lemon color, lemon perfumed.
8L4982 — 5-oz. oval cakes. Shipping weight, 4 pounds.
12 cakes for.59c
8L4900 — 4-oz. oblong cakes. Shipping wt., 3½ lbs.
12 cakes for.49c
THE PALMOLIVE CO. LEMON MILWAUKEE, U.S.A.
Lilac.
Fine white soap, lilac perfumed.
8L4989 — 5-oz. oval cakes. Shipping weight, 4 pounds.
12 cakes for.59c
8L4974 — 4-oz. oblong cakes. Shipping wt., 3½ lbs.
12 cakes for.49c
THE PALMOLIVE CO. LILAC MILWAUKEE, U.S.A.
Transparent.
Light amber color, nicely perfumed.
8L4977 — 5-ounce cakes. Shpg. wt., 4 lbs.
12 cakes for.....59c
8L4949 — 4-ounce cakes. Shpg. wt., 3½ lbs.
12 cakes for.....49c
TRANSPARENT SOAP
White Almond Cocoa.
Heavy lather soap, almond perfume.
8L4979 — 5-ounce cakes. Shpg. wt., 4 lbs.
12 cakes for.....59c
8L4973 — 4-ounce cakes. Shpg. wt., 3½ lbs.
12 cakes for.....49c
ALMOND COCOA HARDWATER
Pink and White.
Popular lather soap for hard water.
8L4976 — 5-ounce cakes. Shpg. wt., 4 lbs.
12 cakes for.....59c
8L4988 — 4-ounce cakes. Shipping weight, 3½ pounds.
12 cakes for.....49c
COCOA MOTTLED
Fine Toilet Soaps
Lady Janis Complexion Soap.
For those who desire a fine quality perfumed complexion soap.
Everyone admires a beautiful complexion; a clear, fresh, youthful skin is the greatest of all charms. We owe it to ourselves to improve our appearance as much as possible, and the skin can be materially improved by a little care and attention. The skin is changing constantly. As the old skin dies, new skin is forming to take its place, and this new skin can be kept clear and soft if we will only do our part.
Begin now to cleanse your skin with a soap suited to its special needs. Use it daily, follow the simple directions and massage, and you will be pleased with the improvement. Enlarged pores, blackheads, etc., can be eliminated by following the simple directions wrapped around each cake. Shipping weight, one cake, 8 ounces; 3 cakes, 1 pound.
8L4932—1 cake.....19c
8L4933—3 cakes.....55c
LADY JANIS COMPLEXION SOAP
Imported Spanish Castile Soap.
For Those Desiring a Pure Imported Olive Oil Castile Soap of Extra Fine Quality.
This fine quality White Castile Soap is imported direct from Spain. Made of a selected quality edible olive oil. There are many so called "Castile Soaps" on the market and we have examined many kinds, but here is a real quality fine Castile soap that we can recommend to you.
We have priced this soap exceptionally low so that every family may use it. Also note the extra large size cakes. Weight, 6 ounces when cut.
Shipping weight, 3 cakes, 1¼ pounds.
8L4996—3 cakes for.....59c
PURE OLIVE OIL MADE IN SPAIN CASTILE SOAP
30-CAKES-FOR 98¢
MASTERPIECE CARNATION
MASTERPIECE GERANIUM
MASTERPIECE ELDERFLOWER
MASTERPIECE LEMON
Big Soap Value.
Thirty 3-ounce cakes of good quality perfumed toilet or bath soap. Four popular odors: Geranium, Carnation, Lemon and Elder Flower. Shipping weight, 6 pounds.
8L4965
30 cakes.....98c
12-Cake Assortment, 49c.
PURE REFRESHING ROSE LILAC LEMON TOILET SOAP
12 Big 4-Ounce Cakes.
Try this big value assortment. Twelve 4-ounce cakes of good quality toilet or bath soap. Three popular odors: Rose, Lilac and Lemon. Note the exceptionally low price. Shipping weight, 12 cakes, 4 pounds.
8L4991—12 cakes.....49c
Advertised Toilet Soaps
Antoinette Donnelly. 3 cakes. 8L4962..59c
Olive Cream. 12 cakes. 8L4901..69c
My Baby's Castile. 3 cakes. 8L4997..49c
Tarola Shampoo. 3 cakes. 8L4908..55c
Woodbury's. 3 cakes. 8L4927..59c
Resinol. 3 cakes. 8L4928..63c
Packer's Tar. 3 cakes. 8L4954..57c
Djer Kiss. 1 cake. 8L4950..39c
Cuticura. 3 cakes. 8L4903..59c
Pumex. 12 cakes. 8L4941..55c
Industrial Tar. 12 cakes. 8L4939..55c
Lifebuoy. 12 cakes. 8L4970..89c
Ivory. 12 cakes. 8L4916..89c
Palmolive. 12 cakes. 8L4912..89c
Jap Rose. 12 cakes. 8L4983..89c
Pear's Scented. 3 cakes. 8L4980..59c
Pear's Unscented. 3 cakes. 8L4981..39c
Lana Oil. 12 cakes. 8L4956..69c
Shpg. wt., 3 cakes, 1 lb.; 12 cakes, 4 lbs.
SEARS, ROEBUCK AND CO. 487

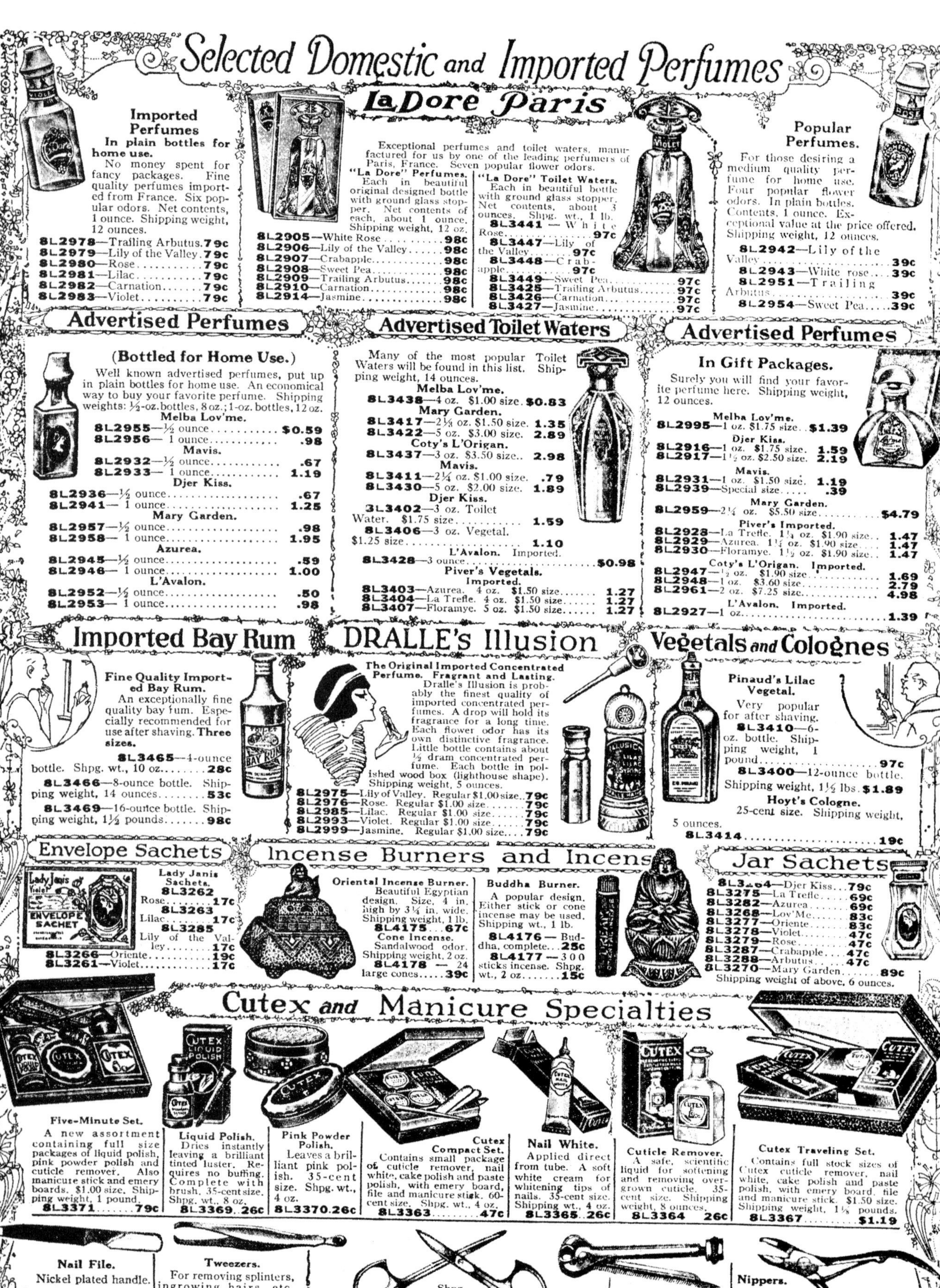

Selected Domestic and Imported Perfumes
La Dore Paris
Imported Perfumes
In plain bottles for home use.
No money spent for fancy packages. Fine quality perfumes imported from France. Six popular odors. Net contents, 1 ounce. Shipping weight, 12 ounces.
8L2978—Trailing Arbutus. 79c
8L2979—Lily of the Valley. 79c
8L2980—Rose. 79c
8L2981—Lilac. 79c
8L2982—Carnation. 79c
8L2983—Violet. 79c
Exceptional perfumes and toilet waters, manufactured for us by one of the leading perfumers of Paris, France. Seven popular flower odors.
"La Dore" Perfumes. Each in beautiful original designed bottle with ground glass stopper. Net contents of each, about 1 ounce. Shipping weight, 12 oz.
8L2905—White Rose. 98c
8L2906—Lily of the Valley. 98c
8L2907—Crabapple. 98c
8L2908—Sweet Pea. 98c
8L2909—Trailing Arbutus. 98c
8L2910—Carnation. 98c
8L2914—Jasmine. 98c
"La Dore" Toilet Waters. Each in beautiful bottle with ground glass stopper. Net contents, about 3 ounces. Shpg. wt., 1 lb.
8L3441—White Rose. 97c
8L3447—Lily of the Valley. 97c
8L3448—Crabapple. 97c
8L3449—Sweet Pea. 97c
8L3425—Trailing Arbutus. 97c
8L3426—Carnation. 97c
8L3427—Jasmine. 97c
Popular Perfumes.
For those desiring a medium quality perfume for home use. Four popular flower odors. In plain bottles. Contents, 1 ounce. Exceptional value at the price offered. Shipping weight, 12 ounces.
8L2942—Lily of the Valley. 39c
8L2943—White rose. 39c
8L2951—Trailing Arbutus. 39c
8L2954—Sweet Pea. 39c
Advertised Perfumes
(Bottled for Home Use.)
Well known advertised perfumes, put up in plain bottles for home use. An economical way to buy your favorite perfume. Shipping weights: ½-oz. bottles, 8 oz.; 1-oz. bottles, 12 oz.
Melba Lov'me.
8L2955—½ ounce. $0.59
8L2956—1 ounce. .98
Mavis.
8L2932—½ ounce. .67
8L2933—1 ounce. 1.19
Djer Kiss.
8L2936—½ ounce. .67
8L2941—1 ounce. 1.25
Mary Garden.
8L2957—½ ounce. .98
8L2958—1 ounce. 1.95
Azurea.
8L2945—½ ounce. .59
8L2946—1 ounce. 1.00
L'Avalon.
8L2952—½ ounce. .50
8L2953—1 ounce. .98
Advertised Toilet Waters
Many of the most popular Toilet Waters will be found in this list. Shipping weight, 14 ounces.
Melba Lov'me.
8L3438—4 oz. $1.00 size. $0.83
Mary Garden.
8L3417—2⅓ oz. $1.50 size. 1.35
8L3422—5 oz. $3.00 size. 2.89
Coty's L'Origan.
8L3437—3 oz. $3.50 size. 2.98
Mavis.
8L3411—2¼ oz. $1.00 size. .79
8L3430—5 oz. $2.00 size. 1.89
Djer Kiss.
3L3402—3 oz. Toilet Water. $1.75 size. 1.59
8L3406—3 oz. Vegetal. $1.25 size. 1.10
L'Avalon. Imported.
8L3428—3 ounce. $0.98
Piver's Vegetals. Imported.
8L3403—Azurea. 4 oz. $1.50 size. 1.27
8L3404—La Trefle. 4 oz. $1.50 size. 1.27
8L3407—Floramye. 5 oz. $1.50 size. 1.27
Advertised Perfumes
In Gift Packages.
Surely you will find your favorite perfume here. Shipping weight, 12 ounces.
Melba Lov'me.
8L2995—1 oz. $1.75 size. $1.39
Djer Kiss.
8L2916—1 oz. $1.75 size. 1.59
8L2917—1½ oz. $2.50 size. 2.19
Mavis.
8L2931—1 oz. $1.50 size. 1.19
8L2939—Special size. .39
Mary Garden.
8L2959—2¼ oz. $5.50 size. $4.79
Piver's Imported.
8L2928—La Trefle. 1¼ oz. $1.90 size. 1.47
8L2929—Azurea. 1¼ oz. $1.90 size. 1.47
8L2930—Floramye. 1½ oz. $1.90 size. 1.47
Coty's L'Origan. Imported.
8L2947—½ oz. $1.90 size. 1.69
8L2948—1 oz. $3.60 size. 2.79
8L2961—2 oz. $7.25 size. 4.98
L'Avalon. Imported.
8L2927—1 oz. 1.39
Imported Bay Rum
Fine Quality Imported Bay Rum.
An exceptionally fine quality bay rum. Especially recommended for use after shaving. Three sizes.
8L3465—4-ounce bottle. Shpg. wt., 10 oz. 28c
8L3466—8-ounce bottle. Shipping weight, 14 ounces. 53c
8L3469—16-ounce bottle. Shipping weight, 1½ pounds. 98c
DRALLE's Illusion
The Original Imported Concentrated Perfume. Fragrant and Lasting.
Dralle's Illusion is probably the finest quality of imported concentrated perfumes. A drop will hold its fragrance for a long time. Each flower odor has its own distinctive fragrance. Little bottle contains about ½ dram concentrated perfume. Each bottle in polished wood box (lighthouse shape). Shipping weight, 5 ounces.
8L2975—Lily of Valley. Regular $1.00 size. 79c
8L2976—Rose. Regular $1.00 size. 79c
8L2985—Lilac. Regular $1.00 size. 79c
8L2993—Violet. Regular $1.00 size. 79c
8L2999—Jasmine. Regular $1.00 size. 79c
Vegetals and Colognes
Pinaud's Lilac Vegetal.
Very popular for after shaving.
8L3410—6-oz. bottle. Shipping weight, 1 pound. 97c
8L3400—12-ounce bottle. Shipping weight, 1½ lbs. $1.89
Hoyt's Cologne.
25-cent size. Shipping weight, 5 ounces.
8L3414. 19c
Envelope Sachets
Lady Janis Sachets.
8L3262 Rose. 17c
8L3263 Lilac. 17c
8L3285 Lily of the Valley. 17c
8L3266—Oriente. 19c
8L3261—Violet. 17c
Incense Burners and Incense
Oriental Incense Burner.
Beautiful Egyptian design. Size, 4 in. high by 3¼ in. wide. Shipping weight, 1 lb.
8L4175. 67c
Cone Incense.
Sandalwood odor. Shipping weight, 2 oz.
8L4178—24 large cones. 39c
Buddha Burner.
A popular design. Either stick or cone incense may be used. Shipping wt., 1 lb.
8L4176—Buddha, complete. 25c
8L4177—300 sticks incense. Shpg. wt., 2 oz. 15c
Jar Sachets
8L3264—Djer Kiss. 79c
8L3275—La Trefle. 69c
8L3282—Azurea. 69c
8L3268—Lov'Me. 83c
8L3277—Oriente. 83c
8L3278—Violet. 47c
8L3279—Rose. 47c
8L3287—Crabapple. 47c
8L3288—Arbutus. 47c
8L3270—Mary Garden. 89c
Shipping weight of above, 6 ounces.
Cutex and Manicure Specialties
Five-Minute Set.
A new assortment containing full size packages of liquid polish, pink powder polish and cuticle remover. Also manicure stick and emery boards. $1.00 size. Shipping weight, 1 pound.
8L3371. 79c
Liquid Polish.
Dries instantly leaving a brilliant tinted luster. Requires no buffing. Complete with brush. 35-cent size. Shpg. wt., 8 oz.
8L3369. 26c
Pink Powder Polish.
Leaves a brilliant pink polish. 35-cent size. Shpg. wt., 4 oz.
8L3370. 26c
Cutex Compact Set.
Contains small package of cuticle remover, nail white, cake polish and paste polish, with emery board, file and manicure stick. 60-cent size. Shpg. wt., 4 oz.
8L3363. 47c
Nail White.
Applied direct from tube. A soft white cream for whitening tips of nails. 35-cent size. Shipping wt., 4 oz.
8L3365. 26c
Cuticle Remover.
A safe, scientific liquid for softening and removing overgrown cuticle. 35-cent size. Shipping weight, 8 ounces.
8L3364 26c
Cutex Traveling Set.
Contains full stock sizes of Cutex cuticle remover, nail white, cake polish and paste polish, with emery board, file and manicure stick. $1.50 size. Shipping weight, 1⅛ pounds.
8L3367. $1.19
Nail File.
Nickel plated handle. Length, 6 inches. Shipping weight, 1 ounce.
8L3901. 17c
Tweezers.
For removing splinters, ingrowing hairs, etc. Length, 3½ inches. Shipping weight, 2 ounces.
8L3905. 19c
Shpg. weight, 2 oz.
8L3928—Nail Scissors. Length, 3½ inches. 67c
8L3929—Cuticle Scissors. Length, 4 inches. 67c
Nail Clipper.
Comprises nail clipper, file and cleaner. Fits in vest pocket. Length, 2¼ in. Shipping wt., 3 oz.
8L3930. 25c
Nippers.
Used for toe or finger nails. Size, about 4⅜ in. Shipping wt., 3 oz.
8L3934. 98c

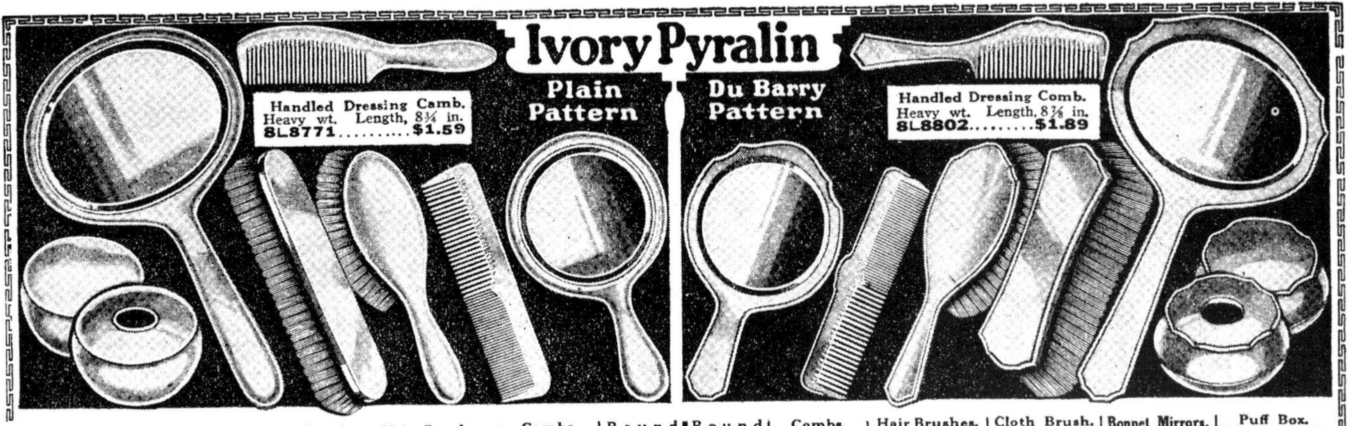

Ivory Pyralin
Plain Pattern
Du Barry Pattern
Handled Dressing Camb. Heavy wt. Length, 8¾ in. 8L8771..........$1.59
Handled Dressing Comb. Heavy wt. Length, 8⅞ in. 8L8802..........$1.89

Hair Receiver. Diameter, 4 inches; height, 2¼ inches. **8L8735 98c**
Puff Box. Diameter, 4 inches; height, 2¼ inches. **8L8736 98c**

Bonnet Mirrors. Bevel plate glass. Length, 14¾ in.; width, 8¾ in. **8L8712 $5.98**
Length, 12⅝ in. width, 7⅜ in. **8L8713 $4.98**

Hat Brush. Length, 5⅞ in Six rows white bristles. **8L8708 $1.87**
Cloth Brush. Length, 6⅞ inches. Eight rows white bristles. 8L8709. **$2.69**

Hair Brushes. Solid backs. Length, 8⅞ in. Eleven rows 1⅛-inch stiff white bristles. **8L8766 $4.98**
Lgth., 8⅞ in. Nine rows ⅞-in. white bristles. 8L8705.. **$2.98**

Combs. Heavy weight. Length, 8⅝ inches. **8L8761** Coarse and fine....**98c**
8L8760 All coarse. **98c**

Round Mirror. Bevelplate glass. Length, 10¼ in.; width, 5⅝ inches. **8L8715 $3.98**

Round Mirror. Bevelplate glass. Length, 10¾ in.; width, 6⅜ inches. **8L8811 $5.69**

Combs. Medium weight. Length, 7¾ inches. **8L8800** Coarse and fine....**98c**
8L8801 All coarse. **98c**

Hair Brushes. Solid backs. Length, 8¾ in. Thirteen rows 1⅛-inch stiff white bristles. 8L8846 **$6.98**
Length, 8½ in. Thirteen rows 1-inch white bristles. 8L8804 **$4.98**

Cloth Brush. Lgth., 6¾ in. Nine rows white bristles. **8L8854 $4.98**
Hat Brush. Lgth., 6 in. Seven rows white bristles. **8L8855 $2.98**

Bonnet Mirrors. Bevel plate glass. Length, 15⅞ in.; width, 8¾ in. **8L8848 $9.98**
Lgth., 14 in.; width, 8⅜ in. **8L8847 $7.69**

Puff Box. Diameter, 4⅞ inches; height, 2 inches. **8L8869 $2.39**
Hair Receiver. Diameter, 4⅞ inches; height, 2 inches. **8L8870 $2.39**

Shipping weight of puff boxes and hair receivers, 8 ounces; mirrors, 3 pounds; brushes, 9 ounces; combs, 6 ounces.

Dresser and Manicure Tray.
8L8746 —11½x8½ inches..........**$2.98**
8L8719 —10x6¼ inches.........**$1.87**

8L8894—Buffer with Boat; reversible chamois. Length, 5¼ inches.**$1.39**
8L8895—Buffer, as above. Length, 6½ inches.........**$1.89**

Heavy Handle Pattern.
8L8726—Nail File. Lgth., 7⅜ in.**33c**
8L8727—Button Hook. 7¼ in. **29c**
8L8728—Cuticle Knife. Length, 5⅛ inches.........................**29c**

HeavyHandle Pattern Scissors.
Finger Nail. 8L8788...**98c**
Cuticle. 8L8789...**98c**

Du Barry Pattern Scissors.
Finger Nail. 8L8892..**$1.19**
Cuticle. 8L8893..**$1.19**

Genuine Du Barry Pattern.
8L8861—Buffer with Boat; reversible chamois. Length, 5 inches..............**$1.67**
8L8866—Same as above. Lgth., 6⅛ in. **2.39**
8L8822—Button Hook. Lgth., 7⅝ in. **.79**
8L8823—Cuticle Knife. Lgth., 5 in. **.87**
8L8824—Nail File. Lgth., 7¼ in... **.87**

Genuine DuBarry Pattern. De Luxe Trays.
Dresser Tray, 12x7½ inches. **8L8820....$3.98**
Manicure Tray, 7½x4⅞ inches. **8L8821....$1.79**

Shipping weight of trays, 14 ounces; manicure articles, 6 ounces.

Fancy Pattern
Grooved Edge

8L8906 Puff Box. Diameter, 4½ in.; height, 1⅝ in. **$1.39**

8L8901 Round Mirror. Bevel plate glass. Length, 10¾ in.; width, 6 in...**$3.79**

8L8902 Hair Brush. Thirteen rows stiff white bristles. Lgth., 8½ in. **$2.39**

Manicuring Implements.
8L8924—Nail File. 7⅜ inches long............**59c**
8L8925—Button Hook. 7 inches long..........**59c**
8L8926—Cuticle Knife. 4¾ inches long.**59c**

Dresser or Manicure Trays.
8L8920—12x8¼ inches.........**$1.89**
8L8921—9¼x6¼ inches.........**$1.39**
8L8922—7½x4¾ inches............**98c**

Buffer. **8L8923** 5⅞ inches long. **$1.39**

Combs. **8L8904** 8 in. Coarse and fine..**59c**
8L8905 All coarse. **59c**

8L8903 Cloth Brush. Length, 6¾ inches. Nine rows white bristles. **$2.59**

8L8900 Bonnet Mirror. Bevelplate glass. Lgth., 13 in.; width, 8 in. **$4.69**

8L8907 Hair Receiver. Diameter, 4½ inches; height, 1⅝ inches. **$1.39**

Shipping weight of puff boxes and hair receivers, 8 ounces; trays, 14 ounces; mirrors, 3 pounds; brushes, 9 ounces; combs, 6 ounces; manicure articles, 6 ounces.

Special Combs

Medium Weight. Length, 8¾ in. **8L8754** Coarse and fine. **47c**
8L8753 All coarse. **47c**

Heavy Weight. Length, 8¾ in. **8L8750** Coarse. **69c**
8L8751 Coarse and fine. **69c**

Shipping weight, 6 oz.

Miscellaneous Articles to Complete Sets

8L8832 Puff Box. Diam., 3¾ in.; height, 2¼ in..**98c**
8L8833 Hair Receiver. Diam., 3¾ in.; height, 2¼ in......**98c**
Shipping wt., 8 oz.

Boudoir Candle Lamp. Average height, about 10 in. Ivory finish metal holder. Cretonne shade. Furnished with candle and adjustable shade support. Shpg. wt., 1⅛ lbs. **8L8782..$1.19**

8L8774 Soap Box. Holds average cake of soap. Size, 3¾x2¾ in. Shpg. wt., 2 ounces. **89c**

Fancy Picture Frames. Du Barry Pattern. Easel backs. Sizes given over all. Shpg. wt., 10 ounces.
8L8867–Abt. 5½x7...**$3.39**
8L8868–Abt. 4¾x5⅞.**$1.98**

Pincushion and Jewel Case.
8L8790—Fancy pattern as illustrated. Cushion raises up for jewels. Diam., 4¼ in.; height, 1⅞ in.$2.39
8L8791—Plain pattern, Diam., 3⅞ in.; height, 1⅞ in. **$1.98**
Shpg. wt., 12 oz.

Boudoir Oil Lamp. **8L8784** Height, 10¾ inches. Complete with ivory finish metal holder, fancy cretonne shade, chimney and wick complete. Shpg. wt., 1⅛ lbs. **$1.39**

Ring and Jewel Box. **8L8794** Size, 5⅜x3⅞ in. Height, 1¾ in. Velvet lined. Shipping wt., 11 oz. **$3.39**

Special Brushes

8L8706 Hair Brush. Solid back, 9 rows of ⅞-inch white bristles. Length, 8¼ in. **$2.39**

8L8701 Hair Brush. Solid back. 11 rows ⅞-inch white bristles. Length, 9¼ in. **$3.98**

Shipping weight, 9 ounces.

FOR OTHER TOILET ARTICLES SEE PAGE 422. FOR OTHER HAIR RECEIVERS SEE PAGE 422.

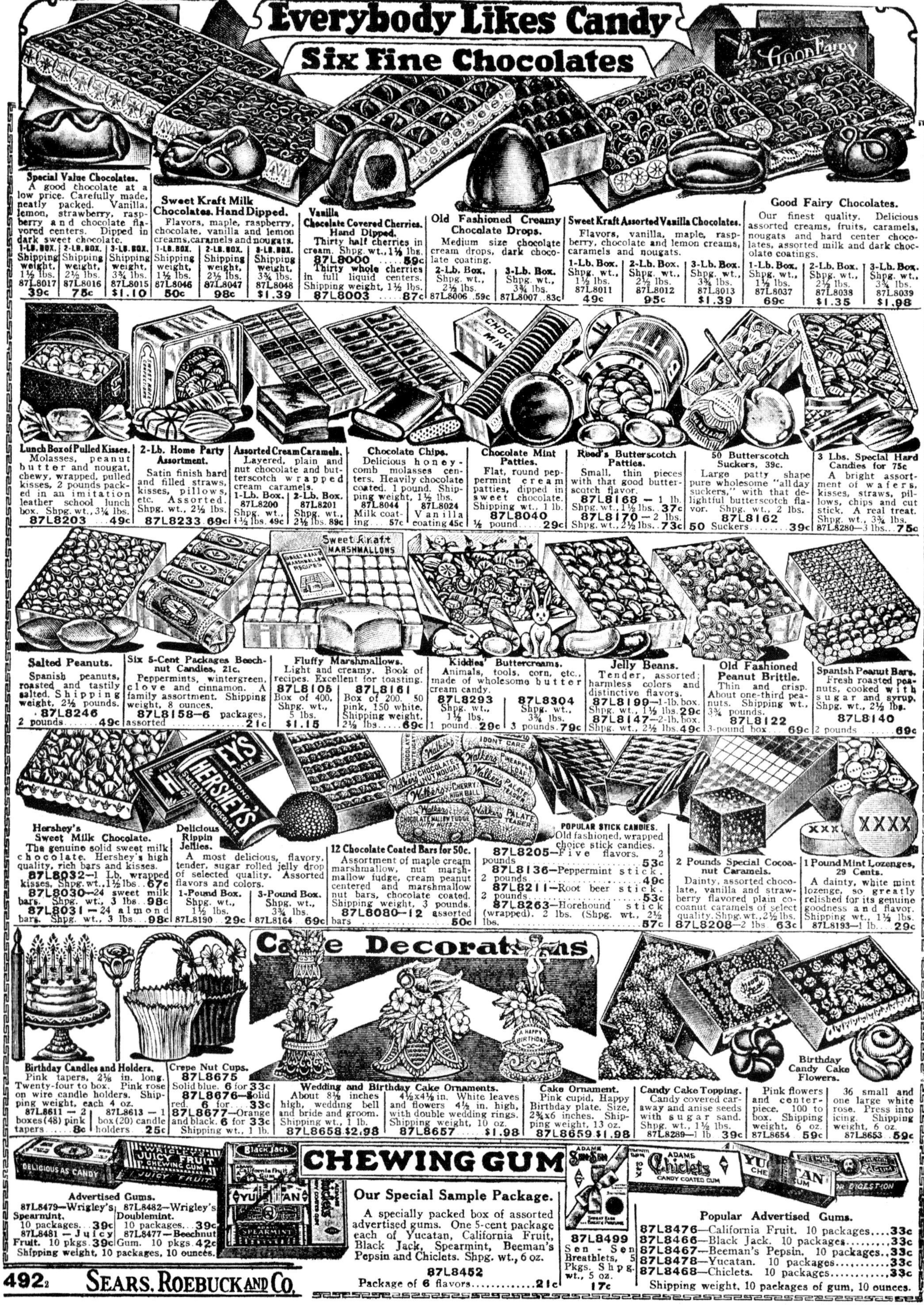

Everybody Likes Candy
Six Fine Chocolates
Good Fairy
Special Value Chocolates. A good chocolate at a low price. Carefully made, neatly packed. Vanilla, lemon, strawberry, raspberry and chocolate flavored centers. Dipped in dark sweet chocolate.
1-LB. BOX. Shipping weight, 1½ lbs. 87L8017 39c | 2-LB. BOX. Shipping weight, 2½ lbs. 87L8016 75c | 3-LB. BOX. Shipping weight, 3¾ lbs. 87L8015 $1.10
Sweet Kraft Milk Chocolates. Hand Dipped. Flavors, maple, raspberry, chocolate, vanilla and lemon creams, caramels and nougats.
1-LB. BOX. Shipping weight, 1½ lbs. 87L8046 50c | 2-LB. BOX. Shipping weight, 2½ lbs. 87L8047 98c | 3-LB. BOX. Shipping weight, 3¾ lbs. 87L8048 $1.39
Vanilla Chocolate Covered Cherries. Hand Dipped. Thirty half cherries in cream. Shpg. wt., 1½ lbs. 87L8000 ..... 59c
Thirty whole cherries in full liquid centers. Shipping weight, 1½ lbs. 87L8003 ..... 87c
Old Fashioned Creamy Chocolate Drops. Medium size chocolate cream drops, dark chocolate coating.
2-Lb. Box. Shpg. wt., 2½ lbs. 87L8006 .. 59c | 3-Lb. Box. Shpg. wt., 3¾ lbs. 87L8007 .. 83c
Sweet Kraft Assorted Vanilla Chocolates. Flavors, vanilla, maple, raspberry, chocolate and lemon creams, caramels and nougats.
1-Lb. Box. Shpg. wt., 1½ lbs. 87L8011 49c | 2-Lb. Box. Shpg. wt., 2½ lbs. 87L8012 95c | 3-Lb. Box. Shpg. wt., 3¾ lbs. 87L8013 $1.39
Good Fairy Chocolates. Our finest quality. Delicious assorted creams, fruits, caramels, nougats and hard center chocolates, assorted milk and dark chocolate coatings.
1-Lb. Box. Shpg. wt., 1½ lbs. 87L8037 69c | 2-Lb. Box. Shpg. wt., 2½ lbs. 87L8038 $1.35 | 3-Lb. Box. Shpg. wt., 3¾ lbs. 87L8039 $1.98
Lunch Box of Pulled Kisses. Molasses, peanut butter and nougat, chewy, wrapped, pulled kisses, 2 pounds packed in an imitation leather school lunch box. Shpg. wt., 3¼ lbs. 87L8203 ..... 49c
2-Lb. Home Party Assortment. Satin finish hard and filled straws, kisses, pillows, etc. Assorted. Shpg. wt., 2½ lbs. 87L8233 69c
Assorted Cream Caramels. Layered, plain and nut chocolate and butterscotch wrapped cream caramels.
1-Lb. Box. 87L8200 Shpg. wt., 1½ lbs. 49c | 2-Lb. Box. 87L8201 Shpg. wt., 2½ lbs. 89c
Chocolate Chips. Delicious honeycomb molasses centers. Heavily chocolate coated. 1 pound. Shipping weight, 1½ lbs.
87L8044 Milk coating ... 57c | 87L8024 Vanilla coating 45c
Chocolate Mint Patties. Flat, round peppermint cream patties, dipped in sweet chocolate. Shipping wt., 1 lb. 87L8040 ½ pound 29c
Reed's Butterscotch Patties. Small, thin pieces with that good butterscotch flavor. 87L8168 — 1 lb. Shpg. wt., 1½ lbs. 37c 87L8170 — 2 lbs. Shpg. wt., 2½ lbs. 73c
50 Butterscotch Suckers, 39c. Large patty shape pure wholesome "all day suckers," with that delightful butterscotch flavor. Shpg. wt., 2 lbs. 87L8162 50 Suckers 39c
3 Lbs. Special Hard Candies for 75c. A bright assortment of wafers, kisses, straws, pillows, chips and cut stick. A real treat. Shpg. wt., 3¾ lbs. 87L8280—3 lbs. .. 75c
Sweet Kraft Marshmallows
Salted Peanuts. Spanish peanuts, roasted and tastily salted. Shipping weight, 2½ pounds. 87L8246 2 pounds 49c
Six 5-Cent Packages Beechnut Candies, 21c. Peppermints, wintergreen, clove and cinnamon. A family assortment. Shipping weight, 8 ounces. 87L8158—6 packages, assorted 21c
Fluffy Marshmallows. Light and creamy. Book of recipes. Excellent for toasting.
87L8105 Box of 400. Shpg. wt., 5 lbs. $1.15 | 87L8161 Box of 200. 50 pink, 150 white. Shipping weight, 2½ lbs. 69c
Kiddies' Buttercreams. Animals, tools, corn, etc., made of wholesome butter cream candy.
87L8293 Shpg. wt., 1½ lbs. 1 pound 29c | 87L8304 Shpg. wt., 3¾ lbs. 3 pounds 79c
Jelly Beans. Tender, assorted; harmless colors and distinctive flavors. 87L8199—1-lb. box. Shpg. wt., 1½ lbs. 29c 87L8147—2-lb. box. Shpg. wt., 2½ lbs. 49c
Old Fashioned Peanut Brittle. Thin and crisp. About one-third peanuts. Shipping wt., 3¾ pounds. 87L8122 3-pound box 69c
Spanish Peanut Bars. Fresh roasted peanuts, cooked with sugar and syrup. Shpg. wt., 2½ lbs. 87L8140 2 pounds 69c
Hershey's Sweet Milk Chocolate. The genuine solid sweet milk chocolate. Hershey's high quality, rich bars and kisses. 87L8032—1 Lb. wrapped kisses. Shpg. wt., 1½ lbs. 67c 87L8030—24 sweet milk bars. Shpg. wt., 3 lbs. 98c 87L8031—24 almond bars. Shpg. wt., 3 lbs. .. 98c
Delicious Rippin Jellies. A most delicious, flavory, tender, sugar rolled jelly drop of selected quality. Assorted flavors and colors.
1-Pound Box. Shpg. wt., 1½ lbs. 87L8190 29c | 3-Pound Box. Shpg. wt., 3¾ lbs. 87L8164 69c
12 Chocolate Coated Bars for 50c. Assortment of maple cream marshmallow, nut marshmallow fudge, cream peanut centered and marshmallow nut bars, chocolate coated. Shipping weight, 3 pounds. 87L8080—12 assorted bars 50c
Popular Stick Candies. Old fashioned, wrapped choice stick candies. 87L8205—Five flavors. 2 pounds ..... 53c 87L8136—Peppermint stick. 2 pounds ..... 49c 87L8211—Root beer stick. 2 pounds ..... 53c 87L8263—Horehound stick (wrapped). 2 lbs. (Shpg. wt., 2½ lbs. 57c
2 Pounds Special Cocoanut Caramels. Dainty, assorted chocolate, vanilla and strawberry flavored plain cocoanut caramels of select quality. Shpg. wt., 2½ lbs. 87L8208—2 lbs. 63c
1 Pound Mint Lozenges, 29 Cents. A dainty, white mint lozenge, so greatly relished for its genuine goodness and flavor. Shipping wt., 1½ lbs. 87L8193—1 lb. 29c
Cake Decorations
Birthday Candles and Holders. Pink tapers, 2⅛ in. long. Twenty-four to box. Pink rose on wire candle holders. Shipping weight, each 4 oz.
87L8611 — 2 boxes (48) pink tapers ..... 8c | 87L8613 — 1 box (20) candle holders. 25c
Crepe Nut Cups. 87L8675 Solid blue. 6 for 33c 87L8676—Solid red. 6 for ..... 33c 87L8677—Orange and black. 6 for 33c Shipping wt., 1 lb.
Wedding and Birthday Cake Ornaments. About 8½ inches high, wedding bell and bride and groom. Shipping wt., 1 lb. 87L8658 $2.98 | 4½x4½ in. White leaves and flowers 4½ in. high, with double wedding rings. Shipping weight, 10 oz. 87L8657 ..... $1.98
Cake Ornament. Pink cupid. Happy Birthday plate. Size, 2¾x6 inches. Shipping weight, 13 oz. 87L8659 $1.98
Candy Cake Topping. Candy covered caraway and anise seeds with sugar sand. Shpg. wt., 1½ lbs. 87L8289—1 lb 39c
Birthday Candy Cake Flowers.
Pink flowers and centerpiece. 100 to box. Shipping weight, 6 oz. 87L8654 59c
36 small and one large white rose. Press into icing. Shipping weight, 6 oz. 87L8653 ..... 59c
CHEWING GUM
Advertised Gums. 87L8479—Wrigley's Spearmint. 10 packages ... 39c 87L8482—Wrigley's Doublemint. 10 packages ... 39c 87L8481 — Juicy Fruit. 10 pkgs. 39c 87L8477—Beechnut Gum. 10 pkgs. 42c Shipping weight, 10 packages, 10 ounces.
Our Special Sample Package. A specially packed box of assorted advertised gums. One 5-cent package each of Yucatan, California Fruit, Black Jack, Spearmint, Beeman's Pepsin and Chiclets. Shpg. wt., 6 oz. 87L8452 Package of 6 flavors 21c
87L8499 Sen - Sen Breathlets, 5 Pkgs. Shpg. wt., 5 oz. 17c
Popular Advertised Gums. 87L8476—California Fruit. 10 packages .... 33c 87L8466—Black Jack. 10 packages ..... 33c 87L8467—Beeman's Pepsin. 10 packages .. 33c 87L8478—Yucatan. 10 packages ..... 33c 87L8468—Chiclets. 10 packages ..... 33c Shipping weight, 10 packages of gum, 10 ounces.

Shipping weight of 25 cigars, 1 lb.; 50 cigars, 1¾ lbs.

CIGARS AND TOBACCO

Shipping weight, 50 stogies or 50 small cigars, 10 ounces.

Berriman's Havana Specials. Made from cuttings from very high grade clear Havana cigars. Perfecto shape. Medium short filler. 4¾ inches.
87L590—Per can of 25...$1.25

Davenport. An old reliable smoke. A fine combination of imported Havana and domestic tobaccos. Imported Sumatra wrapper. 4¾ inches. Fifty to box.
87L570—Perfecto shape. Box of 50..$3.98
87L572—Londres shape. Box of 50..$3.98

Flor de Madrid. Made from clear Havana cuttings from high quality cigars. Havana wrapper. Handmade. Excellent value at the price. 4⅝ inches.
87L514—Per box of 25...$2.19

Kisara. Long Havana filler. Connecticut binder. Imported Sumatra wrapper. This combination makes a mild smoke.
87L555—Panetela shape, 5 inches. 50 to box...$3.98
87L557—Perfecto shape, 4½ in. 25 to box...$2.69

Sir Francis Grant. A favorite with our smokers. Carefully selected long Havana and Porto Rican filler. Connecticut binder, imported Sumatra wrapper. A pleasing smoke. Twenty-five to box.
87L581—Perfecto shape, 4¾ in..$2.49
87L582—Londres shape, 5 in....$2.39

Rio Santos. Our most popular medium priced cigar. Genuine Spanish handmade Porto Rican cigar. Fine imported tobaccos. Fifty to box.
87L567—Corona shape, 5½ in.$5.69
87L565—Cabinet shape, 5¼ inches......$4.69

Royal Club House. A large club shape cigar, containing a fine blend of Havana and domestic long filler tobaccos, imported Sumatra wrapper. 4½ inches. Fifty to box.
87L575—Per box..........$3.39

El Escorto. Just the package for the pocket and car. Wisconsin and Pennsylvania tobaccos. Connecticut wrapper. 5 inches. Londres shape. Shipping weight, 8 oz.
87L592—Per can of 10.......50c

Our Special Leader Value.

Shipping weight, 50 stogies or 50 small cigars, 10 ounces.

Shpg. wt., 25 cigars, 1 lb.; 50 cigars, 1¾ lbs.

William Beal. A perfecto shape cigar. Connecticut wrapper, domestic short filler tobaccos. 4½ inches.
87L593—Per can of 25..........75c

The New Currency. A nationally advertised smoke. Perfecto shape cigar, made of domestic tobacco. 4½ inches.
87L550—Per box of 50...$2.29

Lillian Russell. An advertised cigar of known quality. High grade domestic tobaccos. Perfecto shape, 4⅜ in.
87L617—Per box of 50...$2.29

50 Pony Post Cigars for $1.89. Our leading 5-cent cigar. A blend of Ohio, Wisconsin and Pennsylvania tobaccos. Connecticut wrapper. Banded. Perfecto shape, 4¼ inches.
87L506—Per box of 50.......$1.89

James G. Blaine. Advertised cigar of pleasing blend. Only domestic tobaccos used. Perfecto shape. 4¾ inches.
87L613—Per box of 50...$2.29

Reina Bella. A blend of Pennsylvania and Ohio tobaccos with a Connecticut wrapper. A mild smoke. Perfecto shape. 4⅝ inches.
87L509—Per box of 50. Special..........$2.49

Quality Dutch Stogies. Mild and sweet. All long filler, selected Pennsylvania tobacco. An extra high grade stogie. 6 inches long.
87L701—Per box of 50....$1.79

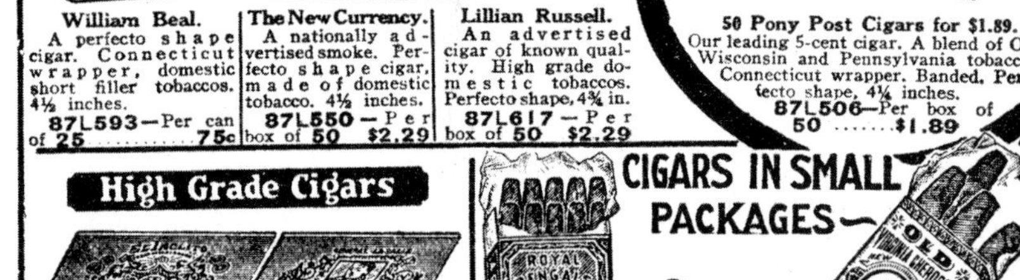

High Grade Cigars

CIGARS IN SMALL PACKAGES

El Inclito. Our highest quality, long filler, clear Havana cigar. Only finest selected tobacco used.
87L585—Rothchild's extra shape, 4¾ inches long. 25 in box.$3.49
87L589—Brevas shape, 4¾ inches long. 50 in box...........$6.98

General La Salle. A selected, high grade smoke. Handmade long clear Havana filler. Connecticut binder. Imported Sumatra wrapper. Invincible shape, 5½ in.
87L552—Per box of 25......$2.98

Royal Bengals. A nationally known small cigar. Shpg. wt., ½ lb.
87L825 3 boxes, 30 cigars....69c

Between the Acts. Those advertised, popular, mild small cigars.
87L817 5 boxes, 50 cigars.....87c

Your Favorites. Old Virginia Cheroots. Genuine Old Virginia Cheroots. Packed in handy pocket package of four cheroots each.
87L578—Per box of 50...$1.10

Congress Whiffs. A mild little cigar. Special pocket size tins.
87L822 5 boxes, 50 cigars.....87c

Rio Santos. A high grade, selected, mild, short smoke. Shpg. wt., 8 oz.
87L830 3 boxes, 30 cigars.......79c

High Grade Cigars

Belle Claire. Mild domestic tobaccos with selected Connecticut wrapper. A delightfully satisfying smoke. Fifty to box.
87L525 — Londres. 5 inches..........$3.69
87L527 — Perfecto. 4¾ inches..........$3.79

Sardou. Domestic long filler and binder, imported Sumatra wrapper. Fifty to box.
87L520—Invincibles, 5½ in. $4.39
87L521 — Perfectos, 4¾ in. $4.35

Killickinnick. 1 Pound Granulated Tobacco. A mixture of medium fine granulated Virginia tobaccos. Mild and full of flavor. A favorite with pipe smokers for years. A trial will convince you of its fragrance and blend. Shpg. wt., 1¼ lbs.
87L49—1-pound bag...49c

Choice Cigar Clippings. A high grade cigar clipping of Havana and broadleaf tobaccos. Free from stems and dust. Shpg. wt., 1¾ lbs.
87L401—14-oz. bag 67c

Legal Tender Plug. 79c 10 Cuts.

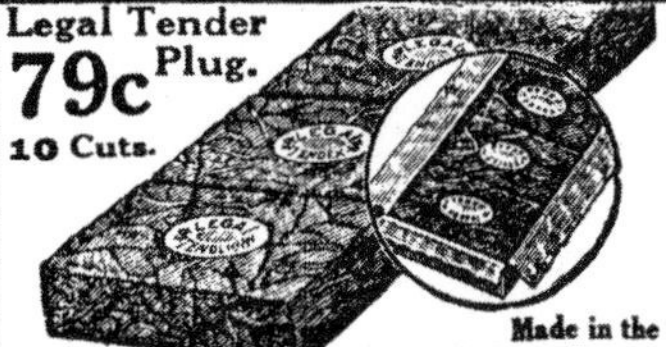

Made in the Piper Heidseik Factory.

An exceptionally fine grade plug tobacco. Made by the Piper Heidseik factory in Louisville, Ky. The plug with the delightful CHAMPAGNE flavor. A very satisfying chew. Net wt., about 13 oz. Shpg. wt., 1¼ lbs.
87L100—Per plug of 10 cuts.......79c

Advertised Plug Tobaccos.
87L102—Spear Head Plug. Eight 10c cuts..............69c
87L103 — Horse Shoe Plug (Smooth). Nine 10c cuts.....79c
87L104—Climax Smooth Plug. Nine 10c cuts..............77c
87L105—Standard Navy. Eight 10c cuts..................69c
87L106—Red Cross Plug. Six 10c cuts...................54c
Shipping weight, 1¼ pounds.

Old Favorite Long Cut Tobacco.

A choice mixture of popular, old fashioned Kentucky Burley and Virginia tobaccos. Very free from "shorts." This high grade mixture makes a cool, fragrant smoke, and a sweet, lasting chew. Packed in a tin pail. Shipping weight, 1½ pounds.
87L41—14-ounce pail.....75c

Columbia Cut Plug. Kentucky Burley Tobacco.
87L322—Tin lunch box, containing 14 oz. Shpg. wt., 1½ lbs.$1.00
87L320—Large size tin, net weight, 1 pound. Shpg. wt., 1¾ lbs.$1.00
87L321—2-oz. sacks. Shipping weight, 13 oz. 3 sacks for......39c

Advertised Burley Tobaccos.

Prince Albert.
87L211 1-lb. tin. Shpg. wt., 1¾ lbs. $1.25
87L212 Glass humidor. Shpg. wt., 4 lbs. $1.39
87L210 ½-lb. tin. Shpg. wt., 1⅛ lbs. 65c

Tuxedo.
87L206 1-lb. tin. Shpg. wt., 1¾ lbs. $1.25
87L207 Glass humidor. Shpg. wt., 4 lbs. $1.39
87L205 ½-lb. tin. Shpg. wt., 1⅛ lbs. 65c

Velvet.
87L216 1-lb. tin. Shpg. wt., 1¾ lbs. $1.25
87L217 Glass humidor. Shpg. wt., 4 lbs. $1.39
87L215 ½-lb. tin. Shpg. wt., 1⅛ lbs. 65c

Sun Kist. Sun Kist has a wonderful richness, a wealth of fragrance and a smack and relish not found in other tobaccos. Try a tin and be convinced.
87L411—1-lb. tin. Shpg. wt., 1¾ lbs.$1.33
87L410—½-lb. tin. Shpg. wt., 1⅛ lbs..69c

Quality Twist Tobaccos.

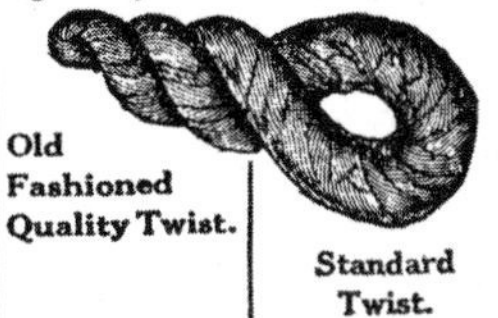

Old Fashioned Quality Twist. A natural, selected leaf twist, unsweetened, for chewing and smoking. Shipping weight, 10 ounces.
8L124—For 6 twists...54c

Standard Twist. A twist of pleasing blend and flavor. Unsweetened. Shipping wt., 10 ounces.
8L107—For 6 twists..49c

Advertised Tobaccos

Osterloh's Quality Mixture.
87L44—1-pound tin (humidor can). Shipping weight, 2 pounds..........83c
87L43—4-ounce pocket package. Shipping weight, ½ pound..........21c

Plow Boy (Long Cut).
87L29—14-ounce tin pail. Shipping weight, 1½ pounds..........67c

Beechnut Scrap.
87L122—6 10c size, pocket size bags. Shipping weight, 1½ pounds..........54c

Sterling (Fine Cut).
87L405—6 10c wax paper, pocket size. Shipping weight, 1½ pounds..........54c

Tip Top (Long Cut).
87L406—14-ounce tin pail. Shipping weight, 1½ pounds..........69c
87L407—6 10c pocket packages. Shipping weight, 1½ pounds..........54c

FOR THE LITTLE TOTS

Screw the hanger hook anywhere. Put baby in swing. He can't fall out. The slightest move sets him swinging. Seat and straps made of white duck. Frames are enameled steel. Size of seat, 13x13 inches.

"Rock-a-Bye" Swing With Back Rest. Taped holes for baby's legs. Has duck back rest with steel wire frame. Height, 52 in. Spring can be used with this swing.
69L9139 Shipping wt., 2 lbs.... **$1.25**

"Rock-a-Bye" Swing. Duck seat, with taped holes for baby's legs. Complete with hook to attach to doorway or ceiling. Spring can be used with this swing. Size of seat, 13x13 in.; height, 40 in.
69L9140 Shpg. wt., 1½ lbs.. **79c**

High Grade Steel Spring.
Use with the "Rock-a-Bye" swings. Size, 8½x1 inch. Shipping weight, 1½ lbs.
69L9183 .... **29c**

Genuine Toddler-Cars. Very Fine Quality.

Disc Steel Wheels.

With ½-inch rubber tires. Sturdy construction; high grade materials throughout and high class workmanship. Finished a brilliant red and yellow and nicely varnished. Foot rest on front part of seat for child's feet for coasting.

	Ht. to Top of Seat, In.	Lgth., In.	Shpg. Wt., Lbs.	
79L7510¼—1 to 2 yrs.	8½	16¼	6	$2.39
79L7511¼—2 to 3 yrs.	10	18¾	8	2.98
79L7512¼—3 to 4 yrs.	11½	20	10	3.69

Same as above, except with bright red wood wheels, no rubber tires.

79L7546¼—1 to 2 yrs.	8½	16¼	6	$1.98
79L7547¼—2 to 3 yrs.	10	18¾	8	2.25
79L7548¼—3 to 4 yrs.	11½	20	10	2.47

Kiddie's Velocipede, For Youngsters from 2 to 5 Years.
Beautifully finished bright red enamel frame, yellow seat. Practically indestructible. Smooth wood seat is about 10 inches from the floor and is large enough to make a comfortable and safe riding pedal car for the beginners. Steel frame, axles, etc. Metal handle bars with wooden grips. ⅜-inch rubber tires. 9-inch front, 7-inch rear wheels. Saddle to pedal at lowest point, 16 inches. Shipping weight, 12 pounds.
79L8362¼ **$2.98**

Babies' Bassinet and Play Swing.
Makes a swinging bed 13¼ by 30 in. and 10 in. deep; or a seat is formed deep enough for baby to be secure against falling. Seat has two holes for legs, and a table in front for toys. Pictures above show swing for both uses. Enameled steel frame. Bed and seat of strong duck. Four duck strap supports with a steel hook. When bed is in use, the tape strings can be tied over top to hold baby in. Height of swing over all, 50 inches. Spring can be used. Shpg. wt., 4¾ lbs.
79L9175¼ **$1.79**

Baby Developer.
With baby sitting comfortably spring the seat up and down a few times. After a few times in the swing adjust the seat a little higher and baby attains a walking position in order to touch the floor, then he starts walking forward until his feet are nearly off the floor, then swings back. Adjustable from 54 to 66 in. Size of seat, 9½x10¾ in. **Spring furnished.** Shipping weight, 2½ pounds.
69L9174 Complete.... **$1.47**

Sand Box Toys. Bright Enamel Colors.
Six metal sand molds (not light gauge tin), all different shapes and covered with heavy enamel coat of bright colors; also a shovel 6¾ inches long and a sieve, 3⅜-inch diameter. All sewed in neat box, size, 8¾x11 inches. Shpg. wt., 1 lb.
49L1800 **39c**

Perfection Baby Walker.
A comfortable seat for baby while learning to walk. Frame made of heavy steel. Edges rounded and smooth; wide base prevents tipping. Diameter, about 26 inches; stands about 15 inches high.
Shipping weight, 8 pounds........ **$1.98**
79L9149¼

For other Baby Walkers see page 669.

$1.98

Babies' Nursery Seat.
Clamps on regular toilet seat. Made of smooth varnished wood. Metal sides. Strong duck strap to hold child in. Sides can be folded up when not in use. Saves using a vessel. Shipping weight, 3¼ lbs.
69L9181 .... **$1.98**

For other Nursery Seats see page 698.

Tom Tinker the Ballman.
He is made of nice, smooth wooden balls strung on strong cords which extend to a handle ball above his head. His many colors are made from harmless dyes that do not come off. About 7 in. high. Packed in pretty box. Shipping weight, 10 ounces.
49L7223 .... **43c**

For outdoor or indoor exercise this slide cannot be beaten. Youngsters must have plenty of exercise to develop their tender muscles. Get one for your little youngster and notice how he develops.

Well seasoned hardwood just like the large ones. Aluminum finish steel pipe railing and 7-step ladder, about 3 feet high. Slide and side rails are selected maple, smoothly sanded and finished in natural color; balance painted red.

Not a Toy, But a Real Miniature Size Slide.
Length, 7½ feet over all. Shpg. wt., 40 lbs.
79L9147¼ .. **$12.98**

Soft and Lovable Teddy Bears.
Made out of heavy cinnamon brown plush. These bears all have movable arms and legs and head turns. Can sit down or be placed in a number of lifelike poses. Glass eyes. The two larger bears have squeaker voices. Three sizes.
49L4322—Full height, 10 inches. Shipping weight, 1 lb. **59c**
49L4319—Full height, 12 in. Shipping weight, 1⅛ lbs..... **79c**
49L4323 — Full height, 14 in. Shpg. wt., 1¼ lbs.. **98c**

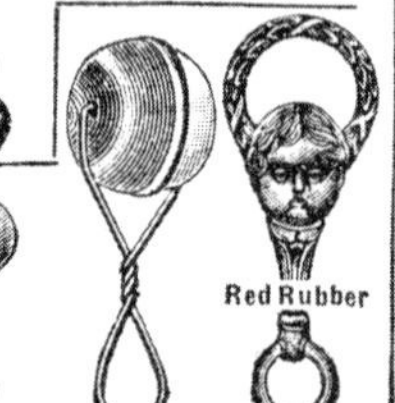

White Baby Head Rattle With Teething Ring. Length, 5 in. Shpg. wt., 2 oz.
49L4415 **10c**

Chick in Egg Rattle. White rubber with whistle and teething ring. Length, 4¼ inches over all. Shpg. wt., 3 oz.
49L4400 **12c**

Celluloid Rattle. Two-color or celluloid. Lgth., 6¾ in. Diam. of ball, 3 in. Shpg. wt., 2 oz.
49L4405 **10c**

Red Rubber

Extra Large RED Rubber Two Faces. Teething rings. Excellent quality red rubber. Length, 7½ inches. Whistle. Shpg. wt., 4 oz.
49L4403 **39c**

Little Fairy Hay Wagon.
Painted bright red. Wood, with metal wheels; 24-inch twisted wire handle. Size of wagon, 10¼x5½x5¾ inches. Shipping weight, 1¾ pounds.
49L5458 .... **39c**

All Steel Red Wagon.
Plenty big for the little toddler. The front wheels turn under just like on larger wagons. Size, 10¼x6¼ in. Enameled red. Wire handle. Shpg. wt., 1¼ lbs.
49L5457 ...... **29c**

Red Rubber Animals.
Best Quality.

Cat, about 4¾ in. long, 2 in. high.
49L4433 **21c**
Dog, about 5 in. long, 4¼-in. high.
49L4436 **39c**
Shipping wt., 5 oz.

White Rubber Animals.
Assorted dogs, horses, etc. About 4¾x2¾ inches. Shipping weight, 5 oz.
49L4406—Each **15c**

Colored Glass Beads to String.
A variety of beautiful colors. Furnishes endless amusement stringing them. Two sizes.

Our 19c String. About 144 beads. ¼-inch diam. Shpg. wt., 8 oz.
49L3803 **19c**

Our 39c String. About 84 beads. ⅝-inch diam. Shpg. wt., 1 lb.
49L3825 .. **39c**

Low Priced Roller Chime.
To pull over the floor and as it rolls along a continuous tinkling sound is heard. Diameter of wood wheel, 5 inches. Colored pictures on barrel, wire handle. Shipping weight, 1 pound.
49L2406 **39c**

Rubber Balls.
Real colored rubber, not painted. Assorted colors and designs.
4 - Inch Diameter. Shipping weight, 8 ounces.
69L7734 **39c**
3 - Inch Diameter. Shipping weight, 7 ounces.
69L7739 **27c**

30 Blocks, 39c.
Thirty A B C Blocks. 1¼-in. cubes. Two sides embossed with painted letters; pictures on four sides. Round corners are safe for baby. Shipping weight, 1⅛ pounds.
49L3658 30 blocks **39c**

Large Teething Ring.
Strong, white, tubular celluloid teething ring on silk cord. ½ inch thick. Crimped surface. Diameter of ring, 3⅜ in. Shpg. wt., 2 oz.
49L4407...... **19c**

Baby Reins.
Strong oilcloth reins, colors. Nickel plated, 1 inch. Fit over the shoulders and under the arms. Five 1-inch bells. Brother or sister, or even mother, will drive and baby trots along like a pony, making music with the bells. Shipping weight, 4 oz.
49L2322 .... **19c**

Aluminum Set for Babies.
Real cup, saucer and A B C plate of satin finish bright aluminum. Plate 6⅞ inches in diameter. Other pieces in proportion. Light weight. Shipping wt., 10 ounces.
49L1859 **25c**

PLAYING CARDS

Well Known Standard Quality.

Same as we have always listed. Shpg. wt., 8 oz.
Touring...49L150—55c
Flinch....49L151—59c
Bunco....49L152—59c
Gypsy Witch...49L153—59c
Rook....49L148—59c
Pit....49L149—59c
100 Bunco Score Cards and 3 Bunco Dice............49L206—25c

Boys and Girls, PLAY TENNIS!

Fun, excitement and lively exercise for the youngsters. The two rackets are 18¾ inches long and have cords stretched taut in frame, nicely varnished and shaped wood handle. The balls are 2½ inches in diameter, felt covered. Shipping weight, complete, 1½ pounds.
49L176 .... 98c

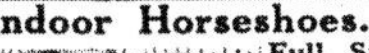

Indoor Horseshoes.

Full Size Rubber Horseshoes. Will Not Injure Furniture or Floors. Four horseshoes, made of rubber, and two pegs firmly fastened in enameled broad metal plates to set on floor. You can get ringers and "leaners" just like outdoors. Shpg. wt., 3⅛ lbs.
49L128—Per set....... $1.39

Chessmen.

Hardwood French pattern finished white and black, varnished. Height of king, 3 in. Packed in wood box. Shpg. wt., 12 oz.
49L202........98c

Hardwood Staunton pattern. Smooth, waxed luster finish. Wood box. King, 2¼ inches high. Shipping wt., 14 oz.
49L203.................. $1.39

Donkey Party.

Starched cloth sheet, 18¾x30 inches. With picture of donkey without tail and 24 detached tails. Blindfold and give each a donkey tail and have him pin it on the donkey. Shipping weight, 5 ounces.
49L135.......................... 19c

FOR ALL AGES

Crokinole, Carroms, Parlor Pool, Etc.

The Old Reliable Carrom Board—57 Games.

The same old standby board that was used by Dad when he was a boy. Remember how the whole family used to gather around it in the evening after supper and play crokinole, carroms, parlor pool, checkers, tenpins, etc. Reversible net pockets. Equipment to play **fifty-seven games.** The panel is made of three-ply maple veneer, natural wood finish. Highly varnished. Size, 29 inches square. Shipping weight, 11½ pounds. Cannot be shipped by parcel post.
79L489¼.............................. $4.39

Ouija Boards.

Have you tried one? Do spirits communicate through them? Intensely interesting and your family and friends will enjoy one.
49L112—Large size, genuine Fuld board, 22x15 in. Shipping wt., 3¼ lbs. 98c
49L129—Genuine Fuld Ouija Board. Size, 12½x18 inches. Shipping weight, 3 pounds..... 79c
49L110—Small size, 15x12½ inches. Shipping weight, 3 pounds..... 33c

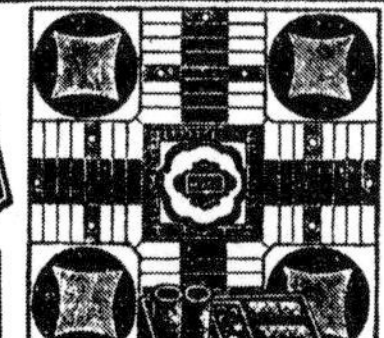

Parcheesi.

It is still the most popular of all board games. 18¼x18¼-inch folding board, dice, cups, colored counters and directions. Shpg. wt., 2¼ lbs.
49L100........ 89c

Dominoes.

49L165—Double-Twelve. Size, 1⅝x¾x¼ inches.................. 59c
49L123—Double-Nine. Size, 1⅝x¾x¼ inches.................. 39c
49L232—Double-Nine. Size, 1⁷⁄₁₆x¾x¼ inches.................. 25c
49L233—Double-Six. Size, 1⅝x⅞x⁵⁄₁₆ inches.................. 19c
Shipping weight, 1 pound.

Tell Tomorrow's Weather

These wonderfully popular weather prophets forecast the weather eight to twenty-four hours in advance. Merely hang up outside on the window frame or side of the house or even put on mantel in the home and the little figures will do the rest. When the weather is going to be fine the woman comes out and man stays in, but if a storm if brewing or rain threatens, the man comes out and the woman stays in. When the weather is changeable, both man and woman stand at the door. Of course, it is not to be expected this would be as accurate as a high grade barometer, but it really does a very good job of forecasting, considering it is a curio as well as a weather prophet.

A Big Value Two-Room Shack.

Made of wood and heavy cardboard. Nicely finished and decorated. Two artificial trees in front of house. Three dummy windows and thermometer. Man and woman beautifully finished. Height to tip of roof, about 7 inches; width, 7 inches; depth, about 4 inches. Similar to above picture. Shipping wt., 1 lb.
69L9196 .... 59c

The One We Recommend.

An exceptionally well made large weather house. Decorated with pine cones, fancy crosspieces and railing. Two dummy windows. Reindeer's head. Height, about 7¼ inches; width, 5¾ inches; depth, about 3 inches.
Shipping weight, 1 pound.
69L9148...................... 98c

CARD GAMES

Standard Brands at Reduced Prices.

P. E. means Plain Edge. G. E. means Gilt Edge. Shpg. wt., 4 oz.
Congress G. E. Picture Back. 49L252—63c
Bicycle, P. E. 49L250..........45c
American Beauty, P. E. 49L275..........39c
Auction Bridge, G. E. (Whist Size.) 49L280..........59c
Radium Pinochle, P. E.....49L276—39c
Uncle Sam, P. E..........49L277—25c
Peerless, G. E..........49L279—59c
Society G. E. Picture Back. 49L278—48c

Both for 19 Cents.
Authors and Old Maid Games.

An exceptional purchase enables us to sell these two games at this price. Boxes, size, 5x6½ inches. Splendid quality, smooth, white enameled, round cornered cards. Printed in colors. No matter what else you buy, include these two games in your order. Shipping weight, two games, 10 oz.
49L244.................. 19c

LOTTO.

Any number can play. Twenty-four cards, embossed counters, glass markers. Wood box, 6x9 inches. Shipping weight, 1 lb.
49L143................ 39c

Our 19-Cent Trunk Lotto. Twelve cards, wood counters, no glass markers. Cardboard box, 6¾x4 inches. Shipping weight, 1⅛ pounds.
49L145................ 19c

Cootie Game. Nearly seven million sold last year. Fun and laughter by the hour. For all ages. Metal frame box, 4x5 inches, glass top. Persons getting all four cooties in wire cage in shortest time wins. A dandy game. Shipping weight, 10 oz.
49L134...... 19c

Tiddledy Winks.

A great game for children. Place glass in center of table and flip the small discs into it. Sixteen colored bone discs, four bone shooting discs and glass (not wood cup). Shipping weight, 12 ounces.
49L147—Complete........... 25c

KINDERGARTEN and EDUCATIONAL TOYS

Story Sewing Cards. Twelve Picture Sewing Cards, 7½x6 inches, plainly perforated for sewing. Pretty box. Five skeins assorted colored threads and needle. Shipping wt., 10 oz.
49L3814—Per set..... 25c

Water Color Brushes. Four wood handle, camels' hair brushes, ranging in size from No. 2 to No. 7. Shpg. wt., 1 oz.
49L3886—4 brushes..... 19c

Prang's Water Colors. These are the standard Prang's quality school paints bearing the maker's brand. Two sizes.

Sixteen-Paint Size, 87 Cents. Sixteen pans semi-moist colors, enameled metal box, two brushes. Size box, 8⅝x3⅛ inches. Shipping weight, 10 ounces.
49L3879.................. 87c

Eight-Paint Size, 44 Cents. Eight pans semi-moist colors, enameled metal box, one brush. Size box, 8x2 inches. Shipping weight, 8 ounces.
49L3876.................. 44c

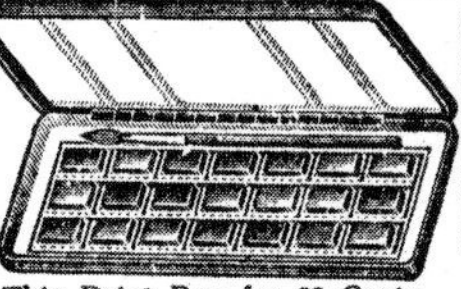

This Paint Box for 29 Cents. Twenty-one cakes dry water colors and camels' hair brush with wood handle in enameled metal hinged cover box, size 8x3 inches. Shipping weight, 8 oz.
49L3860.................. 29c

A Special Value for 15 Cents.

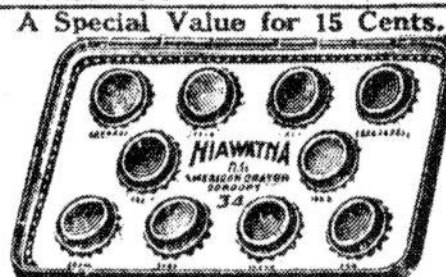

Ten large round cakes water color paints and one brush in metal box, with compartment top for mixing. Size, 6¼x3¼ inches. Shipping weight, 9 oz.
49L3895.................. 15c

Kroma Quality, Semi-Moist Water Colors. Hinged cover metal boxes.
49L3878—Sixteen colors and two brushes, 8⅛x3⅛ inches. Shpg. wt., 10 oz.... 79c
49L3877—Eight colors, one brush, 8x2 inches. Shipping weight, 8 ounces...... 39c

Our Wonder Stencil Outfit. Amusing and educational. Consists of twelve animals, six fruit and six flower cardboard stencils, 24 in all. The child merely lays the stencil on the paper, and with lead pencil, included, makes pictures for coloring or cutting out. This is one of our best and most popular toys. In pretty gift box. Size, 10½x8 inches. Shipping wt., 1 pound.
49L3827.................. 39c

Registering Banks. Nicely made, beautifully finished.

69L8701 For pennies, nickels, dimes, quarters and halves, 7x6⅞ x 4¾ inches. Shipping wt., 3 lbs $2.39
69L8702—For pennies, nickels, dimes and quarters. 5x4¼x4¾ in. Shpg. wt., 2½ lbs. $1.79
69L8700—Registers nickels, dimes and quarters. 5x5⅝x4 in. Shpg. wt., 2¼ lbs.... $1.37
69L8705—For dimes only. 3¾x3¾x3 inches. Shipping weight, 2 lbs........ 87c

For Enlarging Any Picture. Any child can do it—endless enjoyment.

A small size pantograph like those used by artists for enlarging. With it a child can reproduce or enlarge pictures. Extremely simple. Box size, 15¼x9¼ inches. Includes a pad of drawing paper, water color paints, colored wax crayons, steel pantograph, and instructions. Shipping wt., 1½ lbs.
49L3843.................. 39c

Paper Weaving for Little Fingers. Numerous assorted colored, glazed paper sheets, 7¼x5 inches, perforated. Also a large bundle of colored strips of paper to weave through the perforation and make pretty and pleasing designs. Several design sheets to show how. Box, 8¼x6½ inches. Shipping wt., 10 ounces.
49L3847 19c

Colored Wood Beads and Stringing Cords.
In Wooden Boxes.

Teaches child colors and counting.
49L3810—About 350 Beads. Assorted shapes. Round, oval, cylinders and cubes. Shipping weight, 1¼ lbs.... 79c
49L3813—About 175 Beads. Assorted round and oval. Shipping weight, 1 pound... 47c
49L3831—About 100 Beads. Round. Assorted colors. Shipping weight, 8 ounces. Per set.................. 25c

Three Bubbler Outfits for 25c. Every child enjoys blowing beautifully colored bubbles. Three boxes, each containing bubbler and small cakes of soap. Dip the bubbler into water and blow. Shipping weight, 11 ounces.
69L9152 25c
3 boxes for..

OUTDOOR TOYS

High Grade Velocipedes

Adjustable handle bars. Leather wound wood grips. Bicycle style front fork, coil spring padded saddle. Seat can be raised or lowered. Frame is painted, striped and strongly reinforced with bicycle style head. Strong wheels, ⅝-inch rubber tires. **Give measurement of child from crotch to heel.**

Sizes are with seat as low as it will go. **Unmailable.**

Catalog No.	Wheels Front, In.	Wheels Rear, In.	Saddle to Lower Pedal	Shpg. Wt., Lbs.	Price
79L8363¼	14	10	Abt. 17 in.	24	**$6.98**
79L8364¼	16	12	Abt. 20½ in.	28	**7.59**
79L8365¼	20	14	Abt. 23 in.	33	**7.98**
79L8366¼	23½	16	Abt. 25 in.	38	**8.98**

Same as above, only with ball bearing wheels and pedals. Tires are ¾-inch cushion style.

Catalog No.	Wheels Front, In.	Wheels Rear, In.	Seat to Lower Pedal	Shpg. Wt., Lbs.	Price
79L8379¼	14	10	17 inches	28	**$9.98**
79L8380¼	16	12	21½ inches	32	**10.98**
79L8381¼	20	14	23½ inches	34	**11.98**
79L8382¼	23½	16	25½ inches	40	**12.98**

Quality Velocipedes

Sturdy, Best Construction and Most Practical Ball Bearing Velocipede.

Channel steel frame, bicycle style fork heavily nickel plated from top to bottom. Adjustable handle bars of nickel plated steel tubing, about ⅝-inch in diameter. Leather saddle on nickel plated coil springs. Rat trap ball bearing pedals and nickel plated cranks and large hub caps. Frame and wheels finished with sapphire blue bicycle enamel, striped in gold. 1-inch cushion tires. **Unmailable.**

Catalog No.	Wheels Front	Wheels Rear	Saddle to Lower Pedal	Shpg. Wt.	Price
79L8384¼	14 in.	10 in.	17 in.	27 lbs.	**$10.98**
79L8385¼	16 in.	12 in.	19 in.	30 lbs.	**11.98**
79L8386¼	20 in.	14 in.	21 in.	38 lbs.	**13.67**
79L8387¼	24 in.	16 in.	23 in.	40 lbs.	**14.98**

Serviceable Velocipedes

Quality Throughout—Not Built for Price.

Excellent quality velocipedes, ½-inch rubber tires on front wheel and ⅜-inch tire on rear wheels. Rigid handle bars. Half oval steel frame. Black enameled. Wheels painted; spring saddle has about 2½ inch adjustment. One-piece crank. Sizes below measured with seat as low as it will go. **Measure from crotch to heel.**

Catalog No.	Wheels Front, In.	Wheels Rear, In	Seat to Lower Pedal	Shpg. Wt.	Price
79L8356¼	16	12	19 in.	17 lbs.	**$3.98**
79L8357¼	20	14	21 in.	22 lbs.	**4.59**
79L8358¼	23½	16	23 in.	23 lbs.	**4.98**

Same as above, but without rubber tires.

Catalog No.	Wheels Front	Wheels Rear	Seat to Lower Pedal	Shpg. Wt.	Price
79L8351¼	16 in.	12 in.	20 in.	16 lbs.	**$3.39**
79L8360¼	20 in.	14 in.	22 in.	18 lbs.	**3.69**
79L8361¼	24 in.	16 in.	23 in.	20 lbs.	**3.98**

The Ford Auto

$5.98

All steel construction. No castings are used, eliminating worry of parts snapping off. Frame reinforced with steel cross members supporting body and gears. Rigid front axle with steel knuckle joints for easy steering. Smooth 6-inch steering wheel and nickel plated hub caps. Body and seat painted a brilliant red with chrome yellow striping, other parts black. 8-inch wheels with ⅜-inch rubber tires. Measures 33x18½ inches over all and 18½ inches from center to lowest pedal. Shipping weight, 27 lbs.

79L8900¼.................. **$5.98**

Harvard Farm Wagons

Built on specifications of big farm wagons, only smaller. Seat, sides and box bed can be removed. The Harvard Wagon is built to withstand hard knocks—the wheels are staggered spoked, with metal bushings, the tires are 5⁄32-inch metal; metal hub caps. 9⁄16-inch steel axle and wagon is hardwood throughout.

Harvard Junior

$8.98

79L7675¼—Body over all, 18x36 in. Complete with handle and one extra extension reach. Shipping wt., crated, 62 lbs. **Unmailable**.. **$8.98**

The Harvard Wagon.

79L7676¼—Body over all, 18x40 in. Complete with handle and one extra extension reach. Shipping weight, 64 pounds. **Unmailable**.. **9.98**

25 Imitation Onyx Glass Marbles.

Pretty mottled brown, blue and green colors; 22 are ⅝-inch diameter and 3 are 1 1⁄16-inch. Each marble perfectly smooth and nicely polished. Shipping weight, 1 pound.

69L7743.................................. **33c**

Our Best Marbles.

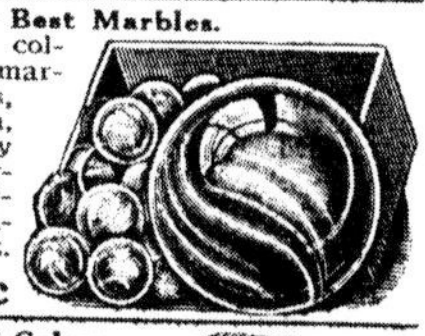

Twenty assorted colors, imitation onyx marbles. Made of glass, all perfectly smooth, polished and brightly colored. Each marble size of large marble shown in illustration. Shpg. wt., 8 oz.

69L7740 20 for.... **33c**

Marbles—Assorted Colors.

Some shooters, some medium size and balance in smaller marbles.

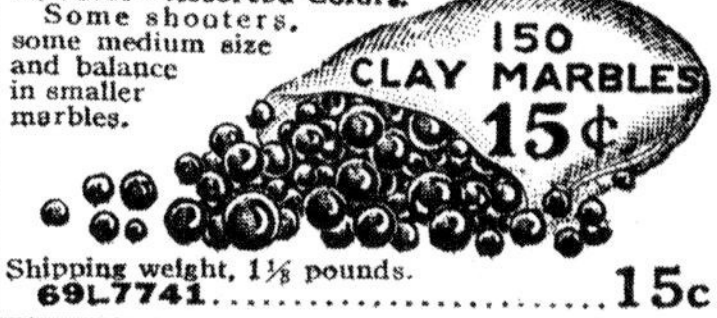

150 CLAY MARBLES 15¢

Shipping weight, 1⅛ pounds.

69L7741.................................. **15c**

Mysterious Gyroscope Top.

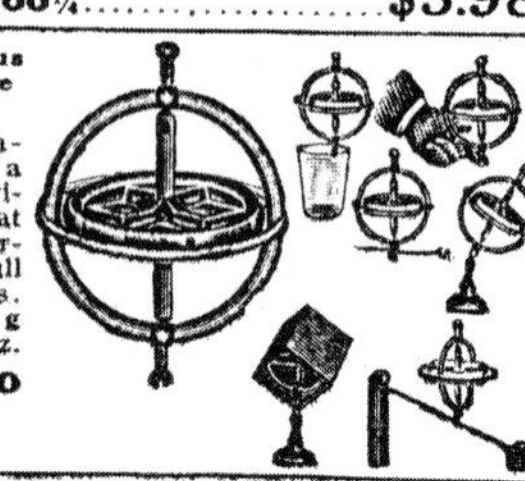

Illustration shows a few experiments that can be performed. Full directions. Shipping weight, 8 oz.

69L7760

19c

Toy Steam Engine.

Real steam cylinder; revolving dummy governor. Steam dome, safety valve, water gauge and whistle. Pulley for running toys. Brass boiler. Size, 6¾x5½x5¾ inches. Shipping weight, 4½ pounds.

49L5330 **$3.98**

Best Quality Pop Guns

Fine blued steel barrels; solid walnut stock. They shoot harmless corks with extra loud report. Look like real guns.

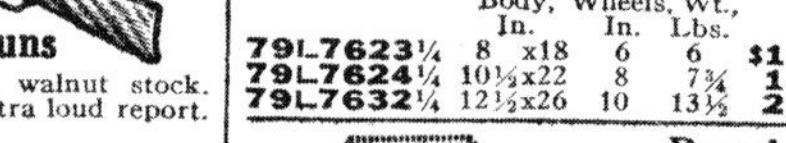

Large. Break action. Length over all, 29½ in. Shpg. wt., 1½ lbs.

49L5600 **67c**

Small. Lever action. Length over all, 17 in. Shpg. wt., 8 oz.

49L5647 **25c**

Ocean Liner.

Height to top of funnel, 3 inches; 8 inches long. Water tight. See it run in water. Shipping wt., 1½ pounds.

49L5744

59c

Buy a Real Wagon

EXPRESS WAGON

Steel Express.

Olympic Wagons.

The strong, substantial metal wagons. Stronger bracing, better gears and wheels, curved handle bar and made to stand wear.

	Size of Body, In.	Wheels, In.	Shpg. Wt. Lbs.	
79L7641¼	12½x26¼	10	18½	**$2.79**
79L7643¼	14½x30	11	21½	**2.98**

Faultless Wagons.

Lighter weight wagons, straight wooden tongue, and not so strongly made as our better grade Olympic wagons.

	Size of Body, In.	Wheels, In.	Shpg. Wt., Lbs.	
79L7623¼	8 x18	6	6	**$1.47**
79L7624¼	10½x22	8	7¾	**1.79**
79L7632¼	12½x26	10	13½	**2.25**

Boys! Make Your Own Kites.

These are the high flyers you want. Enough sticks, paper and cord to make **three strong kites**, size about 18x16 inches. Nice colored paper. Will make good appearance. Complete directions enclosed in each package. Shpg. wt., 10 oz.

69L7786—1 pkg. of **3** kites............ **15c**

Scoot-A-Way

A great muscle developer for children 4 to 15 years of age. One foot on the platform and one on the sidewalk, and watch him go. Will cover from 14 to 18 feet on good pavement with no more effort than an ordinary step. 8-inch disc steel wheels with rubber tires. All steel frame and fork, enameled bright maroon. All steel platform, 4½x14 inches. Light running and high grade construction. Height, 30 inches. Shpg. wt., 18 lbs.

79L8803¼........... **$2.39**

Balloons

6 for 25c

A large watermelon shape balloon, bagpipe, airship, sausage squawker and two round balloons. Shipping weight, 4 ounces.

69L9120—Complete assortment, **6** balloons.......... **25c**

4 Different Tops 19¢

One large size rubber neck top, another a ball point, the third a split top and fourth a good spinner. Four cords included. Shipping weight, 12 ounces.

69L7771—1 pkg. of **4**..... **19c**

Toy Garden Set.

33c

When daddy makes garden the kiddies always want to dig, too. Rake, spade and hoe about 25 inches long. Heads stamped out of strong steel (not cast). Good trowel. Wood handles. Shipping weight, 2¼ pounds.

69L9129.................................. **33c**

The Electric Thriller.

Many interesting stunts can be worked with this thriller, such as a cat trap on the back fence, or picking a coin out of a pan of water, etc. Simply turn crank. Mild or strong thrill, will not hurt anyone. The faster you turn the more thrills you get. Metal frame, wood base, 2½x4 inches. Shipping weight, 1¼ pounds.

69L5903—Complete.... **$1.39**

FOR GIRLS and BOYS

Extraordinary Values. $5.98

Genuine Reed.

Fancy bulge sides. Roll on hood and body. Strong 7-inch wheels with rubber tires. Body reinforced with hardwood dowels. Full lined hood and seat. Reclining back, adjustable hood; nickel plated hub caps; wood handle, nicely enameled, 25 inches high. Body, 21½x10½ inches over all. Hood, 29 inches high. Shipping weight, 15 pounds.

Royal Blue Body With Cream Color Wheels and Black Gear. 79L8240¼ $5.98

Gray Body With Black Wheels and Gear. 79L8241¼ $5.98

Dolly's Sulky. Only 89c. Substantially made of metal, enameled black with yellow wheels and striping on seat. 6-in. wheels. ¼-in. rubber tires. Folding handle, about 20 in. long. Seat, abt. 7x7 in. Back of seat, abt. 5 in. high. Big value at this price. Shpg. wt., 3 lbs. 79L8250¼ 89c

Semi-Collapsible Metal Go-Carts.

WITH THREE-BOW FOLDING HOOD.

Holds 22-inch doll. 7-inch wheels. ¼-inch rubber tires. Reclining back. Shipping weight, 6¼ pounds. 79L8260¼ $2.39

WITH FLAT FOLDING HOOD.

Holds 18-inch doll. 6-inch wheels. ¼-inch rubber tires. Shpg. wt., 4 lbs. 79L8253¼ $1.39

Perambulator With Yellow Enameled Rubber Tired Wheels, $2.69.

Strong metal body, nicely enameled in black with neat yellow stripes. Will stand hard knocks. Wood handle, about 23 inches from floor. Three-bow folding hood, covered with artificial leather. Body measures over all, 17x7½ inches. 6-inch wheels with ¼-inch rubber tires. Will hold an 18-inch doll. Shpg. wt., 6 lbs. 79L8213¼ $2.69

Big Values $3.47

Fiber Reed.

Large body, 17¾x8½ inches, bulge sides. Handle, 20½ in. high. 6-inch double spoke wheels with ¼-in. tires. Well made gear. Will hold an 18-in. doll. Cream color. Shipping weight, 10 pounds. 79L8201¼ $3.47

Same size body as above, only made of genuine reed. Fancy roll on body and hood. 7-inch wheels. Shipping weight, 10 pounds.

79L8238¼—Gray enamel $4.47
79L8239¼—Cream enamel 4.47

Canary Songster.

Your pet canary will pout with envy when this bird sings. When singing it opens its mouth and moves its tail just like a real bird. The bird is made of brass, lacquered to look like gold. Length of bird, 3 in. Length over all, 4¼ in. Shipping weight, 3 ounces. 49L2327 19c

Racket and Return Ball.

Lots of fun for little tots, as the ball doesn't get lost. Hit it hard as you want to. Racket, 18 in. long. Cords stretched tight in frame. Round wood handle. The 2-in. ball is fastened securely with two elastic cords, each decorated with a small bell. Shipping weight, racket and ball, 8 ounces. 49L138 39c

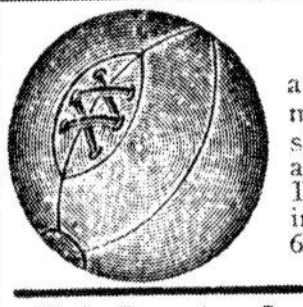

Toy Basket Ball.

Cover made of heavy artificial leather. Hand made rubber round shape bladder, like in a regulation size ball. Diameter, about 7½ inches. Shipping wt., 6 ounces. 69L7738 79c

DISC WHEEL COASTER WAGONS

BOYS!! BE UP TO DATE.

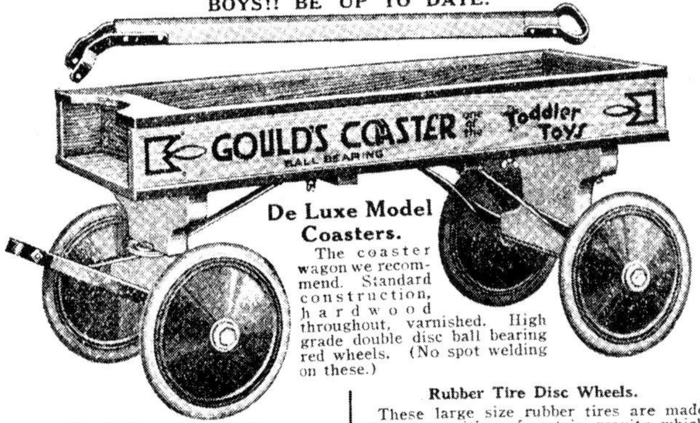

De Luxe Model Coasters.

The coaster wagon we recommend. Standard construction, hardwood throughout, varnished. High grade double disc ball bearing red wheels. (No spot welding on these.)

Metal Tire Disc Wheels.

This wheel is reinforced with a convex surface, increasing the wearing qualities of the rim and makes easy steering.

32-Inch Body. 79L7666¼ 8-inch wheels; box size, 14x32 in.; ht., 14½ in. Shpg. wt., 36 lbs. $6.98

38-Inch Body. 79L7667¼ 10-inch wheels; box size, 16x38 in.; ht., 15½ in. Shpg. wt., 40 lbs. $7.98

Rubber Tire Disc Wheels.

These large size rubber tires are made of a composition of certain gravity which will stand hard service. Noiseless and the most popular coaster wagon.

32-Inch Body. 79L7668¼ 8-inch wheels; box size, 14x32 in.; ht., 14½ in. Shpg. wt., 36 lbs. $7.98

38-Inch Body. 79L7669¼ 10-inch wheels; box size, 16x38 in.; ht., 15½ in. Shpg. wt., 40 lbs. $8.98

For Other Playground Equipment, Teeter-Totter, Slides, Giant Strides, Etc. See Page 767.

Jordan.

The Very Latest in Children's Auto.

Equipped with 10-inch disc steel, ball bearing wheels, enameled a brilliant red, with ½-inch rubber tires; sloping wind shield; dummy gearshift; steel bumper; 7-inch wood steering wheel with dummy gas control; speedometer and clock stenciled on dashboard. All steel body enameled a pretty coach green with carmine and yellow striping. Car measures, over all, about 44 inches in length and 19 inches in width. From center of seat to lowest pedal, 21½ inches. Shipping weight, 50 pounds. 79L8904¼ $9.98

High Bouncing Sponge Rubber Ball.

Every child will enjoy this soft ball. It is made of sponge rubber, very light in weight. About 2½ inches in diameter. Shipping weight, 10 oz. 69L7722 10c

Little Girls' Toy Wrist Watch.

Made of bright yellow metal. Spring sliding link bracelet like mother's watch. Octagon shape, transparent crystal. Winding stem turns hands. Each in box. Weight, 2 ounces. 69L9122 25c

Jump Rope, Also Jacks and Ball.

A dandy jump rope with enameled wood handles, also set of ten metal jacks and a solid rubber ball. Shpg. wt., 12 oz. 69L9156—Complete set for 33c

Four Bags Glass Beads for Stringing, 25c.

Four bags of imported colored glass beads. About 225 assorted color glass beads; each bag different size beads. Size of each bag, about 2x3 inches. Shpg. wt., 8 oz. 49L3815 25c

JOKES—Laughs for Everyone—JOKES

The Joke Mouse, 17 Cents.

Same size and color as ordinary mouse. Has friction movement. No spring. Merely push on floor or carpet, release and the mouse runs over floor with very natural motion. Shipping weight, 4 oz. 69L6305 17c

Old Maid Mirror, 19 Cents.

By looking in mirror one way it makes the tall person short and fat, and by reversing, makes the stout person tall and slim. You can have lots of fun showing friends their funny pictures. Size, 4½x3 inches. Shpg. wt., 5 oz. 69L6301 19c

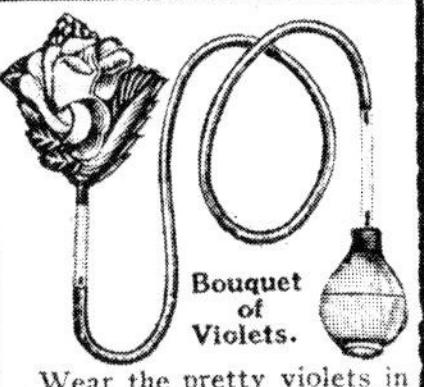

Bouquet of Violets.

Wear the pretty violets in coat lapel, fill the bulb with water and then ask your friend to smell the sweet violets. Press bulb and see some fun start. Shipping weight, 3 ounces. 69L6319 25c

Trick Pencils, 19 Cents.

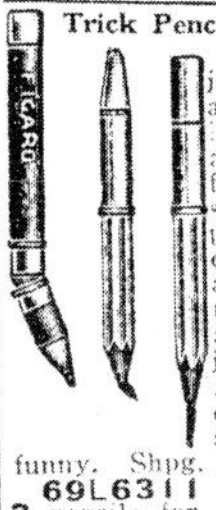

Offer your joke pencils to a person wanting to write and see the fun. Three assorted trick pencils, one each rubber and steel imitation leads, and one double jointed pencil. Bends in middle on pressure. Very funny. Shpg. wt., 3 oz. 69L6311 3 pencils for 19c

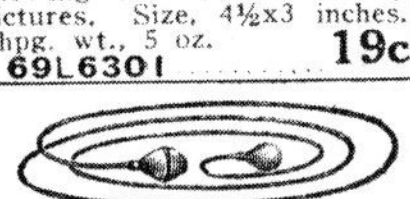

Magic Plate Lifter.

Merely place small bulb beneath tablecloth under any small dish or plate. Long rubber tubing. By pressing the large bulb the article will jump. Can also be used to exaggerate the beats of the heart. Shipping weight, 3 ounces. 69L6302 19c

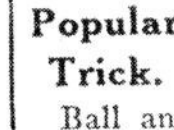

Popular Trick.

Ball and vase trick with which you can make the ball disappear and reappear. Shpg. wt., 3 oz. 69L6016 10c

Funny View Box.

Stout people look thin and thin people look stout. By getting a focus on passing pedestrians, horses, cars, etc., the most ludicrous pictures are witnessed. Shpg. wt., 5 oz. 69L6300 25c

Three Very Good Tricks for 25 Cents.

Three boxes of matches that look real; scratch but will not light. Chair squawker to place on chair or pillow. Also a big lifelike spider on string with long wire legs. Shipping weight, 8 ounces. 69L6320 25c

Aluminum Tea Set, 21 Pieces, 98 Cents.

Beautiful satin finish aluminum tea set. Teapot, 3$\frac{1}{16}$-inch diameter base and 2$\frac{9}{16}$ inches high; four plates, 4¾ inches in diameter. Other pieces in proportion. Very dainty and clean looking. Rolled rims. No sharp edges. Shpg. wt., 1 lb. 49L1873 98c

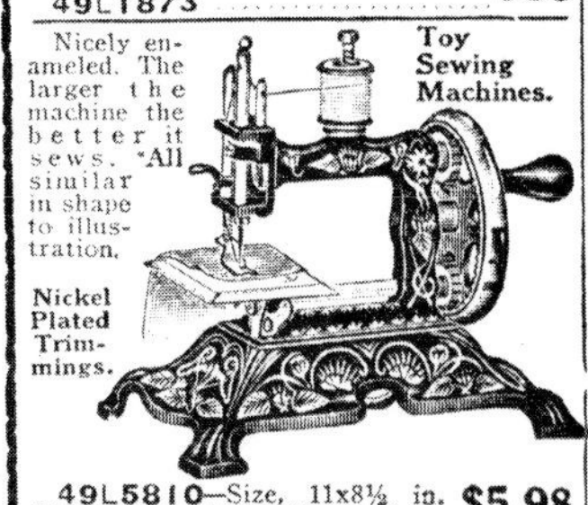

Toy Sewing Machines.

Nicely enameled. The larger the machine the better it sews. All similar in shape to illustration.

Nickel Plated Trimmings.

49L5810—Size, 11x8½ in. Shipping weight, 11½ pounds $5.98
49L5800—Size, 8x7½ in. Shipping weight, 6 pounds 4.69
49L5809—Size, 7¾x7 in. Shipping weight, 3⅛ pounds 3.39
49L5801—Size, 6¾x6¾ in. Shipping weight, 2¾ pounds 2.39
49L5803—Size, 4¼x5 in. Shipping weight, 1⅛ pounds .79

BIG VALUES in Fine Dolls and Doll

This Doll Walks-Talks and Goes to Sleep

Special Low Prices
CELEBRATED
Horsman Quality

Every little girl these days wants one of these new soft body "Mama" Dolls. These are the celebrated Horsman quality and have the Lloyd patented voice which calls "Mama!" most distinctly. Doll can be made to toddle (walk) like a real baby by holding it under arms and guiding it as you would a child just learning to walk. These lovely baby dolls have bright lifelike moving eyes which with the "Mama!" voice make a combination sure to delight. Handsome party style dress of sheer white organdy is lace trimmed and is very effective, having a colored underslip and bloomers. Bonnet matches dress and has ribbon tie. Fine mohair wig with pretty curls. Beautiful Horsman composition head. Fancy socks and fine slippers. Our dolls are real bargains considering the quality and price.

18L2927—Ht., abt. 17 in. Shpg. wt., 3¼ lbs. **$4.59**
18L2929—Ht., abt. 14 in. Shpg. wt., 2¼ lbs. **3.68**

BIG BARGAIN! Full Ball Jointed Doll-Finely Dressed

A Beauty

23-Inch Imported Doll
Extra Fine White Organdy Dress

Special Price Only $4.98

18L2909—If you want a wonderful doll we strongly recommend this number. The gorgeous dress is of very sheer extra fine organdy, and is handsomely embroidered with chenille. Large wired picture hat is of same material with a wreath of flowers on brim. Exceptionally high grade composition body, finished with glossy flesh colored lacquer and jointed at neck, shoulders, elbows, wrists, hips and knees. Pretty bisque head; moving glass eyes, which open and shut. Lovely mohair wig with many curls. Sheer lawn, lace trimmed underwear; fancy hose and slippers. A quality doll which is only found in the best large city stores and at very, very much higher figures than our special price. Height, about 23 inches. Shipping weight, 6¾ pounds.

READY for DRESSING

WALKS TALKS SLEEPS

	Ht. Abt.	Shpg. Wt.	
18L3155	20 in.	3½ lbs.	**$4.98**
18L3157	18 in.	3¼ lbs.	**4.25**
18L3159	15 in.	2¼ lbs.	**3.48**

You will want to dress these new and most charming character dolls of celebrated Horsman quality. They have moving eyes so they can go to sleep, and when picked up will say "Mama" most distinctly, being fitted with patented B. E. Lloyd Mama voice. Lovely wig of fine quality curled mohair sewed on net foundation. Body shaped like a real child and stuffed plump with soft cotton. Doll can be made to toddle like a real baby. Beautiful composition breast plate head permits using charming low neck style dresses. Dolly has a close fitting union suit of good quality knitted pink material, also socks and slippers. The larger the doll, the better the proportions.

Such a Darling!

Special Price $1.98

18L3193—Dandy Play Doll, and a bargain. Her well shaped body is stuffed plump with soft cotton and she has a voice which calls "Mama." Head of strong yet light composition, with painted hair, eyes and features. Good quality cotton romper dress in printed check pattern with white collar and cuffs; sash with large bow in back. Clever hat to match dress. White socks and imitation patent leather slippers. Height, about 18½ inches. Shipping weight, 3 pounds.

Walking Doll With Sleeping Eyes.

ONLY $1.35

18L2919—Walking doll (just take her arm) with strong composition head. Moving eyes and mohair wig. No mechanism to get out of order. Cloth covered stuffed body, jointed at hips and shoulders. White lawn dress and bonnet. Height, about 15 inches. Shipping wt., 2 lbs.

Chime Doll.

48c

18L3196—This doll has a chime inside body so when doll is shaken it gives out a pleasing musical sound. Chime does not get out of order. Doll has durable composition head and arms; painted hair and features. Stuffed body. Jointed at hips and shoulders. Height, about 14 inches. Shipping wt., 1¼ lbs.

Special Values in Better Grade Celluloid Dolls

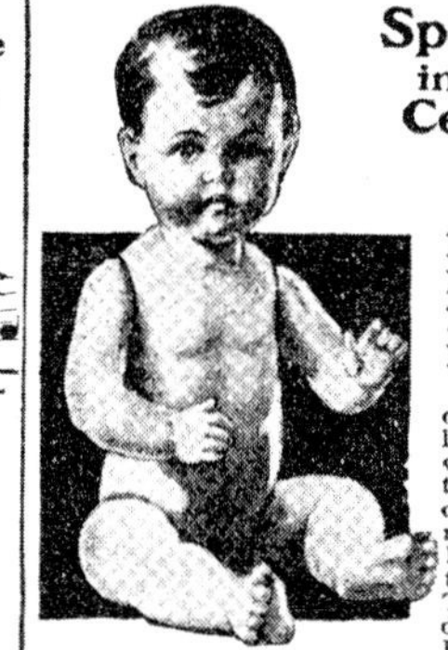

	Ht. In.	Shpg. Wt. Oz.	
18L3192	12	12½	$1.45
18L3190	9	9	.95
18L3188	7	7½	.43
18L3186	5	6¾	.21

These dolls are of the better quality heavy celluloid with lifelike tinting that will not come off. We do not handle the cheaper grades of flimsy construction because they do not give satisfaction. Celluloid dolls make a fine floating toy when baby is being bathed. They are also pretty when dressed in most any costume. Painted eyes, hair and features. Jointed at shoulders and hips.

Two Big Values

Imported Doll. 79c

18L3429—Fully dressed imported doll. Has good quality lifelike bisque head with sleeping eyes and curled mohair wig. Jointed at elbows, shoulders, hips and knees. Medium quality papier mache body. Pretty costume. Neat shoes, socks and underwear. Fine value at this low price. Height, about 12 inches. Shipping weight, 1½ pounds.

98c Character Doll.

18L3212

Durable composition head and soft mohair wig. Painted eyes and features. Plump body stuffed with cork. Smooth joints at shoulders and hips. Pretty dress is of colored cotton material in crossbar pattern, and is lace trimmed. Neat underwear, socks and slippers. Height of doll, 12¾ inches. Shipping weight, 1½ lbs.

Jointed Baby Dolls

Most Charming Dress

THREE SIZES

	Height, About, In.	Shpg. Wt., Lbs.	
18L2955	15	3	**$1.95**
18L2954	12½	2¼	**1.35**
18L2953	9½	1½	**.98**

All girls love these "sweet as sugar" baby character dolls. Natural baby shaped body and bent legs of nicely tinted papier mache. Bisque head has mohair wig and moving glass eyes. White flannel bonnet and removable jacket to match, fastening with pearl buttons. Dress underneath coat of good white cotton material; lawn underwear and diapers. Slippers and stockings. Big value dolls at our prices.

Imported Doll
Bisque Head
Moving Eyes

Only $1.69

A Dandy

15¾ In. High

18L2937

This doll has beautifully tinted bisque head with sleeping glass eyes and mohair ringlet wig. Full jointed papier mache body; wood thighs. Lovely dress of good material, with bonnet to match, as pictured. Neat underclothes, socks and slippers. Shipping weight, 2¾ pounds.

Accessories

POPULAR "Sunshine" Grade.

	Height, Inches	Shipping Weight		
18L3058	25	8	lbs.	$4.98
18L3056	23	4	lbs.	4.39
18L3048	21	3½	lbs.	3.98
18L3044	18	34	oz.	3.29
18L3040	16	29	oz.	2.45

These are fine quality dolls at very low prices. Do not confuse them with cheaply made, shapeless dolls sometimes offered in the market at similar prices. "Sunshine" bodies are of good, clean, white kid leather. Riveted joints at hips and knees. Full leather arms riveted at shoulders, with bisque hands and forearms. The beautiful bisque heads have bright lifelike moving glass eyes and natural looking hair eyelashes. Dainty curled mohair wigs with ribbon bow. Openwork stockings, good quality sandals.

OUR FAMOUS GUARANTEED UNBREAKABLE DOLLS With MINERVA METAL HEADS

Both "Wearwell" and "Knock Out" Dolls are made with strong pink silesia cloth covered bodies, stuffed nice and plump. The forearms and hands are of carved wood, enameled in natural flesh tint. We guarantee the wearing qualities of these dolls and are positive you will find them the greatest values offered for such low prices. The larger the doll, the better its proportions.

Our "Wearwell" Brand. Has "Minerva" metal head with parted, curled mohair wig and bright fixed glass eyes.

	Height, In.	Shpg. Wt., Lbs.	
18L3131	22½	3	$1.65
18L3128	20	2¾	1.48
18L3124	17	2	1.19
18L3120	15	1¾	.85

Our "Knock Out" Grade. Has "Minerva" metal head with painted eyes and molded painted hair.

	Height, In.	Shpg. Wt., Lbs.	
18L3108	15¾	1½	79c
18L3104	13¾	1	63c
18L3100	11¾	¾	45c

Ready to Dress

"Violet" Dolls Are Great Favorites.

	Height, Inches	Shpg. Wt., Lbs.	
18L3130	23½	3¾	$3.48
18L3127	20	3	2.98
18L3126	18	2½	2.45
18L3125	16	2	1.98

These popular dressing dolls have the celebrated, long lasting "Minerva" metal heads, fitted with bright lifelike sleeping eyes. Bodies are splendidly proportioned and covered with durable white imitation kid leather. Riveted joints at hips and knees. Molded celluloid forearms and hands. Full sewed curled wigs parted and ribbon tied. Fitted with good shoes and stockings. Sleeping eyes prevent our guaranteeing these dolls as absolutely unbreakable, but extreme care has been taken to make them as nearly unbreakable as possible. Our prices represent great value. The larger the doll, the better the proportions.

BEAUTIFUL DOLL HEADS

Measure the body from shoulder to shoulder across the top. After you have this measurement it is very easy for you to tell what number to order.

Finest Quality Bisque Heads.

	Height	Across Shldr.	Shpg. Wt.	
18L3404	4¼ in.	3 in.	7 oz.	$0.75
18L3405	4¾ in.	3¼ in.	8 oz.	.89
18L3406	5¼ in.	3¾ in.	10 oz.	1.15
18L3407	5¾ in.	4½ in.	14 oz.	1.45
18L3408	6¾ in.	5¼ in.	20 oz.	1.98
18L3410	7¼ in.	5¾ in.	25 oz.	2.75

Quality considered, the prices we quote are exceptionally low. The features are most lifelike and the heads have bright moving glass eyes. Lips are parted, showing pretty teeth. The lovely mohair wig is hand curled, parted and tied with ribbon bow.

Genuine "Minerva" Metal Heads. With Moving Eyes—Fine Wigs.

	Height	Across Shoulder	Shpg. Wt.	
18L3446	3¾ in.	2⅝ in.	6 oz.	$1.19
18L3448	4¾ in.	3½ in.	8 oz.	1.55
18L3450	5½ in.	3⅞ in.	13 oz.	1.95
18L3452	6½ in.	5¼ in.	15 oz.	2.59

"Minerva" metal doll heads combine the durability of metal and the beauty of bisque. The natural tinted enamel finish is baked on by a terrific heating process. Good quality wigs of mohair, nicely curled and parted. Beautiful sleeping eyes and open lips, showing teeth.

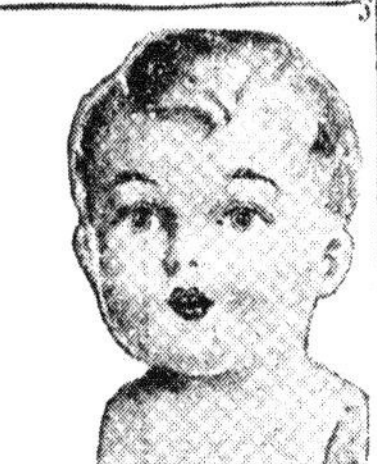

Rubber Dolls

19c Each

18L3461 Girl Doll.

18L3459 Boy Doll.

Good grade red rubber dolls, with whistle. Safe playthings in the hands of infants. Height, about 5 inches. Shipping weight, 3 ounces.

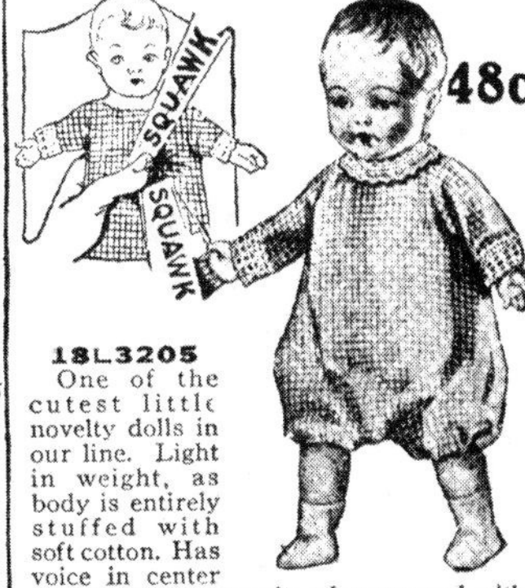

48c

18L3205 One of the cutest little novelty dolls in our line. Light in weight, as body is entirely stuffed with soft cotton. Has voice in center of body which squawks when pressed with fingers. Head of durable composition with painted eyes and features. Rompers of good quality colored cotton material trimmed with lace and ruching at neck. Cotton socks. Height, about 12 inches. Shipping weight, 1 pound.

Composition Heads With Wigs.

	Height	Across Shldr.	Shpg. Wt.	
18L3503	7¼ in.	4½ in.	1½ lbs.	$1.19
18L3505	6¼ in.	4 in.	1¼ lbs.	.89
18L3509	5½ in.	3⅜ in.	1 lb.	.68
18L3521	4¾ in.	2¾ in.	¾ lb.	.49

Strong composition fiber, light yet durable. Bobbed style wigs of good quality mohair. Finely tinted, painted eyes. From best American makers.

Unbreakable Metal Heads With Painted Hair.

	Height	Across Shoulder	Shpg. Weight	
18L3432	3¼ in.	2½ in.	6 oz.	$0.23
18L3434	3⅝ in.	2¾ in.	6 oz.	.33
18L3438	4⅝ in.	3¼ in.	7 oz.	.44
18L3440	5⅛ in.	3½ in.	8 oz.	.59
18L3442	6 in.	4¾ in.	12 oz.	.98
18L3468	6¼ in.	5 in.	14 oz.	1.29

Celebrated "Minerva" heads at low prices. Safe for the smallest child; no hair to get in baby's mouth. Two largest sizes have fixed glass eyes.

Composition Heads With Painted Features.

	Height	Across Shldr.	Shpg. Wt.	
18L3525	7¼ in.	4½ in.	1½ lbs.	67c
18L3529	6¼ in.	4 in.	1¼ lbs.	55c
18L3533	5½ in.	3⅜ in.	1 lb.	35c
18L3537	4¾ in.	2¾ in.	¾ lb.	29c

Painted hair and features. Made by the best American makers. We do not sell the cheaper grade heads, because they do not give satisfaction.

DOLL WIGS

Nicely made Wigs of human hair, thoroughly sanitary, in beautiful curls which are easily cared for and recurled. Sizes, 10 to 15 inches. **State inches.** Shipping weight, 4½ ounces.

18L3556 Light shades....**$1.25**

18L3560 Dark shades....**$1.25**

How to Measure for Wigs. Measure the circumference of the doll's head and **state the inches.** Use strong glue in fastening wigs.

DOLL ACCESSORIES

14c

18L3585 Good quality. Artificial patent leather shoes with tie and pompon. **Give length, heel to toe.** Sizes, 1¾ inches to 3½ in. Shpg. wt., 3 oz.

19c

18L3460 One-piece dress of fine cotton material. Has belt, buttons and button holes. Will fit dolls from 14 to 20 in. **In ordering give height of doll to be dressed.** Shipping weight, 6 oz.

8c

18L3574 — Fancy openwork white cotton doll stockings. Comes in sizes from 1¼ inches up to 3¾ inches. **In ordering give length of foot from heel to toe.** Shpg. wt., 2 oz.

Cut-Out Doll 21c

Shpg. wt., 2 oz.

18L3278—The dolls are lithographed upon a sheet, about 21x36 inches. Height of large doll, about 23½ inches; height of two little dolls, about 7 inches each. Cut out, sew and stuff and you have ideal unbreakable rag dolls for the youngsters.

Ribbons.

A first class ribbon made of the best materials. Will give clear sharp impressions and will not fill the type or fade. In spite of the low price, the strength and durability of these ribbons are equal to any of the high cost ribbons on the market. Shipping weight, each, 4 ounces. Made for Harris, Underwood, Oliver, Royal, L. C. Smith, Remington or Corona typewriters.

3L9876—Black Color Ribbon 58c
3L9874—Blue Color Ribbon 58c
3L9875—Purple Color Ribbon 58c

Be sure to state make of typewriter and color of ribbon.

Stenographers' Pencils.

Fine quality pencils. Try them for the wearing quality of the lead. Length, 7 in. Pointed at both ends. Six pencils in box with metal pointprotector. Shpg. wt., 8oz.

3L13601—Box of 6, with protector 25c

Typewriter Paper.

"Dorado" Bond. A light weight white bond paper of good quality. Shipping weight, per box of 500 sheets, 4½ pounds.

3L9700—Size, 8½x11 in. 500 sheets...$1.18
3L9701—Size, 8½x13 in. 500 sheets... 1.35

"Dorado" Bond. A medium weight white bond paper that is very suitable for general use. Shipping weight, per box of 500 sheets, 6 pounds.

3L9702—Size, 8½x11 in. 500 sheets...$1.19
3L9703—Size, 8½x13 in. 500 sheets... 1.39

"Dorado" Bond. Our best grade of white heavy bond paper, suitable for the most particular kind of work. Shipping wt., per box of 500 sheets, 7 lbs.

3L9704—Size, 8½x11 in. 500 sheets...$1.38
3L9705—Size, 8½x13 in. 500 sheets... 1.55

"Saranac" Linen Finish. A water marked, linen laid paper of exceptional quality. Shipping weight, per box of 500 sheets, 6 pounds.

3L9706—Size, 8½x11 in. 500 sheets...$3.39
3L9707—Size, 8½x13 in. 500 sheets... 3.98

Second Sheets.

A fair grade light weight white sheet, especially designed for making carbon copies. Shpg. wt., 5 lbs.

3L9708—Size, 8½x11 in. Per package of 1,000 sheets $1.48

Carbon Paper.

A good quality purple carbon paper which will make many clear, clean cut copies. Will not blur or smear. Size, 8½x11 inches. Shipping weight, per roll of 25 sheets, 2 ounces.

3L9709—25 sheets 45c

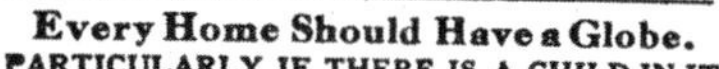

Every Home Should Have a Globe.

PARTICULARLY IF THERE IS A CHILD IN IT.

Our globes are of the very highest grade. The maps are made by George Philip & Son, Ltd., London Geographical Institute, engraved and beautifully lithographed in colors and are up to date in every particular. The balls can be washed without injury. The metal mountings in oxidized copper finish are attractive and substantial. Diameter of ball, 12 inches. Height on stand, 20 inches. Weight, packed for shipment, 10 lbs.

Geographical Globe.

This globe which is illustrated here is beautifully printed in different colors which define the boundaries of each country clearly. The principal cities and towns are printed in black and all water areas, including lakes and rivers, are printed in blue. The principal railways and ocean routes with distances are shown. Ideal for home or school. Regular price, $13.20.

3L12618 .... $9.95

Physical Globe.

This globe shows the height of the land and depth of the water by seven different tints of brown and green and five tints of blue. River courses, lakes, principal railways, cities and towns, ocean distance and currents, etc., are shown. Boundaries are distinctly overprinted in carmine. Regular price, $14.20.

3L12619 .... $10.75

Rubber Type Outfits

Make Your Own Rubber Stamps.

Illustrations show exact style and size of letters.

Deep Type Outfits.

Made of good quality solid rubber in full depth. Useful for getting up from one to four lines of printed matter on short notice. Each outfit in cardboard box. **Each outfit has holder, self inking pad and tweezers.**

AAaaaBbCcDdEeFfGgHh11234

3L16112—Set of Type, containing over 165 pieces; large and small letters, two sets of figures, punctuation marks, fancy ornaments, "$" and "£" marks, etc.; 3⅝-inch three-line metal holder. Weight, 10 ounces. Complete outfit 69c

AAaaa Bb Cc Dd Ee 12345

3L16127—Set of Type, containing over 200 pieces; both large and small letters, two sets of figures, fractions, "¼," "½" and "¾," punctuation marks, fancy ornaments, words "Price," "No." and "Co."; "$," "£" and "&" marks, etc.; 3¼-inch four-line metal holder. Weight, 1 pound. Complete outfit $1.25

AAAAAaaaaaaB3

3L16116—Set of Type, solid rubber, over 310 pieces, consisting of large and small letters, three sets of figures, six fancy ornaments, punctuation marks, three dashes and the words "For Sale By," "From," "Return in 10 days to," and "$," "%" and "£" marks; 3¼-inch four-line metal holder. Weight, 1⅛ pounds. Complete outfit $1.25

AAAAAaaaaaa Bb Cc Dd Ee Ff 12345

3L16104—Set of Type, containing over 275 pieces; large and small letters, three sets of figures, fancy ornaments, punctuation marks, etc., with three-line wood holder. Weight, 10 ounces. Complete 45c

Hinge Top Box File.

Wood frame. Strong colored cardboard top and bottom, complete with index. Size, 11¾x12¼x2⅞ in. Shipping weight, 2 pounds.

3L18155
59c

Automatic Dustproof Inkwell.

Ink cannot evaporate. Size, 2½x3 inches. Weight, packed, either style, 1⅛ pounds.

3L15110—Black top.. 58c
3L15109—Red top... 58c

Convenient Dater.

Metal holder with revolving type. Will print month, date and year. Rec'd, Ans'd, Paid. Shpg. weight, 3 ounces.

3L16425 ... 24c

Metal Valuable Paper or Bond Box.

A place to put valuable papers. Made of heavy gauge sheet metal, is lacquered in brown with gilt marking, with lock and two keys. Length, 11¼ inches. Height, 3 inches. Width, 5½ in. Shpg. wt., 2 lbs.

3L17505 .... $1.18

DRAFTING SUPPLIES

Venus Pencil, 8 Cents.

Furnished in fifteen degrees of hardness: 4-B, 3-B, 2-B, B, HB, F, 1-H, 2-H, 3-H, 4-H, 5-H, 6-H, 7-H, 8-H and 9-H. **State kind.** Shipping weight, 1 dozen, 5 ounces.

3L24405—Each, 8c; doz., 90c; gross $9.80

Koh-I-Noor Pencil, 10 Cents.

Furnished in fifteen degrees of hardness: 4-B, 3-B, 2-B, B, HB, F, 1-H, 2-H, 3-H, 4-H, 5-H, 6-H, 7-H, 8-H and 9-H. **State kind.** Shipping weight, 1 dozen, 5 ounces.

3L24401—Each, 10c; doz., $1.15; gross, $12.90

White Pencil.

Standard length, 7⅛ inches. Writes in white on all dark, rough surfaces, including blue prints, photo albums, photo mounts, black, brown or gray drawing paper or dark cloth. Shipping weight, 1 dozen, 3 oz.

3L24473—Each, 8c; per dozen 85c

For Other Pencils and Erasers See Page 453.

Artgum.

Invaluable to draftsmen, artists, photographers and show card writers. It cleans well without wearing the surface. **State size.**

3L24407
Size, 1⅛x1⅛x1⅛ in. Shpg. wt., 2 oz 4c
Size, 2x1x1 in. Shipping wt., 2 oz 8c

Venus Eraser.

Made of soft, pliable live rubber. Especially suited to the exacting draftsman. Size, 1¼x2x½ in. Shipping weight, 2 ounces.

3L24408 .... 9c

Liquid Drawing Inks.

3L24380—Higgins' Black Waterproof Ink. Put up in ¾-ounce bottles. Stopper fitted with quill for filling pen. Shipping weight, 7 ounces.

Per bottle $0.19
Doz. bottles. (Shpg. wt., 4½ lbs.) ... 2.25

3L24385—Dietzgen's Black Waterproof Ink. Meets all requirements of the most particular draftsman. ¾-ounce bottles. Stopper fitted with quill for filling pen. Shipping weight, 7 ounces.

Per bottle $0.20
Dozen bottles ...... (Shpg. wt., 4½ lbs.) 2.30

3L24386—Colored Drawing Inks. Indelible, put up in the same style bottle as 3L24385, with quill for filling pen; yellow, orange, scarlet, carmine, green, blue or brown. **State color.**

Per bottle .... (Shpg. wt., 7 oz.) $0.20
Per doz. bottles, asstd. colors. (Shpg. wt., 4½ lbs.) 2.30

Erasing Shield, 10 Cents.

Metal Erasing Shield, 2⅛x3¼ in., fourteen openings of various shapes. Shipping weight, 2 ounces.

3L24417 .... 10c

Whatman's Drawing Paper.

We believe Whatman's Drawing Paper is one of the very finest drawing papers in the world. We furnish the paper in two styles, the hot pressed and the cold pressed. The hot pressed paper has a smooth surface and is suitable for very fine line drawings in either pencil or ink. The cold pressed paper has a finely grained surface and is suitable for general drawing and water color painting. **Order by catalog number and be sure to state size.**

3L24431—**Cold Pressed.** See sizes and prices below.
3L24432—**Hot Pressed.** See sizes and prices below.

	Size, In.	Shpg. Wt.		
Cap	13x17	1 lb. 5 oz.	Per quire (24 sheets)	$0.80
Demy	15x20	2 lbs. 1 oz.	Per quire (24 sheets)	1.35
Royal	19x24	2 lbs. 15 oz.	Per quire (24 sheets)	2.40
Imperial	22x30	4 lbs. 6 oz.	Per quire (24 sheets)	4.30

Cream Drawing Paper.

Drawing Paper of a light cream color with a smooth and hard drawing surface. It is extremely tough and suitable for detail drawings, both pencil and ink. Stands rough handling well. **State width.**

3L24435

Width, inches	30	36	42
Shipping wt., per yard	1 lb. 5 oz.	1 lb. 10 oz.	2 lbs.
Per yard	11c	14c	16c

School Drawing Paper.

This Drawing Paper is an excellent white paper with a slightly grained surface. It is strong, stands pencil erasing well and is suitable for work in pencil, ink or color. **State size.**

3L24425

Size, 10x13½ inches. Shipping weight, per quire, 2 lbs. 5 oz. Per quire (24 sheets) 25c
Size, 13x20 inches. Shipping weight, per quire, 4½ lbs. Per quire (24 sheets) 50c

Tracing Paper in 15x20-Inch Sheets.

Very convenient for Home and School Use.

A dull finish, tough, transparent paper in very convenient form for use in both home and school. Just the thing for tracing maps, pictures, embroidery patterns, etc., and for numerous other uses. Put up in packages of fifteen sheets, 15x20 inches in size. Shipping weight, 8 ounces.

3L24426—Per package 22c

Tracing Paper.

This tracing paper is pure white, very thin, tough and transparent, with dull finish. It takes either pencil or ink and stands erasing well. Put up only in 20-yard rolls, 42 inches wide. Shipping weight, 2 pounds 13 ounces.

3L24445—Per roll $1.65

Tracing Paper.

This tracing paper is of medium thickness and oil finished, making it very transparent, and is a fine paper for making tracings from blue prints. Takes either ink or pencil and stands erasing well. Comes only in 20-yard rolls, 42 inches wide. Shipping weight, 3 pounds.

3L24450—Per roll 96c

Tracing Paper.

A rough detail sketching paper. Medium thickness. Oil finished, making it very transparent. Takes either ink or pencil and stands erasing well. Put up only in 50-yard rolls, 40 inches wide. Shipping weight, 6 pounds.

3L24446—Per roll $1.65

Tracing Cloth.

A fine imported vellum Tracing Cloth. **One side glazed and the other side dull. State width.**

3L24440

Width, inches	30	36	42
Shp. wt., per yard	1 lb. 2 oz.	1 lb. 7 oz.	1 lb. 9 oz.
Per yard	89c	98c	$1.12
Shpg. wt., per 24-yd. roll	4 lbs. 15 oz.	5 lbs. 11 oz.	6 lbs. 7 oz.
Per 24-yard roll	$17.82	$20.79	$25.66

"Union Satin" Blue Print Paper.

This grade of "Union Satin" Blue Print Paper is the best we can buy. The stock is heavy; the coating of the best. The paper is especially suited to architects' use. It gives a rich, deep blue color and clear white lines—fine, even lines showing sharp and clear. Furnished only in 10-yard rolls. **State width.**

3L24455

Width, inches	30	36	42
Shpg. wt., per 10-yd. roll	2 lbs. 1 oz.	2 lbs. 7 oz.	2 lbs. 15 oz.
Per 10-yard roll	$1.18	$1.41	$1.64

"Commercial Satin" Blue Print Paper.

Medium weight blue print paper, produced to meet the demand for a dependable paper at a low price. The stock is not quite so good as in our "Union Satin" Blue Print Paper, but it has the same high grade coating. Furnished in 10-yard rolls. **State width.**

3L24456

Width, inches	30	36	42
Shpping wt., per 10-yard roll	2 lbs.	2¼ lbs.	2¾ lbs.
Per 10-yard roll	67c	79c	90c

Protractors.

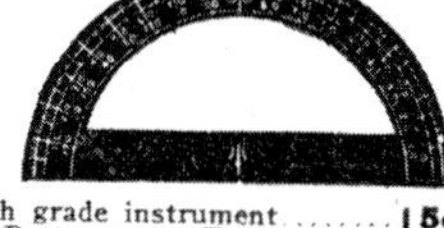

For dividing circles into any number of equal parts and determining angles. Shipping weight, any protractor, 3 ounces.

3L24261—Brass Protractor. Diameter, 3¾ in., half circle; 1 degree graduations. A high grade instrument 15c

3L24262—Celluloid Protractor. Transparent half circle; diameter, 6 in.; ½ degree graduations 48c

Irregular Curve.

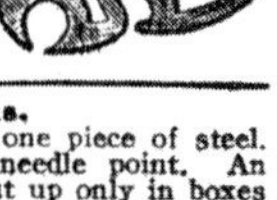

Irregular Curve, accurately made of transparent celluloid. Invaluable for drawing in irregular curves. Shipping weight, 4 ounces.

3L24372 .... 52c

Steel Thumb Tacks.

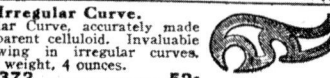

Steel Thumb Tacks, stamped from one piece of steel. Have finely tempered and finished needle point. An excellent tack at a very low price. Put up only in boxes of 50. **State size.** Shipping weight, per box, 3 ounces.

3L24360—Diameter, inch	5/16	3/8	7/16
Per box of 50	20c	24c	29c

DRAFTING SUPPLIES

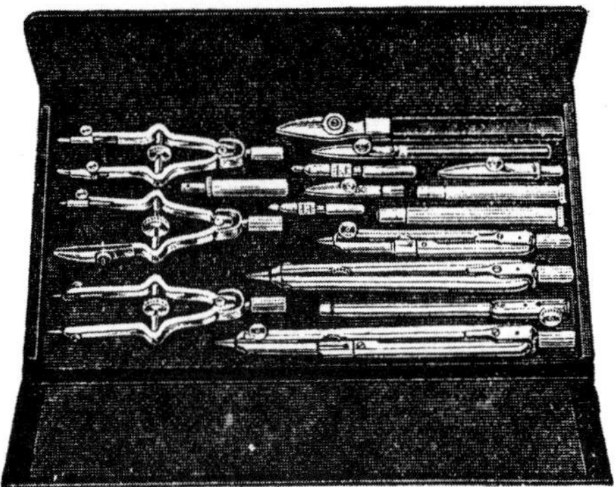

GRADE A Instruments, $7.25 to $18.90.

The GRADE A instruments are made from the finest nickel silver (formerly known as German silver), of great elasticity, density and hardness, with the tongues of the joints, needle points and spring parts of finely tempered and hardened steel. The finish throughout is remarkably fine and so bright that the least flaw or fault in the workmanship, form or quality may be instantly detected. These instruments are made with improved pivot joint and straightening device, and the spring bow instruments are of a very desirable design with center screw adjustment.

GRADE A INSTRUMENTS—Large Assortment. Contains the following:
6¼-Inch Compass with needle, pencil and pen points, lengthening bar and metal handle with reserve needle points. 4½-Inch Compass with needle, pencil and pen points and metal handle with reserve points. 5¾-Inch Hairspring Divider. 4-Inch Spring Bow Divider with central adjusting screw and circular spring bow. 4-Inch Spring Bow Pencil with central adjusting screw and circular spring bow. 4-Inch Spring Bow Pen with central adjusting screw and circular spring bow. 4¼-Inch Ruling Pen, spring blade. 5½-Inch Swedish Ruling Pen, spring blade. Combination screwdriver and lead case with extra leads. Furnished in artificial leather, pocketbook style case with velveteen lining. Shipping weight, 1¼ pounds.
3L24126....................$18.90

GRADE A INSTRUMENTS—Medium Assortment—Contains the following:
6¼-Inch Compass with needle, pencil and pen points, lengthening bar and metal handle with reserve needle points. 5¾-Inch Hairspring Divider. 5-Inch Ruling Pen, spring blade. 4-Inch Spring Bow Divider with central adjusting screw and circular spring bow. 4-inch Spring Bow Pencil with central adjusting screw and circular spring bow. 4-Inch Spring Bow Pen with central adjusting screw and circular spring bow. 4¼-Inch Ruling Pen, spring blade. Combination screwdriver and lead case with extra leads. In artificial leather, pocketbook style case with velveteen lining. Shipping wt., 1 lb. 1 oz.
3L24116....................$12.25

GRADE A INSTRUMENTS—Small Assortment—Contains the following:
6¼-Inch Compass with needle, pencil and pen points, lengthening bar and metal handle with reserve needle points. 4-Inch Spring Bow Pen with central adjusting screw and circular spring bow. 5-Inch Ruling Pen, spring blade. Combination screwdriver and lead case with extra leads. In artificial leather, pocketbook style case with velveteen lining. Shpg. wt., 10 oz.
3L24104....................$7.25

GRADE B Instruments, $5.25 to $14.90.

The GRADE B instruments are also very carefully constructed of fine nickel silver and perfectly tempered steel, and very well finished. They are made with pivot joints. The circular spring bow instruments are made with side screw adjustment.

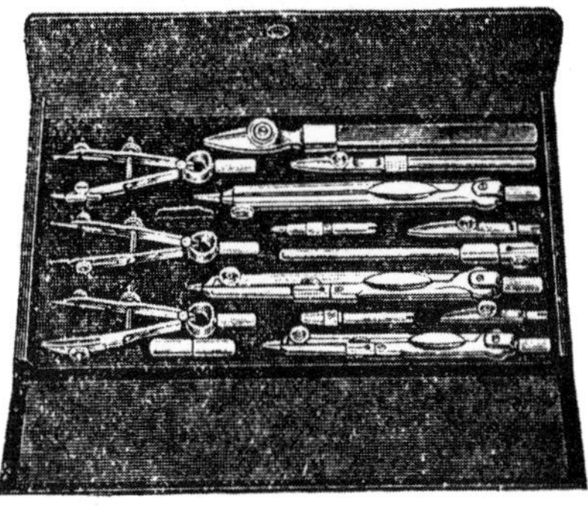

GRADE B INSTRUMENTS—Large Assortment—Contains the following:
5¾-Inch Compass with needle, pen and pencil points and lengthening bar. 4½-Inch Compass with needle, pen and pencil points. 5¾-Inch Hairspring Divider. 3½-Inch Circular Spring Bow Divider. 3½-Inch Circular Spring Bow Pencil. 3½-Inch Circular Spring Bow Pen. 4-Inch Ruling Pen, spring blade. 5½-Inch Swedish Ruling Pen. Box with extra leads. In artificial leather, pocketbook style case with velveteen and cotton moire lining. Shipping weight, 1 pound.
3L24121....................$14.90

GRADE B INSTRUMENTS—Medium Assortment—Contains the following:
5¾-Inch Compass with needle, pen and pencil points and lengthening bar. 5¾-Inch Hairspring Divider. 3½-Inch Circular Spring Bow Divider. 3½-Inch Circular Spring Bow Pencil. 3½-Inch Circular Spring Bow Pen. 5-Inch Ruling Pen, spring blade. 4-Inch Ruling Pen, spring blade. Box with extra leads. In artificial leather, pocketbook style case with velveteen and cotton moire lining. Shpg. wt., 15 oz.
3L24111....................$9.20

GRADE B INSTRUMENTS—Small Assortment—Contains the following:
5¾-Inch Compass with needle, pen and pencil points and lengthening bar. 3½-Inch Circular Spring Bow Pen. 5-Inch Ruling Pen, spring blade. Box with extra leads. In artificial leather, pocketbook style case, velveteen and cotton moire lining. Shipping weight, 8 ounces.
3L24102....................$5.25

Triangular Boxwood Scales.

3L24275—Architects' Triangular Boxwood Scale, divided 3/32, 1/8, 3/16, 1/4, 3/8, 1/2, 3/4, 1, 1½, 3 inches to the foot, 1/16 inch. High grade seasoned boxwood, engine divided. U. S. standard, 12 in. long.....(Shpg. wt., 6 oz.)......63c
3L24277—Architects' Triangular Boxwood Scale, engine divided, same as 3L24275, but with white edges, 12 inches long......(Shipping weight, 6 ounces)......$2.25
3L24280—Engineers' Triangular Boxwood Scale, divided 10, 20, 30, 40, 50 and 60 parts to inch. High grade seasoned boxwood, engine divided. U. S. standard, 12 inches long.........(Shipping weight, 6 ounces)............63c
3L24282—Engineers' Triangular Boxwood Scale, engine divided, same as 3L24280, but with white edges, 12 inches long.........(Shipping weight, 6 ounces)............$2.25

Double Bevel White Edge Pocket Scales.

Architects' Pocket Scale. Made of boxwood, with scale divisions on white celluloid veneered edges. 6-inch length, divided 1/8, 1/4, 3/8, 1/2, 3/4, 1, 1½ and 3 inches to the foot. Scale measures over all 6¾ inches long, 5/32 in. thick and 1 inch wide, a very convenient size for the vest pocket, and has neatly beveled edges on both faces of scale. Guaranteed accurate. Shpg. wt., either scale, 2 oz.
3L24284—With leather sheath....................$1.25
3L24285—Engineers' Pocket Scale. Same as 3L24284, but divided 10, 30, 40 and 50 parts to inch.
With leather sheath....................$1.25

Transparent Celluloid Triangles.

These triangles allow rapid and accurate work owing to their transparency, do not collect dust and keep their edges almost like metal tools. Open center. **State size and catalog number.**

Size, Inches	3L24305 30x60 degrees	3L24306 45 degrees	Shipping Weight
4	17c	18c	2 ounces
6	24c	30c	3 ounces
8	38c	40c	4 ounces
10	48c	60c	5 ounces
12	59c	80c	8 ounces

Keuffel & Esser Polyphase Slide Rule.

Made of mahogany with divisions on white facings. Engine divided. Improved "frameless" glass indicator, so constructed that every figure on the rule is clearly visible at all times. Patent slide adjustment insures smooth working of slide under all conditions. Ten inches long. Put up in a case. Shipping weight, 1 pound.
3L24259—With book of instructions............$6.65

The Ready Drafting Instrument.

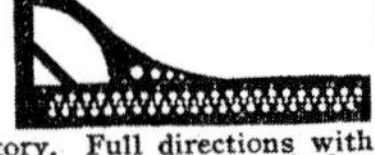

This instrument can be used as a rule, T square, triangle, square, tractor or compass. With it you can draw anything from an ordinary straight line to the parts of a machine or the plan of an entire factory. Full directions with each one. Made of aluminum, 6 5/16 inches long and 2 5/16 inches wide. Shipping weight, 3 ounces.
3L24145....................68c

Learn to Draw at Less Than 9c a Lesson!

Famous Drawing Course by Joseph Cummings Chase Now Offered, Complete, for Only 98c.

Think of this marvelous offer! For only 98 cents you can now obtain the complete drawing course by Joseph Cummings Chase, famous artist with our Expeditionary Forces in France during the war. You can now learn to draw under this master artist, obtaining the benefit of his valuable teachings for only 98 cents. This is exactly the same complete drawing course for which thousands of people paid a greatly higher price. This course contains twelve illustrated lessons, setting forth the great principles of art so that even a person who has never drawn a line before will obtain a knowledge of the principles of this fascinating profession.

Artists in Great Demand

The demand for artists has never before been so great. Newspapers, business houses, advertising concerns need people who can draw and do original work. If you like to draw, there is nothing so delightful, that will increase your earning capacity.

It is not difficult to learn to draw. The simplicity of Mr. Chase's course will astonish you. You will be inspired by your rapid progress in learning to draw under the instructions of one of America's foremost artists. Learning to draw in this new way is a delightful and profitable pastime.

Many young fellows just starting out are able to fatten their weekly pay envelope by the ready sale of little pictures and illustrated jokes drawn during their spare hours.

The ability to draw is a desirable asset in every field. If parents will allow their children to draw, aided by this course, it may reveal hidden talents and shape for the child a brilliant career as an artist.

Grasp This Amazing Offer!

Act today and get this famous drawing course at this marvelously low price. It is not possible to compare the extraordinary value that anyone can obtain from this course with the small sum which we ask for it. Thousands of these sets have been sold at the higher price. Take advantage of this remarkable offer now—it will be withdrawn in a short time.

The course is complete in twelve comprehensive lessons. Each lesson is complete in a separate booklet and is bound in Art Craft paper covers, making it easy and convenient to carry with you. The lessons are fully illustrated so as to show the actual carrying out of the instructions given in the text. These twelve lessons are put up in a slip case with pasted art label. Weight, packed in neat box, 2 pounds.
60L640—Joseph Cummings Chase's Course in Drawing, complete................98c

School Drawing Set, $1.98.

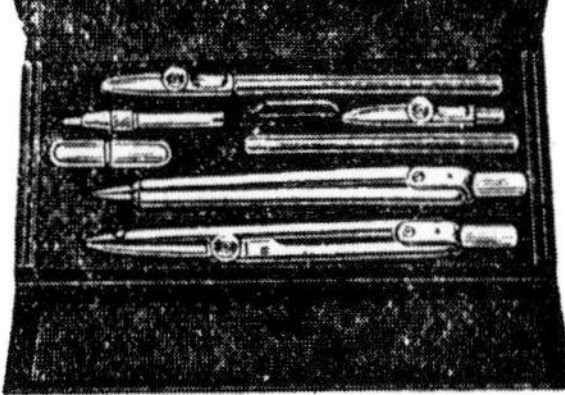

In imitation leather, bar lock style case with cotton twill lining. Contains the following:
4¾-Inch Ruling Pen, black hollow steel handle. 5⅛-Inch Nickel Silver Compass with fixed needle point, divider point pen, pencil point, key and box with extra leads. 5⅛-Inch Nickel Silver Divider with fixed steel needle points. 3⅛-Inch Hollow Steel Handle for use with compass pen.
3L24134 (Shipping weight, 6 ounces)......$1.98

Drawing Boards.

3L24335—Drawing Board, medium grade, with two drawing surfaces and side ledges. **State size.**

Size, inches	12x17	16x21	20x26	23x31
Shipping wt.	3¼ lbs.	4¼ lbs.	8¾ lbs.	11 lbs.
Each	98c	$1.35	$1.76	$2.10

3L24337—Drawing Board made of thoroughly seasoned pine with hardwood ledges dovetailed into the board to allow contraction and expansion. **State size.**

Size, inches	23x31	31x43
Shpg. wt.	19 lbs.	33 lbs.
Each	$2.75	$3.70

Cannot be shipped by parcel post.

T Squares.

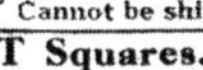

3L24328—Our best T square. Amber lined maple blade. Curved walnut fixed head. **State length.**

Length, inches	24	30	36	42
Shipping wt., oz.	15	20	23	31
Each	$1.76	$2.25	$2.75	$3.13

3L24324—T Square, mahogany, ebony lined blade and fixed head. A very fine square. **State length.**

Length, inches	24	30	36	42
Shipping weight, oz.	15	20	23	31
Each	99c	$1.17	$1.48	$1.65

3L24318—T Square, ash wood blade, maple lined and black walnut fixed head, shellac finish. **State length.**

Length, inches	24	30	36	42
Shipping weight, oz.	15	20	23	31
Each	78c	98c	$1.18	$1.38

3L24315—T Square, with maple blade and fixed head.

Length, inches	15	18	24	30	36	42
Shipping weight, oz.	12	13	15	20	23	31
Each	23c	34c	45c	53c	59c	69c

School Drawing Outfits.

Outfit consists of well made basswood board, T square and two triangles. Board so constructed that T square and triangles are held firmly to under side of it when not in use. Compact, inexpensive and durable.

	Size of Board	Shpg. Wt.	Price
3L24350	10x12 inches	1 lb. 5 oz.	$0.68
3L24351	13x19 inches	2⅜ lbs.	1.09
3L24352	17x22 inches	4 lbs.	1.95

ARTISTS' MATERIALS

Prepared Canvas for Oil Painting.
Smooth Finish Linen.
3L24460—30-Inch.
Per yard **$3.20**
3L24461—36-Inch.
Per yard **3.60**
Shipping weight, per yard, 2 pounds.

Smooth Cotton Sketching Canvas.
3L24462—Single prime, gray tint, white back, 40-inch. Per yard **$1.40**
Shipping weight, per yard, 2 pounds.

Medium Rough Cotton Sketching Canvas.
3L24463—Single prime, gray tint, 40-inch. Per yard **$1.85**
Shipping weight, per yard, 2 pounds.

Academy Boards.
Smooth or Rough Surface.
3L23194—6x9 inches **9c**
Shipping weight, 10 ounces.
3L23195—9x12 inches **17c**
Shipping weight, 1 pound.
3L23196—12x18 inches **28c**
Shipping weight, 1¼ pounds.
Be sure to specify whether smooth or rough surface.

Canvas Sketching Boards.
Smooth or Rough Surface.
3L23197—9x12 inches **24c**
Shipping weight, 1 pound.
3L23198—12x18 inches **42c**
Shipping weight, 1¼ pounds.
Be sure to specify whether smooth or rough surface.

Artists' Oils, Varnishes and Mediums.
In 2½-Ounce Bottles.
(Unmailable).
3L23106—Linseed Oil **22c**
3L23107—Poppy Oil **30c**
3L23108—Pale Drying Oil **27c**
3L23109—Turpentine **20c**
3L23110—Genuine Mastic Varnish **60c**
3L23111—Picture Mastic Varnish **40c**
3L23112—Picture Copal Varnish **30c**
3L23113—White Damar Varnish **35c**
3L23114—Retouching Varnish **40c**
3L23115—White Shellac **35c**
3L23116—Japan Dryer **30c**
3L23117—Fixatif **25c**
Shipping weight, 8 ounces.

Winsor and Newton's Prepared Moist Water Colors.
Series No. 1.

Burnt Sienna.	Mauve.
Burnt Umber.	Naples Yellow.
Chinese White.	New Blue.
Chrome Lemon.	Olive Green.
Chrome Yellow.	Permanent Blue.
Chrome Deep.	Prussian Blue.
Emerald Green.	Raw Sienna.
Gamboge.	Raw Umber.
Hooker's Grn, No. 1	Sap Green.
Hooker's Grn, No. 2	Terre Verte.
Indian Red.	Vandyke Brown.
Ivory-Black.	Yellow Ochre.

3L23118—Half Pan **24c**
3L23119—Whole Tube **35c**

Series No. 2.

Alizarin Crimson.	Scarlet Lake.
Brown Madder.	Sepia.
Crimson Lake.	Vermilion.

3L23120—Half Pan **35c**
3L23121—Whole Tube **68c**

Series No. 3.

Cadmium Yellow, pale.	Cobalt Blue.
Cadmium Yellow.	French Ultra-marine.
Cadmium Yellow, deep.	Indian Yellow.
	Lemon Yellow.
Cerulean Blue.	Viridian.

3L23122—Half Pan **46c**
3L23123—Whole Tube **85c**

Series No. 4.

Aurora Yellow.	Madder Lake.
Carmine.	Rose Madder.

3L23124—Half Pan **$0.68**
3L23125—Whole Tube **1.23**
Shipping weight, half pans, each, 1 ounce.
Shipping weight, tubes, each, 2 ounces.

French Charcoal.
50 sticks, 6 inches long, to a box.
3L23126—Ordinary Grade.
Per box **60c**
3L23127—Berville Venetian.
Per box **95c**
Shipping weight, either style, 6 ounces.

Charcoal Paper.
Size, 19x25 Inches.
3L23199—French Ordinary White.
Per dozen sheets **85c**
3L23200—American Ordinary White.
Per dozen sheets **48c**
Shipping weight, per dozen sheets, 2 lbs.

Palette Crayon Box.
For Art School Students, Etc. The lid of the box is covered on the inside with chamois leather for stumping. The thumbhole is so arranged as to allow the box being held on the hand. Contains twelve black and four white crayons, one tube of crayon sauce, two sticks charcoal, an assortment of stumps and a crayon holder. Shipping weight, 1 pound.
3L23152—Complete **$2.25**

Artists' Oil Colors in Tubes.

3L23103—American make. Single tube **10c**
3L23104—American make. Double tube **18c**
3L23105—Winsor and Newton make. Oil colors. Single tubes only **19c**
Shipping weight, single tube, 2 ounces; double tubes, 4 ounces.
Always Specify Colors and Size of Tubes Desired.

*American Vermilion.	Flake White.	Purple Lake.
Antwerp Blue.	Ivory-Black.	Prussian Blue.
Burnt Sienna.	Indian Red.	Raw Sienna.
Burnt Umber.	Indigo.	Raw Umber.
Cremnitz White.	King's Yellow.	Sap Green.
Chrome Green, light.	Light Red.	Scarlet Lake.
Chrome Green, medium.	Lamp Black.	Silver White.
Chrome Green, deep.	Magenta.	Terre Verte.
Chrome Yellow, light.	Mauve.	Vandyke Brown.
Chrome Yellow, medium.	Megilp.	Venetian Red.
Chrome Yellow, deep.	Naples Yellow.	Verdigris.
Chrome Yellow, orange.	New Blue.	Yellow Ochre.
Crimson Lake.	Payne's Gray.	Zinc White.
Emerald Green.	Permanent Blue.	

*Note—American Vermilion is American make only; the rest of colors can be had in either Winsor and Newton or American make.
Tubes vary in size from ½x2 inches to ¾x4 inches, according to the color.

Special Oil Colors.
3L23101—American make. Single tubes, ½x2 inches **28c**
3L23102—American make. Double tubes, ½x4 inches **48c**
3L23100—Winsor and Newton make. Single tubes **47c**

Carmine No. 2.	Cerulean Blue.	Madder Lake.	Ultramarine.
Cobalt Blue.	Lemon Yellow.	Rose Madder.	*Vermilion.

Shipping weight, single tubes, 2 ounces; double tubes, 4 ounces.
*Winsor and Newton make only.

The Beginner's Oil Color Outfit.
A nicely polished wood case, 9 inches long, 5½ inches wide and 1¾ inches deep. Contains 10 assorted oil colors, a mahogany palette, palette knife, palette cup, two brushes, badger blender and a bottle of pale drying oil. Shipping weight, 1½ pounds.
3L23153 .... **$3.48**

The Amateur Oil Color Outfit.
A polished wood box, size, 10¾x7x2 inches. Contains an assortment of thirteen high grade artists' oil colors, mahogany palette, palette knife, palette cup, three sable and bristle brushes, badger blender, bottle of pale drying oil, bottle of turpentine, Academy board, tracing paper, crayons and impression paper. Shipping weight, 2½ pounds.
3L23154 .... **$4.18**

The Professional Oil Color Outfit.

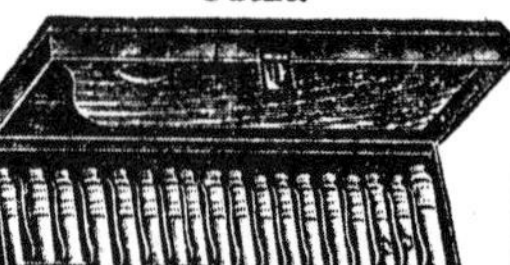

Japanned metal box, size, 13x9x3¼ inches. Contains twenty high grade artist's oil colors, mahogany palette, artist's palette knife, palette cup, eight sable and bristle brushes, badger blender, bottle of pale drying oil, bottle of turpentine, 5-inch crayon holder and charcoal. Shipping weight, 5¼ pounds.
3L23155 .... **$10.80**

Campana's "Student" China Painting Outfit.
Put up in a polished wooden box. Contains twelve China colors, one small vial liquid bright gold, five quill brushes and handles, palette knife, 6x6-inch tile, mixing medium, graphite and tracing paper, set of ring dividers, new process black outlines, wax for molding, three studies and instruction book. Shipping weight, 2½ pounds.
3L23156—Complete set for **$4.50**
3L23128—Separate colors as in outfit:

Yellow Red	**18c**	Rose Color	**18c**
Banding Blue	**13c**	Finishing Brown	**18c**
Shading Green	**13c**	Dark Violet, No. 2	**12c**
Yellow Green	**13c**	Primrose Yellow	**11c**
Yellow Brown	**11c**	Ruby Purple, No. 2	**40c**
Best Black	**11c**	**Be sure to specify color wanted.**	
Albert Yellow	**13c**		

Fine Water Color Outfits for Students and Artists.

Imported German Outfits.

The Large Rembrandt. Contains 12 half pans semi-moist colors and brushes. Shpg. wt., 8 oz.
3L23157 .... **55c**

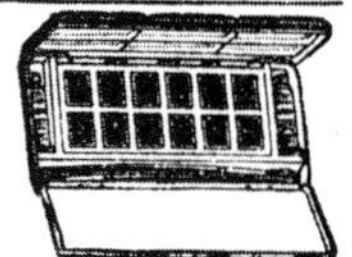

The Small Murillo. Contains 12 pans of semi-moist colors, a tube each of Chinese white and sepia, and brushes. Shipping weight, 12 ounces.
3L23158
$1.10

The Large Murillo. Contains 16 pans of semi-moist colors and a tube each of Chinese white and sepia, and brushes. Shpg. wt. 15 oz.
3L23159 .... **$1.35**

Winsor and Newton "Ideal" Outfit.

A high grade set of the finest imported English water colors. Contains 12 pans semi-moist colors and 3 brushes. Shpg. wt. 12 oz.
3L23163 .... **$4.89**

Prang's No. 16 School Water Colors.

Contains 16 half pans semi-moist water colors and a No. 7 brush. Shipping wt., 11 ounces.
3L23162 .... **89c**

Titian Imported French Outfit.

Contains 15 pans of semi-moist colors, a tube of Chinese white and 2 brushes. A high grade imported paint for particular artists. Shpg. wt., 12 ounces.
3L23161 .... **$2.89**

Students' Moist Water Color Outfit.
Colors put up in tubes, ½x3 inches. This box is beautifully enameled with a black exterior and white interior. The cover is divided into compartments for mixing trays. Contains twelve tubes of assorted colors. Shipping weight, 1¼ pounds.
3L23160 .... **$2.58**

Tourists' Sketch Books.
Stiff canvas cover, twenty-four leaves of white paper for sketching.
3L23164—4x7 inches **55c**
Shipping weight, 10 ounces.
3L23165—6x8¾ inches **70c**
Shipping weight, 1 pound.
3L23166—8x10 inches **90c**
Shipping weight, 1¼ pounds.

Wooden Palettes.
Oiled Mahogany.
Oval.
3L23169—9-inch **30c**
3L23170—10-inch **40c**
3L23171—12-inch **55c**
3L23172—14-inch **70c**
Oblong.
3L23173—5½x10¾ in. **30c**
3L23174—8x11 inches **45c**
3L23175—10x14 inches **70c**
Shipping weight, each, 12 oz.

Japanned Tin Palette Cups.
3L23177—Single, no cover **9c**
3L23178—Double, no cover **18c**
3L23179—Single, with cover **18c**
3L23180—Double, with cover **30c**
Shipping weight, each, 5 ounces.

Artists' Steel Palette Knives.
3L23181—2¼ inches **45c**
3L23182—3 inches **50c**
3L23183—3½ inches **50c**
3L23184—4 inches **55c**
Shipping weight, each 6 ounces.

Atomizers.
Japan Tin.
3L23185 .... **18c**
Shipping weight, 4 ounces.

Bronze Powders for Decorative Purposes.
Highest Luster Grade.

Pale Gold.	Red.
Rich Gold.	Crimson.
Imitation Gold.	Fire.
Aluminum.	Copper.
Antique Statuary.	Lilac.
	Lavender.
Green Gold.	Dark Blue.
Lemon.	Blue Green.
Orange.	Dark Green.

3L23130—Per 1-oz. pkg. **21c**
Shipping weight, 3 ounces.

Bronzing Liquid.
Used for Mixing Bronze Powders.
(Not Mailable.)
3L23131—Per 2-ounce bottle ... **12c**
Shipping weight, 6 ounces.

Bronzing Brush.
3L23132—½-inch soft camel hair.
Each **18c**
Shipping weight, 3 ounces.

Brushes.
3L23133—English Artists' Bristle Brushes, for oil painting. With metal ferrules and polished handles. **Round or flat style. State style wanted.**

Size		Size	
2	**10c**	8	**16c**
4	**11c**	10	**18c**
6	**13c**	12	**22c**

3L23134—Artists' Bristle Brushes, thin brights.

Size		Size	
1	**13c**	4	**17c**
2	**14c**	5	**18c**
3	**16c**	6	**21c**

3L23135—Red Sable Artists' Brushes, round style, nickel ferrules and polished handles.

Size		Size	
2	**17c**	8	**25c**
4	**19c**	10	**30c**
6	**21c**	12	**38c**

3L23136—'Brights' Russia Sable Artists' Brushes, flat style with polished handles and nickel ferrules.

Size		Size	
2	**18c**	8	**25c**
4	**20c**	10	**32c**
6	**22c**		

3L23137—Round Badger Blenders.

Size		Size	
2	**59c**	6	**$1.25**
4	**80c**		

3L23138—Paris Red Sable Brushes for Water Color Painting. Polished handles, nickel ferrules.

Size		Size	
1	**21c**	4	**32c**
2	**23c**	5	**40c**
3	**25c**	6	**55c**

Shipping weight, each, 3 ounces.

Extra Soft Pastel Crayon Outfits.
3L23201
Contains thirty assorted crayons.
Each **$1.18**
Shipping weight, 12 ounces.
3L23203
Contains forty-four assorted crayons.
Each **$1.89**
Shipping weight, 1 pound.
3L23204—Contains fifty-six assorted crayons **$2.85**
Shipping weight, 1¼ pounds.
3L23205—Contains sixty-four assorted crayons **$4.28**
Shipping weight, 1¾ pounds.

SUPERTONE Player Rolls

FOR PLAYER PIANO

SUPERTONE (Our Own Trade Mark, Reg. U. S. Patent Office) ROLLS are mechanically perfect and musically correct. They embody all the good features which enter into the making of rolls of the highest quality.

They are very carefully arranged, both as to melody and accompaniment, and are so mathematically perfect in their construction that the rendition of any composition will have the effect of being performed by an artist of reputation instead of a mechanical apparatus.

Our Music Rolls are put up in neat and serviceable strawboard boxes covered in imitation of leather.

Word Rolls

Often when you play a song with a catchy melody in it, don't you have an inspiration to sing, and isn't it provoking that you are unable to do so because you do not know the words? Then, again, would it not be a pleasure at your parties and gatherings to be able to group around the player and all join in singing the best and the latest songs from the words plainly printed at the right hand side of the roll?

Supertone Player Word Rolls afford these enjoyments, and we feel sure that you will derive great pleasure from them.

It is just as natural to sing to these rolls as it is to play them on your player piano.

However, if you do not care to sing, these rolls are arranged so they may be played as piano solos, and many of them can be used for dancing.

NOTE—All Supertone Player Word Rolls are hand played reproductions by famous pianists. They come in three classes, priced according to the royalty paid the publishers, for the use of the words and music; for example, on standard and old time songs there is no royalty. On the second class rolls (regular popular song hits) there is a single royalty charge, and on the third class (numbers from musical plays and extremely big hits) there is a double royalty charge.

Most of these rolls are also arranged for dancing. Those marked with an asterisk (*) are for waltzes. Those marked with a single dagger (†) are for one-steps. Those marked with a double dagger (‡) are for fox trots.

Order by Catalog number and roll number. Shipping weight, each, 12 ounces.

12L5901 STANDARD AND OLD TIME SONG ROLLS, EACH 37c

(NON-ROYALTY)

Roll No.	Title
153	A Dream—Bartlett
101	All Hail the Power of Jesus' Name
104	Aloha Oe (Farewell to Thee)
102	America
124	Annie Laurie
133	Battle Cry of Freedom
126	Battle Hymn of the Republic
105	Beautiful Isle of Somewhere
154	Ben Bolt
145	Columbia, the Gem of the Ocean
128	Coming Thro' The Rye
132	Darling Nellie Gray
106	Face to Face
107	Home, Sweet Home
157	I Know That My Redeemer Liveth
108	I Need Thee Every Hour
109	In the Evening by the Moonlight
155	Jesus, Lover of My Soul
141	Juanita
142	Just Before the Battle, Mother
110	La Marseillaise
129	Last Rose of Summer
151	Lead, Kindly Light
139	Marching Through Georgia
156	Marie, Marie
131	Massa's in the Cold, Cold Ground
112	My Old Kentucky Home
148	Nearer, My God, to Thee
130	Oh, Promise Me
144	O Sole Mio (My Sunshine)
138	Old Black Joe
134	Old Oaken Bucket, The
115	Onward, Christian Soldiers
152	Rock of Ages
116	Rocked in the Cradle of the Deep
111	Rosary, The—Nevin
143	Santa Lucia (Over the Rippling Sea)
117	Silent Night
114	Sing Me to Sleep
113	Star Spangled Banner, The
118	Swanee River
135	Sweet Genevieve
120	Sweet Hour of Prayer
146	The Palms
127	Then You'll Remember Me
137	Way Down Yonder in the Cornfield
125	We're Tenting Tonight
121	What a Friend We Have in Jesus
136	When Johnny Comes Marching Home Again
122	When You and I Were Young, Maggie
123	Where Is My Wandering Boy Tonight?
158	Where the Silvery Colorado Wends Its Way
149	Whispering Hope
140	Yankee Doodle

12L5903 SONG ROLLS, EACH 45c

(SINGLE ROYALTY)

Roll No.	Title
1001	A Perfect Day
1127	After the Ball
*1002	Alabama Lullaby
‡1003	Alcoholic Blues
*1010	Beautiful Ohio
1013	Break the News to Mother
‡1017	Chong (He Came From Hong Kong)
1035	Holy City, The
1045	In the Baggage Coach Ahead
1133	Just as the Ship Went Down
*1066	Missouri Waltz
1073	My Wild Irish Rose
*1078	One, Two, Three, Four
1126	Pretty Kitty Kelly
†1084	Rose of Honolulu
1086	Silver Threads Among the Gold
1096	The Rose of No Man's Land
*1131	There's a Mother Old and Gray
*1099	Till We Meet Again
1101	Turkey in the Straw
1104	When the Sunset Turns the Ocean's Blue to Gold
1121	Where the River Shannon Flows

12L5905 SONG ROLLS FROM MUSICAL PLAYS AND EXTREMELY BIG HITS, EACH 49c

(DOUBLE ROYALTY)

Roll No.	Title
5178	After the Rain
5134	Ain't We Got Fun
5137	All By Myself
‡5243	All for the Love of Mike
‡5228	All Over Nothing at All
5213	Angel Child
5179	April Showers
‡5229	Are You Playing Fair?
5152	Bimini Bay
5180	Birds of a Feather
5249	Blue
5198	Blue Danube Blues
5118	Bright Eyes
5181	By the Old Ohio Shore
‡5224	By the Sapphire Sea
5214	California
5225	Call MeBack,Palo'Mine
5252	Carolina in the Morning
5182	Carolina Rolling Stone
5246	Carry Me Back to Old Virginny
5006	Casey Jones
5139	Cherie
5253	Coal Black Mammy o'Mine
5201	Cutie
‡5219	Dancing Fool
5171	Dapper Dan
‡5008	Dardanella
5202	Dear Old Southland
‡5240	Didn't Love Him Anyhow Blues
5251	Don't Bring Me Posies
5203	Don'tLeaveMe Mammy
*5101	Down the Trail to Home, Sweet Home
5135	Do You Ever Think of Me?
5184	Everybody Step
‡5108	Feather Your Nest
‡5238	Gallagher and Shean
‡5423	Gee But I Hate to Go Home Alone
‡5230	Georgette
5185	Granny
*5011	Hawaiian Dreams
*5012	Hawaiian Nights
5115	Hiawatha's Melody of Love
5186	High Brown Blues
5121	Home Again Blues
‡5244	Homesick
‡5231	Hot Lips
5132	Humming
5167	I Ain't Nobody's Darling
5133	I Never Knew
5117	I Used to Love You, But It's All Over Now
5164	I Want My Mammy
5254	I Wish I Knew
5232	I Wish There Was a Wireless to Heaven
‡5087	I'd Love to Fall Asleep and Wake Up in My Mammy's Arms
5107	I'll Be With You in Apple Blossom Time
*5017	I'm Forever Blowing Bubbles
5089	I'm in Heaven When I'm in My Mother's Arms
5143	I'm Nobody's Baby
‡5070	Indiana
*5019	In the Shade of the Old Apple Tree
5153	I've Got the Blues for My Old Kentucky Home
5154	Jazz Me, Blues
5155	Jealous of You
5187	Just a Little Love Song
‡5233	Just Because You're You
5199	Ka-Lu-a
*5023	Kentucky Dreams
‡5234	Kitten on the Keys
5169	Leave Me With a Smile
5025	Let the Rest of the World Go By
‡5221	Lovable Eyes
‡5086	Love Nest
5145	Ma
5116	Make Believe
5109	Margie
5188	Marie
5215	Melancholy Moon
5149	Mello Cello
5026	Memories
5206	Mickey O'Neill
5194	Mississippi Cradle
5027	Mother Macree
5241	My Cradle Melody
5125	My Mammy
5189	My Mammy Knows
*5029	My Isle of Golden Dreams
5142	My Sunny Tennessee
‡5220	Nobody Lied
‡5242	Old Fashioned Girl
‡5103	Old Pal, Why Don't You Answer Me?
5190	On the Gin Gin Ginney Shore
*5033	Oh, What a Pal Was Mary!
‡5035	Patches
5140	Peggy O'Neill
5191	Pick Me Up and Lay Me Down in Dear Old Dixieland
5207	Play That Song of India Again
†5041	Red Wing
5080	Rose of Washington Square
‡5043	Sand Dunes
‡5235	Say It While Dancing
5156	Say It With Music
5157	Second-Hand Rose
5196	She's Mine, All Mine
‡5044	Silver Bell
‡5222	Some Sunny Day
*5045	Sometime
5072	Somewhere a Voice Is Calling
5046	Star of the East
5192	Stealing
5151	Stolen Kisses
‡5218	Stumbling
5073	Sunshine of Your Smile
5227	Swanee Bluebird
5197	Swanee River Moon
5250	Sweet Bye and Bye
*5083	Sweet Hawaiian Moonlight
‡5226	Sweet Indiana Home
5208	Tell Her at Twilight
‡5050	Tell Me
5168	Ten Little Fingers and Ten Little Toes
‡5245	The Sneak
*5049	That Naughty Waltz
5051	There's a Long, Long Trail
*5099	There's a Vacant Chair at Home, Sweet Home
5163	The Sheik of Araby
‡5052	The Vamp
5209	Three O'Clock in the Morning
‡5239	Tomorrow
5255	Tomorrow Morning
5217	Too Many Kisses Mean Too Many Tears
5248	Toot! Toot! Tootsie Goodbye
*5096	Tripoli
5150	Tuck Me to Sleep in My Old Tucky Home
5211	Virginia Blues
5158	Wabash Blues
5141	Wang Wang Blues
5172	When Francis Dances With Me
5131	When I'm Gone You'll Soon Forget
5057	When My Baby Smiles at Me
5166	When Shall We Meet Again?
5247	When the Leaves Come Tumbling Down
5130	When You're Gone I Won't Forget
5212	While Miami Dreams
5236	While the Years Roll By
‡5091	Whispering
5146	Who'll Be the Next One to Cry Over You?
‡5237	Why Should I Cry Over You?
5144	Wyoming
5161	Yoo Hoo
5065	You're a Million Miles From Nowhere When You're One Little Mile From Home

12L5921 EACH 33c INSTRUMENTAL PLAYER ROLLS 12L5921 EACH 33c

You Can't Go Wrong If You Order Our Recommendations. When Ordering State Roll Numbers and Catalog Numbers.

MARCHES

Roll No.	Title
10062	Across the Border
10193	America Forever
10090	American Patrol
10003	America Over All
10044	Army and Navy March
10194	Battery A
10045	Battle Line of Liberty
10195	Black Opal
10196	Blaze of Honor
10197	Boy Scouts of America
10007	Brass Buttons
10186	Buck-Eye State
10094	Clayton's Grand March
10010	Clover Blossoms
10011	College Life
10198	Commander-in-Chief
10182	Coronation March
10012	Creole Belles
10199	Dawn of the Century
10015	Dixie Darlings
10200	Drum Major
10150	El Capitan (Sousa)
10201	Emblem of Peace
10152	General Grant March
10202	Gladiators
10187	Great American
10103	Hands Across the Sea
10178	Heaven's Artillery
10105	High School Cadets
10203	Italian Royal March
10110	King Cotton (Sousa)
10204	Lamb's March
10170	La Sorella
10205	Lincoln Centennial March
10155	Lohengrin Wedding March
10158	Manhattan Beach March (Sousa)
10206	March Lorraine
20006	Martial Medley
10207	Midnight Flyer
10244	Military March
10160	Military Parade
10208	National Army
10031	**National Emblem March** Irresistible March Swing
10209	National Fencibles (Sousa)
10060	National Honor March
10180	On the Square
10163	**Our Director's March** One of the Best
10179	Palace of Peace
10032	Panama Pacific March and Two-Step
10210	Paul Revere's Ride
10184	Rapid Fire March
10165	**Repasz Band March.** Very harmonious and stirring
10181	Rotary March
10211	Salute the Flag
10064	Seventh Regiment March, I. N. G.
20058	Sousa March Medley
10066	Spirit of Independence March
10119	**Stars and Stripes Forever** Sousa's popular march
10212	Storm King
10213	Thunderer, The
10214	Tipperary Guards
10183	Under the Banner of Victory
10255	Under the Double Eagle
10040	U. S. Field Artillery
10141	Washington Post (Sousa)
10239	Wedding March

(See Next Page for Waltzes, Reveries, Overtures, Etc.)

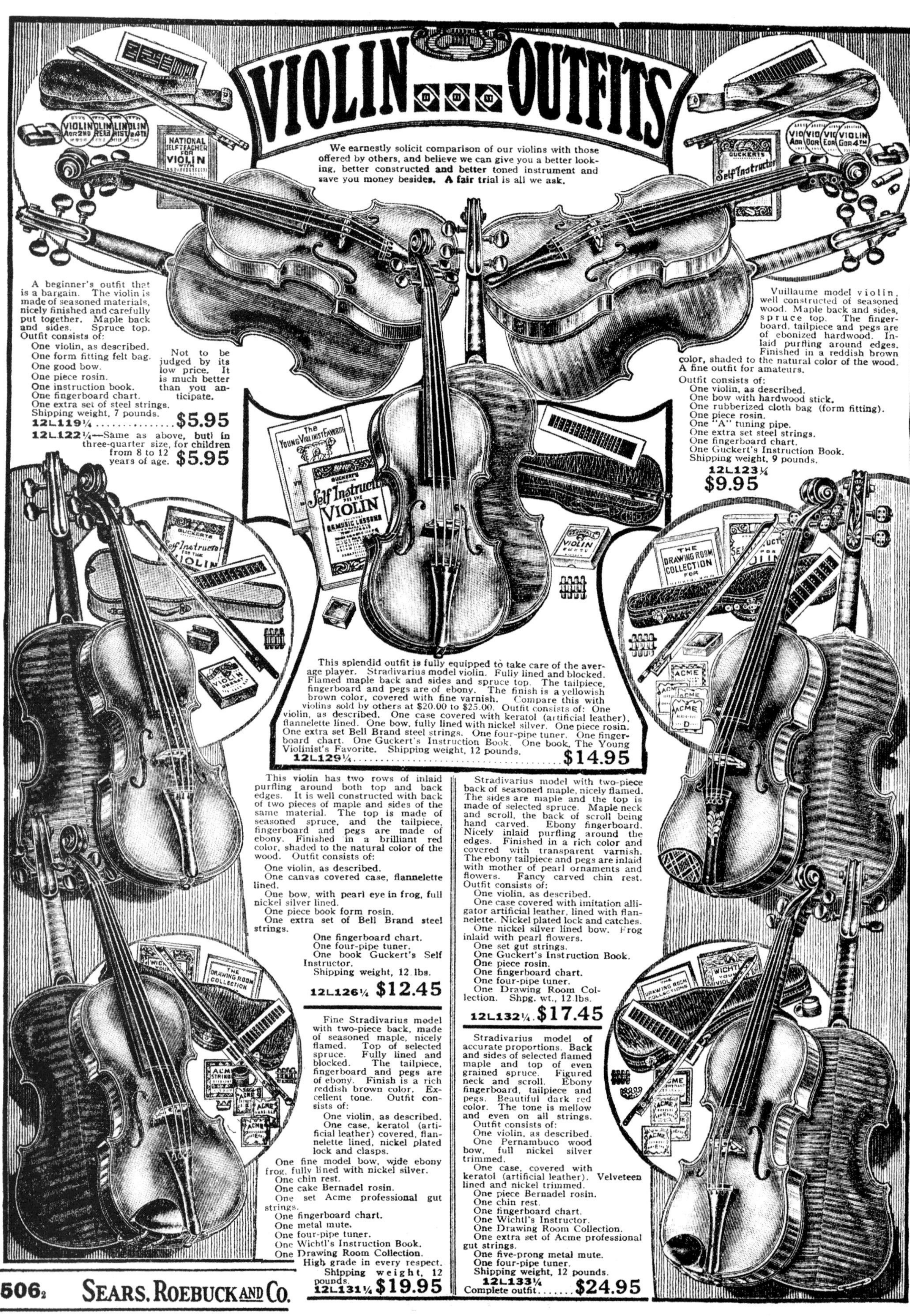
VIOLIN OUTFITS
We earnestly solicit comparison of our violins with those offered by others, and believe we can give you a better looking, better constructed and better toned instrument and save you money besides. A fair trial is all we ask.
A beginner's outfit that is a bargain. The violin is made of seasoned materials, nicely finished and carefully put together. Maple back and sides. Spruce top. Outfit consists of:
One violin, as described.
One form fitting felt bag.
One good bow.
One piece rosin.
One instruction book.
One fingerboard chart.
One extra set of steel strings.
Shipping weight, 7 pounds.
Not to be judged by its low price. It is much better than you anticipate.
12L119¼ $5.95
12L122¼—Same as above, but in three-quarter size, for children from 8 to 12 years of age. $5.95
Vuillaume model violin, well constructed of seasoned wood. Maple back and sides, spruce top. The fingerboard, tailpiece and pegs are of ebonized hardwood. Inlaid purfling around edges. Finished in a reddish brown color, shaded to the natural color of the wood. A fine outfit for amateurs.
Outfit consists of:
One violin, as described.
One bow with hardwood stick.
One rubberized cloth bag (form fitting).
One piece rosin.
One "A" tuning pipe.
One extra set steel strings.
One fingerboard chart.
One Guckert's Instruction Book.
Shipping weight, 9 pounds.
12L123¼ $9.95
This splendid outfit is fully equipped to take care of the average player. Stradivarius model violin. Fully lined and blocked. Flamed maple back and sides and spruce top. The tailpiece, fingerboard and pegs are of ebony. The finish is a yellowish brown color, covered with fine varnish. Compare this with violins sold by others at $20.00 to $25.00. Outfit consists of: One violin, as described. One case covered with keratol (artificial leather), flannelette lined. One bow, fully lined with nickel silver. One piece rosin. One extra set Bell Brand steel strings. One four-pipe tuner. One fingerboard chart. One Guckert's Instruction Book. One book, The Young Violinist's Favorite. Shipping weight, 12 pounds.
12L129¼ $14.95
This violin has two rows of inlaid purfling around both top and back edges. It is well constructed with back of two pieces of maple and sides of the same material. The top is made of seasoned spruce, and the tailpiece, fingerboard and pegs are made of ebony. Finished in a brilliant red color, shaded to the natural color of the wood. Outfit consists of:
One violin, as described.
One canvas covered case, flannelette lined.
One bow, with pearl eye in frog, full nickel silver lined.
One piece book form rosin.
One extra set of Bell Brand steel strings.
One fingerboard chart.
One four-pipe tuner.
One book Guckert's Self Instructor.
Shipping weight, 12 lbs.
12L126¼ $12.45
Stradivarius model with two-piece back of seasoned maple, nicely flamed. The sides are maple and the top is made of selected spruce. Maple neck and scroll, the back of scroll being hand carved. Ebony fingerboard. Nicely inlaid purfling around the edges. Finished in a rich color and covered with transparent varnish. The ebony tailpiece and pegs are inlaid with mother of pearl ornaments and flowers. Fancy carved chin rest. Outfit consists of:
One violin, as described.
One case covered with imitation alligator artificial leather, lined with flannelette. Nickel plated lock and catches.
One nickel silver lined bow. Frog inlaid with pearl flowers.
One set gut strings.
One Guckert's Instruction Book.
One piece rosin.
One fingerboard chart.
One four-pipe tuner.
One Drawing Room Collection. Shpg. wt., 12 lbs.
12L132¼ $17.45
Fine Stradivarius model with two-piece back, made of seasoned maple, nicely flamed. Top of selected spruce. Fully lined and blocked. The tailpiece, fingerboard and pegs are of ebony. Finish is a rich reddish brown color. Excellent tone. Outfit consists of:
One violin, as described.
One case, keratol (artificial leather) covered, flannelette lined, nickel plated lock and clasps.
One fine model bow, wide ebony frog, fully lined with nickel silver.
One chin rest.
One cake Bernadel rosin.
One set Acme professional gut strings.
One fingerboard chart.
One metal mute.
One four-pipe tuner.
One Wichtl's Instruction Book.
One Drawing Room Collection.
High grade in every respect.
Shipping weight, 12 pounds.
12L131¼ $19.95
Stradivarius model of accurate proportions. Back and sides of selected flamed maple and top of even grained spruce. Figured neck and scroll. Ebony fingerboard, tailpiece and pegs. Beautiful dark red color. The tone is mellow and even on all strings.
Outfit consists of:
One violin, as described.
One Pernambuco wood bow, full nickel silver trimmed.
One case, covered with keratol (artificial leather). Velveteen lined and nickel trimmed.
One piece Bernadel rosin.
One chin rest.
One fingerboard chart.
One Wichtl's Instructor.
One Drawing Room Collection.
One extra set of Acme professional gut strings.
One five-prong metal mute.
One four-pipe tuner.
Shipping weight, 12 pounds.
12L133¼
Complete outfit....... $24.95

VIOLINS
Excepting Where Stated There Are No Extras Included With the Violins on This Page.
A Very Popular Style.
12L161¼
$14.95
Appearance and Tone Will Please.
An exceptionally well constructed Stradivarius model violin with a two-piece back of nicely flamed maple and spruce top. The wood used has been well seasoned and is of selected quality. The fingerboard, tailpiece and pegs are made of ebony wood, dull finish. It is finished in a reddish color and nicely varnished, bringing out the flaming in the wood. It is superior to instruments sold by many other dealers at higher prices, so don't judge it by our low price. Shipping weight, 10 pounds.
12L161¼ $14.95
A Copy of Amati.
12L163¼
$19.95
Carved Scroll and One-Piece Back.
Amati model with one-piece back of good quality flamed maple. Sides of the same material. The top is of fine spruce. The purfling is set with care. Neck of figured maple with the scroll artistically hand carved on back and sides. The color is a reddish amber shaded at the center of the body to a golden yellow. The tailpiece and pegs and dull finish fingerboard are of ebony. Pegs and tailpiece are fluted. The tone is sure to please. Shipping weight, 10 pounds.
12L163¼ $19.95
At This Price You Are Getting a Wonderful Violin.
12L146¼
$24.95
Fine Orchestra Violin.
You will be surprised at the quality of workmanship, beauty and finish of this instrument when you consider the price. It is a true Stradivarius model, very carefully constructed. Maple back and sides nicely flamed. Flamed maple scroll and neck. The fingerboard, tailpiece and pegs are made of ebony in dull finish, the tailpiece and pegs being richly hand engraved. You will note particularly the clean and excellent workmanship. It is finished in a reddish brown color, covered with a deep transparent varnish. The tone is particularly suited to orchestral work. Shipping weight, 10 pounds.
12L146¼ $24.95
Copied From a Fine Stradivarius.
12L165¼
$29.95
A Smooth and Mellow Tone.
This violin is made from the measurements of a famous Stradivarius and is carefully planned and worked out in every detail. The wood used is selected for its sonority as well as beauty. The back, sides, neck and scroll are made of finely flamed maple. Well matched spruce top of even grain. Ebony fingerboard, tailpiece and pegs. The body is finished in a deep red color and covered with a rich varnish which is semi-dull rubbed. The tone of this violin is mellow and smooth, even on all strings, and is suitable for either solo or orchestral playing. Shipping weight, 10 pounds.
12L165¼ $29.95
Old Violins Are Very Valuable. This Is a Fine Copy.
12L166¼
$34.95
A Perfect Copy of an Old Stradivarius.
We have entered into an agreement with a famous maker who reproduces old violins perfectly. For this instrument he used the measurements of an old Stradivarius that is worth thousands of dollars. We are more than pleased with the result of our agreement, as the violins are reproduced perfectly, even to the marks of age. They have been so carefully constructed that the tone is as mellow and even as the violins which have been in use a good many years. The maple back, sides and neck are of beautifully figured stock, which has been thoroughly seasoned. Well matched, even grained top made from old spruce. Needless to say, the fingerboard, tailpiece and pegs are of the finest ebony, beautifully finished. Shipping weight, 10 pounds.
12L166¼ $34.95
A Fine Violin With Hand Carved Trimmings.
12L170¼
$49.95
A Master's Best Effort.
Made by one of the foremost makers of Europe and is a perfect copy of a fine Stradivarius. Two-piece back, made of beautifully flamed maple. Top of the best spruce obtainable. Neck of fine curly maple and the scroll purely Stradivarius in style. Finish of rich golden amber, with edges brought to a natural color. The varnish used is of the finest quality, deep and transparent. It is fitted with an ebony tailpiece and chin rest, carved in a beautiful and elaborate design. The pegs are also of carved ebony and have solid gold tips. The fingerboard is of first quality ebony, dull in finish. Mellow tone of great volume and carrying power. Shipping weight, 10 pounds.
12L170¼ $49.95
Made by a Famous Master.
12L172¼
$64.75
A Violin for the Soloist.
A special contract forbids our using the maker's name in connection with the sale of this violin. This instrument, produced under his own name, is well known throughout the world as the best that money can buy, and sells for considerably more than what we ask. The very best of materials enter into its construction. Stradivarius model, two-piece back of beautifully flamed maple. Figured maple neck and scroll. Selected and matched spruce top. Finest ebony fingerboard, tailpiece and pegs. Light red finish covered with a rich transparent varnish which brings out the beautiful high lights of the wood. Has a tone that will appeal to the most exacting soloist. Shpg. wt., 10 lbs.
12L172¼ $64.75
SEARS, ROEBUCK AND CO. 507

SUPERTONE GUITARS

TRADE MARK REG. U.S. PAT. OFF.

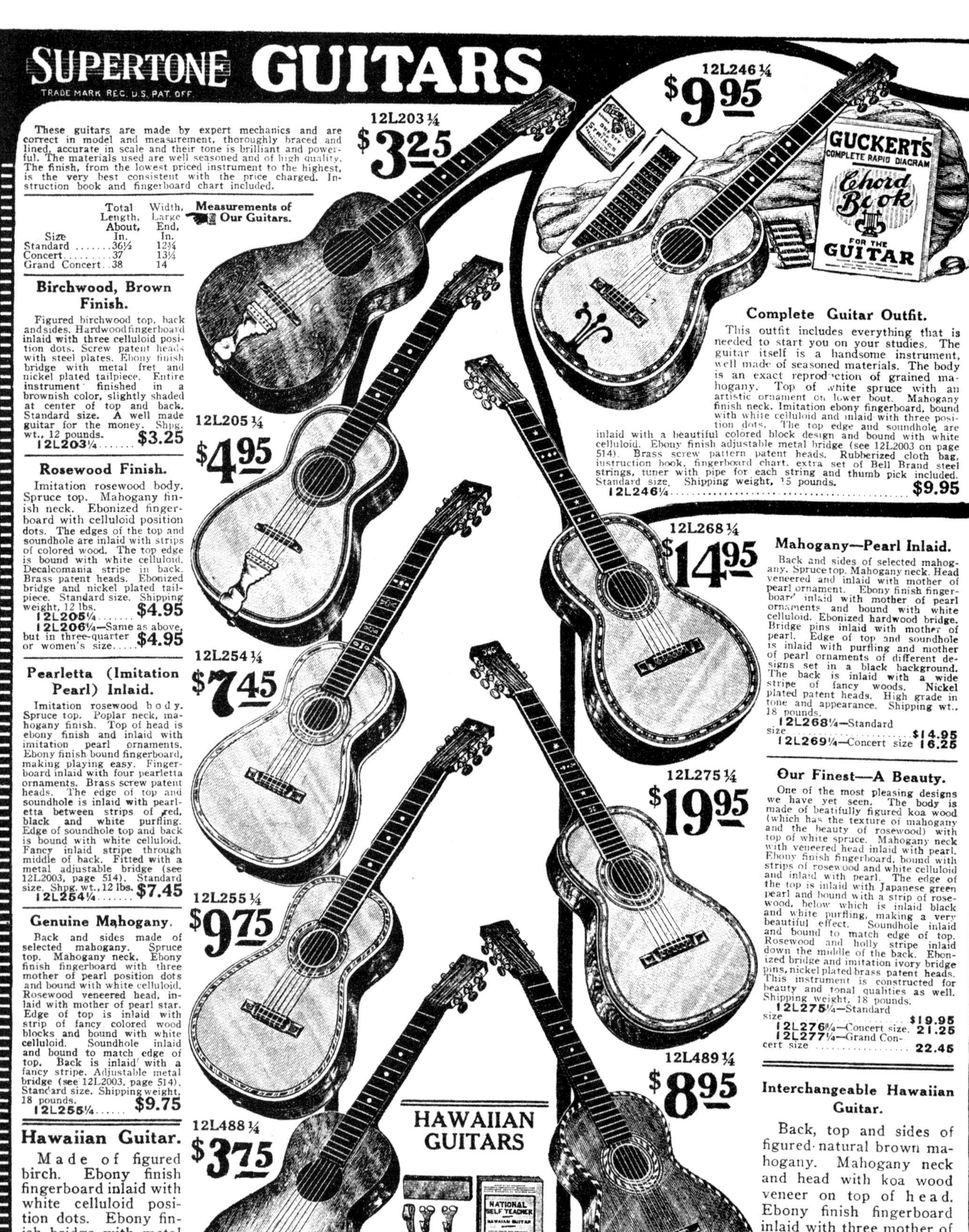

These guitars are made by expert mechanics and are correct in model and measurement, thoroughly braced and lined, accurate in scale and their tone is brilliant and powerful. The materials used are well seasoned and of high quality. The finish, from the lowest priced instrument to the highest, is the very best consistent with the price charged. Instruction book and fingerboard chart included.

Size	Total Length, About, In.	Width, Large End, In.
Standard	36½	12¾
Concert	37	13¼
Grand Concert	38	14

Measurements of Our Guitars.

12L203¼ $3.25

Birchwood, Brown Finish.

Figured birchwood top, back and sides. Hardwood fingerboard inlaid with three celluloid position dots. Screw patent heads with steel plates. Ebony finish bridge with metal fret and nickel plated tailpiece. Entire instrument finished in a brownish color, slightly shaded at center of top and back. Standard size. A well made guitar for the money. Shpg. wt., 12 pounds.
12L203¼ $3.25

12L205¼ $4.95

Rosewood Finish.

Imitation rosewood body. Spruce top. Mahogany finish neck. Ebonized fingerboard with celluloid position dots. The edges of the top and soundhole are inlaid with strips of colored wood. The top edge is bound with white celluloid. Decalcomania stripe in back. Brass patent heads. Ebonized bridge and nickel plated tailpiece. Standard size. Shipping weight, 12 lbs.
12L205¼ $4.95
12L206¼—Same as above, but in three-quarter or women's size $4.95

12L254¼ $7.45

Pearletta (Imitation Pearl) Inlaid.

Imitation rosewood body. Spruce top. Poplar neck, mahogany finish. Top of head is ebony finish and inlaid with imitation pearl ornaments. Ebony finish bound fingerboard, making playing easy. Fingerboard inlaid with four pearletta ornaments. Brass screw patent heads. The edge of top and soundhole is inlaid with pearletta between strips of red, black and white purfling. Edge of soundhole top and back is bound with white celluloid. Fancy inlaid stripe through middle of back. Fitted with a metal adjustable bridge (see 12L2003, page 514). Standard size. Shpg. wt., 12 lbs.
12L254¼ $7.45

12L255¼ $9.75

Genuine Mahogany.

Back and sides made of selected mahogany. Spruce top. Mahogany neck. Ebony finish fingerboard with three mother of pearl position dots and bound with white celluloid. Rosewood veneered head, inlaid with mother of pearl star. Edge of top is inlaid with strip of fancy colored wood blocks and bound with white celluloid. Soundhole inlaid and bound to match edge of top. Back is inlaid with a fancy stripe. Adjustable metal bridge (see 12L2003, page 514). Standard size. Shipping weight, 18 pounds.
12L255¼ $9.75

12L488¼ $3.75

Hawaiian Guitar.

Made of figured birch. Ebony finish fingerboard inlaid with white celluloid position dots. Ebony finish bridge with metal fret. Nickel plated tailpiece. Patent heads with composition buttons. Body, including top and neck, is finished in brownish color, slightly shaded at center of top and back. Shipping weight, 12 lbs.
12L488¼ $3.75

12L246¼ $9.95

Complete Guitar Outfit.

This outfit includes everything that is needed to start you on your studies. The guitar itself is a handsome instrument, well made of seasoned materials. The body is an exact reproduction of grained mahogany. Top of white spruce with an artistic ornament on lower bout. Mahogany finish neck. Imitation ebony fingerboard, bound with white celluloid and inlaid with three position dots. The top edge and soundhole are inlaid with a beautiful colored block design and bound with white celluloid. Ebony finish adjustable metal bridge (see 12L2003 on page 514). Brass screw pattern patent heads. Rubberized cloth bag, instruction book, fingerboard chart, extra set of Bell Brand steel strings, tuner with pipe for each string and thumb pick included. Standard size. Shipping weight, 15 pounds.
12L246¼ $9.95

12L268¼ $14.95

Mahogany—Pearl Inlaid.

Back and sides of selected mahogany. Spruce top. Mahogany neck. Head veneered and inlaid with mother of pearl ornament. Ebony finish fingerboard inlaid with mother of pearl ornaments and bound with white celluloid. Ebonized hardwood bridge. Bridge pins inlaid with mother of pearl. Edge of top and soundhole is inlaid with purfling and mother of pearl ornaments of different designs set in a black background. The back is inlaid with a wide stripe of fancy woods. Nickel plated patent heads. High grade in tone and appearance. Shipping wt., 18 pounds.
12L268¼—Standard size $14.95
12L269¼—Concert size 16.25

12L275¼ $19.95

Our Finest—A Beauty.

One of the most pleasing designs we have yet seen. The body is made of beautifully figured koa wood (which has the texture of mahogany and the beauty of rosewood) with top of white spruce. Mahogany neck with veneered head inlaid with pearl. Ebony finish fingerboard, bound with strips of rosewood and white celluloid and inlaid with pearl. The edge of the top is inlaid with Japanese green pearl and bound with a strip of rosewood, below which is inlaid black and white purfling, making a very beautiful effect. Soundhole inlaid and bound to match edge of top. Rosewood and holly stripe inlaid down the middle of the back. Ebonized bridge and imitation ivory bridge pins, nickel plated brass patent heads. This instrument is constructed for beauty and tonal qualities as well. Shipping weight, 18 pounds.
12L275¼—Standard size $19.95
12L276¼—Concert size. 21.25
12L277¼—Grand Concert size 22.45

HAWAIIAN GUITARS

Steel bar, three picks, instruction book and fingerboard chart included.

The Hawaiian (sometimes called the steel) Guitar, with its beautiful quavering and sharp staccato tones, has done much to make the Hawaiian music so extremely popular. Owing to the fact that the strings when open or when barred form natural chords, it is comparatively easy to learn to play this instrument.

12L489¼ $8.95

Interchangeable Hawaiian Guitar.

Back, top and sides of figured natural brown mahogany. Mahogany neck and head with koa wood veneer on top of head. Ebony finish fingerboard inlaid with three mother of pearl position dots. Fancy corded inlaying around soundhole and edge of top. Brass screw pattern patent heads with composition buttons. Ebony finish bridge and heavy nickel plated fancy tailpiece. Patented removable nut. (By removing nut it can be played in regular style.) A handsome, beautifully toned instrument. Shipping weight, 12 pounds.
12L489¼ $8.95

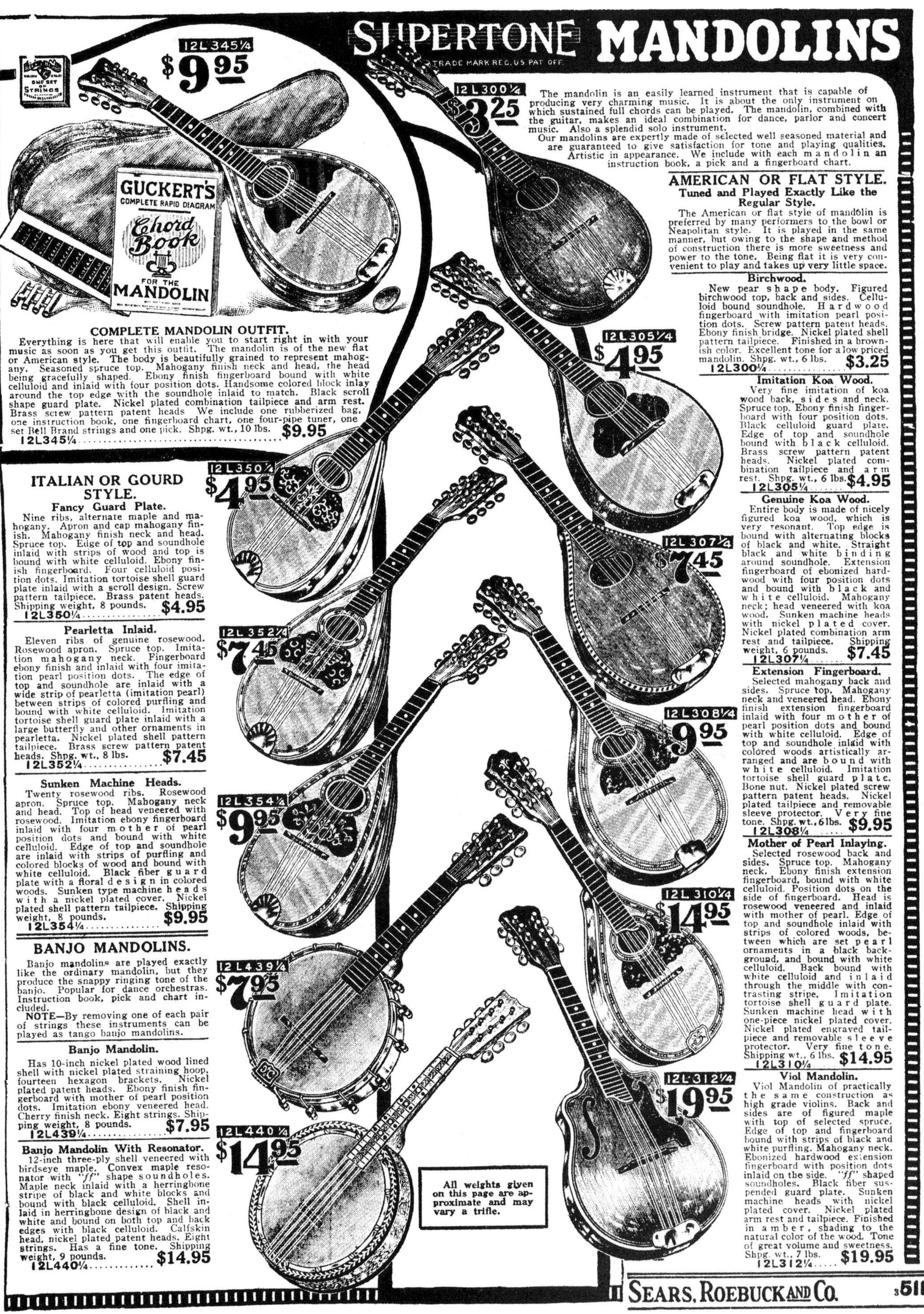

SUPERTONE MANDOLINS

TRADE MARK REG. U.S. PAT. OFF.

The mandolin is an easily learned instrument that is capable of producing very charming music. It is about the only instrument on which sustained full chords can be played. The mandolin, combined with the guitar, makes an ideal combination for dance, parlor and concert music. Also a splendid solo instrument.

Our mandolins are expertly made of selected well seasoned material and are guaranteed to give satisfaction for tone and playing qualities. Artistic in appearance. We include with each mandolin an instruction book, a pick and a fingerboard chart.

12L345¼ $9.95

GUCKERT'S COMPLETE RAPID DIAGRAM Chord Book FOR THE MANDOLIN

ONE SET OF STRINGS

COMPLETE MANDOLIN OUTFIT.

Everything is here that will enable you to start right in with your music as soon as you get this outfit. The mandolin is of the new flat or American style. The body is beautifully grained to represent mahogany. Seasoned spruce top. Mahogany finish neck and head, the head being gracefully shaped. Ebony finish fingerboard bound with white celluloid and inlaid with four position dots. Handsome colored block inlay around the top edge with the soundhole inlaid to match. Black scroll shape guard plate. Nickel plated combination tailpiece and arm rest. Brass screw pattern patent heads. We include one rubberized bag, one instruction book, one fingerboard chart, one four-pipe tuner, one set Bell Brand strings and one pick. Shpg. wt., 10 lbs.
12L345¼ $9.95

AMERICAN OR FLAT STYLE.

Tuned and Played Exactly Like the Regular Style.

The American or flat style of mandolin is preferred by many performers to the bowl or Neapolitan style. It is played in the same manner, but owing to the shape and method of construction there is more sweetness and power to the tone. Being flat it is very convenient to play and takes up very little space.

Birchwood.

New pear shape body. Figured birchwood top, back and sides. Celluloid bound soundhole. Hardwood fingerboard with imitation pearl position dots. Screw pattern patent heads. Ebony finish bridge. Nickel plated shell pattern tailpiece. Finished in a brownish color. Excellent tone for a low priced mandolin. Shpg. wt., 6 lbs.
12L300¼ $3.25

Imitation Koa Wood.

Very fine imitation of koa wood back, sides and neck. Spruce top. Ebony finish fingerboard with four position dots. Black celluloid guard plate. Edge of top and soundhole bound with black celluloid. Brass screw pattern patent heads. Nickel plated combination tailpiece and arm rest. Shpg. wt., 6 lbs.
12L305¼ $4.95

Genuine Koa Wood.

Entire body is made of nicely figured koa wood, which is very resonant. Top edge is bound with alternating blocks of black and white. Straight black and white binding around soundhole. Extension fingerboard of ebonized hardwood with four position dots and bound with black and white celluloid. Mahogany neck; head veneered with koa wood. Sunken machine heads with nickel plated cover. Nickel plated combination arm rest and tailpiece. Shipping weight, 6 pounds.
12L307¼ $7.45

Extension Fingerboard.

Selected mahogany back and sides. Spruce top. Mahogany neck and veneered head. Ebony finish extension fingerboard inlaid with four mother of pearl position dots and bound with white celluloid. Edge of top and soundhole inlaid with colored woods artistically arranged and are bound with white celluloid. Imitation tortoise shell guard plate. Bone nut. Nickel plated screw pattern patent heads. Nickel plated tailpiece and removable sleeve protector. Very fine tone. Shpg. wt., 6 lbs.
12L308¼ $9.95

Mother of Pearl Inlaying.

Selected rosewood back and sides. Spruce top. Mahogany neck. Ebony finish extension fingerboard, bound with white celluloid. Position dots on the side of fingerboard. Head is rosewood veneered and inlaid with mother of pearl. Edge of top and soundhole inlaid with strips of colored woods, between which are set pearl ornaments in a black background, and bound with white celluloid. Back bound with white celluloid and inlaid through the middle with contrasting stripe. Imitation tortoise shell guard plate. Sunken machine head with one-piece nickel plated cover. Nickel plated engraved tailpiece and removable sleeve protector. Very fine tone. Shipping wt., 6 lbs.
12L310¼ $14.95

Viol Mandolin.

Viol Mandolin of practically the same construction as high grade violins. Back and sides are of figured maple with top of selected spruce. Edge of top and fingerboard bound with strips of black and white purfling. Mahogany neck. Ebonized hardwood extension fingerboard with position dots inlaid on the side. "ff" shaped soundholes. Black fiber suspended guard plate. Sunken machine heads with nickel plated cover. Nickel plated arm rest and tailpiece. Finished in amber, shading to the natural color of the wood. Tone of great volume and sweetness. Shpg. wt., 7 lbs.
12L312¼ $19.95

ITALIAN OR GOURD STYLE.

Fancy Guard Plate.

Nine ribs, alternate maple and mahogany. Apron and cap mahogany finish. Mahogany finish neck and head. Spruce top. Edge of top and soundhole inlaid with strips of wood and top is bound with white celluloid. Ebony finish fingerboard. Four celluloid position dots. Imitation tortoise shell guard plate inlaid with a scroll design. Screw pattern tailpiece. Brass patent heads. Shipping weight, 8 pounds.
12L350¼ $4.95

Pearletta Inlaid.

Eleven ribs of genuine rosewood. Rosewood apron. Spruce top. Imitation mahogany neck. Fingerboard ebony finish and inlaid with four imitation pearl position dots. The edge of top and soundhole are inlaid with a wide strip of pearletta (imitation pearl) between strips of colored purfling and bound with white celluloid. Imitation tortoise shell guard plate inlaid with a large butterfly and other ornaments in pearletta. Nickel plated shell pattern tailpiece. Brass screw pattern patent heads. Shpg. wt., 8 lbs.
12L352¼ $7.45

Sunken Machine Heads.

Twenty rosewood ribs. Rosewood apron. Spruce top. Mahogany neck and head. Top of head veneered with rosewood. Imitation ebony fingerboard inlaid with four mother of pearl position dots and bound with white celluloid. Edge of top and soundhole are inlaid with strips of purfling and colored blocks of wood and bound with white celluloid. Black fiber guard plate with a floral design in colored woods. Sunken type machine heads with a nickel plated cover. Nickel plated shell pattern tailpiece. Shipping weight, 8 pounds.
12L354¼ $9.95

BANJO MANDOLINS.

Banjo mandolins are played exactly like the ordinary mandolin, but they produce the snappy ringing tone of the banjo. Popular for dance orchestras. Instruction book, pick and chart included.

NOTE—By removing one of each pair of strings these instruments can be played as tango banjo mandolins.

Banjo Mandolin.

Has 10-inch nickel plated wood lined shell with nickel plated straining hoop, fourteen hexagon brackets. Nickel plated patent heads. Ebony finish fingerboard with mother of pearl position dots. Imitation ebony veneered head. Cherry finish neck. Eight strings. Shipping weight, 8 pounds.
12L439¼ $7.95

Banjo Mandolin With Resonator.

12-inch three-ply shell veneered with birdseye maple. Convex maple resonator with "ff" shape soundholes. Maple neck inlaid with a herringbone stripe of black and white blocks and bound with black celluloid. Shell inlaid in herringbone design of black and white and bound on both top and back edges with black celluloid. Calfskin head, nickel plated patent heads. Eight strings. Has a fine tone. Shipping weight, 9 pounds.
12L440¼ $14.95

12L300¼ $3.25 · 12L305¼ $4.95 · 12L307¼ $7.45 · 12L308¼ $9.95 · 12L310¼ $14.95 · 12L312¼ $19.95 · 12L350¼ $4.95 · 12L352¼ $7.45 · 12L354¼ $9.95 · 12L439¼ $7.95 · 12L440¼ $14.95

All weights given on this page are approximate and may vary a trifle.

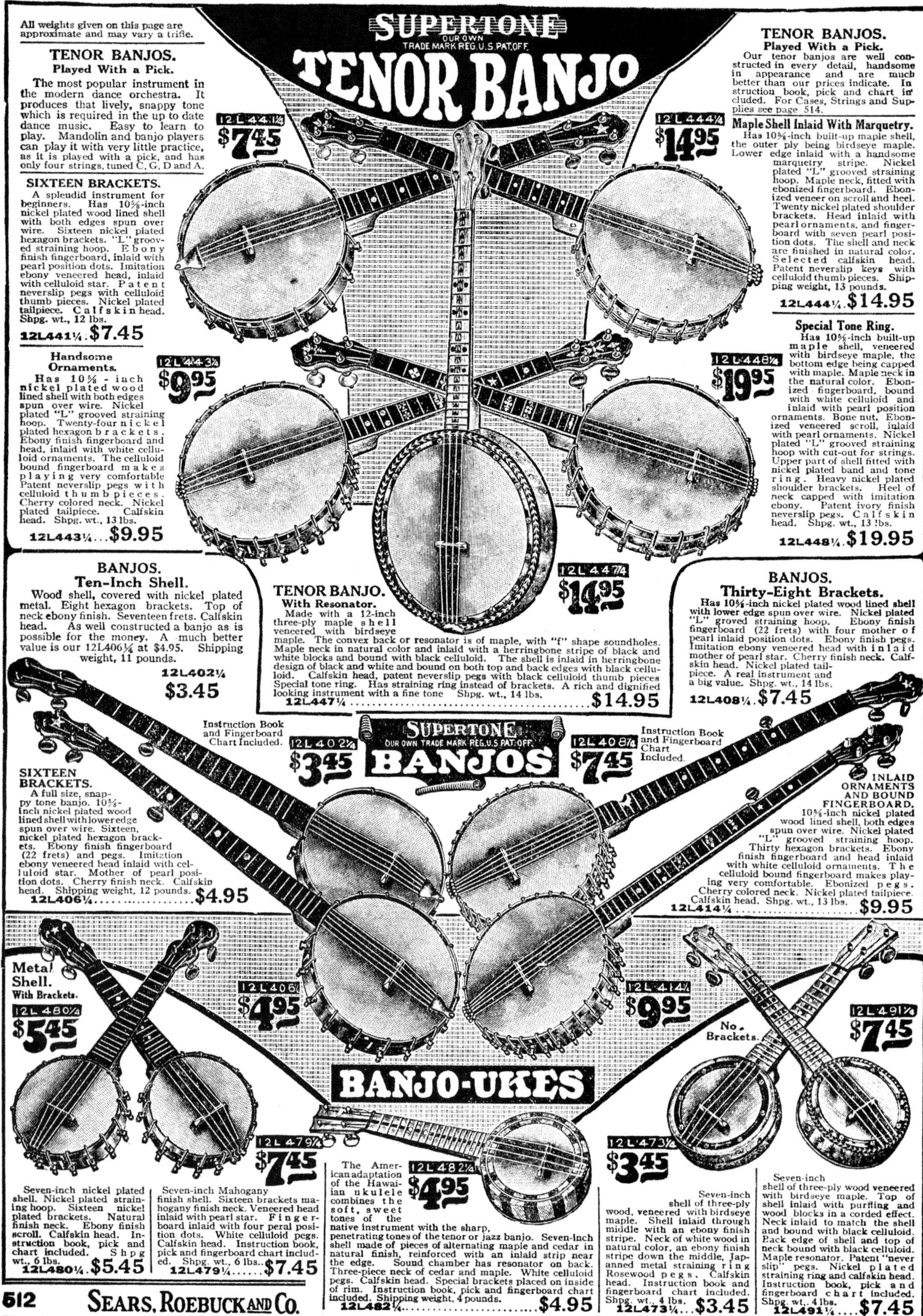
All weights given on this page are approximate and may vary a trifle.
SUPERTONE
OUR OWN TRADE MARK REG. U.S. PAT. OFF.
TENOR BANJO
TENOR BANJOS.
Played With a Pick.
The most popular instrument in the modern dance orchestra. It produces that lively, snappy tone which is required in the up to date dance music. Easy to learn to play. Mandolin and banjo players can play it with very little practice, as it is played with a pick, and has only four strings, tuned C, G, D and A.
12L441¼ $7.45
SIXTEEN BRACKETS.
A splendid instrument for beginners. Has 10⅝-inch nickel plated wood lined shell with both edges spun over wire. Sixteen nickel plated hexagon brackets. "L" grooved straining hoop. Ebony finish fingerboard, inlaid with pearl position dots. Imitation ebony veneered head, inlaid with celluloid star. Patent neverslip pegs with celluloid thumb pieces. Nickel plated tailpiece. Calfskin head. Shpg. wt., 12 lbs.
12L441¼..$7.45
12L443¾ $9.95
Handsome Ornaments.
Has 10⅝-inch nickel plated wood lined shell with both edges spun over wire. Nickel plated "L" grooved straining hoop. Twenty-four nickel plated hexagon brackets. Ebony finish fingerboard and head, inlaid with white celluloid ornaments. The celluloid bound fingerboard makes playing very comfortable. Patent neverslip pegs with celluloid thumb pieces. Cherry colored neck. Nickel plated tailpiece. Calfskin head. Shpg. wt., 13 lbs.
12L443¼...$9.95
TENOR BANJOS.
Played With a Pick.
Our tenor banjos are well constructed in every detail, handsome in appearance and are much better than our prices indicate. Instruction book, pick and chart included. For Cases, Strings and Supplies see page 514.
12L444¼ $14.95
Maple Shell Inlaid With Marquetry.
Has 10⅝-inch built-up maple shell, the outer ply being birdseye maple. Lower edge inlaid with a handsome marquetry stripe. Nickel plated "L" grooved straining hoop. Maple neck, fitted with ebonized fingerboard. Ebonized veneer on scroll and heel. Twenty nickel plated shoulder brackets. Head inlaid with pearl ornaments, and fingerboard with seven pearl position dots. The shell and neck are finished in natural color. Selected calfskin head. Patent neverslip keys with celluloid thumb pieces. Shipping weight, 13 pounds.
12L444¼.$14.95
12L448¼ $19.95
Special Tone Ring.
Has 10⅝-inch built-up maple shell, veneered with birdseye maple, the bottom edge being capped with maple. Maple neck in the natural color. Ebonized fingerboard, bound with white celluloid and inlaid with pearl position ornaments. Bone nut. Ebonized veneered scroll, inlaid with pearl ornaments. Nickel plated "L" grooved straining hoop with cut-out for strings. Upper part of shell fitted with nickel plated band and tone ring. Heavy nickel plated shoulder brackets. Heel of neck capped with imitation ebony. Patent ivory finish neverslip pegs. Calfskin head. Shpg. wt., 13 lbs.
12L448¼.$19.95
BANJOS.
Ten-Inch Shell.
Wood shell, covered with nickel plated metal. Eight hexagon brackets. Top of neck ebony finish. Seventeen frets. Calfskin head. As well constructed a banjo as is possible for the money. A much better value is our 12L406¼ at $4.95. Shipping weight, 11 pounds.
12L402¼
$3.45
12L447¼ $14.95
TENOR BANJO.
With Resonator.
Made with a 12-inch three-ply maple shell veneered with birdseye maple. The convex back or resonator is of maple, with "f" shape soundholes. Maple neck in natural color and inlaid with a herringbone stripe of black and white blocks and bound with black celluloid. The shell is inlaid in herringbone design of black and white and bound on both top and back edges with black celluloid. Calfskin head, patent neverslip pegs with black celluloid thumb pieces. Special tone ring. Has straining ring instead of brackets. A rich and dignified looking instrument with a fine tone. Shpg. wt., 14 lbs.
12L447¼....$14.95
BANJOS.
Thirty-Eight Brackets.
Has 10⅝-inch nickel plated wood lined shell with lower edge spun over wire. Nickel plated "L" groved straining hoop. Ebony finish fingerboard (22 frets) with four mother of pearl inlaid position dots. Ebony finish pegs. Imitation ebony veneered head with inlaid mother of pearl star. Cherry finish neck. Calfskin head. Nickel plated tailpiece. A real instrument and a big value. Shpg. wt., 14 lbs.
12L408¼.$7.45
Instruction Book and Fingerboard Chart Included.
12L402¼ $3.45
SUPERTONE
OUR OWN TRADE MARK REG. U.S. PAT. OFF.
BANJOS
12L408¼ $7.45
Instruction Book and Fingerboard Chart Included.
SIXTEEN BRACKETS.
A full size, snappy tone banjo. 10⅝-inch nickel plated wood lined shell with lower edge spun over wire. Sixteen, nickel plated hexagon brackets. Ebony finish fingerboard (22 frets) and pegs. Imitation ebony veneered head inlaid with celluloid star. Mother of pearl position dots. Cherry finish neck. Calfskin head. Shipping weight, 12 pounds.
12L406¼.........$4.95
INLAID ORNAMENTS AND BOUND FINGERBOARD.
10⅝-inch nickel plated wood lined shell, both edges spun over wire. Nickel plated "L" grooved straining hoop. Thirty hexagon brackets. Ebony finish fingerboard and head inlaid with white celluloid ornaments. The celluloid bound fingerboard makes playing very comfortable. Ebonized pegs. Cherry colored neck. Nickel plated tailpiece. Calfskin head. Shpg. wt., 13 lbs.
12L414¼.........$9.95
12L406¼ $4.95
12L414¼ $9.95
Metal Shell.
With Brackets.
12L480¼ $5.45
No Brackets.
12L491¼ $7.45
BANJO-UKES
12L479¼ $7.45
12L482½ $4.95
12L473¼ $3.45
Seven-inch nickel plated shell. Nickel plated straining hoop. Sixteen nickel plated brackets. Natural finish neck. Ebony finish scroll. Calfskin head. Instruction book, pick and chart included. Shpg wt., 6 lbs.
12L480¼.$5.45
Seven-inch Mahogany finish shell. Sixteen brackets mahogany finish neck. Veneered head inlaid with pearl star. Fingerboard inlaid with four peral position dots. White celluloid pegs. Calfskin head. Instruction book, pick and fingerboard chart included. Shpg. wt., 6 lbs.
12L479¼......$7.45
The American adaptation of the Hawaiian ukulele combines the soft, sweet tones of the native instrument with the sharp, penetrating tones of the tenor or jazz banjo. Seven-inch shell made of pieces of alternating maple and cedar in natural finish, reinforced with an inlaid strip near the edge. Sound chamber has resonator on back. Three-piece neck of cedar and maple. White celluloid pegs. Calfskin head. Special brackets placed on inside of rim. Instruction book, pick and fingerboard chart included. Shipping weight, 4 pounds.
12L482½..........$4.95
Seven-inch shell of three-ply wood, veneered with birdseye maple. Shell inlaid through middle with an ebony finish stripe. Neck of white wood in natural color, an ebony finish stripe down the middle. Japanned metal straining ring. Rosewood pegs. Calfskin head. Instruction book and fingerboard chart included. Shpg. wt., 4 lbs.
12L473¼...$3.45
Seven-inch shell of three-ply wood veneered with birdseye maple. Top of shell inlaid with purfling and wood blocks in a corded effect. Neck inlaid to match the shell and bound with black celluloid. Back edge of shell and top of neck bound with black celluloid. Maple resonator. Patent "never slip" pegs. Nickel plated straining ring and calfskin head. Instruction book, pick and fingerboard chart included. Shpg. wt., 4 lbs.
12L491¼....$7.45

SUPERTONE UKULELES

Complete Ukulele Outfit.

$2.75

Everything that's needed to play the ukulele. The ukulele is made of selected birchwood. Brownish color. Dull finish. Substantially built and of good tone. With this ukulele we include one felt bag, one instruction book, one fingerboard chart and one felt pick. Don't judge this outfit by its low price. It is made to give service. Shipping weight, 5 pounds.
12L460¼........................... $2.75

BIRCHWOOD UKULELE.

Birch body and neck carefully made and finished. Top edge is bound with white celluloid in contrast to the brownish finish. A better instrument than the price would indicate. Shipping weight, 4 pounds.
12L452¼ $1.95

MAHOGANY UKULELE.

Body of figured natural brown mahogany. Mahogany neck and head. Edge of top inlaid with purfling and wood of contrasting colors in corded effect. Soundhole inlaid to match edge. Finished in the natural color and dull rubbed. Accurately fretted. Excellent tone. Shipping weight, 4 pounds.
12L457¼.... $2.95

THE "EUKELET" A Pocket Ukulele. Can Be Carried in Coat Pocket.

Just what is wanted for parties, picnics, camps and canoes. Takes up very little space. Weighs only 5 ounces; 17½ inches long, 5 inches wide and 1 inch deep. Body is made of real koa wood. Mahogany neck. The tone is soft and sweet (just right for accompanying the voice). A wonderful little instrument for the money. Shpg. wt., 2 lbs.
12L459........................ $1.63

KOA WOOD UKULELE.

Natural color body and neck, carefully made and finished. The top edge, fingerboard and soundhole are bound with white and black celluloid which, in contrast to the brownish finish of the body, gives a rich and dignified effect. Extension fingerboard. Sweet toned. Shpg. wt., 4 pounds.
12L458¼.... $4.45

KOA WOOD UKULELE.

Finely figured genuine koa wood body, neck and head. Top edge inlaid in beautiful corded effect. Soundhole inlaid to match edge of top with additional circles of purfling. Extension fingerboard inlaid with a stripe in corded effect and six dots. Natural color. Patent "never-slip" keys. Beautiful tone. Shpg. wt., 4 lbs.
12L465¼ $6.45

AUTOHARPS

37 Strings. 12 Bars. $8.95

23 Strings. 5 Bars. $4.95

Thirty-seven strings. The twelve bars produce twelve chords, as follows: G major, D seventh, C major, A minor, G seventh, E seventh, F major, D minor, C seventh, A seventh, B flat major and G minor. It has a full chromatic scale, therefore capable of great musical possibilities. Size, 11½x22 inches. Body of highly polished imitation ebony with decalcomania decorations. Instruction book included. Shipping weight, 11 pounds.
12L506¼........................ $8.95

Same as 12L506¼ but with only 23 strings and five chords, C major, F major, G major, C seventh and A sharp major. Size, 9½x18½ inches. Shipping weight, 6 pounds.
12L502¼........................ $4.95

SUPERTONE PHONOHARP

Easiest of Them All to Learn to Play.

46 Strings. 4 Chords.

Patent Device For Playing Chords.

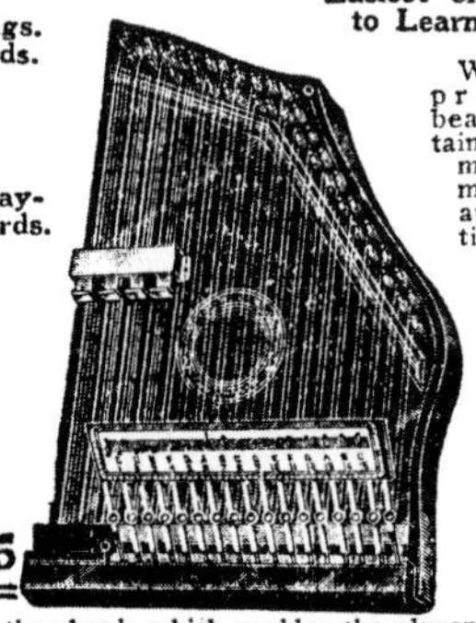

$6.95

With it you can produce the beautiful sustained tremolo melody of the mandolin and at the same time the harmonizing chords of the guitar. It gives a clear, trilling, repeated tone like a mandolin or a single tone at the will of the player. It is fitted with a special device for playing the chords which enables the player to strike all four strings at once without using thumb pick. Top is finished in imitation rosewood, with ebony finish sides. Fancy decalcomania ornaments around soundhole and on top. Highly polished. Size, 13¾x20 inches. Instruction book, tuning wrench, pick and six charts of music included. Shipping weight, 12 pounds.
12L550¼.................. $6.95

ZITHER

$3.95

Mahogany finish top, thirty-one strings (four chords and fifteen melody strings). Chords are C major, G major, F major and A minor. A simplified zither. Wrench, pick, instruction book and music rack included. Size, 10x18 inches. Shipping wt., 9 lbs.
12L520¼ $3.95

"LITTLE JOE"

98c

"Little Joe" is a complete musical instrument. Both melody and accompaniments can be played upon it. The harmonica takes care of the air or melody, the strings of the chords or accompaniments. The instrument is made of wood, finished dark red. Has imitation guard plate of fancy design. Nickelplated tailpiece. It is fitted with four strings which are tuned to produce a chord when they are struck with fingers or pick. By the aid of the fingerboard additional chords can be produced. "Little Joe" comes complete with a ten-hole, twenty-reed harmonica of good grade. Packed in a cardboard box, which contains complete instructions on "how to play," also tuning wrench and pick. Shpg. wt., 2½ lbs.
12L514¼.................. 98c

OCARINAS

Imported and made of special clay. We guarantee them to be in perfect tune, easy to blow and bird-like in tone. A sheet of instructions with each instrument. **Be sure to give catalog number and key of instrument desired.**

12L5222

Key of	Sopranos	Shpg. Wt.,
C	17c	6 oz.
B flat	19c	6 oz.
A	21c	7 oz.
G	23c	7 oz.
F	25c	9 oz.
E	27c	9 oz.
E flat	29c	9 oz.

12L5223

Key of	Altos,	Shpg. Wt.
D	32c	13 oz.
C	35c	14 oz.

12L5223

Key of	Altos	Wt. Shpg.
B flat	38c	15 oz.
A	39c	15 oz.
G	42c	24 oz.
F	49c	38 oz.
E flat	56c	38 oz.

12L5224

Key of	Basses	Shpg. Wt.
D	$0.71	47 oz.
C	.93	47 oz.
B flat	1.07	56 oz.
A	1.18	56 oz.
G	1.38	5 lbs.

12L5225—QUARTET. Key of C. Consists of 1st and 2d Tenor, 1st and 2d Bass. These are carefully selected instruments and accurately tuned to harmonize with each other. Shipping weight, 2½ pounds. Quartet of four ocarinas............ $1.25

12L5227—Same as 12L5225, but in the key of G. Shpg. wt., 2½ lbs. Quartet of four ocarinas ... $1.65

Picks for Zither and Autoharp. Celluloid Thumb Pick. For Autoharp or Zither. Shipping weight, 1 ounce.
12L5512.................. 3c

Tuning Pins for Phonoharp, Zither and Autoharp. Nickel plated. Shipping weight, 4 ounces.
12L5516—Dozen.................. 13c

Song-O-Phone Minstrel or Fun Producing Instruments.

SLIDE TROMBONE — CORNETINA — ALTO — VALVE CORNET — BASS OR TUBA — CORNET — BUGLE — SAXOPHONE

Organize a Song-O-Phone Band. Any one who can sing or hum a tune can play without studying and produce various amusing imitations. The skin diaphragm carries the tone which is amplified by the bell or sound chamber.

12L5360—Cornetina, 7 in. long, 3⅝-inch bell. Shpg. wt., 10 oz........ $0.45
12L5361—Cornet, 9 in. long, 4¼-inch bell. Shpg. wt., 1 lb. 3 oz........ 69
12L5362—Alto, 12 in. long, 5¼-inch bell. Shpg. wt., 1 lb. 7 oz........ 89
12L5363—Valve Cornet, 9 in. long, 4½-in. bell. Shpg. wt., 1 lb. 9 oz.... 1.48
12L5364—Slide Trombone. Bell part, 12 inches; slide, 20 inches. Shipping weight, 2½ pounds.................. $1.99
12L5365—Tuba, 12 inches long, 6-inch bell. Shpg. wt., 2 lbs.......... $2.17
12L5366—Bugle, 12 in. long, 4½-in. bell, with cord. Shp. wt., 1 lb. 9 oz... 1.46
12L5367—Saxophone, 17 in. long, 4½-inch bell. Shpg. wt., 2⅛ lbs...... 1.98
12L5369—Extra Vibrators for above. 1 doz. in pkg. Shpg. wt., 1 oz. Per pkg. .04

All Song-o-phones are made of metal in imitation brass.

AUTOHARP, PHONOHARP AND SIMPLIFIED ZITHER STRINGS.

Autoharp Strings.
Shpg. wt., 4 oz.
12L5522 Set for No. 23 (12L502¼) 49c
12L5526 Set for No. 73 (12L506¼) 89c

Simplified Zither Strings.
12L5532—Set for 12L520¼.................. 98c

Phonoharp Strings.
12L5541—Per set.... 98c

The "Jazzbo."

A new instrument capable of producing many amusing imitations. Hum, talk or imitate the calls of animals, tones of instruments, etc., and the skin vibrator will reproduce them in a natural manner. Old and young will get lots of fun playing them and listening to the amusing imitations. Made of metal. Shpg. wt., 4 oz.
12L5145.................. 7c

ACCORDIONS

Excepting the violin and piano, the accordion is probably the most popular of instruments. It is played the world over. It is generally played by ear and most people can learn to play a melody in half an hour or so. The accordion can also be played by note, which requires a little more time to learn. Instruction books included with all our accordions.

Milano Organetto—Italian Model.

The name Milano Organetto was given to these accordions because of their organlike tone. They are made for us by the manufacturers of our Beaver line and are correspondingly superior in construction and tone to most other accordions of similar style.

Beaver Brand—German Model.

Beaver Brand Accordions are made exclusively for us by one of the leading manufacturers of Germany. They are extra well made and finished, accurately tuned and have a strong and pleasing tone.

ITALIAN STYLE

Milano Organetto. 10 Keys, 4 Basses.

Body imitation rosewood, highly polished. Panels cut in fancy scrollwork, net lined. Cloth bound bellows of ten folds with metal corner protectors. Two sets of reeds. Ten keys, four basses, mother of pearl buttons. Fitted with new pattern lyre shape thumbscrew clasps, permitting the instant tightening of the frame to the bellows or removal for cleaning and repairs. Nickel plated trimmings. A very neat instrument with a strong tone. Size, 6x11 inches. Weight, boxed for shipment, 10 lbs.

12L631¼ **$4.95**

Milano Organetto. 21 Keys, 8 Basses.

Same finish and construction as 12L631¼, but has 21 treble and 8 bass keys and four sets of reeds, mounted on removable reed blocks. Bellows of fourteen folds. The basses are arranged in both major and minor chords. A first class instrument in every respect and one of our most popular styles. Size, 6x11 inches. Shipping wt., 12 lbs.

12L633¼ **$8.95**

Fancy Milano Organetto. 21 Keys, 12 Basses.

Frame, perfect imitation, of rosewood, inlaid on front and on keyboard with various colored woods. The eight corners of the frame are bound with nickel plated metal. Bellows of sixteen folds bound with imitation leather. Four sets of reeds, each two reeds mounted on a separate plate and the plates mounted on removable blocks. Twenty-one treble and twelve bass keys. Pearl buttons. Treble panel, imitation rosewood, cut in fancy scroll design. Nickel plated trimmings. Size, 6x11 in. Shipping weight, 14 pounds.

12L635¼ **$12.45**

Milano Organetto. 31 Keys, 12 Basses.

This is our largest Milano Organetto and is exceptionally powerful. Same construction and finish as 12L631¼, but is larger in size and has bellows with fourteen folds. Three rows of keys for the melody, thirty-one keys, twelve basses, six sets of reeds, mounted on removable reed blocks. Air valve almost the entire length of the panel. This accordion can be played in three different keys. Basses are tuned in both major and minor chords. Size, 6¼x11¾ inches. Shpg. wt., 16 lbs.

12L637¼ **$14.95**

GERMAN STYLE

"Beaver Brand."

Two Stops, Two Basses.

Frame, ebony finish, with shaped moldings. Double bellows of eight folds with metal corner protectors. Two sets of reeds, ten keys with nickel plated buttons, two basses, nickel plated ornaments. Size, 5½x 10¼ in. Shpg. wt., 5½ lbs.

12L601¼ **$2.95**

"Beaver Brand."

Two Stops, Triple Bellows.

Frame, imitation ebony. Panels, bright colored. Bellows, triple pattern, nine folds with corner protectors. Two sets of reeds, two stops, ten keys with metal buttons, two basses. Nickel plated polished trimmings. This instrument is very durably made and reliable. Size, 5½x10¾ inches. Shipping weight, 6 pounds.

12L603¼ **$3.45**

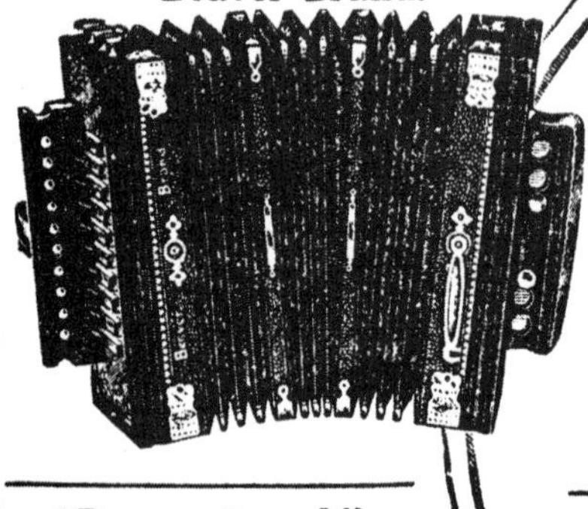

"Beaver Brand."

Three Stops—Large Size.

This is a large accordion and therefore powerful in tone. Frame, ebony finish, with shaped moldings. Panels, finished in very bright colors with embossed gilt ornaments. Bellows, triple style, with nine folds fully protected by metal corners. Three full sets of reeds. Three stops, ten nickel plated keys with metal buttons, nickel plated clasps and trimmings. Size, 6x12 inches. Shipping weight, 9 pounds.

12L606¼ **$4.45**

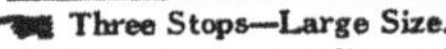

"Beaver Brand."

Four Sets of Reeds.

Frame, imitation ebony. Panels, bright color, stamped with gilt ornaments. Bellows, double style of ten folds, with corner protectors. Four sets of reeds, twenty-one keys, four stops, four basses. Nickel plated polished trimmings. This instrument is large and powerful in tone. Size, 6¾x12½ inches. Shipping weight, 10 pounds.

12L609¼ **$6.95**

"Beaver Brand."

Fancy Panels.

Moldings, fluted and finished in imitation mahogany. Bass panel, maple, finished in natural color. Part of the treble panel is in mahogany, the sunken part being in maple surrounded by nickel plated moldings. Bellows, double, with ten folds and metal protectors. Bass keys and other metal parts are nickel plated and polished. Two sets of steel bronzed reeds very powerful and sweet, two stops of ornamental metal, ten beautifully decorated keys with brass rods. Mother of pearl buttons. The treble panel is ornamented with four gilt and enameled corners of a neat and pleasing design. Size, 6½x13 inches. Shipping weight, 9 pounds.

12L612¼ **$5.45**

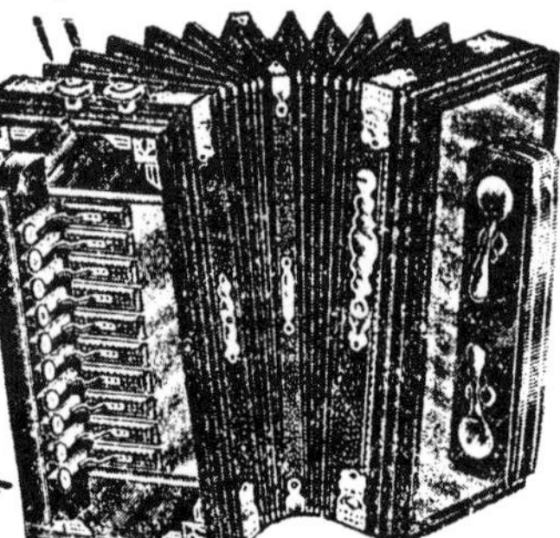

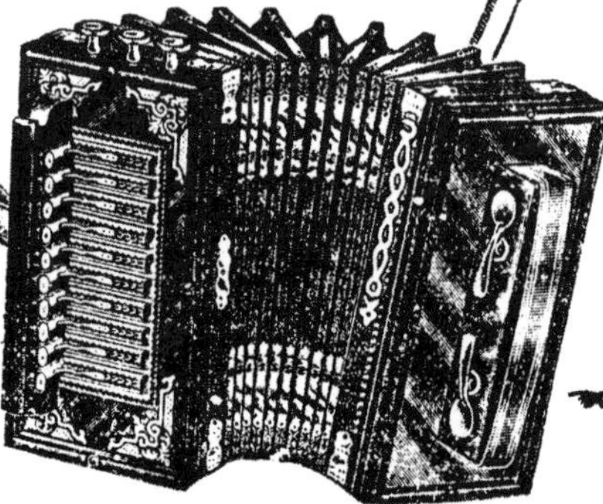

"Beaver Brand."

Octave Tuning. Very Fancy.

Frame, ebony finish, with treble panel of the sunken type. The sunken part is finished in imitation of silver and the balance in imitation ebony with pressed ornaments. Extra deep bellows made of ten wooden ribs, each rib covered by nickel plated band of metal surmounted by a gilt ornament and held in place by fancy gilt studs. Three sets of bronze reeds, three fancy metal stops, ten ebony finish and fancy ornamented keys with brass rods. Mother of pearl buttons. Nickel plated and polished trimmings. Beautifully varnished and highly finished. The combination of colors used harmonizes most exquisitely. Size, 7x13¼ in. Shpg. wt., 11 lbs.

12L614¼ **$7.85**

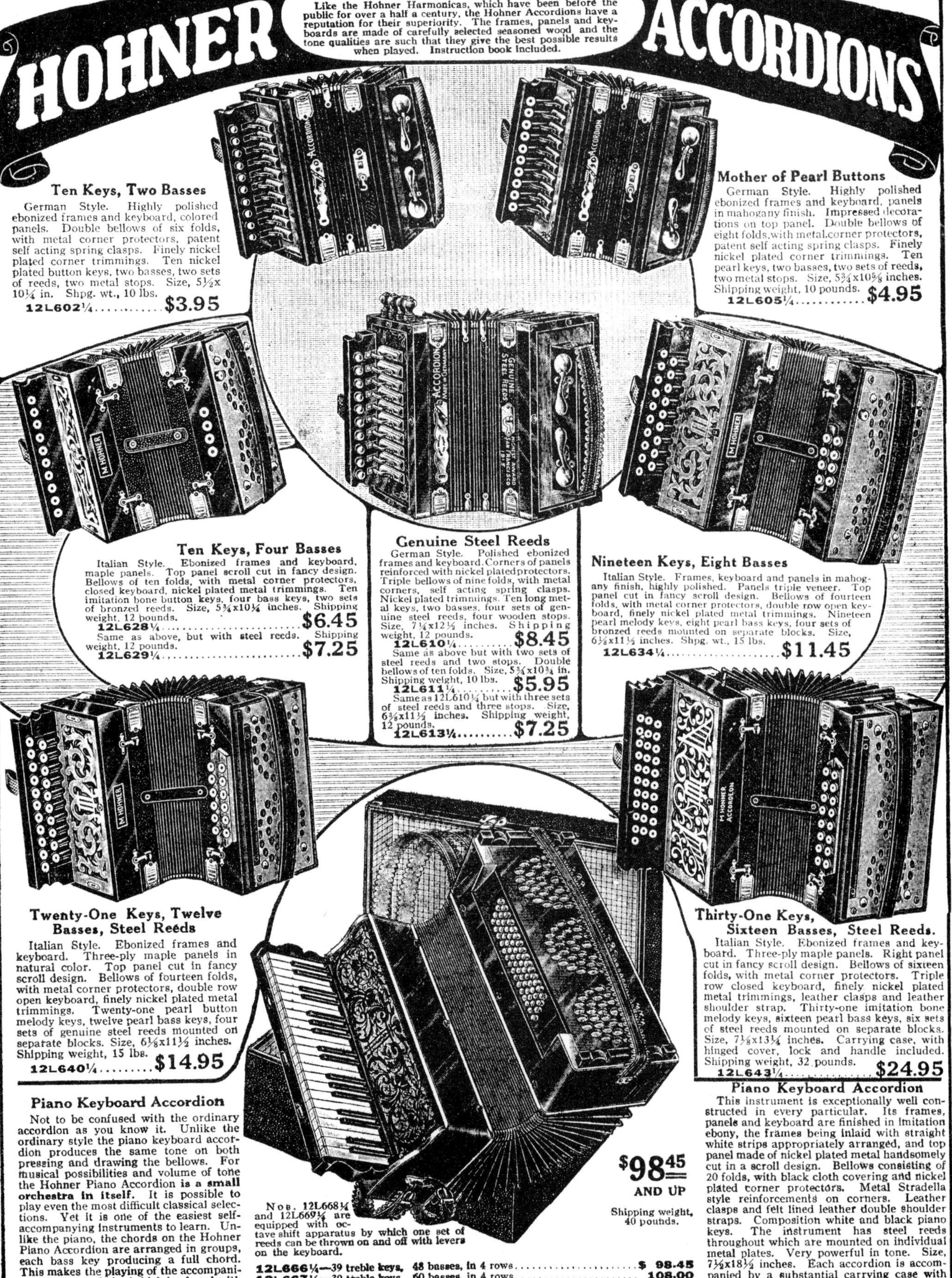
HOHNER ACCORDIONS
Like the Hohner Harmonicas, which have been before the public for over a half a century, the Hohner Accordions have a reputation for their superiority. The frames, panels and keyboards are made of carefully selected seasoned wood and the tone qualities are such that they give the best possible results when played. Instruction book included.
Ten Keys, Two Basses
German Style. Highly polished ebonized frames and keyboard, colored panels. Double bellows of six folds, with metal corner protectors, patent self acting spring clasps. Finely nickel plated corner trimmings. Ten nickel plated button keys, two basses, two sets of reeds, two metal stops. Size, 5½x 10¼ in. Shpg. wt., 10 lbs.
12L602¼..........$3.95
Mother of Pearl Buttons
German Style. Highly polished ebonized frames and keyboard, panels in mahogany finish. Impressed decorations on top panel. Double bellows of eight folds, with metal corner protectors, patent self acting spring clasps. Finely nickel plated corner trimmings. Ten pearl keys, two basses, two sets of reeds, two metal stops. Size, 5¾x10⅝ inches. Shipping weight, 10 pounds.
12L605¼..........$4.95
Ten Keys, Four Basses
Italian Style. Ebonized frames and keyboard, maple panels. Top panel scroll cut in fancy design. Bellows of ten folds, with metal corner protectors, closed keyboard, nickel plated metal trimmings. Ten imitation bone button keys, four bass keys, two sets of bronzed reeds. Size, 5¾x10¼ inches. Shipping weight, 12 pounds.
12L628¼..........$6.45
Same as above, but with steel reeds. Shipping weight, 12 pounds.
12L629¼..........$7.25
Genuine Steel Reeds
German Style. Polished ebonized frames and keyboard. Corners of panels reinforced with nickel plated protectors. Triple bellows of nine folds, with metal corners, self acting spring clasps. Nickel plated trimmings. Ten long metal keys, two basses, four sets of genuine steel reeds, four wooden stops. Size, 7¼x12½ inches. Shipping weight, 12 pounds.
12L610¼..........$8.45
Same as above but with two sets of steel reeds and two stops. Double bellows of ten folds. Size, 5¾x10¾ in. Shipping weight, 10 lbs.
12L611¼..........$5.95
Same as 12L610¼ but with three sets of steel reeds and three stops. Size, 6⅜x11½ inches. Shipping weight, 12 pounds.
12L613¼..........$7.25
Nineteen Keys, Eight Basses
Italian Style. Frames, keyboard and panels in mahogany finish, highly polished. Panels triple veneer. Top panel cut in fancy scroll design. Bellows of fourteen folds, with metal corner protectors, double row open keyboard, finely nickel plated metal trimmings. Nineteen pearl melody keys, eight pearl bass keys, four sets of bronzed reeds mounted on separate blocks. Size, 6⅜x11½ inches. Shpg. wt., 15 lbs.
12L634¼..........$11.45
Twenty-One Keys, Twelve Basses, Steel Reeds
Italian Style. Ebonized frames and keyboard. Three-ply maple panels in natural color. Top panel cut in fancy scroll design. Bellows of fourteen folds, with metal corner protectors, double row open keyboard, finely nickel plated metal trimmings. Twenty-one pearl button melody keys, twelve pearl bass keys, four sets of genuine steel reeds mounted on separate blocks. Size, 6⅛x11½ inches. Shipping weight, 15 lbs.
12L640¼..........$14.95
Thirty-One Keys, Sixteen Basses, Steel Reeds.
Italian Style. Ebonized frames and keyboard. Three-ply maple panels. Right panel cut in fancy scroll design. Bellows of sixteen folds, with metal corner protectors. Triple row closed keyboard, finely nickel plated metal trimmings, leather clasps and leather shoulder strap. Thirty-one imitation bone melody keys, sixteen pearl bass keys, six sets of steel reeds mounted on separate blocks. Size, 7⅛x13¼ inches. Carrying case, with hinged cover, lock and handle included. Shipping weight, 32 pounds.
12L643¼..........$24.95
Piano Keyboard Accordion
Not to be confused with the ordinary accordion as you know it. Unlike the ordinary style the piano keyboard accordion produces the same tone on both pressing and drawing the bellows. For musical possibilities and volume of tone the Hohner Piano Accordion is a small orchestra in itself. It is possible to play even the most difficult classical selections. Yet it is one of the easiest self-accompanying instruments to learn. Unlike the piano, the chords on the Hohner Piano Accordion are arranged in groups, each bass key producing a full chord. This makes the playing of the accompaniment to the melody (which is played with the right hand) simple and easy. The piano keys for playing the melody are arranged exactly as on a piano to produce the chromatic scale.
Nos. 12L668¼ and 12L669¼ are equipped with octave shift apparatus by which one set of reeds can be thrown on and off with levers on the keyboard.
$98.45 AND UP
Shipping weight, 40 pounds.
12L666¼—39 treble keys, 48 basses, in 4 rows..........$ 98.45
12L667¼—39 treble keys, 60 basses, in 4 rows..........108.00
12L668¼—42 treble keys, 80 basses, in 5 rows..........125.00
12L669¼—42 treble keys, 120 basses, in 6 rows..........$162.45
12L01987—Piersanti Method for the Piano Keyboard and Chromatic Accordions. Shipping weight, 1⅛ pounds..........$1.89
Piano Keyboard Accordion
This instrument is exceptionally well constructed in every particular. Its frames, panels and keyboard are finished in imitation ebony, the frames being inlaid with straight white strips appropriately arranged, and top panel made of nickel plated metal handsomely cut in a scroll design. Bellows consisting of 20 folds, with black cloth covering and nickel plated corner protectors. Metal Stradella style reinforcements on corners. Leather clasps and felt lined leather double shoulder straps. Composition white and black piano keys. The instrument has steel reeds throughout which are mounted on individual metal plates. Very powerful in tone. Size, 7½x18½ inches. Each accordion is accompanied by a substantial carrying case with hinged cover, lock and handle.
SEARS, ROEBUCK AND CO.
2517

HOHNER—MOUTH ORGANS—HOHNER

"Up to Date Tremolo."

A large double sided instrument representing the very finest in harmonica construction. Forty-eight double holes, ninety-six reeds, brass plates, nickel plated covers. Fancy gilt stamped extension ends with frame in mahogany finish. Two harmonicas in one in different keys. Comes in a very handsome paper covered wooden case with metal clasps. Length, 9 inches. Shipping weight, 1 pound.
12L5144 .... $2.39

Hohner "Auto Valve."

This instrument is fitted with a wind saving arrangement which is one of the greatest harmonica improvements produced. This makes the concert harp as easy blowing as any single reed mouth organ. Ten double holes, forty reeds, brass plates, nickel covers, full concert style. 4½ inches long. In fine box with hinged cover. Shipping weight, 5 ounces.
12L5139 .... $1.23

Hohner Vest Pocket Chimes.

A compact double sided harmonica with twelve double holes and twenty-four reeds on each side mounted on brass plates; tremolo tuning. Nickel plated rounded covers with fancy perforations. Measures 4¼ inches in length, and is packed in strong hinged box. Shipping weight, 7 ounces.
12L5130 .... 77c

Hohner Full Concert.

The recognized standard harmonica used by players throughout the world. Ten double holes, forty reeds, brass plates. Covers finely nickel plated and have turned in ends. 4⅜ inches long. Packed in a strong hinged box. Shipping weight, 5 ounces.
12L5131 .... 83c

Hohner "Harmonette."

A decidedly novel and practical instrument comprising a harmonica with fourteen double holes, twenty-eight tremolo tuned reeds, genuine brass plates and nickel plated covers. To the harmonica is attached a harp resonator or sound box of wood, reinforced with metal back. A beautiful effect can be produced with it. Length, 4¾ inches. Packed in a durable telescope container. Shipping weight, 8 ounces.
12L5147 .... $1.23

We are showing a variety of good mouth organs, including a fine line of genuine Hohner styles. The Hohner harmonicas have always been a favorite with players on account of their durability, easy blowing qualities and accurate tuning.

The "Beaver" harmonicas are made expressly for us by one of the best manufacturers of Europe and sold under our own name. They all have brass plates, heavy nickel plated covers, bronze metal reeds, hardwood frames, and are especially tuned and tested.

"MarineBandTremolo"

A double sided harmonica in two different keys, measuring 8½ inches in length. Forty double holes, eighty reeds, brass plates, fancy nickel plated covers, mahogany finished frame, extension ends with gilt decorations. Each instrument packed in a heavy telescope box. Shipping weight, 14 ounces.
12L5141 .... $1.89

"Hohner Goliath."

A large size single side tremolo harmonica, measuring 7½ in. in length. Twenty-four double holes, forty-eight reeds, brass plates. Heavy convex and flaring nickel plated covers. An ideal harmonica for concert playing as it has an exceptionally high range. Each instrument is packed in a heavy wooden case, covered in imitation alligator skin. Shipping weight, 11 ounces.
12L5140 .... $1.49

"Hohner Sportsman."

Sixteen double holes, thirty-two reeds, brass plates, nickel plated covers, tremolo tuned, extension ends. A very attractive instrument measuring 5¼ inches in length. Comes in a fancy hinged box with lithographed design. Shpg. wt., 6 oz.
12L5137 .... 85c

Hohner Band.

Ten holes and twenty reeds, mounted on brass plates. Nickel plated flaring covers. Hardwood frame, nicely varnished and polished. A sturdy little instrument. 4 inches long. Telescope box. Shipping weight, 4 oz.
12L5132 .... 39c

The "Hohnerphone."

The finest tone effect can be produced by keeping a slight movement of the hand over the mouth of the horn. Horn also increases volume of tone. Ten single holes, twenty reeds, brass plates, nickel plated covers, with highly polished detachable horn in brass finish. Shipping weight, 12 ounces.
12L5135 .... $1.18

"Hohner Trumpet Call."

Full concert. Ten double holes, forty reeds, producing an organ-like tone, brass plates. The reeds of this instrument are directly connected with a wooden sound box into which the tone passes and finds an outlet through five brass trumpet horns which protrude from the box. For volume of tone production there is no other harmonica like it. It is in a class by itself. Size of harmonica, 4⅞ inches. Furnished in fine hinged box. Shipping weight, 14 ounces.
12L5148 .... $2.45

BEAVER MOUTH ORGANS

"Beaver" University Chimes.

Double Harmonica, tuned in two harmonizing keys and with the assistance of the bells a varied and catchy class of music can be produced. Hardwood frame; forty-eight holes on each side, four brass plates with twenty-four reeds on each or ninety-six reeds in all. Four separate nickel plated covers. The bells are mounted on a special bridgelike frame and are very easily operated. In heavy cardboard box. 9 inches long. Shipping weight, 1⅛ pounds.
12L5118 $1.48

"Beaver" Magic Organ.

Brass plates, double covers with nickel plated imitation organ pipes. Tuned to produce the tremolo or wavy effect in tone. Sixteen double holes, thirty-two reeds with organ-like tone. Length, 5½ inches. Neat cardboard telescope box. Shipping weight, 5 ounces.
12L5126 .... 39c

"Beaver" Regimental Band.

A favorite little instrument with ten holes, twenty reeds, heavy covers flaring at back and with extension ends. Brass plates. Four inches long. Comes in substantial telescope cardboard box. Shipping weight, 4 ounces.
12L5121
23c

"Beaver" Concert Regimental Band.

Full Concert Harmonica. Has brass plates, ten double holes, forty fine reeds. Covers of metal, flaring in model and with extension ends. Beautifully tuned, and rich in tone and carrying power. 4¾ inches long. Complete in fine hinged cover cardboard box. Shipping weight, 5 oz.
12L5122
48c

"Beaver" Celestial Echoes.

Our most popular "Beaver" harmonica.

Tremolo tuned, has hardwood frame, twenty-four double holes, forty-eight reeds, heavy nickel plated covers and brass plates, in a very handsome paper covered wooden box with metal clasp. Sweet, powerful tone, tuned to produce the tremolo or wavy effect. 7 inches long. Shipping weight, 10 ounces.
12L5114 .... 89c

"Beaver" Silvery Sounds.

A double side harmonica with forty-eight holes and forty-eight reeds on each side, brass plates, hardwood frame, beautifully finished, and nickel plated covers, neatly chased and perforated. 7 inches long. Tremolo tuned. Comes in a neat telescope cardboard box. Shipping weight, 12 ounces.
12L5116
$1.15

The Belfry.

Two silver toned bells. Ten double holes, twenty reeds, brass plated plates and nickel plated covers. Two bells mounted on bridge style attachment. Charming effects can be produced with the bells. In cardboard box. Shipping weight, 6 oz.
12L5159 .... 35c

The Magnet.

A full Concert Harp, for the price ordinarily asked for a 20-reed. Compact in size but strong in tone. Ten double holes, forty reeds, brass plates, nickel plated covers with rounded edges. Hardwood enameled frame. In fancy hinged cover box. Very durable. Shipping weight, 4 ounces.
12L5109
39c

Harmonica Holder.

Indispensable to players who desire to have free hands with which to play mandolin, guitar or zither in connection with the harmonica. Will fit any harmonica not more than 4¾ inches in length. When not in use it may be folded into small compass. Shipping weight, 5 ounces.
12L5124 ... 39c

Fourteen Trumpets.

Sixteen double holes and thirty-two reeds. Tremolo tuned. Nickel plated covers, fancy design, mounted with little trumpets, through which the tone passes. Brass plated plates. One of our best sellers. Shipping weight, 5 oz.
12L5152 .... 27c

The Espera.

Ten holes, twenty reeds, genuine brass plates, nickel plated covers. A full size harmonica priced as low as most harmonicas with imitation brass plates. Paper covered pasteboard box. Shipping weight, 3 ounces.
12L5101 .... 17c

CLARINETS, FLUTES, PICCOLOS, FIFES, ETC.

In our better grade of clarinets we carry both wood and ebonite, as quite a few players prefer ebonite (a composition made principally of hard rubber) because they do not check or split and trouble with keys sticking on account of expansion is minimized. The tone is mellow and sonorous and very responsive in blowing quality.

As wood clarinets are susceptible to climatic conditions and sometimes require careful readjustment of the keys there may be a delay of two or three days in shipping.

Practically all organizations are now using low pitch, consequently we have discontinued handling high pitch clarinets.

Lafayette Clarinets.

Albert System. Grenadilla wood, dull finish. Bored and finished with care. Nickel silver keys, highly polished. These instruments are much better than our low price would indicate. They have always given satisfactory results. Mouthpiece cap and instruction book included. Shipping weight, 5 pounds.

12L800¼ — 13 keys, 2 rings, in the keys of A, B flat or C, low pitch. **State key wanted** . . . **$14.95**

12L801¼ — 15 keys, 2 rings, in the keys of A, B flat or C, low pitch. **State key wanted** **$17.45**

12L802¼ — 15 keys, 4 rings and 4 roller keys, in the keys of A, B flat or C, low pitch. **State key wanted** **$19.95**

Supertone C Melody Clarinet.

With the C Melody Clarinet you can play all the popular songs written for the piano without transposing the music. Gives practically the same result as the C Melody Saxophone. It is essential for the modern syncopated or jazz music, and yet its tones are soft and mellow for ballads, love songs and music of similar nature.

12L806¼—Albert System. 15 keys, 4 rings, 4 rollers. Made of grenadilla wood in dull finish. Carefully adjusted. Polished nickel silver keys. Rich and powerful in tone. Mouthpiece cap and instruction book included. **Low pitch only.** Shipping weight, 5 pounds **$24.95**

12L814¼—Same as above, but made of ebonite **$24.95**

12L811¼—Same as 12L814¼, but Boehm System, 17 keys, 6 rings **47.95**

Dupont Boehm System Clarinet.

It is the ambition of nearly all clarinet players to own a Boehm System clarinet, but a player of ordinary means is unwilling to pay $65.00 to $90.00, the price which a reliable instrument of this kind is usually sold for. Their desire can now be fulfilled because at our remarkably low price we offer an instrument that will meet all professional requirements. Fitted with nickel silver keys, ferrules and rings. Shipping weight, 5 pounds.

12L809¼—Grenadilla wood, dull finish. 17 keys, 6 rings, key of A or B flat, low pitch only. **State key wanted.** Mouthpiece cap and instruction book included **$47.95**

12L810¼—Same as above, but made of ebonite **$47.95**

Dupont Clarinets.

Albert System. Made of grenadilla wood, which has been oil treated. Dull finish. Keys and trimmings are of nickel silver, highly polished. The action of the keys is perfect, the intervals correct and the tone brilliantly rich. The keys are elegantly shaped and easy to manipulate. Mouthpiece cap and instruction book included. Shipping weight, 5 pounds. **State key wanted.**

12L805¼ — 15 keys, 2 rings, in the keys of A, B flat or C, low pitch **$22.45**

12L812¼ — Same as 12L805¼, but made of ebonite . . **$22.45**

12L807¼ — 15 keys, 4 rings and 4 roller keys, in the keys of A, B flat or C, low pitch **$24.95**

12L813¼ — Same as 12L807¼, but made of ebonite . . **$24.95**

12L830—One Key.

12L833—Six Keys.

Flute. 12L835—Eight Keys. Flute.

12L830—Coco wood, ebony finished. One nickel silver key. Key of C. Tuning slide. Shpg. wt., 1¼ lbs. . . . **$3.95**

12L833—Grenadilla wood. Six nickel silver keys. Tuning slide. Key of C. Shpg. wt., 1½ lbs. **$4.95**

12L835—Grenadilla wood. Eight keys. Nickel silver embouchure or lip plate. Tuning slide. Key of C. Shpg. wt., 1 lb. 9 oz. **$7.45**

Flutes and Piccolos

Our flutes and piccolos are carefully inspected and the keys adjusted before leaving our store. The wooden instruments are made of as thoroughly seasoned material as can be procured and with proper care will not check or split.

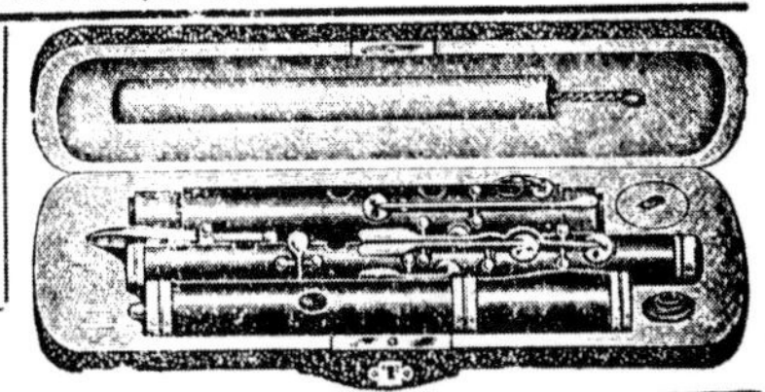

Meyer System Flutes.

Key of C. Low Pitch.

Selected grenadilla wood, cork joints; nickel silver keys, kid pads. Fine lined case, with grease box, screwdriver, swab, pads, lock and key. Shipping weight, 3 pounds.

Cat. No.	No. of Keys	Kind of Head	
12L837	8	Ebonite	**$ 9.95**
12L841	10	Ebonite	**12.45**
12L843	13	Ebonite	**14.95**
12L845	13	Ivory	**24.95**

Piccolos.

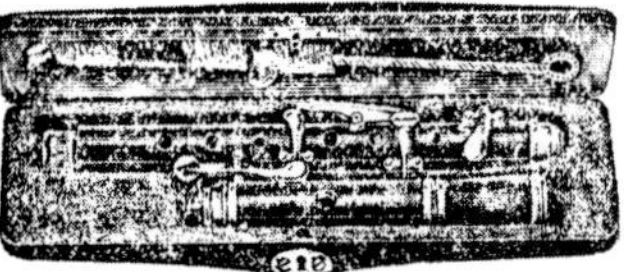

Meyer system, grenadilla wood, hard rubber head, 6 keys, with tuning slide, cork joints and nickel silver trimmed, in velveteen lined cloth covered case, with lock and key. Shipping weight, 15 ounces.

12L858—C. Low pitch, for orchestra.
12L859—D flat. Low pitch, for band. **$3.95**

Grenadilla wood, six keys, with tuning slide, cork joints and nickel silver trimming. Shipping weight, 15 ounces.

12L854—C. Low pitch, for orchestra.
12L857—D flat. Low pitch, for band. **$2.45**

12L850—Coco wood, ebony finished, with one key and no tuning slide. Without cork joints. Key of C. **1.25**

Piccolo Flageolet.

12L870—A combination instrument with two heads. Can be played as a piccolo or flageolet. Grenadilla wood, with six nickel silver keys and tuning slide. Key of C, low pitch. Shipping weight, 15 ounces **$2.95**

FIFES

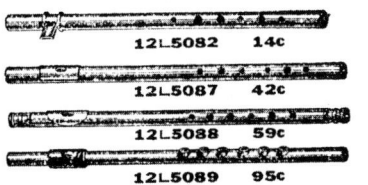

12L5082 14c

12L5087 42c

12L5088 59c

12L5089 95c

12L5091 $1.18

Fifes are made in B flat or C. **State key wanted.** Shipping weight, 9 ounces.

12L5082—Nickel plated metal, with mouthpiece adjusted all ready for playing. Key of B flat or C. **State key** **14c**

12L5087—Brass nickel plated metal, with brass lip plate. B flat or C. **State key** **42c**

12L5088—Heavy nickel plated brass, with brass lip plate and fancy ferrule at ends. B flat or C. **State key** **59c**

12L5089—Nickel plated brass, with raised finger holes and hard rubber lip band. Key of B flat or C. **State key** **95c**

12L5091—Brass, nickel plated, for professional players. Made in two pieces. Key of C or B flat. **State key.** Shipping weight, 10 oz. **$1.18**

Combination B Flat and C Fife.

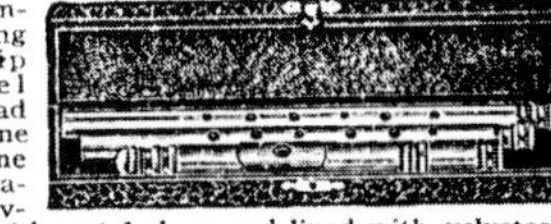

Made of seamless brass tubing with raised lip plate. Nickel plated. One head and two stocks, one for B flat and one for C. Paper imitation leather covered wood case with metal clasp and lined with velveteen. Shipping weight, 1⅜ pounds.

12L5095 . . . . **$2.48**

Fife Mouthpiece.

Composition metal, adjustable 2⅜ to 2⅝ inches in circumference. Shpg. wt., 2 oz.

12L5099 . . . . **9c**

Fife Instruction Book.

Contains simple instructions. Also rudiments of music and chart for fingering, exercises and a number of selections for the fife. Shpg. wt., 2 oz.

12L02011 . . . . **13c**

Multiflute.

Is a combination instrument and it can be played as fife, flageolet or piccolo. Has three detachable mouthpieces. Made of cast metal. Tuned in key of F. Very easy to play. Shipping weight, 1 pound.

12L5174 . . . . **98c**

Chromatic Metal Flageolet.

Made of cast metal. Very easy to blow and to play. Instruction chart included. Shipping weight, 10 oz.

12L5175 . . . . **49c**

ORCHESTRA BELLS.

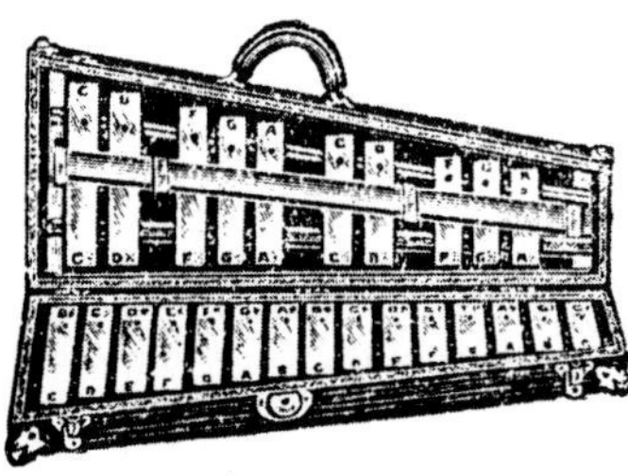

Strictly professional Bells in every sense of the word. Heavy and of superior quality steel, perfectly tuned; loud, penetrating and pure in tone. Nickel plated polished bars, 1¼ inches wide by 5/16 inch thick. Bells are reversible and can be played in either high or low pitch. The case is made of thoroughly seasoned wood with dovetailed corners (no nails used) and is covered with seal grain artificial leather. Nickel plated spring lock, clasps and corner protectors. A feature of this case is the arrangement for holding bars in position, as illustrated. Instruction book and two sets of hammers (one with rubber and one with brass heads) included.

12L5233¼—Thirty-one bars, 2½ octaves, G to C, chromatic. Length, including case, 28 inches. Shipping weight, 34 pounds. . **$24.95**

12L5232¼—Same as 12L5233¼, but with twenty-six bars, 2 octaves, C to C, chromatic. Length, including case, 24 inches. Shipping weight, 30 pounds **$19.95**

These bells are of the same quality steel as 12L5233¼, to the left, but smaller and lighter. The bars are 1 inch wide and ¼ inch thick. Substantial wood case in imitation oak, with dovetailed corners, has nickel plated clasps and name plate. Bells are reversible and can be played in either high or low pitch. Instruction book and two sets of hammers (one with brass heads and one with rubber heads) included.

12L5231¼—Thirty-one nickel plated bars. 2½ octaves, G to C, chromatic. Length, including case, 23½ inches. Shipping wt., 25 lbs. **$14.95**

12L5230¼—Same as 12L5231¼, but with twenty-six nickel plated bars, 2 octaves, C to C, chromatic. Length, including case, 20 in. Shipping weight, 22 pounds **$9.95**

Orchestra Bell Stand.

Made of nickel plated steel with nickel plated brass tubing. Adjustable and folds into small compass. A sturdy and dependable stand. Shipping weight, 3½ pounds.

12L5235
$3.45

Hammers.

Shipping weight, per pair, 4 ounces.

12L5398—For orchestra bells, with brass heads for general playing. Per pair **29c**

12L5400—With wooden heads, for orchestra bells, xylophone or tubephone. Per pair **17c**

12L5399—For orchestra bells with soft rubber heads for practicing. Per pair **23c**

Brass Band

Marceau B Flat Cornet Outfit.

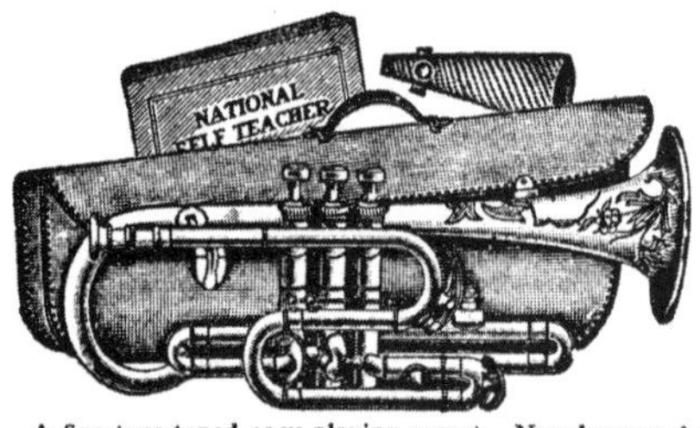

A fine, true toned, easy playing cornet. New long model which is graceful in style and accurate in proportions. Has elaborate engraving on the bell. Mother of pearl finger buttons and two water keys. High and low pitch. 15½ inches long. Comes in canvas covered flannelette lined case with shoulder strap, mute and instruction book. Shipping weight, 8 pounds.

12L735¼—Brass finish only............**$14.95**

B Flat Army Bugle.

$3.45

On account of its convenient size and full, round tone, the U. S. Government has adopted this style of bugle for the army. Made of high grade brass and finished in the new lacquered style called khaki. Shipping weight, 2½ pounds.

12L5069¼............................**$3.45**

Cavalry Trumpet.

U. S. Army Specifications.

Key of G with F slide. Made of fine quality brass. Graceful model. Fine loud tone. Complete with nickel plated mouthpiece. Very popular with Boy Scouts. Weight, boxed, 8 pounds.

12L5070¼—Brass, polished............ **$3.75**

12L5071¼—Nickel plated.............. **4.75**

12L5072½—Silver plated, satin finish..... **6.75**

MARCEAU CORNETS

Long Model B Flat With Quick Change to A.

High and low pitch. No shanks. Fitted with water key. Length, 16 inches. Shipping weight, 7 pounds.

12L705¼—Brass........................**$10.95**
12L706¼—Nickel plated.................. **12.25**
12L707½—Silver plated, satin finish; gold plated bell.................................... **17.95**

Long Model B Flat With Quick Change to A.

Elaborate Engraving—Pearl Buttons.

High and low pitch. No shanks. Fitted with two water keys. Mother of pearl valve buttons. Elaborate engraving on bell. Length, 15½ inches. Shipping weight, 7 pounds.

12L725¼—Brass........................**$16.75**
12L726¼—Nickel plated.................. **17.95**
12L727½—Silver plated, satin finish; gold plated bell.................................... **23.75**

HENRI GAUTIER "VIRTUOSO" CORNETS

New Long American Model B Flat Cornet.

Quick Change to A. Mother of Pearl Buttons. High and Low Pitch.

Adjustment rod on quick change slide. Extra low pitch slide. Engraved wreath on bell. Length, 16 inches. Shipping weight, 7 pounds.

12L751¼—Brass, polished................**$18.75**
12L752¼—Nickel plated, polished.......... **19.95**
12L753½—Silver plated, satin finish; gold plated bell.................................... **26.75**

B Flat Trumpet Model Cornet.

Mother of Pearl Buttons. High and Low Pitch.

Many players prefer this to the regular model B flat cornet. Engraving on bell. Length, 18½ inches. Shipping weight, 8 pounds.

12L741¼—Brass, polished................**$22.45**
12L742¼—Nickel plated, polished.......... **23.75**
12L743½—Silver plated, satin finish; gold plated bell.................................... **30.45**

Right Proportions Long Model Cornet.

Engraving on Bell. High and Low Pitch. Mother of Pearl Buttons.

B flat with quick change to A. Length, 16¼ inches. Complete with low pitch slide, two mouthpieces, music lyre and swab holder. Shipping weight, 8 pounds.

12L744¼—Brass, polished................**$24.75**
12L745¼—Nickel plated, polished.......... **26.05**
12L746½—Silver plated, satin finish; gold plated bell.................................... **32.75**
12L7035—Handsome keratol covered, sateen and plush lined case for above.................. **7.95**

Gautier C Melody or Three-Key Cornet.

In C, B flat and A, high and low pitch. Length, 14¼ inches. Engraving on bell. With it you may play in a band or orchestra or play with the piano or organ without transposing. Especially adapted to church and home playing. Shipping weight, 8 lbs.

12L747¼—Brass, polished................**$24.95**
12L748¼—Nickel plated, polished.......... **26.25**
12L749½—Silver plated, satin finish; gold plated bell.................................... **32.95**
12L7037—Handsome keratol covered, sateen and plush lined case for above.................. **7.95**

The above illustration represents the style of the Alto, Baritone and Bass. For prices see opposite page.

OUR MARCEAU INSTRUMENTS

Are of proper proportions and good models, made of a fine quality brass, have graceful tapering bells and nickel silver piston valves. They are carefully braced and reinforced where necessary. They can be adjusted to either high or low pitch. Mouthpiece, music lyre and instruction book included with all instruments.

We allow a ten days' trial of any instrument you buy, and if not found perfectly satisfactory return the instrument and we will return your money, including transportation charges.

NOTE

SILVER PLATED INSTRUMENTS are not carried in stock. They are specially plated upon receipt of order, thus insuring you a bright, clean horn. On all orders for silver plated instruments allow ten to twelve days' time for shipment to reach you.

All weights given on this page are approximate and may vary a trifle.

Instruments

The above illustration represents the style of the Alto, Baritone and Bass.

HENRI GAUTIER

Instruments are the result of years of patient study and repeated tests by experienced workmen in a factory well equipped with modern machinery. They are built to stand the exacting requirements of professional musicians and are perfect in tune, tone and intonation. Also graceful in model and of accurate proportions. If you are a member of a band or an orchestra, or if you are a soloist and want an instrument that will meet every demand made upon it, order a Henri Gautier. You will be satisfied with it. If they do not prove up to the standard which you require, return them at our expense and we will return your money, including transportation charges.

NOTE

SILVER PLATED INSTRUMENTS are not carried in stock. They are specially plated upon receipt of order, thus insuring you a bright, clean horn. On all orders for silver plated instruments allow ten to twelve days' time for shipment to reach you.

SLIDE TROMBONES

Marceau.

For Bags and Cases to fit our Slide Trombones see page 523.

Length, 44 inches. Bell, 7 inches. Shipping weight, 12 pounds.

12L765¼—Brass........ **$10.45**
12L766¼—Nickel plated........ **11.95**
12L767½—Silver plated, satin finish, gold plated bell........ **18.95**

Henri Gautier "Virtuoso."

Total length, about 44 inches. Bell, 7 inches. Shipping weight, 12 lbs.

12L781¼—Brass, polished........ **$19.45**
12L782¼—Nickel plated, polished........ **20.95**
12L783½—Silver plated, satin or dull finish, with gold plated bell........ **27.95**

For case to fit order 12L7002¼ or 12L7012¼, page 523.

VALVE TROMBONES

Marceau.

Preferred by band men to the upright tenor, which is practically obsolete.

Length, 41½ inches. Bell, 7 inches. Shipping weight, 15 pounds.

12L768¼—Brass........ **$19.95**
12L769¼—Nickel plated........ **22.75**

Henri Gautier "Virtuoso."

Length, about 42 inches. Bell, 7 inches. Shipping weight, 15 pounds.

12L791¼—Brass, polished........ **$34.25**
12L792¼—Nickel plated, polished........ **37.25**

MARCEAU UPRIGHT HORNS

Catalog No.	Instrument	Brass Polished	Nickel Plated	Length, Inches	Diameter of Bell, In.	Shpg. Wt., Lbs.
12L736¼	E flat Alto	$17.45	$19.45	21	8	13
12L755½	B flat Baritone	23.45	26.95	23	9½	27
12L758½	E flat Bass	39.75	44.75	28	11½	38

HENRI GAUTIER "VIRTUOSO" UPRIGHT HORNS

Catalog No.	Instrument	Brass Polished	Nickel Plated	Length, Inches	Diameter of Bell, In.	Shpg. Wt., Lbs.
12L760½	E flat Alto	$29.95	$32.25	20	9	15
12L777½	B flat Baritone	37.95	41.75	24	11	25
12L780½	E flat Bass (Large)	77.45	83.25	30	16	45

For Bags, Cases and Supplies for Brass Instruments See Page 523.

All weights given on this page are approximate and may vary a trifle.

(Our Own Trade Mark, Copyrighted U. S. Pat. Office.)

C Melody Saxophone

For the Home, Band or Orchestra.

$69.75

Perhaps the most popular of all instruments today is the C Melody Saxophone. This is due to the fact that it can be used as a home instrument for playing the melody parts of songs with the piano accompanying it, there being no necessity for transposing the music. It can be used in band and orchestra, taking the oboe and 'cello parts. Contrary to the general belief, the saxophone is an instrument very easy to master, it being arranged so that the keys come in easy reach of the fingers. Our Supertone Saxophone is fitted with modern improvements such as the automatic octave key, roller keys, etc. It is made of high grade brass. The key system is very accurately and securely situated. Handsome engraving on bell. Comes fitted with a hard rubber mouthpiece, nickel plated reed holder and mouthpiece cap. Music lyre. Braided strap with metal snap. We also include an illustrated saxophone chart. **Low pitch only.** Shipping weight, 20 pounds.

12L895¼—Brass........ **$69.75**
12L896½—Silver plated, satin finish with gold plated bell. Burnished keys and engraving........ **$89.45**
12L897½—Silver plated, satin finish, gold plated bell, keys and engraving gold plated, burnished **$113.25**

Orchestra and Street Snare Drums

SUPERTONE
TRADE MARK REG. U.S. PATENT OFFICE

Junior Snare Drum. Not a toy, but a real drum for the youngsters. Nickel plated shell, size, 3x12 inches. Ebony finish hoops with inlaid metal bands, six thumbscrew rods, adjustable snare strainer. Two calfskin heads, four snares. Includes webbed sling with nickel plated snap, a pair of sticks and instruction book. Shipping weight, 8 pounds.
12L901.......... $4.75

All weights given on this page are approximate and may vary a trifle.

Tango Banjo (Jazz) Orchestra Snare Drum. Twelve-inch two-ply maple shell, finished in the natural color, 2 inches high. Twenty-two brackets, fine calfskin head, nickel plated adjustable snare strainer and twelve braided snares. Hickory sticks and instruction book included. A snappy drum. Shipping weight, 6 lbs.
12L910¼........ $7.95

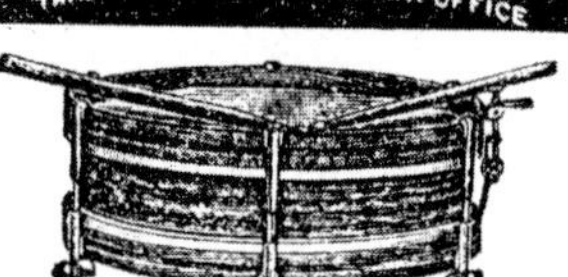

Dance Orchestra Snare Drum. Solid mahogany shell, size, 3x13 in. Natural maple hoops inlaid with nickel plated metal band. Six nickel plated thumbscrew rods. Eight wire and silk snares, nickel plated snare strainer with patent snare release, enabling the player to instantly change to a tom-tom effect. Fine calfskin heads. Very sharp and snappy drum. Selected hickory sticks and instruction book included. Shipping weight, 9 pounds.
12L905¼.............. $9.95

New Style Tango Snare Drum. Very sensitive, brilliant and snappy. Made with two metal counter hoops, the rods passing through the lower and fastening to the upper hoop, enabling the performer to tighten heads evenly and quickly. Finely finished throughout. Fine hickory sticks and instruction book included. Thirteen-inch maple shell. Three inches high. Six thumbscrew rods, ten wire snares, calfskin heads. Shipping weight, 6 lbs.
12L906¼ $11.45

Orchestra Snare Drum. Just the drum with enough snap and volume for all around work. Solid mahogany shell, size, 4x15 inches. Natural maple hoops inlaid with nickel plated metal band. Eight nickel plated thumbscrew rods. Ten wire and silk snares, nickel plated snare strainer with patent snare release, enabling the player to instantly change to a tom-tom effect. Fine calfskin heads. Selected hickory sticks and instruction book included. Shpg. wt., 9 lbs.
12L907¼ $12.45

BASS DRUMS

Rod Pattern. Regulation Sizes.

Mahogany shell, 10x24 inches, varnished finish. Natural maple hoops, ten nickel plated thumbscrew rods with strong center support. (No key or wrench required.) Two fine white calfskin heads. Well finished and well built in every way. A drum with a big tone. Stick, sling and instruction book included. Shpg. wt., 50 pounds.
12L946¼ $19.45

Same as above, but with shell size 12x28 in., and twelve nickel plated rods. Shipping weight, 50 lbs.
12L947¼.............. $24.95

Same as above, but with shell size 14x30 inches. Thirteen rods. Shipping weight, 50 pounds.
12L948¼.............. $29.75

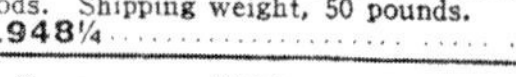

Street or Military Snare Drum.

Instruction Book, Sling and Sticks Included.

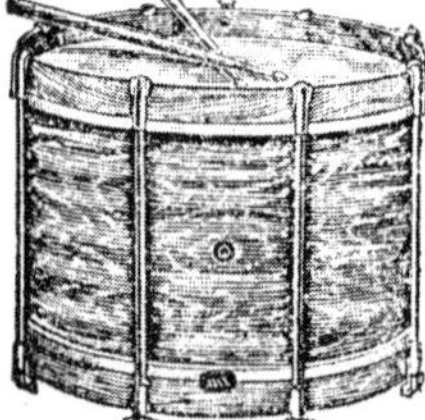

Fifteen-inch mahogany shell, 9½ inches high, about 11½ inches including hoops. Eight nickel plated thumbscrew rods (no key or wrench required.) Calfskin heads. Eight woven snares. Nickel plated snare strainer. The drum for fife and drum corps. Shipping weight, 15 pounds.
12L917¼.............. $14.95

A Drum for Professionals

Patent Snare Release.

Heads Tightened Separately.

All Metal Separate Tension Snare Drum. Not to Be Confused With Cheap Drums of This Type. The separate tension rods enable the drummer to adjust each head separately to any desired tension. With the patent snare release he can change to the tom-tom effect instantaneously. Nickel plated heavy shell of spun brass with center reinforcement. Very sensitive transparent calfskin heads. Wrench, fine hickory sticks and instruction book included. Size of shell, 4x14 inches (about 5½x14 inches, including hoops). The popular size. Shipping weight, 12 pounds.
12L920¼ $19.95

JUNIOR DRUM OUTFIT

$19.95

It is the ambition of nearly every youngster to become a drummer, but the price of the regular outfit is too high for the average lad to pay. To overcome this, we are listing a complete outfit at a very low price. The outfit consists of: One junior snare drum, nickel plated shell, 3x12 inches, six thumbscrew rods, two calfskin heads, hickory sticks; one single head bass drum, 6½x22-inch maple shell, six thumbscrew rods; one nickel plated folding snare drum stand; one 10-inch brass cymbal; one 8-inch brass cymbal; cymbal clamp; one cymbal arm. Shipping weight, 55 pounds.
12L957¼ $19.95

TANGO BASS DRUMS

Dance Orchestra (or Jazz) Bass Drum.

Twenty-four inch mahogany shell, 6 inches high (8 inches high including hoops), natural finish maple hoops, eight nickel plated thumbscrew rods (no wrench or key required), and two fine quality calfskin heads. Just the thing for jazz bands. As the drum is used with pedal beater no stick or sling is included. Instruction book included. See page 523 for pedal beater. Shipping weight, 48 pounds.
12L944¼........ $14.95

Tambourines.

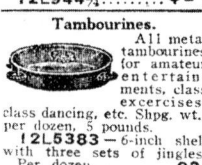

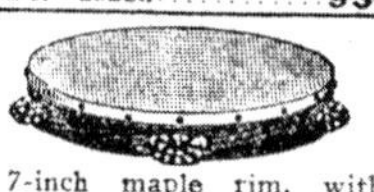

All metal tambourines for amateur entertainments, class excercises, class dancing, etc. Shpg. wt., per dozen, 5 pounds.
12L5383—6-inch shell with three sets of jingles. Per dozen.......... 93c

7-inch maple rim, with tacked head and three sets of jingles. Shpg. wt., 8 oz.
12L5384 .... 69c

Maple rim, 8-inch tacked calfskin head, nine sets of jingles. Shpg. wt., 12 oz.
12L5387 .... 89c

Maple rim, 10-in. tacked calfskin head, twelve sets of jingles. Shipping wt., 1 lb.
12L5388 ...... $1.18

Salvation Army Tambourine.

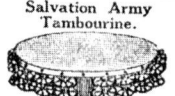

10-inch maple hoop, twenty-eight sets of metal jingles, calfskin head fastened with brass tacks. Shipping weight, 1⅝ lbs.
12L5390 $1.89

Drummers' Tambourine.

10-inch reinforced shell, veneered with highly polished birdseye maple, twelve sets of heavy nickel silver jingles. Transparent head waterproofed by a special process. Shipping weight, 1¼ pounds.
12L5391 $2.75

Snare Drum Case. Wooden frame covered with hard fiber board, reinforced with metal corners. Brass lock and clasps, hand fitting handle. Fiber board cover for drum. A case that will last and give service. For drums up to 18 in. in diameter and 6 in. high. Shpg. wt., 10 lbs.
12L9035 $4.95

State size of shell not including hoops. Rubberized cloth bound edges. Handle, pocket and patent clasp.

Bags, Bass Drum.

	Diameter	Height	Price
12L9001	24 in.	10 in.	$3.25
12L9003	28 in.	12 in.	3.95
12L9005	30 in.	12 in.	4.25

Shipping weight, 2 pounds 1 ounce.

Bags, Snare Drum.

	Diameter	Height	Price
12L9007	15 in.	4 in.	$1.45
12L9009	16 in.	4 in.	1.49
12L9011	16 in.	6 in.	1.59

Shipping weight, 1 pound.

Adjustable Folding Drum Stand.

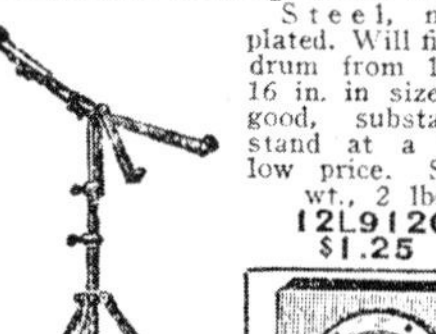

Steel, nickel plated. Will fit any drum from 14 to 16 in. in size. A good, substantial stand at a very low price. Shpg. wt., 2 lbs.
12L9120 $1.25

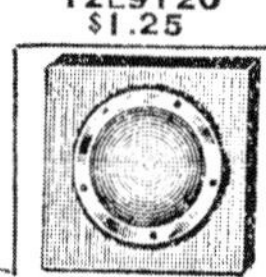

Practice Pad. Hardwood frame. Fiber covered pad held by a nickel plated steel ring. Very responsive and can be heard only a few yards away. Shipping weight, 1½ pounds.
12L9088 .... 85c

The Song Whistle.

An instrument that has become very popular. The tone of the lower octaves is somewhat similar to the human voice. Can be played with your talking machine, the piano, with voices, jazz bands or even as a solo instrument. Makes very delightful music, and it is easy to learn to play. The tones are produced by a sliding rod. It is made of heavy brass, highly nickel plated and polished. The slide is self-lubricating. Shpg. wt., 15 oz.
12L5368 $1.35

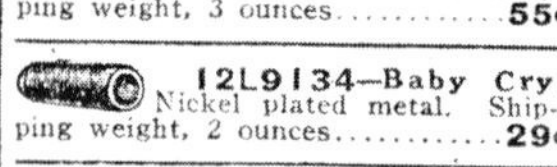

12L9145—Bird Whistle. Nickel plated metal. Shipping weight, 3 ounces.......... 55c

12L9134—Baby Cry. Nickel plated metal. Shipping weight, 2 ounces.......... 29c

12L9158 Cuckoo. Nickel plated metal. Shpg. wt., 4 oz..... 59c

12L9171—Locomotive Whistle. Nickel plated metal. Shipping weight, 1¼ lbs.......... 98c

12L9147 Calf Bawl. Nickel plated metal. Shipping weight, 7 ounces.......... 49c

Cymbals (Brass).

Twelve-inch is the standard size of cymbals. Smaller ones do not give desirable effects, as the tone is shallow and the vibrations are not lasting. Shipping weight, 3 pounds 7 ounces.
12L9041—12-in. Per pair $2.45

Cymbol, Chinese Crash.

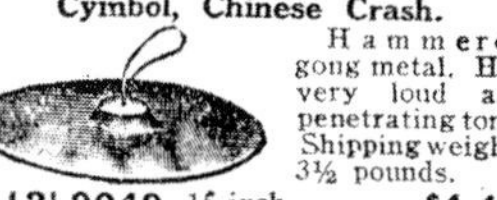

Hammered gong metal. Has very loud and penetrating tone. Shipping weight, 3½ pounds.

12L9049—15-inch $4.45
12L9047—13-inch 2.65
12L9048—14-inch 3.65

12L9182—Sleigh Bells. 13 bells on handle. Shipping weight, 1½ pounds.......... $1.65

Chinese Wood Block Drum.
12L9152
Chinese redwood, which gives a very loud tone. Shipping wt., 1⅛ lbs. Without holder..... 79c
12L9153—Nickel Plated Holder for same. Shpg. wt., 3 oz. 39c

Two-Tone Wood Block. Resonant. Wood cylinder body. Two different tones. Nickel plated adjustable holder with strong clamp. Shipping weight, 2½ pounds.
12L9156 .... $2.25

Special Professional Triangles. Extra heavy tool steel of a remarkably loud, penetrating sound. Vibration is lasting and clear. Complete with hammer.
12L9193—6½-inch. Shipping wt., 1¼ lbs... 63c
12L9194—8-inch. Shipping weight, 1½ pounds.......... 72c
12L9195—10-inch. Shipping weight, 2 pounds.......... 89c

Triangle and Cymbal Holder. Made of nickel plated metal. Rubber covered triangle holder. Cymbal holder for top of drum. Sure clamp. Shipping wt., 6 oz.
12L9198 .... 59c

Chinese Drum or Tom-Tom.

12L9151—Shell of a special composition with skin heads decorated with a Chinese design. 10-in. diam. Shpg. wt., 2½ lbs. $2.25
12L9150—Holder for Chinese Drum. Nickel plated metal. Two parts. Shpg. wt., 12 oz. 63c

"Jazerup" Bells. 12L9172
Bronze finish metal. Four tones tuned in chime effect. With nickel plated holder. Shipping weight, 3 pounds. Without holder.......... $3.29

12L9177 Rattle. Steel frame. Shipping wt., 18 oz. .. $1.98

12L9155 Cow Bell. Shipping wt., 15 oz... 39c

Cornet Cases.

Covered with seal grain keratol (artificial leather). Lined with velveteen. Trimmings, protectors, lock and clasps are nickel plated and highly burnished. Spring jaw for holding the cornet. For cornets, 14 to 16 in. long. Shipping weight, 6 pounds.
12L7031 .... $7.25

12L7021—Strawboard, canvas covered, flannelette lined, with shoulder strap. For cornets up to 17 inches long. Shipping wt., 2 lbs. $1.39

12L7022—Split leather sides and top, flannelette lined. Shoulder strap. For cornets up to 17 in. long. Shpg. wt., 2¾ lbs. $2.39

12L7026—Same as 12L7022, but 21 inches long, to fit trumpet cornet. Shipping weight, 3⅝ pounds. $2.75

Green Felt Bags for Band Instruments.

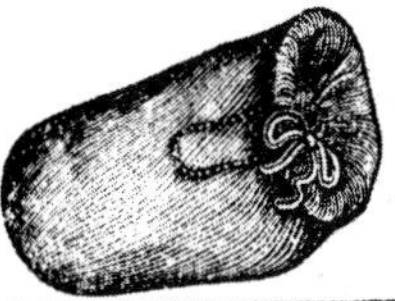

High grade, close texture green felt. Each bag is fitted with a pocket for holding mouthpiece, music lyre and other parts. **State diameter of bell and height of instrument** for which bag is wanted.

Catalog No.	Instrument		Shpg. Wt.
12L7100	For Cornet	$0.59	4 oz.
12L7101	For Alto	.75	5 oz.
12L7104	For Tenor	.98	6 oz.
12L7106	For Baritone	1.17	7 oz.
12L7108	For E Flat Bass	1.38	12 oz.
12L7110	For Slide Trombone	1.10	9 oz.
12L7111	For Tenor Valve Trombone	1.10	9 oz.

Slide Trombone Cases.

Made Only for Our Instruments. Give Catalog Number.

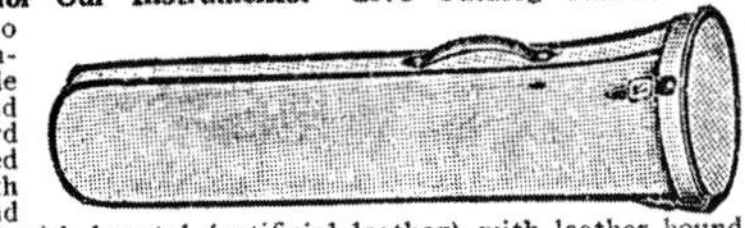

We furnish two qualities of trombone cases, made with heavy and strong strawboard bodies, one covered with canvas with bound edges and the other covered with keratol (artificial leather) with leather bound edges. They are flannelette lined and are fitted with artificial leather handles. Nickel plated buckles. End openings. We guarantee our cases will retain their shape and protect your instrument. Shipping weight, 7 pounds.

12L7002¼—Canvas $3.45

12L7012¼—Same as 12L7002¼, but covered with artificial leather 4.75

Slide Trombone Cases. Side Opening.

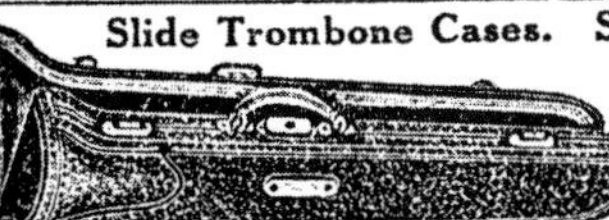

Veneer frame covered with a fine artificial leather of seal grain, lined with velveteen. It is fitted with brass nickel plated lock, patent clasps and trimmings. Shipping weight, 8 pounds.

12L7039¼—Tenor Slide Trombone Case $11.45

Ideal B Flat Cornet Mouthpiece. Brass, silver plated. Wide rim. Invaluable for long marches and severe playing. Shpg. wt., 3 oz.
12L7050 .... 67c

Cornet Mutes.

New Style Mute, made of specially treated material finished in gilt. Shipping weight, 4 oz.
12L7055 .... 33c

Regular model, cork holders, brass, nickel plated. Shipping weight, 5 oz.
12L7056 .... 87c

Music Racks for Band Instruments.

Shipping Wt., 5 Ounces	Catalog No.	Brass	Nickel Plated
For Cornets, All Upright Horns and Saxophones	12L7070	23c	33c
For Clarinet	12L7074	63c	73c
For Slide Trombone	12L7080	48c	58c

Saxophone Cases.

12L8950¼—Saxophone shape. Made of keratol (artificial leather) in imitation of seal grain. Velveteen lining. Brass nickel plated lock and clasps. Has compartment for mouthpiece, mouthpiece joint and music holder. Shipping weight, 10 pounds $11.45

12L8951—Oblong shaped. Lined with flannelette. Made of cheaper construction and trimmings than the above. Shipping weight, 10 pounds $6.95

Saxophone Reeds.

12L8960—Fine quality. For C Melody Saxophone only. Shpg. wt., 3 oz. One-half dozen 95c

Saxophone Mouthpieces.

12L8956—Solid rubber with metal band. Shipping weight, 5 ounces $2.95

12L8957—Nickel silver, heavily silver plated. Reed holder included. Shipping weight, 8 ounces $5.95

Saxophone Mouthpiece Cap.

12L8959—Brass, nickel plated. Shpg. wt., 4 oz. 59c

Saxophone Reed Holders.

12L8961—Brass, nickel plated, with adjusting screws. Shpg. wt., 3 oz. 37c

Saxophone Reed Case.

12L8963—Covered with artificial leather. Lined. Glass plate will hold six reeds. Shipping weight, 6 ounces 65c

Slide Trombone Position Indicator.

12L7065—With this attachment, and in conjunction with the book of instructions included, one can learn the positions easily and quickly. Shpg. wt., 14 oz. $1.45

Clarinet Reeds.

Each reed is sterilized and waterproofed and enclosed in an envelope. A, B flat or C. Shpg. wt., 3 oz.

12L8029—Per doz. reeds, packed in metal box $1.75

Clarinet Reeds.

12L8027

NOTE—B flat reeds are used on A, B flat and C clarinets. Reeds shipped on metal plates. Shipping weight, 8 ounces.

12L8021—Marceau. Each 3c
12L8023—Lafayette. Each 6c
12L8024—Dupont Superior Quality. Each 9c
12L8025—Carl Schubert Waterproof. Each 13c
12L8027—Carl Schubert "Artist." Each 16c

Clarinet Cases

Buffalo grained split leather; flannelette lined; metal catch. Opens at end. For clarinets in A, B flat and C. Shipping weight, 1¼ pounds.
12L8001 .... $1.95

Covered with black keratol (artificial leather). Trimmings are nickel plated. The inside is lined with flannelette. Will hold two clarinets. Shipping weight, 6 pounds.
12L8005 .... $5.75

For one clarinet. Body covered with keratol (artificial leather) and lined with flannelette. Nickel plated lock, hinges and spring clasps. Shipping wt., 3 lbs.
12L8002 .... $4.75

Lined. Keratol (artificial leather) covered. Glass plate with elastic band for holding reeds. Nickel plated spring clasp. Will hold 8 reeds. Shpg. wt., 4 oz.
12L8031 .... 55c

Clarinet Mouthpieces.

12L8014 Grenadilla wood with cork joint. Fine shape. A, B flat or C. **State key.** Shpg. wt., 4 oz. 12L8014 .... 73c

Genuine hard rubber with cork joint. Fine shape and lay. A, B flat or C. **State key.** Shpg. wt., 4 oz.
12L8015 .... $1.75

Mouthpiece Cap.

Nickel plated, for A, B flat or C mouthpiece. **State key wanted.** Shipping weight, 3 ounces.
12L8016 .... 18c

Clarinet, Flute and Piccolo Cleaners.

12L8010

Shipping weight, 2 ounces.
12L8010—Clarinet or Flute Cleaner 15c
12L8048—Piccolo Cleaner 6c

Reed Holder.

Nickel silver with adjustable screws, for A, B flat or C. **State key.** Shpg. wt., 3 oz.
12L8037 .... 17c

Clarinet Adjustable Tuning Joint.

Hard rubber. Screw adjustment enables player to lengthen the clarinet about half a tone. **For B flat clarinets only. Give inside diameter of lower end of barrel joint.** Fine for band and orchestra players. Shipping weight, 4 oz.
12L8039 .... $3.45

Clarinet Key Pads.

Fine quality; to fit clarinet. Fifteen pads to set. Not illustrated. Shipping wt., 2 oz. 12L8012—Per set 32c

Drum and Cymbal Beaters.

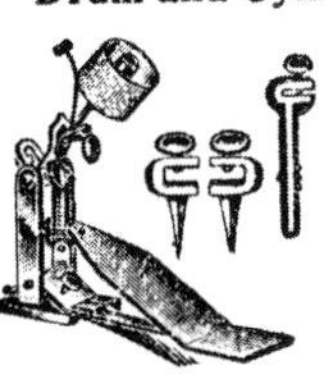

12L9018

Adjustable. Quick, sure action. Set screw, enabling player to adjust spring. Adjustable cymbal striker and beater. Plated metal with fiber footboard. Cymbal arm and spurs. Folds into space 3x4x12 inches. Shpg. wt., 3 lbs. $3.25

Fraser "Direct Stroke."

12L9020

Made entirely of metal. Easy and quick of operation. No side strain or lost motion. Complete with felt head, spurs and cymbal holder. **Front action.** Shpg. wt., 3¼ lbs. $6.95

Crash Cymbal Holder.

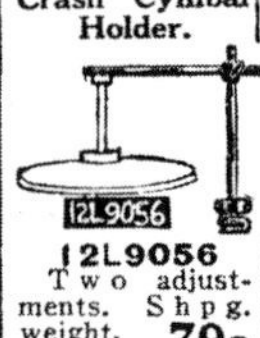

12L9056

Two adjustments. Shpg. weight, 11 ounces. 79c
Cymbal not included.

Cymbal Arm.

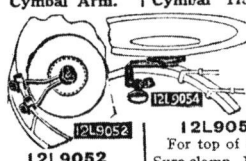

12L9052

For side of drum. Sure clamp. Shpg. wt., 10 oz. 65c
Cymbals not included.

Cymbal Holder.

12L9054

For top of drum. Sure clamp. Plated metal. Shpg. wt., 4 oz. 39c

Triangle and Cymbal Holder.

12L9198

Rubber covered triangle holder. Cymbal holder for top of drum. Sure clamp. Shpg. wt., 6 oz 59c

Drum Spurs.

12L9114

Thumbscrew to fasten to hoop. Shipping wt., 6 ounces. Per pair 43c

Drum Heads.

Soak in lukewarm water until soft before placing on instrument.

No.	Size	For	
12L9072	18	15-in. drum	$1.39
12L9073	19	16-in. drum	1.59
12L9074	20	17-in. drum	1.85
12L9076	28	24-in. bass drum	3.95
12L9078	32	28-in. bass drum	5.25
12L9079	34	30-in. bass drum	6.25

Shipping wt., 18 to 20-inch head, 8 oz.
Shipping wt., 28 to 34-inch head, 1¼ lbs.

Rods.

Thumbscrew style. Light and strong. Nickel plated metal.

	For	Shpg. Wt.	
12L9090	4-in. drum	4 oz.	33c
12L9091	6-in. drum	5 oz.	35c
12L9092	8-in. drum	6 oz.	39c
12L9093	12-in. drum	7 oz.	45c
12L9094	14-in. drum	8 oz.	49c

Snares.

Shipping weight, 3 oz.

12L9105—Twelve braided and waterproofed snares with fiber holder 34c
12L9106—Same as above, but wound with wire 68c
12L9107—Twelve closely wound wire snares with fiber holder 59c

Head Tucker.

12L9085—For putting new heads on drums. Shipping weight, 3 ounces 19c

Sticks, Snare Drum.

Shipping weight, 11 ounces.
12L9130—15½-inch hickory. Orchestra size. Per pair 33c
12L9133—15½-inch. Turtle ebony, for orchestra. Fine shape. Per pair 79c

Sticks, Bass Drum.

12L9125—Hardwood stick. Shipping weight, 10 ounces $0.75
12L9126—Felt head, hardwood stick. Extra fine quality. Shipping weight, 12 ounces 1.25

Combination Piano and Player Piano Bench.

Strongly built and beautifully finished. Broad base. The top is veneered with genuine wood according to finish, 15x36 inches, is hand rubbed and polished. Can be used as ordinary bench or slanted for use as player piano bench. Large music compartment. Height, 20 inches. Shipping weight, 50 pounds. **Shipped from factory in INDIANA.**

Square Tapering Legs.

12L5239⅓—Mahogany finish $9.95
12L5241⅓—Solid oak 9.95
12L5243⅓—Walnut finish 9.95

Round Legs.

12L5236⅓—Mahogany finish $9.90
12L5237⅓—Solid oak 9.90
12L5238⅓—Walnut finish 9.90

Piano Bench Cushion.

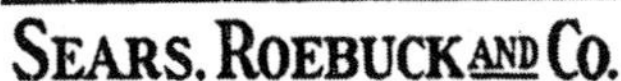

Covered with flowered velour, lined with a heavy felt and interlined with genuine hair cloth. Three straps across the bottom to go over the cover of the bench which keeps it from slipping off when the cover is opened. Very serviceable and a great protection for your piano bench. Comes in either **blue or mulberry. State color.** Shipping weight, 2½ pounds.
12L5242 .... $4.45

Metronomes (Maelzel System).

Used by students of music, especially of the piano, to indicate the tempo or time. The time is indicated both to eye and ear, the movement being in sight and ticking similar to a clock, but much louder. They are made of wood, finished in mahogany. The mechanism is similar to clockwork, is carefully made and as accurately adjusted. All are examined and tested before being shipped. Instructions with each. Shipping weight, 2½ pounds.

12L5165—Pyramid style. Good reliable movement $3.75

12L5166—Same as 12L5165, but with bell attachment striking on the first beat of every measure $4.95

The Famous Beckwith Organs

Noted for Their True Pipe Organ Tones
Lowest Prices ~ Easy Payments

Style F.

Parlor Organs

Once more we are able to offer our customers a complete line of Beckwith organs, the instrument which received the highest award at the St. Louis World's Fair, and which for over 30 years has been considered one of the finest and sweetest tone organs that money could buy. And today, with its improved pipe organ tone, the Beckwith is unquestionably superior to any reed organ on the market, regardless of name, make or price. When you buy a Beckwith, you are getting the very finest, the most up to date, and absolutely the strongest reed organ to be had. We are glad to offer an organ of such fine quality to our customers, for we know they will give continual satisfaction.

We illustrate two very popular styles of parlor organs. Each represents the greatest possible value for the price. These organs are made of selected oak, finished a light golden color. Both are trimmed with daintily executed carvings, the one shown at the right being more ornate having three bevel plate mirrors and besides is fitted with our **grand orchestral action.**

Beckwith Organs are guaranteed for 25 years against defects in material or workmanship.

Size: 82 inches high, 24 inches deep. The 5-octave cases are 46 inches wide and the 6-octave cases, 52½ inches wide. Both organs are fitted with knee swells and the latest type stops. **Price includes fine stool and instruction book.**

Shipped only from factory in KENTUCKY. Shipping weight, Style F, 350 pounds; Style H, 400 pounds

	Reeds	Stops	Octaves	
Style F				
46L585	122	11	5	**$ 98.00**
46L685	146	11	6	**109.00**
Style H				
46L392	122	11	5	**$119.00**
46L492	244	17	5	**139.00**
46L692	146	11	6	**127.00**
46L792	292	17	6	**148.00**

Price payable $10.00 with order, balance $5.00 a month. No interest or extras to be added. Use Order Blank enclosed.

Style H.

Small Monthly Payments

The Organs on this page are sold on payments of $5.00 monthly. Select the one wanted, fill in and mail time payment order blank enclosed and include a deposit of $10.00.

30 Days' Trial

We will ship the organ for 30 days' trial in your home. At the end of this time, if satisfied, send us your first monthly payment of $5.00; balance at the rate of $5.00 per month. The $10.00 deposit will then be applied on your account. If at the end of 30 days' trial for any reason you are not satisfied, we will take the organ back and immediately return your deposit.

Freight Prepaid

We will prepay the freight charges and add the amount to the price. In this way it will cost you nothing to get the organ from your local freight depot.

Piano Organs

This is the finest Beckwith Organ action in a beautiful piano case. It closely resembles a piano in size, design and finish. It is easy to play, as it has 7⅓ octaves the same as a piano. The case is veneered in mahogany, hand rubbed and polished.

We furnish this high grade instrument in two different actions. 46L1170 contains 35 Diapason, 53 Melodia, 35 Viola and 53 Celeste reeds. **46L1172** is the same with the addition of 35 Harp Aeolian and 53 Cello reeds.

The two outer pedals are used to pump the organ, while the middle pedal controls the swells, giving easy and instant control over the volume of tone. An added feature is the octave coupler which couples two keys together so that both will play by pressing only one of them.

Size: 4 feet 8 inches high, 5 feet 2 inches long, 2 feet 2 inches wide. Weight, boxed for shipment, 450 pounds. **Shipped from factory in KENTUCKY.**

Furnished in mahogany only. Order blank enclosed.

46L1170—176 Reeds, 7⅓ Octaves. Payable at $5.00 a month................ **$149.00**

46L1172—264 Reeds, 7⅓ Octaves. Payable at $5.00 a month................ **159.00**

A fine stool and instruction book are furnished with each instrument without extra charge.

Church Organs

This Organ was designed particularly for small churches or chapels requiring a sweet toned organ at a low price. The case is very neat and attractive, durably constructed of selected oak and nicely finished in a light golden color. As the workmanship is of the best and all of the materials are of high quality, you can order this instrument with a full assurance that it will give the service and satisfaction you have a right to expect.

An organ for church use should have a full, rich and resonant tone and be susceptible of the most delicate variations. Therefore, we use an action having reeds which are specially voiced for choir and solo work. The organ illustrated is fitted with two knee swells, the grand organ and the swell organ, and will permit of delightfully soft music as well as a great volume of tone when the full organ is played. The bellows are extra large, but easy to pedal.

For more impressive and larger church organs write for prices and descriptions.

Made in one size only. 52 inches high, 24 inches deep and 46 inches wide. **Shipped from factory in KENTUCKY.** Shipping weight, 275 pounds. Sold on easy payments of $5.00 a month and on our 30-day trial offer, as explained above. **Use Order Blank enclosed.**

46L590—122 Reeds, 11 Stops, 5 Octaves.................................. **$89.00**

Price Includes Fine Stool and Instruction Book.

Every Home should have a Piano

Beckwith Instruments Offer Biggest Values at Prices You Can Afford

Play as You Pay

Small Monthly Payments

30 Days Trial

When you buy a Beckwith you are getting a high grade instrument, guaranteed to give lasting satisfaction and made and sold in such a way you pay the lowest possible price for it.

Every Beckwith Is an Instrument of Fine Quality

THERE is a great variation in the quality of pianos and player pianos, as in most articles which you use and wear, but in none is durability so important as in a piano. **It represents a lifetime investment and should be built accordingly.**

Quality is built into every Beckwith piano and player piano from the ground up. Every part is carefully made of first class material by experienced piano makers who have devoted their lives to making high quality pianos. Only those principles of construction are used which our long experience in building fine pianos has taught us to be the best and most satisfactory.

We could fill pages with a technical description of Beckwith pianos, but, unless you are experienced in piano construction, it would be of little interest to you. The illustrations at the left show three of the most important features of Beckwith pianos. Without them, no piano would be considered a strictly first grade instrument. We also use a pin block made of crossbanded rock maple; hammers and dampers which contain only long fiber felt; copper wound bass strings, ivory keys and other refinements, all of which combine to make Beckwith one of the **leading quality pianos of the day.**

Sounding board made of mountain spruce.

Scale showing full length metal plate.

Showing the substantial back construction of the Beckwith.

Our Manufacturing and Selling Plan Insures Big Savings

The high quality of Beckwith pianos should make them expensive, and they would be if it were not for our economical manufacturing and selling methods: We make Beckwith pianos in our own factory, in which we have installed the very latest labor saving devices in order to reduce the cost of the different operations. **We make thousands of pianos yearly,** which enables us to buy raw materials in large quantities and thus get rock bottom prices. Instead of selling through dealers or middlemen, we sell from factory **direct to your home.** By keeping down the cost of manufacturing on the one hand and selling expense on the other we are able to make big savings for purchasers of Beckwith pianos. You pay for quality and quality only when you buy from us; not one cent for the upkeep of expensive showrooms, dealers' and salesmen's profits and commissions and other expenses. **All the quality is built into a piano at the factory.** Nothing can be added later except expense, and this would have to be included in the selling price. When you buy a Beckwith you eliminate unnecessary expense and get an instrument which is technically and musically perfect, beautiful in design and finish, vibrant and sweet in tone, and **sold at the lowest possible price** at which anyone can furnish you an instrument of similar quality.

It Pays to Buy From a Reliable House

There are but few people who really know what constitutes a fine piano. As it represents a lifetime purchase for the average home at a comparatively big investment, everyone, unless he thoroughly understands piano construction, should **choose the firm he buys from** as carefully as he does the piano. **In fact, the selection of a reliable firm is probably of more importance than the selection of the instrument itself.**

You can safely put your piano purchase in our hands. Our nation wide reputation for fair dealing, the result of over 30 years of honest merchandising, is your assurance that we will live up to our word and that the instrument you buy will give you the satisfaction and service you have a right to expect.

A Guarantee That Really Guarantees

A guarantee is composed of so many words, of a certain meaning. Of themselves they insure nothing. They find their true value only when the integrity, strength and reputation of the firm or individual responsible for the guarantee are considered.

If you want real protection it is well to consider the financial responsibility, the past business history and the honesty of the firm you are dealing with. When you buy a guaranteed article from us, be it a garment or a piano, you know that the guarantee will be lived up to. Our reputation for square dealing is so firmly established that today more than six million families supply their daily needs from the pages of our catalogs, buying on the strength of our guarantee, secure in their knowledge that every transaction must be made satisfactory before we consider it closed. When you buy a Beckwith, you buy it under a **25-year guarantee** against defects in material or manufacture, **a guarantee which really protects you.** Behind the guarantee, we stand with our reputation, willing to back it to the limit of our resources. No greater protection to its customers can possibly be given by any firm.

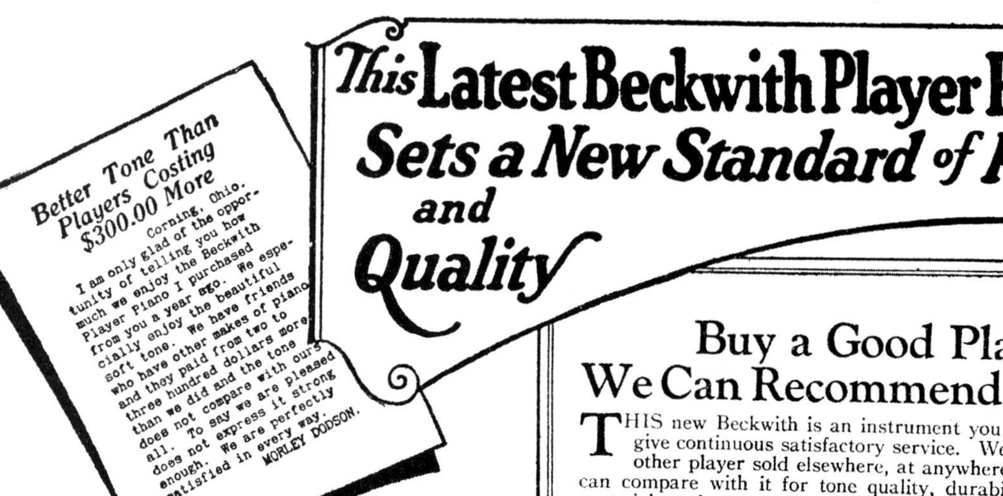

This Latest Beckwith Player Piano
Sets a New Standard of Price
and
Quality
Better Tone Than Players Costing $300.00 More
Corning, Ohio.
I am only glad of the opportunity of telling you how much we enjoy the Beckwith Player Piano I purchased from you a year ago. We especially enjoy the beautiful soft tone. We have friends who have other makes of piano and they paid from two to three hundred dollars more than we did and the tone does not compare with ours all. To say we are pleased does not express it strong enough. We are perfectly satisfied in every way.
MORLEY DODSON.

Buy a Good Player We Can Recommend This One

THIS new Beckwith is an instrument you can depend upon to give continuous satisfactory service. We do not know of any other player sold elsewhere, at anywhere near our price that can compare with it for tone quality, durability or excellence of materials and construction.

You will like this new Beckwith. Its handsome design, hand polished finish and beautiful tone quality will surely appeal to you. It is easy to play, and the player action is of such simple construction that very little attention is required to keep it in good playing condition.

We are glad to recommend to our customers a player of this grade, for we know it will meet every requirement and fulfill every claim we make for it. Many of the highest priced player pianos on the market are no better built than this sturdy Beckwith Player. The extra heavy plate, hammers and strings of extra strength, spruce sounding board, ivory covered keys and other essentials of a high grade instrument, insure years of satisfactory service.

This instrument has an automatic tracking device, pneumatic pedals, extra heavy control wires and a full complement of levers and buttons which will enable anyone to secure the best musical results. It plays all standard 88-note music rolls. The construction is such that the player parts can be readily concealed and the instrument is then ready for hand playing, the same as any upright piano.

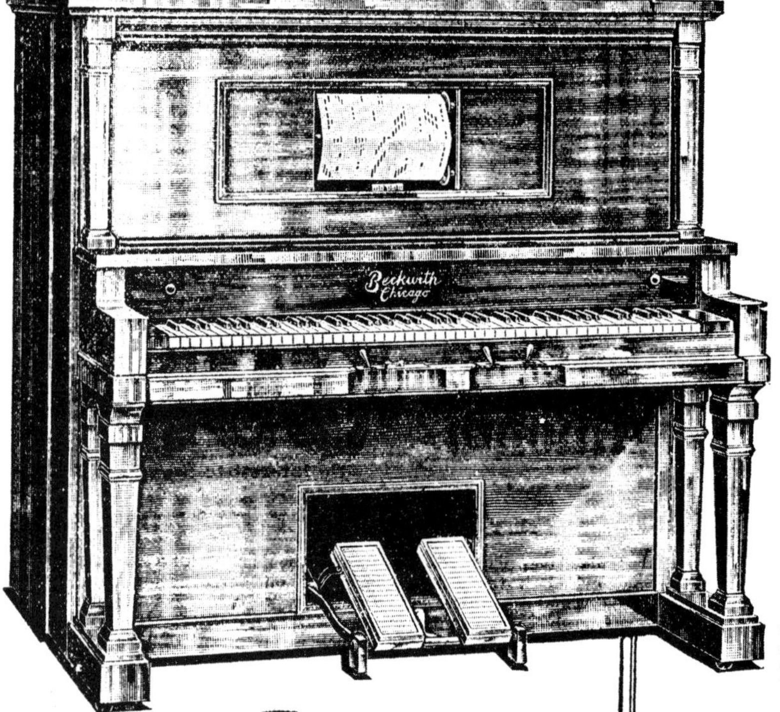

Anyone Can Play

No skill and no experience is required to play a Beckwith Player. Anyone, young or old, can play. Even a child can play well.

The price quoted is the total cost. There is no interest—no extras—no mortgage notes to sign.

Size: 5 feet 3 inches long, 4 feet 8 inches high, 2 feet 4 inches wide. Shipping weight, about 1,000 pounds. Shipped from factory.

46L275—Beckwith Style 75 Player Piano, including cabinet, combination bench and 20 rolls of music. In mahogany at $12.00 per month . **$447.00**

For oak or walnut add $10.00.

$12.00 a month

NO INTEREST CHARGES.

When you buy a Beckwith, you do not have to consider interest. All you pay is the price quoted and the freight from the factory to your station. Bear this in mind when comparing our prices with others. Remember, there is no interest to be added and no extra charges of any kind except the freight.

30 DAYS' TRIAL.

Fill in the order blank enclosed in this catalog and mail it to us, together with a deposit of $10.00, and we will ship the player for a thirty-day trial in your home. If entirely satisfied, send your first monthly payment of $12.00, at which time we will apply the $10.00 deposit to your account. If for any reason you are not entirely satisfied, we will give you instructions for returning the instrument and will immediately send back your deposit.

FREIGHT PREPAID.

To make it easy for you to try this instrument in your home, we will prepay the freight charges with the understanding that the freight will be added to the price. No matter where you buy you will have to pay freight, as this is always added to the price by the dealer making the sale.

No Extra Charge for Cabinet, Bench and Rolls

The cabinet illustrated is made expressly for player rolls but is also an ideal rack for magazines and newspapers. It is 40 inches high, 18 inches wide and 14 inches deep. Holds 90 music rolls of average size. Comes in three finishes to match our players and is durably constructed to give long hard service.

Combination Player and Duet Bench

The bench used with a player should have a sloping top. When the piano is to be played by hand, such a bench is not desirable. Therefore, we furnish a combination type on which the top can be leveled when the piano is to be played by hand, or raised to a sloping position when used with the player. The bench is 30 inches long and 14 inches wide and has a compartment for sheet music.

We also include 20 rolls of music, these being specially selected to bring out the different musical effects so the purchaser can give the player a thorough trial.

Page 527 will interest every piano buyer.

Our Customers Say:

"This Beckwith Concert Grand is The World's Greatest Piano Value"

The Best Money Could Buy

Woonsocket, R. I.

Sears, Roebuck and Co.

I enclose check for the last payment on the Beckwith Upright Piano bought from you.

I want to say that the piano has been a real surprise to us from the time we first saw it, not only for its beauty and fine workmanship, but also for its excellent tone quality. Even some agents selling the highest grade pianos had to admit that my piano was the best at any price within two or three hundred dollars of what I paid.

WILFRED DOYON.

Mr. Doyon bought a Style 9 Concert Grand Piano.

The Most Popular Piano We Have Ever Sold

The Style 9 has always been our most popular design. Its massive proportions and elegant simplicity of detail appeal to all who see it. We have sold thousands of these pianos to our customers in all parts of the country, and the universal praise of their owners is the best evidence we could offer of their high quality. We have hundreds of letters like the one at the right in our files which convince us that our customers think the Beckwith Style 9 is the **world's greatest piano value.**

Read Page 527.

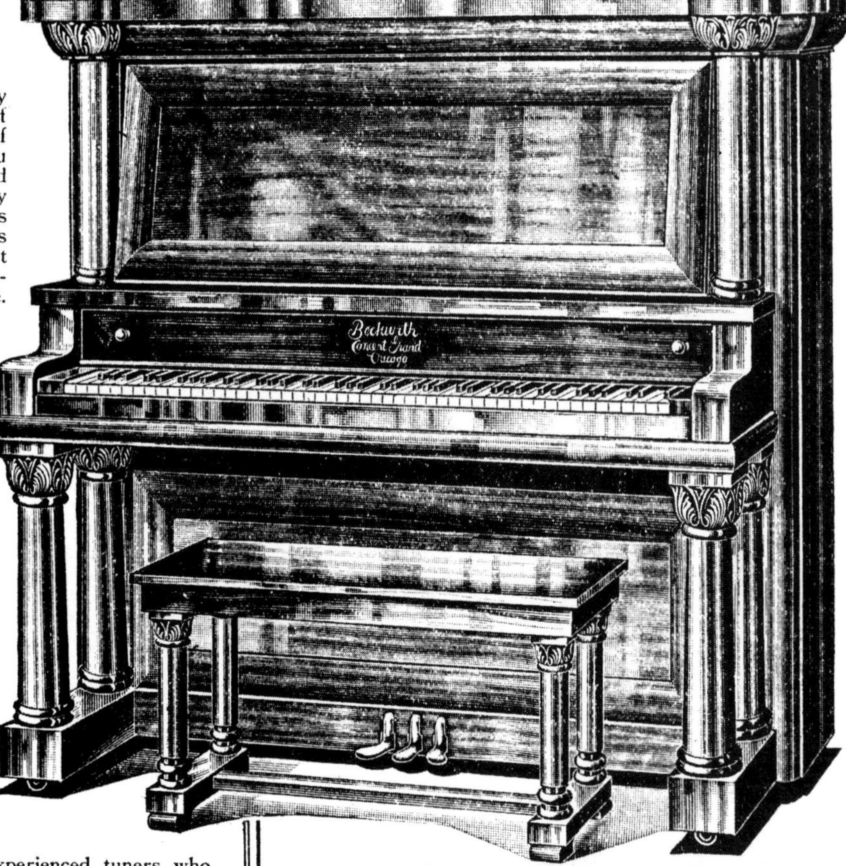

What More Will Money Buy?

If you want the finest piano your money will buy—if you want the most in beauty of design, elegance of finish and perfection of tone—if you want a piano that will give you the maximum in service and satisfaction, and that will compare favorably with the very finest instruments on the market, regardless of make, name or price, then you want this handsome Beckwith Concert Grand, our finest piano and, in the opinion of its many satisfied owners, the world's greatest piano value.

Every detail of design has been worked out with the idea of making this piano the finest looking instrument of its kind on the market—one that will be the pride of any home in which it is placed—and that will stand as testimony to the discrimination and good taste of its owner.

Full Mellow Concert Tone

In keeping with the handsome design of the case, we have placed in this piano the finest piano action we know of. From keys to sounding board, every detail of construction represents the very highest attainment in modern piano development. Only the finest materials are used, and no more care could be lavished on the building of this instrument if it were to sell at twice the price we ask.

The tone of the Style 9 is just what one would expect from a piano of this grade—full, mellow and with that resonance and purity usually associated only with grand pianos, and instruments used for concert and recital work. Expert pianists and experienced tuners who have tried and examined this Beckwith Style 9 agree that it is a piano worthy of the highest praise, for beauty of design, finish and musical excellence.

Our price plus freight from the factory is the total cost. There is nothing extra to be paid for the bench—there are no interest charges to be added. Remember this when comparing prices.

SIZE: 5 feet 4 inches long, 4 feet 10 inches high, 2 feet 3 inches wide. Furnished in mahogany, oak or walnut, polished finish. Weight, boxed, including bench to match, 950 pounds. Shipped direct from factory.

46L209—Beckwith Concert Grand Piano, Style 9, with duet bench and instruction book. Price, in mahogany, at **$10.00** per month. **$395.00**

For oak or walnut, add **$10.00.**

Use the order blank enclosed in this catalog.

$10.00 a month

NO INTEREST TO PAY.

There is no interest to be added or extra payments of any kind except freight.

30 DAYS' TRIAL.

You can try the piano in your home for 30 days to satisfy yourself it is all we claim for it.

FREIGHT CHARGES PREPAID.

We will prepay the freight and add it to the price as explained on the opposite page.

Silvertone

Sold Only by Sears, Roebuck and Co.

Columbia Records Included Without Extra Charge

Every SILVERTONE Phonograph on these two pages is in itself a most extraordinary bargain. But the value is even more amazing when you consider that with these SILVERTONES *we give 25 to 40 genuine Columbia double disc records without any extra charge whatever.*

Records Mean Big Extra Saving

These records at the regular Columbia retail price mean **an extra saving of from $18.75 to $30.00!** Each record has two selections—*fifty to eighty selections* in all—which we will carefully choose from our big list of titles by famous singers, orchestras and entertainers. They include vocal selections, dance hits, band, Hawaiian and other selections, giving you *a complete assortment of music without spending a cent extra for records.*

Five records are shipped with the phonograph. The balance are sent, postpaid, on receipt of first monthly payment.

Why We Can Save You Money

We can offer phonographs of unexcelled quality at such astonishingly low prices because we sell to you direct, at factory cost plus only our one profit. We have cut out the big profits to dealer, wholesaler and jobber—that part of the price which gives you no better music and no better phonograph. In musical quality and mechanical perfection, the SILVERTONE is the equal of any phonograph on the market, regardless of name, make or price. In workmanship and finish, SILVERTONE cabinets are of that superior quality usually associated only with furniture of the better grade.

Many Records FREE

The "Majestic"

Plays All Disc Records.

$8.00 a month

35 Columbia Records Included

Louis Fifteenth Period Art Design

Gold Plated Metal Parts.

Study the dimensions and note the graceful sweeping lines and tasteful decorations of this handsome SILVERTONE with its **gold plated fittings.** Then find out what other dealers ask for a phonograph of this size and excellence of design and finish, *without records.* This will show you what a remarkable bargain the MAJESTIC is.

With its sweet, extraordinarily deep and mellow tone, marvelous volume of sound and quiet running motor, the MAJESTIC is indeed a triumph mechanically. And the elegance of its design, excellent workmanship and fine finish make it suitable for the finest home.

Be sure to read page 535.

48 in. high, 23¾ in. wide, 23¾ in. deep. Net weight, 120 lbs. Silvertone Reproducer; Convertible Tone Arm to play any disc record. Tone modulator. Extra powerful silent running double spring precision motor. Twelve-inch velvet covered turntable. Visible metal parts **gold plated.** Steel needles included. Shipping weight, 190 pounds.

20L4736—The MAJESTIC. Mahogany, American Walnut or Fumed Oak. **State finish.**

Including thirty-five double disc Columbia Records **$148.00**

Terms: $10.00 deposit with order; 2 weeks' trial; $8.00 A MONTH

Use the order blank enclosed in this catalog.

Says "Excellent" After Almost Seven Years' Use

Sears, Roebuck and Co.

We received the phonograph O. K. in every way. It is a splendid instrument. We consider this phonograph to be one of the very best on the market. I purchased a smaller SILVERTONE about seven years ago from you. It has proved to be an excellent phonograph. I am still using it in my school work.

H. J. SCOTT,
South Zanesville, Ohio.

The "Victoria"

$8.00 a month

40 Columbia Records Included

Jacobean Art Design

If you want the very best, both in furniture and in musical qualities, at a remarkable saving in price, you will be delighted with this fine SILVERTONE. The VICTORIA is one of the finest and most gracefully designed phonographs we have ever offered. In harmony of proportion, in beauty of design and finish, in excellence of cabinet work and materials, no less than in musical qualities, it will meet the requirements of the most particular buyer. Its wonderfully clear, mellow, rounded tone and accuracy of reproduction represent the highest degree of phonograph development.

Be sure to read page 535.

36¼ inches high, 41¼ inches wide, 23¾ inches deep. Net weight, 110 pounds. Silvertone Reproducer and Convertible Tone Arm to play any disc record. Tone modulator. Extra powerful double spring silent running precision motor. Twelve-inch turntable. Steel needles included. Shipping weight, 225 pounds.

20L4750—The VICTORIA CONSOLE, *including forty double disc Columbia Records.* **State finish.**

Mahogany or American Walnut, polished, with gold plated metal parts **$165.00**

Cathedral Oak, dull finish, with nickel plated parts, more in keeping with the rich, dull oak finish **145.00**

Terms: $10.00 deposit with order; 2 weeks' trial; $8.00 A MONTH

Use the order blank enclosed in this catalog.

Two Weeks Trial
Easy Monthly Payments

Convince Yourself of Silvertone Quality

We do not ask you to take our word for the high quality of the SILVERTONE. We want you to try it for yourself, in your own home, without any obligation to buy. Just fill out the order blank enclosed and send it with the small deposit, as evidence of good faith.

Compare the SILVERTONE with any other phonograph, regardless of price. Judge if the tone is not fully equal to the highest priced phonograph you have ever heard. Examine the fine cabinet work. Note the perfection of the playing parts. See what you would have to pay elsewhere for a phonograph of the same high quality and beauty of design. If after two weeks' trial you are not fully satisfied with the SILVERTONE simply ship it back at our expense and we will return your deposit, with all transportation charges.

Pay a Little Each Month

If the SILVERTONE is satisfactory, pay for it in small monthly installments, beginning at end of two weeks' trial. Our payment plan is so liberal that you can buy one of our finest SILVERTONES, which will give you a lifetime of pleasure and musical satisfaction, and scarcely feel the expense. Get your SILVERTONE now. Start sharing the joy and happiness which the SILVERTONE has already brought to hundreds of thousands of families all over the country.

No Interest or Extra Charges

Bear in mind, the price we quote is all you pay for the SILVERTONE. There is no extra charge for the records, nor any interest or other charges of any kind.

The "Duchess"

Plays All Disc Records.

Many Records FREE

The "Marquis"

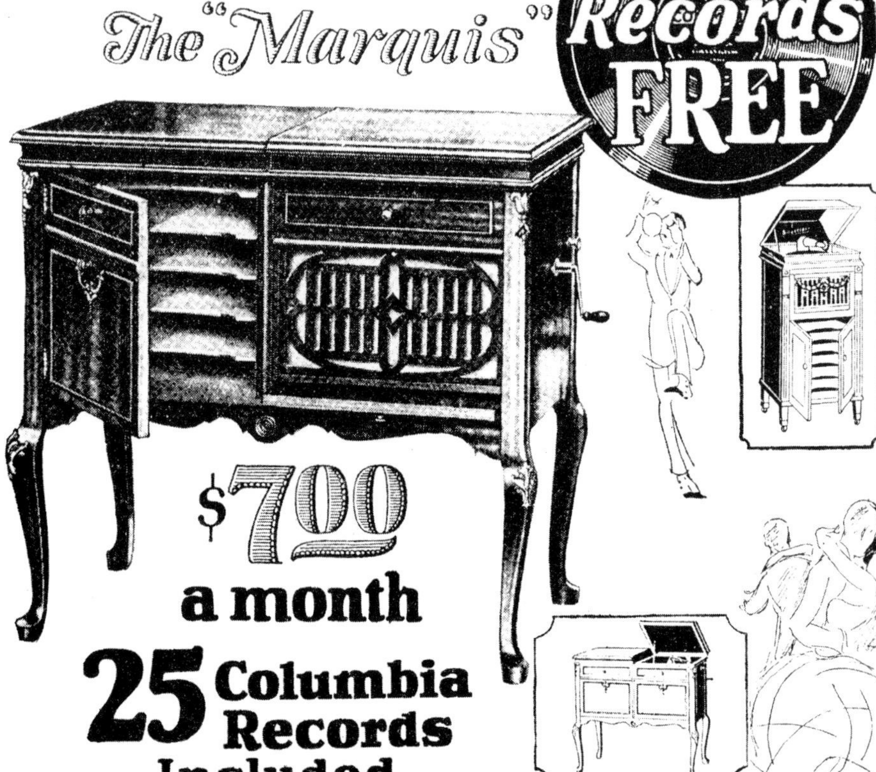

$7.00 a month

25 Columbia Records Included

$7.00 a month

30 Columbia Records Included

Louis Sixteenth Period Design

Gold Plated Metal Parts.

Here is a phonograph value we do not believe you can duplicate. Even our handsome illustration cannot do justice to its beauty. When you see it and hear it play, you will realize what an extraordinary value it is. It is massive, yet so skillfully proportioned that it is extremely graceful. The fine furniture woods used in its construction are rubbed and finished by hand to bring out their full beauty. Visible metal parts are beautifully **gold plated.**

In keeping with the handsome design and finish, the DUCHESS has exceptionally beautiful quality of tone and unusual volume of sound. It will afford you the highest musical and artistic satisfaction.

Be sure to read page 535.

47½ inches high, 22¼ inches wide, 20¾ inches deep. Net weight, 98 pounds. Silvertone Reproducer. Silvertone Convertible Tone Arm to play any disc record Tone modulator. Extra powerful double spring silent running precision motor. Twelve-inch velvet covered turntable. Visible metal parts **gold plated.** Steel needles included. Shipping weight, 160 pounds.

20L4734—The DUCHESS. Mahogany, Walnut or Fumed Oak. **State finish.**

Including thirty double disc Columbia Records.. .. .. **$129.00**

Terms: $10.00 deposit with order; 2 weeks' trial;

$7.00 A MONTH

Use the order blank enclosed in this catalog.

Queen Anne Period Console Design

This beautiful console SILVERTONE, with the twenty-five Columbia records Free, is an exceptional value at our low price. Not only is it an instrument of extraordinary sweetness and mellowness of tone and faithfulness of reproduction, but its beautiful design and fine workmanship appeal immediately to every lover of the beautiful.

Left half of cabinet is an elaborate, spacious record compartment. Panel which protects the silk covered grille in right half of cabinet may be dropped down when playing and used as a shelf for light objects, or may be slid out of the way, as shown in the large illustration.

We know you will be more than pleased with this popular model. It is the very latest in design, first class in workmanship and materials, and of unexcelled musical quality. And you can't match the price anywhere.

Be sure to read page 535.

34¼ inches high, 35¼ inches wide, 20¾ inches deep. Net weight, about 90 pounds. Silvertone Reproducer. Silvertone Convertible Tone Arm to play all disc records. Tone modulator. Powerful silent running double spring precision motor; 12-inch felt covered turntable Nickel plated metal parts. Steel needles included. Shipping weight, 180 pounds.

20L4765—The MARQUIS. Mahogany or Cathedral Oak. **State finish.**

Including twenty-five double disc Columbia Records............ **$98.00**

Terms: $5.00 deposit with order; 2 weeks' trial;

$7.00 A MONTH

Use the order blank enclosed in this catalog.

Musical Qualities Unsurpassed

Sears, Roebuck and Co.

I am well pleased with the SILVERTONE. I have priced the ——, —— and ——, which I think are good machines, but they are priced almost double what I paid for the SILVERTONE. I agree with you that the fine musical qualities are not surpassed by any phonograph on the market at any price.

J. C. NICHOLS,
Moberly, Mo.

COLUMBIA

All Columbia Record

Vocal Selections

All with orchestra accompaniment unless otherwise noted.

Achin' Hearted Blues. Struttin' Blues. Leona Williams, comedienne, and Jazz Band.	A 3599 10-in. 75c
Ain't We Got Fun? Van and Schenck, comedians. **Oh! Dear.** Furman, tenor, and Nash, baritone.	A 3412 10-in. 75c
Alabama Jubilee. Memphis Blues. Collins, baritone; Harlan, tenor.	A 1721 10-in. 75c
Alcoholic Blues. Murray, tenor. **I'm Goin' to Settle Down Outside of London Town.** Murray and Peerless Quartet.	A 2702 10-in. 75c
April Showers. Al Jolson. **Weep No More (My Mammy).** Vernon Dalhart, tenor.	A 3500 10-in. 75c
As I Sat Upon My Dear Old Mother's Knee. Will Oakland. **With All Her Faults I Love Her Still.** Oakland, counter-tenor.	A 1306 10-in. 75c
Atta Baby. Cow Bells. Nora Bayes, comedienne.	A 3633 10-in. 75c
Beautiful Hawaii. Campbell and Burr, tenors. **Rose of My Heart.** Ash, tenor.	A 3363 10-in. 75c
Beautiful Hawaiian Love. Hawaiian Hours With You. Tenor duets, Campbell and Burr.	A 2893 10-in. 75c
Beautiful Ohio. Burr, tenor. **I'm Forever Blowing Bubbles.** Campbell and Burr, tenors.	A 2701 10-in. 75c
Birmingham Blues. Wicked Blues. Edith Wilson and Jazz Band.	A 3558 10-in. 75c
Break the News to Mother. Burr, tenor, and Columbia Stellar Quartet. **Just as the Sun Went Down.** Peerless Quartet.	A 2436 10-in. 75c
Broadway Blues, The. Singin' the Blues Nora Bayes, comedienne.	A 3311 10-in. 75c
Broadway Rose. Peerless Quartet. **Mother's Lullaby.** Sterling Trio.	A 3333 10-in. 75c
Bye-Low. Campbell and Burr. **I'll Always Be Waiting for You.** Charles Harrison, tenor.	A 2827 10-in. 75c
Call Me Back, Pal o' Mine. Lewis James, tenor. **While the Years Roll By.** Criterion Male Quartet.	A 3686 10-in. 75c
Carry Me Back to Old Virginny. (Bland.) Unaccompanied. **Old Oaken Bucket.** (Geibel.) Both Columbia Stellar Quartet.	A 1820 10-in. 75c
Casey Jones. Irving and Jack Kaufman. Tenor duet. **Steamboat Bill.** Irving Kaufman.	A 2809 10-in. 75c
Casey Jones Went Down on the Robert E. Lee. Harlan and Collins. **Whistling Jim.** Peerless Quar.	A 1271 10-in. 75c
Climbing Up the Golden Stairs. Browne. Banjo and piano acc. **Johnny, Get Your Gun.** Browne, baritone. Banjo and orch. acc.	A 2430 10-in. 75c
Crazy Blues. Royal Garden Blues. Mary Stafford and Jazz Band.	A 3365 10-in. 75c
Cuddle Up Blues. I've Got the Wonder Where He Went and When He's Coming Back Blues. Marion Harris, comedienne.	A 3555 10-in. 75c
Dapper Dan. Crumit, tenor. **Ten Little Fingers and Ten Little Toes.** Irving Kaufman.	A 3477 10-in. 75c
Darktown Strutters' Ball. Collins and Harlan. **I'm All Bound 'Round With the Mason-Dixon Line.** Jolson.	A 2478 10-in. 75c
Darling Nellie Gray. You're the Flower of My Heart, Sweet Adeline. Alice Nielsen, soprano.	A 1143 10-in. $1.00
Dixie. Stanley and Harlan. Fife and drum effect. **De Little Old Log Cabin in de Lane.** C. C. Clark. Banjo acc.	A 696 10-in. 75c
Down in Sunshine Valley. Campbell and Burr. **I Want a Girl Just Like the Girl That Married Dear Old Dad.** Columbia Male Quartet.	A 1034 10-in. 75c
Dreams. Sterling Trio. **Alabama Lullaby.** Tenor duet. Campbell and Burr.	A 2717 10-in. 75c
Dreamy Alabama. Hawaiian Lullaby. Tenor duets, Campbell and Burr.	A 2781 10-in. 75c
Driftin' Along on a Blue Lagoon. On Miami Shore. Tenor duets, Campbell and Burr.	A 3302 10-in. 75c
Freckles. Ev'rybody Calls Me Honey. Nora Bayes, comedienne.	A 2816 10-in. 75c
Give Me the Moonlight, Give Me the Girl and Leave the Rest to Me. Ash, tenor. **Give Me the Right to Love You All the While.** Sterling Trio.	A 2415 10-in. 75c

THE PHONO-BRETTO

Contains the words of over 600 favorite sung and spoken phonograph records of the various makes.

It will enable you to thoroughly enjoy your sung and spoken records. In addition, it is one of the most extensive collections of favorite songs, recitations, addresses, verses, etc., ever published.

Printed on good quality paper, handsomely bound in cloth with gilt lettering. Contains over 400 pages. Shipping weight, 12 ounces.

12L6197................$1 25

Good Morning, Mr. Zip-Zip-Zip. Buckley and Peerless Quar. **K-K-K-Katy.** Buckley, baritone.	A 2530 10-in. 75c
Hear Dem Bells. Keemo Kimo. Browne, baritone, and Peerless Quar. Orch. and banjo acc.	A 2853 10-in. 75c
He Comes Up Smiling. Fields, baritone. **Cows May Come, Cows May Go, but the Bull Goes on Forever.** Peerless Quartet.	A 1696 10-in. 75c
Hesitating Blues, The. Adele Rowland. **I'm Goin' to Break That Mason-Dixon Line.** Harry Fox.	A 2769 10-in. 75c
Hiawatha's Melody of Love. Lewis James, tenor. **Underneath the Southern Skies.** James and Harrison, tenors.	A 2914 10-in. 75c
Hi, Jenny, Ho, Jenny Johnson. Razors in the Air. Harry Browne and Peerless Quar.	A 2922 10-in. 75c
Hi Le, Hi Lo. Yodle Song. **Snyder, Does Your Mother Know You're Out?** Yodle Song.	A 572 10-in. 75c
How 'Ya Gonna Keep 'Em Down on the Farm? When Yankee Doodle Sails Upon the Good Ship Home, Sweet Home. Nora Bayes, comedienne.	A 2687 10-in. 75c
I Ain't Got Nobody. Everybody's Crazy 'Bout the Dog-Gone Blues, but I'm Happy. George H. O'Connor, tenor.	A 2481 10-in. 75c
I Can't See the Good in Good-Bye. Lewis James, tenor. **That Wonderful Mother of Mine.** Henry Burr, tenor.	A 2711 10-in. 75c
I Could Have Had You, but I Let You Get By. Nora Bayes and Hickman's Orchestra. **Love Nights.** Nora Bayes.	A 3347 10-in. 75c
I Know What It Means t Be Lonesome. George Meader, tenor. **I Never Knew.** George Meader.	A 2826 10-in. 75c
I'll Be With You in Apple Blossom Time. If I Wait Till the End of the World. Tenor duets, Campbell and Burr.	A 2967 10-in. 75c
I'll Say She Does. Al Jolson. **Just as We Used to Do.** Billy Murray, tenor.	A 2746 10-in. 75c
I'll See You in C-u-b-a. Jack Kaufman. **That Wonderful Kid From Madrid.** Al Jolson, comedian.	A 2898 10-in. 75c
I Love a Lassie. (Harry Lauder.) **He Was Very Kind to Me.** (Harry Lauder.) Both by Sandy Shaw.	A 639 10-in. 75c
I'm Always Chasing Rainbows. Harry Fox. **I Wonder What They're Doing Tonight.** Fields and Peerless Quartet.	A 2557 10-in. 75c
I'm Hungry for Beautiful Girls. I Love Her, She Loves Me. Eddie Cantor, comedian.	A 3624 10-in. 75c
I'm in Heaven When I'm in My Mother's Arms. Henry Burr. **There's a Vacant Chair at Home, Sweet Home.** Campbell and Burr, tenors.	A 2978 10-in. 75c
In My Home Town. The 19th Hole. Frank Crumit, tenor.	A 3666 10-in. 75c
In Sweet September. Al Jolson. **Early in the Morning (Down on the Farm).** Frank Crumit, tenor.	A 2946 10-in. 75c
In Your Arms. Just Like a Gypsy. Nora Bayes, comedienne.	A 6138 12-in. $1.25
It's a Long, Long Way to Tipperary. Stanley Kirkby, baritone. **Old Comrades March.** Band.	A 1608 10-in. 75c
It's Nice to Get Up in the Mornin', but It's Nicer to Lie in Bed. (Lauder.) **Breakfast in My Bed on Sunday Mornin'.** (Lauder.) Evan Davies, baritone.	A 2289 10-in. 75c
I Wish There Was a Wireless to Heaven. Billy Jones, tenor. **Mary, Dear.** Shaw, baritone.	A 3655 10-in. 75c
Japanese Sandman. You're Just as Beautiful at Sixty as You Were at Sweet Sixteen. Both by Nora Bayes.	A 2997 10-in. 75c
Just Before the Battle, Mother. My Own United States. Both by Columbia Stellar Quar.	A 2246 10-in. 75c
Just Snap Your Fingers at Care. Why Worry? Nora Bayes, comedienne.	A 3360 10-in. 75c

The Popular

Stumbling. Frank Crumit, tenor. **Coo-Coo.** Al Jolson, comedian.	A 3626 10-in. 75c
Tomorrow (I'll Be in My Dixie Home Again). Homesick. Both by Nora Bayes, comedienne.	A 3711 10-in. 75c
Tuck Me to Sleep in My Old Kentucky Home. Edwin Dale, baritone; George Reardon, tenor. **My Sunny Tennessee.** Broadway Quartet. (Male Quartet.)	A 3465 10-in. 75c
Wabash Blues. Got to Have My Daddy Blues. Dolly Kay, comedienne.	A 3534 10-in. 75c
Mr. Gallagher and Mr. Shean. When Those Finale Hoppers Start Hopping Around. Furman, tenor; Nash, baritone.	A 3609 10-in. 75c
Toot, Toot, Tootsie! (Goo' Bye). Al Jolson, comedian. **True Blue Sam (The Traveling Man).** Frank Crumit, tenor.	A 3705 10-in. 75c
Virginia Blues. Carolina Rolling Stone. Van and Schenck, comedians.	A 3577 10-in. 75c

12L6200

Order by catalog number (12L6200) and give selection number of each record wanted. It is not necessary to write out the names of selections; just give the numbers.

Ka-Lu-A. Shannon Four. Male Quartet. **Lalawana Lullaby.** Jones, tenor; Hare, baritone.	A 3552 10-in. 75c
Keep in de Middle ob de Road. Oh, Dem Golden Slippers. Browne, baritone, and Knickerbocker Quartet.	A 2116 10-in. 75c
Keep the Home Fires Burning. Pack Up Your Troubles in Your Old Kit Bag. Seagle, baritone, and Columbia Stellar Quartet.	A 6028 10-in. $1.50
Kentucky. Benny Davis, tenor. **I'm Coming Back to Dixie and You.** Frank Crumit, tenor.	A 3320 10-in. 75c

Sacred

VOCAL

Abide With Me. George Alexander. **Where Is My Wandering Boy Tonight?** Burr, tenor.	A 236 10-in. 75c
Beautiful Isle of Somewhere. Columbia Stellar Quartet. **Home of the Soul.** Columbia Mixed Quartet. Double string quartet accompaniment.	A 2048 10-in. 75c
Beautiful Isle of Somewhere. Columbia Mixed Quartet. **Wonderful Words of Life.** Burr.	A 935 10-in. 75c
Brighten the Corner Where You Are. If Your Heart Keeps Right. Homer Rodeheaver, baritone.	A 1990 10-in. 75c
Calling Thee. Transformed. Asher, contralto; Rodeheaver, baritone.	A 3340 10-in. 75c
Face to Face. Henry Burr. **Palm Branches.** Baritone and tenor duet.	A 256 10-in. 75c
God Be With You. Columbia Male Quartet. **Heaven Is My Home.** Burr, tenor. Organ accompaniment.	A 757 10-in. 75c
Hark! The Herald Angels Sing. Burr, tenor. Organ acc. **Tell Mother I'll Be There.** Columbia Male Quartet.	A 264 10-in. 75c
Holy City, The. Columbia Mixed Quintet. **The Lost Chord.** Columbia Stellar Quartet.	A 5744 12-in. $1.25
I Heard the Voice of Jesus Say. Croxton, bass. **When the Roll Is Called Up Yonder.** Peerless Quartet.	A 1305 10-in. 75c
In the Sweet Bye and Bye. Onward, Christian Soldiers. Columbia Stellar Male Quartet.	A 2220 10-in. 75c
Jesus, Lover of My Soul. Face to Face. Henry Burr, tenor.	A 2323 10-in. 75c
Jesus, Saviour, Pilot Me. Seagle and Columbia Stellar Quartet. **Will There Be Any Stars in My Crown?** Seagle, baritone.	A 2808 10-in. $1.00
Lead, Kindly Light. Columbia Male Quartet. **Just as I Am.** George Alexander.	A 249 10-in. 75c
Let the Lower Lights Be Burning. (Bliss.) **Softly Now the Light of Day.** Chautauqua Preachers' Quar.	A 1584 10-in. 75c
Lord's Prayer and 23rd Psalm. Recitation Len Spencer. **March Religioso.** Band.	A 1035 10-in. 75c
Love Divine, All Love Excelling. Lawrence, boy soprano, and Miller, tenor. **Holy City, The.** Lawrence.	A 5453 12-in. $1.25
Meet Me on de Golden Shore. Keep Those Golden Gates Wide Open. Browne, baritone, and Peerless Quar. Banjo and orchestra acc.	A 2992 10-in. 75c
Meet Mother in the Skies. My Mother's Bible. William McEwan, tenor.	A 2498 10-in. 75c
Memories of Mother. Will the Circle Be Unbroken? William McEwan, tenor.	A 1364 10-in. 75c
My Mother's Prayer. Sometime We'll Understand. William McEwan, tenor.	A 1362 10-in. 75c
My Mother's Songs. One by one We're Passing Over. William McEwan, tenor.	A 2881 10-in. 75c
Nearer, My God, to Thee. (Mason.) Baritone solo. **Let the Lower Lights Be Burning.** Baritone and tenor.	A 247 10-in. 75c

RECORDS

Prices Are Postpaid

Vocal Selections

All with orchestra accompaniment unless otherwise noted.

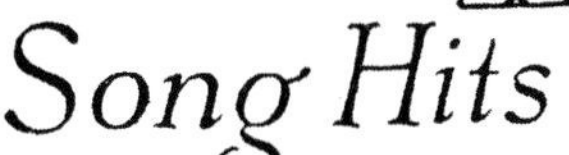

Song Hits

Oh! Is She Dumb? **Susie.** Eddie Cantor, comedian. — A 3682, 10-in., 75c

Swanee River Moon. Columbia Stellar Quartet. **Held Fast in a Baby's Hands.** Reardon, tenor; Mellor, baritone. — A 3432, 10-in., 75c

Why Should I Cry Over You? Billy Jones, tenor. **Sleepy Little Village (Where the Dixie Cotton Grows).** Hart Sisters, harmonizers. — A 3650, 10-in., 75c

Three o'Clock in the Morning. **Moonlight.** Frank Crumit, tenor. — A 3431, 10-in., 75c

Angel Child. Al Jolson, comedian. **Angel Child.** Fox Trot. The Columbians. — A 3568, 10-in., 75c

The Sheik. Male Trio. **Granny.** Male Trio. Hart, Shaw and Clark. — A 3556, 10-in., 75c

I'm Askin' Ye, Ain't It the Truth? Ruth Roye, comedienne. **Georgette.** Ruth Roye. — A 3714, 10-in., 75c

Play on Any Disc Phonograph.

Columbia Records play on any disc phonograph, no matter what kind, no special attachment of any kind being necessary except with the Edison.

Lazy Mississippi. Tenor duet. Campbell and Burr. **Rose of Virginia.** Henry Burr. — A 2909, 10-in., 75c

Left All Alone Again Blues. **Everybody But Me.** Marion Harris, comedienne. — A 2939, 10-in., 75c

Let the Rest of the World Go By. Campbell and Burr, tenors. **Rings.** James and Harrison, tenors. — A 2829, 10-in., 75c

Little Ford Rambled Right Along, The. Fields, baritone. **Si's Been Drinking Cider.** Collins and Harlan. — A 1754, 10-in., 75c

Look for the Silver Lining. **I'm Gonna Do It if I Like It.** Marion Harris, comedienne. — A 3367, 10-in., 75c

Look! What You've Done With Your Dog-Gone Dangerous Eyes. **Love, Honor and O-Baby!** Benny Davis, tenor. — A 3348, 10-in., 75c

Macushla Asthore. **'Tis an Irish Girl I Love and She's Just Like You.** Both by Chauncey Olcott, tenor. — A 2988, 10-in., 75c

Mandy. Van and Schenck. **I'll Be Happy When the Preacher Makes You Mine.** Irving and Jack Kaufman, tenors. — A 2780, 10-in., 75c

Mickey. Sterling Trio. **Mickey.** Medley. Prince's Orchestra. — A 2662, 10-in., 75c

Mighty Lak' a Rose. Robinson. **When You and I Were Young, Maggie.** Robinson, soprano. — A 2571, 10-in., 75c

Missouri Waltz. Campbell and Burr. Tenor duet. **Sing Me Love's Lullaby.** Burr. — A 2358, 10-in., 75c

My Isle of Golden Dreams. Charles Harrison, tenor. **Venetian Moon.** James and Harrison. Tenor duet. — A 2954, 10-in., 75c

My Little Bimbo Down on the Bamboo Isle. Crumit, tenor. **She Gives Them All the Ha-Ha-Ha.** Crumit and Brown, tenors. — A 2981, 10-in., 75c

Nestle in Your Daddy's Arms. **Pucker Up and Whistle.** Both by Frank Crumit, tenor. — A 3406, 10-in., 75c

Never Let No One Man Worry Your Mind. Marion Harris, comedienne. **I'm a Jazz Vampire.** Harris. — A 3328, 10-in., 75c

Oh! By Jingo. **So Long, Oolong.** Both by Frank Crumit, tenor. — A 2935, 10-in., 75c

Oh! How I Laugh When I Think That I Cried Over You. **Snoops, the Lawyer.** Nora Bayes, comedienne. — A 2852, 10-in., 75c

Oh, Judge (He Treats Me Mean). **He Done Me Wrong.** Marion Harris, comedienne. — A 2968, 10-in., 75c

Oh! What a Pal Was Mary. Burr, tenor. **Waiting.** Charles Harrison, tenor. — A 2786, 10-in., 75c

Old Folks at Home. **Massa's in de Cold, Cold Ground.** Seagle, baritone, and Columbia Stellar Quartet. — A 6082, 12-in., $1.50

Out Where the West Begins. **When the Shadows Softly Come and Go.** Both by Charles Harrison, tenor. — A 3315, 10-in., 75c

Peggy O'Neil. Harrison, tenor. **If Shamrocks Grew Along the Swanee Shore.** Broadway Qt. — A 3438, 10-in., 75c

Perfect Day, A. **Rosary, The.** Charles Harrison, tenor. — A 2212, 10-in., 75c

Whistling Selections

Alice, Where Art Thou? **Song Without Words.** Sybil Sanderson Fagan. — A 2919, 10-in., 75c

Boy and the Birds. Band. **In the Valley of Sunshine and Roses.** Burr. Bird imitations by Sybil Sanderson Fagan. — A 2494, 10-in., 75c

Sybil Sanderson Fagan.

Flower Song. **Simple Confession.** Sybil Sanderson Fagan. — A 3549, 10-in., 75c

Nightingale and the Frogs. Fagan. Orchestra acc. **Whistling Rufus.** Prince's Orch. Whistling by Fagan. — A 2838, 10-in., 75c

Senora. **Song of the Wood Bird.** Both by Guido Gialdini. — A 934, 10-in., 75c

Whistler and His Dog. **Warbler's Serenade.** Prince's Band; whistling by Fagan. — A 2654, 10-in., 75c

Perfect Day, A. Columbia Mixed Quartet. **Oh, Fair, Oh, Sweet and Holy.** Violin, flute and harp trio. — A 1622, 10-in., 75c

Pinkie. The Flapper Song. **By the Riverside.** Frank Crumit, tenor. — A 3651, 10-in., 75c

Red Wing. Stanley, baritone, and Burr, tenor. **Virginia Song.** Myers, baritone. — A 468, 10-in., 75c

Rockabye, Lullaby Mammy. **I'd Love to Fall Asleep and Wake Up in My Mammy's Arms.** Both by Harry Fox. — A 2964, 10-in., 75c

Rock Me in My Swanee Cradle. Shannon Four Male Quartet. **Gee! But I Hate to Go Home Alone.** Billy Jones, tenor. — A 3641, 10-in., 75c

Roll On, Silvery Moon. **Sleep, Baby, Sleep.** Yodle Songs by Matt Keefe. — A 2378, 10-in., 75c

Rosary, The. Henry Burr. **Silver Threads Among the Gold.** Henry Burr, tenor. — A 2308, 10-in., 75c

Scandinavia. Jolson, comedian. **Funeral Blues (Eat Custard and You'll Never Break a Tooth).** Blossom Seeley, com. — A 3382, 10-in., 75c

Silver Threads Among the Gold. **Those Songs My Mother Used to Sing.** Harry McClaskey, tenor. — A 5658, 12-in., $1.25

Skeeter and the June Bug. H. C. Browne, baritone. **Dar's a Lock on de Chicken Coop Door.** Prowne and male quartet. — A 3622, 10-in., 75c

Sleep, Baby, Sleep. **Emmett's Yodles.** Medley. Yodle songs, George P. Watson. — A 573, 10-in., 75c

Somewhere a Voice Is Calling. Kerns, soprano, and Stuart, baritone. **Whispering Hope.** Kearns, soprano, and Potter, contralto. — A 1686, 10-in., 75c

Springtime. **With the Coming of Tomorrow.** Grant Stephens, tenor. — A 3362, 10-in., 75c

Star Spangled Banner. **Battle Hymn of the Republic.** Harrison, tenor, and Columbia Stellar Quartet. — A 2367, 10-in., 75c

Sweet Mama (Papa's Getting Mad). Marion Harris, comedienne. **I Told You So.** Marion Harris. — A 3300, 10-in., 75c

Taxation Blues. **Prohibition Blues.** Nora Bayes, comedienne. — A 2823, 10-in., 75c

Tenting Tonight on the Old Camp Ground. **The Vacant Chair.** Columbia Stellar Quartet. — A 1808, 10-in., 75c

That Tumble Down Shack in Athlone. Sterling Trio. **You're Still an Old Sweetheart of Mine.** Sterling Trio. — A 2698, 10-in., 75c

There's a Long, Long Trail. Burton, tenor; Stuart, baritone. **There's a Little Lane Without a Turning.** Henry Burr, tenor. — A 1791, 10-in., 75c

Till We Meet Again. Campbell and Burr. Tenor duet. **Dreaming of Home, Sweet Home.** Sterling Trio. — A 2668, 10-in., 75c

Trail of the Lonesome Pine. Campbell and Burr, tenors. **A Little Bunch of Shamrocks.** Burr, tenor; Stoddard, baritone. — A 1315, 10-in., 75c

Wait Till You Get Them Up in the Air, Boys. Murray, tenor. **I've Got My Captain Working for Me Now.** Jolson, comedian. — A 2794, 10-in., 75c

When Francis Dances With Me. **Da, Da, Da, My Darling.** Both by Frank Crumit, tenor. — A 3521, 10-in., 75c

When My Baby Smiles. Burr. **Daddy, You've Been a Mother to Me.** Lewis James, tenor. — A 2894, 10-in., 75c

When Shall We Meet Again? Hart, tenor; Shaw, baritone. **Just a Little Love Song.** Howard Marsh, tenor. — A 3529, 10-in., 75c

When the Autumn Leaves Begin to Fall. Fred Hughes, tenor. **Like We Used to Be.** Fred Hughes, tenor. — A 3344, 10-in., 75c

When You and I Were Young, Maggie. Oscar Seagle, baritone. **Believe Me, if All Those Endearing Young Charms.** Oscar Seagle, baritone. — A 3619, 10-in., $1.00

When You and I Were Young, Maggie. McClaskey, tenor. **Gypsy's Warning.** McClaskey. — A 1913, 10-in., 75c

When You Wore a Tulip and I Wore a Big Red Rose. Columbia Stellar Quartet. **Sweet Kentucky Lady.** Coombs and Aldwell, tenors. — A 1683, 10-in., 75c

Where the River Shannon Flows. Broadway Quartet. **A Little Bit of Heaven.** Columbia Stellar Quartet. — A 1916, 10-in., 75c

Where the Silvery Colorado Wends Its Way. Harrison, tenor. **In the Evening by the Moonlight.** Columbia Stellar Quartet. — A 2683, 10-in., 75c

Wicked Blues. **Birmingham Blues.** Edith Wilson and Jazz Band. — A 3558, 10-in., 75c

With His Hands in His Pockets and His Pockets in His Pants. **I'm a Twelve o'Clock Fellow in a Nine o'Clock Town.** Byron G. Harlan, tenor. — A 2219, 10-in., 75c

You Can Have Every Light on Broadway (Give Me That Little Light at Home). Billy Jones, tenor. **Time After Time.** Dale, tenor. — A 3574, 10-in., 75c

You'd Be Surprised. Irving Kaufman. **Just Leave It to Me.** Irving and Jack Kaufman. Tenor duet. — A 2815, 10-in., 75c

You Didn't Want Me When You Had Me, So Why Do You Want Me Now? I. Kaufman. **That's Worth While Waiting For.** Irving Kaufman, tenor. — A 2796, 10-in., 75c

You Remind Me of My Mother. Charles Hart, tenor. **Nellie Kelly, I Love You.** Waltz. Prince's Dance Orchestra. — A 3698, 10-in., 75c

You're a Million Miles From Nowhere When You're One Little Mile From Home. **Once Upon a Time.** Both by Fred Hughes, tenor. — A 2862, 10-in., 75c

You're the Only Girl That Made Me Cry. Henry Burr, tenor. **Drifting.** Peerless Quartet. — A 2984, 10-in., 75c

You Tell 'Em. Van and Schenck. **After You Get What You Want You Don't Want It.** Van and Schenck. — A 2966, 10-in., 75c

Selections

Nearer, My God, to Thee. **Lead, Kindly Light.** Columbia Stellar Quartet. — A 3469, 10-in., 75c

Ninety and Nine. Henry Burr. **Throw Out the Life Line.** Henry Burr, tenor. — A 2352, 10-in., 75c

Oh! Reign, Massa Jesus, Reign! **Most Done Traveling.** Fiske University Jubilee Singers. — A 2901, 10-in., 75c

One Sweetly Solemn Thought. **I Love to Tell the Story.** Cyrena Van Gordon, contralto. — A 3561, 10-in., $1.00

Onward, Christian Soldiers. Columbia Male Quartet. **Safe in the Arms of Jesus.** Burr. — A 244, 10-in., 75c

River of Jordan. **Couldn't Hear Nobody Pray.** Fiske University Jubilee Singers. — A 1932, 10-in., 75c

Saved by Grace. Henry Burr. **I'll Go Where You Want Me to Go.** Henry Burr, tenor. — A 722, 10-in., 75c

Shepherd, Show Me How to Go. Weld, baritone. Organ acc. **Nearer, My God, to Thee.** Columbia Male Quartet. — A 250, 10-in., 75c

Shout All Over God's Heaven. **Swing Low, Sweet Chariot.** Fiske University Jubilee Singers. — A 1883, 10-in., 75c

Silent Night, Hallowed Night. Mixed chorus. **Oh, Come All Ye Faithful.** Columbia Mixed Quartet. — A 1859, 10-in., 75c

Since Jesus Came Into My Heart. Rodeheaver, baritone. **Mother's Prayers Have Followed Me.** Rodeheaver. — A 2175, 10-in., 75c

Some o' These Days. **Heab'n.** Asher and Rodeheaver. — A 3559, 10-in., 75c

Sweet Hour of Prayer. **Rescue the Perishing.** Henry Burr, tenor. — A 3385, 10-in., 75c

Tell Mother I'll Be There. **Work for the Night Is Coming.** Earle F. Wilde, evangelist. — A 2772, 10-in., 75c

Throw Out the Life Line. Henry Burr, tenor. **What a Friend We Have in Jesus.** Stanley, baritone; Burr, tenor. — A 266, 10-in., 75c

When the Roll Is Called Up Yonder. Chautauqua Preachers' Quartet. **In the Garden.** Asher, contralto, and Rodeheaver, baritone. — A 2667, 10-in., 75c

When the Roll Is Called Up Yonder. Wilde, evangelist. **Softly and Tenderly.** Wilde. — A 2873, 10-in., 75c

Where Is My Wandering Boy Tonight? Burr, tenor. **Jesus, Lover of My Soul.** Burr. — A 2498, 10-in., 75c

Ye Olden Yuletide Hymns. Parts I and II. Columbia Stellar Quartet. — A 2993, 10-in., 75c

INSTRUMENTAL.

Cathedral Chimes. Orchestra. **Christmas Chimes.** Orchestra. Chimes by Howard Kopp. — A 2644, 10-in., 75c

Jesus, Lover of My Soul, and Rock of Ages. **Onward, Christian Soldiers.** Chimes solos, Howard Kopp. — A 2304, 10-in., 75c

Lead, Kindly Light. Chimes. **Rock of Ages.** Chimes. — A 889, 10-in., 75c

Lost Chord, The. Gatty Sellars. **Largo.** Gatty Sellars. Pipe organ selections. — A 6004, 12-in., $1.25

Nearer, My God, to Thee. Creatore's Band. **The Last Hope.** Prince's Symphony Orchestra. — A 5881, 12-in., $1.25

Safe in the Arms of Jesus. Chimes. **Saviour, Lead Me Lest I Stray.** Henry Burr, tenor. — A 239, 10-in., 75c

Kewpie Kameras

Get the Pictures

$2.25 to $4.95

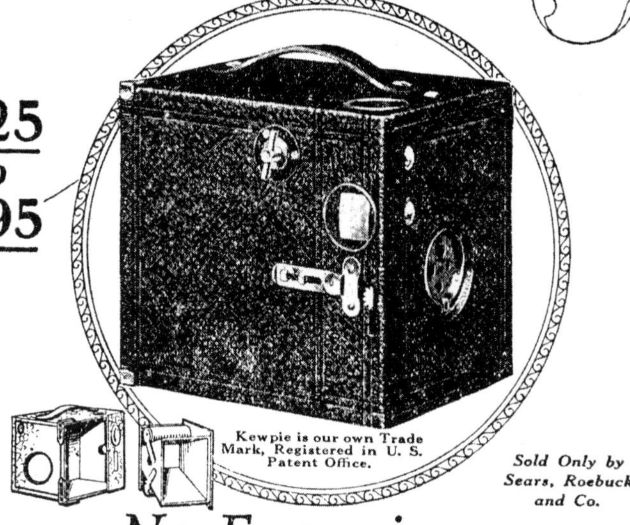

Kewpie is our own Trade Mark, Registered in U. S. Patent Office.

Sold Only by Sears, Roebuck and Co.

Low Priced—Easy to Use

REAL CAMERAS—Reliable, Simple in Construction and Surprisingly Low in Price. That's what those who have bought and used them think of KEWPIE KAMERAS.

That a camera be reliable is perhaps the most important consideration—and you can always depend on a KEWPIE to get the picture. That is why it has been such a favorite with amateurs and the surprisingly large number of owners of high priced cameras who have also bought them.

KEWPIES **need not be focused.** They are "fixed focus" cameras. Just push the lever and the picture is taken.

No Focusing—No Guessing at Distances

Kewpie Kameras are equipped with fine single achromatic lenses. Each lens is carefully fitted and adjusted to the camera in which it will be used. That accounts for the sharpness of detail in KEWPIE pictures.

Provided with four diaphragm openings or stops, by means of which the depth of focus may be regulated or exposures adapted to varying degrees of light. Equipped with automatic rotary shutters, making both time and instantaneous exposures; very simple in construction, positive in action and easy to operate.

Made throughout of kiln dried wood, covered with good grade keratol. Metal trimmings nickel plate or fine black enamel.

Equipped with two finders, one for horizontal and one for vertical pictures.

No. 3 and No. 3A Kameras are each provided with two tripod sockets, for horizontal and vertical pictures.

Illustrated instruction book and Conley Photographic Exposure Guide included with each Kewpie Kamera.

PRICES

Kewpie Kamera	Catalog No.	Size of Picture, Inches	Size of Kamera, Inches	Net Weight	Shipping Weight	For Film See Following on Page 554	Price
No. 2	3L41200	2¼x3¼	3¼x4⅛x5⅜	14 oz.	2 lbs.	3L42716 to 3L42718	$2.25
No. 2A	3L41220	2½x4¼	3½x5¼x5⅞	18 oz.	2¼ lbs.	3L42728 to 3L42733	3.15
No. 3	3L41240	3¼x4¼	4½x5¼x5⅞	22 oz.	2¾ lbs.	3L42760 to 3L42764	4.10
No. 3A	3L41260	3¼x5½ Post card size.	4¾x6½x6¾	31 oz.	3 lbs.	3L42776 to 3L42780	4.95

Conley Junior Film Cameras

Four Sizes—$9.85 to $13.75

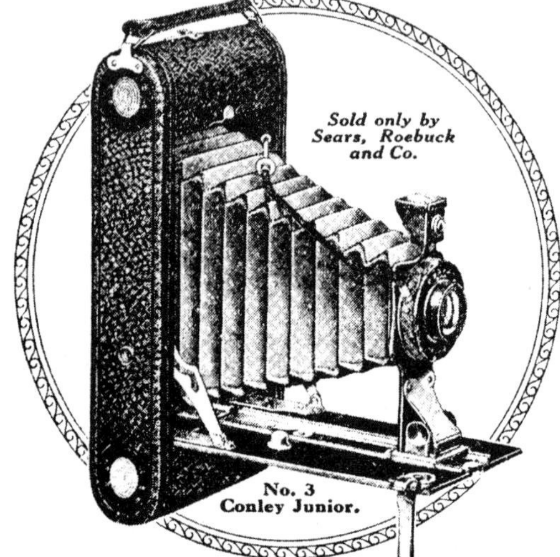

Sold only by Sears, Roebuck and Co.

No. 3 Conley Junior.

CONLEY Junior Roll Film Cameras—there are four of them, made in the sizes that are most popular with amateur photographers—are daylight loading, moderately priced, beautiful in finish and design and thoroughly practical and efficient. They are marvels of compactness, the smaller sizes slipping easily into one's coat pocket. Yet they are all so complete in their equipment as to meet every essential requirement.

An exclusive feature of these cameras is our depth of field focus scale, which shows at a glance both the maximum and minimum distances at which objects are in focus. The ordinary focus scale shows definitely only one distance at which objects are in focus, and the user must guess as to whether objects at greater or less distance will also be sharp, often resulting in pictures which are wholly or in part out of focus, or "blurred." This difficulty is overcome when you use a Conley Junior.

Made chiefly from aluminum, reinforced with ebony finish hardwood. Covering is genuine seal grain leather. All wearing parts of hard brass or steel. Bellows of "Insted" Leather, lined with lightproof gossamer cloth. Brilliant reversible view finder. Two tripod sockets. Rigid all metal lens, standard and folding legs.

Nos. 2A and 3A Conley Junior Cameras are like the No. 3 Conley Junior illustrated. The manner of closing and the lens standard and folding leg of the No. 2 are different. In these details the No. 2 Junior is like the No. 2 Fixed Focus Camera illustrated on opposite page.

Conley Junior Cameras are all equipped with Extra Rapid Rectilinear Lenses, designed expressly for them. They work at F:8 (U. S. 4), which makes them just four times as fast as single achromatic lenses, and twice the speed of the F:11 rapid rectilinear lenses, usually furnished with moderate priced cameras. Each has the Victo Shutter, making time and bulb exposures of any desired length and automatically controlled exposures of 1/10, 1/25, 1/50, and 1/100 of a second; it is provided with wire push release and may also be operated by finger release and is equipped with iris diaphragm.

Very complete, illustrated instruction book and Conley Photographic Exposure Guide included with each Conley Junior Camera.

Keratol Carrying Cases for Conley Junior. With Shoulder Strap.

3L42300—For No. 2 Camera. Shipping weight, 12 ounces.......**$1.70**
3L42310—For No. 2A Camera. Shipping weight, 1¼ pounds.......**$1.80**
3L42320—For No. 3 Camera. Shipping weight, 1½ pounds.......**$1.95**
3L42330—For No. 3A Camera. Shipping weight, 1⅝ pounds.......**$2.05**

PRICES

CONLEY JUNIOR FILM CAMERAS.

Camera	Catalog No.	Size of Picture, Inches	Size of Camera, Closed, Inches	Weight	Shipping Weight	Uses Following Film, Page 554	Price
No. 2	3L42210	2¼x3¼	1½x3½x6⅞	1⅛ lbs.	2 lbs.	3L42716 to 3L42718	$ 9.85
No. 2A	3L42230	2½x4¼	1¾x3¾x8½	1 lb. 7 oz.	3½ lbs.	3L42728 to 3L42733	10.75
No. 3	3L42250	3¼x4¼	1¾x4½x8¼	1⅝ lbs.	3¾ lbs.	3L42752 to 3L42757	12.50
No. 3A	3L42270	3¼x5½ Post card size	1⅞x4⅝x9¾	2 lbs. 3 oz.	4¾ lbs.	3L42768 to 3L42773	13.75

Conley Fixed Focus Folding Cameras

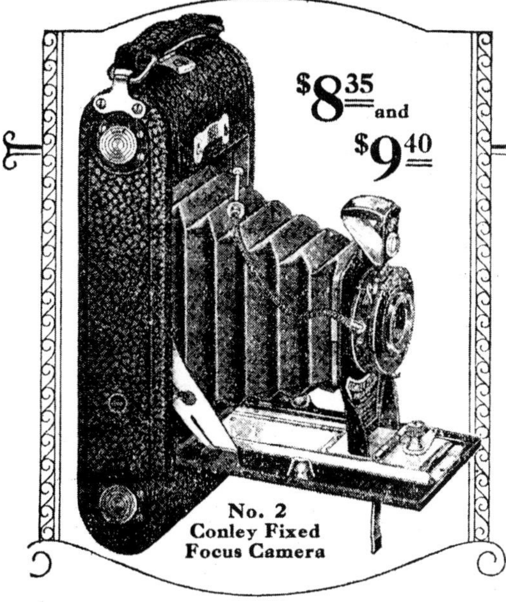

No. 2 Conley Fixed Focus Camera

No Focusing—No Guessing at Distances—Snapshots Always Sharp

PICTURE TAKING with the Conley Fixed Focus Folding Cameras is simplicity itself. Snapshots made with these wonderful little cameras are always sharp. No focusing is required, no guessing at distances—whether the subject is six feet or a hundred feet away, it makes no difference; your entire picture will be sharp. All there is to taking snapshots is to open the camera, pull out the front until it automatically stops in focus, point the camera at the subject and snap the picture. These cameras possess all the desirable features of box cameras with the big additional advantage of compactness, the No. 2 size readily slipping into the coat pocket.

Conley Fixed Focus Folding Cameras are made chiefly from aluminum, the sides reinforced with ebony finish hardwood, and covered with artificial seal grain leather. Wearing parts are of hard brass or steel, nicely finished in black or polished nickel. Bellows of "Insted" Leather, a wonderful substitute for leather. Provided with brilliant reversible view finder, two tripod sockets, rigid all metal lens standard and folding leg. The No. 2 camera is illustrated. The No. 2A is identical in design with the exception of the lens standard, the folding leg and the manner of closing the camera. In these details the No. 2A Fixed Focus Camera is like the No. 3 Conley Junior Camera illustrated on opposite page.

Conley Fixed Focus Folding Cameras are equipped with fine single achromatic lenses, designed especially for them, and Ultex shutters, making time and bulb exposures of any desired length and automatically controlled exposures of 1/25, 1/50 and 1/100 second. They are provided with iris diaphragms and with wire push release and may also be operated by finger release.

Very complete, illustrated instruction book and Conley Photographic Exposure Guide included with each Conley Fixed Focus Folding Camera.

PRICES

With Single Achromatic Lens in Ultex Shutter.

No. 2 Conley Fixed Focus Folding Camera.

For 2¼x3¼ pictures.

Size of camera, closed, 1⅜x3⅛x6⅞ inches. Net weight, 1⅛ pounds. Shipping weight, 2 pounds.

3L42800 .... **$8.35**

For film see 3L42716 to 3L42718, page 554.

3L42300—Keratol Carrying Case, with shoulder strap, for No. 2 Camera. Shipping weight, 12 ounces. **$1.70**

No. 2A Conley Fixed Focus Folding Camera.

For 2½x4¼ pictures.

Size of camera, closed, 1¾x3¾x8½ inches. Net weight, 1 pound 7 ounces. Shipping weight, 3½ pounds.

3L42810 .... **$9.40**

For film see 3L42728 to 3L42733, page 554.

3L42310—Keratol Carrying Case, with shoulder strap, for No. 2A Camera. Shipping weight, 1¼ pounds.......... **$1.80**

Conley De Luxe Roll Film Cameras

Sold only by Sears, Roebuck and Co.

CONLEY De Luxe Roll Film Cameras are beautifully made in every detail, handsomely finished in black enamel, with nickel plated trimmings. They are graceful in appearance, compact and handy, and wonderfully simple in operation. And simplicity in operation always means a greater percentage of good pictures, especially for the amateur. The focusing device is unusually convenient in operation.

The No. 2A camera is furnished with extra long bellows extension, thus being especially suitable for making portraits of satisfactory size without employing any auxiliary or supplementary lens, a feature not to be found in many film cameras.

A desirable feature of the Conley De Luxe Cameras is the unusually rigid and accurate construction of the U Shape metal lens standard, insuring perfect alignment of lens and film. This is very essential in order to utilize to the fullest extent the special advantages of fast anastigmat lenses.

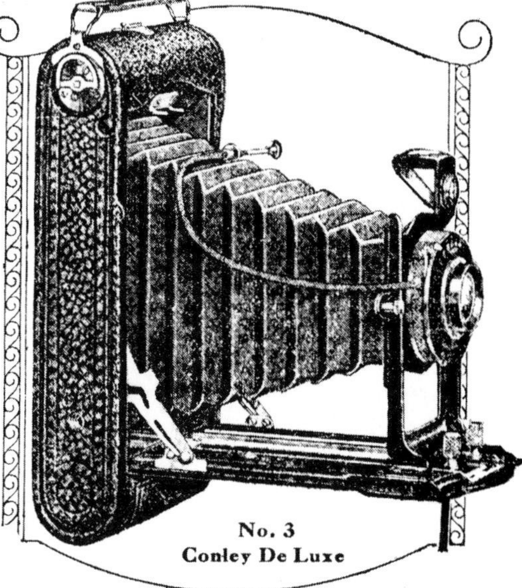
No. 3 Conley De Luxe

Fitted With Anastigmat Lenses at Very Low Prices.

The increased speed and greater covering power of anastigmat lenses has long been appreciated by camera users. The Luxar and Citar Anastigmat Lenses are of exceptional quality. And if you will compare our prices with those generally asked for cameras equipped with F:7.7 and F:6.3 anastigmat lenses of this superior quality, especially F:6.3 lenses, you will find the Conley De Luxe Cameras extraordinarily low priced.

The rapid rectilinear lenses furnished with these cameras are the equal of any rapid rectilinear lens sold.

The Acme shutter makes automatically controlled exposures of 1/500, 1/200, 1/100, 1/50, 1/25, 1/5, 1/2 and 1 second. Victo shutter makes automatically controlled exposures of 1/100, 1/50, 1/25 and 1/10 second. Both make time and bulb exposures of any desired length, are equipped with iris diaphragm and provided with wire push and finger releases.

Details—Conley De Luxe Cameras.

Combination aluminum and ebony finished wood body, covered with black genuine leather. Rising and falling front. Reversible, brilliant finder. Metal parts black enamel or nickel plated. Automatic focus stop at infinity. Very accurate focus scale, locking automatically at each graduation. Black "Insted" leather bellows with lightproof gossamer lining. Two tripod sockets. Improved film holding device, very easy to load or unload. Exposure guide and complete, illustrated instruction book included.

PRICES FOR CONLEY DE LUXE CAMERAS.

Camera	Size of Pictures, Inches	Size of Camera, Closed, In.	Weight	Shipping Weight	Uses Film	Catalog No.	Lens and Shutter Equipment	Price
No. 2A	2½x4¼	1¾x3⅞x 8¼	1 lb. 14 oz.	4 lbs.	3L42728 to 3L42733, page 554.	**3L41038**	F:8 Rapid Rectilinear Lens in Victo Shutter	**$18.25**
						3L41036	F:7.7 Luxar Anastigmat Lens in Victo Shutter	**22.75**
						3L41039	F:6.3 Citar Anastigmat Lens in Acme Shutter	**39.50**
No. 3	3¼x4¼	1¾x4⅝x 8¼	1 lb. 15 oz.	4 lbs.	3L42752 to 3L42757, page 554.	**3L41003**	F:8 Rapid Rectilinear Lens in Victo Shutter	**18.50**
						3L41006	F:7.7 Luxar Anastigmat Lens in Victo Shutter	**23.10**
						3L41005	F:6.3 Citar Anastigmat Lens in Acme Shutter	**40.00**
No. 3A	3¼x5½ Post card size	1⅞x4¾x 9¼	2 lbs. 9 oz.	5 lbs.	3L42768 to 3L42773, page 554.	**3L41018**	F:8 Rapid Rectilinear Lens in Victo Shutter	**21.25**
						3L41021	F:7.7 Luxar Anastigmat Lens in Victo Shutter	**26.25**
						3L41027	F:6.3 Citar Anastigmat Lens in Acme Shutter	**43.00**

Sig. 14.

Carrying Cases for Conley De Luxe Cameras.

3L41041—Plush Lined Leather Carrying Case, with shoulder strap, for No. 2A Camera. Shipping weight, 1½ pounds. **$2.80**

3L41016—Plush Lined Leather Carrying Case, with shoulder strap, for No. 3 Camera. Shipping weight, 1½ pounds... **$2.95**

3L41031—Plush Lined Leather Carrying Case, with shoulder strap, for No. 3A Camera. Shipping weight, 1⅝ pounds. **$3.15**

Photographic Papers and Supplies

Improved Darko Papers and Post Cards

Free From Abrasion Marks With Any Developer—Anti-Friction Developer Not Required.

IMPROVED DARKO produces brilliant, pleasing prints, practically free from stain and fog, even with prolonged development.

IMPROVED DARKO Papers and Post Cards are usually printed by artificial light and developed and fixed much as a negative is. There are three grades—**a grade for every negative.** They are known as Red Label, or Contrast Emulsion, for use with soft or flat negatives; Green Label, or Medium Emulsion, for use with average negatives, and Blue Label, or Soft Emulsion, for use with very hard or contrasty negatives.

The Glossy surface is very smooth and has a high polish. Prints on this paper should be squeegeed to bring out the full degree of glossiness. The Velvet surface is of a velvety texture, having a slight gloss.

All IMPROVED DARKO, except the India Tint Darko and Darko Post Cards, is furnished in regular weight stock. The India Tint Darko and Post Cards come in double weight.

Red Label Papers and Post Cards.

Contrast Emulsion for Soft Negatives.

Red Label Darko will make good prints from negatives which are flat on account of overexposure and underdevelopment. It will make extra strong and contrasty prints from normal or hard negatives.

Order by catalog number.

	Papers	Post Cards
Glossy surface	**3L42641**	**3L42660**
Velvet surface	**3L42642**	**3L42661**

See price list at right.

Blue Label Papers and Post Cards.

Soft Emulsion for Contrasty Negatives.

Blue Label Darko is suited to very hard or contrasty negatives. Underexposed plates which have been forced in development to bring out as much detail as possible are usually harsh and contrasty, and give the best results with this grade.

Order by catalog number.

	Papers	Post Cards
Glossy surface	**3L42651**	**3L42667**
Velvet surface	**3L42652**	**3L42668**

See price list at right.

Price List for Improved Darko Papers and Post Cards

Order by catalog number and state size.

Size	Shpg. Wt.	Per Doz.	Shpg. Wt.	Per 2 Doz.	Shpg. Wt.	Per One-Half Gross	Shpg. Wt.	Per Gross
2¼x3¼	Not furnished		2 oz.	**13c**	These sizes not furnished in one-half gross packages. Less than one gross furnished only at 2-dozen rate.		8 oz.	**$0.56**
2½x4¼	Not furnished		2 oz.	**17c**			10 oz.	**.68**
3¼x4¼	Not furnished		3 oz.	**17c**			12 oz.	**.90**
2⅞x4⅞	Not furnished		3 oz.	**17c**			13 oz.	**.90**
4 x5	Not furnished		4 oz.	**25c**			1 lb.	**1.15**
3¼x5½	Not furnished		4 oz.	**21c**			1 lb.	**1.07**
4 x6	3 oz.	**13c**	These sizes not furnished in 2-dozen packages. Less than one-half gross furnished only at the dozen rate.		9 oz.	**$0.68**	1¼ lbs.	**1.30**
5 x7	6 oz.	**21c**			1 lb.	**1.02**	1¾ lbs.	**1.95**
6½x8½	7 oz.	**34c**			1½ lbs.	**1.70**	2¾ lbs.	**3.20**
8 x10	8 oz.	**47c**			1¾ lbs.	**2.38**	3¾ lbs.	**4.50**
Post Cards	3½ oz.	**17c**	5 oz.	**29c**	12 oz.	**.77**	1¾ lbs.	**1.44**

Green Label Papers and Post Cards.

Medium Emulsion for Normal Negatives.

Green Label Darko is suited to negatives of medium or average strength. Negatives which have been correctly exposed and correctly developed usually yield the best prints when this grade is used.

Order by catalog number.

	Papers	Post Cards
Glossy surface	**3L42646**	**3L42663**
Velvet surface	**3L42647**	**3L42664**

See price list at left.

India Tint Papers and Post Cards.

India Tint Darko is coated upon a special extra heavy velvet surface stock of soft India color, so the high lights in the picture are of a delicate sepia or cream color. The paper is handled just the same as the other grades of Darko, and the India tint of the paper in contrast with the rich blacks of the picture gives an effect that is unique and exceedingly artistic. Unusually pleasing effects are obtained by using the Darko Sepia Toner with this paper. Furnished only in medium emulsion for normal negatives.

3L42656—India Tint Darko Papers.
3L42666—India Tint Darko Post Cards.

See price list at left.

Professional Photographers' Scale.

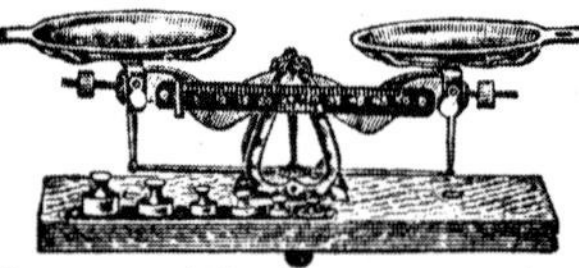

Thoroughly practical, accurate and durable, adjusted and guaranteed sensitive to 1 grain. Large, interchangeable pans. Adjusting screws maintain a perfect balance at all times Beam registers from 1 to 50 grains. Loose weights run from 2 ounces down to 50 grains. Weights and all metal work nickel plated and polished. Beam is black enameled, with white lines and figures. Base quarter sawed oak. Shipping weight, 1⅞ pounds.

3L41275 .... **$3.75**

Good Photo Scale, Only 58c.

One of the best low priced scales devised. Answers all the requirements in making up solutions, etc. Simple, nothing to get out of order, accurate, clean and convenient, no loose weights. Weighs up to 12 drams or 720 grains. Glass pan easily cleaned. Shpg. wt., 6 oz.

3L41265 .... **58c**

3L41267—Extra Glass Pan for 3L41265 Scale. Shipping wt., 3 oz. Each **15c**

Thermometers.

Tray thermometer, made with two clips to hold it on the edge of tray so temperature of solution may be watched. Shipping weight, 4 ounces.

3L41399 .... **16c**

3L41399

Tank Thermometer, designed specially for use with developing tanks. Has hook for fastening to edge of tank. Shipping weight, 6 ounces.

3L41400 .... **23c**

3L41400

Negative Racks.

Well made rack. Holds 24 negatives 8x10 inches or smaller. Shipping weight, 2¼ pounds.

3L41367 .... **25c**

Engraved Graduates.

Cone shape graduates with all lines and figures engraved by hand. Among the most carefully made and accurate graduates on the market. **Be sure to state size.**

3L41244

Capacity	Shpg. Wt.		Capacity	Shpg. Wt.	
120-minim	1 lb.	**45c**	8-oz.	2 lbs. 12 oz.	**$0.49**
2-oz.	1 lb. 3 oz.	**34c**	16-oz.	4 lbs. 4 oz.	**.82**
4-oz.	1 lb. 13 oz.	**45c**	32-oz.	6 lbs. 6 oz.	**1.42**

Measuring Glasses.

Tumbler shaped, for liquids; 2 and 4 ounces, graduated in ounces and drams; the 8-ounce in ounces and ½ and ¼ pints. Not quite so convenient as the cone shaped graduate, but very low priced. **State size wanted.**

3L41241

Capacity	Shipping Weight	
2-oz.	1 lb. 3 oz.	**10c**
4-oz.	1 lb. 13 oz.	**13c**
8-oz.	2 lbs. 12 oz.	**17c**

Fluted Glass Funnels.

Glass Funnels, fluted for filtering. More desirable than plain funnels because filtering is much more rapid. **Be sure to state size wanted.**

3L41255

Size	Shpg. Wt.		Size	Shpg. Wt.	
½ pt.	2 lbs. 8 oz.	**20c**	1 qt.	5 lbs. 10 oz.	**42c**
1 pt.	3 lbs. 8 oz.	**27c**	2 qts.	9 lbs. 13 oz.	**64c**

Oil Ruby Lamp.

A medium size high class oil burning darkroom lamp. Provided with special burner giving unusual volume of light. Perfect combustion without smoke or odor. Made with hinged metal front which can be placed at any angle to regulate volume of light, and fitted with both orange and deep ruby glasses, insuring a perfectly safe and nonactinic light. Light can be turned up without opening lamp. Height of lamp, 8⅛ inches; size of glasses, 3⅝x4⅝ inches. Shipping weight, 2 pounds.

3L41211 .... **90c**

Waterproofed Apron.

Made of rubber sheeting. Protects the clothing from chemical stains and dirt. Made for hard usage. An excellent value which will give long service. Length, 42 inches. Shipping weight, 13 ounces.

3L41312 .... **95c**

Film Developing Tray.

Designed for developing roll films in the strip. Supplied with smooth glass rod 6¼ inches long, which revolves in a socket in each end of tray. Film is passed under the roller and see-sawed up and down through the developer. We advise the purchase of two; one for developing, the other for fixing. Shipping weight, each, 3¼ pounds.

3L41290—2 trays **84c**
Each **45c**

Glass Trays.

Made of good quality molded glass, with ribbed bottom. Very easy to keep clean. **Be sure to state size.**

3L41286

For Plates, Inches	Shpg. Wt., Lbs.		For Plates, Inches	Shpg. Wt., Lbs.	
4 x5	2½	**24c**	6½x 8½	4¾	**55c**
3¼x5½	2½	**25c**	8 x10	5	**75c**
5 x7	4	**35c**			

White Enameled Steel Trays.

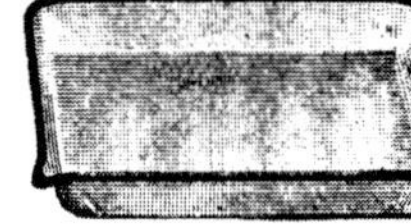

Very satisfactory trays. Made in one piece and are easily cleaned, are practically unbreakable and stand the action of any photographic chemical.

3L41285

For Plates, Inches	Shpg. Wt., Lbs.		For Plates, Inches	Shpg. Wt., Lbs.	
4 x5	¾	**47c**	8x10	2	**$1.07**
3¼x5½	¾	**51c**	10x12	2¾	**1.54**
5 x7	1	**67c**	11x14	4½	**2.37**
6½x8½	1½	**91c**	14x17	5¾	**3.30**

Photographic Clips.

Brass, Nickel plated. Heavy jaws with sharp points prevent film slipping. 3¼ inches wide. Shipping weight, 3 oz. per pair.

3L48560

Per dozen (six pairs) **$1.63**
Per pair .... **.28**

Photographic Chemicals and Supplies

Get All the Fun of Finishing Your Own Pictures and Save Money Too

Developing and Printing Outfits

The Junior Outfit

The Senior Outfit

Junior Outfits for Film Cameras contain everything necessary for finishing pictures complete, except film, which must be ordered extra. See page 554. Each outfit contains the following items:

1 **Candle Ruby Lamp.**
1 **Tray,** for developing film.
1 **Tumbler Shaped Graduate,** for measuring liquids.
1 **Glass Stirring Rod.**
1 **Photographic Thermometer,** for testing temperature of developer.
2 **Tubes Pyro Powders** (make 36 ounces of developer for films).
1 **Package Acid Hypo** (makes 32 ounces of fixing bath for negatives or prints).
1 **Piano Hinged Amateur Printing Frame with glass.**
2 **Dozen Green Label Velvet IMPROVED DARKO Paper.**
2 **Dozen Red Label Velvet IMPROVED DARKO Paper.**
2 **Tubes Developing Powders for paper** (make 16 ounces of developer for prints).
1 **Large Tray,** for developing prints.
1 **Large Tray,** for fixing prints.
1 **Cloth Bound Album,** with 100 pages (50 leaves).
1 **Package (100) Rexo Mounting Corners.**
Instructions for developing and printing.
No film is included with these outfits.

3L45100—Junior Outfit for 2¼x3¼ Film Cameras. Shpg. wt., 7½ lbs...**$2.45**
3L45110—Junior Outfit for 2½x4¼ Film Cameras. Shpg. wt., 8 lbs... **2.65**
3L45130—Junior Outfit for 3¼x4¼ Film Cameras. Shpg. wt., 9 lbs... **2.75**
3L45140—Junior Outfit for 3¼x5½ Film Cameras. Shpg. wt., 9½ lbs... **2.85**

Senior Outfits for Film Cameras are unusually large, of extra quality throughout and complete in every detail, except film, which must be purchased extra. See page 554. Each outfit contains the following items:

1 **Oil Darkroom Lamp,** with ruby and orange glasses.
1 **Special Glass Tray with Roller,** for developing film.
1 **Cone Shaped Graduate,** for measuring liquids.
1 **Glass Stirring Rod.**
1 **Photographic Thermometer,** for testing temperature of developer.
1 **Package Pyro Developing Powders.** Six tubes (make 108 oz. of developer).
1 **Package Acid Hypo** (makes 64 ounces of fixing bath for negatives or prints).
1 **Pair Film Clips,** for hanging up negatives to dry.
1 **Professional Printing Frame with glass.**
2 **Dozen Red Label Velvet IMPROVED DARKO Paper.**
2 **Dozen Green Label Velvet IMPROVED DARKO Paper.**
2 **Dozen Blue Label Velvet IMPROVED DARKO Paper.**
1 **Large Tray,** for developing prints.
1 **Large Tray,** for fixing prints.
1 **Package Developing Powders for paper.** Six tubes (make 48 ounces of developer for prints).
1 **Cloth Bound Album,** with 100 pages (50 leaves).
1 **Jar Photo Paste,** for mounting prints in album.
1 **Package (100) Rexo Mounting Corners.**
Instructions for developing and printing.
No film is included with these outfits.

3L45150—Senior Outfit for 2¼x3¼ Film Cameras. Shpg. wt., 15 lbs...**$4.35**
3L45160—Senior Outfit for 2½x4¼ Film Cameras. Shpg. wt., 16 lbs... **4.65**
3L45180—Senior Outfit for 3¼x4¼ Film Cameras. Shpg. wt., 17 lbs... **4.80**
3L45190—Senior Outfit for 3¼x5½ Film Cameras. Shpg. wt., 18 lbs... **5.15**

Red Seal Developers for Plates or Films

Red Seal Developing Powders for plates or films are of exceptional quality, put up in tightly sealed tubes, six tubes to the package. Only the purest chemicals are used and we recommend these powders as very satisfactory in results, very simple and convenient to use, and economical.

Pyro Developing Powders.

For Developing Plates or Films.

Pyro is absolutely the best developer for plates or films. Each tube makes 18 oz. of solution for six-minute tray development, or 36 oz. for twenty-minute tank development at 65 degrees. Some workers object to pyro because of its tendency to stain the hands, but so far as results are concerned we believe pyro is the best developer that can be used. Shipping weight, 6 ounces.

3L41949—Per package of **6** tubes.....**21c**

Hydro-Metol Developing Powders.

Contrast Formula for Plates or Films.

Yields negatives of great brilliancy and a marked degree of contrast. Each tube makes 8 oz. of developer for tray development, complete package making 48 oz. Shpg. wt., 5 oz.

3L41951—Per package of **6** tubes.....**22c**

NOTE—Metol sometimes has an irritating or poisonous effect on the skin of some individuals, and those who are susceptible to it should select a developer which does not contain metol.

Hydro-Metol Developing Powders.

Soft Formula for Plates or Films.

Yields soft negatives with abundant detail. Each tube makes 8 ounces of developer for tray development, complete package making 48 ounces. Shipping weight, 5 ounces.

3L41953—Per package of **6** tubes.....**22c**

Neg-Dry, 23 Cents.

Hardening solution for either plates or films. Negatives treated with Neg-Dry may be dried in less than five minutes after they have been washed by placing them in this preparation and drying them by artificial heat. Neg-Dry also eliminates all traces of hypo that may not have been washed out of the film. Preparation may be used over and over again. Shpg. wt., 13 oz.

3L41980—Per bottle....................**23c**

PHOTOGRAPHIC CHEMICALS

3L42008—**Genuine Monomethyl Paramidophenol Sulphate** (formerly known as **METOL**). Per ounce, **41c**; ¼ pound, **$1.50**; ½ pound, **$2.84**; 1 pound....................**$5.33**

3L42002—**Pyro (Pyrogallic Acid).** Crystals. Just the same chemically as the resublimed and cleaner to handle. Does not blow about the workroom to cause spots on prints and negatives. We do not handle resublimed pyro. Per ounce, **20c**; ¼ pound, **53c**; ½ pound, **$1.00**; 1 pound....................**$1.87**

3L42005—**Hydrochinon.** The best grade pure white hydrochinon, put up in cartons. Per ounce, **16c**; ¼ pound, **46c**; ½ pound, **88c**; 1 pound....................**$1.65**

3L42020—**Acetic Acid,** No. 8, 16-ounce bottle..........**26c**

3L42043—**Common Alum,** pure pulverized. 16-ounce carton..........**19c**

For other Alum see page 484.

3L42105—**Potassium Bromide.** 16-oz., **45c**; per oz...**9c**

3L42138—**Sodium Carbonate,** dry. A chemically pure, anhydrous carbonate. 16-ounce carton..........**16c**

3L42144—**Sodium Sulphite,** dry. A chemically pure, anhydrous sulphite. 16-ounce carton..........**22c**

Reducing Powders, 27 Cents.

Negatives which from over-development or other causes are too dense can be quickly reduced to the proper density for printing by the use of this preparation. Twelve powders to the package, each powder making 10 ounces of reducing solution. For use they are simply dissolved in water. Shipping weight, 3 ounces.

3L41972—Per box..........**27c**

Intensifying Powders, 27 Cents.

A weak, thin negative lacking in contrast may be greatly improved by intensification, and this powder is a very convenient means of preparing a good intensifier. It requires only to be dissolved in water to be made ready for use. Each package makes 24 ounces of solution. Shipping weight, 5 ounces.

3L41970—Per package.....**27c**

Acid Hypo, 19 Cents a Pound.

A special combination, put up under the most approved formula, of pure granular hypo and other ingredients which, when dissolved in water, makes a clean, clear fixing solution for either plates, films or papers. A very convenient means of preparing the acid fixing bath. It gives unusually good results, as it not only fixes to perfection but also acts as a clearing solution and a hardener. Put up only in 1-pound packages, sufficient to make 64 ounces acid fixing bath. Full instructions included. Shipping wt., 1½ lbs.

3L41985—Per pound....................**19c**

Red Seal Hypo.

(Hyposulphite of Sodium.)

Pea Crystal Hypo, as the name implies, is simply small crystals. The granular is made by crushing or grinding the crystal hypo until it is about like coarse granulated sugar. This form dissolves very readily and is convenient to use.

Pea Crystal Hypo.

3L42156—In 1-lb. carton. Shpg. wt., 1½ lbs. Per lb..........**9c**
3L42157—In 5-lb. friction top can. Shpg. wt., 7½ lbs. Per can..**40c**
3L42158—In 10-lb. screw top can. Shpg. wt., 13 lbs. Per can..**75c**

Granular Hypo.

3L42160—In 1-lb. carton. Shpg. wt., 1½ lbs. Per lb..........**10c**
3L42161—In 5-lb. friction top can. Shpg. wt., 7½ lbs. Per can..**42c**
3L42162—In 10-lb. screw top can. Shpg. wt., 13 lbs. Per can..........**80c**

Red Seal Developers for Developing Papers

Red Seal Developing Powders for papers are very simple and easy to use. It is only necessary to dissolve the contents of one tube in 8 ounces of water and the developer is ready for use. Packed in boxes, as illustrated, six tubes to box, so each box makes 48 ounces of developer ready for use.

Hydro-Metol Powders for Papers.

Regular. For Bluish Black Tones.

A combination of hydrochinon and metol which gives bluish black tones. We recommend this developer in preference to any other for velvet and matte surface papers. Shipping weight, 5 ounces.

3L41936—Per box of **6** tubes..........**22c**

NOTE—Metol sometimes has an irritating or poisonous effect on the skin of some persons. Those who are susceptible to it should select a developer which does not contain metol.

Hydro-Metol Powders for Papers.

Anti-Friction. For Olive Black Tones.

A combination of hydrochinon and metol which gives olive black tones. Put up under a special formula which prevents the dark lines or irregular markings known as friction or abrasion marks, which often disfigure prints on glossy papers. Shipping weight, 5 ounces.

3L41937—Per box of **6** tubes..........**22c**

Amidol Powders for Papers.

Yield very fine blue black tones suitable for use with any standard developing papers or post cards. A very satisfactory developer for prints. Shipping weight, 5 ounces.

3L41938—Per package of **6** tubes.....**22c**

Red Seal Darko Sepia Toner, 35 Cents.

Contains five tubes of bleaching powder for bleaching solution and one tube of powder for redeveloping solution, with full directions. With this toner, prints on any of the various styles or grades of Darko may be toned to a beautiful light brown or sepia color. Especially fine results are obtainable with India Tint Darko. Prints to be toned are first developed in the regular way, the toning constituting an additional operation. Shipping weight, 5 ounces.

3L41948....................**35c**

Eastman, Vulcan and Rexo Film

When ordering film, measure the length of the empty spool in your camera. Spool lengths are stated for each style. As a further guide to the style you require note the maker's number as stated under the description of each kind. You will find this numbering both on the roll itself and on the carton.

Vulcan and Eastman N. C. Roll Film

Both Made by the Eastman Kodak Co.

Both brands are non-curling. Both are orthochromatic. Both are free from halation. Both are extremely rapid. **We have made repeated tests of the two and can see no difference in the speed of the film or in the quality of the negatives.**

Both Vulcan and Eastman N. C. Film may be used in Autographic Kodaks and Brownies, but no record or title can be made on the film at the time of exposure.

FOR 1⅝x2½ PICTURES.

Length of spool, 1 15/16 inches. Makers' numbers: Vulcan, 254; Eastman, 127; Rexo, 407. Fits cameras using 1⅝x2½ film, with spool length of 1 15/16 inches.

3L42708—Vulcan. 8-exp. roll..**21c**
3L42709—Eastman. 8-exp. roll..**22c**
3L42710—Rexo. 8-exp. roll..**20c**
Shipping weight, 1½ ounces.

FOR 2¼x2¼ PICTURES.

Length of spool, 2⅝ inches. Makers' numbers: Vulcan, 234; Eastman, 117; Rexo, 410. Fits cameras using 2¼x2¼ film, with spool length of 2⅝ inches.

3L42712—Vulcan. 6-exp. roll..**17c**
3L42713—Eastman. 6-exp. roll..**18c**
3L42714—Rexo. 6-exp. roll..**16c**
Shipping weight, 2 ounces.

FOR 2¼x3¼ PICTURES.

Length of spool, 2 9/16 inches. Makers' numbers: Vulcan, 240; Eastman, 120; Rexo, 415. Fits:

No. 2 Kewpie
No. 2 Folding Kewpie
No. 2 Conley Junior
No. 2 Conley Fixed Focus

and any other camera using 2¼x3¼ film, with a spool length of 2 9/16 inches.

3L42716—Vulcan. 6-exp. roll..**22c**
3L42717—Eastman. 6-exp. roll..**23c**
3L42718—Rexo. 6-exp. roll..**21c**
Shipping weight, 2 ounces.

FOR 2½x4¼ PICTURES.

Length of spool, 2⅞ inches. Makers' numbers: Vulcan, 232; Eastman, 116; Rexo, 425 and 426. Fits:

No. 2A Kewpie Kamera
No. 2A Folding Kewpie Kamera
No. 2A Model A Conley
No. 2A Conley Junior
No. 2A Conley Fixed Focus
No. 2A Conley De Luxe
Model E Conley

or any other camera using 2½x4¼ film, with a spool length of 2⅞ inches.

3L42728—Vulcan. 6-exp. roll.**26c**
3L42729—Vulcan. 12-exp. roll.**52c**
3L42730—Eastman. 6-exp. roll.**27c**
3L42731—Eastman. 12-exp. roll.**54c**
3L42732—Rexo. 6-exp. roll.**24c**
3L42733—Rexo. 12-exp. roll.**48c**
Shipping weights, 6-exposure roll, 2½ ounces; 12-exposure roll, 3 ounces.

FOR 2⅞x4⅞ PICTURES.

Length of spool, 3 9/16 inches. Makers' numbers: Vulcan, 236; Eastman, 130;
Continued in next column.

REXO ROLL FILM

Rexo Film possesses in the highest degree every quality desirable in a roll film. We are offering it to our customers only after repeated tests, under a variety of conditions and over a good length of time have satisfied us that it is a high class product, a film we can fully recommend.

Try a roll of Rexo alongside any of the other popular brands, under the same conditions, and convince yourself that Rexo Film will produce negatives of highest quality at a considerable saving in cost.

Rexo Film is extremely rapid, non-curling, orthochromatic and free from halation. It can be used in Autographic Cameras, but no record can be made on the film at time of exposure.

Continued from column 1

Rexo, 436 and 438. Fits No. 2C Kewpie and any other camera using 2⅞x4⅞ film, with a spool length of 3 9/16 inches.

3L42736—Vulcan. 6-exp. roll.**40c**
3L42737—Vulcan. 10-exp. roll.**66c**
3L42738—Eastman. 6-exp. roll.**41c**
3L42739—Eastman. 10-exp. roll.**68c**
3L42740—Rexo. 6-exp. roll.**38c**
3L42741—Rexo. 10-exp. roll.**62c**
Shipping weights, 6-exposure roll, 3 ounces; 10-exposure roll, 3½ ounces.

FOR 3½x3½ PICTURES AND STEREOS.

Length of spool, 3¾ inches. Makers' numbers: Vulcan, 202; Eastman, 101; Rexo, 440. Fits any camera using 3½x3½ film, with a spool length of 3¾ inches.

3L42744—Vulcan. 6-exp. roll.**31c**
3L42746—Eastman. 6-exp. roll.**32c**
3L42748—Rexo. 6-exp. roll.**29c**
Shipping weight, 3½ ounces.

FOR 3¼x4¼ PICTURES AND STEREOS.

Length of spool, 3½ inches. Makers' numbers: Vulcan, 236; Eastman, 118; Rexo, 430. Fits:

No. 3 Model C Conley
No. 3 Model B Conley
Model B-1 Conley
No. 3 Conley Junior
No. 3 Conley De Luxe

and any other camera using 3¼x4¼ film, with a spool length of 3½ inches.
Continued in next column.

Continued from column 2

3L42752—Vulcan. 6-exp. roll.**40c**
3L42754—Eastman. 6-exp. roll.**41c**
3L42756—Rexo. 6-exp. roll.**38c**
Shipping weight, 3½ ounces.

CAUTION—Be careful in ordering film for 3¼x4¼ pictures. We list two kinds. Note length of spool and makers' numbers.

FOR 3¼x4¼ PICTURES.

Length of spool, 3¾ inches. Makers' numbers: Vulcan, 248; Eastman, 124; Rexo, 435. Fits:

No. 3 Kewpie
No. 3 Folding Kewpie
Model A-1 Conley
No. 3 Model A Conley
Conley Jr., old style with square ends

and any other camera using 3¼x4¼ film, with a spool length of 3¾ inches.

3L42760—Vulcan. 6-exp. roll.**40c**
3L42761—Vulcan. 12-exp. roll.**79c**
3L42762—Eastman. 6-exp. roll.**41c**
3L42763—Eastman. 12-exp. roll.**81c**
3L42764—Rexo. 6-exp. roll.**38c**
Shipping weights, 6-exposure roll, 3½ ounces; 12-exposure roll, 4½ ounces.

FOR 3¼x5½ PICTURES.

Length of spool, 3¾ inches. Makers' numbers: Vulcan, 244; Eastman, 122; Rexo, 445 and 446. Fits:

No. 3A Folding Kewpie
No. 3A Conley Junior

Continued in next column.

Continued from column 3

Conley Junior, old style with square ends

No. 3A Model B Conley
No. 3A Model C Conley
No. 3A Model D Conley
Model A-2 Conley
No. 3A Conley De Luxe

and any other camera using 3¼x5½ film, with a spool length of 3¾ inches. (Does not fit Kewpie Box Kamera; see 3L42776 to 3L42780.)

3L42768—Vulcan. 6-exp. roll.**48c**
3L42769—Vulcan. 10-exp. roll.**78c**
3L42770—Eastman. 6-exp. roll.**50c**
3L42771—Eastman. 10-exp. roll.**81c**
3L42772—Rexo. 6-exp. roll.**44c**
3L42773—Rexo. 10-exp. roll.**73c**
Shipping weights, 6-exposure roll, 4 ounces; 10-exposure roll, 4½ ounces.

CAUTION—Be careful in ordering film for 3¼x5½ pictures. We list two kinds. Note length of spool and makers' numbers.

FOR 3¼x5½ PICTURES AND STEREOS.

Length of spool, 4 inches. Makers' numbers: Vulcan, 250; Eastman, 125; Rexo, 450. Fits **No. 3A Kewpie Kamera** and any other camera using 3¼x5½ film, with a spool length of 4 inches.

3L42776—Vulcan. 6-exp. roll.**48c**
3L42777—Vulcan. 10-exp. roll.**78c**
3L42778—Eastman. 6-exp. roll.**50c**
3L42779—Eastman. 10-exp. roll.**81c**
3L42780—Rexo. 6-exp. roll.**44c**
Shipping weights, 6-exposure roll, 4 ounces; 10-exposure roll, 4½ ounces.

FOR 4x5 PICTURES AND 3½x12 PANORAMS.

Length of spool, 4 inches. Makers' numbers: Vulcan, 206; Eastman, 103; Rexo, 460. Fits cameras using 4x5 film, with a spool length of 4 inches.

3L42784—Vulcan. 6-exp. roll.**48c**
3L42786—Eastman. 6-exp. roll.**50c**
3L42788—Rexo. 6-exp. roll.**44c**
Shipping weight, 4 ounces.

Note—When used in Panoramic Cameras, the "6-exposure" roll makes two 3½ x12 exposures.

Eastman Autographic Film

This film is designed for use in Autographic Kodaks and Autographic Brownies. It may be used in other cameras, but the autographic feature—that is, the recording of writing on the film at time exposure is made—can be made use of only when this film is used in an Autographic Camera.

For 1⅝x2½ Pictures.

Fits Vest Pocket Autographic Kodaks. Eastman's number A127.

3L42820—8-exp. roll......**23c**
Shipping weight, 1½ ounces.

For 2¼x3¼ Pictures.

Fits No. 1 Autographic Kodaks or No. 2 Folding Autographic Brownie. Eastman's number A120.

3L42823—6-exp. roll,.....**23c**
Shipping weight, 2 ounces.

For 2½x4¼ Pictures.

Fits No. 1A Autographic Kodaks or No. 2A Folding Autographic Brownie. Eastman's number A116.

3L42826— 6-exp. roll.**27c**
3L42827—12-exp. roll.**54c**
Shipping weights, 6-exposure roll, 2½ ounces; 12-exposure roll, 3 ounces.

For 2⅞x4⅞ Pictures.

Fits No. 2C Autographic Kodaks or No. 2C Folding Autographic Brownie. Eastman's number A130.

3L42830— 6-exp. roll.**41c**
3L42831—10-exp. roll.**68c**
Shipping weights, 6-exposure roll, 3 ounces; 10-exposure roll, 3½ ounces.

For 3¼x4¼ Pictures.

Fits No. 3 Autographic Kodaks. Eastman's number A118.

3L42834— 6-exp. roll.**41c**
3L42835—12-exp. roll.**81c**
Shipping weights, 6-exposure roll, 3½ ounces; 12-exposure roll, 4 ounces.

For 3¼x5½ Pictures.

Fits No. 3A Autographic Kodaks or No. 3A Folding Autographic Brownie. Eastman's number A122.

3L42838— 6-exp. roll.**50c**
3L42839—10-exp. roll.**81c**
Shipping weights, 6-exposure roll, 4 ounces; 10-exposure roll, 4½ ounces.

FILM PACKS, ADAPTERS AND PLATES.

Kodak Film Packs.

Use Films in Your Plate Camera.

By the use of a film pack adapter you can instantly convert any plate camera, up to the 5x7 size, into a daylight loading film camera. Instead of two exposures with a plate holder, a film pack adapter occupying no more space gives you twelve exposures without reloading camera. Each pack contains twelve films. Any number of the films may be exposed and removed without exposing the remainder.

Kodak Film Packs also fit film pack cameras without an adapter.

Catalog No.	Size	Shpg. Wt.	Each
3L41892	2¼x3¼	4 ounces	**$0.45**
3L41893	2½x4¼	5 ounces	**.54**
3L41894	3¼x4¼	6 ounces	**.81**
3L41895	4 x5	7 ounces	**.99**
3L41896	*3 x5¼	7 ounces	**.90**
3L41897	*3¼x5½	7 ounces	**.99**
3L41898	5 x7	12 ounces	**1.71**

*The 3x5¼ Film Pack fits all 3¼x5½ cameras except the Premo and Seneca models of 1913 and later. These cameras take the full 3¼x5½ size.

Film Pack Adapters for Plate Cameras.

A device the same size and shape as a plate holder which enables one to use film with any ordinary glass plate camera. Fit any style of Conley or Seroco cameras, also the Century, Premo, Seneca and Black Beauty cameras.

3L45880

Size	Shpg. Wt.	Each
3¼x4¼	5 oz.	**$1.38**
4 x5	6 oz.	**1.45**
*3 x5¼	6 oz.	**1.45**
*3¼x5½	7 oz.	**1.45**
5 x7	10 oz.	**2.30**

*All 3¼x5½ cameras, except Premos and Senecas made since 1913, take the 3x5¼ adapter.

Hammer "Extra Fast" Orthochromatic Plates.

Suitable for landscape views, portraits and all subjects in which color values are important, such as flowers and paintings. For full color correction, a ray filter should be used.

3L41764

Size	Weight, per Dozen	Shipping Weight, per Dozen	Per Dozen
3¼x4¼	17 oz.	23 oz.	**$0.58**
4 x5	1½ lbs.	1⅞ lbs.	**.81**
3¼x5½	1⅜ lbs.	1¾ lbs.	**.81**
5 x7	2⅝ lbs.	3½ lbs.	**1.30**
6½x8½	4 lbs.	5¾ lbs.	**1.98**
8 x10	5¾ lbs.	8 lbs.	**2.88**

Hammer "Extra Fast" Plates.

Probably the most popular of all the Hammer plates for general all around photography.

3L41760

Size	Weight, per Dozen	Shipping Weight, per Dozen	Per Dozen
3¼x4¼	17 oz.	23 oz.	**$0.52**
4 x5	1½ lbs.	1⅞ lbs.	**.72**
3¼x5½	1⅜ lbs.	1¾ lbs.	**.72**
4¼x6½	2 lbs.	2⅝ lbs.	**1.04**
5 x7	2⅝ lbs.	3½ lbs.	**1.16**
6½x8½	4 lbs.	5¾ lbs.	**1.76**
8 x10	5¾ lbs.	8 lbs.	**2.56**

Stanley "Regular" Dry Plates.

Meet the requirements for general all around photography. Extremely rapid, yielding crisp, clean, snappy negatives of excellent printing quality. We recommend them highly for landscapes, architectural subjects, general instantaneous work, portraits in the studio or in the home, groups and for general commercial work.

3L41778

Size	Weight, per Dozen	Shipping Weight per Dozen	Per Dozen
3½x3½	16 oz.	21 oz.	**$0.44**
3¼x4¼	17 oz.	23 oz.	**.52**
4 x5	1½ lbs.	1⅞ lbs.	**.72**
3¼x5½	1⅜ lbs.	1¾ lbs.	**.72**
4¼x6½	2 lbs.	2⅝ lbs.	**1.04**
5 x7	2⅝ lbs.	3½ lbs.	**1.16**
6½x8½	4 lbs.	5¾ lbs.	**1.76**
8 x10	5¾ lbs.	8 lbs.	**2.56**

Folders, Mounts and Photo Supplies

Iris Slip-In Post Card Folders.

A beautiful slip-in post card folder consisting of a single piece of heavy rough surfaced stock, folded to produce a cut-out mat. Cover is embellished with printed double line border and embossed, printed corner design. Double lines, one wide, one narrow, extend around the picture opening and a double line border is embossed around the outside edge of the mat. The panel shape of this mounting lends distinction to the post card. Shipping weight, per dozen, 1 pound 5 ounces.

Colors: Gray or brown. **State color wanted.**

3L43206—Size, closed, 4½x9 inches; opening, 2¾x5 inches; for photos, 3¼x5½ inches. Per ½ gross, **$2.68**; per dozen....**47c**

Sold only in original sealed packages.

Datura Slip-In Folders.

A double flap slip-in folder, especially designed to meet the needs of the amateur photographer. Cover has embossed surface. Mat is of lighter tint than cover and has a printed and embossed border line around opening, which is square. Sold only in original sealed packages.

Colors: Brown or gray. **Be sure to state color.**

	Size, Closed	Opening	For Photos	Shpg. Wt., per Doz.	Per ½ Gross	Per Doz.
3L43209	2⅞x4¼	1 7/16x2 5/16	1⅝x2½	6 oz.	$2.11	37c
3L43210	3⅝x5¼	2 1/16x3⅛	2¼x3¼	10 oz.	2.28	40c
3L43211	3⅞x6¼	2 5/16x4 1/16	2½x4¼	13 oz.	2.46	43c
3L43212	4⅝x6½	2 15/16x4	3¼x4¼	13 oz.	2.91	51c
3L43213	4⅞x7 5/16	2 15/16x5	3½x5½	1 lb.	3.02	53c

Printing Made Easy With Our Electric Printer and Safelight.

Makes printing certain, as exposures are under accurate control. Made of wood with ventilated metal panels. Prints from films or glass negatives of any size up to and including 5x7. A hinged door operated from outside the box divides box into two compartments. Socket for electric lamp is in lower compartment. Exposure is made by raising and lowering the door. There is a safelight by which the operator can see to place the paper on the negative. A yellow window in the side forms a safelight by which to develop prints. Replacing this with a ruby glass, 3¼x5¼ inches, the printer becomes a darkroom lamp.

3L41423—Electric Printer and Safelight, complete with 6 feet of cord, plug and socket, without lamp. Shipping weight, 5 pounds....**$3.40**

Ground Glass.

Ground glass, 5x7 inches. This glass fits the Electric Printer and Safelight. Shipping wt., 2 pounds 2 ounces.

3L41230....**25c**

A Properly Trimmed Print Mounts Well.

3L41465—Print Trimming Board. Blade of tempered steel; board of polished hardwood, graduated measure and guide. For all prints 4x5 and smaller. Shpg. weight 1½ lbs.

Each....**59c**

3L41466—Trimming Board, same as 3L41465, but for prints up to 5x7. Shpg. wt., 2¼ lbs....**76c**

3L41467—Trimming Board, same as 3L41465, but with 10½-inch blade, for prints up to 8x10. Shipping weight, 4 pounds....**$1.15**

Our Best Trimming Boards.

Best Print Trimming Boards. Blade of good quality tempered steel. Hardwood, polished. Board will not warp. **Spring joint** keeps the two cutting edges always in contact.

3L41472				
Length of blade, in.	6½	8½	10½	12½
Shipping weight	3¼ lbs.	4½ lbs.	6 lbs.	8¾ lbs.
Each	**$1.80**	**$2.40**	**$2.80**	**$3.50**

Coleus Slip-In Easel Mounts.

Just the Thing for Amateur Photos!

A dainty easel folder mount which may be used either as an easel or as a folder, as shown in the illustrations. The picture is held by the slip-under corners, no pasting required. Sold only in original sealed packages.

Made of fine grade stock in two colors, gray or brown. **State color wanted.**

	For Photos	Size of Mount, Closed	Shpg. Wt., per Dozen	Per ½ Gross	Per Doz.
3L43220	2¼x3¼	2½x3½	7 oz.	**$1.77**	**31c**
3L43221	2½x4¼	2¾x4½	9 oz.	**2.11**	**37c**
3L43222	3¼x4¼	3½x4½	10 oz.	**2.28**	**40c**
3L43223	3¼x5½	3½x5¾	15 oz.	**2.57**	**45c**

Azura Mounts.

Made from fine quality heavy stock, with heavily embossed artistic design. Square corners. Sold only in original sealed packages.

Colors: Gray or brown. **State color wanted.**

	For Photos	Shpg. Wt., per 25	Per 100	Per 25
3L43224	1⅝x 2½	13 oz.	**$0.81**	**$0.22**
3L43225	2¼x 3¼	15 oz.	**1.20**	**.32**
3L43226	2½x 4¼	1 lb. 2 oz.	**1.39**	**.37**
3L43227	3¼x 4¼	1 lb. 4 oz.	**1.44**	**.38**
3L43228	4 x 5	2 lbs.	**1.57**	**.42**
3L43229	3¼x 5½	2 lbs.	**1.62**	**.43**
3L43230	5 x 7	3 lbs.	**2.15**	**.57**
3L43231	6½x 8½	4 lbs. 6 oz.	**3.34**	**.88**
3L43232	8 x10	6 lbs.	**4.71**	**1.24**

Rainier Slip-In Mounts.

An excellent quality slip-in view mount. Card of good weight cloth grained stock. Mat of printed, embossed stock. Decoration is simple and effective. The printed, embossed border around the opening and the printed, embossed line around the outside of the mat give a most pleasing effect. Sold only in original sealed packages.

Colors: Brown or gray. **Be sure to state color.**

For photos, in.	**3L43233**	**3L43234**	**3L43235**
Opening, inches	5x7	6½x8½	8x10
Size of card, in.	4½x6½	6x8	7x9½
	7⅜x9⅜	9x11	10½x12½
Shipping weight, per dozen, lbs.	1⅜	1⅞	2
Per ½ gross	**$3.53**	**$5.30**	**$7.47**
Per dozen	**.62**	**.93**	**1.31**

Amateur Printing Frames.

High grade light weight printing frames with piano hinge, usually found only on high priced frames. **Be sure to state size.**

Size, Inches	Shipping Weight	3L41410 Without Glass	Shipping Weight	3L41411 With Glass
2¼x3¼	8 oz.	**22c**	1 lb. 1 oz.	**23c**
2½x4¼	9 oz.	**23c**	1 lb. 3 oz.	**24c**
3¼x4¼	9 oz.	**23c**	1 lb. 4 oz.	**25c**
4 x5	10 oz.	**24c**	1 lb. 5 oz.	**27c**
3¼x5½	10 oz.	**25c**	1 lb. 12 oz.	**28c**
3⅝x6	11 oz.	**26c**	1 lb. 14 oz.	**30c**
5 x7	14 oz.	**27c**	2 lbs. 2 oz.	**31c**

NOTE—For printing post cards from negatives 3¼x5½ or smaller, use a 3⅝x6 frame with glass.

Glass is necessary when printing from glass negatives smaller than frame and when printing from film negatives.

Professional Printing Frames.

Heavy weight printing frames, strongly made. Springs slide under steel plates instead of grooves in the wood. Corners mortised; back in three pieces to prevent warping. **State size wanted.**

Size, Inches	Shpg. Wt.	3L41415 Without Glass	Shpg. Wt.	3L41416 With Glass
3¼x 4¼	15 oz.	**50c**	1 lb. 8 oz.	**52c**
4 x 5	1 lb.	**53c**	2 lbs. 2 oz.	**56c**
3¼x 5½	1 lb.	**55c**	2 lbs. 2 oz.	**58c**
3⅝x 6	1 lb. 1 oz.	**60c**	2 lbs. 2 oz.	**63c**
5 x 7	1 lb. 5 oz.	**65c**	2 lbs. 8 oz.	**69c**
6½x 8½	1 lb. 11 oz.	**83c**	3 lbs. 3 oz.	**88c**
8 x10	2 lbs. 2 oz.	**91c**	3 lbs. 4 oz.	**98c**

See note under 3L41410 and 3L41411.

Cleome Slip-In Post Card Folders.

A double flap slip-in folder of rich design and good quality mottled stock. Cover has all around border decoration and is embellished with an artistic design embossed across it. Mat is of same stock as folder but in fancy embossed grain. Opening has printed and embossed borders and a narrow, printed embossed line extends around outside edge of mat. A truly beautiful mount. Sold only in original sealed packages. Size, closed, 4¾x8½ in.; opening, 3x4½ in., for post card prints. Shipping weight, per dozen, 1 pound 4 ounces.

Colors: Gray or brown. **State color wanted.**

	Opening	Per ½ Gross	Per Doz.
3L43207	Oval	**$3.88**	**68c**
3L43208	Square	**3.88**	**68c**

Torenia Slip-In Folders

An exceptional quality slip-in double flap folder of strikingly handsome design. Card and mat both of extra high grade clouded, embossed stock. The gracefully rounded flaps are embellished with artistic printed and embossed designs. A fine embossed, printed line, surrounded by a narrow border of stipple embossing on the outside edges of the flaps, gives them a most pleasing finish. Mat has either square or oval opening, surrounded by beautiful printed and stipple embossed border, narrow embossed, printed line extends around outer edge of mat. Opening is ¼ to ½ inch smaller on each dimension than size print for which it is listed. Sold only in original sealed packages. **Brown only.**

Square Opening	Oval Opening	Size, Closed	For Photos	Shpg. Wt., Per Doz.	Per ½ Gross	Per Dozen
3L43214	3L43217	4¼x6	3 x4	14 oz.	**$3.36**	**59c**
3L43215	3L43218	5⅛x7⅛	3½x5	1⅛ lbs.	**3.70**	**65c**
3L43216	3L43219	5⅞x9¼	4 x6	1¾ lbs.	**4.62**	**81c**

Improve Your Negatives by Retouching.

Contains one retouching pencil, one small bottle retouching varnish, one small bottle opaque, one etching knife, one spotting brush and one spotting pencil. Shipping weight, 4 ounces.

3L41578—Complete Retouching Outfit....**89c**

Kensington Film Negative Books.

Very convenient for filing film negatives. Keeps them clean, flat and readily accessible. Book contains 50 numbered envelopes, made of tough, translucent paper, and an index with corresponding numbers. Shipping weight, 4 ounces.

3L47580—For negatives 2½x4¼ or smaller....**35c**

3L47590—For negatives 4x5 or 3¼x5½....**45c**

Squeegee Plates.

Good quality light weight ferrotype plates for squeegeeing glossy prints to produce a highly polished surface. Size, 10x14 inches. Shipping weight, per dozen, 4¼ pounds; each, 12 ounces.

3L41546—Per dozen, **$1.53**; 3 for....**45c**

Polish for Squeegee Plates.

A dry preparation for polishing ferrotype or squeegee plates to prevent the prints sticking to them. Shipping weight, 2 ounces.

3L41548—Per box....**12c**

Print Roller.

For smoothing down prints after mounting and for squeegeeing prints on ferrotype plates. Rubber covered 4-inch roller. Black enameled metal handle. Shipping weight, 12 ounces.

3L41492....**22c**

Perfection Blotter Book.

For drying prints flat. Has twelve sheets of lintless blotting paper, 9x12 inches, interleaved with fine quality wax paper, all bound in heavy manila covers. Shipping weight, 13 ounces.

3L41528....**23c**

Photographers' Blotting Paper.

3L41526—For drying or mounting prints, free from lint. **State size.**

Size, 9x12 in. Shipping weight, 10 oz. Dozen....**14c**

Size, 19x24 in. Shipping wt., 2 lbs. 3 oz. Dozen....**45c**

High in Quality! Groceries Low in Price!

Special Sample Selection of PIE FRUITS

Mother's pies! A fitting ending to the delicious meals that mother provides. With perfect pie fruits mother's task is made easy and her desserts more appetizing than ever.

We want you to try our groceries, and this special sample selection of pie fruits is listed here just for that purpose. It will show you, as only a trial will, that we have reason to be proud of the quality of our foods.

Another illustration of the money you can save and the better groceries you secure when you order from our Grocery Catalog.

12 cans of our most popular pie fruits.

2 cans No. 2 Rivera Brand Pumpkin.
1 can No. 2 Kingston Brand Red Sour Cherries.
2 cans No. 2 Montclair Brand Crushed Pineapple.
2 cans No. 2 Kingston Brand Blackberries.
1 can No. 2 Montclair Brand Wet Mince Meat.
2 cans No. 2½ Kingston Brand Pie Peaches.
1 can No. 2½ Kingston Brand Apricots.
1 can No. 2 Montclair Brand Blueberries.

7L12697—12 cans...**$2.45**

Contents No. 2 cans, 1 pound 4 ounces. Contents No. 2½ cans, 1 lb. 14 oz.

Shipping weight, 12 cans, 20 pounds.

Pie fruits are only one of the many food products that we offer at bargain prices in our Grocery Catalog. SEND FOR A COPY.

Special Sample Selection of CANNED VEGETABLES

There's nothing quite so much appreciated as good vegetables. And there are no better vegetables canned than those we sell under our Montclair Brand!

Just to acquaint you with the superiority of Montclair foods we have made up this special sample selection of the best canned vegetables on earth!

There's a treat in store for you if you've never tasted Montclair corn or beans or any of the others listed in this assortment.

Order this assortment today and we are confident that you'll always serve our canned vegetables. They're so good and so reasonably priced!

Montclair Brand

12 No. 2 cans of our highest quality canned vegetables.

1 can Montclair Golden Bantam Corn.
1 can Montclair Maine Corn.
1 can Montclair Kidney Beans.
1 can Montclair Red Beans.
1 can Montclair Pork and Beans.
1 can Montclair Succotash.
1 can Montclair Pumpkin.
1 can Montclair Spaghetti.
1 can Montclair Brussels Sprouts.
1 can Montclair Soup Vegetables.
1 can Montclair Milk Hominy.
1 can Montclair Spinach.

7L13517—12 cans...**$1.60**

All No. 2 cans. Contents, 1 pound 4 ounces.

Shipping weight, 12 cans, 19 pounds.

This is only one of the many bargains we have for you in our Grocery Catalog. SEND FOR A COPY.

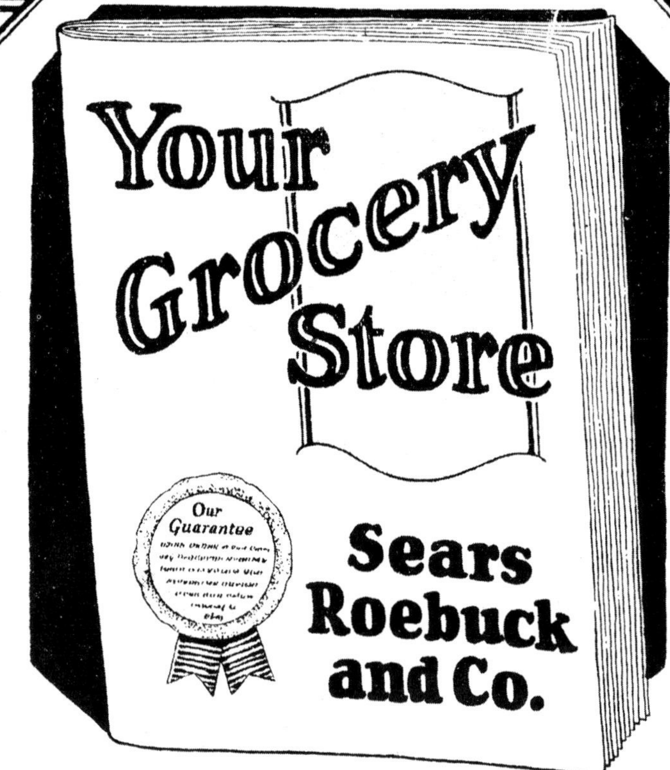

Your
Grocery
Store
Our Guarantee
Sears
Roebuck
and Co.

Full of good things to eat and at money saving prices our Grocery Catalog—YOUR GROCERY STORE—is yours for the asking.

Just a few of the hundreds of articles that await your selection are shown on these pages. We want you to get this book. But we want you most of all to test the quality of our groceries and to test the saving that you can make by using this catalog.

Many of our customers tell us that they save as much as one-fourth. You can do the same, and get better groceries.

Our grocery guarantee tells what confidence we have in our products. Here it is:

You may return at our expense any groceries you buy of us which are not satisfactory in every way, or on which you think you have not saved money. We will then exchange them for exactly what you want or return your money, together with any transportation charges you paid.

Order from this page, or ask for our Grocery Catalog 7GCL606. DO IT TODAY!

Dried Fruits

Our dried fruits are simply delicious! Meaty, flavory, appetizing, they are the pick of the season's crops.

Hundreds of thousands of housekeepers testify to the superiority of our dried fruits. Grade for grade, they are the best fruits that money will buy, and when you remember that we guarantee a saving on the quality you prefer, you will realize that we have the utmost confidence in what we are selling.

Ask for our Grocery Catalog. Send us a test order. Prove to your own satisfaction how much better and how much cheaper our fruits are.

Coffee Headquarters

2 lbs. Fulton Brand 59¢

Nowhere else can you get the quality, the wonderful blends, the big values that we offer in coffees! When we started our coffee business, over twenty-five years ago, we were determined *we would give our customers such extraordinary values that they would always look to us to furnish them with coffee.*

Beginning with the source of supply, we selected only coffees grown in those climates especially adapted for producing fine flavored berries. A careful system of grading, sorting, blending and roasting under the direct supervision of our own experts was developed and perfected. Many economies in handling enabled us to quote prices for quality coffees which brought these within the reach of all! Our limited space here does not permit us to display our remarkable facilities for blending and roasting coffee, but we extend a cordial invitation to all of our customers to visit this department whenever they come to Chicago.

For years thousands of families have been enjoying the world's finest blended coffees at prices usually charged for ordinary grades. Now we want you to give our coffees a trial. You can order a 2½-pound canister of Montclair or a 2-pound package of Fulton Coffee by parcel post if you wish. Then try it out in your own home, serve it at your meals. We'll rest our case right there. If you don't think you've enjoyed the best coffee, at the price, that you've ever tasted, return it and we will cheerfully return your money, also transportation charges.

Montclair Coffee is a blend of the highest grade coffees grown. Rich, fragrant, smooth, it is the choice of experts.

Fulton Brand Coffee has become very popular with our many customers who have proclaimed it the Greatest Coffee Value Offered!

Montclair Brand Coffee.

5-Pound Canister, $2.00

7L183—2½-pound canister...**$1.03**
Shipping weight, 3¼ pounds.

7L185—5-pound canister.....**$2.00**
Shipping weight, 6½ pounds.

Fulton Brand Coffee.
"The World's Greatest Coffee at the Price."

5 Pounds, $1.46

7L152—2-pound package.......**$0.59**
Shipping weight, 2¼ pounds.

7L153—5-pound package....... **1.46**
Shipping weight, 6½ pounds.

Coffee is only one of the many articles offered in our Grocery Catalog at money saving prices. SEND FOR A COPY!

We Have Many Other Splendid Bargains Like This in Our Grocery Catalog. Send for a Copy.

Special Sample Selection Pacific Coast Canned Fruits

12 Cans Kingston Brand $3.40

The whole world knows Pacific Coast Fruits. Large, luscious, palatable, appetite-creating, our canned fruits are the best that the most famous orchards produce.

So that you may taste the goodness of these wonderful canned fruits, we have made two special selections of our **Kingston** and **Montclair** Brands. The former, our lowest priced, is standard grade, while the latter is a super grade, for which we use only the largest and most perfect fruits. In each selection are twelve large cans of these delicious Pacific Coast fruits.

Just let us send you one of our selections. If you don't think these the best fruits you ever tasted—regardless of price—if you don't think you've saved money, return them and your money with transportation charges will be returned without question.

Kingston Brand.
Fruit Assortment.
12 Cans of Fruit for $3.40.

Standard grade, perfect fruit in lighter syrup than our Montclair Brand. No. 2½ cans.

- 2 cans Yellow Cling Peaches, Halves.
- 2 cans Yellow Cling Peaches, Sliced.
- 2 cans Apricots.
- 2 cans Bartlett Pears.
- 2 cans Egg Plums.
- 1 can White Cherries.
- 1 can Muskat Grapes.

Contents of each can, 1 pound 14 ounces.

7L12687—12 cans........**$3.40**
Shipping weight, 28 pounds.

Montclair Brand.
Fruit Assortment.
12 Cans of Fruit for $4.23.

Largest, most perfect fruits packed, in extra heavy syrup. No. 2½ cans.

- 2 cans Yellow Cling Peaches, Halves.
- 2 cans Yellow Cling Peaches, Sliced.
- 1 can Apricots.
- 1 can White Cherries.
- 1 can Bartlett Pears.
- 1 can Egg Plums.
- 1 can Muskat Grapes.
- 1 can Sliced Pineapple.
- 1 can Crushed Pineapple.
- 1 can Assorted Fruits for Salad.

Contents of each can, 1 pound 14 ounces.

7L12657—12 cans **$4.23**
Shipping weight, 28 pounds.

TEAS We Guarantee Quality and Saving!

62¢ a lb.

Only the Orient can produce teas such as we sell under our **Rivera Brand** and only the choice grade, the first picking, is used by us. Tender, fragrant and aromatic, Rivera Teas are sure to please you.

You can depend upon saving if you order your teas from us and you can depend upon full weight, too. We import direct and the tea comes to you in airtight canisters. Just send a trial order! For only 62 cents a pound we will send you any of the following wide selection. Order at least a pound, today, to be sent by parcel post if you wish, and let's get acquainted! State catalog number.

1-Pound Canister, 62c.

7L50—Uncolored Japan. GREEN—A beautiful light liquor.
7L52—Basket Fired Japan. GREEN—A very pale liquor.
7L54—Imperial Gunpowder. GREEN—A very aromatic liquor.
7L56—Young Hyson. GREEN—A medium light liquor.
7L61—English Breakfast. BLACK—A fine amber liquor.
7L63—Oolong. BLACK—A medium amber liquor.
7L65—India-Ceylon. Orange-Pekoe. BLACK—An amber liquor.
7L68—Green and Black Mixed. A blend of Oolong and Gunpowder teas. A medium amber liquor.
7L70—Special Breakfast. A mixture of choice teas. A medium dark liquor.

Shipping weight, each pound canister, 1¼ pounds. Contents, 1 pound.

Teas are only one of the hundreds of bargains we offer in our Grocery Catalog. SEND FOR A COPY!

Special Sample Selection Sea Foods

12 Cans $2.20

Here's a selection of tasty, healthy, freshly packed goods from our fish department—a "get acquainted" lot that we've made up so it would be easy for you to judge for yourself the superior quality of our goods and the economy of prices. **Remember our guarantee.**

Choice, Fresh Packs of 12 Favorite Sea Foods.

- 1 can No. 1 Montclair (Wet) Shrimp. 5¾ oz.
- 1 can Montclair Fish Roe. 10 ounces.
- 1 can No. 3 Montclair Clam Chowder. 2 pounds 1 ounce.
- 1 can No. 1 Doxee Neptune Whole Clams. 10 ounces.
- 1 can Montclair Fish Flakes. 6 ounces.
- 1 can No. 1 Crown Kippered Silver Hake. 14½ ounces.
- 1 can No. 1 Marshall's Kippered Herring. 1 lb.
- 1 can No. 1½ Gorton Fish Balls. 1 pound.
- 1 can No. 1 Rivera Sardines in Tomato Sauce. 15 ounces.
- 1 can No. 1 Kingston Alaska Red Salmon. 1 lb.
- 1 can ¼ Dupont Imported Portuguese Sardines. 3¾ ounces.
- 1 can No. 1 Pilchards. 1 lb.

7L16947—12 cans.......(Shipping weight, 12 cans, 17 lbs.).........**$2.20**

This is only one of the many bargains in our Grocery Catalog. SEND FOR A COPY!

Soaps and Laundry Supplies

We're famed all over the country for our soaps and laundry supplies. And rightly so. For we exercise the greatest care to see that only the best ingredients money can buy go into our products.

Because of our great volume, we are able to sell these needed articles much below regular market prices. Note, too, that our packages are large, full weight. Whatever you need for laundry or bath, we have it in our big grocery department and at prices that mean a decided saving. Send for a copy of our Grocery Catalog!

Look at These Extraordinary Values In Popular Luxurious Parlor Sets

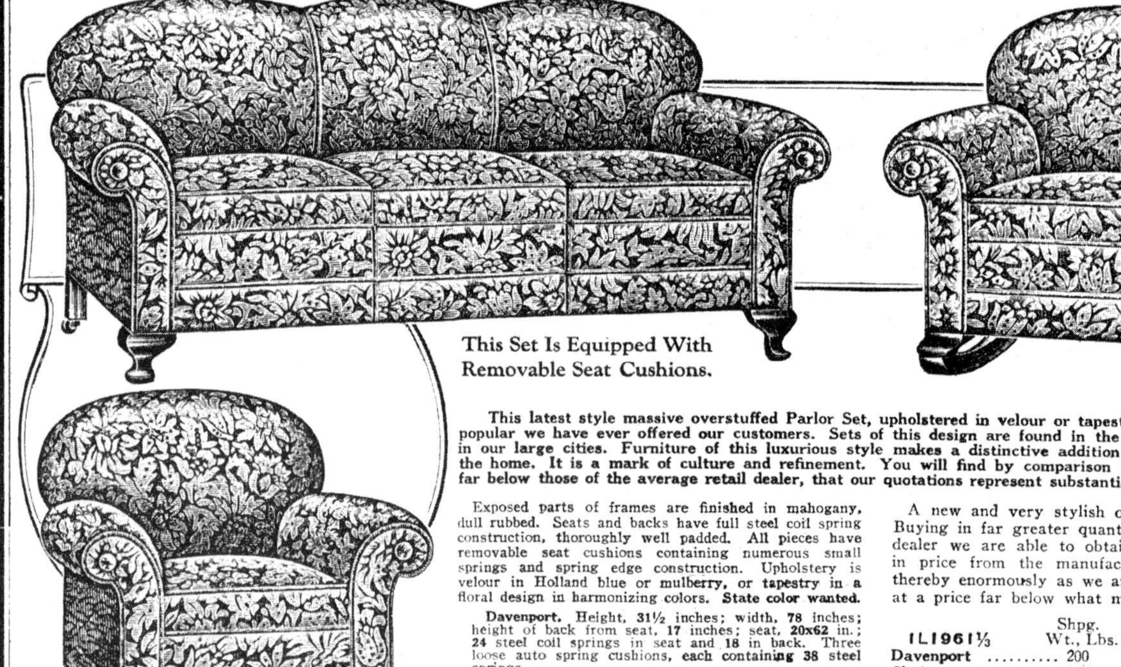

This Set Is Equipped With Removable Seat Cushions.

This latest style massive overstuffed Parlor Set, upholstered in velour or tapestry, is one of the most popular we have ever offered our customers. Sets of this design are found in the homes of the wealthy in our large cities. Furniture of this luxurious style makes a distinctive addition to the furnishings of the home. It is a mark of culture and refinement. You will find by comparison that our prices are so far below those of the average retail dealer, that our quotations represent substantial money savings.

Exposed parts of frames are finished in mahogany, dull rubbed. Seats and backs have full steel coil spring construction, thoroughly well padded. All pieces have removable seat cushions containing numerous small springs and spring edge construction. Upholstery is velour in Holland blue or mulberry, or tapestry in a floral design in harmonizing colors. **State color wanted.**

Davenport. Height, 31½ inches; width, **78** inches; height of back from seat, 17 inches; seat, **20x62** in.; 24 steel coil springs in seat and 18 in back. Three loose auto spring cushions, **each containing 38** steel springs.

Chair and Rocker. Height, 31½ inches; height of back from seat, **17** inches; seat, **17x20** inches. Seat contains 9 steel coil springs and back contains 6 pillow springs. Loose seat cushions contain **36** small seat **springs.**

A new and very stylish overstuffed Parlor Set. Buying in far greater quantities than the average dealer we are able to obtain marked concessions in price from the manufacturer. YOU benefit thereby enormously as we are able to list this set at a price far below what most others would ask.

1L1961⅓	Shpg. Wt., Lbs.	Velour	Tapestry
Davenport	200	**$68.00**	**$65.00**
Chair	105	**33.75**	**31.65**
Rocker	110	**33.85**	**31.75**

Shipped set up from factory near CHICAGO.

Any of these pieces may be purchased separately.

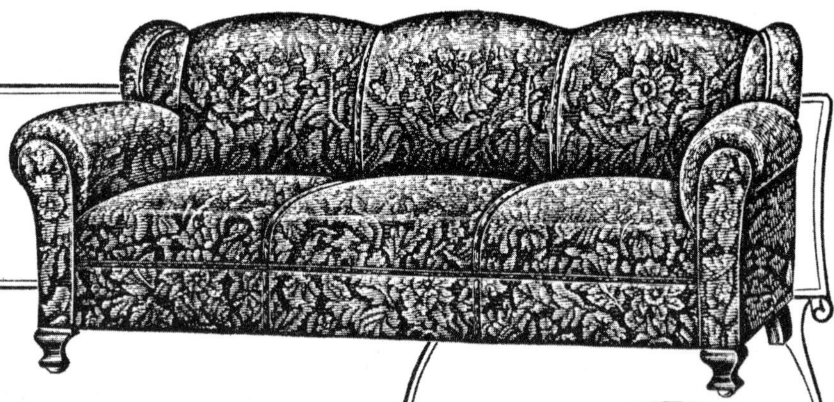

Priced 40 per Cent Below Usual Retail Price

You will be proud to have a set like this in your home. Not only on account of its very fine appearance, but because of its excellent quality and luxurious comfort. Special features of this set are the full steel coil spring seat construction, which are made with firm spring edges and the steel coil springs in the back, also. All the pieces are large and roomy, full size. The upholstering, either in velour or tapestry, is a very good grade. Our prices on this set are about 40 per cent below what most retailers usually ask.

Frames constructed of good grade of hardwood. Exposed parts nicely finished in mahogany, dull rubbed. All pieces made spring edge construction. This set is well padded and expertly upholstered. Your choice of three coverings: mulberry or Holland blue velour, with a floral design in a harmonizing color; or tapestry, floral design. **State kind and color of upholstering wanted.**

Any of These Pieces May Be Purchased Separately.

1L1968⅓	Shpg. Wt., Lbs.	**Upholstered in Velour**	**Upholstered in Tapestry**
Davenport	240	**$49.75**	**$46.00**
Rocker	100	**26.75**	**24.85**
Chair	100	**26.65**	**24.75**

Shipped from factory near CHICAGO or factory in CENTRAL NEW YORK.

Davenport. Has spring edge. **Thirty** springs in seat, **eighteen** steel coil springs in back. Length, **77** inches. Length between arms, **61** inches. Height of back from seat, **17** inches. Height, **34** inches. Depth of seat, **20** inches. Shipped set up.

The springs of this set are covered with a good grade filling, which makes it very comfortable.

Chair and Rocker. Six steel coil springs in each seat, supported on steel crimp wires and twine tied on top. Height of back from seat, **22½** inches. Backs are comfortably padded. Height, **34** inches. Size of seat, **17x20½** inches. Shipped set up.

These are the Talk of the Country

Three-Piece Set in Mahogany or Oak.

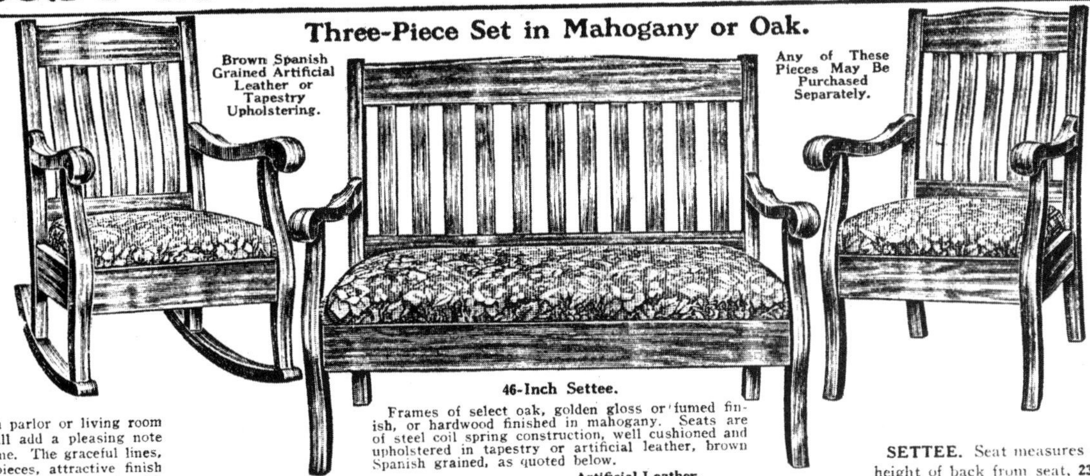

Brown Spanish Grained Artificial Leather or Tapestry Upholstering.

Any of These Pieces May Be Purchased Separately.

46-Inch Settee.

Here is a parlor or living room set that will add a pleasing note to your home. The graceful lines, full sized pieces, attractive finish and upholstering, combined with strong construction and a remarkably low price, make this one of the outstanding values of the season. **Shipped set up from our store.**

Frames of select oak, golden gloss or fumed finish, or hardwood finished in mahogany. Seats are of steel coil spring construction, well cushioned and upholstered in tapestry or artificial leather, brown Spanish grained, as quoted below.

1L1293

	Artificial Leather, Brown Spanish Grained: Golden Oak	Artificial Leather, Brown Spanish Grained: Fumed Oak	Tapestry Mahogany
Settee. Shipping weight, 85 pounds	$19.65	$19.85	$21.45
Chair. Shipping weight, 56 pounds	11.45	11.65	12.65
Rocker. Shipping weight, 52 pounds	11.55	11.75	12.75
Complete Set. Shipping weight, 193 pounds	42.65	43.25	46.85

SETTEE. Seat measures **44x21** inches; height of back from seat, **23** inches.

CHAIR AND ROCKER. Seat, **19x19½** inches; height of back from seat, **22½** in.

Four-Piece Parlor Set.

Walnut or Mahogany Finish.

Attractive Cane Panels.

Velour Upholstering.

Shop where you will, you cannot duplicate this set for attractive appearance, quality of upholstering and finish, sturdy construction and pleasing style for even considerably more money than we ask.

Queen Anne Period design. Frames are select hardwood veneered with genuine walnut, American walnut finish, or finished in mahogany. Seats have full steel spring construction and are well upholstered in velour, in choice of mulberry or blue color. **State color. Shipped set up from our store.**

Any of These Pieces May Be Purchased Separately.

1L1295

	Velour Mahogany	Velour Walnut
Settee. Shipping weight, 55 pounds	$29.65	$29.75
Chair. Shipping weight, 35 pounds	16.75	16.80
Rocker. Shipping weight, 35 pounds	16.85	16.90
Table. Shipping weight, 90 pounds	27.75	27.85

TABLE. Top measures **24x42** inches.

SETTEE. Seat, **18x50** inches; height of back from seat, **21** inches.

CHAIR AND ROCKER. Seat, **18x19½** inches; height of back from seat, **20½** inches.

Cane Back and Sides—Velour or Mohair Upholstery

Three Pillows, as Shown, Included With Each Set.

An Unusual Value in a Three-Piece Parlor Set.

CHAIR AND ROCKER. Seat, **22x24** inches. **Fifty-six** springs in each cushion. **Twelve** steel coil springs securely anchored to frame.

Shipped from factory near CHICAGO. Shipping weights: Chairs, each, 125 pounds; Davenport, 200 lbs.

All Pieces Have Removable Seat Cushions.

Exposed parts of frames are finished in mahogany, dull rubbed. Attractive features are the spring edge and the spring cushion seats. Your choice of two kinds of upholstering: Velour in mulberry or blue floral design in harmonizing colors, or blue or taupe mohair. **State color wanted.**

DAVENPORT. Height of back from seat, **19** inches. Inside measurements, **66** inches wide, **24** inches deep. Length over all, **69** inches. Entire height, **36** inches. Made spring edge with **thirty-six** steel coil springs securely anchored to frame. Three loose cushions, each having **fifty-six small steel coil springs.**

Any of These Pieces May Be Purchased Separately.

1L1971⅓

	Velour	Mohair
Davenport	$68.65	$98.50
Chair	31.75	37.85
Rocker	31.85	37.95

These are the Talk of the Country They Set a New Mark in Value Giving

Golden or Fumed. Oak

Mission Library Set.

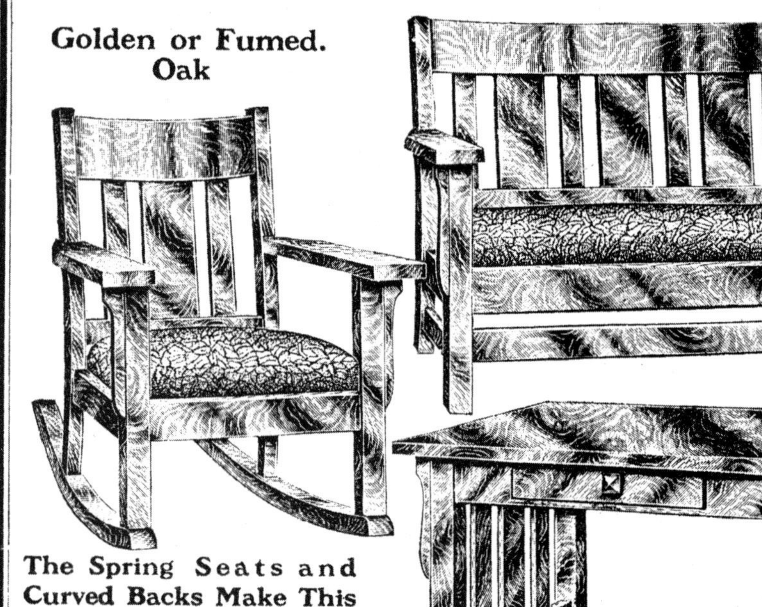

The Spring Seats and Curved Backs Make This a Very Comfortable Set.

The trim, neat design of this pretty set is well shown in the illustration. We know you will be pleased with this furniture, not alone because of the design, but with the quality of material, workmanship and finish. The prices speak for themselves.

MATERIAL AND FINISH.

Each piece is made of oak, with the arms, front rails and backs of settee, rocker and chair and the top of table made of quarter sawed oak. Choice of golden or fumed finish. (Fumed finish is a rich nut brown color.) The seats of settee, chair and rocker are covered with genuine leather, brown Spanish grained.

1L1279⅓

	Genuine Leather, Brown Spanish Grained Golden Oak	Fumed Oak
Settee. Shipping weight, 70 pounds	**$13.75**	**$13.85**
Rocker. Shipping weight, 45 pounds	**8.75**	**8.85**
Arm Chair. Shipping weight, 45 pounds	**8.65**	**8.75**
Library Table. Shipping weight, 68 pounds	**9.85**	**9,95**
Complete set (four pieces). Shpg. wt., 230 lbs.	**41.00**	**41.40**

All pieces shipped from our CHICAGO store or factory in WESTERN NEW YORK.

Complete Set Only $41.00 Golden Oak

ROCKER AND CHAIR.

have full spring seats with steel coil springs filled on top with tow and a layer of cotton felt. Curved bent backs. Size of seats, **18x19** inches. Square front posts. Height of backs from seats, **22** inches. Entire height, rocker, **36** inches; chair, **36** inches. Chair fitted with steel sliding casters.

SETTEE.

Curved bent back. Full box spring seat, size **18x40½** inches, contains steel coil cone shape springs. Height of back from seat, **22 inches. Length** outside, **48** inches. **Height, 38** inches. Steel sliding casters.

TABLE.

has top **24x38** inches. Broad box rim. Large center drawer with knob. Broad lower shelf. Height, **29½** inches. Shipped knocked down.

A Popular Arts and Crafts Design

Complete Set, Only $14.95

ROCKER AND CHAIR

have full spring seats, covered with artificial leather, brown Spanish grained. Large roomy seats, 19x19 inches. Medium high backs, 21½ inches high from seats. Entire height, chair, **35** inches. Chair has sliding casters.

A Well Made Inexpensive Library Set.

All pieces are made of select oak, fumed finish. The construction is strong and durable. You will be surprised that we are able to sell furniture as good as this for so little money. Chair and rocker are full size and covered with a good grade of upholstering.

TABLE

is **34** inches long and **24** inches wide. It has a wide box rim. Convenient lower shelf, 7 inches wide. Sliding casters. Three pieces packed knocked down in one crate. Shipped from factory in INDIANA. Shipping weight, 100 pounds.

1L1265⅓—Complete, **3** pieces **$14.95**

LARGE ARM ROCKER.

A great favorite. Roomy and comfortable. Has spring seat construction and is well cushioned. Seat measures **18x19½** inches. Height of back from seat, **23½** inches.

LIBRARY TABLE.

Made of oak, fumed finish. Top measures **24x35** inches. Paneled ends. Broad lower shelf adds to appearance and strength. Strongly and solidly built.

LARGE ARM CHAIR.

Spring seat construction. Noiseless, gliding casters. Seat measures **18x19½** inches. Height of back from seat, **23½** inches; entire height, 38 inches.

This Seven-Piece Library Set is an example of the wonderful furniture bargains we are offering. If you should price furniture of this quality and style in retail stores throughout the country you would find you can save by sending your order to us. This set is made of solid oak throughout with backs and arms of chairs of quarter sawed oak. The fumed oak finish is a rich nut brown color and goes well with the brown Spanish grained artificial leather upholstering. The backs are nicely padded. The seats are steel spring construction and very comfortable. This set is crated and shipped knocked down from factory in WESTERN NEW YORK or from factory near CHICAGO. Shipping weight, 180 pounds.

1L1278⅓—Complete set............................**$29.45**

COMFORT OR SEWING ROCKER AND CHAIR.

The sewing rocker is ideal for the purpose. Chair and rocker are used for reception chairs. Have comfortable spring cushion seat construction. Size of seat, **14½x15** in. Height of back from seat, **19** in.; entire height, chair, **36** in.; rocker, **33** in.

BOOK BLOCKS.

These popular book ends are of oak, fumed finish, to match set. They measure **4½** inches wide by **6¾** inches high.

TABOURET.

A convenient article. Fumed oak. Top, **11¾** inches in diameter; stands **17** inches high.

A Beautiful Sofa By Day — DAVENPORT BED-SOFAS — A Large Comfortable Bed by Night

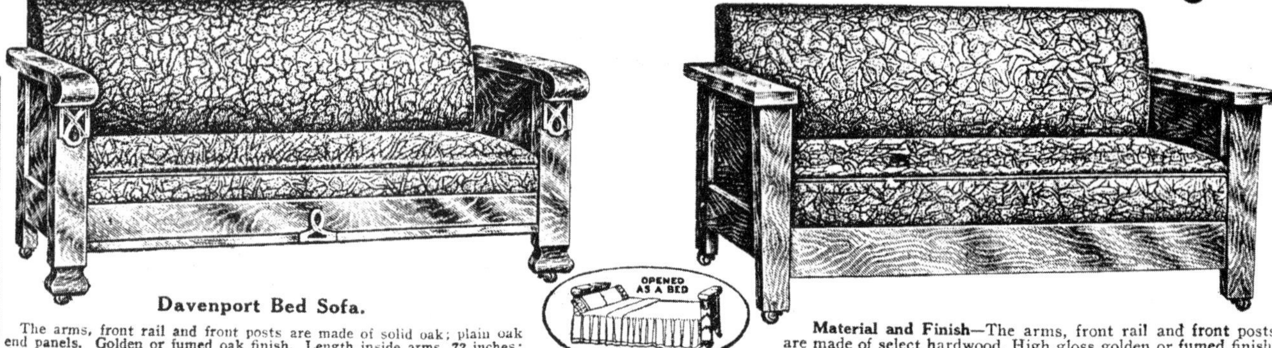

Davenport Bed Sofa.

The arms, front rail and front posts are made of solid oak; plain oak end panels. Golden or fumed oak finish. Length inside arms, **72** inches; depth of seat, **20** inches; length over all, **79½** inches; height, **36** inches. When made into a bed it is **72x42** inches. Spring seat and back; seat contains **24** and back **16** steel coil springs. Between springs and covering is a soft, even filling. Shipped from factory near CHICAGO. Shipping weight, 150 pounds.

1L1704⅓	Black Artificial Leather	Artificial Leather, Brown Spanish Grained
Golden oak	$28.75	$29.25
Fumed oak	28.85	29.35

Mattress for Davenport Bed Sofa.

Much more comfort is obtained when a mattress is used on these Davenport Bed Sofas. We list here a good, serviceable pad mattress, soft and comfortable. Made with 1½-inch square box edges, with strong felt binding. Covered with good quality floral art pattern ticking. Diamond shape tufts. Size **42x72** in. Wt., 18 lbs.

1L7208 .... **$4.85**

Material and Finish—The arms, front rail and front posts are made of select hardwood. High gloss golden or fumed finish. **Construction**—Spring seat and back; seat contains **24** and back **16** steel coil springs. Between springs and covering is a soft, even filling. Length, inside arms, **72** inches. When made into a bed it is **72x42** inches. Shipped from factory near CHICAGO. Shpg. weight, 150 pounds.

1L1701⅓	Black Artificial Leather	Artificial Leather, Brown Spanish Grained
Golden oak	$26.35	$26.85
Fumed oak	26.45	26.95

EASIFOLD BED DAVENPORTS

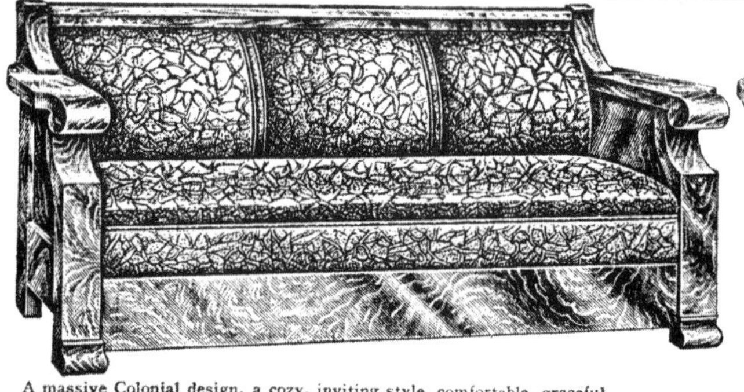

A massive Colonial design, a cozy, inviting style, comfortable, graceful and clean cut. Frame is made of selected oak finished with quarter sawed oak, except panels, back legs and side rails, which are hardwood, and front rail, which is hardwood finished with quarter sawed oak. Comes either in golden, rubbed or fumed finish. (Fumed finish is a dull nut brown color.) Length, outside, **84¾** inches; length between arms, **75½** inches; depth of seat, **20** inches; height of back from seat, **19** inches; entire height, **36** inches; size, opened as bed, **48x72** inches.

For Mattress to fit see page 607E.

The seat contains **twenty-seven** heavy deep coil springs resting on steel crimp wire supports and securely fastened on top. The filling over the springs is soft and even, forming a level foundation for the upholstery covering. Shipped from factory near CHICAGO or in CENTRAL NEW YORK. Shipping weight, 290 pounds.

1L1765⅓	Artificial Leather, Black	Artificial Leather, Brown Spanish Grained	Genuine Leather, Black	Genuine Leather, Brown Spanish Grained
Golden oak	$45.75	$46.25	$55.35	$55.85
Fumed oak	45.85	46.35	55.45	55.95

This Easifold or long bed davenport is a distinctive plain design which will appeal strongly to discriminating home furnishers. The frame is of select oak, quarter sawed oak veneered with plain oak panels. Finish, fumed or golden gloss; also made of birch, finished in brown mahogany. The seat has an all steel coil spring construction, containing **36** wide coil springs, supported on steel crimp wires, securely fastened to the frame. Steel interlocking top fastening forms a soft flexible top. Very well padded with a filling of fine tow and felted cotton. Length outside, **84** inches; depth of seat, **21½** inches; height of back from seat, **17** inches; entire height, **35** inches; size, open as a bed, **48x72** inches. Shipped knocked down from factory near CHICAGO or in CENTRAL NEW YORK. Shpg. wt., 265 lbs.

1L1763	Artificial Leather Brown Spanish Grained	Genuine Leather Brown Spanish Grained
Golden oak	$41.75	$51.75
Fumed oak	41.85	51.85
Mahogany finish	41.95	51.95

De Luxe Bed Davenports

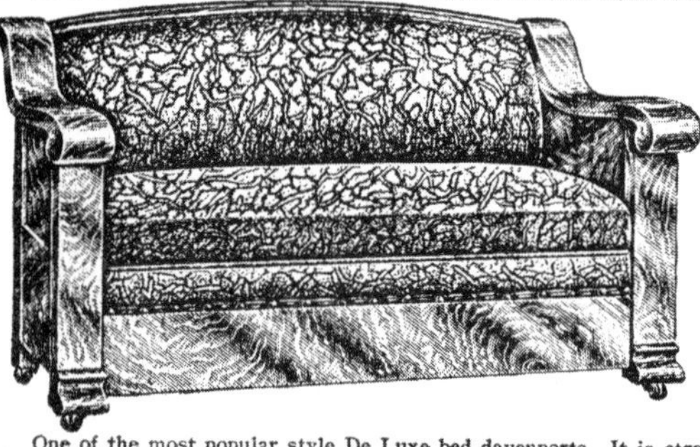

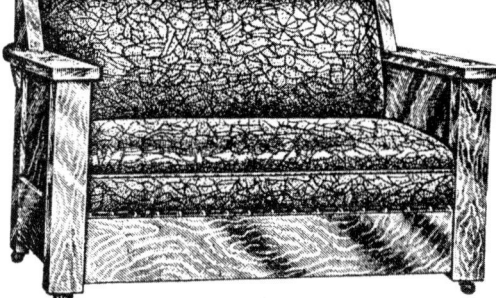

One of the most popular style De Luxe bed davenports. It is strongly constructed of select oak, top rails, arms and entire front quarter sawed oak veneered. Plain panels, nicely finished in golden oak or fumed. Also made of birch, veneered with mahogany, birch panels, brown mahogany finish. This is a very comfortable bed davenport, made with all steel spring construction in seat. Contains **24** springs, supported on steel crimped wires. Over springs is a soft, even filling. This forms a smooth, firm foundation for the outer covering. Length over all, **60½** inches; length between arms, **52½** inches; depth of seat, **21** inches; entire height, **34** inches. Shipped knocked down from our factory near CHICAGO or in CENTRAL NEW YORK. Shpg. wt., 230 lbs.

1L1738⅓	Artificial Leather, Black	Artificial Leather, Brown Spanish Grained	Genuine Leather, Black	Genuine Leather, Brown Spanish Grained
Golden oak	$38.25	$38.75	$46.25	$46.75
Fumed oak	38.35	38.85	46.35	46.85
Mahogany finish	38.45	38.95	46.45	46.95

A plain design popular style De Luxe bed davenport. Well made of select oak, top rails, arms and entire front quarter sawed oak veneered. Plain panels, golden oak or fumed finish. All steel spring construction in seat. Contains **18** springs, supported on steel crimped wires. Over springs is a soft, even filling. This forms a smooth, firm foundation for the outer covering. Length over all, **59** inches; length between arms, **52** inches; depth of seat, **20½** inches. Shipped knocked down from our factory near CHICAGO or in CENTRAL NEW YORK. Shipping weight, 220 pounds.

1L1737⅓	Artificial Leather Brown Spanish Grained	Genuine Leather Brown Spanish Grained
Golden oak	$31.85	$39.75
Fumed oak	31.95	39.85

Why Our Cedar Chests Are the Best

ALL TRAYS ARE THREE-QUARTER LENGTH.

Material—Nothing but selected fragrant Tennessee red cedar is used in our chests, every piece of which is thoroughly dried and properly seasoned in our modern kilns. It is absolutely necessary to do this, since without properly seasoned material it is impossible to produce quality work. After seasoning and machining, a great deal of care is required in selection, as Tennessee cedar is full of defects, mostly streaks of white sap, large black knots and dry rot. These are all eliminated by us, which necessarily means a large percentage of waste. Small knots are not a defect, in fact they are a necessity, as most of the cedar oil that produces the fragrance is contained in the knots. They also add to the beauty of the grain.

Construction—In constructing we use all fragrant solid red cedar finished ⅞ inch thick.

The Corner Construction—The corner construction has been acknowledged the best self locking corner ever used in the manufacture of cedar chests. No nails, dowels or screws to mar the beauty on the side. Joints are locked together and glued. This makes it impossible for them to open.

Workmanship—Our chests are the product of skilled cabinetmakers—genuine mechanics of the old school, to whom perfection has become a habit. Every chest is accorded the same individual care in both workmanship and selection of materials to make it a genuine piece of art. Every joint is carefully fitted. All lids are made to fit snug and tight and overlap on the joint, insuring a chest that will be permanent proof against moths, vermin and dust.

Finishing—The natural non-fading red cedar color is maintained throughout in finishing, after being sanded perfectly smooth; no stains, acids or other preparations are used. These are unnecessary when the cedar is well selected as to color. Each chest is given a soft, velvety finish with high grade varnish finish. We give some of our chests a rubbed finish on the top only and others a rubbed finish throughout. This rubbing is done by hand and the varnish is rubbed down to a beautiful polish, making each chest a very fine piece of furniture.

All our Cedar Chests are equipped with roller casters, wood handles and substantial locks. (Two keys.) The hinges and patent lid stays are plain metal, polished. We use only genuine 10-ounce copper for trimming our cedar chests.

OUR NEW CORNER CONSTRUCTION
POSITIVELY PREVENTS JOINTS FROM PARTING OR OPENING

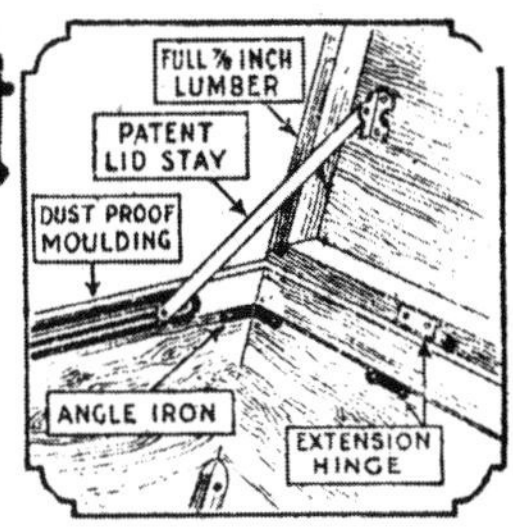

A Cedar Chest Is Like An Insurance Policy

It insures your furs, linens and clothes against damage or destruction by moths, mice, dust, etc.

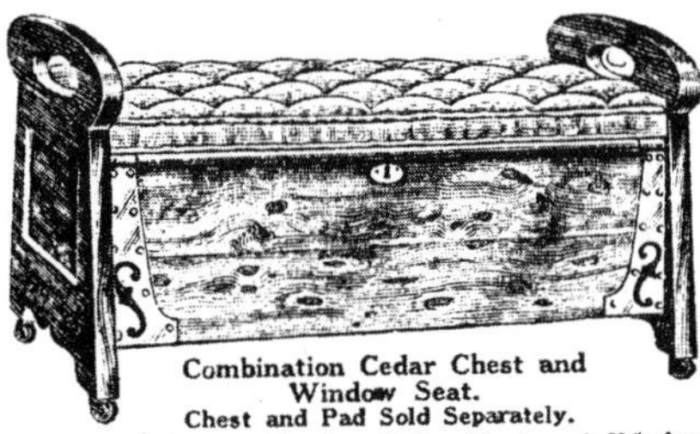

Combination Cedar Chest and Window Seat. Chest and Pad Sold Separately.

An attractive combination Cedar Chest and Window Seat. The ends are 1½-inch red cedar (paneled). Top, dull satin rubbed finish; sides and ends gloss finish. Artistically cut metal straps on each end of front panel, extending from lid to base. Metal used is copper with a frosted satin finish, lacquered to prevent tarnishing, and studded with round copper head nails. Shipping weight, 125 pounds.

1L3765

Size, 48x24x18½ inches. Chest only, without pad **$33.95**

Pad is made of felted cotton covered with a good grade of green brocaded covering, nicely tufted.

For pad only. Shipping weight, 10 pounds.....**$3.95**

Note the Paneled Effect Front.

Lid is hand rubbed to a dull satin finish. Sides and front are given an attractive gloss finish. Ornamented with 3-inch metal bands extending over lid. Metal used is copper having frosted satin finish, lacquered to prevent tarnishing, and studded with round copper head nails.

1L3760

Size, Inches	Height, Inches	Shpg. Wt., Lbs.	
45x20	19	87	$24.45
48x21	20	95	27.75

Very Popular Design. Furnished With or Without Tray.

Lid is hand rubbed to a dull satin finish, balance of chest having an attractive gloss finish. The beauty of this chest is greatly enhanced by the gracefully rounded corners on lid. Has fancy cut metal straps over each end of front, giving chest a massive appearance. Metal used is copper with frosted satin finish, lacquered to prevent tarnishing, and studded with round copper head nails.

1L3763

Size, Inches	Height, Inches	Shpg. Wt., Lbs.	Without Tray	With Tray
42x19	18	70	$22.95	$24.85
48x21	20	90	27.85	29.95

For Moth Bags See Page 367.

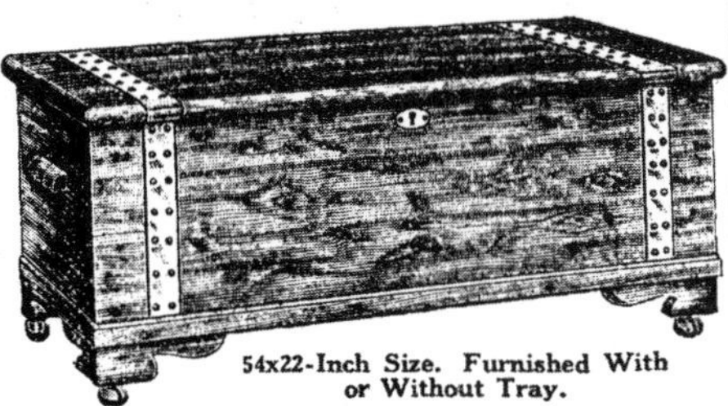

54x22-Inch Size. Furnished With or Without Tray.

The lid has a beautiful rubbed dull satin finish, the balance of the chest having an attractive gloss finish. Has 2-inch straps extending over complete width of lid and down front panel to base. All metal used is copper with a frosted satin finish, lacquered to prevent tarnishing, and studded with round copper head nails.

1L3755

Size, Inches	Height, Inches	Shpg. Wt., Lbs.	Without Tray	With Tray
38x18	17	70	$14.85	Not Furnished
45x20	19	87	21.00	
54x22	21	115	28.95	$31.65

Utility Boxes

Inside made of ½-inch seasoned hardwood. Covered with high grade Japanese white matting, glued on. Trimmings around top and base, ¾-inch bands of white birch. (Panel effect.) Lid is trimmed with 1¼-inch panel frame, which gives it a neat, trim, finished appearance. Equipped with strong extension hinges, substantial lid stays and wood handles.

1L3770

Size, Inches	Height, Inches	Shpg. Wt., Lbs.	
36x16	15½	35	$7.65
40x17	15½	45	8.45
45x19	15½	55	8.95

45x20-Inch Size. Furnished With or Without Tray.

Lid has hand rubbed dull satin finish. Balance of chest given an attractive gloss finish. Lid ornamented with 2-inch metal bands of copper having a frosted satin finish and lacquered to prevent tarnishing. Studded with round copper head nails.

1L3756

Size, Inches	Height, Inches	Shpg. Wt., Lbs.	Without Tray	With Tray
38x18	21	90	$18.95	Not Furnished
45x20	23	110	23.85	$25.95

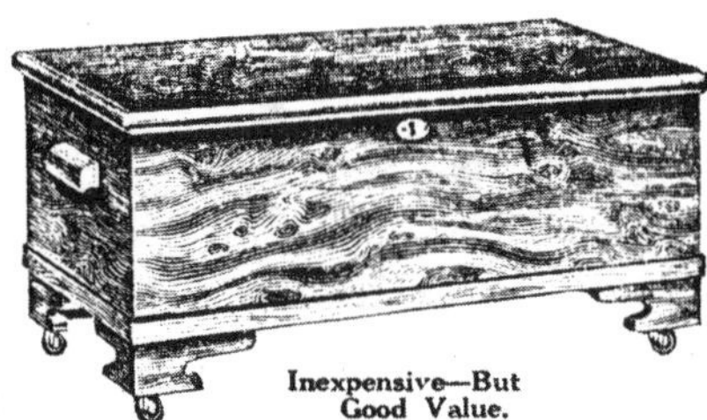

Inexpensive—But Good Value.

Lid is hand rubbed to a dull satin finish. Sides and front have an attractive gloss finish. Lid is edged with tapered molding which fits snugly over the chest, making it air-tight and dustproof.

1L3753

Size, Inches	Height, Inches	Shpg. Wt., Lbs.	
34x17	16	70	$9.65
42x19	18	85	14.95

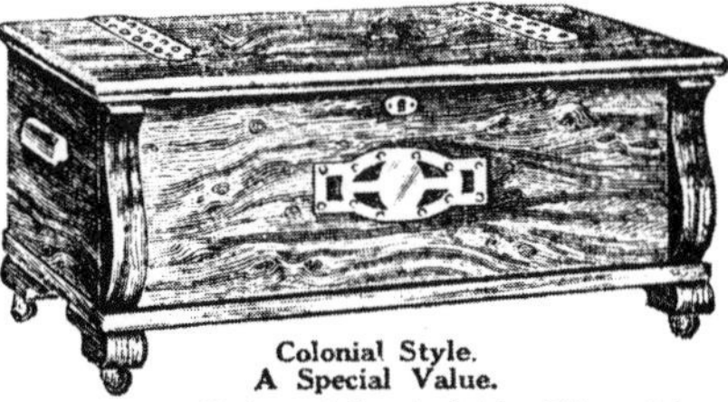

Colonial Style. A Special Value.

Lid is hand rubbed to a dull satin finish. Sides and front have an attractive gloss finish. Trimmed with fancy metal straps above each hinge and metal plate in center of front panel. Metal used is copper with frosted satin finish and lacquered to prevent tarnishing. Studded with round copper head nails.

1L3759

Size, Inches	Height, Inches	Shpg. Wt., Lbs.	
42x19	18	85	$17.95
48x21	20	95	23.85

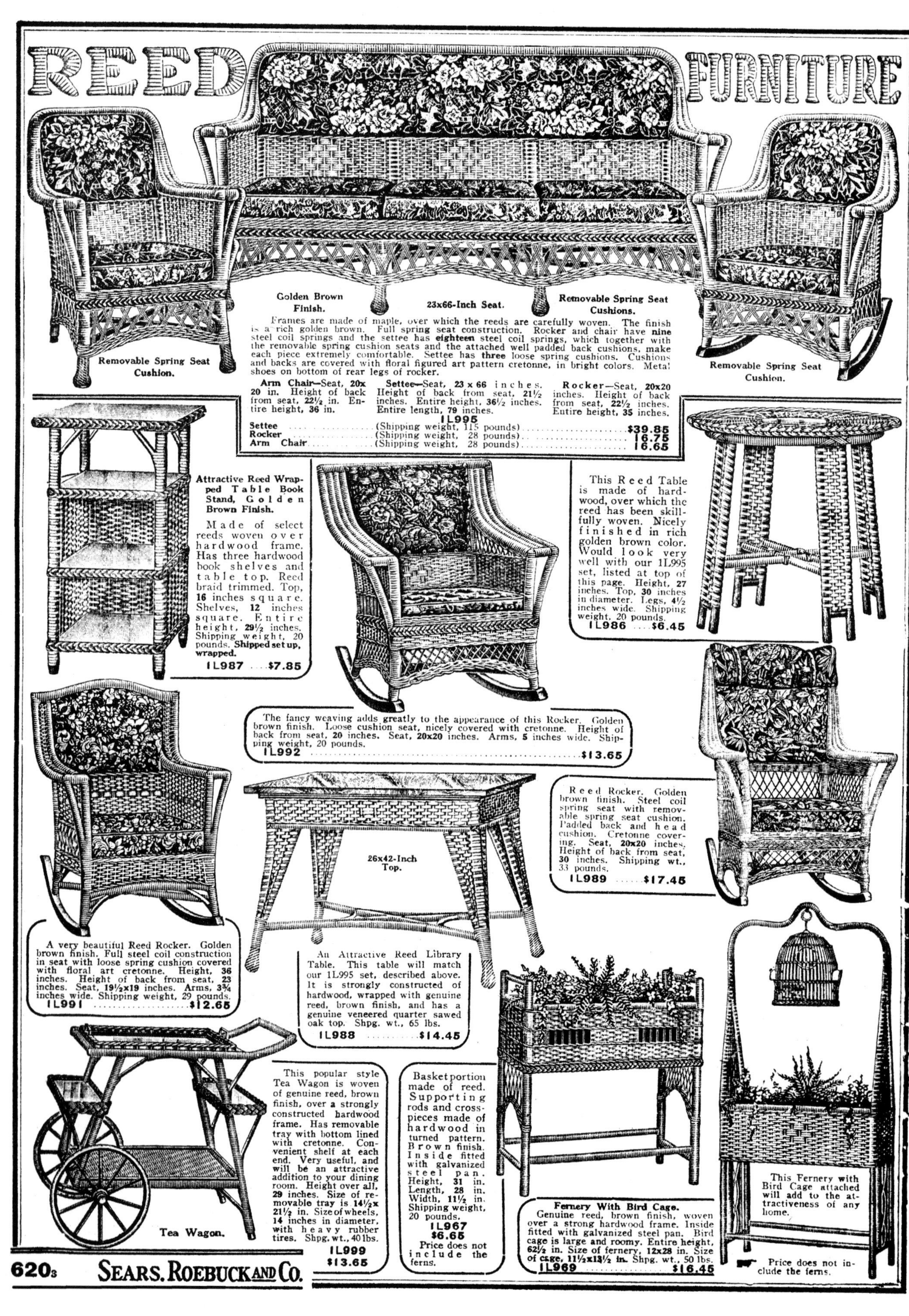

REED FURNITURE

Golden Brown Finish.

23x66-Inch Seat.

Removable Spring Seat Cushions.

Removable Spring Seat Cushion.

Removable Spring Seat Cushion.

Frames are made of maple, over which the reeds are carefully woven. The finish is a rich golden brown. Full spring seat construction. Rocker and chair have **nine** steel coil springs and the settee has **eighteen** steel coil springs, which together with the removable spring cushion seats and the attached well padded back cushions, make each piece extremely comfortable. Settee has **three** loose spring cushions. Cushions and backs are covered with floral figured art pattern cretonne, in bright colors. Metal shoes on bottom of rear legs of rocker.

Arm Chair—Seat, **20x20 in.** Height of back from seat, **22½** in. Entire height, **36** in.

Settee—Seat, **23 x 66** inches. Height of back from seat, **21½** inches. Entire height, **36½** inches. Entire length, **79** inches.

Rocker—Seat, **20x20** inches. Height of back from seat, **22½** inches. Entire height, **35** inches.

1L995

Settee	(Shipping weight, 115 pounds)	**$39.85**
Rocker	(Shipping weight, 28 pounds)	**16.75**
Arm Chair	(Shipping weight, 28 pounds)	**16.65**

Attractive Reed Wrapped Table Book Stand, Golden Brown Finish.

Made of select reeds woven over hardwood frame. Has three hardwood book shelves and table top. Reed braid trimmed. Top, **16** inches square. Shelves, **12** inches square. Entire height, **29½** inches. Shipping weight, 20 pounds. **Shipped set up, wrapped.**

1L987 ....$7.85

This Reed Table is made of hardwood, over which the reed has been skillfully woven. Nicely finished in rich golden brown color. Would look very well with our 1L995 set, listed at top of this page. Height, **27** inches. Top, **30** inches in diameter. Legs, **4½** inches wide. Shipping weight, 20 pounds.

1L986 ....$6.45

The fancy weaving adds greatly to the appearance of this Rocker. Golden brown finish. Loose cushion seat, nicely covered with cretonne. Height of back from seat, **20** inches. Seat, **20x20** inches. Arms, **5** inches wide. Shipping weight, 20 pounds.

1L992 ....$13.65

Reed Rocker. Golden brown finish. Steel coil spring seat with removable spring seat cushion. Padded back and head cushion. Cretonne covering. Seat, **20x20** inches. Height of back from seat, **30** inches. Shipping wt., 33 pounds.

1L989 ....$17.45

26x42-Inch Top.

A very beautiful Reed Rocker. Golden brown finish. Full steel coil construction in seat with loose spring cushion covered with floral art cretonne. Height, **36** inches. Height of back from seat, **23** inches. Seat, **19½x19** inches. Arms, **3¾** inches wide. Shipping weight, 29 pounds.

1L991 ....$12.65

An Attractive Reed Library Table. This table will match our 1L995 set, described above. It is strongly constructed of hardwood, wrapped with genuine reed, brown finish, and has a genuine veneered quarter sawed oak top. Shpg. wt., 65 lbs.

1L988 ....$14.45

This popular style Tea Wagon is woven of genuine reed, brown finish, over a strongly constructed hardwood frame. Has removable tray with bottom lined with cretonne. Convenient shelf at each end. Very useful, and will be an attractive addition to your dining room. Height over all, **29** inches. Size of removable tray is **14½x21½** in. Size of wheels, **14** inches in diameter, with heavy rubber tires. Shpg. wt., 40 lbs.

Tea Wagon.

1L999
$13.65

Basket portion made of reed. Supporting rods and cross-pieces made of hardwood in turned pattern. Brown finish. Inside fitted with galvanized steel pan. Height, **31** in. Length, **28** in. Width, **11½** in. Shipping weight, 20 pounds.

1L967
$6.65

Price does not include the ferns.

Fernery With Bird Cage.

Genuine reed, brown finish, woven over a strong hardwood frame. Inside fitted with galvanized steel pan. Bird cage is large and roomy. Entire height, **62½** in. Size of fernery, **12x28** in. Size of cage, **11½x13½** in. Shpg. wt., 50 lbs.

1L969 ....$16.45

This Fernery with Bird Cage attached will add to the attractiveness of any home.

Price does not include the ferns.

SEARS, ROEBUCK AND CO.

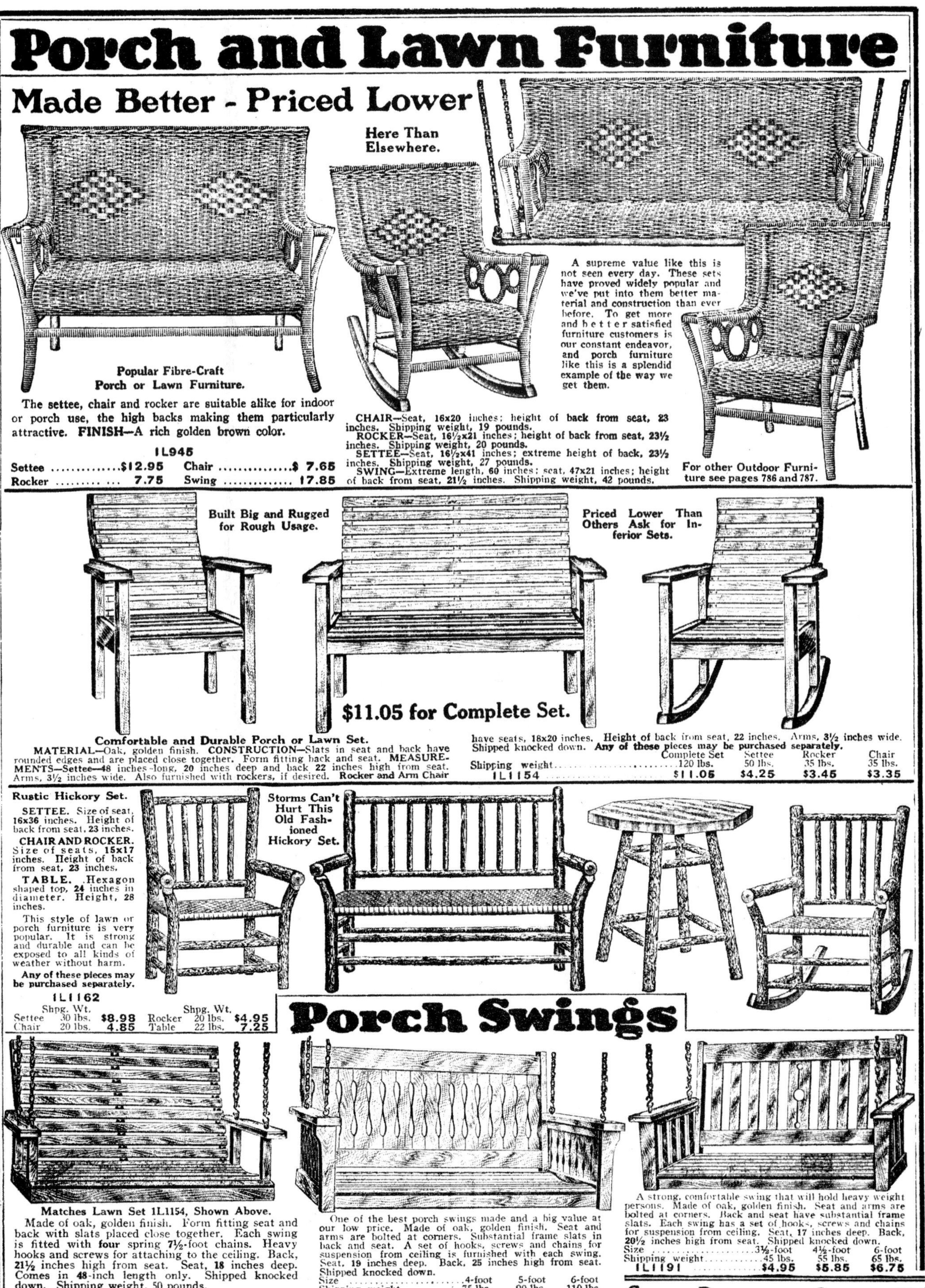

Porch and Lawn Furniture
Made Better - Priced Lower
Here Than Elsewhere.
Popular Fibre-Craft Porch or Lawn Furniture.
The settee, chair and rocker are suitable alike for indoor or porch use, the high backs making them particularly attractive. FINISH—A rich golden brown color.
1L945
Settee ....$12.95 Chair ....$ 7.65
Rocker .... 7.75 Swing 17.85
A supreme value like this is not seen every day. These sets have proved widely popular and we've put into them better material and construction than ever before. To get more and better satisfied furniture customers is our constant endeavor, and porch furniture like this is a splendid example of the way we get them.
CHAIR—Seat, 16x20 inches; height of back from seat, 23 inches. Shipping weight, 19 pounds.
ROCKER—Seat, 16½x21 inches; height of back from seat, 23½ inches. Shipping weight, 20 pounds.
SETTEE—Seat, 16½x41 inches; extreme height of back, 23½ inches. Shipping weight, 27 pounds.
SWING—Extreme length, 60 inches; seat, 47x21 inches; height of back from seat, 21½ inches. Shipping weight, 42 pounds.
For other Outdoor Furniture see pages 786 and 787.
Built Big and Rugged for Rough Usage.
Priced Lower Than Others Ask for Inferior Sets.
$11.05 for Complete Set.
Comfortable and Durable Porch or Lawn Set.
MATERIAL—Oak, golden finish. CONSTRUCTION—Slats in seat and back have rounded edges and are placed close together. Form fitting back and seat. MEASUREMENTS—Settee—48 inches long, 20 inches deep and back 22 inches high from seat. Arms, 3½ inches wide. Also furnished with rockers, if desired. Rocker and Arm Chair have seats, 18x20 inches. Height of back from seat, 22 inches. Arms, 3½ inches wide. Shipped knocked down. Any of these pieces may be purchased separately.
Complete Set Settee Rocker Chair
Shipping weight .... 120 lbs. 50 lbs. 35 lbs. 35 lbs.
1L1154 $11.05 $4.25 $3.45 $3.35
Rustic Hickory Set.
SETTEE. Size of seat, 16x36 inches. Height of back from seat, 23 inches.
CHAIR AND ROCKER. Size of seats, 15x17 inches. Height of back from seat, 23 inches.
TABLE. Hexagon shaped top, 24 inches in diameter. Height, 28 inches.
This style of lawn or porch furniture is very popular. It is strong and durable and can be exposed to all kinds of weather without harm.
Any of these pieces may be purchased separately.
1L1162
Shpg. Wt. Shpg. Wt.
Settee 30 lbs. $8.98 Rocker 20 lbs. $4.95
Chair 20 lbs. 4.85 Table 22 lbs. 7.25
Storms Can't Hurt This Old Fashioned Hickory Set.
Porch Swings
Matches Lawn Set 1L1154, Shown Above.
Made of oak, golden finish. Form fitting seat and back with slats placed close together. Each swing is fitted with four spring 7½-foot chains. Heavy hooks and screws for attaching to the ceiling. Back, 21½ inches high from seat. Seat, 18 inches deep. Comes in 48-inch length only. Shipped knocked down. Shipping weight, 50 pounds.
1L1186 ....$3.98
One of the best porch swings made and a big value at our low price. Made of oak, golden finish. Seat and arms are bolted at corners. Substantial frame slats in back and seat. A set of hooks, screws and chains for suspension from ceiling is furnished with each swing. Seat, 19 inches deep. Back, 25 inches high from seat. Shipped knocked down.
Size 4-foot 5-foot 6-foot
Shipping weight .... 75 lbs. 90 lbs. 110 lbs.
1L1193 .... $5.95 $6.95 $7.95
A strong, comfortable swing that will hold heavy weight persons. Made of oak, golden finish. Seat and arms are bolted at corners. Back and seat have substantial frame slats. Each swing has a set of hooks, screws and chains for suspension from ceiling. Seat, 17 inches deep. Back, 20½ inches high from seat. Shipped knocked down.
Size 3½-foot 4½-foot 6-foot
Shipping weight .... 45 lbs. 55 lbs. 65 lbs.
1L1191 .... $4.95 $5.85 $6.75
SEARS, ROEBUCK AND CO. 621

Telephone Table with Stool Costumers

PARLOR TABLES

Folding Card Table Folding Sewing Table

Pedestals

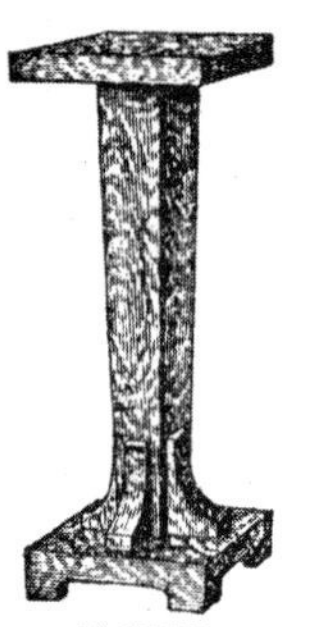

1L3785
Golden oak..$4.85
Mahogany .. 4.90

Hardwood, imitation quarter sawed oak, golden gloss or imitation mahogany. Height, 36 inches. Shipping weight, 34 pounds.

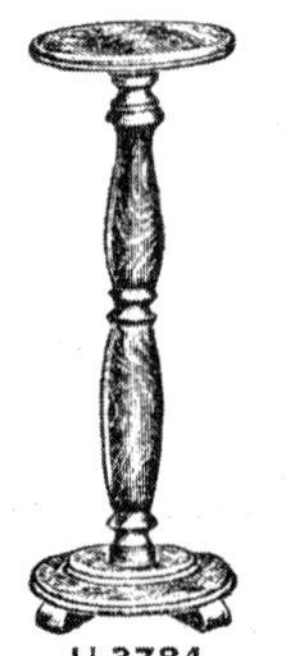

1L3784
Golden oak. $3.95
Mahogany .. 4.00

Made of solid quarter sawed oak, golden gloss finish on base and column, polished top, or solid birch, mahogany gloss finish on base and column, polished top. Height, 36 inches. Top, 12 inches in diameter. Column, 3 inches in diameter at widest part. Shipping weight, 20 lbs.

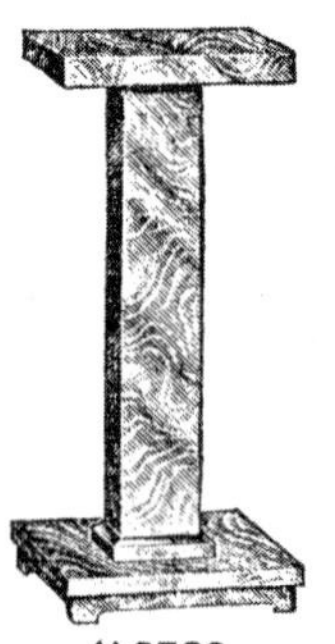

1L3789
Golden oak. $6.65
Fumed oak.. 6.70
Mahogany .. 6.75
Walnut 6.85

Made of hardwood, veneered with quarter sawed oak, golden or fumed finish, mahogany or walnut. Height, 36 inches. Top, 13x13 in. Column, 4½ in. square. Shipping weight, 25 pounds.

1L3521
Golden oak..$6.95
Fumed 7.00

A pretty pedestal style Parlor Table of strong construction. Made of quarter sawed oak, golden or fumed finish. Top, hand rubbed, polished finish; base and column gloss finish. Top has rounded corners. Shipped knocked down. Top, 24x24 inches. Base, 14x14 inches. Column, 4⅝ inches square. Shipping weight, 45 lbs.

Glass Ball Foot Table. Exceptional Value.

1L3529....$3.98

Made of hardwood in imitation quarter sawed oak, golden gloss finish. Shipped knocked down. Top, 24x24 inches. Shipping wt., 30 pounds.

Bedroom Table.

Made of hardwood. Finished in imitation quarter sawed oak, golden gloss, white enamel or dull ivory, as desired. Full box rim with center drawer. Strong knock-down construction. Legs securely fastened to the top and firmly braced, with broad lower shelf. Shipped knocked down. Top, 18x24 inches. Height, 28 inches. Shipping weight, 35 pounds.

1L3538
Golden oak..........$4.95
White enamel.......... 5.45
Dull ivory.......... 5.50

Telephone Table with Stool

Has shelf with ample space for directory. Stool fits between legs of table. Made of solid oak, finished in golden or fumed finish, or birch, mahogany finish or hardwood finished in walnut. Shipped knocked down. Top, 14x15 in. Shipping weight, 34 pounds.

1L3548
Golden oak..........$4.65
Fumed oak.......... 4.70
Mahogany 4.75
Walnut 4.85

COSTUMERS

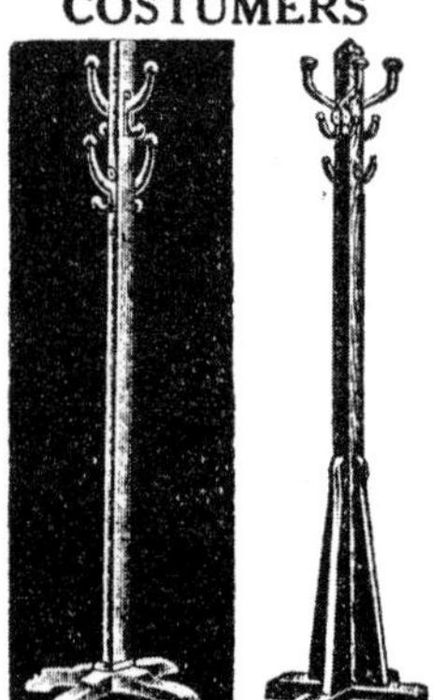

Made of hardwood, ivory or walnut finish. Has four two-prong hooks. Height, 66¾ inches. Shipping weight, 10 pounds.

1L1218
Walnut $3.85
Ivory.. 3.95

Made of solid oak, dull golden or fumed finish. Has two four-prong hooks and two single prong hooks. Height, 73 inches. Shipping weight, 12 pounds.

1L1217
Golden oak $3.45
Fumed oak. 3.50

1L3528
Golden gloss finish $4.85
Fumed finish 4.90

A beautifully finished and substantially constructed Parlor Table. Made of quarter sawed oak, finished in golden or fumed. Top and shelf rubbed polished finish, legs gloss finish. Large, roomy lower shelf. Shipped knocked down. Top, 24x24 inches. Shipping weight, 38 pounds.

A Good Popular Priced Table.

1L3523....$2.95

Made of hardwood in imitation quarter sawed oak, golden gloss finish. Shipped knocked down. Top, 24 inches square. Shipping weight, 28 pounds.

Parlor Stand.

Inexpensive, but nevertheless a very good value and a better table than you would expect for the price. Substantially made of hardwood in imitation quarter sawed oak, golden gloss finish. Thoroughly substantial construction and well finished. Shipped knocked down. Size of top, 15x15 inches. Shipping weight, 14 pounds.

1L3514 ....$1.95

Folding Card Table.

Especially desirable for card parties, tea parties and writing. When not in use this table can be folded, as shown in illustration. Framework made of birch, finished in imitation mahogany. Top covered with dark olive green felt or dark green artificial leather. Top, 28¾ inches square. Height, 26½ inches. Shipping wt., 14 lbs.

1L3748
Felt top..........$2.65
Leather top.......... 2.70

Folding Sewing Table.

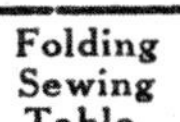

May be used for a variety of purposes, such as measuring, cutting and sewing. Made of hardwood, finished in golden gloss. Has yard measure stamped on top of table. Size of top, 19x36 inches. Height, about 25¾ inches. Shipping weight, 20 pounds.

1L3738 ....$1.78

For other Folding Tables see pages 784 and 785.

Pedestals

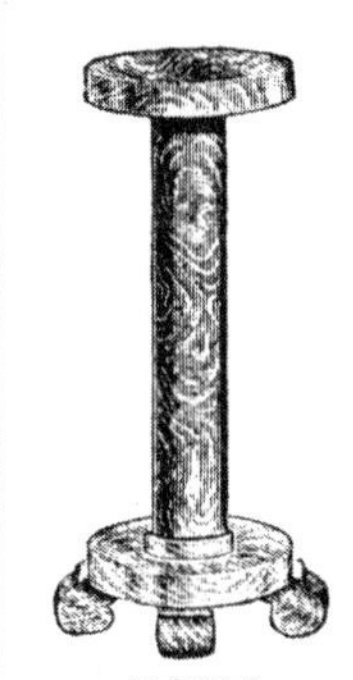

1L3786
Golden oak. $4.65
Mahogany .. 4.70

Made of hardwood, imitation quarter sawed oak gloss finish or mahogany gloss finish. Height, 33¾ inches. Top, 12 inches in diameter. Column, 3¾ inches in diameter. Shipping weight, 25 pounds.

1L3783
Golden oak..$4.65
Fumed oak. 4.70
Mahogany .. 4.75

Quarter sawed oak in polished golden or fumed finish, or of birch, imitation mahogany finish, polished. Height, 36½ inches. Top, 12 inches square. Column, 4 inches square. Shipping weight, 31 pounds.

1L3782
Golden oak..$4.35
Mahogany .. 4.40

Made of hardwood, golden oak gloss or mahogany gloss finish. Top, 12x12 in. Shelf, 10x10 in. Height, 34 inches. Shipping weight, 20 pounds.

Better Values in Bookcases and Writing Desks

Made of oak, with quarter sawed oak front, in high gloss golden or fumed finish, as desired. (Fumed finish is a dull nut brown color.) Has four adjustable shelves; will hold about 190 average size books. Glass paneled doors. Height, **54** inches. Width, **44** inches. Shipping weight, 150 pounds. Shipped from factory in ROCKFORD, ILL.

1L3246⅓—Golden oak........................**$19.85**
Fumed oak........................**19.95**

Combination Book Rack and Table

This Revolving Bookstand is a great convenience. It also makes a very nice table. Holds about 50 books, all within easy reach. Strongly made of select elm, imitation oak, golden gloss finish. Top section is attached to base by steel plates with steel ball bearings. Top, **20x20** inches; shelves, **16x16** inches; height over all, **35½** inches. Shipping weight, 60 pounds. Shipped from our store.

1L3205........................**$7.45**

Colonial style. Made of oak with quarter sawed oak front. Polished golden or fumed finish, as desired; also in birch, mahogany finish, polished. Has four adjustable shelves. Will hold about 200 average size books. Glass paneled doors. Height, **53** inches. Width, **44** inches. Shpg. wt., 150 pounds. Shipped from factory in ROCKFORD, ILL.

1L3248⅓—Polished golden oak........**$24.65**
Fumed oak........................**24.75**
Mahogany finish........................**24.85**

$16.35 Golden Oak.

Made of oak, with quarter sawed oak front. Golden gloss or fumed finish. Glass paneled doors. Four adjustable shelves. Will hold about 150 average size books. Height, **54** inches. Width, **32** inches. Shipping weight, 110 pounds. Shipped from factory in ROCKFORD, ILL.

1L3220⅓—Golden oak........................**$16.35**
Fumed oak........................**16.45**

$19.75 Golden Oak.

An Arts and Crafts Design. Will hold about 70 average size books. Made of oak. Golden gloss or fumed finish. Four adjustable shelves. Desk section has pigeonhole case with drawer. Above desk is cupboard with door. Height, **64** inches. Width, **36¾** inches. Mirror, **10x14** inches. Shipping weight, 130 pounds. Shipped from factory in ROCKFORD, ILL.

1L3112⅓—Golden oak........................**$19.75**
Fumed oak........................**19.85**

$11.35 Golden Oak.

Made of oak, high gloss golden or fumed finish; also birch, imitation mahogany finish. Holds about 75 average size books. Height, **52** inches. Width, **24** inches. Depth, **12** inches. Shipping weight, 80 pounds. Shipped from factory in ROCKFORD, ILL.

1L3210⅓—Golden oak........................**$11.35**
Fumed oak........................**11.45**
Mahogany finish........................**11.55**

Convenient — WRITING DESKS — *Useful*

$16.65 Golden Oak.

A Very Useful Writing Desk.

Made of oak, with quarter sawed oak veneer. Golden gloss or fumed finish. Below desk are three extra deep drawers and a handy cupboard with private drawer and box letter file. File is included at price quoted. Wood knobs. Height, **46** inches; width, **32** inches; depth, **16** inches. Shipping weight, 145 pounds. Shipped from factory near CHICAGO.

1L3332⅓—Golden oak........................**$16.65**
Fumed oak........................**16.75**

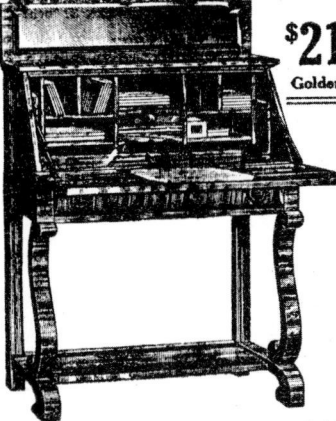

$21.85 Golden Oak.

Made of quarter sawed oak, polished golden finish, or birch, imitation mahogany finish, with drawer, front and outside of desk veneered in genuine mahogany. Wood knobs. Height, **48** inches; width, **30** inches; depth, **17** inches. Beveled mirror, **6x28** inches. Shipping weight, 90 pounds. Shipped from factory in ROCKFORD, ILL.

1L3822⅓—Golden oak........................**$21.85**
Mahogany finish........................**21.95**

$28.65 Golden Oak.

Made of quarter sawed oak, polished golden finish, or birch, imitation mahogany finish, with front of fall lid and drawer pulls veneered in genuine mahogany, highly polished. Bevel plate mirror, **6x30** inches. Wood knobs. Height, **50** inches. Width, **32** inches. Depth, **16½** inches. Shipping weight, 110 pounds. Shipped from factory in ROCKFORD, ILL.

1L3831⅓—Golden oak........................**$28.65**
Mahogany finish........................**28.75**

Music Cabinets - Tea Wagon - Desk - Sewing Cabinet

Music Cabinet.

Made of birch with genuine mahogany veneered doors, or quarter sawed oak, polished golden finish, or finished in American walnut. Will hold 90 player piano rolls that are not over 2½ inches wide. Height, 46 inches. Width, 24¾ inches. Depth, 14 inches. Shpg. wt., 110 pounds. Shipped from factory in ROCKFORD, ILL.

1L3876⅓
Golden oak. $19.75
Mahogany. 19.85
Walnut finish ... 19.95

Can be used either as a table or desk. Made of oak, dull golden finish. Pedestal has three drawers, 16 inches wide. Top drawer is divided into eight compartments. Can be folded into compact form, as shown in small illustration. Top, 23x48 inches. Height, 27 inches. When closed the top is 23x23¾ inches. Shpg. wt., 90 lbs. Shipped from factory in ROCKFORD, ILL.

1L3725⅓ $13.45

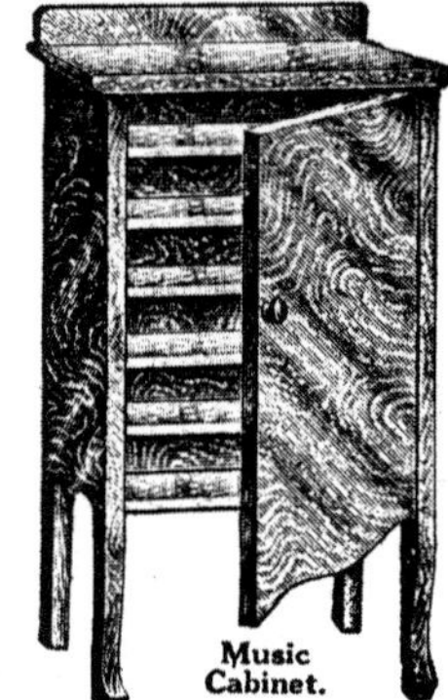

Music Cabinet.

Made of birch, mahogany finish, or in oak with quarter sawed oak veneered front and top, polished golden finish. A well made, nicely polished, inexpensive music cabinet. Height, 40 inches. Width, 20 inches. Depth, 12 inches. Shipping weight, 70 pounds. Shipped from our store.

1L3857
Golden oak... $10.75
Mahogany finish 10.85

Serviceable Screens

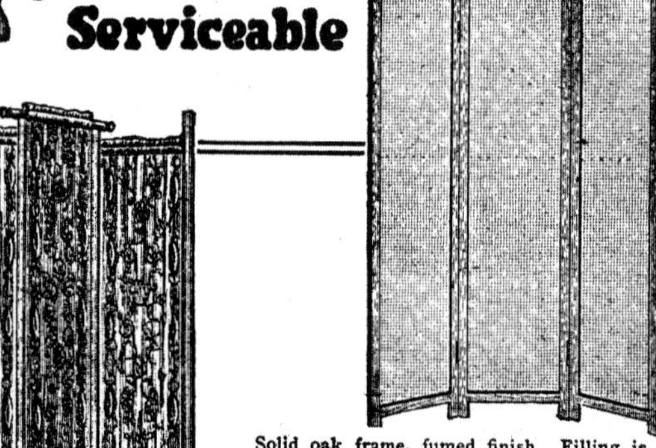

Frame made of hardwood, golden oak gloss finish. Has an attractive floral pattern cretonne filling in green with red flowers, shirred on wood rods at the top and base. Height, 61 inches. Width, 49 inches, open. Shipping weight, 9 pounds.

1L3704 .... $2.48

Solid oak frame, fumed finish. Filling is dark green burlap. Has three 17¾-inch wings. Height, 67 inches. Total width, opened, 53¼ inches. Shipping weight, 13 lbs.

1L3717 .... $5.45

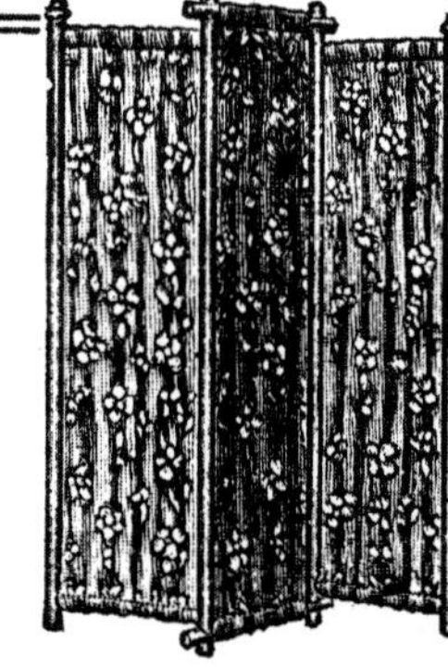

Hardwood frame, high gloss golden finish. Fitted with floral pattern cretonne in green with red flowers, shirred on rods at top and base. Height, 61 inches. Width, open, three wings, 55½ in. Five wings, 91½ in. Shipping weight, 12 pounds and 16 pounds, respectively.

1L3713
Three wings... $3.65
Five wings.... 5.95

Combination Tea Wagon and Serving Table.

Made of gumwood, mahogany or walnut, dull rubbed satin finish. Has loose serving tray top. Artillery wheels, 14 inches in diameter, with heavy rubber tires. Legs equipped with 3-inch faultless casters. Height, 28 inches. Top, 25½x15½ inches with side leaves down. Top, 35½x25½ inches with side leaves raised. Shipping weight, 42 pounds.

1L3623
Mahogany finish $18.85
Walnut finish 18.95

Convenient, Attractive. Gate Leg Table.

Made of gumwood. Choice of brown mahogany finish with genuine mahogany veneered top, or dull satin American walnut finish with genuine walnut veneered top. The two legs on each side are joined together with hinges to base proper to support the side leaves when raised. Has large roomy drawer. Top, open, is 34x42 inches. Size with both sides dropped, 14 inches wide and 34 inches long. Height, 28 inches. Shipping weight, 78 pounds.

1L3633
Mahogany $23.65
Walnut 23.75

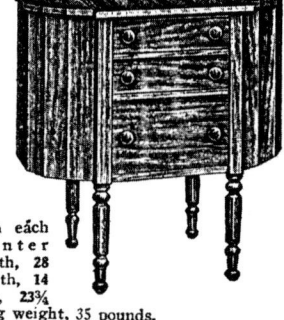

Martha Washington Sewing Cabinet.

Has solid mahogany top and gumwood base, brown mahogany finish, or made of gumwood finished in dull American walnut. Sewing compartment in each end. Three center drawers. Length, 28 inches. Width, 14 inches. Height, 23¾ inches. Shipping weight, 35 pounds.

1L3614
Mahogany finish............................ $15.85
Walnut finish............................ 15.95

Very Latest Styles and Best Values in End Tables

This Davenport End Table will prove to be very serviceable and add to the attractiveness of a room.

This table is of a very neat design, strong construction and nicely finished.

Made of gumwood with genuine walnut veneered or genuine mahogany veneered top. Mahogany or walnut finish. Size of top, 12x24 inches. Height, 26 inches. Shpg. wt., 17 lbs.

1L3636—Mahogany $5.45
Walnut 5.55

Tables in these fashionable designs are found in the big city stores at much higher prices.

Parlor Table. Genuine mahogany and black walnut top, mahogany or walnut dull rubbed satin finish. The two-tone effect of the mahogany and black walnut is the very latest style in exclusive home furnishings. This table is splendidly made and finished and priced so low as to be a tremendous bargain. Octagonal top measures 35 inches in diameter. Shipping wt., 85 lbs.

1L3635—Mahogany $21.65
Walnut 21.75

These companion tables with tops of genuine mahogany and black walnut are widely popular.

Davenport End Table. A new design and very attractive. The new two-tone effect is well carried out in the genuine mahogany and black walnut top. Mahogany or walnut dull rubbed satin finish. Davenport end tables are greatly in demand and you will find them selling elsewhere for quite a bit more than our price. Top measures 14x26 inches. Height, 24 inches. Shipping weight, 40 pounds.

1L3631—Mahogany $10.85
Walnut 10.95

Polychrome Mirrors

The finish on the frames of these mirrors is polychrome. Polychrome is the mingling or blending of many colors harmoniously. These mirrors have become very popular, as they add greatly to the attractiveness of a living room, bedroom or hall.

We have taken great care to select only the most popular designs and sizes. They are all well constructed, have good grade of plate glass, and are exceptional values at our low prices. Securely packed for shipping.

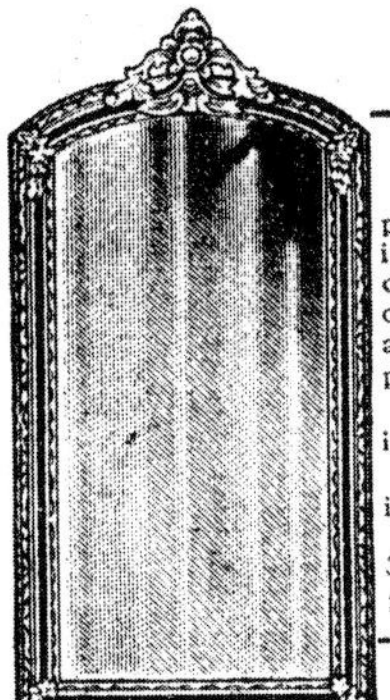

MAYFIELD

The frame of this pretty mirror is finished in blue polychrome with artistically colored flowers and ornaments. Plain plate mirror.

Frame—17½x34¼ inches over all.

Mirror—14x28 inches.

Shipping weight, 30 pounds.

1L4122...$8.45

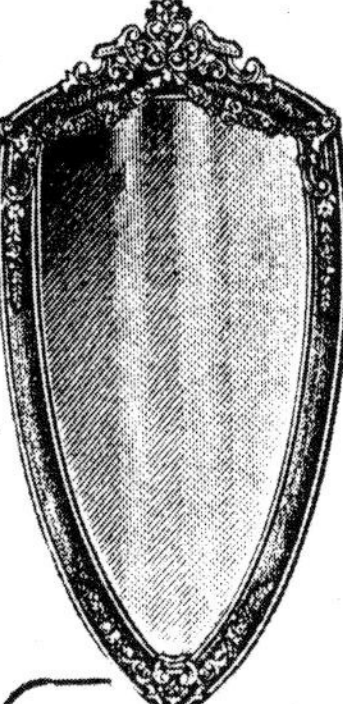

CRESTFIELD

This crest shape mirror is a very popular style. Frame finished in grayish blue polychrome, decorated with flowers in harmonizing colors. Plain plate mirror.

Frame—19½x36½ inches over all.

Mirror—16x30 in.

Shpg. wt., 29 lbs.

1L4125.$11.45

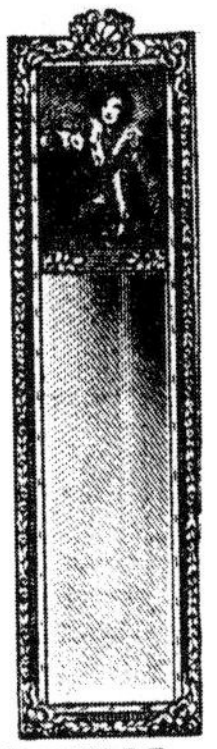

PANELLE

Frame beautifully toned in gold polychrome, decorated with artistically colored flowers and ornaments. Plain plate mirror. Assorted pictures at top.

Frame — 7¾x29 in. over all.

Mirror—6x18 inches.

Picture—6x8 inches.

Shipping weight, 12 pounds.

1L4126...$2.58

SUNBURST

Frame is polychrome finish, gold and blue predominating, ornamented in harmonizing colors. Plain plate mirrors. Mirror at top has a sunburst cut in glass.

Frame — 15x35 inches over all.

Top Mirror—12x9 in.

Bottom Mirror—12x20 inches.

Shipping wt., 25 lbs.

1L4134.......$9.65

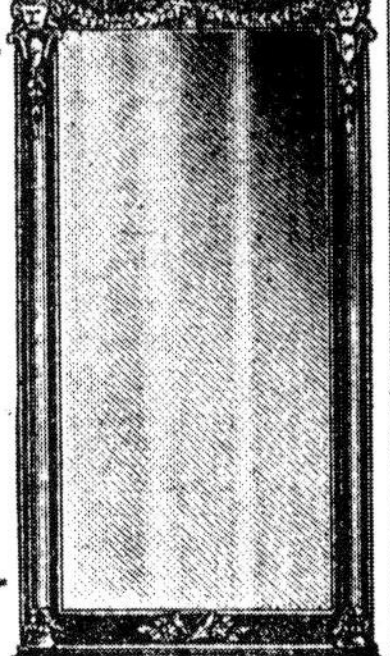

BOUDOIR

The design of this frame is sure to make it very popular. Beautifully toned in brown polychrome, with the flowers in various colors to harmonize. Frame is 21x34½ inches. Mirror, 14x28 inches. Shipping wt., 30 pounds.

1L4129..$9.95

GOREAU

The frame is green polychrome, gold burnished trimming, artistically decorated with colored flowers and ornaments. Plain plate mirror.

Frame—14x28 inches over all.

Mirror—12x22 in.

Shipping weight, 22 pounds.

1L4119.....$4.95

Frame is made of wood, nicely finished in gold color (not polychrome) and decorated with floral ornaments and picture in colors. Plain plate mirror.

Frame—13½x28¾ inches.

Mirror—10x18 in.

Picture—6x10 in.

Shipping weight, 15 pounds.

1L4127....$3.95

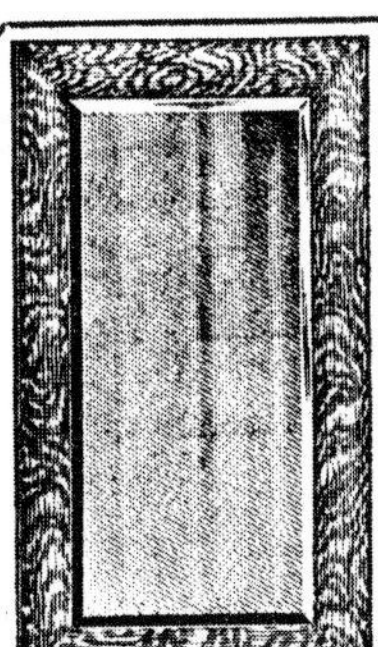

Plain wood frame, finished in imitation quarter sawed oak, imitation mahogany or imitation Circassian walnut, gloss finish. **State finish wanted.** Bevel plate mirror.

Frame—About 21x43 inches.

Width of Frame—3 inches.

Mirror—18x40 in.

Shpg. wt., 40 lbs.

1L4113

Quarter sawed oak....... **$9.85**

Mahogany.... **9.90**

Circassian walnut.... **9.95**

A polychrome finish wood frame plate Mirror in a very attractive design. Gold and brown predominate in the frame. The decorative carvings are in harmonizing colors. Plain plate mirrors.

Frame—15½x40½ inches over all. Mirrors—11¾ inches square. Shipping weight, 25 pounds.

1L4136......................................$7.65

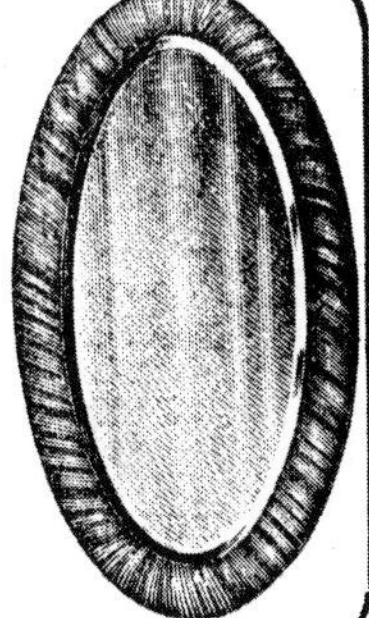

BEVEL PLATE MIRROR

Frame, 3 inches wide, in imitation quarter sawed oak, golden finish or imitation walnut, high gloss finish. Bevel plate mirror, 16x28 inches. Shipping wt., 56 pounds.

1L4114

Imitation quarter sawed oak...**$8.40**

Walnut finish. **8.45**

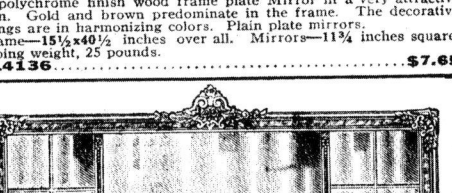

FOXHALL MANOR

This mirror is so richly colored and decorated as to make it an exceptionally beautiful ornament. The frame is polychrome finish, gold and brown predominating. Ornaments and flowers in harmonizing colors. Plain plate mirror.

Frame—18¼x48 inches over all. Center Mirror—12x24 inches. End Mirrors—10x12 inches. Shipping weight, 35 pounds.

1L4133..................................$11.95

Frame of oak, golden gloss finish. Side mirror frames attached to center mirror frame with strong hooks, permitting adjusting to any position desired. Bevel plate mirrors.

Length, over all—About 47 inches.

Height—15 inches.

Center Mirror—12x20 inches.

Each End Mirror—9x12 inches.

Shipping weight, 28 inches.

1L4135......................$8.45

Frame made of oak, polished golden finish. Fitted with four double hooks, made of metal, oxidized finish. Bevel plate mirror.

Frame—About 17x24 inches.

Width of Frame—About 3 inches.

Mirror—14x24 inches.

Shipping weight, 30 pounds.

1L4132......................................$7.65

Frame made of solid oak, high gloss finish, or hardwood, white enamel finish. Carefully fitted corners. Plain or bevel plate mirror.

1L4102

Width Frame, Inches	Size Plate, Inches	Oak Frame: Plate, Plain Edges	Oak Frame: Plate, Bevel Edges	White Enamel, Plate Bevel Edges	Shpg. Wt., Lbs.
1½	7x 9	$0.88	......	....	5
1½	9x12	1.08	$1.48	....	6
1½	10x14	1.45	1.95	$2.25	7½
2	12x20	2.45	2.95	2.95	14
2	14x24	3.65	......	....	17
3	16x28	5.25	......	....	22½
3	18x36	......	8.85	....	33
3	18x40	8.95	......	....	35

Steel and Wood Medicine Cabinets

These Medicine Cabinets are well made and exceptionally well finished. Mirrors are genuine plate glass. These values are unbeatable.

Made of sheet steel, outside and inside finished in white gloss enamel. Has two shelves with partition. Plain plate mirror.

Height—21 inches.

Width—17 inches.

Mirror—6½x 11½ inches.

Shipping weight, 29 pounds.

1L4152..................$7.45

Made of sheet steel, outside and inside finished in white gloss enamel. Plain plate mirror in door. Two stationary steel shelves.

Height—20 inches.

Width—13½ in.

Depth—4¾ inches.

Mirror—7x12 in.

Shpg. wt., 20 lbs.

1L4146...$4.95

A very good wood cabinet, inside and outside finished in white enamel. Has two shelves and one drawer. Door equipped with catch. Plain plate mirror.

Height—21¼ inches.

Width—14⅜ inches.

Depth—5¾ inches.

Shpg. wt., 21 lbs.

1L4145.....$5.45

Wood cabinet. White enameled inside and outside. Has two shelves. Door equipped with catch. Plain plate mirror.

Height—16½ in.

Width—12½ in.

Depth—6 inches.

Shipping weight, 19 pounds.

1L4143..........................$4.48

Do You Know Values? Here's a Big One!

Queen Anne Period Style Dining Room Set

Genuine Walnut Veneer.

Queen Anne Period Style—that means the most popular style in dining room furniture. People who appreciate pleasant surroundings, particularly in dining room furnishings, have given their approval to the Queen Anne Period Design.

The angle-brace construction of this set insures long service. The genuine walnut veneer is usually found only on much higher priced furniture.

Our price on this set positively cannot be equalled elsewhere. We are proud of this value, as you will be proud of the furniture itself.

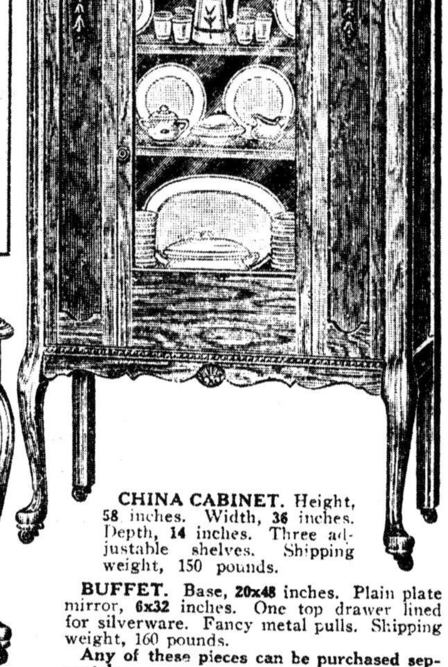

The buffet mirror is genuine heavy plate glass, not the greatly inferior so-called "crystal" or shock glass.

Here is where each of your dollars buys considerably more than one hundred cents' worth of furniture.

Tops, fronts and end panels genuine walnut veneered. Chairs are constructed of gumwood, walnut finish.

EXTENSION TABLE. Top, **45x54** inches. Furnished only in one size, **8**-foot extension. Shipping weight, 185 pounds.

CHAIRS. Removable slip seats, size **16½x15** inches, upholstered in **genuine leather,** brown or blue Spanish grained. **State color.** Height of back from seat, **21** inches. Shipping weight, 15 pounds.

CHINA CABINET. Height, **58** inches. Width, **36** inches. Depth, **14** inches. Three adjustable shelves. Shipping weight, 150 pounds.

BUFFET. Base, **20x48** inches. Plain plate mirror, **6x32** inches. One top drawer lined for silverware. Fancy metal pulls. Shipping weight, 160 pounds.

Any of these pieces can be purchased separately.

1L2325⅓

Buffet	**$27.45**
China Cabinet	**24.85**
Extension Table, 8-foot	**23.95**
Dining Chair, brown Spanish grained	**5.65**
Dining Chair, blue Spanish grained..	**5.85**

Chairs shipped from our CHICAGO or PHILADELPHIA store; all other pieces shipped from factory in INDIANA.

Attention! Where Can You Buy a Better Set for $80.65 with 6 Chairs and 6 Foot Table

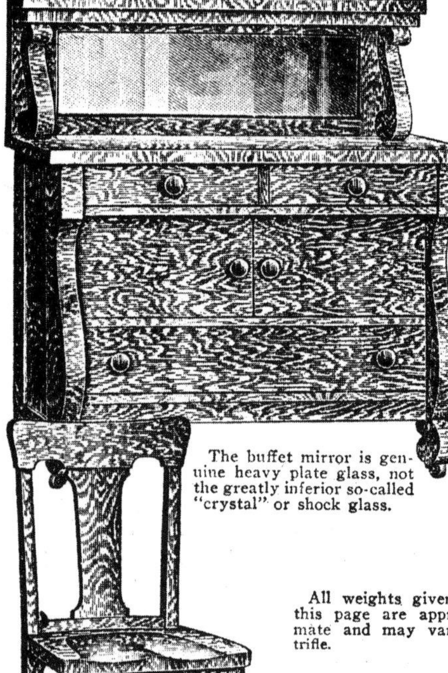

The buffet mirror is genuine heavy plate glass, not the greatly inferior so-called "crystal" or shock glass.

All weights given on this page are approximate and may vary a trifle.

Shop right here if you want an inexpensive Dining Room Set. We've priced it far below the usual retail price. Style, quality and construction are much better than you would expect for the price.

Made of hardwood in **imitation of quarter sawed oak, high gloss golden finish.**

EXTENSION TABLE. Top, **45** inches in diameter. Pedestal, **10** inches in diameter, is non-dividing in all lengths and in all except **6**-foot length has patent drop leg construction.

CHAIRS. Saddle shape seats. Height of back from seat, **20** inches. Entire height, **37½** inches.

CHINA CABINET. 58 inches high, **40** inches wide and **14** inches deep. Glass paneled doors with lock. Three wood grooved adjustable shelves. Glass paneled ends. Casters.

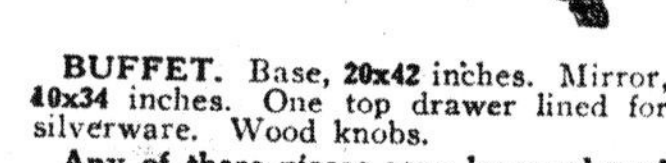

BUFFET. Base, **20x42** inches. Mirror, **10x34** inches. One top drawer lined for silverware. Wood knobs.

Any of these pieces may be purchased separately.

1L2311⅓—Extension Table.

Size	Shipping Weight	
6-foot	165 pounds	**$17.65**
8-foot	195 pounds	**21.65**
10-foot	225 pounds	**25.65**
12-foot	255 pounds	**29.65**
Buffet. Shipping wt., 175 lbs....		**24.45**
China Cabinet. Shipping weight, 135 pounds		**22.65**
Chairs. Shipping weight, each, 13 pounds		**2.65**

Buffet, China Cabinet and Table shipped from factory in WISCONSIN. Chairs from our CHICAGO or PHILADELPHIA store.

A Choice Selection of the Most Popular Styles.

Buffets

Our Buffets Are Famous for High Quality and Low Price.

$49.85

22x54-Inch Size.

Genuine Walnut Veneered Buffet. A most attractive addition to your dining room. Graceful Queen Anne Period design which is so popular today in the big cities. Top, front and ends are genuine walnut veneer. Finish is dull satin American walnut. There are six compartments in the top drawer, five of them lined for silverware. Lock and key. Easy rolling casters. Choice of two sizes as noted below. Shipped from factory in INDIANA.

1L2667⅓

Base, 22x54 inches. Mirror, 8x52 inches. Shipping wt., 200 lbs....**$49.85**
Base, 22x60 inches. Mirror, 8x56 inches. Shipping wt., 230 lbs....**59.65**

$49.65

Golden Oak.

Material—Quarter sawed oak. **Finish**—Polished golden or fumed. Interior finished with two coats of varnish. Right top drawer lined for silverware. Shelf 7 inches wide extends all the way across lower cupboard section. Wood knobs. **Measurements**—Base, 24x60 inches. Bevel plate mirror, 12x52 inches. Shipped from factory in INDIANA. Shipping weight, 300 pounds.

1L2670⅓—Golden finish....**$49.65**
Fumed finish....**49.75**

The Buffets on this page are equipped with heavy plate mirrors, not the greatly inferior mirrors made of so called crystal or shock glass.

$33.75

Golden Oak.

Material—Quarter sawed oak. **Finish**—Golden gloss or fumed. Finished inside with two coats of varnish. Bevel plate mirror, 10x42 inches. Right top drawer lined for silverware. Cupboard is not divided inside. Square metal drawer pulls. **Measurements**—Base, 22x48 inches. Shipped from factory in INDIANA. Shipping weight, 170 pounds.

1L2673⅓—Golden oak....**$33.75**
Fumed oak....**33.85**

$37.75

Golden Oak.

Material—Quarter sawed oak. **Finish**—Golden gloss or fumed. Inside finished with two coats of varnish. Bevel plate mirror. Right top drawer lined for silverware. Cupboard is not divided inside. Wood knobs. **Measurements**—Base, 22x48 inches. Beveled mirror, 8x40 inches. Shipped from factory in INDIANA. Shipping weight, 210 lbs.

1L2664⅓—Golden oak....**$37.75**
Fumed oak....**37.85**

$24.85

Golden Oak.

Material—Oak, with front, top of base and mirror standards veneered with quarter sawed oak. **Finish**—Golden gloss or fumed. Bevel plate mirror. Right top drawer lined for silverware. Cupboard is not divided inside. Wood knobs. **Measurements**— Base, 20x42 inches. Mirror, 8x36 inches. Shipped from factory in INDIANA. Shipping weight, 170 pounds.

1L2672⅓

Golden oak...**$24.85**
Fumed oak... **24.95**

$27.45

Made of hardwood with top, front and end panels veneered with genuine walnut. Dull satin American walnut finish. Base measures 20x48 inches. Plain plate mirror, 6x32 inches. Has one top drawer lined for silverware. Fancy metal drawer pulls. Shipped from factory in INDIANA. Shipping weight, 160 pounds.

1L2645⅓

Walnut.**$27.45**

Queen Anne Style Buffet. This stylish period design is found in the fashionable homes of the big cities. Our low price enables you to have the same design to adorn your dining room.

China Cabinets

Material—Quarter sawed oak. **Finish**—Golden gloss or fumed. **Construction**—Full bent glass ends and door. Wood lattice work at top of panel in door. Four shelves. **Measurements**—Height, **68** inches; width, **39** inches. Mirror in top, **6x 34** inches. Shipped from factory in ROCKFORD, ILL. Shipping weight, 150 pounds.

1L2723⅓—Golden oak........**$29.85**
Fumed oak........................**29.95**

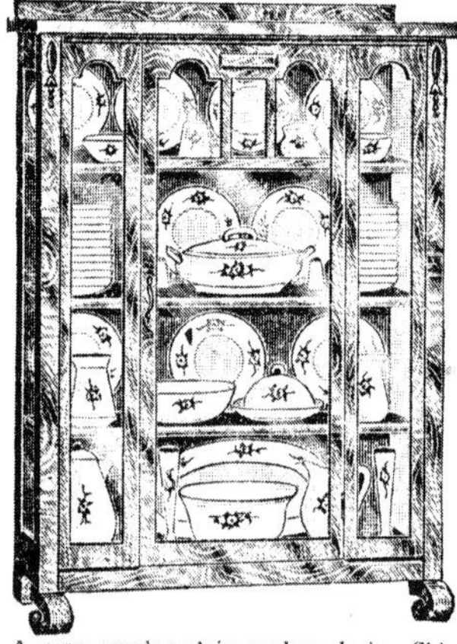

A very popular plain modern design China Cabinet. Made of quarter sawed oak in golden gloss or fumed finish. Has glass side and end panels and glass paneled door with wood lattice work in the design. Three adjustable shelves.

Measurements—Height, **58** inches; width, **42** inches. Shipped from factory in ROCKFORD, ILL. Shipping weight, 170 pounds.

1L2718⅓—Golden oak........**$31.75**
Fumed oak........................**31.85**

Material—Quarter sawed oak. **Finish**—Golden gloss or fumed. **Construction**—Bent glass ends and wide glass paneled front door with handle. Four adjustable shelves have molded edges and grooves near the back edge for plates. Steel sliding casters. **Measurements**—Height, **62** inches; width, **36** inches. Shipped from factory in ROCKFORD, ILL. Shipping weight, 140 pounds.

1L2716⅓—Golden oak........**$22.65**
Fumed oak........................**22.75**

Refrigerators

Our Refrigerators Are Proved Best by Actual Use.

There are no better refrigerator values to be had than we offer on these two pages. The large assortment shown covers a variety of sizes and prices that will meet all requirements. Every refrigerator here listed is constructed by expert workmen and is built according to the most modern scientific principles. We do not experiment. These refrigerators have been proved best by actual use. We guarantee these values cannot be equaled by any retail dealer in the country.

All Refrigerators on this and opposite page are shipped from factory near GRAND RAPIDS, MICH., or direct from our PHILADELPHIA store.

Outside Case—Made of ash, high gloss golden oak finish. Swinging baseboard. **Trimmings**—Solid metal nickel plated hinges and swing lever fasteners on doors. Self retaining socket casters. Automatic air trap fitted in bottom of provision chamber. **Ice Chamber**—Single door. Galvanized steel lining. All metal ice rack. **Food Chambers**—White enameled. Steel wire shelves. Removable drain pipe.

1L2854⅓—**Ice Chamber**—Width, **12** inches; depth, **15** inches; height, **19¾** inches. Capacity, 100 pounds. **Large Provision Chamber**—Width, **12¾** inches; depth, **15 inches**; height, **30¼** inches. **Small Provision Chamber**—Width, **12** inches; depth, **15** inches; height, **7½** inches. **Outside Measures**—Width, **33** inches; depth, **19** inches; height, **46** inches. Shipping weight, 195 pounds........**$28.45**

1L2856⅓—**Ice Chamber**—Width, **12¾** inches; depth, **16** inches; height, **22⅜** inches. Capacity, 125 pounds. **Large Provision Chamber**—Width, **13¾** inches; depth, **16** inches; height, **34¾** inches. **Small Provision Chamber**—Width, **12¾** inches; depth, **16** inches; height, **9¼** inches. **Outside Measures**—Width, **35** inches; depth, **20** inches; height, **50** inches. Shipping weight, 225 pounds........**$32.95**

Water Cooler. This shows the water cooler of Refrigerator 1L2819⅓. Made of steel, with white enameled lining, fitted at the left hand side of ice chamber. A cup stand and faucet are also furnished, as shown in large illustration.

Refrigerator With Water Cooler.

How convenient it is to have cold drinking water handy in the house—ready whenever you want it. Our very low price makes this refrigerator equipped with water cooler an exceptional value.

Outside Case—Made of ash, high gloss golden oak finish. Swinging baseboard. **Trimmings**—Solid metal lever, nickel plated hinges and swing lever fasteners on doors. Self retaining casters. Air trap fitted to bottom of provision chamber. **Ice Chamber**—Opens from the top. Galvanized steel lining. All metal ice rack. Removable flue. **Provision Chamber**—White enameled, steel lined. Floor on level with bottom of door. Steel wire shelves, tinned finish. Removable drain pipe.

1L2819⅓—**Ice Chamber**—Width, **20** inches; depth, **12⅛** inches; height, **11** inches. Capacity, 85 pounds. **Provision Chamber**—Width, **19½** inches; depth, **12¾** inches; height, **17** inches. **Outside Measures**—Width, **26** inches; depth, **17** inches; height, **44** inches. Shipping weight, 150 pounds..**$26.75**

Equipped with water cooler as illustrated above.

White Porcelain

White Enameled Lined Refrigerators are the most desirable because they are the most sanitary. The smooth white enameled surface is hard and polished, the easiest surface to clean and keep clean. Buy refrigerators with white porcelain enameled lining both for looks and health. And our prices are low. In fact, our prices on porcelain enameled lined refrigerators are no higher than many dealers ask for ordinary and inferior grades.

One of the best refrigerators on the market. Outside case, northern ash, high gloss golden oak finish. Has paneled ends, swinging baseboard. Trimmings—Nickel plated metal hinges and latest type lever fasteners hold doors into airtight joint. Has retaining spring socket casters. Air trap is fitted to bottom of provision chamber. Ice chamber. Single door. Galvanized steel lining. All metal ice rack. Provision chamber, made of one piece of sheet metal, finished on inner side in pure white porcelain enamel. This porcelain enamel lining has the same hard durable polished surface as on porcelain cooking ware and will last indefinitely. It is an ideal lining for refrigerators, as it is clean and sanitary, having no cracks or crevices to hold anything that may be spilled. Easily cleaned and kept clean.

Outside Dimensions—Width, **36** inches; depth, **20** inches; height, **52** inches. Ice capacity, 125 pounds. Shipping weight, 265 pounds. **Ice Chamber**—Width, **13⅝** inches; depth, **15⅝** inches; height, **22⅛** inches. **Large Provision Chamber**—Width, **13⅝** inches; depth, **15⅝** inches; height, **22⅛** inches. **Small Provision Chamber**—Width, **13⅝** inches; depth, **15⅝** inches; height, **10¾** inches.

1L2893⅓..............................**$48.00**

Best Made-Lowest Priced

Read Here the Description of Refrigerators Listed.

Insulation is a composition, felted into sheets. It does not settle in the wall leaving the ice chamber unprotected, nor can it sift through the inner walls to the provision chamber.

Ice Chamber. The ice rests on a removable rack. The water flows into the waste pipe and fills the drip cup. The end of pipe is always immersed in the water in cup. No foul air can enter ice chamber through waste pipe.

Provision Chamber. All our refrigerators, except 1L2832⅓ and 1L2889⅓, have the top, sides and bottom of the provision compartment lined with heavy galvanized steel. The shelves are made of steel wire, tinned finish.

Ice Capacity. A cake of ice cut the exact dimensions of the ice chamber will weigh the number of pounds given in the description. We recommend selection of refrigerator with ice capacity large enough to hold amount of ice you wish to use, the unmelted portion of previous cake, as well as milk bottles, etc.

Outside Case—Ash, high gloss golden finish. Swinging baseboard. **Trimmings**—Stamped steel, nickel plated hinges and swing lever fasteners on doors. Self retaining socket casters. Air trap on bottom of provision chamber. **Ice Chamber**—Galvanized steel lining. All metal ice rack. **Provision Chambers**—White enamel finish lining. Wire shelves. Removable drain pipe.

1L2833⅓—Ice Chamber—Width, **10⅛** in.; depth, **12** in.; height, **15⅝** in. Capacity, 50 pounds. **Large Provision Chamber**—Width, **10⅛** in.; depth, **12** in.; height, **25⅜** in. **Small Provision Chamber**—Width, **10⅛** in.; depth, **12** in.; height, **7¼** in. **Outside Measures**—Width, **28** in.; depth, **16** in.; height, **41** in. Shipping weight, 150 lbs. . **$19.85**

1L2834⅓—Ice Chamber—Width, **11⅜** in.; depth, **14** in.; height, **17¾** in. Capacity, 75 pounds. **Large Provision Chamber**—Width, **11⅜** in.; depth, **14** in.; height, **28¼** in. **Small Provision Chamber**—Width, **11⅜** in.; depth, **14** in.; height, **7½** in. **Outside Measures**—Width, **31** in.; depth, **13** in.; height, **44** in. Shipping weight, 170 lbs. . . . **$22.95**

Outside Case—Ice Chamber and Provision Chamber are constructed and finished as described above on 1L2833⅓ and 1L2834⅓.

1L2827⅓—Ice Chamber—Width, **17** in.; depth, **9¾** in.; height, **10¼** in. Capacity, 50 pounds. **Provision Chamber**—Width, **15¾** in.; depth, **10¼** in.; height, **14⅜** in. **Outside Measures**—Width, **21** in. depth, **14** in.; height, **39** in. Shpg. wt., 105 lbs. . **$15.85**

1L2828⅓—Ice Chamber—Width, **19** in.; depth, **10¾** in.; height, **11¼** in. Capacity, 70 pounds. **Provision Chamber**—Width, **17¾** in.; depth, **11¼** in.; height, **15⅜** in. **Outside Measures**—Width, **23** in. depth, **15** in.; height, **41** in. Shpg. wt., 120 lbs. . **$16.95**

1L2839⅓—Ice Chamber Width, **21** in.; depth, **11¾** in.; height, **12¼** in. Capacity, 90 pounds. **Provision Chamber**—Width, **19¾** in.; depth, **12¼** in.; height, **16¾** in. **Outside Measures** — Width, **25** in.; depth, **16** in.; height, **43** in. Shipping weight, 130 pounds **$19.45**

Case of seasoned northern elm, golden finish. The inner walls are built in the same perfect manner as the refrigerators. Economical in the use of ice. They are lined with a good quality galvanized steel and have galvanized steel shelves and wood rack. Equipped with drain pipe and drip cup.

	Outside Measures. Width, In.	Ht., In.	Depth, In.	Shpg. Wt., Lbs.	
1L2804⅓	23	25	16	90	**$ 9.95**
1L2806⅓	26	27	18	95	**12.85**
1L2808⅓	32	31	22	145	**16.95**

Enameled Lining

Outside Case—Northern ash, high gloss golden oak finish. Paneled ends. Swinging baseboard. **Trimmings**—Nickel plated metal hinges and swing lever fastener hold doors into an airtight joint. Self **retaining** spring socket casters. Air trap fitted to bottom of provision chamber. **Ice Chamber** — Single door. Galvanized steel lining. All metal ice rack. **Provision Chamber**—Made of one piece of sheet steel, the inner side of which has a pure white porcelain enamel finish, as described on the opposite page.

1L2889⅓

Ice Chamber — Width, **12⅝** in.; depth, **14⅝** in.; height, **19⅝** in. Capacity, 100 pounds. **Large Provision Chamber**—Width, **12⅝** in.; depth, **14⅜** in.; height, **32¾** in. **Small Provision Chamber**—Width, **12⅝** in.; depth, **14⅜** in.; height, **9¾** in. **Outside Measures**—Width, **34** in.; depth, **19** in.; height, **47** in. Shipping weight, 266 pounds **$44.00**

1L2886⅓—Ice Chamber—Width, **10⅝** in.; depth, **12½** in.; height, **15¾** in. Capacity, 60 pounds. **Large Provision Chamber**—Width, **10⅝** in.; depth, **12½** in.; height, **26⅞** in. **Small Provision Chamber**—Width, **10⅝** in.; depth, **12½** in.; height, **7¾** in. **Outside Measures**—Width, **30** in.; depth, **17** in.; height, **43** in. Shipping weight, 185 pounds **$36.75**

Outside Case—Northern ash, golden gloss oak finish. **Trimmings**—Nickel plated metal hinges and drop lever fasteners. Has self retaining spring socket casters. **Ice Chamber**—Single door. Galvanized steel lining. All metal ice rack. **Provision Chamber**—Made of one piece of sheet steel, finished in pure white porcelain enamel.

1L2832⅓

Ice Chamber—Width, **21¼** in.; depth, **13⅝** in.; height, **13½** in.; capacity, 100 pounds. **Provision Chamber**—Width, **21¼** in.; depth, **13⅜** in.; height, **17¾** in.; three shelves. **Outside Measures**—Width, **27⅝** in.; depth, **18** in.; height, **49** in. Shipping weight, 100 lbs. **$34.85**

Cupboards

1L2236⅓—Made of hardwood, with solid oak front. **Finish**—Golden gloss. Stands **80½** inches high, **41** inches wide and **16** inches deep, outside measurements. Top section doors have glass panels, **12x28** inches. Inside has three removable and adjustable shelves. Wood knobs. Shipped knocked down from factory in INDIANA. Shipping weight, 150 pounds **$17.75**

1L2223⅓—Made of hardwood. **Finish**—Golden gloss, **71½** in. high, **38½** in. wide and **16¾** in. deep, extreme outside measurements. Top section has glass panel doors and the inside fitted with two shelves. Inside below is fitted with a shelf. Two roomy drawers. Wood knobs. Shipped knocked down from factory in INDIANA. Shipping weight, 120 pounds **$14.65**

1L2218⅓—Made of hardwood. **Finish**—Golden gloss. Stands **71½** inches high and is **38½** inches wide and **16¾** inches deep, extreme outside measurements. Upper compartment has two inside shelves and the lower compartment has one. Drawers have knobs. Shipped knocked down from factory in INDIANA. Shipping weight, 115 pounds **$12.45**

The World's Greatest

\$43.85 \$48.85

Golden Oak White Enamel

Where Can You Buy a Better Kitchen Cabinet for the Money.

We are particularly proud of our ability to offer our customers this really remarkable Kitchen Cabinet value. We doubt if there has been anything like it ever offered the American public before. No better cabinet is made in America at the price. Cabinets of similar size, style, equipment, construction and quality are generally sold for at least 40 to 50 per cent more than we ask.

Note the full size, the white porcelain enameled table top, the large roomy shelves, all white enameled, in the upper section. All of the best labor saving devices and conveniences are included in this cabinet. The construction known as the improved angle brace lock joint, framed-in type, is the best known to the cabinet trade, as it is a positive insurance against warping or shrinkage from kitchen heat or moisture. The light natural oak color finish harmonizes nicely with other kitchen furnishings, or if you prefer the white enamel finish, you are assured of one of the handsomest cabinets ever made. We invite a most rigid comparison of this cabinet with any other cabinet on the market, as we are sure such a comparison will quickly convince you that in buying this cabinet you will effect a saving of many dollars.

1L2169⅓
In Golden Oak or White Enamel.

White Porcelain Enameled Top

So numerous are the advantages and benefits to be derived from kitchen cabinets that many people believe them to be as much a household necessity as a bed, a chair or a table. This cabinet is so constructed and arranged that the things most needed in the preparation of meals are READY TO YOUR HAND, thereby saving thousands of steps and so systematizing the work of the kitchen that many additional leisure hours are found for each busy day.

This cabinet is one of superior quality, workmanship and design and at our price insures your entire satisfaction as well as a substantial saving of money.

MATERIAL, CONSTRUCTION AND FINISH.

Made of { Oak, light natural oak, dull rubbed satin finish.
Oak, finished all over in white enamel.

Strong lock joint framed-in type construction which resists the kitchen steam and heat.

Interior is all white enameled, sanitary and easily cleaned.

Porcelain enameled top is of especially high quality, smooth, clean and highly polished.

SOME OF THE SPECIAL FEATURES ARE:

Easy filling flour bin, window ont and patent sifter. Capacity, about 50 pounds.
Swinging glass sugar container.
Cutlery drawer.
Meal bin. Capacity, 20 pounds.
Sliding pan tray in bottom of cupboard in base section.
Bracket support for food chopper.
Dustproof base top.
Mouse proof cake and bread box drawer with perforated lid.
Sliding drop curtain.
New type catches, insuring accurately closed, dustproof doors.

THIS CABINET IS ALSO EQUIPPED WITH

Menu cards. Bill hooks. Recipe card rack. Metal coin tray. Cook book holder. Crystal glass coffee, tea and spice jars and many other conveniences and labor saving devices.

Measurements—72 inches high, 48 inches wide and 27 inches deep, when top of base section is closed; the sliding extension porcelain table top when pulled forward increases the working surface to 38x48 inches, Height of table top, 34 inches; depth of base section, 23 inches; depth of top section, 11 inches; pan compartment in base, 22x27x16½ inches; metal bread drawer, 13½x9x18½ inches; tilting metal bin, 10x9x13 inches. Shipped from factory in INDIANA. Shipping weight, 325 pounds.

1L2169⅓—Golden oak \$43.85
White enamel 48.85

A Few of the Special Features Which Make This Cabinet a Supreme Value.

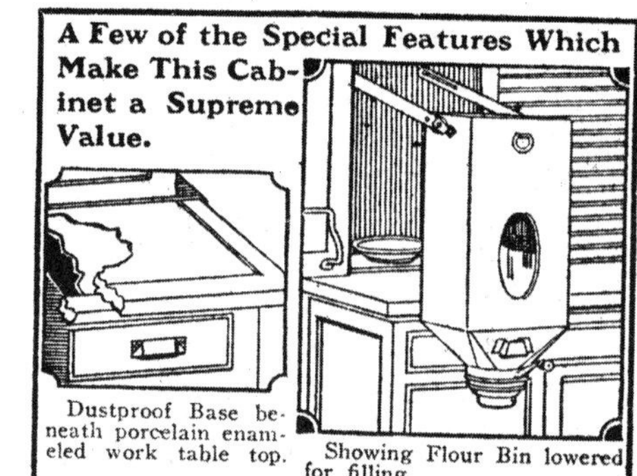

Dustproof Base beneath porcelain enameled work table top.

Showing Flour Bin lowered for filling.

New Type Catches, insuring accurately closed, dustproof doors.

Bracket Support for food chopper.

Only those articles specified in description are included with cabinets.

Kitchen Cabinet Values

WITH WHITE PORCELAIN ENAMELED TOP

10 SPECIAL FEATURES

The ten special features of this cabinet will eliminate a great many of the unpleasant kitchen duties. It places everything necessary to prepare a meal within easy reach. It saves many steps, which are the most tiresome and unpleasant part of a housewife's work. Read the description below and we know you will be fully convinced that we are offering one of the biggest kitchen cabinet values on the market today.

$34.65 GOLDEN OAK.

$38.65 WHITE ENAMEL.

Easily filled vermin proof 50-pound flour bin with leak proof patent sifter. Glass window shows level of flour.

Sanitary and moisture proof crystal glass containers for coffee, tea and spices.

Disappearing wood curtain door. Opens without disturbing contents of cupboard or removal of things from work table.

Crystal glass sugar jar with patent sifter dispenser swings conveniently in and out of cabinet, always at hand, never in the way.

Permanently attached rigid bracket for food chopper. No interference with sliding top or doors.

Removable metal trays for pots and pans, trays slide out for easy access and easy cleaning.

Scientifically aerated vermin proof metal bread and cake box, which effectively prevents mold. Easily cleaned.

Open sanitary base, high enough for easy cleaning and free circulation of air.

Porcelain enameled sliding table top is reinforced with wood, proof against warping or bulging. Slides from under top section for larger working space.

Bread board slides into receptacle under shelf in the dustproof cupboard, easily removed without disturbing other contents.

1L2167⅓
In Golden Oak or White Enamel.

Here is another striking example of our ability to offer our customers unequaled values. You would pay at least 40 or 50 per cent more than our price elsewhere for a cabinet of similar design, construction and equipment. Not quite as large a cabinet as shown on the opposite page, but having the very same superior construction and made of the same selected materials. The white porcelain table top alone—in a cabinet at this low price—makes this an outstanding bargain.

MATERIAL, CONSTRUCTION AND FINISH.

Made of oak, in light natural oak color, dull satin finish.
Made of oak, finished all over in white enamel.
Strong lock joint frame in type construction which resists the kitchen steam and heat.
Interior is white enameled, sanitary and easily cleaned.
The legs allow sufficient room for sweeping beneath.

MEASUREMENTS.

Height over all, 72 in. Height of table top, 33½ inches. Work base, closed, 25x42 inches; with extension pulled forward, 34x42 inches. Pan compartment in base, 22x22x16½ inches. Metal bread drawer, 13½x9x18½ inches. Capacity of flour bin, 50 pounds. Shipped from factory in INDIANA. Shipping weight, 250 pounds.

1L2167⅓
Golden oak.......$34.65
White enamel.... 38.65

THE EQUIPMENT ALSO INCLUDES:

Metal coin tray.
Removable metal shelf in top section.
Bill hooks.
Recipe card rack.
Cook book holder.
Table of weights and measures.

Kitchen Cabinet Stool

It is so much easier to do your work when seated.

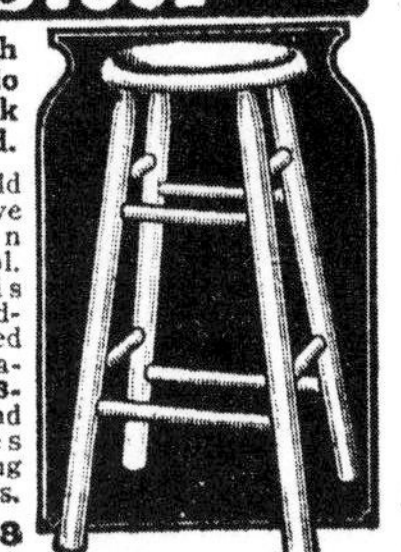

You should certainly have a Kitchen Cabinet Stool. This one is made of hardwood, finished in white enamel, has a 13-inch seat, and is 24 inches high. Shipping weight, 6 lbs.
1L307 $1.98

Only those articles specified in description are included with cabinets.

Why Buy Elsewhere

Now You Can Buy This Popular Queen Anne Style Bedroom Set for a Very Low Price

People everywhere are buying PERIOD furniture. The Queen Anne Period design here shown will harmonize with the best of home furnishings. Special features are the genuine walnut veneered tops, fronts and ends, the stylish bow end bed, drawers smoothly finished inside, and the mirrors of heavy genuine plate glass.

The latest style in bedroom furniture in a quality you cannot obtain elsewhere at **so low a price.** You will be more than pleased with this stylish Queen Anne Style Bedroom Set in your home. Exclusive furniture stores handle this kind of design at much higher prices. While **our price is far less than what others ask,** we have not sacrificed quality. Material, construction and finish are far superior to what you would expect for so little money. The **Bow End Bed** is very popular and usually comes only in very high priced furniture. **Genuine walnut veneered** tops, fronts and ends; American walnut, dull satin finish, or made of select oak with quarter sawed oak veneered tops, posts and ends, golden gloss finish. Drawers and doors are equipped with metal pulls.

BED.

Head end, **50** inches high; foot end, **30½** inches high; width, **54** inches. Reversible steel side rails. **When ordering a spring for this bed select one quoted for bow end metal bed, in size 54 inches.** See pages 662 and 663 for springs.

CHAIRS AND ROCKER.

Solid saddle shape seats, size, **16x17** inches. Height of back from seat: Chair, **19¾** inches; rocker, **21½** inches. Entire height of chair, **36½** inches. Height of dressing table chair seat from floor, **19** inches.

DRESSER.

Two sizes—Base, **19x36** inches. Plain mirror, **18x22 inches.** Or, base, **19x40** in.; mirror, **22x26** inches.

CHIFFORETTE.

Base, **19x32** inches. Two drawers, one large shelf and two removable trays in cupboard.

SEMI-VANITY DRESSER.

Base, **18x40** inches. Plain mirrors. Large center mirror is **18x36** inches, and swinging wing mirrors are each **8x24** inches.

1L4354⅓

All mirrors are heavy plate glass.

Any of these pieces may be purchased separately.

	Dresser, 36-Inch Base	Dresser, 40-Inch Base	Chifforette	Semi-Vanity Dresser	Bed	Rocker	Chair	Dresser Chair
Shipping wt....	135 lbs.	150 lbs.	120 lbs.	115 lbs.	115 lbs.	14 lbs.	14 lbs.	14 lbs.
Walnut	$18.95	$21.85	$16.95	$22.85	$19.60	$6.35	$5.85	$5.95
Oak	18.85	21.75	16.85	22.75	19.50	6.25	5.75	5.85

Chairs are shipped from our CHICAGO or PHILADELPHIA store. Other pieces from factory in INDIANA.

Buy Here and Save

Beautify Your Bedroom With This Colonial Design Bedroom Set

Always dignified, always in perfect taste, the Colonial Period design has won a high place in the esteem of those who appreciate furniture of the better sort. The set shown here, in oak or walnut, is at once dignified and graceful. The heavy bevel plate mirrors, and strong angle brace construction—a patented feature—are not usually found in furniture at these low prices, and assure you of an unequaled value.

Colonial Style Bedroom Set. The very best that can be bought anywhere at this price. Our unequaled buying power and enormous business reduces the cost so much that our furniture prices remain unequaled by any other retailer. **You save money when you buy furniture from us.**

Bed—Height head end, **49** inches; foot end, **35** inches. Length, inside, **76** inches; width, **54** inches. Reversible metal bed rails. When ordering spring for this bed select one for metal bed, in size **54** inches.

Dresser—Bevel plate mirror, **24x26** inches. Base, **19x42** inches. Wood knobs.

Semi-Vanity Dresser—Equipped with bevel plate mirrors. Swinging wing mirrors, each, **8x22** inches. Center mirror, **16x34** inches. Has two drawers; wood knobs. Top, **18x40** inches.

Chiffonier—Has roomy, convenient drawers. Wood knobs. Bevel plate mirror, **14x16** inches. Base, **18x32** inches.

Chairs and Rocker—Full box, saddle shape seats. Height of back from seat: Rocker and chair, **20** inches. Height of dressing table chair seat from floor, **19** inches.

Each piece is securely packed in a strong wood crate and shipped from factory in INDIANA.

Material—Made of oak, with upright matched fronts beautifully finished in golden gloss, or of select hardwood, dull satin American walnut finish. **Construction**—Strong angle brace construction throughout. Has large roomy drawers equipped with wood pulls, and the top drawers of chiffonier and dresser are equipped with good grade lock and key. All large pieces, with the exception of semi-vanity dresser, equipped with easy rolling casters.

Mirrors are genuine heavy plate glass. Any of these pieces may be purchased separately.

1L4340⅓

	Chiffonier	Bed	Semi-Vanity Dresser	Dresser	Rocker	Chair	Semi-Vanity Dressing Chair
Shipping weight......	145 lbs.	120 lbs.	110 lbs.	150 lbs.	20 lbs.	20 lbs.	20 lbs.
Oak	**$21.45**	**$14.85**	**$23.75**	**$24.85**	**$6.10**	**$5.60**	**$5.70**
Walnut	**22.95**	**16.35**	**25.25**	**26.85**	**6.20**	**5.70**	**5.80**

Chairs are shipped from our CHICAGO or PHILADELPHIA store; other pieces from factory in INDIANA.

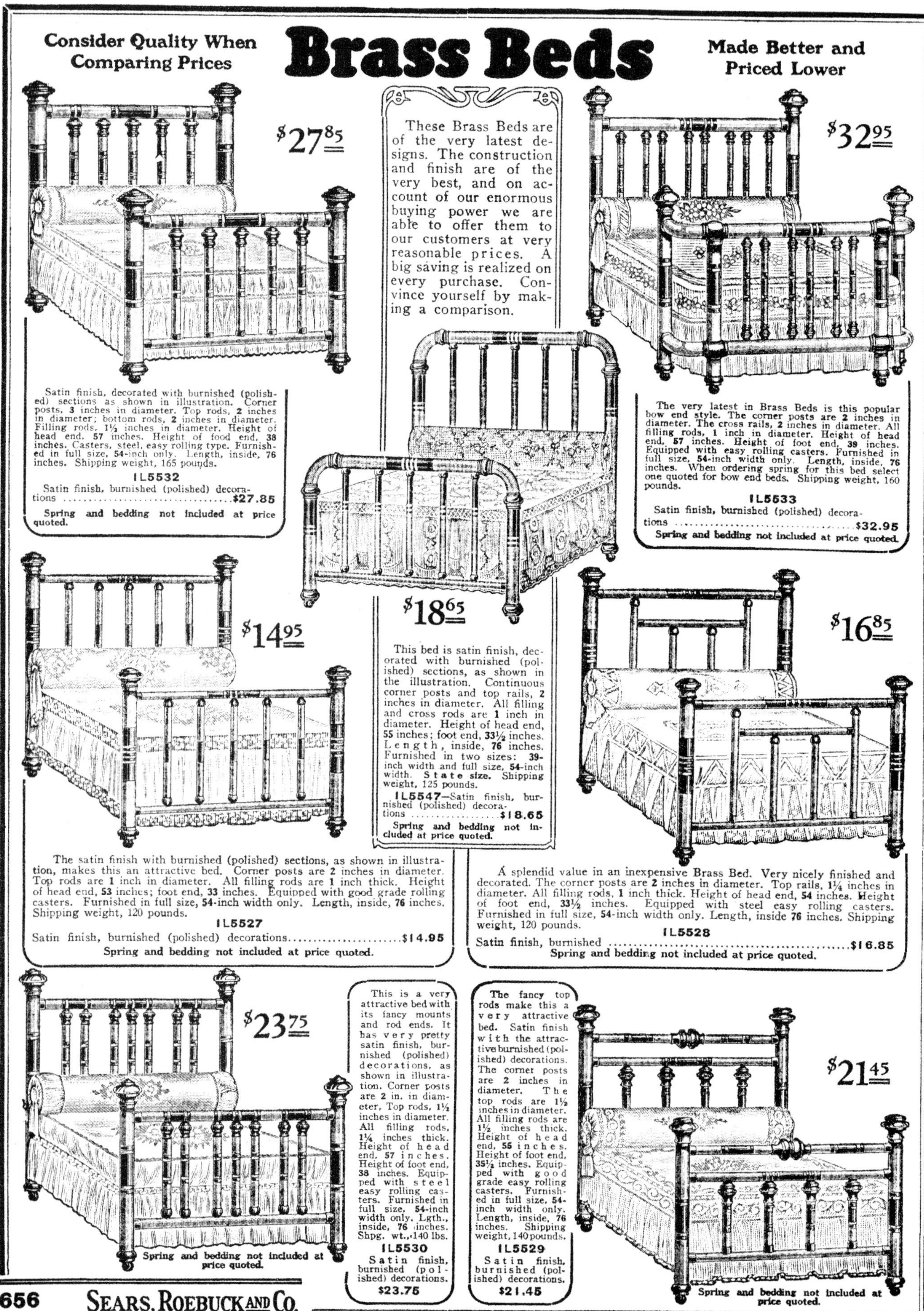
Consider Quality When Comparing Prices
Brass Beds
Made Better and Priced Lower
These Brass Beds are of the very latest designs. The construction and finish are of the very best, and on account of our enormous buying power we are able to offer them to our customers at very reasonable prices. A big saving is realized on every purchase. Convince yourself by making a comparison.
$27.85
Satin finish, decorated with burnished (polished) sections as shown in illustration. Corner posts, 3 inches in diameter. Top rods, 2 inches in diameter; bottom rods, 2 inches in diameter. Filling rods, 1½ inches in diameter. Height of head end, 57 inches. Height of food end, 38 inches. Casters, steel, easy rolling type. Furnished in full size, 54-inch only. Length, inside, 76 inches. Shipping weight, 165 pounds.
1L5532
Satin finish, burnished (polished) decorations $27.85
Spring and bedding not included at price quoted.
$32.95
The very latest in Brass Beds is this popular bow end style. The corner posts are 2 inches in diameter. The cross rails, 2 inches in diameter. All filling rods, 1 inch in diameter. Height of head end, 57 inches. Height of foot end, 39 inches. Equipped with easy rolling casters. Furnished in full size, 54-inch width only. Length, inside, 76 inches. When ordering spring for this bed select one quoted for bow end beds. Shipping weight, 160 pounds.
1L5533
Satin finish, burnished (polished) decorations $32.95
Spring and bedding not included at price quoted.
$18.65
This bed is satin finish, decorated with burnished (polished) sections, as shown in the illustration. Continuous corner posts and top rails, 2 inches in diameter. All filling and cross rods are 1 inch in diameter. Height of head end, 55 inches; foot end, 33½ inches. Length, inside, 76 inches. Furnished in two sizes: 39-inch width and full size, 54-inch width. State size. Shipping weight, 125 pounds.
1L5547—Satin finish, burnished (polished) decorations $18.65
Spring and bedding not included at price quoted.
$14.95
The satin finish with burnished (polished) sections, as shown in illustration, makes this an attractive bed. Corner posts are 2 inches in diameter. Top rods are 1 inch in diameter. All filling rods are 1 inch thick. Height of head end, 53 inches; foot end, 33 inches. Equipped with good grade rolling casters. Furnished in full size, 54-inch width only. Length, inside, 76 inches. Shipping weight, 120 pounds.
1L5527
Satin finish, burnished (polished) decorations $14.95
Spring and bedding not included at price quoted.
$16.85
A splendid value in an inexpensive Brass Bed. Very nicely finished and decorated. The corner posts are 2 inches in diameter. Top rails, 1¼ inches in diameter. All filling rods, 1 inch thick. Height of head end, 54 inches. Height of foot end, 33½ inches. Equipped with steel easy rolling casters. Furnished in full size, 54-inch width only. Length, inside 76 inches. Shipping weight, 120 pounds.
1L5528
Satin finish, burnished $16.85
Spring and bedding not included at price quoted.
$23.75
This is a very attractive bed with its fancy mounts and rod ends. It has very pretty satin finish, burnished (polished) decorations, as shown in illustration. Corner posts are 2 in. in diameter, Top rods, 1½ inches in diameter. All filling rods, 1¼ inches thick. Height of head end, 57 inches. Height of foot end, 38 inches. Equipped with steel easy rolling casters. Furnished in full size, 54-inch width only. Lgth., inside, 76 inches. Shpg. wt., 140 lbs.
1L5530
Satin finish, burnished (polished) decorations.
$23.75
Spring and bedding not included at price quoted.
$21.45
The fancy top rods make this a very attractive bed. Satin finish with the attractive burnished (polished) decorations. The corner posts are 2 inches in diameter. The top rods are 1½ inches in diameter. All filling rods are 1½ inches thick. Height of head end, 55 inches. Height of foot end, 35½ inches. Equipped with good grade easy rolling casters. Furnished in full size, 54-inch width only. Length, inside, 76 inches. Shipping weight, 140 pounds.
1L5529
Satin finish, burnished (polished) decorations.
$21.45
Spring and bedding not included at price quoted.
656
SEARS, ROEBUCK AND CO.

Feathers and Pillows

GUARANTEED ALL NEW FEATHERS

INVEST IN RESTFUL SLEEP, which is essential to good health, by purchasing our Hygienically Treated Feathers and Feather Pillows. We unqualifiedly guarantee that all feathers used in the manufacture of our pillows or bulk grades of feathers are absolutely new, never used before. The original stock is carefully selected. Our modern machinery and exclusive processes treat these feathers in a thoroughly scientific manner, eliminating all dust, odor, quills and animal matter (which sticks to the stems of the feathers, if not properly treated), leaving the feathers fluffy and scrupulously clean. Every pillow and every sack of feathers bears our guarantee that the feathers are all new feathers, thoroughly sterilized, deodorized, and that they fully comply with the legislative and board of health requirements of states with sanitary bedding laws.

Our descriptions make it easy for you to select the grade you want.

We are the largest retailers of feathers and pillows in the world and are therefore able to give you the finest qualities at prices which, by careful comparison, cannot be duplicated elsewhere.

Our prices for feathers in muslin sacks are for net weight of feathers, put up in 1, 2, 3, 5 and 10-pound neatly stenciled muslin sacks. Each sack contains the exact weight of feathers purchased. You readily see what a big saving this means in purchasing feathers.

We unreservedly guarantee "Purity Brand" Feathers and Pillows to be unusual in their resiliency and buoyancy, extra fine quality and hygienically clean. Look for the "Purity Brand" silk label on every pillow; it stands for the best you can get and is the undisputed standard for pillows and feathers. "Purity Brand" means permanent pillow satisfaction.

BUY THE PURITY BRAND

FEATHERS AND PILLOWS

IT IS A GUARANTEE OF HIGH QUALITY

2¼ LBS. $2.95 EACH — THE VERY FINEST QUALITY PRODUCED

1L7194—Purity Brand Pillows. Our finest grade of choice white goose feathers, scientifically blended. The acme of perfection encased in 8-ounce satin finish art ticking; newest designs in patterns.

Size, Inches	Weight, Each Pillow	Each Pillow	The Pair
20x27	2¼ lbs.	$2.95	$5.90
21x27	2½ lbs.	3.45	6.90
22x28	3 lbs.	3.95	7.90
23x29	3½ lbs.	4.45	8.90

2 LBS. $2.65 EACH — AN EXTRA HIGH GRADE PILLOW

1L7193—Purity Brand Pillows. Filled with choice white and gray goose feathers. The lowest priced all goose feather pillow that can be purchased, resiliency and loftiness considered. Encased in 8-ounce best quality satin finish art Gobelin ticking.

Size, Inches	Weight, Each Pillow	Each Pillow	The Pair
18x26	2 lbs.	$2.65	$5.30
21x26	2½ lbs.	3.15	6.30
22x27	3 lbs.	3.65	7.30

2 LBS. $2.15 EACH — A VERY FINE QUALITY PILLOW

1L7185—Purity Brand Pillows. Made of one-quarter choice white goose and three-quarters choice duck feathers, selected for its excellent filling capacity. Light, buoyant and a very desirable grade. Encased in best quality 8-ounce art Gobelin ticking; beautiful design.

Size, Inches	Wt., Each Pillow	Each Pillow	The Pair
18x26	2 lbs.	$2.15	$4.30
21x26	2½ lbs.	2.65	5.30
22x27	3 lbs.	3.15	6.30
22x28	3½ lbs.	3.65	7.30

FEATHERS AND DOWN

OUR PRICES ARE FOR *ACTUAL WEIGHT* OF FEATHERS WITHOUT CONTAINERS. PUT UP IN 1, 2, 3, 5 AND 10-POUND SACKS

Description.	Per Lb.
1L7103—New dry picked western hen feathers, double dusted, guaranteeed sanitary and odorless..	$0.27
1L7108—Prime selected dry picked hen feathers, one-third white, two-thirds colored; considerably better than 1L7103 in color and filling capacity.....	.35
1L7112—About one-third choice goose and duck feathers and two-thirds of white and colored hen (equal proportions) feathers. A very good mixture; light color..............	.49
1L7116—An excellent grade, especially adapted for pillows and feather beds, about one-fourth choice white goose and three-fourths gray duck feathers..	.72
1L7120—Purity Brand. A choice white feather, about one-half white goose and one-half white hen. Recommended for color..............	.88
1L7124—Purity Brand. Choice white and gray goose feathers, about equal proportions. Recommended for its excellent filling qualities..........	.98
1L7128—Purity Brand. All white goose feathers. Good quality with excellent color and filling qualities; unusual value..............	1.28
1L7133—Purity Brand. Our best grade of choicest downy white goose feathers, selected to please the most critical buyer. We cannot recommend this grade too highly where the best in feathers is desired..............	1.42
1L7145—Purity Brand choice down; "white as snow." One-half choice goose down and one-half choice white hen down. Free from feathers. Soft and buoyant..............	1.68

2½ LBS. $1.98 EACH — A VERY FINE GRADE PILLOW

1L7181—Purity Brand Pillows. Made of about one-third choice light goose feathers, one-third choice duck and one-third prime curled turkey. An excellent grade. Covered in 8-ounce ticking; beautiful design.

Size, Inches	Weight, Each Pillow	Each Pillow	The Pair
19x26	2½ lbs.	$1.98	$3.96
21x27	3 lbs.	2.38	4.76
22x28	3½ lbs.	2.78	5.56

2½ LBS. $1.58 EACH — AN EXCELLENT VALUE

1L7177—Purity Brand Pillows. About one-third choice goose and duck and two-thirds prime curled turkey, slightly blended with curled hen feathers. Buoyant and elastic, and a very low price. Covered in newest creation in satin finish border ticking.

Size, Inches	Weight, Each Pillow	Each Pillow	The Pair
20x26	2½ lbs.	$1.58	$3.16
21x27	3 lbs.	1.88	3.76
22x28	3½ lbs.	2.18	4.36

3 LBS. $1.45 EACH — A VERY GOOD VALUE

1L7173—Pillows. Made of about one-quarter gray duck feathers, three-quarters prime curled western hen feathers. Good medium priced pillow. Covered in attractive fancy ticking; good quality.

Size, Inches	Weight, Each Pillow	Each Pillow	The Pair
20x26	3 lbs.	$1.45	$2.90
21x27	3½ lbs.	1.65	3.30
22x28	4 lbs.	1.85	3.70

Patent Open Bolster Rolls.

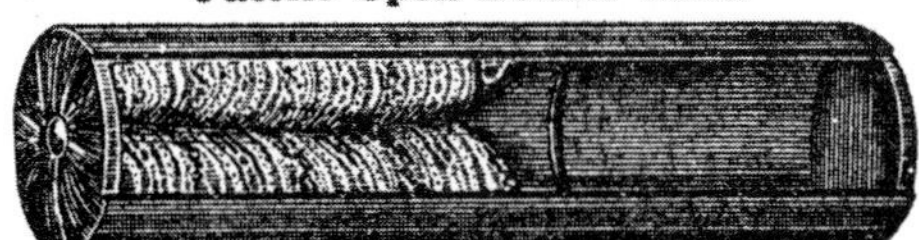

A useful and ornamental receptacle for pillows in the daytime, as well as artistically dressing the bed. Made of extra heavy cardboard with wooden ends. Durability guaranteed. Covered and interlined with good quality cambric. 54-inch size in white, blue or pink. State color. Sizes 39, 42 and 48-inch in blue only. State size. Packed in individual cartons. Shipping weight, 20 pounds.

1L7247 .... $2.95

Each Pair of Pillows Encased in a Paper Envelope Without Additional Charge.

For Blankets, Sheets and Pillowcases, see pages 344 to 347.

2½ LBS. 95c EACH — A VERY POPULAR PILLOW

1L7165—Pillows. Made of one-half prime curled turkey and one-half prime western hen feathers, dry picked. Unusually well filled. Covered in fancy art ticking.

Size, Inches	Weight, Each Pillow	Each Pillow	The Pair
18x25	2½ lbs.	$0.95	$1.90
19x26	3 lbs.	1.15	2.30
21x27	3½ lbs.	1.35	2.70
22x28	4 lbs.	1.55	3.10

2 LBS. 58c EACH — A VERY LOW PRICE

1L7164—Pillows. Made of prime curled western hen feathers, thoroughly deodorized and double dusted. The best cheap feather pillow made. Narrow blue and white striped ticking.

Size, Inches	Weight, Each Pillow	Each Pillow	The Pair
17x24	2 lbs.	$0.58	$1.16
18x25	2½ lbs.	.74	1.48
19x26	3 lbs.	.90	1.80
20x27	3½ lbs.	1.06	2.12

12 OZ. 35c EACH

1L7161—Choice turkey feathers, half down cushion, thoroughly renovated. Covered in good quality white cambric.

Size, inches	16x16	18x18	20x20	22x22
Weight	12 oz.	16 oz.	23 oz.	32 oz.
Each	35c	45c	65c	85c

7 OZ. 32c EACH

1L7158—100 per cent prime No. 1 Java kapok, a silky vegetable fiber, non-absorbent, very resilient and elastic. An ideal soft cushion. Covered with good quality white cambric.

Size, inches	16x16	18x18	20x20	22x22	24x24	26x26
Weight	7 oz.	10 oz.	13 oz.	17 oz.	20 oz.	24 oz.
Each	32c	42c	58c	74c	90c	$1.06

Folding Steel Couch and Cots

This Comfortable Steel Couch will make a very pretty Day Bed

This steel couch with comfortable pad, cretonne covered, is being sold for an unusually low price. **Material**—Frame is made of 1½-inch angle steel, strongly riveted together, so that it will stand hard usage and give long service. The spring is constructed of **55** steel coil springs, securely fastened together and supported on 1-inch steel bands. The fabric has anti-rust tin coating. The frame is finished in light gray enamel. Length over all, about **72** inches; width, about **29** inches. The comfortable pad which is furnished with this couch is a heavy felted cotton pad, nicely tufted and covered on top with a good grade of floral pattern cretonne. The ruffle on pad hangs down, covering the steel construction of the couch, making this a very pretty day bed. This pad is equipped with tapes which enable you to fasten it securely to spring. The small illustration below shows this couch with pad attached and head rest elevated. Shipping weight, 85 pounds.

1L5831—Steel Couch with Pad, complete....... **$17.75**
Cot only........(Shipping weight, 65 pounds)....... **9.00**

$17.75 Cot and Pad Complete.

Strongly Constructed.

This large illustration shows the construction of this comfortable steel couch. The head rest may be lowered. It is easy to operate and is strongly constructed so that it is not easy to break or get out of order. The legs of this couch can be folded under, making it easy to move or store away when not in use. We believe this to be one of the most strongly constructed and comfortable couches on the market today.

$9.00 Cot Only.

Steel Folding Cot Finished in Gray Enamel.

This is an especially attractive steel cot which has curved head and foot ends. Can be made into a comfortable day bed. Height of ends, **31¾** inches. Continuous side posts and top rails of steel tubing **1** inch in diameter. Straight vertical filling rods ⅜ inch in diameter. Steel bottom cross rods ⅝ inch thick. Has a good steel wire fabric spring, fastened to ends with helical springs. The fabric spring is elevated about **4** inches above the side rails. Length over all, **76** inches. Inside measurements, **30x72** inches or **36x72** inches. Furnished in light gray enamel. **State size wanted.** Shipping weight, 70 and 75 pounds, respectively.

1L5898
30-inch width **$7.95**
36-inch width **8.70**

Makes a Comfortable Bed.

Continuous steel corner posts, **1 1/16** inches in diameter. Straight vertical filling rods, **5/16** inch in diameter. Bottom cross rod, ⅝ inch in diameter. Height of head end, **36** inches; foot end, **30** inches. Has comfortable steel fabric spring, securely fastened to angle ends with helical steel springs. The steel fabric spring is elevated above the side rails. Length over all, **76** inches. Widths, **30**-inch and **36**-inch. Length of spring, **72** inches. **State width desired.** Shipping weights, 80 and 85 pounds, respectively.

1L5896
30-inch width **$7.45**
36-inch width **7.95**

Make into a bed at night. Fold up and put away in the daytime.

Folds flat, as shown. **Material and Construction**—Continuous steel corner posts, **1 1/16** in. in diameter. Straight vertical filling rods, ¼ inch thick. Steel bottom cross rods, ⅜ inch thick. Has steel wire fabric spring fastened to the angle ends with helical steel springs. Steel fabric spring elevated **4** inches above side rails. **Finish**—Light gray enamel. **Measurements**—Height, head and foot ends, **24** inches. Length over all, **76** inches. Spring fabric, is **17½** inches from floor. Made in **30**-inch and **36**-inch widths; **72** inches long. **State width desired.** Shipping weights, 60 and 65 pounds, respectively.

1L5891
30-inch width **$6.85**
36-inch width **7.35**

For Other Folding Cots See Page 785.

An Inexpensive Steel Cot.

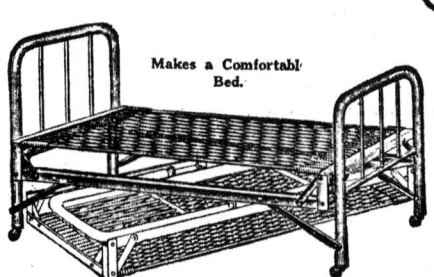

Folds flat. **Material**—Continuous posts and top rail of steel tubing, **1 1/16** in. thick. Fitted with double prong fabric and steel helical springs. Angle steel ends and side rails. Finished in light gray enamel. **Measurements**—Head and foot ends, **24** inches high. Length over all, **79** inches. Spring fabric, **14½** inches from floor. Made in **30**-inch and **36**-inch widths. **72** inches long. **Be sure to state width desired.** Shipping weights, 45 and 50 pounds, respectively.

1L5889
30-inch width **$4.45**
36-inch width **4.95**

A Specially Low Priced Cot.

One of the best low priced Folding Cots on the market. **Material**—Hard maple. **Finish**—Varnish. **Construction**—Upright posts firmly braced. Single weave, closely woven wire fabric. Length over all, **75** inches. Shipping weight, 30 pounds for **30**-inch width cot and 32 pounds for **36**-inch width cot.

1L5854
30-inch width **$2.65**
36-inch width **3.10**

A Specially Big Value.

Material—1¼-inch angle Bessemer steel. Finished in light gray enamel. Fitted with double prong steel fabric with steel helical end springs. **Measurements**—Height of head and foot ends, **24½** inches. Length over all, **76** inches. Furnished in three sizes. **State width.** Shipping weight, 40, 50 and 60 pounds, respectively.

1L5884
30-inch width **$3.75**
36-inch width **4.25**
42-inch width **4.75**

Filled With Felted Cotton Stock.

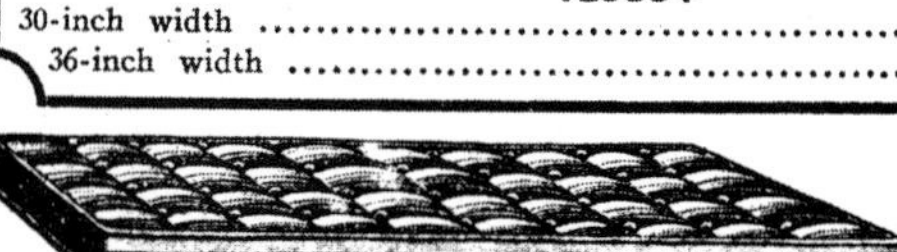

A medium priced pad, extremely comfortable. **Material and Construction**—Made of built-up layers of cotton felted stock, the same as our regular high grade cotton felted mattresses. Tufts are evenly placed and strongly bound. Strongly woven twilled ticking. Has 1½-inch square box edges.

1L7230
Size 30x72 inches, actual weight, 14 pounds....... **$4.38**
Size 36x72 inches, actual weight, 16 pounds....... **4.98**
Size 42x72 inches, actual weight, 19 pounds....... **5.58**

Pad Mattresses for Cots

When you order a cot be sure to include one of these box edge mattresses. These mattresses are made especially for our cots and are of the same good value. Each one comes packed in heavy paper and new burlap. For shipping weights add 2 pounds to actual weights given.

Felted Cotton Cot Mattress. **Material and Construction**—Filled with guaranteed quality cotton stock. Fleeced, worked and fitted into loose, fluffy layers and encased in the ticking. **Construction**—Same as our regular high grade cotton felted mattresses. Full tufted with biscuit shape tufts. 3-inch box edge. Fancy art pattern ticking.

1L7237
Size, 30x72 inches, actual weight, 17 pounds....... **$5.35**
Size, 36x72 inches, actual weight, 20 pounds....... **6.15**
Size, 42x72 inches, actual weight, 23 pounds....... **6.95**

BABIES' CRIBS AND CRADLES

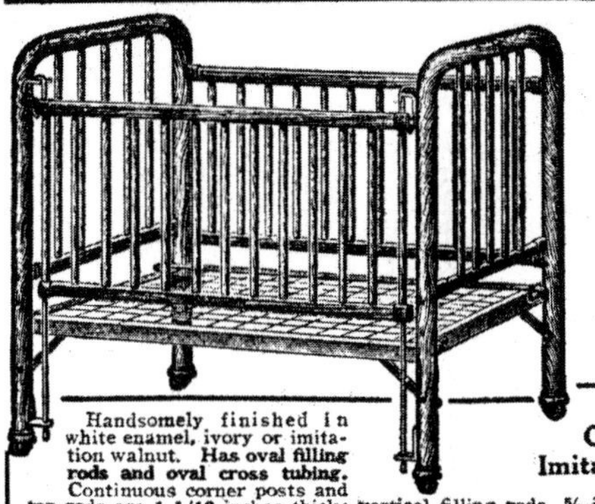

Mattresses to fit these cribs and cradle are shown on page 661

Latest Style Sanitary Extra Strong Steel Cribs. Sliding Drop Sides. Furnished in Imitation Walnut, Ivory or White Enamel Finish.

Handsomely finished in white enamel, ivory or imitation walnut. **Has oval filling rods and oval cross tubing.** Continuous corner posts and top rods are **1 1/16** inches thick; vertical filling rods, ⅝-inch; cross rods are 1 1/16-inch oval tubing. Height, **46½** inches. Sides, each, **23½** inches high. Has strong wire fabric spring, firmly attached to ends with spiral helical steel springs. Easy working sliding drop side. Finished as listed below. Shipped knocked down. Shipping weights, 65 and 75 pounds, respectively.

1L5988	White Enameled	Ivory	Imitation Walnut
Size, 30x54 inches	$11.85	$11.95	$13.35
Size, 36x60 inches	12.45	12.55	13.95

Square corner posts, top, end and side rails. Corner posts, 1¼ inches square; top rails, ends, 1¼ inches square; sides, 1 inch square. Bottom rails, ends, 1 inch square; sides, 1 inch square. Filling rods are ⅜x1 inch. Easy rolling steel casters. Height, **46** inches. Depth of sides, **22** inches. Has strong double prong wire fabric spring attached to ends with steel spiral helicals. Finished in white enamel, ivory or imitation walnut. Note sizes and finish in price list. Shipping weights, 65 and 80 pounds, respectively.

1L5978	White Enameled	Ivory	Imitation Walnut
Size, 30x54 inches	$14.35	$14.45	$15.85
Size, 36x60 inches	14.95	15.05	16.45

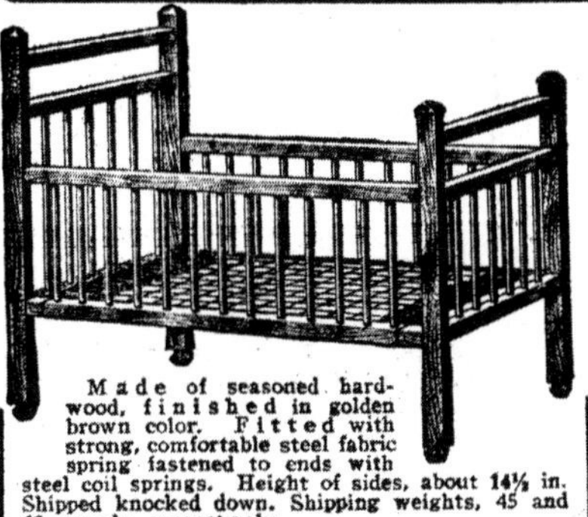

Made of seasoned hardwood, finished in golden brown color. Fitted with strong, comfortable steel fabric spring fastened to ends with steel coil springs. Height of sides, about 14½ in. Shipped knocked down. Shipping weights, 45 and 63 pounds, respectively.

1L5940—Size, 30x60 inches..........$6.45
1L5949—Size, 40x60 inches..........7.25

Inexpensive Down Pillows.

Just the Thing for Baby's Crib.

The Dependo.

Half-Down Crib Pillow

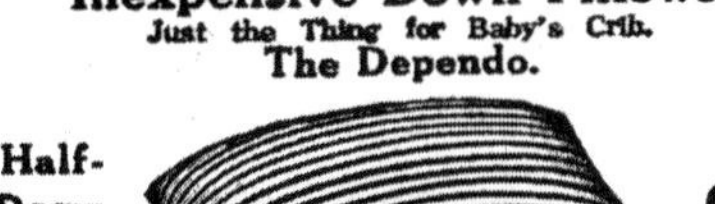

95c

A splendid half-down pillow, containing 25 per cent goose, 50 per cent turkey and 25 per cent duck half-down. Covered with best quality satin finished ticking. Size, **14x18** inches. Shipping weight, 18 ounces.

1L7155..........95c

The Regent.

Half-Down Crib Pillow

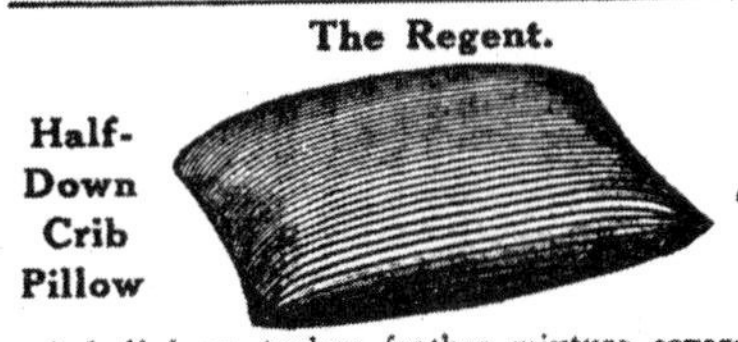

48c

A half-down turkey feather mixture covered with a good grade of striped ticking. Size, **10x15** inches. Shipping weight, 7 ounces.

1L7151..........48c

Sliding Drop Side. Continuous corner posts and top rods, ⅞ inch thick. Vertical filling rods, ¼ inch thick. Top and bottom rods on sides, ½ inch thick. Height of head and foot ends, **45** in. Sides, each, **23½** inches high. Fitted with strong fabric spring. Furnished in white enamel, ivory or Vernis Martin (gold color bronze) finish. Shipped knocked down. Shipping wts., 65 and 75 lbs., respectively.

1L5989	White Enameled	Ivory	Vernis Martin
Size, 30x54 in.	$7.85	$7.95	$8.25
Size, 36x60 in.	8.45	8.55	8.85

Made of hardwood, finished either in golden brown finish or in white enamel, as desired. Has one stationary and one sliding drop side, each 18¾ in. high. Height of head and foot ends, 35¼ in. Fitted with strong, comfortable steel fabric spring, fastened to ends with steel coil springs. **State finish desired.** Shipped knocked down. Shipping wts., in golden brown, about 52 and 63 lbs., and in white enamel finish, which is crated, 55 and 65 lbs., respectively.

	Golden Brown	White Enameled
1L5967—Size, 30x54 in.	$6.95	$7.70
1L5969—Size, 36x60 in.	7.75	8.50

Hardwood frame, white enameled. The sides, ends and top are of durable wire screen. **Length, 37** inches; width, **21** inches; height, **30** inches. Wheels, **6** inches in diameter, with rubber tires. Wire spring bottom. Shipped knocked down. Shipping weight, 25 pounds.

1L5909..........$8.95
1L7276—Felted Cotton Pad Mattress to fit. Actual weight, 5 pounds..........$2.15
1L7277—Java Kapok (Silk Floss) Pad Mattress to fit. Actual weight, 3¼ lbs...$2.95

Corner posts and top rods made of steel tubing, ⅞ in. in diameter. Top tubes on sides, **7/16** inch; vertical filling rods, ¼ inch in diameter. One stationary and one sliding drop side, about **16½** inches high. Height, **39½** inches. No projections on which baby might injure itself. Strong removable steel fabric spring. Shipped knocked down. Shipping weights, 55 and 65 pounds, respectively. Although similar in design, this crib is not so large as 1L5989, shown above.

1L5976	White Enameled	Vernis Martin
Size, 30x54 inches	$7.25	$7.65
Size, 36x60 inches	7.85	8.25

Frame of hardwood, white enameled. Length, **35** inches; width, **19¾** inches; height, **31** inches; height of sides, **12½** inches. Wheels, **6** inches high, with rubber tires. Wire spring bottom. Shipped knocked down. Shipping wt., 21 pounds.

1L5907..........$4.95
1L7270—Felted Cotton Pad Mattress to fit. Actual weight, 3 pounds..........$1.68
1L7271—Java Kapok (Silk Floss) Pad Mattress to fit. Actual weight, 2 pounds....$2.28

Cradle—Made of thoroughly seasoned hardwood, golden brown finish, which presents a bright, cleanly appearance. Size, **24x44** inches. Height of sides, about **14** inches. Woven wire spring. Shipped knocked down. Shipping weight, 26 pounds.

1L5908..........$3.95

Crib—Same as above, but without rocker runners. Shipping weight, 24 pounds.

1L5904..........$3.45

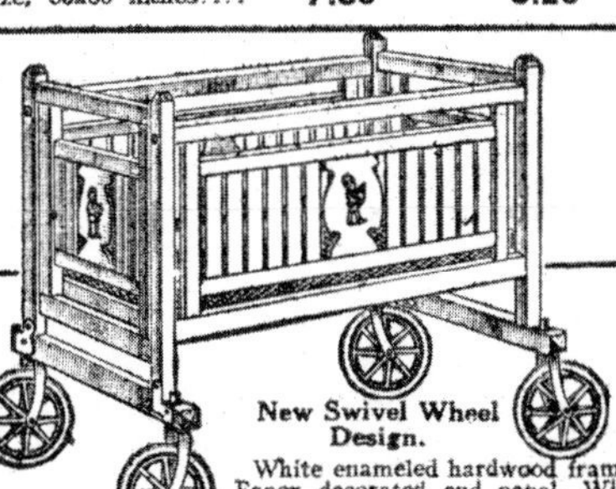

New Swivel Wheel Design.

White enameled hardwood frame. Fancy decorated end panel. Wire spring bottom. Length, **39** inches; height, **34** inches; width, **22** inches; height of sides, **15¼** inches; 6-inch swivel wheels with rubber tires. Shipped knocked down. Shipping weight, 27 pounds.

1L5914—White Enamel.....$9.65; Ivory.....$9.75
1L7276—Felted Cotton Pad Mattress to fit. Actual weight, 5 pounds..........$2.15
1L7277—Java Kapok (Silk Floss) Pad Mattress to fit. Actual weight, 3¼ pounds..........$2.95

Collapsible Go-Carts

$7.45

1L7704—A Collapsible Go-Cart of this design and quality usually retails for at least 40 per cent more than our price. One of our most popular go-carts. Solid flat steel pusher handles with one-piece black enameled hand grips. Black enameled gear frame. Wheels are **10** inches in diameter, with ½-inch rubber tires. Adjustable reclining back and dash. Size of body inside with back down, **14½x32** inches. Seat supported underneath with our special steel coil compression springs. Covering is a serviceable grade of black artificial leather. Shipping weight, 30 lbs.

1L7704..................$7.45

1L7716—A value like this cannot be equaled elsewhere. Sides, front and back of fiber board. Steel gear frame, black enameled. Long flat steel pusher handles, with one-piece black enameled hand grips. Adjustable back and dash. Seat supported underneath with steel coil compression springs. Covering is a serviceable grade of black artificial leather. Folds into compact form. **10**-inch wheels with ½-inch rubber tires. Size of body inside with back down, **31** inches long and **13** inches wide. Shipping weight, 35 lbs.

1L7716............$8.95

$8.95

Twin Go-Cart.

1L7738—You can't buy a better collaspible Twin Go-Cart anywhere for the price we ask. The high adjustable backs and folding dashes are separate and independent of each other. Width of body between arms, **22** inches. Total width over all, **27½** inches. Has **10**-inch wheels with ½-inch rubber cushion tires. Seat supported by four compression springs which absorb the jar. Black enameled hand grips. Flat steel pusher handles. Covering is a serviceable grade of black artificial leather. Frame black enameled. This go-cart folds and unfolds easily. Shipping weight, 55 lbs.

1L7738........$13.85

$13.85

Very Popular.

1L7721—Big values like this collapsible Go-Cart for only $11.95 have made us famous. Framework is steel, black enameled. Dash adjustable to three positions, either dropped as shown, or raised to bring sides of dash level with seat, or up to make level bed. Wheels, **12** inches in diameter, ½-inch rubber tires. Size of body with back down, **13½x36** inches. Foot brake. Seat supported with steel coil compression springs. Covering is a serviceable grade of black artificial leather. Shipping weight, 40 pounds.

1L7721........$11.95

$11.95

Collapsible Sulkies

$5.45

Sturdy Collapsible Sulky. Framework of baby carriage steel, securely riveted and braced, finished in black enamel. Strong steel handle, wood turned hand grip. Back, seat and hood of serviceable black artificial leather. Wide foot rest. Protection strap across front. Seat, **9½x11½** inches. Has **10**-inch wheels with ⅜-inch rubber tires. Folds into compact form. Shipping weight, 20 pounds.

1L7763..................$5.45

Folds into compact form. The sides and back are fibre-craft (a fiberlike material resembling reeds). Framework, black enameled steel. The long curved handle is of flat steel with black wood hand grip. Size of seat, **11x12** inches, supported underneath by two steel coil compression springs. Full three-bow hood adjusts to any position and is covered with a serviceable grade of black artificial leather, as is the seat. The wheels are **10** inches in diameter, with ½-inch rubber tires. Small steel rear wheels, **2½** inches high. Shipping wt., 31 lbs.

1L7784 ....$8.85

$8.85

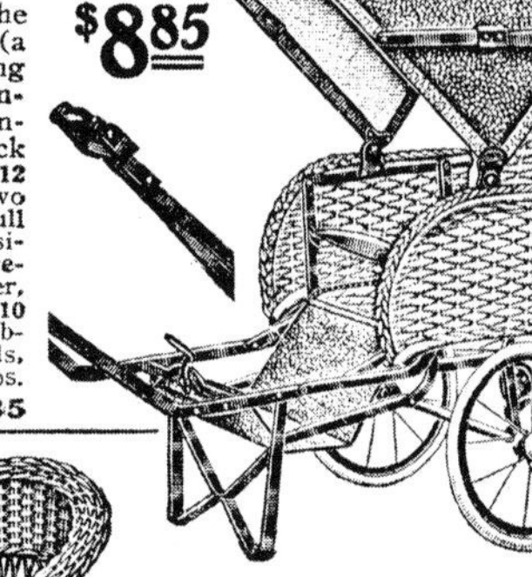

$2.95

Framework made of steel, securely riveted and braced, finished in black enamel. Seat, **8½x13** inches, covered with a serviceable grade of black artificial leather. Wheels, **10** inches in diameter, with ⅜-inch rubber tires. Foot rest. Long pusher handle, wood turned hand grip. Where pusher connects to frame is a simple locking device which allows the cart to fold into compact form. Shipped knocked down. Shipping weight, 18 pounds.

1L7749 .... $2.95

$4.95

Black Enameled Steel Gear.

Body made of maple fiber stakes with fibre-craft and reed. Seat, **8x13** inches, covered with a serviceable grade of black artificial leather. Has **10**-inch wheels with ⅜-inch rubber tires. Foot rest. Long pusher handle, wood turned hand grip. A simple locking device allows cart to fold into compact form. Shipped knocked down. Shpg. wt., 18 lbs.

1L7764 .... $4.95

$6.95

A lower price and better quality than you can get elsewhere. Body made of fibre-craft material (fiberlike material resembling reeds), finished in ecru color. Black enameled steel framework. Steel pusher handle with wood turned hand grip. Back, seat and hood of serviceable black artificial leather. Wide foot rest. Protection strap across front. Seat measures **9½x11½** inches. Has **10**-inch wheels with ⅜-inch rubber tires. Folds into compact form. Shipping weight, 20 pounds.

1L7757 ......$6.95

All weights given on this page are approximate and may vary a trifle.

$7.45

A very comfortable Sulky, the seat being supported by two sensitive steel coil compression springs. Size of body, **13x32** inches with dash up and back down. Framework made of flat steel, black enameled. Seat, back, sides, dash and hood covered with a serviceable grade black artificial leather. Long steel handle with round wood hand grip. Adjustable back and dash. Wheels are **10** inches in diameter with ½-inch rubber tires. Small steel rear wheels are **2½** inches high. Folds up into compact form. Shipping weight, 30 pounds.

1L7775 ....$7.45

We Are Famous for Big Baby Carriage Values

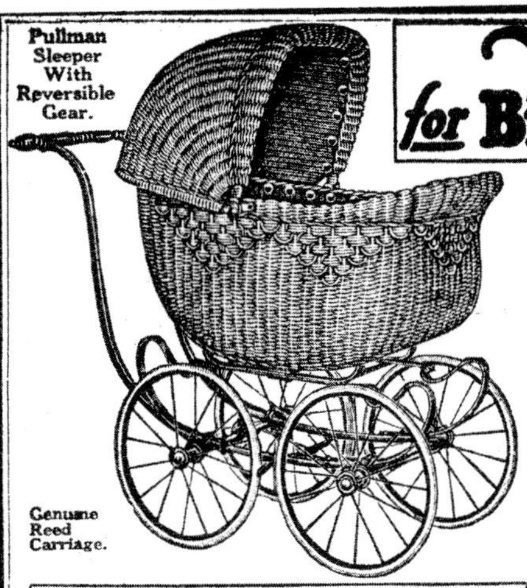

Pullman Sleeper With Reversible Gear.

Genuine Reed Carriage.

Equipped with reversible gear turntable with automatic lock on side which holds carriage in place. Has deep body equipped with foot well. Sides, seat, back, well flap and inside of hood nicely upholstered with genuine corduroy. Body measures 14½ inches wide, **24** inches long with back up and **34** inches long with back reclining; **12** inches deep. Gear, baby carriage spring steel, nicely enameled and gold color striped. Wheels, **16** inches in diameter with ⅝-inch rubber tires. **Colors:** Frosted ebony finish on body with blue corduroy upholstering, black gear and white wheels, or cream color ivory finish on body with brown corduroy upholstering, black gear and black wheels, or gray body, upholstering and gear. **State color.** Shipping weight, 75 pounds.

1L7880
Cream color ivory finish......**$31.75**
Gray enamel finish......**31.85**
Frosted ebony finish......**32.25**

Storm Curtain.

Made of corduroy in color to match body of carriage. Has elastic cord around edges to hold in position and fastener to attach to hood. Furnished to fit Carriage 1L7868, 1L7880 and 1L7870. **State color and catalog number of carriage to which it is to be fitted.** Shipped by parcel post. Shipping weight, 2 pounds.

1L7795......**$1.65**

New Style Stroller.

Body made of Fibre-Craft, resembling reed. Upholstering is a serviceable fabric resembling corduroy. Black enameled tubular steel pusher handles. Body measures 13 inches wide. Seat, 10 inches long, increased to 32 inches with dash up; dash adjustable to three positions. 12-inch wheels in rear and 8-inch front wheels with ½-inch rubber tires. **Colors:** Cream color ivory finish body with brown upholstering and black gear and wheels, or frosted ebony finish on body with gray upholstering, black gear and cream color wheels. **State color.** Shipping weight, 50 lbs.

1L7779—Cream color ivory finish......**$15.65**
Frosted ebony finish......**16.15**

Charming New Design.

Has deep body of Fibre-Craft (a fiberlike material resembling genuine reeds) equipped with foot well. **Upholstering is genuine corduroy.** Body measures 14½ inches wide, **24** inches long with back up and **34** inches long with back reclining; **14** inches deep. Gear, baby carriage spring steel, nicely enameled and gold color striped. Wheels, **16** inches in diameter with ½-inch rubber tires. **Colors:** Frosted ebony finish on body with blue corduroy upholstering, black gear and white wheels, or cream color ivory finish on body with brown corduroy upholstering, black gear and black wheels, or gray body, upholstering and gear. **State color.** Shipping weight, 80 pounds.

1L7882—Cream color ivory finish......**$34.45**
Gray enamel finish......**34.55**
Frosted ebony finish......**35.95**

A Value Hard to Equal.

Made of Fibre-Craft (a fiberlike material resembling genuine reeds). Upholstered in a serviceable fabric resembling corduroy. Body measures 14½ inches wide, 25 inches long with back up and 36 inches long with back reclining; depth is 9 inches. 14-inch wheel with ½-inch rubber tires. **Colors:** Frosted ebony finish on body with gray upholstering, black gear and cream color wheels, or cream color ivory finish on body with brown upholstering and black gear and wheels. **State color.** Shipping weight, 65 pounds.

1L7870
Cream color ivory finish......**$19.45**
Frosted ebony finish......**19.95**

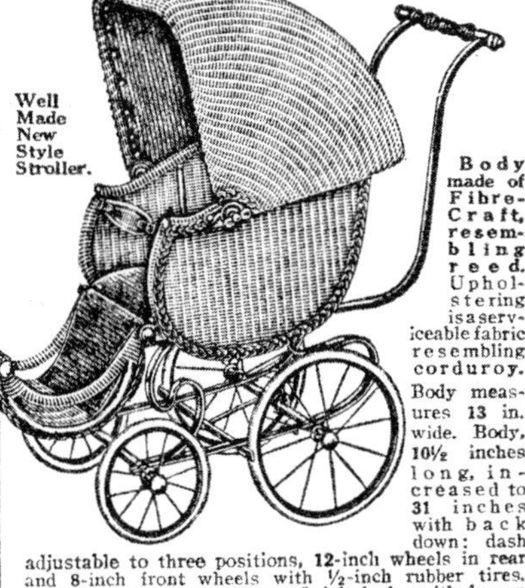

Well Made New Style Stroller.

Body made of Fibre-Craft, resembling reed. Upholstering is a serviceable fabric resembling corduroy. Body measures 13 in. wide. Body, 10½ inches long, increased to 31 inches with back down; dash adjustable to three positions, 12-inch wheels in rear and 8-inch front wheels with ½-inch rubber tires. **Colors:** Cream color ivory finish body with brown upholstering and black gear and wheels, or frosted ebony finish on body with gray upholstering, black gear and cream color wheels, or gray body, upholstering and gear. **State color.** Shipping wt., 35 lbs.

1L7780—Cream color ivory finish......**$13.85**
Gray enamel finish......**13.95**
Frosted ebony finish......**14.35**

You Will Be Proud of This Carriage.

Body made of Fibre-Craft (a fiberlike material resembling genuine reeds). **Upholstered in genuine corduroy.** Body measures 14½ inches wide, **24** inches long, extending to **34** inches with back down, and **12** inches deep. Has 14-inch wheels with ½-inch rubber tires. Gear made of baby carriage steel, nicely enameled. **Colors:** Cream color ivory enameled body with brown upholstering, black gear and wheels; frosted ebony body with blue upholstering, black gear and cream color wheels, or gray enameled body with gray upholstering, gray gear and wheels. **State color wanted.** Shipping weight, 75 pounds.

1L7878—Cream color ivory finish......**$26.85**
Gray enamel finish......**26.95**
Frosted ebony finish......**27.35**

Twin Baby Carriage, Pullman Sleeper Style.

Half Round Genuine Reed Carriage.

Equipped with two foot wells. Each section is **12x34** inches with back down and **23** inches long with back up. **Upholstered with genuine corduroy.** Gear made of baby carriage spring steel, nicely enameled and black and gold color striped. Wheels, **14** inches in diameter with ½-inch rubber tires. Furnished in cream color ivory finish with brown upholstering, black gear and wheels, or in frosted ebony finish with gray upholstering and ivory finished wheels. **State color.** Shipping weight, 80 pounds.

1L7886
Cream color ivory finish......**$38.95**
Frosted ebony finish. **39.45**

1L7799—Special Corduroy Storm Curtain to fit this twin carriage, similar to storm curtain shown at top of page. Shipping weight, 2 lbs **$1.95**

One of Our Most Popular Styles.

Body made of half round genuine reed. Upholstering is a serviceable fabric resembling corduroy. Body measures **15** inches wide, **29** inches long and **8** inches deep. Has 12-inch wheels with ½-inch rubber tires. Gear made of baby carriage steel, nicely enameled. **Colors:** Cream color ivory enameled body with brown upholstering, black gear and wheels; frosted ebony body with blue upholstering, black gear and cream color wheels, or gray enameled body with gray upholstering, gray gear and wheels. **State color.** Shipping weight, 55 pounds.

1L7868
Cream color ivory finish......**$16.65**
Gray enamel finish......**16.75**
Frosted ebony finish......**17.15**

$3.00 Brings this Typewriter on trial!

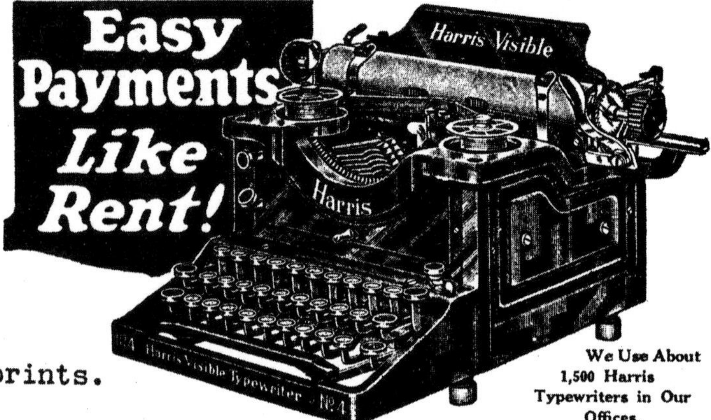

$3.00 deposit with your order brings you the genuine Harris, a typewriter of guaranteed high quality and exactly the same as used by Sears, Roebuck and Co. in their enormous business, on ten days' trial.

Test the Harris out in your own office or home for ten days and compare it with the best and most expensive typewriters used in your vicinity, then if fully satisfied pay the balance in easy monthly payments of $5.00 each, starting after thirty days. If, for any reason, you do not wish to keep the Harris after ten days' trial, just box it up and send it back to us and we will at once return your full deposit and pay all transportation charges both ways, so the trial costs you nothing.

This is a specimen of type the Harris prints.

This liberal offer makes it very easy for the student, teacher, lawyer, minister or business man to become the owner of a practical, standard size high quality writing machine on terms so easy that the typewriter is paid for and is your property almost before you realize it.

$5.00 a month is only about 17 cents a day. You can easily put by that much. Then why be without a good writing machine for another week? Why continue to write in slow, tedious and tiresome longhand, when with the Harris you can turn out a beautifully clear and legible printed page with far less time and effort?

The Harris Typewriter makes writing a pleasure. A single light touch of the key takes the place of two to five or six carefully directed strokes necessary to form each letter in writing longhand. Longhand wastes the time of both writer and reader and is frequently misread or misunderstood.

To be businesslike use the Harris. It enables you to make one or more carbon copies when writing letters, making bids, proposals, quoting prices, etc., that are exact duplicates of the original. By keeping these for reference you can frequently prevent disputes or law suits which may easily arise through misunderstanding, faulty memory, etc., if you have no accurate record of what has been written.

You will find it very simple and easy to handle, requiring no previous experience of any kind. Our complete instruction booklet, sent with each machine gives simple directions which will enable almost anyone to write at satisfactory speed after a little practice.

We Use About 1,500 Harris Typewriters in Our Offices.

The Harris is a front strike visible writer of standard size and shape, beautifully finished in bright nickel and polished black enamel, and will ornament the finest office or study.

It is an always visible writer. Every character is in plain sight from the moment it is printed. Errors are instantly seen and easily corrected. It has twenty-eight keys, writing eighty-four letters, figures and other characters. The type is regular pica size, as illustrated, and as used in about 95 per cent of all business typewriters; universal keyboard with arrangement of letters the same as on other standard machines. Light, lively touch and speedy action. Has tabulator, back spacer, shift lock and release, shift keys at both ends of keyboard and marginal release. Nine-inch writing line with extra wide carriage, taking in paper up to 11 inches wide, and other valuable features. Comes to you complete with black record ribbon, rubber cover, cleaning outfit and complete instructions.

Money Saving Prices.

58L285—Harris Typewriter with regular keyboard.
58L287—Harris Typewriter with keyboard containing fractions ¼, ½ and ¾ instead of the ∧, * and = signs **$66.75**

Terms: **$3.00** deposit with order and balance in monthly payments of **$5.00** each, beginning at end of thirty days.
If full amount of cash is sent with order **$61.75**

Harris Typewriters are always shipped by express. For express rates see page 457. Shipping weight, 48 pounds.

58L234—Extra Ribbons for Harris Typewriter. State color wanted.
Shipping weight, 3 ounces **48c**

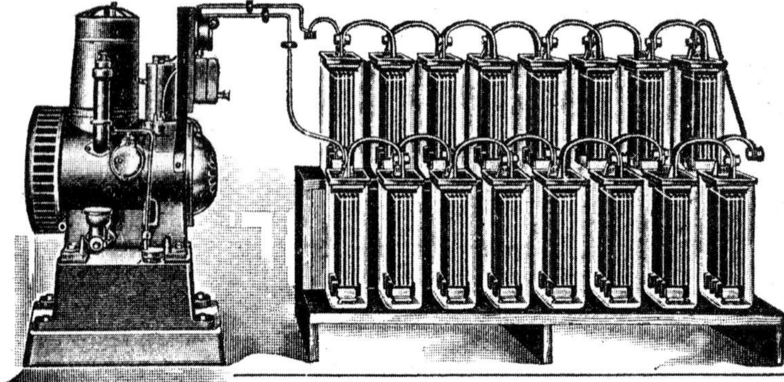

ELECTRIC LIGHT and POWER

Send for This Catalog ☛

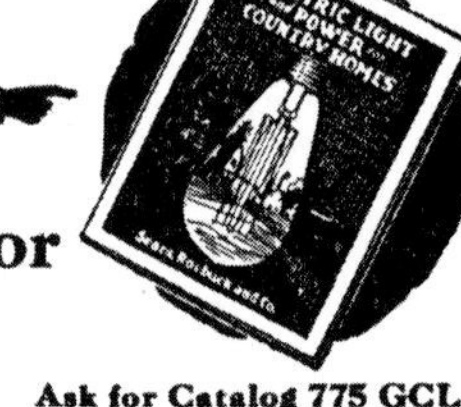

Tungsten Lamps for Farm Plants

Ask for Catalog 775 GCL.

NEW BATTERIES FOR YOUR PLANT

Batteries Guaranteed for a Definite Number of Years. Liberal Allowance for Your Old Battery.

Give full information asked on form below and mail to us at Chicago, Ill.

(Answer each question.)

Name.......... Postoffice..........

Rural Route.......... Box No.......... County.......... State..........

What make of plant have you?..........

The battery jars measure..........inches high, ...inches wide, ...inches across.

The battery plates measure..........inches high, ...inches wide.

There are..........plates in each cell **(not including the wood separators).** How many cells or jars make up the battery?..........

Is it a Sealed Glass Jar Battery?..........Open Glass Jar?..........or Rubber Jar?.......... It is rated as..........ampere-hour capacity battery.

The battery has been in service about..........years..........months. Is the battery now completely worn out?..........

If not, how much longer do you think it will last?.......... At how many amperes have you been charging your present battery?..........

About how many hours per week do you now charge it?..........Are you interested in a battery of larger capacity?..........

MAYFLOWER

The Range With a Sure Oven

QUALITY First and Foremost From Top to Bottom. The acme of value and our most popular range. Quality first and foremost is the slogan that has put it in the front rank. Built up to the highest standard, not down to a price. Equal to the best ranges made anywhere, it has no superiors in quality and but few equals. You cannot match it anywhere for the money; in fact, our prices save you at least $20.00 and probably more.

QUALITY goes clear through from top to bottom in this cast iron range. The more you use it, the better you like it. Extra big, strong and lasting, just the range for farmers and other homes where a great amount of cooking is done. We know it will make good every time and all the time.

You will take pride in its beauty and joy in its satisfaction. The many splendid advantages of this range make every buyer enthusiastic. The solid cast iron body is made extra strong and durable. Good for twenty-five or thirty years' work.

The Sure Oven Is the Supreme Feature.

Guaranteed to give satisfaction all the time. The housewife who has developed a reputation for cooking and baking excellence recognizes her dependence on her stove. The MAYFLOWER oven never fails. It always bakes and roasts to perfection.

Enjoy the Advantages of a Polished Cooking Top.

Always bright and smooth, one of the beauty features of this range. No blackening or polishing required; just wash it with a dishcloth or wipe with an oily rag. A labor saver. Be sure to mark order very plainly if polished top is wanted. Plain black cooking top is also furnished for those who prefer it.

The left section of top, over the fire, is hinged and can be raised by the handle; a catch holds it at any height wanted. You can toast or broil right over the hot coals or spread the fuel evenly. Height to cooking top, 31 to 33 inches, according to size of range.

Other Features of Interest and Convenience.

The nickel plating is handsome and easily kept bright. Fire boxes do quick baking and cooking without waste of fuel. Fire box length for wood, 20 to 22 inches, according to size of range. When range is ordered for burning wood only, the coal feed door is not furnished.

Reservoir is white porcelain enameled inside. If range is wanted without reservoir, deduct $5.25 from prices. Water front to fit in fire box to heat 30-gallon hot water boiler, $3.75 extra; for larger boiler, $4.75 extra. If you order water front, be sure to state if range is wanted with or without reservoir.

Shipped from Springfield, Mass., Philadelphia, Penna., Harrisburg, Penna., Atlanta, Ga., Newark, Ohio, Chicago, Ill., St. Louis Mo., Kansas City, Mo., or St. Paul, Minn.

Made in Four Sizes. Order by Number	Size of Lids, No.	Oven Measures, Inches	Stove With Black Cooking Top			Stove With Polished Cooking Top			Cooking Top, Including Reservoir, Inches	Capacity of Reservoir, Quarts	Size Pipe, Inches	Shpg. Wt., Lbs.
			For Soft Coal and Wood	For Hard (Anthracite) Coal	For Wood Only	For Soft Coal and Wood	For Hard (Anthracite) Coal	For Wood Only				
22L3117	8	16x17	$64.25	$64.70	$64.05	$69.25	$69.70	$69.05	42½x25	13½	7	560
22L3118	8	18x18	68.50	68.95	68.30	73.50	73.95	73.30	46½x27	17	7	628
22L3120	8	20x20	72.75	73.20	72.55	77.75	78.20	77.55	49 x27	22	7	667
22L3121	9	20x20	73.15	73.60	72.95	78.15	78.60	77.95	49 x27	22	7	667

We Offer the Greatest Variety of Sizes and Styles

No other store anywhere offers such a variety as we do, so it is naturally a very easy matter for you to choose just the stove best suited to your own needs and wishes, and at a price much lower than you would pay elsewhere for equal value. Look over our complete line and be convinced.

Our Stoves Are Made in Our Own Factories. We Guarantee Quality.

Our stoves are made in very large foundries using the latest and best machine equipment, and our enormous production eliminates wastes and losses frequent in smaller factories. This enables us to manufacture better stoves at lower costs than foundries not so large or so well equipped as ours.

Our big sales make possible the low costs; low costs make possible our unusually low prices; and our unusually low prices account for the big sales. Every-one benefits.

We Guarantee to Save You Money.

Size and quality being equal, we guarantee to save you 15 to 30 per cent in prices, sometimes more, even after you have paid the freight charges from one of our stove warehouses near you. The benefits of enormous factory production and our method of selling direct from factory to you are evident in our big stove values that emphasize the savings we make you without skimping one bit on quality.

How to Compare Stove Values When Buying.

You must consider more than appearance and price. You must compare size and weight, quality and style. One of the most important points of a stove, that is often overlooked when buying from a catalog, is the weight. If you were to compare two stoves standing side by side you would quickly satisfy yourself as to the relative value by looking into the oven and fire pot, by lifting the lids, etc. When buying from a catalog always be sure to look at the weight as well as the other measurements; the heavier the stove is, the more value there is bound to be.

All weights given on this page are approximate and may vary a trifle.

These Cast Iron Stoves for Small Families Are Real Bargains for Bargain Hunters

Made of cast iron, every piece and part the right weight and thickness to give good wear. The baking and roasting oven does splendid work, quickly and without waste of fuel. Has two oven doors. Four-lid cooking top is made the proper thickness.

Fire box has flat shaking and dumping grates and built-in **cast iron fireback.** Coal feed door and ash pan are furnished on stove when ordered for **burning coal and** wood; neither is furnished on stove for burning wood only. Takes 6-inch **stovepipe.**

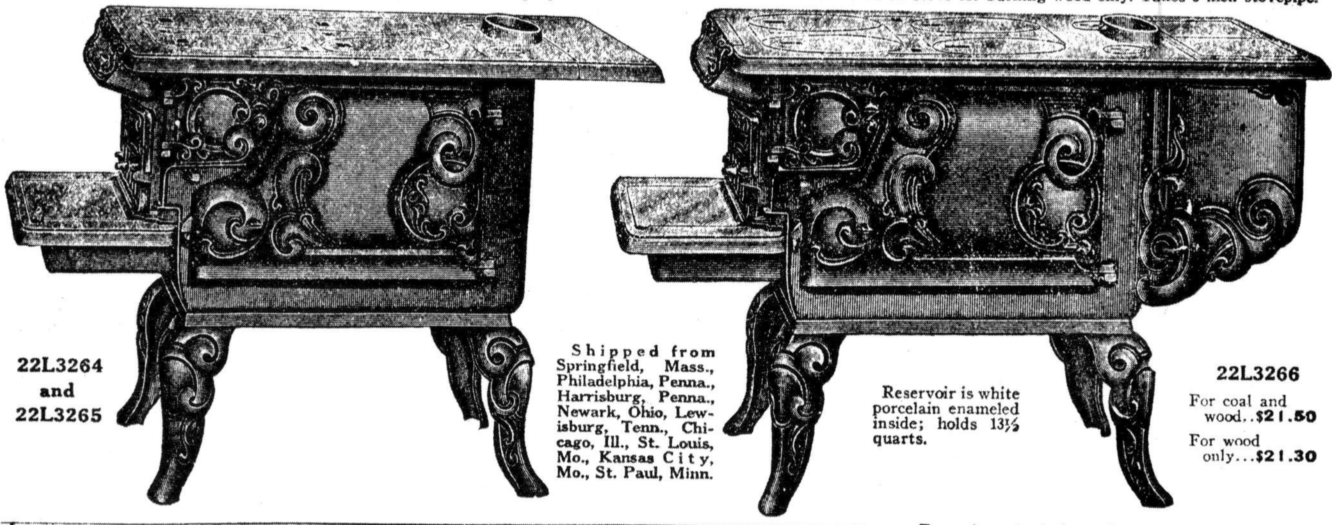

22L3264 and 22L3265

Shipped from Springfield, Mass., Philadelphia, Penna., Harrisburg, Penna., Newark, Ohio, Lewisburg, Tenn., Chicago, Ill., St. Louis, Mo., Kansas City, Mo., St. Paul, Minn.

Reservoir is white porcelain enameled inside; holds 13½ quarts.

22L3266

For coal and wood..**$21.50**

For wood only...**$21.30**

Order by Number	Style	Size of Lids, Inches	Oven Measures, Inches	Stove for Coal and Wood	Stove for Wood Only	Cooking Top, Inches	Fire Box Length for Wood, Inches	Shpg. Wt., Lbs.
22L3264	Without reservoir	8	15x13½x10	**$17.35**	**$17.15**	21x28½	17½	190
22L3265	Without reservoir	8	17x14¼x10½	**19.00**	**18.80**	22x30½	18½	215
22L3266	With reservoir	8	17x14¼x10½	**21.50**	**21.30**	22x37¼	18½	288

Bear in mind that these stoves are suitable only for small families, or where only a small size stove is needed. Be sure to order a stove large enough to do your work properly.

Common Sense High Oven Stoves, Convenient and Compact

It has not been necessary to sacrifice one bit of efficiency in making these new type stoves. But they have added a great deal of comfort and convenience to the use of the oven. You do not have to stoop over or get down on your knees when putting pans into the oven or taking them out. The oven is at the most comfortable height—right in front of you. These stoves will do the same satisfactory baking and cooking as all the other stoves we sell. The oven heats very quickly; can also be used for heating dishes and keeping cooked food warm until ready to serve.

Another advantage is the small floor space needed for a complete stove of this type. Especially recommended for small or crowded rooms.

Good Value.

Large enough for almost any family. Made heavy and solid to stand hard wear and last for years. Full cast iron body. Oven section is made of steel. The fire box is the same kind we use in our regular style ranges. Has duplex grate for burning any kind of coal, corn cobs and short wood. Ash pan is furnished. Roomy cooking top; two side shelves.

Oven measures 16x16 inches. Four 8-inch lids. Top, 28¾x21½ inches. Fire box length for wood, 14 inches. Takes 7-inch stovepipe. Height to cooking top, 27 inches. Total height, 63 inches. Shipping weight, 338 pounds.

Shipped from Newark, Ohio, or Chicago, Ill.

22L6476 ....**$35.00**

Low Priced.

This stove is not so large or heavy as the stove illustrated at left, but it is large enough to do all the cooking, baking and laundry work for a medium size family.

The stove is made of cast iron with steel oven. Round dump style grate. Burns any kind of coal, corn cobs and short wood. Will heat a room comfortably in really cold weather and hold fire all night.

Oven measures 16x16 inches. Cooking top, 21x21 inches. Four 8-inch lids. Height to cooking top, 22 inches. Total height, 58 inches. Diameter of fire pot at top, 12 inches. Takes 6-inch stovepipe. Shipping weight, 200 pounds.

Shipped from Springfield, Mass., Philadelphia, Penna., Harrisburg, Penna., Newark, Ohio, Chicago, Ill., or Kansas City, Mo.

22L3477 ....**$16.98**

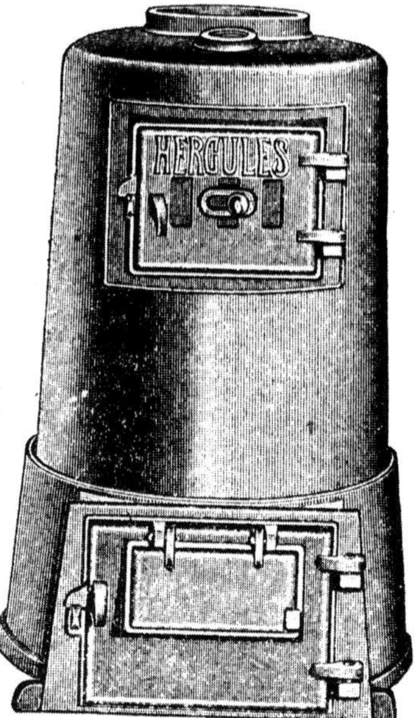

Big Capacity Coal Burning Water Heater.

This stove burns hard and soft coal, coke and short wood. Gas Water Heaters are listed on page 703.

This water heater is to be used in connection with hot water tank or range boiler for running water systems having constant pressure through pipes and must be connected to boiler or tank before starting fire. For range boilers to be used with these heaters see page 703.

A very powerful and efficient heater. Will heat 100 gallons an hour to temperature suitable to ordinary household purposes.

This heater will furnish the great amount of hot water required in large homes, flat and apartment buildings, for bath, shower, kitchen and laundry; can also be used for small hot water heating systems in garages, greenhouses, poultry buildings, etc. Capacity of radiation, 160 square feet. Quick and economical.

Made of cast iron, extra thick, strong and lasting. Draw center shaking grate. Inside diameter of fire pot, 12 inches. Height, 27 inches. 1¼-inch water pipe connection. Takes 6-inch stovepipe. Shipping weight, 276 pounds.

22L6481 ....**$28.50**

Shipped from Springfield, Mass., Philadelphia, Penna., Harrisburg, Penna., Newark, Ohio, or Chicago, Ill.

All weights given on this page are approximate and may vary a trifle.

Four 8-Inch Lid Stove With Sadiron Heaters.

22L3490

$7.38

$2.75

Stovepipe Baking Oven.

You should have one of these Stovepipe Baking Ovens attached to your laundry stove, for often you will find it convenient to get a quick meal on the laundry stove. It is a quick and satisfactory baker, uses the waste heat passing up the pipe. We especially recommend that in ordering a laundry stove or a small heating stove you order a Stovepipe Baking Oven. (This oven will not work satisfactorily with water heating laundry stoves, such as 22L6480 and 22L1499 listed at bottom of page.)

The oven is strongly made of steel. Furnished with one joint of stovepipe to connect oven to stove, as illustrated. Wire oven rack is removable. Oven measures 9 inches wide and 17¾ inches long inside; will hold two 8-inch pie plates or an 8x17-inch baking or roasting pan. Pipe collar on top and bottom to fit 6-inch stovepipe. Shipping weight, 21 pounds.

The small illustration shows the big capacity of four-lid laundry stove and Stovepipe Baking Oven. Our prices do not include articles shown.

Shipped from the same cities as our laundry stoves. See list below.

22L5500—Stovepipe Baking Oven, with one joint of stovepipe.......... **$2.75**

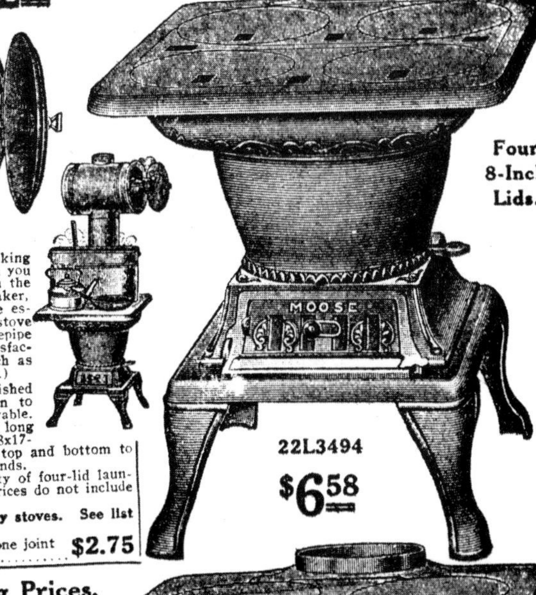

Four 8-Inch Lids.

22L3494

$6.58

Standard Laundry Stoves at Money Saving Prices.

Our complete line represents an unusual opportunity to select a style and size exactly suited to your needs. Made with two or four lids, full standard sizes and full weight castings. Strong and lasting. Shaking and dumping style grates. Burn any kind of coal, coke, corn cobs, short wood and rubbish.

The styles with sadiron heaters are especially desirable and convenient when you want to heat sadirons at the same time that the stove top is being used for other purposes. The irons heat very quickly. Sadirons illustrated are not included in prices.

Our prices show you a good money saving compared with prices asked elsewhere for stoves of similar size and quality.

Height of stove, 22 to 24 inches. All sizes take 6-inch stovepipe.

We recommend that you order a Stovepipe Baking Oven with laundry stove, for often you may find it convenient to get a quick meal on the laundry stove. The oven is a quick and satisfactory baker, using the waste heat going up the pipe. See description above.

These stoves and stovepipe baking ovens are shipped from Springfield, Mass., Philadelphia, Penna., Harrisburg, Penna., Atlanta, Ga., Newark, Ohio, Chicago, Ill., St. Louis, Mo., Kansas City, Mo., or St. Paul, Minn.

All weights given on this page are approximate and may vary a trifle.

Made in Five Sizes. Order by Number	Number of Lids	Diameter of Fire Pot, Inches	Stove Only	With Stovepipe Baking Oven	Top Measures, Inches	Shpg. Wt. of Stove, Lbs.
22L5496	Two No. 7	10½	**$5.20**	**$ 7.95**	19 x13¼	73
22L5498	Two No. 8	12	**5.80**	**8.55**	21 x13½	82
22L3493	Two No. 8	12	**6.50**	**9.25**	21 x13½	90
22L3494	Four No. 8	12	**6.58**	**9.33**	21½x20¼	116
22L3490	Four No. 8	12	**7.38**	**10.13**	21½x20¼	116

Two 8-Inch Lid Stove With Sadiron Heaters.

PRIDE

22L3493

$6.50

Two-Lid Stove With Coal Feed Door.

PET

22L5496—Two 7-Inch Lids.......... $5.20

22L5498—Two 8-Inch Lids.......... $5.80

Small Laundry Stove.

If a small stove will serve your needs and do your work, here is the one to buy. Be sure to notice the measurements and order a stove large enough to do your work properly. Has two No. 7 lids, front feed door, flat style grate. Burns coal, coke, corn cobs and short wood. Top measures 18x15 inches. Height, 17½ inches. Fire pot diameter at top, 9¼ in. Takes 6-in. stovepipe. Shipping weight, 64 pounds.

Shipped from the same cities as our other laundry stoves listed above.

22L5495 DANDY stove **$4.15**

22L5495—DANDY stove with STOVEPIPE BAKING OVEN..... **$6.90**

DANDY

22L5495

$4.15

Four 8-Inch Lid Laundry Stove and Tank Heater.

WIZARD

THE WEHRLE CO. NEWARK OHIO

22L6480

$15.35

Coal Burning Water Heaters.

These stoves burn hard and soft coal, coke and short wood. Gas Water Heaters are listed on page 703.

These water heaters are to be used in connection with hot water or range boiler for running water systems having constant pressure through pipes and must be connected to boiler or tank before starting fire. For range boilers to be used with these heaters see page 703.

These stoves are particularly desirable for laundry and kitchen. Will heat more than 40 gallons per hour to a temperature suitable for ordinary household purposes.

Will supply plenty of hot water for household needs and can be used for cooking and heating wash boiler at the same time. The water heats very quickly as it circulates through the hollow fire pot.

Made of cast iron, heavy and strong. Draw center shaking grate. Inside diameter of fire pot, 9½ inches. Top measures 21x21 inches, and has four No. 8 lids. Height, 26¼ inches. ¾-inch water pipe connection. Takes 6-inch stovepipe. Shipping weight, 199 pounds.

22L6480.....**$15.35**

Made of cast iron, strong and durable. Dump style grate. Inside diameter of fire pot, 10¼ inches. Top measures 13½x21 inches. Two No. 8 lids. Height, 23½ inches. ¾-inch water pipe connection. Takes 6-inch stovepipe. Has coal feed door. Shipping weight, 108 pounds. **Shipped from Newark, Ohio.**

22L1499.........**$9.35**

Shipped from Springfield, Mass., Philadelphia, Penna., Harrisburg, Penna., Atlanta, Ga., Newark, Ohio, or Chicago, Ill.

Two 8-Inch Lid Laundry Stove and Tank Heater.

22L1499

$9.35

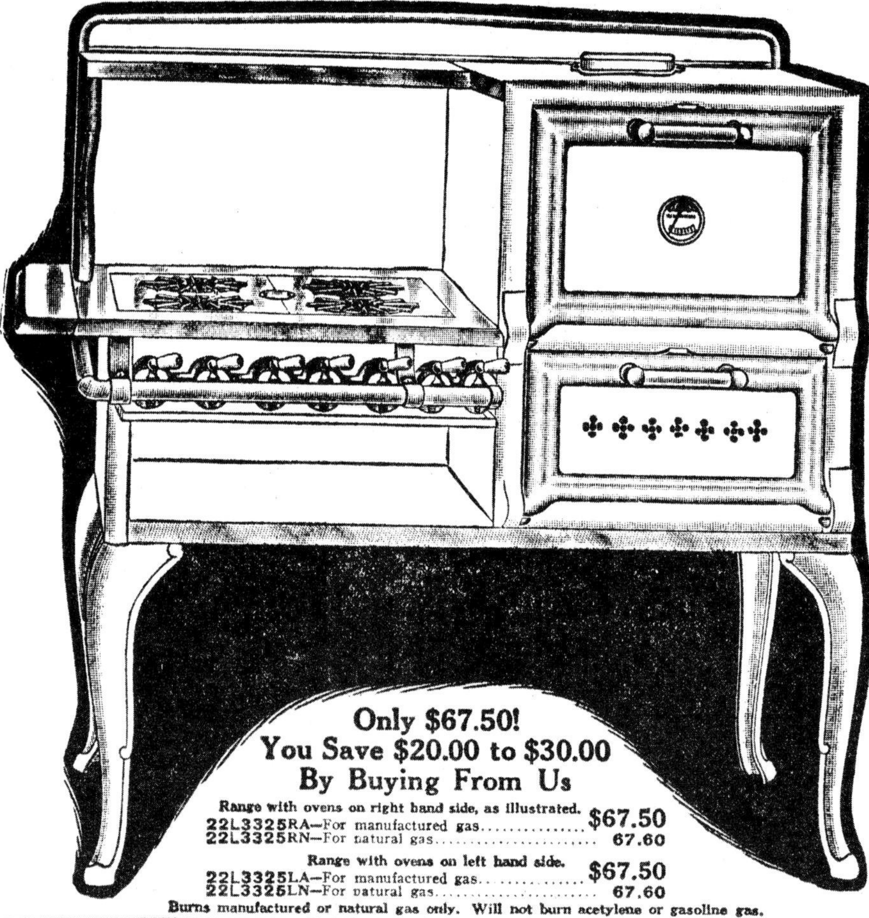

Only $67.50!
You Save $20.00 to $30.00
By Buying From Us

Range with ovens on right hand side, as illustrated.
22L3325RA—For manufactured gas................$67.50
22L3325RN—For natural gas........................67.60

Range with ovens on left hand side.
22L3325LA—For manufactured gas...............$67.50
22L3325LN—For natural gas........................67.60

Burns manufactured or natural gas only. Will not burn acetylene or gasoline gas.

White Porcelain Enamel— The Everlasting Finish Without a Fault

Enameled Kitchen Equipment Is Dominant Today.

There is an ever increasing demand for full porcelain enameled ranges to complete the modern, up to date, sanitary and work saving kitchens equipped with enameled sinks, tables and kitchen cabinets. One of the most beautiful and showy ranges ever designed. It will add a wonderful charm to your kitchen. The enamel is fused onto the steel and practically becomes a part of it, so that it will last as long as the range. The whole range is porcelain enameled, front and back, inside and outside, except the oven linings, which are zinc plated, and the cooking top and end shelf which are polished metal. The enamel on the inside of the range body is the really important feature, for it protects the steel from rust. Most ranges wear out from the inside, not the outside, and that is what we have guarded against by enameling the range both inside and outside. There is practically no wear out to such a range. The enamel is heatproof and rustproof and we guarantee it not to crack or chip from heat. Will not spot or discolor. Can be cleaned with a damp cloth as easily as a china dish. It forever eliminates the work and dirt of blackening. Nickel plated door frames and trimmings are brilliantly polished.

Enjoy the Advantages of a Polished Cooking Top.

Always bright and smooth, one of the beauty features of this range. No blackening or polishing required; just wash it with a dishcloth or wipe with an oily rag. A work saver.

The ovens are good size and a big amount of baking can be done at a time. Cast iron oven bottom will not rust out as steel oven bottoms so often do. This bottom has a removable lid so that vegetables can be cooked in the oven and have the odor go up the stovepipe instead of out into the room. Broiling oven is very convenient for broiling meat and fish, toasting, etc. Baking and broiling ovens are heated with two burners that heat both ovens at the same time. These burners are fitted with safety lighter.

Cooking top burners are easily lifted out for cleaning. Has three standard size burners, one giant and one combination simmer burner and lighter for other burners. You can light any burner instantly without matches and without reaching over the top or moving utensils. **Be sure to state whether you burn manufactured or natural gas.**

Measurements—Baking oven, 18 inches wide, 20 inches deep. Cooking top, 28x23 inches. Height to cooking top, 32 inches. Length over all, 51 inches. Floor space, 46x24 inches. Size pipe, 5 inches. Shipping weight, 335 pounds.

Shipped from Springfield, Mass., Philadelphia, Penna., Harrisburg, Penna., Newark, Ohio, Chicago, Ill., Kansas City, Mo.

Unusual Value—Popular Style—Porcelain Enameled Range

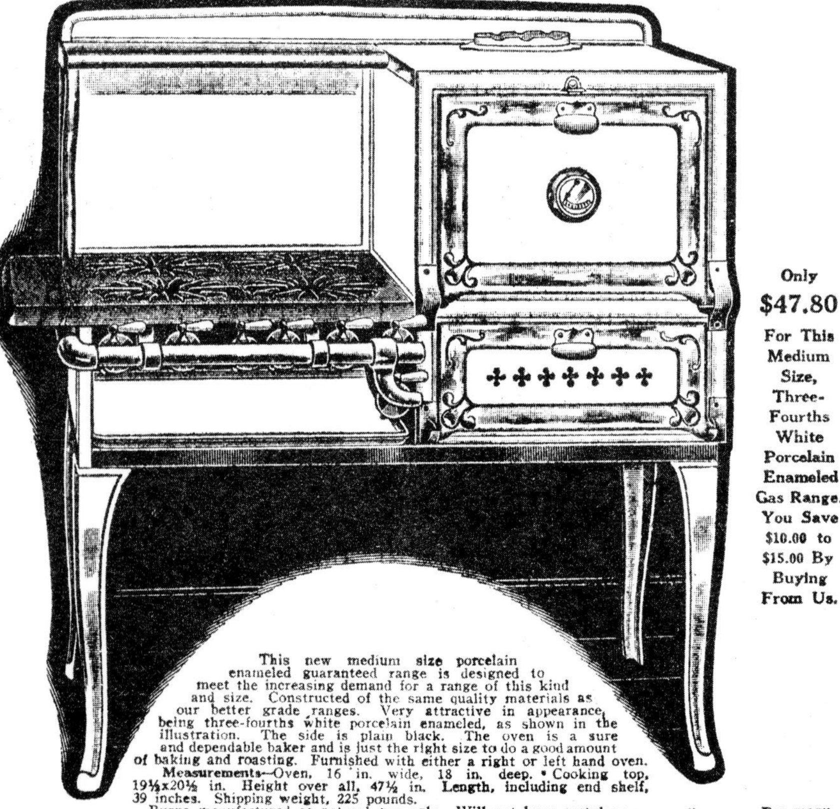

Only $47.80 For This Medium Size, Three-Fourths White Porcelain Enameled Gas Range. You Save $10.00 to $15.00 By Buying From Us.

This new medium size porcelain enameled guaranteed range is designed to meet the increasing demand for a range of this kind and size. Constructed of the same quality materials as our better grade ranges. Very attractive in appearance, being three-fourths white porcelain enameled, as shown in the illustration. The side is plain black. The oven is a sure and dependable baker and is just the right size to do a good amount of baking and roasting. Furnished with either a right or left hand oven.

Measurements—Oven, 16 in. wide, 18 in. deep. Cooking top, 19½x20½ in. Height over all, 47½ in. Length, including end shelf, 39 inches. Shipping weight, 225 pounds.

Burns manufactured or natural gas only. Will not burn acetylene or gasoline gas. For manufactured gas, open top grates are furnished, as illustrated. For natural gas, we furnish closed top.

Shipped from PHILADELPHIA, PENNA., NEWARK, OHIO, or CHICAGO, ILL.

Range with ovens on right hand side, as illustrated.
22L3326RA—For manufactured gas.......$47.80
22L3326RN—For natural gas.............47.90

Range with ovens on left hand side.
22L3326LA—For manufactured gas.......$47.80
22L3326LN—For natural gas.............47.90

Unbeatable Value

White Porcelain Enameled Three-Burner Gas Cooker With Combination Baking and Broiling Oven.

This is positively the greatest stove value we have ever offered. Just think! A three-burner cooker, with white porcelain enameled cooking top, main front, legs and drip pan—nickel plated trimmings—18-inch baking and broiling oven combined, for only $22.50.

Especially designed for small or crowded kitchens, that will not accommodate a large cabinet range. Cleans easy, without effort. Eliminates the most unpleasant work in the kitchen, that of blacking the stove. Enamel guaranteed not to crack or chip from heat and will not discolor.

Burns manufactured or natural gas only. Will not burn acetylene or gasoline gas. For manufactured gas, open top grates are furnished. For natural gas, closed tops are furnished.

Measurements—Oven, 18 inches wide, 13 inches deep. Cooking top, 13x29½ inches. Shipping weight, 130 pounds.

Shipped from Philadelphia, Penna., Newark, Ohio, or Chicago, Ill.

22L3363A—For manufactured gas......$22.50
22L3363N—For natural gas............22.60

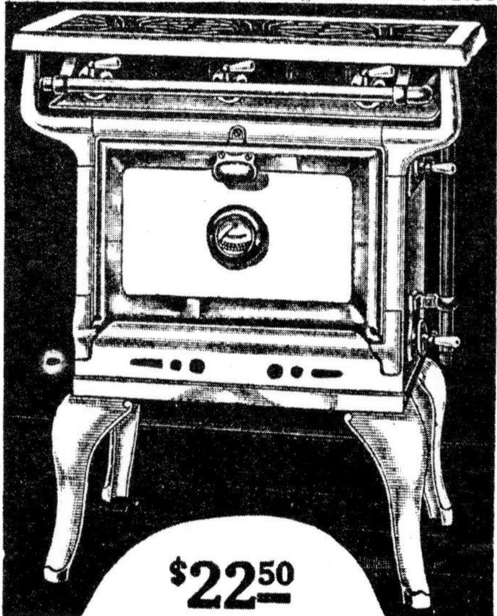

$22.50

Outstanding Gas Range Value

The tremendous number of orders that we are receiving daily from all over the country indicates that the majority of gas stove users regard this model as the greatest gas range value of today.

This is by far the most popular style gas range now on the market, because cabinet gas ranges have so many conveniences that make your cooking and baking real pleasure and comfort, instead of just work. But you want your stove to be something more than a cooking machine. It should be a piece of kitchen furniture as well. You can show this range to your friends and neighbors with a great deal of pride and satisfaction. It's so easy to keep spick and span. The enamel is hard and durable, will not crack or chip from heat and will not spot or discolor from grease spatterings. Sanitary and rustproof, easily cleaned with a damp cloth.

You cannot buy such a high class range elsewhere at anywhere near our money saving prices. This range is very standard in style and size and can be readily compared with nationally advertised brands of about equal value that retail for $50.00 to $65.00. Our prices save you $10.00 or more. Made of good grade steel, with strong cast iron frame and legs. Good enough for any kitchen.

The ovens are good size and so arranged that a big amount of baking can be done at a time. Cast iron oven bottom will not rust out as steel oven bottoms so often do. This bottom has a removable lid so that vegetables can be cooked in the oven and have the odor go up the stovepipe instead of out into the room. Broiling oven is very convenient for broiling meat and fish, toasting, etc. Baking and broiling ovens are heated with two burners that heat both ovens at the same time. These burners are fitted with safety lighter.

Cooking top burners are easily lifted out for cleaning. Has three standard size burners, one giant and one combination simmer burner and lighter for the other burners. You can light any burner instantly without matches. Cooking top for natural gas is closed. Range for manufactured gas has large open top grates, as shown in illustration. Be sure to state whether you burn manufactured or natural gas. **Cannot be used for acetylene or gasoline gas.**

Measurements—Baking oven, 18 inches wide, 20 inches deep. Cooking top, 28x23 inches. Height to cooking top, 32 inches. Length over all, 51 inches. Floor space, 46x24 inches. Size pipe, 5 inches. Shipping weight, 335 pounds.

Shipped from Springfield, Mass., Philadelphia, Penna., Harrisburg, Penna., Newark, Ohio, Chicago, Ill., Kansas City, Mo.

All weights given on this page are approximate and may vary a trifle.

Only $41.25! You Save $10.00 to $20.00 by Buying From Us

Range with ovens on right hand side, as illustrated.

22L3329RA—For manufactured gas.................. **$41.25**
22L3329RN—For natural gas.............................**$41.35**

Range with ovens on left hand side.

22L3329LA—For manufactured gas.................................**$41.25**
22L3329LN—For natural gas.......................................**41.35**

Burns manufactured or natural gas only. Will not burn acetylene or gasoline gas.

White Porcelain Enameled Four-Burner Gas Cooker

With White Porcelain Mantel Shelf

$33.25

Values are determined not by price alone, but by what you receive for the money that you spend. By actual comparison with other brands of dependable stoves, our prices are lower.

A most beautiful and stylish gas cooker, designed to meet the ever increasing demand for porcelain enameled stoves. This model is especialy adapted to use in small kitchens. Cast iron legs, main front, door frame and cooking top. All these parts are white porcelain enameled. The sparkling white finish and the nickel plated trimings harmonize charmingly with any kitchen furniture. White porcelain enameled splasher and high shelf add much to the appearance and convenience of this lovely stove.

The enamel is guarantee not to crack or chip from the heat and will not discolor. Forever eliminates the most unpleasant work of the kitchen, that of blackening the stove. Easy to clean by just wiping with a damp cloth.

Large and roomy oven that bakes quickly and satisfactorily. Oven bottom is made of cast iron that will not rust as many steel oven bottoms so often do. All burners are equipped with adjustable gas cocks and stove is also furnished with a patented simmer and lighter burner.

Measurements: Oven, 18 inches wide, 17¼ inches deep. Cooking top, 22x28 inches, including end shelves. Shipping weight, 220 lbs. **Shipped from Philadelphia, Penna., Chicago, Ill., or Newark, Ohio.**

Burns manufactured or natural gas only. Will not burn acetylene or gasoline gas.

22L3341A — Stove for manufactured gas.......... **$33.25**
22L3341N — Stove for natural gas **$33.35**

Good Value Low Priced Gas Range

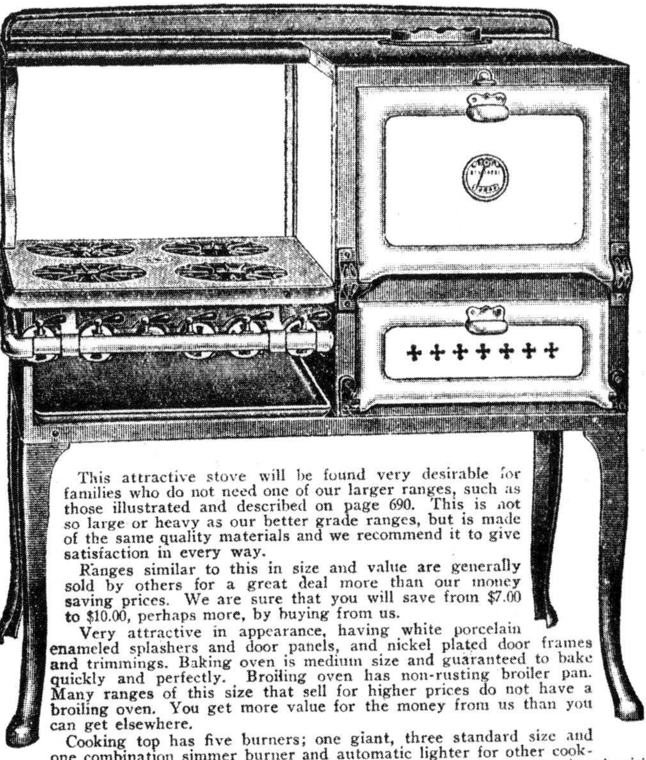

This attractive stove will be found very desirable for families who do not need one of our larger ranges, such as those illustrated and described on page 690. This is not so large or heavy as our better grade ranges, but is made of the same quality materials and we recommend it to give satisfaction in every way.

Ranges similar to this in size and value are generally sold by others for a great deal more than our money saving prices. We are sure that you will save from $7.00 to $10.00, perhaps more, by buying from us.

Very attractive in appearance, having white porcelain enameled splashers and door panels, and nickel plated door frames and trimmings. Baking oven is medium size and guaranteed to bake quickly and perfectly. Broiling oven has non-rusting broiler pan. Many ranges of this size that sell for higher prices do not have a broiling oven. You get more value for the money from us than you can get elsewhere.

Cooking top has five burners; one giant, three standard size and one combination simmer burner and automatic lighter for other cooking burners. Baking and broiling ovens are heated by one burner, equipped with safety oven lighter.

Measurements—Baking oven, 16x18 inches. Cooking top, 21x21 inches. Length, over all, 40 inches. Shipping weight, 199 pounds.

Shipped from Philadelphia, Penna., Newark, Ohio, or Chicago, Ill.

Burns manufactured or natural gas only. Will not burn acetylene or gasoline gas.

We make this range with ovens on right hand side only, as illustrated.

22L6333A—For range to burn manufactured gas.................. **$29.75**
22L6333N—For range to burn natural gas.......................... **$29.85**

Hercules The Pipeless Furnace *of* Superior Quality

$66.95 Cash Buys This High Grade Pipeless Furnace. 18-Inch Fire Pot Size

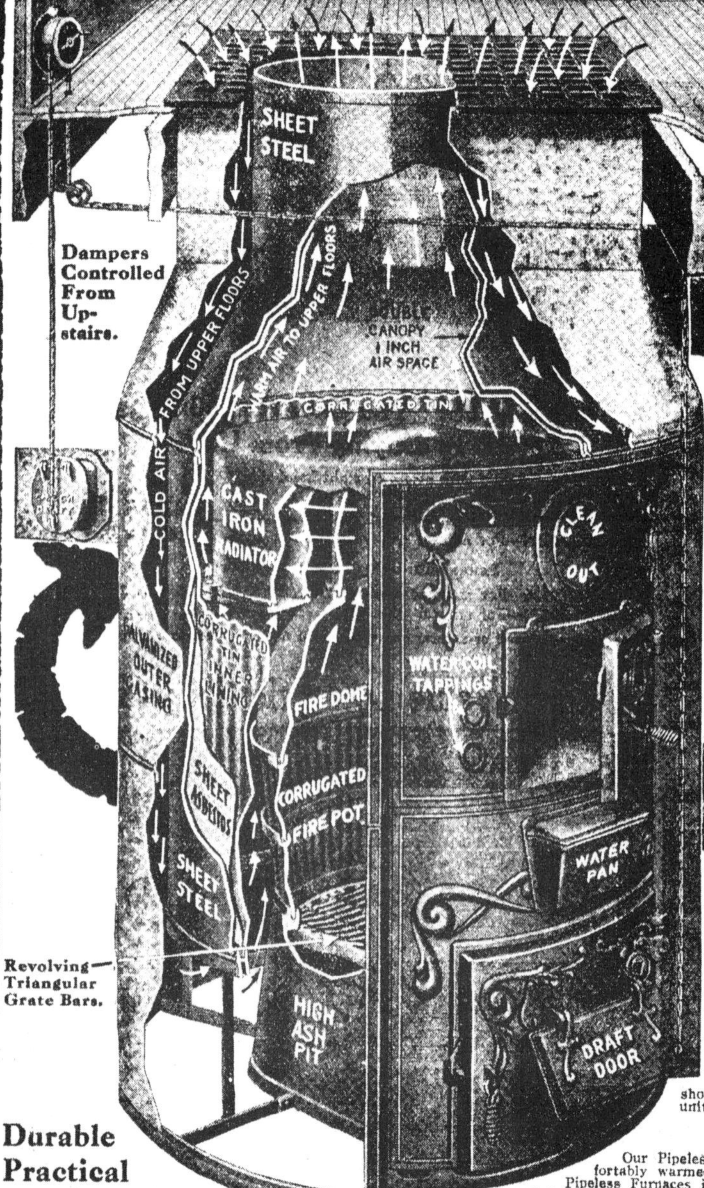

Also Sold on Easy Payments

Terms and Prices Quoted Below

No Interest or Other Extras to Be Added

There is an advantage, of course, in paying cash in full for your furnace as you get the benefit of a lower price. However, if it is more convenient for you to purchase on our easy payment plan you will find our special easy payment terms and prices quoted below. You pay down the amount stated below and pay each month the amount stated for ten months. **There is no interest or other extras to be added to the monthly payment prices.** If you wish to purchase on easy payments, please fill out our special easy payment order blank inclosed with this catalog. **To take advantage of our easy payment offer it will be necessary for you to have title to the building in which furnace is to be placed.**

Simplest and Most Economical of All Heating Systems.

Our Pipeless Furnaces are the simplest of all systems of heating and, by reason of their extreme simplicity, they are without any question the most economical in fuel. The heat is delivered to your rooms almost instantaneously, without passing through any long system of pipes, so there is practically no heat wasted. Your basement will always be cool enough for vegetables, and for every shovelful of coal you put into this furnace you will get big returns in actual heat units in your rooms.

Keeps Entire House Comfortably Warmed.

Our Pipeless Furnaces not only keep the room in which the large register is placed comfortably warmed, but experience has proved that in any ordinary residence where one of our Pipeless Furnaces is installed (if the doors between adjoining rooms are left open, or registers installed in the ceiling, so as to allow a reasonable amount of circulation between the rooms) every room will be comfortably warmed.

Durable Practical Economical Maximum Efficiency

Our Hercules Pipeless Furnace illustrated on this page embodies the most advanced improvements in pipeless furnace design. Cold air is taken from upstairs, it enters the large register at the sides, drops down, and turns up under the inner or sub-casing near the base of the furnace. It then strikes the hot surface of the furnace and is carried up through the sub-casing and is discharged through the center of the large register above.

No Heat Lost in Basement.

Just observe that this furnace provides its own heat insulation. It is entirely encased in a jacket of cold air, which is in constant circulation and which absorbs every unit of heat which might otherwise escape through the furnace casing, and it carries this heat right back into the heating chamber and up into your living rooms upstairs. That is why you get the full benefit out of every shovelful of coal you put into this furnace. There is no heat wasted. The warm air rising out of the large register above circulates to all parts of the building, delivering heat to every room in the house.

Size to Order.

Figure up the total cubic contents of your building and select a furnace of corresponding capacity from the table on this page.

If you live in a real cold climate, or your house is unusually hard to heat; if you want to keep your house unusually warm, or if you will burn wood or soft coal, we would recommend ordering one size larger furnace. It is always better to have a furnace a little larger than is actually required. **The 18-inch fire pot furnace makes an excellent furnace for small houses, bungalows, etc., but is not recommended for houses having more than five rooms.**

If you prefer we will estimate the size furnace you require for your building. Write for one of our Special Heating Information Blanks, fill it out carefully and return it to us. Our engineers will advise you the correct size furnace needed to heat your building, and we will also advise you what the freight will amount to if you wish us to do so.

Prices of Hercules Pipeless Furnaces.

Prices are for furnace complete with register, casing, shaker handle, poker, check draft, chains, pulleys, damper chain plate and cement. **Shipped from factory in OHIO.**

	42L3899⅓	42L3900⅓	42L3901⅓	42L3902⅓	42L3904⅓
Diameter of fire pot, inches	18	20	22	24	28
Diameter of inside casing, inches	34	35	38	41	52
Diameter of outer casing, inches	40	42	45¼	49	60
Size of feed door, inches	10½x13	11x13	11½x13½	11½x13½	11x13
Depth of fire pot, inches	11	11½	12	12½	14
Diameter of smoke pipe, inches	7	8	8	8	9
Heating capacity, cubic feet	10,000	14,000	20,000	28,000	40,000
Shipping weight, pounds	950	1,060	1,250	1,450	2,070
Cash price	$66.95	$75.75	$88.75	$108.95	$159.25
Price on easy payments	77.00	87.00	102.00	125.00	183.00
Payment down	12.00	12.00	12.00	15.00	18.00
Payment, per month, for ten months	6.50	7.50	9.00	11.00	16.50

If ordering on easy payments please fill out special easy payment order blank enclosed with this catalog.

Smoke pipe not included at above prices. When ordering, state amount of smoke pipe needed and we will include it with your order at our net catalog price.

Prices are subject to market changes.

Hot Water Coil for **Hercules Furnace**, $1.75 extra. **If wanted with Gas Burner in addition to coal grate, write for prices.**

Fuel Economy Unsurpassed.

Here you have a furnace which positively cannot waste heat. The cold air coming down from upstairs surrounds the entire heating chamber and carries all stray heat units right back into the heating chamber, from which they are delivered up into your rooms. We positively cannot point to a single heating device of any kind now on the market, barring none, that will deliver more actual heat units to your rooms per pound of coal consumed than our Hercules Pipeless Furnace illustrated here. The warm air rises through the center of the large register and circulates to all parts of the house. By placing combination registers in the ceiling the warm air will circulate to the bedrooms, bathroom, etc., on the second floor, and you will be able to maintain an even and comfortable warmth throughout every room in the house.

Modern Heating Systems

Cash or Easy Payments

CASH $51.95 PRICE

To take advantage of our easy payment offer it will be necessary that you own the building in which the furnace is to be installed. If ordering on easy payments please fill out special easy payment order blank which you will find enclosed with this catalog.

Volcano Pipeless Furnace.

Cash Price $64.95

Burns Hard Coal, Soft Coal or Wood.

This is one of the simplest, most efficient pipeless furnaces on the market and is recommended with the assurance that it will give you highly satisfactory service. Like all of our pipeless furnaces, it is made of high quality material throughout. The castings are heavy and well designed for greatest efficiency. Casing is of heavy galvanized iron, well insulated to prevent loss of heat. Upper half is lined with bright corrugated tin and sheet asbestos, as shown in illustration at left. Cold air is taken in at the bottom of the casing through screened cold air intakes. This gives the full capacity of the large register for warm air circulation. The furnace has heavy cast iron radiator, two-piece corrugated fire pot, corrugated fire dome and revolving triangular grate bars. Draft dampers can be controlled from upstairs. It is economical in fuel and easily operated. Furnace is adapted for any height of basement up to 7 feet. Prices include furnace complete with large steel register, check damper, shaker, pulleys and cement, but no smoke pipe. State amount of smoke pipe you need when ordering and we will add same to your order at our net catalog price. **Shipped from factory in OHIO.** Read information on page 692 regarding correct size to order. **Prices are subject to market changes.**

PRICES AND DIMENSIONS OF VOLCANO PIPELESS FURNACE.

Catalog No.	Diameter Fire Pot, Inches	Heating Capacity, Cubic Ft.	Diam. Smoke Pipe, In.	Shpg. Wt., Lbs.	Cash Price	Easy Payment Price	Payment Down	Payment per Month
42L3920⅓	20	14,000	8	825	$ 64.95	$ 75.00	$10.00	$ 6.50
42L3921⅓	22	20,000	8	1,050	77.75	90.00	10.00	8.00
42L3922⅓	24	28,000	8	1,275	91.85	106.00	11.00	9.50
42L3923⅓	26	35,000	9	1,480	105.95	122.00	12.00	11.00
42L3924⅓	28	40,000	10	1,650	126.95	146.00	16.00	13.00
42L3925⅓	30	48,000	10	1,850	156.95	181.00	21.00	16.00

Hummer Pipeless Furnace.

A Red Hot Rapid Heater. Burns Fuel of Any Kind.

Burns any kind of fuel, and especially adapted for burning wood. This furnace has a large, heavy steel drum combustion chamber which is covered with a heavy cast iron top plate, all parts being well made and carefully cemented, bolted or riveted, making sound, gastight joints. Grates are of the revolving triangular type. This furnace is a rapid heater, giving almost instantaneous results from a very small fire. The large volume of heat delivered by this furnace from a very small amount of fuel is really remarkable. Fire pot is made in two parts so as to allow for expansion and contraction to prevent cracking. Adapted for any height of basement up to 7 feet. Price includes furnace complete with large steel register, check damper, shaker, pulleys and cement, but no smoke pipe. When ordering state amount of smoke pipe needed and we will include same with your order at our net catalog price. **Shipped from factory in OHIO.** Read information on page 692 regarding correct size to order. **Prices are subject to market changes.**

PRICES AND DIMENSIONS OF HUMMER PIPELESS FURNACE.

Catalog No.	Diam. Fire Pot, Inches	Heating Capacity, Cubic Feet	Diam. Smoke Pipe, Inches	Shpg. Wt., Lbs.	Cash Price	Easy Payment Price	Payment Down	Payment per Month
42L3910⅓	20	10,000	8	625	$51.95	$60.00	$10.00	$5.00
42L3911⅓	22	15,000	8	800	63.00	72.50	12.50	6.00
42L3912⅓	24	21,000	8	1,000	75.00	88.00	13.00	7.50

Hot Water Coil for any size furnace, $1.75 extra. If wanted with Gas Burner in addition to coal grate, write for prices.

Warm Air Heating Plants CASH OR EASY PAYMENT TERMS.

For real practical heating results, durability, efficiency of design, workmanship, finish and material, our Warm Air Furnaces are equal to the best on the market. We can furnish you a complete Modern Warm Air Heating System, including all necessary registers, warm air pipes, tin wall stacks, etc., at a very moderate price. All of this material is fully illustrated and described in our Special Heating Catalog 498GCL, which we will gladly send you on request. Send us a sketch of your building and get our estimate on a complete warm air heating system for your home.

The prices given below are our cash prices. Our warm air pipe furnaces are not sold separately on easy payments. However, if you wish to install a complete warm air heating system you may take advantage of our easy payment plan.

Hummer Warm Air Pipe Furnace Best for Burning Wood.

Burns hard or soft coal, but especially well adapted for burning wood. Throws a large volume of heat. It is remarkably economical in fuel and gives instant response to firing. Large fuel door and combustion chamber permits burning of logs. The radiator is made of sheet steel with a cast iron top and bottom. Diving flue spreads heat to all parts of radiator. Fire pot, ash pit and casing are same construction as our Volcano Furnaces. Pipes, registers or collars, etc., not included. **Shipped from foundry in OHIO. Prices are subject to market changes. Hot Water Coil for any size furnace, $1.75 extra. If wanted with Gas Burner in addition to coal grate, write for prices.**

Catalog No.	Diam. Fire Pot, In.	Heating Capac'y, Cubic Feet	Size of Feed Door, In.	Total Height, Inches	Shpg. Wt., Lbs.	Cash Price
42L3952⅓	20	10,000	10x14	56	600	$47.00
42L3953⅓	22	15,000	10x14	59	750	58.00
42L3954⅓	24	21,000	12x15	62	875	69.00

Volcano Warm Air Pipe Furnace.

Burns hard coal, soft coal, wood or coke. Casing is galvanized sheet steel. Circular cast iron radiator. Fire pot is made in two sections, heavily corrugated and with deep cup joints. Upper casing has asbestos felt lining and bright tin reflector. Deep ash pit and four triangular revolving grate bars.

Prices, complete with casing, shaker handle, poker, check draft, chains, pulleys, regulator plate and cement, but with no pipes of any kind or collars or other fittings. **Shipped direct from factory in OHIO. Prices are subject to market changes.**

Catalog No.	Diam. Fire Pot, In.	Heating Capacity, Cubic Feet	Size Feed Door, Inches	Height, With Casing, Inches	Shpg. Wt., Lbs.	Cash Price
42L3940⅓	20	12,000	9 x11¾	56	775	$ 57.45
42J 3941⅓	22	16,000	9 x11¾	59	910	70.45
42L3942⅓	24	23,000	9⅜x12⅝	62	1,115	81.95
42L3943⅓	26	32,000	9⅜x12⅝	65	1,330	97.95
42L3944⅓	28	38,000	10¼x12¾	68	1,565	115.95
42L3945⅓	30	45,000	10¼x12¾	73	1,710	135.95

Hot water coil to fit any size furnace, $1.75 extra.

Our complete line of warm air floor and wall registers and borders, warm air furnace pipe and fittings fully illustrated and quoted in our Special Heating Catalog.

Combination Floor and Ceiling Register and Ventilator.

$3.90

10x12 In.

Used with Pipeless Furnaces. Carries heat from room below to heat room above. Lets heat circulate to second floor. Adjustable to fit any floor or ceiling from 7 to 12 inches apart.

Catalog No.	Size Opening in Floor, In.	Size of Register Face, In.	Shpg. Wt., Lbs.	
99L2118	8x10	10x12	11	$2.80
99L2119	10x12	12x14	13	3.90
99L2120	12x15	14x17	18	6.40

Any Handy Man Can Easily Install Our Furnaces.

The Pleasure of Doing.

All of us like to do things. No greater satisfaction can come to a man than the pride he takes in something he has built or made for himself. The small boy enjoys tearing a thing to pieces to see what it is made of, and in after years one still delights in taking a piece of machinery apart to discover the "why" of it. But far greater pleasure lies in putting something together, beginning with the smallest parts, and then seeing it work perfectly.

An Honest Pride.

Now think of doing something really worth while; something you never tried before, and perhaps never thought you could do; something mighty interesting in its theory and construction; finally, something that you would always feel a pride in having done yourself, a well deserved pride. Think of saying to your friends and neighbors: "Yes, I put this furnace in practically all by myself. My brother Jack helped a little, which enabled me to get through a little sooner, but that was all. And I have learned a lot about furnaces. Mighty interesting they are, too."

A Most Interesting Job for You.

Are You Handy?

It is a little work, of course. But, much more than that, it is really fun to see the furnace grow under your hands to completion, and then work perfectly when you have it finished. It is just common sense, after all, combined with a little practical knowledge, and the real pleasure comes in seeing the thing accomplished.

Save Needless Expense.

Even if you do not intend to do the work yourself you do not need an expert. We repeat that any reasonably handy man can install one of our pipeless furnaces with the directions we supply. But do it yourself if you possibly can. Not only is it interesting and not difficult, but by doing it yourself you will save a tidy sum. You can use that money as well as the other fellow, if not better.

Steam and Hot Water Heating Systems

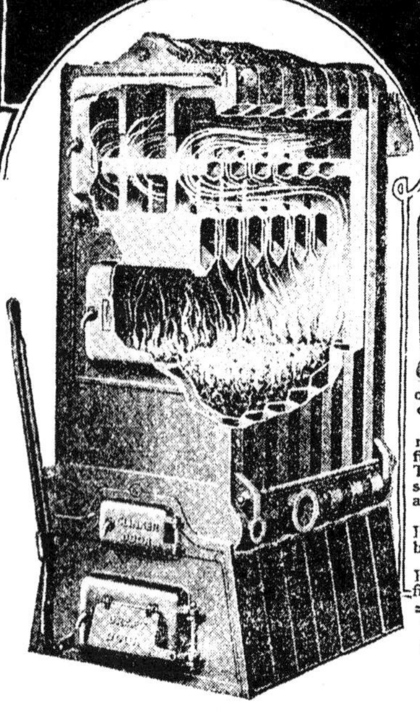

Hercules Air Moistener.

For Health and Comfort.

Filled with water and hung over the back of a radiator, or placed above a hot air register, it keeps the air properly moistened by the evaporation of the water. Consists of a galvanized sheet metal tank finished in aluminum or gold bronze. **Hangs behind radiator, out of the way and out of sight.** Shpg. wt., 4 lbs.

NOTE—These air moisteners will not fit five-column radiators. They will fit only the single column, two-column and three-column patterns.

42L156—Hercules Humidifier, aluminum bronze finish **60c**

42L157—Hercules Humidifier, gold bronze finish **65c**

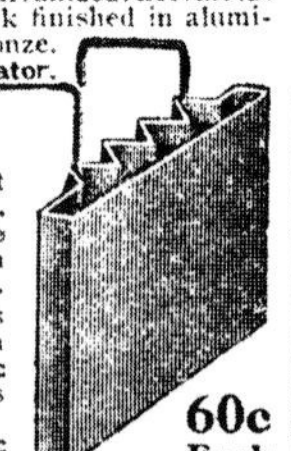

60c Each

Write for Our Special Heating Catalog Sent Postpaid on Request

Cash or Easy Payment Terms

The prices given in this catalog are our cash prices. There is an advantage, of course, in paying cash for your material. Our heating boilers and radiators are not sold separately on easy payments. However, if you wish to purchase a complete heating system, including boiler, radiators, pipes, valves, etc., and if it is more convenient for you to pay for your system on monthly payments, provided you own the property in which the material is to be installed, we will make special arrangements so that you can purchase your material on easy payment terms.

By taking advantage of this offer you need not wait until you have accumulated sufficient funds to pay for the material in full. You can make this great improvement in your home right now and begin at once to enjoy this great comfort and convenience while paying for the material in moderate monthly payments easily within your means.

If you will send us a sketch of your building and advise us just what kind of heating system you would like to install, we will prepare a special estimate for you and will advise you just what terms we will make.

Hercules Boilers For Home Heating.

Hercules Is Our Own Registered Trade Mark.

Efficient—Durable—Economical.

When selecting a heating system for your home, the boiler should have your most careful consideration. No matter how well the system is planned, or how good the material may be, if the boiler is not designed for best efficiency you will not get the results that you should from the fuel you burn.

Illustration on this page shows clearly the very efficient construction of our 42L3972⅓ and 42L3973⅓ Hercules Home Heating Boilers for steam and hot water heating. These boilers burn hard coal, soft coal, wood or coke and embody the latest improvements in home heating boiler design. The V shaped construction of the sections in the fire pot presents the greatest amount of heating surface in proportion to the volume of fuel. Our Hercules boilers 42L3970⅓ and 42L3971⅓ are of very similar design to the boiler shown above except that on account of their smaller sizes they have not as many flue spaces.

Hercules Boilers are equipped with latest improved type of rocking grate bars. Grates are of heavy construction and the mechanical action is perfect.

Prices on steam boilers include steam trimmings, complete. Prices on hot water boilers are for boilers only, without trimmings. Firing tools, including hoe, poker, rake and flue brush with handle, furnished with all boilers, both steam and hot water. Our Hercules Home Heating Boilers are, without a doubt, among the most efficient home heating boilers on the market. **Thousands now in use on our Hercules Home Heating Systems are giving satisfactory service in all parts of the country. For more detailed information about these boilers see our Special Heating Catalog, 498GCL. Shipped from factory in WESTERN NEW YORK. All boiler prices quoted are subject to market changes.**

Boilers With Fire Pot 17 Inches Wide.
42L3972⅓—Steam Boilers. 42L3973⅓—Hot Water Boilers.

No.	Size of Fire Pot, Inches	Size of Smoke Pipe, Inches	Height Over All, Inches	Flow and Return, Tappings	Rating, Hot Water	Rating, Steam	Hot Water	Steam
174	17x12	9	58	2-2½ in.	650	400	$ 78.95	$ 92.30
175	17x16½	9	58	2-2½ in.	750	475	87.50	100.85
176	17x20½	9	58	2-2½ in.	925	550	100.95	114.45
177	17x25	9	58	3-2½ in.	1,100	650	118.75	132.25
178	17x29	9	58	3-2½ in.	1,275	750	135.65	149.00

Boilers With Fire Pot 13½ Inches Wide.
42L3970⅓—Steam Boilers. 42L3971⅓—Hot Water Boilers.

No.	Size of Fire Pot, Inches	Size of Smoke Pipe, Inches	Height Over All, Inches	Flow and Return, Tappings	Rating, Hot Water	Rating, Steam	Hot Water	Steam
134	13x10	7	49½	2-2 in.	300	150	$47.70	$60.45
135	13x13½	7	49½	2-2 in.	400	225	59.65	72.45
136	13x17	7	49½	3-2 in.	500	300	71.75	84.45

For prices on larger size Boilers see our Special Heating Catalog.

You Can Easily Install Your Own Steam or Hot Water Heating System and Make a Big Saving.

Others Have Done It—So Can You.

By taking advantage of our low prices and installing your own heating plant you will make a big saving in the labor cost as well as in the purchase price of the material. The great majority of our customers install their own heating systems. It is not nearly as difficult as most people imagine. It is really surprising how simple and easy the work becomes after the job is started. "Irresistibly interesting" is the way one customer put it when telling us about his experience in installing his own heating system. It amounts to nothing more than a little substantial exercise, and after the job is finished you will feel the better for it. One man and a boy can easily install any of our Hercules Heating Systems. The big saving you can make well warrants your looking after the work yourself.

Experience Not Necessary.

It does not require an experienced plumber or steamfitter by any means to install our heating systems. Any man who can cut and thread ordinary iron pipe and who has a little mechanical ability can easily do this work by following our simple plans and instructions. We cut and thread all the larger size pipes for you so that all you have to do is follow the plans and screw them together.

Write for our estimate.

Why Be Without This Great Home Comfort?

If you haven't a Hercules Heating System in your home you are missing one of the greatest of all modern comforts. Just think of what it means to have your entire house heated to a comfortable and even temperature throughout the coldest winter weather; only one fire to attend to, no smoke, dust or gas in your living rooms; no carrying of coal and ashes over your rugs and carpets. House cleaning a real pleasure. You can get up on the coldest winter morning in a nice warm bedroom, have the bathroom as warm as any other room in the house, and you can use a modern oil or gas stove for cooking in the kitchen if you want to. This great comfort and convenience is easily within your means.

Cast Iron Radiators for Steam or Hot Water Heating Systems.

Good quality cast iron, properly machined and tested. Graceful, plain design.

Steam Radiators are tapped as follows, unless otherwise requested:
Up to 24 feet, 1-inch supply; 24 feet to 50 feet, 1¼-inch supply; 50 feet to 100 feet, 1½-inch supply; over 100 feet, 2-inch supply.

Hot Water Radiators are tapped as follows, unless otherwise requested:
Up to 75 feet, ¾-inch feed and return; 75 to 120 feet, 1-inch feed and return; over 120 feet, 1½-inch feed and return.

List of Sizes, Hercules Three-Column Radiators

	Square Feet of Heating Surface		
	38-Inch Height, 5 Sq. Ft. per Section	26-Inch Height, 3¾ Sq. Ft. per Section	22-Inch Height, 3 Sq. Ft. per Section
42L3998⅓ Steam. Per square ft.	32½c	39c	41½c
42L3999⅓ Hot Water. Per square ft.	33½c	40c	42½c

List of Sizes, Hercules Five-Column Radiators

	Square Feet of Heating Surface		
	22-Inch Height, 6 Sq. Ft. per Section	18-Inch Height, 5 Sq. Ft. per Section	14-Inch Height, 4 Sq. Ft. per Section
42L4002⅓ Steam. Per square ft.	43c	48½c	51½c
42L4003⅓ Hot Water. Per square ft.	43c	48½c	51½c

Shipped from factory in WESTERN NEW YORK. Prices subject to market changes.

Hercules Thermostat Heat Regulator.

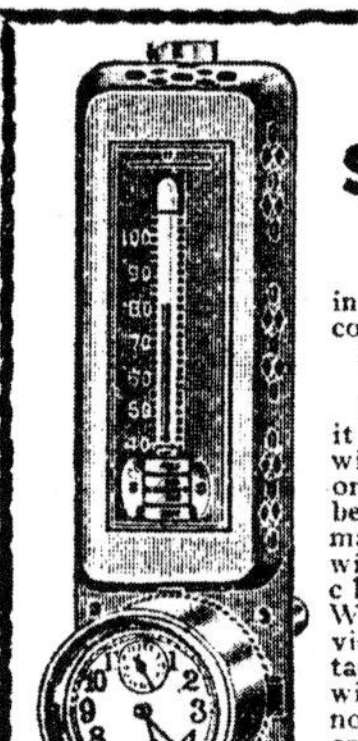

$31.45

This device will not only keep the temperature in your rooms regulated to an even and exact degree throughout the winter, but it will open up the dampers on your heating system for you before you arise in the morning so that your rooms will be warm and cozy when you get up.

It will maintain any temperature you want to set it for.

PAYS FOR ITSELF IN ONE WINTER.

It makes a big saving in your fuel bills. Night or day it is always on the job and it will pay for itself the first winter used. You will need to give your heating system only about one-half the attention, as the draft doors will be regulated automatically. Furnished with or without clock attachment. We recommend device with clock attachment, as outfit without clock does not have feature of opening draft before you arise in the morning.

42L1770¼—Without clock attachment. Shipping wt., 27 lbs **$26.75**

42L1771¼—Outfit complete with clock attachment. Shipping weight, 30 pounds **$31.45**

Ever Ready Automatic Electric Water Supply System

$99.85 Gives You Running Water in Your Home

Why Not Enjoy This Great Convenience

It Starts Itself—It Stops Itself Night or Day It Never Forgets to Pump When Water Is Needed

A High Grade Outfit, Built for Service. The Modern Water Supply System is the foundation of all plumbing conveniences in the home. After you once have a good, reliable water supply system installed, it is a very simple matter to extend the pipes and have running water wherever you want it on your premises.

Our Ever Ready Automatic Electric Water Supply System, illustrated here, will automatically supply running water to your bathroom, kitchen sink, laundry, etc. These outfits may be used for pumping from wells or cisterns not over 20 feet deep. It is a strictly high grade, practical outfit and is built for service. You can extend the piping to your barn for watering stock, if you wish, and have hydrants conveniently placed for lawn or garden sprinkling, fire protection, etc. All of these conveniences are yours as soon as you have one of these modern outfits installed in your home. There is nothing complicated or difficult to understand about this outfit. It is easy to connect up. Any handy man can easily install it by following our simple instructions.

Figuring cost of current at 10 cents per kilowatt hour, it will cost you only 3 cents to pump 180 gallons of water.

Outfit may be placed any distance away from the well or cistern but pump should not be more than 20 feet above lowest water level.

Water in Abundance When and Where You Want It.

A suction pipe is, of course, extended to your well, and discharge pipe from the pump is connected to the water supply pipe in your home. Aside from the occasional supplying of oil, it needs no attention. It starts and stops automatically and will always keep a liberal supply of water under pressure in the tank, and the moment you open a faucet anywhere on the piping system you will get a plentiful supply of running water.

Important—Please Note.

These outfits are furnished for four types of electric current only, as follows: No. 1—110-volt, A. C., single phase, 60-cycle. No. 2—115-volt, A. C., single phase, 60-cycle. No. 3—110-volt, D. C., and No. 4—32-volt, D. C., same as furnished with our farm electric lighting plants. **When ordering be sure to state which is wanted.**

You can get this information from the people who supply you with electric current. We cannot furnish motors for current other than the four types stated above.

Wire from motor can be attached direct to the wires on any electric lighting system having current mentioned above. There are no difficult electrical connections to make and you will not require the services of an experienced electrician or plumber for any part of the work. The outfit is very simple and the installation and operation is all very clearly explained in our special instructions sent with each outfit. Complete outfit includes pump, tank, electric motor, with automatic switch and all necessary valves, gauges and connections. ¾-inch galvanized suction pipe to well or ½-inch supply pipe to plumbing fixtures not included.

Stroke of pump, 2 in.; bore, 1¼ in. Capacity, 180 gallons per hour. Size of motor, ¼ H.-P. Cylinder is brass lined. Piston rod is solid brass. Piston is leather packed. Maximum working pressure, 50 pounds. Range of pressure for which automatic cut-out switch can be adjusted, 15 to 45 lbs.

PRICES FOR EVER READY AUTOMATIC ELECTRIC WATER SUPPLY SYSTEM.

	Diameter, Tank, Inches	Length, Tank, Feet	Total Capacity, Gallons	Working Capacity, Gallons	Shpg. Wt., Lbs.	
Outfits With Black Tanks.						
42L3600¼	18	4	53	42	225	$ 99.85
42L3602¼	20	5	82	65	290	107.85
Outfits With Galvanized Tanks.						
42L3601¼	18	4	53	42	225	$107.95
42L3603¼	20	5	82	65	290	113.25

42L3610⅓—**Pump, switch and motor only,** without tank or connections. Shipping weight, 100 pounds **$69.95**

Steel tank, motor and pump on all above outfits shipped from CENTRAL ILLINOIS, or from our PHILADELPHIA store; gauges and valves shipped from our CHICAGO or PHILADELPHIA store.

We also sell Automatic Electric Pumping Outfits for Deep Wells. Write for Our Special Water Supply Circular 6375GCL.

"Water Boy" Pneumatic Water Supply System.

Easy to Operate.

$49.75 Gives You Running Water in Your Home at Small Expense.

It is a very much improved type of system. In this outfit the physical effort required to pump the water is greatly reduced. The momentum built up in turning the heavy balance wheel makes the pump work smoothly and easily over the load or pumping stroke, so that the tank is rapidly brought up to pressure with surprisingly little effort. The great advantages of this outfit are strikingly apparent. You can place it in your basement or elsewhere and you can locate it any distance away from the house, the well, or the cistern. The outfit will still furnish a plentiful supply of water to your bathroom fixtures on the second floor as well as to the kitchen sink on the first floor, or to any other plumbing fixtures in your home. Tank can be pumped up with ease in ten minutes and pumping once or twice a day will fill all the requirements of the average family. A water pressure as high as 35 pounds may be developed. It is one of the most scientific pumping outfits ever placed on the market for the home. Pump has brass rod and 2-inch brass cylinder; stroke of pump, 3 inches; tank size, 20x60 inches; total capacity, 82 gallons.

42L3630¼—**"Water Boy" Pneumatic Water Supply System,** complete with black tank, all valves, gauges, etc., as illustrated, ready for service. Shipping weight, 325 pounds **$49.75**

42L3631¼—**"Water Boy" Outfit,** same as above, with tank galvanized inside and out. Shipping weight, 325 pounds **56.75**

Septic Tanks for Sewage Disposal.

Purifies Sewage by Bacterial Action.

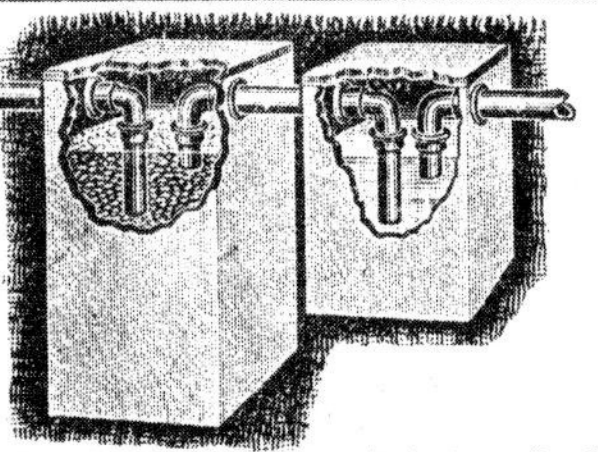

We furnish the two cement tanks or compartments only. Each compartment has a removable cover.

The sewage is discharged into the large tank with ordinary cast iron soil pipe or glazed sewer tile. Tanks should be interconnected at top with sewer pipe. Outlet discharges into a system of ordinary porous field drain tile for final purification. Tanks are made of cement composition, well reinforced with steel rods. They are made in one solid piece with removable covers; no seams, no chance for leakage.

Size, large tank, 24x24x52 inches. Small tank, 24x24x36 inches. Thickness of walls, 1½ inches. Drilled for connections as shown.

CAPACITY. This outfit fills all requirements of average family of about 6 persons. Capacity may be increased by installing a battery of two or more units. Tile disposal system should cover a ground area of about 50x50 feet and should be located not less than 100 feet from house. Shipping weight, both tanks, 1,350 pounds. **Shipped from factory in NORTHERN ILLINOIS.**

Any handy man can easily install it. Full instructions furnished.

42L3560⅓—Two tanks only, for sanitary sewage disposal system as described, with covers; no pipe fittings, tile or pipe included **$29.75**

For cast iron soil pipe and fittings see page 704.

Hand Power Pneumatic Water Supply System.

For Wells or Cisterns Up to 20 Feet Deep.

Vertical Tank Outfits. $52.75

Our Ever Ready Hand Power Water Supply Systems quoted below are equipped with one of the best hand power pumps on the market. Easy working geared head drive, brass cased piston rod, brass air cylinder attachment. Pumps air and water at the same time. Large air chamber, well made and carefully machined. These pumps are up to the highest standard in workmanship, finish and material. Pump will draw water from any well or cistern up to 20 feet deep.

These outfits will force water wherever you wish to extend the pipe and connect a faucet. With one of these outfits you can have all the conveniences of modern plumbing. A sink, bathtub, lavatory or water closet, with hot and cold water always on tap. **The first cost is the only cost.** Fifteen minutes' pumping each day supplies sufficient water for the ordinary household. **Complete** outfit includes one vertical black tank, one double acting geared head force pump with 3-inch brass lined cylinder and automatic air pump attached; also all necessary valves, gauges and connections, etc., as shown. **Suction pipe to well or cistern or water supply pipe to plumbing fixtures not included. Order 1¾-inch galvanized pipe to connect pump with well and ¾-inch galvanized pipe to make connection to plumbing fixtures.**

"Ever Ready" Is Our Own Trade Mark.

Prices and Dimensions—Vertical Tank Outfits

Diam. Tank, In.	Lgth. Tank, Feet	Total Capac., Gallons	Working Capac., Gallons	Shpg. Wt., Lbs.	With Black Tank		With Galvanized Tank	
					Catalog No.	Outfit	Catalog No.	Outfit
24	6	140	105	470	42L3612⅓	$52.75	42L3613⅓	$ 82.65
30	6	230	165	660	42L3614⅓	59.95	42L3615⅓	100.95
36	6	315	215	860	42L3616⅓	72.45	42L3617⅓	117.25

Tanks on all these outfits are shipped from factory in CENTRAL ILLINOIS or from our PHILADELPHIA store; pumps and trimmings from our CHICAGO or PHILADELPHIA store. Shipment will be made from point nearer you.

Any of above outfits can easily be converted into a power outfit for gasoline engine or electric motor drive by simply attaching one of our No. 312 Pump Jacks as listed on page 818.

Modern Bathroom Outfits

$71.35 Buys This High Grade Outfit

Why be without this great convenience in your home? It is not difficult to install a modern bathroom outfit. By following our simple plans and instructions you can install your own plumbing material and make a big saving. Send us a sketch showing how you would like to locate the plumbing fixtures and let us send you an estimate on a complete modern plumbing system.

Our prices are remarkably low. Remember all our plumbing material is strictly first quality. We do not sell seconds or "B" grade plumbing fixtures of any kind.

Easily Installed in the Home

Our Complete Plumbing Systems Also Sold on Easy Payment Terms. Write for Our Estimate.

The prices given in this catalog are our cash prices. There is an advantage, of course, in paying cash for your material. However, if it is more convenient for you to pay for your plumbing system on monthly payments, provided you own the property in which the material is to be installed, we will make special arrangements so that you can purchase your material on easy payment terms.

By taking advantage of this offer you need not wait until you have accumulated sufficient funds to pay for the material in full. You can make this great improvement in your home right now and begin at once to enjoy this great comfort and convenience while paying for the material in moderate monthly payments easily within your means.

Fairview Bathroom Outfit.

A very popular combination. Modern, of neat appearance and durable. Suitable for the finest residence. A good serviceable outfit at a moderate price. You can easily install it yourself. Why not enjoy this great comfort in your home now?

BATHTUB is 5 feet long, 30 inches wide over rim at top and 16½ inches deep, sufficient to prevent water from splashing over sides. Height from floor to top of tub, 22 inches. Made of cast iron, coated inside with white porcelain enamel, painted red outside and has 3-inch roll rim. Standard Fuller pattern bath cock, 1⅜-inch connected waste and overflow and two ⅜-inch offset supply pipes to floor, all of brass, nickel plated and polished.

LAVATORY is 18 inches from front to back, 21 inches wide and has 8-inch high back and roll or turnover rim 4 inches deep. Soap cup cast in top directly above overflow. 1¼ inch trap with outlet to wall, two compression faucets with china tops, one marked "hot," the other "cold," and two ⅜-inch supply pipes to wall. All faucets, traps, supply pipes, etc., made of brass, in the latest designs, nicely nickel plated.

CLOSET TANK—White vitreous china, artistically designed. A beautiful sanitary closet outfit that is easily kept clean and white, and will add greatly to the appearance of your bathroom. **Closet shipped from our store.** Shipping weight, 130 pounds. **Bathtub and lavatory shipped from LAYTON PARK, WIS.** Shipping weight, 395 pounds.

42L3664¼—Fairview Bathroom Outfit (bathtub, lavatory and closet), with supply and waste pipes to wall, as illustrated. **$71.35**

42L3665¼—Fairview Bathroom Outfit (bathtub, lavatory and closet), with supply and waste pipes to floor. **$71.80**

If wanted with waste and supply pipes threaded for iron pipe, allow $1.50 extra.

Prices for above outfits do not include paper holders, towel bars, etc. **For Bathroom Trimmings see page 934.**

Iroquois Bathroom Outfit.

White vitreous china closet tank, Siphon jet noiseless closet bowl and large square lavatory are special features of this outfit.

THE BATHTUB—Same as furnished with Fairview Bathroom Outfit described above, complete with all trimmings to floor.

LAVATORY is 18 inches from front to back, 24 inches wide and has a 10-inch high back and deep apron. Made of cast iron with soap cup cast in top. Inside of bowl, top, back and outside of apron glazed with white porcelain enamel, 1¼-inch trap with outlet to wall, two compression cocks with china tops, marked "hot" and "cold," and two ⅜-inch air chamber supply pipes to wall. All trimmings of brass, nicely nickel plated.

CLOSET has siphon jet bowl and chinaware tank. Seat of seasoned birchwood, highly polished mahogany finish, with nickel plated brass hinges. China handle flush lever, nickel plated ⅜-inch supply pipe to floor.

Closet shipped from our store. Shipping weight, 130 pounds. **Bathtub and lavatory shipped from LAYTON PARK, WIS.** Shipping weight, 410 pounds.

42L3672¼—Iroquois Bathroom Outfit (bathtub, lavatory and closet), with supply and waste pipes of lavatory to wall as illustrated. **$83.10**

42L3673¼—Iroquois Bathroom Outfit (bathtub lavatory and closet), with waste and supply pipes of lavatory to the floor. **$83.60**

If wanted with waste and supply pipes threaded for connecting to iron pipe, allow $1.50 extra.

Prices do not include paper holder, towel bar, stool, bath seat, etc. **For Bathroom Trimmings see page 934.**

Chippendale Bathroom Outfit.

Here is an outfit at a moderate price that would be appropriate for the finest mansion. What a pride it will be for you to have this beautiful outfit installed in your home.

BATHTUB is 5 feet long, 30 inches wide over rim at top and about 17 inches deep inside. Cast iron, with a 3-inch roll rim and base, all cast in one piece. Tub is white porcelain enameled inside and painted red outside. Has Fuller bath cock, with china handles, marked "hot" and "cold," 1½-inch Model waste with china top marked "waste," and two ½-inch supply pipes, all brass, nickel plated.

LAVATORY is cast iron white porcelain enameled, 20 inches from front to back and 24 inches wide. 1¼-inch trap with outlet to wall; Model waste; two Fuller pattern cocks with china handles, marked "hot" and "cold," and two ⅜-inch bottle air chamber supply pipes to wall. All trimmings nickel plated.

CLOSET has siphon jet bowl of vitreous earthenware and chinaware tank. Seat is of solid birchwood with highly polished ivory white finish, fitted with nickel plated brass hinges. China handle flush lever, nickel plated; ⅜-in. supply tank from tank to floor.

Closet shipped from our store. Shipping weight, 130 pounds. **Bathtub and lavatory shipped from LAYTON PARK, WIS.** Shipping weight, 670 pounds.

42L3668¼—Chippendale Bathroom Outfit (bathtub, lavatory and closet), with supply and waste pipes of lavatory to wall, as illustrated. **$107.65**

42L3669¼—Chippendale Bathroom Outfit (bathtub, lavatory and closet), with supply and waste pipes of lavatory to floor. **$108.25**

If wanted with waste and supply pipes threaded for iron pipe, allow $1.50 extra.

Prices of outfits above do not include paper holder, towel bar, bath stool, etc. **For Bathroom Trimmings see page 934.**

Bathtub Bargains

Look Over These Rock Bottom Prices

Improve Your Home Now!

4½ ft Size
$22.95
Without Fittings

All Bathtubs on This Page Are Strictly First Quality. We Do Not Sell "Seconds" or "B Grade" Plumbing Fixtures of Any Kind.

4½ ft Size
$22.90
Without Fittings

Brookside Cast Iron Roll Rim Bathtub With Legs.

Heavy Cast Iron One-Piece Bathtub, detachable cast iron legs, 3-inch roll rim. Heavily coated inside and over rim with white porcelain enamel. Outside painted one coat of iron filler paint. Furnished with or without trimmings as quoted below. Trimmings include Fuller bath cock, connected waste and overflow and supply pipes to floor, all of nickel plated brass. Width, over rim, 30 inches. Depth, 16 inches. Height, 22 inches. **Shipped from LAYTON PARK, WIS.**

Length, Feet	Shipping Weight, Pounds	42L3680⅓ Enameled Inside, Painted Outside, Complete	42L3681⅓ Same as 42L3680⅓ With Legs But No Trimmings
4½	280	$31.45	$22.95
5	340	31.90	23.95
5½	350	34.75	26.95
6	380	39.95	33.60

NOTE—Trimmings are necessary to connect to pipes. **If wanted for iron pipe connection, add 75 cents to price of tub with trimmings.**

Sheet Steel Bathtub.

Made of sheet steel, painted with paint enamel. Inside white, outside blue. Has 3-inch wood rim, varnished. Nickel plated connected waste and overflow threaded for 1-inch iron pipe. Drilled for bath cock, but bath cock or supply pipes not included at prices quoted. **Shipped from DETROIT, MICH.** Shipping wt., 90 lbs.

	42L3706⅓	42L3707⅓	42L3708⅓	42L3709⅓
Size	4 ft. 6 in.	5 ft.	5 ft. 6 in.	6 ft.
Each	$10.30	$10.70	$11.10	$11.70

Irving Bathtub for Small Bathrooms.

Only 26 Inches Wide Over All.

Same as 42L3680⅓, except narrower width, for small bathrooms where space is limited. Width over all, 26 inches. Width of rim, 2 inches. Trimmings include Fuller bathtub faucet, supply pipes to floor and connected waste and overflow, all brass, nickel plated. **Shipped from LAYTON PARK, WIS.**

L'gth, Feet	Shpg. Wt., Lbs.	42L3692⅓ Complete With Trimmings	42L3693⅓ With Legs, No Trimmings
4½	270	$31.40	$22.90
5	300	31.85	23.90
5½	330	34.70	26.90

For iron pipe connection add 75¢ to price of tub with trimmings.

For Folding Bathtubs See Page 702.

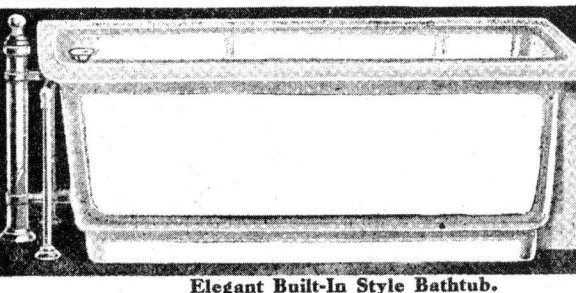

4½ ft Size
$54.35
Complete

4½ ft Size
$41.65
Complete

Elegant Built-In Style Bathtub.

This tub is cast iron with three coats of porcelain enamel inside and outside. Artistic and sanitary. Has nickel plated brass model waste and overflow with a china lifting knob, nickel plated brass supply pipes with ball offset connections and a nickel plated Fuller bath faucet with china handles. This tub sets up against the back and end walls and fits tight to the floor. It is therefore strictly sanitary and easily kept clean. Dirt or other foreign substances cannot collect at the end, beneath or behind the tub. Furnished for either right or left hand corner. **Illustration shows tub for right hand corner. State which is wanted. Shipped from LAYTON PARK, WIS. For iron pipe connection add 75 cents.**

Length of tub, feet	4½	5	5½
Shipping weight, pounds	390	450	500
42L3694⅓—For Right Hand Corner	$54.35	$56.70	$60.60
42L3695⅓—For Left Hand Corner	54.35	56.70	60.60

Marquette Bathtub. Stylish Design With Base.

Tub and base are cast in one solid piece. White porcelain enameled inside, painted one coat outside. Also furnished with white porcelain enameled finish inside and outside. Fitted complete with nickel plated china lifting waste, nickel plated Fuller bath cock and nickel plated supply pipes to floor, as illustrated. Dust or other foreign substances cannot collect beneath this tub. A beautiful, sanitary fixture.

Bathtub, complete as described, porcelain enameled inside, painted outside.

Length, feet	4½	5	5½
Shpg. wt., lbs.	400	460	510
42L3696⅓	$41.65	$41.75	$45.20

Bathtub, complete as described, porcelain enameled inside and outside.

Length, feet	4½	5	5½
Shpg. wt., lbs.	400	460	510
42L3697⅓	$55.90	$58.55	$62.55

For iron pipe connection add 75 cents. Shipped from LAYTON PARK, WIS.

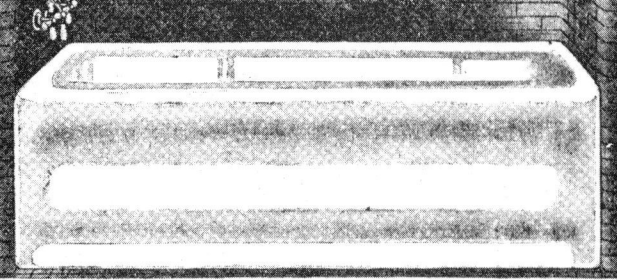

Venetian Built-In Corner Bathtub.

5-ft. Size,
$85.50
Complete

This beautiful massive cast iron white porcelain enameled corner bathtub is truly a masterpiece of the designer's art. It is the last word in sanitary bathroom fixture design. The entire tub is coated both inside and out with highest quality genuine white porcelain enamel. Tub is furnished complete with latest pattern solid brass nickel plated fixtures, standing waste and supply pipes with individual shut off cocks and china handles. **Shipped from LAYTON PARK, WIS.** Furnished for right or left hand corner. Illustration shows tub for right hand corner. **State which is wanted. For iron pipe connecton add 75 cents.**

42L3700⅓—For Left Hand Corner. Size, 5 feet. Shipping wt., 470 lbs. **$85.50**
42L3702⅓—For Right Hand Corner. Size, 5 feet. Shipping wt., 470 lbs. **85.50**

Alhambra Recess Built-In Bathtub.

5-ft. Size,
$85.00
Complete

This elegant tub is an exact counterpart of our Venetian Built-In-Tub described at left, except that it is designed to be built in at both ends and one side as shown. The waste and supply fittings are concealed in the wall and the controlling knobs only protrude.

This style of tub is now in great popular favor and is to be found in the most modern and fashionable residences in the country. Tub is furnished complete with concealed fixtures, nickel plated wall plates and controlling knobs with china handles for setting in wall as shown.

42L3704⅓—Size, 5 feet. Shipping weight, 450 pounds **$85.00**

For iron pipe connection add 75 cents.
Shipped from LAYTON PARK, WIS.

Handy Portable Bathtubs and Heaters

No Pipes. No Sewers. No Running Water. No Plumbing Work

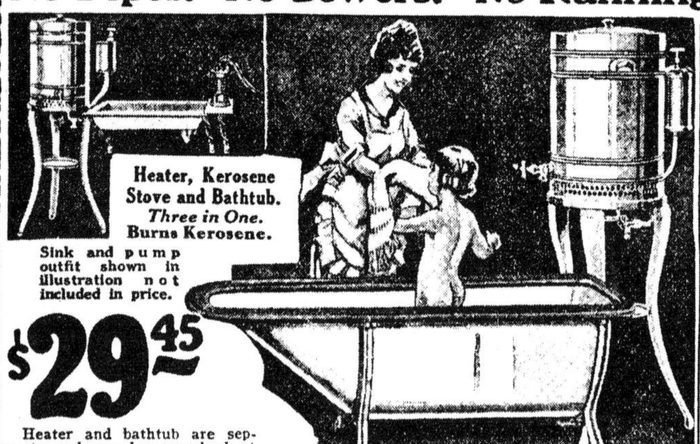

Heater, Kerosene Stove and Bathtub. *Three in One.* **Burns Kerosene.**

Sink and pump outfit shown in illustration not included in price.

$29.45

Heater and bathtub are separate and can be moved about independently. You can use the heater for heating water, cooking, etc., while the bathtub is not in use. Both the heater and tub are light and easily handled. When not in use tub can easily be lifted up and stood on end in some closet or corner where it will be out of the way, and the heater can be moved into the kitchen and placed alongside of the sink, as shown in the illustration at the left, and you will always have hot water for cooking, washing dishes, etc., by just opening the faucet on the tank. During hot summer days you can remove the tank and use the stove part of heater as a kerosene stove for cooking, frying, heating irons, etc., so that you will not have to keep a hot coal or wood stove burning in your kitchen.

An Ideal Outfit for the Summer Cottage.

This outfit fulfills three big needs in every home, as shown by the illustrations. Has 6-foot rubber hose to let water run outdoors. Tub is galvanized sheet steel nicely painted, 5½ feet long and 28 inches wide with varnished oak rim. Tank holds 12 gallons. It is made of galvanized sheet steel and is covered with bright nickel plated sheet metal, giving it a beautiful finish. Entire outfit is exceptionally well made and elegantly finished throughout. Just the thing for the country home which has no running water. As a vacation outfit for the summer cottage it can't be beat. Shipping weight, 185 pounds. **Shipped from factory at DETROIT, MICH.**

42L3711⅓—Outfit complete as described.............$29.45

No Pipes or Plumbing Work Necessary.

$36.40

You will be impressed with the appearance, fine workmanship and practical utility of this outfit. Just fill the heater tank with water and light the burner. Water will be heated sufficiently for bath in thirty minutes. The water runs right into the tub by simply opening the faucet. To drain the water out of the tub, a 6-foot length of hose is provided which you can attach to the water outlet and let the water run outdoors. Water heater holds 12 gallons.

Tub is 5 feet long and 30 inches wide. It is made of galvanized sheet steel. Entire tub is heavily coated with paint enamel, giving the outfit a handsome and sanitary appearance. The upright stand is also white enameled to match the tub. Hardwood rim around entire top of tub, white enameled painted to match. Heater has galvanized steel inner casing covered with an outer casing of light nickel plated sheet steel, highly polished.

Light, Portable, Easily Handled.

The entire outfit is on roller bearing rollers and can be pushed out of the way into some convenient storeroom or clothes closet. Water circulates through a copper tube in the heater directly over the burner, so that it heats quickly. Heater may be provided with gasoline or kerosene oil burner, as desired. **State which is wanted.** This outfit fulfills a real necessity in every rural home where there is no modern plumbing installed. Shipping weight, 165 pounds. **Shipped from factory at DETROIT, MICH.**

42L3710⅓—Outfit complete as described.............$36.40

Sheet Steel Sinks Will Not Break.

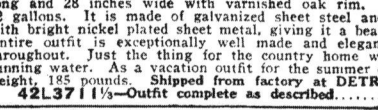

$2.10 and Up

Sheet Steel Sinks with flat rim. Made from one piece of steel. No seams. Strong, light and durable. Fitted for 1½-inch lead pipe. **If wanted for 1½-inch iron pipe connection, add 25c.**

Size, over all, inches	16x24	18x30	20x30
Shipping weight, pounds	10	14½	16½
42L1630¼—Painted	**$2.10**	**$2.70**	**$2.98**
42L1632¼—Galvanized	**2.35**	**2.95**	**3.48**

Roll Rim Sheet Steel Sinks.

$2.35 and Up

Size, over all, inches	16x24	18x30	20x30
Shipping weight, pounds	20	28	32
42L1634¼—Painted	**$2.35**	**$2.80**	**$3.48**
42L1636¼—Galvanized	**2.58**	**3.48**	**3.95**

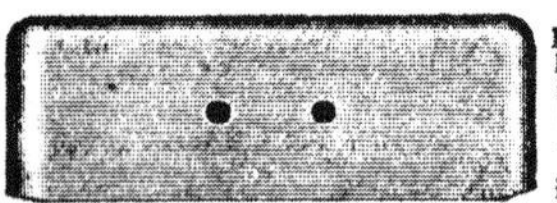

42L1644¼

Painted Steel Sink Back—Length, 24 in., 15 in. high. Shipping weight, 12 lbs. ....**$1.78** Length, 30 in., 15 in. high. Shipping wt., 16 lbs. **$2.10**

42L1646¼—Galvanized Steel Sink Back.

Length, 24 inches, 15 inches high................**$2.15**

Length, 30 inches, 15 inches high................**2.35**

Drain Boards.

42L1910¼

Reversible cast iron enameled drain board. With iron bracket. For roll rim sinks.

Size, 18x20 inches. Shipping weight, 44 pounds **$5.25**

Size, 18x24 inches. Shipping weight, 48 lbs. **$5.50**

Size, 20x20 inches. Shipping weight, 45 lbs. **$5.80**

Size, 20x24 inches. Shipping weight, 54 lbs. **$5.90**

Sink Brackets.

Steel Sink Brackets.

Can be used with any sink on this page. Very neat in appearance. Shipping weight, 3 pounds per pair.

42L350—No. 1 Plain, for 16 and 18-in. sinks. Per pair............**32c**

42L351—No. 2 Plain, for 20-inch sinks and larger. Per pair.......**45c**

Genuine White Porcelain Enameled Flat Rim Kitchen Sinks.

$4.75 and Up

Cast Iron Kitchen Sinks.

The Old Reliable Cast Iron Flat Rim Sink. Furnished with genuine white porcelain enamel finish inside and painted finish outside; also with painted finish inside and outside. Both styles quoted below. Regularly furnished for lead pipe connection. **Threaded for 1¼ or 1½-inch iron pipe, 25c extra.**

Size, over all	16x24	18x30	18x36	20x30	20x36	20x40
Shpg. wt., lbs.	40	48	55	56	68	80
42L1479¼—Painted inside and out	**$3.15**	**$3.55**	**$4.20**	**$3.95**	**$4.75**	**$ 5.45**
42L1480¼—Porcelain enameled inside, painted outside	**4.75**	**4.90**	**7.35**	**5.65**	**7.40**	**10.75**

Kitchen Sink Outfit With Cistern Pump.

Includes cast iron flat rim sink, porcelain enameled inside, 3-inch pitcher spout pump with iron cylinder, 1½-inch cast iron sink trap to floor or wall fitted for iron pipe connections, pump board, three brackets. Pump threaded for 1¼-inch iron suction pipe. Waste pipe or suction pipe not included.

State if trap is wanted to floor or to wall.

42L2029¼—Outfit complete as described.

Size of sink, inches	18x30	18x36	20x30	20x36	20x40
Shpg. wt., pounds	95	105	97	110	115
With trap to wall.	**$8.10**	**$ 9.10**	**$8.80**	**$10.30**	**$12.70**
With trap to floor.	**9.30**	**10.30**	**9.95**	**11.50**	**13.90**

If pump with brass lined cylinder is wanted instead of iron, add 60c to above prices.

Sinks, Backs and Brackets.

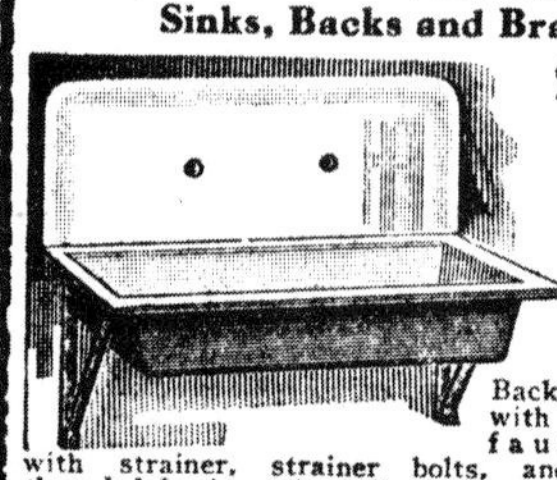

$8.95 and Up Complete

Cast Iron Flat Rim Sink, white enameled inside, enameled back, steel brackets. Back, 12 inches high with holes for two faucets. Furnished with strainer, strainer bolts, and trap coupling threaded for iron pipe. Faucets or trap not included.

42L2034¼

Size, inches	18x30	18x36	20x30	20x36	20x40
Shpg. wt., lbs.	88	100	88	101	110
	$8.95	**$12.65**	**$9.85**	**$12.80**	**$17.20**

Above with waste pipe threaded for iron pipe, 25c extra.

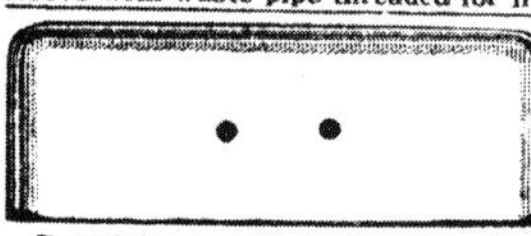

Cast Iron Sink Backs.

$3.50 and Up

Porcelain enameled, for flat rim sinks, 12 inches high and 2¼ inches deep.

42L1481¼

Length, inches	24	30	36	40
Shipping weight, lbs.	30	38	45	50
	$3.50	**$3.75**	**$4.95**	**$6.00**

Perfection Kitchen Sink and Pump Outfit.

18x30 Size. $13.10

Water may be drawn from the pump spout, forced into an attic tank or used for sprinkling, etc.

State if trap is wanted to floor or to wall.

Includes cast iron flat rim sink, porcelain enameled inside, 3-inch brass body cistern force pump with cock spout threaded for hose connection, three sink brackets, oak pump board, 1½-inch cast iron trap, fitted for iron pipe connection. Pump has 1¼-inch suction and 1-inch discharge for iron pipe connection. No pipe included.

42L2031¼

Size of sink, in.	18x30	18x36	20x30	20x36	20x40
Shpg. wt., lbs.	92	96	97	101	111
With trap to floor.	**$13.10**	**$15.55**	**$14.05**	**$15.80**	**$19.15**
With trap to wall.	**11.90**	**14.35**	**12.85**	**14.60**	**17.95**

Complete Plumbing Systems

Sold for Cash or on Easy Payments

Write for Our Estimate

The prices given in this catalog are our cash prices. There is an advantage, of course, in paying cash for your material. Plumbing fixtures such as, sinks, bathtubs, closets, lavatories, range boilers, gas water heaters, laundry tubs, etc., are not sold separately on easy payments. However, if you wish to purchase a complete plumbing system and it is more convenient for you to pay for your system on monthly payments, provided you own the property in which the material is to be installed, we will make special arrangements so that you can purchase your material on easy payment terms.

10 Months to Pay

By taking advantage of this offer you need not wait until you have accumulated sufficient funds to pay for the material in full. You can make this great improvement in your home right now and begin at once to enjoy this great comfort and convenience while paying for the material in moderate monthly payments easily within your means. You can take ten months to pay.

If you will send us a sketch of your building and advise us just what plumbing fixtures you would like to install, we will prepare a special estimate for you and will advise you just what terms we will make. Install your own plumbing system and save money.

Simplex Range Boiler Covering.

Saves gas or coal, pays for itself in a very short time. Water heats up quicker and holds the heat. Made of two layers of canvas with hair felt insulation between. Made to fit boiler, sizes as quoted. Has two rows of brass lacing hooks. Just wrap it around boiler and lace up the seam. Lacing furnished. Range boiler not included.

42L1810—For Boiler, capacity, 30 gallons; height, 5 feet; diameter, 12 inches. Shipping wt., 6 pounds........**$6.35**

42L1811—For Boiler, capacity, 40 gallons; height, 5 feet; diameter, 14 inches. Shpg. wt., 7 lbs....**$6.85**

42L1812—For Boiler, capacity, 52 gallons; height, 5 feet; diameter, 16 inches. Shpg. wt., 8 lbs.......**$7.45**

Hercules Kerosene Heater.

42L1834¼—Will heat about 10 gallons of water per hour to a temperature suitable for ordinary household purposes. Will burn twelve hours on one gallon of kerosene. An inexpensive means of supplying hot water in the country home having running water but no city gas. Four spiral copper coils, cast iron jacket. Tapped for ¾-inch iron pipe. Removable reservoir holds 1 gallon of kerosene and is fitted with glass tube showing height of oil. Shipping weight, 47 pounds **$20.95**

Instantaneous Water Heater.

State if wanted for natural or artificial gas.

42L1816¼ Instantaneous Gas Water Heater. Connected direct to city pressure. No boiler necessary. Water flows out of side of heater, but water cannot be forced to fixture above heater. Copper, nickel plated. Diameter, 11 inches; height, 30 inches; heats 1½ to 2 gallons per minute to a temperature suitable for all household purposes. ½-inch water pipe and ½-inch gas pipe. Shipping weight, 65 pounds...**$32.75**

Kerosene Water Heater.

Just the Thing Where Neither City Gas Nor Water Pressure Are Available.

42L1836¼—Kerosene Water Heater. Made of galvanized iron, nickel plated outside. Reservoir holds about 9 gallons of water. Draft chimney is entirely surrounded by water. This makes the water heat quicker. Will heat enough water for bath in fifteen to twenty minutes. Shipping wt., 28 pounds..........**$10.98**

Above cannot be connected with water pressure. No range boiler is required.

Laundry Tub Faucet.

42L303—Combination laundry tub faucet and mixing valve. Made of brass. By turning handles marked hot and cold you can regulate temperature of water to exact degree you want. Spout swings to right and left so that two tubs may be supplied with hot and cold water from one faucet. Spout threaded for hose connection. Clamps over back of tub, saves drilling holes for pipes. Threaded for ½-inch iron pipe. Only two connections necessary. Shipping weight, 10 pounds...**$3.50**

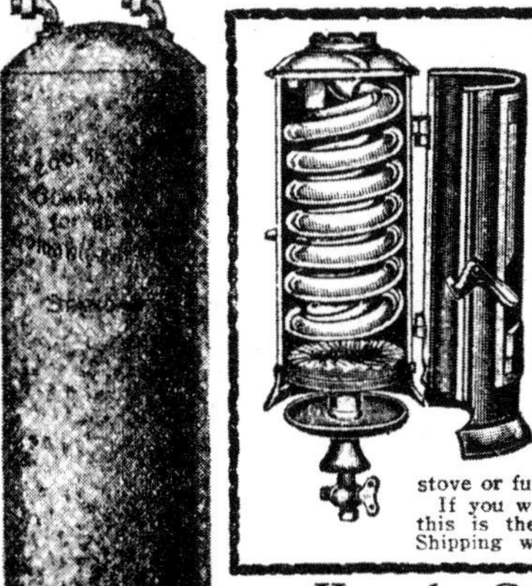

$10.25 and up Complete

Double Copper Coil Large Capacity Gas Water Heater

$6.25

42L1824¼ Large Capacity Double Copper Coil Gas Water Heater. Heats 30 gallons of water to temperature suitable for ordinary household purposes in 40 minutes at a gas consumption of from 50 to 60 cubic feet per hour. Has two coils of ¾-inch copper tubing as shown, cast iron jacket with door, ¾-inch water connection and ½-inch gas connection. Iron pipe threads. One of the most efficient gas heaters on the market. Being constructed entirely of cast iron and copper, it is very durable.

In summer, when you are not using your stove or furnace, one of these heaters is a necessity.

If you want plenty of hot water and want it quickly, this is the heater to buy. Artificial or natural gas. Shipping weight, 28 pounds..........**$6.25**

Hercules Galvanized Range Boilers.

Our Hercules Range Boilers are well made of steel plate heavily galvanized inside and out, and are fully tested and guaranteed for pressures stated.

Prices quoted below on vertical boilers complete include stand, tube and couplings. Prices on horizontal boilers complete include tube and couplings. Stand not required with horizontal boilers. Prices quoted for range boilers only do not include anything except range boilers.

Shipping weights given are for boilers complete. If boiler only is ordered, deduct 10 pounds.

Couplings furnished with complete boilers are for lead pipe; if couplings for iron pipe connections are wanted, add **20 cents** to prices of boilers complete.

Standard Weight Range Boilers.

Tested to 200 pounds and guaranteed for 85 pounds working pressure.

Approximate capacity, gallons	30	40	52
Size, inches	12x60	14x60	16x60
Shipping weight, pounds	90	100	125
42L2036¼—Vertical Range Boilers, complete	**$10.25**	**$12.45**	**$20.85**
42L2037¼—Horizontal Range Boilers, complete	**10.35**	**12.55**	**20.95**
42L1802¼—Vertical Range Boilers only	**8.25**	**10.35**	**17.20**
42L1803¼—Horizontal Range Boilers only	**8.35**	**10.45**	**17.30**

Extra Heavy Range Boilers.

Tested to 250 lbs. pressure and guaranteed for working pressure up to 150 lbs.

Approximate capacity, gallons	30	40	52
Size, inches	12x60	14x60	16x60
Shipping weight, pounds	95	110	145
42L2038¼—Vertical Range Boilers, complete	**$11.45**	**$14.25**	**$24.45**
42L2039¼—Horizontal Range Boilers, complete	**11.55**	**14.35**	**24.55**
42L1804¼—Vertical Range Boilers only	**9.65**	**12.25**	**20.95**
42L1805¼—Horizontal Range Boilers only	**9.75**	**12.35**	**21.25**

Single Coil Gas Water Heater.

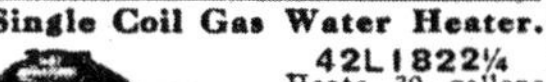

42L1822¼ Heats 30 gallons in one hour. Cast iron jacket. Note the conical shape of the copper coil, which adds greatly to the efficiency of the heater. Very efficient and satisfactory where a moderate amount of hot water is required. If a great amount of hot water is used, we recommend Heater 42L1824¼, described at the left. ¾-inch water pipe, ½-inch gas pipe connections. Artificial or natural gas. Shipping weight, 24 pounds....**$5.60**

Hercules Hot Water Marvel.

Range Boiler and Water Heater Combined.

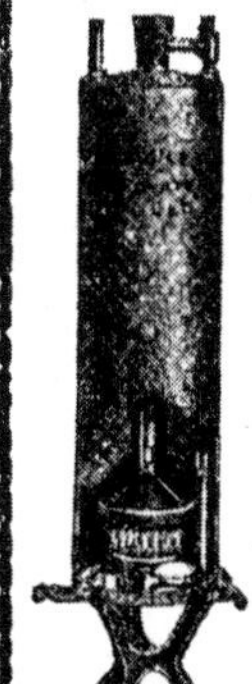

Illustrated at left. Burns gas. Has gas water heater built right into boiler as shown. Heater is connected by circulation pipe to top and bottom of boiler. Water circulates constantly from bottom to top of boiler through heater. Heat travels up through central flue entirely surrounded by water. No waste of heat. One of the quickest and most economical water heaters on the market. Entirely self contained, takes up little space in kitchen. Eliminates extra labor and expense in connecting up heater and boiler separately. Simple, safe, efficient and clean. May also be connected to coil in range, stove or furnace, if desired. **For natural or artificial gas. State which is wanted.**

42L2040¼—Size, 30 gallons. Shipping weight, 140 pounds.......**$25.85**

42L2041¼—Size, 40 gallons. Shipping weight, 150 pounds **$32.50**

Stoneware Laundry Tubs

These stationary laundry tubs do away with all lifting and carrying of water, spilling and splashing, etc. You can have an abundant supply of running hot and cold water right at the tubs by simply opening the faucets.

Molded in one solid piece from a special granitine composition material. No seams to leak or catch grease or dirt. Tubs have metal rims and guards. Ornamental cast iron legs. Tubs measure 24 inches from front to back and are 14 inches deep. **Prices quoted on laundry tubs on this page do not include faucets or traps.** Two-part tubs require four faucets and one trap. Three-part tubs require six faucets and one trap. Two-part tubs have one waste outlet. Three-part tubs have two waste outlets. These may be connected together so that only one trap is necessary. Use Drum Trap 42L436, shown on page 704. For laundry tub faucets, see page 706. Also see special laundry tub faucet shown on this page at left. Shipped from factory in PHILADELPHIA, PENNA., ST. PAUL, MINN., CINCINNATI, OHIO, or CHICAGO, ILL.

For Laundry Stoves and Coal Burning Water Heaters See Pages 684 and 685.

Two-Compartment Tubs.

	42L3873⅓	42L3874⅓	42L3875⅓
Length, over all, inches	48	54	60
Size, each tub, over all, in.	24x24x15½	24x27x15½	24x30x15½
Shipping weight, pounds	380	400	450
	$11.48	**$13.10**	**$14.60**

If waste connection is wanted for iron pipe, add **25 cents.**

Two-Compartment Tubs With High Backs.

All Laundry Tubs quoted on this page are shipped from factory in Philadelphia, Penna., St. Paul, Minn., Cincinnati, Ohio, or Chicago, Ill. Shipment will be made from point nearest you.

	42L3870⅓	42L3871⅓	42L3872⅓
Length over all, inches	48	54	60
Size of each tub over all, inches	24x24x15½	24x27x15½	24x30x15½
Shipping weight, lbs.	450	500	550
	$15.65	**$18.25**	**$20.95**

If waste outlet on above tubs is wanted for iron pipe connections, add **25 cents.**

Three-Compartment Tub. (Not illustrated.) Same as 42L3873⅓ Above, Except With Three Compartments. Made in one size only, 72 inches long. Each tub measures 24x24x15½ inches over all. The washing can be done in one tub, the clothes rinsed in the middle tub and the third tub used for bluing. From this last tub the clothes are wrung out into the basket. This greater convenience is well worth extra cost. If wanted for iron pipe connections, add 50 cents.

42L3879⅓—Shipping weight, 550 pounds........**$18.25**

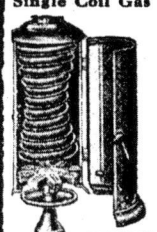

Cast Iron Soil Pipe and Fittings

Quarter Bends. For Soil Pipe.

42L1514¼

Standard. 2-inch. Shipping weight, 3½ pounds 33c

Standard. 4-inch. Shipping weight, 8 pounds 60c

42L1515¼

Extra Heavy. 2-inch. Shipping weight, 6½ pounds 46c

Extra Heavy. 4-inch. Shipping weight, 16 pounds 73c

Quarter Bends. With Heel Inlet.

42L1586¼

Standard.

Size, In.	Shpg. Wt., Lbs.	
2x2	6	$1.04
4x2	12	1.05

42L1587¼

Extra Heavy.

Size, In.	Shpg. Wt., Lbs.	
2x2	8	$1.27
4x2	18	1.32

42L404

Clean Out Ferrule. Cast iron body, brass screw cover. Placed in soil pipe for clean out.

Size, In.	Shpg. Wt., Lbs.	
2	1½	37c
4	2½	60c

Clean Out Y With Brass Screw Cover.

Placed in pipe so that it may be cleaned out through opening at side.

42L1594¼

Standard. Size, 2 inches. Shipping weight, 5 pounds....$1.15

Standard. Size, 4 inches. Shipping weight, 15 pounds....$1.98

42L1595¼

Extra Heavy. Size, 2 inches. Shipping wt., 7 lbs...$1.55

Extra Heavy. Size, 4 inches. Shipping wt., 18 lbs......$2.55

Eighth Bends.

42L1512¼

Standard.

Size, In.	Shpg. Wt., Lbs.	
2	3	29c
4	7	53c

42L1513¼

Extra Heavy.

Size, In.	Shpg. Wt., Lbs.	
2	6	33c
4	11	66c

Soil Pipe Offsets.

For offsetting soil pipe stack over foundation wall of building.

42L1572¼

Standard.

Size, In.	Shpg. Wt., Lbs.	
2x12	8	$0.95
4x12	18	1.32

42L1573¼

Extra Heavy.

Size, In.	Shpg. Wt., Lbs.	
2x12	14	$1.15
4x12	24	1.65

The cast iron soil pipe and fittings shown on this page are of excellent quality. The joints of this pipe are easily made up with calking lead and oakum. Full instructions for connecting pipe and much other valuable plumbing information will be found in our **Special Plumbing Instruction Book 7777GCL, sent postpaid on request. All prices on soil pipe and soil pipe fittings are subject to market changes.**

Single Hub Soil Pipe.

Sold in 5-ft. lengths only.

42L1500¼—Standard. 2-inch size. Shipping weight, 17½ pounds. Per 5-ft. length $1.00

42L1502¼—Standard. 4-inch size. Shipping wt., 35 lbs. Per 5-foot length $1.75

42L1501¼ — Extra Heavy. 2-inch size. Shipping weight, 27 pounds. Per 5-ft. length $1.35

42L1503¼ — Extra Heavy. 4-inch size. Shipping wt., 61 lbs. Per 5-foot length....$2.75

Double Hub Soil Pipe.

Sold in 5-ft. lengths only.

42L1504¼—Standard. 2-inch size. Shipping weight, 22 pounds. Per 5-ft. length $1.05

42L1506¼—Standard. 4-inch size. Shipping weight, 38½ lbs. Per 5-foot length......$1.80

42L1505¼—Extra Heavy. 2-inch size. Shipping wt., 29 lbs. Per 5-foot length $1.40

42L1507¼—Extra Heavy. 4-inch size. Shipping wt., 65 lbs. Per 5-foot length $2.90

Prices Subject to Market Changes.

Soil Pipe Rests.

Used to brace soil pipe stack firmly in wall and to keep it from slipping down.

42L456—Standard.

Size, In.	Shpg. Wt., Lbs.	
2	2¼	24c
4	3	33c

42L457—Extra Heavy.

Size, In.	Shpg. Wt., Lbs.	
2	3½	33c
4	5¾	36c

Y Branches.

42L1564¼

Standard.

Size, In.	Shipping Weight	
2x2	7 lbs.	45c
4x2	12½ lbs.	70c
4x4	16½ lbs.	95c

42L1565¼

Extra Heavy.

Size, In.	Shpg. Weight	
2x2	10 lbs.	$0.62
4x2	16 lbs.	.90
4x4	23 lbs.	1.15

Sanitary T Branch.

Used to connect closet to soil pipe stack.

42L1528¼

Standard.

Size, In.	Shipping Weight	
2x2	7 lbs.	44c
4x2	12 lbs.	70c
4x4	15 lbs.	95c

42L1529¼

Extra Heavy.

Size, In.	Shpg. Weight	
2x2	9 lbs.	$0.62
4x2	16 lbs.	.91
4x4	23 lbs.	1.15

T Branch Tapped for Iron Pipe.

42L1560¼

Standard.

Size, In.	Shipping Weight	
2x1½	5 lbs.	58c
4x1½	10 lbs.	82c
4x2	10 lbs.	86c

42L1561¼

Extra Heavy.

Size, In.	Shipping Weight	
2x1½	7 lbs.	$0.70
4x1½	12 lbs.	1.05
4x2	12 lbs.	1.05

Sanitary Tapped T Branch.

42L1548¼

Standard.

Size, In.	Shipping Weight	
2x1½	5½ lbs.	$0.82
4x1½	11 lbs.	1.05
4x2	11 lbs.	1.05

42L1549¼

Extra Heavy.

Size, In.	Shipping Weight	
2x1½	8 lbs.	$0.74
4x1½	13 lbs.	1.27
4x2	13 lbs.	1.32

Sanitary T With Side Inlet.

Used in stack for waste of closet, with side opening for waste of lavatory and bathtub. Size, 4 inches. Either right or left side inlet. Illustration shows left hand side inlet. **State which is wanted.** Shpg. wts., standard, 22 lbs.; extra heavy, 24 pounds.

42L1532¼

Standard.

With 2-in. side inlet for soil pipe $1.80

With 1½-in. tapped side inlet for iron pipe $1.98

42L1533¼

Extra Heavy.

With 2-in. side inlet for soil pipe $2.05

With 1½-in. tapped side inlet for iron pipe $2.15

Adjustable Roof Flange.

Easily adjustable to fit any pitch of roof. Furnished in copper or galvanized iron. Has lead collar at top. This is calked in tight around soil pipe, making watertight joint. Shipping weight, 4-inch, 2¾ pounds; 5-inch, 3¼ pounds.

Size	42L184¼ Galvanized	42L186¼ Copper
4-inch	$1.05	$1.95
5-inch	1.40	2.90

Cast Iron Adjustable Closet Bend With Flexible Gasket.

Size, 4x15½ inches. Used to connect closet bowl to soil pipe stack. Has adjustable floor flange, bolts and flexible composition gasket. Tapped on both sides for 1½-inch iron pipe. Makes watertight and gastight joint without using putty. Shipping weight, 20 lbs.

42L1511¼ $1.98

42L1510¼—Closet Bend only, same except without gasket........1.50

Half S Traps.

42L1540¼ — Standard. Shipping weight, 17 pounds........$1.20

42L1541¼—Extra Heavy. Shipping wt., 23¾ lbs..$1.55

Size, 4 In.

Cast Iron Soil Pipe Increasers.

Required at top of soil stack where going through roof to increase opening and prevent clogging with frost.

42L1556¼—Standard. Size, 2 to 4x24 in. long. Shipping wt., 14 pounds........$1.05

Standard. Size, 4 to 5x30 in. long. Shpg. wt., 24 lbs..$1.65

42L1557¼ — Extra Heavy. Size, 2 to 4x24 inches long. Shipping weight, 14 pounds $1.25

Extra Heavy. Size, 4 to 5x30 in. long. Shpg. wt., 31 lbs.$2.05

Lead Bend and Ferrule.

42L428

For connecting soil pipe with closet bowl. Flanged over floor, and joint made to bowl with putty.

Size, In.	Shpg. Wt., Lbs.	
4x12	10	$2.30
4x14	11	2.50
4x18	13	2.95

Ventilating Cap.

Used for ventilating sewer outside of building. Size, 4x6 inches.

42L1552¼ Standard. Shipping weight, 8 pounds 70c

42L1553¼ Extra Heavy. Shipping weight, 9 pounds....90c

Lead Pipe.

42L1620

Prices subject to market changes. Sizes given are inside diameter.

Size, Inches	Wt., per Foot	Per Foot
½	10 oz.	8c
⅝	1 lb.	13c
1	1½ lbs.	19c
1¼	2 lbs.	26c
1½	3 lbs.	38c
2	4 lbs.	51c
4	5 lbs.	63c

Rope Oakum.

For calking iron soil pipe and fittings. A 4-inch soil pipe joint requires ¼ pound of oakum; a 2-inch joint, ⅛ pound.

42L1469¼

Per pound........$0.10½

Per 50-lb. coil. 5.10

42L1472¼—Pig lead for making calked lead joints, etc. Come in 8-pound ingots.

Per pound........10½c

Price subject to market changes.

Medium Weight Lead Pipe.

42L1622

Size, Inches	Wt., per Foot	Per Foot
½	1½ lbs.	19c
⅝	2 lbs.	25½c
¾	2¼ lbs.	28½c
1	3¼ lbs.	41c

Plumbers' Solder.

42L486 — Bars are about 1¼ lbs. each. Less than a bar not sold.

Per bar........30c

Price subject to market changes. For Tinners' Solder and Block Tin see page 853.

Short Lead Bend.

42L432 — Used to connect closet to soil pipe stack. Size, 4x12 inches. Shipping weight, 7½ pounds........$1.85

42L408

Combination Lead and Iron Ferrules.

Size, In.	Shpg. Wt., Lbs.	
2x4	1½	$0.30
4x6	5	.88
4x8	5½	1.10
4x12	7½	1.50

Extra Long S Lead Traps.

42L414 Size, 1¼ in. Shipping wt., 4½ lbs. 97c

Size, 1½ in. Shipping weight, 6¼ lbs........$1.45

42L416 Same as above, for iron pipe connection.

Size, Inches	
1¼	$1.50
1½	1.98

Sheet Lead.

Used to make watertight joint where soil pipe goes through roof, lining chemical vats, etc. **Thickness,** 3/32 and 1/16 inch; 3/32-inch weighs 2 pounds per square foot; 1/16-inch weighs 4 pounds per square foot. **State thickness wanted.**

42L1474—3/32-inch. Per square foot....26½c

42L1475—1/16-inch. Per square foot....53c

Basement or Garage Floor Drain Trap.

With bell trap. Placed in floor and connected to 4-inch soil or sewer pipe. Just the thing for the garage or basement floor.

42L616—Size, about 6x6 inches. Shipping weight, 4 pounds........50c

42L618—Size, about 9x9 inches. Shipping weight, 11½ pounds........94c

42L620—Size, about 12x12 inches. Shipping weight, 16 pounds $1.40

Full S Standard Lead Traps.

For connecting sinks, laundry tubs, etc., to pipe in floor.

42L412 — Size, 1¼ inches. Shipping weight, 2½ lbs 60c

Size, 1½ inches. Shipping weight, 4¼ pounds........95c

42L410 — Same as above, except with iron pipe connection on outlet. Size, 1¼ inches. Shipping wt., 4 lbs...$1.15

Size, 1½-in. Shipping weight, 5½ lbs $1.50

Lead Half S Traps.

42L420

Size, 1¼ in. Shipping weight, 2¼ lbs........53c

Size, 1½ in. Shipping weight, 4½ lbs........79c

42L422

Same as above, except with iron pipe connection on outlet.

1¼-inch. Shipping weight, 3¼ pounds . $1.10

1½-inch. Shipping weight, 4 pounds $1.38

Plumbers' Blast Furnaces and Calking Tools

Plumbers' Gasoline Blast Furnace.

$4.59

A very high grade furnace, strong and substantially made. Used by plumbers for melting lead, heating soldering irons, etc. By removing top it may be used as a torch for thawing frozen water pipes, burning off old paint, heating glue pots, etc. In fact, it may be used for any purpose for which a small portable stove with real hot flame is desired. Many campers use this furnace for cooking, frying, heating water, etc. Easy to light. Produces a hot blue flame. Melts lead or solder quickly. Heavy, durable construction. Reservoir is galvanized. Shipping weight, 10¼ pounds.

42L1720¼........$4.59

Melting Pot not included in above price.

Large Capacity Combination Gasoline Blast Furnace.

$7.98

Serves every purpose for which Blast Furnace 42L1720¼ is used and has larger flame, giving greater heating capacity. Much used by linemen and electricians, etc., as well as by plumbers. Gives intense heat. Heats soldering irons and melts lead or solder at the same time. Has new type burner, designed for outdoor or indoor use. Burner is attached with swivel joint so it can be turned to horizontal or vertical position without changing position of furnace. Hood easily removed so tank with burner may be used separately. Reservoir holds 2¾ quarts of gasoline and is made of heavy gauge pressed steel with welded leakproof seams. Fitted with copper filler plug and improved brass pump with needle stop valve. Hood has cover, soldering iron rest and wind shield. Height, 12 inches; diameter, 8¼ inches. Shipping weight, 11 pounds.

42L1722¼........$7.98

Melting Pot.

Made of cast iron. Very durable. To fit 42L1720¼ and 42L1722¼ Furnaces. Shipping weight, 2 lbs.

42L528........37c

Plumbers' Melting or Pouring Ladle.

Wrought iron. 3-inch ladle fits Melting Pot 42L528.

Size, across bowl, in.	3	4	5
Shipping weight, lbs.	1	1¾	2½
42L1460	18c	24c	42c

Asbestos Joint Runner

For pouring lead into horizontal soil pipe joints. Fits up close to mouth of hub. Lead cannot leak out. Easily attached. Made of strong fiber asbestos with cast iron pouring lip and clamp. For 2, 4, 5 and 6-inch pipe. Shipping weight, 2 pounds.

42L484........$2.10

Calking Chisel.

42L490 Straight Calking Chisel. Shipping wt., 12 ounces........21c

42L492—Right Hand Offset Calking Chisel........22c

42L494—Left Hand Offset Calking Chisel........23c

Yarning Iron.

For forcing oakum into soil pipe joint before calking. Shipping weight, 1 lb.

42L488........27c

42L443—Drum Trap, same as 42L442 (below), with offset outlet for straight line connection. Shipping weight, 10½ pounds........$2.30

42L442—Swivel Base Cast Iron Drum Trap. Base is separate from body and can be turned in any direction, making the trap easy to connect and saving one elbow. When trap is set in proper position base is then connected to body of trap with a calked joint. Shipping weight, 8¾ pounds........$2.25

Cast Iron Sink Traps.

Cast iron traps will not fit our cast iron roll rim sinks. They are used only with our cast iron flat rim sinks 42L1479¼ and 42L1480¼, and our flat rim and roll rim steel sinks shown on page 702.

42L349 Cast Iron Sink Trap. Size, 1½-inch, for iron pipe connection. Long sweep prevents clogging. Shpg. wt., 4½ lbs........55c

Cast Iron S Sink Trap.

42L347 Size, 1½-inch, for iron pipe connection. Shipping weight, 10 lbs........$1.75

Iron Drum Trap.

42L436 — Size, 4¼x8¼ inches. Tapped both sides for 1½-inch pipe. Nickel plated brass cover. Shipping weight, 8½ pounds........98c

Iron Drum Trap.

42L438—Size, 4¼x5¼ in. Nickel plated brass cover. Tapped 1½ inches side and bottom. Shipping weight, 6¼ pounds........80c

Save Money Do Your Own Pipefitting

Quick Action Pipe Vise. Side latch and hinge permit taking pipe out sideways. Holds pipe, rods and bars. Malleable iron with steel jaws. **$1.55 AND UP**

42L1668¼—No. 0. Holds pipe from ⅛ to 1¼ inches. Shipping weight, 4¼ pounds.........**$1.55**

42L1669¼—No. 1. Holds pipe from ¼ to 2½ inches. Shipping weight, 12¼ pounds.........**$2.75**

42L1670¼—No. 2. Holds pipe from ¼ to 3 inches. Shpg. wt., 16¼ lbs. **$3.95**

42L1410—Steel jaws only for No 0. vise. Shipping weight, 1 pound. Per set.........**88c**

42L1411—Steel jaws only for No. 1 vise. Shipping weight, 1 pound. Set.........**$1.10**

42L1412—Steel jaws only for No. 2 vise. Shipping weight, 1 pound. Set.........**$1.60**

Combination Bench and Pipe Vise. A Combination Bench and Pipe Vise on a swivel base. Made extra strong. Jaws tempered tool steel. **$6.50 AND UP**

42L1660¼—Width of jaw, 3⅛ inches. Takes pipe ⅛ to 1½ inches. Shipping weight, 22½ pounds. **$6.50**

42L1661¼—Width of jaw, 3½ inches. Takes pipe ⅛ to 2 inches. Shipping weight, 58¼ pounds.........**$9.90**

42L1662¼—Width of jaw, 4¾ inches. Takes pipe ⅛ to 3 inches. Shipping weight, 74 pounds...**$12.50**

When ordering repair parts give casting number of part wanted.

Hercules Pipe Grip. 27c Consists of two malleable iron jaws mounted on steel spring loop. Fits between jaws of vise. Changes your bench vise into a pipe vise instantly. Shipping weight, 8 ounces.

42L954.........**27c**

Blow Torch Bargains!

Gasoline Blue Blast Blow Torch. **$3.65**

Saves Plumbers' Bills. This is a strictly high grade torch made of strong gauge polished brass and is very durable. Not to be compared to the light weight flimsy torches on the market. Used for soldering, burning off old paint, thawing frozen water pipes, tempering tools, vulcanizing automobile tires, etc. It is a portable blast furnace, giving intense heat wherever you need it. For the plumbing repair jobs around the home it pays for itself the first time used. Has soldering copper attachment. High quality brass.

42L1450—Capacity, 1 pint. Shipping weight, 2½ pounds.........**$3.65**

42L1451—Capacity, about 1 quart. Shipping weight, 4 pounds.........**$4.20**

Gasoline Flat Blow Torch. **$6.25** With soldering copper attachment. Takes but little space in your automobile tool box or your plumbers' tool kit. Made flat for use in narrow or close places. Made of brass, with steel conductor tube and needle valve. Has double check pump and bracket to prevent tipping. Size, about 7x9x2 inches. Capacity, 1 pint. Shipping weight, 4¼ pounds.

42L1453.........**$6.25**

Gasoline Brazing Torch. Gives Intense Heat. **$4.90** With soldering copper attachment. No pump. Pressure created by gasoline vaporizing. Gives 1,800 degrees heat under proper conditions. Can be held in any position. Copper plated steel container, welded joints. Capacity, ⅔ pint. Shipping weight, 3 pounds.

42L1452.........**$4.90**

Automatic Blow Torch. **$1.60** Handy self blowing alcohol torch; requires no pumping. Gives highly concentrated, very intense heat at point of flame. Made of brass. Height over all, 5½ inches. A very practical and serviceable small torch for the dentist, jeweler, electrician or mechanic. Brass highly nickel plated. Has wood handle on side. Shipping weight, 6 ounces.

42L1455.........**$1.60**

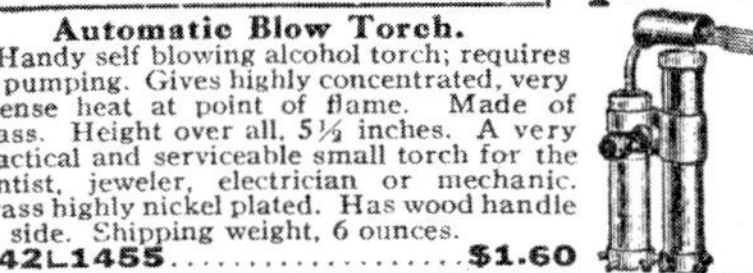

Pipe Taps and Reamers.

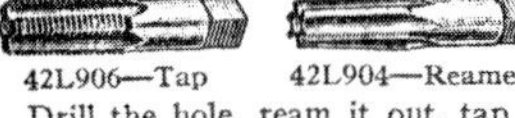

42L906—Tap 42L904—Reamer

Drill the hole, ream it out, tap it and then screw in the pipe. Reamers and taps are both made of tempered tool steel.

Size, In.	Shpg. Wt.	42L906 Tap	42L904 Reamer
⅛	1 oz.	$0.31	$0.47
¼	3 oz.	.37	.55
⅜	4 oz.	.50	.74
½	5 oz.	.64	.93
¾	9 oz.	.87	1.30
1	14 oz.	1.35	2.10
1¼	1¾ lbs.	1.85	2.85
1½	2¼ lbs.	2.45	3.75
2	3¼ lbs.	3.65	5.75

Burring Reamer.

Tool steel. Fits common bit brace. Removes burr from pipe after cutting.

42L908—For pipe size ⅛ to 1 in. Shipping weight, 4 ounces.......**68c**

42L910—For pipe size ¼ to 1¼ inches. Shipping weight, 6 oz...**90c**

42L912—For pipe size ½ to 2 in. Shipping weight, 1 pound....**$1.65**

A Quick and Easy Pipe Threader.

Stocks are made of malleable iron, cutters in dies are of tempered tool steel. Cuts clean, sharp threads. Nos. 2 and 3 furnished with lead screw. Lead screw not required on sizes smaller than No. 2.

Catalog No.	Size No	Threads Pipe Size	Size Dies	Shpg. Wt. Lbs.	Complete Set	Extra Dies, Each	Extra Gds. Each
42L860¼	0	⅛, ¼, ⅜, ½	2 x½	7¼	$4.40	$0.81	$0.17
42L862¼	1	¼, ⅜, ½, ¾, 1	2½x¾	13	5.80	.92	.23
42L864¼	1½	¾, 1, 1¼	3 x¾	14½	5.25	1.15	.35
42L863¼	1⅞	⅜, ½, ¾, 1, 1¼, 1½	3 x¾	20	9.75	1.16	.35
42L866¼	2	1¼, 1½, 2	4 x⅞	25	8.00	1.44	.43

Armstrong Pattern Stocks and Dies.

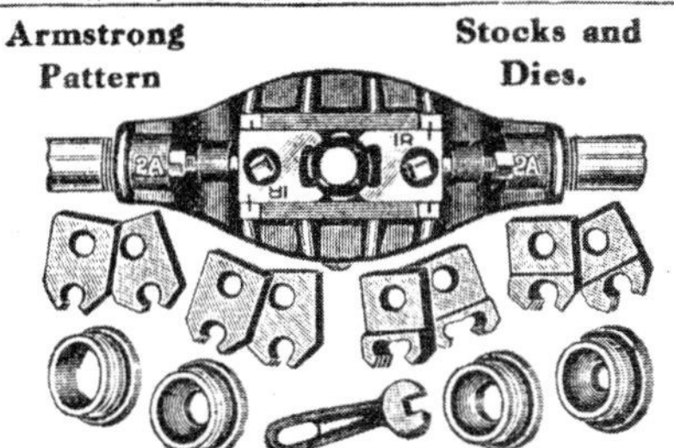

Armstrong Pattern Adjustable Stocks, fitted with dies that can be adjusted to cut a deep or shallow thread. The dies are quick and easy cutters and can be sharpened on grindstone or emery wheel when worn. Easy to start die on the pipe without filing, even when there is a swell or burr. **Parts interchangeable.**

Catalog No.	Threads Pipe Size, In.	Shpg. Wt. Lbs.	Complete Set	Extra Dies Each Size, Two Pieces	Extra Guides
42L880¼	¼ to 1 inclusive	13	$ 6.55	$1.15	23c
42L882¼	½ to 1¼ inclusive	15¾	7.70	1.45	35c
42L884¼	¼ to 1¼ inclusive	19	10.95	1.50	36c
42L886¼	1¼ to 2 inclusive	29	9.95	2.60	43c

White Cotton Waste.

For cleaning and wiping mechanical parts, machinery etc. Highly absorbent, pure white. **No. 1 quality.**

42L1462¼

1 lb.	5 lbs.	10 lbs.	25 lbs.	50 lbs.	100 lbs.
21c	$1.05	$2.05	$4.87	$9.50	$18.75

Monarch Drill Ratchet.

Drills Holes in Iron, Steel, Marble or Wood.

42L914—Made of steel. By attaching sleeve, listed below, round shank drill may be used.

10-inch. Shipping weight, 3¼ pounds.........**$3.95**
12-inch. Shipping weight, 5½ pounds.........**4.45**
15-inch. Shipping weight, 7 pounds.........**5.15**

Ratchet Drill Sleeve. For Round Shank Drills. Fits 42L914 Ratchet.

42L916—For ½-inch round shank drills. Shipping weight, 8 ounces..**$1.20**

42L918—For ⅝-inch round shank drills. Shipping weight, 12 ounces.........**$1.25**

Roller Tube Expander. Steel with hardened rollers. Guaranteed satisfactory. Give outside diameter (not inside) of tube the expander is to be used in.

42L742

Size	Shpg. Wt.		Size	Shpg. Wt.	
1½ in.	2½ lbs.	$3.40	2¼ in.	5¾ lbs.	$4.10
1¾ in.	3¾ lbs.	3.45	2½ in.	7¼ lbs.	4.85
2 in.	5 lbs.	3.50	3 in.	10½ lbs.	6.25

Steel Wire Flue Brush. For cast iron heating boilers, warm air furnaces, etc. Size, 2x4½x6 inches. You will greatly increase the efficiency of your boiler or furnace and save coal by keeping the flues clean. Tapped for ¼-inch pipe connection. Shipping weight, 1 pound.

42L752.........**65c**

Stillson Pipe Wrench. Tool Steel. 69c AND UP A very satisfactory pipe wrench used by best tradesmen. Sizes 6 and 8 in. have wood covered handles. All other sizes have solid steel handles.

42L1148

Lgth., Open, In.	Takes Pipe, In.	Shpg. Wt., Lbs.	
6	⅛ to ½	½	$0.69
8	⅛ to ¾	¾	.80
10	⅛ to 1	1¼	.90
14	¼ to 1½	2½	1.25
18	¼ to 2	4¼	1.79
24	¼ to 2½	8	2.58
36	½ to 3½	14	4.85

Genuine Trimo Pipe Wrench. 84c AND UP A well known reliable pipe wrench, made of steel with hardened jaws. Strong, well made and durable.

42L1146

Lgth., Open, In.	Takes Pipe In.	Shpg. Wt., Lbs.	
6	⅛ to ½	½	$0.84
8	⅛ to ¾	¾	.95
10	⅛ to 1	1¼	1.05
14	¼ to 1½	3	1.40
18	¼ to 2	5	1.98
24	¼ to 2½	8¼	3.05
36	½ to 3½	15¾	5.70

Automatic Pipe Wrench. Drop Forged Steel. 95c AND UP Always ready. Grips pipe firmly and instantly. No adjusting necessary. A big time saver.

42L1150

10-inch. Takes pipe ⅛ to 1 inch. Shipping weight, 1¼ pounds.........**95c**

14-inch. Takes pipe ¼ to 1½ in. Shipping weight, 2½ pounds.........**$1.35**

18-inch. Takes pipe ⅜ to 2 inches. Shipping weight, 4½ pounds.........**$1.90**

24-inch. Takes pipe ½ to 2½ inches. Shipping weight, 7 pounds.........**$2.75**

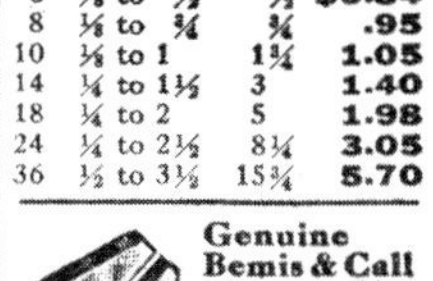

Genuine Bemis & Call Nut and Pipe Wrench. $1.78 AND UP One side takes bolts rods or pipe; other side for square or hexagonal nuts. Very handy. Wrought bar and casehardened jaws and adjusting nut.

42L1152

10-inch. Takes pipe ⅛ to 1 inch. Shipping weight, 2 pounds.........**$1.78**

12-inch. Takes pipe ⅛ to 1¼ inches. Shipping weight, 3¼ pounds.........**$2.14**

15-inch. Takes pipe ½ to 2¼ inches. Shipping weight, 4½ pounds.........**$2.86**

Handy Adjustable Wrenches.

For gripping pipe, bolts, rods, etc. Well made. Steel jaws.

42L1142

Lgth., In.	Takes Pipe, In.	Shpg. Wt., Lbs.	
6	⅛ to ⅜	½	$0.55
8	⅛ to ¾	¾	.70
10	⅛ to 1	1¾	.80
12	⅛ to 1½	2½	1.20

Same as 42L1142, except has jaws at both ends. One for bolts or pipe, the other for squares or hexagons. Takes pipe ⅛ to ½ inch; length, 6 inches. Shipping weight, 11 ounces.

42L1144.........**85c**

Brock's Chain Pipe Wrench.

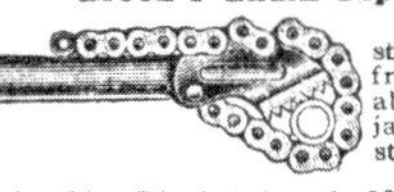

Holds tight and withstands great strain. Forged from bar steel. Has removable milled and tempered jaw. Chain made of special steel.

42L1680¼—Size 1. Holds pipe ¼ to 2 inches; length, 20 inches. Shipping weight, 5½ pounds.........**$1.98**

Size, 2. Holds pipe ¼ to 3 inches; length, 27 inches. Shipping weight, 9¼ pounds.........**$2.85**

Size, 3. Holds pipe ¾ to 6 inches; length, 37 inches. Shipping weight, 19 pounds.........**$3.95**

Saunders' Pipe Cutter.

Wheel cuts groove in pipe and the two rollers roll edges down smooth so that there is no burr left which would prevent die taking hold easily when threading pipe. Malleable iron tool steel cutting wheel.

42L852

	No. 1	No. 2
Cuts pipe from	⅛ to 1 in.	1 to 2 in.
Shipping weight, pounds	3½	6½
Each	$1.55	$2.50
Extra wheels, each	.14	.19

Barnes' Three-Wheel Pipe Cutter.

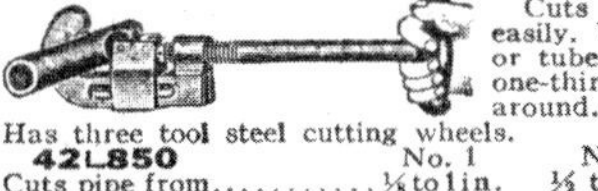

Cuts pipe quickly and easily. Will cut off a pipe or tube by turning only one-third of the way around. Malleable iron. Has three tool steel cutting wheels.

42L850

	No. 1	No. 2	No. 3
Cuts pipe from	⅛ to 1 in.	½ to 2 in.	1½ to 3 in.
Shipping weight, pounds	3	5	8¼
Each	$1.92	$2.65	$4.25
Extra wheels, each	.12	.15	.20

Hercules Spring Flue Cleaner.

Very effective. Spiral spring knives, kept tight against flue and remove soot with little effort. 3-inch size tapped for ⅜-inch pipe, other sizes for ¼-inch pipe.

42L746

Size	Shpg. Wt.		Size	Shpg. Wt.	
1¾ in.	1½ lbs.	$1.10	2½ in.	2 lbs.	$1.50
2 in.	1½ lbs.	1.20	2¾ in.	2¼ lbs.	1.65
			3 in.	2½ lbs.	1.80

Serviceable Pumps for the Farm

High Spout Set Length Anti-Freezing Lift Pump for Wells Up to 60 Feet Deep.

Strong and substantially made; has small hole just above cylinder which lets water run out so pump cannot freeze. Pump is all made up, with pump rod, pipe and iron cylinder, as shown. Stroke, 6 inches.

$5.60 2½x10-Inch.

42L2350¼ Set Length Pump, as described, with 2½x10-inch iron cylinder. Shipping wt., 62 pounds ... $5.60
With brass body cylinder ... $7.50

42L2351¼—Same Pump, with 3x10-inch iron cylinder. Shipping weight, 65 pounds ... $5.65
With brass body cylinder, each ... $7.75

42L2352¼—Same Pump, with 3½x10-inch iron cylinder. Shipping weight, 70 pounds ... $6.40
With brass body cylinder, each ... $8.45

For galvanized pipe between pump and cylinder instead of black add 25c to above prices and allow for cost of galvanized pipe to reach water in well.

42L2355¼—Pump Standard only, without cylinder, pipe or pump rod. Shipping weight, 40 pounds ... $3.15

Takes 1¼-Inch Pipe.

Bottom of Cylinder Comes 4 Feet Below Pump Platform.

High Spout Set Length Anti-Freezing Lift Pump With Closed Top.

$6.40 3x10-In.

Same as 42L2350¼, except that it is of heavier construction. Also has oscillating link connection between pump rod and handle and closed top.

42L2360¼—Set Length Pump, with 3x10-inch iron cylinder. Shpg. wt., 70 lbs. Each ... $6.40
With brass body cylinder, each ... $8.50

42L2361¼—Same Pump, with 3½x10-in. iron cylinder. Shpg. wt., 75 lbs. ... $7.15
With brass body cylinder ... $9.20

For galvanized pipe between pump and cylinder instead of black add 25c to above prices and allow for cost of galvanized pipe to reach water in well.

42L2364¼—Pump Standard only; same as above, without cylinder, pipe or pump rod. Shipping wt., 48 lbs. ... $3.90

Takes 1¼-Inch Pipe.

Extra Heavy Set Length Hand Force Pump.

Spout has hose and shut off clevis, so that water can be forced to elevated tank or through hose for sprinkling, washing windows, buggies, automobiles, etc.

$7.95 2½x10-In.

Bottom of Cylinder Comes 4 Feet Below Pump Platform.

Pump Takes 1¼-Inch Pipe. 6-Inch Stroke.

42L2409¼ Set Length Hand Force Pump, with 2½x10-in. iron cylinder. Shipping wt., 76 lbs. ... $7.95
With brass body cylinder ... $9.85

42L2410¼ Same Pump, with 3x10-in. iron cylinder. Shpg. wt., 85 lbs. ... $8.10
With brass body cylinder ... $10.20

42L2411¼—Same Pump, with 3½x10-inch iron cylinder. Shpg. wt., 90 lbs. Each ... $8.70
With brass body cylinder ... $10.75

For galvanized pipe add 25c to above prices and allow for cost of galvanized pipe to reach water in well.

Back Attachment, $1.50 Extra.

Cock Spout, $1.65 Extra.

How to Order Your Pump Outfit.

If the distance from pump platform to the lowest water level is less than 25 feet, you can use one of our set length pumps, quoted on this page, and order sufficient extra pipe to reach to about 3 feet below water. In that case the cylinder need not be let down. Just screw the pipe into the bottom of the cylinder and extend the pipe down below the water. This simplifies the installation very much.

If the distance to the lowest water level is over 25 feet, then you will have to extend the cylinder down, and in that case we recommend that the cylinder be immersed in the water and sufficient extra pipe and pump rod ordered to place it in that position. We call your attention to our complete pump outfits, ready cut and fitted, ready to assemble for various depths of wells, as quoted on page 709. Lift pumps draw water to the pump spout only. Force pumps will force water to an elevated tank. If you just want to pump water to a bucket or tank at the pump spout, a lift pump will do. If you want to force it to an elevated tank or a pressure tank, order a force pump.

We charge 2 cents per foot in addition to cost of pump rod and pipe for cutting and fitting pump outfits to order.

Note Our Complete Ready Made Pump Outfits for Various Depths of Wells as Quoted on Page 709. For Drive Well Points, Pump Rod Cylinders, Etc., See Page 710.

NOTE—Prices quoted on pump standards only do not include any pipe, pump rod or cylinder.

Pitcher Spout Iron Lift Pump.

$1.79 3-In. Size.

One of the most popular and serviceable Kitchen Sink Pumps on the market. For wells or cisterns 20 feet deep or less. Has revolving top so handle may be turned as desired. Cylinders of all pumps listed below tapped for 1¼-inch suction pipe. Order pipe from water to pump. Shipping weight, 28 pounds.

42L2370¼—3-in. iron cylinder ... $1.79
42L2371¼—3-inch brass lined cylinder ... $2.40
42L2372¼ 3½-inch iron cylinder ... $2.10
42L2373¼—3½-inch brass lined cylinder ... $2.80

Simplex Cistern Force Pump.

$4.55 2½-In. Size.

For wells or cisterns 20 feet deep or less. Made of high quality cast iron with brass cylinders, valve set and stuffing nut. Gooseneck spout fitted with hose coupling or faucet cock, as illustrated. Tapped for 1¼-inch suction and 1-inch top outlet for forcing water to elevated tank. Shipping weight, 20 lbs.

42L2376¼—With plain spout and 2½-inch cylinder ... $4.55
42L2377¼—With plain spout and 3-inch cylinder ... $4.70
42L2378¼—With faucet spout as illustrated and 2½-inch cylinder ... $5.15
42L2379¼—With faucet spout and 3-inch cylinder ... $5.25

Suction Pipe From Well Not Included in Above Prices.

See Also Our Complete Sink and Pump Outfits on Page 702.

Extra Heavy Windmill Set Length Force Pump.

$8.95 2½x10-In.

Adapted for wells up to 150 feet deep. Can be operated by hand, windmill or pump jack. Six-inch stroke. Tapped back of spout for 1-inch pipe. Spout has hose and shut off clevis, so that water can be forced to elevated tank or through hose for sprinkling, washing windows, buggies, automobiles, etc. Bottom of cylinder, 4 feet below pump platform.

42L2392¼—Set Length Pump with 2½x10-inch iron cylinder. Shipping weight, 100 pounds ... $8.95
With brass body cylinder ... $10.85

42L2393¼—Same Pump, with 3x10-inch iron cylinder. Shpg. wt., 105 lbs. ... $9.10
With brass body cylinder ... $11.20

42L2394¼—Same Pump with 3½x10-inch iron cylinder. Shipping wt., 110 lbs. ... $9.75
With brass body cylinder, each ... $11.80

For galvanized pipe add 25 cents to above prices and allow for cost of galvanized pipe to reach water in well.

Takes 1¼-Inch Pipe.

6-Inch Stroke.

Cock Spout, $1.65 Extra.

Back Attachment, $1.50 Extra.

Please Note — When ordering repair parts for any of our pumps, give casting number of part wanted.

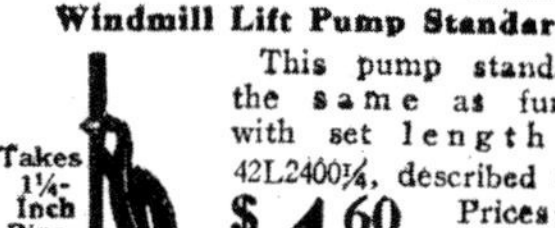

High Spout Set Length Anti-Freezing Windmill Lift Pump.

$7.15 3x10-Inch.

Operates by hand, windmill or pump jack. A good strong pump for farm service. Six-inch stroke. Bottom of cylinder comes 4 feet below platform.

42L2400¼—Windmill Pump with 3x10-in. iron cylinder. Shpg. wt., 70 lbs. Each ... $7.15
With brass body cylinder, each ... $9.25

42L2401¼ Same Pump with 3½x10-inch iron cylinder. Shpg. weight, 75 lbs. Each ... $7.90
With brass body cylinder, each ... $9.95

For galvanized pipe add 25 cents to above prices and allow for cost of galvanized pipe to reach water in well.

42L2404¼—Windmill Pump Standard only, without cylinder, pipe or pump rod. Shipping weight, 50 lbs. ... $4.60

Windmill Lift Pump Standard.

This pump standard is the same as furnished with set length pump 42L2400¼, described above. Prices include standard only, as illustrated, but no pipe, pump rod or cylinder.

$4.60 6-In. Stroke.

Takes 1¼-Inch Pipe.

42L2404¼ Windmill Lift Pump Standard only as illustrated, 6-inch stroke. Tapped for 2-inch pipe or smaller. Shipping weight, 50 pounds. $4.60

42L2405¼—Windmill Lift Pump Standard only as illustrated. 10-inch stroke. Tapped for 2-inch pipe or smaller. Shipping weight, 55 pounds ... $5.20

Windmill Force Pump Standard.

This pump standard is the same as furnished with set length pump 42L2392¼ described above. Prices are for standard only as illustrated, but no pipe, pump rod or cylinder.

$6.50 6-In. Stroke.

Takes 1¼-Inch Pipe.

42L2396¼ Windmill Force Pump Standard only as illustrated, 6-inch stroke. Tapped for 2-inch pipe or smaller. Shipping weight, 65 pounds ... $6.50

42L2397¼ Windmill Force Pump Standard only as illustrated, 10-inch stroke. Tapped for 2-inch pipe or smaller. Shipping weight, 70 lbs. ... $7.15

42L2413¼—Pump Standard only as illustrated, except without windmill extension. This pump standard is same as standard furnished with pumps 42L2409¼, etc., as listed at the left. For hand pumping only. No windmill attachment. 6-inch stroke. Tapped for 1¼-inch pipe. Shipping weight, 45 pounds ... $5.40

Extra Heavy Set Length Anti-Freezing Lift Pump.

Stroke, 8 Inches

$7.95 3½x10-In.

Large pumping capacity and extra heavy construction. Distance from pump platform to bottom of cylinder, 4 feet.

42L2382¼—Set Length Pump with 3½x10-in. iron cylinder. Shpg. wt., 80 lbs. Each ... $7.95
With brass body cylinder ... $9.95

42L2383¼—Same Pump with 4x10-inch iron cylinder. Shipping weight, 85 pounds ... $8.80
With brass body cylinder ... $11.25

For galvanized pipe add 25 cents to above prices and allow for cost of galvanized pipe to reach water in well.

42L2386¼—Pump Standard only, without cylinder, pipe or pump rod. Shipping weight, 50 pounds ... $4.70

Takes 1½-Inch Pipe.

Windmill Force Pump for Deep Wells.

Underground Pipe. Anti-Freezing.

$13.95 2½x10-Inch.

For wells up to 150 feet deep. This pump is single acting. Underground valve is operated by wheel handle on spout. Furnished for 6 or 10-inch stroke.

42L2424¼—Single Acting Windmill Force Pump, 2½x10-inch iron cylinder, 6-inch stroke. Shipping wt., 135 lbs. ... $13.95
With brass body cylinder, each ... $15.85

42L2425¼—Same Pump with 3x10-inch iron cylinder. Shpg. wt., 135 lbs ... $14.25
With brass body cylinder, each ... $16.35

42L2428¼—Same Pump with 2½x16-in. iron cylinder. 10-inch stroke. Shipping weight, 135 lbs. ... $15.50
With brass body cylinder, each ... $17.45

42L2429¼—Same as 42L2428¼, with 3x16-inch iron cylinder. Shipping wt., 140 pounds ... $15.85
With brass body cylinder, each ... $17.90

42L2430¼—Same as 42L2428¼, with 3½x16-inch iron cylinder. Shipping wt., 145 pounds ... $16.90
With brass body cylinder, each ... $18.80

Takes 1¼, 1½ and 2-Inch Pipe.

Distance From Pump Platform to Bottom of Cylinder, About 6 Feet.

Easy Working Wood Farm Lift Pumps.

Light acting, easy working. Has large pumping capacity. For wells or cisterns up to 30 feet deep. Has 3¼-inch cylinder. Made from first grade Washington spruce, measuring about 5½ inches square. Nicely painted, striped and varnished. Iron handle brackets and iron spout. When ordering include tubing listed below to reach from bottom of pump to bottom of well. When more than 94 feet of tubing is required order coupling for each joint to connect tubing. Shpg. wt., 38 lbs.

$6.80 6-Foot Size.

42L2192¼—Length, 6 feet, for wells up to 20 feet deep. Plain pump ... $6.80
With porcelain lined cylinder ... $8.40

42L2193¼—Length, 7 feet, for wells 20 to 25 feet deep. Plain pump ... $7.40
With porcelain lined cylinder ... $9.00

42L2194¼—Length, 8 feet, for wells 25 to 30 feet deep. Plain pump ... $8.00
With porcelain lined cylinder ... $9.60

42L2197¼—Wood Tubing to fit above pumps, 4 inches square, with 1¾-inch bore, in random lengths from 6 to 12 feet. Shpg. wt., per ft., 1¼ lbs. Per foot ... 24c

42L2198¼—Couplings for Tubing. Shipping weight, each, 5 pounds ... 72c

42L832—Pioneer Leathers, size, 3¼ inches, for above pumps. Shipping weight, 8 ounces. Each ... 18c

42L833—Special Iron Pipe Connection for attaching iron pipe to above pumps. Shipping weight, 10 pounds ... $1.25

Note—Iron pipe can be used with wood pumps on drive wells only. For other wells use wooden pump tubing quoted above to extend down to water in well.

Ready Made Pumping Outfits

High Spout Anti-Freezing Windmill Lift Pumping Outfit for Dug or Drilled Wells.

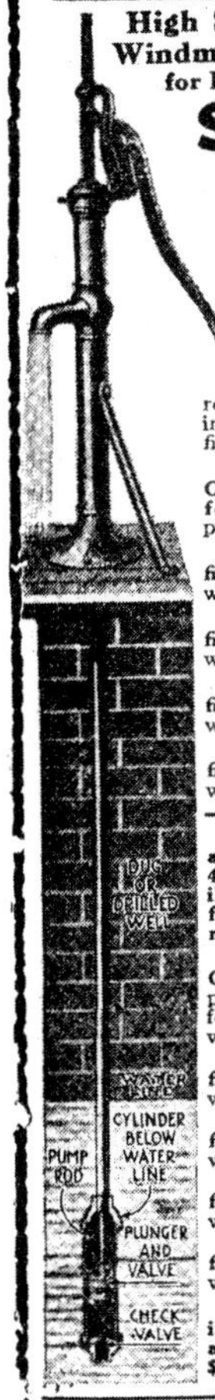

$12.40

COMPLETE OUTFIT FOR 30-FT. WELL.

Complete Outfit includes 42L2400¼ pump an described on page 708, with cylinder galvanized pipe and galvanized pump rod all cut and fitted ready to install in depths of wells specified.

42L2464¼—Pumping Outfit as described, for **30-foot well.** Shipping weight, 150 pounds............**$12.40**

42L2466¼—Same Outfit, for **40-foot well.** Shipping weight, 175 pounds...**$14.35**

42L2468¼—Same Outfit, for **50-foot well.** Shipping weight, 202 pounds...**$16.30**

42L2470¼—Same Outfit, for **60-foot well.** Shipping weight, 230 pounds...**$18.25**

42L2472¼—Same Outfit, for **70-foot well.** Shipping weight, 256 pounds...**$20.20**

Outfits quoted below same as above, except Hand Pump 42L2360¼, as described and illustrated on page 708, is furnished instead of windmill pump.

42L2440¼—Pumping Outfit as described, with hand pump instead of windmill pump for **30-foot well.** Shipping weight, 145 pounds...**$11.65**

42L2442¼—Same Outfit, for **40-foot well.** Shipping weight, 170 pounds.. **$13.80**

42L2444¼—Same Outfit, for **50-foot well.** Shipping weight, 197 pounds...**$15.55**

42L2446¼—Same Outfit, for **60-foot well.** Shipping weight, 225 pounds...**$17.50**

42L2448¼—Same Outfit, for **70-foot well.** Shipping weight, 251 pounds...**$19.45**

For brass body cylinder instead of iron cylinder on any of above outfits, add $2.10 to the price quoted.

Extra Heavy Windmill Force Pumping Outfit for Dug or Drilled Wells.

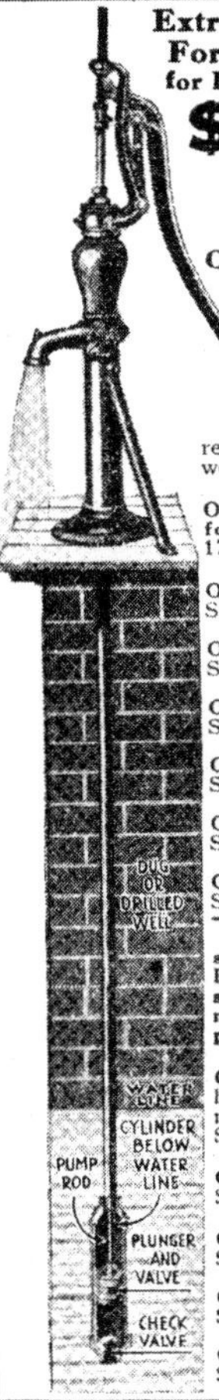

$14.40

COMPLETE OUTFIT FOR 30-FT. WELL.

Complete Outfit includes 42L2393¼ pump as described on page 708, with cylinder galvanized pipe and galvanized pump rod all cut and fitted ready to install in depths of wells specified.

42L2474¼—Pumping Outfit as described, for **30-foot well.** Shipping weight, 172 pounds........**$14.40**

42L2476¼ — Same Outfit, for **40-foot well.** Shpg. wt., 200 lbs..**$16.35**

42L2478¼ — Same Outfit, for **50-foot well.** Shpg. wt., 227 lbs..**$18.30**

42L2480¼ — Same Outfit, for **60-foot well.** Shpg. wt., 254 lbs..**$20.25**

42L2482¼ — Same Outfit, for **70-foot well.** Shpg. wt., 282 lbs..**$22.20**

42L2484¼ — Same Outfit, for **80-foot well.** Shpg. wt., 310 lbs..**$24.15**

42L2486¼ — Same Outfit, for **90-foot well.** Shpg. wt., 337 lbs..**$26.10**

Outfits quoted below are same as above, except Hand Pump 42L2410¼, as described on page 708, is furnished instead of windmill pump.

42L2450¼—Pumping Outfit as described, with hand pump instead of windmill pump for **30-foot well.** Shpg. wt., 150 lbs..**$13.40**

42L2452¼ — Same Outfit, for **40-foot well.** Shpg. wt., 176 lbs..**$15.35**

42L2454¼ — Same Outfit, for **50-foot well.** Shpg. wt., 203 lbs..**$17.30**

42L2456¼ — Same Outfit, for **60-foot well.** Shpg. wt., 231 lbs..**$19.25**

42L2458¼ — Same Outfit, for **70-foot well.** Shpg. wt., 258 lbs..**$21.20**

Drive Well Pump Outfit With Heavy Windmill Force Pump.

$11.60

COMPLETE OUTFIT FOR 10-FT. WELL.

Same as Outfits 42L2474¼, etc., as described at left, except that it is furnished with 1¼-inch 60-gauze well point, 30 inches long, for drive well.

42L2504¼
Drive Well Pumping Outfit as described, for **10-foot well.** Shipping weight, 120 pounds.........**$11.60**

42L2505¼
Same Outfit, for **15-foot well.** Shipping wt., 130 lbs..**$12.25**

42L2506¼
Same Outfit, for **20-foot well.** Shpg. wt., 140 lbs..**$12.90**

42L2507¼
Same Outfit, for **25-foot well.** Shpg. wt., 150 lbs..**$13.55**

Outfits quoted below are same as above, except Hand Force Pump Standard 42L2410¼, illustrated and described on page 708, furnished instead of windmill pump.

42L2510¼
Drive Well Pumping Outfit as described, with hand pump instead of windmill pump for **10-foot well.** Shpg. wt., 110 lbs..**$10.60**

42L2511¼
Same Outfit, for **15-foot well.** Shpg. wt., 120 lbs..**$11.25**

42L2512¼
Same Outfit, for **20-foot well.** Shpg. wt., 130 lbs..**$11.90**

42L2513¼
Same Outfit, for **25-foot well.** Shpg. wt., 140 lbs..**$12.55**

For brass body cylinder instead of iron cylinder on any of above outfits, add $2.10.

Anti-Freezing Windmill Lift Pumping Outfit for Drive Well.

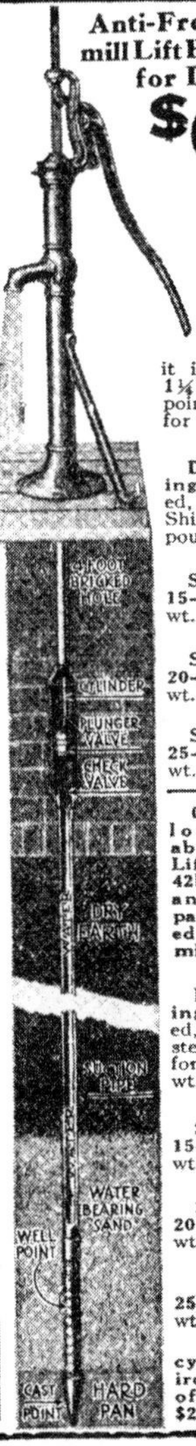

$9.65

COMPLETE OUTFIT FOR 10-FT. WELL.

Same as Outfits 42L2464¼, etc., as described at left, except that it is furnished with 1¼-inch 60-gauze well point, 30 inches long, for drive well.

42L2492¼
Drive Well Pumping Outfit, as described, for **10-foot well.** Shipping weight, 97 pounds.......**$9.65**

42L2493¼
Same Outfit, for **15-foot well.** Shpg. wt., 110 lbs..**$10.30**

42L2494¼
Same Outfit, for **20-foot well.** Shpg. wt., 120 lbs..**$10.95**

42L2495¼
Same Outfit, for **25-foot well.** Shpg. wt., 130 lbs..**$11.60**

Outfits quoted below are same as above, except Hand Lift Pump, same as 42L2360¼, described and illustrated on page 708, is furnished instead of windmill pump.

42L2498¼
Drive Well Pumping Outfit as described, with hand pump instead of windmill pump for **10-foot well.** Shpg. wt., 87 lbs.....**$8.90**

42L2499¼
Same Outfit, for **15-foot well.** Shpg. wt., 100 lbs...**$9.55**

42L2500¼
Same Outfit, for **20-foot well.** Shpg. wt., 110 lbs..**$10.20**

42L2501¼
Same Outfit, for **25-foot well.** Shpg. wt., 120 lbs..**$10.85**

For brass body cylinder instead of iron cylinder on any of above outfits, add $2.10.

Sure Tight Cistern Cover for Safety and Cleanliness.

$3.95
21-INCH SIZE

42L1846¼—For 21-inch opening. Made of cast iron with ring for lifting and two catches, making it necessary to turn cover slightly before it can be raised. Not easily opened by children, so it relieves danger of their falling into cistern or catch basin. Shipping weight, 50 pounds..............................**$3.95**

42L1847¼—Same as above, except with holes in cover for siphon chamber of septic tank...........................**$4.20**

Challenge Steel Curb Purifying Pumps.

$10.98

COMPLETE FOR 10-FOOT WELL.

Turning of crank gives large, continuous steady flow of water at pump spout. A child can easily operate it. Water is raised by means of an endless chain with buckets which pass around bearings in pump and well. Buckets carry air to bottom of well, where it is released and works its way to the surface in small bubbles. These bubbles liberate the gases the water contains, while the oxygen in the air helps to purify the water. Order outfit that will reach to within 2 feet of well bottom, as water is cooler and fresher at this point. Pumps consist of galvanized steel curb, nicely painted; galvanized endless chain with buckets and lower bearing all ready to place in well. Shipping weight, 90 pounds.

42L2232¼—For 10-foot well..**$10.98**
42L2233¼—For 15-foot well.. **12.58**
42L2234¼—For 20-foot well.. **13.98**
42L2235¼—For 25-foot well.. **15.48**
42L2236¼—For 30-foot well.. **16.98**

Challenge Wood Curb Bucket Chain Pumps.

Same as our 42L2232¼, etc., illustrated above, except that wood curb or top casing illustrated at right, nicely painted, is furnished instead of galvanized steel. Shipping weight, 90 pounds.

42L2211¼—For 10-foot well..**$ 9.35**
42L2212¼—For 15-foot well.. **10.85**
42L2213¼—For 20-foot well.. **12.30**
42L2214¼—For 25-foot well.. **13.75**
42L2215¼—For 30-foot well.. **15.25**

42L2231—**Extra Galvanized Buckets and Chain** to fit any of above pumps, with either wood or steel curb. Shipping weight, per foot, 12 ounces. Per foot..........**15c**

Wood Curb Tubular Chain Pump.

$6.59

COMPLETE FOR 10-FOOT WELL.

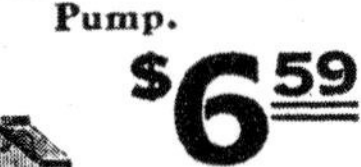

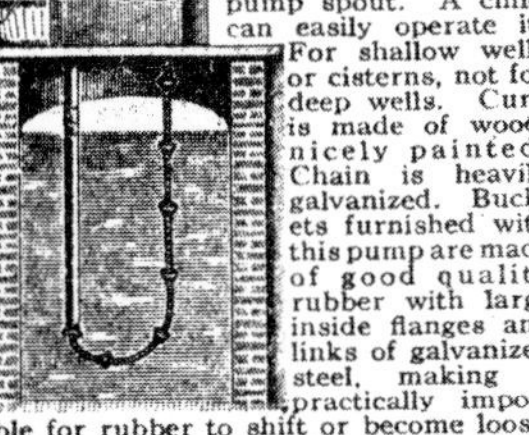

Turning of crank gives large, continuous steady flow of water at pump spout. A child can easily operate it. For shallow wells or cisterns, not for deep wells. Curb is made of wood, nicely painted. Chain is heavily galvanized. Buckets furnished with this pump are made of good quality rubber with large inside flanges and links of galvanized steel, making it practically impossible for rubber to shift or become loose. We advise the use of this pump for wells no deeper than 20 feet. For wells of greater depth, purchase one of our iron pumps. Shipping weight, 64 pounds.

42L2201¼—For 10-foot well.**$6.59**
42L2202¼—For 12-foot well. **6.69**
42L2203¼—For 15-foot well. **7.15**
42L2204¼—For 18-foot well. **7.98**
42L2205¼—For 20-foot well. **8.30**

Steel Curb Tubular Chain Pump.

Same as our 42L2201¼, etc., illustrated above, except that galvanized steel curb or top casing illustrated at left is furnished instead of wood. Shpg. wt., 75 lbs.

42L2221¼—For 10-foot well.**$7.98**
42L2222¼—For 12-foot well. **8.45**
42L2223¼—For 15-foot well. **8.90**
42L2224¼—For 18-foot well. **9.75**
42L2225¼—For 20-foot well. **9.95**

Let Our Simplex Hydraulic Ram Pump Water for You.

$10.95 AND UP

No engine, electric motor or other artificial source of power necessary to drive this pump. It makes the water pump itself by its own power.

Our Simplex Hydraulic Ram, illustrated at the right, will pump water anywhere from 20 to 120 feet above point where it is placed, depending upon height of fall of water to ram. It works continuously night and day and never gets tired.

Will elevate water to a high tank or reservoir or will force it a long distance from the source of supply. Set the ram below a spring or body of water at a distance of from 25 to 50 feet. Will lift water 4 to 10 feet for every foot of fall (difference in level) between water and ram. The ram lifts about one-seventh of the water which runs into it; the balance of the water runs out of the ram at the valve in the operation of lifting. All you have to do is to connect the supply or feed pipe to ram and place one end in the water. Connect the discharge pipe to the other side of ram and run it to your storage tank. Ram will not work with less than 2 feet of fall. We furnish all directions for installing. Better write and tell us just what conditions you have to operate a ram and we will prepare a special estimate. Ask for our Hydraulic Ram Circular No. 7429GCL, sent postpaid on request.

Catalog No.	Gallons of Water per Minute Necessary to Operate Ram	Size of Supply Pipe	Size of Discharge Pipe	Shipping Weight, Lbs.	
42L2300¼	½ to 2	¾ in.	½ in.	32	**$10.95**
42L2301¼	1½ to 4	1 in.	½ in.	41	**13.00**
42L2302¼	3 to 7	1¼ in.	¾ in.	58	**16.95**
42L2303¼	6 to 11	2 in.	1 in.	76	**20.75**
42L2304¼	11 to 25	2½ in.	1¼ in.	98	**39.75**

Galvanized Steel Tubing.

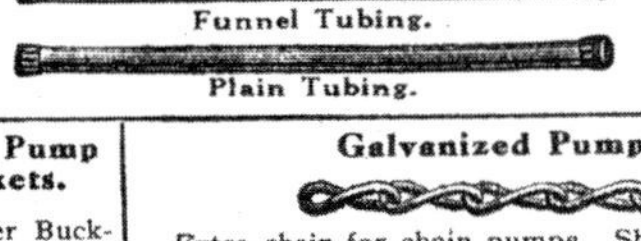

Reservoir Tubing.
Funnel Tubing.
Plain Tubing.

Used in connection with 42L2221¼ and 42L2201¼ Chain Pumps as listed at the left. Furnished in 5, 6, 7 or 8-foot lengths. Reservoir tubing is used in connection with the pump curb. Funnel tubing goes at bottom of pump and plain tubing connects reservoir and funnel tubing where the two are not sufficient. Shpg. wt., per foot, 12 oz. **State length and kind wanted.**

42L2256¼
Per foot............**15c**

Chain Pump Buckets.

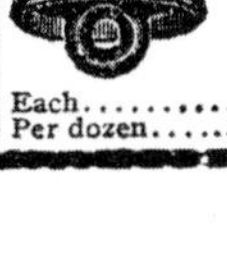

Rubber Buckets for chain pumps. Shipping weight, 3 ounces.

42L368
Each.......................**3½c**
Per dozen...................**40c**

Galvanized Pump Chain.

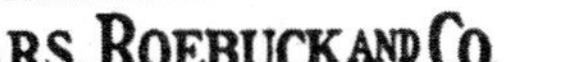

Extra chain for chain pumps. Shipping weight, per foot, 5 ounces.

42L370—Per foot..................................**4c**

ICELESS "ICE BOX"

No Spoiled Food. No Expense for Ice.

Keeps Your Provisions Cool During Hottest Weather Without Ice.

This Sanitary Iceless Cooler serves every purpose of an ice box with no expense for ice. It is a practical necessity on every farm.

It may be installed in any open or dug well, or if you have a drilled or driven well which cannot accommodate it you can make an excavation in your cellar or other convenient place 8 to 10 feet deep and lined with 18-inch tile. At this depth the temperature is as low as 55 degrees. You can make this excavation directly under the kitchen if you wish and the outfit can then be installed right in your kitchen, where it will be handy at all times. It is simply a "dumb waiter," and anything you want may be obtained in a moment. Saves expense for ice. First cost is only cost. Any handy man can easily install it. Provision chamber is bug and animal proof. It protects your foodstuffs from insects and prowling animals as well as keeping them cool and preventing their spoiling even in hottest summer weather. Provision chamber is 40 inches high and 14 inches in diameter with three adjustable shelves. **Shipped from factory in IOWA.**

42L3635⅓—Outfit complete with windlass, galvanized steel cable, and guide rods extending 8 feet into well as shown. Shipping weight, 95 pounds..... **$16.95**

Note—If guide rods longer than 8 feet are wanted, add 25c for each additional foot.

For Refrigerators see pages 640 and 641.

Rotary Barrel Pump.

Hand Rotary Barrel Pump. For pumping gasoline, motor oil, syrup or other liquids out of barrels. Much used for filling gasoline tanks on traction engines, automobiles, etc. Forces liquids 15 to 20 feet above pump. Works same as Acme Rotary Power Force Pump shown on this page. Made of iron. Furnished to fit steel barrel or wooden barrel. Both types quoted below. Has 3-foot suction pipe, 5-foot discharge hose and gooseneck. Pumps 13 gallons per minute. Shipping weight, 65 pounds.

Barrel Not Included.

42L2259¼—With attachment for steel barrel as shown........ **$12.45**

42L2260¼—With attachment for wooden barrel.................. **12.25**

Large Capacity Tank Pump.

For filling thresher tanks, stock watering tanks, etc.; pumping from shallow wells or cisterns. Used also as a deck pump, trench pump or for emptying cesspools, etc., or for any purpose where it is desired to pump a large volume of water. Pumps 50 gallons per minute. This is a strictly high grade pump. Has brass valve seats. Cylinder, 5 inches in diameter, 5-inch stroke; suction and discharge, 2 inches. Has extra connection for discharge, to connect 1-inch hose for sprinkling, etc. Shipping weight, 78 pounds.

42L2266¼—Pump as described............ **$10.65**

42L828—5-Inch Crimped Plunger Leathers. Shipping weight, 8 ounces. Set of two **85c**

For Suction Hose see page 854.

Galvanized Malleable Strainers.

42L826 Placed at end of pump suction pipe in well to keep out foreign matter. Covered with brass wire cloth.

Size, In.	Shpg. Wt., Lbs.	
1	1	32c
1¼	1¼	36c
1½	1½	42c
2	2¼	56c

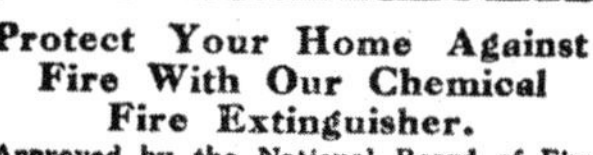

Protect Your Home Against Fire With Our Chemical Fire Extinguisher.

Approved by the National Board of Fire Underwriters.

"Hercules" is our own registered trade mark.

One of the most effective fire extinguishers on the market. Puts fire out almost instantly. How much would you be willing to pay for a reliable extinguisher in case of an actual fire starting in your home? How long would it take a fire to burn $13.85 worth of clothing, property, etc., to say nothing of a possible loss of life? Think it over. To operate, turn extinguisher upside down and direct stream to base of fire. Pressure is automatically generated, and liquid chemical when coming in contact with fire generates gas which envelops the fire and quickly smothers flame. Capacity, 2½ gallons. Shipping weight, 23 pounds.

42L1776¼ **$13.85**

Well Boring Outfit.

$7.20

Bores wells 25 feet deep, 8, 9, 10, 11, 12, 13 or 14 inches in diameter, or digs post holes quickly and with little effort. Requires no special machinery for boring wells; simply place auger on ground where you want well and turn handle. Bores through sand, gravel, sticky clay, mud or hard pan. Will dig under water; works easily and rapidly. When auger is full, draw it out of hole, pull up release catch and it opens, allowing earth to drop out without shaking or pounding. Can be used for digging holes for foundation piling for barns, etc. Outfit consists of a new type earth auger, extension rods with malleable couplings for boring down 25 feet, extension blade for increasing diameter of hole and a smooth finish hardwood handle. Shipping weight, 50 lbs.

42L1778¼—Hercules Well Boring Outfit and Post Hole Digger, complete as described above..... **$7.20**

For complete line of Post Hole Implements see page 843.

Air and Water Force Pump.

Easy working. Geared head drive. High efficiency. For water supply systems. Pumps air and water at same time.

Highly efficient for pumping water from cisterns or shallow wells into pneumatic pressure tanks. Has all brass valves and geared head drive. 1¼-inch suction and 1-inch discharge. 4-inch stroke; 3-inch brass lined cylinder. Shipping weight, 76 pounds.

42L2188¼

Pump, as described.... **$13.45**

Double Acting Pump.

Same as 42L2188¼, illustrated at left and described above, except that it does not have brass air pump attached at rear of cylinder. Pumps liquid only; no air. Much used for pumping water to elevated tanks, cleaning, filling and testing boilers and tanks, for fire protection and as a deck pump. Shipping weight, 65 pounds.

42L2186¼ With 3-inch iron cylinder, **$11.45**

Suction limit, 20 feet.

42L2187¼ With 3-inch brass lined cylinder....... **$12.45**

Acme Rotary Power Force Pump.

Will force water a horizontal distance of 200 feet and throw a solid stream 20 feet. An excellent pump for raising a large amount of water with minimum amount of power where lift is not over 15 feet and elevated tank into which water is forced not over 50 feet above pump. Reliable protection in case of fire. Driving shaft made long enough to allow use of balance wheel with handle so pump may be worked by hand when desired. Spout threaded for iron pipe at end and also at top where it connects to pump. Can be run at 200 revolutions per minute without injury, although 100 is recommended. One of the simplest and most satisfactory power pumps on the market. High speed permits belting direct to engine. If wanted with balance wheel and handle for hand power, allow $5.50 extra. Prices quoted do not include pipe or belt.

Catalog No.	Gals. per Minute	H.-P.	Pipe Size, Inches: Inlet	Pipe Size, Inches: Outlet	Pulley, Inches	Shipping Weight	
42L2270¼	13	1	1¼	1	10x3	80 lbs.	$13.25
42L2271¼	14	1	1¼	1	10x3	86 lbs.	14.55
42L2272¼	17	1	1½	1¼	10x3	98 lbs.	16.95
42L2273¼	27	2	2	1½	12x3½	145 lbs.	30.95
42L2274¼	36	3	2	2	12x3½	157 lbs.	34.95

For Gasoline Engines see pages 818 and 819.

Water Conductors.

42L856—Hang on pump spout and connect with iron pipe for conducting water to tank or trough. **State size wanted.** Shipping weight, 5¼ pounds.

1¼-inch pipe.................. **38c**

1½-inch pipe.................. **48c**

Little Giant Pipe Holder.

42L1780¼—Harder the pull tighter the grip. For 1, 1¼ and 1½-inch pipe. Dog has corrugated chilled surface. Shipping weight, 35 lbs. **$2.48**

Pipe Lifting Clevis.

42L854—Used to prevent pipe from slipping when being taken from well. For 1¼ and 1½-inch pipe. Shipping weight, 1½ lbs. **95c**

Malleable Iron Driving Caps.

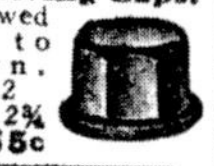

42L834 — Screwed on top of pipe to protect threads when driving it down.

Size, inches	1¼	1½	2
Shipping weight, lbs	1½	2	2¾
	25c	30c	55c

Iron and Brass Body Pump Cylinders.

10-inch cylinders have 6-inch stroke, 12-inch an 8-inch stroke, and 16-inch a 10-inch stroke. Cylinders 3½ inches in diameter fitted for 1½-inch pipe; 4 inches in diameter for 2-inch pipe; all others fitted for 1¼-inch pipe. 10-inch cylinders have one leather on plunger. 12 and 16-inch cylinders have two. Brass body cylinders have a brass cage and valve in plunger.

IRON BODY CYLINDERS.

	Diam., Inches	Shpg. Wt., Lbs.	10 In. Long	Wt., Lbs.	12 In. Long	Wt., Lbs.	16 In. Long
42L1100¼	2	9	$1.25	10	$1.85	12	$2.05
42L1101¼	2½	11	1.45	12	2.05	16	2.35
42L1102¼	3	12	1.70	14	2.35	18	2.70
42L1103¼	3½	15	2.35	19	3.05	24	3.80
42L1104¼	4	21	3.05	24	3.85	30	4.85

BRASS BODY CYLINDERS. Weights same as Iron Body.

	Diam., In.	10 In. Long	12 In. Long	16 In. Long
42L1112¼	2	$3.15	$3.35	$3.80
42L1113¼	2½	3.35	3.60	4.30
42L1114¼	3	3.80	3.95	4.75
42L1115¼	3½	4.40	4.75	5.70
42L1116¼	4	5.50	5.95	7.40

42L838

Float Valve to regulate supply of water in watering troughs, tanks, etc. Float not included.

Size, inches	¾	1	1¼	1½
Shpg. wt., lbs.	3	3	3¼	7
Valve without float	85c	98c	$1.15	$2.50

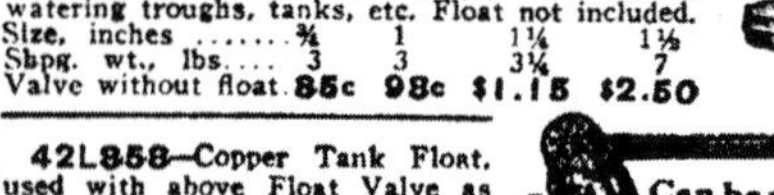

42L858—Copper Tank Float, used with above Float Valve as shown. Size, 9½x2¾ inches. Shipping weight, 2 pounds.... **$1.40**

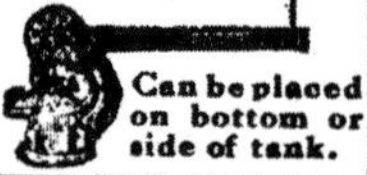

Can be placed on bottom or side of tank.

Spiral Earth Augers.

42L1704¼—Boring wells, prospecting, etc. Tool steel auger with pipe shank. Works inside of pipe. **State size.** Order auger one size smaller than inside diameter of pipe.

Outside Diam., Auger	Shpg. Wt.	Shank Pipe, Size	
2 in.	7 lbs.	1 in.	$4.95
2½ in.	12 lbs.	1¼ in.	5.50
3 in.	14 lbs.	1¼ in.	5.90

Sand Pump and Drill.

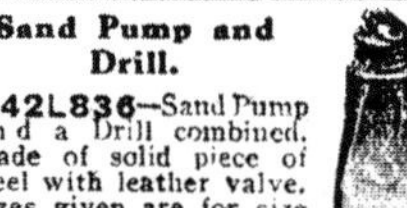

42L836—Sand Pump and a Drill combined. Made of solid piece of steel with leather valve. Sizes given are for size pipe drills will pass through. Shipping weight, 4½ pounds.

Size, inches	2	2½	3
Threaded for pipe size	1 in.	1 in.	1¼ in.
	$1.85	$2.95	$3.95

Foot Valves.

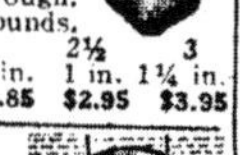

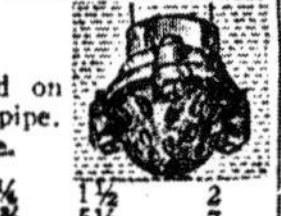
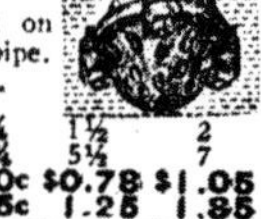

42L894 — Screwed on end of pump suction pipe. Acts as a check valve.

Pipe size, inches	1¼	1½	2
Shpg. wt., lbs.	3¾	5½	7
Black	60c	$0.78	$1.05
Galvanized	95c	1.25	1.85

Drive Well Points.

Made of wrought iron pipe, galvanized inside and out after holes are punched. Covered with brass gauze, and gauze is covered and protected by a perforated brass jacket. No. 60 gauze is used for coarse sand and No. 100 gauze for fine sand. Post maul or drive cap shown in illustration not included.

42L1696¼—60-Gauze Drive Well Point.

42L1698¼—100-Gauze Drive Well Point.

Diameter, In.	Lgth., In.	Shpg. Wt., Lbs.	60-Gauze	100-Gauze
1¼	24	5	$1.25	$2.68
1¼	30	6¼	1.55	3.39
1¼	36	7½	1.89	3.98
1¼	42	8¾	2.25	4.68
1½	30	7½	1.98	3.89
1½	36	9	2.45	4.68
2	36	14	3.49	6.29
2	48	16	4.45	8.25

Galvanized pump rod couplings. Shipping wt., 2 oz.

42L820—Size, ⅜ inch, 14 threads to the inch........ **4c**

42L822—Size, 7/16 inch, 12 threads to the inch........ **5c**

42L824—Size, ½ inch, 14 threads to the inch........ **6c**

New Process Pump Leathers.

42L828—Made of good quality leather, saturated with a special compound which makes it waterproof and very tough. Shipping weight, 8 ounces.

Sizes are inside diameter of cylinders in which leathers are to fit.

Size, inches	2	2¼	2½	2¾	
Each	11c	13c	14c	16c	
Size, in.	3	3¼	3½	3¾	4
Each	18c	21c	24c	30c	35c

42L830—Made of very tough good quality leather. Shpg. wt., 4 oz.

Diam., cylinder, in.	2	2¼			
Each	4c	5c			
Diam., cylinder, in.	2½	2¾			
Each	6c	7c			
Diam., cyl'r, in.	3	3¼	3½	3¾	4
Each	8c	10c	11c	13c	14c

State size wanted.

When ordering 42L830 for iron cylinder give inside diameter of cylinder and when for brass body cylinders specify leathers ¼ inch less than inside diameter of cylinder.

Round Steel Pump Rod.

42L2290—For connecting pump cylinders to pump heads. Furnished in any length desired up to 20 feet. Size, 7/16 inch. **Not threaded.** Shipping weight, per foot, 8 oz.

Per foot.......................... **4c**

Threads, extra, per cut, 2 cents.

Five Hours Ironing in Less than Two

That's what you can do with an
Allen Ironer

Saves three to four hours' time every ironing day. For the benefit of those who are not familiar with ironing machines, we will state that the plain flat pieces, heavy or light, such as tablecloths, napkins, handkerchiefs, sheets, pillow slips, bedspreads, blankets, towels, doilies, centerpieces, curtains and similar articles which take so much time and strength, can be ironed in the Allen in about one-fifth the time it takes to do them up by hand, and the result will b much more pleasing. The design in table linens will stand out like new and the finish will be more perfect than you can possibly get by hand. The embossing in a bedspread and the embroidery of doilies, centerpieces and similar articles, stand out beautifully. Wearing apparel, such as shirts, kitchen aprons, house dresses, underwear, nightgowns, pajamas, children's rompers, dresses and all similar articles, can be almost completely ironed on the Allen. Ruffles and frills may require a slight finishing with a hand iron.

The Allen Ironer is the result of a long series of experiments, made with a view of providing an ironing machine that is easily controlled, simple enough to be sold without personal demonstration and at a price that is reasonable. If you will do one ironing on an Allen Ironer, that is all you need to prove that we have succeeded in our efforts. So sure are we of this that we will sell it with the understanding that you can try it out in comparison with any other ironer on the market, and if **you can find a machine that is easier to handle, that will do better work, a greater variety of work, or which you consider a more desirable machine than the Allen, you can return it at our expense and we will return your money.**

The highly polished shoe of the Allen Ironer is 46 inches long, curved to fit the padded roll, having an actual contact of about 7 inches, the full length of the roll, making an ironing surface of more than 250 square inches. The movement of the roll is controlled by a foot pedal arrangement extending the entire length of the machine, so it can be controlled from any position. Machine can be used in a sitting position as well as standing. Pressure of roll against the shoe is maintained by two oil tempered steel springs, with means for adjusting the pressure as desired. Gearing is very simple, consisting of a steel worm, bronze worm wheel and a pair of spur gears. Drive is by belt. No complicated adjustments to make. One end of machine is open as illustrated for ironing collars, cuffs and the like.

Electrically Operated—Gas Heated.

The machine is furnished complete with high grade electric motor, cord and plug for connection to any electric light socket and flexible metal gas hose with rubber ends for connection with gas supply. Gas burner is fitted with control valve and is covered with a sheet metal shield. Shipping weight, 350 pounds.

Shipped from factory in NEWARK, OHIO.

44L7800—Allen Ironer. Complete, as described......... **$86.75**

Gasoline Heating Equipment.

We can also supply the Allen Ironer with equipment using gasoline as fuel for heating. The gasoline attachment is self contained and consists of a brass tank holding about 2 quarts, which will operate the burner for about five hours, so one filling is more than sufficient to do a large ironing.

44L7804—Allen Ironer with electric drive and gasoline heating equipment, as described. Complete **$105.00**

This is an exceptionally well made machine of a very popular type. Measures 19x28½x14 inches inside. **Shipped from CHICAGO or PHILADELPHIA, whichever is nearest.** Shipping weight, 50 lbs.

44L7530

$5.25

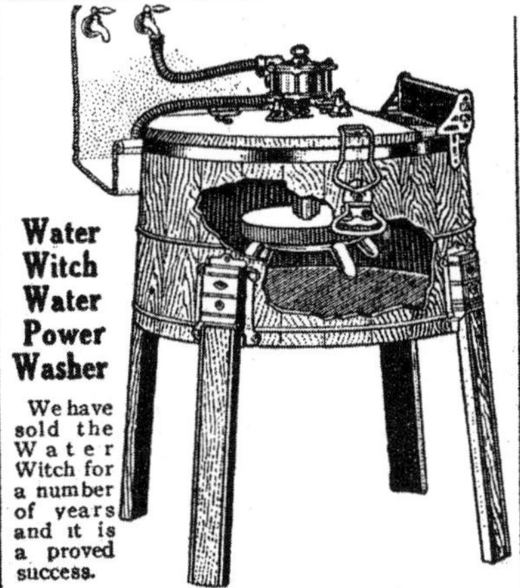

Water Witch Water Power Washer

We have sold the Water Witch for a number of years and it is a proved success.

To use a water power washer it is necessary that the water be supplied from a pumping system of some kind at a pressure of not less than 15 to 20 pounds to the square inch and a flow of 4 to 6 gallons per minute.

The Water Witch motor is of the gearless type, made entirely of brass and is very simple. Inlet hose is provided with standard hose coupling.

Full directions for use furnished. Shipping weight, 70 pounds. **Shipped from CHICAGO, PHILADELPHIA or factory in CENTRAL OHIO, whichever is nearest.**

44L7518—Water Witch Water Power Washing Machine.......... **$16.45**

44L7519—Smooth Faucet Hose Attachment. A device that enables you to connect hose to any standard smooth faucet. Weight, 1 pound........................ **45c**

High Speed Wizard

This is the most popular and successful hand operated washer we have ever sold. It will wash the clothes clean, runs easily, and will give good service. The design of our High Speed gearing is such that it requires very little effort to keep the machine in operation. All parts are much heavier than you will usually find in a machine of this type, so that it will give long service. Handle socket is arranged so that you can operate it from either a sitting or standing position. Sold subject to our well known "money back if not satisfied" guarantee, and you have full thirty days' trial before you decide whether you will keep it. Shipping weight, 90 pounds. **Shipped from CHICAGO, PHILADELPHIA or factory in CENTRAL OHIO, whichever is nearest you.**

44L7513—High Speed Wizard Washer........ .... **$16.95**

Golden Crown Washer.

The Golden Crown Washer is a high grade full size rotary machine. Turning the flywheel in one direction causes the dolly inside the tub to turn a three-quarter revolution and then automatically reverse. A very satisfactory low priced rotary washer. **Shipped from CHICAGO, PHILADELPHIA or factory in CENTRAL OHIO, whichever is nearest.** Shipping weight, 85 pounds.

44L7524—Golden Crown Washer.......................... **$12.25**

The New Improved Allen Electric Washer

All the Latest Improvements

High Sanitary Base :: Swinging Wringer

Oscillating Wood Cylinder

Enclosed Gears

In our 1923 Model Improved Allen Washer we have incorporated every feature that years of experience has proved desirable. **We have adopted the high sanitary base,** so it is easy to clean under the machine instead of forming a dirt catcher, as is the case where cabinet extends to the floor.

The Swinging Wringer is the latest power type with safety release; can be operated in either direction and swung to various positions as illustrated at the right.

The gearing is enclosed by a metal panel, locked shut by simple catches, but is easily opened to expose mechanism for inspection.

The Oscillating Wood Cylinder has been in use for years and is no experiment. The clothes are placed in the wood cylinder which rocks back and forth and are cleaned in a remarkably short time. The wood cylinder will not stain the clothes and is easily kept sweet and clean.

The Allen is a well constructed machine and will give years of satisfactory service.

The outer casing and water container are built of "ARMCO" Iron, which is a special grade of galvanized iron and practically rustproof. The frame is securely supported by the heavy corner angles, which also form the legs, and machine is mounted on large, easy rolling casters.

A brass drain faucet is provided. Motor is full ¼ horse-power and is powerful enough to operate both washer and wringer at same time. Machine comes complete ready to connect to any electric light socket and is fully guaranteed.

Our trial offer. Send us your order for an Allen Washer, enclosing our price of $89.50. Use it in your home for 30 days. Compare it with any other machine.

If you can find any washer that will do better work or which you think is better constructed, or a better value than the Allen, we want you to return the Allen at our expense and all your money will be returned to you.

The Allen Electric Washer measures 37½ inches high, not including wringer, 24½ inches wide and 27 inches long, outside measurements. It is handsomely finished in white enamel with gray trimming. Extra tubs, bench and basket illustrated are not included. **Shipped from factory in SOUTHERN MICHIGAN.** Shipping weight, 300 pounds.

44L7565—Improved Allen Washer.................................... **$89.50**

Wood Tub Electric Washers

Angle Steel Base. — **Swinging Reversible Wringer.**

Dolly Type. — **30 Days' Trial.**

These machines operate on the time proved and popular dolly (sometimes called peg or dasher type) principle with which most everyone is familiar.

The tubs are full size, measuring 22¾ inches in diameter and 12 inches deep inside, and have a washing capacity of six sheets or their equivalent in other clothes, and are made of specially selected washing machine lumber, which is practically rot and warp proof. Brass faucet is provided for draining tubs and it is threaded so hose can be attached.

The gearing is designed for heavy power work and is not hand machine gearing changed slightly to be operated by power.

Thirty days' trial allowed. Order either one of these machines, enclosing our catalog price. Use it in your home for 30 days. Compare it with any other machine from the standpoint of mechanical construction or washing efficiency, and if for any reason you are not perfectly satisfied with it, you can send it back and every cent of your money will be returned to you. Judge for yourself whether you want to keep the machine.

The Liberty Electric Washer.

We have sold this machine for a number of years and the thousands in use have demonstrated its success. If you want a single tub dolly type electric washer, the Liberty will suit you in every respect and our price will save you money.

The mechanism is very simple. The drive pulley and gears are mounted on a single large casting securely attached to the steel base. Raising the lid stops the washing without the use of clutches or other device. The power driven wringer can be swung into position to wring the clothes from the washer, rinsing tubs or bluing tubs. Wringer drive is simple, a slight movement of the lever being sufficient to start, stop or reverse the movement of the rolls, which are made of extra grade white rubber, securely vulcanized on the shafts.

The steel stand is fitted with casters and as it measures but 24 inches square, it is easily stored away. It can be used in connection with stationary laundry tubs, and in this case the swinging wringer is a big convenience. If you use ordinary tubs for rinsing and bluing, these can be placed on any bench you may have. Motor will operate washer and wringer at the same time. The Liberty Electric Washer is furnished complete with cord and plug ready to connect with any electric light socket, and full directions for operating. **Shipped from factory in CENTRAL OHIO.** Shipping weight, 255 pounds.

44L7555—Liberty Electric Washing Machine, complete with swinging wringer and electric motor, as described.. **$59.95**

Double Tub Electric Washer.

This double tub machine is especially recommended. The advantages of a double tub machine are many. Both tubs can be used for washing and, together with the wringer, can be operated at the same time, or either tub or wringer can be operated independently of one another. You can wash twice as fast as with a single tub machine; one tub can be used for slightly soiled light articles, while the other can be used for heavier or colored clothes which require different treatment. While washing in one tub, the other can be used for rinsing or bluing, and when machine rinsed, the rinsing is more thoroughly taken care of than when done in the old way. **Each double tub machine is fitted with a folding basket rack strong enough and large enough to hold two more ordinary tubs,** or a tub and basket. The gearing is extra heavy, very simple and fully enclosed. The main gears are packed in grease, and all the bearings are very generous in size and made with grease retainers so that very little attention is required.

The wringer is the latest type full size power wringer, mounted so it can be swung in various positions.

The electric motor is of standard make and is guaranteed to operate both washer tubs and wringer at the same time.

Machine is furnished complete with all necessary connections and directions ready to run. **Shipped from factory in CENTRAL OHIO.** Shipping weight, 375 pounds.

44L7572—Double Tub Electric Washer, complete as described........................ **$93.75**

Engine Driven Power Washers

All washers on this page are sold subject to thirty days' trial with our usual return privilege and money back if not satisfied.

Here is a large double capacity washing machine which is just the thing for the home having no electric power.

The washer is made up of two full size standard power washer tubs which are made of lumber specially adapted for the purpose. **These are mounted on a steel stand,** securely braced and fitted with a folding rack large enough and strong enough to carry two more tubs or a tub and basket. Large, easy rolling casters are provided, so machine is easily moved.

The gearing is very simple and entirely covered or enclosed. The main gears run in a greasetight case and are packed with grease. The bearings are designed with grease retainers, so little attention is required.

The wringer is a full size standard power wringer and swings to various positions, as illustrated below.

The engine is mounted under the tub out of the way, but is easily accessible. We are particularly proud of the engine on this machine. It will develop enough power to operate both washer tubs and wringer at the same time. The engine can be started by an easy pull on the strap of the pull starter, as illustrated below. The engine on this washer is a very simple, yet highly efficient, air cooled four-cycle type, developing in excess of ½ horse-power. When you see this engine you will be just as enthusiastic about it as we are. The construction is exceptionally good and is on the same lines as a high grade automobile engine. For example, the crankshaft is a drop forging, ⅞ inch in diameter, and is mounted in two large bronze bearings. The connecting rod is strong enough for an engine twice the size. Valves are in the removable head and of large size. Exhaust valve is operated by the usual adjustable rocker arm and push rod. Inlet valve is automatic. Gasoline tank is in the base and carburetor is automatic, requiring no adjustment. Jump spark ignition is used and each outfit includes a high grade spark coil, four-cell dry battery sealed in waterproof case, and necessary connections.

We especially recommend this outfit for its many advantages.

Both tubs can be used for washing and, together with the wringer, can be operated independently of one another. You can wash twice as fast as with a single tub machine: one tub can be used for slightly soiled light articles, while the other can be used for heavier or colored clothes which require different treatment. Where clothes are badly soiled they can be first washed in one tub with a warm suds until they are practically clean, and a second washing in the other tub with very hot suds finishes the job. While washing in one tub, the other can be used for rinsing or bluing and, when machine rinsed, the rinsing is more thoroughly taken care of than when done in the old way.

For the most efficient kind of a washing machine, order this one if you have no electric power. **Shipped from factory in CENTRAL OHIO.** Shipping wt., 400 lbs.

44L7573—Double Tub Washer with attached engine, complete as described **$117.50**

$117.50

These small illustrations show the advantages of the swinging wringer.

A pull on the strap of our mechanical starter starts the engine quickly and easily.

Wringing clothes from one washer tub to the other.

While one lot is being washed, the operator is wringing another lot from the other tub into the rinse tub placed on folding tub stand.

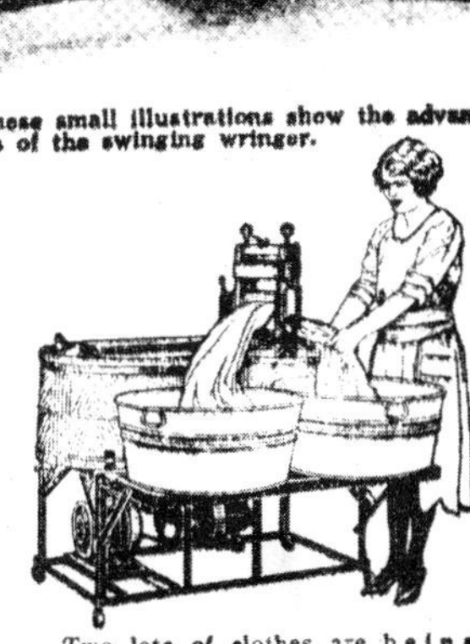
Two lots of clothes are being washed in the washing tubs while another is wrung from the rinse water to the bluing water.

Double Tub Power Washer

$68.25

This is the same washer as described above, but is fitted with plain power pulley, 8 inches in diameter, 1½-inch face and must be driven about 300 revolutions per minute.

If you already have an engine or line shaft outfit and do not want a machine with attached engine, this is just the washer for you. It has all the advantages of the power equipped machine shown above and we are sure it will please you. **Shipped from factory in CENTRAL OHIO.** Shipping weight, 350 pounds.

44L7574—Double Tub Power Washer with pulley, as described **$68.25**

Liberty Power Washer

This is the same Liberty Washer as described to the right under 44L7576, but without power, and is recommended to those who already have an engine or line shaft outfit and want a power washing machine. No casters are furnished. Washer is fitted with drive pulley, 10 inches in diameter, 1¾-inch face, and should be driven 200 revolutions per minute. **Shipped from factory in CENTRAL OHIO.** Shipping weight, 200 lbs.

44L7543—Liberty Power Washer with plain power pulley.... **$39.85**

The Farwell Power Washer

We have had a great many requests for a good power washer without a wringer, and the Farwell was designed to meet that demand. It is a well built, full size power machine, made up of material which years of experience has proved to be the best for the purpose. It has a plain power pulley, 8 inches in diameter, 2¼-inch face, which should be driven at approximately 225 revolutions per minute. **Shipped complete from factory in CENTRAL OHIO.** Shipping weight, 250 pounds.

44L7580—Farwell Power Washer **$22.95**

The Liberty Power Washer With Attached Engine.

$87.50

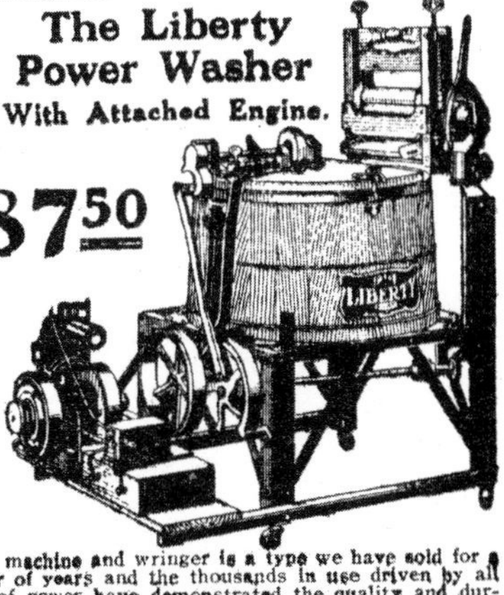

This machine and wringer is a type we have sold for a number of years and the thousands in use driven by all kinds of power have demonstrated the quality and durability of the machine. The engine and washer are compactly mounted together on a substantial steel base with casters, making it portable and easy to put away between washdays.

Place your rinsing and bluing tubs on any bench or stand you may have.

The engine furnished is the same highly efficient engine furnished with our double tub outfit as described above.

Direction card sent with each outfit describing how to operate and care for it. **Shipped from factory in CENTRAL OHIO.** Shipping weight, 315 pounds.

44L7576—Liberty Power Washer with attached engine **$87.50**

FINE QUALITY PLAIN CUT AND ETCHED CRYSTAL (CLEAR) GLASS TABLEWARE

Pressed Colonial Stemware.

Made of medium weight clear crystal pressed glass, highly polished in a Colonial shape. This Colonial glassware is in the best of taste and most desirable for anyone wanting a neat, plain line of tableware at a low price.

35L2300—Water Goblet. Height, 6¾ in. Weight, per dozen, 7 lbs. Per dozen. **$1.75**

35L2301—Tall Footed Sherbet. Ht., 4½ in. Weight, per doz., 4 lbs. Per doz **$1.70**

35L2302—Low Footed Sherbet. Ht., 3⅝ in. Wt., per dozen, 3 lbs. Per dozen **$1.35**

35L2303—Standard Table Tumbler. Ht., 4 in. Wt., per dozen, 5 lbs. Per doz.. **$1.45**

35L2304—Tall Lemonade or Iced Tea Tumbler. Height, 5 inches. Weight, per doz., 6 pounds. Per dozen.................**$1.80**

Plain Blown Stemware.

Made of thin blown crystal glass, highly polished, with one-piece drawn stems, in the popular and handsome Fifth Avenue shape. This plain blown tableware is always in good taste and will look well on any table.

35L2000—Water Goblet. Height, 7 in. Weight, per doz., 3 lbs. Per dozen **$3.48**

35L2001—Tall Footed Sherbet. Height, 4⅝ in. Wt., per doz., 3 lbs. Per doz **$3.45**

35L2002—Low Footed Sherbet. Height, 3⅜ in. Wt., per doz., 3 lbs. Per doz **$2.98**

35L2003—Standard Table Tumbler. Ht., 3¾ in. Wt., per doz., 3 lbs. Per doz **$1.10**

35L2004—Tall Lemonade or Iced Tea Tumbler. Height, 5½ in. Weight, per doz., 3 pounds. Per dozen....................**$1.98**

Needle Etched Stemware.

Made of thin blown crystal glass, highly polished, in a pleasing new shape. Solid one-piece drawn stems. Decorated with a wide needle etched border in a standard and ever popular design. A very pretty pattern.

35L2200—Water Goblet. Height, 6½ in. Weight, per doz., 5 lbs. Per doz.. **$4.98**

35L2008—Tall Footed Sherbet. Height, 5 in. Wt., per doz., 3½ lbs. Per doz **$4.95**

35L2009—Low Footed Sherbet. Height, 3 in. Wt., per doz., 2½ lbs. Per doz **$4.65**

35L2214—Standard Table Tumbler. Ht., 3¾ in. Wt., per doz., 3 lbs. Per doz **$2.35**

35L2216—Tall Lemonade or Iced Tea Tumbler. Height, 5½ in. Weight, per doz., 4½ pounds. Per dozen............**$2.95**

Star Cut Stemware.

Made of thin blown crystal glass, highly polished, with one-piece drawn stems. Each piece is cut with three 6-point polished stars with cut silver gray rays. Cut on the handsome and popular Fifth Avenue shape.

35L2900—Water Goblet. Height, 6¼ in. Weight, per doz., 4½ lbs. Per doz **$4.95**

35L2901—Tall Footed Sherbet. Ht., 4½ in. Wt., per doz., 3½ lbs. Per doz **$4.90**

35L2902—Low Footed Sherbet. Height, 3⅜ in. Wt., per doz., 3 lbs. Per doz.**$4.65**

35L2903—Standard Table Tumbler. Ht., 4⅛ in. Wt., per doz., 3 lbs. Per doz.**$1.75**

35L2904—Tall Lemonade or Iced Tea Tumbler. Height, 5⅝ in. Weight, per doz., 3½ pounds Per dozen............**$2.65**

Optic Blown Stemware.

Made of thin blown crystal glass, highly polished, with one-piece drawn stems, in a very attractive fancy optic (fluted) shape. This glassware is a happy medium between the plain and decorated glass and will appeal to those wanting a simple pattern.

35L3000—Water Goblet. Height, 6⅞ in. Weight, per doz., 3½ lbs. Per doz **$4.48**

35L3001—Tall Footed Sherbet. Height, 4⅞ in. Wt., per doz., 3 lbs. Per doz **$4.45**

35L3002—Low Footed Sherbet. Height, 3⅝ in. Wt., per doz., 3 lbs. Per doz **$4.18**

35L3003—Standard Table Tumbler. Ht., 3¾ in., Wt., per do 3 lbs. Per doz **$2.10**

35L3004—Tall Lemonade or Iced Tea Tumbler. Height, 5½ in. Weight, per doz., 4½ pounds. Per dozen..............**$2.95**

Daisy Cut Stemware.

Made of thin blown crystal glass, highly polished, with strong solid one-piece drawn stems. Each glass is cut in silver gray finish with daisies and their stems and leaves on both sides of a new, attractive shape. This daisy pattern is very pleasing.

35L2800—Water Goblet. Height, 6½ in. Wt., per doz., 5 lbs. Per doz..**$6.98**

35L2811—Tall Footed Sherbet. Height, 4½ in. Wt., per doz., 3 lbs. Per doz..**$6.95**

35L2810—Low Footed Sherbet. Height, 3 in. Wt., per doz., 2½ lbs. Per doz **$6.75**

35L2805—Standard Table Tumbler. Ht., 3¾ in. Wt., per doz., 2½ lbs. Per doz **$3.68**

35L2806—Tall Lemonade or Iced Tea Tumbler. Height, 5½ in. Weight, per doz., 4½ pounds. Per dozen............**$4.60**

Plate Etched Stemware.

Made of thin blown crystal glass, highly polished, in a new aristocratic optic (fluted) shape. Solid one-piece drawn stems. Beautifully decorated with a genuine plate etched border in fuchsia design. This is a very high grade line. A rich and handsome glassware.

35L2700—Water Goblet. Height, 7 in. Weight, per dozen, 5 lbs. Per dozen...**$8.98**

35L2708—Tall Footed Sherbet. Height, 4½ in. Wt., per doz., 3 lbs. Per doz**$8.95**

35L2709—Low Footed Sherbet. Height, 3½ in. Wt., per doz., 2½ lbs. Per doz.**$8.75**

35L2714—Standard Table Tumbler. Ht., 3¾ in. Wt., per doz., 3 lbs. Per doz **$5.48**

35L2716—Tall Lemonade or Iced Tea Tumbler. Height, 5½ in. Weight, per doz., 4½ pounds. Per dozen..............**$6.45**

Floral Cut Stemware.

Made of thin blown crystal glass, highly polished, with one-piece-drawn stems. Each glass is cut with silver gray daisies and a beautiful garland border in silver gray on a new shape which is very handsome and pleasing. This is a high grade line of cut tableware.

35L3100—Water Goblet. Height, 7 in. Wt., per doz., 4½ lbs. Per dozen. **$8.50**

35L3101—Tall Footed Sherbet. Height, 5⅝ in. Wt., per doz., 4 lbs. Per doz.**$8.45**

35L3102—Low Footed Sherbet. Ht., 4⅛ in. Wt., per doz., 4½ lbs. Per doz **$8.25**

35L3103—Standard Table Tumbler. Ht., 3⅞ in. Wt., per doz., 3 lbs. Per doz **$4.98**

35L3104—Tall Lemonade or Iced Tea Tumbler. Height, 5¼ in. Weight, per dozen, 4½ pounds. Per dozen..............**$5.75**

Floral Cut Sherbets. Made of thin blown crystal glass. Cut on both sides with floral spray in silver gray finish. **Low Footed Sherbet.** Height, 3¼ inches. Weight, per dozen, 3 lbs.

35L2818—Per dozen **$4.68**

Tall Footed Sherbet. Height, 4½ inches. Weight, per dozen, 4 lbs.

35L2819—Per dozen..............**$4.98**

Etched Footed Sherbet.

Made of thin blown crystal glass, highly polished, with one-piece drawn stems in a new attractive shape. The decoration is a deep plate etching in a Dresden border design. This handsome plate etching is exceedingly graceful and dainty. Height, 3½ inches. Weight, 2½ pounds.

35L2813—Per dozen..............**$9.75**

15-Piece Iced Tea or Lemonade Set

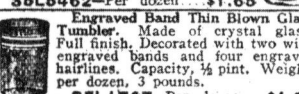

This Fifteen-Piece Iced Tea or Lemonade Set is one of the most practical sets offered. Its appeal is instant. It is an ideal set for use in the parlor or on the porch. It will also make a handsome and pleasing gift. This set contains all the pieces necessary for serving cool and refreshing beverages. The 10½-inch covered pitcher and six tall bell shape iced tea or lemonade glasses are made of cut thin blown crystal glass. The decoration consists of an allover silver gray floral cutting with deep cut and polished leaves and stems. The tray is made of wood in mahogany finish with cut-out handles. The bottom of the tray is decorated with conventional border and center design. The hollow glass spoon sippers are made of clear thin blown glass with assorted colored bowl. This very practical and pleasing set contains the following pieces: One large 10½-inch covered pitcher; six tall iced tea or lemonade glasses; six glass spoon sippers; one mahogany finish serving tray with glass bottom, 13¾x19¾ inches. Weight, packed, 13 pounds.

35L134—Iced Tea or Lemonade Set with Tray..........................**$6.75**

Gold Band Colonial Tumbler Set.

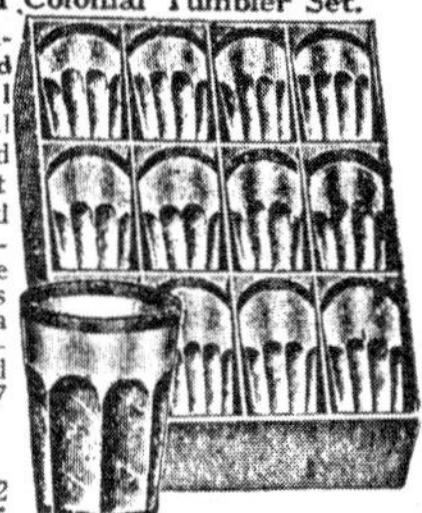

Twelve tumblers of good quality crystal glass in Colonial style. Decorated with wide bright gilt band around the edge. Capacity, ½ pint. The twelve tumblers are put up in a neat compartment pasteboard box. Weight, 7 pounds.

35L5488

Per box (12 tumblers) **$1.45**

Colonial Footed Sherbet.

For ice cream or sundaes. Made of pressed crystal glass, highly polished. Heavy pressed crystal bottom. Ht., 3 in. Wt., doz., 4 lbs.

35L5543—Per doz..**$1.18**

Colonial Footed Sherbet or Sundae.

Made of pressed crystal glass, highly polished, with an attractive fluted or rib design around the lower part. Fancy stem. Height, 3 inches. Weight, per dozen, 4 lbs.

35L5463—Per dozen.....**$2.45**

Needle Etched Footed Sherbet.

Made of thin pressed crystal glass ornamented with needle etched band. Height, 3½ inches. Weight, per dozen, 5¼ pounds.

35L5418

Per dozen **$2.98**

Plain Footed Sherbet.

For everyday use. Made of pressed crystal glass, highly polished. For sundaes, ice creams, etc. Height, 3 inches. Weight, per dozen, 3 lbs.

35L5542—Per dozen..**$1.35**

Colonial Table Tumbler.

Made of pressed crystal glass, highly polished, with an attractive fluted or rib design around the lower part. Pressed star and ground bottom. Capacity, ½ pint. Ht., 3¼ in. Weight, per dozen, 6½ pounds.

35L5460—Per dozen..............**$1.98**

Colonial Water Goblet.

Made of pressed crystal glass, highly polished, with an attractive fluted or rib design around the lower part. Fancy stem. Height, 5½ inches. Weight, per dozen, 13½ pounds.

35L5461—Per dozen..**$2.95**

Needle Etched Handled Custard.

Made of thin pressed crystal glass ornamented with needle etched band. Wt., per doz., 4 pounds.

35L5419—Per doz **$2.78**

PLAIN AND FANCY TABLE TUMBLERS.

Medium Weight Full Finished Pressed Glass Colonial Tumbler. Made of full finish crystal glass, polished bottom. Height, 4 inches. Capacity, ½ pt. Weight, doz., 6 lbs.

35L5470—Per dozen.................**79c**

Extra Heavy Full Finished Pressed Glass Colonial Tumbler. Fancy shape. Clear crystal glass. Finished and polished bottom of extra heavy weight. Cap., ⅝ pint. Wt., per doz., 7½ lbs.

35L5462—Per dozen.....**$1.65**

Engraved Band Thin Blown Glass Tumbler. Made of crystal glass. Full finish. Decorated with two wide engraved bands and four engraved hairlines. Capacity, ½ pint. Weight, per dozen, 3 pounds.

35L1767—Per dozen....**$1.38**

Plain Thin Blown Glass Tumbler. Straight shape. Made of crystal glass. Full finish. Light and thin. Capacity, ½ pint. Weight, per dozen, 3 pounds.

35L1755—Per dozen..............**95c**

Needle Etched Thin Blown Glass Tumbler. Bell shape. Wide needle etched band in a standard and popular design. Full finish. Capacity, ½ pint. Weight, per dozen, 3 lbs.

35L5496—Per dozen..**$1.95**

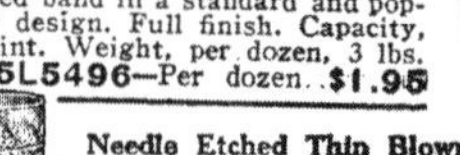

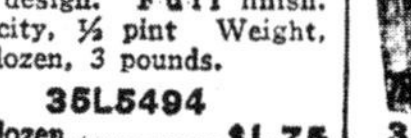

Needle Etched Thin Blown Glass Tumbler. Straight shape. Wide needle etched band in a standard and popular design. **Full finish.** Capacity, ½ pint Weight, per dozen, 3 pounds.

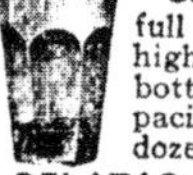

35L5494

Per dozen **$1.75**

Plain Thin Blown Glass Tumbler. Bell shape. Made of crystal glass. Full finish. Light and thin. Good quality. Capacity, ½ pint. Weight, per dozen, 5½ pounds.

35L1765—Per dozen.......**98c**

Grape Border Decorated Thin Blown Glass Tumbler. Decorated with an enamel border composed of grapes with stems and leaves. Height, 3¾ inches. Capacity, ½ pint. Weight, per dozen, 3 pounds.

35L1751—Per dozen..**$1.39**

Colonial Iced Tea Tumbler. A full finished tall pressed glass, highly polished. Heavy ground bottom. Height, 5 inches. Capacity, ¾ pint. Weight, per dozen, 13 pounds.

35L1710—Per dozen..........**$1.75**

9-Ounce Heavy Ground Bottom Hotel Tumblers. Made of pressed crystal glass. Full finish. Non-nesting. Ground heavy bottom. A strong, first class hotel tumbler. Weight, per dozen, 10 pounds.

35L6095—Per doz..**$1.48**

9-Ounce Fancy Barrel Shape Hotel Tumbler. A nice tumbler. Made of pressed crystal glass. Full finish. Non-nesting, with fluted and ground heavy bottom. Weight, per dozen, 7 pounds.

35L6092—Per dozen.........**$1.45**

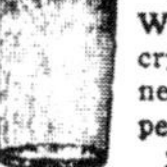

12-Ounce Heavy Bottom Milk or Water Glass. Made of pressed crystal glass. Full finish. Non-nesting. Height, 4¾ in. Weight, per dozen, 15 pounds.

35L6093—Per doz..**$1.89**

GIFT SHOP

Polychrome Book Ends. Made of composition and carved in a fancy design. Finished in polychrome and antique gold color bronze. Size, 6x7 inches. Weight, packed, per pair, 9 pounds.
35L1251—Per pair $2.85

8½-Inch Colored Glass Fruit or Flower Bowl. Made of blue iridescent glass in a very attractive shape and mounted on an ebony glass detachable base. The blue bowl and black base make a very pleasing combination. Height, 6⅓ inches. Weight, packed, 6½ pounds.
35L1252 .... $1.75

10-Inch Blue Iris Glass Fruit Bowl with Artificial Fruit. Made of iridescent glass in a popular shape and mounted on an ebony glass detachable base. Height, 5 inches. Weight, packed, 10 lbs.
35L1253 .... $3.25

Complete Table Decoration, consisting of a 5¼-inch hammered brass footed fern dish on 8-inch metal edge mirror plateau, filled with Japanese green air ferns, which do not need watering. Shipping weight, 4½ lbs.
35L1198 $1.98

Mirror Plateau.
Made of heavy plate glass, highly polished, with notched cut beveled edge. Mounted in a fancy embossed silver plated brass frame. Height, 2¼ inches.
35L201—10-in. Wt., 8½ lbs $3.98
35L202—12-in. Wt., 9½ lbs 4.98
35L204—14-in. Wt., 13 lbs 6.75

Polychrome Electric Lamp. The 11-in. base is made of composition and carved in a fancy design. Finished in polychrome and antique gold color bronze. The 10-inch shade is made of parchment paper, hand decorated. Height, 16 in. Weight, packed, 6 lbs. Complete with bulb and cord.
35L1283
$4.95

Iridescent Glassware.

Made of Thin Blown Crystal Glass, highly polished, with one-piece drawn stems in a very attractive fancy optic (fluted) shape. Iridescent glassware is very popular and will look well and harmonize with any dinnerware.
35L3030—Water Goblet. Height, 7 inches. Weight, per dozen, 13 pounds. Per dozen $5.98
35L3031—Tall Footed Sherbet. Height, 4⅝ inches. Weight, per dozen, 8 pounds. Per dozen $5.95
35L3032—Low Footed Sherbet. Height, 3⅜ inches. Weight, per dozen, 7 pounds. Per dozen $5.45
35L3033—Standard Table Tumbler. Height, 3¾ inches. Weight, per dozen, 6 pounds. Per dozen $2.98
35L3034—Tall Lemonade or Iced Tea Tumbler. Height, 5½ inches. Wt., per dozen, 8 pounds. Per dozen $3.85

35L915
7-Inch Handled Cut Glass Vase. Made of crystal glass in a very attractive shape. Cut in a floral and fern design in silver gray finish. Weight, 3 lbs. ... $1.65

35L1288
Polychrome Electric Candle. Made of composition, in bayberry style. Finished in polychrome and antique gold color bronze. Height, 20 in. Weight, 10 lbs. Complete with cord and bulb.
$3.25

35L1257
10½-Inch Cut Glass Bud Vase. Made of thin blown crystal glass with silver gray floral cutting. Fitted with artificial color rosebuds. Weight, 2 lbs.
98c

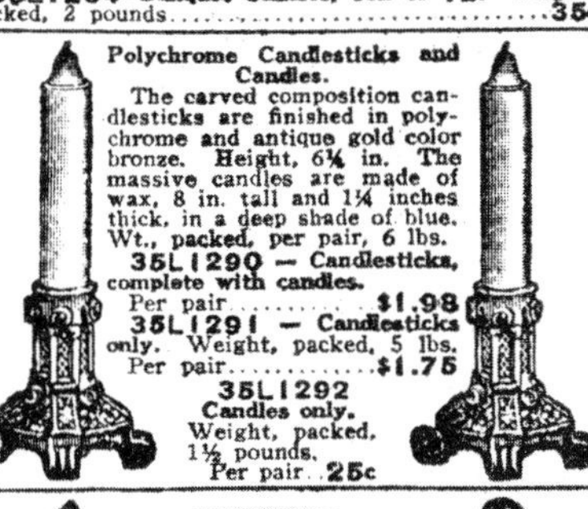

Three-Piece Polychrome Table Set.

Consists of one 8½-Inch Footed Compote and two 11½-Inch Candlesticks. Made of composition and carved in fruit and floral design. Finished in polychrome and gold color bronze with blue and red trimmings. The candlesticks, and compote, filled with artificial fruit, make an attractive table decoration. Wt., pkd., 12 lbs.
35L1262—3-Piece Set (without fruit) $4.98
35L1263—Artificial Fruit, illustrated in compote as follows: Apples, oranges, bananas, pears and peaches. Weight, packed, each, 1 pound $0.60
Grapes, per bunch. Weight, packed, 1½ lbs. 1.40
35L1264—Banquet Candles, box of 12. Weight, packed, 2 pounds 35c

Polychrome Candlesticks and Candles.
The carved composition candlesticks are finished in polychrome and antique gold color bronze. Height, 6¼ in. The massive candles are made of wax, 8 in. tall and 1¼ inches thick, in a deep shade of blue. Wt., packed, per pair, 6 lbs.
35L1290 — Candlesticks, complete with candles.
Per pair $1.98
35L1291 — Candlesticks only. Weight, packed, 5 lbs. Per pair $1.75
35L1292
Candles only. Weight, packed, 1½ pounds. Per pair. 25c

35L1265
Glass Salt and Pepper Set, in plated stand. Silver tops. Weight, pkd., 1 lb $1.78

35L1271
10-Inch Iridescent Glass Salad Bowl and 8½-Inch Plate. Optic effect. Wt., pkd., 4½ lbs. $1.18

35L1269
Blue Iris Color Glass Jelly or Bonbon Dish. Height, 6 inches. Weight, pkd., 3 lbs ... 95c

35L1267—Iridescent Glass Sherbet Set. Optic effect. Six footed sherbets and six 6-inch plates. Wt., packed, 5 pounds $1.95

35L2836
Cut Glass Candy Jar. Made of thin crystal glass and cut with large wild roses and rosebuds, leaves and stems in silver gray finish. Notched cut cover. Height, 10 inches. Capacity, 1 pound. Weight, 5 pounds $1.85

35L1287
14-Inch Tiffany Style Finish Boudoir Lamp with Silk Shade. The base is made of wood, and the 8-inch shade is made of silk with fancy braid around the top and bottom. Height, 14 in. Wt., 4 lbs. Complete with bulb and cord $3.98

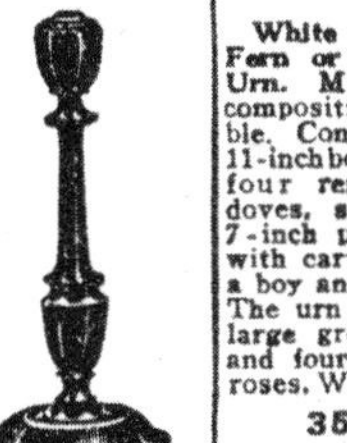

35L1280
12-Inch Candlesticks. Made of birch wood in mahogany finish in a dainty and graceful Colonial shape. Weight, 1½ lbs.
Each $0.98
Per pair 1.95

Cut Glass Covered Marmalade or Jam Jar with colored spoon. Made of thin blown crystal glass with silver gray floral cutting. Covers have assorted colored knobs. Height, 4¾ inches. Weight, packed, 2 pounds.
35L909 .... 69c

Colored Glass Whipped Cream or Mayonnaise Set. Made of topaz color glass. Set contains 5-inch dish and 8-inch plate and colored glass spoon. Weight, packed, 3½ pounds.
35L1255 .... $1.48

8-Inch Bulb or Flower Bowl. Made of mottled blue earthenware, waterproof. Just the thing for raising natural bulbs. Can also be fitted with a removable block and used for natural cut flowers or artificial flowers. Weight, packed, 4 pounds.
35L1286—Bowl and block $1.35

10½-Inch Fruit or Flower Bowl. Made of green iridescent glass in a very attractive shape mounted on a removable black base. Weight, packed, 8 pounds.
35L1259 .... $1.98

Kalo-Chrome Candles. Fancy candles are all the rage. Here is a variety to select from:

E D C B A

35L1274—(A) 6-Inch Twisted Banquet Candle. Blue, red or lavender. **State color wanted.** Weight, 1 lb. Box, 12 candles. 35c
35L1275—(B) 8¼-Inch Twisted Banquet Candle. Blue, red or lavender. **State color wanted.** Wt., 1⅛ lbs. Box, 8 candles. 35c
35L1276—(C) 10-Inch Handmade Candle. Blue, red or lavender. **State color wanted.** Wt., 1⅛ lbs. Box, 6 candles. 98c
35L1277—(D) 9-Inch Kalo-Chrome Tiffany Finish. Mottled blue, pink and green. Weight, 1 lb. Per pair 45c
35L1278—(E) 10-Inch Bayberry Style, "Etruscan" Finish. Weight, 1⅛ pounds. Per pair $1.25

White Marble Fern or Flower Urn. Made of composition marble. Consists of 11-inch bowl with four removable doves, set on 7-inch pedestal, with carved figures of a boy and girl at play. The urn is filled with large green air fern and four artificial red roses. Wt., pkd., 16 lbs.
35L1282
Complete $4.98

Aquarium or Fish Globe. Made of heavy crystal glass in a low squat shape and mounted in a wrought iron stand, decorated in verde green and antique gold color bronze. The frame, besides being ornamented, also prevents the bowl from tipping or falling over. Capacity, 2 gallons. Height, in frame, 10 in. Weight, packed, 10 pounds.
35L1284 .... $1.98

English Jet Teapot. Made of English earthenware in a popular shape. Decorated with a green design with colored enamel beading. The handle, spout and cover are traced with bright gold. Capacity, 3 pts. Wt., packed, 4 lbs.
35L570 .... $1.25

23-Piece Luster Tea or Lunch Set.

Tea sets in bright luster colors have become the very latest rage. The tea set shown here is made of semi-porcelain in a plain edge shape, which is very attractive. It is decorated all over with a pink luster. Around the edges of the cups and plates there is a narrow black band and a black hairline inside the cup. The handles, knobs and spout are decorated with black. The set consists of: 6 Tea Cups; 6 Tea Saucers; 6 Pie Plates; 1 Teapot, 1⅜-pint; 1 Creamer, ⅝-pint; 1 Sugar Bowl. Weight, packed, 18 pounds.
35L5660 .... $10.95

Hand Decorated Jardiniere.
Made of stoneware in a very attractive shape. Hand decorated all over with woodland scenes. Colors on a full tinted ivory green and brown background. Full glazed inside and out, making it waterproof.
35L1260—Size, 7½ in. Weight, packed, 8 lbs. $3.25
35L1261—Size, 8½ in. Weight, packed, 10 lbs $4.45
For Pedestals see page 628.

Children's Two-Piece Sets.
Made of cream color pottery and decorated with four small "bunnies" and a wide green band **and black hairline around the edge.**

35L1270
Children's Two-Piece Set. Rolled edge baby plate and handled mug. Weight, 4 pounds ... $1.25

35L1272—Children's Two-Piece Set. 6-inch mush and milk bowl and cream pitcher. Weight, 3 pounds 98c

Footed Brass Jardinieres.
Made of solid spun brass in a new Colonial shape. Finished in brush (satin) brass finish. **Three heavy brass ball feet.**
35L1201
Size, inside, 7¼ inches. Height, 7⅛ in. Shpg. wt., 3 lbs. ... $2.25
35L1202—Size, inside, 10¼ in. Height, 8⅛ in. Shipping. wt., 4½ lbs. $4.25

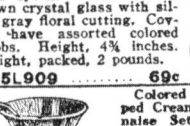

GLASS BRACKET OIL LAMPS

Fitted with Improved Burner and Chimney.

SWINGING KITCHEN BRACKET LAMP. | SWINGING DINING ROOM OR HALL BRACKET LAMP.

Solid metal swinging bracket with wall plate in bronze finish. Fitted with No. 2 Improved brass burner and newly designed straight top chimney, fully described on this page, wick and removable crystal glass fount. Back of the lamp, attached to the bracket frame, is a 7-in. mirrored glass reflector. Capacity, 1½ pints. Weight, packed, 6½ pounds.

35L722 .. $1.35

Fancy ornamental heavy solid metal swinging bracket with wall plate and fount holder in bronze gilt finish. Fitted with No. 2 Improved brass burner and newly designed straight top chimney, fully described on this page, wick and removable crystal glass fount. Back of the lamp, fastened to the upper part of the frame, is an adjustable 8-inch mirrored glass reflector. Capacity, 1½ pts. Weight, packed, 7½ lbs.

35L724 .... $1.68

NEW IMPROVED BURNER AND CHIMNEY

No Smoke. No Flicker. More Light.

The small feeble light from the old style burner and crimp top chimney.

The steady white light from the new burner and straight top chimney.

Exhaustive scientific experiments by the Standard Oil Company demonstrated the fact that the users of oil lamps fitted with the ordinary No. 2 Brass Burner and Crimp Top Chimney are not getting the maximum amount of light from this combination. The Standard Oil Company working with the manufacturers of burners and chimneys have developed an all brass No. 2 burner and a lead glass chimney with straight top which increases the efficiency of the oil and produces a far greater volume of light than has ever been obtained before from a No. 2 brass burner; and this without increasing the consumption of the oil.

You cannot use satisfactorily the new Burner and the old Crimp Top Chimney, neither can you use an old Burner with a new approved Chimney. These combinations will not give the desired result. To obtain this bigger and better light, **it is necessary to use both the new improved No. 2 Brass Burner and the new approved Straight Top Chimney.**

35L730—No. 2 Brass Burner and Macbeth-Evans Lead Glass Chimney. Weight, packed, 1¼ lbs. For both........ **$0.35**
35L731—Chimney only. (Wt., 2¾ lbs.) Per ½ dozen.. **1.20**
35L732—Burner only. Each......(Wt., 2 oz.)...... **.15**

BRASS BRACKET OIL LAMPS

SWINGING CENTER DRAFT BRASS BRACKET LAMP.

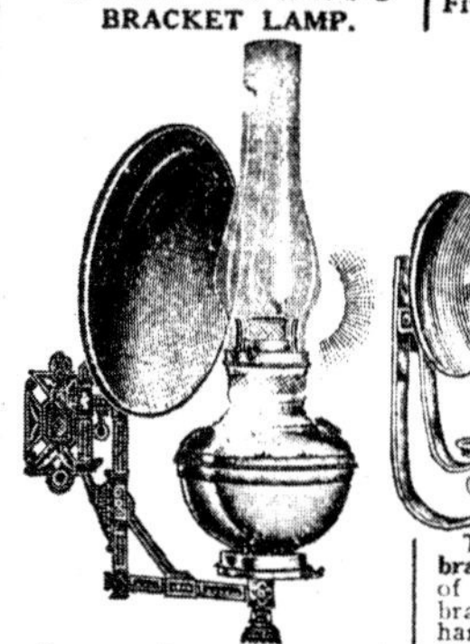

Large, strong ornamental solid metal swinging bracket with wall plate in full gilt finish. Fitted with polished brass removable oil fount and No. 2 center draft burner, chimney and wick. Back of lamp, fastened to top of frame, is an adjustable 10-inch mirrored glass reflector. A splendid light giver. Capacity, 2½ pints. Weight, packed, 8 lbs.

35L750 ...... $3.95

COMBINATION BRACKET AND TABLE LAMP.
Fitted With Improved Burner and Chimney.

The lamp is made of polished brass and the bracket is made of cold rolled steel, heavily brass plated. **The lamp has a handle by which it can be lifted off the bracket and makes** it easy to carry about as a hand lamp. Back of the lamp and fastened to the top of the frame, is a 6-inch mirrored glass reflector. Fitted with the No. 2 Improved Queen Anne brass burner, and newly designed straight top chimney described on this page, and wick. Capacity, 1½ pints. Height of lamp, 13½ inches. Weight, packed, 8 pounds.

35L721 .... $2.75

PLAIN GLASS OIL LAMPS.

Fitted with Improved Burner and Chimney.

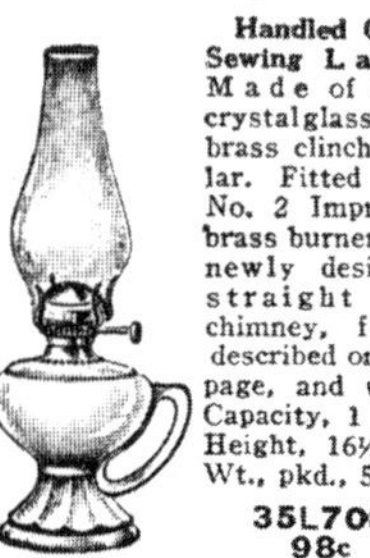

Handled Glass Sewing Lamp. Made of clear crystal glass with brass clinch collar. Fitted with No. 2 Improved brass burner and newly designed straight top chimney, fully described on this page, and wick. Capacity, 1 qt. Height, 16½ in. Wt., pkd., 5 lbs.

35L708
98c

Footed Stand Lamp. Made of clear crystal glass with brass clinch collar. Fitted with No. 2 Improved brass burner and newly designed straight top chimney, fully described on this page, and wick. Height, 19 in. Capacity, 24 oz. Wt., packed, 5 lbs.

35L711
78c

Footed Glass Sewing Lamp. Made of clear crystal glass with brass clinch collar. Fitted with No. 2 Improved brass burner and newly designed straight top chimney, fully described on this page, and wick. Height, 17½ in. Capacity, 24 oz. Wt., pkd., 5 lbs.

35L710
85c

FANCY GLASS LAMPS.

With Embossed Base and Chimney.

Extra Large Fancy Crystal Table or Sewing Lamp. Fitted with large No. 3 burner and wick. Richly embossed base and chimney. Clinch collar. Gives a fine light. Even a person with very weak eyes can use this lamp. Very ornamental. Oil capacity, 2½ pints. Height, 19 inches. Wt., pkd., 6 lbs.

35L728
$1.25

Richly Embossed Large Ruby and Gold Colored Sewing Lamp. Chimney decorated with ruby color flowers. Solid gilded base with embossing tinted ruby. Complete with No. 3 burner, chimney and wick. Oil capacity, 2½ pints. Clinch collar. Height, 19 in. Wt., packed, 6 pounds.

35L715
$1.48

KITCHEN, WALL OR TABLE LAMP.

Fitted With Improved Burner and Chimney.

Black enameled metal frame. Fitted with polished tin reflector, removable glass fount, No. 2 Improved brass burner and newly designed straight top chimney fully described on this page, and wick. Capacity, 1 pint. Height, 12¼ inches. Weight, packed, 6 pounds.

35L705 ...... 75c

"EVER-BRITE" Gasoline Table Lamp

$8.25 Complete | **$9.25 Complete**

Lights With a Match

The "Ever-Brite" Lamp is all that the name implies. It produces a light of unusually high candle power, which is of greater volume than that produced by twenty ordinary No. 2 flat wick kerosene lamps. Its consumption of fuel is very small, being only 4 pints of gasoline in twenty-four hours' continuous burning. In other words, you can burn this lamp three hours a night for a month on about 1½ gallons of gasoline. It is fitted with two mantles, and for this reason it casts no shadows, which is a very great advantage, as users of ordinary kerosene lamps will quickly realize.

The quality of light approaches sunlight, which makes it easy on the eyes and shows colors in their true values. The light is absolutely steady—no flicker or smoke, which makes reading and sewing under it a pleasure.

It is simple, convenient, safe, clean and economical. It lights almost instantly with an ordinary parlor match. Anyone can operate it.

There are no complicated parts to get out of order and no trouble. It requires practically no attention.

It requires filling only once a week, and a few strokes with a small hand pump every two or three days to keep up the air pressure.

Absolutely safe. The fuel has only one outlet—a tiny hole in the generator—**so there is absolutely no danger.** The lamp can be turned over and over and will burn safely in any position. **If filled by daylight, there is no chance of accident.**

The lamp gives off no smoke, soot or odor. It has no wick to trim, no chimney to clean or break. It cannot leak or become oily; it is absolutely air-tight and clean as a dish to handle.

Built for lifelong service and everyday satisfaction. All exposed parts of the "Ever-Brite" Lamp, base and burner, are made of heavy brass, nickel plated. The standard is made of brass and is covered with heavy black wood, which adds to the appearance of the lamp and forms a cool handle for carrying. The height of the lamp is 22 inches and the diameter of the base, which contains the fount, is 8 inches. Capacity of fount, 2 quarts.

"EVER-BRITE" GASOLINE TABLE LAMP WITH WHITE SHADE.

This lamp is fitted with a white ribbed opal glass shade and is the ideal lamp for living room, dining room, bedroom and for general all around purposes. Complete with two mantles, air pump, wrench, cleaning pick, extra generator and extra gas tip. Weight, packed, 11 pounds.

35L840—Complete **$8.25**

35L2498—"Ever-Brite" Gasoline Mantles.
Per half dozen................(Weight, 4 ounces).................. **43c**

"EVER-BRITE" GASOLINE TABLE LAMP WITH GREEN SHADE.

This lamp is fitted with a green ribbed glass shade lined with white opal glass. It is the ideal lamp for the library and for reading, as it concentrates the light on the table or book. Complete with two mantles, air pump, wrench, cleaning pick, extra generator and extra gas tip. Weight, packed, 11 pounds.

35L842—Complete **$9.25**

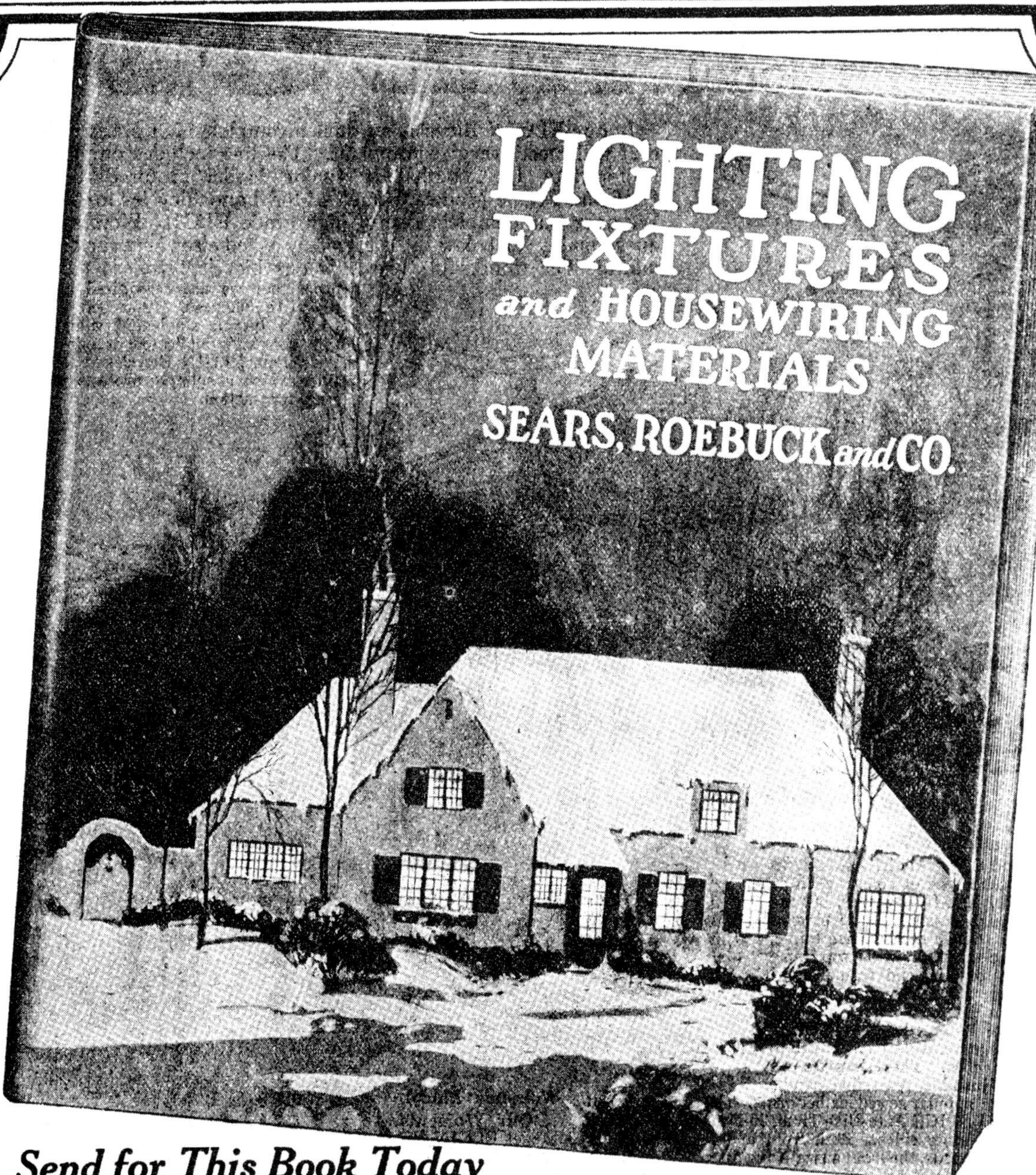

Send for This Book Today

THE lighting is a most important feature of a well appointed home. Our new Lighting Fixture Catalog illustrates a comprehensive line of Lighting Fixtures, embodying the latest ideas in modern illumination. It also contains a complete selection of standard electrical house wiring materials, lighting glassware and portable lamps. It will pay you to send for this catalog. Quality should be considered when buying lighting fixtures. Quality is the foundation of our line, and our liberal guarantee protects you.

Our Lighting Fixture Catalog also lists:
Art Domes
Brackets
Chandeliers
Cleats
Clusters
Conduits
Desk Lamps
Domes
Drop Cords
Floor Lamps
Glassware

Have you been thinking of putting a new lighting fixture in your living room or dining room, something to replace the old one that's an eyesore to you? Or is the lighting of the bedroom bad, making it hard to see how to dress? Why not put that new fixture in right away?

Perhaps you have a house or an apartment to rent or sell. Why not make it more beautiful and valuable by installing attractive Lighting Fixtures? The increase in the rent or sale price will pay for the fixtures many times over.

Our electric fixtures are properly constructed and wired for use with either city current or any home lighting plant.

Our fixtures are shipped to you completely assembled, wired ready to hang, fitted with "NOTORCH" attachments, which enables you to install them without the use of a soldering torch or any tools other than a screwdriver.

Don't delay—write today for your copy of our new Lighting Fixture Catalog 416GCL. We will send it to you postpaid. Take advantage of our low prices and order your lighting fixtures at your earliest opportunity.

Our Lighting Fixture Catalog also lists:
Insulating Joints
Outlet Boxes
Plugs
Receptacles
Rubber Covered Wire
Shades
Sockets
Switches
Table Lamps
Tungsten Lamps

Our fixtures are approved by the Underwriters' Laboratories, Inc., under the direction of the National Board of Fire Underwriters, and each and every fixture bears their stamp of approval in the form of a serial numbered label.

Boys' and Girls' Models Women's Model ELGIN

In our ELGIN Bicycles we offer a complete high grade range of models that will meet every requirement. The line includes our famous ELGIN Motor-Bike, the Youths' or Junior Model Motor-Bike, an attractive Diamond Frame Model, Women's Model and Bicycles for Boys and Girls. We offer our complete line of high grade ELGIN Bicycles at prices unquestionably far below those regularly asked elsewhere for bicycles of equal merit. The wonderful volume of our bicycle business today is striking proof that we are recognized as leaders in the bicycle business. You will find every model ELGIN Bicycle a splendid value, one which you will not find duplicated at our price elsewhere. We sell our ELGIN Bicycles under a binding guarantee which absolutely protects you from risk in every transaction.

Boys' and Girls' Models.

Our Boys' and Girls' Model ELGIN Bicycles have the same type of construction as our adults' model ELGIN Bicycles, but are furnished in a smaller size. These bicycles are made with a 16-inch combination frame, with 26-inch wheels. A dip in the top frame bar of the boys' or diamond frame model permits adjusting the seat post for comfortable riding on the part of any boy who can ride a 16 or 18-inch model bicycle, which will include leg measurements from crotch to heel of 18 to 26 inches. These bicycles are equipped with adjustable reversible handle bars fitted with the popular sewed leather grips. The tire equipment is our celebrated JUSTICE Auto-Bike Tires, 26-inch size, with black studded tread and white sides. Both front and rear wheels are equipped with mud guards, the front guard being fitted with splasher, the rear guard with stand lock clip. Stand comes in dull nickel finish. Equipped with Juvenile Troxel Saddle with beehive cushion springs. Special rubber pedals, comfortable and durable, with adjustable pedal rubbers. Has a 1-inch pitch roller chain. Complete with tool bag with bicycle wrench, hand pump, tube of tire repair cement and oiler.

The girls' model is furnished with drop style frame instead of diamond and is equipped with skirt guard. All models are furnished in attractive cherry red, with black striping.

We can also furnish our Boys' and Girls' ELGIN Model Bicycles in khaki brown and cream color, with cream color frame head, khaki brown seat post mast, khaki brown top and lower frame bars and fork sides (khaki brown drop frame bars on the Girls' model), with contrasting color darts. Rims and mud guards enameled with cream color centers and khaki brown sides to match frame.

Prices of Our Boys' and Girls' Model ELGIN Bicycles.

28L1357¼—Boys' Model ELGIN Bicycle, equipped with New Departure Coaster Brake. Furnished in cherry red with black striping or in khaki brown and cream finish. **State finish wanted**.................................**$22.45**

28L1361¼—Girls' Model ELGIN Bicycle, equipped with New Departure Coaster Brake. Furnished in cherry red with black striping or in khaki brown and cream finish. **State finish wanted**.................................**$23.95**

Shipping weight, 45 pounds.

Women's Model.

Our Women's Model ELGIN Bicycle is a high grade, well equipped model, attractively designed and finished. It is a worthy companion to our men's model ELGIN bicycles. The equipment has been carefully selected to include those features that make for comfort, durability and attractiveness.

Features of Our Women's Model ELGIN Bicycle.

FRAME—Reinforced 1-inch steel tubing, drop style instead of diamond. **Size, 20-inch only.**

Women's Model Troxel Saddle, comfortable and easy riding. Has beehive cushion springs, finished in black. We furnish a 1-inch roller chain on this model. Equipment includes substantial metal chain guard, laced skirt guard, mud guards, with splasher fitted to front guard and stand lock clip to rear guard, a regulation type dull nickel finish stand, special rubber pedals, adjustable reversible handle bars with about 6-inch sewed leather bulldog grips and tool bag with bicycle wrench, hand pump, tube of tire repair cement and oiler. The tire equipment is our splendid JUSTICE Auto-Bike Tires with black studded tread and white side walls. (For description and illustration of these tires see page 735.)

FINISH—Our Women's Model ELGIN Bicycle is furnished in cherry red with black striping. The many nickel plated metal parts make an impressive contrast with the rich red finish of the frame, mud guards and rims.

We can also furnish our Women's Model ELGIN Bicycle in blue and white finish. Furnished with white frame head, blue seat post mast and blue drop frame bars and fork sides, with contrasting color darts. Rims and mud guards enameled with white centers and blue sides to match frame.

28L1351¼—Women's Model ELGIN Bicycle, equipped with New Departure Coaster Brake. Furnished in cherry red and black striping, or in blue and white finish. **State finish wanted.** Shipping weight, 60 pounds.......................**$26.95**

BICYCLES Diamond Frame Model Youths' Motor Bike

Our ELGIN Diamond Frame Model Bicycle combines a high grade construction with well selected equipment, representing up to date ideas in bicycle design.

Frame, 1-inch steel tubing, with thoroughly reinforced flush joints. Furnished in 20 or 22-inch sizes only. **In ordering be sure to state size.** Maple rims, laced with 36 spokes, both front and rear, for 28-inch tires. Hubs are one-piece spindle type. Well known box type handle bars, forward extension style, with about 6-inch sewed leather bulldog grips. These bars offer a comfortable riding position similar to a motor-bike. Popular Troxel Tip Top Saddle with good grade black leather top and beehive type black finish springs. Length of top, about 10 inches; width, about 7½ inches. Arched and corrugated light weight type mud guards, with black supporting braces. Front guard is fitted with rubber splasher, rear guard with stand lock clip. Substantial bicycle stand of dull nickel finish; can be securely engaged to stand clip when bicycle is being ridden.

High grade JUSTICE Auto-Bike Tires with black studded tread and white side walls. 28-inch size. (For complete description and illustration of these splendid tires see page 735.)

Comfortable riding motor-bike type corrugated rubber pedals, with pedal rubbers adjustable or removable. 5/16-inch Diamond Roller Chain. One-piece drop forged crank complete with 7-inch tapering pedal cranks. Sprocket has 26 teeth and is of light weight attractive type. Substantial heavy leather tool bag, reinforced with black metal ends. Tool equipment comprises wrench, hand pump, tube of tire repair cement and oiler.

Furnished in attractive cherry red, with black striping. Rims and mud guards are enameled to match the finish of the frame. We can also furnish our ELGIN Diamond Frame Bicycle in up to date Arizona brown, black striped.

Our Youths' Model ELGIN Motor-Bike is especially designed for riders of regulation style bicycles of 20 or 22-inch diamond frame and will therefore interest the many youths who want their own size motor-bike. Furnished in 20-inch size only. The illustration of this model shows its attractiveness from the standpoint of design, equipment and general appearance.

The frame is 1-inch bicycle tubing with flush joint reinforcements. The dip in top frame bar permits wide adjustment of the seat post. The fork crown is of up to date late type triple truss keystone arch design.

The tire equipment is our splendid JUSTICE Auto-Bike Tires with black studded tread and white side walls. (See page 735.) De Luxe handle bars with reinforcing bar; 6-inch sewed leather bulldog grips. Troxel Tip Top saddle, with black leather top and beehive cushion springs, black finish. The mud guards are of drop side style, with flat braces. The front mud guard is fitted with splasher, the rear guard with clip for fastening stand, which is furnished in dull nickel finish. Comfortable, durable rubber pedals, with adjustable removable pedal rubbers. 5/16x1-inch pitch Diamond Roller Chain, drop forged one-piece crank complete with 7-inch tapering pedal cranks and a light weight 26-tooth sprocket. Rims are maple, of crescent cement type, and laced with 36 spokes. A substantial leather tool bag with our regulation tool equipment. Furnished in cherry red, black striped.

We can also furnish our Youths' Model ELGIN Motor-Bike in black and green finish. Furnished with dark green frame head and seat post mast, black enameled top and lower bars with contrasting color darts, gold color edged. Rims and mud guards are enameled to match the frame.

The term "Motor-Bike" has reference only to the type of frame, meaning that it is built on the order of a motorcycle.

28L1368¼—Our Youths' Model ELGIN Motor-Bike, equipped with New Departure Coaster Brake. Furnished in attractive cherry red with black striping or in black and green finish. **State finish wanted.** Shipping weight, 63 pounds.......... **$24.45**

28L1366¼—Our ELGIN Diamond Frame Bicycle, equipped with New Departure Coaster Brake. Furnished in attractive cherry red with black striping, or in Arizona brown, black striped. Shipping weight, 63 pounds.......... **$23.95**

NOTE—Furnished in 20 or 22-inch frame. Order 20-inch **frame** for leg measurement of 28 to 32 inches from crotch to heel, and 22-inch frame for 32 to 35 inches. **In ordering be sure to state size.**

ELGIN Motor Bike

The Term "Motor-Bike" has reference only to the type of frame, meaning that it is built on the order of a motorcycle.
$27.45
Coaster Model

The Features of ELGIN Motor-Bike Model Quality Include the Following:

FRAME.

Up to date motor-bike frame of approved truss type. Made of 1-inch steel bicycle tubing, with thoroughly reinforced flush joints. Patterned after standard motorcycle frame design.

SIZE.

Furnished in 22-inch size only, with dip in top frame bar, making it practical to raise or lower the seat post to permit comfortable riding for practically anyone who can ride a 20, 22 or 24-inch diamond frame bicycle.

TIRES.

Famous JUSTICE Auto-Bike Tires, 28-inch size, with black studded tread and white side walls. (See page 735 for more complete description of these tires.).

HANDLE BARS.

Famous De Luxe Motor-Bike Bars of forward extension type. Complete with substantial reinforcing bar with diamond tapered ends. Up to date type sewed leather grips, about 6 inches long.

FRONT FORK.

Approved motor-bike type.

MUD GUARDS AND STAND.

Mud guards are of drop side style, front guard fitted with rubber splasher and rear guard with stand lock clip. Substantial motor-bike type stand fastening to clip on rear mud guard when bicycle is being ridden.

SADDLE.

High grade comfortable riding Troxel saddle of motor-bike type, with good grade leather top. Has beehive type cushion springs, black finish. Size of top, length over all, about 10¾ inches; width, about 8½ inches.

PEDALS.

Motor-bike type corrugated rubber pedals, with pedal rubbers removable or adjustable.

CHAIN.

³⁄₁₆x1-inch Diamond Roller Chain.

HANGER

One-piece drop forged crank with hanger lock ring. Well designed sprocket, 26 teeth, of light weight. Has 7-inch tapering pedal crank.

WHEELS AND RIMS.

Maple rims, crescent cement type, for 28-inch tires. Both front and rear wheels have 36 spokes. Front hub is spindle type; rear hub New Departure Coaster Brake Hub.

TOOL EQUIPMENT.

Heavy leather tool bag, reinforced with black metal ends. Has nickel plated clasp and ring. Equipment includes telescope type bicycle pump, bicycle wrench, tube of tire repair cement and oiler.

FINISH.

A distinctive finish that is quickly recognized and everywhere appreciated. Our ELGIN Motor-Bike Model frame comes in attractive cherry red, black striped. Rims and mud guards finished to match the frame. The many nickel plated metal parts complete a most attractive appearance.

We can also furnish our ELGIN Motor-Bike in cherry red and ivory color. Frame head, crossbar and seat post mast are ivory color, balance of frame cherry red with contrasting color darts on front fork sides and frame bars. Rims and mud guards have ivory color centers and cherry red sides, matching the frame finish.

Our ELGIN Motor-Bike Model combines the latest ideas in up to date motor-bike designing with sturdy, reliable construction, a splendid equipment and a distinctive finish that is quickly recognized and universally appreciated.

28L1370¼—Our ELGIN Motor-Bike Model, equipped with New Departure Coaster Brake. Furnished in attractive cherry red with black striping or in cherry red and ivory color. **State finish wanted.** Shipping weight, 70 pounds.... **$27.45**

America Red or Blue Studded Tread Bicycle Tires

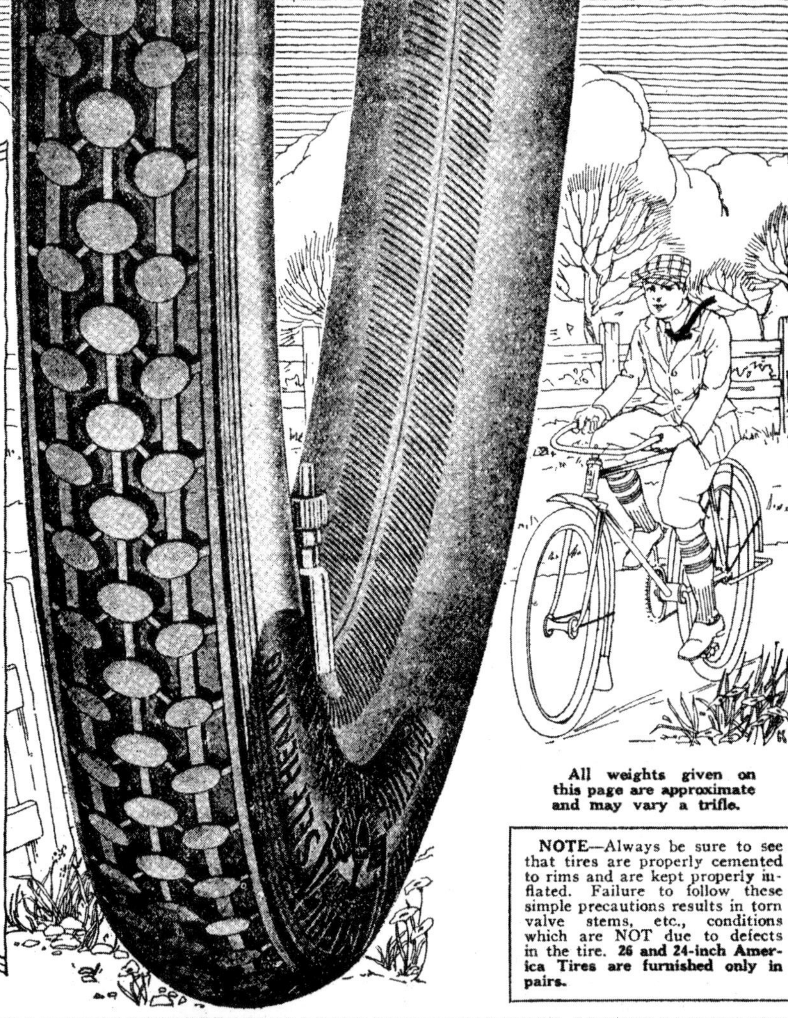

28, 26 and 24-Inch Sizes.

America Bicycle Tires stand the test of time. America Studded Tread Tires, although introduced only a few years ago, proved their worth so quickly that today they are our fastest selling bicycle tires. America Studded Tread Tires owe their splendid reputation to their many excellent features. A number of closely woven fabric layers are united into a one-piece several-ply construction by layers of frictioned rubber. The tough anti-skid studded tread with its large, thick studs offers a combination of long wear, protection from skidding and attractive appearance. America tires are noted for their resiliency or liveliness. An inside rubber air chamber is treated to make most small punctures self healing.

We GUARANTEE to replace or repair without extra charge any America Tire which develops defects in use. This does not cover punctured or cut tires or tires worn out in actual service.

We furnish America Studded Tread Tires in either 28, 26 or 24-inch sizes with either blue tread and white side walls or red tread and white side walls.

Order 28-inch tires for adults' model bicycles. 26 and 24-inch tires are for use on 18 and 16-inch frame juvenile bicycles. (Elgin bicycles for boys and girls have 26-inch tires.) Shipping weight, 28-inch tires, each, 3⅛ pounds; per pair, 6¼ pounds. 26-in. size, per pair, 5 pounds; 24-in. size, per pair, 4¾ pounds.

AMERICA STUDDED TREAD TIRES.

Blue Tread and White Side Walls.

28L1810—Size, 28x1½ inches. Per pair.......... **$3.95**

28L1811—Size, 28x1½ inches. Each.............. **$2.00**

28L1814—Size, 26x1½ inches. Per pair.......... **$3.85**

28L1816—Size, 24x1½ inches. Per pair.......... **$3.80**

AMERICA STUDDED TREAD TIRES.

Red Tread and White Side Walls.

28L1803—Size, 28x1½ inches. Per pair.......... **$3.95**

28L1804—Size, 28x1½ inches. Each.............. **$2.00**

28L1822—Size, 26x1½ inches. Per pair.......... **$3.85**

28L1824—Size, 24x1½ inches. Per pair.......... **$3.80**

All weights given on this page are approximate and may vary a trifle.

NOTE—Always be sure to see that tires are properly cemented to rims and are kept properly inflated. Failure to follow these simple precautions results in torn valve stems, etc., conditions which are NOT due to defects in the tire. **26 and 24-inch America Tires are furnished only in pairs.**

Other Favorite Bicycle Tires

UNITED STATES TIRES

Chain Tread. G. & J. **$6.98 A PAIR**

28L2029 United States Chain Tread G. & J. Clincher Tires. Sizes, 28x1⅛ or 28x1⅝ inches. **State size.** Shipping weight, 5⅝ lbs. Per pair, two casings and two inner tubes **$6.98**

28L2031 Same. Casing only. Shipping weight, 2½ pounds. Each ... **$2.75**

28L1875 United States Corrugated Tread G. & J. Clincher Tires. Sizes, 28x1⅛ or 28x1⅝ inches. **State size.** Shpg. wt., 5 lbs. Per pair, two casings and two inner tubes...... **$6.25**

28L1877—Same. Casing only. Shipping weight, 2½ pounds. Each.......................... **2.48**

Chain Tread. Single Tube Style.

28L1818 United States Chain Tread Tires, Single Tube Style. Size, 28x1½ in. only. Shipping weight, per pair, 4½ pounds. Per pair. **$4.45**

28L1819 Same, one tire only. Shipping weight, each, 2¼ pounds. Each, **$2.25**

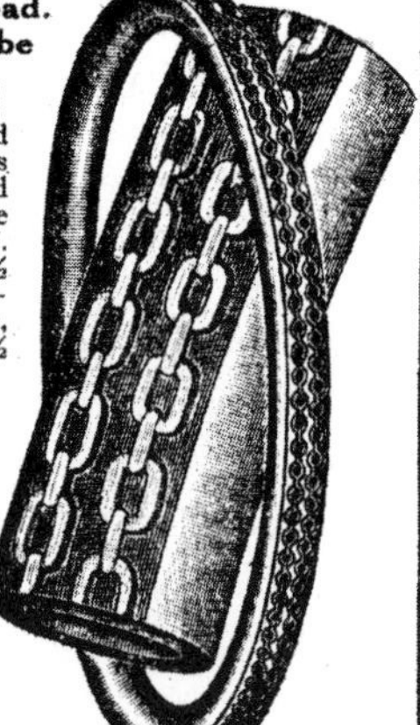

VITALIC TIRES

Well known high grade single tube tires. Furnished in one size only, 28x1½ inches.

28L1840—Vitalic Tires, 28x1½ inches. Shipping weight, per pair, 5½ lbs. Per pair...... **$5.98**

28L1841—Same as above, single tire only. Shipping weight, each, 2¾ pounds. Each....... **3.05**

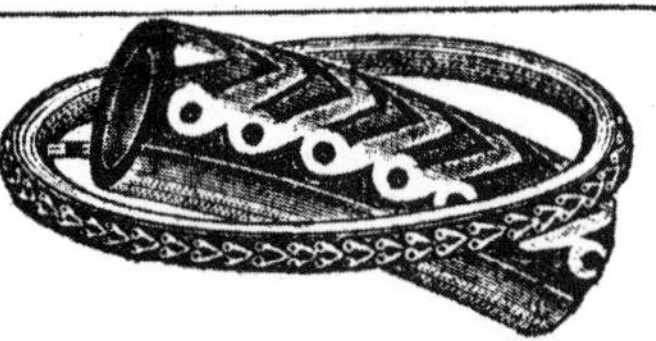

JUSTICE AUTO-BIKE TIRES

$4.45 A PAIR

JUSTICE Is Our Own Trade Mark, Registered in U. S. Patent Office.

We furnish JUSTICE Auto-Bike Tires as regular equipment on every ELGIN Bicycle. They are a high grade tire with a studded tread noted both for its anti-skid and wear resisting qualities. They are of puncture healing type, a feature responsible in no small measure for their great popularity. They have a two-ply construction, being heavy enough to give splendid wear on any bicycle and lively enough for comfortable riding. They are furnished in red tread with white side walls, in 28-inch size only, and with black tread and white side walls in either 28 or 26-inch sizes. Shipping weight, per pair, 28-inch size, 5⅝ lbs.; 26-inch size, 5¼ lbs.

28L1854—JUSTICE Auto-Bike Tires, black tread and white side walls. Size, 28x1½ inches. Per pair....... **$4.45**

28L1855—JUSTICE Auto-Bike Tires, with red tread and white side walls. Size, 28x1½ in. only. Per pair.. **$4.45**

28L1848—JUSTICE Auto-Bike Tires, black tread and white side walls. Size, 26x1⅜ in. Per pair **4.35**

We guarantee to replace or repair without charge any JUSTICE Auto-Bike Tire which develops defects in use. This does not cover punctures or cuts or tires worn out in actual service.

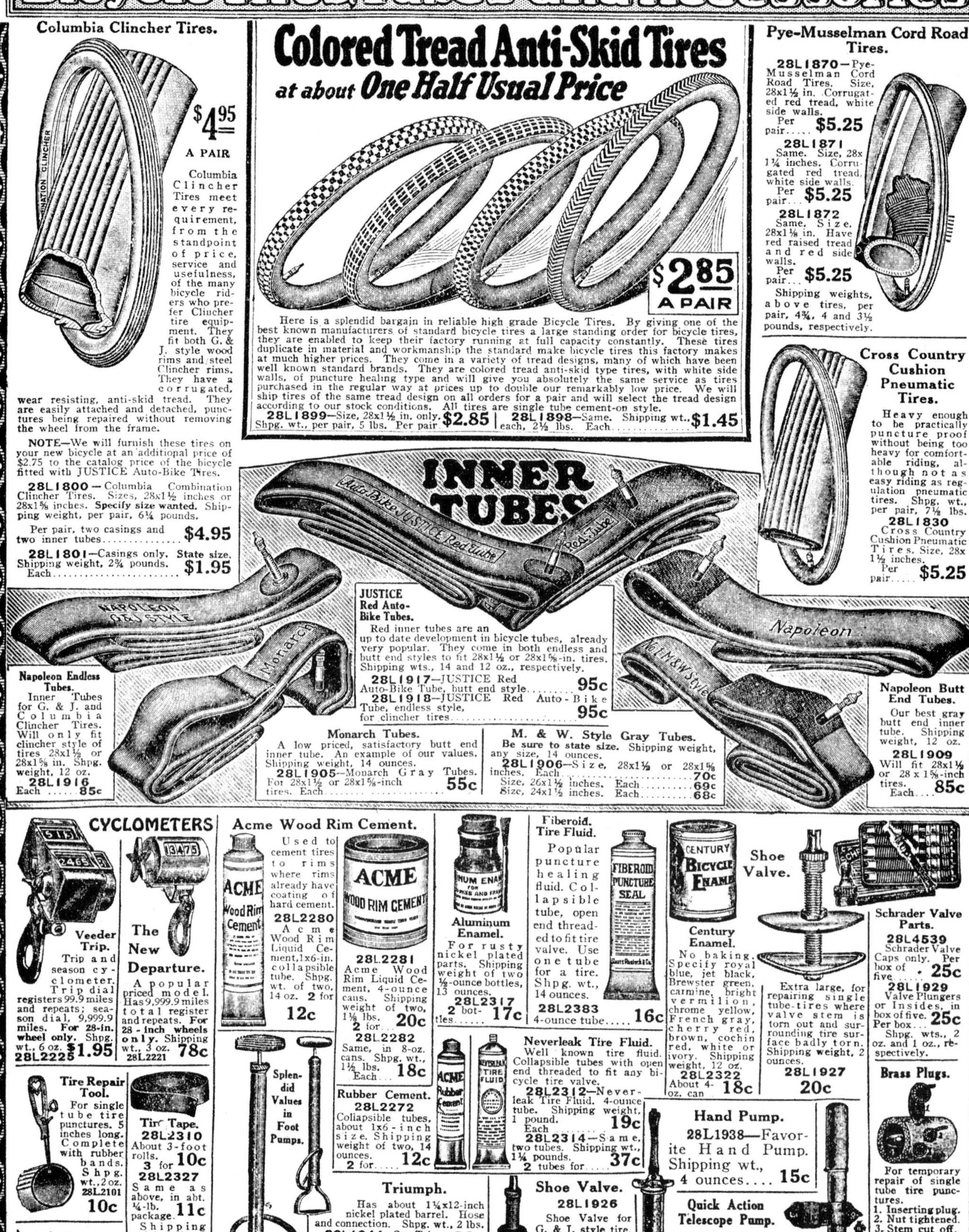

Bicycle Tires, Tubes and Accessories
Columbia Clincher Tires.
$4.95 A PAIR
Columbia Clincher Tires meet every requirement, from the standpoint of price, service and usefulness, of the many bicycle riders who prefer Clincher tire equipment. They fit both G. & J. style wood rims and steel Clincher rims. They have a corrugated, wear resisting, anti-skid tread. They are easily attached and detached, punctures being repaired without removing the wheel from the frame.
NOTE—We will furnish these tires on your new bicycle at an additional price of $2.75 to the catalog price of the bicycle fitted with JUSTICE Auto-Bike Tires.
28L1800—Columbia Combination Clincher Tires. Sizes, 28x1½ inches or 28x1⅝ inches. Specify size wanted. Shipping weight, per pair, 6¼ pounds.
Per pair, two casings and two inner tubes.............. $4.95
28L1801—Casings only. State size. Shipping weight, 2¾ pounds. Each........................ $1.95
Colored Tread Anti-Skid Tires
at about One Half Usual Price
$2.85 A PAIR
Here is a splendid bargain in reliable high grade Bicycle Tires. By giving one of the best known manufacturers of standard bicycle tires a large standing order for bicycle tires, they are enabled to keep their factory running at full capacity constantly. These tires duplicate in material and workmanship the standard make bicycle tires this factory makes at much higher prices. They come in a variety of tread designs, many of which have been well known standard brands. They are colored tread anti-skid type tires, with white side walls, of puncture healing type and will give you absolutely the same service as tires purchased in the regular way at prices up to double our remarkably low price. We will ship tires of the same tread design on all orders for a pair and will select the tread design according to our stock conditions. All tires are single tube cement-on style.
28L1899—Size, 28x1½ in. only. Shpg. wt., per pair, 5 lbs. Per pair $2.85
28L1898—Same. Shipping wt., each, 2½ lbs. Each.............. $1.45
Pye-Musselman Cord Road Tires.
28L1870—Pye-Musselman Cord Road Tires. Size, 28x1½ in. Corrugated red tread, white side walls. Per pair..... $5.25
28L1871 Same. Size, 28x1¾ inches. Corrugated red tread, white side walls. Per pair... $5.25
28L1872 Same. Size, 28x1⅝ in. Have red raised tread and red side walls. Per pair... $5.25
Shipping weights, above tires, per pair, 4¾, 4 and 3½ pounds, respectively.
Cross Country Cushion Pneumatic Tires.
Heavy enough to be practically puncture proof without being too heavy for comfortable riding, although not as easy riding as regulation pneumatic tires. Shpg. wt., per pair, 7½ lbs.
28L1830 Cross Country Cushion Pneumatic Tires. Size, 28x1½ inches. Per pair..... $5.25
INNER TUBES
Auto-Bike JUSTICE Red Tube
Red Tube
NAPOLEON G&J STYLE
Monarch
Napoleon
No. 1 M&W Style
JUSTICE Red Auto-Bike Tubes.
Red inner tubes are an up to date development in bicycle tubes, already very popular. They come in both endless and butt end styles to fit 28x1½ or 28x1⅝-in. tires. Shipping wts., 14 and 12 oz., respectively.
28L1917—JUSTICE Red Auto-Bike Tube, butt end style.......... 95c
28L1918—JUSTICE Red Auto-Bike Tube, endless style, for clincher tires.......... 95c
Napoleon Endless Tubes.
Inner Tubes for G. & J. and Columbia Clincher Tires. Will only fit clincher style of tires 28x1½ or 28x1⅝ in. Shpg. weight, 12 oz.
28L1916 Each 85c
Monarch Tubes.
A low priced, satisfactory butt end inner tube. An example of our values. Shipping weight, 14 ounces.
28L1905—Monarch Gray Tubes. For 28x1½ or 28x1⅝-inch tires. Each.................. 55c
M. & W. Style Gray Tubes.
Be sure to state size. Shipping weight, any size, 14 ounces.
28L1906—Size, 28x1½ or 28x1⅝ inches. Each.................. 70c
Size, 26x1½ inches. Each.......... 69c
Size, 24x1½ inches. Each.......... 68c
Napoleon Butt End Tubes.
Our best gray butt end inner tube. Shipping weight, 12 oz.
28L1909 Will fit 28x1½ or 28 x 1⅝-inch tires. Each.... 85c
CYCLOMETERS
Veeder Trip.
Trip and season cyclometer. Trip dial registers 99.9 miles and repeats; season dial, 9,999.9 miles. For 28-in. wheel only. Shpg. wt., 6 oz. 28L2225 $1.95
The New Departure.
A popular priced model. Has 9,999.9 miles total register and repeats. For 28-inch wheels only. Shipping wt., 3 oz. 28L2221 78c
Acme Wood Rim Cement.
ACME Wood Rim Cement
Used to cement tires to rims where rims already have coating of hard cement.
28L2280 Acme Wood Rim Liquid Cement, 1x6-in. collapsible tube. Shpg. wt. of two, 14 oz. 2 for 12c
ACME WOOD RIM CEMENT
28L2281 Acme Wood Rim Liquid Cement, 4-ounce cans. Shipping weight of two, 1⅛ lbs. 2 for... 20c
28L2282 Same, in 8-oz. cans. Shpg. wt., 1½ lbs. Each.... 18c
Aluminum Enamel.
For rusty nickel plated parts. Shipping weight of two ½-ounce bottles, 13 ounces.
28L2317 2 bottles...... 17c
Fiberoid Tire Fluid.
FIBEROID PUNCTURE SEAL
Popular puncture healing fluid. Collapsible tube, open end threaded to fit tire valve. Use one tube for a tire. Shpg. wt., 14 ounces.
28L2383 4-ounce tube..... 16c
CENTURY BICYCLE ENAMEL
Century Enamel.
No baking. Specify royal blue, jet black, Brewster green, carmine, bright vermilion, chrome yellow, French gray, cherry red, brown, cochin red, white or ivory. Shipping weight, 12 oz.
28L2322 About 4-oz. can 18c
Shoe Valve.
Extra large, for repairing single tube tires where valve stem is torn out and surrounding tire surface badly torn. Shipping weight, 2 ounces.
28L1927 20c
Schrader Valve Parts.
28L4539 Schrader Valve Caps only. Per box of five..... 25c
28L1929 Valve Plungers or Insides, in box of five. Per box... 25c
Shpg. wts., 2 oz. and 1 oz., respectively.
Tire Repair Tool.
For single tube tire punctures. 5 inches long. Complete with rubber bands. Shpg. wt., 2 oz. 28L2101 10c
Tire Tape.
28L2310 About 3-foot rolls. 3 for 10c
28L2327 Same as above, in abt. ¼-lb. package. 11c
Shipping weights, 5 and 6 ounces, respectively.
Trousers Guard.
28L2232—Shipping wt., 2 oz. Per pair...... 5c
Splendid Values in Foot Pumps.
Rubber Cement.
ACME Rubber Cement
28L2272 Collapsible tubes, about 1x6-inch size. Shipping weight of two, 14 ounces. 2 for...... 12c
NEVERLEAK TIRE FLUID
Neverleak Tire Fluid.
Well known tire fluid. Collapsible tubes with open end threaded to fit any bicycle tire valve.
28L2312—Neverleak Tire Fluid. 4-ounce tube. Shipping weight, 1 pound. Each........ 19c
28L2314—Same, two tubes. Shipping wt., 1¼ pounds. 2 tubes for...... 37c
Hand Pump.
28L1938—Favorite Hand Pump. Shipping wt., 4 ounces.... 15c
Brass Plugs.
For temporary repair of single tube tire punctures.
1. Inserting plug.
2. Nut tightened.
3. Stem cut off.
4. Plug in position.
Shpg. wt., per set, 3 oz.
28L2266 Puncture Plugs. Per set, two each large, medium and small size plugs 18c
Triumph.
Has about 1¼x12-inch nickel plated barrel. Hose and connection. Shpg. wt., 2 lbs.
28L1944—Our Triumph Foot Pump........ 39c
Quick Action Foot Pump.
Size of barrel, about 1¼x20 inches. Brass finish heavy tubing. Extra long rubber pump hose. Shipping wt., 3½ lbs.
28L1956—Quick Action Foot Pump... 80c
Shoe Valve.
28L1926 Shoe Valve for G. & J. style tire. Shpg. wt., 2 oz. Each.. 20c
Inner Tube Patches.
Twelve patches, about 1½ inches in diameter. Shipping weight, 2 ounces.
28L2297 Per package...... 10c
Quick Action Telescope Pump.
Quick working and handy; having clamp for attaching to bicycle frame. Length, closed, about 10½ inches; open, without tubing, about 18 in. Shipping wt., 12 ounces.
28L1954—Telescope Bicycle Pump, complete.................. 42c

Accessories and Equipment Priced Very Low

Electric Head-Tail Lamp Outfit.

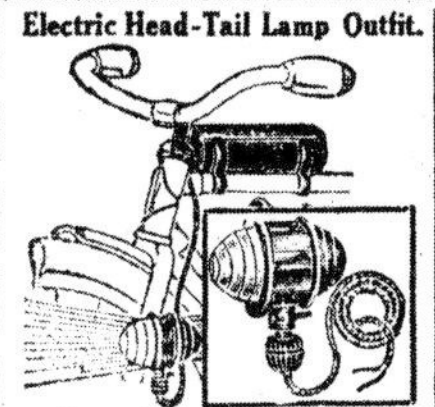

Comprises up to date lamp with one each red and white oval dome style lens (the latter of etched type), single cell battery container (without battery), together with necessary electric cord. Lamp has about 1½-volt double contact bulb. Ediswan base, on and off switch and bracket for attaching to front fork side. Throws white light ahead and red light behind. Shipping weight, complete outfit, 2 lbs. **$1.68**
28L2234

The Searchlight Gas Lamp. $2.95

"IMPROVED SEARCH-LIGHT"

An old time favorite bicycle gas lamp giving a splendid light. Has about 2½-inch convex lens, protected reflector, red and green side lights and large water reservoir. Comes nickel plated, about 8 inches high. Fitted with universal adjustable bracket for bicycle fork or head. Shipping weight, 2½ pounds.
28L2005—Searchlight Gas Lamp....... **$2.95**

Electric Headlight Outfit.

Includes lamp with about 3½-inch front diameter, handy on and off switch at back, 2.8-volt bulb, double contact Ediswan base, bracket for attaching lamp to handle bar stem and two-cell battery container, about 13 inches long, with clamps for attaching to bicycle frame and necessary wire, without batteries. Lamp and container come in black finish. Shipping weight, 2¾ pounds.
28L2135 **$1.65**
28L2126—Lamp, with single-cell container and 1½-volt bulb, without dry batteries. Shpg. wt., 2 lbs. **1.45**
28L2017—Extra Headlight Bulbs, 2.8 volts, for 28L2135 Lighting Outfit. Shipping weight, 4 ounces...... **.18**
28L2137—Extra Headlight Bulbs, 1½ volts, for 28L2126 Lighting Outfit. Shipping weight, 4 ounces **.18**

Light Weight Oil Lamp.

Light, but well made. Has about 2½-inch lens, green and ruby color side lights. Nickel plated finish. Height, about 5¾ inches. Rigid adjustable bracket furnished. Shipping weight, 1⅞ pounds.
28L2018—Light Weight Oil Lamp...... **98c**

Pirate All Metal Pedals.

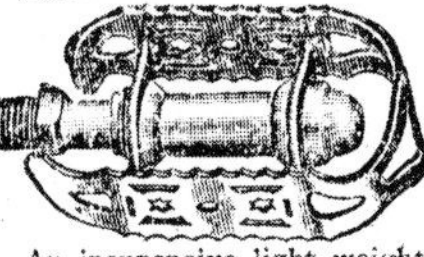

An inexpensive light weight pedal. Shipping weight, per pair, 1¼ pounds.
28L2064—Pirate Pedals. Per pair...... **72c**

Classy Hand Horn.

A reliable, effective warning signal, operated by hand. Height, over all, about 4½ inches; diameter, at front, about 3 inches. Clamps to handle bar. Shipping weight, 1⅜ lbs.
28L2335
Classy Hand Horn. **72c**

Ready Ringing Bell.

Push striker to ring bell as long as desired. Has ratchet mechanism, gong about 2¼ inches in diameter. Top has color decoration. Shipping wt., 10 ounces.
28L2124—Ready Ringing Bell.............. **62c**

Favorite Siren.

Gives the familiar siren warning. Roller is brought into contact with front tire by pulling chain. Has ball bearing construction. Outer drum nickel plated. Cannot be used on bicycles where front mud guard extends beyond fork. Shipping weight, 1⅛ lbs.
28L2132—Favorite Bicycle Siren.............. **75c**

Triumph Rubber Pedals.

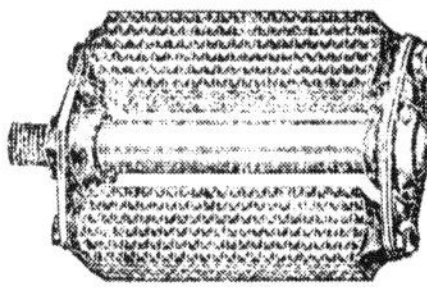

Shipping weight, per pair, 2¼ pounds.
28L2059—Pair... **98c**
28L2073—Pedal Rubbers for above. Shipping weight, per pair, 7 ounces. Pair, for one pedal... **15c**

New Departure Coaster Hub.

28L1986—New Departure Coaster Hub. Shpg. wt., 3 lbs... **$4.25**
28L1994¼—Built-Up Coaster Hub Wheel. Crescent Cement Natural Wood Rim. Size, 28 inches. Shipping weight, 4⅝ pounds.............. **$6.75**
28L1995¼—Built-Up Coaster Hub Wheel. G. & J. style or Columbia Clincher Natural Wood Rim. Shipping weight, 4⅞ pounds.............. **$6.85**
28L1996¼—Built-Up Coaster Hub Wheel. Steel Crescent Cement Rim, Black. Shpg. wt., 5½ lbs. **$6.75**
28L1997¼—Built-Up Coaster Hub Wheel, Steel Clincher Rim, Black. Shipping weight, 6 pounds...... **$6.65**
NOTE—28L1994¼ and 28L1995¼ Rear Wheels can be furnished with enameled rims for 22 cents extra. For colors see note under 28L2178¼, at the right. All Steel Rims are furnished in black only. We do not furnish steel lined wood rims.
Unless otherwise specified, for 1x3⁄16-inch chain we will ship you a hub for thirty-six spokes with a 9-tooth sprocket (7, 8 or 10-tooth also furnished). When used with ½x⅛-in. chain we will ship you an 18-tooth sprocket unless 14, 16, 20 or 22-tooth is specified.

Bicycle Wheels.

Cement-on type wheels have a coat of hard cement to be moistened with gasoline or benzine before putting tire on. All wheels have spindle hubs. Sprockets for rear wheels are furnished 9-tooth for use with 1x3⁄16-inch chains (7, 8 or 10-tooth also furnished). Hubs are furnished with 18-tooth sprocket for use with ½-inch chains (14, 16, or 20-tooth also furnished). **In ordering state number of sprocket teeth.** Always order wheels by tire size, not by rim size. Adults' bicycles (20, 22 and 24-inch frames, also motor-bikes) always have 28-inch tires. Children's bicycles (16 and 18-inch frames) take 24 and 26-inch tires, respectively. Wheels for children's bicycles come in black, red centers, or in cochin red, black striped. Elgin Bicycles for Boys and Girls require wheels for 26-inch tires and come in khaki brown, cream color centers; also cochin red.

WHEELS FOR 28-INCH TIRES, CRESCENT CEMENT NATURAL WOOD RIMS.
28L2141¼—Rear Wheel only. Crescent cement wood rim. Shipping weight, 3 pounds.............. **$3.45**
28L2142¼—Front Wheel only. Crescent cement wood rim. Shipping weight, 2½ pounds.............. **$2.65**

NOTE—Steel rim wheels furnished only in black.
All wheels with clincher rims, either wood or steel rims, are adapted for Columbia or G. & J style clincher tires.

BUILT-UP WHEELS FOR 28-INCH G. & J. STYLE CLINCHER TIRES. NATURAL WOOD RIMS.
28L2143¼—Rear Wheel only. G. & J. style clincher wood rim. Shipping weight, 3⅜ pounds **$3.60**
28L2144¼—Front Wheel only. G. & J. style clincher wood rim. Shipping weight, 2⅝ lbs...... **$2.75**

BUILT-UP WHEELS FOR 28-INCH TIRES, CRESCENT AND CLINCHER STEEL RIMS.
28L2152¼—Rear Wheel only, equipped with Crescent cement steel rim. Shipping wt., 4 lbs. **$3.35**
28L2153¼—Front Wheel only, equipped with Crescent cement steel rim. Shpg. wt., 3½ lbs... **$2.35**
28L2154¼—Rear Wheel only, equipped with clincher steel rim. Shipping weight, 4⅝ lbs...... **$3.30**
28L2155¼—Front Wheel only, equipped with clincher steel rim. Shipping weight, 3½ pounds... **$2.35**

BUILT-UP WHEELS FOR 24 AND 26-INCH TIRES, CRESCENT CEMENT NATURAL WOOD RIMS. STATE SIZE.
28L2150¼—Rear Wheel only. Shipping weight, 2¾ lbs...... **$3.45**
28L2151¼—Front Wheel only. Shipping weight, 2½ lbs...... **$2.65**
NOTE—For New Departure coaster brake built-up wheels see column at left.
NOTE—For enameled rims add 22 cents. See "Bicycle Rims" at right.

Bicycle Rims.

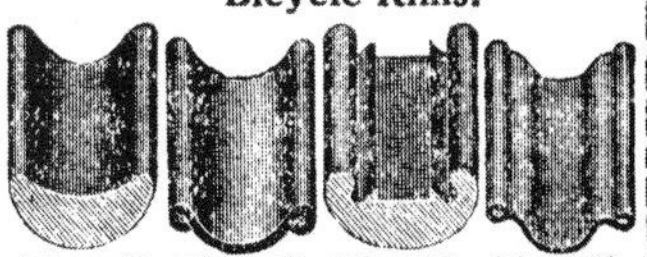

28L2171¼ 28L2183¼ 28L2178¼ 28L2184¼

Wood rims are maple. Rims come in 36-hole drill, in 28-inch size only. Rims fitting 28x1⅝-inch tires also fit 28x1½-inch tires.

WOOD RIMS, NATURAL.
28L2171¼—Crescent Cement Rim, 36 holes. Shpg. wt., 1¾ lbs... **80c**
28L2178¼—G. & J. Style Rim, 36 holes. Shpg. wt., 1¾ lbs... **98c**
NOTE—Add 22 cents for enameled wood rims. Colors: Cherry red with ivory color center; motor-bike red; Arizona brown with black panel, white striped; black with green center; white centers with blue sides; Elgin cochin red with black stripes. **In ordering, state color.**

STEEL RIMS, BLACK.
28L2183¼—Crescent Cement Rim, 36 holes, 28-inch size only. Shipping weight, 2¼ pounds.............. **65c**
28L2184¼—Clincher Rim, 36 holes, 28-inch size only. Shipping weight, 2⅜ pounds. **70c**
We do not furnish steel lined wood rims.

Sprocket Lock.

Also fits rear hub. Comprises keyless combination padlock with long steel shackle. Shipping weight, 13 ounces.
28L2354 Keyless Combination Sprocket Lock. **72c**

Repair Hanger.

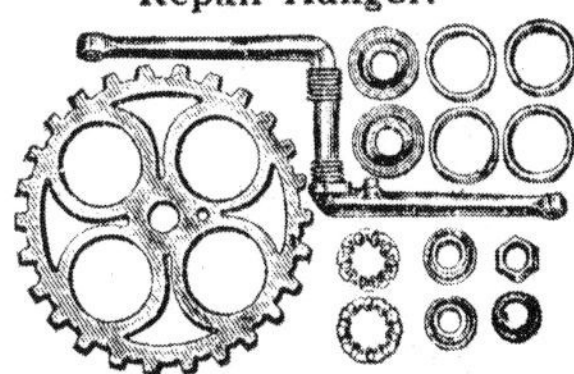

Comprises shaft and crank in one piece, cranks each 7 inches long, drilled and tapped for pedal shaft, 26-tooth sprocket, complete with balls, ball retainers, key washer, lock nut, cups 25⁄32 inch in diameter and two sets of extra bushings to make cups 1 11⁄16 and 2 1⁄32 inches in diameter. Shipping weight, complete outfit 4¾ pounds.
28L2358—Bicycle Repair Hanger. Complete as illustrated **$3.65**

Steel Balls.

	Size, In.	Shpg. Wt.	2 Doz.
28L2200	⅛	2 oz.	5c
28L2201	5⁄32	2 oz.	7c
28L2202	3⁄16	2 oz.	9c
28L2203	¼	2 oz.	13c
28L2204	5⁄16	2 oz.	15c

3⁄16 ¼ 5⁄16

Spokes.

Complete, with nipples and washers. Lengths, about 10¼, 11⅛ and 12⅛ inches. **State size.**
28L2180—Spokes. Shipping weight, per dozen, 8 ounces. Per dozen.............. **$0.15**
28L2181—Same, per 100. Shipping weight, 2 pounds. Per 100.............. **1.05**

Spindle Hubs.

One-piece, nickel plated. Sprockets for rear hubs furnished 7, 8, 9 or 10-tooth and 14, 16, 18 or 20-tooth. **State sprocket size.** All hubs drilled 36 holes.
28L2157—Rear Hub and Sprocket. Shpg. wt., 2 lbs...... **$1.50**
28L2158—Spindle Front Hub only. Shpg. wt., 10 oz...... **55c**

All weights given on this page are approximate and may vary a trifle.

BICYCLE SADDLES

Tip Top Saddle.

Motor-bike type. Light weight, easy riding type. Has black beehive type cushion springs, good quality leather top. Size of top, about 10 in. long and about 7⅛ in. wide. Shipping wt., 5⅝ lbs.
28L2095.............. **$1.55**

Favorite Juvenile Saddle.

An excellent Saddle for boys' and girls' bicycles, similar in type to 28L2389 Saddle at right, but with size of top about 6⅝x9 inches. (Not illustrated.) Black beehive type cushion springs. Shipping wt., 2¾ lbs.
28L2388.............. **$1.20**

Troxel Motor-Bike Saddle.

A splendid large saddle of motorcycle type, well padded and shaped leather top, neatly finished. Size of top, about 10¾ inches long, about 8⅛ inches wide. Double truss spring frame and popular beehive type cushion springs; black finish. Universal saddle clamp. Shipping weight, 6⅝ lbs.
28L2045—Troxel Motor-Bike Saddle **$1.85**

Peerless Troxel Saddle.

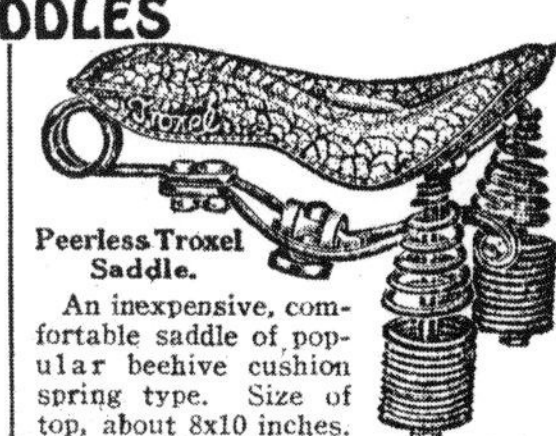

An inexpensive, comfortable saddle of popular beehive cushion spring type. Size of top, about 8x10 inches. Substantial wire frame. Shipping weight, 4 pounds.
28L2389—Peerless Troxel Saddle.... **$1.38**

Nipple Grip.

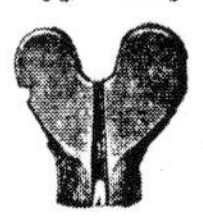

For tightening spoke nipples. Nickel plated. Shipping weight, 3 ounces.
28L2249 Nipple Grip....... **7c**

Bicycle and Motorcycle Equipment

Luggage Carrier-Stand.

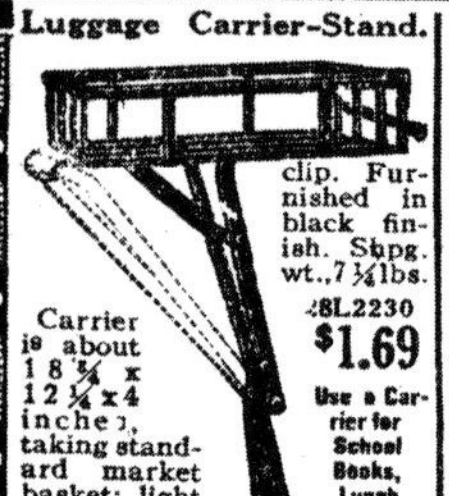

Carrier is about 18¾ x 12¾ x 4 inches, taking standard market basket; light but strong. Stand attaches to rear axle pin. Carrier has stand lock clip. Furnished in black finish. Shpg. wt., 7¼ lbs.
28L2230
$1.69

Use a Carrier for School Books, Lunch, Errands, Camping Trips, etc.

Bicycle Tool.

Screwdriver shape end; nipple grip openings and three hex nut openings; 4 inches long. Shipping weight, 2 ounces.
28L2237.....**15c**

Diamond Wrench.

Nickel plated, about 5½ inches long. Shpg. wt., 8 oz.
28L2241—Diamond Bicycle Wrench..........**30c**

GRIPS.

Grip-Well.

Give unusually secure grip. Note arrangement of knobs. Come about 5 inches long. Shipping wt., per pair, 6 oz.
28L2352—Pair...**19c**

Bulldog.

Sewed leather grips, about 5¾ in. long, with wooden core shaped to fit hand. Open ends have nickel plated ferrule. Shipping wt., per pair, 5 oz.
28L2248—Pair...**27c**

MUDGUARDS

For 28-inch wheels only.

Strong, light weight. Arched and corrugated. Black enamel finish. Shipping weight, per set, 4⅝ lbs.
28L2098—Per set, for front and rear wheels.......**69c**

Mud Guard Attaching Outfit.

Saves drilling or tapping frame. Outfit includes front bridge, necessary clamps, bolts and nuts. (Not illustrated.) Shpg. wt., 3 oz.
28L2105—Per set..........**8c**

GRIPS.

Motor-Bike Junior.

Rubber. Shock absorbing. Come about 4 inches long. Shipping wt., per pair, 4 oz.
28L2380 White. Per pair.....**19c**
28L2381 Black. Per pair.....**13c**

Motor-Bike.

Popular type motor-bike rubber grips, 6 inches long. Absorb jolts and jars. Shipping weight, per pair, 9 oz.
28L2116 Per pair..........**18c**

Favorite Parcel Carrier.

Parcel carrier. ½-inch flat cross-pieces. Length over all, about 17½ inches; carrier size, about 7x13 inches. Attach to rear fork stays and rear axle. Shpg. wt., 3 lbs.
28L2227
33c

A popular type Carrier at a remarkably low price. You need one!

Bicycle Repair Fork.

Fork crown and sides nickel plated. Long stem, 1-inch diameter, 24 threads to inch. Triple truss keystone crown. Shipping weight, 3 pounds.
28L2094
Bicycle Repair Fork.**$1.98**

Chief Tool Bag.

Popular type, with black metal ends and nickel plated clasp and ring. Brown leather. Comes about 6 inches long.
28L2081
Chief Tool Bag....**32c**

28L2076—Master Tool Bag. Black leather, with metal ends. (Not illustrated.) Shpg. wt., either, 10 oz...**32c**

Extension Adjustable Bar.

Has about 2¼-inch forward extension, 3¾ inch drop. Bar is about 20¾ inches wide. Shipping weight, 3 lbs.
28L2086—Extension Bar..........**98c**
28L2087—Extra Stem, ⅞-inch diameter. Shipping weight, 1 pound...**39c**

HANDLE BARS

Chief de Luxe.

A high grade up to date handle bar, of motor-bike type. Bar comes about 21½ inches wide. Has reinforcing bar with ends tapered to permit ready removal of stem. Nickel plated, with 6-inch CHIEF red grips, motor-bike type. Complete with substantial expander stem. Shipping weight, 3¾ pounds.
28L2085..........**$1.75**

Napoleon Handle Bar.

A motor-bike bar of popular V type. Comes about 22 inches wide, with about 7-inch rise and 2¾-inch forward extension stem. Has 4-inch rubber grips. Shipping weight, 2⅛ pounds.
28L2139—Napoleon Handle Bars..........**$1.18**

Offset Wrench.

Handy Offset Wrench with four socket openings, the assortment of openings fitting any nut on bicycle. Hex opening sizes are ½, ⅜, ²¹⁄₃₂ and ²³⁄₃₂-inch. Furnished in steel, natural finish. Shipping weight, 12 ounces.
28L1970—Offset Wrench..........**48c**

MOTORCYCLE SUPPLIES.

Federal Motorcycle Tires.

Well known Federal Motorcycle Tires, in sizes as listed below. All casings furnished single clinch only. 28L306¼ is a B. B. style rim and 28L308¼ is a C. C. style rim.

	Size, Inches	Shpg. Wt., Lbs.	Single Casing	Single Tube	Shpg. Wt. Lbs.
28L306¼	28x2½	7½	**$7.25**	**$1.15**	1¼
28L308¼	28x3	10	**7.45**	**1.25**	1½

Diamond Roller Chain.

Furnished ¼ inch wide, ⅝-inch pitch. Shipping weight, per foot, 11 ounces.
28L617—Diamond Motorcycle Roller Chain. Per foot..........**60c**

28L618—Roller Chain Single Connecting Link; one link and one connector. (Not illustrated.) Shipping weight, 2 ounces..........**16c**
28L619—Roller Chain Double Connecting Link; two connectors and one link. (Not illustrated.) Shipping weight, 3 ounces..........**25c**

Roller Chains.

28L2195—Roller Chain, ⁵⁄₁₆x1-inch size. Shipping weight, 1 pound 9 ounces..**$1.40**
28L2196—Extra Combination Repair Link for above chain. Shpg. wt., 2 oz. **.10**
28L2197—Roller Chain, ⅛x½-inch size. Shipping weight, 1⅜ pounds.......**1.50**
28L2199—Extra Combination Repair Link for above chain. Shpg. wt., 2 oz. **.10**

"B" Block Chain.

A well known block chain. Furnished 60 links, ⁵⁄₁₆x1-inch size. Shipping weight, 1¾ pounds.
28L2190
"B" Block Chain..........**$1.58**

Repair Links.

Diamond Block Chain Repair Links in box of two links. Shipping weight, 2 ounces.
28L2192...**15c**

All weights given on this page are approximate and may vary a trifle.

Auto Springs

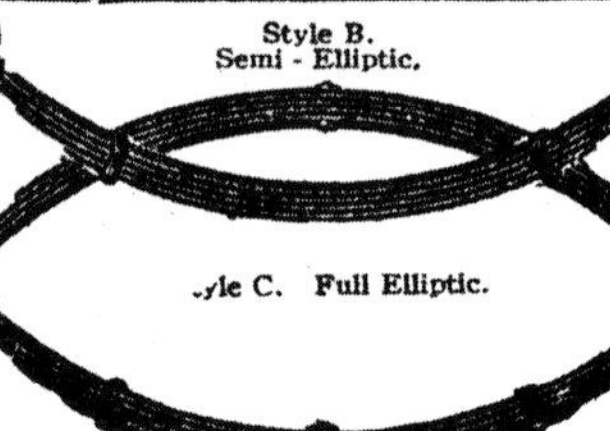

Style B. Semi - Elliptic.

Style C. Full Elliptic.

High grade steel springs with leaves hand fitted, oil tempered and graphited. All springs are black finish.

We can furnish springs for practically any car. If the spring you want is not shown here, ask for our complete Catalog of Automotive Supplies, 512GCL, sent postpaid on request.

Style D. (½-Cantilever.) (Not illustrated.)

Catalog No.	Model	Style	Location	W'th, In.	Lgth., In.	Wt., Lbs.	
	BRISCOE.						
28L13305½	4-24—1917-18-19.....	C	Front	1¾	29¾	29	**$2.95**
28L13306½	4-24—1917-18-19.....	C	Rear	1¾	32⅜	38	**3.95**
28L15218½	4-34, 1920-21........	B	Front	1¾	36	23	**2.75**
	BUICK.						
28L13320½	D-45-47—1916........	B	Front	2	35½	28	**2.95**
28L13321½	D-45-47—1916........	B	Rear	2½	45¼	58	**6.45**
28L13324½	D-34-35—1917........	B	Front	1¾	33½	21	**2.25**
28L15226½	D-34, 1917..........	B	Rear	2	46⅞	28	**3.25**
28L13325½	D-35—1917..........	B	Rear	2	48	34	**3.75**
28L15227½	E-34-35, 1918........	B	Front	1¾	33½	21	**2.65**
28L13476½	E-44—1918..........	B	Front	2	35¾	33	**3.95**
28L13477½	E-45, 1918..........	B	Rear	2½	45	58	**6.95**
	CHEVROLET.						
28L13358½	H-2-H-4............	B	Front	1¾	35½	22	**2.65**
28L13359½	490—1916...........	D	Front	1¾	21½	11	**1.45**
28L15259½	490—1917...........	D	Front	1¾	22½	21	**2.65**
28L13361½	490—1918-19........	D	Front	1¾	22½	21	**2.25**
28L13362½	490—1920...........	D	Front	1¾	23½	19	**2.45**
28L13360½	490—1916-19........	D	Rear	2	30	29	**3.45**
28L15260½	490—1920...........	D	Rear	2	30	29	**3.45**
	DODGE.						
28L13365½	1915-20.............	B	Front	1¾	35	21	**2.25**
28L13366½	1915-20.............	B	Rear	2	41¾	30	**3.45**
28L13481½	1915-20.............	A	U. Q. R.	2	16¼	13	**1.50**
	DORT.						
28L15281½	1919................	B	Front	2	37	28	**3.45**
	ELGIN.						
28L15289½	1917................	B	Front	1¾	33¼	19	**2.45**
28L15290½	1917................	B	Rear	2¼	45¾	45	**5.45**
	ESSEX.						
28L15299½	1917-18-19-20-21....	B	Front	2	36	24	**3.25**
	GRANT.						
28L15301½	1915-16-17..........	B	Rear	2	37½	41	**4.75**

Catalog No.	Model	Style	Location	W'th, In.	Lgth., In.	Wt., Lbs.	
	HUPMOBILE.						
28L13329½	K & N—1915-16....	B	Front	1¾	36½	25	**$2.95**
28L13389½	K & N—1915-16....	B	Rear	2	50	42	**4.25**
28L13338½	R—1918..........	B	Front	1¾	35½	22	**2.65**
28L13339½	R—1918..........	B	Rear	2	49⅜	38	**4.95**
28L15381½	Model R, '21 Nib type	B	Front	1¾	35½	22	**2.65**
	JEFFREY.						
28L15383½	6-71............	B	Front	2	38¼	32	**3.95**
	MAXWELL.						
28L13395½	1915-16-17-18-19....	B	Front	1¾	31½	18	**1.95**
28L13396½	1915-16..........	B	Rear	1¾	38	25	**2.65**
28L13397½	1917............	B	Rear	1¾	44	25	**2.95**
28L13398½	1918-19..........	B	Rear	2	47½	32	**3.95**
	OAKLAND.						
28L13405½	32B—1916........	B	Rear	1¾	38½	25	**2.65**
28L13407½	34S. S.—1917.....	B	Rear	1¾	49¼	35	**3.95**
28L13404½	32B—1916........	B	Front	1¾	34½	22	**2.25**
28L13406½	34S. S.—1917.....	B	Front	1¾	34½	23	**2.65**
28L13408½	34B—1918-19.....	B	Front	1¾	34½	21	**2.75**
28L13409½	34B—1918-19.....	B	Rear	1¾	50¾	35	**3.95**
28L15412½	34-C—1920-21.....	B	Front	1¾	35½	24	**2.75**
28L15413½	34-C—1920-21.....	B	Rear	1¾	49¼	36	**4.45**
	OLDSMOBILE.						
28L13415½	8-45............	B	Front	2	36	26	**2.95**
28L13416½	37..............	B	Front	2	35½	23	**2.85**
	OVERLAND.						
28L13424½	81-83...........	B	Front	1¾	35½	23	**2.85**
28L13432½	90..............	B	Front	1¾	36	24	**2.45**
28L13433½	90..............	B	Rear	2	40¾	37	**3.75**
28L13434½	4—1920..........	D	Front	2	20⅞	15	**1.95**
28L13435½	4—1920..........	D	Rear	2	24¼	20	**2.35**
28L13428½	75..............	B	Front	1¾	36	21	**2.45**
28L13429½	75..............	B	Rear	2	41	37	**4.45**
28L13425½	81-83...........	B	Rear	1¾	45	29	**3.45**
28L13427½	84..............	B	Front	1¾	35¼	28	**3.35**
28L13430½	85-4, 85-6.......	B	Front	1¾	36	29	**3.45**
	REO.						
28L13449½	5th–1913-14-15-16-17	B	Rear	2	43¾	34	**3.45**
28L13450½	6th–1915-16-17-18..	B	Front	2	37½	34	**3.95**
	SAXON.						
28L13455½	4 Cyl............	D	Front	1½	21	14	**1.45**
28L13456½	4 Cyl............	D	Rear	1½	22¼	14	**1.45**
28L13459½	6 Cyl.—1915-16....	D	Rear	2	30	34	**3.95**
	STUDEBAKER.						
28L13466½	6 Cyl.—1917-18....	B	Front	2	37½	28	**3.35**

NOTE—All springs shipped from factory in NORTHEASTERN ILLINOIS or EASTERN PENNSYLVANIA, whichever is nearer you, or from our store, according to stock conditions.

JUSTICE INNER TUBES

Guaranteed for Two Years

We do not know of a better inner tube than our popular high grade JUSTICE Heavy Red or Gray Inner Tubes. Hundreds of thousands of these tubes are in use today and giving complete satisfaction. They are a splendid foundation for satisfactory tire service. Tough, heavy, high grade rubber materials are combined by excellent workmanship into inner tubes that will give exceptionally long wear. They can be used in casing after casing. One JUSTICE Inner Tube in one of your casings will pave the way for more of the same make.

JUSTICE Inner Tubes Are Guaranteed for Two Years Against Any Defects in Material or Workmanship!

JUSTICE Inner Tube Prices Save You One-Third or More.

We Believe There Are No Better Inner Tubes Made Regardless of Name, Make or Price

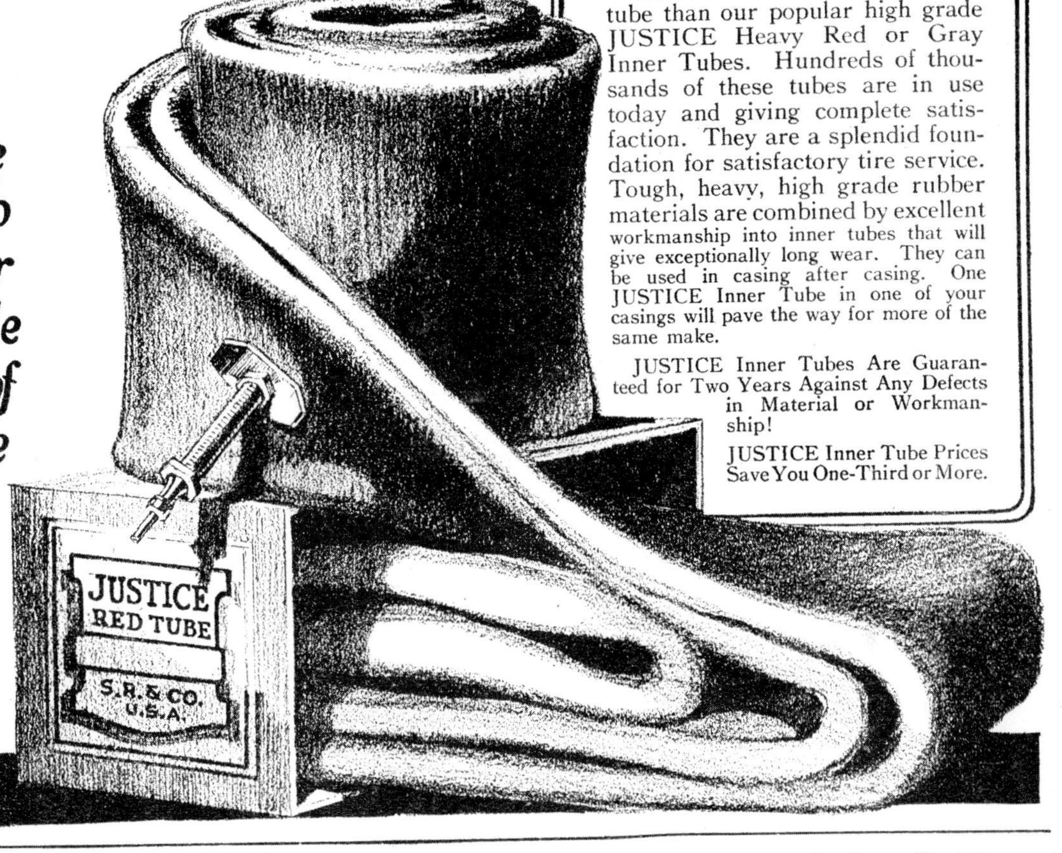

JUSTICE RED TUBE S.R.&CO. U.S.A.

Prices of JUSTICE Inner Tubes

HEAVY RED

Catalog No.	Tube Size, Inches	Shipping Wt., Lbs.	
28L3204	28x3	2	$1.05
28L3205	30x3	2⅛	1.12
28L3212	30x3½	2½	1.45
28L3213	31x3¾	2⅝	1.50
28L3214	32x3½	2⅞	1.55
28L3220	31x4	3¼	1.70
28L3221	32x4	3¼	1.75
28L3222	33x4	3½	1.80
28L3223	34x4	3½	1.85
28L3225	36x4	3¾	1.90
28L3230	32x4½	3⅞	2.25
28L3231	33x4½	4	2.30
28L3232	34x4½	4	2.35
28L3233	35x4½	4¼	2.40
28L3234	36x4½	4¼	2.45
28L3239	33x5	4⅞	2.55
28L3241	35x5	5⅛	2.75
28L3243	37x5	5⅛	2.85

HEAVY GRAY

Catalog No.	Tube Size, Inches	Shipping Wt., Lbs.	
28L12804	28x3	2	$0.99
28L12805	30x3	2	1.05
28L12812	30x3½	2⅜	1.25
28L12813	31x3¾	2½	1.35
28L12814	32x3½	2½	1.35
28L12820	31x4	2⅞	1.45
28L12821	32x4	3	1.50
28L12822	33x4	3	1.59
28L12823	34x4	3⅛	1.65
28L12830	32x4½	3¾	1.95
28L12831	33x4½	4	2.00
28L12832	34x4½	4	2.10
28L12833	35x4½	4¼	2.15
28L12834	36x4½	4¼	2.20

28L13065—JUSTICE Combination Red Inner Tubes for either 30x3 or 30x3½-inch fabric tires. Shipping weight, 2¼ pounds....................**$1.23**

EXTRA HEAVY RED

Catalog No.	Tube, Size Inches	Shipping Wt., Lbs.	
28L4100	30x3½	3	$1.80
28L4101	32x3½	3½	1.98
28L4102	31x4	3⅞	2.38
28L4103	32x4	4	2.45
28L4104	33x4	4¼	2.50
28L4105	34x4	4¼	2.59
28L4106	32x4½	5	2.85
28L4107	33x4½	5	2.90
28L4108	34x4½	5¼	2.95
28L4109	35x4½	5¼	3.00
28L4110	36x4½	5½	3.10
28L4111	33x5	5⅞	3.25
28L4112	35x5	6	3.35
28L4113	37x5	6⅜	3.45

Extra Heavy De Luxe Red Inner Tubes.

10 Per Cent Larger, 25 Per Cent Heavier.

Especially Adapted for Cord Tires.

JUSTICE Extra Heavy Red Inner Tubes offer the finest example of high quality inner tube construction that we have ever seen. They are made by the makers of our famous JUSTICE Heavy Red Inner Tubes, who have justly earned the reputation of being leaders in the manufacture of high class inner tubes. The splendid satisfaction JUSTICE Heavy Tubes have invariably given during the ten years we have sold them marks them as tubes of exceptional quality.

In JUSTICE Extra Heavy Red Inner Tubes we offer a wonderful new super-tube, built of the very finest rubber materials obtainable and made 10 per cent larger and of 25 per cent heavier gauge. This splendid tube will prove its worth in any size casing. Its unusual thickness makes it practically free from pinching. Its heavy gauge safeguards the driver to a large extent from puncture troubles. It is especially well adapted for use in cord tires, its size and thickness adapting it perfectly for use in oversize cord tires without being stretched out of shape.

Get More Mileage from Your Tires

Showing Sectional View.

INNER CASINGS.

Help Keep Your Old Tires in Service.

Don't remove a tire when it begins to show signs of considerable wear—an Inner Casing will usually help you get worth while extra mileage service from the tire. By using Inner Casings you can avoid the average puncture due to small nails or tacks, the extra thickness inside the casing helping to keep these objects away from the inner tube. Inner casings are built up of several layers of fabric and an outside layer of rubber. They are shaped to fit the tire. They are practically an endless reliner, but of heavier construction and with the sides extending well down past the beads to cover nearly the entire tube. Furnished in all popular tire sizes.

Prices of Inner Casings.

Catalog Number	Inner Casing Size, Inches	Shpg. Wt., Lbs.	
28L9482	30x3	3½	$1.95
28L9484	30x3½	3¾	2.40
28L9486	32x3½	4	2.45
28L9488	31x4	4½	2.85
28L9490	32x4	4¾	2.95
28L9492	33x4	5¼	3.10
28L9494	34x4	5¼	3.25
28L5304	33x4½	5¾	3.70
28L13206	34x4½	6¼	3.90
28L13207	35x4½	6⅜	4.05

TRIUMPH RELINERS.

Make Your Tires Last Longer.

Don't Throw Away Partly Worn Tires.

A Reliner will usually help you get considerably more mileage from them.

Many car owners put Triumph Reliners in almost new tires to help avoid punctures. Triumph Reliners are made of several plies of frictioned tire fabric. Wipe out inside of casing and moisten dark (cemented) side of reliner with gasoline saturated cloth, press reliner into place, a small section at a time.

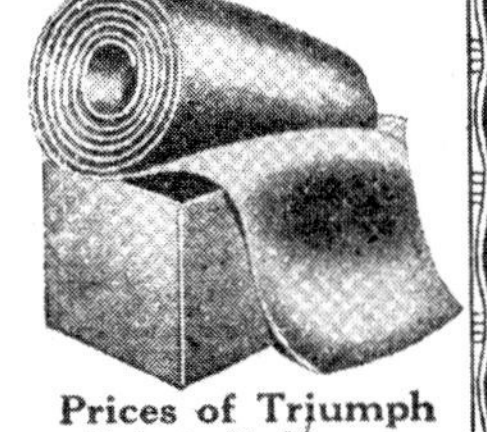

Prices of Triumph Fabric Reliners.

Catalog No.	For Tire Size, In.	Shipping Weight, Pounds	
28L3804	28x3	2⅜	$0.95
28L3805	30x3	2⅞	1.08
28L3812	30x3½	2⅞	1.23
28L3814	32x3½	3½	1.47
28L3820	31x4	4¼	1.63
28L3821	32x4	4½	1.84
28L3822	33x4	4⅝	1.98
28L3823	34x4	4⅝	2.06
28L3831	33x4½	4⅝	2.10
28L3832	34x4½	4⅝	2.20
28L3833	35x4½	4¾	2.33
28L3834	36x4½	5	2.47

JUS

JUSTICE Tire Prices

(JUSTICE Is Our Own Trade Mark, Registered in U. S. Patent Office.)

We are offering a standard brand tire under our own name JUSTICE, at the lowest prices ever quoted on a tire of standard quality. JUSTICE Tires are made side by side with a well known tire manufacturer's own tires, the materials and tire building processes used throughout being identical. They are finished in molds bearing the name JUSTICE, with our trade marked JUSTICE tread instead of the maker's own brand name and tread, the sole points of difference. These are the only important respects in which they differ from any of the widely advertised brands of standard tires that are offered to-day. We sell JUSTICE Tires under the most economical selling plan known, direct by mail to the user. The cost of handling by middlemen is eliminated. You get standard tire quality, reliability, appearance and service at prices which save you one-third or more over prices asked elsewhere on tires of similar construction. You take absolutely no risk in sending us a trial order. Our liberal policy protects you in every transaction. We furnish JUSTICE Tires in a complete line of sizes, in both FABRIC and CORD construction.

TICE Tires

A Standard Tire in Every Respect Our Method of Selling Saves You 1/3 or More

Cords and Fabrics

More and more car owners are equipping their cars with cord tires as the real solution of the tire problem. Their experience has proved most important advantages from the use of cord tires—economy in gasoline consumption, greater tire mileage and increased tire resiliency, resulting in easier riding.

JUSTICE CORDS, with their big, massive, oversize construction and their great strength and resiliency will prove a revelation to you in the complete satisfaction their use uniformly insures.

JUSTICE CORDS are not an experiment. They duplicate the standard product of a celebrated tire maker except in name and tread design. The same excellent materials and the same degree of skilled workmanship are employed in their construction. They do not differ in a single important respect from any of the better known standard brand cord tires sold today. Tire names and tread designs are principally a means of identification. Try JUSTICE CORDS under our well known guarantee of satisfaction on every transaction.

We furnish JUSTICE CORDS in regulation ribbed tread, also in our regular JUSTICE anti-skid tread, in the sizes listed below, straight side style only, except 30x3½-inch size, which is also furnished in clincher style.

Small Car Owners!

JUSTICE CORDS for Ford, Chevrolet and all other cars using 30x3½-inch rims offer you the same class of tire equipment that owners of larger cars are choosing today in greater and greater numbers.

JUSTICE CORDS are furnished in 30x3½-inch size, in either Clincher or Straight Side styles. Order a set or pair NOW and learn what real tire satisfaction awaits you.

28L3912¼—JUSTICE SUPER-OVERSIZE CORDS, ribbed tread, 30x3½-inch size. Shipping weight, 14 pounds.
Clincher style only........ **$9.95**

28L4012¼—JUSTICE SUPER-OVERSIZE CORDS, anti-skid tread, 30x3½-inch size. Shipping weight, 15 pounds.
Clincher style only........ **$9.95**

28L4010¼—JUSTICE SUPER-OVERSIZE CORDS, anti-skid tread, 30x3½-inch size. Shipping weight, 15 pounds.
Straight Side style only........ **$10.95**

Prices
JUSTICE FABRIC Tires
Ribbed Tread.

Catalog No.	Tire Size, Inches	Shipping Weight, Lbs.	Price
28L3304¼	28x3	8½	*$ 5.85
28L3305¼	30x3	9½	*5.95
28L3312¼	30x3½	12½	*6.95
28L3314¼	32x3½	14½	9.39
28L3320¼	31x4	15½	*10.45
28L3321¼	32x4	18½	11.95
28L3322¼	33x4	19	12.25
28L3323¼	34x4	19½	12.95
28L3332¼	34x4½	26	17.95
28L3333¼	35x4½	27½	18.25

***Clincher style only. All other sizes furnished in straight side style only.**

Anti-Skid Tread.

Catalog No.	Tire Size, Inches	Shipping Weight, Lbs.	Price
28L3404¼	28x3	8½	*$ 5.85
28L3405¼	30x3	9½	*5.95
28L3412¼	30x3½	12½	*6.95
28L3018¼	31x3¾**	15	*8.95
28L3414¼	32x3½	14½	9.39
28L3420¼	31x4	15½	*10.45
28L3421¼	32x4	19	11.95
28L3422¼	33x4	19½	12.25
28L3423¼	34x4	20	12.95
28L3425¼	36x4	21½	14.95
28L3431¼	33x4½	25½	17.45
28L3432¼	34x4½	26	17.95
28L3433¼	35x4½	27½	18.25
28L3434¼	36x4½	28½	18.75
28L3441¼	35x5	33	18.95

***Clincher style only. All other sizes furnished in straight side style only.**
****Use on 30x3½-inch clincher rims.**

Special JUSTICE MAMMOTH Tire.

A splendid fabric tire in clincher anti-skid style only; size, 31x3¾ inches, to fit 30x3½-inch clincher rims. Their use takes your Ford, Chevrolet, Briscoe, Dort, Maxwell, Overland cars, etc., out of the small car class in appearance, gives you easier riding and reduces your tire cost per mile. Shipping weight, 15 pounds.

28L3018¼—Special JUSTICE MAMMOTH Tire........ **$8.95**

We guarantee JUSTICE FABRIC Tires against defects in material or workmanship on the basis of 6,000 miles' service. We will repair or replace a DEFECTIVE casing on the above basis, charging only for the mileage received from the tire.

We guarantee JUSTICE CORD Tires against defects in material or workmanship on the basis of 10,000 miles' service. We will repair or replace a DEFECTIVE casing on the above basis, charging only for the mileage received from the tire.

NOTE—We can only furnish JUSTICE FABRIC and CORD Tires in the sizes and styles listed.

JUSTICE Tires can be shipped by parcel post. Tire shipments can be made up to 70-pound packages in local zone and zones 1, 2 and 3, and up to 50-pound packages in all other zones.

Prices
JUSTICE OVERSIZE CORDS
FURNISHED IN STRAIGHT SIDE STYLE ONLY, EXCEPT 30x3½-INCH SIZE.

Ribbed Tread.

Catalog No.	Tire Size, Inches	Shipping Weight, Lbs.	Price
28L3912¼	30x3½	16	*$ 9.95
28L3914¼	32x3½	21	15.45
28L3921¼	32x4	25	17.25
28L3922¼	33x4	26	17.75
28L3923¼	34x4	26	18.95
28L3932¼	34x4½	30	23.95
28L3933¼	35x4½	32	24.95
28L3941¼	35x5	41½	29.95

***Clincher style only.**

Anti-Skid Tread.

Catalog No.	Tire Size, Inches	Shipping Weight, Lbs.	Price
28L4012¼	30x3½	16	*$ 9.95
28L4010¼	30x3½	16	†10.95
28L4014¼	32x3½	21	15.45
28L4020¼	31x4	22	16.75
28L4021¼	32x4	25	17.25
28L4022¼	33x4	26	17.75
28L4023¼	34x4	26	18.95
28L4030¼	32x4½	27½	22.45
28L4031¼	33x4½	28	22.95
28L4032¼	34x4½	30	23.95
28L4033¼	35x4½	32	24.95
28L4034¼	36x4½	32½	25.25
28L4039¼	33x5	35	28.95
28L4041¼	35x5	41½	29.95
28L4043¼	37x5	42	31.45

***Clincher style only.**
†New 30x3½-inch straight side style.

For Auto Beds and other Camping Supplies see page 784.

All weights given on this page are approximate and may vary a trifle.

ACCESSORIES FOR TIRE REPAIRS

Self Curing Tire Repair Kit.

Quick, Lasting, Inexpensive.

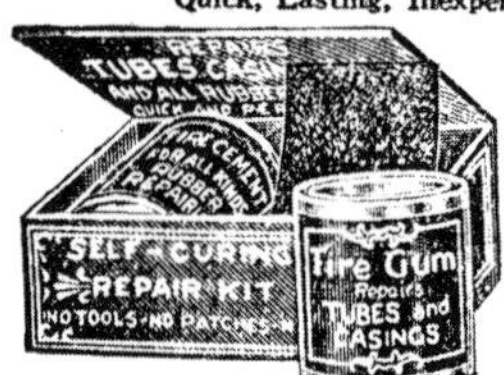

For making quick, lasting repairs of inner tube punctures or tears and cuts or holes in casings without vulcanizing. Repairs can be quickly made and the tube put into use at once. Repairing casing cuts or holes lengthens the life of the tire, preventing dirt and water from causing early ruin to tire. Outfit comprises can of Tire Gum containing sufficient material for repairing from thirty to fifty tube punctures or small casing holes, and can of Cement. Complete with directions for using. Shipping weight, 10 ounces.

28L9384 .... **32c**

Rubber Cement.

28L8231 For inner tube patches, etc. About 4-ounce can. Shipping weight, 9 oz.... **12c**

Tire Tape.

Comes in about 4-oz. package. Shpg. wt., 5 oz. **28L9378** Per package...... **11c**

Tire Flaps.

To keep tube from pinching, chafing against rough, rusty rim, etc. Replace badly worn flaps before they injure the tube. Made of several fabric plies, with beveled edges.

28L7554—For 30x3½-inch clincher cord tires **55c**

28L7556 For 32x3½ or 33x4-in. tires **55c**

28L7560 Same, for 32x4 or 33x4½-in. tires........ **55c** Shipping weight, any of above, 14 ounces.

28L7564 Same, for 34x4 or 35x4½-inch tires...... **55c** Shipping weight, 1 lb.

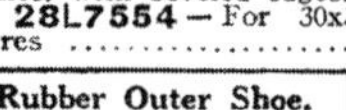

Rubber Outer Shoe.

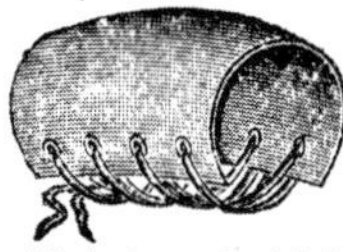

Made from tire fabric and rubber tread stock. Has rawhide lace. Comes in three sizes only.

Prices of Outer Shoes.

	For Tire Size, Inches	
28L4740	3 and 3½	48c
28L4742	4	62c
28L4744	4½ and 5	68c

Length of above shoes, about 9, 10 and 11 inches, respectively. Shipping weights, 1, 1½ and 1¾ pounds, respectively.

Acme Cut Healer.

For filling cuts in casings. A heavy rubber compound. Comes in about 1x6-inch tubes. Shpg. wt., 7 ounces.

28L13087 **20c**

Five-Minute Vulcanizer.

For Inner Tubes.

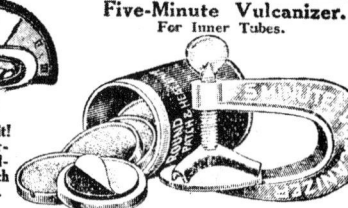

Think of it! A neat, permanent vulcanized patch in 5 minutes.

A thoroughly vulcanized patch on your inner tube in five minutes! Heat is produced by igniting disc. Apply rubber patch to tube surface, clamp container, with disc in place, over patch and ignite heat unit. Outfit includes twelve patches and heat units. At our remarkably low price this splendid little outfit should be a part of every auto owner's equipment. Shipping weight, 1 pound.

28L10556—Complete outfit.................. **85c**

28L10558—Box of Twelve Patches and Heat Units. Shipping weight, 8 ounces............... **55c**

Quick-Fix No-Cement Rubber Patching Outfit.

For Inner Tubes.

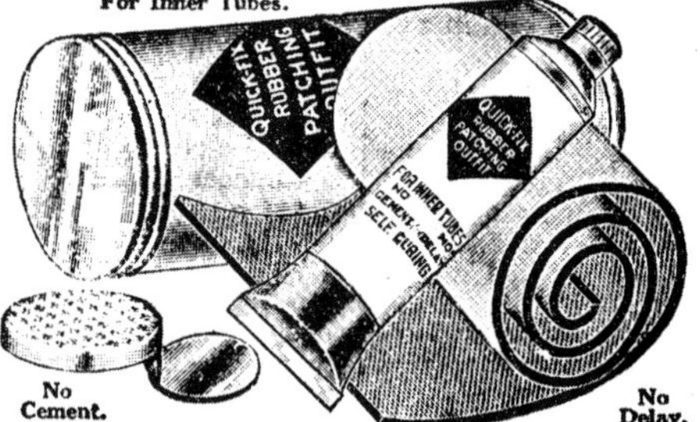

No Cement. No Delay.

You can make any inner tube repair, either puncture or blowout tear, quickly and successfully, without either vulcanizing or using cement. No waiting for a vulcanized patch to cool or for a "cold" patch to "set." The repaired tube can be put into the casing AT ONCE and the repair will last indefinitely.

Outfit comprises patching material, tube of cleaning compound and tube buffer. Instructions furnished for making quick, satisfactory, permanent repairs. Comes in two sizes.

28L10481—Outfit with about 30 square inches of patching rubber. Shipping weight, 8 ounces...... **29c**

28L10561—Outfit with about 72 square inches of patching rubber. Shipping weight, 10 ounces...... **49c**

Cactus Rubber Patching Outfit.

For Inner Tubes.

An efficient low priced patching outfit for inner tubes. Outfit comprises patching material, rubber cement and tube buffer. Instructions furnished for making a simple, sure repair. Heat generated from friction vulcanizes patch onto tube. Made of high grade rubber, stretches with tube and will make any size repair from a puncture to a blowout.

28L11226—Outfit with about 30 square inches patching rubber. Shipping weight, 8 ounces.......... **17c**

28L11221—Outfit with about 72 square inches patching rubber. Shipping weight, 10 ounces.......... **23c**

Adamson Vulcanizer.

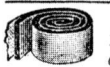

For either casings or tubes. Complete with repair stock, measuring cup and scissors. Instructions furnished. Shipping wt., 3¾ lbs.

28L9073—Complete outfit **$2.15**

28L13075—Repair Stock for Adamson Vulcanizer, about ¼-pound rolls. Shipping weight, 6 ounces.................. **25c**

Double Flap Inner Shoe.

For fabric breaks, etc. Tire beads and rim hold flaps in place.

REGULAR LENGTH	For Tire Size	Length, About	
28L5296	3 in.	8 in.	18c
28L5297	3½ in.	8½ in.	20c
28L5298	4 in.	9 in.	26c
28L5299	4½ in.	9½ in.	37c
EXTRA LONG	**For Tire Size**	**Length, About**	
28L5300	3 in.	15 in.	40c
28L5301	3½ in.	15 in.	45c
28L5302	4 in.	15 in.	50c

Shpg. wts.: 1, 1, 1⅜, 1⅝, 1⅜, 1⅝ and 1⅝ lbs.

Mica Tire Powder.

Comes in can about 8x2 in., with sifter top. Shpg. wt., 14 oz. **28L10705** Per can, **10c**

Triumph Lever Handle Lift Jack.

A practical, durable lever handle jack, remarkably low priced. Has reliable spiral gear action, with 4⅜-inch rise. Has about 28-inch T type folding and jointed bar handle. Use on cars weighing up to 3,000 pounds. Note ease and convenience of operation as shown in the illustration. Lowest height, about 8¾ inches. Has 3-inch hinge extension top for cars with high axles. Made of malleable casting. Shipping weight, 10 lbs.

28L6309.......................... **$1.95**

Tire Saver Jacks.

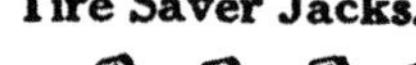

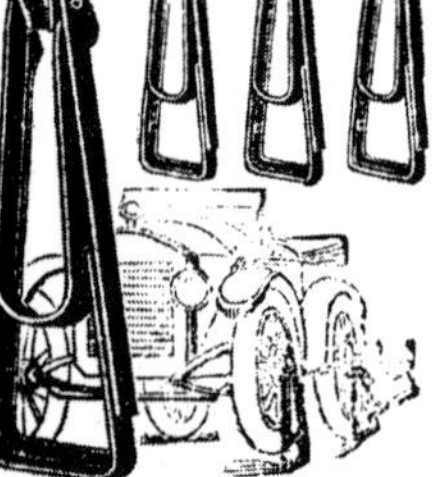

Keep your tires off an oily, damp wood or cement floor and keep weight of car off tires. Wheel is lifted and secured in place in one operation. Come in sets of four. Black finish.

28L15067¼—Adjustable type for cars having 32x3½-inch tires or larger. Shipping weight, per set, 16 pounds. Per set of 4... **$2.90**

28L13204¼—Non-adjustable type for cars having 30x3 or 3½-inch tires. (Not illustrated.) Shipping wt., per set, 13 lbs. Per set of 4.... **$1.98**

JACKS

A complete line showing various types of efficient jacks, all splendid values.

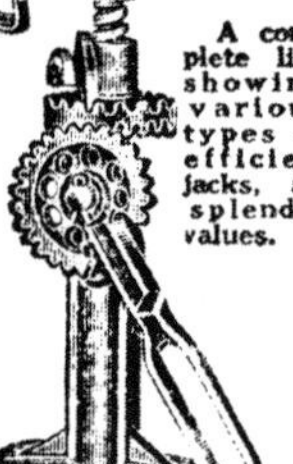

Screw Jack.

An inexpensive light weight Jack. Use on cars weighing up to 3,500 pounds. Operate catch to raise or lower jack. Height, about 10 inches; rise, about 5⅜ inches. Shipping weight, 5 pounds. **28L5024**....... **85c**

Steel Jack.

Practical, light weight inexpensive Jack. Use on cars weighing up to 2,000 pounds. Height, about 10¼ inches; rise, about 5¼ inches. Has corrugated top for holding securely when placed under axle. Complete with 15-inch bent bar handle. Shipping wt., 7¼ pounds. **28L8052**.. **95c**

Light Weight Jack.

A practical light weight malleable Jack of quick working ratchet type. Side hook gives it wide adjustability. Use on cars weighing up to 3,500 lbs. Comes about 9¼ in. high, with 5½-in. rise. Has wood handle. Shipping wt., 5 lbs. **28L11019**.. **$1.00**

Hercules Jack.

A big sturdy Jack for any weight of car, or height of axle, high or low.

Side hook adjustable from about 4½ to 16½ in. high. Use on cars weighing up to 4,500 pounds. Has steel rack bar with milled or machine cut teeth. Red enamel finish. Furnished with wood handle. Height, about 11½ inches; rise, about 6 in. Shpg. wt., 12½ lbs. **28L5150**...... **$2.45**

Ball Bearing Lever Handle Jack.

A splendid high grade lever handle Jack of ball bearing swivel gear type. Operates unusually smoothly and easily. Ball bearings are enclosed in race. Has convenient folding handle; length of handle, open, 32 inches; closed, 16 inches. Has 2-inch hinge extension top for cars with high axles. Lowest height, 9¼ inches; has 5¾-inch rise. Use on cars weighing up to 5,000 pounds. Shipping weight, 10 pounds.

28L8332.................. **$2.75**

All weights given on this page are approximate and may vary a trifle.

CUT PRICES ON THESE TIRE ACCESSORIES

Emergency Tire Chain.

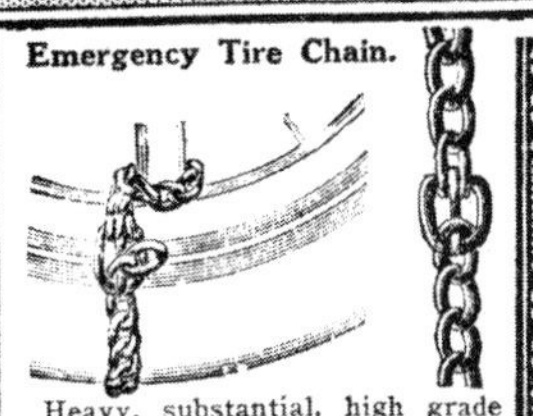

Heavy, substantial, high grade tire chain, in sets of four. Comes about 30 inches long, each section. Can be wrapped around spoke and tire and fastened with buckle and tongue to make a practical skid chain in mud or sand. Buckle together in series to make a strong emergency tow chain. Use with a good lock to lock car or spare tires. Bright finish. Come in cloth bag. Shipping weight, set of four, 7 pounds.
28L4653—Set of **4**. **$2.35**

Chain Pliers.

For removing and replacing broken cross chains. About 10 in. long. Instructions for using. Shipping weight, 1¼ lbs.
28L4292—Per pair. **49c**

Repair Links for Skid Chains.

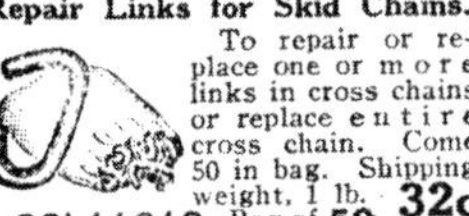

To repair or replace one or more links in cross chains or replace entire cross chain. Come 50 in bag. Shipping weight, 1 lb.
28L11618—Bag of **50**. **32c**

Weed De Luxe Cross Chains.

Catalog No.	For Tire Size, Inches	Shpg. Wt., Half Dozen	Half Doz.
28L4256	3½	1¾ lbs.	33c
28L4258	4	2 lbs.	39c
28L4260	4½	2¼ lbs.	44c
28L4262	5	2⅞ lbs.	50c

Catalog No.	Size, Inches	Shpg. Wt.	Each
28L4255	3½	6 oz.	6c
28L4257	4	7 oz.	7c
28L4259	4½	8 oz.	8c
28L4261	5	9 oz.	9c

Schrader Valve Parts.

28L4539

28L4539—Caps, in tin box of five with sliding cover. Per box of **5**. **25c**
28L4571—Valve Plungers or insides. Box of **5**. **25c**
Shpg. wt., either of above, 3 oz.

GENUINE WEED DE LUXE CHAINS AT CUT PRICES

Improved Lever Locking Connecting Hook.

Note New Reinforcing Link Feature.

A new Weed chain of the latest improved type. A new construction makes it stronger than ever and gives it greatly increased wearing qualities. Each cross chain has several reinforcing links firmly gripping it, permitting hardening cross chains to much greater depth and greatly lengthening their life. Reinforcing links also give better traction by affording more points of road contact. A new style lever locking connecting hook gives a positive lock, unaffected by chain tension. Our cut prices on this splendid chain offer you a money saving opportunity you cannot afford to miss.

Cut Prices on Genuine Weed De Luxe Tire Chains.

Catalog No.	Order Same Size for Either Fabric or Cord Tires	Shipping Weight, Pounds	Per Pair
28L4152	30x3½	16	$3.45
28L4154	32x3½	17½	3.95
28L4156	31x4	17½	4.25
28L4158	32x4	18½	4.35
28L4160	33x4	18¾	4.65
28L4162	34x4	19	4.95
28L4163	32x4½	20½	4.95
28L4164	33x4½	20¾	4.95
28L4166	34x4½	21½	5.25
28L4168	35x4½	22½	5.75
28L4171	33x5	26½	6.00
28L4172	35x5	27¼	6.45
28L4174	36x5	27½	6.45
28L4176	37x5	27¾	6.85

Mud Hooks.

Will often pull you out of deep mud, etc., when chains will not. Web straps for fastening.
28L6953 Mud Hooks for 3 and 3½-inch tires. Per pair. **$1.29**
28L6955—Same, for 4 and 4½-inch tires. Per pair. **$1.29**
Shpg. wt., either of above, 4½ lbs.

Lawco Rim Tool.

For quickly removing and replacing tires on practically any demountable rim that splits across. Tool both contracts and expands rim. Comprises set of heavy, strong steel arms with claws for hooking over rim and lever for locking tool in place. Folds like a jackknife for storing in car. Instructions for using and illustrations showing method of operation on Kelsey, Baker, Detroit, Stanweld, etc., rims. Shipping weight, 6¼ pounds.
28L5114—Lawco Rim Tool. **$2.65**

Han-D Tire Tool.

For quickly taking off or putting on any 3, 3½ or 4-inch tire without injuring tube or casing. Both side wall and bead of tire are forced over rim at the same time, shoulder of tool fitting bead. About 16½ inches long. Shipping weight, 2½ pounds.
28L11377—Han-D Tire Tool. **49c**

Tire Locking Cable.

For locking spare tire, wheel, etc. Made of steel cable with leather covering. Eyes on end securely fastened. Comes 36 in. long. For padlock see below. Shipping weight, 13 ounces.
28L8083—Tire Locking Cable. **64c**
28L11195—Keyless Combination Padlock for tire locking chains. Shipping weight, 7 ounces. **69c**

Puncture Proof RELINERS

MAKE YOUR TIRES PUNCTURE AND BLOWOUT PROOF

An excellent, highly efficient Puncture Proof Reliner, made of extra heavy, tough, resilient rubber with two layers of overlapping brass discs or plates. These metal discs, separated by rubber, make your tire puncture proof; nails, tacks and other sharp objects penetrating to the plates being turned back. Reliner is built with long tapered ends and feather edge tapered sides, protecting tube from being pinched and fitting tire snugly. They conform perfectly to shape of casing when tube is inflated. **We guarantee Puncture Proof Reliners to be both puncture and blowout proof for the life of the tire.** Made in all popular tire sizes as listed below.

Catalog No.	Size, Inches	Shpg. Wt., Lbs.	Each
28L2650	30x3	4⅝	$2.45
28L2651	30x3½	4⅞	2.95
28L2652	32x3½	5¼	3.55
28L2653	31x4	5¼	3.65
28L2654	32x4	5½	3.75
28L2655	33x4	5⅝	4.25
28L2656	34x4	5⅝	4.35
28L2657	32x4½	6⅝	4.25
28L2658	33x4½	6⅝	4.35
28L2659	34x4½	6¾	4.65
28L2660	35x4½	6⅞	4.70
28L2661	33x5	6⅞	5.45
28L2662	35x5	7	5.50

Puncture Healing Inner Tubes

5,000-Mile Guarantee Against Puncture.

Avoid punctures with their annoying features and get all the pleasure from your automobile trips that freedom from worry over possible tire trouble will bring you. Twentieth Century Inner Tubes are guaranteed for 5,000 miles against punctures. A special process gives these tubes puncture resisting features. Nails, tacks and other small objects that penetrate the tube can be pulled out from the tire without losing the air in the tube. Twentieth Century Tubes are made of high grade rubber materials and will give long, satisfactory wear. Try them under our liberal guarantee of freedom from puncture.

Catalog No.	Size, Inches	Shipping Weight	Each
28L6007	30x3	3¼ lbs.	$1.70
28L6009	30x3½	4¼ lbs.	1.88
28L6017	32x3½	4½ lbs.	1.93
28L6019	31x4	5¼ lbs.	2.26
28L7036	32x4	5½ lbs.	2.34
28L7038	33x4	5¾ lbs.	2.38
28L7040	34x4	6½ lbs.	2.43
28L5213	32x4½	6 lbs.	2.83
28L5216	33x4½	6⅛ lbs.	2.87
28L5218	34x4½	6¼ lbs.	2.93
28L5220	35x4½	7¾ lbs.	2.98

Handy Combination Puncture Healing Inner Tube for Ford Cars.

Ford Car Owners! We have a special size Twentieth Century Inner Tube to use in either 30x3 or 30x3½-inch fabric tires. Shipping weight, 4¼ pounds.
28L7042. **$1.78**

All weights given on this page are approximate and may vary a trifle.

TIRE ACCESSORIES AT MONEY SAVING PRICES

Tire Pumps

Samson Single Cylinder Pump. A high grade, single action foot pump, with about 1½x16½-inch extra heavy non-rusting seamless brass barrel. Complete with extra long, heavy rubber pump hose and thumb lock valve connection. Has heavy malleable base, brass finish. Large comfortable handle. Shipping weight, 4⅞ pounds.
28L4841—Samson Single Cylinder Foot Pump. **$2.10**

Ideal Cmpound Pump.

Length of barrels, about 16 inches; diameter, about 1½ inches and ⅞ in. Brass finish. Shipping weight, 5 lbs.
28L4860 — Ideal Compound Pump.......... **$1.45**

Crown.

A low priced single cylinder pump, non-rusting brass barrel. Length of barrel, about 17½ inches; diameter of barrel, 1⅛ inches. Malleable base. Shipping weight, 3 pounds.
28L4866—Crown Pump.................. **85c**

Inland Running Board Pump.

A splendid, powerful, easy working running board pump. Folds compactly in about 3x4½x13½ inches. Has many original exclusive features of design and construction. It delivers up to 150 pounds of air pressure. Has leakproof seamless drawn steel cylinder about 3x6 inches, high grade oil treated, long lasting leather plunger. Malleable iron frame and handle, screw steel thumbscrews. Patent folding handle closes in contact with cylinder; clamp folds under pump. Hose screws to pump and is quickly detached. Complete with about 7 feet of hose.
28L7790—Inland Running Board Pump, without pressure gauge...... **$3.75**
28L7792—Inland Running Board Pump, with pressure gauge.................. **4.25**
Shipping weight, above pumps, 8½ and 8⅝ pounds, respectively.

Triple Cylinder Pump.

Has three cylinders, brass finish, about 18 inches long. Diameter 1¾, 1¼ and ¾ inch. Shipping weight, 9 pounds.
28L4885 Triple Cylinder Pump.......... **$3.25**

Rubber Pump Tubing.

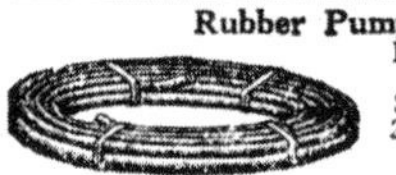

Inside Diameter Given.
28L4950—5/16-inch. Shipping weight, per foot, 2 ounces. Per foot.......... **7c**
28L4951—¼-inch. Shipping weight, per foot, 3 ounces. Per foot.................................... **8c**

Apex Pump Connection.

Fasten to free end of foot pump hose. Turn arm to quickly and securely attach over tire valve stem. Shipping weight, 2 ounces.
28L4890—Apex Pump Connection .. **19c**

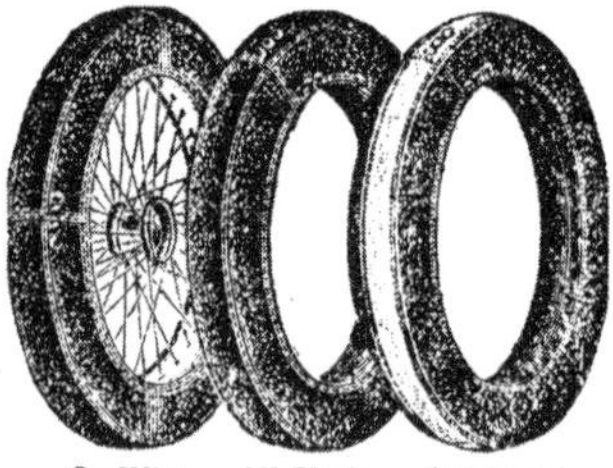

On Wire Wheel. All Black. Black and White.

In ordering be sure to specify size of tire and whether plain or anti-skid type tread.

Tire Covers.

Your spare tires will last longer if you protect them from dust, rain and sun with a tire cover. This helps prevent the rubber from checking, splitting or rotting from exposure, also adds to the appearance of the car. Made of black enameled cloth. Fastened with ball and socket type fasteners.

Black and White Universal Tire Covers for Fabric Tires.

Black and White Universal Tire Covers, made of black enameled cloth with white center stripe; very attractive in appearance. Use them for demountable tires for any make of car, including special holder type, covers only extending to demountable rim and not covering it. Also fit any make of wire wheels.

Black and White Universal Fabric Tire Covers.

	Tire Size	Shipping Weight	
28L5284	30x3½ in.	1¾ lbs.	$1.55
28L5289	32x3½ in.	1¾ lbs.	1.60
28L5290	31x4 in.	1¾ lbs.	1.65
28L5291	32x4 in.	1¾ lbs.	1.65
28L5292	33x4 in.	1¾ lbs.	1.65
28L5293	34x4 in.	1¾ lbs.	1.65
28L5136	33x4½ in.	1¾ lbs.	1.70
28L5294	34x4½ in.	1¾ lbs.	1.85
28L5295	35x4½ in.	1¾ lbs.	1.85

All Black Universal Fabric Tire Covers.
Same as above, except all black.

	Tire Size	Shipping Weight	
28L5274	30x3½ in.	1¾ lbs.	$1.35
28L5275	32x3½ in.	1¾ lbs.	1.40
28L5276	31x4 in.	1¾ lbs.	1.45
28L5277	32x4 in.	1¾ lbs.	1.50
28L5278	33x4 in.	1¾ lbs.	1.50
28L5279	34x4 in.	1¾ lbs.	1.50
28L5280	33x4½ in.	1⅞ lbs.	1.65
28L5281	34x4½ in.	1⅞ lbs.	1.70
28L5282	35x4½ in.	1⅞ lbs.	1.70

Universal Cord Tire Covers, All Black.
All black tire covers, made of black enameled cloth, for cord tires. Furnished in sizes listed below.

	Tire Size	Shipping Wt.	
28L5117	30x3½ inches	1⅝ pounds	$1.45
28L5118	32x3½ inches	1⅝ pounds	1.50
28L5119	31x4 inches	1¾ pounds	1.60
28L5120	32x4 inches	1¾ pounds	1.65
28L5121	33x4 inches	1¾ pounds	1.65
28L5122	34x4 inches	1¾ pounds	1.65
28L5123	33x4½ inches	1⅞ pounds	1.75
28L5124	34x4½ inches	1⅞ pounds	1.75
28L5126	35x4½ inches	1⅞ pounds	1.75

Black and White Cord Tire Covers.
Same as above, except with white center stripe.

	Tire Size	Shipping Wt.	
28L5127	30x3½ inches	1½ pounds	$1.65
28L5128	32x3½ inches	1¾ pounds	1.65
28L5129	31x4 inches	1¾ pounds	1.80
28L5130	32x4 inches	1¾ pounds	1.80
28L5131	33x4 inches	1¾ pounds	1.80
28L5132	34x4 inches	1¾ pounds	1.80
28L5133	33x4½ inches	1¾ pounds	1.90
28L5134	34x4½ inches	1¾ pounds	1.90
28L5135	35x4½ inches	1¾ pounds	1.90

Handy Car Equipment

Running Board Mats.

Cocoa. Rubber.

Two clips hold mat to running board. Size, about 14x8 inches. Shipping wt., 1¾ pounds.
28L13170 Cocoa Running Board Mat **79c**

A well made corrugated black rubber mat for running board use. Size, about 8x11½ inches. Attach with bolts furnished. Shipping weight, 1¾ pounds.
28L10569—Running Board Mat. Rubber.......... **30c**

3-In-1 Wind Shield Cleaner.

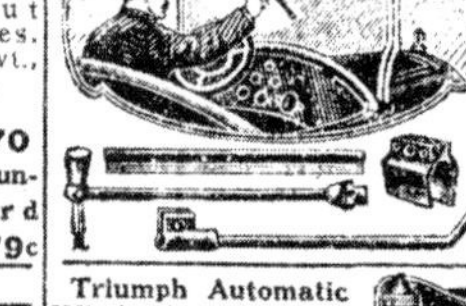

Fits any make or style car in any one of three positions. Wipes quickly and effectively, rotating on bearing at stroke end. Has adjustable wiping bar tension, renewable rubber wiping bar and nickel plated handle with cushion, allowing moving cleaner practically without effort. Shipping weight, 14 oz.
28L6608 .... **95c**

Tire Gauges.

Keep the air pressure in your tires to the figure shown on the tire and get more mileage.

Every mile run on soft or under inflated tires shortens the tire service many times that mile.

Schrader Tire Pressure Gauge.

Complete with case. Shipping weight, 3 ounces.
28L4902 Schrader Tire Pressure Gauge.. **$1.25**

Twitchell Air Gauge.

Indicator is held by catch at pressure registered. Release catch to return indicator to position. Nickel plated. Shipping weight, 3 ounces.
28L4900 — Twitchell Air Gauge........... **$1.25**

Step Plates for Any Car.

Rubber Insert.

The latest type step plates, with aluminum frame. Rubber insert pad comes about 6¾x8 inches in size between frame rims. Over all size of step plate, about 9½ inches square. Bolt to running board. Bolts furnished. Shpg. wt., pair, 4⅝ lbs.
28L12077 Per pair............... **$1.55**

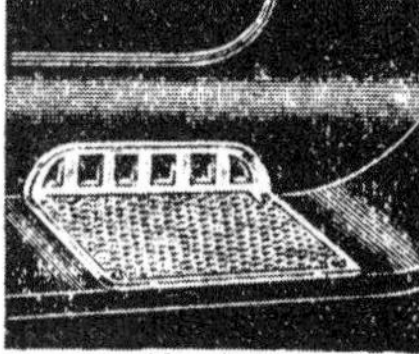

Aluminum.

Up to date and attractive. Prevents marring splash apron when getting in or out of car. Bolts to running board. Comes about 9½ inches long and 9 inches wide over all. Bolts furnished. Shipping weight, per pair, 4⅜ pounds.
28L5330—Pair... **$1.45**

Triumph Automatic Wind Shield Cleaner.

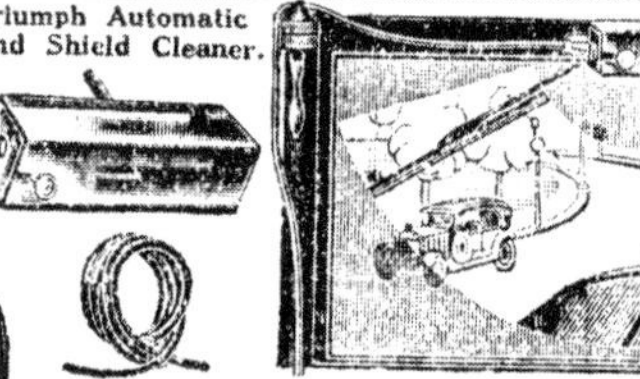

A splendid unusually quick operating, effective wind shield cleaner, working from engine suction. Install on upper wind shield frame, as illustrated, attach rubber hose furnished, drilling hole in intake manifold. Operates by turning button. Furnished in polished aluminum. Complete with cleats for securely attaching hose to car body. Wiping arm is complete with high grade rubber strip. Will positively remove sleet or snow from the shield, as well as rain. Guarantees you clear vision in stormy weather. Shipping weight 1⅝ pounds.
28L5014—Triumph Automatic Wind Shield Cleaner.................. **$3.68**

Air Line Gauge.

Use gauge to determine pressure in tire and when inflating tire with pump or free air line having no gauge. Registers up to 150 pounds pressure. Attached to two short lengths of pump hose, one of which has locking style slip-on connection at free end, the other a tire valve. Shipping weight, 5 ounces.
28L10816—Air Line Gauge............ **$1.15**

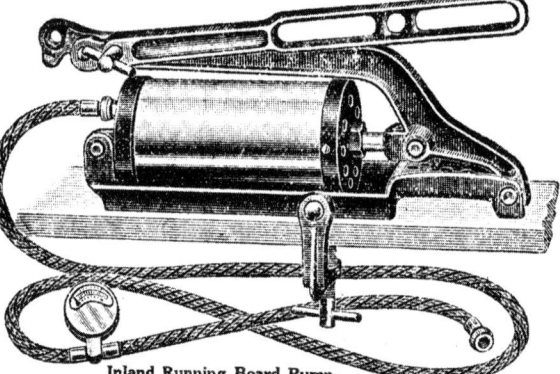

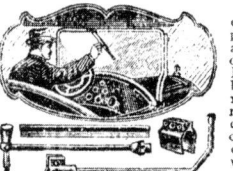

COVERINGS FOR STANDARD CARS

Top Coverings for Well Known Touring Cars

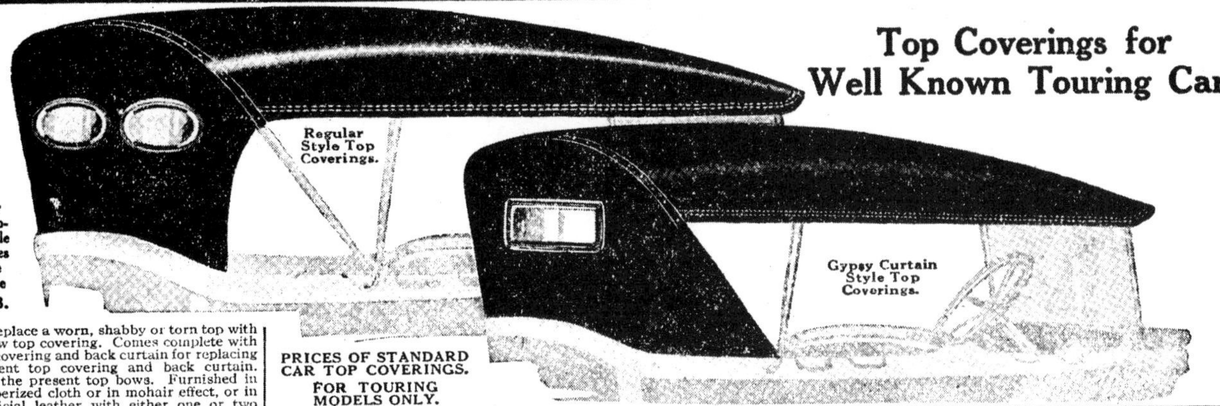

For Automobile Robes See Page 933.

Replace a worn, shabby or torn top with a new top covering. Comes complete with top covering and back curtain for replacing present top covering and back curtain. Use the present top bows. Furnished in rubberized cloth or in mohair effect, or in artificial leather with either one or two 6x12-inch oval bevel plate glass lights, or with one oblong bevel plate glass light, 7¼x13½ inches. Mohair effect material is made in dark gray cotton cloth, in imitation of mohair. Artificial leather covers are made of double texture long grain rubber material. **Specify whether black or khaki color inside lining is wanted on mohair effect or artificial leather top coverings. Rubberized cloth top coverings come only in black color inside lining. In ordering be sure to state year and model of car.** FURNISHED ONLY FOR MAKES, YEARS AND MODELS OF CARS LISTED.

NOTE—Gypsy back curtain is furnished with top covering only on models whose regular equipment includes gypsy back curtain.

*1919 models not equipped with Gypsy back curtain.

†Specify on 1919-21 models if regular equipment has regular or Gypsy style back curtain.

‡Specify on 1919-90T if regular equipment has regular or Gypsy style back curtain.

PRICES OF STANDARD CAR TOP COVERINGS. FOR TOURING MODELS ONLY.

Catalog No.	Car and Model
28L12750⅓	Buick D-45-1916-1917, E-45-1918
28L12752⅓	Buick H45, 1919, K45, 1920, both with Gypsy Back Cur'n.
28L12754⅓	Chevrolet H4, 1916, Baby Grand, 490-1916, 490-1919-1921, F. A. 1918
28L12756⅓	Chevrolet 490-1917, 1918
28L12758⅓	Chevrolet Baby Grand F. B. 1919-1921, Gypsy Back Curtain
28L12760⅓	Dodge 1916-1919* Touring
28L12762⅓	Dodge 1919-1921 Touring, Gypsy Back Curtain
28L12764⅓	Maxwell 25, 1916-1921†
28L12766⅓	Oakland 4-38, 1916, 34-1917, 34B-1918-19
28L12767⅓	Oakland 34C, 1920-21, Gypsy Back Curtain
28L12768⅓	Overland 90 C. C., 1917-1918, 90T, 1918-1919‡
28L12770⅓	Overland 75-1916, 83-1916, 75B-1917, B85 4 cyl. 1917, 4-85-1918, 90-1919
28L12771⅓	Overland Light 4, 1920-21, Gypsy Back Curtain
28L12780⅓	Saxon 6 cyl., 1916, 1917, 1918, 1919
28L12782⅓	Studebaker L6, 1919-1921 5 pass., Gypsy Back Curtain
28L12783⅓	Studebaker Special Six, 1920-21, Gypsy Back Curtain

	Rubberized Cloth	Mohair Effect	Artificial Leather
One Oval Glass Light	$11.65	$12.55	$13.45
Two Oval Glass Lights	$12.75	$13.65	$14.65
One Oblong Glass Light	$13.85	$14.75	$15.75

NOTE—Shipped from factory in NORTHEASTERN ILLINOIS, in about fifteen days' time. Shipping weight, one-glass light type, 11 pounds; two-glass light type, 15 pounds.

"Stik-Tite" Roof Patch Strips.

For repairing holes or breaks in rubberized cloth, mohair effect or artificial leather tops. Mohair, artificial leather and rubberized cloth strips, come 6x18 inches, large size; mohair effect and artificial leather strips, come 3x12 inches, small size; rubberized cloth strips, about 5x9 inches, small size. Under surface has cement coating, protected by sheeting. Apply like an inner tube patch, following instructions given.

28L11047—Stik-Tite Roof Patch Strip, small size, for rubberized cloth tops.......... **27c**

28L9636—Same, large size, for rubberized cloth tops..... **79c**

28L11048—Same, small size, for mohair effect tops........ **27c**

28L9632—Same, large size, for mohair effect tops........ **79c**

28L9630—Same, small size, for artificial leather tops...... **27c**

28L9634—Same, large size, for artificial leather tops...... **79c**

Shipping weights, above outfits, small size, 4 ounces; large size, 6 oz.

Celluloid Cement Outfit.

You can repair small cracks or breaks in celluloid curtains. Outfit comprises about 1-ounce bottle of transparent celluloid cement, brush for applying to edges of crack and strips of celluloid. Shipping weight, 7 ounces.

28L7654—Celluloid Cement Outfit.......... **25c**

Glaroscope.

Protects driver from glare of dazzling headlights. Made of dark green celluloid. Attach to wind shield by tight gripping rubber suction cup. Shipping weight, 4 ounces.

28L5694—Glaroscope. **15c**

Back Curtains With Glass Lights for Popular Model Touring Cars.

FURNISHED ONLY FOR MAKES, YEARS AND MODELS OF CARS LISTED.

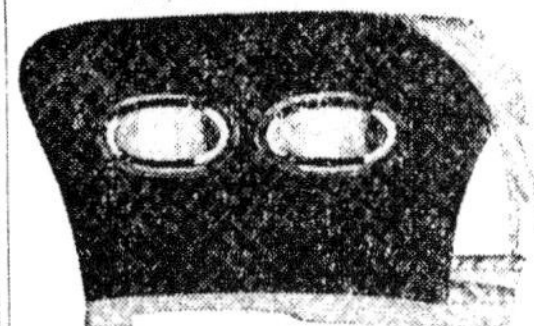

Artificial Leather Back Curtains With Two Oval Glass Lights.

Mohair Style Back Curtains With Single Oblong Glass Light.

Bring your car up to date and give the car rear an attractive appearance by equipping it with one of these popular late model bevel plate oval or oblong glass light type back curtains. These new curtains are especially worth while where the back curtain is shabby or where cracked or broken celluloid lights give the car rear an unsightly appearance. These curtains are furnished in either rubberized cloth or in mohair effect or in artificial leather and with either one or two 6x12-inch oval bevel plate glass lights, or one 7¼x13½-inch oblong bevel plate glass light. **Specify whether black or khaki color inside lining is wanted on mohair effect or artificial leather back curtains. Rubberized cloth back coverings come only in black color inside lining. Gypsy style curtain is furnished only where it is regular equipment on the car. In ordering be sure to state year and model of car.**

Catalog No.	Car and Model
28L12964⅓	Buick D-45-1916-1917, E-45-1918
28L12966⅓	Buick H45, 1919, K45, 1920, both with Gypsy Back Curtain
28L12968⅓	Chevrolet H4, 1916 Baby Grand, 490-1916, 490-1919-1921. F. A. 1918
28L12970⅓	Chevrolet 490-1917, 1918
28L12972⅓	Chevrolet Baby Grand F. B. 1919-1921, Gypsy Back Curtain
28L12974⅓	Dodge 1916-1919* Touring
28L12976⅓	Dodge 1919-1921 Touring, Gypsy Back Curtain
28L12978⅓	Maxwell 25, 1916-1921†
28L12982⅓	Oakland 4-38, 1916, 34-1917, 34B-1918-19
28L12991⅓	Oakland, 34C, 1920-21, Gypsy Back Curtain
28L12984⅓	Overland 90 C. C., 1917, 1918, 90T, 1918-1919‡
28L12986⅓	Overland 75, 1916, 83-1916, 75B-1917, B85 4 cyl. 1917, 4-85-4-1918, 90-1919
28L12993⅓	Overland Light 4, 1920-21, Gypsy Back Curtain
28L12996⅓	Saxon 6 cyl., 1916, 1917, 1918, 1919
28L12998⅓	Studebaker L6, 1919-1921 5 pass., Gypsy Back Cur'n
28L12995⅓	Studebaker Special Six, 1920-21, Gypsy Back Curtain

*1919 models not equipped with Gypsy Back Curtain.

†Specify on 1919-21 models if regular equipment has regular or Gypsy style back curtain.

‡Specify on 1919-90T if regular equipment has regular or Gypsy style back curtain.

	Rubberized Cloth	Mohair Effect	Artificial Leather
One Oval Glass Light	$5.45	$6.25	$6.55
Two Oval Glass Lights	$6.15	$6.95	$7.25
One Oblong Glass Light	$6.85	$7.65	$7.95

For Touring Models Only.

Mohair effect material is made in dark gray cotton cloth, in imitation of mohair.

Artificial leather covers are made of double texture long grain rubber material.

NOTE—Shipped from factory in NORTHEASTERN ILLINOIS, in about fifteen days' time. Shipping weight, 5 pounds.

Seat Pads.

Protect your clothes and seat cushions with these attractive, serviceable, two-piece straw seat covers, bound and reinforced with strong khaki material. Size of each piece, about 17x17 inches. Pieces are fastened together to make combination back and seat pad. Clean them with sponge or damp cloth. Almost a necessity for hot weather, but valuable at all times, both for auto and for widespread outdoor use. Shipping weight, 1¼ pounds.

28L6969—Seat Pads. Per Two-Piece Pad........ **57c**

Auto Top Mending Outfit.

For repairing rubberized or mohair style tops, top boots, seat covers, side curtains, etc. Comprises 2-oz. bottle of cement, patching material and swab for applying. Shipping weight, 12 ounces.

28L9395—Mending Outfit for rubberized tops. **30c**

28L9397—Mending Outfit for mohair style tops **30c**

CELLULOID.

28L7745¼

Celluloid in rolled sheets about 20x36 inches.

Per sheet...... **$1.29**

28L13155¼

Same in sheets about 20x50 inches.

Per sheet...... **$1.55**

Shipping weight, either of above, 1⅝ pounds.

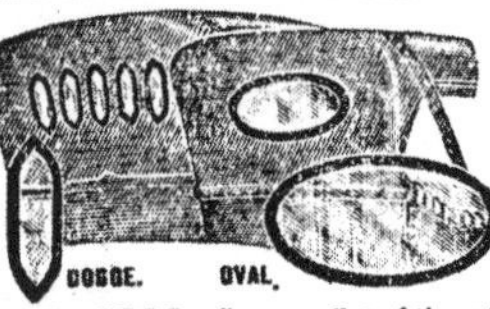

"Stik-Tite" Celluloid Curtain Lights.

Stick Them On Like a Postage Stamp.

28L4294

For Dodge rear or side curtains. Use cement furnished to make quick, satisfactory, permanent repair. Shpg. wt., 6 oz. Each.. **35c**

28L9223—Same. Set of three lights. Shipping weight, 12 ounces. Set.................. **$1.00**

28L5116—Stik-Tite Oval Curtain Light for Overland and Chevrolet cars, also Buick and Maxwell cars through 1917 only. Use on any car having oval back curtain light up to 8½x18½-inch opening size. Cement furnished for applying. Shipping weight, 12 ounces.............. **95c**

HANDY CAR EQUIPMENT AT LOW PRICES

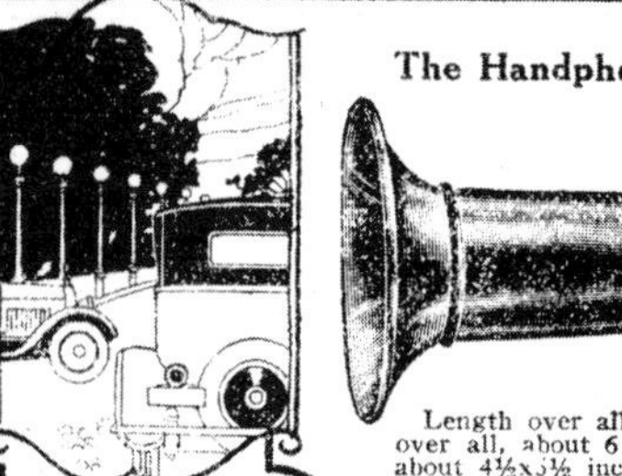

The Handphone.

Length over all, about 9 in.; height over all, about 6 inches; size of bell, about 4½x5½ inches. Furnished with bracket for any standard car.

28L6755—The Handphone **$2.25**

28L13202—Same, with bracket for attaching to Ford car **$2.48**

Shipping weight, either of above, 3¾ pounds.

All weights given on this page are approximate and may vary a trifle.

CLOCKS

FRISCO. An inexpensive auto clock with about 2⅛-inch white dial. Width, over all, about 3½ inches; depth (front to back), about 1½ in. Furnished in black finish, nickel plated rim. Has thirty-six hour movement. Screws and bolts furnished. Shipping wt., 1¼ lbs.

28L7150 **$3.75**

DENVER. Popular priced 8-day clock. Has about 2⅛-inch dial, black face and black case with nickel plated rim. Width over all, about 3⅛ in.; depth (front to back), about 2 in. Complete with screws and bolts. Shpg. wt., 1¾ lbs.

28L7143 **$5.45**

To install either of above clocks, cut hole in dash about 2½ inches in diameter.

Tru-Tone Under Hood Horn.

Use with storage battery. Mount under hood on engine. Diameter at mouth, 4 inches; length over all, 10 inches; over all height, 6 inches. Cord and push button not furnished.

28L12980—For use with 6-8-volt battery **$4.48**

28L5660—Same, for use with 12-16-volt battery **$4.48**

28L12983—Same as 28L12980 Horn with bracket for Fordengine **$4.48**

Shipping wt., any of above horns, 4¾ pounds.

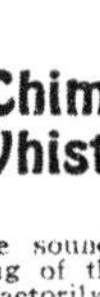

These pages offer a splendid variety of car equipment items at prices which make them wonderful values.

Chime Whistle

A high grade, attractive sounding signal, producing a blending of three tones. Can be used satisfactorily on any car having priming cups. Barrel size, about 2 inches in diameter, 4¼ inches long. Place on compression or exhaust valve side of engine. Shipping weight, 1¼ pounds.

28L9075 Chime Whistle **$2.98**

Stiles' Explosion Whistle.

Whistle is aluminum, nickel plated finish. Barrel size, about 4 inches long, 2 inches in diameter. Adjustable to any tone, regardless of strength of motor compression, a valuable feature. Has fitting for replacing priming cup, also spark plug adapter for use on cars not equipped with priming cups.

28L5303—With fitting replacing priming cup, but without spark plug adapter. Shipping weight, 1½ pounds **$2.10**

For Cars Not Equipped with Priming Cups.

28L9284—Same, complete with ½-inch spark plug adapter **$2.40**

28L9286—Same, complete with ⅞-inch adapter **2.40**

Shipping wts. of above, 1¼ and 1⅜ lbs., respectively.

Cut-Out Outfits

Leader Cut-Out Outfit. Cut V in exhaust pipe to install. Outfit shown includes valve, pedal, pulley and cable. Fits any car with outside diameter exhaust pipe the same as any of the valve sizes given.

28L12555—Leader Cut-Out Outfit for Maxwell cars. Valve size, 1⅝ inches **75c**

28L12558—Same, for Saxon up to and including 1917 models, Allen, Apperson, Briscoe, Buick, Cadillac, Grant, Hupmobile and Oldsmobile cars. Valve size, 1¾ inches **80c**

28L12560—Same, for Chandler, Dodge, Essex, Liberty, Lexington and Studebaker cars. Valve size, 2 inches **90c**

28L11715—Same, for Ford cars. Valve size, 1½ inches **65c**

Shipping weights of above, 3, 3, 3 and 3¼ lbs., respectively.

Master Cut-Out Outfit. A well made, extra heavy cut-out. Has 1½-inch opening; top of valve is open, valve being attached to exhaust pipe by clamps. Outfit is complete with substantial locking pedal and cable. No pulley is needed, action of pedal opening cut-out. Valve and pedal furnished in black finish. Shipping weight, 2¼ pounds.

28L12584—Master Cut-Out Outfit........ **$1.20**

Goodrich Motor Testing Valve.

A Popular, Up to Date, High Grade Cut-Out. Has bell mouth opening, intensifying explosion sound, allows extra space for gas explosion and relieves motor of all back pressure. Unusually durable. Instructions for installing.

28L13249—Valve size, 1⅝ inches. Shipping weight, 5 pounds **$2.95**

28L13247—Valve size, 1¾ inches. Shipping weight, 4½ pounds **$2.95**

28L13252—Valve size, 2 inches. Shipping weight, 4½ pounds **$2.95**

28L13254—Valve size, 2¼ inches. Shipping weight, 6¾ pounds **$3.38**

You can afford to make your car equipment complete. These prices prove it.

Folding Chair

Popular type folding auto chair. Has padded artificial leather seat and back with round metal frame; black. Height of seat from floor, about 16 inches. Folds compactly. Shipping weight, 9¼ pounds.

28L10311—Standard Folding Auto Chair **$2.39**

Imitation Tortoise Shell Goggles (Celluloid).

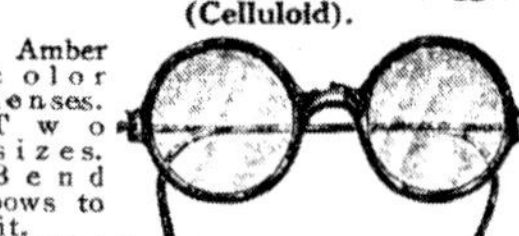

Amber color lenses. Two sizes. Bend bows to fit.

28L10478—Imitation Tortoise Shell (Celluloid) Goggles. Lens diameter, about 1⅞ inches **39c**

28L10557—Same, lens diameter, about 1¾ inches..... **39c**

Shipping weight, either size, 7 oz.

Side Shield Amber Goggles.

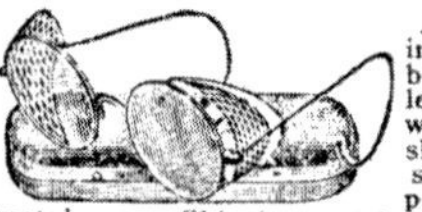

About 2-inch amber color lens with wire dust shields at side. Complete in metal case. Shipping weight, 8 oz.

28L4622—Side Shield Amber Goggles. Per pair **35c**

Adjustable Luggage Carrier

A Splendid Value. **A Touring Necessity.** Adjustable luggage carrier, quickly clamped to running board with three thumb fasteners. Greatest inside length, about 50 inches, about 15 inches high. Complete with end brackets. Adjustable feature permits holding securely suitcase, bags or other packages carried. Shipping weight, 14⅝ pounds.

28L15041—Adjustable Luggage Carrier **$1.35**

LUGGAGE CARRIER COVER.

Substantial enameled cloth or artificial leather Luggage Covers for protecting luggage from rain, mud, etc. Furnished about 51 inches long (fitting snugly and without folding two average suitcases placed end to end in luggage carrier), 12 inches wide and 16 inches high. Open at bottom.

28L13241—Running Board Luggage Cover, enameled cloth. Shpg. wt., 2 lbs. **$1.25**

28L13243—Same, artificial leather. Shipping weight, 2⅝ pounds **1.75**

Siren.

A melodious warning signal of penetrating tone. Plays a variety of notes. Attach to exhaust manifold and operate by simply pulling cord. Attach cord to dash or steering wheel column. Shipping weight, 1¼ pounds.

28L9163—Siren **$1.49**

Rear View Mirrors

Up to date popular type. Size, about 10x2½ inches. Has bevel edge. Complete with bracket fitting any standard open car wind shield frame. Bracket is adjustable in four positions. Shipping weight, 1⅜ pounds.

28L9628—For open model standard cars **$1.05**

The above mirror with bracket fitting wind shield frame on any standard model closed car. (Not illustrated.) Mirror can be tilted up or down, or to either side. Shipping weight, 1 pound.

28L9626 .... **85c**

For similar mirrors for Ford cars, see page 759.

Auto Visors

Leather—For Sun, Rain or Bright Lights.

Popular type. Adjustable to shield driver's eyes from sun's rays or headlight glare; keeps upper wind shield clearer in stormy weather. Black artificial leather with metal framework. Will not readily sag or rattle. Size, about 9½x40 inches. Fits all but V shape wind shields. On open cars, attach brackets to front top bow, shield brackets being adjustable to fit any size car. On closed cars screw brackets to under side of roof. Screws furnished. Shipping weight, 5½ pounds.

28L10630¼ **$1.70**

Shipping wt., either style, 7¼ lbs.

Metal—For protection against Sun, Rain or Bright Lights.

All metal, black enameled finish. Has bracket for clamping to any open car wind shield frame, except with V shape wind shields, with thumbnut adjustment for tilting to angle desired. Size, about 9 inches wide and 42 inches long over all.

28L11588¼ Metal Auto Visor, open car type **$2.80**

Closed Car Type. 28L5264¼ Same as above, for use on any closed model car **$2.80**

Rear View Mirrors

A 5-inch round mirror of reducing type, black. Attaches to wind shield frame. Adjustable. Shpg. wt., 1⅜ lbs.

28L6795 .... **78c**

A round mirror; 5-inch front, beveled edge glass, black frame. Adjustable. (Not illustrated.)

28L6767 .... **$1.15**

Shipping weight, 1⅝ pounds.

FENDER TYPE.

An inexpensive mirror of reducing type for attaching to fender to show approach of cars from rear. Comes in all black finish, with about 5-inch front and upright bracket. Complete with bolts for attaching. Shipping wt., 1 pound.

28L7456 Fender Mirror **74c**

Ribbed Green Glass Sun Shield.

ADJUSTABLE BRACKET

A substantial, attractive, green ribbed sun shield with splendid, durable, highly polished nickel plated brackets, fitting any open or closed model Ford or standard type cars. Glass size, about 10 inches wide, 44 inches long. Glass has no frame, thus giving shield no unnecessary weight. Edges of glass are ground. Glass design effectively kills all sun glare. Shield also keeps upper wind shield from being covered by rain in stormy weather. One trip will prove its value when driving against the sun. Will not rust or rattle. When properly attached, shield is as solid as any part of the car. Shipping weight, 21¼ pounds.

28L9987¼—Complete with brackets **$9.95**

SOCKET WRENCH SETS AND TOOLS

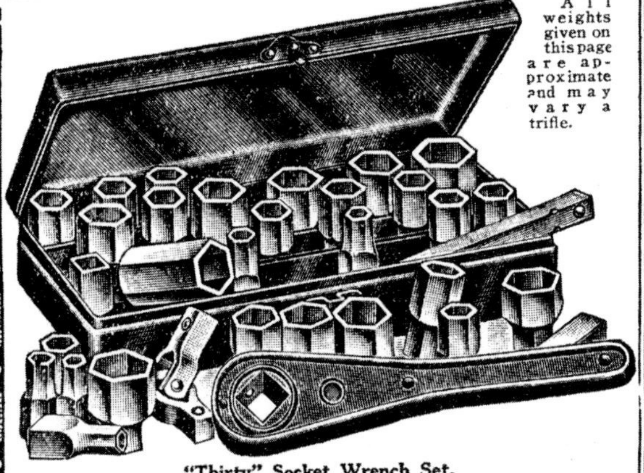

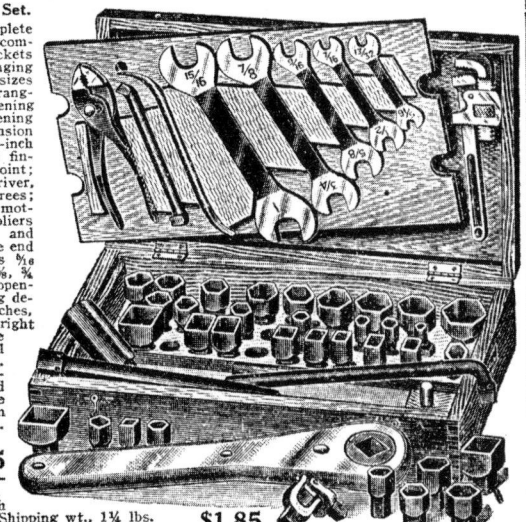

All weights given on this page are approximate and may vary a trifle.

"Thirty" Socket Wrench Set.

A high grade socket wrench set, fitting practically every nut and bolt on the car. Comprises twenty-seven steel hex sockets with opening sizes ranging from 5/16 to 1 9/32 inches and three square sockets, 13/32, 17/32 and 21/32 inch opening sizes; spark plug socket, long extension bar, universal joint and substantial ratchet wrench handle. Comes in substantial metal case, black finish. Shipping weight, 9¼ pounds.

28L15048—"Thirty" Socket Wrench Set........ **$5.75**

Universal Socket Wrench Set.

Our finest and most complete socket wrench set. Outfit comprises thirty-four steel sockets (twenty-three hex sockets ranging from 5/16 to 1 9/32-inch opening sizes and eleven square sockets ranging from 13/32 to 1 9/32-inch opening sizes), spark plug socket, opening size 31/32-inch; 7-inch extension bar; 10-inch offset handle; 7-inch Stillson style wrench, bright finish; drop forged Universal joint; about 5¾-inch offset screwdriver, 3/8-inch blade, offset 90 degrees; 7½-inch cotter pin extractor, mottled finish; one pair 6-inch pliers with milled handle, bright and lacquered jaw tips; five double end wrenches with opening sizes 5/16 and 13/32, 7/16 and ½, 9/16 and 5/8, ¾ and 7/8, and 15/16 and 1-inch openings offset 15 degrees (giving desirable features of S wrenches, yet considerably stronger), bright lacquered ends, and high grade substantial 10-inch steel nickel plated ratchet wrench handle. All sockets and tools black finish unless otherwise stated and of anti-rust type. Complete with tray in substantial wooden box. Shipping wt., 16⅛ lbs.

28L5312 Universal Socket Wrench Set...... **$9.95**

Ratchet Wrench Handle.

Same as furnished with 28L5312 Wrench Set above. Shipping wt., 1¾ lbs.

28L10958........................ **$1.85**

Our "Fourteen" Leader Tool Kit.

A popular low price kit containing fourteen tools and accessories most in demand. Has case with individual pockets. Shipping weight, 5½ pounds.

28L11290—Leader Tool Kit.. **$2.75**

"Thirty" Tool Kit.

A well selected tool kit, including four machinists' wrenches, 9-inch auto wrench, monkey wrench, spark plug wrench, 9¼-inch Stillson style wrench, alligator wrench, ball pein hammer, three screwdrivers, offset screwdriver, three files, file handle, two pair pliers, two cold chisels, cape chisel, solid punch, center punch, two bundles copper wire, etc. (Not illustrated.) Shipping weight, 12 pounds.

28L11250—"Thirty" Tool Kit **$9.45**

Tappet Wrench Set.

Comprises six double end wrenches, two of each size for adjusting valve tappets. Opening sizes, 5/8x11/16, ½x9/16 and 3/8x7/16-inch; this range being desirable for almost any size of valve tappet. Wrenches are casehardened steel, complete in artificial leather case. Shipping wt., 10 oz.

28L4868—Tappet Wrench Set.... **68c**

Socket Brace.

Comes about 18 inches over all. All metal. Ball and spring friction holds sockets in place. Has about 9-inch shank. Fits any regulation hex or square socket. Use with socket fitting rim lugs to make a satisfactory rim brace wrench. Shipping weight, 1⅝ pounds.

28L9303—Socket Brace......... **45c**

Han-D Wrench Set.

Twelve wrenches in one. Comprises six double end wrenches with opening sizes from 3/8 to ¾-inch, graduated in 1/32-inch sizes (except 19/32-inch). Shipping weight, 1 pound. Three sizes of wrenches; smallest size, about 5⅛ inches long; largest size, about 7 in. long. Knurled thumbscrew holds wrenches together when not in use. Come in blued finish.

28L11057—Han-D Wrench Set............ **69c**

Sexto Wrench Set.

Similar to above set, but smaller. Opening sizes from 7/32 to 9/16 inch in 1/32-inch graduated sizes. Readily taken apart for using any one wrench. Knurled thumbscrew holds wrenches together. Nickel plated. Length over all, about 5 inches. (Not illustrated.) Shipping weight, 7 ounces.

28L11894...................... **38c**

Bearing Scrapers.

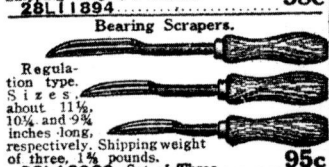

Regulation type. Sizes, about 11⅛, 10¾ and 9¾ inches long, respectively. Shipping weight of three, 1¾ pounds.

28L10909—Set of Three........ **95c**

Aristocrat No. 1 Socket Wrench Set.

For removing practically any nut or cap screw on any car by ratchet motion. Permits tightening nuts at points hard to reach. Set comprises substantial universal ratchet wrench handle, extension bar, offset screwdriver, universal joint, spark plug socket and thirty steel sockets, including twenty-seven hexagon wrench sockets, having average opening sizes as follows: 5/16, 11/32, 3/8, 13/32, 7/16, 15/32, ½, 17/32, 9/16, 19/32, 5/8, 21/32, 11/16, 23/32, ¾, 25/32, 13/16, 27/32, 7/8, 29/32, 15/16, 31/32, 1, 1 1/32, 1 1/16, 1 3/32, 1 9/32 inches, and three square wrench sockets having opening sizes 13/32, 17/32 and 21/32 inch. Complete with hardwood box. Opening size of spark plug wrench, about 29/32 inch. Shipping weight, 9⅛ pounds.

28L11203—Aristocrat No. 1 Socket Wrench Set.................. **$6.70**

Lock Washers.

Average thirty-nine, assorted sizes. Shipping weight, 6 ounces.

28L11318 **10c**

"Fifteen" Socket Wrench Set.

A well selected moderate price socket wrench set meeting all requirements of many car owners. Comprises fourteen hexagon steel sockets, 7-inch extension bar, about 8½-inch ratchet wrench handle, white metal finish, and spark plug socket, 31/32-inch opening size. Socket opening sizes, 13/32, 15/32, 17/32, 19/32, 5/8, 21/32, 23/32, 25/32, 13/16, 27/32, 29/32, 1, 1 1/32 and 1 3/32 inches. Complete in substantial wooden box. Shipping weight, 4⅞ pounds.

28L5313—"Fifteen" Wrench Set..... **$3.45**

Campbell Cotter Pins.

Can be locked in place without spreading. Insert pin and drive on head until points are same length. Average 100 pins to a box. Shpg. wt., 6 oz.

28L11328... **14c**

T Handle Socket Wrench

For reaching nuts and bolts in places ordinarily hard to reach, for use as rim brace wrench, etc. Length over all, about 12 inches; length of handle, 7½ inches. Nickel plated.

	Hex Socket Opening Size, Inches	
28L9542	7/16	39c
28L9544	½	39c
28L9546	9/16	39c
28L9548	5/8	39c

Shpg. wt., any of above, 1⅝ lbs.

Mechanics' Assortment.

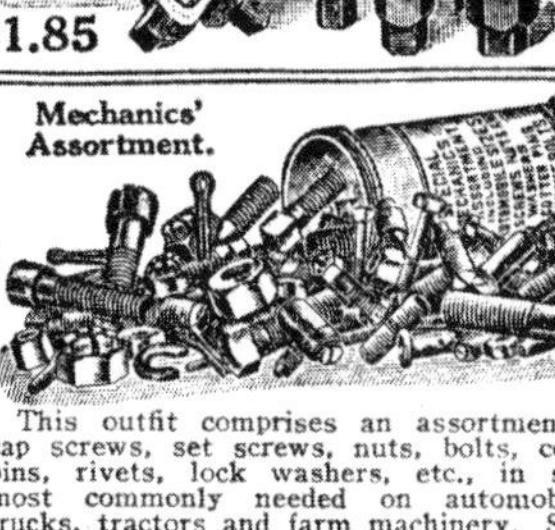

This outfit comprises an assortment of cap screws, set screws, nuts, bolts, cotter pins, rivets, lock washers, etc., in sizes most commonly needed on automobiles, trucks, tractors and farm machinery. Shipping weight, 3⅞ pounds.

28L12574........................ **98c**

Spring Lubricating Tool.

For opening spring leaves to insert lubricant. Avoid rusty, squeaky springs. Also used as a temporary repair clamp. Black finish. Shipping weight, 1 pound.

28L8671 Spring Lubricating Tool........................ **98c**

Hex Steel Sockets.

Shipping weight, 3 ounces. Furnished in sizes as listed below.

	Opening Size, In.			Opening Size, In.	
28L5752	5/16	12c	28L5763	19/32	12c
28L5753	11/32	12c	28L5764	5/8	12c
28L5754	3/8	12c	28L5765	21/32	12c
28L5757	13/32	12c	28L5766	11/16	12c
28L5758	7/16	12c	28L5767	23/32	12c
28L5759	15/32	12c	28L5768	¾	12c
28L5760	½	12c	28L5769	25/32	12c
28L5761	17/32	12c	28L5770	13/16	12c
28L5762	9/16	12c	28L5771	27/32	12c
			28L5772	7/8	12c

Spring Cotters.

Average 100, assorted sizes, ¾ to about 1¼-in. Shpg. wt., 6 ounces.

28L11325 **9c**

4-In-1 Screwdriver.

Four screwdrivers in one. Sizes, about 5¾, 3, 1¾ and 1⅛ inches long, respectively. Each size screwdriver fits into next larger size. All metal, nickel plated blades, upper end knurled. Largest blade has sheath. Screw cap. Shipping weight, 5 ounces.

28L7875—4-In-1 Screwdriver..................... **45c**

Speed Wrench.

High grade speed wrench with bell shape steel swivel top handle. Comes about 20 in. long in satin nickel finish, in sizes as listed below.

	Fits Nut Size, In.			Fits Nut Size, In.	
28L4628	7/16	54c	28L4638	11/16	54c
28L4630	½	54c	28L4640	¾	54c
28L4632	9/16	54c	28L4642	25/32	54c
28L4634	19/32	54c	28L4644	13/16	54c
28L4636	5/8	54c	28L4646	7/8	54c

Shipping weight, any size, 2 pounds.

Castellated Nuts.

Fifteen, assorted sizes, S. A. E. (A. L. A. M.) thread. Shipping weight, 8 ounces.

28L11310 **28c**

Steel Carbon Scrapers.

For reaching practically any point where carbon accumulates. Length of scrapers, about 13½ inches. Shipping weight of three, 1⅛ pounds.

28L10915—Set of Three Carbon Scrapers.................. **37c**

We Save You From \$8.00 to \$20.00 on Every Battery

THE Peerless Battery we ship on your order will duplicate your present battery in these important respects—voltage, terminals, size and shape. IN ORDERING A BATTERY BE SURE TO GIVE US THE MAKE, YEAR AND MODEL OF YOUR CAR. If battery is wanted for car not shown below, write for price, giving us the same information. We can furnish Peerless Batteries to fit any make of car.

Catalog No.	Car	Year and Model	Shpg. Wt. Lbs.	Regular Wood Box Type	Solid Rubber Case Type
28L13706⅓	Briscoe★	1916, 4-38, 8-38	77	\$19.25	\$22.75
28L13708⅓	Briscoe	1916-17, 4-24	58	16.45	....
28L13709⅓	Briscoe	1918-19-20, ALL	58	16.45	....
28L13718⅓	Buick	1914, B24, 25, 36, 37	67	16.45	....
28L13719⅓	Buick	1914-15, B36, 37, 54, 55, C24, 25	67	16.45	....
28L13713⅓	Buick	1916, D-54, 55	67	16.45	19.45
28L13714⅓	Buick	1916, D-44, 45, 46, 47	58	13.75	15.45
28L13715⅓	Buick	1917, D6-44, 45, 46, 47, E6-49; 1918, ALL	58	13.75	15.45
28L13717⅓	Buick	1919-20-21, ALL; 1922, 6 cyl	67	16.45	19.45
28L13725⅓	Chalmers	1916, 32; 1917-18, 6-30	67	16.45	....
28L13729⅓	Chandler	1916, Model 17	58	16.45	19.45
28L13723⅓	Chandler	1917-18, ALL	58	13.75	15.45
28L13731⅓	Chandler	1919-20, ALL	67	16.45	19.45
28L13727⅓	Chandler	1921-22, ALL	67	16.45	19.45
28L13724⅓	Chevrolet	1916-17-18, 490	58	13.75	15.45
28L13739⅓	Chevrolet	1919-20-21-22, 490	58	13.75	15.45
28L13735⅓	Chevrolet	1917, Series F, Baby Grand	67	16.45	19.45
28L13736⅓	Chevrolet	1917-18, Series FR, Baby Grand	67	16.45	19.45
28L13737⅓	Chevrolet	1917-18, D-Series, D-4 and 5	75	19.25	....
28L13738⅓	Chevrolet	1919-20, FB5; 1921, FB and T	67	16.45	19.45
28L13740⅓	DodgeBros★	1915-16-17-18-19-20-21, ALL	77	19.25	22.75
28L13741⅓	Essex	1919-20, ALL	67	16.45	19.45
28L13615⅓	Ford	1919-20-21-22, Liberty Starting System	50	12.45	14.45
28L13742⅓	Franklin★	1916-17-18, Series 9; 1919-20, ALL	77	20.95	....
28L13751⅓	Hudson	1916, 6-40	67	16.45	19.45
28L13752⅙	Hudson	1917-18, Super Six	67	16.45	19.45
28L13757⅓	Hupmobile	1916-17, N, NL, NQ, NR, NU	67	16.45	19.45
28L13761⅓	Hupmobile	1918, R, RR; 1919, ALL	58	13.75	15.45
28L13760⅓	Hupmobile	1920-21, ALL	58	13.75	15.45
28L13767⅓	Jeffrey-Nash	1915-16, Jeffrey 4	67	16.45	19.45
28L13768⅓	Jeffrey-Nash	1917, ALL	67	16.45	19.45
28L13782⅓	Maxwell★	1915-16-17, 25 (4 Cables)	77	19.25	22.75
28L13783⅓	Maxwell★	1918, 25; 1919, ALL; 1920, 25, up to car No. 300,000 (2 Cables)	77	19.25	22.75
28L13784⅓	Maxwell	1920 cars numbered over 300,000; 1921-22, ALL	67	16.45	19.45
28L13791⅓	Mitchell	1917-18, D-40	58	13.75	15.45
28L13793⅓	Mitchell	1919-20, E40, E42	58	13.75	15.45
28L13794⅓	Mitchell	1921, F40, F42; 1922, ALL	67	16.45	19.45
28L13801⅓	Nash	1917-18, 681, 682	58	13.75	15.45
28L13803⅓	Nash	1919-20, ALL, except truck	67	\$16.45	\$19.45
28L13804⅓	Nash	1921, 641	67	13.75	15.45
28L13807⅓	Nash	1921-22, 691	67	16.45	19.45
28L13812⅓	Oakland	1916, 32	58	13.75	15.45
28L13813⅓	Oakland	1916, 38	58	13.75	15.45
28L13815⅓	Oakland	1917, 34	58	13.75	15.45
28L13819⅓	Oakland	1918, 34B; 1919, 38, 34, 34B	58	13.75	15.45
28L13808⅓	Oakland	1920, 34C	67	16.45	19.45
28L13821⅓	Oldsmobile	1916, 43, 44, and 1918, 45A	58	13.75	15.45
28L13822⅓	Oldsmobile	1917, 45; 1919-20, 37A; 1922, 6 cyl. 37A	58	13.75	15.45
28L13828⅓	Oldsmobile	1919-20, 45B	67	16.45	19.45
28L13829⅓	Oldsmobile	1921, ALL; 1922, 8-cyl	67	16.45	19.45
28L13825⅓	Overland	1914, 65, 79	67	16.45	19.45
28L13826⅓	Overland	1915, 80, 81, 82; 1916, 86, 83, 83B, 83BOE; 1917, 85, 85-6	67	16.45	....
28L13827⅓	Overland	1916, 75; 1917-18, 90, 90B, 90T, 90R; 1919, 90CC and Del.; 1920-21, ALL	58	13.75	15.45
28L13830⅓	Paige	1916, 6-38, 6-46	58	16.45	19.45
28L13831⅓	Paige	1917, K-6-17, J-6-17; 1917-18, 6-39, 6-40, 6-51, 6-55; 1919-20, ALL	67	16.45	19.45
28L13833⅓	Paige	1921-22, ALL	67	16.45	19.45
28L13840⅓	Reo	1916-20, ALL	67	16.45	19.45
28L13845⅓	Saxon	1915, S; and 1916, S-2	58	13.75	15.45
28L13847⅓	Saxon	1917, S-4, B-5, B-14; 1918, B-5, 60, 61, 62, 63	58	13.75	15.45
28L13848⅓	Saxon	1919-20, ALL	58	13.75	15.45
28L13855⅓	Studebaker	1914, EB, SC	67	16.45	....
28L13856⅓	Studebaker	1915, EC, SD	67	16.45	....
28L13857⅓	Studebaker	1916-17, ED, SF; 7; 1918, EG, EH, SH	67	16.45	19.45
28L13858⅓	Studebaker	1919, ALL	67	16.45	19.45
28L13859⅓	Studebaker	1920, ALL	67	16.45	19.45
28L13853⅓	Studebaker	1921-22, Big 6 and Special 6	67	16.45	19.45
28L13854⅓	Studebaker	1921-22, Light Six	58	13.75	15.45
28L13866⅓	Velie	1915-16, 22 Series 22; 1916-17, Biltwell 28 Series; 1917, Biltwell 38 Series	58	13.75	15.45
28L13867⅓	Velie	1917, Biltwell 37 Series; 1918, 39	67	16.45	19.45
28L13868⅓	Velie	1919, ALL	58	13.75	15.45
28L13869⅓	Velie	1920, 38	67	16.45	19.45
28L13870⅓	Willys-Kn't	1916-17, 84, 88-4; 88-8, 1918, 88-8B; 1919, ALL	75	19.60	....
28L13871⅓	Willys-Six	1917-18, 86-B, 89-6	67	16.45	....

NOTE—All starting batteries shipped from factory in NORTHEASTERN ILLINOIS or SOUTHEASTERN NEW YORK, whichever is nearer you. All weights given on this page are approximate and may vary a trifle. ★ 12-volt. All batteries other than those specially marked are 6-volt type.

A Peerless Battery

Always Give Us the Make, Year and Model of Car

(Rubber Case Type Battery.)

For Dry Batteries, see page 813.

WHY YOU SHOULD BUY A PEERLESS BATTERY.

PEERLESS Starting Batteries are guaranteed for eighteen months' satisfactory service. With proper care they will give you good service for a much longer period.

PEERLESS Batteries are sold on sixty days' trial. Use your PEERLESS Battery for sixty days. At the end of that time, if you are not fully satisfied with it in every respect, send it back to us and we will return the purchase price, together with any transportation charges you may have paid.

PEERLESS Batteries are the standard product of a well known manufacturer of high grade batteries. There is not a better made battery on the market today, even at much higher prices. This is proved by PEERLESS construction and everyday performance. PEERLESS Batteries have ten years of satisfactory service behind them. Thousands and thousands of motorists can testify as to their excellence.

PEERLESS Starting Batteries are shipped to you promptly direct from the factory, fully charged and ready for use. You save up to half the price you would ordinarily pay a dealer for a starting battery of similar type, capacity and quality. PEERLESS Battery prices include only one profit above actual manufacturing cost. Buying direct by mail is the most economical method of buying. You save all middlemen's profit. Every PEERLESS Battery is sold under a binding guarantee of satisfaction or your money back. You take no risk in sending us your order for a PEERLESS Battery. PEERLESS Battery quality coupled with their amazingly low prices make them the best buy for your battery dollars.

Why We Can Guarantee PEERLESS Starting Batteries for Eighteen Months' Satisfactory Service.

PEERLESS Battery construction includes the following excellent features: Hand pasted plates filled with correctly varied lead oxides. Selected cedar separators. Rubber jars of highest quality. Molded rubber covers sealed around posts with pure gum gaskets. Terminal posts surrounded by well to permit thorough bushing and proper sealing. Terminals interchangeable with those on battery originally supplied with car. Heavy molded lead connectors connecting cells. Convenient firmly imbedded handles. Hardwood boxes, covered with acidproof paint, with dovetail corners dowel locked.

Order a Rubber Battery.

Solid rubber case batteries are the latest and best development in battery construction. The solid rubber case is made in one piece, complete with jar compartments. It does away with leaky cells. It is practically unbreakable. It cannot be damaged by mud, water or oil thrown up from the road or from stones and other objects that the wheels may throw against the battery box. We can furnish PEERLESS Batteries with this splendid rubber case in almost every type. (See list on opposite page.) At the left we show an illustration of our PEERLESS Rubber Case Battery. On the opposite page we list PEERLESS Rubber Case Batteries for both standard type and Ford cars.

HYDROMETER OUTFITS.

Leader Hydrometer.

A practical, inexpensive battery acid testing outfit, comprising hydrometer, rubber bulb, glass and rubber tube. Instructions for using. Shipping weight, 8 ounces.

28L10192 Leader Hydrometer.. **65c**

Leader Battery Filler Jar Outfit.

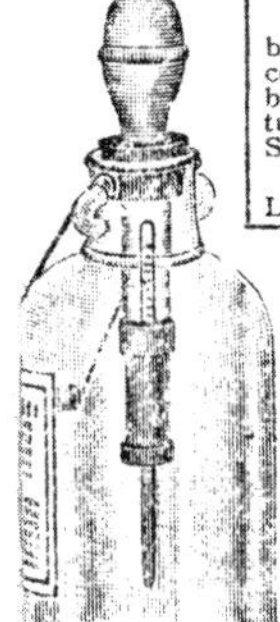

Outfit comprises heavy glass jar, one-half gallon size, for storing distilled water for battery, hydrometer, rubber bulb, glass tube and rubber tubing. Jar has bail or handle for handy carrying and cork top with opening for inserting hydrometer when not in use. Label on jar has spaces for recording battery readings systematically. Shipping weight, 4½ pounds.

28L10824 — Leader battery Filler Jar Outfit, complete with hydrometer.................. **$1.35**

Challenge Hydrometer.

A low priced practical battery hydrometer. Comprises rubber bulb and tube, glass tube and reliable, accurate hydrometer float for registering specific gravity of acid in battery cells. Shipping weight, 8 oz.

28L8672 Challenge Hydrometer... **40c**

De Luxe Hydrometer.

Our highest grade battery hydrometer. Has excellent quality durable rubber bulb and rubber tube, desirable straight barrel tube and high grade float accurately registering cell gravity. Complete with shoulders holding float away from walls of glass tube, insuring more accurate reading. Shipping weight, 12 ounces.

28L4851 De Luxe Hydrometer............ **65c**

De Luxe Hydrometer-Jar Outfit.

Inexpensive, popular type outfit. Comprises heavy glass one-half gallon jar for storing distilled water for battery and our No. 28L4851 De Luxe hydrometer as listed above. Keep hydrometer in jar when not in use. Shipping weight, 4⅝ pounds.

28L7168—De Luxe Hydrometer-Jar Outfit, complete with De Luxe Hydrometer.................... **98c**

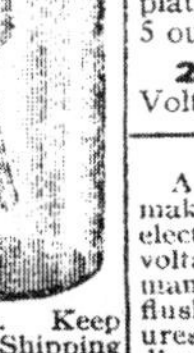

IGNITION COILS.

High Quality Replacement Transformer Coils. For Many Standard Cars.

28L15016 28L15017

Bracket A. Bracket C.

High quality Transformer Coils for many standard cars, replacing the equipment coil. These Replacement Coils are of unusually high quality and we guarantee them to give satisfactory service for the life of the car. Furnished as listed below in 6-volt type.

28L15016—Replacement Coil for Connecticut System. Has special watertight porcelain top. For use on Allen 1914-19, Briscoe 1915-19, Chevrolet 1914-17, Crow Elkhart 1915-19, Dort 1915-18, Lexington 1915-19, Mitchell 1914-17, Overland 1916-18 and Willys-Knight 1916-21. (Shpg. wt., 2¼ lbs.)....... **$5.65**

28L15017—Replacement Connecticut Coil. Use on Allen 1920, Briscoe 1920-21, Crow Elkhart 1920-22, Dort 1919-22 and Overland 1919-22. (Shpg. wt., 2¼ lbs.).......... **$5.75**

28L15023—Remy Replacement Two-Post Universal Coil, Bracket A Type. (Coil not illustrated.) Has bakelite tube, enameled metal top, silicon steel laminated core and rustproof terminals, base and brackets. Use for Apperson 1916-22, Auburn 1919-22, Briscoe 1915-16, Chalmers 1916-22, Chevrolet 1917-21, Elgin 1917-19, Grant 1917-19, Mitchell 1917-22, Paige 1916-19, Reo 1915-19, Studebaker 1918-19 and Velie 1918. (Shpg. wt., 2¼ lbs.). **$5.45**

28L15028—Remy Replacement Two-Post Universal Coil, Bracket C Type. (Coil not illustrated.) Use for Chevrolet 490 1918-22, Reo 1914-15 and Studebaker 1914-17. (Shpg. wt., 2¼ lbs.) **5.45**

Magneto Cable.

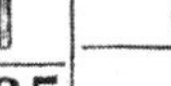

High Tension (secondary) Magneto or Spark Plug Cable, rubber covered. Comes 11/32 inch in diameter (9 M.M.). Widely used on Briscoe, Buick, Chalmers, Chandler, Chevrolet, Dodge, Ford, Hudson, Hupmobile, Maxwell, Nash, Oakland, Oldsmobile, Saxon and Studebaker Cars. Shipping weight, per foot, 2 ounces.

28L6358—Per foot.............. **6c**

Armored Primary Cable.

Armored Primary or Low Tension Ignition Cable. Shipping weight, per foot, 2 oz.

28L8310—Armored Primary Cable. Per foot........ **4c**

High Tension Cable.

High Tension Cable with braid covering. Comes 14-gauge, 5/16 inch diameter. Especially adapted for cars listed under 28L6358 cable above (not illustrated). Shipping weight, per foot, 2 ounces.

28L6372—Per foot........... **5c**

Low Tension Cable.

Low Tension and Lighting Cable. Comes 14-gauge, 3/32 inch diameter, with double braid covering (not illustrated). Shipping weight, per foot, 2 ounces.

28L6380—Per foot.......... **3c**

Battery Testing Meter.

Combination. Registers from 0 to 35 amperes and from 0 to 8 volts. Use as voltmeter to test voltage of dry cells or storage batteries. Nickel plated. Shipping weight, 5 ounces.

28L5204 Volt-Ammeter.... **64c**

Universal P. V. Ammeters.

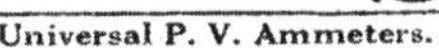

A high grade ammeter adapted for all makes of cars and all makes and types of electric lighting and starting systems of all voltages. Used as standard equipment on many cars. Comes in black enameled finish, flush type; has black dial with silvered figures. Range is 30-0-30 amperes. Over all diameter, about 2¾ inches. Complete with fittings for mounting on cowl dash. Shipping weight, 10 ounces.

28L10490—P. V. Ammeter.................. **$2.15**

Peerless Storage Batteries. Guaranteed for 18 Months.

Fully charged. Adapted for ignition, lighting and all general storage battery purposes, except automobile starting. Length of batteries, 7¼, 9½, 10 and 11½ inches, respectively. The small size is for ignition, and is 6¾ inches wide and 8¾ inches high; the larger sizes are for lighting or ignition, and are 7½ inches wide and 9 inches high. Battery sizes given above are over all. One ampere lamp load is equivalent to 6-candle power at 6 volts.

Sizes and Prices of Peerless Storage Batteries.

	Type of Battery	Shpg. Wt., Lbs.	
28L8521⅓	6-V. 40-Am.	31½	$10.25
28L8510⅓	6-V. 80-Am.	45	14.15
28L8507⅓	6-V. 100-Am.	51	15.75
28L8505⅓	6-V. 120-Am.	57	16.95

NOTE—All lighting batteries shipped from factory in NORTHEASTERN ILLINOIS or SOUTHEASTERN NEW YORK, whichever is nearer to you.

BARGAIN PRICES ON STANDARD SPARK PLUGS

STYLES OF SPARK PLUGS

Be Sure to Order Right Size.

⅞-inch (A. L. A. M. or S. A. E.) size plug of long shank or extension type. (See 28L7691 plug just below for style.)

½-inch size. Have no collar or gasket. (See 28L7861 plug at right for style.)

Regulation type, ⅞-inch (A. L. A. M. or S. A. E.) size plug with shoulder or collar. All ⅞-inch plugs have gasket. (See 28L7226, plug at left for style.)

NOTE—Porcelains cannot be furnished separately for any of the plugs listed on this page.

Spark Plug Amplifier.

Use on spark plugs to secure a stronger spark and to permit a more economical carburetor adjustment without sacrificing smooth running qualities. Fit only screw type terminals. Shipping weight, 2 ounces.

28L10664 **15c**

Spark Plug Tester.

For testing spark plug to see if it is firing properly, spark jumping across gap in tester. Black composition handle. Shipping weight, 3 ounces.

28L10662 Spark Plug Tester...... **25c**

A. C. Titan. Regular ⅞-Inch Type.

Regular equipment on Cadillac, Case, Chalmers, Cleveland, Cole, Dort, Elgin, Hudson through 1920, Hupmobile, Maxwell, Oakland up to 1918, Oldsmobile Eight, Paige-Detroit and Essex cars and fits Allen, Briscoe, Lincoln and Velie cars. Shipping weight, 6 ounces.

28L7226—A. C. Titan, regular ⅞-inch type...... **65c**

A. C. Titan. ⅞-Inch Type, S. A. E. Small Body, Extra Long Shank.

Regular equipment on 1920-23 Nash cars. (Not illustrated.) Shipping weight, 5 ounces.

28L7403—A. C. Titan, ⅞-inch S. A. E. small body, extra long shank...... **65c**

A. C. Titan. Two-Piece Plug, ⅞-Inch Long Type.

Fits Buick, Chevrolet, Kissel Kar, Scripps-Booth 4-cylinder, Sheridan and Stephens cars. (Not illustrated.) Shipping weight, 6 ounces.

28L7764...... **75c**

A. C. Titan. Long Shank.

Regular equipment for Buick, Chevrolet and Kissel Kar, and fits Sheridan and Stephens cars. Furnished in ⅞-inch size only, with long shank. Will not fit cars using ⅞-inch plugs with shank of regular length. Shipping weight, 5 ounces.

28L7691 A. C. Titan Long Shank Spark Plug...... **65c**

A. C. Titan. Dodge Special.

Regular equipment on Dodge cars. Slightly longer than ⅞-in. regular plugs, and also fits Cunningham and Hanson Six cars. Shipping weight, 5 ounces.

28L13135 A. C. Titan, Dodge Special...... **65c**

A. C. Titan. ⅞-Inch S. A. E. Small Body, Long Shank.

Regular equipment on Durant Four, Nash 1918-19, Oakland, 4 and 6-cylinder Oldsmobile, Saxon and 6-cylinder Scripps-Booth cars, and fits Franklin cars. Shipping weight, 5 ounces.

28L7222—A. C. Titan, ⅞-inch type S. A. E. Small body, long shank...... **65c**

A. C. Titan. Regular ⅞-Inch S. A. E. Type, Small Body.

Regular equipment on Apperson, Durant Six, Mitchell and Stearns-Knight cars and fits Studebaker cars. Shipping weight, 5 ounces.

28L7224—A. C. Titan, Regular S. A. E. small body type...... **65c**

A. C. Titan. Two-Piece Plug, ⅞-Inch Size, Regular Body.

Fits Allen, Briscoe, Cadillac, Chalmers, Cleveland, Dort, Hudson (except 1921-23), Hupmobile, Lincoln, Maxwell, Oldsmobile Eight, Paige, Velie and Westcott cars. Shipping weight, 6 ounces.

28L7741 **75c**

A. C. Titan. ½-Inch Type, Carbon Proof.

Porcelain has ribs with saw tooth edges (see cross section) to burn oil deposits. Fits Ford, Overland, Studebaker through 1919. Fordson and other tractors. Shipping wt., 5 oz.

28L7861...... **65c**

A. C. Titan. ½-Inch Long Type.

Fits Reo cars. (Not illustrated.) Shipping weight, 5 ounces.

28L7443—A. C. Titan, ½-inch long type.. **65c**

Regular Metric Type.

Regular equipment for Duesenberg, Essex, Hudson 1921-23 and Wills-Sainte Claire cars. (Not illustrated.) Shipping weight, 5 ounces.

28L7762—A. C. Titan, regular metric type. **65c**

Magneto Parts
(Platinum Points.)

For Bosch. 1001 1002

For Bosch. **28L7880**—(1001) Long Contact Screw. Each **$1.95**

For Remy. 2312 2313

28L7881 — (1002) Short Contact Screw. Each **$1.95**

For Splitdorf. 1810 1809

For Remy. **28L7901**—(2313) Interrupter Screw, platinum point. Each **$1.59**

28L7902—(2312) Interrupter Lever. Each...... **$1.79**

For Splitdorf.

28L7903—(1809) Interrupter Lever, Model Dixie 40. Each **$2.10**

28L7904—(1810) Interrupter Screw, Model Dixie 40. Each **$1.74**

Shipping weight, any of above, 2 ounces.

Spark Plug Intensifier.

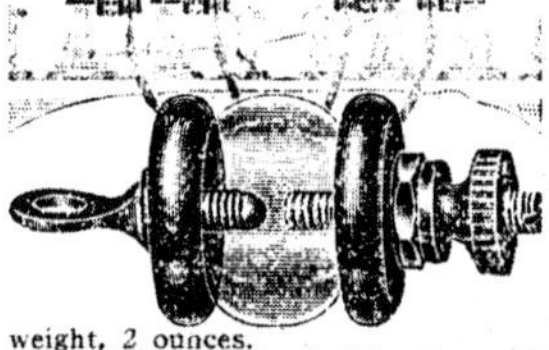

Produces hot spark, tending to explode entire gas charge at one time. Strengthens spark, aids it in burning through and removing carbon and oil deposits on sparking points, and indicates plug which is not sparking. Attach to spark plug terminal and wire. Shipping weight, 2 ounces.

28L11493—Spark Plug Intensifier...... **26c**

Parts for Connecticut Ignition System.

28L7915 — (2700) Interrupter Lever with Tungsten point. Each **34c**

28L7956 — (2701) Spring for 28L7915. Each **6c**

28L7914—(2706) Contact Screw with Tungsten point and locknut. Each...... **18c**

28L7966—(2708) Distributer Arm for Model 15. Each **15c**

28L7967—(2709) Distributer Arm for Model 16. Each **15c**

Shipping weight, any of above, 2 ounces.

28L11982—(2717) Distributer Cap, 4-Cylinder, Model 15. Each...... **98c**

28L11986—(2720) Distributer Cap, 4-Cylinder, Model 16. Each...... **98c**

28L11988—(2721) Distributer Cap, 6-Cylinder, Model 16. Each...... **$1.19**

Shipping weight, distributer caps, 8 ounces.

2706 2720 2717 2700 2721 2708 2701 2709

Firing Tubes.

Made of high copper alloy, heating readily to high temperature for burning oil and carbon off plug firing points.

For making plugs fire effectively in cylinder whose piston passes oil. Screw into cylinder in place of plug and mount plug into tube. Keeps plug firing points free from oil and carbon soot, mixture being fired inside tube. Shipping weight, 4 ounces.

28L11404—Firing Tube, ⅞-inch size, with three holes in bottom of tube...... **50c**

28L6604—Firing Tube, ⅞-inch extension type, with three holes in bottom of tube. (Not illustrated)...... **50c**

28L11402—Same, ½-inch size, with single hole at the bottom. (Not illustrated.) **50c**

Splitdorf Plugs

⅞-Inch Late Buick Type.

For all 1918-1923 model Buick cars. Equipped with special slip-on terminal. (Not illustrated.) Shipping weight, 6 ounces.

28L7200 — Splitdorf Plug, ⅞-inch, late Buick type...... **65c**

P179, ½-Inch Type.

For Ford, Fordson Tractor, 1915 Overland Four, Regal, Reo cars up to and including 1916 models, and Studebaker cars up to and including 1919. Shipping weight, 6 oz.

28L7204 Splitdorf Plug, P179, ½-inch type...... **59c**

P179, ½-Inch Type. P163, Regular ⅞-Inch Type.

P163, Regular ⅞-Inch Type.

For Briscoe, Cadillac, Case, Chalmers, Chevrolet 490, Cole, Dodge, Hudson, Hupmobile, Jeffery, Maxwell, Saxon, Stearns-Knight, 1920-23 Studebaker, Velie and Westcott cars. Shipping weight, 6 ounces.

28L7202—Splitdorf Plug, regular ⅞-inch type...... **65c**

For a more complete line of spark plugs and ignition accessories, ask for a copy of our latest catalog of Automotive Supplies, 512GCL, sent POSTPAID ON REQUEST.

P165, ⅞-Inch S. A. E. Type.

For Holmes, Nash 1920-23, Oakland 32, Oldsmobile, R. & V. Knight, Saxon cars, etc. A long open end style plug. (Not illustrated.) Shipping weight, 6 ounces.

28L7350—Splitdorf Plug, P165, S. A. E. type...... **65c**

Champion Regular.

Furnished in two sizes, as listed. Shipping weights, 5 and 6 oz., respectively.

28L7485 ½-in. size. **60c**

28L7487 S. A. E. (A.L.A.M. or ⅞x18) size **69c**

Our 4-for-1 Spark Plug Outfit.

$1.05 Per Set.

Four regulation type spark plugs, complete with fabric case. Open end porcelain plugs, furnished in ½-in. or S. A. E. small body, (A. L. A. M. or ⅞x18) sizes. **In ordering specify size desired.** Shipping wt., 1⅛ lbs.

28L7435 Our 4-for-1 Spark Plug Set. Per set.... **$1.05**

P785, Heavy Duty.

An excellent spark plug of heavy duty type. Used for Chalmers, Hupmobile, Grant, Elgin cars, etc. Shipping weight, 5 ounces.

28L13133 — Splitdorf Heavy Duty Type Spark Plug, ⅞-inch size...... **65c**

Star Jump Spark Coils.

An inexpensive, efficient, economical, popular box type coil. Furnished in polished oak case for one or two-cylinder high speed marine or stationary engines.

28L7813—Star Jump Spark Coil, 1-cylinder model. Shipping weight, 3⅛ pounds...... **$2.63**

28L7814—Same, 2-cylinder model. Shipping weight, 6⅛ pounds...... **5.25**

LOW PRICES ON LAMPS AND LIGHTING ACCESSORIES

Parking Lamps

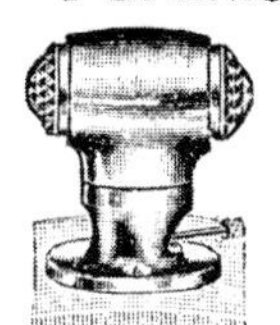

Beauty Junior.

A very attractive little parking light made of polished aluminum. Has 7/8-inch cut glass lens both front and rear, crystal in front and red in rear. Complete with on and off switch and 6-8-volt bulb. Height over all, 2½ in.; length over all, 1¾ inches. Complete with wire leads; screws, nuts and washers furnished for attaching to fender. Shipping weight, 1 pound.
28L8700....**$1.39**

Challenge.

A small, inexpensive, light weight attractive lamp in black enameled finish with nickel plated lens rims. Has about 2-inch beehive type ruby lens and frosted white lens. 6-8-volt bulb, on and off switch. Length over all, about 3¾ in.; height, about 3 inches; diameter of barrel, about 2⅛ inches. Wire leads. Shipping weight, 1¼ lbs.
28L9065
98c

Hard Rubber Screw Cap Plugs.

28L8166—Double contact base.................... **9c**
28L8165—Single contact base. (Not shown)........................ **9c**
Shipping weight, either style, 2 ounces.

Auto Lamp Dimmers

McKee Lens.

A well known and efficient dimmer lens which has been approved in all states having "anti-headlight glare" laws. Gives good driving light at side of road close to the car as well as straight ahead. Instructions for focusing. **Order according to the measurements of present headlight glass.**

	Present Glass Diam. Over All	Glass Diam. Between Door Rim	Shpg. Wt., per Pair	Per Pair
28L8677	8⅝ in.	7¼ in.	4 lbs.	$1.59
28L8679	8½ in.	7½ in.	4 lbs.	1.59
28L8681	8¾ in.	7½ in.	4 lbs.	1.89
28L8683	8¾ in.	8 in.	4¼ lbs.	1.89
28L8685	9 in.	8½ in.	4¾ lbs.	1.89
28L8687	9¼ in.	8½ in.	5 lbs.	1.89
28L8689	9½ in.	8¼ in.	5 lbs.	1.89
28L8691	9½ in.	8⅜ in.	5 lbs.	1.89

28L10289
McKee Lens for all Ford cars, including late models equipped with green visor lens. Shipping weight, per pair, 3⅝ pounds.
Per pair...**79c**

Classy Trouble Lamp.

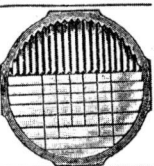

Attractive, unusual type. Comes about 6 in. high, closed, with black wood handle and metal reflector top. Has 6-8-volt bulb, with hinged nickel plated reflector folding like a globe and fastening at top to protect bulb when lamp is not in use. Complete with about 10 feet of cord. Shipping wt., 1 lb.

Closed.

28L11554—Double contact type... **98c**
28L11552—Same, single contact type.............................. **98c**

Universal Tail Lamp.

The type tail lamp used on a great many popular standard cars, including Buick, Chalmers, Dort, Durant, Elgin, Maxwell through 1921, Nash, Oldsmobile, etc. Complete with 2-candle power, 6-8-volt single contact bulb. Over all depth, front to back, about 4 inches. Has about 3-inch ruby lens. Black finish. Shipping weight, 1⅝ pounds.
28L10686.............................. **48c**

Automobile Spot Lights

Junior Master Spot Light.

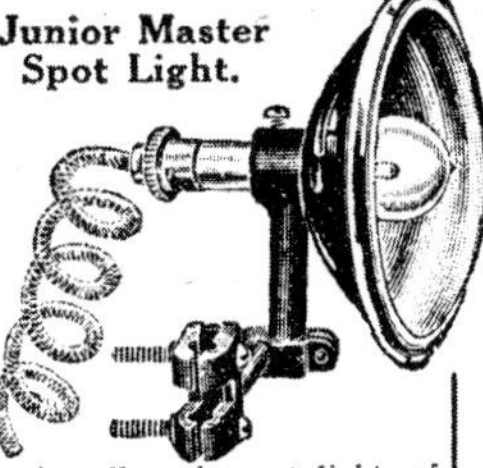

A well made spot light, of unusually attractive design. Has about 4½-inch front, with brass body and reflector combined in one. Has 21-candle power double contact bulb, handy on and off switch, about 5 feet of cord and outside focusing device. Furnished in black enameled finish with nickel plated trimmings. Universal type bracket fits all types of open car wind shields. Furnished as listed below. Shipping weight, 2 pounds.
28L15008—Junior Master Spot Light, 6-8 volt bulb **$2.15**
28L15010 Same, 12-16-volt bulb **2.15**
28L15012—Same, 18-24-volt bulb, for use with magneto on Ford cars **$2.15**

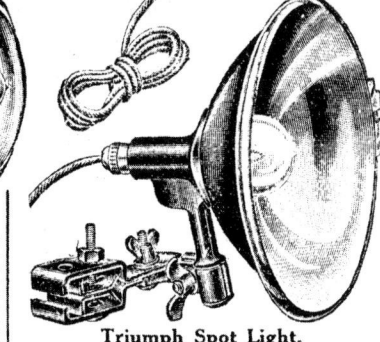

Triumph Spot Light.

Has 7-inch front and brass reflector forming body of lamp, with nitrogen bulb. Has swivel action and conforms to state laws regulating height of spot light rays from ground. Fitted with about 3½ feet of lighting cord, terminals, on and off switch at handle end and focusing device. Universal bracket jaws fit practically any wind shield frame. For use on Ford cars order 28L5234 Spot Light if spot light is to be operated from magneto, and 28L5230 if for use with 6-8-volt battery. Shpg. wt., 3 lbs.
28L5230—With 6-8-volt bulb....... **$2.95**
28L5232—Same, with 12-16-volt bulb........................ **2.95**
28L5234—Same, with 18-24-volt bulb for Ford cars **2.95**

28L4651—6-8-volt type, with closed car bracket........................ **$2.15**
28L4654—12-16-volt type, with closed car bracket......................... **2.15**
28L4656—18-24-volt type, with closed car bracket......................... **2.15**

Armored Lamp Wire.

For spot light use, etc. Furnished in two-strand type. Shipping weight, foot, 2 ounces.
28L8312—Per foot **8c**

Lighting Switch.

In 1, 2 or 3-gang type, push button style, black finish. For use on wood dash only, screws being furnished.

	Gangs	Shpg. Wt.	
28L8104	1	3 oz.	22c
28L8106	2	4 oz.	43c
28L8108	3	6 oz.	65c

Lamp Cord.

New code twisted green and yellow cotton lamp cord. No. 18. Shipping weight, 1 ounce.
28L6356—Per foot **3c**

Spot Light Bracket, for Closed Cars.

Screw to car body outside and clamp spot light to bracket. Shipping weight, 4 ounces.
28L9110.......... **29c**

Drum Headlights.

The up to date popular type. Have universal bracket for attaching to fenders and adjustable to throw light up and down or to left or right. Have about 9-inch front, 21-candle power, 6-8-volt single contact bulbs. Have black enameled body with nickel plated brass rims. Shipping weight, per pair, either style, 13½ pounds.

For type for Fords see page 756.

28L15001—Per pair...... **$7.45**
28L7024—Two-bulb (dim and bright) type drum headlights, 6-8-volt single contact type. Per pair.............. **$8.75**

For Dodge Cars.

28L15002—Drum headlights with special bracket and 12-16-volt bulb, for Dodge cars. Shipping wt., per pair, 11 lbs. Per pair..... **$7.45**

Automobile Bulbs

Double Contact.

Be sure to order the correct style base.

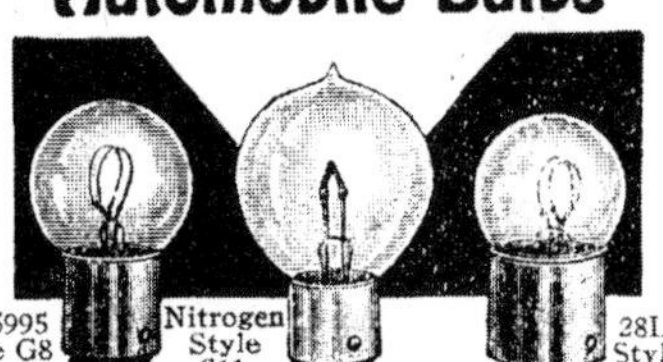

28L5995 Style G8 — Nitrogen Style S11 — 28L5990 Style G6

Single Contact.

Double contact style base is shown at left.

Lamps shown about ¼ size.

	C. P.	Volts	Contact	
Tungsten Headlight Bulbs. Style S11.				
28L6048	21	9	Double	22c
28L12037★	2¼–20	6–8	Double	39c
Nitrogen (Type C) Headlight Bulbs. Style S11.				
28L5970★	21	9	Double	22c
28L5974	32	6–8	Double	33c
28L5975	32	6–8	Single	33c
28L6075	21	6–8	Double	22c
28L6076	21	6–8	Single	22c
28L6991	27	6–8	Double	30c
28L6992	27	6–8	Single	30c
28L6515★	27	9	Double	24c
28L6993	32	12–16	Double	39c
28L6994	32	12–16	Single	39c
28L6044	21	12–16	Single	24c
28L6046	21	12–16	Double	24c
28L6995	27	18–24	Double	39c
28L13072	27	18–24	Single	39c

Tungsten Tail Light Bulbs. Style G6.

	C. P.	Volt	Contact	
28L6021	2	3–4	Double	15c
28L6022	2	3–4	Single	15c
28L5993	2	18–24	Double	19c
28L5994	2	18–24	Single	19c
28L5987	2	6–8	Double	15c
28L5990	2	6–8	Single	15c
28L5991	2	12–16	Double	15c
28L5992	2	12–16	Single	15c

Tungsten Side Light Bulbs. Style G8.

	C. P.	Volt	Contact	
28L5995	4	6–8	Double	[illegible]
28L5998	4	6–8	Single	[illegible]

★ For Ford lighting systems operated from magneto. †Tulite bulb for Ford cars equipped with Ford starting system. Shipping weights: S11 bulbs, 5 ounces; G8 and G6 bulbs, 4 ounces.

Dash Lamps

Complete With On and Off Switch.

6-8-Volt Single and Double Contact Bulbs. No Wire Leads Furnished.

Nickel plated, with 2-C.P. tungsten bulb. Ediswan fittings. Distance from center of bulb to back of flange, about 2 inches.

28L8086 — 28L8089

28L8089 Double contact type. For metal dash.......................... **49c**
28L6039—Same, with single contact bulb.......................... **49c**
28L8086—Double contact type. For wood dash......................... **37c**
28L6042—Same, with single contact bulb.......................... **49c**
Shipping weight, any of above lamps, 6 oz.

Triumph Dash Lamp and Wire.

For FORD Cars.

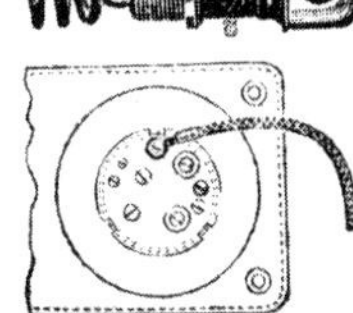

A neat, handy Dash Lamp. Comes complete with insulated wire attached to lamp, with terminal on other end of wire to be fastened direct to lighting switch. Has 6-8-volt single contact bulb, metal dash connector. Nickel plated. Shipping weight, 7 ounces.
28L8330........................ **57c**

Triumph Stop Signal

A Practical Need for Safe Driving.

You cannot afford to drive without a stop signal. The instant pressure is put on the brake pedal a powerful red light flashes up at the rear of the car, warning traffic close behind to slow down and be watchful.

Our Triumph Stop Signal is an attractive, well made, inexpensive lamp of neat design. Furnished with about 4-inch lens, with nickel plated rim; balance of lamp black enameled. Lamp has rust resisting all brass shell. Has bracket for attaching to fender or license bracket. Has high grade enclosed type knife switch (illustrated and described below), which will not short circuit from dampness, etc., necessary lamp cord connecting lamp, switch and battery and wire with clamp for attaching to brake rod.
28L10304—With 6-8-volt bulb **$1.29**
28L10316 — Same, with suitable bulb for cars having 12-16-volt battery, and for Ford cars having lights operated from magneto **$1.29**
Shipping weight, either of above, 1⅝ lbs.

TELL TALE FOR STOP LIGHT

Tells you instantly if stop light lights up when brake is applied. A dash light with red celluloid window, 3-volt bulb and resistance coil for reducing battery current to 3 volts. Mount on dash or attach vertically to bottom edge of instrument board. Wire and instructions furnished. Shipping weight, 8 ounces.
28L9988................................ **72c**

Leader Stop Signal Switch.

A positive working, durable Stop Signal Switch of enclosed type, built to prevent short circuiting from dampness, etc. Has strong, reliable knife contact of type similar to power house electric switches. Shipping wt., 4 oz.
28L10658............ **24c**

BIG VALUES ENGINE AND RADIATOR ACCESSORIES LOW PRICES

Master Valve Grinder.

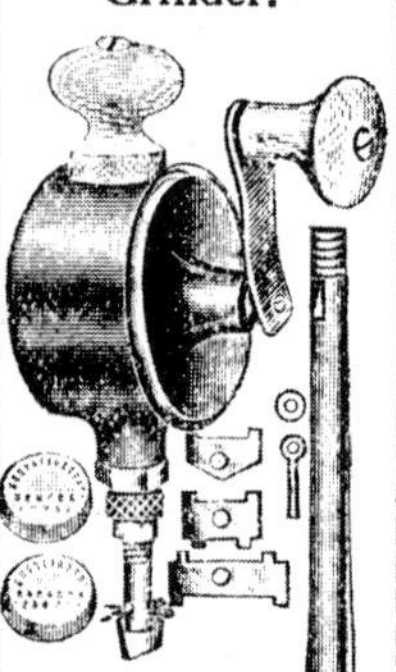

An excellently designed and well made valve grinder. Turning handle steadily in one direction gives the reciprocating or forward and backward movement needed for satisfactory valve grinding. Insures even grinding, alternating a full turn in one direction with a three-quarter turn in the opposite direction. Complete with long extension shank, screwdriver bit and three driving points, fitting valves on Fords and most other small cars. Has ball bearing action, insuring easy operation. Shipping weight, 2½ pounds.
28L12960 **$2.55**

Quick Seating Oil and Compression Rings.

A well made one-piece piston ring of step-cut type with correctly designed oil groove and quick seating face. Designed to stop oil leakage; quick seating feature insures extra power. Adapted for any engine of internal combustion type.

Catalog No.	Size, In.	Price
28L5746	2 3/8x3/16	20c
28L5745	2 1/4x3/16	20c
28L9502	2 3/4x3/16, .010 oversize	20c
28L5743	2 13/16x3/16	20c
28L5827	2 13/16x3/16, .0025 oversize	20c
28L8160	2 13/16x3/16, .010 oversize	20c
28L5744	2 7/8x3/16	20c
28L9504	2 7/8x3/16, .010 oversize	20c
28L5742	3 x3/16	20c
28L5741	3 1/8x3/16	20c
28L5739	3 1/4x3/16	20c
28L5597	3 1/4x3/16, .0025 oversize	20c
28L5596	3 1/4x3/16, .010 oversize	20c
28L5723	3 1/4x1/4	20c
28L5738	3 3/8x3/16	20c
28L5829	3 3/8x3/16, .0025 oversize	20c
28L5722	3 3/8x1/4	20c
28L5737	3 1/2x3/16	20c
28L5599	3 1/2x3/16, .0025 oversize	20c
28L5598	3 1/2x3/16, .010 oversize	20c
28L5717	3 1/2x1/4	20c
28L5608	3 9/16x3/16	20c
28L5735	3 5/8x3/16	20c
28L5716	3 5/8x1/4	20c
28L5733	3 11/16x3/16	20c
28L9506	3 11/16x3/16, .010 oversize	20c
28L5731	3 3/4x3/16	20c
28L5714	3 3/4x1/4★	20c
28L5610	3 3/4x1/4★ .0025 oversize	20c
28L5618	3 3/4x1/4★ .005 oversize	20c
28L5621	3 3/4x1/4★ .010 oversize	20c
28L5624	3 3/4x1/4★ .015 oversize	20c
28L5713	3 3/4x1/4★ .025 oversize	20c
28L5712	3 3/4x1/4★ .03125 oversize	20c
28L5705	3 7/8x5/32	20c
28L5729	3 7/8x3/16	20c
28L9508	3 7/8x3/16, .010 oversize	20c
28L5728	4 x3/16	20c
28L5711	4 x1/4	20c
28L5689	4 x5/16	20c
28L5727	4 1/8x3/16	20c
28L5710	4 1/8x1/4	20c
28L5663	4 1/8x3/8	20c
28L9514	4 1/4x3/16	20c
28L5709	4 1/4x1/4	20c
28L5708	4 1/2x1/4	20c
28L5669	4 1/2x3/16	20c

★For Ford Cars. Shipping weight, any of above, 3 ounces; 28L5710 and following sizes, 4 ounces.

Quick Seating Plain Piston Rings.

A high grade individually cast piston ring, accurately made, seating quickly and designed to increase engine power.

Catalog No.	Size, In.	Price
28L5924	2 5/8x3/16	14c
28L5926	2 3/4x3/16	14c
28L9495	2 3/4x3/16, .010 oversize	14c
28L5928	2 13/16x3/16	14c
28L5837	2 13/16x3/16, .0025 oversize	14c
28L5929	2 13/16x3/16, .010 oversize	14c
28L5931	2 7/8x3/16	14c
28L9497	2 7/8x3/16, .010 oversize	14c
28L5933	3 x3/16	14c
28L5936	3 1/8x3/16	14c
28L5939	3 1/4x3/16	14c
28L5841	3 1/4x3/16, .0025 oversize	14c
28L5839	3 1/4x3/16, .010 oversize	14c
28L5942	3 1/4x1/4	14c
28L5944	3 25/64x3/16	14c
28L5946	3 3/8x3/16	14c
28L5948	3 3/8x1/4	14c
28L5951	3 1/2x3/16	14c
28L5604	3 1/2x3/16, .015 oversize	14c
28L5953	3 1/2x1/4	14c
28L5958	3 5/8x3/16	14c
28L5960	3 5/8x1/4	14c
28L5962	3 11/16x3/16	14c
28L9498	3 11/16x3/16, .010 oversize	14c
28L12385	3 3/4x3/16	14c
28L11728	3 3/4x1/4★	11c
28L11174	3 3/4x1/4★ in .0025 oversize	12c
28L5616	3 3/4x1/4★ .005 oversize	12c
28L12528	3 3/4x1/4★ .010 oversize	12c
28L5623	3 3/4x1/4★ .015 oversize	12c
28L5966	3 3/4x1/4★ .025 oversize	12c
28L11176	3 3/4x1/4★ .03125 oversize	12c
28L5968	3 7/8x5/32	14c
28L5969	3 7/8x3/16	14c
28L9499	3 7/8x3/16, .010 oversize	14c
28L12386	4 x3/16	14c
28L12387	4 x1/4	14c
28L12388	4 x5/16	14c
28L5971	4 1/8x3/16	14c
28L5973	4 1/8x3/8	14c
28L12389	4 1/8x1/4	14c
28L9512	4 1/4x3/16	14c
28L12391	4 1/4x1/4	14c
28L12393	4 1/2x1/4	14c
28L5667	4 1/2x3/16	14c

★For Ford Cars. Shipping weight, any of above, 3 oz.

Boyce Motometers.

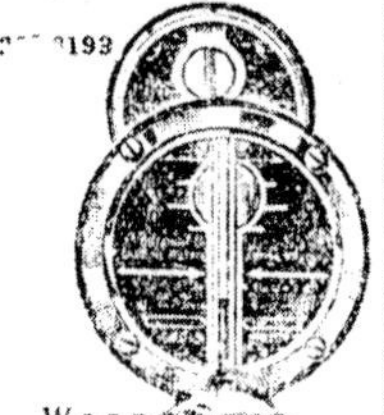

Warns you when motor is overheating from lack of water, oil, etc. Furnished without radiator cap.

We do not furnish motometers to fit Willys-Overland cars.

Standard Model. Has 3¼-inch front, heavy beveled crystals both front and rear. Comes in nickel plated finish with black dial. A splendid model with latest improved features. Shipping weight, 1¼ pounds.
28L4828 **$7.75**

Universal Model. Of similar type to the above, but somewhat smaller. Comes in nickel plated finish with black dial, about 2¼ inches in diameter. (Not illustrated.) Shpg. wt., 1 lb.
28L6319... **$5.45**

Boyce Midget Model. Front and rear beveled crystals; black and nickel plated dial, about 2¼ in. in diameter. Shipping weight, 11 ounces.
28L8199... **$2.98**

Premier Valve Grinding Compound.

Comes in cans about 1 inch high, containing equal divisions fine and coarse compound. Shipping weight, 8 ounces.
28L10730.. **19c**

Chain Valve Lifter.

Attach as illustrated for compressing valve spring and lifting out valve for grinding, etc., excellent leverage being obtained. Outfit comprises spring compressor-valve lifter complete with chain and hook. Shipping weight, 1¾ pounds.
28L12568—Chain Valve Lifter............ **39c**

Valve Lifter-Spring Compressor.

Combination valve lifter-spring compressor. Shpg. wt., 1¼ lbs.
28L10899. **36c**

Radiator Moto-Wings.

A popular new radiator ornament. Held in place by motometer. Furnished nickel plated in two sizes as listed below. Use with any type motometer. (Motometer not furnished.)
28L6071—Moto-Wings, 10-inch size.......... **45c**
28L6073—Moto-Wings, 14-inch size.......... **60c**
Shpg. wt. of above, 12 oz. and 1¼ lbs., respectively.

Worko Carbon Remover.

Can of about twenty-four tablets. Shpg. weight, 6 oz.
28L10678
70c

Spring Hood Clips for Chevrolet Cars.

Replace the wing nut type hood fasteners on Chevrolet cars. Use present hood bracket and bolt, attaching spring clip as shown in illustration. Shipping weight, per set, 14 ounces.
28L7105—Spring Hood Clips, black enameled finish, for model 490 Chevrolet cars. Set of four. **70c**
28L7021—Same, nickel plated finish. Set of four..... **$1.05**

Poor compression and loss of power

Poor compression and loss of power are often due to the worn condition of piston rings. The excess oil passing worn rings forms carbon and fouls the spark plugs, causing irregular firing.

NOTE—In ordering be sure to state make, year and model of car and size of ring. Measure diameter of ring with the points of ring separated by about the thickness of a sheet of paper. Also state width (height) of ring.

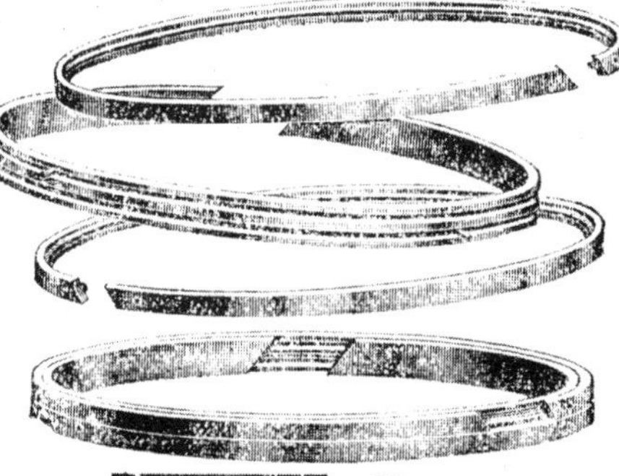

PistonTITE Rings

Three-piece piston rings, individually cast.

Catalog No.	Size, Inches	Price
28L12536	3 3/4x1/4★	60c
28L9104	3 3/4x1/4★ .025 oversize	60c
28L5852	2 5/8x3/16	70c
28L5854	2 3/4x3/16	70c
28L5858	2 13/16x3/16	70c
28L5862	2 7/8x3/16	70c
28L5864	3 x3/16	70c
28L5867	3 1/8x3/16	70c
28L5869	3 1/4x3/16	70c
28L5872	3 1/4x1/4	70c
28L5875	3 3/8x3/16	70c
28L5879	3 1/2x3/16	70c
28L5881	3 1/2x1/4	70c
28L5885	3 5/8x3/16	70c
28L5887	3 5/8x1/4	70c
28L5889	3 11/16x3/16	70c
28L5891	3 3/4x3/16	70c
28L5899	3 7/8x3/16	70c
28L5903	4 x3/16	70c
28L5905	4 x1/4	70c
28L5893	4 x5/16	70c
28L5907	4 1/8x3/16	70c
28L5911	4 1/8x1/4	70c
28L5909	4 1/8x3/8	70c
28L5913	4 1/4x1/4	70c

★For Ford Cars. Shipping weight, any of above, 3 ounces; 28L5905 and larger sizes, 4 ounces.

Mossberg Valve Grinder.

An adjustable valve grinder that will fit all valve tops. All metal, substantially made. Length, over all, about 21 inches. Shipping weight, 2 pounds.
28L10940—Mossberg Valve Grinder....... **69c**

Mossberg Valve Spring Lifter-Compressor.

A handy, well made tool. Readily inserted under the spring washer, raised by finger pressure and locked where you wish it by ratchet. Ratchet can be quickly released, driving levers together for further use. Shipping weight, 1½ pounds.
28L9496—Mossberg Valve Spring Lifter-Compressor....... **98c**

S. R. Spring Leaf Oilers.

For Ford Front Springs.

Standard Type.

Their use eliminates spring rust and squeaks, insures uniformly perfect spring action, reduces road shocks and spring breakage. They make your car ride easier and lengthen the life of the tires. Easily attached, being adjustable both to width and depth of spring. Have oil reservoir at top and felt packing inside, through which oil is fed to spring leaves. Use two oilers for each semi-elliptic spring. Shipping weight, each, 1 lb.
28L6706—S. R. Spring Leaf Oiler for any standard car spring, front or rear. Each......... **78c**
28L6710—Set of 4 Oilers, for Ford springs (two each for front and rear springs). Shipping weight, per set, 1¾ pounds. Per set of 4...... **$2.75**

Reserve Gasoline Tank.

Substantial heavy galvanized can for carrying one gallon of gasoline in reserve. Can is 10½ in. high, 8¼ inches wide and 3 inches from front to back. Painted red. Stores under rear seat. Shipping wt., 2 lbs.
28L13192..... **65c**

Radiator Hose.

Comes in 1, 2 or 3-ft. lengths only, in inside diameter sizes as listed.

Catalog No.	In.	Shpg. Wt., per Ft., Oz.	Per Ft.
28L6615	1	10	15c
28L6616	1¼	12	18c
28L6617	1½	13	21c
28L6618	1¾	14	24c
28L6619	2	16	27c
28L6629	2¼	17	30c

Radiators for Chevrolet 490, Dodge and Maxwell Cars.

High grade radiators with brass cores and tanks. Connections are made of rust and corrosion proof alloy. Shells are pressed steel and black enameled. **In ordering state year and model of car.**
28L15105⅓—Radiator for 1915-1917 model 490 Chevrolet. Thermo-Siphon Type. State year........ **$28.50**
28L15107⅓—Radiator for 1918-1923 model 490 Chevrolet, pump system.......... **$22.50**
28L15109⅓—Radiator for 1917-1922 Dodge cars... **28.50**
28L15111⅓—Radiator for 1916-1919 Maxwell cars........................ **28.50**
28L15117⅓—Radiator for 1920-1921 Maxwell cars........................ **28.50**
Shipping weight, any of above, 45 pounds; 28L15109⅓, 50 lbs.
NOTE—Shipped promptly from factory in NORTHEASTERN ILLINOIS.

All weights given on this page are approximate and may vary a trifle.

Tin Funnel.

28L7850 1-quart size.... **12c**
28L7851 2-qt. size. **16c**
28L7852 3-qt. size. **20c**
Shpg. wts., 1½, 1¾ and 2¼ pounds respectively.

Iron Cement.

For repairing cracks in engine cylinder castings, water jackets, leaky iron radiators, etc. Withstands heat, steam, oil, etc. Instructions furnished for using. Comes in about ¼-lb. cans. Shpg. wt., 12 oz.
28L8100 Iron Cement..... **20c**

Measure Funnels.

Have both pouring lip and funnel.

Catalog No.	Size, Qts.	
Copper Plated.		
28L10085	1	$0.73
28L10086	2	.85
28L10087	4	1.05
Galvanized.		
28L10088	1	68c
28L10089	2	80c
28L10090	4	9[illegible]c

Shipping weights, 2, 2½ and 3½ lbs.

AUTO REPAIRS AND SUPPLIES. ALL BARGAINS

BRAKE BAND LINING

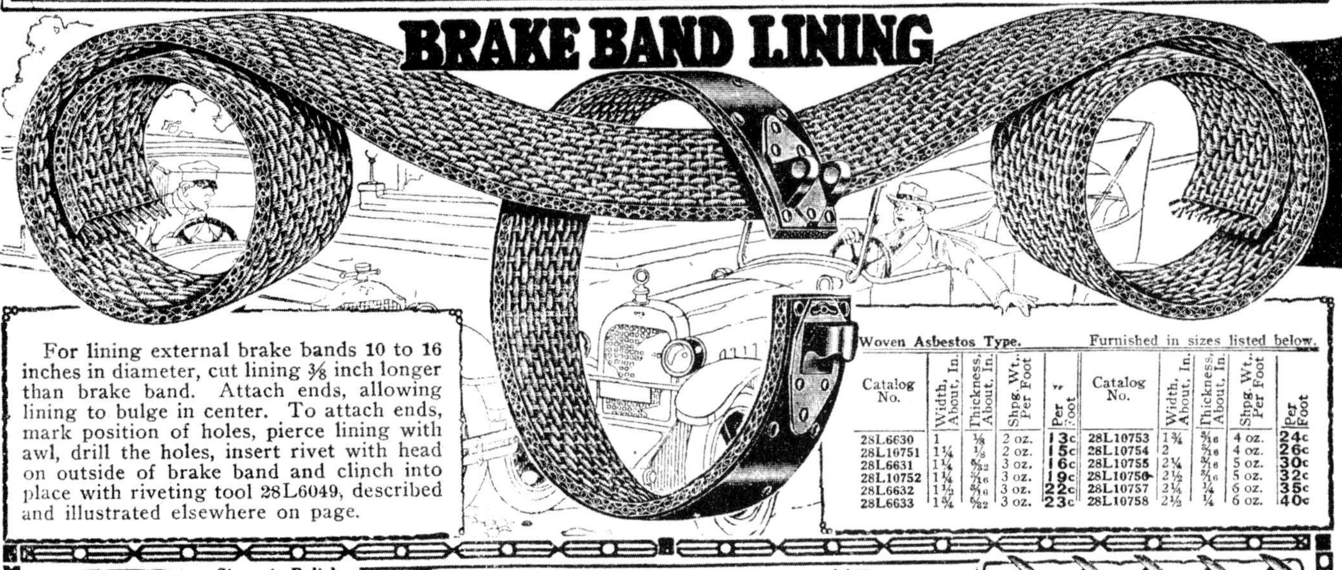

For lining external brake bands 10 to 16 inches in diameter, cut lining ⅜ inch longer than brake band. Attach ends, allowing lining to bulge in center. To attach ends, mark position of holes, pierce lining with awl, drill the holes, insert rivet with head on outside of brake band and clinch into place with riveting tool 28L6049, described and illustrated elsewhere on page.

Woven Asbestos Type. Furnished in sizes listed below.

Catalog No.	Width, About, In.	Thickness, About, In.	Shpg. Wt., Per Foot	Per Foot	Catalog No.	Width, In.	Thickness, About, In.	Shpg. Wt., Per Foot	Per Foot
28L6630	1	⅛	2 oz.	13c	28L10753	1¾	3/16	4 oz.	24c
28L10751	1¼	⅛	2 oz.	15c	28L10754	2	3/16	4 oz.	26c
28L6631	1¼	5/32	3 oz.	16c	28L10755	2¼	3/16	5 oz.	30c
28L10752	1¼	3/16	3 oz.	19c	28L10756	2½	3/16	5 oz.	32c
28L6632	1½	3/16	3 oz.	22c	28L10757	2¼	¼	6 oz.	35c
28L6633	1¾	5/32	3 oz.	23c	28L10758	2½	¼	6 oz.	40c

Simoniz Polish.

Simoniz gives a lustrous finish that protects the car body surfaces splendidly against wind, rain and sun. You can remove dust, dirt and road splashes from a "Simonized" car without scratching the finish or even washing the body surface. Apply Simoniz to a new car to keep the finish perfect.

Use Simon's Kleener to clean the car body before applying Simoniz. It removes greasy, dirty spots and stains.

28L13106—Simoniz (Paste Polish). Per 5-ounce can............ **39c**
28L13107—Simon's Kleener. Per 9-ounce can.................. **39c**
Shipping wts., 1 and 1⅜ lbs., respectively.

Carshine Wax.

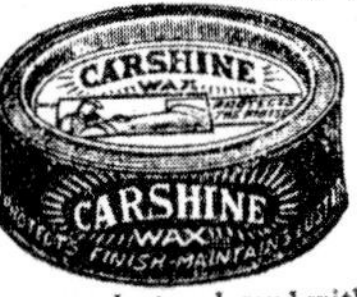

A high grade wax and cleaner, giving the car body surfaces a lasting, lustrous finish, proof against rain, wind and sun. Occasionally wipe waxed surface with a cloth to remove dust and mud without injuring finish.

28L13104—Carshine Wax, about 5-ounce can... **30c**
28L13105—Carshine Cleaner, about 9-oz. can.. **30c**
Shipping weight, 1 and 1⅜ pounds, respectively.

Oil-Grease Guns.
Quick Action Oil-Grease Gun.

An unusually substantial brass grease gun. About 9¼ inches long and 2 inches in diameter. Has positive spiral worm action. Small tip at spout end is for oil use. Loosen thumb nut near handle to secure pushing action for plunger in oiling. Spout, about 5 inches long. Shpg. wt., 2⅛ lbs.

28L10930—Quick Action Oil-Grease Gun **$1.58**

Leader Oil-Grease Gun.

Has brass barrel, about 1¼ inches in diameter and 7 inches long. Has tip for spout for use as oil gun. A practical, inexpensive gun. Shipping weight, 1 pound.

28L13188 Leader Oil-Grease Gun **35c**

Chamois and Sponges.
Sponges.

Especially adapted for automobile use. Commercially called Rock Island Wool. Diameter, about 5½ inches.

28L10855 Auto Sponge... **55c**
28L10858—High Duty Sponge. Known as a Velvet Sponge. (Not illustrated.)........ **39c**
Shpg. wt., either, 7 oz.

Sheepskin Chamois.

Sheepskin Chamois in pieces about 26x28 or 24x27 inches. Shipping weight, 6 and 8 ounces, respectively.

28L11858 Size, 24x27 inches........... **$1.19** | **28L10850** Size, 26x28 inches........... **$1.25**

For Sponges in other sizes see pages 478 and 952. For other sizes of Chamois see page 478.

Axle and Drive Shafts for Standard Cars

High grade carbon steel axle and drive shafts for the following cars:

Axle Shaft.

Drive Shaft.

Axle Shaft.

Catalog No.	Car and Model	Shpg. Wt., Lbs.	
28L15125¼	Buick, B-54-55, C54-55, Right, 1914-15	18¼	$3.35
28L15127¼	Buick, B54-55, C54-55, Left, 1914-15	17	3.35
28L15129¼	Buick, D-E-H-K-44-5-6-7-9-50, 1917-18-19-20	13¼	2.95
28L15131¼	Buick, D-E, 34-35, B-35, 1916-17-18	9⅝	2.65
28L15133¼	Buick, D54-55, Right, 1916	17⅝	3.75
28L15135¼	Buick, D-54-55, Left, 1916	16⅝	3.75
28L15069¼	Chevrolet 490	7⅝	1.25
28L15075¼	Chevrolet Baby Grand, 1918-20	11⅛	2.70
28L15077¼	Dodge, All Models	10¼	1.55
28L15137¼	Dort, 5-5A-6-8-9-10-C11-29	9⅞	2.70
28L15139¼	Dort, Model 15, 56-inch Tread	10⅛	3.25
28L15085¼	Maxwell, Model 25, Right	8⅜	1.35
28L15083¼	Maxwell, Model 25, Left	8⅜	1.35
28L15141¼	Oakland, 1916-17-18-19-20, R. H., Square End	9⅝	2.65
28L15097¼	Overland, Models 69, 79, 80, 81, 83, Right, 33¼ inches long	11¼	1.98
28L15095¼	Overland, Models 69, 79, 80, 81, 83, Left, 32½ inches long	11	1.98
28L15089¼	Overland, 90 Touring, Right	11⅝	1.98
28L15087¼	Overland, 90 Touring, Left	10⅝	1.98
28L15093¼	Overland, 75 or 90 C. C. Right	9½	1.98
28L15091¼	Overland, 75 or 90 C. C. Left	8⅝	1.98
28L15123¼	Studebaker, 4 cyl., 1919-20 Light 4	11⅞	2.10
28L15121¼	Studebaker, 6 cyl., 1919-20 Big 6	15½	2.10
28L15099¼	Studebaker, 4 and 6 cyl., 1914-18 models	12⅛	2.10

Drive Shaft.

Catalog No.	Car and Model	Shpg. Wt., Lbs.	
28L15071¼	Chevrolet 490, 1917	14¼	$2.40
28L15073¼	Chevrolet 490, 1918-19	14⅛	2.40
28L15145¼	Dodge, All Models	18½	5.25
28L15147¼	Dort, 11-15, 1919-20	15¼	6.75
28L15079¼	Maxwell, Model 25, 1917	13¼	1.95
28L15081¼	Maxwell, Model 25, 1918-19	14⅛	2.05
28L15143¼	Overland, 90 T	13	2.65

Brass Split Rivets.
Shown Actual Size.

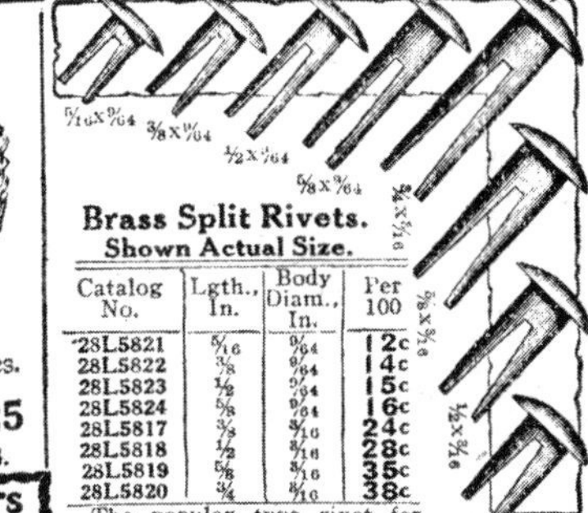

Catalog No.	Lgth., In.	Body Diam., In.	Per 100
28L5821	5/16	9/64	12c
28L5822	⅜	9/64	14c
28L5823	½	9/64	15c
28L5824	⅝	9/64	16c
28L5817	⅜	3/16	24c
28L5818	½	3/16	28c
28L5819	⅝	3/16	35c
28L5820	¾	3/16	38c

The popular type rivet for fastening brake lining. Quickly applied without punching holes or using special tools; heads fit snugly without noticeable projection. Furnished as listed above.

Shipping weight, per 100, 4, 4, 5, 6, 6, 7, 8, 9 ounces, respectively.

Brake Rivet Tool.

Substantial, practical tool for riveting brake band lining in place. Furnished about 12½ in. long, black finish. Shipping weight, 2⅞ pounds.

28L6049—Brake Rivet Tool.. **$2.25**

Jim-Dandy Fire Extinguisher.

A low priced, efficient, well known fire extinguisher. Direct stream at base of flame and operate pump handle. Liquid contents of extinguisher form air excluding blanket above flame, quickly smothering it. Contains enough fluid for a number of pump strokes. All brass. Complete with bracket for attaching. Use for car, garage, house, barn, etc. Contents will not leak nor deteriorate quickly. Shipping weight, 4 pounds.

28L9079............ **$2.95**

All weights given on this page are approximate and may vary a trifle.

Swivel Joints.

Shpg. wt., first two Nos., 1⅜ lbs.; last two Nos. 1¼ lbs.

28L13024

A high grade swivel joint, unusually well made, with dust tight washer. Bolt can be adjusted for wear or swivel taken apart.

28L13024—Swivel Joint, right hand, for 1911-1923 Stewart Speedometers...... **$1.85**
28L13026—Swivel Joint, left hand, for 1911-1923 Stewart Speedometers.............. **1.85**

An unusually low priced, very satisfactory, swivel joint. (Not illustrated.)

28L4648—Swivel Joint, right hand, for 1911-1923 Stewart Speedometers............ **$1.50**
28L4650—Swivel Joint, left hand, for 1911-1923 Stewart Speedometers............ **1.50**

Right hand wheel is wheel at your right hand when sitting in the car.

SPEEDOMETER SHAFTS.

Complete with end fittings. Housing is flexible tubing.

28L13030—Shaft for Stewart Speedometers. Specify 50, 54, 60, 66, 72 or 78-inch shaft.................... **$2.25**
Shipping weight, 3¼, 3½, 3¾, 4¼, 4½ and 4⅝ pounds, respectively.

28L13031—Shaft for Standard or Sears Cross Speedometers for 1915-1923 Ford cars, or for Jones Speedometer for 1915 Ford cars. Specify 54, 60, 63 or 66-inch shaft. Shipping weights, 2⅞, 3, 3⅛ and 3¼ pounds, respectively.............. **$2.25**

Clutches and Shaft Fittings for Chain Drive Stewart Speedometer.

28L13032 1. Upper Clutch... **8c**
28L13033 2. Lower Sliding Bar Clutch **15c**
Shpg. wt., any of these parts, 2 oz.
28L13034—3. Repair Links. Half dozen links.... **8c**

1 3 2

Speedometer Parts

Diameter, No. 1 pinions, 2¼ in.; No. 2 pinions, 2 inches; No. 3 pinions, 1¾ inches.

Fiber Gears.
1 2 3

28L13035—No. 1 Cotter Pin Fiber Pinion for Stewart Speedometer.... **10c**
28L13036—No. 2 Cotter Pin Fiber Pinion for Stewart Speedometer.... **10c**
28L13037—No. 3 Cotter Pin Fiber Pinion for Stewart Speedometer............ **10c**
28L13041—Fiber Pinion with oblong hole for Sears Cross or Standard Speedometer for 1915-1923. Ford cars, or for Jones Speedometer for 1915 Ford cars. (Not illustrated.)............ **18c**
Shipping weight, any of the above, 2 ounces.

Miscellaneous Equipment for Ford Cars

New Boyce Motometer.

Warns you when motor is overheating from lack of water, oil, etc. Improved construction, having beveled crystals and easily readable broad thermometer tube. Shipping weight, 15 ounces.
28L6333—Complete with radiator cap $3.19

Winged Radiator Cap.

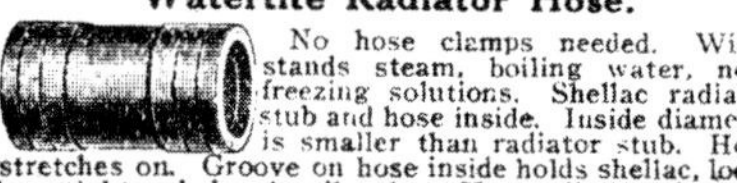

Substantial and attractive. Made of heavy white brass, nickel plated. Length, over all, about 6½ in. Top inside cut for tapping out with center punch for attaching motometer. Shipping weight, 13 oz.
28L4829—Winged Radiator Cap 38c
28L12315—Challenge Cap. Similar type to above, but of cheaper construction. Nickel plated. Shpg. wt., 1¼ lbs. (Not illustrated) 26c

Watertite Radiator Hose.

No hose clamps needed. Withstands steam, boiling water, non-freezing solutions. Shellac radiator stub and hose inside. Inside diameter is smaller than radiator stub. Hose stretches on. Groove on hose inside holds shellac, locks hose tight and absorbs vibration. Unusually long lasting.
28L11332—Inlet Hose, about 1¾x2⅞ in. Shipping weight, 7 ounces 25c
28L11462—Outlet Hose, about 2x4⅛ in. Shipping weight, 10 ounces 32c

Water Circulator.

Starts instant circulation with first turn of motor, keeping motor from overheating in summer weather and freezing up in cold weather. Replaces upper water casting. Readily installed. Complete with substantial oak tanned leather belt. Fits 1917-1923 Ford cars. Shipping wt., 3¼ lbs.
28L11097—Water Circulator for 1917-1923 Ford cars. Complete with belt $2.35

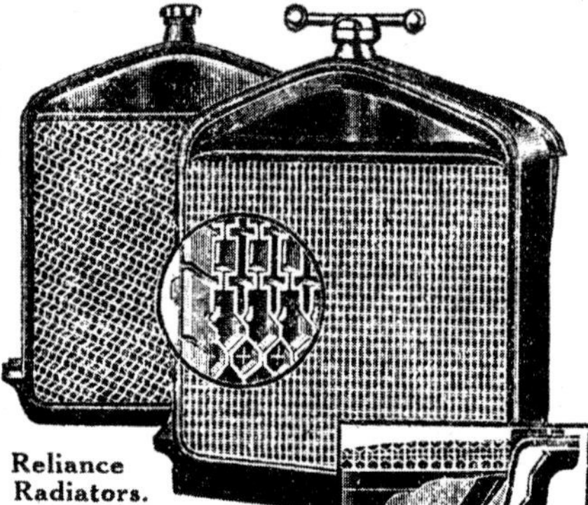

Reliance Radiators.

Honeycomb style radiators in black finish, for 1909-1916 Ford cars; also for 1917-1923 Ford cars.
28L13250¼ Reliance Radiator for 1909-1916 Ford cars. (Not illustrated.) $10.95
28L13255¼ Reliance Radiator for 1917-1923 Ford cars. $10.95
Shipping wt., 37 and 45 lbs., respectively.

Master Radiator.

A splendid radiator, with many special features. Has bronze water channels made of one-piece material and practically seamless, eliminating chance of leaks. Has necessary toughness and resisting qualities to prevent ready corrosion from alkali and lime water. Core will not burst from freezing, cell construction permitting expansion and having no seams to open. Has spring steel bracket, removing the strain on radiator and core produced by spreading action of car frame, a cause of leaks. Complete with pressed steel shell with black baked enamel finish. Furnished only for 1917-1923 models. Shipping weight, 53 pounds.
28L15102¼—Master Radiator for 1917-1923 Ford cars. Complete with bar Cap $12.25

Fan Belts.

Carry an Extra One With You.

An Oak Tanned Leather Endless Fan Belt. Shipping weight, 4 ounces.
28L11770—For Ford cars up to and including 1916 models. Size, about 22¾x1 inch 19c
28L11243—For 1917-1920 Ford cars. Size, about 26¾x1 in. 19c
28L9215—For 1921-1923 Ford cars. Size, about 27⅛x1 inch 19c
Fabric Fan Belt. (Not illustrated.) Shipping weight, 4 ounces.
28L11771—For Ford cars up to and including 1916 models 15c
28L12544—For 1917-1920 Ford cars 15c
28L9213—For 1921-1923 Ford cars 15c

Radiator Cap.

Regulation type, black finish Radiator Cap. Shpg. wt., 5 oz.
28L13271 .... 12c

Radiator Hose.

28L11793—Outlet (Upper) Hose, about 2 inches inside diameter, 3½ inches long 9c
28L12086—Outlet Hose, about 2x4 inches 11c
28L11794—Inlet (Lower) Hose, about 1¾ inches inside diameter, 2¾ inches long. Shipping weight, 8, 10 and 7 ounces, respectively 8c

Red Arrow Circulating Pump.

This splendid pump prevents overheating in summer and reduces danger of freezing in cold weather, keeping water circulating rapidly with engine running. Requires no oiling or greasing, having bronze bearings lubricated by water action. Complete with heavy oak tanned leather belt for any model Ford car or truck. Instructions for installing.
28L9153—Shpg. wt., 7¾ lbs. Complete $6.25

Oil and Grease Gun.

Has brass finish barrel, about 8 in. long and 1³⁄₁₆ in. in diameter. Has about 3¼-in. grease spout, with tip for oil use. Threaded to fit filling hole in differential. Shipping weight, 1 pound.
28L12080—Oil and Grease Gun 37c

Muffler Cut-Outs

Master. Extra heavy cut-out. Has 1½-inch opening; top of valve is open, valve being attached to exhaust pipe by clamps. Complete with substantial locking pedal and cable. No pulley is needed, action of pedal opening cut-out. Black finish. Shipping weight, 2¼ pounds.
28L12584 Complete $1.20

Leader. Outfit includes cut-out valve, 1½-in. size, for steel tubing, locking pedal, cable and pulley. Cut a V in exhaust pipe and clamp valve on. Shipping wt., complete, 3 pounds.
28L11715 Complete 65c

Interchangeable Fenders for Ford Passenger Cars.

All weights given on this page are approximate and may vary a trifle.

Sold Only in Sets.

Furnished for 1909-1916 Ford cars; also for 1917-1923 Ford cars, as illustrated, duplicating the Ford fenders. Japanned finish. (Running boards and shields are not furnished with fenders. See below.) Shipping weight, per set, either style, 80 pounds.
28L13064¼—Interchangeable Fenders for 1909-1916 Ford cars. Per set of four (fenders only) $9.95
28L13066¼—Interchangeable Fenders, for 1917-1923 Ford cars. Per set of four (fenders only) 9.95

Running Boards and Shields.

Well made and furnished in black baked enamel finish.
28L11764¼—Running Boards for 1917-1923 Model T Ford cars. Sold only in pairs. Shipping weight, per pair, 15½ pounds. Per pair $2.35
28L15068¼—Running Board Shield for 1917-1923 Ford cars. Sold only in pairs. Shipping weight, per pair, 22¾ lbs. Per pair $3.85

Washers and Gaskets.

Felt Washers. Complete Set of Twenty Felt Washers, including two each Nos. 2809, 2510B, 3111B, 3451, 3363 long, 3377B and one each 2580, 3012, 3070, 3071, 3102, 3279 and 3544. Shipping weight, per set, 8 ounces.
28L11891—Per set 25c

Cork Gaskets for Ford Cars.

Set of Cork Gaskets, including two each Ford Nos. 3111B, 3377B and one each 2580, 3070, 3071, 3102, 3279, 3363 short, 3363 long and 3379. Shipping wt., per set, 1¼ lbs.
28L4737—Per set 35c

Felt Washers for Axle.

A widely used felt grease retainer washer to prevent leakage of grease from Ford rear axles on to wheels, brake drums, etc. Use three on each outside axle end. Shipping weight, per set of six, 8 ounces.
28L4905—Felt Washers for axle. Set of 6 .... 28c

Fero-O-Lock.

A high grade steering wheel lock which locks steering wheel in disengaged or free spinning position. Has heavy bronze housing, nickel plated. Half turn of key locks or unlocks gear shaft. Has 16-tumbler lock which cannot be picked. Approved by the Underwriters' Laboratories. Can be installed with wrench. Shipping weight, 2 lbs.
28L11154 Fero-O-Lock $3.95

Steering Wheel.

Popular 17 inch size steering wheel. Furnished with walnut finish rim, corrugated on inner side for giving secure grip. Attractive aluminum spider. Readily installed. Shipping weight, 3½ pounds.
28L6070 .... $2.30

Four-Tone Chime Whistle.

A high grade well known chime whistle with four distinct notes. Gives an unfailing, powerful, pleasing warning signal. Install in front of muffler by cutting out a piece of exhaust pipe. Complete with cut-out pedal, cable, valve with double butterfly valve, with 1½-inch opening, whistle, etc. Whistle comes nickel plated with length of tubes about 11, 9, 7¾ and 5¾ inches. Shipping wt., 5¼ lbs.
28L10566—Complete $5.25

Horns

Magneto. Operates from magneto. Length over all, about 7 in. Diaphragm is about 5½ in. in diameter. Complete with push button, wire and bracket for mounting under hood. Shipping weight, 2 pounds.
28L6324 Complete $1.89

For Other Repair Parts for Ford Cars See Page 761.

Battery. A compact sturdy 6-8 volt Electric Horn of diaphragm type, operated from storage battery. An under hood type horn readily fitting in place of magneto horn. Comes about 9 inches long; diaphragm, 4½ inches in diameter and about 3-inch opening. Shipping weight, 3 pounds.
28L6757—Electric Horn $2.65

Wheel Pulls.

Adjustable for clamping tightly to hub. Has heavy set screw. Shpg. wt., 1¾ lbs.
28L11734 Adjustable Wheel Pull 45c

A very efficient all steel wheel pull, substantially made. Has heavy loose plunger aiding in starting off wheel "stuck" or "frozen" to axle and is adjustable for clamping tightly to hub. Shipping weight, 2¼ pounds.
28L8195 Wheel Pull $1.05

Hub Parts

Front Hub Roller Bearings.

Worn bearings result in wabbly wheels and undue tire wear. These roller bearings take end thrust and side strain successfully on entire length of rollers. Bearings are self aligning; rollers are concave, cup and cone convex. Installed by simply substituting in place of ball bearings regularly furnished. Shipping weight, per set, for two hubs, 3⅛ pounds.
28L7877—Per set, for 2 hubs $3.75

Standard Speedometer.

Latest model Standard Speedometer. Sets flush on instrument board, having no overhang or loose corners. Complete, as illustrated, with metal frame. Attached to instrument board by four machine screws and nuts furnished. **Adapted only for late model Ford cars regularly equipped with cowl board.** Black dial with white figures. Registers 9,999.9 season mileage with 100-mile trip register readily reset to zero. Shpg. wt., 7¾ lbs.
28L13180—Complete with necessary fittings $11.85

Jiffy Wind Shield Cleaner.

Cleans and drys both sides of upper wind shield at one time. Made of aluminum and rubber with tempered steel spring. On or off in a jiffy. Shipping weight, 1 lb.
28L6311 .... 98c

Popular Needs Engine and Carburetor Accessories Priced Low

Valve Grinder. Operate handle in quarter circle. Metal, with wood handle and grinder top, about 4½ in. high. Shpg. wt., 8 oz.
28L11815 Valve Grinder.... 35c

Victor Gaskets.

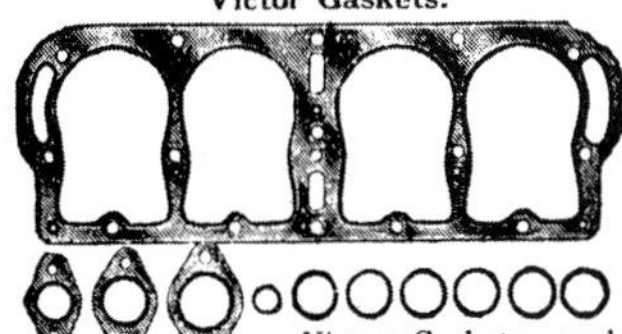

Victor Gaskets, made in one piece, ready for placing over all engine cylinders at one time, to prevent loss of compression. Shpg. wt., 1¼ lbs.
28L11750 .... 24c

Victor Gasket Set, containing eleven gaskets, in all sizes needed around the Ford engine, including cylinder head gasket. Shipping weight, 1⅜ pounds.
28L11751—Per set 36c

Crank Shaft.

28L14266—(Ford No. 3030) Crank Shaft, Chrome Vanadium steel. Shipping wt., 18½ lbs $6.50

Counter Balanced Crank Shaft. A high grade counter balanced crank shaft made of one piece drop forged alloy crank steel. Counter balances are a part of shaft itself. Has perfect running balance, eliminating motor noise and vibration. (Not illustrated.) Shipping weight, 20 pounds.
28L15056¼ $13.95

Piston Guide.

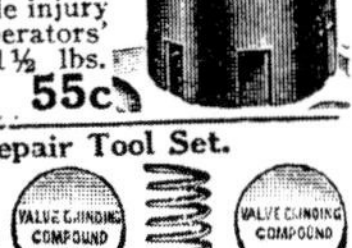

Place guide on top or bottom of cylinder and push piston through. Guide has correctly tapered spring steel jaws which automatically compress the rings evenly as they pass through. A quick, efficient device which saves time and prevents possible injury to piston rings or operators' fingers. Shipping wt., 1½ lbs.
28L11692 Piston Guide.... 55c

Master Vale Repair Tool Set.

A splendid set of high grade tools, comprising valve grinder, valve seat reamer, valve refacing tool, handy valve spring and small can each of fine and coarse valve grinding compound, complete with excellent instruction book clearly illustrating and describing every operation from draining the radiator to the replacing of cylinder head. Practically any Ford owner by following the simple and complete directions furnished can satisfactorily ream out the valve seats, reface and grind the valves of his Ford engine and keep his engine running smoothly. Shpg. wt., complete, 4¾ lbs.
28L10631—Master Valve Repair Tool Set.... $5.45

Master Valve Refacer. Valve Refacer from above set. Smooths face of valve quickly, working like a small lathe. Requires only a few turns to make valve fit perfectly to valve seat. Shipping weight, 1¼ lbs.
28L11609—Master Valve Refacer $1.30

Valve. Shpg. wt., 5 ounces.
28L11813 Valve 8c

Filler Cap Gauge.

Replaces filler cap. White dial, marked as illustrated. An inexpensive, serviceable gauge.
28L10778—Filler Cap Gauge for open car models, including 1919 .... 65c
28L10779—Same, for 1920-1923 open car models 65c
28L9052—Same, for Sedan models 65c
Shipping weight, any of the above, 14 ounces.

Moto-Oiler. Insures Correct Oiling.

The Moto-Oiler's operation insures the oil level being constantly maintained half way between the pet cocks, the admittedly correct level. It feeds oil automatically from a one-gallon reservoir. Complete with splendid substantial 3-inch oil regulator with heavy glass sides and large copper float, maintaining correct amount of oil at all times in crank case. Keeping the correct oil level in the Ford crank case will prevent spark plug fouling (from over oiling), scoring of cylinders and burning out of bearings and transmission band linings through lack of oil and clogging of oil pipe from particles of burnt out band linings getting into oil pipes. Occasional inspection soon tells you how long the gallon of oil in the reservoir lasts. Once this is determined, your oiling needs no further attention than refilling your gallon tank at intervals. Economical oil consumption is insured by systematic automatic oiling. Regulator bracket is spot welded to regular and firmly anchored to crank case bolt. Complete with all necessary fittings. Full instructions furnished for attaching, as illustrated. Shipping weight, complete outfit, 7½ pounds.
28L9992—Moto-Oiler, complete with fittings $7.45

Quick Seating One-Piece Piston Rings.

28L11728—Individually cast. Size, 3¾x¼ inch 11c
28L11174—Same as above, but .0025 inch oversize. 12c
28L5616—Same, .005 inch oversize 12c
28L12528—Same as above, but .010 inch oversize.. 12c
28L5623—Same, .015 inch oversize 12c
28L5966—Same as above, but .025 inch oversize 12c
28L11176—Same as above, but .03125 inch oversize. 12c
Shipping wt., any of above, 4 oz.

28L12536—Pistontite Piston Ring. A three-piece ring. Size, about 3⅜x¼ inch 60c
28L9104—Same as above, but .025 inch oversize, for Ford cars with worn cylinders 60c
Shipping weight, either of above, 4 ounces.

Myles Standish Connecting Rod Bearing Bolts.

Automatic take-up connecting rod bearing bolts for keeping connecting rod bearings tight. Used without shims. Only polished surface of bolt comes in contact with bearing cap. No threads exposed to bind bearing. Major thread on bearing bolt is milled, insuring free, perfect thread, and follower nut moves forward readily and perfectly, regardless of how little bearing wears. Furnished in sets of 8. Shipping weight, per set, 1½ lbs.
28L14267—Myles Standish Connecting Rod Bearing Bolts. Per set, complete $3.95

Sturdy Valve Lifter. An inexpensive, practical tool for compressing Ford valve springs to remove pin and lift out valve. Works without spring action. Shipping weight, 12 oz.
28L10888—Sturdy Valve Lifter 52c

Connecting Rod. Connecting rod for Fords. Many Ford car owners make this repair themselves. Shpg. wt., 2 lbs.
28L11010—Connecting Rod for Ford cars 90c

Rebabbitting Jig for Connecting Rod. A practical, durable rebabbitting jig for use on Ford connecting rods. Shipping wt., 2½ lbs.
28L15035 Rebabbitting Jig only, for Ford connecting rod. 78c

Light Weight Pistons. Popular type light weight pistons in standard size and five oversizes. Complete with rings, bushings and wrist pins. Shipping wt., 3⅛ lbs.
28L14029—Light Weight Piston, standard size $1.15
28L14030—Same, .0025 oversize $1.15
28L11558—Same, .005 oversize $1.15
28L11557—Same, .010 oversize 1.15
28L11556—Same, .025 oversize 1.15
28L14031—Same, .03125 oversize 1.15

Oil Gauge. Well made. Has drain cock. Note glass tube protected by metal guard. Heavier construction than regulation type. Replaces lower pet cock. Shipping weight, 7 ounces.
28L11655 Oil Gauge 45c

ACCELERATORS

Use an accelerator and have both hands free for steering, etc. Also permits regulating speed better according to road conditions, pressure of foot being readily varied.

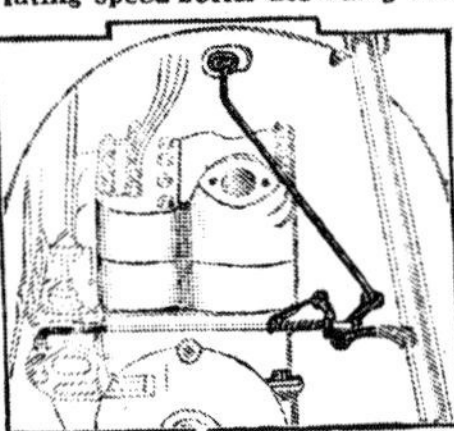

E-Z Action Accelerator. A well known efficient foot accelerator of improved type. Instructions furnished for attaching. Has adjustable foot rest. Shpg. wt., 1¼ lbs.
28L9077 .... $1.20

Get-A-Way Accelerator. Fits both Kingston and Holly Carburetors. Substantially made, easily attached without bending or shortening pedal pull rod. Permits regulating speed through throttle hand lever regardless of pressure on foot pedal. Shipping weight, 1 pound.
28L4624 .... 58c

U. and J. Accelerator. Has adjustable foot rest and guide. An excellent, widely known accelerator with adjustable foot rest and guide permitting feeding gas steadily under all driving conditions. Gives splendid control of motor. Made with hardened bushings and reinforced joints. Nickel plated. Works without any change of carburetor throttle rod. Easily attached, being clamped to steering column. Shipping weight, 1⅜ lbs.
28L11726 .... $2.32

New Kingston Model L4K Carburetor. Produces more power, greater gasoline mileage and greater flexibility. Has spray nozzle carburetion principle, insuring quick get away and smooth "idling." Equipped with an improved float, special metal fuel valve, practically indestructible bronze air valve and a good strainer, invaluable for use with the ordinary gasoline available. Shipping wt., 3 lbs.
28L13258—New Kingston Model L4K Carburetor.. $4.85

For a more complete line of Engine and Carburetor Accessories, ask for a copy of our latest complete catalog of Automotive Supplies, 512GCL, sent POSTPAID on request.

Carburetor Choke-Control. Replaces present choker on board. Use present carburetor rod, directions for installing. Graduated dial shows position of needle valve, making correct mixture easily obtained under all driving conditions. Use as choker when engine is cold. As motor warms up reduce richness of mixture by operating knob on dial. Shipping weight, 1 pound.
28L5557 .... 73c

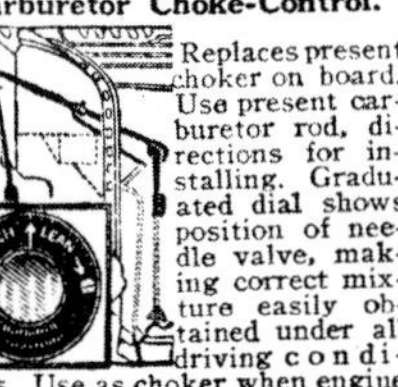

Triumph Accelerator. Simple, substantial and positive in action. Is furnished complete with foot rest and instructions for attaching. Clamps to steering column bracket. Working parts steel, machined to fit well and to permit installing easily. Will not readily rattle or get out of order. Shipping weight, 1 pound.
28L11288 .... 98c

Standard Carburetor Control. For convenient and efficient adjustment of carburetor from instrument board on cowl type Ford cars having instrument board. Dial markings permit close adjustment of carburetor, by operating knob indicator. Outfit includes universal joints, adjustable connectors, and special fork piece for carburetor. Replaces present carburetor control.
28L13280—For open model Ford cars. Shipping weight, 1⅛ pounds $1.50
28L10918—Same for Sedan model. Shipping weight, 1⅜ pounds 1.45

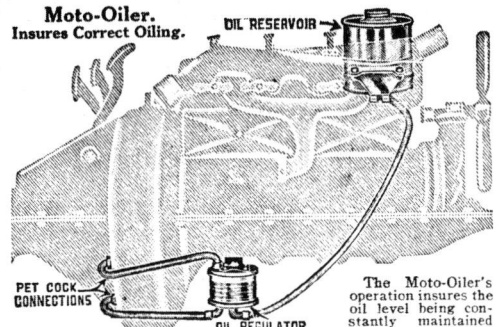

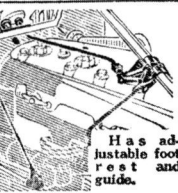

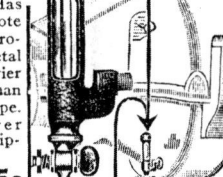

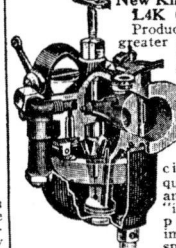

Big Values Lighting Accessories - Tools Low Prices

Sunlight Lens

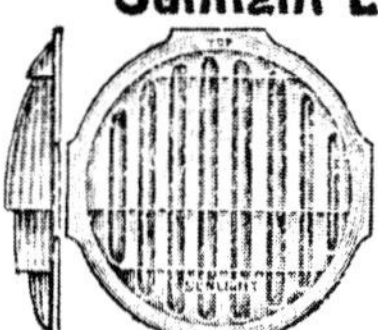

A new lens of step cut prismatic type diffusing light properly and focusing it on road at proper height. Throws enough side light to outline edges of road. Meets legal requirements in every state. Shipping weight, per pair, 3½ lbs.
28L10194—Sunlight Lens for Ford Cars. Per pair **50c**

Green Headlight Lens.

Use only on Ford cars equipped with green headlight lens. Shpg. weight, 2 pounds.
28L6086 Each........ **18c**

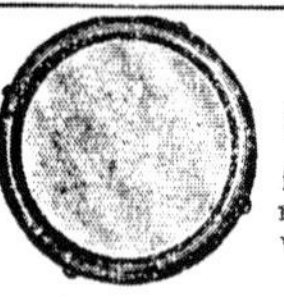

Headlight Door.

28L10312 (Ford No. 6594X) Headlight Door for 1915 to 1923 models. Shipping weight, 2¾ lbs. **35c**

Reflector.

28L14252 (Ford No. 6585X) Reflector for 1915 to 1923 Ford Electric Headlights. Shpg. wt., 1½ lbs......... **49c**

Spotlight Regulator.

For 1916-1923 Ford cars. Transforms magneto current supplied to spotlight to 6-volt, current being run through transformer. Use 6-volt bulbs. Gives a practically steady, brilliant light at any engine speed. Complete with necessary wires. Instructions furnished. Shipping weight, 6 oz.
28L8192—Spotlight Regulator **69c**

ELECTRIC HEADLIGHTS

Electric Headlight Assembly.

For Ford cars through 1914. Outfit comprises two electric headlights, switch, lighting cable, staples, etc. Lamps are black, with about 9-inch door; width between prop centers, about 7 in. Have parabolic reflector and 6-volt bulb. Shipping weight, complete outfit, 8½ pounds.
28L5565—For early model Ford cars **$4.85**

Junior Master Spotlight.

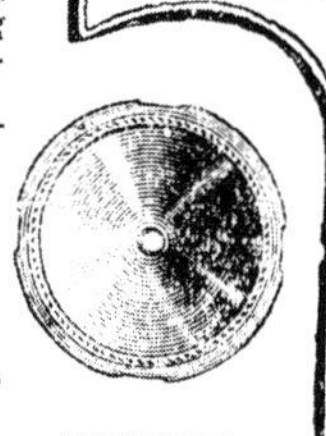

An attractive spotlight of popular small type. Has 4½-inch front. For complete description and larger illustration see page 751.
28L15008 With 6-8 volt, 21 candle power, double contact bulb for Ford cars with lights run from battery **$2.15**
28L15012—Same, with 18-24 volt bulb for Ford cars with lights run from magneto. **$2.15**
Shpg. wt., either of above, 2 lbs.

Electric Headlights, Drum Type.

Has special bracket support. For all 1915-1923 models. Bracket adjustable for throwing light up or down. Have 9-in. front, nickel plated parabolic reflector, 21-candle power, 6-8-volt double contact bulb. Have black enameled body, nickel plated brass rims. Shipping weight, per pair, 11¾ pounds.
28L15004 Per pair **$6.45**
28L15006—Same as above, with 9-volt bulbs. Per pair **$6.45**

Electric Headlights, Center Bracket.

For 1915 to 1923 models. Lamps are black enameled; ht. over all, including bracket, abt. 13 in. Switch lighting cord, etc., not furnished. 9-volt bulbs. Shpg. wt., pair, 9¾ lbs.
28L5567—Pair **$3.80**

Victor Electric Tail Lamp.

For 1915-1923 models. Has 3-inch front with bullseye lens, 6-volt 2-candle power tungsten lamp with Ediswan base and on and off connector. Use with 6-volt battery. Fitted with threaded lug and nut.
28L5583 ...... **45c**
28L7028—Same, 18-24-volt bulb for magneto use. Shpg. wt., either style, 1¼ lbs.... **55c**

Dual System Electric Tail Lamp.

Electric Tail Lamp.

For lighting tail lamp from magneto with engine running and from dry batteries at other times. Includes switch, wire, terminals, tail lamp with red semaphore lens and two bulbs, without dry cells. Wire one bulb to headlight switch and other to switch furnished, operating two dry cells, as per instructions. Shpg. wt., 1¾ lbs.
28L11702—Dual System Electric Tail Lamp **$1.49**

Brite-Lite Coil

Gives you full, clear driving light at slow engine speeds and acts as a choke coil at high engine speeds to prevent excess current from burning out bulbs. Use your regular bulbs. Use with all Ford cars where lights are operated from magneto. Shipping weight, 1¼ pounds.

Dimmer Switch.

28L8553 ...... **$1.15**
28L8252—Brite-Lite Coil complete with dimmer switch, as illustrated above. Shipping weight, 1¾ pounds **$1.95**
28L6515—Style S11 Nitrogen Bulb, 27-C. P., 9-volt, double contact base. Shpg. wt., 4 oz. **24c**
For sizes of Style S11 bulbs see page 751.

Bright Light Socket.

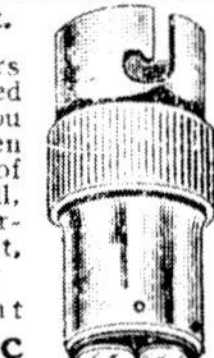

For use on Ford cars where lights are operated from magneto. Gives you bright light even when driving at a slow rate of speed. Easy to install, instructions being furnished. Shipping weight, 2 ounces.
28L8190—Bright Light Socket......... **34c**

Battery Box.

Protects your starting battery from dirt, water, road oil, etc. Use with regular Ford starting system battery attached to car frame. Box fastens to outside of battery cradle and is quickly attached. Lid is slotted for battery terminals and can be lifted off by operating suit case catch at front of box. Well made and furnished in black enamel finish. Shipping weight, 7⅜ pounds.
28L11824¼ Battery Box **$1.05**

Tools, Tool Boxes and Socket Sets

Triumph Socket Wrench Set.

Specially designed to meet practically every need of Ford owners. Comprises five hex steel sockets and 19/32-inch square socket. 17/32 and 19/32-inch hex sockets are tapered to properly fit rear axle housing bolts, engine base bolts, etc.; 21/32-inch for cylinder cap screws; 25/32-inch and 31/32-inch for general purposes. Complete with substantial ratchet wrench handle, white metal finish, about 8½ inches long, in durable wood box. Shipping weight, 2¾ pounds.
28L5311—Triumph Socket Wrench Set **$1.78**

Triple End Socket Wrench.

Has hexagon socket openings about 41/64 and 33/64 inch at double end and 41/64 inch at single end. Fits cylinder head bolts and rear axle housing nuts. Shipping weight, 1¼ pounds.
28L11823—Triple End Socket Wrench........... **34c**

Flywheel Cap Screw Wrench.

Drop forged offset wrench for removing flywheel cap screw, having 41/64-inch opening size. Shipping weight, 1 pound.
28L11021—Flywheel Cap Screw Wrench **43c**

Spark Plug Cylinder Head Wrench.

Open end fits Champion X spark plug; socket fits cylinder head bolts. A one-piece wrench with both ends hardened. Shipping weight, 1⅜ pounds.
28L6024—Spark Plug Cylinder Head Wrench **24c**

Connecting Rod Wrench.

For removing the back connecting rod easily. Length, about 12 inches; opening size, about 41/64 inch. Shpg. wt., 1 lb.
28L5796—Connecting Rod Wrench.................. **27c**

Superior Socket Wrench Set.

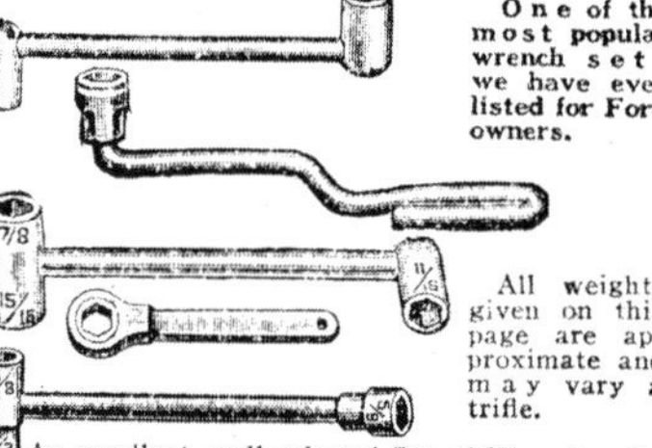

One of the most popular wrench sets we have ever listed for Ford owners.

All weights given on this page are approximate and may vary a trifle.

An excellent, well selected Set of Wrenches for Ford cars. Comprises connecting rod wrench (indispensable for cylinder next to dash), ratchet transmission wrench, triple end offset wrench with 11/16, ⅞ and 15/16-inch hex openings fitting many nuts and bolts on the car, triple end cylinder head and rear axle housing wrench, and double end offset socket wrench with hex openings sizes 9/16 and ¾ inch. All steel, nickel finish. Shipping weight, 5⅝ pounds.
28L13307—Superior Wrench Set **$2.45**

Tool Box.

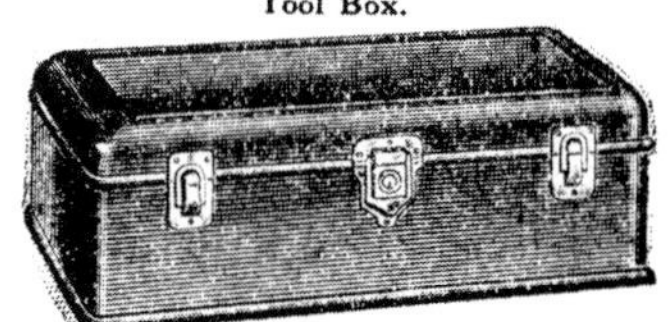

Shallow height permits Ford door opening over closed box. Fitted with good lock and end suit case catches. Inside measurements with box closed: Length, about 22 inches; width, about 9 inches; over all height, about 7 inches. Shipping weight, 10½ pounds.
28L11742¼—Tool Box.......... **$1.95**
28L15042¼—Similar to the above box, but of somewhat lighter construction. Has end suit case catches and inexpensive center lock. Size, inside length, about 21½ inches; width, about 9 inches; over all height, about 6½ inches. Shipping weight, 10 pounds.. **$1.65**

Superior Wrench Set.

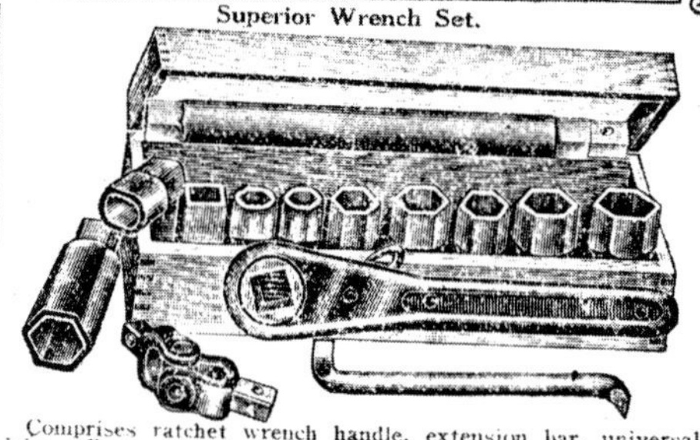

Comprises ratchet wrench handle, extension bar, universal joint, offset screwdriver, spark plug socket, seven hex sockets, 17/32, 19/32, 21/32, 23/32, 25/32, 29/32 and 31/32 inch, one square socket, 19/32-inch size, and one oval socket, about 19/32-inch, fitting crankshaft bearing bolt. 17/32 and 19/32-inch hex sockets are tapered. Steel sockets and tools. Substantial hardwood box. Shipping weight, 4⅝ pounds.
28L11821 .... **$3.68**

Hex Socket Wrench Set.

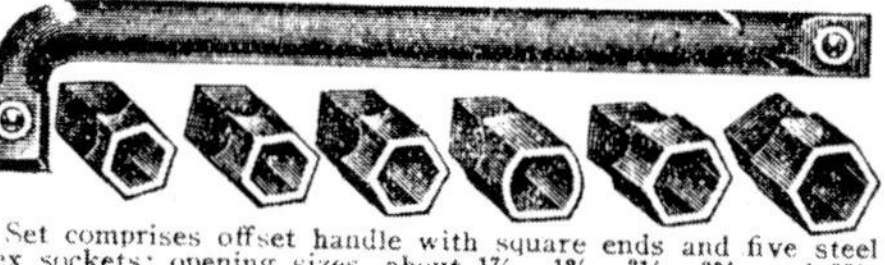

Set comprises offset handle with square ends and five steel hex sockets; opening sizes, about 17/32, 19/32, 21/32, 25/32 and 29/32 inch, and one oval socket fitting main bearing bolt. Sockets are held by ball friction. Shipping weight, 1⅝ pounds.
28L11820—Hex Socket Wrench Set **98c**

Combination Bushing Reamer.

Lower part is for reaming out Ford No. 2713 spindle body bushings and part above shoulder for reaming out Ford No. 2714 spindle arm bushings.
28L9157 .... **$0.95**
28L9106—Reamer for use with Ford No. 3022½ Piston Pin Bushing. (Not illustrated.) **1.15**
28L9108—Reamer for use with Ford No. 3314½ Transmission Gear Bushing. (Not illustrated.) **1.15**
Shipping weight, above, 10, 13 and 12 ounces respectively.

Spindle Body Bushing Remover.

Specially designed for quickly removing spindle body bushings. Push tool through until split ends reach inside edge of bushing. Sleeve compresses split ends. Bushing is then easily removed by tapping handle of tool with hammer. Used for both upper and lower bushings. Shipping weight, 14 ounces.
28L8341—Spindle Body Bushing Remover....... **65c**

Lowest Prices on Ignition Needs

A. C. Titan 1075 Spark Plug.

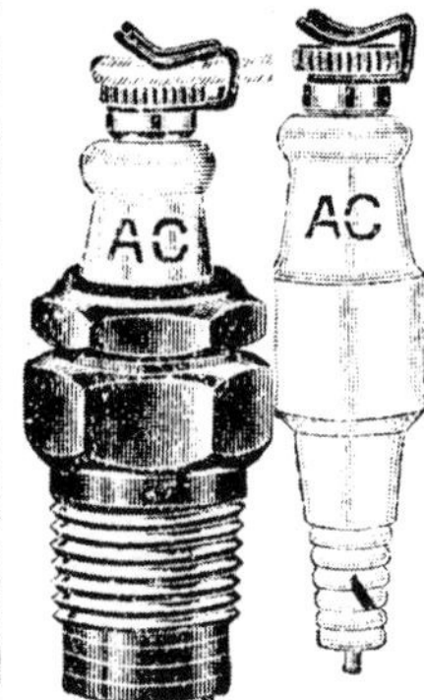

A new A. C. "carbon proof" plug with spring terminal clip for quickly attaching and detaching wires with motor running, for testing plugs and coils. Has sturdy porcelain with saw tooth edges to burn oil deposits. Special type electrode to prevent oil lodging in spark gap. Quickly taken apart; has only one gasket. Shipping weight, 5 ounces.

28L7660 — A. C. Titan 1075 Plug, ½-inch size only **43c**

A. C. Titan 1075 Porcelain. "Carbon proof" porcelain only for above plug. Complete with spring terminal clip. A big, substantial porcelain. Shipping weight, 2 ounces.

28L7352—Porcelain for A. C. Titan 1075 Plug **25c**

(Bridge) Vibrator Points.

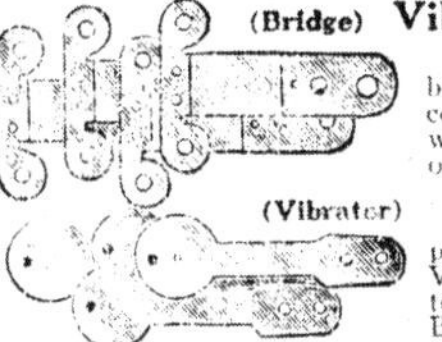

(Vibrator)

Vibrator and Vibrator Bridge for coil units. Shipping weight, per set of one each, 2 ounces.

28L11932 Tungsten points, per pair (one each Vibrator and Vibrator Bridge) **10c**

Extra Life Reversible Timer.

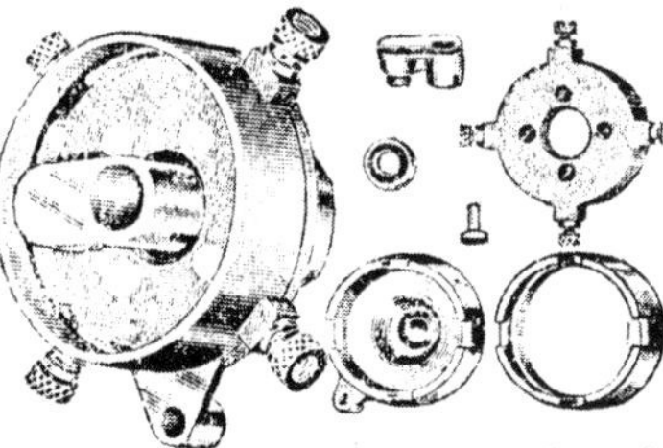

A substantial, well made timer with die casting case, white metal finish. The commutator is made of one piece fiber. Contact plugs extend clear through fiber, permitting reversing commutator for double wear. Commutator will not short circuit from oil or water and will not warp, expand or contract from dampness or dryness. Timer requires no oil. Design permits replacing any worn part. Shipping weight, 1⅛ pounds.

28L8320—Extra Life Reversible Timer **98c**

Regulation Roller Contact Timer.

Roller contact timer for late 1912 and 1913-1923 model Ford cars. A well made, inexpensive timer of regulation type. Shipping weight, 14 ounces.

28L5593 .... **57c**

Ignition Outfits

Primary Wiring Set, Lighting Wire and Spark Plug Wiring Set.

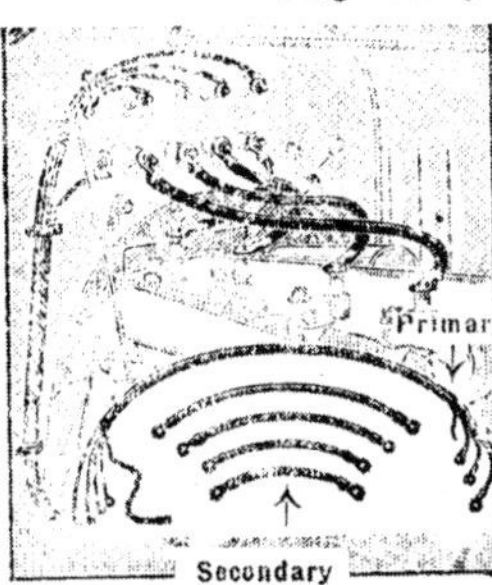

Comprises set of four primary timer wires, lighting wire and set of four secondary or spark plug wires for any model Ford car. Wires cut correct length and complete with terminals. Come in contrasting colors for identifying and replacing properly. Primary wiring set is protected by braid insulation against short circuiting due to oil, water, etc.

28L11712 Outfit, complete as above. Shipping weight, 1 pound **48c**

28L12363—Set of Four Primary Wires and Lighting Wire only. Shipping weight, 12 oz. **32c**

28L12364—Set of Four Secondary or Spark Plug Wires only. Shipping weight, 7 oz. **18c**

28L12088—Six-Wire Primary Set, for models equipped with regular Ford electric self starting system. Shipping weight, 1 pound **42c**

HORSESHOE

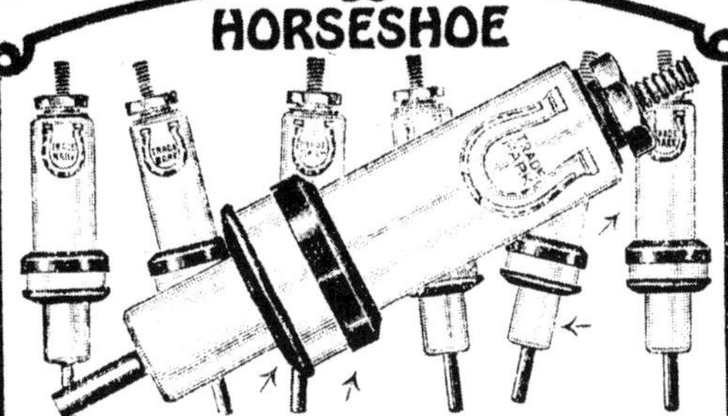

Porcelains

Gaskets. Copper, asbestos lined. Fit perfectly.

Porcelains. Made from high grade baked clay. Will withstand tremendous heat.

High grade Horseshoe Porcelains, capable of withstanding tremendous heat, fitting Champion X, Horseshoe and both regular and extension shank Bethlehem spark plugs used as equipment on many Ford cars. Complete with gaskets. New porcelains will make your plugs as good as new at a trifling cost. Our prices make it poor economy to clean old porcelains. Furnished as listed below.

28L7373—Horseshoe Porcelain. Shipping weight, 2 ounces. Each **10c**

28L7378—Same, per one-half dozen. Shipping weight, 9 ounces **57c**

Horseshoe Spark Plugs duplicate Champion X Plugs in every important respect, the porcelains fitting perfectly and all parts being interchangeable. Buy them and save one-half the usual price on your next set of spark plugs.

A high grade spark plug with well made heat resisting porcelain, compression-tight asbestos lined copper gaskets, cold rolled steel shell and bushing, oil drip type electrode and perfectly cut threads. Furnished as listed below.

28L5214—Horseshoe Spark Plug. Shipping weight, 5 ounces. Each **25c**

28L5594—Same, in handy clip of four for convenient carrying. Shpg. wt., 1 lb. Per set of 4 .... **95c**

Spark Plugs

All weights given on this page are approximate and may vary a trifle.

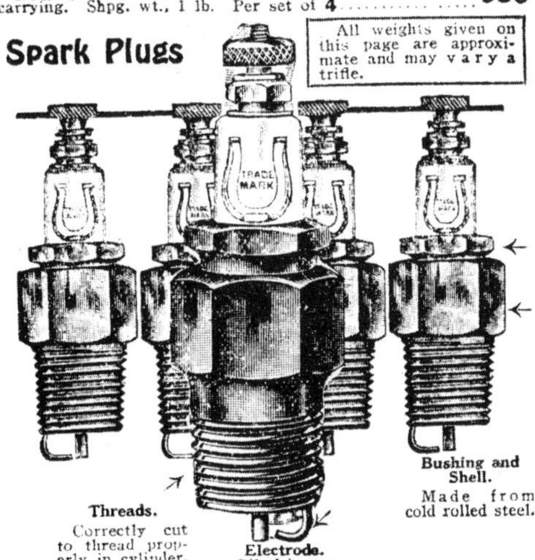

Threads. Correctly cut to thread properly in cylinder.

Electrode. Oil drip type.

Bushing and Shell. Made from cold rolled steel.

Battery Leads.

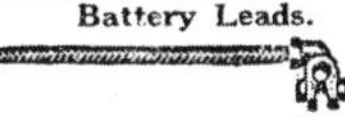

28L6754 Positive lead from battery to generator, 34 inches long. Excellently insulated. Complete with necessary connections. Shipping weight, 1⅝ pounds **60c**

28L6753—Same, about 28 inches long. Shipping weight, 1⅜ pounds **55c**

Overhead Ignition Outfit.

Note flexible metal tubing housing.

An outfit similar to 28L9170 Protected Timer Wire Assembly, shown at right above, in an up to date overhead form. Housing has two clips to be fastened to radiator hood rod and clip to be fastened under cylinder head outlet connection nut, preventing housing from coming in contact with fan. Keeps wires in handy position, away from oil, grease and water of drip pan. Shipping weight, 1 pound.

28L9168 .... **83c**

Brushes.

28L11990 28L11994 28L11992

28L11990—(Ford No. 5104) Starter Brush. Each **21c**

28L11992—(Ford No. 5126) Generator Brush (small). Each **12c**

28L11994—(Ford No. 5125) Generator Brush (large). Each **15c**

Shipping weight, any of above brushes, 2 ounces.

Champion X Spark Plug.

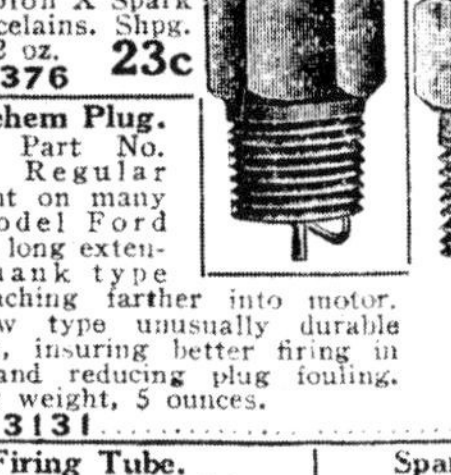

Regular equipment on many Ford cars and made especially for the Ford motor. ½-inch size only. Shpg. wt., 5 oz.

28L11818 .... **43c**

Porcelains. Champion X Spark Plug Porcelains. Shpg. weight, 2 oz.

28L7376 .... **23c**

Bethlehem Plug. (Ford Part No. 5200.) Regular equipment on many late model Ford cars. A long extension shank type plug reaching farther into motor. Has new type unusually durable porcelain, insuring better firing in engine and reducing plug fouling. Shipping weight, 5 ounces.

28L13131 .... **43c**

Firing Tube.

For making plugs fire effectively in cylinder whose piston passes oil. Screw into cylinder in place of plug and mount plug into tube. Keeps plug firing points free from oil and carbon soot, mixture being fired inside tube and flames passing out through hole in bottom of tube. Shipping weight, 4 ounces.

28L11402—Firing Tube, ½-inch size, with single hole at bottom **50c**

Spark Coils.

28L14251 (Ford No. 5007) Coil Unit. Shipping wt., 2¼ pounds. Each **$1.25**

Everlasting Ball Bearing Timer.

Practically indestructible timer, giving unfailing automatic ball bearing contact. Bearing outside acts as roller. Operates from shaft as positively as a car wheel. Reduces wear on case. Should give several years' satisfactory service. Shipping weight, 1 pound.

28L7868 — Everlasting Ball Bearing Timer **$3.45**

Elevated Ignition Outfit.

Keeps ignition wires away from oil, grease and water in drip pan. Wires are fastened to bracket which bolts to forward end of motor and drop straight to timer. Lighting wire is left in regular position under hood frame. Complete with terminals. Shipping weight, 12 oz.

28L13080 .... **73c**

Protected Timer Wire Assembly.

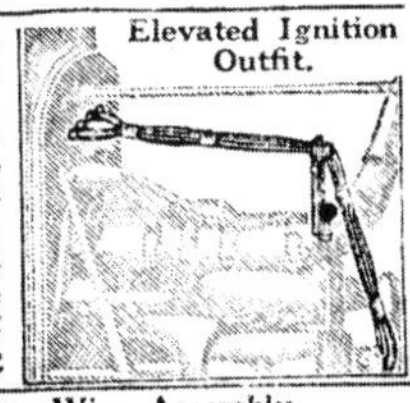

Comprises substantial flexible metal tubing housing with necessary wires, cut correct lengths, and complete with terminals, to run from dash coil units to timer. Keeps oil and water off wires, thereby preventing short circuiting; protects user against confusing wires, fits any Ford timer. Wires are covered with varnished insulated braid at timer end. Shipping weight, complete, 1 pound.

28L9170 .... **85c**

Peerless Starting Battery.

Peerless Starting Battery for Ford cars equipped with the regular Ford (Liberty) Starting System. Comes in 6-volt type. All Peerless Batteries are guaranteed for 18 months. Shipping wt., 50 pounds.

28L13615⅓

Peerless Starting Battery for Ford (Liberty) Starting System. Regular wood box type **$12.45**

With rubber case. (See the large illustration of this excellent type battery on page 749.) Shpg. wt., 50 pounds **$14.45**

Rubber Case Type.

NOTE—Both of above batteries are shipped from factory in NORTHEASTERN ILLINOIS or SOUTHEASTERN NEW YORK, whichever is nearer you.

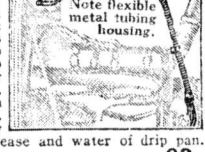

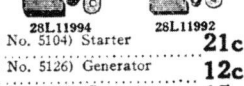

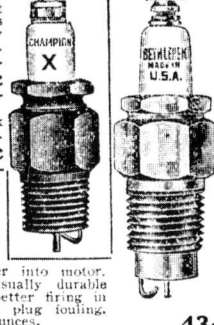

You Can Afford New Coverings

Rubberized Cloth Top Coverings

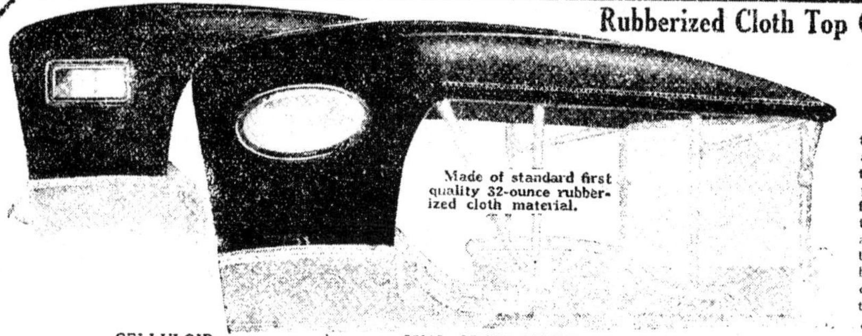

For 1915-1922 Ford Cars.

Set includes top covering and back curtain only, complete with fasteners and trimmings for attaching. Use the present bows. In ordering be sure to specify year of car.

Made of standard first quality 32-ounce rubberized cloth material.

CELLULOID.

28L12890—Rubberized Cloth Top Covering for roadster cars. Has back curtain with three celluloid lights. Shipping weight, 6 lbs.
Per set, complete........ **$3.85**

28L12895—Same as above, but for touring cars. Shpg. wt., 10 lbs.
Complete set.............. **$4.95**

OVAL GLASS LIGHT.

28L13354—Rubberized Cloth Top Covering for roadster cars. Has back curtain with bevel plate oval glass light, about 6x12 inches in size, with nickel plated rim. Shipping weight, 8½ pounds.
Per set, complete.......... **$6.15**

28L13355—Same as above, but for touring cars. Shpg. wt., 11½ lbs.
Per set, complete.......... **$7.95**

OBLONG GLASS LIGHT.

28L5320—Rubberized Cloth Top Covering for roadster cars. Has back curtain with bevel plate **oblong** glass light, about 7x13½ inches in size, with nickel plated rim. Shipping weight, 8½ pounds.
Per set, complete.......... **$7.35**

28L5322—Same as above, but for touring cars. Shpg. wt., 11½ lbs.
Per set, complete.......... **$8.95**

Special Back Curtains

With Glass Light.
For 1915-1922 Models Only.

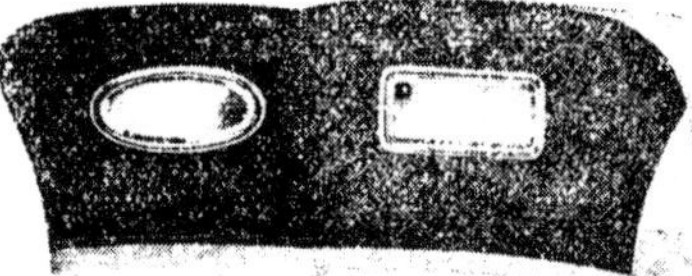

Attractive back curtains of popular glass light type. The light furnished in these special back curtains is bevel plate, about 6x12 inches in size (oval) and has nickel plated rim. (Oblong light is about 7x13½ inches.) Fit 1915-1922 Ford touring cars or roadsters. **In ordering state year of car.**

	Style Car	Finish	Style Glass	
28L13240	Touring	Rubberized	Oval	**$3.45**
28L13242	Roadster	Rubberized	Oval	**3.45**
28L13232	Touring	Mohair Effect	Oval	**3.95**
28L13234	Roadster	Mohair Effect	Oval	**3.95**
28L9200	Touring	Rubberized	Oblong	**3.95**
28L9202	Roadster	Rubberized	Oblong	**3.95**
28L9206	Touring	Mohair Effect	Oblong	**5.85**
28L9204	Roadster	Mohair Effect	Oblong	**5.85**

Shipping weight, above curtains, 4½, 4½, 3¾, 3¾, 5, 4⅝, 4¼ and 4¼ pounds respectively.

NOTE—Rubberized Cloth Back Curtains with celluloid lights are listed lower in the column.

"One-Man" Top.

Up to date standard car type. A real "One-Man" Top, strongly made with reinforced sockets. Furnished with popular gypsy back curtain, in rubberized cloth. Complete with side curtains and fittings for attaching. Has bevel plate 6x12-inch oval glass light with nickel plated rim. Fit 1915-1922 models. **State year.**

28L13227⅓ — "One-Man" Top with gypsy back curtain............ **$26.75**

28L9999⅓—Same, with one 7x13½-inch oblong bevel plate glass light............ **28.65**

28L13228⅓—"One-Man" Top with gypsy back curtain, oval glass light and "open with door" side curtains (see illustration at right) instead of side curtains furnished with "One-Man" Tops listed above. Complete........ **$29.95**

28L9998⅓—Same as 28L13228⅓, with **oblong** light replacing oval glass light. Complete.............. **31.85**

Shipping weight, any of above, 100 pounds.

NOTE—Shipped promptly from factory in NORTHEASTERN ILLINOIS or from store, as stock conditions permit.

"Open With Door" Side Curtains.

An up to date equipment that gives you the convenience of a closed car in getting in and out. Comprises complete set of side curtains, the same kind of material as Ford rubberized cloth curtains. Curtains on three doors open with door. Fits 1915-1922 models. **State year of car and for 1922 touring model also give motor number.** Eyelets furnished fit fasteners used for present curtains. Curtains complete with celluloid lights. Furnished in two sizes.

28L7154¼—Set of "Open With Door" Curtains for Ford touring car. Shipping weight, 12½ pounds. Per set.......... **$14.65**

28L7152¼—Set of "Open With Door" Curtains for Ford roadster. Shipping weight, 8½ pounds. Per set.......... **7.85**

Back Light Sets

Wooden Frame Glass.

Attractive and durable. Replaces celluloid lights. Glass size between frame openings, 5x9 inches. Furnished with nickel plated metal lining strips and upholsterers' tacks, curtain edges being securely held in place by inserting them between frame and metal strips and tacking strips to frame. Glass is securely held in channel in frame and kept from rattling. Shipping weight, per set of three, 2⅝ pounds.

28L5147—Per set of 3...... **$1.35**

Glass Back Light Set.

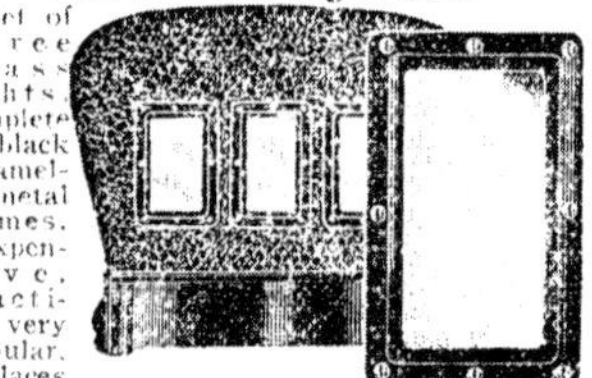

Set of three glass lights complete in black enameled metal frames. Inexpensive, practical, very popular. Replaces celluloid lights regularly furnished in rear curtains for 1917-1922 Ford cars. Readily installed. Very durable and permit a better rear view. Shipping weight, per set of three, 3¾ pounds.

28L10069—Per set of 3...... **98c**

Metal Frame Celluloid Replace Light.

Inexpensive and attractive, for back curtain of 1917-1922 cars. Celluloid has bevel plate effect around edges and metal frame with eyelets. Insert fasteners furnished through eyelets and turn down on inside of curtain. Over all size, each light, about 6½x10 inches. Shipping weight, 13 ounces.

28L13099—Per set of 3...... **95c**

Replace Curtain Lights.

28L11043 28L11041 28L12350 28L12351

Installed without sewing. Furnished with fasteners.

28L12350—Replace Back Curtain Light for 1914-1916 cars. Size, about 10½x18 in. Shpg. wt., 8 oz.
Each.............. **48c**

28L11041—Replace Back Curtain Light for 1917-1922 cars. Size, about 6½x10½ in. Shpg. wt., 4 oz.
Per single light.......... **19c**

28L13309—Same, per set of 3 lights. Shpg. wt., 6 oz........ **56c**

28L12351—Large Replace Side Curtain Light for 1914-1922 cars. Size, about 7½x20½ in. Shpg. wt., 5 oz.
Each.............. **37c**

28L11043—Replace Side Curtain Light for 1915-1922 touring cars. Size, about 10½x14½ in. Shpg. wt., 7 oz.
Each.............. **38c**

28L12352—Small Replace Side Curtain Light for 1914-1922 cars. (Not illustrated.) Size, about 7½x10½ inches. Shipping weight, 4 ounces.
Each.............. **20c**

Special Back Curtains With Celluloid Lights.

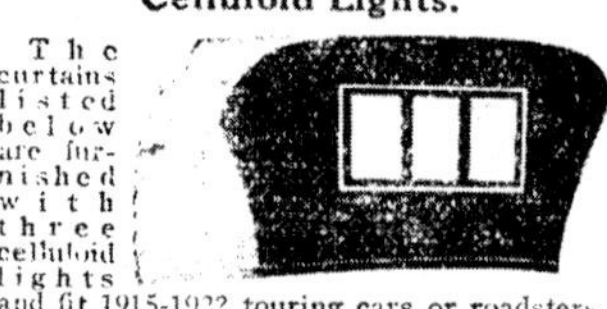

The curtains listed below are furnished with three celluloid lights and fit 1915-1922 touring cars or roadsters. **In ordering state year.**

28L13244 — Special Back Curtain. Rubberized cloth top material with three celluloid lights, for touring cars. Shipping weight, 2½ pounds..... **$1.95**

28L13246—Same, for roadster model cars. Shipping weight, 2½ pounds............ **$1.95**

Side Curtains.

Furnished in same kind of material as Ford rubberized cloth curtains, in set of four pieces. Complete with eyelets to attach to curtains and extra fasteners. **In ordering be sure to state year of car, and for 1922 touring model also give motor number.**

28L12910—Set of Side Curtains for 1915-1922 roadsters. Shipping weight, per set, 5 pounds.
Per set.............. **$4.65**

28L12912—Set of Side Curtains for 1915-1922 touring cars. Shpg. wt., per set, 7¼ lbs.
Per set.............. **$6.95**

Stick-Tite Curtain Light.

"Stick Them On Like a Postage Stamp."

For rear curtain of 1917-1922 Ford cars. Cement into place, no sewing or metal fasteners. Complete with cement. Shipping weight, 7 oz.

28L12958 **35c**

28L9224 Shipping wt., 14 oz. Set of 3 **$1.00**

Mats and Carpets.

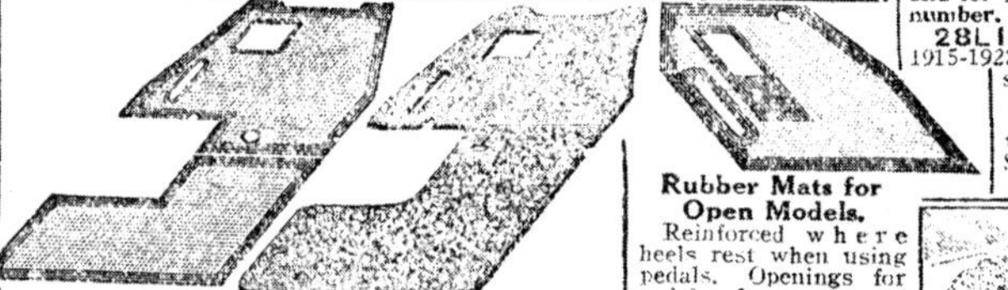

Rubber Mats for Closed Models.

Sedan Mat.

Two-Piece, fits all Ford sedans. Easily cleaned; no fasteners or tacks needed. Shpg. wt., 10½ lbs.

28L15038
Each.............. **$2.19**

28L15040—Mat, one-piece, for Ford coupe. (Not illustrated.) Shipping wt., 7 pounds.
Each.............. **$1.45**

Closed Car Carpets.

Attractive, durable gray carpet rugs. Bound edges.

28L4743—Coupe Carpet or Rug. (Not illustrated.) Shipping weight, 3¼ pounds.......... **$2.10**

28L4745—Sedan Carpet or Rug. Two Piece. Shpg. wt., 3¾ lbs...... **$3.60**

Rubber Mats for Open Models.

Reinforced where heels rest when using pedals. Openings for pedals, levers and speedometer shaft.

28L11782—For 1915-1920 cowl dash type open model Ford cars....... **80c**

28L15026—Same for 1921-22-23 Ford cars. State year.......... **80c**

28L11783—Same for Ford cars not equipped with cowl dash. (Not illustrated.)..... **80c**

Shpg. wt., 4½, 4⅜ and 4 lbs. respectively.

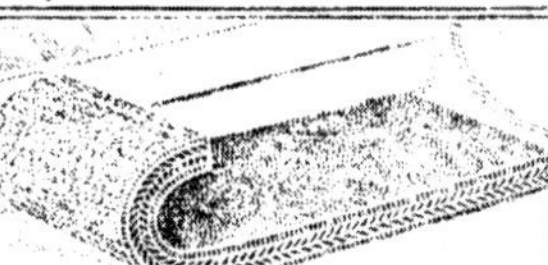

Cocoa Tonneau Mat.

Attractive, substantial Cocoa Mat for touring rear floor. Size, about 21x28 inches; about 1¼ inches thick. Gives comfortable footing and will last a long time. Shipping weight, 6 lbs.

28L5009.............. **$2.25**

All weights given on this page are approximate and may vary a trifle.

For Your Ford At These Prices

Mackintosh Cloth Seat Covers.

For 1915-1923 touring cars or roadsters. Made of khaki color mackintosh cloth. Well bound. Roadster door and touring rear door covers fitted with pockets. Furnished with or without top cover. **In ordering be sure to specify year of car. Top covers of same material as seat covers.**

28L11776—Set of Seat Covers only, for touring car. Shipping weight, 8¾ pounds. Per set. **$10.25**
28L11775—Set of Seat Covers complete with Top Cover, for touring car. Shpg. wt., 11¾ lbs. Per set.. **11.80**
28L11778—Set of Seat Covers only, for roadster. Shipping weight, 4½ pounds. Per set........ **5.55**
28L11777—Set of Seat Covers complete with Top Cover, for roadster. Shpg. wt., 7¼ lbs. Per set...... **6.15**

Striped Upholstery Coverings for Closed Cars.

For 1916-1923 Models.

An attractive, durable high grade striped seat covering material adding greatly to the appearance of the car and protecting both clothes and the car upholstery. Covering material furnished for doors and car side panels as well as for seats and cushions. One door covering is furnished with pocket. Complete with stud fasteners, eyelets and upholsterers' tacks for attaching.

28L13381—Striped Upholstery Coverings for Ford coupe models. **State year.** Shipping weight, 3¾ pounds. Per set........ **$6.85**
28L13383—Same, for Ford sedan models. **State year.** Shipping weight, per set, about 5¾ pounds. Per set.............. **$10.85**

Striped Seat Covers.

A high grade **striped seat cover material, very durable** and attractive in appearance, for Ford touring cars. Seat covers are furnished with or without top cover. **Furnished as illustrated for 1915-1923 touring models only.** Rear door covers fitted with pockets. **In ordering be sure to state year of car.**

28L5252—Striped Seat Covers only, for Ford touring cars. Shipping weight, 6 pounds. Per set.. **$8.60**
28L5253—Striped Seat Covers, complete with top cover for Ford touring cars. Shipping weight, 8 pounds. Per set........................ **$13.40**

Slip-On Seat Covers.

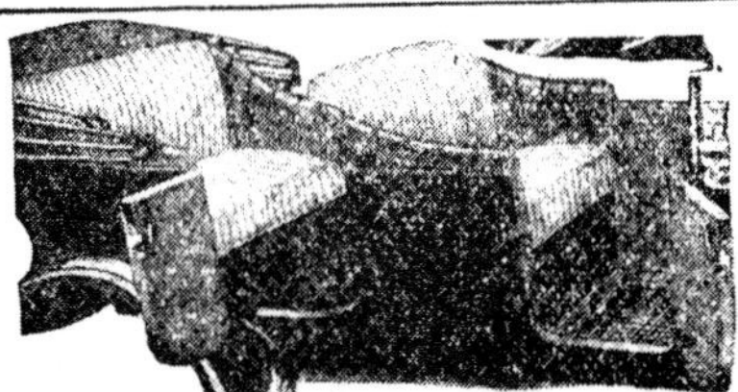

Showing Slip-On Covers on Sedan Seats.

Protect your clothes from dirty, greasy cushions by a set of these attractive appearing seat covers of striped washable material. Readily put on or taken off, being fitted with buttonholes, which button over tack buttons furnished for attaching to top of seat back. Front seat covers on sedan models have cap or hood covering entire back of seat and require no fasteners.

28L5062—Slip-On Seat Cover for Ford roadsters. Each................ **$1.95**
28L5064—Slip-On Seat Covers for Ford touring cars. Set of two...... **3.95**
28L5066—Slip-On Seat Covers for Ford sedans. Set of three.......... **4.75**

NOTE—Furnished only as listed above. Shpg. wt., 1⅛, 2¾ and 3⅜ lbs., respectively.

Seat Cushions.

Divided Type.

Replace a worn, damaged or lumpy seat cushion with a pair of these well made artificial leather covered seat cushions of handy divided type for front seat of 1913-23 touring models, also for roadsters. Permits raising right hand cushion for filling gas tank, etc., without handling entire cushion. (Unmailable.) Shipping weight, 20½ pounds.

28L13294¼—Divided Seat Cushions for front seat of touring car, or for roadster seat. **State year.** Per pair.................................. **$6.75**

One-Piece Type.

One-piece high grade artificial leather covered seat cushion for front or rear seat of 1913-1922 touring models, or for roadsters. An unusually well made cushion with wooden frame and inside burlap reinforcement for taking strain off leather covering of cushion. Spring construction includes wire trusses for holding springs in position. Has larger number of springs than usually used. (Unmailable.)

28L13262¼—One-Piece Type Seat Cushion for touring rear seat. **State year.** Shipping weight, 26½ pounds.......... **$5.60**
28L13261¼—Same, for front touring seat, or for roadster. **State year.** Shipping weight, 22 pounds.......... **5.35**

Ball Grip Handles.

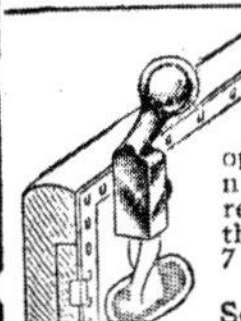

Make opening and closing of doors a simple, convenient operation. Well made, nickel plated. Can be readily attached. Sets of three. Shipping weight, 7 ounces.

28L10942 Set of 3.......... **35c**

Challenge Door Handles.

A handy, neat, inexpensive door handle, in sets of three. Well made; black finish. Shipping weight, per set, 1 pound.

28L6058 Per set of 3........ **24c**

Pedal Rubbers.

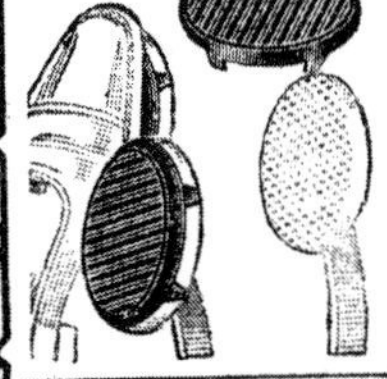

Give sure and more comfortable footing on pedals. Furnished in sets of three. Shpg. wt., per set, 9 ounces.

28L11832 Pedal Rubbers. Set of 3 **30c**

Upholstery.

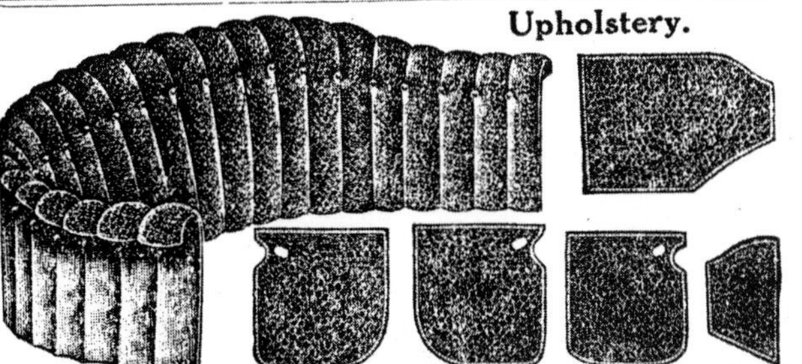

Replace worn out or torn upholstery with this made up ready to install upholstery. Made of well padded artificial leather complete with side arm rests, as illustrated.

28L13489¼—Upholstery for front of touring car or for roadsters, complete with tacks and binding for attaching............ **$5.35**
28L13491¼—Same, for rear of touring car........................ **5.35**
Shipping weight of above, 8½ and 10 pounds.
28L13174¼—Upholstery for both front and rear of touring car. Shipping weight, set, 14⅞ pounds. Per set......................... **9.95**

Door Cover Set.

Similar to present door covers. Complete with upholsterers' tacks.

28L13235—Door Cover Set of three pieces, complete, for Ford roadster. Shipping weight, per set, 4 pounds. Per set.......... **$0.75**
28L13239—Same, in set of five pieces, as illustrated, for Ford touring car. Shipping weight, per set, 5⅜ pounds. Set of five......... **1.35**

Safety Mirror.

Up to date popular type rear view mirror. Size, about 7x2½ inches. Has beveled edge. Complete with bracket fitting sedan or cope wind shield frame. Mirror can be tilted up or down or to either side. Shipping weight, 1 pound.

28L9622—Rear View Mirror for closed model Ford cars........ **65c**

Open Car Type.

The above mirror with bracket fitting wind shield frame on Ford touring cars or roadsters. (Not illustrated.) Bracket is adjustable in four positions. Shipping weight, 1¼ pounds.

28L9624—Rear View Mirror for open model Ford cars......... **95c**

Adjustable Top Springs.

Handy, long lasting, Adjustable Top Springs, requiring no straps. Replace top straps. Readily adjusted and adjustment locked in place. Will not become uncoiled or stretched in use. Shpg. wt., per pair, 8 oz.

28L11563—Per pair......... **48c**

Rives' Extension Pedal Pads.

Two outside pads have flat rubber surface of about 2¼x3¼ inches; inner pad is corrugated. Substantial clamps for attaching to pedals. Shipping weight, per set of three, 1⅛ pounds.

28L5718—Rives' Extension Pedal Pads. Set of 3........ **75c**

Starting Crank Holder.

28L5316—Leather Starting Crank Holder with closed end. Shipping weight, 5 oz.. **20c**

Linoleum Covered Wood Running Boards.

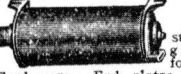

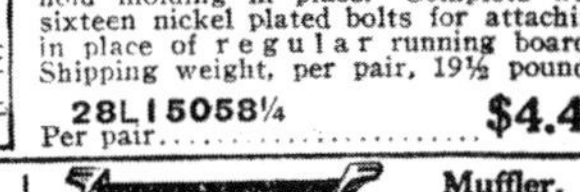

Make Your Ford Look Like a Higher Price Standard Car.

Seasoned, non-warping, ⅞-inch oak running boards. Covered with dark brown automobile cork linoleum and bound with aluminum molding with ½-inch corrugated top. Nickel plated screws hold molding in place. Complete with sixteen nickel plated bolts for attaching in place of regular running boards. Shipping weight, per pair, 19½ pounds.

28L15058¼ Per pair........................ **$4.45**

Black and Nickel Robe Rail.

28L11835—Robe Rail with folding ends. Nickel plated crossbar, about 27x⅝ inch; black enameled ends. Shipping weight, 2¼ pounds........... **57c**

Muffler.

An unusually substantial, high grade muffler for all Model T Ford cars. End plates are castings. Shipping weight, 7⅜ pounds.

28L10051—Muffler........ **$1.35**

Door Hand Pad.

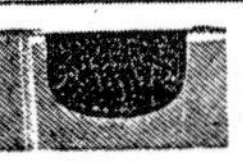

Prevents finger print marks or scratches showing on door. Metal body, artificial leather cover. Shipping weight, pair, 10 ounces.

28L7009—Pair.............. **20c**

All Black Robe Rail.

Length of crossbar, about 28 inches. Has curved ends. Shipping weight, 3 pounds.

28L11836—All Black Robe Rail.............................. **34c**

All weights given on this page are approximate and may vary a trifle.

You Need These Riding and Driving Aids

Transmission Lining

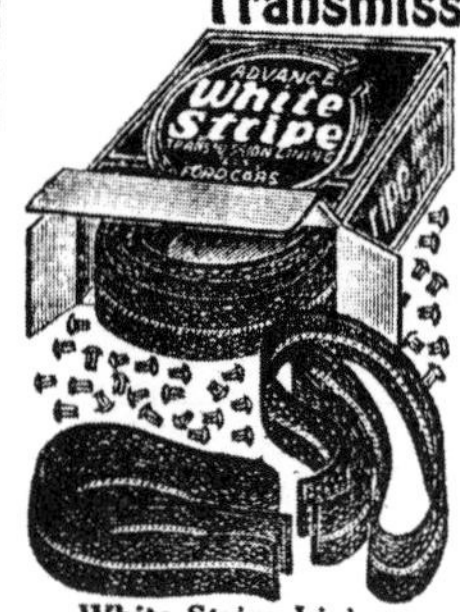

White Stripe Lining.
A splendid widely advertised brake lining which we believe is the strongest and most durable plain lining made for Ford transmission bands. Woven from tough long fiber hose cord yarn. Has more cotton to inch and more yarn than any other lining. A special process for treating the threads keeps them soft and pliable, protecting them from hot oil, and giving them unusual wearing qualities. Comes in set of three pieces, complete with rivets. Shipping weight, per set, 14 ounces.
28L4731—Set... **$1.48**

Advance Cork Feltback Lining.
Improved type. Felt back cushions braking action, equalizes grip around transmission drum and floods lining surfaces with oil during braking action, preventing lining burning. Felt absorbs oil thrown up by flywheel, acts as oil reservoir and holds cork inserts in place. Cork inserts grip smoothly, instantly and firmly. Shipping weight, per set of three, 14 ounces.
28L9222—Set **$2.15**

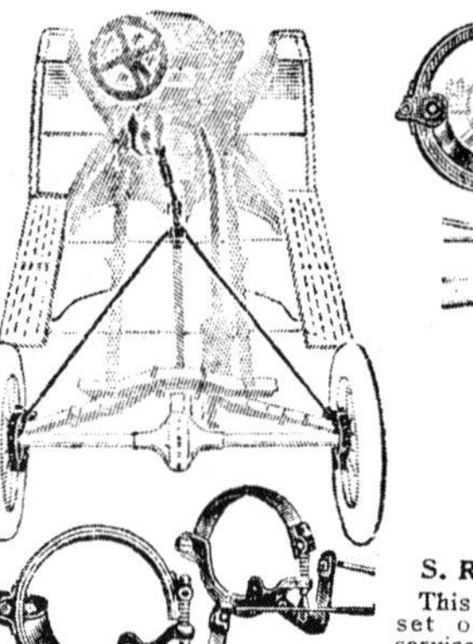

S. R. Brakes.
This well known set of external service brakes delivers braking power direct to rear wheels, eliminating strain on axle shafts, gears, keys and transmission bearings. Enables driver to stop car without jarring or chattering of transmission brake band, reduces skidding and wear on transmission. Brake bracket and band are attached to rear axle housing and connected by cable and equalizer, using present brake pedal. Does not interfere with present emergency brake. Instructions furnished for installing. Shipping weight, 14⅛ pounds.
28L13347—S. R. Brakes, per set.... **$11.75**

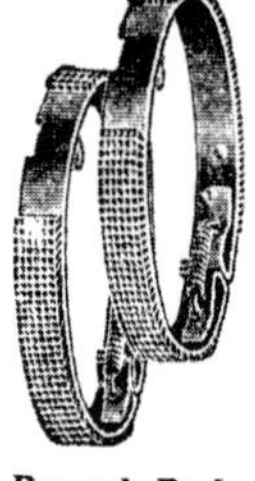

Pressed Brake Shoe.
Pressed from one piece. Flexible and efficient. Worn linings can be quickly replaced. Furnished in sets of two, complete with brake springs. Shipping wt., per set, 3 pounds.
28L11710
Per set........ **$1.25**

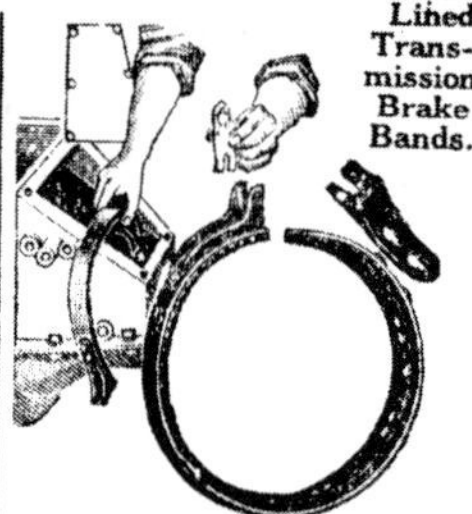
Lined Transmission Brake Bands.

Lined Transmission Brake Bands.
Put in a set when old band linings are worn instead of relining old bands. To take out present bands, remove upper half of transmission case and install lined bands, which are fitted with removable end lug. When new bands require relining take them out through top plate, reline and replace them without removing upper half of case, removable end lug making easy the removing and relining of bands. Shipping weight, 4¼ pounds.
28L12089
Set of three........... **$3.60**

All weights given on this page are approximate and may vary a trifle.

Regulation Transmission Band Lining.
Comes in a set of three pieces, complete with rivets. A well made brake lining, chemically treated to insure long wear and smooth operation. Used for low gear, reverse and transmission brake. (Not illustrated.) Shpg. wt., 13 oz.
28L11761—Per set............ **38c**

Asbestos Transmission Band Lining.
High grade long wearing Asbestos Transmission Band Lining. (Not illustrated.) Comes in sets of three pieces. Complete with rivets. Shipping weight, 1 pound.
28L10932—Set **95c**

Reverse and Brake Wrench.

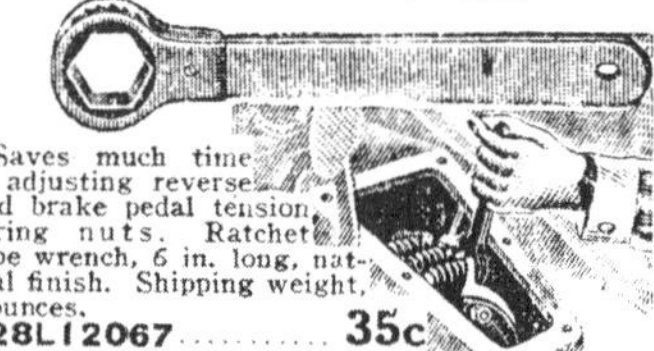

Saves much time in adjusting reverse and brake pedal tension spring nuts. Ratchet type wrench, 6 in. long, natural finish. Shipping weight, 5 ounces.
28L12067.......... **35c**

Double Arm Lever Shock Absorbers. For Any Model Ford Car.
Installed on Rear. Installed on Front.
Popular type double arm lever shock absorbers. Installed without machining. Furnished in sets of four only, for open or closed model Ford cars. Instructions for attaching. Shipping wt., per set, 17¼ lbs.
28L13161—Double Arm Lever Shock Absorbers. Per set of 4 **$5.00**

Springs.
Front and rear springs fitting all model T Ford cars. Japanned.
28L13052¼ (Ford No. 3800) Front Spring. Shipping weight, 17 pounds. **$1.25**
Front. Rear.
28L11796¼—Rear Spring, standard tread. Shipping weight, 43 pounds.............. **$4.15**
28L13452¼—Rear Spring, 10-leaf, for closed models and Ford trucks. Complete with special clips. Shipping weight, 54 pounds..................... **$6.50**

Running Board and Fender Brace Set.
Strengthens running boards and fenders, reducing vibration and rattling. Comprises substantial metal brace rod, with hooks fastening over frame near center and blocks fitting under running board. Shipping weight, 10¼ lbs.
28L10112¼ **85c**

Angle Radius Rod Support.
A substantial, practical one-piece angle iron auxiliary radius rod support which braces the radius rod successfully, reducing front axle and steering wheel vibration and wear on ball and socket at rear of rod. Black finish. **For all models up to and including 1919.** Shipping weight, 5⅛ pounds.
28L10077¼—Angle Radius Rod Support **40c**

Anti-Rattling Ball Sockets.

Take up wear and prevent rattling.

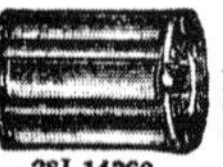

Replaces cap that holds front axle radius rod in position. Shipping weight, 12 ounces.
28L11755 **20c**

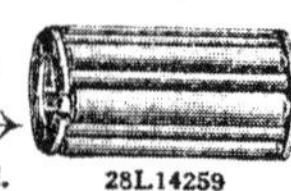

Replace cap on end of steering gear connecting rod. Shpg. wt., per pair, 12 oz.
28L11752 **20c**
Per pair

Shafts

Rear Axle Shaft.
28L9955¼—Rear axle shaft, replacing Ford No. 2505D rear axle shaft. Shipping weight, 16¾ pounds **95c**

Drive Shaft.
28L13173¼—(Ford No. 2595B) D. S. only (tapered). (Unmailable.) Shipping weight, 13⅜ pounds **$1.65**

Special Ratio Gears.
The 3-1 Ratio Gears give considerable increase in speed. Fit Ford rear axle without special adjustment, replacing regular drive shaft pinion and differential gear. Widely used in level country and with speedster bodies, etc. Gears are made from 3½ per cent nickel steel with correctly designed teeth; durable and quiet. Gear has thirty-nine teeth; pinion, thirteen teeth.
28L8542—Special Ratio Gears, 3-1. Per set of 2 gears **$5.25**

4.2-1 Ratio Gears.
The 4.2-1 Ratio Gears give considerable increase in power. Especially well adapted for sandy or hilly country and for use on Ford trucks and closed models. Gear has forty-two teeth; pinion, ten teeth.
28L8544—Special Ratio Gears, 4.2-1. Per set of 2 gears **$4.98**
Shpg. wt., above, per set, 4 and 4⅞ lbs., respectively.

Ideal Roll-Ball Bearings for Axle and Drive Shaft.
28L14260 — Showing Six Steel Rollers Mounted on Steel Balls at Either Thrust End. — 28L14259
A new type Ball-Roller Bearing, practically eliminating side play and friction. Mechanically perfect as regards axle and thrust pressure. Frictionless, dependable and long lived.
28L14259—Replaces Ford No. 2508. Shipping weight, 2¼ pounds......... **92c**
28L14260—Replaces Ford No. 2587. Shipping weight, 1¾ pounds......... **98c**

Straightaway Steersman.
A practical aid to easier driving, making desirable tension between steering rod and front axle. Comprises two strong springs with hooks for attaching at steering rod connection and clamp for attaching other end of spring at center of axle. Helps prevent rattling of steering rod connections. Complete with instructions. Shipping weight, 3¾ pounds.
28L12576—Straightaway Steersman.................. **$1.45**

Shaw's Automatic Stabilizer.

Attach between radius rod and front tie rod to hold front wheels straight ahead and prevent steering wheel vibration and steering rod rattling. Illustration clearly shows method of attachment. Shipping weight, 1¼ pounds.
28L13330—Shaw's Automatic Stabilizer................. **75c**

Ball Thrust Bearings.

Replace plain flat washers regularly used in rear axle and keep gears in proper mesh, preventing breaking of gear teeth, grinding and jerking. Hardened steel balls between outer plates take pressure on a rotating, frictionless, wear resisting surface. Complete as illustrated. Shipping weight, 12 ounces.
28L11724—Per set............... **$1.20**

For a more complete line of Riding and Driving Aids, ask for a copy of our latest catalog of Automotive Supplies, 512GCL. Sent POSTPAID on request.

Micalene Anti-Chatter Compound.
Remove transmission cover plate, loosen bands and pour contents on bands, allowing it to soak in. 8-ounce can. Shipping wt., 1¼ pounds.
28L9620 **52c**

Anti-Rattlers

Brake Rod Silencers.
An efficient device for preventing brake rod rattling. Come in sets of four. Illustration shows methods of attaching. Shipping weight, per set, 3 oz.
28L11454 **22c**
Per set of 4......

Steering Rod Silencers.
Silencer and brace for steering rod connections. They prevent undue wear. Made of wire, japanned. Shipping weight, per pair, 9 oz.
28L11236—Steering Rod Silencers.
Per pair........... **22c**

Clutch Release.

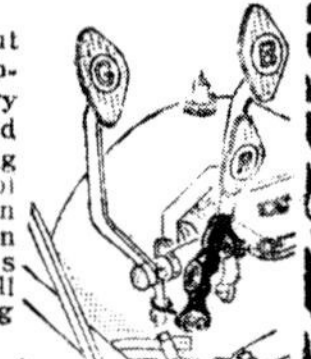

Automatically throws out clutch when foot brake is applied, making it unnecessary to operate clutch pedal to avoid killing engine when slowing down or stopping. The use of this device will also lengthen the life of the transmission bands. Illustration shows method of attaching. Fits all 1913-1923 Ford cars. Shipping weight, 12 ounces.
28L10890
Clutch Release......... **45c**

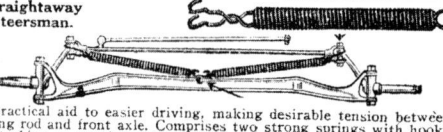

Repair Parts for Ford Cars

Wheel and Hub Parts.

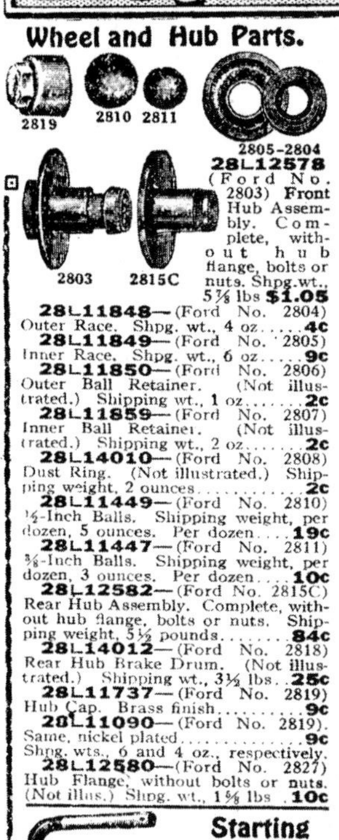

28L12578—(Ford No. 2803) Front Hub Assembly. Complete, without hub flange, bolts or nuts. Shpg.wt., 5⅜ lbs **$1.05**

28L11848—(Ford No. 2804) Outer Race. Shpg. wt., 4 oz. **4c**

28L11849—(Ford No. 2805) Inner Race. Shpg. wt., 6 oz. . . . **9c**

28L11850—(Ford No. 2806) Outer Ball Retainer. (Not illustrated.) Shipping wt., 1 oz. **2c**

28L11859—(Ford No. 2807) Inner Ball Retainer. (Not illustrated.) Shipping wt., 2 oz. **2c**

28L14010—(Ford No. 2808) Dust Ring. (Not illustrated.) Shipping weight, 2 ounces. **2c**

28L11449—(Ford No. 2810) ½-Inch Balls. Shipping weight, per dozen, 5 ounces. Per dozen. . . . **19c**

28L11447—(Ford No. 2811) ⅜-Inch Balls. Shipping weight, per dozen, 3 ounces. Per dozen. . . . **10c**

28L12582—(Ford No. 2815C) Rear Hub Assembly. Complete, without hub flange, bolts or nuts. Shipping weight, 5½ pounds. **84c**

28L14012—(Ford No. 2818) Rear Hub Brake Drum. (Not illustrated.) Shipping wt., 3½ lbs. . **25c**

28L11737—(Ford No. 2819) Hub Cap. Brass finish. **9c**

28L11090—(Ford No. 2819). Same, nickel plated. **9c**

Shpg. wts., 6 and 4 oz., respectively.

28L12580—(Ford No. 2827) Hub Flange, without bolts or nuts. (Not illus.) Shpg. wt., 1⅝ lbs . **10c**

Starting Crank.

3900 3901 3906

28L14231 (Ford No. 3900) Starting Crank only. Shipping weight, 1¾ pounds. **37c**

28L14232—(Ford No. 3901) Starting Crank Handle. Shipping weight, 7 ounces **10c**

28L14235—(Ford No. 3906) Starting Crank Ratchet. Shipping weight, 9 ounces. **12c**

Spring Parts.

3808 3848 3818B 3843 3813 3833

28L14218—(Ford No. 3808) Front Spring Clip. Shpg. wt., 11 oz. **13c**

28L14221—(Ford No. 3813) Front Spring Hanger. Shpg. wt., 6 oz. **9c**

28L14222—(Ford No. 3815.) Front and Rear Spring Hanger Nut. (Not illustrated.) Shpg. wt., 2 oz. **2c**

28L13045—(Ford No. 3818B) Front Spring Perch, Right, 1919-23 models. Shpg. wt., 1⅜ lbs. **37c**

28L13043—(Ford No. 3819B) Front Spring Perch, Left, 1919-23 models. (Not illustrated.) Shipping weight, 1⅜ lbs. **37c**

28L1514—(Ford No. 3833) Rear Spring Clip. Shpg. wt., 13 oz. . . . **16c**

28L14227—(Ford No. 3840) Rear Spring Hanger. (Not illustrated.) Shipping weight, 7 ounces. **8c**

28L14228—(Ford No. 3843) Rear Spring Perch. Shpg. wt., 14 oz. **20c**

28L14229—(Ford No. 3845) Rear Spring Perch Nut. (Not illustrated.) Shipping weight, 3 oz. **4c**

28L14256—(Ford No. 3844) Rear Spring Perch Bushing. (Not illustrated.) Shipping weight, 3 oz. **3c**

28L14230—(Ford No. 3848) Rear Spring Clip Assembly with Bolt and Nut. Shipping wt., 5 oz. **4c**

Steering Gear.

3517 3519 3547

28L14205—(Ford No. 3506) Steering Gear Cover Bushing. (Not illustrated.) Shipping wt., 2 oz. . . **2c**

28L14208—(Ford No. 3517) Steering Gear Pinion. Shpg. wt., 2 oz. **9c**

28L14210—(Ford No. 3519) Steering Gear Drive Pinion. Shipping weight, 8 ounces. **19c**

28L14212—(Ford No. 3534B) Commutator Pull Rod. (Not illustrated.) Shipping weight, 8 oz. **4c**

28L11727—(Ford No. 3536) Ball and Socket Joint, with nut. (Not illustrated.) Shipping weight, 2 oz. **5c**

28L14215—(Ford No. 3545) Steering Post Bracket Bushing. (Not illustrated.) Shpg. wt., 6 oz. 2 pieces. **7c**

28L14216—(Ford No. 3547) Steering Gear Ball Arm. Shipping weight, 1 pound **28c**

Rear Axle Parts

2508 2509 2587 2589B 2597B

The repair parts for Ford cars shown on this page duplicate the original Ford parts. They are guaranteed perfect in fit and workmanship. The Ford part number given in each case serves to identify the part.

Note the remarkably low money saving prices we quote on these parts in every instance.

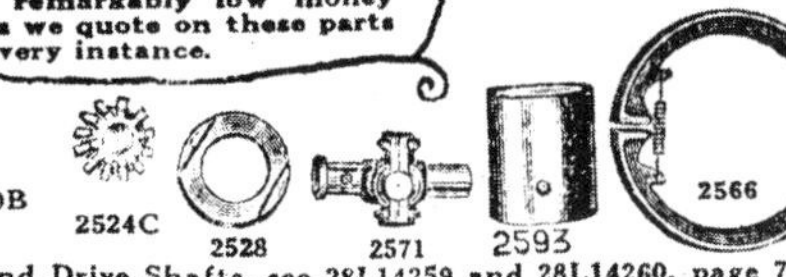

For Ideal Roll-Ball Bearings for Axle and Drive Shafts, see 28L14259 and 28L14260, page 760.

28L11504—(Ford No. 2508) Shaft Roller Bearing. Shipping weight, 1¼ pounds. **88c**

28L11505—(Ford No. 2509) Right Sleeve for above. Shipping weight, 8 ounces. **11c**

28L11508—(Ford No. 2509B) Left Sleeve for above. (Not illustrated.) Shpg. wt., 8 oz. **11c**

28L11510—(Ford No. 2510A) Outer Roller Bearing Washer. (Not illus.) Shpg. wt., 2 oz. . **2c**

28L11080—(Ford No. 2512C) Differential Case, Left. (Not illustrated.) Shpg. wt., 4½ lbs. **92c**

28L12575—(Ford No. 2518) Differential Drive Gear. Shipping weight, 3¼ lbs. **$1.84**

28L11512—(Ford No. 2519) Screw for same. (Not illustrated.) Shipping weight, 2 oz. **3c**

28L11078—(Ford No. 2520B) Differential Gear. Shipping weight, 1⅜ pounds. **84c**

28L11517—(Ford No. 2524C) Differential Pinion. Shipping weight, 8 ounces. **23c**

28L11518—(Ford No. 2528) Differential Thrust Washer. Shipping weight, 6 ounces . . **10c**

28L12535—(Ford No. 2528B) Differential Thrust Washer. Bronze. (Not illustrated.) Shipping weight, 8 ounces. **23c**

28L11519—(Ford No. 2529) Differential Thrust Plate. Shpg. wt., 3 oz. **6c**

28L11520—(Ford No. 2531) Differential Thrust Plate Pin. (Not illustrated.) Shpg. wt., 2 oz. **2c**

28L10074¼—(Ford No. 2547) Rear Radius Rod. (Not illustrated.) Shpg. wt., 4½ lbs. **$1.35**

28L11528—(Ford No. 2566) Hub Brake Shoe. Shipping weight, 1¾ pounds. **19c**

28L10969—(Ford No. 2571) Universal Joint Assembly. Shipping weight, 2¾ pounds. . . **$1.09**

28L11544—(Ford No. 2583B) D. S. Roller Bearing Housing. (Not illus.) Shpg. wt., 2¾ lbs. **98c**

28L11082—(Ford No. 2587) D. S. Roller Bearing. Shipping weight, 1¼ pounds. **79c**

28L11084—(Ford No. 2589B) D. S. Ball Bearing Assembly. Shipping weight, 9 oz. . . . **36c**

28L10963—(Ford No. 2593) D. S. Roller Bearing Sleeve. Shipping weight, 13 oz. **84c**

28L10964—(Ford No. 2596) D. S. Sleeve. (Not illustrated.) Shipping weight, 6 ounces. . **29c**

28L10965—(Ford No. 2597B) D. S. Pinion. Shipping weight, 12 ounces. **50c**

Front Axle Parts.

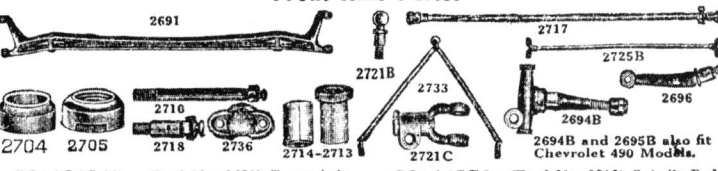

2691 2721B 2717 2725B 2733 2696 2704 2705 2716 2718 2736 2714-2713 2721C 2694B

2694B and 2695B also fit Chevrolet 490 Models.

28L10124¼—(Ford No. 2691) Front Axle only. Shpg. wt., 18½ lbs. (Unmailable.) **$5.60**

28L10634—(Ford No. 2694B) Spindle Body, Right, 1911-1923 models. **90c**

28L10643—(Ford No. 2695B) Spindle Body, Left, 1911-1923 models. (Not illustrated.) **90c** Shipping weight, either of above, 3½ pounds.

28L10641—(Ford No. 2696) Spindle Arm, Right or Left, up to and including 1919 models. Shipping weight, 1 pound. **25c**

28L13494—(Ford No. 2696C) Spindle Arm, Right, for 1920-23 models. Shpg. wt., 1 lb. . **27c**

28L13495—(Ford No. 2696D) Spindle Arm, Left, for 1920-23 models. Shpg. wt., 1 lb. . **27c**

28L11839—(Ford No. 2704) Stationary Cone. Shipping weight, 5 ounces. **8c**

28L10954—(Ford No. 2705) Adjusting Cone, Right Thread. Shpg. wt., 3 oz. **6c**

28L10956—(Ford No. 2706) Adjusting Cone, Left Thread. (Not illus.) Shpg. wt., 3 oz. **6c**

28L11892—(Ford No. 2710) Spindle Bolt with Oiler. Shipping weight, 10 ounces. **9c**

28L10676—(Ford No. 2711) Spindle Bolt Nut. (Not illustrated.) Shpg. wt., 2 oz. **3c**

28L14254—(Ford No. 2713) Spindle Body Bushing (pair). Shipping weight, 5 ounces. **10c**

28L11658—(Ford No. 2714) Spindle Arm Bushing. Shipping weight, 2 ounces. **3c**

28L10856¼—(Ford No. 2717) Spindle Connecting Rod. Shpg. wt., 3½ lbs. (Unmailable.) **$1.25**

28L11893—(Ford No. 2718) Spindle Connecting Rod Bolt with Oiler. Shpg. wt., 4 oz. . **9c**

28L14001—(Ford No. 2719) Spindle Connecting Rod Bolt Nut. (Not illus.) Shpg. wt., 2 oz. **2c**

28L1502—(Ford No. 2721B) Spindle Connecting Rod Yoke Ball. Shpg. wt., 5 oz. **13c**

28L14003—(Ford No. 2721C) Spindle Connecting Rod Yoke. Shipping weight, 8 oz. . **26c**

28L11867—(Ford No. 2725B) Steering Gear Connecting Rod. Shpg. wt., 2 lbs. . . . **55c**

28L10076¼—(Ford No. 2733) Front Radius Rod. (For models up to and including 1919.) Shipping weight, 7½ lbs. (Unmailable.) **$1.10**

28L10062¼—(Ford No. 2733B) Same for 1920-23 models. (Not illustrated.) Shipping weight, 7¾ pounds. (Unmailable.) **$1.00**

28L14007—(Ford No. 2736) Front Radius Rod Ball Cap. Shipping weight, 4 ounces. **6c**

Motor Parts

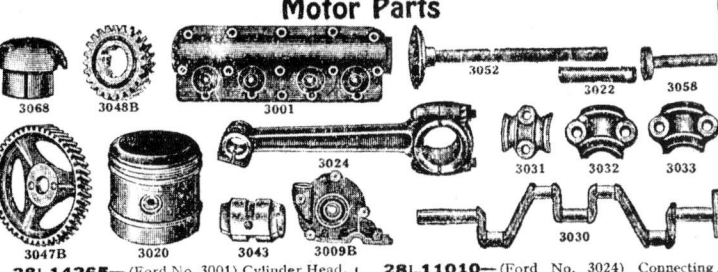

3068 3048B 3001 3052 3022 3058 3024 3031 3032 3033 3047B 3020 3043 3009B 3030

28L14265—(Ford No. 3001) Cylinder Head. Shipping wt., 30¾ pounds. **$4.00**

28L14019—(Ford No. 3004B) Cylinder Head Outlet Connection. For 1917-1923 models. (Not illustrated.) Shpg. wt., 2 lbs. **22c**

28L14023—(Ford No. 3009B) Cylinder Front Cover, 1909-19 models. Shpg. wt., 4½ lbs. **60c**

28L13497—(Ford No. 3009C) Cylinder Front Cover, 1919-23 models, starter type. (Not illustrated.) Shipping weight, 4¼ pounds. . . **60c**

28L13496—(Ford No. 3014) Cylinder Cover and Crank Case Bolt and Nut. (Not illustrated.) Shipping weight, 2 ounces **4c**

28L14026—(Ford No. 3015) Cylinder Water Inlet Connection. (Not illus.) Shpg. wt., 1¼ lbs. **18c**

28L14029—(Ford No. 3020.) Light Weight Piston, standard size **$1.15**

28L14030—(Ford No. 3020C). Same, .0025 oversize **$1.15**

28L11558—Same, .005 oversize. . . . **1.15**

28L11557—Same, .010 oversize. . . . **1.15**

28L11556—Same, .025 oversize. . . . **1.15**

28L14031—(Ford No. 3020D). Same, .03125 oversize **$1.15**

Shpg. wt., any of above, 3⅛ lbs.; 28L14029, 3 lbs.; 28L14031, 3⅜ pounds.

28L14032—(Ford No. 3022) Piston Pin. Shipping weight, 5 ounces. **11c**

28L14255—(Ford No. 3022½) Piston Pin Bushing (pair). (Not illustrated.) Shipping weight, 4 ounces. Per pair. **8c**

28L11010—(Ford No. 3024) Connecting Rod. Shipping weight, 2 pounds. **90c**

28L14266—(Ford No. 3030) Crank Shaft, Chrome Vanadium steel. Shpg. wt., 18½ lbs. **$6.50**

28L14035—(Ford No. 3031) C. S. Rear Bearing Cap. Shipping weight, 1⅛ lbs. **37c**

28L14036—(Ford No. 3032) C. S. Front Bearing Cap. Shipping weight, 12 ounces. . . **25c**

28L14037—(Ford No. 3033) C. S. Center Bearing Cap. Shipping weight, 12 ounces. . . **25c**

28L14040—(Ford No. 3043) Cam Shaft Center Bearing. Shipping weight, 8 oz. . . . **20c**

28L14041—(Ford No. 3044) Cam Shaft Rear Bearing. (Not illustrated.) Shipping weight, 4 ounces. **7c**

28L1504—(Ford No. 3047B) Time Gear, large. Shipping weight, 2 pounds. **63c**

28L1506—(Ford No. 3048B). Same, small. Shipping weight, 1 pound. **51c**

28L11813—(Ford No. 3052) Valve. Shipping weight, 5 ounces. **8c**

28L14045—(Ford No. 3058) Push Rod. Shipping weight, 3 ounces. **7c**

28L14047—(Ford No. 3063) Inlet and Exhaust Pipe Gasket. (Not illustrated.) Shipping weight, 1 ounce **2c**

28L12564—(Ford No. 3068) Breather Pipe Cap. Shipping weight, 3 ounces. **6c**

For other Engine Accessories and Repair Parts see page 755.

Transmission Parts.

3301 3306 3314½ 3304 3309 3311

28L14058—(Ford No. 3301) Transmission Reverse Plate Assembly. Shpg. wt., 6 lbs. . . **$1.89**

28L1508—(Ford No. 3304) Transmission Reverse Gear Bushing. Shipping weight, 11 ounces. **33c**

28L14059—(Ford No. 3306) Transmission Slow Speed Plate Assembly. Shpg. wt., 5½ lbs. **$1.73**

28L1510—(Ford No. 3309) Transmission Slow Speed Gear Bushing. Shipping weight, 7 ounces. **24c**

28L14060—(Ford No. 3311) Transmission Brake Drum Assembly. Shipping weight, 8 pounds. . . **$2.00**

28L14061—(Ford No. 3314½) Transmission Triple Gear Flanged Bushing. Shipping wt., 5 oz. . . . **16c**

28L14067—(Ford No. 3328) Transmission Clutch Disc, small. (Not illus.) Shpg. wt., 8 oz. . . . **5c**

28L14068—(Ford No. 3329) Transmission Clutch Disc, large. (Not illus.) Shpg. wt., 8 oz. . . . **5c**

28L14074—(Ford No. 3362) Transmission Cover Bolt. (Not illustrated.) Shipping wt., 2 oz. . . . **3c**

28L14084—(Ford No. 3413) Transmission Band Assembly. (Not illustrated.) Shpg. wt., 1¼ lbs. . **51c**

28L14087—(Ford No. 3425) Transmission Band Spring. (Not illustrated.) Shpg. wt., 2 oz. . . . **3c**

28L14099¼—(Ford No. 3468) Hub Brake Pull Rod, Right, and (Ford No. 3469) Hub Brake Pull Rod, Left. (Not illus.) Shpg. wt., 1¼ lbs. **16c**

Hood.

Duplicates the 1917-23 Ford hood. Comes in baked black enamel finish. Shipping weight, 25 pounds.

28L15045¼—Hood for 1917-1923 Model Ford cars. **$3.75**

Fan Parts.

3960 3962 3961C 3963 3962B

28L14238—(Ford No. 3960) Fan and Pulley Assembly, 1909-1920. Shipping weight, 3¼ pounds. . . . **61c**

28L14263—(Ford No. 3960B) Fan and Bracket Assembly, 1920-23 models. (Not illus.) Shpg. wt., 3½ lbs. **78c**

28L14239—(Ford No. 3961) Fan Blade, 1909-20. (Not illustrated.) Shipping weight, 4 ounces. **3c**

28L14240—(Ford No. 3961B) Fan Blade Rivets, 1 dozen. (Not illustrated.) Shipping weight, 2 oz. . . . **3c**

28L14264—(Ford No. 3961C) Fan Blade Assembly, 1920-23 models. Shipping weight, 1 pound **10c**

28L14241—(Ford No. 3962) Driven Fan Pulley, 1909-20. Shipping weight, 1⅜ pounds. **50c**

28L13498—(Ford No. 3962B) Driven Fan Pulley with Bushing, 1920-23 models, starter type. Shipping weight, 2 pounds. **45c**

28L14242—(Ford No. 3963) Fan Drive Pulley, 1909-20. Shipping weight, 11 ounces. **13c**

28L14261—(Ford No. 3963B) Fan Drive Pulley, 1920-23 models. (Not illustrated.) Shpg. wt., 1 pound. . . . **16c**

28L14243—(Ford No. 3966) Fan Shaft, 7⁄16x4⁹⁄16 inches, 1909-20. (Not illustrated.) Shipping wt., 5 oz. . . . **10c**

28L14262—(Ford No. 3966B) Fan Shaft, 1920-23 models. (Not illustrated.) Shipping weight, 6 oz. . . . **12c**

28L14244—(Ford No. 3967B) Fan Bracket, 1917-23. (Not illustrated.) Shipping weight, 12 oz. . . **18c**

28L14245—(Ford No. 3970) Fan Grease Cup. (Not illustrated.) Shipping weight, 2 ounces. **6c**

Muffler.

4037B

28L14247¼—(Ford No. 4037B) Long Exhaust Pipe, 1909-20 models. Shpg. wt., 2½ lbs. (Unmailable.) **65c**

Baseball

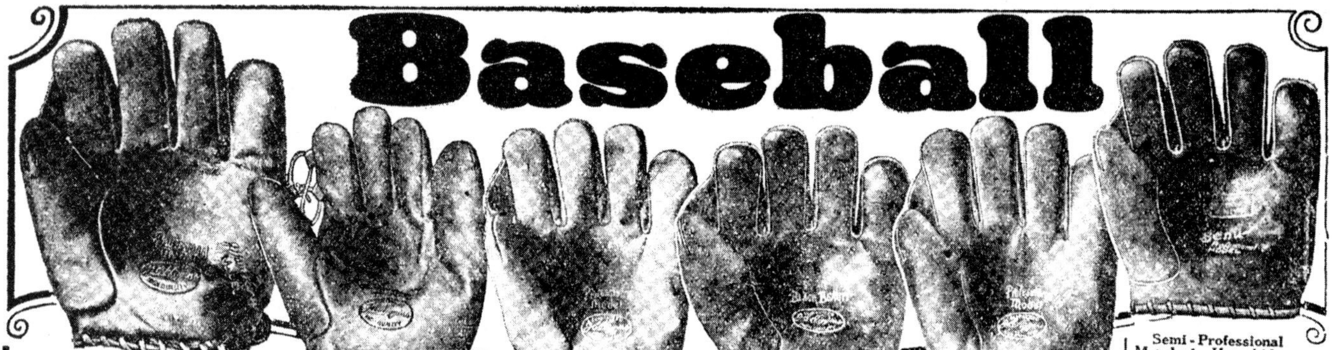

"J. C. Higgins" Professional "Wagner" Model Glove. Highest quality tan color oil treated horsehide, soft and pliable. Extra large pattern, first quality felt padding, a deep natural pocket, ready broke in; leather bound, leather welted seams, rawhide leather lace at wrist. Full leather lined. Shpg. wt., 1½ lbs.
6L1666 To wear on left hand **$4.20**
6L1667 To wear on right hand; for left handed throwers **$4.20**

"J. C. Higgins" Laced Thumb Fielders' Glove. Professional short fingered model. Laced between thumb and forefinger. Made of excellent quality tan horsehide, felt padded, full lined with soft glove leather, welted seams, leather bound, rawhide lace at wrist. Shpg. wt., 1½ lbs.
6L1676—To wear on left hand **$3.60**
6L1677—To wear on right hand; for left handed throwers **$3.60**

"J. C. Higgins" Professional Model Glove. Standard in quality and workmanship. One of the lastest professional model gloves. Excellent quality buffed drab color pliable horsehide, leather lined. Leather bound, welted seams, rawhide lace at wrist, strap and button at wrist, natural deep pocket. Shpg. wt., 1½ lbs.
6L1664—To wear on left hand **$3.30**
6L1665—To wear on right hand, for left handed throwers **$3.30**

"J. C. Higgins" "Black Beauty" Laced Fielders' Glove. "Professional model." Made of the best quality black flexible horsehide throughout, full lined with soft glove leather, welted seams, first quality felt padding, leather bound, rawhide lace at wrist. Shpg. wt., 1½ lbs.
6L1668—To wear on left hand **$3.10**
6L1669—To wear on right hand; for left handed throwers **$3.10**

Professional Model Laced Glove. Made of good quality light tan horsehide, lined with soft glove leather, felt padded, welted seams, leather bound, well stitched, deep pocket, rawhide lace at wrist. Shpg. wt., 1½ lbs.
6L1670—To wear on left hand **$2.75**
6L1671—To wear on right hand; for left handed throwers **$2.75**

Semi-Professional Model Horsehide Glove. Full size black horsehide glove, felt padding, full leather lined, deep pocket, leather welted seams, strap and button at wrist, fabric bound, leather laced wrist. Exceptional value. Shipping weight, 1½ pounds.
6L1672 To wear on left hand **$2.15**
6L1673—To wear on right hand; for left handed throwers **$2.15**

The J.C.Higgins League Ball

Guaranteed for a full game up to eighteen innings against ripping, tearing or losing its shape, if not played with when wet. Rubber center, wool yarn wound, specially tanned selected horsehide covering, sewed with strong thread. It carries our own J. C. Higgins trade mark as an absolute guarantee. Each ball wrapped in tisue paper and tinfoil and packed in individual sealed box. Sold only by us. Shipping weight, each, 8 ounces.
6L1600—Each **$1.25**
Per half dozen, **$7.25**; per dozen **14.25**

Professional League Ball. Fine quality horsehide cover; rubber center, yarn wound. Guaranteed for nine innings. Size, 9 inches. Weight, 5 oz. Shpg. wt., 8oz.
6L1605
98c

Boys' League Ball. Good grade horsehide cover; rubber center, yarn wound. Guaranteed for nine innings. Size, 8½ inches. Weight, 4½ oz. Shpg. wt., 8 oz.
6L1606
80c

Pitchers' Pride Ball. Good grade horsehide cover, well sewed; rubber center. An excellent ball for boys. Size, 9 in. Weight, 5 oz. Shpg. wt., 8 oz.
6L1607
55c

Boys' Winner Ball. Genuine leather covered. A splendid ball for small boys. Shipping weight, 7 ounces.
6L1609 ..... **35c**

Junior Ball. Artificial leather cover, well stitched. A soft ball for small boys and girls. Shpg. wt., 6 oz.
6L1610 ..... **14c**

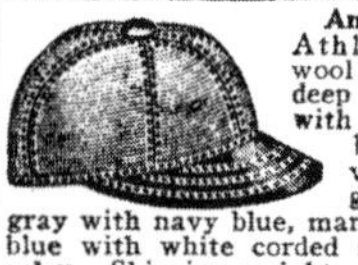

Amateur Baseball Cap. Athletic flannel, one-third wool and two-thirds cotton; deep crown, Brooklyn style, with corded seams and buckram unbreakable visor. Colors: Light gray with maroon, light gray with navy blue, maroon with white or navy blue with white corded seams. **State size and color.** Shipping weight, 4 ounces.
6L1778 ..... **65c**

Professional Baseball Cap. One-half wool and one-half cotton athletic flannel. New York style, with silk corded seams and buckram unbreakable visor; deep ventilated crown. Colors: Light gray with maroon, light gray with navy blue, maroon with white or navy blue with white corded seams. **State size and color wanted.** Shipping wt., 4 ounces.
6L1779 ..... **98c**

Sun Vision Visor Baseball Cap. Latest style baseball cap. Gives ample room to properly see through visor and still protect the eyes from the sun glare. Adaptable to any player's position. Can be readily adjusted for the catching of fly balls or resting the eyes. Cutout is fitted with green celluloid set into an unbreakable buckram visor. Extra deep crown, professional model, of fine quality athletic flannel, three-fourths wool. Ventilated sewed eyelets. Furnished in plain navy blue and plain black only. **State size.** Shipping weight, 4 ounces.
6L1775—Plain navy blue **$1.50**
6L1776—Plain black **1.50**

READY TO WEAR

Baseball Uniforms

Carried in Stock for Immediate Shipment.

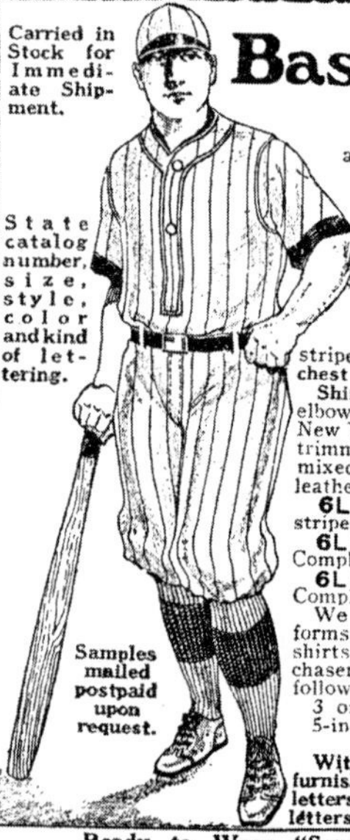

State catalog number, size, style, color and kind of lettering.

Samples mailed postpaid upon request.

Major League Baseball Uniforms.

To enable us to give our customers lower prices and quick service, we have discontinued our Made-to-Order Uniforms and carry in stock ready for immediate shipment, high grade uniforms.

Professional in every respect, in material, style and workmanship. Extra heavy athletic flannel, 13 ounces to the double width, about half wool, thereby materially adding to the wearing qualities. Furnished in three patterns: Oxford gray with a green and wine outline stripe, Yale gray with a maroon and navy blue outline stripe, and white with navy blue stripe. **Sizes, 34 to 44 inches chest measure. Give chest measure and size of cap.** Shpg. wt., 3 lbs.

Shirt—Professional, V neck, button front, trimmed elbow sleeves. **Pants**—Unlined, tunnel loops. **Cap**—New York style, deep crown, visor and button to match trimmings, unbreakable buckram visor. **Hose**—Wool mixed in colors to match uniform. **Belt**—1¼-inch leather with nickel buckle and loop.

6L1782—Yale gray with maroon and navy blue stripes. Complete **$8.95**
6L1783—Oxford gray with green and wine stripes. Complete **$8.95**
6L1785—White with navy blue stripes. Complete **$8.95**

We can furnish felt letters only when baseball uniforms are ordered from us. We do not sew letters on shirts, but they can easily be sewed on by the purchaser. Colors, navy blue, green or maroon, in the following styles:
3 or 4-inch plain block letters. Each **5c**
5-inch Old English letters. Each **15c**

Special Offer on Club Orders.
With an order for nine or more uniforms we will furnish, without extra charge, any name in felt letters, not to exceed 12 block letters or 2 Old English letters to a uniform.

Ready to Wear "Semi-Pro League" Baseball Uniforms.
Made of especially woven athletic flannel, two-thirds cotton, one-third wool, with broad woven stripe. Furnished in two patterns, oxford gray with green stripes, Yale gray with navy blue stripes. Made strictly along professional lines, embodying every feature needed to insure comfort. **Sizes, 34 to 42 inches chest measure. Give chest measure and size of** g. wt., 2½ lbs.
Shirt—V neck with trimmed insert, two rows cordage on front and elbow sleeves. **Pants**—Tunnel loops, peg style. **Cap**—Deep crown, Brooklyn style, corded seams, with unbreakable buckram visor. **Cotton Hose**—Oxford gray with stripe to match trimmings. **Belt**—1¼-inch leather.
6L1786—Yale gray with navy blue stripes. Complete **$6.50**
6L1781—Oxford gray, with green stripes. Complete **6.50**

Boys' Ready to Wear Baseball Uniforms.
Made of a special athletic flannel, 15 per cent wool, strong and durable, along professional lines. Oxford gray with blue stripes. Consists of shirt, pants, leather belt, hose and cap. Same style as our "Semi-Pro League" Uniforms. Will please the youngsters. **State chest measure and size of cap.** Shpg. wt., 2½ lbs.
6L1787—Complete sizes, 28 to 36 chest measure **$4.95**

Juvenile Baseball Uniform. Big League Suits for Little Fellows.
Made of cotton baseball flannel, oxford gray with navy blue stripes, along professional lines, embodying every feature of big league suits, in style, workmanship and appearance. Consists of shirt, pants, leather belt and cap. **Ages, 4 to 12 years. State age and size of cap.** Shipping weight, 2 pounds.
6L1784—Complete as above **$3.25**

Athletic Footless Stockings. Suitable for athletes of all kinds. Sanitary, and cotton hose can be used under them. Shpg. wt., per pair, 8 oz.
Heavy Ribbed Cotton Stockings. Colors: Black, navy blue or maroon.
6L2053—Pair **62c**
Same as 6L2053, with a 3-inch white calf stripe; also black with 3-in. orange stripe or navy blue with 3-in. orange stripe, oxford gray with either navy, maroon or green stripe. **State color.**
6L2054—Per pair **64c**
Half Wool and Half Cotton Double Ribbed Stockings. Colors: Black, navy blue or maroon. **State color.**
6L2062—Pair **$1.10**
6L2063—Same as 6L2062, with 3-in. white calf stripe. Also black with 3-in. orange stripe and oxford gray with either navy blue, maroon or green stripe. **State color.** Per pair **$1.12**
All Wool Worsted Heavy Ribbed Stockings, Professional style. Colors: Navy blue, with 5-inch white shoe top; black, with 5-in. white shoe top; maroon, with 5-inch white shoe top, and plain oxford gray.
6L2058—Per pair **$1.60**
For Athletic Stockings with Feet refer to page 769.

Professional Shoe Plates. Made with tempered steel prongs and beveled edge; light weight, very strong. Complete with screws. Shpg. wt., per pair, 3 oz.
6L1730—Professional League Steel Heel Plates. Per pair. **25c**
6L1731—Professional League Steel Heel Plates. Per pair. **25c**

Amateur Shoe Plates, complete with screws.
6L1732—Per pair **15c**
Amateur Heel Plates.
6L1733—Per pair **15c**
Shipping wt., per pair, 3 oz.

Pitchers' Toe Plates.

Pitchers' Aluminum Toe Plate with screws, for right or left foot. Shpg. wt., 2 oz.
6L1734—For right foot. **22c**
6L1735—For left foot. **22c**

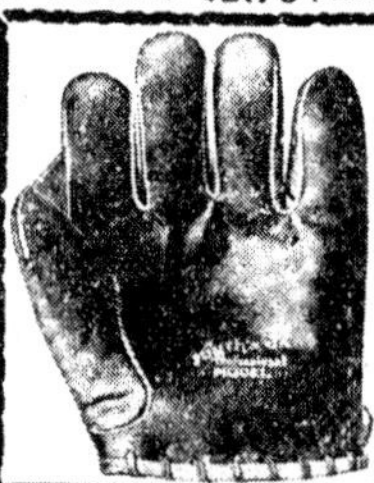

Junior Professional Laced Glove. Tan color, soft glove leather, good quality felt padding; full leather lined, welted seams, laced wrist, fabric bound, well padded. Full size. Shipping weight, 9 oz.
6L1674—To wear on left hand **$1.75**
6L1675—To wear on right hand; for left handed throwers **$1.75**

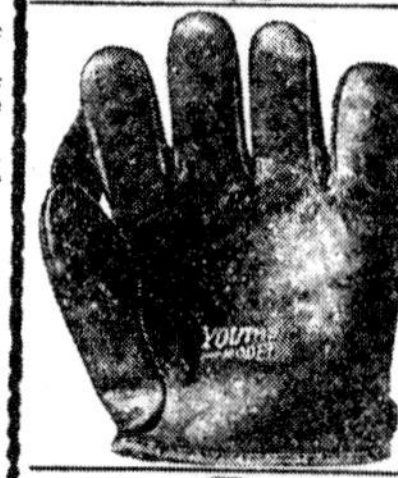

Youths' Glove. Large size, napa tanned soft glove leather, good quality felt padding, palm and fingers leather lined, fabric bound, very serviceable glove for the young fellow. Shipping weight, 8 ounces.
6L1680
To wear on left hand **$1.25**
6L1681 — Right hand; for left handed throwers **$1.25**

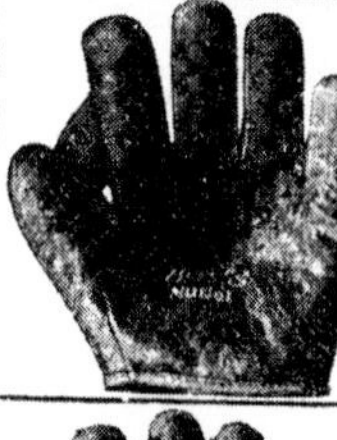

Boys' Large Size Glove. Soft glove leather, felt padding, palm leather lined, fabric bound edge. Excellent value. Shpg. wt., 6 oz.
6L1684
To wear on left hand **69c**
6L1685
To wear on right hand; for left handed throwers **69c**

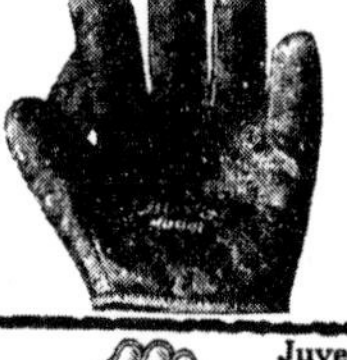

Boys' All Leather Glove. Good quality soft glove leather, felt padding, palm leather lined. Shpg. wt., 6 oz.
6L1686
To wear on left hand **48c**
6L1687
To wear on right hand for left handed throwers **48c**

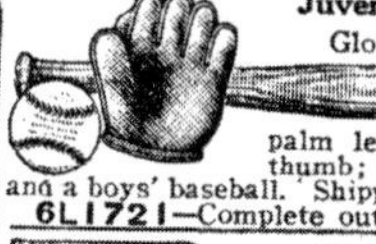

Juvenile Baseball Outfit. Glove is made of fine soft leather, felt padded; palm leather lined, with web thumb; a fine hardwood bat and a boys' baseball. Shipping weight, 2 lbs. 3 oz.
6L1721—Complete outfit **$1.25**

Official Baseball Score Book for 22 games; cloth covered. Shipping weight, 2 ounces.
6L1743 ..... **45c**

Official Baseball Rules and Records.
Not issued before April 1st of each year. Shipping weight, 6 ounces.
6L1765 ..... **14c**

HANDBALLS.

Leather Covered Handball. Rubber center, yarn wound, horsehide cover, hand stitched. Regulation size and weight. Shipping weight, 5 oz.
6L1297 ..... **33c**

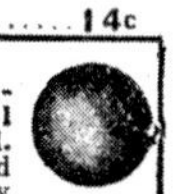

Professional Handball. Good quality black rubber. Shipping weight, 3 ounces.
6L1298 .. **29c**

Handball Rule Book. Shipping wt., 4 ounces.
6L1839 ..... **14c**

Supplies

Boys' All Leather Catcher's Mitt. Made of soft glove leather throughout; well padded and well stitched. A good serviceable mitt for the youngsters. Shipping weight, 12 ounces.

6L1650—To wear on left hand..........98c

6L1651—To wear on right hand..........98c

Youths' Large Size Laced Mitt. Improved model. Good quality soft glove leather throughout; well padded; deep pocket; laced edge; well stitched throughout. Adjustable thumb strap and buckle at wrist. Shpg. wt., 1⅝ lbs.

6L1648—To wear on left hand..........$1.75

6L1649—To wear on right hand..........$1.75

Amateur Model Mitt. Improved this season with black horsehide palm, back and fingers of soft glove leather, well padded; deep pocket, full bound laced edge, adjustable thumb strap and buckle at wrist. Will give excellent service. Shpg. wt., 2¼ lbs.

6L1642—To wear on left hand..........$2.40

6L1643—To wear on right hand; for left handed throwers..........$2.40

Semi-Professional Model Mitt. High quality brown color cowhide palm; back and fingers are made of soft tan glove leather, special felt and hair padding, deep pocket, leather laced, wrist protector. Well sewed. Shipping weight, 2⅝ pounds.

6L1638—To wear on left hand..........$3.85

6L1639—To wear on right hand; for left handed throwers..........$3.85

The J. C. Higgins Professional Black Mitt. High quality black horsehide used throughout, leather bound edge, full leather laced, patent wrist protector. Felt padding is molded into shape, giving the mitt a deep pocket. Shpg. wt., 3 lbs.

6L1634—To wear on left hand..........$5.45

6L1635—To wear on right hand; for left handed throwers..........$5.45

Our J. C. Higgins Best Professional Model Mitt.

First quality, dark tan color, pliable cowhide leather used throughout. Padded with standard felt, hand molded, natural pocket, welted and leather bound, full leather laced edge, double stitched throughout, patent wrist protector. Shpg. wt., 3 lbs.

6L1632—To wear on left hand..........$7.25

6L1633—To wear on right hand; for left handed throwers..........7.25

1st Basemen's Mitt

"J. C. Higgins" Professional Model Mitt. Excellent quality dark tan pliable horsehide, palm and back leather lined, standard felt padding, leather bound and leather laced, well sewed, natural deep pocket, patent wrist protector. Shipping weight, 1½ pounds.

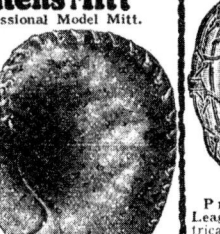

6L1656 To wear on left hand....$4.50

6L1657 To wear on right hand $4.50

"Black Beauty" First Basemen's Mitt. High quality black horsehide leather palm, soft glove leather back, leather lined, leather bound and well stitched, felt padded. Rawhide laced edge. Patent wrist protector. Shipping weight, 1 lb. 9 oz.

6L1658 To wear on left hand..........$3.20

6L1659 To wear on right hand..........$3.20

Youths' Large Size First Basemen's Mitt. Good quality soft and pliable glove leather finger back, laced edge, well padded. A good, serviceable mitt for youths. Shipping weight, 1¼ pounds.

6L1660 To wear on left hand..........$1.75

6L1661 To wear on right hand..........$1.75

-Mask-

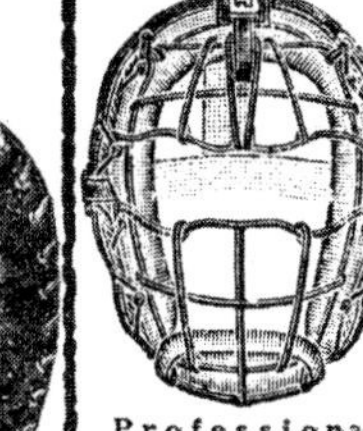

Professional League Mask. Electrically welded clear vision frame, dull black enameled finish. Hair stuffed head, cheek and chin pads laced to frame. Shipping wt., 2 lbs.

6L1694..$2.45

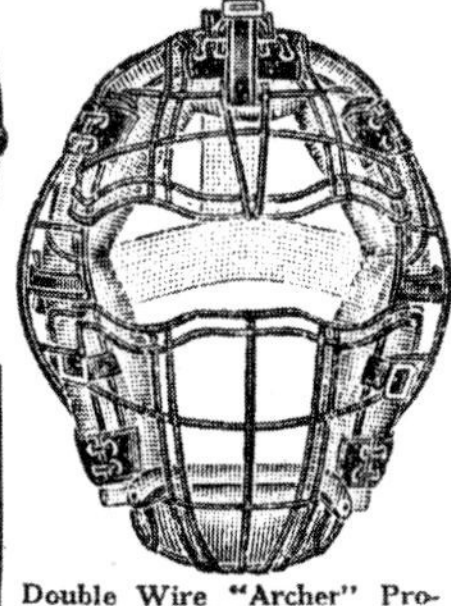

Double Wire "Archer" Professional Model Mask. This is a duplicate of the model used by the big league catchers, close to the face, full open vision and ear protectors. Made of high quality dull black enameled wire. All joints securely electrically welded. Hair filled leather side, forehead and cheek pads laced to frame. Adjustable elastic head strap. Shipping weight, 3 lbs.

6L1691..........$4.48

Sun Shield Mask. Made of dull black enameled wire, electrically welded joints, clear vision frame, hair stuffed head, cheek and chin pads laced to frame. Fiber shield to protect eyes from sun and bright light is laced to frame. Shipping weight, 3 pounds.

6L1692..........$3.65

Catchers' and Umpires' Neck Protecting Mask. Made of dull black enameled wire, electrically welded joints, clear vision frame, hair stuffed head, cheek and chin pads laced to the frame, elastic head strap. Shipping weight, 3¾ pounds.

6L1690..$3.35

Umpires' Indicator.

Made of white celluloid. Size, 3x1½ in. Indorsed by the league umpires. Shipping wt., 3 oz.

6L1745..........55c

Winding Tape.

For taping handles of baseball bats and tennis rackets. One roll will tape either bat or racket handle. Shpg. wt., 2 oz.

6L1744—3 rolls..........10c

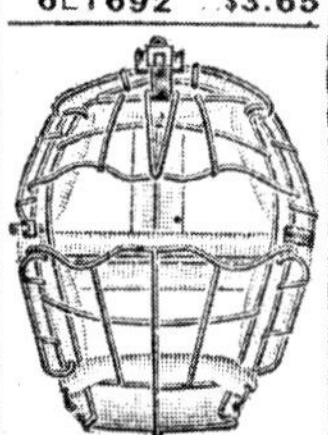

Semi-Professional Mask. Electrically welded clear vision frame, dull black enameled finish, leather forehead and cheek pads; padded chin piece laced to frame, elastic head strap. Shipping weight, 1½ pounds.

6L1695....$2.00

Our Bats are made specially for us by the makers of the famous Louisville Slugger Bats.

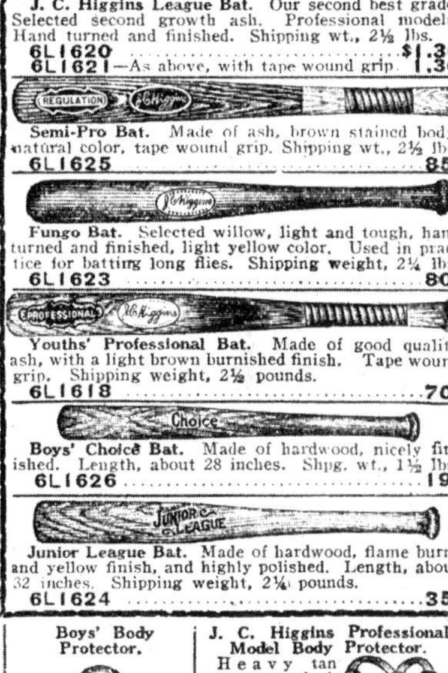

J. C. Higgins "Slugger" Professional Models. Our best grade bat, same as used in big leagues. Thoroughly seasoned second growth ash. Hand rubbed and finished in oil. A light yellow color. Shipping weight, 2⅝ pounds.

6L1619..........$1.60

J. C. Higgins "Slugger" Bat. Same as above, but natural finish. Highly polished. Shpg. wt., 2½ lbs.

6L1617..........$1.60

J. C. Higgins League Bat. Our second best grade. Selected second growth ash. Professional models. Hand turned and finished. Shipping wt., 2½ lbs.

6L1620..........$1.30

6L1621—As above, with tape wound grip. 1.30

Semi-Pro Bat. Made of ash, brown stained body, natural color, tape wound grip. Shipping wt., 2½ lbs.

6L1625..........85c

Fungo Bat. Selected willow, light and tough, hand turned and finished, light yellow color. Used in practice for batting long flies. Shipping weight, 2¼ lbs.

6L1623..........80c

Youths' Professional Bat. Made of good quality ash, with a light brown burnished finish. Tape wound grip. Shipping weight, 2½ pounds.

6L1618..........70c

Boys' Choice Bat. Made of hardwood, nicely finished. Length, about 28 inches. Shpg. wt., 1½ lbs.

6L1626..........19c

Junior League Bat. Made of hardwood, flame burnt and yellow finish, and highly polished. Length, about 32 inches. Shipping weight, 2¼ pounds.

6L1624..........35c

J. C. Higgins League Official Indoor Ball. Excellent quality horsehide cover, well stitched, hand sewn. Filled with genuine kapok. Three sizes.

6L1611—17-Inch Ball. Shipping weight, 1 pound....$1.65

6L1612—14-Inch Ball. Shipping weight, 13 ounces....$1.30

6L1613—12-Inch Ball. Shipping weight, 12 ounces....$1.05

Indoor Ball Rules.

6L1839—Shipping weight, 2 ounces......14c

Regulation Indoor Bat.

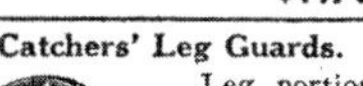

Made of second growth white ash, taped handle, highly polished. Shipping weight, 1⅝ pounds.

6L1627..........65c

Official Outseam Playground Ball. Finest quality horsehide cover. Filled with genuine kapok. Well made, hand sewn. Will stand lots of hard use. Shipping weight, 13 oz.

6L1608—14-inch ball..........$1.33

6L1615—12-inch ball..........$1.10

Playground Ball. Genuine leather cover, well sewed. Very soft, about 11 in. in circumference. Excellent value. Shpg. wt., 6 oz.

6L1614......38c

Junior Playground Ball.

Boys' soft ball with split leather cover, well stitched. Size, about 10½ inches. Shipping weight, 7 ounces.

6L1604..........20c

Youths' League Mask. Electrically welded, dull black enameled frame, leather head and cheek pads, elastic head strap and padded chin piece. Shipping weight, 1⅝ lbs.

6L1696 $1.75

Youths' Mask. Electrically welded frame, dull black enameled finish, leather temple and cheek pads, nicely finished. Size, 8¾x6½ in. Shipping wt., 1¼ lbs.

6L1697......78c

Small Boys' Mask. Size, 8¼x6¼ inches. Shipping weight, 6 oz.

6L1698..........30c

Catchers' Leg Guards.

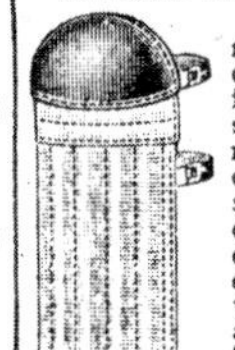

Leg portion made of heavy canvas, reinforced with strong, round reeds. Knee cap made of special molded fiber, padded with layer felt. Straps to buckle around legs. Shipping wt., 2 pounds.

6L1760 Per pair $4.40

Professional Model Catchers' Leg Guards.

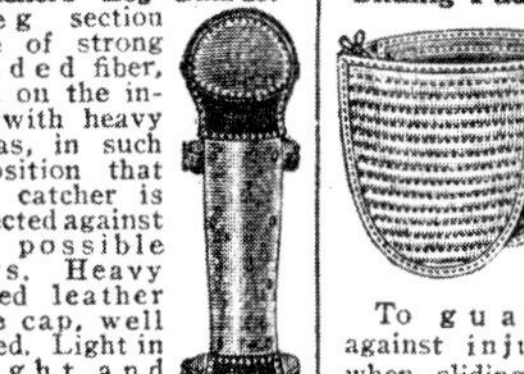

Leg section made of strong molded fiber, lined on the inside with heavy canvas, in such a position that the catcher is protected against any possible blows. Heavy molded leather knee cap, well padded. Light in weight and quickly adjusted. Shpg. wt., 3 lbs.

6L1761—Pair..$5.95

Sliding Pads.

To guard against injury when sliding in bases. Made of cotton, covered and quilted; adjustable to any size waist. Worn inside of pants. Shpg. wt., 10 oz.

6L1749 Pr. $1.98

Regulation League Bases.

Made of extra heavy canvas, quilted top, furnished complete with leather straps and stakes. Shpg. wt., 14 lbs.

6L1754—Set of 3 bases...$7.00

6L1753—Not stuffed Canvas Bases with straps and stakes. Can easily be stuffed by purchaser. Shipping weight, 3 pounds. Per set of 3 bases..........$3.85

Boys' Body Protector.

Made of olive tan drill, front and back well padded and quilted. Leather neck and body straps. Shipping weight, 1⅝ pounds.

6L1716 $1.35

J. C. Higgins Professional Model Body Protector.

Heavy tan color cotton duck stuffed with good quality hair. No inflating required. Full leather bound edge. Adjustable leather body strap, elastic shoulder strap. Shipping weight, 4 lbs.

6L1710 $4.80

Youths' Size Professional Model Body Protector.

Same style as above, but a trifle smaller and made of fine quality tan drill.

6L1711..........$2.80

"Save Hide" Sliding Pad.

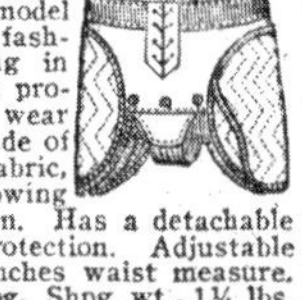

Latest improved professional model sliding pad, worn in pants like fashion, thereby constantly staying in comfortable position. The best protection a player can possibly wear and still be non-interfering. Made of fine quality quilted cotton fabric, padding in two leaf sections allowing for the necessary sliding friction. Has a detachable elastic supporter for added protection. Adjustable lacing front. Sizes, 30 to 40 inches waist measure. **Give waist measure when ordering.** Shpg. wt., 1¼ lbs.

6L1748..........$3.20

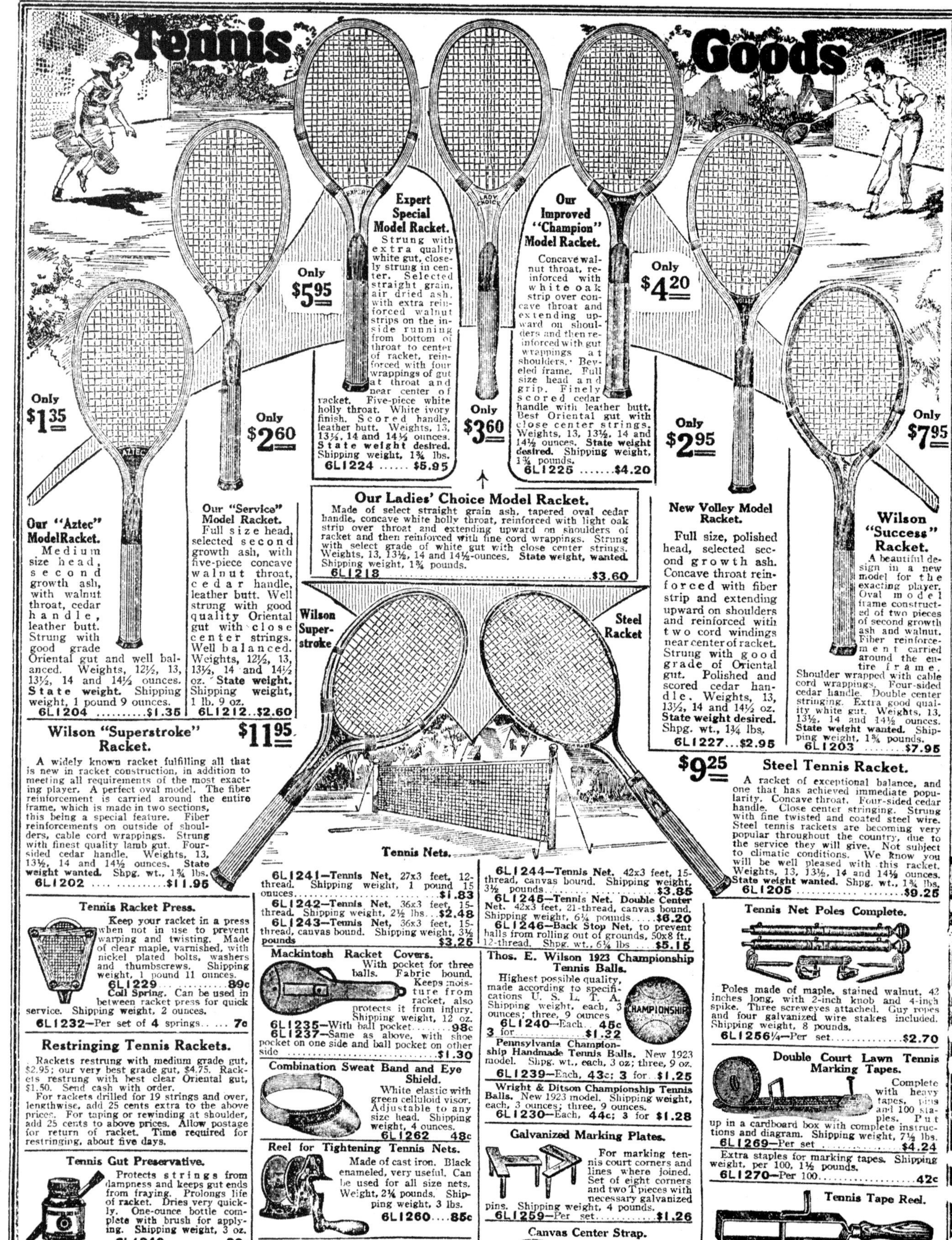

Tennis Goods

Our "Aztec" Model Racket. Medium size head, second growth ash, with walnut throat, cedar handle, leather butt. Strung with good grade Oriental gut and well balanced. Weights, 12½, 13, 13½, 14 and 14½ ounces. **State weight.** Shipping weight, 1 pound 9 ounces.
6L1204 .... $1.35
Only $1.35

Our "Service" Model Racket. Full size head, selected second growth ash, with five-piece concave walnut throat, cedar handle, leather butt. Well strung with good quality Oriental gut with close center strings. Well balanced. Weights, 12½, 13, 13½, 14 and 14½ oz. **State weight.** Shipping weight, 1 lb. 9 oz.
6L1212..$2.60
Only $2.60

Expert Special Model Racket. Strung with extra quality white gut, closely strung in center. Selected straight grain, air dried ash, with extra reinforced walnut strips on the inside running from bottom of throat to center of racket, reinforced with four wrappings of gut at throat and near center of racket. Five-piece white holly throat. White ivory finish. Scored handle, leather butt. Weights, 13, 13¼, 14 and 14½ ounces. **State weight desired.** Shipping weight, 1¾ lbs.
6L1224 ...... $5.95
Only $5.95

Only $3.60

Our Improved "Champion" Model Racket. Concave walnut throat, reinforced with white oak strip over concave throat and extending upward on shoulders and then reinforced with gut wrappings at shoulders. Beveled frame. Full size head and grip. Finely scored cedar handle with leather butt. Best Oriental gut with close center strings. Weights, 13, 13¼, 14 and 14½ ounces. **State weight desired.** Shipping weight, 1¾ pounds.
6L1225 .... $4.20
Only $4.20

Our Ladies' Choice Model Racket. Made of select straight grain ash, tapered oval cedar handle, concave white holly throat, reinforced with light oak strip over throat and extending upward on shoulders of racket and then reinforced with fine cord wrappings. Strung with select grade of white gut with close center strings. Weights, 13, 13½, 14 and 14½-ounces. **State weight, when wanted.** Shipping weight, 1¾ pounds.
6L1218 .... $3.60

New Volley Model Racket. Full size, polished head, selected second growth ash. Concave throat reinforced with fiber strip and extending upward on shoulders and reinforced with two cord windings near center of racket. Strung with good grade of Oriental gut. Polished and scored cedar handle. Weights, 13, 13½, 14 and 14½ oz. **State weight desired.** Shpg. wt., 1¾ lbs.
6L1227...$2.95
Only $2.95

Wilson "Success" Racket. A beautiful design in a new model for the exacting player. Oval model frame constructed of two pieces of second growth ash and walnut. Fiber reinforcement carried around the entire frame. Shoulder wrapped with cable cord wrappings. Four-sided cedar handle. Double center stringing. Extra good quality white gut. Weights, 13, 13½, 14 and 14½ ounces. **State weight wanted.** Shipping weight, 1¾ pounds.
6L1203 .... $7.95
Only $7.95

Wilson Superstroke

Steel Racket

Wilson "Superstroke" Racket. $11.95

A widely known racket fulfilling all that is new in racket construction, in addition to meeting all requirements of the most exacting player. A perfect oval model. The fiber reinforcement is carried around the entire frame, which is made in two sections, this being a special feature. Fiber reinforcements on outside of shoulders, cable cord wrappings. Strung with finest quality lamb gut. Four-sided cedar handle. Weights, 13, 13½, 14 and 14½ ounces. **State weight wanted.** Shpg. wt., 1¾ lbs.
6L1202 .... $11.95

Steel Tennis Racket. $9.25

A racket of exceptional balance, and one that has achieved immediate popularity. Concave throat. Four-sided cedar handle. Close center stringing. Strung with fine twisted and coated steel wire. Steel tennis rackets are becoming very popular throughout the country, due to the service they will give. Not subject to climatic conditions. We know you will be well pleased with this racket. Weights, 13, 13½, 14 and 14½ ounces. **State weight wanted.** Shpg. wt., 1¾ lbs.
6L1205 .... $9.25

Tennis Nets.

6L1241—Tennis Net, 27x3 feet, 12-thread. Shipping weight, 1 pound 15 ounces $1.83
6L1242—Tennis Net, 36x3 feet, 15-thread. Shipping weight, 2½ lbs...$2.48
6L1243—Tennis Net, 36x3 feet, 15-thread, canvas bound. Shipping weight, 3½ pounds $3.25
6L1244—Tennis Net. 42x3 feet, 15-thread, canvas bound. Shipping weight, 3½ pounds $3.85
6L1245—Tennis Net. Double Center Net. 42x3 feet, 21-thread, canvas bound. Shipping weight, 6¼ pounds..... $6.20
6L1246—Back Stop Net, to prevent balls from rolling out of grounds, 50x8 ft., 12-thread. Shpg. wt., 6¼ lbs $5.15

Tennis Racket Press.

Keep your racket in a press when not in use to prevent warping and twisting. Made of clear maple, varnished, with nickel plated bolts, washers and thumbscrews. Shipping weight, 1 pound 11 ounces.
6L1229 .... 89c
Coil Spring. Can be used in between racket press for quick service. Shipping weight, 2 ounces.
6L1232—Per set of 4 springs.... 7c

Restringing Tennis Rackets.

Rackets restrung with medium grade gut, $2.95; our very best grade gut, $4.75. Rackets restrung with best clear Oriental gut, $1.50. Send cash with order.

For rackets drilled for 19 strings and over, lengthwise, add 25 cents extra to the above prices. For taping or rewinding at shoulder, add 25 cents to above prices. Allow postage for return of racket. Time required for restringing, about five days.

Tennis Gut Preservative.

Protects strings from dampness and keeps gut ends from fraying. Prolongs life of racket. Dries very quickly. One-ounce bottle complete with brush for applying. Shipping weight, 3 oz.
6L1249 .... 30c

Tennis Rules and Records. Shipping weight, 4 ounces.
6L1289 .... 14c

Mackintosh Racket Covers.

With pocket for three balls. Fabric bound. Keeps moisture from racket, also protects it from injury. Shipping weight, 12 oz.
6L1235—With ball pocket 98c
6L1237—Same as above, with shoe pocket on one side and ball pocket on other side $1.30

Combination Sweat Band and Eye Shield.

White elastic with green celluloid visor. Adjustable to any size head. Shipping weight, 4 ounces.
6L1262 .... 48c

Reel for Tightening Tennis Nets.

Made of cast iron. Black enameled, very useful. Can be used for all size nets. Weight, 2¼ pounds. Shipping weight, 3 lbs.
6L1260....85c

Dry Court Marker.

Fill with marble dust or air slaked lime, no mixing material required. Comes fitted with handle and steel bracket. Weight, 2¼ pounds. Shpg. wt., 2¾ lbs.
6L1268 .... $1.29

Thos. E. Wilson 1923 Championship Tennis Balls.

Highest possible quality, made according to specifications U. S. L. T. A. Shipping weight, each, 3 ounces; three, 9 ounces.
6L1240—Each.. 45c
3 for $1.32

Pennsylvania Championship Handmade Tennis Balls. New 1923 model. Shpg. wt., each, 3 oz; three, 9 oz.
6L1239—Each, 43c; 3 for $1.25

Wright & Ditson Championship Tennis Balls. New 1923 model. Shipping weight, each, 3 ounces; three, 9 ounces.
6L1230—Each, 44c; 3 for $1.28

Galvanized Marking Plates.

For marking tennis court corners and lines where joined. Set of eight corners and two T pieces with necessary galvanized pins. Shipping weight, 4 pounds.
6L1259—Per set $1.26

Canvas Center Strap.

Heavy Canvas Center Strap, with brass turnbuckle and galvanized stake for holding center of net at regulation height. Will not chafe the net. Shipping weight, 1½ pounds.
6L1250—Price, each $1.10

Tennis Net Poles Complete.

Poles made of maple, stained walnut, 42 inches long, with 2-inch knob and 4-inch spike. Three screweyes attached. Guy ropes and four galvanized wire stakes included. Shipping weight, 8 pounds.
6L1256¼—Per set $2.70

Double Court Lawn Tennis Marking Tapes.

Complete with heavy tapes, pins and 100 staples. Put up in a cardboard box with complete instructions and diagram. Shipping weight, 7½ lbs.
6L1269—Per set $4.24
Extra staples for marking tapes. Shipping weight, per 100, 1½ pounds.
6L1270—Per 100 .... 42c

Tennis Tape Reel.

Made of metal with wood handle. For winding up marking tapes. Especially handy for players who use the public courts. By its use a set of tapes can be rolled up and put in a box in a very few minutes. Length, 13 inches; width, 1¾ inches; depth, 1 inch. Shipping weight, 8 ounces.
6L1255 .... 58c

Beginners' Golf Outfit.

$9.25

An exceptional golf outfit for a beginner or one who does not want an expensive set. Excellent value. Consists of four good quality clubs—brassie, mid iron, mashie and putter, of popular models; a 4½-in. brown canvas bag, 35 inches long, three steel stays, imitation black leather trimmings, leather bottom. For right hand players only. Shipping wt., 11 pounds.

6L1374¼ $9.25

A good low priced golf bag. Bag only as described above. Shpg. wt., 2 lbs.

6L1380 .... $2.60

Golf Bags With Hood.

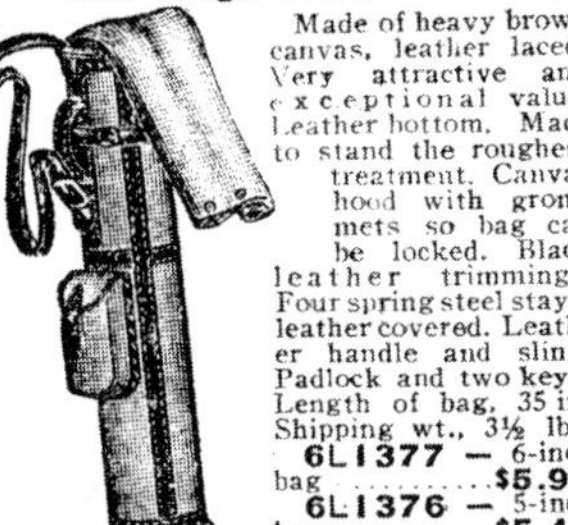

Made of heavy brown canvas, leather laced. Very attractive and exceptional value. Leather bottom. Made to stand the roughest treatment. Canvas hood with grommets so bag can be locked. Black leather trimmings. Four spring steel stays, leather covered. Leather handle and sling. Padlock and two keys. Length of bag, 35 in. Shipping wt., 3½ lbs.

6L1377 — 6-inch bag **$5.95**
6L1376 — 5-inch bag **$5.40**

Five-Inch Golf Bags.

Made of high grade canvas, with genuine leather trimmings. Three spring steel stays, leather covered. Leather bottom, adjustable leather sling. A medium size bag, used extensively by women. A bag that will please. Length of bag, 34 inches. Shipping weight, 3 pounds.

6L1379—Brown color canvas, black leather trimmed **$4.45**
6L1375—White color canvas, brown leather trimmed **$4.45**

Sunday Golf Bag.

Made of white canvas with a canvas sling. Has double bottom. Ball pocket with leather strap. 4½ inches in diameter; length, 34 inches. A very light, handy bag. Shipping weight, 1¼ pounds.

6L1382 .... 85c

GOLF GOODS

Order by catalog number and name of club and state length.

Glencoe Golf Clubs With Hickory or Steel Shafts

Hickory Shafts.

These clubs are branded especially for us; in all other respects they are the same as one of the most popular makes of clubs usually sold elsewhere for considerably more money. Patterned after the most popular professional models. First class in every respect. The heads of wood clubs are made of selected persimmon, turned, polished and well finished. The shafts are of selected second growth hickory with calf leather grips. Our irons are made of fine quality steel, drop forged, highly finished and have second growth hickory shafts with calf leather grips.

Catalog Numbers Right Hand	Catalog Numbers Left Hand	Kind of Club	Length, Inches		Shipping Weight
6L1351	6L1352	Driver	42 to 43	$2.50	1½ lbs.
6L1353	6L1354	Brassie	42 to 43	2.50	1½ lbs.
6L1355	6L1356	Driving Iron	38 to 39	2.50	1½ lbs.
6L1357	6L1358	Mid Iron	37 to 38½	2.50	1½ lbs.
6L1359	6L1360	Mashie	36 to 37	2.50	1½ lbs.
6L1361	6L1362	Niblick	35 to 36½	2.50	1½ lbs.
6L1363	6L1364	Putter	33 to 35	2.50	1½ lbs.
6L1369	6L1370	Bakspin Mashie	36 to 37	2.50	1½ lbs.
6L1365	6L1366	Aluminum Putter	33 to 35	3.70	1½ lbs.

Steel Shafts.

Steel Shafted Clubs have become popular with amateurs and professionals everywhere, and they give absolute satisfaction. The steel shaft is lighter than hickory. This throws the weight down near the head, giving a perfectly balanced club. They are uniform in balance, will not warp or break, and are exceptionally durable. Calf leather grips, iron and wood heads are of the late improved model. Well finished. Right hand clubs carried in stock only; if left handed are wanted, can ship from factory within ten days.

Catalog No.	Kind of Club	Length in Inches		Shipping Weight
6L1344	Driver	42 to 43	$6.75	1½ lbs.
6L1345	Brassie	42 to 43	6.75	1½ lbs.
6L1346	Driving Iron	38 to 39	6.15	1½ lbs.
6L1347	Mid Iron	37 to 38½	6.15	1½ lbs.
6L1348	Mashie	36 to 37	6.15	1½ lbs.
6L1349	Putter	33 to 35	6.15	1½ lbs.

Our Own "162" Golf Balls.

High grade ball. Regular 75c value. Made for us by a manufacturer of a popular golf ball. Standard and official in size and weight. Recessed markings. Sinker. Shipping weight, half dozen, 14 ounces.

6L1309
Each **$0.55**
Half dozen **3.20**

Victor "75" Golf Balls.

6L1329—No. 75 Sinking Ball. Official size and weight. Recessed markings. Shipping wt., half dozen, 14 ounces.
Each, **68c**; per half dozen **$4.00**

St. Mungo Colonel No. 162 Golf Balls.

Standard and official in size and weight. Sinker. Recessed markings. Shipping weight, half dozen, 14 ounces.
6L1332—Each, **68c**; half dozen .. **$4.00**

Radio Standard Golf Balls.

Standard and official in size and weight. Recessed markings. Sinker. Shipping weight, half dozen, 14 ounces.
6L1311—Each, **68c**; half dozen ... **$4.00**

Wilson's Success No. 162 Golf Balls.

Standard and official in size and weight. Recessed markings. Sink in water. Shipping weight, per half dozen, 14 ounces.
6L1312—Each, **68c**; half dozen . **$4.00**

Wilson's Floater.

6L1313—Mesh markings.
Each, **45c**; per half dozen **$2.60**
Shipping weight, per half dozen, 14 ounces.

Golf Hose.

High quality Golf Hose in two grades, medium weight, ribbed, fancy color cuff. Sizes, 10, 10½, 11 and 11½. **State size.** Shpg. wt., 8 ounces.

Wool Hose.

6L1302 — Brown heather color.
Per pair.... **$1.75**
6L1303 — Camel hair color (tan).
Per pair .. **$1.75**

Wool and Artificial Silk Hose.

6L1304—Brown heather color.
Per pair **$1.30**
6L1305—Gray color. Pair....... **1.30**

For Golf Oxfords see page 766.

Golf Knickers.

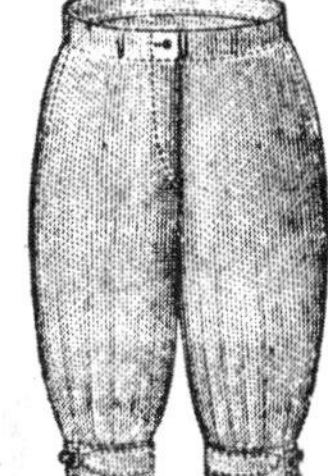

In two grades of high quality material, linen crash and Palm Beach. Have usual pockets and belt loops, removable buckle at cuff. Easily laundered. Ideal for all outdoor sports. Sizes, 30, 31, 32, 33, 34, 36, 38, 40 and 42 in. waist measure. **State size.** Shipping wt., 2 lbs.

6L1300 — Linen. Natural color **$3.85**
6L1301—Palm Beach. Light brown color **$4.95**

American Made

Vacuum Bottles

Each Bottle Carries Manufacturer's Guarantee Tag.
Will keep liquids hot twenty-four hours or cool three full days.

Our Favorite Lunch Kit.

Popular with factory, office and outdoor workers, school children and all who carry lunch. Black enameled metal case, leather handle, hinged lid, brass plated clasps. Good quality pint vacuum bottle, enameled body, aluminum cup and shoulder, held in upper compartment, provides hot coffee or tea or cold drinks. Lower compartment keeps lunch clean. Shipping wt., 3½ pounds.

6L4917 .... $2.43

Our Highest Grade Vacuum Bottle.

Made of seamless brass, heavily nickel plated. Removable filler and screw off top with handle. Shipping weight: Pint size, 2 lbs.; quart size, 4½ pounds.

6L4930—Pint size.. **$2.95**
6L4931—Quart size. **4.25**

An Added Feature.

This detachable handle included free with Vacuum Bottles 6L4903, 6L4904, 6L4930, 6L4931, 6L4909 and 6L4910; also Lunch Kits 6L4913 and 6L4917. Attaching handle to top of bottle makes a fine drinking cup as illustrated.

Corrugated Vacuum Bottle.

Corrugated seamless brass, nickel plated case, with removable filler and screw off top. A well built and popular model. The corrugations prevent slipping of bottle from hands. Shpg. wt.: Pint, 2 lbs.; quart, 4½ lbs.

6L4903
Pint size...... **$1.90**
6L4904
Quart size... **$2.75**

Metal Lunch Kit With One-Half Pint Vacuum Bottle.

New and ideal kit for school children and all who carry their lunch. Metal carrying case; black finish, equipped with a high quality one-half pint vacuum bottle, nickel plated screw off top, enameled body. Removable filler. Case is partitioned so that bottle will stay in place and lunch can be kept clean. Shipping weight, 3½ pounds.

6L4913 .... $2.33

Vacuum Carafe.

Shpg. wt., 4½ lbs.

Ideal for dining room, bedroom or guest room. Capacity, 1 quart. Height, 11 inches. Diameter of bottle, 5 inches. Nickel plated and well finished. Filler is removable.

6L4939
$4.50

Extra Stoppers.
Shipping weight, each, 8 oz.
6L4924 .... 60c

Extra Fillers or Inside Bottles.

Can easily be fitted. Will also fit Icy Hot Bottles.

6L4944
Pint size, for bottles 6L4903, 6L4909, 6L4917 and 6L4930. Shipping wt., 2 lbs **$1.05**
6L4946
Quart size, for bottles 6L4904, 6L4910 and 6L4931. Shipping weight, 3 pounds... **$1.55**
6L4945
Filler for carafe 6L4939. Shipping wt., 3lbs **$3.00**
6L4947—Filler for pint size food jar 6L4934. Shpg. wt., 2¼ lbs. **$1.55**
6L4948—Filler for quart size food jar 6L4935. Shipping weight, 4½ pounds... **$2.10**
6L4950—Filler for ½-pint food jar 6L4933. Shipping weight, 3 pounds **$1.45**
6L4943—Filler for ½-pint Vacuum Bottle 6L4913. Shipping weight, 1½ lbs **$1.00**

$1.25 Pint Size. Our Leader. Shipping weight, pint size, 2 pounds.

$2.10 Quart Size. Green Enameled. Shipping weight, quart size, 4½ lbs.

Vacuum Bottle.

Green enameled body and aluminum shoulder and cup, with removable filler and screw off top. Top serves as a drinking cup.

6L4909—Pint size... **$1.25**
6L4910—Quart size **2.10**

Carrying Cases for Vacuum Bottles.

Made of imitation leather. Strongly constructed and nicely finished. Leather strap fastening over covers.

6L4953—For one 1-pint bottle. Shipping wt., 14 oz.... **$1.95**
6L4957—For one 1-quart bottle. Shpg. wt., 2 lbs...... **$2.40**
6L4959—For two 1-quart bottles. Shipping weight, 2½ pounds **$2.70**

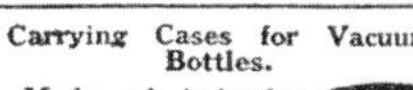

Vacuum Jar.

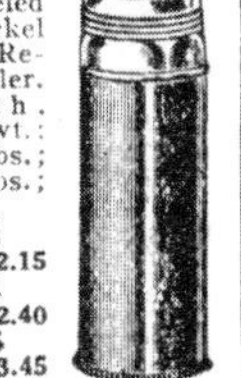

Keeps solid foods hot or cold. Fine for ice cream, etc. Black enameled body and nickel plated top. Removable filler. Wide mouth. Shipping wt.: ½-pint, 2 lbs.; pint, 2½ lbs.; quart, 5 lbs.

6L4933
½-pt. size **$2.15**
6L4934
1-pint size **$2.40**
6L4935
1-qt. size **$3.45**

Extra Corks.

To fit bottles 6L4930, 6L4931, 6L4903, 6L4904, 6L4909 and 6L4910; also 6L4913 and 6L4917 Lunch Kits. Shipping weight, 3 ounces.
6L4922—Each **5c**
To fit 6L4933, 6L4934 and 6L4935 Food Jars. **State which is wanted.** Shipping wt., 5 oz.
6L4923—Each **15c**

ATHLETIC SHOES

Feather Weight Professional Baseball Shoes. Our best Baseball Shoes. Genuine kangaroo leather uppers, leather reinforced; oak leather soles; English welt, flexible shank; fitted with hardened league toe and heel plates; solid rivets. Leather laces. Shoes strongly sewed throughout. Sizes and half sizes, 5 to 11. **State size.** Shipping weight, 2¼ pounds.
6L1963 .... $6.35

Professional Baseball Shoes. High quality gun metal calf uppers, chrome tanned, reinforced vamp. English welt. Strong and durable and has the celebrated sprinting sole. Latest cut shank, giving extreme flexibility. Hardened steel heel and toe plates, solid rivets, leather laces. Furnished in sizes and half sizes, 5 to 11. **State size wanted.** Shpg. wt., 2½ lbs.
6L1945—Per pair. $4.60; 9 pairs for $40.50

Golf Oxfords.

High quality soft and pliable cowhide leather, pearl elk tanned uppers, brown saddle strap across instep, latest pattern. Goodyear stitched; rubber suction cup sole. A very attractive shoe for all outdoor wear. Sizes and half sizes, 6 to 11. **State size.** Shipping weight, 3 pounds.
6L1961 .... $5.75

Same style as above, but with oak leather sole fitted with steel screw-in calks. Sizes and half sizes, 6 to 11.
6L1962. $6.40

How to measure for shoes, see Order Blank in back of catalog.

Amateur League Baseball Shoes.

Box side leather uppers. Oak tanned leather sole, McKay sewed. Steel toe and heel plates. A shoe of exceptional value at this price and one that will render complete satisfaction. Sizes and half sizes, 5 to 11. Full widths only. **State size wanted.** Shipping weight, 2½ pounds.
6L1947 .... $2.75
9 pairs for 23.65

Juvenile Baseball Shoes.

Made on the same style last as our Amateur League Shoes 6L1947, but in juvenile sizes. Fitted with steel toe and heel plates. Furnished in sizes and half sizes, 13 to 4½. **State size wanted.** Shipping weight, 1¾ pounds.
6L1958 .... $2.60

Klaykort Tennis Shoes.

Heavy white canvas uppers, reinforced vamp and eyelet stays, high quality rubber sole, leather insole, jar resisting felt cushion between inner and outer soles. Especially constructed for dirt and clay courts. **State size.** Shipping weight, 2½ pounds.
6L1969—Men's. Sizes, 6 to 11, $2.50
6L1974—Women's. Sizes, 2½ to 7. $2.30
6L1975 Boys'. Sizes, 2½ to 6. $2.30

Boys' Tennis Shoes.

Made with white canvas uppers and rubber trimmings, suction rubber sole, rubber toe cap, kool-foot insole. This shoe is also adapted for outdoor sports. Sizes and half sizes, 2½ to 6. **State size wanted.** Shipping wt., 2 pounds.
6L1960 $1.35

Bowling Shoes.

Laced to Toe Style.

For bowling, boxing and wrestling. Black chrome tanned leather, soft and pliable. Fine quality imitation buck sole. English welt. Strongly stitched. High cut laced to toe style, giving good protection to ankles. Sizes and half sizes, 5 to 11. **State size wanted.** Shipping wt., 2 lbs.
6L1949 $3.60

Gymnasium Oxfords.

Low cut. For indoor ball, handball, volley ball, fencing or sparring. Soft and pliable leather strongly sewed chrome sole. Black only. **State size wanted.** Shipping weight, 1¼ pounds.
6L1954—For boys. Sizes, 9 to 5½ $1.20
6L1955—For women. Sizes, 2 to 7 .... 1.25
6L1956—For men. Sizes, 6 to 12 .... 1.30

Gymnasium Shoes.

High cut. For indoor ball volley ball, handball, fencing or sparring. Soft and pliable leather. Medium weight, turned sole (commonly called the "never-slip"), strongly sewed. Very comfortable. Black only. **State size wanted.**
6L1957—For boys. Sizes, 12 to 5½ $1.45
6L1959—For men. Sizes, 6 to 12 .... 1.48

Camping and Outing Shoes.

Extra high cut shoes, in two heights. Dark brown oil tanned cowhide uppers, moccasin style hand sewed vamp; oil tanned heavy belt leather sole. An ideal shoe for fishing, hunting and trailing. A very comfortable fitting shoe. Sizes and half sizes, 5 to 11. **State size wanted.** Shpg. wt.: 9-in., 4 lbs.; 14-in., 6 lbs.
6L1946 9 in. high. $5.85
6L1948—14 inches high $7.25

Canvas Basket Ball Shoes.

Laced to Toe Style.

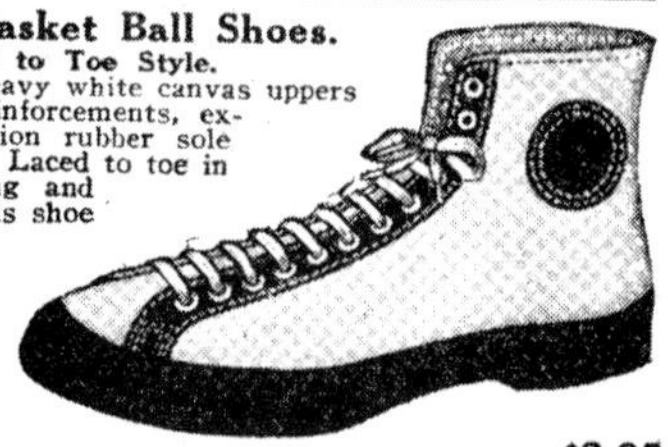

Made with heavy white canvas uppers with rubber reinforcements, extra heavy suction rubber sole leather insole. Laced to toe in sures a snug and smooth fit. This shoe is also adapted for golf, tennis or camping. Sizes and half sizes, 5 to 11. **State size wanted.** Shpg. wt., 3 lbs.
6L1968 .... $2.95

Klaykort Tennis Oxfords.

White canvas uppers, reinforced vamp and eyelet stays, combination red and gray rubber sole, jar resisting felt cushion, leather insole. Sizes and half sizes. **State size wanted.** Shipping weight, 2½ pounds.
6L1980—Men's. Sizes, 6 to 11 .. $2.35
6L1981—Women's. Sizes, 2½ to 7 2.15
6L1982—Boys'. Sizes, 2½ to 6 2.15

Leather Moccasins.

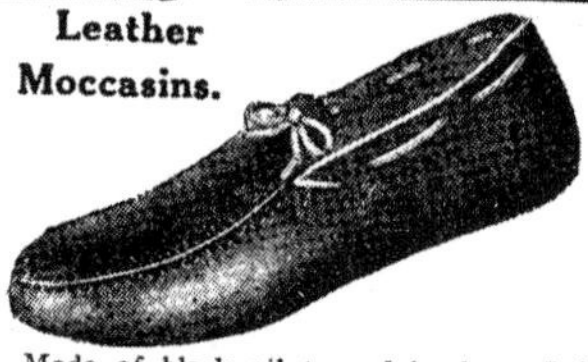

Made of black oil tanned leather. Soft and pliable. Laced front with draw string around ankle. Snug fitting. Used for camping, trailing, fishing and canoeing. Also ideal for indoor wear. Sizes and half sizes. **State size.** Shipping weight, 2½ lbs.
6L1937—For women. Sizes, 2 to 7. $2.40
6L1938—For men. Sizes, 5 to 11 2.65

Jumping Shoes.

Outdoor Jumping, Hurdling and Vaulting Shoes. Made of horsehide leather, soft and flexible. Reinforced heel and eyelet stays. Oak leather tap with six patented hand forged steel spikes and heel fitted with one spike. Sizes and half sizes, 3 to 11. **State size wanted.** Shipping wt., 1½ lbs.
6L1932 .... $5.25

Running Shoes.

Horsehide leather, soft and pliable. Seamless toe, flexible shank, reinforced heel stay, strongly sewed oak leather tap. Fitted with six good patented hand forged steel spikes. Sizes and half sizes, 3 to 11. **State size wanted.** Shipping weight, 1 pound 9 ounces.
6L1944 .... $4.85

Our Improved Swinging Rings Trapeze Bar and Swing.

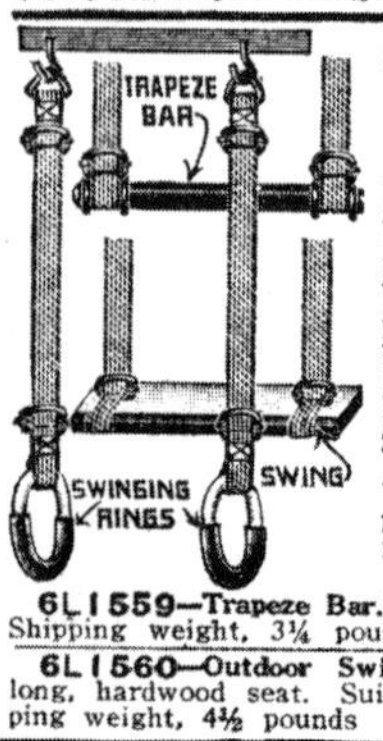

TRAPEZE BAR — SWINGING RINGS — SWING

Made of good quality webbing, very strong. Adjustable steel buckles, steel rings and hooks. All wood parts nicely smoothed and well finished. Strong, safe and simple to set up. Equip a small home gymnasium at little expense. Suitable for adults as well as youngsters. Can be used outdoors as well. Sold separately.
6L1558—Swinging Rings With Rubber Grips. Webbing, about 5 feet long. Shipping weight, 3 pounds. Per pair $2.60
6L1559—Trapeze Bar. Webbing, about 5 feet long. Shipping weight, 3¼ pounds $2.20
6L1560—Outdoor Swing. Webbing, about 9 feet long, hardwood seat. Suitable for outdoor use. Shipping weight, 4½ pounds $2.20

HOME EXERCISING APPARATUS

Peerless Combination Wire Exerciser.

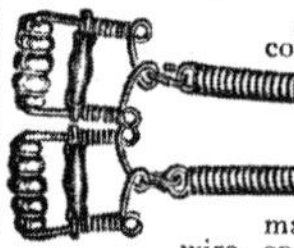

A universal exerciser for a complete course of exercise controlling all the principal muscles of the body. Can be instantly adjusted for use as a wall exerciser or as a chest pull. Combination grip and massage roller handles. Tempered music wire springs. All metal parts nickel plated. Chart of instructions included. Shipping weight, 2½ lbs.
6L1564—Medium tension $2.40
6L1563—Heavy tension 2.50
6L1565—Extra heavy tension 2.90
6L1562—Light tension 2.10

Strong Grip Wrist Machine.

Develops the muscles of the wrist and arm. Strengthens the grip. Nickel plated steel wire spring. Fitted with polished black handles. Shipping weight, 1 pound.
6L1572—Per pair 28c

Special Athletic Exerciser.

Medium light weight for women or children, full size, elastic cable, wood pulleys and wood handles, nickel plated trimmings. Chart of illustrations included, showing many forms of exercise that can be brought into play by the use of an elastic wall exerciser. Shipping weight, 1¾ lbs.
6L1577 .... $1.75

Elastic Cord with japanned wood handles, wood pulleys, nickel plated trimmings. Chart of illustrations included. Shipping wt., 1½ lbs.
6L1573 .... $2.10

Elastic Cord Chest Pulls.

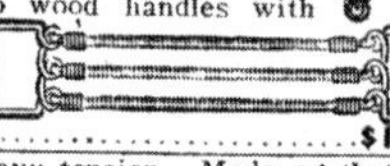

Ideal for developing shoulders and strengthening muscles of back and arms. Elastic cables, attached to two wood handles with nickel plated fittings. Medium tension for women and youths; 30 in. Shpg. wt., 15 oz.
6L1567 .... $1.15

Five strands, extra heavy tension. Made so that one or two strands may be removed to change tension. Shipping weight, 1½ pounds.
6L1569 .... $1.90

ROLLER SKATES

Ball Bearing Extension Skates.

Harris & Reed.

Extra strong construction. For sidewalk. Double truss brace, heavy stock, rubber cushions at front and rear axles, extra large ¼-inch ball bearings, casehardened braces. Ball bearings and cones are enclosed and cannot come out. Sizes, 8 to 11½ inches. Shipping weight, per pair, 5¾ pounds.

6L4832—Men's Skates. Per pair..**$2.30**
6L4833—Women's Skates, with high leather heel band. Per pair..**$2.40**

6L4846—Self Contained Steel Ball Bearing Rolls for above skates. Shipping weight, 5 ounces. Each....**12c**

Union Hardware Co. Rink Skates.

Improved self contained ball bearing 2-inch fiber rolls, similar to 6L4823 and 6L4824, but heavier constructed and have "U" shape bar of steel extending full length of bottom of foot plate, which means additional strength. Shipping weight, 5 pounds.

6L4861—Men's and Boys' Adjustable Skates. Per pair....**$3.90**
6L4862—Women's and Girls' Adjustable Skates. Per pair....**$4.20**

Extra Rolls for Above.
6L4863—Fiber Self Contained Ball Bearing Rolls, for Skates 6L4861 and 6L4862. Shipping weight, 5 oz. Each....**24c**

Boys' and Girls' Extension Skates.

Plain bearings. Rubber cushions. Trucks, clamps and stamping of good grade cold rolled steel, finely finished. Plain bearing pressed steel rolls. The skate extends to fit all sizes from 8½ to 11½ inches. Shipping weight, 3¾ pounds.

6L4822
Per pair....**$1.20**

6L4844—Steel Plain Bearing Rolls for above. Shipping weight, 5 ounces. Each....**4c**

Union Hardware Co. Ball Bearing Extension Skates With Self Contained Ball Bearing Steel Rollers.

Tops, trucks and clamps made of good quality cold rolled steel, nickel plated and polished. Oscillating trucks with rubber cushions; steel, self contained, ball bearing rolls. Ball bearings and cones are enclosed and cannot come out. Will extend to fit shoes 8½ to 11½ inches. Shipping weight, 4⅝ pounds.

6L4823—Men's and Boys' Skates, with toe clamp and high steel heel band with strap. Per pair..**$1.65**

6L4824—Women's and Girls' Skates, with toe clamp and high leather heel band with strap. Per pair....**$1.71**

6L4857—Steel Self Contained Ball Bearing Rolls for above. Shipping weight, 5 ounces. Each....**12c**

Extra Parts.

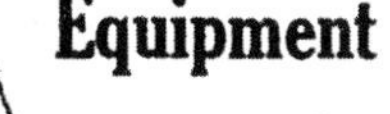

6L4846—Rolls, complete with ¼-inch ball bearings and cones. So made that cones or bearings will not drop out, require no adjusting and will fit any plain or ball bearing skate. Shipping weight, each, 5 ounces. Each....**12c**

6L4848—3/16-Inch Steel Balls. Shipping weight, per dozen, 1 ounce. Per 100, **60c**; per two dozen....**18c**

6L4850—Roller Skate Keys. Shipping weight, each, 2 ounces. Per dozen, **42c**; each....**4c**

Richardson Rink Skates.

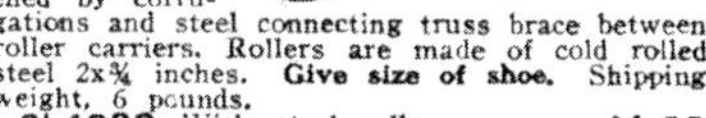

Richardson high grade, truss frame, anti-jar, ball bearing rink skates, nickel plated finish. Foot plates made of high quality steel. Strengthened by corrugations and steel connecting truss brace between roller carriers. Rollers are made of cold rolled steel 2x¾ inches. **Give size of shoe.** Shipping weight, 6 pounds.

6L4828—With steel rolls....**$4.60**
6L4830—With 2⅛x⅞-inch fiber rolls....**4.65**

Extra Rolls for Above.
6L4852—Steel Rolls for skates 6L4828. Shipping weight, 5 ounces. Each....**19c**
6L4853—Fiber Rolls, for skates 6L4830. Shipping weight, 5 ounces. Each....**20c**

Juvenile or Small Children's Skates.

Improved extension sidewalk skates. Steel foot plates, plain bearing, pressed steel rolls. Fit shoes from 6 to 10 inches long. Shpg. wt., 2⅝ lbs.

6L4821
Per pair....**79c**

Extra Rolls for Above.
6L4840—Plain maple. Shipping weight, 8 oz. Per set of **8**....**12c**

Skate Strap Pad.

Used with either roller or ice skates. Soft, pliable leather, well lined, with loop on top for inserting strap. Relieves pressure of skate strap on shoe buttons or lace hooks, as well as preventing chafing of the shoe. Shipping weight, 4 ounces.

6L4819—Per pair....**28c**

Union Hardware Co. Children's Extension Skates With Steel Self Contained Ball Bearing Rolls.

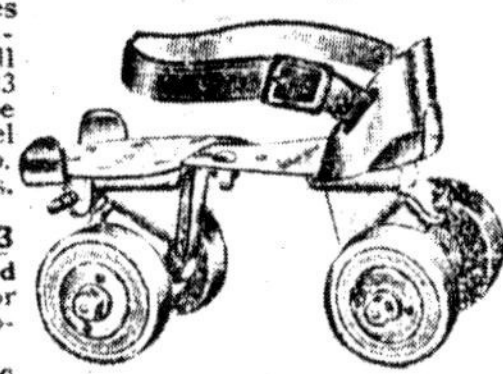

To fit children's shoes from sizes 6 to 8. Construction similar to full size Skates 6L4823 above. Nickel plated toe clamp and high heel band with leather strap. Shipping weight, 4 lbs.

6L4827
Per pair....**$1.63**

Steel Self Contained Ball Bearing Rolls, for Skates 6L4827. Shipping weight, 5 oz.
6L4857—Each **12c**

Playground Equipment

For our complete line of Playground equipment, write for Sporting Goods Catalog, 568GCL.

Help Make the Children Happy and Healthy.

Playground Swing Outfit.

A strong, well built, three-swing outfit. The frame is 10 feet high above ground. The upright supports and the top piece are of 2-inch painted iron pipe. The swings are supported by heavy galvanized chains, the seats being of selected hardwood set in malleable iron brackets in such a way as to insure strength and safety. The frame should be embedded in concrete or cement and permitted to harden thoroughly before being used. Shipped from factory in NORTHERN INDIANA. Shpg. wt., 300 lbs.

6L6263⅓—Complete, as illustrated....$49.50

Protective Feature Playground Slide.

→ Protective Hand Rail.

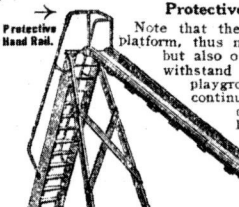

Note that the hand rails entirely surround the platform, thus not only adding a factor of safety but also offering more exercise. Designed to withstand very severe strain in large public playgrounds or parks where subjected to continued use by a great number of children. It is built throughout of heavy hardwood stock. The steps, rails and all braces are assembled with strength and durability foremost in mind. The bottom of the chute is constructed with a substantial tongue and groove wood flooring, and then lined with galvanized "Armco" iron, thus making the slide "fast" and capable of withstanding severe weather conditions. Can also furnish Slides with chute made entirely of hard maple. **State kind of chute wanted.** It is well painted and finished with great care throughout. Shipped from factory in NORTHERN INDIANA.

Catalog No.	Length Chute	Height Ladder	Shipping Weight	Each
6L6105⅓	16 feet	8 feet	350 lbs.	$61.40
6L6106⅓	20 feet	10 feet	550 lbs.	82.15

Giant Stride.

Vertical post of 4-inch iron pipe 18 feet long, 4 feet of which should be embedded in concrete. Six all steel chain strides suspended from a revolving weatherproof malleable iron ball bearing head. Can be furnished with fittings only, without the upright pipe, if desired, as the fittings can be readily fastened on any 4-inch pipe. No threading of pipe required.

6L6108⅓—Complete. Shipping weight, 245 pounds. Shipped from factory in NORTHEASTERN INDIANA....**$49.00**

6L6109⅓—Without pipe, fittings only. Shipping weight, 75 lbs. Shipped from factory in NORTHERN INDIANA....**$34.00**

Giant Stride Head or Pivot. Revolves on bearings. Heavy malleable iron, to be driven into end of 4-inch pipe. Accomodates six or eight strides. Shipped from factory in NORTHERN INDIANA. Shipping weight, 28 lbs.

6L6110⅓....**$15.00**

Extra Strides or Ladders. Length over all, 80 inches. Galvanized chain, all steel. Shipping weight, 6 pounds.

6L6113⅓....**$3.25**

Playground Slides.

A portable slide, constructed throughout of well seasoned hardwood, thoroughly coated with green paint. The ladder is built with steps of hardwood, strongly put together, and provided with a steel pipe railing. The bottom or bed of the slide is sheet steel. This type of slide is "fast" and will withstand the weather. A special feature is the hinge arrangement connecting the slide and ladder. This makes it compact and easy to set up or take down. Suitable for either private or public use.

Catalog No.	Height to Top of Ladder	Length of Slide, Feet	Shpg. Wt., Lbs.	Each
6L6100⅓	5 ft.	10	90	$17.25
6L6101⅓	6 ft.	12	100	21.00
6L6102⅓	7 ft.	14	105	24.00
Extra Heavy Slide.				
6L6103⅓	8 ft.	16	290	$51.05

Shipped from factory in NORTHERN INDIANA

Junior Teeter Totter.

For youngsters from 4 to 8 years of age. Made of hardwood throughout. The board is 10 feet long, 8 inches wide, and is substantially mounted on a steel pivot. Height, about 24 inches.

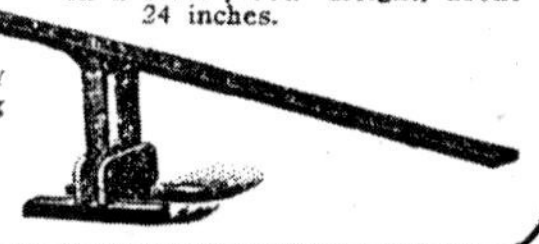

Shipped immediately from stock. Shipping weight, 30 pounds.

6L6120¼
$5.75

Bathing Suits and Accessories

Aviators' Style Rubber Cap. Suitable for women as well as for men. Natural color. Pure Para rubber. Shipping weight, 3 oz.
6L2109...**49c**

Divers' Plain Style Rubber Cap. Natural color. Pure Para rubber. Shpg. wt., 3 ounces.
6L2105...**42c**

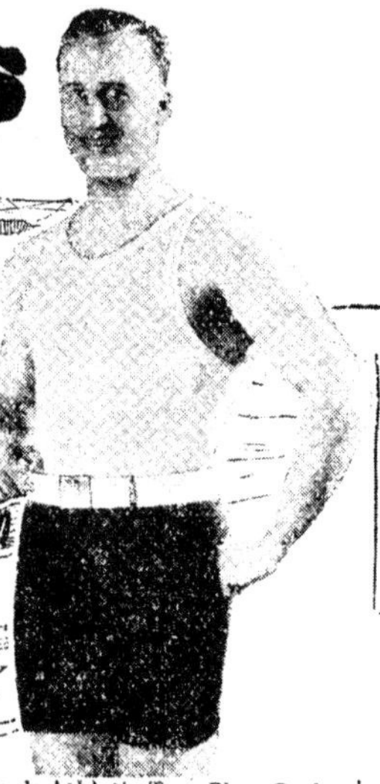

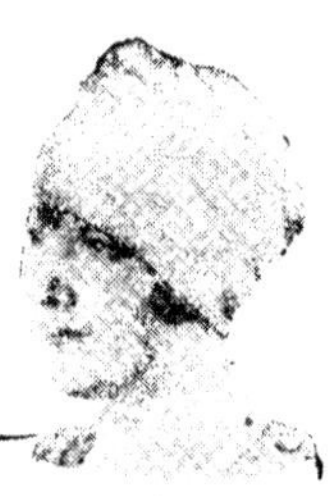

Divers' Style Rubber Cap. For both women and men. Plain colors. Shipping weight, 3 oz.
6L2100—Black.**19c**
6L2101—Blue..**19c**
6L2102—Green.**19c**
6L2103—Brown.**19c**
6L2104—Red..**19c**

Men's Medium Weight All Wool Worsted Bathing Suit. Brown heather with orange and royal blue chest stripes. Sizes, 36 to 46 inches chest measure. **State size.** Shipping weight, 1¼ pounds.
6L2125..............**$3.45**

Men's Heavy Weight All Wool Worsted Bathing Suit. A very high grade suit. Navy blue with white trimming. No chest stripes. V shape neck. Sizes, 36 to 46 inches chest measure. **State size.** Shipping weight, 1¼ pounds.
6L2141..............**$3.80**

NOTICE
All our bathing suits are California style, with the exception of our two-piece suits 6L2143 and 6L2146.
They are all exceptionally well made, extra full sizes and are unusual value for the prices we ask. Not to be compared with cheaper suits that are skimped in size.

Men's Athletic Two-Piece Bathing Suit. The very latest style for men. All wool worsted knitted white shirt with combination supporter attached. Knitted navy blue fly front pants. White web belt with nickel plated buckle. Sizes, 34 to 46 in. chest measure. **State size.** Shpg. wt., 1½ lbs.
6L2143—White shirt, with navy blue knitted pants and belt...**$3.95**

6L2146—**Same as above,** but with white shirt and all wool navy blue flannel pants and web belt. **State size**....................**$3.95**

Men's Medium Weight All Wool Worsted Bathing Suit. Plain navy blue. Excellent value for the money. Sizes, 36 to 46 inches chest measure. **State size.** Shipping weight, 1¼ pounds.
6L2140....................**$2.85**

Men's All Cotton Bathing Suit. Medium weight. Style as above. Navy blue with trimming of contrasting color. **State size.** Shipping weight, 1¼ pounds.
6L2139....................**95c**

Women's All Wool Worsted Bathing Suit. Sizes, 36 to 46 inches bust measure. **State size.** Shipping weight, 1½ pounds.
6L2164—Color, brown with tan trimming. V shape neck. **State size**.....**$4.35**

Same style as shown above. Color, Kelly green with white trimming. V shape neck. **State size.**
6L2169....................**$4.35**

6L2163—Navy blue with white trimming. Round neck....................**$3.75**

Women's Cotton Bathing Suit. Medium weight. Navy blue with trimming of contrasting colors. **State size.**
6L2160....................**$1.10**

Women's All Wool Worsted Bathing Suit. Medium weight. California style, with belt. Trimmed with gold braid, lustrous finish in two beautiful color combinations. A very attractive suit. Sizes, 36 to 46 inches chest measure. **State size.** Shpg. wt., 1½ lbs.
6L2167—Celestial blue with gold braid trimmings..............**$4.95**
6L2166—Kelly green with gold braid trimmings...**$4.95**

Men's Medium Weight All Wool Worsted Suit. Myrtle green with cardinal trimming. Round neck. Sizes, 36 to 46 inches chest measure. **State size.** Shipping weight, 1¼ pounds.
6L2142....................**$3.00**

Men's Medium Weight All Wool Worsted Suit. Brown with tan trimming. Sizes, 36 to 46 inches chest measure. **State size.** Shipping weight, 1¼ pounds.
6L2127....................**$3.10**

Women's Bow Style Rubber Diving Cap. Very attractive. Shipping weight, 3 ounces.
6L2126—Blue, gold trimmed.....**40c**
6L2128—Red, blue trimmed.....**40c**
6L2129—Green, black trimmed....**40c**

Women's Butterfly Style Rubber Diving Cap. Shpg. wt., 4 oz.
6L2107—Red, blue trimmed..........**40c**
6L2110—Green, black trimmed......**40c**
6L2111—Blue, white trimmed......**40c**

Jockey Style Rubber Cap. Visor acts as a sun shade. Shpg. wt., 5 oz.
6L2136—Green, white bow.........**60c**
6L2137—Red, white bow.........**60c**
6L2138—Blue, white bow.........**60c**

Ayvad's Water Wings. When inflated will support an adult of 200 pounds at the proper level for comfortable swimming. An excellent support to use when learning to swim. Shipping weight, 4 oz.
6L2121—Pair.**35c**

Juvenile All Wool Bathing Suit. Old rose color, no trimming. Sizes, 22 to 28 inches chest measure. **State size.** Shipping weight, 1 pound.
6L2124......**$1.25**

Rubber Surf Ball. Inflated rubber ball. Raised stars prevent ball from slipping when hands are wet. 13 inches in circumference. Can be used in water as well as on the beach. Shipping weight, 6 oz.
6L2117.....**25c**

Cork Surf Ball. A great fun producer for bathers. Measures 9 inches in circumference. Shipping weight, 4 oz.
6L2115.....**25c**

Women's Trimmed Style Rubber Diving Cap. Shipping weight, 3 ounces.
6L2106—Blue, white trimmed.....**40c**
6L2108—Green, red trimmed.......**40c**
6L2120—Red, blue trimmed..........**40c**
6L2122—Black, orange trimmed....**40c**

Bathers' Web Belts. Color, white. Nickel plated buckle. Non-tarnishable. Sizes, 26 to 42 in. waist measure. **State size.** Shpg. wt., 5 oz.
6L2195.......**18c**

Misses' All Wool Worsted Bathing Suit. Copenhagen blue with scarlet and black chest stripes. Sizes, 28 to 34 inches bust measure. **State size.** Shpg. wt., 1¼ lbs.
6L2194.........**$2.50**

Misses' All Cotton Bathing Suit. Navy blue with trimming of contrasting colors. **State size.**
6L2187............**98c**

Boys' All Wool Worsted Bathing Suit. Maroon with combination orange and blue chest trimming. Sizes, 28 to 34 inches chest measure. **State size.** Shipping weight, 1 pound.
6L2123..........**$2.48**

Boys' All Cotton Suit. Navy blue with trimming of contrasting colors. **State size.**
6L2131............**88c**

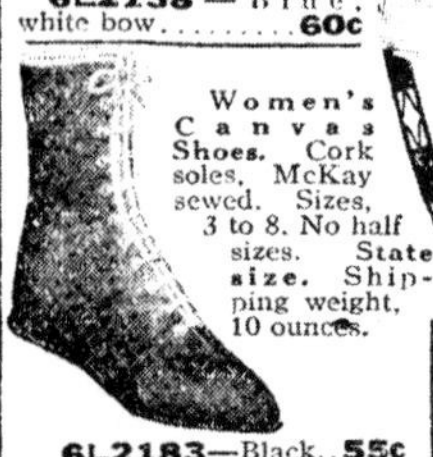

Women's Canvas Shoes. Cork soles, McKay sewed. Sizes, 3 to 8. No half sizes. **State size.** Shipping weight, 10 ounces.
6L2183—Black..**55c**
6L2184—White..**55c**

Women's Sateen Shoes. Rubber soles, McKay sewed. Sizes, 3 to 7. No half sizes. **State size.** Shipping weight, 1 pound.
6L2173—Black..**75c**
6L2174
Navy blue..........**75c**
6L2175—Green.**75c**

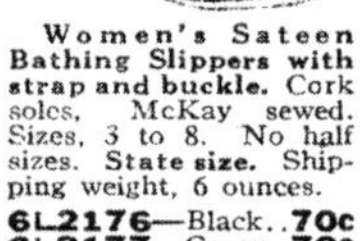

Women's Sateen Bathing Slippers with strap and buckle. Cork soles, McKay sewed. Sizes, 3 to 8. No half sizes. **State size.** Shipping weight, 6 ounces.
6L2176—Black..**70c**
6L2177—Green..**70c**
6L2178—Red...**70c**

Neptune Safety Swimming Suspenders.
Consists of inflated rubber tubes, front and back, comfortably fitted, out of the way, easily filled through two good valves. Beginners will find they are an aid in swimming, with but little exertion. This is not a life preserver, but will enable a non-swimmer to keep head above water with just a slight movement of the feet and keep afloat for a long time without tiring. Adult size. Shpg. wt., 2 lbs.
6L2208............**$3.25**

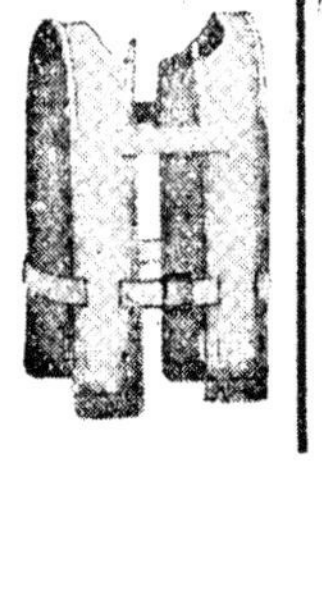

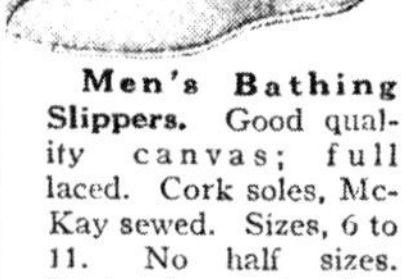

Men's Bathing Slippers. Good quality canvas; full laced. Cork soles, McKay sewed. Sizes, 6 to 11. No half sizes. **State size and color.** Shipping weight, 8 oz.
6L2186
Black...........**75c**
6L2185
White...........**75c**

Cork soles. McKay sewed.
Shipping weight, 6 ounces.
Women's Canvas Bathing Slippers. Sizes, 3 to 8. No half sizes. **State size.**
6L2156—Black..**54c**
6L2157—White..**54c**

Women's Sateen Bathing Slippers. Sizes, 3 to 8. No half sizes. **State size.**
6L2200—Black..**65c**
6L2201
Navy blue..........**65c**
6L2202—Green.**65c**
6L2203—Brown.**65c**

Children's Canvas Bathing Slippers. Sizes, 11 to 2. No half sizes. **State size.**
6L2154—Black..**50c**
6L2155—White..**50c**

ATHLETIC CLOTHING

All Wool Worsted Striped Athletic Shirts.

Sleeveless style with plain chest stripe. Medium weight. Sizes, 30 to 42 inches chest measure. **State size wanted.** Shipping wt., 12 oz.

6L2043—Oxford gray with maroon stripes **$1.95**
6L2044—Navy blue with white stripes **$1.95**
6L2045—Maroon with white stripes **$1.95**
6L2046—Black with orange stripes **$1.95**

Medium Weight Cotton Striped Athletic Shirt. Sleeveless style, with stripe around body. Sizes, 30 to 42 inches chest measure. **State size.** Shipping weight, 10 oz.

6L2031—Navy blue with white stripe **80c**
6L2020—Maroon with white stripe **80c**
6L2019—Black with orange stripe **80c**

All Wool Worsted Athletic Shirts. Solid Colors.

Made in sleeveless style only. Medium weight. Plain colors. Sizes, 30 to 42 in. chest measure. **State size.** Shpg. wt., 12 oz.

6L2032—Oxford gray **$1.85**
6L2033—Maroon **1.85**
6L2034—Navy blue **1.85**
6L2038—White **1.85**

Cotton Athletic Shirts. Solid Colors.

Sleeveless style, medium weight cotton, well sewed. Sizes, 26 to 44 inches chest measure. **State size.** Shipping weight, 10 ounces.

6L2025—Black **60c**
6L2026—Navy blue **65c**
6L2027—White **50c**

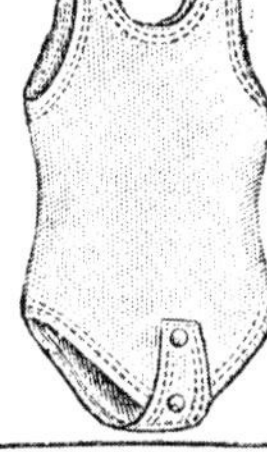

Cotton Athletic Shirt With Supporter.

Made of medium weight cotton in white color only. Supporter attachment is a real feature. In addition to a supporter, it will prevent shirt from pulling out of trunks. Sizes, 28 to 42 chest measure. **State size wanted.** Shipping weight, 10 ounces.

6L2021—White only **55c**

Athletic Stockings.

Adapted for athletic use, especially for basket ball use, by rolling down hose, making same into a cuff to suit players' convenience. Elastic band will keep them in proper place. Furnished in sizes 9 to 11. Shipping weight, 8 ounces.

Heavy Ribbed Cotton Stockings. Colors: Black, navy blue, white or maroon. **State size and color.**

6L2060 .... **65c**

Same as 6L2060, with 3-in. white calf stripe; also black with 3-in. orange stripe, or navy blue with 3-in. orange stripe, oxford gray with navy, maroon or green stripe. **State size and color.**

6L2061 .... **70c**

All Wool Double Ribbed Stockings. Cotton feet. Colors: Black, navy blue, white or maroon. **State size and color.**

6L2064 .... **$1.28**

Same grade as 6L2064, but with 3-inch single white calf stripe. State color.

6L2065 .... **$1.33**

Same grade as 6L2064, but with triple varsity stripe (one wide and two narrow stripes). Navy blue with white stripes, maroon with white stripes or black with orange stripes. **State color and size.**

6L2071 .... **$1.35**

Wrestlers' Supporter.

Pfister model; high quality black elastic waistband with knitted jersey cloth pouch and under-strap. Sizes, 28 to 44 in. waist measure. **Give waist measure.** Shpg. wt., 8 oz.

6L2094 .... **$1.23**

Athletes' Morton Style Supporter.

Made of Canton flannel, lace front. Sizes, 24 to 46 in. waist measure. **Give waist measure.** Shipping weight, 5 ounces.

6L2039 .... **50c**

Supporter with elastic gore on each side, otherwise made same as above. Shipping weight, 5 oz.

6L2040 .... **68c**

For full line of Supporters see page 474.

Women's and Girls' Athletic Clothing.

These bloomers and middy blouses are made for gymnasium use, and are endorsed by instructors everywhere. When you buy these, you may be sure that you are buying the proper garments for the purpose.

Middy Blouse.

Made of excellent quality, heavy cotton twill cloth. Square sailor collar, lace front, one pocket. Furnished with long or short sleeves. White only. Sizes, 30 to 44 inches bust measure. **State size.** Shipping weight, 12 ounces.

6L2051—With long sleeves ... **$1.90**
6L2052—With short sleeves ... **1.60**

Cotton Bloomers.

Full plaited black twill, good weight, extra full. Two-button fastening at waist, elastic band around knees. Black only. Sizes, 24 to 34 inches waist measure. **State waist measure.** Shipping wt., 12 oz.

6L2083 .... **$1.65**

Children's Bloomers.

Same style as 6L2083. Sizes, 6 to 14 years. **State age.** Shipping weight, 10 oz.

6L2096 .... **$1.20**

Serge Bloomers.

Good quality, one-half wool. Extra full, plaited to knee, high grade, well made garment. Two-button fastening at waist, elastic band around knees. Sizes, 24, 26, 28, 30, 32 and 34 inches waist measure only. **State waist measure.** Shpg. wt., 1 lb.

6L2081—Black **$3.35**
6L2082—Navy blue **3.35**

Sweat Shirt.

Made of a good grade of cotton, fleece lined, gray color, with low collar and long sleeves. Highly recommended for all kinds of sports. Protects the athlete from chill. Keeps the body warm. Worn with or without a top shirt. Sizes, 34 to 42 inches chest measure. **State size.** Shipping weight, 2½ pounds.

6L2015 ..... **98c**

All Wool Worsted Wrestling Tights With Double Knees.

Good Quality All Wool Worsted Knit Full Length Tights with double knees; for wrestling, hockey, skating or gymnasium. Sizes, 30 to 40 inches waist measure and 30 to 36 inches inseam measure. **State waist and inseam measures.** Black color only. Shipping wt., 1 lb.

6L2093 .... **$4.50**

Cotton Wrestling Tights.

Same style and sizes as above, but without double knees. Color, black only. **State size.**

6L2092 .... **$1.98**

All Wool Worsted Striped Athletic Shirts.

Sleeveless style with stripes of latest pattern around body; V trimmed neck. Medium weight. The very latest style. Sizes, 30 to 42 in. chest measure. **State size wanted.** Shipping weight, 12 ounces.

6L2016—Navy blue with white stripes **$2.10**
6L2017—Maroon with white stripes **$2.10**
6L2018—Black with orange stripes **$2.10**

Basket Ball Pants.

Olive drab drill, heavy quality material, padded hips; fly front, belt loops, full seat, short inseam. 26 to 42 inches waist. **State waist measure.** Shipping weight, 9 ounces.

6L2072 .... **$1.10**

Flannel Basket Ball Pants.

Same style as above. Medium weight flannel, about one-fourth wool, with one stripe on side seams, as listed below. Padded at hips. Sizes, 28 to 40 inches waist measure. **State waist measure and color.** Shpg. wt., 12 oz.

6L2074—Navy blue with white **$1.75**
6L2075—Oxford gray with navy blue **1.75**
6L2077—Black with orange **1.75**

Basket Ball Pants With Removable Pads.

Olive drab, heavy quality drill with latest improved removable pads; adjustable at waistband; fly front has strap and buckle at waist; double stitched seams. Can also be used as **track, running and camping pants** by removing the pads. Sizes, 28 to 40 inches waist measure. **State size.** Shipping weight, 12 ounces.

6L2078 .... **$1.65**

"Prep" Basket Ball Pants. Used in All Match Games.

Our Best Grade Flannel Basket Ball Pants with loose hanging hip pads. Made of suitable weight flannel, about one-half wool, with latest style strap and buckle piece. Prominent stripe on side seams. Sizes, 28 to 40 in. waist measure. **State size.** Shpg. wt., 1 lb.

6L2048—Navy blue with white stripes **$2.60**
6L2049—Maroon with white stripes **$2.60**
6L2050—Orange with black stripes **$2.60**

Knitted Athletic and Boxing Trunks.

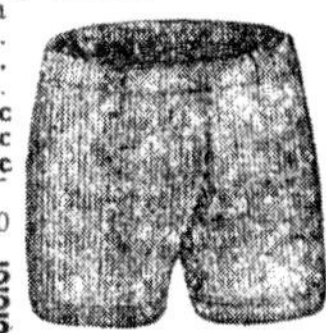

Knitted Cotton Trunks. Fitted with draw strings. No opening in front. Sizes, 26 to 40 inches waist measure. **State size.** Shipping weight, 6 ounces.

6L2035—White **55c**
6L2036—Black **60c**
6L2037—Navy blue **68c**

All Wool Worsted Knit Trunks.

Same style as above. Sizes, 26 to 40 in. waist measure. Shpg. wt., 8 oz.

6L2097—Black **$1.25**
6L2095—Navy blue **1.25**
6L2099—Green **1.25**

Track, Gymnasium and Soccer Pants.

Flannel Athletic Pants with belt loops and hip pocket. About one-third wool. Unpadded. Sizes, 26 to 40 in. waist measure. Shipping weight, 8 oz.

6L2085—Navy blue **$1.75**
6L2088—Gray **1.76**

Plain White Muslin Track and Gymnasium Pants.

Fly front. Made same as above, but without belt loops and pocket. Sizes, 26 to 40 inches waist measure. Shipping weight, 5 ounces.

6L2089 .... **55c**

Cotton Web Belts.

Adapted for athletic use because of its flexibility. Leather strap and polished buckle, 1¼ inches wide. **State size.** Shipping weight, 2 oz.

6L2066—Black .... **30c** | **6L2068**—White **30c**
6L2067—Navy blue .... **30c** | **6L2069**—Maroon **30c**

Dumbbells and Indian Clubs.

Fine quality rock maple, polished. Weight given is the approximate weight of each club or dumbbell. For shipping weight add about 1 pound to actual weight of a pair of dumbbells or clubs. **State weight wanted.**

Dumbbells. 6L1557 Per Pair	Dumbbells. 6L1557 Dozen Pairs	Weight, Pounds	Indian Clubs. 6L1556 Per Pair	Indian Clubs. 6L1556 Dozen Pairs
$0.54	$ 6.36	½	$0.66	$ 7.76
.65	7.68	¾	.70	8.28
.75	8.88	1	.85	10.08
.90	10.68	1¼	.97	11.52
1.15	13.68	2	1.25	14.88
1.47	17.40	3	1.52	18.00

Iron Quoits.

Will stand hard usage. Plainly stamped as illustrated.

6L5956—About 2 pounds each. Shipping weight, 11 lbs. Set of 4 .... **80c**

6L5957—About 3 pounds each. Shipping weight, 14 lbs. Set of 4 .... **$1.20**

6L5958—Iron Pegs. Shipping weight, 1 pound 5 ounces. Per pair **12c**

SPORT ACCESSORIES

Regulation Pitching Horseshoes.

Made of good grade steel, unbreakable, regulation weight. One pair stamped number 1, painted blue color, and the other pair stamped number 2, painted red color, so that each pair can be easily distinguished. A set of official rules included with each pair. Shipping weight, 7 pounds.

6L5967—Blue shoes, number 1. Per pair **$1.25**
6L5968—Red shoes, number 2. Per pair **1.25**
6L5959—Regulation size steel stakes. Shipping weight, 12 pounds. Per pair **1.25**

Bamboo Vaulting Poles.

Selected bamboo, spike riveted to end. For outdoor use.

6L5963¼—10 feet. Shipping wt., 4 lbs. **$6.95**
6L5964¼—12 feet. Shipping wt., 4½ lbs. **7.40**
6L5965¼—14 feet. Shipping wt., 5 lbs. **7.95**
6L5966¼—16 feet. Shipping wt., 7 lbs. **8.40**

Iron Dumbbells.

If only one dumbbell is wanted in the 10, 15, 20 or 25-pound weights, take one-half of the price listed below. For shipping weights, add 1 pound for each dumbbell. **State weight wanted.**

6L1516—**Black finish.**

Wt.	Per Pr.	Wt.	Per Pr.	Wt.	Per Pr.
1 lb.	18c	5 lbs.	$0.84	15 lbs.	$2.52
2 lbs.	34c	8 lbs.	1.33	20 lbs.	3.25
3 lbs.	52c	10 lbs.	1.70	25 lbs.	4.10

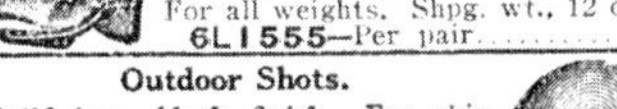

Dumbbell and Indian Club Hangers.

Made of iron, black enamel finish. For all weights. Shpg. wt., 12 oz.

6L1555—Per pair **10c**

Outdoor Shots.

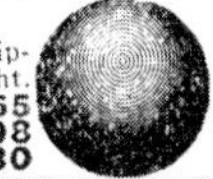

Solid iron, black finish. For shipping weight, add 1 lb. to actual weight.

6L5960—8-pound weight ... **$0.65**
6L5961—12-pound weight ... **.98**
6L5955—16-pound weight ... **1.30**

Boxing Gloves

Pupils' Special Style Boxing Gloves.
Corbett pattern, with double length padded cuffs, finger grips and laced wrists. High quality tan color soft leather, drill lined and stuffed with good quality hair. Made strong and durable to withstand the severe usage given boxing gloves by amateurs. Ventilated palms. Designed with extra long cuffs to give all the protection possible. Weight, each, about 9 oz. Shipping weight, 4½ pounds.
6L1409—Set of 4 gloves **$6.10**

Latest Corbett Pattern Gloves.
Fine quality olive tanned soft leather, stuffed with good quality hair, double stitched, drill lined, leather bound, laced wrists, padded cuffs, finger grips, ventilated palms. Excellent gloves for instructors. Weight, each glove, about 8 oz. Shipping weight, 3¼ lbs.
6L1407—Set of 4 gloves **$5.45**

Championship Model Gloves.
Corbett Pattern 8-Ounce Boxing Gloves. Full heel pads below the lacing, outside palm grips, full padded cuffs; laced wristbands; made of high quality tan color leather, stuffed with excellent quality curled hair. Shipping weight, 3¼ lbs.
6L1411—Set of 4 gloves **$6.80**

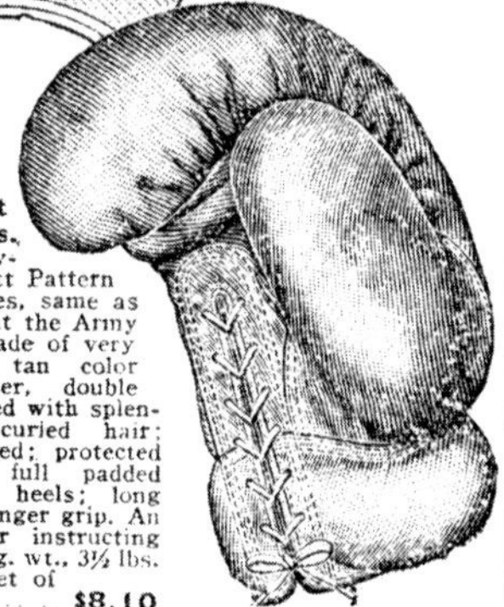

Government Pattern Gloves.
Regular Government Corbett Pattern 10-Ounce Gloves, same as used throughout the Army and Navy. Made of very high quality tan color glove leather, double stitched; stuffed with splendid quality curled hair; strong drill lined; protected thumbs and full padded cuffs; padded heels; long laced wrists; finger grip. An ideal glove for instructing purposes. Shpg. wt., 3½ lbs.
6L1412—Set of 4 gloves **$8.10**

Approved Battling Gloves.
Men's Approved Battling Pattern Gloves. Made of good quality soft glove leather, with finger grips and toe pads, ventilated palms, padded wrists, full lined, leather binding, laced wrists, stuffed with good quality hair, double stitched throughout. About 6-ounce sparring gloves. Shipping weight, 3 pounds.
6L1413—Set of 4 gloves **$4.65**

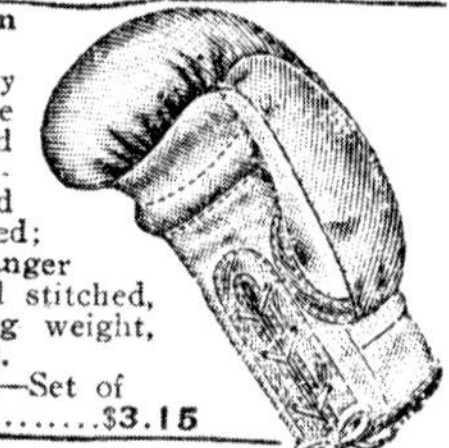

Youths' Corbett Pattern All Leather Gloves.
Youths' Good Quality All Leather 5-Ounce Gloves. Made of good quality soft glove leather. Stuffed with good quality hair; full lined; ventilated palm; finger grips. Full laced; well stitched, full bound. Shipping weight, 2¼ pounds.
6L1404—Set of 4 gloves **$3.15**

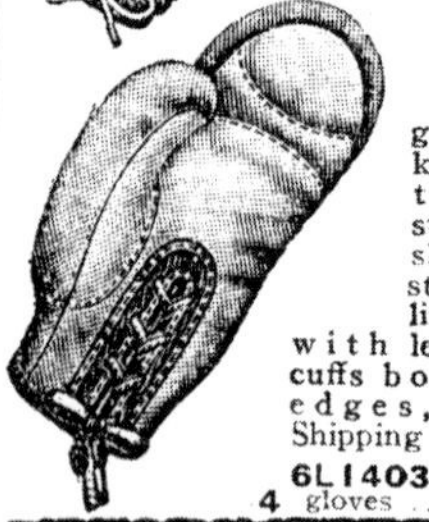

Boys' Favorite Gloves.
Made of good quality khaki drill throughout, stuffed with short hair, well stitched, full lined, laced with leather lacing, cuffs bound around edges, finger grips. Shipping weight, 2 lbs.
6L1403—Set of 4 gloves **$1.60**

Juvenile All Leather Boxing Gloves. 
Excellent gloves for youngsters, ages 4 to 8 years. Made of tan color soft glove leather; stuffed with good quality hair, canvas lined. Well stitched throughout. Standard pattern. Shipping wt., 1½ lbs.
6L1401—Set of 4 gloves **$1.75**

Men's Corbett Pattern Gloves. 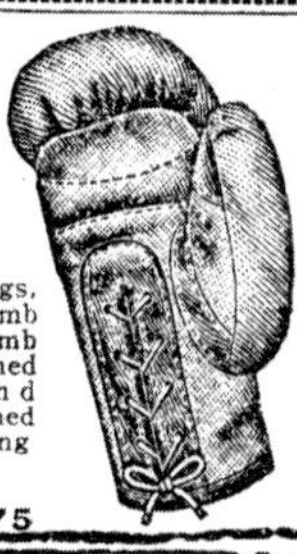
Corbett Pattern 6-Ounce All Leather Gloves. Made of soft tanned glove leather. Stuffed with good quality hair, full lined, deep lacing with leather facings, leather reinforced thumb tips, also where thumb joins glove, stitched finger grips, bound edges. Well stitched throughout. Shipping weight, 2¼ pounds.
6L1402—Set of 4 gloves **$3.75**

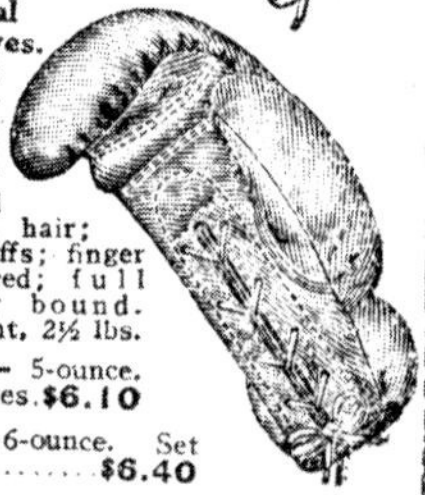

Professional Fighting Gloves.
Standard Pattern Professional Model Gloves. Olive color leather, stuffed with curled hair; padded cuffs; finger grip; full laced; full lined; leather bound. Shipping weight, 2½ lbs.
6L1417 — 5-ounce. Set of 4 gloves. **$6.10**
6L1418—6-ounce. Set of 4 gloves **$6.40**

Striking Bags and Accessories

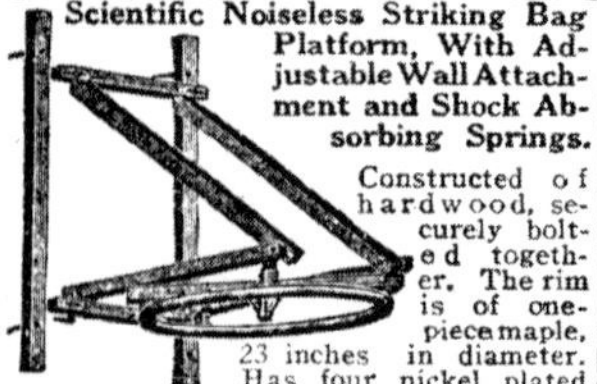

Scientific Noiseless Striking Bag Platform, With Adjustable Wall Attachment and Shock Absorbing Springs.
Constructed of hardwood, securely bolted together. The rim is of one-piece maple, 23 inches in diameter. Has four nickel plated vibrating cushion springs between the rim and the frame which absorb all vibration. Weight, packed for shipment, 15 pounds.
6L5815¼ **$4.50**

Elastic Floor Attachment.
Elastic covered with braided cotton. Used for attaching the bottom of a double end bag to the floor. Shipping weight, 3 ounces.
6L1442 .... **25c**

Striking Bag Mitts. 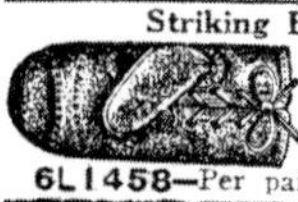
Soft leather, grip in center, padded backs; laced wrists. Shpg. wt., 6 oz.
6L1458—Per pair **$1.25**

Pear Shape Striking Bags.

Youths' Size All Leather Bag. Made of soft napa tanned leather. Well stitched, taped seams, drill lined, strong leather loop top. Shipping wt., 1¼ lbs.
6L1435
With bladder **$2.65**

Tan Color Napa Tanned Leather Bag. Well stitched and taped seams; drill lined; strong loop top; fine quality bladder. Shipping wt., 1¼ lbs.
6L1438—With bladder **$3.00**

Excellent Quality Olive Green Color Napa Tanned Leather Bag. Welted and triple seams, stitched with strong thread; full lined; strong leather loop top; 31 inches in circumference when inflated. Shpg. wt., 1¼ lbs.
6L1439—With bladder **$3.60**

Professional Bag. High quality tanned horsehide, tan color. Hand sewed leather loop top; welted seams; full lined; 32 inches in circumference when inflated. Shpg. wt., 1¾ lbs.
6L1440—With bladder **$4.75**

Striking Bag Bladder. For pear shaped bag, fine quality rubber. This style can only be used in pear shaped bags. Shpg. wt., 5 oz. 
6L1454 .... **55c**

Oval Shape Striking Bags.

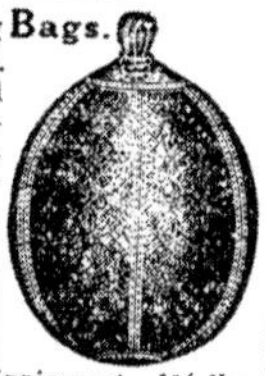

Boys' Size All Leather Bag. Soft napa tanned leather, full canvas lined. Reinforced bottom and top; leather loops at both ends. Rubber bladder. Shipping weight, 1 lb.
6L1424 .... **$1.90**

Youths' Size Bag. Soft napa tanned leather, full lined; 30 inches in circumference. Complete with rope, elastic and bladder. Shipping wt., 1⅛ lbs.
6L1426 .... **$2.40**

Full Size, Tan Color, Soft Tanned Leather Bag, full lined. Triple stitched. Complete with rope, elastic and bladder. Shpg. wt., 1¼ lbs.
6L1429 .... **$3.20**

Expert Bag, made of selected tanned horsehide; very strong and tough. Drill lined, triple seams, welted; strong loop; high class in every respect; 32 inches in circumference when inflated. Complete with rope, elastic and bladder. Shipping weight, 1½ pounds.
6L1436 .... **$5.45**

Striking Bag Bladder. For oval and bell shape striking bags. Fine quality pure rubber. Shpg. wt., 5 oz. 
6L1453 .... **50c**

Bell Shape Striking Bag. Double End. 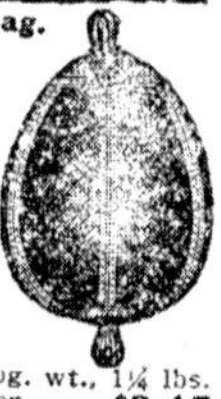
Fine quality black tanned leather, full lined, welted and taped seams. Triple stitched with strong linen thread. Strong loops at top and bottom, so it can be used either as single or double end bag. Furnished complete with rubber bladder and elastic, rope and screw eyes. Shpg. wt., 1¼ lbs.
6L1441—With bladder **$3.45**

Professional Kno-Knot Bag Swivel. 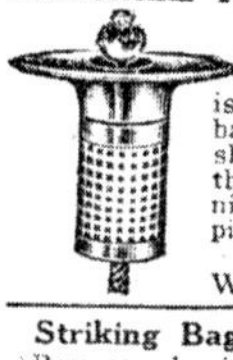
Made of malleable iron, removable sockets, full nickel plated. Rope is fastened into a steel ball, as illustrated. No slipping or twisting of the rope. Rope is furnished with swivel. Shipping weight, 2 pounds.
6L1443
With rope **90c**

Striking Bag Swivel. 
Bag can be instantly removed or a new rope inserted by unscrewing the projecting stem from the round disc. Nickel plated. Shpg. wt., 10 oz.
6L1450 .... **53c**

Flags and Wool Felt

United States Silk Flags. 
Good quality printed silk taffeta with hemmed edges. Mounted on dark stained sticks fitted with gilt spearheads. Furnished in quantities only as listed. A very beautiful flag for decorating or use on almost any occasion.
6L4418—One only, silk flag, size 24x36 inches. Shipping weight, 2½ pounds **$1.85**

Catalog No.	Size, Inches	Set of 3	Shpg. Wt.	Per Doz.	Shpg. Wt.
6L4415	5x 8	$0.57	8 oz.	$ 2.25	1¾ lbs.
6L4416	8x12	.82	1 lb.	3.20	2 lbs.
6L4417	12x18	1.42	1½ lbs.	5.60	2½ lbs.
6L4418	24x36	5.48	3 lbs.	21.85	6 lbs.

United States Muslin Flags.
Mounted on plain sticks without spearheads. We do not furnish in smaller quantities than quoted below.

Catalog No.	Size, Inches	Per Doz.	Shpg. Wt.	Per Gross	Shpg. Wt.
6L4410	3½x 6	$0.10	4 oz.	$ 1.10	1 lb.
6L4411	6 x 9½	.24	1 lb.	2.70	8 lbs.
6L4412	12 x22	1.00	2 lbs.	11.75	14 lbs.
6L4413	18 x27½	1.75	3 lbs.	20.00	26 lbs.

United States Bunting Flags.
Good grade soft cotton bunting; hemmed edges. Mounted on wood staffs fitted with gilt spearheads.

Catalog No.	Size, Inches	Per Doz.	Shpg. Wt.	Per Gross	Shipping Wt., Lbs.
6L4435	7½x11	$0.66	1 lb.	$ 7.25	7½
6L4436	12 x17	1.00	2 lbs.	10.98	19
6L4437	15 x24	1.75	2½ lbs.	19.15	24
6L4438	24 x36	3.50	4½ lbs.	38.25	50

Wool Felt.
Fine Quality All Wool Felt in piece for making pennants, pillow tops, etc. 72 inches wide. Comes in black, white, purple, maroon, orange, red, yellow, green, old gold or tan. **State color wanted.** We do not sell less than ¼ yard. Shipping weight, per yard, about 1¼ pounds.
6L5871—Any color above except white. Per yard **$2.40**
6L5872—White only. Per yard **2.55**

United States Flags. Not Mounted. 
Sewed bunting flags. Forty-eight stars, sewed on both sides of field, and placed according to Government regulation. Stripes sewed with double seams. Regulation width of flags is 10/19's of length.
Size recommended for average schoolhouse flag is 8 feet long. A 4-foot flag has but thirteen stars.

ALL WOOL BUNTING FLAGS—U. S. War Department Standard. Fast Color.

Catalog No.	L'gth, Feet	Each	Shpg. Wt.	Catalog No.	L'gth, Feet	Each	Shpg. Wt., Lbs.
6L4440	4	$1.85	6 oz.	6L4449	10	$ 7.74	1⅞
6L4441	5	2.73	10 oz.	6L4451	12	10.38	2⅞
6L4443	6	3.44	12 oz.	6L4452	14	13.83	3¾
6L4445	7	4.34	16 oz.	6L4453	16	17.45	5
6L4447	8	5.34	21 oz.	6L4455	20	24.23	7

SEWED COTTON BUNTING FLAGS—Fast Color. Imitating Standard Bunting.

Catalog No.	L'gth, Feet	Each	Shpg. Wt.	Catalog No.	L'gth, Feet	Each	Shpg. Wt.
6L4423	5	$1.74	8 oz.	6L4425	8	$3.03	19 oz.
6L4424	6	2.06	13 oz.	6L4427	10	4.37	26 oz.

BASKET SOCCER VOLLEY AND FOOT BALLS

"J. C. Higgins" Official Playground Outseam Basket Ball.
Our best grade improved outseam basket ball is especially constructed for outdoor use. Made on the four-piece pattern. High quality American pebbled grain leather with raised seams, which will prevent wear on the stitching. Regulation size, complete with rubber bladder, lace and needle. Shpg. wt., 1¾ lbs.
6L1822 .... **$6.75**

"Scholastic" Outseam Basket Ball.
Another special offer of a low priced cowhide outseam basket ball of reegulation size. Special grade of cowhide, heavy canvas lined, waxed linen thread. Made on the four-piece pattern. Raised seams prevent wear on stitching. For use on rough playgrounds. Complete with rubber bladder, leather lace and needle. Shipping weight, 1¾ pounds.
6L1817 .... $4.40

"Amateur" Outseam Basket Ball.
Good quality pebbled grain sheepskin leather, canvas lined, full size. Made on the four-piece pattern. Raised leather seams make it suitable for rough playgrounds. Will give good service, but cannot be compared with cowhide balls above. Complete with bladder, lace and needle. Shipping wt., 1¼ lbs.
6L1819 .... $3.10

Playground Outseam Volley Ball.
Made of fine quality pebbled grain cowhide with raised seams to prevent wear from rough surfaces. For outdoor use. Heavy canvas lined and sewed with waxed thread. Made on the four-piece pattern. Regulation size and perfect shape. Complete with fine quality bladder, leather lace and lacing needle. Shipping weight, 2 pounds.
6L1840 .... $4.75

"J. C. Higgins" Official Volley Ball.
Our new official volley ball is made of high grade pearl color horsehide, heavy canvas lined and sewed with waxed thread. Official in size, shape and weight. Complete with fine quality bladder, leather lace and lacing needle. Shipping weight, 1½ pounds.
6L1831 .... $3.80

"Service" Volley Ball.
Made of tan color pebbled grain cowhide leather. Regulation size. Canvas lined and sewed with waxed thread. Adapted for hard usage. Furnished with rubber bladder, lace and needle. Shpg. wt., 1¾ lbs.
6L1836 .... $4.45

"Amateur" Volley Ball. Good quality tan color soft leather, canvas lined. Otherwise the same as our "Service" ball. Complete with bladder and lace. Shipping weight, 1½ lbs.
6L1837 .... $2.35

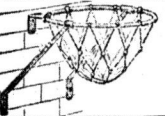

School Playball.
Used in place of the old style round black rubber football. Made on the order of a soccer football, but smaller and lighter in weight. Good quality leather. Canvas lined, well sewed. Popular at playgrounds. Complete with rubber bladder and lace.
6L1801—Boys' size, about 19 inches in circumference. Shipping weight, 5 ounces $1.38
6L1805—Extra rubber bladder for Playball 6L1801. Shpg. wt., 4 oz. .. 37c
6L1802—Youths' size, about 27 inches in circumference. Shipping weight, 9 ounces $1.85
6L1806—Extra rubber bladder for Playball 6L1802. Shipping weight, 4 ounces 40c

Official Basket Ball Goals.
Drop forged iron rim and braces. Made with 6-inch extension; furnished complete with screws and handmade net with draw string at bottom. Shipping weight, per pair, 7 pounds.
6L1832—Per pair $5.60

Regulation Size Basket Ball Goals.
Iron frame, fitted with a cotton net. Draw string bottom which can be left open in practice and closed for match games. Shipping weight, per pair, 12 pounds.
6L1828—Per pair $2.90

Volley Ball Net.
Regulation size, 27 feet long, 3 feet wide, made of No. 12 white cotton twine, same as used in our tennis nets. Shpg wt., 1 pound 15 ounces.
6L1838 .... $1.85

Official Basket Ball Score Book.
To score 20 games. Cloth covered cardboard cover. Shipping weight, 4 ounces.
6L1878 .... 45c

Basket Ball Rules.
Basket Ball Rules for men and women, embodied in one book for the present season. Shipping weight, 4 ounces.
6L1829 .... 14c

Volley Ball Rules. Complete instructions and rules on how to play the game for the present season. Shpg. wt., 5 oz.
6L1839 .... 14c

Club Size Inflater.
Made of polished tube brass. Length, 13 in.; ⅝ inches in diameter. Shpg. wt., 6 oz.
6L1895 .... 35c

"J. C. Higgins" Official Rugby Football.
Our highest grade football. Excellent quality pebbled grain cowhide leather of English tanning and tempering process whereby the stretch is removed. Reinforced seams, sewed with lockstitch waxed linen thread. Hand finished. Each ball is carefully inspected and tested for perfection of weight, size and shape. Fully guaranteed. Fine quality bladder, rawhide lace and lacing needle. Shpg. wt., 1¾ lbs.
6L1810 .... **$5.75**

"College" Rugby Football.

Made of fine quality pebbled grain cowhide leather; stitched with heavy waxed thread. Canvas lined. Second in quality to our J. C. Higgins official ball. Fine for practice. Extra strong. Official size. Furnished with rubber bladder, leather lace and lacing needle. Shpg. wt., 1 lb. 11 oz.
6L1800 .... $4.15

"Prep" Rugby Football.
We are now able to offer a low priced cowhide football of regulation size. Made of a special good grade cowhide leather, canvas lined, strongly stitched with waxed linen thread. A good practice football. Furnished complete with strong rubber bladder, leather lace and lacing needle. Shipping weight, 1½ pounds.
6L1807 .... $2.85

"Leader" Rugby Football.
Regulation size ball. Made of high quality grain pebbled sheepskin leather, canvas lined, full size. Strongly stitched with thread. A well finished ball. Leather lace and rubber bladder included. A very strong ball and one that represents excellent value. Shipping weight, 1¼ pounds.
6L1809 .... $1.85

"Junior" Rugby Football.
Medium size ball. It possesses a good quality pebbled grained sheepskin leather cover and is canvas lined and well made. Stitched with strong thread. Lace and rubber bladder included. A genuine bargain and a ball that will please the boys. Shpg. wt., 15 oz.
6L1813 .... $1.55

Boys' All Leather Rugby Football.
A high grade ball for boys. Well made of good quality sheepskin leather, canvas lined and strongly stitched. Good pure rubber bladder included. Just a little smaller than regulation size and made as good as the larger balls. Not to be confused with the cheap imitation leather balls. Shipping weight, 12 ounces.
6L1818 .... **$1.18**

Extra Bladder for footballs, 6L1813 and 6L1818. Shipping weight, 4 ounces.
6L1811 .... 32c

RUBBER BLADDERS

Basket Ball Bladder.
Good quality rubber. Regulation size, heavy weight. Shpg. wt., 5 oz.
6L1827 .... 63c

Rugby Football Bladder.
Good quality rubber bladder for regulation Rugby footballs. Shipping weight, 5 ounces.
6L1820 .... 48c

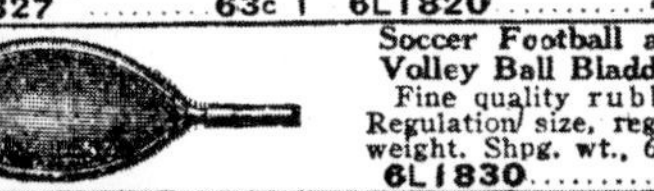

Soccer Football and Volley Ball Bladder.
Fine quality rubber. Regulation size, regular weight. Shpg. wt., 6 oz.
6L1830 .... 48c

Pocket Size Inflater.
Nickel plated football, basket ball, soccer ball, volley ball and striking bag inflater. Shipping weight, 4 ounces.
6L1894 .... 18c

"J. C. Higgins" Official Basket Ball.
Guaranteed to Comply With Official Rules.
Our highest grade basket ball. Made on the official four-piece pattern. Excellent quality pebbled grain cowhide leather of English tanning and tempering process whereby the stretch is removed. Lined with heavy canvas. Sewed with heavy waxed thread, reinforced at seams. Each ball is carefully calipered for size and tested for weight and perfection of shape. **For indoor use only.** Furnished complete with fine quality rubber bladder, leather lace and lacing needle. Shipping weight, 2 pounds.
6L1824 .... **$8.85**

"College" Basket Ball.
Fine quality pebbled grain cowhide leather, canvas lined, sewed with heavy waxed thread. Regulation size. Made on the four-piece pattern. Second in quality to our Official Basket Ball and guaranteed to satisfy. **For indoor use only.** This ball is not adapted to outdoor play. Furnished with lace, lacing needle and high quality rubber bladder. Shipping weight, 1⅞ pounds.
6L1821 .... $6.50

"Scholastic" Basket Ball.
We are now able to offer a low priced cowhide basket ball of regulation size. Made of a special good grade of cowhide leather, canvas lined, strongly stitched with waxed linen thread. Made on the four-piece pattern. A strong ball and highly recommended for practice. **For indoor use only.** Complete with good rubber bladder, leather lace and lacing needle. Shipping weight, 1⅞ pounds.
6L1825 .... $4.15

"Amateur" Basket Ball.
Made of good grade pebbled grain sheepskin leather, full lined; sewed with good quality thread. **For indoor use only.** Lace, lacing needle and full size rubber bladder. This basket ball is made on the four-piece pattern and will give good service. Shipping weight, 1⅛ pounds.
6L1826 .... $2.98

"College" Soccer Football.
Made of good quality American tanned grain cowhide leather, hand sewed ends, canvas lined. Regulation size. Furnished with leather lace, needle and good quality bladder. Shipping wt., 1½ lbs.
6L1815 .... $4.48

"Practice" Soccer Football.
Made of good quality pebbled sheepskin leather, full canvas lined, leather lace and full size rubber bladder. Regulation size. Shpg. wt., 1¼ lbs.
6L1816 .... $2.30

Leather Wrist Strap.
Soft tan color leather, chamois lined, stitched edges. Fitted with clinch buckle which allows adjustment to the smallest fraction of an inch. Shipping weight, 5 oz. **Single Strap**, 2¼ inches.
6L1885 .... 26c
Double strap style, 2¾ in. wide, made the same as above.
6L1887 .... 42c

Supporter and Protector for Basket Ball, Football and Baseball.
A combination protector and supporter. Made with 3-inch elastic abdominal band and two 3-inch leg bands. Jersey knit pouch contains a light aluminum guard, felt padded around the edge. Waist, 26 to 42 inches. **Give waist measure.** Shipping weight, 7 oz.
6L1890 .... $1.98

Basket Ball Elbow and Knee Pads.
Made of tan color khaki drill, with padding on inside, with elastic web at each end to hold pad in place. This style pad affords protection to basket ball and indoor ball players. Shipping weight, 6 ounces.
6L1876—Elbow Pads. Pair 93c
6L1877—Knee Pads. Pair 94c

Knee and Elbow Protectors.
Combined protector and elastic bandage with felt padding, for football or basket ball. Shipping weight, 12 ounces.
6L1880—Knee protector. Per pair. $1.28
6L1881—Elbow protector. Per pair 75c

Referees' Whistle.
Loud and shrill, full nickel plated, with ring for chain. Also makes an excellent dog call. Shipping weight, 2 ounces.
6L1872 .... 18c

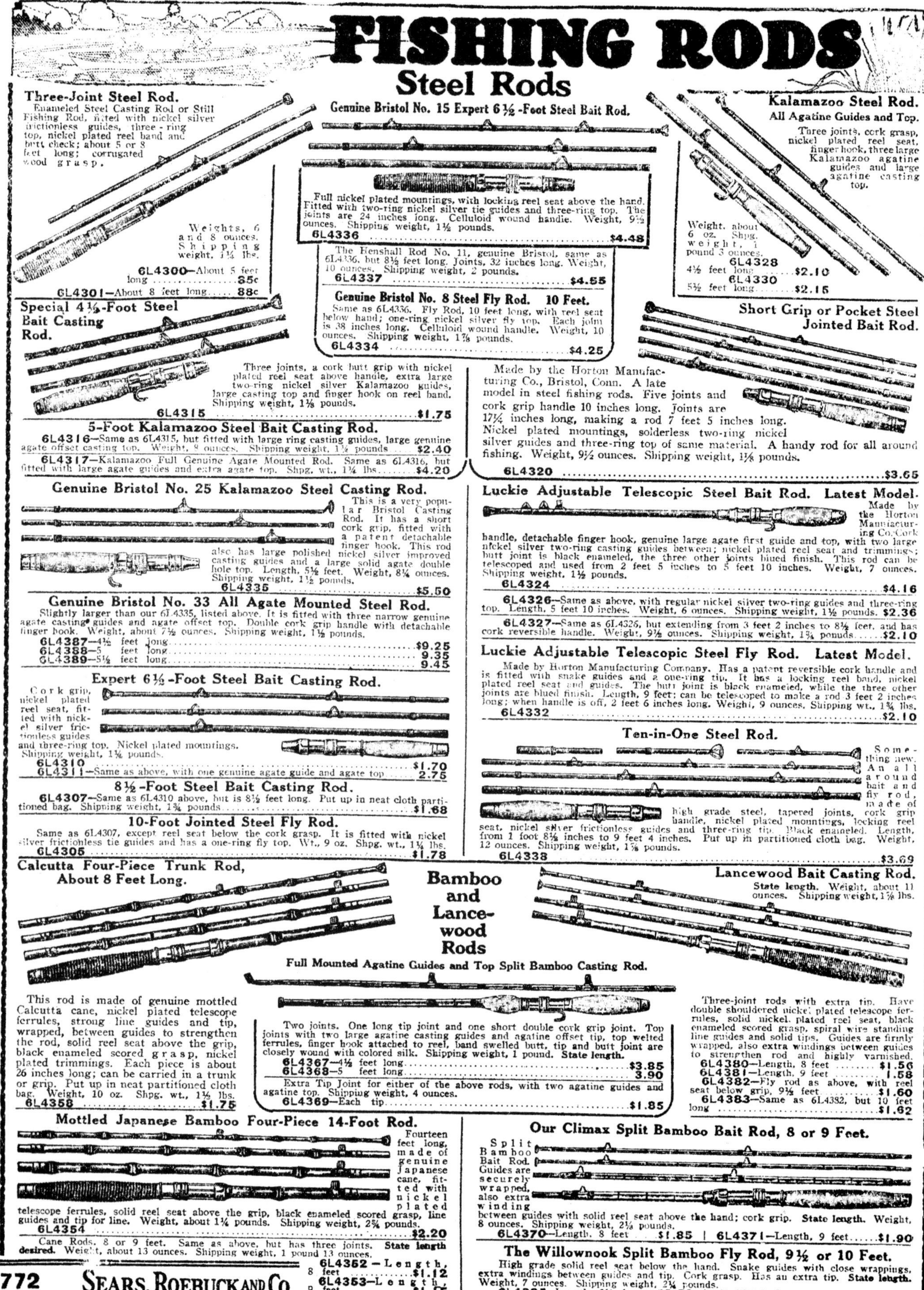

FISHING RODS

Steel Rods

Three-Joint Steel Rod.
Enameled Steel Casting Rod or Still Fishing Rod, fitted with nickel silver frictionless guides, three-ring top, nickel plated reel band and butt check; about 5 or 8 feet long; corrugated wood grasp.

Weights, 6 and 8 ounces. Shipping weight, 1¼ lbs.

6L4300—About 5 feet long **35c**
6L4301—About 8 feet long **88c**

Genuine Bristol No. 15 Expert 6½-Foot Steel Bait Rod.

Full nickel plated mountings, with locking reel seat above the hand. Fitted with two-ring nickel silver tie guides and three-ring top. The joints are 24 inches long. Celluloid wound handle. Weight, 9½ ounces. Shipping weight, 1½ pounds.
6L4336 ..... **$4.48**

The Henshall Rod No. 11, genuine Bristol, same as 6L4336, but 8½ feet long. Joints, 32 inches long. Weight, 10 ounces. Shipping weight, 2 pounds.
6L4337 ..... **$4.55**

Genuine Bristol No. 8 Steel Fly Rod. 10 Feet.
Same as 6L4336. Fly Rod, 10 feet long, with reel seat below hand; one-ring nickel silver fly top. Each joint is 38 inches long. Celluloid wound handle. Weight, 10 ounces. Shipping weight, 1⅞ pounds.
6L4334 ..... **$4.25**

Kalamazoo Steel Rod.
All Agatine Guides and Top.
Three joints, cork grasp, nickel plated reel seat, finger hook, three large Kalamazoo agatine guides and large agatine casting top.

Weight, about 6 oz. Shpg. weight, 1 pound 3 ounces.
6L4328
4½ feet long **$2.10**
6L4330
5½ feet long **$2.15**

Special 4½-Foot Steel Bait Casting Rod.
Three joints, a cork butt grip with nickel plated reel seat above handle, extra large two-ring nickel silver Kalamazoo guides, large casting top and finger hook on reel band. Shipping weight, 1⅛ pounds.
6L4315 ..... **$1.75**

5-Foot Kalamazoo Steel Bait Casting Rod.
6L4316—Same as 6L4315, but fitted with large ring casting guides, large genuine agate offset casting top. Weight, 9 ounces. Shipping weight, 1¼ pounds **$2.40**
6L4317—Kalamazoo Full Genuine Agate Mounted Rod. Same as 6L4316, but fitted with large agate guides and extra agate top. Shpg. wt., 1¼ lbs. **$4.20**

Short Grip or Pocket Steel Jointed Bait Rod.
Made by the Horton Manufacturing Co., Bristol, Conn. A late model in steel fishing rods. Five joints and cork grip handle 10 inches long. Joints are 17¼ inches long, making a rod 7 feet 5 inches long. Nickel plated mountings, solderless two-ring nickel silver guides and three-ring top of same material. A handy rod for all around fishing. Weight, 9½ ounces. Shipping weight, 1⅛ pounds.
6L4320 ..... **$3.65**

Genuine Bristol No. 25 Kalamazoo Steel Casting Rod.
This is a very popular Bristol Casting Rod. It has a short cork grip, fitted with a patent detachable finger hook. This rod also has large polished nickel silver improved casting guides and a large solid agate double hole top. Length, 5½ feet. Weight, 8¼ ounces. Shipping weight, 1½ pounds.
6L4335 ..... **$5.50**

Genuine Bristol No. 33 All Agate Mounted Steel Rod.
Slightly larger than our 6L4335, listed above. It is fitted with three narrow genuine agate casting guides and agate offset top. Double cork grip handle with detachable finger hook. Weight, about 7½ ounces. Shipping weight, 1½ pounds.
6L4387—4½ feet long **$9.25**
6L4388—5 feet long **9.35**
6L4389—5½ feet long **9.45**

Expert 6½-Foot Steel Bait Casting Rod.
Cork grip, nickel plated reel seat, fitted with nickel silver frictionless guides and three-ring top. Nickel plated mountings. Shipping weight, 1¼ pounds.
6L4310 ..... **$1.70**
6L4311—Same as above, with one genuine agate guide and agate top **2.75**

8½-Foot Steel Bait Casting Rod.
6L4307—Same as 6L4310 above, but is 8½ feet long. Put up in neat cloth partitioned bag. Shipping weight, 1¾ pounds **$1.68**

10-Foot Jointed Steel Fly Rod.
Same as 6L4307, except reel seat below the cork grasp. It is fitted with nickel silver frictionless tie guides and has a one-ring fly top. Wt., 9 oz. Shpg. wt., 1¼ lbs.
6L4305 ..... **$1.78**

Luckie Adjustable Telescopic Steel Bait Rod. Latest Model.
Made by the Horton Manufacturing Co. Cork handle, detachable finger hook, genuine large agate first guide and top, with two large nickel silver two-ring casting guides between; nickel plated reel seat and trimmings; butt joint is black enameled, the three other joints blued finish. This rod can be telescoped and used from 2 feet 5 inches to 5 feet 10 inches. Weight, 7 ounces. Shipping weight, 1½ pounds.
6L4324 ..... **$4.16**
6L4326—Same as above, with regular nickel silver two-ring guides and three-ring top. Length, 5 feet 10 inches. Weight, 6 ounces. Shipping weight, 1½ pounds. **$2.36**
6L4327—Same as 6L4326, but extending from 3 feet 2 inches to 8½ feet, and has cork reversible handle. Weight, 9½ ounces. Shipping weight, 1¾ pounds **$2.10**

Luckie Adjustable Telescopic Steel Fly Rod. Latest Model.
Made by Horton Manufacturing Company. Has a patent reversible cork handle and is fitted with snake guides and a one-ring tip. It has a locking reel band, nickel plated reel seat and guides. The butt joint is black enameled, while the three other joints are blued finish. Length, 9 feet; can be telescoped to make a rod 3 feet 2 inches long; when handle is off, 2 feet 6 inches long. Weight, 9 ounces. Shipping wt., 1¾ lbs.
6L4332 ..... **$2.10**

Ten-in-One Steel Rod.
Something new. An all around bait and fly rod, made of high grade steel, tapered joints, cork grip handle, nickel plated mountings, locking reel seat, nickel silver frictionless guides and three-ring tip. Black enameled. Length, from 1 foot 8½ inches to 9 feet 4 inches. Put up in partitioned cloth bag. Weight, 12 ounces. Shipping weight, 1⅝ pounds.
6L4338 ..... **$3.69**

Bamboo and Lancewood Rods

Calcutta Four-Piece Trunk Rod, About 8 Feet Long.
This rod is made of genuine mottled Calcutta cane, nickel plated telescope ferrules, strong line guides and tip, wrapped, between guides to strengthen the rod, solid reel seat above the grip, black enameled scored grasp, nickel plated trimmings. Each piece is about 26 inches long; can be carried in a trunk or grip. Put up in neat partitioned cloth bag. Weight, 10 oz. Shpg. wt., 1½ lbs.
6L4358 ..... **$1.75**

Full Mounted Agatine Guides and Top Split Bamboo Casting Rod.
Two joints. One long tip joint and one short double cork grip joint. Top joints with two large agatine casting guides and agatine offset tip, top welted ferrules, finger hook attached to reel, band swelled butt, tip and butt joint are closely wound with colored silk. Shipping weight, 1 pound. **State length.**
6L4367—4½ feet long **$3.85**
6L4368—5 feet long **3.90**
Extra Tip Joint for either of the above rods, with two agatine guides and agatine top. Shipping weight, 4 ounces.
6L4369—Each tip **$1.85**

Lancewood Bait Casting Rod.
State length. Weight, about 11 ounces. Shipping weight, 1⅛ lbs.
Three-joint rods with extra tip. Have double shouldered nickel plated telescope ferrules, solid nickel plated reel seat, black enameled scored grasp, spiral wire standing line guides and solid tips. Guides are firmly wrapped, also extra windings between guides to strengthen rod and highly varnished.
6L4380—Length, 8 feet **$1.56**
6L4381—Length, 9 feet **1.58**
6L4382—Fly rod as above, with reel seat below grip, 9½ feet **$1.60**
6L4383—Same as 6L4382, but 10 feet long **$1.62**

Mottled Japanese Bamboo Four-Piece 14-Foot Rod.
Fourteen feet long, made of genuine Japanese cane, fitted with nickel plated telescope ferrules, solid reel seat above the grip, black enameled scored grasp, line guides and tip for line. Weight, about 1¼ pounds. Shipping weight, 2¾ pounds.
6L4354 ..... **$2.20**
Cane Rods, 8 or 9 feet. Same as above, but has three joints. **State length desired.** Weight, about 13 ounces. Shipping weight, 1 pound 13 ounces.
8 **6L4352**—Length, 8 feet **$1.12**
9 **6L4353**—Length, 9 feet **$1.15**

Our Climax Split Bamboo Bait Rod, 8 or 9 Feet.
Split Bamboo Bait Rod. Guides are securely wrapped, also extra winding between guides with solid reel seat above the hand; cork grip. **State length.** Weight, 8 ounces. Shipping weight, 2¼ pounds.
6L4370—Length, 8 feet ..... **$1.85** | **6L4371**—Length, 9 feet **$1.90**

The Willownook Split Bamboo Fly Rod, 9½ or 10 Feet.
High grade solid reel seat below the hand. Snake guides with close wrappings, extra windings between guides and tip. Cork grasp. Has an extra tip. **State length.** Weight, 7 ounces. Shipping weight, 2¼ pounds.
6L4385—Length, 9½ feet ..... **$2.50** | **6L4386**—Length, 10 feet **$2.95**

Fishing Reels

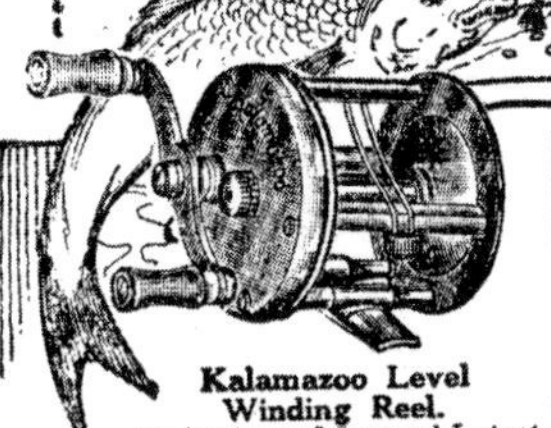

Kalamazoo Level Winding Reel.
Shakespeare Improved Latest Model Kalamazoo Level Winding Quadruple Reel with the new level winding crossbar and attachments. Has jeweled tension oil caps. Double handles on crank. Has click with thumb button on tail plate. Length of pillars, 1¾ inches. Diameter of spool, 1½ inches. 100-yard size. Weight, 8¼ oz. Shipping weight, 12 ounces.
6L4109 .... $9.45

South Bend Casting Reel—Anti-Back Lash.

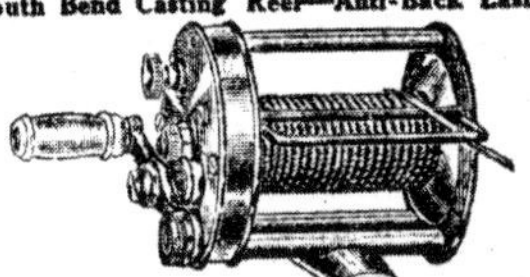

Made of nickel silver, satin finish; solid frame. Full quadruple gear ratio 4 to 1. Balanced crank; imitation ivory grip. End plates, 2 inches in diameter; spool, 1⅞ inches in diameter; spool end, 1½ inches. Will hold 100 yards of No. 5 standard size line. No thumbing required. It automatically stops the line on reel when your bait strikes the water, so you do not get back lashes or snarls. Drag is adjustable to any size bait. Sliding click. Shipping weight, 12 oz.
6L4184 .... $9.95

South Bend Level Winding Anti-Back Lash Casting Reel.

Late model double grip balance handle, two adjustable screw off jeweled oil caps. No thumbing required. Full quadruple gear ratio 4 to 1. Made of nickel silver, satin finish. End plates, 2¼ inches in diameter, 1¾-inch spool. Will hold 100 yards No. 5 standard size silk line. Shipping weight, 14 ounces.
6L4185 .... $19.85

Double Multiplying Reel.

Raised pillar, made of brass, nickel plated, with two screw off oil caps and patent adjustable side drag and back sliding click; polished bearings; wide spool. Holds No. 5 line.
6L4143—40-yard size. Shipping wt., 8 ounces 90c
6L4144—60-yard size. Shipping wt., 9 ounces 95c

Jeweled Wonder Reel.

Raised pillar, wide spool 1⅜ in., white balance handle with adjustable click and drag, two large jeweled screw off oil caps; double multiplying; made of brass, nickel plated and polished.
6L4148—60-yard size. Shipping wt., 8 ounces $1.32
6L4149—80-yard size. Shipping wt., 10 ounces $1.52

Winner Jeweled Casting Reel.
A medium priced casting reel. A 60-yard size, wide spool pattern; oil caps are fitted with jewel; steel pivots and pinions, steel cog post, large metal bushings, white handle; fitted with adjustable click and drag. Gear ratio 3½ to 1. Known as quadruple. Shipping weight, 13 ounces.
6L4159 .... $3.48

Beaver Nickel Silver Jeweled Reel.
Kentucky pattern wide spool fitted with large fancy imitation ivory balance handle. Gear ratio 3½ to 1. Known as quadruple. Adjustable front sliding drag and back sliding click. Pinions and pivots of English steel; bushings of high grade phosphor bronze. The spool has a dead center bearing, agates at each end. Shipping weight, 1 pound.
6L4182—60-yard size $5.50
Same as above, 80-yard size.
6L4183 .... 6.10

Shakespeare Service Reel.
Solid takedown frame, 1⅞-inch spool. Circle finish. Adjusted screw off improved oil caps, set with agates. Shakespeare patent graduated adjustable drag, which permits of accurate and long distance casting. Click is made of hardened tool steel, operated by a thumb button on the tail plate. 80-yard size. Gear ratio 3½ to 1. Known as quadruple. Weight, 7 ounces. Shipping weight, 10 ounces.
6L4107 .... $3.85
We do not believe this reel can be bought elsewhere for less than $5.00.

Blue Grass Reel.

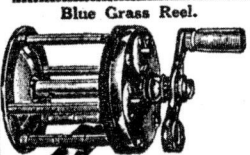

Made of nickel silver. Full quadruple gear ratio 4 to 1. With click and drag, two screw off oil caps, large balance handle, spiral gears and pinions. Diameter of end plate, 2 in.; diameter of spool head, 1½ in.; length of spool, 1⅝ in. Shpg. wt., 14 oz.
6L4116—80-yard. Plain Oil Caps, No. 3B $19.00
6L4117—80-yard. Jeweled Pivot Bearings, No. 3 J. B. $23.75
6L4118—100-yard. Plain Oil Caps. No. 4B $22.80
6L4119—100-yard. Jeweled Pivot Bearings, No. 4 J. B. $27.55

Improved Blue Grass Simplex Reel No. 33.
Can take apart without tools. Sliding click, two screw off oil caps. 80-yard size. Shipping weight, 12 ounces.
6L4112 .... $11.95

Go-ite Bait Casting Reel.
(Anti-Back Lash.)

Made of aluminum with brass bushing which revolves on steel axle with brass tension nut and spring housing, which can be adjusted to suit the size bait you are using. Has two wood winding handles, agate line guide which can be adjusted or set to any angle to suit guide on rod you are using to insure level winding on reel. Diameter of spool, 5¼ in. Wt., 5 oz. Shpg. wt., 12 oz.
6L4127 .... $4.65

Meisselbach Tri-Part Reel.

An 80-yard size, with wide nickel silver spool. Taken apart in a few seconds by simply unscrewing the metal bands. The pivots are turned on the solid steel shaft, which extends the entire length of the spool. The bronzed gear wheel is securely braced and bridged. Fitted with a friction cap, which enables the fisherman to regulate the speed of the reel and prevents back lashings; also sliding click. Gear ratio 3¼ to 1. Known as quadruple. Shipping weight, 14 oz.
6L4130 .... $4.28

Meisselbach Free Spool Tri-Part Reel.
Similar to above, but, when casting, the handle automatically releases and does not turn, thereby giving you a greater casting distance. No fear of back lashing. 80-yard size. Gear ratio 3½ to 1. Known as quadruple. Shpg. wt., 14 oz.
6L4131 .... $5.63

Royal Blue Jeweled Reel.
Quadruple multiplying 3½ to 1 ratio reel with click and drag. Entire reel, except handle, which is white, is blued finish, highly polished. Has jeweled bearing caps and balanced handle. Capacity, 60 yards. Shipping weight, 14 ounces.
6L4135 .... $2.95

Union Hardware Samson Take Apart Reel.

Quadruple multiplying nickel plated reel with removable spool. Can be taken apart in a few seconds by pressing latch button which releases hinged head. All parts are fastened and cannot become lost. Has click button and button on shaft to regulate casting and prevent back lashing. Capacity, 80 yards. This is a very sturdy reel with practically no parts to get out of order. Shipping weight, 15 oz.
6L4164 .... $3.25

Stubby Rod and Reel.
For casting, trolling and still fishing. Very handy to carry in your pocket or car. Reel has adjustable drag. The two-piece rod is 2 feet long and has agatine tip. Shipping weight, 2 pounds.
6L4302 .... $2.25

Shakespeare Improved Marhoff Level Winding Reel.
Latest model with the new level winding crossbar and attachment. Nickel plated, head box made of vulcanized rubber, inlaid with metal agate jeweled oil caps, white double grip handle, click and drag combined in one member situated in the head of reel. Length of pillars, 1¾ inches; diameter of spool, 1½ inches. Capacity, 100 yards. Packed in sheepskin chamois bag with screwdriver. Shipping weight, 1 lb. 3 oz.
6L4105 .... $14.85

Our Ideal Reel.

Round disc, wide spool, screw off oil caps. The oil caps, discs and post are milled, giving the reel a handsome appearance. Fitted with steel axle and steel pinion; also with click and drag. Gear ratio 3½ to 1. Known as quadruple. Holds No. 5 line. Shpg. wt., 13 oz.
6L4150—40-yard size $2.50
6L4151—60-yard size 2.63
6L4152—80-yard size 2.75

Shakespeare Standard-Professional Reel.

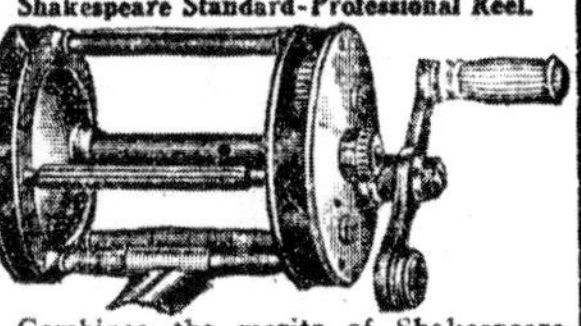

Combines the merits of Shakespeare Standard and Professional designs. Nickel silver frame takedown pattern, with hard rubber head and tail discs, metal bound. 80-yard size. Gear ratio 3½ to 1. Known as quadruple. Spool, 1⅞ inches wide. Has two adjustable screw off oil caps set with agates with click and drag. Shipping weight, 10 ounces.
6L4108 .... $8.25

Kentucky Pattern Jeweled Satin Finish Reel.
80-yard size, extra wide, 1⅞-in. spool, steel pinion and steel axle. Pillars extend through front and rear plates, securing great strength and rigidity. Steel axle bears on jewels at each end, which are fitted by hand with sliding click and drag. Gear ratio 3½ to 1. Known as quadruple. Mechanism is quickly accessible by removing two small screws. Shipping weight, 13 oz.
6L4157 .... $3.85

Surf Casting Reels.
For tarpon, tunny and all salt water fishing. Double multiplier, except 6L4171, has a gear ratio 1¾ to 1, hard rubber disc, metal bound. Has click and drag and leather thumb brake. Steel pivots, spiral tooth gear, steel ratchet and click, and two screw off oil caps.
6L4171—150-yard size. Shipping weight, 1 pound 1 ounce $5.80
6L4174—250-yard size. Shipping weight, 1 pound 5 ounces $6.85
6L4172—300-yard size. Shipping weight, 1 pound 9 ounces $7.25
6L4173—400-yard size. Shipping weight, 2 pounds $8.85

Trout, Fly and Single Action Reels.

Gem Single Action Reel.

Made of metal, gunmetal finish, very light and strong. Can be taken apart by removing one screw. The open or perforated spool makes it a line drier. Sliding click. Double line guides with movable eyelet sliding back and forward. For fly or bass fishing. 60-yard size. Shpg. wt., 6 oz.
6L4170 .... 55c

Nickel Silver Reel.
Single action. The double tone shape nickel silver spool takes in the line as fast as an ordinary multiplying reel with click.
6L4160—80-yard size. Shipping weight, 8 oz. $1.20
6L4162—80-yard size with sliding jewel guide. Shipping weight, 8 ounces $1.35

The Utica Automatic Reel.
"The Little Finger Does It."
One of the latest and best light weight automatic trout fly casting reels. Made of aluminum and brass. Brass is nickel plated. All wearing parts of hard metal, reinforced edges. Compact, lays flat on reel seat and has a release to let down tension. Holds 50 yards of trout line. Weight, 8⅛ oz. Shipping weight, 12 oz.
6L4142 .... $4.50

Fly Casting Reel.

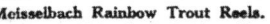

Made with nickel silver frame and pillars. Black hard rubber side plates. Single action, with back sliding click and pivots; balanced handle. Shipping weight, 7 ounces.
6L4168—60-yard size $1.95
6L4169—80-yard size. Shpg. wt., 10 ounces. $2.35

Meisselbach Rainbow Trout Reels.

Meisselbach Rainbow Trout Reels. Take-A-Part of the English design. Light in construction and very strongly built. Made of alloy in dull black finish with nickel silver trimmings. Ivoryloid handle.
6L4128—Diameter, 2⅞ inches; width of spool, ¾ inch; holds 35 yards D line. Weight, 3¼ ounces. Shipping weight, 6 ounces $3.98
6L4129—As above, but larger; diameter, 3¼ inches; width of spool, ¾ inch. Holds 50 yards D line. Weight, 4¼ oz. Shipping weight, 7 ounces $4.55

Fish Lines and Tackle

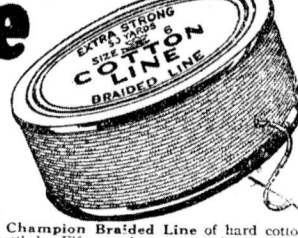

Pure Linen Lines.

Natural Linen Cuttyhunk Line. An old reliable twisted line made of Irish linen. The requirements for a line of this type call for the greatest possible strength and to be properly twisted and free from defects of all kinds. Two spools connected if desired. Shpg. wt., 4 oz.

Catalog No.	Size	Breaking Strength	Per 150-Ft. Spool
6L4280	No. 9	18 lbs.	$0.49
6L4281	No. 12	24 lbs.	.59
6L4282	No. 15	30 lbs.	.68
6L4283	No. 18	36 lbs.	.78
6L4284	No. 21	42 lbs.	.98
6L4285	No. 24	48 lbs.	1.05
6L4286	No. 27	54 lbs.	1.18

Same as above, but dyed green color before twisting, to insure a fast color. Shpg. wt., 4 oz.

Catalog No.	Size	Breaking Strength	Per 150-Ft. Spool
6L4186	No. 9	18 lbs.	$0.58
6L4187	No. 12	24 lbs.	.66
6L4188	No. 15	30 lbs.	.79
6L4189	No. 18	36 lbs.	.90
6L4190	No. 21	42 lbs.	1.20
6L4191	No. 24	48 lbs.	1.30
6L4192	No. 27	54 lbs.	1.50

Green Linen Cuttyhunk Line for Sea Fishing. Comes on 200-yard spool. Shipping weight, 12 ounces.

Catalog No.	Size	Breaking Strength	Per 200-Yd. Spool
6L4291	No. 9	18 lbs.	$2.00
6L4292	No. 15	30 lbs.	3.00
6L4293	No. 21	42 lbs.	4.50
6L4294	No. 27	54 lbs.	6.00

Lake Green Linen Line. Made of selected quality Scotch flax. Is green color and especially adapted for game fish. Fifty feet on coil, six coils connected if desired. Shpg. wt., 2 oz.

Catalog No.	Size	Breaking Strength	Per 50-Ft. Coil
6L4255	No. 9	18 lbs.	17c
6L4256	No. 12	24 lbs.	20c
6L4257	No. 15	30 lbs.	23c
6L4258	No. 18	36 lbs.	28c

Hard Braided Linen Line. Put up 25 yards in a coil and may be had four coils connected. Made from high grade fine Scotch linen fiber, evenly braided over a flexible cotton core and well finished. Shipping weight, 4 ounces.

Catalog No.	Size	Breaking Strength	25-Yard Coil
6L4238	No. 5	23 lbs.	35c
6L4239	No. 4	28 lbs.	38c
6L4240	No. 3	30 lbs.	41c
6L4241	No. 2	43 lbs.	44c
6L4242	No. 1	46 lbs.	48c
6L4243	No. 1–0	52 lbs.	52c
6L4244	No. 2–0	58 lbs.	56c
6L4245	No. 3–0	64 lbs.	63c

Patent Snap Wire Leader. Just the thing for bait casting. Will prevent fish from biting the line and will not kink or break. Brass swivel and patent snap. Snap can be instantly detached from leader. Length, about 7 inches. Shipping weight, each, 1 ounce; six, 4 ounces.
6L3793—6 for 39c; each..........7c
6L3794—Same as above, but 3 feet long. 6 for 55c; each..........10c

Twisted Wire Leaders. Braided rustproof flexible wire gimp leaders with wire wrapped over loops and soldered. Cooper snap and swivel attached. Very strong. Handy for quick changing of baits or hooks. Shipping weight, 1 ounce.
6L3798— 6 inches long. Each....10c
6L3799—10 inches long. Each....13c

Short Gut Casting Leaders. Five inches long; 3-ply machine twist casting leaders with No. 3 swivel attached, and snood loop; for bass, pike and pickerel casting. Shpg. wt., 1 oz.
6L3792—5 in. long. 3 for..........18c
6L3795—9 in. long. 3 for..........24c

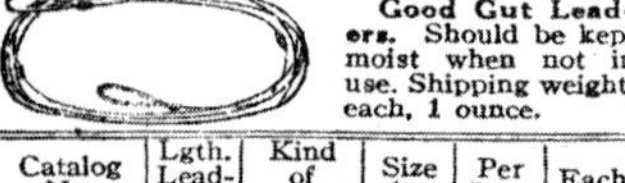

Good Gut Leaders. Should be kept moist when not in use. Shipping weight, each, 1 ounce.

Catalog No.	Lgth. Leaders	Kind of Gut	Size for	Per Doz.	Each
6L3782	3 feet	Single	Trout	$0.43	4c
6L3783	6 feet	Single	Trout	.86	8c
6L3784	3 feet	Double	Bass	.87	8c
6L3785	6 feet	Double	Bass	1.72	16c

High Grade Selected Gut Leaders.

Catalog No.	Lgth. Leaders	Kind of Gut	Size for	Per Doz.	Each
6L3779	3 feet	Single	Salmon	$1.10	10c
6L3781	6 feet	Single	Salmon	2.08	19c
6L3786	3 feet	Double	Salmon	2.20	20c
6L3787	6 feet	Double	Salmon	4.16	38c

Dropper Loop Leaders. High grade, stout Marana selecto, single straight round gut, one on a card. Shipping weight, each, 2 ounces.

Cat. No.	Length of Leaders	Each	Per Doz.
6L3788	3 ft., extra loop	20c	$2.19
6L3789	6 ft., 2 extra loops	40c	4.38

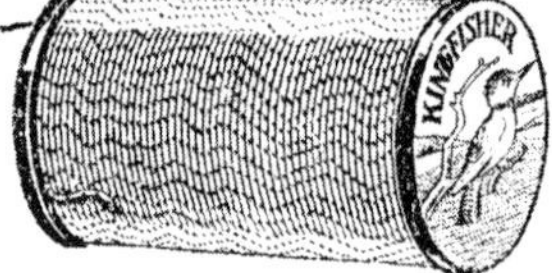

Pure All Silk Lines.

Genuine Kingfisher Bait Casting Line. A well known tested braided silk casting line of white silk with black markings. Put up on 50-yard spools, two spools connected if desired. Shipping weight, 3 ounces.

Catalog No.	Size	Breaking Strength	Per 50-Yd. Spool
6L4208	No. 5	12 lbs.	$0.98
6L4209	No. 4	16 lbs.	1.35
6L4210	No. 3	23 lbs.	1.60
6L4211	No. 2	28 lbs.	1.98

Expert Black Silk Waterproof Line. Made of selected Japan long silk, hard braided over a pure silk twisted center, making it a very easy running line. Fifty-yard spools, two spools connected if desired. Shipping weight, 4 ounces.

Catalog No.	Breaking Strength	Per 50-Yard Spool
6L4271	14 lbs.	$1.20
6L4272	18 lbs.	1.30
6L4273	24 lbs.	1.40
6L4274	28 lbs.	1.50

Mottled Brown Silk Casting Line. Made of selected grade long Japan silk, evenly and carefully braided over a silk core, which insures a round perfect running line. Fifty-yard spools, two connected if desired. Shipping weight, 4 ounces.

Catalog No.	Breaking Strength	Per 50-Yard Spool
6L4194	14 lbs.	$0.80
6L4195	17 lbs.	1.00
6L4196	20 lbs.	1.20
6L4197	25 lbs.	1.40
6L4198	30 lbs.	1.60

White and Black Mottled Silk Line. A good grade pure silk line for all kinds of fishing. Evenly braided and easy running. Put up on 25-yard spools, or 100 yards connected if desired. Shipping weight, 2 oz.

Catalog No.	Breaking Strength	Per 25-Yard Spool
6L4251	12 lbs.	29c
6L4253	16 lbs.	32c
6L4254	20 lbs.	36c

The Beaver Braided Silk Casting Line. Braided of high quality genuine silk of long fiber over silk core and possesses exceptional finish, smoothness and hardness. 50 yards on spool and in a sealed transparent tube. Two spools connected if desired. Shipping weight, 6 ounces.
6L4227—18-lb. breaking strength. 50 yards..........$1.25
6L4228—23-lb. breaking strength. 50 yards..........$1.40

South Bend Anti-Back Lash Brand Silk Casting Line. Soft braided over silk center; color, mottled white and black. Fifty-yard spool, two spools connected if desired. For bass, pike, pickerel or musky. Shpg. wt., 3 oz.

Cat. No.	Breaking Strength	50-Yd. Spool
6L4223	12 lbs.	$1.35
6L4224	19 lbs.	1.80
6L4225	25 lbs.	2.10

Italian Enameled Silk Fly Casting Line. Our highest grade enameled silk waterproof fly casting line. Strong, flexible and will not crack. 25-yard coil, four coils connected if desired. Shipping weight, 3 ounces.

Cat. No.	Breaking Strength	25-Yd. Coil
6L4216	12 lbs.	48c
6L4217	16 lbs.	58c
6L4218	20 lbs.	70c
6L4219	23 lbs.	85c

Expert Enameled Silk Fly Casting Line. Our very highest grade of enameled silk line, made of selected long fiber silk and specially enameled. A suitable line for the exacting caster. Comes in 50-yard coils, four coils connected if desired. Shipping weight, 3 ounces.

Catalog No.	Breaking Strength	Per 50-Yard Coil
6L4298	13 lbs.	$1.00
6L4299	18 lbs.	1.25

Champion Braided Line of hard cotton, mottled. Fifty yards on a spool; two spools connected if desired. Shipping weight, 5 oz.

Catalog No.	Size	Per 50-Yard Spool
6L4295	No. 6	25c
6L4296	No. 5	26c
6L4297	No. 4	27c

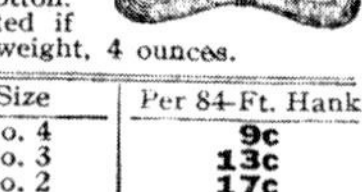

Braided Cotton Line. Strong and durable; made of good grade cotton. Two hanks connected if desired. Shipping weight, 4 ounces.

Catalog No.	Size	Per 84-Ft. Hank
6L4230	No. 4	9c
6L4231	No. 3	13c
6L4232	No. 2	17c
6L4233	No. 1	20c
6L4234	No. 1-0	24c
6L4235	No. 2-0	29c
6L4236	No. 3-0	34c
6L4237	No. 4-0	43c

Braided Cotton Line. Mottled effect, having the appearance of silk. Put up on 50-foot coils, four coils connected if desired. Shipping weight, 2 ounces.
6L4246—For trout and small game fish. 50 feet..........10c
6L4247—For bass, pickerel and other large fish. 50 feet..........12c

Braided Bronze Wire Fishing Line. Bronze wire closely braided over a cotton center. Particularly adapted for deep water trolling, no sinker being required; also for long pole spatting. Not to use on reel. Put up in spools of 100 feet, three spools connected if desired. Shipping weight, 5 ounces.
6L4202—Per spool of 100 feet.....68c

Rigged Silk Braided Line. Good quality silk complete with gut, hook, fancy barrel cork float and adjustable sinker. For still fishing. Shipping weight, 3 ounces.
6L4252..........22c

Rigged Linen Braided Line. Good linen line rigged complete with gut hook, adjustable sinker and painted barrel shaped float. Shipping weight, 3 ounces.
6L4226..........20c

Select Silkworm Gut. Put up 100 in a bunch. For fishermen who make their own snelled hooks, flies and leaders. Shipping weight, per bunch, 2 ounces.
6L3796—About 10 inches long. Good Padron, superior quality. Per bunch....60c
6L3797—About 12 inches long. Select Marana, extra quality, heavy and very strong. Per bunch..........$1.32

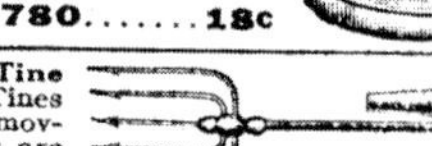

Aluminum Leader Box. Diameter, 4 inches. Two felt pads. Shipping weight, 3 ounces.
6L3780.......18c

Salt Water and Muskellunge Fishing Tackle.

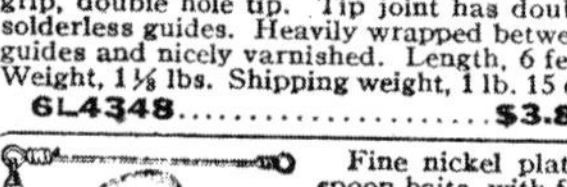

Salt Water Rod. Made of good heavy split bamboo. Thickness at tip end, ¼ inch; at winding check, ¾ inch. Two joints, 1⅛-inch reel seat, extra long enameled scored grip, double hole tip. Tip joint has double solderless guides. Heavily wrapped between guides and nicely varnished. Length, 6 feet. Weight, 1⅛ lbs. Shipping weight, 1 lb. 15 oz.
6L4348..........$3.83

Fine nickel plated spoon baits, with full feathered treble hooks. Used for muskellunge and large pickerel. Shipping weight, 5 ounces.
6L3257—2⅝-inch spoon, for 10 to 20-pound fish..........36c
6L3258—3¼-inch spoon, for 20 to 100-pound fish..........45c

O'Shaughnessy Ringed Hook. High quality, forged, tapered, tested and bronzed. Tapered eye. Dublin point. We do not sell less than 1 dozen of a size. State size.

6L3606

Size	3	2	1	1-0	2-0
Shpg. wt., 2 oz.		2 oz.	3 oz.	3 oz.	4 oz.
Per dozen..8c		9c	10c	11c	12c
Size		3-0	4-0	5-0	6-0
Shipping wt.		4 oz.	5 oz.	5 oz.	6 oz.
Per dozen		12c	14c	16c	18c
Size		7-0	8-0	9-0	10-0
Shipping wt.		7 oz.	7 oz.	8 oz.	8 oz.
Per dozen		20c	24c	30c	36c

Gaff Hook. Illustration shows hook open, ready for action. Japanned blue. Size, 8½ inches between points of jaws when open. Shipping weight, 1⅜ pounds.
6L3508..........$1.38

Beaver Hand Split Bamboo Fly Rod, 9½ or 10 Feet.

A high grade split Bamboo Fly Rod. Solid sectional cork grip. All trimmings nickel silver. The wrappings are of colored silk, closely wound. Put up on a handsome green velveteen covered wood form with nickel plated cap ends, in a cloth bag. State length. Weight, about 7 ounces. Shipping weight, 2 pounds 5 ounces.
6L4372—Length, 9½ feet..........$7.45
6L4373—Length, 10 feet..........7.48
6L4374—Same as 6L4372, but fitted with one agate guide and two agate tips. Length, 9½ feet..........$9.30
6L4375—Same as 6L4374. Length, 10 feet..........9.35

Beaver Special Handmade Split Bamboo Bait Rod.

High grade. Reinforced at butt above reel seat with strips of cedar inlaid between the bamboo. Fitted with one genuine agate guide and two agate tips, the remaining guides being nickel silver trumpet guides; all very closely wrapped with red and black silk. Fitted with a short cork grip. All trimmings nickel silver. Velveteen covered wood form. Length of rod, 5½ feet. Weight, 5½ ounces. Shipping weight, 1½ pounds.
6L4376..........$9.65

Shakespeare Expert Casting Rod. With extra tip. Expert, flexible, combination 5-foot casting rod, with two tips, one extra flexible and light for casting light bait and the other slightly heavier for heavier bait casting. Made of finest possible selected split bamboo closely wrapped and fitted with genuine agate guides and tip. Double cork grip butt with special locking reel band. Tip joint is in one piece, 4 feet long, which makes rod more flexible than a rod with more joints. Weight, 5 ounces. Put up in a strong, serviceable bag. Shipping weight, 2 pounds.
6L4359..........$14.85

Five-Tine Spear. Tines not removable but are set in heavy head of brass. Threaded to screw into socket of handle. Total length, 15 inches; length of tines, 4 inches; width, across tines, 3½ inches. Shipping weight, 1½ pounds.
6L3505..........$1.13

Fish Spear. Made of tempered steel. Hole in ferrule 1 inch in diameter, barbs on each tine made on solid shank. By using a larger wedge the outside tines can be removed at any time to convert into a three-tine spear. Width, 4½ inches; length of tines, 5½ inches; 15 inches long. Shipping weight, 2 pounds 3 ounces.
6L3504..........$1.50

Fish Spear. Five tines. Length of tines, about 3⅝ in.; width of spear, about 3½ in. Total length, 7½ in. Socket for pole. Shpg. wt., 1 lb.
6L3503..........49c

Fish and Frog Spear. Four 2¾-inch tines. Socket for pole. Shipping weight, 6 ounces.
6L3502..........15c

Steel Gaff Hook. Has three 7-inch steel joints that screw together. For pickerel and muskellunge. Can be carried in pocket or tackle box. Weight, 6 ounces. Shipping weight, 9 oz.
6L3507..........42c

Fishing Reel Bag. Protects your reel from dust and dampness. Made of soft sheepskin with draw strings. For reel up to 100-yard size. Shipping weight, 2 ounces.
6L3490..........20c

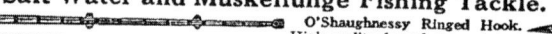

Fishing Tackle

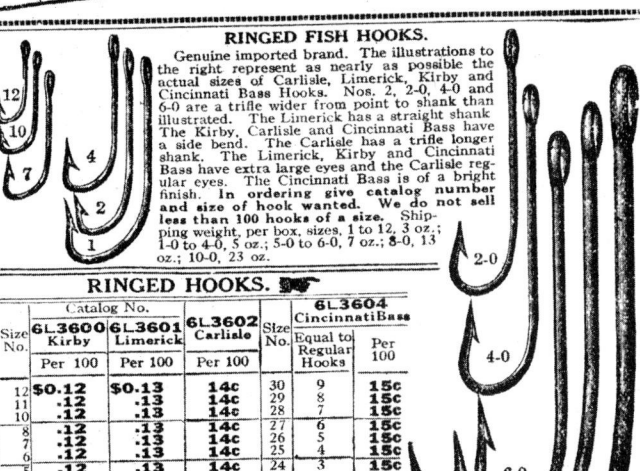

Chatillion Sportsman Scale. Weighs from 1 to 15 lbs. by ¼ lb. Equipped with tare allowance. Shpg. wt., 3 oz.
6L3485....................42c

Chain Fish Stringer. Strong links, nickel plated. Length, 42 in. Shpg. wt., 3 oz.
6L3477..........12c

Keystone Fish Stringer. Brass needle at one end for stringing and ring at other end to loop first fish. Six feet long. Shipping wt., 4 oz.
6L3479..........15c

Fish Scaler. Stamped out piece of sheet steel. Shpg. wt., 3 oz.
6L3483......7c

Aluminum Hook Disgorger. Double end. Very light. Will not rust. Shipping weight, 3 oz.
6L3495....................18c

Fish Knife and Scaler. 9 inches long; 4½-inch steel blade, hardwood handle. Shipping wt., 8 oz.
6L3499....................35c

Liquid Mosquito Chaser. An effective, harmless, pleasant smelling preventive against mosquitos and other insects. Few drops on the hands and face or on towel hung over head of bed will keep away mosquitos and other insects. Shipping weight, 5 ounces.
6L3449—1-oz. bottle.....29c

Jungle Mosquito Protector. Fine, close mesh netting, with large, square, soft copper wire front. Used with any style hat. Shpg. wt., 2 oz.
6L3512....................84c

Fish Bait Oil. When still fishing put one or two drops of this oil on your bait. Patent screw top set in cork. 1-oz. bottle. Shpg. wt., 5 oz.
6L3451.....29c

"Drifl." A waterproofing solution to use on flies and lines to keep them from sinking when casting. Shpg. wt., 3 oz.
6L3453.....24c

Kingfisher Rod Varnish. High grade. Will not crack or chip. Remove all old varnish from rod before putting on new varnish. Put up in 4-oz. cans. Shpg. wt., 6oz.
6L3493—Per can......20c

Agatine Offset Stirrup Casting Tips. Agatine is very strong and durable, red in color, set in nickel silver. Made for rod tips of the following diameters. Shpg. wt., 2 oz.
6L4017—⁵⁄₃₂ inch....................26c
6L4018—⁶⁄₃₂ inch....................27c
6L4019—⁷⁄₃₂ inch....................28c
6L4020—⁸⁄₃₂ inch....................29c
6L4021—⁹⁄₃₂ inch....................30c

Rod Shortener. Fits the grip of any jointed steel rod. Takes the two smallest joints, leaving out the joint which fits in the grip. With this shortener you can make an 8½-foot rod 6 ft. long. Shpg. wt., 2 oz.
6L4014....................10c

Emergency Tip. Nickel silver. Fits any jointed steel rod. Can be used in second joint of rod in case of accident to regular tip point or to stiffen rod for trolling. Shipping weight, 2 oz.
6L4012....................10c

Large Agatine Casting Guides. Agatine is very strong and durable, red in color, set in nickel silver mountings. Shipping weight, 2 ounces.
6L4022—Small....................30c
6L4023—Medium....................32c
6L4024—Large....................34c
6L4025—Extra large....................38c

Feathered Treble Hook. Well tied. High quality hook. **State size.** Shipping weight, 3 ounces.
6L3655—Sizes, 1, 2, 3, 4 and 5. **3** for....................19c
6L3656—Sizes, 1-0, 2-0, 3-0 and 4-0. **3** for....................23c
6L3657—Sizes, 6-0 and 8-0. **3** for..35c

Bucktail Treble Hook. Made from genuine deer tail hair. Can be attached to almost any spoon bait. **State size.** Shipping weight, 4 oz.
6L3660—Sizes, 2, 4 and 6..15c
6L3661—Sizes, 1-0, 2-0 and 3-0....................19c
6L3662—Size, 8-0....................34c

Red Bucktail Treble Hook. Large size No. 8-0. For muskellunge and pickerel. Full tied and long red bucktail. Shpg. wt., 4 oz.
6L3663....................42c

Marvel Automatic Fish Hook. For use on a pole or set line. The slightest nibble will make it close, catching the fish. Shipping wt., 2 oz.
6L3644—Size, 1-0, for small fish....................9c
6L3645—Size, 3-0, for large fish....................10c

Steel Split Rings. For attaching hooks and flies to spoons and baits. Come in ⅜-in. ⁵⁄₁₆-in. and ¾-in. **State size.** Shpg. wt., 1 oz.
6L3720—Per dozen......8c

RINGED FISH HOOKS.

Genuine imported brand. The illustrations to the right represent as nearly as possible the actual sizes of Carlisle, Limerick, Kirby and Cincinnati Bass Hooks. Nos. 2, 2-0, 4-0 and 6-0 are a trifle wider from point to shank than illustrated. The Limerick has a straight shank The Kirby, Carlisle and Cincinnati Bass have a side bend. The Carlisle has a trifle longer shank. The Limerick, Kirby and Cincinnati Bass have extra large eyes and the Carlisle regular eyes. The Cincinnati Bass is of a bright finish. **In ordering give catalog number and size of hook wanted. We do not sell less than 100 hooks of a size.** Shipping weight, per box, sizes, 1 to 12, 3 oz.; 1-0 to 4-0, 5 oz.; 5-0 to 6-0, 7 oz.; 8-0, 13 oz.; 10-0, 23 oz.

12 10 7 4 2 1 2-0 4-0 6-0 8-0 10-0

RINGED HOOKS.

Size No.	6L3600 Kirby	6L3601 Limerick	6L3602 Carlisle	Size No.	6L3604 Cincinnati Bass	
	Per 100	Per 100	Per 100		Equal to Regular Hooks	Per 100
12	$0.12	$0.13	14c	30	9	15c
11	.12	.13	14c	29	8	15c
10	.12	.13	14c	28	7	15c
8	.12	.13	14c	27	6	15c
7	.12	.13	14c	26	5	15c
6	.12	.13	14c	25	4	15c
5	.12	.13	14c	24	3	15c
4	.12	.13	14c	23	2	15c
3	.12	.13	14c	22	1	15c
2	.12	.13	14c	21	1-0	18c
1	.12	.13	14c	20	2-0	22c
1-0	.14	.15	19c	19	3-0	27c
2-0	.18	.19	24c	18	4-0	34c
3-0	.21	.22	32c	17	5-0	38c
4-0	.25	.26	35c	16	6-0	46c
5-0	.32	.33	38c	15	7-0	63c
6-0	.36	.37	43c	....	...	....
8-0	.60	.61	65c	....	...	....
10-0	1.05	1.08	....	....	...	....

Carlisle Snelled Hooks. Carlisle Spring Steel Hooks, tied to gut with loop. Not sold less than one-half dozen of a size. Shipping weight, 3 ounces.

Catalog No.	Kind	Size No.	Per Doz.
6L3738	Single Gut	10 to 1	22c
6L3739	Double Gut	6 to 1	29c
6L3740	Double Gut	1-0	31c
6L3741	Double Gut	2-0	34c
6L3742	Double Gut	3-0	36c
6L3743	Double Gut	4-0	38c
6L3744	Double Gut	5-0	40c
6L3745	Double Gut	6-0	42c

Carlisle "A" Grade Hollow Point Select Double Gut Hooks. Tied with silk to 7-inch gut. Not sold less than one-half dozen. Shipping weight, per dozen, 3 ounces.

Catalog No.	Kind	Size No.	Per Doz.
6L3618	Double Gut	6 to 1	48c
6L3619	Double Gut	1-0	50c
6L3620	Double Gut	2-0	52c
6L3621	Double Gut	3-0	54c
6L3622	Double Gut	4-0	56c
6L3623	Double Gut	5-0	58c
6L3624	Double Gut	6-0	60c

Cincinnati Bass Snelled Hooks. Bright steel, tied to double gut with loop. Not sold less than one-half dozen of a size. Shpg. wt., per doz., 3 oz.

Catalog No.	Size	Equal to Reg. No.	Per Doz.
6L3750	No. 26	5	31c
6L3751	No. 25	4	31c
6L3752	No. 24	3	31c
6L3753	No. 23	2	31c
6L3754	No. 22	1	31c
6L3755	No. 21	1-0	33c
6L3756	No. 20	2-0	35c
6L3757	No. 19	3-0	37c
6L3758	No. 18	4-0	39c
6L3759	No. 17	5-0	41c
6L3760	No. 16	6-0	43c

Hercules Pennell Hooks. Wire Lead. Turndown eyes neck hollow point hook, tied with waterproof silk wrapping on phosphor bronze cable flexible wire, extra whipped-over silk with nickeled steel wire at loop and hook. Not sold less than one-half dozen of a size. Shipping weight, 3 ounces.
6L3771—Size, 2-0. Per ½ doz....52c
6L3772—Size, 3-0. Per ½ doz....55c
6L3774—Size, 4-0. Per ½ doz....58c
6L3776—Size, 5-0. Per ½ doz....61c
6L3778—Size, 6-0. Per ½ doz....64c

Hensel Steel Leader Hook. Made of bright steel piano wire Cincinnati bass hook wired and soldered to wire. Full length, 7 inches. Three sizes, 20, 19 and 18 Cincinnati bass. **Mention size.** Shipping weight, 2 oz.
6L3773—6 for **36c**; each......7c

Limerick Spear Point Snelled Hooks. Tied to high quality gut with loop. Not sold less than one-half dozen of a size. Shipping weight, per dozen, 3 oz. **State size.**

Catalog No.	Kind	Size No.	Per Doz.
6L3730	Single Gut	8 to 1	20c
6L3731	Double Gut	4 to 1	26c
6L3732	Double Gut	1-0	28c
6L3733	Double Gut	2-0	31c
6L3734	Double Gut	3-0	33c

Weedless Weighted Hook. Hollow point sneck hook with long shank and weight. Double flexible guard of steel spring wire with correct tension. A first class frog hook for weed casting. Shpg. wt., 2 oz.
6L3635—Size, 3-0 **3** for........25c
6L3636—Size, 4-0 **3** for........25c

Stanley Pork Rind Tandem Weedless Hook. Can be used with frog, pork rind, etc. Hollow point sneck hooks. Weed guards are steel spring wire. Shipping weight, 2 ounces.
6L3329—Size, 3-0....................28c
6L3330—Size, 4-0....................29c
6L3331—Size, 5-0....................30c

Lucky Worm Gang. Consists of three No. 8 hollow point hooks tied to gut. Twist worm around hooks. Shipping weight, each, 1 oz.
6L3777—6 for **30c**; each........6c

Henzel's Weedless Hook, with double guard. Hollow point long shank Cincinnati bass hook. Can be used with frog, pork rind or minnow. Shipping weight, 2 ounces.
6L3632—Size 19, for bass, **3** for....19c
6L3633—Size 18, pickerel, **3** for....20c
6L3634—Size 17, for pike, **3** for....21c

Improved Greer's Lever Hooks. Adapted to all kinds of fishing by sliding the little clamp on the hook. Made with Carlisle hooks in the following sizes. Shpg. wt., 2 oz.
6L3647—Size, 1-0, small..........9c
6L3648—Size, 3-0, medium..........10c
6L3649—Size, 5-0, large..........11c

Plain Treble Hook. Bright finish high quality. Sizes compare with regular hooks. **State size.** Shpg. wt., 4 oz.
6L3650—Sizes, 1 to 4. **3** for..8c
6L3651—Sizes, 1-0, 2-0 and 3-0. **3** for....................10c
6L3652—Sizes, 4-0, 5-0, 6-0 and 7-0. **3** for......16c
For wood minnows use size No. 1.

Safety Snap with Swivels. Blued brass barrel swivel with steel safety snap device; very strong and easy to open and close.

Catalog No.	Size No.	Full Lgth.	Shpg. Wt.	6 for
6L3695	10	1 in.	2 oz.	15c
6L3696	6	1⅕ in.	2 oz.	16c
6L3697	4	2 in.	2 oz.	17c
6L3698	2	2¼ in.	2 oz.	19c
6L3699	1-0	2½ in.	3 oz.	20c
6L3700	3-0	2¾ in.	3 oz.	26c

Brass Barrel Swivels. For attaching to spoon baits, etc. The length mentioned is the entire length.

Catalog No.	Size No.	Lgth., In.	Shpg. Wt., Doz.	Per Doz.
6L3665	10	½	1 oz.	12c
6L3666	7	¾	1⅕ oz.	13c
6L3667	4	⅞	1½ oz.	14c
6L3668	2	1⅛	2 oz.	15c
6L3669	1-0	1⅜	2 oz.	18c
6L3670	3-0	1⅝	3 oz.	26c

Adjustable Cork Float. Good grade cork. Painted in two colors.

Catalog No.	Length	Weight	Each
6L3723	2 in.	¼ oz.	8c
6L3724	3 in.	½ oz.	14c
6L3725	4 in.	¾ oz.	22c

New Adjustable Round Cork Ball Floats. Striped and varnished. With adjustable wires at top and bottom.

Catalog No.	Length	Weight	Each
6L3735	1⅜ in.	¼ oz.	10c
6L3736	1⅝ in.	½ oz.	15c
6L3737	1⅞ in.	¾ oz.	18c

Adjustable Lead Sinkers. Can be attached or detached by a few turns of the line.

Catalog No.	Size No.	Length, Inches	Shpg. Wt., per Dozen	Per Doz.
6L3679	1	¾	2 oz.	5c
6L3680	2	1	2 oz.	6c
6L3681	3	1	3 oz.	7c
6L3682	4	1¼	4 oz.	10c
6L3683	5	1⅓	5 oz.	13c
6L3684	6	1¾	6 oz.	17c

Six Assorted Adjustable Lead Sinkers, same style as above. One of each, sizes, 1 to 6. Shipping weight, 4 ounces.
6L3685—Per package............8c

Dipsey Swivel Lead Sinkers. Will not kink the line. No. 6 sinker is excellent for bait casting practice.

Catalog No.	Size	Shpg. Weight	3 for
6L3688	10	2 oz.	9c
6L3689	8	4 oz.	10c
6L3690	6	6 oz.	11c
6L3691	4	8 oz.	12c
6L3692	2	10 oz.	15c
6L3693	1-0	12 oz.	20c
6L3694	3-0	12 oz.	32c

Rangely Cinch Sinkers. Soft lead, split in center, soft ends. Line is put in slot and the ends bent over with fingers, which fastens sinker to line. Can be removed and used again.

Catalog No.	Size No.	Shipping Wt., Doz.	Per Doz.
6L3710	0	3 oz.	5c
6L3711	1	4 oz.	6c
6L3712	2	5 oz.	7c
6L3713	3	5 oz.	10c
6L3714	4	6 oz.	13c
6L3715	5	7 oz.	14c
6L3716	6	10 oz.	19c

Seven Assorted Rangely Lead Sinkers. Same style as above. One of each, sizes 0 to 6. Shipping weight, 4 ounces.
6L3717—Per package............9c

Split Shot. Round lead split. Put up in small wood boxes. Shipping weight, 2 oz.
6L3706—Size BB. **2** dozen for...4c
6L3707—Size 2B. **1** dozen for...5c

Set Line Complete.

Trot line with hooks and staging, complete, ready to attach to line. Line is 150 feet long, 60-thread twine. Fifty Kirby hooks, size No. 3-0, with 12-inch staging, attached by looping staging to line. Leaves a space of 3 feet between the hooks. Shipping weight, 1 pound.
6L3548....................90c

Same as above, but with stronger line, 84-thread twine, and No. 4-0 hooks. Shipping weight, 1 pound 3 ounces.
6L3549....................$1.05

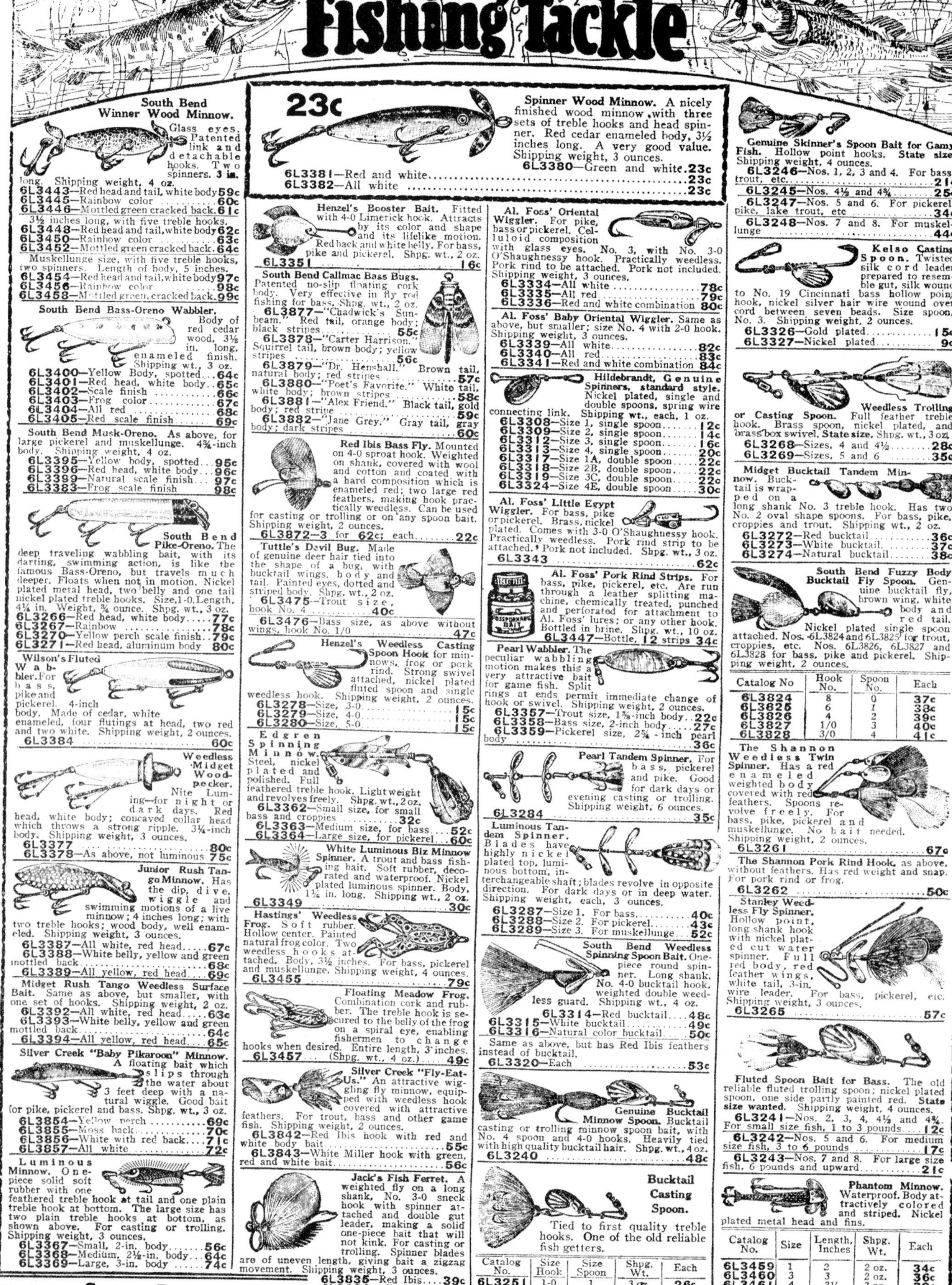

Fishing Tackle

South Bend Winner Wood Minnow. Glass eyes. Patented link and detachable hooks. Two spinners. 3 in. long. Shipping weight, 4 oz.
6L3443—Red head and tail, white body 59c
6L3445—Rainbow color 60c
6L3446—Mottled green cracked back 61c
3½ inches long, with five treble hooks.
6L3448—Red head and tail, white body 62c
6L3450—Rainbow color 63c
6L3452—Mottled green cracked back 64c
Muskellunge size, with five treble hooks, two spinners. Length of body, 5 inches.
6L3454—Red head and tail, white body 97c
6L3456—Rainbow color 98c
6L3458—Mottled green, cracked back 99c

South Bend Bass-Oreno Wabbler. Body of red cedar wood, 3½ in. long, enameled finish. Shipping wt., 3 oz.
6L3400—Yellow Body, spotted 64c
6L3401—Red head, white body 65c
6L3402—Scale finish 66c
6L3403—Frog color 67c
6L3404—All red 68c
6L3405—Red scale finish 69c

South Bend Musk-Oreno. As above, for large pickerel and muskellunge. 4¾-inch body. Shipping weight, 4 oz.
6L3395—Yellow body, spotted 95c
6L3396—Red head, white body 96c
6L3399—Natural scale finish 97c
6L3383—Frog scale finish 98c

South Bend Pike-Oreno. The deep traveling wabbling bait, with its darting, swimming action, is like the famous Bass-Oreno, but travels much deeper. Floats when not in motion. Nickel plated metal head, two belly and one tail nickel plated treble hooks. Size, 1-0. Length, 4¼ in. Weight, ¾ ounce. Shpg. wt., 3 oz.
6L3266—Red head, white body 77c
6L3267—Rainbow 78c
6L3270—Yellow perch scale finish 79c
6L3271—Red head, aluminum body 80c

Wilson's Fluted Wabbler. For bass, pike and pickerel. 4-inch body. Made of cedar, white enameled, four flutings at head, two red and two white. Shipping weight, 2 ounces.
6L3384 60c

Weedless Midget Wood-pecker. Nite Luming—for night or dark days. Red head, white body; concaved collar head which throws a strong ripple. 3¼-inch body. Shipping weight, 3 ounces.
6L3377 80c
6L3378—As above, not luminous 75c

Junior Rush Tango Minnow. Has the dip, dive, wiggle and swimming motions of a live minnow; 4 inches long; with two treble hooks; wood body, well enameled. Shipping weight, 3 ounces.
6L3387—All white, red head 67c
6L3388—White belly, yellow and green mottled back 68c
6L3389—All yellow, red head 69c

Midget Rush Tango Weedless Surface Bait. Same as above, but smaller, with one set of hooks. Shipping weight, 2 oz.
6L3392—All white, red head 63c
6L3393—White belly, yellow and green mottled back 64c
6L3394—All yellow, red head 65c

Silver Creek "Baby Pikaroon" Minnow. A floating bait which slips through the water about 3 feet deep with a natural wiggle. Good bait for pike, pickerel and bass. Shpg. wt., 3 oz.
6L3854—Yellow perch 69c
6L3855—Moss back 70c
6L3856—White with red back 71c
6L3857—All white 72c

Luminous Minnow. One-piece solid soft rubber with one feathered treble hook at tail and one plain treble hook at bottom. The large size has two plain treble hooks at bottom, as shown above. For casting or trolling. Shipping weight, 3 ounces.
6L3367—Small, 2-in. body 56c
6L3368—Medium, 2½-in. body 64c
6L3369—Large, 3-in. body 74c

23c **Spinner Wood Minnow.** A nicely finished wood minnow, with three sets of treble hooks and head spinner. Red cedar enameled body, 3½ inches long. A very good value. Shipping weight, 3 ounces.
6L3380—Green and white 23c
6L3381—Red and white 23c
6L3382—All white 23c

Henzel's Booster Bait. Fitted with 4-0 Limerick hook. Attracts by its color and shape and its lifelike motion. Red back and white belly. For bass, pike and pickerel. Shpg. wt., 2 oz.
6L3351 16c

South Bend Callmac Bass Bugs. Patented no-slip floating cork body. Very effective in fly rod fishing for bass. Shpg. wt., 2 oz.
6L3877—"Chadwick's Sunbeam." Red tail, orange body; black stripes 55c
6L3878—"Carter Harrison." Squirrel tail, brown body; yellow stripes 56c
6L3879—"Dr. Henshall." Brown tail, natural body; red stripes 57c
6L3880—"Poet's Favorite." White tail, white body; brown stripes 58c
6L3881—"Alex Friend." Black tail, gold body; red stripe 59c
6L3882—"Jane Grey." Gray tail, gray body; dark stripes 60c

Red Ibis Bass Fly. Mounted on 4-0 sproat hook. Weighted on shank, covered with wool and cotton and coated with a hard composition which is enameled red; two large red feathers, making hook practically weedless. Can be used for casting or trolling or on any spoon bait. Shipping weight, 2 ounces.
6L3872—3 for 62c; each 22c

Tuttle's Devil Bug. Made of genuine deer hair tied into the shape of a bug, with bucktail wings, body and tail. Painted eyes, dotted and striped body. Shpg. wt., 2 oz.
6L3475—Trout size, hook No. 4 40c
6L3476—Bass size, as above without wings, hook No. 1/0 47c

Henzel's Weedless Casting Spoon Hook for minnows, frog or pork rind. Strong swivel attached, nickel plated fluted spoon and single weedless hook. Shipping weight, 2 ounces.
6L3278—Size, 3-0 15c
6L3279—Size, 4-0 15c
6L3280—Size, 5-0 15c

Edgren Spinning Minnow. Steel, nickel plated and polished. Full feathered treble hook. Light weight and revolves freely. Shpg. wt., 2 oz.
6L3362—Small size, for small bass and croppies 32c
6L3363—Medium size, for bass 52c
6L3364—Large size, for pickerel 60c

White Luminous Biz Minnow Spinner. A trout and bass fishing bait. Soft rubber, decorated and waterproof. Nickel plated luminous spinner. Body, 1¼ in. long. Shipping wt., 2 oz.
6L3349 30c

Hastings' Weedless Frog. Soft rubber. Hollow center. Painted natural frog color. Two weedless hooks attached. Body, 3½ inches. For bass, pickerel and muskellunge. Shipping weight, 4 ounces.
6L3455 79c

Floating Meadow Frog. Combination cork and rubber. The treble hook is secured to the belly of the frog on a spiral eye, enabling fishermen to change hooks when desired. Entire length, 3 inches.
6L3457 (Shpg. wt., 4 oz.) 49c

Silver Creek "Fly-Eat-Us." An attractive wiggling fly minnow, equipped with weedless hook covered with attractive feathers. For trout, bass and other game fish. Shipping weight, 2 ounces.
6L3842—Red Ibis hook with red and white body bait 55c
6L3843—White Miller hook with green, red and white bait 56c

Jack's Fish Ferret. A weighted fly on a long shank, No. 3-0 sneck hook with spinner attached and double gut leader, making a solid one-piece bait that will not kink. For casting or trolling. Spinner blades are of uneven length, giving bait a zigzag movement. Shipping weight, 3 ounces.
6L3835—Red Ibis 39c
6L3836—Brown body, brown and red feathers 40c
6L3837—Bucktail 35c

Al. Foss' Oriental Wiggler. For pike, bass or pickerel. Celluloid composition with glass eyes. No. 3, with No. 3-0 O'Shaughnessy hook. Practically weedless. Pork rind to be attached. Pork not included. Shipping weight, 3 ounces.
6L3334—All white 78c
6L3335—All red 79c
6L3336—Red and white combination 80c

Al. Foss' Baby Oriental Wiggler. Same as above, but smaller; size No. 4 with 2-0 hook. Shipping weight, 3 ounces.
6L3339—All white 82c
6L3340—All red 83c
6L3341—Red and white combination 84c

Hildebrandt, Genuine Spinners, standard style. Nickel plated, single and double spoons, spring wire connecting link. Shipping wt., each, 1 oz.
6L3308—Size 1, single spoon 12c
6L3309—Size 2, single spoon 14c
6L3312—Size 3, single spoon 16c
6L3313—Size 4, single spoon 20c
6L3317—Size 1A, double spoon 22c
6L3318—Size 2B, double spoon 22c
6L3319—Size 3C, double spoon 22c
6L3324—Size 4E, double spoon 30c

Al. Foss' Little Egypt Wiggler. For bass, pike or pickerel. Brass, nickel plated. Comes with 3-0 O'Shaughnessy hook. Practically weedless. Pork rind strip to be attached. Pork not included. Shpg. wt., 3 oz.
6L3343 62c

Al. Foss' Pork Rind Strips. For bass, pike, pickerel, etc. Are run through a leather splitting machine, chemically treated, punched and perforated for attachment to Al. Foss' lures; or any other hook. Bottled in brine. Shpg. wt., 10 oz.
6L3447—Bottle, 12 strips 34c

Pearl Wabbler. The peculiar wabbling motion makes this a very attractive bait for game fish. Split rings at ends permit immediate change of hook or swivel. Shipping weight, 2 ounces.
6L3357—Trout size, 1⅞-inch body 22c
6L3358—Bass size, 2-inch body 27c
6L3359—Pickerel size, 2¾-inch pearl body 36c

Pearl Tandem Spinner. For bass, pickerel and pike. Good for dark days or evening casting or trolling. Shipping weight, 6 ounces.
6L3284 35c

Luminous Tandem Spinner. Blades have highly nickel plated top, luminous bottom, interchangeable shaft; blades revolve in opposite direction. For dark days or in deep water. Shipping weight, each, 3 ounces.
6L3287—Size 1. For bass 40c
6L3288—Size 2. For pickerel 43c
6L3289—Size 3. For muskellunge 52c

South Bend Weedless Spinning Spoon Bait. One-piece round spinner. Long shank. No. 4-0 bucktail hook, weighted double weedless guard. Shipping wt., 4 oz.
6L3314—Red bucktail 48c
6L3315—White bucktail 49c
6L3316—Natural color bucktail 50c
Same as above, but has Red Ibis feathers instead of bucktail.
6L3320—Each 53c

Genuine Bucktail Minnow Spoon. Bucktail casting or trolling minnow spoon bait, with No. 4 spoon and 4-0 hooks. Heavily tied with high quality bucktail hair. Shpg. wt., 4 oz.
6L3240 48c

Bucktail Casting Spoon. Tied to first quality treble hooks. One of the old reliable fish getters.

Catalog No.	Size Hook	Size Spoon	Shpg. Wt.	Each
6L3251	1-0	1	3 oz.	26c
6L3252	2-0	2	3 oz.	27c
6L3253	3-0	3	3 oz.	28c
6L3254	4-0	4	3 oz.	29c

Genuine Skinner's Spoon Bait for Gamy Fish. Hollow point hooks. State size. Shipping weight, 4 ounces.
6L3246—Nos. 1, 2, 3 and 4. For bass, trout, etc. 21c
6L3245—Nos. 4½ and 4¾ 25c
6L3247—Nos. 5 and 6. For pickerel, pike, lake trout, etc. 34c
6L3248—Nos. 7 and 8. For muskellunge 44c

Kelso Casting Spoon. Twisted silk cord leader prepared to resemble gut, silk wound to No. 19 Cincinnati bass hollow point hook, nickel silver hair wire wound over cord between seven beads. Size spoon, No. 3. Shipping weight, 2 ounces.
6L3326—Gold plated 15c
6L3327—Nickel plated 9c

Weedless Trolling or Casting Spoon. Full feather treble hook. Brass spoon, nickel plated, and brass box swivel. State size. Shpg. wt., 3 oz.
6L3268—Sizes, 4 and 4½ 28c
6L3269—Sizes, 5 and 6 35c

Midget Bucktail Tandem Minnow. Bucktail is wrapped on a long shank No. 3 treble hook. Has two No. 2 oval shape spoons. For bass, pike, croppies and trout. Shipping wt., 2 oz.
6L3272—Red bucktail 36c
6L3273—White bucktail 37c
6L3274—Natural bucktail 38c

South Bend Fuzzy Body Bucktail Fly Spoon. Genuine bucktail fly, brown wing, white body and red tail. Nickel plated single spoon attached. Nos. 6L3824 and 6L3825 for trout, croppies, etc. Nos. 6L3826, 6L3827 and 6L3828 for bass, pike and pickerel. Shipping weight, 2 ounces.

Catalog No	Hook No.	Spoon No.	Each
6L3824	8	0	37c
6L3825	6	1	38c
6L3826	4	2	39c
6L3827	1/0	3	40c
6L3828	3/0	4	41c

The Shannon Weedless Twin Spinner. Has a red enameled weighted body covered with red feathers. Spoons revolve freely. For bass, pike, pickerel and muskellunge. No bait needed. Shipping weight, 2 ounces.
6L3261 67c

The Shannon Pork Rind Hook, as above, without feathers. Has red weight and snap. For pork rind or frog.
6L3262 50c

Stanley Weedless Fly Spinner. Hollow point, long shank hook with nickel plated cut water spinner. Full red body, red feather wings, white tail, 3-in. wire leader. For bass, pickerel, etc. Shipping weight, 3 ounces.
6L3265 57c

Fluted Spoon Bait for Bass. The old reliable fluted trolling spoon; nickel plated spoon, one side partly painted red. State size wanted. Shipping weight, 4 ounces.
6L3241—Nos. 2, 3, 4, 4½ and 4¾. For small size fish, 1 to 3 pounds 12c
6L3242—Nos. 5 and 6. For medium size fish, 3 to 6 pounds 17c
6L3243—Nos. 7 and 8. For large size fish, 6 pounds and upward 21c

Phantom Minnow. Waterproof. Body attractively colored and striped. Nickel plated metal head and fins.

Catalog No.	Size	Length, Inches	Shpg. Wt.	Each
6L3459	1	2	2 oz.	34c
6L3460	3	3	2 oz.	36c
6L3461	4	3¼	3 oz.	39c
6L3462	5	3½	3 oz.	41c
6L3463	6	4	4 oz.	46c
6L3464	8	4½	4 oz.	58c

Fishing Tackle

TROUT FLIES.

We have selected only such flies as we know from experience to be killers. Our flies are excellent in beauty, quality and workmanship. We do not carry the very cheap grade of trout flies.

A Grade, high quality, with Pennell turned eye which prevents the feathers and silk wrappings from becoming loose and whipped off. Hollow point sneck hook, full silk body with divided wings, and silk tied to first quality selected full length clear round gut.

B Grade, a good grade silk body fly, plain wings, well finished, tied to full length clear gut on file point sneck hook.

NOTICE—When ordering give number of fly, also size of hooks wanted, and whether A or B Grade. We do not sell less than one-half dozen flies nor less than two flies of one kind and size of hook. Shipping weight, per dozen, 2 ounces.

Catalog No.	Name	Size of Hooks	Per Dozen A Grade	Per Dozen B Grade
6L3801	Rube Wood	6, 8, 10 or 12	$1.12	78c
6L3802	Professor	6, 8, 10 or 12	1.12	78c
6L3803	Governor	6, 8, 10 or 12	1.12	78c
6L3804	Golden Spinner	6, 8, 10 or 12	1.12	78c
6L3805	Silver Doctor	6, 8, 10 or 12	1.12	78c
6L3806	Seth Green	6, 8, 10 or 12	1.12	78c
6L3808	Cow Dung	6, 8, 10 or 12	1.12	78c
6L3809	Queen of Waters	6, 8, 10 or 12	1.12	78c
6L3810	King of Waters	6, 8, 10 or 12	1.12	78c
6L3811	Grizzly King	6, 8, 10 or 12	1.12	78c
6L3814	Brown Hackle	6, 8, 10 or 12	1.12	78c
6L3815	Gray Hackle	6, 8, 10 or 12	1.12	78c
6L3816	Coachman	6, 8, 10 or 12	1.12	78c
6L3817	Royal Coachman	6, 8, 10 or 12	1.12	78c
6L3819	Parmachenee Belle	6, 8, 10 or 12	1.12	78c
6L3820	Montreal	6, 8, 10 or 12	1.12	78c
6L3821	Black Gnat	6, 8, 10 or 12	1.12	78c
6L3823	White Miller	6, 8, 10 or 12	1.12	78c
6L3829	Yellow May	6, 8, 10 or 12	1.12	78c

High Grade Split Double Wing Divided Dry Flies. These flies are high grade in material and workmanship and are made from water fowl feathers as far as possible. The divided wings cause the fly to drop onto the water in a lifelike and natural way. The hooks are tested, and the gut absolutely perfect; tied to Pennell sneck hollow point hooks, turndown eye. Shipping weight, 2 ounces.

Catalog No.	Name of Fly	Size of Hook	Per Doz.	Catalog No.	Name of Fly	Size of Hook	Per Doz.
6L3860	Red Quill	8, 10 or 12	$1.80	6L3866	Royal Coachman	8, 10 or 12	$1.80
6L3861	Olive Quill	8, 10 or 12	1.80	6L3867	Grizzly King	8, 10 or 12	1.80
6L3862	Ginger Quill	8, 10 or 12	1.80	6L3868	Western Bee	8, 10 or 12	1.80
6L3863	Brown Hackle	8, 10 or 12	1.80	6L3869	Black Gnat	8, 10 or 12	1.80
6L3864	March Brown	8, 10 or 12	1.80	6L3870	Cow Dung	8, 10 or 12	1.80
6L3865	Rube Wood	8, 10 or 12	1.80	6L3871	Professor	8, 10 or 12	1.80

Fly Book. Selected quality pigskin, light tan color; leather lining; aluminum box for dry flies; fitted with dampening pad and cork hook holder. Book has three felt drying pads, two celluloid leaves with Bray style springs and clips on each side, one leather pocket. Box holds 8 dozen flies. Celluloid leaves hold 4 dozen flies. Size, 6½x3½x1¼ in. Shpg. wt., 9 oz.
6L3569 ....$3.98

Vest Pocket Fly Book. 6x2¼ inches when folded. Has two envelope compartments, and four pages with metal tacks for hooks or flies. Double drying pad. Black leatherette case and metal snap button. Shipping weight, 3 oz.
6L3565 ....45c

Tackle Book. Waterproof brown canvas. Eleven pockets, bound with artificial leather. Size, 10x4½ inches. Single strap fastening under leather loops.
6L3570—Shpg. wt., 8 oz.......$1.20

Tackle and Fly Book. Black artificial leather covered; 3½x6 inches when folded. Has two large envelope compartments and four pages of Bray style fasteners, with double flannel drying pads. Fitted with snap button fastener.
6L3563—Shpg. wt., 8 oz.......59c

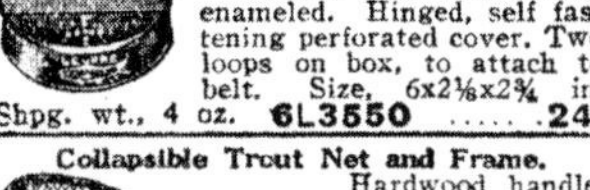

Crescent Bait Box. Nicely enameled. Hinged, self fastening perforated cover. Two loops on box, to attach to belt. Size, 6x2⅛x2¾ in. Shpg. wt., 4 oz. **6L3550** 24c

Collapsible Trout Net and Frame. Hardwood handle, 13 inches long, with hook. 10½x10½-inch steel ring. Patent closing top with 18-inch brown waterproof net, ¾-inch mesh. Shpg. wt., 10 oz.
6L3543 ....$1.10

Minnow and Dip Net Complete. Solid round ring, 14x14 inches, fitted with 20-inch net, ½-inch top mesh, ¼-inch bottom mesh. Used for landing large fish or as minnow net. 4-foot jointed bamboo handle. Shpg. wt., 1¾ lbs.
6L3545 ....$1.20

Compact Folding Dip Net and Frame. No. 1 ring, 12½x14 inches, mounted with a 20-inch square bottom brown waterproof net. ¾-inch mesh, 4-foot jointed handle. Shpg. wt., 1¼ lbs.
6L3544 ....$1.75

Minnow Dip Net. Strong wire handle. Ring, 4¼ inches in diameter; net, ¼-inch mesh at top and ⅛-inch mesh at bottom. Shpg. wt., 8 oz.
6L3546 ....20c

South Bend Moth-Oreno Fly. (Robert Page Lincoln Design.) Tied on number 8 turndown eye hollow point bronzed Limerick Hook. Fuzzy cork body with suitable coloring. It is a splendid imitation of a natural moth. Supplied in six colors of feathered wings. Shipping wt., per half dozen, 2 oz.
6L3885—White color wings.
6L3886—Gray color wings.
6L3887—Tan color wings.
6L3888—Speckled color wings.
6L3889—Yellow color wings.
6L3890—Brown color wings.
Per half dozen.... **$1.72** | Each **30c**

Spinning Coachman with 1⅛-inch highly nickel plated spoon. No. 1 hook, tied with double gut; with swivel attached. Each bait comes with a Coachman and two other popular bass flies mounted on white card. Shipping weight, 3 ounces.
6L3346—Per card......**32c**

TROUT BAIT.

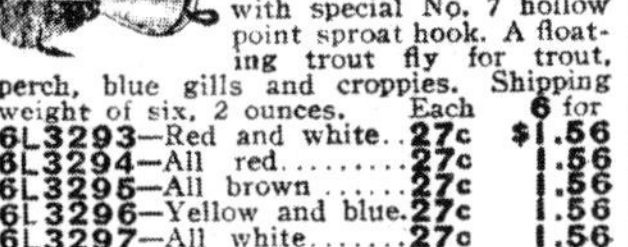

Jamison Coaxer Floating Trout Fly. Made of cork with special No. 7 hollow point sproat hook. A floating trout fly for trout, perch, blue gills and croppies. Shipping weight of six, 2 ounces.

		Each	6 for
6L3293	Red and white	27c	$1.56
6L3294	All red	27c	1.56
6L3295	All brown	27c	1.56
6L3296	Yellow and blue	27c	1.56
6L3297	All white	27c	1.56
6L3298	All black	27c	1.56

Luminous Trout Spinner. Has two luminous spinners. Feathers and spinners bright and attractive. Shpg. wt., 2 oz.
6L3311 ....**28c**

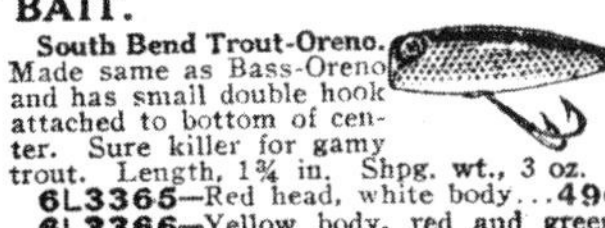

South Bend Trout-Oreno. Made same as Bass-Oreno and has small double hook attached to bottom of center. Sure killer for gamy trout. Length, 1¾ in. Shpg. wt., 3 oz.
6L3365—Red head, white body....**49c**
6L3366—Yellow body, red and green spotted **50c**
6L3370—Red body, blended head.**51c**
6L3371—Natural scale finish.....**52c**

Trout Bait. Improved Colorado spinner, nickel plated. Hollow point hooks, two small brass swivels, ¾-inch spoon. Shpg. wt., 2 oz.
6L3310 ....**11c**

Natural Preserved Minnows. Lifelike appearance. Toughened and will stay on hook longer than live minnows. Will keep in any climate.
6L3435—Small Shiners, 1 to 1½ inches. About six dozen to a jar. Shipping wt., 1 lb.........**25c**
6L3436—Medium Shiners, about 3 in. About 5 dozen to a jar. Shipg. wt., 1⅛ lbs.....**28c**
6L3437—Large Shiners, 3½ inches or over. About 3 dozen to a jar. Shipping weight, 1¼ pounds.................**32c**

Preserved Pork Bait With Rind. Minnow shape. Six pieces, 3 inches long. Tough, yet soft and pliable to give a minnow tail wiggle. Each piece has rind on top. For bass, pike, pickerel and muskellunge; 4-oz. jar. Shpg. wt., 1 lb.
6L3440—Per jar **31c**

Chunk Shape Pork With Rind, preserved with full top rind. Used mostly for surface bait casting. About 1½ inches long, 1 inch wide, ⅜ inch thick. Cut to wabble when reeling in. Six pieces in 4-ounce jar. Shipping weight, 1 lb.
6L3442—Per jar **34c**

For Al Foss Pork Rind Refer to Page 776.

"Gitz-Em" Fish Worm Extractor. Makes fish worms, angle worms and night crawlers come out on top, day or night, in one to three minutes. Simply mix with water and pour on ground. Will not injure worms, ground or grass. Always keep a can on hand to avoid the tedious task of digging for worms. Results are surprising. Shipping weight, 2 pounds.
6L3480 ....**79c**

Salmon Eggs. Preserved for bait. Large and firm enough to make them stay on the hook. Come in 2-ounce glass jar containing about 140 eggs. Weight, 8 ounces.
6L3444—Per jar.......**32c**

Single Salmon Egg Pennell Sneck Hook attached to gut, 10 inches long, with knot under eye. Three sizes. Nos. 8, 10 and 12. **State size.** Shpg. wt., 2 oz.
6L3775—Per dozen hooks **45c**

Soft Rubber Minnow. Length, 1¾ in. For still fishing. Shpg. wt., 2 oz.
6L3466 ....**22c**

Same style as 6L3466, 2 inches long, chub shape, with gimp leader and swivel. Red striped gills and fins. For bass, pike or pickerel. Shpg. wt., 2 oz.
6L3468 ....**33c**

Rubber Grasshopper Bait. Size, 1¾ in. long. Very natural. Shipping weight, 2 ounces.
6L3474 ....**20c**

Red Rubber Angle Worm. A good imitation of a live worm. About 2½ in. long. Shpg. wt., 2 oz.
6L3470 ....**29c**

Soft Rubber Helgamite or Dobson. Length, 2¼ inches, with hook, natural color, 2-inch gimp leader and swivel. Shipping weight, 3 ounces.
6L3472 ....**36c**

FISHING TACKLE OUTFITS.

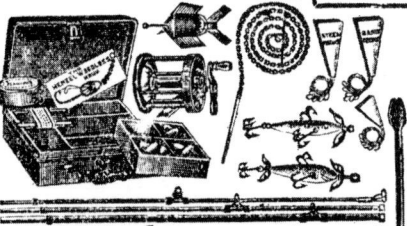

Consists of one 5½-foot three-jointed steel casting rod, extra large ring nickle silver guides and tip, cork grasp, nickle plated reel seat with finger hook; one nickel plated emergency tip; one rod shortener, one rubber plate jeweled oil cap reel, 60-yard, click and drag; large balance handle, gear ratio, 3¼ to 1, known as quadruple; 50 yards Italian silk bass casting line; 25 yards standard quality hard braided linen line; 1½ dozen assorted high quality double gut hooks; two popular spoon baits; one weedless spoon hook; two wood minnows; one disgorger; one chain fish stringer; ½ dozen assorted sinkers; one wire leader, 7 inches long, fitted with snap and swivel at ends, and one Henzel Booster bait. Packed in metal tackle box, 9½ inches long, 3½ inches high and 7 inches wide. Weight, 4½ pounds. Shipping weight, complete, 6 lbs.
6L4400—Outfit complete with box **$9.35**

Consists of one 6-foot three-jointed steel casting rod, fitted with cork grasp, nickel plated reel seat with finger hook; extra large ring nickel silver guides and large tip; excellent 60-yard wide spool reel, fitted with two screw off oil caps, click and drag, full nickel plated fancy bone balance handle, gear ratio, 3¼ to 1, known as quadruple; 75 feet of hard braided silk casting line; 75 feet hard braided linen line suitable for trolling or still fishing, two wood minnows, 1½ dozen assorted gut hooks, two popular attractive spoon baits, one weedless casting spoon, one Booster bait, ½ dozen assorted sinkers, one 3-foot double gut leader, one worm gang, one chain fish stringer, one fancy cork float, one fish scaler, one disgorger. Shipping weight, 2 pounds 1 ounce.
6L4396—Outfit **$5.95**

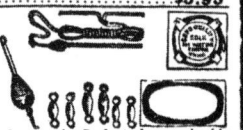

Consists of a good grade two-piece jointed Japanese bamboo rod about 7 feet long, nickel plated ferrules and nickel plated reel seat, black enameled scored grasp with nickel plated butt cap; 50-yard high frame nickel plated, free running reel, with sliding click; 50 feet of braided line, one weedless casting spoon, one dozen gut hooks, 3-foot gut leader, ½ dozen adjustable sinkers, fancy adjustable float, strong nickel plated chain fish stringer with pointed needle, one wood minnow with treble hooks and spinner, nicely enameled. Shipping weight, 2⅝ pounds.
6L4398—Outfit **$2.28**

Consisting of one two-piece jointed Japanese bamboo rod, 7 feet long, double ferrules, one Simplicity reel, 50-yard size, heavily nickel plated, 60 feet good quality braided fish line, 1 dozen assorted gut hooks, one 3-foot gut leader, one one-half dozen adjustable painted float, one wood minnow, one three-hook worm gang; one-half dozen adjustable lead sinkers, assorted sizes; one weedless casting spoon, one fish hook extractor and one fish stringer of braided line with needle and wire end. Shpg. wt., 1⅝ lbs.
6L4397—Outfit **$1.09**

Tackle Boxes.

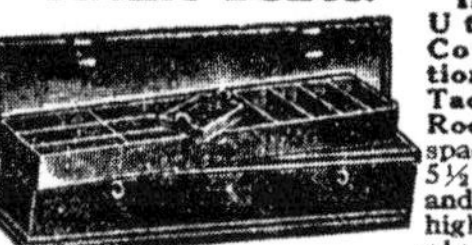

Improved Utility Combination Box for Tackle and Rod. Inside space, 23½x5½ inches and 5¼ in. high. Enameled light green with gold color decorations. Two front clasps. One large tray with twelve compartments and one small tray. Bottom has four large compartments. Holds eight 80-yard reels or four large salt water reels. Space for short jointed rods, 23½ inches long, 2 inches wide and 3 inches high. Shpg. wt., 7 lbs.
6L3551........................**$3.68**

Tackle Box. Size, 9½ in. long, 3½ in. high, 7 in. wide. One tray with four compartments and one compartment large enough for a 100-yard reel or two smaller reels; one compartment the full length of the box for snelled hooks and flies. Three smaller compartments for spoon baits, etc. Shipping weight, 2 lbs. 13 oz.
6L3554........................**$1.38**

Tackle Box. American Anglers' Improved Tackle Box with two patent sliding trays. Made of heavy metal, highly enameled in dark green with gold color decorations. Size, 13½ in. long, 5¼ in. wide, 5 in. high. Two sliding trays with eight compartments which slide out of box and hold fast as shown. Bottom has three large compartments for reels, lines, baits, etc. Shipping wt., 4 lbs.
6L3556........................**$2.18**

Tackle Box. Size, 10½ inches long, 3¾ inches high and 5 inches wide. Has one tray with five compartments suitable for sinkers, swivels, hooks, floats, spoons, etc. Three compartments large enough to hold one 100-yard reel, lines, wood, minnow, gut hooks, etc. Shpg. wt., 2½ lbs.
6L3555........................**$1.47**

Tackle Box. Made of steel, finished in hard baked dull enamel; 11 in. long, 5¼ in. wide, 2¼ in. deep. With lock and two keys. Compartments for reel, lines, spoons, baits, etc. Dividing lid to keep tackle in place. Shipping weight, 2¼ lbs.
6L3564........................**$1.07**

Never Sink Floating Minnow Bucket. Outside bucket made of heavy tin plate, bottom of galvanized iron, highly enameled in light green color with gold color decorations. Inside bucket made of heavy tin plate and galvanized bottom. Water and air pass through meshes, giving minnows plenty of fresh air. Air chamber at top allows inner bucket to float about 2 inches above water. Galvanized wire hinged top lid with clasp, to get minnows out easily. Strongly made, and nicely finished. Shipping weight, 4 pounds.
6L3567—8-quart size **$1.78**
6L3568—10-quart size **1.92**

Trout Baskets. Select split willow, strong and light. Clasp of leather and metal.
6L3580—Small size No. 1; capacity, 10 pounds. Shipping weight, 15 ounces..........**$1.84**
6L3581—Medium size No. 2; capacity 13 pounds. Shipping wt., 1¼ lbs. **$2.15**
6L3582—Large size No. 3; capacity, 19 pounds. Shipping weight, 1½ lbs. **$2.75**
6L3579—Extra large size No. 4; capacity, 25 pounds. Shpg. wt., 2 lbs. **$2.95**

Trout Basket Strap. Made of webbing attached to leather straps with buckles, rings and snap. Leaves the arms free. Shipping weight, 4 ounces.
6L3583........................**34c**

Folding Canvas Creel or Trout Basket. Folds quickly and can be used for carrying game or as a pack bag. Olive tan drill and lined. Shoulder sling has sliding buckle, snap and ring; leather shoulder pad. Has ventilating eyelets at bottom and spring steel spreader at top; ribs sewed in to hold shape. Shipping weight, 14 ounces.
6L3575—No. 1. Small size. Capacity, 12 pounds........................**95c**
6L3576—No. 2. Medium size. Capacity, 20 pounds........................**$1.05**
6L3577—No. 3. Large size. Capacity, 25 pounds........................**$1.15**

"Keep Alive" Net. Has wire hoops and knotted netting. Weight at bottom to keep net spread open, and draw string at top, to keep fish alive and fresh.

Catalog No.	Diameter	Shpg. Wt.	
6L3540	10 in.	14 oz.	**50c**
6L3541	12 in.	1 lb.	**65c**
6L3542	14 in.	1½ lbs.	**80c**

Row Boats and Canoes

Build your own boat and save money on labor, freight and hauling. We furnish the material. All you need is a hammer and screwdriver. The boats, when put together, are safe, comfortable and easy rowing. When you buy a boat ready made, you pay three or four times first class freight or express charges. Buying your boat knocked down, as above, means that all you have to pay is second class freight rate, which in itself is a very considerable saving.

The lumber we furnish is selected, seasoned material from our own yards, of the proper thickness, with strong reinforcements. The stem is hardwood; the stern is reinforced with extra brackets and has a plate board attached for outboard motor. The rear seat is extra wide. Bottom is in three sections, calked and reinforced with shaped strips. Everything necessary, including nails, screws, calking cotton and oar sockets, with one pair of copper tipped oars with plates and oarlocks attached. The 14, 15 and 16-foot boats are fitted with two pairs of sockets. The materials are all smoothly finished, with one coat of paint. Full directions for assembling the boat accompany each shipment. **Shipped from factory in OHIO.**

Catalog No.	Length, Feet	Width Beam, Inches	Height, Inches	No. of Seats	Weight, With Oars, Lbs.	Shpg. Wt., Crated, Lbs.	
6L5690⅓	13	40	15	3	168	241	**$29.95**
6L5691⅓	14	42	15	4	188	264	**31.48**
6L5692⅓	15	44	15	4	209	288	**32.97**
6L5693⅓	16	44	15	4	229	311	**34.46**

Extra pair of oars, painted, with horns and plates attached, extra..........**$3.23**

Galvanized Steel Flat Bottom Fishing and Pleasure Boat. Built of standard quality 20-gauge galvanized steel. All joints double seamed and heavily soldered. Coated with two coats of paint in battleship gray color. Two large air chambers, one in bow and one in stern, will keep boat afloat when filled with water. Well braced on inside with angle iron 1x1x⅛ inch thick. Also reinforced all around top with angle iron. Not affected by sun or weather. Specially braced for outboard motor. Has a steel keel which is securely fastened to bottom of boat from stem to stern, which helps to guide boat and also prevents bottom from being scraped when pulled up on stone or gravel shores. One pair of painted copper tipped oars, fitted with oar plates and North River oarlocks. The 14 and 15-foot boats are fitted with two pairs of sockets. Pulley at bow of boat for anchor rope. When storing away turn boat upside down. **Shipped from factory in INDIANA.**

Catalog No.	Length, Feet	Width Beam, Inches	Height, Inches	Number Seats	Weight, Pounds	
6L5700⅓	13	39	15	3	192	**$31.64**
6L5701⅓	14	41	15	4	203	**33.95**
6L5702⅓	15	43	15	4	213	**35.20**

Extra for section wood flooring in boat. Shipping weight, 30 pounds..........**$2.20**
Extra for additional coat of paint, any color..........**1.50**
Extra pair of oars, painted and fitted with oar plates and horns..........**3.55**
Lettering on boats in 2-inch letters, extra per letter, 5 cents.

Hiawatha Canoe.

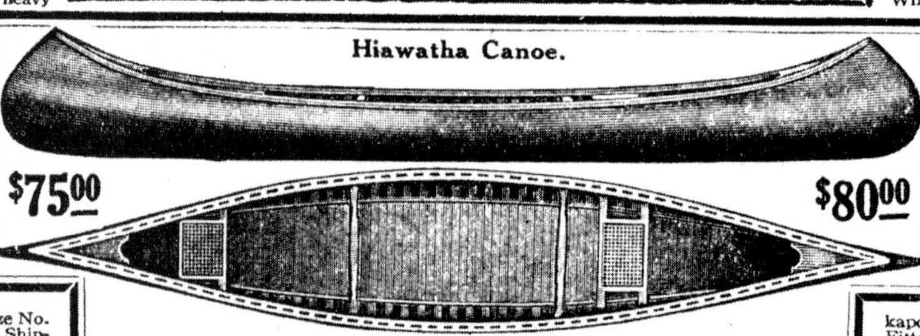

$75.00 **$80.00**

The Hiawatha Model Canoe is the latest creation in the line of canvas covered canoes. No effort has been spared to surpass the other models in refinement and distinction. Ribs are of finest selected cedar, 5/16-inch thick, with half ribs of same thickness placed between full length ribs on bottom of canoe. Planking of 5/16-inch selected cedar in full length strips with bevel edge, insuring tight joints and smooth surface. All fastenings are of copper or brass. Open gunwales of full length straight grained spruce. Thwarts and keel are of selected white oak. Seat frames of selected white oak with caned center. Canvas is one piece of closely woven duck filled with flexible waterproof filler and covered with two coats of color paint and two coats of varnish. Inside of canoe covered with three coats of elastic marine varnish. Comes in two colors: Dark green and Tuscan red. Do not confuse this with cheaper canoes, as this canoe is in a class by itself. Shipped from factory in Northern Wisconsin.

Catalog No.	Length	Beam	Depth Amid	Weight	
6L5703½	16 feet	33 inches	13 inches	70 lbs.	**$75.00**
6L5704½	17 feet	34 inches	13 inches	75 lbs.	**80.00**

If Sponsons are wanted allow $20.00 extra.

Galvanized Oval Floating Minnow Buckets. Striped. Strongly made. Inside bucket is also galvanized and galvanized wire mesh around inner bucket floats one inch above water; has hinged cover with clasp. Shipping weight, 4 pounds.
6L3573—8-qt. size **$3.16**
6L3574—10-qt. size **3.35**

Duplex Folding Minnow Bucket. For fishermen who do not care to carry the large tin bucket. Folds very small and can be put in hand bag or suit case. Made of heavy canvas, waterproofed. Metal parts spring steel, rustproof. Aeration is secured by metal eyelets inserted in top. Top sunk to hold ice if desired. Shpg. wt., 2¼ lbs. Size, In.

6L3586	6-quart	10x7½	**$1.88**
6L3587	8-quart	12x7½	**2.26**
6L3588	10-quart	12x9½	**2.64**

Boat Cement. An excellent watertight calking for boats, canoes, launches, tanks and troughs. Far superior to tar or ordinary pitch. Simple to use with directions; simply pour into the seam or opening and allow to dry. Shipping weight, 2½ pounds.
6L5761—Per quart can..........**98c**

Mushroom Boat Anchors. Iron Boat Anchor in mushroom shape, which prevents dragging, but will not snag when pulled up. Has holes in base to allow mud and water to drain off. The shape of anchor makes weight heavy enough to hold any ordinary boat in a current. Weight, 15 lbs.
6L5714........................**$1.50**

Sure Catch Minnow and Crawfish Trap. Galvanized sheet steel and wire screening. May be separated in the center and one end telescoped into the other. Length, 17 inches; diameter, 8½ inches at center, and at ends 7½ inches. Shipping weight, 2¾ pounds.
6L3566........................**88c**

Northern Fir Oars.

Copper Tipped Northern Fir Oars. Well made; lighter than ash oars and are very strong. **Cannot be shipped by parcel post.**

Catalog No.	Length	Shpg. Wt.	Per Pair
6L5730¼	6 feet	5½ lbs.	**$2.16**
6L5731¼	6½ feet	6 lbs.	**2.34**
6L5732¼	7 feet	6½ lbs.	**2.52**
6L5733¼	7½ feet	7 lbs.	**2.70**
6L5734¼	8 feet	7¾ lbs.	**2.88**
6L5735¼	8½ feet	9 lbs.	**3.06**

Selected White Ash Oars. Copper Tipped White Ash Oars. Unmailable.

Catalog No.	Length	Shpg. Wt.	Per Pair
6L5740¼	6 feet	6 lbs.	**$2.34**
6L5741¼	6½ feet	7 lbs.	**2.54**
6L5742¼	7 feet	8 lbs.	**2.78**
6L5745¼	7½ feet	9 lbs.	**2.95**
6L5743¼	8 feet	10 lbs.	**3.12**
6L5744¼	8½ feet	11 lbs.	**3.31**

Northern White Fir Paddles.

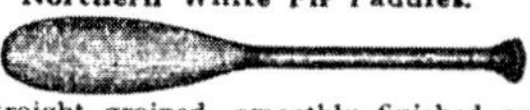

Straight grained, smoothly finished and varnished. Properly shaped and balanced.

Catalog No.	Length	Shpg. Wt.	Each
6L5747¼	4½ feet	1¾ lbs.	**$1.44**
6L5748¼	5 feet	2 lbs.	**1.45**
6L5749¼	5½ feet	2¾ lbs.	**1.46**

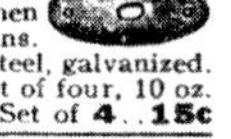

Oar Plates. Will double the serviceability of your oars when used in oarlocks with pins. Stamped from heavy gauge steel, galvanized. Easily attached. Wt., per set of four, 10 oz.
6L5728—Oar Plates. Set of 4. **15c**

Safety Oarlocks with spring which holds the shank of the horns; makes it impossible to pull the horns out of the socket accidentally. Malleable iron, 2 inches between horns. Weight, per pair, 2 pounds.
6L5726—Per pair... **55c**

Malleable Iron Oarlocks. With pin; 2 inches between horns. Good weight. Weight, per pair, 1½ lbs.
6L5724
Pair... **36c**

Round Socket Malleable Iron Oarlocks. 2¼-inch hole for oars. Shpg. wt., 2 lbs.
6L5725
Per pair..... **38c**

Folding Boat Seat With Back Rest. Frame is made of heavy round steel. Padded seat and back rest of artificial leather securely riveted. Will fit any boat seat and make boating a pleasure. Folds up into flat package when not in use. Shipping weight, 4½ pounds.
6L5713........................**$1.45**

Boat Cushion and Life Preserver, filled with prime Java kapoc and covered with artificial leather. Fitted with two handles. Size, 15x15 inches. Will sustain a 250-pound person afloat in water for 48 hours. Shipping wt., 3 lbs.
6L5709—Color, brown........**$1.40**
6L5710—Color, green........ **1.40**

Life Preserver [illegible] belt, made of [illegible] blocks of cork covered with [illegible] for use only [illegible] and rivets. Shpg. wt., 8 lbs.
6L5[illegible]......**$1.75**

Jeffery's Special Marine Canoe Glue. A quick repair for wood or canvas covered canoes. It is absolutely waterproof. In case of a puncture or leak, melt a piece of this glue with a candle and apply to puncture. You can then continue on your journey. Directions how to use on each can.
6L5719—6-ounce friction top emergency can. Shpg. wt., 8 oz........**32c**
6L5720—1-lb. can. Shpg. wt., 1¾ lbs........**76c**
6L5721—3-lb. can. Shpg. wt., 4¾ lbs........**$1.75**

The Baker Fishing Torch. Forms brilliant jets of flame about 8 in. long. Wind will not blow it out. Flow of oil governed by a valve. Can be used on boats, piers or on land. Galvanized tank holds 5 quarts. One gallon of gasoline will burn about four hours. Weight, 4¾ pounds. Shipping weight, 7 pounds.
6L5753¼—One-Burner Torch. **$3.10**
6L5754¼—As above, but with two burners........**$4.98**

For other Gasoline Torches see page 825.

AWNINGS and CURTAINS

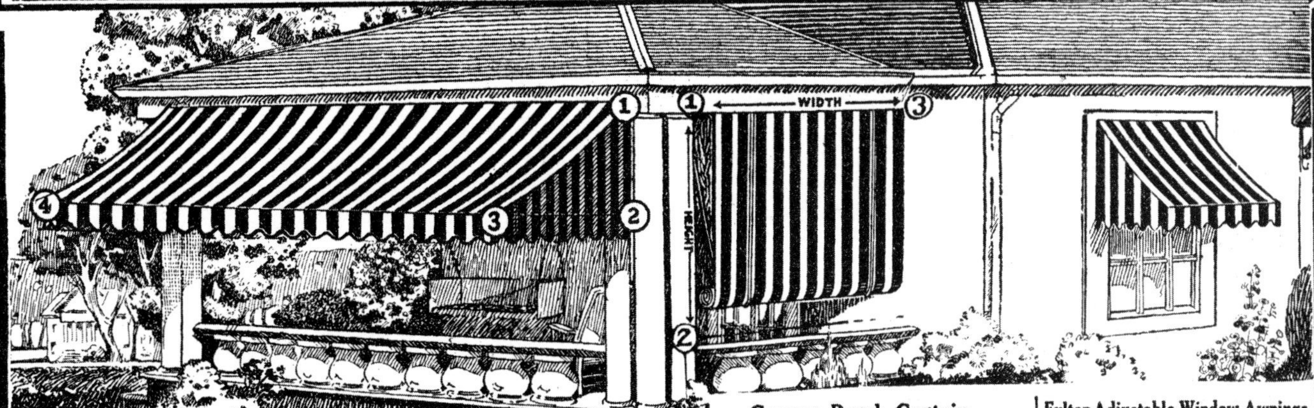

Porch and Store Awnings.

Following prices are for awnings made of plain white, blue and white or brown and white striped duck. Good weight and not to be confused with lighter lining material regularly sold. **When ordering give height from number 1 to 2, projection from number 2 to 3 and width of front from 3 to 4.** Allow four to seven days to make; in June or July allow seven to ten days. Not shipped C. O. D. For lettering awnings allow four days extra.

All exposed metal parts are japanned or galvanized. Awnings 7 feet or less in width are made with a one-piece iron frame across front; over 7 feet, wood front. We will be pleased to quote on any sizes not shown in the following list. The prices quoted herewith are for awnings complete, ready to be set up. Shipping weight given below is for the 10-foot size. 12-foot size weighs one-fifth more; 14-foot, two-fifths more; 16-foot, three-fifths more; 18-foot, four-fifths more, and 20-foot twice the 10-foot size weight.

State size; also if to be fastened to wood or brick building.

6L7653½—Blue and white striped duck.
6L7654½—Brown and white striped duck.
6L7655½—Plain white 10-ounce duck.

Height, From 1 to 2	Projection From 2 to 3	Width of Front—From 3 to 4						Shpg. Wt. for 10-Foot Size
		10 Feet	12 Feet	14 Feet	16 Feet	18 Feet	20 Feet	
4 ft.	4 ft.	$10.63	$12.35	$14.47	$15.60	$17.49	$18.73	22 lbs.
4 ft. 6 in.	4 ft. 6 in.	11.54	13.24	15.59	16.82	18.84	20.17	27 lbs.
5 ft.	5 ft.	12.40	14.20	16.63	18.15	20.16	21.63	29 lbs.
5 ft. 6 in.	5 ft. 6 in.	13.44	15.24	17.97	19.55	21.68	23.20	32 lbs.
6 ft.	6 ft.	14.63	16.63	19.37	21.01	23.30	25.04	34 lbs.

If cover only is wanted, deduct 20 per cent, or one-fifth, from the above prices, and state whether for wood or iron front bar. Lettering roof of awning, 15 cents a letter extra; lettering on curtain, 12 cents a letter extra.

Canvas Porch Curtain.

Plain white or striped duck. Furnished complete with pulleys, ropes, roller, etc., ready to attach. By pulling the rope you revolve roller, the curtain being raised or lowered as desired. When ordering give height and width as illustrated. We sell these curtains by the square foot, the price per square foot including all accessories. A curtain 7 feet high and 10 feet wide contains 70 square feet, and at 7¾ cents per square foot would cost $5.44 when **made to order.** Allow four to six days to make, and **be sure to state height from 1 to 2 and width from 1 to 3.** See illustration. **Also state color.**

Made to Order.

6L7677½—8-ounce white duck. Per square foot.............. **7¾c**
6L7664½—Blue and white striped duck. Per square foot...... **9½c**
6L7665½—Brown and white striped duck. Per square foot.... **9½c**

Carried in Stock.

Following sizes carried in stock for immediate shipment. **State size.**
6L7673¼—Plain white only. 8-ounce duck.

Height	Width	Each	Shpg. Wt.	Height	Width	Each	Shpg. Wt.
6 ft.	6 ft.	**$2.77**	9 lbs.	7 ft.	9 ft.	**$4.86**	14 lbs.
6 ft.	8 ft.	**3.70**	12 lbs.	8 ft.	8 ft.	**4.94**	14 lbs.
7 ft.	7 ft.	**3.78**	10 lbs.	8 ft.	10 ft.	**6.17**	21 lbs.

Wood Porch Shades.

Made of smooth sawed bass wood strips, oil stained dark green color only. Strong and durable, woven with strong cord. Equipped with easy hanging device and cords and are easily raised or lowered. Allows free circulation of air. Ideal for the sleeping porch. Can only be furnished in sizes listed below. **Unmailable.**

Catalog No.	Hght.	Width	Each	Shpg. Wt.
6L5972¼	7 ft.	6 ft.	**$4.50**	13 lbs.
6L5973¼	7 ft.	8 ft.	**6.15**	18 lbs.
6L5974¼	7 ft.	10 ft.	**7.95**	22 lbs.

Fulton Adjustable Window Awnings.

Manufactured and sold by us exclusively. Can be shipped immediately from stock. The particular points of merit are the quick adjustment (in width only at the side arms), the rigidity of the frame, the quality of the material used and the workmanship. Made of blue and white awning stripe duck with scalloped curtain. Complete, ready to hang up. Prices include rope, screws and cleat. These awnings are very simple in construction and so made that anyone can readily put them up. Five sizes, to fit windows 2 feet 4 inches to 4 feet 6 inches wide. Height, 3½-foot; projection, 3-foot. **Give width of your windows from center to center of casing on each side, and specify catalog number of awning wanted. Be sure to order awning wide enough.**

6L5981¼—No. 1 Awning will fit any window from 2 feet 4 inches to 2 feet 7 in. Weight, 6¼ lbs... **$3.05**

6L5982¼—No. 2 Awning will fit any window from 2 feet 8 inches to 3 feet. Weight, 6¾ pounds.... **$3.12**

6L5983¼—No. 3 Awning will fit any window from 3 feet 1 inch to 3 feet 6 inches. Weight, 7 pounds.. **$3.30**

6L5984¼—No. 4 Awning will fit any window from 3 feet 7 inches to 4 feet. Weight, 7½ pounds.... **$3.56**

6L5985¼—No. 5 Awning will fit any window from 4 feet 1 inch to 4 feet 6 inches. Weight, 8 pounds.. **$3.74**

Canvas Covers

FOR AUTO

FOR WAGON

FOR THRASHING MACHINE

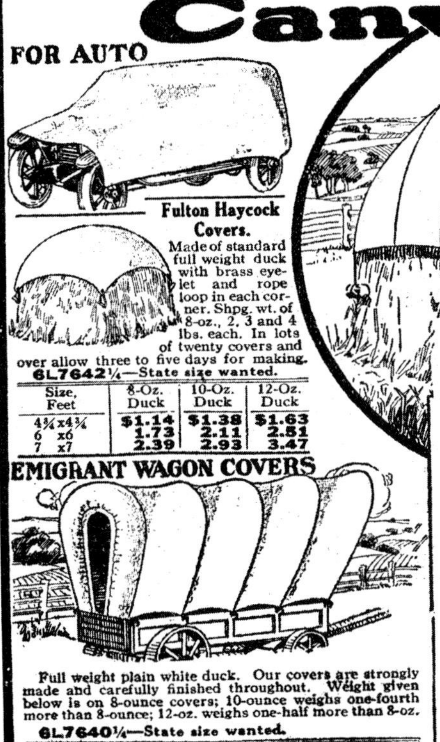

Fulton Haycock Covers.

Made of standard full weight duck with brass eyelet and rope loop in each corner. Shpg. wt. of 8-oz., 2, 3 and 4 lbs. each. In lots of twenty covers and over allow three to five days for making.

6L7642¼—State size wanted.

Size, Feet	8-Oz. Duck	10-Oz. Duck	12-Oz. Duck
4¾x4¾	**$1.14**	**$1.38**	**$1.63**
6 x6	**1.73**	**2.11**	**2.51**
7 x7	**2.39**	**2.93**	**3.47**

EMIGRANT WAGON COVERS

Full weight plain white duck. Our covers are strongly made and carefully finished throughout. Weight given below is on 8-ounce covers; 10-ounce weighs one-fourth more than 8-ounce; 12-oz. weighs one-half more than 8-oz.

6L7640¼—State size wanted.

Size, Feet	8-Oz. S. F. Duck	10-Oz. S. F. Duck	12-Oz. S. F. Duck	Wt., Lbs.	Size, Feet	8-Oz. S. F. Duck	10-Oz. S. F. Duck	12-Oz. S. F. Duck	Wt., Lbs.
10x10	$4.78	$5.90	$7.04	8	11x15	$7.73	$9.57	$11.40	13
10x12	5.62	6.97	8.30	9	11x16	8.46	10.47	12.48	15
10x14	6.70	8.28	9.86	11	12x16	9.29	11.52	13.69	15
10x16	7.77	9.62	11.49	13	12x20	10.97	13.58	16.17	18
11x13	6.76	8.36	9.98	11					

Stack and Machine Covers.

Also makes an excellent cover for your automobile, wagon or farm machinery.

Made of standard 29-inch white duck, finished around edges with brass eyelets and short ropes, thoroughly reinforced, no poles. Rope easily removed if not needed. Write for samples of canvas. Carried in stock up to size 16x20; size 16x24 and larger made to order. Allow six days for making. Weights given are for 8-oz. covers. 10-oz. weighs one-fourth more and 12-oz. one-half more than 8-oz.

6L7630¼—State size wanted.

Size, Feet	8-Oz. Duck	10-Oz. Duck	12-Oz. Duck	Weight of 8-Oz.
10x16	**$ 7.50**	**$ 9.27**	**$11.03**	12 lbs.
10x18	**8.60**	**10.61**	**12.60**	14 lbs.
12x14	**7.83**	**9.65**	**11.48**	13½ lbs.
12x16	**8.93**	**11.03**	**13.10**	13 lbs.
12x18	**10.00**	**12.34**	**14.66**	16 lbs.
12x20	**11.07**	**13.65**	**16.23**	17 lbs.
14x16	**10.67**	**13.13**	**15.62**	16 lbs.
14x18	**11.98**	**14.76**	**17.55**	18 lbs.
14x20	**13.24**	**16.34**	**19.43**	20 lbs.
14x24	**15.78**	**19.49**	**23.21**	25 lbs.
16x16	**11.97**	**14.75**	**17.54**	18 lbs.
16x18	**13.34**	**16.44**	**19.56**	19 lbs.
16x20	**14.78**	**18.25**	**21.71**	23 lbs.
6L7631½—State size wanted.				
16x24	**$17.65**	**$21.79**	**$25.92**	29 lbs.
18x20	**16.49**	**20.33**	**24.18**	26 lbs.
18x24	**19.72**	**24.34**	**28.97**	32 lbs.
18x28	**22.85**	**28.22**	**33.60**	37 lbs.
18x30	**24.50**	**30.24**	**36.02**	40 lbs.
20x24	**21.53**	**26.60**	**31.67**	34 lbs.
20x36	**32.08**	**39.69**	**47.30**	51 lbs.
24x30	**32.42**	**40.12**	**47.83**	53 lbs.
24x40	**43.05**	**53.30**	**63.53**	67 lbs.
24x50	**53.66**	**66.44**	**79.22**	82 lbs.

Fulton Black Waterproof Covers or Tarpaulins.

The preparation used to waterproof our covers is mildew and water proof, and the covers will remain soft and pliable. Made of an extra heavy duck. Shipping weights: 6x8 size, 8 lbs.; 6x9 size, 10 lbs.; 6x10 size, 11 lbs.; 6x12 size, 13 lbs.; 6x14 size, 14 lbs.; 7x9 size, 11 lbs.; 7x10 size, 12 lbs.; 7x12 size, 13 lbs.; 7x14 size, 16 lbs.; 8x10 size, 13 lbs.; 8x12 size, 16 lbs.; 8x14 size, 18 lbs.; 8x16 size, 20 lbs.; 9x14 size, 23 lbs.

6L7635¼—State size wanted.

Size, Ft.	Each	Size, Ft.	Each	Size, Ft.	Each
6x 8	**$4.76**	7x 9	**$6.24**	8x10	**$ 7.94**
6x 9	**5.35**	7x10	**6.96**	8x12	**9.52**
6x10	**5.96**	7x12	**8.34**	8x14	**11.10**
6x12	**7.14**	7x14	**9.73**	8x16	**12.70**
6x14	**8.34**	....	.....	9x14	**12.80**

FULTON

FULTON BRAND TENTS are made only from full weight standard 29-inch duck. The standard way of measuring the weight of duck is to take the weight of a yard 29 inches wide. Thus our 8-ounce duck will weigh 8 ounces to the yard, 29 inches wide, and our 10-ounce duck, 10 ounces to the yard, 29 inches wide, etc. Occasionally you may find that some firms will specify the weight of duck measured by the square yard.

For the purpose of comparison with our qualities and prices, it is well for you to know that if we measured our duck by the square yard, our 8-ounce duck would be rated as weighing over 9½ ounces; our 10-ounce duck over 12 ounces, and our 12-ounce duck over 14½ ounces.

Fulton Brand Canvas goods are sold exclusively by us, and by reason of their being made in our own factory and under close supervision, we are enabled to rigidly maintain the quality.

GARAGE TENTS

Especially constructed for the smaller models of automobiles, such as Ford, Maxwell, Dodge, etc. Made with 6-foot wall, 10½-foot center pole at rear, and four corner poles. Roll door operated by a rope; size of door, 6½ feet wide, 7 feet high. Wood frame above door supporting ridge pole. Furnished complete with guy ropes and stakes. Allow 4 to 8 days to make.

6L7644¼—State size wanted.

Length	Width	8-Ounce S. F. Duck	10-Ounce S. F. Duck	12-Ounce S. F. Duck	Weight, Pounds
12 ft.	9½ ft.	**$30.38**	**$35.91**	**$41.43**	120
14 ft.	9½ ft.	**33.62**	**39.83**	**46.00**	130
16 ft.	9½ ft.	**36.96**	**43.74**	**50.50**	150

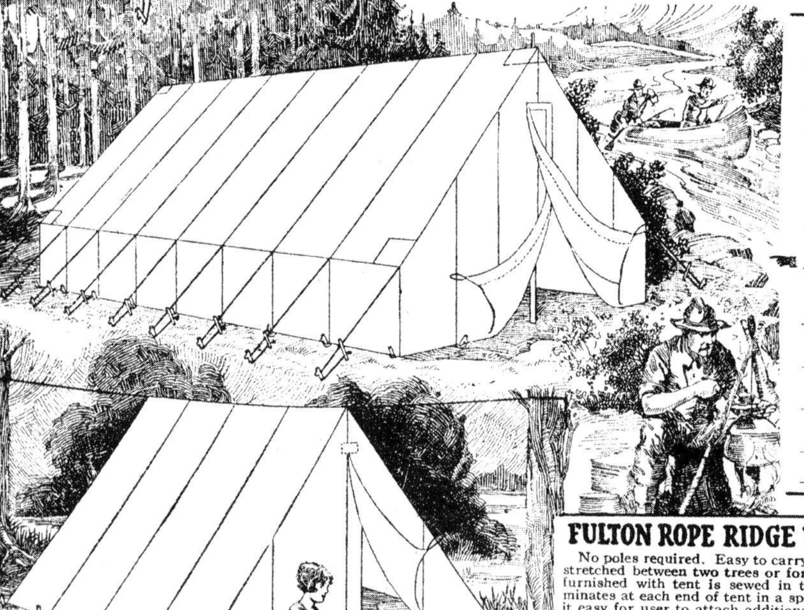

FULTON WALL TENTS

Width and Length	Height of Wall	H'ght of Tent	Our Price Includes Poles, Pegs, Guys, Guy Ropes, etc. Tent complete, ready to set up.			Wt., 8-Oz.
Feet	Feet	Feet	8-Ounce S. F. Duck	10-Ounce S. F. Duck	12-Ounce S. F. Duck	Lbs.
6L7600¼						
7 x 7	3	7	**$11.48**	**$13.73**	**$16.04**	39
7 x 9	3	7	**13.40**	**16.04**	**18.73**	47
9 x 9	3	7⅓	**16.07**	**19.35**	**22.63**	53
9½x12	3	7½	**18.64**	**22.43**	**26.22**	62
9½x14	3	7½	**20.95**	**25.20**	**29.49**	66
12 x12	3½	8	**22.22**	**26.79**	**31.37**	71
12 x14	3½	8	**24.93**	**30.08**	**35.26**	78
12 x16	3½	8	**28.15**	**33.84**	**39.56**	92
12 x18	3½	8	**30.92**	**37.20**	**43.50**	98
6L7610½						
14 x14	4	9	**30.40**	**36.66**	**42.92**	87
14 x16	4	9	**33.61**	**40.44**	**47.25**	105
14 x18	4	9	**35.66**	**42.83**	**50.00**	115
14 x20	4	9	**38.91**	**46.77**	**54.63**	125
14 x24	4	9	**44.56**	**53.43**	**62.30**	141
16 x16	5	11	**39.62**	**47.70**	**55.81**	111
16 x18	5	11	**41.66**	**50.14**	**58.60**	122
16 x20	5	11	**47.78**	**57.60**	**67.44**	143
16 x24	5	11	**53.48**	**64.45**	**75.43**	162
16 x30	5	11	**61.48**	**73.98**	**86.50**	210
18 x20	5	11	**49.40**	**59.55**	**69.70**	153
18 x24	5	11	**55.58**	**66.94**	**78.30**	167
18 x30	5	11	**65.76**	**79.20**	**92.64**	212
18 x35	5	11	**75.56**	**91.17**	**106.76**	233

FULTON ROPE RIDGE WEDGE TENT

No poles required. Easy to carry about and can be stretched between two trees or forked sticks. Rope furnished with tent is sewed in the ridge and terminates at each end of tent in a spliced loop, making it easy for user to attach additional ropes according to his requirements.

6L7621¼—State size wanted.

Width and Lgth., Feet	Ht., Feet	8-Ounce Duck	10-Ounce Duck	12-Ounce Duck	Wt., 8-oz. Lbs.
5x7	6	**$ 6.52**	**$ 7.99**	**$ 9.48**	13
7x7	7	**8.52**	**10.47**	**12.44**	15
7x9	7	**10.44**	**12.84**	**15.23**	20

HUNTERS SHELTER TENTS

A fire built in front will keep the tent warm. Furnished complete with poles. However, new poles can be cut whenever a change of camp is made, and if poles are not wanted with tent, deduct 5 per cent of 8-ounce price. Height of wall, 3 feet.

6L7629¼—State size wanted.

Width and Lgth., Feet	Ht. of Center	8-Ounce Duck	10-Ounce Duck	12-Ounce Duck	Shpg. Wt., 8-oz. Lbs.
7x 9	7 ft.	**$14.97**	**$17.90**	**$20.83**	54
9x 9	8 ft.	**18.20**	**21.88**	**25.56**	59
9x12	8 ft.	**21.03**	**25.30**	**29.59**	65

FULTON REFRESHMENT TENTS

Made of plain white duck. Prices include poles pins and guy ropes, complete, ready to set up. We furnish double corner guy ropes made of high quality manila. Illustration shows front open with canvas drawn to the side; the front may be closed or stretched out as an awning, or taken off altogether, as it is fitted with snaps for these changes.

Be sure to be very careful in giving us the correct measurements. Allow five to eight days for making; ten days during July and August. Not shipped C. O. D.

6L7614½—State size wanted.

Width and Length in Feet	Ht., Side Walls	Ht., Center, Ft.,	8-Oz. White Duck	10-Oz. White Duck	12-Oz. White Duck	Wt., 8-Oz., Lbs.
9x14	6 ft.	10	**$32.11**	**$37.83**	**$43.52**	85
12x19	6 ft.	11	**45.92**	**54.27**	**62.62**	140
14x21½	6 ft.	11	**50.40**	**59.64**	**68.87**	155
14x23½	6 ft.	11	**56.28**	**66.80**	**77.30**	160

WEIGHT OF TENTS is given above in 8-ounce with poles; 10-ounce will weigh about one-quarter more than 8-ounce; 12-ounce about one-half more than 8-ounce. The weight may vary slightly, as the poles do not always run alike.

WE CARRY IN STOCK all sizes of wall tents from 7x7 to 12x18 feet, and immediate shipment can be made. On all other sizes allow three to five days' time to make; in June and July allow from six to twelve days. If a stock tent is ordered with a fly, the order requires special handling. Allow four to eight days for making.

Extras for Fulton Wall Tents.

A TENT FLY is an extra removable roof spread over the top of tent and is staked down, leaving an air space between it and tent. The extra roof provides additional protection from cold and dampness. Buy a tent fly if the roof of tent leaks. The cost of a fly is one-half the price of a wall tent of corresponding size and weight of duck. For example, a fly for a $20.00 wall tent would cost $10.00. Necessary stakes and ropes always included without extra charge. Allow three to six days for making. **In ordering specify 6L7658½.**

EXTENSION FLY covers not only the roof of the tent, but extends in front, forming a shade or awning. The cost of an extension fly is in the same proportion as the length of the extension is to the length of the regular fly. A 6-foot extension on a fly 12 feet long would increase the length of the fly one-half and would increase the cost a like amount; in other words, if the cost of a fly 12 feet long were $10.00, the cost of a 12-foot fly with a 6-foot extension would be $15.00. Allow three to six days for making. **In ordering specify 6L7674½.**

WHEN HIGHER WALL than regularly quoted is wanted, add 5 per cent of the cost of the tent for each additional 6 inches. Allow three to six days for making. **When ordering specify 6L7604½.**

EXTRA DOOR at back end of tent is furnished without additional charge on all tents over 16x24 feet if wanted. On all tents under this size where extra door is wanted we make an additional charge of 60 cents. **In ordering specify 6L7615½.**

IF POLES ARE NOT WANTED deduct 5 per cent of the price of 8-ounce tent.

We do not ship made to order tents C. O. D.

A SOD CLOTH on a tent is a strip of canvas 9 inches wide, sewed to the bottom of the ends and sides of the tent, upon which are placed stones or earth to keep out wind, flies, etc. In measuring amount of sod cloth required, measure distance around tent. A 12x16-foot tent would require 56 lineal feet of sod cloth. **Can be had sewed on or separate as desired.**

6L7617½—Per lineal foot **5c**

TENTS

By contracting for our duck in large quantities, as we do, directly with large mills in the heart of the cotton growing country in the south, we are enabled—combined with our economical selling methods—to offer the remarkable values in canvas goods that have made the Fulton brand so widely known.

Fulton Brand Tents possess many features of special merit. Two doors, one at each end, are furnished, without additional charge, on all tents 16x24 feet or larger, if desired. If you order a Wall Tent, this size or larger, mention whether you want the extra door. All doors are provided with a large, full flap of the same weight of canvas as is used throughout the tent. Pure manila and sisal rope only are furnished. All guy ropes are provided with a hand worked sailor splice. On all tents 12x12 feet or larger we provide two additional top guy ropes, one at each end. Our improved steel ring grommets are sewed into all tents at points of greatest tension. **Notice the height of wall and center shown on page 782.**

TOURIST TENTS

Can be set up along the side of any car. Has a front piece that goes over the roof of the car and ties to wheels on opposite side or can be stretched out straight. No poles required. A very serviceable, compact tent, amply large for all practical purposes. It is 7 feet wide (side against car) and 7 feet in depth. Rear wall 3 feet high. When folded makes a package 28 inches long by 15 inches wide. If larger quarters are wanted we recommend the purchase of two tents, using one on each side of car. Carried in stock ready for immediate shipment. Weight, 8-oz., 22 lbs., 10-oz., 28 lbs., 12-oz., 33 lbs., 10-oz. Khaki, 28 lbs.

6L7643¼—State kind of duck wanted.

Height	8-Ounce White Duck	10-Ounce White Duck	12-Ounce White Duck	10-Ounce Khaki Duck
6 feet 6 inches	**$10.50**	**$12.80**	**$15.07**	**$14.68**

CHILDRENS PLAY TENT

Just what the children like. Invites health, joyful outdoor play Ideal for yard or lawn. The front of the tent, as illustrated, can be raised as an awning or let down, closing it on all four sides. Made without wall, the roof sloping direct from ridge to ground, as shown. Material is a good weight, plain white drill; well finished throughout, the same as a regular tent. Size, 6 feet wide by 6 feet long. Height, 5½ feet. Height of extra poles used to support the awning, 4½ feet. Carried in stock for immediate shipment. Weight, packed for shipment, 26 pounds.

6L7641¼—Including poles, ropes and stakes............**$6.27**

WINDOW AND VENTILATOR

Can be instantly attached to any style tent by slitting seam open about 10 inches and clamping the ventilator in place. This ventilator has an oval opening 8½ inches long and 5¼ inches wide. Made with a removable mosquito screen and removable transparent window. These ventilators are usually placed in the front and rear gable ends of tent.

6L7651½—Attached to any tent......................**$1.05**

We can furnish ventilator separate, not attached to tent, at the above price. Shpg wt., 1 lb. 5 oz.

THE FULTON BRAND MINERS TENTS

Used largely by miners and prospectors, as they can be very conveniently carried and require but one pole for erecting. May also be used as play tents for children. The weights which we give include pole. If wanted without pole deduct 20 cents from price quoted. Carried in stock ready for immediate shipment.

6L7622¼—State size wanted.

Size of Base	Height	8-Oz. Duck	10-Oz. Duck	12-Oz. Duck	Wt., 8-Oz. Lbs.
7x7 feet	7 feet	**$6.33**	**$7.68**	**$ 9.03**	17
9x9 feet	8 feet	**9.70**	**11.81**	**13.90**	24

WATERPROOF COATING

When applied on canvas will make it water repellent. The canvas can be folded without danger of cracking; it acts like rubber. For covers, tents, etc. One gallon will cover about 100 square feet. Comes in white or brown color. **State which is wanted.** Shipping weight, 1-gallon can, 10 pounds; 5-gallon can, 42 pounds.

6L5950—White. 1-gallon can......................**$1.55**

6L5952¼—White. 5-gallon can......................**$7.75**

6L5951 — Brown. 1-gallon can......................**$1.48**

6L5953¼—Brown. 5-gallon can... **7.40**

FULTON BRAND-A-WEDGE TENTS

The weights which we give include poles. When poles are not wanted with tents, deduct 5 per cent of the price of 8-ounce tent. Every tent guaranteed as to workmanship and quality of material. Carried in stock ready for immediate shipment.

6L7620¼—State size wanted.

Width and Length	Height	8-Oz. Duck	10-Oz. Duck	12-Oz. Duck	Wt., 8-Oz. Lbs.
7 x 7 ft.	7 ft.	**$ 9.44**	**$11.39**	**$13.32**	30
7 x 9 ft.	7 ft.	**11.38**	**13.76**	**16.12**	40
9 x 9 ft.	7 ft.	**13.59**	**16.44**	**19.30**	45
9½x12 ft.	7½ ft.	**15.68**	**19.00**	**22.35**	50

"Can-Va-Sek" Waterproofing Preparation.

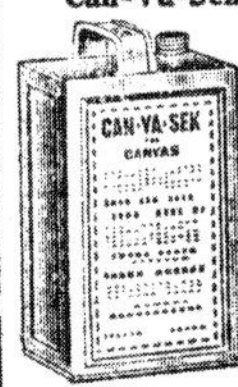

For tents, awnings and canvas of all kinds. Waterproofs thoroughly. Increases tensile strength and prevents mildew. A thin, clear liquid, easily applied with spray, sponge or sprinkling can. Does not discolor. Can be used on any color canvas; will not rub or wash out. Allows free ventilation through canvas. One gallon will cover from 100 to 125 square feet, depending upon weight of duck.

6L5949—1-gallon can. Shpg. wt., 10 lbs.......**$1.70**

6L5954¼—5-gallon can. Shipping weight, 40 pounds......................**$7.95**

PALMETTO LAWN TENTS

Palmetto Lawn Tents are intended as playhouses for children, for lawn parties, fairs, etc. They are made of awning stripe and set up with one pole and a light iron frame sewed into the tent around the eaves. Every tent guaranteed full size. Carried in stock ready to ship.

6L7633¼—State size wanted.

Size of Base, Feet	Size of Top	Height at Center	Height at Side	Price	Shpg. Wt., Lbs.
7x 7	2 ft. 4 in.	7 ft. 6 in.	6 ft.	**$10.25**	22
8x 8	2 ft. 4 in.	8 ft.	6 ft. 6 in.	**11.16**	25
10x10	3 ft. 6 in.	9 ft.	7 ft. 6 in.	**16.53**	42

Stovepipe Hole Protector.

Galvanized sheet metal. Can be placed in any seam of the tent without special fitting or sewing. Comes in 4, 5 and 6-inch hole. **State which is wanted.** Sold separately at same price. Shipping weight, 1 pound.

6L7650½—Attached to any wall tent......**55c**

6L5481—Same as above. Size, 3-in. hole...**55c**

Iron Tent Pegs.

They last a lifetime. Not easily broken.

6L5930—Short Peg. 8¾ inches long. Weight, 4½ ounces each. Shipping weight, 1 dozen, 5 pounds......................**65c**

6L5931—Long Peg. 13½ inches long. Weight, 7¼ ounces each. Shipping weight, 1 dozen, 8 pounds. Per dozen............**$1.15**

CAMPING

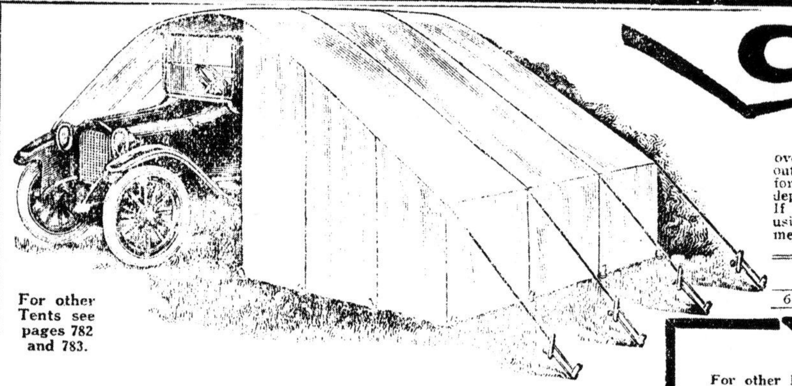

For other Tents see pages 782 and 783.

Automobile Tourists' Tent

Can be set up along the side of any car. Has a front piece that goes over the roof of car and ties to wheels on opposite side or can be stretched out straight. No poles required. A very serviceable, compact tent, amply large for all practical purposes. It is 7 feet wide (side against car) and 7 feet in depth. When folded, makes a package 28 inches long by 15 inches wide. If larger quarters are wanted we recommend the purchase of two tents, using one on each side of car. Carried in stock ready for immediate shipment. Shipping weight, 8-oz., 24 lbs.; 10-oz., 27 lbs.; 12-oz., 30 lbs.

6L7643¼—State weight wanted.

Height	8-Ounce Duck	10-Ounce Duck	12-Ounce Duck	10-Ounce Khaki Duck
6 feet 6 inches	**$10.50**	**$12.80**	**$15.07**	**$14.68**

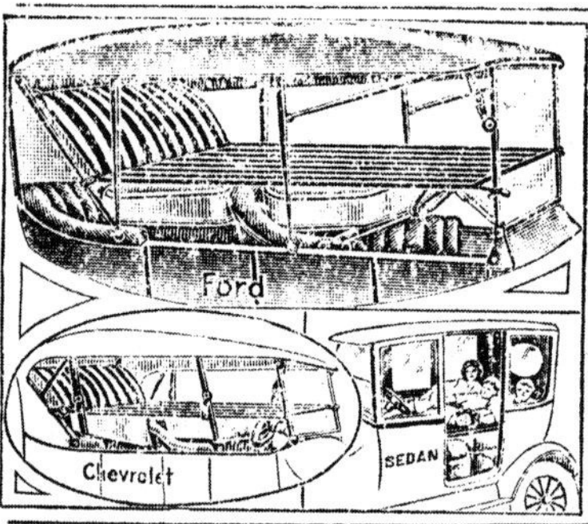

New Universal Car Bed for Ford Cars and Others.

For cars of the 100-inch wheel base size only, such as Fords, Overland 4, Chevrolet 490, Durant, Star, etc. Made in two styles, for touring cars and sedans. Easily and quickly installed without leaving a scar on car. Bed part is made of heavy washable awning duck, edges are doubled with a wide overlap to meet any strain. Hems are reinforced with heavy trunk web, and special tape where needed. Two pillow pockets are provided in case pillows are wanted. Weight, only 7½ pounds. Instructions furnished with each bed. Shpg. wt., 8 lbs.

6L5443¼—For touring cars. $8.95
6L5444¼—For sedans 8.95

Auto Bed.

For use in any five to seven-passenger touring car and the larger styles of sedans and closed cars. Very practical and comfortable and can be put in place or taken down in a short time. Plenty of room for two grown persons. Weight is supported by legs of bed. Folds very compactly into a package 4 feet long and 5 inches in diameter. Length, 74 inches. Width, over all, 48 inches. Frame made of hardwood. Bed is of heavy brown duck. Has adjustable end bar to regulate tension to suit individual requirements. Complete instructions furnished with each bed. Shipping weight, 23 lbs.

6L5442¼ $11.25

For other Folding Furniture see pages 624 and 637.

"Gold Medal" Folding Camp Chair.

Hardwood, very strong and light weight. Seat made of heavy brown duck. Legs reinforced by steel braces. Back rest folds with lower parts very compactly. Shpg. wt., 4½ lbs.

6L5429¼ .. 90c

"Gold Medal" Folding Chair.

Frame made of selected hardwood, finished in natural color. Canvas seat and back. Folds up compactly and is especially recommended for camping, outdoor use and motor boats. Ideal for the home, club or hotel. Shipping wt., 15 lbs.

6L5480¼ $3.60

"Gold Medal" Folding Chair.

Instantly adjusts itself to the body. Seat and back of heavy striped duck; the frame of hardwood in natural finish, reinforced with iron braces. Size, when folded, 3 inches by 4 inches by 3 feet long. Shipping wt., 8 lbs.

6L5434¼ $2.30

"Gold Medal" Roll Top Camp Table.

An excellent table for the tourist or camper, as it is strong and rigid but extremely light and folds compactly. Simple in construction and easily folded. Made of selected hardwood, varnished, steel plates riveted at joints. Legs fold, top rolls up. When folded makes a package about 6 inches in diameter. Equipped with handle for carrying. Stands 28 inches high and top measures 36x27 inches. Weighs 14 pounds. Will comfortably seat four persons. Shipping weight, 18 pounds.

6L5424¼ $3.40

In selecting our line of camping equipment great care was taken to select only items that are necessary for the comfort and pleasure of the tourist or camper. The items shown on these two pages are of a high grade and are guaranteed to give satisfactory service. We are in position to give you the best equipment for a very reasonable price. Make a comparison and you will be satisfied that on each purchase you make a worth while saving.

American Pedometer.

Hang the pedometer in your watch pocket or on your belt, and every step you take will register. The figures on the face of the pedometer indicate the miles or fraction of a mile you walk. Registers 100 miles and repeats. Directions included. Shipping weight, 4 ounces.

6L472 .... $1.70

Insulated Jug.

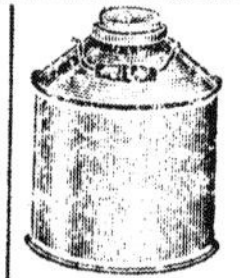

Ideal for touring, camping, fishing and for the farm or shop. Will keep liquids hot or cold from 7 to 8 hours. Outer casing of sheet steel, and inner casing of earthenware, cork stopper, wide mouth. Capacity, about 1 gallon. Sanitary, and can be very easily cleaned. Shipping weight, 15 pounds.

6L4936 .... $2.90

Dunnage Bag.

Made of white medium weight duck; 36 inches high, 12 inches in diameter; round bottom. Cotton rope draw string top. Well sewed throughout. Just the thing for carrying clothes, blankets, etc. Shipping weight, 2 pounds.

6L1000 .... 99c

Duluth Pack Sack for Hunters, Explorers or Campers.

Made of brown duck with 2¼-inch double thickness canvas head straps, 2-inch double thickness canvas shoulder straps, grain leather billets and buckle stays; lapover top fastened with three leather straps. Size of sack, 24x26 inches. Shipping weight, 2½ pounds.

6L997 .... $3.45

Hunting and Sticking Knife.

Strong 5-inch tempered blade. Length over all, 8½ in. Leather handle, brass and fiber trimmings. Finger bilt to give firm grasp. Durable leather sheath. Shpg. wt., 14 oz.

6L7126¼—With sheath $2.35

Sheffield Pattern Bowie Hunting Knife.

6-inch blade. Stag pattern handle. Nickel silver guard. Leather sheath. Shipping weight, 15 ounces.

6L7135¼- $1.85

CAMPERS' OR HUNTERS' KNIFE.

Swaged clip blade, 5 inches long. Baked enameled handle with polished ferrule. Entire length, 9¾ inches. Shipping weight, 6 ounces.

6L7137¼—With sheath 60c

Stanolind Camp Nite-Lite.

For campers, fishermen, hunters, tourists and general use. A handy, quick, clean and safe light for use in tents, cottages, summer homes or garages. Each light will burn about fifteen hours, and can be blown out and lit in a second. Packed twelve lights in a carton with one strong, thick red glass container, guaranteed not to crack from heat of the candle. When one light is burned out, place another in the glass. Shipping weight, 2⅜ pounds.

6L4647—Per box of twelve lights, with red glass container 62c

Canvas Wall Pocket.

Good weight white canvas with three brass eyelets at top. Eight pockets, about 6 in. wide by 5 in. deep, and two large pockets, about 12½ in. wide by 10 in. deep. Shipping weight, 1½ pounds.

6L5938 .... 86c

Ideal Tent Light.

Stamped metal base with finger hook, as illustrated. Shipping weight, 1¾ pounds.

6L4638
With six candles 15c

Extra Candles.

For the above or regular use. Shpg. wt., 1¼ lbs.

6L4639—Per doz. 15c

Waterproof Match Box.

Absolutely water and moisture proof. Made of seamless drawn brass, nickel plated. Fitted with a rubber gasket in cover. Very convenient size to carry in pocket. Your matches are always dry. Shpg. wt., 3 oz.

6L470 .... 48c

Waterproof Bed Sheet.

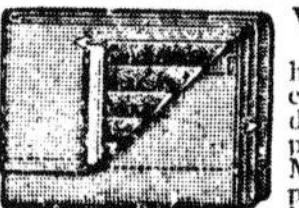

For campers, tourists, herders and all who are compelled to sleep outdoors. Often used in place of sleeping bag. Made of brown waterproof canvas, lined with blanket containing a mixture of cotton with a very small percentage of wool. Fitted with snaps and rings, permitting the sheet to be fastened around the body like a sleeping bag. Can be spread open and aired.

6L5970¼—Size, 6x12 feet. Weight, 5 pounds $10.32
6L5971¼—Size, 7x16 feet. Weight, 20 pounds $15.87

Camp Blankets.

Always remain soft and pliable. Made of very closely woven duck, dark color. Treated with a waterproof solution, making the duck as waterproof as is possible to produce. No seams; has eyelets in corners and sides. Is also used as a ground cloth.

6L5816—58x72 inches. Shipping weight, 4½ pounds $3.93
6L5817—58x96 inches. Shipping weight, 6 pounds $5.06

For complete line of Camping Blankets see page 346.

Waterproof Ponchos.

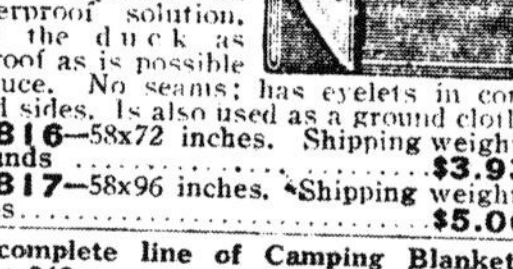

Poncho, made of slicker cloth, olive tan color, with hole and fly, to be used as a cape. For fishing, camping, touring, etc. Can be used as a blanket. Shipping weight, 3½ pounds.

6L5936—Men's size, 66x90 inches $3.45
6L5937—Boys' size, 45x72 inches $1.95

Nickel Plated Cup.

Size, open, 2⅝ in. high, 2⅜ in. in diameter. Folds compactly in snug fitting nickel plated cover. Shipping weight, 5 ounces.

6L4965 .... 33c

Duplex Folding Canvas Water Pail.

Very useful article around camp and popular with automobilists. Made of brown color waterproofed duck, handle for carrying, hinged braces, top and bottom. Size, when open, 11x9½ in. Capacity, 10 quarts. Shipping wt., 2⅛ lbs.

6L5989 ..... $1.80

Same as above, but smaller. Size, 10x7½ inches. Capacity, 6 quarts. Shipping weight, 1¾ pounds.

6L5988 .... $1.42

Double Folding Canvas Wash Basin.

Made of brown color waterproofed duck, two handles, hinged rim at top and bottom, very handy around a camp. Size, when open, 12x7½ inches. Shipping weight, 2 pounds.

6L5990 .... $1.79

Same as above, but smaller. Size, 10x5½ inches. Shipping weight, 1½ pounds.

6L5991 .... $1.40

African Water Bags.

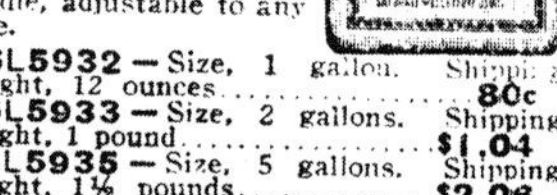

Made of a specially constructed heavy flax canvas, which has the peculiar property of holding water and exuding just enough to the surface to keep up a continual evaporation. Bag fitted with a mouthpiece, cork attached. Rope handle, adjustable to any size.

6L5932—Size, 1 gallon. Shipping weight, 12 ounces 80c
6L5933—Size, 2 gallons. Shipping weight, 1 pound $1.04
6L5935—Size, 5 gallons. Shipping weight, 1½ pounds $2.06

One-Piece Solid Aluminum Cup.

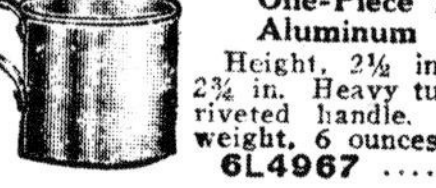

Height, 2½ in.; width, 2¾ in. Heavy turned rim, riveted handle. Shipping weight, 6 ounces.

6L4967 .... 10c

EQUIPMENT

For Vacuum Goods refer to page 765.

"Gold Medal" Folding Camp Bed.

One of the most popular Camp Cots on the market. Size, 6 feet 6 inches long, 2 feet 3 inches wide and 16¾ inches high. May be folded into a parcel 39 inches long and 6 inches in diameter. Frame of selected hardwood, steel plates riveted on at all joints. Covered with 12-ounce double filled brown canvas, with pillow casing, which may be stuffed with straw, hay or clothing to serve as a pillow. Strongly made and folds very compactly.

6L5438¼—Shipping weight, 20 pounds **$3.65**

6L5439¼—Same as above, but extra wide; 36 inches wide, 18 inches high, 6 feet 6 inches long. Folds compactly into a package 39 inches long and 8 inches in diameter. Shipping weight, 24 pounds **$5.25**

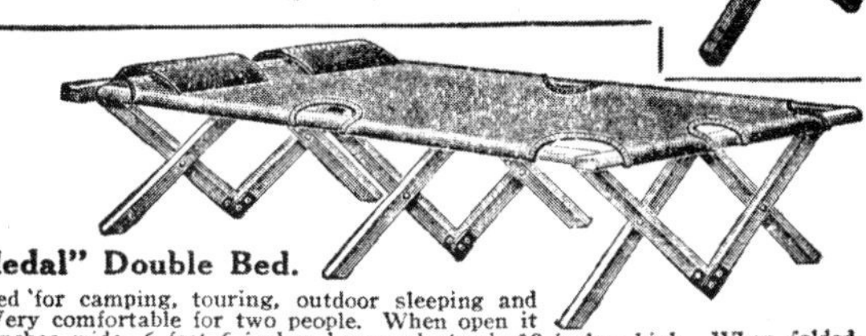

"Gold Medal" Double Bed.

A double bed for camping, touring, outdoor sleeping and home uses. Very comfortable for two people. When open it measures 52 inches wide, 6 feet 6 inches long and stands 18 inches high. When folded, measures 3 feet 3 inches long, 5 inches thick and 10 inches wide. Frame of selected hardwood stock and strongly braced throughout, the same as all other Gold Medal brand cots. Constructed with center rail which folds with the bed. Covered with extra heavy duck. Shipping weight, 35 pounds.

6L5437¼ **$8.50**

Camp Stools and Table.

"Boko" Folding Steel Stool.

For automobiling, fishing, boating, camping, etc. Light, strong, durable and comfortable. All steel frame, heavy canvas seat. Weighs only 2 pounds. Folds flat into a space of 7½x10 inches. Height, when open, 15 inches. Shipping wt., 2½ lbs.

6L5436 .... **98c**

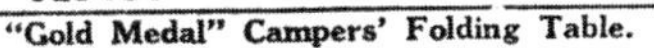

"Gold Medal" Folding Camp Stool.

Covered with brown canvas, reinforced at the corners. Hardwood legs. When folded it makes a package 2 feet long and 2½ inches square. Shipping weight, each, 2 pounds.

6L5477¼

Each **$0.55**

Dozen **6.20**

"Gold Medal" Folding Camp Stool.

Very strong and folds very compactly. Light weight and easy to carry about. Seat made of 10-ounce brown duck. Hardwood frame. Legs reinforced by steel braces. Shipping weight, 3½ pounds.

6L5430¼

Each **70c**

"Gold Medal" Campers' Folding Table.

Popular with campers and tourists. Hardwood throughout, natural finish. Accommodates four persons. Can be folded in a package 3 feet by 5x7 inches. Top of the table measures 2 feet 3 inches wide by 3 feet long and will comfortably seat four people. Height, 28 inches. Legs strongly reinforced with iron braces. It is so constructed that there are no separate parts to either top or legs. Shipping weight, 22 pounds.

6L5433¾ **$4.30**

"Gold Medal" Ever-Level Single Folding Cot.

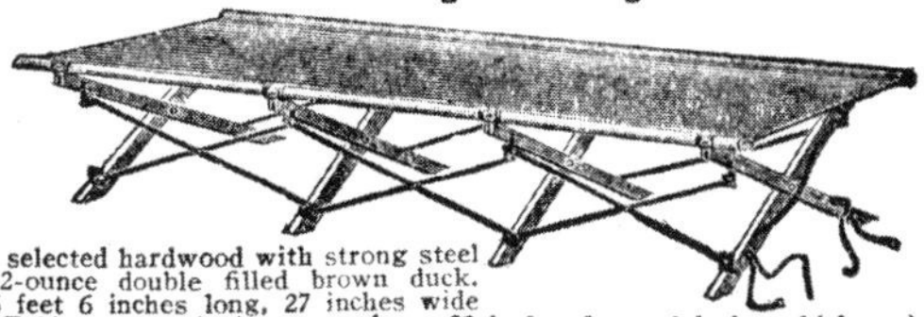

Constructed in such a way that it will automatically adjust itself to uneven surfaces. Always rigid and comfortable in spite of irregularities in the ground, making it ideal for campers and tourists. Frame made of selected hardwood with strong steel braces. Cover is 12-ounce double filled brown duck. Size, when opened, 6 feet 6 inches long, 27 inches wide and 18 inches high. Folds compactly into a package 32 inches long, 4 inches thick and 7½ inches wide. Shipping weight, 18 pounds.

6L5427¼ **$4.20**

"Gold Medal" Ever-Level Double Folding Cot.

Has the same features as single cot above and folds very compactly. Hardwood frame with strong steel braces. Cover is 17-ounce brown duck. Size, when opened, 6 feet 6 inches long, 50 inches wide, 18 inches high. When folded, 34 inches long, 8 inches thick and 10 inches wide. Shipping weight, 28 pounds.

6L5428¼ **$8.75**

For other Folding Cots see pages 665 and 666.

Kapok Cot Mattress.

Made for use with camp cots, also makes a very comfortable bed when sleeping on the ground. Filled with kapok, well tufted and covered with khaki colored denim. About 2 inches thick and folds compactly. Two mattresses, size 27x78 inches, will be suitable for Cot 6L5437¼.

6L5446¼—Size, 27x78 inches. For Cots 6L5438¼ and 6L5427¼. Shipping weight, 6 pounds **$4.20**

6L5447¼—Size, 36x78 inches. For Cots 6L5439¼ and 6L5428¼. Shipping weight, 7½ pounds **$5.25**

Folding Camp Grid.

For campers, tourists, surveyors, automobilists, etc. Made of standard grade tinned steel wire, electrically welded. Cross wires are 1 inch apart. Can burn wood, leaves, etc. Illustration shows grid with charcoal grate. Shipping weights: No. 1, 3¼ pounds; No. 2, 5¼ pounds.

6L5466—No. 1, without charcoal grate. Size, 10x14 inches **40c**

6L5467—No. 2, without charcoal grate. Size, 13x22 inches **64c**

6L5468—No. 1, with charcoal grate. Size 10x14 inches **65c**

6L5469—No. 2, with charcoal grate. Size, 13x22 inches **93c**

Upton Kamp Kook Kit, Better Known as Stopple Mess Kit.

Complete camping outfit for small party; two cups, two frying pans, a boiler for coffee, etc., and a grid. Folded, 9½ inches long, 4¼ inches wide, 2¼ inches high. Weight, about 2 pounds. Shipping weight, 3 pounds.

6L5450—Complete **$2.38**

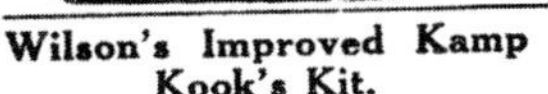

Wilson's Improved Kamp Kook's Kit.

Fifty-three pieces. Fire jacks, two boilers suitable for use as an oven, frying pan, coffeepot and all utensils and tableware for a party of six. Boilers are made of 26-gauge smooth steel. The entire kit nests in small space, and when packed ready for shipment makes a package of 14½x10½x8 inches, all nested together, and can be firmly locked up with an ordinary padlock. Weight, complete, 25 pounds.

6L5455¼—Complete **$8.98**

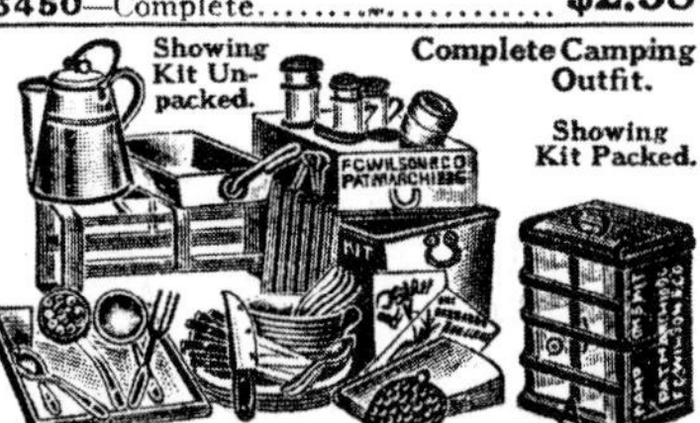

Showing Kit Unpacked.

Complete Camping Outfit.

Showing Kit Packed.

Campers' Table Set.

A very practical set for campers, consisting of four knives, four forks, four teaspoons. All are coated with pure block tin to prevent rust. Put up in a canvas partitioned cover with two eyelets, so that it can be hung on wall. Shipping wt., 2 lbs.

6L7395 .... **$1.29**

Handy Camp Stove for Heating and Cooking. Burns Gasoline.

Made of metal, strongly constructed. Burns gasoline, producing an intense heat, generates easily, lights quickly and can be regulated to suit. Capacity of fount, one quart. Operates six hours on one filling. Height, 7½ inches. Diameter, 6 inches. Shipping weight, 5 pounds.

6L5474—Complete with pump, funnel and lighter **$6.10**

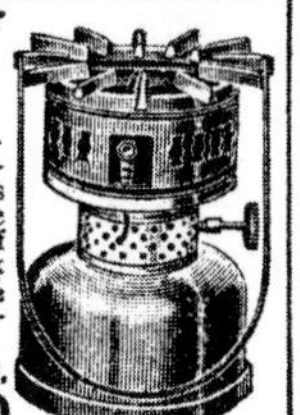

Comfort Camp Cook Stove.

Made of a specially prepared steel and enameled black. When collapsed, measures 4x10x18 inches. Weight, 12 pounds. Cover and lids used as a shield in windy weather and as a warming and serving table. Main burner when lighted produces gas for second burner. Use one or both. Burns common motor gasoline. No funnel needed for filling. Shipping weight, 15 pounds.

6L5462¼ **$9.75**

Portable Folding Cast Iron Stove.

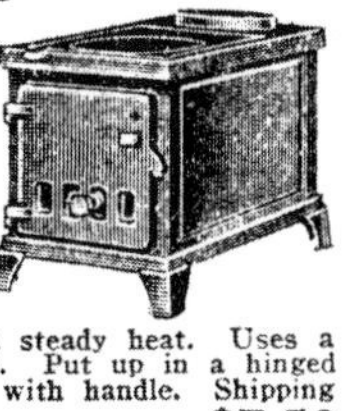

A strong cast iron stove which will not rust or warp like sheet iron. No bolts or screws required. Folds up into a flat package, 18x10x4 in. Full size when set up is 12 inches high, 17 inches long and 11 inches wide. Will burn coal or wood and give a steady heat. Uses a standard 5-inch pipe. Put up in a hinged wood carrying case with handle. Shipping weight, 40 pounds.

6L5464¼ **$7.50**

The "Comfort" Heater and Camp Stove. Burns Gasoline.

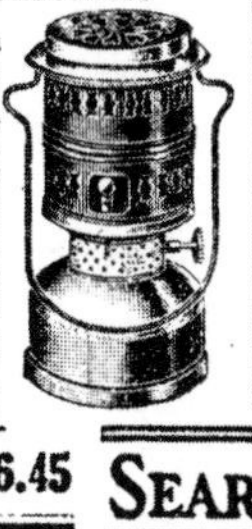

For campers, sportsmen and hunters' camps. Made of metal, fount nickel plated and polished. Holds one quart of gasoline; burns about eight hours. The heat drum is of finished blued steel. The top rim or cover at top of drum can be removed for frying and cooking purposes. Equipped with automatic gas tip cleaner, pump, funnel and asbestos torch for lighting. Height, 12¼ inches; diameter, 6 in. Weight, 3½ pounds. Shipping weight, 4¾ lbs.

6L5470¼—Complete **$6.45**

Improved Folding Camp Stove.

Made of blued steel. Width, 14 inches; length, 17¾ inches; height, 12 inches. Folds 14x17¾x2 inches. 7-inch lids and lifter. Has no bottom. Collar attached to top; telescope pipe, about 4 feet long, 4 inches wide at bottom and 3 inches at top. Used inside tent by building fire on a little sand or dirt. Weight, with pipe, 11¼ pounds. Shipping weight, 20 pounds.

6L5471¼ **$3.20**

6L5475—Extra elbow, for 3-inch pipe. Shipping weight, 1½ pounds .... **.10**

6L5476—Extra length 3-inch pipe. Shipping weight, 2 pounds .... **.12**

6L5481—Stovepipe Hole Protector. Galvanized sheet metal. Placed in any seam of tent, for 3-in. pipe. Shpg. wt., 1 lb. **55c**

Winner Collapsible Camp Stove.

For cooking, baking and heating. Made of blued steel, 18 inches wide, 27½ inches long, 12 inches high, or, with cast iron legs attached, 18 inches high; size, folded, 18x27x4 inches. Oven is 11 inches wide; 16 inches long, 8 inches high. Even temperature inside for baking. Four holes with 7-inch lids and lifter. Furnished with telescope pipe, 4 feet long, 4 inches wide at bottom, 3 inches at top. Shipping weight, 62 pounds.

6L5472¼ **$8.75**

Genuine Optimus Swedish Kerosene Oil Stove.

For outdoor or camp use. Made partly of brass. Width at base, 6½ inches. Height, when set up, 8 inches. Generates its own gas from common kerosene. One quart of kerosene is sufficient for six hours. A strong, safe, well built stove. Shipping weight, 4¾ pounds.

6L5473 **$4.98**

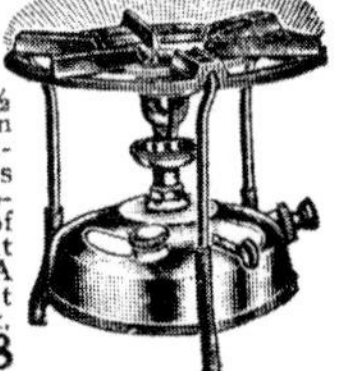

Outdoor Furnishings

NOTICE—If couch is to hang from ceiling it is necessary to use Chains 6L5122, also Hooks 6L5123, shown below.

"Ideal" Hammock Couch.

Couch is 72 inches long and 24 inches wide. Constructed with angle steel frame, fitted with wire link fabric spring, helical spring at both ends. Covered with extra quality dark tan drill. Ends of couch are constructed so they will not pull away from canvas. Fitted with full tufted covered excelsior and cotton mattress and fringed curtain. Another feature is the padded top rail or back rest and adjustable head rest. Exceptional value for the money. Shipping weight, 55 pounds. For stand and canopy see 6L5918¼ and 6L5915¼, listed below.

6L5914¼ $9.95

"Comfort" Hammock Couch.

Covered with 8-ounce army duck, gray with blue stripes, fast color, attractive pattern. Constructed with a steel frame fitted with wire link fabric springs, helical springs at both ends. Full tufted covered cotton mattress with fringed curtain. Padded back rail, which can be easily taken off and put on as desired. Ends of couch are constructed so they will not pull away from canvas. Adjustable head rest. First class in every respect. Length of couch, 72 inches; width, 24 inches. Shipping weight, 55 pounds.

6L5919¼ $12.75

"Luxury" Hammock Couch.

One of the latest styles. A hammock couch with both seat and back made of wire link fabric springs with helical extensions and padded cushions with cotton filling, making couch invitingly comfortable. All angle steel frame. Canvas covering is made of 8-ounce army duck, gray color with blue stripes. Fast color. A very striking pattern. Length of couch is 72 inches; seat is 22 inches wide and back is 21 inches high. Shipping weight, 70 pounds.

6L5916¼—Hammock Couch only.......$16.75
6L5918¼—Stand only. (Shpg.wt., 45 lbs.) 3.95
6L5917¼—Canopy only. (Shpg.wt., 10 lbs.) 6.00

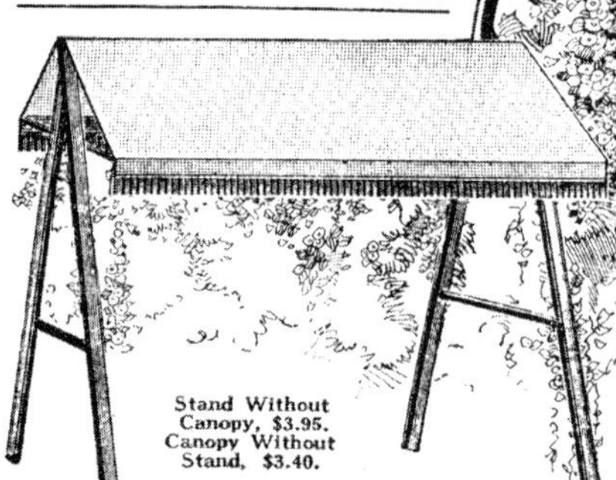

Stand Without Canopy, $3.95. Canopy Without Stand, $3.40.

Hammock Couch Stand.

Made of angle iron, nicely finished. Strong and durable. Very easily set up without tools, no bolts or nuts required. Can be used with our couches or any measuring 72 inches long. Shipping weight, 45 pounds.

6L5918¼—Couch Stand only......$3.95

Hammock Couch Canopy.

Made of good grade dark tan drill to match either Couch 6L5914¼ or Couch 6L5913¼. Trimmed with fringe. Fitted over wood and steel frame, can be thrown up and back as desired. Shipping weight, 12 pounds.

6L5915¼—Canopy only............$3.40

"Leader" Hammock Couch.

A low priced couch of good value. All angle steel frame, fitted with wire link fabric springs, helical springs at both ends. Ends of couch are constructed so they will not pull away from canvas. Fitted with well tufted excelsior mattress covered with good quality dark tan color drill. Couch is 72 inches long and 24 inches wide. Shipping weight, 50 pounds. For stand and canopy see 6L5918¼ and 6L5915¼ listed at left.

6L5913¼ $7.55

Steel Coil Springs.

To connect to supporting chains on hamock couches, porch swings, etc. Black enamel finish. Shipping weight, medium, 3¼ pounds; large, 5 pounds.

6L5126—Medium size, hold about 600 pounds. Per pair......45c
6L5127—Large size, hold about 1,000 lbs. Per pair 60c

Hammock Couch and Porch Swing Hooks.

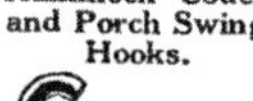

Tinned, ⅜ inch in diameter. Length, over all, 2½ inches. To screw in ceiling. Shipping weight, 10 oz.

6L5123
Per pair..........10c

Porch Swing Chains.

A weather resisting chain that can be used on most styles swinging porch settees or swings, but not adapted for hammock couches. Easily attached. Length, 8 feet. Furnished with two ceiling hooks. Shipping weight, 5 lbs.

6L5128
Per pair..........80c

Hammock Couch Chains.

Weather resisting hammock couch chains with hook at both ends. Length, 6 feet. Shipping weight, 3 pounds.

6L5122—Per pair.........43c

Folding Wood Lawn Settee.

Selected hardwood; frame painted in red and well varnished in natural finish. The seat and back pieces are securely fastened to frame. A strongly built settee. Length, 40 inches; height of back, 15¼ inches. Shipping weight, 17 lbs.

6L5484¼ $1.75

Steel-Slat Settees.

Ideal for outdoor use. Full size, well built, strongly braced, nicely finished in green paint. Built of ⅞x⅜-inch steel. Braces, 1x¼ in. flat steel. **Shipped knocked down from factory in OHIO VALLEY.**

6L5488⅓—4 feet long. Shipping weight, 70 pounds....................$6.35
6L5489⅓—5 feet long. Shipping weight, 80 pounds....................$8.30

Croquet Sets

Rules and Instructions With All Croquet Sets.

Our Amateur Croquet Sets.

Made of good quality hardwood, furnished in both four and eight balls. The mallet handles and balls are nicely varnished and striped. Wire arches. Each set put up in strong wooden box with hinged cover.

6L5942¼ — Eight-Ball Set. Shipping weight, 22 pounds.
Per set....................$2.30
6L5941¼—Four-Ball Set. Shipping weight, 18 lbs. Set.... 1.70

Favorite Eight-Ball Croquet Set.

Consists of eight nicely painted and varnished mallets with 5-inch heads, eight striped and varnished balls, two large fancy striped stakes, ten heavy wire arches. An excellent set at a low price. Shipping weight, 24 pounds.

6L5944¼—Per set....................$3.35

Champion Six-Ball Croquet Set.

Consists of six nicely finished striped mallets with 8-inch heads, six hard maple striped and varnished balls, two striped fancy stakes, heavy wire arches; put up in a strong wooden box with hinged cover. Shpg. wt., 22 lbs.

6L5945¼—Per set....................$3.45

Expert Croquet Set.

Eight balls and eight mallets. Eight-inch mallets, scored and beaded handles, well painted, nicely striped; well seasoned hardwood balls, painted and striped; two fancy beaded stakes, painted and beautifully striped. Ten heavy wire arches. A very handsome set. Shipping weight, 31 pounds.

6L5948¼—Per set....$6.45

Folding Lawn Chair.

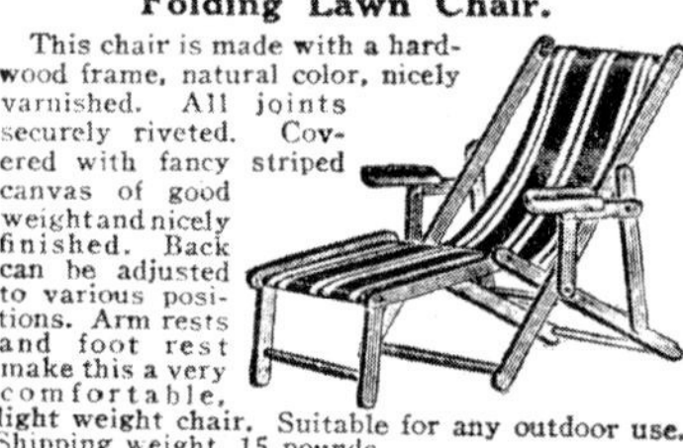

This chair is made with a hardwood frame, natural color, nicely varnished. All joints securely riveted. Covered with fancy striped canvas of good weight and nicely finished. Back can be adjusted to various positions. Arm rests and foot rest make this a very comfortable, light weight chair. Suitable for any outdoor use. Shipping weight, 15 pounds.

6L5479¼—With foot rest..........$2.10
6L5478¼—Without foot rest........ 1.70

Park or Lawn Settee.

Hardwood slats and channel steel frame. Very strong. Just the thing for parks and lawns. Frame painted black, slats natural finish. Packed flat. **Shipped from factory in OHIO VALLEY.**

	Length	Shpg. wt.	
6L5485⅓	4 feet	40 lbs.	$4.25
6L5486⅓	5 feet	45 lbs.	4.90
6L5487⅓	6 feet	50 lbs.	6.20

SEARS, ROEBUCK AND CO.

Repeating and Single and Double Barrel Shotguns

Special Model Single Barrel Gun.

BARREL—Blued steel, choke bored. Strongly built. FRAME—Solid steel, mottled finish. Top thumb lever. All action parts are of good grade steel, assembled by hand. STOCK—Plain pistol grip and fore-end; rubber butt plate.
Weight, about 6¾ pounds. Shipping weight, 10 pounds.
6L100¼—12-gauge, 30 or 32-inch barrel. **State length of barrel wanted**........ **$7.98**
410-Caliber Shotgun, as above, but with automatic ejector. Shoots 44-caliber X. L. and 410-caliber shells. Weight, 4½ pounds. Shipping weight, 10 pounds.
6L142—410-caliber, 26-inch barrel........ **$9.18**

A Long Range Single Barrel Gun.

A long range 36-inch barrel gun, used for geese, turkeys, jack rabbits, etc. Strongly built.
BARREL—Blued steel, fitted with a heavy lug; full choke; 36 inches long.
FRAME—Solid steel, mottled finish; made extra heavy and reinforced.
STOCK—Plain pistol grip; rubber butt plate; snap hinged fore-end.
The 12-gauge weighs from 7 to 7¼ pounds, and 16-gauge from 6¾ to 7 pounds. Shipping weight, 12 lbs.
6L129¼—12-gauge, 36-inch barrel........ **$9.68** | **6L130¼**—16-gauge, 36-inch barrel........ **$9.70**

Single Barrel Shotgun Outfit.

This outfit consists of a single barrel shotgun, either 12, 16 or 20-gauge, adapted to either black or smokeless powder, fitted with a blued steel barrel, strongly built; rebounding hammer, top snap lever, and plain pistol grip stock with rubber butt plate. Snap hinged fore-end; mottled finish, solid steel frame. Also one box of 25 Smokeless Powder Shells; one wood duck call; one bottle gun oil; one Tomlinson cleaner (the most practical cleaner for gun barrels); one hardwood cleaning rod with swab, scratch brush and wiper. Every article in this outfit guaranteed to give excellent service. Shipping weight, 13 pounds.
6L101¼—12-gauge outfit, 30 or 32-inch barrel........ **$9.20**
6L102¼—16-gauge outfit, 30-inch barrel........ **9.25**
6L103¼—20-gauge outfit, 28-inch barrel........ **9.30**
Cannot be sent by parcel post.

Stevens Hammerless Repeating Shotgun.

STEVENS

Stevens No. 520, 12-gauge, takedown, six-shot full coke hammerless repeating shotgun. Walnut stock, 13¾ inches long, with 2⅝-inch drop at heel. Full pistol grip and rubber butt plate. This gun has a solid breech with independent safety lock, and cannot be discharged before it is tightly closed. Weight, about 7¾ pounds. Packed for shipment, 12 pounds. Comes in 12-gauge, 30 or 32-inch barrel. **State length.**
6L209¼........ **$41.85**

Winchester Repeating Takedown Shotgun.

WINCHESTER

Winchester Repeating Takedown Shotgun, 1897 Model. Made in 12-gauge and is six-shot. Has a 30-inch steel barrel fitted with a solid blued frame, the shell being ejected entirely from the side. Plain pistol grip stock, not checkered, 13¾ inches long, 1¾-inch drop at the comb and 2⅝-inch drop at the heel. The takedown is strong and simple. Is made with a full choke bored barrel, great care being taken that none goes out which will not make a good target. Will shoot black or smokeless shells and accommodate shells 2¾ inches or 2⅝ inches in length. Weight, about 7¾ pounds. Packed for shipment, 12 pounds.
6L180¼—Winchester 12-Gauge Repeating Shotgun........ **$41.25**

Remington Model 10A Repeating Shotgun.

REMINGTON

Six-shot, takedown model. Made in 12-gauge only. It has a 30-inch blued barrel and a matted top frame; solid breech; American walnut pistol grip stock. The stock is 13¾ inches long and has a 2¾-inch drop. The gun is of the hammerless type, a favorite feature with many shooters. The operation is smooth and positive and the gun can be fired very rapidly. Ample safety devices are provided. It is a takedown model, so arranged that the magazine and barrel can be taken from the receiver without the aid of any tools; simply a quarter turn. It is also fitted with an adjustable bushing which takes up any looseness that might develop from wear. Weight, about 7½ pounds. Packed for shipment, 12 pounds.
6L190¼—Remington Repeating Takedown Shotgun. 12-gauge only; 30-inch barrel. **$47.10**

Double Barrel Hammerless Shotgun.

This American Shotgun is manufactured for us by a well known Eastern firm of fire arm manufacturers.
BARRELS—Blued steel, matted top rib, left barrel full choke, right barrel slightly modified; positive extractor. Locking lug is solid extension from barrel.
ACTION—Hammerless, snap top lever; automatic thumb safety and casehardened frame.
STOCK—Pistol grip, nicely checkered, rubber butt plate; length, 14 inches; drop, 3 to 3¼ inches; snap fore-end checkered. Packed for shipment, 14 pounds.
6L10¼—12-gauge, 30 or 32-inch barrels. **State length of barrels wanted.** Weight, 7½ to 8¼ pounds........ **$19.50**
6L11¼—16-gauge, 30-inch barrels only. Weight, 7¼ to 7¾ pounds........ **19.55**
6L12¼—20-gauge, 28-inch barrels only. Weight, 7 to 7¼ pounds........ **19.60**

410-Caliber Double Barrel Hammerless Gun.
BARRELS—Blued finish, 26 inches long. Chambered for both 44 XL shot cartridges and the 410 smokeless powder loaded shells. Positive extractor.
STOCK—Pistol, checkered grip. Snap checkered fore-end. A good grade, light weight gun, very effective for squirrels, rabbits and small game. Weight, about 6 lbs. Shipping weight, 10 lbs.
6L19¼........ **$22.50**

Complete Double Barrel Hammerless Gun Outfit.

American made hammerless gun. Fitted with 12-gauge, 30 or 32-inch blued steel barrels, 16-gauge, 30-inch barrels or 20-gauge, 26 or 28-inch barrels. Positive extractor, taper choke bored, full pistol grip checkered stock and fore-end. Outfit consists of gun, 25 Pointer shells, 1 bottle gun oil, 1 Tomlinson cleaner, 1 cleaning rod with swab, 1 scratch brush and wiper and 1 improved duck call. Weight, packed for shipment, 22 pounds. **Cannot be sent by parcel post.**
6L15¼—**12-gauge** outfit, 30 or 32-inch barrels. **State length**........ **$21.35**
6L16¼—Same as 6L15¼, but in **16-gauge**, 28 or 30-inch barrels. **State length**........ **21.40**
6L17¼—Same as above, but in **20-gauge**, 26 or 28-inch barrels. **State length**........ **21.45**

Pointer SMOKELESS Shells

Shipping Weight.
Box of 25 Shells.. 5 lbs.
Box of 50 Shells.. 7½ lbs.
Box of 100 Shells.. 14 lbs.
Box of 500 Shells.. 65 lbs.

(Unmailable) 12-GAUGE. Loaded With Drop Shot.

Catalog No.	Grains of Smokeless Powder equal to	Oz. of Shot	Size of Drop Shot	Per Box of 25 Shells	Per 100 Shells	Per Case of 500 Shells of One Load Only	Per 1,000 Shells
6L238¼ 6L239¼ 6L240¼ 6L241¼	3 Drams	1	No. 4 No. 6 No. 8 No. 10	$0.86	$3.37	$16.60	$33.20
6L242¼ 6L243¼ 6L244¼ 6L245¼ 6L246¼ 6L247¼	3 Drams	1⅛	No. 2 No. 4 No. 5 No. 6 No. 7 No. 8	.91	3.55	17.50	35.00
6L290¼ 6L248¼ 6L249¼	3¼ Drams	1⅛	No. 2 No. 4 No. 6	.93	3.63	17.90	35.80
6L295¼	3¼ Drams	1⅛	BB	1.05	4.13	20.40	40.80

10-Gauge Pointer Smokeless Shells. Loaded With Drop Shot.

Catalog No.	Grains of Smokeless Powder equal to	Oz. of Shot	Size of Drop Shot	Per Box of 25 Shells	Per 100 Shells	Per Case of 500 Shells of One Load Only	Per 1,000 Shells
6L253¼ 6L254¼	3¼ Drams	1⅛	No. 6 No. 8	$0.97	$3.80	$18.75	$37.50
6L255¼ 6L256¼ 6L257¼	3½ Drams	1¼	No. 2 No. 4 No. 6	1.00	3.91	19.30	38.60

28-Gauge Pointer Smokeless Shells. Loaded With Drop Shot.

Catalog No.	Grains of Smokeless Powder equal to	Oz. of Shot	Size of Shot	Per Box of 25 Shells	Per 100 Shells	Per Case of 500 Shells of One Load Only	Per 1,000 Shells
6L259¼ 6L260¼ 6L261¼	1¾ Drams	⅝	6 ch. 8 drop 10 dr.	88c	$3.43	$16.95	$33.90

Always give catalog number and state size of shot load wanted

Compare Our Prices With Others

The long brass cup protects the shells, keeps out moisture and makes them better, stronger and safer.

Guaranteed high quality in velocity, pattern and penetration.

All Pointer Shells are loaded with a high grade smokeless bulk powder of a hard, clean grain. Primed with a powerful nitro primer set in a gas-tight battery cup. Instantaneous ignition. Loaded by automatic machinery, guaranteeing uniformity. Pointer shells are the ideal shells to use in magazine, double or single guns for trap or field use. Sold exclusively by us.

If you want to determine the freight charges on a case of 500 shells weighing 65 pounds, to various central points, refer to the list of cities shown below. By taking the city nearest to where you live, you can approximately determine what the charges would be on a case of shells shipped to your town.

Case prices on full cases (500) of one load only. We do not furnish other loads than those specified.

410-Caliber Smokeless Shells.

Catalog No.	Grains of Smokeless Powder equal to	Oz. of Shot	Size of Ch'l'd Shot No.	Per Box of 25 Shells	Per 100 Shells	Per Case of 500 Shells of One Load Only
6L277¼ 6L278¼	⅝ Dram	⅓	6 7½	67c	$2.51	$12.30

Always give catalog number and state size of shot load wanted.

(Unmailable) 12-GAUGE. Loaded With Chilled Shot.

Catalog No.	Grains of Smokeless Powder equal to	Oz. of Shot	Size of Chilled Shot	Per Box of 25 Shells	Per 100 Shells	Per Case of 500 Shells of One Load Only	Per 1,000 Shells
6L264¼ 6L265¼ 6L263¼	3 Drams	1⅛	No. 4 No. 6 No. 7½	$0.96	$3.77	$18.60	$37.20
6L267¼ 6L268¼ 6L269¼	3¼ Drams	1⅛	No. 4 No. 5 No. 6	.98	3.84	18.95	37.90
6L270¼ Trap Load	3½ Drams	1¼	No. 7½	1.01	3.98	19.65	39.30
6L279¼	3 Drams	1¼	No. 7½	.99	3.91	19.30	38.60

16-Gauge Pointer Smokeless Shells. Loaded With Drop Shot.

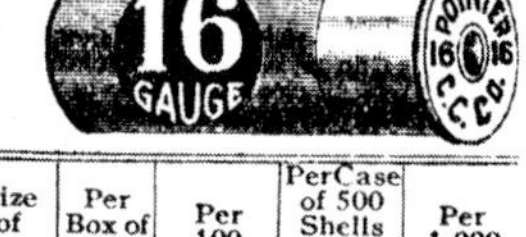

Catalog No.	Grains of Smokeless Powder equal to	Oz. of Shot	Size of Drop Shot	Per Box of 25 Shells	Per 100 Shells	Per Case of 500 Shells of One Load Only	Per 1,000 Shells
6L271¼ 6L272¼ 6L273¼	2½ Drams	1	No. 4 No. 6 No. 8	86c	$3.34	$16.45	$32.90

20-Gauge Pointer Smokeless Shells. Loaded With Drop Shot.

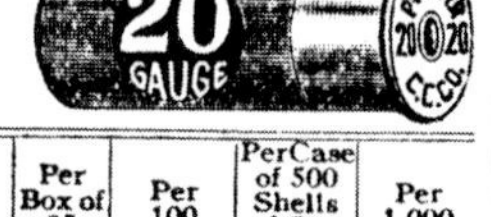

Catalog No.	Grains of Smokeless Powder equal to	Oz. of Shot	Size of Drop Shot	Per Box of 25 Shells	Per 100 Shells	Per Case of 500 Shells of One Load Only	Per 1,000 Shells
6L274¼ 6L275¼ 6L276¼	2¼ Drams	⅞	No. 4 No. 6 No. 8	85c	$3.30	$16.25	$32.50

Shells Cannot Be Shipped by Parcel Post.

A good way is to buy in case lots and have them come to you by freight. You then effect a great saving in carrying charges.

See approximate freight charges at bottom of page for case lots.

Loaded Black Powder Shotgun Shells

An excellent grade. Very popular with many shooters.

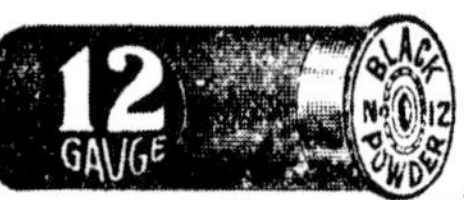

Buy by the case. You save in cost price and freight charges.

We guarantee every black powder shell against misfire, hangfire or blowback. We use a special black powder of a hard grain. It burns rapidly and leaves very little residue. Case prices on full cases (500) of one load only. A case of 500 shells weighs approximately as follows: 12-gauge, 65 lbs.; 10-gauge, 75 lbs.; 16-gauge, 53 lbs.; 20-gauge, 48 lbs.

Shells Cannot Be Shipped by Parcel Post.

(Unmailable) 12-Gauge (Black Powder Shells)

Catalog No.	Drams of Powder	Ounces of Shot	Size of Drop Shot	Per Box of 25 Shells	Per 100 Shells	Per Case of 500 Shells	Per 1,000 Shells
6L215¼ 6L216¼	3	1	6 8	73c	$2.90	$14.35	$28.70
6L226¼ 6L217¼ 6L218¼	3	1⅛	2 4 6	74c	2.94	14.45	28.90
6L219¼ 6L220¼ 6L221¼	3¼	1⅛	4 5 6	75c	2.98	14.65	29.30
6L222¼ 6L223¼	3½	1⅛	4 6	76c	3.02	14.85	29.70
6L224¼	3½	1⅛	BB	81c	3.22	15.85	31.70
6L225¼	3½	1	4 Buck	80c	3.18	15.65	31.30

(Unmailable) 10-Gauge (Black Powder Shells)

Catalog No.	Drams of Powder	Ounces of Shot	Size of Drop Shot	Per Box of 25 Shells	Per 100 Shells	Per Case of 500 Shells	Per 1,000 Shells
6L227¼ 6L228¼ 6L229¼	4½	1¼	2 4 6	91c	$3.60	$17.75	$35.50
6L233¼	4½	1¼	BB	94c	3.75	18.50	37.00

16-Gauge (Black Powder Shells).

Catalog No.	Drams of Powder	Ounces of Shot	Size of Drop Shot	Per Box of 25 Shells	Per 100 Shells	Per Case of 500 Shells	Per 1,000 Shells
6L234¼ 6L235¼	2¾	1	4 6	74c	$2.94	$14.45	$28.90

20-Gauge (Black Powder Shells).

Catalog No.	Drams of Powder	Ounces of Shot	Size of Drop Shot	Per Box of 25 Shells	Per 100 Shells	Per Case of 500 Shells	Per 1,000 Shells
6L236¼ 6L237¼	2½	⅞	6 8	70c	$2.78	$13.65	$27.30

Mallard Smokeless Shells

12-Gauge Only. A case of 500 weighs about 65 pounds.

A good grade smokeless powder shell. These shells are loaded with a good grade bulk smokeless powder, quick, clean and powerful. Mallard shells are guaranteed to be uniform and have a low breech pressure. They are primed with a powerful No. 3 primer and will be found highly satisfactory for all kinds of game shooting. Sold in 12-gauge only. Buy them by the case and effect a saving in freight charges. A case weighs about 65 pounds. Note the list below giving freight rates to various points.

Shells Are Unmailable.

Catalog No.	Dr'ms of Powder	Oz. of Shot	Size of Shot	Per Box of 25 Shells	Per 100 Shells	Per Case of 500 Shells	Per 1,000 Shells
6L280¼ 6L281¼ 6L282¼	3	1	4 6 8	$0.82	$3.26	$16.05	$32.10
6L283¼ 6L284¼	3	1⅛	4 6	.85	3.38	16.60	33.20
6L285¼ 6L286¼	3¼	1⅛	4 6	.86	3.42	16.95	33.90
6L287¼	3	1⅛	6 Chilled	.90	3.58	17.70	35.40
6L288¼	3	1⅛	7½ Chilled	.90	3.58	17.70	35.40

FREIGHT CHARGES ON A CASE OF 500 SHELLS.

From Chicago to—		From Chicago to—		From Chicago to—	
Denver, Colo.	$2.21	Detroit, Mich.	$0.67	Memphis, Tenn.	$1.47
Indianapolis, Ind.	.61	Minneapolis, Minn.	.76	Fargo, N. Dak.	1.50
Dubuque, Iowa	.59	Helena, Mont.	3.40	Milwaukee, Wis.	.50
New Orleans, La.	1.98	Kansas City, Mo.	.99	Springfield, Ill.	.61

From Phila'phia to—		From Phila'phia to—		From Phila'phia to—		From Phila'phia to—		From Phila'phia to—	
Hartford, Conn.	$0.50	Bangor, Me.	$0.88	Atlantic City, N. J.	$0.50	Chambersburg, Pa.	$0.57	Providence, R. I.	$0.61
Georgetown, Del.	.52	Houlton, Me.	1.59	Trenton, N. J.	.50	Harrisburg, Pa.	.50	Newport, Vt.	.79
Atlanta, Ga.	1.89	Baltimore, Md.	.50	Albany, N. Y.	.57	Easton, Pa.	.50	Montpelier, Vt.	.79
Portland, Me.	.70	Boston, Mass.	.61	Pittsburgh, Pa.	.70	Towanda, Pa.	.57		

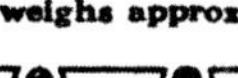
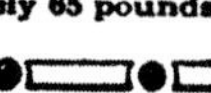
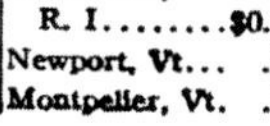

A case of 500 12-gauge shells weighs approximately 65 pounds.

Metallic Ammunition

COMPARE OUR PRICES WITH OTHERS!

Ammunition Cannot Be Sent by Parcel Post

Rim Fire Black Powder Cartridges.

Rim fire cartridges, loaded with black powder, with the exception of 22 short, 22 long, 22 long rifle and 22 Special for Winchester Model 1890 Rifle, which are loaded with Lesmok powder. If in doubt about the caliber, send with your order a sample shell that has been shot, or send the cover of the box. **Cartridges cannot be shipped by parcel post.**

Catalog No.	Cannot Be Sent by Parcel Post — Caliber	For 50	For 100	For 1,000	Good for Yards	Gr'ns of Powder	Bullet Wt., Gr'ns	Wt., per 100
6L315	B. B. Caps	......	$0.36	$ 3.45	15	..	20	7 oz.
6L316	22 Short Lesmok	18c	.35	3.35	30	3	29	10 oz.
6L317	22 Long Lesmok	24c	.46	4.50	30	5	35	11 oz.
6L318	22 Long Rifle Lesmok	28c	.54	5.30	100	5	40	14 oz.
6L319	22 Special for Mod. 1890 Win.	44c	.86	8.50	125	7	45	18 oz.
6L320	25 Stevens Rim Fire	68c	1.34	13.30	150	11	65	29 oz.
6L321	32 Short Rim Fire	45c	.88	8.70	100	9	80	27 oz.
6L322	32 Long Rim Fire	54c	1.06	10.50	125	13	90	30 oz.
6L323	41 Short Rem'gton Derringer	69c	1.36	13.50	125	13	130	41 oz.

Black Powder Center Fire Cartridges.

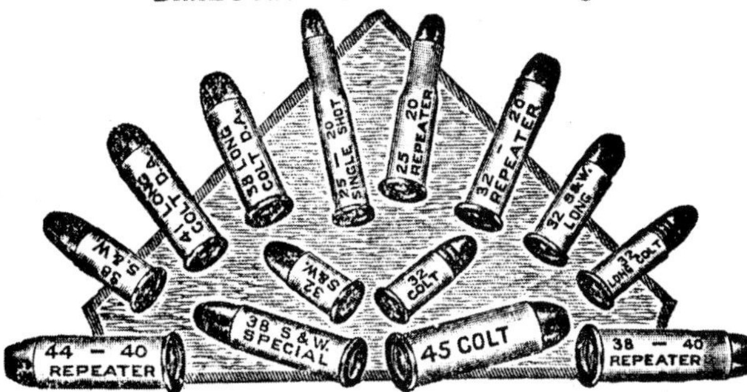

Pistol and rifle cartridges loaded with black powder. The illustrations above represent one-half actual lengths of cartridges. If in doubt about the caliber, send with your order a sample shell that has been shot, or send the cover of the box. **Cartridges cannot be shipped by parcel post.**

Catalog No.	Cannot be sent by Parcel Post — Caliber	For 50	For 100	Good for Yards	Grains of Powder	Bullet Wt., Gr'ns	Weight, per 100
6L330	25-20 Single Shot Rifles	$1.44	$2.86	200	20	86	2½ lbs.
6L331	25-20 Repeating Rifles	1.19	2.36	200	17	86	2 lbs.
6L332	32 Smith & Wesson Long	.90	1.79	125	13	98	2¼ lbs.
6L333	32 Smith & Wesson	.77	1.52	75	10	85	1¾ lbs.
6L334	32 Short Colt's Revolver	.84	1.66	75	9	80	1¾ lbs.
6L335	32 Long Colt's Revolver	.90	1.79	125	13	82	2 lbs.
6L336	32-20 Repeating Rifles	1.19	2.36	200	20	115	3 lbs.
6L337	38 Smith & Wesson	.89	1.75	100	14	145	2½ lbs.
6L338	38 Smith & Wesson Special	1.23	2.45	200	21	158	3½ lbs.
6L339	38 Long Colt's Revolver	1.09	2.16	175	19	150	3¼ lbs.
6L340	38-40 Repeating Rifles	1.42	2.82	300	40	180	4½ lbs.
6L341	41 Long Colt's Revolver	1.33	2.65	175	21	200	4 lbs.
6L342	44-40 Repeating Rifles	1.42	2.82	300	40	200	5 lbs.
6L343	45 Colt's Revolver	1.65	3.28	300	40	255	5½ lbs.

Black Powder Center Fire Military and Sporting Cartridges.

Catalog No.	Cannot be sent by Parcel Post — Caliber	For 20	For 100	Good for Yards	Grains of Powder	Bullet Wt., Gr'ns	Wt., per 100
6L344	32-40 Winchester and Marlin	$0.88	$4.38	400	40	165	5½ lbs.
6L345	38-55 Winchester and Marlin	1.08	5.39	500	55	255	7 lbs.
6L346	45-70 405 Government	1.20	5.95	700	70	405	10 lbs.

Rim Fire Smokeless Cartridges.

Rim fire cartridges loaded with smokeless powder. The above illustrations represent one-half actual lengths of cartridges. If in doubt about the caliber, send with your order a sample shell that has been shot, or send the cover of the box. **Cartridges cannot be shipped by parcel post.**

Catalog No.	Cannot be sent by Parcel Post — Caliber	For 20	For 50	For 100	For 1,000	Bullet Wt., Gr'ns	Weight, per 100
6L350	22 Short Rim Fire	......	20c	$0.39	$ 3.80	30	10 oz.
6L351	22 Short Hollow Point	......	22c	.42	4.15	28	9 oz.
6L352	22 Long Rim Fire	......	30c	.58	5.70	30	11 oz.
6L353	22 Long Rifle Rim Fire	......	34c	.66	6.50	40	14 oz.
6L354	22 Winchester Automatic	......	46c	.90	8.90	45	1 lb.
6L355	22 Remington Auto. Loading	......	47c	.92	9.10	45	1 lb.
6L356	22 Spec. Winch., Mod. 1890	......	46c	.90	8.90	45	1¼ lbs.
6L357	41 Swiss Rim Fire	$1.08	......	5.39	53.80	300	7 lbs.
6L349	22 Short Rim Fire Spotlight	......	38c	.74	7.30	30	10 oz.

Center Fire Smokeless Cartridges.

M. P. means metal patched bullet.

S. P. means soft point bullet.

Center fire smokeless pistol and rifle cartridges. Above illustrations represent one-half actual lengths of cartridges. If in doubt about caliber, send a sample shell that has been shot, or send the cover of the box. **Cartridges cannot be shipped by parcel post.**

Catalog No.	Kind of Bullet	Cannot be sent by Parcel Post — Caliber	For 50	For 100	Bullet Wt., Grains	Weight, per 100
6L360	M. P.	25-20 For Repeating Rifles	$1.49	$2.97	86	2¼ lbs.
6L361	S. P.	25-20 For Repeating Rifles	1.50	2.99	86	2¼ lbs.
6L362	M. P.	32 Colt's Automatic Pistol	1.34	2.66	74	1¾ lbs.
6L363	S. P.	32 Colt's Automatic Pistol	1.35	2.68	74	2 lbs.
6L364	M. P.	25 Colt's Automatic Pistol	1.23	2.45	50	1¼ lbs.
6L365	M. P.	32-20 For Repeating Rifles	1.49	2.97	115	2½ lbs.
6L366	S. P.	32-20 For Repeating Rifles	1.50	2.99	115	3 lbs.
6L358	Lead	38 Smith & Wesson	1.16	2.30	145	2½ lbs.
6L359	Lead	38 Long Colt's Revolver	1.23	2.45	150	3¼ lbs.
6L367	M. P.	38 Colt's Automatic Pistol	1.99	3.97	130	3 lbs.
6L374	M. P.	380 Colt's Automatic Pistol	1.96	3.90	95	3 lbs.
6L368	M. P.	45 Colt's Automatic Pistol	2.24	4.43	230	6½ lbs.
6L369	S. P.	30 Luger	2.07	4.13	93	2½ lbs.
6L370	M. P.	30 Luger	2.08	4.15	93	2½ lbs.
6L371	M. P.	8 m m Mannlicher (in clips)	5.00	9.98	227	7½ lbs.
6L372	M. P.	351 Self Loading	2.75	5.48	180	4½ lbs.
6L373	S. P.	351 Self Loading	2.76	5.49	180	4½ lbs.

Smokeless Sporting Rifle Cartridges.

M. P. means metal patched bullet.

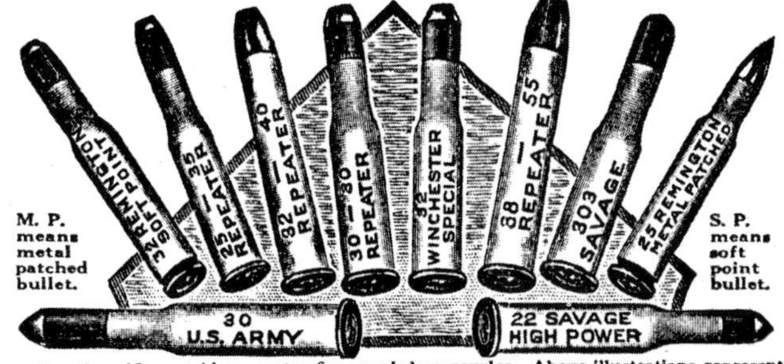

S. P. means soft point bullet.

Sporting rifle cartridges, center fire, smokeless powder. Above illustrations represent one-half actual lengths of cartridges. If in doubt about caliber, send a sample shell that has been shot, or send cover of box. **Cartridges cannot be shipped by parcel post.**

Catalog No.	Kind of Bullet	Cartridges cannot be sent by Parcel Post — Caliber	For 20	For 100	Grains of Powder	Bullet Wt., Gr'ns	Weight, per 100
6L375	M. P.	22 Hi-Power for Savage Rifles	$1.29	$6.40	12	70	3¾ lbs.
6L376	S. P.	22 Hi-Power for Savage Rifles	1.30	6.45	12	70	3¾ lbs.
6L377	M. P.	25-35 for Winchester Rifles	1.11	5.50	19	117	4¼ lbs.
6L378	S. P.	25-35 for Winchester Rifles	1.12	5.55	19	117	4¼ lbs.
6L398	M. P.	250-3000 for Savage Rifles	1.43	7.10	..	87	4 lbs.
6L399	S. P.	250-3000 for Savage Rifles	1.44	7.15	..	87	4 lbs.
6L379	M. P.	30-30 for Repeating Rifles	1.20	5.95	23	170	5 lbs.
6L380	S. P.	30-30 for Repeating Rifles	1.21	6.00	23	170	6 lbs.
6L406	M. P.	30 Springfield Rimless, 1906	1.94	9.70	..	150	8 lbs.
6L407	S. P.	30 Springfield Rimless, 1906	1.95	9.75	..	150	8 lbs.
6L381	M. P.	303 Savage Repeating Rifles	1.20	5.95	27	190	6¼ lbs.
6L382	S. P.	303 Savage Repeating Rifles	1.21	6.00	27	190	6¼ lbs.
6L383	S. P.	30 U. S. Army	1.70	8.45	35	220	7½ lbs.
6L384	M. P.	32-40 for Repeating Rifles	1.07	5.30	24	165	5½ lbs.
6L385	S. P.	32-40 for Repeating Rifles	1.08	5.35	24	165	5½ lbs.
6L386	M. P.	32 Winchester Special	1.20	5.95	..	170	5¾ lbs.
6L387	S. P.	32 Winchester Special	1.21	6.00	..	170	5¾ lbs.
6L388	M. P.	38-55 for Repeating Rifles	1.33	6.60	26	255	6 lbs.
6L389	S. P.	38-55 for Repeating Rifles	1.34	6.65	26	255	6½ lbs.

Rimless High Power Smokeless Cartridges.

Catalog No.	Kind of Bullet	Cartridges cannot be sent by Parcel Post — Caliber	For 20	For 100	Bullet Weight, Grains	Weight, per 100
6L390	M. P.	25 Remington Rimless	$1.14	$5.65	117	4¼ lbs.
6L391	S. P.	25 Remington Rimless	1.15	5.70	117	4¼ lbs.
6L392	M. P.	30 Remington Rimless	1.27	6.30	160	5½ lbs.
6L393	S. P.	30 Remington Rimless	1.28	6.35	170	6 lbs.
6L394	M. P.	32 Remington Rimless	1.27	6.30	170	6 lbs.
6L395	S. P.	32 Remington Rimless	1.28	6.35	170	6 lbs.
6L396	M. P.	35 Remington Rimless	1.43	7.10	200	7 lbs.
6L397	S. P.	35 Remington Rimless	1.44	7.15	200	7 lbs.

Shot Cartridges.

Cannot be sent by parcel post

22-caliber loaded with No. 12 shot; 44-caliber with No. 8 shot.

Catalog No.	Caliber	For 50	For 100	Wt., 100, Lbs.
6L408	22 Long, R. F.	$0.43	$0.83	¾
6L409	44 XL. C.F.	1.28	2.54	5

Blank Cartridges.

Cannot be sent by parcel post. Primed with regular powder charges, but without bullets.

Catalog No.	Caliber	For 50	For 100	For 1,000	Wt., 100, Oz.
6L410	22 Rim.	11c	$0.20	$ 1.95	4
6L411	32 S. & W.	48c	.94	9.30	10
6L412	38 S. & W.	61c	1.20	11.90	15

Ammunition Cannot Be Sent by Parcel Post.

1000 Shot Repeating Air Rifle

1 Tube of Shot Included

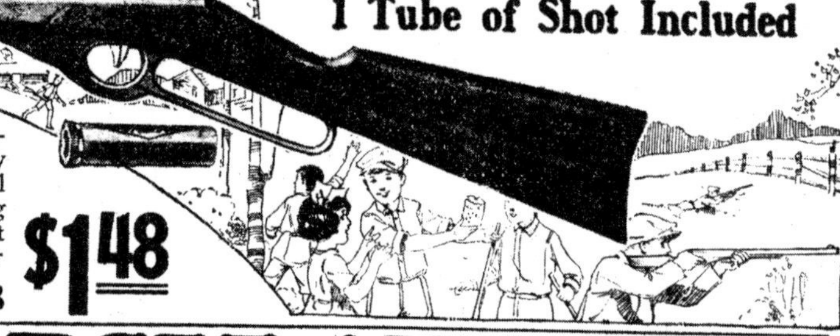

Lever Action Repeating Air Rifle

Lever action repeating air rifle. Will shoot 1,000 times without reloading. Blued steel barrel and walnut stock. It is very strongly constructed and neatly finished. The opening in barrel for loading is gauged so as to prevent any oversize shot being used which would jam the mechanism. Also has improved shot race and inner barrel. We include a tube of shot with every air rifle. Shipping weight, 4½ pounds.

6L800—With tube of shot............................**$1.48**

$1.48

AIR RIFLES, TARGET AND SHOT

$1.15

Upton Single-Shot Air Rifle.

1 Tube of Shot Included.

Lever action. Shoots air rifle shot. The frame is gunmetal finished and strongly made. Length of barrel, 19 inches; length over all, 31 inches. Weight, 1¼ pounds. Shipping weight, 3⅛ pounds.

6L804—With tube of shot........................**$1.15**

Upton Special Air Rifle.

1 Tube of Shot Included.

Single shot, break action, gunmetal finish. Substantial construction. Walnut finished stock, reinforced and rigidly attached to barrel holder. Strong and accurate shooter. Chambers air rifle shot only. Length, 29 inches. Shipping weight, 2 pounds.

6L802—With tube of shot..............................**68c**

68c

Air Rifle Shot.

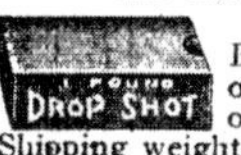

One-pound box. For use in any of the air rifles on this page. Shipping weight, 1⅛ pounds.

6L515—Per pound.......**14c**

Darts for Target Practice.

For use in King Air Rifles. Caliber, 17-100. Shipping weight, 2 ounces.

6L805—Per dozen...**18c**

Self Setting Air Rifle Target.

Our very latest design self setting Air Rifle Target. All steel. Very compact. Can be used anywhere. Size, 4½x6½ inches. Can be used for indoor practice. Shipping weight, 1 lb.

6L808..........................**19c**

Air Rifle Shot.

In tube container, patent top. Very convenient to carry, also for loading magazines of repeating rifles. Contains 4½ ounces, or about 250 pellets. Shipping weight, 7 ounces.

6L516..........................**5c**

RIFLE AND SHOTGUN SIGHTS

Catalog No.	Name of Sight	Kind of Sight	Each	Shipping Weight	
6L830	Ivory Bead Sight.	Front	**$0.90**	2 ounces	**Always mention caliber, name and model of rifle when ordering sights.**
6L831	Sheard Gold Color Bead Hunting Sight.	Front	**1.35**	3 ounces	
6L832	Marble's Reversible Ivory and Gold Color Sight.	Front	**1.35**	3 ounces	
6L833	Improved Ivory Sight.	Front	**.90**	3 ounces	
6L834	Adjustable Folding Leaf Sight.	Rear	**.90**	3 ounces	
6L835	Blank Piece to replace rear sight.	Rear	**.22**	3 ounces	
6L836	Sporting Rear.	Rear	**.80**	2 ounces	
6L837	Nickel Silver. Rocky Mountain.	Front	**.60**	2 ounces	

Marble's Flexible Rear Sight.

The Marble Flexible Rear Sight is one of the best rear sights we sell. It is furnished with two interchangeable discs with large and small apertures. **When ordering state the name of your rifle, also the caliber and model of same**, as these sights are made to fit each particular model and caliber of rifle. When using this sight the regular rear sight should be removed and blank piece 6L835 should be used. Shipping weight, 6 ounces.

6L840..**$2.98**

Marble's Vickers-Maxim Front Sight.

This is a very popular front sight. Made of hard steel. The face and lining of the aperture are made of an alloy of copper commonly known as Pope's Island gold. It is visible in the dimmest light. For quick shooting we recommend this sight. **When ordering be sure to give name, model and caliber of rifle.** Shipping weight, 2 ounces.

6L841..**$1.35**

Marble's Simplex Rear Sight.

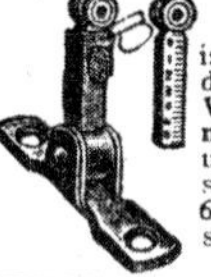

Made for 22-caliber rifles only. It is furnished with two interchangeable discs, with large and small apertures. **When ordering state the name and model of your 22-caliber rifle.** When using this sight the regular rear sight should be removed and blank piece 6L835 or folding leaf sight 6L834 should be used. Shipping wt., 4 oz.

6L842........................**$1.60**

Discs for Marble's Flexible and Simplex Rear Sights.

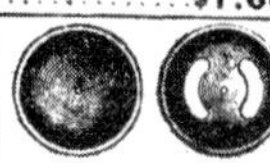

These cup discs can be attached to Marble's Flexible and Simplex rear sights as listed above. The target disc greatly assists in accuracy. It is especially adapted for target shooting. In dim light and for failing eyesight the side light disc is invaluable. **Order by catalog number.** Shipping weight, 2 oz.

6L843—Target Disc........................**48c**

6L844—Side Light Disc....................**48c**

Our Own Patent Globe Front Sight for double barrel shotguns. For 10, 12 and 16-gauge breech loaders. Shipping weight, 2 ounces.

6L847—For double barrel shotguns...**30c**

MISCELLANEOUS POLICE GOODS

NOTICE—We sell revolvers and pistols only to such persons as are legally entitled to carry or possess them. We, therefore, cater exclusively to the wants of policemen, sheriffs, deputies, marshals, game wardens, express and bank messengers and officers of the militia and national guard, and ask that intending purchasers send with order a permit or recommendation signed by the local Mayor, Chief of Police, Justice of the Peace or Sheriff.

Bean's Pattern Handcuffs.

Lock automatically. Unlocked by key. Light weight and popular with detectives and other officers of the law. Shipping weight, 14 ounces.

6L436—Polished. Per pair..........................**$3.35**

6L438—Extra Keys for above handcuffs. Shipping weight, 2 ounces. Each..........................**28c**

Heavy Police Whistle.

A very loud, shrill whistle of heavy construction. Nickel plated. Also used as a dog call or referee's signal. Shipping weight, 3 ounces.

6L452..........................**30c**

Police Whistle.

A heavily made regulation police whistle. Nickel plated; strong stationary ring at end. Length, 2½ inches. Very loud and shrill. Shipping weight, 3 ounces.

6L451..........................**29c**

Police Whistle.

Heavy metal whistle, 2⅛ inches long. A very loud, shrill whistle; suitable for referee's signal or dog call. Shipping weight, 3 ounces.

6L4787..........................**18c**

Colt's Automatic 32-Caliber Pistol.

Hammerless pocket model. Eight-shot. Fancy rubber stock, safety on grip. Blued finish. Entire length, 6¼ inches. Each shot throws out shell and puts in new cartridge. Shoots Cartridges 6L362 and 6L363. Length of barrel, 3¾ inches. Weighs 23 ounces. Shipping wt., 1⅞ lbs.

6L416¼—32-Caliber..........**$20.50**

6L402—Extra Magazine for above..........**95c**

Colt's 38-Caliber Automatic. Pocket Model.

Hammer model. Eight-shot. Blued finish. Cocks itself by own recoil, same as above automatic. Length of barrel, 4½ inches. Weighs 31 ounces. Shoots Cartridge 6L367. Shipping weight, 2⅜ lbs.

6L417¼..........................**$42.25**

6L403—Extra Magazine for above..........**1.20**

Colt's 25-Caliber Automatic.

A small and compact automatic. Shoots 25-caliber rimless, and smokeless center fire Cartridge 6L364. Blued finish; seven-shot. Fitted with a slide lock safety, also grip safety. Length over all, 4½ inches. Length of barrel, 2 inches. Weighs 13 ounces. Shipping weight, 1 pound.

6L415¼..........................**$17.00**

6L401—Extra Magazine for above..........**.95**

Colt's Automatic Target Pistol.

22-caliber. Designed, with a long barrel, for target work. Capacity of magazine, ten shots. Blued finish; checkered wood stocks. Bead front sight, adjustable for elevation. Rear sight with adjusting screw, adjustable for windage. Length over all, 10½ inches. Length of barrel, 6½ inches. Weighs 28 ounces. Shoots Cartridges 6L318. Shipping weight, 2 pounds.

6L414¼..........**$32.00**

6L400—Extra Magazine for above..........**$1.90**

Colt's Police Positive Special Revolver.

Large size frame for accurate shooting. Blued finish; six-shot. Shoots Cartridges 6L338 and 6L339. Shipping weight, 2 pounds.

6L422¼—38-Special. Length of barrel, 4 inches..........................**$28.50**

Colt's Pocket Positive Revolver.

Double action revolver, small size frame and handle. Suitable for pocket carrying. Blued finish; six-shot. Shoots 32 S. & W. regular and 32 S. & W. long, 6L332 and 6L333. Weighs 17 ounces. Length of barrel, 3½ inches. Shipping wt., 1¼ lbs.

6L425¼..........................**$26.50**

Colt's Army Special Revolver.

Double action, jointless solid frame, simultaneous ejection, center fire, blued finish; six-shot. Length of barrel, 5 inches. Shipping weight, 2¼ pounds.

6L426¼ — 38-caliber. Shoots Cartridges 6L338 and 6L339..........................**$30.00**

Colt's 45-Caliber Automatic.

The automatic adopted by the United States Government. Hammer model, automatic grip and slide lock safety. Blued finish; checkered walnut stocks. Length over all, 8½ inches. Length of barrel, 5 inches. Capacity, seven shots. Shoots Cartridge 6L368. Weight 39 ounces. Shipping weight, 2 pounds 15 ounces.

6L418¼..........................**$36.75**

6L404—Extra Magazine for above..........**1.50**

Remington Automatic Pistol, Model 51.

380-Caliber Hammerless Automatic Pistol. Magazine holds seven cartridges. Hard rubber handles, dull black finish. Automatic grip safety prevents accidental discharge. Entire length, 6⅝ inches. Weight, 21 ounces. Shape of the stock insures a perfect grip. The 380-caliber steel cased bullet gives accuracy and stopping power above police requirements. Shoots Cartridge 6L374. Shipping weight, 2 pounds.

6L424¼..........................**$19.50**

Dark Lantern.

Slide is thrown off or on by means of a thumb latch at the top of the handle, requiring the use of but one hand to operate. Fitted with a 3-inch heavy bullseye. Burns for hours with one filling. Use signal oil only. Shipping weight, 1⅞ pounds.

6L453..........................**$1.68**

6L454—1-Quart Can Signal Oil. Shipping weight, 3 pounds..........**35c**

6L455—1-Gallon Can Signal Oil. Shipping weight, 11 pounds..........**95c**

Police Stars.

We sell police stars and badges only to persons authorized to wear them. Kindly furnish evidence when ordering.

Police and Officers' Five-Ball Pointed Star. Nickel silver. Furnished lettered as follows only: Police, Special Police, Marshal, City Marshal, Constable, Detective, Deputy Sheriff, Sheriff, Watchman or Game Warden. **State plainly which is wanted.** We sell police stars only to persons authorized to wear them. Kindly furnish evidence when ordering. We do not furnish any other lettering. Shipping weight, 3 ounces.

6L445¼..........................**60c**

Police and Watchmen's Clubs.

Made of solid sole leather on a spring steel core, turned to a hard, smooth polish. 1¼ inches in diameter. As solid as wood and far more serviceable, as it will not crack or chip.

	Length	Shpg. Wt.	Each
6L464	10 in.	12 oz.	**$1.69**
6L465	12 in.	14 oz.	**1.98**
6L466	14 in.	1 lb.	**2.19**

Shooters' Equipment

Improved Duck Call.

Seasoned wood with nickel plated brass ferrule. Tongue of very flexible nickel silver. Our largest and easiest blowing duck call. Shipping weight, 5 ounces.
6L700 .... 47c

Wood Duck Call.

Will not check or crack and is not affected by weather conditions. Has good, strong tone and is easy to blow. Shipping weight, 4 oz.
6L702 .... 33c

Crow Call.

Well seasoned wood; fine nickel silver reed. With practice you can soon learn to call crows successfully. Shipping weight, 2 ounces.
6L701 .... 33c

Turkey Call.

Hold the caller in the left hand, and with the right hand rub the slate on the side of the caller. 4½ in. long, 2½ in. wide. Shpg. wt., 3 oz.
6L705 .... 55c

Indian Game Call.

Made from a hollow bone. For plover, quail, snipe, rail birds and hawks. Also an excellent dog call. Length, 4 inches. Shpg. wt., 3 oz.
6L703 .... 48c

Barnum's Game Carrier.

For carrying ducks and other birds. Holds about eighteen ducks. Shipping weight, 7 ounces.
6L989 .... 15c

Cedar Wood Decoy Ducks.

Light in Weight, Substantial and Naturally Colored. Will Not Sink if Shot.

They come in mallard, canvasback, redhead, black duck, bluebill, teal or pintail. **State which species you wish.** Shipping weight, 40 pounds per dozen. We furnish only eight drakes and four hens in each dozen.

No. 1 decoy ducks, with glass eyes. Nicely painted and well shaped decoys. **State species wanted.**

6L5995¼
Each $0.88
Per dozen, all one species 9.80

Bar Lead.

By tacking lengthwise on decoy will keep it well balanced. One strip is enough for one decoy. Shipping weight, 1 lb.
6L513—Price for 3 bars. 15c

Folding Decoys.

Made of wood throughout. Will not sink if shot. Float like live birds. Handsomely painted; glass eyes. One dozen can be made into a small package weighing only 15 pounds. Furnished in mallard, canvasback and bluebill only (8 drakes and 4 hens to each dozen). **State species wanted.** Shipping weight, each, 1¼ pounds; per dozen, 17 pounds.
6L5999—Each $0.85
Per dozen, all one species 9.75

Decoy Anchor.

One-piece iron, mushroom shape. Wt., each anchor, 1¼ lbs. Shipping weight, each, 1¼ lbs.; per dozen, 17 lbs.
6L5998
Each $0.17
Per doz. $1.95

Hunting or Driving Gloves.

One-Finger Gloves, made of soft pliable glove leather, fleece lined, close fitting knit wrists. Shpg. wt., 8 oz.
6L1001—Per pair $1.98

Hunters' Ax.

Solid steel, ground and tempered. Has 14-inch hickory handle. Total length, 16 in. Shpg. wt., 2½ lbs.
6L990 .... 98c

For complete line of Axes see page 856.

Hunters' or Scouts' All Steel Ax.

Made of tempered steel. The handle is formed of hollow steel, strongly reinforced and strongly riveted to the head; baked black enamel finish. Width of blade, 3¼ inches; length of handle, 11¼ in. Shipping wt., 2¼ lbs.
6L993—Ax and sheath $1.10

Ax Sheaths.

Made of heavy russet grain leather, to fit Hunters' Ax 6L990. Sheath 6L991 has adjustable shoulder strap. Sheath 6L992 made to carry on belt. Shipping weight, each, 7 ounces.
6L992 .... 50c | 6L991 .... 72c

Plain Face Gallery Target.

Made with solid steel plate 12 inches in diameter, intended for 22 or 32-caliber rim fire cartridges. Bell rings when bullseye is hit. The bullseye can be had in two sizes, ½ inch or ¾ inch in diameter. **State size wanted.** Shipping weight, 15 pounds.
6L685 .... $1.95

Rough Turned Walnut Stock.

Thoroughly seasoned, turned to shape, leaving the square end 1⅞ in. wide and 2¼ inches from top to bottom; length, 17¼ inches; butt measure, 5x1¼ inches. Made of good American walnut. Not fitted, just shaved. Suitable for double barrel breech loading guns.
6L687—Medium quality. Shipping weight, 2¼ pounds 75c
6L688—Selected quality. Shipping weight, 2¼ pounds 98c

Canvas Shell Bag.

Olive Tan Canvas Bag, canvas bound, with pocket and adjustable leather carrying strap. Shipping weight, 12 ounces.
6L995—Holds 75 shells $0.95
6L996—Holds 150 shells 1.00

Victoria Gun Case With Bag.

Heavy tan color canvas, reinforced with leather lock and muzzle protector and pocket for cleaning rod; also shell bag to hold fifty shells. For single, double or pump guns with 26, 28, 30 or 32-inch barrels. **State length.** Shipping weight, 1 pound 11 ounces.
6L914 .... $1.90

Same style as 6L914, to fit Remington or Winchester Automatic Shotguns. **State style wanted.** Shipping weight, 1 pound 11 ounces.
6L915 .... $2.05

Cowhide Shotgun Case.

Good quality dark russet color cowhide leather, embossed to represent pigskin, strawboard reinforced. Canton flannel lined, reinforced bottom seam, leather handle and sling, brass plated trimmings, rod pocket attached to the inside partition. For single, double and repeating shotguns only, 26, 28, 30 or 32-inch barrels. **State barrel length.** Shpg. wt., 4 lbs.
6L905 .... $5.70

Takedown Rifle Case.

Same as 6L905, but made of black imitation leather on the outside. Inside rod pocket, leather billets, handle and sling, brass plated lock buckle and trimmings. For rifles only. **Mention make, model and length of barrel of rifle.** Shpg. wt., 3¼ lbs.
6L906 .... $4.60

Straight Style Case for Pump or Repeating Shotguns Only.

Good quality leather, oak tan color, strawboard reinforced. Lined with napped cheesecloth. Two brass plated lock buckles, brass name plate and brass plated trimmings; leather handle and shoulder sling. Will fit any pump or repeating gun. Furnished for 26, 28, 30 or 32-inch barrels. **State length.** Shipping weight, 4¼ pounds.
6L907 .... $8.95

Straight Case for Remington and Winchester Automatics.

Same as above, but for automatic shotguns. **Give catalog number.**
6L903—To fit Remington Automatic Shotgun $9.90
6L904—To fit Winchester Automatic Shotgun $9.95

Our Best Cowhide Gun Case.

Good quality oak tanned cowhide, strawboard reinforced; reinforced bottom seam; rod pocket on outside with two straps and buckles; handle and returned shoulder sling; brass plated lock buckle, name plate and trimmings. Made to fit double, single and repeating guns with 26, 28, 30 and 32-inch barrels. Shpg. wt., 5½ lbs. **State length of barrel and style of gun.**
6L900 .... $8.50

Full Length Duck Cover.

Tan Duck Cover for rifles or shotguns. Full canvas bound, with heavy leather lock and muzzle protector, handle and sling. **State whether cover is wanted for rifle or shotgun, and give make, model and length of barrel.** Shipping wt., 15 ounces.
6L910 .... $1.60

Same style as above, to fit Remington or Winchester **Automatic Shotguns. State style.**
6L912 .... $1.70

Saddle Rifle Sheath.

Heavy oak tanned russet grain leather. For carrying rifle on saddle, leaving stock of rifle exposed so it may be easily grasped. For 24, 26 and 28-inch barrel rifles only. **Give make of rifle, model and length of barrel.** Shipping weight, 1⅞ pounds.
6L923 .... $3.45

Carbine Sheath, same as above, for carbines only. Furnished for 20 or 22-inch barrel. **State make of carbine and length of barrel.** Shipping weight, 1 pound 9 ounces.
6L924 .... $3.05

Folding Gun Case.

Tan color canvas, reinforced ends, leather muzzle protector, with sling strap and handle. For single, double or pump guns with 26, 28, 30 or 32-inch barrels. **Mention length of barrel.** Shpg. wt., 1¼ lbs.
6L917 .... $1.63

Light Weight Duck Gun Case.

Tan color case for take-down shotgun. Has inside rod pocket. Lined with napped cheesecloth. For single, double or pump guns with 26, 28, 30 or 32-inch barrels. **Give length of barrel.** Shipping weight, 12 ounces.
6L916 .... 82c

Takedown Rifle Cover.

Tan canvas; folding style for takedown rifle. Reinforced ends, leather protectors. **Mention make, model and length of barrel.** Shipping weight, 12 ounces.
6L918 .... $1.55

Supplemental Chambers.

Used in 30-30 and 32-40 rifles, enabling you to shoot a short range cartridge in a high power rifle. Resemble shells with the heads cut off, and are chambered to take a pistol cartridge. By placing one of these chambers in your rifle you can, with a 30-30 rifle, shoot a 32 Smith & Wesson cartridge, and with a 32-40 rifle a 32 Colt cartridge. Made of brass, nicely nickel plated.

Supplemental Chamber for 30-30 rifle. Takes the 32 Smith & Wesson Cartridge 6L333. Shipping weight, 2 ounces.
6L681 .... 58c

As above for 32-40 Marlin or Winchester, takes the 32 Colt Cartridges 6L334 and 6L335. Shipping weight, 2 ounces.
6L682 .... 59c

"Corol" Anti-Rust Compound.

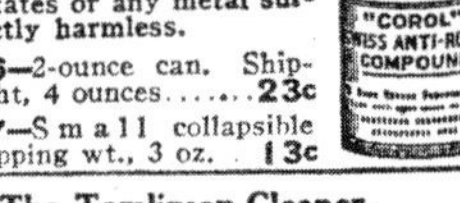

An effective aid in preventing rusting of gun barrels, rifles, knives, skates or any metal surface. Strictly harmless.
6L586—2-ounce can. Shipping weight, 4 ounces 23c
6L587—Small collapsible tube. Shipping wt., 3 oz. 13c

The Tomlinson Cleaner.

Made of brass. Takes out all burnt powder. Polishes inside of the barrels. Will fit any jointed rod. Shipping weight, 3 ounces.

6L639—12-ga. 23c | 6L641—20-ga. 25c
6L640—16-ga. 24c | 6L638—10-ga. 26c

Brass Wire Brush.

Wire brush for removing lead, powder caking and rust spots from gun barrels. Shipping weight, 3 ounces.

6L646—12-ga. 32c | 6L645—10-ga. 35c
6L647—16-ga. 33c | 6L649—28-ga. 36c
6L648—20-ga. 34c

Gun Cleaning Implements.

Our Jointed Cleaning Rods, made of beech or maple wood; patent brass joints and three implements, swab, scratch brush and wiper. **State gauge.**
6L650—36 inches long, 10, 12 and 16-gauge. Shipping wt., 10 oz. Per set 22c
6L651—36 inches long, 20 and 28-gauge. Shipping weight, 7 ounces. Per set 24c
6L652—35 inches long, 410-caliber. Shipping weight, 7 ounces. Per set 24c
6L653—48 inches long, 10, 12 and 16-gauge. Shipping weight, 11 oz. Per set, 35c

Marble's Anti-Rust Rope.

When saturated with oil this rope excludes air and moisture, preventing barrels from becoming rusted or pitted. Rope gives a constant pressure of oil against entire inner surface of barrel. Shpg. wt., 4 oz.
6L609—For shotguns. State gauge. Each 48c
6L610—For rifles. State caliber. Each 48c

Shotgun Cleaning Outfit.

Outfit consists of one hardwood three-joint cleaning rod, one wool swab, one slotted wiper, one wire scratch brush, one Tomlinson wire gauze spring center gun cleaner, one bottle gun oil and one box of gun grease. Shipping weight, 1¾ pounds.
6L630—10-ga. 65c | 6L632—16-ga. 67c
6L631—12-ga. 66c | 6L633—20-ga. 68c

Pocket Style Gun Cleaner.

Consists of a bristle brush and slotted wiper, with detachable cord and weight for dropping through barrel; a separate slotted wiper for drawing through a dry cloth and for oiling. Shipping wt., 4 oz.

6L672—22-cal. 19c | 6L676—38-cal. 19c
6L673—25-cal. 19c | 6L677—45-cal. 19c
6L674—30-cal. 19c | 6L678—50-cal. 19c
6L675—32-cal. 19c

Brass Cleaning Rod.

One-Piece Brass Cleaning Rod for rifles; 30 inches long, swivel handle; loop end screws off so that 6L662 to 6L670 Brush may be used. Shipping weight, 10 oz.
6L660—22-caliber 32c
6L661—For 25, 30, 32, 38, 44, 45 and 50-caliber. **State caliber** 34c

Brass Rifle Brush.

Brass Wire Brush to fit 6L658, 6L659, 6L660 and 6L661 Cleaning Rods. Brass shank. Especially made for cleaning rust and burnt powder out of rifles. Shipping weight, 2 ounces.

6L662—22-cal. 9c | 6L666—38-cal. 9c
6L663—25-cal. 9c | 6L667—40-cal. 9c
6L664—30-cal. 9c | 6L669—45-cal. 9c
6L665—32-cal. 9c | 6L670—50-cal. 9c

Four-Piece Brass Cleaning Rod.

Each joint is about 8½ in. long and when put together the entire rod is about 33 in. long. Has revolving handle. Shpg. wt., 10 oz.
6L658—22-caliber 34c
6L659—For 25, 30, 32, 38, 44, 45 and 50-caliber. **State caliber** 39c

Shipping weight, 8 ounces.

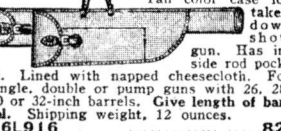

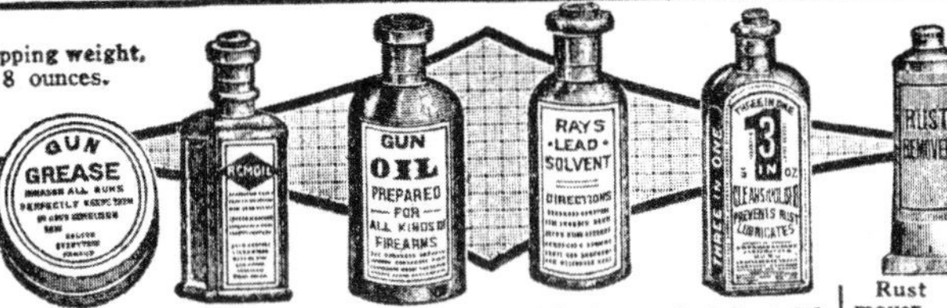

Gun Grease. Prevents rust on gun or rifle barrels, cutlery, razors. Wt., 2 ounces.
6L580 Per box, 8c

Remoil. Excellent lubricant and powder solvent, for all fire arms and machines of all kinds. Put up in 2½-ounce bottles.
6L581 Per bottle, 24c

Gun Oil. High quality, for guns, gun locks and fine machinery. Prevents rust and will not gum. We recommend this oil.
6L582 Per 2-ounce bottle 8c

Lead Solvent Cleaner. Removes all residue from barrels. One bottle is sufficient for a 22-caliber rifle; a 38-55 rifle and similar sizes require two bottles; a 12-gauge shotgun barrel, three bottles.
6L583 1 bottle .... 24c; 3 bottles 62c

3-In-1 Oil. The celebrated 3-In-1 Oil for fire arms, reels, razor strops, hones, sewing machines, clocks, etc.
6L584 Per 1-ounce bottle 13c

Rust Remover. For removing rust from tools, knives, skates, etc., or polishing any metal surface. 2-oz. tube with screw-off top.
6L585 14c

Hunters' Clothing

6L5130 — Army Shelter Tent Duck Hunting Coat. $5.75

Full Lined, Double Stitched, Olive Tan Coat, made of army shelter tent duck. Corduroy collar and corduroy lined adjustable cuffs. Reinforced shoulders and gussets under arms. Large inside game pocket, full width of coat, with front and side openings. Two large shell and one double breast pocket, as illustrated. All pockets are buttoned, to prevent game or shells from dropping out. Match scratcher on inside. **Comes in even sizes—30 to 48 inches chest measure. State chest measure.** Shipping wt., 3¾ lbs.

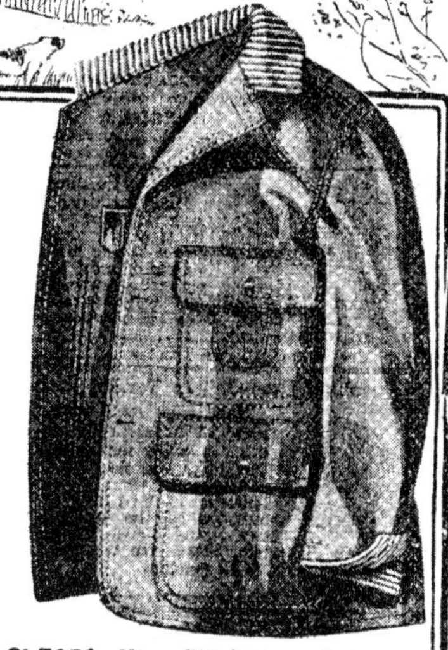

6L5133—Medium Weight Duck Hunting Coat. $4.25

Medium Weight, Khaki Color Coat. Drill lined, corduroy collar and corduroy adjustable cuffs. Double stitched and reinforced shoulders. Three large inside game pockets and two large shell and one double breast pocket with flaps. **Comes in even sizes—30 to 48 inches chest measure. State chest measure.** Shipping weight, 3¼ pounds.

6L5141—Heavy Weight Duck Hunting Coat. $5.60

A Strong, Warm, Full Lined, Double Stitched Khaki Hunting Coat. Reinforced shoulders, corduroy collar and corduroy lined adjustable cuffs. Has gussets under arms. Three large inside game pockets. Two large shell pockets and one double breast pocket with buttoned flaps. A serviceable coat, sure to please. Match scratcher on inside. **Comes in even sizes —30 to 48 inches chest measure. State chest measure.** Shipping weight, 3¼ pounds.

Field Favorite Hunting Coat

6L5135 — Field Favorite Hunting Coat. Not Lined. $2.20

A very practical garment for both hunters and fishermen. Made of khaki tan duck. Cut over large and roomy pattern. Has two large double outside lower pockets and one upper pocket. Large inside game pocket, about 11 inches deep in two divisions, giving ample room for game. All pocket corners are bar tacked and seams are double stitched. Buttons are securely fastened. **Comes in even sizes —30 to 48 inches chest measure. State chest measure.** Shipping wt., 2¾ lbs.

6L5139—Duck Hunting Pants. $1.98

Hunting Pants to match above coat. Long plain bottoms and cut full in hips and knees. Has two front pockets and one hip pocket. Seams are double stitched. **SIZES—30 to 44 inches waist measure and 30 to 36 inches inseam measure. State waist and inseam measures.** Waist measure in even sizes only. Shipping weight, 2¼ pounds.

6L5145—Corduroy Suit (coat, vest, pants and cap). Shipping weight, 7¼ lbs.. **$16.50**

6L5146—Corduroy Coat, 34 to 46 inches chest measure. Shipping weight, 4⅛ pounds.... **$8.00**

6L5147—Corduroy Vest, 34 to 46 inches chest measure. Shipping weight, 2¼ lbs... **$3.35**

6L5148—Corduroy Plain Bottom Long Pants, 30 to 44 inches waist measure; 30 to 36 inches inseam measure. Shipping weight, 3 pounds................. **$4.50**

6L5149—Corduroy Cap. Sizes, 6¾ to 7¾. **Give size.** Shipping weight, 8 ounces...... **$1.40**

Entire suit made of a thickset, good weight corduroy in dead grass shade. Cotton drill full lining makes inside game pocket large as body of coat, with front and back openings. Two very large outside pockets, one breast and one whistle pocket. Adjustable tab on cuffs. Vest has two very large breast patch pockets, which button, covered with loops for shells of 10 or 12 gauge. Lined with good strong drill. Pants can be worn with belt or suspenders. Cap has warm lined inside turndown band to cover ears. **Give measurements.**

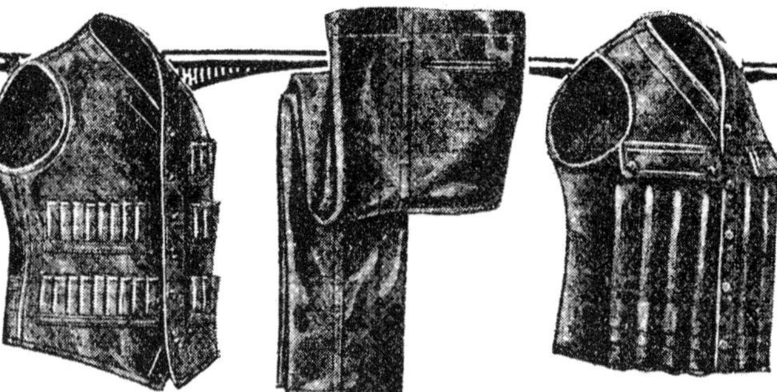

6L5155 — For 10 or 12-gauge shells.

6L5156 — For 16 or 20 gauge shells. $1.55

Hunting Vests. Good weight khaki color duck. At least 32 to 36 shell loops of same material. **SIZES—34 to 48 inches chest measure. State chest measure.** Shipping weight, 1¼ pounds.

6L5159—Long Plain Bottom Duck Hunting Pants. $2.75

Good Quality Heavy Weight Khaki Color Army Duck Hunting Pants. Cut over improved patterns; full in hips and knees. Has usual pockets and can be worn with belt or suspenders. **SIZES—30 to 44 inches waist measure and 30 to 36 inches inseam measure. State measurements.** Shipping weight, 2½ pounds.

6L5162—For 12-gauge shells. $4.25
6L5163—For 16-gauge shells.
6L5164—For 20-gauge shells.

Automatic Shell Hunting Vest. Khaki color medium weight duck. Four shell compartments on each side fitted with brass clips at bottom. Forty shells, or five to each compartment. When one is pulled another drops into place. **SIZES —34 to 46 inches chest measure. State chest measure.** Shipping weight, 2 pounds.

6L5191 Duck Cap. 95c

A Medium Weight Hunting Cap of closely woven duck. Reinforced shade and stitched rim which can be turned either up or down. All seams double stitched. Has top button and leather sweatband. A fine cap for all outdoor sports. **SIZES—6¾ to 7¾. State size.** Shipping weight, 1 pound.

Ventilated Hat.

Crusher Style Ventilated Hat. Suitable for hunting, fishing or camping. Double sewed khaki crown with stitched brim. Has two brass wire gauze ventilators. An ideal hat for warm weather. **SIZES—6¾ to 8. State size wanted.** Shipping weight, 12 ounces.
6L5195.............................. **95c**

SPORT-SEK Waterproofing for Hunters' Clothing.

A clean, sanitary waterproofing for hunters' clothing, etc. Textiles so treated are rendered waterproof and appreciably strengthened; also proof against mildew and mold. Fabric is not changed in appearance, has no odor and is not greasy. Dries clean and remains flexible. One quart is enough for an entire suit. Shipping weight, 3 pounds.
6L5192—Per quart can....... **$1.20**

6L5189 Duck Cap. 75c

Made of khaki color duck. Body of cap and large cape cotton flannel lined. Excellent rough or cold weather cap. Cape can be neatly folded around body of cap in good weather. See small illustration. **SIZES—6¾ to 7¾. State size.** Shipping weight, 9 ounces.

6L5165—Hunters' and Fisherman's Waterproof Storm Slicker. $4.50

For hunters and fishermen. Coat made of slicker cloth. Is pliable and will not crack. Shoulders and sleeves are lined. Has corduroy collar and adjustable cuffs. Double buttoned, overlapping front. **SIZES—32 to 46 inches chest measure. State chest measure when ordering.** Shipping weight, 3½ pounds.

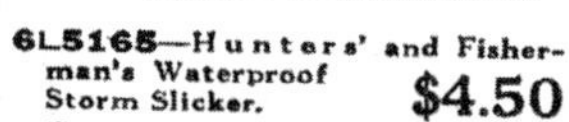

Holsters, Belts and Recoil Pads

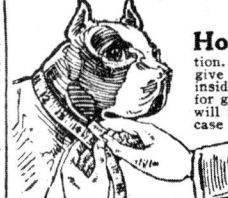

Gun Recoil Pad.

Grain Leather Recoil Pad, sewed end; laced side. The cushion is made of sponge rubber and will take up the recoil. Not made for rifles. **Give length of butt plate and catalog number.** Shipping weight, 6 ounces.

6L972—4¾-inch butt plate........**77c**
6L973—5 -inch butt plate........**78c**
6L974—5¼-inch butt plate........**79c**
6L975—5½-inch butt plate........**80c**
6L976—5¾-inch butt plate........**81c**
6L977—6 -inch butt plate........**82c**

Combination Cheek and Recoil Pad.

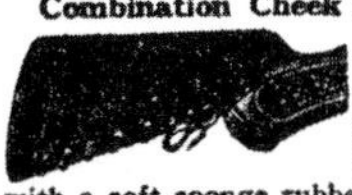

Soft leather. Protects top and sides of stock; also affords protection to shoulder as regular recoil pad. Fitted with a soft sponge rubber cushion. Not made for rifles. **Give length of butt plate and state catalog number.** Shipping wt., 4 oz.

6L980—4¾ and 5-inch butt plate...**90c**
6L981—5¼-inch butt plate........**92c**
6L982—5½ and 5¾-inch butt plate..**94c**

Sponge Rubber Recoil Pads.

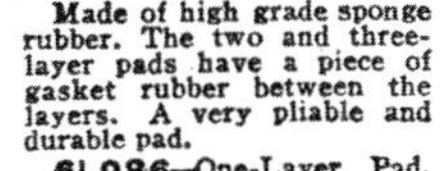

Made of high grade sponge rubber. The two and three-layer pads have a piece of gasket rubber between the layers. A very pliable and durable pad.

6L986—One-Layer Pad, ¾ inch thick. Shipping wt., 12 ounces..........**$1.20**

6L987—Two-Layer Pad, 1 inch thick. Shipping wt., 12 ounces. Each..........**$1.55**

6L986 6L987

6L988—Three-Layer Pad, 1⅛ in. thick. Shipping weight, 14 ounces..........**$2.00**

These pads will fit any model gun. Full directions for fitting furnished.

Combination Cartridge and Money Belt.

Made of heavy grain leather. Width, about 2¼ inches. Strong and durable, nicely embossed, edges double stitched. **Mention caliber and waist measure.** Shipping weight, 1⅞ pounds.

6L930—32-caliber **$2.30**
6L931—38-caliber **2.31**
6L932—44 and 45-caliber........**2.32**

Plain Leather Belt.

Made about 1 in. wide (no loops). Length, 28 to 46 inches. **State length.** Shipping weight, 4 ounces.

6L933 ....**45c**

NOTICE—**When ordering holsters always give the make, length of barrel and caliber of your revolver.**

Leather Flap and Open Top Holsters.

Made of high quality grain leather, nicely embossed, with loop for belt. **When ordering state make, caliber and length of barrel of your revolver. Order by catalog number.**

Flap Holster. Open Top Holster.

Flap Holsters.

To fit 22, 32 and 38-caliber revolvers, 3, 3½ and 4-inch barrels. Shipping weight, 5 oz.
6L1085 ....**65c**

To fit 22, 32 and 38-caliber revolvers, 5 and 6-inch barrels. Shipping weight, 7 oz.
6L1086 ....**73c**

To fit Stevens' 22-caliber target pistols, 8 and 10-inch barrels. Shipping wt., 5 oz.
6L1087 ....**83c**

To fit Colt's New Navy, New Army and S. & W. military and police revolvers. Shipping weight, 6 ounces.
6L1088 ....**83c**

To fit all large revolvers on 44 or 45 frame. Shipping weight, 7 ounces.
6L1089 ....**$1.10**

To fit Colt's 25, 32 and 38-caliber automatic pistols. Shipping weight, 5 ounces.
6L1090 ....**$1.00**

To fit 22 and 45-caliber Colt's and Luger automatic pistols. Shipping weight, 7 oz.
6L1091 ....**$1.05**

Open Top Holsters.

To fit 22, 32 and 38-caliber revolvers, 3, 3½ and 4-inch barrels. Shipping weight, 5 oz.
6L1092 ....**50c**

To fit 22, 32 and 38-caliber revolvers, 5 and 6-inch barrels. Shipping weight, 7 oz.
6L1093 ....**55c**

To fit Stevens' 22-caliber target pistols, 8 and 10-inch barrels. Shipping wt., 5 oz.
6L1094 ....**57c**

To fit Colt's New Navy, New Army and S. & W. military and police revolvers. Shipping weight, 6 ounces.
6L1095 ....**60c**

To fit all large revolvers on 44 or 45 frame. Shipping weight, 7 ounces.
6L1096 ....**78c**

To fit Colt's 25, 32 and 38-caliber automatic pistols. Shipping weight, 5 ounces.
6L1097 ....**60c**

To fit 22 and 45-caliber Colt's and Luger automatic pistols. Shipping wt., 7 oz.
6L1098 ....**72c**

Texas Shoulder Holsters.

Made of soft leather, nicely embossed, with web strap to pass around the chest to hold holster on shoulder, as shown in illustration. **When ordering always give catalog number in full and state the make and style of your revolver; give the length of barrel and we will fit your revolver.** Shipping weight, 6 ounces.

To fit 22, 32 and 38-caliber revolvers, 3, 3½ and 4-in. barrels. **Give length of barrel.**
6L1078 ....**98c**

To fit 22, 32 and 38-caliber revolvers, 5 or 6-inch barrel. **Give length of barrel.**
6L1079 ....**$1.00**

To fit Colt's New Navy, New Army and S. & W. military and police revolvers.
6L1080 ....**$1.20**

To fit all large revolvers on 44 or 45 frame.
6L1081 ....**$1.30**

To fit all Colt's 25, 32 and 38-caliber automatic pistols.
6L1082 ....**$1.20**

To fit 22 and 45-caliber Colt's and Luger automatic pistols.
6L1083 ....**$1.35**

Money Belts.

Soft, pliable, leather; double stitched, with three compartments, 3¼ in. wide; center pocket, 8 inches long; two end pockets, 5 inches long; outside cover folds over closely and is fastened by snap buttons. Shpg. wt., 6 oz.
6L958 ....**$1.35**

Soft sheepskin, generally known as chamois. Soft and comfortable and made in three compartments. Shpg. wt., 3 oz.
6L957 ....**92c**

Cartridge Belt.

Cartridge Belt, grain leather, with loops for cartridges. Width, about 1½ inches. Plain buckle. Length, 28 to 46 inches. **Mention waist measure and caliber.** Shipping wt., 6 oz.

6L925—22-caliber **68c**
6L926—32-caliber **68c**
6L927—38-caliber **68c**
6L928—44 and 45-caliber........**68c**

Anson Mills Woven Shell Belt.

Genuine Anson Mills Woven Web Belt, with loops woven into the belt. Adjustable to any size waist. **Mention gauge.** Shipping weight, 1¼ pounds.

6L940—12-gauge **$2.10**
6L941—16-gauge **2.15**

Anson Mills Hunters' Belt.

Loops are woven, closed at bottom, protecting shells; no sewing on belt. Adjustable to any size waist; no shoulder straps. **Mention gauge.** Shipping wt., 12 oz.

6L944—12-gauge **$1.75**
6L945—16-gauge **1.76**
6L946—20-gauge **1.77**

Shell Belts for Shotgun Shells.

Web or russet grain leather and with shoulder strap. **Order by number and give gauge and waist measure.**

6L952-6L955 6L950-6L951

Catalog No.	Made of	Size Gauge	Each	Shpg. Wt.
6L950	Web	12	$0.88	8 oz.
6L951	Web	16	.88	8 oz.
6L952	Rus. Leather	10	1.05	10 oz.
6L953	Rus. Leather	12	1.06	10 oz.
6L954	Rus. Leather	16	1.07	10 oz.
6L955	Rus. Leather	20	1.08	10 oz.

Patented Combination Shell and Game Carrier.

Made of olive tan duck. Front part has shell compartments with automatic shell clips. Holds forty 12-gauge shells. Bag is 19 in. long, 17 in. wide. Suitable for carrying supplies or game. Shell pocket on back which will hold 25 additional shells. Shpg. wt., 1¾ pounds.
6L994 ....**$4.95**

Automatic Shell Bag.

For either trap or field shooting. Made of olive tan duck, with adjustable neck and body straps. Buttoned flap at top. Holds twenty-five 12-gauge shells only. Always in position, ready for quick loading. Shipping weight, 1 pound.
6L999 ....**$2.25**

Braided Leather Dog Whip.

Made of black four-plait leather, strong center. Braided cord whip snap. Full length, 36 inches. Can be fastened to collar or harness and used as a lead. 1½-inch snap at butt end. Shipping wt., 6 oz.
6L4785 ....**40c**

Dog Call.

Heavy Metal Whistle, 2⅛ in. long. A very loud, shrill whistle; suitable as police or referees' signal. Shpg wt., 3 oz.
6L4787 ....**18c**

Kennel Dog Chains.

Polished steel round wire. Niagara style safety links; two swivel snap hooks. Will not kink.

Catalog No.	Size Links	Lgth.	Each	Shpg. Wt.
6L4790	Med.	4½ ft.	30c	10 oz.
6L4791	Heavy	6 ft.	40c	19 oz.

Braided Leather Dog Lead.

Made of black four-plait leather braided full length, leather center. 1½-inch snap at one end and braided loop handle at other end. Full length, 54 inches. Shipping weight, 6 oz.
6L4786 ....**42c**

Dog Lead.

Four-Foot Nickel Plated Steel Lead, with swivel snap and handle. Shipping weight, 3 ounces.
6L4792 ....**20c**

DOG COLLARS

How to Measure. Do not measure the old collar, but measure the dog's neck with tape measure as shown in illustration. Draw the tape measure tight around the dog's neck, as illustrated, and give us the exact measure. The measurements of our collars are first hole or inside hole measurements, with two extra holes, or 1½ to 2 inches allowance for growing dogs. If your dog's neck measures 11 or 12 inches, our 11-inch collars will fit, etc. This illustration shows a dog's neck measuring 14 inches, in which case our 14-inch collar will fit, with plenty of room for enlarging. **Prices on collars do not include locks. Name engraved on collar for 5 cents a letter. Write plainly name to be engraved.**

Nickel Plated Studded Collars.

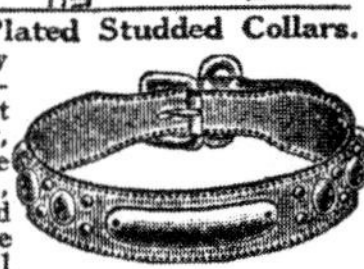

Heavy oak tanned russet leather, double thickness, trimmed with large and small round nickel plated studs, alternating around collar, nickel plated name plate, "D" ring and nickel plated lock buckle. **Give exact measurement of dog's neck and order according to catalog number.**

Catalog No.	Neck Measure.	Width Collar	Each	Shpg. Wt.
6L4690	13 in.	1 in.	$0.70	3 oz.
6L4691	15 in.	1¼ in.	.85	4 oz.
6L4692	17 in.	1¼ in.	.88	5 oz.
6L4693	19 in.	1¼ in.	.98	6 oz.
6L4694	21 in.	1½ in.	1.15	7 oz.
6L4695	23 in.	1½ in.	1.28	8 oz.

Jewel Studded Collars.

Fine black grain leather, double row of round brass plated studs with imitation turquoise studs in polished mountings; sheepskin chamois lined. Fitted with brass plated name plate, ring and lock buckle. **When ordering give measurement of dog's neck and give catalog number of the size collar nearest to length wanted.**

Catalog No.	Neck Measure	Width of Collar	Each	Shipping Wt.
6L4680	9 in.	½ in.	30c	3 oz.
6L4681	11 in.	⅝ in.	35c	4 oz.
6L4682	13 in.	¾ in.	41c	4 oz.
6L4683	15 in.	1 in.	53c	5 oz.
6L4684	17 in.	1 in.	60c	7 oz.
6L4685	19 in.	1 in.	66c	9 oz.

Bull Terrier Spike Collars.

Heavy black leather, leather lined. 13 and 15-in. collars have each a single row of pointed brass spikes, ⅝ inch long, and a double row of ⅝-inch brass plated studs. 17, 19 and 21-inch collars have each a double row of spikes and a triple row of studs. Brass plated name plate, "D" ring and lock buckle. **When ordering give size of collar that is nearest to the measurement around your dog's neck.**

Catalog No.	Neck Measure	Width of Collar	Each	Shipping Wt.
6L4650	13 in.	1½ in.	$1.30	8 oz.
6L4651	15 in.	1½ in.	1.50	12 oz.
6L4652	17 in.	2 in.	2.49	17 oz.
6L4653	19 in.	2¼ in.	2.60	19 oz.
6L4654	21 in.	2¼ in.	2.80	21 oz.

Dog Harness.

Strongly built harness that will hold any dog. Made of black heavy leather, double thickness. All connections are hand sewed. Trimmed with large round brass plated studs. Name plate and "D" ring. **Give measurement around body at back of front legs and give catalog number of harness nearest to size desired.**

Catalog No.	Body Measure	Width Straps	Each	Shpg. Wt.
6L4710	14 in.	½ in.	$1.45	5 oz.
6L4711	16 in.	⅝-in.	1.70	6 oz.
6L4712	18 in.	⅝ in.	1.76	7 oz.
6L4713	20 in.	¾ in.	2.03	8 oz.
6L4714	22 in.	¾ in.	2.10	9 oz.
6L4715	24 in.	1 in.	2.70	10 oz.
6L4716	26 in.	1¼ in.	3.60	14 oz.
6L4717	28-32 in.	1¼ in.	4.00	18 oz.

Leather Muzzles for Dogs.

Adjustable Skeleton Style. Russet strap leather. Buckles for adjustment around neck, nose and head.

Catalog No.	Neck	Each	Shpg. Wt.
6L4730	12 in.	20c	3 oz.
6L4731	14 in.	24c	3 oz.
6L4732	16 in.	33c	4 oz.
6L4733	18 in.	38c	4 oz.
6L4734	21 in.	68c	5 oz.
6L4735	26 in.	80c	5 oz.

Small Dog and Cat Collars.

First quality wine color tanned leather, sheepskin chamois lined, fitted with two nickel plated bells, nickel plated round studs, name plate and "D" ring.

Catalog No.	Neck Measure	Width of Collar	Each	Shpg. Wt.
6L4674	7 in.	½ inch	37c	2 oz.
6L4675	9 in.	½ inch	40c	2 oz.
6L4676	11 in.	½ inch	43c	2 oz.
6L4677	13 in.	¾ inch	49c	2 oz.

NOTICE—**When ordering give measurement of neck and give catalog number of the size collar that is nearest to the length wanted.**

Round Style Collars.

For collies and all long haired dogs; also used as a training collar. Made of double thickness good quality black leather, ⅜ inch round, fitted with brass plated name plate, lock buckle and "D" ring.

Catalog No.	Neck Measure	Each	Shpg. Wt.
6L4700	13 inches	$1.12	4 oz.
6L4701	15 inches	1.17	5 oz.
6L4702	17 inches	1.26	6 oz.
6L4703	19 inches	1.35	7 oz.

Dog Collar Lock.

Padlock 1x1¼ inch, all nickel plated, with key. Shipping weight, 3 oz.
6L4784 **26c**

Spratt's Dog and Puppy Cakes.

Spratt's dog foods are favorably known. Used in large dog kennels and dog shows. Two sizes.

Catalog No.	Kind Cakes	Size Boxes, Lbs.	Each	Shpg. Wt., Lbs.
6L4770	Dog	2	$0.35	3
6L4771	Dog	5	.80	5¾
6L4772¼	Dog	25	3.65	30
6L4773	Puppy	1¾	.40	2½
6L4774	Puppy	4½	.86	4¾

Sportsmen's and Trappers' Lamps

Campers', Hunters' and Sportsmen's Hand Carbide Light. Made of metal, nickel plated and highly polished, with handles and hook and self lighter attachment. Windproof and rainproof shield over lava tip. Will not smoke; has non-clog water feed; burns about six hours with one filling of carbide. Has 3-in. deep concave reflector, ht., abt., 5¾ in. Throws light about 100 feet. No glass, oil or wick. Use miners' size carbide as listed below.
6L4642— (Shpg. wt., 2 lbs.) **$2.35**

SEARS, ROEBUCK AND CO. Improved Match Lighting High Power Gasoline Lantern. One large match or two small matches will light this lantern—no alcohol or spirits needed. Used by sportsmen, campers, etc. Also excellent for lighting carnivals, boat landings, yards and large pavilions. Holds 2½ pints of gasoline and will burn 12 to 15 hours. Can be carried in the severest gale or storm. Gives a white, penetrating light. Is safe, clean and will not explode if dropped or upset. 14 inches high, exclusive of handle, and 6 inches wide at base. Fount made of heavy brass, other parts of brass and metal, nickel plated. Mica globe with reflector. Packed complete with pump, mantles and directions. Shpg. wt., 6 lbs.
6L4600—Complete with double burner and mantles, as illustrated **$6.40**
6L4602—Extra Mantles for above. Shipping weight, 4 oz. Per half dozen.... **.42**
6L4603—Large Mantles for single burners similar to above lantern. Per half doz. **.54**
6L4604—Extra Mica Globes with reflector. Shipping weight, 8 ounces. Each.. **.65**
6L4606—Extra Coil Generator for above. Shipping weight, 2 ounces.......... **.30**

Sportsmen's and Hunters' Searchlight. With Belt, Self Lighter Attachment and Extra Lens.
For sportsmen, campers, miners, etc. Carbide container carried on belt, attached by rubber tube to burner. Headlight made of brass, nickel finish. Generator made of steel, black finish. Strong glass lens, diameter, 2½ inches, will focus light about 300 feet. Headpiece weighs 5 ounces. Will not blow out. Burns 9 to 10 hours. Can be fastened on almost any cap or hat. With this light we give an extra No. 49 Concentrated Lens for spotlight purposes, which increases the power and distance. Cap not included. Shipping weight, 4 pounds.
6L4616—Complete.............. **$5.05**
6L4619—Extra tip. Shipping weight, 2 ounces. 3 for.............. **23c**
6L4620—Extra Brass Tube and Tip. Shipping weight, 3 ounces. Each.... **31c**

Campers', Hunters' and Sportsmen's Hand Carbide Light. Made of brass and metal, nickel plated and highly polished, with handles and hook and self lighter attachment. Has windproof and rainproof shield over lava tip. Will not smoke; has non-clog water feed; burns about four hours with one filling of carbide. Throws light about 75 feet. Height, 4⅝ in. Shpg. wt., 2 lbs.
6L4645 .... **$1.59**

Campers', Hunters' and Sportsmen's Light. Same make, style and finish as above, but 4 inches high. Burns about 2½ hours and throws light about 50 feet. Shpg. wt., 2 lbs.
6L4646 .... **$1.22**

Carbide. Can be used in any style carbide lamp. Miners' size, 2-pound can. Shipping weight, 2½ pounds.
6L4640 .... **25c**
Ten-pound can. Shipping weight, 12 pounds.
6L4641 .... **$1.18**

Metal Carbide Container. Concave in shape, to fit pocket. Has a tight fitting sliding cover. Will hold 10 ounces of carbide. Carbide not included. Size, 4 inches high, 3½ inches wide. Shipping weight, 6 ounces.
6L4634 .... **12c**

Campers', Miners' and Sportsmen's Light Weight Carbide Lamp. Made of brass, nickel plated and polished, with self lighter attachment. Has windproof and rainproof shield over lava tip, also improved non-clogging water feed valve which insures uniform burning. Fitted with a 2½-inch reflector. Burns about 2½ hours with one filling. Height, about 3⅞ inches. Throws light about 50 feet.
6L4648.....(Shpg. wt., 1 lb.)...... **98c**

Felt Holders. To hold felt packing 6L4645, 6L4646 and 6L4648. Shipping weight, each, 3 ounces.
6L4614—Each **4c**
Per half dozen **20c**
For 6L4616 and 6L4642 lamps.
6L4607—Per half dozen............ **20c**

Felt Packing. To fit 6L4645, 6L4646 and 6L4648 or similar lights. Shipping weight, 4 ounces.
6L4623—Half dozen............ **8c**
For 6L4616 and 6L4642 lamps.
6L4605—Per half dozen............ **8c**

Carbide Light Tips. Made of metal outside, with a lava center. Will not break or crack. For lights 6L4642, 6L4645, 6L4646 and 6L4648 only. Shpg. wt., each, 2 oz.; half dozen, 4 oz.
6L4615—Each, 4c; half dozen..... **20c**

"Justrite" Carbide Lantern. For sportsmen, campers, fishermen and trappers; also railway and mine use. Constructed of brass and metal, highly nickel plated. Gives an 18 to 20 candle power light. Penetrates a distance of 100 feet. Burns 3 to 5 hours. Has a bullseye lens. Will not explode or blow out. Height, 9 inches, exclusive of handle; width at base, 4¼ in. Shipping wt., 3 pounds.
6L4631 ...... **$4.85**
6L4626—Extra Glass Globe. Shpg. wt., 6 oz. **20c**

The "Justrite" Automatic Lighter. For lighting carbide lights, fuses, gas, etc. Produces a strong spark. Safe and dependable. Length, 4¼ inches. Average shipping weight, about 3 ounces.
6L4621 .... **25c**
6L4622—Extra Flints for above lighter. 3 for.............. **11c**

Lighter Attachment. Complete for 6L4642, 6L4645, 6L4646, 6L4648, or similar lights. Shipping weight, 2 ounces; per ¼ dozen, 5 ounces.
6L4617—Each, **10c**; 3 for...... **25c**

Extra Flints. For self lighter attachment 6L4617 and all lights with self lighter attachment. Shpg. wt., 2 oz.
6L4628—Half dozen **15c**

Headlight Caps. Plain white canvas, with a shield fitted in front. Sizes, 6½ to 7½. State size wanted. Shipping weight, 12 ounces.
6L4627............ **24c**

Tip Cleaner and Brush. Made of steel wires, with rust resisting coating; has sliding cover. Pocket size. Shipping weight, 3 ounces.
6L4632 .... **9c**

Rubber Gaskets. To fit 6L4645, 6L4646 and 6L4648 or similar lights. Diameter, 1⅛ in. Shipping weight, 5 ounces.
6L4625—Half dozen **10c**
For 6L4616 and 6L4642 lamps.
6L4624—Half dozen.............. **10c**

SMALL ANIMAL TRAPS AND TRAPPERS' SUPPLIES

Lightning Tanner. For quickly tanning furs and skins such as mink, muskrat, raccoon, dog, beaver, opossum, fox, wolf and other small fur bearing animals; also for land and water fowl. Directions with package.
6L5350—Box with powder sufficient to tan two raccoon skins in 36 hours. Shipping weight, 3 ounces......... **15c**
6L5351—Box containing three times the above. Shipping weight, 5 ounces...... **28c**
6L5352—Box containing about 1 pound. Shipping weight, 1 pound 7 ounces **75c**

Newhouse Fur Stretcher. Very light and durable. Made of a 6-gauge galvanized wire, 22 inches long. Spurs 3⅜ inches, which lock automatically where set. Skins are stretched lengthwise as well as crosswise. For muskrat, skunk, opossum, etc. Weight, 6 ounces. Shipping weight, ½ dozen, 3 pounds.
6L5327—Each **15c**
Per half dozen........ **82c**

Self Spreading Gambrel. Made of iron, with adjustable hooks to hold any size animal. Simply fold together and hook into the tendons. When animal is raised the gambrel spreads automatically, enabling one to split and clean animal very easily. Height, 17 inches; width, when folded, 18⅛ inches; width, when spread, 38⅛ inches. Shipping weight, 13 pounds.
6L7592 .... **$2.50**

Folding Skinning Gambrel. All metal, for any animal from a rabbit to a wolf. Spreads legs of animal to a width of 13 inches. Length of arms, when open, 14 inches; length, folded, 7 inches. Shipping weight, 1½ lbs.
6L7595 .... **95c**

Oneida's Latest Quick and High Catching Trap. No. 1. With chain. A new light weight trap with strong spring and sure grip jaws. Has a very wide jaw spread and lies very flat. For muskrat, skunk, opossum, marten, rabbits, etc. Shipping weight, 10 ounces.
6L5242—Each **$0.13**
Per dozen **1.49**

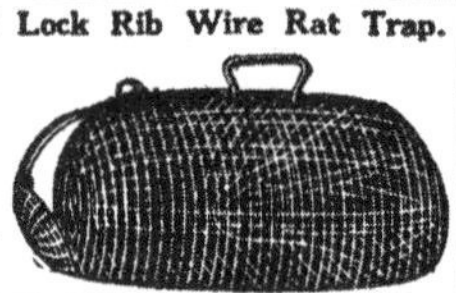

Improved Lock Rib Wire Rat Trap. Has safety door to prevent hands from being bitten. Length, 17 in.; width, 9½ in.; height, 7 inches; weight, 2½ pounds; capacity, about twelve rats. Well coated with a rustproof lacquer. Shipping weight, 2½ pounds.
6L5304 .... **95c**

Mouse Trap. Similar to 6L5304, but smaller, 5 inches high, 8 inches long. Shipping wt., 10 oz.
6L5303 ... **49c**

Improved Model Animal or Game Smoker. For driving mink, skunk, opossum, fox, rabbit and other game out of their dens. Operated by working two cylinders up and down. Galvanized steel. About 18 inches long, 4 inches in diameter, with hole tapered to 1 inch. Wire mesh 2 inches from hole to prevent cartridges, etc., from falling out and stopping smoke. Box of charcoal and sulphur compound, with directions. Shipping weight, 5 pounds.
6L5325 .... **$1.95**

Extra Boxes Sulphur and Charcoal Compound. One box sufficient to make about 15 cartridges. Shipping weight, each, 8 oz.
6L5326—Per box **$0.15**
Per dozen boxes... **1.65**

Victor Traps. Size No. 0. Spread of jaws, 3½ inches. Shipping weight, 12 ounces.
6L5200
Each, **17c**; doz. **$1.78**

Size No. 1. Spread of jaws, 3⅞ inches. Shipping weight, 1 pound.
6L5201—Each, **20c**; dozen...... **$2.13**

Size No. 1½. Mink trap. Spread of jaws, 4⅝ in.; single spring, with chain. Shipping weight, 1⅝ pounds.
6L5202—Each, **31c**; doz. **$3.20**

"Out o' Sight" Mole Trap. Standard for many years. A sure catch where moles travel. Steel, light in weight. Directions. Shipping wt., 1½ lbs.
6L5321 .... **$1.05**

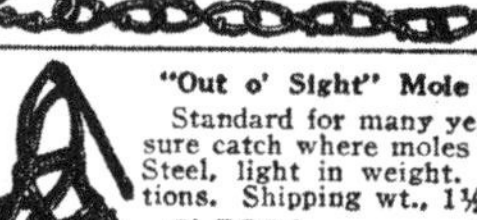

Newhouse Gopher and Salamander Trap. Made of tempered steel. Spring easily set, very lively and strong. Trap stamped "Newhouse." Weight, 6 oz. Shpg. wt., 9 oz.
6L5324
Each **$0.19**
Per dozen.... **2.13**

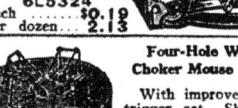

Four-Hole Wood Choker Mouse Trap. With improved loop trigger set. Shipping weight, 6 ounces.
6L5305 .... **9c**

Burbank's Animal Scents. The well known Burbank Natural Animal Scents. Shpg. wt., per bot., 60 sets, 4 oz.; 120 sets, 6 oz.
Burbank's Trout Oil, especially recommended for mink. One bottle sufficient for 120 sets.
6L5360—Per bottle..... **70c**
Burbank's Muskrat Scent. One bottle contains 60 sets.
6L5362—Per bottle ... **32c**
Burbank's Opossum Scent, for skunk and opossum. 60 sets.
6L5365—Per bottle ... **26c**
Burbank's Rat and Mouse Scent, for house rats and mice; four or five drops will entice them into trap. Bottle contains about 60 sets.
6L5366—Per bottle.............. **24c**
Burbank's Raccoon Scent, for raccoons and weasels. 60 sets.
6L5367—Per bottle.............. **25c**

Improved Little Giant Self Setting Mole Trap. Very sensitive. No danger in setting; set by pulling up plunger rod. Trigger catches itself. Made of heavy tinned steel. One of the simplest and surest mole traps made. Shipping weight, 3 pounds.
6L5320—Each **$0.30**
Per ½ dozen **3.50**

Official Mouse Trap. With double acting trigger released by either downward or upward pressure on bait holder. Sure to catch. Strongly made of wood and spring wire. Shipping weight of three, 10 ounces.
6L5307-3 for **10c**; per dozen.... **38c**

Victor Rat Trap. Extra thick heavy hardwood base; short bait trigger; heavy powerful spring. Shpg. wt., 6 oz.
6L5301—Each **10c**
6L5302—6 for **55c**
Victor Mouse Trap. Shpg. wt. of six, 14 oz.
6L5300—Per half doz., **12c**; doz. **20c**

Gopher and Salamander Trap. With wood sides. Can be easily set. Weight, 12 ounces. Shipping weight, 1 pound 1 ounce.
6L5315
Each **$0.18**
Per dozen...... **2.05**

BILLIARD AND POOL SUPPLIES

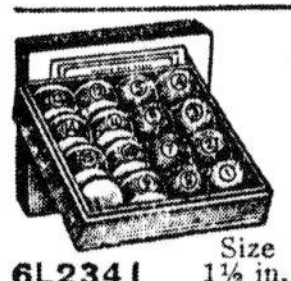

Balls for Small Tables. Sixteen to set. High grade composition, mottled effect, known as agate. Accurately made, highly polished.

	Size	Per Set	Shpg. Wt.
6L2341	1½ in.	$4.10	4 lbs.
6L2342	1¾ in.	4.25	5 lbs.
6L2343	1⅞ in.	4.50	6 lbs.
6L2344	2 in.	5.20	7 lbs.

Cushions for Small Tables. Come in two styles, as shown. The illustrations represent exact size. Made of a high grade plain rubber. Sold by the running foot. Not recommended for full size tables.

6L2650 6L2649

6L2649—Per foot 43c
Shipping weight, per foot, 3 ounces.
6L2650—Per foot 23c
Shipping weight, per foot, 2 ounces.

Wood Triangles, well finished. Sizes given represent sizes of balls for which triangles are intended.

	Size	Each	Shpg. Wt.
6L2505	2¼ in.	42c	1½ lbs.
6L2506	2 in.	40c	1½ lbs.
6L2507	1⅞ in.	37c	1¼ lbs.
6L2508	1¾ in.	37c	1¼ lbs.
6L2509	1½ in.	35c	1 lb.
6L2510	1 in.	35c	1 lb.

Bridge Heads. Design as illustrated. **6L2618**—Aluminum. Shipping weight, 2 ounces. Each 45c
6L2617—Solid hardwood. Shipping wt. of two, 6 ounces. 2 for 13c

Billiard Markers. Varnished maple. 100 to a set. Two sides, 50 black and 50 white. Shipping weight, 2 pounds.
6L2530—Per set. $1.10
Thirty-four markers to a side. Shipping weight, 1½ pounds.
6L2529—Per set 95c

Green Court Plaster. For repairing tears in cloth. Comes in strips 2½ inches wide by 18 inches long. Shipping weight, 3 oz.
6L2609—Per strip 23c

Repair Leathers. Short black leathers to place over broken or cut pocket iron leathers. Shipping weight, 6 ounces.
6L2559-3 for 23c

Bent Needles. High quality, polished steel. For mending cloth on tables. Shipping weight, 2 ounces.
6L2612-3 for 10c

Leather Table Pockets. Good size. Well finished and made to last. Colors, green or tan. Shipping weight, 9 ounces.
6L2583—Green.
6L2584—Tan.
Set of **6** $2.45

Our "Jumbo" Pocket. An extra large, full pocket, made of heavy thick leather. Length, over all, about 9 inches. Made to stand extremely hard use. Comes in green or tan. Shipping weight, 14 ounces.
6L2585—Green. Set of **6**......$3.35
6L2586—Tan. Set of **6**......3.35

The "Junior" Pocket. An excellent lighter weight leather pocket. Length over all, about 8 inches. Not so full as above numbers, but of similar construction. Shipping weight, 7 ounces.
6L2587—Green Set of **6**......$1.65
6L2588—Tan. Set of **6**......1.65

Worsted Pockets. With worsted covered fringe. Dark green. Standard size. Shipping weight, 8 ounces.
6L2603—Set of **6** $2.75

Trade Checks. Raised letter checks, round or octagon design. Price includes special lettering on one side and "Good for 1c, 2½c, 5c, 10c, 12½c, 25c or $1.00 in Trade," or "Good for 1 Pint or 1 Quart of Milk" on other side. There will be an additional charge of $1.00 for any other denomination than mentioned above. Sold in lots not less than 100. **State plainly lettering wanted for both sides, also whether round or octagon design.** Allow about ten days to fill order. Shipping weight, 100, 1⅛ lbs.

6L2484⅓—Brass, ¾ inch in diameter.
100 for ....$ 3.50
500 for 10.00
6L2486⅓—Aluminum, ¾ inch in diam.
100 for $3.25
500 for 9.25
6L2487⅓—Brass. 1 inch in diameter.
100 for ....$ 4.25
500 for 14.00
6L2488⅓—Aluminum. 1 inch in diam.
100 for ....$ 3.75
500 for 11.50

Rubber Chalk Holders. Shipping weight, 2 ounces.
6L2418—Square 9c
6L2419—Round 10c
6L2420—DoubleEndRound..11c

Selected Billiard Cues.

Plain, highly polished maple cue, natural finish. Full size, 57 inches. A strong, well made cue for all around use. Shipping weight, each, 3 lbs.; half dozen, 10 lbs.
6L2210¼—Each, **65c**; per half dozen **$3.75**

Fancy Inlaid Two-Prong Cue. Length, 57 inches. White bone ferrule; butt of handsome dark grained wood; top of highly polished maple, natural finish. A well balanced, nicely finished cue. Shipping weight, each, 3 lbs.; half dozen, 10 lbs.
6L2211¼—Each, **$1.30**; per half dozen **$7.50**

Fancy Four-Prong Cue. Length, 57 inches. White bone ferrule; butt of highly finished dark wood; top of plain maple, natural finish. Comes plain or with twine wound butt. Shipping weight, each, 3 pounds; half dozen, 10 pounds.
6L2213¼—Wound butt. Each, **$1.99**; per half dozen............**$11.60**
6L2212¼—Plain butt. Each, **1.55**; per half dozen............**9.00**

Four-Prong Cue with machine carved butt, affording a firm, easy grip. Length, 57 inches. Selected grained dark wood, highly polished. Clear maple top fitted with white bone ferrule. Shipping weight, each, 3 pounds; half dozen, 10 pounds.
6L2217¼—Each, **$1.85**; per half dozen............**$10.70**

Cues for Small Tables. Made from clear, seasoned stock, natural finish. Length, 45 inches. Tapered nicely and fitted with good tip. Shipping weight, each, 2½ pounds; half dozen, 8 pounds.
6L2214¼—Each, **45c**; per half dozen............**$2.60**

Two-Prong Cue. Length, 48 inches. Butt of selected dark wood, shaft of plain maple. Polished. Shipping weight, each, 2½ pounds; half dozen, 8 pounds.
6L2216¼—Each, **75c**; per half dozen............**$4.25**

Spink's Cue Tips, Without Glue Well made, nicely finished and of high quality. Come in three heights and three widths. The tip illustrations indicate exact heights and circles the exact diameter or widths in millimeters. **State whether 13 m/m, 14 m/m or 15 m/m width is wanted.** The prices are for a box of 100 tips. Shipping weight, per box, 4 oz.

Catalog No.	Height	Shape	13 ½	14 9/16	15 5/8
6L2253	1		96c	$1.05	$1.12
6L2254	2		92c	.99	1.07
6L2255	3		86c	.95	1.02

Assorted tips as above. Come three widths to a box in Nos. 1, 2 or 3 height. **State which height is wanted.** Shipping weight, 4 ounces.
6L2256—Box of **100** **$1.10**

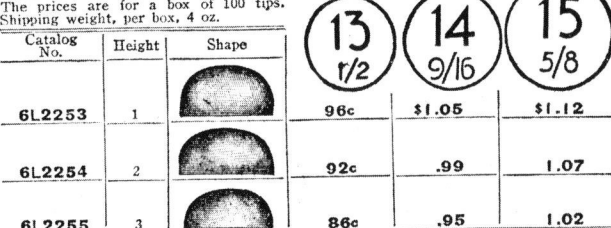

Our Tuxedo Chalk. A high grade chalk. Colors, green or blue. Square or round shape and nicely labeled. Good quality; will not break. Shipping weight, per dozen, 10 oz.; per gross, 7 lbs.

	Color	Style	Per Doz.	Per Gross
6L2401	Green	Square	19c	$2.10
6L2402	Green	Round	19c	2.10
6L2403	Blue	Square	19c	2.10
6L2404	Blue	Round	19c	2.10

TUXEDO CHALK

The "Knickerbocker" Brand Billiard Cloth.

Our "Knickerbocker" brand of billiard cloth is a grade of cloth we highly recommend for first class use. It is an all wool 18-ounce cloth of an excellent shade of green, and is adapted for either billiard or pocket billiard use. It is evenly and firmly woven and possesses a very high finish. It is a quality cloth in all respects.

For 4x8-Foot Table.

			Shpg. Wt.
6L2661	For Bed	$14.70	3¼ lbs.
6L2662	For Cushion	4.50	1⅜ lbs.
6L2663	For Both	19.20	4¼ lbs.

For 4½x9-Foot Table.

			Shpg. Wt.
6L2664	For Bed	$17.65	4 lbs.
6L2665	For Cushion	4.80	1⅝ lbs.
6L2666	For Both	22.45	5 lbs.

The "Knickerbocker" Brand Billiard Cloth in the piece, 56 inches wide. Shipping weight, per yard, 1⅛ pounds.
6L2660—Per yard **$5.90**

The "Emeraltex" Brand Billiard Cloth. An excellent lower grade cloth that is used very extensively. Intended for cheaper grades of tables and for use where it is not desired to use a high quality cloth. It is an all wool, 16-ounce cloth and comes in a good shade of green. It possesses a good finish and will be found very satisfactory.

For 4x8-Foot Table.

			Shpg. Wt.
6L2673	For Bed	$11.35	3¼ lbs.
6L2674	For Cushion	3.45	1⅜ lbs.
6L2675	For Both	14.80	4¼ lbs.

For 4½x9-Foot Table.

			Shpg. Wt.
6L2676	For Bed	$13.60	4 lbs.
6L2677	For Cushion	3.80	1⅜ lbs.
6L2678	For Both	17.40	5 lbs.

The "Emeraltex" Billiard Cloth in the piece, 56 inches wide. Shipping weight, per yard, 1⅛ pounds.
6L2672—Per yard **$4.55**

Enameled Back Billiard Cloth. A good weight dark green drill, covered on one side with a black composition. Excellent for use on tables subject to abuse or rough playing and where it is not desired to use the better grades of regular cloth. This cloth will prove highly serviceable.

For 4x8-Foot Table.

			Shpg. Wt.
6L2681¼	For Bed	$3.60	2¾ lbs.
6L2682	For Cushion	1.15	1⅜ lbs.
6L2683¼	For Both	4.75	4¼ lbs.

For 4½x9-Foot Table.

			Shpg. Wt.
6L2684¼	For Bed	$3.90	4 lbs.
6L2685	For Cushion	1.22	1⅜ lbs.
6L2686¼	For Both	5.12	5 lbs.

Above cloth in the piece. Width, 56 inches. Shipping weight, per yard, 1⅜ lbs.
6L2680—Per yard **$1.30**

Billiard Table Covers. A composition, black cloth cover, dust and moisture proof.
6L2690—For 3x6-foot table. Shipping weight, 2½ pounds............**$1.50**
6L2687—For 4x8-foot table. Shipping weight, 3⅞ pounds............**$2.00**
6L2688—For 4½x9-foot table. Shipping weight, 4⅞ pounds **$2.35**

Pocket Billiard or Pool Balls. The regulation 2¼-inch composition solid stripe and inlaid number balls, 16 to set, as used generally in billiard halls. Our price is exceptionally low. Strictly new balls, quality guaranteed. Shipping weight, 8⅝ pounds.
6L2306—Per set **$19.50**

Single Balls. One numbered ball only. Any number. **State which is wanted.** Grade as above, 2¼ inches in diameter. Shipping weight, 14 ounces.
6L2307—Each **$1.30**

The "Gold Crown" Cue Ball. Our highest grade. Guaranteed not to chip, crack or dent. 2¼-inch size only. Light brown color. Shipping weight, 15 ounces.
6L2329—Each **$4.00**

Baseball Pocket Billiards. Partial set to add to any regular 15-ball pocket billiard set. Striped balls numbered from 16 to 21. With large triangle and score pad. Shipping weight, 10 pounds.
6L2336—Per set **$12.75**

Extra Score Sheets. Pad of ninety-six extra score sheets for above. Shipping weight, 2 pounds.
6L2337 ....**98c**

Composition Billiard Balls. Excellent grade. Sizes, 2⅜ and 2^7⁄$_{16}$ inches in diameter. Two white and one red. Shipping weight, 3½ pounds.
6L2349—Size, 2⅜ in. Set of **3** **$7.00**
6L2350—Size, 2^7⁄$_{16}$ in. Set of **3** **8.00**

Triangle, made of maple, brass reinforced, for 2¼-inch balls. A high grade triangle. Shipping weight, 1½ pounds.
6L2504 ....**$1.18**

The "White Wonder" Cue Ball. A good grade, plain white hard ball, that will give good satisfaction. Diameter, 2¼ inches. Shipping weight, 14 ounces.
6L2328—Each **95c**

Our "Purple Star" Cue Ball is a very hard, lively white ball with an inlaid purple star. Will stand hard abuse and guaranteed not to crack or break. Size, 2¼ inches. Shipping weight, 14 ounces.
6L2324—Each **$1.45**

Shake or Tally Balls. Flat on one side. Come numbered 1 to 16 to set. Shipping weight, 5 oz.
6L2371—Maple. Per set **17c**
6L2372—Composition. Per set. **30c**

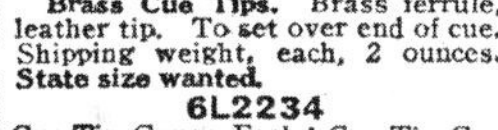

Brass Cue Tips. Brass ferrule, leather tip. To set over end of cue. Shipping weight, each, 2 ounces. **State size wanted.**

6L2234

Cue Tip	Gauge	Each	Cue Tip	Gauge	Each
^{11}⁄$_{32}$-in.	No. 6	9c	½-in.	No 10	9c
⅜-in.	No. 7	9c	^{9}⁄$_{16}$-in.	No. 11	9c
^{13}⁄$_{32}$-in.	No. 8	9c	^{19}⁄$_{32}$-in.	No. 12	9c
^{7}⁄$_{16}$-in.	No. 9	9c			

Assorted Brass Cue Tips, as above. Come one-half dozen in a package. An assortment that will accommodate various sizes of cues. Shipping weight, 4 ounces.
6L2233—Package **50c**

Cue Points. White bone. Three-quarters inch high. Come ^{19}⁄$_{32}$ inch to ^{13}⁄$_{32}$ inch outside diameter. All drilled with ¼-inch holes. **State outside measurements wanted.**

6L2241
Each ......(Shpg. wt., 2 oz.)......**$0.18**
Dozen ......(Shpg. wt., 6 oz.)......**1.18**

Lightning Cue Clamps. All metal. To hold new tips until glue is dry. Instantly put on or removed. Shipping wt.. 14 ounces.
6L2630—Set of **3** **19c**

Rail Bolt Caps. Brass or oxidized finish. **State kind wanted.** Screws included. Shipping weight, 3 ounces.
6L2623—Dozen **20c**

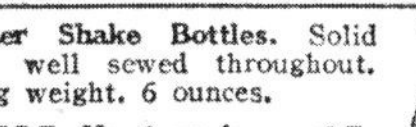

Leather Shake Bottles. Solid leather, well sewed throughout. Shipping weight, 6 ounces.
6L2525—No. 1 grade.....**65c**
6L2524—No. 2 grade.....**56c**

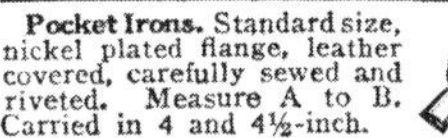

Pocket Irons. Standard size, nickel plated flange, leather covered, carefully sewed and riveted. Measure A to B. Carried in 4 and 4½-inch.
6L2545—Set of six, 4-inch, complete with bolts. Shpg. wt., 5 lbs. Per set **$3.70**
6L2546—Set of six, 4½-inch, complete with bolts. Shpg. wt., 5 lbs. **$3.85**
6L2547—One only, 4-inch. **State if corner or side iron is wanted.** Shipping weight, 1 pound 5 ounces. Each............**65c**
6L2548—One only, 4½-inch. **State if corner or side iron is wanted.** Shipping weight, 1 pound 5 ounces. Each ... **68c**

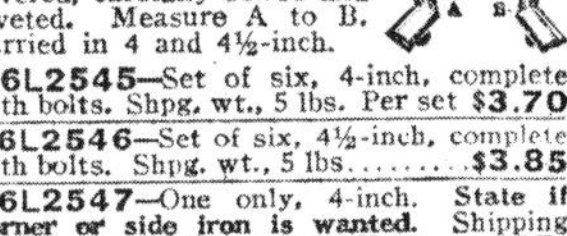

Pocket Knives

Dakota Cowboys' Knife. Stag pattern handle, 3⅝ in. long. Double bolsters. Spear blade, 2⅞ in. long; sheep's foot blade, 2¼ in. long; pen blade, 2⅛ in. long. Shipping weight, 6 oz.
6L7065.......... 89c

Stag Pattern Handle 2-Blade Knife. 3⅝ inches long. Double bolsters. Spear blade, 2¾ inches long; pen blade, 2 inches long. A substantial knife. Shipping weight, 5 oz.
6L7058.......... 59c

Four-Blade Cattlemen' Knife. Stag pattern handle, 3⅝ inches long. Nickel silver bolsters. Clip blade, 2⅞ inches long; sheep's foot blade, 2⅛ inches long; spaying blade, 2⅛ inches long; awl or punch blade, 2 inches long. Shipping weight, 6 oz.
6L7074.......... $1.15

Stockmen's Lock Blade Awl Knife. Celluloid handle, 3¾ inches long. Clip blade, 3 inches long; spaying blade, 2¾ inches long; lock awl blade, 2¾ inches long, which cannot be released unless large blade is pressed down. Shipping weight, 6 oz.
6L7066.......... $1.38

Shipping wt., 4 ounces. **Slide Button Fly Lock Knife.** Will not open by accidental pressure against the button while in your pocket, but must be pushed sideways to be opened. Remains locked when opened or closed. Celluloid imitation rosewood handle, 3⅝ in. long, brass lined. Large blade, 2¼ in. long; small blade, 2 in. long.
6L7082..... $1.10
6L7083—Same as above, with jet black handle..... 1.10

Easy Opener Knife. Polished red wood handle, 3¼ in. long. Large spear blade, 2⅝ in.; small blade, 1⅞ inches long. Strong chain, securely fastened, with attachment to fasten to button of clothing. Shipping weight, 6 ounces.
6L7012..... 35c

Boys' Pocket Knife. Rosewood handle pocket knife, with 2½-inch spear blade. Single bolster; length of handle, 3¼ inches. Shpg. wt., 4 oz.
6L7005..... 18c

Pocket Knife. Stag pattern handle, 3⅝ inches long. Clip blade, 2½ inches long. Small blade, 2 inches long. Single bolster. A very strong, practical knife for any boy. Shipping wt., 5 oz.
6L7015..... 33c

Barlow Pattern Jackknife. Is the well known Barlow pattern jackknife. Bone handle, 3⅜ inches long. Heavy single steel bolster, 1⅛ in. long, making the knife extra strong. Spear blade, 2¾ in. long; pen blade, 2 in. long. Shipping wt., 5 oz.
6L7016..... 31c

Ebony Handle Easy Opener Knife. Length, 3⅝ inches. Double bolsters and fancy nickel silver shield, with easy opener feature. Large blade, 2¾ inches long; pen blade, 2 inches long. Blades and handle are extra wide. Shipping wt., 5 ounces.
6L7008..... 65c

Easy Opener Knife. Stag pattern handle, 3⅝ inches long. Double bolsters. Spear blade, 2⅞ inches long; pen blade, 2 inches long. A strong, all around, practical knife. Shipping wt., 5 oz.
6L7030..... 80c

Cattlemen's Knife. Stag pattern handle, 3⅞ inches long; with clip point blade open, 7 inches long. Spaying blade is 3 inches long from bolster. Nickel silver bolsters. Shipping weight, 5 oz.
6L7034.......... 85c

Three-Blade Knife. Stag pattern handle, 3⅝ inches long, swelled in middle to insure a firm hold. Long double bolsters, brass lined. Large spear blade, 2¾ inches long; two pen blades, each 1¾ inches long. Shipping weight, 4 ounces.
6L7057..... 98c

Stockmen's Four-Blade Knife. Stag pattern handle, 3¾ inches long. Nickel silver bolsters. Clip blade, 2⅝ inches long; sheep's foot blade, 2¼ in. long; spaying blade, 2¼ in. long; pen blade, 1⅞ in. long. Shpg. wt., 5 oz.
6L7070.. $1.20

Penknives

Watch Chain Knife. A very beautiful flat knife, 2⅜ inches long, handle resembling abalone pearl. Has pen blade, 1½ inches long, and file and nail blade, 1¾ inches long. Nickel silver lined. Shipping weight, 3 ounces.
6L7025.......... 75c

Flat Penknife or Vest Pocket Knife. Stag pattern handle, 3 inches long. Nickel silver tips and shield, brass lined. The wide flat shape makes this a convenient vest pocket knife. Very strongly made. Spear blade, 2¼ inches long; small blade, 1¾ inches long. Shpg. wt., 4 oz.
6L7055.......... 65c

Celluloid Handle Penknife. Handle of brown celluloid in artistic design resembling shell, very thin and flat, 2¾ inches long, with nickel silver tips and shield. Large blade, 2⅛ inches; pen blade, 1½ inches long. Can be carried in vest pocket. Shipping weight, 3 ounces.
6L7103.......... 68c

Combination Penknife and Cigar Cutter. Nickel silver handle, 2⅞ inches long. Pen blade, 1⅝ inches long. Press down pen blade to cut cigar end. Has linked end to fasten to watch chain or key ring. Flat shape. A good vest pocket size. Shpg. wt., 2 oz.
6L7108.......... 33c

Leather Knife Purse. For pocket knives having handles not longer than 4 inches. Give length of knife you intend to carry in purse. Keeps knife from rusting. Shipping weight, 1 ounce.
6L7124.......... 8c

Perfection Pocket Knife Desk Hone. Will keep pocket knife or small cutting tool sharp. Every pocket knife owner should have one. 3 inches long by ⅞ inch wide. Shipping weight, 4 oz.
6L7142.......... 10c

Three-Blade Cattle Knife. Stag pattern handle, 3⅝ inches long. Double bolsters. Spear blade, 2¾ in. long; sheep's foot blade, 2¼ inches long; awl blade, 2 inches long. Shipping weight, 5 ounces.
6L7064..... 95c

Stag Pattern Handle Knife. Two blades. Handle, 3½ inches long. Large blade, 2⅞ inches long. Small blade, 2 inches long. Double bolsters and shield of nickel silver. Strong chain, securely fastened, with attachment to fasten to button of clothing. Shipping wt., 6 oz.
6L7010..... 55c

Texas Three-Blade Stock Knife. Stag pattern handle, 3⅞ inches long. Nickel silver bolsters. Clip blade, 3¼ inches long; sheep's foot blade, 2¼ inches long; spaying blade, 2¼ inches long. Shipping weight, 5 ounces.
6L7068..... $1.05

Vermilion Handle Knife. Polished genuine vermilion wood handle, 3 inches long. Single bolster. Blades are of good steel, nicely polished and well tempered. Spear blade, 2⅝ inches long; pen blade, 2⅛ inches long. Shipping weight, 4 ounces.
6L7001.......... 32c

Ebony Handle Jackknife. Genuine ebony handle, 3¾ inches long. Double bolsters and shield of nickel silver. A very strong knife. Clip blade, 3 inches long; pen blade, 2¼ inches long. Shipping weight, 5 ounces.
6L7013.......... 85c

Stag Pattern Handle Knife. Three and a half inches long. Double bolsters and nickel silver shield. Clip blade, 2¾ inches long; pen blade, 1¾ inches long. Good quality steel, well tempered. Shipping weight, 4 ounces.
6L7018..... [illegible]c

Horn Handle Jackknife. Handle of selected horn, 3⅜ inches long. Double fluted bolsters and shield of nickel silver. Clip blade, 2⅝ inches long; pen blade, 2 inches long. Shipping weight, 5 oz.
6L7047.......... 92c

Premier Stockmen's Three-Blade Knife. Beautiful dark red celluloid handle, closely resembling mahogany, 3½ inches long. Nickel silver bolsters. Clip blade, 2⅜ inches long; sheep's foot blade, 2¼ inches long; pen blade, 2 inches long. Shipping weight, 5 ounces.
6L7061...... $1.25

Texas Toothpick Knife. Stag pattern handle, 3⅞ inches long. Double bolsters. Clip point saber blade, 3⅛ inches long; pen blade, 2⅛ inches long. Shipping wt., 5 oz.
6L7032..... 78c

Stag Pattern Handle Jackknife. Length, 3¼ inches. Two blades. Double bolsters. Clip blade, 2⅝ inches long; pen blade, 2 inches long. A well made knife. Nicely polished and ground. Shipping weight, 5 ounces.
6L7020.......... 53c

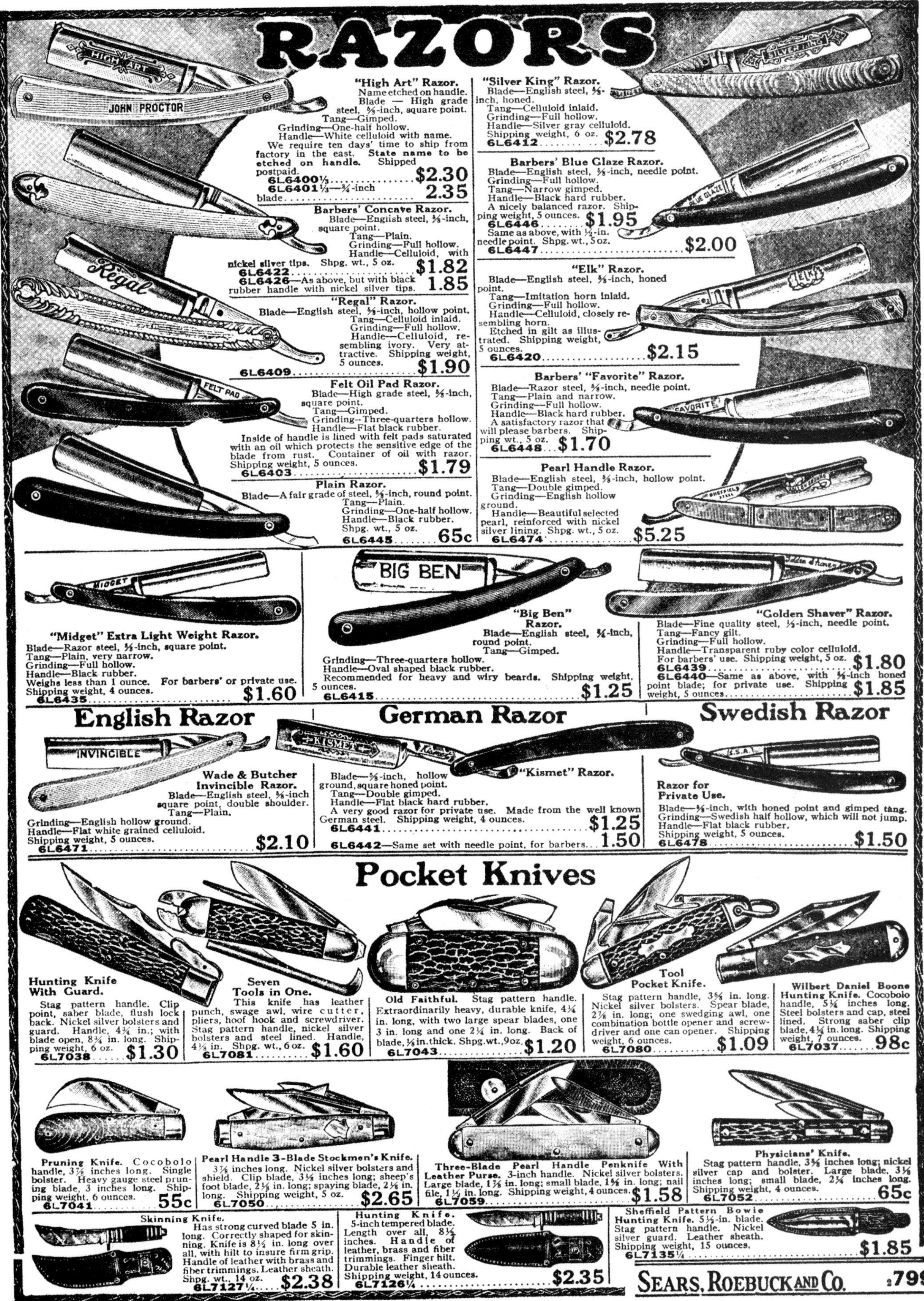
RAZORS
HIGH ART
JOHN PROCTOR
"High Art" Razor. Name etched on handle. Blade — High grade steel, ⅝-inch, square point. Tang—Gimped. Grinding—One-half hollow. Handle—White celluloid with name. We require ten days' time to ship from factory in the east. State name to be etched on handle. Shipped postpaid.
6L6400½ $2.30
6L6401½—¾-inch blade 2.35
Barbers' Concave Razor. Blade—English steel, ⅝-inch, square point. Tang—Plain. Grinding—Full hollow. Handle—Celluloid, with nickel silver tips. Shpg. wt., 5 oz.
6L6422 $1.82
6L6426—As above, but with black rubber handle with nickel silver tips. 1.85
Regal
"Regal" Razor. Blade—English steel, ⅝-inch, hollow point. Tang—Celluloid inlaid. Grinding—Full hollow. Handle—Celluloid, resembling ivory. Very attractive. Shipping weight, 5 ounces.
6L6409 $1.90
FELT PAD
Felt Oil Pad Razor. Blade—High grade steel, ⅝-inch, square point. Tang—Gimped. Grinding—Three-quarters hollow. Handle—Flat black rubber. Inside of handle is lined with felt pads saturated with an oil which protects the sensitive edge of the blade from rust. Container of oil with razor. Shipping weight, 5 ounces.
6L6403 $1.79
Plain Razor. Blade—A fair grade of steel, ⅝-inch, round point. Tang—Plain. Grinding—One-half hollow. Handle—Black rubber. Shpg. wt., 5 oz.
6L6445 65c
SILVER KING
"Silver King" Razor. Blade—English steel, ⅝-inch, honed. Tang—Celluloid inlaid. Grinding—Full hollow. Handle—Silver gray celluloid. Shipping weight, 6 oz.
6L6412 $2.78
BLUE GLAZE
Barbers' Blue Glaze Razor. Blade—English steel, ⅝-inch, needle point. Grinding—Full hollow. Tang—Narrow gimped. Handle—Black hard rubber. A nicely balanced razor. Shipping weight, 5 ounces.
6L6446 $1.95
Same as above, with ½-in. needle point. Shpg. wt., 5 oz.
6L6447 $2.00
ELK
"Elk" Razor. Blade—English steel, ⅝-inch, honed point. Tang—Imitation horn inlaid. Grinding—Full hollow. Handle—Celluloid, closely resembling horn. Etched in gilt as illustrated. Shipping weight, 5 ounces.
6L6420 $2.15
FAVORITE
Barbers' "Favorite" Razor. Blade—Razor steel, ⅝-inch, needle point. Tang—Plain and narrow. Grinding—Full hollow. Handle—Black hard rubber. A satisfactory razor that will please barbers. Shipping wt., 5 oz.
6L6448 $1.70
SHEFFIELD STEEL
Pearl Handle Razor. Blade—English steel, ⅝-inch, hollow point. Tang—Double gimped. Grinding—English hollow ground. Handle—Beautiful selected pearl, reinforced with nickel silver lining. Shpg. wt., 5 oz.
6L6474 $5.25
MIDGET
"Midget" Extra Light Weight Razor. Blade—Razor steel, ⅝-inch, square point. Tang—Plain, very narrow. Grinding—Full hollow. Handle—Black rubber. Weighs less than 1 ounce. For barbers' or private use. Shipping weight, 4 ounces.
6L6435 $1.60
BIG BEN
"Big Ben" Razor. Blade—English steel, ¾-inch, round point. Tang—Gimped. Grinding—Three-quarters hollow. Handle—Oval shaped black rubber. Recommended for heavy and wiry beards. Shipping weight, 5 ounces.
6L6415 $1.25
Golden Shaver
"Golden Shaver" Razor. Blade—Fine quality steel, ½-inch, needle point. Tang—Fancy gilt. Grinding—Full hollow. Handle—Transparent ruby color celluloid. For barbers' use. Shipping weight, 5 oz.
6L6439 $1.80
6L6440—Same as above, with ⅝-inch honed point blade; for private use. Shipping weight, 5 ounces $1.85
English Razor
INVINCIBLE
Wade & Butcher Invincible Razor. Blade—English steel, ⅝-inch square point, double shoulder. Tang—Plain. Grinding—English hollow ground. Handle—Flat white grained celluloid. Shipping weight, 5 ounces.
6L6471 $2.10
German Razor
KISMET
"Kismet" Razor. Blade—⅝-inch, hollow ground, square honed point. Tang—Double gimped. Handle—Flat black hard rubber. A very good razor for private use. Made from the well known German steel. Shipping weight, 4 ounces.
6L6441 $1.25
6L6442—Same set with needle point, for barbers 1.50
Swedish Razor
S.S.A.
Razor for Private Use. Blade—⅝-inch, with honed point and gimped tang. Grinding—Swedish half hollow, which will not jump. Handle—Flat black rubber. Shipping weight, 5 ounces.
6L6478 $1.50
Pocket Knives
Hunting Knife With Guard. Stag pattern handle. Clip point, saber blade, flush lock back. Nickel silver bolsters and guard. Handle, 4¾ in.; with blade open, 8¾ in. long. Shipping weight, 6 oz.
6L7038 $1.30
Seven Tools in One. This knife has leather punch, swage awl, wire cutter, pliers, hoof hook and screwdriver. Stag pattern handle, nickel silver bolsters and steel lined. Handle, 4⅛ in. Shpg. wt., 6 oz.
6L7081 $1.60
Old Faithful. Stag pattern handle. Extraordinarily heavy, durable knife, 4¼ in. long, with two large spear blades, one 3 in. long and one 2¼ in. long. Back of blade, ⅛ in. thick. Shpg. wt., 9 oz.
6L7043 $1.20
Tool Pocket Knife. Stag pattern handle, 3⅝ in. long. Nickel silver bolsters. Spear blade, 2⅞ in. long; one swedging awl, one combination bottle opener and screwdriver and one can opener. Shipping weight, 6 ounces.
6L7080 $1.09
Wilbert Daniel Boone Hunting Knife. Cocobolo handle, 5¼ inches long. Steel bolsters and cap, steel lined. Strong saber clip blade, 4⅛ in. long. Shipping weight, 7 ounces.
6L7037 98c
Pruning Knife. Cocobolo handle, 3½ inches long. Single bolster. Heavy gauge steel pruning blade, 3 inches long. Shipping weight, 6 ounces.
6L7041 55c
Pearl Handle 3-Blade Stockmen's Knife. 3⅞ inches long. Nickel silver bolsters and shield. Clip blade, 3⅛ inches long; sheep's foot blade, 2½ in. long; spaying blade, 2½ in. long. Shipping weight, 5 oz.
6L7050 $2.65
Three-Blade Pearl Handle Penknife With Leather Purse. 3-inch handle. Nickel silver bolsters. Large blade, 1⅞ in. long; small blade, 1⅝ in. long; nail file, 1½ in. long. Shipping weight, 4 ounces.
6L7059 $1.58
Physicians' Knife. Stag pattern handle, 3⅝ inches long; nickel silver cap and bolster. Large blade, 3⅛ inches long; small blade, 2¼ inches long. Shipping weight, 4 ounces.
6L7052 65c
Skinning Knife. Has strong curved blade 5 in. long. Correctly shaped for skinning. Knife is 8½ in. long over all, with hilt to insure firm grip. Handle of leather with brass and fiber trimmings. Leather sheath. Shpg. wt., 14 oz.
6L7127¼ $2.38
Hunting Knife. 5-inch tempered blade. Length over all, 8½ inches. Handle of leather, brass and fiber trimmings. Finger hilt. Durable leather sheath. Shipping weight, 14 ounces.
6L7126¼ $2.35
Sheffield Pattern Bowie Hunting Knife. 5½-in. blade. Stag pattern handle. Nickel silver guard. Leather sheath. Shipping weight, 15 ounces.
6L7135¼ $1.85
SEARS, ROEBUCK AND CO.
2799

Hair Cutting and Shaving Supplies

Waldorf Roller Bearing Clipper.

Has a fine fluted curved bottom plate. No rough edges. Cuts hair 1/8 inch long. Three-coil tempered wire spring. Teeth carefully beveled and finely finished. Roller bearing. A very easy operating clipper. Nickel plated and polished. Shipping weight, 13 ounces.
6L6708 .... $2.20

Extra spring for Waldorf Hair Clipper. Shipping weight, 1 ounce.
6L6709 .... 9c

The Clippers on this page to be used for human hair only. For Dog and Horse Clippers turn to our Harness pages.

Neck Shaver Clipper.

Cuts hair almost as close as a razor; also for cutting beards. The blades are made of tool steel, accurately machined and ground. The teeth are finely finished and beveled with just the right taper. Nickel plated throughout. A high quality clipper. Shipping weight, 9 ounces.
6L6716 .... $1.90

Extra spring for Neck Shaver Clipper. Shipping weight, 2 ounces.
6L6717 .... 9c

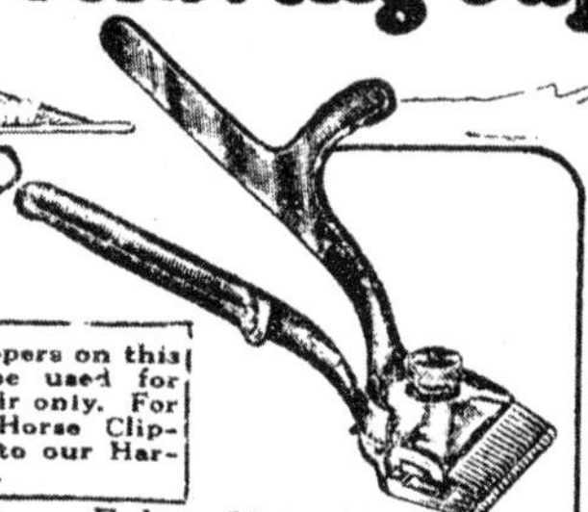

Fulton Hair Clipper.

This clipper is full size, well made of good steel, properly tempered, finely finished and nickel plated. Cuts hair 1/8 inch long. Although low in price, this clipper will give excellent satisfaction. For private use we strongly recommend it. Shipping weight, 12 ounces.
6L6712 .... 98c

Extra spring for the Fulton Hair Clipper. Shipping weight, 1 ounce.
6L6713 .... 9c

Browne & Sharpe's Barbers' Clippers, Bressant Pattern.

Shipping weight, each, 15 ounces.
6L6720—No. 0, cuts 1/16 inch $3.48
6L6722—No. 000 Shaver, cuts extra close $3.60
6L6723—No. 00 Shaver, cuts 1/32 inch $3.55
6L6724—Spring for Bressant pattern clippers. Shipping weight, 2 ounces 7c
6L6725—Spring for B. & S. improved pattern. Shipping weight, 2 ounces 4c

Safetee Shaving Stick.

Made of fine ingredients, with a cocoa butter center. Lathers very quickly. The cocoa butter center gives a pleasing odor and a soothing effect. Shipping weight, 5 ounces.
6L6646 25c

Colgate's Shaving Stick.

In nickel plated box. Shipping weight, 5 ounces.
6L6643 32c

Extra refills for above.
6L6647 23c

Williams' Shaving Stick.

In nickel plated box. No waste; always clean and ready for use. Shipping weight, 5 oz.
6L6637 30c

Extra reloads for above.
6L6640 21c

Williams' Quick and Easy Shaving Powder.

Quicker than stick or mug and brush. Sprinkle a little powder on a wet brush and apply to the face. It forms a rich lather. Shipping weight, 5 ounces.
6L6636 30c

Wilbert Face Creams.

In 16-ounce glass jars. Shipping weight, 2 pounds.
6L6628—Camphor Cream 62c
6L6629—Menthol Cream 75c
6L6630—Vanishing Cream ... 55c
6L6631—Rolling Massage Cream, 60c

Lemon Vanishing Cream.

A delightful vanishing cream, lemon odor. In 8-oz. jars. Shpg. wt., 12 oz.
6L6632
Jar 45c

Williams' and Colgate's Barbers' Shaving Soap.

Well known Standard Barbers' Bar Shaving Soaps. Also very popular with the private user.
6L6639—Williams' Barbers' Bar Soap. Shipping weight, 1¼ lbs. Per pound (six cakes) 50c
6L6638—Shipping weight, 12 ounces. Per ½ pound (three cakes) 27c
6L6642—Colgate's Barbers' Bar Soap. (Eight cakes.) Shipping weight, 1¼ pounds. Per pound 59c

Kolax the "Speed Shave."

A shaving cream requiring no soap, brush or mug. Simply dampen face with either hot or cold water and apply Kolax. No rubbing required, as beard is softened very quickly. Will not irritate or inflame and is refreshing and soothing. Shpg. wts.: ½-pound jar, 12 ounces; 1-pound jar, 1½ pounds.
6L6648
½-lb. jar 39c
6L6649
1-lb. jar 70c

Williams' Shaving Cream.

Nicely perfumed. A small quantity on a wet brush will work up a good lather. In large size tubes. Shpg. wt., 5 oz.
6L6635
Per tube, 30c

Mennen's Shaving Cream.

Nicely perfumed. In large size tubes. Shipping weight, 8 ounces.
6L6644
Per tube, 40c

Johnson's Shaving Cream.

High grade, nicely perfumed. In large tubes. Shipping weight, 5 ounces.
6L6645
Per tube, 24c

Palmolive Shaving Cream.

A well known cream of pleasant odor. Lathers very quickly. Shipping weight, 5 oz.
6L6623
27c

Cocoa Butter Cream.

Most delightful for after shaving. Cool and soothing. In large tubes. Shipping weight, 6 ounces.
6L6608
Per tube 25c

In 8-ounce jars. Shipping weight, 12 ounces.
6L6609—Per jar. 62c

6L6608 6L6609

For full line of Toilet Preparations see pages 488 to 490.

Razor Handles.

Black Hard Rubber Oval Razor Handle. Shipping weight, 1 ounce.
6L6663—Complete with rivets 15c
White Celluloid Handle with artistic design.
6L6664—Complete with rivets 33c

Williams' Tonsorial Shaving Soap.

Put up especially for barbers' use. Five cakes to the pound, of a size and shape economical for the barbers' stand. Barbers have endorsed this "no waste" cake as the best for shop work. Shipping weight, 1¼ pounds.
6L6619—Per pound package 50c
6L6614—Per 10-pound box. (Shpg. wt., 11 lbs.) $4.95

Williams' Shaving Cream.

In 1-lb. jars. Especially put up for barbers' use. Shipping weight, 2 pounds.
6L6615 .... 60c

For Men's Wigs see page 129.

Razor Strop Holder.

Nickel Plated Razor Strop Holder with swivel. Shipping weight, 4 ounces.
6L6683
Each 8c
Dozen 80c

Leather Chair Strap.

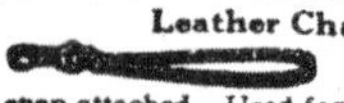

Leather chair strap with nickel plated snap attached. Used for attaching razor strop to barbers' chair or any furniture without marring same. Does away with the use of hooks and swivels. Shipping weight, 2 ounces.
6L6684 .... 12c

Razor Strop Dressing.

A compound possessing just the right amount of hardness to put an edge on a razor. Comes in cakes 2 inches long, ¾ inch wide. Shipping weight, 1 ounce.
6L6727—Per cake 9c

Razor Strop Softening Compound.

Contains no grit or emery. Softens the leather of any strop which has become hard and dry. Also makes a good filler for canvas. Shipping weight, 2 ounces.
6L6726—Small jar 12c

Barbers' Tweezers.

Tweezers with blackhead remover at other end. Shipping weight, 2 ounces.
6L6666 .... 15c

Shaving Mug.

Semi-Porcelain Mug. Nicely decorated. Gilt rimmed top and heavy base flange. About 3¾ inches high. Shipping weight, 1¾ pounds.
6L6739 ..... 30c

Aluminum Mug.

Cast aluminum handle strongly riveted to cup. Satin finish body, neatly engraved. Size, 3¼x3¾ inches. Shipping weight, 8 ounces.
6L6736 .... 85c
Engraving initials on aluminum cups, extra, per letter, 5 cents. Be sure to state letters wanted.

Barbers' Shaving Mug.

Glass mug. With fluted sides. Size, 3½x3½ inches. Shipping weight, 1 pound.
6L6731 .... 20c

Barbers' Razor Box.

Barbers' wood box, covered with black fiber artificial leather, with snap lock and key. Holds 9 razors. Shpg. wt., 10 ounces.
6L6668—$1.15

Barbers' Razor Pockets.

Made of artificial leather, flannelette lined. Shpg. wt., 6 oz.
6L6670 — Holds three razors 50c
6L6650 — Holds six razors. Shipping weight, 7 ounces .. 55c

Razor Purse.

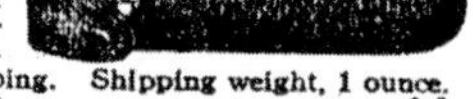

Made of good quality leather. Prevents razor from damage by dropping. Shipping weight, 1 ounce.
6L6658 .... 14c

Steel Unbreakable Mirror.

Nickel plated and highly polished on both sides. Can be hung anywhere.
6L6695—Size, 3x4 inches. Shipping weight, 3 ounces 10c

Circular Extension Mirror.

Shipping wt., 3½ pounds.

A French bevel glass, 8 inches in diameter set in solid black burnished nickel plated frame. Can be attached to the wall and extended to 24 inches, being also adjustable to any angle desired. An unusually handy, convenient article for general household use, but especially suitable for attaching near window where plenty of light is desired.
6L6675 .... $2.67

Adjustable Shaving Mirror.

Shaving Mirror, mounted on nickel plated brass frame. Fine 6-inch French bevelglass. Can be set on table with glass at any angle, or used as a hand mirror. Shipping weight, 2 lbs.
6L6677 ... 90c

"Shavezy" Adjustable Razor Guard.

Reversible and adjustable to any razor from ⅛ inch to ¾ inch wide. Makes a practical safety razor. Nickel plated. Shpg. wt., 2 oz.
6L6729 .... 20c

Styptic Pencil.

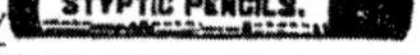

A necessary article for barbers. Heals small cuts quickly and stops flow of blood. Shpg. wt., each, 1 oz.; per dozen, 7 oz.
6L6634
Each 4c; 3 sticks for 10c; dozen 37c

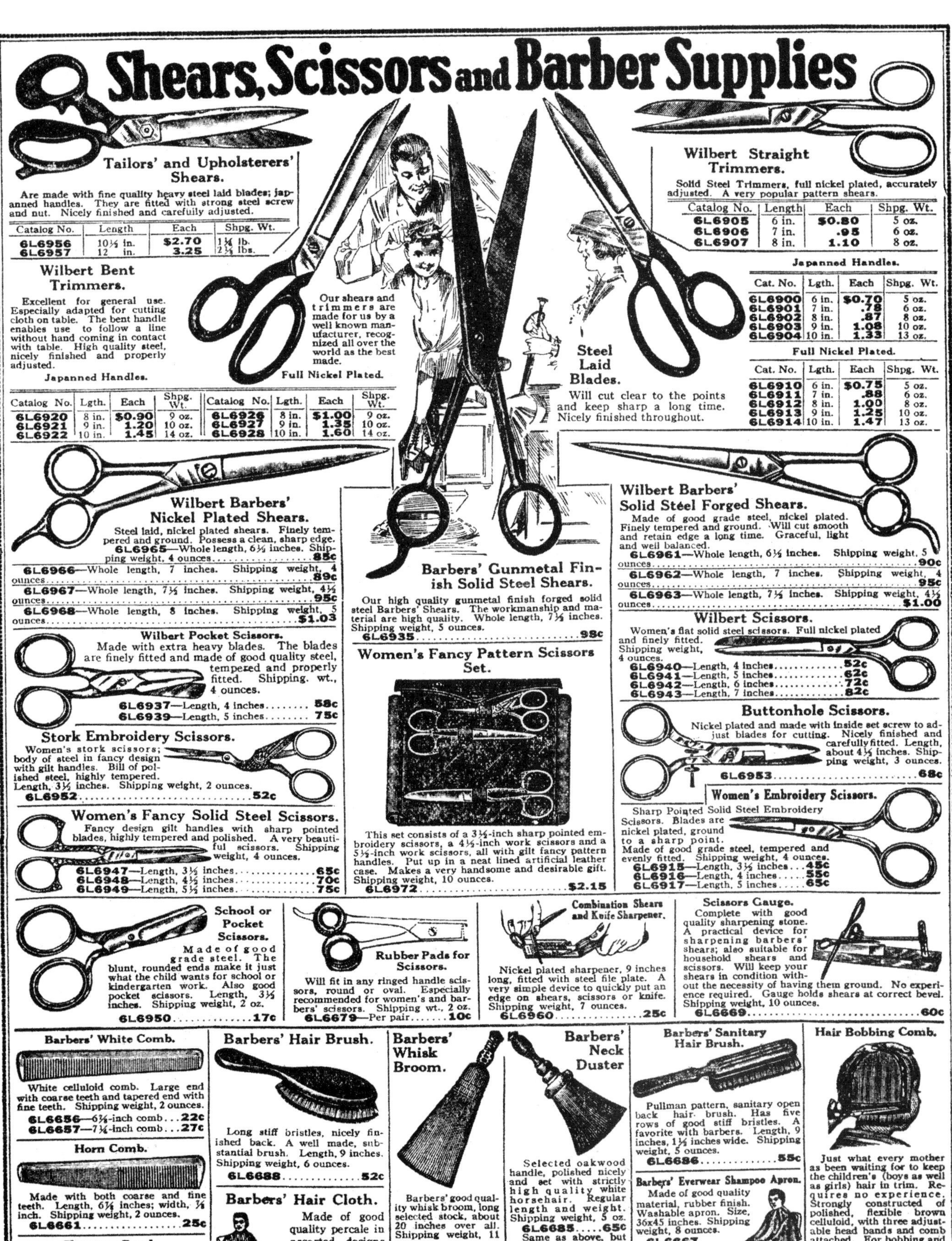

Shears, Scissors and Barber Supplies

Tailors' and Upholsterers' Shears.

Are made with fine quality heavy steel laid blades; japanned handles. They are fitted with strong steel screw and nut. Nicely finished and carefully adjusted.

Catalog No.	Length	Each	Shpg. Wt.
6L6956	10½ in.	$2.70	1¾ lb.
6L6957	12 in.	3.25	2½ lbs.

Wilbert Bent Trimmers.

Excellent for general use. Especially adapted for cutting cloth on table. The bent handle enables use to follow a line without hand coming in contact with table. High quality steel, nicely finished and properly adjusted.

Japanned Handles. / Full Nickel Plated.

Catalog No.	Lgth.	Each	Shpg. Wt.	Catalog No.	Lgth.	Each	Shpg. Wt.
6L6920	8 in.	$0.90	9 oz.	6L6926	8 in.	$1.00	9 oz.
6L6921	9 in.	1.20	10 oz.	6L6927	9 in.	1.35	10 oz.
6L6922	10 in.	1.45	14 oz.	6L6928	10 in.	1.60	14 oz.

Our shears and trimmers are made for us by a well known manufacturer, recognized all over the world as the best made.

Steel Laid Blades.

Will cut clear to the points and keep sharp a long time. Nicely finished throughout.

Wilbert Straight Trimmers.

Solid Steel Trimmers, full nickel plated, accurately adjusted. A very popular pattern shears.

Catalog No.	Length	Each	Shpg. Wt.
6L6905	6 in.	$0.80	5 oz.
6L6906	7 in.	.95	6 oz.
6L6907	8 in.	1.10	8 oz.

Japanned Handles.

Cat. No.	Lgth.	Each	Shpg. Wt.
6L6900	6 in.	$0.70	5 oz.
6L6901	7 in.	.78	6 oz.
6L6902	8 in.	.87	8 oz.
6L6903	9 in.	1.08	10 oz.
6L6904	10 in.	1.33	13 oz.

Full Nickel Plated.

Cat. No.	Lgth.	Each	Shpg. Wt.
6L6910	6 in.	$0.75	5 oz.
6L6911	7 in.	.88	6 oz.
6L6912	8 in.	1.00	8 oz.
6L6913	9 in.	1.25	10 oz.
6L6914	10 in.	1.47	13 oz.

Wilbert Barbers' Nickel Plated Shears.

Steel laid, nickel plated shears. Finely tempered and ground. Possess a clean, sharp edge.
6L6965—Whole length, 6½ inches. Shipping weight, 4 ounces 85c
6L6966—Whole length, 7 inches. Shipping weight, 4 ounces 89c
6L6967—Whole length, 7½ inches. Shipping weight, 4½ ounces 95c
6L6968—Whole length, 8 inches. Shipping weight, 5 ounces $1.03

Barbers' Gunmetal Finish Solid Steel Shears.

Our high quality gunmetal finish forged solid steel Barbers' Shears. The workmanship and material are high quality. Whole length, 7½ inches. Shipping weight, 5 ounces.
6L6935 .... 98c

Wilbert Barbers' Solid Steel Forged Shears.

Made of good grade steel, nickel plated. Finely tempered and ground. Will cut smooth and retain edge a long time. Graceful, light and well balanced.
6L6961—Whole length, 6½ inches. Shipping weight, 5 ounces 90c
6L6962—Whole length, 7 inches. Shipping weight, 4 ounces 95c
6L6963—Whole length, 7½ inches. Shipping weight, 4½ ounces $1.00

Wilbert Pocket Scissors.

Made with extra heavy blades. The blades are finely fitted and made of good quality steel, tempered and properly fitted. Shipping. wt., 4 ounces.
6L6937—Length, 4 inches 58c
6L6939—Length, 5 inches 75c

Stork Embroidery Scissors.

Women's stork scissors; body of steel in fancy design with gilt handles. Bill of polished steel, highly tempered. Length, 3½ inches. Shipping weight, 2 ounces.
6L6952 .... 52c

Women's Fancy Solid Steel Scissors.

Fancy design gilt handles with sharp pointed blades, highly tempered and polished. A very beautiful scissors. Shipping weight, 4 ounces.
6L6947—Length, 3½ inches 65c
6L6948—Length, 4½ inches 70c
6L6949—Length, 5½ inches 75c

Women's Fancy Pattern Scissors Set.

This set consists of a 3½-inch sharp pointed embroidery scissors, a 4½-inch work scissors and a 5½-inch work scissors, all with gilt fancy pattern handles. Put up in a neat lined artificial leather case. Makes a very handsome and desirable gift. Shipping weight, 10 ounces.
6L6972 .... $2.15

Wilbert Scissors.

Women's flat solid steel scissors. Full nickel plated and finely fitted. Shipping weight, 4 ounces.
6L6940—Length, 4 inches 52c
6L6941—Length, 5 inches 62c
6L6942—Length, 6 inches 72c
6L6943—Length, 7 inches 82c

Buttonhole Scissors.

Nickel plated and made with inside set screw to adjust blades for cutting. Nicely finished and carefully fitted. Length, about 4½ inches. Shipping weight, 3 ounces.
6L6953 .... 68c

Women's Embroidery Scissors.

Sharp Pointed Solid Steel Embroidery Scissors. Blades are nickel plated, ground to a sharp point. Made of good grade steel, tempered and evenly fitted. Shipping weight, 4 ounces.
6L6915—Length, 3½ inches 45c
6L6916—Length, 4 inches 55c
6L6917—Length, 5 inches 65c

School or Pocket Scissors.

Made of good grade steel. The blunt, rounded ends make it just what the child wants for school or kindergarten work. Also good pocket scissors. Length, 3½ inches. Shipping weight, 2 oz.
6L6950 .... 17c

Rubber Pads for Scissors.

Will fit in any ringed handle scissors, round or oval. Especially recommended for women's and barbers' scissors. Shipping wt., 2 oz.
6L6679—Per pair 10c

Combination Shears and Knife Sharpener.

Nickel plated sharpener, 9 inches long, fitted with steel file plate. A very simple device to quickly put an edge on shears, scissors or knife. Shipping weight, 7 ounces.
6L6960 .... 25c

Scissors Gauge.

Complete with good quality sharpening stone. A practical device for sharpening barbers' shears; also suitable for household shears and scissors. Will keep your shears in condition without the necessity of having them ground. No experience required. Gauge holds shears at correct bevel. Shipping weight, 10 ounces.
6L6669 .... 60c

Barbers' White Comb.

White celluloid comb. Large end with coarse teeth and tapered end with fine teeth. Shipping weight, 2 ounces.
6L6656—6¾-inch comb ... 22c
6L6657—7¾-inch comb ... 27c

Horn Comb.

Made with both coarse and fine teeth. Length, 6¾ inches; width, ⅞ inch. Shipping weight, 2 ounces.
6L6661 .... 25c

Shampoo Brush.

Made of hardwood with good quality stiff bristles. 6¼ inches long by 2⅛ inches wide. Shipping weight, 12 ounces.
6L6694 .... 50c

Barbers' Hair Brush.

Long stiff bristles, nicely finished back. A well made, substantial brush. Length, 9 inches. Shipping weight, 6 ounces.
6L6688 .... 52c

Barbers' Hair Cloth.

Made of good quality percale in assorted designs and colors. Represents a value that any barber will immediately appreciate. Size, 48x48 inches. Shipping weight, 8 ounces.
6L6665 .... 55c

Barbers' Whisk Broom.

Barbers' good quality whisk broom, long selected stock, about 20 inches over all. Shipping weight, 11 ounces.
6L6691 ... 45c

Barbers' Neck Duster

Selected oakwood handle, polished nicely and set with strictly high quality white horsehair. Regular length and weight. Shipping weight, 5 oz.
6L6685 ..... 65c
Same as above, but with black hair.
6L6689 ..... 45c

Barbers' Everwear Bib.

To be used when cutting hair. Made of good quality material, rubber finish; adjustable to fit any size neck. Shipping weight, 2 ounces.
6L6659
Each $0.32
Per dozen 3.85

Barbers' Sanitary Hair Brush.

Pullman pattern, sanitary open back hair brush. Has five rows of good stiff bristles. A favorite with barbers. Length, 9 inches, 1½ inches wide. Shipping weight, 5 ounces.
6L6686 .... 55c

Barbers' Everwear Shampoo Apron.

Made of good quality material, rubber finish. Washable apron. Size, 36x45 inches. Shipping weight, 8 ounces.
6L6667
$1.18

Massage Apron.

As above, but 24x30 inches. Shpg. wt., 5 oz.
6L6662
55c

Hair Bobbing Comb.

Just what every mother as been waiting for to keep the children's (boys as well as girls) hair in trim. Requires no experience. Strongly constructed of polished, flexible brown celluloid, with three adjustable head bands and comb attached. For bobbing and trimming bangs; will fit any child's head. Eliminates the trouble of taking frightened children into public barber shops and trying to keep them quiet. Shipping weight, 6 ounces.
6L6655 .... 68c

BUTCHER SUPPLIES

For Trade and Household Use

The Genuine Diamond "A" Butcher Knives.

These knives are made specially for the butcher trade, where a knife is required to be not only of good material, but also of just the right shape and proper balance to do good work. Every Diamond "A" Knife is hand forged and ground by experienced workmen and honed to a keen edge. They are inspected before leaving the factory and are guaranteed to give satisfaction and to be of a very good quality.

Lion Brand Sticking Knife. Fitted with beechwood handle. A very keen, sharp knife. Solidly riveted.

Catalog No.	Length Blade	Each	Shipping Weight
6L7413	6 in.	25c	6 oz.
6L7415	6½ in.	35c	7 oz.

Lion Brand Butcher Knife. Fitted with beechwood handle. A very strongly built knife.

Catalog No.	Length Blade	Each	Shipping Weight
6L7408	6 in.	25c	7 oz.
6L7410	7 in.	30c	8 oz.
6L7411	8 in.	35c	9 oz.

Lion Brand Butcher Knife. Fitted with ebony handle. Ground to a very keen, sharp edge.

Catalog No.	Length Blade	Each	Shipping Weight
6L7400	6 in.	45c	7 oz.
6L7402	7 in.	50c	8 oz.
6L7404	8 in.	70c	9 oz.
6L7406	10 in.	95c	11 oz.

Lion Brand Skinning Knife. Beechwood handle. Well tempered and possesses excellent cutting qualities.

Catalog No.	Length Blade	Each	Shipping Weight
6L7418	6 in.	30c	7 oz.
6L7419	7 in.	35c	8 oz.

Butcher Knife. Represents a strong, well finished, low priced knife.

Catalog No.	Length Blade	Each	Shipping Weight
6L7422	6 inches	20c	7 ounces
6L7424	8 inches	30c	9 ounces
6L7426	10 inches	50c	11 ounces

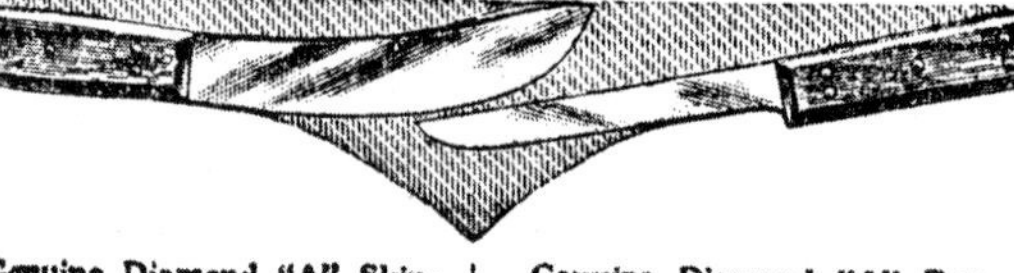

Genuine Diamond "A" Skinning Knife. Beechwood handle.

6L7462—Length of blade, 6 inches. Shipping wt., 7 oz...35c

6L7464—Length of blade, 7 inches. Shipping wt., 8 oz..45c

Genuine Diamond "A" Boning Knife. Fitted with beechwood handle. Solidly riveted. Length of blade, 6 inches. Shipping weight, 7 ounces.

6L7472 ....30c

Genuine Diamond "A" Sticking Knife. Fitted with beechwood handle. Very strongly made.

Catalog No.	Length Blade	Each	Shpg. Wt.
6L7466	6 in.	30c	7 oz.
6L7468	7 in.	40c	8 oz.

Genuine Diamond "A" Steak Knife. Scimitar blade. Fitted with beechwood handle. Solidly riveted.

Catalog No.	Length Blade, Inches	Each	Shpg. Wt., Lbs.
6L7476	10	$0.70	¾
6L7478	12	.95	⅞
6L7480	14	1.25	1⅛

Genuine Diamond "A" Market Cleaver. Forged by hand and fitted with walnut handle. Solidly riveted. Very strongly made.

Catalog No.	Length Blade, Inches	Each	Shpg. Wt., Lbs.
6L7560	7	$3.10	3
6L7562	8	3.40	3½
6L7564	9	3.70	4
6L7566	10	4.10	5

Genuine Diamond "A" Butcher Knife. Fitted with beechwood handle. Very strongly made.

Catalog No.	Length Blade, Inches	Each	Shpg. Wt., Oz.
6L7450	6	$0.30	7
6L7452	7	.40	8
6L7454	8	.50	9
6L7456	9	.60	10
6L7458	10	.80	11
6L7460	12	1.00	12

Crackajack Slicer or general household knife. Cocobolo handle, three large head brass rivets, single bolster, clip point, quality steel blade. Length of blade, 9¼ inches. Shipping weight, 8 oz.

6L7420 ....60c

Utility Slicer. A carver, slicer, butcher and bread knife combined in one. Blade of high grade tempered steel, ground thinner than a regular butcher knife, which gives it a very sharp edge. Handle of cocobolo, with three brass rivets. Length of blade, 9 inches. Shipping weight, 9 ounces.

6L7441 ....55c

Butcher Knife. Ground and finished by hand. Cocobolo handle.

Catalog No.	Length Blade	Each	Shipping Weight
6L7469	6 inches	50c	7 ounces
6L7470	7 inches	60c	8 ounces
6L7471	8 inches	70c	9 ounces

Butcher Knife. Fine quality steel blade, three large brass rivets, cocobolo handle; a high grade, serviceable knife.

Catalog No.	Length Blade	Each	Shipping Weight
6L7436	7 inches	48c	8 ounces
6L7438	8 inches	58	9 ounces

Butcher Knife. Solid cocobolo handle, fancy bolster, steel blade. No joints or places for grease to collect.

Catalog No.	Length Blade	Each	Shipping Weight
6L7430	6 inches	50c	7 ounces
6L7431	7 inches	60c	8 ounces
6L7432	8 inches	70c	9 ounces

Our 55-Cent Kitchen Saw.

Flat steel frame. Good steel blade. Beechwood handle with two screws. Length, 16 inches. Weight, 13 ounces. Shipping weight, 1 pound.

6L7554 ....55c

Hoffman Magnetic Butcher Steel.

Natural color handle with steel swivel ring. Nickel plated diamond shape guard.

Catalog No.	Length of Blade	Shpg. Wt.	Each
6L7587	12 in.	1½ lbs.	$0.90
6L7588	14 in.	1⅝ lbs.	1.10
6L7589	16 in.	1¾ lbs.	1.35

Lion Brand Butcher Steel.

Oriental rosewood handle with steel swivel ring. Nickel plated guard to prevent rolling. Strong construction throughout.

No.	Length	Shpg. Wt.	Each
6L7580	10 in.	1⅛ lbs.	$0.80
6L7581	12 in.	1¼ lbs.	1.00
6L7582	14 in.	1½ lbs.	1.25

Boss Kitchen Saw.

Flat steel frame, beechwood handle and attachment for tightening blade. Blade, 14 inches; width, ½ in. Weight, 13 oz. Shpg. wt., 1 lb.

6L7556—With one blade.............65c

6L7558—Extra Blade for above. Shipping weight, 3 ounces..........................10c

Fulton Butcher Saw.

Flat polished steel frame, ¾ inch wide, ¼ inch thick; beechwood handle, varnished, fastened with three large brass plated screws; ¾-inch blade.

Catalog No.	Length Blade	Shpg. Wt.	Each
6L7502	18 in.	2⅛ lbs.	$1.15
6L7504	20 in.	2¼ lbs.	1.20
6L7506	22 in.	2½ lbs.	1.25
6L7508	24 in.	2¾ lbs.	1.30

6L7510—Handle and Screws for above saw. Shipping weight, 11 ounces . 35c

Kelso Butchers' Saw.

Steel frame, 1 inch wide, ¼ inch thick, fitted with polished beechwood handle, fastened by three large nickel plated screws. All come fitted with ¾-inch spring steel butcher saw blades.

Catalog No.	Length Blade	Shpg. Wt.	Each
6L7514	18 in.	2⅞ lbs.	$1.45
6L7516	20 in.	3 lbs.	1.55
6L7518	22 in.	3⅛ lbs.	1.65
6L7520	24 in.	3½ lbs.	1.75
6L7522	26 in.	3¾ lbs.	1.85

6L7524—Handle and Screws for above saw. Shipping weight, 11 ounces....35c

Beef Splitting Saw.

Steel frame, wood handles. Weight, about 6 pounds. 32-inch blade, inches wide. Shipping weight, 7½ pounds.

6L7550¼ $3.25

Extra Blade, 32x2 inches. Weight, 10 ounces. Shipping weight, 12 ounces.

6L7552 ....48c

Double End Scraper for Hogs.

Heavy gauge galvanized steel; hardwood handle. Shipping weight, 9 ounces.

6L7585 ....9c

Household Saw.

Strong, well made, intended for household use. Flat 20-inch frame, bright finish; beechwood handle, varnished; three brass plated screws; a finely tempered blade. Made of durable material and is guaranteed to give satisfaction. Size, 20-inch only. Weight, 1¾ pounds. Shipping weight, 2¼ pounds.

6L7500 ....$1.15

Butchers' Cleaver.

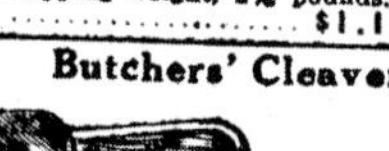

Hand forged from a good grade of steel. Properly tempered and very sharp. Natural beechwood handle.

Catalog No.	Length	Shipping Weight	Each
6L7568	7 inches	1⅞ lbs.	$2.40
6L7570	8 inches	2 lbs.	2.75
6L7572	9 inches	2¾ lbs.	3.20
6L7574	10 inches	4 lbs.	3.50

Folding Skinning Gambrel.

All metal, for any animal from a rabbit to a wolf. Spreads legs of animal to a width of 18 inches. Length of arms, when open, 14 in.; length, folded, 7 inches. Shipping weight, 1⅛ lbs.

6L7595 ....95c

Hog Hook.

Made from steel. Diameter, ⁵⁄₁₆ inch. Length of hook, 7¼ inches; length of handle, 9 inches. Weight, 12 ounces. Shipping weight, 1 pound.

6L7586........35c

Butcher Saw Blades.

Butcher Saw Blades. Made of good quality spring steel, finely ground and brightly polished. Filed and set, ready for immediate use. Width, ¾ inch. Shipping weight, each, 5 oz.

Catalog No.	Length	Each	Per Dozen
6L7528	18 inches	12c	$1.35
6L7530	20 inches	13c	1.45
6L7532	22 inches	14c	1.60
6L7534	24 inches	15c	1.70
6L7536	26 inches	16c	1.80

Self Spreading Gambrel.

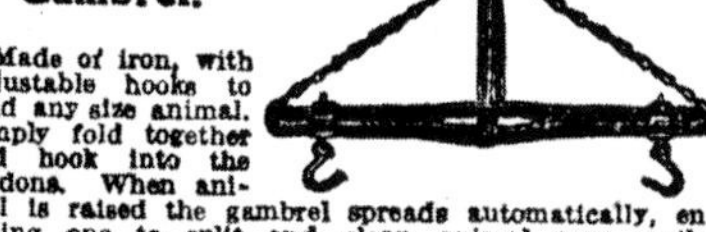

Made of iron, with adjustable hooks to hold any size animal. Simply fold together and hook into the tendons. When animal is raised the gambrel spreads automatically, enabling one to split and clean animal very easily. Height, 17 inches; width, when folded, 18⅛ inches; width, when spread, 38½ inches. Shipping weight, 18 pounds.

6L7592 ....$2.50

Universal Amplifying Transformer

Designed for use with any tube on the market, including the new types. High grade construction throughout. Specifications: Primary resistance, 900 ohms (approximately). Secondary winding, 15,000 turns No. 40 B. & S. gauge enameled wire. Secondary resistance, about 5,000 ohms (approximately). All insulation is varnished cambric and silk. Binding posts are put on Bakelite top. Cores are made of .007 silicon steel. Frame is heavy brass, stamped; brushed finish. Shipping weight, 2½ pounds.

6L9732—Universal Amplifying Transformer **$3.04**
6L9733—Coil and Core only. Shipping weight, 2 pounds **$2.28**

Special Amplifying Transformer

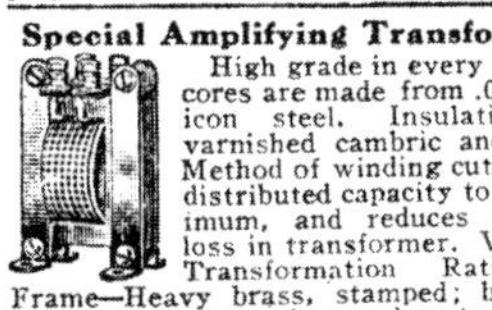

High grade in every detail; cores are made from .007 silicon steel. Insulation is varnished cambric and silk. Method of winding cuts down distributed capacity to a minimum, and reduces energy loss in transformer. Voltage Transformation Ratio—3.1. Frame—Heavy brass, stamped; brushed finish. Primary and secondary terminals mounted on Bakelite on top of transformer. Over All Size—3 inches high; 1⅞ inches wide. Shipping weight, 1 pound.

6L9503—Special Amplifying Transformer **$5.10**
6L9504—Coil and Core only. Shipping weight, 2 pounds **$4.33**

Panel Mounting Rheostat

A specially made rheostat for back panel mounting only. Resistance is special non-corrosive alloy, mounted around Bakelite insulation, ¼ in. thick, 2¾ inches in diameter, and will not creep. The bolt is 1¼ in. long, which permits mounting on a panel of any thickness from ⅛ to ¼ in. Knob is molded Bakelite, 1⅜ inches in diameter. Pointer and bearing collar are heavily nickel plated. Contact to the resistance is made by laminated lever which is remarkably smooth running. Gives unusually close filament temperature adjustment on either 4 or 6-volt vacuum tubes, and is suitable for filament control of 5-watt transmitter tubes. Resistance, 2 ohms; capacity, 3 amperes continually. Shipping weight, 8 ounces.

6L9422—Panel Mounting Rheostat, complete with bushings and screws. **$1.34**

Junior Panel Rheostat

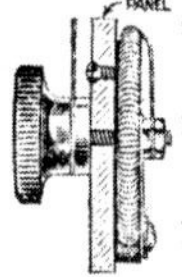

Designed for compact panel mounting. Resistance unit mounted on Bakelite disc 2 inches in diameter; 4 ohms resistance; carrying capacity, 1½ amperes. An "off" position is provided, obviating the necessity of a filament switch. Furnished complete with model 1⅝-inch Bakelite knob, shaft, nickel plated pointer and supporting screw.

6L9371—Shpg. wt., 5 oz **77c**

Porcelain Base Rheostat, New Model

Used to regulate battery current. Used with small motors, miniature lamps, detector batteries, etc. Resistance, 10 ohms, capacity, 3 amperes continually; 4 inches in diameter, 1⁵⁄₁₆ inch thick. Shipping weight, 1½ pounds.

6L9277 .... **69c**

Radio For a more complete line of high grade apparatus. Write for our Radio Catalog 459GCL Sent postpaid on request

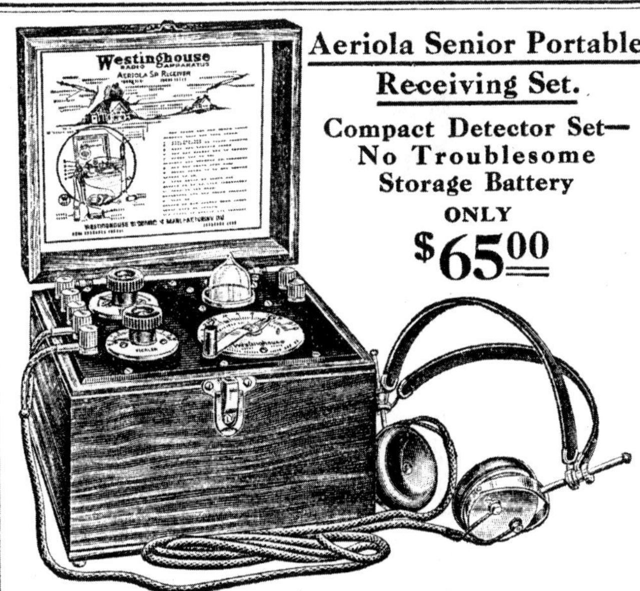

Aeriola Senior Portable Receiving Set.

Compact Detector Set— No Troublesome Storage Battery

ONLY **$65.00**

An ideal radio outfit for campers, tourists and persons situated in the rural districts. The Aeriola will be found especially useful to the farmer for the daily reception of market and weather reports. These messages are sent out by the U. S. Government stations on wave lengths of 400 and 485 meters, and are received like regular telephone conversations. It is not necessary for the operator to know the telegraph codes. Thus, this instrument proves of value to the great farming centers of the United States which are served by local radiophone broadcasting stations.

This set is particularly adapted to the rural districts, as the filament circuit is operated by an ordinary dry cell, thus doing away with the expense and inconvenience of a storage battery.

This set is also recommended for hunters, scouts and campers. It may be carried by one man without overburdening him, whether he is on a long hike or not. The complete outfit with the necessary small "B" battery and dry cell for its operation, the insulators and wire for the antenna, as well as the wire used for the ground connection, may be placed in a haversack. The total weight is less than 15 pounds. Shipping weight, 22 pounds.

6L9315—Aeriola Senior Portable Receiving Set **$65.00**

Detector and Amplifier Tubes.

These tubes, as listed below, are the standard four-prong tubes for use with 6-volt battery. The detector tubes require about 22½ volts on the plate, and the amplifier, 45 volts or more, for maximum signal sensitivity. Shipping weight, 1 pound.

Audiotron Tubes.

6L9650—Audiotron Detector Tube C-300. Each **$4.60**
6L9651—Audiotron Amplifier Tube C-301. Each **6.15**

Radiotron Tubes.

6L9438—Radiotron Detector U. V. 200. Each **$4.60**
6L9540—Radiotron Amplifier U. V. 201. Each **6.15**

Standard Double Slide Tuning Coil

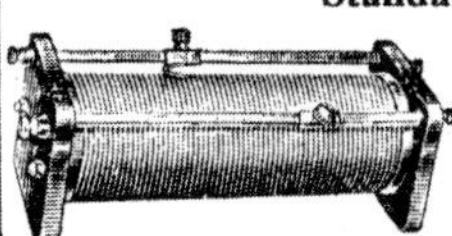

With suitable aerial this coil will respond to wave lengths up to 1,000 meters. Coil is bare copper wire wound, with two sliding contacts, which are nickel plated and polished. Ends are of molded insulation. Slider rods and binding posts are polished brass and lacquered. Length, 8¾ inches. Shipping weight, 3 pounds.

6L9246—Standard Double Slide Tuning Coil **$2.98**

Audiotron Bakelite Molded Variometer

Molded in two pieces from genuine Bakelite, primary winding on the inside, insuring an absolutely rigid stator winding. Rotor is molded from Bakelite and permits of minimum clearance. Brass supports permit of ready table mounting. Bearings are of brass, fitted with a spring, insuring perfect contact. Over all height, 5¾ inches; length, except the protruding end of the shaft, is 5¼ inches. The total width of the variometer across the stator is 3 inches. Not to be compared with the small wooden types now on the market.

6L9781—Audiotron Variometer. Shipping weight, 5 pounds **$5.65**
6L9782—Audiotron Variometer, with 6L9646 Dial and Knob. Shipping weight, 5 pounds **$6.20**

Audiotron Variocoupler

This variocoupler has coupling range of 180 degrees instead of 90 degrees (as is the case with many other variocouplers). Primary winding is green silk covered wire on a tube, 4 inches in diameter by 2¾ inches deep. Ten taps are provided. The secondary rotor molded Bakelite. This coupler will tune over a range of 150 to 500 meters with secondary variometer, and with secondary condenser of .001 MFD, will tune to 700 meters.

6L9784 — Audiotron Variocoupler. Shipping weight, 5 pounds **$4.15**
6L9785 — Audiotron Variocoupler, with 6L9646 Bakelite Dial and Knob. Shipping weight, 5 pounds **$4.72**

Variocoupler

Variocoupler winding is made over formica tubing 3½ in. in diameter, wire having raised points for taps. Bearing Shafts with spacing shoulders turned from the shaft itself are used, assuring good contacts between rotor windings and bearing standards without "pig-tailing." Variocoupler may be mounted on back of panel or directly on table, as desired. Variocoupler is completely assembled. Primary wire is No. 20 gauge and rotor wire is No. 22 gauge. Shpg. wt., 2 lbs.

6L9685—Variocoupler **$3.42**

Variometer

Wood parts are of thoroughly kiln dried stock, accurately turned and carefully finished. Stator and rotor windings are secured with special cement, which is colorless, extremely adhesive and has no capacity effect. Bearing parts are of brass, 1 inch in width, sunken flush with wood forms, allowing variometers to be mounted flat on back of panel without spacers, and also insuring rigidity and permanency of spacing between rotor and stator windings. Stator blocks measure 4¾x4¾ inches. Shpg. wt., 3 lbs.

6L9684—Variometer **$3.38**

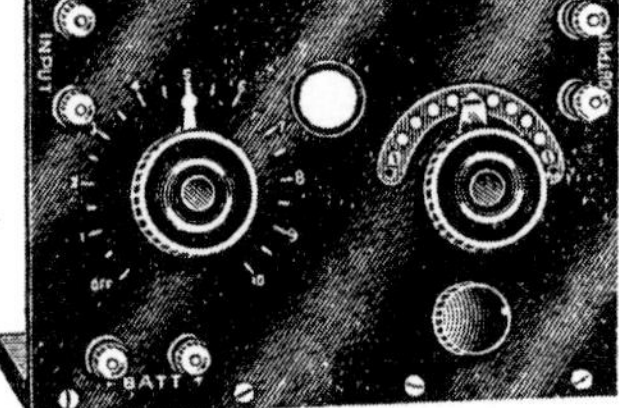

Molded Bakelite Panel Is 7¼ Inches Long, 5 Inches High and ³⁄₁₆ Inch Thick

This panel is of genuine molded Bakelite, 5x7¼x³⁄₁₆ inch. Surface is highly polished glossy black, and lettering and scales are molded in and filled with white enamel. Filament current controlled by Panel Rheostat, back mounted, and provided with an off position. Audiotron Potentiometer, connected across the storage battery, provides the close adjustment of plate potential necessary for sensitive detector action. The grid leak is variable; grid condenser is back mounted and is the correct capacity for the new gas content detector tubes. Molded Bakelite Socket is used and supports the tube vertically, insuring maximum filament life. Two or more panels can be mounted in a line to form detector-amplifier combination. Terminals at the back of the panel are provided with flexible leads for the B battery connection. Shipping weight, 5 pounds.

6L9652—Audiotron Bakelite Control Panel. **$6.42**

Audiotron Amplifier Units

Audiotron Amplifier

This amplifier unit is designed to work in connection with Audiotron Detector Panel and is the same size as the detector unit. Panel is molded Bakelite and is fitted with rheostat control, tube socket and binding posts. All wiring is of the bus bar type. We furnish this unit both with and without transformer, as many experimenters already have transformers which they could use in this set. This unit is especially designed for Cunningham Audiotron Amplifier Tube, Type C-301, and is also suitable for any standard four-prong tube on the market. Shpg. wt., 8 lbs.

6L9570—Audiotron Amplifier Panel, without transformer **$4.53**
6L9571—Audiotron Amplifier Panel, complete with Transformer 6L9503. Shipping weight, 8 pounds **$9.64**

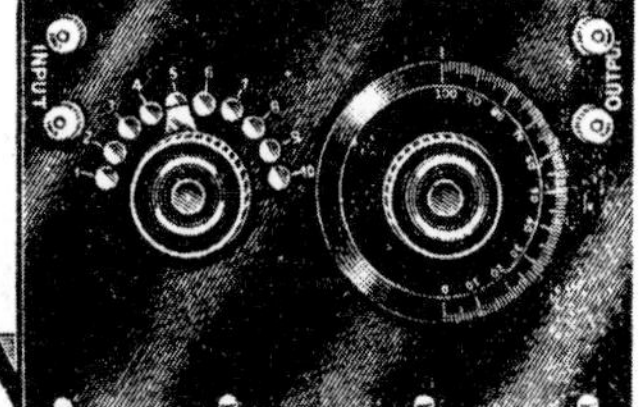

Audiotron Variocoupler Panel

Variocoupler 6L9785 is also supplied mounted on a Bakelite panel, 5x7¼x³⁄₁₆ inch. This panel is finished in glossy black and all lettering is white filled. Input and output binding posts are at opposite ends of the panel; special switch lever with bakelite knob to match the dial knob is provided for varying the primary inductance. The wiring to the panel is of the approved bus bar type and all connections from the primary taps to the panel contacts are soldered. Hardwood base, 7¼x4x⅜ inch, is finished in black. Shipping weight, 6 pounds.

6L9786
Audiotron Variocoupler Panel **$10.12**

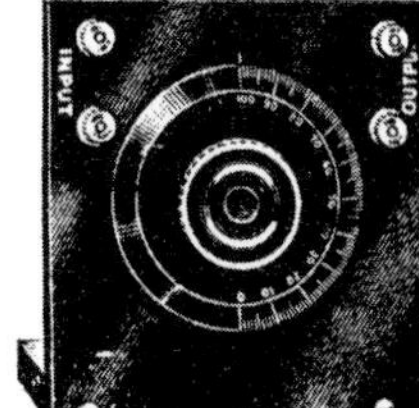

Audiotron Variometer Panel

Variometer 6L9782 is supplied mounted on a Bakelite molded panel, 5x5x³⁄₁₆ inch, with hardwood base, finished in black, 5x3x⅜ inch. The panel is the same height as the Audiotron Variocoupler and 6L9652 Detector Panel and the arrangement of the binding posts permits ready interconnection of panels to form any desired receiving combination. Shipping weight, 6 pounds.

6L9783 — Audiotron Variometer Panel with Dial **$8.22**

Small Well Finished Jacks.

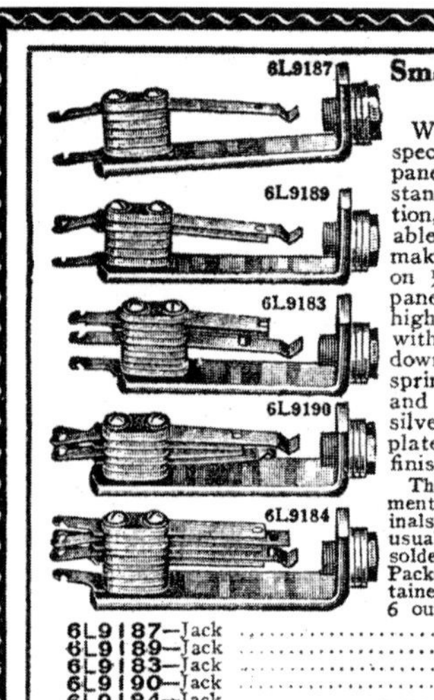

Well finished jacks, specially designed for panel work. Are of standardized construction, to be interchangeable with other standard makes. Can be mounted on ⅛, 3/16 or ¼-inch panels. Insulation is of high grade and will withstand 110-volt breakdown test. Contact springs are nickel silver and contact points pure silver. Frame, nickel plated and highly buffed finish.

The "spread" arrangement of the spring terminals allows twice the usual amount of space for soldering to the wiring. Packed in individual containers. Shipping weight, 6 ounces.

6L9187—Jack 44c
6L9189—Jack 47c
6L9183—Jack 67c
6L9190—Jack 54c
6L9184—Jack 81c

High Grade Head Sets
$4.40
2,000 Ohms

A high grade set of receivers offered at a very reasonable price. These phones compare favorably with the mica diaphragm and amplifying head sets listed at much higher prices and can be used to good advantage in loud speakers. Resistance is of double magnet type and metal diaphragm is carefully adjusted in relation to the magnets. Workmanship is thoroughly tested after each step in construction, to insure most sensitive results. The phones are fitted with Army-Navy style headband, covered with heavy webbing. Also a 6-foot connecting cord with round tip terminals. Shipping weight, 1½ pounds.

6L9216—Double Set. 2,000 ohms $4.40
6L9217—Double Set. 3,000 ohms $5.30

Radio Plug.

A very practical and efficient plug. Will fit all standard jacks; finished in hard rubber insulation and best non-conductor bushings. Requires no tools for connecting. The cord tips are brought through the handle and inserted in the screw adjustment. A small loop is provided for anchoring the cords. Shipping wt., 8 oz.

6L9194—Radio Plug 89c

Cord Tip Radio Plug.

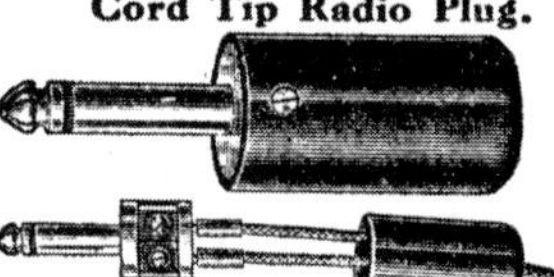

With this type of plug the cord tips are attached directly to plug. No longer necessary to remove tips, form loops and fasten under head of screws. This plug is handsomely finished, polished hard rubber sleeve. Can be used with any standard make of jack. Cord tips slip into the base of plug and are held firmly in place by simply tightening the two screws, as shown in the illustration of plug with sleeve removed. Shipping weight, 8 ounces.

6L9182—Plug 82c

Fixed Condenser.

Made of hard rubber composition with nickeled binding posts. Used to shunt across the receivers. Terminals are fitted with special cord tip clamps. Capacity, .0016 M.F.D. Size, over all, 2⅝x1½ inches. Shipping weight, 5 ounces.

6L9264—Fixed Condenser.... 65c

Mica Grid Condenser and Grid Leak Combined.

Made from formica sheet, grained finish and best India mica as insulation. Unit is held together by two screws which also act as binding posts and are fitted with molded knobs. Condenser capacity is .00025 M.F.D. Adjustable grid leak of about 2 megohms, made of gray fiber. Very convenient for panel mounting. Size, 2½x1x½ inches. Shipping weight, 4 ounces.

6L9658—Mica Grid Condenser and Grid Leak Combined.. 52c

Tubular Fixed Condenser.

Can be used with great success in the receiving circuit. A high grade fixed condenser mounted on a nickel plated tube. Base and top of hard rubber composition. Capacity is .003 M.F.D. Shipping weight, 10 ounces.

6L9400—Tubular Fixed Condenser 90c

The Leader Binding Posts.

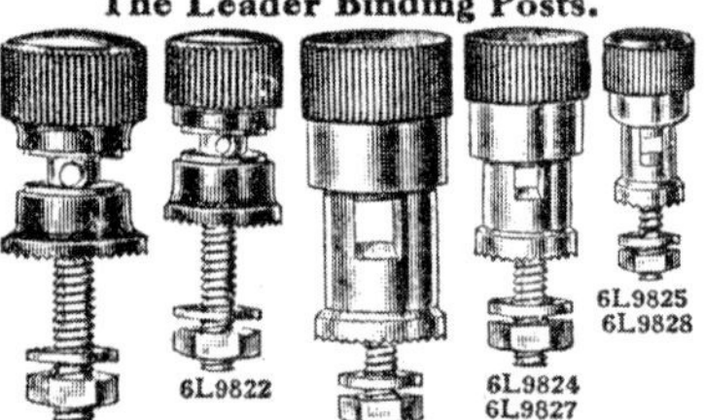

The latest development in binding post manufacture; heads will not come off and yet will allow plenty of room for wires. All sizes will take standard telephone cord tip. Special knurled base makes excellent contact and prevents turning. Built for long service. Positive contact for fine wires or solid terminals is assured by lock grip. Furnished complete as shown. Illustrations show two-thirds size.

6L9821—Black Molded Insulated Post. Shipping weight, 2 ounces 20c
6L9822—Black Molded Insulated Post. Shipping weight, 2 ounces 19c
6L9823—Polished Nickel Binding Post. Shipping weight, 3 ounces 35c
6L9824—Polished Nickel Binding Post. Shipping weight, 2 ounces 16c
6L9825—Polished Nickel Binding Post. Shipping weight, 1 ounce 11c
6L9826—Polished Brass Binding Post. Shipping weight, 3 ounces 33c
6L9827—Polished Brass Binding Post. Shipping weight, 2 ounces 19c
6L9828—Polished Brass Binding Post. Shipping weight, 1 ounce 9c

Switch Contact Points—Polished Nickel.

Solid brass, nickel plated; two nuts, ¼-in. diameter, ¼-in. head, ⅝-in. length of thread. Shipping wt., per doz., 12 oz.; 100, 5 lbs.	Size, ¼-in. diameter, 5/16-in. head; otherwise same as 6L9472. Shipping wt., per doz., 10 oz.; 100, 5 pounds.	Solid brass head, nickel plated, size, 5/16x5/16 in., otherwise same as 6L9251. Shipping wt., per doz., 10 ounces; 100, 5 lbs.	Solid brass head, ¼x5/16 in., nickel plated, otherwise same as 6L9251. Shipping wt., per dozen, 10 oz.; 100, 5 lbs.	Solid brass head, ¼x¼ inch, nickel plated. Fitted with brass screw and soldering lug. Shipping wt., per dozen 10 oz.; 100, 5 lbs.
6L9472	6L9473	6L9304	6L9271	6L9251
Doz. $0.31	Doz.. $0.30	Doz.. $0.29	Doz.. $0.30	Doz.... $0.31
100 2.48	100 2.40	100 2.35	100 2.40	100 2.48

Aerial Lightning Arrester.

Designed especially for protecting radio receiving apparatus from atmospheric lightning disturbances. Besides efficiently protecting the receiving apparatus, it practically eliminates any fire hazard of an unprotected aerial lead. Porcelain weatherproof body with metal discharge plates, separated by an air gap of 1/50 inch. May be suspended either by heavy set screw connections or attached to any suitable support by means of a steel band and screws, which are supplied. Shipping weight, 1½ pounds.

6L9417—Lightning Arrester........ $1.48

Knob.

Used on many instruments shown in this catalog. Fine for detectors, condensers, small switches, etc. Has 8/32 bushing. Shipping wt., each, 2 ounces; dozen, 1 pound.

6L9469—Knob. Each 5c
Per dozen 18c

Unit Switch Lever.

Switch lever used on our Progressive units. Very popular for all kinds of panel sets. Lever of brass, nickel plated and polished, fitted with coil tension spring and two nuts. Knob of black composition and knurled. 1½-inch radius. Bearing is ⅜ inch from panel to lever. Shipping weight, 5 ounces.

6L9628 .... 33c

Popular Switch Lever.

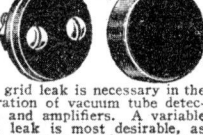

Improved spring type. Bearing collar fits directly against panel. Fitted with 1-inch lever, coil spring, terminal lug and nuts. Shpg. wt., each, 5 oz.; ½ dozen, 1¼ pounds.

6L9465—Each $0.31
Per half dozen 1.70

New Style Government Knobs.

Late design used extensively on government and high grade experimental apparatus. Large and small sizes shown. These knobs are used with our Bakelite dials. Top is concave. Each knob has a brass bushing, 10/32 thread, and two holes for stay pins.

6L9303
Each .... (Shipping wt., 3 oz.) 9c
Per dozen .... (Shipping wt., 1 lb.) 98c
6L9470
Each .... (Shipping wt., 4 oz.) $0.11
Per dozen .... (Shipping wt., 1½ lbs.) 1.24

Army-Navy Polished Nickel Plated Panel Switch Lever.

Army-Navy Panel Switch Lever. Knob of genuine formica and has knurled edge. Switch blade is of spring brass, 2-inch radius. Switch bolt extends through heavy brass bushing, fitted with large nut used to mount the lever on panel. Furnished complete with two nuts. All metal parts nickel plated. Shpg. wt., 8 oz.

6L9463—Army-Navy Panel Switch Lever. Each $0.54
Per half dozen 3.10

Audiotron Variable Grid Leak.

A grid leak is necessary in the operation of vacuum tube detectors and amplifiers. A variable grid leak is most desirable, as the necessary resistance may vary from ½ to 5 megohms. The base of this Grid Leak is molded from bakelite and a pencil mark between the contact studs provides the variable resistance. Metal cap is finished in black celluloid enamel and the studs are provided with washers and nuts for panel mounting. 6L9660 Grid Condenser will fit directly on screws. Shipping weight, 2 ounces.

6L9659—Audiotron Variable Grid Leak 30c

Marconi Knob.

Marconi Knob, for large panels, switchboards, variometers, transformers, etc. No bushing. Drilled for 5/16-inch rod at bottom and has 7/16-inch hole in top. Highly polished.

6L9460
Marconi Knob.
Each .... (Shipping wt., each, 4 oz.) $0.18
Per dozen .... (Shipping wt., 3 lbs.) 2.03

Binding Posts.

Polished Nickel Plated Binding Posts.

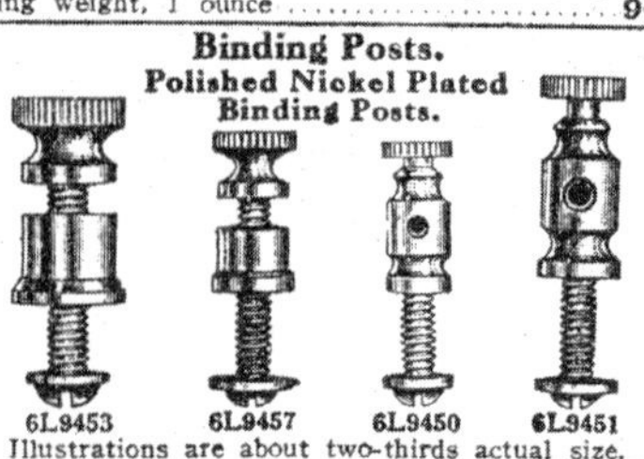

Illustrations are about two-thirds actual size. Made from brass stock, nickel plated and buffed; high grade in every respect. Each post fitted with brass screw and washer. Two styles, two sizes each style.

6L9453—Each $0.13
Per dozen 1.45
Shipping weight, each, 3 ounces; dozen, 1½ lbs.
6L9457—Each $0.09
Per dozen 1.04
Shipping wt., each, 3 oz.; dozen, 1½ lbs.
6L9450—Each 8c
Per dozen 87c
Shipping wt., each, 1 oz.; doz., ½ lb.
6L9451—Each $0.10
Per dozen 1.09
Shipping wt., each, 2 oz.; doz., 1 lb.

Audiotron Fixed Grid Condenser.

Stamped from copper sheet and insulated with paraffin paper. Entire unit is encased and impregnated and the terminals are spaced so as to mount on the back of the panel directly on the 6L9659 Grid Leak. Capacity is .00025 M.F.D., the correct value for the type C-300 Detector tube. Shipping weight, 1 ounce.

6L9660—Fixed Grid Condenser 15c

Molded Navy Key Knob.

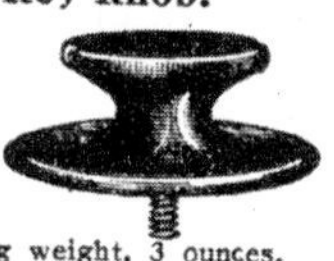

Molded Navy Key Knob. The latest and most approved type. Adds speed and accuracy to operating. Construction is flameproof. Used on our navy type key, and on a great many of the best keys. Has 8/32-inch stem, which will fit most all keys. Shipping weight, 3 ounces.

6L9381—Navy Key Knob 21c

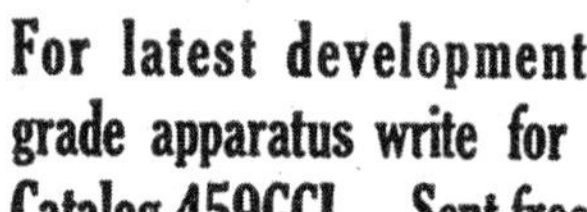

Universal Helix Clip.

Used for making connections on the Helix and Oscillation Transformer. Nickel plated. Shipping weight, 1 oz.

6L9409—Universal Helix Clip... 7c

Bradleystat Filament Control.

May be used where the finest filament adjustment is necessary. Resistance controlled by varying contact pressure of graphite discs and screw adjustment permits a critical current regulation. Very strong construction and will handle a 1½ to 2½-ampere current. Porcelain case encloses graphite discs, shaft fitted with Bakelite knob. Shipping weight, 1 pound.
6L9275—Bradleystat **$1.55**

Laboratory Variable Air Condensers.

Plates made from No. 22-gauge hard steel aluminum. Spacers are turned from brass stock and gauged for accuracy. All insulation cut from grade M Formica sheet, 3/16 inch thick. Panel mounting type. We list them in two sizes, suitable for all classes of radio, telegraph and telephone work. Can be used on ⅛, 3/16 or ¼-inch panel. Panel mounting, 43 plate; 21 rotary, 22 stationary. Capacity, .001 MFD. Shipping weight, 2¼ pounds.
6L9298 ....**$2.75**
Panel mounting, 21 plate; 10 rotary, 11 stationary. Capacity, .0005 MFD. Shipping weight, 1¾ pounds.
6L9299 ....**$2.15**

Crystal Detector.

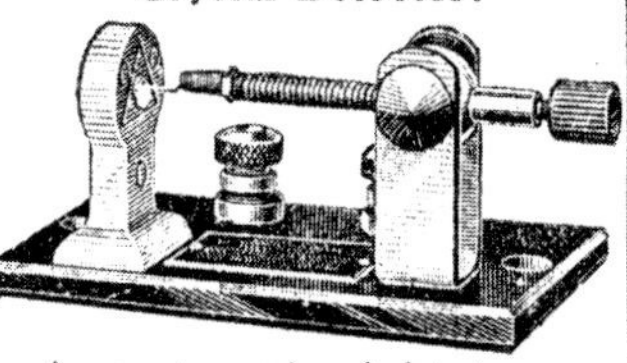

A new type mineral detector, constructed with a view to long service and most sensitive results. All parts are nickel plated and detector is fitted with tested galena crystal neatly mounted in soft metal on new style mineral cup. The detector is equipped with a cat whisker and ball socket rod adjustment, which provides for a wide range over the mineral and insures most sensitive contact. Shpg. wt., 1 lb.
6L9261 — Mineral Detector, complete with Crystal..............**$1.60**

Arlington Tested Minerals.

Each Arlington Tested Mineral has been individually tested, and unless it has shown extraordinary results it is discarded. They must bring in distant stations loud and clear. Individually wrapped and packed and sealed in a box. Shipping weight, per crystal, 2 ounces.
6L9285 — Arlington Tested Galena. Per crystal..........................**21c**
6L9286 — Arlington Tested Silicon. Per crystal..........................**21c**

"NAA" Receiving Transformer.

Bakelite Coil Head. Capacity, 2,500 Meters.

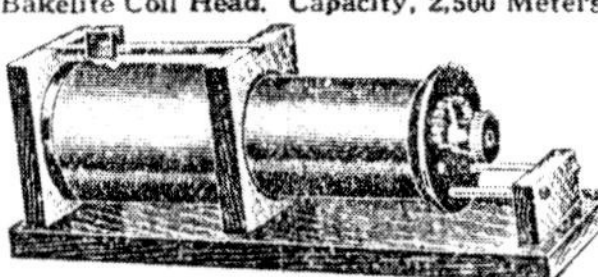

An efficient, well made, long wave length tuner at a low price. With it you can tune in with the big wireless stations using wave lengths of 2,500 meters and more. Coil head is of turned Bakelite. All binding posts mounted on primary coil end. The windings of both coils are of green silk-cotton covered wire. The slider is very selective, as it will make contact on a single turn. The secondary inductance is varied by means of a 10-point switch. All metal parts are of brass, nickel plated. Size of base, 18⅛x6 inches. Shipping weight, 14 pounds.
6L9333 ....**$6.55**

Radio Magnavox

Only $38.75

Radio Magnavox is a new type of sound magnifier, made on the electro-dynamic principle. It will reproduce radiophone speech or music to such a degree that the sound may be heard a considerable distance away from the receiving set. Equipped with a 14-inch horn. Should be used with 6-volt storage battery. Metal parts are finished in black enamel. Shipping weight, 20 pounds.
6L9779—Radio Magnavox. Each................**$38.75**

For latest developments in high grade apparatus write for our Radio Catalog 459GCL. Sent postpaid on request.

Antenna and Ground Outfit.

A complete outfit with all the necessary parts for installing aerial and ground connections to the home receiving set.

The set consists of 125 feet of seven strand copper aerial cable, with insulators, which will make an aerial capable of receiving all of the amateur radiophone wave lengths. Also 50 feet of No. 14-gauge weatherproof wire, with cleats, for the lead-in and ground wires. An approved lightning arrester and ground clamp are also furnished with the set. Shipping weight, 15 pounds.
6L9435—Antenna and Ground Outfit....................................**$3.18**

High Grade Indicating Dials.

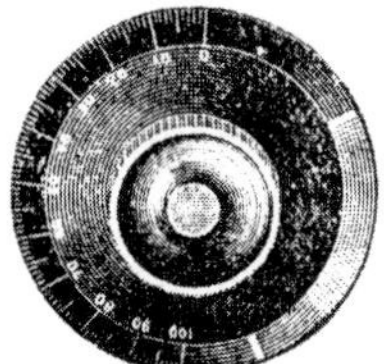

This 3-inch beveled edge dial is molded from genuine black Bakelite and will not warp or discolor. The surface is highly polished and will add to the appearance of any panel. The edge of each dial is ground true. The knob is molded Bakelite. The bushing is drilled for 3/16-inch shaft and the set screw passes through both knob and bushing. The construction insures an absolutely true running dial. Shipping weight, 4 ounces.
6L9646—3-Inch Bakelite Dial with knob and bushing, 3/16-inch **57c**
6L9649—Same as 6L9646, with ¼-inch bushings..**57c**

Radio Storage Batteries at Reduced Prices.

A storage battery developed for radio service under the exacting specifications of the United States Government during the recent war. Extra thick plates and the highest grade insulation are the basis of an unusually high ampere-hour capacity and long life. Assembled in neat, clean wood case. Highest grade rubber jars tested under 20,000 volts give absolute protection against leakage. Lead headed knurled binding posts make quick connections easy.

Shipped direct from depots in CHICAGO, ILL., PHILADELPHIA, PENNA., ATLANTA, GA., MINNEAPOLIS, MINN., or SAN FRANCISCO, CALIF.
6L9519⅓—Storage Battery, 6-volt, 40-ampere hour. Shpg. wt., 35 lbs....**$9.08**
6L9520⅓—Storage Battery, 6-volt, 60-ampere hour. Shpg. wt., 40 lbs..**$10.90**
6L9521⅓—Storage Battery, 6-volt, 90-ampere hour. Shpg. wt., 50 lbs..**$13.68**
6L9522⅓—Storage Battery, 6-volt, 120-ampere hour. Shpg. wt., 55 lbs..**$16.76**
6L9523⅓—Storage Battery, 6-volt, 150-ampere hour. Shpg. wt., 65 lbs..**$19.82**

Meteor Rectifier.

Used to charge any 6-volt storage battery. Rectifier is automatic. Screw the plug into any electric light socket on a 110-volt 60-cycle circuit, connect the leads to the battery and the charging will immediately commence. There is no danger of overcharging, as the rate of charging automatically tapers. The battery connections may be reversed without danger to battery. Will fully charge any 6-volt storage battery at a cost of from 4 to 10 cents. Furnished complete with book of directions. Shipping weight, 10 pounds.
6L9683—Meteor Rectifier.....**$13.45**

Beginners' Wireless Practice Set.

This set provides an excellent method of quickly learning the code. The key may also be used with a spark coil and the buzzer for testing detectors. Set consists of a wireless key and buzzer mounted on a polished wood base. Buzzer is nickel plated and reproduces the high piched sound of the wireless sation. Complete with one dry cell, 3 feet insulated wire, code chart and diagram of connections and instructions. Size of base, 7x4½ inches. Shipping weight, 4⅛ pounds.
6L9200 ....**$2.40**

Meteor "B" Batteries.

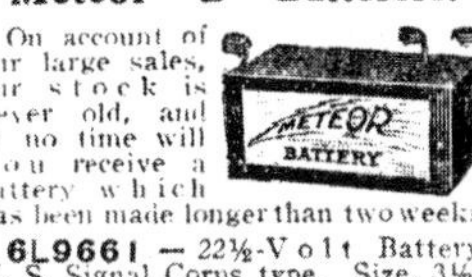

On account of our large sales, our stock is never old, and at no time will you receive a battery which has been made longer than two weeks.
6L9661 — 22½-Volt Battery, U. S. Signal Corps type. Size, 3½x2⅛x2½ inches. Shpg. wt., 4 lbs **93c**
6L9662—22½-Volt Battery with 18-volt tap. U. S. Navy type. Recommended for use with Audiotron and Radiotron Tubes. Size, 6½x4x3 in. Shpg. wt., 9 lbs **$1.59**
6L9601—45-Volt Battery developed to meet the requirements of the Audiotron and Radiotron Amplifier Tubes. Size, 6⅝x2¾x7¾ in. Shipping wt., 12 lbs **$3.20**

Molded Bakelite Socket.

Designed for all standard four-prong based vacuum tubes. The socket is molded from black Bakelite, and the bayonet lock is reinforced with metal insert. Screws are provided for panel mounting. Base is 2¼x2¾ inches and the height is 1⅛ inches. Shpg. wt., 8 oz.
6L9542—Bakelite Molded Tube Socket, panel or table mounting..........**77c**

Porcelain Socket.

Made in one piece of porcelain, the same material that is used in the base of the vacuum tube to insulate the four prongs. Designed to prevent short circuiting the high voltage B Battery current across the filament contacts. Will fit any standard four-prong tube base. Shpg. wt., 8 oz.
6L9533—Porcelain Socket **43c**

G-R Tube Socket.

A high grade tube socket fitted with positive contact springs. Base is of molded Bakelite, and the tube and terminal screws are of brass with polished nickel finish. This socket is adapted to any of the standard four-prong tubes. Transmitting tubes can be used by simply changing two screws. Shipping weight, 8 ounces.
6L9655—G-R Tube Socket ..**$1.37**

Porcelain Strain Insulator.

Made of brown glazed porcelain, heavily ribbed. Has smooth end holes. Excellent for small aerials. Over all size, 2¾ inches long by 1¼ inches in diameter.
6L9273
Each. Shpg. wt., 5 oz.........**7c**
Per dozen. Shpg. wt., 3 lbs...**76c**

Electrose Insulators.

6L9337 6L9338 6L9339-6L9340

Standard with the U. S. Army and Navy. Natural color, brown with galvanized eyebolts.
6L9337—Ball Insulator. Length, 3½ inches. Shipping wt., 8 oz.........**26c**
6L9338—Strain Insulator. Length, 5 inches. Shipping wt., 12 oz. **35c**
6L9339—Strain Insulator. Length, 10 inches. Shpg. wt., 1 lb. **60c**
6L9340—Strain Insulator. Length, 16 inches. Shipping wt., 2 lbs ..**$1.55**

Aerial Connector Block.

The weakest point in most stations is where the wires from the aerial join the lead-in. This brass aerial connector block does away with soldered joints and loose connections. Size, 2 inches high by 1¾ inches wide by ¾ inch thick. Shipping weight, 6 ounces.
6L9272—Aerial Connector Block **21c**

Triple Braid Weatherproof Wire.

Covered with three saturated braids. For outside work only. Made to meet weather conditions better than rubber covered.
6L9975¼—No. 14-gauge, 25 ft **$0.23**
Per 100 ft...(Shpg. wt., 3 lbs.).. **.74**
Per 1,000 ft...(Shpg. wt., 28 lbs.).. **6.75**

Litzendraht Wire.

Consists of twenty strands of No. 38 special Belden enameled wire, twisted and covered with a double serving of white silk. Shipping weight, 8 ounces per 100 feet.
6L9942—Per 100 feet **65c**

Annunciator or Bell Wire.

Annunciator Wire No. 18, in ½ or 1 pound coils (about 150 feet to the pound).
6L9900—Per pound **47c**

Office Wire.

Office Wire No. 18, in 1-pound coils (about 95 feet to the pound). Same as annunciator wire, but with heavier and thicker insulation.
6L9902—Per pound **54c**

Seven-Strand No. 22 Copper Aerial Cable.

Composed of seven strands of No. 22 B. & S. gauge copper wire. Used extensively by experimenters, amateurs, etc. Put up in standard coils as listed below. Not sold any other way.
6L9979¼—Stranded Copper Aerial Cable.
Per 100 ft. (Shpg. wt., 1½ lbs.) **$0.64**
Per 500 ft. (Shpg. wt., 8 lbs.) **3.00**
Per 1,000 ft. (Shpg. wt., 15 lbs.) **5.65**

Bare Copper Aerial Wire.

Put up in standard coils as listed below. Not sold any other way.
6L9989¼—No. 14-Gauge Bare Copper Wire.
Per 100 ft. (Shpg. wt., 1½ lbs.) **$0.39**
Per 500 ft. (Shpg. wt., 9 lbs.) **1.65**
Per 1,000 ft. (Shpg. wt., 20 lbs.) **3.20**
6L9990¼—No. 12-Gauge Bare Copper Wire.
Per 100 ft. (Shpg. wt. 2½ lbs.) **$0.57**
Per 500 ft. (Shpg. wt., 12 lbs.) **2.55**
Per 1,000 ft. (Shpg. wt., 30 lbs.) **4.95**

Electric Accessories for Home Comfort

Standby Percolator

Useful and decorative. Made of aluminum, highly polished. Specially designed heat storage plates keep liquids hot for a long time after current is turned off. Easily cleaned and sanitary. Height, 9¾ inches. Capacity, 6 cups. Cool ebony finish handle. Six feet of cord with plug. Operates on usual city current of 105 to 115 volts. Shipping weight, 4 pounds.

6L8004—Standby Percolator.................. **$5.75**

For other Coffee Percolators see pages 826 and 829.

Combination Electric Stove and Flat Top Toaster.

One of the daintiest and most serviceable table stoves made, all metal construction, beautifully nickel plated. Stove is 9 inches long, 5 inches wide and 3 inches high. Complete with cord and universal plug. Operates on usual city circuit of 110 to 115 volts. Shipping weight, 3 pounds.

6L8022.................. **$2.65**

Electric Curling Iron.

A very neat, compact curling iron that heats quickly. Made of steel, nickel plated and polished. 6L8062 for usual city lighting circuit of 105 to 115 volts; 6L8063 for 32-volt private lighting plants. Be sure to select the correct one for your circuit. Shipping weight, 8 oz.

6L8062—105 to 115-volt circuit.................. **$2.75**

6L8063—For use on 32-volt circuit.................. **3.15**

For other Hair Curlers and Wavers see pages 131 and 486.

Electric Stove or Hot Plate.

Heating surface, 5½ inches in diameter. 660-watt size and is equipped with a three-heat switch, giving full, medium or low heat. 5½ inches high, rigidly constructed, nickel plated and polished. Equipped with 6 feet of cord and separable plug. Suitable for usual city circuit of 105 to 115 volts. Shipping weight, 16 pounds.

6L8027¼.............. **$10.67**

Electric Warming Pad.

For cases of illness where heat applications are required. Switch controls heat to 125, 160 or 190 degrees. Will not overheat. Size, 12x15 inches. Covered with tan eiderdown. 10-foot connecting cord with heat control switch and separable attachment plug to fit any lamp socket. Shpg. wt., 2½ lbs.

6L8054—110-volt service circuit type.................. **$8.14**

6L8055—32-volt farm plant type.................. **8.52**

Electric Toaster.

A high grade Toaster which will take any size bread. Toasts quickly and evenly. Low current consumption. Nickel plated finish. Furnished with 6 feet of cord and separable attachment plug. Shipping weight, 3 pounds.

6L8020—Toaster for 105 to 115-volt current....... **$5.31**

6L8021—Toaster for 32-volt lighting plants....... **5.80**

Cords and Plugs.

Fit all electric irons and heating devices using a two-conductor plug of similar construction. Shipping weight of cord, 12 ounces; plug, 6 ounces.

6L8087—Six-foot cord with separable socket plug and "Fitzall" heating device plug........ **82c**

6L8088—"Fitzall" plug with protecting spring only.................. **22c**

Electric Table Stove.

This Stove broils, boils, steams, fries or toasts. Bakes biscuits, muffins and cup cakes. Handsomely finished in polished nickel plate. Three pans are furnished; also broiler rack, cups and cup rack. The pans are 7¾ inches square and ½ inch, 1¼ inches and 1¾ inches deep, respectively. The three-heat switch permits the use of full, medium or low heat. Furnished with 6 feet of cord and separable attachment plug. Shipping weight, 10 pounds.

6L8033—Electric Table Stove for 105 to 115-volt circuit.................. **$11.30**

6L8034—Same as above, but constructed for 32-volt circuit.................. **11.90**

ELECTRIC IRON

A good looking, efficient iron which will be found very satisfactory. The heating part of this iron is absolutely guaranteed to give years of service. Furnished with steel stand, the upper shelf having a groove in which you can place a curling iron for heating. Complete with 6 feet of cord, feed through switch and plug. Suitable for usual city circuit of 105 to 115 volts. Weight of iron, 6 pounds. Shpg. wt., 10 lbs.

6L8037.................. **$3.92**

Junior Electric Iron.

A practical light weight iron, weighing 1½ pounds, suitable for lingerie, laces and all delicate articles. Its compactness and light weight make it especially desirable when traveling. Operates on city circuit of 105 to 115 volts. Shipping weight, 2 pounds.

6L8040 **$3.28**

Iron With Tip Up Stand.

Size, 6 pounds. Ample heat storage capacity for all average work. Blued base. Nickel plated hood. Ebonized handle. Separate heat proof stand and tip up stand on heel of iron. With 6-foot connecting cord and separable attachment plug to fit any lamp socket. Shipping weight, 7 lbs.

6L8045—110-volt service circuit type......... **$5.27**

6L8046—32-volt farm plant type.......... **5.76**

MEDICAL BATTERIES

Medical Batteries are in extensive use for the relief of certain diseases and disorders. Their advocates are enthusiastic over the possibilities for relief resulting from the proper use of these machines. We advise consultation with a physician before placing an order, so that the purchaser may be sure to order the particular machine best suited to the purpose for which it is to be used. We guarantee these appliances to be high grade in every respect.

Triple Cell Medical Battery.

Polished oak case, about 9¾x7¾x8¼ inches. Three dry cells. Faradic coil, 1¼ inches in diameter. Circular carbon rheostat regulates current. Four-point switch permits one cell, two cells or three cells to be used at a time. Wheel rheotome for interrupted current, adjustable for slow or rapid interruptions. Metal parts all nickel plated. One pair conducting cords with tips, pair insulating wooden handles, pair nickel plated metal handles, pair sponge electrodes, one nickel plated foot plate, hair brush and an instruction pamphlet included. Otherwise same as 6L9105. Shipping wt., 20 lbs.

6L9110¼—Complete with accessories and three dry cells, **$13.55**

Double Cell Medical Battery.

Polished oak case, about 8¾x6¼x8 inches, with compartment in cover for accessories and compartment in base for two dry cells. Faradic coil, 1⅛ inches in diameter, with locknut, spring vibrator and adjustable ball attachment for slow vibration. Intensity of current regulated by withdrawing shield. Three-point switch. Pole changer for changing the polarity or direction of the current. Metal parts all nickel plated. One pair conducting cords with tips, pair insulating wooden handles, pair nickel plated metal handles, pair sponge electrodes, nickel plated foot plate, hair brush and instruction pamphlet included. Shipping wt., 13 lbs.

6L9105—Complete with accessories and two dry cells.................. **$9.15**

For extra batteries see 6L8635 on page 813.

Single Cell Medical Battery.

Polished oak case, about 10x6¾x5 inches, opening top and bottom. Lower compartment for dry cell and battery mechanism, upper compartment for accessories. Faradic coil, 1¹⁄₁₆ inch in diameter, with locking device and spring vibrator. Intensity of current regulated from mild to strong by withdrawing shield from core of coil. Metal parts nickel plated. One pair conducting cords with tips, one pair insulating wooden handles, pair nickel plated metal handles, pair sponge electrodes, nickel plated foot plate and an instruction pamphlet included. Shipping weight, 7¾ pounds.

6L9100—Complete with accessories and one dry cell.. **$5.80**

Sweep and Sew the Electric Way

$29.85 NOTE THE EXCEEDINGLY LOW PRICE.

Order an ENERGEX MODEL 5 **Vacuum Cleaner**

It Combines Quality and Price

It is easy to connect attachments to the Energex. You do not have to turn the machine upside down.

THIS new, bigger and better Energex Cleaner embraces the twofold cleaning principle of powerful suction and gentle sweeping of a positive rotating brush operating independent of the motor.

The strong suction extracts all deep down, germ laden dust and grit. The revolving full bristle brush picks up all clinging lint, litter, hair, threads, ravelings, etc., without injuring the most delicate rug or fabric. The horizontal underslung motor permits of the use of the cleaner under furniture, radiators, etc. Its light weight and swivel rear caster make it easy to operate. Mounted on rubber tired wheels throughout, it will not mar or scratch the most highly polished floor. Wheels inside the nozzle prevent breakage and permit of thorough cleaning in corners. Specifications are as follows:

NOZZLE AND FRAME. Made of highly polished cast aluminum, nozzle is 13 inches long.

MOTOR. Motor is 110-volt Universal type (will operate on A. C. or D. C.), suitable for use on any circuit of 105 to 120 volts. Fan cooled and equipped with phosphor bronze bearings. It is practically indestructible and with proper care, cleaning and a few drops of oil at intervals, should last many years. Requires little current to operate. Motor housing is of cast aluminum polished to match frame.

SUCTION FAN. Suction fan is attached directly to motor shaft, five-blade cast aluminum, accurately balanced.

BRUSH. The soft bristle brush is driven by gears enclosed in a metal dustproof housing. These gears, directed to a supporting wheel in the nozzle, operate the spiral bristle brush in a ratio of four to one to the small rubber tired wheel. Brush is readily removed for cleaning.

DUST BAG. Made of heavy specially prepared vacuum cleaner cloth, chemically treated. This bag retains all dirt and grit and does not permit of any dust seepage. Easily disconnected from exhaust by simple bayonet bag ring. Opens at top for easy removal of dirt.

HANDLE AND SWITCH. Strong wooden handle, ebony finish, fitted with special trigger switch and easy to grip ball knob. Furnished with 22 feet black mercerized cord and separable attachment plug. Shipping weight, 16 pounds.

6L9800¼ **$29.85**

6L9801¼—Same as above, except for use on 30 to 32-volt farm lighting plants **$30.65**

We can also ship, direct from factory, cleaners for special voltages. **Be sure to state voltage required.**

6L9803⅓—Special Cleaner **$31.50**

FOR OTHER VACUUM CLEANERS SEE PAGE 824.

ENERGEX MODEL 5

Note the Exceedingly Low Price.

Attachments for Energex Model 5.

These attachments are priced at a very low figure to enable our customers to get the full benefit of the cleaners at the least expense. Their many features, such as the cleaning of mattresses, walls, draperies, upholstery, etc., make them particularly useful. Metal parts are aluminum. Hose is rubber, cotton covered. Other parts are hard fiber. Shpg. wt., 4 lbs.

6L9802—Per set of 5 pieces **$8.70**

VIBRATORS

Large Vibrator.

Vibrator for 104 to 115-volt direct or alternating circuit. Gives a rubbing motion. Switch and speed regulator conveniently located. Constructed for more various uses than the smaller outfits and more applicators furnished, as shown: One facial, one scalp, one button, one ball, one large hard disc, one soft rubber sponge, one soft rubber ball, one hard rubber applicator, one spinal, and one curved applicator; also one jar of massage cream and one oil can. In attractive leatherette case with silk lining. Size of case, about 13¾x10¼x4½ inches, equipped with a lock. Shipping weight, 9 lbs.

6L9127 .... **$24.50**

Same as above, except for 32-volt circuit. Amount of current consumed is .5 ampere.

6L9128 .... **$24.60**

General Purpose Vibrator.

For 104 to 110-volt circuit. Gives heavy vibration. Speed regulator on top of machine. Complete with six applicators, as shown: One button, one ball, one soft rubber scalp, one soft rubber face, one large rubber disc, one soft sponge applicator, one oil can. In imitation leather covered case, with an electric conductor cord. Shpg. wt., 6 lbs.

6L9125 .... **$13.35**

Same as above, except for 32-volt direct circuit. Amount of current consumed is .5 ampere.

6L9126 .... **$14.85**

Excellent Vibrator for Home Use.

Popular priced vibrator, packed in a neat case. Size, about 9¼x6½x4 inches. Especially recommended for home use. You will find it a very satisfactory applicator. The motor and equipment are good quality. Furnished with four applicators, as shown: One facial, one scalp, one button and one ball applicator. Operates on 104 to 115-volt direct or alternating circuit. Shipping weight, 4½ pounds.

6L9121 .... **$9.12**

Low Volt Vibrator, same as 6L9121 and constructed to operate on 32-volt private lighting plant circuit. Amount of current consumed is .5 ampere.

6L9123 .... **$10.76**

Electric Sewing Machine Motor.

This well made Motor can be quickly attached to your sewing machine without screws, bolts or clamps and without harming the machine in any way. The operation of the friction pulley against the balance wheel operates the machine and the speed can be easily regulated to meet your requirements. It produces a way for you to do a great deal more sewing without being worn out from running the machine. This motor is recommended for use in the general household or domestic sewing and any woman who has ever operated a sewing machine by foot power will appreciate the great advantage of having this motor on her machine. It is a motor built to meet the requirements of sewing done at home, but cannot be used in homes that do not have electricity and is not constructed to operate from dry batteries. The motor and control need never be in the way, because they can be removed when the machine is not in use. It operates on city current of 105 to 115 volts and the cost for the electric current is very small. A suitable cord and plug are furnished so they may be conveniently attached to the lamp socket. Shipping weight, 10 pounds.

6L9820—Electric Sewing Machine Motor **$14.95**

SHURLITE Flashlights and Supplies

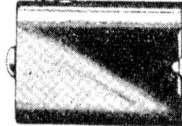

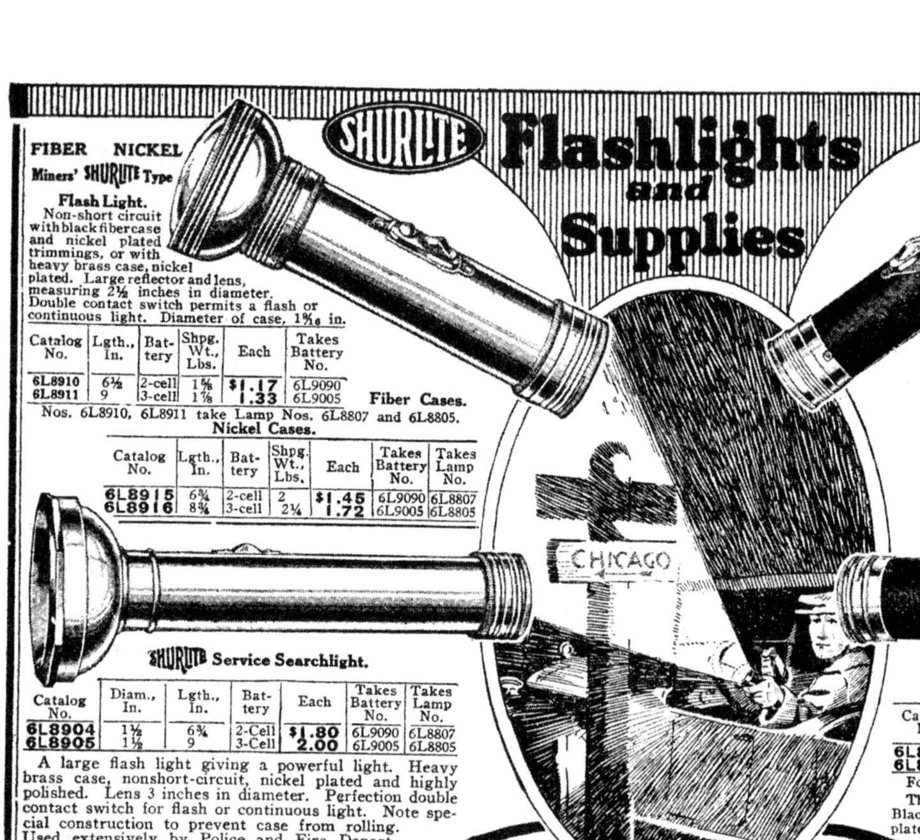

FIBER NICKEL Miners' SHURLITE Type Flash Light.

Non-short circuit with black fiber case and nickel plated trimmings, or with heavy brass case, nickel plated. Large reflector and lens, measuring 2½ inches in diameter. Double contact switch permits a flash or continuous light. Diameter of case, 1$\frac{9}{16}$ in.

Fiber Cases.

Catalog No.	Lgth., In.	Battery	Shpg. Wt., Lbs.	Each	Takes Battery No.
6L8910	6½	2-cell	1⅝	$1.17	6L9090
6L8911	9	3-cell	1⅞	1.33	6L9005

Nos. 6L8910, 6L8911 take Lamp Nos. 6L8807 and 6L8805.

Nickel Cases.

Catalog No.	Lgth., In.	Battery	Shpg. Wt., Lbs.	Each	Takes Battery No.	Takes Lamp No.
6L8915	6¾	2-cell	2	$1.45	6L9090	6L8807
6L8916	8¾	3-cell	2¼	1.72	6L9005	6L8805

FIBER Tubular SHURLITE NICKEL Flash Lights.

Non-short circuit tubular lights, made with black fiber case and nickel plated trimmings, or with heavy brass case, nickel plated and polished. Double contact switch for flash or continuous light. Very popular size. Diameter of cases, 1⅛, 1⅓ and 1½ inches.

Fiber Cases.

Catalog No.	Lgth., In.	Battery	Shpg. Wt., Lbs.	Each	Takes Battery No.
6L8907	5¼	2-cell	1	$0.75	6L9091
6L8908	6¾	2-cell	1⅜	1.02	6L9090
6L8909	9	3-cell	1⅝	1.15	6L9005

6L8907, 6L8908, 6L8909 take Lamp Nos. 6L8806, 6L8807 and 6L8805, respectively.

Nickel Cases.

Catalog No.	Lgth., In.	Battery	Shpg. Wt., Lbs.	Each	Takes Battery No.	Takes Lamp No.
6L8912	5½	2-cell	⅞	$0.95	6L9091	6L8806
6L8913	6¼	2-cell	1⅜	1.14	6L9090	6L8807
6L8914	9	3-cell	1¾	1.40	6L9005	6L8805

For extra batteries and lamps see below.

SHURLITE Service Searchlight.

Catalog No.	Diam., In.	Lgth., In.	Battery	Each	Takes Battery No.	Takes Lamp No.
6L8904	1½	6¾	2-Cell	$1.80	6L9090	6L8807
6L8905	1½	9	3-Cell	2.00	6L9005	6L8805

A large flash light giving a powerful light. Heavy brass case, nonshort-circuit, nickel plated and highly polished. Lens 3 inches in diameter. Perfection double contact switch for flash or continuous light. Note special construction to prevent case from rolling. Used extensively by Police and Fire Departments, sportsmen and railroad men. Two lengths. Shipping weight, 2 pounds and 3 pounds, respectively.

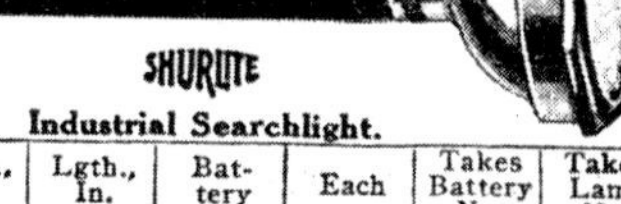

SHURLITE Industrial Searchlight.

Catalog No.	Diam., In.	Lgth., In.	Battery	Each	Takes Battery No.	Takes Lamp No.
6L8918	1½	9	3-Cell	$2.00	6L9005	6L8805
6L8919	1½	14	5-Cell	2.40	6L9015	6L8816

For extra batteries and lamps see below.

Throws a powerful light farther than the ordinary lens. Black vulcanized fiber non-short circuiting case, polished nickel plated trimmings. Perfection double contact switch gives flash or continuous light. Note special construction to prevent case from rolling. Large beveled lens, 3 inches in diameter. Two lengths. Shipping weight, 3 pounds and 4 pounds, respectively.

Penlite

Handy for the vest pocket or hand bag. Neatly made and finished in black enamel. Fitted with pocket clip. Length, 5⅛ inches; greatest diameter, ⅝ inch. Use battery 6L9006 and lamp 6L8800. Shipping weight, 8 ounces.

6L8921—Penlite $1.00

Handy Pocket Size Flash Light.

Flat nickel plated, highly polished pocket case. Very handy for pocket or woman's hand bag. Length, 3 inches; width, 1⅝ inches; ⅝ inch thick, or 3 inches long, 2¾ inches wide and ¾ inch thick.

	Width	Battery	Shpg. Wt., Lbs.	Each	Takes Battery	Takes Lamp
6L8922	1⅝	2-cell	½	$1.00	6L9050	6L8800
6L8923	2¾	3-cell	¾	1.25	6L9051	6L8802

Mazda SHURLITE Flash Light Lamps.

We offer genuine MAZDA Flash Light Lamps which are very high grade. Mazda lamps give a white light. The actual candle power of the lamps, as stated below, is greatly magnified by the reflector and lens of the flash lights. It is advisable to have an extra lamp available. Shipping weight, any one of these lamps, 2 oz.

6L8800

Catalog No.	Mazda No.	Rating	Candle Power
6L8800	Mazda 1	2.2 volts	¾
6L8802	Mazda 2	3.3 volts	1½
6L8803	Mazda 11	2.3 volts	1
6L8804	Mazda 12	3.5 volts	1½
6L8805	Mazda 17	3.8 volts	1½
6L8806	Mazda 3	2.3 volts	1
6L8807	Mazda 16	2.5 volts	1
6L8809	Mazda 18	4 volts	1½
6L8810	Mazda 19	1.25 volts	¾
6L8816	Mazda 31	6.2 volts	2

Each .... 15c; 10 for $1.35

Be sure to give catalog number of lamp wanted.

Miniature Electric Lamp Outfit.

An interesting toy which is also useful for lighting dark closets, etc. Consists of three cells Red Label dry battery, a 1½-candle power miniature Mazda tungsten electric lamp, a receptacle, switch and 30 feet of insulated wire. Shipping weight, 8½ lbs.

6L9030 ... $1.55

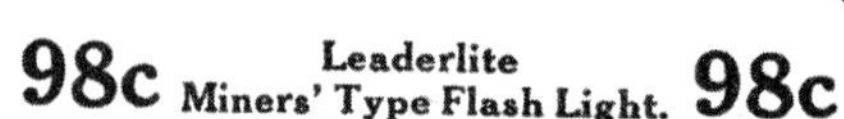

98c Leaderlite Miners' Type Flash Light. 98c

Black fiber case, with nickel plated trimmings. The reflector and lens measure 2½ inches in diameter. An excellent flash light at a very low price. Case 1½ inches in diameter, 6½ inches long over all. Takes two-cell battery 6L9090 and lamp 6L8807. Shipping weight, 1⅝ lbs.

6L8920—Leaderlite. Complete with battery...... 98c

"Giant" Utility Double Cell Lantern.

Strong steel case. Has both bail and grip handles. Latter fold flush with body in back. Heavy lens gives bullseye effect. Reflector is heavily plated and polished. Switch located conveniently in back. Black enamel finish with nickel plated trimmings. 7½ in. high, 5½ in. wide and 4 inches thick. Requires two standard 2½x6-inch dry cells.

6L8993—"Giant" Utility Double Cell Lantern, complete, without batteries. Shipping weight, 3 pounds...... $1.76

6L8994—"Giant" Utility Double Cell Lantern, complete, with two dry cells. Shipping weight, 7 pounds...... $2.44

For extra batteries see 6L8635 on page 813.

For extra lamps see 6L8807 at left.

"Safety First" Electric Vehicle Light.

When driving at night see the road ahead and warn those behind. Outfit consists of steel battery case and high grade lamp, having 3¼-inch reflector, 1¾-inch deep red lens in rear. Black enamel finish, nickel plated trimmings. Switch controls light. Brackets for mountings. Case, 7x5⅝x2¾ inches. Will hold two No. 6 dry cells. Ten feet of wire furnished.

6L8999—Vehicle Light, complete, without batteries. Shipping wt., 2½ lbs.. $2.10

6L8978—Vehicle Light, complete, with two dry batteries. Shpg. wt., 9 lbs. $2.78

For extra batteries see 6L8635 on page 813.

For extra lamps see 6L8807 at left.

The Spotlight

This standard dry cell light is made with a reflector by means of which it throws a narrow, bright shaft of light. Uses a regular size dry battery. Black metal case with handle and bail. Diameter of case, 2⅞ inches. Height, over all, 8 inches. Diameter of lens, 2⅛ in.

6L8988—Spotlight. Complete without battery. Shipping weight, 3 pounds.................. $1.12

6L8989—Spotlight. Complete with battery. Shipping weight, 5 pounds.. $1.46

For extra lamp see 6L8810 at left. For extra battery see 6L8635 on page 813.

Flash Light Batteries SHURLITE Flash Light Batteries

Our flash light batteries are very high grade. They store an unusual amount of energy, and have great recuperative power. Our large volume of sales insures fresh stock at all times. Purchase an extra battery and have one available when the first one has served its usefulness.

6L9000—Two-Cell Battery for Flat Pocket Case formerly handled by us; $\frac{9}{16}$ inch thick, 1⅛ inches wide, 1⅞ inches high. Shipping weight, 2 ounces...... 20c

6L9003—Three-Cell Battery for House Lamp formerly handled by us; ¾ inch thick, 2$\frac{7}{16}$ inches wide, 2⅝ inches high. Shipping weight, 6 ounces............ 27c

6L9050—Two-Cell Battery for Flat Pocket Case formerly handled by us; ⅝ inch thick, 1$\frac{9}{16}$ inches wide, 2¼ inches high. Shipping weight, 3 ounces 20c

6L9051—Three-Cell Battery for Flat Pocket Case formerly handled by us; $\frac{11}{16}$ inch thick, 2 inches wide, 2¼ inches high. Shipping weight, 4 ounces..... 27c

6L9010—Five-Cell Battery for Search Lantern formerly handled by us. Diameter, 3⅝ inches; height, 3 inches. Shipping wt., 1½ lbs 78c

6L9031—Five-Cell Battery, 1⅜ inches thick, 3 inches high, 6½ in. wide. Shipping wt., 1⅝ lbs.... 77c

6L9034—Three-Cell Battery for Cap Light and Electric Lantern formerly handled by us; 1⅜ inches thick, 4 inches wide, 3 inches high. Shipping weight, 1 pound...... 45c

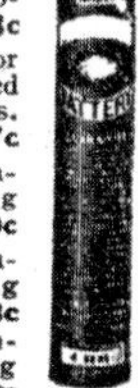

6L9005—Three-Cell Tubular Battery. Diameter, 1⅜ inches; length, 7 inches. Shipping weight, 12 ounces.................. 33c

6L9006—Two-Cell Tubular Battery for Pen Light and Pistol Light formerly handled by us. Diameter, $\frac{9}{16}$ inch; length, 4 inches. Shipping weight, 3 ounces................ 17c

6L9015—Five-Cell Tubular Battery. Diameter, 1⅜ inches; length, 11¾ inches. Shipping weight, 1¼ pounds.......................... 60c

6L9090—Two-Cell Tubular Battery. Diameter, 1⅜ inches; length, 4¾ inches. Shipping weight, 8 ounces.......................... 23c

6L9091—Two-Cell Tubular Battery. Diameter, 1 inch. Length, 3¾ inches. Shipping weight, 4 ounces 20c

Power Motors at Low Prices.

Can be used to operate washing machines, churns, ice cream freezers and grinders. Iron frame, good bearings. Steel shaft, felt wick oil cups. Nicely finished V or grooved pulley. **Furnished only for current and voltage shown below.** Prices do not include cord or plug. See Flexible Cord 6L9690 below. Flat face pulleys, 1½ or 2 inches in diameter by 2-inch face, $1.00 extra. **State diameter wanted.**

60-Cycle, 105 to 115-Volt Alternating Current Motors.					32-Volt Direct Current
Catalog No.	6L9672¼	6L9673¼	6L9674¼	6L9638¼	6L9675¼
Horse-power	⅛	⅙	¼	½	¼
Shpg. wt., lbs.	21	25	35	65	35
Speed, R.P.M.	1,725	1,725	1,720	1,800	1,740
Height, inches	5⅝	5⅝	6 5/16	9¼	6 5/16
Diam., inches	6	5½	6⅛	8¾	6⅛
Length over all, inches	8	8 15/16	8⅜	13¾	8¾
Diameter, pulley, inches	⅞	1½	1⅝	3½x2	1⅝
Round belt, in.	¼	5/16	⅜	Flat	⅜
Each	**$13.60**	**$14.40**	**$16.50**	**$54.32**	**$21.00**

Flexible Cord With Terminals and Plug.

6L9690—To use with power motors. Cord of standard quality. Length, 15 feet. Shipping weight, 1½ lbs. Complete...**95c**

Jump Spark Coil.

High grade, suitable for any one-cylinder engine with spark plug, regardless of horse-power. Gives a hot, fat spark with very little current consumption; effective and economical. Best results with four cells of battery. Shipping weight, 3½ pounds.

6L9500 ....$2.60

For Automobile Engine Ignition see page 757.

Gasoline Engine Magnetos.

Made with two large powerful magnets and completely enclosed armature, affording protection from dirt and moisture. Equipped with governor, which regulates the speed, preventing burning out the spark coil or contact points. A uniform spark at all times, insuring full engine efficiency. Once mounted on the engine, they require practically no attention aside from oiling and renewal of brushes about once in from twelve to eighteen months. Not suitable for use on automobiles or motor boats. Shpg. wt. of all make and break magnetos, 18 lbs.

6L9524¼—Friction Drive Magneto, for make and break ignition, without coil...........**$8.95**

6L9525¼—Friction Drive Magneto, for make and break ignition, with coil mounted in the magnets...........**$9.70**

6L9526¼—Belt Drive Magneto, for make and break ignition, without coil...........**$9.50**

6L9527¼—Belt Drive Magneto, make and break ignition, with coil mounted in the magnets...........**$10.45**

6L9528¼—Friction Drive Magneto, for jump spark ignition...........**$9.25**

6L9529¼—Belt Drive Magneto, for jump spark ignition...........**$9.75**

Shipping weight, jump spark magnetos, 20 pounds.

NOTE—We do not furnish jump spark magnetos with coil mounted in the magnets.

Utility Motors for Farm Lighting Plants, 32-Volt.

Intended for operating many of the household and dairy machines which you now operate by hand. Will run cream separators, grindstones, churns, washing machines, small pumps, etc.

Motor is a ¼ horse-power back geared type, mounted on tripod and furnished with about 19 feet of flexible cord and plug to attach to lamp sockets. Tripod feet can be quickly removed for bench mounting.

The motor proper runs at a speed of 2,000 R. P. M., and for running high speed machines there is a grooved pulley for ⅜-inch round belt; effective diameter, 1¾ inches. Three other pulleys are furnished, geared down to run at approximately 350 R. P. M. One for ⅜-inch round belt having an effective diameter of 2¼ inches, a larger one, 3⅝ inches in diameter, and a pulley 6½ inches in diameter for using a flat belt. Over all height with tripod, 23½ in.; length of the motor shaft, about 12½ in. Wt. of motor, complete, 36 lbs. Shpg. wt., 44 lbs.

6L9643¼—Utility Motor.........**$35.50**

Red Label Dry Batteries.

Our Red Label Dry Battery is suitable for telephone use, door bells, annunciators, burglar alarms, medical batteries, toy motors or other similar work. It is made with high internal resistance and medium amperage, averaging 18 amperes. Owing to its high internal resistance it has lasting qualities. Size, 2½ inches diameter by 6 inches high.

6L8635

Each......(Shpg. wt., 2⅝ lbs.)....**$0.34**
Dozen ...(Shpg. wt., 30 lbs.).... **4.00**

Stand-By "Special" Dry Batteries.

The Stand-By "Special" is a dry battery of low internal resistance and is designed especially for ignition work. It is suitable for use with stationary gasoline engines, automobiles and launches. It may also be used for operating any motor, lighting miniature incandescent lamps, etc. Each cell is guaranteed to test 25 amperes. Size, 2½x6 inches. Shipping wt., 2¾ pounds.

6L8645 ....**41c**

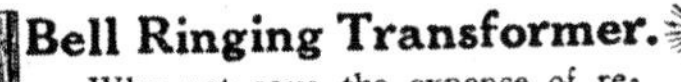

Bell Ringing Transformer.

Why not save the expense of renewing from time to time the batteries that ring your door bell or operate your buzzer or door opener?

A bell ringing transformer connected into your electric lighting circuit will do the work beautifully, uses no current to speak of, and once installed, will last indefinitely without further attention. Easily installed. We furnish full directions.

Transformer suitable for ordinary city current (105 to 115-volt, 60 to 133-cycle, alternating). Has range of three voltages, 6, 8 and 14. Strong metal case, black enameled, 2½ inches wide, 4 inches high, 1⅝ inches thick over all. Shpg. wt., 2¼ lbs.

6L8516 ....**95c**

(Price does not include Bell, Push Button or Wire.)

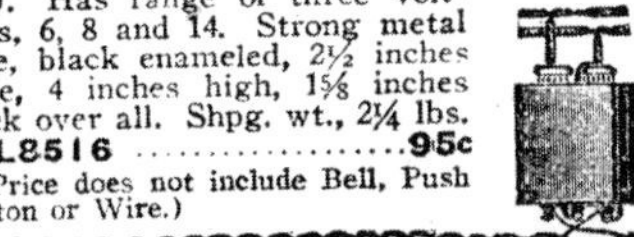

Electric Bell Outfit.

For Door and Call Bell Service.

Consists of one Red Label dry battery, one push button, one 2½-inch bell, 75 feet of annunciator wire and necessary staples. Directions for installing furnished. Shipping weight, 4¼ pounds.

6L8522 ....**$1.35**

If more than 75 feet of wire is needed to connect as pictured, order wire 6L9900 shown on page 809.

Make and Break Coil.

For use with any engine of make or break ignition. Mounted in strong steel case, 2¾x2¼x6½ in. Takes up about same space as a dry battery. Easily inserted in battery box. Shipping weight, 3 pounds.

6L9508 — Without switch **74c**

Make and Break Coil, shown above, equipped with switch for making and breaking circuit.

6L9509 ....**81c**

Trojan Double-Life Dry Battery.

Waterproof—High Amperage—Long Life.

Especially designed for continuous or heavy duty. Average amperage is 30. Renders much better service and has longer life than the ordinary round cell type battery. The zinc (the active metal) is placed on the inside of the cell in the form of ribbon strips and cannot spot out. Measures 6½x3⅝x2 inches. Can be operated entirely under water except for the terminals, for regular life of battery. **Shipping weight, 2¾ lbs.**

6L8648—Each **$0.55**
Per dozen **6.35**

Trojan Double-Life Multiple Batteries.

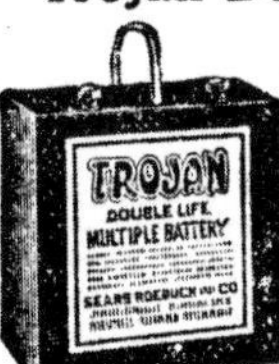

Made from Trojan double-life dry cells combined into compact units. Entirely waterproof and embodying all the double life features. Each size, except 6L8652, is fitted with handle to facilitate carrying.

6L8649—Trojan 6-Volt Multiple Battery made up of 4 cells. Size over all, 4¼x6¾x7 inches. Shipping weight, 12 pounds......**$2.60**

6L8650—Trojan 7½-Volt Multiple Battery made up of 5 cells. Size over all, 3½x10⅝x7 inches. Shipping weight, 15 pounds......**$2.90**

6L8651—Trojan 9-Volt Multiple Battery made up of 6 cells. Size over all, 6¾x6¾x7 inches. Shipping weight, 18 pounds......**$3.65**

6L8652—Trojan 7½-Volt Multiple Battery made up of 10 cells, as 2 multiples of 5 cells in series. Size over all, 6¾x10⅝x7 inches. Shipping weight, 30 pounds...........**$6.00**

Gravity Batteries.

The Gravity Battery is a closed circuit battery used almost entirely for telegraph work. The 5x7 battery requires 1½ pounds of blue vitriol for a charge, the 6x8 battery requires 3 pounds. High grade batteries.

Gravity Battery. Size, 5x7 inches, consisting of jar, copper and zinc. Shipping wt., 7¾ lbs.

6L8610—Without blue vitriol..........**85c**

Gravity Battery. Size, 6x8 inches, consisting of jar, copper and zinc. Shipping weight, 9⅝ lbs.

6L8611—Without blue vitriol..........**98c**

NOTE—Blue Vitriol is not furnished with these batteries. It is always sold extra.

Battery Jar, glass, 5x7 inches. Shipping weight, 5⅝ pounds.
6L8612 ....**45c**

Battery Jar, glass, 6x8 inches. Shipping weight, 5⅞ pounds.
6L8613 ....**50c**

Zinc for 5x7 battery. Shipping weight, 2⅝ pounds.
6L8614 ....**30c**

Zinc for 6x8 battery. Shipping weight, 4⅛ pounds.
6L8615 ....**36c**

Copper for 5x7 battery. Shipping weight, 3 ounces.
6L8616 ....**16c**

Copper for 6x8 battery. Shipping weight, 4 ounces.
6L8617 ....**18c**

Blue Vitriol. Price subject to market change. Shipping weight of 1 pound, 1½ pounds. Not mailable.
6L8618—Per pound.........**14c**

Rectangular Zinc Stick, ¼x⅜x6½ inches. Shipping weight, 2 ounces.
6L8601—3 zincs.........**22c**
Per dozen..................**80c**
Shipping weight, 3 pounds.

Sal Ammoniac. Shipping weight, 1½ pounds.
6L8604—Per pound......**18c**

6L8605—Sal Ammoniac sufficient for one charge. Shipping weight, 7 oz. Per package.........**8c**

Electric Door Bells.

High grade bell, 2½-inch gong; cover, stamped sheet steel, finished black. Guaranteed to ring clearly through 200 feet of No. 18 annunciator wire (100 feet from bell to battery) on one battery cell. To get maximum sound, two batteries should be used. Add one cell for each additional 50 feet of wire. Suitable for use with transformer. Shipping weight, 1⅛ pounds.

6L8501 ....**49c**

Same as above, but with 3-inch gong. Shipping weight, 1⅛ pounds.
6L8500 ....**53c**

Same as above, but with 4-inch gong. Not suitable to use with transformer. Shipping weight, 1⅝ pounds.
6L8502 ....**71c**

NOTE—These bells cannot be used with telephones. Send for our Electrical Goods Catalog No. 520GCL for Telephone Bells.

Push Buttons.

Wood Push Buttons. High grade springs, porcelain center. Shipping weight, each, 3 ounces; dozen, 2½ pounds.
6L8535—Each **7c**
Per dozen **80c**

Anti-Wood Push Buttons. A metal push button. Shipping weight, 3 ounces.
6L8538—Each **$0.12**
Per dozen **1.38**

For other Push Buttons see page 834.

Buzzer.

Makes a comparatively low buzzing sound and is used in place of bell where loud ringing is not desirable. Directions for installing furnished. Shipping weight, 12 oz.
6L8517 ....**53c**

Battery Switches.

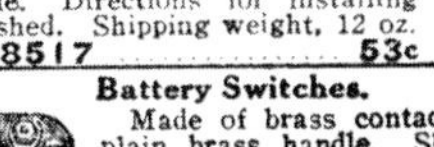

Made of brass contacts and plain brass handle. Shipping weight, 3 ounces.
6L8541—Single Throw Switch **16c**

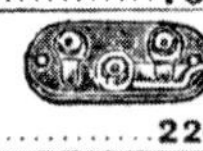

6L8542—Double Throw Switch for use with two sets of batteries or one set of batteries and magneto **22c**

Wood Base Switches.

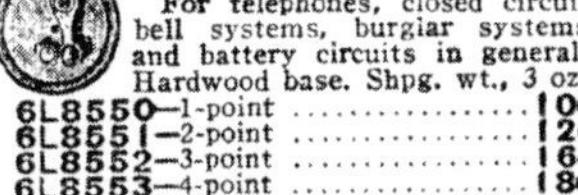

For telephones, closed circuit bell systems, burglar systems and battery circuits in general. Hardwood base. Shpg. wt., 3 oz.

6L8550—1-point **10c**
6L8551—2-point **12c**
6L8552—3-point **16c**
6L8553—4-point **18c**

Telephones and Accessories

Description of Magneto Telephones. What Kind to Order.

Bridging magneto telephones as here listed are the standard for party lines. If you need series phones for private line and city exchange work we will gladly quote prices, but we recommend bridging phones for this work also. Party lines can use only bridging phones.

All telephones on one line should be of the same ohms resistance. We guarantee all our telephones to be a standard make and high grade in every respect. Every part that goes into the telephone is new. We do not sell rebuilt phones.

Every telephone is equipped with solid back transmitter, bi-polar receiver, adjustable ringers with 2½-inch brass gongs, lightning arresters and generators with laminated magnets, each magnet consisting of three laminations, tempered and magnetized separately, producing a powerful generator. In other words, our six-bar generator has eighteen of these magnets.

These phones are shipped from factory near CHICAGO. Shipping weight, each, 43 pounds. If you do not want batteries, deduct 76 cents per telephone.

Compact Bridging Telephones With 6-Bar Generator, Including Two Dry Batteries per Phone.

Catalog No.	Type of Ringer	Each	Six for
6L8112⅓	1,000-ohm	$13.95	$82.50
6L8113⅓	1,600-ohm	14.04	83.04
6L8114⅓	2,000-ohm	14.11	83.46
6L8115⅓	2,500-ohm	14.15	83.70

Compact Bridging Telephones With 5-Bar Generator, Including Two Dry Batteries per Phone.

Catalog No.	Type of Ringer	Each	Six for
6L8100⅓	1,000-ohm	$13.79	$81.54
6L8101⅓	1,600-ohm	13.88	82.08
6L8102⅓	2,000-ohm	13.95	82.50
6L8103⅓	2,500-ohm	13.99	82.74

Series Magneto Phones. For Distances to 5 Miles.

Can only be used in pairs. Will operate satisfactorily on lines 5 miles in length. Can be used with copper or iron wire; see 6L9900 on page 809, or 6L9915¼ shown to the right. Only one wire required between phones and enough extra to run into ground at each phone. Good grade transmitter, receiver and generator. Cases of oak, exposed metal parts nickel plated. **Shipped from factory near CHICAGO.** Shipping weight, per pair, 46 lbs.

6L8143⅓—Per pair (two telephones)...**$21.98**

Four dry batteries included with each pair.

Wall Telephone for Two-Station Line.

A very efficient and well appearing telephone built for single line service with a station at each end. Material, workmanship and finish are strictly high grade throughout. Will not detract from the appearance of any room in which you may install it. Will ring satisfactorily over line 600 feet long, using 6L9919¼ or 6L9920¼ Twisted Pair Wire shown to the right (only one length needed), and four dry batteries (longer distances require larger gauge wire). Shipping weight, 5½ pounds.

6L8150

Per pair, without batteries or wire.......**$13.50**

Receivers and Receiver Shells.

Bi-Polar Receiver. Can be used on any telephone. Furnished with cord having spade tips at free end, but cord can be reversed to give straight terminals at free end. Shipping weight, 1 pound 5 ounces.

6L8165..................................**$1.35**

Receiver Shell only, for all standard type phones. Shipping weight, 8 ounces.

6L8167..**47c**

Mouthpieces.

Solid Back Transmitter. Button type granular carbon transmitter, fits both old and new type compact, desk and Southwestern telephones. Shpg. wt., 1¼ lbs.

6L8159..**$1.89**

Composition Mouthpiece. Fits our old style phones, Bell and Automatic, having thread size 30 per inch, 1⅝ inch in diameter. Shipping weight, 5 oz.

6L8237.................................**9c**

Composition Mouthpiece. Fits style A, B, C, etc., telephones, and Kellogg, Chicago and Dean phones having thread size 20 per inch, 1⁵⁄₁₆ inch in diameter. Shipping weight, 5 ounces.

6L8238.................................**9c**

Receiver Cord.

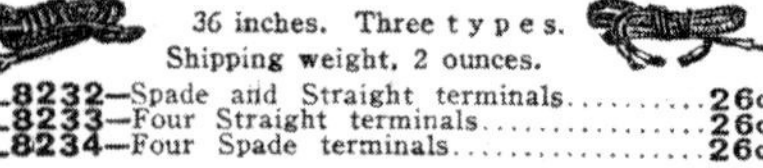

36 inches. Three types. Shipping weight, 2 ounces.

6L8232—Spade and Straight terminals..........**26c**
6L8233—Four Straight terminals...............**26c**
6L8234—Four Spade terminals..................**26c**

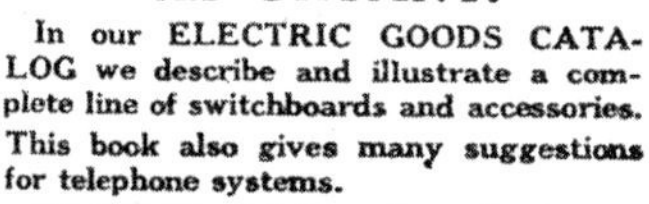

IMPORTANT.

In our ELECTRIC GOODS CATALOG we describe and illustrate a complete line of switchboards and accessories. This book also gives many suggestions for telephone systems.

We will be glad to make estimates on switchboards and to advise on any of your telephone problems.

Write today for Catalog 520GCL. Sent postpaid.

Single and Double Groove Pony Glass Insulators.

No. 9 for telephone, telegraph and fire alarm work, 400 in barrel. Weight, per barrel, 300 pounds.

6L8390¼

400 insulators..........................**$21.40**
Each, less than barrel lots.........**.06**

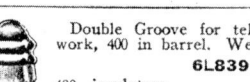

Double Groove for telephone transposition work, 400 in barrel. Weight, barrel, 300 lbs.

6L8391¼

400 insulators..........................**$22.35**
Each, less than barrel lots..........**.06**

Fuses.

Shipping weight, per dozen, 4 ounces.

	Per 100	Per Doz.
6L8342—Western Union, copper tip.	$2.75	34c
6L8343—Postal, copper tip........	2.98	37c

Ground Rod.

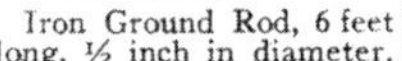

Iron Ground Rod, 6 feet long, ½ inch in diameter. Galvanized. Shipping wt., each, 3½ pounds.

6L8350¼

Each **$0.38**
Per dozen..........**4.50**

Iron Guy Rod.

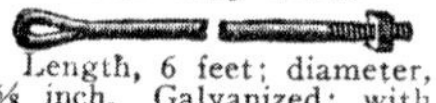

Length, 6 feet; diameter, ⅝ inch. Galvanized; with square nut and one square washer. Shipping weight, each, 6¼ pounds.

6L8407¼

Dozen, **$8.90**; each...**75c**

Pony Oak Brackets.

For telephone and telegraph line construction. Can be fastened to side of pole or house. Painted. Weight, each, 8 ounces.

6L8400¼

Each **$0.03**
20 for..**.52**
100 for...**2.40**

Size, 1¼x8 inches. For use on crossarms. Painted. Weight, each, about 8 ounces.

6L8401¼

Per sack, containing 250 pins. (Shpg. wt., 62 lbs.)..**$4.75**
Each, in less than sack lots.........................**.03**

Powerful Electro Magnet.

Operates by batteries. With one dry cell this magnet has a lifting power of 2 pounds, with two cells will lift about 5 pounds, with four cells will lift over 10 pounds. Shipping weight, 1 pound.

6L8560—With two 2½-foot conducting cords, but without batteries...........**$2.16**

Our Desk Telephones—Bridging Type.

Combines excellent quality with a very low price. It is a standard bridging magneto telephone and can be used on any magneto telephone line with bridging phones. Our prices include a complete desk telephone, and generator and ringer box with a lightning arrester mounted on the box. We also furnish two dry batteries with each instrument. The generator and ringer box is highly polished oak and all exposed metal parts are finished in black enamel and nickel plate. The generators have ¾-inch laminated magnets. See further description at left.

In ordering for a new line it is always best to select telephones all with the same ohms resistance. When adding telephones to a line, order those of the same ohms resistance as the ones already installed. **Shipped from factory near CHICAGO.** Shipping weight of each, 45 pounds. If you do not want batteries, deduct 76 cents per telephone.

Desk Type Bridging Telephone With 6-Bar Generator, Including Two Dry Batteries per Phone.

Catalog No.	Type of Ringer	Each	Six for
6L8132⅓	1,000-ohm	$15.56	$92.16
6L8133⅓	1,600-ohm	15.65	92.70
6L8134⅓	2,000-ohm	15.72	93.12
6L8135⅓	2,500-ohm	15.76	93.36

Desk Type Bridging Telephone With 5-Bar Generator, Including Two Dry Batteries per Phone.

Catalog No.	Type of Ringer	Each	Six for
6L8136⅓	1,000-ohm	$15.40	$91.20
6L8137⅓	1,600-ohm	15.49	91.74
6L8138⅓	2,000-ohm	15.56	92.16
6L8139⅓	2,500-ohm	15.60	92.40

Linemen's Tool Belt.

Made of high grade leather. Width of strap, 2 inches. Length, 46 inches. Shipping weight, 1½ lbs.

6L8453...**$1.51**

Climber Straps.

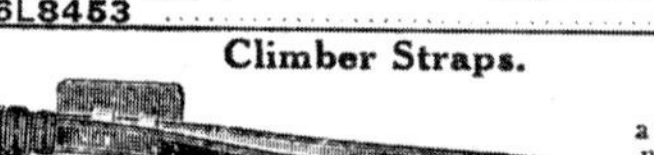

Straps are furnished with a leather pad which prevents the climber from digging into the knee. Can be used with any style of climber, either eastern or western. Shipping weight, 2 pounds.

6L8450—Per set of four; 2 upper, 2 lower **$1.42**

Linemen's Safety Strap.

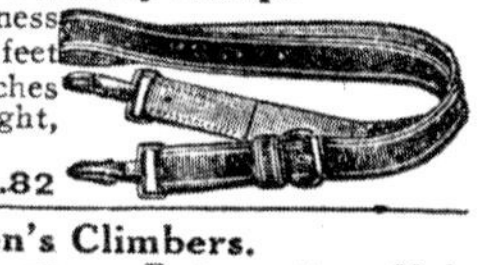

Made of prime harness leather, single strap, 5 feet 6 inches long, 1¾ inches wide. Shipping weight, 2 pounds.

6L8452.........**$1.82**

Linemen's Climbers.

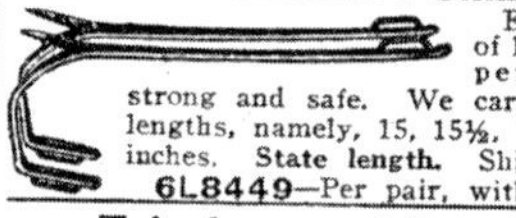

Eastern pattern. Made of high grade steel, tempered, finely finished, strong and safe. We carry them in standard lengths, namely, 15, 15½, 16, 16½, 17 and 17½ inches. **State length.** Shipping weight, 4 lbs.

6L8449—Per pair, without straps **$2.52**

Telephone and Telegraph Wire.

Has a heavy coating of zinc covering every part of the wire, protecting it against rust and corrosion. **Sold only in coils of ½ mile each. Shipped from factory in CHICAGO.**

		Weight of Coil	Per ½ Mile Coil
6L9913¼	No. 12 Steel	86 lbs.	$5.77
6L9914¼	No. 14 Steel	50 lbs.	3.78
6L9915¼	No.12 B. B. Iron	86 lbs.	6.30
6L9916¼	No.14 B. B. Iron	50 lbs.	3.95

Rubber Covered Twisted Pair.

For outside work. Each conductor a No. 19 copper wire with insulation of black rubber compound and saturated braid covering. Shipping weight, per 100 feet, 3½ pounds.

6L9919¼—Per 100 feet.........................**$1.26**

Same as above, but for inside work, with rubber insulation of black rubber compound and dry braid of yellow cotton. Shipping weight, per 100 feet, 3½ pounds.

6L9920¼—Per 100 feet.........................**$1.26**

Neither of the above is suitable for lighting circuit use.

Baby Knife Switches.

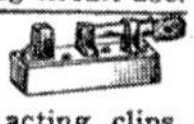

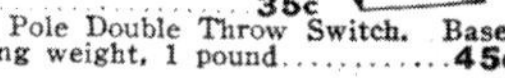

Porcelain base telephone, telegraph and battery switches, with return bend, self adjusting, smooth acting clips. Furnished only in 15-ampere size for 125 volts or less.

6L8353—Single Pole Single Throw Switch. Base, 1⅛x3⅛ inches. Shipping weight, 6 ounces..........**21c**

6L8354—Single Pole Double Throw Switch. Base, 1⅛x4 inches. Shipping weight, 12 ounces..........**28c**

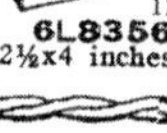

6L8355—Double Pole Single Throw Switch. Base, 2x2⅝ inches. Shipping weight, 12 ounces..........**35c**

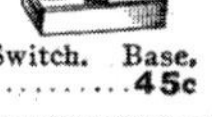

6L8356—Double Pole Double Throw Switch. Base, 2½x4 inches. Shipping weight, 1 pound..........**45c**

Sport Bodies for Ford Cars

A low, racy Speedster Body for Ford car, designed after a well-known racing car, consisting of body with torpedo back, hood and bullet nosed radiator shell. Made of automobile steel, securely fastened to substantial wood frame, hardwood floor boards, steel wear plates for pedals, and instrument board.

Bullet nosed radiator shell fits over regular Ford radiator. (Can furnish Fiat radiator shell if desired at same price.) Body can be installed on any Ford chassis. Special fenders as shown in upper right hand corner, or may be used without fenders. Regular Ford headlights attach to standard brackets. Ford tank fits in back of seat.

Spring seat and back upholstered in a high grade artificial leather, both removable, very comfortable, with plenty of leg room. Body, hood and shell painted vermilion red, canary yellow, chrome green or battleship gray. (Other standard colors, **$5.00** extra.) **State color.**

Body, 141 inches long; **Cowl,** 32 inches wide, 25 inches high. **Seat,** 34 inches wide, 6 inches from floor, 17 inches deep. Actual weight of body complete, 275 pounds. Shipping weight, 320 pounds.

Shipped from factory near CHICAGO, ILL.

11L25—Speedster Body with Hood, Bullet Nosed Radiator Shell, Upholstered Spring Seat and Back, and Hardwood Instrument Board. Shipping weight, 320 pounds $69.95
11L3—Two-Piece Ventilating Wind Shield (black enameled). Weight, 25 lbs ... 12.95
11L9—Collapsible Khaki Top with Side Curtains. Shipping weight, 15 pounds 13.95
11L10—Aluminum Military Steps. Shipping weight, 6 pounds. Per pair 4.35
11L13—Aluminum Cowl Ventilator with Dash Control. Shipping weight, 2 lbs ... 3.50
11L7—Individual Fenders (black enameled). Weight, 160 lbs. Per set of four ... 26.75
11L8—Nickel Plated Electric Lights (two cowl and one parking light). Weight, 2 pounds 6.75
11L6—Four Outside Wheel Discs. (State if wheels have demountable rims with 4 or 5 lugs). Shipping weight, 20 pounds 7.35
11L11—Four Inside Wheel Discs, with space cut out for inflating tires. Weight, 20 pounds 7.35

The above illustration shows the special fenders that we can furnish for this job. They are made of heavy fender steel, well braced to the frame of the car.

The illustration in the upper left hand corner shows the collapsible khaki top and the two-piece ventilating wind shield with rubber strip across the bottom; finished in black enamel.

We recommend the use of the Underslung Parts as illustrated below with this body, and the Bucket Seat Speedster, although they are not absolutely necessary. They lower the body about 3½ inches, so that the car holds the road better and rides easier.

Wheel discs bolt on the outside of the regular wood wheel, as illustrated below, and are painted same color as body. **Be sure to tell us if the wheels on your car have demountable rims with four or five lugs.**

Runabout Body for Ford Cars.

For Wheel Discs see prices above.

A high grade Runabout Body complete, consisting of body, special hood and radiator shell, two-piece wind shield and bow top with side curtains, that fits on any Ford chassis in place of the regular body. Price does not include chassis, fenders or running board.

Body made of automobile steel over hardwood frame. Radiator shell fits over regular Ford radiator. Ford gasoline tank goes in back of car. Regular Ford or drum type headlights can be used. Artificial leather upholstered spring seat and padded back. Holds two people comfortably with plenty of leg room.

Body, 116 inches long. **Cowl,** 34 inches wide, 24½ inches from floor. **Seat,** 34 inches wide, 8½ inches from floor. **Rear Compartment,** 30 inches wide, 16 inches high, with 10x30 inch hinged lid that can be locked. **Painted** vermilion red, chrome green or battleship gray. **Be sure to state color desired. Shipped from factory near CHICAGO, ILL.**

11L21—Runabout Body, as illustrated and described above, including wind shield and top. Shipping weight, 290 pounds **$84.80**
11L22—Runabout Body only, without wind shield and top. Shipping weight, 240 pounds **$52.50**
11L13—Two-Piece Ventilating Wind Shield (black enameled). Shipping weight, 25 pounds **$12.95**
11L14—Bow Top with Side Curtains. Weight, 25 lbs **19.35**

Underslung Parts.

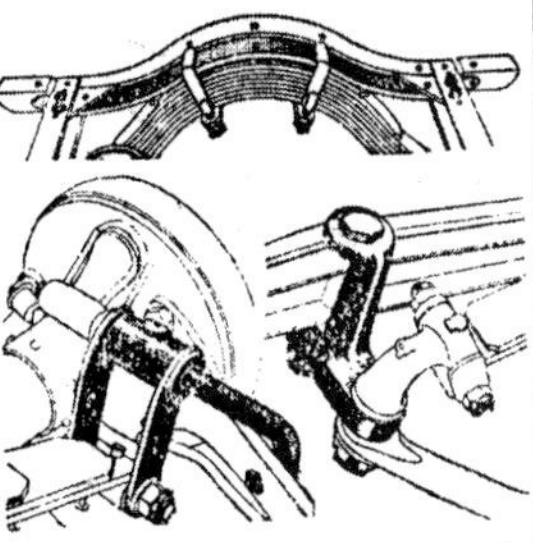

Used on regular Ford chassis with speedster body which lowers the body about 3½ inches, so the car rides easier and can be handled better. No change is necessary in frame or rear axle. Above illustrations show parts installed. Full instructions accompany each set. **Shipped from factory near CHICAGO, ILL.** Weight, 12 pounds.

11L5—Set of Underslung Parts **$7.85**

Bucket Seat Speedster.

Consisting of hardwood frame with cowl, hood, radiator shell and two bucket seats with artificial leather cushions, all ready to put on any Ford chassis. Mount the regular Ford gasoline tank just back of the seats with a tool box or carrying compartment on the rear and you have a sporty looking speedster at a very low price. Price does not include tank or tool box. Painted vermilion red, chrome green or battleship gray. **State color wanted. Shipped from factory near CHICAGO, ILL.** Shipping weight, 150 lbs.

11L18—Speedster Body with cowl, hood, radiator shell and bucket seats **$38.95**
11L3—Two-Piece Ventilating Wind Shield (black enameled). Shipping weight, 25 pounds **$12.95**
11L6—Four Outside Wheel Discs, painted same color as body. **(State if wheels have demountable rims with four or five lugs.)** Shipping weight, 20 pounds **$7.35**
11L11—Four Inside Wheel Discs, with space cut out for inflating tires. Weight, 20 pounds **$7.35**

Slip-On Body for Ford Runabout.

A good substantial Body that we guarantee will give you years of satisfactory service, made by one of the best body builders in the country of the best materials for the purpose. 34 inches wide, 52 inches long. Painted a plain gloss black. Heavy sills and rail on top of side panels; outside brace irons in rear; drop endgate with patent fasteners; irons on top of endgate; 5-inch flare boards attached to body; 8½-inch side panels. Furnished complete with rear lamp bracket and with bolts for attaching. Weight, 95 lbs.

11L350—Slip-On Body for Ford Runabout. Shipping wt., 140 lbs. **Shipped from EVANSVILLE, IND., or factory in EASTERN PENNSYLVANIA. $9.95**

Standard Wagon Box.

A well made, substantial wagon box that we guarantee to give you satisfactory service.

Hardwood cross and bolster cleats; bevel edge irons on top of side and ends; ⅝-inch box straps; ½-inch side braces; ⅜-inch end rods. Painted green, striped and varnished. Complete with hinged endgate, side box fasteners, grain cleats and spreader chains. 10½ feet long by 26 inches high. Furnished 38 inches wide only. **Shipped from KANSAS CITY, MO., or ST. PAUL, MINN.**

11L1903—Standard Wagon Box, 38 inches wide. Weight, 350 pounds. **$22.50**
11L1907—10-Inch Tip Top Box for box 38 inches wide. Shipping weight, 25 pounds **$7.95**
11L1909—Spring Seat for 38-inch Wagon Box. Shipping weight, 10 lbs ... **$4.50**

Standard Wood Wheel Farm Truck.

4,000-Pound Capacity.

11L1871—Wood Wheel Farm Truck, complete with standard drop tongue. Weight, 650 pounds **$49.95**
11L1800—Set of Doubletrees and Neckyoke. Weight, 34 pounds **$4.35**
11L1805—Brake complete with box attachment. Weight, 75 pounds **7.35**
11L1806—Brake without box attachment. Weight, 65 pounds **5.70**

Trucks are built of high grade well seasoned material, 4,000 pounds capacity. Standard wagon track, 4 feet 8 inches between wheel centers. Axles have truss rods, fitted with 3¼x10-inch skeins, set in red lead. Wood hounds with drop tongue; 10-foot reach; bolsters securely ironed, 38 inches between stakes. Well made trucks that we guarantee will give you excellent service for farm work.

The 11L1871 Truck is fitted with high grade wood wheels made of selected material, 36-inch front, 40-inch rear, 3x⅜ inch tires. Painted dark orange color.

The 11L1872 Truck is fitted with steel wheels, 28-inch front, 32-inch rear, 4x⅜-inch grooved steel tires. Spokes riveted in cast hubs and steel tires, are guaranteed never to come loose. Truck painted dark orange, wheels black.

Both trucks shipped direct from factory near CHICAGO, ILL., or ST. PAUL, MINN.

Standard Metal Wheel Farm Truck.

4,000-Pound Capacity.

11L1872—Metal Wheel Farm Truck, complete with standard drop tongue. Weight, 540 pounds **$39.95**
11L1800—Set of Doubletrees and Neckyoke. Weight, 34 pounds **$4.35**
11L1805—Brake complete with box attachment. Weight, 75 pounds **$7.35**
11L1806—Brake without box attachment. Wt., 65 lbs **5.70**

Triple Panel Auto Seat American Beauty Buggy

$84.75

2½-Bow Top with Southern Style Drop Back Seat furnished on our American Beauty Buggy in place of regular, no extra charge.

Special Features

Metal Auto Seat With Triple Panel Back.
Waterproof Skeleton Auto Top.
12-Inch Wrought Fifth Wheel.
Second Growth Hickory Shafts and Wheels.
High Grade Auto Finish.
Shipped on Thirty Days' Trial.
Guaranteed Against Defect in Material and Workmanship.

11L3508—Triple Panel Auto Seat American Beauty Buggy, with triple braced shafts and steel tires......... **$84.75**

This illustration shows the back and side of the Triple Panel Auto Seat as regularly furnished on our American Beauty Top Buggy.

DESCRIPTION.

SEAT—Triple panel steel automobile style, 32½ inches across top of cushion. (Can furnish all wood seat if specified, no extra charge.) **UPHOLSTERY**—Black Chase Leatherwove (a high grade artificial leather), spring cushion and back, nicely tufted. Padded and lined seat ends. **TOP**—Three-bow, automobile skeleton style, black auto rubber top with tan inside. Auto rubber quarters, stays and back curtain. With side curtains and storm apron; black auto fasteners. **BODY**—Piano style, 23 inches wide, 56 inches long; heavy hardwood frame and corner posts; well ironed and braced; steel corner irons. **GEAR**—1⁵⁄₁₆-inch long distance 2½-inch true sweep arch axles; hickory axle caps; double hickory reaches, ironed full length; 34-inch end springs, three-leaf front, four-leaf rear; center bearing body loops, 12-inch full wrought fifth wheel. **WHEELS**—Selected, Sarven's patent style, 39 inches front, 43 inches rear. ⅞-inch screwed rims bolted between spokes, fitted with ⅝-inch oval edge steel tires; hickory spokes and felloes. **SHAFTS**—Selected hickory, triple braced; flat straps; neatly trimmed; quick shifting anti-rattler shaft couplers. **SUNDRIES**—Curved patent leather padded dash with hand holds and line rail; rubber mat; fiber boot and storm apron. **PAINTING**—Body, plain black; gear, Brewster green, neatly striped. **TRACK**—4 feet 8 inches, narrow, or 5 feet 2 inches, wide. **State width wanted.**

Weight, 360 pounds. Shipping weight, crated under 34 inches, 500 pounds. **Shipped from factory at EVANSVILLE, IND.**

Changes We Can Make in American Beauty Top Buggy or Runabout.

PAINTING—Carmine red, black, wine or yellow gear, no extra charge.

TOP—Four-bow top or 2½-bow with southern style drop back seat as shown in the upper right hand corner, no extra charge.

BODY—18 or 20-inch body, seat 29 inches across top of cushion, no extra charge.

WHEELS—37-inch front, 41-inch rear, or 41-inch front, 45-inch rear wheels. ¾-inch rim with ¼-inch tire, no extra charge.

Pole in place of shafts, add.................................... **$3.55**
Pole and shafts, add.................................... **8.90**
⅞-inch high grade rubber tires in place of steel, add.......... **8.90**

American Beauty Runabout

Panel Stick Seat as furnished on 11L3513 American Beauty Runabout.

$59.00

Shipped From EVANSVILLE, IND.

11L3515

DESCRIPTION.

SEAT—Bent panel seat; 31 inches across top of cushion. **UPHOLSTERY**—Black artificial leather; tufted panel back and spring cushion. **BODY**—Piano style, 23 inches wide by 56 inches long; hardwood sills and corner posts with steel corner irons. **GEAR**—1⁵⁄₁₆-inch long distance 2½-inch true sweep arch axles; hickory axle caps; double hickory reaches, ironed full length; three-leaf front and four leaf rear open head springs, 34 inches long; center bearing body loops; 12-inch full wrought fifth wheel. **WHEELS**—Selected hickory, Sarven's paten style; ⅞-inch screwed rims bolted between spokes, fitted with ⁵⁄₁₆-inch ova edge steel tires; 39 inches front and 43 inches rear. **PAINTING**—Body and seat, plain black; gear, Brewster green, neatly striped. **SHAFTS**—Selected hickory; triple braced; flat straps; neatly trimmed; quick shifting anti-rattler shaft couplers. **SUNDRIES**—Braced padded patent leather dash and rubber mat. **TRACK**—4 feet 8 inches, narrow, or 5 feet 2 inches, wide. **State width wanted.**

Weight, 290 pounds. Shipping weight, crated under 34 inches, 435 pounds. **Shipped from factory at EVANSVILLE, IND.**

11L3515—American Beauty Runabout, with bent panel seat, triple braced shafts and steel tires........................ $59.00

11L3513—American Beauty Runabout, with panel stick seat, triple braced shafts and steel tires........................ 60.50

For changes we can make see above.

A Good Serviceable Standard Buggy at a Low Price

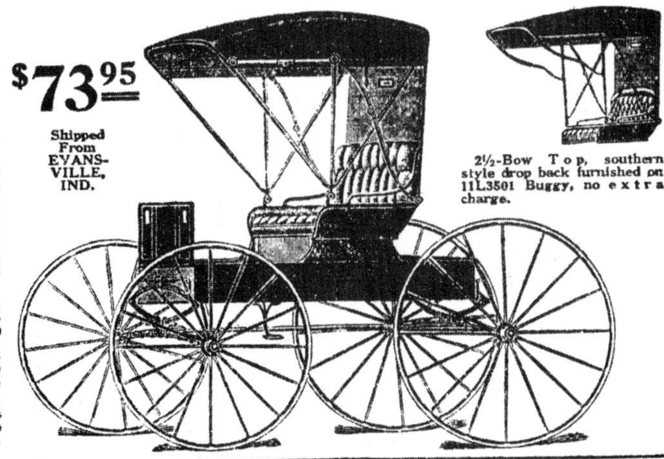

$73.95

Shipped From EVANSVILLE, IND.

2½-Bow Top, southern style drop back furnished on 11L3501 Buggy, no extra charge.

SEAT—Solid panel back with arm rails; regular buggy style; 30½ inches across top of cushion. **UPHOLSTERY**—Black artificial leather; solid panel tufted back and spring cushion; seat ends padded and lined. **TOP**—Three-bow skeleton auto style; auto rubber quarters, stays, roof and back curtain; waterproof side curtains; black knob fasteners. **BODY**—Piano style, 23 inches wide by 56 inches long, heavy hardwood frame and corner posts with steel corner irons. **GEAR**—1 5/16-inch long distance 2½-inch true sweep arch axles; hickory axle caps; double hickory reaches, ironed full length; three-leaf front and four-leaf rear end springs, 34 inches long; center bearing body loops; 12-inch full wrought fifth wheel. **WHEELS**—Sarven's patent style, 39 inches front and 43 inches rear; ⅞-inch screwed rims, bolted between spokes, fitted with 5/16-inch oval edge steel tires; hickory spokes and felloes. **PAINTING**—Body, plain black; gear, Brewster green, neatly striped. **SHAFTS**—Selected hickory; triple braced; flat straps; neatly trimmed; quick shifting anti-rattler shaft couplers. **SUNRDIES**—Patent leather padded dash; rubber mat; fiber boot and storm apron. **TRACK**—4 feet 8 inches, narrow, or 5 feet 2 inches, wide. **State width wanted.**

Weight, 335 pounds. Shipping weight, crated under 34 inches, 465 pounds.
Shipped from factory at EVANSVILLE, IND.

11L3501—With triple braced shafts and steel tires.......... **$73.95**

Pole in place of shafts, add.......... **$3.55**
Pole and shafts, add.......... **8.90**
⅞-inch high grade rubber tires, add.......... **8.90**

Changes We Can Make Without Extra Charge. **Body**—18 or 20-inch, 27½ inches across top of cushion. **Top**—4-bow or 2½-bow with drop back seat as illustrated in upper right hand corner. **Painting**—Gear, black, carmine red, wine or yellow. **Wheels**—37-inch front, 41-inch rear, or 41-inch front, 45-inch rear. ¾-inch rim with ¼-inch tire.

A High Grade Three-Spring Market Wagon

$68.25

Shipped From EVANSVILLE, IND.

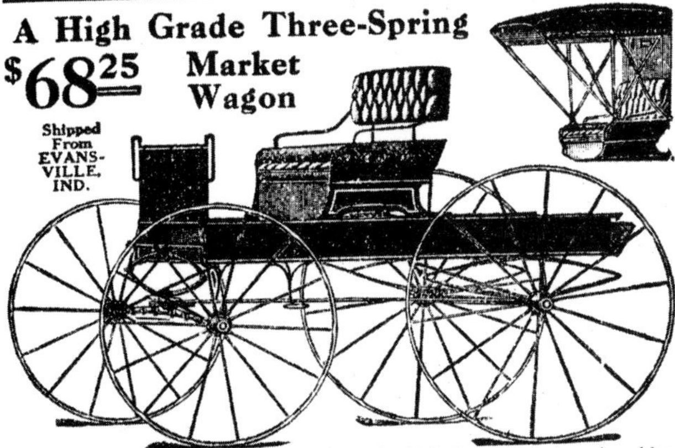

SEAT—Special panel seat with lazyback; 34 inches across top of cushion; open risers. **UPHOLSTERY**—Black artificial leather; box spring tufted cushion; back padded and tufted. Seat ends padded and lined. **BODY**—Hardwood frame and panels, ironed and braced; 76 inches long by 32 inches wide; drop endgate; three-prong steps. **GEAR**—1 1/16-inch straight long distance axles with wide washer bearing, hickory cap on front axle; single reach, ironed full length and braced; two three-leaf springs in rear, 1¼ inches wide by 34 inches long; one four-leaf spring in front, 1⅜ inches wide by 34 inches long; wood body loop in front; rear circle fifth wheel. **WHEELS**—Sarven's patent style; selected hickory; 1-inch screwed rims, fitted with 5/16-inch oval edge steel tires, bolted between spokes; 39 inches front and 43 inches rear. **PAINTING**—Body, dark green, striped, with black molding; gear, green, striped. (Will paint wheels red or yellow in place of green if desired.) **SHAFTS**—Triple braced hickory shafts; flat straps; anti-rattlers. **SUNDRIES**—Wood dash with black rail. **TRACK**—4 feet 8 inches, narrow, or 5 feet 2 inches, wide. **State width wanted.**

Weight, 385 pounds. Shipping weight, crated under 34 inches, 515 pounds.
Shipped from factory at EVANSVILLE, IND.

11L5021—With triple braced shafts and steel tires, one seat....... **$68.25**
11L5023—With triple braced shafts and steel tires, two-seats..... **79.95**

Pole in place of shafts, add.......... **$ 3.50**
Pole and shafts, add.......... **9.00**
1-inch high grade rubber tires in place of steel, add.......... **10.25**
Four-bow auto rubber skeleton top with panel back seat, add.......... **15.50**
Hand brake, add.......... **5.75**

Pony Runabout

BODY—Piano style; hardwood frame; poplar panels on the first two sizes 21x44 inches; on the large size, 21x50 inches. **GEAR**—¾-inch medium arch axles; full length axle caps; double reaches, ironed and braced; wood body loops; elliptic springs. **WHEELS**—Sarven's patent style; ¾-inch rims. (For height see quotation below.) **PAINTING**—Body, black; gear, carmine red, striped; or we will furnish gear painted black, Brewster green, or wine if ordered, at no extra charge. **SHAFTS**—Extra strong and well made. Bradley couplers. **TRIMMING**—Upholstered in heavy tan whipcord; or we can furnish black or tan artificial leather; spring cushion tufted; back padded and tufted; plain padded dash; seat 31 inches across top of cushion. **TRACK**—3 feet 9 inches.

Weight, 185 pounds; shipping weight, crated under 34 inches, 320 pounds.
Shipped from factory in SOUTHERN OHIO.

11L2000—For pony 33 to 42 inches high; shafts, 54 inches long; wheels, 26 inches front and 30 inches rear. With steel tires and shafts... **$65.00**
11L2001—For pony 42 to 50 inches high; shafts, 60 inches long; wheels, 30 inches front and 34 inches rear. With steel tires and shafts... **65.00**
11L2002—For pony 50 inches high; shafts, 66 inches long; wheels, 34 inches front and 38 inches rear. With steel tires and shafts.......... **65.00**

EXTRAS WE CAN FURNISH.

Pole in place of shafts, add.... **$ 6.75** | ¾-in. guaranteed rubber tires, add. **$ 8.40**
Both pole and shafts, add...... **14.75** | Square canopy umbrella top, add. **12.35**

Skeleton Road Cart

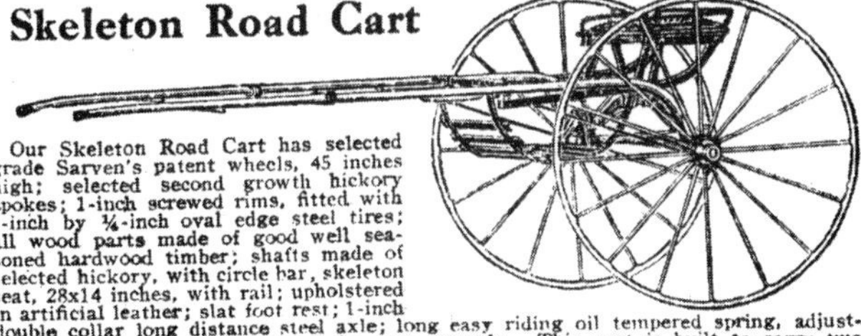

Our Skeleton Road Cart has selected grade Sarven's patent wheels, 45 inches high; selected second growth hickory spokes; 1-inch screwed rims, fitted with 1-inch by ¼-inch oval edge steel tires; all wood parts made of good well seasoned hardwood timber; shafts made of selected hickory, with circle bar, skeleton seat, 28x14 inches, with rail; upholstered in artificial leather; slat foot rest; 1-inch double collar long distance steel axle; long easy riding oil tempered spring, adjustable and hung so as to balance the seat properly. This cart is built to carry two passengers, but the adjustment of the spring is such that it will ride very easily with only one passenger. Painted carmine red with black striping.

TRACK—4 feet 8 inches, narrow, or 5 feet 2 inches, wide. **State width wanted.**
Shipping wt., 150 lbs. **Shipped from factory at EVANSVILLE, IND.**

11L34—Skeleton Road Cart.......... **$22.65**

Phaeton Body Road Cart

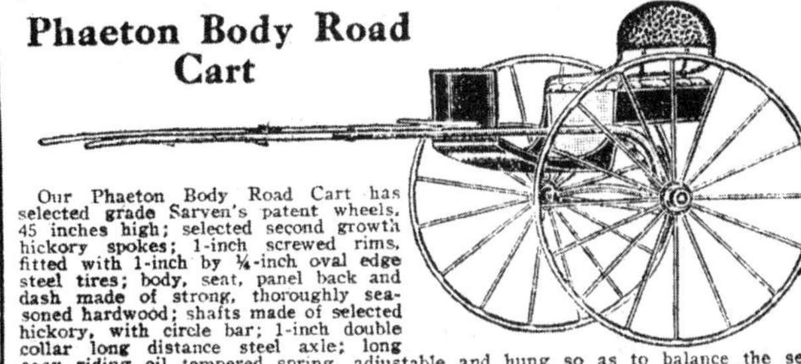

Our Phaeton Body Road Cart has selected grade Sarven's patent wheels, 45 inches high; selected second growth hickory spokes; 1-inch screwed rims, fitted with 1-inch by ¼-inch oval edge steel tires; body, seat, panel back and dash made of strong, thoroughly seasoned hardwood; shafts made of selected hickory, with circle bar; 1-inch double collar long distance steel axle; long easy riding oil tempered spring, adjustable and hung so as to balance the seat perfectly; seat, 28x15 inches, and lazyback as shown in illustration, upholstered in artificial leather. The seat is hinged so that it can be raised and small articles carried in the box under the seat. Built to carry two passengers, although the adjustment of the spring is such that it will ride very easily with only one passenger. The body of this cart is painted black, with a rich blood carmine gear, neatly striped.

TRACK—4 feet 8 inches, narrow, or 5 feet 2 inches, wide. **State width wanted.**
Shipping wt., 180 lbs. **Shipped from factory at EVANSVILLE, IND.**

11L35—Phaeton Body Road Cart.......... **$27.95**

Pony Cart

BODY—Regular cart body with full size springs; wood dash; seat, 15 inches deep, 28 inches wide, will hold two people comfortably. **WHEELS**—Sarven's patent style with ¾-inch rims and steel tires to suit height of pony. (See below.) **AXLE**—¾-inch coach style. **PAINTING**—Body, black; gear and shafts, red. **TRIMMING**—Dark green artificial leather. **TRACK**—3 feet 9 inches only.

Weight, 110 pounds. Shipping weight, crated under 34 inches, 155 pounds.
Shipped from factory in SOUTHERN OHIO.

11L2020—For pony 33 to 42 inches high; 56-inch shafts; 30-inch wheels.......... **$26.75**
11L2021—For pony 42 to 50 inches high; 60-inch shafts; 34-inch wheels.......... **26.75**
11L2022—For pony over 50 inches high; 66-inch shafts; 38-inch wheels.......... **26.75**

If ¾-inch guaranteed rubber tires are wanted, add.......... **$4.10**

Economy Gasoline Engines

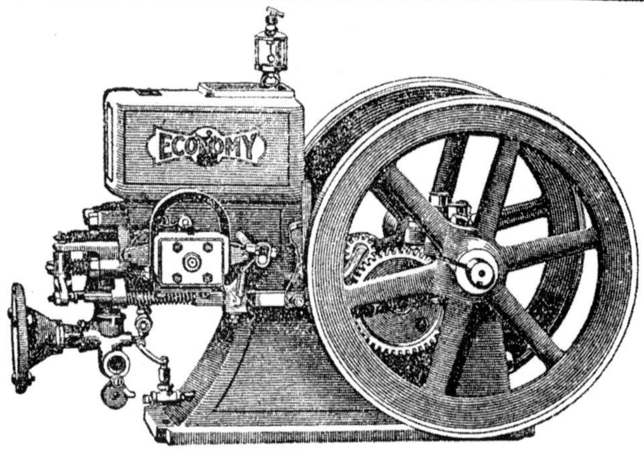

60 DAYS TRIAL

We will ship you any engine or outfit shown on this and the opposite page, on receipt of the full purchase price with the understanding that you are to try it sixty days before you decide to keep it. If, at the end of that time, you are not entirely satisfied, you may return it to us and we will exchange it for an outfit that will give you satisfactory service, or we will return your money, together with any freight charges you have paid.

Economy Gasoline Engines are guaranteed against defect in material and workmanship as long as they last. We will replace defective parts at any time free of expense to you.

Quick Delivery From City Near You.

We carry the engines and equipment listed on this and the opposite page in stock and can ship them at once either direct **from the factory at EVANSVILLE, IND.,** at the factory price, or from the warehouse in any of the following cities at the warehouse price.

ST. PAUL, MINN., OMAHA, NEB., HARRISBURG, PENNA.

Mail your order to our Chicago or Philadelphia store, allowing either factory or warehouse price according to whichever point is nearest to you and we will make shipment immediately.

Guaranteed Horse-Power	Cylinder Bore, Inches	Stroke of Piston, Inches	Engine Speed, R.P.M.	FLYWHEELS Diameter, Inches	FLYWHEELS Weight, Pounds	Crank-shaft Diameter, Inches	Actual Engine Weight, Pounds
1½	3¼	5	550	18	43	1¼	246
2	3⅝	5	550	19¾	52	1⅜	295
3	4¼	6	450	22	88	1⅝	483

1½, 2 and 3 Horse-Power Economy Gasoline Engines.

	Price From Factory	Price From Warehouse
47L115—1½ Horse-Power Economy Gasoline Engine with Webster Magneto and 4x4-Inch Pulley. Shipping weight, 278 pounds	**$49.95**	**$52.00**
47L12—2 Horse-Power Economy Gasoline Engine with Webster Magneto and 4x4-Inch Pulley. Shipping weight, 330 pounds	**63.00**	**65.50**
47L23—3 Horse-Power Economy Gasoline Engine with Webster Magneto and 8x4-Inch Pulley. Shipping weight, 520 pounds	**79.95**	**83.75**

Pump Jacks.

A pump jack is used with an engine to operate a pump. Jack 47L337, on the right, can be used on any hand or windmill force pump so that pump can be operated by hand or with engine. Will handle wells up to 300 feet deep. The Jack, 47L312, as shown above, is used to operate a horizontal pump, a clamp being furnished to fasten around the handle of the pump. It can also be used to operate a three-way pump or as an overhead jack. Both jacks are back geared four to one, with three strokes, 4½, 7 and 9½ inches, equipped with 13-inch tight and loose pulleys, 2¼-inch face. Should run 160 revolutions per minute operating the pump forty strokes a minute.

47L337

	Price From Factory	Price From Warehouse
47L337—Double Gear Jack, for wells up to 300 feet deep. Shipping weight, 90 pounds	**$6.50**	**$7.15**
47L312—Double Gear Horizontal Jack, with hand clamp and stand support. Shipping wt., 95 lbs.	**7.15**	**7.75**

Pumping Outfits.

These outfits as shown at the left consist of an Engine with pulley, a 47L337 Double Gear Pump Jack and an 11-foot belt. Can be used on any hand or windmill force pump. Pump can be run with engine or by hand. Will handle wells up to 300 feet deep.

	Price From Factory	Price From Warehouse
47L15337—1½ Horse-Power Pumping Outfit, including Engine, Jack and Belt. Shipping weight, 380 pounds	**$58.20**	**$60.95**
47L12337—2 Horse-Power Pumping Outfit, including Engine, Jack and Belt. Shipping weight, 420 pounds	**71.25**	**74.45**

Direct Connected Gear Driven Pumping Outfits.

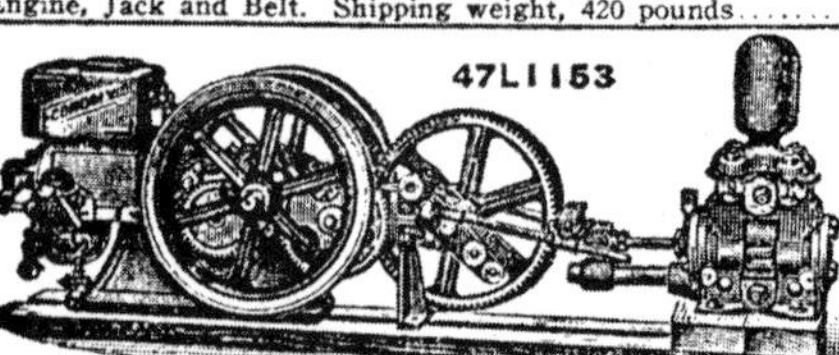

Price does not include the pump.

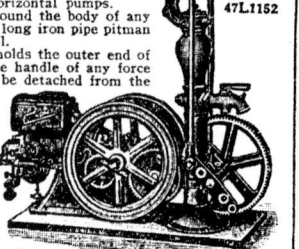

These outfits consist of a 1½ Horse-Power Economy Gasoline Engine with **Webster Magneto,** 4x4-Inch Pulley and a Pump Jack, fastened by four cap screws to the base of the engine so that the jack is driven by gears instead of a belt. Jacks furnished for vertical or horizontal pumps.

The vertical jack shown below is built to clamp around the body of any ordinary hand or windmill force pump, furnished with long iron pipe pitman rods and crosshead. Will handle up to 300-foot well.

The horizontal jack shown above has a stand that holds the outer end of the jack and short rods with a clamp to fasten to the handle of any force pump. When engine is not used for pumping it can be detached from the jack and used for other work. Shipping weight, 390 pounds. **Shipped only from factory at EVANSVILLE, IND.**

47L1152

47L1152—1½ Horse-Power Economy Gasoline Engine with Webster Magneto, 4x4-Inch Pulley and Direct Connected Vertical Pump Jack **$63.95**

47L1153—1½ Horse-Power Economy Gasoline Engine with Webster Magneto, 4x4-Inch Pulley and Direct Connected Horizontal Pump Jack ... **$63.95**

Hand Trucks for Small Engines.

47L1 Truck, as shown in the illustration, has an angle steel frame, 26 inches long; ¾-inch pipe axle, 18 inches wide; cast wheels, 9-inch diameter with 2-inch tires; for use with our 1½ and 2 horse-power Economy or Thermoil engines, or may be used with any 1½ or 2 horse-power engine.

47L35 Truck has an angle steel frame, 36 inches long; 1⅜-inch solid steel axle, 30 inches wide; steel wheels, 14 inches in diameter with 2½-inch tires; for use with our 3 or 5 horse-power Economy engines, or may be used with any engine from 2½ to 5 horse-power.

	Price From Factory	Price From Warehouse
47L1—All Steel Hand Truck for any 1½ and 2 horse-power engines. Shipping weight, 45 pounds	**$ 4.65**	**$ 5.30**
47L35—All Steel Hand Truck for any 2½ to 5 horse-power engines. Shipping weight, 118 pounds	**11.25**	**12.15**

Small Grain Grinding Outfits.

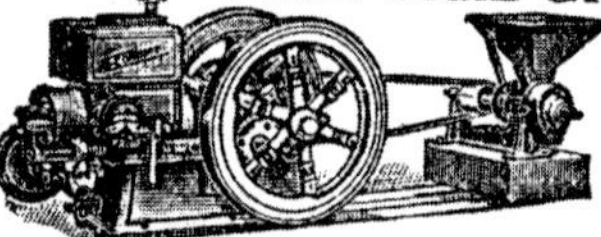

Outfits consist of an Economy Gasoline Engine with pulley, a Little Wonder Small Grain Grinder and 20 feet of 3-inch 4-ply rubber belt. Will grind small quantities of mixed grain, chicken feed, etc., and will make a good grade of table meal. Two sets of 5½-inch burrs for coarse and fine grinding.

	Price From Factory	Price From Warehouse
47L1156—Grinding Outfit, including 1½ Horse-Power Economy Gasoline Engine, Small Grain Grinder and Belt. Shipping weight, 361 pounds	**$63.05**	**$65.65**
47L126—Grinding Outfit, including 2 Horse-Power Economy Gasoline Engine, Small Grain Grinder and Belt. Shipping weight, 415 pounds	**76.10**	**79.15**
47L236—Grinding Outfit, including 3 Horse-Power Economy Gasoline Engine, Small Grain Grinder and Belt. Shipping weight, 603 pounds	**93.05**	**97.40**

Run Your Machines From a Line Shaft.

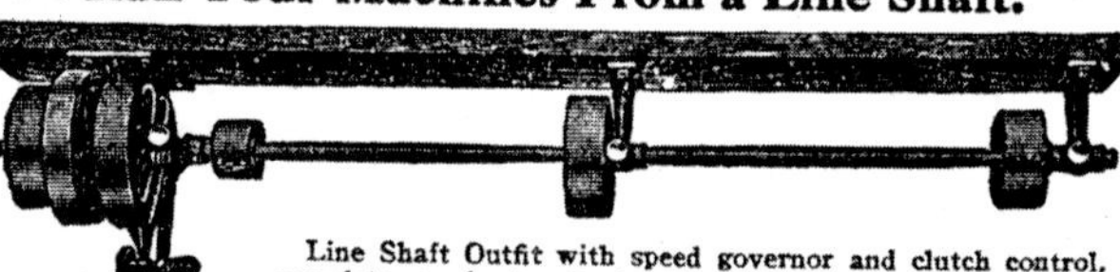

Line Shaft Outfit with speed governor and clutch control, complete, ready to attach to ceiling or side wall. Can be used with any engine up to 3 horse-power, to run the cream separator, churn, washing machine, pump jack or any small machine, as the speed governor takes care of variations in speed of engine and size of engine pulley.

A special adjustment on the speed governor regulates the speed of the line shaft. It can be set permanently for any desired speed, or the speed can be increased or decreased as desired. Governor also acts as a clutch. Full instructions with each outfit.

	Price From Factory	Price From Warehouse
47L321—8-Foot Line Shaft Outfit consists of an 8-foot shaft, the speed governor, one 4-inch, one 6-inch, two 8-inch pulleys and two hangers, mounted on 8-foot wood base. Weight, 115 pounds	**$22.50**	**$23.20**
47L322—12-Foot Line Shaft Outfit consists of a 12-foot shaft, the speed governor, one 4-inch, two 6-inch, three 8-inch pulleys and three hangers, mounted on 12-foot wood base. Weight, 150 pounds	**27.85**	**28.85**
47L323—50 feet of 2-Inch, 3-Ply Rubber Belt, with six belt fasteners for use with above outfits. Wt., 6 lbs.	**8.00**	**8.00**

And Farm Power Outfits

SATISFACTORY SERVICE GUARANTEED

Economy Gasoline Engines are made of high grade material by engine experts who have made a life study of the engine business. There are over 150,000 in use on all kinds of work and we honestly believe that there is no better engine made or sold at any price.

Economy Gasoline Engines operate on the four-cycle principle of the hit and miss type, the speed being controlled by a very sensitive fly ball governor. For ignition we furnish at the price quoted the well known **Webster Magneto** that gives a big hot spark for starting and running the engine.

The engines are complete with cast iron belt pulley, starting crank, lubricator and grease cups. They are water cooled, so that all you have to do is to fill the water reservoir on top of the engine with water, put gasoline in the tank in the base of the engine and start the engine according to the simple instructions given in a complete instruction book that we send with each engine.

5, 7, 9 and 12 Horse-Power Economy Gasoline Engines.

	Price From Factory	Price From Warehouse
47L25—5 Horse-Power Economy Gasoline Engine with Webster Magneto and 12x6-Inch Pulley. Shipping wt., 860 lbs.	**$108.85**	**$114.75**
12x6-Inch Friction Clutch Pulley in place of regular, add	**13.75**	**13.75**
47L27—7 Horse-Power Economy Gasoline Engine with Webster Magneto and 16x6-Inch Pulley. Shipping wt., 1,200 lbs.	**157.50**	**167.00**
16x8-Inch Friction Clutch Pulley in place of regular, add	**16.45**	**16.45**
47L29—9 Horse-Power Economy Gasoline Engine with Webster Magneto and 20x8-Inch Pulley. Shipping wt., 1,985 lbs.	**212.00**	**228.00**
20x8-Inch Friction Clutch Pulley in place of regular, add	**24.20**	**24.20**
47L212—12 Horse-Power Economy Gasoline Engine with Webster Magneto and 24x8-Inch Pulley. Shipping wt., 2,600 lbs.	**269.00**	**289.00**
24x8-Inch Friction Clutch Pulley in place of regular, add	**27.40**	**27.40**

Guaranteed Horse-Power	Cylinder Bore, Inches	Stroke of Piston, Inches	Engine Speed, R.P.M.	FLYWHEELS		Crank-shaft Diameter, Inches	Actual Engine Weight, Pounds
				Diameter, Inches	Weight, Pounds		
5	5	7½	425	28	168	2	777
7	5¾	9	375	34	242	2¼	1,107
9	6½	11	325	38	386	2½	1,823
12	7½	12	300	44	530	2¾	2,433

Feed Grinding Outfits.

These outfits consist of an engine with pulley, an 8-Inch David Bradley Corn and Cob Crusher and Feed Grinder and a 30-Foot 5-Inch 4-Ply Rubber Belt complete, **all ready to grind feed**. **Grinder furnished** with one set of coarse burrs for crushing or cracking ear and shelled corn for rough feed and one set of medium burrs for finer grinding of shelled corn and small grain. For complete description of Grinder see page 889.

	Price From Factory	Price From Warehouse
47L258—5 Horse-Power Engine, with 12x6-Inch Pulley, Grinder and Belt. Shipping wt., 1,146 pounds	**$143.40**	**$151.50**
47L278—7 Horse-Power Engine, with 16x6-Inch Pulley, Grinder and Belt. Shipping wt., 1,485 pounds	**192.05**	**203.75**

Friction Clutch Pulley for Any Make of Engine.

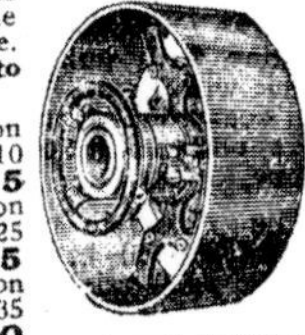

We recommend the use of Friction Clutch Pulleys on all engines of 5 horse-power or larger, as they enable you to start the engine alone or you can stop the machine without stopping the engine. Prices are for shipment from factory or warehouse. **Be sure to give name and horse-power of engine.**

47L510—10x6-Inch Friction Clutch Pulley. Weight, 83 pounds **$14.95**

47L512—12x6-Inch Friction Clutch Pulley. Weight, 83 pounds **$16.35**

47L516—16x8-Inch Friction Clutch Pulley. Weight, 100 pounds **$19.95**

47L520—20x8-Inch Friction Clutch Pulley. Weight, 210 pounds **$29.95**

47L524—24x8-Inch Friction Clutch Pulley. Weight, 225 pounds **$34.65**

47L528—28x8-Inch Friction Clutch Pulley. Weight, 235 pounds **$38.50**

Drag Saw Outfit.

Shipped From Evansville, Ind., Only.

47L2232—**Economy Portable Drag Saw Outfit, complete with Engine and 5-foot saw, as shown.** . **$87.50**

47L229—5-Foot Extra Saw Blade **5.60**

47L330—4x4-Inch Pulley for power purposes, extra **1.25**

The Economy Drag Saw Outfit is complete with a 1½ horse-power Economy gasoline engine with Webster magneto for ignition, mounted on a 9-foot hardwood frame. A multiple disc lever clutch enables the operator to start the engine without starting the saw or to stop the saw without stopping the engine and prevents damage to saw blade in case it binds in the cut.

The two all steel wheels, 18 inches in diameter, with 3-inch tires, are mounted on swivel axles that can be easily changed, so the outfit may be drawn from one place to another or moved from cut to cut along the log. Saw blade is 5 feet long, made of high grade saw steel and is filed and set, ready for use. Length of stroke, 20 inches; saw operates at 150 strokes per minute.

The saw is held by a guide, which also holds the saw rigid while the outfit is being moved. A truss rod extends the full length of the frame, strengthening the outfit and preventing vibration. The anchor hook holds the outfit to the log, is easily handled and prevents slipping.

Standard binder chain is used for driving the saw, as chain drive is very flexible, easy to handle and repairs can be purchased at any hardware store.

Engine can be removed from the saw frame and used for pumping water or any other work. Pulley furnished at extra price shown above. Engine furnishes plenty of power to drive the saw and we guarantee the outfit to give you satisfactory service. Weight, complete, 490 pounds; crated for shipment, 525 pounds. Full instructions accompany the outfit, telling how to set it up, how to start and run the engine. **Shipped complete from factory at EVANSVILLE, IND.**

We cannot furnish saw frame without engine, as engine is made special for this outfit. Any engine that you might have, even an Economy, will not work.

Engine and Steel Truck.

These outfits consist of an engine with regular cast iron pulley and an all steel truck with pole. The truck is of special construction, with bent channel steel frame. The channel steel frame is well braced to solid steel axles, reinforced with I beams. The wheels are all steel, 24-inch front and 32-inch rear. A high grade outfit that we guarantee will give you satisfactory service.

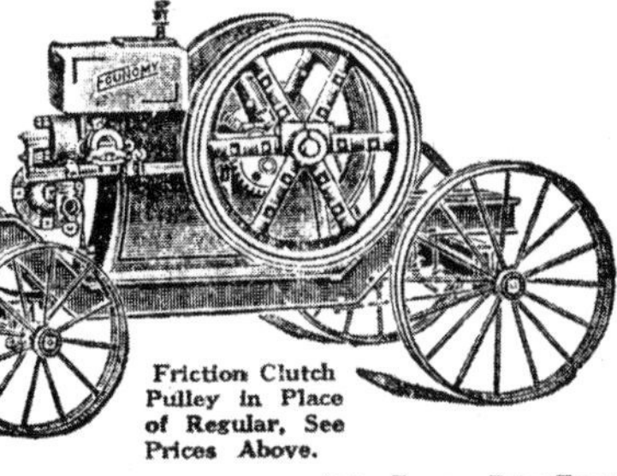

Friction Clutch Pulley in Place of Regular, See Prices Above.

	Price From Factory	Price From Warehouse
47L257—5 Horse-Power Engine, 12x6-Inch Pulley, Truck and Pole. Shipping weight, 1,447 pounds	**$148.85**	**$162.75**
47L277—7 Horse-Power Engine, 16x6-Inch Pulley, Truck and Pole. Shipping weight, 1,787 pounds	**197.50**	**215.00**
47L2978—9 Horse-Power Engine, 20x8-Inch Pulley, with Truck and Pole. Shipping weight, 2,633 pounds	**259.75**	**284.50**
47L21278—12 Horse-Power Engine, 24x8-Inch Pulley, with Truck and Pole. Shipping weight, 3,248 pounds	**316.75**	**345.50**
47L7—Truck only for any make of engine up to 7 horse-power. Shipping weight, 587 pounds	**40.00**	**48.00**
47L78—Truck only for any make of engine up to 12 horse-power. Shipping weight, 648 pounds	**47.75**	**56.50**

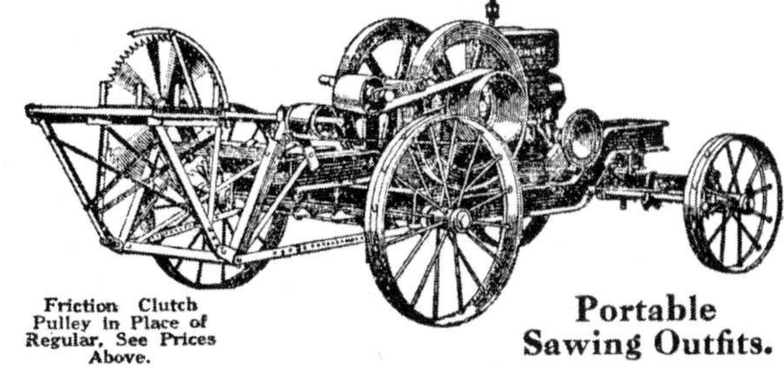

Friction Clutch Pulley in Place of Regular, See Prices Above.

Portable Sawing Outfits.

These saw rigs are complete, ready to saw wood, consisting of an Economy gasoline engine with regular cast iron pulley, an all steel truck with pole, and a steel saw frame fastened and braced to the back of the truck. Saw frame consists of an all steel tilting or sliding table, a saw with guard, a belt tightener and rubber belt. Saw table can be taken off the truck when not in use. **Be sure to tell us if you want the tilting or sliding table.** Illustration shows tilting table. Shipped on trial; satisfactory service guaranteed.

	Price, From Factory	Price, From Warehouse
47L2573—Portable Sawing Outfit, with 5 Horse-Power Economy Engine, 12x6-Inch Pulley, Saw Table, 26-Inch Saw and Belt. Shipping wt., 1,703 lbs.	**$179.00**	**$196.15**
47L2773—Portable Sawing Outfit with 7 Horse-Power Economy Engine, 16x6-Inch Pulley, Saw Table, 30-Inch Saw and Belt. Shipping wt., 2,043 lbs.	**229.00**	**249.75**

Thermoil Cheap Fuel Engines

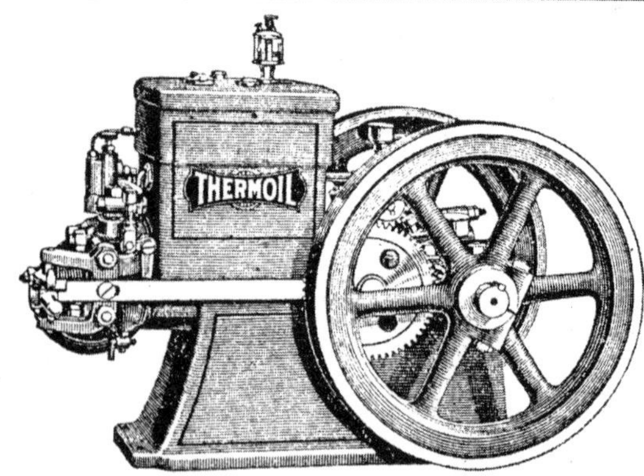

1½ Horse-Power Thermoil Engine.

47L18—1½ Horse-Power Thermoil Engine with 4x4-Inch Pulley. Shipping weight, 365 pounds. Shipped from COLUMBUS, IND., only.................. **$65.75**

Thermoil Engines Start and Run on Cheap Fuel.

Thermoil Cheap Fuel Engines were designed to start and run on all grades of kerosene, as this is the fuel that is now most readily obtainable in all parts of the country; the lowest grades, costing about one-half as much as gasoline, will give the best results. You do not have to start on gasoline and then switch to kerosene. All that is necessary is to turn on the fuel by opening the speed control wheel, relieve the compression and crank the engine. The instruction book that we send with each engine tells how to do it.

Thermoil Engines are complete with belt pulley, fuel tank, starting crank, lubricator and grease cups. They operate on the four-cycle throttling governor principle, both intake and exhaust valves are mechanically operated by an all steel cam shaft, piston is extra long, fitted with fine special ground rings, insuring good compression.

Thermoil Engines have no carburetor, batteries or electrical ignition of any kind. The fuel is fired by the heat of compression. The piston compresses the air in the cylinder to 480 pounds per square inch, heating it to about 1,000 degrees Fahrenheit, which sets fire to the fuel in the cylinder.

Shipped on 60 Days' Trial.

We will ship you any engine or outfit shown on this page on receipt of the full purchase price with the understanding that you are to try it sixty days before you decide to keep it. If, at the end of that time, you are not entirely satisfied, you may return it to us and we will exchange it for an outfit that will give you satisfactory service, or we will return your money, together with any freight charges you have paid.

Thermoil Engines are guaranteed against defect in material and workmanship as long as they last. We will replace defective parts at any time free of expense to you.

7 and 9 Horse-Power Thermoil Engines.

	Shipped From Evansville, Ind.	Shipped From St. Paul, Minn., Omaha, Neb., or Harrisburg, Penna.
47L78—7 Horse-Power Thermoil Engine with 10x6-Inch Pulley. Shipping weight, 1,250 pounds....	**$219.85**	**$228.50**
10x6-Inch Friction Clutch Pulley in place of regular	**12.30**	**12.30**
47L98—9 Horse-Power Thermoil Engine with 14x8-Inch Pulley. Shipping weight, 1,745 pounds...	**269.50**	**283.00**
14x8-Inch Friction Clutch Pulley in place of regular	**15.90**	**15.90**

Save Two-Thirds of Your Fuel Cost.

Thermoil Cheap Fuel Engines are so constructed that they use only 60 per cent as much fuel as the gasoline or throttling governor kerosene engines, and the larger sizes, 7 and 9 horse-power, will operate on the lower grade fuels. If you are in position to buy fuel oil, crude oil or any of the lower grade fuels, you can take advantage of this additional saving, as the 7 and 9 horse-power engines will operate on any fuel that is thin enough to flow through the fuel pipe.

Fuel Consumption of Thermoil Engines.

For an Eight-Hour Run on One-Fourth, One-Half, Three-Fourths and Full Load, Using Kerosene as Fuel.

	ON ¼ LOAD	ON ½ LOAD	ON ¾ LOAD	ON FULL LOAD
1½ HORSE POWER	6/10 OF A GALLON	¾ OF A GALLON	9/10 OF A GALLON	1 GALLON
7 HORSE POWER	2½ GALLONS	3 GALLONS	4 GALLONS	5 GALLONS
9 HORSE POWER	3 GALLONS	3¾ GALLONS	5 GALLONS	6½ GALLONS

Feed Grinding Outfits.

These outfits consist of a Thermoil Engine with pulley, an 8-inch David Bradley Corn and Cob Crusher and Feed Grinder and a 30-Foot 5-Inch 4-Ply Rubber Belt complete, all ready to grind feed. Grinder regularly furnished with one set of coarse burrs for crushing or cracking ear and shelled corn for rough feed and one set of medium burrs for finer grinding of shelled corn and small grain.

For complete description of Grinder see page 889.

	Shipped From Evansville, Ind.	Shipped From St. Paul, Minn., Omaha, Neb., or Harrisburg, Penna.
47L788—7 Horse-Power Thermoil Engine, with 10x6-Inch Pulley Grinder and Belt. Shpg. wt., 1,536 lbs.	**$254.40**	**$265.25**
47L988—9 Horse-Power Thermoil Engine, with 14x8-Inch Pulley, Grinder and Belt. Shpg. wt., 2,035 lbs.	**304.05**	**319.75**
47L1700—Pair of Special Oats Burrs. Postpaid from factory		**1.50**

Portable Engines.

These outfits consist of a Thermoil Engine with pulley and an all steel truck with pole. The truck is of special construction, with bent channel steel frame.

The channel steel frame is well braced to solid steel axles, reinforced with I beams. The wheels are all steel, 24-inch front and 32-inch rear.

A high grade outfit that we guarantee will give you satisfactory service.

	Shipped From Evansville, Ind.	Shipped From St. Paul, Minn., Omaha, Neb., or Harrisburg, Penna.
47L787—7 Horse-Power Engine, with 10x6-Inch Pulley, with Truck and Pole. Shipping wt., 1,837 lbs..	**$259.35**	**$276.50**
47L9878—9 Horse-Power Engine, with 14x8-Inch Pulley, with Truck and Pole. Shipping wt., 2,393 lbs...	**317.25**	**339.50**

Portable Sawing Outfits.

These saw rigs are complete, ready to saw wood, consisting of a Thermoil Kerosene Engine with pulley, an all steel truck with pole, an all steel tilting or sliding table saw frame fastened and braced to the back of the truck, a saw with guard, a belt tightener and rubber belt. Saw table can be taken off the truck when not in use.

Be sure to tell us if you want tilting or sliding table. Illustration shows tilting table. Shipped on trial; satisfactory service guaranteed.

	Shipped From Evansville, Ind.	Shipped From St. Paul, Minn., Omaha, Neb., or Harrisburg, Penna.
47L7873—Portable Sawing Outfit, with 7 Horse-Power Thermoil Engine, Saw Frame, 30-Inch Saw and Belt. Shipping weight, 2,093 pounds	**$290.85**	**$311.25**
47L9873—Portable Sawing Outfit, with 9 Horse-Power Thermoil Engine, 14x8-Inch Pulley, Saw Frame, 30-Inch Saw and Belt. Shipping weight, 2,649 pounds	**348.75**	**374.25**

Motorgo Row Boat Engine $79.95

Easy Starter
Guaranteed Magneto
Safety Clutch Propeller
Tilting Device

Fastens on the back of any ordinary row boat and will push it from 6 to 8 miles an hour any place that a row boat will go—through streams with barely enough water to float the boat; through weeds and over rocks, stumps or other obstructions that would put the ordinary rowboat engine out of commission.

The Motorgo is equipped with a tilting device so that if you strike a rock or log or run into a bunch of weeds the propeller tilts just enough to let it go by. If the propeller blade strikes the obstruction it is equipped with an automatic safety clutch to release the propeller, allowing it to slip until you have passed over the object, when it takes hold again and continues to push the boat.

The Motorgo steers with the propeller and is equipped with a compensating spring that takes care of the side strain on the steering handle, which makes the outfit very flexible and easy to handle, particularly in rough weather. The engine is reversible—can be stopped and started instantly—and with a little practice you will be able to dock your boat or get away from the dock just as quickly and easily as with the rudder steered outfit.

It does just what we say it will and we guarantee it.

The Motorgo is equipped with a high grade guaranteed magneto that insures easy starting and constant running. We also furnish a rope starter as shown in the illustration to the right. One pull turns the engine over several times which, together with the high grade magneto, eliminates all starting trouble. (We have discontinued the battery equipped outfit entirely.)

Some manufacturers are making a feature of light weight, but we prefer to furnish an outfit built of material that we know will stand up and give years of satisfactory, dependable service. We guarantee every part of the Motorgo against defect in material or workmanship and will replace defective parts free of charge at any time.

The fuel tank, holding 3 quarts of gasoline, enough for about a four hours' run, is made of aluminum, as are also the flywheel, exhaust manifold and crankcase. The engine is accurately balanced so that it will give the minimum amount of vibration and we guarantee the Motorgo will give you just as good service as any outfit you can buy regardless of price.

The bracket that holds the engine to the back of the boat is made of malleable iron and is adjustable to any angle of the stern. Malleable iron is heavier than aluminum, but it gives you the strength where it is needed and we believe that is what you want.

The lower part of the Motorgo is made of bronze, so that it can be used in either fresh or salt water. The gears on the inside of the housing are of steel, packed in grease. The automatic safety clutch propeller is weedless, 9 inches in diameter with 10-inch pitch. The cylinder and piston are made of close grained gray iron, ground to a perfect fit. The connecting rod is made of high grade phosphor bronze. The crank shaft is of carbon steel. The cylinder is 2⅝-inch bore, 2½-inch stroke, running 850 to 1,000 revolutions per minute, commonly called 2 horse-power.

The exhaust side, showing method of using the rope starter.

The tilting device and automatic safety clutch propeller make it possible for you to go through any water that is deep enough to float a boat; over rocks, stumps, weeds or any obstructions without fear of damage to the propeller or engine.

Shipped from factory in SOUTHERN MICH.

The automatic safety clutch propeller is simple in construction, there are no complicated parts. It is always on the job when you need it to prevent damage to propeller or engine.

47L812—Motorgo Rowboat Engine, with magneto ignition. Actual weight, 65 pounds. Shipping weight, 95 pounds. **Shipped from factory in SOUTHERN MICHIGAN**.......... **$79.95**

47L813—Under Water Exhaust. Weight, 2 pounds.......... **$2.40**

47L814—Waterproof Canvas Cover that goes over the entire top of the outfit to protect it from rain. Weight, 1 pound.......... **$1.65**

Two Cycle Boat Engines

We have been selling Motorgo Two-Cycle Boat Engines for over ten years. They are giving satisfactory service in thousands of boats in all kinds of waters.

Every one of these engines was shipped on a thirty days' trial, guaranteed against defect in material or workmanship. Very few ever come back, which is conclusive proof that they are a superior engine and will give years of satisfactory service.

Furnished with single and double cylinders. Painted battleship gray enamel, nicely balanced to run your boat with the least vibration. Economical in the use of fuel and are guaranteed to give you satisfactory service.

Send us your order, enclose the price listed below, we will make immediate shipment from factory in **SOUTHERN MICHIGAN** with the understanding that you may try the engine thirty days. If at the end of that time you are not entirely satisfied—if you don't feel that you have a high grade engine at a much lower price than you would have to pay elsewhere, you may return the engine and we will send you back your money together with any freight charges that you have paid. You must be satisfied with a Motorgo Engine or we don't want you to keep it.

Engines.

47L803—2½ horse-power single cylinder with mixing valve, muffler and battery ignition. Shipping weight, 148 lbs..... **$71.95**

47L804—4 horse-power single cylinder with Schebler carburetor, muffler and battery ignition. Shipping weight, 200 pounds.......... **$85.50**

47L806—6 horse-power double cylinder with Schebler carburetor, muffler and battery ignition. Shipping weight, 218 pounds.......... **112.95**

47L808—8 horse-power double cylinder with Schebler carburetor, muffler and battery ignition. Shipping weight, 288 pounds.......... **131.00**

Bosch Magneto Ignition in place of Battery on any of the above engines, add.......... **15.00**

At the prices quoted above engines are complete with timer, Schebler carburetor (mixing valve on the 2½ horse-power), bronze plunger pump, grease cups, a spark plug and priming cup for each cylinder, thrust bearing, rear coupling for attaching a propeller shaft, battery ignition, consisting of six dry cell batteries, a high grade vibrating spark coil and the necessary wiring with terminals, muffler, starting crank and book of instructions.

Boat Equipment.

Consisting of 6 feet of propeller shaft (steel for fresh water, bronze for salt water), a bronze stuffing box with lag screws and a two-blade bronze speed propeller (or will furnish weedless if specified on your order). Propellers specified are of the proper size to give the best results considering speed of engine and horse-power developed.

	Fresh Water	Salt Water
47L8031—12-Inch Two-Blade Bronze Speed Propeller and Stuffing Box with 6 feet of ¾-Inch Propeller Shaft for 2½ Horse-Power Engine. Weight, 38 pounds	**$8.00**	**$11.70**
47L8041—14-Inch Two-Blade Bronze Speed Propeller and Stuffing Box with 6 Feet of ⅞-Inch Propeller Shaft for 4 and 6 Horse-Power Engines. Weight, 42 pounds	**$9.10**	**$13.45**
47L8081—15-Inch Two-Blade Bronze Speed Propeller and Stuffing Box with 6 Feet of 1-Inch Propeller Shaft for 8 Horse-Power Engine. Weight, 62 pounds	**$9.85**	**$15.00**

5 Horse Power Four Cycle Boat Engine $99.75

A four-cycle single cylinder boat engine developing 5 full horse-power with only 160 pounds of weight.

All movable parts subject to wear are interchangeable with the same parts on the Ford engine, except the oil and water pump, crankshaft, magneto and cylinder. It is even equipped with the Holly carburetor and a ½-inch spark plug so that replacements can be purchased wherever parts for the Ford engine are sold. This is a new addition to our famous line of Motorgo engines—one that we are proud of and that we guarantee will give you just as satisfactory service as the other Motorgo Engines that we have been selling for a good many years.

The engine is complete at the price quoted below and is shipped on thirty days' trial with the understanding that if it is not entirely satisfactory it may be returned for exchange or we will send you back your money together with any freight charges you have paid—it is guaranteed against defect in material or workmanship as long as it lasts and, being made of high grade material all the way through, it should give you a lifetime of satisfactory service.

47L805—5 Horse-Power Four-Cycle Boat Engine, complete with Bosch magneto and impulse starter, regular Ford carburetor, and rear coupling. Shipping weight, 200 lbs. **Shipped from factory in SOUTHERN MICHIGAN**... **$99.75**

47L8051—14-Inch Two-Blade Bronze Speed Propeller and Stuffing Box with 6 Feet of ⅞-Inch Propeller Shaft. Weight, 42 pounds. Fresh Water **$9.10** Salt Water **$13.45**

47L8052—Muffler or Expansion Chamber for the above engine. Weight, 25 pounds.......... **2.60**

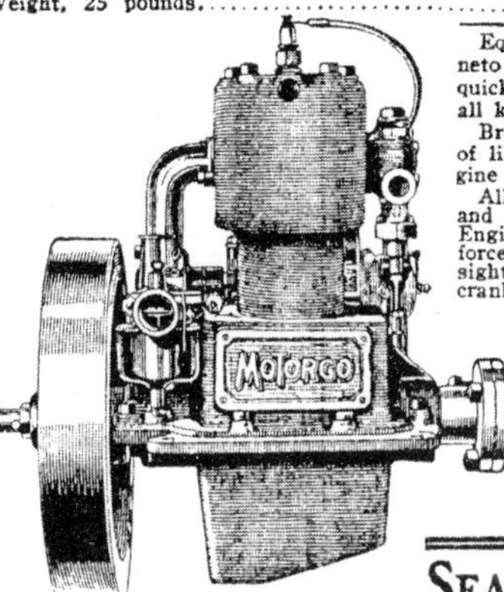

Equipped with Bosch High Tension Magneto with Impulse Starter, guaranteeing a quick, easy start and dependable running in all kinds of weather.

Bronze water and oil plunger type pumps of liberal size, accurate workmanship. Engine can be used in either salt or fresh water.

All bearings and gaskets are removable and interchangeable with those on the Ford Engine. Lubrication is taken care of by force feed pump and splash system with sight feed oil glass. The oil pump in the crank case holds 2 quarts of oil.

The crankshaft is drop forge steel, counterweighted and accurately balanced. The flywheel is taper bored, held on with a nut.

The cylinder is made of a special grade gray iron, 3¾-inch bore, 4-inch stroke, fitted with Ford piston and rings. Speed is 350 to 1,000 revolutions per minute. Two-blade bronze propeller, 12-inch diameter, 12-inch pitch.

"Protection" Ball Bearing Bench Wringer. $7.95

Guaranteed for Three Years. Enclosed Gears Mean Protection.

Rolls measure 10x1¾ in. and are guaranteed for three years. Steel pressure spring, improved wheel top screws. Hardwood frame. Enclosed gears protect fingers and clothes.

Holds two tubs. Folds into small space. Reversible drain board. Shipping weight, 46 pounds. Not mailable.

99L2035 $7.95

The New Modern Improved Protection Wringer ONLY $4.90

NEWEST DESIGN — WELL MADE — LATEST FEATURES

Quick Acting Reversible Drain Board.

Best Quality 11x1¾-Inch Rubber Rolls—Guaranteed for 3 Years.

"Oil-Less" Wood Bearings Need No Attention.

No Gears—Direct Even Pressure on Rolls at All Times. Large Central Adjusting Hand Wheel Regulates Pressure on Rolls.

Selected Hardwood Frame.

For Laundry Tubs See Page 703.

Shipping weight, 28 pounds.

99L2112 .... $4.90

OUR "Best Made" Ball Bearing Bench Wringer.

Guaranteed for Five Years.

Large Rolls, Enclosed Gears. The Reversible Drain Board can be set to drain on either side of wringer.

Rolls, guaranteed for five years, measure 11x1¾ in. Enclosed gears protect fingers and clothes. Have steel pressure spring and improved wheel top screws. Hardwood frame. Holds two tubs. Folds into small space. Not mailable. $8.90

Shipping weight, 48 pounds.

99L2036 $8.90

$3.80 "Acme Star" Wringer

An Excellent Quality Low Priced Wringer.

Guaranteed for One Year.

Rolls measure 10x1¾ inches and are guaranteed for one year, but with ordinary care will last much longer. Spiral steel springs and improved wheel top screws. Galvanized metal frame.

99L2026..(Shpg. wt., 22 lbs.)..$3.80

Solid Brass Washboard With Truss Back. 56c

Large rubbing surface, size 10¾x11 inches of sheet brass, with improved corrugated cable crimp, protector top and open back.

99L2052 56c

Double Zinc Washboard. Globe crimp on one side, plain crimp on the other.

99L2051 54c

Single Zinc Washboard. Globe crimp.

Shipping weight, 3 pounds.

99L2050 .... 46c

$2.65 Set of 4 Household Baskets.

A durable and serviceable set, comprising the four baskets most commonly used about the home, includes one large size clothes basket, one large clothes hamper, one market basket and one farm or vegetable basket, as illustrated. Shipping weight, 26 pounds. Not Mailable.

99L2079 — For 4 baskets, as illustrated $2.65

Our "Best Made" Ball Bearing Wringer $6.50

Rolls guaranteed for five years.

The Reversible Drain Board and Rigid Clamps Make These Wringers Especially Suitable for Stationary Laundry Tubs. Enclosed gears protect fingers and clothes. Have steel pressure spring and improved wheel top screws. Hardwood frame. Rolls measure 11x1¾ inches. Shipping wt., 34 lbs.

99L2033 .... $6.50

Vacuum Clothes Washer. 29c

Washes a tubful of clothes at once. Forces steam, suds and water through the clothes, quickly removing the dirt. Made of tin. Weight, 10 ounces.

9L2851 Without handle 29c

$4.85 Our Best Made Solid Copper Wash Boiler

Why Pay $6.50 to $7.00 Elsewhere?

Copper Plated Steel Cover.

Extra well made. The kind of boiler to buy when service and durability are desired. Extra heavy, solid copper wash boiler.

Large size, 23¾x12x12⅞ inches high, inside measurements. Tinned on the inside with block tin.

Heavy steel wire rod around top and corrugation of sides add to the strength of boiler. Double seamed and guaranteed not to leak. Stationary wood grip handles. Cover is unusually strong and is made from one piece of copper plated steel with rigid wood handle. Shipping weight, 17 pounds. Not mailable.

99L2045............ $4.85

$1.30 Extra Grade Galvanized Washtubs.

Heavy sheet steel, galvanized after tub is formed. Note wringer lugs. Will outwear two of ordinary kind. Shipping weight, 10 lbs. Not mailable.

24¾ In.

	Top Diam.	Depth	
99L2086	22 in.	10¾ in.	$1.15
99L2087	24¾ in.	11 in.	1.30

65c Standard Quality Galvanized Washtubs.

21½ In.

Shipping weight, 8 pounds. Not Mailable.

	Top Diam.	Depth	
99L2084	21½ in	10½ in	65c
99L2085	24 in	11 in.	80c

Extra Grade Galvanized Pails. 55c 14-Qt.

Heavy sheet steel, galvanized after pail is formed. Shipping weight, 4 pounds.

	Cap., Qts.	
99L2096	12	50c
99L2097	14	55c

Standard Quality Galvanized Pails. 25c 12-Qt.

Shipping weight, 2 pounds.

	Cap., Qts.	
99L2094	12	25c
99L2095	14	28c

Solid Copper Wash Boiler $3.95

Tin Covers.

22x11⅝x13 In.

Standard high quality solid copper wash boiler. Made of heavy polished sheet copper, tinned on the inside. Drop handles, wired around the top and double seamed to prevent leaking.

99L2043 — Size, 20⅜x10¾x13 inches high, inside measurements. Shipping weight, 13 pounds....$3.65

99L2044 — Size, 22x11⅝x13 inches high, inside measurements. Shipping weight, 14 pounds....$3.95

Galvanized SHEET STEEL Wash Boiler $2.35

NOTE—COVER IS GALVANIZED SAME AS BOILER.

Large Size—Strong and Durable.

Sides and bottom of heavy sheet steel, galvanized after boiler is formed. Measures 23x12x13 inches high, inside measurements. Shipping weight, 18 lbs. Not mailable.

99L2040 ....$2.35

Copper Rim and Bottom Tin Wash Boilers. $2.20

20¼x10¼x13 Inches.

A serviceable boiler. Low priced. Full size. Stationary wood grip handles.

99L2041 Size, 20¼x10¾x13in. high, inside measurements. Shipping weight, 12 lbs....$2.20

99L2042 Size, 22x11¼x13 in. high, inside measurements. Shipping weight, 14 lbs....$2.45

Copper Bottom Tin Wash Boilers. $1.70

Full size, double seamed. Drop handles.

99L2038—Size, 20¼x10¼x13 inches high, inside measurements. Shipping weight, 12 pounds....$1.70

QUITE NATURALLY OUR PRICES ARE LOW. Our great purchases and economical selling methods assure it. But add to this the quality our scientific tests assure you, and you must admit our values are extraordinary.

All weights and measurements given on this page are approximate and may vary a trifle.

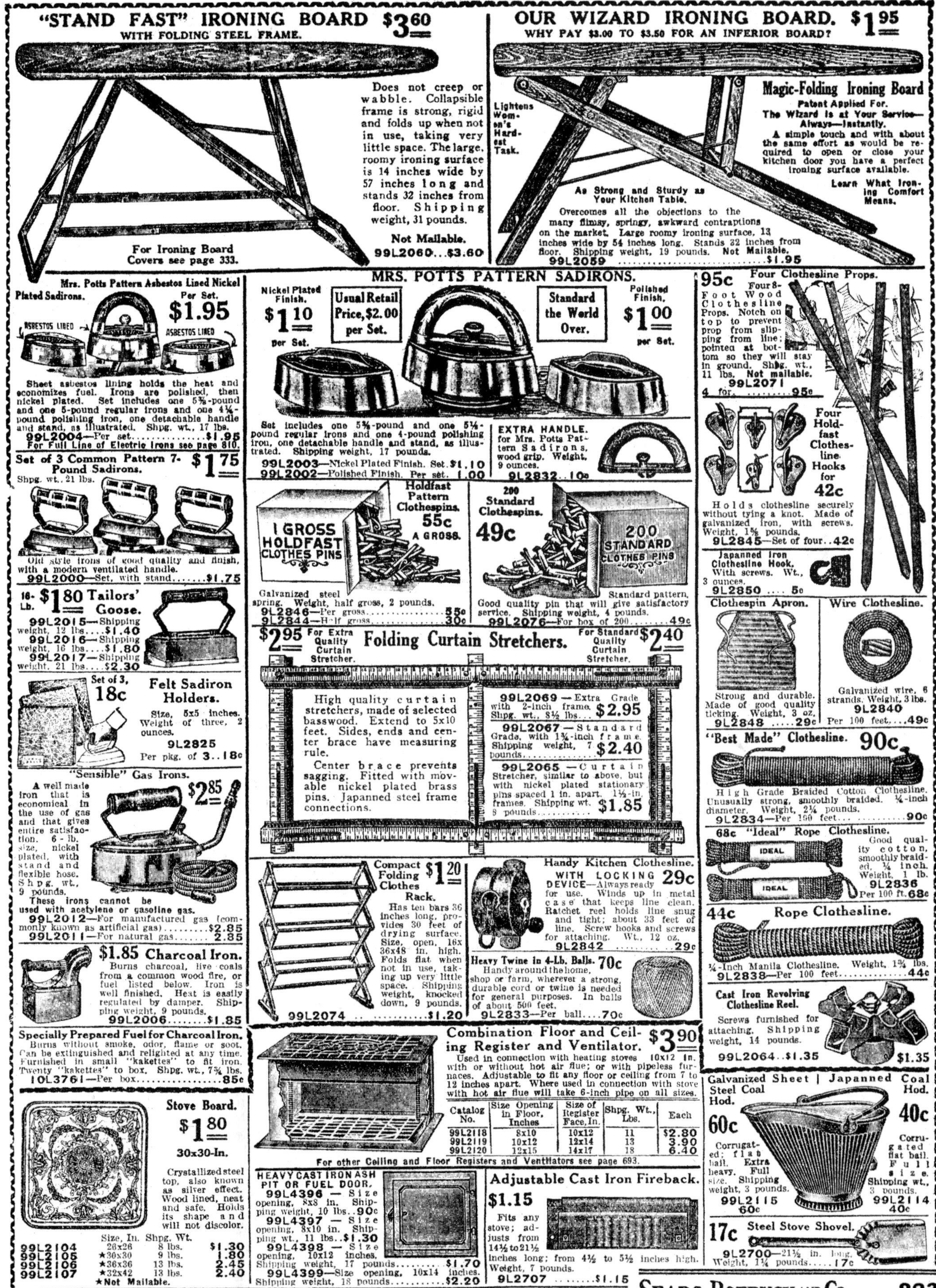

"STAND FAST" IRONING BOARD $3.60
WITH FOLDING STEEL FRAME.

Does not creep or wabble. Collapsible frame is strong, rigid and folds up when not in use, taking very little space. The large, roomy ironing surface is 14 inches wide by 57 inches long and stands 32 inches from floor. Shipping weight, 31 pounds.

Not Mailable.
99L2060...$3.60

For Ironing Board Covers see page 333.

OUR WIZARD IRONING BOARD. $1.95
WHY PAY $3.00 TO $3.50 FOR AN INFERIOR BOARD?

Lightens Women's Hardest Task.

Magic-Folding Ironing Board
Patent Applied For.
The Wizard is at Your Service—Always—Instantly.
A simple touch and with about the same effort as would be required to open or close your kitchen door you have a perfect ironing surface available.

Learn What Ironing Comfort Means.

As Strong and Sturdy as Your Kitchen Table.
Overcomes all the objections to the many flimsy, springy, awkward contraptions on the market. Large roomy ironing surface, 13 inches wide by 54 inches long. Stands 32 inches from floor. Shipping weight, 19 pounds. Not Mailable.
99L2059 ....$1.95

Mrs. Potts Pattern Asbestos Lined Nickel Plated Sadirons.
Per Set. $1.95

ASBESTOS LINED

Sheet asbestos lining holds the heat and economizes fuel. Irons are polished, then nickel plated. Set includes one 5⅝-pound and one 5-pound regular irons and one 4¼-pound polishing iron, one detachable handle and stand, as illustrated. Shpg. wt., 17 lbs.
99L2004—Per set..............$1.95
For Full Line of Electric Irons see page 810.

MRS. POTTS PATTERN SADIRONS.
Nickel Plated Finish. $1.10 per Set.

Usual Retail Price, $2.00 per Set.

Standard the World Over.

Polished Finish. $1.00 per Set.

Set includes one 5⅝-pound and one 5¼-pound regular irons and one 4-pound polishing iron, one detachable handle and stand, as illustrated. Shipping weight, 17 pounds.
99L2003—Nickel Plated Finish. Set. $1.10
99L2002—Polished Finish. Per set. 1.00

EXTRA HANDLE. for Mrs. Potts Pattern Sadirons, wood grip. Weight, 9 ounces.
9L2832...10c

95c Four Clothesline Props.
Four 8-Foot Wood Clothesline Props. Notch on top to prevent prop from slipping from line; pointed at bottom so they will stay in ground. Shpg. wt., 11 lbs. Not mailable.
99L2071
4 for..........95c

Set of 3 Common Pattern 7-Pound Sadirons. $1.75
Shpg. wt., 21 lbs.

Old style irons of good quality and finish, with a modern ventilated handle.
99L2000—Set, with stand.......$1.75

Holdfast Pattern Clothespins. 55c A GROSS.
1 GROSS HOLDFAST CLOTHES PINS

Galvanized steel spring. Weight, half gross, 2 pounds.
9L2846—Per gross.......55c
9L2844—Half gross......30c

200 Standard Clothespins. 49c
200 STANDARD CLOTHES PINS

Standard pattern. Good quality pin that will give satisfactory service. Shipping weight, 4 pounds.
99L2076—For box of 200.......49c

Four Holdfast Clothes-line Hooks for 42c
Holds clothesline securely without tying a knot. Made of galvanized iron, with screws. Weight, 1⅝ pounds.
9L2845—Set of four..42c

Japanned Iron Clothesline Hook. With screws. Wt., 3 ounces.
9L2850 .... 5c

16-Lb. $1.80 Tailors' Goose.
99L2015—Shipping weight, 12 lbs....$1.40
99L2016—Shipping weight, 16 lbs....$1.80
99L2017—Shipping weight, 21 lbs....$2.30

$2.95 For Extra Quality Curtain Stretcher. Folding Curtain Stretchers. For Standard Quality Curtain Stretcher. $2.40

High quality curtain stretchers, made of selected basswood. Extend to 5x10 feet. Sides, ends and center brace have measuring rule.

Center brace prevents sagging. Fitted with movable nickel plated brass pins. Japanned steel frame connections.

99L2069—Extra Grade with 2-inch frame. Shpg. wt., 8½ lbs....$2.95
99L2067—Standard Grade, with 1¾-inch frame. Shipping weight, 7 pounds....$2.40
99L2065—Curtain Stretcher, similar to above, but with nickel plated stationary pins spaced 1 in. apart. 1½-in. frames. Shipping wt. 8 pounds..........$1.85

Clothespin Apron.
Strong and durable. Made of good quality ticking. Weight, 3 oz.
9L2848 .....29c

Wire Clothesline.
Galvanized wire, 6 strands. Weight, 3 lbs.
9L2840
Per 100 feet....49c

Felt Sadiron Holders.
Set of 3, 18c
Size, 5x5 inches. Weight of three, 2 ounces.
9L2825
Per pkg. of 3..18c

"Best Made" Clothesline. 90c
High Grade Braided Cotton Clothesline. Unusually strong, smoothly braided. ¼-inch diameter. Weight, 2¼ pounds.
9L2834—Per 100 feet..........90c

"Sensible" Gas Irons. $2.85
A well made iron that is economical in the use of gas and that gives entire satisfaction. 6-lb. size, nickel plated, with stand and flexible hose. Shpg. wt., 9 pounds.
These irons cannot be used with acetylene or gasoline gas.
99L2012—For manufactured gas (commonly known as artificial gas).........$2.85
99L2011—For natural gas.......2.85

68c "Ideal" Rope Clothesline.
IDEAL
Good quality cotton, smoothly braided, ¼ inch. Weight, 1 lb.
9L2836
Per 100 ft. 68c

Compact Folding Clothes Rack. $1.20
Has ten bars 36 inches long, provides 30 feet of drying surface. Size, open, 16x36x48 in. high. Folds flat when not in use, taking up very little space. Shipping weight, knocked down, 9 pounds.
99L2074 ....$1.20

Handy Kitchen Clothesline. WITH LOCKING DEVICE 29c
Always ready for use. Winds up in metal case that keeps line clean. Ratchet reel holds line snug and tight; about 33 feet of line. Screw hooks and screws for attaching. Wt., 12 oz.
9L2842 ....29c

44c Rope Clothesline.
¼-Inch Manila Clothesline. Weight, 1¾ lbs.
9L2838—Per 100 feet..............44c

$1.85 Charcoal Iron.
Burns charcoal, live coals from a common wood fire, or fuel listed below. Iron is well finished. Heat is easily regulated by damper. Shipping weight, 9 pounds.
99L2006..............$1.85

Heavy Twine in 4-Lb. Balls. 70c
Handy around the home, shop or farm, wherever a strong, durable cord or twine is needed for general purposes. In balls of about 500 feet.
9L2833—Per ball....70c

Cast Iron Revolving Clothesline Reel.
Screws furnished for attaching. Shipping weight, 14 pounds.
99L2064..$1.35

Specially Prepared Fuel for Charcoal Iron.
Burns without smoke, odor, flame or soot. Can be extinguished and relighted at any time. Furnished in small "kakettes" to fit iron. Twenty "kakettes" to box. Shpg. wt., 7¾ lbs.
10L3761—Per box..............85c

Combination Floor and Ceiling Register and Ventilator. $3.90
Used in connection with heating stoves 10x12 in. with or without hot air flue; or with pipeless furnaces. Adjustable to fit any floor or ceiling from 7 to 12 inches apart. Where used in connection with stove with hot air flue will take 6-inch pipe on all sizes.

Catalog No.	Size Opening in Floor, Inches	Size of Register Face, In.	Shpg. Wt., Lbs.	Each
99L2118	8x10	10x12	11	$2.80
99L2119	10x12	12x14	13	3.90
99L2120	12x15	14x17	18	6.40

For other Ceiling and Floor Registers and Ventilators see page 693.

Galvanized Sheet Steel Coal Hod. | Japanned Coal Hod.
60c Corrugated; flat bail. Extra heavy. Full size. Shipping weight, 3 pounds.
99L2115 60c

40c Corrugated flat bail. Full size. Shipping wt., 3 pounds.
99L2114 40c

Stove Board. $1.80 30x30-In.
Crystallized steel top, also known as silver effect. Wood lined, neat and safe. Holds its shape and will not discolor.

	Size, In.	Shpg. Wt.	
99L2104	26x26	8 lbs.	$1.30
99L2105	★30x30	9 lbs.	1.80
99L2106	★36x36	13 lbs.	2.45
99L2107	★32x42	13 lbs.	2.40

★Not Mailable.

HEAVY CAST IRON ASH PIT OR FUEL DOOR.
99L4396—Size opening, 8x8 in. Shipping weight, 10 lbs..90c
99L4397—Size opening, 8x10 in. Shipping wt., 11 lbs..$1.30
99L4398—Size opening, 10x12 inches. Shipping weight, 17 pounds........$1.70
99L4399—Size opening, 10x14 inches. Shipping weight, 18 pounds........$2.20

Adjustable Cast Iron Fireback.
$1.15
Fits any stove; adjusts from 14½ to 21½ inches long; from 4½ to 5½ inches high. Weight, 7 pounds.
9L2707 ....$1.15

17c Steel Stove Shovel.
9L2700—21½ in. long. Weight, 1¼ pounds.....17c

All weights and measurements given on this page are approximate and may vary a trifle.

2 Brooms for the Usual Price of 1

Why Pay More?

Our Price, $1.25 For 2 Brooms.

Includes one extra fancy parlor broom for sweeping fine quality carpets, rugs, etc., and one good quality house broom for general use around the home. Shipping weight, 4½ pounds.

99L2142 2 brooms......$1.25

Floor Brush. 14-In. $1.35

Complete With Handle.

99L2147—Size, 12 inches. Shipping weight, 2½ pounds......$1.20
99L2148—Size, 14 inches. Shipping weight, 2½ pounds......$1.35
99L2149—Size, 16 inches. Shipping weight, 3 pounds......$1.50

Dusting Brush. 52c

9L2874—Standard quality. Length over all, 13¾ inches. Weight, 7 oz. ..52c

Japanned Heavy Tin Dust Pan. 30c

Weight, 1¼ lbs.
9L2880 ....30c

Combination Mop and Brush Holder Set. 95c

The Mop Holder is adjustable to hold either mop or brush. Can also be used for holding sponges, etc. Set includes one mop, one brush and one holder. Shpg. wt., 2½ lbs.
99L2165 ......95c

Self Wringing Cotton Mop. 85c

Keeps the hands out of the dirty, grimy and skin cracking mop water. To wring mop, just pull out the handle and twist it. Has long wood handle. Shipping weight, 2¼ pounds.
99L2162 ......85c

Extra Mop Cloth for above mop. Weight, 8 oz.
9L2858 ......30c

"High Grade" Mop Wringer Bucket.

Wrings out mop easily and quickly; 16-quart galvanized bucket. Will outlast two or three wood bucket wringers. Guaranteed to be satisfactory. Shipping wt., 15 lbs.
99L2098 $2.40

Cotton Mop Head. 19c

Made of cotton strand or yarn, stitched in the middle to strong tape forming the head for fastening in any ordinary mop stick or handle. Weight, 1 pound.
9L2862—Mop Head only, without handle. 19c

Our "Colonial" Carpet Sweeper

$3.95

Why Pay $5.00 to $6.00 Elsewhere?

Lever Adjustment for Taking Up Wear of Brush and Adding to Life of Sweeper. Mahogany Finish Hardwood Case.

A high grade Carpet Sweeper in every respect, being carefully made throughout. Adjustable to long or short nap carpet by pressure on handle. Wheels have rubber tires and patent roller bearings. Protecting band of woven braid encircles the sweeper to prevent marring of furniture. Strongly constructed and guaranteed to give the best possible service in everyday use. Shipping weight, 7 pounds.
99L2132......$3.95

BIG 3-IN-1 COMBINATION MOP OUTFIT.

Use the Polish Mop Once a Week. Our Price, $1.30. Regular $2.00 Outfit. Use the Dry Dust Mop Daily.

MAKES HOMES BRIGHTER AND HOUSEWORK LIGHTER.

A YEAR'S SAVING in buying all the family wearing apparel, dry goods, hardware and general supplies from us makes a goodly sum for a rainy day or to buy something extra.

Adjustable handle permits keeping the mop set square on the floor.
99L2153—Big 3-In-1 Combination Outfit, consisting of one Oil Mop, one Dry Dust Mop, two handles and one can of Mop Oil for renewing oil mop. Shipping weight, 5 pounds $1.30

Requires No Electricity!

The "Sanitary Special" VACUUM CLEANER

Non-Electric.

Powerful suction gets dust and dirt. Revolving brush gathers up threads and lint. All you do is put it down on the floor and push. No work to it at all. Any child can push it around the room and it cleans in just the same manner as electrically operated vacuum cleaners.

Equal in Value to Similar Cleaners Sold at From $5.00 to $10.00 More Than We Ask.

Costs Nothing to Operate

Is gear driven. The wheels drive gears propelling a fan, which in turn draws the dust, dirt and lint by SUCTION, just as do electric cleaners.

It is sturdy, well made, and will give you long service. The brush is removable, permitting you to operate the machine for suction cleaning alone, without brushing the surface to be cleaned.

Body is cast aluminum. Wheels are rubber tired. Bag is made of dust-proof fabric and readily removed for emptying. Actual weight, 8½ pounds. Shipping weight, 11 pounds.

99L2141

$18.75

For Electric Vacuum Cleaners see page 811.

only $18.75

Fiber Scrub Brush. Length, 8½ inches. Palmyra fiber. Weight, 10 ounces.
9L2868 ......15c

Horsehair Window Brush. 65c
9L2872 Length, 8 inches. Weight, 12 ounces. Without handle......65c

"Collector" Carpet Sweeper. $3.65

A serviceable carpet sweeper at a moderate price. Mahogany finish case, fitted with band of woven braid to prevent marring of furniture. Guaranteed to give good service. Shipping weight, 7 lbs.
99L2131 ......$3.65

"Champion" Carpet Sweeper. $2.95

Has woven braid furniture protector and hardwood case. Shpg. wt., 6½ lbs.
99L2130 ......$2.95

FURNI-FLOR Mop Oil

Furni-Flor is our own Trade Mark, Registered in U. S. Patent Office.

The Universal Polish.

For Renewing Dustless Mops and Dust Cloths.

An oil mop needs frequent renewing. Reoil your mop or dust cloth frequently and it will give far better service. To renew mop, wash with soap and warm water, rinse well and add a quantity of Furni-Flor Mop Oil.

30L3403
1-pt. can.....(Shipping wt., 2 lbs.)....23c
1-qt. can.....(Shipping wt., 3½ lbs.)....29c

Oil Polish Mop. 65c

Use once a week to keep floors in first class condition. Adjustable handle permits keeping mop set square on the floor. Shipping weight, 2½ lbs.
99L2152 With handle and can Mop Oil for renewing mop. 65c

Dustless Dry Mop. 60c

Use daily to keep floors looking spick and span. Adjustable handle permits keeping mop set square on the floor. Shipping weight, 2 pounds.
99L2151—With handle 60c

Spick and Span Oil Mop. 40c

For Use on All Hardwood Floors.

Good quality. Use an old mop handle with this mop. Wt., 1½ pounds.
9L2863—Without handle 40c

Lightning Oil Mop, Without Handle.

Used in any ordinary mop stick. Weight, 12 ounces.
9L2864 — Without handle 45c

Dustless Floor Sweeper. 75c

Used dry only. Size, 10 inches. Shipping weight, 2¼ pounds.
99L2154 With handle. 75c

Dustless Hand Duster. 50c

Length over all, 20 inches. Weight, 14 oz.
9L2854 50c

Dustless Dust Cloth. 24x36-Inch Size. 15c

Dustless dust cloths remove dust from woodwork, bric-a-brac, walls, etc.
9L2855—Size, 24x36 inches. Weight, 2½ oz. 15c
9L2856—Size, 36x48 inches. Weight, 3 oz...30c

All weights and measurements given on this page are approximate and may vary a trifle.

$1.45 Gray Enameled Seamless Combinet.

Made of sheet steel, well enameled inside and out. This combinet has a broad roll top, 11 in. in diameter. Capacity, 11 quarts. Shpg. wt., 5¾ lbs.

99L2221 .... $1.45

Gray Enameled Seamless Chamber Pail. $1.40

Very durable. Sheet steel covered with two coats of tough enamel. Top, 9½ inches in diameter. Holds 12 quarts. Shipping weight, 5½ pounds.

99L2220 .... $1.40

Gray Enameled Chamber Covers.

18c Medium.

9L2108—Medium size. Weight, 8 ounces 18c
9L2109—Large size. Weight, 10 ounces 20c

Gray Enameled Chambers.

40c Medium.

9L2105—Medium size. Weight, 1 pound 40c
9L2106—Large size. Weight, 1¼ pounds ... 50c

White Enameled Combinet.

$1.95

With white enameled cover. Easily cleaned. Can be used as chamber or pail. Capacity 12 quarts; 11 inches top diameter. Shipping weight, 6½ pounds.

99L2200
$1.95

White Enameled Chamber Cover. 30c Large.

Weight, 10 ounces.

	Size	
9L2482	Medium	25c
9L2483	Large	30c

White Enameled Chamber. 65c Large.

Weight, 2 pounds.

	Size	
9L2480	Medium	55c
9L2481	Large	65c

Gray Enameled Wash Basins.

26c Medium.

9L2097—Small size. Diameter, 10⅛ inches. Weight, 12 ounces 22c
9L2098—Medium size. Diameter, 11½ inches. Weight, 14 ounces 26c
9L2099—Large size. Diameter, 12½ inches. Weight, 1 pound 30c

Gray Enameled Foot Bathtub.

$1.30

9L2095
Size, 18½ inches. Capacity, 20 quarts. Weight, 3¼ lbs....$1.30

SEAMLESS WHITE ENAMELED WASH BASINS.

9L2455—Medium. Diameter, 11½ in. Wt., 1¼ lbs. 35c
9L2456—Large. Diameter, 12½ in. Wt., 1½ lbs. 45c
9L2457 — Extra large. Diameter, 13 inches. Weight, 1¾ pounds.... 55c

White Enameled Foot Bath Tub.

$1.50

9L2445
Size, 18½ in. Capacity, 20 quarts. Wt., 3¼ lbs.
$1.50

Gray Enameled Water Pail.

99L2232
Capacity, 12 qts. Shipping wt., 4¼ lbs..... 95c
99L2233
Capacity, 14 qts. Shipping wt., 5¼ lbs.. $1.10

Gray Enameled Soap Dish.

10c

9L2010 — Size, 4x6 inches. Weight, 4 ounces 10c

$3.35 4-Piece Galvanized Toilet Set.

Serviceable and durable. Made of sheet steel, galvanized after being formed. Includes 12-qt. combinet, 24-qt. foot tub, 4½-quart water pitcher and 14-inch wash basin. Shipping weight, 16 pounds.

99L2099 .... $3.35

Nickel Plated Cuspidor. 30c

9L2626—Loaded bottom. Weight, 1 pound 30c

White Enameled Washstand Set. $5.95

Strong, durable and sanitary. Angle steel stand 30 in. high with soap dish and towel bar. White enameled seamless pitcher and wash basin. Pitcher holds 4 quarts. Wash basin 14 in. in diameter. Shipping wt., 17 lbs. Not mailable.

99L2204
$5.95

Complete Washstand Set. $3.60

Steel stand 30 in. high, 12-inch wash basin, 4-quart water dish. Bowl and pitcher are made of heavy sheet steel, covered inside and out with white enamel. Soap dish and stand are white japanned. Shipping weight, 12 pounds. Not mailable.

99L2205 .. $3.60

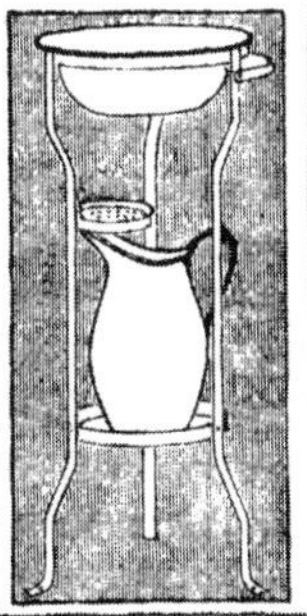

For other Washstands, see page 652.

White Enameled Wash Set.

$2.45

Durable seamless white enameled ware. Looks like china, but is more durable. Set includes 6-quart water pitcher, 14-inch wash basin and soap dish with grate. Shipping wt., 7½ lbs.

99L2202 .... $2.45

White Enameled Toilet Set. $5.65

High quality seamless white enameled ware. Set consists of 14-inch wash basin, 6-quart water pitcher, 12-quart combinet and cover, large size chamber and cover and soap dish with grate. Each made from one piece of heavy sheet steel. Shipping weight, 21 pounds. Not mailable.

99L2207 .... $5.65

A Very Handy Wash Set. 65c

Basin holder, soap rack and 12-inch white enameled basin with screws for attaching to wall. Weight, 1¾ pounds.

9L2466 .... 65c

"Handy" White Enameled Wash Set. $2.20

A handy set for the back porch or kitchen. High quality seamless white enameled ware. Includes 12-quart water pail, 13-in. wash basin and ¾-quart water dipper. Shipping weight, 8¼ lbs.

99L2206 .... $2.20

THE THRIFTY BUYER always considers **quality** as well as **price**, because it requires both to measure **value**. Our experienced merchandising assures you the biggest possible value for every dollar.

Gasoline Torch. $3.65

NOTE: Joints in Gasoline Torches should be tightened before using.

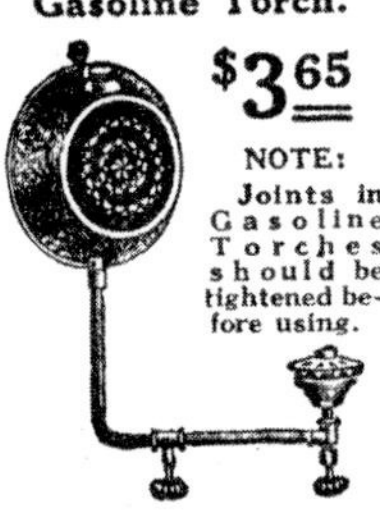

Gives a strong light, for indoor or outdoor lighting. Screw furnished for hanging torch. Mammoth Single Burner Gasoline Torch with 9½-qt. galvanized reservoir and large size improved burner. Shipping weight, 8¾ lbs.

99L2177 ..... $3.65

Standard Gasoline Torches with 3½-quart japanned iron reservoir and regular burners.

99L2175—Single Burner. Shipping weight, 4½ pounds........ $1.95
99L2176—Double Burner. Shipping weight, 7 pounds........ $2.90

Galvanized Oil Can. 90c

5-Gallon.

Made of sheet steel, galvanized after being put together, making it absolutely tight. Has corrugated rounded top, which adds strength to the can. Every can tested.

99L2100 — 5-gallon size. Shipping weight, 7 lbs... 90c
9L2608 — 1-gallon size, with tin top part. Weight, 1 pound 33c

Drivers' Special Cold Blast Dash Lantern. $2.40

Large reflector throws a strong, bright light. Clamps on dash by means of a spring clamp at back of lantern. No. 1 burner with ⅝-in. wick. Well made and fully guaranteed. Weight, 2½ pounds.

9L2508 .. $2.40

9L2513—Extra Cold Blast Globe to fit 9L2508 Lantern. Weight, 10 ounces. 10c

Junior Cold Blast Wagon Lantern. $1.55

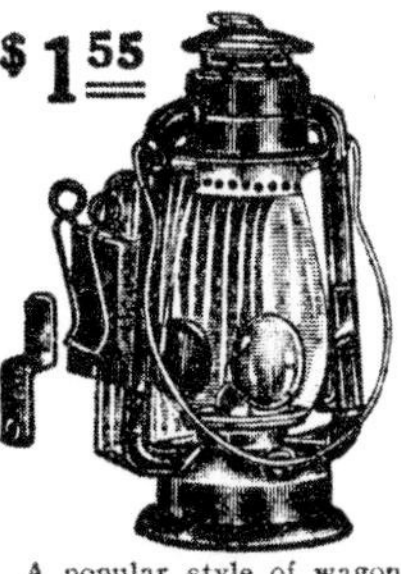

A popular style of wagon lantern that complies with all night driving laws. Fitted with 2¼-inch red danger lens in rear. Attaches to wagon by means of brackets furnished or by spring clamp. Nicely finished in black enamel. Height, 12 inches. ⅝-inch wick. Weight, 3 pounds.

9L2504 .... $1.55

9L2518—Extra globe for above lantern. Weight, 10 ounces 10c

For Kerosene see page 957.
For Electric Vehicle Lamps see page 812.

"Little Wonder" Lantern. 55c

A well made, small size lantern that will burn steadily in wind or storm. No. 1 burner, ⅝-in. wick. Height, 7¾ inches. Weight, 1½ pounds.

9L2505..... 55c

9L2506 — Extra Globe for 9L2505 "Little Wonder" Lantern. Weight, 8 ounces 10c

Junior Cold Blast Lantern. 85c

Well made. Same high quality as our High Grade Cold Blast Lantern. No. 1 burner, ⅝-in. wick. Weight, 2 pounds.

9L2509...... 85c

9L2510 — Extra Globe for 9L2509 Junior Cold Blast Lantern. Weight, 10 ounces 10c

High Grade Cold Blast Lantern. $1.30

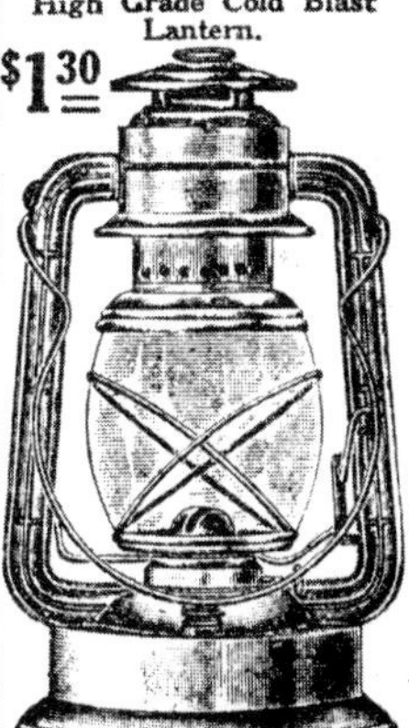

Extra large and well made. Gives a strong, bright light. Outlasts the ordinary kind. No. 2 burner, 1-inch wick, extra large fount. Weight, 2¾ pounds.

9L2511 .... $1.30

9L2515—Extra Globe for 9L2511 High Grade Lantern. Weight, 11 ounces....... 10c

New Jersey Pattern R. R. Milk Cans

5-Gal. **$3.40** 8-Gal. **$4.50** 10-Gal. **$5.30**

Our Prices Save You $1.00 to $2.00 According to Size.

Latest Improved Construction. One of the Best and Most Popular Patterns on the Market.

Made of heavy steel plate, seamed and riveted, heavily tinned and soldered so as to leave no crevices or corners to collect germs and dirt.

Body and bottom rolled together, forming a rigid joint of great durability.

New Jersey Pattern Cans are rigid and stiff and will stand hard usage. Breast and bottom are full rounded, which makes cleaning easy.

New Jersey Pattern Cans are guaranteed to give satisfaction.

99L2506 5-GALLON SIZE. Shipping wt., 12 pounds.... **$3.40**

99L2507 8-GALLON SIZE. Shipping wt., 18 pounds.... **$4.50**

99L2508 10-GALLON SIZE. Shipping wt., 20 pounds.... **$5.30**

Milwaukee Pattern Riveted Milk Cans

5-Gal. **$2.80** 8-Gal. **$3.60** 10-Gal. **$4.30**

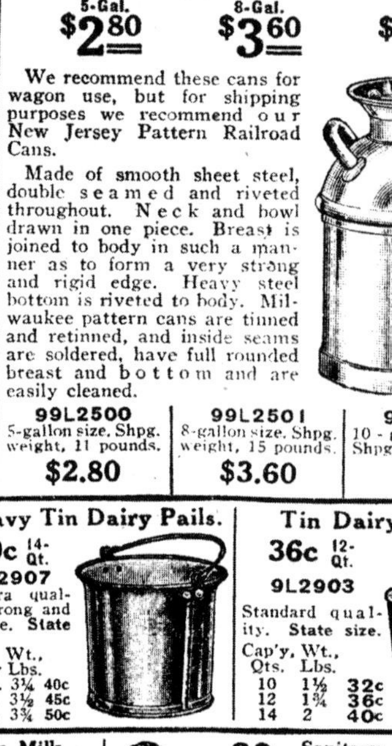

We recommend these cans for wagon use, but for shipping purposes we recommend our New Jersey Pattern Railroad Cans.

Made of smooth sheet steel, double seamed and riveted throughout. Neck and bowl drawn in one piece. Breast is joined to body in such a manner as to form a very strong and rigid edge. Heavy steel bottom is riveted to body. Milwaukee pattern cans are tinned and retinned, and inside seams are soldered, have full rounded breast and bottom and are easily cleaned.

99L2500 5-gallon size. Shpg. weight, 11 pounds. **$2.80**

99L2501 8-gallon size. Shpg. weight, 15 pounds. **$3.60**

99L2502 10-gallon size. Shpg. wt., 17 lbs. **$4.30**

BUTTER in Three Minutes! The "Holstein" Butter Reaper

by Actual Tests Makes Butter From Sweet or Ripened Cream in From 3 to 5 Minutes. Satisfaction Guaranteed or Your Money Returned.

The peculiar construction of the dasher gives a violent action to the cream, without breaking their granular form, quickly extracts all butter fat particles, forming solid first quality butter in from three to five minutes' time.

$4.95

Save Your Time! Save Your Strength!

Why waste valuable time using the old time slow and hard method of churning when by using modern methods you can obtain better results in but a few minutes' time?

Churn is light in weight and easy running. Can made of heavy tin plate and holds $3\frac{1}{2}$ gallons. Churns up to $1\frac{1}{4}$ gal. Shipping wt., 23 lbs.

99L2591 .... **$4.95**

Our "Best Made" Aluminum Pail.

99L2424 Capacity, 10 quarts. Shpg. wt., 3 lbs. **$1.55**

99L2425 Capacity, 12 quarts. Shipping wt., $3\frac{1}{2}$ lbs. **$1.75**

Sanitary Strainer Pails.

$1.00 14-Qt.

9L2900 Strong, substantial tin pail with brass strainer. State size.

Cap.	Wt., Lbs.	
12 qts.	$3\frac{1}{2}$	**$0.90**
14 qts.	$3\frac{1}{2}$	**1.00**

Heavy Tin Dairy Pails.

50c 14-Qt.

9L2907 Extra quality strong and durable. State size.

Cap'ty	Wt., Lbs.	
10 qts.	$3\frac{1}{4}$	**40c**
12 qts.	$3\frac{1}{2}$	**45c**
14 qts.	$3\frac{3}{4}$	**50c**

Tin Dairy Pails.

36c 12-Qt.

9L2903 Standard quality. State size.

Cap'y, Qts.	Wt., Lbs.	
10	$1\frac{1}{2}$	**32c**
12	$1\frac{3}{4}$	**36c**
14	2	**40c**

"High Speed" Rotary Wood Churns.

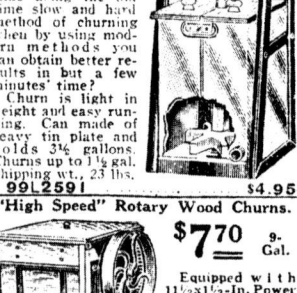

$7.70 9-Gal.

Equipped with $11\frac{1}{2}$x$1\frac{1}{2}$-In. Power Pulley and Detachable Crank, as illustrated, for Hand or Power. A churn that makes, gathers, works and salts butter without removing from churn. Reinforced with four truss rods, which makes it extra strong and rigid. Hardwood body.

Not mailable.	Holds	Churns About	Shpg. Wt.	
99L2605	9 gal.	5 gal.	50 lbs.	**$7.70**
99L2606	11 gal.	7 gal.	57 lbs.	**8.15**
99L2607	13 gal.	10 gal.	59 lbs.	**8.55**
99L2608	15 gal.	12 gal.	61 lbs.	**8.95**

Milk Can Strainer.

Heavy tin. Bowl is seamless, $10\frac{1}{2}$ in. across top with $4\frac{1}{2}$-in. brass strainer cloth. Wt., 15 oz.

9L2896 .... **55c**

55c

15c Bristle Milk Bottle Brush.

Bristles securely fastened. Stiff tampico tufts for cleaning corners and bottom. Length, 16 inches. Weight, 2 ounces.

9L2876 .... **15c**

30c Sanitary Milk Can Brush.

Bristles held securely between the twisted wire frame and guaranteed not to pull out. Length, 21 inches. Weight, 8 ounces.

9L2878 .... **30c**

Lock Cover Cream Setting Cans.

Self Locking Covers. May be completely submerged in water without leaking. We quote a special price on lots of six cans and do not sell less quantities. Not mailable.

$3.00 for 6 14-Qt.

	Cap., Qts.	Shpg. Abt., Wt., Lbs.	6 for
99L2524	14	21	**$3.00**
99L2525	18	24	**3.40**
99L2526	20	25	**3.80**

Old Reliable Star Barrel Churns.

$5.95 10-Gal. | **$6.45** 15-Gal.

Why Pay More?

Easy to operate and keep clean. Churns quickly and gets all the butter. Barrels made of oak. Cover fits tight and will not leak. Fasteners are attached to outside of churn and clamp the cover with a compound lever action. Full directions furnished with each churn. **Not mailable.**

	Holds	Churns Up to	Shpg. Weight	
99L2592	6 gal.	3 gal.	29 lbs.	**$5.20**
99L2593	10 gal.	5 gal.	34 lbs.	**5.95**
99L2594	15 gal.	7 gal.	41 lbs.	**6.45**
99L2595	20 gal.	9 gal.	50 lbs.	**7.20**

"Success" Power Barrel Churns.

$10.85 15-Gal.

Well made of seasoned oak. Tight fitting cover. Full directions furnished. Furnished with 12-inch tight and loose pulleys, which take a 2-inch belt. **Shipped direct from factory near CHICAGO.** Shipment usually made in 5 to 10 days after order is received. **Not mailable.**

	Holds	Churns Up to	Shpg. Wt., Lbs.	
99L8348⅓	15 gal.	7 gal.	60	**$10.85**
99L8349⅓	20 gal.	9 gal.	70	**11.80**
99L8350⅓	25 gal.	12 gal.	90	**12.95**
99L8351⅓	35 gal.	16 gal.	100	**14.40**

Cedar Cylinder Churns.

$4.70 7-Gal. Size.

Easy to turn, agitates cream violently, makes butter quickly. Made of clear straight grained cedar. ★Not mailable.

	No.	Holds About	Churns About	Shipping Weight	
99L2613	1	3 gal.	2 gal.	15 lbs.	**$3.40**
99L2614	2	4 gal.	$2\frac{1}{2}$ gal.	18 lbs.	**4.05**
99L2615	★3	7 gal.	4 gal.	22 lbs.	**4.70**
99L2616	★4	10 gal.	5 gal.	27 lbs.	**5.30**

$5.40 14x23 In. Standard Butter Worker.

Easy to operate. Works a batch of butter in three to five minutes. **Not mailable.**

	Size	Works	Shpg. Wt.	
99L2540	14x23 in.	10 lbs.	29 lbs.	**$5.40**
99L2542	20x30 in.	30 lbs.	39 lbs.	**6.60**

Flint Glass Milk Bottles.

Good quality clear glass. Uniform in size and properly annealed to insure toughness and strength. Smoothly finished inside and out. Will stand a great deal of rough handling.

70c Per Dozen. Half pints. Shpg. wt., per 6 dozen, 53 pounds.

99L2577 Per doz. **70c**

99L2580 Per crate of 6 doz. Not mailable. **$3.40**

90c Per Dozen. Pints. Shpg. weight, per 6 dozen, 81 lbs.

99L2578 Per doz. **90c**

99L2581 Per crate of 6 doz. Not mailable. **$4.20**

$1.15 Per Dozen. Quarts. Shpg. wt., per 6 doz., 132 lbs.

99L2579 Per dozen, **$1.15**

99L2582 Per crate of 6 doz. Not mailable. **$5.60**

Our "Improved" Milk Cooler and Aerator. $5.90 18-Qt.

Draws out the animal odors and helps to prevent the growth of bacteria in new milk. Used with either cold running water or with ice water. Heavy tin plate with galvanized steel bottom, painted inside. Prices include double cheesecloth strainer, spring pins and stirring ladle. **Not mailable.**

	Size No.	Receiver Holds	For Cows	Shpg. Wt., Lbs.	Each
99L2532	2	18 qts.	5 to 25	27	**$5.90**
99L2533	3	34 qts.	25 to 50	35	**7.10**
99L2534	4	52 qts.	50 to 100	38	**8.40**

Dilution Cream Separators.

$3.90 10-Gallon. **$4.60** 14-Gallon. **$5.20** 18-Gallon.

Separates cream from milk in three to four hours; gives sweet diluted milk, which is far superior to sour milk as stock food. Saves several hours' waiting for cream to rise. As water mixes with the milk, cream separates and rises to top; this can be watched through gauges. Heavy tin, securely seamed and soldered. Enameled on outside. Prices include tin tubes, strainer faucet and legs.

99L2550—Capacity, 10 gallons. Shipping weight, 14 pounds.... **$3.90**

99L2551—Capacity, 14 gal. Shipping wt., 16 lbs. **Not mailable**.... **$4.60**

99L2552—Capacity, 18 gal. Shipping wt., 20 lbs. **Not mailable**.... **$5.20**

For Dairy Thermometers see page 427. For Dairy Paper see page 478.

Cast Aluminum Butter Ladle. 48c

Size, $3\frac{1}{4}$x$8\frac{3}{4}$ in. Weight, $3\frac{1}{2}$ oz.

9L2310 **48c**

"Babcock Pattern" Standard Milk Testing Outfit. 4-Bottle Size for Milk and Cream.

$6.95

These Milk Testers tell the exact quality of each cow's milk.

Price includes bottle of acid, test bottles, brush, acid measure, pipette and simple directions for making tests. Shipping wt., 19 lbs. **Not mailable.**

99L2573 4-bottle size for milk and cream.... **$6.95**

99L2571 2-bottle size for milk and cream.... **$6.70**

Extra Glassware for Babcock Testers.

9L2882—50 per cent Cream Bottles. **30c**

9L2884—10 per cent Milk Bottles. **28c**

9L2886—$\frac{5}{100}$ of 1 per cent Skim Milk Bottles.... **65c**

9L2888—17.6-18-C. C. Pipettes.... **24c**

9L2890—17.5-C. C. Acid Measures. **12c**

$15.50 WATERPROOF MILK BOTTLE CAPS.

First quality, paraffined both sides. Fit any standard size bottle. Barrel of 50,000.

6,000 FOR **$2.40**

We can only furnish plain caps without printing of any kind.

36c

99L2585 In barrels containing about 50,000 caps. Shpg. wt., 133 lbs. Not mailable. **$15.50**

99L2587 In 20-qt. galvanized pail containing abt. 6,000 caps. Shipping wt., 17 lbs. Not mailable. **$2.40**

9L2892 In packages containing about 1,000 caps. Shpg. wt., $2\frac{1}{4}$ lbs. **36c**

Double Can Creamer.

$5.60 6-Gal.

Separates cream from milk in four to six hours in warm or cold weather. Removable inner can makes cleaning easy. Has glass gauges. Gives pure, undiluted milk. Inner can heavy tin. Outer can galvanized iron.

★Not Mailable.	Capacity of Inner Can	Shpg. Wt.	
99L2562	4 gal.	14 lbs.	**$4.90**
99L2563	★ 6 gal.	17 lbs.	**5.60**
99L2564	★ 8 gal.	20 lbs.	**6.20**
99L2565	★10 gal.	22 lbs.	**6.65**
99L2566	★12 gal.	$23\frac{1}{2}$ lbs.	**7.10**

Guaranteed Waterproof Wash Aprons. $1.95 With Legs.

Made of black oiled cotton duck. To protect clothing when stooping or bending, apron is provided with legs that hold it close to body. Apron measures 36x45 inches. Weight, 2 pounds.

9L3521 .... **$1.95**

9L3522 — Wash Aprons, same as above, but without legs. Weight, $1\frac{1}{2}$ lbs **$1.55**

ALL WEIGHTS AND MEASUREMENTS GIVEN ON THIS PAGE ARE APPROXIMATE AND MAY VARY A TRIFLE.

"High Grade" Double Beam Counter Scale.

$14.90

Capacity of platform, 240 pounds by ½ pounds.

Platform measures about 9½x12¾ inches. Capacity of scoop, 30 pounds by ½ ounces. Poises are equipped with adjusting screws. Shipping wt., 57 lbs.
99L2775.................$14.90

"High Grade" Single Beam Counter Scale.

Similar in construction to the double beam scale listed above; same capacity. Shipping weight, 53 pounds.
99L2774.........................$12.85

OUR PRICES ARE LOW but that is only part of the benefit of buying from us. The big advantage is in the QUALITY of our goods. This is assured by our constant testing.

Household or Family Scales

Scales of this type are not legal for use in trade, but for ordinary weighing they will be found very handy and entirely satisfactory for use in the home. Made of sheet steel. We sell many thousands every every year.

$3.45

General Purpose 60-Pound Spring Scale. Large strong scale with tin scoop. Body is heavy sheet steel, black finish. Springs are tempered steel. Has double platform supports. Capacity, 60 pounds by 2 ounces. Shipping weight, 9¼ pounds.
99L2785..........$3.45

$1.95

Family Spring Scale. With Tin Scoop and Extra Tare Hand. 25-Pound Capacity. Gives weight separately of scoop and articles weighed. Scale can also be used in the ordinary way. Has steel body and platform. Capacity, 25 pounds by ounces. Shipping wt., 5½ lbs.
99L2786..........$1.95

$1.75

Family Spring Scale. 25-Pound Capacity. Has steel body and platform, steel spring, white finished dial. Capacity, 25 pounds by ounces. Not made with scoop. Shpg. wt., 4¾ lbs.
99L2781.....$1.75

$2.55 25-POUND CAPACITY.

Family Spring Scale. Good quality scale, made of sheet steel, light blue finish; dial has glass front; white tile platform. Capacity, 25 pounds by ounces. Shipping weight, 6 lbs.
99L2784..........................$2.55

ATLAS U. S. STANDARD SCALES

Guaranteed to meet the requirements of inspection by any Government authorized sealer of weights in any state of the United States and to be free from defective material and workmanship. **All Atlas Scales are shipped from factory in SOUTHEASTERN MICHIGAN.**

Atlas Portable Scales.

Atlas Scale Beams are solid brass. The beams for 600 and 1,000-pound capacity are marked by ¼ pounds to 50 pounds. On the larger sizes by ½ to 100 pounds.

Atlas Test Weight. 99L8401⅓ 50-Pound Sealed Test Weight. Guaranteed accurate. Used for testing any scale. $5.60

Atlas Weightless Scale.

99L8403⅓ Weightless Type. $28.00

99L8421⅓ Double Beam Type. $45.00 Capacity, 2,000 Lbs.

99L8407⅓ Single Beam Type. $25.00 Capacity, 1,000 Lbs.

Portable and Platform Scales Are Not Mailable.

Capacity, 600 Pounds. This scale has no loose weights. Upper beam marked to 100 pounds by ½-pound marks; lower beam to 500 pounds by 100-pound notches. Platform measures 16½x24 inches. Shpg. wt., 160 lbs.
99L8403⅓.$28.00

Single Beam Type Catalog No.		Capacity, Lbs.	Size of Platform, Inches	Shpg. Wt., Lbs.	Double Beam Type Catalog No.	
99L8405⅓	$22.00	600	16½x24	165	99L8415⅓	$24.00
99L8407⅓	25.00	1,000	17½x25	175	99L8417⅓	27.00
99L8409⅓	36.00	1,600	20½x29¾	275	99L8419⅓	38.00
99L8411⅓	43.00	2,000	25¼x34½	355	99L8421⅓	45.00
99L8413⅓	46.00	2,500	25¼x34½	365	99L8423⅓	47.00

Scale Beams. With Two Poises.

$4.50 400-Lb. Capacity.

Not Legal for Use in Trade.

Heavy strong beam for rough weighing. Plainly marked on both sides. Will weigh to full capacity.

	Capacity, Lbs.	Shpg. Wt., Lbs.	
99L2791	200	8	$2.35
99L2793	400	19	4.50
99L2794	600	32	5.95

Atlas Pit Scale.

$95.00 to $185.00

Requires a pit 22 inches deep. Beam boxes painted and fitted with lock and key. Building plans included. Weight beams, solid brass. Upper bar of double beam marked to 1,000 pounds by 5-pound marks; lower bar by 500-pound notches to 9,000 pounds. Triple beam has upper bar marked to 2,000 pounds by 10-pound marks, lower bar to 500 pounds by 5-pound marks and middle bar to 9,000 pounds by 500-pound notches. The 8-ton has one loose weight for weighing between 5 and 8 tons. Platform size, 8x14 feet. Lumber for platform and frame not included.

99L8425⅓—5-Ton Double Beam Atlas Pit Scale. Shipping weight, 565 pounds. **Not mailable**..........$95.00

99L8427⅓—5-Ton Triple Beam Atlas Pit Scale. Shipping weight, 600 lbs. **Not mailable**...$105.00

99L8429⅓—8-Ton Triple Beam Atlas Pit Scale. Shipping weight, 800 lbs. **Not mailable**.....$185.00

Atlas Pitless Scale.

$110.00 AND $137.50 Not Mailable.

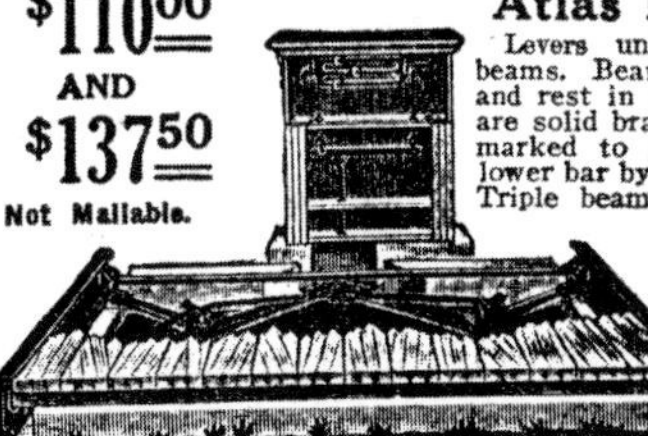

Levers under platform are T shape steel beams. Bearing pivots are hardened tool steel and rest in hard bearing loops. Weigh beams are solid brass. Upper bar of the double beam marked to 1,000 pounds by 5-pound marks; lower bar by 500-pound notches to 9,000 pounds. Triple beam has upper bar marked to 2,000 pounds by 10-pound marks, lower bar to 500 pounds by 5-pound marks and middle bar to 9,000 pounds by 500-pound notches; steel I beams securely tied together by iron pipe and locknuts. Furnished with painted beam box and directions. Lumber for frame and platform not included.

99L8431⅓—5-Ton Pitless Scale. Furnished for wood ends with double beams. Shipping weight, 900 lbs. **Not mailable**........$110.00

99L8433⅓—5-Ton Pitless Scale. Furnished with steel ends and triple beam. We recommend this scale. Shipping weight, 1,160 pounds. **Not mailable**..........$137.50

NOTE OUR PRICES ON STANDARD WIRE NAILS SEE WHAT YOU SAVE

We handle nails only for the convenience of our regular customers buying other merchandise. We will not fill orders for nails alone.

60d	20d	8d
$1.95	$2.85	$3.50
Per Keg (100 lbs.), Abt. 1,100 Nails.	Per Keg (100 lbs.), Abt. 2,800 Nails.	Per Keg (100 lbs.), Abt. 9,800 Nails.

$1.95 Per Keg for 60d Size. Which is less than carload price to dealers at time this catalog was printed.

STANDARD WIRE NAILS IN 100-POUND KEGS.

We do not break kegs at full keg prices. Where less than a keg of one size is wanted, order by the 5-pound package as priced below.

99L1150—State size wanted.

Size	60d	40d	30d	20d	16d	12d
Length	6 in.	5 in.	4½ in.	4 in.	3½ in.	3¼ in.
★Per 100-pound keg	$1.95	$2.30	$2.60	$2.85	$3.05	$3.20
Size	10d	8d	6d	4d	3d	2d
Length	3 in.	2½ in.	2 in.	1½ in.	1¼ in.	1 in.
★Per 100-pound keg	$3.35	$3.50	$3.65	$3.90	$4.20	$4.60

Size	10d	8d	6d	4d	3d	2d
99L1145—Galvanized Shingle Nails—★Per 100-lb. keg	....	....	....	$4.90	$5.30	....
99L1147—Smooth Box Nails—★Per 100-lb. keg	....	....	$3.80	....	....	....
99L1151—Wire Lathing Nails—★Per 100-lb. keg	....	....	....	....	4.60	$4.90
99L1152—Wire Finishing Nails—★Per 100-lb. keg	$3.60	$3.75	3.95	4.25	....	....
99L1153—Wire Flooring Nails—★Per 100-lb. keg	3.50	3.65	3.85	....	....	....
99L1154—Wire Casing Nails—★Per 100-lb. keg	3.50	3.65	3.85	4.10	....	....
99L1156—Wire Fencing Nails—★Per 100-lb. keg	3.45	3.60	....			

(★ Not Mailable.)

STANDARD WIRE NAILS IN 5-POUND PACKAGES.

Size	60d	40d	30d	20d	16d	12d	10d	8d	6d	4d	3d	2d
Length, inches	6	5	4½	4	3½	3¼	3	2½	2	1½	1¼	1
9L4284—Per 5-lb. pkg.	15c	16c	17c	18c	19c	20c	21c	22c	23c	24c	25c	26c

Size	10d	8d	6d	4d	3d	2d
9L4281—Galvanized Shingle Nails—Per 5-lb. package	.....	.....	.....	30c	33c	.....
9L4283—Smooth Box Nails—Per 5-lb. package	.....	.....	24c	.....	.....	.....
9L4285—Wire Lathing Nails—Per 5-lb. package	.....	.....	.....	.....	31c	32c
9L4286—Wire Finishing Nails—Per 5-lb. package	27c	28c	29c	30c	.....	.....
9L4287—Wire Flooring Nails—Per 5-lb. package	26c	27c	28c	.....	.....	.....
9L4288—Wire Casing Nails—Per 5-lb. package	26c	27c	28c	29c		
9L4290—Wire Fencing Nails—Per 5-lb. package	22c	23c	.....	....	.....	....

All weights and measurements given on this page are approximate and may vary a trifle.

Big Value! 16c

Order One or More Packages! One Pound Assorted Wood Screws, in Assorted Sizes, Lengths and Finishes. Each package contains about 200 screws.
9L4271—Per 1-pound package....16c

100 Lbs. $2.20 Assorted Wire Nails. 25 Lbs. 73c

Well Assorted—Many Sizes. Order Now and Have the Nails You Want When You Want Them.
99L1166—Per package of 25 pounds..........73c
99L1160—Per keg of 100 lbs. Not mailable.$2.20

FLAT HEAD BRIGHT IRON SCREWS.

Order by catalog number and give length, size number and quantity wanted.

Lgth., in., and size No.	¼-1	⅜-2	½-2	½-3	½-4	½-6	⅝-4	⅝-6	⅝-8
9L4272—Pkg. 3 doz.	6c	6c	6c	7c	7c	7c	8c	8c	8c
9L4273—Per gross	19c	20c	21c	22c	23c	24c	25c	26c	27c
Lgth., in., and size No.	¾-5	¾-6	¾-7	¾-8	¾-9	¾-10	⅞-5	⅞-6	⅞-7
9L4272—Pkg. 3 doz.	7c	7c	8c	8c	9c	9c	7c	7c	8c
9L4273—Per gross	24c	25c	26c	27c	28c	29c	25c	26c	27c
Lgth., in., and size No.	⅞-8	⅞-9	⅞-10	1-6	1-7	1-8	1-9	1-10	1-12
9L4272—Pkg. 3 doz.	8c	8c	8c	9c	9c	9c	10c	10c	10c
9L4273—Per gross	28c	29c	30c	31c	32c	33c	34c	35c	36c
Lgth., in., and size No.	1¼-8	1¼-9	1¼-10	1¼-12	1½-8	1½-9	1½-10	1½-12	2-10
9L4272—Pkg. 3 doz.	9c	9c	10c	10c	11c	11c	12c	12c	13c
9L4273—Per gross	33c	34c	35c	36c	37c	38c	40c	42c	42c
Lgth., in., and size No.	2-12	2-14	2½-12	2½-14	3-14	4-16	....	....	....
9L4272—Pkg. 3 doz.	14c	16c	18c	20c	24c	28c	....	....	....
9L4273—Per gross	46c	50c	56c	64c	78c	96c	....	....	....

ROUND HEAD BLUED IRON SCREWS FOR WOOD.

Lgth., in., and size No.	½-3	½-4	¾-4	¾-6	⅞-5	⅞-6	1-6	1-8	....
9L4275—Pkg. 3 doz.	7c	7c	8c	8c	9c	9c	10c	10c	....
9L4276—Per gross	24c	25c	26c	27c	28c	30c	32c	34c	....
Lgth., in., and size No.	1-10	1¼-6	1¼-8	1¼-10	1½-8	1½-10	2-8	2-10	....
9L4275—Pkg. 3 doz.	10c	10c	11c	11c	12c	13c	14c	15c	....
9L4276—Per gross	36c	38c	40c	42c	44c	46c	48c	50c	....

ASSORTED SCREWS. 21c PER GROSS.

Assorted Flat Head Bright Iron Wood Screws, ⅜, ½, ⅝, ¾, ⅞, 1, 1¼ and 1½ in. long. Weight, 12 ounces.
9L4270
Gross (144 screws)....21c

Wire Brads. 8c Per Lb. and Up.

9L4292—In 1-lb. packages. State length.

Length, inches	½	⅝	¾	⅞	1	1¼	1½
Gauge wire	19	18	18	17	16	15	15
Per pound	18c	16c	14c	12c	10c	9c	8c

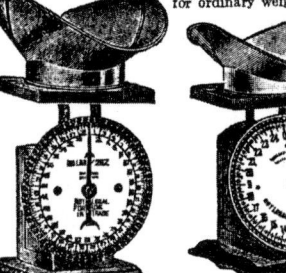

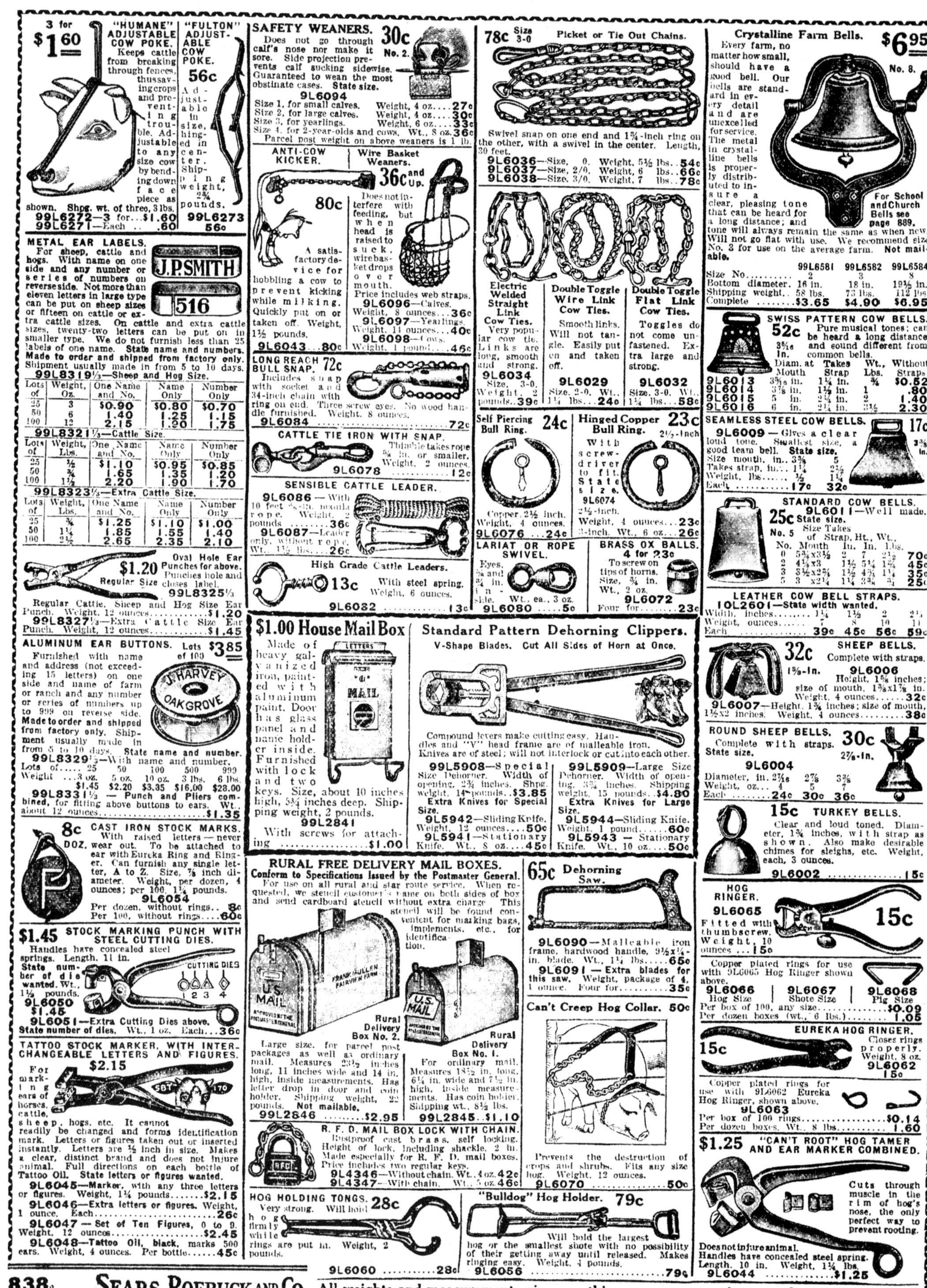

3 for $1.60

"HUMANE" ADJUSTABLE COW POKE. Keeps cattle from breaking through fences, thus saving crops and preventing trouble. Adjustable to any size cow by bending down face piece as shown. Shpg. wt. of three, 3 lbs.
99L6272—3 for...$1.60
99L6271—Each ... .60

"FULTON" ADJUSTABLE COW POKE. 56c
Adjustable in size, hinged in center. Shipping weight, 2¾ pounds.
99L6273 56c

METAL EAR LABELS.
For sheep, cattle and hogs. With name on one side and any number or series of numbers on reverse side. Not more than eleven letters in large type can be put on sheep sizes or fifteen on cattle or extra cattle sizes. On cattle and extra cattle sizes, twenty-two letters can be put on in smaller type. We do not furnish less than 25 labels of one name. State name and numbers. Made to order and shipped from factory only. Shipment usually made in from 5 to 10 days.

J.P.SMITH
516

99L8319⅓—Sheep and Hog Size.

Lots of	Weight, Oz.	One Name and No.	Name Only	Number Only
25	3	$0.90	$0.80	$0.70
50	6	1.40	1.25	1.15
100	12	2.15	1.20	1.75

99L8321⅓—Cattle Size.

Lots of	Weight, Lbs.	One Name and No.	Name Only	Number Only
25	½	$1.10	$0.95	$0.85
50	¾	1.65	1.35	1.20
100	1½	2.20	1.90	1.70

99L8323⅓—Extra Cattle Size.

Lots of	Weight, Lbs.	One Name and No.	Name Only	Number Only
25	¾	$1.25	$1.10	$1.00
50	1¼	1.85	1.55	1.40
100	2½	2.65	2.35	2.10

$1.20 Regular Size
Oval Hole Ear Punches for above. Punches hole and closes label.
99L8325⅓
Regular Cattle, Sheep and Hog Size Ear Punch. Weight, 12 ounces...........$1.20
99L8327⅓—Extra Cattle Size Ear Punch. Weight, 12 ounces...........$1.45

ALUMINUM EAR BUTTONS. Lots of 100 $3.85
Furnished with name and address (not exceeding 15 letters) on one side and name of farm or ranch and any number or series of numbers up to 999 on reverse side. Made to order and shipped from factory only. Shipment usually made in from 5 to 10 days. State name and number.

J. HARVEY OAKGROVE

99L8329⅓—With name and number.

Lots of	25	50	100	500	999
Weight	3 oz.	5 oz.	10 oz.	3 lbs.	6 lbs.
	$1.45	$2.20	$3.35	$16.00	$28.00

99L8331⅓ — Punch and Pliers combined, for fitting above buttons to ears. Wt., about 12 ounces.....................$1.35

8c DOZ.
CAST IRON STOCK MARKS. With raised letters—never wear out. To be attached to ear with Eureka Ring and Ringer. Can furnish any single letter, A to Z. Size, ⅞ inch diameter. Weight, per dozen, 4 ounces; per 100, 1¼ pounds.
9L6054
Per dozen, without rings.. 8c
Per 100, without rings....60c

$1.45 **STOCK MARKING PUNCH WITH STEEL CUTTING DIES.**
Handles have concealed steel springs. Length, 11 in. State number of die wanted. Wt., 1½ pounds.
CUTTING DIES 1 2 3 4
9L6050 $1.45
9L6051—Extra Cutting Dies above. State number of dies. Wt., 1 oz. Each.....36c

TATTOO STOCK MARKER, WITH INTERCHANGEABLE LETTERS AND FIGURES. $2.15
For marking ears of horses, cattle, sheep, hogs, etc. It cannot readily be changed and forms identification mark. Letters or figures taken out or inserted instantly. Letters are ⅜ inch in size. Makes a clear, distinct brand and does not injure animal. Full directions on each bottle of Tattoo Oil. State letters or figures wanted.
9L6045—Marker, with any three letters or figures. Weight, 1¼ pounds.......$2.15
9L6046—Extra letters or figures. Weight, 1 ounce. Each..............26c
9L6047 — Set of Ten Figures, 0 to 9. Weight, 12 ounces.................$2.45
9L6048—Tattoo Oil, black, marks 500 ears. Weight, 4 ounces. Per bottle......45c

SAFETY WEANERS. 30c No. 2.
Does not go through calf's nose nor make it sore. Side projection prevents calf sucking sidewise. Guaranteed to wean the most obstinate cases. State size.
9L6094
Size 1, for small calves. Weight, 4 oz....27c
Size 2, for large calves. Weight, 4 oz....30c
Size 3, for yearlings. Weight, 6 oz....33c
Size 4, for 2-year-olds and cows. Wt., 8 oz.36c
Parcel post weight on above weaners is 1 lb.

ANTI-COW KICKER. 80c
A satisfactory device for hobbling a cow to prevent kicking while milking. Quickly put on or taken off. Weight, 1½ pounds.
9L6043...80c

Wire Basket Weaners. 36c and Up.
Does not interfere with feeding, but when head is raised to suck, wire basket drops over mouth. Price includes web straps.
9L6096—Calves. Weight, 8 ounces...36c
9L6097—Yearlings. Weight, 11 ounces...40c
9L6098—Cows. Weight, 1 pound....46c

LONG REACH BULL SNAP. 72c
Includes snap with socket and 34-inch chain with ring on end. Three screw eyes. No wood handle furnished. Weight, 8 ounces.
9L6084 ....72c

CATTLE TIE IRON WITH SNAP.
Thimble takes rope ⅝ in. or smaller. Weight, 2 ounces.
9L6078 ....12c

SENSIBLE CATTLE LEADER.
9L6086 — With 10 feet ⅝-in. manila rope. Weight, 2 pounds.......36c
9L6087—Leader only, without rope. Wt., 1½ lbs.....26c

High Grade Cattle Leaders. 13c
With steel spring. Weight, 6 ounces.
9L6032 ....13c

78c Size 3-0 **Picket or Tie Out Chains.**
Swivel snap on one end and 1¾-inch ring on the other, with a swivel in the center. Length, 30 feet.
9L6036—Size, 0. Weight, 5½ lbs..54c
9L6037—Size, 2/0. Weight, 6 lbs..66c
9L6038—Size, 3/0. Weight, 7 lbs..78c

Electric Welded Straight Link Cow Ties.
Very popular cow tie. Links are long, smooth and strong.
9L6034
Size, 3-0. Weight, 2 pounds..39c

Double Toggle Wire Link Cow Ties.
Smooth links. Will not tangle. Easily put on and taken off.
9L6029
Size, 2-0. Wt., 1¼ lbs....24c

Double Toggle Flat Link Cow Ties.
Toggles do not come unfastened. Extra large and strong.
9L6032
Size, 3-0. Wt., 1¼ lbs....58c

Self Piercing Bull Ring. 24c
Copper, 2½ inch. Weight, 4 ounces.
9L6076 ....24c

Hinged Copper Bull Ring. 23c 2½-inch
With screwdriver to fit. State size.
9L6074
2½-inch. Weight, 4 ounces...23c
3-inch. Wt., 6 oz....26c

LARIAT OR ROPE SWIVEL.
Eyes, ⅝ and ¾ in. inside. Wt. ea., 3 oz.
9L6080 .....5c

BRASS OX BALLS. 4 for 23c
To screw on tips of horns. Size, ¾ in. Wt., 2 oz.
9L6072
Four for.........23c

Crystalline Farm Bells. $6.95 No. 8.
Every farm, no matter how small, should have a good bell. Our bells are standard in every detail and are unexcelled for service. The metal in crystalline bells is properly distributed to insure a clear, pleasing tone that can be heard for a long distance; and tone will always remain the same as when new. Will not go flat with use. We recommend size No. 3 for use on the average farm. Not mailable.
For School and Church Bells see page 889.

	99L6581	99L6582	99L6584
Size No.	2	3	8
Bottom diameter	16 in.	18 in.	19½ in.
Shipping weight	58 lbs.	73 lbs.	112 lbs.
Complete	$3.65	$4.90	$6.95

SWISS PATTERN COW BELLS. 52c 3⁹⁄₁₆ in.
Pure musical tones; can be heard a long distance and sound different from common bells.

	Diam. at Mouth	Takes Strap	Wt., Lbs.	Without Straps
9L6013	3⁹⁄₁₆ in.	1¼ in.	¾	$0.52
9L6014	3⅞ in.	1½ in.	1	.80
9L6015	5 in.	2¼ in.	2	1.40
9L6016	6 in.	2¼ in.	3½	2.30

SEAMLESS STEEL COW BELLS. 17c 3⅜ in.
9L6009 — Gives a clear loud tone. Smallest size, a good team bell. State size.

Size mouth, in..	3⅜	5
Takes strap, in..	1¼	2½
Weight, lbs.....	½	1¾
Each	17c	32c

STANDARD COW BELLS. 25c No. 5
9L6011—Well made. State size.

No.	Size Mouth	Takes of Strap, In.	Ht., In.	Wt., Lbs.	
0	5¾x3½	2	7	2½	70c
2	4¼x3	1½	5¼	1½	45c
3	3½x2¾	1½	4¾	1¼	35c
5	3 x2¼	1¼	3¾	¾	25c

LEATHER COW BELL STRAPS.
10L2601—State width wanted.

Width, inches	1¼	1½	2	2½
Weight, ounces	7	8	10	11
Each	39c	45c	56c	59c

SHEEP BELLS. 32c 1⅝-in.
Complete with straps.
9L6006
Height, 1⅝ inches; size of mouth, 1⅝x1⅞ in. Weight, 4 ounces......32c
9L6007—Height, 1¾ inches; size of mouth, 1¾x2 inches. Weight, 4 ounces.........38c

ROUND SHEEP BELLS. 30c 2⅞-in.
Complete with straps. State size.
9L6004

Diameter, in.	2⁷⁄₁₆	2⅞	3⅜
Weight, oz.	4	5	7
Each	24c	30c	36c

15c **TURKEY BELLS.**
Clear and loud toned. Diameter, 1¾ inches, with strap as shown. Also make desirable chimes for sleighs, etc. Weight, each, 3 ounces.
9L6002 ....15c

$1.00 House Mail Box
Made of heavy galvanized iron, painted with aluminum paint. Door has glass panel and name holder inside. Furnished with lock and two keys. Size, about 10 inches high, 5¾ inches deep. Shipping weight, 2 pounds.
LETTERS MAIL
99L2841
With screws for attaching $1.00

Standard Pattern Dehorning Clippers.
V-Shape Blades. Cut All Sides of Horn at Once.
Compound levers make cutting easy. Handles and "V" head frame are of malleable iron. Knives are of steel; will not interlock or cut into each other.
99L5908—Special Size Dehorner. Width of opening, 2¾ inches. Shpg. weight, 14 pounds..$3.85
Extra Knives for Special Size.
9L5942—Sliding Knife. Weight, 12 ounces....50c
9L5941—Stationary Knife. Wt., 8 oz....45c
99L5909—Large Size Dehorner. Width of opening, 3¾ inches. Shipping weight, 15 pounds..$4.80
Extra Knives for Large Size.
9L5944—Sliding Knife. Weight, 1 pound........60c
9L5943 — Stationary Knife. Wt., 10 oz....50c

RURAL FREE DELIVERY MAIL BOXES.
Conform to Specifications Issued by the Postmaster General.
For use on all rural and star route service. When requested, we stencil customer's name on both sides of box and send cardboard stencil without extra charge. This stencil will be found convenient for marking bags, implements, etc., for identification.
U.S. MAIL
FRANK J. ALLEN FAIRVIEW FARM
Rural Delivery Box No. 2.
Large size, for parcel post packages as well as ordinary mail. Measures 23½ inches long, 11 inches wide and 14 in. high, inside measurements. Has letter drop in door and coin holder. Shipping weight, 22 pounds. Not mailable.
99L2846 ....$2.95
Rural Delivery Box No. 1.
For ordinary mail. Measures 18½ in. long, 6¼ in. wide and 7½ in. high, inside measurements. Has coin holder. Shipping wt., 8½ lbs.
99L2845..$1.10

R. F. D. MAIL BOX LOCK WITH CHAIN.
Rustproof cast brass, self locking. Height of lock, including shackle, 2 in. Made especially for R. F. D. mail boxes. Price includes two regular keys.
9L4346—Without chain. Wt., 4 oz.42c
9L4347—With chain. Wt., 5 oz.46c

65c **Dehorning Saw.**
9L6090—Malleable iron frame, hardwood handle, 9½x¼-in. blade. Wt., 1¼ lbs.....65c
9L6091 — Extra blades for this saw. Weight, package of 4, 1 ounce. Four for..........35c

Can't Creep Hog Collar. 50c
Prevents the destruction of crops and shrubs. Fits any size hog. Weight, 12 ounces.
9L6070 ....50c

HOG HOLDING TONGS. 28c
Very strong. Will hold hog firmly while rings are put in. Weight, 2 pounds.
9L6060 ....28c

"Bulldog" Hog Holder. 79c
Will hold the largest hog or the smallest shote with no possibility of their getting away until released. Makes ringing easy. Weight, 4 pounds.
9L6056 ....79c

HOG RINGER. 15c
9L6065
Fitted with thumbscrew. Weight, 10 ounces ...15c
Copper plated rings for use with 9L6065 Hog Ringer shown above.

9L6066	9L6067	9L6068
Hog Size	Shote Size	Pig Size

Per box of 100, any size...............$0.09
Per dozen boxes (wt., 6 lbs.)...........1.05

EUREKA HOG RINGER. 15c
Closes rings properly. Weight, 8 oz.
9L6062 15c
Copper plated rings for use with 9L6062 Eureka Hog Ringer, shown above.
9L6063
Per box of 100 rings..................$0.14
Per dozen boxes, Wt., 8 lbs..........1.60

$1.25 **"CAN'T ROOT" HOG TAMER AND EAR MARKER COMBINED.**
Cuts through muscle in the rim of hog's nose, the only perfect way to prevent rooting.
Does not injure animal. Handles have concealed steel spring. Length, 10 in. Weight, 1¼ lbs.
9L6044 ....$1.25

 All weights and measurements given on this page are approximate and may vary a trifle.

Acme Power Grist Mill.

This is our 99L6135 large size Acme Grist Mill, illustrated to the right, equipped with pulley for use with belt power. Size of pulley, 7½x2½ inches. Should be run about 200 revolutions per minute. Capacity, about 2 lbs. of corn per minute. Shipping wt., 83 lbs. **Not mailable.** $7.65

99L6136 $7.65

$3.90 Poultry Raisers' Special Grist Mill.

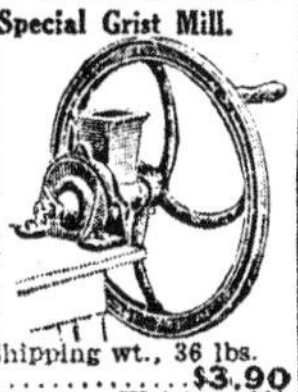

For grinding coarse grains, hominy, dry shells, dry bones, charcoal, etc. Cracked grain digests quicker. Dry shells, bones and charcoal are necessary for every fowl. Enables you to prepare your own poultry food. Grinds coarse, medium or fine. Screws furnished for attaching. Shipping wt., 36 lbs.

99L6130 .... $3.90
9L6144 — Extra Burrs. Weight, 8¼ pounds. Per pair........ $1.35

Our Hustler Hand Grist Mill. $3.85

For rough, coarse grinding. Will grind salt, peas, corn, wheat, rye and other grains, and is especially recommended for grinding chicken feed, hominy, etc. No chicken raiser should be without a grist mill. Heavy flywheel makes it easy running. Screws furnished for attaching. Shipping wt., 35 lbs.

99L6129 .... $3.85
9L6143 — Extra Burrs. Weight, 1¼ pounds. Per pair........ 75c

FULTON IMPROVED ADJUSTABLE CORN SHELLER.

$2.15 Usual $3.50 Retail Value.

Complete with butting and tipping attachment with which, in shelling seed corn, the imperfect grains can be shelled off end of ear before shelling the ear. Shells rapidly all sizes of field corn. Adjustable spring tension and thumbscrew make it suitable for small or large size ears; also equipped with guide for discharging cobs outside of box. A sheller that is guaranteed to be satisfactory in every respect. Can be clamped on any ordinary box. Shipping weight, 22 pounds.

99L6141—Complete $2.15

OUR HUSTLER CORN SHELLER.

You cannot afford to do without a corn sheller at our price. This sheller is made of good quality material, has hardened teeth, adjustable spring tension regulated by thumbscrew and will take any size ear. Shells clean from butt to tip and does not crack the grains. Fitted with attachment for shelling popcorn. Quickly attached to any box or board. Shipping weight, 16 pounds.

$1.30

99L6140 .... $1.30

For larger Corn Shellers see page 888.

$2.80 FOR 10 "SECURITY" FOLDING VEGETABLE CRATES.

Capacity, abt. 1 bushel. Recommended for field use. **Easily set up or collapsed. Folds into small space when not in use.** Packed in bundles of ten crates, eight folded and two set up. Crates measure 18 inches long, 12 inches wide, and 13½ inches high. Made of hardwood. Corners are wired, as shown in illustration. Shipping weight, bundle, 60 lbs. **Not mailable.**

99L6250—Per bundle of ten. $2.80

Galvanized Measure.

Capacity, ½ bushel. Shipping weight, 3 pounds.

99L2170 50c

Wire Chain Seed Corn Hangers. $1.45 Doz.

Each hanger holds 20 ears of corn. Made of galvanized wire. Lgth., 18 in. Wt., per doz., 7 lbs.

9L6131 Per dozen. $1.45

Acme Adjustable HAND Grist Mills

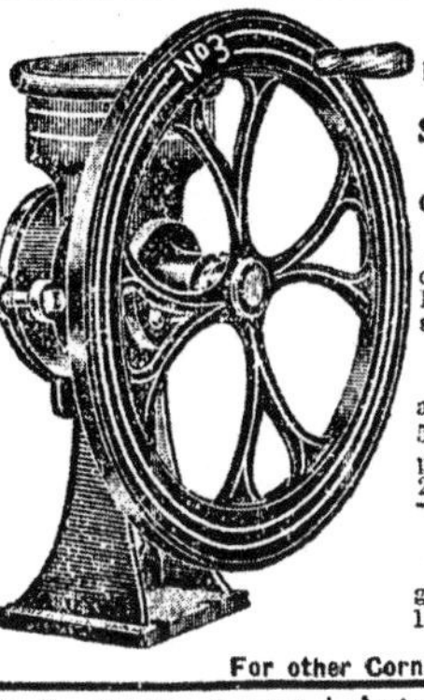

For Grinding Stock and Poultry Feed. May Also Be Used for Grinding Coffee.

Small Size $3.65 | Medium Size $4.95 | Large Size $6.30

OUR PRICES ON GRIST MILLS SHOW A DECIDED SAVINGS UNDER USUAL RETAIL PRICES.

Grinds corn, small grains, roots, bark, salt, dried bones, dried shells, etc. Adjustable to grind coarse, medium or fine. Furnished with screws for attaching. **These mills are not suitable for grinding flour, green bones or shells.**

99L6133 Small size. Grinds about 2 lbs. corn in 5 minutes. Shipping wt., 25 lbs.... $3.65

99L6134 Medium size. Grinds about 4 lbs. corn in 5 min. Shpg. wt., 42 pounds. $4.95

99L6135 Large size. Grinds about 6 lbs. corn in 5 minutes. Shipping wt., 74 lbs. $6.30 Not Mailable.

EXTRA BURRS.

9L6140 For small size grist mill. Wt., 1¼ lbs. Per pair.... 65c

9L6141 For medium size grist mill. Wt., 2¾ lbs. Per pair $1.10

9L6142 For large size grist mill. Wt., 6½ lbs. Per pair. $1.45

For other Corn, Feed and Grain Grinders see pages 818, 819, 820 and 889.

Poultry Drinking Fountain.

50c 1-Gal.

Well made of galvanized iron.

99L6075 — Cap., 1 gal. Shpg. wt., 2¼ lbs. 50c
99L6076 — Cap., 2 gal. Shpg. wt., 3¼ lbs. 70c

Automatic Air Lock Poultry Fountain. $5.65

With Heater to Keep Water From Freezing in Cold Weather.

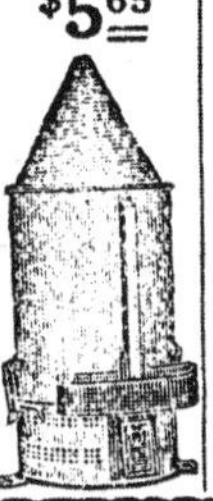

Includes 5-gallon reservoir equipped with air lock siphon pipe which automatically keeps water at proper level in the cast iron drinking cup. Base equipped with kerosene burner to keep water from freezing in extremely cold weather. Height over all, 26½ inches. Shipping weight, 15 pounds.

99L6073 .... $5.65

Poultry Drinking Fountain. 13c

Waters smallest chicks or largest fowls. Any Mason jar fits. Weight, 4 oz.

9L6182 — Without glass jar........ 13c

Our "Chickadee" Galvanized Dry Mash Poultry Feeders.

Prevents Waste and Keeps Feed Clean.

With Rain Top. $1.80 For Outdoor Use. | Without Rain Top. $1.45 For Indoor Use.

Body of feeder measures 4¼ inches in diameter by 27 inches long, with out turned flanges at opening. Has guard rod above opening to prevent fowls from getting into the feed. Lengthwise in the basin of feeder are two vibrator rods which prevent fowls from hooking or billing feed out of feeder. Capacity of feeder, about 10 pounds.

99L6070—With rain top, as illustrated, for outdoor use. Shipping weight, 7 pounds........ $1.80

99L6071—Without rain top, as illustrated, for indoor use. Shipping weight, 5 pounds........ $1.45

Our "CHICKADEE" Galvanized Chick Feeder.

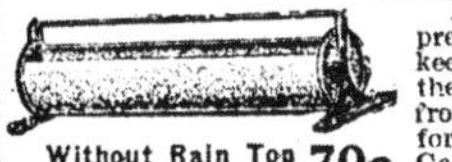

Keeps the feed in, the chick out, prevents waste, and with top keeps older birds from eating the chick feed. Protects feed from rain and sun. Can be used for feed, water or sour milk. Galvanized trough, cast iron end supports. Length, 20 inches.

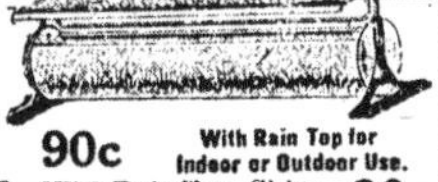

Without Rain Top for Indoor Use. 70c | With Rain Top for Indoor or Outdoor Use. 90c

99L6081—Without Rain Top. Shipping weight, 1¾ pounds........ 70c

99L6082—With Rain Top. Shipping weight, 2¼ pounds........ 90c

Spring Poultry Punch.

24c

For marking chicks by punching holes through web between toes. Used on fowls of any age. Weight, 2 ounces.

9L6178 24c

Aluminum Poultry Leg Band.

20c for 25

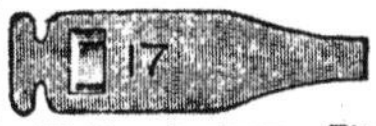

Pliable aluminum. Fits any size fowl. Once adjusted will not come off. Weight, per package of 25, 1 ounce. **State numbers.** Furnished numbered consecutively as follows:

9L6183— 1 to 25.. 20c
9L6184—26 to 50.. 20c
9L6185—51 to 75.. 20c
9L6186—76 to 100.. 20c

Celluloid Spiral Leg Bands.

Weight of fifty, 2 ounces. **Colors: Red, white, blue, green or black. State color wanted.**

½-In. 20c for 50

9L6179 — Diameter inside, ½ inch. Per package of **50**, all one color........ 20c

9L6181 — Diameter inside, ⅝ inch. Per package of **50**, all one color........ 24c

$4.45 Lice Killing Machine.

Put two tablespoonfuls of lice powder in machine and three old or six or eight young fowls, close it up and turn it around about four revolutions, then release the fowls. Wood frame, sheet steel cylinder 30 in. long and 12 in. in diameter. Shpg. wt., 23 lbs. **Not mailable.**

99L6099—Including ½ pound of best lice powder........ $4.45

Sanitary Wire Hens' Nests.

Keep Down Vermin. $1.95 A DOZEN.

Intended to fasten to wall by hooks or screws. Should be lined with straw and supported at bottom. Heavy steel wire, copper plated, reinforced. Diam., 13½ in. Shpg. wt., per doz., 8 lbs.

99L6121—Per dozen........ $1.95

HUSKING GLOVE. 39c

Curved palm plate, two malleable spur hooks. Made of leather with double strapped wristband. Right hand only. Weight, 4 oz.

9L6138 .... 39c

HUSKING GLOVE.

Made of leather. Corrugated steel hook. Two straps. Right hand only. Weight, 2 ounces. 16c

9L6136 16c

HUSKING PIN. 17c

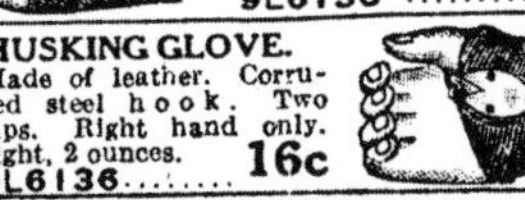

Steel pin with steel guard curved in front of wide forefinger protector. Made of leather with strap. Weight, 1 ounce.

9L6134 .... 17c

HUSKING PIN. 10c

Steel pin, mounted on good quality leather with laced thong adjustment. Weight, 1 oz.

9L6132 .... 10c

Humpty Dumpty Egg Case. 50c 12-Doz.

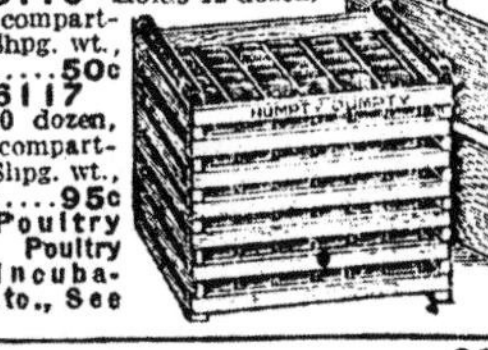

Folds together when not in use. Made of wood. Cardboard fillers.

99L6116—Holds 12 dozen, single compartment. Shpg. wt., 6¼ lbs.... 50c

99L6117 Holds 30 dozen, double compartment. Shpg. wt., 13 lbs...... 95c

For Poultry Houses, Poultry Feeds, Incubators, Etc., See Index.

Six-Hole Egg Tester. 90c

Made of steel. Enables you to determine the live eggs after six days' setting. No lamp or dark-room necessary. Wt., 1½ lbs.

99L6125 .... 90c

Galvanized Steel Brooder Coops. $1.95

Mother hens crowd their chicks into corners and sometimes smother them. Round brooder coops prevent this. Has ventilator top and two sliding doors. Slotted door gives ventilation and protects chicks from rats, etc., at night. Solid door keeps out cold. Bottom of coop is removable for cleaning. Height, 18½ inches. Bottom diameter, 21 inches. Shipping weight, each, 20 pounds. Shipping weight, per crate of 3, 65 pounds. **Not mailable.**

99L6093—Each $1.95
99L6094—Per crate of 3........ 5.40

"Chickadee" Automatic Poultry Feeder and Exerciser.

85c Keeps Your Chickens Working.

Poultry, when kept busy scratching, are healthier, more vigorous and lay more eggs. Crossarm is baited with corn or other food. By pecking bait, agitator is moved, allowing a few grains to drop on the ground. Capacity, 7 qts. Shipping wt., 3¼ lbs.

99L6089 .... 85c

Our "Chickadee" Automatic Dry Mash Feeder. $1.20

To get more eggs it is absolutely necessary to feed dry mash. Our Chickadee Automatic Dry Mash Feeder is sanitary, cannot clog and prevents waste. Fourteen hens can feed at one time. Made of galvanized steel, body is 12 inches in diameter by 5 inches in length and holds 7 quarts. Cover can be dropped down at night to keep out rats, mice, etc. Shpg. wt., 2 lbs.

99L6079 .... $1.20

"4-In-1" Dry Feeder. $1.00

Made of galvanized steel with four separate compartments for feeding dry mash, charcoal, oyster shells, grit, etc. Size, 7x7x11½ inches high. Shipping weight, 3½ pounds.

99L6080 ... $1.00

Triple Grit or Shell Box. $1.15

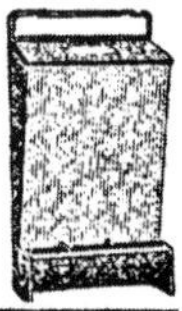

Prevents waste, keeps grit and shells out of dirt. Grit runs down automatically, but will not run out. Galvanized steel; hangs on wall. Width, 8 in.; height, 10 inches. Shipping weight, 3¼ pounds.

99L6087 .... $1.15

Galvanized Feed Basket. 1-Bu. 74c

99L2173—Capacity, 1 bushel. Shipping weight, 5¼ lbs. **Not mailable** 74c

99L2174—Capacity, 1½ bushels. Shipping weight, 5½ pounds. **Not mailable**... 92c

"Safety First" Parcel Post Egg Carriers.

$4.60 for 25 4-Doz. Size.

Light in weight; safe in shipping.

99L6104—Package of 10 carriers, each holding 10 dozen eggs. Shpg. wt., 42 lbs... $3.80
99L6105—Package of 25 carriers, each holding 4 dozen eggs. Shpg. wt., 52 lbs.... $4.60
99L6106—Package of 25 carriers, each holding 3 dozen eggs. Shpg. wt., 42 lbs.... $3.70
99L6107—Package of 25 carriers, each holding 2 dozen eggs. Shpg. wt., 28 lbs.... $2.70
99L6108—Package of 50 carriers, each holding 1 dozen eggs. Shpg. wt., 30 lbs.... $3.30

We Do Not Sell Less Than a Full Package. A full package of Parcel Post Egg Carriers, on account of its size, is not mailed.

Egg Case Cartons or Carriers.

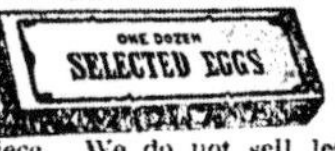

Printed as illustrated. Shipped knocked down, all complete in one piece. We do not sell less than 100. Shipping weight, per 100, 13 pounds.

	Bundle of	Per Bundle
99L6101	100	$1.75
99L6102	200	3.40
99L6103	500	8.40

All weights and measurements given on this page are approximate and may vary a trifle.

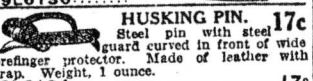

Our "Best Made" Garden Hose Outfits

COMPLETE HOSE OUTFITS SPELL HOSE ECONOMY.

$13.00 For 50-Foot Outfit.

Do you realize that, aside from the great convenience and cleanliness of keeping garden hose on a reel, the life of the hose is more than doubled thereby? The reel drains the water from the hose, thus allowing it to dry inside and pays for itself many times over.

"An Outfit You Will Be Proud Of"

This "De Luxe" outfit comprises our "Best Made" Red Corrugated Seamless Garden Hose, our "Best Made" Rubber Tired Stand Fast Hose Reel and a heavy cast brass hose nozzle. One of the best Garden Hose Outfits ever offered.

99L6601	99L6602	99L6603
50-Foot ¾-Inch Hose, Reel and Nozzle. Shpg. wt., 57 lbs. Complete...$13.00	75-Foot ¾-Inch Hose, Reel and Nozzle. Shpg. wt., 63 lbs. Complete...$17.25	100-Foot ¾-Inch Hose, Reel and Nozzle. Shpg. wt., 75 lbs. Complete...$21.50

Not Mailable.

FOR LONG SERVICE WE RECOMMEND THE PURCHASE OF

MOLDED RUBBER GARDEN HOSE

Seamless, Non-Kinkable and Durable.

Made with two seamless double braided jackets over a tube of high grade rubber incased in a heavy molded rubber outside cover. As nearly leakproof as garden hose can be made. Our price includes standard ¾-inch patent brass couplings properly put on.

Sold in Any Length Desired From 10 Feet Up to 500 Feet.

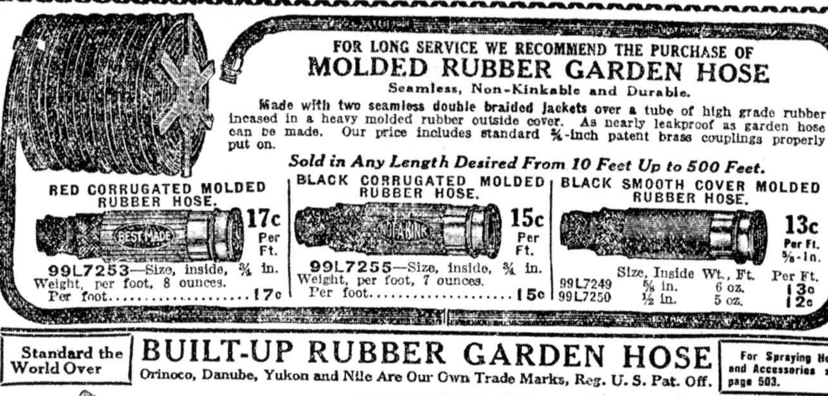

RED CORRUGATED MOLDED RUBBER HOSE. 17c Per Ft.
99L7253—Size, inside, ¾ in. Weight, per foot, 8 ounces. Per foot....17c

BLACK CORRUGATED MOLDED RUBBER HOSE. 15c Per Ft.
99L7255—Size, inside, ¾ in. Weight, per foot, 7 ounces. Per foot....15c

BLACK SMOOTH COVER MOLDED RUBBER HOSE. 13c Per Ft. ⅝-In.

	Size, Inside	Wt., Ft.	Per Ft.
99L7249	⅝ in.	6 oz.	13c
99L7250	½ in.	5 oz.	12c

Standard the World Over

BUILT-UP RUBBER GARDEN HOSE

Orinoco, Danube, Yukon and Nile Are Our Own Trade Marks, Reg. U. S. Pat. Off.

For Spraying Hose and Accessories see page 503.

ORINOCO PLY — DANUBE PLY — YUKON PLY — NILE PLY

1-INCH 18c Per Foot in 50-Ft. Lengths. 7-Ply 1-Inch "Orinoco" Hose. Has heavy rubber inner tube, seven plies of strong cotton fabric and a heavy rubber outer coating. Size, inside, 1 inch, complete with 1-inch couplings properly put on. Shipping wt., 25 feet, 15 lbs.
99L7238 Length, 25 ft. $4.75
99L7239 Length, 50 ft. $9.00

¾-INCH 12c Per Foot in 50-Ft. Lengths. 6-Ply ¾-Inch "Danube" Hose. Has rubber inner and outer coating and six plies of strong cotton fabric. Size, inside, ¾ inch. Complete with ¾-inch couplings properly put on. Shipping weight, 25 feet, 11 pounds.
99L7234 Length, 25 ft. $3.25
99L7235 Length, 50 ft. $6.00

⅝-INCH 10c Per Foot in 50-Ft. Lengths. 5-Ply ⅝-Inch "Yukon" Hose. Good quality solid walls. Rubber inner tube, five plies cotton fabric and good rubber outer cover. Size, inside, ⅝ inch. Complete with ¾-inch couplings properly put on. Shipping weight, 25 feet, 8 pounds.
99L7224 Length 25 feet $2.75
99L7225 50 feet 5.00

½-INCH 8½c Per Foot in 50-Ft. Lengths. 4-Ply ½-Inch "Nile" Hose. Rubber inner tube, four plies cotton fabric and rubber outer coating, with ¾-inch couplings. Size, inside, ½ inch. Shipping weight, 25 feet, 4½ pounds.
99L7190—Length, 25 ft. $2.35
99L7191—Length, 50 ft. 4.25

BUCKEYE PRUNING SHEARS. $1.40
Cuts good size limbs. Steel blades. Handles, 22 in. long; 2½-inch steel cutters. Shipping wt., 2¾ lbs.
99L5593 ....$1.40

HEDGE SHEARS. $1.65 and Up.
Steel blades with improved notch for cutting twigs. State length.

Length of blade, in.	8	10	12
Shpg. wt., lbs.	3	3¼	3¾
99L5595	$1.65	$2.15	$2.70

DOUBLE EDGE PRUNING SAW. $1.10
Has 18-inch steel blade, toothed fine on one edge, coarse on other. Hardwood handle. Weight, 8 ounces.
9L5042 ....$1.10

TREE PRUNER.
$1.95 12-Ft. Not Mailable.
No ladders required, as pruning can be done while standing on the ground. Blade of tempered steel.

	Length	Shpg. Wt.	
99L6673	8 ft.	3½ lbs.	$1.60
99L6674	10 ft.	5 lbs.	1.75
99L6675	12 ft.	6 lbs.	1.95
9L5940—Extra Knife. Wt., 2 oz.			.12

TELEGRAPH TREE PRUNER. $1.10
Can be screwed to pole of any length. Price, without pole or rope.
9L5934—Wt., 1¼ lbs. $1.10
9L5935—Extra Spring. Weight, 1 ounce....12c
9L5936—Extra Blade. Wt., 2 ounces....12c

Parrot Bill Pattern Pruning Shear. $1.55
This Pruning Shear is well made and finished. Constructed so the cutter pushes the twig into the hook instead of away from it. Both cutter and hook are laid the entire length with cutlery steel. The handle is made to fit the hand and the spring is constructed and placed so that the shear works most easily. Weight, 10 ounces.
9L5939 ....$1.55

Our "Best Made" Hose Reel. $4.50 Holds 150 Feet of Hose.
A Really Practical Household Reel.
Built low to the ground, our "Best" Made" reel stands solid and firm. Rolls quietly on wire wheels with rubber tires. Steel construction throughout. Shipping weight, 32 pounds. Not Mailable.
99L6608 ....$4.50

Fulton Hose Reel Rigid and Strong.
Steel construction throughout. Made to last. Will stand hard service. Shpg. wt., 21 lbs.
5-in. iron wheels. Height over all, 37 inches.
$1.90 Holds 150 ft. of Hose.
99L6609 $1.90

Turbine Revolving Lawn Sprinkler. $1.85
Shpg. wt., 4 lbs.
Works well and covers a large area. Solid brass arms and head. Takes ¾-in. hose couplings, is 11½ in. high and arms have a spread of 13 inches.
99L6615 ....$1.85

Aurora Lawn Sprinkler. 75c

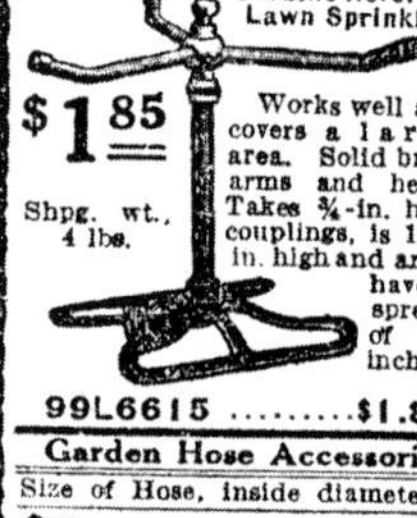

Distributes water evenly over large area. For ¾-inch hose coupling. Weight, 3 lbs.
9L7084 ....75c

Garden Hose Accessories. State Size Wanted.

Size of Hose, inside diameter, inch	½	¾	1
9L7092 Gem Pattern Solid Brass Hose Nozzle. Weight, 8 ounces (9L7092—Nozzle)	....	55c	75c
9L7096—Galvanized Steel Hose Clamp. Weight, 2 ounces (9L7096 Clamp)	3c	4c	5c
9L7098 Brass Hose Coupling. Wt., 5 ounces (9L7098—Coupling)	15c	18c	26c
9L7102—Malleable Iron Hose Splicer. Weight, 2 ounces (9L7102—Splicer)	5c	6c	7c
9L7094—Rubber Hose Washers. Weight, 1 ounce. Per dozen (9L7094 Washers)		4c	

BRASS HOSE NIPPLES. 22c ¾-In.
For connecting hose to iron pipe. State size.

Pipe size, inch	½	¾	1
Fits hose coupling, in.	¾	¾	1
Weight, ounces	2	3	4
9L7100	18c	22c	28c

Household Soldering Outfit.
Small soldering iron, scraper, bar of solder and box of rosin, as illustrated. Wt., 1 lb.
9L3581 ....34c

Guaranteed Waterproof Wash Aprons With Legs.
Made of black oiled cotton duck. To protect clothing when stooping or bending, apron is provided with legs that hold it close to body. Apron measures 36x45 in. Wt., 2 lbs.
9L3521 $1.95
Same as above, but without legs. Weight, 1½ lbs.
9L3522 $1.55

"WHY DON'T YOU SPEAK FOR YOURSELF JOHN?", said Priscilla to John Alden. No need to tell a Sears-Roebuck shipment that. The quality is there.

BRASS STENCIL LETTERS AND FIGURES.

Complete font of 70 pieces consisting of letters, figures, periods, etc.
9L6250—State size.

Size Letters, Inches	Wt., Oz.	
½	6	$0.95
¾	8	1.05
1	10	1.20
1½	12	1.45

BLACK STENCIL INK
With sponge cup and sponge. Weight, 15 oz.
9L6252..39c

STENCIL BRUSH.
1-inch ferrule. Weight, 4 ounces. For other Stencil brushes, see page 952.
9L6253 ..23c

Steel Grain Scoops.
Not Mailable.
Seams riveted, handles securely put on.
99L6735—Half bushel size. Shpg. wt., 7 lbs. $2.20
99L6736—Bushel size. Shpg. wt., 9½ lbs. $2.60

$3.20 Handy Bag or Box Truck.
Not Mailable

For farmers, grain dealers, etc. Made of selected hardwood with 3-foot 10-inch bent handles. Measures 17 in. between handles and 12 in. across nose. Has 11-inch straps, ¾-inch square axle and 6-inch wheels. Shipping wt., 21 pounds.
99L6740 ....$3.20

SWISS PATTERN PRUNING SHEARS. $2.15
A high grade shear, well known to orchard men. Blade is steel, ground convex. Length, 8½ in. Wt., 1¼ lbs.
9L5933 ....$2.15

CALIFORNIA PATTERN PRUNING SHEARS. 80c $1.20
9L5930—Standard Grade. Length, 9 in.; 2½-in. blade. Weight, 14 oz....80c
9L5931—Extra High Grade. Length, 9 in.; 2½-in. blade. Wt., 14 oz....$1.20

EASY SPRING CALIFORNIA PATTERN PRUNING SHEARS. $1.50
Full polished, new patent easy brass spring and improved locknut. Length, 9 in.; 2½-inch tempered steel blade. Weight, 1 lb.
9L5932 ....$1.50

All weights and measurements given on this page are approximate and may vary a trifle.

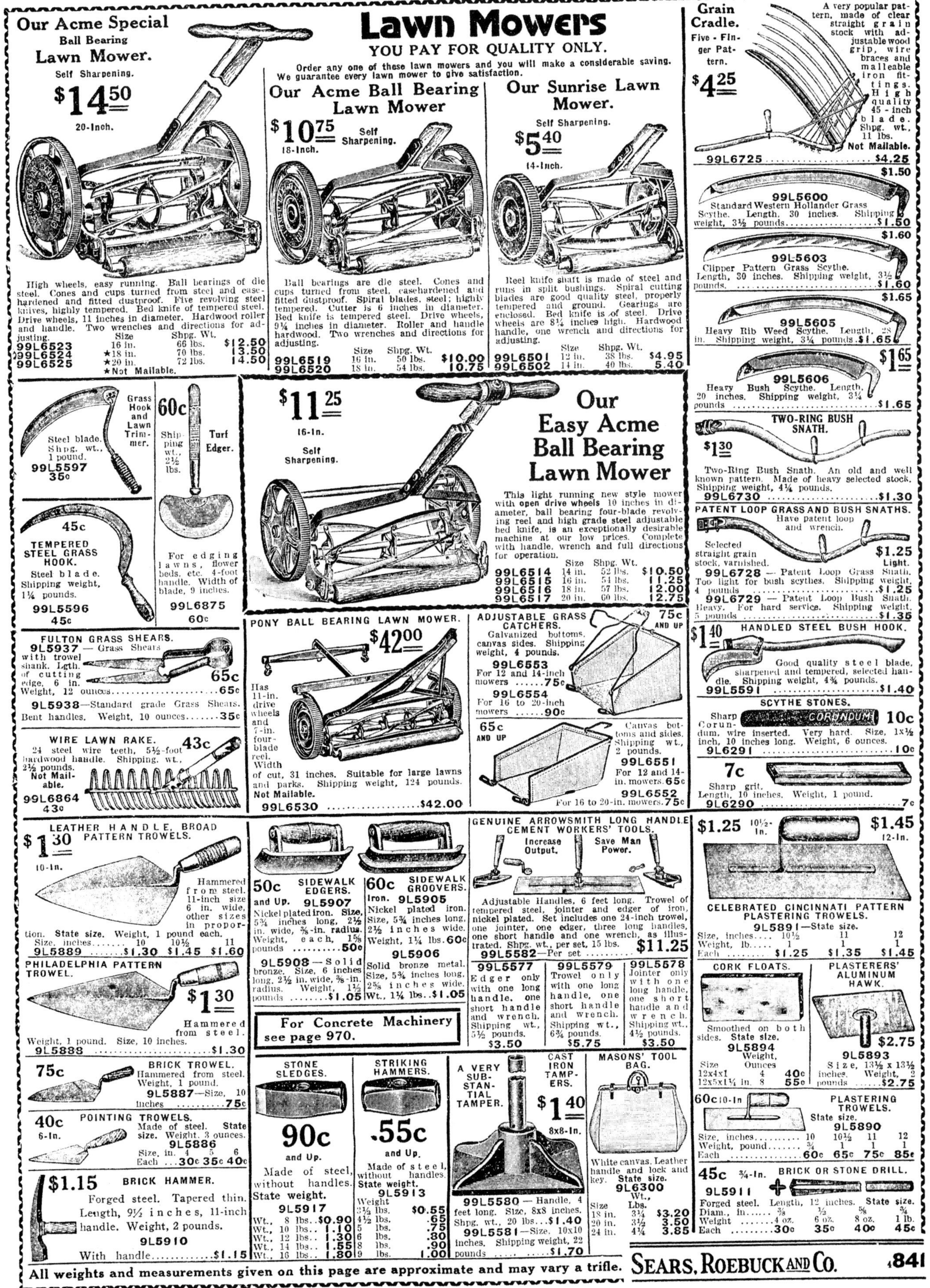

Lawn Mowers

YOU PAY FOR QUALITY ONLY.

Order any one of these lawn mowers and you will make a considerable saving. We guarantee every lawn mower to give satisfaction.

Our Acme Special Ball Bearing Lawn Mower.

Self Sharpening.

$14.50 20-Inch.

High wheels, easy running. Ball bearings of die steel. Cones and cups turned from steel and casehardened and fitted dustproof. Five revolving steel knives, highly tempered. Bed knife of tempered steel. Drive wheels, 11 inches in diameter. Hardwood roller and handle. Two wrenches and directions for adjusting.

	Size	Shpg. Wt.	
99L6523	16 in.	66 lbs.	$12.50
99L6524	★18 in.	70 lbs.	13.50
99L6525	★20 in.	72 lbs.	14.50

★Not Mailable.

Our Acme Ball Bearing Lawn Mower

$10.75 Self Sharpening. 18-Inch.

Ball bearings are die steel. Cones and cups turned from steel, casehardened and fitted dustproof. Spiral blades, steel; highly tempered. Cutter is 6 inches in diameter. Bed knife is tempered steel. Drive wheels, 9¼ inches in diameter. Roller and handle hardwood. Two wrenches and directions for adjusting.

	Size	Shpg. Wt.	
99L6519	16 in.	50 lbs.	$10.00
99L6520	18 in.	54 lbs.	10.75

Our Sunrise Lawn Mower.

Self Sharpening.

$5.40 14-Inch.

Reel knife shaft is made of steel and runs in split bushings. Spiral cutting blades are good quality steel, properly tempered and ground. Gearings are enclosed. Bed knife is of steel. Drive wheels are 8¼ inches high. Hardwood handle, one wrench and directions for adjusting.

	Size	Shpg. Wt.	
99L6501	12 in.	38 lbs.	$4.95
99L6502	14 in.	40 lbs.	5.40

Grain Cradle.

Five - Finger Pattern.

$4.25

A very popular pattern, made of clear straight grain stock with adjustable wood grip, wire braces and malleable iron fittings. High quality 45-inch blade. Shpg. wt., 11 lbs. Not Mailable.

99L6725 .... $4.25

$1.50

99L5600 Standard Western Hollander Grass Scythe. Length, 30 inches. Shipping weight, 3½ pounds $1.50

$1.60

99L5603 Clipper Pattern Grass Scythe. Length, 30 inches. Shipping weight, 3½ pounds $1.60

$1.65

99L5605 Heavy Rib Weed Scythe. Length, 28 in. Shipping weight, 3¼ pounds $1.65

$1.65

99L5606 Heavy Bush Scythe. Length, 20 inches. Shipping weight, 3¼ pounds $1.65

TWO-RING BUSH SNATH.

$1.30

Two-Ring Bush Snath. An old and well known pattern. Made of heavy selected stock. Shipping weight, 4¼ pounds.

99L6730 .... $1.30

PATENT LOOP GRASS AND BUSH SNATHS.

Have patent loop and wrench.

Selected straight grain stock, varnished.

$1.25 Light.

99L6728 — Patent Loop Grass Snath. Too light for bush scythes. Shipping weight, 4 pounds $1.25

99L6729 — Patent Loop Bush Snath. Heavy. For hard service. Shipping weight, 5 pounds $1.35

Grass Hook and Lawn Trimmer.

Steel blade. Shpg. wt., 1 pound.

99L5597 35c

Turf Edger.

60c Shipping wt., 2½ lbs.

For edging lawns, flower beds, etc. 4-foot handle. Width of blade, 9 inches.

99L6875 60c

TEMPERED STEEL GRASS HOOK.

45c

Steel blade. Shipping weight, 1¼ pounds.

99L5596 45c

Our Easy Acme Ball Bearing Lawn Mower

$11.25 16-In. Self Sharpening.

This light running new style mower with open drive wheels 10 inches in diameter, ball bearing four-blade revolving reel and high grade steel adjustable bed knife, is an exceptionally desirable machine at our low prices. Complete with handle, wrench and full directions for operation.

	Size	Shpg. Wt.	
99L6514	14 in.	52 lbs.	$10.50
99L6515	16 in.	54 lbs.	11.25
99L6516	18 in.	57 lbs.	12.00
99L6517	20 in.	60 lbs.	12.75

FULTON GRASS SHEARS.

9L5937 — Grass Shears with trowel shank. Lgth. of cutting edge, 6 in. Weight, 12 ounces 65c

9L5938—Standard grade Grass Shears. Bent handles. Weight, 10 ounces....... 35c

WIRE LAWN RAKE.

43c

24 steel wire teeth, 5½-foot hardwood handle. Shipping wt., 2½ pounds. Not Mailable.

99L6864 43c

PONY BALL BEARING LAWN MOWER.

$42.00

Has 11-in. drive wheels and 7-in. four-blade reel. Width of cut, 31 inches. Suitable for large lawns and parks. Shipping weight, 124 pounds. Not Mailable.

99L6530 .... $42.00

ADJUSTABLE GRASS CATCHERS.

75c and up

Galvanized bottoms, canvas sides. Shipping weight, 4 pounds.

99L6553 For 12 and 14-inch mowers 75c

99L6554 For 16 to 20-inch mowers 90c

65c and up

Canvas bottoms and sides. Shipping wt., 2 pounds.

99L6551 For 12 and 14-in. mowers. 65c

99L6552 For 16 to 20-in. mowers. 75c

HANDLED STEEL BUSH HOOK.

$1.40

Good quality steel blade, sharpened and tempered, selected handle. Shipping weight, 4¾ pounds.

99L5591 .... $1.40

SCYTHE STONES.

10c

Sharp Corundum, wire inserted. Very hard. Size, 1x½ inch, 10 inches long. Weight, 6 ounces.

9L6291 .... 10c

7c

Sharp grit. Length, 10 inches. Weight, 1 pound.

9L6290 .... 7c

LEATHER HANDLE, BROAD PATTERN TROWELS.

$1.30 10-In.

Hammered from steel. 11-inch size 6 in. wide, other sizes in proportion. State size. Weight, 1 pound each.

Size, inches	10	10½	11
9L5889	$1.30	$1.45	$1.60

PHILADELPHIA PATTERN TROWEL.

$1.30

Hammered from steel. Weight, 1 pound. Size, 10 inches.

9L5888 .... $1.30

BRICK TROWEL.

75c

Hammered from steel. Weight, 1 pound.

9L5887—Size, 10 inches 75c

POINTING TROWELS.

40c 6-In.

Made of steel. State size. Weight, 3 ounces.

9L5886			
Size, in.	4	5	6
Each	30c	35c	40c

BRICK HAMMER.

$1.15

Forged steel. Tapered thin. Length, 9½ inches, 11-inch handle. Weight, 2 pounds.

9L5910

With handle $1.15

SIDEWALK EDGERS.

50c and Up.

9L5907 Nickel plated iron. Size, 5¾ inches long, 2½ in. wide, ⅝-in. radius. Weight, each, 1⅝ pounds 50c

9L5908 — Solid bronze. Size, 6 inches long, 2½ in. wide, ⅜-in. radius. Weight, 1½ pounds $1.05

SIDEWALK GROOVERS.

60c Iron.

9L5905 Nickel plated iron. Size, 5¾ inches long, 2½ inches wide. Weight, 1¼ lbs. 60c

9L5906 Solid bronze metal. Size, 5¾ inches long, 2⅝ inches wide. Wt., 1¼ lbs.. $1.05

For Concrete Machinery see page 970.

GENUINE ARROWSMITH LONG HANDLE CEMENT WORKERS' TOOLS.

Increase Output. Save Man Power.

Adjustable Handles, 6 feet long. Trowel of tempered steel, jointer and edger of iron, nickel plated. Set includes one 24-inch trowel, one jointer, one edger, three long handles, one short handle and one wrench, as illustrated. Shpg. wt., per set, 15 lbs.

99L5582—Per set $11.25

99L5577 Edger only with one long handle, one short handle and wrench. Shipping wt., 5½ pounds. $3.50

99L5579 Trowel only with one long handle, one short handle and wrench. Shipping wt., 6¾ pounds. $5.75

99L5578 Jointer only with one long handle, one short handle and wrench. Shipping wt., 4½ pounds. $3.50

CELEBRATED CINCINNATI PATTERN PLASTERING TROWELS.

$1.25 10½-In. $1.45 12-In.

9L5891—State size.

Size, inches....	10½	11	12
Weight, lb.....	1	1	1
Each	$1.25	$1.35	$1.45

CORK FLOATS.

Smoothed on both sides. State size.

9L5894

Size	Weight, Ounces	
12x4x1	4	40c
12x5x1¼ in.	8	55c

PLASTERERS' ALUMINUM HAWK.

$2.75

9L5893 Size, 13½ x 13½ inches. Weight, 2 pounds $2.75

PLASTERING TROWELS.

60c 10-In.

State size.

9L5890

Size, inches..........	10	10½	11	12
Weight, pound.......	¾	1	1	1
Each	60c	65c	75c	85c

STONE SLEDGES.

90c and Up.

Made of steel, without handles. State weight.

9L5917

Wt.,	8 lbs..	$0.90
Wt.,	10 lbs..	1.10
Wt.,	12 lbs..	1.30
Wt.,	14 lbs..	1.55
Wt.,	16 lbs..	1.80

STRIKING HAMMERS.

.55c and Up.

Made of steel, without handles. State weight.

9L5913

Weight	
3½ lbs.	$0.55
4½ lbs.	.65
5 lbs.	.75
6 lbs.	.80
8 lbs.	.90
9 lbs.	1.00

A VERY SUBSTANTIAL TAMPER.

CAST IRON TAMPERS.

$1.40 8x8-In.

99L5580 — Handle, 4 feet long. Size, 8x8 inches. Shpg. wt., 20 lbs...$1.40

99L5581—Size, 10x10 inches. Shipping weight, 22 pounds $1.70

MASONS' TOOL BAG.

White canvas. Leather handle and lock and key. State size.

9L6300

Size	Wt., Lbs.	
18 in.	3¼	$3.20
20 in.	3½	3.50
24 in.	4¼	3.85

BRICK OR STONE DRILL.

45c ¾-In.

9L5911 Forged steel. Length, 12 inches. State size.

Diam., in......	⅜	½	⅝	¾
Weight	4 oz.	6 oz.	8 oz.	1 lb.
Each	30c	35c	40c	45c

All weights and measurements given on this page are approximate and may vary a trifle.

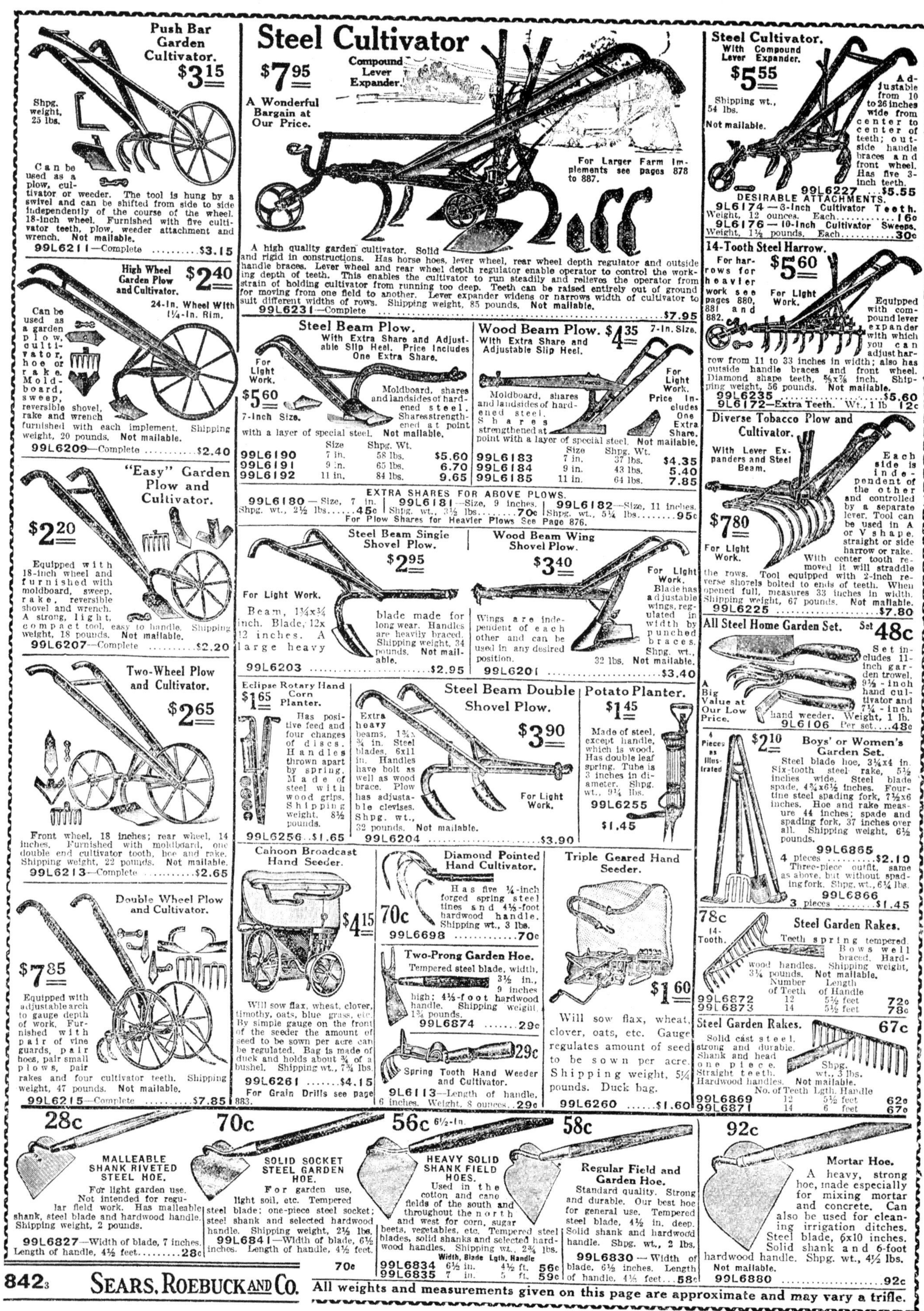

Push Bar Garden Cultivator.

$3.15

Shpg. weight, 25 lbs.

Can be used as a plow, cultivator or weeder. The tool is hung by a swivel and can be shifted from side to side independently of the course of the wheel. 18-inch wheel. Furnished with five cultivator teeth, plow, weeder attachment and wrench. Not mailable.

99L6211—Complete $3.15

Steel Cultivator

$7.95

Compound Lever Expander.

A Wonderful Bargain at Our Price.

For Larger Farm Implements see pages 878 to 887.

A high quality garden cultivator. Solid and rigid in constructions. Has horse hoes, lever wheel, rear wheel depth regulator and outside handle braces. Lever wheel and rear wheel depth regulator enable operator to control the working depth of teeth. This enables the cultivator to run steadily and relieves the operator from strain of holding cultivator from running too deep. Teeth can be raised entirely out of ground for moving from one field to another. Lever expander widens or narrows width of cultivator to suit different widths of rows. Shipping weight, 85 pounds. Not mailable.

99L6231—Complete $7.95

Steel Cultivator.

With Compound Lever Expander.

$5.55

Shipping wt., 54 lbs.

Not mailable.

Adjustable from 10 to 26 inches wide from center to center of teeth; outside handle braces and front wheel. Has five 3-inch teeth.

99L6227 ....$5.55

DESIRABLE ATTACHMENTS.

9L6174—3-Inch Cultivator Teeth. Weight, 12 ounces. Each 16c

9L6176—10-Inch Cultivator Sweeps. Weight, 1½ pounds. Each 30c

High Wheel Garden Plow and Cultivator.

$2.40

24-In. Wheel With 1¼-In. Rim.

Can be used as a garden plow, cultivator, hoe or rake. Moldboard, sweep, reversible shovel, rake and wrench furnished with each implement. Shipping weight, 20 pounds. Not mailable.

99L6209—Complete $2.40

Steel Beam Plow.

With Extra Share and Adjustable Slip Heel. Price Includes One Extra Share.

For Light Work.

$5.60

7-Inch Size.

Moldboard, shares and landsides of hardened steel. Shares strengthened at point with a layer of special steel. Not mailable.

	Size	Shpg. Wt.	
99L6190	7 in.	58 lbs.	$5.60
99L6191	9 in.	65 lbs.	6.70
99L6192	11 in.	84 lbs.	9.65

Wood Beam Plow. $4.35 7-In. Size.

With Extra Share and Adjustable Slip Heel.

For Light Work. Price Includes One Extra Share.

Moldboard, shares and landsides of hardened steel. Shares strengthened at point with a layer of special steel. Not mailable.

	Size	Shpg. Wt.	
99L6183	7 in.	37 lbs.	$4.35
99L6184	9 in.	43 lbs.	5.40
99L6185	11 in.	64 lbs.	7.85

EXTRA SHARES FOR ABOVE PLOWS.

99L6180—Size, 7 in. Shpg. wt., 2½ lbs......45c | 99L6181—Size, 9 inches. Shpg. wt., 3½ lbs.......70c | 99L6182—Size, 11 inches. Shpg. wt., 5¼ lbs.......95c

For Plow Shares for Heavier Plows See Page 876.

14-Tooth Steel Harrow.

$5.60

For harrows for heavier work see pages 880, 881 and 882.

For Light Work.

Equipped with compound lever expander with which you can adjust harrow from 11 to 33 inches in width; also has outside handle braces and front wheel. Diamond shape teeth, ⅝x⅝ inch. Shipping weight, 56 pounds. Not mailable.

99L6235 ....$5.60

9L6172—Extra Teeth. Wt., 1 lb 12c

"Easy" Garden Plow and Cultivator.

$2.20

Equipped with 18-inch wheel and furnished with moldboard, sweep, rake, reversible shovel and wrench. A strong, light, compact tool, easy to handle. Shipping weight, 18 pounds. Not mailable.

99L6207—Complete $2.20

Steel Beam Single Shovel Plow.

$2.95

For Light Work.

Beam, 1¾x¾ inch. Blade, 12x12 inches. A large heavy blade made for long wear. Handles are heavily braced. Shipping weight, 34 pounds. Not mailable.

99L6203 ....$2.95

Wood Beam Wing Shovel Plow.

$3.40

For Light Work.

Blade has adjustable wings, regulated in width by punched braces. Shpg. wt., 32 lbs. Not mailable.

Wings are independent of each other and can be used in any desired position.

99L6201 ....$3.40

Diverse Tobacco Plow and Cultivator.

With Lever Expanders and Steel Beam.

$7.80

For Light Work.

Each side is independent of the other and controlled by a separate lever. Tool can be used in A or V shape, straight or side harrow or rake. With center tooth removed it will straddle the rows. Tool equipped with 2-inch reverse shovels bolted to ends of teeth. When opened full, measures 33 inches in width. Shipping weight, 67 pounds. Not mailable.

99L6225 ....$7.80

Two-Wheel Plow and Cultivator.

$2.65

Front wheel, 18 inches; rear wheel, 14 inches. Furnished with moldboard, one double end cultivator tooth, hoe and rake. Shipping weight, 22 pounds. Not mailable.

99L6213—Complete $2.65

Eclipse Rotary Hand Corn Planter.

$1.65

Has positive feed and four changes of discs. Handles thrown apart by spring. Made of steel with wood grips. Shipping weight, 8½ pounds.

99L6256 .. $1.65

Steel Beam Double Shovel Plow.

$3.90

Extra heavy beams, 1¾x¾ in. Steel blades, 6x11 in. Handles have bolt as well as wood brace. Plow has adjustable clevises. Shpg. wt., 32 pounds. Not mailable.

For Light Work.

99L6204 ....$3.90

Potato Planter.

$1.45

Made of steel, except handle, which is wood. Has double leaf spring. Tube is 3 inches in diameter. Shpg. wt., 9¼ lbs.

99L6255

$1.45

All Steel Home Garden Set.

Set 48c

A Big Value at Our Low Price.

Set includes 11-inch garden trowel, 9¼-inch hand cultivator and 7¼-inch hand weeder. Weight, 1 lb.

9L6106 Per set ... 48c

Boys' or Women's Garden Set.

$2.10

4 Pieces as Illustrated

Steel blade hoe, 3¼x4 in. Six-tooth steel rake, 5½ inches wide. Steel blade spade, 4¾x6½ inches. Four-tine steel spading fork, 7½x6 inches. Hoe and rake measure 44 inches; spade and spading fork, 37 inches over all. Shipping weight, 6½ pounds.

99L6865

4 pieces $2.10

Three-piece outfit, same as above, but without spading fork. Shpg. wt., 6¼ lbs.

99L6866

3 pieces $1.45

Double Wheel Plow and Cultivator.

$7.85

Equipped with adjustable arch to gauge depth of work. Furnished with pair of vine guards, pair hoes, pair small plows, pair rakes and four cultivator teeth. Shipping weight, 47 pounds. Not mailable.

99L6215—Complete $7.85

Cahoon Broadcast Hand Seeder.

$4.15

Will sow flax, wheat, clover, timothy, oats, blue grass, etc. By simple gauge on the front of the seeder the amount of seed to be sown per acre can be regulated. Bag is made of duck and holds about ¾ of a bushel. Shipping wt., 7¾ lbs.

99L6261 ....$4.15

For Grain Drills see page 883.

Diamond Pointed Hand Cultivator.

70c

Has five ¼-inch forged spring steel tines and 4½-foot hardwood handle. Shipping wt., 3 lbs.

99L6698 ....70c

Two-Prong Garden Hoe.

Tempered steel blade, width, 3½ in., 9 inches high; 4½-foot hardwood handle. Shipping weight, 1¾ pounds.

99L6874 ....29c

Spring Tooth Hand Weeder and Cultivator.

29c

9L6113—Length of handle, 6 inches. Weight, 8 ounces .. 29c

Triple Geared Hand Seeder.

$1.60

Will sow flax, wheat, clover, oats, etc. Gauge regulates amount of seed to be sown per acre. Shipping weight, 5¼ pounds. Duck bag.

99L6260 ....$1.60

Steel Garden Rakes.

78c

14-Tooth.

Teeth spring tempered. Bows well braced. Hardwood handles. Shipping weight, 3¼ pounds. Not mailable.

	Number of Teeth	Length of Handle	
99L6872	12	5½ feet	72c
99L6873	14	5½ feet	78c

Steel Garden Rakes.

67c

Solid cast steel, strong and durable. Shank and head one piece. Straight teeth. Hardwood handles. Shpg. wt., 3 lbs. Not mailable.

	No. of Teeth	Lgth. Handle	
99L6869	12	5½ feet	62c
99L6871	14	6 feet	67c

Malleable Shank Riveted Steel Hoe.

28c

For light garden use. Not intended for regular field work. Has malleable shank, steel blade and hardwood handle. Shipping weight, 2 pounds.

99L6827—Width of blade, 7 inches. Length of handle, 4½ feet 28c

Solid Socket Steel Garden Hoe.

70c

For garden use, light soil, etc. Tempered steel blade; one-piece steel socket; steel shank and selected hardwood handle. Shipping weight, 2½ lbs.

99L6841—Width of blade, 6½ inches. Length of handle, 4½ feet.

70c

Heavy Solid Shank Field Hoes.

56c 6½-In.

Used in the cotton and cane fields of the south and throughout the north and west for corn, sugar beets, vegetables, etc. Tempered steel blades, solid shanks and selected hardwood handles. Shipping wt., 2¾ lbs.

	Width, Blade	Lgth. Handle	
99L6834	6½ in.	4½ ft.	56c
99L6835	7 in.	5 ft.	59c

Regular Field and Garden Hoe.

58c

Standard quality. Strong and durable. Our best hoe for general use. Tempered steel blade, 4½ in. deep. Solid shank and hardwood handle. Shpg. wt., 2 lbs.

99L6830—Width of blade, 6½ inches. Length of handle, 4½ feet ... 58c

Mortar Hoe.

92c

A heavy, strong hoe, made especially for mixing mortar and concrete. Can also be used for cleaning irrigation ditches. Steel blade, 6x10 inches. Solid shank and 6-foot hardwood handle. Shpg. wt., 4½ lbs. Not mailable.

99L6880 ....92c

All weights and measurements given on this page are approximate and may vary a trifle.

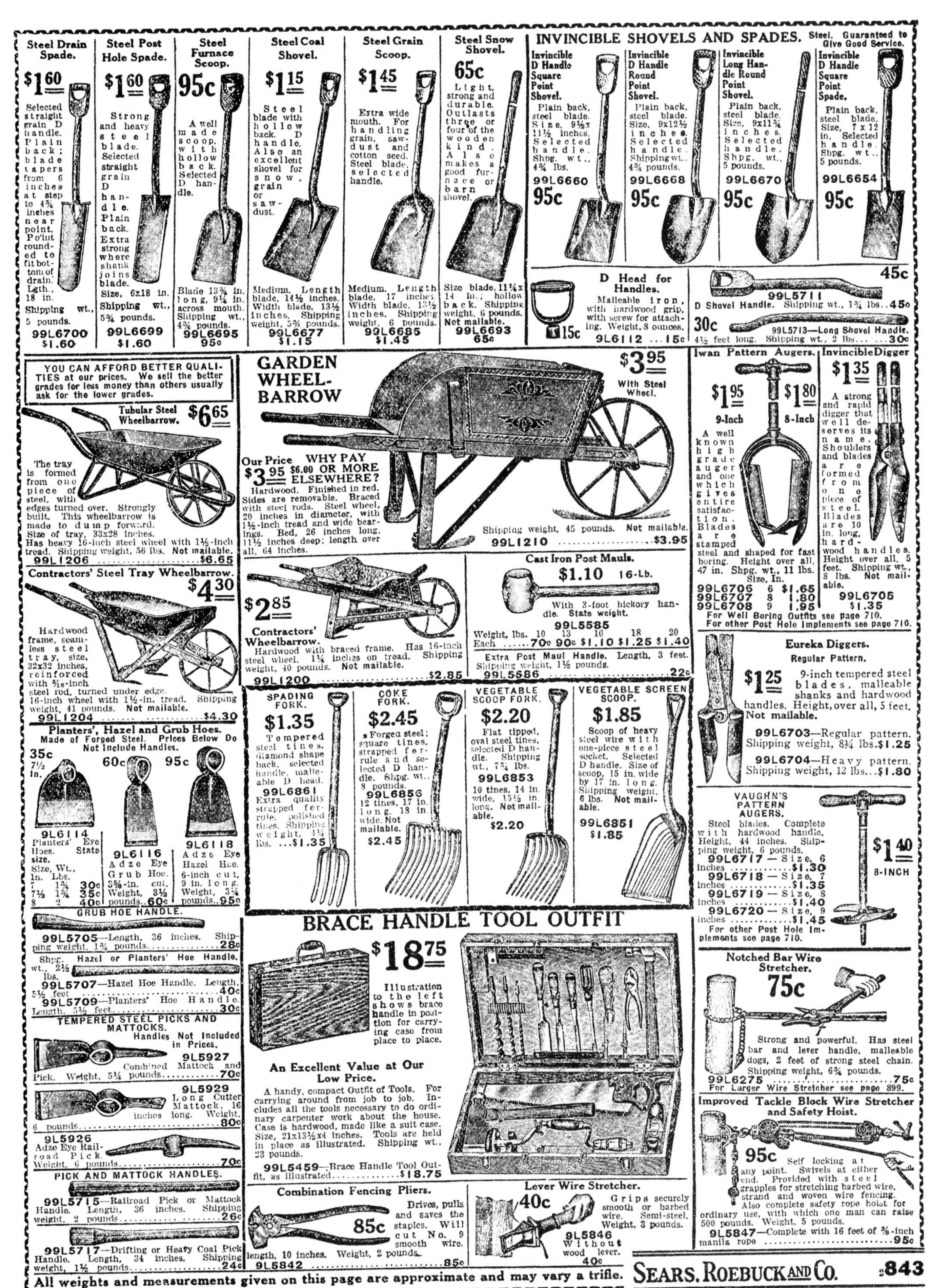

Steel Drain Spade. $1.60

Selected straight grain D handle. Plain back; blade tapers from 6 inches at step to 4¾ inches near point. Point rounded to fit bottom of drain. Lgth., 18 in. Shipping wt., 5 pounds.
99L6700 $1.60

Steel Post Hole Spade. $1.60

Strong and heavy steel blade. Selected straight grain D handle. Plain back. Extra strong where shank joins blade. Size, 6x18 in. Shipping wt., 5¾ pounds.
99L6699 $1.60

Steel Furnace Scoop. 95c

A well made scoop, with hollow back. Selected D handle. Blade 13¾ in. long, 9¼ in. across mouth. Shipping wt., 4¾ pounds.
99L6695 95c

Steel Coal Shovel. $1.15

Steel blade with hollow back. D handle. Also an excellent shovel for snow, grain or sawdust. Medium. Length blade, 14½ inches. Width blade, 13½ inches. Shipping weight, 5¾ pounds.
99L6677 $1.15

Steel Grain Scoop. $1.45

Extra wide mouth. For handling grain, sawdust and cotton seed. Steel blade, selected handle. Medium. Length blade, 17 inches. Width blade, 13½ inches. Shipping weight, 6 pounds.
99L6685 $1.45

Steel Snow Shovel. 65c

Light, strong and durable. Outlasts three or four of the wooden kind. Also makes a good furnace or barn shovel. Size blade, 11¼x14 in.; hollow back. Shipping weight, 6 pounds. Not mailable.
99L6693 65c

INVINCIBLE SHOVELS AND SPADES.

Steel. Guaranteed to Give Good Service.

Invincible D Handle Square Point Shovel. Plain back, steel blade. Size, 9½x11½ inches. Selected handle. Shpg. wt., 4¾ lbs.
99L6660 95c

Invincible D Handle Round Point Shovel. Plain back, steel blade. Size, 9x12½ inches. Selected handle. Shipping wt., 4¾ pounds.
99L6668 95c

Invincible Long Handle Round Point Shovel. Plain back, steel blade. Size, 9x11¾ inches. Selected handle. Shpg. wt., 5 pounds.
99L6670 95c

Invincible D Handle Square Point Spade. Plain back, steel blade. Size, 7 x 12 in. Selected handle. Shpg. wt., 5 pounds.
99L6654 95c

D Head for Handles. Malleable iron, with hardwood grip, with screw for attaching. Weight, 3 ounces. 15c
9L6112 ... 15c

45c
99L5711
D Shovel Handle. Shipping wt., 1¾ lbs. .45c

30c
99L5713—Long Shovel Handle. 4½ feet long. Shipping wt., 2 lbs. ... .30c

YOU CAN AFFORD BETTER QUALITIES at our prices. We sell the better grades for less money than others usually ask for the lower grades.

Tubular Steel Wheelbarrow. $6.65

The tray is formed from one piece of steel, with edges turned over. Strongly built. This wheelbarrow is made to dump forward. Size of tray, 33x28 inches. Has heavy 16-inch steel wheel with 1½-inch tread. Shipping weight, 56 lbs. Not mailable.
99L1206 ... $6.65

Contractors' Steel Tray Wheelbarrow. $4.30

Hardwood frame, seamless steel tray, size, 32x32 inches, reinforced with ⁵⁄₁₆-inch steel rod, turned under edge. 16-inch wheel with 1½-in. tread. Shipping weight, 41 pounds. Not mailable.
99L1204 ... $4.30

GARDEN WHEELBARROW

$3.95 With Steel Wheel.

Our Price $3.95 WHY PAY $6.00 OR MORE ELSEWHERE? Hardwood. Finished in red. Sides are removable. Braced with steel rods. Steel wheel, 20 inches in diameter, with 1½-inch tread and wide bearings. Bed, 26 inches long, 11½ inches deep; length over all, 64 inches.

Shipping weight, 45 pounds. Not mailable.
99L1210 ... $3.95

$2.85 **Contractors' Wheelbarrow.** Hardwood with braced frame. Has 16-inch steel wheel. 1¼ inches on tread. Shipping weight, 40 pounds. Not mailable.
99L1200 ... $2.85

Cast Iron Post Mauls.

$1.10 16-Lb.

With 3-foot hickory handle. State weight.
99L5585

Weight, lbs.	10	13	16	18	20
Each	70c	90c	$1.10	$1.25	$1.40

Extra Post Maul Handle. Length, 3 feet. Shipping weight, 1½ pounds.
99L5586 ... 22c

Iwan Pattern Augers.

$1.95 9-Inch; $1.80 8-Inch

A well known high grade auger and one which gives entire satisfaction. Blades are stamped steel and shaped for fast boring. Height over all, 47 in. Shpg. wt., 11 lbs.

	Size, In.	
99L6706	6	$1.65
99L6707	8	1.80
99L6708	9	1.95

Invincible Digger $1.35

A strong and rapid digger that well deserves its name. Shoulders and blades are formed from one piece of steel. Blades are 10 in. long, hardwood handles. Height over all, 5 feet. Shipping wt., 8 lbs. Not mailable.
99L6705 $1.35

For Well Boring Outfits see page 710.
For other Post Hole Implements see page 710.

Eureka Diggers. Regular Pattern.

$1.25 9-inch tempered steel blades, malleable shanks and hardwood handles. Height, over all, 5 feet. Not mailable.

99L6703—Regular pattern. Shipping weight, 8¾ lbs. $1.25
99L6704—Heavy pattern. Shipping weight, 12 lbs. .. $1.80

Planters', Hazel and Grub Hoes. Made of Forged Steel. Prices Below Do Not Include Handles.

35c 7½ In. — 60c — 95c

9L6114 Planters' Eye Hoes. State size.

Size, In.	Wt., Lbs.	
7	1¾	30c
7½	1¾	35c
8	2	40c

9L6116 Adze Eye Grub Hoe. 3⅝-in. cut. Weight, 3½ pounds .. 60c

9L6118 Adze Eye Hazel Hoe. 6-inch cut, 9 in. long. Weight, 3¼ pounds .. 95c

SPADING FORK. $1.35 Tempered steel tines, diamond shape back, selected handle, malleable D head.
99L6861 Extra quality strapped ferrule, polished tines. Shipping weight, 4¼ lbs. ... $1.35

COKE FORK. $2.45 Forged steel; square tines, strapped ferrule and selected D handle. Shpg. wt., 8 pounds.
99L6856 12 tines, 17 in. long, 18 in. wide. Not mailable. $2.45

VEGETABLE SCOOP FORK. $2.20 Flat tipped, oval steel tines, selected D handle. Shipping wt., 7¾ lbs.
99L6853 10 tines, 14 in. wide, 15½ in. long. Not mailable. $2.20

VEGETABLE SCREEN SCOOP. $1.85 Scoop of heavy steel wire with one-piece steel socket. Selected D handle. Size of scoop, 15 in. wide by 17 in. long. Shipping weight, 6 lbs. Not mailable.
99L6851 $1.85

VAUGHN'S PATTERN AUGERS. $1.40 8-INCH

Steel blades. Complete with hardwood handle. Height, 44 inches. Shipping weight, 6 pounds.
99L6717—Size, 6 inches ... $1.30
99L6718—Size, 7 inches ... $1.35
99L6719—Size, 8 inches ... $1.40
99L6720—Size, 9 inches ... $1.45

For other Post Hole Implements see page 710.

GRUB HOE HANDLE.
99L5705—Length, 36 inches. Shipping weight, 1¾ pounds ... 28c

Shpg. wt., 2½ lbs. Hazel or Planters' Hoe Handle.
99L5707—Hazel Hoe Handle. Length, 5½ feet ... 40c
99L5709—Planters' Hoe Handle. Length, 5½ feet ... 30c

TEMPERED STEEL PICKS AND MATTOCKS. Handles Not Included in Prices.

9L5927 Combined Mattock and Pick. Weight, 5¼ pounds ... 70c

9L5929 Long Cutter Mattock, 16 inches long. Weight, 6 pounds ... 80c

9L5926 Adze Eye Railroad Pick. Weight, 6 pounds ... 70c

PICK AND MATTOCK HANDLES.

99L5715—Railroad Pick or Mattock Handle. Length, 36 inches. Shipping weight, 2 pounds ... 26c

99L5717—Drifting or Heavy Coal Pick Handle. Length, 34 inches. Shipping weight, 1½ pounds ... 24c

BRACE HANDLE TOOL OUTFIT

$18.75

Illustration to the left shows brace handle in position for carrying case from place to place.

An Excellent Value at Our Low Price.

A handy, compact Outfit of Tools. For carrying around from job to job. Includes all the tools necessary to do ordinary carpenter work about the house. Case is hardwood, made like a suit case. Size, 21x13½x4 inches. Tools are held in place as illustrated. Shipping wt., 23 pounds.

99L5459—Brace Handle Tool Outfit, as illustrated ... $18.75

Notched Bar Wire Stretcher. 75c

Strong and powerful. Has steel bar and lever handle, malleable dogs, 2 feet of strong steel chain. Shipping weight, 6¾ pounds.
99L6275 ... 75c

For Larger Wire Stretcher see page 899.

Improved Tackle Block Wire Stretcher and Safety Hoist.

95c Self locking at any point. Swivels at either end. Provided with steel grapples for stretching barbed wire, strand and woven wire fencing. Also complete safety rope hoist for ordinary use, with which one man can raise 500 pounds. Weight, 5 pounds.
9L5847—Complete with 16 feet of ⅜-inch manila rope ... 95c

Combination Fencing Pliers. 85c

Drives, pulls and saves the staples. Will cut No. 9 smooth wire. Length, 10 inches. Weight, 2 pounds.
9L5842 ... 85c

Lever Wire Stretcher. 40c

Grips securely smooth or barbed wire. Semi-steel. Weight, 3 pounds.
9L5846 Without wood lever. 40c

All weights and measurements given on this page are approximate and may vary a trifle.

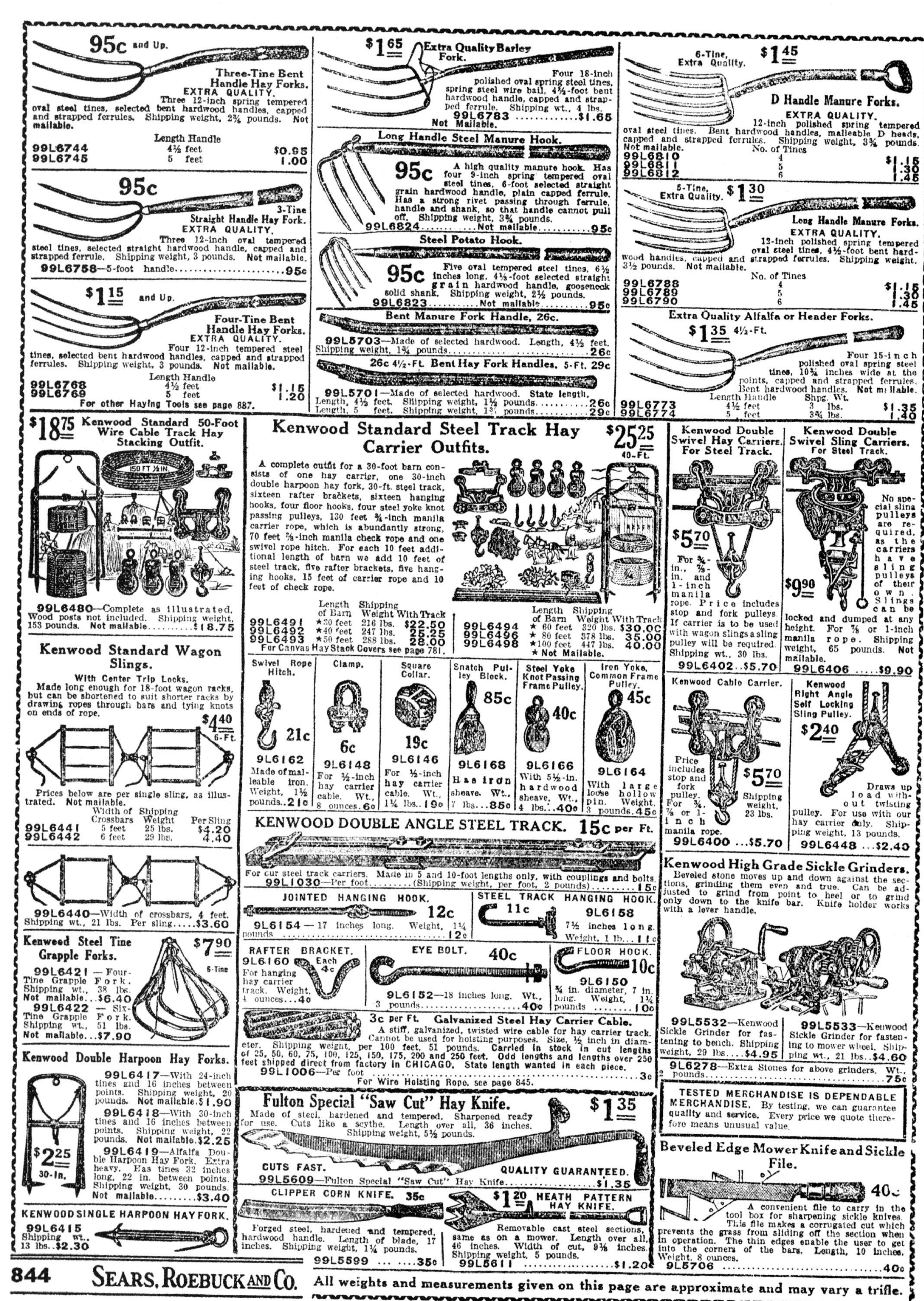

All weights and measurements given on this page are approximate and may vary a trifle.

Note Our Low Prices on Manila Rope.

Pure Manila Rope

Long Fiber, Fresh New Stock, Full Size and Strength.

1-In. $5.20 per 100 feet.

Guaranteed all pure manila, no jute or sisal mixed with it. High grade in every respect. For hay carriers it is unexcelled.

Manila rope is never measured exact diameter, one-third of the circumference being considered the diameter. Can furnish in one piece any length up to 1,200 feet.

PURE MANILA LARIAT ROPE. 2c Per Foot.

7/16-Inch Pure Manila Lariat Rope. Made of the same quality of stock as our regular manila rope. Four-strand, hard laid. A high grade lariat rope. Will not kink, runs freely and maintains a perfect loop. An excellent rope for tying out cattle, etc. Shpg. wt., per 100 feet, 6¼ lbs.

99L7826—Per foot........ **2c**

For Rope Lariats with Hondas see page 924.

NOTE—Rope for our hay carriers should not be ordered larger than 1 inch in diameter.

99L7825—State size wanted when ordering.

Size	3/16 in.	¼ in.	5/16 in.	⅜ in.	½ in.	⅝ in.	¾ in.	⅞ in.	1 in.	1⅛ in.	1¼ in.	1½ in.
Wt., per 100 ft.	1½ lbs.	2 lbs.	2½ lbs.	4 lbs.	7½ lbs.	13½ lbs.	17 lbs.	23 lbs.	27 lbs.	36 lbs.	42 lbs.	60 lbs.
Per 100 feet	35c	45c	60c	90c	$1.55	$2.70	$3.40	$4.30	$5.20	$6.40	$7.60	$10.40
Per foot	4/10c	6/10c	8/10c	1c	1¾c	3c	4c	5c	6c	7c	8c	11c

STANDARD WIRE ROPE 6c to 24c Per Foot.

Made of six strands of nineteen wires each, twisted around a fiber center. **Shipment made only from factory in CHICAGO. When ordering give catalog number, size of rope and length wanted in each piece.**

99L8381⅓—Iron Rope. A general purpose hoisting rope for use where great strain is not required.

Diameter	¼-in.	5/16-in.	⅜-in.	½-in.	⅝-in.	¾-in.	⅞-in.	1-in.
Min. Size of Drum or Sheave	1½ ft.	2 ft.	2¾ ft.	3 ft.	4 ft.	4½ ft.	5½ ft.	6 ft.
Proper Working Load, lbs.	440	600	1,960	1,560	2,400	3,400	4,720	5,800
Approx. Breaking Strain, lbs.	2,200	3,000	4,800	7,800	12,000	17,000	23,600	29,000
Shipping wt., per 100 feet	10 lbs.	15 lbs.	22 lbs.	39 lbs.	62 lbs.	89 lbs.	120lbs.	158lbs.
Per foot	6c	7c	8c	9c	11c	14c	18c	24c

99L8382⅓—Crucible Cast Steel Rope. A high grade general purpose hoisting rope for all moderate loads.

Diameter	¼-in	5/16-in.	⅜-in.	½-in.	⅝-in.	¾-in.	⅞-in.	1-in.
Min. Size of Drum or Sheave	1 ft.	1¼ ft.	1½ ft.	2 ft.	2½ ft.	3 ft.	3½ ft.	4 ft.
Proper Working Load, lbs.	880	1,240	1,920	3,360	5,000	7,000	9,200	12,000
Approx. Breaking Strain, lbs.	4,400	6,100	9,600	16,800	25,000	35,000	46,000	60,000
Shipping wt., per 100 feet	10 lbs.	15 lbs.	22 lbs.	39 lbs.	62 lbs.	89 lbs.	120lbs.	158lbs.
Per foot	7c	8c	9c	10c	12c	15c	19c	25c

99L8390⅓—Plow Steel Rope. An extra high grade general purpose hoisting rope for heavy loads and where great strength is desired. Especially recommended for stump pulling.

Diameter	⅝-in.	¾-in.	⅞-in.	1-in.	1¼-in.
Min. Size of Drum or Sheave	2½ ft.	3 ft.	3½ ft.	4 ft.	5 ft.
Proper Working Load, lbs.	6,200	9,200	11,600	15,200	24,000
Approx. Breaking Strain, lbs.	31,000	46,000	58,000	76,000	116,000
Shipping wt., per 100 feet	62 lbs.	89 lbs.	120 lbs.	158 lbs.	245 lbs.
Per foot	12c	16c	21c	26c	39c

Galvanized Iron Rope.

Composed of six strands (seven wires to a strand), with a hemp center. Each wire is galvanized, making rope rustproof. This rope is not suitable to run over drums or pulleys, but can be used for guys, etc. **State diameter wanted.**

99L8385⅓

Diameter, Inch	Shpg. Wt., per 100 Ft.	Per Ft.
5/16	15 lbs.	4c
⅜	22 lbs.	5c
½	39 lbs.	6c
⅝	62 lbs.	8c
¾	89 lbs.	10c

3c and Up. Wire Rope Clips and Clamps.

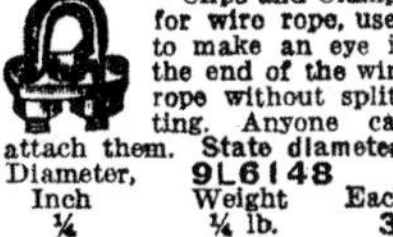

Clips and Clamps for wire rope, used to make an eye in the end of the wire rope without splitting. Anyone can attach them. **State diameter.**

9L6148

Diameter, Inch	Weight	Each
¼	¼ lb.	3c
5/16	¼ lb.	4c
⅜	¼ lb.	5c
½	½ lb.	6c
⅝	¾ lb.	8c
¾	1¼ lbs.	11c
⅞	1½ lbs.	15c
1	1¾ lbs.	18c

Transmission Rope 9c Per Ft. 1-In.

Four-strand. Made from selected manila fiber for transmission of power. Manila core, graphite and tallow laid. Can furnish any length up to 1,200 feet in one piece.

	Size	Shipping Weight, per 100 Feet	Per Foot
99L7840	⅞ in.	25 lbs.	6c
99L7841	1 in.	32 lbs.	9c
99L7842	1¼ in.	44 lbs.	12c

Strapped Steel Tackle Blocks.

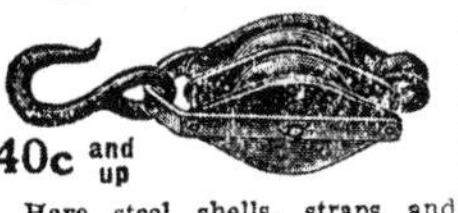

40c and up

Have steel shells, straps and pins and iron hooks. Edges are rounded to prevent wearing of rope; straps extend below the pins.

Single Steel Tackle Block.

Catalog No.	Size Shell	For Rope	Wt., Lbs.	Each
9L7046	4 in.	½ in.	1½	$0.40
9L7047	6 in.	¾ in.	3	.55
99L7991	8 in.	1 in.	7¼	.85
99L7992	10 in.	1⅛ in.	13	1.90

Double Steel Tackle Block.

Catalog No.	Size Shell	For Rope	Wt., Lbs.	Each
9L7048	4 in.	½ in.	2	$0.75
9L7049	6 in.	¾ in.	4	.95
99L7993	8 in.	1 in.	9¾	1.60
99L7994	10 in.	1⅛ in.	19	3.40

Triple Steel Tackle Block.

Catalog No.	Size Shell	For Rope	Wt., Lbs.	Each
9L7050	4 in.	½ in.	2½	$1.00
9L7051	6 in.	¾ in.	6	1.45
99L7995	8 in.	1 in.	13	2.30
99L7996	10 in.	1⅛ in.	26	4.80

Strapped Wood Tackle Blocks.

50c and up

Iron strap with iron sheaves and steel pins. Strap extends below the pins, making them unusually strong. Edges are rounded.

Single Wood Tackle Block.

Catalog No.	Size Shell	For Rope	Wt., Lbs.	Each
9L7022	4 in.	½ in.	1½	$0.50
9L7023	6 in.	¾ in.	4	.70
99L7973	8 in.	1 in.	8	1.00
99L7974	10 in.	1⅛ in.	16	1.80

Double Wood Tackle Block.

Catalog No.	Size Shell	For Rope	Wt., Lbs.	Each
9L7024	4 in.	½ in.	2½	$0.95
9L7025	6 in.	¾ in.	7	1.30
99L7977	8 in.	1 in.	12	1.80
99L7978	10 in.	1⅛ in.	25	2.90

Triple Wood Tackle Block.

Catalog No.	Size Shell	For Rope	Wt., Lbs.	Each
9L7026	4 in.	½ in.	3	$1.40
9L7027	6 in.	¾ in.	9	1.80
99L7981	8 in.	1 in.	17	2.65
99L7982	10 in.	1¼ in.	29	3.95

Big Value! 16c Order One or More Packages!

One Pound Assorted Wood Screws, in Assorted Sizes, Lengths and Finishes.

Each package contains about 200 screws.

9L4271—Per 1-pound package.... **16c**

$1.80 Security Automatic Hoists.

1,500 Lbs.

Hoists, lowers, locks and unlocks without the bother of a trip rope. Malleable iron. **Prices do not include rope.**

	Takes Rope, Inch	Cap'y, Lbs.	Shpg. Wt., Lbs.	Without Rope
9L7064	⅜	1,000	2	$0.85
99L7853	½	1,500	7	1.80
99L7854	¾	2,500	14	3.10

Safety Steel Hoist. $2.90 1,500 Lbs.

Prices Quoted Do Not Include Rope.

For contractors and builders. Have no wedge, eccentric springs, etc., to get out of order.

	Pulleys	Takes Rope, In.	Cap'y, Lbs.	Shpg. Wt., Lbs.	Without Rope
9L7066	Double	⅜	800	4	$1.70
99L7855	Double	½	1,500	5½	2.90
99L7856	Double	⅝	2,000	8¼	3.80
99L7857	Double	⅝	2,500	14	4.70
99L7858	Triple	½	3,000	8½	5.60
99L7859	Triple	⅝	3,500	12	6.50
99L7860	Triple	¾	5,000	21	7.40

Differential Chain Hoists. $16.50 2,000 Pounds.

Weston pattern. A powerful, low priced chain hoist. Automatically holds load in any position.

	Capacity, Tons	Lift, Feet	Chain Pull to Lift Full Load, Lbs.	Shpg. Wt., Lbs.	Differential Chain Hoist.
99L7880	½	7	122	41	$13.00
99L7881	1	8	215	59	16.50
99L7882	★1½	8½	245	95	21.50
99L7883	★2	9	308	150	27.00

★ Not Mallable.

100 Lbs. $2.20 Assorted Wire Nails. 25 Lbs. 73c

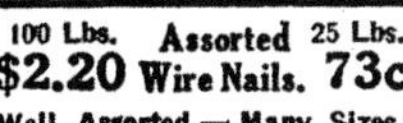

Well Assorted — Many Sizes.

Order Now and Have the Nails You Want When You Want Them.

99L1166—Per package of 25 pounds........... **73c**

99L1160—Per keg of 100 lbs. Not mallable. **$2.20**

8c Galvanized Awning Pulleys. 12c

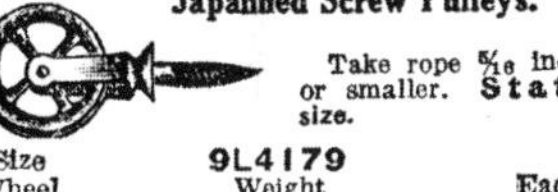

Take 5/16-in. rope, 1½-inch wheels.

9L4181—Single. Weight, 7 oz..... **8c** | **9L4183**—Double. Weight, 10 oz... **12c**

Japanned Screw Pulleys.

Take rope 5/16 inch or smaller. **State size.**

9L4179

Size Wheel	Weight	Each
2 in.	4 oz.	8c
2⅞ in.	8 oz.	13c

Japanned Well Wheel.

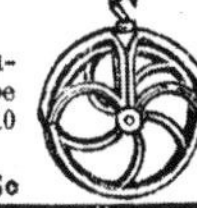

Takes ½-inch rope. Regular well chain can also be used. Width of frame, 10 inches. Weight, 4 pounds.

9L6220 .... **95c**

Samson Malleable Wagon Jacks.

$6.60 $4.95 $3.90 Built for Heavy Service.

It raises and trips or may be lowered notch by notch automatically. No springs or complex parts to get out of order.

	To Lift Corner of	Shpg. Wt., Lbs.	
99L5516	2-ton load or less	16	$3.90
99L5517	6-ton load or less	25	4.95
99L5518	8-ton load or less	31	6.60

BELL BOTTOM JACK SCREWS $1.80 AND UP.

Screws are made of cold rolled steel, lathe turned threads and cast iron stands. Diameter of screw given is actual measurement of bar before thread is cut. Diameter of finished screw is slightly less. We do not furnish levers. **Be sure to state size wanted.**

Catalog No.	Di'm'ter of Screw	Height, Stand	Length of Screw	Height Over All	Cap'ty, Tons	Shpg. Wt.	Each
99L5320	1¼ in.	8 in.	7¾ in.	10½ in.	10	12 lbs.	$1.80
99L5322	1½ in.	10 in.	9½ in.	13 in.	12	17 lbs.	2.45
99L5325	1¾ in.	12 in.	11½ in.	15 in.	16	26 lbs.	3.30
99L5328	2 in.	12 in.	11½ in.	15½ in.	20	29 lbs.	3.95
99L5330	2 in.	16 in.	15½ in.	19¼ in.	20	40 lbs.	4.90

OUR PRICES ARE LOW, but that is only part of the benefit in buying from us. The big advantage is in the QUALITY of our goods. This is assured by laboratory test.

15½-Inch $5.90 Acme Ratchet Jack Screws.

Screw is made of cold rolled steel, threads are lathe turned. Stand and cap are cast iron. Lever socket is malleable iron. Capacity, 25 tons. Diameter of base, 9¾ in.; diameter of screw, 2 inches. We do not furnish levers.

	Ht., Closed, In.	Ht., Open, In.	Shpg. Wt., Lbs.	
99L5346	15½	24	42	$5.90
99L5347	19	30	57	7.20

$3.90 Cast Iron Jack Screws.

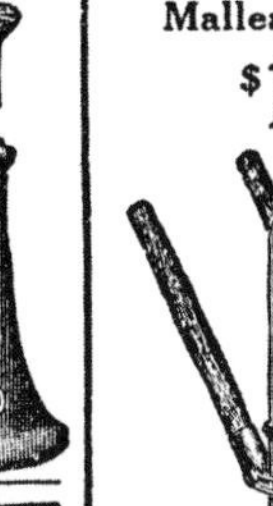

24-In.

Recommended for house movers, contractors, builders, etc. Price includes cap as shown. Trade size of screw is 3 inches; actual size is 2 11/16 inches. We do not furnish levers.

	Height Over All, Inches	Shpg. Wt., Lbs.	
99L5340	20	59	$3.40
99L5341	24	65	3.90
99L5342	30	73	4.50
99L5343	★36	80	5.10

★ Not Mallable.

Malleable Wagon Jacks.

$1.30 $1.95 $2.85

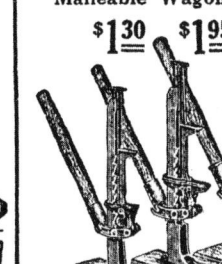

A low priced easily adjusted jack. Simple in construction and positive in its action. Locks when handle passes center. Raising handle will lower load without a jar.

	For	Shpg. Wt., Pounds	
99L5510	Buggies	6	$1.30
99L5511	Light Wagons	12	1.95
99L5512	Heavy Wagons	14	2.85

All weights and measurements given on this page are approximate and may vary a trifle.

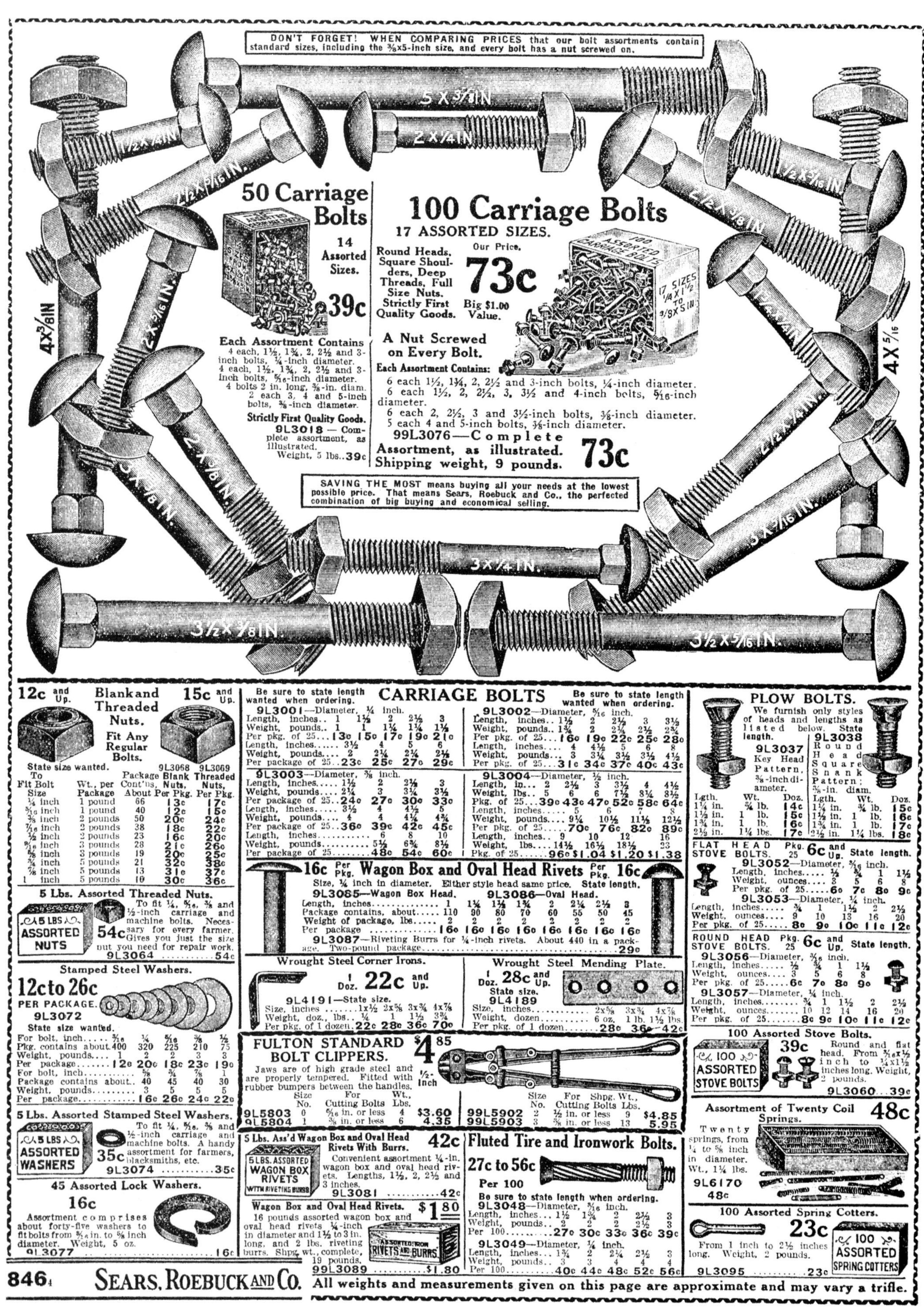

DON'T FORGET! WHEN COMPARING PRICES that our bolt assortments contain standard sizes, including the ⅜x5-inch size, and every bolt has a nut screwed on.

5 X ⅜ IN. — 2 X ¼ IN. — 1½ X ¼ IN. — 2½ X 5/16 IN. — 2 X 5/16 IN. — 4 X ⅜ IN. — 2 X ⅜ IN. — 3 X ⅜ IN. — 3 X ¼ IN. — 3½ X ⅜ IN. — 3½ X 5/16 IN. — 1½ X 5/16 IN. — 2½ X ⅜ IN. — 4 X 5/16 — 2½ X ¼ IN. — 3 X 7/16 IN.

50 Carriage Bolts

14 Assorted Sizes. **39c**

Each Assortment Contains
4 each, 1½, 1¾, 2, 2½ and 3-inch bolts, ¼-inch diameter.
4 each, 1½, 1¾, 2, 2½ and 3-inch bolts, 5/16-inch diameter.
4 bolts 2 in. long, ⅜-in. diam.
2 each 3, 4 and 5-inch bolts, ⅜-inch diameter.

Strictly First Quality Goods.
9L3018 — Complete assortment, as illustrated. Weight, 5 lbs..39c

100 Carriage Bolts

17 ASSORTED SIZES.

Round Heads, Square Shoulders, Deep Threads, Full Size Nuts. Strictly First Quality Goods.

Our Price. **73c** Big $1.00 Value.

A Nut Screwed on Every Bolt.

Each Assortment Contains:
6 each 1½, 1¾, 2, 2½ and 3-inch bolts, ¼-inch diameter.
6 each 1½, 2, 2½, 3, 3½ and 4-inch bolts, 5/16-inch diameter.
6 each 2, 2½, 3 and 3½-inch bolts, ⅜-inch diameter.
5 each 4 and 5-inch bolts, ⅜-inch diameter.

99L3076—Complete Assortment, as illustrated. Shipping weight, 9 pounds. **73c**

SAVING THE MOST means buying all your needs at the lowest possible price. That means Sears, Roebuck and Co., the perfected combination of big buying and economical selling.

Blank and Threaded Nuts.

12c and Up. 15c and Up.

Fit Any Regular Bolts.

State size wanted.

To Fit Bolt Size	Wt., per Package	Package Cont'ns, About	9L3068 Blank Nuts, Per Pkg.	9L3069 Threaded Nuts, Per Pkg.
¼ inch	1 pound	66	13c	17c
5/16 inch	1 pound	40	12c	15c
⅜ inch	2 pounds	50	20c	24c
7/16 inch	2 pounds	38	18c	22c
½ inch	2 pounds	23	16c	20c
9/16 inch	3 pounds	28	21c	26c
⅝ inch	3 pounds	19	20c	25c
¾ inch	5 pounds	21	32c	38c
⅞ inch	5 pounds	13	31c	37c
1 inch	5 pounds	10	30c	36c

5 Lbs. Assorted Threaded Nuts.

5 LBS ASSORTED NUTS

54c To fit ¼, 5/16, ⅜ and ½-inch carriage and machine bolts. Necessary for every farmer. Gives you just the size nut you need for repair work.
9L3064 .... 54c

Stamped Steel Washers.

12c to 26c PER PACKAGE.
9L3072
State size wanted.

For bolt, inch	3/16	¼	5/16	⅜	½
Pkg. contains about	400	320	225	210	75
Weight, pounds	1	2	2	3	3
Per package	12c	20c	18c	23c	19c

For bolt, inch	⅝	¾	⅞	1
Package contains about	40	45	40	30
Weight, pounds	3	5	5	5
Per package	16c	26c	24c	22c

5 Lbs. Assorted Stamped Steel Washers.

5 LBS ASSORTED WASHERS

35c To fit ¼, 5/16, ⅜ and ½-inch carriage and machine bolts. A handy assortment for farmers, blacksmiths, etc.
9L3074 .... 35c

45 Assorted Lock Washers.

16c

Assortment comprises about forty-five washers to fit bolts from 3/16 in. to ⅝ inch diameter. Weight, 5 oz.
9L3077 .... 16c

CARRIAGE BOLTS

Be sure to state length wanted when ordering.

9L3001—Diameter, ¼ inch.

Length, inches	1	1½	2	2½	3
Weight, pounds	1	1	1¼	1¼	1⅓
Per pkg. of 25	13c	15c	17c	19c	21c
Length, inches	3½	4	5	6	
Weight, pounds	2	2¼	2¼	2⅓	
Per package of 25	23c	25c	27c	29c	

9L3002—Diameter, 5/16 inch.

Length, inches	1½	2	2½	3	3½
Weight, pounds	1¾	2	2¼	2½	2¾
Per pkg. of 25	16c	19c	22c	25c	28c
Length, inches	4	4½	5	6	8
Weight, pounds	3	3¼	3½	3½	4½
Per pkg. of 25	31c	34c	37c	40c	43c

9L3003—Diameter, ⅜ inch.

Length, inches	1½	2	2½	3
Weight, pounds	2¼	3	3¼	3½
Per package of 25	24c	27c	30c	33c
Length, inches	3½	4	4½	5
Weight, pounds	4	4	4¼	4¾
Per package of 25	36c	39c	42c	45c
Length, inches		6	8	10
Weight, pounds		5½	6¾	8½
Per package of 25		48c	54c	60c

9L3004—Diameter, ½ inch.

Length, in.	2	2½	3	3½	4	4½
Weight, lbs.	5	6	6	7½	8¼	8½
Pkg. of 25	39c	43c	47c	52c	58c	64c

Length, inches	5	6	7	8
Weight, pounds	9¼	10½	11½	12½
Per pkg. of 25	70c	76c	82c	89c
Length, inches	9	10	12	16
Weight, lbs.	14½	16½	18½	23
Pkg. of 25	96c	$1.04	$1.20	$1.38

16c Per Pkg. Wagon Box and Oval Head Rivets Per Pkg. 16c

Size, ¼ inch in diameter. Either style head same price. State length.
9L3085—Wagon Box Head. 9L3086—Oval Head.

Length, inches	1	1¼	1½	1¾	2	2¼	2½	3
Package contains, about	110	90	80	70	60	55	50	45
Weight of package, lbs.	2	2	2	2	2	2	2	2
Per package	16c	16c	16c	16c	16c	16c	16c	16c

9L3087—Riveting Burrs for ¼-inch rivets. About 440 in a package. Two-pound package 29c

Wrought Steel Corner Irons.

1 Doz. 22c and Up.
9L4191—State size.

Size, inches	1x2⅞	2x5⅜	3x3¾	4x7⅞
Weight, doz., lbs.	¼	1	1½	3¾
Per pkg. of 1 dozen	22c	28c	36c	70c

Wrought Steel Mending Plate.

1 Doz. 28c and Up.
State size.
9L4189

Size, inches	2x⅝	3x¾	4x⅞
Weight, dozen	6 oz.	1 lb.	1½ lbs.
Per pkg. of 1 dozen	28c	36c	42c

FULTON STANDARD BOLT CLIPPERS.

$4.85 ½-Inch

Jaws are of high grade steel and are properly tempered. Fitted with rubber bumpers between the handles.

	Size No.	For Cutting Bolts	Wt., Lbs.	
9L5803	0	5/16 in. or less	4	$3.60
9L5804	1	⅜ in. or less	6	4.35

	Size No.	For Cutting Bolts	Shpg. Wt., Lbs.	
99L5902	2	½ in. or less	9	$4.85
99L5903	3	⅝ in. or less	13	5.95

5 Lbs. Ass'd Wagon Box and Oval Head Rivets With Burrs. 42c

5 LBS. ASSORTED WAGON BOX RIVETS WITH RIVETING BURRS

Convenient assortment ¼-in. wagon box and oval head rivets. Lengths, 1½, 2, 2½ and 3 inches.
9L3081 .... 42c

Wagon Box and Oval Head Rivets. $1.80

16 pounds assorted wagon box and oval head rivets ¼-inch in diameter and 1½ to 3 in. long, and 2 lbs. riveting burrs. Shpg. wt., complete, 19 pounds.

ASSORTED IRON RIVETS AND BURRS

99L3089 .... $1.80

Fluted Tire and Ironwork Bolts.

27c to 56c Per 100

Be sure to state length when ordering.

9L3048—Diameter, 5/16 inch.

Length, inches	1½	1¾	2	2¼	3
Weight, pounds	2	2	2	2½	3
Per 100	27c	30c	33c	36c	39c

9L3049—Diameter, ¼ inch.

Length, inches	1¾	2	2¼	2½	3
Weight, pounds	3	3	4	4	4
Per 100	40c	44c	48c	52c	56c

PLOW BOLTS.

We furnish only styles of heads and lengths as listed below. State length.

9L3037 Key Head Pattern, ⅜-inch diameter.

Lgth.	Wt.	Doz.
1¼ in.	¾ lb.	14c
1½ in.	1 lb.	15c
1¾ in.	1 lb.	16c
2½ in.	1¼ lbs.	17c

9L3038 Round Head Square Shank Pattern; ⅜-in. diam.

Lgth.	Wt.	Doz.
1¼ in.	¾ lb.	15c
1½ in.	1 lb.	16c
1¾ in.	1 lb.	17c
2½ in.	1¼ lbs.	18c

FLAT HEAD STOVE BOLTS. Pkg. 25 6c and Up. State length.

9L3052—Diameter, 3/16 inch.

Length, inches	½	¾	1	1½
Weight, ounces	3	5	6	8
Per pkg. of 25	6c	7c	8c	9c

9L3053—Diameter, ¼ inch.

Length, inches	¾	1	1½	2	2½
Weight, ounces	9	10	13	16	20
Per pkg. of 25	8c	9c	10c	11c	12c

ROUND HEAD STOVE BOLTS. Pkg. 25 6c and Up. State length.

9L3056—Diameter, 3/16 inch.

Length, inches	½	¾	1	1½
Weight, ounces	3	5	6	8
Per pkg. of 25	6c	7c	8c	9c

9L3057—Diameter, ¼ inch.

Length, inches	¾	1	1½	2	2½
Weight, ounces	10	12	14	16	20
Per pkg. of 25	8c	9c	10c	11c	12c

100 Assorted Stove Bolts.

100 ASSORTED STOVE BOLTS

39c Round and flat head. From 3/16x½ inch to ¼x1½ inches long. Weight, 2 pounds.
9L3060 .... 39c

Assortment of Twenty Coil Springs. 48c

Twenty springs, from ¼ to ⅝ inch in diameter. Wt., 1¼ lbs.
9L6170 48c

100 Assorted Spring Cotters.

23c From 1 inch to 2½ inches long. Weight, 2 pounds.

100 ASSORTED SPRING COTTERS

9L3095 .... 23c

All weights and measurements given on this page are approximate and may vary a trifle.

DON'T FORGET! WHEN COMPARING PRICES that our bolt assortments contain standard sizes, including the 3/8x5-inch size, and every bolt has a nut screwed on.

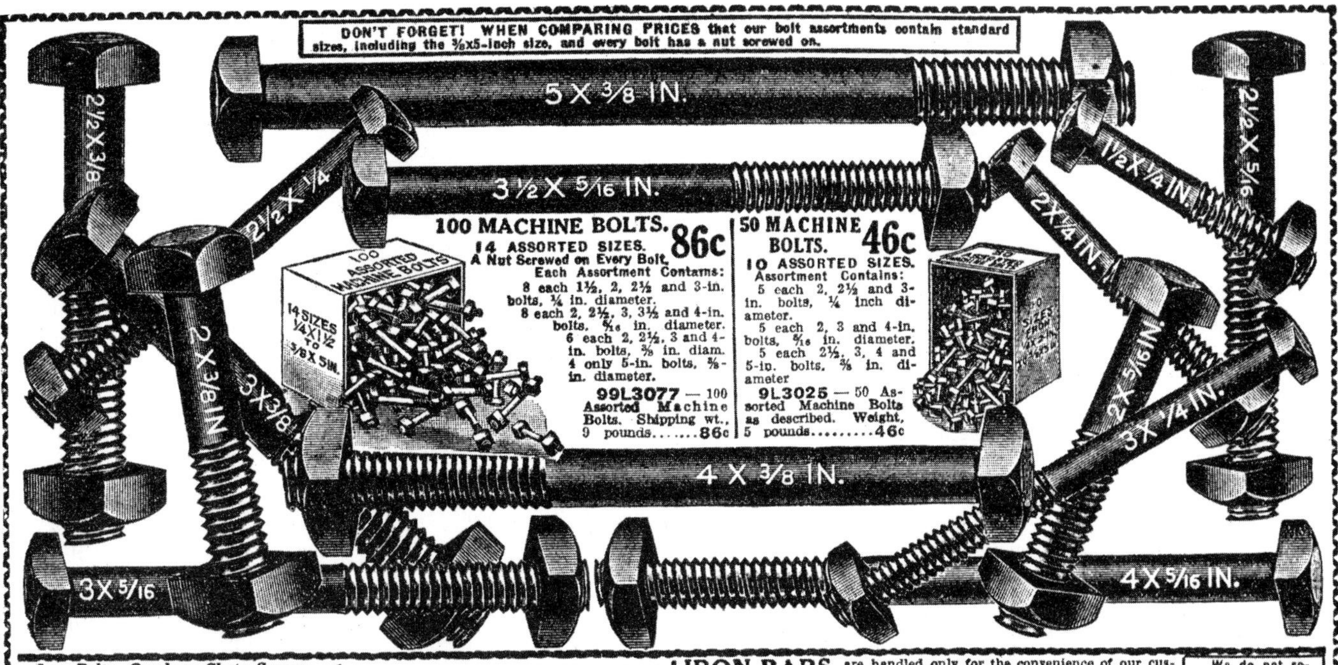

100 MACHINE BOLTS. 86c

14 ASSORTED SIZES.
A Nut Screwed on Every Bolt.
Each Assortment Contains: 8 each 1½, 2, 2½ and 3-in. bolts, ¼ in. diameter. 8 each 2, 2½, 3, 3½ and 4-in. bolts, 5/16 in. diameter. 6 each 2, 2½, 3 and 4-in. bolts, ⅜ in. diam. 4 only 5-in. bolts, ⅜-in. diameter.

9L3077 — 100 Assorted Machine Bolts. Shipping wt., 9 pounds.......**86c**

50 MACHINE BOLTS. 46c

10 ASSORTED SIZES.
Assortment Contains: 5 each 2, 2½ and 3-in. bolts, ¼ inch diameter. 5 each 2, 3 and 4-in. bolts, 5/16 in. diameter. 5 each 2½, 3, 4 and 5-in. bolts, ⅜ in. diameter.

9L3025 — 50 Assorted Machine Bolts as described. Weight, 5 pounds.........**46c**

Lag Bolts, Coach or Skein Screws.

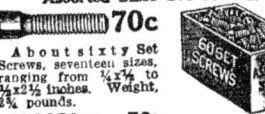

12c to 36c Per Doz. State length.

9L3041—Diameter, 5/16 inch.

Length, inches	2	2½	3	4
Weight, pounds	½	⅝	¾	15/16
Per dozen	**12c**	**14c**	**16c**	**18c**

9L3042—Diameter, ⅜ inch.

Length, inches	2½	3	4	5
Weight, pounds	⅞	1	1¼	1½
Per dozen	**15c**	**17c**	**20c**	**23c**

9L3043—Diameter, ½ inch.

Length, in.	4	5	6	7	8
Weight, lbs.	2¾	3	3½	4¼	5
Per dozen	**24c**	**27c**	**30c**	**33c**	**36c**

Assorted Sizes Cap Screws. 95c

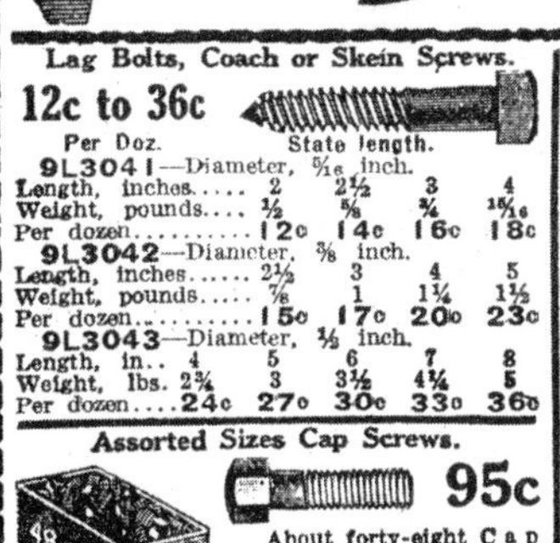

About forty-eight Cap Screws, ten sizes, ranging from ¼x¾ to ½x2 inches. Hexagon head. Weight, 5 pounds.

9L3033 ....**95c**

Assorted Sizes Set Screws. 70c

About sixty Set Screws, seventeen sizes, ranging from ¼x¼ to ½x2½ inches. Weight, 2¾ pounds.

9L3029 .....**70c**

MACHINE BOLTS

9L3009—State length. Diameter, ¼ inch.

Lgth., In.	Wt.	Per Pkg. of 25
1½	1 lb.	**16c**
2	1¼ lbs.	**19c**
2½	1¼ lbs.	**22c**
3	1½ lbs.	**25c**
4	2 lbs.	**28c**
5	2¼ lbs.	**31c**
6	2¼ lbs.	**34c**

9L3010—State length. Diameter, 5/16 inch.

Lgth., In.	Wt.	Per Pkg. of 25
1½	1¾ lbs.	**21c**
2	2 lbs.	**24c**
2½	2¼ lbs.	**27c**
3	2½ lbs.	**30c**
4	3 lbs.	**33c**
5	3½ lbs.	**36c**
6	5 lbs.	**39c**

9L3011—State length. Diameter, ⅜ inch.

Lgth., In.	Wt.	Per Pkg. of 25
1½	2½ lbs.	**24c**
2	3 lbs.	**27c**
2½	3½ lbs.	**30c**
3	3½ lbs.	**33c**
3½	4 lbs.	**36c**
4	4½ lbs.	**39c**
4½	4¾ lbs.	**42c**
5	5 lbs.	**46c**
6	6 lbs.	**50c**
8	8 lbs.	**54c**

For Complete Line of Wrenches see page 871.

9L3012—State length. Diameter, ½ inch.

Lgth., In.	Wt.	Per Pkg. of 25
1½	5 lbs.	**$0.38**
2	5½ lbs.	**.42**
2½	6¼ lbs.	**.47**
3	6¼ lbs.	**.53**
3½	7¾ lbs.	**.60**
4	8½ lbs.	**.66**
4½	10 lbs.	**.72**
5	10 lbs.	**.78**
6	10¾ lbs.	**.84**
7	13½ lbs.	**.90**
8	14½ lbs.	**.98**
10	18 lbs.	**1.10**
12	20 lbs.	**1.22**
16	26 lbs.	**1.40**
20	33 lbs.	**1.60**

IRON BARS

are handled only for the convenience of our customers who buy other merchandise from us as well.

We Will Not Fill Orders for Iron Bars Alone.

Orders must be for five bars or more, all one size or assorted as desired, on account of expense in bundling for shipment.

DO NOT ORDER BY THE POUND. WE SELL IRON BY THE BAR ONLY. ORDER NOT LESS THAN FIVE BARS.

99L3600—★Round Iron, 14-foot bars. State size wanted.

Size, inches	⅜	7/16	½	9/16	⅝	¾	⅞	1
Weight, pounds	5¼	7½	9½	12	15	31	28	37
Per bar	**12c**	**17c**	**22c**	**29c**	**34c**	**50c**	**68c**	**86c**

99L3610—★Flat Iron, ⅛ inch thick, 14-ft. bars. State size wanted.

Width, inches	¾	⅞	1	1¼	1½	1¾	2
Weight, pounds	4½	5	7	7	9	11	12
Per bar	**14c**	**16c**	**19c**	**23c**	**27c**	**31c**	**36c**

99L3615—★Flat Iron, 3/16 inch thick, 14-ft. bars. State size wanted.

Width, inches	¾	⅞	1	1¼	1½	2
Weight, pounds	7	8	9	11	13	18
Per bar	**22c**	**24c**	**27c**	**31c**	**[illegible]**	**50c**

99L3620—★Flat Iron, ¼ inch thick, 14-ft. bars. State size wanted.

Width, inches	¾	⅞	1	1⅛	1¼	1½	2
Weight, pounds	8	10	12	13	15	18	23
Per bar	**23c**	**25c**	**28c**	**32c**	**38c**	**44c**	**58c**

99L3630—★Flat Iron, ⅜ inch thick, 14-ft. bars. State size wanted.

Width, inches	1	1¼	1½	2
Weight, pounds	18	22	26	35
Per bar	**43c**	**53c**	**63c**	**83c**

99L3635—★Flat Iron, ½ inch thick, 14-ft. bars. State size wanted.

Width, inches	1	1½	2	3
Weight, pounds	23	35	46	70
Per bar	**56c**	**86c**	**$1.16**	**$1.66**

99L3640—★Round Edge Tire Iron. Sold only in full sets consisting of four bars 12½ feet long. Weights and prices are for full sets of four bars. State size.

Size, inches	⅞x¼	1x¼	1⅛x5/16	1¼x⅜	1½x½	1¾x½
Weight, pounds	45	47	65	87	140	164
Per set	**$1.00**	**$1.15**	**$1.55**	**$2.05**	**$3.25**	**$3.85**

★ Not Mailable.

NOTE—Our complete line of blacksmiths' tools shown on page 873.

We do not solicit inquiries for large quantities of iron bars for the use of dealers and manufacturers.

SOLID COPPER RIVETS AND BURRS.

Made of solid copper with burrs. State length wanted.

Size 8 30c Per Lb.

9L6329

Length	Per 1 Lb.
⅜ inch	**30c**
½ inch	**30c**
⅝ inch	**30c**
¾ inch	**30c**

Assorted, ⅜ to ¾ inch. Per 1 lb..**32c**

Size 9 32c Per Lb.

9L6330

Length	Per 1 Lb.
⅜ inch	**32c**
½ inch	**32c**
⅝ inch	**32c**
¾ inch	**32c**

Assorted, ⅜ to ¾ inch. Per 1 lb..**34c**

Size 10 35c Per Lb.

9L6331

Length	Per 1 Lb.
⅜ inch	**35c**
½ inch	**35c**
⅝ inch	**35c**
¾ inch	**35c**

Assorted, ⅜ to ¾ inch. Per 1 lb..**37c**

Size 12 38c Per Lb.

9L6332

Length	Per 1 Lb.
⅜ inch	**38c**
½ inch	**38c**
⅝ inch	**38c**
¾ inch	**38c**

Assorted, ⅜ to ¾ inch. Per 1 lb..**40c**

Copper Plated Iron Rivets and Burrs.

8c ½-Lb. No. 8.

In ½-lb. pkg. Size No. 8, assorted lengths, ⅜ to ¾ in.

9L6335

Per ½-pound package **8c**

Per dozen packages ....**[illegible]**

Japanned Tubular Steel Rivets.

For use with riveting machines shown on this page and other similar riveting machines. State size wanted.

¾ in. 11/16 in. ⅝ in. 9/16 in. ½ in. 7/16 in. ⅜ in. 5/16 in. ¼ in. 3/16 in.

9L6323—Japanned Tubular Steel Rivets, per package of about 100.

Length, inch	¾	11/16	⅝	9/16	½	7/16	⅜	5/16	¼	3/16
Weight, ounces	4	4	4	4	4	4	4	4	2	2
Per box	**24c**	**23c**	**22c**	**21c**	**20c**	**19c**	**18c**	**17c**	**16c**	**15c**

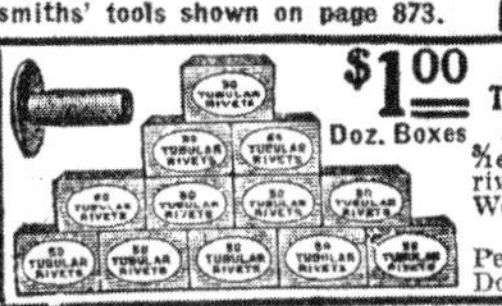

$1.00 Doz. Boxes

Assorted Japanned Tubular Steel Rivets. Assorted lengths, 5/16 to ½ in. About 50 rivets in each box. Weight, each, 2 oz.

9L6320

Per box.....**$0.10**

Dozen boxes..**$1.00**

Lever Riveting Machine. 60c

Strong and substantially made of cast iron. Takes tubular rivets 9L6320 and 9L6323 shown on this page. Will not take solid rivets. Wt., 4 lbs.

9L5850 ....**60c**

Combination Tubular Riveting Machine and Leather Punch. 95c

For riveting tubular rivets listed at right. Made of cast iron, japanned finish. Wt., about 7 pounds.

9L5849 ...**95c**

Steel Rivet Sets.

Sets suit rivets of same number. Sizes, either 7, 8, 9 or 10. State size wanted. Weight, 6 oz.

9L5861 ....**12c**

Handy Rivet Assortment.

Contains about 200 assorted rivets. Six boxes, one each, No. 5½ section, ½ to 1 inch; No. 6 section, ½ to 1 inch; No. 8 coppered iron, ⅜ to ¾ inch, with burrs; tinners' tinned assorted sizes; tubular, 5/16 to ½ inch; clinch, ¼ to ½ inch. Weight, 12 ounces.

9L6337 ....**33c**

Slotted Steel Clinch Rivets. Pkg. of 100 Rivets. 9c

Made of copper plated annealed steel. No tool required except a hammer. Assorted lengths. Weight, per package, 4 ounces.

9L6326—Pkg. of about 100 rivets..**$0.09**

Per dozen packages...................**1.00**

Riveting Hammer. 55c SIZE No. 1.

Forged steel. Weight includes handle. State weight.

Size No.	1	2	3	4
Weight, ounces	10	12	16	18
9L5748	**55c**	**60c**	**65c**	**70c**

Hollow Tube Revolving Spring Punches.

An extra high grade punch for harness makers' use and for those who desire a punch that will stand severe service. Drop forged handles, screw tubes.

9L5884 — 6-Tube. Weight, 12 oz..**$1.35**

9L5883 — 4-Tube. Weight, 12 oz..**$1.20**

Saddlers' Hollow Drive Punch. No. 6 16c

9L5862—Made of steel with tempered ends. Wt., 2 ounces. State size.

Size No.	2	4	6	8	10
Each	12c	14c	16c	18c	20c

Fulton High Grade 6-Tube Spring Punch.

Does the Work of Six Single Tube Punches.

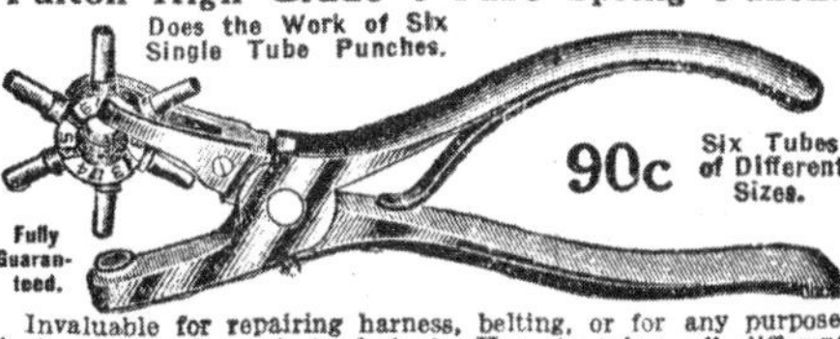

90c Six Tubes of Different Sizes.

Invaluable for repairing harness, belting, or for any purpose where a good punch is desired. Has six tubes, all different sizes, quickly set for any size. Tubes and spring are of tempered steel. Length, 8 inches. Weight, 12 ounces.

9L5865 ....**90c**

Fulton High Grade 4-Tube Spring Punch.

Four tubes, all different sizes. Tubes and spring are tempered steel. Length, 8 inches. Weight, 12 ounces.

9L5864 ....**70c**

All weights and measurements given on this page are approximate and may vary a trifle.

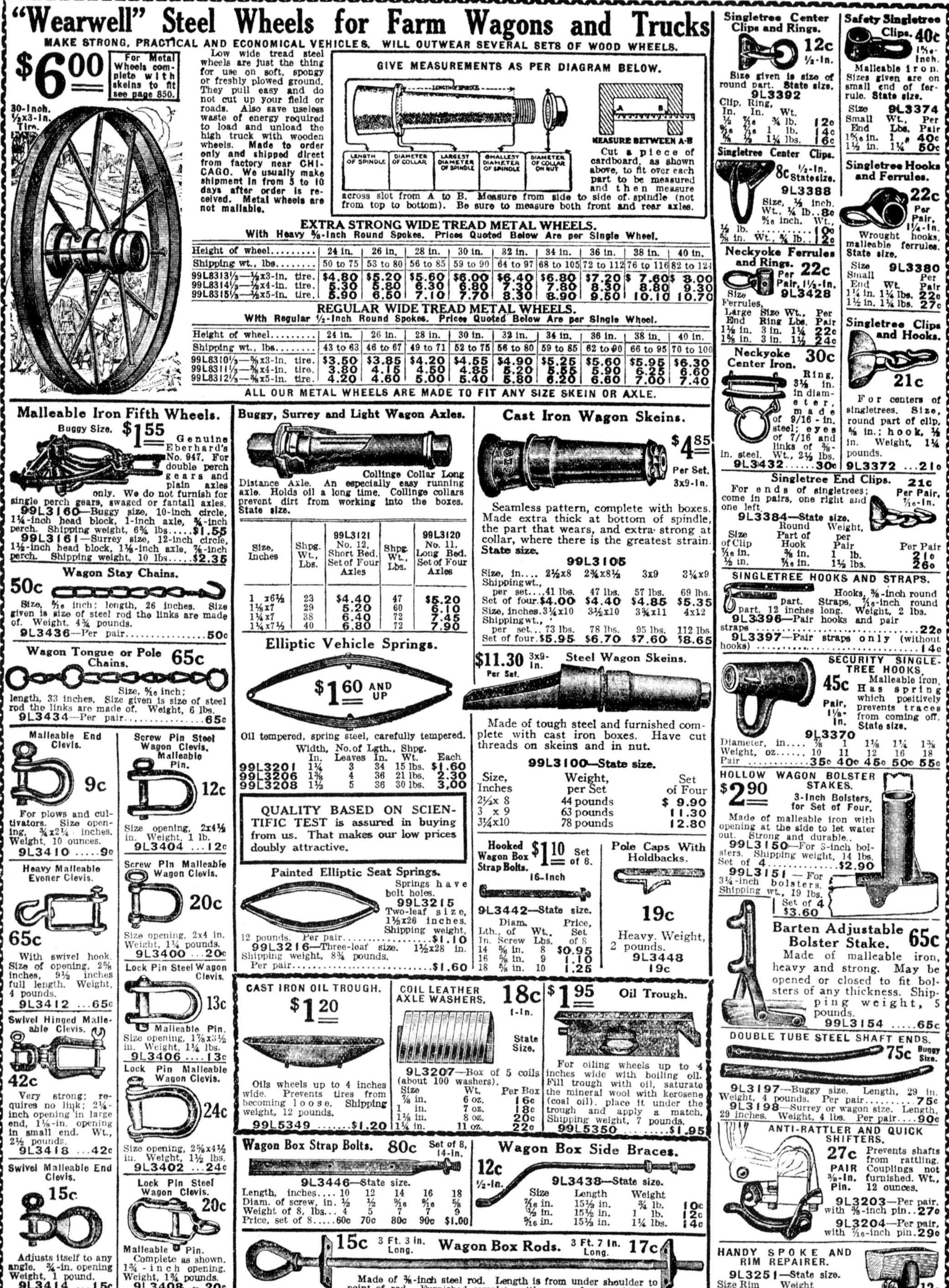

"Wearwell" Steel Wheels for Farm Wagons and Trucks

MAKE STRONG, PRACTICAL AND ECONOMICAL VEHICLES. WILL OUTWEAR SEVERAL SETS OF WOOD WHEELS.

$6.00 For Metal Wheels complete with skeins to fit see page 850.

30-Inch. ½x3-In. Tire.

Low wide tread steel wheels are just the thing for use on soft, spongy or freshly plowed ground. They pull easy and do not cut up your field or roads. Also save useless waste of energy required to load and unload the high truck with wooden wheels. Made to order only and shipped direct from factory near CHICAGO. We usually make shipment in from 5 to 10 days after order is received. Metal wheels are not mailable.

GIVE MEASUREMENTS AS PER DIAGRAM BELOW.

LENGTH OF SPINDLE | DIAMETER OF COLLAR | LARGEST DIAMETER OF SPINDLE | SMALLEST DIAMETER OF SPINDLE | DIAMETER OF COLLAR ON NUT

MEASURE BETWEEN A-B

Cut a piece of cardboard, as shown above, to fit over each part to be measured and then measure across slot from A to B. Measure from side to side of spindle (not from top to bottom). Be sure to measure both front and rear axles.

EXTRA STRONG WIDE TREAD METAL WHEELS.

With Heavy ⅝-Inch Round Spokes. Prices Quoted Below Are per Single Wheel.

Height of wheel	24 in.	26 in.	28 in.	30 in.	32 in.	34 in.	36 in.	38 in.	40 in.
Shipping wt., lbs.	50 to 75	53 to 80	56 to 85	59 to 90	64 to 97	68 to 105	72 to 112	76 to 116	82 to 124
99L8313⅓—½x3-in. tire.	$4.80	$5.20	$5.60	$6.00	$6.40	$6.80	$7.20	$ 7.60	$ 8.00
99L8314⅓—½x4-in. tire.	5.30	5.80	6.30	6.80	7.30	7.80	8.30	8.80	9.30
99L8315⅓—½x5-in. tire.	5.90	6.50	7.10	7.70	8.30	8.90	9.50	10.10	10.70

REGULAR WIDE TREAD METAL WHEELS.

With Regular ½-Inch Round Spokes. Prices Quoted Below Are per Single Wheel.

Height of wheel	24 in.	26 in.	28 in.	30 in.	32 in.	34 in.	36 in.	38 in.	40 in.
Shipping wt., lbs.	43 to 63	46 to 67	49 to 71	52 to 75	56 to 80	59 to 85	62 to 90	66 to 95	70 to 100
99L8310⅓—⅜x3-in. tire.	$3.50	$3.85	$4.20	$4.55	$4.90	$5.25	$5.60	$5.95	$6.30
99L8311⅓—⅜x4-in. tire.	3.80	4.15	4.50	4.85	5.20	5.55	5.90	6.25	6.60
99L8312⅓—⅜x5-in. tire.	4.20	4.60	5.00	5.40	5.80	6.20	6.60	7.00	7.40

ALL OUR METAL WHEELS ARE MADE TO FIT ANY SIZE SKEIN OR AXLE.

Malleable Iron Fifth Wheels.

Buggy Size. $1.55

Genuine Eberhard's No. 947. For double perch gears and plain axles only. We do not furnish for single perch gears, swaged or fantail axles.

99L3160—Buggy size, 10-inch circle, 1¼-inch head block, 1-inch axle, ⅞-inch perch. Shipping weight, 6¾ lbs. $1.55

99L3161—Surrey size, 12-inch circle, 1½-inch head block, 1⅛-inch axle, ⅞-inch perch. Shipping weight, 10 lbs. $2.35

Wagon Stay Chains.

50c

Size, 5/16 inch; length, 26 inches. Size given is size of steel rod the links are made of. Weight, 4¾ pounds.

9L3436—Per pair 50c

Wagon Tongue or Pole Chains. 65c

Size, 5/16 inch; length, 33 inches. Size given is size of steel rod the links are made of. Weight, 6 lbs.

9L3434—Per pair 65c

Malleable End Clevis.

9c

For plows and cultivators. Size opening, ¾x2¼ inches. Weight, 10 ounces.

9L3410 ....9c

Screw Pin Steel Wagon Clevis. Malleable Pin.

12c

Size opening, 2x4½ in. Weight, 1 lb.

9L3404 ...12c

Heavy Malleable Evener Clevis.

65c

With swivel hook. Size of opening, 2⅝ inches, 9½ inches full length. Weight, 4 pounds.

9L3412 ...65c

Screw Pin Malleable Wagon Clevis.

20c

Size opening, 2x4 in. Weight, 1¼ pounds.

9L3400 ...20c

Lock Pin Steel Wagon Clevis.

13c

Malleable Pin. Size opening, 1⅞x3½ in. Weight, 1½ lbs.

9L3406 ....13c

Swivel Hinged Malleable Clevis.

42c

Very strong; requires no link; 2¼-inch opening in large end, 1⅛-in. opening in small end. Wt., 2½ pounds.

9L3418 ...42c

Lock Pin Malleable Wagon Clevis.

24c

Size opening, 2⅝x4½ in. Weight, 1½ lbs.

9L3402 ...24c

Swivel Malleable End Clevis.

15c

Adjusts itself to any angle. ¾-in. opening Weight, 1 pound.

9L3414 ...15c

Lock Pin Steel Wagon Clevis.

20c

Malleable Pin. Complete as shown. 1¾-inch opening. Weight, 1¾ pounds.

9L3408 ...20c

Buggy, Surrey and Light Wagon Axles.

Collinge Collar Long Distance Axle. An especially easy running axle. Holds oil a long time. Collinge collars prevent dirt from working into the boxes. State size.

Size, Inches	Shpg. Wt., Lbs.	99L3121 No. 12, Short Bed. Set of Four Axles	Shpg. Wt., Lbs.	99L3120 No. 11, Long Bed. Set of Four Axles
1 x6½	23	$4.40	47	$5.20
1⅛x7	29	5.20	60	6.10
1¼x7	38	6.40	72	7.45
1¼x7½	40	6.80	72	7.90

Elliptic Vehicle Springs.

$1.60 AND UP

Oil tempered, spring steel, carefully tempered.

	Width, In.	No. of Leaves	Lgth., In.	Shpg. Wt.	Each
99L3201	1¼	3	34	15 lbs.	$1.60
99L3206	1⅜	4	36	21 lbs.	2.30
99L3208	1½	5	36	30 lbs.	3.00

QUALITY BASED ON SCIENTIFIC TEST is assured in buying from us. That makes our low prices doubly attractive.

Painted Elliptic Seat Springs.

Springs have bolt holes.

99L3215 Two-leaf size, 1½x26 inches. Shipping weight, 12 pounds. Per pair $1.10

99L3216—Three-leaf size. 1½x28 in. Shipping weight, 8¾ pounds. Per pair $1.60

Cast Iron Wagon Skeins.

$4.85 Per Set. 3x9-In.

Seamless pattern, complete with boxes. Made extra thick at bottom of spindle, the part that wears, and extra strong at collar, where there is the greatest strain. State size.

99L3105

Size, in.	2½x8	2¾x8½	3x9	3¼x9
Shipping wt., per set	41 lbs.	47 lbs.	57 lbs.	69 lbs.
Set of four	$4.00	$4.40	$4.85	$5.35
Size, inches	3¼x10	3½x10	3¾x11	4x12
Shipping wt., per set	73 lbs.	78 lbs.	95 lbs.	112 lbs.
Set of four	$5.95	$6.70	$7.60	$8.65

Steel Wagon Skeins.

$11.30 Per Set. 3x9-In.

Made of tough steel and furnished complete with cast iron boxes. Have cut threads on skeins and in nut.

99L3100—State size.

Size, Inches	Weight, per Set	Set of Four
2½x 8	44 pounds	$ 9.90
3 x 9	63 pounds	11.30
3¼x10	78 pounds	12.80

Hooked Wagon Box Strap Bolts.

$1.10 Set of 8. 16-Inch

9L3442—State size.

Lth., In.	Diam. of Screw	Wt., Lbs.	Price, Set of 8
14	⅝ in.	8	$0.95
16	⅝ in.	9	1.10
18	⅝ in.	10	1.25

Pole Caps With Holdbacks.

19c

Heavy. Weight, 2 pounds.

9L3448 19c

CAST IRON OIL TROUGH.

$1.20

Oils wheels up to 4 inches wide. Prevents tires from becoming loose. Shipping weight, 12 pounds.

99L5349 ....$1.20

COIL LEATHER AXLE WASHERS.

18c 1-In. State Size.

9L3207—Box of 5 coils (about 100 washers).

Size	Wt.	Per Box
⅞ in.	6 oz.	16c
1 in.	7 oz.	18c
1⅛ in.	8 oz.	20c
1¼ in.	11 oz.	22c

$1.95 Oil Trough.

For oiling wheels up to 4 inches wide with boiling oil. Fill trough with oil, saturate the mineral wool with kerosene (coal oil), place it under the trough and apply a match. Shipping weight, 7 pounds.

99L5350 ....$1.95

Wagon Box Strap Bolts. 80c

Set of 8, 14-In.

9L3446—State size.

Length, inches	10	12	14	16	18
Diam. of screw, in.	½	½	9/16	9/16	⅝
Weight of 8, lbs.	4	5	7	7	9
Price, set of 8	60c	70c	80c	90c	$1.00

Wagon Box Side Braces.

12c ½-In.

9L3438—State size.

Size	Length	Weight	
7/16 in.	15½ in.	¾ lb.	10c
½ in.	15½ in.	1 lb.	12c
9/16 in.	15½ in.	1¼ lbs.	14c

Wagon Box Rods.

15c 3 Ft. 3 In. Long. 17c 3 Ft. 7 In. Long.

Made of ⅜-inch steel rod. Length is from under shoulder to point of rod. Furnished complete with screws for attaching.

99L3144—3 ft. 3 in. long. For narrow bed. Shipping wt., 1½ lbs. 15c

99L3145—3 ft. 7 in. long. For wide bed. Shipping wt., 1¾ lbs. 17c

Singletree Center Clips and Rings.

12c ½-In.

Size given is size of round part. State size.

9L3392

Clip, In.	Ring, In.	Wt.	
½	7/16	¾ lb.	12c
9/16	7/16	1 lb.	14c
⅝	½	1¼ lbs.	16c

Safety Singletree Clips. 40c

1 5/16-Inch.

Malleable iron. Sizes given are on small end of ferrule. State size.

Size Small End	9L3374 Wt., Lbs.	Per Pair
1 5/16 in.	1	40c
1½ in.	1¼	50c

Singletree Center Clips.

8c ½-In. State size.

9L3388

Size, ½ inch. Wt., ¼ lb. ..8c

9/16 inch. Wt., ½ lb. 10c

⅝ in. Wt., ¾ lb. ..12c

Singletree Hooks and Ferrules.

22c Per Pair, 1¼-In.

Wrought hooks, malleable ferrules. State size.

Size Small End	9L3380 Wt.	Per Pair
1¼ in.	1¼ lbs.	22c
1½ in.	1¼ lbs.	27c

Neckyoke Ferrules and Rings.

22c Per Pair, 1½-In.

9L3428

Size Ferrules, Large End	Size Ring	Wt., Lbs.	Per Pair
1½ in.	3 in.	1¼	22c
1⅝ in.	3 in.	1½	24c

Singletree Clips and Hooks.

21c

For centers of singletrees. Size, round part of clip, ⅝ in.; hook, ½ in. Weight, 1¼ pounds.

9L3372 ...21c

Neckyoke Center Iron. 30c

Ring, 3½ in. in diameter, made of 9/16-in. steel; eyes of 7/16 and links of ⅜-in. steel. Wt., 2½ lbs.

9L3432 ....30c

Singletree End Clips.

21c Per Pair, 7/16-In.

For ends of singletrees; come in pairs, one right and one left.

9L3384—State size.

Size of Clip	Round Part of Hook	Weight, per Pair	Per Pair
7/16 in.	⅜ in.	1 lb.	21c
½ in.	7/16 in.	1½ lbs.	26c

SINGLETREE HOOKS AND STRAPS.

Hooks, ⅜-inch round part. Straps, 7/16-inch round part, 12 inches long. Weight, 2 lbs.

9L3396—Pair hooks and pair straps 22c

9L3397—Pair straps only (without hooks) 14c

SECURITY SINGLETREE HOOKS.

45c Pair, 1⅛-In.

Malleable iron. Has spring which positively prevents traces from coming off. State size.

9L3370

Diameter, in.	⅞	1	1⅛	1¼	1⅜
Weight, oz.	10	11	12	16	18
Pair	35c	40c	45c	50c	55c

HOLLOW WAGON BOLSTER STAKES.

$2.90 3-Inch Bolsters, for Set of Four.

Made of malleable iron with opening at the side to let water out. Strong and durable.

99L3150—For 3-inch bolsters. Shipping weight, 14 lbs. Set of 4 ....$2.90

99L3151—For 3¼-inch bolsters. Shipping wt., 19 lbs. Set of 4 $3.60

Barten Adjustable Bolster Stake. 65c

Made of malleable iron, heavy and strong. May be opened or closed to fit bolsters of any thickness. Shipping weight, 5 pounds.

99L3154 ....65c

DOUBLE TUBE STEEL SHAFT ENDS.

75c Buggy Size.

9L3197—Buggy size. Length, 29 in. Weight, 4 pounds. Per pair 75c

9L3198—Surrey or wagon size. Length, 29 inches. Weight, 4 lbs. Per pair 90c

ANTI-RATTLER AND QUICK SHIFTERS.

27c PAIR ⅜-In. Pin.

Prevents shafts from rattling. Couplings not furnished. Wt., 12 ounces.

9L3203—Per pair, with ⅜-inch pin ..27c

9L3204—Per pair, with 7/16-inch pin ..29c

HANDY SPOKE AND RIM REPAIRER.

12c 1-In.

9L3251—State size.

Size Rim	Weight	
¾ in.	2 oz.	11c
1 in.	3 oz.	12c
1¼ in.	3 oz.	13c

All weights and measurements given on this page are approximate and may vary a trifle.

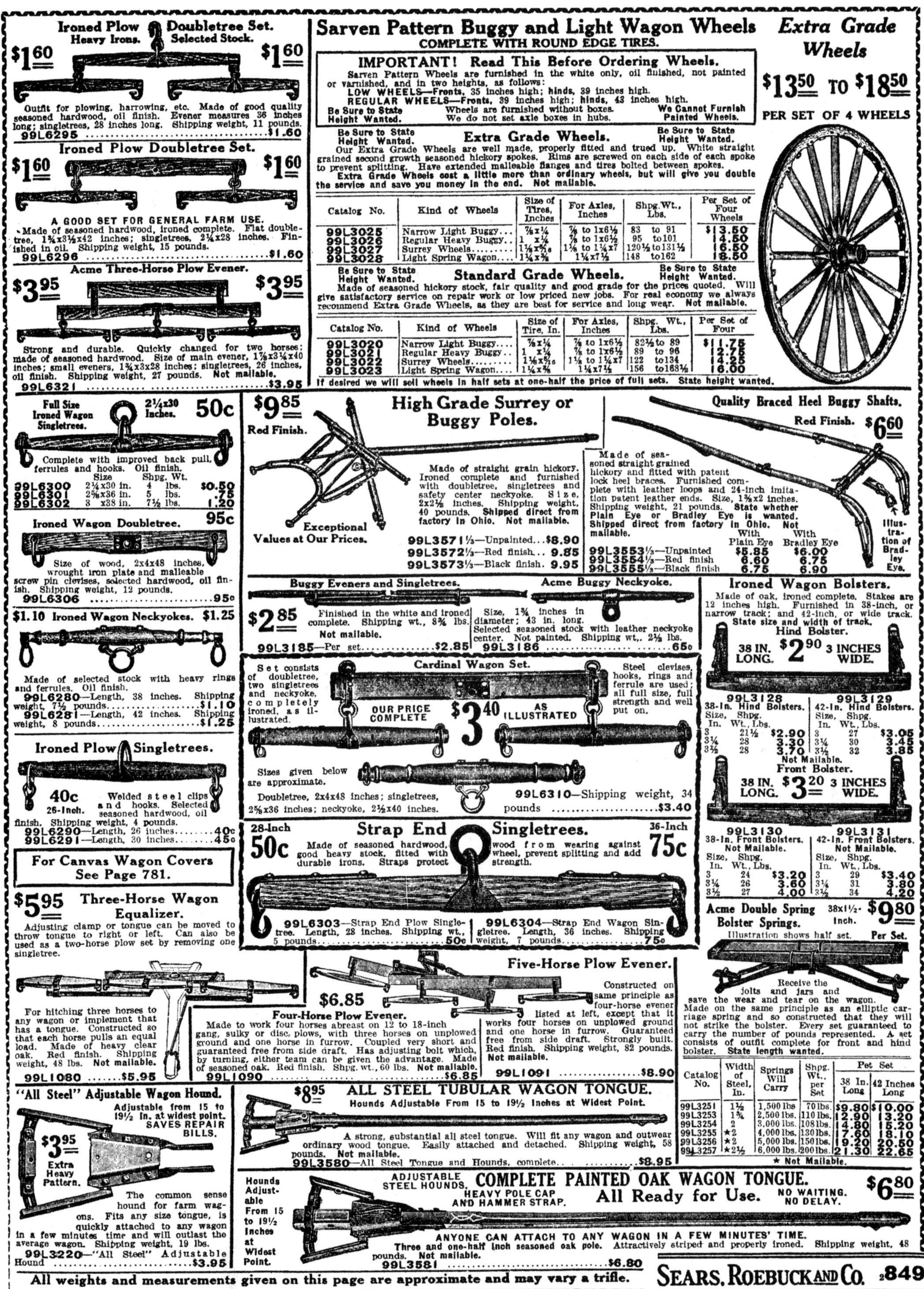

Ironed Plow Doubletree Set.
Heavy Irons. Selected Stock.

$1.60 $1.60

Outfit for plowing, harrowing, etc. Made of good quality seasoned hardwood, oil finish. Evener measures 36 inches long; singletrees, 28 inches long. Shipping weight, 11 pounds.
99L6295 ....$1.60

Ironed Plow Doubletree Set.

$1.60 $1.60

A GOOD SET FOR GENERAL FARM USE.
Made of seasoned hardwood, ironed complete. Flat doubletree, 1¾x3½x42 inches; singletrees, 2¼x28 inches. Finished in oil. Shipping weight, 15 pounds.
99L6296 ....$1.60

Acme Three-Horse Plow Evener.

$3.95 $3.95

Strong and durable. Quickly changed for two horses; made of seasoned hardwood. Size of main evener, 1⅞x3¼x40 inches; small eveners, 1¾x3x28 inches; singletrees, 26 inches, oil finish. Shipping weight, 27 pounds. Not mailable.
99L6321 ....$3.95

Sarven Pattern Buggy and Light Wagon Wheels
COMPLETE WITH ROUND EDGE TIRES.

IMPORTANT! Read This Before Ordering Wheels.
Sarven Pattern Wheels are furnished in the white only, oil finished, not painted or varnished, and in two heights, as follows:
LOW WHEELS—Fronts, 35 inches high; hinds, 39 inches high.
REGULAR WHEELS—Fronts, 39 inches high; hinds, 43 inches high.
Be Sure to State Height Wanted. Wheels are furnished without boxes. We do not set axle boxes in hubs. We Cannot Furnish Painted Wheels.

Extra Grade Wheels
$13.50 TO $18.50
PER SET OF 4 WHEELS

Extra Grade Wheels.
Be Sure to State Height Wanted. Be Sure to State Height Wanted.

Our Extra Grade Wheels are well made, properly fitted and trued up. White straight grained second growth seasoned hickory spokes. Rims are screwed on each side of each spoke to prevent splitting. Have extended malleable flanges and tires bolted between spokes.
Extra Grade Wheels cost a little more than ordinary wheels, but will give you double the service and save you money in the end. Not mailable.

Catalog No.	Kind of Wheels	Size of Tires, Inches	For Axles, Inches	Shpg. Wt., Lbs.	Per Set of Four Wheels
99L3025	Narrow Light Buggy...	⅞x¼	⅞ to 1x6½	83 to 91	$13.50
99L3026	Regular Heavy Buggy..	1 x¼	⅞ to 1x6½	95 to 101	14.50
99L3027	Surrey Wheels.........	1⅛x5/16	1⅛ to 1¼x7	120½ to 131½	16.50
99L3028	Light Spring Wagon....	1¼x⅜	1¼x7½	148 to 162	18.50

Standard Grade Wheels.
Be Sure to State Height Wanted. Be Sure to State Height Wanted.

Made of seasoned hickory stock, fair quality and good grade for the prices quoted. Will give satisfactory service on repair work or low priced new jobs. For real economy we always recommend Extra Grade Wheels, as they are best for service and long wear. Not mailable.

Catalog No.	Kind of Wheels	Size of Tire, In.	For Axles, Inches	Shpg. Wt., Lbs.	Per Set of Four
99L3020	Narrow Light Buggy....	⅞x¼	⅞ to 1x6½	82½ to 89	$11.75
99L3021	Regular Heavy Buggy..	1 x¼	⅞ to 1x6½	89 to 96	12.75
99L3022	Surrey Wheels.........	1⅛x5/16	1⅛ to 1¼x7	122 to 134	14.25
99L3023	Light Spring Wagon....	1¼x⅜	1¼x7½	156 to 168½	16.00

If desired we will sell wheels in half sets at one-half the price of full sets. State height wanted.

Full Size Ironed Wagon Singletrees.
2¼x30 Inches. 50c

Complete with improved back pull, ferrules and hooks. Oil finish.

	Size	Shpg. Wt.	
99L6300	2¼x30 in.	4 lbs.	$0.50
99L6301	2⅝x36 in.	5 lbs.	.75
99L6302	3 x38 in.	7½ lbs.	1.20

Ironed Wagon Doubletree. 95c

Size of wood, 2x4x48 inches, wrought iron plate and malleable screw pin clevises, selected hardwood, oil finish. Shipping weight, 12 pounds.
99L6306 ....95c

$1.10 Ironed Wagon Neckyokes. $1.25

Made of selected stock with heavy rings and ferrules. Oil finish.
99L6280—Length, 38 inches. Shipping weight, 7½ pounds....$1.10
99L6281—Length, 42 inches. Shipping weight, 8 pounds....$1.25

Ironed Plow Singletrees.

40c 26-Inch.

Welded steel clips and hooks. Selected seasoned hardwood, oil finish. Shipping weight, 4 pounds.
99L6290—Length, 26 inches........40c
99L6291—Length, 30 inches........45c

For Canvas Wagon Covers See Page 781.

$5.95 Three-Horse Wagon Equalizer.

Adjusting clamp or tongue can be moved to throw tongue to right or left. Can also be used as a two-horse plow set by removing one singletree.

For hitching three horses to any wagon or implement that has a tongue. Constructed so that each horse pulls an equal load. Made of heavy clear oak. Red finish. Shipping weight, 48 lbs. Not mailable.
99L1080 ....$5.95

High Grade Surrey or Buggy Poles.
$9.85 Red Finish.

Made of straight grain hickory. Ironed complete and furnished with doubletree, singletrees and safety center neckyoke. Size, 2x2⅝ inches. Shipping weight, 40 pounds. Shipped direct from factory in Ohio. Not mailable.

Exceptional Values at Our Prices.

99L3571⅓—Unpainted...$8.90
99L3572⅓—Red finish... 9.85
99L3573⅓—Black finish. 9.95

Quality Braced Heel Buggy Shafts.
Red Finish. $6.60

Made of seasoned straight grained hickory and fitted with patent lock heel braces. Furnished complete with leather loops and 24-inch imitation patent leather ends. Size, 1⅜x2 inches. Shipping weight, 21 pounds. State whether Plain Eye or Bradley Eye is wanted. Shipped direct from factory in Ohio. Not mailable.

Illustration of Bradley Eye.

	With Plain Eye	With Bradley Eye
99L3553⅓—Unpainted	$5.85	$6.00
99L3554⅓—Red finish	6.60	6.75
99L3555⅓—Black finish	6.75	6.90

Buggy Eveners and Singletrees.

$2.85 Finished in the white and ironed complete. Shipping wt., 8¾ lbs. Not mailable.
99L3185—Per set....................$2.85

Acme Buggy Neckyoke.

Size, 1¾ inches in diameter; 43 in. long. Selected seasoned stock with leather neckyoke center. Not painted. Shipping wt., 2½ lbs.
99L3186 ....65c

Cardinal Wagon Set.

Set consists of doubletree, two singletrees and neckyoke, completely ironed, as illustrated.

Steel clevises, hooks, rings and ferrule are used; all full size, full strength and well put on.

OUR PRICE COMPLETE $3.40 AS ILLUSTRATED

Sizes given below are approximate.
Doubletree, 2x4x48 inches; singletrees, 2⅝x36 inches; neckyoke, 2½x40 inches.
99L6310—Shipping weight, 34 pounds $3.40

Strap End Singletrees.
28-Inch 50c 36-Inch 75c

Made of seasoned hardwood, good heavy stock, fitted with durable irons. Straps protect wood from wearing against wheel, prevent splitting and add strength.

99L6303—Strap End Plow Singletree. Length, 28 inches. Shipping wt., 5 pounds....................50c
99L6304—Strap End Wagon Singletree. Length, 36 inches. Shipping weight, 7 pounds....................75c

Ironed Wagon Bolsters.

Made of oak, ironed complete. Stakes are 12 inches high. Furnished in 38-inch, or narrow track; and 42-inch, or wide track. State size and width of track.

Hind Bolster.
38 IN. LONG. $2.90 3 INCHES WIDE.

99L3128 38-In. Hind Bolsters. Size, In.	Shpg. Wt., Lbs.		99L3129 42-In. Hind Bolsters. Size, In.	Shpg. Wt., Lbs.	
3	21½	$2.90	3	27	$3.05
3¼	28	3.30	3¼	30	3.45
3½	28	3.70	3½	32	3.85

Not Mailable.

Front Bolster.
38 IN. LONG. $3.20 3 INCHES WIDE.

99L3130 38-In. Front Bolsters. Not Mailable. Size, In.	Shpg. Wt., Lbs.		99L3131 42-In. Front Bolsters. Not Mailable. Size, In.	Shpg. Wt., Lbs.	
3	24	$3.20	3	29	$3.40
3¼	26	3.60	3¼	31	3.80
3½	27	4.00	3½	34	4.20

Acme Double Spring Bolster Springs.
38x1½-Inch. $9.80 Per Set.

Illustration shows half set.

Receive the jolts and jars and save the wear and tear on the wagon. Made on the same principle as an elliptic carriage spring and so constructed that they will not strike the bolster. Every set guaranteed to carry the number of pounds represented. A set consists of outfit complete for front and hind bolster. **State length wanted.**

Catalog No.	Width of Steel, In.	Springs Will Carry	Shpg. Wt., per Set	Pet Set 38 In. Long	42 Inches Long
99L3251	1½	1,500 lbs	70 lbs.	$9.80	$10.00
99L3253	1¾	2,500 lbs.	110 lbs.	12.90	13.20
99L3254	2	3,000 lbs.	108 lbs.	14.80	15.20
99L3255	★2	4,000 lbs.	130 lbs.	17.60	18.10
99L3256	★2	5,000 lbs.	150 lbs.	19.20	20.50
99L3257	★2½	6,000 lbs.	200 lbs.	21.30	22.65

★ Not Mailable.

Five-Horse Plow Evener.

Constructed on same principle as four-horse evener listed at left, except that it works four horses on unplowed ground and one horse in furrow. Guaranteed free from side draft. Strongly built. Red finish. Shipping weight, 82 pounds. Not mailable.
99L1091 ....$8.90

Four-Horse Plow Evener.
$6.85

Made to work four horses abreast on 12 to 18-inch gang, sulky or disc plows, with three horses on unplowed ground and one horse in furrow. Coupled very short and guaranteed free from side draft. Has adjusting bolt which, by turning, either team can be given the advantage. Made of seasoned oak. Red finish. Shpg. wt., 60 lbs. Not mailable.
99L1090 ....$6.85

"All Steel" Adjustable Wagon Hound.

Adjustable from 15 to 19½ In. at widest point. SAVES REPAIR BILLS.

$3.95 Extra Heavy Pattern.

The common sense hound for farm wagons. Fits any size tongue, is quickly attached to any wagon in a few minutes time and will outlast the average wagon. Shipping weight, 19 lbs.
99L3220—"All Steel" Adjustable Hound $3.95

$8.95 ALL STEEL TUBULAR WAGON TONGUE.
Hounds Adjustable From 15 to 19½ Inches at Widest Point.

A strong, substantial all steel tongue. Will fit any wagon and outwear ordinary wood tongue. Easily attached and detached. Shipping weight, 58 pounds. Not mailable.
99L3580—All Steel Tongue and Hounds, complete....................$8.95

COMPLETE PAINTED OAK WAGON TONGUE.
ADJUSTABLE STEEL HOUNDS, HEAVY POLE CAP AND HAMMER STRAP. All Ready for Use. NO WAITING. NO DELAY. $6.80

Hounds Adjustable From 15 to 19½ Inches at Widest Point.

ANYONE CAN ATTACH TO ANY WAGON IN A FEW MINUTES' TIME.
Three and one-half inch seasoned oak pole. Attractively striped and properly ironed. Shipping weight, 48 pounds. Not mailable.
99L3581 ....$6.80

All weights and measurements given on this page are approximate and may vary a trifle.

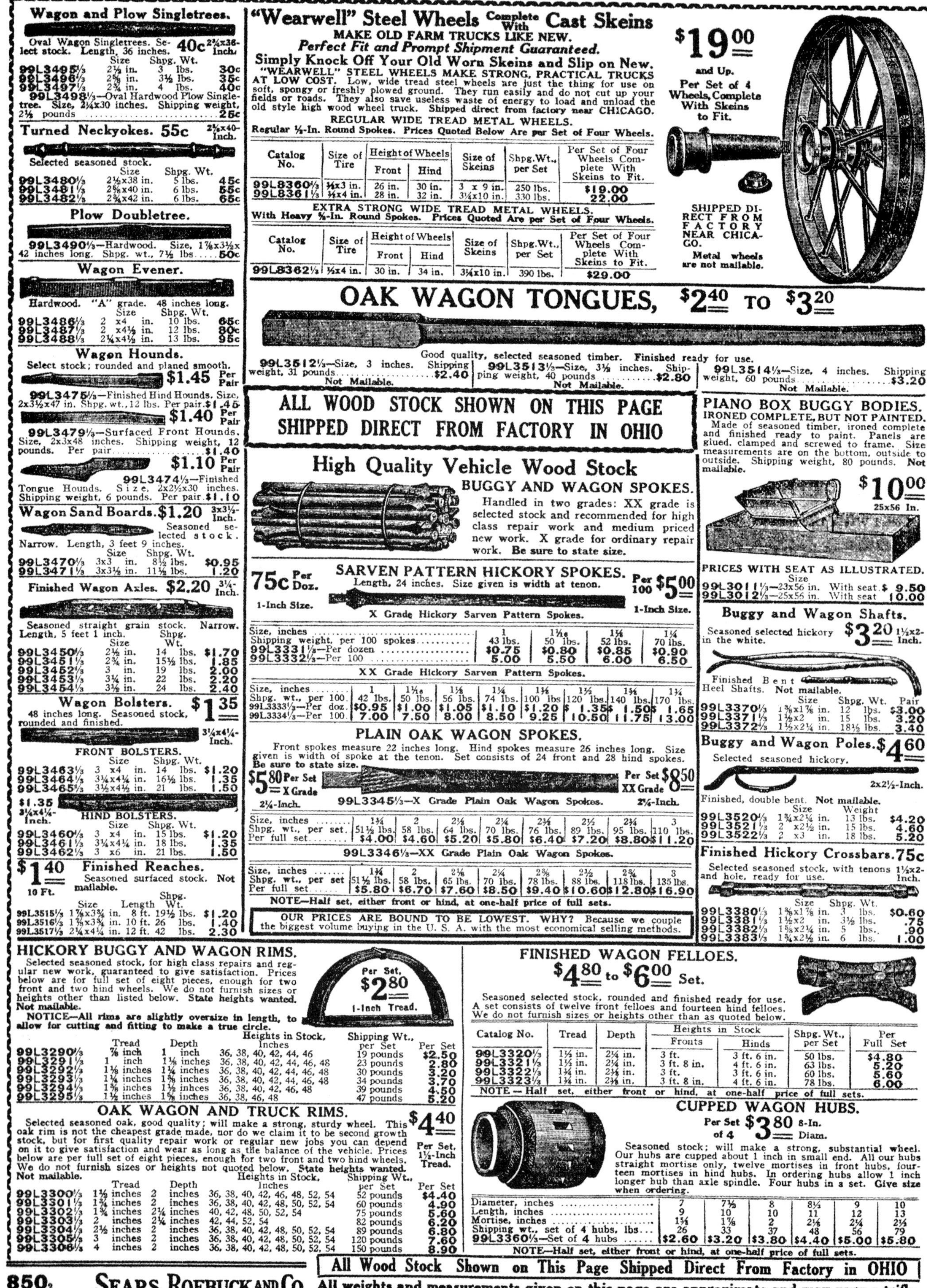

Wagon and Plow Singletrees.

Oval Wagon Singletrees. Select stock. Length, 36 inches. **40c** 2¾x36-Inch.

	Size	Shpg. Wt.	
99L3495⅓	2½ in.	3 lbs.	30c
99L3496⅓	2⅝ in.	3½ lbs.	35c
99L3497⅓	2¾ in.	4 lbs.	40c

99L3498⅓—Oval Hardwood Plow Singletree. Size, 2¼x30 inches. Shipping weight, 2½ pounds 25c

Turned Neckyokes. 55c 2⅝x40-Inch.

Selected seasoned stock.

	Size	Shpg. Wt.	
99L3480⅓	2½x38 in.	5 lbs.	45c
99L3481⅓	2⅝x40 in.	6 lbs.	55c
99L3482⅓	2¾x42 in.	6 lbs.	65c

Plow Doubletree.

99L3490⅓—Hardwood. Size, 1⅞x3½x42 inches long. Shpg. wt., 7½ lbs. 50c

Wagon Evener.

Hardwood. "A" grade. 48 inches long.

	Size	Shpg. Wt.	
99L3486⅓	2 x4 in.	10 lbs.	65c
99L3487⅓	2 x4½ in.	12 lbs.	80c
99L3488⅓	2¼x4½ in.	13 lbs.	95c

Wagon Hounds.

Select stock; rounded and planed smooth.

$1.45 Per Pair

99L3475⅓—Finished Hind Hounds. Size, 2x3½x47 in. Shpg. wt., 12 lbs. Per pair. $1.45

$1.40 Per Pair

99L3479⅓—Surfaced Front Hounds. Size, 2x3x48 inches. Shipping weight, 12 pounds. Per pair $1.40

$1.10 Per Pair

99L3474⅓—Finished Tongue Hounds. Size, 2x2½x30 inches. Shipping weight, 6 pounds. Per pair. $1.10

Wagon Sand Boards. $1.20 3x3½-Inch.

Seasoned selected stock. Narrow. Length, 3 feet 9 inches.

	Size	Shpg. Wt.	
99L3470⅓	3x3 in.	8½ lbs.	$0.95
99L3471⅓	3x3½ in.	11½ lbs.	1.20

Finished Wagon Axles. $2.20 3¼-Inch.

Seasoned straight grain stock. Narrow. Length, 5 feet 1 inch.

	Size	Shpg. Wt.	
99L3450⅓	2½ in.	14 lbs.	$1.70
99L3451⅓	2¾ in.	15½ lbs.	1.85
99L3452⅓	3 in.	19 lbs.	2.00
99L3453⅓	3¼ in.	22 lbs.	2.20
99L3454⅓	3½ in.	24 lbs.	2.40

Wagon Bolsters. $1.35

48 inches long. Seasoned stock, rounded and finished. 3¼x4¼-Inch.

FRONT BOLSTERS.

	Size	Shpg. Wt.	
99L3463⅓	3 x4 in.	14 lbs.	$1.20
99L3464⅓	3¼x4¼ in.	16½ lbs.	1.35
99L3465⅓	3½x4½ in.	21 lbs.	1.50

$1.35 3¼x4¼-Inch.

HIND BOLSTERS.

	Size	Shpg. Wt.	
99L3460⅓	3 x4 in.	15 lbs.	$1.20
99L3461⅓	3¼x4¼ in.	18 lbs.	1.35
99L3462⅓	3 x6 in.	21 lbs.	1.50

Finished Reaches.

$1.40 10 Ft.

Seasoned surfaced stock. Not mailable.

	Size	Length	Shpg. Wt.	
99L3515⅓	1⅞x3⅜ in.	8 ft.	19½ lbs.	$1.20
99L3516⅓	1⅞x3⅜ in.	10 ft.	26 lbs.	1.40
99L3517⅓	2¼x4¼ in.	12 ft.	42 lbs.	2.30

"Wearwell" Steel Wheels Complete With Cast Skeins

MAKE OLD FARM TRUCKS LIKE NEW.

Perfect Fit and Prompt Shipment Guaranteed.

Simply Knock Off Your Old Worn Skeins and Slip on New.

"WEARWELL" STEEL WHEELS MAKE STRONG, PRACTICAL TRUCKS AT LOW COST. Low, wide tread steel wheels are just the thing for use on soft, spongy or freshly plowed ground. They run easily and do not cut up your fields or roads. They also save useless waste of energy to load and unload the old style high wood wheel truck. Shipped direct from factory near CHICAGO.

$19.00 and Up. Per Set of 4 Wheels, Complete With Skeins to Fit.

REGULAR WIDE TREAD METAL WHEELS.

Regular ½-In. Round Spokes. Prices Quoted Below Are per Set of Four Wheels.

Catalog No.	Size of Tire	Height of Wheels Front	Height of Wheels Hind	Size of Skeins	Shpg. Wt., per Set	Per Set of Four Wheels Complete With Skeins to Fit.
99L8360⅓	⅜x3 in.	26 in.	30 in.	3 x 9 in.	250 lbs.	$19.00
99L8361⅓	⅜x4 in.	28 in.	32 in.	3¼x10 in.	330 lbs.	22.00

EXTRA STRONG WIDE TREAD METAL WHEELS.

With Heavy ⅝-In. Round Spokes. Prices Quoted Are per Set of Four Wheels.

Catalog No.	Size of Tire	Height of Wheels Front	Height of Wheels Hind	Size of Skeins	Shpg. Wt., per Set	Per Set of Four Wheels Complete With Skeins to Fit.
99L8362⅓	½x4 in.	30 in.	34 in.	3¼x10 in.	390 lbs.	$29.00

SHIPPED DIRECT FROM FACTORY NEAR CHICAGO.

Metal wheels are not mailable.

OAK WAGON TONGUES, $2.40 TO $3.20

Good quality, selected seasoned timber. Finished ready for use.

99L3512⅓—Size, 3 inches. Shipping weight, 31 pounds .... $2.40 Not Mailable.

99L3513⅓—Size, 3½ inches. Shipping weight, 40 pounds .... $2.80 Not Mailable.

99L3514⅓—Size, 4 inches. Shipping weight, 60 pounds .... $3.20 Not Mailable.

ALL WOOD STOCK SHOWN ON THIS PAGE SHIPPED DIRECT FROM FACTORY IN OHIO

High Quality Vehicle Wood Stock

BUGGY AND WAGON SPOKES.

Handled in two grades: XX grade is selected stock and recommended for high class repair work and medium priced new work. X grade for ordinary repair work. **Be sure to state size.**

SARVEN PATTERN HICKORY SPOKES.

Length, 24 inches. Size given is width at tenon.

75c Per Doz. 1-Inch Size. — Per 100 **$5.00** 1-Inch Size.

X Grade Hickory Sarven Pattern Spokes.

Size, inches	1	1$\frac{1}{16}$	1⅛	1¼
Shipping weight, per 100 spokes	43 lbs.	50 lbs.	52 lbs.	70 lbs.
99L3331⅓—Per dozen	$0.75	$0.80	$0.85	$0.90
99L3332⅓—Per 100	5.00	5.50	6.00	6.50

XX Grade Hickory Sarven Pattern Spokes.

Size, inches	1	1$\frac{1}{16}$	1⅛	1¼	1⅜	1½	1⅝	1¾
Shpg. wt., per 100	42 lbs.	50 lbs.	56 lbs.	74 lbs.	100 lbs	120 lbs.	140 lbs.	170 lbs.
99L3333⅓—Per doz.	$0.95	$1.00	$1.05	$1.10	$1.20	$1.35	$1.50	$1.65
99L3334⅓—Per 100	7.00	7.50	8.00	8.50	9.25	10.50	11.75	13.00

PLAIN OAK WAGON SPOKES.

Front spokes measure 22 inches long. Hind spokes measure 26 inches long. Size given is width of spoke at the tenon. Set consists of 24 front and 28 hind spokes. **Be sure to state size.**

$5.80 Per Set X Grade 2¼-Inch. — Per Set **$8.50** XX Grade 2¼-Inch.

99L3345⅓—X Grade Plain Oak Wagon Spokes.

Size, inches	1¾	2	2⅛	2¼	2⅜	2½	2¾	3
Shpg. wt., per set	51½ lbs.	58 lbs.	64 lbs.	70 lbs.	76 lbs.	89 lbs.	95 lbs.	110 lbs.
Per full set	$4.00	$4.60	$5.20	$5.80	$6.40	$7.20	$8.80	$11.20

99L3346⅓—XX Grade Plain Oak Wagon Spokes.

Size, inches	1¾	2	2⅛	2¼	2⅜	2½	2¾	3
Shpg. wt., per set	51½ lbs.	58 lbs.	65 lbs.	70 lbs.	78 lbs.	88 lbs.	118 lbs.	135 lbs.
Per full set	$5.80	$6.70	$7.60	$8.50	$9.40	$10.60	$12.80	$16.90

NOTE—Half set, either front or hind, at one-half price of full sets.

OUR PRICES ARE BOUND TO BE LOWEST. WHY? Because we couple the biggest volume buying in the U. S. A. with the most economical selling methods.

PIANO BOX BUGGY BODIES.

IRONED COMPLETE, BUT NOT PAINTED.

Made of seasoned timber, ironed complete and finished ready to paint. Panels are glued, clamped and screwed to frame. Size measurements are on the bottom, outside to outside. Shipping weight, 80 pounds. Not mailable.

$10.00 25x56 In.

PRICES WITH SEAT AS ILLUSTRATED.

	Size		
99L3011⅓	23x56 in.	With seat	$ 9.50
99L3012⅓	25x56 in.	With seat	10.00

Buggy and Wagon Shafts.

Seasoned selected hickory in the white. **$3.20** 1½x2-Inch.

Finished Bent Heel Shafts. Not mailable.

	Size	Shpg. Wt.	Pair
99L3370⅓	1⅜x1⅞ in.	12 lbs.	$3.00
99L3371⅓	1½x2 in.	15 lbs.	3.20
99L3372⅓	1½x2¼ in.	18½ lbs.	3.40

Buggy and Wagon Poles. $4.60

Selected seasoned hickory. 2x2½-Inch.

Finished, double bent. Not mailable.

	Size	Weight	
99L3520⅓	1¾x2¼ in.	13 lbs.	$4.20
99L3521⅓	2 x2½ in.	15 lbs.	4.60
99L3522⅓	2 x3 in.	18 lbs.	5.20

Finished Hickory Crossbars. 75c

Selected seasoned stock, with tenons and hole, ready for use. 1½x2-Inch.

	Size	Shpg. Wt.	
99L3380⅓	1⅜x1⅞ in.	3 lbs.	$0.60
99L3381⅓	1½x2 in.	3½ lbs.	.75
99L3382⅓	1⅝x2¼ in.	5 lbs.	.90
99L3383⅓	1¾x2½ in.	6 lbs.	1.00

HICKORY BUGGY AND WAGON RIMS.

Selected seasoned stock, for high class repairs and regular new work, guaranteed to give satisfaction. Prices below are for full set of eight pieces, enough for two front and two hind wheels. We do not furnish sizes or heights other than those listed below. State heights wanted. Not mailable.

Per Set, **$2.80** 1-Inch Tread.

NOTICE—All rims are slightly oversize in length, to allow for cutting and fitting to make a true circle.

	Tread	Depth	Heights in Stock, Inches	Shipping Wt., per Set	Per Set
99L3290⅓	⅞ inch	1 inch	36, 38, 40, 42, 44, 46	19 pounds	$2.50
99L3291⅓	1 inch	1⅛ inches	36, 38, 40, 42, 44, 46, 48	23 pounds	2.80
99L3292⅓	1⅛ inches	1¼ inches	36, 38, 40, 42, 44, 46, 48	30 pounds	3.20
99L3293⅓	1¼ inches	1⅜ inches	36, 38, 40, 42, 44, 46, 48	34 pounds	3.70
99L3294⅓	1⅜ inches	1½ inhces	36, 38, 40, 42, 46, 48	39 pounds	4.50
99L3295⅓	1½ inches	1⅝ inches	36, 38, 46, 48	47 pounds	5.20

OAK WAGON AND TRUCK RIMS.

Selected seasoned oak, good quality; will make a strong, sturdy wheel. This oak rim is not the cheapest grade made, nor do we claim it to be second growth stock, but for first quality repair work or regular new jobs you can depend on it to give satisfaction and wear as long as the balance of the vehicle. Prices below are per full set of eight pieces, enough for two front and two hind wheels. We do not furnish sizes or heights not quoted below. State heights wanted. Not mailable.

$4.40 Per Set. 1½-Inch Tread.

	Tread	Depth	Heights in Stock, Inches	Shipping Wt., per Set	Per Set
99L3300⅓	1½ inches	2 inches	36, 38, 40, 42, 46, 48, 52, 54	52 pounds	$4.40
99L3301⅓	1¾ inches	2 inches	36, 38, 40, 42, 48, 50, 52, 54	60 pounds	4.90
99L3302⅓	1¾ inches	2¼ inches	40, 42, 48, 50, 52, 54	75 pounds	5.60
99L3303⅓	2 inches	2¼ inches	42, 44, 52, 54	82 pounds	6.20
99L3304⅓	2½ inches	2 inches	36, 38, 40, 42, 48, 50, 52, 54	89 pounds	6.80
99L3305⅓	3 inches	2 inches	36, 38, 40, 42, 48, 50, 52, 54	120 pounds	7.60
99L3306⅓	4 inches	2 inches	36, 38, 40, 42, 48, 50, 52, 54	150 pounds	8.90

FINISHED WAGON FELLOES.

$4.80 to $6.00 Set.

Seasoned selected stock, rounded and finished ready for use. A set consists of twelve front felloes and fourteen hind felloes. We do not furnish sizes or heights other than as quoted below.

Catalog No.	Tread	Depth	Heights in Stock Fronts	Heights in Stock Hinds	Shpg. Wt., per Set	Per Full Set
99L3320⅓	1½ in.	2¼ in.	3 ft.	3 ft. 6 in.	50 lbs.	$4.80
99L3321⅓	1½ in.	2¼ in.	3 ft. 8 in.	4 ft. 6 in.	63 lbs.	5.20
99L3322⅓	1¾ in.	2⅜ in.	3 ft.	3 ft. 6 in.	60 lbs.	5.60
99L3323⅓	1¾ in.	2⅜ in.	3 ft. 8 in.	4 ft. 6 in.	78 lbs.	6.00

NOTE—Half set, either front or hind, at one-half price of full sets.

CUPPED WAGON HUBS.

Per Set of 4 **$3.80** 8-In. Diam.

Seasoned stock; will make a strong, substantial wheel. Our hubs are cupped about 1 inch in small end. All our hubs straight mortise only, twelve mortises in front hubs, fourteen mortises in hind hubs. In ordering hubs allow 1 inch longer hub than axle spindle. Four hubs in a set. **Give size when ordering.**

Diameter, inches	7	7½	8	8½	9	10
Length, inches	9	10	10	11	12	13
Mortise, inches	1⅝	1⅞	2	2⅛	2¼	2½
Shipping wt., set of 4 hubs, lbs.	26	33	37	48	56	79
99L3360⅓—Set of 4 hubs	$2.60	$3.20	$3.80	$4.40	$5.00	$5.80

NOTE—Half set, either front or hind, at one-half price of full sets.

All Wood Stock Shown on This Page Shipped Direct From Factory in OHIO

All weights and measurements given on this page are approximate and may vary a trifle.

$17.75 Heavy Rubber Drill Top.

Shipped direct from factory near CHICAGO.

Will wear longer than our Reliable Full Rubber Drill Top, as roof, quarters, stays and back curtains are made of a heavy weight rubber drill. Furnished in four-bow style, as illustrated, complete with rubber drill side and back curtains. Shpg. wt., 53 lbs. **Not mailable.** See diagram below for taking measurements.

99L3711⅓—For Plain Panel Seats, without shifting rail.......**$17.75**

$17.25 Reliable Full Rubber Drill Top.

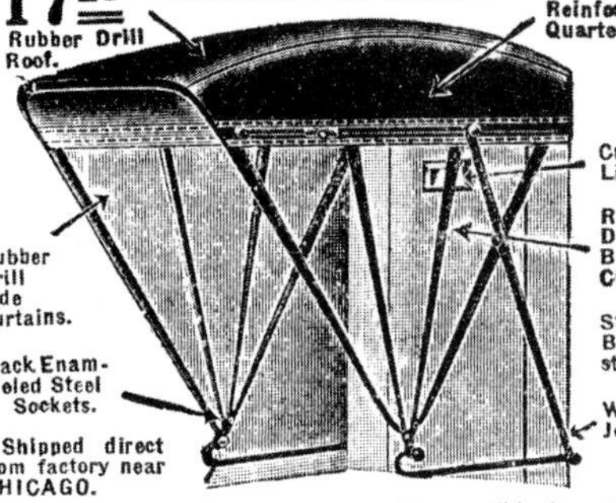

Rubber Drill Roof. Reinforced Quarters. Curtain Light. Rubber Drill Back Curtain. Stiffened Back-stays. Wrought Joints. Rubber Drill Side Curtains. Black Enameled Steel Sockets.

Shipped direct from factory near CHICAGO.

Complete with side and back curtains. Shipping weight, 52 pounds. **Not mailable. See diagram below for taking measurements.**

99L3708⅓—For Plain Panel Seats, without shifting rail, four-bow style, as illustrated.......**$17.25**

$16.80 Competition Enameled Drill Top.

Shipped direct from factory near CHICAGO.

Made of black enameled drill. Furnished in four-bow style, complete with side and back curtains, for plain panel seats only. Shipping weight, 53 lbs. **Not mailable. See diagram below for taking measurements.**

99L3705⅓—For Plain Panel Seats only, without shifting rail.......**$16.80**

$3.65 Biscuit Tufted Seat Cushion.

99L3771⅓

Black Morocco Grain Artificial Leather Cushion with fall. To fit seats 34 inches in width (S to T) or less...**$3.65** Shpg. wt., 5 lbs.

Shipped direct from factory near CHICAGO. See diagram below for taking measurements.

Biscuit Tufted Full Back.

99L3772⅓ **$3.65** Full Back to match above cushion. To fit seats 39 inches in width (A to B) or less...**$3.65**

Shipped direct from factory near CHICAGO. See diagram below for taking measurements.

Directions for Taking Measurements for Cushions and Backs.

Cushions and backs are shipped direct from factory near CHICAGO, usually within five days after order is received.

MEASUREMENTS FOR CUSHIONS.

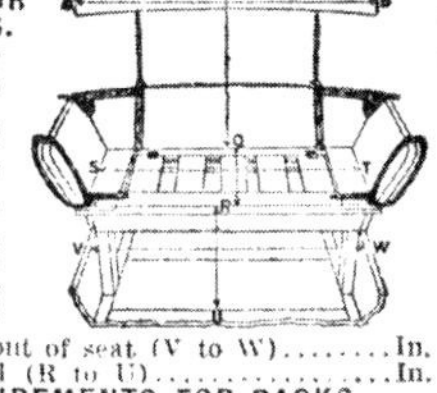

Width of seat (S to T).......In.
Depth of seat (Q to R).......In.
Width of fall. Give width of inside of buggy body at front of seat (V to W).......In.
Depth of fall (R to U).......In.

MEASUREMENTS FOR BACKS.

Distance across top of back (A to B)....In.
Distance from top of back to bottom of seat (P to Q).......In.

EXTRAS TO PRICES ON CUSHIONS AND BACKS.

Buggy Cushions, over 34 inches wide (S to T), each additional inch, extra. **8c**
Buggy Cushions, with extra cushion springs, extra.......**75c**
Buggy Backs, over 39 inches wide (A to B), each additional inch, extra... **8c**
Buggy Backs, with coil springs, extra..**60c**

Directions for Taking Measurements for Tops to Fit Plain Panel Seats.

Buggy Tops are shipped direct from factory near CHICAGO, usually within five days after order is received.

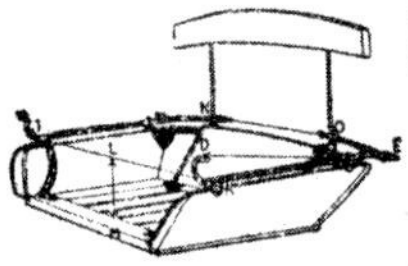

MEASUREMENTS FOR TOPS. Inches

Give distance from hole to hole (J to K).......
Depth of seat, from a line parallel with top of seat to bottom of seat (L to M).......
Distance from center of gooseneck to center of prop (D to E).......

EXTRAS TO PRICES ON TOPS.

Tops, built 3½-bow Handy Style, extra.......**35c**
Tops, 43 to 50 inches wide, extra.......**75c**
Shifting Rail for Plain Panel Seats, per pair, extra.......**90c**
99L3740⅓—Wood and Irons for Backs to attach to seat, complete, extra.......**$1.95**

MEASUREMENTS FOR WOOD AND IRONS FOR BACKS. Inches

Give distance across front of seat (J to K).......
Give distance from hole to hole on back of seat (N to O).......

Note

Buggy Tops, Wagon Tops, Buggy Cushions and Buggy Backs shown on this page are shipped direct from factory near Chicago, shipment usually being made within five days after order is received.

For Canvas Wagon Covers see page 781.

$10.80 Brown Duck Wagon Top.

Shipped direct from factory near CHICAGO.

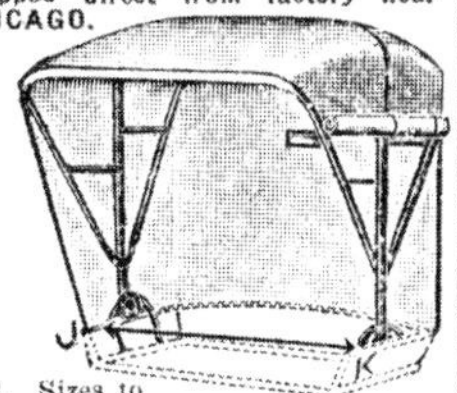

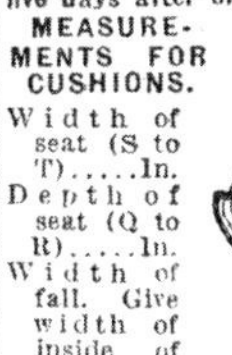

Covered with heavy weight closely woven brown duck. Furnished complete with side and back curtains, necessary fixtures and bolts for attaching to seat. Sizes to fit seats 32, 34, 36, 38, 40, 42 and 44 inches wide (J to K), as shown in illustration, are carried in stock at factory. Other sizes are made up special and are usually shipped within five days after order is received. **Give width outside to outside of flare on top of seat (J to K).** Shipping weight, 21 lbs. **Not mailable.**

99L3701⅓—Without seat......**$10.80**

ALL ITEMS SHOWN ABOVE SHIPPED DIRECT FROM FACTORY NEAR CHICAGO.

Acme Storm Shields and Aprons.

Adjustable Storm Aprons. $1.85 AND UP

Strong elastic cord across dash with hooks at each end which hold apron securely.

99L3751 28-oz. extra quality rubber drill. Size, 50 by 60 in. Fits dashes 23 to 28 in. Shipping wt., 4 lbs.**$2.70**
99L3752—28-ounce extra quality rubber drill. Size, 50 by 70 inches. Fits dashes 29 to 34 inches. Shipping wt., 4¾ lbs..**$2.80**
99L3749—20-ounce good quality rubber drill. Size, 50 by 52 inches. Fits dashes 21 to 28 inches. Shpg. wt., 3 lbs.......**$1.85**
99L3750—24-ounce good quality rubber drill. Size, 50 by 60 inches. Fits dashes 23 to 28 inches. Shpg. wt., 3½ lbs.......**$2.15**
99L3754—Black oiled drill with plaid back. Will outwear rubber drill. Size, 50 by 60 inches. Fits dashes 23 to 28 inches. Shipping weight, 2¾ pounds.......**$2.95**

Acme High Grade Storm Shields. $4.90

For buggy tops measuring 40 to 44 inches across top of bows.

This high grade storm shield is made of heavy waterproof material, fitted with large celluloid front light, side lights and a 12-inch line pocket. Has connections for attaching to dash, body and bows. Shipping weight, 2½ pounds.

99L3743.......**$4.90**

Standard quality storm shields, similar to above, made of rubber drill, with celluloid front and side lights, line pocket and fasteners for attaching.

99L3755—For buggy tops measuring 36 to 40 inches across top of bows. Shipping weight, 3½ pounds.......**$3.60**
99L3756—For buggy tops measuring 41 to 48 inches across top of bows. Shipping weight, 3½ pounds.......**$4.60**

Folding Third Seats. 82c Brussels Carpet.

Make a buggy as comfortable for three persons as for two. Fit between the two on the seat and take up little room. Weight, 5 pounds.

Covered With

9L3125—Heavy brown duck.......**78c**
9L3126—Brussels carpet.......**82c**
9L3127—Velvet carpet.......**86c**

KEEP QUALITY IN MIND AS WELL AS PRICE when you are trying to make your money do its full duty. Of course our prices are low, but the quality our constant testing assures, makes our values real.

THE IDEAL MATERIAL FOR AUTO AND BUGGY COVERINGS.

Morocco Grain Artificial Leather. $1.40 A YARD

50 Inches Wide.

Made on tough cloth back, will not peel or crack, is not affected by heat or cold and is embossed by the same process as genuine leather. Made on heavy colored back. If interested, write for free samples of Morocco Grain Artificial Leather.

Extra quality. Shpg. wt., per yd., 1⅛ lbs.

99L7511—Black. Per yard......**$1.40**
99L7512—Green. Per yard...... **1.40**
99L7514—Tan. Per yard...... **1.40**
99L7515—Special for upholsterings. Black, dull finish. Extra grade only. Shpg. wt., per yard, 3½ pounds. Per yard.......**$1.40**

Standard quality. Shpg. wt., per yard, 1 lb.

99L7516—Black. Per yard.......**95c**
99L7517—Green. Per yard.......**95c**
99L7519—Tan. Per yard.......**95c**

Spanish Grain Artificial Leather. $1.55 A YARD

50 Inches Wide.

Very attractive and serviceable. Heavy sateen back. Coating will not crack or peel. Will stand hard usage and all weather conditions. Recommended for tops, cushions, pillows, etc. If interested, write for free samples of artificial leather. Shpg. wt., per yard, 1¼ lbs.

99L7500—Green. Per yard......**$1.55**
99L7501—Red. Per yard...... **1.55**
99L7502—Brown. Per yard...... **1.55**
9L3196—Gimp to match. In about 25-yard rolls. Colors, green, red or brown. **State color desired.** Weight, 3 oz. Per roll....**22c**
9L3188—Buttons to match. Colors, green, red or brown. **State color desired.** Weight, 4 ounces. Per pkg. of about 100.......**18c**
9L3194—Nails to match. Colors, green, red or brown. **State color desired.** Weight, 4 ounces. Per pkg. of about 100.......**12c**

For Artificial Leather for Furniture Upholstery see page 561.

"BRONCO LEATHER" $2.20 A YARD

Exceptionally Strong Artificial Leather. Will Outwear Most Genuine Upholstering Leather. Positively Will Not Crack or Peel. No Waste. Easily Applied. The Ideal Material for High Class Work. 50 Inches Wide. Shipping weight, per yard, 2 pounds.

99L7506—Black Leather Grain. Per yard.......**$2.20**
99L7507—Tan. Beautiful Spanish Grain. Per yard....... **2.20**

Black Face Rubber Drill Carriage Cloth. 70c A YARD AND UP

50 Inches Wide.

Leather grain finish. Excellent for tops, cushions, full backs, curtains, aprons, boots, etc. Does not harden or crack and will outlast enameled cloth two to one. Prices below are for our best grade rubber drill.

99L7540—20-ounce white back. Shipping weight, per yard, 1¼ pounds. Per yard..**70c**
99L7541—24-ounce white back. Shipping weight, per yard, 1¾ pounds. Per yard..**75c**
99L7542—28-ounce white back. Shipping weight, per yard, 2 pounds. Per yard....**85c**
99L7550—22-ounce dark back. Shipping weight, per yard, 1½ pounds. Per yard..**85c**
99L7551—28-ounce dark back. Shipping weight, per yard, 2 pounds. Per yard.....**95c**

Black Enameled Carriage Cloth. 40c A YARD AND UP

50 Inches Wide.

A low priced material of attractive appearance which may be used for tops, cushions, curtains, etc. Shpg. wt., per yard, 1¼ lbs.

99L7555—Leather Grain Muslin, white back. Per yard.......**40c**
99L7556—Leather Grain Drill, white back. Per yard.......**50c**
99L7557—Leather Grain Drill, dark back. Per yard.......**60c**
99L7559—Leather Grain Duck, white back. Per yard.......**70c**
99L7560—Leather Grain Duck, dark back. Per yard.......**80c**

Special Quality 32-Oz. Morocco Grain Rubber Drill. 75c A YARD

50 Inches Wide.

High grade. Especially adapted for automobile tops, particularly for Ford cars, being of the same quality and appearance as the top material furnished with this popular car. Black back. Shpg. wt., 1¾ lbs.

99L7565—Per yard.......**75c**

Leakproof Double Fabric Auto Top Material. $1.35 A YARD

54 Inches Wide.

A durable rubber treated material possessing remarkable weather resisting properties. Suitable for the highest class work. The high grade drab drill back is combined with the outer fabric by a thoroughly waterproof coating of rubber. The outer rubber surface is then finished in imitation of long grain leather. Shipping weight, 2¼ pounds per lineal yard.

99L7570—Per yard.......**$1.35**

Metalene Nails. 10c Per 100

Package of about 100 nails. Colors black, green, maroon or tan. **State color.** Weight, 4 ounces.

9L3193

Per package.......**10c**

Extra Heavy White Cotton Duck or Canvas.

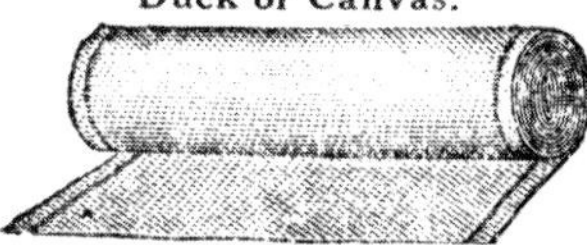

70c A YARD AND UP

Our No. 8 white cotton duck is made from strictly first quality cotton. Twisted hard and woven very close to shed water. Recommended for wagon covers, tops and cushions, tents, stack covers, etc.

Note—This is a high quality extra heavy duck, not the light weight so called 8, 10 and 12-ounce duck.

99L7580

Width, inches	36	54	72
Shpg. wt., yard, oz.	18	27	36
Per yard	70c	$1.05	$1.40

For other Canvas Duck see page 331.

Black Oiled Cotton Duck. $1.40 A YARD

FOR WAGON COVERS.

Absolutely waterproof. Made of extra quality No. 10 black cotton duck, heavily coated on one side with thick black oil which when dried becomes very tough and smooth. Is flexible; oil will not crack or peel off. Width, 50 inches. Shipping weight, per yard, 2 lbs.

99L7585—Per yard.......**$1.40**

Wagon Umbrella. $2.65

Shipping wt., 8 lbs.

Protects driver from sunshine, rain or snow. Has heavy duck (not drill) cover, and will give long and satisfactory service. This umbrella has six stiff steel ribs and a spread of about 5 feet 8 inches. Price includes fixture and screws for attaching.

99L3078—Covered brown duck.**$2.65**
99L3079—Covered white duck. **2.65**

Riding Plow Umbrella Holder. 55c

A strong malleable iron umbrella holder or arm which can be attached to riding plows, mowers, binders and other implements. Holds umbrella securely. Properly finished. Weight, 2 pounds.

9L3456.......**55c**

Trimmers' Gimp. 25c PER ROLL

In rolls of about 25 yards. Colors, black, green, maroon or tan. **State color.** Width, ½ inch. Weight, 6 ounces.

9L3195

Per roll.......**25c**

For other Furniture Gimp see page 561.

All weights and measurements given on this page are approximate and may vary a trifle.

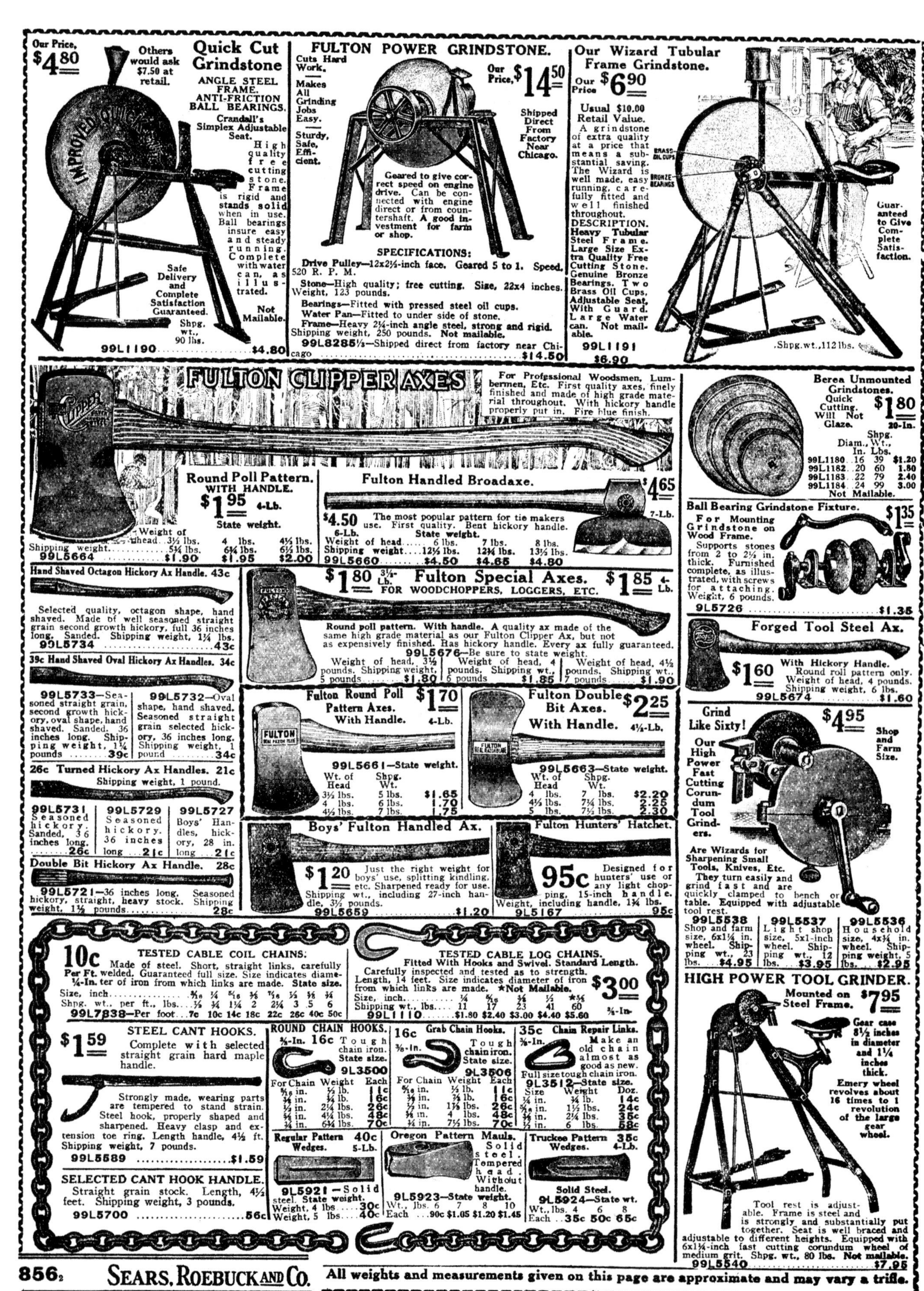

Our Price, $4.80
Others would ask $7.50 at retail.
Quick Cut Grindstone
ANGLE STEEL FRAME. ANTI-FRICTION BALL BEARINGS. Crandall's Simplex Adjustable Seat.
IMPROVED QUICK CUT
High quality free cutting stone. Frame is rigid and stands solid when in use. Ball bearings insure easy and steady running. Complete with water can, as illustrated.
Safe Delivery and Complete Satisfaction Guaranteed.
Not Mailable.
Shpg. wt., 90 lbs.
99L1190 ....$4.80
FULTON POWER GRINDSTONE.
Cuts Hard Work.
Makes All Grinding Jobs Easy.
Sturdy, Safe, Efficient.
Our Price, $14.50
Shipped Direct From Factory Near Chicago.
Geared to give correct speed on engine drive. Can be connected with engine direct or from countershaft. A good investment for farm or shop.
SPECIFICATIONS:
Drive Pulley—12x2½-inch face. Geared 5 to 1. Speed, 520 R. P. M.
Stone—High quality; free cutting. Size, 22x4 inches. Weight, 123 pounds.
Bearings—Fitted with pressed steel oil cups.
Water Pan—Fitted to under side of stone.
Frame—Heavy 2¼-inch angle steel, strong and rigid. Shipping weight, 250 pounds. Not mailable.
99L8285⅓—Shipped direct from factory near Chicago $14.50
Our Wizard Tubular Frame Grindstone.
Our Price $6.90
Usual $10.00 Retail Value.
A grindstone of extra quality at a price that means a substantial saving. The Wizard is well made, easy running, carefully fitted and well finished throughout.
DESCRIPTION. Heavy Tubular Steel Frame. Large Size Extra Quality Free Cutting Stone. Genuine Bronze Bearings. Two Brass Oil Cups. Adjustable Seat, With Guard. Large Water can. Not mailable.
BRASS OIL CUPS
BRONZE BEARINGS
Guaranteed to Give Complete Satisfaction.
Shpg.wt.,112 lbs.
99L1191 $6.90
FULTON CLIPPER AXES
CLIPPER
For Professional Woodsmen, Lumbermen, Etc. First quality axes, finely finished and made of high grade material throughout. With hickory handle properly put in. Fire blue finish.
Round Poll Pattern. WITH HANDLE.
$1.95 4-Lb. State weight.
Weight of head .. 3½ lbs. 4 lbs. 4½ lbs.
Shipping weight ....5¾ lbs. 6¾ lbs. 6½ lbs.
99L5664 ....$1.90 $1.95 $2.00
Fulton Handled Broadaxe.
$4.65 7-Lb.
$4.50 6-Lb. The most popular pattern for tie makers use. First quality. Bent hickory handle. State weight.
Weight of head ...... 6 lbs. 7 lbs. 8 lbs.
Shipping weight ....12½ lbs. 12¾ lbs. 13½ lbs.
99L5660 ....$4.50 $4.65 $4.80
Berea Unmounted Grindstones.
Quick Cutting. Will Not Glaze.
$1.80 20-In.
Diam., In. / Shpg. Wt., Lbs.
99L1180 .16 39 $1.20
99L1182 .20 60 1.80
99L1183 .22 79 2.40
99L1184 .24 99 3.00
Not Mailable.
Ball Bearing Grindstone Fixture.
$1.35
For Mounting Grindstone on Wood Frame.
Supports stones from 2 to 2½ in. thick. Furnished complete, as illustrated, with screws for attaching. Weight, 6 pounds.
9L5726 ....$1.35
Hand Shaved Octagon Hickory Ax Handle. 43c
Selected quality, octagon shape, hand shaved. Made of well seasoned straight grain second growth hickory, full 36 inches long. Sanded. Shipping weight, 1¼ lbs.
99L5734 ....43c
39c Hand Shaved Oval Hickory Ax Handles. 34c
99L5733—Seasoned straight grain, second growth hickory, oval shape, hand shaved. Sanded. 36 inches long. Shipping weight, 1¼ pounds 39c
99L5732—Oval shape, hand shaved. Seasoned straight grain selected hickory, 36 inches long. Shipping weight, 1 pound 34c
26c Turned Hickory Ax Handles. 21c
Shipping weight, 1 pound.
99L5731 Seasoned hickory. Sanded, 36 inches long. 26c
99L5729 Seasoned hickory. 36 inches long ...21c
99L5727 Boys' Handles, hickory, 28 in. long ...21c
Double Bit Hickory Ax Handle. 28c
99L5721—36 inches long. Seasoned hickory, straight, heavy stock. Shipping weight, 1½ pounds 28c
$1.80 3½-Lb. Fulton Special Axes. $1.85 4-Lb.
FOR WOODCHOPPERS, LOGGERS, ETC.
FULTON
Round poll pattern. With handle. A quality ax made of the same high grade material as our Fulton Clipper Ax, but not as expensively finished. Has hickory handle. Every ax fully guaranteed.
99L5676—Be sure to state weight.
Weight of head, 3½ pounds. Shipping weight, 5 pounds $1.80
Weight of head, 4 pounds. Shipping wt., 6 pounds $1.85
Weight of head, 4½ pounds. Shipping wt., 7 pounds $1.90
Forged Tool Steel Ax.
$1.60 With Hickory Handle. Round poll pattern only. Weight of head, 4 pounds. Shipping weight, 6 lbs.
99L5674 ....$1.60
Fulton Round Poll Pattern Axes. With Handle.
$1.70 4-Lb.
FULTON
99L5661—State weight.
Wt. of Head / Shpg. Wt.
3½ lbs. 5 lbs. $1.65
4 lbs. 6 lbs. 1.70
4½ lbs. 7 lbs. 1.75
Fulton Double Bit Axes. With Handle.
$2.25 4½-Lb.
FULTON
99L5663—State weight.
Wt. of Head / Shpg. Wt.
4 lbs. 7 lbs. $2.20
4½ lbs. 7¼ lbs. 2.25
5 lbs. 7½ lbs. 2.30
Boys' Fulton Handled Ax.
$1.20 Just the right weight for boys' use, splitting kindling, etc. Sharpened ready for use. Shipping wt., including 27-inch handle, 3½ pounds.
99L5659 ....$1.20
Fulton Hunters' Hatchet.
95c Designed for hunters' use or any light chopping. 15-inch handle. Weight, including handle, 1¾ lbs.
9L5167 ....95c
Grind Like Sixty!
$4.95 Shop and Farm Size.
Our High Power Fast Cutting Corundum Tool Grinders.
Are Wizards for Sharpening Small Tools, Knives, Etc.
They turn easily and grind fast and are quickly clamped to bench or table. Equipped with adjustable tool rest.
99L5538 Shop and farm size, 6x1¼ in. wheel. Shipping wt., 23 lbs. $4.95
99L5537 Light shop size, 5x1-inch wheel. Shipping wt., 12 lbs. $3.95
99L5536 Household size, 4x¾ in. wheel. Shipping weight, 5 lbs. $2.95
10c Per Ft. ⅛-In.
TESTED CABLE COIL CHAINS.
Made of steel. Short, straight links, carefully welded. Guaranteed full size. Size indicates diameter of iron from which links are made. State size.
Size, inch .... 3/16 ¼ 5/16 ⅜ 7/16 ½ ⅝ ¾
Shpg. wt., per ft., lbs. ...⅓ ¾ 1¼ 2 2¾ 3 5 6
99L7838—Per foot ...7c 10c 14c 18c 22c 26c 40c 50c
TESTED CABLE LOG CHAINS.
Fitted With Hooks and Swivel. Standard Length.
Carefully inspected and tested as to strength. Length, 14 feet. Size indicates diameter of iron from which links are made. ★Not Mailable.
$3.00 ⅜-In.
Size, inch .... ¼ 5/16 ⅜ ½ ★⅝
Shipping wt., lbs. ... 11 17 23 41 60
99L1110 .... $1.80 $2.40 $3.00 $4.40 $5.60
$1.59 STEEL CANT HOOKS.
Complete with selected straight grain hard maple handle.
Strongly made, wearing parts are tempered to stand strain. Steel hook, properly shaped and sharpened. Heavy clasp and extension toe ring. Length handle, 4½ ft. Shipping weight, 7 pounds.
99L5589 ....$1.59
SELECTED CANT HOOK HANDLE.
Straight grain stock. Length, 4½ feet. Shipping weight, 3 pounds.
99L5700 ....56c
ROUND CHAIN HOOKS. ⅜-In. 16c
Tough chain iron. State size.
9L3500
For Chain / Weight / Each
5/16 in. ½ lb. 11c
⅜ in. ½ lb. 16c
½ in. 2½ lbs. 26c
⅝ in. 4¼ lbs. 48c
¾ in. 6¼ lbs. 70c
16c Grab Chain Hooks. ⅜-In.
Tough chain iron. State size.
9L3506
For Chain / Weight / Each
5/16 in. ½ lb. 11c
⅜ in. ½ lb. 16c
½ in. 1⅞ lbs. 26c
⅝ in. 4 lbs. 48c
¾ in. 7½ lbs. 70c
35c Chain Repair Links. ⅜-In.
Make an old chain almost as good as new. Full size tough chain iron.
9L3512—State size.
Size / Weight / Doz.
¼ in. ¾ lb. 14c
5/16 in. 1½ lbs. 24c
⅜ in. 2¼ lbs. 35c
½ in. 6 lbs. 58c
Regular Pattern Wedges. 40c 5-Lb.
9L5921—Solid steel. State weight.
Weight, 4 lbs 30c
Weight, 5 lbs 40c
Oregon Pattern Mauls.
Solid steel. Tempered head. Without handle.
9L5923—State weight.
Wt., lbs. 6 7 8 10
Each ...90c $1.05 $1.20 $1.45
Truckee Pattern Wedges. 35c 4-Lb.
Solid Steel.
9L5924—State wt.
Wt., lbs. 4 6 8
Each ..35c 50c 65c
HIGH POWER TOOL GRINDER.
Mounted on Steel Frame. $7.95
Gear case 8½ inches in diameter and 1¼ inches thick.
Emery wheel revolves about 16 times to 1 revolution of the large gear wheel.
Tool rest is adjustable. Frame is steel and is strongly and substantially put together. Seat is well braced and adjustable to different heights. Equipped with 6x1¼-inch fast cutting corundum wheel of medium grit. Shpg. wt., 80 lbs. Not mailable.
99L5540 ....$7.95

 All weights and measurements given on this page are approximate and may vary a trifle.

Superior in Quality.

Fulton Crosscut Saws

Our Prices Save You Money.

FULTON SAWS ARE MADE OF HIGH GRADE SAW STEEL, properly tempered and carefully ground, filed and set by hand. All Fulton Saws are sent out sharp and ready for use. Satisfaction guaranteed.

FULTON CLIPPER TOOTH WIDE CROSSCUT SAWS.

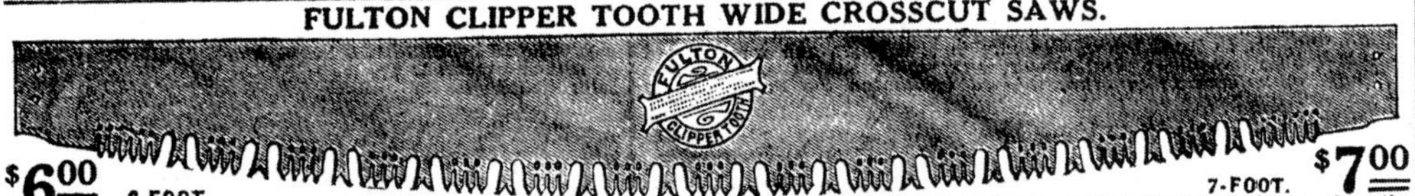

$6.00 6-FOOT. 7-FOOT. $7.00

99L5928—A popular saw for hard service. Fulton Clipper Tooth Saws, at middle of back, are ground six gauges thinner than the toothed edge. Fast cutting. Easy running. If you want high quality, this is the saw to buy. Without handles. For handles to fit see below. State length.

5 FEET LONG. Shipping weight, 6 pounds...$5.00 | 6 FEET LONG. Shipping weight, 9 pounds. Not mailable...$6.00 | 7 FEET LONG. Shipping weight, 11 pounds. Not mailable..$7.00

FULTON PERFORATED LANCE TOOTH WIDE CROSSCUT SAWS.

$4.25 5-FOOT. 6-FOOT. $5.10

99L5927—A popular and most satisfactory saw. Fulton Perforated Lance Tooth Saws, at middle of back, are ground five gauges thinner than the toothed edge. Easy to keep in order. Without handles. For handles see below. State length.

5 FEET LONG. Shipping weight, 7 pounds...$4.25 | 6 FEET LONG. Shipping weight, 9 pounds. Not mailable...$5.10

$3.50 5-FOOT. Fulton Diamond Tooth Wide Crosscut Saws. 6-FOOT. $4.20

Fulton Diamond Tooth Saws, at middle of back, are ground four gauges thinner than the toothed edge. Without handles. For handles see below. State length.

Length	5 ft.	★6 ft.
Shipping weight	6½ lbs.	8¾ lbs.
99L5926 (★Not Mailable.)	$3.50	$4.20

$3.25 5½-FOOT. Fulton Champion Tooth Wide Crosscut Saw.

Fulton Champion Tooth Wide Crosscut Saw with thin back. Well known and a favorite for many years. Without handles. For handles see below.

Length	5½ ft.
Shipping weight	8 lbs.
99L5922	$3.25

$2.50 5-FOOT. Fulton Peg Tooth Wide Crosscut Saws.

Fulton Peg or V Tooth Wide Crosscut Saw with thin back. Standard quality. Without handles. For handles see below.

Length	5 ft.
Shipping weight	6 lbs.
99L5921	$2.50

$3.25 5-FOOT. Fulton Perforated Lance Tooth Felling Saws. 6-FOOT. $3.90

Fulton Perforated Lance Tooth Thin Back Felling Saw. A smooth, quick cutter. Without handles. For handles see below. State length.

Length	5 ft.	★6 ft.
Shipping weight	5 lbs.	7 lbs.
99L5918 (★Not Mailable.)	$3.25	$3.90

$2.50 5-FOOT. Fulton Champion Tooth Felling Saws. 6-FOOT. $3.00

Fulton Champion Tooth Felling Saw. Without handles. For handles see below. State length.

Length	5 ft.	6 ft.
Shipping weight	5½ lbs.	6¾ lbs.
99L5917	$2.50	$3.00

$2.00 5-FOOT. Fulton Champion Tooth Narrow Crosscut Saws. 6-FOOT. $2.40

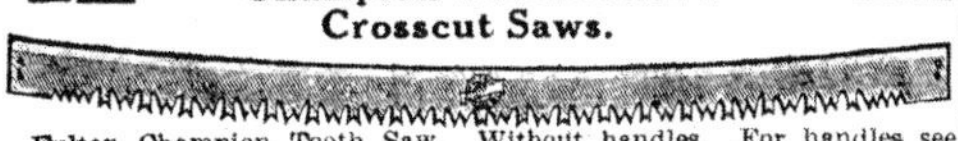

Fulton Champion Tooth Saw. Without handles. For handles see below. State length.

Length	5 ft.	6 ft.
Shipping weight	5 lbs.	6 lbs.
99L5916	$2.00	$2.40

18c One-Man Crosscut Saw Handles.

9L5085 — Supplementary Handle. Complete with rivet. Wt., 4 ounces...18c

9L5086 32c Good stock, varnished edges, without screws. Weight, 8 oz...32c

Farmers' and Mechanics' One-Man Crosscut Saw. $2.00

ONE MAN SAW

99L5912—For sawing down trees, cutting logs, firewood, timbers, etc. Made of good steel, sharpened and set ready for use. Length of blade, 3 feet. Complete with handles, as shown. Shipping weight, 4 pounds....$2.00

Crosscut Saw Handles.

34c 9L5089—Reversible pattern. Wt., 1¼ pounds. Per pair...34c

46c 9L5090—Heavy, extra quality, loop pattern. Weight, 2 pounds. Per pair...46c

FULTON PERFORATED LANCE TOOTH ONE-MAN STRAIGHT BACK CROSSCUT SAWS.

$3.60 4-FOOT. 5-FOOT. $4.50

99L5915—Fulton Perforated Lance Tooth, Taper Ground, Thin Back One-Man Saw. Complete with handles, as illustrated. State length.

4 FEET LONG. Shipping weight, 6¾ pounds. $3.60 | 5 FEET LONG. Shipping weight, 8½ pounds. $4.50

$2.80 3½-FOOT. Fulton Champion Tooth, One-Man, Straight Back Crosscut Saws. 4½-FOOT. $3.60

Fulton Champion Tooth, Taper Ground, Thin Back One-Man Saw. Complete with handles, as illustrated. State length.

Length	3½ ft.	4½ ft.
Shipping weight	5 lbs.	7 lbs.
99L5914	$2.80	$3.60

Fulton Peg or V Tooth One-Man Saw. 3-FOOT. $2.25

Fulton Peg or V Tooth, Taper Ground, Thin Back One-Man Saw. Complete with handles, as illustrated.

Length	3 ft.
Shipping weight	4 lbs.
99L5913	$2.25

FULTON CLIPPER BUCK SAW. $2.00

A High Grade, Fast Cutting Saw. Blade is made of special saw steel, correctly tempered. Champion pattern coarse teeth, which cut free and easily without choking or wedging tight. Width of blade, 2¼ inches. Selected hardwood frame, red finish. Saw comes sharpened and set ready for use. Weight, 4 pounds.

9L5007—Complete...$2.00
9L5011—Extra Blades. Weight, 8 oz. Each...50c

$1.75 Fulton Special Buck Saw.

FULTON SPECIAL

Frame made of hardwood and varnished. Steel blade properly tempered and fully guaranteed. Teeth are V shape, full bevel, filed, set and ready for use. Blade, 1½ inches wide. Weight, 3¼ pounds.

9L5006—Complete...$1.75
9L5010—Extra Blades. Weight, 6 oz. Each...45c

$1.50 Fulton Buck Saw.

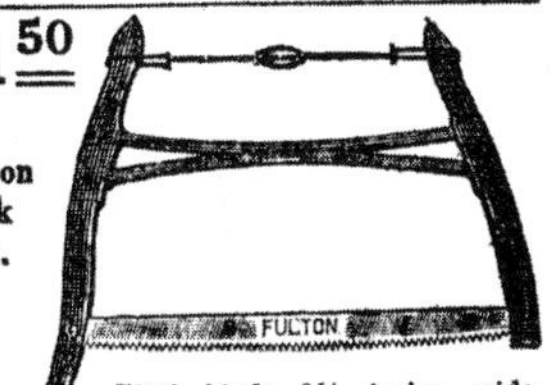

FULTON

Steel blade 2¼ inches wide. V shape teeth. Hardwood frame, red finish. Weight, 3¼ pounds.

9L5005—Complete...$1.50
9L5009—Extra Blades. Weight, 10 oz. Each...40c

Fulton Pond Ice Saws. $3.95 4-Ft.

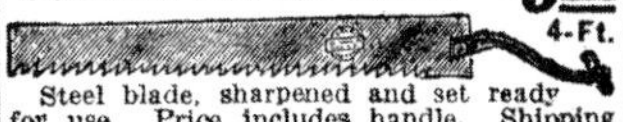

Steel blade, sharpened and set ready for use. Price includes handle. Shipping weight, 15 pounds.

9L5969—State length.

Length blade, feet	4	4½	5
Including handle	$3.95	$4.45	$4.95

$1.45 Fulton Wagon Ice Saws. 24-In.

Steel blade, with malleable iron handle. Sharpened and set ready for use. Fully guaranteed. State length. Shipping wt., 4 lbs.

Length, in.	24	26	28	30
99L5967	$1.45	$1.65	$1.85	$2.15

Steel Ice Tongs.

Medium size. Drop forged steel. Will easily handle a 200-pound block. Shipping weight, 4 lbs. $1.20

99L2770..$1.20

"Hold Fast" Screw Calks for Boots.

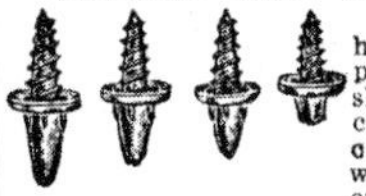

For bottoms and heels of shoes to prevent wearer from slipping. In boxes containing fifty calks and one wrench. Weight, 4 ounces.

9L6000—State size.

Size	Large	Medium	Small	Blunt
Length over all	1 in.	13/16 in.	11/16 in.	⅝ in.
Per box	42c	41c	40c	39c

$1.20 16-LB. STEEL CROWBARS. 12-LB. 90c

Solid steel, tempered points, well proportioned, strong and substantial.

Wedge Point Crowbar.
- 99L5501—Shipping weight, 12 lbs. Length, about 4 feet......$0.90
- 99L5502—Shipping weight, 16 lbs. Length, about 4½ feet......1.20
- 99L5503—Shipping weight, 20 lbs. Length, about 5 feet......1.40

Pinch Point Crowbar.
- 99L5504—Shipping weight, 12 lbs. Length, about 4 feet......$0.90
- 99L5505—Shipping weight, 16 lbs. Length, about 4½ feet......1.20
- 99L5506—Shipping weight, 20 lbs. Length, about 5 feet......1.40

WRECKING BARS.

35c 30-INCH. Forged steel. Forged from ¾-inch stock. 30c 24-INCH.

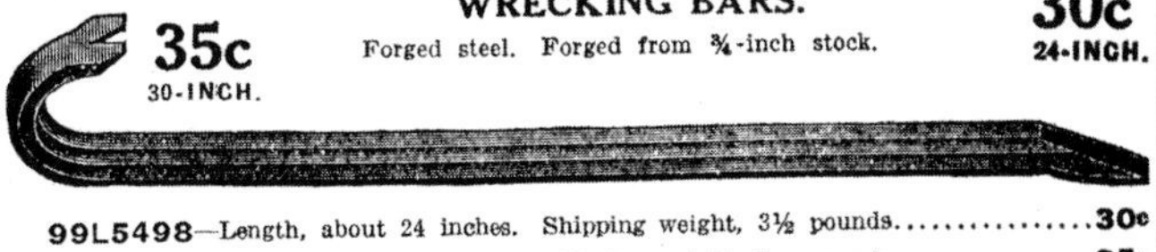

99L5498—Length, about 24 inches. Shipping weight, 3½ pounds...30c
99L5499—Length, about 30 inches. Shipping weight, 5 pounds...35c

All weights and measurements given on this page are approximate and may vary a trifle.

FULTON SPECIAL HAND SAWS
Fulton Special Hand Saws are made of high grade saw steel. Ground extra thin on back, have a true taper and perfect balance, are hand smithed, hand filed, set and finished with a high polish. Fitted with a carved handle of correct shape, properly attached. The Fulton Special is as nearly perfect as modern skill and equipment can make it and is guaranteed to be satisfactory in every way on the finest class of work.
Made in both straight and skew back patterns, in lengths and points as listed below. Be sure to give length and point wanted when ordering. For cabinet work, miter sawing and all uses where accuracy is demanded, we recommend the Fulton Special Hand Saw.
A good mechanic needs a good saw. Buy our Fulton Special and you will have a saw capable of doing the finest cabinet work.
$2.70 26-INCH
Guaranteed Full Value in Quality for Every Cent You Pay.
Note Our Low Prices on Our Highest Quality (Fulton Special) Hand Saws.
9L5069—Skew Back Panel Saw. 9L5070—Straight Back Panel Saw.
9L5071—Skew Back Hand Saw. 9L5072—Straight Back Hand Saw.
9L5073—Skew Back Rip Saw. 9L5074—Straight Back Rip Saw.
Panel Saw. Length, 20 inches. 10 or 12 Points. Weight, 1¼ pounds. $2.25
Panel Saw. Length, 22 inches. 10 or 12 Points. Weight, 1½ pounds. $2.40
Hand Saw. Length, 24 inches. 7, 8, 9 or 10 Points. Weight, 1¾ pounds. $2.55
Hand Saw. Length, 26 inches. 6, 7, 8, 9, 10 or 12 Points. Weight, 2¼ pounds. $2.70
Hand Saw. Length, 28 inches. 7 or 8 Points. Weight, 2½ pounds. $2.85
Rip Saw. Length, 24 inches. 5½ or 6 Points. Weight, 1¾ pounds. $2.55
Rip Saw. Length, 26 inches. 5, 5½ or 6 Points. Weight, 2¼ pounds. $2.70
Rip Saw. Length, 28 inches. 5, 5½ or 6 Points. Weight, 2¼ pounds. $2.85
FULTON HAND SAWS
Fulton Hand Saws are of standard grade, and for all ordinary work where an extra fine saw is not required they will give satisfactory service. Are made of good grade saw steel, ground four gauges thinner on the back; are hand filed, carefully set and polished. Handle is of correct shape and is properly attached.
If you buy a Fulton Saw and are not satisfied with it in every way we will return your money on request. Straight and skew back patterns, as below, in lengths and points as given. Be sure to mention length and point when ordering.
A high grade saw at a price which insures a big saving to you.
$1.90 24-INCH Usual Retail Price, $3.50
9L5063 Skew Back Panel Saw. 9L5064 Straight Back Panel Saw.
9L5065 Skew Back Hand Saw. 9L5066 Straight Back Hand Saw.
9L5067 Skew Back Rip Saw. 9L5068 Straight Back Rip Saw.
Panel Saw. Length, 18 in. 10 or 12 Points. Weight, 1 pound. $1.45
Panel Saw. Length, 20 in. 10 or 12 Points. Weight, 1¼ lbs. $1.60
Panel Saw. Length, 22 in. 10 or 12 Points. Weight, 1½ lbs. $1.75
Hand Saw. Length, 24 in. 7, 8, 9 or 10 Points. Weight, 1¼ lbs. $1.90
Hand Saw. Length, 26 in. 6, 7, 8, 9, 10 or 12 Points. Weight, 2 pounds. $2.05
Hand Saw. Length, 28 in. 7 or 8 Points. Weight, 2½ lbs. $2.25
Hand Saw. Length, 30 in. 7 or 8 Points. Weight, 2½ lbs. $2.60
Rip Saw. Length, 24 in. 5½ or 6 Points. Weight, 1¾ lbs. $1.90
Rip Saw. Length, 26 in. 5, 5½ or 6 Points. Weight, 2 lbs. $2.05
Rip Saw. Length, 28 in. 5 or 5½ Points. Weight, 2¼ lbs. $2.25
HOW MANY MERCHANTS REALLY KNOW the exact quality of every article they sell? We know, by laboratory test. You are safer buying here.
Mechanics' C. E. Jennings & Co.'s "ARROW HEAD BRAND." Fine Tools "ARROW HEAD BRAND."
$2.30 All Jennings "Arrow Head Brand" Saws Are Made in the Good Old Way, by Hand. $2.35
26-In. Hand. Jennings No. A7½ Narrow Blade Hand Saw 26-In. Rip.
Runs as easy and smooth as a fine saw worn narrow by use. The 26-inch size is 6 inches wide at butt and tapers to 1½ in. at point (other sizes in proportion). Ground thin on back. Have carved handles and are fully guaranteed. Be sure to give length and point wanted.
9L5050—Jennings No. A7½ Panel or Hand Saws.
Length, 22-inch panel. 10, 11 or 12 points to inch. Weight, 1¼ pounds $2.00
Length, 24 inches. Hand, 8, 9 or 10 points to inch. Weight, 1½ pounds $2.15
Length, 26 inches. Hand, 7, 8, 9 or 10 points to inch. Weight, 1¾ pounds $2.30
9L5051—Jennings No. A7½ Rip Saws.
Length, 26 inches. Rip, 5 or 6 points to inch. Weight, 1¾ pounds $2.35
Length, 28 inches. Rip, 4½, 5 or 5½ points to inch. Weight, 2 pounds $2.50
Jennings No. A70½ Narrow Blade Hand Saw.
Guaranteed to run without set. Ground thin on back. Made especially for high class mechanics who know how to take care of and appreciate a fine tool. Teeth are regular shaped. Have carved handles and are full concave taper ground. Especially useful where rapid and smooth cutting is required. Be sure to give length and point wanted.
$3.40 26-Inch Hand.
9L5052—Jennings No. A70½ Panel or Hand Saws.
Length, 22-inch panel. 10, 11 or 12 points to inch. Weight, 1 pound $2.95
Length, 24 inches. Hand, 8, 9 or 10 points to inch. Weight, 1¼ pounds $3.20
Length, 26 inches. Hand, 8, 10, 11 or 12 points to inch. Weight, 2 pounds $3.40
9L5053—Jennings No. A70½ Rip Saws.
Length, 26 inches. Rip, 5 or 6 points to inch. Weight, 1½ pounds $3.45
Length, 28 inches. Rip, 4½, 5 or 5½ points to inch. Weight, 2 pounds $3.70
Jennings No. 212 Skew Back Hand Saw.
Taper ground, thin back, selected blades with carved handles. For carpenters' general use this saw will give satisfaction. Fully guaranteed. Be sure to give length and point wanted.
$3.20 26-Inch Hand.
9L5054—Jennings No. 212 Panel or Hand Saws.
Length, 22-inch panel. 10, 11 or 12 points to inch. Weight, 1¼ pounds $2.70
Length, 24 inches. Hand, 8, 9 or 10 points to inch. Weight, 1½ pounds $2.95
Length, 26 inches. Hand, 7, 8, 9 or 10 points to inch. Weight, 2 pounds $3.20
9L5055—Jennings No. 212 Rip Saws.
Length, 26 inches. Rip, 5 or 6 points to inch. Weight, 2 pounds $3.25
Length, 28 inches. Rip, 4½, 5 or 5½ points to inch. Weight, 2 pounds $3.60
Fulton Nail Cutting Compass Saws.
60c
Cuts wood, nails, sheet metal, etc. May be filed when dull, requires no setting. Selected handle, two screws. Length, 12 in. Wt., 5 oz.
9L5037 .... 60c
Fulton Combined Keyhole, Compass and Nail Cutting Saw, Complete.
$1.05 A combination of saws that every carpenter should have. Set consists of handle, 12-in. keyhole blade, 14-inch regular compass blade and 14-inch nail cutting compass blade, all interchangeable. Weight, 12 ounces.
9L5041 .... $1.05
Fulton Nail Cutter Saw
$1.90
Mechanics—Don't spoil your high grade saws on repair work, house wrecking or second hand lumber. BUY ONE OF OUR NAIL CUTTER SAWS. They are especially suited for carpenters, plumbers, electricians, tinners or anyone who requires a saw that will actually cut right through wood, nails, sheet metal, wire, gas pipe, etc. This Fulton Nail Cutter Saw is made of special saw steel, hand filed and tempered by a process which makes it exceedingly hard, but very tough. The Fulton Nail Cutter Saws can be filed when they become dull, but do not require setting, being taper ground. Made in one size only, with 18-inch blade. Weight, 1¼ pounds.
9L5057 .... $1.90
Fulton Interchangeable Compass Saws.
Very convenient, only one handle is necessary for the different length blades.
9L5038—Handle only. Weight, 4 ounces 20c
Blades for above handle. State length. Weight, 2 ounces.
Length, inches..... 10 12 14 16
9L5039 .... 25c 30c 35c 40c
Fulton Compass Saws.
30c 10-Inch State length. Weight, 10 ounces.
Length, inches..... 10 12 14 16
9L5040 .... 30c 35c 40c 50c
$1.50 SPRINGFIELD HAND SAWS.
Four screws on large sizes only.
24-Inch Size.
Made of a good grade saw steel, ground nearly three gauges thinner on back. Carefully smithed and blocked. Has varnished hardwood handle and hollow back. It is full bevel, hand filed, set and fully guaranteed. State length and number of points wanted.
9L5060—Springfield Panel Saws.
Panel. Length, 18 in. 10 or 12 points to inch. Weight, 1 pound........ $1.20
Panel. Length, 20 in. 10 or 12 points to inch. Weight, 1¼ pounds........ 1.30
Panel. Length, 22 in. 9 or 10 points to inch. Weight, 1½ pounds........ 1.40
Hand. Length, 24 in. 8, 9, 10 or 12 points to inch. Weight, 1¾ pounds........ 1.50
Hand. Length, 26 in. 7, 8, 9, 10 or 12 points to inch. Weight, 2 pounds........ 1.60
9L5061—Springfield Rip Saws.
Rip. Length, 26 inches. 5, 5½ or 6 points to inch. Weight, 2 pounds........ 1.60
Rip. Length, 28 inches. 5, 5½ or 6 points to inch. Weight, 2½ pounds..... $1.75
90c ODD JOBS HAND SAWS.
24-Inch Size.
This is a very good saw for the price. Especially recommended for the home, for odd jobs about the house or barn, but not intended for mechanics' use. Made of good steel, has a hardwood handle, well put on, is set and sharpened ready for use and is guaranteed to give satisfactory service for the purpose for which it is intended. Odd Jobs saws are furnished only in points as listed below. State length and number of points wanted.
9L5058—Odd Jobs Panel Saws.
Panel. Length, 20 inches. 9, 10 or 12 points to the inch. Weight, 1¼ pounds..... $0.85
Hand. Length, 24 inches. 7, 8, 9 or 10 points to the inch. Weight, 1¾ pounds..... .90
Hand. Length, 26 inches. 7, 8, 9 or 10 points to the inch. Weight, 1¾ pounds..... .95
9L5059—Odd Jobs Rip Saws.
Rip. Length, 26 inches. 5, 5½ or 6 points to the inch. Weight, 2 pounds..... .95
Rip. Length, 28 inches. 5, 5½ or 6 points to the inch. Weight, 2¼ pounds..... 1.10

All weights and measurements given on this page are approximate and may vary a trifle.

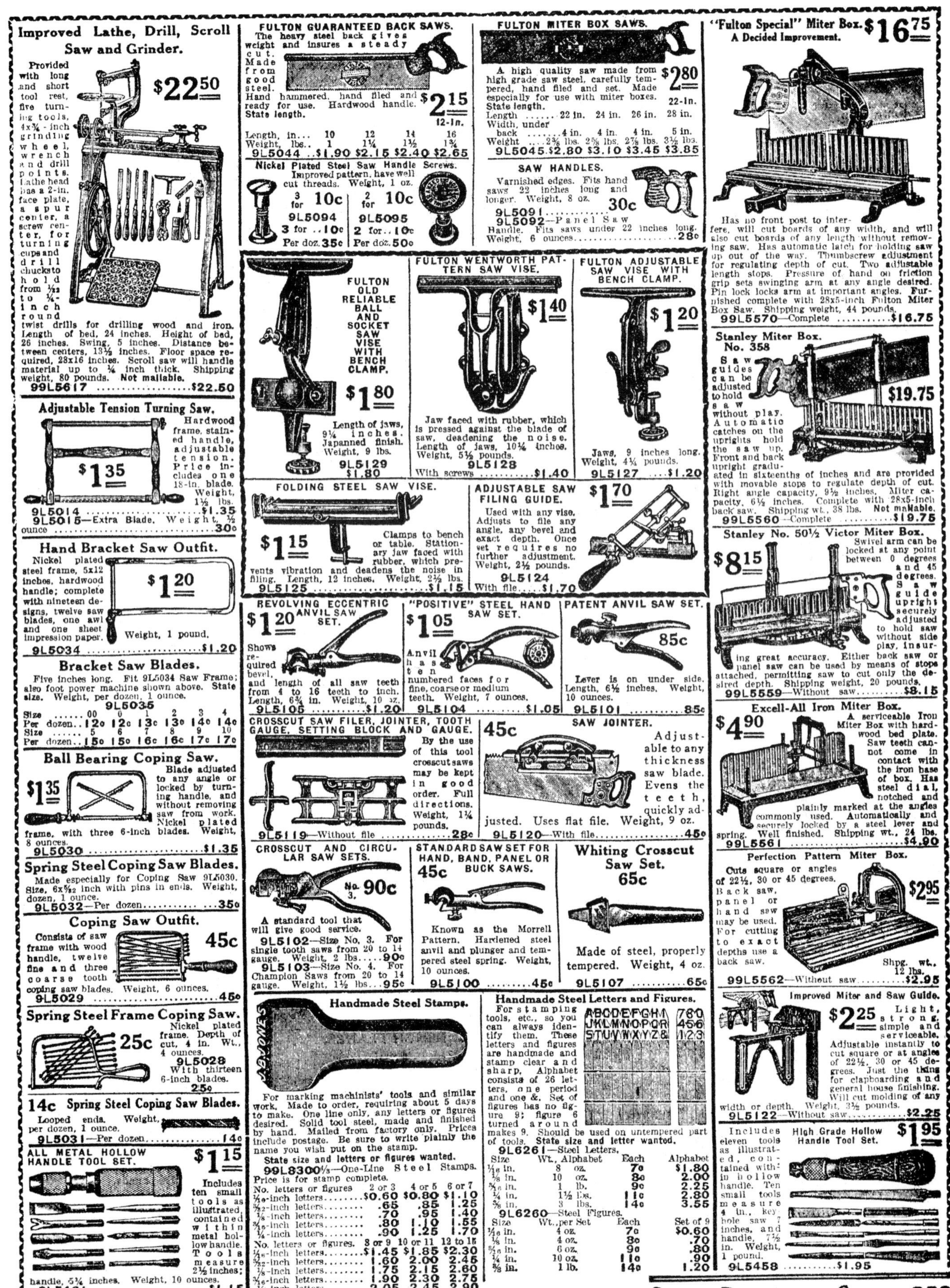

Improved Lathe, Drill, Scroll Saw and Grinder.

$22.50

Provided with long and short tool rest, five turning tools, 4x¾-inch grinding wheel, wrench and drill points. Lathe head has a 2-in. face plate, a spur center, a screw center, for turning cups and drill chucks to hold from 1/32 to ¼-inch round twist drills for drilling wood and iron. Length of bed, 24 inches. Height of bed, 26 inches. Swing, 5 inches. Distance between centers, 13½ inches. Floor space required, 28x16 inches. Scroll saw will handle material up to ¼ inch thick. Shipping weight, 80 pounds. **Not mailable.**

99L5617 ....$22.50

Adjustable Tension Turning Saw.

$1.35

Hardwood frame, stained handle, adjustable tension. Price includes one 18-in. blade. Weight, 1½ lbs.

9L5014 ....$1.35

9L5015—Extra Blade. Weight, ½ ounce 30c

Hand Bracket Saw Outfit.

$1.20

Nickel plated steel frame, 5x12 inches, hardwood handle; complete with nineteen designs, twelve saw blades, one awl and one sheet impression paper. Weight, 1 pound.

9L5034 ....$1.20

Bracket Saw Blades.

Five inches long. Fit 9L5034 Saw Frame; also foot power machine shown above. State size. Weight, per dozen, 1 ounce.

9L5035

Size	00	0	1	2	3	4
Per dozen	12c	12c	13c	13c	14c	14c
Size	5	6	7	8	9	10
Per dozen	15c	15c	16c	16c	17c	17c

Ball Bearing Coping Saw.

$1.35

Blade adjusted to any angle or locked by turning handle, and without removing saw from work. Nickel plated frame, with three 6-inch blades. Weight, 8 ounces.

9L5030 ....$1.35

Spring Steel Coping Saw Blades.

Made especially for Coping Saw 9L5030. Size, 6x5/32 inch with pins in ends. Weight, dozen, 1 ounce.

9L5032—Per dozen 35c

Coping Saw Outfit.

45c

Consists of saw frame with wood handle, twelve fine and three coarse tooth coping saw blades. Weight, 6 ounces.

9L5029 ....45c

Spring Steel Frame Coping Saw.

25c

Nickel plated frame. Depth of cut, 4 in. Wt., 4 ounces.

9L5028

With thirteen 6-inch blades. 25c

14c Spring Steel Coping Saw Blades.

Looped ends. Weight, per dozen, 1 ounce.

9L5031—Per dozen 14c

ALL METAL HOLLOW HANDLE TOOL SET.

$1.15

Includes ten small tools as illustrated, contained within metal hollow handle. Tools measure 2½ inches; handle, 5¾ inches. Weight, 10 ounces.

9L5461 ....$1.15

FULTON GUARANTEED BACK SAWS.

$2.15 12-In.

The heavy steel back gives weight and insures a steady cut. Made from good steel. Hand hammered, hand filed and ready for use. Hardwood handle. State length.

Length, in.	10	12	14	16
Weight, lbs.	1	1¼	1½	1¾
9L5044	$1.90	$2.15	$2.40	$2.65

Nickel Plated Steel Saw Handle Screws.

Improved pattern, have well cut threads. Weight, 1 oz.

3 for 10c — 9L5094 — 3 for ..10c — Per doz. 35c

2 for 10c — 9L5095 — 2 for ..10c — Per doz. 50c

FULTON MITER BOX SAWS.

$2.80 22-In.

A high quality saw made from high grade saw steel, carefully tempered, hand filed and set. Made especially for use with miter boxes. State length.

Length	22 in.	24 in.	26 in.	28 in.
Width, under back	4 in.	4 in.	4 in.	5 in.
Weight	2⅜ lbs.	2⅝ lbs.	2⅞ lbs.	3½ lbs.
9L5045	$2.80	$3.10	$3.45	$3.85

SAW HANDLES.

30c

Varnished edges. Fits hand saws 22 inches long and longer. Weight, 8 oz.

9L5091

9L5092—Panel Saw Handle. Fits saws under 22 inches long. Weight, 6 ounces 28c

"Fulton Special" Miter Box. $16.75

A Decided Improvement.

Has no front post to interfere, will cut boards of any width, and will also cut boards of any length without removing saw. Has automatic latch for holding saw up out of the way. Thumbscrew adjustment for regulating depth of cut. Two adjustable length stops. Pressure of hand on friction grip sets swinging arm at any angle desired. Pin lock locks arm at important angles. Furnished complete with 28x5-inch Fulton Miter Box Saw. Shipping weight, 44 pounds.

99L5570—Complete $16.75

FULTON OLD RELIABLE BALL AND SOCKET SAW VISE WITH BENCH CLAMP.

$1.80

Length of jaws, 9¼ inches. Japanned finish. Weight, 9 lbs.

9L5129 $1.80

FULTON WENTWORTH PATTERN SAW VISE.

$1.40

Jaw faced with rubber, which is pressed against the blade of saw, deadening the noise. Length of jaws, 10¼ inches. Weight, 5½ pounds.

9L5128

With screws $1.40

FULTON ADJUSTABLE SAW VISE WITH BENCH CLAMP.

$1.20

Jaws, 9 inches long. Weight, 4¼ pounds.

9L5127 ....$1.20

FOLDING STEEL SAW VISE.

$1.15

Clamps to bench or table. Stationary jaw faced with rubber, which prevents vibration and deadens the noise in filing. Length, 12 inches. Weight, 2½ lbs.

9L5125 ....$1.15

ADJUSTABLE SAW FILING GUIDE.

$1.70

Used with any vise. Adjusts to file any angle, any bevel and exact depth. Once set requires no further adjustment. Weight, 2½ pounds.

9L5124

With file $1.70

REVOLVING ECCENTRIC ANVIL SAW SET.

$1.20

Shows required bevel, and length of all saw teeth from 4 to 16 teeth to inch. Length, 6¾ in. Weight, 10 oz.

9L5105 ....$1.20

"POSITIVE" STEEL HAND SAW SET.

$1.05

Anvil has ten numbered faces for fine, coarse or medium teeth. Weight, 7 ounces.

9L5104 ....$1.05

PATENT ANVIL SAW SET.

85c

Lever is on under side. Length, 6½ inches. Weight, 10 ounces.

9L5101 ....85c

Stanley Miter Box. No. 358

$19.75

Saw guides can be adjusted to hold saw without play. Automatic catches on the uprights hold the saw up. Front and back upright graduated in sixteenths of inches and are provided with movable stops to regulate depth of cut. Right angle capacity, 9½ inches. Miter capacity, 6½ inches. Complete with 28x5-inch back saw. Shipping wt., 38 lbs. **Not mailable.**

99L5560—Complete $19.75

Stanley No. 50½ Victor Miter Box.

$8.15

Swivel arm can be locked at any point between 0 degrees and 45 degrees. Saw guide uprights securely adjusted to hold saw without side play, insuring great accuracy. Either back saw or panel saw can be used by means of stops attached, permitting saw to cut only the desired depth. Shipping weight, 20 pounds.

99L5559—Without saw $8.15

CROSSCUT SAW FILER, JOINTER, TOOTH GAUGE, SETTING BLOCK AND GAUGE.

By the use of this tool crosscut saws may be kept in good order. Full directions. Weight, 1¾ pounds.

9L5119—Without file 28c

SAW JOINTER.

45c

Adjustable to any thickness saw blade. Evens the teeth, quickly adjusted. Uses flat file. Weight, 9 oz.

9L5120—With file 45c

Excell-All Iron Miter Box.

$4.90

A serviceable Iron Miter Box with hardwood bed plate. Saw teeth cannot come in contact with the iron base of box. Has steel dial, notched and plainly marked at the angles commonly used. Automatically and securely locked by a steel lever and spring. Well finished. Shipping wt., 24 lbs.

99L5561 ....$4.90

CROSSCUT AND CIRCULAR SAW SETS.

No. 3. 90c

A standard tool that will give good service.

9L5102—Size No. 3. For single tooth saws from 20 to 14 gauge. Weight, 2 lbs. 90c

9L5103—Size No. 4. For Champion Saws from 20 to 14 gauge. Weight, 1½ lbs. 95c

STANDARD SAW SET FOR HAND, BAND, PANEL OR BUCK SAWS.

45c

Known as the Morrell Pattern. Hardened steel anvil and plunger and tempered steel spring. Weight, 10 ounces.

9L5100 ....45c

Whiting Crosscut Saw Set.

65c

Made of steel, properly tempered. Weight, 4 oz.

9L5107 ....65c

Perfection Pattern Miter Box.

$2.95

Cuts square or angles of 22½, 30 or 45 degrees. Back saw, panel or hand saw may be used. For cutting to exact depths use a back saw. Shpg. wt., 12 lbs.

99L5562—Without saw $2.95

Handmade Steel Stamps.

For marking machinists' tools and similar work. Made to order, requiring about 5 days to make. One line only, any letters or figures desired. Solid tool steel, made and finished by hand. Mailed from factory only. Prices include postage. Be sure to write plainly the name you wish put on the stamp.

State size and letters or figures wanted.

99L8300½—One-Line Steel Stamps. Price is for stamp complete.

No. letters or figures	2 or 3	4 or 5	6 or 7
1/16-inch letters	$0.60	$0.80	$1.10
3/32-inch letters	.65	.85	1.25
⅛-inch letters	.70	.95	1.40
3/16-inch letters	.80	1.10	1.55
¼-inch letters	.90	1.25	1.70
No. letters or figures	**8 or 9**	**10 or 11**	**12 to 15**
1/16-inch letters	$1.45	$1.85	$2.30
3/32-inch letters	1.60	2.00	2.45
⅛-inch letters	1.75	2.15	2.60
3/16-inch letters	1.90	2.30	2.75
¼-inch letters	2.05	2.45	2.90

Handmade Steel Letters and Figures.

For stamping tools, etc., so you can always identify them. These letters and figures are handmade and stamp clear and sharp. Alphabet consists of 26 letters, one period and one &. Set of figures has no figure 9; figure 6 turned around makes 9. Should be used on untempered part of tools. **State size and letter wanted.**

9L6261—Steel Letters.

Size	Wt., Alphabet	Each	Alphabet
1/16 in.	8 oz.	7c	$1.80
⅛ in.	10 oz.	8c	2.00
3/16 in.	1 lb.	9c	2.25
¼ in.	1½ lbs.	11c	2.80
⅜ in.	3 lbs.	14c	3.55

9L6260—Steel Figures.

Size	Wt., per Set	Each	Set of 9
1/16 in.	4 oz.	7c	$0.60
⅛ in.	4 oz.	8c	.70
3/16 in.	6 oz.	9c	.80
¼ in.	10 oz.	11c	.90
⅜ in.	1 lb.	14c	1.20

Improved Miter and Saw Guide.

$2.25

Light, strong, simple and serviceable. Adjustable instantly to cut square or at angles of 22½, 30 or 45 degrees. Just the thing for clapboarding and general house finishing. Will cut molding of any width or depth. Weight, 3½ pounds.

9L5122—Without saw $2.25

High Grade Hollow Handle Tool Set.

$1.95

Includes eleven tools as illustrated, contained within hollow handle. Ten small tools measure 4 in., key hole saw 7 inches, and handle, 7½ in. Weight, 1 pound.

9L5458 ....$1.95

All weights and measurements given on this page are approximate and may vary a trifle.

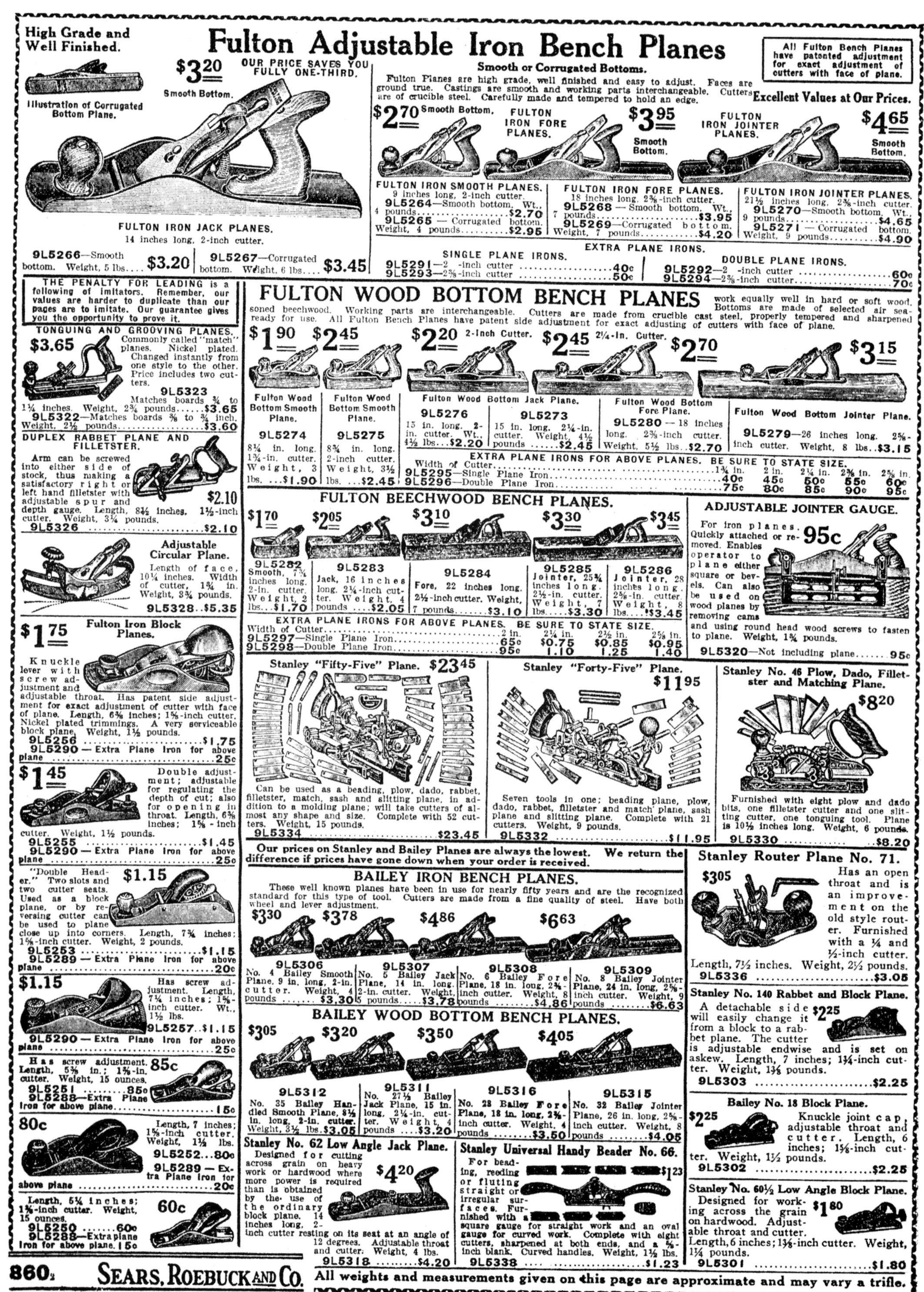

High Grade and Well Finished.

Fulton Adjustable Iron Bench Planes

All Fulton Bench Planes have patented adjustment for exact adjustment of cutters with face of plane.

$3.20 OUR PRICE SAVES YOU FULLY ONE-THIRD. Smooth Bottom.

Illustration of Corrugated Bottom Plane.

Smooth or Corrugated Bottoms.

Fulton Planes are high grade, well finished and easy to adjust. Faces are ground true. Castings are smooth and working parts interchangeable. Cutters are of crucible steel. Carefully made and tempered to hold an edge.

Excellent Values at Our Prices.

FULTON IRON JACK PLANES. 14 inches long, 2-inch cutter.

9L5266—Smooth bottom. Weight, 5 lbs.... $3.20

9L5267—Corrugated bottom. Weight, 6 lbs.... $3.45

$2.70 Smooth Bottom. FULTON IRON SMOOTH PLANES. 9 inches long, 2-inch cutter.
9L5264—Smooth bottom. Wt., 4 pounds $2.70
9L5265 — Corrugated bottom. Weight, 4 pounds $2.95

$3.95 Smooth Bottom. FULTON IRON FORE PLANES. 18 inches long, 2⅜-inch cutter.
9L5268 — Smooth bottom. Wt., 7 pounds $3.95
9L5269—Corrugated bottom. Weight, 7 pounds $4.20

$4.65 Smooth Bottom. FULTON IRON JOINTER PLANES. 21½ inches long, 2⅜-inch cutter.
9L5270—Smooth bottom. Wt., 9 pounds $4.65
9L5271 — Corrugated bottom. Weight, 9 pounds $4.90

EXTRA PLANE IRONS.

SINGLE PLANE IRONS.
9L5291—2 -inch cutter 40c
9L5293—2⅜-inch cutter 50c

DOUBLE PLANE IRONS.
9L5292—2 -inch cutter 60c
9L5294—2⅜-inch cutter 70c

THE PENALTY FOR LEADING is a following of imitators. Remember, our values are harder to duplicate than our pages are to imitate. Our guarantee gives you the opportunity to prove it.

FULTON WOOD BOTTOM BENCH PLANES

work equally well in hard or soft wood. Bottoms are made of selected air seasoned beechwood. Working parts are interchangeable. Cutters are made from crucible cast steel, properly tempered and sharpened ready for use. All Fulton Bench Planes have patent side adjustment for exact adjusting of cutters with face of plane.

$1.90 Fulton Wood Bottom Smooth Plane. 9L5274 8¼ in. long. 1¾-in. cutter. Weight, 3 lbs. ... $1.90

$2.45 Fulton Wood Bottom Smooth Plane. 9L5275 8¾ in. long. 2-inch cutter. Weight, 3½ lbs. ... $2.45

$2.20 2-Inch Cutter. $2.45 2¼-In. Cutter. Fulton Wood Bottom Jack Plane.
9L5276 15 in. long. 2-in. cutter. Wt., 4½ lbs... $2.20
9L5273 15 in. long. 2¼-in. cutter. Weight, 4½ pounds $2.45

$2.70 Fulton Wood Bottom Fore Plane. 9L5280 — 18 inches long. 2⅜-inch cutter. Weight, 5½ lbs.. $2.70

$3.15 Fulton Wood Bottom Jointer Plane. 9L5279—26 inches long. 2⅝-inch cutter. Weight, 8 lbs.. $3.15

EXTRA PLANE IRONS FOR ABOVE PLANES. BE SURE TO STATE SIZE.

Width of Cutter	1¾ in.	2 in.	2¼ in.	2⅜ in.	2⅝ in.
9L5295—Single Plane Iron	40c	45c	50c	55c	60c
9L5296—Double Plane Iron	75c	80c	85c	90c	95c

TONGUING AND GROOVING PLANES.

$3.65 Commonly called "match" planes. Nickel plated. Changed instantly from one style to the other. Price includes two cutters.

9L5323 Matches boards ¾ to 1¼ inches. Weight, 2¾ pounds...... $3.65
9L5322—Matches boards ⅝ to ¾ inch. Weight, 2½ pounds $3.60

DUPLEX RABBET PLANE AND FILLETSTER.

Arm can be screwed into either side of stock, thus making a satisfactory right or left hand filletster with adjustable spur and depth gauge. Length, 8½ inches. 1½-inch cutter. Weight, 3¼ pounds. $2.10

9L5326 .... $2.10

Adjustable Circular Plane. Length of face, 10¼ inches. Width of cutter, 1¾ in. Weight, 3¾ pounds.

9L5328.. $5.35

$1.75 Fulton Iron Block Planes.

Knuckle lever with screw adjustment and adjustable throat. Has patent side adjustment for exact adjustment of cutter with face of plane. Length, 6⅝ inches; 1⅝-inch cutter. Nickel plated trimmings. A very serviceable block plane. Weight, 1½ pounds.

9L5256 .... $1.75
9L5290 — Extra Plane Iron for above plane 25c

$1.45 Double adjustment; adjustable for regulating the depth of cut; also for opening in throat. Length, 6⅝ inches; 1⅝-inch cutter. Weight, 1½ pounds.

9L5255 .... $1.45
9L5290 — Extra Plane Iron for above plane 25c

"Double Header." Two slots and two cutter seats. Used as a block plane, or by reversing cutter can be used to plane close up into corners. Length, 7¾ inches; 1⅝-inch cutter. Weight, 2 pounds. $1.15

9L5253 .... $1.15
9L5289 — Extra Plane Iron for above plane 20c

$1.15 Has screw adjustment. Length, 7¼ inches; 1⅝-inch cutter. Wt., 1½ lbs.

9L5257.. $1.15
9L5290 — Extra Plane Iron for above plane 25c

Has screw adjustment. Length, 5⅝ in.; 1⅜-in. cutter. Weight, 15 ounces. 85c

9L5251 .... 85c
9L5288—Extra Plane Iron for above plane 15c

80c Length, 7 inches; 1⅝-inch cutter. Weight, 1½ lbs.

9L5252... 80c
9L5289 — Extra Plane Iron for above plane 20c

Length, 5¼ inches; 1⅜-inch cutter. Weight, 15 ounces. 60c

9L5250 .... 60c
9L5288—Extra plane iron for above plane. 15c

FULTON BEECHWOOD BENCH PLANES.

$1.70 9L5282 Smooth, 7¾ inches long. 2-in. cutter. Weight, 2 lbs... $1.70

$2.05 9L5283 Jack, 16 inches long. 2¼-inch cutter. Weight, 4 pounds $2.05

$3.10 9L5284 Fore, 22 inches long. 2½-inch cutter. Weight, 7 pounds $3.10

$3.30 9L5285 Jointer, 25¾ inches long. 2½-in. cutter. Weight, 7 lbs.... $3.30

$3.45 9L5286 Jointer, 28 inches long. 2⅝-in. cutter. Weight, 8 lbs.... $3.45

EXTRA PLANE IRONS FOR ABOVE PLANES. BE SURE TO STATE SIZE.

Width of Cutter	2 in.	2¼ in.	2½ in.	2⅝ in.
9L5297—Single Plane Iron	65c	$0.75	$0.85	$0.95
9L5298—Double Plane Iron	95c	1.10	1.25	1.40

ADJUSTABLE JOINTER GAUGE.

For iron planes. Quickly attached or removed. Enables operator to plane either square or bevels. Can also be used on wood planes by removing cams and using round head wood screws to fasten to plane. Weight, 1¾ pounds. 95c

9L5320—Not including plane...... 95c

Stanley "Fifty-Five" Plane. $23.45

Can be used as a beading, plow, dado, rabbet, filletster, match, sash and slitting plane, in addition to a molding plane; will take cutters of almost any shape and size. Complete with 52 cutters. Weight, 15 pounds.

9L5334 .... $23.45

Stanley "Forty-Five" Plane. $11.95

Seven tools in one; beading plane, plow, dado, rabbet, filletster and match plane, sash plane and slitting plane. Complete with 21 cutters. Weight, 9 pounds.

9L5332 .... $11.95

Stanley No. 46 Plow, Dado, Filletster and Matching Plane. $8.20

Furnished with eight plow and dado bits, one filletster cutter and one slitting cutter, one tonguing tool. Plane is 10½ inches long. Weight, 6 pounds.

9L5330 .... $8.20

Our prices on Stanley and Bailey Planes are always the lowest. We return the difference if prices have gone down when your order is received.

BAILEY IRON BENCH PLANES.

These well known planes have been in use for nearly fifty years and are the recognized standard for this type of tool. Cutters are made from a fine quality of steel. Have both wheel and lever adjustment.

$3.30 9L5306 No. 4 Bailey Smooth Plane, 9 in. long, 2-in. cutter. Weight, 4 pounds $3.30

$3.78 9L5307 No. 5 Bailey Jack Plane, 14 in. long, 2-in. cutter. Weight, 5 pounds $3.78

$4.86 9L5308 No. 6 Bailey Fore Plane, 18 in. long, 2⅜-inch cutter. Weight, 8 pounds $4.86

$6.63 9L5309 No. 8 Bailey Jointer Plane, 24 in. long, 2⅝-inch cutter. Weight, 9 pounds $6.63

BAILEY WOOD BOTTOM BENCH PLANES.

$3.05 9L5312 No. 35 Bailey Handled Smooth Plane, 8½ in. long, 2-in. cutter. Weight, 3½ lbs. $3.05

$3.20 9L5311 No. 27½ Bailey Jack Plane, 15 in. long, 2¼-in. cutter. Weight, 4 pounds $3.20

$3.50 9L5316 No. 28 Bailey Fore Plane, 18 in. long, 2⅜-inch cutter. Weight, 4 pounds $3.50

$4.05 9L5315 No. 32 Bailey Jointer Plane, 26 in. long, 2⅝-inch cutter. Weight, 8 pounds $4.05

Stanley Router Plane No. 71. $3.05

Has an open throat and is an improvement on the old style router. Furnished with a ¼ and ½-inch cutter. Length, 7½ inches. Weight, 2½ pounds.

9L5336 .... $3.05

Stanley No. 140 Rabbet and Block Plane. $2.25

A detachable side will easily change it from a block to a rabbet plane. The cutter is adjustable endwise and is set on askew. Length, 7 inches; 1¼-inch cutter. Weight, 1⅛ pounds.

9L5303 .... $2.25

Bailey No. 18 Block Plane. $2.25

Knuckle joint cap, adjustable throat and cutter. Length, 6 inches; 1⅝-inch cutter. Weight, 1½ pounds.

9L5302 .... $2.25

Stanley No. 62 Low Angle Jack Plane. $4.20

Designed for cutting across grain on heavy work or hardwood where more power is required than is obtained by the use of the ordinary block plane. 14 inches long. 2-inch cutter resting on its seat at an angle of 12 degrees. Adjustable throat and cutter. Weight, 4 lbs.

9L5318 .... $4.20

Stanley Universal Handy Beader No. 66. $1.23

For beading, reeding or fluting straight or irregular surfaces. Furnished with a square gauge for straight work and an oval gauge for curved work. Complete with eight cutters, sharpened at both ends, and a ⅛-inch blank. Curved handles. Weight, 1½ lbs.

9L5338 .... $1.23

Stanley No. 60½ Low Angle Block Plane. $1.80

Designed for working across the grain on hardwood. Adjustable throat and cutter. Length, 6 inches; 1⅜-inch cutter. Weight, 1¼ pounds.

9L5301 .... $1.80

All weights and measurements given on this page are approximate and may vary a trifle.

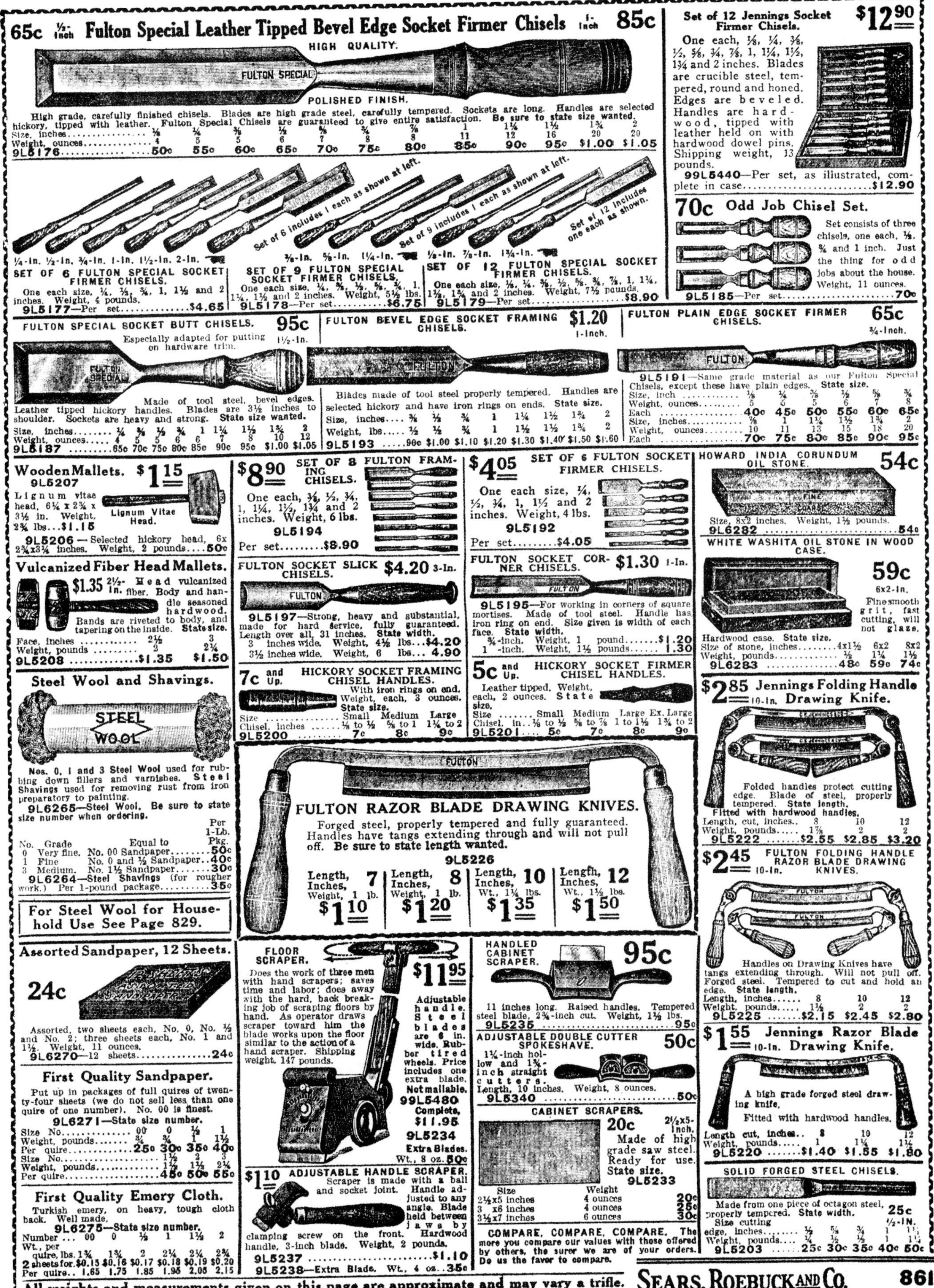

65c ½-inch Fulton Special Leather Tipped Bevel Edge Socket Firmer Chisels 1-inch 85c

HIGH QUALITY. POLISHED FINISH.

High grade, carefully finished chisels. Blades are high grade steel, carefully tempered. Sockets are long. Handles are selected hickory, tipped with leather. Fulton Special Chisels are guaranteed to give entire satisfaction. Be sure to state size wanted.

Size, inches	⅛	¼	⅜	½	⅝	¾	⅞	1	1¼	1½	1¾	2
Weight, ounces	4	5	5	6	7	8	8	11	12	16	20	20
9L5176	50c	55c	60c	65c	70c	75c	80c	85c	90c	95c	$1.00	$1.05

SET OF 6 FULTON SPECIAL SOCKET FIRMER CHISELS. One each size, ¼, ½, ¾, 1, 1½ and 2 inches. Weight, 4 pounds.
9L5177—Per set....$4.65

SET OF 9 FULTON SPECIAL SOCKET FIRMER CHISELS. One each size, ¼, ⅜, ½, ⅝, ¾, 1, 1¼, 1½ and 2 inches. Weight, 5½ lbs.
9L5178—Per set....$6.75

SET OF 12 FULTON SPECIAL SOCKET FIRMER CHISELS. One each size, ⅛, ¼, ⅜, ½, ⅝, ¾, ⅞, 1, 1¼, 1½, 1¾ and 2 inches. Weight, 7½ pounds.
9L5179—Per set....$8.90

Set of 12 Jennings Socket Firmer Chisels. $12.90

One each, ⅛, ¼, ⅜, ½, ⅝, ¾, ⅞, 1, 1¼, 1½, 1¾ and 2 inches. Blades are crucible steel, tempered, round and honed. Edges are beveled. Handles are hardwood, tipped with leather held on with hardwood dowel pins. Shipping weight, 13 pounds.
99L5440—Per set, as illustrated, complete in case....$12.90

70c Odd Job Chisel Set.

Set consists of three chisels, one each, ½, ¾ and 1 inch. Just the thing for odd jobs about the house. Weight, 11 ounces.
9L5185—Per set....70c

FULTON SPECIAL SOCKET BUTT CHISELS. 95c 1½-In.

Especially adapted for putting on hardware trim.

Made of tool steel, bevel edges. Leather tipped hickory handles. Blades are 3½ inches to shoulder. Sockets are heavy and strong. State size wanted.

Size, inches	¼	⅜	½	¾	1	1¼	1½	1¾	2
Weight, ounces	4	5	5	6	6	7	8	10	12
9L5187	65c	70c	75c	80c	85c	90c	95c	$1.00	$1.05

FULTON BEVEL EDGE SOCKET FRAMING CHISELS. $1.20 1-Inch.

Blades made of tool steel properly tempered. Handles are selected hickory and have iron rings on ends. State size.

Size, inches	⅜	½	¾	1	1¼	1½	1¾	2
Weight, lbs.	½	½	¾	1	1½	1½	1¾	2
9L5193	90c	$1.00	$1.10	$1.20	$1.30	$1.40	$1.50	$1.60

FULTON PLAIN EDGE SOCKET FIRMER CHISELS. 65c ¾-Inch.

9L5191—Same grade material as our Fulton Special Chisels, except these have plain edges. State size.

Size, inch	⅛	¼	⅜	½	⅝	¾
Weight, ounces	5	5	5	6	7	8
Each	40c	45c	50c	55c	60c	65c
Size, inches	⅞	1	1¼	1½	1¾	2
Weight, ounces	10	11	13	15	18	20
Each	70c	75c	80c	85c	90c	95c

Wooden Mallets. $1.15

9L5207
Lignum vitae head, 6¼ x 2¾ x 3½ in. Weight, 2¾ lbs...$1.15

9L5206—Selected hickory head, 6x2¾x3¼ inches. Weight, 2 pounds....50c

Vulcanized Fiber Head Mallets.

$1.35 2½-In. Head vulcanized fiber. Body and handle seasoned hardwood. Bands are riveted to body, and tapering on the inside. State size.

Face, inches	2½	3
Weight, pounds	2	2¼
9L5208	$1.35	$1.50

Steel Wool and Shavings.

Nos. 0, 1 and 3 Steel Wool used for rubbing down fillers and varnishes. Steel Shavings used for removing rust from iron preparatory to painting.
9L6265—Steel Wool. Be sure to state size number when ordering.

No.	Grade	Equal to	Per 1-Lb. Pkg.
0	Very fine.	No. 00 Sandpaper	50c
1	Fine	No. 0 and ½ Sandpaper	40c
3	Medium.	No. 1½ Sandpaper	30c

9L6264—Steel Shavings (for rougher work.) Per 1-pound package....35c

For Steel Wool for Household Use See Page 829.

Assorted Sandpaper, 12 Sheets.

24c

Assorted, two sheets each, No. 0, No. ½ and No. 2; three sheets each, No. 1 and 1½. Weight, 11 ounces.
9L6270—12 sheets....24c

First Quality Sandpaper.

Put up in packages of full quires of twenty-four sheets (we do not sell less than one quire of one number). No. 00 is finest.
9L6271—State size number.

Size No.	00	0	½	1
Weight, pounds	¾	¾	1	1½
Per quire	25c	30c	35c	40c
Size No.	1½	2	3	
Weight, pounds	1½	1½	2¼	
Per quire	45c	50c	55c	

First Quality Emery Cloth.

Turkish emery, on heavy, tough cloth back. Well made.
9L6275—State size number.

Number	00	0	½	1	1½	2
Wt., per quire, lbs.	1¾	1¾	2	2¼	2¼	2¾
2 sheets for	$0.15	$0.16	$0.17	$0.18	$0.19	$0.20
Per quire	1.65	1.75	1.85	1.95	2.05	2.15

$8.90 SET OF 8 FULTON FRAMING CHISELS.

One each, ⅜, ½, ¾, 1, 1¼, 1½, 1¾ and 2 inches. Weight, 6 lbs.
9L5194
Per set....$8.90

$4.05 SET OF 6 FULTON SOCKET FIRMER CHISELS.

One each size, ¼, ½, ¾, 1, 1½ and 2 inches. Weight, 4 lbs.
9L5192
Per set....$4.05

FULTON SOCKET SLICK CHISELS. $4.20 3-In.

9L5197—Strong, heavy and substantial, made for hard service, fully guaranteed. Length over all, 31 inches. State width.
3 inches wide. Weight, 4½ lbs...$4.20
3½ inches wide. Weight, 6 lbs... 4.90

FULTON SOCKET CORNER CHISELS. $1.30 1-In.

9L5195—For working in corners of square mortises. Made of tool steel. Handle has iron ring on end. Size given is width of each face. State width.
¾-inch. Weight, 1 pound....$1.20
1-inch. Weight, 1½ pounds....1.30

7c and Up. HICKORY SOCKET FRAMING CHISEL HANDLES.

With iron rings on end. Weight, each, 3 ounces. State size.

Size	Small	Medium	Large
Chisel, inches	⅛ to ½	⅝ to 1	1¼ to 2
9L5200	7c	8c	9c

5c and Up. HICKORY SOCKET FIRMER CHISEL HANDLES.

Leather tipped. Weight, each, 2 ounces. State size.

Size	Small	Medium	Large	Ex. Large
Chisel, in.	⅛ to ½	⅝ to ⅞	1 to 1½	1¾ to 2
9L5201	5c	7c	8c	9c

HOWARD INDIA CORUNDUM OIL STONE. 54c

Size, 8x2 inches. Weight, 1½ pounds.
9L6282....54c

WHITE WASHITA OIL STONE IN WOOD CASE.

59c 6x2-In.

Fine smooth grit, fast cutting, will not glaze.

Hardwood case. State size.

Size of stone, inches	4x1½	6x2	8x2
Weight, pounds	½	1¼	1½
9L6283	48c	59c	74c

$2.85 Jennings Folding Handle 10-In. Drawing Knife.

Folded handles protect cutting edge. Blade of steel, properly tempered. State length. Fitted with hardwood handles.

Length, cut, inches	8	10	12
Weight, pounds	1⅞	2	2
9L5222	$2.55	$2.85	$3.20

FULTON RAZOR BLADE DRAWING KNIVES.

Forged steel, properly tempered and fully guaranteed. Handles have tangs extending through and will not pull off. Be sure to state length wanted.

9L5226

Length, Inches, 7	Length, Inches, 8	Length, Inches, 10	Length, Inches, 12
Weight, 1 lb.	Weight, 1 lb.	Wt., 1¼ lbs.	Wt., 1½ lbs.
$1.10	$1.20	$1.35	$1.50

$2.45 FULTON FOLDING HANDLE RAZOR BLADE DRAWING KNIVES. 10-In.

Handles on Drawing Knives have tangs extending through. Will not pull off. Forged steel. Tempered to cut and hold an edge. State length.

Length, inches	8	10	12
Weight, pounds	1½	2	2
9L5225	$2.15	$2.45	$2.80

FLOOR SCRAPER. $11.95

Does the work of three men with hand scrapers; saves time and labor; does away with the hard, back breaking job of scraping floors by hand. As operator draws scraper toward him the blade works upon the floor similar to the action of a hand scraper. Shipping weight, 147 pounds.

Adjustable handle. Steel blades are 6 in. wide. Rubber tired wheels. Price includes one extra blade. Not mailable.
99L5480 Complete, $11.95
9L5234 Extra Blades. Wt., 8 oz. 50c

$1.10 ADJUSTABLE HANDLE SCRAPER.

Scraper is made with a ball and socket joint. Handle adjusted to any angle. Blade held between jaws by clamping screw on the front. Hardwood handle, 3-inch blade. Weight, 2 pounds.
9L5237....$1.10
9L5238—Extra Blade. Wt., 4 oz..35c

HANDLED CABINET SCRAPER. 95c

11 inches long. Raised handles. Tempered steel blade, 2¾-inch cut. Weight, 1½ lbs.
9L5235....95c

ADJUSTABLE DOUBLE CUTTER SPOKESHAVE. 50c

1¾-inch hollow and 1¾-inch straight cutters. Length, 10 inches. Weight, 8 ounces.
9L5340....50c

CABINET SCRAPERS.

20c 2½x5-Inch. Made of high grade saw steel. Ready for use. State size.
9L5233

Size	Weight	
2½x5 inches	4 ounces	20c
3 x6 inches	4 ounces	25c
3½x7 inches	6 ounces	30c

COMPARE, COMPARE, COMPARE. The more you compare our values with those offered by others, the surer we are of your orders. Do us the favor to compare.

$1.55 Jennings Razor Blade 10-In. Drawing Knife.

A high grade forged steel drawing knife.

Fitted with hardwood handles.

Length cut, inches	8	10	12
Weight, pounds	1	1¼	1½
9L5220	$1.40	$1.55	$1.80

SOLID FORGED STEEL CHISELS.

Made from one piece of octagon steel, properly tempered. State width. 25c ½-IN.

Size cutting edge, inches	½	⅝	¾	1	1¼
Weight, pounds	¼	½	½	1	1¼
9L5203	25c	30c	35c	40c	50c

All weights and measurements given on this page are approximate and may vary a trifle.

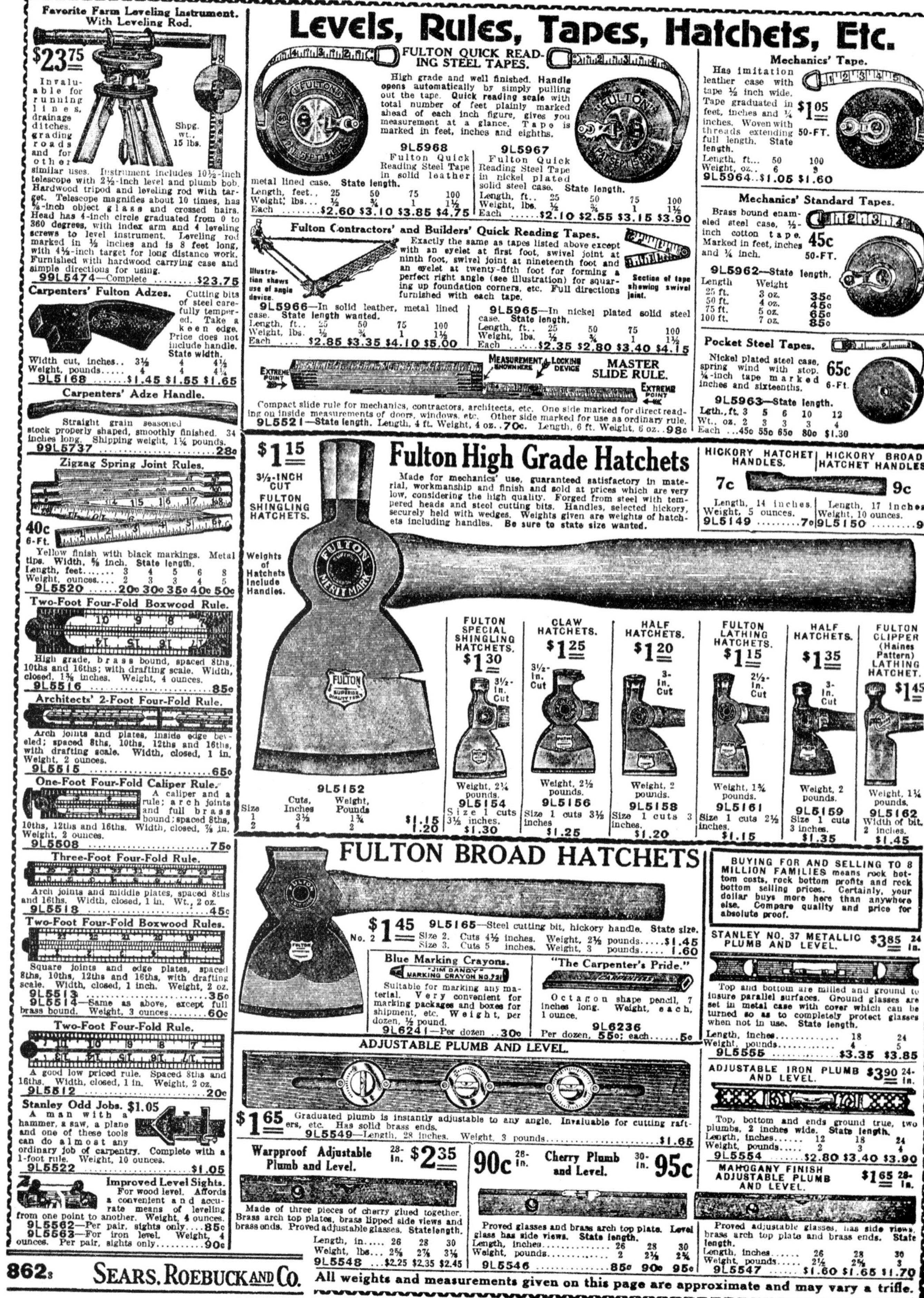

Levels, Rules, Tapes, Hatchets, Etc.

Favorite Farm Leveling Instrument. With Leveling Rod.

$23.75

Invaluable for running lines, drainage ditches, grading roads and for other similar uses. Instrument includes 10½-inch telescope with 2½-inch level and plumb bob. Hardwood tripod and leveling rod with target. Telescope magnifies about 10 times, has ⅝-inch object glass and crossed hairs. Head has 4-inch circle graduated from 0 to 360 degrees, with index arm and 4 leveling screws to level instrument. Leveling rod marked in ½ inches and is 8 feet long, with 4½-inch target for long distance work. Furnished with hardwood carrying case and simple directions for using. Shpg. wt., 15 lbs.

99L5474—Complete $23.75

Carpenters' Fulton Adzes.

Cutting bits of steel carefully tempered. Take a keen edge. Price does not include handle. State width.

Width cut, inches..	3½	4	4½
Weight, pounds.....	4	4	4¼
9L5168	$1.45	$1.55	$1.65

Carpenters' Adze Handle.

Straight grain seasoned stock properly shaped, smoothly finished. 34 inches long. Shipping weight, 1¼ pounds.

99L5737 ....28c

Zigzag Spring Joint Rules.

40c 6-Ft.

Yellow finish with black markings. Metal tips. Width, ⅝ inch. State length.

Length, feet.......	3	4	5	6	8
Weight, ounces....	2	3	3	4	5
9L5520	20c	30c	35c	40c	50c

Two-Foot Four-Fold Boxwood Rule.

High grade, brass bound, spaced 8ths, 10ths and 16ths; with drafting scale. Width, closed, 1⅜ inches. Weight, 4 ounces.

9L5516 ....85c

Architects' 2-Foot Four-Fold Rule.

Arch joints and plates, inside edge beveled; spaced 8ths, 10ths, 12ths and 16ths, with drafting scale. Width, closed, 1 in. Weight, 2 ounces.

9L5515 ....65c

One-Foot Four-Fold Caliper Rule.

A caliper and a rule; arch joints and full brass bound; spaced 8ths, 10ths, 12ths and 16ths. Width, closed, ⅞ in. Weight, 2 ounces.

9L5508 ....75c

Three-Foot Four-Fold Rule.

Arch joints and middle plates, spaced 8ths and 16ths. Width, closed, 1 in. Wt., 2 oz.

9L5518 ....45c

Two-Foot Four-Fold Boxwood Rules.

Square joints and edge plates, spaced 8ths, 10ths, 12ths and 16ths, with drafting scale. Width, closed, 1 inch. Weight, 2 oz.

9L5513 ....35c

9L5514—Same as above, except full brass bound. Weight, 3 ounces........60c

Two-Foot Four-Fold Rule.

A good low priced rule. Spaced 8ths and 16ths. Width, closed, 1 in. Weight, 2 oz.

9L5512 ....20c

Stanley Odd Jobs. $1.05

A man with a hammer, a saw, a plane and one of these tools can do almost any ordinary job of carpentry. Complete with a 1-foot rule. Weight, 10 ounces.

9L5522 ....$1.05

Improved Level Sights.

For wood level. Affords a convenient and accurate means of leveling from one point to another. Weight, 4 ounces.

9L5562—Per pair, sights only....85c

9L5563—For iron level. Weight, 4 ounces. Per pair, sights only..........90c

FULTON QUICK READING STEEL TAPES.

High grade and well finished. Handle opens automatically by simply pulling out the tape. Quick reading scale with total number of feet plainly marked ahead of each inch figure, gives you measurement at a glance. Tape is marked in feet, inches and eighths.

9L5968 Fulton Quick Reading Steel Tape in solid leather metal lined case. State length.

Length, feet..	25	50	75	100
Weight, lbs...	½	¾	1	1½
Each	$2.60	$3.10	$3.85	$4.75

9L5967 Fulton Quick Reading Steel Tape in nickel plated solid steel case. State length.

Length, ft..	25	50	75	100
Weight, lbs.	½	¾	1	1½
Each	$2.10	$2.55	$3.15	$3.90

Fulton Contractors' and Builders' Quick Reading Tapes.

Exactly the same as tapes listed above except with an eyelet at first foot, swivel joint at ninth foot, swivel joint at nineteenth foot and an eyelet at twenty-fifth foot for forming a perfect right angle (see illustration) for squaring up foundation corners, etc. Full directions furnished with each tape.

Illustration shows use of angle device.

Section of tape showing swivel joint.

9L5966—In solid leather, metal lined case. State length wanted.

Length, ft..	25	50	75	100
Weight, lbs.	½	¾	1	1½
Each	$2.85	$3.35	$4.10	$5.00

9L5965—In nickel plated solid steel case. State length.

Length, ft..	25	50	75	100
Weight, lbs.	½	¾	1	1½
Each	$2.35	$2.80	$3.40	$4.15

MASTER SLIDE RULE.

Extreme Point — Measurement shown here — Locking device — Extreme Point

Compact slide rule for mechanics, contractors, architects, etc. One side marked for direct reading on inside measurements of doors, windows, etc. Other side marked for use as ordinary rule.

9L5521—State length. Length, 4 ft. Weight, 4 oz..70c. Length, 6 ft. Weight, 6 oz..98c

Mechanics' Tape.

$1.05 50-FT.

Has imitation leather case with tape ½ inch wide. Tape graduated in feet, inches and ¼ inches. Woven with threads extending full length. State length.

Length, ft...	50	100
Weight, oz..	6	9
9L5964..	$1.05	$1.60

Mechanics' Standard Tapes.

45c 50-FT.

Brass bound enameled steel case, ½-inch cotton tape. Marked in feet, inches and ¼ inch.

9L5962—State length.

Length	Weight	
25 ft.	3 oz.	35c
50 ft.	4 oz.	45c
75 ft.	5 oz.	65c
100 ft.	7 oz.	85c

Pocket Steel Tapes.

65c 6-Ft.

Nickel plated steel case, spring wind with stop. ¼-inch tape marked inches and sixteenths.

9L5963—State length.

Lgth., ft.	3	5	6	10	12
Wt., oz.	2	3	3	3	4
Each ...	45c	55c	65c	80c	$1.30

Fulton High Grade Hatchets

Made for mechanics' use, guaranteed satisfactory in material, workmanship and finish and sold at prices which are very low, considering the high quality. Forged from steel with tempered heads and steel cutting bits. Handles, selected hickory, securely held with wedges. Weights given are weights of hatchets including handles. Be sure to state size wanted.

$1.15 — 3½-INCH CUT FULTON SHINGLING HATCHETS.

Weights of Hatchets Include Handles.

9L5152

Size	Cuts, Inches	Weight, Pounds	
1	3½	1¾	$1.15
2	4	2	1.20

FULTON SPECIAL SHINGLING HATCHETS. $1.30 — 3½-In. Cut. Weight, 2¼ pounds. 9L5154 Size 1 cuts 3½ inches. $1.30

CLAW HATCHETS. $1.25 — 3½-In. Cut. Weight, 2½ pounds. 9L5156 Size 1 cuts 3½ inches $1.25

HALF HATCHETS. $1.20 — 3-In. Cut. Weight, 2 pounds. 9L5158 Size 1 cuts 3 inches. $1.20

FULTON LATHING HATCHETS. $1.15 — 2½-In. Cut. Weight, 1¾ pounds. 9L5161 Size 1 cuts 2½ inches. $1.15

HALF HATCHETS. $1.35 — 3-In. Cut. Weight, 2 pounds. 9L5159 Size 1 cuts 3 inches. $1.35

FULTON CLIPPER (Haines Pattern) LATHING HATCHET. $1.45 — Weight, 1¼ pounds. 9L5162 Width of bit, 2 inches. $1.45

HICKORY HATCHET HANDLES.

7c

Length, 14 inches. Weight, 5 ounces.

9L5149 ....7c

HICKORY BROAD HATCHET HANDLES.

9c

Length, 17 inches. Weight, 10 ounces.

9L5150 ....9c

FULTON BROAD HATCHETS

No. 2 $1.45

9L5165—Steel cutting bit, hickory handle. State size.

Size 2. Cuts 4½ inches. Weight, 2½ pounds.....$1.45

Size 3. Cuts 5 inches. Weight, 3 pounds..... 1.60

Blue Marking Crayons.

"JIM DANDY" MARKING CRAYON NO. 721

Suitable for marking any material. Very convenient for marking packages and boxes for shipment, etc. Weight, per dozen, ½ pound.

9L6241—Per dozen ..30c

"The Carpenter's Pride."

Octagon shape pencil, 7 inches long. Weight, each, 1 ounce.

9L6236

Per dozen, 55c; each......5c

BUYING FOR AND SELLING TO 8 MILLION FAMILIES means rock bottom costs, rock bottom profits and rock bottom selling prices. Certainly, your dollar buys more here than anywhere else. Compare quality and price for absolute proof.

STANLEY NO. 37 METALLIC PLUMB AND LEVEL. $3.85 24 In.

Top and bottom are milled and ground to insure parallel surfaces. Ground glasses are set in metal case with cover which can be turned so as to completely protect glasses when not in use. State length.

Length, inches..............	18	24
Weight, pounds............	4	5
9L5555	$3.35	$3.85

ADJUSTABLE IRON PLUMB AND LEVEL. $3.90 24-In.

Top, bottom and ends ground true, two plumbs, 2 inches wide. State length.

Length, inches......	12	18	24
Weight, pounds.....	2	3	4
9L5554	$2.80	$3.40	$3.90

ADJUSTABLE PLUMB AND LEVEL.

$1.65

Graduated plumb is instantly adjustable to any angle. Invaluable for cutting rafters, etc. Has solid brass ends.

9L5549—Length, 28 inches. Weight, 3 pounds.........................$1.65

Warpproof Adjustable Plumb and Level. 28-In. $2.35

Made of three pieces of cherry glued together. Brass arch top plates, brass lipped side views and brass ends. Proved adjustable glasses. State length.

Length, in.....	26	28	30
Weight, lbs...	2⅝	2⅞	3⅛
9L5548 ...	$2.25	$2.35	$2.45

Cherry Plumb and Level. 28-In. 90c — 30-In. 95c

Proved glasses and brass arch top plate. Level glass has side views. State length.

Length, inches..............	26	28	30
Weight, pounds............	2	2⅛	2¼
9L5546	85c	90c	95c

MAHOGANY FINISH ADJUSTABLE PLUMB AND LEVEL. $1.65 28-In.

Proved adjustable glasses, has side views, brass arch top plate and brass ends. State length.

Length, inches......	26	28	30
Weight, pounds.....	2½	2⅝	3
9L5547	$1.60	$1.65	$1.70

All weights and measurements given on this page are approximate and may vary a trifle.

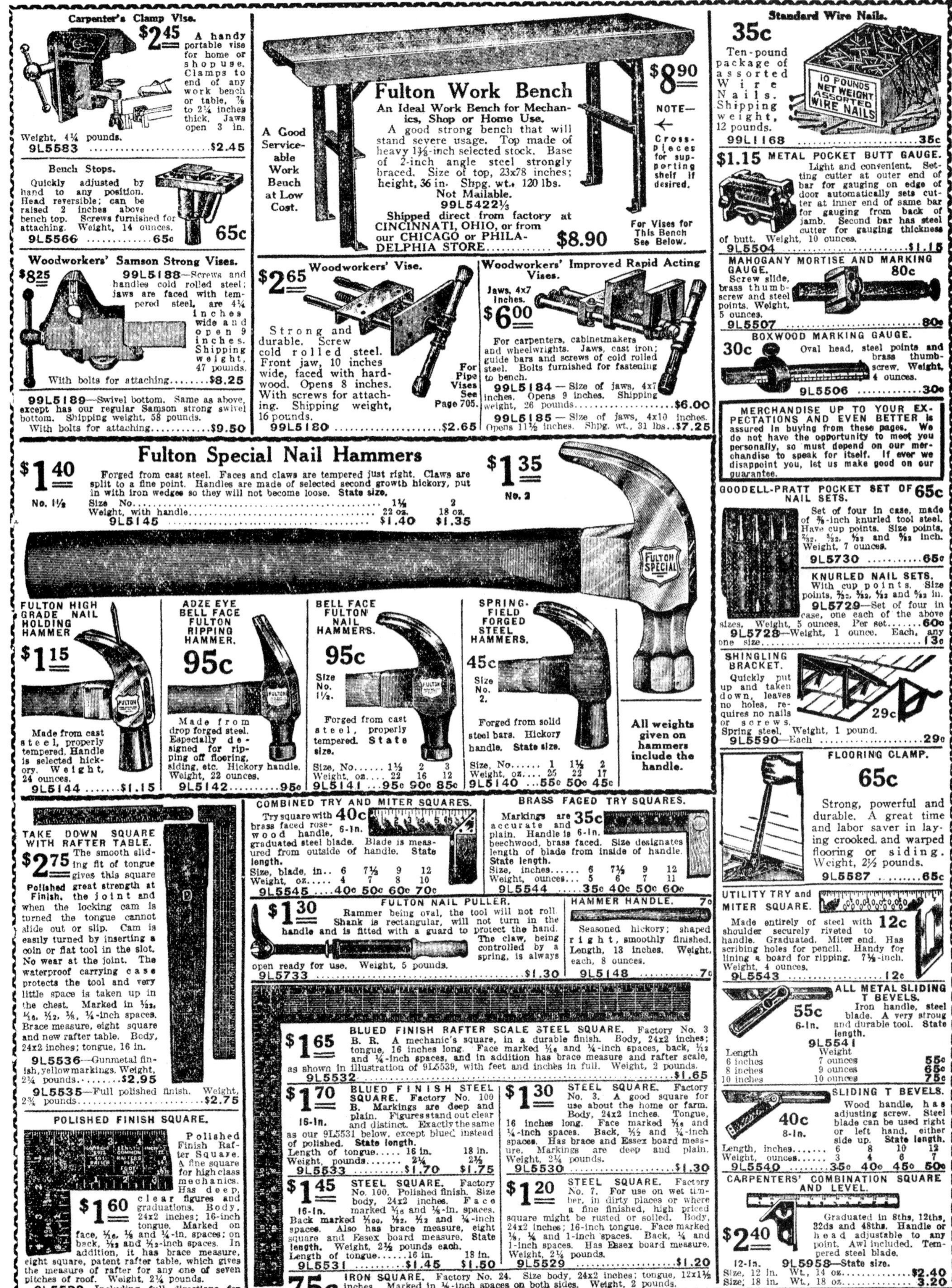

Carpenter's Clamp Vise.

$2.45 A handy portable vise for home or shop use. Clamps to end of any work bench or table, ⅞ to 2¼ inches thick. Jaws open 3 in. Weight, 4¼ pounds.
9L5583 .... $2.45

Bench Stops.

Quickly adjusted by hand to any position. Head reversible; can be raised 2 inches above bench top. Screws furnished for attaching. Weight, 14 ounces.
9L5566 .... 65c

65c

Woodworkers' Samson Strong Vises.

$8.25 99L5188—Screws and handles cold rolled steel; jaws are faced with tempered steel, are 4¼ inches wide and open 9 inches. Shipping weight, 47 pounds. With bolts for attaching $8.25

99L5189—Swivel bottom. Same as above, except has our regular Samson strong swivel bottom. Shipping weight, 58 pounds. With bolts for attaching $9.50

Fulton Work Bench

A Good Serviceable Work Bench at Low Cost.

An Ideal Work Bench for Mechanics, Shop or Home Use.

A good strong bench that will stand severe usage. Top made of heavy 1⅜-inch selected stock. Base of 2-inch angle steel strongly braced. Size of top, 23x78 inches; height, 36 in. Shpg. wt., 120 lbs. Not Mailable.
99L5422⅓
Shipped direct from factory at CINCINNATI, OHIO, or from our CHICAGO or PHILADELPHIA STORE $8.90

$8.90

NOTE— ← Cross-pieces for supporting shelf if desired.

For Vises for This Bench See Below.

Standard Wire Nails.

35c Ten-pound package of assorted Wire Nails. Shipping weight, 12 pounds.
99L1168 .... 35c

$1.15 METAL POCKET BUTT GAUGE.

Light and convenient. Setting cutter at outer end of bar for gauging on edge of door automatically sets cutter at inner end of same bar for gauging from back of jamb. Second bar has steel cutter for gauging thickness of butt. Weight, 10 ounces.
9L5504 .... $1.15

$2.65 Woodworkers' Vise.

Strong and durable. Screw cold rolled steel. Front jaw, 10 inches wide, faced with hardwood. Opens 8 inches. With screws for attaching. Shipping weight, 16 pounds.
99L5180 .... $2.65

For Pipe Vises See Page 705.

Woodworkers' Improved Rapid Acting Vises.

Jaws, 4x7 inches. $6.00

For carpenters, cabinetmakers and wheelwrights. Jaws, cast iron; guide bars and screws of cold rolled steel. Bolts furnished for fastening to bench.
99L5184 — Size of jaws, 4x7 inches. Opens 9 inches. Shipping weight, 26 pounds $6.00
99L5185 — Size of jaws, 4x10 inches. Opens 11½ inches. Shpg. wt., 31 lbs. .. $7.25

MAHOGANY MORTISE AND MARKING GAUGE. 80c

Screw slide, brass thumbscrew and steel points. Weight, 5 ounces.
9L5507 .... 80c

BOXWOOD MARKING GAUGE.

30c Oval head, steel points and brass thumbscrew. Weight, 4 ounces.
9L5506 .... 30c

MERCHANDISE UP TO YOUR EXPECTATIONS AND EVEN BETTER is assured in buying from these pages. We do not have the opportunity to meet you personally, so must depend on our merchandise to speak for itself. If ever we disappoint you, let us make good on our guarantee.

Fulton Special Nail Hammers

$1.40 No. 1½ — $1.35 No. 2

Forged from cast steel. Faces and claws are tempered just right. Claws are split to a fine point. Handles are made of selected second growth hickory, put in with iron wedges so they will not become loose. State size.

Size No.	1½	2
Weight, with handle	22 oz.	18 oz.
9L5145	$1.40	$1.35

GOODELL-PRATT POCKET SET OF NAIL SETS. 65c

Set of four in case, made of ⅜-inch knurled tool steel. Have cup points. Size points, $\frac{2}{32}$, $\frac{3}{32}$, $\frac{4}{32}$ and $\frac{5}{32}$ inch. Weight, 7 ounces.
9L5730 .... 65c

KNURLED NAIL SETS.

With cup points. Size points, $\frac{2}{32}$, $\frac{3}{32}$, $\frac{4}{32}$ and $\frac{5}{32}$ in.
9L5729—Set of four in case, one each of the above sizes. Weight, 5 ounces. Per set 60c
9L5728—Weight, 1 ounce. Each, any one size 13c

FULTON HIGH GRADE NAIL HOLDING HAMMER

$1.15 Made from cast steel, properly tempered. Handle is selected hickory. Weight, 24 ounces.
9L5144 .... $1.15

ADZE EYE BELL FACE FULTON RIPPING HAMMER.

95c Made from drop forged steel. Especially designed for ripping off flooring, siding, etc. Hickory handle. Weight, 22 ounces.
9L5142 .... 95c

BELL FACE FULTON NAIL HAMMERS.

95c Size No. 1½.

Forged from cast steel, properly tempered. State size.

Size, No.	1½	2	3
Weight, oz.	22	16	12
9L5141	95c	90c	85c

SPRINGFIELD FORGED STEEL HAMMERS.

45c Size No. 2.

Forged from solid steel bars. Hickory handle. State size.

Size, No.	1	1½	2
Weight, oz.	25	22	17
9L5140	55c	50c	45c

All weights given on hammers include the handle.

SHINGLING BRACKET.

Quickly put up and taken down, leaves no holes, requires no nails or screws. Spring steel. Weight, 1 pound. 29c
9L5590—Each 29c

FLOORING CLAMP.

65c Strong, powerful and durable. A great time and labor saver in laying crooked and warped flooring or siding. Weight, 2½ pounds.
9L5587 .... 65c

TAKE DOWN SQUARE WITH RAFTER TABLE.

$2.75 The smooth sliding fit of tongue gives this square great strength at the joint and when the locking cam is turned the tongue cannot slide out or slip. Cam is easily turned by inserting a coin or flat tool in the slot. No wear at the joint. The waterproof carrying case protects the tool and very little space is taken up in the chest. Marked in $\frac{1}{32}$, $\frac{1}{16}$, $\frac{1}{12}$, ⅛, ¼-inch spaces. Brace measure, eight square and new rafter table. Body, 24x2 inches; tongue, 16 in.

Polished Finish.

9L5536—Gunmetal finish, yellow markings. Weight, 2¼ pounds $2.95
9L5535—Full polished finish. Weight, 2¾ pounds $2.75

COMBINED TRY AND MITER SQUARES.

Try square with brass faced rosewood handle, graduated steel blade. Blade is measured from outside of handle. State length. 40c 6-In.

Size, blade, in.	6	7½	9	12
Weight, oz.	4	7	8	10
9L5545	40c	50c	60c	70c

BRASS FACED TRY SQUARES.

Markings are accurate and plain. Handle is beechwood, brass faced. Size designates length of blade from inside of handle. State length. 35c 6-In.

Size, inches	6	7½	9	12
Weight, ounces	5	6	7	11
9L5544	35c	40c	50c	60c

UTILITY TRY and MITER SQUARE.

Made entirely of steel with shoulder securely riveted to handle. Graduated. Miter end. Has scribing holes for pencil. Handy for lining a board for ripping. 7¼-inch. Weight, 4 ounces. 12c
9L5543 .... 12c

FULTON NAIL PULLER.

$1.30 Rammer being oval, the tool will not roll. Shank is rectangular, will not turn in the handle and is fitted with a guard to protect the hand. The claw, being controlled by a spring, is always open ready for use. Weight, 5 pounds.
9L5733 .... $1.30

HAMMER HANDLE. 7c

Seasoned hickory; shaped right, smoothly finished. Length, 13 inches. Weight, each, 8 ounces.
9L5148 .... 7c

ALL METAL SLIDING T BEVELS.

55c 6-In. Iron handle, steel blade. A very strong and durable tool. State length.

9L5541 Length	Weight	
6 inches	7 ounces	55c
8 inches	9 ounces	65c
10 inches	10 ounces	75c

POLISHED FINISH SQUARE.

$1.60 Polished Finish Rafter Square. A fine square for high class mechanics. Has deep, clear figures and graduations. Body, 24x2 inches; 16-inch tongue. Marked on face, $\frac{1}{16}$, ⅛ and ¼-in. spaces; on back, $\frac{1}{12}$ and $\frac{1}{32}$-inch spaces. In addition, it has brace measure, eight square, patent rafter table, which gives the measure of rafter for any one of seven pitches of roof. Weight, 2¼ pounds.
9L5539—Including full directions for using $1.60

$1.65 BLUED FINISH RAFTER SCALE STEEL SQUARE. Factory No. 3 B. R. A mechanic's square, in a durable finish. Body, 24x2 inches; tongue, 16 inches long. Face marked $\frac{1}{16}$ and ¼-inch spaces, back, $\frac{1}{12}$ and ¼-inch spaces, and in addition has brace measure and rafter scale, as shown in illustration of 9L5539, with feet and inches in full. Weight, 2 pounds.
9L5532 .... $1.65

$1.70 BLUED FINISH STEEL SQUARE. 16-In. Factory No. 100 B. Markings are deep and plain. Figures stand out clear and distinct. Exactly the same as our 9L5531 below, except blued instead of polished. State length.

Length of tongue	16 in.	18 in.
Weight, pounds	2¼	2½
9L5533	$1.70	$1.75

$1.30 STEEL SQUARE. Factory No. 3. A good square for use about the home or farm. Body, 24x2 inches. Tongue, 16 inches long. Face marked $\frac{1}{16}$ and ¼-inch spaces. Back, $\frac{1}{12}$ and ¼-inch spaces. Has brace and Essex board measure. Markings are deep and plain. Weight, 2¼ pounds.
9L5530 .... $1.30

$1.45 STEEL SQUARE. 16-In. Factory No. 100. Polished finish. Size body, 24x2 inches. Face marked $\frac{1}{16}$ and ⅛-in. spaces. Back marked $\frac{1}{100}$, $\frac{1}{32}$, $\frac{1}{12}$ and ¼-inch spaces. Also has brace measure, eight square and Essex board measure. State length. Weight, 2½ pounds each.

Length of tongue	16 in.	18 in.
9L5531	$1.45	$1.50

$1.20 STEEL SQUARE. Factory No. 7. For use on wet timber, in dirty places or where a fine finished, high priced square might be rusted or soiled. Body, 24x2 inches; 16-inch tongue. Face marked ⅛, ¼ and 1-inch spaces. Back, ¼ and 1-inch spaces. Has Essex board measure. Weight, 2¼ pounds.
9L5529 .... $1.20

75c IRON SQUARE. Factory No. 24. Size body, 24x2 inches; tongue, 12x1½ inches. Marked in ⅛-inch spaces on both sides. Weight, 2 pounds.
9L5528 .... 75c

SLIDING T BEVELS.

40c 8-In. Wood handle, has adjusting screw. Steel blade can be used right or left hand, either side up. State length.

Length, inches	6	8	10	12
Weight, ounces	3	4	6	7
9L5540	35c	40c	45c	50c

CARPENTERS' COMBINATION SQUARE AND LEVEL.

$2.40 12-In. Graduated in 8ths, 12ths, 32ds and 48ths. Handle or head adjustable to any point. Awl included. Tempered steel blade.
9L5958—State size.
Size, 12 in. Wt., 14 oz. $2.40
Size, 18 in. Wt., 18 oz. 3.20

All weights and measurements given on this page are approximate and may vary a trifle.

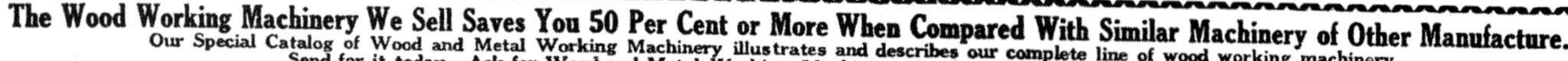

The Wood Working Machinery We Sell Saves You 50 Per Cent or More When Compared With Similar Machinery of Other Manufacture.

Our Special Catalog of Wood and Metal Working Machinery illustrates and describes our complete line of wood working machinery. Send for it today. Ask for Wood and Metal Working Machinery Catalog, No. 528GCL.

PLANING MILL SPECIAL.

8 Machines in One.

$510.00

Shipped Direct From Factory at CINCINNATI, OHIO.

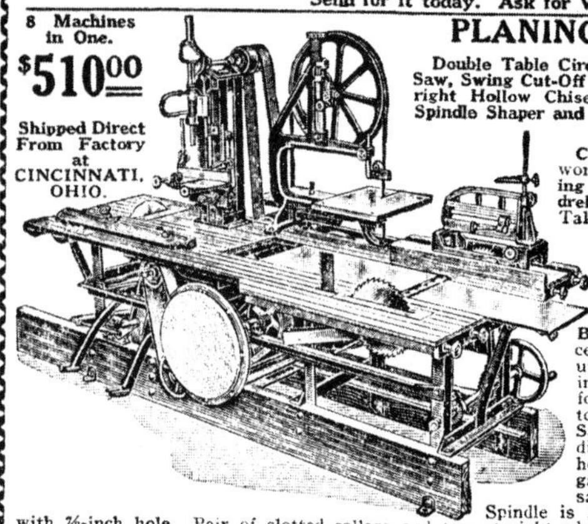

Double Table Circular, Rip and Crosscut Saw, Band Saw, Swing Cut-Off Saw, 12-Inch Jointer, Tenoner, Upright Hollow Chisel Mortiser and Borer, Reversible Spindle Shaper and Sanding Machine.

SPECIFICATIONS.

Circular Saw Table—Has double working surface. Width, 44 in. **Raising and Lowering Circular Saw Mandrel**—Diameter of mandrel, 1⁹⁄₁₆ inches. Takes saws, etc., with 1-inch hole. Has friction clutch to start and stop. **Jointer**—12 inches wide, with round safety head and four knives. **Tilting Guide** for jointer tilts to angle of 45 degrees and is adjustable up to 17 inches in width. **Band Saw**—22-inch swing, cuts to center of 44-inch circle. Table measures 14½x17½ inches, stands 40 inches from floor and can be tilted for sawing bevels. Has belt shifter to start and stop. **Swing Cut-Off Saw**—Saw mandrel 1³⁄₁₆ inches in diameter takes saws with 1-inch hole. Throwing out lever which engages bevel frictions, stops swing saw. **Reversible Spindle Shaper**—Spindle is 1 inch in diameter, takes cutters with ⅞-inch hole. Pair of slotted collars and two straight edge knives furnished. **Upright Hollow Chisel Mortiser and Borer**—Bed plate tilts to angle of 45 degrees, is adjustable and will take up to 3 inches to center of bit. Foot treadle forces chisels and bits into the work. Has belt shifter to start and stop. **Tenoner**—Furnished with two 9-inch grooving saws and sliding carriage. Carriage has vertical stop for making tenons of exact lengths. **Sand Disc**—18 inches in diameter. **Rip Guide**—Is adjustable for any width up to 22 inches. **Crosscut Guide**—Has angle adjustment of 45 degrees and adjustable stop for cutting to length. **Countershaft**—1⁵⁄₁₆ inches in diameter with 10x4-inch tight and loose pulleys. Floor space, 4 feet 10 in. x 9 feet 6 in. Tight and loose drive pulleys, 10x4-inch face. Speed, 550 R. P. M. Power, 5 to 7½ H.-P. Shipping weight, 1,800 lbs.

99L8685⅓—Planing Mill Special, complete with one 12-inch cut-off saw, one 12-inch rip saw, one 10-inch crosscut saw, one ⅜-inch band saw and five belts. Shipped direct from factory at CINCINNATI, OHIO **$510.00**

WAGON SHOP SPECIAL.

Shipped Direct From Factory at Cincinnati, Ohio.

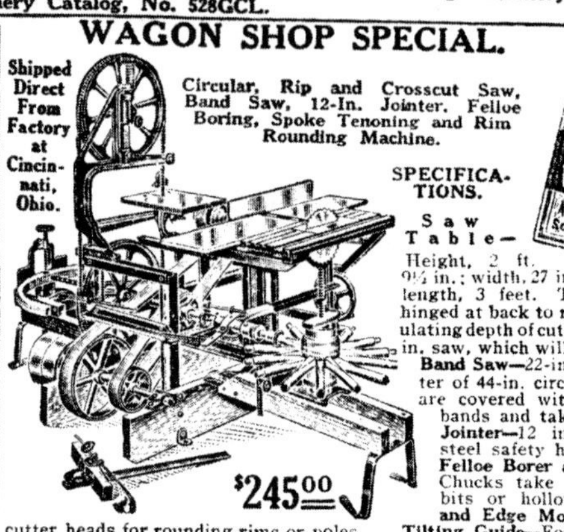

Circular, Rip and Crosscut Saw, Band Saw, 12-In. Jointer, Felloe Boring, Spoke Tenoning and Rim Rounding Machine.

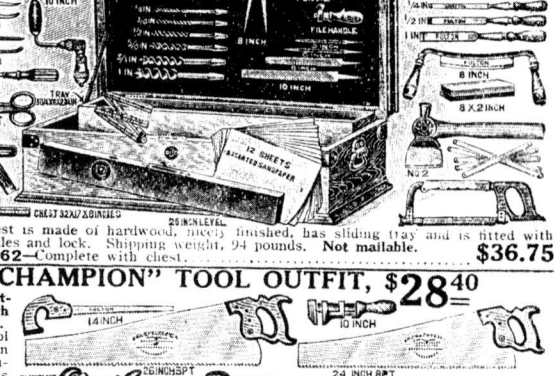

$245.00

SPECIFICATIONS.

Saw Table—Height, 2 ft. 9½ in.; width, 27 in.; length, 3 feet. Table hinged at back to raise or lower for regulating depth of cut. Has opening for 14-in. saw, which will cut up to 5 in. deep. **Band Saw**—22-in. swing, cuts to center of 44-in. circle. Band saw wheels are covered with endless rubber bands and take saws 12 feet long. **Jointer**—12 in. wide, with round steel safety head and four knives. **Felloe Borer and Spoke Tenoner**—Chucks take ½-inch round shank bits or hollow augers. **Rounder and Edge Molder**—Furnished with cutter heads for rounding rims or poles. **Tilting Guide**—For jointer or rip saw, tilts to angle of 45 degrees. **Crosscut Guide**—Slides in milled groove. Has angle adjustment of 45 degrees. **Saw Mandrel**—1⁵⁄₁₆ inches in diameter. Takes saws, cutter heads or emery wheels with 1-inch hole. **Countershaft**—1³⁄₁₆ inches in diameter; size of tight and loose drive pulleys, 10x4-inch face. **Bearings**—4 inches wide, full babbitted with high quality babbitt. Floor space, 7 feet 9 inches x 6 feet 5 inches. Tight and loose drive pulleys, 10x4 inches. Speed, 550 R. P. M. Power, 5 H.-P. Shipping weight, 950 pounds.

99L8710⅓—Wagon Shop Special, complete with one 12-in. rip saw, one 10-in. crosscut saw, one ⅜-in. band saw, one ⅝-in. bit, one 1⅛-in. hollow auger with ⅝-in. and ⅞-in. bushings and one 4-in. and one 3-in. belt. Shipped direct from factory at CINCINNATI, OHIO **$245.00**

$47.80 OUR "CARPENTERS' SPECIAL" TOOL SET.

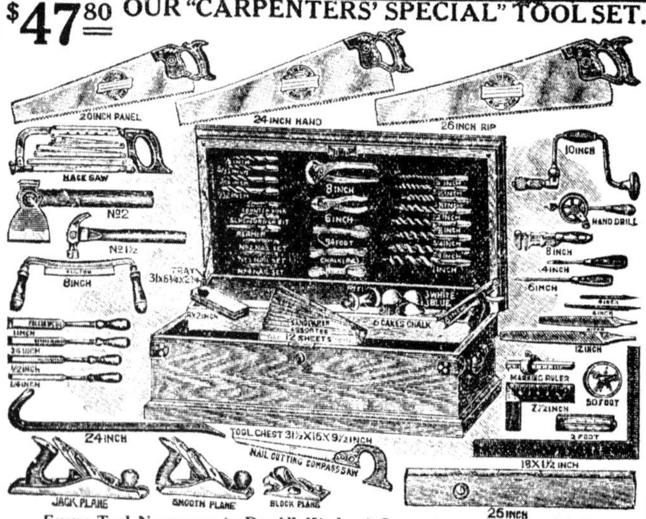

Every Tool Necessary to Do All Kinds of Carpenter or Cabinet Work. Nothing Unnecessary; Nothing Left Out.

Each piece of our "Carpenters' Special" Tool Set has been carefully selected for its utility from our regular stock of high grade tools, making this a most complete and serviceable tool set. Shipping weight, 114 pounds. **Not mailable.**

99L5365—Our "Carpenters' Special" Tool Set, complete as illustrated, in substantial hardwood chest **$47.80**

OUR "VICTOR" TOOL OUTFIT. $36.75

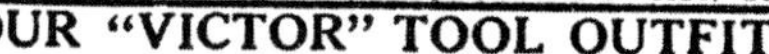

All the Tools in Our Victor Outfit Are High Grade, Selected From Our Regular Stock and Fully Guaranteed.

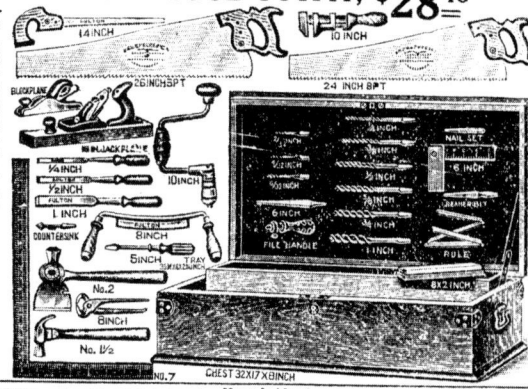

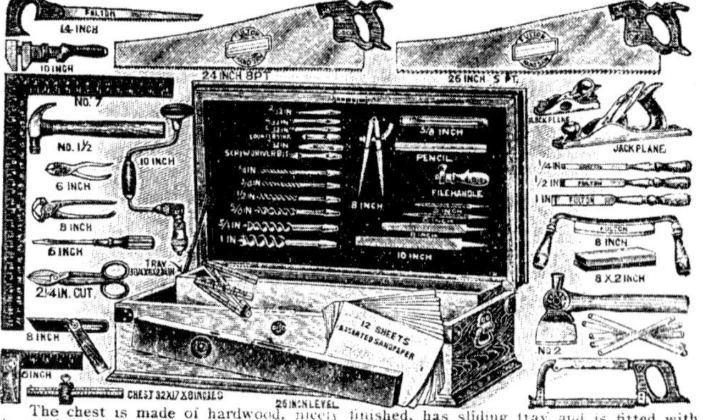

The chest is made of hardwood, nicely finished, has sliding tray and is fitted with hinges, handles and lock. Shipping weight, 94 pounds. **Not mailable.**

99L5362—Complete with chest **$36.75**

OUR "CHAMPION" TOOL OUTFIT, $28.40

A Big Assortment of High Grade Tools. Every tool included in this Champion Outfit is selected from our regular stock and fully guaranteed. The chest we furnish with our Champion Outfit is made of hardwood, nicely finished, has lock and hinges. Shpg. wt., 86 lbs. **Not mailable.**

99L5361 Complete with chest. **$28.40**

$18.50 THE "MANUAL TRAINER" FOOT POWER Band Saw Machine

Not a toy—but a real honest-to-goodness saw that can be readily operated by any ordinary boy of high school age. Also makes a handy machine for small carpenter shops, etc.

DESCRIPTION:

Swing—10½ inches.
Saw Guides are adjustable. Upper guide can be raised to saw material up to 2 inches thick.
Table Top of hardwood, measures 18x19 in. and stands 31 inches from floor.
Frame—1¼-in. angle steel, riveted and bolted together.
Cabinet for holding tools, drawings, etc., 11x12x20 inches high inside.
Height over all, 44 inches.
Foot Power—Drive wheel, 11x1½ inches with pedals and complete with belt.
Shipping weight, 165 lbs. **Not mailable.**

99L5645⅓ — Complete with three ³⁄₁₆-in. band saws. Shipped direct from factory at CINCINNATI, OHIO, or from our CHICAGO or PHILADELPHIA store **$18.50**

Extra Band Saw Blades for above. Size, ³⁄₁₆ in. by 66 in. Wt. of three, 4 oz.
99L5647⅓ — Per package of 3 blades **$1.35**

STOOL. For "Manual Trainer" Band Saw Machine. Adjustable from 20 to 25 inches in height. Shpg. wt., 36 pounds.
99L5646⅓ $2.40

Band Saw Blades, and Stool shown above, shipped direct from factory at CINCINNATI, OHIO, or from our CHICAGO or PHILADELPHIA store.

Jennings HOME TOOL Outfit $14.90

Handy assortment of small size tools, for ordinary use around the house. Size of case, 17x8½x6½ in. Shpg. wt., 18 lbs.

99L5458 Complete **$14.90**

Householders' "Odd Jobs" Carpenters' Tool Outfit, $5.95

Includes 18 Useful Tools for Doing Any Ordinary Job Around the House.

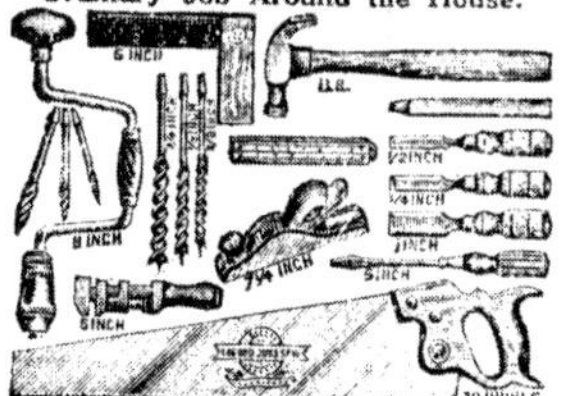

Every householder should own at least a few tools. The "Odd Jobs" gives you a complete outfit of serviceable tools at very little cost. Shipping weight, 10 pounds.

99L5360—Complete outfit, as illustrated **$5.95**

 All weights and measurements given on this page are approximate and may vary a trifle.

$79.50 OUR BIG MASTERWORKMAN BLACKSMITHS' OUTFIT

READ THESE DESCRIPTIONS CAREFULLY.

This outfit will take care of the heaviest work, and contains all the tools needed by the blacksmith and horseshoer. Especially for high class iron workers, men who demand the best.

ANVIL. Fulton All-Steel One-Piece Anvil. Wt., 100 pounds. Long well shaped horn and heel. Wide spread of base prevents tipping when forging heavy work on extreme ends.

BLOWER. Our Tiger Blower complete with piping and tuyere. Has flat, straight cut gears; no belts, clutches or ratchets to get out of order, steel shafts, composition bearings, and turns easily. 8⅞-inch fan; 11¼-inch fan case; height, over all, 43 inches.

DRILL. Our Acme Ball Bearing Two-Speed Self Feed Third Gear Drill. Drills holes up to 1¼ inches and to center of 14¾-inch circle, takes ½-inch round shank drill bits and has 3½-inch run of spindle.

SCREW PLATE. Fulton Double Die Screw Plate, cuts five sizes, ¼[20], 5/16[18], ⅜[16], ½[13] and ⅝[11] inch. Complete with 22-inch stock and 16-inch tap wrench, in hardwood case.

VISE. Our blacksmiths' 60-pound solid box wrought Vise with 4½-inch tempered steel jaws.

BOLT CLIPPERS. One pair Fulton Bolt Clippers will cut up to and including ½-inch bolts. Adjustable to take up wear and have rubber bumpers.

Outfit also includes complete assortment of good quality tools, as illustrated.

99L5018—Complete outfit, as illustrated and described. Shpg. wt., 460 lbs. Not mailable...... **$79.50**

DO YOU KNOW

That we are the largest distributors of Blacksmiths' Supplies?

WHY?

OUR PRICES are considerably lower than those charged by the ordinary retail store.

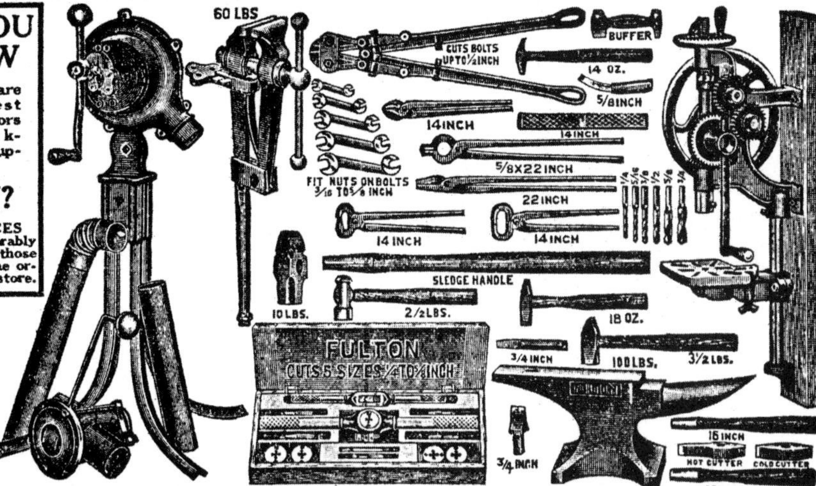

$68.75 IMPROVED VULCAN OUTFIT OF BLACKSMITHS' TOOLS

Each article included in this outfit is fully guaranteed in material and workmanship. The tools are heavy and strong, built for regular everyday shop use, and will give long and satisfactory service.

ANVIL. Fulton All-Steel One-Piece Anvil. Weight, 100 pounds. Horn and heel are long and well shaped. Base has a wide spread, insuring great stability, and prevents tipping when forging heavy work on the extreme ends of the anvil.

FORGE. Our "Eclipse" Forge with half hood. 8-inch blower, 25x36-inch hearth, 30 inches high. Crank turns either forward or backward. Will produce a welding heat on 3-inch iron in five minutes.

DRILL. Our Acme Ball Bearing Self Feed Third Gear Post Drill. Drills up to 1¼-inch holes and to center of 14¾-inch circle; has 3½-inch run of spindle, takes ½-inch round shank drills.

SCREW PLATE. Invincible Screw Plate, with five taps and five dies cutting ¼[20], 5/16[18], ⅜[16], ½[13] and ⅝[11] inch. Set has one stock, 16 inches long, and one tap wrench, 16 inches long in hardwood box.

VISE. Our Blacksmiths' 60-pound wrought Vise with solid box and 4½-inch steel jaws. Outfit also includes complete assortment of good quality hand tools, as illustrated.

99L5013—Complete outfit, as illustrated and described. Shipping weight, 471 pounds. Not mailable........ **$68.75**

$56.50 OUR ACME OUTFIT OF BLACKSMITHS' TOOLS

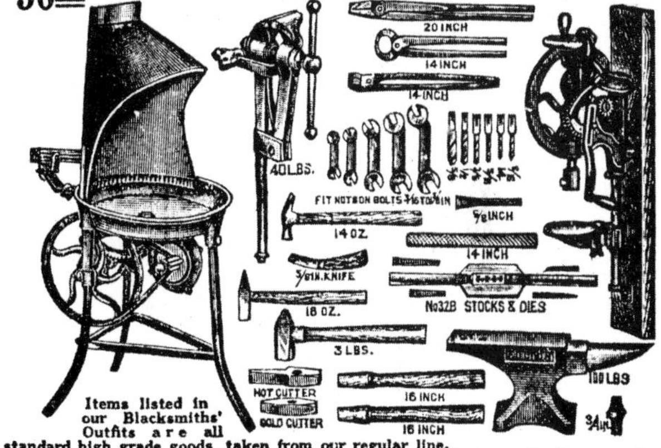

Items listed in our Blacksmiths' Outfits are all standard high grade goods taken from our regular line.

Especially recommended for repair men, farmers and others who have use for a reliable outfit of blacksmiths' tools. Every item we put into this set is of good quality, fully guaranteed and meets the requirements of any workman.

READ THESE DESCRIPTIONS.

ANVIL. Fulton All-Steel One-Piece Anvil. Weight, 100 pounds. Has long well shaped horn and heel. Wide spread of base prevents tipping when forging heavy work on extreme ends.

FORGE. Our "Challenge" Forge. Has 8½-inch blower, 18x21-inch hearth, and measures 30 inches from floor to top of hearth.

DRILL. Our Acme Two-Speed Self Feed Post Drill. Takes ½-inch round shank drill bits, has 2¾-inch run of spindle and drills holes up to ¾ inch and to center of 12-inch circle.

VISE. 40-Pound Blacksmiths' Wrought Iron Vise. Has solid box and 4-inch hardened steel jaws.

STOCK AND DIES No. 32B. Cut ¼ to ¾ inch, 10, 12, 14 and 16 threads to the inch; four taps and four dies.

Outfit also includes complete assortment of good quality tools, as illustrated.

Shipping weight, complete outfit, 319 pounds. Not mailable.

99L5011—Complete outfit, as illustrated and described............ **$56.50**

$45.40 OUR ECLIPSE OUTFIT OF BLACKSMITHS' TOOLS

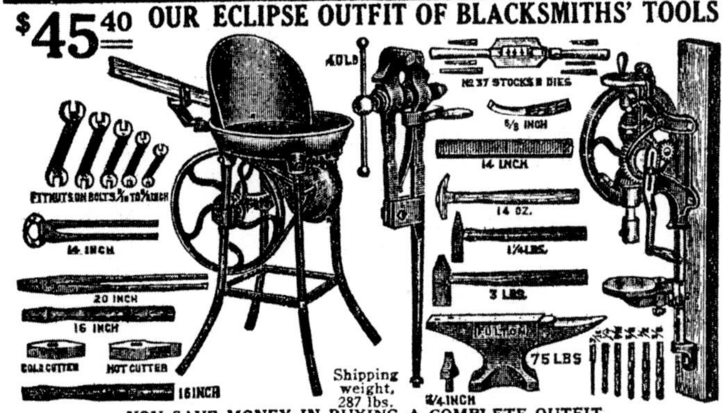

Shipping weight, 287 lbs.

YOU SAVE MONEY IN BUYING A COMPLETE OUTFIT.

Includes the most important tools, each one of high quality, the same that we furnish in our other outfits. **READ THESE DESCRIPTIONS CAREFULLY.**

ANVIL. Fulton All-Steel One-Piece Anvil, wt., 75 lbs. Has long well-shaped horn and heel. Wide spread of base prevents tipping when forging heavy work on extreme ends.

FORGE. Invincible Lever Forge. Has 18-inch hearth, which stands 30 inches from floor, and 8-inch blower.

VISE. Blacksmiths' 40-lb. wrought Vise. Has solid box and 4-inch tempered steel jaws.

DRILL. Acme Two-Speed Self Feed Post Drill. Takes ½-inch round shank drills, has 2¾-inch run of spindle and drills holes up to ¾ inch and to center of 12-inch circle.

STOCK AND DIES No. 37. Cut ¼ to 9/16 inch, 14, 18 and 22 threads to the inch, six taps and three dies.

Outfit also includes assortment of good quality tools, as illustrated.

99L5010—Complete outfit, as illustrated and described. Not mailable.... **$45.40**

CHALLENGE OUTFIT OF BLACKSMITHS' HIGH CLASS TOOLS

$28.90

Every tool selected from our regular stock and amply strong and heavy enough for all kinds of light and medium work. Quality not sacrificed in this outfit.

Shpg. weight, 230 lbs.

READ THESE DESCRIPTIONS.

ANVIL. 70-pound cast iron Anvil.

FORGE. Our Invincible Portable Lever Forge has 18-inch hearth, 8-inch blower and is 30 inches high.

DRILL. Our horizontal Bench Drill has screw feed chuck bored for ½-inch round shank drills.

VISE. Blacksmiths' cast iron Vise with hinged jaws. Jaws, 4 inches wide and faced with tempered steel. Weight, 44 pounds.

STOCK AND DIES, No. 41C. Cut ¼ to 9/16 inch, 12, 14 and 16 threads to the inch, six taps and three dies.

Outfit also includes assortment of good quality hand tools, as illustrated.

99L5009—Compltee outfit, as illustrated and described. Not mailable... **$28.90**

All weights and measurements given on this page are approximate and may vary a trifle.

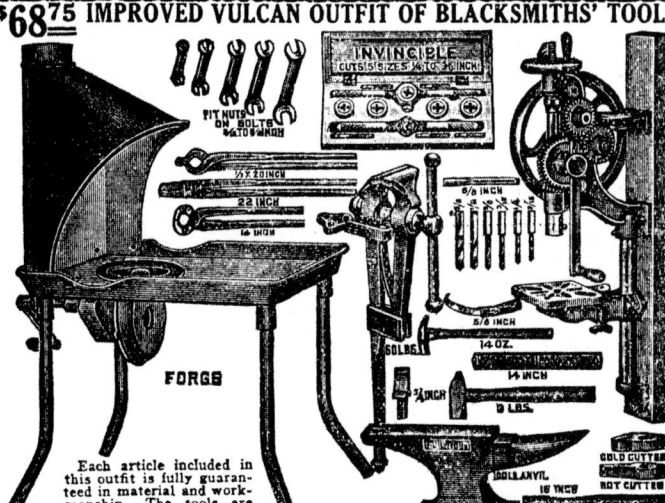

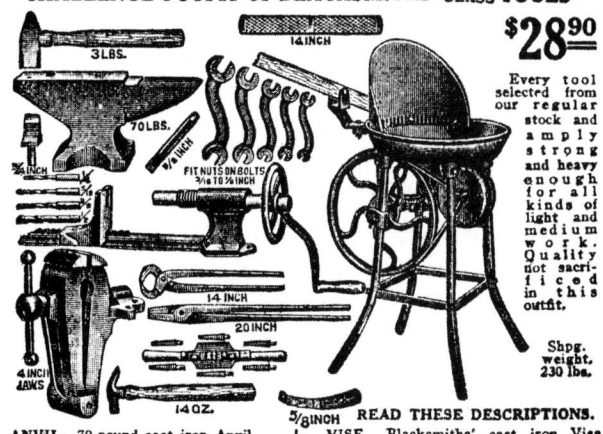

BLACKSMITHS' HIGH GRADE POST DRILLS

Blacksmiths' Acme Ball Bearing Two-Speed Self Feed Third Gear Post Drill.

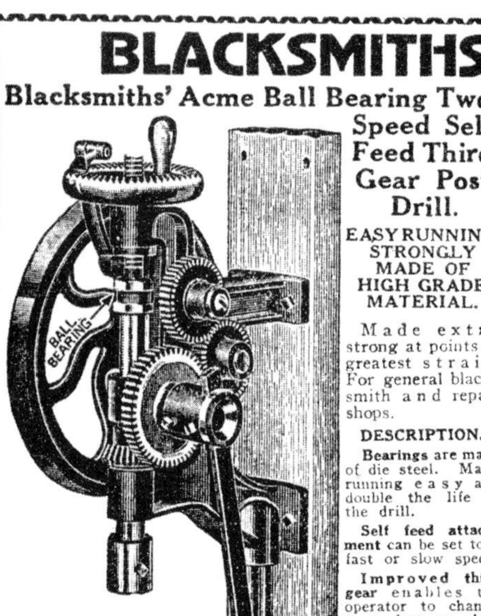

EASY RUNNING. STRONGLY MADE OF HIGH GRADE MATERIAL.

Made extra strong at points of greatest strain. For general blacksmith and repair shops.

DESCRIPTION.

Bearings are made of die steel. Make running easy and double the life of the drill.

Self feed attachment can be set to a fast or slow speed.

Improved third gear enables the operator to change to a fast or slow speed by simply changing the crank from one shaft to another on the same side of the machine.

Drill table is extra heavy, can be raised or lowered 10½ inches, swung around or removed entirely.

Shaft and Spindle are made of high grade steel. Drills up to 1¼-inch holes and to center of 14⅝-inch circle. Has up and down run of 3¼ inches and is bored for ½-inch round shank drills. Shpg. wt., 124 lbs. Not mailable.

$13.90

99L5081—With lag screws for fastening to post **$13.90**

Acme Two-Speed Self Feed Post Drill.

$8.95

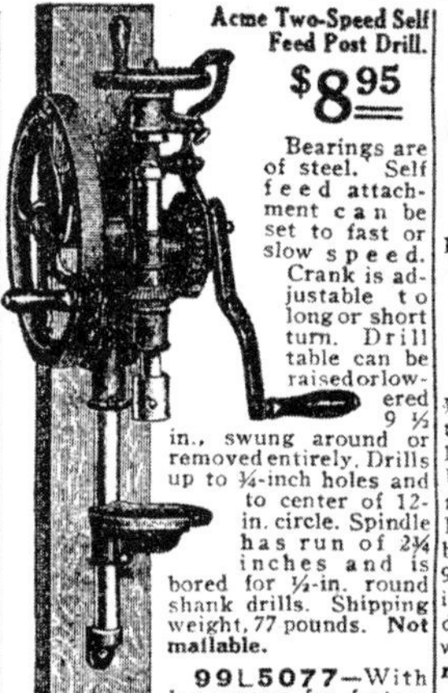

Bearings are of steel. Self feed attachment can be set to fast or slow speed. Crank is adjustable to long or short turn. Drill table can be raised or lowered 9½ in., swung around or removed entirely. Drills up to ¾-inch holes and to center of 12-in. circle. Spindle has run of 2¾ inches and is bored for ½-in. round shank drills. Shipping weight, 77 pounds. **Not mailable.**

99L5077—With lag screws for fastening to post......**$8.95**

Our Challenge Post Drill.

$7.90

For Farm, Shop or Garage Use.

Drills up to ¾-in. hole and to center of 12-inch circle. Up and down run of 2¼ in. Drill table can be raised or lowered 9½ inches. Takes ½-inch round shank drills. Shipping weight, 85 pounds. Not **mailable.**

99L5073 ......**$7.90**

Acme Special Drill

For Hand or Power.

$14.80

Belt and Hand Power.

Has both flywheel and crank. Spindle has up and down run of 3 inches. Table has up and down run of 10¾ inches. Drills up to 1¼-inch holes and to center of 15-inch circle. 8x2⅛-inch tight and loose pulleys. Takes ½-inch round shank drills. Should be run at about 200 revolutions per minute. Shipping weight, 126 pounds. **Not mailable.**

99L5083—With lag screws for fastening to post..............................**$14.80**

99L5082—Same as above, but without flywheel. For power use only. Shipping weight, 109 pounds.

With lag screws for fastening to post **$13.80**

"Garage Special" Power Drill.

$29.00

This drill is made extra heavy throughout. The materials which are used for its manufacture are the very best for the different parts.

Bearings are extra long, insuring easy running and long life. Steel ball thrust bearings at the end of the spindle. Spindle is 1⅛-inch steel, bored for standard ½-inch round shank drills. Feed adjustment, 3¼ in.

Table is adjustable up and down on a cold rolled steel column. Distance from center of drill to column, 7¼ inches. Vertical adjustment of table, 17 inches. Size of table, 7x8 inches.

Main frame of drill is fastened to a pipe column. Distance from floor to center of pulley shaft, 5 feet. Distance from lower base to end of spindle, 4 feet. Base, or floor plate, 13½x17 inches.

Furnished with 10x2-inch tight and loose pulleys; also a hand lever—when used without power.

A substantially made machine, suitable for any kind of drilling within its capacity.

Automatic feed attachment is very simple. Drill can also be used with hand feed if desired.

Drill weighs 200 pounds. Weight, crated for shipment, 225 pounds.

99L8280⅓—**"Garage Special" Power Drill. Shipped direct from factory near CHICAGO** **$29.00**

Goodell-Pratt Drill Chucks.

$2.30

⅜-Inch ½in. round shank.

9L5651 Holds straight shank drills up to ⅜-inch. Weight, 10 oz. **$1.70**

9L5652 — Holds straight shank drills up to ½-inch. Wt., 1 pound **$2.30**

Drill Chuck.

½-In. Round Shank.

70c

Has square socket, takes square shank bit stock drills. Weight, 6 ounces.

9L5654 ....**70c**

LATHE ACCESSORIES.

Shipped Direct From Factory Near Chicago.

INDEPENDENT LATHE CHUCK.

With Four Independent Reversible Jaws.

	Rated Size of Chuck	No.	Will Hold About	
99L88233⅓	5 in.	300	7 in.	**$21.00**
99L88234⅓	6 in.	301	7½ in.	**22.00**
99L88235⅓	7½ in.	302	8¾ in.	**25.00**
99L88236⅓	8 in.	302½	9½ in.	**26.00**

For fitting Independent Chuck to lathe before it leaves factory, using Machined Chuck Back furnished with regular equipment, extra...............................**$1.00**

"STANDARD" DRILL CHUCK.

Made to fit taper arbor, which will fit both head and tail spindle of lathe.

	Capacity, Inches	Diameter, Inches	
99L88268⅓	0 to ¼	1⅜	**$ 6.00**
99L88269⅓	0 to ⅜	1¹¹⁄₁₆	**6.50**
99L88270⅓	0 to ½	2³⁄₁₆	**7.00**
99L88271⅓	0 to ¾	2⅞	**8.00**
99L88272⅓	0 to 1	3⁷⁄₁₆	**10.00**

For fitting Drill Chuck to lathe, including arbor, extra **$2.00**

TURNING TOOL.

99L88254⅓—No. 00-S. Size of shank, ⁵⁄₁₆x¼x4½ inches. Size of cutter, ³⁄₁₆ inch square. Complete **$1.80**

CUTTING-OFF TOOL.

99L88255⅓—No. 29-R. Size of shank, ⁵⁄₁₆x¾ inch. Size of blades, ⁵⁄₆₄x½ inch. Complete...............................**$1.90**

BORING TOOL.

Each set consists of Holder and Bar, with straight and 45-Degree End Caps, two High Speed Cutters (ground for boring) and a Double End Wrench.

99L88256⅓—No. 00-B. Size of shank, ⁵⁄₁₆x¾ in. Size of bar, ½ in. dia. Size of cutter, ³⁄₁₆ in. square. Complete **$3.25**

LATHE DOGS.

99L88260⅓—Size, ¼ in.............**40c**
99L88261⅓—Size, ½ in.............**50c**
99L88262⅓—Size, ¾ in.............**60c**
99L88263⅓—Size, 1 in.............**70c**
99L88264⅓—Size, 1¼ in............**80c**
99L88265⅓—Size, 1½ in............**95c**

Our "Garage Special" Produces Profits

Only to Be Compared in Value With Lathes Sold at $200.00 or More.

$125.00

As Illustrated.

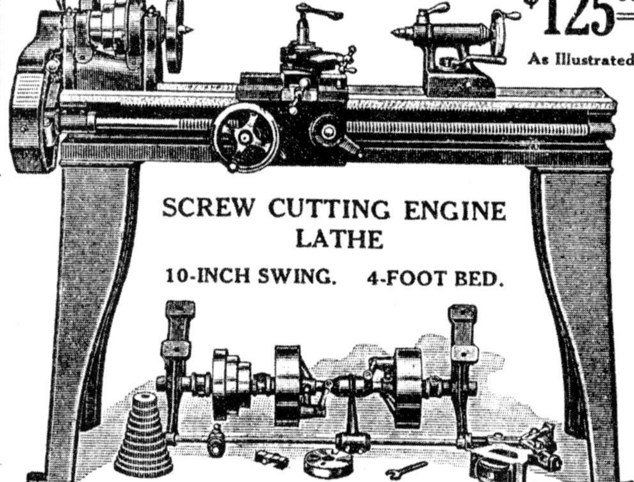

SCREW CUTTING ENGINE LATHE
10-INCH SWING. 4-FOOT BED.

Designed and manufactured exclusively for us to meet the demand for a sturdy, practical, accurate and economical lathe capable of handling the thousand and one metal working jobs that come to garages, machine and repair shops.

A GUARANTEED QUALITY LATHE AT A REMARKABLE PRICE.

A screw cutting engine lathe adaptable for the garage, machine and repair shop, for electrical work, or any place where fine, accurate machine work is required. The headstock is equipped with an improved reverse, the spindle cone has 3-step for 1-inch belt, spindle has ¾-inch hole and the centers are No. 2 Morse Taper. The bearings are the best phosphor bronze and are adjustable for wear.

The tailstock is offset to allow compound rest to swivel parallel with the bed and is provided with set over for turning taper.

The carriage is fitted with a graduated compound rest which may be set on any angle for turning or boring. The feed of the carriage is operated by clamping the split nuts on the lead screw.

The lathe will cut threads 4 to 40, right or left, including 11½-inch pipe thread, and by compounding gears many other threads may be cut.

The equipment as shown in the illustration is included in the price of the lathe and consists of large and small face plates, compound rest, two steel centers, center rest, change gears for screw cutting, chuck back fitted to spindle nose, adjustable stop for screw cutting and double friction countershaft.

The lathe swings 10¼ inches over the bed, length of bed is 4 feet, takes between centers 29½ inches and swings over the carriage 7¾ inches. Tool post takes ⅜x⅞-inch tool, countershaft speed is 290 revolutions per minute. Weight, crated ready for shipment, 500 pounds.

Shipped direct from factory near CHICAGO.

99L8275⅓—"Garage Special," complete as illustrated............ **$125.00**

99L8276⅓—Bench Lathe, same as above, with short legs. Shipping weight, 450 pounds.................. **117.00**

"Garage Special" Draw Cut Power Saw.

$14.75

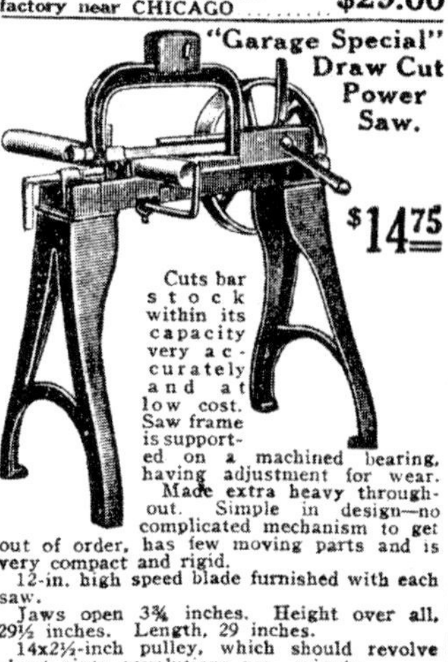

Cuts bar stock within its capacity very accurately and at low cost. Saw frame is supported on a machined bearing, having adjustment for wear. Made extra heavy throughout. Simple in design—no complicated mechanism to get out of order, has few moving parts and is very compact and rigid.

12-in. high speed blade furnished with each saw.

Jaws open 3¾ inches. Height over all, 29½ inches. Length, 29 inches.

14x2½-inch pulley, which should revolve about sixty revolutions per minute.

Weight, 110 pounds net. Weight, crated for shipment, about 125 pounds.

99L8282⅓ — **"Garage Special" Draw Cut Power Saw.**

Shipped direct from factory near CHICAGO **$14.75**

"GARAGE SPECIAL" GRINDER.

$29.50

Manufactured especially for use in garages and repair shops. The combination of disc and wheel makes it especially desirable for this class of work. Sandpaper for woodwork or emery paper for metal work can be clamped on the disc.

Rests for disc and wheel are adjustable.

Has extra large and long bearings.

Spindle, 1½ inches in diameter; disc, 12 in. in diameter; pulley measures 4 in. in diam. by 2¾-in. face. Actual wt., 245 lbs. Wt., crated for shipment, 270 pounds.

99L8281⅓ **"Garage Special" Grinder. Shipped direct from factory near CHICAGO** **$29.50**

All weights and measurements given on this page are approximate and may vary a trifle.

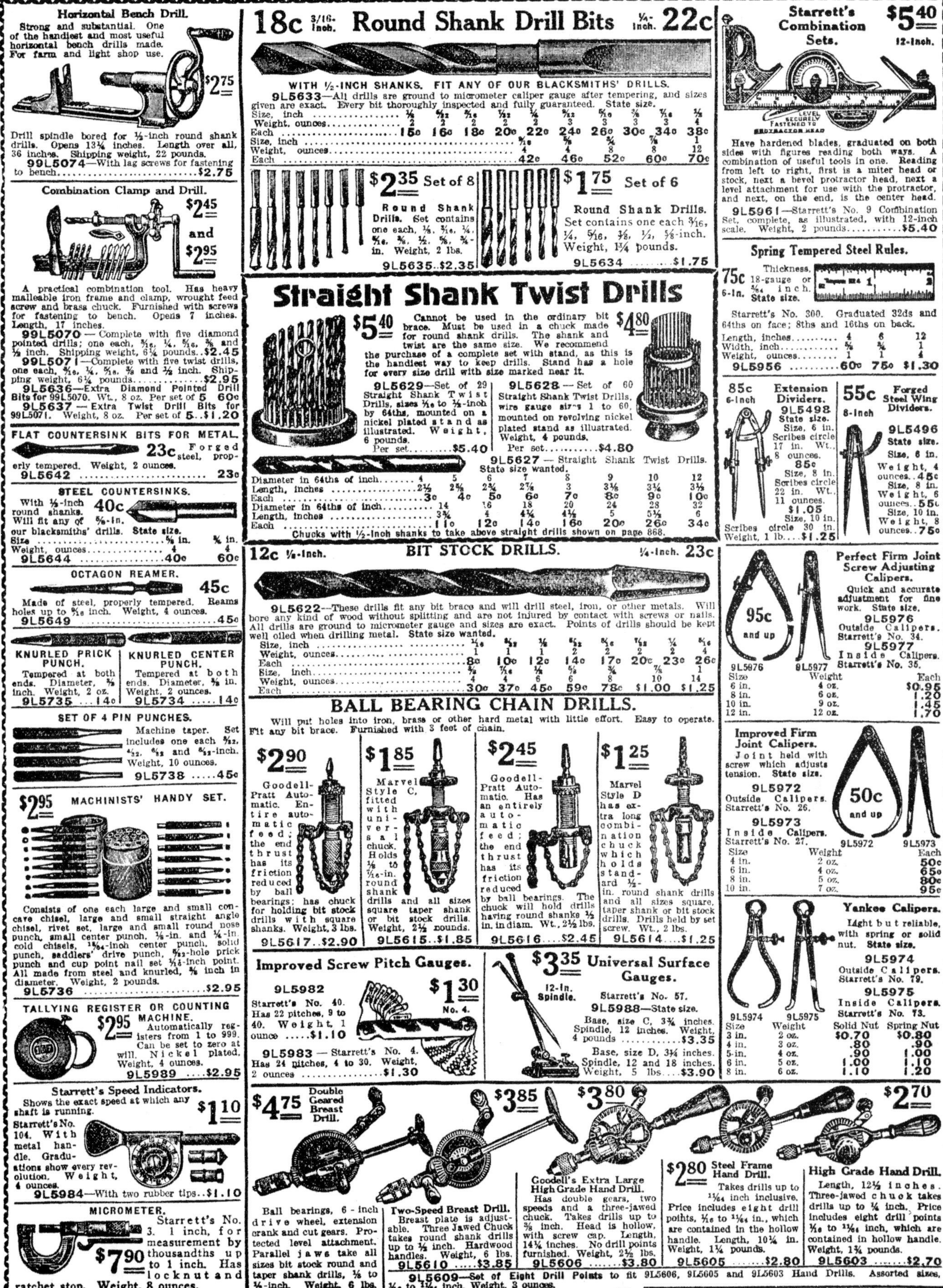

Horizontal Bench Drill. Strong and substantial. One of the handiest and most useful horizontal bench drills made. For farm and light shop use. $2.75

Drill spindle bored for ½-inch round shank drills. Opens 13¾ inches. Length over all, 36 inches. Shipping weight, 22 pounds.
99L5074—With lag screws for fastening to bench $2.75

Combination Clamp and Drill. $2.45 and $2.95

A practical combination tool. Has heavy malleable iron frame and clamp, wrought feed screw and brass chuck. Furnished with screws for fastening to bench. Opens 7 inches. Length, 17 inches.
99L5070 — Complete with five diamond pointed drills; one each, 3/16, ¼, 5/16, ⅜ and ½ inch. Shipping weight, 6¼ pounds.. $2.45
99L5071—Complete with five twist drills, one each, 3/16, ¼, 5/16, ⅜ and ½ inch. Shipping weight, 6¼ pounds $2.95
9L5636—Extra Diamond Pointed Drill Bits for 99L5070. Wt., 8 oz. Per set of 5 60c
9L5637 — Extra Twist Drill Bits for 99L5071. Weight, 8 oz. Per set of 5..$1.20

FLAT COUNTERSINK BITS FOR METAL. 23c Forged steel, properly tempered. Weight, 2 ounces.
9L5642 .... 23c

STEEL COUNTERSINKS. 40c ⅝-in. With ½-inch round shanks. Will fit any of our blacksmiths' drills. State size.

Size	⅝ in.	¾ in.
Weight, ounces	4	4
9L5644	40c	60c

OCTAGON REAMER. 45c Made of steel, properly tempered. Reams holes up to 9/16 inch. Weight, 4 ounces.
9L5649 .... 45c

KNURLED PRICK PUNCH. Tempered at both ends. Diameter, ⅜ inch. Weight, 2 oz.
9L5735 ...14c

KNURLED CENTER PUNCH. Tempered at both ends. Diameter, ⅜ in. Weight, 2 ounces.
9L5734 .....14c

SET OF 4 PIN PUNCHES. Machine taper. Set includes one each 3/32, 4/32, 6/32 and 8/32-inch. Weight, 10 ounces.
9L5738 .....45c

$2.95 MACHINISTS' HANDY SET.

Consists of one each large and small concave chisel, large and small straight angle chisel, rivet set, large and small round nose punch, small center punch, ⅛-in. and ¼-in. cold chisels, 15/64-inch center punch, solid punch, saddlers' drive punch, 5/32-hole prick punch and cup point nail set 1/16-inch point. All made from steel and knurled, ⅜ inch in diameter. Weight, 2 pounds.
9L5736 .... $2.95

TALLYING REGISTER OR COUNTING MACHINE. $2.95 Automatically registers from 1 to 999. Can be set to zero at will. Nickel plated. Weight, 4 ounces.
9L5989 ....$2.95

Starrett's Speed Indicators. Shows the exact speed at which any shaft is running. $1.10
Starrett's No. 104. With metal handle. Graduations show every revolution. Weight, 4 ounces.
9L5984—With two rubber tips..$1.10

MICROMETER. Starrett's No. 3. 1 inch, for measurement by thousandths up to 1 inch. Has locknut and ratchet stop. Weight, 8 ounces. $7.90
9L5980 .... $7.90

18c 3/16-inch Round Shank Drill Bits ¼-inch 22c

WITH ½-INCH SHANKS. FIT ANY OF OUR BLACKSMITHS' DRILLS.
9L5633—All drills are ground to micrometer caliper gauge after tempering, and sizes given are exact. Every bit thoroughly inspected and fully guaranteed. State size.

Size, inch	⅛	5/32	3/16	7/32	¼	9/32	5/16	⅜	7/16	½
Weight, ounces	2	2	2	2	2	3	3	3	3	4
Each	15c	16c	18c	20c	22c	24c	26c	30c	34c	38c

Size, inch	9/16	⅝	¾	⅞	1
Weight, ounces	4	4	8	8	12
Each	42c	46c	52c	60c	70c

$2.35 Set of 8 Round Shank Drills. Set contains one each, ⅛, 3/16, ¼, 5/16, ⅜, ½, ⅝, ¾-in. Weight, 2 lbs.
9L5635..$2.35

$1.75 Set of 6 Round Shank Drills. Set contains one each 3/16, ¼, 5/16, ⅜, ½, ⅝-inch. Weight, 1¼ pounds.
9L5634 ....$1.75

Straight Shank Twist Drills

$5.40 $4.80 Cannot be used in the ordinary bit brace. Must be used in a chuck made for round shank drills. The shank and twist are the same size. We recommend the purchase of a complete set with stand, as this is the handiest way to keep drills. Stand has a hole for every size drill with size marked near it.

9L5629—Set of 29 Straight Shank Twist Drills, sizes 1/16 to ½-inch by 64ths, mounted on a nickel plated stand as illustrated. Weight, 6 pounds. Per set.........$5.40

9L5628 — Set of 60 Straight Shank Twist Drills, wire gauge sizes 1 to 60, mounted on revolving nickel plated stand as illustrated. Weight, 4 pounds. Per set..........$4.80

9L5627 — Straight Shank Twist Drills. State size wanted.

Diameter in 64ths of inch	4	5	6	7	8	9	10	12
Length, inches	2½	2⅝	2¾	2⅞	3	3⅛	3¼	3½
Each	3c	4c	5c	6c	7c	8c	9c	10c

Diameter in 64ths of inch	14	16	18	20	24	28	32
Length, inches	3¾	4	4¼	4½	5	5½	6
Each	11c	12c	14c	16c	20c	26c	34c

Chucks with ½-inch shanks to take above straight drills shown on page 868.

12c ⅛-inch. BIT STOCK DRILLS. ¼-inch. 23c

9L5622—These drills fit any bit brace and will drill steel, iron, or other metals. Will bore any kind of wood without splitting and are not injured by contact with screws or nails. All drills are ground to micrometer gauge and sizes are exact. Points of drills should be kept well oiled when drilling metal. State size wanted.

Size, inch	1/16	3/32	⅛	5/32	3/16	7/32	¼	5/16
Weight, ounces	1	1	1	2	2	2	2	4
Each	8c	10c	12c	14c	17c	20c	23c	26c

Size, inch	⅜	7/16	½	⅝	¾	⅞	1
Weight, ounces	4	4	6	6	8	10	14
Each	30c	37c	45c	59c	78c	$1.00	$1.25

BALL BEARING CHAIN DRILLS.

Will put holes into iron, brass or other hard metal with little effort. Easy to operate. Fit any bit brace. Furnished with 3 feet of chain.

$2.90 Goodell-Pratt Automatic. Entire automatic feed; the end thrust has its friction reduced by ball bearings; has chuck for holding bit stock drills with square shanks. Weight, 3 lbs.
9L5617..$2.90

$1.85 Marvel Style C, fitted with universal chuck. Holds ⅛ to 7/16-in. round shank drills and all sizes square taper shank or bit stock drills. Weight, 2½ pounds.
9L5615..$1.85

$2.45 Goodell-Pratt Automatic. Has an entirely automatic feed; the end thrust has its friction reduced by ball bearings. The chuck will hold drills having round shanks ½ in. in diam. Wt., 2½ lbs.
9L5616....$2.45

$1.25 Marvel Style D has extra long combination chuck which holds standard ½-in. round shank drills and all sizes square, taper shank or bit stock drills. Drills held by set screw. Wt., 2 lbs.
9L5614....$1.25

Improved Screw Pitch Gauges. $1.30 No. 4.
9L5982 Starrett's No. 40. Has 22 pitches, 9 to 40. Weight, 1 ounce $1.10
9L5983 — Starrett's No. 4. Has 24 pitches, 4 to 30. Weight, 2 ounces $1.30

$3.35 Universal Surface Gauges. 12-In. Spindle. Starrett's No. 57.
9L5988—State size. Base, size C, 3¾ inches. Spindle, 12 inches. Weight, 4 pounds $3.35
Base, size D, 3¼ inches. Spindle, 12 and 18 inches. Weight, 5 lbs....$3.90

$4.75 Double Geared Breast Drill. Ball bearings, 6-inch drive wheel, extension crank and cut gears. Protected level attachment. Parallel jaws take all sizes bit stock round and taper shank drills, ⅛ to ½-inch. Weight, 6 lbs.
9L5612......$4.75

$3.85 Two-Speed Breast Drill. Breast plate is adjustable. Three Jawed Chuck takes round shank drills up to ½ inch. Hardwood handles. Weight, 6 lbs.
9L5610 .....$3.85

$3.80 Goodell's Extra Large High Grade Hand Drill. Has double gears, two speeds and a three-jawed chuck. Takes drills up to ⅜ inch. Head is hollow, with screw cap. Length, 14¼ inches. No drill points furnished. Weight, 2½ lbs.
9L5606 ....$3.80

$2.80 Steel Frame Hand Drill. Takes drills up to 11/64 inch inclusive. Price includes eight drill points, 1/16 to 11/64 in., which are contained in the hollow handle. Length, 10¼ in. Weight, 1¼ pounds.
9L5605 ....$2.80

$2.70 High Grade Hand Drill. Length, 12½ inches. Three-jawed chuck takes drills up to ¼ inch. Price includes eight drill points 1/16 to 11/64 inch, which are contained in hollow handle. Weight, 1¾ pounds.
9L5603 ....$2.70

9L5609—Set of Eight Drill Points to fit 9L5606, 9L5605 and 9L5603 Hand Drills. Assorted sizes, 1/16 to 11/64 inch. Weight, 3 ounces. Per set of 8.......... 25c

Starrett's Combination Sets. $5.40 12-Inch.

Have hardened blades, graduated on both sides with figures reading both ways. A combination of useful tools in one. Reading from left to right, first is a miter head or stock, next a bevel protractor head, next a level attachment for use with the protractor, and next, on the end, is the center head.
9L5961—Starrett's No. 9 Combination Set, complete, as illustrated, with 12-inch scale. Weight, 2 pounds..........$5.40

Spring Tempered Steel Rules. 75c 6-In. Thickness, 18-gauge or 3/64 inch. State size.

Starrett's No. 300. Graduated 32ds and 64ths on face; 8ths and 16ths on back.

Length, inches	4	6	12
Width, inch	⅝	¾	1
Weight, ounces	1	1	4
9L5956	60c	75c	$1.30

85c 6-Inch Extension Dividers. 9L5498 State size. Size, 6 in. Scribes circle 17 in. Wt., 8 ounces. 85c
Size, 8 in. Scribes circle 22 in. Wt., 11 ounces. $1.05
Size, 10 in. Scribes circle 30 in. Weight, 1 lb....$1.25

55c 8-Inch Forged Steel Wing Dividers. 9L5496 State size. Size, 6 in. Weight, 4 ounces..45c
Size, 8 in. Weight, 6 ounces..55c
Size, 10 in. Weight, 8 ounces..75c

Perfect Firm Joint Screw Adjusting Calipers. 95c and up. Quick and accurate adjustment for fine work. State size.
9L5976 Outside Calipers. Starrett's No. 34.
9L5977 Inside Calipers. Starrett's No. 35.

Size	Weight	Each
6 in.	4 oz.	$0.95
8 in.	6 oz.	1.20
10 in.	9 oz.	1.45
12 in.	12 oz.	1.70

Improved Firm Joint Calipers. 50c and up. Joint held with screw which adjusts tension. State size.
9L5972 Outside Calipers. Starrett's No. 26.
9L5973 Inside Calipers. Starrett's No. 27.

Size	Weight	Each
4 in.	2 oz.	50c
6 in.	4 oz.	65c
8 in.	5 oz.	80c
10 in.	7 oz.	95c

Yankee Calipers. Light but reliable, with spring or solid nut. State size.
9L5974 Outside Calipers. Starrett's No. 79.
9L5975 Inside Calipers. Starrett's No. 73.

Size	Weight	Solid Nut	Spring Nut
3 in.	2 oz.	$0.70	$0.80
4 in.	3 oz.	.80	.90
5-in.	4 oz.	.90	1.00
6 in.	5 oz.	1.00	1.10
8 in.	6 oz.	1.10	1.20

All weights and measurements given on this page are approximate and may vary a trifle.

Blacksmiths' Taper Taps.

25c AND UP.

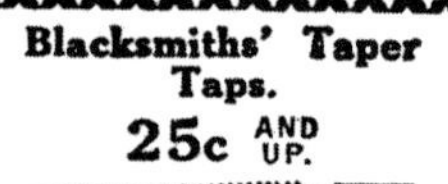

9L5679—Right hand Taper Tap. **State size and number of threads.**

Size, In.	No. Threads to Inch	Wt., Oz.	
¼	20	1	**25c**
5/16	18	1	**30c**
⅜	16	1	**35c**
7/16	14, 16	2	**40c**
½	12, 14	2	**45c**
⅝	12	2	**55c**
¾	10	6	**70c**

BLACKSMITHS' STOCKS AND DIES

Stocks made of malleable iron. Dies and taps made of die steel. Perfect fit and adjustment guaranteed. All parts of these tools are guaranteed against defective material or workmanship. Should any part prove defective we will replace it free of charge.

$5.45 As Illustrated. **$5.45** As Illustrated.

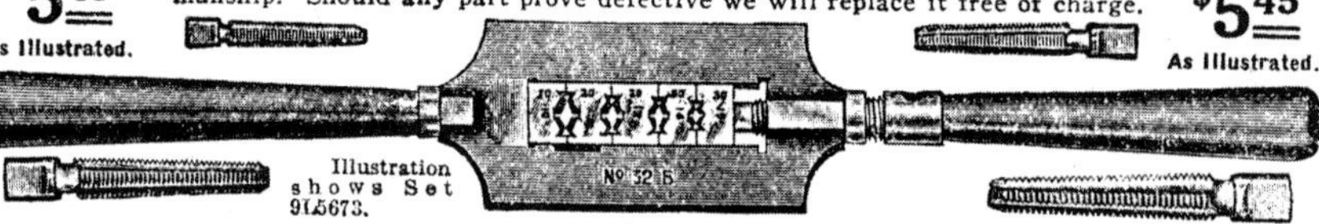

Illustration shows Set 9L5673.

Factory No. 37. 6 taps, 3 dies, 14, 18 and 22 threads to inch. Cuts ⅝ to 5/16 inch. Weight, 4 pounds. **9L5672** Per set........................**$4.95**

Factory No. 32B. 4 taps, 4 dies, 10, 12, 14 and 16 threads to inch. Cuts ¾ to ¼ inch. Weight, 6 pounds. **9L5673** Per set........................**$5.45**

Factory No. 60. 6 taps, 4 dies, 10, 12, 14 and 18 threads to inch. Cuts ¾ to 5/16 inch. Weight, 6 pounds. **9L5675** Per set........................**$6.85**

T Tap and Drill Wrench.

60c

Combination Tap and Drill Wrench. Furnished with two high carbon jaws, one for holding taps, sizes 1/16 to ⅜, and one for holding drills 1/16 to ¼. T handle is 4⅝ inches **long**. Weight, 4 ounces.
9L5686........**60c**

INVINCIBLE ADJUSTABLE DIE SCREW PLATES.

Dies are cut through on one side and a screw inserted in the opening. By turning screw, dies are opened or closed. U. S. standard thread, which fits regular carriage and machine bolts.

INVINCIBLE ADJUSTABLE DIE FULL MOUNTED SCREW PLATE. $16.75

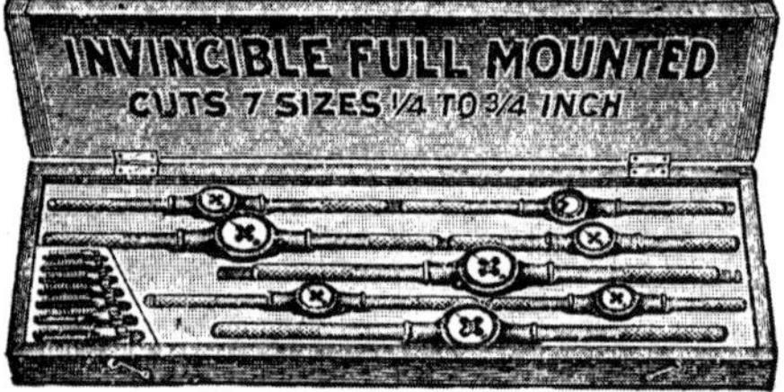

Illustration shows Set 99L5134.

U. S. Standard Thread.

99L5133—Cuts five sizes, ¼^{20}, 5/16^{18}, ⅜^{16}, 7/16^{14} and ½^{13}; complete with five taps and five dies, and one 10-inch tap wrench; stocks measure from 12 to 16 inches long. In hardwood case. Shipping weight, 21 lbs. Per set....**$11.50**

99L5134—Cuts seven sizes, ¼^{20}, 5/16^{18}, ⅜^{16}, 7/16^{14}, ½^{13}, ⅝^{11} and ¾^{10}; complete with seven taps and seven dies, and one 16-inch tap wrench; stocks measure from 12 to 24 inches long. In hardwood case. Shipping weight, 34 pounds. Per set........................**$16.75**

99L5135—Cuts nine sizes, ¼^{20}, 5/16^{18}, ⅜^{16}, 7/16^{14}, ½^{13}, ⅝^{11}, ¾^{10}, ⅞^{9} and 1^{8}; complete with nine taps and nine dies, and two tap wrenches, 12 and 20 inches; stocks measure from 12 to 27 inches long. In hardwood case. Shipping weight, 51 pounds. Per set........................**$28.50**

INVINCIBLE ADJUSTABLE DIE SCREW PLATE. $16.00

Illustration shows Set 99L5131.

U. S. Standard Thread.

99L5130—With six taps and six dies, cutting ¼^{20}, 5/16^{18}, ⅜^{16}, ½^{13}, ⅝^{11} and ¾^{10}. Furnished with two stocks, 14 and 26 inches long, and one tap wrench, 16 inches long. In hardwood case. Shipping weight, 20 lbs. Per set....**$13.00**

99L5131—With seven taps and seven dies and collets, cutting ¼^{20}, 5/16^{18}, ⅜^{16}, 7/16^{14}, ½^{13}, ⅝^{11} and ¾^{10}. Has two stocks, 14 and 26 inches long, and tap wrench, 16 inches long. In hardwood case. Shipping wt., 21 lbs Per set....**$16.00**

99L5132—With nine taps and nine dies and collets, cutting ¼^{20}, 5/16^{18}, ⅜^{16}, 7/16^{14}, ½^{13}, ⅝^{11}, ¾^{10}, ⅞^{9} and 1^{8}. Has two stocks, 14 and 26 inches long, and tap wrench, 15 in. long. In hardwood case. Shipping wt., 25 lbs. Per set..**$24.50**

9L5691—Extra Taps, Dies and Collets for Invincible Screw Plates. Price includes die and collets with tap to match.

Size, inch	¼	5/16	⅜	7/16	½
Weight	7 oz.	8 oz.	9 oz.	10 oz.	14 oz.
State size	**$1.45**	**$1.60**	**$1.90**	**$2.25**	**$2.75**
Size, inch		⅝	¾	⅞	1
Weight		1¼ lbs.	1½ lbs.	2 lbs.	2¼ lbs.
State size		**$3.00**	**$3.25**	**$3.75**	**$4.25**

INVINCIBLE ADJUSTABLE DIE SCREW PLATE. $10.25

U. S. Standard Thread.

99L5129—With five taps and five dies, cutting ¼^{20}, 5/16^{18}, ⅜^{16}, ½^{13} and ⅝^{11}. Furnished with one stock, 16 inches long, and one tap wrench, 16 inches long. In hardwood case. Shipping weight, 11 pounds. Per set........................**$10.25**

Blacksmiths' Separate Plate Stocks and Dies. $9.85

Right Hand Only.

As Illustrated.

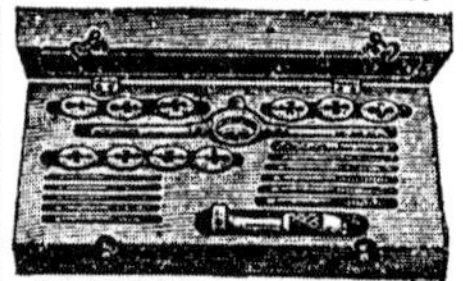

9L5667—Factory No. 51B, cuts ½ to 5/16 inch, number of threads to inch, 12, 14, 16 and 18; number of taps in set, 4; number of dies in set, 4 pairs. Weight, 3 pounds. Per set........................**$4.95**

9L5668—Factory No. 27D, cuts ¾ to 5/16 inch, number of threads to inch, 10, 11, 12, 14, 16, 18 and 20; number of taps in set, 7; number of dies in set, 7 pairs. Weight, 6 pounds. Per set........................**$9.85**

Special Right and Left Hand Stock and Dies. $4.80

Factory No. 42, cuts ½ to 5/16 inch right hand, 14 and 20 threads to inch, and ½ to 5/16 inch left hand, 14 threads to inch, 6 taps and 3 dies. Weight, 2½ pounds.
9L5677—Per set........................**$4.80**

Small Size Little Giant Screw Plates. $8.50

Round adjustable dies, 13/16 inch in diameter. These plates are furnished in varnished hardwood cases with plush lined tops. Small figures to right of sizes indicate number of threads to inch. U. S. standard thread. Weight, 1½ lbs.

9L5659—Factory No D10. Cutting screw gauge sizes Nos. 2^{56}, 3^{48}, 4^{30}, 5^{36}, 6^{32}, 8^{32}, 10^{24}, 10^{32}, 12^{24}, 14^{20} inch. 10 taps, 10 dies, stock and adjustable tap wrench. Set..**$8.50**

9L5658—Factory No. DD6. Cutting 7/64^{48}, ⅛^{40}, 5/32^{36}, 3/16^{24}, 7/32^{24}, ¼^{20} inch. 6 taps, 6 dies, stock and adjustable tap wrench. Per set........................**$6.25**

9L5657—Factory No. D5. Cutting screw gauge sizes Nos. 4^{30}, 6^{32}, 8^{32}, 10^{24}, 12^{24} inch, 5 taps, 5 dies, stock and adjustable tap wrench. Per set........**$5.80**

9L5656—Factory No. DD5. Cutting ⅛^{40}, 5/32^{36}, 3/16^{24}, 7/32^{24}, ¼^{20} inch, 5 taps and 5 dies, stock and adjustable tap wrench. Per set........................**$5.85**

Little Giant Screw Plate. $8.25

Factory No. F6. Round adjustable dies, 1 inch in diameter. Cutting ⅛^{40}, 3/16^{24}, ¼^{20}, 5/16^{18}, ⅜^{16}, 7/16^{14} inch. Stocks, 9 inches long; 6 dies, 6 taps and adjustable tap wrench. U. S. standard thread. In hardwood case. Wt., 2½ lbs.
9L5661
Per set........**$8.25**

Automobile Screw Plate. $12.75

Complete with roll. Set includes 14-inch stock, seven sizes round adjustable dies and seven plug taps, cutting ¼^{28}, 5/16^{24}, ⅜^{24}, 7/16^{20}, ½^{20}, 9/16^{18} and ⅝^{18} in. (Small figures to right of sizes indicate number of threads to the inch.) S. A. E. thread. Wt., 4 pounds.

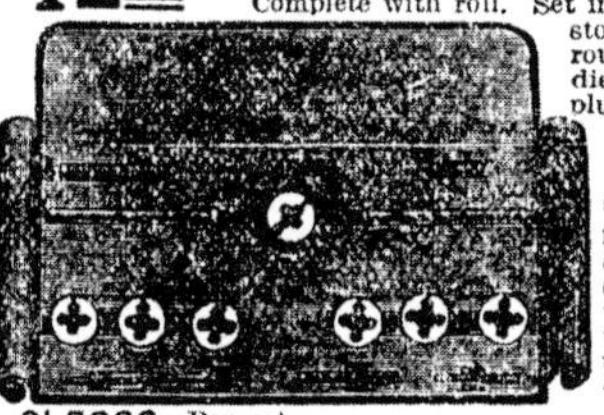

9L5666—Per set........................**$12.75**

Fulton Utility Solid Die Screw Plates. $8.75

Nothing to Adjust or Get Out of Order.

A low priced screw plate with the same actual working features of the higher priced sets. Cuts right hand U. S. standard thread. Stock has black body and polished handles.

99L5102—With 4 dies and 4 taper taps to match and one wrench plate. Cuts four sizes, ¼^{20}, 5/16^{18}, ⅜^{16} and ½^{13}. Stock, 18 inches long. Shipping wt., 4¼ lbs...**$8.75**

FULTON FULL MOUNTED DOUBLE DIE SCREW PLATES. $19.50

Illustration shows Set 99L5114.

A high grade screw plate having a stock for each die. Two-piece adjustable double dies are made of tool steel and have two sets of cutting edges, cutting equally from either face, giving twice the service of the ordinary die. Collets are hardened steel. Taper taps are of tempered tool steel. Right hand U. S. standard thread, which fits regular carriage and machine bolts. Furnished complete in hardwood case.

99L5114—With 7 taper taps and 7 dies and collets, cutting ¼^{20}, 5/16^{18}, ⅜^{16}, 7/16^{14}, ½^{12}, ⅝^{11} and ¾^{10} inch. Stocks, 11 to 24 inches long. Tap wrench, 16 inches long. Shipping weight, 27 pounds. Per set........................**$19.50**

99L5115—With 9 taper taps and 9 dies and collets, cutting ¼^{20}, 5/16^{18}, ⅜^{16}, 7/16^{14}, ½^{12}, ⅝^{11}, ¾^{10}, ⅞^{9} and 1^{8} inch. Stocks, 11 to 28 inches long. Tap wrench, 21 inches long. Shipping weight, 46 pounds. Per set........................**$30.00**

FULTON SPECIAL OPENING DIE SCREW PLATES. $18.50

Illustration shows Set 99L5107.

Patent opening die enables you to lift the tool clear of the work after cutting the thread without running back over the finished thread. A wonderful time saver. Furnished with a special patent stock and improved tap wrench. Right hand U. S. standard thread, which fits regular carriage and machine bolts. Furnished complete in hardwood case.

99L5107—With 6 taper taps and 6 dies and collets cutting ¼^{20}, 5/16^{18}, ⅜^{16}, ½^{13}, ⅝^{11} and ¾^{10} inch. Stock, 26 inches long. Tap wrench, 16 inches long. Shipping weight, 19 pounds. Per set........................**$18.50**

99L5108—With 7 taper taps and 7 dies and collets, cutting ¼^{20}, 5/16^{18}, ⅜^{16}, 7/16^{14}, ½^{13}, ⅝^{11} and ¾^{10} inch. Stock, 26 inches long. Tap wrench, 16 inches long. Shipping weight, 20 pounds. Per set........................**$19.50**

FULTON DOUBLE DIE SCREW PLATES. $10.90

Illustration shows Set 99L5103.

Double dies cut from both faces and are so milled and chamfered that they will cut close to the shoulder and will thread a very short piece. Cuts right hand U. S. standard thread, which fits regular carriage and machine bolts. Furnished complete in hardwood case.

99L5103—With 5 taper taps and 5 dies and collets, cutting ¼^{20}, 5/16^{18}, ⅜^{16}, ½^{13} and ⅝^{11} inch. Stock, 22 inches long. Tap wrench, 16 inches long. Shipping weight, 14 lbs. Per set........................**$10.90**

99L5104—With 6 taper taps and 6 dies and collets, cutting ¼^{20}, 5/16^{18}, ⅜^{16}, ½^{13}, ⅝^{11} and ¾^{10} inch. Stock, 22 inches long. Taper wrench, 16 inches long. Shipping weight, 15 pounds. Per set........................**$12.85**

Gunsmiths' or Jewelers' Screw Plate.

$2.35 Cuts seven sizes, from ⅛ to ¼ inch. Cuts threads exact sizes. Weight, 8 ounces.
9L5663—Per set........................**$2.35**

All weights and measurements given on this page are approximate and may vary a trifle.

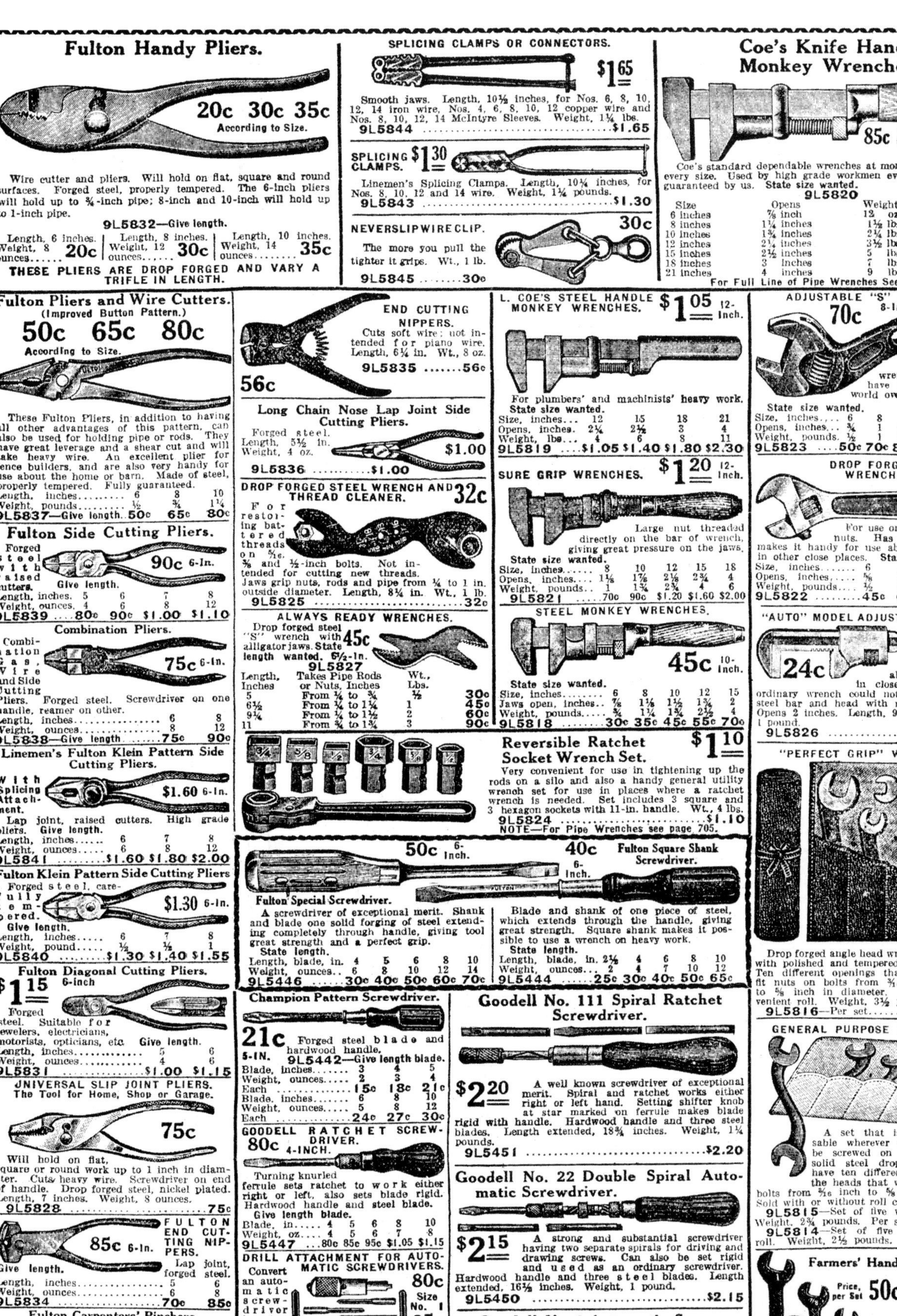

Fulton Handy Pliers.

20c 30c 35c According to Size.

Wire cutter and pliers. Will hold on flat, square and round surfaces. Forged steel, properly tempered. The 6-inch pliers will hold up to ¾-inch pipe; 8-inch and 10-inch will hold up to 1-inch pipe.

9L5832—Give length.

Length, 6 inches. Weight, 8 ounces...... **20c** | Length, 8 inches. Weight, 12 ounces...... **30c** | Length, 10 inches. Weight, 14 ounces...... **35c**

THESE PLIERS ARE DROP FORGED AND VARY A TRIFLE IN LENGTH.

SPLICING CLAMPS OR CONNECTORS.

$1.65

Smooth jaws. Length, 10½ inches, for Nos. 6, 8, 10, 12, 14 iron wire, Nos. 4, 6, 8, 10, 12 copper wire and Nos. 8, 10, 12, 14 McIntyre Sleeves. Weight, 1¼ lbs.

9L5844 ....$1.65

SPLICING CLAMPS. **$1.30**

Linemen's Splicing Clamps. Length, 10¼ inches, for Nos. 8, 10, 12 and 14 wire. Weight, 1¼ pounds.

9L5843 ....$1.30

NEVERSLIPWIRECLIP. **30c**

The more you pull the tighter it grips. Wt., 1 lb.

9L5845 ....30c

Coe's Knife Handle Monkey Wrenches.

85c 10-Inch. **$1.05** 12-Inch.

Coe's standard dependable wrenches at money saving prices on every size. Used by high grade workmen everywhere and fully guaranteed by us. State size wanted.

9L5820

Size	Opens	Weight	
6 inches	⅞ inch	12 oz.	$0.55
8 inches	1¼ inches	1½ lbs.	.70
10 inches	1¾ inches	2¼ lbs.	.85
12 inches	2¼ inches	3½ lbs.	1.05
15 inches	2½ inches	5 lbs.	1.40
18 inches	3 inches	7 lbs.	1.80
21 inches	4 inches	9 lbs.	2.30

For Full Line of Pipe Wrenches See Page 705.

Fulton Pliers and Wire Cutters. (Improved Button Pattern.)

50c 65c 80c According to Size.

These Fulton Pliers, in addition to having all other advantages of this pattern, can also be used for holding pipe or rods. They have great leverage and a shear cut and will take heavy wire. An excellent plier for fence builders, and are also very handy for use about the home or barn. Made of steel, properly tempered. Fully guaranteed.

Length, inches	6	8	10
Weight, pounds	½	¾	1¼
9L5837—Give length	50c	65c	80c

Fulton Side Cutting Pliers.

90c 6-In.

Forged steel with raised cutters. Give length.

Length, inches	5	6	7	8
Weight, ounces	4	6	8	12
9L5839	80c	90c	$1.00	$1.10

Combination Pliers.

75c 6-In.

Combination Gas, Wire and Side Cutting Pliers. Forged steel. Screwdriver on one handle, reamer on other.

Length, inches	6	8
Weight, ounces	8	12
9L5838—Give length	75c	90c

Linemen's Fulton Klein Pattern Side Cutting Pliers.

$1.60 6-In.

With Splicing Attachment. Lap joint, raised cutters. High grade pliers. Give length.

Length, inches	6	7	8
Weight, ounces	6	8	12
9L5841	$1.60	$1.80	$2.00

Fulton Klein Pattern Side Cutting Pliers

$1.30 6-In.

Forged steel, carefully tempered. Give length.

Length, inches	6	7	8
Weight, pound	½	½	1
9L5840	$1.30	$1.40	$1.55

Fulton Diagonal Cutting Pliers.

$1.15 6-inch

Forged steel. Suitable for jewelers, electricians, motorists, opticians, etc. Give length.

Length, inches	5	6
Weight, ounces	4	6
9L5831	$1.00	$1.15

UNIVERSAL SLIP JOINT PLIERS.

The Tool for Home, Shop or Garage.

75c

Will hold on flat, square or round work up to 1 inch in diameter. Cuts heavy wire. Screwdriver on end of handle. Drop forged steel, nickel plated. Length, 7 inches. Weight, 8 ounces.

9L5828 ....75c

FULTON END CUTTING NIPPERS.

85c 6-In.

Lap joint, forged steel. Give length.

Length, inches	5	6
Weight, ounces	6	8
9L5834	70c	85c

Fulton Carpenters' Pinchers.

45c 8-In.

Claw on one handle, screwdriver on other. Forged steel.

Length, inches	6	8	10	12
Weight, pounds	½	¾	1¼	1½
9L5833—Givelength	35c	45c	55c	65c

END CUTTING NIPPERS.

56c

Cuts soft wire; not intended for piano wire. Length, 6¼ in. Wt., 8 oz.

9L5835 ....56c

Long Chain Nose Lap Joint Side Cutting Pliers.

$1.00

Forged steel. Length, 5½ in. Weight, 4 oz.

9L5836 ....$1.00

DROP FORGED STEEL WRENCH AND THREAD CLEANER. **32c**

For restoring battered threads on ⁵⁄₁₆, ⅜ and ½-inch bolts. Not intended for cutting new threads. Jaws grip nuts, rods and pipe from ¼ to 1 in. outside diameter. Length, 8¼ in. Wt., 1 lb.

9L5825 ....32c

ALWAYS READY WRENCHES.

Drop forged steel "S" wrench with alligator jaws. State length wanted. 6½-In. **45c**

9L5827

Length, Inches	Takes Pipe Rods or Nuts, Inches	Wt., Lbs.	
5	From ¼ to ¾	½	30c
6½	From ¼ to 1¼	1	45c
9¼	From ¾ to 1½	2	60c
11	From ¾ to 1¾	3	90c

L. COE'S STEEL HANDLE MONKEY WRENCHES. **$1.05** 12-Inch.

For plumbers' and machinists' heavy work. State size wanted.

Size, inches	12	15	18	21
Opens, inches	2¼	2½	3	4
Weight, lbs	4	6	8	11
9L5819	$1.05	$1.40	$1.80	$2.30

SURE GRIP WRENCHES. **$1.20** 12-Inch.

Large nut threaded directly on the bar of wrench, giving great pressure on the jaws. State size wanted.

Size, inches	8	10	12	15	18
Opens, inches	1⅛	1⅞	2⅛	2¾	4
Weight, pounds	1	1¾	2¾	4	6
9L5821	70c	90c	$1.20	$1.60	$2.00

STEEL MONKEY WRENCHES.

45c 10-Inch.

State size wanted.

Size, inches	6	8	10	12	15
Jaws open, inches	⅞	1⅛	1½	1¾	2
Weight, pounds	¾	1¼	1¾	2½	4
9L5818	30c	35c	45c	55c	70c

ADJUSTABLE "S" WRENCHES.

70c 8-Inch.

Malleable iron wrenches of this style have been standard the world over for many years. State size wanted.

Size, inches	6	8	10	12	14
Opens, inches	¾	1	1⅜	1⅝	2
Weight, pounds	½	1	1¾	3	5
9L5823	50c	70c	85c	$1.20	$1.60

DROP FORGED WRENCH. **60c** 8-Inch.

For use on hexagon or square nuts. Has angle head that makes it handy for use about machinery and in other close places. State size wanted.

Size, inches	6	8	10	12
Opens, inches	⅝	⅞	1	1⅜
Weight, pounds	½	¾	1¼	1¾
9L5822	45c	60c	75c	90c

"AUTO" MODEL ADJUSTABLE WRENCH.

24c

A good, serviceable wrench for use in close places where an ordinary wrench could not be used. Forged steel bar and head with malleable iron jaw. Opens 2 inches. Length, 9¼ inches. Weight, 1 pound.

9L5826 ....24c

Reversible Ratchet Socket Wrench Set. **$1.10**

Very convenient for use in tightening up the rods on a silo and also a handy general utility wrench set for use in places where a ratchet wrench is needed. Set includes 3 square and 3 hexagon sockets with 11-in. handle. Wt., 4 lbs.

9L5824 ....$1.10

NOTE—For Pipe Wrenches see page 705.

"PERFECT GRIP" WRENCH SET.

Drop forged angle head wrenches with polished and tempered jaws. Ten different openings that will fit nuts on bolts from ³⁄₁₆ inch to ⅝ inch in diameter. Each set in convenient roll. Weight, 3½ pounds. **$1.20**

9L5816—Per set $1.20

Fulton Special Screwdriver. **50c** 6-Inch.

A screwdriver of exceptional merit. Shank and blade one solid forging of steel extending completely through handle, giving tool great strength and a perfect grip. State length.

Length, blade, in.	4	5	6	8	10
Weight, ounces	6	8	10	12	14
9L5446	30c	40c	50c	60c	70c

Fulton Square Shank Screwdriver. **40c** 6-Inch.

Blade and shank of one piece of steel, which extends through the handle, giving great strength. Square shank makes it possible to use a wrench on heavy work. State length.

Length, blade, in.	2¼	4	6	8	10
Weight, ounces	2	4	7	10	12
9L5444	25c	30c	40c	50c	65c

Champion Pattern Screwdriver.

21c 5-IN. Forged steel blade and hardwood handle.

9L5442—Give length blade.

Blade, inches	3	4	5
Weight, ounces	2	3	4
Each	15c	18c	21c
Blade, inches	6	8	10
Weight, ounces	5	8	12
Each	24c	27c	30c

GOODELL RATCHET SCREWDRIVER.

80c 4-INCH.

Turning knurled ferrule sets ratchet to work either right or left, also sets blade rigid. Hardwood handle and steel blade. Give length blade.

Blade, in.	4	5	6	8	10
Weight, oz.	4	5	6	7	8
9L5447	80c	85c	95c	$1.05	$1.15

DRILL ATTACHMENT FOR AUTOMATIC SCREWDRIVERS.

80c Size No. 1 **85c** Size No. 2

Convert an automatic screwdriver into an automatic drill.

9L5453—Size 1. Fits screwdriver 9L5449. Weight, 4 ounces. Chuck and eight drill points, as illustrated 80c

9L5454—Size 2. Fits screwdrivers 9L5451 and 9L5450. Weight, 4 ounces. Chuck and eight drill points, as illustrated 85c

Goodell No. 111 Spiral Ratchet Screwdriver.

$2.20

A well known screwdriver of exceptional merit. Spiral and ratchet works either right or left hand. Setting shifter knob at star marked on ferrule makes blade rigid with handle. Hardwood handle and three steel blades. Length extended, 18¾ inches. Weight, 1¼ pounds.

9L5451 ....$2.20

Goodell No. 22 Double Spiral Automatic Screwdriver.

$2.15

A strong and substantial screwdriver having two separate spirals for driving and drawing screws. Can also be set rigid and used as an ordinary screwdriver. Hardwood handle and three steel blades. Length extended, 16½ inches. Weight, 1 pound.

9L5450 ....$2.15

Goodell No. 1 Automatic Screwdriver.

$1.30

A good moderate priced automatic screwdriver which can be used as a spiral or plain screwdriver. Furnished with hardwood handle and three steel blades. Extended, 14½ inches. Weight, 10 ounces.

9L5449 ....$1.30

GENERAL PURPOSE WRENCH SET.

90c

A set that is almost indispensable wherever there are nuts to be screwed on or off. The five solid steel drop forged wrenches have ten different size openings in the heads that will fit the nuts on bolts from ³⁄₁₆ inch to ⅝ inch in diameter. Sold with or without roll case as desired.

9L5815—Set of five wrenches with roll. Weight, 2¾ pounds. Per set............90c

9L5814—Set of five wrenches without roll. Weight, 2½ pounds. Per set.......70c

Farmers' Handy Wrench Set.

Price, per Set **50c**

Steel wrenches fit nuts on bolts from ⁵⁄₁₆ inch to ⅝ in. Thread cleaner restores battered threads on ⁵⁄₁₆, ⅜ and ½-in. bolts. Grips up to 1 in. Wt., 2¼ pounds.

9L5830..50c

All weights and measurements given on this page are approximate and may vary a trifle.

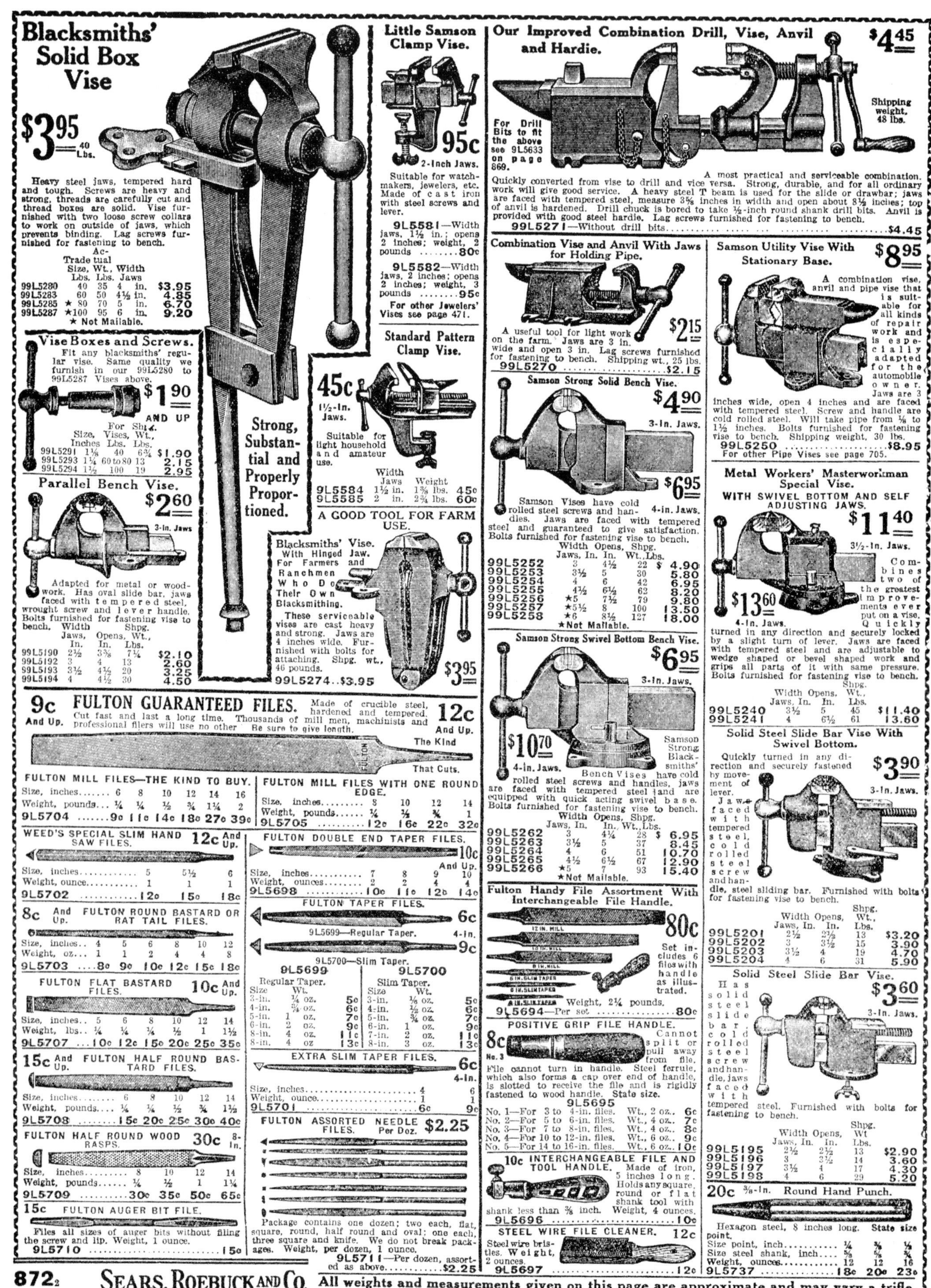

Blacksmiths' Solid Box Vise

$3.95 40 Lbs.

Heavy steel jaws, tempered hard and tough. Screws are heavy and strong, threads are carefully cut and thread boxes are solid. Vise furnished with two loose screw collars to work on outside of jaws, which prevents binding. Lag screws furnished for fastening to bench.

	Trade Size, Lbs.	Actual Wt., Lbs.	Width Jaws	
99L5280	40	35	4 in.	$3.95
99L5283	60	50	4½ in.	4.85
99L5285	★ 80	70	5 in.	6.70
99L5287	★100	95	6 in.	9.20

★ Not Mailable.

Strong, Substantial and Properly Proportioned.

Vise Boxes and Screws.

Fit any blacksmiths' regular vise. Same quality we furnish in our 99L5280 to 99L5287 Vises above.

$1.90 AND UP

	Size, Inches	For Vises, Lbs.	Shpg. Wt., Lbs.	
99L5291	1⅛	40	6¾	$1.90
99L5293	1¼	60 to 80	13	2.15
99L5294	1½	100	19	2.95

Parallel Bench Vise.

$2.60 3-In. Jaws

Adapted for metal or woodwork. Has oval slide bar, jaws faced with tempered steel, wrought screw and lever handle. Bolts furnished for fastening vise to bench.

	Width Jaws, In.	Opens, In.	Shpg. Wt., Lbs.	
99L5190	2½	3⅜	7¼	$2.10
99L5192	3	4	13	2.60
99L5193	3½	4½	20	3.25
99L5194	4	4½	30	4.50

Little Samson Clamp Vise.

95c 2-Inch Jaws.

Suitable for watchmakers, jewelers, etc. Made of cast iron with steel screws and lever.

9L5581—Width jaws, 1½ in.; opens 2 inches; weight, 2 pounds 80c

9L5582—Width jaws, 2 inches; opens 2 inches; weight, 3 pounds 95c

For other Jewelers' Vises see page 471.

Standard Pattern Clamp Vise.

45c 1½-In. Jaws.

Suitable for light household and amateur use.

	Width Jaws	Weight	
9L5584	1½ in.	1⅜ lbs.	45c
9L5585	2 in.	2¾ lbs.	60c

A GOOD TOOL FOR FARM USE.

Blacksmiths' Vise. With Hinged Jaw. For Farmers and Ranchmen Who Do Their Own Blacksmithing.

These serviceable vises are cast heavy and strong. Jaws are 4 inches wide. Furnished with bolts for attaching. Shpg. wt., 46 pounds.

99L5274..$3.95

$3.95

Our Improved Combination Drill, Vise, Anvil and Hardie.

$4.45

For Drill Bits to fit the above see 9L5633 on page 869.

Shipping weight, 48 lbs.

A most practical and serviceable combination. Quickly converted from vise to drill and vice versa. Strong, durable, and for all ordinary work will give good service. A heavy steel T beam is used for the slide or drawbar; jaws are faced with tempered steel, measure 3⅝ inches in width and open about 8½ inches; top of anvil is hardened. Drill chuck is bored to take ½-inch round shank drill bits. Anvil is provided with good steel hardie. Lag screws furnished for fastening to bench.

99L5271—Without drill bits $4.45

Combination Vise and Anvil With Jaws for Holding Pipe.

$2.15

A useful tool for light work on the farm. Jaws are 3 in. wide and open 3 in. Lag screws furnished for fastening to bench. Shipping wt., 25 lbs.

99L5270 .... $2.15

Samson Strong Solid Bench Vise.

$4.90 3-In. Jaws.

$6.95 4-In. Jaws.

Samson Vises have cold rolled steel screws and handles. Jaws are faced with tempered steel and guaranteed to give satisfaction. Bolts furnished for fastening vise to bench.

	Width Jaws, In.	Opens, In.	Shpg. Wt., Lbs.	
99L5252	3	4½	22	$ 4.90
99L5253	3½	5	30	5.80
99L5254	4	6	42	6.95
99L5255	4½	6½	62	8.20
99L5256	★5	7½	79	9.80
99L5257	★5½	8	100	13.50
99L5258	★6	8½	127	18.00

★Not Mailable.

Samson Strong Swivel Bottom Bench Vise.

$6.95 3-In. Jaws.

$10.70 4-In. Jaws.

Samson Strong Blacksmiths' Bench Vises have cold rolled steel screws and handles, jaws are faced with tempered steel and are equipped with quick acting swivel base. Bolts furnished for fastening vise to bench.

	Width Jaws, In.	Opens, In.	Shpg. Wt., Lbs.	
99L5262	3	4½	28	$ 6.95
99L5263	3½	5	37	8.45
99L5264	4	6	51	10.70
99L5265	4½	6½	67	12.90
99L5266	★5	7	93	15.40

★Not Mailable.

Samson Utility Vise With Stationary Base.

$8.95

A combination vise, anvil and pipe vise that is suitable for all kinds of repair work and is especially adapted for the automobile owner. Jaws are 3 inches wide, open 4 inches and are faced with tempered steel. Screw and handle are cold rolled steel. Will take pipe from ⅛ to 1½ inches. Bolts furnished for fastening vise to bench. Shipping weight, 30 lbs.

99L5250 .... $8.95

For other Pipe Vises see page 705.

Metal Workers' Masterworkman Special Vise.

WITH SWIVEL BOTTOM AND SELF ADJUSTING JAWS.

$11.40 3½-In. Jaws.

$13.60 4-In. Jaws.

Combines two of the greatest improvements ever put on a vise. Quickly turned in any direction and securely locked by a slight turn of lever. Jaws are faced with tempered steel and are adjustable to wedge shaped or bevel shaped work and grips all parts of it with same pressure. Bolts furnished for fastening vise to bench.

	Width Jaws, In.	Opens, In.	Shpg. Wt., Lbs.	
99L5240	3½	5	45	$11.40
99L5241	4	6½	61	13.60

Solid Steel Slide Bar Vise With Swivel Bottom.

$3.90 3-In. Jaws.

Quickly turned in any direction and securely fastened by movement of lever. Jaws faced with tempered steel, cold rolled steel screw and handle, steel sliding bar. Furnished with bolts for fastening vise to bench.

	Width Jaws, In.	Opens, In.	Shpg. Wt., Lbs.	
99L5201	2½	2½	13	$3.20
99L5202	3	3½	15	3.90
99L5203	3½	4	19	4.70
99L5204	4	6	31	5.90

Solid Steel Slide Bar Vise.

$3.60 3-In. Jaws.

Has solid steel slide bar, cold rolled steel screw and handle, jaws faced with tempered steel. Furnished with bolts for fastening to bench.

	Width Jaws, In.	Opens, In.	Shpg. Wt, Lbs.	
99L5195	2½	2½	13	$2.90
99L5196	3	3½	14	3.60
99L5197	3½	4	17	4.30
99L5198	4	6	29	5.20

9c And Up. FULTON GUARANTEED FILES. 12c And Up.

Cut fast and last a long time. Thousands of mill men, machinists and professional filers will use no other. Be sure to give length. Made of crucible steel, hardened and tempered.

The Kind That Cuts.

FULTON MILL FILES—THE KIND TO BUY.

Size, inches	6	8	10	12	14	16
Weight, pounds	¼	¼	½	¾	1¼	2
9L5704	9c	11c	14c	18c	27c	39c

FULTON MILL FILES WITH ONE ROUND EDGE.

Size, inches	8	10	12	14
Weight, pounds	¼	½	¾	1
9L5705	12c	16c	22c	32c

WEED'S SPECIAL SLIM HAND SAW FILES. 12c And Up.

Size, inches	5	5½	6
Weight, ounce	1	1	1
9L5702	12c	15c	18c

FULTON DOUBLE END TAPER FILES. 10c And Up.

Size, inches	7	8	9	10
Weight, ounces	2	2	4	4
9L5698	10c	11c	12c	14c

8c And Up. FULTON ROUND BASTARD OR RAT TAIL FILES.

Size, inches	4	5	6	8	10	12
Weight, oz.	1	1	2	4	4	8
9L5703	8c	9c	10c	12c	15c	18c

FULTON TAPER FILES.

6c 4-In. 9L5699—Regular Taper.

9c 9L5700—Slim Taper.

9L5699 Regular Taper. Size	Wt.		9L5700 Slim Taper. Size	Wt.	
3-in.	¼ oz.	5c	3-in.	⅛ oz.	5c
4-in.	⅜ oz.	6c	4-in.	½ oz.	6c
5-in.	1 oz.	7c	5-in.	¾ oz.	7c
6-in.	2 oz.	9c	6-in.	1 oz.	9c
8-in.	4 oz.	11c	7-in.	2 oz.	11c
8-in.	4 oz.	13c	8-in.	3 oz.	13c

FULTON FLAT BASTARD FILES. 10c And Up.

Size, inches	5	6	8	10	12	14
Weight, lbs.	¼	¼	¼	½	1	1½
9L5707	10c	12c	15c	20c	25c	35c

15c And Up. FULTON HALF ROUND BASTARD FILES.

Size, inches	6	8	10	12	14
Weight, pounds	¼	¼	½	¾	1½
9L5708	15c	20c	25c	30c	40c

EXTRA SLIM TAPER FILES. 6c 4-In.

Size, inches	4	6
Weight, ounce	1	1
9L5701	6c	9c

FULTON HALF ROUND WOOD RASPS. 30c 8-In.

Size, inches	8	10	12	14
Weight, pounds	¼	½	1	1¼
9L5709	30c	35c	50c	65c

15c FULTON AUGER BIT FILE.

Files all sizes of auger bits without filing the screw and lip. Weight, 1 ounce.

9L5710 .... 15c

FULTON ASSORTED NEEDLE FILES. Per Doz. $2.25

Package contains one dozen; two each, flat, square, round, half round and oval; one each, three square and knife. We do not break packages. Weight, per dozen, 1 ounce.

9L5711—Per dozen, assorted as above $2.25

Fulton Handy File Assortment With Interchangeable File Handle.

80c

12 IN. MILL, 10 IN. MILL, 8 IN. MILL, 6 IN. SLIM TAPER, 5 IN. SLIM TAPER, 4 IN. SLIM TAPER

Set includes 6 files with handle as illustrated. Weight, 2¼ pounds.

9L5694—Per set 80c

POSITIVE GRIP FILE HANDLE.

8c No. 3

Cannot split or pull away from file. File cannot turn in handle. Steel ferrule, which also forms a cap over end of handle, is slotted to receive the file and is rigidly fastened to wood handle. State size.

9L5695

No. 1—For 3 to 4-in. files.	Wt., 2 oz.	6c
No. 2—For 5 to 6-in. files.	Wt., 4 oz.	7c
No. 3—For 7 to 8-in. files.	Wt., 4 oz.	8c
No. 4—For 10 to 12-in. files.	Wt., 6 oz.	9c
No. 5—For 14 to 16-in. files.	Wt., 6 oz.	10c

10c INTERCHANGEABLE FILE AND TOOL HANDLE.

Made of iron, 5 inches long. Holds any square, round or flat shank tool with shank less than ⅜ inch. Weight, 4 ounces.

9L5696 .... 10c

STEEL WIRE FILE CLEANER. 12c

Steel wire bristles. Weight, 2 ounces.

9L5697 .... 12c

20c ⅜-In. Round Hand Punch.

Hexagon steel, 8 inches long. State size point.

Size point, inch	¼	⅜	½
Size steel shank, inch	⅝	⅝	¾
Weight, ounces	12	12	16
9L5737	18c	20c	23c

All weights and measurements given on this page are approximate and may vary a trifle.

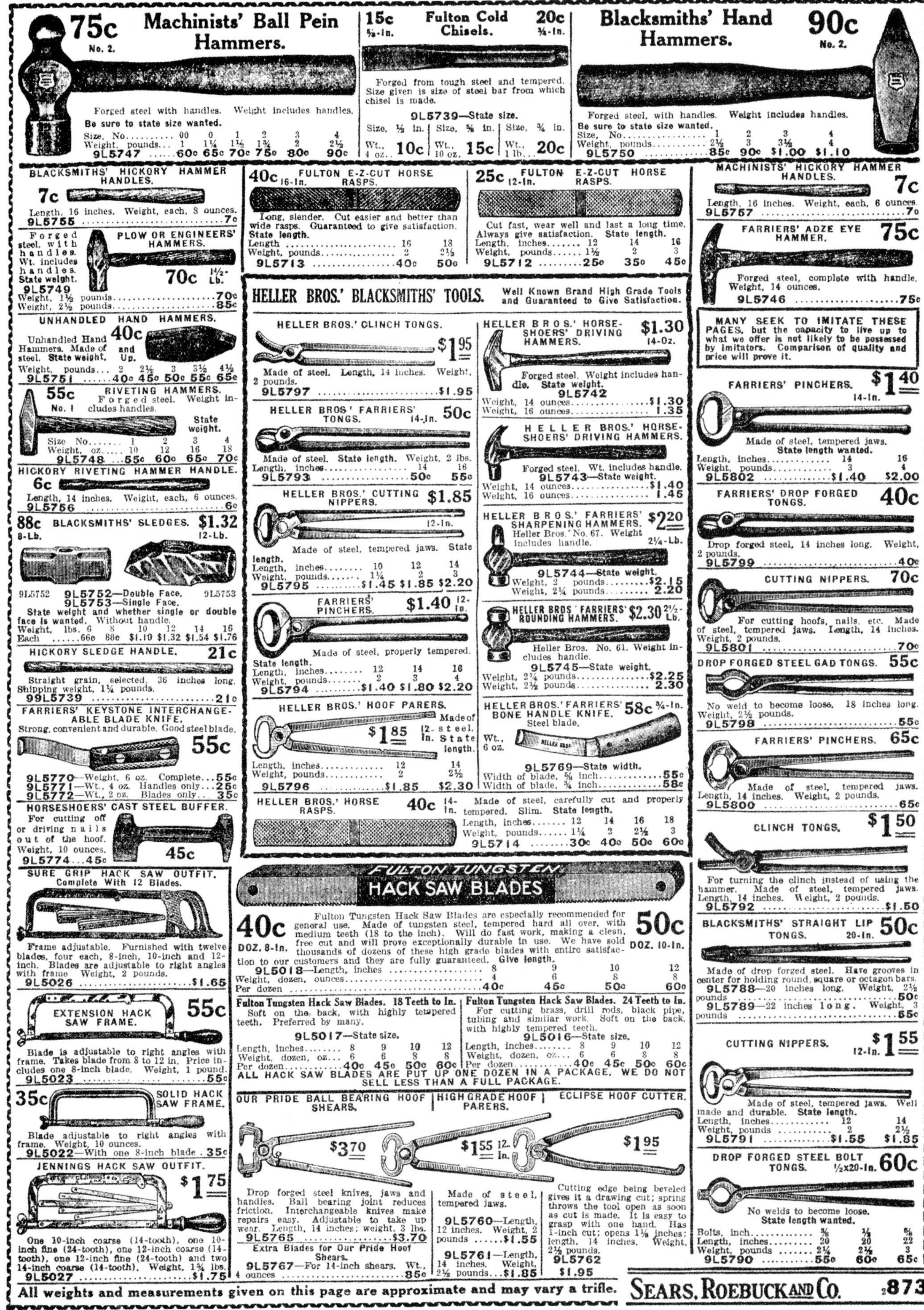

Machinists' Ball Pein Hammers. 75c No. 2.

Forged steel with handles. Weight includes handles.
Be sure to state size wanted.

Size, No.	00	0	1	2	3	4
Weight, pounds	1	1¼	1½	1¾	2	2½
9L5747	60c	65c	70c	75c	80c	90c

Fulton Cold Chisels. 15c ⅝-In. 20c ¾-In.

Forged from tough steel and tempered. Size given is size of steel bar from which chisel is made.

9L5739—State size.

Size, ½ in. Wt., 4 oz. **10c** | Size, ⅝ in. Wt., 10 oz. **15c** | Size, ¾ in. Wt., 1 lb. **20c**

Blacksmiths' Hand Hammers. 90c No. 2.

Forged steel, with handles. Weight includes handles.
Be sure to state size wanted.

Size, No.	1	2	3	4
Weight, pounds	2½	3	3½	4
9L5750	85c	90c	$1.00	$1.10

BLACKSMITHS' HICKORY HAMMER HANDLES. 7c

Length, 16 inches. Weight, each, 8 ounces.
9L5755 ....7c

PLOW OR ENGINEERS' HAMMERS. 70c 1½-Lb.

Forged steel, with handles. Wt. includes handles. State weight.
9L5749
Weight, 1½ pounds 70c
Weight, 2½ pounds 85c

UNHANDLED HAND HAMMERS. 40c and Up.

Unhandled Hand Hammers. Made of steel. State weight.

Weight, pounds	2	2½	3	3½	4½
9L5751	40c	45c	50c	55c	65c

RIVETING HAMMERS. 55c No. 1

Forged steel. Weight includes handles. State weight.

Size No.	1	2	3	4
Weight, oz.	10	12	16	18
9L5748	55c	60c	65c	70c

HICKORY RIVETING HAMMER HANDLE. 6c

Length, 14 inches. Weight, each, 6 ounces.
9L5756 ....6c

BLACKSMITHS' SLEDGES. 88c 8-Lb. $1.32 12-Lb.

9L5752 9L5753
9L5752—Double Face.
9L5753—Single Face.
State weight and whether single or double face is wanted. Without handle.

Weight, lbs.	6	8	10	12	14	16
Each	66c	88c	$1.10	$1.32	$1.54	$1.76

HICKORY SLEDGE HANDLE. 21c

Straight grain, selected. 36 inches long. Shipping weight, 1¼ pounds.
99L5739 ....21c

FARRIERS' KEYSTONE INTERCHANGEABLE BLADE KNIFE. 55c

Strong, convenient and durable. Good steel blade.
9L5770—Weight, 6 oz. Complete...55c
9L5771—Wt., 4 oz. Handles only...25c
9L5772—Wt., 2 oz. Blades only...35c

HORSESHOERS' CAST STEEL BUFFER. 45c

For cutting off or driving nails out of the hoof. Weight, 10 ounces.
9L5774...45c

SURE GRIP HACK SAW OUTFIT. Complete With 12 Blades.

Frame adjustable. Furnished with twelve blades, four each, 8-inch, 10-inch and 12-inch. Blades are adjustable to right angles with frame. Weight, 2 pounds.
9L5026 ....$1.65

EXTENSION HACK SAW FRAME. 55c

Blade is adjustable to right angles with frame. Takes blade from 8 to 12 in. Price includes one 8-inch blade. Weight, 1 pound.
9L5023 ....55c

SOLID HACK SAW FRAME. 35c

Blade adjustable to right angles with frame. Weight, 10 ounces.
9L5022—With one 8-inch blade .35c

JENNINGS HACK SAW OUTFIT. $1.75

One 10-inch coarse (14-tooth), one 10-inch fine (24-tooth), one 12-inch coarse (14-tooth), one 12-inch fine (24-tooth) and two 14-inch coarse (14-tooth). Weight, 1¾ lbs.
9L5027 ....$1.75

FULTON E-Z-CUT HORSE RASPS. 40c 16-In.

Long, slender. Cut easier and better than wide rasps. Guaranteed to give satisfaction. **State length.**

Length	16	18
Weight, pounds	2	2⅓
9L5713	40c	50c

FULTON E-Z-CUT HORSE RASPS. 25c 12-In.

Cut fast, wear well and last a long time. Always give satisfaction. **State length.**

Length, inches	12	14	16
Weight, pounds	1½	2	3
9L5712	25c	35c	45c

HELLER BROS.' BLACKSMITHS' TOOLS.

Well Known Brand High Grade Tools and Guaranteed to Give Satisfaction.

HELLER BROS.' CLINCH TONGS. $1.95

Made of steel. Length, 14 inches. Weight, 2 pounds.
9L5797 ....$1.95

HELLER BROS.' FARRIERS' TONGS. 14-In. 50c

Made of steel. **State length.** Weight, 2 lbs.

Length, inches	14	16
9L5793	50c	55c

HELLER BROS.' CUTTING NIPPERS. $1.85 12-In.

Made of steel, tempered jaws. State length.

Length, inches	10	12	14
Weight, pounds	1¼	2	3
9L5795	$1.45	$1.85	$2.20

FARRIERS' PINCHERS. $1.40 12-In.

Made of steel, properly tempered. **State length.**

Length, inches	12	14	16
Weight, pounds	2	3	4
9L5794	$1.40	$1.80	$2.20

HELLER BROS.' HOOF PARERS. $1.85 12-In.

Made of steel. State length.

Length, inches	12	14
Weight, pounds	2	2½
9L5796	$1.85	$2.30

HELLER BROS.' HORSESHOERS' DRIVING HAMMERS. $1.30 14-Oz.

Forged steel. Weight includes handle. **State weight.**
9L5742
Weight, 14 ounces $1.30
Weight, 16 ounces 1.35

HELLER BROS.' HORSESHOERS' DRIVING HAMMERS.

Forged steel. Wt. includes handle.
9L5743—State weight.
Weight, 14 ounces $1.40
Weight, 16 ounces 1.45

HELLER BROS.' FARRIERS' SHARPENING HAMMERS. $2.20 2¼-Lb.

Heller Bros.' No. 67. Weight includes handle.
9L5744—State weight.
Weight, 2 pounds $2.15
Weight, 2¼ pounds 2.20

HELLER BROS.' FARRIERS' ROUNDING HAMMERS. $2.30 2½-Lb.

Heller Bros. No. 61. Weight includes handle.
9L5745—State weight.
Weight, 2¼ pounds $2.25
Weight, 2½ pounds 2.30

HELLER BROS.' FARRIERS' BONE HANDLE KNIFE. 58c ¾-In.

Steel blade. Wt., 6 oz.
9L5769—State width.
Width of blade, ⅝ inch 55c
Width of blade, ¾ inch 58c

HELLER BROS.' HORSE RASPS. 40c 14-In.

Made of steel, carefully cut and properly tempered. Slim. **State length.**

Length, inches	12	14	16	18
Weight, pounds	1¼	2	2½	3
9L5714	30c	40c	50c	60c

FULTON TUNGSTEN HACK SAW BLADES

40c DOZ. 8-In. — 50c DOZ. 10-In.

Fulton Tungsten Hack Saw Blades are especially recommended for general use. Made of tungsten steel, tempered hard all over, with medium teeth (18 to the inch). Will do fast work, making a clean, free cut and will prove exceptionally durable in use. We have sold thousands of dozens of these high grade blades with entire satisfaction to our customers and they are fully guaranteed. **Give length.**

9L5018—Length, inches	8	9	10	12
Weight, dozen, ounces	4	6	8	8
Per dozen	40c	45c	50c	60c

Fulton Tungsten Hack Saw Blades. 18 Teeth to In. Soft on the back, with highly tempered teeth. Preferred by many.
9L5017—State size.

Length, inches	8	9	10	12
Weight, dozen, oz.	6	6	8	8
Per dozen	40c	45c	50c	60c

Fulton Tungsten Hack Saw Blades. 24 Teeth to In. For cutting brass, drill rods, black pipe, tubing and similar work. Soft on the back, with highly tempered teeth.
9L5016—State size.

Length, inches	8	9	10	12
Weight, dozen, oz.	6	6	8	8
Per dozen	40c	45c	50c	60c

ALL HACK SAW BLADES ARE PUT UP ONE DOZEN IN A PACKAGE. WE DO NOT SELL LESS THAN A FULL PACKAGE.

OUR PRIDE BALL BEARING HOOF SHEARS. $3.70

Drop forged steel knives, jaws and handles. Ball bearing joint reduces friction. Interchangeable knives make repairs easy. Adjustable to take up wear. Length, 14 inches; weight, 3 lbs.
9L5765 ....$3.70
Extra Blades for Our Pride Hoof Shears.
9L5767—For 14-inch shears. Wt., 4 ounces 85c

HIGH GRADE HOOF PARERS. $1.55 12-In.

Made of steel, tempered jaws.
9L5760—Length, 12 inches. Weight, 2 pounds $1.55
9L5761—Length, 14 inches. Weight, 2½ pounds...$1.85

ECLIPSE HOOF CUTTER. $1.95

Cutting edge being beveled gives it a drawing cut; spring throws the tool open as soon as cut is made. It is easy to grasp with one hand. Has 1-inch cut; opens 1⅛ inches; length, 14 inches. Weight, 2½ pounds.
9L5762
$1.95

MACHINISTS' HICKORY HAMMER HANDLES. 7c

Length, 16 inches. Weight, each, 6 ounces.
9L5757 ....7c

FARRIERS' ADZE EYE HAMMER. 75c

Forged steel, complete with handle. Weight, 14 ounces.
9L5746 ....75c

MANY SEEK TO IMITATE THESE PAGES, but the capacity to live up to what we offer is not likely to be possessed by imitators. Comparison of quality and price will prove it.

FARRIERS' PINCHERS. $1.40 14-In.

Made of steel, tempered jaws. **State length wanted.**

Length, inches	14	16
Weight, pounds	3	4
9L5802	$1.40	$2.00

FARRIERS' DROP FORGED TONGS. 40c

Drop forged steel, 14 inches long. Weight, 2 pounds.
9L5799 ....40c

CUTTING NIPPERS. 70c

For cutting hoofs, nails, etc. Made of steel, tempered jaws. Length, 14 inches. Weight, 2 pounds.
9L5801 ....70c

DROP FORGED STEEL GAD TONGS. 55c

No weld to become loose. 18 inches long. Weight, 2½ pounds.
9L5798 ....55c

FARRIERS' PINCHERS. 65c

Made of steel, tempered jaws. Length, 14 inches. Weight, 2 pounds.
9L5800 ....65c

CLINCH TONGS. $1.50

For turning the clinch instead of using the hammer. Made of steel, tempered jaws. Length, 14 inches. Weight, 2 pounds.
9L5792 ....$1.50

BLACKSMITHS' STRAIGHT LIP TONGS. 20-In. 50c

Made of drop forged steel. Have grooves in center for holding round, square or octagon bars.
9L5788—20 inches long. Weight, 2½ pounds 50c
9L5789—22 inches long. Weight, 3 pounds 55c

CUTTING NIPPERS. $1.55 12-In.

Made of steel, tempered jaws. Well made and durable. **State length.**

Length, inches	12	14
Weight, pounds	2	2½
9L5791	$1.55	$1.85

DROP FORGED STEEL BOLT TONGS. ½x20-In. 60c

No welds to become loose. **State length wanted.**

Bolts, inch	⅜	½	⅝
Length, inches	20	20	22
Weight, pounds	2¼	2¼	3
9L5790	55c	60c	65c

All weights and measurements given on this page are approximate and may vary a trifle.

Tiger Blower. $16.90

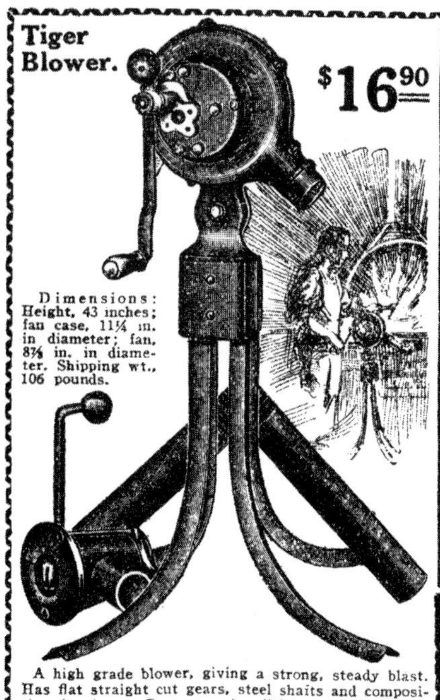

Dimensions: Height, 43 inches; fan case, 11¼ in. in diameter; fan, 8⅞ in. in diameter. Shipping wt., 106 pounds.

A high grade blower, giving a strong, steady blast. Has flat straight cut gears, steel shafts and composition bearings. Gear case is oiltight and dustproof, permitting gears to run in continuous bath of oil. Blower case can be raised or lowered. Crank turns forward or backward. **Not mailable.**

99L5001—Complete with piping and heavy anticlinker tuyere, as illustrated **$16.90**

Farmers' Combination Blacksmith Repair Outfit.

$16.50 As Illustrated.

A BIG OUTFIT OF 16 USEFUL TOOLS. OUTFIT INCLUDES 15 Serviceable Tools, as illustrated, and Our Big Four-In-One Combination Forge, Drill, Anvil and Vise. **Forge**—Pan measures 12x15 inches and stands 32 inches from floor. Fan case measures 7 inches in diameter. **Drill**—Chuck takes drills with ½-inch round shank. Vise lever regulates feed. **Anvil**—Face measures, 3¾x5¼ inches. Furnished with ⅝-inch steel hardie. **Vise**—Jaws are 3½ inches wide and are faced with tempered steel. Movable jaw operates on heavy steel T beam, which can be adjusted by means of steel pin for opening jaws up to 8½ inches.

1 LB. 4 INCH 6 INCH 6 INCH 6 INCH 6 INCH 8 INCH 10 INCH 8 INCH BLADE ¼ IN. ⅜ IN. ⅝ IN. 5 INCH ⅞ INCH

ONLY **$13.75** For 4-in-1 Combination Outfit Only. Without Tools.

99L5004—Farmers' Big Combination Repair Outfit, as illustrated and described. Shipping weight, 123 pounds. Not mailable **$16.50**

99L5003—Combination Vise, Anvil, Drill, Forge and Blower only, without any tools. Shpg. wt., 114 pounds. Not mailable **$13.75**

Niagara Spiral Geared Ball Bearing Blower. $21.80

Fitted with iron cut spur gear, steel worm gear and bronze spiral gear.

Dimensions: Tuyere, 10x 8¾x4½ in. (inside measurement); 12-inch fan case adjustable from 44 to 48 inches in height.

A high grade well finished easy running hand power blower. Develops a steady, powerful blast. Gear case is dustproof and oiltight, permitting gears to run in a continuous bath of oil. Crank turns either forward or backward. Fan case can be raised or lowered or turned in any direction. Shipping weight, 144 lbs.

99L5002—Complete with piping and tuyere, as illustrated. Not mailable **$21.80**

Adjustable Blast Tuyere Iron. $2.45

Blast regulated by turning large rod; cinders and ashes removed by pulling small rod. Levers and spring can be changed to either side for right or left hand use. Shpg. wt., 28 lbs.

99L5315 .... **$2.45**

Tuyere Irons. $1.95 Medium.

With dump and shaker. Made of cast iron.

99L5311 Medium size. Shpg. wt. 22 lbs. **$1.95**

99L5312—Large size. Shpg. wt., 26 lbs **$2.30**

Our Challenge Forge. $10.75

A well made forge that will give good service.

Description. 18x21-in. fire pan, 30 inches high, 8½-inch fan case, adjustable lever handle and ratchet clutch.

Shpg. wt., 102 lbs. Not mailable.

99L5021 .. **$10.75**

HOW MUCH IS A DOLLAR? It is worth ninety cents, just a dollar, or a dollar and a quarter according to where you invest it. Volume buying, scientific testing and economical selling gives your dollar a premium in our catalog.

Challenge Outfit of Blacksmiths' High Class Tools, $28.90

3 LBS. 14 INCH 70 LBS. ¾ INCH ⅝ INCH FIT NUTS ON BOLTS 5/16 TO ¾ INCH 14 INCH 20 INCH 4 INCH JAWS 14 OZ. ⅝ INCH

Every tool selected from our regular stock and amply strong and heavy enough for all kinds of light and medium work. Quality not sacrificed in this outfit.

Shpg. wt., 230 lbs. **Not mailable.**

READ THESE DESCRIPTIONS.

ANVIL. 70-pound cast iron Anvil.
FORGE. Our Invincible Portable Lever Forge has 18-inch hearth, 8-inch blower and is 30 inches high.
DRILL. Our horizontal Bench Drill has screw feed chuck bored for ½-inch round shank drills.
VISE. Blacksmiths' cast iron Vise with hinged jaws. Jaws, 4 inches wide and faced with tempered steel. Weight, 44 pounds.
STOCK AND DIES. 41C. Cut ¼ to 5/16 inch, 12, 14 and 16 threads to the inch, six taps and three dies.

Outfit also includes assortment of good quality hand tools, as illustrated.

99L5009—Complete outfit as illustrated and described **$28.90**

Blacksmiths' Apron With Split Center. $2.95

Made of selected tough hides. Split center makes apron more serviceable for blacksmiths and horseshoers than the one-piece apron. Size, 28x36 inches.

Weight, 3 pounds.

9L3527 .... **$2.95**

Leather Apron for Blacksmiths and Horseshoers. $2.60

Split leather. Made of tough hides. With bib and strings. Strings are put on with rivets. Size, 28x38 inches. Weight, 2 pounds.

9L3525 .... **$2.60**

Eclipse Forge and Blower. $16.40 With Half Hood.

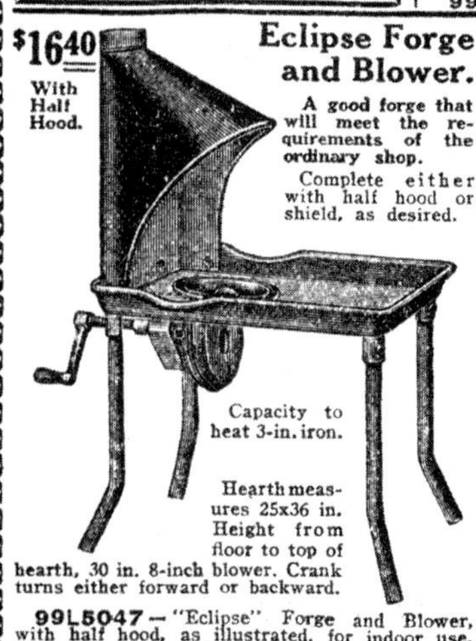

A good forge that will meet the requirements of the ordinary shop. Complete either with half hood or shield, as desired.

Capacity to heat 3-in. iron.

Hearth measures 25x36 in. Height from floor to top of hearth, 30 in. 8-inch blower. Crank turns either forward or backward.

99L5047—"Eclipse" Forge and Blower, with half hood, as illustrated, for indoor use. Shpg. wt., 140 lbs. **Not mailable** **$16.40**

99L5048—"Eclipse" Forge and Blower, with shield. Shpg. wt., 136 lbs. **Not mailable.** **$15.60**

Acme Combination Forge and Blower. $35.50

A Strictly High Grade Large Size Forge for Heavy Work.

It pays to buy the best. This is the largest and best forge that we sell and we cannot recommend it too highly. It is well made of good material and is of heavy construction throughout. Length over all, 53 inches.

Hearth measures 45x31½ inches. From top of hearth to floor measures 30 inches. Solid fire pot with tuyere ball furnishing side and center blast.

Blower is powerful and easy to operate; gear case is oiltight and dustproof. Crank turns either forward or backward. Fan case is 11½ inches in diameter.

Capacity to heat 4-inch iron. Plenty of surplus heat to weld the largest size wagon axle. The forge for the shop where the work is heavy.

99L5049—Acme Combination Crank Forge, with coal box, half hood and tool rest, as illustrated. Shipping weight, 290 lbs. **Not mailable** ... **$35.50**

Peerless Combination Forge and Blower. $25.50 With Half Hood.

For the Farmer, Blacksmith, Ranchman, Horseshoer and Boiler Maker.

A forge that will require but little attention and will give you years of satisfactory service.

Capacity to heat 3-inch iron. Size of hearth, 35½x24½ inches; duck nest fire pot and anticlinker tuyere. 9-inch blower. Crank turns either forward or backward. Gear case is oiltight and dustproof. Height from floor to top of hearth, 30 inches.

99L5034—Peerless Combination Forge and Blower, with half hood as illustrated. Shpg. wt., 167 lbs. **Not mailable** . **$25.50**

99L5033—Peerless Combination Forge and Blower, with shield. Shipping weight, 165 pounds. Not mailable **$24.70**

 All weights and measurements given on this page are approximate and may vary a trifle.

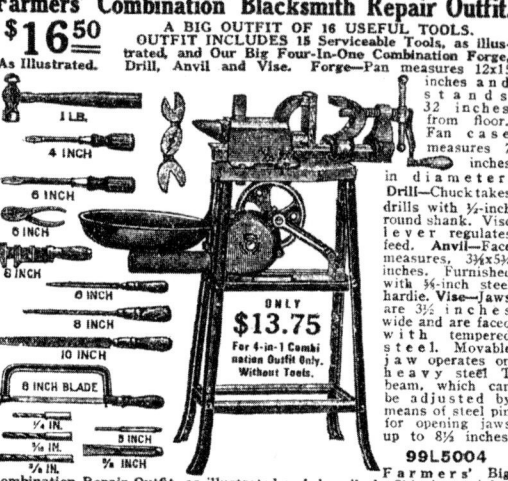

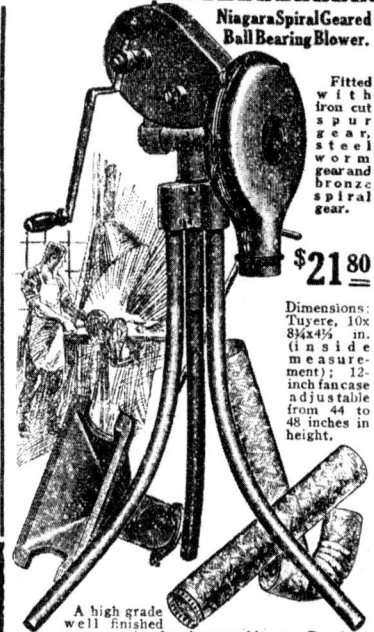

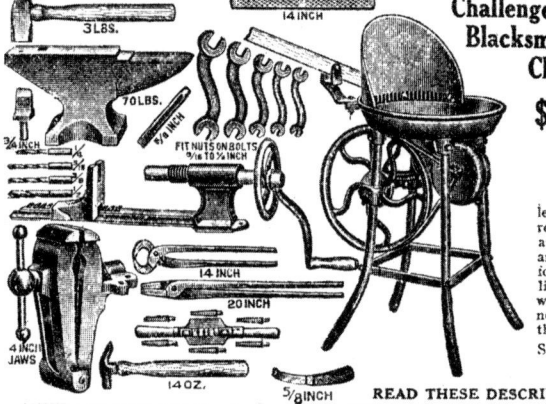

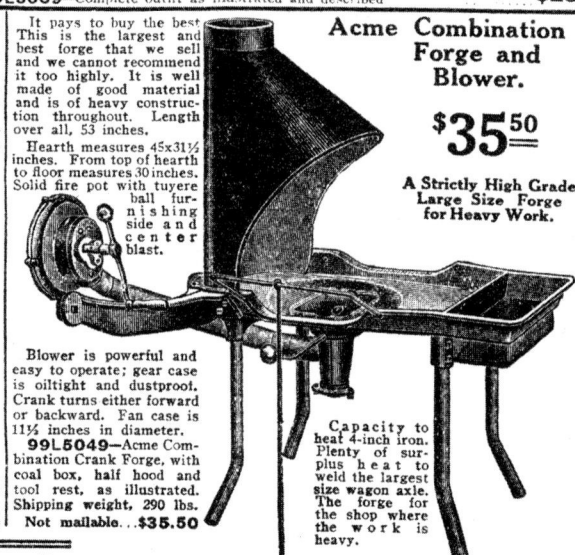

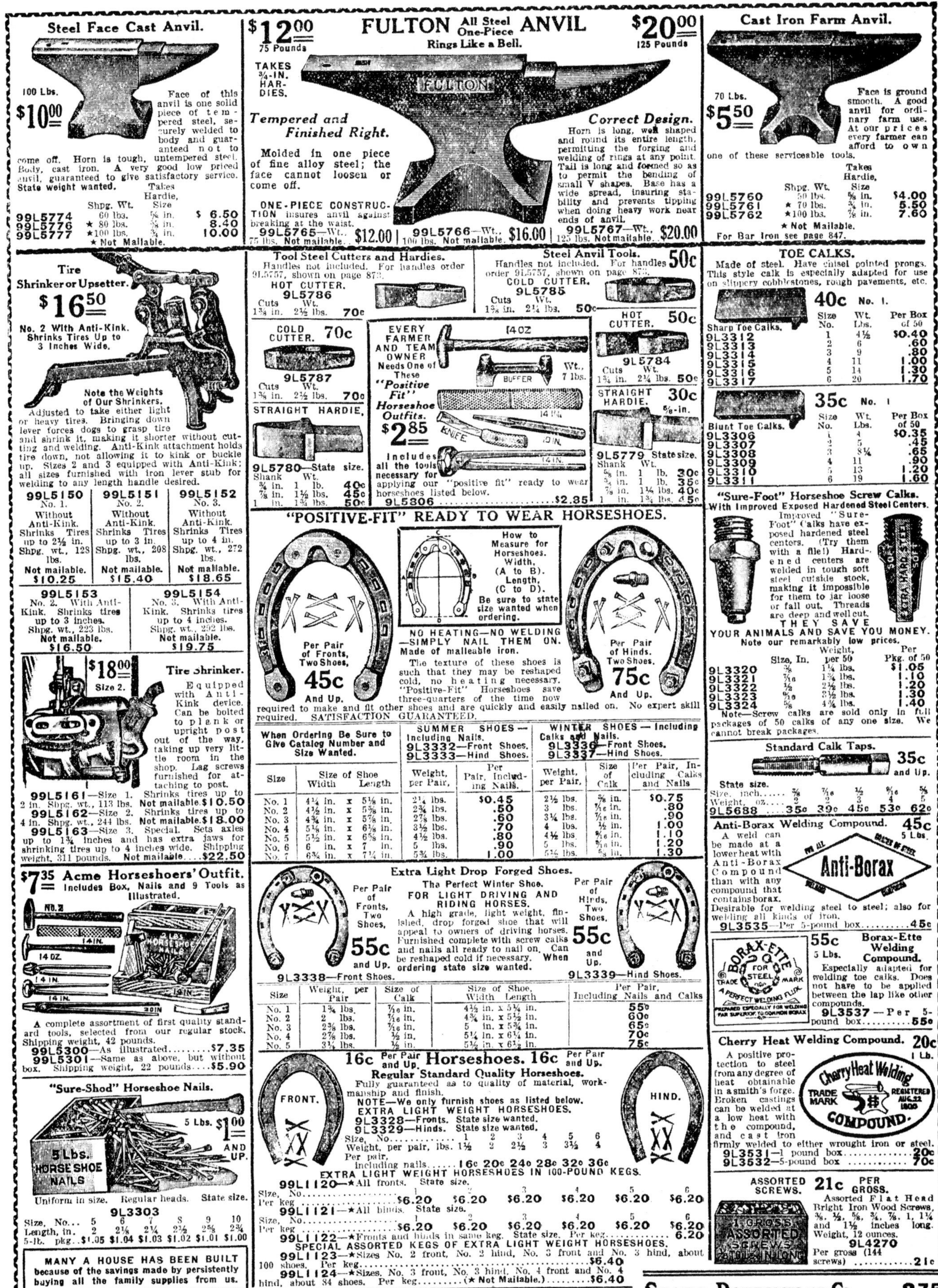

Steel Face Cast Anvil.

100 Lbs. **$10.00**

Face of this anvil is one solid piece of tempered steel, securely welded to body and guaranteed not to come off. Horn is tough, untempered steel. Body, cast iron. A very good low priced anvil, guaranteed to give satisfactory service. **State weight wanted.**

	Shpg. Wt.	Takes Hardie, Size	
99L5774	60 lbs.	5/8 in.	$ 6.50
99L5776	★ 80 lbs.	3/4 in.	8.40
99L5777	★100 lbs.	3/4 in.	10.00

★ Not Mailable.

$12.00 FULTON All Steel One-Piece ANVIL $20.00

75 Pounds — Rings Like a Bell. — 125 Pounds

TAKES 3/4-IN. HARDIES.

Tempered and Finished Right.

Molded in one piece of fine alloy steel; the face cannot loosen or come off.

ONE-PIECE CONSTRUCTION insures anvil against breaking at the waist.

Correct Design. Horn is long, well shaped and round its entire length, permitting the forging and welding of rings at any point. Tail is long and formed so as to permit the bending of small V shapes. Base has a wide spread, insuring stability and prevents tipping when doing heavy work near ends of anvil.

99L5765—Wt., 75 lbs. Not mailable. **$12.00** | 99L5766—Wt., 100 lbs. Not mailable. **$16.00** | 99L5767—Wt., 125 lbs. Not mailable. **$20.00**

Cast Iron Farm Anvil.

70 Lbs. **$5.50**

Face is ground smooth. A good anvil for ordinary farm use. At our prices every farmer can afford to own one of these serviceable tools.

	Shpg. Wt.	Takes Hardie, Size	
99L5760	50 lbs.	5/8 in.	$4.00
99L5761	★ 70 lbs.	3/4 in.	5.50
99L5762	★100 lbs.	7/8 in.	7.60

★ Not Mailable.

For Bar Iron see page 847.

Tire Shrinker or Upsetter. $16.50

No. 2 With Anti-Kink. Shrinks Tires Up to 3 Inches Wide.

Note the Weights of Our Shrinkers. Adjusted to take either light or heavy tires. Bringing down lever forces dogs to grasp tire and shrink it, making it shorter without cutting and welding. Anti-Kink attachment holds tire down, not allowing it to kink or buckle up. Sizes 2 and 3 equipped with Anti-Kink; all sizes furnished with iron lever stub for welding to any length handle desired.

99L5150 No. 1.	99L5151 No. 2.	99L5152 No. 3.
Without Anti-Kink. Shrinks Tires up to 2½ in. Shpg. wt., 128 lbs. Not mailable. $10.25	Without Anti-Kink. Shrinks Tires up to 3 in. Shpg. wt., 208 lbs. Not mailable. $15.40	Without Anti-Kink. Shrinks Tires up to 4 in. Shpg. wt., 272 lbs. Not mailable. $18.65

99L5153	99L5154
No. 2. With Anti-Kink. Shrinks tires up to 3 inches. Shpg. wt., 223 lbs. Not mailable. $16.50	No. 3. With Anti-Kink. Shrinks tires up to 4 inches. Shpg. wt., 292 lbs. Not mailable. $19.75

Tire Shrinker. $18.00 Size 2.

Equipped with Anti-Kink device. Can be bolted to plank or upright post out of the way, taking up very little room in the shop. Lag screws furnished for attaching to post.

99L5161—Size 1. Shrinks tires up to 2 in. Shpg. wt., 113 lbs. Not mailable. $10.50
99L5162—Size 2. Shrinks tires up to 4 in. Shpg. wt., 244 lbs. Not mailable. $18.00
99L5163—Size 3. Special. Sets axles up to 1¾ inches and has extra jaws for shrinking tires up to 4 inches wide. Shipping weight, 311 pounds. Not mailable....$22.50

$7.35 Acme Horseshoers' Outfit.

Includes Box, Nails and 9 Tools as Illustrated.

A complete assortment of first quality standard tools, selected from our regular stock. Shipping weight, 42 pounds.

99L5300—As illustrated........$7.35
99L5301—Same as above, but without box. Shipping weight, 22 pounds....$5.90

"Sure-Shod" Horseshoe Nails.

5 Lbs. $1.00 AND UP.

Uniform in size. Regular heads. State size.

9L3303

Size, No...	5	6	7	8	9	10
Length, in.	2	2⅛	2¼	2½	2⅝	2¾
5-lb. pkg..	$1.05	$1.04	$1.03	$1.02	$1.01	$1.00

MANY A HOUSE HAS BEEN BUILT because of the savings made by persistently buying all the family supplies from us.

Tool Steel Cutters and Hardies.

Handles not included. For handles order 9L5757, shown on page 873.

HOT CUTTER. 9L5786 — Cuts 1¾ in. Wt. 2½ lbs. **70c**

COLD CUTTER. **70c** 9L5787 — Cuts 1¾ in. Wt. 2½ lbs. **70c**

STRAIGHT HARDIE. 9L5780—State size.

Shank	Wt.	
¾ in.	1 lb.	40c
⅞ in.	1½ lbs.	45c
1 in.	1¾ lbs.	50c

Steel Anvil Tools. 50c

Handles not included. For handles order 9L5757, shown on page 873.

COLD CUTTER. 9L5785 — Cuts 1⅜ in. Wt. 2¼ lbs. **50c**

HOT CUTTER. **50c** 9L5784 — Cuts 1¾ in. Wt. 2¼ lbs. **50c**

STRAIGHT HARDIE. **30c** ⅝-in. 9L5779 State size.

Shank	Wt.	
⅝ in.	1 lb.	30c
¾ in.	1 lb.	35c
⅞ in.	1¼ lbs.	40c
1 in.	1¾ lbs.	45c

EVERY FARMER AND TEAM OWNER Needs One of These *"Positive Fit" Horseshoe Outfits.* **$2.85**

Includes all the tools necessary for applying our "positive fit" ready to wear horseshoes listed below.

9L5806..........................$2.85

"POSITIVE-FIT" READY TO WEAR HORSESHOES.

Per Pair of Fronts, Two Shoes, **45c** And Up. — Per Pair of Hinds, Two Shoes, **75c** And Up.

How to Measure for Horseshoes. Width, (A to B). Length, (C to D). Be sure to state size wanted when ordering.

NO HEATING—NO WELDING —SIMPLY NAIL THEM ON. Made of malleable iron.

The texture of these shoes is such that they may be reshaped cold, no heating necessary. "Positive-Fit" Horseshoes save three-quarters of the time now required to make and fit other shoes and are quickly and easily nailed on. No expert skill required. SATISFACTION GUARANTEED.

When Ordering Be Sure to Give Catalog Number and Size Wanted.

SUMMER SHOES—Including Nails. 9L3332—Front Shoes. 9L3333—Hind Shoes.

WINTER SHOES—Including Calks and Nails. 9L3336—Front Shoes. 9L3337—Hind Shoes.

Size	Width	Length	Summer: Weight, per Pair	Summer: Per Pair, Including Nails	Winter: Weight, per Pair	Winter: Size of Calk	Winter: Per Pair, Including Calks and Nails
No. 1	4¼ in.	x 5¼ in.	2¼ lbs.	$0.45	2½ lbs.	⅜ in.	$0.75
No. 2	4½ in.	x 5⅝ in.	2¾ lbs.	.50	3 lbs.	7/16 in.	.80
No. 3	4¾ in.	x 5⅞ in.	2⅞ lbs.	.60	3¼ lbs.	7/16 in.	.90
No. 4	5⅛ in.	x 6⅛ in.	3½ lbs.	.70	4 lbs.	½ in.	1.00
No. 5	5½ in.	x 6⅝ in.	4½ lbs.	.80	4½ lbs.	9/16 in.	1.10
No. 6	6 in.	x 7 in.	5 lbs.	.90	5 lbs.	9/16 in.	1.20
No. 7	6¾ in.	x 7¼ in.	5¾ lbs.	1.00	5½ lbs.	⅝ in.	1.30

Extra Light Drop Forged Shoes.

The Perfect Winter Shoe. FOR LIGHT DRIVING AND RIDING HORSES.

Per Pair of Fronts, Two Shoes, **55c** and Up. — Per Pair of Hinds, Two Shoes, **55c** and Up.

A high grade, light weight, finished, drop forged shoe that will appeal to owners of driving horses. Furnished complete with screw calks and nails all ready to nail on. Can be reshaped cold if necessary. When ordering state size wanted.

9L3338—Front Shoes. 9L3339—Hind Shoes.

Size	Weight, per Pair	Size of Calk	Size of Shoe, Width Length	Per Pair, Including Nails and Calks
No. 1	1¾ lbs.	7/16 in.	4½ in. x 5¼ in.	55c
No. 2	2 lbs.	7/16 in.	4¾ in. x 5½ in.	60c
No. 3	2⅜ lbs.	7/16 in.	5 in. x 5¾ in.	65c
No. 4	2⅞ lbs.	½ in.	5¼ in. x 6¼ in.	70c
No. 5	3¼ lbs.	½ in.	5½ in. x 6½ in.	75c

16c Per Pair and Up. Horseshoes. 16c Per Pair and Up.

Regular Standard Quality Horseshoes.

Fully guaranteed as to quality of material, workmanship and finish.

NOTE—We only furnish shoes as listed below.

EXTRA LIGHT WEIGHT HORSESHOES.

9L3328—Fronts. State size wanted.
9L3329—Hinds. State size wanted.

Size, No.	1	2	3	4	5	6
Weight, per pair, lbs.	1½	2	2½	3	3½	4
Per pair, including nails	16c	20c	24c	28c	32c	36c

EXTRA LIGHT WEIGHT HORSESHOES IN 100-POUND KEGS.

99L1120—★All fronts. State size.

Size, No.	1	2	3	4	5	6
Per keg	$6.20	$6.20	$6.20	$6.20	$6.20	$6.20

99L1121—★All hinds. State size.

Size, No.	1	2	3	4	5	6
Per keg	$6.20	$6.20	$6.20	$6.20	$6.20	$6.20

99L1122—★Fronts and hinds in same keg. State size. Per keg.............. 6.20

SPECIAL ASSORTED KEGS OF EXTRA LIGHT WEIGHT HORSESHOES.

99L1123—★Sizes No. 2 front, No. 2 hind, No. 3 front and No. 3 hind, about 100 shoes. Per keg...............................$6.40

99L1124—★Sizes. No. 3 front, No. 3 hind, No. 4 front and No. 4 hind, about 84 shoes. Per keg.............(★ Not Mailable.)..........$6.40

TOE CALKS.

Made of steel. Have chisel pointed prongs. This style calk is especially adapted for use on slippery cobblestones, rough pavements, etc.

40c No. 1.

Sharp Toe Calks.	Size No.	Wt. Lbs.	Per Box of 50
9L3312	1	4½	$0.40
9L3313	2	6	.60
9L3314	3	9	.80
9L3315	4	11	1.00
9L3316	5	14	1.30
9L3317	6	20	1.70

35c No. 1

Blunt Toe Calks.	Size No.	Wt. Lbs.	Per Box of 50
9L3306	1	4	$0.35
9L3307	2	5	.45
9L3308	3	8¼	.65
9L3309	4	11	.90
9L3310	5	13	1.20
9L3311	6	19	1.60

"Sure-Foot" Horseshoe Screw Calks.

With Improved Exposed Hardened Steel Centers.

Improved "Sure-Foot" Calks have exposed hardened steel centers. (Try them with a file!) Hardened centers are welded in tough soft steel outside stock, making it impossible for them to jar loose or fall out. Threads are deep and well cut.

THEY SAVE YOUR ANIMALS AND SAVE YOU MONEY. Note our remarkably low prices.

	Size, In.	Weight, per 50	Per Pkg. of 50
9L3320	⅜	1¼ lbs.	$1.05
9L3321	7/16	1¾ lbs.	1.10
9L3322	½	2¼ lbs.	1.20
9L3323	9/16	3½ lbs.	1.30
9L3324	⅝	4¾ lbs.	1.40

Note—Screw calks are sold only in full packages of 50 calks of any one size. We cannot break packages.

Standard Calk Taps. 35c and Up.

State size.

Size, inch	⅜	7/16	½	9/16	⅝
Weight, oz.	2	2	3	4	5
9L5688	35c	39c	45c	53c	62c

Anti-Borax Welding Compound. 45c 5 Lbs.

A weld can be made at a lower heat with Anti-Borax Compound than with any compound that contains borax. Desirable for welding steel to steel; also for welding all kinds of iron.

9L3535—Per 5-pound box.............45c

Borax-Ette Welding Compound. 55c 5 Lbs.

Especially adapted for welding toe calks. Does not have to be applied between the lap like other compounds.

9L3537—Per 5-pound box..............55c

Cherry Heat Welding Compound. 20c 1 Lb.

A positive protection to steel from any degree of heat obtainable in a smith's forge. Broken castings can be welded at a low heat with the compound, and cast iron firmly welded to either wrought iron or steel.

9L3531—1 pound box..............20c
9L3532—5-pound box..............70c

ASSORTED SCREWS. 21c PER GROSS.

Assorted Flat Head Bright Iron Wood Screws, ⅜, ½, ⅝, ¾, ⅞, 1, 1¼ and 1½ inches long. Weight, 12 ounces.

9L4270 Per gross (144 screws)..............21c

All weights and measurements given on this page are approximate and may vary a trifle.

David Bradley Garden City Clipper Walking Plows

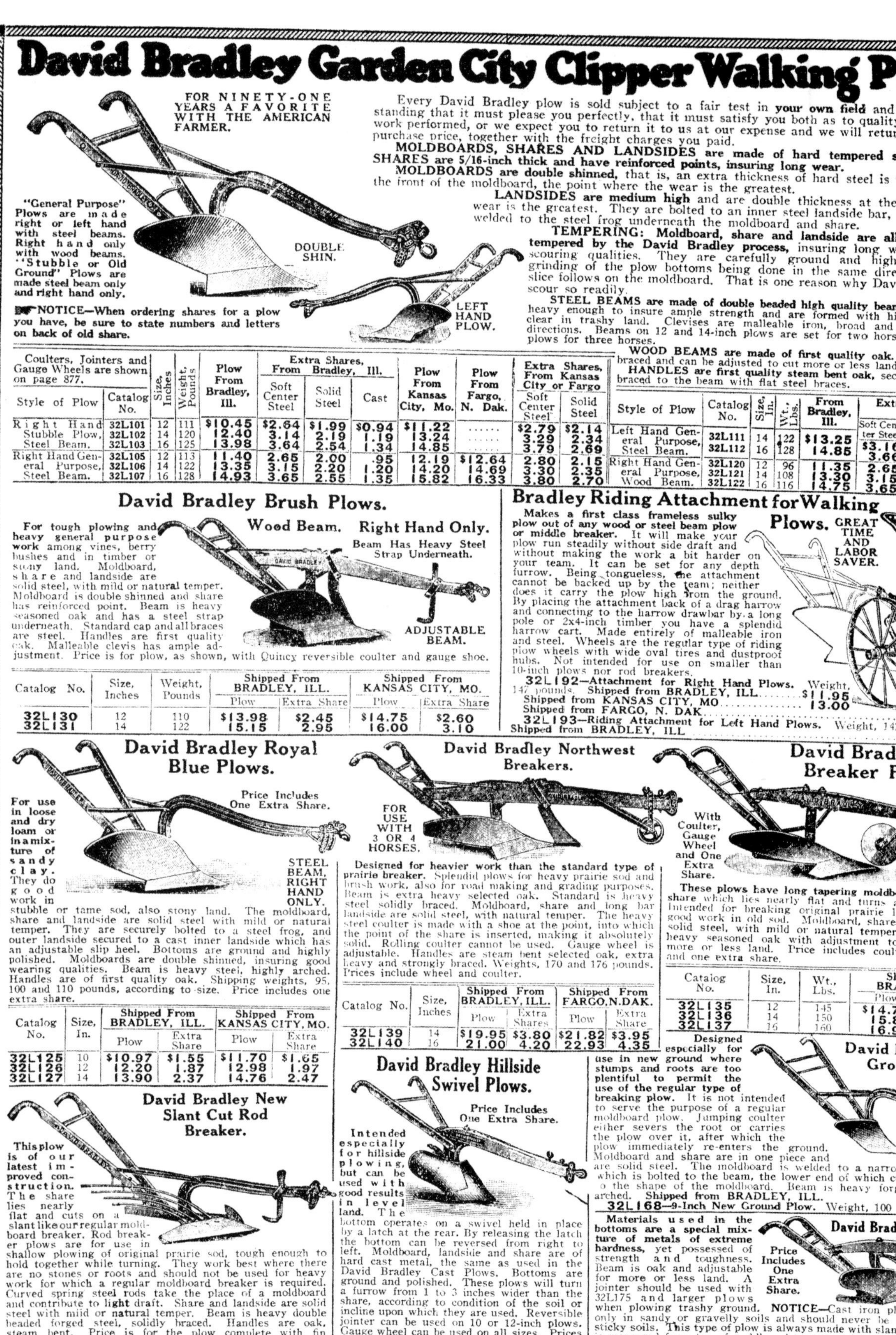

FOR NINETY-ONE YEARS A FAVORITE WITH THE AMERICAN FARMER.

"General Purpose" Plows are made right or left hand with steel beams. Right hand only with wood beams. "Stubble or Old Ground" Plows are made steel beam only and right hand only.

DOUBLE SHIN.

LEFT HAND PLOW.

NOTICE—When ordering shares for a plow you have, be sure to state numbers and letters on back of old share.

Every David Bradley plow is sold subject to a fair test in **your own field** and with the understanding that it must please you perfectly, that it must satisfy you both as to quality of material and work performed, or we expect you to return it to us at our expense and we will return to you the full purchase price, together with the freight charges you paid.

MOLDBOARDS, SHARES AND LANDSIDES are made of hard tempered soft center steel. SHARES are 5/16-inch thick and have reinforced points, insuring long wear.

MOLDBOARDS are double shinned, that is, an extra thickness of hard steel is welded on top of the front of the moldboard, the point where the wear is the greatest.

LANDSIDES are medium high and are double thickness at the heel, where the wear is the greatest. They are bolted to an inner steel landside bar, which is securely welded to the steel frog underneath the moldboard and share.

TEMPERING: Moldboard, share and landside are all uniformly hard tempered by the David Bradley process, insuring long wearing and easy scouring qualities. They are carefully ground and highly polished, the grinding of the plow bottoms being done in the same direction the furrow slice follows on the moldboard. That is one reason why David Bradley plows scour so readily.

STEEL BEAMS are made of double beaded high quality beam steel. They are heavy enough to insure ample strength and are formed with high arch so as to clear in trashy land. Clevises are malleable iron, broad and adjustable in all directions. Beams on 12 and 14-inch plows are set for two horses and on 16-inch plows for three horses.

WOOD BEAMS are made of first quality oak. They are rigidly braced and can be adjusted to cut more or less land. **Right hand only.**

HANDLES are first quality steam bent oak, securely attached and braced to the beam with flat steel braces.

Coulters, Jointers and Gauge Wheels are shown on page 877.

Style of Plow	Catalog No.	Size, Inches	Weight, Pounds	Plow From Bradley, Ill.	Extra Shares, From Bradley, Ill.: Soft Center Steel	Solid Steel	Cast	Plow From Kansas City, Mo.	Plow From Fargo, N. Dak.	Extra Shares, From Kansas City or Fargo: Soft Center Steel	Solid Steel
Right Hand Stubble Plow, Steel Beam.	32L101	12	111	$10.45	$2.64	$1.99	$0.94	$11.22	......	$2.79	$2.14
	32L102	14	120	12.40	3.14	2.19	1.19	13.24	......	3.29	2.34
	32L103	16	125	13.98	3.64	2.54	1.34	14.85	......	3.79	2.69
Right Hand General Purpose, Steel Beam.	32L105	12	113	11.40	2.65	2.00	.95	12.19	$12.64	2.80	2.15
	32L106	14	122	13.35	3.15	2.20	1.20	14.20	14.69	3.30	2.35
	32L107	16	128	14.93	3.65	2.55	1.35	15.82	16.33	3.80	2.70

Style of Plow	Catalog No.	Size, In.	Wt., Lbs.	From Bradley, Ill.	Extra Shares, From Bradley, Ill.: Soft Center Steel	Solid Steel	Cast
Left Hand General Purpose, Steel Beam.	32L111	14	122	$13.25	$3.16	$2.21	$1.21
	32L112	16	128	14.85	3.66	2.56	1.36
Right Hand General Purpose, Wood Beam.	32L120	12	96	11.35	2.65	2.00	.95
	32L121	14	108	13.30	3.15	2.20	1.20
	32L122	16	116	14.75	3.65	2.55	1.35

David Bradley Brush Plows.

Wood Beam. Right Hand Only.

Beam Has Heavy Steel Strap Underneath.

For tough plowing and heavy general purpose work among vines, berry bushes and in timber or stony land. Moldboard, share and landside are solid steel, with mild or natural temper. Moldboard is double shinned and share has reinforced point. Beam is heavy seasoned oak and has a steel strap underneath. Standard cap and all braces are steel. Handles are first quality oak. Malleable clevis has ample adjustment. Price is for plow, as shown, with Quincy reversible coulter and gauge shoe.

ADJUSTABLE BEAM.

Catalog No.	Size, Inches	Weight, Pounds	Shipped From BRADLEY, ILL.: Plow	Extra Share	Shipped From KANSAS CITY, MO.: Plow	Extra Share
32L130	12	110	$13.98	$2.45	$14.75	$2.60
32L131	14	122	15.15	2.95	16.00	3.10

Bradley Riding Attachment for Walking Plows.

GREAT TIME AND LABOR SAVER.

Makes a first class frameless sulky plow out of any wood or steel beam plow or middle breaker. It will make your plow run steadily without side draft and without making the work a bit harder on your team. It can be set for any depth furrow. Being tongueless, the attachment cannot be backed up by the team; neither does it carry the plow high from the ground. By placing the attachment back of a drag harrow and connecting to the harrow drawbar by a long pole or 2x4-inch timber you have a splendid harrow cart. Made entirely of malleable iron and steel. Wheels are the regular type of riding plow wheels with wide oval tires and dustproof hubs. Not intended for use on smaller than 10-inch plows nor rod breakers.

Dustproof Hubs.

32L192—Attachment for Right Hand Plows. Weight, 147 pounds. **Shipped from BRADLEY, ILL.** $11.95
Shipped from KANSAS CITY, MO 13.00
Shipped from FARGO, N. DAK $13.57
32L193—Riding Attachment for Left Hand Plows. Weight, 147 pounds. **Shipped from BRADLEY, ILL** 11.96

David Bradley Royal Blue Plows.

Price Includes One Extra Share.

For use in loose and dry loam or in a mixture of sandy clay. They do good work in stubble or tame sod, also stony land. The moldboard, share and landside are solid steel with mild or natural temper. They are securely bolted to a steel frog, and outer landside secured to a cast inner landside which has an adjustable slip heel. Bottoms are ground and highly polished. Moldboards are double shinned, insuring good wearing qualities. Beam is heavy steel, highly arched. Handles are of first quality oak. Shipping weights, 95, 100 and 110 pounds, according to size. Price includes one extra share.

STEEL BEAM, RIGHT HAND ONLY.

Catalog No.	Size, In.	Shipped From BRADLEY, ILL.: Plow	Extra Share	Shipped From KANSAS CITY, MO.: Plow	Extra Share
32L125	10	$10.97	$1.55	$11.70	$1.65
32L126	12	12.20	1.87	12.98	1.97
32L127	14	13.90	2.37	14.76	2.47

David Bradley Northwest Breakers.

FOR USE WITH 3 OR 4 HORSES.

Designed for heavier work than the standard type of prairie breaker. Splendid plows for heavy prairie sod and brush work, also for road making and grading purposes. Beam is extra heavy selected oak. Standard is heavy steel solidly braced. Moldboard, share and long bar landside are solid steel, with natural temper. The heavy steel coulter is made with a shoe at the point, into which the point of the share is inserted, making it absolutely solid. Rolling coulter cannot be used. Gauge wheel is adjustable. Handles are steam bent selected oak, extra heavy and strongly braced. Weights, 170 and 176 pounds. Prices include wheel and coulter.

Catalog No.	Size, Inches	Shipped From BRADLEY, ILL.: Plow	Extra Shares	Shipped From FARGO, N. DAK.: Plow	Extra Share
32L139	14	$19.95	$3.80	$21.82	$3.95
32L140	16	21.00	4.20	22.93	4.35

David Bradley Prairie Breaker Plows.

With Coulter, Gauge Wheel and One Extra Share.

ADJUSTABLE BEAM.

These plows have long tapering moldboard and slanting share which lies nearly flat and turns a smooth furrow. Intended for breaking original prairie land, but will do good work in old sod. Moldboard, share and landside are solid steel, with mild or natural temper. Beam is extra heavy seasoned oak with adjustment to make plow cut more or less land. Price includes coulter, gauge wheel and one extra share.

Catalog No.	Size, In.	Wt., Lbs.	Shipped From BRADLEY, ILL.: Plow	Extra Share
32L135	12	145	$14.75	$2.15
32L136	14	150	15.80	2.55
32L137	16	160	16.95	2.95

David Bradley New Ground Plow.

Complete With Jumping Coulter.

Designed especially for use in new ground where stumps and roots are too plentiful to permit the use of the regular type of breaking plow. It is not intended to serve the purpose of a regular moldboard plow. Jumping coulter either severs the root or carries the plow over it, after which the plow immediately re-enters the ground. Moldboard and share are in one piece and are solid steel. The moldboard is welded to a narrow steel landside which is bolted to the beam, the lower end of which curves to conform to the shape of the moldboard. Beam is heavy forged steel, highly arched. **Shipped from BRADLEY, ILL.**

32L168—9-Inch New Ground Plow. Weight, 100 pounds .. **$9.98**

David Bradley New Slant Cut Rod Breaker.

This plow is of our latest improved construction. The share lies nearly flat and cuts on a slant like our regular moldboard breaker. Rod breaker plows are for use in shallow plowing of original prairie sod, tough enough to hold together while turning. They work best where there are no stones or roots and should not be used for heavy work for which a regular moldboard breaker is required. Curved spring steel rods take the place of a moldboard and contribute to light draft. Share and landside are solid steel with mild or natural temper. Beam is heavy double beaded forged steel, solidly braced. Handles are oak, steam bent. Price is for the plow complete with fin cutter, gauge shoe, malleable clevis and one extra share.

Catalog No.	Size, In.	Wt., Lbs.	Shipped From BRADLEY, ILL.: Plow	Extra Share	Shipped From KANSAS CITY, MO.: Plow	Extra Share
32L141	12	85	$9.87	$2.15	$10.46	$2.25
32L142	14	90	10.92	2.55	11.55	2.65
32L143	16	95	11.97	2.95	12.63	3.05

David Bradley Hillside Swivel Plows.

Price Includes One Extra Share.

Intended especially for hillside plowing, but can be used with good results in level land. The bottom operates on a swivel held in place by a latch at the rear. By releasing the latch the bottom can be reversed from right to left. Moldboard, landside and share are of hard cast metal, the same as used in the David Bradley Cast Plows. Bottoms are ground and polished. These plows will turn a furrow from 1 to 3 inches wider than the share, according to condition of the soil or incline upon which they are used. Reversible jointer can be used on 10 or 12-inch plows. Gauge wheel can be used on all sizes. Prices include one extra share. **Shipped from BRADLEY, ILL.**

Catalog No.	Size, In.	Wt., Lbs.	Plow	Extra Share
32L180	6	75	$7.98	$0.75
32L181	8	80	8.85	.95
32L182	10	115	11.25	1.15
32L183	12	145	13.97	1.65

32L184—Reversible Jointer. Weight, 18 pounds. Extra **$1.85**
32L185—Gauge Wheel. Weight, 11 pounds. Extra **95c**

David Bradley Cast Plows.

Price Includes One Extra Share.

Materials used in the bottoms are a special mixture of metals of extreme hardness, yet possessed of strength and toughness. Beam is oak and adjustable for more or less land. A jointer should be used with 32L175 and larger plows when plowing trashy ground. **NOTICE—**Cast iron plows are for use only in sandy or gravelly soils and should never be used in black or sticky soils. This type of plow is always made with slanting landside and is measured from top of moldboard to outer edge of share. The bottom of share measures about 2 inches less than full cut of plow. Price includes one extra share, wrench and adjustable malleable clevis. Jointer and gauge wheel are extra. **Shipped from BRADLEY, ILL.**

Catalog No.	Will Cut Furrow From	Share Measures on Bottom	Wt., Lbs.	Plow	Extra Share
32L170	6 to 8 in.	6 in.	50	$5.98	49c
32L172	8 to 10 in.	8 in.	70	7.88	58c
32L175	10 to 12 in.	10 in.	108	10.90	79c
32L176	11 to 14 in.	11½ in.	135	13.50	98c

32L178—Jointer. Weight, 12 pounds. Extra **$1.73**
32L179—Gauge Wheel. Weight, 11 pounds. Extra .... **.92**

David Bradley Riding Plows

Write for Our
TIME PAYMENT OFFER
on David Bradley Farm Implements.

X-Rays Sulky Plow

$42.25 AND UP

High Lift. Foot Lift.

Every David Bradley implement is sold with the understanding that the purchaser may subject it to a fair trial in his own field before he decides to keep it, and if he is not perfectly satisfied and will notify us promptly, we guarantee to immediately make the implement entirely satisfactory to him, or will instruct him to return it to our factory to be exchanged for a satisfactory implement, at our expense, or, if he desires, we will return the money he paid for it and freight charges.

These famous plows are the product of the David Bradley factory's ninety-one years of successful plow building experience. Consequently, they need little, if any, introduction to farmers. They are made from the best materials procurable and designed to meet the hardest plowing conditions with the least possible draft. So perfect is the design of their foot lift and lever arrangement that a small boy can operate them with remarkable ease.

Frames are formed from extra heavy first quality bar steel, 2¼ inches wide by ⅝ inch thick, making one of the heaviest frames put on any riding plow.

Beams are highest quality double headed beam steel, 1¾ inches thick, formed with high arch for ample clearance in trashy ground. Beam braces on gang plows are 1¾-inch round steel bars, flattened at the ends and hot pressed to fit perfectly the channels of the beams, and for additional strength extend clear down to the frog.

Bottoms are the famous David Bradley Garden City Clipper Bottoms, with double shin moldboards made of hard tempered soft centered steel and polished in the same direction the furrow slice follows over their surfaces. They take the soil polish more quickly because it is a continuation of the factory process.

Shares are hard tempered soft center steel, $\frac{5}{16}$ inch thick, and are reinforced by welding a ¼-inch slab of steel on top, the point where wear is greatest. Uniformly tempered, finely ground and polished.

Wheels are steel, with staggered spokes set into long dustproof "long distance" hubs, and have wide half oval tires. Land wheels are 34 inches high, front furrow wheels 24 inches high and rear wheels 20 inches high. Wheel bearing surfaces are 9 inches long. An adjustable steel rod connects furrow wheels to pole. The pole plate lever permits quick adjustment of the furrow wheel from the seat. Rear wheel casters automatically and is set to relieve landside of friction, thus the draft is distributed on the thoroughly lubricated wheels.

Coulters are made of finest coulter steel, 15 inches in diameter, and revolve on dustproof chilled bearings which are adjustable for wear. Yokes swivel on long standards which are adjustable. **Jointers will be furnished in place of coulters if so ordered.**

The hitch on these plows is a valuable feature, adjustment being accomplished by simply loosening a bolt and sliding clevis to right or left. Adjustment to the finest point is possible, affording greater accuracy than when limited to a number of holes spaced a given distance apart.

David Bradley Garden City Clipper Bottoms.

Furnished on All David Bradley Riding Plows.
Your selection of a plow bottom should be governed by the nature of your soil.

Stubble or Old Ground.

Right Hand Only.

Hard Tempered Soft Center Steel. For use in stubble or ground which has been frequently cultivated. The moldboard is comparatively short with a bluff turn which pulverizes this particular soil and puts it in better shape than any other style. The bottom we recommend for the Red River Valley and most parts of Kansas, Nebraska, Minnesota and the Dakotas. Not intended for general sod plowing, for which our General Purpose shape described below is best adapted.

General Purpose or Stubble and Sod.

Right or Left Hand.

Hard Tempered Soft Center Steel. The General Purpose shape has a longer turn moldboard which turns timothy, clover, tame sod, etc., clear over. We recommend this shape for plowing very weedy ground, clover, timothy or blue grass sod and for deep stubble plowing.

David Bradley New Extra High Lift No. 6 Gang Plow

$64.95 AND UP

Double Bail.

Shipped from BRADLEY, ILL., KANSAS CITY, MO., or FARGO, N. DAK.

Prices of X-Rays Sulky Plows.

Prices are for the plows complete with three-horse hitch, pole, neckyoke, Bradley 15-inch rolling coulter, weed hook and wrench. Jointer will be furnished in place of coulter if so ordered. Sulky plows shipped from FARGO are equipped with special Northwest Bottoms.

Catalog No.	Size, In.	Style of Garden City Clipper Bottom. Right Hand Plows Marked R. H. Left Hand Marked L. H.	Wt., Lbs.	Plow Shipped From Bradley, Ill.	Plow Shipped From Kansas City, Mo.	Plow Shipped From Fargo, N. Dak.
32L225	12	Stubble, Right Hand	510	$42.25	$45.82	....
32L226	12	General Purpose, R. H.	512	42.75	46.33	....
32L227	14	Stubble, Right Hand	515	43.25	46.85	....
32L228	14	General Purpose, R. H.	517	43.75	47.37	....
32L229	14	General Purpose, L. H.	518	43.80	....	....
32L230	16	Stubble, Right Hand	520	44.25	47.89	....
32L231	16	General Purpose, R. H.	525	44.75	48.43	....
32L232	16	General Purpose, L. H.	526	44.80	....	....
32L233	14	Northwest Special	520	....	....	$49.98
32L234	16	Northwest Special	530	....	....	51.08

Extra price for four-horse abreast hitch in place of three-horse hitch...... $2.10
Extra price for five-horse string out hitch in place of three-horse hitch.... 3.50

Prices of High Lift No. 6 Gang Plows.

Plows are furnished complete with rolling coulters, pole, four-horse evener, neckyoke, weed hooks and wrench. Will furnish a four-horse string out hitch instead, if wanted, without extra charge. Jointers furnished in place of coulters if so ordered. Gang plows shipped from FARGO are equipped with Special Northwest Bottoms and extra heavy 2-inch beams.

Catalog No.	Size, In.	Style of Garden City Clipper Bottom Right Hand Plows Marked R. H. Left Hand Marked L. H.	Wt., Lbs.	Plow Shipped From Bradley, Ill.	Plow Shipped From Kansas City, Mo.	Plow Shipped From Fargo, N. Dak.
32L280	12	Stubble, Right Hand	735	$64.95	$70.10	....
32L281	12	General Purpose, R. H.	740	65.45	70.63	....
32L284	12	General Purpose, L. H.	742	65.50	....	....
32L282	14	Stubble, Right Hand	745	66.45	71.67	....
32L283	14	General Purpose, R. H.	750	66.95	72.20	....
32L285	14	General Purpose, L. H.	752	67.00	....	....
32L297	12	Northwest Special	750	....	....	$75.45
32L298	14	Northwest Special	760	....	....	77.05

5-horse string out hitch in place of 4-horse hitch. Extra.................. $2.50
6-horse string out hitch in place of 4-horse hitch. Extra.................. 3.75

Extra Prairie Breaker Bottoms.

With hard tempered soft center steel moldboard.

Right Hand Only.

Intended for breaking prairie sod. They turn the furrow to perfection and are remarkably light of draft. Moldboard is of hard tempered soft center steel, share is solid steel of mild temper, and each bottom is furnished with one extra share. Buy extra breaker bottoms for your Bradley Sulkies or Gangs. Order two bottoms for gang plows.

Catalog No.	Sizes, Inches	Weight, Pounds	Shipped From Bradley, Ill.	Shipped From Fargo, N. Dak.
32L237	12	55	$ 9.50	$10.10
32L238	14	60	10.00	10.66
32L239	16	65	10.50	11.22

Extra Shares for David Bradley Sulky and Gang Plows.

If extra shares should be ordered for plows previously purchased, then we must know whether they are for Stubble or General Purpose Plows, and the number and letters appearing on the back of the old moldboard, as well as the marks appearing on the back of the share.	Weight, Pounds	Shipped From Bradley, Ill.	Shipped From Kansas City, Mo., or Fargo, N. Dak.
12-Inch Hard Tempered Soft Center Steel	12	$2.65	$2.80
14-Inch Hard Tempered Soft Center Steel	14	3.15	3.30
16-Inch Hard Tempered Soft Center Steel	16	3.65	3.80
12-Inch Mild Solid Steel	12	2.00	2.15
14-Inch Mild Solid Steel	14	2.20	2.35
16-Inch Mild Solid Steel	16	2.55	2.70
12-Inch Cast Iron	12	.95	....
14-Inch Cast Iron	14	1.20	....
16-Inch Cast Iron	16	1.35	....

Bradley Power Lift Tractor Plow

$79.50 UP

Guaranteed to Work Satisfactorily With All Successful Makes of Tractors.

A proved plow that has successfully met the most difficult plowing conditions in all territories and recommended by hundreds of satisfied users throughout the country.

Guaranteed to work satisfactorily with any tractor. A tractor has neither sense nor temperament—it knows nothing of trade names, it pulls just so many pounds at the drawbar and handles a plow easier only when that plow is of lighter draft, regardless of the trade mark on it, and when you can buy a better plow than the one which happens to be offered with a tractor you should be privileged to use your judgment in selecting the plow as well as the tractor, especially when equal or better quality can be had for less money. We guarantee the Bradley Tractor Plow to be as stanchly constructed, as conveniently handled, to do as good work, regardless of conditions, and pull with as light draft as any other plow cutting the same number of furrows of same width and depth, and in the event this plow for any other reason does not perfectly satisfy the purchaser he may return it to us and we will return all money paid for it and the freight charges. Dealers desiring to recommend this plow for use with their tractors can assure their customers of the protection of our guarantee.

Lifts to full height in less than 3 feet of travel. The plow has an extremely quick lift, so essential for plowing close to the ends of the row. However, this quickness of lift has not increased the power demand on the land wheel, as advantage is taken of the natural tendency of a plow. The first movement of the lifting device turns the plow points upward, causing them to **automatically run out of the ground,** power being required only from the ground up and from which point two heavy counterbalancing springs assist in the work. So easy is this action that **wheel lugs are seldom used.** Likewise, immediate return, **points first,** to full depth is accomplished by another slight jerk on the single trip rope which serves to both raise and lower the plows.

Powerful, Simple Lift. The unique lifting device consists only of two cog toothed quadrant gears and a simple clutch. The lower gear is supported on the land wheel axle and when the clutch is engaged by pull rope, rotates with the wheel, lifting with it the upper gear, which is part of the frame, and at the same time bringing the arch of the front axles to a vertical position, thus lifting the plow to full height in less than 3 feet of travel.

BRADLEY QUICK DETACHABLE SHARES.

Patented.

A Truly Quick Detachable Share. Simply loosen the handwheel and it is unlocked from frog without removing a single bolt. With equal ease the share is drawn up and locked into place as the handwheel is tightened.

A compound lever connecting the front bail with the toggle joint on rear wheel gives a uniform lift to all bottoms, and automatically locks in either raised or lowered position.

Levers extend well forward in easy reach of operator on tractor. Their closely spaced locking notches permit the finest adjustment and a plowing depth of from 2 to 10 inches.

Garden City Clipper Tractor Plow Bottoms hold a pleasant surprise for those who have yet to try them. The peculiar roll of these moldboards permits much greater traveling speed without affecting the lay of the slice, which makes a graceful turn at any reasonable speed, working equally well in tame sod and old ground stubble plowing. **Moldboards and Shares** are made of hard tempered first quality soft center steel. Moldboards are double shinned with an extra thickness of steel welded to the front which stands the greater wear. Shares are made from 5/16-inch soft center steel, and are reinforced at the point by welding a 1/4-inch slab of steel on top where the wear is greatest.

BEAMS.

Beams are heavy double beaded steel I beams, 2 1/8 inches thick and arched to give a clearance of 22 inches from the bottom of the furrow. This liberal clearance, together with the 24-inch clearance from moldboard to moldboard, is evidence that the plow will not clog even in the most foul plowing.

WHEELS.

Front wheels are 26 inches in diameter with 4-inch oval tires; staggered steel spokes are welded into long greasetight hubs which are provided with large grease cups. The rear caster wheel, 20 inches in diameter, is pressed from solid steel to prevent the gathering of trash.

DRAWBAR.

Drawbar is made with length and up and down adjustments to fit any tractor. Is provided with wood break pins to protect plows from hidden obstructions. The drawbar permits backing the plows by power.

COULTERS.

The 15-inch self castering coulters are made of the finest coulter steel and mounted on adjustable, chilled cone, dustproof bearings with large receptacles for hard grease. **Steel Moldboard Combination Jointers** to attach to coulters are furnished at an extra price and may be ordered at any time.

Plows are furnished complete with Quick Detachable Shares and Rolling Coulters. Jointers are extra.

NOTICE—When ordering shares for a plow you have, always tell us the numbers and letters which appear on the back of old share.

Catalog No.	Size of Plow and Shares	Weight, Pounds	Shipped From Bradley, Ill.	From Kansas City, Mo.	From Fargo, N. Dak.
32L266	**Two 12-Inch Bottom Tractor Plow**	857	$ 79.50	$ 85.50	$ 88.92
32L267	**Three 12-Inch Bottom Tractor Plow**	1,038	99.75	107.00	111.17
32L274	**Two 14-Inch Bottom Tractor Plow**	875	81.00	87.12	90.62
32L275	**Three 14-Inch Bottom Tractor Plow**	1,065	102.00	109.45	113.70
32L276	**Combination Jointer,** each, extra	13	1.25	1.35	1.40
32L248	Extra Q. D. Share, 12-Inch Soft Center Steel	13	2.75	2.85	2.90
32L249	Extra Q. D. Share, 12-Inch Solid Steel	13	2.00	2.10	2.15
32L252	Extra Q. D. Share, 14-Inch Soft Center Steel	15	3.25	3.35	3.40
32L253	Extra Q. D. Share, 14-Inch Solid Steel	15	2.20	2.30	2.35

Bradley Tractor Disc Harrow

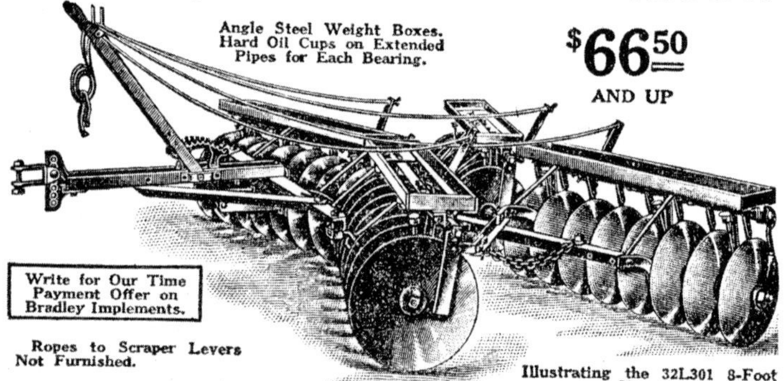

Angle Steel Weight Boxes. Hard Oil Cups on Extended Pipes for Each Bearing.
$66.50 AND UP
Write for Our Time Payment Offer on Bradley Implements.
Ropes to Scraper Levers Not Furnished.

Illustrating the 32L301 8-Foot Harrow.

Catalog No.	Without Scrapers	Wt., Lbs.	Bradley, Ill.	Kansas City, Mo.	Fargo, N. Dak.	Prices With Scrapers. Wt., Lbs.	Bradley, Ill.	Kansas City, Mo.	Fargo, N. Dak.
32L300	7-Foot Harrow, 28 16-inch discs	690	$66.50	$71.33	$74.10	792	$71.60	$77.15	$ 80.30
32L301	8-Foot Harrow, 32 16-inch discs	780	72.00	77.46	81.58	890	78.45	84.68	88.25
32L303	10-Foot Harrow, 40 16-inch discs	970	81.50	88.29	92.17	1,110	89.75	97.49	101.95

One needs but to consider the meritorious features of this David Bradley Heavy Duty Tractor Harrow to be convinced that the Bradley factory has scored another triumph in having overcome the well known faults that have been common to the many makes of tractor harrows.

ALL BEARINGS RELIEVED OF SIDE THRUST.

Disc blades do their work entirely by side thrust. The end strain of a disc gang is in the same proportion as the forward pull, and all of this strain would be borne by end thrust on the bearings if the "bumper" principle were not used. This is the way horse drawn harrows are made and the way the forward gangs of tandem harrows are made, but with others when it comes to making the trailer or rear gangs of tandem harrows, they abandon the very principle that has added so many years to the life of disc harrows.

SWIVEL CHAIN END THRUST EQUALIZER.

Flexible swiveling chain connection between rear gangs equalizes the end pull of the gang just the same as when the forward gangs roll together on bumper washers. The only strain on the bearings is the forward pull. Though the gangs usually revolve at the same speed, the chain revolving with them, provision is made for either gang rolling independently by chilled swivel connections at each end of the chain.

CROSS ARM COMPENSATING TRAILER CONNECTIONS.

Our rear gangs are guided and rear discs held to splitting the furrow of forward discs by compensating connections or drawbars. Every farmer knows the tendency of a disc harrow to "buckle up" in the center and the inner discs to run shallow; in a tandem harrow the discs of rear gangs being reversed, it is the tendency of inner discs to run too deep, and by the manner of crossing the Bradley compensating drawbars these forces are equalized. Thus the deep running discs of rear gangs exert their force on the shallow running discs of front gangs and all are kept cutting at a uniform depth. Straight line draft is accomplished for the rear gangs. The draft is direct from the tractor to the bearings on the front gangs and direct from the same front draw irons through **the compensating connections** to the rear bearings. The front bearings carry none of the weight of rear gangs. BEARINGS.

Here again Bradley construction scores an advantage. We use oil soaked hard maple liners in all bearing boxes. Both front and rear boxes are interchangeable, a big advantage in repairs.

FRAME AND WEIGHT BOXES.

The forward frame and stub pole is made into one solid unit from heavy angle steel. Weight boxes are formed from angle steel and are regular equipment. DISC BLADES.

These are of the highest quality disc steel, fully tempered and highly polished. The oscillating spring steel scrapers have throw-off levers which instantly free them from the discs when not needed.

David Bradley Perfection Sulky Plows

An excellent, light, medium high lift frameless Sulky Plow. All parts subject to strain are steel or malleable iron. The beam is high quality steel beam stock with high arch and ample clearance. Bottoms are the same as used on other David Bradley sulky plows. Levers can be easily operated by a boy. The rear wheel casters and locks in line with the plow after turning corners. Rear wheel lock casting is adjusted by two set screws, enabling the operator to set plow to carry the landside away from the land, reducing friction and insuring light draft. The short lever adjusts the front furrow wheel for more or less land while plow is in motion. Strong wheels have staggered spokes. Land wheel is 30 inches, rear and front furrow wheels are 20 inches in diameter. Rear wheel is of solid pressed steel and will not gather trash. Wheel boxings are dustproof with 7 1/4-inch wearing surface and have large receptacles for hard grease. The rolling coulter can be adjusted as required. The plow is furnished with three-horse evener, coulter, wrench and weed hook. **Shipped from BRADLEY, ILL., or HARRISBURG, PENNA.**

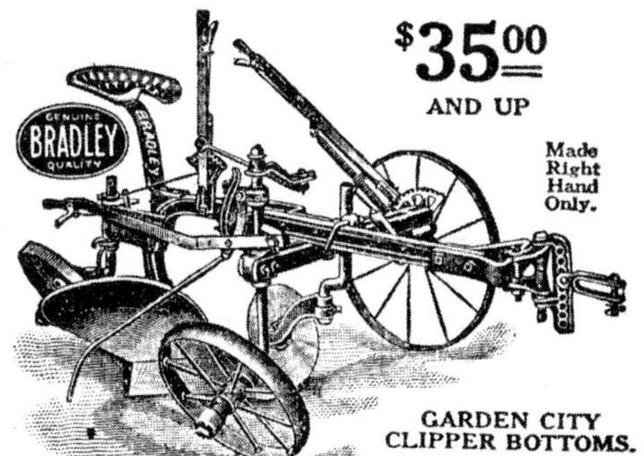

$35.00 AND UP
BRADLEY
Made Right Hand Only.

GARDEN CITY CLIPPER BOTTOMS.

Catalog No.	Size, In.	Style of Bottom, Right Hand Only	Wt., Lbs.	Shipped From Bradley, Ill. Plow	Extra Shares. Soft Center Steel	Solid Steel	Cast	Shipped from Harrisburg, Penna. Plow	Extra Shares. Soft Center Steel	Solid Steel	Cast
32L216	12	General Purpose	432	$35.00	$2.75	$2.00	$0 95	$38.25	$2.90	$2.15	$1.10
32L218	14	General Purpose	436	36.00	3.25	2.20	1.20	39.27	3.40	2.35	1.35
32L221	16	General Purpose	442	37.00	3.75	2.55	1.35	40.32	3.90	2.70	1.50

Quality based on Scientific Test is assured in buying from us. That makes our low prices doubly attractive.

David Bradley Ideal Tongueless Disc Harrows

These machines possess every important improvement in disc harrow construction, and we guarantee them to be unusually strong and perfect working implements. **The frame is made of heavy steel throughout.** It is rigidly braced in all directions. The gangs have bumper washers on their inner ends and are operated by independent levers. **The steel bars connecting the levers with the disc gangs are placed under a heavy steel yoke, which is adjustable up and down on the frame. Hitch is low** and the pull is direct from the gang bearings.

Weight boxes are heavy steel. Discs are carefully tempered steel and highly polished. **Dustproof bearings run in oil soaked maple boxes** and are provided with large compression grease cups. Axles are steel, ⅞ inch square. **The oscillating spring steel scrapers** are operated by slight foot pressure on a foot lever at each side of seat post. They are also provided with **a throw off lever** at rear. This feature not only saves wear on scrapers and discs, but lightens the draft when scrapers are not required. **The spring seat post,** fully covered by Bradley patents, can be used on no other disc harrow. This feature, together with the large, roomy seat, makes the David Bradley Ideal an extremely easy riding disc harrow.

A David Bradley adjustable oscillating tongue truck is secured to the stub tongue. The advantages of a tongueless disc harrow are so readily appreciated by farmers that it has come into almost universal use.

The truck relieves your horses of worry and neck weight and makes it possible for you to work clear out to the edge of plowed ground, close to fences, and to turn corners easily. The tongue truck can be easily detached and used on numerous other implements. A forward pole can be used on the truck if desired. If you wish us to furnish a pole with the harrow, we will do so for $2.10 extra. If you intend to use a forward pole, you should order extra either our 32L348 Pole Plate, price extra, 50 cents, or 32L349 Offset Pole Attachment, price extra, $1.35. **Buy this harrow and you will have a disc harrow as satisfactory and up to date as you could purchase at any price,** a machine which does perfectly uniform work. **Prices are for the disc harrows complete with eveners, scrapers, weight boxes and tongue truck.** We furnish a two-horse hitch with harrows having eight and ten discs, a three-horse hitch with harrows having twelve discs and a four-horse hitch with harrows having fourteen, sixteen or eighteen discs and five-horse hitch with harrows having twenty discs.

Shipped from BRADLEY, ILL., KANSAS CITY, MO., FARGO, N. DAK., or HARRISBURG, PA.

Bearing With Oil Soaked Maple Box and Hard Oil Cup. No Dust Can Get In.

Illustration Shows Harrow 32L334.

Hard Oil Cups on Tongue Truck.

If disc harrows are wanted with pole, and without tongue truck, deduct $3.35 from prices at left.

Prices of David Bradley Ideal Tongueless Disc Harrows, Complete With Weight Boxes, Scrapers and Eveners.

Catalog No.	No. of Discs	Width of Cut	16-Inch Discs					18-Inch Discs	
			Wt., Lbs.	Shipped From Bradley, Ill.	Shipped From Kansas City, Mo.	Shipped From Fargo, N. Dak.	Shipped From Harrisburg, Pa.	Wt., Lbs.	Shipped From Bradley, Ill.
32L330	8	4 ft.	400	$28.50	$31.30	$32.90	$31.50	415	$30.50
32L331	10	5 ft.	430	30.55	33.55	35.28	33.77	450	33.05
32L332	12	6 ft.	490	33.95	37.38	39.34	37.63	525	36.95
32L333	14	7 ft.	535	36.50	40.24	42.38	....	575	40.00
32L334	16	8 ft.	590	38.75	42.88	45.24	....	635	42.75
32L322	18	9 ft.	650	41.85	46.40	49.00	....	700	46.35
32L319	20	10 ft.	722	44.75	49.80	52.69	....	762	49.75

Amount to Deduct When Ideal Disc Harrows Are Wanted Without Eveners, Weight Boxes or Scrapers.

Deduct from harrows having	8 Discs	10 Discs	12 Discs	14 Discs	16 Discs	18 Discs	20 Discs
Without **eveners,** deduct	$1.50	$1.50	$2.75	$4.00	$4.00	$4.00	$6.00
Without **weight boxes,** deduct	1.00	1.00	1.50	1.50	2.00	2.00	2.00
Without **scrapers,** deduct....	2.25	2.50	2.75	3.00	3.25	3.50	3.75

SEND FOR OUR TIME PAYMENT OFFER. Although it is to your advantage to buy at these cash prices, no worthy farmer need do without a needed implement simply because he has not the ready money. Our offer will enable him to buy **David Bradley Implements** now at low prices and pay for them later. Write and say—**"Send me your Time Payment Offer on David Bradley Implements."**

David Bradley Tongue Truck.

OFFSET POLE ATTACHMENT.

POLE PLATE.

For use with disc harrows, seeders, land rollers, etc.

The same truck furnished with the David Bradley Tongueless Disc Harrow. Made with cast iron wheels, 15 inches high, with flanged tires 2 inches wide. They have long, oil retaining dustproof hubs and hard grease cups. Distance between wheels is 24 inches. The high arch steel axle allows the wheels to pass clear under the tongue. The casting which attaches to the stub pole is adjustable up or down on the standard, so as to accommodate all heights of hitches and insure direct draft. The construction of this casting is such as to permit the wheels to oscillate and adjust themselves to uneven ground. We furnish at extra prices quoted a pole plate, which can be used with two or four-horse hitches whenever it is desired to use a forward pole with the truck, or an offset pole attachment for forward pole with three or five-horse hitch. The truck is easily attached. It is only necessary to saw off but about 4 feet of the tongue and bolt the truck to the stub with bolts we furnish.

Dustproof Magazine Hubs and Hard Oil Cups.

Catalog No. and Description	Weight, Pounds	Shipped From Bradley, Ill.	Shipped From Kansas City, Mo.	Shipped From Fargo, N. Dak.	Shipped From Harrisburg, Pa.
32L347B—Tongue Truck..	65	$5.45	$5.90	$6.16	$5.93
32L348—Pole Plate for two or four horses...........	4	.50	.53	.55	.53
32L349—Offset Pole Attachment for three or five horses.	18	1.35	1.47	1.55	1.50

David Bradley Reversible Disc Harrow and Cultivator.

It serves the purpose of a disc harrow or a disc cultivator. The discs can be set at any angle desired and can be reversed to throw the dirt in or out. Can be set close together for harrowing or separated so as to straddle the row. Distance between gangs adjustable up to 18 inches. When set close together inside discs will be 6 inches apart. When discs are close together the six-disc machine cuts about 3 and the eight-disc machine cuts about 4 feet. Discs are 16 inches in diameter, tempered and polished. Aside from cultivating grain, hilling potatoes, cultivating tobacco, etc., this implement can be used to run between rows of berries, grapes, etc., for which work the ordinary disc harrow is too wide. Discs can be tilted either way so as to hill or trench. They have oil soaked maple bearings. Neckyoke and pole with two-horse hitch furnished regularly. Scrapers furnished at extra prices shown.

Catalog No.	Number of Discs	Weight, Pounds	Shipped From Bradley, Ill.	Shipped From Kansas City, Mo.	Shipped From Harrisburg, Pa.
32L345	6	217	$18.95	$20.47	$20.58
32L346	8	238	21.25	22.92	23.05

32L345B—Extra for Scrapers for 6-Disc Harrows. Weight, 8 pounds.......... **$1.15**
32L346B—Extra for Scrapers for 8-Disc Harrows. Weight, 10 pounds.......... **1.30**

Bradley Rotary Harrow Attachment.

Plow and Harrow at One Operation.

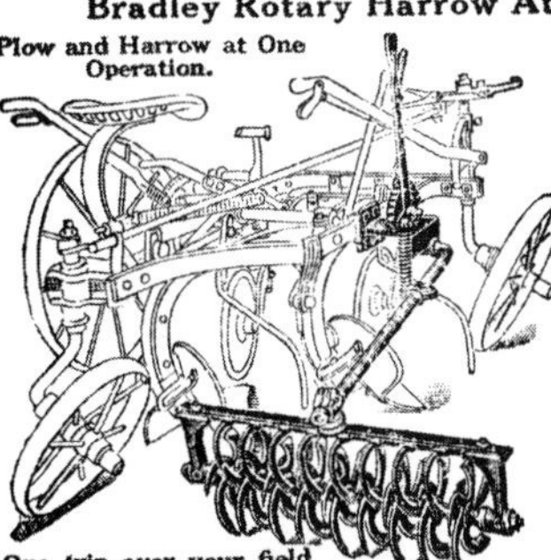

One trip over your field instead of two or three.

The cutting blades are curved. They cut throughout their entire length, not simply at the point, as do straight blades. They are self cleaning, will not clog, and they bury the trash well under the soil. Blades are malleable iron, cast in one piece and will not break or bend. Dustproof roller bearings. Blades are stationary on axle and all turn together. Conveniently placed lifting lever, assisted by tension spring, holds harrow to its work. We guarantee it to do satisfactory work in any soil. Attachment for sulky plows has six blades and cuts 18 inches. Attachment for gang plows or 2 bottom tractor plow has nine blades and cuts 28 inches. Attachment for 3 bottom tractor gang has 14 blades and cuts 42 inches. Easily attached to the frame of any riding plow. **Shipped from BRADLEY, ILL.**

	Wt. Lbs.	
32L580—Harrow Attachment for Right Hand Sulky Plow..........	80	**$10.95**
32L582—Harrow Attachment for Right Hand Gang Plow..........	100	**12.85**
32L583—Harrow Attachment for Left Hand Gang Plow..........	105	**12.90**
32L585—Harrow Attachment for Two-Bottom Tractor Plow..........	118	**14.50**
32L584—Harrow Attachment for Three-Bottom Tractor Plow..........	133	**15.50**

X-L-All Lime and Fertilizer Sower.

Strongly built to handle all kinds of commercial fertilizers, ashes, lime, marl and pulverized limestone.

The agitator and force feed consists of heavy cast paddles or propeller wheels mounted on the 1¼-inch square steel shaft axles and driven from both wheels. Drive shaft has a center bearing and each wheel drives half the shaft independently of the other half. A clutch on each wheel, operated by a hand lever, enables operator to throw out one or both drives.

The feed wheels are spaced 6 inches apart the entire length of hopper, one directly over each discharge opening in hopper bottom. Two steel rods extending full length pass through each feed wheel and in addition to strengthening the feed they serve as agitators. **This arrangement of feed wheels and agitators** gives a steady, uniform feed of any material, light or heavy. **Hopper holds 10 bushels, is strongly built** of good lumber, well braced, with hinged lid and sheet iron bottom. A removable galvanized screen keeps out obstructions which would not pass through discharge openings. A deflecting board underneath distributes fertilizer uniformly. Size of openings regulated by hand lever to sow any desired quantity. Width of sowing, 8 feet. Steel wheels, 30 inches in diameter, 4-inch concave tires. Price includes pole, without neckyoke or eveners. **Shipped from factory in WESTERN OHIO.**

32L1133—Lime and Fertilizer Sower. Weight, 385 pounds.......... **$42.75**

David Bradley Transport Truck for Disc Harrows.

Distance between wheels, 3 feet 2 inches. Wheels, 11¾ inches in diameter. Tires, 2 inches wide.

A very practical and useful device. Intended for use with Bradley Tongueless Disc Harrows or others with similar frame construction. The forward arms are placed under the disc axles of both gangs and hooked on to the forward frame. When the lever is thrown up the disc gangs are automatically raised from the ground and the harrow is ready for transporting on the road or from field to field, while the driver rides in the seat without danger of tipping over. Saves your discs and makes the work easier for horses. Strongly made entirely of iron and steel and should last a lifetime.

32L340—Transport Truck. Weight, 80 pounds.
Shipped from BRADLEY, ILL..........$6.00
Shipped from KANSAS CITY, MO...... 6.55
Shipped from FARGO, N. DAK.......... 6.88

David Bradley Steel U-Bar Lever Harrows.

For Garden Harrows for light work see page 842.

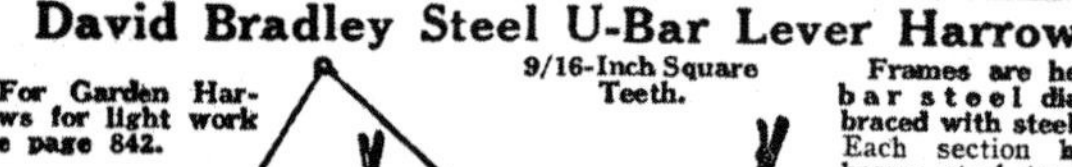

9/16-Inch Square Teeth.

Frames are heavy U-bar steel diagonally braced with steel braces. Each section has five heavy steel tooth bars. Teeth are attached by steel clips with set screws and lock nuts. Tooth clips are movable to any place on the bars, so you can use as many teeth as you want on each bar. The Bradley tooth clip is most satisfactory; when the teeth are properly set they will not become loose. The teeth are 9/16 inch square, 8 inches long, headed, and can be easily adjusted for depth. Made in sections of either 25 or 30 teeth. Each 30-tooth section cuts 5 feet wide and the 25-tooth section 4 feet 3 inches. Each section has runner teeth at corners. The teeth can be reversed to renew the cutting edge. Illustration shows the 60-tooth two-section harrow. Wood drawbar and irons furnished with two, three and four-section harrows. Draw irons only furnished with one-section harrows.

	Shipped From Bradley, Ill.	Shipped From Kansas City, Mo.
32L350— 25-Tooth One-Section Harrow. Weight, 89 lbs.	$ 5.17	$ 5.79
32L351— 30-Tooth One-Section Harrow. Weight, 102 lbs.	5.75	6.46
32L352— 50-Tooth Two-Section Harrow. Weight, 190 lbs.	11.25	12.58
32L353— 60-Tooth Two-Section Harrow. Weight, 220 lbs	12.75	14.29
32L354— 75-Tooth Three-Section Harrow. Weight, 285 lbs.	16.15	18.15
32L355— 90-Tooth Three-Section Harrow. Weight, 330 lbs.	19.45	21.76
32L356—100-Tooth Four-Section Harrow. Weight, 430 lbs	22.38	25.38
32L357—120-Tooth Four-Section Harrow. Weight, 470 lbs.	25.90	29.19

32L375—Extra Teeth. Weight, each, 6 ounces. Each 5c

32L376—Extra Clips. Weight, each, 5 ounces. Each 5c

32L379—End Guard Rails, similar to those used with our regular Guarded End Harrows, can be furnished for these harrows. Easily attached. One pair sufficient for each complete harrow. Weight, 13 pounds. Per pair 90c

David Bradley Wood Frame Lever Harrows.

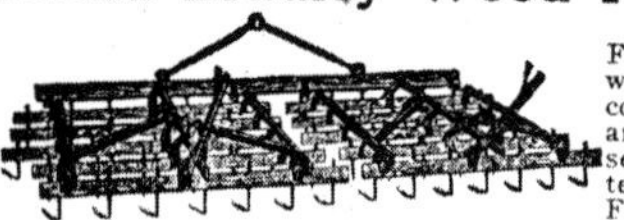

Strong and durable, lighter than steel. Frames are first quality seasoned hardwood, strongly riveted at each tooth and connected by iron and steel pivot plates and crosspieces. Made in sections, each section having thirty-five ½-inch square teeth and cutting about 5 feet wide. Frames are painted with high grade oil paint. Teeth can be set at any angle by adjusting the ratchet levers, or harrow can be tilted to run on the runners. Price includes drawbar.

	Shipped From Bradley, Ill.	Shipped From Fargo, N. Dak.
32L380— 35-Tooth One-Section Lever Harrow. Wt., 105 lbs.	$ 7.45	$ 8.60
32L381— 70-Tooth Two-Section Lever Harrow. Wt., 210 lbs.	15.20	17.50
32L382—105-Tooth Three-Section Lever Harrow. Wt., 330 lbs.	23.40	27.03
32L383—140-Tooth Four-Section Lever Harrow. Wt., 440 lbs.	31.50	36.34

Steel Frame Spring Tooth Harrows.

Favorites in timber countries. Of latest design with inner bar of one section cut off at forward hinge and bridged over the top to rear hinge, thus preventing clogging between the bars. Frames are of heavy angle steel. Teeth of standard design. One-section harrow cuts 3 feet. Illustration of single-section harrow shows steel handles, which are furnished at the extra price shown below. The two-section 6 feet and sections are hinged at center. For $1.00 extra we furnish the two-section harrow with special pointed teeth for cultivating alfalfa.

32L388—1-Section 9-Tooth Harrow. Wt., 135 pounds.
Shipped From Bradley, Ill .... $ 9.90
Shipped from Harrisburg, Penna. 10.91

32L389—2-Section 17-Tooth Harrow. Wt., 270 pounds.
Shipped from Bradley, Ill $17.85
Shipped from Harrisburg, Penna 19.87

32L390—Extra Spring Teeth. Weight, 5 pounds.
Shipped from Bradley, Ill. Each 48c
Shipped from Harrisburg, Penna. Each 53c

32L391—Extra Alfalfa Teeth. Weight, 5 pounds.
Shipped from Bradley, Ill. Each 57c
Shipped from Harrisburg, Penna. Each 62c

32L399—Steel Handles for 9-Tooth Harrow. Weight, 15 pounds.
Shipped from Bradley, Ill. Each $1.05
Shipped from Harrisburg, Penna. Each 1.15

David Bradley Steel Guarded End Lever Harrows.

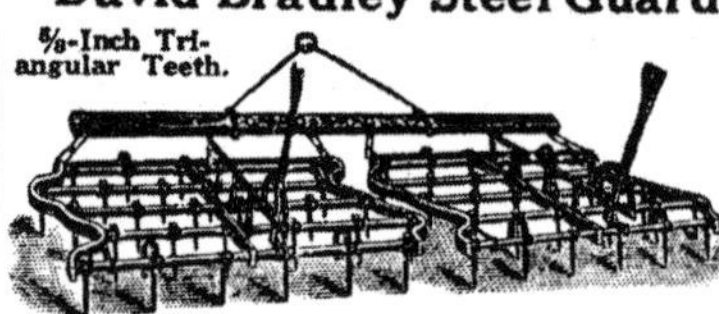

⅝-Inch Triangular Teeth.

Popular for use in new and stumpy ground. The sides and front end of the harrow are protected by a heavy steel I beam which forms the frame. The U shaped steel tooth bars are fastened to this frame at each end, making a very rigid construction. This outer frame serves to prevent the harrow from catching on trees, stumps, stones or other obstructions. The teeth are triangular in shape. They are 7 inches long and connected to the bars with the celebrated **Bradley steel tooth clip, the same as used on the Bradley U-bar harrow.** This clip enables one to place the teeth as far apart as desired or to use as many teeth on a bar as wanted. Teeth are swedged. Harrows are made in sections having either 25 or 30 teeth; each section has five tooth bars and runner tooth at each corner. The 25-tooth sections cut about 4 feet and the 30-tooth sections about 4¾ feet wide. Illustration shows a 60-tooth two-section harrow. Wood drawbar and irons **furnished with two, three and four-section** harrows. Draw irons only with one-section harrows.

	Shipped From dley, Ill.	Shipped From Harrisburg, Penna.
32L362— 25-Tooth One-Section Harrow. Weight, 90 lbs.	$ 5.37	$ 6.05
32L363— 30-Tooth One-Section Harrow. Weight, 97 lbs.	5.98	6.70
32L364— 50-Tooth Two-Section Harrow. Weight, 185 lbs.	11.65	13.05
32L365— 60-Tooth Two-Section Harrow. Weight, 210 lbs.	13.10	14.67
32L366— 75-Tooth Three-Section Harrow. Weight, 275 lbs.	17.60	....
32L367— 90-Tooth Three-Section Harrow. Weight, 310 lbs.	19.75	....
32L368—100-Tooth Four-Section Harrow. Weight, 365 lbs.	22.95	....
32L369—120-Tooth Four-Section Harrow. Weight, 440 lbs.	26.25	....

David Bradley Wood Frame Spring Tooth Harrows.

Especially adapted for use in timber countries. Made in two sections, hinged and flexible at center. Frames are first quality seasoned hardwood painted with high grade oil paint and fitted with stump guard around the sides and front and **lined with steel on the bottoms.** Teeth are made of tempered steel, firmly secured to the bars. The 16-tooth harrow cuts 5 feet; the 18-tooth, 6 feet, and 20-tooth, 7 feet wide. Price includes drawbar. Don't buy an unlined harrow; it will cost you nearly as much as a lined harrow and wear out in a much shorter time. **Shipped from BRADLEY, ILL.**

32L384—16-Tooth Lined Harrow. Weight, 175 pounds $13.96
32L385—18-Tooth Lined Harrow. Weight, 195 pounds 15.68
32L386—20-Tooth Lined Harrow. Weight, 215 pounds 16.87
32L387—Extra Harrow Tooth. Weight, 5 pounds .... .49

David Bradley Oak Frame Boss Harrows.

Stock at FARGO, N. D.

Frames are made of seasoned oak, strongly riveted at each tooth and firmly braced. All wood parts are well painted with a **first quality oil paint.** Teeth are ½ inch square. Sections are independently connected to drawbar with drop link clevises, allowing flexibility. The 48-tooth harrow cuts 8½ feet, has two side sections and two-horse drawbar; the 60-tooth harrow cuts 10½ feet, has one center and one side section and a two-horse drawbar; the 78-tooth harrow cuts 13 feet, has one center and two narrow side sections, and a two-horse drawbar; the 102-tooth harrow cuts 16½ feet, has one center and two wide side sections, and a three-horse drawbar; the 150-tooth harrow cuts 26 feet, has one center, two wide and two narrow side sections, and a four-horse drawbar which is fitted with a **sheave pulley and roller chain draft equalizer.** If you purchase a 150-tooth harrow you can use it as a 78-tooth or as a 102-tooth harrow by separating the sections and using a shorter drawbar.

	Shipped From Bradley, Ill.	Shipped From Fargo, N. Dak.
32L374— 48-Tooth Two-Section Boss Harrow. Wt., 135 lbs.	$ 7.97	$ 9.46
32L370— 60-Tooth Two-Section Boss Harrow. Wt., 145 lbs.	9.90	11.49
32L371— 78-Tooth Three-Section Boss Harrow. Wt., 190 lbs.	11.40	13.49
32L372—102-Tooth Three-Section Boss Harrow. Wt., 245 lbs.	15.32	18.02
32L373—150-Tooth Five-Section Boss Harrow. Wt., 375 lbs.	22.45	26.57

Universal Binder Tongue Truck.

Especially designed for use on binders and harvesters, but can also be successfully used with disc harrows and other implements. Enables one to do away with side draft on any harvester. Easily attached to any machine, either right or left hand. Wheels set 36 inches apart; truck will not skid. Oscillating frame permits truck to follow line of draft over rough ground without affecting tongue or implement. Truck can be used with or without forward pole. Strong steel wheels, 22 inches in diameter, have 3-inch tires. **Shipped from BRADLEY, ILL.**

32L555—Universal Binder for Tongue Truck. Weight, 132 pounds... **$11.45**

David Bradley All Steel Harrow Cart.

Can be attached to any style of drag harrow and will save many a day's hard tramp. It is strongly built and simple in construction. The driver is always in line with and facing his team. The seat is mounted on a steel **seat spring** which is attached to the frame and axle. As the axle is pivoted to the frame there is no cramping nor straining in turning. By tripping a foot lever when turning, the axle pivots to allow the cart to follow around after the harrow, and when straightened it automatically locks itself. The cart rides steadily and does not drift on side hill work. Furnished with either 24 or 42-inch wheels with 2-inch oval tires. Distance between wheels on 24-inch cart is 25¼ inches, on the 42-inch cart 43 inches. The shafts or connecting bars are heavy angle iron, attached to harrow evener by means of steel clasps.

Catalog No.		Size, Wheels	Weight, Pounds	Shipped From Bradley, Ill.	From Kansas City, Mo.
32L358	Harrow Cart	24-Inch	95	$7.60	$ 8.25
32L359	Harrow Cart	42-Inch	152	9.40	10.45

Bradley Combined Land Roller, Pulverizer, Crusher and Packer.

Illustration shows how split collars can be used in place of wheels to straddle rows.

Write for Our Easy Payment Offer on Bradley Implements.

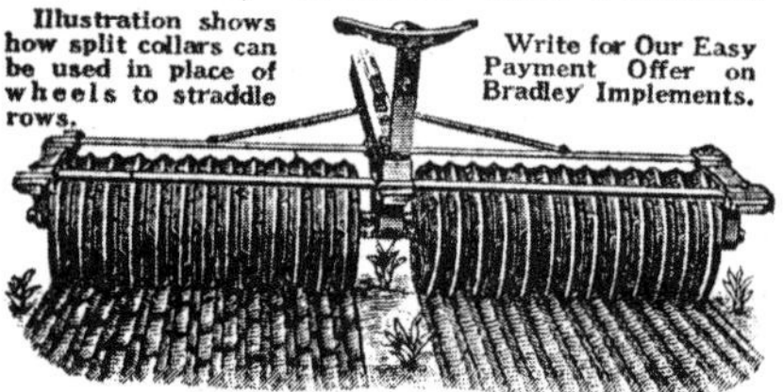

USED AS A SUBSURFACE PACKER. USED AS A ROLLER.

For thorough preparation of seed bed, pulverizing and packing the soil about the roots of growing plants this machine is without a superior. The 7-foot roller is particularly adapted for rolling corn, it being just the right size to use among standard widths of rows. The heavy cast iron discs turn independently on a 1½-inch steel shaft. Discs are 18 inches in diameter, 3 inches wide and have crushing lugs. These catch and pulverize the clods instead of pushing them along. Run the smooth face of the lugs forward and you have an ideal smoothing roller. Transfer the pole to the other side and run in opposite direction and the machine acts as a pulverizer and subsurface packer. Works the clods down and leaves small loose ridges on the surface so rain can soak away. For rolling young corn, wheat, clover and meadow in the spring, preparing seed bed for wheat in the fall, and for use in irrigated land and alfalfa culture it is invaluable and will often pay for itself in a single season on a 50-acre crop, because ground will hold moisture better and cause grain to germinate and mature earlier. Frame is heavy bar steel to which are attached the end and center hangers through which shaft runs. By removing one or two of the discs and using the two split collars furnished, roller will straddle rows if desired. If more than two split collars are wanted, we can furnish them at the extra price quoted. **Price does not include eveners or neckyoke. We will furnish with stub pole for tractor hitch in place of regular pole if so desired.**

Catalog No.	Length	Rolling Surface	No. of Discs	Weight	Shipped From Bradley, Ill.	Shipped From Kansas City, Mo.
32L324	7 feet	6 feet 4 inches	24	1,000 lbs.	$52.50	$59.50
32L342	9½ feet	8 feet 10 inches	34	1,280 lbs.	69.75	78.70

32L344—Extra Split Collars. Weight, 4 pounds. Each 38c

For Garden Harrows for light work see page 842.

David Bradley No. 1 Corn Planter

The David Bradley No. 1 Planter has stood the most severe tests in all sections of the country for more than thirty years. It has given excellent satisfaction under the most trying conditions. We therefore do not hesitate to recommend it to our customers as a planter in which the utmost dependence may be placed.

It has all the convenient features found in the most up to date planters of this type. It will successfully handle all kinds and sizes of seed, and if one uses the same care in grading and sorting the seed as is required with the more complicated edge drop style of planter, equal accuracy of drop can be obtained and a far greater degree of reliability. The plates of the David Bradley No. 1 are revolved by a pinion, which is operated by the check rod, and deposits the corn in the boots, where it is held by valves until the check is made. Then the plunger automatically forces the lower valve open, depositing the corn in the furrow without scattering. It is of strong, rigid construction, yet light enough to be easily handled by any team. The simplicity of mechanism, ease of adjustment and uniformity of drop are its most striking features. The frame is made of heavy channel steel, all connections are steel or malleable iron; it will stand the hardest use without danger of breakage. The runners are large and strong and their depth can be regulated independently. Seed boxes are large and are hinged and tilt forward, making it an easy matter to change plates without emptying seed. Wheels are steel and of strong construction. An automatic axle clutch shifter stops the drilling device when the front frame is raised, and when the frame is lowered the sprocket is automatically thrown in gear. The planter fitted for drilling only does not have the check rower and its attachments. It is furnished with four plates to drill single kernels of corn 12, 17 or 21 inches apart, and one blank plate. Check rower planter is complete with anchor stakes, 80 rods wire and automatic reel, which distributes the wire evenly. With the check rower planter we furnish seven plates to drill single kernels of corn, 12, 17 or 21 inches apart, and to drop two, three or four small, medium or large kernels, and one blank drill plate, which can be drilled as desired. Special plates can be furnished if you will send samples of seed. Can furnish special plates for kaffir corn, shoe peg and pop corn; also for large Maryland, Kentucky and Tennessee corn. The new automatic reel is operated by friction. The reel is lowered by foot lever and the friction wheel fits outer rim of planter wheel, which revolves it and winds the wire. An anchor stake is placed through the eye on seat bracket and the wire is placed on the hook at the end of anchor stake which, if shifted back and forth, insures even winding of wire. The spring anchor stake removes the strain on the check wire and adds to its life.

Planters are adjustable for rows 3 feet 4 inches, 3 feet 6 inches or 3 feet 8 inches apart. But all planters are set at 3 feet 6 inches when shipped. Can furnish check rower wire with buttons 3 feet, 3 feet 4 inches, 3 feet 6 inches, 3 feet 8 inches, 3 feet 10 inches or 4 feet apart.

Shipped From Bradley, Ill., Kansas City, Mo., or Fargo, N. Dak.

Accurate and Reliable Under All Conditions.

Illustration Shows Check Rower Planter 32L466.

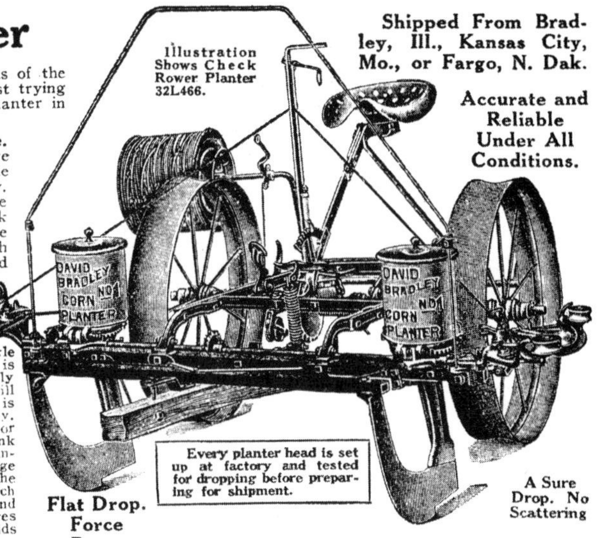

Every planter head is set up at factory and tested for dropping before preparing for shipment.

Flat Drop. Force Drop.

A Sure Drop. No Scattering

At extra price quoted we furnish a steel fertilizer attachment which attaches to the rear frame between seed boxes and wheels. When used with check rower planter this attachment drops the fertilizer 2 inches to the rear of the hill and when used with the drill planter it drills the fertilizer continuously in the rows.

Prices of Bradley No. 1 Planters and Attachments.

Prices include pole, disc marker and arch for marker rope to pass over.

Write for Our Time Payment Offer on Bradley Implements.

Important—Don't fail to state whether you want concave or open wheels, and height; also distance between buttons on check rower wire. Concave tires shipped unless otherwise ordered.	30-Inch Concave Wheels			38-In. Concave Wheels		30-Inch Open Wheels			38-Inch Open Wheels	
	From Bradley, Ill.	From Kansas City, Mo.	From Fargo, N. D.	From Bradley, Ill.	From Kansas City, Mo.	From Bradley, Ill.	From Kansas City, Mo.	From Fargo, N. D.	From Bradley, Ill.	From Kansas City, Mo.
32L465—No. 1 Corn Planter, fitted for drilling only. Weight, 334 pounds	$36.95	$39.28	$40.62	$39.90	$42.23	$37.45	$39.78	$41.12	$40.40	$42.73
32L466—No. 1 Check Rower Planter, with 80 rods of wire. Weight, 465 pounds	45.75	49.00	50.87	48.70	51.95	46.25	49.50	51.37	49.20	52.45
32L468—Fertilizer Attachment, extra. Weight, 120 pounds	13.75	....	....	....	....	....	....	....	....	....

When ordering Fertilizer Attachment **only**, advise whether your planter has 30 or 38-inch wheels.

Planter Wire.

If you want more than the 80 rods of wire furnished with our planter, or if you wish to purchase wire for an old planter of any make, order as many extra rods of this wire as you require. It is the same as is used on all standard planters and is sold in bundles of 10, 20, 40 and 80 rods, and with buttons 3 feet, 3 feet 4 inches, 3 feet 6 inches, 3 feet 8 inches, 3 feet 10 inches or 4 feet apart. **Be sure to state distance desired between buttons.**

Shipped from BRADLEY, ILL.

32L469—Check Rower Wire. Weight, 8 ounces per rod. Per rod........ **5½c**

David Bradley No. 3 Champion Combined Cotton and Corn Planter.

STEEL FRAME ONE-HORSE DRILL.

A highly satisfactory combination cotton and corn planter. It has the reliable Bradley Champion force feed which has made David Bradley cotton planters so successful. A slide in the bottom of cotton box regulates the amount of cotton seed dropped. Machine can be regulated to plant from 1 peck to 2 bushels of cotton seed to the acre. Will not clog, does not injure or bunch the seed and plants it as it comes from the gin, without any previous preparation. Machine can be thrown out of gear at end of rows by means of throwout rod attached to handle. For planting corn, the same seed box is used, and we furnish plates for drilling one kernel of corn every 17 or 25 inches and a blank plate which can be drilled for small seed or grain to suit. The change from a cotton to a corn planter is quickly made. Shovel shanks have a friction break and a depth gauge runner is at the end of each shovel. The machine is made of iron and steel excepting handles, which are of oak.

Shipped from BRADLEY, ILL.

32L459—No. 3 Combined Cotton and Corn Planter. Wt., 83 lbs....... **$9.65**

Kenwood Wheelbarrow Grass Seeder.

For sowing small seeds broadcast. Made with hoppers 12, 14 or 16 feet long. The wheel governs the feed and the index plate and the speed of travel governs the quantity sown. Has iron wheel and force feed which can be accurately adjusted. Can be instantly thrown out of gear. The single hopper seeder is for sowing smooth seeds only, such as clover, timothy, alfalfa, millet, flax, etc. The double hopper seeder will sow all kinds of smooth seeds and also light and chaffy seeds, such as red-top, orchard grass, blue grass, etc. Weight, 45, 50 and 55 pounds, according to size. **Shipped from factory in NORTHERN OHIO.**

FAVORITE IN ALFALFA COUNTRY.

	12-foot	14-foot	16-foot
32L1160—Single Hopper Seeder	$8.35	$8.60	$9.20
32L1161—Double Hopper Seeder	9.67	10.00	10.50

Two-Cylinder Rotary Seed Corn Grader.

Shipped from factory in SOUTHWESTERN OHIO.

This grader is a wonder for capacity and precision of grading. The inner cylinder, to which corn is first delivered by positive screw feed in the supply passage, handles the coarser grades first, getting them out of the way without going into the second or larger cylinder, which handles the smaller grades, thus relieving it of unnecessary volume. Overloading is positively prevented by the screw feed. Its big capacity is accounted for by the deep corrugation of both cylinders, which forces the kernel to stand on edge or on end for different grades ready to pass out the desired opening. Unlike a flat grader, the rotary motion prevents clogging. By raising the feed end ½ inch and running through again, two extra grades are obtained, making five grades possible. The grader is 30 inches high, 32 inches long and 15 inches wide. Is simple to operate and has a capacity of 10 to 12 bushels per hour. It is crated, with legs detached, to insure the best freight rate. Price does not include measures shown in illustration.

32L7075—Two-Cylinder Corn Grader. Weight, 45 pounds............ **$9.25**

David Bradley Nos. 2 and 3 Presser Wheel One-Horse Corn Drills.

A high grade machine in every respect. This machine has rotary force feed dropping device which is controlled by sprockets made to drop one kernel of corn every 10, 15, 20 or 26 inches. We furnish four sprockets and two plates suitable for dropping large or small corn and blank plate which can be drilled to plant peas, beans, cane or other seed. An opening in the boot enables the operator to see the corn at all times as it leaves the hopper. By means of a slip clutch the drill can be backed without throwing out of gear. The wide presser wheel is provided with a steel scraper. It properly covers every hill. No. 3 drill is shown in illustration at the right. No. 2 drill is the same, excepting that it is not provided with a fertilizer attachment. It sows any quantity of fertilizer desired. Machine is strongly made.

These Drills Cannot Be Used for Planting Cotton Seed.

Prices are for the machine complete, as illustrated.

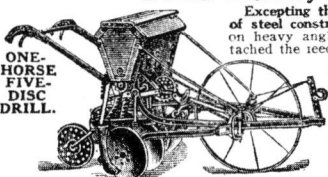

32L452—No. 2 One-Horse Corn Drill. Weight, 90 pounds.
Shipped from BRADLEY, ILL......... **$11.98**
Shipped from KANSAS CITY, MO........ **12.60**

32L453—No. 3 One-Horse Corn and Fertilizer Drill. Weight, 128 pounds.
Shipped from BRADLEY, ILL........ **$15.05**
Shipped from KANSAS CITY, MO........ **15.95**
Shipped from HARRISBURG, PA........ **16.02**

32L454—Marker and Pole. Weight, 7 pounds.
Shipped from BRADLEY, ILL., 50c; from KANSAS CITY, MO........ **55c**

David Bradley Grain Drill.

ONE-HORSE FIVE-DISC DRILL.

Excepting the sides and lid of seed box, this drill is of steel construction throughout. The box is mounted on heavy angle bars and to these bars are also attached the feed cups, making a solid unit in which true alignment is assured. The feed is of the well known fluted force feed type which will sow any small grain or larger seeds, such as cowpeas or corn, without cracking. Feed is chain driven from front wheel and amount sown is shown on gauge, which is changeable while in motion; agitator furnished regular, hence their adaptability to handling long bearded oats.

The 13-inch polished discs run on dustproof bearings, having large receptacles for hard grease. Distance between discs is adjustable and they cut an extreme width of 34 inches.

Weight of drill is carried on front wheel and two rear caster wheels, allowing discs to float at any desired depth, which is maintained by the pressure springs above each disc. Depth adjustment is instantly regulated by releasing hand latch and raising or lowering the handles.

The two outside discs are pivoted at the clevis and swing against compression springs which allow them to give when striking obstructions.

Drill is equipped with steel ribbon tubes, covering chains and hand guards.

Shipped from factory at BRADLEY, ILL. **$27.50**

32L1113—Five-Disc Drill. Weight, 350 pounds.

Kenwood Endgate Broadcast Seeder.

A good Agitator Endgate Broadcast Seeder. Sows broadcast all kinds of small seeds and dry fertilizers. Feed is adjustable, no loose plates to lose or mislay. Will distribute seed evenly 12 to 40 feet in width, depending upon weight and kind of seed. Seeder is fitted with a spring clutch and in starting or stopping there is no sudden jar or strain. Machine is attached to board, taking the place of the rear endgate. Price includes large sprocket wheel and clips for fastening it on the wagon wheel; also chain for driving the machine and full directions for attaching and operating. Will sow 100 acres of wheat per day with team traveling at the rate of 2½ miles per hour. **Shipped from factory in SOUTHEASTERN WISCONSIN.**

32L1150—Kenwood Endgate Seeder. Weight, 95 pounds...... **$12.70**

For Hand Seeders see page 842.

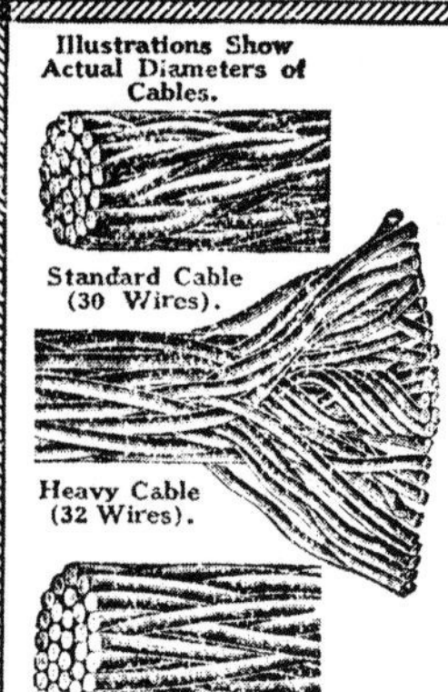

Illustrations Show Actual Diameters of Cables.

Standard Cable (30 Wires).

Heavy Cable (32 Wires).

Extra Heavy Cable (34 Wires).

Copper Cable Lightning Rods—Furnished in Three Sizes

In sections where electrical storms are unusually severe the heavier cables have been extensively adopted, especially for large buildings or those requiring three or more groundings. For all ordinary buildings, however, our Standard cable is sufficient.

Our two heaviest cables are so woven as to permit the greatest possible circulation of air around and between the wires. This reduces the danger of overheating and thus melting the cable.

Vanes are mounted on brass swivel to turn easily with the wind.

In measuring for cable allow enough so that all bends or turns over eaves or around corners will be gradual, and where cable passes close to metal cornices, water spouts, ornaments, gutters, etc., it should be connected to such points. At least two ground connections are needed for every building, and on large buildings with more than five tops or points, three groundings. Allow 10 feet for each ground connection.

Tops or points should not be over 25 feet apart. Each cupola and tower should have a point or top, and each gable or wing extending from main structure 10 feet or more should also have a point or top. There should also be a top or point at the side of each chimney.

An expert is not needed for measuring your building nor for putting on the rods. You can easily do this yourself and save this extra expense. **If your building is of unusual construction, or you want more information, write us, stating what you need, and make a rough sketch of the building and we will send you a special catalog, giving full instruction for rodding buildings; also showing illustrations of our complete line of fixtures not shown on this page.**

Prices include sufficient copper clips and nails for fastening cable to the building.

A complete lightning rod outfit with cable and complete tops will weigh, packed for shipment, from ¾ to 1 pound to the foot. **Shipped from factory near CHICAGO.**

32L3700—Standard Copper Lightning Rod Cable. Per foot........ **$0.06¾**
32L3696—Heavy Copper Lightning Rod Cable. Per foot........ **.07½**
32L3695—Extra Heavy Copper Lightning Rod Cable. Per foot........ **.09¾**
32L3698—Complete Plain Top........ **1.00**
32L3702—Complete Top With Animal Figure and Glass Ball. (State figure, whether horse, cow, pig, sheep or rooster, and color of ball.)........ **2.35**
32L3712—Complete Top With Arrow Vane and Glass Ball. (State color of glass ball wanted)........ **1.75**

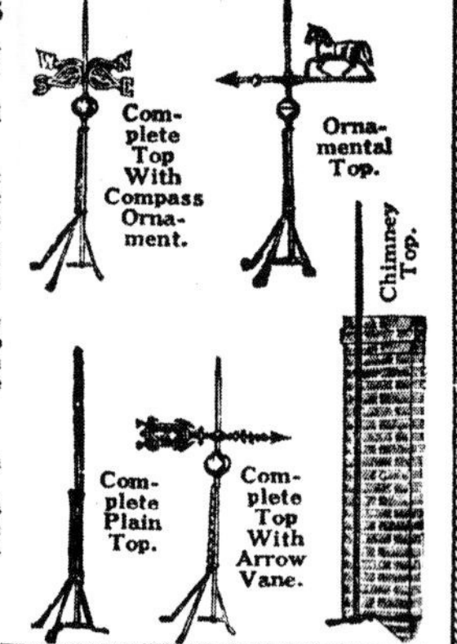

Complete Top With Compass Ornament.

Ornamental Top.

Chimney Top.

Complete Plain Top.

Complete Top With Arrow Vane.

High Quality Road Scrapers.

Made of Heavy Gauge Steel.

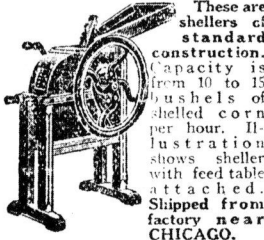

Bowls are pressed from a single sheet of special high carbon scraper steel, and by a **special method of stamping the full thickness of steel is preserved at the point of greatest wear.** Corners, sides and back are formed on a gradual curve, causing scrapers to fill and clean easily. Sides are stiffened by a projecting flange all around upper edge. Nose is rounded and enters ground easily. Bail and swivel are heavy forged steel. Hooks and handle sockets are heavy steel securely riveted in place. Hardwood handles. **Double runner scrapers** are the favorite, having two runners of hardened steel riveted to bottom. **Double bottom** scrapers have an extra bottom plate of hard steel, 13x21 inches, riveted to bottom. Our scrapers are unsurpassed for the heaviest work of farmers, contractors, townships and railroads. **Shipped from factory in WESTERN OHIO.**

DIMENSIONS OF SCRAPERS.

Size	Length, Inches	Width, Inches	Depth, Inches	Capacity, Cubic Feet	Gauge, Steel	Weight, Pounds
No. 3	32½	27	13	3½	10	75 to 85
No. 2	32½	30	13¾	5	10	85 to 95
No. 1	35	32½	13¾	7	9	95 to 105

	No. 3	No. 2	No. 1
32L2500—Smooth Bottom Scraper	**$6.65**	**$6.95**	**$7.35**
32L2503—Double Runner Scraper	**7.30**	**7.60**	**7.98**
32L2506—Double Bottom Scraper	**8.00**	**8.35**	**8.70**

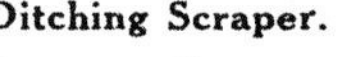

Ditching Scraper.

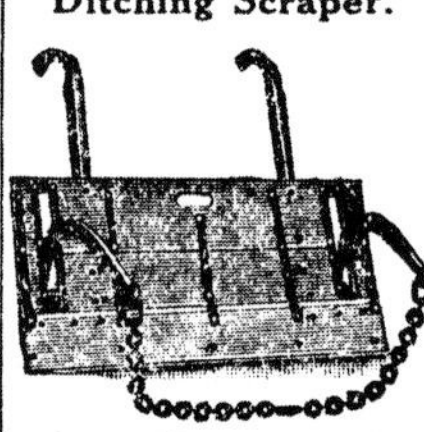

An excellent scraper for cleaning out and filling ditches, and for leveling roads and uneven places. Is well made of seasoned hardwood. Steel bit is 48 inches long, 7 inches wide and ¼ inch thick. This scraper is well ironed, has 1¾ by ¾-inch steel hounds, with ⅜-inch cable chain. **Shipped from factory in WESTERN OHIO.**

32L2525—Ditching Scraper. Weight, 75 pounds........ **$8.48**

Improved Ideal Farm Road Grader.

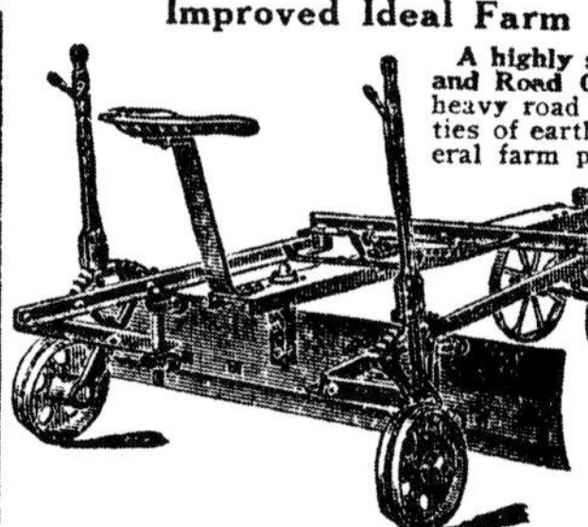

A highly successful light two-horse Farm and Road Grader. It is not intended for heavy road work where very large quantities of earth are to be moved, but for general farm purposes and for light and frequent working of roads. The blade can be reversed merely by pressing on foot latch and turning team to right or left. Will cut deep or shallow. High carbon steel blade is 6 inches wide, moldboard with blade attached is 12 inches wide and 6 feet long. Blade is bolted to curved steel moldboard and can be replaced when worn. Wheels reverse automatically with blade. Machine is made entirely of iron and steel, and measures 6 feet long from center to center of trucks. Rear wheels are 16 inches in diameter and have steel roller bearings. Front trucks have both caster and ball and socket action, so machine can be worked either with or without tongue. As the grader can be worked where very heavy machines could not be used, it is of great value to farmers for leveling fields and barnyards, keeping lanes and driveways in order, etc. Generally used for tongueless, but we can furnish a tongue for $3.50 extra. Price does not include hitches. **Shipped from factory in CENTRAL OHIO.**

32L2552—Improved Farm and Road Grader. Weight, 600 pounds........ **$61.50**
32L2553—Extra Blade. Weight, 30 pounds........ **4.90**

One-Hole Corn Shellers.

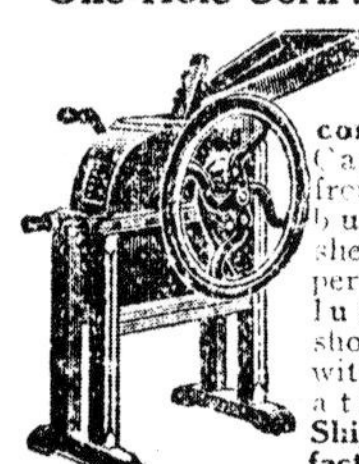

These are shellers of standard construction. Capacity is from 10 to 15 bushels of shelled corn per hour. Illustration shows sheller with feed table attached. **Shipped from factory near CHICAGO.**

32L1400—One-Hole Sheller, without table or fan. Weight, 130 pounds........ **$10.30**

32L1401—One-Hole Sheller, with table only. Weight, 137 pounds........ **$10.90**

32L1402—One-Hole Sheller, with table and fan. Weight, 145 pounds........ **$11.23**

32L1403—Clamp Pulley, 8x2½ inches. Weight, 6 pounds.... **6Cc**

For Small Hand Sheller see page 839.

We handle a complete line of **Grain Bins** and **Corn Cribs.** If interested write for prices.

David Bradley Two-Hole Corn Sheller.

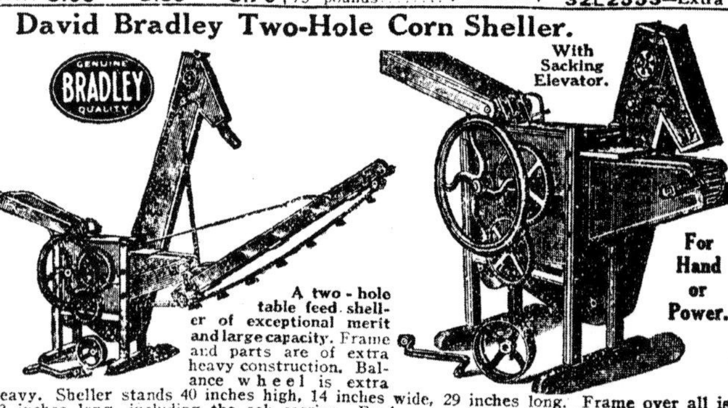

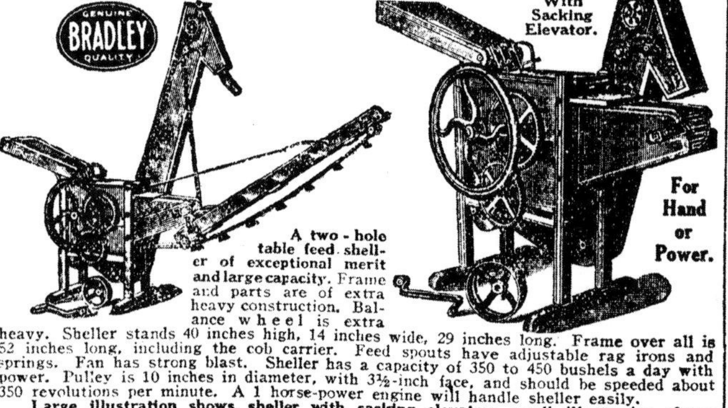

With Sacking Elevator.

For Hand or Power.

A two-hole table feed sheller of exceptional merit and large capacity. Frame and parts are of extra heavy construction. Balance wheel is extra heavy. Sheller stands 40 inches high, 14 inches wide, 29 inches long. Frame over all is 52 inches long, including the cob carrier. Feed spouts have adjustable rag irons and springs. Fan has strong blast. Sheller has a capacity of 350 to 450 bushels a day with power. Pulley is 10 inches in diameter, with 3½-inch face, and should be speeded about 350 revolutions per minute. A 1 horse-power engine will handle sheller easily.

Large illustration shows sheller with sacking elevator; small illustration shows sheller with wagon elevator and cob stacker, but price on sheller does not include any of these attachments, which are furnished only at the extra prices quoted below. Price on sheller includes crank, pulley, fan and feed table.

32L1408—Bradley Two-Hole Sheller. Weight, 290 pounds.
Shipped from BRADLEY, ILL........ **$24.95**
Shipped from KANSAS CITY, MO........ **26.98**
32L1409—6-Foot Cob Stacker. Weight, 35 pounds.
Shipped from BRADLEY, ILL........ **7.85**
Shipped from KANSAS CITY, MO........ **8.10**
32L1410—5-Foot Sacking Elevator. Weight, 65 pounds.
Shipped from BRADLEY, ILL........ **12.45**
Shipped from KANSAS CITY, MO........ **12.90**
32L1411—8-Foot Wagon Elevator. Weight, 90 pounds.
Shipped from BRADLEY, ILL........ **17.25**
Shipped from KANSAS CITY, MO........ **17.88**

Triple Geared Ball Bearing Sweep Feed Mill.

For grinding corn and cobs, shelled corn, oats and other small grain for feeding, but will not grind ear corn with the husks.

Easy running and has good capacity. Ball bearings support the internal revolving mechanism. The gearing is enclosed. The cob breaker breaks the ear corn and forces it into the grinding rings. Grinding rings make three revolutions to one round of the horse. Average capacity, based on coarse grinding of dry corn and cobs, is from 8 to 15 bushels an hour when two horses are used. Grinding rings regularly furnished with the mills are for coarse grinding of corn and cobs or shelled corn for feeding purposes. For finer grinding of shelled corn or for grinding wheat, oats, rye or other small grain, fine grinding rings are required. Capacity on shelled corn is from 10 to 25 bushels an hour, oats, 4 to 8 bushels an hour. Grinding medium or fine with fine rings reduces the capacity one-third to one-half. The mill is mounted on a platform. Price is for the mill complete with one pair of grinding rings only, sweep and hitch hook. **When ordering mill or extra rings, state whether you want coarse or fine grinding rings. Shipped from factory near CHICAGO.**

32L1620—Triple Geared Sweep Feed Mill. Weight, 635 pounds........ **$37.50**
32L1621—Pair of Extra Grinding Rings. **(State whether coarse or fine is wanted.)** Weight, 45 pounds........ **$4.20**

X-L-ALL Self Feed Two-Hole Corn Sheller.

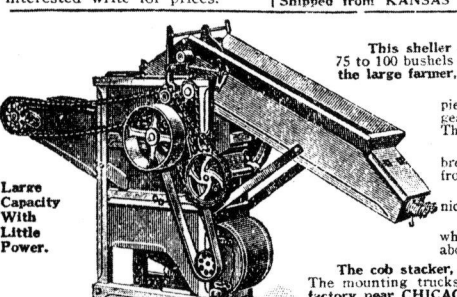

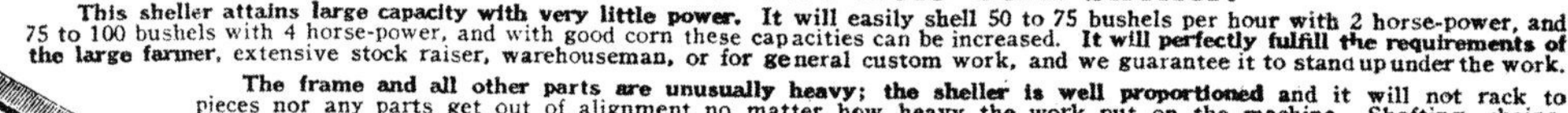

Large Capacity With Little Power.

This sheller attains large capacity with very little power. It will easily shell 50 to 75 bushels per hour with 2 horse-power, and 75 to 100 bushels with 4 horse-power, and with good corn these capacities can be increased. **It will perfectly fulfill the requirements of the large farmer,** extensive stock raiser, warehouseman, or for general custom work, and we guarantee it to stand up under the work.

The frame and all other parts are unusually heavy; the sheller is well proportioned and it will not rack to pieces nor any parts get out of alignment no matter how heavy the work put on the machine. Shafting, chains, gears and all mechanism are very heavy. Running parts are accurately fitted, insuring smoothness of operation. The interior of machine is easily accessible from the top through a hinged lid covering the adjusting springs.

The sheller is quickly and easily adjusted for different kinds of corn, and is guaranteed to shell clean without breaking the cobs or grinding the corn. Having an extra large and powerful fan, it delivers the shelled corn free from cobs and dirt.

Sheller is 5 feet high, 30 inches wide and 55 inches long over all. Frame is selected hardwood and machine is nicely painted, striped and varnished.

Gears are protected by guards and a shifting clutch throws feeder in or out of gear instantly. Solid web balance wheel, 22 inches in diameter, weighing 65 pounds, insures steady operation. Pulley is 12x4 inches and should run about 700 revolutions per minute.

The cob stacker, wagon box elevator and the 5-foot sacking elevator are furnished only when ordered at the extra prices shown. The mounting trucks, 32L1585, are especially useful when sheller is used for custom work. **Sheller and attachments are shipped from factory near CHICAGO.**

32L1430—Self Feed Two-Hole Sheller. Weight, 705 pounds........ **$86.50**
32L1431—Wagon Box Elevator. Weight, 165 pounds........ **27.40**
32L1432—5-Foot Sacking Elevator. Weight, 108 pounds........ **17.95**
32L1433—8-Foot Straightaway Cob Stacker. Weight, 65 pounds........ **$11.90**
32L1434—8-Foot Swivel Cob Stacker. Weight, 140 pounds........ **21.50**
32L1585—Set of Mounting Trucks. Weight, 500 pounds........ **54.75**

David Bradley Cob Crusher Feed Grinder

Ear Corn will not bridge in this large capacity hopper.

Dust Tight Burr Cover.

Instantaneous hand lever adjustment.

Large Heavy Flywheel.

Angle Steel Legs, strongly braced.

Sacking Elevator is furnished only when ordered at the extra price.

Successfully grinds corn on the cob (without the shuck), shelled corn, oats or other small grain, and to most any degree of fineness from cracked feed to table meal.

The cob breakers which cut and crush the cobs are set spirally on the heavy shaft and force the grain to the corrugated ring crusher and worm feed, which set next to the burrs and keep them working to their full capacity. Gate between hopper and burrs regulates the amount according to the power available. Burrs are 8-inch, self aligning, are protected by a cushion spring, easily changed, and are instantly thrown out of gear by a handy safety lever. The frame bed is one solid casting with a hinged bottom held to place by wood brake pins, which protect the grinder against solid obstructions that may be in the grain. The stanch construction of this grinder, with the heavy angle steel legs solidly braced, large, true running bearings with hard grease cups and the balanced heavy flywheel, insure long years of service.

Grinder should run 300 to 350 revolutions per minute and requires 4 to 6 horse-power. Pulley is 12 inches in diameter with 6-inch face, or we will furnish either 8x6, 10x6 or 14x6-inch pulley instead without extra charge. **Please state size of your engine pulley and its speed.** Capacity depends on the power, the speed, the kind of grain and its condition. The range of capacities is from 10 to 25 bushels of dry ear corn and 10 to 40 bushels of shelled corn, wheat, oats or barley per hour. Regularly furnished with one set of coarse burrs for crushing or cracking ear and shelled corn for rough feed and one set of medium burrs for finer grinding of shelled corn and small grain. At the extra price we furnish special oat burrs, which are also suitable for grinding table meal.

The sacking elevator is furnished only when ordered at the extra price shown.

Illustrating the "Slice Cut" Burrs furnished as extras at extra price quoted.

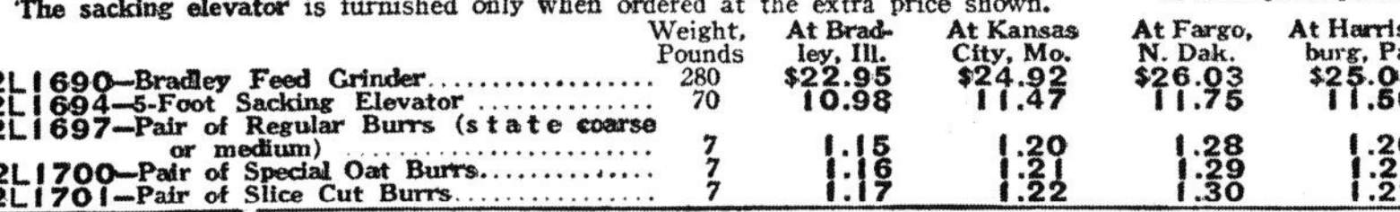

	Weight, Pounds	At Bradley, Ill.	At Kansas City, Mo.	At Fargo, N. Dak.	At Harrisburg, Pa.
32L1690—Bradley Feed Grinder	280	$22.95	$24.92	$26.03	$25.05
32L1694—5-Foot Sacking Elevator	70	10.98	11.47	11.75	11.50
32L1697—Pair of Regular Burrs (state coarse or medium)	7	1.15	1.20	1.28	1.20
32L1700—Pair of Special Oat Burrs	7	1.16	1.21	1.29	1.21
32L1701—Pair of Slice Cut Burrs	7	1.17	1.22	1.30	1.22

David Bradley Small Grain Grinder

A feed grinder and family meal and flour mill combined.

No. 2L Grinder.

We have developed this mill to meet the demand for a general purpose mill which can be used for grinding meal and flour for table use as well as the coarser classes of work. It is designed for use with gasoline engines, 1 horse-power and larger. Its capacity, of course, depends upon the power furnished and the class of grinding desired, but in any event we believe you will be surprised at the results and highly satisfied. The self aligning 5¼-inch burrs are mounted in a dust tight case with hand wheel adjustment for fineness of grinding, and hand lever locknut which securely holds the adjustment desired. The removal of two bolts gives access to the burrs. A slide in the hopper regulates the quantity of grain admitted to the spiral FORCE FEED, which keeps the burrs working to their full capacity for the power available. The perforated bolting screen bolts corn meal, graham flour, etc., ready for table use and is operated by the wabble cam wheel on main shaft. The main shaft runs in long babbitted bearings which are lubricated by large grease cups. Pulley is 4x4 inches and should run 700 to 750 revolutions per minute. Capacity is from 5 to 15 bushels per hour, depending upon fineness of grinding, condition of grain and horse-power used. Furnished with one set of coarse burrs for grinding shelled corn, oats and other small grains for feed, and one set of fine burrs for grinding meal and flour for table use. We regard this mill as one of the best values we have ever offered in a small grinder and guarantee it to satisfy you in every respect. Each grinder is furnished with two sets of burrs and the bolting screen.

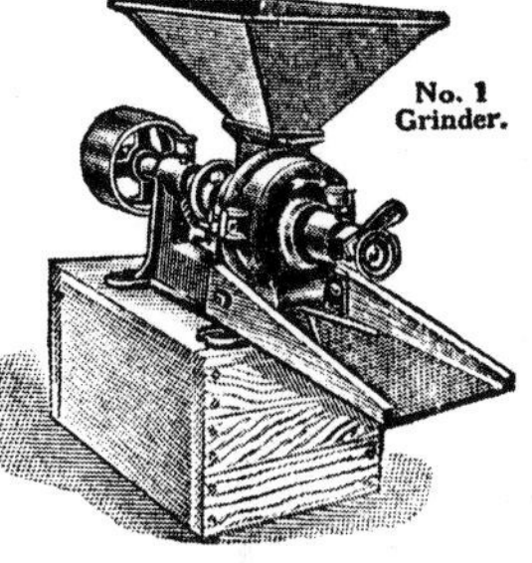

No. 1 Grinder.

Illustrating one method of mounting the grinders without legs on box or bench. Box not furnished.

For Other Grinders see pages 818 to 820 and 839.

	Weight, Pounds	At Bradley, Ill	At Philadelphia	At Kansas City, Mo.	At Fargo, N. Dak.
32L1660—No. 1 Grinder without Legs or Flywheel	72	$ 8.98	$ 9.59	$ 9.48	$ 9.77
32L1661—No. 1L Grinder with Legs and No Flywheel	83	10.28	10.98	10.86	11.19
32L1662—No. 2 Grinder with Flywheel and No Legs	90	10.95	11.72	11.58	11.95
32L1663—No. 2L Grinder with Flywheel and Legs	103	12.25	13.12	12.97	13.38
32L1664—Extra Burrs (state coarse or fine)	4	.76	.80	.79	.81

No. 6 Model X-L-All Fanning Mill and Grain Grader

Cleans, Grades and Separates Any Small Grain All in One Operation.

This double-shoe mill, with its regular equipment of screens and riddles, will clean and grade wheat, oats, barley, beans, peas, corn, buckwheat, rice, cotton seed, clover, alfalfa, timothy, etc. The regular equipment consists of four riddles and four sieves and covers all ordinary work. For special separation and grading see list of extra attachments. Riddles and screens are all the same size, 25x27½ inches; this permits screens being used in upper shoe as riddles when desired and is a marked advantage over fixed riddles. Screens are kept clean and prevented from bagging by scrubbing bars. Upper shoe has side shake and lower shoe has end shake, the length of shake being adjustable. The **side spout** is the outlet for all small seeds before they strike the air blast. This saves for recleaning such seeds as timothy, which are ordinarily blown out with the chaff.

The extra large hopper, holding about 2 bushels, has two distinctive and valuable features, the **feed regulator** and **agitator.** The slide regulator admits only the desired amount of grain to the riddle and the agitator keeps the grain moving evenly without the possibility of clogging. Another valuable feature is the screening box or drawer shown in the bottom of the mill, which prevents the screenings becoming mixed again with the grain.

Mill is 3 feet 7 inches long, 3 feet 7 inches high and 3 feet wide. All lumber used in its construction is selected high quality and thoroughly kiln dried. Capacity, 60 bushels of grain per hour.

Shipped from factory in NORTHERN INDIANA.

32L1727—No. 6 Mill without sacking elevator. Weight, 225 pounds........**$32.50**

32L1754—Sacking Elevator for No. 6 Mill. Weight, 60 pounds........**$8.95**

32L1728—Succotash Attachment. For separating wild or tame oats from wheat. Weight, 15 pounds........**$6.95**

32L1730—Flax Attachment. For ordinary clean threshed flax. Takes mustard, wheat, oats, weeds, etc., from flax. Weight, 10 pounds........**$3.70**

32L1737—Lespedeza or Japanese Clover Attachment. Consists of five zinc riddles and two screens. Weight, 20 pounds........**$7.45**

32L1732—Barley Attachment. For separating barley from oats. Weight, 14 pounds........**$6.40**

32L1755—Timothy Attachment. Takes out sand, clover, alfalfa, millet, alsike, red top; also pepper grass, plantain, daisy and other weeds. Weight, 15 pounds........**$6.95**

32L1751—Power Attachment. Consists of 8x2-inch tight and loose pulleys with socket wheels and chains; should run about 350 revolutions per minute. Weight, 15 pounds........**$2.25**

NOTE—These mills are not intended for hulling or shelling peas, but with the equipment regularly furnished they will clean and grade peas already hulled. We do not handle pea hullers.

Eureka Stone Burr Mills

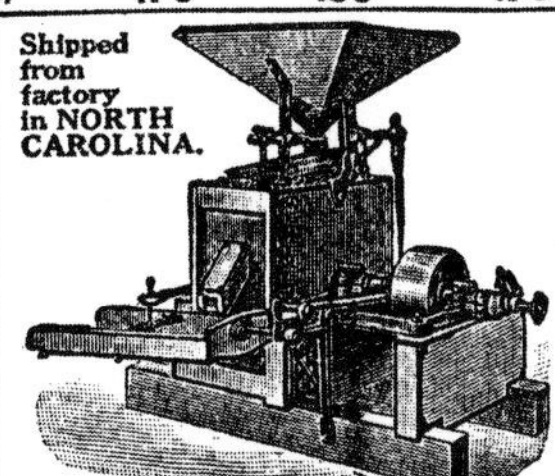

Shipped from factory in NORTH CAROLINA.

These mills are fitted with native stone burrs which make a soft meal. Are preferred in some sections, especially the southern states, for making table meal. They also do good work on grinding shelled corn and small grains for feeding purposes, and are suitable for custom work, as well as for individual farmers' use.

The wood frame is made of extra heavy selected lumber. Entire construction is very strong and the mill is well finished throughout and nicely painted. Hopper is wood. Feed is adjustable for regulating flow of grain and is self locking. A fan cleaner blows out light trashy matter before reaching the burrs.

The bolter or sifter is driven from an eccentric on main shaft. Bolter can be attached or detached while mill is in motion. These mills will grind rapidly, coolly and evenly. Instructions are furnished for operating mills and for redressing stones. **Shipped from factory in NORTH CAROLINA.**

Catalog No.	Size of Burrs, Inches	Speed, R.P.M.	Bushels of Table Meal, per Hour	Bushels of Feed Meal, per Hour	Horse-Power Required	Size Pulley, Inches	Wt., Lbs.	Each
32L1710	14	900	4 to 6	5 to 10	5 to 6	10x6	600	$ 72.50
32L1711	16	800	5 to 8	6 to 12	6 to 7	12x6	700	79.75
32L1712	18	700	6 to 9	7 to 14	7 to 8	12x6	800	90.00
32L1713	20	675	7 to 10	9 to 15	7 to 10	12x6	900	97.50
32L1714	22	650	8 to 12	10 to 20	8 to 10	14x6	1,100	113.00
32L1715	24	625	10 to 12	12 to 20	10 to 12	14x6	1,200	124.50
32L1716	26	625	12 to 15	15 to 25	12 to 15	16x8	1,500	142.50
32L1717	30	550	15 to 20	20 to 35	15 to 20	16x8	1,800	169.00

OUR FAMOUS CRYSTALLINE BELLS.

Guaranteed for Five Years Against Breakage.

Cast from a special mixture of metal of high quality. Thousands of schools, churches and factories use our Crystalline bells with perfect satisfaction.

We allow sixty days' trial on every Crystalline bell we sell. During this period you can give it a thorough trial. Compare it in tone, volume and quality with composition bells of any other make, and if you find any reason to be dissatisfied you can return the bell to us at our expense and we will return to you every cent you paid for it and freight charges.

For Small Farm Bells see page 838.

Thirty-eight inch and larger bells are mounted on **roller bearings.** These bearings enable one person to ring our largest bells with great ease. Bells 24 inches and larger are fitted with our **improved springs and clapper,** which insures a full stroke of the clapper without the possibility of a second stroke. Iron rope wheels are furnished with 20 to 36-inch bells and sectional wood rope wheels with 38 to 48-inch bells. **Tolling hammer is furnished with church bells.**

Prices are for the bells complete with frame, wheel and wood sills. **Shipped from factory in CENTRAL OHIO.**

School and Factory Bells.

	Diameter, Inches	Weight, Pounds	Each
32L2800	20	165	$10.21
32L2801	22	205	13.18
32L2802	24	250	16.25
32L2803	26	350	25.15
32L2804	28	450	31.34
32L2805	30	555	40.66
32L2806	32	630	47.15
32L2807	34	745	57.17
32L2808	36	930	71.20
32L2809	38	985	79.00
32L2810	40	1,275	95.55

Church Bells.

	Diameter, Inches	Weight, Pounds	Each
32L2822	24	260	18.42
32L2823	26	365	27.51
32L2824	28	465	35.17
32L2825	30	570	45.09
32L2826	32	640	49.98
32L2827	34	765	59.80
32L2828	36	950	73.40
32L2829	38	1,010	83.74
32L2830	40	1,300	99.59
32L2832	44	1,790	127.05
32L2834	48	2,280	162.75

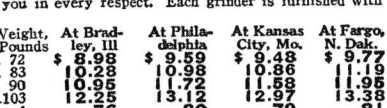

David Bradley Light Draft Auto-Steer Manure Spreader

Roller Bearings.
Write for Our Time Payment Offer on Bradley Implements.
Two horses handle it easily under ordinary conditions.
DAVID BRADLEY
Auto-Steer Front Wheels, permitting same tread as rear without cut under box and all tongue whip eliminated.
All Steel Frame and Spiral Beater.
Top of box 43 inches from ground.

Compare each feature of this new improved spreader with the best you know in spreader construction and we believe you will agree that no other spreader offers so many real worth while improvements, or so much quality in every square inch of its makeup. Honest capacity, stanch construction with minimum weight, extreme simplicity, the fewest parts possible, thorough pulverization with the least power demand, elimination of torsion strains and ease of handling for both man and team are accomplished facts in this superior spreader which we offer under the broadest guarantee possible: It must satisfy you perfectly after a fair trial, or we accept its return at our expense and cheerfully return your money.

All Steel Frame—One-piece channel steel side sills bound together by steel channels, crossed braced with tie rods and **bridge trussed** by angle steel apron guides underneath, form a solid unit that defies the most severe strains and maintains true alignment. Heavy beater supports rest directly on the sills and are braced by the heavy angle steel rake arch, thus relieving the box sides of any strain. Box sides are supported by steel bars extending from top to bottom and secured to main sills by bolts. Top edges of box are iron bound.

All Steel Spiral Beater—Formed of steel U-bars mounted spirally on three malleable heads of large diameter. This distinctive Bradley feature of setting the bars spirally has a fourfold purpose. It stiffens and renders them proof against side twist by the teeth. It relieves the bars of the usual strain of all teeth on a bar engaging the load at the same instant, the load or torque being taken on one tooth at a time. Spirals prevent the throwing out of chunks because at no time is there a wide opening where the load can fall between two bars, the open space between bars running crosswise from an upper to opposite lower corner. This also relieves the driving mechanism of the usual strain of a jerky, uneven load because with this overlap of the bars, before the last tooth of one bar leaves the solid part of the load the first tooth of the following bar is engaging it. Self aligning bearings prevent binding.

The All Steel Leveling Rake is hinged to the heavy angle steel arch, has long compression relief springs, permitting unyielding obstructions to pass through without detriment, and distributes the load evenly to the beater.

Showing the simple, accessible and powerful drive. Large spur gear permanent on axle, large sprocket permament on countershaft, small pinion gear is always in mesh with large gear and slides on the countershaft to or from the large sprocket, which it engages with heavy jaw teeth when thrown in gear. Beater chains encircle both sprockets.

Powerful, Simple Double Drive.

The heavy rear axle is the driving member for all moving parts and receives with equal distribution the full tractive force of both rear wheels. Unequal demand on the drive wheels is the cause of wheels choking down and skidding, a condition that is impossible with the Bradley, regardless of the frozen or slippery condition of the ground. The differential, a simple three-dog clutch in the hub of one wheel (see illustration), permits the spreader to run full tilt when turning corners and without strain. The countershaft, which drives both the beater and apron, receives its power through the heavy spur gear mounted solidly to rear axle and the sliding pinion gear on the countershaft. The large beater driving sprocket is mounted solidly to the countershaft and has clutch teeth on its outer face which, when the left hand lever is thrown, engages the clutch teeth on the inner face of the sliding pinion gear. The valuable feature of this simple drive is that the large gear and sliding gear are always in mesh. The beater sprocket chain encircles both sprockets, thus distributing the strain around the sprockets, not on one side; a much better method than running a chain on one side of a sprocket with fewer teeth engaged and the enormous pressure necessary to keep it from jumping off. The apron cannot be started without the beater.

With the chain encircling the sprockets, gears always in mesh, the clutch gear removable independently of any other part and each drive wheel carrying equal share of the load, we believe this drive to be the simplest, stanchest and most accessible ever devised.

Silent, Positive Ratchet Feed.

Note the simplicity: The lever on the countershaft crank (see large illustration at top) works the curved lever up and down on a bearing midway between the ratchet fingers, and imparts to them alternately a pushing motion that turns the feed wheel with smoothness and regularity. The feed control lever at the operator's right slides the top of the crank lever in or out on the curved lever for different lengths of stroke, thus regulating the speed of the apron and consequently the thickness of spreading. Merely sliding this lever, without a gear, clutch, spring or intricate part, changes the thickness of spreading from a light top dressing to a heavy coat of manure. A highly efficient trouble proof apron feed. Few parts, all in plain sight, any part removable without disturbing the others, and no adjustment necessary or possible.

Pivot Wheel, Auto-Steer Front Truck.

The Auto-Steer front trucks eliminate all tongue whip, permit wide tread with high wheels and short turning without cut under wheels.

This construction keeps the load sitting solidly on all four wheels, eliminates the twist and strain caused by fifth wheels permitting the load to rock over center and so much of the time out of balance. The front wheels pivot the same as on an automobile, and make short turns without cutting under the box. This makes possible the valuable feature of a wide front tread, the same as rear wheels, without the objectionable long axle with its tongue whip and its cutting under the box. In the Bradley every bit of tongue whip is eliminated; in fact, these spreaders, fully loaded, have been driven over rough level ground with the tongue removed.

Full Slatted Endless Apron; narrow wooden slats fitted closely together and mounted on endless link chain belts with bearing surfaces that slide on smooth steel tracks. Apron chains run over sprocket shafts at each end, which insures true alignment. The closely spaced slats and tight fitting box slides form a box that retains liquid manure. Apron always starts easily because the box is wider at the rear than in front.

Roller Bearings—All bearings are self aligning with hard grease cups. Large roller bearings carry the rear axle and are used in the front wheels also.

Wide Spread Pulverizing Attachment.

Peculiar to the development of anything mechanical, and strange as it may seem, the most complicated devices are tried out first, real progress being marked by the degree it is simplified.

One look at the powerful, smooth running Bradley Wide Spread in action evokes the thought: "Why didn't some one think of that long before?" It consists of only four round, flat discs mounted slantingly on a shaft running through their centers. Discs being round and mounted at their centers, the attachment is in balance, all jerkiness eliminated; discs being mounted on a slant, the revolving shaft imparts to them a smooth but powerful side slap motion that thoroughly pulverizes the manure and spreads both ways a fine even coat well beyond the wheel tracks.

Rear view showing the Wide Spread Attachment, which consists of four flat steel discs mounted slantingly on a revolving shaft. This we believe to be the most efficient and simplest Wide Spread and Pulverizer ever produced.

Capacity—Box is 132 inches long, 45 inches wide and 15 inches deep, giving a larger capacity than the usual 65-bushel spreader, which will be most evident if you will compare with other makes. When comparing capacities, compare the measurements of spreader boxes and be sure the length does not include the beater and that there is no waste at the bottom for cut under wheels.

A Two-Horse Spreader—So light is the draft that under ordinary conditions two horses handle the Bradley with ease. Only in very heavy going are three horses necessary and we provide a combination two and three-horse hitch to meet such emergency.

High Wheels, Wide Tires, Low Down Box—Rear wheels have traction cleats and are 42 inches high. Front wheels are 25 inches high. Top of box is 43 inches high at the center. Width of spreader over all is 74 inches.

PRICES.

Shipped from BRADLEY, ILL.

32L815—David Bradley 65-Bushel Manure Spreader. Weight, 1,935 pounds........ **$107.50**

32L816—Wide Spread Attachment. Weight, 90 pounds........ **8.75**

For a more complete description and larger illustrations send for our Manure Spreader Circular 9243GCL.

Trussed Frame 2-Wheel Sweep Rake and Power Lift 4-Wheel Push Rake.

Our Sweep Rakes and Stackers are of improved design, made of high quality materials and stanchly constructed throughout. As between this and cheaper lines we could offer we believe our customers will appreciate the quality for small additional cost.

The trussed frame of the two-wheel rake prevents sagging under heavy loads and the loose tongues allow the rake to follow unevenness of the ground freely. Rake is 12 feet wide, has 13 selected teeth and is carried on large wheels 16 inches in diameter with 3-inch faces. **The Power Lift 4-Wheel Rake** is of extra strong construction to meet the heaviest haying conditions. A special feature is the "power lift" which carries the teeth high from the ground, the horses supplying the lifting power. Teeth are doubly braced by the heavy axle, which is carried on extra large wheels. Only selected high grade material is used in both rakes. They are strongly built, nicely painted and varnished.

2-Wheel Sweep Rake.

Shipped from SOUTHERN MINNESOTA.

32L5504—2-Wheel Sweep Rake. Weight, 300 pounds........ **$22.95**

32L5509—4-Wheel Push Rake. Weight, 635 pounds........ **$45.85**

4-Wheel Push Rake.

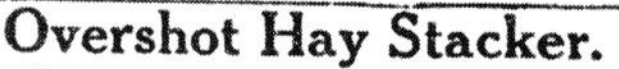

Overshot Hay Stacker.

This stacker is made on a strong ground frame that is provided with extension pieces at the back for making the stack higher when so desired.

The "A" frame is provided with metal boxes working on a steel pipe at the bottom, and is strongly made and braced.

The lifting arms are made double, forming a truss brace on each side.

The pitcher teeth are adjustable to any desired angle.

The spring attachment of this stacker is a most desirable feature. It is easy to adjust, and works automatically. In elevating a load, the horse travels but 35 feet when the load is dumped in the middle of a stack.

The teeth are 9 feet long and are provided with metal points, well fitted and securely fastened.

Stacker is equipped with 3 pulleys, 50 feet of ¾-inch rope and 38 feet of ⅝-inch rope.

This stacker is made of well selected material, is manufactured in a first class, workmanlike manner, is nicely painted and varnished.

Shipped from SOUTHERN MINNESOTA.

32L5502—Overshot Stacker. Weight, 725 pounds........ **$55.75**

Peerless Dairy Barn Equipment

Simple, strong, practical. No complicated parts to get out of order, yet contain all the features necessary for cleanliness and convenience. Made from high carbon brazed steel tubing, 1⅝-inch outside diameter, 1⅜-inch inside diameter, very stiff and strong. Put together with heavy malleable clamp fittings. No threads in tubing or fittings, everything bolted together. Upright posts stand 5 feet 5 inches high above floor with 5 inches additional for imbedding in the concrete.

The triple bend stall partitions extend back 42 inches from uprights and are 42 inches high from floor. Partitions have a cast flange riveted to bottom end to hold them securely in the concrete.

Stalls are finished in a durable blue gray enamel paint. Shipped from factory near CHICAGO, ILL.

When ordering stalls be sure to give us the following information:

Into how many rows the stalls are to be divided.

The desired width of stalls.

Do the rows join the building wall at one or both ends?

Also furnish us with a rough pencil drawing showing the number of stalls in each row, the location and size of any posts that may be in the way of the rows, and the distances between posts from center to center in both directions.

For Barns and Silos see pages 965 and 971.

Each stall has two uprights and crosspiece for supporting partition. Prices include top rail, two side uprights, one bent partition, one stanchion with anchor and necessary clamps and bolts. Furnished 3 feet 6 inches to 4 feet wide, as ordered. **State width.** Weight, 75 pounds. Illustration shows two stalls with partitions and one extra end section.

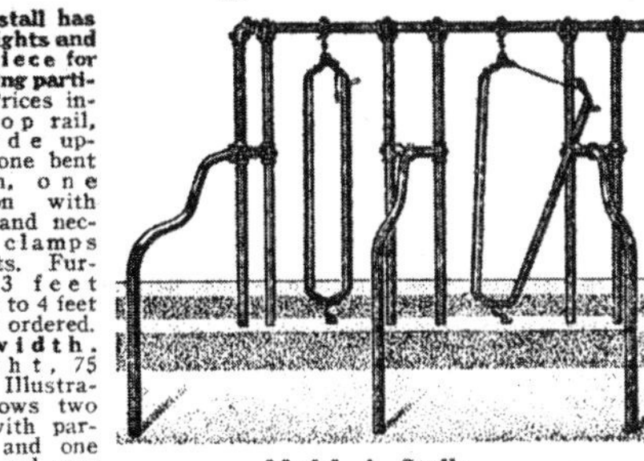

Model A Stalls.

The Stanchion furnished is our 32L2214 Peerless Oval Steel Stanchion described below to the left.

32L2263—Model A Steel Stall **$7.90**

Illustration shows two stalls and one extra end section. Price is for stall complete, consisting of top rail, one side upright and partition, stanchion with hanger and bottom anchor and necessary clamps and bolts. Furnished any width, 3 to 4 feet. **State width desired.** Weight, 56 pounds.

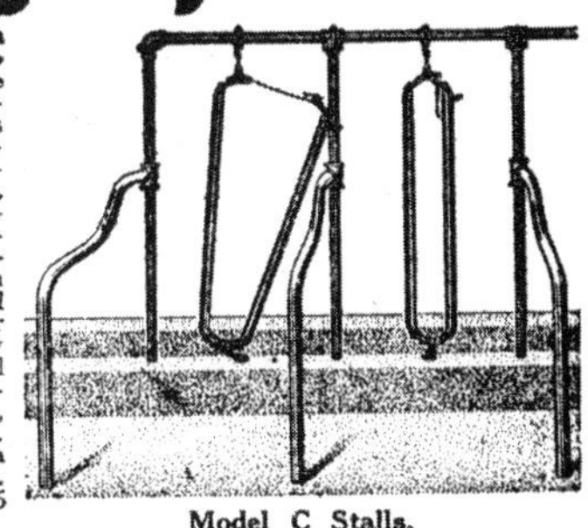

Model C Stalls.

The Stanchion furnished is our 32L2214 Peerless Oval Steel Stanchion described below to the left.

32L2267—Model C Steel Stall **$6.10**

Where both ends of a row of stalls are independent and do not join the wall of building, one extra section composed of an upright and bent partition is required for finishing off the end of row.

32L2268—Extra End Section for finishing end of row. Shipping weight, 25 pounds **$2.88**

Peerless Oval Steel Stanchions.

Our Most Popular Pattern.

Side bars are heavy U bar steel, lined with smooth, rounded hardwood strips. The channel of the steel U bars, being turned inward, the wood linings are firmly supported by the steel.

Hinge, latch and top are malleable castings.

Adjustable top and bottom for 5½, 6½ and 7½ inches neck space, and are 48 inches high inside. Chains, 6 inches long. Durably finished in black japan.

32L2214—Oval Steel Stanchion, Plain. Weight, 21 pounds. **Shipped from factory near Chicago, Ill.** **$2.29**

Shipped from Philadelphia store **$2.45**

32L2220—Bottom Anchor for fastening stanchion to concrete curb. (Suitable for any chain hanging stanchion.) Weight, 1½ pounds **23c**

32L2221—Malleable Clamp with bolts for hanging stanchion to pipe frame (1⅝ inches outside diameter). Weight, 8 ounces **23c**

Peerless Two-Way Steel Stanchion.

Sides are heavy steel U bars, lined with hardwood strips. This stanchion opens both ways from the center, giving extra wide head space when open. A pinion or cog wheel pivoted in the center engages with the two malleable top bars, causing them to move in or out equally on both sides. Mechanism is firmly supported by strong malleable frame. Stanchion is about 48 inches high inside. Neck space is adjustable. Spring latch locks stanchion when closed. The top bar construction provides a rigid frame all around and relieves the bottom hinge from strain.

32L2211—Peerless Two-Way Stanchion. Weight, 27 pounds. Shipped from **factory near Chicago, Ill.** **$3.42**

Shipped from Philadelphia store 3.65

Model A Stall, regularly furnished with Stanchion 32L2214, will be furnished with the Two-Way Stanchion 32L2211 at an extra price of $1.10 per stall.

Peerless Wood Bar Stanchions.

32L2215

Low priced, but strong and satisfactory. Sold by us for many years, and in their present improved stage are still in great demand.

Side bars are clear white hardwood, rounded, smoothed and oiled. Tops and bottoms heavy malleables. Adjustable for 5, 5½, 6, 6½, 7, 7½ or 8 inches neck space. Cowproof lock.

Pivot swinging or chain hanging. Pivoted stanchion has a tumbler which prevents stanchion turning in frame when open. Stanchions are 49 inches high inside. Chains are 6 in. long, pivots are 2½ inches long.

32L2215—Wood Stanchion, Pivot Hanger. Weight, 17 pounds. **Shipped from factory near Chicago, Ill.** **$1.99**

Shipped from Philadelphia store **$2.15**

32L2216

32L2216—Wood Stanchion, Chain Hanger. Weight, 17 pounds. **Shipped from factory near Chicago, Ill** **$2.15**

Shipped from Philadelphia store **2.30**

Write for This Special Catalog of Dairy Barn Equipment No. 504GCL.

Besides showing larger illustrations with more complete descriptions of the items shown on these pages, it shows our complete line of Dairy Barn Equipment, including cow, calf and bull pens, mangers, etc.; also our complete line of Litter and Feed Carriers.

Peerless Quick Detachable Water Bowls.

Can be attached to either steel or wood posts. A storage tank elevated a few feet will provide all the pressure necessary for these bowls. The valve consists of a rubber Fuller ball closing against a brass seat and held in place by a brass spring. The water can be piped from either above or below. When piped from below two ordinary elbows and a short nipple are required. These are not included in the price of the bowl, nor are the pieces of pipe and the coupling shown above the bowl in the illustration. The clamps for supporting the supply pipe and a clamp with lever for attaching the bowl to the stall post are furnished. Peerless bowls can be quickly detached for cleaning merely by moving the small lever, which locks the bowl securely in place. Bowls are made of malleable iron and are 10x10 inches in diameter and 4¼ inches deep, with rolled edges. Weight, 12 pounds.

32L2242—Water Bowl for Steel Stalls.
Shipped from factory near Chicago, Ill **$2.10**
Shipped from Philadelphia store **2.25**

32L2243—Water Bowl for Wood Stalls.
Shipped from factory near Chicago, Ill. **2.24**
Shipped from Philadelphia store **2.40**

Steel Stall Partitions for Wood Frames.

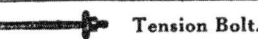

Any of our stanchions can be used. If chain hanging stanchions are used, they can be set forward or backward of frame by using the 6-inch Hook Bolts 32L2273. When wanted for wood floor one extra heavy malleable Floor Flange, 32L2272 must be ordered with each partition. With each partition we include one flange with lag screws for uprights. Prices do not include stanchion or woodwork, and are for one partition only.

32L2270—Stall Partition. Weight, 15 lbs. **Shipped from factory near Chicago, Ill.** **$1.65**

Shipped from Philadelphia store **$1.75**

32L2272—Flange for Wood Floors. Wt., 3 lbs. **.34**

32L2273—6-Inch Hook Bolt. Weight, 1 pound .. **.07**

How to Order a Litter Carrier

For each line of rod track, two Tension Bolts (32L1282) and two End Loop Clamps (32L1281) are required, and if the track extends out to a post an Anchor Loop (32L1284) is necessary. If the rod track turns a square corner, two independent lines of rod track are necessary, each with tension bolts and end loop clamps; also a Curve or Switch (32L1285). Carriers or Track shipped from factory near CHICAGO, ILL.

If there is anything you do not understand, write to us for information.

Plain Carrier, Without Automatic Lift.

Best for barns with low ceilings. Bucket, 42 in. long, 24 inches wide, 16 in. deep. No. 18-gauge reinforced galvanized steel body, double hardwood ends, 2 in. thick. Channel steel bail. Large wheels with roller bearings. Malleable castings throughout. Weight, 125 pounds.

32L1278—Carrier only .. **$18.75**

Carrier With Automatic Lift.

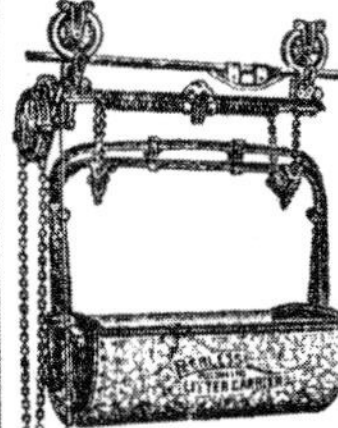

All galvanized steel bucket, body No. 18-gauge, ends heavier No. 16-gauge, reinforced all around with steel angles. Heavy 2-inch channel bail. Roller bearing wheels. Bucket locks at both ends. Works either way from trip. Automatic trip and hand trip bucket, 42 inches long, 24 inches wide, 16 inches deep. Quick lift, self lowering, automatic clutch and brake. Weight, 215 pounds.

32L1274—Carrier only .. **$33.85**

Complete 100-Foot Straightaway Rod Track Outfits.

Include carrier, 100 feet rod track, two end loop clamps, two tension bolts, one anchor loop with turnbuckle and anchor rod.

32L1275—Complete Outfit with Carrier 32L1278 and 100 feet No. 0000 rod track. Weight, 200 pounds **$26.95**

32L1277—Complete Outfit with Carrier 32L1274 and 100 feet No. 000000 rod track. Weight, 250 pounds **$44.45**

If longer track is wanted, order as much extra Rod Track (32L1280 or 32L1293) **as required. State whether wanted all in one piece, or in what lengths if more than one piece.**

Rod Track.

32L1280 32L1293

Polished round steel rod, can be looped like ordinary wire. Rolls in a coil for shipment.

32L1280—No. 0000 Rod Track, 13/32 **inch in diameter.** Weight, per foot, ⅓ pound. Per foot **4c**

32L1293—No. 000000 Rod Track, 15/32 **inch in diameter.** Weight, per foot, ½ pound. Per foot **5c**

Tension Bolt.

One required at each end of rod track. Welded eye. Size, ⅞x30 inches, with nut and washer. Weight, 7 pounds.

32L1282—Tension Bolt **85c**

End Loop Clamp.

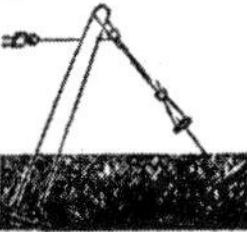

One required at each end of rod track. They hook into tension bolt for stretching the track.

32L1281—End Loop Clamp. Weight, 2 pounds **26c**

Curve or Switch.

For turning a square corner with rod track. Made of angle steel with malleable points. Weight, 25 pounds.

32L1285 ...... **$3.38**

Latest Pattern I-Beam Track.

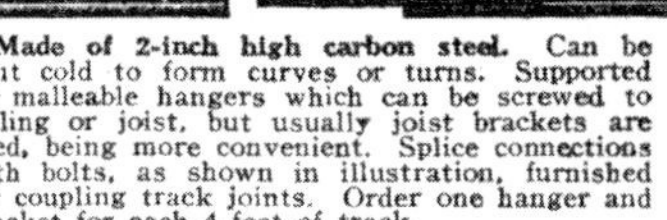

Made of 2-inch high carbon steel. Can be bent cold to form curves or turns. Supported by malleable hangers which can be screwed to ceiling or joist, but usually joist brackets are used, being more convenient. Splice connections with bolts, as shown in illustration, furnished for coupling track joints. Order one hanger and bracket for each 4 feet of track.

32L1214—Rail Track. Weight, per foot, 1½ pounds. Per foot **9½c**

32L1215—Malleable Hanger, 11½ inches long. Weight, 2 pounds **30c**

32L1216—Malleable Hanger, 9 inches long. Weight, 1⅝ pounds **23c**

32L1217—Joist Bracket for track, running across joist. Weight, 1 pound. Bracket only. **7c**

32L1218—Joist Bracket for track running parallel with joist. Weight, 1 pound.

Bracket only **7c**

We furnish carriers with slightly smaller wheels when Hanger 32L1216 is used.

We also furnish track hangers with extension rods where track must pass below cross timbers, plain, two and three-way switches and, in fact, have a complete line of Barn Equipment. Send for Special Catalog 504GCL.

For Other Carrier Track Brackets See Page 844.

Anchor Loop.

Necessary where rod track attaches to post. Includes 20 feet No. 000000 rod, two loop clamps, turnbuckle and ¾ inch by 6 feet anchor rod. Weight, 20 pounds.

32L1284 ...... **$3.17**

Handy Feed Cookers.

Full Capacity.

Boiler is made entirely of heavy galvanized sheet steel, strongly bound at top and bottom, and has a close fitting hinged cover. Fire box is No. 16-gauge blue annealed steel. Fire flue is 4 in. deep and extends entire length of boiler. A partition in the center deflects the heat from the fire box to the opposite end and then back to the smoke pipe. Can be used for cooking feed, boiling water and many other purposes. Has heavy cast iron grate and will burn coal, wood or cobs. Price includes one joint of 6-inch pipe. Shipped from factory in NORTHERN INDIANA.

	Size, Gal.	Wt., Lbs.	Each
32L1970	60	180	$13.65
32L1971	90	190	14.70
32L1972	115	206	16.40
32L1973	160	245	19.85

Bradley Automatic Hog Feeder.

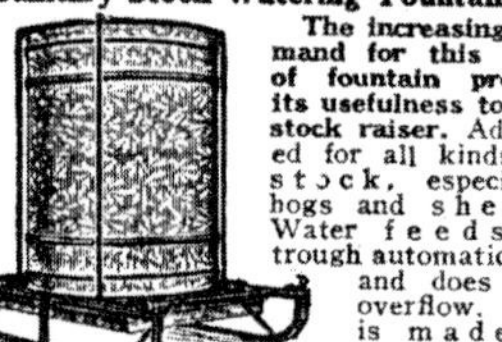

Let your hogs feed themselves and grow fat.

Thousands of large hog raisers have installed this feeder with satisfactory results. It has a large compartment to hold about twenty-five bushels shelled corn and two small compartments to hold two bushels each of mineral feed and tankage. It is built in sections, to be bolted together, and is shipped knocked down flat, taking low freight rate. Dimensions, about 6 feet long, 3½ feet wide, 3½ feet high. Ends, sides and bottom are ⅞-inch dressed lumber with lapped joints. Trough has 2-inch sides. A removable board above trough provides a larger opening for feeding ear corn. Painted with oil paint. Runners on bottom with hook for moving. We believe this is the best made feeder of this type on the market, and it should not be confused with inferior feeders made of light material. Shipped from BRADLEY, ILL.

32L1875 — Automatic Hog Feeder. Weight, 300 pounds $13.95

Sanitary Stock Watering Fountain.

The increasing demand for this type of fountain proves its usefulness to the stock raiser. Adapted for all kinds of stock, especially hogs and sheep. Water feeds to trough automatically and does not overflow. Sled is made of heavy angle steel, well braced throughout. Tank is made of heavy galvanized steel, both top and bottom being double seamed by special machinery and not held in place by solder. Top is pressed into a funnel shape with large brass filling plug in center, which makes it quick and easy to fill. By using this construction there are no rivets through the tank and no solder around head to crack and break and cause leaks. Shipped from factory in OHIO.

	Size, Gal.	Wt., Lbs.	Each
32L1860	65	100	$10.95
32L1861	85	105	11.55
32L1862	100	110	12.80

Kenwood Agricultural Boilers.

FULL CAPACITY GUARANTEED.

Design patented.

A strictly high grade boiler, and our own exclusive design. Popular among both farmers and butchers. Can be used for any purpose where a fine, smooth kettle is required, such as rendering lard, cooking feed for stock or boiling sap. **Caldrons are made of fine grain smooth iron, with black lead finish inside, and we guarantee them to be full capacity. Furnaces are cast iron throughout,** put together as well as any stove made, and the design is pleasing and ornamental. Price is for the furnace and caldron complete as shown, but does not include pipe. For coal they have iron grate and heavy firebrick lining, which can be taken out and replaced through the door. **Shipped from factory in CENTRAL OHIO.**

Wood Burning Boilers.

Catalog No.	Size, Gals.	Size, Pipe	Wt., Lbs.	Each
32L1900	15	5 in.	244	$15.98
32L1901	22	6 in.	295	19.90
32L1902	30	6 in.	365	24.35
32L1903	45	7 in.	444	28.85
32L1904	60	7 in.	646	39.30
32L1905	75	8 in.	737	44.50

Coal Burning Boilers.

Catalog No.	Size, Gals.	Size, Pipe	Wt., Lbs.	Each
32L1907	15	5 in.	275	$18.75
32L1908	22	6 in.	327	21.35
32L1909	30	6 in.	425	26.20
32L1910	45	7 in.	494	31.45
32L1911	60	7 in.	698	42.25
32L1912	75	8 in.	804	46.85

Farmers' Friend Feed Cooker and Caldron Furnace.

Elbow, Damper and Pipe With Each Cooker.

The jacket is heavy rolled steel plate, supported at the bottom by heavy iron bands. **Kettles are made of smooth, fine grained iron with black lead finish inside, and we guarantee them to be full capacity.** The rim of the kettle rests on top of the jacket and the kettle can be easily removed. Cooker is intended to set on the ground, as it has no bottom. For indoor use, set on a base made of brick and sand. Can be used for cooking feed, rendering lard, or for any other purpose where an ordinary kettle can be used. Prices are for the cooker for burning wood, complete with elbow, damper and one joint of pipe. Coal grates and cover are extra. **Shipped from factory in CENTRAL OHIO.**

32L1930—15-Gallon Farmers' Friend Cooker. Weight, 113 pounds....$ 7.85
32L1931—22-Gallon Farmers' Friend Cooker. Weight, 150 pounds.... 9.45
32L1932—30-Gallon Farmers' Friend Cooker. Weight, 174 pounds.... 12.45
32L1933—45-Gallon Farmers' Friend Cooker. Weight, 245 pounds.... 14.75
32L1934—60-Gallon Farmers' Friend Cooker. Weight, 288 pounds.... 18.70
32L1935—75-Gallon Farmers' Friend Cooker. Weight, 341 pounds.... 21.50
32L1937—Coal Grate for 15 and 22-Gallon Cookers. Weight, 41 lbs... 2.98
32L1938—Coal Grate for 30, 45, 60 and 75-Gal. Cookers. Wt., 63 lbs... 4.00
32L1940—Wood Hinged Cover for 15, 22 and 30-Gal. Cookers. Wt., 10 lbs. .95
32L1941—Wood Hinged Cover for 45, 60 and 75-Gal. Cookers. Wt., 16 lbs. 1.25

The All Year Sanitary Hog Watering Fountains.

Will not freeze in winter and will keep the water cool and clean in summer. Will give hogs, sheep and other small animals all the water they can drink at a comfortable temperature in **coldest weather,** thereby contributing to their growth and saving feed. The water compartment being airtight, the water is automatically fed to the drinking pan by the partial vacuum, which releases it only as used. Trough is always full, but never overflows. No mud hole around the drinking place. Made with either one or two drinking troughs in each size and **are protected so stock cannot get their feet into them.** The air space in the top of the tank helps prevent freezing in winter and keeps the water cool in summer. One lamp furnished with single trough and two lamps with double trough. **Lamps set directly under drinking pans, and when properly cared for will protect troughs against freezing with temperature 35 degrees below zero. Waterers with troughs outside cannot stand such extreme cold.** Tank and outer casing are separate, with air space between. Heat from lamp passes around drinking pans and up through casing around tank, and keeps water at right temperature. Lamps burn kerosene. Waterers are made of galvanized steel. The 55-gallon size is 31¼ inches in diameter, 41½ inches high. The 90-gallon size is 31¼ inches in diameter, 52½ inches high. **Shipped from factory in CENTRAL IOWA.**

32L1884—55-Gallon Single Trough Fountain. Weight, 93 pounds....$16.95
32L1885—90-Gallon Single Trough Fountain. Weight, 138 pounds.... 23.75
32L1882—55-Gallon Double Trough Fountain. Weight, 95 pounds.... 20.40
32L1883—90-Gallon Double Trough Fountain. Weight, 140 pounds.... 27.20

Extra Strong Galvanized One-Piece Hog Trough.

Shipped nested at a saving of one-half in freight charges.

Trough is formed from one piece of heavy galvanized steel, which makes a liquid tight trough without soldered or riveted joints. Legs are separate, and slotted to slip over the folded end, and creased at the top for easy bending over the trough end. Thus they are held there securely and cannot become loosened. Galvanized crossbars are spaced 12 inches apart. The troughs are shipped with legs and crossbars off, nested together, thus saving about half the freight you would pay on troughs which cannot be shipped nested. Troughs of same or different sizes can be nested. It requires but a moment to attach the ends and braces. Troughs are 11 inches across top, 6 inches deep. **Shipped from our store.**

	Weight	Each
32L4400—2-Ft. Trough.	8 lbs.	$0.68
32L4401—4-Ft. Trough.	17 lbs.	1.53
32L4402—6-Ft. Trough.	24 lbs.	1.95
32L4403—8-Ft. Trough.	30 lbs.	2.60

Cast Iron Hog Troughs.

On account of their weight and durability these troughs are very popular among hog raisers. They have smooth bottoms inside and can be easily kept clean. Made of a good quality of close grained iron and are sufficiently heavy so they are not easily tipped over. The 3-foot trough has one iron crossbar, the 4-foot trough has two, and the 5-foot trough has three crossbars. Troughs are 12 inches wide. 2-foot and 3-foot troughs **shipped from our store; other sizes from factory near CHICAGO or PHILADELPHIA.**

32L4495 — 2-Foot Trough. Weight, 25 pounds $1.58
32L4496 — 3-Foot Trough. Weight, 35 pounds $2.65
32L4497 — 4-Foot Trough. Weight, 55 pounds $3.80
32L4498 — 5-Foot Trough. Weight, 70 pounds $4.73

Caldron Kettles.

The same kettles as are used in our Farmers' Friend Food Cookers. Made of fine grain smooth iron, and can be used for cooking feed, rendering lard, boiling syrup, etc. Shipped from CENTRAL OHIO.

Full Capacity.

	Size, Gal.	Wt., Lbs.	Each
32L1950	15	62	$ 3.55
32L1951	22	87	4.60
32L1952	30	105	6.35
32L1953	45	139	8.45
32L1954	60	209	11.95
32L1955	75	254	14.90

For Copper and Other Kettles see page 829.

Marvel Feed Cookers.

The boiler is made of heavy galvanized steel, with hinged cover and heavy band iron around top so that the weight of feed will not bend it out of shape when being lifted about. The furnace is made of black steel. Front, back and hearth are heavy castings. 32L1956, 32L1957 and 32L1958 have cast iron grate and will burn either coal or wood, and the furnace in these sizes has an extra steel inner lining with air space between it and the outer wall.

The cookers for burning wood have no grate, nor inner lining, and the furnace is made without the ash pit and ash pit door. Illustration shows the cooker for burning coal or wood. All cookers substantially made and well crated.

Price includes one length 6-inch pipe and elbow. **Shipped from factory in WESTERN ILLINOIS.**

32L1956 — 25-Gallon Cooker for coal. Weight, 87 pounds $12.98
32L1957 — 50-Gallon Cooker for coal. Weight, 120 pounds $13.95
32L1958 — 100-Gallon Cooker for coal. Weight, 177 pounds $21.50
32L1959 — 25-Gallon Cooker for wood. Weight, 81 pounds $10.95
32L1960 — 50-Gallon Cooker for wood. Weight, 110 pounds $12.45
32L1961 — 100-Gallon Cooker for wood. Weight, 165 pounds $19.95

"Can't Clog" Rotary Hog Feeder.

A feeder that can't clog. Particularly desirable for feeding tankage or other feeds which bridge in ordinary feeders. Hog feeds himself without waste and gets all he wants to eat. Will pay for itself in short time in the saving of feed. **Cone inside rotates as hogs naturally push against partitions inside of trough and feed comes down just fast enough to supply their needs.** Base and feed trough made of heavy lumber and trough is steel clad. Balance of feeder made of heavy galvanized steel. Cover is hinged. Adjusting screw under cover regulates space between base and drum, according to kind of feed used or the speed of feed flow wanted. Feeder is 47 inches high and 40 inches in diameter, with a capacity of about 13 bushels. Can be used for feeding any kind of ground feed or small grain, shorts, middlings, shelled corn, tankage, etc. **Shipped from factory in WISCONSIN.**

32L3048 "Can't Clog" Rotary Hog Feeder. Weight, 160 pounds........$22.50

Rotary Hog Oiler.

Over 30,000 in use.

One of the most simple, durable and economical oilers on the market. Made of heavy cast iron. Animal cannot tip it. When the animal rubs wheels rotate and pick up oil from basin; sufficient to spread on the hog and balance runs back. No waste. Requires practically no attention and as there are no springs or valves it will never wear out. Will give lasting service under the most exacting use. 13½ inches high and 13½ inches wide. **Shipped from factory in IOWA.**

32L1768—Rotary Hog Oiler. Weight, 87 lbs...$7.48

Three-Bar Hog Oiler.

A three-bar oiler, very simple in construction and automatic in operation. The only moving parts are the rubbing posts. When the animal rubs one of the bars, the plunger formed top of the bar is forced into a small cup, displacing the few drops of oil the cup contains and causing the oil to flow down over that rubbing bar. There are no valves to become clogged or held open and no half way point where a continuous flow of oil would be possible. The oil is contained in and supplied from an inverted tank clamped on the top of the post. Vacuum maintains a full supply chamber into which the plunger works and at the same time seals the only opening in the supply tank. Any oil draining from the rubbing post is caught in the recessed cup in the pedestal, and from there is applied to the under parts of the animal. Economical in handling of the oil, of simple and stanch construction. **Shipped from our store.**

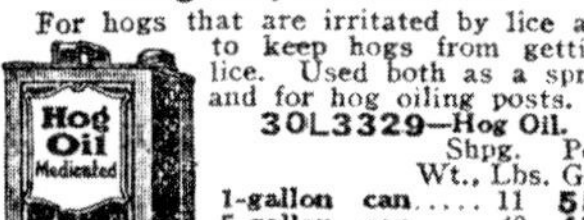

32L1766 — Three-Bar Hog Oiler. Weight, 47 pounds $4.50

Hog Oil, Medicated.

For hogs that are irritated by lice and to keep hogs from getting lice. Used both as a spray and for hog oiling posts.

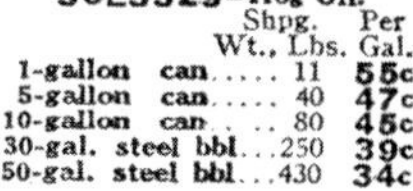

30L3329—Hog Oil.

	Shpg. Wt., Lbs.	Per Gal.
1-gallon can	11	55c
5-gallon can	40	47c
10-gallon can	80	45c
30-gal. steel bbl.	250	39c
50-gal. steel bbl.	430	34c

Home Meat Smoker

Kill and Cure Your Own Meat Supply.

The Home Portable Smokehouse—Can be used any place on the farm. It can be placed on the porch, in an outbuilding or in the house, as it has an outlet for connection to any ordinary chimney, which permits the escape of excess smoke. It is safe and fireproof.

The low cost makes the **Home Portable Smokehouse** indispensable wherever pork is cured for home or market.

Built in Two Sizes—Three-hog and five-hog capacity. Just the right sizes for farm use, or for meat markets where a limited amount of meat is smoked. Shops that smoke all of their meat frequently use two or more outfits.

Sturdy Construction—The casing, fire chamber and smoke pipe are heavily galvanized iron, strong and well made. There are four hardwood 2x2-inch posts with strap iron hoops on which the meat is hung on hooks. Full set of hooks for hanging meat furnished.

Scientific Method of Smoking—In going from the fire chamber to the smokehouse proper the smoke must pass through a long pipe, thus being thoroughly cooled and avoiding any possibility of overheating the meat. The smoke enters at the bottom of the meat chamber and has thorough circulation through it.

An Ideal Storage House—As a storage chamber alone the **Home Portable Smokehouse** is worth more than its cost. You need not remove the meat from it when smoked. It has two screened ventilators for air circulation which keeps the meat in perfect condition. The smokehouse is vermin and insect proof. **Shipped from factory in IOWA.**

32L4650—Three-hog size, as described. Shpg. wt., 115 lbs....... **$21.00**
32L4651—Five-hog size, as described. Shpg. wt., 135 lbs....... **24.00**

Dairy Boiler and Steam Feed Cooker

Burn cobs, wood or coal. Are very quick steamers and very economical in use of fuel. Used for cooking feed, scalding hogs, scalding milk cans, thawing out water tanks, and for all steam cooking, washing or renovating purposes. Used indoors or out of doors. Require no experience. No flues to clog or burn out. Made of heavy boiler plate steel with steel head. Shell and head are welded together by a special process; there are no riveted or patched joints or seams to leak, and the welded seams are stronger than riveted seams. Has cast iron base and dump grate. Has a cone shaped fire box extending full height of boiler inside. This fire box is surrounded entirely by water and will not burn out. The water surface is least where the heating surface is greatest, making the boiler a quick steamer. Boilers are equipped with safety valve, adjusted to 15 pounds pressure, sufficient for all agricultural purposes. No. 2 boiler holds 30 gallons and No. 3 boiler holds 40 gallons. Boilers are furnished with two sets of steam pipes, shut off valves, one try cock, water glass, safety valve, steam gauge, hand pump and hand pump hose. Hose from steam pipe not furnished. Each boiler when used for cooking will heat to the boiling point about four times as much water as it converts into steam. **Shipped from SOUTHWESTERN MICHIGAN.**

32L1998—No. 2 Boiler. Height, 56 inches; diameter, 17 inches. Weight, 270 pounds.... **$54.95**
32L1999—No. 3 Boiler. Height, 68 inches; diameter, 17 inches. Weight, 290 pounds.... **60.45**

Genuine Dandy Green Bone Cutters

Known everywhere for their large capacity, durability and easy running qualities. The bones are placed in an oblong box, across the end of which revolves a large geared disc on which are placed knives. These revolve across the end of the boxes, reducing the bones to meal. The knives are easily removed and sharpened.

Illustrates 32L3200 Cutter.

The small size cutter is back geared, operated by handle on rim of large balance wheel; suitable for a flock of 50 to 75 fowls. The bone box is 3¼x4⅝x6 inches inside. It is fed by a hand feed screw, which has a split nut that swings entirely out of the way when filling the bone box. Cutter has one straight and two corrugated knives.

Illustrates 32L3241 Cutter.

The large size cutter is equipped with automatic screw feed follower block and split nut. Bone box is 4x4x20 inches inside measurement. Equipped with one straight and three corrugated knives. Has capacity to feed 400 fowls by hand power or 1,500 fowls by engine power. **Shipped from factory in WESTERN PENNSYLVANIA.**

32L3200—No. OB Bone Cutter. Weight, 75 pounds.......... **$12.85**
32L3241—No. 13 Cutter for hand or power, equipped with crank and pulley. Weight, 170 pounds.......... **29.85**

For other sizes and styles of Bone Cutters write for prices.

Clover Cutter

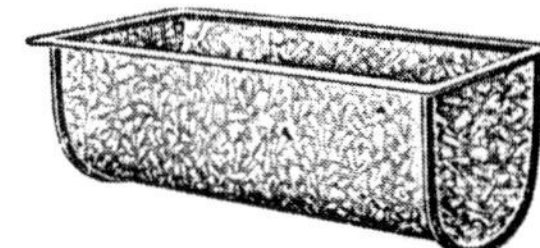

Will cut green or dry clover, alfalfa or vegetable tops into ⅛-inch lengths, suitable for poultry. The cutting head has four 7-inch steel knives bolted to malleable heads. Knives can be easily removed, sharpened and replaced. They cut against a hardened cutting bar which is adjustable. Has two corrugated feed rollers with spring tension, the upper roll raising and lowering as the machine is fed. Cutter head makes about twenty revolutions to one turn of the crank. Machine is made entirely of iron and steel. Has three short legs for bolting to bench or box. **Shipped from factory in PENNSYLVANIA.**

32L3127—Clover Cutter. Weight, 70 pounds.......... **$12.95**

Hog Scalder

Can be used for scalding hogs, cooking feed and other purposes. Entire scalder made of 20-gauge steel throughout. Top frame and end frames are steel angles. **Shipped from factory in CENTRAL OHIO.**

32L1975—Length, 5 feet; width, 30 inches; depth, 18 inches; capacity, 100 gallons. Weight, 60 pounds.... **$12.15**
32L1976—Length, 5 feet; width, 30 inches; depth, 24 inches; capacity, 145 gallons. Weight, 92 pounds.... **$13.20**
32L1977—Length, 6 feet; width, 30 inches; depth, 24 inches; capacity, 174 gallons. Weight, 106 pounds... **$13.95**

One-Horse Dump Cart

A light one-horse cart—suitable for use in fence building, for gathering fruit and truck crops, hauling manure for gardens, light loads of dirt, stone, fodder, etc. Can be taken in places inaccessible to a wagon. Strongly built of good sound lumber, well painted. Box is 5 feet long, 3 feet 4 inches wide, 11½ inches deep, inside measurements, and made of ⅞-inch material, well braced. Bottom is strongly framed. Wheels are heavy steel, 24 inches in diameter. 3-inch tires, and the tread is 48 inches. Axle is solid steel, 1⅜ inches in diameter. Top of box is about 3 feet high from the ground. Box can be dumped by pulling out steel pin in front. **Shipped from factory at BRADLEY, ILL.**

32L2231—Handy Farm Cart. Weight, 300 pounds.......... **$22.75**

Cast Iron Stone Boat Head

Don't pay freight on the wood part of a stone boat. With this iron stone boat head you can make a stone boat and use any straight plank you happen to have. It will outwear several sets of plank. The head is 29 inches wide, heavy and strong. It will last a lifetime. **Shipped from BRADLEY, ILL., or our PHILADELPHIA Store.**

32L2550—Cast Iron Stone Boat Head. Weight, 70 pounds.......... **$3.98**

Steel Frame Barrel and Spray Cart

Useful for carrying swill for feeding; also as an orchard spray cart, and for numerous other purposes. Made entirely of steel and iron. The steel wheels are 36 inches high, with 1¼x¼-inch tires. The frame is bolted solidly to the axle castings, and these castings are made to fit the sides of a barrel to which they are to be bolted. A kerosene, molasses or vinegar barrel can be used. Price includes bolts to attach to barrel and one bracket or rest for bottom of barrel.

32L2225—Barrel Cart. Weight, 66 pounds.
At BRADLEY, ILL.......... **$4.40**
At PHILADELPHIA, PENNA.......... **4.95**

Handy Platform Cart

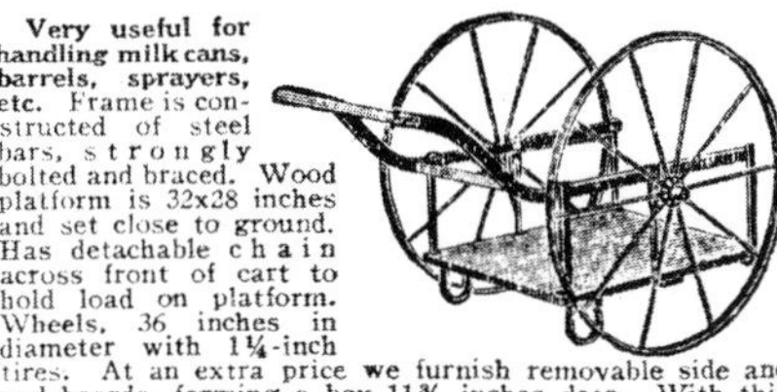

Very useful for handling milk cans, barrels, sprayers, etc. Frame is constructed of steel bars, strongly bolted and braced. Wood platform is 32x28 inches and set close to ground. Has detachable chain across front of cart to hold load on platform. Wheels, 36 inches in diameter with 1¼-inch tires. At an extra price we furnish removable side and end boards, forming a box 11¾ inches deep. With this box the cart is adapted for almost any use to which a hand cart could be put.

32L2236—Handy Platform Cart. Weight, 100 lbs.
From BRADLEY, ILL.......... **$6.95**
From PHILADELPHIA, PENNA.......... **7.80**
32L2237—Removable Box. Weight, 27 pounds.
From BRADLEY, ILL.......... **$1.45**
From PHILADELPHIA, PENNA.......... **1.69**

Bradley Hand Cart

A strongly constructed hand cart. For farm work and other rough usage. Has large and deep hardwood box, 1-inch steel axle. Wheels are 36 inches high with 1¼x¼-inch tires. The box is 36 inches long, 21 inches wide and 9½ inches deep, inside. End boards are held in place by steel rods. Ends and sides can be removed, leaving bottom flat. We can furnish a top extension box, 9½ inches deep, to be set on top of the regular box, as shown in illustration.

REMOVABLE TOP SECTION

32L2227—Bradley Hand Cart. Weight, 100 pounds.
From BRADLEY, ILL.......... **$7.95**
From PHILADELPHIA, PENNA.......... **8.80**
32L2228—Top Extension Box, for above cart. Wt., 15 pounds. From BRADLEY, ILL.......... **$1.50**
From PHILADELPHIA, PENNA.......... **1.65**

Standard Galvanized Steel Tanks

Order a tank from us with the understanding that it must be exactly as we represent it and perfectly satisfactory to you or you can return it at our expense and we will return your money and freight charges. Prompt shipment of all sizes. When tanks are shipped knocked down, all holes are punched, every part is fitted together at the factory and sufficient solder and rivets are sent with which to put the tank together. All tanks made of No. 20-gauge steel unless otherwise specified. Will make regular No. 20-gauge tanks of No. 18-gauge at price 30 per cent higher or No. 16-gauge at price 60 per cent higher.

Certain Standard tanks which are regularly shipped set up can and will be shipped knocked down if so ordered.

All seams in No. 20-gauge tanks are lock seams and are carefully soldered, no edges coming in contact with the water. Stock tanks made of No. 18 and No. 16-gauge steel have riveted and soldered seams. Tops are bound with angle steel. Bottoms of all tanks over 1 foot in height are secured between two pieces of flat steel or are bound with angle steel, depending upon size and shape of tank. (See illustration.) Sides of tanks 6 feet or longer are firmly braced with angle steel bars. Measurements in all cases are outside, over all.

Tanks in these two columns are shipped from factory in SOUTHWESTERN MICHIGAN or KANSAS CITY, MO.

WAGON TANKS.

In our wagon tanks a 1x1x⅛-inch galvanized steel angle is riveted around the inner side at the top and the cover is flanged over this angle and closely riveted between the angle and a strip of ⅞x⅛-inch band steel, extending all around the tank. A three-quarter partition is fitted across the center of 8-foot tanks and two partitions fitted in 10-foot tanks. All made 2 feet high, and with watertight top, having 14-inch round manhole. Have 1-inch pipe connection in rear end.

Shipped set up.	Lgth., Ft.	Width, Ft.	Cap., Gal.	Wt., Lbs.	From Fcty., Michigan	From Fcty., Kans. City
32L4440	6	2	144	125	$13.58	$16.20
32L4441	8	2	197	170	16.98	18.65
32L4442	8	2½	245	185	18.85	20.90
32L4443	8	3	295	205	20.95	23.98
32L4444	10	3	378	255	24.50	26.75

ROUND STORAGE TANKS.

Tanks 32L4388, 32L4389, 32L4392 and 32L4393 are made of No. 20-gauge galvanized steel. Tank 32L4394 is made of No. 18-gauge steel, but can be made of No. 16-gauge at a price 30 per cent higher. Tanks 32L4395 and 32L4398 are made of No. 16-gauge steel, because lighter gauge is not strong enough for such large tanks. **These tanks are always shipped knocked down.** They do not have lock seams, but are punched for rivets. Prices include sufficient solder and rivets.

	Diam., Ft.	Ht., Ft.	Cap., Gal.	Wt., Lbs.	From Fcty., Michigan	From Fcty., Kans. City
32L4388	6	6	1,200	265	$ 28.50	$ 29.95
32L4389	6	8	1,600	330	35.45	37.25
32L4392	8	5	1,800	330	39.60	40.70
32L4393	8	6	2,133	375	43.35	45.75
32L4394	8	8	2,854	590	67.85	70.25
32L4395	8	10	3,592	910	91.75	94.60
32L4398	10	8	4,580	985	106.50	109.50

OVAL TROUGHS.

Shipped set up.

	L'gth., Ft.	Width, Ft.	Depth, In.	Cap., Gal.	Wt., Lbs.	From Fcty., Michigan	From Fcty., Kans. City
32L4410	8	1½	14	72	75	$ 6.60	$ 8.90
32L4411	8	2	12	72	75	6.65	8.95
32L4412	8	2	20	152	100	9.95	13.35
32L4417	10	1½	8	50	65	5.05	6.75
32L4418	10	1½	14	90	85	7.45	9.65
32L4419	10	2	12	90	90	7.50	9.70
32L4420	10	2	20	190	130	11.60	14.62

ROUND TANKS.

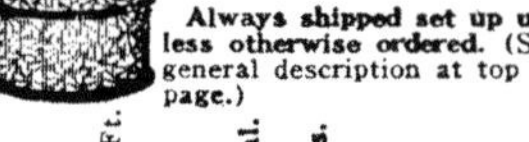

Always shipped set up unless otherwise ordered. (See general description at top of page.)

	Diam., Ft.	Ht., Ft.	Cap., Gal.	Wt., Lbs.	From Fcty., Michigan	From Fcty., Kans. City
32L4310	4	2	166	80	$ 7.25	$ 7.95
32L4311	4	2½	215	90	8.30	9.50
32L4312	4	3	254	100	10.15	10.98
32L4313	4	4	338	125	12.25	13.75
32L4314	5	2	262	110	9.90	10.60
32L4316	5	3	411	135	12.46	13.70
32L4317	5	4	548	160	15.85	17.50
32L4318	5	5	675	185	20.90	21.80
32L4320	6	2	384	140	12.65	13.40
32L4322	6	3	583	170	15.80	17.25
32L4323	6	4	768	200	20.40	23.50
32L4324	6	5	966	235	25.50	28.45

ROUND END TANKS.

Always shipped set up unless otherwise ordered. (See general description at top of page.)

	Lgth., Ft.	Width, Ft.	Ht., Ft.	Cap., Gal.	Wt., Lbs.	From Fcty., Michigan	From Fcty., Kans. City
32L4330	4	2	1	45	45	$ 5.19	$ 6.28
32L4331	6	2	1	70	65	6.22	7.95
32L4332	8	2	1	100	85	7.80	10.15
32L4334	4	2	2	91	61	5.45	6.97
32L4336	6	2	2	144	90	7.75	9.78
32L4337	8	2	2	197	120	9.65	12.60
32L4339	8	2½	2	245	135	10.65	14.09
32L4340	8	2½	2½	310	145	12.44	16.40
32L4342	8	3	2½	375	155	13.26	17.65
32L4343	8	4	2	386	150	14.25	17.90
32L4347	10	3	2	384	170	14.60	18.75
32L4350	10	4	2	496	195	16.85	21.80
32L4351	10	4	2½	625	220	20.39	24.85
32L4354	10	6	2	813	230	22.80	27.90
32L4356	16	4	2	826	300	27.15	33.40
32L4357	16	5	2	1,072	335	33.10	37.80

SQUARE END TANKS.

Always shipped set up unless otherwise ordered. (See general description at top of page.)

	Lgth., Ft.	Width, Ft.	Ht., Ft.	Cap., Gal.	Wt., Lbs.	From Fcty., Michigan	From Fcty., Kans. City
32L4365	4	2	1	50	50	$ 4.95	$ 5.80
32L4366	6	2	1	75	70	6.35	8.15
32L4368	4	2	2	101	75	6.40	8.17
32L4369	6	2	2	152	100	8.00	10.95
32L4370	8	2	2	202	130	11.97	13.98
32L4374	8	3	2	318	155	14.20	17.20
32L4379	10	3	2	397	195	16.88	20.90
32L4382	10	4	2	530	225	19.68	24.20

Kenwood Ball Bearing Windmill

Extra heavy and strong and will prove more satisfactory than light built windmills.

Ball Bearing Turntables and Ball Bearing End Thrust back of the wind wheel permit the Kenwood to respond readily to changes in direction of the wind, and pump in the lightest breezes.

Self governing. The adjustable weight and lever governor will hold the windmill into any wind in which it is safe to run. Should the wind become too high, the windmill will swing quietly out of the wind and then resume pumping when wind abates.

Workmanship and materials are first class. Large shaft bearings lined with hard engine babbitt.

The steel wheels are made of heavy gauge steel and galvanized after all the parts are made. Each wheel section is thus practically soldered into one solid piece by the galvanizing. There are no raw edges or bolt holes exposed. All bolts on wheel and rudder are galvanized and have double nuts. Steel wheels and rudders are nicely striped with red paint.

A 6-Foot Windmill is intended for light service. For wells over 25 feet in depth and where considerable water is required we recommend a windmill 8 feet or larger. Even for light service they are preferable as a permanent investment.

Wood wheels and rudders are made from clear straight grain material, thoroughly painted with white lead paint and trimmed with red.

Complete instructions for erecting are furnished, with which any farmer without expert help can easily erect his windmill.

Prices include pump pole, pull-out wire, reefing gear, bedplate and truing spider with which the mill is attached to tower, but prices do not include tower or platform. Windmill ordered without tower is equipped with bedplate and truing spider for a four-post **wood tower,** and with sufficient pump pole and pull-out wire for a tower 40 feet high, unless otherwise ordered.

Graphite Bearing.

Windmills fitted with graphite bearings require no oiling. The graphite mixture is baked into grooves in the inner surface of the bearing, and fine particles of graphite are distributed over the entire surface. Will last as long as the bearing itself and as long as any oil lubricated bearing. When finally worn out a new bearing can be slipped on over the end of the shaft without taking down the windmill head.

Catalog No.	Kind of Mill	Size of Mill	Wt., Lbs.	From Factory in Indiana: With Regular Bearings	From Factory in Indiana: With Graphite Bearings	From Warehouse at Kansas City, Mo. Regular Bearings
32L4006	Back Geared Steel Windmill	6 feet	305	$27.65	$32.50	$30.40
32L4008	Back Geared Steel Windmill	8 feet	425	38.50	44.85	42.32
32L4010	Back Geared Steel Windmill	10 feet	630	56.25	66.50	61.92
32L4018	Direct Stroke Steel Windmill	8 feet	400	37.75	41.50	.....
32L4020	Direct Stroke Steel Windmill	10 feet	565	49.90	55.25	.....
32L4028	Direct Stroke Wood Windmill	8 feet	390	37.60	41.35	41.10
32L4030	Direct Stroke Wood Windmill	10 feet	485	49.25	54.75	53.60
32L4032	Direct Stroke Wood Windmill	12 feet	650	65.75	72.50	.....

For complete line of windmill pumps, pump cylinders and well boring tools see pages 708 to 710.

Kenwood Four-Post Windmill Towers.

Our Steel Towers are strong and substantial. Every corner post, brace, band girth, bolt and nut is **heavily galvanized after all machine work is done. Our towers are braced diagonally as well as crosswise at every corner post joint** and will withstand the most severe storms. No. 1 towers have one set of bands for each 10 feet of their height; also a band at the platform and two bands above the platform which serve as steps. No. 1 towers are intended for use with 6-foot windmills. Nos. 2 and 3 towers have bands 5 feet apart for the entire height of the tower; also a band at the platform and two extra bands above the platform which serve as steps. No. 2 towers are suitable for 8-foot and 10-foot windmills. No. 3 towers are suitable for 10-foot or 12-foot windmills at any height. **Our towers are full height;** every corner post section is 10 feet 6 inches long, the extra 6 inches being allowed for the lap of one post over the one below it. This feature makes a stronger and better finished tower and also serves to prevent water from running into the corner post joints. **Illustration shows a 40-foot No. 2 or No. 3 Tower. Towers are built only for our windmills and will not support tanks for storage purposes.** Prices are for towers complete with platform, ladder, rod guides, anchor posts, anchor plates and instructions for erecting, but do not include windmill, bedplate, truing spider, pump pole, pullout wire nor reefing gear, these being parts of the windmill.

	Tower	Wt., Lbs.	From Factory in Indiana	From Kansas City, Mo.
32L4112	20-Foot No. 1	345	$23.95	....
32L4113	30-Foot No. 1	485	33.67	....
32L4122	20-Foot No. 2	355	25.45	$28.65
32L4123	30-Foot No. 2	520	36.95	41.63
32L4124	40-Foot No. 2	700	49.60	55.90
32L4132	20-Foot No. 3	405	26.55	....
32L4133	30-Foot No. 3	580	38.80	....
32L4134	40-Foot No. 3	785	52.00	....
32L4135	50-Foot No. 3	1,085	73.10	....
32L4136	60-Foot No. 3	1,410	95.25	....

OIL WAGON TANKS.

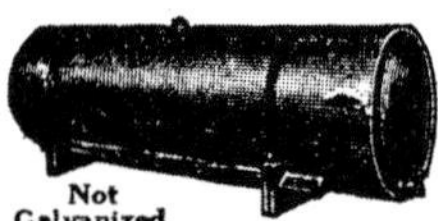

Not Galvanized.

Especially designed for hauling kerosene and gasoline for tractors and other purposes, but can be used for other liquids as well. They are made of extra heavy black sheet steel with welded seams, making one solid steel shell. No riveted nor soldered seams. Have 2-inch threaded and plugged opening in top and 1-inch threaded opening on rear end close to bottom with 1-inch brass lever handle faucet. Prices are for the tank only and do not include the wooden bolsters and bands shown in illustration. Tanks are nicely painted. **Shipped set up from factory in SOUTHWESTERN MICHIGAN or KANSAS CITY, MO.**

	Lgth., Feet	Diam., Inches	Gauge Steel	Cap'y, Gal.	Wt., Lbs.	From Factory, Michigan	From Factory, Kansas City
32L4423	6	24½	16	142	125	$19.45	$19.95
32L4424	8	24½	16	190	155	21.95	23.45
32L4425	8	30	14	290	255	28.90	32.25
32L4426	8	34½	12	380	410	40.00	46.30
32L4427	10	34½	12	475	500	46.35	51.75

32L4428—Wooden Bolster and Bands. Weight, 60 to 75 pounds. From factory, Michigan **$5.35**
From factory, Kansas City **6.25**
32L4429—Foot Rest and Rack, for attaching spring seat. (Seat not furnished.) From factory, Michigan **$4.80**
From factory, Kansas City **4.95**
Foot Rest and Rack can be furnished only when ordered with the bolsters.

Steel Tank Covers.

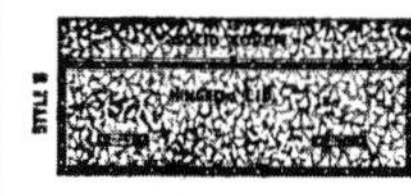

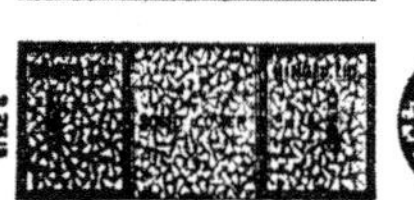

Covers are sold by the square foot, measuring round tanks as though they were square, and round end tanks as though they were square at the ends. Example: A cover for a round end tank 6 feet long and 2 feet wide would measure six times two, or 12 **square feet, and would cost twelve times the price of one square foot.** Made of 20-gauge steel. Weight, about 2 pounds per square foot. **State style wanted. Shipped from tank factory in SOUTHWESTERN MICHIGAN.**

32L4306—Steel Cover. Square foot. **26c**

Thresher Tanks.

Made of No. 20-gauge galvanized steel. Will make of No. 18-gauge at prices 30 per cent higher. Trusses are heavy angle steel and entire tank is built far in excess of necessary strength. Seams are double locked. Fuel box is large and its bottom is lined with 1-inch plank. Rear end of top has a 14-inch round manhole. Has 1-inch feed pipe connection in rear end of tank near the bottom. Just wide enough to fit a 38-inch bolster, but you can bolt a piece of 2x4-inch lumber to each side of each truss to make it fit a 42-inch bolster. A stationary metal bulkhead or splashboard is secured to fastenings on inside walls. Price does not include truck. **Shipped set up from factory in NORTHERN INDIANA.**

	Size No.	Length, Feet	Ht., Feet	Capacity, Barrels	Wt., Lbs.	
32L4445	8	8	2	9½	372	$24.50
32L4446	10	10	2	12	444	27.90
32L4447	12	10	2½	15	490	29.40

Wood Stock and Storage Tanks.

We can furnish cypress wood tanks of most any size in 1½ and 2-inch material. If interested write for prices.

PIPE CONNECTION FOR STEEL TANKS.

We do not cut pipe connection holes in steel tanks, because it is a difficult matter for anyone to tell until he receives his tank just where it is best to cut the hole. You can easily cut the hole by using a cold chisel, and cutting against a block of hardwood. Our galvanized pipe connections consist of one close nipple, two washers, two locknuts and one pipe cap, to close the pipe hole when you wish. **Shipped from our store or from factory.**

32L4300 — Pipe Connection.

Size	Shpg. Wt.	
¾ in.	10 oz.	$0.43
1 in.	1 lb.	.54
1¼ in.	1¾ lbs.	.75
1½ in.	2 lbs.	1.00
2 in.	3 lbs.	1.25

Bee Keepers' Supplies

We can supply nearly everything used by bee keepers. If the articles you want are not listed on this page, write us for prices.

Our entire line of beehives and bee keepers' supplies represents the highest standard of quality. We guarantee them to satisfy any bee keeper, even the most particular professional, and solicit your orders with the understanding that if you do not find our goods fully equal to any others on the market, regardless of price, and that you have saved money by ordering from us, you may send them back at our expense and we will gladly return your money.

The bodies of our hives are made of clear white pine, free from sap. This material takes a smooth finish and cuts easily, permitting the very close accuracy of fit which is so necessary to insure a perfectly satisfactory completed hive.

Beehives are always shipped partly knocked down and are not painted. The small illustrations show how five complete hives are crated for shipment. The large crate contains the five 1-story hives with the covers, bottoms, frames and other parts. Three of the hive bodies are nailed together, and all other parts are packed inside. The small crate contains the supers, two of the super bodies being nailed up and the balance with other parts packed inside. It is a very easy matter for anyone, even without previous experience, to put the hives together. Complete and simple instructions for putting together are sent with each crate of brood chambers and supers, and a sufficient supply of nails is also included.

High Quality Guaranteed.

Five Complete Hives and Supers Crated for Shipment.

Beehives are generally used 1½ stories high, made up of a 1-story hive and the addition of a super or upper half-story. Hives are generally sold in lots of five, ten, etc., but we list single hives so that you can order any quantity.

Shipped from our store, or factory in OHIO.

One-Story Dovetailed Hives.

These are the standard hives—universally used by bee keepers and are fitted with Hoffman frames, which are pierced for wiring, excelsior cover and reversible bottom. No foundation, starters or division board furnished. The metal roofed double cover, which we furnish at an additional price, has telescoping sides which fit down over the hive and an inner wood cover which can be used as an escape board.

Supers, or upper stories, are not included. To make 1½-story hives you must also order the styles of super desired; or you can make full depth 2-story hives by ordering the hive bodies we list below.

Complete 1-Story Hive.

	Wood Cover	Metal Cover
32L3501—Crate of five 8-Frame, 1-Story Hives. Wt., 113 lbs.	**$11.25**	**$13.75**
32L3508—Crate of five 10-Frame, 1-Story Hives. Wt., 130 lbs.	**12.10**	**14.55**
32L3498—Single 1-Story 8-Frame Hive. Weight, 25 lbs.	**2.49**	
32L3499—Single 1-Story 10-Frame Hive. Weight, 28 lbs.	**2.68**	

Hive Bodies.

Complete with frames but without cover and bottom. The Jumbo hive body has the same dimensions as the standard 10-frame except that it is 11$^{5}/_{16}$ inches deep, standard being 9$^{5}/_{16}$ inches deep, and is designed to accommodate those who prefer a brood chamber deeper than the standard 10-frame. No division board or foundation starters furnished. Nails and tin rabbets are included. These bodies may be used as brood chambers or as a full depth upper story to make a full 2-story hive for extracted honey.

32L3480—Crate of five 8-Frame Standard Hive Bodies. Wt., 62 lbs. **$5.65**
32L3481—Crate of five 10-Frame Standard Hive Bodies. Wt., 70 lbs. **6.49**
32L3482—Crate of five 10-Frame Jumbo Hive Bodies. Wt., 78 lbs. **7.28**

Supers for Comb or Extracted Honey.

The supers for comb honeys are 4¾ inches deep and are fitted with scalloped section holders, scalloped separators, follower board and flat springs. No sections or starters are included, except that sections are included with the single supers. 4¼x4¼x1⅞ scalloped sections only can be used with this furniture.

Super for Comb Honey. The sections or honey boxes shown in illustration must be ordered separately.

Super With Extracting Frames.

Supers for extracted honey are 5⅝ inches deep and are complete with 5⅝-inch shallow frames. No wax foundation is included.

32L3504—Crate of five 8-Frame Supers for Comb Honey. Without sections. Weight, 30 pounds. **$3.78**
32L3509—Crate of five 10-Frame Supers for Comb Honey. Without sections. Weight, 35 pounds. **$4.25**
32L3490—Single 8-Frame Super for Comb Honey. Includes twenty-four 4¼x4¼x1⅞ sections. Weight, 8 pounds. **$1.28**
32L3491—Single 10-Frame Super for Comb Honey. Includes twenty-eight 4¼x4¼x1⅞ sections. Weight, 10 pounds. **$1.40**

32L3512—Crate of five 8-Frame Supers for Extracted Honey. With frames. Weight, 37 pounds. **$4.05**
32L3516—Crate of five 10-Frame Supers for Extracted Honey. With frames. Weight, 42 pounds. **$4.50**
32L3492—Single 8-Frame Super for Extracted Honey. With frames. Weight, 7 pounds. **$1.05**
32L3493—Single 10-Frame Super for Extracted Honey. With frames. Weight, 9 pounds. **$1.12**

Section Honey Boxes.

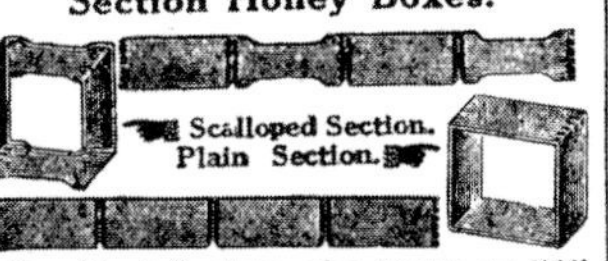

Scalloped Section. Plain Section.

Our No. 1 Sections, also known as "A" grade sections, are exceptionally high in quality. They are made of clear basswood, polished very smooth, are perfect in finish and free from defect. We furnish them plain (no beeway) or scalloped. The scalloped sections are 4¼x4¼x1⅞ inches, with two beeways. The very close sorting to grade of our No. 1 sections leaves many slightly darker or stained or slightly imperfect. These are known as No. 2 or B grade sections. They are polished and perfectly made, and where a strictly clear white section is not required, will serve every purpose. Sections are sold only in full size packages listed. When ordering sections be careful to order the size and style for which your hives are equipped. Weights are as follows: 100 sections, 8 pounds; 250 sections, 18 pounds; 500 sections, 35 pounds.

32L3524—No. 1 Scalloped Sections. Size, 4¼x4¼x1⅞ inches.
Per package of 100. **$1.30**
Per package of 250. **3.15**
Per crate of 500. **6.00**

32L3526—No. 1 Plain Sections. Size, 4¼x4¼x1½ inches.
Per package of 250. **$2.92**
Per crate of 500. **5.64**

32L3522—No. 1 Plain Sections. Size, 4x5x1⅜ inches.
Per package of 250. **$2.90**
Per crate of 500. **5.60**

32L3530—No. 2 Scalloped Sections. Size, 4¼x4¼x1⅞ inches.
Per crate of 500. **$5.49**

Porter Double Bee Escape.

To be placed in the center of a honey board. Saves the work and worry of smoking and brushing bees out of supers. Place the escape in a board and slip between super and hive and the next morning your bees are out of the super.

32L3558—Porter Bee Escape without board. Weight, 2 ounces. **15c**
32L3559—Porter Bee Escape with 8-frame board. Weight, 1¾ lbs. **49c**
32L3560—Porter Bee Escape with 10-frame board. Weight, 2 lbs. **50c**

Hoffman Self Spacing Frame.

The drop frame is the same size used in all of our brood hives. It is 17⅝ inches long by 9⅛ inches deep, with 18¾-inch top piece. The shallow frame is the same as is used in our extracting hive supers. It is 17⅝ inches long, 5⅜ inches deep and has 18½-inch top piece. **Sold only in full crates.** Shipped in flat. Price includes nails.

32L3519—Crate of 50 Shallow Frames. Weight, 15 pounds. **$1.93**
32L3520—Crate of 100 Shallow Frames. Weight, 25 pounds. **$3.69**
32L3518—Crate of 50 Deep Hoffman Frames. Weight, 27 pounds. **$2.95**
32L3521—Crate of 100 Deep Hoffman Frames. Weight, 50 pounds. **$5.70**

Wax Comb Foundation.

Very high quality, genuine Weed Process Foundation, clear and tough, and easily worked by the bees. Will suit the most particular bee keeper. Brood foundation sheets are about 7⅞x16¾ inches. Medium brood foundation runs 8 sheets to the pound, and is used without wiring. Light brood foundation sheets run 9½ sheets to the pound and should always be wired. Super foundation sheets are 3⅞x15½ inches. Thin super foundation runs about 28 sheets to the pound, extra thin super foundation about 32 sheets to the pound. Thin is used for starters, extra thin when full sheets are used. Medium brood and thin super foundation are recommended to the average bee keeper. Shipping weight, 1-pound boxes, 1½ pounds; 5-pound boxes, 6 pounds.

32L3528—Medium Brood Foundation. 5 pounds, **$3.15**; per pound **68c**
32L3533—Light Brood Foundation. 5 pounds, **$3.25**; per pound **70c**
32L3529—Thin Super Foundation. 5 pounds, **$3.55**; per pound **76c**
32L3534—Extra Thin Super Foundation. 5 pounds, **$3.65**; per pound **78c**

Parker Foundation Fastener.

For fastening comb foundation into section boxes.

32L3532—Parker Foundation Fastener. Weight, 12 ounces **32c**

Drone Trap and Swarm Guard.

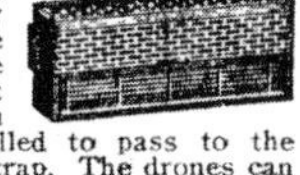

For exterminating drones and preventing loss of swarms. The openings permit the worker bees to pass, but the drones and queen cannot, and are compelled to pass to the upper compartment, or trap. The drones can then be destroyed, or if swarming the trap is placed on the new hive and when swarm returns, as it will, the slide is opened and the queen will go into the hive with the swarm. This is the improved trap with wire spacing, through which the bees can pass without injury.

32L3549—Wire Front Queen and Drone Trap. Weight, 1 pound **66c**

Queen Excluder or Honey Board.

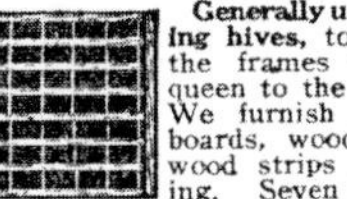

Generally used with extracting hives, to be placed over the frames to confine the queen to the brood chamber. We furnish the improved boards, wood bound with wood strips and wire spacing. Seven wires in each strip. The wood and wire boards are preferred by experienced bee keepers to the perforated zinc boards, as they preserve the bee space more accurately and the smooth wire passages do not injure the bees' wings. 8-frame boards weigh 1½ pounds; 10-frame boards, 1¾ pounds each.

32L3554—Wood and Wire Board. 8-frame size **53c**
32L3555—Wood and Wire Board. 10-frame size **59c**

Honey Extractors.

Latest type of construction, containing all improvements, standard in every respect. Have slip gear which throws crank and pinion out of gear at maximum speed, allowing reel to continue in motion with crank hanging loose. Have ball bearings top and bottom. Frames of comb baskets are heavily tinned. The wire cloth against which the combs rest can be easily removed for cleaning. Improved pocket hinges permit rapid reversing to extract both sides of the comb. Cans are galvanized steel, strongly bound, with handles and are enameled outside. No. 15 is 20 inches in diameter, No. 17 is 23 inches in diameter. **Shipped from factory in OHIO only.**

For larger Hand or Power Extractors, write for prices.

32L3475—No. 15 2-Frame Extractor. Comb pockets, 9⅝x16 inches. Weight, 110 pounds **$31.25**
32L3476—No. 17 2-Frame Extractor. Pockets, 12x16 inches. Weight, 125 pounds **$36.50**

Spur Wire Imbedder.

Used for imbedding the wire of a wired frame into a full sheet of foundation. The teeth of the wheel straddle the wire, thus forcing the wire into the wax.

32L3591—Spur Wire Imbedder. Wt., 3 oz. **29c**

Tinned Steel Wire.

For wiring frames when full sheets of foundation are used; the No. 30 wire has been found to be the most suitable.

32L3595—Tinned Steel Wire.
½-lb. spool. Wt., 11 oz. **22c**
1-lb. spool. Wt., 1¼ lbs. **39c**

Globe Bee Veil With Springs.

Made of French cotton tulle with silk face piece. Five spring steel bars keep the veil away from the face and neck. These bars button to studs on neckbands of veil.

32L3544—Globe Bee Veil. Weight, 5 ounces **$1.10**

Alexander Bee Veil.

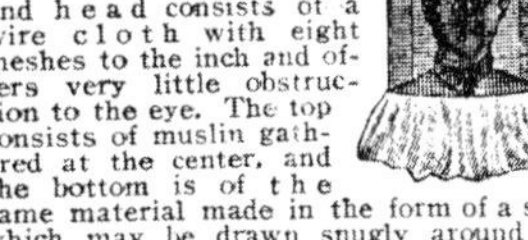

This style is very popular among bee keepers. The portion around the face and head consists of a wire cloth with eight meshes to the inch and offers very little obstruction to the eye. The top consists of muslin gathered at the center, and the bottom is of the same material made in the form of a skirt which may be drawn snugly around the collar by means of a draw string.

32L3542—Alexander Bee Veil. Weight, 1 pound **$1.00**

Hand Section Press.

For putting together one-piece section honey boxes. Presses the dovetailed ends together squarely without breaking corners.

32L3538—Hand Section Press. Weight, 1¼ lbs. **54c**

Bee Gloves.

Furnished in two grades and large, medium or small sizes. Have fingers and long gauntlet with rubber cord to exclude bees. The canvas gloves are heavy and rendered stingproof by special treatment. The rubber gloves are made of a thin elastic material, rubber coated, with sewed seams. We recommend rubber gloves as most comfortable. **State size wanted.**

32L3546—Canvas Bee Gloves. Wt., 8 ounces. Per pair **82c**
32L3543—Rubber Bee Gloves. Wt., 8 oz. Pair **98c**

Honey Shipping Cases.

Cases are made of smooth, sound basswood, and hold 24 sections in single tier, with 2-inch glass front. They have corrugated paper linings and a paper pan to go under the packing to catch any drippings. Made for 4¼x4¼x1⅞-inch scalloped sections, or 4¼x4¼x1½-inch or 4x5x1⅜-inch plain sections. **State size of section used.** Cases are packed in flat and sold only in packages of ten cases. **Shipped from factory in OHIO only.** Price includes nails.

32L3465—Crate of Ten Regular Shipping Cases. Weight, 36 lbs. **$4.95**

Bee Smokers.

Latest Improved Model. Made in three sizes. The two larger sizes have flexible hinge which permits close, tight fit of nozzle over fire chamber. The small smoker has no hinge. All are made of heavy tin and guaranteed perfect.

32L3471—Standard Smoker. 3¼-inch diameter. Weight, 2¼ pounds **$1.00**
32L3472—Jumbo Smoker, 4-inch diameter. Weight, 2½ pounds **$1.29**
32L3470—Junior Smoker, 2½-inch diameter. Weight, 1½ pounds **74c**

Swarm Catcher.

This is a very simple device but is very effective. It is a conical wire basket with four sides and a cover and is to be attached to a long pole or handle. Pole is not furnished.

32L3570—Swarm Catcher. Weight, 8 pounds **$1.98**

Bristle Bee Brush.

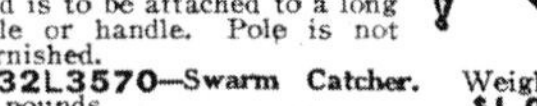

Made of black horsehair bristles, firm enough to easily clear the combs of bees, but soft and pliable so they will not be injured. Can be washed out repeatedly and will usually last several seasons.

32L3564—Bee Brush. Wt., 8 oz. **28c**

Super-Hatcher Double Wall Incubators

FAMOUS FOR THE HEALTHY, VIGOROUS CHICKS THEY PRODUCE

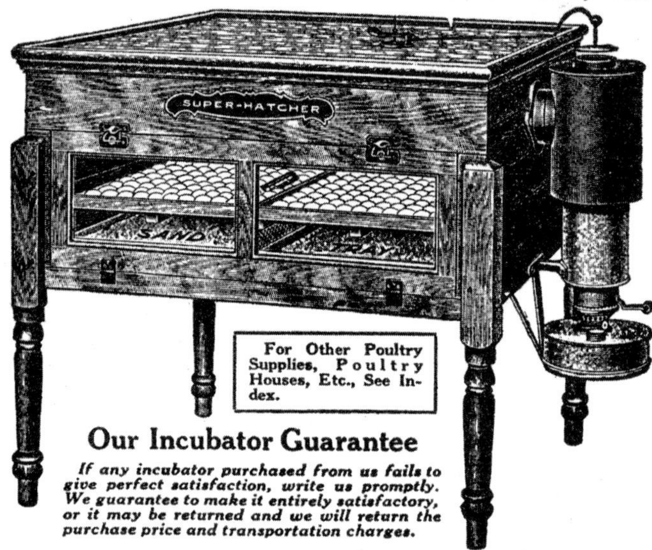

For Other Poultry Supplies, Poultry Houses, Etc., See Index.

Our Incubator Guarantee

If any incubator purchased from us fails to give perfect satisfaction, write us promptly. We guarantee to make it entirely satisfactory, or it may be returned and we will return the purchase price and transportation charges.

Price is for incubator complete with lamp, high grade thermometer, egg tester and full instructions for operating. **Shipped from CHICAGO or factory in PENNSYLVANIA.**

32L3022—125-Egg Hot Air Incubator. Height on legs, 37 inches. Length of case, 28 inches. Width, 27 inches. Weight, 130 pounds **$23.75**

32L3023—240-Egg Hot Air Incubator. Height on legs, 40 inches. Length of case, 38 inches. Width, 36 inches. Weight, 200 pounds **$37.45**

To fully appreciate the value and high quality of the "Super-Hatcher" you must read every word covering the detailed construction.

No matter how much you pay, you cannot obtain a better constructed or a more successful hatcher. Made by one of the oldest and most successful incubator builders in the country. Thousands have been sold with perfect satisfaction to the user. Super-Hatchers are being used by many experimental stations and expert poultry men who are only satisfied with the very best.

Top radiates heat, down draft ventilation and applied moisture. These three great principles, original with the Super-Hatcher, account for its superior hatching qualities, regardless of climatic conditions. The high quality cabinet has wide double walls, properly insulated and put together with screws and nails and corners locked. This extreme care in construction assures you constant, even temperature under all conditions and saves oil expense.

THE HEATER. Powerful heater constructed and insulated so there is absolutely no loss in the transmission of heat. Every joint airtight. Smoke and fumes cannot enter into the hatching chamber. Removable safety lamp made of heavy galvanized steel.

HEAT REGULATOR. Double expanding bars of zinc, reinforced with steel channels with a forceful power thrust, actuate levers balanced on knife edge bearings. Most dependable regulator ever produced. Will not only hold temperature within fraction of a degree, but will last as long as incubator is in use.

HEAT DISTRIBUTION. Dead air space or sluggish circulation is a thing unknown in the Super-Hatcher. The sheet metal distributer, at the top of the egg chamber, receives the heated air current from the horizontal tube above and distributes it evenly to every part of the chamber.

VENTILATION. This was one of many incubators tested out by state experimental stations and it was conclusively demonstrated that it afforded the most successful system of ventilation, far outstripping its nearest competitor. The fresh air is filtered around the heater without contact with the flame and in a separate passage from the smoke or gases. Then it passes down to all parts of the egg chamber just over the eggs, not down through them, but passes out exhaust holes above the eggs and down under the nursery and out below. No heat is applied under the eggs, as they are insulated from the heat below and separated from it by the moist sand tray. The nursery is warmed only enough to remove the chill.

GALVANIZED SAND TRAY. Holds moistened sand and covers entire bottom of egg chamber, giving more even distribution of moisture than is possible with sponges or open pans. Moisture coming from below the eggs prevents evaporation from them. This idea is the nearest approach to the "stolen nest on the ground" ever produced, and is original in this hatcher.

CLEANLINESS. The removal and emptying of the sand tray removes every particle of filth from the hatch. Refill it and the incubator is ready for another hatch without cooling or loss of time.

EGG TRAYS. Well made with wooden frames and galvanized screens, bottoms and removable chick drops.

Built of the most suitable materials, durable, free of any hanging curtains or catch-alls for dust, sanitary in every respect, and a system of incubation which is the closest approach to nature ever produced, adaptable to any climate and with every part superbly fitted and finished, this hatcher offers value unsurpassed.

Electric Incubator. For Either 32-Volt or 110-Volt Current.

Electricity is being so extensively used today that the demand for electric heated incubators is constantly increasing. The expression "You can do it better with electricity" cannot be more effectively used than in connection with artificial incubation. At last, we can offer our customers a dependable electric incubator—an incubator that is absolutely automatic and operates at a very low cost—furnished complete, ready for use for attaching to lamp socket. Made entirely of metal. The case is double walled and packed with wool felt insulation, insuring constant, even temperature with minimum expense. The heating arrangement extends around the entire top of the machine, distributing the heat evenly. The heat is controlled by a high grade regulator, operated by a very sensitive thermostat, and having a dial with index point, making it easy to adjust. This regulator is absolutely automatic, and requires no attention after once being adjusted, and will always maintain heat at the correct temperature. Another convenience is an electric bulb or pilot light inside of the incubator to throw light directly on the thermometer. This lighting arrangement is also automatic—the light goes on when hinged lid is raised and goes out when the lid is put down. The nest or egg tray is made of closely woven, heavy galvanized wire screen, easily kept clean. The bottom of the machine is perforated, providing necessary ventilation. **Can be used on either direct or alternating current.**

Machine is very economical in fuel consumption, as the current is automatically turned off when temperature reaches the required degree, so that current is only used about one-half of the time. The amount will vary somewhat, depending on outside temperature, but in a room of ordinary living temperature, the small size consumes about 15-K. W. H. and the large size about 20 K. W. H. in 21 days. Price includes an egg tester, high grade thermometer and full instructions for operating. **Shipped from our store or factory in INDIANA.**

32L3006— 60-Egg Electric Incubator for 32-Volt Current. Weight, 25 lbs... **$13.75**
32L3007— 60-Egg Electric Incubator for 110-Volt Current. Weight, 25 lbs.... **13.80**
32L3008—100-Egg Electric Incubator for 32-Volt Current. Weight, 40 lbs.... **17.90**
32L3009—100-Egg Electric Incubator for 110-Volt Current. Weight, 40 lbs.... **17.95**

Electric Hover. For Either 32-Volt or 110-Volt Current.

If you have electric current available, you will appreciate the convenience of an electric outfit. **This Electric Hover is made entirely of metal with double outing flannel curtains.** Has large heating surface and will keep the chicks always at a comfortable temperature; burns day and night without any attention. Furnished complete with cord and plug ready to attach to lamp socket. **Sixty-chick size** is 18 inches in diameter and stands 10 inches high; 100-chick size is 22 inches in diameter and stands 10 inches high. **Shipped direct from our store or factory in INDIANA.**

32L3090—60-Chick Electric Hover for 32-Volt Current. Weight, 18 pounds **$9.78**
32L3091—60-Chick Electric Hover for 110-Volt Current. Weight, 18 pounds **$9.83**
32L3092—100-Chick Electric Hover for 32-Volt Current. Weight, 30 pounds **$13.15**
32L3093—100-Chick Electric Hover for 110-Volt Current. Weight, 30 pounds **$13.20**

Little Brown Hen Incubator.

50-Egg Capacity.

Tens of thousands of poultry raisers are using this incubator with perfect success. Simple in construction, convenient and easy to operate. Many large poultry farms use a number of these machines and set fresh eggs as fast as settings are collected. Give the Little Brown Hen a fair trial and if it does not suit you perfectly, send it back and we will return the price you paid, with transportation charges.

The incubator is 18 inches in diameter, stands 15 inches high and holds about fifty average size hen eggs. It is made entirely of metal with double walled nest and top lined with insulating felt. Nest slopes toward center so that by taking out a few eggs the others can be rolled over, a simple method of turning.

Heat radiates above and around nest and is uniformly distributed, the fumes being carried off through side openings.

Regulator is of the expansion disc type, with brass disc.

Thermometer is guaranteed high grade and can be easily read through glass window in top. Lamp has heavy one-piece bowl and burner and chimney of improved safety design. Incubator is finished outside in a durable brown enamel. Complete instructions furnished.

Can be shipped by parcel post, freight or express. When ordered by parcel post be sure to send amount of postage extra. **Shipped from our store.**

32L3011—Little Brown Hen Incubator. Weight, 15 pounds **$4.98**

Copper Tank Hot Water Heated Incubator.

100-Egg Capacity.

We recommend this very successful 100-Egg Hot Water Heated All Metal Incubator because of its simplicity in construction, economy in operation and dependability; this model has become very popular among poultry raisers. Made of strong, durable material throughout, nicely finished in gray enamel—double top and walls lined with insulating material. The circulating water tank is made of **heavy copper** to prevent rusting and extends around the entire top of the machine, insuring even distribution of heat. The lamp sets directly under the water tank—no loss of heat. The thermostat is of the expansion disc type, very sensitive and positive; will hold heat to a fraction of a degree. Thermometer is high grade and can be read through glass window. The egg tray or nest is made of heavy galvanized wire screen, easily kept clean. The bottom of the machine is perforated, providing necessary ventilation. The incubator is 22 inches in diameter and stands 13 inches high and holds about 100 average size hen eggs. Incubator is complete with lamp, egg tester, thermostat and directions for operating. **Shipped from our store or factory in INDIANA.**

32L3015—100-Egg Hot Water Incubator. Weight, 40 pounds...... **$10.85**

Radio Oil Burning Colony Brooder

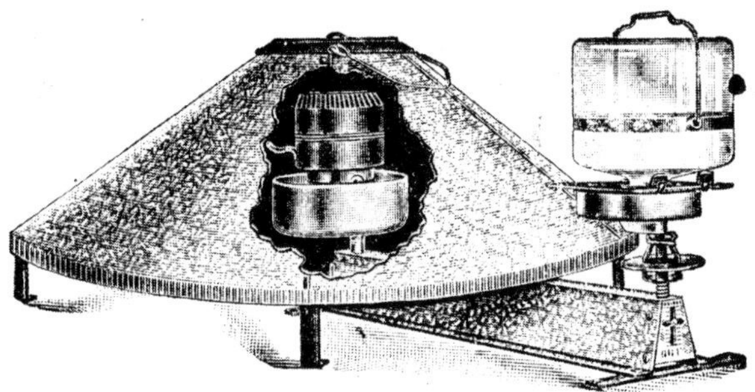

A convenient, economical and practical form of brooder which has gained great popularity. Burns ordinary kerosene oil, gives a wide spread of continuous warmth with ample ventilation and light.

Height of flame and degree of heat are determined by raising or lowering the glass reservoir by means of the handwheel screw adjustment underneath, and the indicator scale guides the operator in determining the adjustment required.

The burner gives a steady blue flame without smoke or odor. It has been successfully used for years by one of the most prominent oil stove makers. It lights easily and generates more rapidly than most others. The asbestos lighting ring is easily cleaned and generally lasts through a season, but can be cheaply renewed. The glass reservoir holds about 1 gallon and requires but one filling a day. The oil pipe from reservoir to burner is continuous, no threaded or packed joints to leak, part of it being a flexible brass tubing which permits up and down adjustment of the reservoir, without use of a stuffing box or gasket joint. The glass fountain locks into position and the burner is protected by a seamless metal pan underneath, while the oil pipes are shielded by the galvanized steel frame which securely holds all parts in alignment. The galvanized sheet steel canopy has ventilation openings and damper at the top. It is shipped knocked down, but can be easily put together with bolts furnished. Complete instructions and two lighting rings with every brooder. **Shipped from our store.**

32L3074—Brooder with 42-inch canopy, for 500 chicks or less. Weight, 55 pounds........**$10.65**

32L3075—Brooder with 52-inch canopy, for 1,000 chicks or less. Weight, 65 pounds........**12.00**

Imperial (Coal Burning Stove) Colony Brooder

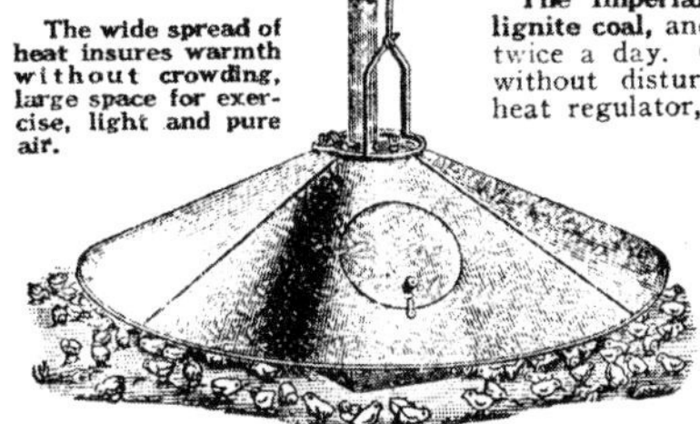

The wide spread of heat insures warmth without crowding, large space for exercise, light and pure air.

The Imperial Brooder burns hard, soft or lignite coal, and requires attention but once or twice a day. Coal is put in through the top without disturbing the canopy hover. The heat regulator, controlling the draft, is automatic and positive in maintaining an even temperature. Canopy is suspended from the ceiling by cord and pulleys furnished, and can be raised while the floor is cleaned or lowered to regulate temperature desired.

Imperial Brooders are made in two sizes. No. 1 stove will care for up to 500 chicks and measures 18 inches high, 11 inches in diameter; grate is 9 inches in diameter, and canopy 42 inches in diameter. No. 2 stove will care for up to 1,000 chicks. It measures 22 inches high, 12 inches in diameter; grate is 10¾ inches in diameter and canopy 52 inches in diameter.

Canopies are made of galvanized sheet steel. Prices include stove, canopy, pulleys and cord and thermostat regulator, but no stovepipe. Stove takes 3-inch pipe. Ordinary galvanized 3-inch drain pipe will serve the purpose. **Shipped from our store or from factory in PENNSYLVANIA.**

32L3000—No. 1 Imperial Stove Brooder. Weight, 80 pounds........**$13.97**

32L3001—No. 2 Imperial Stove Brooder. Weight, 105 pounds........**$17.95**

Mammoth Poultry Feeder.

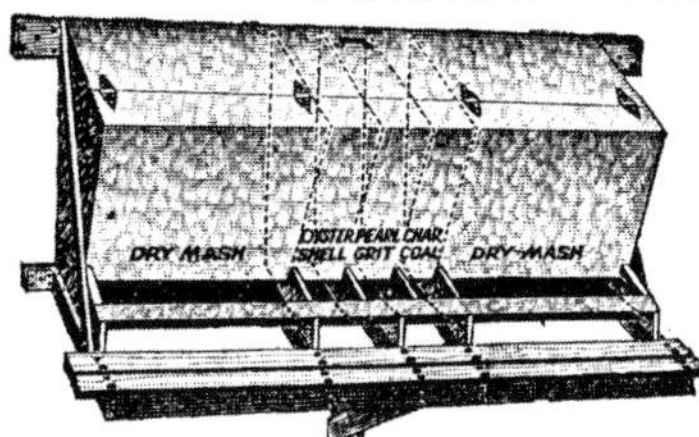

One of the greatest labor saving and sanitary feeders made. Only needs filling about once a month and eliminates daily filling of small hoppers and grit boxes. Its large capacity makes unnecessary extra storage of feed or leaving it in sacks for mice and rats to get at.

Made entirely of galvanized steel. No. 1 size, 45 inches long, holds 100 pounds dry mash, 60 pounds pearl grit, 60 pounds oyster shell, 30 pounds charcoal. No. 2 size, 95 inches long, holds 300 pounds dry mash and same capacity of grit, shells and charcoal as smaller size.

Both sizes are 16 inches wide and 30 inches high. Platform extends 8 inches in front of feeder. Prices include feeder complete with entire platform and hangers for attaching to wall, 18 inches above floor. This allows chickens all floor space. **Shipped crated from factory in ILLINOIS.**

32L3045—No. 1 Mammoth Feeder. Weight, 65 pounds........**$10.85**
32L3046—No. 2 Mammoth Feeder. Weight, 100 pounds........**17.70**

Cozy Hover and Brooder

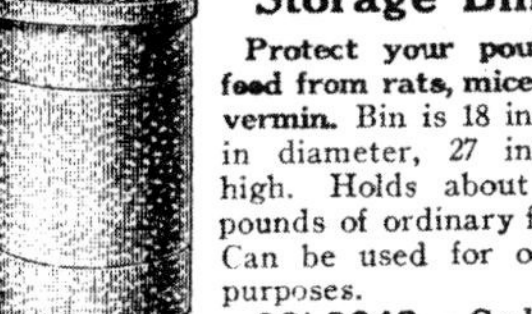

Our Cozy Hover and Cages will give your chicks the necessary warmth and protection and enable you to save many chicks that you would lose otherwise. Made of sheet metal with double outing flannel curtains. Can be used in a room, shed or box. **When set inside a substantial box it makes an excellent outdoor brooder.** Small size hover is 18 inches in diameter, 12 inches high and will care for about fifty chicks. Large size hover is 22 inches in diameter and will care for about 100 chicks.

Lamp compartment is galvanized steel, with brass screen and mica window in door. Lamp and burner are specially designed for safety and economy in use of oil. Lamp fumes are carried off through a galvanized pipe which gives steady draft for lamp flame and prevents fumes from entering hover.

The cage, if wanted, must be ordered extra. It is made of close meshed galvanized wire and sheet steel. It can be spread to allow for exercise or closed tight around the hover to protect chicks from rats, etc. Order a Cozy Hover with your Little Brown Hen Incubator and you will have an ideal outfit. Send extra money for postage if you wish shipped by parcel post. **Shipped from our store.**

Hover With Cage.

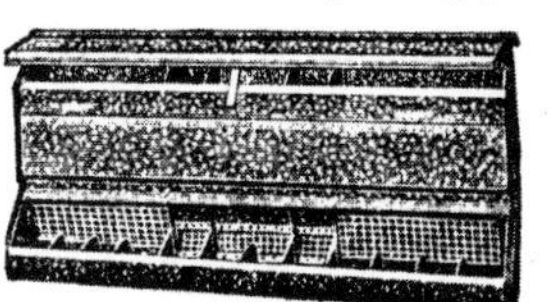

32L3013—50-Chick Size Cozy Hover, without cage. Weight, 15 lbs....**$4.45**
32L3033—100-Chick Size Cozy Hover. Weight, 32 pounds........**6.50**

32L3014—Pest Proof Cage for 50-Chick Cozy Hover. Wt., 11 lbs.....**$2.65**
32L3034—Pest Proof Cage for 100-Chick Cozy Hover. Wt., 13 lbs........**3.35**

Galvanized Steel Feed or Storage Bin.

Protect your poultry feed from rats, mice and vermin. Bin is 18 inches in diameter, 27 inches high. Holds about 100 pounds of ordinary feed. Can be used for other purposes.

32L3042 — Galvanized Feed Bin. Weight, 35 pounds.

Shipped from factory in INDIANA........**$3.10**
Shipped from our store........**3.35**

Round Hopper Feeder.

Ideal hopper for feeding dry mash for large flocks of fowls. Feed pan, 22 inches in diameter, 3½ inches deep and divided by partitions placed close together to prevent fowls from throwing out feed. Cover can be closed over openings on feed trough to keep out mice and rats. Made of heavy galvanized steel and holds about 1½ bushels.

32L3044—Round Hopper Feeder. Weight, 60 pounds.

Shipped from factory in INDIANA........**$6.98**
Shipped from our store........**7.40**

Handy Hopper Feeder.

A popular Hopper Feeder with many poultry raisers. Made of heavy galvanized steel throughout. All parts strongly soldered. Has five partitions provided with adjustable swinging fronts to prevent clogging. Has hinged lids over both hopper and trough, making it mouse, rat and dirt proof. Feeder is 36 inches long, 15 inches high, 14½ inches wide. **Shipped set up, ready to attach to wall.**

32L3050—Feeder. Weight, 20 pounds.
Shipped from factory in INDIANA........**$3.45**
Shipped from our store........**3.65**

Oats Sprouter.

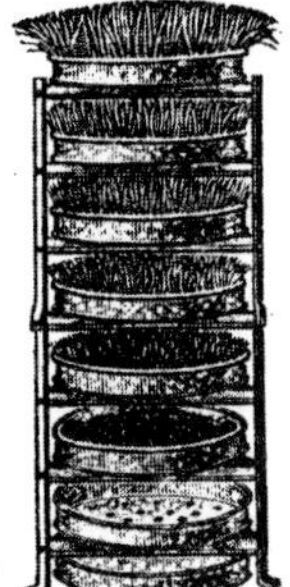

Poultrymen will appreciate the merits of this sprouter. Each pan can be removed without disturbing the other pans. Made entirely of galvanized steel. Sprouter consists of a series of round pans, 12½ inches in diameter. All but lower pan are perforated to allow proper drainage, while bottom or drip pan is solid to catch surplus moisture. Strong galvanized triangular frame, 30½ inches high, braced by truss rods which form shelves for pans.

No lamp or fuel required. Put in the oats and apply a little moisture to top pan daily, and in three or four days you will have sprouted oats. Also suitable for sprouting and raising early plants. Made with seven perforated pans and one drip pan. Can be shipped by parcel post. **Shipped from our store.**

32L7074—Eight-Pan Oats Sprouter. Weight, 23 pounds........**$3.54**

Lamp Heated Sectional Oats Sprouter.

Made of galvanized steel and insulated with lining of rubberoid roofing which retains heat. Sprouter is made in two sections so that for a small flock you need purchase only the base section, which is a complete sprouter. For a larger flock you can double the sprouting capacity by purchasing extra top section. Base section is 20 inches square and 24 inches high and contains the lamp, two grain trays, moisture pan, and has a perforated cover.

Top section is 20 inches square, 14 inches high, and has two grain trays. All trays are 18 inches square, have perforated bottoms for the passage of moisture, and each holds 1 gallon of unsprouted grain. Ventilating spaces provided around the trays and these, with the perforated top, provide ample ventilation to prevent mold. Doors have large glass panes for the admission of light to stimulate growth of the grain. Mouse proof and requires very little heat. Removal of perforated top cover permits watering trays from top. **Shipped crated from factory in INDIANA.**

32L7072—Sprouter (Base Section only.) Capacity, ¼ bushel grain. Shpg. wt., 65 lbs....**$8.00**
32L7073—Extra Top Section. Capacity, ¼ bushel grain. Shipping weight, 35 pounds....**$5.67**

Galvanized Laying and Trap Nests.

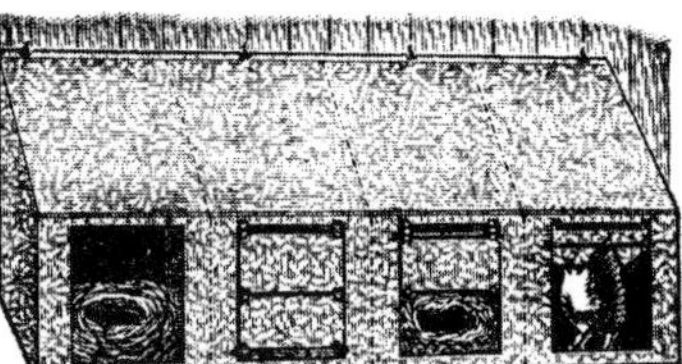

An essential item for every poultry raiser, whether farmer, small flock man, large flock rancher or fancier. Cull your hens and avoid feeding non-layers during winter months when feed is expensive. Made of galvanized iron. Hung at the top to the wall by staples. Nests have no backs. To clean them it is only necessary to take hold at bottom of front and pull upward so contents will fall out. Absolutely sanitary. No place for mites and lice to hide. Nests have sloping top, making it impossible for hens to roost upon it. Trap is simple, but effective. Has no catches, triggers or springs. Trap consists of two shutters, hanging from the top and hinged to center with wire rods, sliding on outside of nest.

When hen enters nest shutters touch her back and close behind her, and she is confined, with ample ventilation, until you wish to release her. Traps can be unhooked when not needed. Furnished in two sizes only. Small size has four nests, is 48 inches long, 13 inches deep, 10 inches high in front and 19 inches high in back. Large size has eight nests and is 94 inches long. Other dimensions same as small size.

32L3040—Four-Hen Trap Nest. Weight, 25 pounds.
Shipped from factory in ILLINOIS........**$2.97**
Shipped from Philadelphia store........**3.20**
32L3041—Eight-Hen Trap Nest. Weight, 60 pounds.
Shipped from factory in ILLINOIS........**5.65**
Shipped from Philadelphia store........**6.30**

For Bone Cutters See Page 893.

$7.70 **Handy Whitewashing and Spraying Machine.**

It forces whitewash into corners and crevices where disease germs lodge and which cannot be reached with brushes. In whitewashing chicken houses infested with lice you can do the work from a door or window, thus avoiding the discomfort of being covered with chicken lice and dirt from head to foot.

For spraying sheep and cattle it is especially efficient as it forces the liquid through to the hide, where it is really wanted.

Has 6-foot iron spray pipe with a ¼-inch brass shut off cock and brass spray nozzle, extra brass nozzle for straight stream-work, rubber suction hose with screen attachment, and 8 feet of ⅜-inch special high pressure spray hose. Shipping wt., 30 lbs.

42L1725¼ **$7.70**

Hercules Power Sprayer $41.75

We furnish, without extra charge, attachments with this sprayer so that it may also be used for hand pumping if desired.

Supplies two nozzles at 125 pounds pressure. Has 2¼-inch brass pump cylinder, with brass ball valves and large cast iron air chamber; 4½-inch stroke; 11¾-inch drive pulley. Back geared 4 to 1. Automatic agitator.

Accessories include pressure gauge, relief or overflow valve, one 20-foot length of ½-inch spray hose with 3 feet of spray pipe, shut off cock and nozzle, 6 feet of suction hose with strainer and 5 feet of overflow hose to lead from relief valve back to tank. Outfit furnished for one lead of hose only. Connections are for standard ¾-inch hose coupling which fits either ½ or ¾-inch hose. If two leads of hose are wanted, order ¾-inch brass twin hose connection 42L667, also extra hose and nozzle. Shipping weight, 175 pounds. Barrel, driving belt or engine not included. For gasoline engines see page 818.

42L2104¼—Hercules Power Sprayer **$41.75**

42L1414¼—High Power Spray Gun only. Shown in operation in above illustration at left. For full description see opposite page. Shipping weight, 3½ pounds **$4.95**

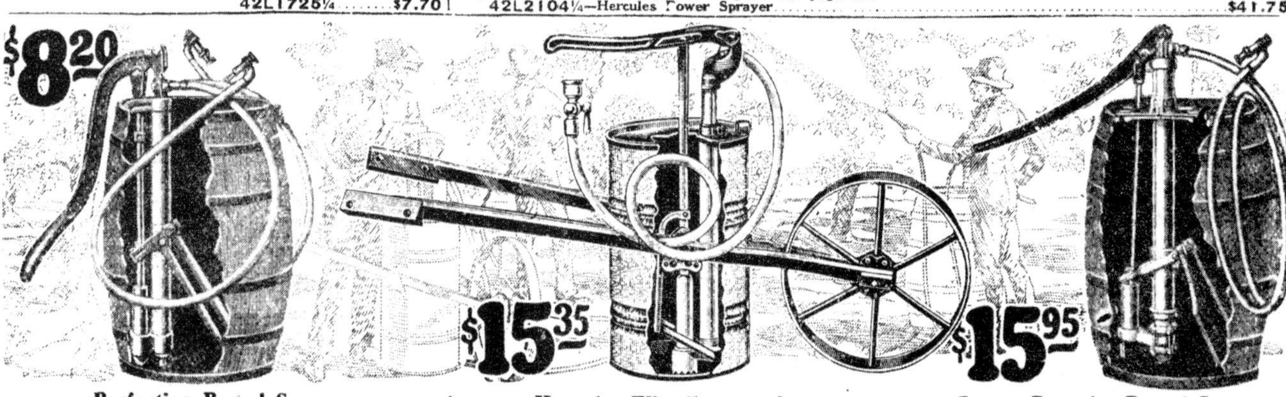

Perfection Barrel Sprayer.

Can also be used for whitewashing, etc. All parts, excepting handle of pump, are inside barrel or tank. Nothing to catch on low hanging branches. Develops 125 pounds pressure. Brass suction cylinder, valves and valve seats. Packing or parts not affected by strong solutions. Two-paddle agitator keeps solution thoroughly stirred up. Easily attached to any barrel. Prices do not include hose, shut offs, nozzles or barrel. Shipping weight, 60 pounds.

42L2130¼—Perfection Barrel Sprayer, with paddle agitator. Fitted for one lead of hose. No hose included **$8.20**

42L2131¼—Perfection Barrel Sprayer. Fitted for two leads of hose. No hose included **$8.95**

Hercules Wheelbarrow Sprayer.

A big labor and time saver for spraying orchards, garden crops, whitewashing or cold water painting chicken houses, hog pens, cattle barns, horse stables, etc. Sprays better, quicker and easier than a small hand sprayer, as it has much larger pumping and carrying capacity. Pump can be removed and tank and barrow used for carting feed, grain, slops, etc. Open top galvanized iron tank holds 10 gallons. Pump cylinder, base castings and valves are all of brass. Produces a pressure of 150 pounds, sufficient to break up solution into a fine mist or fog. Large agitator paddles keep solution thoroughly mixed. Shipping weight, 70 pounds.

42L1726¼—Hercules Barrow Sprayer, complete with 8 feet ½-inch spray hose, shut off cock and spraying nozzle **$15.35**

Large Capacity Barrel Sprayer.

Easily operated. Brass ball valves and valve seat. Brass cylinder, 2¾ inches in diameter. Paddle agitator. For spraying orchards, vineyards, whitewashing and cold water painting. Slow pumping produces heavy, large, powerful mist spray. Adjustable for long or short stroke. Has filler hole with iron cap for filling barrel, brass Vermorel nozzle, twin connection for two leads of ½-inch hose and 15 feet of hose. Barrel not included. Shipping weight, 95 pounds.

42L2108¼—Barrel Spray Pump **$15.95**

$3.95 **Combination Barrel and Bucket Brass Spray Pump.**

May be used with either bucket or barrel for spraying, whitewashing, cold water painting, etc. Cylinders are of heavy seamless brass tubing. Bottom casting and valves are brass. Produces large, powerful continuous spray with slow pumping. Malleable stirrup quickly detachable so pump may be used in barrel by bolting to staves through casting furnished for this purpose, making good barrel pump outfit. Has 3 feet of ⅜-inch hose and two brass nozzles for fine or coarse spray. Agitator in brass base casting keeps liquids thoroughly stirred. Height over all, 31 inches. Shipping weight, 11 pounds.

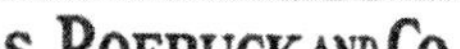

42L1728¼—Combination Barrel and Bucket Brass Spray Pump **$3.95**

$3.65 **Simplex Compressed Air Sprayer.**

For spraying small garden plots, small trees, potatoes and all other low growing vegetation. Extension pipe with nozzle attached directly to tank. No rubber hose required. The automatic lever shut off is attached in such a way that the same hand that carries the sprayer about can operate the shut off, leaving the operator's other hand free to hold back leaves so that the underside can be properly sprayed. Extension pipe is curved and can be turned either up or down. Pressure is pumped up before the sprayer is put in operation, so that all the operator has to do when spraying is to direct the nozzle. The pump is made of brass, the tank will drain dry, pump is of the latest pattern and is adequate for the size of the tank. Unexcelled for greenhouse use. Capacity is 1½ gallons Height over all, 19 inches. Shipping weight, 8 pounds.

42L1733¼ **$3.65**

$4.90 **Marvel Large Capacity Portable Spray and Fire Pump.**

Pump is double acting. This pump will easily develop 180 to 200 pounds pressure. Throws continuous stream or spray. When not used for spraying, keep this spray pump in your house, barn or garage for use in case of fire. Pump and all valves and fittings that come in contact with solution are brass. Two nozzles furnished, one for spraying, other for fire, or washing windows, automobiles, etc. Fire nozzle throws strong stream 25 to 30 feet. Galvanized steel tank and three feet of ⅜-inch hose. Capacity, 6½ gallons. Height over all, 27 inches. Shpg. wt., 22 lbs. Not mailable.

42L1732¼—Marvel Portable Spray and Fire Pump **$4.90**

$8.95 **For Spraying, Whitewashing or Cold Water Painting.**

Furnished complete with 6½-gallon open top reservoir. A strong pressure can be developed with this outfit. It is well adapted for spraying heavy solutions or liquids, such as whitewash, cold water paint, etc. Pump cylinder is brass with iron handle. Agitator keeps solution thoroughly mixed. Pump fitted with 8 feet of three-ply ⅜-inch hose, stop cock and latest pattern large misty spray nozzle. Nozzle has heavy brass strainer and can be easily taken apart for cleaning. Sprayer, filled, is easily carried. Height over all, 33 inches. Shippping weight, 40 lbs. Not mailable.

42L1734¼—Handy Improved Sprayer **$8.95**

On These Finest Quality Sprayers

Handy Bucket Spray Pump.

Bucket sprayer illustrated above has brass body and brass ball valves. Pressure on the handle is entirely on down stroke, making it easy to operate. Spray nozzle can be detached and a solid stream nozzle attached for sprinkling. Shipping weight, 7 lbs.

42L1742¼—Handy Bucket Sprayer, with foot rest, 3 feet of ⅜-inch hose and nozzle, as shown above...**$2.98**

Bucket not included in above price.

For Spraying and Whitewashing.

Bucket Sprayer, Whitewasher, etc. Discharges solution from nozzle in a steady, even spray. Brass, with iron handle and foot rest. Furnished with 3 feet of ⅜-inch hose and two nozzles, one for spraying, the other a straight stream nozzle for sprinkling. Shipping weight, 6 pounds.

42L1744¼—Bucket Whitewash Sprayer **$2.10**

Bucket not included in above price.

For Spraying, Washing Automobiles, Whitewashing, Etc.

Galvanized iron, with brass screw cap at bottom and brass capped wooden plug at top of cylinder for cleaning. For spraying, whitewashing, cold water painting, washing windows, automobiles, etc. Ball check valves in the intake and discharge chambers. Includes 3 feet of ⅜-inch hose, 1 foot brass extension pipe and nozzle for both a coarse and fine spray. Shipping weight, 3½ pounds.

42L1746¼—Bucket Whitewash Sprayer **$1.78**

Bucket not included in above price.

Bucket Spray Pumps illustrated above may be used for spraying vegetables, bushes or small trees, whitewashing, cold water painting, sprinkling, washing windows, automobiles, etc. Hose furnished is ⅜-inch size and sufficiently long for spraying to height of about 6 feet. To reach higher points order extra ⅜-inch hose or extension pipes, quoted on this page.

Gem Crank Duster.

For dusting plants with Paris green, arsenate of lead and other chemicals in powder form. Fan driven by gears, not a belt. This duster will cover two rows at a time as fast as operator can walk and is adjustable for rows of different widths, and for dusting small trees and bushes. Furnished with three tubes, two nozzles, one Y connection, two elbows and carrying strap with snaps. Complete instructions included. Shipping weight, 10 pounds.

42L1740¼—Gem Crank Duster **$7.98**

Powder Duster.

Will dust Paris green, hellebore, dry arsenate of lead and other insect killing powders. Made of light sheet steel, nicely painted. One of the most effective hand operated powder dusters on the market. Throws powder in any direction up or down. Shipping weight, 3 pounds.

42L1036—Dry Powder Duster **$1.30**

Spray Hose for High or Medium Pressure.

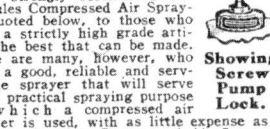

When comparing our prices on sprayer hose remember we furnish two couplings with each length of hose.

Half-inch hose furnished with standard ¾-inch hose connections; ⅜-inch hose furnished with standard ¼-inch connections. Be sure to order correct size hose to fit your sprayer.

42L950¼—Perfection Double Braid Smooth Spray Hose for 150 pounds pressure. Comes only in lengths specified below. **Furnished complete with coupling connections.** Shipping wt., per foot, 5 oz.

Length, ft.	5	10	15	25	50
⅜-inch	$0.75	$1.25	$1.75	$2.75	$5.25
½-inch	.98	1.55	2.10	3.20	5.90

42L952¼—Hercules ½-Inch Double Braid Corrugated Spray Hose for 300 pounds pressure, special seamless braided, high quality. Comes only in lengths specified below. **Furnished complete with coupling connections.** Shipping wt., per foot, 5 oz.

Length, 10 feet	**$1.98**	Length, 25 feet	**$4.10**
Length, 15 feet	**2.65**	Length, 50 feet	**7.75**

Hose Extensions for Barrel Sprayers.

One-Half Inch Five-Ply Spray Hose, complete with Misty Nozzle and Shut Off Cock.

	42L1012¼	42L1014¼	42L1016¼
Length, ft.	5	10	25
Shpg. wt., lbs.	3	6	13
	$2.05	**$2.70**	**$4.20**

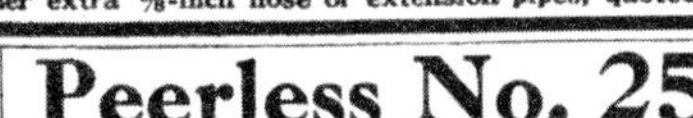

Peerless No. 25

Compressed Air Sprayer $3.95

Screw Lock Type.

For Spraying or Whitewashing.

Showing Screw Pump Lock.

We strongly recommend our Hercules Compressed Air Sprayer, quoted below, to those who want a strictly high grade article, the best that can be made. There are many, however, who want a good, reliable and serviceable sprayer that will serve every practical spraying purpose for which a compressed air sprayer is used, with as little expense as possible. To those we offer our Peerless Compressed Air Sprayer illustrated here. This sprayer operates on the same principle and has same capacity, etc., as our Hercules Sprayer, but the pump is attached to the tank by a screw locking device instead of the quick acting lever-cam lock with which our Hercules Sprayer is equipped. This screw lock is made entirely of brass and threads are machine cut. This sprayer may be used for all general spraying purposes, such as garden spraying, whitewashing, cold water painting, etc. Shipping weight, 9 pounds.

42L1764¼ **$3.95**

$4.45 Hercules

Compressed Air Sprayer

Cam Lock Type.

For Spraying or Whitewashing.

Showing Cam Pump Lock.

Has latest improved quick acting lever cam pump lock. Gives continuous spray for spraying trees, plants, animals, stables and chicken houses, washing windows, automobiles, buggies, whitewashing or cold water painting. A few minutes' pumping charges the tank, and it will then spray without further pumping until pressure in tank gets low again. Three gallons of liquid can be sprayed with three charges of air. Total capacity, 3 gallons. Tank tested to 60 pounds pressure. Pump cylinder is of seamless brass tubing with rubber valve and brass valve spring. Shipping weight, 10 pounds.

42L1760¼—Sprayer with galvanized tank **$4.45**
42L1761¼—Sprayer with brass tank **6.95**
42L1417—Brass Extension Pipes, 2 feet long. Shipping wt., 8 ounces **40c**

Double Acting Tubular Spray Pump.

For spraying, whitewashing and applying disinfectants. Double acting. Gives powerful continuous spray. Operates by working telescoping tubes forward and backward. Sprayer made of brass with wood handle and ½-in. rubber suction hose, with piece of pipe and strainer on end. Four nozzles furnished. One for coarse spray, one for fine spray, one for sprinkling and one for fire stream. Shipping weight, 5 pounds.

42L1032 .... **$2.85**

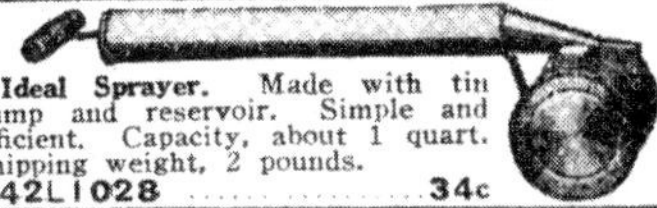

Ideal Sprayer. Made with tin pump and reservoir. Simple and efficient. Capacity, about 1 quart. Shipping weight, 2 pounds.

42L1028 .... **34c**

Little Wonder Sprayer. Has brass ball check valve. Throws continuous spray. Sprays up or down in any position.

42L1018—Sprayer as described, 1 quart size with tin pump and tin reservoir. Shipping weight, 2 pounds **49c**

42L1020—Same as 42L1018, with tin pump and galvanized reservoir **55c**

42L1022—Same as 42L1020, except that it is larger, having capacity of 2 quarts. Shipping weight, 1½ pounds **95c**

Glass Reservoir Sprayer. Plunger has metal expander, keeping leather washer expanded at all times. Capacity, 1 quart. Shipping wt., 3½ lbs.

42L1024—Glass Reservoir Sprayer **44c**

High Power Spray Gun.

With this device you can stand on the ground and spray tall trees. For fast work and thorough spraying this spray gun is unexcelled. One man can do the work of three using old time methods, and there is not half the effort. Sprays liquid most effectively and with least waste. Can be adjusted instantly to suit distance up to twenty-five feet by turning handle at bottom. Owing to the high pressure necessary to operate this gun we recommend it only for use with power sprayers. Shipping weight, 3⅛ lbs.

42L1414¼ **$4.95**

Bamboo Poles.

8 Ft. Long.

For spraying tall trees. Connections are made to a seamless brass tube which extends through center of pole, light and strong. Shut off cock at lower end; brass drip deflector at top. ¼-inch pipe threads. Length, 8 ft. Shpg. wt., 5 lbs.

42L1418¼ **$2.95**

Galvanized Iron Extension Tube, 8 feet long, for spraying tall trees. Threaded both ends, with one coupling. Shipping weight, 7 pounds.

42L1415¼ **48c**

Bordeaux Nozzle. For spraying, whitewashing and cold water painting. Threaded for ¼-inch pipe. Shipping wt., 7 oz.

42L960 .... **39c**

One-Point Vermorel Nozzle. Threaded for ¼-inch pipe. Has coarse and fine discs. Shpg. wt., 4 oz.

42L964 .... **65c**

Brass Elbow. For spraying from beneath. Threaded for ¼-in. iron pipe. Shpg. wt., 6 oz.

42L968 .... **20c**

Brass Strainer. Size, abt. 4¼x 5½ inches. Shpg. wt., 8 oz.

42L1034 .... **79c**

Used to attach two nozzles. Fits ¼-inch pipe. Shipping wt., 5 oz.

42L966 .... **39c**

Cast Brass Shut Off. Threaded for ¼-inch pipe. Shpg. wt., 12 oz.

42L970 .... **63c**

Misty Spray Nozzle.

Has fine and coarse spray discs. Threaded to fit standard ¼-inch pipe connection. Shipping weight, 8 ounces.

42L956—Misty Straight Nozzle **29c**
42L958—Misty Angular Nozzle **30c**

For other Hose Nozzles see page 840.

Popular Sizes At Low Prices

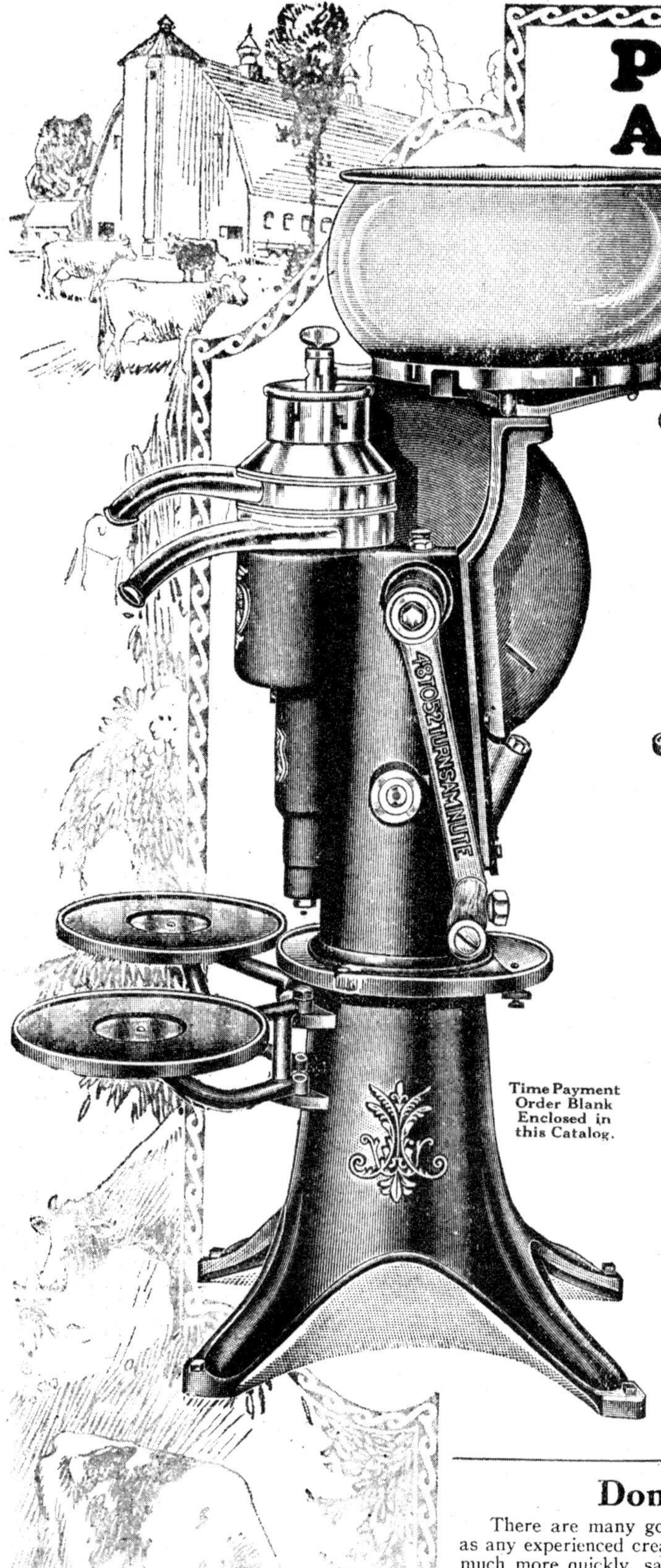

Time Payment Order Blank Enclosed in this Catalog.

Economy King Cream Separator No. 16 With Stand

Skimming Capacity, 600 Pounds (About 290 Quarts) of Milk an Hour.

This is the best size for dairies of from three to twelve or fifteen cows, and is the size selected by most of our customers. It skims at the rate of a milk pailful every two minutes and does a big skimming in short order.

If you have a small or medium size herd, this is the Economy King we recommend. It saves time and labor daily, and because of its big capacity handles the big spring milk flow easily. Should you enlarge your herd in coming seasons, as you will be more than likely to do, it will take care of the extra milk up to a herd of eighteen or twenty cows, if necessary, and save you the expense of buying a separator of larger capacity.

It is a wonderfully close skimmer and takes the cream to a trace from warm or cold milk, gives you a cream of any wished for density, from the thinnest to the heaviest, and cleanses and aerates both the cream and skim milk.

This No. 16 Economy King has all our latest improvements and conveniences. We fully guarantee it and allow you full thirty days for trial, giving you plenty of time to prove its worth in your own dairy.

23L986—Economy King Cream Separator No. 16 with stand. Skimming capacity, 600 pounds of milk an hour. Shipping weight, 240 pounds.

Cash With Order	$5.00 with order and balance in monthly payments of $6.50 each, starting after 30 days.
$62.60	**$70.00**

Economy King Cream Separator No. 18 With Stand

Skimming Capacity, 800 Pounds (About 390 Quarts) of Milk an Hour.

We advise buying this size for herds of ten or twelve to fifty cows or more. It readily takes care of an extra large milk flow when the cows are fresh or on spring pasture, because of its big capacity, and runs so easily that you can handle a large quantity of milk without the use of power, if desired.

It skims at the rate of six and one-half quarts a minute, or a pailful in about one and one-half minutes. City milk dealers and ice cream makers who must separate large quantities of milk without loss of time will find this size just what they need.

It is well adapted for use with power because of its heavy frame and base and substantial construction throughout. The purifying and aerating feature is of great advantage as it removes any dirt or any other foreign matter that may have fallen into the milk, so that both skim milk and cream are in better condition than the whole milk before skimming.

We guarantee this Economy King to please you in every way. Send us your trial order today.

23L988—Economy King Cream Separator No. 18 with stand. Skimming capacity, 800 pounds of milk an hour. Shipping weight, 255 pounds.

Cash With Order	$5.00 with order and balance in monthly payments of $8.00 each, starting after 30 days.
$72.90	**$85.00**

We Ship From a Warehouse Near You.

We insure speedy delivery and low freight charges by shipping the cream separator you order from a warehouse near your home town. See page 904 for list of warehouse points.

Don't Buy Too Small a Size

There are many good reasons for selecting a cream separator of ample capacity, as any experienced cream separator user will tell you. The larger sizes skim the milk much more quickly, saving time and labor every day in the year. They are cleaned about as quickly as the smaller sizes and are practically as easy to run and care for. They easily handle the big milk flow you will have at certain times of the year and save much valuable time in the planting and harvest seasons when every moment counts.

The two larger sizes listed above are extra heavy and substantial, with broad faced gearing, shafting of large diameter and long bearings, giving plenty of wearing surface and thus insuring great durability.

Another advantage of these larger sizes is that they will last longer than the smaller machines. A cream separator, like any other machine, only wears as it is used, and as the larger sizes do any skimming in about one-half the time of the smaller sizes, they are used only about half as long each day and naturally will run about twice as many days before wearing out or needing repairs, so that although the price is a little higher in the first place, in the long run they cost you less per year than the smaller sizes.

Cream Separators With Electric Drive

We supply Economy King Cream Separators in the dairy sizes equipped with electric motors for operation with any form of electric power. If you have a home electric lighting plant, or can purchase electric current, this is one of the most satisfactory and economical methods of operating a cream separator.

Our electrically operated separators are provided with special heavy duty motors especially adapted to stand the heavy starting load. The motors are mounted on a bracket below the drip shelf and drive the separator through an endless belt connecting with a pulley mounted on an extension of the pinion shaft. They do not in any way interfere with hand operation, and if the power is off, the belt can be slipped off the pulley and separating done by turning the hand crank.

In ordering give voltage and say whether you have direct or alternating current. If alternating current is used, find out from your electric company whether single phase, two-phase or three-phase, and the number of cycles, or frequencies. The **32-Volt Direct Current Electric Drive Separator** is complete with 32-volt DIRECT current motor, knife switch, fuse plugs and 40 feet of insulated wire. (Price of extra wire, if wanted, 3 cents per foot.) The **110-Volt Alternating Current Electric Drive Separator** is complete with 110-volt 60-cycle ALTERNATING current motor and 10 feet of cord, with attachment plug for lamp socket. If different from those listed herewith, send us full information and we will quote special prices. We can also supply electric equipment at reasonable prices for the dairy size Economy King you are now using. If interested, write us. Electrically equipped separators are **shipped from CHICAGO, ILL., or BUFFALO, N. Y.**

Electric Drive Economy King Cream Separators

For 32-Volt Direct or 110-Volt Alternating Current.

23L982—Skimming capacity, 250 pounds an hour. Cash with order	$ 80.00	**$10.00** with order and balance in monthly payments of **$7.50** each	$ 92.50
23L984—Skimming capacity, 375 pounds an hour. Cash with order	91.00	**$10.00** with order and balance in monthly payments of **$8.50** each	103.50
23L986—Skimming capacity, 600 pounds an hour. Cash with order	102.50	**$10.00** with order and balance in monthly payments of **$10.00** each	120.00
23L988—Skimming capacity, 800 pounds an hour. Cash with order	112.50	**$10.00** with order and balance in monthly payments of **$11.00** each	131.00

Shipping weights as follows: 23L982, 235 pounds; 23L984, 250 pounds; 23L986, 300 pounds; 23L988, 315 pounds. **See Monthly payment Order Blank enclosed in this Catalog.**

Improved Flexible Drive for Power

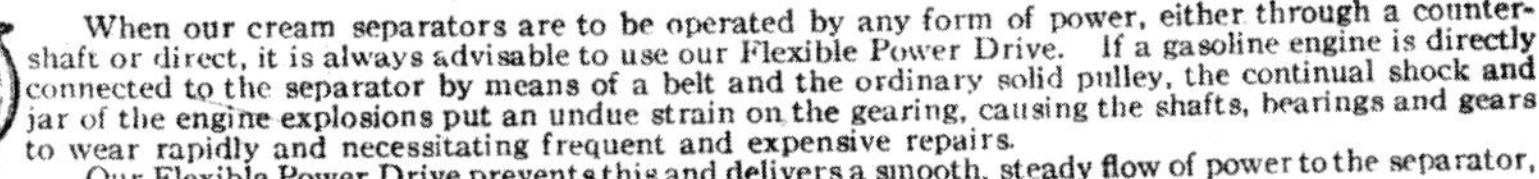

When our cream separators are to be operated by any form of power, either through a countershaft or direct, it is always advisable to use our Flexible Power Drive. If a gasoline engine is directly connected to the separator by means of a belt and the ordinary solid pulley, the continual shock and jar of the engine explosions put an undue strain on the gearing, causing the shafts, bearings and gears to wear rapidly and necessitating frequent and expensive repairs.

Our Flexible Power Drive prevents this and delivers a smooth, steady flow of power to the separator, so that it is under even less strain than with hand operation.

The sudden impulses due to the engine explosions are absorbed by a highly tempered, closely coiled steel spring shaft which transmits the power to the gearing in a smooth, even flow. When the power is applied suddenly a rocker arm swings inward, permitting the belt to slip and applying the power to the separator gear gradually, bringing it up to skimming speed in two or three minutes, without shock or undue strain. Should the engine stop suddenly a ratchet permits the separator to run unchecked until it runs down.

Diameter of driving pulley to be belted to source of power is 3¾ inches and a belt 1 inch wide should be used. Our Flexible Power Drive is shipped complete with special extended pinion shaft to replace the regular shaft used for hand operation. Shipping weight, 50 pounds.

23L772—Flexible Power Drive for Nos. 12 and 14 Economy King.......... **$15.45**

23L776—Flexible Power Drive for Nos. 16 and 18 Economy King.......... **15.75**

The 3¾-inch driving pulley on the flexible power drive should run at the following speeds: For Nos. 12 and 14 Economy King, 707 turns a minute; for No. 16 Economy King, 645 to 696 turns a minute; for No. 18 Economy King, 619 to 671 turns a minute.

Pulleys for Power

Pulleys for use when our cream separators are to be operated by gasoline engine or other power are applied by simply removing the crank and replacing it with the pulley.

Never attempt to belt from gasoline engine direct to these pulleys, as it runs much too fast, and to do so is highly dangerous. Use a countershaft and have the size of the pulleys calculated by some one you know is competent.

23L135—Friction Clutch Pulley for Economy King Cream Separators Nos. 10, 11, 2, 4, 12 or 14. Size, 14 inches in diameter, 1½-inch face. Shipping weight, 25 pounds.......... **$3.90**

23L130—Friction Clutch Pulley for Economy King Cream Separators Nos. 6, 8, 16 or 18. Size 14 inches in diameter, 2-inch face. Shipping weight, 30 pounds.......... **$4.25**

If pulley is ordered separately always state in your order which size separator it is for. We do not furnish pulleys to fit other cream separators.

Cleaning Brushes

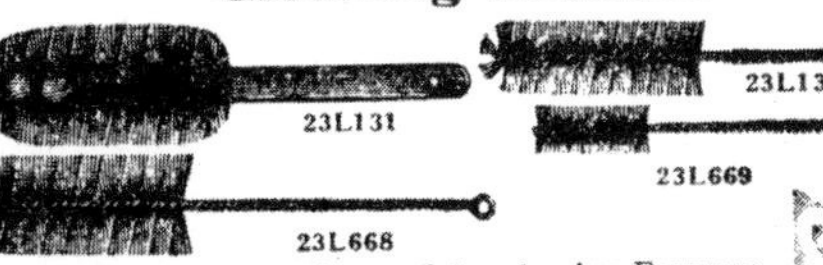

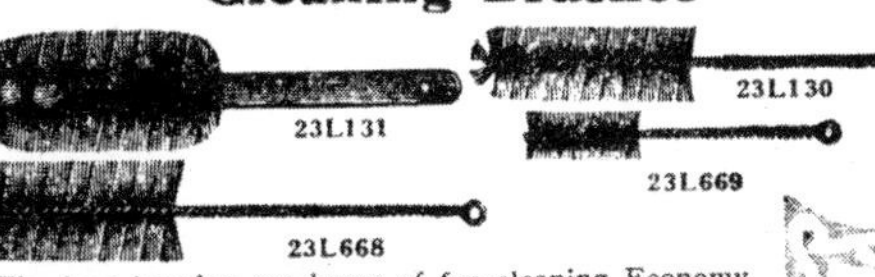

The best brushes we know of for cleaning Economy bowl parts and tinware. Made of extra quality stiff brush stock.

23L131—Wood Handle Brush for Economy King and Economy Chief. Length, 14 inches. Width across brush, 2¾ inches. Shipping weight, 8 ounces.......... **22c**

23L668—Wire Handle Brush for Economy Chief. Length, 13½ inches. Width across brush, 2 inches. Shipping weight, 7 ounces.......... **15c**

23L130—Wire Handle Brush for Economy King. Length, 12 inches. Width across brush, 1½ inches. Shipping weight, 6 ounces.......... **15c**

23L132—Wire Handle Tube Brush for Economy King. Same style as 23L130. Length, 12 inches. Width across brush, ⅞ inch. Shipping weight, 4 oz. **10c**

23L669—Small Wire Handle Brush for Economy Chief. Length, 8½ inches. Width across brush, ⅝ inch. Shipping weight, 2 ounces.......... **5c**

Cream Separator Supplies.

For prices on other supplies and repairs for Economy Cream Separators see direction booklet furnished with the machine. If you have lost or mislaid your direction booklet, write us, giving the name and bowl number of your separator, and we will gladly send you another direction booklet postpaid.

The "JEWEL" Nationally Known Double Farm Harness

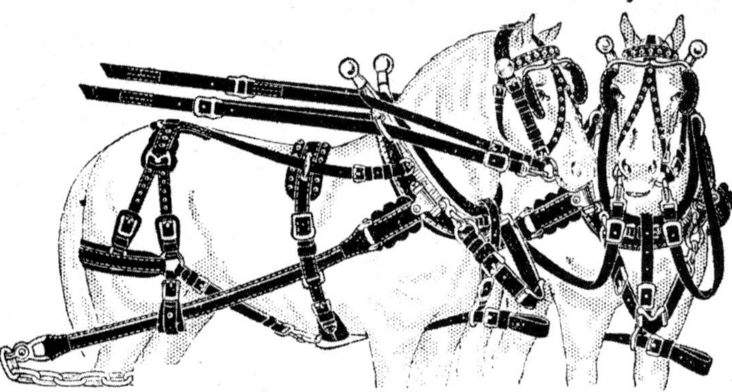

A First Class Harness. Made With Leather Covered Extra Strong Wire Cable Tugs.

$49.75

Satisfies the Wants of the Majority of Farmers.

Buyers of this harness are fully safeguarded because we guarantee the quality. It is to our interest to sell you harness that will give satisfaction not only in appearance but in service as well.

BRIDLES—⅞-inch cheeks; Concord blinds; spotted face pieces; flat reins. **LINES**—1 inch wide, 20 feet long. **HAMES**—Steel Concord bolt; ball top, four hame straps; two spread straps; 1½-inch breast straps and martingales. **PADS**—Felt lined, spotted; adjustable skirts; 1¼-inch bellybands. **BREECHING**—Folded harness leather body; 1-inch hip straps with trace carriers, 1-inch back and side straps. **TRACES**—Leather covered flexible galvanized wire cable. Stronger and more durable than a leather trace. Cable traces, 6 feet 2 inches long, not measuring heel chains. Weight of harness, packed for shipment, 75 pounds.

10L808¼—Double harness, without collars................ **$49.75**

Add extra for lines 1⅛ inches wide.......................... **75c**

You may buy the breeching of this harness separately. If wanted, order:

10L2482—Complete breeching, consisting of hip and back straps, side straps and body, for two horses. Shipping weight, 15 lbs. Per set. **$11.95**

"Clifton" Concord Style Double Farm Harness

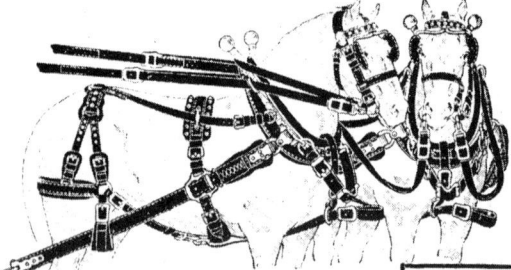

This harness is deserving of special notice. **The heavy running traces are made of two-ply harness leather 6 feet long and have an all leather filling—the popular standard truck style, suitable for** all kinds of teaming and farm work. The other parts are made substantial and strong, assuring a good all around harness. **BRIDLES**—⅞-inch long cheeks; spotted face pieces and fronts; **short checks. LINES—1 inch** wide, 20 feet long, with snaps. **HAMES**—Steel, clip, ball top; four hame straps; two spread straps; folded bellybands. **BREECHING**—Folded body with layer; three-ring, 1-inch double hip straps; 1-inch double backstraps running to the hames. **MARTINGALES AND BREAST STRAPS—1½ in.** wide. Weight, packed for shipment, 75 pounds.

10L695¼—Double harness, with 1½-inch traces, without collars...... **$45.57**

10L696¼—Double harness, with 1¾-inch traces, without collars...... **$47.25**

The "Burton" Harness

Good Leather and Workmanship. Buyers of This Harness Will Be Well Repaid in Satisfactory Service.

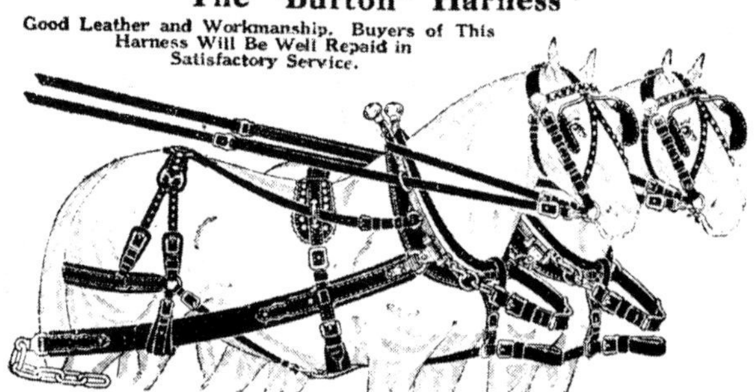

BRIDLES—⅞-inch cheeks; Concord blinds; spotted face pieces; flat check reins.

LINES—1 inch wide, 20 feet long.

HAMES—Steel Concord bolt; ball tops; four hame straps; two spread straps; 1½-inch breast straps and martingales.

PADS—Felt lined; spotted; adjustable skirts; folded bellybands.

BREECHING—Folded harness leather body; 1-inch hip straps with trace carriers; 1-inch back and side straps.

TRACES—Single strap, 3½ inches wide, scalloped layer stitched on both ends; 6 feet long, not measuring heel chain.

Weight of harness, packed for shipment, 75 pounds.

10L707¼—Double harness, without collars........................ **$51.95**

Add extra for lines 1⅛ inches wide.................... **75c**

"Reliance" Farm Harness

BRIDLES — ⅞-inch cheeks; Concord blinds; spotted face pieces; short flat reins. **LINES**—1 inch wide, 20 feet long, with snaps. **HAMES** — Oiled Concord bolt; 1½-inch breast straps; 1½-inch martingales; four hame straps and two spread straps. **TRACES**—Leather, 1½ inches or 1¾ inches wide, 6 feet long, with Concord clevises and heel chains. **PADS**—Flat, felt lined, spotted, metal bridges for double backstraps; 1¼-inch bellybands. **HIP STRAPS**—1 inch; 1-inch double backstraps, with cruppers to buckle on. Weight of harness, packed for shipment, 85 pounds.

10L709¼—Double harness, with 1½-inch traces, without collars...... **$42.75**

10L710¼—Double harness, with 1¾-inch traces, without collars...... **$44.50**

Add extra for lines 1⅛ inches wide........................ **75c**

Add extra for 1-inch bridles........................ **50c**

"Bernard" Wire Cable Trace Harness

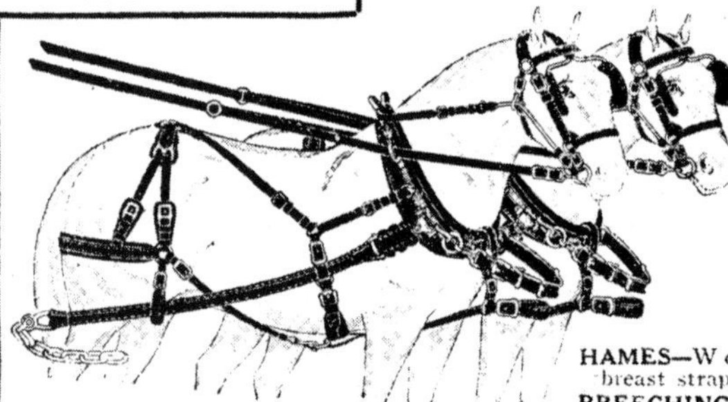

This style of harness appeals to a great many team owners and farmers. It is plain but substantially made and has the strong galvanized wire cable traces, leather covered. A harness that you can order with the feeling that you will not be disappointed in its quality or value.

BRIDLES—⅞-inch short cheek; round reins; Concord blinds.

LINES—1⅛ inches wide, 20 feet long.

HAMES—Wood, overtop; bolt; 1½-inch breast straps and martingales.

BREECHING—Folded leather body with layer, 1-inch hip straps, backstraps and side straps; folded bellyband.

TRACES—Leather covered flexible galvanized wire cable traces, 6 feet 2 inches long, heel chains.

Weight of harness, packed for shipment, 75 pounds.

10L712¼—Double harness, without collars........................ **$42.75**

"Rockwell" Double Farm Harness

Made of genuine bark tanned leather; strong bridles with ⅞-inch cheeks; steel hames; flat pads, felt lined, and heavy breeching. A harness well recommended for all kinds of teaming and one which we know will give long and satisfactory service. **BRIDLES**—⅞-inch short cheeks; flat side checks. **LINES**—1 inch wide, 18 feet long. **HAMES**—Steel, ball top, bolt; four hame straps, two spread straps. **TRACES**—All leather, 6 feet long; heel chains; laced box loop hame tugs. **PADS**—Flat harness leather, felt lined; folded bellybands. **MARTINGALES AND BREAST STRAPS**—1½ inches wide. **BREECHING**—Folded leather body, 1-inch layer, stitched the full length; 1-inch side straps, backstraps and hip straps. Weight of harness, packed for shipment, 80 pounds.

10L816¼ — Double harness with 1½-inch traces, without collars... **$48.95**

10L817¼ — Double harness with 1¾-inch traces, without collars..... **$49.95**

Add extra for lines 1⅛ inches wide............ **75c**

"Hartley" Farm Harness

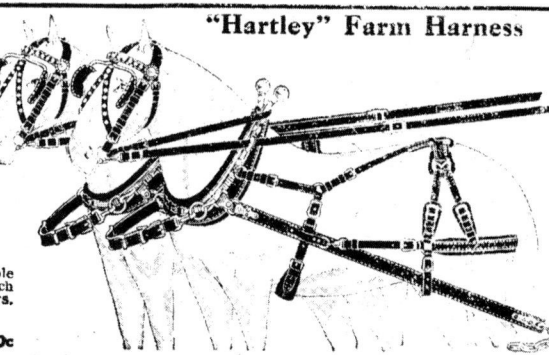

A good leather harness, with double and stitched traces and crotch breeching. This style of harness has an exceptionally big sale. **BRIDLES—⅞-inch** long cheeks with face pieces; sensible blinds; round winker braces, flat side **reins. LINES—1 inch wide, 20 feet long,** with snaps. **TRACES**—6 feet long, fastened to hames with concord jointed clip; six-link heel chains. **HAMES—Steel, ball top,** bolt, four hame straps; two spread straps. **BREAST STRAPS**—1½-inch, with snaps and slides. No martingales. **BELLYBANDS** — 1¼-inch, folded. **BREECHING**—Folded leather body with layer double and stitched; 1-inch side straps, backstraps and hip straps. Weight of harness, packed for shipment, 75 pounds.

10L818¼—Double harness with 1½-inch traces, without collars, **$43.50**

10L819¼—Double harness with 1¾-inch traces, without collars. **$44.50**

Add extra for 1-inch bridles............ **50c**

The "Oakbourn" Heavy Wagon Harness

A Widely Approved Style and a Harness of Select Quality.

$52.75

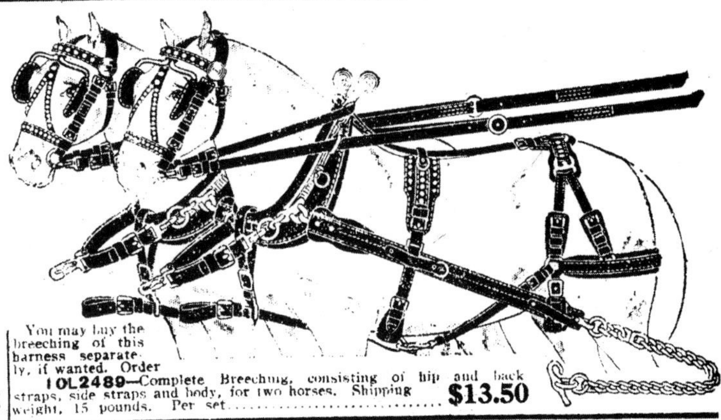

BRIDLES—⅞-inch long cheeks; Concord blinds; combination noseband and face pieces, spotted front; flat reins. **LINES**—Our best and heaviest, 1⅛ inches wide, 20 feet long. **TRACES**—4½ feet long and 2¼ inches wide; three-row stitching; short scalloped safe; 1½-inch bellyband billet; 42-inch chain. **HAMES**—Steel, clip, dandy brass ball top, four hame straps, two spread straps with rings; 1½-inch folded bellybands. **BREECHING**—Folded body, 2½ inches wide with 1¼-inch layer; 1-inch side straps; three-ring style double hip straps, 1 inch wide, with hip strap trace carrier; 1½-inch backstraps to hames; padded rump safe. **PADS**—4 inches wide, tapering to 3 inches; 1¼-inch layer and billet. **MARTINGALES**—1¾ inches wide, with ⅞-inch collar straps. **BREAST STRAPS**—2 inches wide with roller snaps. **TRIMMINGS**—Black japanned; brass spots. Weight of harness, packed for shipment, 90 pounds.

10L860¼—Double harness, without collars.......... **$52.75**

You may buy the breeching of this harness separately, if wanted. Order

10L2489—Complete Breeching, consisting of hip and back straps, side straps and body, for two horses. Shipping weight, 15 pounds. Per set.......... **$13.50**

"Capital" Farm Harness.

BRIDLES—¾-inch cheeks; Concord blinds; face pieces. **LINES**—1 inch wide, 20 feet long. **PADS**—Sewed bottom; 1-inch hip straps; 1-inch backstraps with folded crupper. **TRACES**—All leather, 6 feet long, two rows stitching, clip cockeyes, or leather covered flexible wire cable, 6 feet 2 inches long, with heel chains. **HAMES**—Steel, ball top, bolt. Concord clip attachment; laced hame tugs; four hame straps; two spread straps; folded bellybands. **BREAST STRAPS**—1½-inch, with twin loops. **POLE STRAPS**—1½ inches wide. Weight of harness, packed for shipment, 70 pounds.

10L825¼ Double Harness with 1½-inch leather traces, without collars, **$38.95**

10L826¼ Double Harness with 1¾-inch leather traces, without collars, **$39.95**

10L827¼—Double Harness with leather covered galvanized wire cable traces, without collars.......... **$37.95**

"White Mountain" Eastern Style Truck Harness.

A heavy harness used extensively by the large express and drayage concerns. The leather is of selected quality and the harness well made throughout. **BRIDLES**—⅞-inch cheeks, sensible blinds, face pieces, flat reins. **LINES**—Good heavy leather, sewed and riveted laps, with buckle, billet and snap ends; 1 inch wide, 18 feet long. **HAMES**—Scotch bolt, wood, with steel back, brass ball tops. Heavy hame straps and spread straps. **TRACES**—Three-row stitched front, 2 inches wide; 1¾-inch rear, double and stitched, with dee, hook and 24-inch chain; folded bellybands with return billets. **BREECHING**—Wide folded body with 1¾-inch layer; heavy lead-ups with safes; 1½-inch side straps extending to dee in trace; 1⅛-inch backstraps and double hip straps; padded rump safe. **JOCKEY STRAPS**—1½ inches wide, single strap with hook and ring; 1⅛-inch carrying straps. **TRIMMINGS**—Black japanned. Shipping weight, 100 lbs.

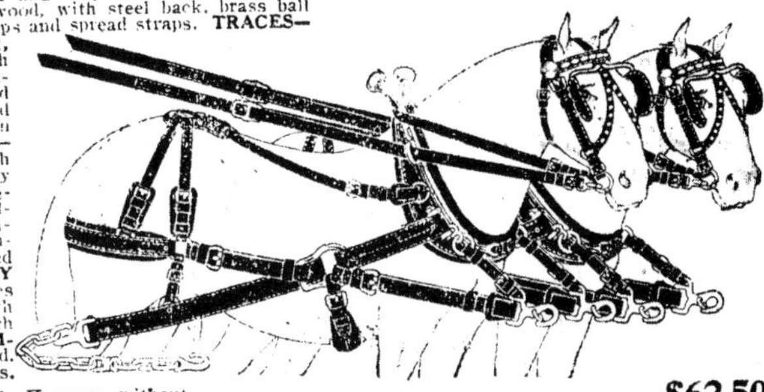

10L858¼—Double Harness, without collars.......... **$62.50**

"Watson" Farm Harness.

The "Watson" harness is one we have sold for a number of years and there are thousands of them in use and giving splendid service. Future purchasers will have the same measure of satisfaction. **BRIDLES**—⅞-inch long cheeks. **LINES**—1 in. wide, 18 feet long. **HAMES**—No. 91 bolt, ball top. **TRACES**—No. 1 stock all leather; 6 feet long, 1¾ inches wide; triangular cockeyes. **HAME TUGS**—1¾-inch, double and stitched. **PADS**—Metal bridge, flat harness leather, felt lined, folded bellybands. **BREAST STRAPS**—1½-inch. **MARTINGALES**—1½-inch. **BREECHING**—Folded leather body; heavy layer; 1-inch side straps, backstraps and hip straps. Weight of harness, packed for shipment, 80 pounds.

10L677¼—Double Harness, without collars.......... **$47.95**

You may buy the breeching of this harness separately, if wanted. Order:

10L2480—Complete Breeching, consisting of hip and back straps, side straps and body for two horses. Shipping weight, 15 pounds. Per set.......... **$11.95**

"Jefferson" Concord Style Truck Harness.

BRIDLES—Heavy 1-inch cheeks. **LINES**—First quality, 1 inch wide, 20 feet long. **HAMES**—Concord bolt, with dandy brass ball tops; 1-inch top hame straps, 1¼-inch bottom hame straps. Spread straps. **TRACES**—Concord truck style, very heavy leather, well made. Connected to hame with our special jointed clip; 6 feet 4 inches long, not measuring the heel chains. Folded bellybands. **BREECHING**—Five-ring style; heavy folded body with layer stitched the full length. 1¼-in. hip, back and side straps, or 1-in. hip and side straps, 1⅛-inch backstraps. **BREAST STRAPS**—1½ or 1¾ inches wide with roller snaps. **POLE STRAPS**—Heavy trucking style, 1½ or 1¾ inches wide. **PADS**—Flat harness leather, felt lined, metal backstrap loops. **TRIMMINGS**—Black japanned buckles, brass spots. Weight, boxed for shipment, 100 pounds.

10L830¼—Double Harness with 1½-inch traces, breast straps and pole straps; breeching with 1-inch hip and side straps, 1⅛-inch backstraps, without collars.......... **$56.85**

10L831¼—Double Harness with 1¾-inch traces, breast straps and pole straps; breeching with 1¼-inch hip, back and side straps, without collars...... **$59.95**

"Rutherford" Double Team Harness.

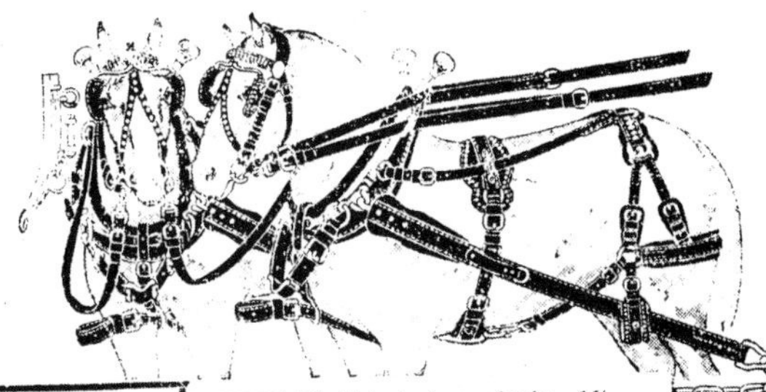

BRIDLES—⅞-inch long cheeks, 1⅝-inch crown, 1-inch fronts, flat reins.

LINES—1⅛ inches wide, 20 feet long.

HAMES—Scotch, long spot, wood, ball top, clip or bolt. Hame straps and spread straps. **TRACES**—1¾-inch Concord truck, double and stitched; 6 feet or 6 feet 4 inches long, not measuring heel chains; folded bellybands. **PADS**—Heavy leather, scalloped, spotted all around, with padded bottom, 5 inches wide at widest part; metal bridges. **BREECHING**—Heavy folded body with 1¾-inch layer, 1-inch hip and side straps, 1⅛-inch backstraps. **MARTINGALES AND BREAST STRAPS**—1½ inches wide. Shipping weight of harness, 100 pounds.

10L910¼—Brass trimmed double harness with clip hames and traces 6 feet long, without collars **$67.50**

10L911¼—Brass trimmed double harness with bolt hames and traces 6 ft. 4 in. long, without collars **$68.95**

The "Diamond" Heavy Truck Harness.

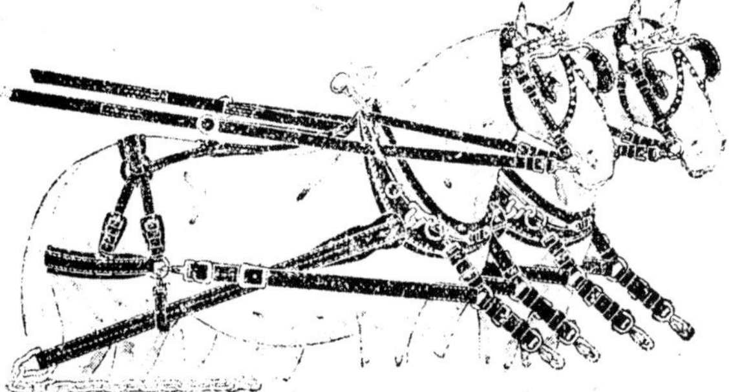

BRIDLES—⅞-inch short cheeks; ⅞-inch flat reins.

LINES—First quality, 1⅛ inches wide, 20 feet long.

TRACES—5 feet 8 inches long, 1¾ inches wide, three rows of stitching; heel chain with hook.

HAMES—Heavy Scotch, clip brass ball top.

BREECHING—Folded body, 2¼-inch layer. New York style backstraps, 1¼-inch hip straps, 1½-inch backstraps.

JOCKEY STRAPS—Boston style, 1¾ inches wide, extending to neckyoke; 1½-inch carrying straps.

TRIMMINGS—Black japanned, brass spots.

Weight of harness, boxed for shipment, 100 pounds.

10L854¼—Double harness, without collars.......... **$61.50**

10L854JS—Jockey Straps with snaps. Shpg. wt., 16 lbs. Set for two horses.......... **10.75**

RELIABLE SADDLES

"Doris" Side Saddle

$14.85

10L1223¼

Russet Leather

TREE—17-inch, low cantle. SEAT—Quilted, hand raised and stitched. SKIRTS—Fancy stamped, 16½ inches wide, 24 in. long on near side; 11 inches wide and 11 inches long on the off side. PAD—Enameled drill top, drill lining, quilted and tufted, very easy on horse. GIRTH—Cotton cord web, buckle on each end, 1¼-inch billets, attached to tree; extra cord surcingle running over tree and skirts, making a double rigged saddle and assuring safety. STIRRUP STRAPS—¾ inch wide, iron stirrups. Weight, about 10 pounds; shipping wt., 16 pounds.

"Walter" Riding Saddle

Pony Saddle for Boys and Girls.

10L1215¼

$10.95

RUSSET LEATHER.

A practical saddle for use around the farm. Well made and strong.
TREE—12-inch, full leather covered, Morgan style, leather covered horn.
SEAT—Half leather covered, tacked to tree, round skirts.
CINCH—Single, corded, 1-inch tie straps.
STIRRUP STRAPS—1-inch, with fenders 6x13 inches; leather hooded stirrups.
Weight, about 9¼ pounds. Shipping weight, 15 pounds.

"Mansfield"

10L1207¼

$5.95

English Style Saddle.

Made in russet leather only. Large enough for a small man or good size boy. Very easy on horse's back. TREE—15-inch, canvas covered, English style.
SEAT—Kip leather.
SKIRTS—Pigskin impression, stitched to seat.
PAD—Cotton serge, stuffed and fitted.
STIRRUP STRAPS—⅞ inch wide, common 3-inch wood stirrups.
GIRTH—Cotton web, buckle on each end, leather billets.
Weight of saddle, about 6 pounds; packed for shipment, 10 pounds.

"Gilbert" Russet Leather Saddle

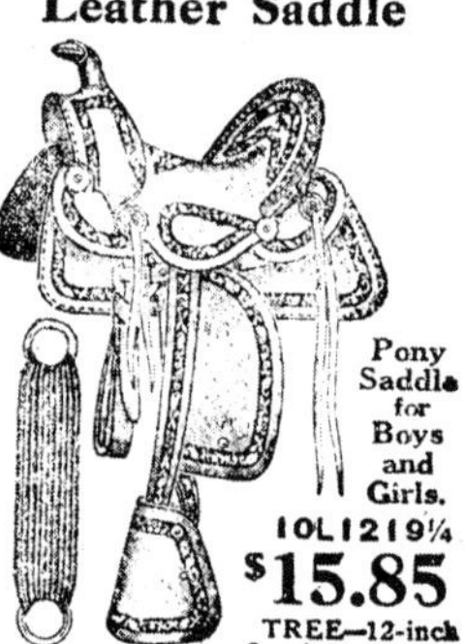

Pony Saddle for Boys and Girls.

10L1219¼

$15.85

TREE—12-inch Omaha, canvas covered, with steel fork. SEAT—Full, leather jockeys, horn and fork full leather covered. CANTLE—Bound, front lined and stitched. HORN—Covered and stitched, with cap. SKIRTS—Round corners and felt lined, good leather. LATIGOS—1¼-inch, single leather. CINCH—4-inch, hard hair. STIRRUP STRAPS—1¼-inch, to buckle. STIRRUPS—2½-inch hooded.
Weight, packed for shipment, 16 pounds.

"Roslyn" Single Rigged Saddle

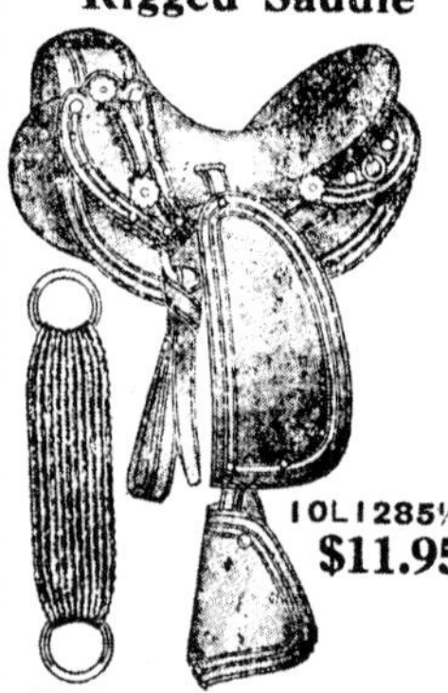

10L1285¼

$11.95

TREE—13-inch, canvas covered, muley style. SEAT—Full leather covered. STIRRUP STRAPS—1⅛ inches wide, to buckle, with fenders 7x14 inches. STIRRUPS—3-inch, wood, leather hooded. CINCH—4 inches wide, with ring. LATIGO STRAPS—1⅛ inches wide, to tie. LEATHER—Russet, good quality embossed border. Flat handhold on near side. Weight, about 11 pounds; packed for shipment, 17 pounds.

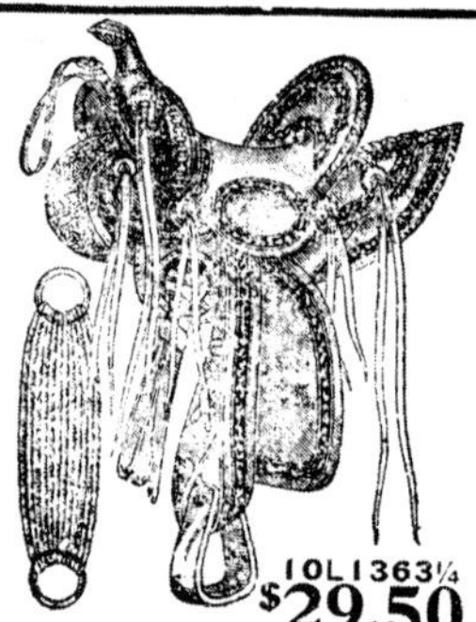

10L1363¼

$29.50

"Seagull" Small Seated Saddle.

For Boys, Girls or Young Men.

LEATHER—Bark tanned russet saddle skirting; embossed border. TREE—12½-inch, Jewel, hide covered, steel fork, 12-inch bulge. SEAT—Full; seat, jockey and cantle in one piece; leather bound cantle and front; leather covered horn. SKIRTS—Round or wing pattern, felt lined. RIGGING—Special feature—Spanish style, large leather covered flat rings; 1⅛-inch latigo tie strap; wide woven hair cinch. STIRRUP STRAPS—2 inches wide to lace, with fenders 6x13 inches, attached; 1½-in. ox bow stirrups, boys' size. Weight of saddle, about 22½ pounds. Shipping weight, 29½ pounds.

Kentucky Style Saddle

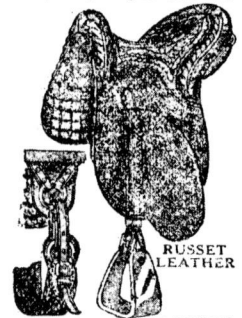

RUSSET LEATHER

10L1229¼ With Fenders.	10L1230¼ Without Fenders.
$23.75	**$21.50**

TREE—17-inch, heavily ironed, canvas covered, double gullet. SEAT—Fine quilted, raised stitching, raised stitched roll cantle and roll front. A very easy riding seat. PAD—Sheepskin top, serge cloth lining, tufted and quilted. SKIRTS—Hogskin impression, 20 inches long from center of seat, 13 inches wide. STIRRUP STRAPS—1¼ inches wide, extra long; hogskin impression, piped fenders, 16½ inches long, 9 inches wide, or without fenders; 4-inch Texas bolt stirrups. CINCH—Williams' improved cinch, heavy web, with ring and buckle; tie strap, to cinch and buckle.
Weight of saddle, about 18 lbs.; packed for shipment, 24 pounds.
Extra for leather hooded stirrups **$1.25**

"Buna Vista" Saddle

$21.95

10L1228¼

TREE—16½-inch, Wilburn style, extended bars, canvas covered. SEAT—Large and roomy, russet kip leather covered. SKIRTS—19 inches long from center of seat, 11 inches wide, embossed russet leather. PAD—Made in two parts, hair stuffed, serge covered, full stitched; very easy on the horse's back. STIRRUP STRAPS—1¼ inches wide with double and stitched piped fenders, 10x16 in. Virginia block stirrups and cord girth.
Weight of saddle, about 16 pounds; packed for shipment, 22 pounds.

A Very Comfortable Saddle for Pleasure Riding or Business, Especially for Large Men.

"Oliver" Morgan Saddle

Pony Saddle for Boys and Girls.

10L1217¼

$11.95

RUSSET LEATHER.

TREE—12-inch, full leather covered, Morgan style, leather covered round horn.
SEAT—Half leather covered, round skirts.
STIRRUP STRAPS—1 inch wide, full length, to buckle; fenders, 6x13 inches, attached; 3-inch wood stirrups, leather hooded.
RIGGING—Double cinch rigged; corded cinches, with rings; 1-inch tie straps.
Weight, about 10⅛ pounds. Shipping weight, 15½ pounds.

"Hoffman" Saddle

Russet Leather.

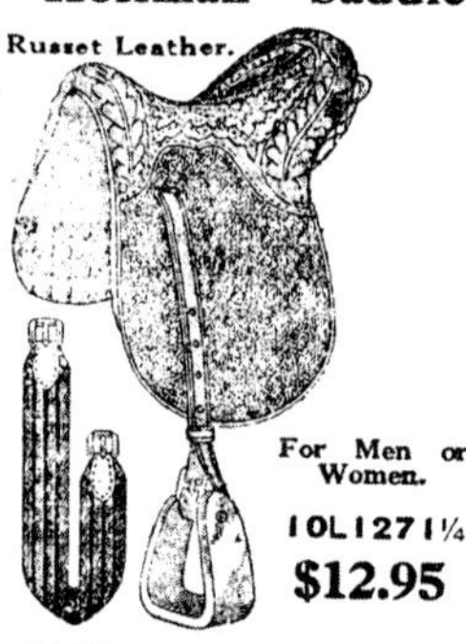

For Men or Women.

10L1271¼

$12.95

TREE—16-inch, canvas covered, Kentucky style. SEAT—Soft, quilted; star stitched. PAD—Sheepskin top, serge cloth lining. SKIRTS—18 inches long from center of seat, 11 inches wide. GIRTH—Cotton web, buckle on each end, 1¼-inch billets. STIRRUP STRAPS—1⅛ inches wide; 4-inch Texas bolt stirrups.
Weight of saddle, about 12 lbs.; packed for shipment 18 pounds.

"Eleanor" Astride Saddle

A Saddle of Good Quality for Women or Girls.

Russet Color.

10L1322¼

$19.50

TREE—14-inch, canvas covered, leather covered horn. SEAT—Buckskin quilted, very comfortable, bound cantle. SKIRTS—Wool lined. TIE STRAPS—1¼-inch latigo; wide cotton cinch. STIRRUP STRAPS—1⅛-inch; women's size pug nose stirrups. Weight of saddle, about 17 pounds; packed for shipment, 23 pounds.

"National" Park Saddle

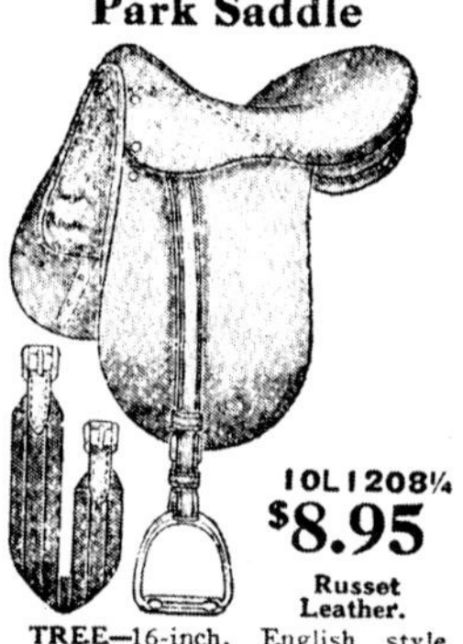

10L1208¼

$8.95

Russet Leather.

TREE—16-inch, English style. SEAT—Seamed seat with jockey in one piece. PAD—Good quality cloth, well stuffed. GIRTH—Corded web. STIRRUP STRAPS—1-inch, to buckle. STIRRUPS—Iron. Weight, packed for shipment, 15 pounds.

"Victoria" French Style Park Saddle

Lght Russet Color.

Made on a 17-inch cut back Colonial French style tree with extended bars, which distribute the weight of the rider over a large surface of the horse's back. This style is the horsewoman's special favorite and is suitable for a rider of almost any weight or build. It will be found very comfortable for both men and women. The seat, jockeys and skirts are of imported pigskin; skirts are 21 inches long from center of seat and 12½ inches wide; imported cowhide stirrup straps, 1¼ inches wide with "Never-Rust" stirrups; built-up calfskin covered pad; folded leather girth. Wt. of saddle, abt. 16½ lbs. Shpg. wt., 22 lbs.

10L1203¼ **$65.50**

The "Ascot" Park Saddle

Light Russet Color.

Designed for those who ride with the stirrup leathers extended to allow for only a slight bend at the knee. The cantle is slightly raised thereby deepening the seat. It assures to the rider the highest degree of comfort and provides added safety, which two features make this particular style of saddle especially desirable for the use of horsewomen. Made with the large 18-inch Colonial style cut back tree; imported pigskin seat and jockeys. Skirts are 21 inches long from center of seat and 12½ inches wide. The calfskin covered pad is an important feature; imported cowhide leather stirrup straps, 1¼ inches wide with "Never-Rust" stirrups. Folded leather girth. Three buckles at each end. Weight of saddle, about 15 pounds. Shipping weight, 21 pounds.

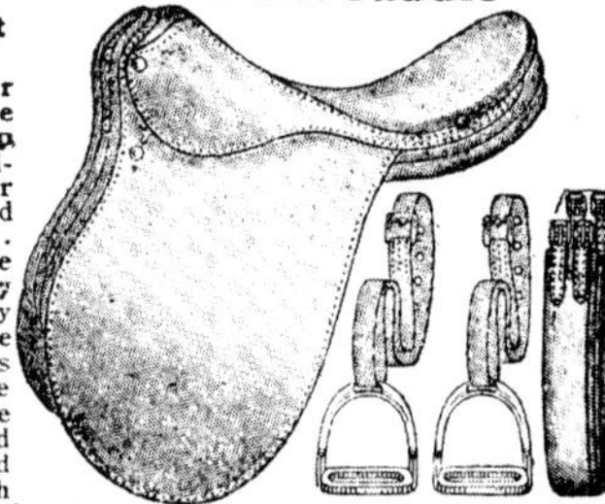

10L1202¼—With imported pigskin skirts.......... **$57.95**

10L1206¼—With Cowhide skirts, hogskin impression... **$52.95**

The "Brighton" Park Saddle

Light Russet Color.

Regular 17-inch Colonial style tree, deep cut back. Seat of imported pigskin, very roomy. Cantle made in 2 different shapes, round or fantail. Skirts, 11½ in. wide, 19½ in. long from center of seat; stirrup straps of imported cowhide leather, 1¼ in. wide, "Never-Rust" stirrups; calfskin covered pad; folded leather girth. Weight of saddle, about 14 pounds. Shipping weight, 20 pounds.

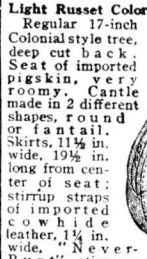

Saddle With Imported Pigskin Skirts.

10L1200¼—With round cantle.............. **$47.85**
10L1201¼—With fantail cantle.............. **47.95**

Saddle With Cowhide Skirts, Pigskin Impression.

10L1204¼—With round cantle.............. **$42.85**
10L1205¼—With fantail cantle.............. **42.95**

Good Leather Riding Bridles—Well Made and Strong

"Rawley" Riding Bridle.

10L1980

95c

Woven russet web; ⅞-inch reins, about 8 feet long; curb bit, light and strong. Shipping weight, 1⅛ pounds.

"Alvira" Riding Bridle.

10L1798 Less Bit. **$2.98**

With Bit. **$3.25**

Russet Leather, smooth finish; ¾-inch sewed cheeks to buckle on near side, with noseband; 7-foot reins to loop in bit. Shipping weight, 2 pounds.

"Anson" Russet Leather Bridle.

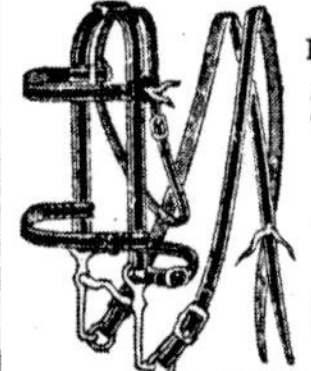

Less Bit. **$2.75**

With Bit. **$3.10**

10L1766

Extra heavy and strong; ⅞-inch double headstall to buckle on top; noseband; ⅞-inch reins, 6 feet long, to buckle in bit; port bit. Shipping weight, 2 lbs.

"Columbian" Riding Bridle.

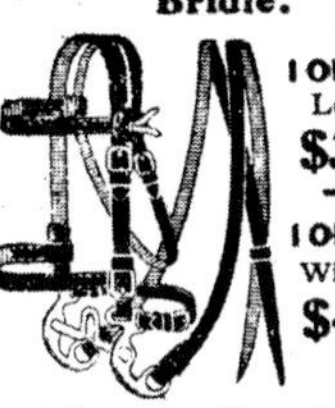

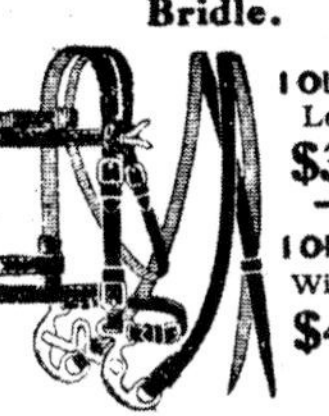

10L1768 Less Bit. **$3.75**

10L1769 With Bit. **$4.15**

A heavy, well made bridle of russet leather, 1-inch throughout. Adjustable on left cheek; long throatlatch; noseband and curb strap; 7-foot reins loop in bit. Shipping weight, 3½ pounds.

"Sioux" Cowboy Riding Bridle

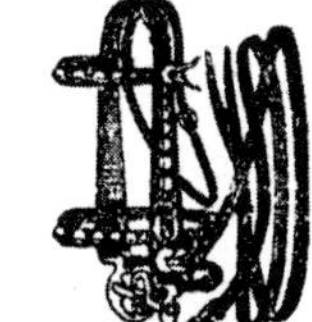

10L1779 Less Bit. **$5.95**

10L1780 With Bit. **$6.35**

Russet leather, embossed cheeks, nickel plated conchas, swedge buckles, tapered front, sunburst spots; 1-inch reins. 6½ feet long. Shipping weight, 3⅜ pounds.

Russet Leather Bridle

Our Most Popular Riding Bridle.

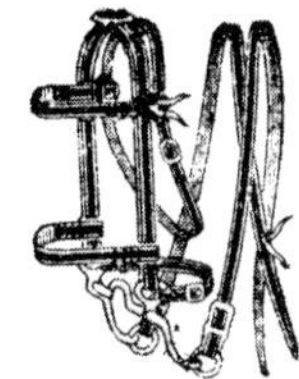

10L1754 **$2.15**

The good quality of this bridle will show after long service. This is the buyer's assurance of a genuine bargain.

Light weight but strong western style bridle for the use of farmers and other riders. Well tanned russet leather, uniformly cut, carefully finished. Adjusts to fit large or small horses, ¾-inch double headstall to buckle on top; noseband; ¾-inch reins, 6 feet long, to loop in bit; port bit and curb strap. Shipping weight, 1⅜ pounds.

White Latigo Riding Bridle.

10L1765 **$3.25**

California style, ⅞-inch double adjustable crown-piece, with buckle on top; ½-inch throatlatch, noseband and curb strap and ⅞-inch white latigo reins; port bit. Shipping weight, 1⅞ pounds.

"Amanda" Riding Bridle.

10L1799 **$4.75**

Russet leather, double and stitched, long tapered spotted cheeks to buckle on crown, nickel plated buckles, conchas tied in; ⅝-inch reins, 7 feet long, to loop in bit; no bit. Shipping weight, 2½ pounds.

Great Western Cowboy Bridle.

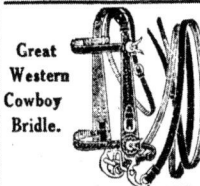

10L1781 Less Bit. **$3.98** | **10L1782** With Bit. **$4.37**

Russet leather, pointed and stamped cheeks, adjustable crown, nickel plated buckles and ornaments. Reins, 6 feet long, ⅞ inch wide; noseband. Shipping weight, 3½ pounds.

Cowboy Bridle.

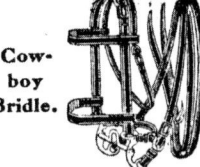

10L1770 Less Bit. **$3.45** | **10L1771** With Bit. **$3.85**

Russet leather, ⅞-inch double headstall, to buckle on top; ⅞-inch reins, 6 feet long, to buckle in bit; noseband. Nickel plated buckles, box loops. Shipping weight, with bit, 2⅜ pounds.

Pony Bridles

Two Styles.

Your Choice of Color.

Russet or black leather. ¾ inch bridle with reins 4 feet long, nickel plated buckles, snaffle bit. Shipping weight, 1½ pounds.

10L1812—Russet.
10L1813—Black.
Each........ **$1.95**

10L1814—Russet leather, ⅝-inch cheeks, double headstall, 4½-foot reins; port bit. Shipping weight, 1½ pounds. **$1.99**

"Weymouth" Park Colonial Style Riding Bridle

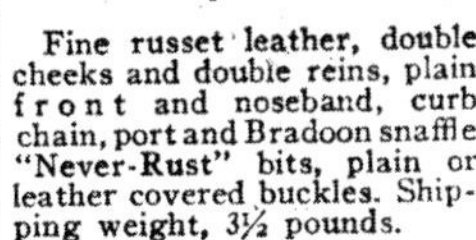

Imported.

Fine russet leather, double cheeks and double reins, plain front and noseband, curb chain, port and Bradoon snaffle "Never-Rust" bits, plain or leather covered buckles. Shipping weight, 3½ pounds.

10L1790—With ⅝-inch cheeks; one ⅝-inch and one ¾-inch rein. **$12.85**

10L1791—With ¾-inch cheeks; one ¾-inch and one ⅞-inch rein. **$13.50**

"Hiawatha" Riding Bridle.

10L1772 Less Bit........ **$6.25**
10L1773 With Bit.......... **6.65**

Russet leather, scalloped and embossed cheeks. Nickel plated conchas tied-in, sunburst spots, swedge buckles; 1-inch reins, 7 feet long. Shipping weight, 3½ pounds.

"Rocky Mountain" Riding Bridle.

10L1776 Headstall. **$4.85**

10L1777 Reins Only. **$2.95**

10L1778 Complete.......... **$8.25**

Russet leather. Spotted cheeks, noseband and front. Nickel plated buckles; ring, link reins, quirt ends. Bit furnished with complete bridle. Shipping weight, 3 pounds.

Team Bridles Without Blinds

"Dan" Open Team Bridle.

Each. **$3.25**

10L2068 XC trimmed.

10L2069 Japan trimmed.

⅞-inch cheeks. XC white metal or japanned roller buckles. Long, round reins. Shipping weight, 3½ pounds.

"Walter" Open Bridle.

Each. **$3.45**

10L2078 XC trimmed.

10L2079 Japan trimmed.

Long reins, 1-inch cheeks, ⅞-inch reins, with bit. XC white metal or japanned buckles. Shipping weight, 5 pounds.

"Alonzo" Team Bridle.

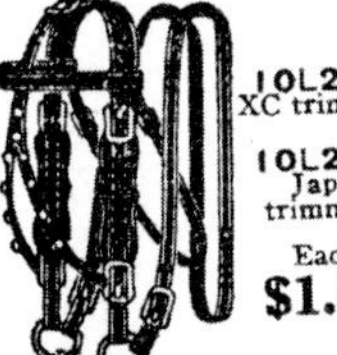

10L2081 XC trimmed.

10L2082 Japan trimmed.

Each. **$1.85**

Open, 1¼-in. sewed cheeks, 1-inch short flat reins. Spotted face piece; XC white metal or japanned bar buckles. Shipping weight, 3⅛ lbs.

Halter or Mule Bridle.

10L1975 **$3.35**

1¼-inch double and stitched cheeks with blinds, 1¼-inch noseband and chin strap, ⅞-inch throatlatch. Bit and snaps. Shipping wt., 3⅜ lbs.

Stallion Bridles

Made of Strong, Heavy Leather, Full Size for Large Animals.

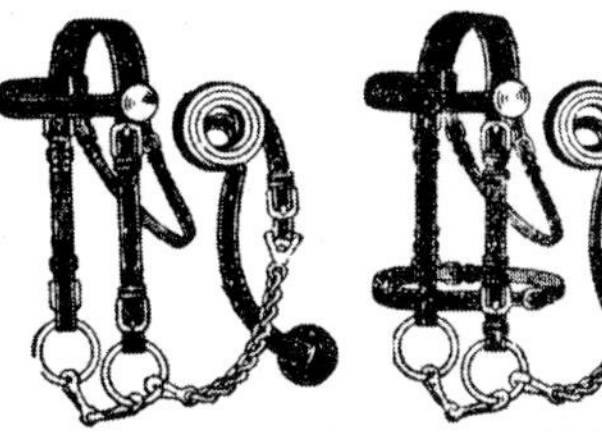

10L1787

Heavy russet leather, 1¾-inch crown, 1¼-inch front and cheeks, 1¼-inch lead reins, 12 feet long, with 18-inch chain, round stopper; nickel plated buckles and rosettes, no bit. Shipping wt., 4 pounds.... **$4.95**

10L1788

Russet leather, solid crown, 1-inch raised front, cheeks and noseband, double stitched laps; 1½-inch lead rein with stopper; 18-inch chain; creased edges, nickel plated buckles. No bit. Shpg. wt., 5 lbs........ **$6.95**

Bridle parts, including crown, rosettes, spotted front, noseband, and face drop for express or team bridles, brass or nickel plated trimmed.

Complete outfit. Shipping wt., 1⅜ pounds.

10L2440—Nickel plated outfit for one bridle.......... **$2.10**
10L2441—Brass trimmed outfit for one bridle.......... **$2.10**

Front only. Shipping weight, 5 ounces.
10L2442—Nickel plated front.............. **45c**
10L2443—Brass trimmed front.............. **45c**

Noseband only. Shipping weight, 6 ounces.
10L2444—Nickel plated trimmed noseband.. **55c**
10L2445—Brass trimmed noseband.......... **55c**

Face drop only. Shipping weight, 7 ounces.
10L2447—Nickel plated trimmed face drop **65c**
10L2448—Brass trimmed face drop.......... **65c**

Braided Leather Bridle

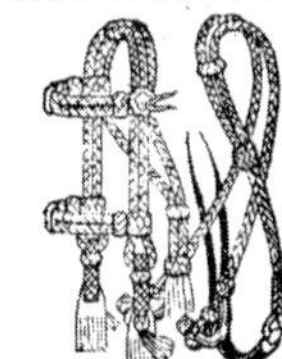

10L1743—Fancy Western Riding Bridle of brown and white leather. Made with double headstall, front and noseband, overhead throat latch and 7-foot reins with romal quirt ends; without bit. Trimmed with braided knots and leather fringe. Shpg. wt., 2 lbs..... **$3.95**

Fine Leather Buggy Bridles

⅝-Inch Box Loop Cheek Buggy Bridle with patent leather blinds, round winker braces, overcheck or side reins; XC buckles; furnished complete with bit. Shipping weight, 4 pounds.

10L1853 Overcheck Bridle.	**10L1854** Side Check Bridle.
$2.97	**$3.35**

⅝-inch sewed flat cheeks (no box loops); leather blinds, overcheck or side rein. Flat winker braces; furnished complete with bit. Shipping weight, 2⅜ lbs.

10L1860 Overcheck Bridle.	**10L1861** Side Check Bridle.
$1.85	**$1.95**

Fine ⅝-Inch Flat Cheek Open Bridle, strong and serviceable, overcheck with noseband, nickel plated or imitation rubber trimmings. Shipping weight, 1 pound 13 ounces.

10L1862 Nickel plated.	**10L1863** Imitation rubber trimmed.
$1.89	**$1.89**

WHIPS

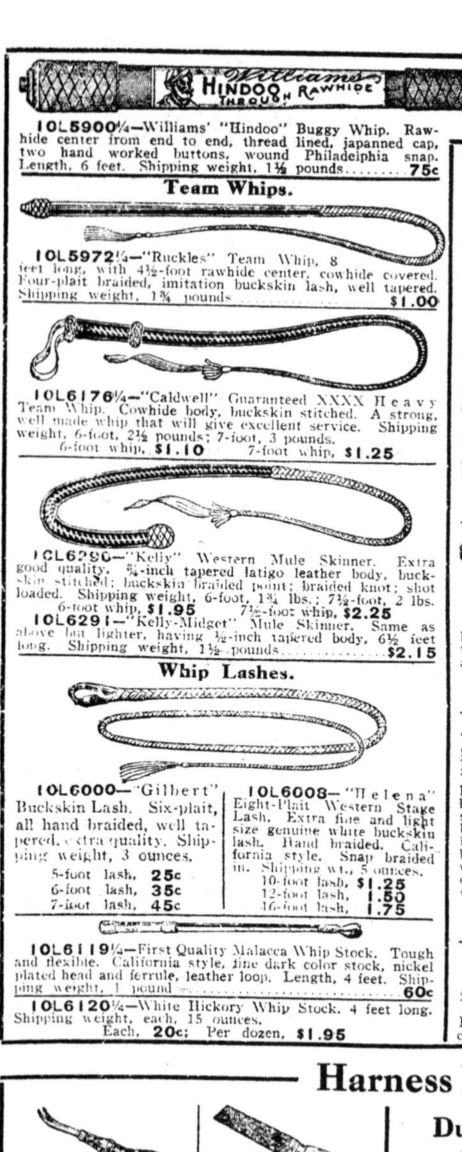

10L5900¼—Williams' "Hindoo" Buggy Whip. Rawhide center from end to end, thread lined, japanned cap, two hand worked buttons, wound Philadelphia snap. Length, 6 feet. Shipping weight, 1½ pounds........**75c**

Team Whips.

10L5972¼—"Ruckles" Team Whip, 8 feet long, with 4½-foot rawhide center, cowhide covered. Four-plait braided, imitation buckskin lash, well tapered. Shipping weight, 1¾ pounds........**$1.00**

10L6176¼—"Caldwell" Guaranteed XXXX Heavy Team Whip. Cowhide body, buckskin stitched. A strong, well made whip that will give excellent service. Shipping weight, 6-foot, 2½ pounds; 7-foot, 3 pounds.
6-foot whip, **$1.10** 7-foot whip, **$1.25**

10L6290—"Kelly" Western Mule Skinner. Extra good quality, ¾-inch tapered latigo leather body, buckskin stitched; buckskin braided point; braided knot; shot loaded. Shipping weight, 6-foot, 1¾ lbs.; 7½-foot, 2 lbs.
6-foot whip, **$1.95** 7½-foot whip, **$2.25**

10L6291—"Kelly-Midget" Mule Skinner. Same as above but lighter, having ½-inch tapered body, 6½ feet long. Shipping weight, 1½ pounds........**$2.15**

Whip Lashes.

10L6000—"Gilbert" Buckskin Lash. Six-plait, all hand braided, well tapered, extra quality. Shipping weight, 3 ounces.
5-foot lash, **25c**
6-foot lash, **35c**
7-foot lash, **45c**

10L6008—"Helena" Eight-Plait Western Stage Lash. Extra fine and light size genuine white buckskin lash. Hand braided. California style. Snap braided in. Shipping wt., 5 ounces.
10-foot lash, **$1.25**
12-foot lash, **1.50**
16-foot lash, **1.75**

10L6119¼—First Quality Malacca Whip Stock. Tough and flexible. California style, fine dark color stock, nickel plated head and ferrule, leather loop. Length, 4 feet. Shipping weight, 1 pound........**60c**

10L6120¼—White Hickory Whip Stock. 4 feet long. Shipping weight, each, 15 ounces.
Each, **20c**; Per dozen, **$1.95**

Drovers' Whips.

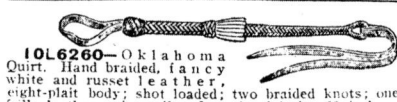

10L6180 "Griffin Special" Boys' Leather Drovers' Whip. Six-plait leather, hand braided California style, well tapered lash, strongly wired to 9-inch wood handle. Shipping weight, 1½ pounds.
6-foot whip........**55c**
8-foot whip........**65c**

10L6203 "Dodson" Improved Jacksonville Drovers' Whip. A great favorite with our western customers. Made strong for hard daily use. California style fine eight-plait leather body, buckskin point. Hand braided, well-tapered and fine swinging whip. Steel front revolving handle. Shot loaded body. Shipping weight, 2⅜ pounds.
8-foot whip........**$1.65**
10-foot whip........**1.85**
12-foot whip........**1.95**

10L6235 "Colorado" Improved Australian Style Cattle Whip. Twelve-plait drumhead rawhide, hand braided, shot loaded whip, revolving handle. Australian knot. The well tapered body and point, together with the revolving handle, assure a well balanced swing. Shipping weight, 2½ pounds.
10-foot whip........**$2.40**
12-foot whip........**2.55**
14-foot whip........**2.75**

10L6243 "Hanley" California Style Drovers' Whip. Twelve-plait latigo leather, hand braided, shot loaded body; buckskin tapered point; eight-plait hand braided white leather covered steel handle, leather wrist loop, and two braided buttons. Handle does not revolve. A strictly western style cattle whip. Many prefer this whip because of the stationary handle. A good swinging whip. Shipping weight, 2¾ pounds.
8-foot whip........**$2.25**
10-foot whip........**2.50**
12-foot whip........**2.75**

Buggy Whip Snaps, 7 inches long. Sold in dozens only. Shipping weight, 2 ounces.

10L6133	**10L6134**	**10L6135**
Half silk, half cotton. Doz **45c**	cotton. Doz **15c**	All silk, first quality. Doz **69c**

10L5925¼—Williams' "Gray Fox" Buggy Whip. Loaded butt, rawhide center from snap to cap. The whip is stocked and loaded to make it a good swinging buggy whip. Length, 6 feet. Shipping weight, 1½ pounds........**98c**

10L6255—San Antonio Quirt. Fancy white and russet leather, hand braided, four-plait body. Two braided knots and one frill. Leather quirt tails. Length of body, 18 inches; full length, about 33 inches. Shipping weight, 9 ounces........**45c**

10L6260—Oklahoma Quirt. Hand braided, fancy white and russet leather, eight-plait body; shot loaded; two braided knots; one frill; leather quirt tails. Length of body, 20 inches; full length, about 33 inches. Shipping weight, 14 ounces........**65c**

10L6270—"Tolby" Mexican Quirt. Eight-plait, hand braided russet leather body. Iron spike, two heavy braided knots, leather hand loop. Length of body, 19 inches. Full length about 34 inches. Shipping weight, 13 ounces........**75c**

10L6280—Twelve plait braided rawhide quirt; full length, 33 inches; body and handle measures 18 inches; handle is heavily weighted and trimmed with braided calfskin; fancy braided knots, leather fringe, quirt tails and wrist loops. Shipping weight, 1 pound........**85c**

10L6285—Brown and white calfskin quirt, 12-plait braided; leather tails and wrist loop; length of body and handle, 18 inches; full length, 33 inches; has weighted handle and is trimmed with braided knots. Shipping weight, 1 pound........**$1.25**

Sixteen plait braided quirts of rawhide or calfskin; tapered body, about 19 inches long; full length, about 33 inches; braided knots, leather tails and wrist loops. Shipping weight, 1 pound.

10L6288—Braided calfskin quirt........**$1.45** | **10L6289**—Braided rawhide quirt........**$1.70**

Harness Leather—Harness Makers' Tools

10L7716
Harness Makers' Edging Tool. For removing sharp corners of new strap work. Five inches long, polished. Shipping weight, 3 ounces...**27c**

10L7605
Square Point Trimming Knife. Round handle, 3½-inch fine tool steel blade. Shipping weight, 4 ounces. **15c**

Harness Makers' Wax. Used with harness thread at right for making wax-ends. Shipping weight of four balls, 5 ounces.
10L7690 4 balls for **5c**
10L7691 30 balls for **30c**

Linen Harness Thread. To be used with Wax 10L7690 for making wax-ends; 2-ounce balls. Shipping weight, 3 oz.
10L7695—"Chicago" Brand, No. 10. **30c**
10L7696—Barbour's No. 10 **35c**

10L7685
Harness Needles. Twenty-five in a paper. Assorted sizes from 0 to 4. Shipping weight, 2 ounces........**9c**

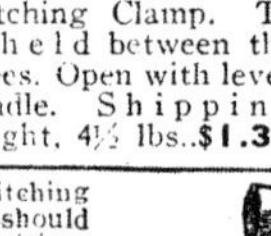

10L7748
"Tim's" Harness Stitching Clamp. To be held between the knees. Open with lever handle. Shipping weight, 4½ lbs..**$1.39**

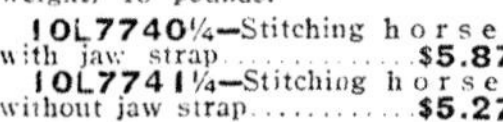

Harness Makers' Wood Stitching Horse. Every horse owner should have one to do his own repairing. Saves time and money. Shipping weight, 18 pounds.
10L7740¼—Stitching horse, with jaw strap........**$5.87**
10L7741¼—Stitching horse, without jaw strap........**$5.27**

Dundee Bark Tanned Harness Leather.

Sold in full sides, or backs only, with the belly cut off.

Bark tanning insures for our harness leather uniform good quality. It is plump, well tanned, thoroughly curried and finished, will wear long and give excellent service. Our harness leather is selected from packers' steer hides and we guarantee it to be the equal of any harness leather sold. Weights run from 16 to 24 pounds per side, and backs from 13 to 19 pounds each. We will send you as near the weight you order as we can. **We do not cut the sides or pieces.**

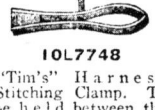

10L7763¼ **Full Sides**, weighing from 16 to 19 pounds each.
Per pound... **53c**

10L7764¼ **Full Sides**, weighing from 20 to 24 lbs. each.
Per pound... **58c**

Black only.

10L7767¼ **Backs only.** Weighing from 13 to 16 pounds.
Per pound... **63c**

10L7768¼ **Backs only.** Heavy trace leather, weighing from 16¼ to 19 pounds.
Per pound... **68c**

10L7771—Harness Leather Bellies, in pieces weighing from 3 to 6 pounds each, used for repairing or making light strap work.
Per pound........**29c**

Special Note—Owing to the uncertainty of the leather and hide market, the prices on leather are subject to change without notice.

Myer's Famous Lockstitch Sewing Awl.

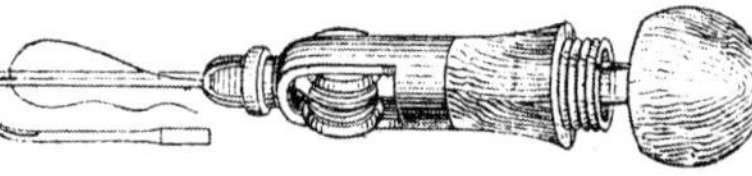

One of the handiest tools you can have for the house or barn. For repairing harness, saddles, shoes, carpets, rugs, tents, awnings, etc., it is invaluable. The needles are grooved to contain the thread. Anyone who has use for a repairing tool should not be without this one. Shipping wt., 5 oz.

10L7742—Complete Awl, with two needles (one medium, one coarse) and spool coarse thread **$0.60**
10L7750—Three Complete Awls, as above **1.75**
10L7751—Spool Waxed Thread. Coarse.. **.15**
10L7752—Spool Waxed Thread. Fine.... **.15**
10L7753—2-Ounce Tube Waxed Thread. Coarse........ **.55**
10L7754—2-Oz. Tube Waxed Thread. Fine **.55**

Extra Needles.
10L7755—**Fine,** straight point........**7c**
10L7756—**Medium,** curved point........**7c**
10L7758—**Coarse,** straight point........**7c**
Shipping wt., needles, 2 oz.; tube, 3 oz.; spool, 2 oz.

10L7550
Leather Gauge Knife. Hollow iron handle. Kind used by practical harness makers. Cuts from ⅛ to 4 inches wide. Shipping wt., 1 lb. 5 oz....**$1.65**
10L7551—**Extra blade** for gauge knife........**19c**

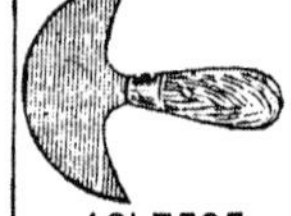

10L7595
Harness Makers' Round Knife. 5-inch tool steel blade; well tempered to take sharp edge. Wood handle. Shipping wt., 7 oz **$1.25**

10L7710
Harness Awl Blades. Straight blades only. Assorted sizes. Shipping weight, 2 ounces.
3 for........**12c**
Per dozen........**45c**

10L7715
Wood Awl Handles. To be used with Blades 10L7710. Made with ferrules. Shipping weight, each, 2 ounces.
Each........**2c**
Per dozen........**21c**

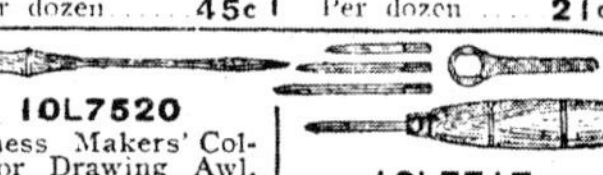

10L7520
Harness Makers' Collar or Drawing Awl. Has large eye for sewing horse collars with leather thongs or whangs. Tool steel, highly tempered. Length, about 9 inches. Shipping weight, 5 ounces........**65c**

10L7717
"Jerry" Quick Setting Handy Awl Haft. With wrench and four awl blades of assorted sizes. Shipping weight, 10 ounces........**35c**

For Other Awls see page 855.

Harness Makers' Hollow Tube Spring Punches.

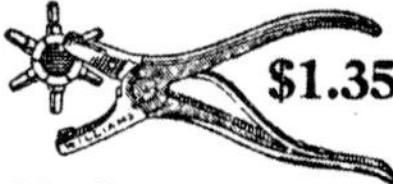

An extra high grade punch for harness makers' use and for those who desire a punch that will stand severe and constant service. Drop forged handles, screw tubes.

$1.35

9L5884—6-Tube Revolving Spring Punch. Shipping weight, 12 ounces........**$1.35**
9L5883—4-Tube Revolving Spring Punch. Shipping weight, 12 ounces........**$1.20**

Bits for Riding, Driving or Heavy Work Bridles

Snaffle and Wire Bits.

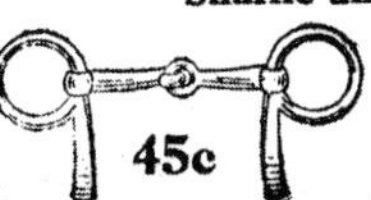

45c

Racine Nickel Plated Driving Bit. Solid cheeks. Shipping weight, 12 oz.
10L8103—Jointed..45c
10L8104—Stiff.....45c

Pony Bit. Same style as above. Shpg. wt., 11 oz.
10L8105—Jointed 27c
10L8106—Stiff 27c

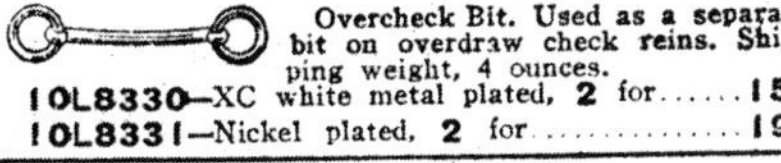

Overcheck Bit. Used as a separate bit on overdraw check reins. Shipping weight, 4 ounces.
10L8330—XC white metal plated, **2** for......**15c**
10L8331—Nickel plated, **2** for **19c**

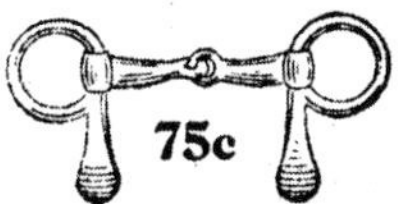

75c

Dexter Driving Bit. Heavy cheeks, tapered mouthpiece, nickel plated. Shipping weight, 15 oz.
10L8133—Jointed..75c
10L8134—Stiff 75c

10L8135—"Goodwood" Nickel Silver Driving Bit. Extra heavy jointed mouthpiece, tapered; Dexter pattern; large heavy rings and half snaffle cheeks. Guaranteed not to rust or corrode. Shpg. wt., 1⅛ lbs. **$1.65**

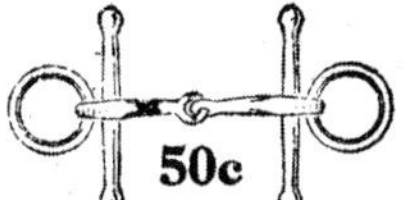

50c

10L8171—Full Snaffle Bridle Bit. Nickel plated. Solid cheeks, loose rings, jointed mouthpiece only. The cheek bars prevent the bit pulling through horse's mouth. Shipping weight, 10 ounces..............**50c**

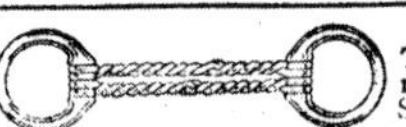

10L8305—Double Twisted Wire Bit. 2¼-inch rings, jointed mouthpiece. Shipping weight, 8 oz...**20c**

10L8306—Single Twisted Wire Bit, 2¼-inch rings, jointed mouthpiece. Shipping weight, 6 ounces.....**17c**

Bits for Weymouth double rein and crown bridles. "Never rust" white metal, highly polished finish. Used singly or together. Price is for one bit of either kind.

10L8110 — Snaffle Bit. Shipping weight, 1 pound **$1.20**

10L8112—Port Bit with curb chain. Shipping wt., 1¾ lbs.**$3.00**

$2.00

10L8117 — Swivel Ring Bar Bit. Made with low port and removable roller; cast steel blued finish. Shipping weight, 1 pound 1 ounce **$2.00**

10L8009—"Eureka" Spurs. Wide steel heel band, nickel plated, 1⅛-in. rowel, nickel plated buttons. Shipping weight, 9 ounces.
Per pair.........**75c**

10L8021 — U. S. Army New Model Regulation Officers' Spurs, with straps. Nickel plated, highly polished; no rowel. Shipping wt., 13 ounces.
Per pair.........**$1.75**

39c

10L8140—Fine Blued Mexican Curb Bit. Short port on mouth bar without roller. Shipping wt., 1 pound.

10L8141—Same as 10L8140, but with roller in mouth bar. Shipping weight, 1 pound **44c**

$1.20

10L8119

Improved Low Port Nickel Plated Bit. Very strong and well made and finely finished. Shipping weight, 14 ounces.

Standard Team Bridle Bits.

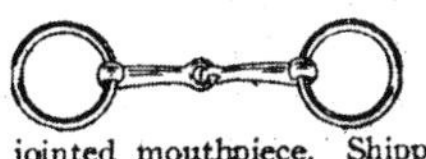

10L8321—Jointed.
10L8322—Stiff.
Team Bridle Bit. Trade No. 47. 2¼-inch ring, stiff or jointed mouthpiece. Shipping weight, 10 ounces.
Each **$0.12**
Per dozen **1.25**

Bits for Vicious Horses or Horses With Hard Mouths.

Success.

Success Driving Bit, has a steel bar. Good controlling power without injuring mouth. Shipping weight, 1 lb. 3 oz.
10L8291—XC white metal plated **$0.79**
10L8292—Nickel plated **1.15**

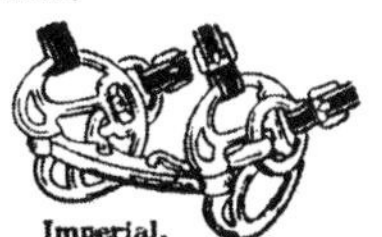

Imperial.

Imperial Driving Bit. Steel bar. Tongue cannot be carried over top of this bit. Shipping weight, 15 ounces.
10L8280—XC white metal plated **$0.64**
10L8281—Nickel plated **1.25**

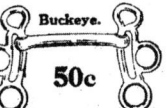

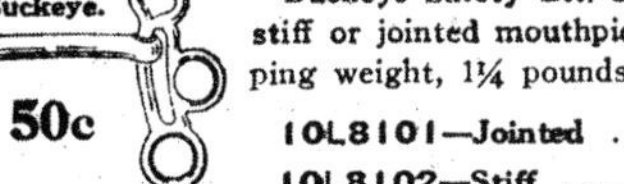

Buckeye.

50c

Buckeye Safety Bit. Loose bar, stiff or jointed mouthpiece. Shipping weight, 1¼ pounds.
10L8101—Jointed 50c
10L8102—Stiff 50c

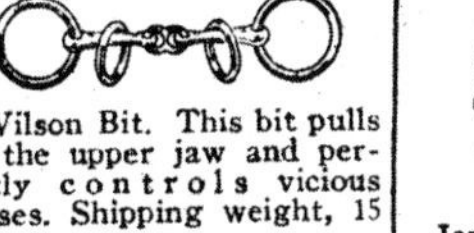

Wilson Bit. This bit pulls on the upper jaw and perfectly controls vicious horses. Shipping weight, 15 ounces.
10L8270—White metal plated **27c**
10L8271—Nickel plated **50c**

29c

Jay-Eye-See Bit. Shipping weight, 15 ounces.
10L8285—XC white metal plated **29c**
10L8286—Nickel plated **55c**

Fine Hand Forged Steel Bits and Spurs.

Hand forging is the surest method known, as well as the oldest, of manufacturing bits and spurs of quality and as near perfect in construction as possible. Very seldom is a flaw found in a piece of hand forged steel, and for that reason we most strongly recommend the use of these goods to owners of spirited or vicious saddle horses.

$1.00

10L8132—"Kelley" Cowboys' Hand Forged Bit; very strong; 5-inch low port mouth bar, 6-inch cheeks, blued finish. Shipping wt., 1½ pounds.

$2.99

10L8138—Polished Hand Forged Steel Port Bit, 5-inch cheeks with three nickel silver ornaments, 5-inch mouthpiece. Shipping weight, 1 lb.

$2.19

10L8139—Low Port Polished Steel Bit, 5-inch mouth bar, 5-inch cheeks, with nickel silver heart ornament on each cheek. Shipping weight, 1 lb.

Bits for Horses With Tender Mouths.

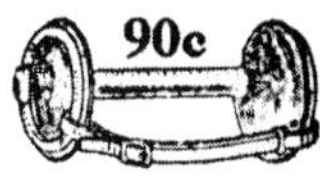

90c

10L8378—"Louvois" Driving Bit. Flexible leather mouthpiece; heavy nickel plated rings with leather shields and curb strap. Shipping weight, 1 pound.

39c

10L8365—"Durham" Rubber Mouthpiece Bit. Nickel plated half snaffle cheeks. Shipping weight, 10 ounces.

69c

10L8367—"Stockton" Leather Mouthpiece Bit. Nickel plated rings and cheeks. Shipping weight, 12 ounces.

70c

10L8376—"Sandow" Humane Driving Bit. Steel nickel plated solid mouthpiece, with leather chin strap and small rings for overcheck; large side rings. Shipping weight, 12 ounces.

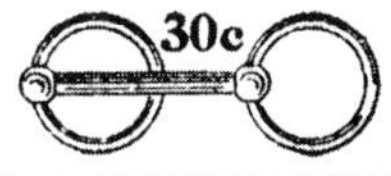

30c

10L8315—Solid Head Bit. No. 2 wire. Trade No. 90, 3-inch loose rings, 5-inch mouthpiece. Shipping weight, 1 pound 1 ounce.

10L8316—Fine Stallion Bit, with large 4-inch rings, stiff or jointed mouth bar (**state choice**). Shipping weight, 1⅛ pounds.......................**95c**

10L8317—Heavy Express Bit. No. 1 wire. Solid head, 6-inch stiff mouthpiece, nickel plated. Shipping weight, 1 pound.........................**40c**

$1.10

10L8245 — California Heavy Cowboy Bit. Patent port, complete with rein chains and roller. A very heavy and strong bit. Shipping weight, 1½ pounds.

90c

10L8184—Four-Ring Port Bit. For use on double rein riding bridles. Malleable iron, nickel plated. Shipping weight, 1 pound.

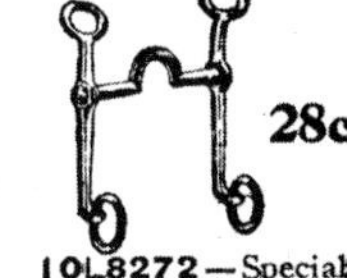

28c

10L8272—Special Port Bit. Malleable iron. Straight posts, low port. One of the most popular riding bridle bits made. Shipping weight, 12 oz.

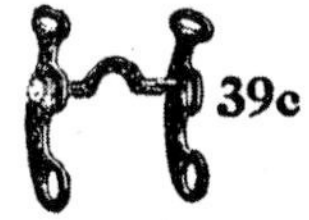

39c

10L8187—Fine Wrought Iron Mouthpiece Port Bit. Japanned finish, strong and heavy bit for hard mouthed horses or mules. Shipping weight, 15 ounces.

English Style Spurs.

10L8025—Malleable iron; XC white metal plated, with spur straps. A neat light spur for dress wear. Shipping weight, 9 ounces.
Per pair..........**59c**

10L8026—Solid brass, highly polished. Steel rowel with spur straps. Shipping weight, 9 ounces.
Per pair........**78c**

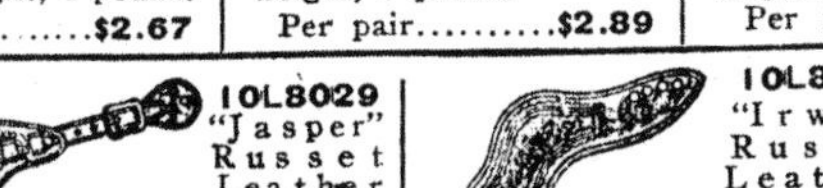

10L8038—Straight Shank Hand Forged Steel Spurs; light; neat; 3-inch shank, 1¼-inch rowel, ½-inch band. Nickel silver ornaments on band, over rowel pin and buttons. Shipping weight, 1 pound.
Per pair..........**$2.67**

10L8040 — Diamond Design Hand Forged Steel Spurs. Two nickel silver buttons, nickel silver diamonds and plate over rowel pins; 2¼-inch shank, 1¼-inch rowel. Shipping weight, 1 pound.
Per pair..........**$2.89**

10L8042—Drop Shank Hand Forged Steel Spurs. Brass dot and two nickel silver hearts on band. Nickel silver over rowel pin and on buttons; 2½-inch shank, 1¼-inch rowel. Shipping weight, 1 pound.
Per pair..........**$3.65**

10L8044—Leg Pattern, 2¼-inch shank, 1¼-inch rowel, ⅝-inch band. Nickel silver stocking, bronze thigh. Nickel silver heart ornaments and bronze diamonds on band. Shipping weight, 1 pound.
Per pair..........**$3.35**

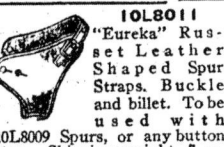

10L8011
"Eureka" Russet Leather Shaped Spur Straps. Buckle and billet. To be used with 10L8009 Spurs, or any button spur. Shipping weight, 7 oz.
Per pair..................**45c**

10L8029
"Jasper" Russet Leather Fancy Stamped Shaped Spur Straps. Adjustable buckle and billet. Can be used on any of our button spurs. Shipping weight, 10 ounces.
Per pair.....................**79c**

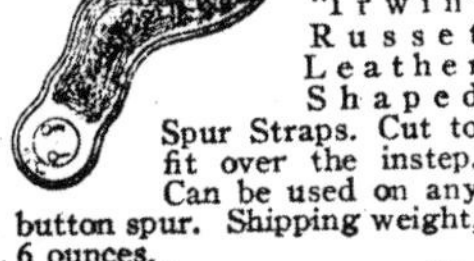

10L8017
"Irwin" Russet Leather Shaped Spur Straps. Cut to fit over the instep. Can be used on any button spur. Shipping weight, 6 ounces.
Per pair...................**47c**

10L8022
Mexican Spurs. Chased and filed malleable iron; single chain; 1¼-inch rowel; two iron knockers. Shipping weight, 8 ounces.
Per pair ..**49c**

10L8045
New Pattern California Spurs. Filed and chased, 1⅛ in. rowel, double chains. A strong, heavy and serviceable spur. Shipping weight, 15 oz.
Per pair...................**$1.50**

Fly Cover and Cord Lashes Combined

Very Big Sellers in the Northwest and Missouri River Territory.

The sheet of closely woven cotton duck over the back, 29 or 40 inches wide, makes a cooling cover and keeps the flies off. The cord lashes on the sides give protection to the lower part of the body and legs. The lashes fasten through a woven web bar stitched to the duck. Full length of covers, 100 and 110 inches. Shipping weight, 3½ pounds.

10L9061—29-Inch Duck.
100-inch cover........**$2.25**
110-inch cover........ **2.50**

10L9062—40-Inch Duck.
100-inch cover........**$2.55**
110-inch cover........ **2.75**

"Missouri Valley" Cotton Cord Team Fly Nets.

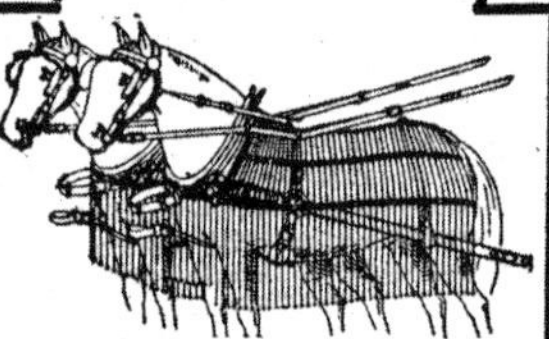

Made with 60 or 100 lashes in the body and extra lashes across breast. Double woven web bars; lashes between the bars sewed together. The body of these nets is 5 feet long, lashes are 7 feet long. Shipping weight, each, 3½ and 4½ pounds, respectively.

10L9082
60-Lash Net.
Each, for one horse....**$1.85**

10L9083
100-Lash Net.
Each, for one horse....**$2.65**

Old Gold Color Woven Net Fly Covers

A Woven Cotton Scrim That Allows Proper Ventilation and Keeps the Flies Out.

These Covers weigh about 32 ounces each. There are four other weights made. One is heavier and three are lighter than those we offer. When ordering it will be well to keep this information in mind.

Highly recommended for hot weather when the flies are bad. Comfortable for the horse, being light weight. These covers will be found very practical, both for excluding the fly and keeping the horse cool. The material is a special weave of cotton scrim netting through which air readily circulates, made specially strong for fly covers. Leather and duck trimmed. Hemmed edges. Shipping weight, each, 2½ pounds.

10L9059—100-Inch Fly Cover.
Each, for one horse..**$2.10**

10L9060—110-Inch Fly Cover.
Each, for one horse..**$2.25**

"Rosser Brand" Cotton Cord Team Fly Nets

WITH HAND BRAIDED BARS AND LASHES WITH WIRE FASTENED ENDS.

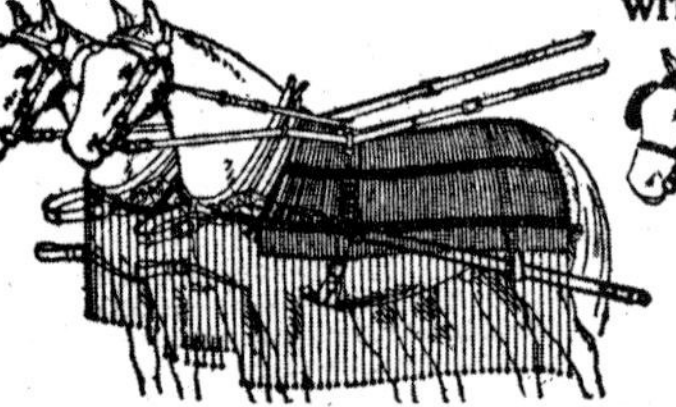

Heavy Cotton Cord Team Nets.

Made of extra quality 3-ply heavy seine twine with hand braided bars. Double the number of lashes over back as there are around legs. Shipping wt., each, 3¾, 4¼ and 5 lbs.

10L9180
With white and black bars. Old gold color lashes. 100 lashes in body, 50 lashes below lower bar.
Each, for one horse ...**$2.75**

10L9181
All brown net, 140 lashes body, 70 lashes below the lower bar.
Each, for one horse, **$3.50**

10L9182
All brown net, 160 lashes in body, 80 lashes below lower bar.
Each, for one horse, **$3.75**

The "Olympus" Fly Nets.

A dark brown net with lashes 7¼ feet long, of 3-ply medium weight cotton fish cord. The bars lie flat on the horse's back and lashes are long enough to keep the flies from the horse's legs and under part of the body. The nets are 5 feet long on the back; breast piece contains enough lashes to keep the flies from the front of the horse. Shipping weight, each, 3¼, 4¾ and 5 pounds.

10L9209
70-Lash Net.
Each, for one horse..**$2.50**

10L9210
84-Lash Net.
Each, for one horse **$2.75**

10L9211
100-Lash Net.
Each, for one horse **$2.95**

The "Colossus" Fly Nets.

EXTRA LARGE NETS FOR BIG HORSES.

Made of cotton fish cord, 5½ feet long on the back and lashes full 9 feet long, to keep the flies from the legs and under part of the body. Lashes are yellow, bars are hand braided in different colors, as mentioned below. The end of each lash is wire fastened to keep it from unraveling. Shipping weight, each, 5 and 5½ pounds.

10L9216
84-Lash Net, with black and purple bars. Each, for one horse **$3.40**

10L9217
100-Lash Net, with white and green bars.
Each, for one horse **$3.75**

"Water-Sweat-Proofed" Fly Nets.

These nets are made of heavy 3-ply cotton fish cord, dyed black, and waterproofed to resist the rotting action of rain and sweat. Body is 5 feet long, extra lashes across the breast; wide, hand braided bars with lashes, 8 feet long, braided through them so that they will not readily pull out; the end of each lash is wire fastened to prevent unraveling. Shipping weight, each, 6, 7 and 8 lbs.

10L9228 — 70-Lash Net. Each, for one horse **$2.75**

10L9229 — 84-Lash Net. Each, for one horse ..**$3.50**

10L9230—100-Lash Net. Each, for one horse **$4.25**

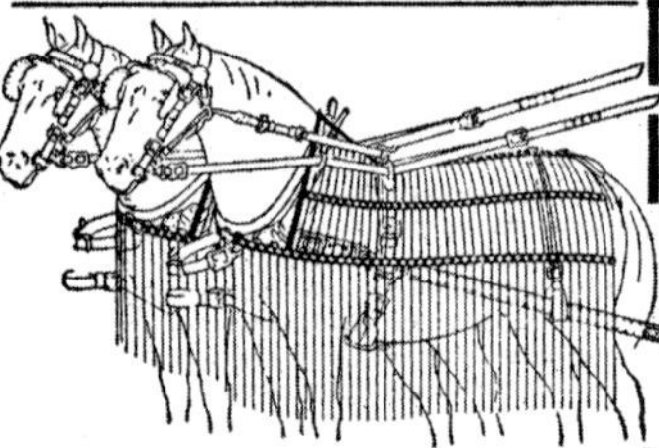

Leather Nets.

Medium Weight Leather Fly Nets, for expressing or teaming; ⅝-inch bars, 5 feet long; round lashes, 7 feet long, pieced on the center bar. Leather nets give long and satisfactory service.

10L9250
60-Lash Net. Shipping weight, 6 lbs.
Each, for one horse........**$3.95**

10L9251
70-Lash Net. Shipping weight, 7 lbs.
Each, for one horse........**$4.50**

Two of the Most Popular Covers Sold.

Plain Burlap Fly Covers.

Burlap makes good light fly excluders at low cost. Have duck trimmed hame and terret holes. Shipping weight, each, 2 lbs.

10L9063
100-Inch Cover.
Each.........**79c**

10L9064
110-Inch Cover.
Each **89c**

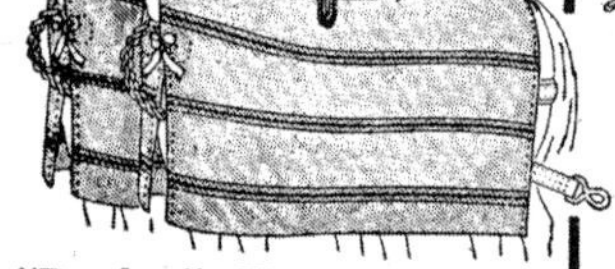

"Dunbar" Fly Sheets.

Made of white cotton Osnaberg woven cloth, striped; a fabric that protects the horse from the intense heat as well as from the flies, making him comfortable and able to do better work. Shipping weight, each, 1¾ pounds.

10L9066—100-Inch Sheet. Each **$1.10**
10L9067—110-Inch Sheet. Each **1.25**

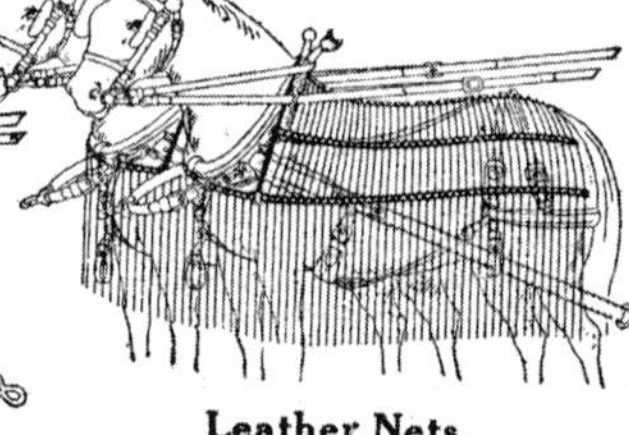

Leather Nets.

Heavy "Jumbo" Leather Team or Express Net. ⅝-inch bars, 5 feet long; round lashes, 7 feet long, pieced on center bar. Extra strong nets for hard usage or heavy work.

10L9255
60-Lash Net. Shipping weight, 6½ lbs.
Each, for one horse.........**$4.65**

10L9256
70-Lash Net. Shipping weight, 7½ lbs.
Each, for one horse **$4.95**

Light Weight Cotton Cord or Leather Net for Buggy or Light Express.

10L9188—Glazed Cord or Shoe String Net, with hand braided bars and lashes braided through the bars; seventy lashes in the body, extra lashes across the breast; black and white bars, black lashes. Shipping weight, 1⅝ pounds...**$2.69**

10L9189—Leather Net, with ½-inch leather bars and sixty round leather lashes in the body. Shipping weight, 5 pounds...**$3.00**

Face Fly Nets.

10L9010

Cotton Cord Face Nets add to your horses' comfort and make them work better. Have one of these face nets for each of your horses. Shipping weight, each, 8 ounces.
Each.....**$0.10**
Per doz.. **1.08**

Belting Leather Net.

10L9124—A low priced, 50-lash leather net, 5 feet long on the back, with ¾-inch bars. Lashes are laced through the bars, pieced in the center, and are 7 feet long. Shipping weight, each, 5½ pounds.
Each, for one horse......**$1.95**

"Keep-Off" Fly Guard.

10L5021—Galvanized Wire Fly Guard. Keeps the flies away from the nostrils of your horses and cattle. Can be fastened over head or tied to bridle or halter. Shpg. wt., each, 1 pound...**35c**

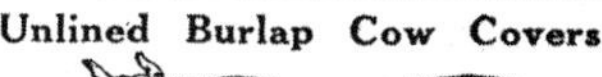

Unlined Burlap Cow Covers.

Your cows will give more milk on the same quantity of feed if you keep the flies off them, and be in better condition than if you tether them without fly covers. Snaps around cow's breast. Wide surcingle and short stays back of fore legs, surcingle going around each hind leg so cover cannot slip off. Three sizes, measured from breast to hind legs. Shipping weight, each, 2½, 3 and 3¼ pounds, respectively.

10L9078—Length, 58 inches...**$1.45**
10L9079—Length, 62 inches... **1.55**
10L9080—Length, 66 inches... **1.65**

For Horse Covers and Other Cattle Covers, see opposite page.

—Sheep Shearing and Horse Clipping Outfits—

Wizard Power Grinder for Grinding Sheep Shearing Combs and Cutters.

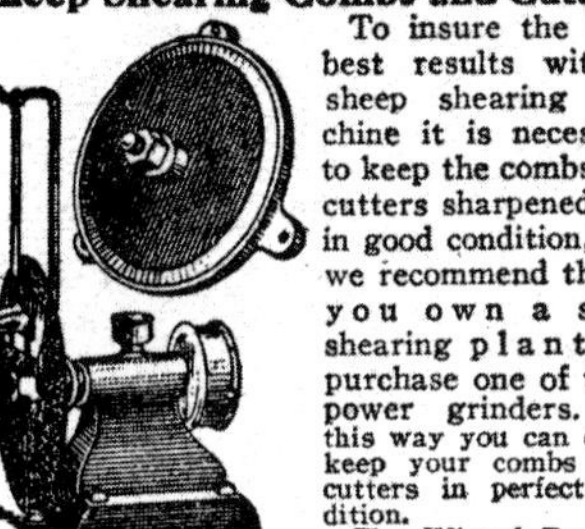

To insure the very best results with a sheep shearing machine it is necessary to keep the combs and cutters sharpened and in good condition, and we recommend that if you own a sheep shearing plant you purchase one of these power grinders. In this way you can easily keep your combs and cutters in perfect condition.

The Wizard Power Grinder has two 11-inch discs with emery paper on each disc, and is complete with comb and cutter holder, wrench and hard grease, fitted for 1½-in. belt. Shpg. wt., complete, 98 lbs.
10L4853¼ **$23.50**

10L4854—Extra Emery Paper. Sheets to fit grinder disc. Shipping weight, each, 3 oz. Each, fine......**18c** Each, coarse......**25c**

The Improved Wizard Narrow Shearing Head.

The Improved Wizard Shearing Head with four narrow combs and four three-point cutters, as furnished with the engine power machines. Shpg. wt., 4 lbs. **$12.95**
10L4851

For extra Combs and Cutters see 10L4883 and 10L4884 below.

The Improved Wizard Wide Shearing Head.

Improved Wizard Shearing Head with four wide combs and four four-point cutters, for engine power machines. Shipping weight, 4½ pounds. **$14.50**
10L4852

10L4885—The Improved Wizard Four-Point Cutter as used on the wide shearing head. Shpg. wt., 4 oz..... **20c**

10L4886—The Improved Wizard Wide Comb as used on the wide shearing head. Shpg. wt., 5 oz...... **65c**

Engine Power Sheep Shearing Machine.

10L4875¼ **$15.95**

Can be run directly from your engine. You do not need line shafting, pulleys or any other accessories except belting. Use flat belting, 1¼, 1½ or 1¾ inches wide. The gears are semi-enclosed; lever quickly throws machine in or out of gear, as desired, while in operation.

The machine is guaranteed to shear sheep regardless of the condition of their wool; it has bee tested and found to work perfectly. All parts are made of steel of high quality, cut and fitted with extreme care, so that every part when kept well oiled will operate smoothly and with the least possible friction. Its construction is simple and it will not easily get out of order. Furnished with a narrow shearing head with three-point type cutter. Four combs and four cutters are supplied with each machine.

Shipping weight, 37 pounds.

Eclipse Hand Power Sheep Shearing Machine.

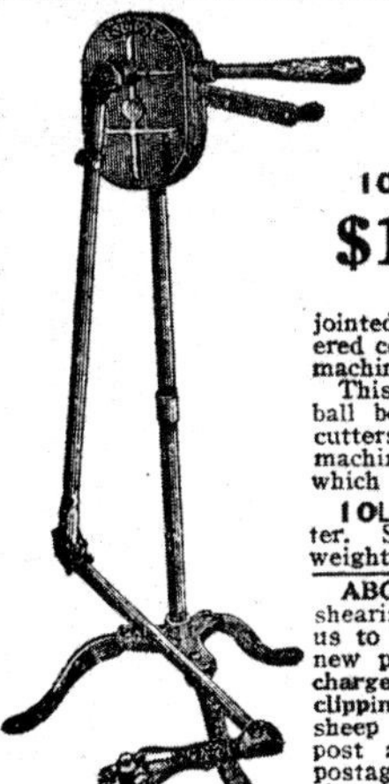

10L4879¼
$14.75

The body and stand of this machine are the same as those used on the Eclipse Clean Cutting Horse and Cattle Clipping Machine, the difference between the two machines being in the shaft, head and knives. The sheep shearing machine has a flexible jointed, ball bearing arm shaft in place of the web covered coil chain drive shaft which is on the horse clipping machine.

This machine is furnished complete with the Eclipse ball bearing sheep shearing head and four combs and cutters. Buy the Eclipse and make sure of getting a machine that for shearing efficiency is most reliable and which is fully guaranteed. Shipping weight, 45 pounds.

10L4883—Extra Cutter. Shipping weight, 3 ounces... **16c**

10L4884—Extra Comb. Shpg. weight, 4 ounces.... **47c**

ABOUT REPAIRS—After you have finished your shearing or clipping, send the cutting head and knives to us to be overhauled. We will charge you only for any new parts needed and for sharpening the knives. **We charge 50 cents per pair for sharpening knives of horse clipping machines** and 35 cents per pair for those of sheep shearing machines. Send parts to us by parcel post and instructions in a separate letter, including postage to return the goods to you. Eight to ten days' time is required for repairing and sharpening combs and cutters.

Eclipse Sheep Shearing Attachment for Horse Clipping Machine.

10L4681¼—This attachment consists of jointed ball bearing arm shaft with sheep shearing head and four pairs of combs and cutters, to attach to the Eclipse horse clipping machine in place of the web covered flexible chain shaft and horse clipping head. If you now have the Eclipse Clean Cutting Horse Clipping Machine you can readily convert it into a sheep shearing machine by the use of this attachment. Shipping weight, 15 pounds... **$9.95**

Eclipse Clean Cutting Horse and Cattle Clipping Machine.

Makes clipping an easy task.

The Eclipse Clean Cutting Horse and Cattle Clipping Machine is small, compact and substantially constructed. Will do its work quickly and efficiently. It has strong power, uniform speed and turns easily, requiring little labor to operate. It is equipped with the guaranteed Eclipse Clean Cutting Head with concaved knives, flexible steel coil shaft, web covered, and flexible chain. Weight of machine, packed for shipment, 48 lbs.

10L4680¼—Eclipse Clean Cutting Horse and Cattle Clipping Machine complete.............. **$9.50**

10L4682—Extra Cutter or Top Plate. Shipping weight, 6 oz.. **$1.00**

10L4684—Extra Comb or Bottom Plate. Shipping weight, 6 oz.. **$1.35**

Waterproof Horse Covers

Oiled cotton duck, unlined. Regular size, 76 inches full length or 60 inches from hame leathers to the tail; extra large size, 84 inches full length or 68 inches from hame leathers to the tail. Both sizes are 72 inches wide.

Have tug leathers to snap around the traces, line pockets and hame leathers; three-hole, "Fit-Rite" adjustable front fastener; unlined; will shed the snow and rain.

10L9898¼
Regular size, 76 inches long. Weight, about 7 pounds. Shipping weight, 8¾ pounds........**$3.98**

10L9899¼
Extra large size, 84 in. long. Weight, about 7¼ pounds. Shipping weight, 8¾ pounds..............**$4.35**

U. S. Government Stable Blanket

A Wonderful Value. Easily Worth Up to $5.00.

Order Early. Don't Miss This Chance to Pick a Sure Winner.

These full lined stable blankets present an unusual opportunity for money saving. We could not offer a blanket of the same quality and value in any other kind at this price. The lot consists of various kinds of blankets not all exactly alike, but every one made to stand hard service, and an unusual bargain. Some are slightly soiled from being stored in Government warehouses. All are sold under our regular catalog guarantee as they have our highest recommendation. Shipping weight, each, 8 to 10 pounds.
10L9750¼.............................. **$2.50**

Cutting Head for Horse Clipping Machine.

10L4685
Eclipse Clean Cutting Horse and Cattle Clipping Machine Head. Shipping wt., 1½ lbs. Furnished complete with one pair of knives.............. **$3.25**

For prices of extra knives, see 10L4682 and 10L4684.

"Crown" Unlined Stable Blankets.

They Keep Horses Clean.

Made of heavy jute, striped canvas, called tarpaulin cloth, a closely woven and strong fabric for making horse blankets; adapted to summer or winter use. These blankets are neatly finished all around, the neck and rump being well reinforced. Made with one body surcingle, two leg surcingles which keep the blanket in position and the "Fit-Rite" three-hole adjustable fastener at the front. A good blanket for a woman's or man's saddle horse.

10L9745¼
Size, 68 inches. Shipping weight, 6 pounds ...**$2.95**

10L9746¼
Size, 72 inches. Shipping weight, 6½ lbs. **$3.15**

10L9747¼
Size, 76 inches. Shipping weight, 7 pounds **$3.35**

"Climax" Fabric Lined Stable Blankets of heavy tarpaulin cloth, striped; bound all around; rounded corners (known as a shaped blanket) fitted with the style of surcingle shown and described with the blankets above and "Fit-Rite" three-hole front fastener; very strong material and a warm blanket for winter use.

10L9752¼
Size, 68 inches. Shipping weight, 7 lbs.....**$4.15**

10L9753¼
Size, 72 inches. Shipping weight, 7½ lbs. **$4.30**

10L9754¼
Size, 76 inches. Shipping weight, 8 lbs.....**$4.45**

"Guernsey" Cow Covers

Made extra wide, of white cotton duck or burlap; hemmed around neck and at rear; adjustable, reinforced front with fastener; front surcingle and stay are stitched on; jute girth and leg surcingles. The surcingle fastening around each hind leg keeps the cover in place.

10L9075—White cotton duck cover. Shipping weight, each, 3½ lbs.
58-inch cover....**$2.35**
62-inch cover.... **2.55**
66-inch cover.... **2.75**

10L9076—Plain burlap cover. Shipping wt., each, 3¾ pounds.
58-inch cover....**$1.75**
62-inch cover.... **1.85**
66-inch cover.... **1.95**

"Little Dan" Unlined Stable Sheets.

Made of mangled burlap in three sizes. Two surcingles with stay extending to first surcingle; hemmed neck and front. The right kind to keep your horse clean in the stable. Shipping weight, 3 pounds.

10L9702—Size, 72 in. Wt., 2¼ lbs...**$1.45**
10L9703—Size, 76 in. Wt., 2½ lbs... **1.55**
10L9704—Size, 80 in. Wt., 2½ lbs... **1.65**

Bathroom and Kitchen Trimmings

All bathroom trimmings shown on this page, unless otherwise specified, are made of brass, nickel plated and polished. Nickel plated screws are included with all fixtures that fasten to the wall. For Other Bathroom Hardware See Page 834.

68c

Shower Bath Spray.

42L260 Spray with Nickel Plated Brass Spray Head. Five feet white rubber hose and faucet connection. Attaches to any bathtub faucet. You can take a nice, cold shower before finishing the bath. Used also for shampooing the hair. Shipping weight, 1 pound. **68c**

42L261—Extra quality red rubber Spray with patent bulb to fit bath cock, otherwise as above. Shipping weight, 1 pound. **90c**

42L280¼—Bathroom Oval Mirror. Has white enameled metal frame. Size, 10x20 inches. Neat and attractive, especially appropriate for the bathroom. Excellent for shaving. Shipping weight, 10 pounds. **$1.95**

42L273—Siphon Bath Spray Brush. Just the thing for taking a refreshing shower bath in the country home without plumbing or running water. Fill bucket with water heated to any desired temperature. Then press bulb and water will flow through spray brush until pail is empty; 6 feet of rubber hose with siphon. Water flows out between rubber teeth of rubber brush. Any ordinary bucket and tub may be used, as shown. Bucket or tub not included. Shipping weight, 3½ pounds. **$2.48**

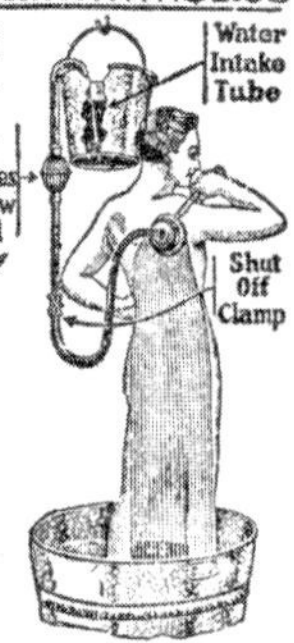

Shower Bath Outfit. 42L1457 Easily attached in a few minutes. Fits bathtub faucet. Made of brass, nickel plated. Has nickel plated brass head, 24-inch curtain ring, white fabric curtain, rubber tube, curtain chain and hook, waterproof cap for protecting hair, and screws to fasten to wall. Shower bath may be taken without wetting the floor or walls of your bathroom. Just the thing for taking a nice, cool, invigorating shower before leaving the tub. Shipping weight, 12 pounds. **$8.95**

Bathtub Seat. Hooks over rim at each side of tub. Made of hardwood in natural wood finish or white enameled. Steel rods nickel plated and rubber covered to protect enamel on tub. Shipping weight, 7½ pounds.

42L264—Natural wood finish **85c**

42L263—White enamel **95c**

Snow White Fixtures

For the Bath Room Beautiful

These beautiful white fixtures will add wonderfully to the beauty and sanitary appearance of your bathroom. Made of steel, except where otherwise mentioned, white coated by a special process with a snow white hard enamel which will not wear or chip off. Easily cleaned and kept clean. Always white and sanitary. Require no polishing. Admired by all.

42L206—Snow White Tooth Brush Holder. Convenient, sanitary. Shipping wt., 5 ounces. **36c**

42L207—Snow White Bath Robe Hook. Useful and very appropriate for the bathroom. Shipping weight, 6 ounces. **15c**

42L1456 White Enameled Bath Stool. Strong and well made of seasoned wood, fine rubbed white finish. Used in or out of tub. Rubber tipped legs. Practical, sanitary and attractive, 15 in. high. 12-in. round top. Shpg. wt., 8 lbs. **$2.25**

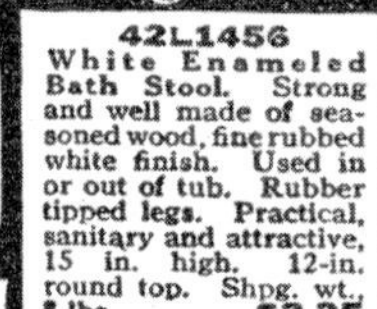

42L209—Snow White Toilet Paper Holder. Spring grip black wood roller. Shpg. wt., 6 oz. **35c**

42L205—Snow White Bathtub Soap Dish. Fits over rim of tub. Size, 3½x6 inches. Shpg. wt., 12 oz. **65c**

42L208—Snow White Tumbler Holder. Shipping weight 8 oz. **50c**

42L203—Snow White Soap Dish. Size, 4½x3¼ inches. Shpg. wt., 8 oz. **57c**

42L231 Sanitary White Opal Tumbler. Fits any holder on page. Shpg. wt., 8 oz. **24c**

42L213—Snow White Tumbler and Tooth Brush Holder. Made of brass, white coated. Shipping wt., 9 ounces. **88c**

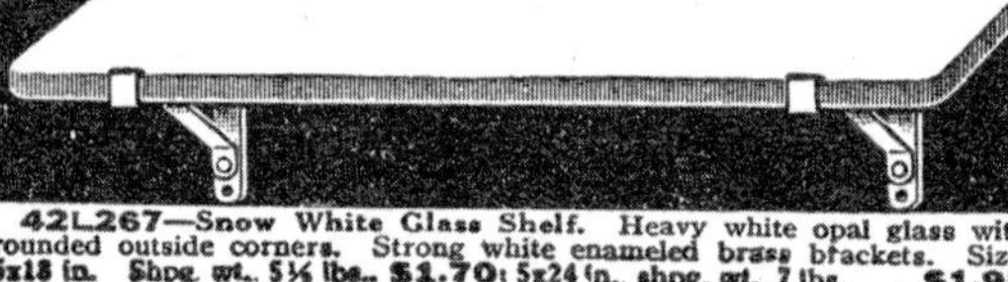

42L267—Snow White Glass Shelf. Heavy white opal glass with rounded outside corners. Strong white enameled brass brackets. Size, 5x18 in. Shpg. wt., 5½ lbs. **$1.70**; 5x24 in., shpg. wt., 7 lbs. **$1.95**

42L211—Snow White Opal Glass Towel Bar. White enameled brass posts, ⅝ inch thick; length, 18 in. Shpg. wt., 3 lbs. **$1.10**
Length, 24 in. Shpg. wt., 4 lbs. **1.25**

42L212—Same as above, 1 inch thick. Length, 18 in. Shpg. wt., 3½ lbs., **$1.70**; length, 24 in. Shpg. wt., 5 lbs. **$1.95**

42L210—Snow White Towel Bar. ⅝ inch diameter; 18 inches long. Shipping weight, 2¼ pounds, **75c**; 24 inches long. Shipping weight, 2½ pounds. **90c**

For Medicine Cabinets See Page 632.

42L262—Combination Massage Brush and Spray. Has 2½-in. perforated brass spray head and detachable rubber perforated massage brush which slips over spray head. Water flows out between rubber teeth, so that thorough shower and massage may be taken at same time. Five feet rubber tubing, aluminum faucet attachment. Shipping weight, 2 lbs. **$1.15**

$1.15

Easy Clean Sanitary Closet Brush. Bristles on heavy wire. Shipping weight, 1 pound.

42L271—First grade palmetto fibers **17c**

42L272—Selected white fibers **18c**

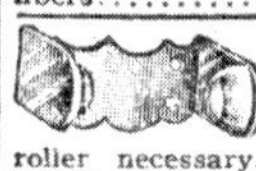

42L275 Toilet Paper Holder. Spring sides. No roller necessary. Shipping weight, 6 ounces. **65c**

42L270 Toilet Paper Holder. Ornamental pattern. Shpg. wt., 9 oz. **44c**

42L227 Tooth Brush and Tumbler Holder. Solid brass, nickel plated. Shpg. wt., 9 oz. **72c**

42L223 Soap Dish for Rim of Bathtub. Solid brass, nickel plated. Size, about 4x5 inches. Shpg. wt., 12 oz. **52c**

42L225 Tumbler Holder. Solid brass, nickel plated. Shpg. wt., 8 oz. **60c**

42L226 Same, without hook. **57c**

42L222 Wall Soap Cup, with ring hook. Solid brass, nickel plated. Shpg. wt., 8 oz. **45c**

42L224 Wall Soap Cup. Solid brass, nickel plated. Has removable opal glass dish. Size, about 3½x5¼ in. Shpg. wt., 2 lbs. **79c**

42L220 Wall Soap Cup. Solid brass, nickel plated. Size, about 2¾x4¼ inches. Shipping weight, 4 ounces. **21c**

42L229 Combination Tumbler and Soap Holder. Made of brass, heavily nickel plated. Shpg. wt., 12 oz. **80c**

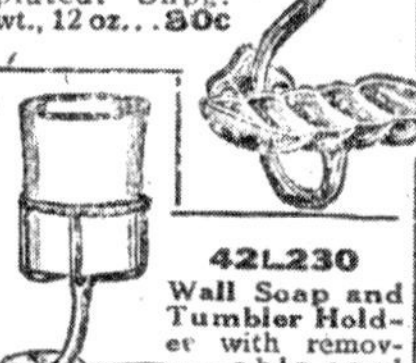

42L232 Wall Soap, Tumbler and Tooth Brush Holder with removable glass dish and tumbler. Solid brass, nickel plated. Shipping weight, 2¼ lbs. **$2.10**

42L230 Wall Soap and Tumbler Holder with removable opal glass soap dish and tumbler. Solid brass, nickel plated. Shipping weight, 2½ pounds. Complete **$1.70**

42L200 Bath Robe Hook. Nickel plated brass. Single. Shpg. wt., 2 oz. Each **$0.10**; Dozen **1.10**

Pipe Kleen Chemical Solvent. A chemical preparation for removing accumulations of grease and filth in waste pipes, drains and closet bowls. Quickly cleans out pipe stoppages and saves plumbers' bills. Should be used frequently. Shipping weight, 1 pound.

42L403 ..... **24c**

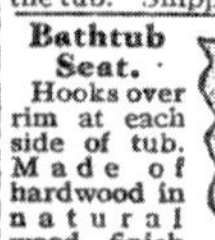

Rubber Force Cup. Saves Plumbers' Bills.

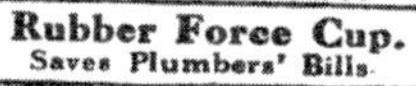

42L326¼—For forcing out stoppages and cleaning waste pipes, closets, sinks, wash bowls and bathtubs. Shpg. wt., 1½ lbs. **25c**

42L328¼—Same size as above, except that cup is made of extra quality red rubber. Shipping weight, 1½ pounds. **30c**

Closet Clean Out Auger. $1.55

42L329—Steel spring clean out auger. Removes stoppages in sink traps, sewer pipes and closet bowls QUICKLY AND EASILY. Length, 55 inches. Shipping weight, 2 pounds. **$1.55**

42L266—¼-Inch Crystal Shelf, rounded edges and bar. About 5x24 inches. Solid brass brackets. Shipping weight, 5½ pounds. **$2.70**

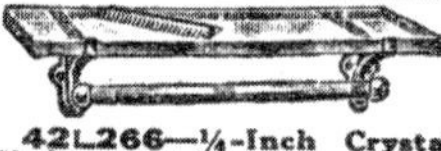

42L265 ¼-Inch Glass Shelf. Rounded edges. Solid brass brackets. About 5x24 inches. Shipping weight, 5½ pounds. **$1.35**

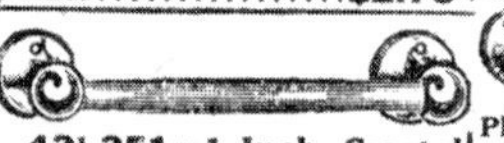

42L251—1-Inch Crystal Glass Towel Bar. Has nickel plated brass posts. Length, 18 in. Wt., 3½ lbs. **$1.05**; length, 24 in. Wt., 4 lbs. **$1.20**; length, 30 in. Wt., 4½ lbs. **$1.35**

42L242—½-Inch Nickel Plated Brass Towel Bar. Strong and well made. Length, 15 in., **42c**; 18 inches, **45c**; 24 in., **53c**

42L243—Size, ⅝ inch. Length, 18 inches, **63c**; 24 inches, **75c**; 30 inches **86c**

Shpg. wts.: 15 in., 2 lbs.; 18 in., 2¼ lbs.; 24 in., 2½ lbs.; 30 in., 3½ lbs.

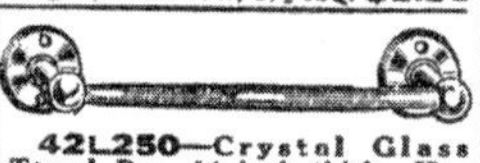

42L250—Crystal Glass Towel Bar, ⅝ inch thick. Has nickel plated brass posts. Shipping weight, 3 pounds.

Length, 18 inches **75c**

Length, 24 inches **88c**

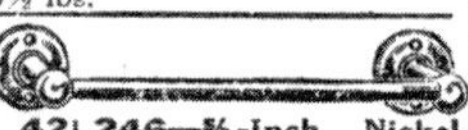

42L246—⅝-Inch Nickel Plated Brass Towel Bar. Has ball end posts. Shipping weight 3 pounds. Length, 18 in., **75c**; length, 24 in., **85c**; length, 30 inches **95c**

42L255 Three-Arm Swinging Towel Rack. Arms are flat, 16 in. long and ⅝ inch wide. Made of steel, nickel plated. Shipping weight, 8 ounces. **56c**

42L256—Double Arm Towel Bar. Solid brass, nickel plated. Length, over all, about 12 in. Shpg. wt., 8 oz. **67c**

42L201 Bath Robe Hook. Nickel plated brass. Double. Shipping weight, 5 ounces. Each, **11c**; Doz. **$1.25**

42L269 ¼-Inch Glass Corner Shelf. Plate Glass Shelf. Rounded edges. About 9x9 in. Shipping weight, 3 lbs. **$1.90**

42L1400 Rubber Plug. Wt., 3 ounces. 1-inch for wash bowl, **15c**; 1½-inch for bathtub, **18c**

Tooth Brush Holder.

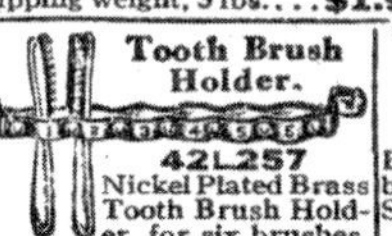

42L257 Nickel Plated Brass Tooth Brush Holder, for six brushes. Shpg. wt., 3 oz. **19c**

42L221—Comb and Brush Tray. Nickel plated brass. Size, about 4x10 in. Shpg. wt., 12 oz. **84c**

For Ivory Dresser Trays see page 491.

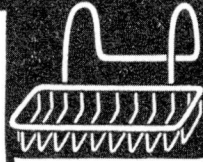

Rufix

Rufix is Our Own Trade Mark, Registered in the U. S. Patent Office.

A high grade, thick, heavy roof paint for the man who wants the best.

For painting roofs and keeping them in good condition. May be used on tin, iron, slate or composition roofing.

A roof paint meets more severe tests than any other paint, for the roof receives more hard wear than any other part of the building. A roof paint must withstand the extremes of weather conditions—the hot summer sun and the ice and snow of winter. It must be adapted to different kinds of roofing. It must be reasonable in price.

Because of these requirements, we have spent much time in an effort to improve the quality of Rufix—our highest grade roof paint. We now have a paint which we feel confident will give you excellent service. Rufix is a combination of weather resisting gums and oils which forms a thick elastic waterproof coating over the surface to which it is applied. There is not an ounce of coal tar in Rufix, nor any other product that will soften during very hot weather.

Rufix is about the consistency of thick paint and should be applied with a brush, preferably a roof paint brush. It both preserves and protects and will make an ordinary roof last many years longer.

We have so much confidence in our Rufix that we have no hesitancy in guaranteeing it to give you the service you have a right to expect from a high grade roof paint. If for any reason it fails to come up to this standard, we agree to furnish you with new paint free of charge.

One gallon will cover approximately 200 square feet, one coat, on smooth surfaces.

	30L1910 Black	30L1920 Maroon
1-gallon can. Per gal. Shpg. wt., 15 lbs.	96c	$1.25
5-gallon can. Per gal. Shpg. wt., 50 lbs.	93c	$1.22
*25-gal. half bbl. Per gal. Shpg. wt., 275 lbs.	91c	$1.20
*50-gal. barrel. Per gal. Shpg. wt., 500 lbs.	81c	$1.10

*The 25 and 50-gallon barrels are shipped from factory in NORTHEASTERN ILLINOIS.

Write for our "How to Paint" Book No. 686GCL, Sent Postpaid on Request.

"Save the surface and you save all"—Paint & Varnish

Asbestos Fiber Roof Coatings

For painting and repairing metal, concrete, tile, stone, slate and composition roofing. Also excellent for repairing skylights, smokestacks, flashings, chimneys, ventilators, and dampproofing the exterior of brick foundations and underground tanks.

You don't have to wait for the tinner or the roofer to repair your roof when it is leaking. Here are two materials that may be used for repairing metal, concrete, tile and composition roofs. You can do the work yourself. These materials come ready to use and are easily applied. Will not harden in the package and can be kept for any length of time. Asbestos Fiber Roof Coatings contain a large percentage of asbestos fibers which prevent these materials from running and spreading and consequently give them more adhesive power so that they can be used practically any place, such as on slanting roofs, sides of buildings, etc., without the least danger of their crawling and running.

Asbestos Fiber Roof Cement.

Asbestos Fiber Roof Cement is of puttylike consistency and while it may be used as an entire roof coating, it is prepared especially for repair work, such as patching large holes and gaps, and repairing gutters, chimney flashings, skylights, cornices, parapets, etc. It should be applied about ⅜ inch thick with a trowel. If, after repairing a badly worn roof with the cement, the roof is given a coat of Liquid Coating it will be as good as new and, if repaired as often as needed, will last for years.

Asbestos Fiber Liquid Roof Coating.

Asbestos Fiber Liquid Roof Coating is a black liquid of paintlike consistency, recommended as a preservative to be applied to the entire roof with a roof brush. It is excellent for dampproofing the exterior of building foundations and underground tanks. It will spread without a single seam or joint exposed in the whole surface. It will not bulge, crack or run.

Which Material to Use.

These compounds are made of exactly the same materials, but in two consistencies. The Cement is the plastic material, like putty, and the Liquid Coating is of paintlike consistency. The nature of the repair work will determine which to use, but in many cases either one will work. When used as a preservative coating on a new roof, the Liquid Coating should be used. When repairing an old roof, go over the surface first and repair the large cracks and holes with the Cement and then give the entire roof a coat of Liquid Coating.

Asbestos Fiber Roof Cement. Apply With a Trowel.

10 pounds will cover about 25 square feet on a smooth surface.

		30L3477—Black.	30L3478—Red.
5-lb. can. Per lb.	(Shpg. wt., 7 lbs.)	10c	15c
10-lb. can. Per lb.	(Shpg. wt., 12 lbs.)	9c	14c
25-lb. pail. Per lb.	(Shpg. wt., 30 lbs.)	8c	13c
50-lb. can. Per lb.	(Shpg. wt., 60 lbs.)	8c	12c
*300-lb. half barrel. Per lb.	(Shpg. wt., 340 lbs.)	5c	10c
*550-lb. barrel. Per lb.	(Shpg. wt., 620 lbs.)	5c	9c

Asbestos Fiber Liquid Roof Coating. Apply With a Roof Brush.

1 gallon will cover about 65 square feet.

		30L3479—Black.	30L3480—Red.
1 gallon	(Shpg. wt., 15 lbs.)	80c	$1.34
5 gallons. Per gallon	(Shpg. wt., 50 lbs.)	75c	1.24
*30-gal. half barrel. Per gal.	(Shpg. wt., 265 lbs.)	50c	1.07
*50-gal. barrel. Per gal.	(Shpg. wt., 425 lbs.)	45c	1.02

*Shipped from factory in NEW JERSEY, CLEVELAND, OHIO, or KANSAS CITY, MO.

6-Inch Trowel.

40c

Made of steel, ground smooth and well shaped. Especially recommended for applying Longlife Elastic Compound. Weight, about 3 ounces.

9L5886 .... 40c

Roof Paint Brush.

Roof Paint Brushes, bound with wire and carefully cemented into a head of hardwood. Will not check or crack. Recommended for heavy tar paints, roof coatings and heavy painting material.

30L3170

Two Knots. Shipping weight, 1 pound.	50c
Three Knots. Shipping weight, 1 lb. 11 oz.	68c
Four Knots. Shipping weight, 2 pounds.	87c

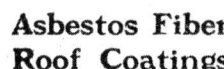

Longlife Creosote Wood Preserver

A Liquid Creosote Oil for Treating Posts, Sills, and All Kinds of Timbers to Prevent Rotting.

Creosote Oil as a wood preservative has been so thoroughly discussed and recommended by the Government that every farmer and construction engineer now knows that it is the best wood preserver known. It will do wonders in prolonging the life of ordinary fence posts, sills, joists, railroad ties, etc. It is invaluable for treating cedar, cottonwood and other soft wood posts.

Creosote Oil, because of its non-evaporating qualities, tends to fill the pores of the wood, killing any bacteria already there and preventing, to a certain extent, the absorption of moisture.

Our Longlife Creosote Wood Preserver is a product which we highly recommend. It is one of the best wood preservers on the market. It is made of pure Creosote Oil and can be depended upon to repay you in every way for its cost and labor of applying. One gallon will treat the butt ends of about twenty ordinary posts to a height of 2½ feet.

Longlife Creosote Wood Preserver gives an artistic dark brown color to new wood that never fades or wears off. It is highly recommended for shingles, as it both preserves and beautifies. One gallon will cover about 80 square feet two coats.

30L1947—Creosote Wood Preserver.		Per Gal.
1-gallon can	(Shipping weight, 10 pounds)	78c
5-gallon can	(Shipping weight, 50 pounds)	76c
10-gallon can	(Shipping weight, 100 pounds)	74c
*25-gallon barrel	(Shipping weight, 550 pounds)	70c

***Shipped from factory in SOUTHEASTERN PENNSYLVANIA or CHICAGO store.**

Asphalt Roof Paint.

For painting felt or composition roofing, iron roofs, trestles, sheds, warehouses and other buildings. Second in quality only to our Rufix.

This Asphalt Roof Paint will give far better service for felt and composition roofing and for painting iron roofs than any of the coal tar preparations.

If you are looking for a good low priced paint, then send us your order for this Asphalt Roof Paint with the guarantee that the paint, considering its low price, will prove satisfactory.

One gallon of this Asphalt Roof Paint will cover about 250 square feet of smooth surface, one coat, or about 150 square feet of rough wood or felt roofing, one coat.

30L1941—Asphalt Roof Paint. Black only.		Per Gal.
1-gallon can	(Shipping weight, 15 pounds)	75c
5-gallon can	(Shipping weight, 50 pounds)	72c
10-gallon can	(Shipping weight, 100 pounds)	70c
*25-gallon half barrel	(Shipping weight, 215 pounds)	69c
*50-gallon barrel	(Shipping weight, 425 pounds)	61c

*The 25 and 50-gal. barrels are shipped from factory in SOUTHEASTERN PENNSYLVANIA or NORTHEASTERN ILLINOIS.

Liquid Coal Tar.

Used for painting exterior of water tanks, fence posts and metal surfaces that are liable to become rusty. Used extensively for recoating gravel roofs and composition roofing. May be applied cold, but is better heated to a boiling temperature.

30L3468	Per Gal.
1-gallon can. Shipping weight, 14 lbs.	57c
5-gallon can. Shipping weight, 51 lbs.	53c
10-gallon can. Shipping weight, 105 lbs.	45c
*25-gallon half barrel. Shipping weight, 250 lbs.	30c
*50-gallon barrel. Shipping weight, 425 lbs.	25c

*Shipped from factory in SOUTHEASTERN PENNSYLVANIA, SOUTHERN WISCONSIN, MISSOURI, ALABAMA, OHIO or NORTHEASTERN ILLINOIS.

Seroco Chromate Paint.

A first coater for metal surfaces. An exceptionally high grade paint for water tanks, sheet iron buildings, metal grain bins, iron fences, bridges, iron work, etc.

It has always been difficult to get paint to stick on metal surfaces. We believe we have a better first coater in our Seroco Chromate Paint than any of the better known red lead or graphite paints that are offered for this purpose. It is made in a pleasing medium brown shade, and while it is made for first coater work it can also be used for the finish coat if you wish. Any other kind of paint can be applied over our Chromate Paint with perfect satisfaction.

30L1908—Medium Brown.		
1-quart can	(Shpg. wt., 5 lbs.)	$0.63
½-gallon can	(Shpg. wt., 9 lbs.)	1.24
1-gallon can	(Shpg. wt., 15 lbs.)	2.33
5-gallon can	(Shpg. wt., 50 lbs.) Per gal.	2.30

Substitute Turpentine.

Our Substitute Turpentine (otherwise known as substitute for oil of turpentine) will prove very satisfactory for thinning paints and varnish and the equal of pure turpentine. It will save you money on your painting bill.

30L3462

½-gal. can..........**40c** Shipping wt., 7 lbs.
1-gal. can..........**62c** Shipping wt., 10 lbs.
5-gal. can, per gallon..**59c** Shipping wt., 45 lbs.
10-gal. can, per gallon..**57c** Shipping wt., 90 lbs.

If you prefer pure turpentine, write for prices.

White Lead.

Science has proved beyond a doubt the superiority of carefully balanced prepared paint over hand mixed white lead and oil. We have decided, therefore, out of fairness to our customers, to discontinue selling white lead. We know that our Seroco Paint will give far better service and cost less than any hand mixed white lead and oil paint, and since we now guarantee our Seroco Paint in stronger terms than anyone can guarantee white lead, we feel that our customers would be no longer interested in mixing their own paint.

Putty.

Will not crumble or crack. Guaranteed a strictly pure linseed oil putty and first quality. We do not handle any other quality.

30L2205

4 pounds..............**32c** Shpg. wt., 5½ lbs.
10 pounds..... (Shpg. wt., 13 lbs.).....**$0.70**
20 pounds..... (Shpg. wt., 25 lbs.).....**1.30**

Brush Renewer.

Don't throw away your old brushes. No matter how dry and hard they are, our Brush Renewer will make them soft and ready to use again. A powder to be dissolved in warm water. One package will clean from six to eight brushes. Full directions on carton.

30L2780

1-lb. carton..........**25c**
Shipping weight, 1⅛ lbs.

Hercules Liquid Glue.

Will cement together wood, leather, ivory, cork, cloth, marble, glass, iron, china, crockery, pearl, metals, porcelain, etc. A high grade hide glue. Thinned out it may be used as a sizing for the back of a rug to prevent sliding.

30L3216

Large tube (Shpg. wt., 4 oz.)..........**8c**
⅛-pint (Shpg. wt., 8 oz.)..........**14c**
½-pint (Shpg. wt., 1¼ lbs.)........**28c**
1-pint (Shpg. wt., 1⅞ lbs.).......**46c**
1-quart (Shpg. wt., 3¼ lbs.)....**78c**

LePage's Tested Ground Glue (Dry.)

A general purpose glue of very good quality, put up in special package for amateur use. May also be used as a kalsominers' glue and for cabinet work, furniture repairing, etc. Shipping wt., 1½ pounds.

30L2791—1-pound carton..........**24c**

LePage's Liquid Glue.

The well-known widely advertised glue for use on anything and everything around the house. Shpg. wt., 8 oz.

30L3218 — ½-pint can..............**38c**

For other Mucilage see page 478.

Extra Quality Cold Water Paint for Interiors. (No Oil.)

We prepare the pigment—You add the water. In powder form, to be mixed with cold water, for coating interiors of barns, basements, chicken houses, factories, factory offices, summer cottages, etc. Used in place of whitewash. Forms a hard, firm and sanitary coating; will not rub, scale or peel off. Not suitable for outside work. Can be applied with a brush or spraying machine. (See Spraying Machine 42L51725¼ listed below.) Ten pounds will cover approximately from 600 to 800 square feet, one coat.

30L2629—White Cold Water Paint.

5-pound package. (Shipping weight, 8 pounds.) Per pound....**10c**
25-pound box. (Shipping weight, 28 pounds.) Per pound....**9c**
***100-pound drum.** (Shipping weight, 110 pounds.) Per pound....**8c**
***320-pound barrel.** (Shipping weight, 385 pounds.) Per pound....**7½c**

30L2631—Cream Color Cold Water Paint.

5-pound package. (Shipping weight, 8 pounds.) Per pound........**11½c**
25-pound box. (Shipping weight, 28 pounds.) Per pound........**10½c**
***100-pound drum.** (Shipping weight, 110 pounds.) Per pound........**9½c**
***320-pound barrel.** (Shipping weight, 385 pounds.) Per pound........**9c**

*Shipped from SOUTHEASTERN NEW YORK or our CHICAGO store.

Outside White Cold Water Paint.

In dry powder form, to be mixed with cold water for painting exteriors of factories, warehouses, docks, stables, farm buildings, sheds, benches, etc. It makes a good interior washable paint, and may be used on sand finished walls, brick, stone, wood, etc. A very good coating for temporary or rough buildings. For permanent buildings of any kind we recommend our Barn Paint as much more durable and protective, and cheaper in the long run. Five pounds will cover approximately from 300 to 400 square feet, one coat. Directions on package.

30L2633—Outside White Cold Water Paint. Per Lb.
5-pound package... (Shipping wt., 8 lbs.)....**11½c**
25-pound box......... (Shipping wt., 28 lbs.)....**10½c**
***100-pound drum**...... (Shipping wt., 110 lbs.)....**9½c**
***320-pound barrel**...... (Shipping wt., 385 lbs.)....**9c**

*Shipped from our CHICAGO store.

Handy Whitewashing and Spraying Machine.

It whitewashes, sprays sheep, cattle, trees and shrubbery and does many other useful things.

It forces whitewash into corners and crevices where disease germs lodge and which cannot be reached with brushes. In whitewashing chicken houses infested with lice you can do the work from a door or window, thus avoiding the discomfort of being covered with chicken lice and dirt from head to foot.

For spraying sheep and cattle it is especially efficient, as it forces the liquid through to the hide, where it is really wanted.

Equipment consists of a 6-foot wrought iron spray pipe with a ¼-inch brass shut off cock and brass spray nozzle, extra brass nozzle for straight stream-work, iron follower wrench (used for tightening plunger ring in bottom pump cylinder), rubber suction hose with screen attachment, and 8 feet of ⅜-inch special high pressure rubber discharge hose. Net weight, 20 lbs. Shipping weight, 30 lbs.

42L51725¼—Complete..........**$7.70**

For full line of Hand and Power Sprayers see pages 902 and 903.

Seroco Colors in Oil.

Used by the painter for coloring or tinting when mixing his own paint. Ground in pure linseed oil, strong and permanent. We exercise great care in grinding these colors, some of the harder pigments being ground several times. The result is that the tinting power is greatly increased. Put up in 1 and 5-pound cans. Shipping weight, 1 lb. 3 oz. and 6 lbs. Per Lb.

30L2230—Lampblack..........**28c**
30L2231—Coach Black..........**23c**
30L2232—Ivory Black..........**23c**
30L2233—Drop Black..........**23c**
30L2235—American (formerly Prussian) Blue..........**46c**
30L2236—Ultramarine Blue..........**46c**
30L2237—Cobalt Blue..........**46c**
30L2240—Raw Umber..........**23c**
30L2241—Burnt Umber..........**23c**
30L2242—Raw Sienna..........**24c**
30L2243—Burnt Sienna..........**24c**
30L2248—Blind or Chrome Green....**27c**
30L2253—English Venetian Red.....**19c**
30L2254—Indian Red..........**31c**
30L2255—Tuscan Red..........**30c**
30L2256—Unfading Red..........**33c**
30L2259—Scarlet Vermilion..........**27c**
30L2262—Chrome Yellow, Light....**28c**
30L2263—Chrome Yellow, Medium..**28c**
30L2264—Chrome Yellow, Orange...**28c**
30L2265—French Yellow Ochre.....**18c**

Lampblack.

30L2343—Lampblack. High quality. 1-pound carton..........**31c**

Paint and Varnish Remover.

One of the best paint and varnish removers on the market. Free from caustics, carbolic acid, lye or any other injurious ingredient. Guaranteed to remove paint, varnish or oil from wood or iron without damage to the wood veneer, glue or filler. Directions on can.

30L2770

	Shpg. wt.	
Pint.....	2½ lbs.....	**$0.33**
Quart.....	4 lbs.....	**.48**
Gallon...	11 lbs.....	**1.60**

Powdered Pumice Stone.

High grade powdered domestic pumice stone. Used for rubbing furniture, woodwork, etc., preparatory to or after varnishing.

30L2673—1 lb.......**8c** Shipping weight, 1⅛ lbs.
25 lbs..........**$1.50** Shipping weight, 30 lbs.

Pure Shellac. (Orange and Clear.)

A high grade standard light body liquid shellac of the proper consistency to be used as a first coater on furniture and woodwork. Clear shellac is used when it is desired to preserve the natural color of the wood. Orange shellac gives the wood a light oak color. It dries in from three to four hours. It is light bodied, cut in the proportion of 2½ pounds of gum to the gallon. One gallon will cover approximately 600 square feet, one coat. We do not recommend using shellac as a first coat on floors. In place of shellac use the varnish itself thinned with 10 per cent turpentine as a first coater. You will find that it wears far better than the shellac. Shellac is recommended only for woodwork and furniture.

30L2656—Clear Shellac.

	Shpg. Wt.	
1 pint	2 lbs.	**$0.56**
1 quart	3 lbs.	**.98**
½ gallon	6½ lbs.	**1.85**
1 gallon	10½ lbs.	**3.56**

30L2657—Orange Shellac.

	Shpg. Wt.	
1 pint	2 lbs.	**$0.51**
1 quart	3 lbs.	**.87**
½ gallon	6½ lbs.	**1.66**
1 gallon	10½ lbs.	**3.17**

Seroco Substitute for Shellac.

This shellac substitute is offered at a low price. It is a product made especially from high grade, selected gums cut in denatured alcohol, has the same consistency as shellac, and dries in about the same time with the characteristic shellac finish. It is made in both orange and clear and can be reduced with denatured alcohol. It can be successfully used instead of shellac in practically every place except on floors.

30L2758—Substitute for Clear Shellac.
30L2759—Substitute for Orange Shellac.

1 pint (Shpg. wt., 2 lbs.)........**38c** | ½ gal. (Shpg. wt., 6½ lbs.)..**$1.05**
1 qt. (Shpg. wt., 3 lbs.)........**60c** | 1 gal. (Shpg. wt., 10½ lbs.)..**1.90**

Wizard Scraper.

Save Your Old Safety Razor Blades.

Use them in this handy tool for removing paint from glass, for removing rust and grease and many other uses that housewives, painters, machinists and farmers will find for this scraper. Furnished with a guard so it can be carried in pocket or tool box. Shipping weight, 5 ounces.

30L2911—Scraper, including 1 blade..**19c**
30L2913—Extra heavy blade..........**5c**

Steel Wool and Shavings.

Nos. 0, 1 and 3 Steel Wool used for rubbing down fillers and varnishes. **Steel Shavings** used for removing rust from iron preparatory to painting. **State size number.**

9L6265—Steel Wool. Per 1-Lb. Pkg.

No.	Grade	Equal to	
0	Very Fine	No. 00 Sandpaper..........	**50c**
1	Fine	Nos. 0 and ½ Sandpaper....	**40c**
3	Medium	No. 1½ Sandpaper..........	**30c**

9L6264—Steel Shavings. (For rougher work.) 1-pound package..........**35c**

Cold Water Dry Paste.

Mixes easily and readily into a smooth, transparent paste. After testing numerous samples of paste submitted for our approval we have selected this one, which has several superior qualities. **It will make more paste per pound than any paste we have ever examined.** It mixes readily and spreads evenly. It will not sour within reasonable length of time and will hang any kind of wall paper. A cold water paste is very convenient, saving time, labor and trouble and gives as good service as anyone can possibly desire. One 2-pound bag will make enough paste for the walls and ceiling of a large room.

30L2787

2-pound bag........(Shpg. wt., 2½ lbs.)...**$0.25**
5-pound bag........(Shpg. wt., 6 lbs.)...**.60**
25-pound drum.....(Shpg. wt., 30 lbs.)...**2.90**

Prepared Flour Paste.

Only Boiling Necessary.

Will make strong adhesive paste. One pound will make 1½ gallons of ready to use paste or 1 gallon solid paste.

30L2786

1-lb. package.......**12c** Shipping weight, 1⅝ lbs.
25-lb. pail.........**$2.60** Shipping weight, 31 lbs.

Seroco Elastic Sizing.

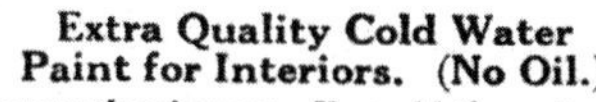

Commonly known as a glue sizing. For use on kalsomined, painted or sand finished walls or ceilings preparatory to hanging wall paper, oilcloth, Lincrusta, burlap, etc. Used on kalsomined walls, paper can be hung without danger of peeling. **Should not be used as a first coater for paint or enamel.** One quart makes enough size for walls of an ordinary room. May be used as a sizing on the back of a rug to prevent sliding.

30L2789

1-quart can..........**35c** Shipping weight, 3 pounds.
1-gallon can..........**$1.20** Shipping weight, 10½ pounds.

Paint Your Own Car!

Auto and Carriage Enamel

Make the Old Car Look New Again!

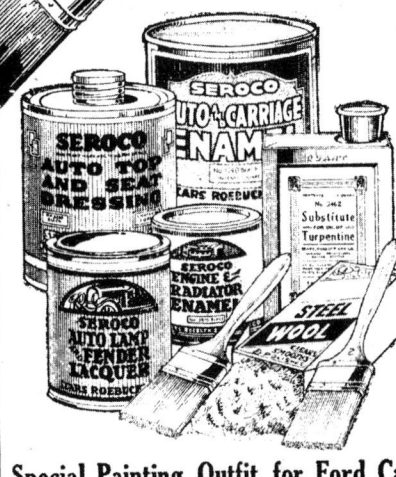

Give it a coat of Auto Enamel and make it shine and glisten like a new car! Thousands of car owners are doing it! You can too!

Here is an enamel of smooth cream like consistency, made of pure non-fading colors and high grade varnish especially for auto refinishing. It flows freely and spreads easily and evenly, so that anybody can apply it. It dries with a smooth mirror like finish and has a beautiful gloss. No other finish need be applied. It is made in our own factory, under our own supervision, and we guarantee it to give perfect satisfaction. It is exactly as good and in many cases better than the auto enamels retailing for over twice as much per gallon.

About three quarts should be enough to refinish the body and gear of a touring car two coats. If the old finish still has a gloss it should be removed, using either steel wool or sandpaper.

For color samples see page 950.

30L1725—Cream.
30L1705—Vermilion.
30L1715—Dark Wine.
30L1720—Mouse Gray.
30L1760—Auto Gray.
30L1735—Brewster Green
30L1755—Dark Blue.
30L1740—Black.

	Shpg. wt.	Price
1-pint can	2½ lbs.	**$0.44**
1-quart can	4¼ lbs.	**.78**
½-gallon can	9 lbs.	**1.48**
1-gallon can	13 lbs.	**2.85**

"Save the surface and you save all" – Paint & Varnish

Auto Top and Seat Dressing.

Jet Black.

For refinishing and waterproofing leather and imitation leather automobile tops and seats. Will dry over night with an excellent finish and positively will not crack or peel. The most delicate fabric will not be affected by coming in contact with the material after it has dried. About 1 quart needed for the top and seats of a runabout, and about 1 quart and 1 pint for the top and seats of a touring car. Apply with a good varnish brush.

30L3420 — Gloss Black.
30L3419—Dull Finish.
1-pt. can..45c
1-qt. can.80c
Shpg. wt.: 1 pt., 2 lbs.; 1 qt., 3 lbs.

For Brushes See Page 940.

Black Touch-Up Varnish.

If you do not want to refinish your entire car but just want to touch up a few scratches, marks, etc., this is the very material you need. May be used on any part of the car for touch up work. Will produce a high quality finish, tough and durable. Apply with a varnish brush. Will dry dust free in two hours and harden in six hours.

30L1850

Shpg. Wt.
1-pint can 2½ lbs.... **50c**
1-qt. can... 4 lbs.... **84c**

Special Painting Outfit for Ford Cars.

Do Your Own Auto Refinishing.

Others Are Doing It! Last Year We Sold Thousands of Painting Outfits to Ford Car Owners.

You can keep your Ford car looking new all the time at a surprisingly small cost. Keep one of our Refinishing Outfits on hand all the time and give it a coat of glossy black enamel when it begins to show signs of wear. Do the painting yourself. It will be dry ready to use in thirty-six hours. This outfit contains everything you will need. Each material is ready to use and easy to apply. They are made especially for the car owner who wants to do his own auto refinishing. No experience is necessary. Just follow the few simple directions given. Enough material to refinish the body and fenders of a Ford car two coats and the top, seats and engine one coat.

Outfit Consists of:

- 1 Quart Black Auto Enamel.
- 1 Quart Black Auto Top and Seat Dressing.
- ½ Pint Black Engine and Radiator Enamel.
- 1 Large Package Steel Wool.
- 1 Pint Lamp and Fender Lacquer.
- 1 Quart Substitute Turpentine.
- 2 Varnish Brushes, 2 inches wide.

Shipping weight, 15 pounds.

30L1699—Complete outfit................**$2.98**

Complete Automobile Refinishing Outfit.

Paint Your Own Car and Save Money.

Others Are Doing It! Last Year We Sold 23,000 Auto Painting Outfits.

You can paint your car and do an excellent job with this outfit. The materials are products which we strongly guarantee. Nothing but the best materials are good enough to repaint a car. Buy the outfit you are sure of—the outfit that will give your car a lasting high gloss, waterproof finish.

30L1773—Outfit for painting cars BLACK.
30L1791—Outfit for painting cars DARK BLUE.
30L1789—Outfit for painting cars BREWSTER GREEN.
30L1785—Outfit for painting cars VERMILION.
30L1787—Outfit for painting cars DARK WINE.
30L1796—Outfit for painting cars MOUSE GRAY.
30L1798—Outfit for painting cars AUTO GRAY.

Outfit Consists of:

- ½ Gallon Can Auto Enamel.
- 1 Quart Substitute Turpentine.
- 1 Package Steel Wool No. 1.
- 1 Package Steel Wool No. 3.
- 1 Quart Auto Top Dressing.
- ½ Pint Black Engine and Radiator Enamel.
- 2-Inch Oval Varnish Brush.
- 3-Inch Varnish Brush, bristles secured in vulcanized rubber.
- 1 Pound Cotton Waste.

Shipping weight, any outfit, 17 pounds.

Complete outfit.............................**$4.98**

Engine and Radiator Enamel.

Paint your engine or radiator with this high grade enamel, and it can be easily kept clean and will not rust, the high gloss of the enamel leaving no surface for the grease and dirt to accumulate. Will not blister, peel or rub off. Apply with a varnish brush.

30L3416—Black.
30L3417—Gray.

	Shpg. Wt.	
½-pint can	1¼ lbs.	**30c**
1-pint can	2¼ lbs.	**51c**
1-quart can	4⅛ lbs.	**93c**

Lamp and Fender Lacquer.

An ideal preparation for refinishing auto hoods and fenders and for painting brass lamps. A high quality finish, tough and durable; will stand hard wear. If the finish on the fenders or radiator hood of your machine is dull and marred, one coat of Seroco Lacquer will give it a new appearance.

30L2185—Black. High gloss.

	Shpg. Wt.	
½-pint can	1¼ lbs.	**30c**
1-pint can	2¼ lbs.	**51c**
1-quart can	4⅛ lbs.	**93c**

Seroco Black Dye for Auto Top Linings.

Apply this dye to the under lining of auto tops and it will make them look like new. It is an indelible black waterproof liquid for dyeing top linings, and is not to be used on leather or imitation leather. For leather or imitation leather auto tops and seats we recommend our Black Auto Top and Seat Dressing listed above.

30L3418

Shpg. Wt.
1-quart can.....3 lbs. **75c**

Black Dressing for Mohair Tops.

Leaves cloth soft and pliable. Mohair has a drill backing held with cement. The cement loses its adhesive qualities through age, thus causing the mohair to readily wear out. This special Seroco Mohair Top Dressing prolongs the life of adhesive qualities of the cement as well as waterproofing the mohair and making the old top look like new. Will dry over night. About 1 pint is needed for a runabout and 1 quart for a touring car.

30L3421

Shpg. Wt.
1-pint can........................2 lbs. **44c**
1-quart can......................3 lbs. **75c**

Auto Body Varnish.

A brilliant, transparent and durable varnish for finishing carriage, buggy and auto bodies. It is very pale and will not darken or injure the lightest shades of body color. If your car does not need repainting, but merely brightening up, our Auto Body Varnish is just what you want. It may also be used over our Auto Enamel to insure a higher gloss and more permanent finish. Easy to apply and dries free from dust in sixteen hours and hardens properly in three days.

30L2740

	Shpg. Wt.	
1-pint can	2½ lbs.	**$0.52**
1-quart can	4 lbs.	**.99**
½-gallon can	6 lbs.	**1.90**
1-gallon can	11 lbs.	**3.60**

Steel Roofing and Siding

2½ and 1¼-Inch Corrugated Steel.

2½-INCH AND 1¼-INCH CORRUGATED SHEETS are furnished in sheets which actually measure 26 inches wide. Both have a covering width of only 24 inches on account of the side lap. When ordering, allow from 4 to 6 inches for end laps, depending upon the pitch or slant of the roof. If used for siding, a 2-inch end lap is sufficient. When ordering remember that it requires 109 square feet to cover 100 square feet of surface; this does not include the end laps.

When Ordering Always Specify Length.

We furnish two thicknesses, United States Standard 28-gauge, which is known the country over as standard weight and thickness, and United States Standard 26-gauge, which we describe as EXTRA HEAVY and which is fully 20 per cent or one-fifth heavier than standard 28-gauge. We recommend the 26-gauge in preference to the standard weight, as it costs but a trifle more, costs no more to lay and gives nearly double the wear. Steel Roofing and Siding are **shipped from our factory in CENTRAL OHIO.**

Two-V and Three-V Crimp Steel.

Two-V and Three-V Crimp Steel Roofing have a covering width of 24 inches after lapping one crimp over the other. **Requires no soldering, folding or hammering of seams or joints;** anyone who has ordinary mechanical ability can put it on.

2½-Inch Corrugated Steel.

Catalog No.	Per 5-Ft. Sheet	Per 6-Ft. Sheet	Per 7-Ft. Sheet	Per 8-Ft. Sheet	Per 9-Ft. Sheet	Per 10-Ft. Sheet	Per 100 Square Feet	Weight, per Square
48L3105—28-Gauge Painted Red.	$0.38	$0.46	$0.54	$0.62	$0.69	$0.76	$3.50	68 lbs.
48L3106—28-Gauge Galvanized.	.54	.65	.76	.87	.98	1.08	4.94	84 lbs.
48L3123—26-Gauge Galvanized.	.59	.71	.83	.95	1.07	1.18	5.39	98 lbs.

1¼-Inch Corrugated Steel.

Catalog No.	Per 5-Ft. Sheet	Per 6-Ft. Sheet	Per 7-Ft. Sheet	Per 8-Ft. Sheet	Per 9-Ft. Sheet	Per 10-Ft. Sheet	Per 100 Square Feet	Weight, per Square
48L3107—28-Gauge Painted Red.	$0.38	$0.46	$0.54	$0.62	$0.69	$0.76	$3.50	68 lbs.
48L3108—28-Gauge Galvanized.	.54	.65	.76	.87	.98	1.08	4.94	84 lbs.
48L3125—26-Gauge Galvanized.	.59	.71	.83	.95	1.07	1.18	5.39	98 lbs.

Two-V Crimp Steel.

Catalog No.	Per 5-Ft. Sheet	Per 6-Ft. Sheet	Per 7-Ft. Sheet	Per 8-Ft. Sheet	Per 9-Ft. Sheet	Per 10-Ft. Sheet	Per 100 Square Feet	Weight, per Square
48L3077—28-Gauge Painted Red.	$0.36	$0.44	$0.51	$0.58	$0.65	$0.72	$3.56	69 lbs.
48L3078—28-Gauge Galvanized.	.50	.60	.70	.80	.90	1.00	5.00	85 lbs.
48L3080—26-Gauge Galvanized.	.54	.65	.76	.87	.98	1.08	5.39	98 lbs.

Three-V Crimp Steel.

Catalog No.	Per 5-Ft. Sheet	Per 6-Ft. Sheet	Per 7-Ft. Sheet	Per 8-Ft. Sheet	Per 9-Ft. Sheet	Per 10-Ft. Sheet	Per 100 Square Feet	Weight, per Square
48L3086—28-Gauge Galvanized.	$0.52	$0.63	$0.74	$0.84	$0.94	$1.04	$5.18	86 lbs.
48L3083—26-Gauge Galvanized.	.57	.69	.81	.92	1.03	1.14	5.61	100 lbs.

Guaranteed Full 28-Gauge.

Pressed Brick Face Steel Siding. Size of single brick, 2¼x8½ inches. Sheets, 60x28 inches.

Pressed Brick Steel Siding after painting can hardly be distinguished from pressed brick. Sold only in full sheets, painted red, or galvanized, as quoted below.

A square consists of 8⁴⁄₇ sheets. Weight, painted, 64 lbs. per square; galvanized, 78 lbs. per square.

Catalog No.	Pressed Brick Face Steel Siding	Per Sheet, 60x28 Inches	Per Square, 100 Square Feet
48L3116	Painted red	43c	$3.64
48L3117	Galvanized	59c	5.00

Guaranteed Full 28-Gauge.

Rock Face Steel Siding. Size of single stone, 7x12 inches. Sheets, 60x28 inches.

An elegant facing for stone fronts; makes a handsome front and is easily applied. A square consists of 8⁴⁄₇ sheets. Weight, painted, 64 pounds; galvanized, 78 pounds per square.

Catalog No.	Rock Face Steel Siding	Per Sheet, 60x28 Inches	Per Square, 100 Square Feet
48L3142	Painted red	43c	$3.64
48L3143	Galvanized	59c	5.00

Galvanized and Black Sheet Steel.

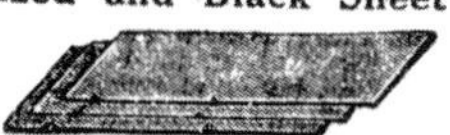

48L3225—Galvanized Sheet Steel, standard grade, absolutely flat. Size of sheet, 28x96 inches. **State gauge wanted.**

No. of gauge	28	26	24
No. of sheets per bundle	10	8	7
Wt., per bundle, lbs.	146	135	151
Per sheet	$0.89	$0.96	$1.18
Per bundle	8.67	7.51	8.12

48L3222—Black Sheet Steel.

No. of gauge	28	26	24
No. of sheets per bundle	10	8	7
Wt., per bundle, lbs.	117	112	131
Per sheet	$0.58	$0.67	$0.87
Per bundle	5.56	5.19	5.97

Valley Tin.

48L3198—Made of a good grade of tin plate in a continuous strip, locked and soldered and painted one side. Full lengths are 50 feet, but we furnish any quantity. **State width. Shipped from CHICAGO, ILL., or PHILADELPHIA, PENNA.**

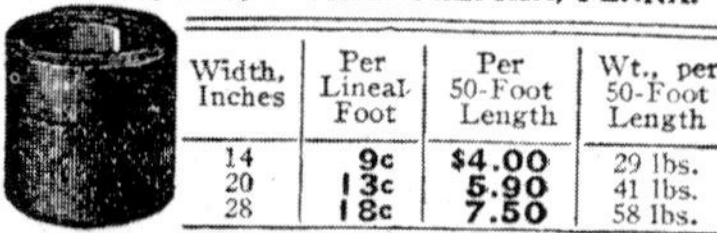

Width, Inches	Per Lineal Foot	Per 50-Foot Length	Wt., per 50-Foot Length
14	9c	$4.00	29 lbs.
20	13c	5.90	41 lbs.
28	18c	7.50	58 lbs.

48L3197—Galvanized Valley in rolls, made of 28-gauge galvanized steel, 14 inches wide.
25-foot roll. Weight, 20 pounds........$1.80
50-foot roll. Weight, 40 pounds........3.55

Beaded Steel Siding or Ceiling.

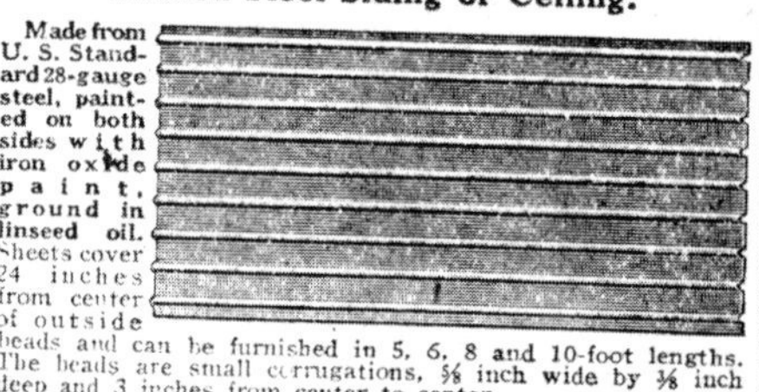

Made from U. S. Standard 28-gauge steel, painted on both sides with iron oxide paint, ground in linseed oil. Sheets cover 24 inches from center of outside beads and can be furnished in 5, 6, 8 and 10-foot lengths. The beads are small corrugations, ⅝ inch wide by ⅜ inch deep and 3 inches from center to center.

When ordering be sure to specify length of sheets desired. Always allow for end lap.

48L3115—Beaded Steel Siding or Ceiling. Weight, painted, 70 pounds per square of 100 square feet.
Per square........ $4.00

Length, feet	5	6	8	10
Sheet	40c	48c	64c	80c

When buying Steel Roofing or Siding compare the weights per square, as more weight means longer service. Steel Roofing, Siding and Ventilators are **shipped from factory in CENTRAL OHIO.**

Have you considered the remarkable values we are now offering on Guaranteed Roofing on the following pages?

Barbed Nails. (For Steel Roofing and Siding only.)

Catalog No.	Length	Per Lb.	Per 100 Lbs.
48L3028	⅞ in.	7c	$6.90
48L3088	1¼ in.	7c	6.90

Galvanized Steel Ventilators.

Extra heavy, 22, 24 and 26-gauge galvanized steel is used in the manufacture of our Majestic Ventilators, according to size.

Each Majestic Ventilator is equipped with four stay rods which hold it securely in place on the roof. A wire screen is provided to keep birds out.

Your choice of gold bronzed horse or cow weather vane.

Shipped from factory in CENTRAL OHIO.

Prices of Majestic Ventilators for large buildings:

Catalog No.	Size Flue	Size Base	Height	Wt., Lbs.	No. Cattle	Each
48L657	18 in.	26 in.	7 ft. 6 in.	115	4	$19.80
48L652	20 in.	29 in.	8 ft. 8 in.	155	6	25.85
48L653	24 in.	36 in.	9 ft. 5 in.	182	8	29.45
48L654	28 in.	40 in.	10 ft. 4 in.	235	12	34.90
48L655	30 in.	44 in.	10 ft. 6 in.	252	14	37.75
48L656	36 in.	55 in.	11 ft. 3 in.	396	20	43.85

Prince Ventilators are intended for small buildings such as small barns, hog houses and poultry houses.

Shipped from factory in CENTRAL OHIO.

Prices of Prince Ventilators for small buildings:

Catalog No.	Size of Flue, In.	Size of Base, In.	Height	Wt., Lbs.	Each
48L650	16	22	3 ft. 5 in.	55	$9.50
48L651	18	24	3 ft. 6 in.	58	10.75

METAL CEILINGS

$7.50 Per 100 Square Feet

FRENCH RENAISSANCE DESIGN. **Deeply embossed, intended for residences, churches, lodge halls or any kind of public or private building.** The cornice drops 12 inches on the side wall. Send us a rough drawing of your ceiling, giving measures of ceiling and of all offsets in wall, and our metal ceiling experts will furnish you with an estimate showing to the penny what this ceiling will cost you. Shipping wt., 65 lbs. to the square. **Shipped from steel mills in CENTRAL OHIO.**

48L2156½—French Renaissance Design.
Per 100 square feet, including nails........$7.50

$3.95 Per 100 Sq. Feet. High Grade Metal Ceiling

Suitable for Large or Small Rooms.

This neat and tastefully designed steel ceiling is used very extensively for stores and moderate priced buildings. It is of a neat, small pattern which will be appropriate for any room whether it be large or small.

48L3114—Steel Siding or Ceiling Covering. Sold at this price only in sheets of 27x96 inches, which includes allowances made for side and end laps as beaded on the sheets. Weight, per square, 56 pounds. Painted in a light drab color on both sides. **Shipped only from steel mills in CENTRAL OHIO.**
Per sheet, **66c**; per 100 square feet........$3.95

48L3118—Egg and Dart Design Border. Width, 3 inches. Per 4-foot length........ .14c

The two illustrations of metal ceilings shown are our two most popular designs. However, our Roofing and Metal Ceiling Catalog shows many other attractive designs. If you want a large assortment to select from write for our Roofing and Metal Ceiling Catalog No. 523GCL. Sent postpaid on request.

No Soldering-The Joints Slip Together

Galvanized Steel, Rust Resisting.

WE FURNISH TWO THICKNESSES. United States Standard 29-gauge, which is known the country over as standard weight and thickness, and United States Standard 26-gauge, which we describe as extra heavy and which is about 33⅓ per cent thicker and heavier than the standard 29-gauge. We recommend the extra heavy grade or 26-gauge in preference to the standard weight 29-gauge, as it costs but little more, costs no more to hang and gives nearly double the wear. All eaves troughs, conductor pipes and flashes are **shipped from our CHICAGO or PHILADELPHIA store.**

OUR SLIP JOINT REQUIRES NO SOLDER. No experience necessary to put up our Slip Joint Eaves Troughs and Conductor Pipes. Directions furnished with order.

Patent Slip Joint Eaves Troughs.
Can Be Put Together Without Soldering Iron.

When ordering state whether right or left hand is wanted. Right hand means that the water flows to the right hand end of trough; left hand that the water flows to the left hand end as you face the building.

STANDARD WEIGHT, U. S. STANDARD 29-GAUGE, GALVANIZED.

48L3148—Right Hand. 48L3149—Left Hand.

Width across top or size, in.	3½	4	4½	5	6
Weight, per length, pounds	4	4½	5	6	7
Per 10-foot length	47c	53c	58c	63c	74c

EXTRA HEAVY WEIGHT, U. S. STANDARD 26-GAUGE GALVANIZED STEEL.

48L3301—Right Hand. 48L3302—Left Hand.

Width across top or size, in.	3½	4	4½	5	6
Weight, per length, pounds	5	6	6½	7	9
Per 10-foot length	54c	60c	67c	72c	85c

ORDER BY SIZE AND NUMBER IN CATALOG.

Extra Heavy Gauge Eaves Trough Corners or Miters.

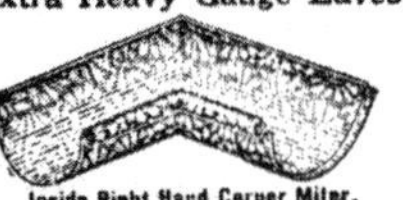
Inside Right Hand Corner Miter.

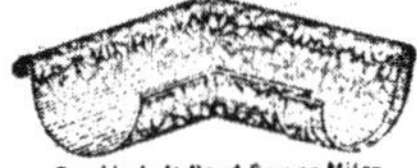
Outside Left Hand Corner Miter.

Make sure whether right or left hand miter is wanted. Furnished in the extra heavy 26-gauge galvanized steel. Corners where two eaves troughs come together require the most strength, as snow and ice collect there in cold weather. **Complete, ready for use for either right or left hand,** inside or outside bend. **Give catalog number and specify size.**

EXTRA HEAVY WEIGHT U. S. Standard 26-Gauge Galvanized Steel Corners or Miters.

48L3317—Inside Corner. Slip Joint. Right Hand.
48L3318—Inside Corner. Slip Joint. Left Hand.
48L3325—Outside Corner. Slip Joint. Right Hand.
48L3326—Outside Corner. Slip Joint. Left Hand.

Width across top or size, in.	3½	4	4½	5	6
Weight, per dozen, pounds	8	10	11	13	14
Each slip joint	29c	31c	33c	34c	38c

Galvanized Steel Elbows and Shoes.

Specify Angle and Number in Catalog.

STANDARD WEIGHT.

U. S. Standard 29-Gauge.
48L3182—Elbow, Angle No. 2.
48L3183—Elbow, Angle No. 3.
48L3184—Conductor Shoe.

No. 2 No. 3 Shoe

Diameter or size, inches	2	3	4
Weight per dozen, pounds	4¼	5	8½
Elbow	16c	20c	30c
Weight, per dozen, pounds	4½	5¾	9
Shoe	27c	30c	42c

EXTRA HEAVY WEIGHT, U. S. STANDARD 26-GAUGE GALVANIZED STEEL.

48L3312—Elbow. Angle No. 2.
48L3313—Elbow. Angle No. 3.
48L3314—Conductor Shoe.

Diameter or size, inches	2	3	4
Weight, per dozen, pounds	4¾	7½	11
Elbow	28c	34c	56c
Weight, per dozen, pounds	5½	9	15
Shoe	35c	44c	69c

Rain Water Cut-Offs.

For Corrugated Conductor. One of the strongest and best rain water cut-offs ever placed on the market.

Diameter of Spouts, Inches		2	3	4
48L3306 Wt., per doz., lbs.		6	9	16
Extra Heavy, U. S. Standard 26-Gauge.	Ea.	81c	86c	$1.19

Corrugated conductor makes a stiff, strong pipe, is easily put together and is most commonly used. Furnished only in 10-foot lengths. They are made to fit the following eaves troughs:

Size of eaves troughs, in.	3½	4	4½	5	6
Size of conductor, in.	2	2	3	3	4

48L3180—Standard Weight, U. S. Standard 29-Gauge.

Size, inches	2	3	4
Wt., per length, lbs.	4	5½	8
Per 10-foot length	52c	62c	77c

48L3303—Extra Heavy Weight, U. S. Standard 26-Gauge Galvanized Steel.

Size, inches	2	3	4
Wt., per length, lbs.	6	7½	10
Per 10-foot length	59c	70c	90c

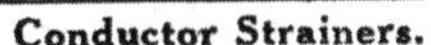

Conductor Strainers.

48L3194 — Galvanized Wire Conductor Strainers, placed in the outlet of eaves trough, prevent leaves, etc., from entering or stopping up the conductor. The size given designates the size outlet strainer will fit.

Size, inches	2	3	4
Wt., per dozen, lbs.	1	1½	2½
Each	9c	10c	13c

Conductor Funnel.

For running two conductors into one. Size indicates size of lower spout.

48L3305—No. 26-Gauge Galvanized Steel.

Size, inches	2	3	4
Wt., per dozen, lbs.	6½	9¼	12½
Each	31c	35c	46c

Adjustable Outlet.

Illustration represents outlet in position, without slip joint end cap. No soldering needed.

48L3308—Extra Heavy 26-Gauge Galvanized Steel Outlets.

Size, inches	3½	4	4½	5	6
Fitted for conductor, size, in.	2	2	3	3	4
Weight, per dozen, pounds	5	6	7	8	9
Each	19c	21c	23c	24c	27c

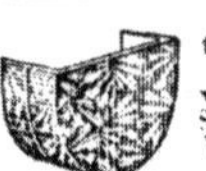

End Cap Slip Joint, for either eaves troughs or our adjustable outlet.

48L3307—Extra Heavy 26-Gauge Galvanized Steel.

Size, inches	3½	4	4½	5	6
Weight, per doz., lbs.	1½	1¾	2	2¼	2½
Each	10c	11c	12c	13c	14c

HOW TO ORDER PIPE AND GUTTER.

First. Be sure to state how much left hand and how much right hand Eaves Trough you need.

Second. Make a list of fittings for Eaves Troughs, such as Miters, Hangers, End Caps and Outlets. These should be exactly the same size as the Eaves Trough.

Third. State how much Conductor Pipe you will need and be sure to order the proper size to fit the Eaves Trough. **Conductor Pipe is always smaller than the Eaves Trough.** For instance, 3-inch Conductor Pipe should be used with 5-inch Eaves Trough.

Fourth. Make a list of fittings which go with Conductor Pipe, such as Elbows, Conductor Hooks, Strainers, Funnels and Cut-Offs. These should be ordered exactly the same size as Conductor Pipe.

Guaranteed Roofing See Pages 960 to 963.

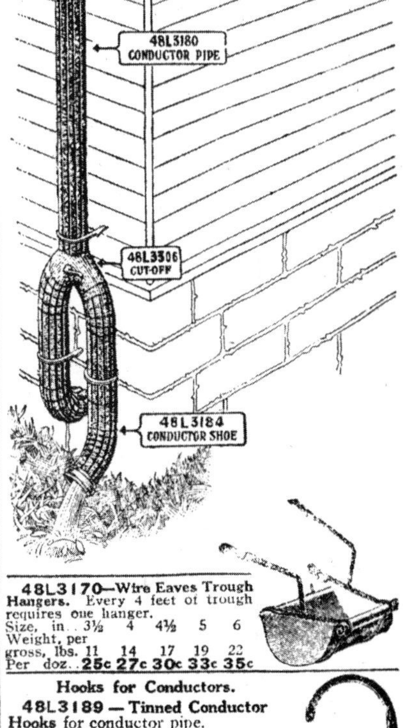

48L3170—Wire Eaves Trough Hangers. Every 4 feet of trough requires one hanger.

Size, in.	3½	4	4½	5	6
Weight, per gross, lbs.	11	14	17	19	22
Per doz.	25c	27c	30c	33c	35c

Hooks for Conductors.

48L3189 — Tinned Conductor Hooks for conductor pipe.

Size, inches	2½	3	4
Weight, per 100 pounds	10	15	20
For wood, dozen	$0.92	$1.15	$1.60
Weight, per 100 pounds	12	17	23
For brick, dozen	1.03	1.26	1.95

Two-inch corrugated conductors require 2½-inch hooks.

Tarred Felt-Building Paper-Deadening Felt

Tarred Felt—Used for Sheathing or General Building Purposes.

Tarred Felt is one of the best kinds of building paper for lining floors and for use between sheathing and siding. Can be used under stucco and brick veneers. It is also used for gravel roofs. **It is made of carefully selected felt, thoroughly saturated with redistilled low heat American coal tar.**

48L3050—TARRED FELT. We recommend this for all first class jobs, because it is the thickest. Used extensively for roofing sheds and temporary buildings. It makes an excellent deadening felt and should be used under all floors where a good job is wanted.

48L3054—TARRED FELT, medium thickness. This thickness of tarred felt is also used for gravel roofing where several layers are specified. It is a lighter weight (not so thick), and where a medium grade of work is required a 48L3054 Tarred Felt will be found very satisfactory.

48L3055—TARRED FELT. This thickness of felt is made with the same care as the two heavier grades, but is much thinner.

	Weight, Per Roll	Number of Sq. Ft. to Roll	W'th, Inches	Per Roll
48L3050	60 lbs.	250	32	**$1.90**
48L3054	60 lbs.	400	32	
48L3055	60 lbs.	500	32	

Homan Brand Deadening Felt.

Deadening Felt is used to deaden the sound between floors and in walls. It is made of a good grade of felt, soft and pliable, and will add much to the warmth of your building. Homan Brand Deadening Felt should be under every floor.

Prices of Homan Brand Deadening Felt.

	Brand	Weight, per Square Yard	Width, Inches	Approximate Wt.	Per Roll
48L3004	Homan	1 pound	36	50 pounds	**$1.98**
48L3006	Homan	1½ pounds	36	75 pounds	**2.95**

About 50 square yards to the roll.
All shipped from CHICAGO or PHILADELPHIA store.

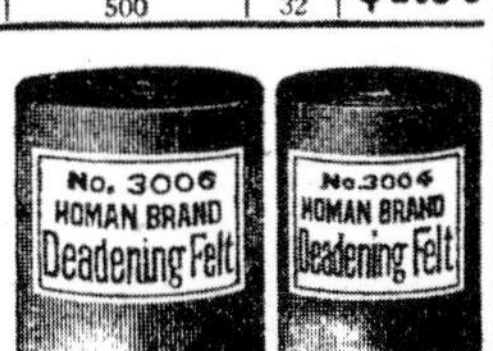

No. 3006 HOMAN BRAND Deadening Felt
No. 3004 HOMAN BRAND Deadening Felt

Red, Rosin Sized Building Paper.

68c and $1.35 Roll.

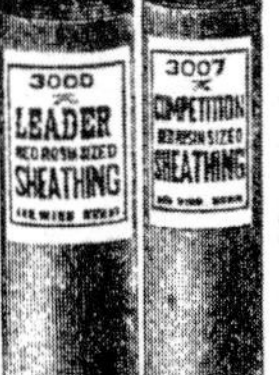

Red, Rosin Sized Building Paper should be used under siding and between floors to exclude wind and moisture. It is rosin sized. We sell two different weights of this paper. We recommend our **Leader Brand,** as it is of sufficient thickness to give entire satisfaction.

Our Competition Brand is a fair grade, light in weight, and runs 20 pounds to the roll.

	Brand	Weight, Pounds	Width, Inches	Per Roll
48L3000	Leader	40	36	**$1.35**
48L3007	Competition	20	36	**.68**

All rolls contain 500 square feet.
All shipped from CHICAGO or PHILADELPHIA store.

Concrete Machinery and Mixers

Our Concrete Machinery Catalog Will Save You Big Money

Now as never before we are in position to save you money on our big line of Concrete Machinery.

While this page shows our very best sellers and quotes our lowest prices on Concrete Machinery and Mixers, we also show other designs of mixers, forms and molds in our Concrete Machinery Catalog, illustrated to the right.

Everyone who uses concrete mixers or block machines needs our book. Experts have stated that it is the best and most complete catalog on concrete machinery ever published. It illustrates, describes and prices a complete line of all concrete machinery, including mixers, tampers, molds for making blocks, bricks, fence posts and porch material.

Our Book of Concrete Machinery shows you how to increase your output and reduce your operating costs. Our new prices mean an EXTRA BIG SAVING.

We have letters from farm owners, concrete products manufacturers and big contractors stating they have increased their profits and built up a big business by using our Concrete Machinery. You can do the same. Our catalog will show you how.

Remember, every price in our new Concrete Machinery Catalog has been reduced. Get these new rock bottom prices. We ship our machinery direct from factory to you, saving you all extra expense. Big stocks on hand. We can make IMMEDIATE SHIPMENT of all machines.

You are sure to find this book interesting, helpful and valuable. It is yours for the asking. Sent postpaid, and without any obligation on your part. Send for Concrete Machinery Catalog 532GCL. You will be glad you did when you receive it.

This Big Book Sent Postpaid. Write for 532GCL.

Five Cubic Foot Ellis Concrete Mixer.

The Ellis Concrete Mixer, with engine and side loader, illustrated directly above, is practical for large or small concrete work. Our price is extremely low for this HIGH GRADE machine, which is guaranteed to be perfect in material and workmanship. Built for service. The kind of service you would expect from a machine of much higher price. 40 to 60 cubic yards of concrete per day can be mixed by two men. It would take three men about four days to mix the same amount of concrete by hand with hoe and shovel. Full directions are furnished so you will have no trouble in starting or taking care of the Economy Gasoline Engine sent with the mixer. Ellis Mixers are **shipped complete from factory in CENTRAL OHIO.**

COMPLETE SPECIFICATIONS.

Mixing Drum—Made of 10-gauge steel; 36 inches in diameter; 30 inches wide. Drum openings, 14 inches.

Capacity—Five cubic feet of unmixed materials, using one-half bag of cement.

Side Loader—Made of heavy sheet steel, properly braced. Holds one batch, 5 cubic feet. Has steel cable and hoist operated by a simple internal expansion ring clutch. Automatically stops when bucket is at top.

Frame—Made of 5x1¾-inch steel channel, with three cross channels of same material.

Trunion Rollers, on which mixing drum is supported, have 2-inch chilled face and measure 10 inches in diameter.

Truck Wheels are all steel with cast hub. Diameter of front wheels, 18 inches; rear wheels, 24 inches; with 4-inch reinforced tires. Axles are cold rolled steel.

Track—Standard 4 feet 8 inches, so wheels will follow the road track.

Power—Equipped with our famous 2½ horse-power Economy Gasoline Engine, with built-in magneto.

All important bearings are provided with compression grease cups.

63L5980—Ellis Concrete Mixer complete with engine and power driven side loader, as illustrated above. Shipping weight, 2,700 pounds **$378.00**

63L5984—Same as above, but with wood loading platform. Shipping weight, 2,600 pounds **$349.75**

63L5982—Same as above, but without wood loading platform. Shipping weight, 2,350 pounds **$331.00**

Write for catalog giving prices on Ten Cubic Foot Ellis Mixer.

Five Cubic Foot Ellis Mixer.

This Ellis Concrete Mixer is of the same high grade construction as the Ellis Mixer shown above, the drum construction of both machines being identical. Mounted on either skids for stationary use or on truck for portable use. Trucks have wheels 16 inches in diameter with 3-inch face. Frame made of 3-inch steel channel iron.

Capacity—When operated by hand, 3 cubic feet. When operated by power, 5 cubic feet of unmixed material per batch, using one-half bag of cement.

Shipped complete from factory in CENTRAL OHIO.

63L5976—Ellis Hand Operated Mixer on Skids. Shipping weight, 860 pounds **$102.75**

63L5978—Ellis Hand Operated Mixer on Truck. Shipping weight, 1,250 pounds **$134.50**

63L5979—Power Pulley. 20 inches in diameter, 3-inch face. Shipping weight, 35 pounds **$6.50**

Wizard Concrete Block Machine.

A Big Capacity Block Making Outfit.

Simple and Speedy.

Anyone Can Operate It.

$83.10

A complete outfit for quantity production of high grade building blocks. Used by many leading block manufacturers. Priced low enough to make its use profitable on farms or other places where concrete products are in demand. Each outfit contains:

One Wizard Machine on heavy stand.	One Face Plate for making inside corner blocks.
Two Face Plates for making whole blocks.	Two Joist Block Attachments.
Two Face Plates for making half and quarter blocks.	One Dividing Plate for making gable blocks.
Two Core Endgates.	Two Wall Plugs for making solid blocks.
Four Return Endgates.	Two Pallet Plugs for making solid blocks.
Two Dividing Plates.	One Strike-Off Tool.
Two Core Dividing Plates.	One Double End Tamper.
One Cast Iron Pallet.	

We furnish outfits with one plain and one rock face plate. Two men can make 250 to 300 blocks a day. **IRON PALLETS** are used. A smooth rounded handle on each end of pallet permits lifting of block with ease. Will not split, warp or swell. Barring accidents, will last forever.

63L5515—8x8x16-Inch Wizard Block Making Outfit. Shipping weight, 425 pounds **$83.10**

63L5537—Pallets for 8x8x16-Inch Blocks. Weight, each, 7 pounds. Each ..... **.66**
Per 100 ..... **64.00**

Shipped direct from factory in CENTRAL OHIO.

WIZARD CONCRETE MIXER ALSO FEED MIXER.

An efficient Concrete Mixer. Made in two sizes, 3 and 4 cubic foot capacities and is especially suited for farm use and for general use on small jobs. The mixing drum is so proportioned that, with the help of three simple mixing blades on the inside, a thorough mix and a quick discharge is obtained. The tilting drum permits loading on one side and discharging on the other, which allows piling material close to machine. Can be operated by hand or power. Will mix concrete, mortar, plaster, wet or dry. It can also be used on the farm as a feed mixer.

SPECIFICATIONS.

Three cubic foot mixer. Capacity, 3 cubic feet unmixed material or 25 to 35 cubic yards concrete per day. Mixing drum 22 inches in diameter by 33 inches deep with a 16-inch opening. A 1½ horse-power engine will drive this mixer; pulley 20 by 3 inches; charging height, 35 inches.

63L5995—Three cubic foot Wizard Concrete Mixer on stand. Shipping weight, 300 pounds **$42.40**

SPECIFICATIONS.

Four cubic foot mixer. Capacity, 4 cubic feet unmixed material or 30 to 40 cubic yards concrete per day. Mixing drum 25 inches in diameter by 33 inches deep, 17¾-inch opening. A 1½ horse-power engine will drive this mixer; pulley 20 by 3 inches; charging height, 37 inches.

63L5996—Four cubic foot Wizard Concrete Mixer on stand. Shipping weight, 390 pounds **$55.00**

Shipped from factory in CENTRAL OHIO.

Triumph Concrete Block Machine.

A durable cast iron machine for making 8x8x16-inch concrete blocks with rock face and double air space. Makes blocks face down. Uses wood pallets. Furnished either with lugs for fastening to a bench as shown or with a well braced cast iron stand. Two men can make from 100 to 125 blocks per day.

We furnish with the machine: One Rock Face Plate for whole blocks. One Rock Face Plate for half and quarter blocks. Two Rock Endgates. Two Core Endgates. Two Dividing Plates. One Gable Block Dividing Plate. Two Joist Block Attachments. One Wood Pallet (others can be made). Plugs, Striker, Tamper.

Shipped complete from factory in CENTRAL OHIO.

63L5704—Triumph Block Machine for 8x8x16-inch blocks, for mounting on bench. Shipping weight, 140 pounds **$23.15**

63L5709—Triumph Block Machine for 8x8x16-inch blocks, on iron stand. Shipping wt., 165 lbs **$30.00**

Triumph Silo Block Mold.

A good, efficient silo can be constructed of concrete blocks and proper reinforcing. Our Silo Block Mold will enable any farm owner with a little practice to make his own blocks.

Mold is made of heavy castings for making 8x8x16-inch blocks of the proper curve for silo. Each block has two hollow cores and space for reinforcing rods. With the mold we send a wooden pallet, face plates for whole blocks and two half blocks, dividing plate and tamper. Additional pallets are very easily made from any smooth lumber. Blocks are suitable for any size silo from 10 to 18 feet in diameter. Full directions sent with each machine.

63L5756 — Triumph Silo Block Mold. Shipping weight, 75 pounds **$19.10**

Shipped direct from factory in CENTRAL OHIO.

Fence Post Mold.

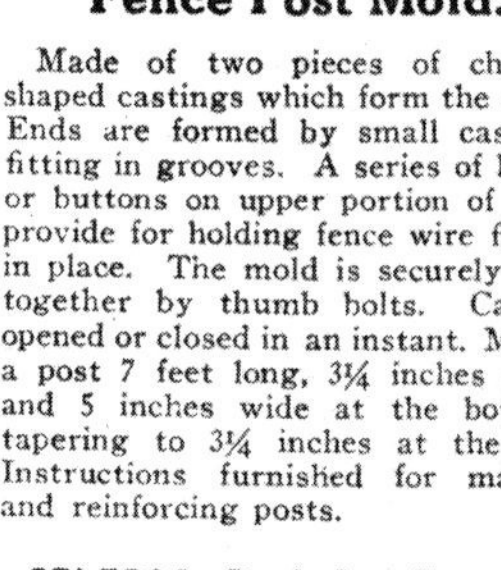

Made of two pieces of channel shaped castings which form the sides. Ends are formed by small castings fitting in grooves. A series of knobs or buttons on upper portion of mold provide for holding fence wire firmly in place. The mold is securely held together by thumb bolts. Can be opened or closed in an instant. Makes a post 7 feet long, 3¼ inches thick and 5 inches wide at the bottom, tapering to 3¼ inches at the top. Instructions furnished for making and reinforcing posts.

63L5896—Single Line Fence Post Mold (for making one post at a time), complete with grooving block and wire tie. Shipping wt., 50 lbs. **$13.00**

Harvard Concrete Mixer.

$33.80

A good power mixer. Cement, sand and gravel are fed into hopper. Water is added through galvanized iron tank. Mixing done by paddles which push mixture forward 4 inches and back 1 inch. Can be used as batch mixer or continuous mixer.

Specifications—Hopper and mixing drum made of 16-gauge rolled sheet steel, securely riveted and held with heavy castings. Fitted with heavy bevel gears with necessary grease cups. Water tank holds 3 gallons and is provided with brass water valve. Capacity, up to 25 cubic yards of concrete per day. **Shipped from factory in CENTRAL OHIO.**

63L5674—Harvard Concrete Mixer on Skids. Shipping weight, 240 pounds **$33.80**

Already Cut FARM BUILDINGS

Build It Yourself!
Pile of Already Cut Lumber, showing how all pieces are numbered to correspond with numbers on plans.
Upper Wall Plate Joists and Truss Cords, Studding and Purlin Posts, showing how they fit together.
Windows ALREADY MADE and Frames Already Cut and Bundled.
Doors ALREADY MADE. Nothing to do but hang them in place.
Studding, Plate Rafter, Truss Principal and Tie, showing how they fit together.
Rafters Already Cut and Fitted.
Studding Already Cut and Fitted.

THINK of all the money you can save when you can dispense with expert labor in putting up your farm buildings. Think of all the money you can save when you can purchase material from us at practically wholesale prices. Think of the satisfaction in knowing that all of the material is guaranteed to be the best of its kind—material that will last you a lifetime.

Only first grade yellow pine and time defying cypress used in our modern farm buildings. We furnish select and clear grade cypress for our barn siding, and cypress has been known to last TWO HUNDRED YEARS.

We are the only concern that we know of that furnishes this high grade material for farm buildings.

The book of **Modern Farm Buildings,** illustrated below, is an exceptional book. It will help you in many ways. It contains, in addition to a complete collection of modern farm buildings, the latest ideas on ventilation, ventilating systems, floor arrangements and barn equipment. To be sure of having a truly modern, sanitary and durable farm building on your property—a building that your children can point to with pride in future years—sit down right now and mail us a letter or a post card asking for our book of **Modern Farm Buildings** 504GCL. It will be sent to you promptly, postage prepaid.

Write for it today, before you turn from this page. It will show you how to get one of the finest barns in your county at a big saving in price. You can build it yourself. No waste material, no time lost in figuring, measuring or guessing.

On this page we show you how the material for our modern farm buildings comes to you already cut and fitted. In addition, all of the doors are ALREADY MADE. All you have to do is to hang them in place.

Our new book of **Modern Farm Buildings** fully describes and illustrates the buildings shown above, together with many others. It contains the very latest ideas in modern farm buildings secured from leading authorities. If you are thinking of building any kind of farm structure, be sure to write for this book of **Modern Farm Buildings** 504GCL today.

Going to Build a Home?

Send for Book of "Honor Bilt" High Grade Modern Homes. 596GCL.

HONOR BILT *Already Cut* HOMES

This illustration shows how well an "HONOR BILT" Home is constructed. Note the double plates over windows with the double studdings at the sides. Three studs at the corners give additional strength. Double floors with heavy building paper between insures against drafty floors.

Yet with this extra heavy construction you can save as high as 40 per cent on your home.

Compare construction when you compare price.

$1,098 Living and dining room, two bedrooms, kitchen and bathroom. Can be built on a lot 27 feet wide. An attractive home for a moderate outlay of money. **$18.00** A MONTH

"HONOR BILT" System Saved 40 Per Cent Carpenter Labor

Our homes are furnished Already Cut and Fitted, according to our "HONOR BILT" SYSTEM. Our new Book of "HONOR BILT" Modern Homes contains certified details of this saving, with photographs, affidavits, etc.

One Order Brings It All

When you purchase a house from us you dispose of the entire transaction in a few minutes. On receipt of your order we ship at factory prices the following materials:

Lumber,
Lath,
Mill Work, Such as Doors, Windows, Molding, Flooring,
Building Paper,
Porch Material, Etc.,
Hardware,
Nails,
Eaves Trough,
Down Spout,
Paint and Varnish.
Medicine Case.
Shingles or Roofing as specified.

At your option: Steam Heating, Furnace Heating, Plumbing Outfit, Electric Wiring, Gas and Electric Fixtures, Wall Paper and Electric Lighting Plants.

Shingles or Fire Chief Shingle Roll Roofing as specified.

No need to shop about in a dozen places.

CATALOG FREE

Ask for Modern Homes No. 596GCL described on opposite page.

Save Time and Money.

1—Note the notches and miters. No use for a saw here.
2—Pieces numbered to correspond with plans.
3—Doors mortised for locks.
4—Every piece cut to fit. A most difficult job made easy.

Easy Payments

$2,093 Five large rooms and bath. This bungalow can be built on a lot 32 feet wide. Real estate dealers ask $6,000.00 for the same house. **$30.00** A MONTH

THE BETTER GRADE OF HOUSES

$1,983 Five rooms and bathroom. Your choice of two floor plans. This house can be built on a lot 40 feet wide. **$30.00** A MONTH

Permanent High Grade Homes

"HONOR BILT" Homes are extra high grade. They have double floors, double walls and 2x4-inch studding. Outside door and window casings are CLEAR CYPRESS, 1⅛ inches thick.

"HONOR BILT" means a more Substantial Home—a more Comfortable Home—a Superior Home—at a big saving in money and time.

Save From $500 to $2,000.

You need our Book on "HONOR BILT" Modern Homes if you are thinking of building. It will save you $500.00 to $2,000.00, depending upon the size house you build. It illustrates over a hundred homes for city, suburb and farm, including the very newest designs in bungalows. Each home is priced, completely illustrated and described. Floor plans are shown for every design.

Sold on Easy Payments

"HONOR BILT" Triple Unit Wardrobe Clothes Closet, fitted with mirror doors, compartments for hats above, shoe drawers below. Plenty of space for suit cases and other heavy articles.

Prices for Modern Homes shown on these pages include the following items:

Lumber, Lath, Mill Work (Such as: Doors, Windows, Molding, Flooring), Porch Material, Building Paper, Hardware, Nails, Eaves Trough, Down Spout, Paint and Varnish, Medicine Case

Shingles or Roofing as specified.

We guarantee enough material to finish the house complete.

"HONOR BILT" Kitchen De Luxe White Tile Sink and Drain Board. White Enamel Cupboards. Plenty of room for kitchen utensils within arm's reach.

"HONOR BILT" Folding Built-In Ironing Board. Strong. Compact. Convenient.

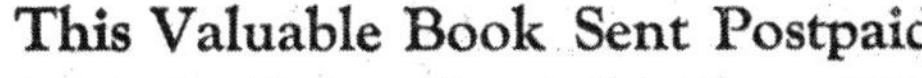

$2,045 Six rooms, bath and porch. A comfortable, well planned home of the Dutch Colonial type. Can be built on a lot 30 feet wide. **$30.00** A MONTH

This Valuable Book Sent Postpaid

Ask for Modern Homes Catalog, 596GCL. It contains 140 pages, and shows homes ranging from cottages at $678.00 to large residences at $4,000.00. Several houses are illustrated in colors. This catalog explains how our "Ready Cut System" saves you money. It contains statements from our customers who have saved from $500.00 to $2,000.00 by trading with us—and are entirely satisfied.

It explains our Easy Payment Plan and shows how you can become the owner of one of these fine homes on small monthly payments.

IT IS FREE.

Send for your copy today.

$1,947 Six rooms and bath. Note the big, roomy porch. Can be built on a lot 30 feet wide. A big, roomy house at a very low price. **$30.00** A MONTH

Sectional Garages and Summer Cottages

Order Right from this Page Bolt It Together Yourself

55L122—Three-Room Summer Cottage With Porch. $462.00
Size, 15 feet 3 inches by 21 feet 3 inches outside. Shipping weight, about 7,800 pounds. Peerless Wall Board for lining walls and ceilings, $86.00 extra.

55L187 —Five-Room Summer Cottage With Porch. $1,058.00
Size, 27 feet 3 inches by 33 feet 3 inches outside. Shipping weight, 19,000 pounds. Peerless Wall Board for lining walls and ceilings, $182.00 extra.

55L337 55L188

55L188—Five-Room Summer Cottage With Porch. Size, 27 feet 3 inches by 21 feet 3 inches outside. Shipping weight, 14,000 pounds. Peerless Wall Board for lining walls and ceilings, $129.00 extra.......................... **$748.00**

55L337—Five-Room Summer Cottage With Bath and Porch. Size, 30 feet 3 inches by 21 feet 3 inches outside. Shipping weight, 17,000 pounds. Peerless Wall Board for lining walls and ceilings, $192.00 extra.................. **$951.00**

MT. VERNON

55L191 — Four-Room Summer Cottage With Bathroom and Breakfast Alcove. $999.00
Size, 30 feet 3 inches by 24 feet 3 inches outside. Shipping weight, 21,000 pounds. Peerless Wall Board for lining walls and ceilings, $180.00 extra.

55L189 — Five-Room Summer Cottage With Porch and Bathroom. $1,089.00
Size, 24 feet 3 inches by 30 feet 3 inches outside. Shipping weight, 18,000 pounds. Peerless Wall Board for lining walls and ceilings, $165.00 extra.

Specifications of "Simplex" Sectional Garages and Summer Cottages

Write for Garage Easy Payment Terms.

Frame summer cottages and garages constructed of No. 1 pine lumber in sections 3 feet wide by 8 feet high, ready to lock together. Walls made of drop siding finished ¾-inch thick; flooring dressed and matched finished 13/16-inch thick; joists, 2x6 inches; sills, 3x6 inches; roof sections and partitions of ⅝-inch beaded ceiling. Doors and windows are hung in their frames. All necessary hardware furnished. Outside of buildings has one coat of gray paint, excepting Mount Vernon, 55L191, which is painted white. 4-foot cedar foundation posts furnished for all cottages and garages. All cottages furnished with galvanized steel chimneys, also 17-Year Guaranteed Fire Chief Shingle Roll Roofing. Avenue Garage furnished with same roofing. Utility Garage furnished with 14-Year Guaranteed Best-of-all Roofing. See page 961 for Fire Chief Shingle Roll Roofing and page 963 for Best-of-all Roofing. All Sectional Frame Buildings on this page shipped from factory in SOUTHERN ILLINOIS. Brick, plaster, cement or stone not furnished.

Extra price of Peerless Wall Board includes ceiling joists, furring strips, baseboard, etc. For description of wall board see page 969.

Chicken Houses.

Chicken Houses furnished with 14-Year Guaranteed Best-of-all Roofing. Walls made of tongued and grooved siding ¾-inch thick; 2x2 and 2x4-inch framing; roof boards tongued and grooved, ⅝-inch thick. Brood Coops made of lighter material than chicken houses, suitable for this purpose, and have red wooden roofs. Shipped from SOUTHERN ILLINOIS.

For other poultry supplies see Index.

Specifications of Monitor Steel Garages.

All steel No. 26-gauge, heavily galvanized, fireproof. Uniform sections 2 feet 6 inches by 8 feet. All hardware included. Steel garages shipped direct from factory in OHIO.

Permanent **Honor Bilt** *Homes Described on Pages 972-973*

55L338 55L106

55L106 — Three-Room Summer Cottage With Porch.......$598.00
Size, 18 feet 3 inches by 21 feet 3 inches outside. Shipping weight, 10,600 pounds. Peerless Wall Board for lining walls and ceilings, $91.00 extra.

55L338—Five-Room Summer Cottage With Bath and Porch...$1,079
Size, 30 feet 3 inches by 21 feet 3 inches outside. Shipping weight, 21,000 pounds. Peerless Wall Board for lining walls and ceilings, $186.00 extra.

55L186 — Four-Room Summer Cottage With Front and Side Porches......$678.00.
Size, 18 feet 3 inches by 24 feet 3 inches. Shipping weight, 12,500 pounds. Peerless Wall Board for lining walls and ceilings, $112.00 extra.

Chicken Houses.
55L300—Size, 10x8 feet. Shipping weight, 1,200 pounds. Without floor. **$67.00**
55L302—Size, 12x10 feet. Shipping weight, 1,800 pounds. Without floor. **$78.00**
55L306—Size, 20x12 feet. Shipping weight, 3,600 pounds. Without floor. **$175.00**

Portable Brood Coops.

Roof ¾ inch thick, floor ¾ inch thick, walls ¾ inch thick, all tongued and grooved.

55L321—Simplex Portable Brood Coop, complete. Shipping weight, 90 pounds.

Each coop...................... **$ 6.95**
Three coops.................... **20.00**
Six coops...................... **38.95**

MONITOR

Catalog No.	Price	Shpg. Weight	Foundation Dimensions Ft. In. Ft. In.
55L193	$110.00	1,100 lbs.	9 4¾x12 7⅞
55L194	130.00	1,215 lbs.	9 4¾x15 1⅞

UTILITY

Catalog No.	Price	Shipping Weight	Foundation Dimensions
55L167	$ 94.00	2,450 lbs	10 ft. x 12 ft.
55L168	107.00	2,600 lbs.	10 ft. x 14 ft.
55L168C	115.00	3,200 lbs.	10 ft. x 16 ft.

Write for Prices of Other Sizes and EASY PAYMENT TERMS.

AVENUE

Catalog No.	Price	Shpg. Weight	Foundation Dimensions Ft. In. Ft. In.
55L19	$210.00	3,600 lbs.	12 3 x 15 3
55L19C	246.00	4,400 lbs.	12 3 x 18 3
55L19D	264.00	5,300 lbs.	18 3 x 18 3
55L19E	318.00	6,400 lbs.	21 3 x 18 3

All weights given on this page are approximate and may vary a little.

$18.75
Requires No Electricity
The "Sanitary Special" Non-Electric Vacuum Cleaner.
Equal in Value to Similar Cleaners Sold at From $5.00 to $10.00 More Than We Ask.
Powerful suction gets dust and dirt. Revolving brush gathers up threads and lint. All you do is put it down on the floor and push, and it cleans in just the same manner as electrically operated vacuum cleaners.
Costs Nothing to Operate.
Is gear driven. The wheels drive gears propelling a fan, which in turn draws the dust, dirt and lint by SUCTION, just as do electric cleaners.
The brush is removable, permitting you to operate the machine for suction cleaning alone, without brushing the surface to be cleaned.
Body is cast aluminum. Wheels are rubber tired. Bag is made of dust-proof fabric and readily removed for emptying. Actual weight, 8½ pounds. Shipping weight, 11 pounds.
99L987
$18.75
Best Grade Aluminum Ware At ½ Usual Retail Prices
9½ IN.
9½ IN.
2 QT.
8 QT.
1 QT.
9 IN.
9 IN.
4 QT.
Eight Big, Full Size Pieces of Best Grade Aluminum Ware.
$2.95 SET
USUAL $4.50 TO $5.00 RETAIL VALUE.
All big, full size pieces. Just the pieces necessary about the kitchen. Shipping weight, 5 pounds.
99L991—Big Eight-Piece Set, as illustrated. $2.95
1½ QT.
2 QT.
3 QT.
6 QT.
4 QT.
2 QT.
90c SET
Set of Three Best Grade Aluminum Saucepans.
Usual $2.00 Retail Value.
Note the capacities in our set, 3, 2 and 1½ quarts. Shpg. wt., 2½ lbs.
99L990—Per set...... 90c
These Three Daily Necessities.
$1.95
Three large pieces for the usual retail price of the 6-quart kettle alone.
Best Grade Aluminum Ware.
99L992—Set of 3 pieces, as illustrated. Shpg. wt., 4 lbs..... $1.95
5 QT.
12 QT.
6 QT.
Best Grade Aluminum Teakettle.
$1.65
Usual Retail Price, $3.00.
Large spout permits being filled direct from any ordinary kitchen faucet without removing cover. Flat bottom fits any stove. Shipping weight, 2½ pounds.
99L988—Capacity, 5 qts.. $1.65
Best Grade Aluminum Dish Pan.
$1.15
Also makes an ideal pan for cooking fruit, etc., for canning. Shipping weight, 4 pounds.
Note our low price on this Best Grade Aluminum Dish Pan.
99L989....... $1.15
For Other Great Aluminum Ware Values See Pages 826 and 827
$1.10
Patent Ears.
Best Grade Aluminum Lipped Convex Kettles.
With patent ears for holding cover in place when pouring hot contents.
99L993—Capacity, 6 quarts. Shipping weight, 2¾ pounds................. $1.10
99L981
99L983
99L984
9L980
99L982
99L985
99L986
PROTECTION
11 x 1¾ IN. ROLLS
IMPROVED GUIDE BOARD
Wash Day Needs At New Low Prices
$1.10 Per Set
Mrs. Potts Pattern Sadirons.
Nickel Plated Finish. Usual Retail Price, $2.00 per Set. Standard the World Over.
Shipping weight, 17 pounds.
99L981—As illustrated, per set........... $1.10
OUR WIZARD IRONING BOARD.
Why Pay $3.00 to $3.50 for an Inferior Board?
$1.95
Large, roomy ironing surface, 13 inches wide by 54 inches long. Stands 32 inches from floor. Shipping weight, 19 pounds.
99L983—Not mailable...... $1.95
Galvanized Wire Clothesline.
43c
Weight, 3 lbs.
9L980
Per 100 feet, 43c
Extra Grade Galvanized Washtubs.
$1.30
Shipping weight, 10 pounds.
Not mailable.
99L985—Top diameter, 24¾ in. Depth, 11 in. $1.30
200 Standard Clothespins.
49c
Standard pattern, good quality pin. Shpg. wt., 4 lbs.
99L984
For box of 200, 49c
Our Best Made Solid Copper Wash Boiler.
Copper Plated Steel Cover.
Why Pay $6.50 to $7.00 Elsewhere?
$4.45
Extra heavy. Large size. Inside measure, 23¼ x 12x12⅜ in. Tinned on the inside with block tin. Shpg wt., 17 lbs. Not mailable
99L982...... $4.45
The New Modern Improved Protection Wringer.
Newest Design. Well Made. Latest Features.
$4.90
Quick acting reversible drain board. Best quality 11x1¾-inch rubber rolls. Guaranteed for three years. "Oil-Less" wood bearings. Selected hardwood frame.
99L986—Shipping weight, 28 pounds.. $4.90

ALL WEIGHTS AND MEASUREMENTS GIVEN ON THIS PAGE ARE APPROXIMATE AND MAY VARY A TRIFLE.